D0208110

SOYBEANS:
Improvement, Production, and Uses

Third Edition

SOYBEANS:
Improvement, Production, and Uses

Third Edition

H. Roger Boerma and James E. Specht, *co-editors*

Editorial Committee
Richard M. Shibles
James E. Harper
Richard F. Wilson
Randy C. Shoemaker

Managing Editor: Lisa K. Al-Amoodi

Editor-in-Chief ASA Publications: Kenneth A. Barbarick

Editor-in-Chief CSSA Publications: Craig A. Roberts

Editor-in-Chief SSSA Publications: Warren A. Dick

Number 16 in the series
AGRONOMY

American Society of Agronomy, Inc.
Crop Science Society of America, Inc.
Soil Science Society of America, Inc.
Publishers
Madison, Wisconsin, USA

2004

American Society of Agronomy, Inc.
Crop Science Society of America, Inc.
Soil Science Society of America, Inc.
677 South Segoe Road, Madison, Wisconsin 53711 USA

Library of Congress Catalog Control Number: 2003114302

Printed in the United States of America

CONTENTS

FOREWORD

On the 50th anniversary of the discovery of the DNA helix by Watson and Crick, the American Society of Agronomy, the Crop Science Society of America, and the Soil Science Society of America are pleased to present the third edition of *Soybeans: Improvement, Production, and Uses*. As in the first two editions, this edition describes the most recent advances in soybean science and technology and the trends in soybean usage that are changing the way we live.

Soybean is a remarkable crop. While long used as a grain crop in its native homeland where it was described as one of the five sacred grains, it came to the USA initially as a forage crop. It evolved to become a grain crop that is the main source of protein in livestock diets. As a grain crop, soybean production was initiated in numerous countries where it has become an integral part of their modern agriculture and food system. These events are recent and many of us remember those pioneers who changed this crop from a forage crop to a grain crop and spread it throughout the world. Now, soybean is emerging as a major protein and oil crop for human consumption. Its increasing importance as food and feed has led to some of the most rapidly accepted genetic modifications in agriculture, and the true coupling of basic and applied science in crop improvement, production, and usage. Soybean is a global crop and the contents of this edition span the breadth of its worldwide significance. This book will be a ready source of useful information for the casual consumer who is interested in knowing more about the foods they eat and part of the world they live in, the student interested in getting a fundamental understanding of a major crop, and the specialist looking for concise information.

We would like to thank the Editors, the organizing committee, the authors, and the reviewers for their efforts in creating this excellent publication. They volunteered their time, skill, creativity, and knowledge to create a publication of the highest quality that will well serve agriculture and those it feeds for years to come.

Robert G. Hoeft, President, 2003
American Society of Agronomy

P. Stephen Baenziger, President, 2003
Crop Science Society of America

Michael J. Singer, President, 2003
Soil Science Society of America

PREFACE

Soybean, *Glycine max* (L.) Merr., remains the world's primary source of protein feed supplement for livestock and accounts for much of the world's vegetable oil supply. During the past 15 yr much has changed in soybean production and soybean research. In the USA soybean production has increased substantially, but significant increases have also occurred in South America. Soybean hectarage has been increasing in the northern USA, especially in the western Great Plains, but has been declining in the Mid-South and southeastern USA. The emergent genetic and agronomic technologies have had a significant impact on the soybean industry. Some newly evolved production practices offer great promise. There are now better genetic and agronomic approaches for dealing with some old as well as new yield-limiting diseases and pests. In addition, we are learning much more about the benefits of soybean consumption on human health.

It is worth noting that 2003 marks the 50th anniversary of the 1953 discovery of the DNA helix by Watson and Crick. Genetic and genomic research has accelerated in the last 15 to 20 yr and has been instrumental in the development of recombinant and other DNA technologies that have revolutionized approaches to soybean improvement in yield, pest protection, and seed quality. Transgenic soybean cultivars occupy a dominant share of the production in several countries, hundreds of thousands of soybean gene transcripts have been partially cloned, and it is probable that in the next 10 yr, we may have in hand the complete DNA sequence of all of the genes in the soybean genome!

Soybean Monographs have been published about every decade and a half. The first edition was published in 1973, and the second in 1987. This 3rd edition of *Soybeans: Improvement, Production, and Uses* is organized into several sections. Chapter 1 provides the reader with recent statistics on USA and world soybean production and trade. The excellent Vegetative and Reproductive Morphology chapters of the 2nd edition are reprinted to keep this information readily accessible for the readers of this new edition of the Monograph. Progress in soybean genetic research and improvement is documented in Chapters 4 through 9. Critical updates on the research advances that have occurred in soybean management, in soybean physiology related to water, carbon, and nitrogen, and soybean seed composition are presented in Chapters 10 through 13. Chapters 14 through 18 present new developments in genetic and management approaches to deal with soybean biotic stresses (i.e., diseases, nematodes, insects, and weeds). Finally, new data and findings in areas of soybean marketing and processing, plus newly emerging information demonstrating the benefits of human soybean consumption are discussed in Chapters 19 through 21.

We would like to draw your attention to the new chapters in this edition. We are pleased to provide readers with current information on soybean genomics (Chapter 6), transgenic soybean development (Chapter 7), and a chapter dedicated to availability and utilization of soybean genetic diversity (Chapter 8). There is also a new chapter describing the impacts of transgenic soybean cultivars and identity preservation systems on soybean marketing and value creation (Chapter 19). Last, but not least, a new chapter was commissioned to cover emerging information on the impact of human soybean consumption on health and nutrition. This chapter will be of significance in terms of understanding the FDA-labeling regulations that have already been enacted, or are being considered, for soy-containing foods in the USA (Chapter 21).

One important driver of soybean research in the USA has been the Soybean Promotion and Research Order authorized by the Soybean Promotion, Research, and Consumer Information Act (7 U.S.C. 6301-6311). For more information, go to this web site http://www.ams.usda.gov/lsg/mpb/soy/soychk.htm. This Congressional act authorized the establishment of a national soybean promotion, research, and consumer information program, to be administered by the 62 members of the United Soybean Board (USB). The goal of this program is to "strengthen the position of soybeans in the marketplace and to maintain and expand domestic and foreign markets and uses for soybeans and soybean products." In so doing, USB funding has been used to stimulate, and coordinate soybean research in the USA in a variety of areas. For example, the USB recently created and funded the Better Bean Initiative, which establishes research targets and goals designed to increase the value of soybean by modifying compositional quality of soybean oil and protein. Such research on oil and protein quality and marketing will eventually bring a "Better Bean" to the consumer.

We would like to highlight numerous macro-changes that have occurred in soybean genetics and production since the publication of the 2nd edition of *Soybeans: Improvement, Production, and Uses.* The introduction of glyphosate-tolerant (Roundup Ready, Monsanto, St. Louis, MO) soybean cultivars was an epic event (i.e., transgenic resistance to a chemical) that significantly modified producer approaches to weed management, not the least of which is more rapid acceptance of conservation tillage practices. In the Mid-South, farmers have readily adopted the Early Soybean Production System (ESPS) to avoid the impact on seed yield of the regular occurrence of late-August and September droughts. The implementation of these production systems was a result of new research developments and has resulted in increases in seed yield and reduction in production costs.

Accompanying the foregoing changes in production practices were shifts in the relative importance of specific weeds, diseases, and pests. The soybean cyst nematode became more clearly recognized as the most important soybean pest as it spread into northern soybean production areas of the USA. New procedures to describe the variation in field populations of the nematode have been developed. These new procedures have created new challenges in classifying the effectiveness of resistance in specific soybean cultivars.

The area of genetic improvement of soybean has experienced many new techniques and accomplishments. Development of genetic markers has provided for the identification of the genomic location of numerous classical genes and many genetic loci conditioning quantitative traits. This research has contributed to the development and release of several private and public soybean cultivars. Other contributions in the area of genetic improvements include development and deposit in public databases of more than 300 000 soybean ESTs, a set of ~ 40 000 unigenes, and the public availability of the first soybean microarrays. These developments should produce significant advances by the publication of the fourth monograph.

In closing, we wish to express our appreciation to the Editorial Committee for this edition of the soybean monograph. This committee consisted of Drs. Jim Harper, Randy Shoemaker, Rich Wilson, and Richard Hussey. Their expertise in the various disciplines helped guide us in the development of chapter content and author selection that eventually resulted in this, the 3rd edition of the monograph. We also wish to express our appreciation and thanks to the many scientists who anonomously provided the three critical reviews of each chapter. Those peer reviews helped the co-editors and authors produce a quality product for delivery to the soybean community. In the end, however, the content of this monograph is the result of the collective efforts of the author(s) of each chapter. We thank all authors for their willingness to contribute their time, creativity, and intellect to this project, and extend our appreciation to all other persons involved in this project for their helpful comments, advice, and quick answers to our questions. Finally, it was a privilege for us to serve as co-editors of this project. The time we spent on the many aspects of this project was long, but it was nevertheless quite rewarding.

H. Roger Boerma and James E. Specht, editors

CONTRIBUTORS

D. Lee Alekel

Associate Professor of Nutrition, Food Science and Human Nutrition, 1127 Human Nutritional Sciences Building, Iowa State University, Ames, IA 50011

Mary S. Anthony

Assistant Professor of Pathology and Public Health Sciences, Health Sciences, Wake Forest University, Medical Center Boulevard, Winston-Salem, NC 27157-1040

Karen L. Bender

Principal Research Specialist, National Soybean Research Lab. and Agricultural & Consumer Economics, 170 Soybean Research Lab., University of Illinois, 1101 W. Peabody Drive, Urbana, IL 61801

Diane F. Birt

Professor and Department Chair, 2312 Food Sciences Building, Iowa State University, Ames, IA 50011-1061

H. Roger Boerma

Distinguished Research Professor, 111 Riverbend Road, Center for Applied Genetic Technologies, University of Georgia, Athens, GA 30602

David J. Boethel

Professor of Entomology, 104 J. Norman Efferson Hall, Louisiana State University Agricultural Center, Baton Rouge, LA 70803

Jason P. Bond

Assistant Professor of Plant Pathology, Plant, Soil and General Agriculture, Southern Illinois University, Agriculture Building, Room 176, 1205 Lincoln Drive, Carbondale, IL 62901

Douglas D. Buhler

Chair and Professor, Department of Crop and Soil Sciences, 286 Plant and Soil Science Building, Michigan State University, East Lansing, MI 48824

Glenn R. Buss

Professor, Crop and Soil Environmental Sciences Department, Virginia Polytechnic Institute and State University, Blacksburg, VA 24061-0404

John B. Carlson

(Reprinted chapters)

Thomas E. Carter, Jr.

Research Geneticist, USDA-ARS, 3127 Ligon Street, Raleigh, NC 27607

Thomas E. Clemente

Associate Professor, Department of Agronomy and Horticulture, Center for Biotechnology, Plant Science Initiative, N308 Beadle Center, University of Nebraska, Lincoln, NE 68588-0665

Perry B. Cregan

Supervisory Research Geneticist, USDA-ARS, Soybean Genomics and Improvement Lab., 10300 Baltimore Avenue, B-006, Room 100, BARC-West, Beltsville, MD 20705

Zhanglin Cui   Research Associate, North Carolina State University, 3127
                                Ligon Street, Raleigh, NC 27607

Brian W. Diers   Associate Professor, 1101 W. Peabody Drive, National Soy-
                                bean Research Center, University of Illinois, Urbana, IL 61801

Anne E. Dorrance   Assistant Professor, Department of Plant Pathology, The Ohio
                                State University/OARDC, 1680 Madison Avenue, Wooster,
                                OH 44691

Roger W. Elmore   Professor of Agronomy and Horticulture and Extension Crops
                                Specialist, Department of Agronomy and Horticulture, Uni-
                                versity of Nebraska, 377 I Plant Science, Lincoln, NE 68583-
                                0724

Donna K. Fisher   Assistant Professor, School of Economic Development, Geor-
                                gia Southern University, P.O. Box 8153, Statesboro, GA 30460

Craig R. Grau   Professor of Plant Pathology, Department of Plant Pathol-
                                ogy, University of Wisconsin, 1630 Linden Drive, Madison,
                                WI 53706-1598

Robert G. Hartzler   Professor, Department of Agronomy, Agronomy Hall, Iowa
                                State University, Ames, IA 50011

Larry G. Heatherly   Research Agronomist, USDA-ARS, P.O. Box 343, Stoneville,
                                MS 38776

Suzanne Hendrich   Professor, Food Science and Human Nutrition, 124 MacKay
                                Hall, Iowa State University, Ames, IA 50011

Theodore Hymowitz   Professor of Plant Genetics, Department of Crop Sciences,
                                University of Illinois, 1102 South Goodwin Avenue, Urbana,
                                IL 61801

Thomas C. Kilen   USDA-ARS, Crop Genetics and Production Research, 141 Ex-
                                periment Station Road, Stoneville, MS 38776

George H. Lacy   Professor of Plant Pathology, Department of Plant Pathol-
                                ogy, Physiology and Weed Science, Virginia Polytechnic In-
                                stitute and State University, Blacksburg, VA 24061-0330

Nels R. Lersten   (Reprinted chapters)

Edmund W. Lusas   Consultant, Ed Lusas, Problem Solvers Incorporated, 3604
                                Old Oaks Drive, Bryan, TX 77802-4743

Randall L. Nelson   Supervisory Research Geneticist, USDA-ARS, National Soy-
                                bean Research Center, University of Illinois, 1101 West
                                Peabody Drive, Urbana, IL 61801

Terry L. Niblack   Professor, Department of Crop Sciences, AW-101 Turner
                                Hall, University of Illinois, 1102 South Goodwin Avenue, Ur-
                                bana, IL 61801-4798

James H. Orf	Professor, Department of Agronomy and Plant Genetics, 411 Borlaug Hall, 1991 Upper Buford Circle, St. Paul, MN 55108
Reid G. Palmer	Research Geneticist, USDA-ARS, G301 Agronomy, Iowa State University, 100 Osborn Drive, Ames, IA 50011
Wayne E. Parrott	Professor, Center for Applied Genetic Technologies, University of Georgia, 111 Riverbend Road, Athens, GA 30602-6810
Todd W. Pfeiffer	Professor of Agronomy, Department of Agronomy, N106 Agricultural Science Building North, University of Kentucky, Lexington, KY 40546-0091
Larry C. Purcell	Professor, Department of Crop, Soil, and Environmental Sciences, University of Arkansas, 1366 West Altheimer Drive, Fayetteville, AR 72704
Robert D. Riggs	University Professor, Department of Plant Pathology, 217 Plant Science, University of Arkansas, Fayetteville, AR 72701
John S. Russin	Professor of Plant Pathology, Department of Plant, Soil and General Agriculture, Southern Illinois University, Carbondale, IL 62901
Randy C. Shoemaker	Research Geneticist, USDA-ARS, G401 Agronomy Hall, Iowa State University, Ames, IA 50011
Thomas R. Sinclair	Plant Physiologist, USDA-ARS, Agronomy Physiology Laboratory, University of Florida, P.O. Box 110965, Gainesville, FL 32611-0965
Clay H. Sneller	Associate Professor of Horticulture and Crop Science, The Ohio State University/OARDC, 1680 Madison Avenue, Wooster, OH 44691
Steven T. Sonka	Soybean Industry Chair in Agricultural Strategy and Director, National Soybean Research Laboratory, University of Illinois, 170 National Soybean Research Lab., 1101 W. Peabody Drive, Urbana, IL 61801
James E. Specht	Professor, Department of Agronomy and Horticulture, 322 Keim Hall, East Campus, University of Nebraska, Lincoln, NE 68583-0915
Sue A. Tolin	Professor of Plant Pathology, Department of Plant Pathology, Physiology and Weed Science, Virginia Polytechnic Institute and State University, Blacksburg, VA 24061-0330
Gregory L. Tylka	Professor, Department of Plant Pathology, 351 Bessey Hall, Iowa State University, Ames, IA 50011
Lila Vodkin	Professor, Department of Crop Sciences, University of Illinois, 384 ERML, 1201 W. Gregory Avenue, Urbana, IL 61801

James R. Wilcox   Professor Emeritus, Department of Agronomy, Purdue University, West Lafayette, IN 47907-1150

Richard F. Wilson   National Program Leader, Oilseeds & Bioscience, USDA-Agricultural Research Service, National Program Staff, Room 4-2214 George Washington Carver Center, 5601 Sunnyside Avenue, Beltsville, MD 20705-5139

Conversion Factors for SI and non-SI Units

Conversion Factors for SI and non-SI Units

To convert Column 1 into Column 2, multiply by	Column 1 SI Unit	Column 2 non-SI Units	To convert Column 2 into Column 1, multiply by
		Length	
0.621	kilometer, km (10^3 m)	mile, mi	1.609
1.094	meter, m	yard, yd	0.914
3.28	meter, m	foot, ft	0.304
1.0	micrometer, μm (10^{-6} m)	micron, μ	1.0
3.94×10^{-2}	millimeter, mm (10^{-3} m)	inch, in	25.4
10	nanometer, nm (10^{-9} m)	Angstrom, Å	0.1
		Area	
2.47	hectare, ha	acre	0.405
247	square kilometer, km^2 (10^3 m)^2	acre	4.05×10^{-3}
0.386	square kilometer, km^2 (10^3 m)^2	square mile, mi^2	2.590
2.47×10^{-4}	square meter, m^2	acre	4.05×10^3
10.76	square meter, m^2	square foot, ft^2	9.29×10^{-2}
1.55×10^{-3}	square millimeter, mm^2 (10^{-3} m)^2	square inch, in^2	645
		Volume	
9.73×10^{-3}	cubic meter, m^3	acre-inch	102.8
35.3	cubic meter, m^3	cubic foot, ft^3	2.83×10^{-2}
6.10×10^4	cubic meter, m^3	cubic inch, in^3	1.64×10^{-5}
2.84×10^{-2}	liter, L (10^{-3} m^3)	bushel, bu	35.24
1.057	liter, L (10^{-3} m^3)	quart (liquid), qt	0.946
3.53×10^{-2}	liter, L (10^{-3} m^3)	cubic foot, ft^3	28.3
0.265	liter, L (10^{-3} m^3)	gallon	3.78
33.78	liter, L (10^{-3} m^3)	ounce (fluid), oz	2.96×10^{-2}
2.11	liter, L (10^{-3} m^3)	pint (fluid), pt	0.473

Mass

To convert Column 1 into Column 2, multiply by	Column 1 SI Unit	Column 2 non-SI Unit	To convert Column 2 into Column 1, multiply by
2.20×10^{-3}	gram, g (10^{-3} kg)	pound, lb	454
3.52×10^{-2}	gram, g (10^{-3} kg)	ounce (avdp), oz	28.4
2.205	kilogram, kg	pound, lb	0.454
0.01	kilogram, kg	quintal (metric), q	100
1.10×10^{-3}	kilogram, kg	ton (2000 lb), ton	907
1.102	megagram, Mg (tonne)	ton (U.S.), ton	0.907
1.102	tonne, t	ton (U.S.), ton	0.907

Yield and Rate

To convert Column 1 into Column 2, multiply by	Column 1 SI Unit	Column 2 non-SI Unit	To convert Column 2 into Column 1, multiply by
0.893	kilogram per hectare, kg ha^{-1}	pound per acre, lb acre^{-1}	1.12
7.77×10^{-2}	kilogram per cubic meter, kg m^{-3}	pound per bushel, lb bu^{-1}	12.87
1.49×10^{-2}	kilogram per hectare, kg ha^{-1}	bushel per acre, 60 lb	67.19
1.59×10^{-2}	kilogram per hectare, kg ha^{-1}	bushel per acre, 56 lb	62.71
1.86×10^{-2}	kilogram per hectare, kg ha^{-1}	bushel per acre, 48 lb	53.75
0.107	liter per hectare, L ha^{-1}	gallon per acre	9.35
893	tonne per hectare, t ha^{-1}	pound per acre, lb acre^{-1}	1.12×10^{-3}
893	megagram per hectare, Mg ha^{-1}	pound per acre, lb acre^{-1}	1.12×10^{-3}
0.446	megagram per hectare, Mg ha^{-1}	ton (2000 lb) per acre, ton acre^{-1}	2.24
2.24	meter per second, m s^{-1}	mile per hour	0.447

Specific Surface

To convert Column 1 into Column 2, multiply by	Column 1 SI Unit	Column 2 non-SI Unit	To convert Column 2 into Column 1, multiply by
10	square meter per kilogram, m^2 kg^{-1}	square centimeter per gram, cm^2 g^{-1}	0.1
1000	square meter per kilogram, m^2 kg^{-1}	square millimeter per gram, mm^2 g^{-1}	0.001

Density

To convert Column 1 into Column 2, multiply by	Column 1 SI Unit	Column 2 non-SI Unit	To convert Column 2 into Column 1, multiply by
1.00	megagram per cubic meter, Mg m^{-3}	gram per cubic centimeter, g cm^{-3}	1.00

Pressure

To convert Column 1 into Column 2, multiply by	Column 1 SI Unit	Column 2 non-SI Unit	To convert Column 2 into Column 1, multiply by
9.90	megapascal, MPa (10^6 Pa)	atmosphere	0.101
10	megapascal, MPa (10^6 Pa)	bar	0.1
2.09×10^{-2}	pascal, Pa	pound per square foot, lb ft^{-2}	47.9
1.45×10^{-4}	pascal, Pa	pound per square inch, lb in^{-2}	6.90×10^3

(continued on next page)

Conversion Factors for SI and non-SI Units

To convert Column 1 into Column 2, multiply by	Column 1 SI Unit	Column 2 non-SI Units	To convert Column 2 into Column 1, multiply by
Temperature			
$1.00\ (K - 273)$	kelvin, K	Celsius, °C	$1.00\ (°C + 273)$
$(9/5\ °C) + 32$	Celsius, °C	Fahrenheit, °F	$5/9\ (°F - 32)$
Energy, Work, Quantity of Heat			
9.52×10^{-4}	joule, J	British thermal unit, Btu	1.05×10^{3}
0.239	joule, J	calorie, cal	4.19
10^{7}	joule, J	erg	10^{-7}
0.735	joule, J	foot-pound	1.36
2.387×10^{-5}	joule per square meter, J m^{-2}	calorie per square centimeter (langley)	4.19×10^{4}
10^{5}	newton, N	dyne	10^{-5}
1.43×10^{-3}	watt per square meter, W m^{-2}	calorie per square centimeter minute (irradiance), cal cm^{-2} min^{-1}	698
Transpiration and Photosynthesis			
3.60×10^{-2}	milligram per square meter second, mg m^{-2} s^{-1}	gram per square decimeter hour, g dm^{-2} h^{-1}	27.8
5.56×10^{-3}	milligram (H$_2$O) per square meter second, mg m^{-2} s^{-1}	micromole (H$_2$O) per square centimeter second, μmol cm^{-2} s^{-1}	180
10^{-4}	milligram per square meter second, mg m^{-2} s^{-1}	milligram per square centimeter second, mg cm^{-2} s^{-1}	10^{4}
35.97	milligram per square meter second, mg m^{-2} s^{-1}	milligram per square decimeter hour, mg dm^{-2} h^{-1}	2.78×10^{-2}
Plane Angle			
57.3	radian, rad	degrees (angle), °	1.75×10^{-2}

Electrical Conductivity, Electricity, and Magnetism

	Column 1 (SI Unit)	Column 2 (non-SI Unit)	
10	siemen per meter, S m^{-1}	millimho per centimeter, mmho cm^{-1}	0.1
10^4	tesla, T	gauss, G	10^{-4}

Water Measurement

	Column 1 (SI Unit)	Column 2 (non-SI Unit)	
9.73×10^{-3}	cubic meter, m^3	acre-inch, acre-in	102.8
9.81×10^{-3}	cubic meter per hour, m^3 h^{-1}	cubic foot per second, ft^3 s^{-1}	101.9
4.40	cubic meter per hour, m^3 h^{-1}	U.S. gallon per minute, gal min^{-1}	0.227
8.11	hectare meter, ha m	acre-foot, acre-ft	0.123
97.28	hectare meter, ha m	acre-inch, acre-in	1.03×10^{-2}
8.1×10^{-2}	hectare centimeter, ha cm	acre-foot, acre-ft	12.33

Concentrations

	Column 1 (SI Unit)	Column 2 (non-SI Unit)	
1	centimole per kilogram, cmol kg^{-1}	milliequivalent per 100 grams, meq 100 g^{-1}	1
0.1	gram per kilogram, g kg^{-1}	percent, %	10
1	milligram per kilogram, mg kg^{-1}	parts per million, ppm	1

Radioactivity

	Column 1 (SI Unit)	Column 2 (non-SI Unit)	
2.7×10^{-11}	becquerel, Bq	curie, Ci	3.7×10^{10}
2.7×10^{-2}	becquerel per kilogram, Bq kg^{-1}	picocurie per gram, pCi g^{-1}	37
100	gray, Gy (absorbed dose)	rad, rd	0.01
100	sievert, Sv (equivalent dose)	rem (roentgen equivalent man)	0.01

Plant Nutrient Conversion

	Elemental	Oxide	
2.29	P	P$_2$O$_5$	0.437
1.20	K	K$_2$O	0.830
1.39	Ca	CaO	0.715
1.66	Mg	MgO	0.602

1   World Distribution and Trade of Soybean

JAMES R. WILCOX

Purdue University
West Lafayette, Indiana

Soybean [*Glycine max* (L.) Merr.] is the leading oilseed crop produced and consumed in the world today. A native of Asia, the soybean was introduced into North America, Europe, then into South and Central America (Hymowitz, 2004, this publication). In each of these production areas, soybean has become a major economic crop.

Current world production of soybean far exceeds that of any other edible oilseed (Plate 1–1). The 176 million megagrams (Mg) of soybean produced in 2001 is 35% of the world total oilseed production. Oil palm (*Elaies guineensis* Jacq.)is second, with 26% of world oilseeds, and coconut (*Cocus nucifera* L.) comprises 10% of world production. Peanut (*Arachis hypogaea* L.), cottonseed (*Gossypium hirsutum* L.), and rapeseed (*Brassica napus* L.) each make up about 7% of world oilseeds. Sunflower (*Helianthus annus* L.) is about 4%, and olive (*Olea europaea* L.) about 3% of world oilseed production.

Soybean has been the dominant oilseed produced since the 1960s (Smith and Huyser, 1987). Since 1985, world soybean production has increased by 75 million Mg (Plate 1–1). During this time, oil palm production has increased 90 million Mg, rapeseed production 15 million Mg, peanut production 15 million Mg, coconut production 12 million Mg, and sunflower about 8 million Mg. Cottonseed and sesame (*Sesamum indicum* L.) seed production have remained relatively unchanged since 1985.

1–1 WORLD SOYBEAN PRODUCTION

About 50 countries in the world grow soybean (Table 1–1). During the past half century, the USA has been the world's leading producer and, in 2000/2001, produced about 77 million Mg of soybean, or 45% of the world total (Plate 1–2). The two largest producers in South America are Brazil, with 35 million Mg (21% of world total) produced on 14 million ha, and Argentina with about 23 million Mg (14% of world total) produced on 9 million ha. The People's Republic of China, with 15 million Mg, (9% of world total) and India, with 5 million Mg, are the two largest soybean producers in Asia and the Middle East. Italy, with 0.9 million Mg, was the largest soybean producer in the European Union. The Russian Federation, with 0.3, and Yugoslavia, with 0.2 million Mg, produced most of the soybean in Eastern Europe.

Table 1–1. World soybean production in specified countries, 2000/2001.†

Country	Area	Yield	Production
	1 000 ha	Mg	1 000 Mg
North and Central America			
Canada	1 035	2.29	2 372
Guatemala	10	3.06	30
Mexico	72	1.66	119
Nicaragua	6	2.22	14
United States	29 423	2.61	76 862
South America			
Argentina	9 478	2.47	23 472
Bolivia	568	1.81	1 033
Brazil	13 788	2.55	35 205
Columbia	21	2.23	47
Ecuador	69	1.67	116
Paraguay	1 193	2.75	3 283
Uruguay	10	1.59	17
Europe			
Austria	16	2.10	33
Croatia	49	1.84	90
Hungary	26	1.78	48
France	99	2.57	254
Italy	246	3.65	894
Romania	54	1.42	70
Russian Federation	373	0.83	302
Ukraine	66	1.05	68
Yugoslavia	116	1.88	200
Africa			
Congo	25	0.48	24
Ethiopia	7	3.57	25
Nigeria	588	0.73	429
Rwanda	28	0.57	16
South Africa	111	1.57	173
Uganda	116	1.13	132
Zambia	13	2.31	30
Zimbabwe	62	2.13	132
Asia and the Middle East			
Burma	69	0.90	62
Cambodia	32	1.05	33
China, People's Republic of	9 003	1.72	15 431
India	6 076	0.88	5 342
Indonesia	774	1.21	940
Iran	90	1.56	140
Japan	124	1.92	238
Korea, Democratic People's	310	1.13	350
Korea, Republic of	82	1.41	115
Myanmar	111	0.95	105
Nepal	20	0.85	17
Thailand	228	1.43	324
Turkey	25	2.13	50
Viet Nam	127	1.20	153
Oceana			
Australia	55	1.89	104
World total	74 821	2.26	168 934

† Data from Food and Agricultural Organization of the United Nations, FAO Statistical tables, Agriculture, 2002.

1–1.1 Soybean Production in the USA

Soybean is grown in the eastern half of the North America, from coastal areas of the Gulf of Mexico north to southern Canada (Plate1–3). These areas have adequate moisture to successfully produce the crop. Soybean production extends west in the USA to the Dakotas, Nebraska, Kansas, Oklahoma, and Texas. The western margins of profitable soybean production in these states have been extended with the use of irrigation.

Soybean is grown as a full-season, spring seeded crop throughout its production area in the USA. From 6 to 9% of the soybean hectarage is planted as a second crop, following winter wheat (*Triticum aestivum* L.), rice (*Oryza sativa* L.), or winter canola, south of 35°N lat. The area planted to double crop soybean increases progressively farther south of 35°N. lat. In these areas, soybean is typically seeded in late June or early July, following fall seeded crops that are harvested in mid-June to early July. Early harvesting of high-moisture winter wheat increases successful production of soybean as a second crop.

Soybean production in the USA increased by 49% since 1985, from 51 to 76 million Mg (Table 1–2). During this time, area planted to soybean increased only 14%, from 25.6 to 29.3 million ha. Seed yields for the entire country increased 28%, from 1.98 to 2.53 Mg ha^{-1}. Leading soybean-producing states are Iowa (13.0 million Mg), Illinois (12.4 million Mg), Minnesota (7.9 million Mg), and Indiana (6.9 million Mg), all in the North-Central USA.

Soybean production in all the North-Central states averaged 67 million Mg during 2000/2002, about 82% of the total for the USA (Table 1–2). This is a change from 15 yr ago when North-Central states produced 71% of the U.S. soybean crop. Since 1985, soybean production decreased in all the southern states except Arkansas and Oklahoma. The decrease in production was due primarily to a decrease in area planted to soybean, from 7.6 million ha in 1983/1985 to 3.7 million ha in 2000/2002.

Production increased in all the North-Central states, with the greatest increases in Iowa, Illinois, Minnesota, Nebraska, and South Dakota (Table 1–2). The increased production is due to increases in both area planted and yields per unit area. States with the greatest increases in area planted are South Dakota (1264 thousand ha), Iowa (1044 thousand ha), Nebraska (970 thousand ha), Minnesota (864 thousand ha), and Illinois (586 thousand ha). Those states with the greatest increases in seed yields since 1985 are Kentucky (0.72 Mg ha^{-1}), Missouri and Nebraska (0.69 Mg ha^{-1}), and Indiana and Iowa (0.65 Mg ha^{-1}).

1–1.2 Soybean Production in Brazil

Brazil is the second largest soybean producer in the world, and the largest producer in South America. Initially introduced into southern Brazil, the soybean opened the Cerrados to agriculture, and more recently contributed to an expanding agriculture in the Northeast (Portugal, 1999). Soybean was first produced on a commercial scale in the state of Rio Grande do Sul, using adapted U.S. cultivars. Production in Brazil increased from 0.5 million Mg in 1964/1965 to 31 million Mg in 1998/1999. During this time, seed yields increased from 1.20 to 2.40 Mg ha^{-1}. Initially, production was confined to the southern states of Rio Grande do Sul, Santa

Table 1–2. Area, yield, and production of soybean by states in the USA, mean of 1983/1985 and 2000/2002.†

State	Area harvested		Yield		Production	
	1983/1985	2000/2002	1983/1985	2000/2002	1983/1985	2000/2002
	——— 1 000 ha ———		——— Mg ———		——— 1 000 Mg———	
Northern states						
Delaware	100	82‡	1.86	2.33	186	193
Illinois	3 652	4 238	2.32	2.91	8 477	12 367
Indiana	1 721	2 259	2.40	3.05	4 155	6 890
Iowa	3 306	4 350	2.34	2.99	7 744	13 029
Kansas	610	1 056	1.44	1.70	870	1 814
Kentucky	557	484	1.79	2.51	986	1 216
Maryland	163	207	1.95	2.31	319	480
Michigan	447	823	2.05	2.33	912	1 913
Minnesota	2 002	2 866	2.20	2.76	4 401	7 900
Missouri	2 115	1 962	1.68	2.37	3 560	4 673
Nebraska	942	1 912	2.03	2.72	1 909	5 225
New Jersey	53	39	1.97	2.15	105	84
New York	--	60	--	2.15	--	129
North Dakota	237	858	1.70	2.23	400	1 920
Ohio	1 473	1 840	2.46	2.58	3 661	4 741
Pennsylvania	67	156	2.11	2.35	144	369
South Dakota	488	1 752	1.83	2.17	890	3 818
Wisconsin	154	606	2.20	2.67	339	1 614
Southern states						
Alabama	526	57	1.52	1.77	786	102
Arkansas	1 538	1 201	1.59	2.07	2 449	2 479
Florida	117	4	1.68	1.08	196	5
Georgia	749	59	1.46	1.64	1 082	97
Louisiana	964	299	1.67	2.02	1 626	596
Mississippi	1 203	556	1.57	1.99	1 874	1 090
North Carolina	693	540	1.55	1.94	1 077	1 051
Oklahoma	86	111	1.32	1.39	113	155
South Carolina	560	173	1.27	1.41	708	244
Tennessee	712	446	1.64	2.04	1 134	908
Texas	151	98	1.71	1.81	260	178
Virginia	275	192	1.58	2.13	442	410
West Virginia	--	6	--	1.99	--	13
USA	25 662	29 296	1.98	2.53	50 809	75 711

† Data from USDA, NASS, 2002.
‡ Values in italics indicate a decline in mean value from 1983/1985 to 2000/2002.

Catarina, Parana, and Sao Paulo. The use of U.S. cultivars, adapted to production in the southern USA, limited successful production to these areas. New cultivars were developed that were adapted to production in lower latitudes. During the 1980s, production expanded in the Cerrado area of the states of Minas Gerais, Mato Grosso do Sul, Mato Grosso, Goias, Bahia, and Maranhao.

The use of improved production technology has contributed to the expansion of profitable soybean production in Brazil. Empresa Brasileira de Pesquisa Agropecuaria (Embrapa) scientists have developed cultivars with increased yield potential, with resistance to pathogens that limited yields, and that are adapted to tropical production areas. The use of reduced tillage, correction of soil acidity, and improved seed quality have all contributed to increased productivity.

1–1.3 Soybean Production in Argentina

Argentina is the second largest producer of soybean in South America. The first planting of soybean in Argentina was in 1862 (Larreche and Firpo Brenta, 1999). During the 1970s, when soybean hectarage expanded rapidly, there was a marked increase in production. Production increased rapidly subsequent to the 1970s, from 1.4 million Mg in 1976/1977 to 19.9 million Mg in 1998/1999. Soybean cultivars in Maturity Groups III to IX, depending upon latitude, are grown from 23 to 39°S lat. Most of the production (85%) is in the Pampean provinces of Sante Fe, Cordoba, and Buenos Aires.

About 65% of the soybean produced in Argentina is grown as a full-season crop, the remaining 35% is double cropped with wheat. Most of the seeding is done using minimum or no tillage. Soybean is commonly rotated with corn (*Zea mays* L.), sorghum [*Sorghum bicolor* (L.) Moench], grassland, peanut, and less commonly with sunflower, oat (*Avena sativa* L.), and flax (*Linum usitatissinum* L.).

Factors that contributed to increased soybean production in Argentina are the end of an export ban on oilseeds in the mid-1970s, and exemptions of soybean from an export tax in the early 1990s. This gave producers access to increasing international prices. The adoption of double-cropping soybean after wheat, and the use of minimum and no-tillage production systems also contributed to increases in productivity. Investments in storage and transportation systems, including updating port structures and road and railway networks provided the infrastructure essential to successfully market the crop.

1–1.4 Soybean Production in China

China has always been the major producer of soybean in Asia. Soybean production in China increased from 7.5 million Mg in 1978 to 15 million Mg in 1999 (Lu and Wang, 1999). Yields increased from 1.06 Mg ha^{-1} in 1978 to 1.83 Mg ha^{-1} in 1999. The largest production areas in China are in the provinces of Heilongjiang, Liaoning, and Inner Mongolia. Soybean is spring seeded in these areas and grown as a full-season crop. Production in these areas is about 45% of the total for the country. About 30% of the soybean crop is produced in the provinces of Henan, Shandong, Hebei, and Anhui, where soybean is seeded in the summer as a second crop following winter wheat. About 25% of the crop is produced in southern China, south of the Yangzee River. Here, soybean may be seeded in the spring, or in the summer or fall following a rice crop.

Increases in soybean yields are attributed to improved cultivars, increased application of fertilizers, better pest control, and improved cultural practices. Mechanized tillage is practiced on about 40% of the soybean production area. Machine harvesting is done on only about 20% of the soybean area. Domestic consumption of soy foods is increasing as the standard of living of the Chinese increases.

1–1.5 Soybean Production in India

India ranks second in soybean production in Asia, and fifth in world production. The rapid expansion of soybean hectarage and increase in production occurred since the 1980s (Paroda, 1999). In 1969, only 3000 ha of soybean were grown. From 1970/1971 to 1980/1981, area increased from 0.03 to 0.61 million ha. During the next decade, area increased to 2.56 million ha, and by 1998/1999 to 6.3 million ha. Total production increased from 0.44 million Mg in 1980/1981 to the current 6 million Mg in 1998/1999. Predominant soybean-growing states are Madhya Pradesh, Maharashtra, and Rajasthan.

Initial large-scale production of soybean in India during the 1960s was based on southern U.S. cultivars, including Bragg, Lee, etc., that were adapted to production areas in India. The rapid expansion of soybean area in the 1970s was due to increased research and development, the establishment of soybean oil extraction plants in central India, and the availability of fallow land during the rainy season.

Soybean is generally grown as a rainy season crop under normal rainfall conditions. Soybean is grown as a single crop following wheat or chickpea (*Cicer arietinum* L.), or in a multiple cropping system of soybean/wheat/chickpea. Intercropping soybean with corn, sorghum, sugarcane (*Saccharum offinarum* L.), and millet [*Pennisetum americanum* (L.) K. Schum] have become profitable production systems. Soybean has replaced some less profitable crops, including sorghum and some millets, and in some areas is replacing cotton.

Of the 6 million Mg of soybean produced, 5% is used for food and feed, 10% for seed, and 85% for oil extraction. Soybean contributes about 8% of the edible oil consumed in India. Most of the meal is exported. Efforts are underway to in-

Table 1–3. Soybean exports and imports by selected areas and countries, 2-yr means, 1985/1986 to 1999/2000.

Country or area	1985/ 1986	1987/ 1988	1989/ 1990	1991/ 1992	1993/ 1994	1995/ 1996	1997/ 1998	1999/ 2000
				1 000 Mg				
Exports								
North and Central America	19 623	19 818	15 524	18 985	19 262	24 966	24 087	25 997
USA	19 472	19 614	15 328	18 746	18 819	24 400	23 380	25 171
Canada	151	204	196	239	440	565	704	824
South America	5 820	5 788	7 688	7 709	8 873	7 476	12 710	15 958
Argentina	2 774	1 740	1 831	3 774	2 669	2 302	1 666	3 594
Bolivia	8	23	74	92	130	238	208	198
Brazil	2 346	2 810	4 348	2 873	4 792	3 570	8 808	10 217
Paraguay	670	1 176	1 406	943	1 274	1 364	2 024	1 922
Europe	129	320	336	484	629	610	1 644	1 463
Belgium-Luxembourg	12	46	26	26	26	51	76	26
France	2	32	22	14	18	33	28	22
Italy	<1	8	25	44	2	6	30	10
Netherlands	100	204	234	320	462	404	1 267	1190
Romania	<1	<1	<1	<1	<1	<1	42	41
Russian Federation	--	--	--	22	42	60	75	32
Asia	1 338	1 704	1 190	978	646	367	271	375
China	1 252	1 594	1 094	884	602	284	178	208
India	8	3	1	<1	<1	1	6	42
Malaysia	2	4	6	13	14	17	19	51
Viet Nam	42	49	36	26	3	41	38	38
World total	26 913	27 634	24 746	28 162	29 452	33 434	38 762	43 836

(continued on next page)

crease the development and acceptance of soy-based food products. This would increase the domestic consumption of soybean as a rich source of protein.

1–2 WORLD TRADE IN SOYBEAN, SOYBEAN OIL, AND SOYBEAN MEAL

1–2.1 World Trade in Soybean

Soybean is a major commodity traded in world markets (Sonka et al., 2004, this publication). Of the 170 million Mg of soybean produced in 2000/2001, about 44 million Mg, or 24% of the crop enters world trade. The major soybean-producing countries are also the main exporters of soybean (Table 1–3). The USA is the leading exporter of soybean, followed by Brazil, Argentina, and Paraguay.

Table 1–3. Continued.

Country or area	1985/ 1986	1987/ 1988	1989/ 1990	1991/ 1992	1993/ 1994	1995/ 1996	1997/ 1998	1999/ 2000
				— 1 000 Mg —				
Imports								
North and Central America	1 485	1 466	1 571	2 301	2 904	3 148	4 138	4 864
Canada	196	154	233	114	132	87	189	406
Mexico	1 160	1 080	1 004	1 795	2 334	2 640	3 450	4 026
USA	--	--	--	--	143	114	222	118
South America	506	668	162	602	836	1 301	2 268	1 689
Brazil	234	273	40	378	544	908	1 139	694
Columbia	87	198	49	100	155	188	190	256
Venezuela	164	172	65	103	122	188	179	166
Europe	14 220	14 936	13 536	14 868	14 120	16 121	16 973	17 046
Belgium-Luxembourg	1 402	1 422	1 092	1 266	1 141	1 233	1 351	1 225
Finland	134	174	144	146	125	159	166	138
France	569	528	325	451	508	704	680	524
Germany	3 002	3 087	2 629	3 044	3 192	2 822	3 283	4 029
Greece	148	184	256	280	293	227	272	264
Italy	1 412	849	719	930	1 350	1 082	811	767
Netherlands	2 853	3 578	3 540	3 994	3 530	4 856	5 150	5 128
Norway	303	299	294	234	112	316	283	387
Portugal	843	870	786	735	581	802	575	613
Romania	310	372	246	199	74	81	42	8
Russian Federation	--	--	--	117	158	33	7	124
Spain	2 173	2 390	2 374	2 488	2 068	2 589	2 950	2 804
United Kingdom	592	625	692	634	686	834	1 007	837
Yugoslavia	224	254	219	100	55	68	105	48
Asia	8 751	9 375	8 950	9 804	10 692	12 080	14 938	21 138
China	1 750	2 240	1 901	2 157	2 489	3 337	5 414	9 697
Indonesia	330	376	466	684	763	677	480	1 290
Israel	432	402	374	412	523	468	547	586
Japan	4 864	4 741	4 514	4 528	4 881	4 842	4 904	4 856
Korea, Republic of	927	1 072	1 035	1 170	1 158	1 469	1 491	1 466
Malaysia	229	322	423	533	490	441	482	565
Thailand	0	16	2	96	72	311	778	1 164
Africa	55	74	92	174	268	356	412	450
World total	26 462	27 978	25 034	28 195	28 876	33 094	38 769	45 199

† Data from Food and Agricultural Organization of the United Nations, 1988 to 1998 and 2002.

Exports of soybean have changed since 1985. In 1985/1986, U.S. exports were 72% of the world total, and exports from South America represented 22% of the world total. Since then, South America has gradually increased soybean exports, and these now represent 36% of the world total. Brazil and Argentina account for 23 and 8%, respectively, of current world soybean exports. In recent years, exports from the USA have increased, but not as rapidly as those from South America, and now are 57% of the world total. Asian exports, primarily from China, have decreased from 1.3 million Mg in 1985/1986 to 0.2 million Mg in 1999/2000.

Soybean has been exported primarily to Europe and Asia. European imports have increased slightly, from 14.2 million Mt in 1985/1986 to 17.0 million Mt in 1999/2000. Imports have been relatively stable among the European countries during this period.

Imports of soybean into Asia have increased considerably since 1985, from 8.8 million Mg in 1985/1986 to 21.1 million Mg in 1999/2000. Japanese imports have been relatively stable, at about 4.9 million Mg, during this period. China has shown the greatest increase in imports of the Asian countries. Chinese imports increased from 1.8 million Mg in 1985/1986 to 9.7 million Mt in 1999/2000. During this same period, Chinese soybean exports decreased from 1.2 to 0.2 million Mg. Thailand became a significant importer of soybean, increasing imports from 0 to 1.2 million Mg since 1985.

Mexico has become the largest soybean importer in the Americas, with imports increasing from 1.2 to 4.0 million Mg from 1985/1986 to 1999/2000.

World trade in soybean has been affected by the widespread production of transgenic soybean cultivars (Parrott and Clemente, 2004, this publication). Currently, about 68% of the U.S. soybean hectarage is planted to soybean resistant to glyphosate, a herbicide with wide spectrum weed control. Glyphosate resistance was incorporated into soybean using genetic transformation, and this genetic modification resulted in widespread concerns about the nutritional value of transgenic soybean cultivars. Most of the soybean grown in the USA and Argentina are glyphosate resistant, and the decrease in exports from these countries in 1997/1998 may be a reflection of concerns about consuming transgenic soybean (Table 1–3.). Current trends indicate that the reluctance to import and consume transgenic soybean with glyphosate resistance have been alleviated.

1–2.2 World Trade in Soybean Oil and Meal

Argentina and Brazil are the two largest exporters of soybean oil, followed by the USA (Table 1–4). Asian countries are the largest importers of soybean oil, with China, India, and Iran the major importers. European countries, with alternative sources of edible oil, imported much less soybean oil than Asian countries. African countries, including Egypt, Morocco, and Tunisia, have increased soybean oil imports significantly over the past 15 yr.

The major soybean-producing countries are the major exporters of soybean meal (Table 1–5). North and South America combined, export nearly 80% of the world's soybean meal. Europe, Germany, and the Netherlands, major importers of soybean, are the leading exporters of soybean meal. India has replaced China as the primary exporter of soybean meal in Asia.

Europe is the major importer of soybean meal, utilizing about half of the world's imports. France has consistently been the largest European importer over the past 15 yr. Other large soybean meal importers in Europe include Belgium-Luxembourg, Denmark, Germany, Italy, and Spain. Soybean meal in Europe is used primarily in livestock rations.

Asia is second to Europe as the largest soybean meal importer. China and Thailand are the largest soybean meal importers in Asia. Indonesia, Japan, and the Republic of Korea, the Philippines, and Thailand have each increased imports to about 1000 Mg of soybean meal annually. Canada and Mexico are the major importers of soybean meal in North and Central America.

1–3 WORLD PRODUCTION TRENDS

Modest increases in soybean production can be expected in the USA over the next 10 yr. Eighty-seven percent of the U.S. crop is produced in the North-Central states. In this area, corn and soybean are typically alternated in a 2-yr rotation on the same land. Because corn is an important grain crop in the North-Central states, it is not anticipated that soybean will replace corn, but that the corn-soybean sequence will continue. Area planted to soybean in the North- Central states was about 21 million ha, slightly less than the 22 million ha planted to corn. This suggests only a 5% potential increase for soybean acreage in the corn-soybean rotation. There are limited opportunities for soybean to replace other major crops in this area.

Area planted to soybean in the southern USA has decreased about 3 million ha since 1985. If this area were again planted to soybean, it would represent about a 10% increase in total U.S. soybean area of production.

Area of soybean production in China is expected to increase very slightly, to about 9 million ha over the next 10 to 20 yr (Lu and Wang, 1999). Land resources that could be committed to soybean are extremely limited. Yields per hectare are anticipated to increase to 2.00 to 2.40 Mg ha^{-1}, for a total production of about 20 million Mg. Soybean production will be consumed domestically, to meet increasing demand for soy foods and livestock feeds.

Soybean area in Argentina is expected to increase to 8 million ha by 2010 (Larreche and Firpo Brenta, 1999). Use of improved cultural practices could increase yields to an estimated 3.10 Mg ha^{-1} by the year 2005. It is anticipated that much of the increased production will be exported.

Brazil has the greatest potential in the world for increased soybean production (Portugal, 1999). The Cerrados, or savannas, has a potential 80 million ha of land that can be used for grain production. Successful expansion of soybean into this area is dependent upon several factors. These include (i) growth in national income in major importing countries, (ii) the construction of crushing plants, and location of these plants, (iii) availability and price of competing vegetable oil products, (iv) the production in, and government policies of, major competing countries, and (v) improvement of transportation and storage facilities, with decreased costs. Currently, most of the Brazilian soybean crop is exported, and this trend is expected to continue.

Table 1–4. Soybean oil exports and imports by selected areas and countries, 2-yr means, 1985/1986 to 1999/2000.†

Country or area	1985/ 1986	1987/ 1988	1989/ 1990	1991/ 1992	1993/ 1994	1995/ 1996	1997/ 1998	1999/ 2000
				1 000 Mg				
Exports								
North and Central America	564	759	629	609	788	865	1 267	794
USA	558	756	622	602	748	799	1 218	732
Canada	6	2	2	4	17	28	33	38
South America	1 306	1 718	1 764	1 961	2 692	3 106	3 542	4 550
Argentina	622	868	903	1 283	1 437	1 403	2 110	2 998
Bolivia	<1	1	6	8	19	56	92	128
Brazil	676	834	843	616	1 140	1 548	1 243	1 312
Paraguay	10	15	14	54	94	97	96	95
Europe	1 315	1 302	1 199	1 229	982	1 216	1 570	1 670
Belgium-Luxembourg	180	163	159	138	110	106	179	114
France	95	66	66	98	62	99	91	60
Germany	212	212	176	189	188	205	358	454
Greece	38	33	24	17	18	12	12	16
Italy	78	53	74	121	60	25	28	33
Netherlands	304	353	418	426	356	422	479	492
Portugal	90	36	19	30	22	73	63	43
Spain	288	330	218	178	114	219	270	214
Asia	61	184	161	132	236	445	1 024	605
China	1	6	17	8	46	113	373	46
Indonesia	--	--	31	23	32	6	14	<1
Iran	--	--	--	--	8	108	118	218
Malaysia	28	62	34	38	68	107	152	158
Singapore	19	98	65	50	48	54	33	32
World total	3 248	3 966	3 755	3 938	4 702	5 638	7 408	7 632

(continued on next page)

The area planted to soybean is expanding in India, and is expected to increase to 10 million ha by the year 2010 (Parada, 1999). Yields are expected to increase from the current 1.25 to 1.5 Mg ha^{-1} by 2010. Soybean oil is largely consumed domestically, with the meal exported. Domestic consumption of soybean and soy products is likely to increase during the next 10 yr.

1–4 POTENTIAL CHANGES IN SEED YIELDS

World average soybean yields were 1.83 Mg ha^{-1} for the period 1983/1985 and 2.26 Mg ha^{-1} for the period 1997/1999. This is a yield increase of 0.425 Mg ha^{-1} over a 15-yr period, or 0.028 Mg ha^{-1} yr^{-1}. If this rate of yield increase can be sustained, it would make a major contribution to world soybean production over the next 10 yr.

Estimates of U.S. seed yield increases for soybean during the past 60 yr provide optimistic evidence for continued increases in productivity for the future. Specht et al. (1999) determined that soybean yields in the USA have increased 0.0232 Mg ha^{-1} yr^{-1} from 1924 to 1997. Rates of yield improvement of publicly developed elite

Table 1–4. Continued.

Country or area	1985/ 1986	1987/ 1988	1989/ 1990	1991/ 1992	1993/ 1994	1995/ 1996	1997/ 1998	1999/ 2000
					1 000 Mg			
Imports								
North and Central America	190	336	220	277	321	388	430	480
Dominican Rep.	42	62	76	85	81	82	95	106
Mexico	43	57	68	61	82	73	101	110
USA	21	156	15	1	36	56	28	41
South America	413	310	295	440	651	627	774	708
Brazil	152	52	30	95	200	186	187	132
Chile	56	32	56	69	97	93	77	48
Columbia	56	43	37	32	71	103	117	144
Ecuador	29	18	27	37	36	43	63	61
Peru	43	74	45	80	108	74	162	99
Venezuela	68	84	94	119	133	125	159	216
Europe	733	718	731	862	722	626	867	1 050
Belgium-Luxembourg	59	66	87	105	78	83	135	226
Denmark	37	40	45	50	32	36	37	30
France	85	73	72	74	69	65	64	44
Germany	144	138	148	142	126	103	45	50
Netherlands	39	35	36	25	19	21	143	124
Poland	46	52	41	52	93	74	107	86
Russian Federation	--	--	--	224	30	22	76	283
Sweden	66	72	82	66	52	37	29	27
United Kingdom	127	168	106	145	79	72	82	28
Asia	1 356	2 014	2 013	1 540	2 032	3 278	3 921	3 962
Bangladesh	60	226	186	262	145	426	484	670
China	108	279	478	278	579	1 398	1 063	604
Egypt	18	16	<1	<1	39	94	66	176
India	303	348	28	42	34	61	243	596
Iran	430	295	600	334	542	478	378	760
Korea, Republic of	<1	<1	5	34	22	49	61	136
Malaysia	2	56	37	27	24	50	118	98
Pakistan	217	375	364	216	222	199	222	267
Singapore	24	109	36	48	58	69	44	53
Turkey	92	164	145	137	185	124	159	163
Africa	315	334	337	420	722	666	669	940
Algeria	<1	1	3	55	178	59	41	12
Egypt	18	19	1	--	39	94	66	176
Morocco	96	98	109	119	168	157	142	272
Senegal	8	1	<1	1	33	32	83	90
Tunisia	52	63	86	81	129	126	131	134
World total	3 218	3 884	3 848	3 678	4 416	5 642	6 706	7 178

† Data from Food and Agricultural Organization of the United Nations, 1988 to 1998 and 2002.

breeding lines in the northern USA averaged 0.030 Mg ha^{-1} yr^{-1}, or about 1% per year over a 60-yr period (Wilcox, 2001). Data from various studies demonstrated that rates of yield increases have been greater in recent than in earlier years, indicating no yield plateaus have been reached in the genetic improvement of yield potential (Specht et al., 1999; Voldeng et al.,1997; Wilcox, 2001).

Table 1–5. Soybean meal exports and imports by selected areas and countries 2-yr means 1985/1986 to 1999/2000.†

Country or area	1985/ 1986	1987/ 1988	1989/ 1990	1991/ 1992	1993/ 1994	1995/ 1996	1997/ 1998	1999/ 2000
					1 000 Mg			
Exports								
North America	5 339	6 156	4 513	5 550	4 631	5 200	7 130	6 126
USA	5 337	6 127	4 468	5 473	4 549	5 099	7 041	6 030
Canada	1	11	1	28	25	45	63	61
South America	10 860	12 369	14 385	14 555	17 035	18 728	20 828	23 922
Argentina	2 974	4 294	4 761	6 251	6 646	6 656	9 732	13 010
Bolivia	17	23	68	99	110	267	463	590
Brazil	7 782	7 965	9 027	8 019	10 029	11 413	10 230	9 910
Paraguay	82	78	62	177	250	370	400	389
Europe	4 406	4 303	3 935	3 532	4 306	4 262	5 130	5 475
Belgium-Luxembourg	1 033	844	786	696	907	793	989	966
France	12	16	12	9	8	12	33	44
Germany	1 148	1 300	869	1 074	1 078	968	1 376	1 218
Italy	128	100	75	84	42	24	26	140
Netherlands	1 541	1 703	1 733	1 918	2 038	2 138	2 267	2 550
Norway	151	147	132	135	67	177	148	152
Portugal	164	101	99	96	27	18	13	54
Spain	193	31	15	12	24	44	130	145
Asia	1 624	2 704	2 980	3 178	3 131	3 329	2 999	2 562
China	1 045	2 060	1 885	1 508	758	485	19	21
India	531	592	854	1 645	2 343	2 787	2 852	2 328
World total	22 274	25 194	26 220	27 460	29 186	31 588	35 114	38 108
Imports								
North and Central America	1 085	1 217	1 364	1 628	1 722	1 819	1 746	2 008
Canada	586	652	571	621	688	748	744	819
Cuba	202	173	223	186	227	229	212	176
Dominican Republic	64	89	101	146	215	240	254	330
Guatemala	30	47	42	44	56	93	114	144
Mexico	70	137	303	402	293	216	79	225
South America	609	1 032	505	803	946	1 359	1 904	2 136
Brazil	--	--	--	--	<1	64	238	88
Chile	29	53	52	117	127	203	318	462
Columbia	4	6	1	54	197	349	436	426
Ecuador	--	1	2	8	4	37	139	140
Peru	72	153	75	93	156	225	277	442
Venezuela	504	814	374	525	435	447	452	526

(continued on next page)

1–5 SUMMARY

Soybean is the major oilseed produced and consumed in the world today. The USA is currently the largest producer of soybean, followed by Brazil, Argentina, the People's Republic of China, and India. Production in the USA is primarily in the North-Central states, where the highest seed yields are attained. The largest producing countries export the greatest amounts of soybean, meal, and oil. Europe and Asia are the major importers of soybean, soy meal, and soy oil.

Production potential for soybean in the future should easily meet demands for soy products. Brazil is the country with the greatest potential for expanding area

Table 1–5. Continued.

Country or area	1985/ 1986	1987/ 1988	1989/ 1990	1991/ 1992	1993/ 1994	1995/ 1996	1997/ 1998	1999/ 2000
				1 000 Mg				
Imports								
Europe	17 856	17 059	16 208	16 073	18 440	17 848	17 850	21 625
Austria	475	467	447	461	455	443	493	474
Belgium-Luxembourg	884	721	811	731	1 071	1 025	1 122	1 320
Bulgaria	390	487	319	123	68	74	52	54
Czechoslovak	413	455	264	400	300	395	404	438
Denmark	1 317	1 318	1 254	1 323	1 432	1 459	1 454	1 554
France	3 616	3 322	3 232	3 492	3 619	3 482	3 550	4 082
Germany	3 367	2 928	2 641	2 194	1 951	1 669	1 814	1 904
Hungary	610	636	649	405	446	527	608	676
Ireland	189	179	182	205	292	279	311	282
Italy	1 396	1 408	1 248	1 425	1 695	1 857	1 849	2 220
Netherlands	1 571	1 198	1 085	852	1 251	1 161	572	1 610
Poland	898	1 051	755	480	506	793	842	868
Romania	64	50	32	183	203	214	131	95
Russian Federation	--	--	--	--	496	92	118	284
Spain	1 115	1 134	1 324	1 653	1 927	1 707	1 742	2 474
Sweden	136	166	150	114	192	292	279	333
United Kingdom	1 205	1 221	1 126	1 292	1 456	1 497	1 354	1 452
Asia	2 116	2 983	3 313	4 746	5 588	7 726	11 022	9 558
China	4	8	47	130	130	973	3 607	587
Indonesia	241	165	60	182	430	812	769	1 084
Iran	243	306	338	481	457	653	446	440
Japan	182	405	542	870	860	795	839	812
Korea Rep.	117	319	466	562	654	1 070	831	1 142
Malaysia	164	157	141	251	383	582	533	584
Philippines	295	457	581	635	739	665	942	806
Saudi Arabia	170	271	248	312	362	432	542	524
Thailand	349	233	256	531	751	740	1 230	1 315
Turkey	2	42	66	235	240	282	378	530
Africa	815	941	1 023	900	1 026	1 040	1 540	2 040
Algeria	229	363	408	288	242	216	222	332
Egypt	280	264	218	294	415	372	590	858
South Africa	53	42	126	73	91	185	359	394
Tunisia	53	70	125	134	147	147	219	250
World total	22 902	26 189	25 814	27 089	27 864	30 007	34 211	37 461

†Data from Food and Agricultural Organization of the United Nations, 1988 to 1998 and 2002.

of soybean production. Seed yields of soybean have increased about 1% per year in major soybean- producing countries. There is no indication that this rate of increase will decrease in the future.

REFERENCES

Food and Agricultural Organization of the United Nations, FAO Trade Yearbooks 1988 to 1998. vol. 41–52.

Food and Agricultural Organizations of the United Nations, FAO statistical tables, agriculture, 2002. http://apps.fao.org/page/collections?subset=agriculture (verified 17 Dec. 2002).

Hymowitz, T. 2004. Speciation and cytogenetics. p. 97–136. In H.R. Boerma and J.E. Specht (ed.) Soybeans: Improvement, production, and uses. 3rd ed. Agron. Monogr. 16. ASA, CSSA, and SSSA, Madison, WI.

Larreche, H.J., and L.M. Firpo Brenta. 1999. State of the soybean industry in Argentina. p. 5–13. *In* H.E. Kauffmann (ed.) Proc. World Soybean Res. Conf. VI, Chicago, IL. 4–7 Aug. 1999. Superior Printing, Champaign, IL.

Lu, M., and L. Wang. 1999. State of the soybean industry in the People's Republic of China. p. 1–5 *In* H.E. Kauffmann (ed.) Proc. World Soybean Res. Conf. VI., Chicago, IL. 4–7 Aug. 1999. Superior Printing, Champaign, IL.

Paroda, S. 1999. Status of soybean research and development in India. p. 13–23. *In* H.E. Kauffmann (ed.) Proc. World Soybean Res. Conf. VI., Chicago, IL. 4–7 Aug. 1999. Superior Printing, Champaign, IL.

Parrott, W.A., and T.E. Clemente. 2004. Transgenic soybean. p. 265–302. *In* H.R. Boerma and J.E. Specht (ed.) Soybeans: Improvement, production, and uses. 3rd ed. Agron. Monogr. 16. ASA, CSSA, and SSSA, Madison, WI.

Portugal, A.D. 1999. State of the soybean agribusiness in Brazil. p. 37–45. *In* H.E. Kauffmann (ed.) Proc. World Soybean Res. Conf. VI., Chicago, IL. 4–7 Aug. 1999. Superior Printing, Champaign, IL.

Smith, K., and W. Huyser. 1987. World distribution and significance of soybean. p. 1–22. *In* J.R. Wilcox (ed.) Soybeans: Improvement, production, and uses. 2nd ed. Agron. Monogr. 16. ASA, CSSA, and SSSA, Madison, WI.

Sonka, S.T., K.L. Bender, and D.K. Fisher. 2004. Economics and marketing. p. 919–948. *In* H.R. Boerma and J.E. Specht (ed.) Soybeans: Improvement, production, and uses. 3rd ed. Agron. Monogr. 16. ASA, CSSA, and SSSA, Madison, WI.

Specht, J.E., D.J. Hume, and S.V. Kumudini. 1999. Soybean yield potential a genetic and physiological perspective. Crop Sci. 39:1560–1570.

United States Department of Agriculture, National Agricultural Statistics Service. 2002. http://www.nass.usda.gov.81/ipedb/ (verified 17 Dec. 2002).

Voldeng, H.D., E.R. Cober, D.J. Hume, C. Gillard, and M.J. Morrison. 1997. Fifty-eight years of genetic improvement of short-season soybean cultivars. Crop Sci. 37:428–431.

Wilcox, J.R. 2001. Sixty years of improvement in publicly developed elite soybean lines. Crop Sci. 41:1711–1716.

2

Vegetative Morphology[1]

NELS R. LERSTEN

JOHN B. CARLSON

The soybean [*Glycine max* (L.) Merr.] is an annual plant 75 to 125 cm in height (Shibles et al., 1974). It may be sparsely or densely branched, depending on cultivar and growing conditions. First-order branching of the main stem is most common; second-order branching is rare (Dzikowski, 1936). Intrinsic genetic factors and environmental effects (e.g., daylength, spacing, and soil fertility) affect branching. The possible branching variations are shown in Fig. 2–1.

The root system is best described as diffuse. It consists of a taproot, which usually cannot be distinguished from other roots of similar diameter, and a large number of secondary roots which in turn support several orders of smaller roots. In addition, multibranched adventitious roots emerge from the lower portion of the hypocotyl. The first bacterial root nodules are visible about 10 d after planting and at maturity the root system is extensively nodulated (Fig. 2–2).

The horizontal and vertical extent of the root system varies depending on cultural conditions. The taproot may reach a depth of 200 cm, and the side roots a length of 250 cm, in plants grown singly in an open field (Dzikowski, 1936), but the root system is less extensive under typical competitive conditions in field plantings. Mitchell and Russell (1971) and Raper and Barber (1970) showed that soybean grown under normal field conditions lacks a distinct taproot. Most lateral roots emerge from the upper 10 to 15 cm of the taproot and remain more-or-less horizontal but some extend obliquely to a depth of 40 to 75 cm, then turn steeply downward, sometimes reaching 180 cm. In the confined area of a rhizotron compartment, varieties differed in rates of growth, but all produced roots capable of reaching the bottom (217 cm) of each compartment (Kaspar et al., 1978). Under almost all conditions, however, most of the roots remained in the upper 15 cm of soil.

There are disadvantages to all extant methods of estimating the total root system (Bohm et al., 1975). An accurate three-dimensional reconstruction of the root system as it exists in the soil is perhaps impossible to attain.

---

[1] This chapter is reprinted from Chapter 3 of *Soybeans: Improvement, Production, and Uses*, 2nd ed. 1987.

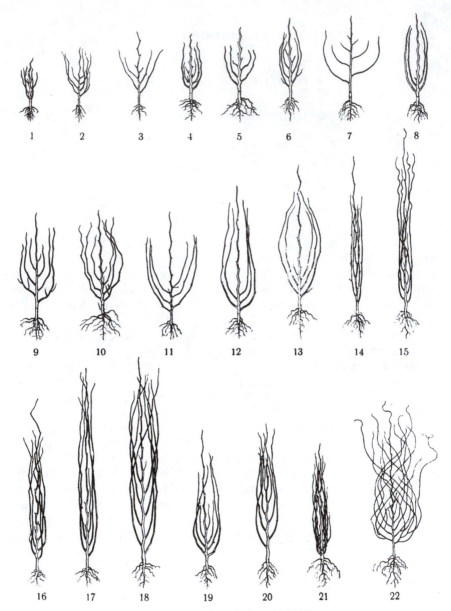

Fig. 2–1. Patterns of branching of the soybean. From Dzikowski (1936).

2–1 LEAF

2–1.1 Gross Morphology

There are four different types of soybean leaves: the first pair of simple cotyledons or seed leaves, the second pair of simple primary leaves, trifoliolate foliage leaves, and the prophylls. Each of the oppositely arranged pair of simple pri-

Fig. 2–2. Portion of mature soybean root system bearing bacterial nodules.

mary leaves is ovate in outline form, with petioles 1 to 2 cm in length and a pair of stipules at its point of attachment to the stem. These leaves occur at the first node above the cotyledons. All leaves produced subsequently are trifoliolate and arranged alternately in two opposite rows (i.e., distichously) (Sun, 1957a). Although individual leaves do not show any preferred planes of orientation, leaves of some cultivars tend collectively to be somewhat more vertically oriented than those of other cultivars (Blad and Baker, 1972).

Leaflets of trioliolate foliage leaves have entire margins and vary in outline from oblong to ovate to lanceolate (Dzikowski, 1936) (Fig. 2–3). Occasionally, four to seven leaflets may occur or lateral leaflets may fuse with the terminal leaflet (Fig. 2–3g and 3h) (Williams, 1950). Leaflets vary from 4 to 20 cm in length and 3 to 10 cm in width. Petiolules of lateral leaflets are 1 cm or less in length, considerably shorter than those of the terminal leaflet. The terminal leaflet has two small subtending stipels, each lateral leaflet a single one. At the base of the petiole there is a pair of stipules. These small, leaflike structures have parallel venation consisting usually of about seven main veins alternating with smaller veins (Fig. 2–4).

A pulvinus occurs at the base of the petiole. Another, but smaller, pulvinus is located at the base of the petiolule of each leaflet. The pulvini act as hinges, allowing an entire leaf and individual leaflets to move. Such movements are diurnal, occurring in response to changes in pulvinar osmotic pressure (Dzikowski,1936, 1937).

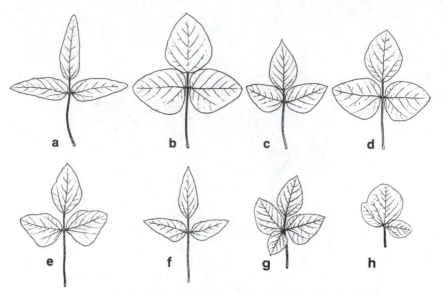

Fig. 2–3. Leaf variation in soybean, including extra leaflet (g) and fused leaflets (h). Stipules and stipels have been omitted. From Dzikowski (1936).

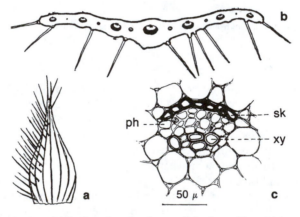

Fig. 2–4. Stipule structure. (a) Surface view showing parallel veins and long trichomes. (b) Transection. (c) Enlarged transection of one stipule bundle. Abbreviations: ph—phloem, sk—sclerenchyma, and xy—xylem. From Dzikowski (1937).

The fourth type of leaf, the prophyll, occurs as a first tiny (rarely over 1 mm long) pair of simple leaves at the base of each lateral branch. Prophylls lack petioles and pulvini.

2–1.2 Petiole and Petiolules

Three vascular traces extend from the stem into the petiole at each node (Watari, 1934; Crafts, 1967). Dzikowski (1937) showed that they immediately

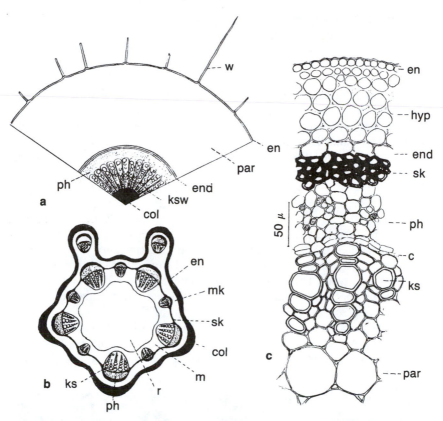

Fig. 2–5. Petiole anatomy. (a) Transection of pulvinus (one sector). (b) Petiole transection. (c) Segment of petiole transection from epidermis to pith. Abbreviations: c—cambium, col—collenchyma, end—endodermis, ep—epidermis, hyp—hypodermis, ks—xylem, ksw—secondary xylem, m—interfascicular parenchyma, mk—cortical parenchyma, par—parenchyma, ph—phloem, r—pith, sk—sclerenchyma, and w—trichomes. From Dzikowski (1937).

merge into one concentric vascular bundle upon entering the basal pulvinus (Fig. 2–5a), which is an enlarged, cushion-like zone with an extensive cortex composed of parenchyma. The vascular tissue in the pulvinus consists of an outer cylinder of phloem and an inner dissected ring of xylem. Several parenchyma rays divide the xylem into wedge-shaped segments, as seen in a transectional view. The pith of the pulvinus appears similar to collenchyma tissue.

Just beyond the pulvinus, vascular tissue expands rather abruptly (Fig. 2–5b) to an eustele consisting of five large vascular bundles (alternating with five small bundles (Dzikowski, 1937; Fisher, 1975). Two prominent vasculated ridges extend along each side of the adaxial surface of the petiole. The single vascular bundle of each ridge becomes part of the vascular tissue of the stipels at the base of each leaflet. Watari (1934) described complex vascular interconnections at the juncture of the three petiolules, beyond which about three bundles extended into each leaflet. According to Bostrack and Struckmeyer (1953), however, seven bundles enter the ter-

minal leaflet; two supply the pair of stipels, the others merge in the leaflet pulvinus and are reduced to one strand.

2–1.3 Epidermis

The leaf epidermis consists mostly of flat, tabular cells with slightly thickened radial walls. A thin cuticle covers all epidermal surfaces (Williams, 1950). Epi-
cuticular wax is a sparse layer of irregular rod-like particles except over guard cells
and some of the immediately adjacent cells (Fig. 2–6) (Flores and Espinoza, 1977).
Dzikowski (1937) noted that upper epidermal cells are somewhat larger than lower
epidermal cells, and the latter also have fewer convolutions or infoldings of the radial
walls.

Stomata occur on both epidermal surfaces (Fig. 2–14), but they are more abundant abaxially (Fig. 2–6). Unpublished data by Carlson indicate that terminal
leaflets of trifóliolate leaves of 'Ottawa-Mandarin' soybean have about three times
as many stomata (17 000 stomata cm^{-2}) on the lower epidermis as on the upper (5400
stomata cm^{-2}). This agrees with data of Ciha and Brun (1975) from field-grown
plants of 43 genotypes. They found variation among genotypes, but overall there
was a mean number of 130 stomata mm^{-2} adaxially and 316 stomata mm^{-2} abaxially.

When closed, the two guard cells are about 12 μm wide and 24 μm long. When
fully opened, the total width, including pore and guard cells, is about 16 μm. The
stomatal pore is, therefore, about 4 μm, similar to measurements from photomi-

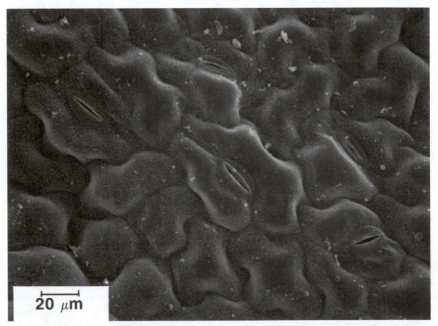

Fig. 2–6. Abaxial leaf epidermis showing stomata. Scanning electron micrograph from Flores and Espinoza (1977).

crographs in Flores and Espinoza (1977). There are no obvious subsidiary cells, although cells immediately adjacent to the guard cells may be more swollen and somewhat different in shape than ordinary epidermal cells (Fig. 2–6).

Trichomes vary in size, density, and color on aerial parts among soybean cultivars commonly grown in the USA. Glabrous strains occur in Japan, but none are grown commercially in North America (Bernard and Singh, 1969). Pubescent cultivars have closely packed uniseriate trichomes on young leaves and the outer surface of stipules (Fig. 2–4, 2–7, 2–10). Woolley (1964) found that such hairs on the upper surface of mature 'Hawkeye' soybean leaves were about 1 mm in length and spaced about 1 mm apart, accounting for perhaps 10% of the total leaf surface. Each hair consists of a long distal cell, 0.5 to 1.5 mm in length, and a short basal cell (Fig. 2–8) arising from a modified epidermal cell which is surrounded by a cushion of epidermal cells (Fig. 2–9). The hairs are slanted slightly toward the tip and edges of the leaflet. Mature hairs dry out and become air-filled or flattened (Dzikowski. 1937).

In addition to these elongate uniseriate trichomes, small five-celled, club-shaped trichomes (Fig. 2–10 and 2–11) are abundant on all young organs (see Fig. 3 of Ali and Fletcher, 1970). These trichomes persist but gradually senesce on the mature leaf. Franceschi and Giaquinta (1983a) reported that the cuticle over the dis-

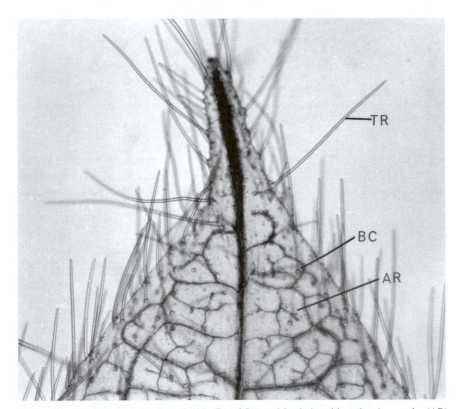

Fig. 2–7. Cleared, unstained tip of terminal leaflet of Ottawa-Mandarin cultivar showing areoles (AR) and trichomes (TR) with basal cells (BC). × 300.

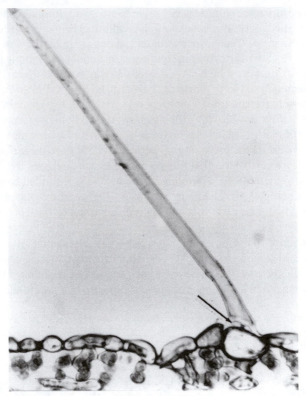

Fig. 2–8. Longisection of trichome showing long terminal cell and a short basal cell (*arrow*) attached to a somewhat enlarged epidermal cell. Compare this with Fig. 2–9.

tal two cells becomes distended, indicating that a secretory product accumulates beneath it. They speculated from ultrastructural evidence that a volatile terpenoid compound is secreted which helps to protect developing leaflets against foraging insects.

2–1.4 Mesophyll

The mesophyll of a soybean leaf consists of five to six interior cell layers, except the vascular bundles (Fig. 2–12 and 2–14). The two layers just below the adaxial epidermis are palisade mesophyll, consisting of columnar cells each containing 15 to 30 chloroplasts (Duane Ford, personal communication, 1983). More insolated upper leaves may develop a third palisade layer (Lugg and Sinclair, 1980). Where stomata occur, one to three palisade cells are lacking, thus forming a substomatal chamber (Fig. 2–14). Spongy mesophyll consists of two or three cell layers interior to the lower epidermis. The cells are irregularly lobed, with conspicuous intercellular spaces and large substomatal chambers (Fig. 2–12 and 2–14), and have fewer chloroplasts than palisade cells.

Lying between palisade and spongy mesophyll is a single layer of flat, horizontally lobed cells at the level of the phloem (Fig. 2–12 and 2–13). This paraveinal

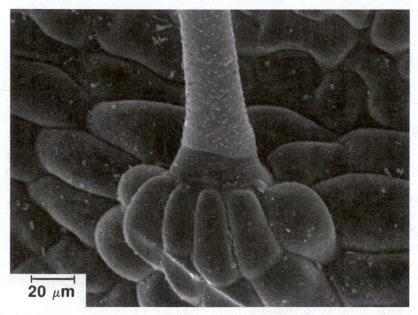

20 μm

Fig. 2–9. Scanning electron micrograph of lower part of setaceous trichome. Short basal cell has thinner cuticle than distal cell. Pillow-like epidermal cells encircle trichome base. From Flores and Espinoza (1977).

mesophyll layer collects and conducts photosynthates laterally to the phloem (Fisher, 1967). It is not restricted to soybean; a similarly modified mesophyll layer occurs in other Phaseoleae and certain other legume tribes, and in a few other families (Lackey, 1978).

Paraveinal mesophyll in soybean is the first leaf tissue to differentiate; mature cells are six to eight times larger than either palisade or spongy mesophyll cells (Franceschi and Giaquinta, 1983b, 1983c). During vegetative growth, paraveinal mesophyll cells have only a few starch-free chloroplasts. The large vacuole in each cell stores a glycoprotein that gradually diminishes during the first 14 d of seed filling.

2–1.5 Vascular Architecture

Each leaflet is richly vasculated (Fig. 2–15). As many as six orders of veins have been recognized (Bán et al., 1981). The midvein in the midrib develops a slight amount of secondary vascular tissue, and there is usually a small accessory bundle adaxial to it (Fig. 2–16).

Second order bundles depart alternately to each side from the midvein and extend as conspicuous lateral veins to the margin, where they form loops that usually connect to the next distal secondary bundle (Fig. 2–15). The progressively smaller orders of vascular bundles form an extensive minor vein retriculum and vein endings between second order veins (Fig. 2–17). All vascular bundles in a leaf are

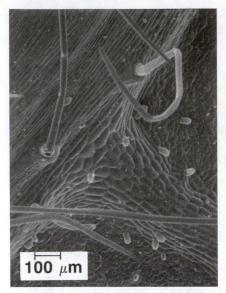

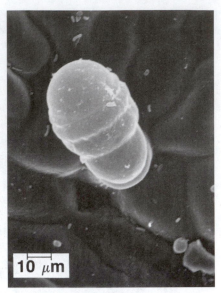

Fig. 2–10. Scanning electron micrograph of abaxial leaf surface at junction of secondary vein (*lower right*) and midvein. Long setaceous hairs and numerous small clavate hairs are present. From Flores and Espinoza (1977).

Fig. 2–11. Scanning electron micrograph of one clavate hair. Outlines of five cells can be seen. From Flores and Espinoza (1977).

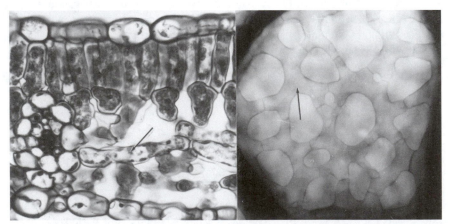

Fig. 2–12. Transection of soybean leaflet. Centrally located minor vein flanked by paraveinal mesophyll (PVM) (*arrow*). Palisade mesophyll is above PVM, spongy mesophyll is below.

Fig. 2–13. Cleared leaflet with focus on paraveinal mesophyll, which consists of flat, lobed cells (*arrow*) with large intercellular spaces.

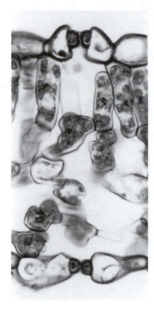

Fig. 2–14. Portion of leaf transection to show stomata on both leaf surfaces. Note substomatal chamber beneath each pair of guard cells.

1 cm

Fig. 2–15. Central leaflet with all veins included. From Bán et al. (1981).

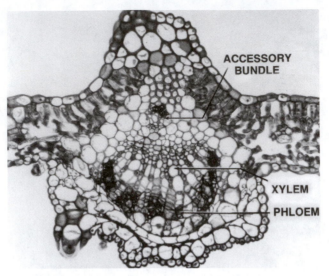

Fig. 2–16. Transection of leaflet midrib.

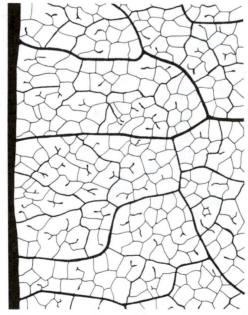

Fig. 2–17. Enlarged view of a small portion of Fig. 3–15 showing minor venation and vein endings. From
   Bán et al. (1981).

collateral, with adaxial xylem and abaxial phloem. The accessory bundle in the
midrib, however, is inverted.

The larger veins are flanked by sclerenchyma; parenchymatous bundle sheath
extensions connect these bundles above and below with each epidermis. Twin pris-

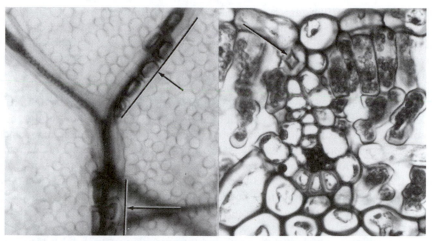

Fig. 2–18. Portion of a cleared leaflet with focus on adaxial bundle sheath cells. Twin prismatic crystals occur in some cells (*arrows*) but not others.

Fig. 2–19. Leaflet transection showing vascular bundle with bundle sheath extension above and below. Arrow indicates crystal in upper bundle sheath cell. Compare with Fig. 2–18.

matic crystals occur in some of these parenchyma cells (Fig. 2–18 and 2–19), which occur in linear patches of various lengths along the third and fourth vein orders. In larger vein sheaths, crystals may occur on both sides, and considerable variation has been found in crystal distribution among cultivars (Bán et al., 1980).

In the smallest two vein orders, sclerenchyma and bundle sheath extensions are replaced by a completely encircling parenchymatous bundle sheath (Fig. 2–12). These ultimate veins and vein endings typically have only one to two xylem vessels and two to three sieve tubes. The latter are greatly reduced in diameter but their associated companion cells and phloem parenchyma cells are greatly enlarged (Fig. 2–12). This size reversal is related to the energy-intensive task of accumulating photosynthates from the surrounding mesophyll and forcing it into the sieve tubes. In many species these enlarged phloem cells also have internal convoluted cell wall ridges or pegs, which greatly increases cell membrane area and enhances their efficiency in this task. Such wall features, however, are said to be lacking in soybean and other members of the tribe Phaseoleae (Watson et al., 1977).

2–2 STEM

2–2.1 Mature Primary Structure

Because the stem is either elongating, or increasing in diameter following internodal elongation, at any particular level it is only briefly at the stage of primary tissue maturation comparable to that of the leaf and much of the root system. For convenience, however, the stem will first be described in its mature primary condition, followed by the stages leading up to it and the subsequent events of secondary growth.

Figure 2–20 illustrates a representative stem internode in which the major regions can be seen: central pith composed of large, thin-walled storage parenchyma cells; zone of vascular bundles arranged in a cylindrical pattern (eustele); and cortex, between eustele and epidermis. The stem appears somewhat irregular in outline because bulging strands of collenchyma occur to the outside of the large vascular bundles.

The stem epidermis has the same cell types and trichomes as described earlier for the leaf. Epidermal cells are elongated and lack radial wall infoldings, and stomata are restricted to narrow strips of epidermis between collenchyma strands (N.R. Lersten, unpublished data, 1984). The cortex includes these collenchyma strands, and below them a cylinder of chlorenchyma three to four cells thick; between collenchyma strands the chlorenchyma extends to the epidermis. Just interior to the chlorenchyma is a single-layered endodermoid layer or starch sheath, lacking chloroplasts but with abundant starch grains (Fig. 2–21). According to Cumbie (1960), these cells also lack casparian strips. The zone of pericyclic fibers (Bell, 1934) lying interior to the starch sheath was interpreted by Miksche (1961) as consisting of primary phloem fibers. The parenchyma tissue separating adjacent bundles constitutes the medullary rays, where cortex merges with pith. The central pith region is composed of large, chloroplast-free parenchyma cells.

In the internodes some vascular bundles are larger than others (Fig. 2–20). According to Curry (1982), the larger bundles are sympodia (bundles which will split off leaf traces at higher levels) whereas the smaller ones are incipient leaf traces. At each node three of these traces depart from the eustele, one from each of three

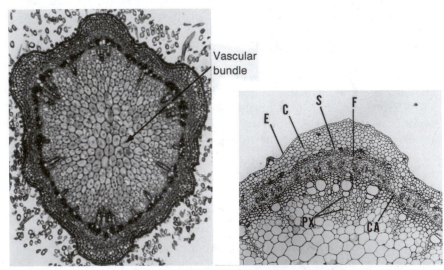

Fig. 2–20. Stem internode transection showing about 16 vascular bundles in a eustelic arrangement, with large pith and narrow cortex. From Curry (1982). × 15.

Fig. 2–21. Enlarged view of a segment of internode transection. Abbreviations: E—epidermis, C—collenchyma rib, CA—cambium, F—phloem fibers, PX—protoxylem, and S—starch sheath. From Miksche (1961).

sympodia. This fits the open type of stem vasculature said by Dormer (1945) to be representative of the Phaseoleae and certain other legume tribes (Fig. 2–22).

Stem vascular bundles are of the common collateral type, with a strand of xylem toward the pith flanked by a strand of phloem to the exterior, and a strip of potential cambial cells in between. Primary xylem consists of protoxylem, the ephemeral portion functioning only during internodal elongation, and metaxylem, the later-maturing permanent water-conducting tissue. Phloem is similarly divided into protophloem and metaphloem, although they are difficult to distinguish at bundle maturity because the thin-walled protophloem cells are largely obliterated by elongation and lateral expansion of surrounding tissue. Also considered to be part of phloem (Miksche, 1961) is the bundle cap, a strand of sclerenchyma fibers just to the exterior of the protophloem (Fig. 2–21).

The ultrastructure of soybean phloem has been described (Fisher, 1975; Wergin and Newcomb, 1970). The most notable feature is the large fusiform P-protein crystal that develops in each sieve tube member.

Soybean vascular bundles have xylem transfer cells with extensive wall ingrowths in the form of ridges or papillae (Kuo et al., 1980). Xylem transfer cells occur at nodes, where they are thought to aid the transfer of water and nutrients to axillary buds. Equally well-developed xylem transfer cells also occur in all vascular bundles in the mid-internode, where they may possibly be involved in the transfer of nitrogenous solutes from xylem to phloem within a bundle.

2–2.2 Secondary Growth

As more stem tissue, leaves and, later, reproductive structures continue to be added as the plant grows, lower portions of the stem must accommodate them by adding more vascular tissue and supporting tissue through the activity of the vas-

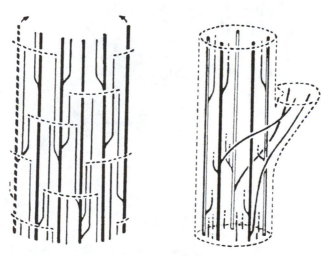

Fig. 2–22. Diagrammatic view of open type of primary stem vasculature found in soybean. Each of the three leaf traces entering the petiole (*right*) arises from a different cauline bundle in the stem (*left*).

cular cambium. This cylindrical meristem is initiated by fascicular cambium located between primary xylem and phloem within each bundle. These meristematic strips become connected later by interfascicular cambium, meristematic strips that form between each pair of bundles. During early cambial activity, secondary xylem and phloem are added only within existing bundles (Fig. 2–20). In nodes and internodes near the ground, where it continues for a long time, a complete cylinder of secondary xylem and phloem is produced (Cumbie, 1960) (Fig. 2–23).

In the lower internodes of stems that have undergone considerable secondary growth, pith cells collapse and the stem becomes hollow. Primary xylem strands are still present but considerable secondary xylem has been added. It consists of large pitted vessels and numerous smaller vessels and tracheids, diffuse-apotracheal axial parenchyma, sclerenchyma fibers scattered in patches, and uniseriate to multiseriate xylem rays (Datta and Saha, 1971). The primary phloem fibers are conspicuous even in these older woody stems. No periderm has been reported (Cumbie, 1960).

A sequence of cross sections from the lower portion of the stem upward would show progressively less secondary growth, especially following the onset of flowering, and the upper internodes would have little or no secondary tissue. Struckmeyer (1941) noted that, prior to flowering, considerable cambial activity is evident in the stem. During flowering, however, cambial activity decreases and cells that had previously been produced by cambial activity develop thickened walls.

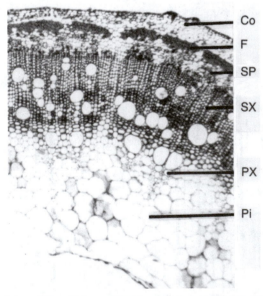

Fig. 2–23. Lower internode in which considerable secondary growth has occurred. Abbreviations: Co—cortex, F—primary phloem fibers, Pi—pith, Px—primary xylem, SP—secondary phloem, SX—secondary xylem.

2–3 ROOT SYSTEM

2–3.1 Primary Structure

The uniseriate epidermis consists of tabular cells elongated along the root axis (Fig. 2–24B). Cell walls are usually thin but outer and radial walls may be somewhat thickened. Intercellular spaces are lacking. Any epidermal cell may form a root hair; these first appear 4 d after germination about 1 cm behind the primary root tip (Anderson, 1961). Branch roots also form root hairs (Fig. 2–24A). Dittmer (1940), in a study of mature field-grown 'Illini' soybean, found root hairs on all roots except the tap root, where secondary growth had removed the epidermis.

Little is known about numbers of root hairs and the total root hair surface of field-grown soybean. Table 2–1 presents data from core samples of field-grown soybean. It is not possible from these data to estimate total root hair number or total area of absorbing surface for the entire root system.

The cortex is the broad zone between epidermis and stele. In the primary root it consists of 8 to 11 layers of slightly elongated cells with much intercellular space (Fig. 2–25). In branch roots the narrower cortex has four to nine layers. Cortical cells close to the epidermis or near the endodermis are considerably smaller than intervening cortical cells. Little or no starch is stored in the cortex, because annual plants store their reserves mostly in seeds. The innermost cortical cell layer is the endodermis, with a continuous suberized casparian strip encircling its radial walls. The casparian strip is recognizable about 2.3 cm from the root apex (Sun, 1955). No other thickenings appear in the endodermis as the root matures.

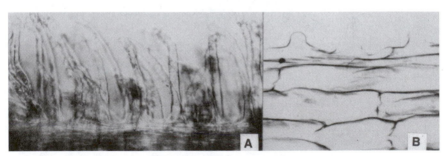

Fig. 2–24. (A) Root hairs extending from epidermis. × 120. (B) Surface view of root epidermis. From Anderson (1961).

Table 2–1. Data on roots and root hairs of mature, field-grown (Illini) soybean. From Dittmer (1940).

	Avg. root diam	Avg. root hair length × diam	No. root hairs mm^{-1} root length
	mm	μ	
Taproot†	2.50	--	--
Secondary roots	0.65	110 × 17	606
Tertiary roots	0.31	90 × 14	210
Quaternary roots	0.23	90 × 14	170

† Secondary growth resulted in loss of most of epidermis and root hairs.

The stele consists of phloem, xylem, associated stelar parenchyma, and surrounding pericycle (Byrne et al., 1977a). Xylem parenchyma cells have transfer cell-like wall ingrowths (Lauchli et al., 1974) which probably allow them to actively secrete ions into the xylem sap or to re-absorb ions from the sap. The pericycle delimits the stele; it has been reported to be two- to three-layers thick in places but Byrne et al. (1977a), described it as only one-layer thick. Pericycle cells have thin walls and then retain their meristematic potential during the maturation of xylem and phloem.

The xylem is almost always tetrarch (Bell, 1934; Byrne et al., 1977a, 1977b). Occasional lateral roots may be triarch, a condition also induced by pathogens (Byrne et al., 1977a, 1977b). Phloem occurs as a single strand midway between each pair of protoxylem ridges (Fig. 2–25).

Lateral roots are protuberances from the stele. They traverse the cortex but have no physiological connection with it. Lateral roots are diarch at their base because they are connected to one protoxylem ridge and adjacent metaxylem. The diarch xylem is surrounded by a cylinder of phloem arising from the two adjacent phloem strands. This stelar arrangement extends about 500 μm into the lateral root, where reorganization into a tetrarch pattern occurs and the root enlarges (Fig.

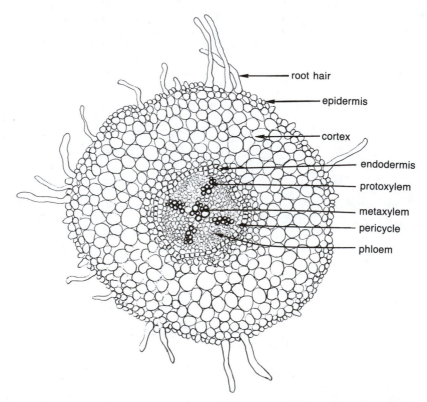

Fig. 2–25. Transection of primary root, 2 cm from apex. Camera lucida drawing of freehand section. × 8.

2–46). The lateral root distally is similar anatomically to comparable levels in the parent root (Byrne et al., 1977a).

The anatomy of individual roots has been shown to change in compacted soils. Baligar et al. (1975) reported these qualitative changes: wavy epidermis with mostly ruptured cells, more spherical cortical parenchyma with increased number of intercellular spaces, change in stelar shape from cylindical to somewhat flattened, and increased wall thickening in xylem cells.

2–3.2 Secondary Growth of Root

Secondary growth in soybean roots follows a common pattern among dicots. Vascular cambium is first visible in 4-d seedlings as bands of tangentially dividing stelar parenchyma cells between phloem and xylem 3 to 5 cm behind the root apex (Anderson, 1961; Sun, 1955). These cambial bands expand laterally by divisions in the pericycle opposite the xylem ridges; this activity results eventually in a continuous cambial layer surrounding the xylem (Fig. 2–26). Acropetal differentiation of the cambium continues as the primary root elongates.

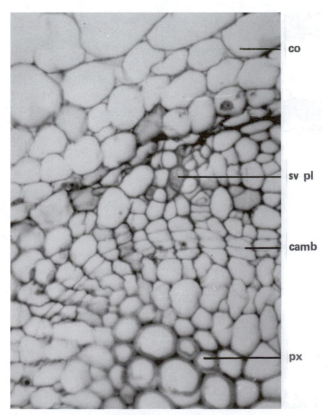

Fig. 2–26. Portion of primary root with newly formed vascular cambium. Abbreviations: camb—vascular cambium, co—cortex, px—protoxylem, and svpl—sieve plate. × 390.

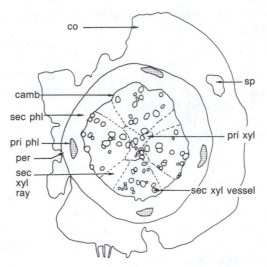

Fig. 2–27. Diagram of root transection with considerable secondary growth. Only largest vessels are indicated. Abbreviations: camb—cambium, co—cortex sloughing off, per—periderm, pri phl—primary phloem, pri xyl—primary xylem, sec phl—secondary phloem, sec xyl—secondary xylem. × 38.

Continued secondary growth eventually produces a cylinder of secondary xylem around the primary xylem, which is in turn surrounded by secondary phloem (Fig. 2–27 and 2–28). The characteristic tetrarch primary xylem pattern is somewhat observed as secondary xylem is added, and the only evidence of primary phloem is the presence of groups of four strands of primary phloem fibers (Fig. 2–27 and 2–28). Secondary xylem consists of pitted vessels, parenchyma cells that may become sclerified, fibers, and parenchymatous rays that are most prominent opposite the original protoxylem ridges. Secondary phloem has parenchyma, sieve

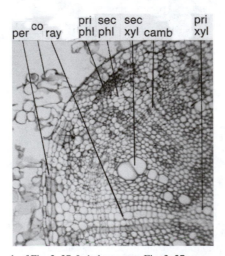

Fig. 2–28. Photomicrograph of Fig. 2–27. Labels same as Fig. 2–27.

tubes, companion cells, and fibers. The cortex develops large intercellular spaces because parenchyma cells are disrupted as the stele increases in diameter.

Where extensive secondary growth occurs, such as in the upper portion of the taproot and the older secondary roots, the epidermis, cortex, and endodermis are disrupted and sloughed off. In these roots, the pericycle forms a cork cambium that divides periclinally and produces a protective periderm (Fig. 2–27 and 2–28) around the vascular tissue.

2–4 GERMINATION AND SEEDLING GROWTH

2–4.1 Establishment

Seeds of most soybean cultivars imbibe water rapidly following planting. Some genotypes, however, especially wild types, have a large proportion of hard seeds that are slow to take up water. The collective results of many studies on legumes indicate that water enters the otherwise impenetrable seed through the lens, a local area of modified palisade epidermal cells that occurs on the opposite side of the hilum from the micropyle (Gunn, 1981).

Under favorable conditions, the radicle emerges from the seed in 1 or 2 d (Dzikowski, 1936; Williams, 1950). Downward growth of the radicle (now called the *primary root*) is rapid and by the 4th or 5th d the first secondary roots emerge 4 to 5 cm behind the primary root apex (Anderson, 1961; Sun, 1955). The cotyledons emerge in 3 to 4 d, and are pulled out of the soil by rapid growth in the doubled-over crook in the upper region of the hypocotyl. Further growth and straightening of the hypocotyl elevates the pale cotyledons, which soon become green and photosynthetic in addition to supplying stored reserves to the seedling. Later, the cotyledons turn yellow and fall off.

Liu et al. (1971) found that microbodies, which they assumed to be glyoxysomes, appear in cotyledon cells about 48 h after planting. As cotyledons emerge and chloroplasts become functional, other microbodies, assumed to be peroxisomes, are associated with the chloroplasts. They speculated that the glyoxysomes help to convert stored lipids to hexose sugar, whereas the peroxisomes indicate that peroxisomal photorespiration occurs in cotyledons. As cotyledons senesce, both chloroplasts and peroxisomes show symptoms of degeneration.

After cotyledons emerge, the two primary leaves expand from the apical bud and mature within a few days. Further growth of the seedling involves the formation of trifoliolate leaves. These early growth stages are shown in Fig. 2–29.

The vascular pattern of the hypocotyl, in the dormant embryo and up to about 36 h after germination, consists principally of a tetrarch arrangement of xylem. Each protoxylem ridge is prominent, with exarch maturation. At the distal end of the hypocotyl the four xylem ridges merge into two—one for each cotyledon (Fig. 2–30).

The characteristic crook in the upper hypocotyl develops because of increased mitotic activity (Bell, 1934). As new cells are added here, more vascular tissue develops. Figure 2–31, of the upper portion of the hypocotyl of Ottawa-Mandarin soybean 48 h after germination, shows additional vascular elements which have arisen from the procambium of the arch.

   Figure 2–32 is a diagrammatic representation of the vascular transition from
the root apex through the hypocotyl and up to the cotyledonary node; the figure legend describes the salient features of the changing vascular pattern. The most sig-

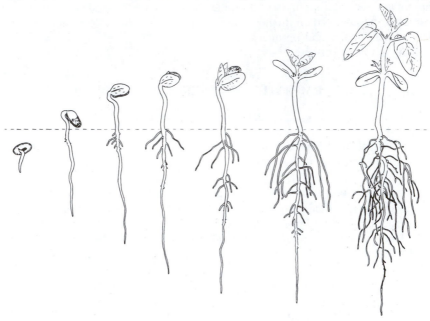

Fig. 2–29. Stages in germination and early seedling growth. Dotted line indicates soil level. Modified
   from Dzikowski (1936).

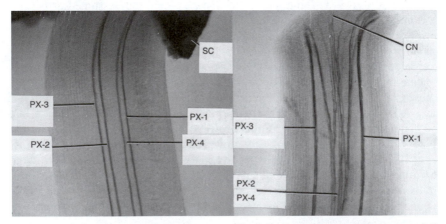

Fig. 2–30. Cleared, unstained upper hypocotyl 36   Fig. 2–31. Cleared, unstained upper hypocotyl 48
   h after planting. PX indicates protoxylem ridges   h after planting. Protoxylem strands PX-1 and
   1–4, SC—seed coat. × 14.   PX-3 are seen from the side and appear as
                                                           single strands each. PX-2 and PX-4 are in face
                                                           view and appear superimposed. They now
                                                           have additional primary xylem elements that
                                                           pass into the cotyledonary node, CN. × 14.

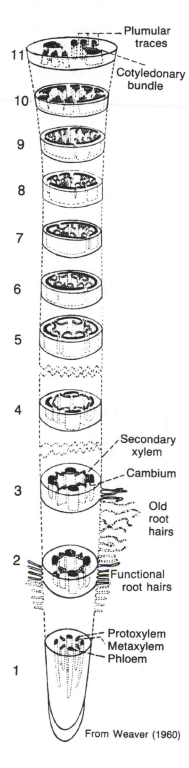

Plumular traces

Cotyledonary bundle

Secondary xylem

Cambium

Old root hairs

Functional root hairs

Protoxylem
Metaxylem
Phloem

From Weaver (1960)

Fig. 2–32. Representation of vascular transition in soybean root (Measurements in parentheses show lengths from each level to next above.)

1. Tetrach, exarch radial arrangement of root (centripetal differentiation). (8 mm.)

2. Vascular cambium internal to primary phloem strands. Xylem strands triangular in cross section. Phloem strands larger in diameter. (5.5 mm.)

3. Xylem strands show doubleness resulting from change in direction of differentiation from centripetal to tangential. Locations of the four points of first primary xylem differentiation have persisted and still occupy positions near the periphery of the axis. (11 mm.)

4. Phloem strands expanded tangentially and reduced radially. There are now eight points of first xylem (all primary) differentiation, that is, a dichotomy of primary xylem strands has occurred so that each tangential wing of primary xylem is now a separate strand. Vascular cambium does not occur at this level or above. (23 mm.) (This pattern was about 30-mm long, over half the distance from level 1 to 11.)

5. Strands differentiated in a concave pattern, with the latest formed vessels indicating a change from exarch to endarch differentiation. The points of initial primary xylem differentiation are still exterior. The convex strands of phloem have been extended tangentially. (2 mm.)

6. Primary xylem has established contact with phloem, the points of initial xylem formation have shifted inward and phloem strands have fused into two broad convex strands, leaving medullary rays in the cotyledonary plane only. Two small endarch xylem strands have differentiated against the phloem strands occupying positions in the intercotyledonary plane. These bundles of xylem are the traces to the plumule. (0.85 mm.)

7. The two endarch plumular traces are distinct from the four primary xylem strands with which they were associated at their origin. In addition to becoming entirely separate the *plumular* traces have increased in magnitude and appear double. This does not result in two strands. (2 mm.)

8. The plumular traces are clearly single collateral bundles. Formation of four new medullary rays at this level has transformed the two phloem strands into six, two of which are associated with plumular xylem strands. Each of the other four is associated with two of the primary xylem strands which have persisted from level 4. These eight xylem strands are almost completely endarch at level 8. (6 mm.)

9. Anastomosis of the phloem strands of the nonplumular bundles results in formation of two opposite collateral bundles, continuous with the cotyledonary bundles. Each exhibits four distinct xylem strands, which are entirely endarch at this level. The xylem strands of the two plumular traces have trifurcated, forming lateral strands of lesser magnitude than the central ones. (2 mm.)

10. Further development of the cotyledonary bundles. The xylem strands have fused in pairs, transforming the bundles into monophloic, bixylary units. Trifurcation of the phloem of the plumular traces has resulted in the establishment of six collateral plumular traces. Central traces are somewhat more massive than the lateral ones. (0.28 mm.)

11. Immediately below the cotyledonary node. Fusion of the xylem strands of the cotyledonary bundles has produced two massive bundles. The plumular traces are of about equal size and form groups of three in the intercotyledonary plane.

nificant fact is that all the primary xylem of the primary root and hypocotyl passes into the two cotyledonary vascular bundles. There is also a gradual transition from the exarch pattern of xylem maturation characteristic of the root to the endarch maturation pattern characteristic of leaves and stem (Level 10, Fig. 2–32).

Weaver (1960) and Compton (1912) showed that xylem passing from the hypocotyl into the epicotyl and the primary leaves arises as a new tissue superimposed upon the preexisting xylem tissue of the root-hypocotyl-cotyledonary axis. These new vascular bundles are endarch in their pattern of maturation. Weaver noted

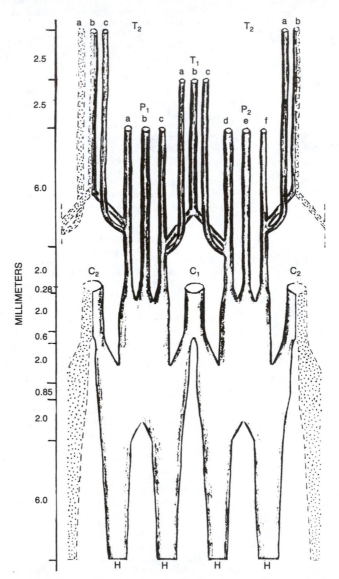

Fig. 2–33. Diagram of phloem distribution in a seedling from hypocotyl to second pair of trifoliolate leaves. From Crafts (1967).

that epicotylar phloem, unlike the xylem, is not superimposed upon the hypocoty-lar phloem. Some of the primary phloem present in the hypocotyl axis of the embryo extends up into the epicotyl during seedling development. Crafts (1967) reconstructed phloem distribution from hypocotyl to second trifoliolate leaf (Fig. 2–33), showing that primary phloem is continuous from the root up through the epicotyl. The upper portion of the embryo axis in the dormant embryo terminates in a short epicotyl consisting of two simple leaves, the primordium of the first trifoliolate leaf, and the stem apex.

Miksche (1961) reported that in the dormant seed there are two unifoliolate leaves with conduplicate vernation and two stipules at the base of each leaf. The first mitotic activity in the embryo occurs about 36 h after planting, in procambial elements within the corpus. After 40 h, divisions were noted in the tunica. By 48 h, mitotic figures were visible in the mesophyll of the unifoliolate leaves, the epidermal hairs, and in the primordium of the first trifoliolate leaf.

Soybean seedlings have red pigment bodies restricted to the subepidermal layer of the hypocotyl and young stem (Nozzolillo, 1973). The pigment is malvadin, known so far only from members of the tribe Phaseoleae.

2–5 PRIMARY MERISTEMS AND ORGAN DIFFERENTIATION

2–5.1 Stem Apical Meristem and Primary Stem

Sun (1957a) reported that the shoot apex of cv. Gibson consists of a two-layered tunica and a rather massive corpus subdivided into three areas: central initiation zone of rather large cells, peripheral zone of somewhat smaller cells, and a rib meristem immediately below the central initiation zone (Fig. 2–34). Grandet (1955) described two tunica layers and a single massive corpus in cv. Rouest 250, as did Bostrack and Struckmeyer (1964) in Hawkeye soybean. All of these investigations studied only seedlings.

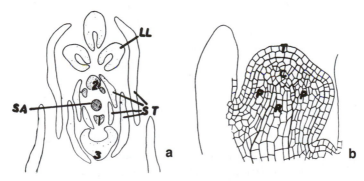

Fig. 2–34. (a) Transection of stem tip showing distichous leaf arrangement in alternate phyllotaxy. 1–3 indicates progressively older leaf primordia. Abbreviations: LL—leaflets of fourth leaf, SA—shoot apex, and ST—stipules. (b) Longisection through primordia 1 and 2 of Fig. 3–34a. Shoot apex shows tunica (T) and three zones within corpus: C—central initiation zone, P—peripheral meristem, r—rib meristem. From Sun (1957a).

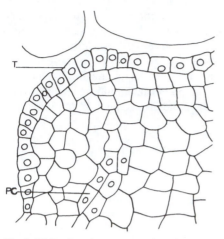

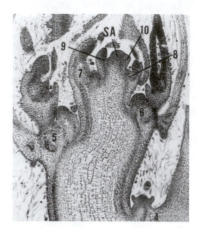

Fig. 2–35. Portion of stem apex region of dormant embryo showing periclinal walls in the layer below the tunica (T), and a procambial strand (PC) extending toward the first trifoliolate leaf primordium. From Miksche (1961).

Fig. 2–36. Median longitudinal section of stem 21 d after planting. First three trifoliolate leaves now shown, other numbered consecutively from oldest (4) to youngest (10). Abbreviations: SA—shoot apex. From Miksche (1961).

Miksche (1961) found only a single-layered tunica and a massive corpus in the dormant embryo. Figure 2–35 shows two periclinal divisions in the layer of cells below the outer tunica layer. These divisions are either associated with the initiation of the first trifoliate leaf, or else they indicate a change from a one-layered to a two-layered tunica as the embryo germinates.

Below the shoot apex, cell divisions virtually cease in the ground meristem as pith and cortical cells enlarge greatly. In the procambium, however, new cells produced by periclinal divisions differentiate into protophloem and protoxylem. These ephemeral conducting strands are obliterated by internodal elongation. As an internode attains maximum length, most of the remaining procambium differentiates into metaphloem and metaxylem. Figure 2–36 illustrates early stages of internodal elongation and concomitant stem differentiation.

Axillary bud primordia are already present in the axil of the second trifoliolate leaf primordium below the stem apex (Sun, 1957a). Each axillary bud primordium is initiated by periclinal divisions of several cells in the outer layer of the corpus, and later involves divisions of the tunica as well.

2–5.2 Apical Meristems and Leaf Development

The initiation and development of trifoliolate leaves and axillary buds were described by Sun (1957a) and Decker and Poslethwait (1960). Figure 2–37 shows stages in the ontogeny of a trifoliolate leaf. The first evidence of leaf initiation occurs 30 to 50 μm below the summit of the shoot apex (Voroshilova, 1964), from periclinal divisions in the second tunica layer and divisions in the corpus just below this second tunica layer. The outer tunica layer divides only anticlinally and thereby maintains itself until it matures as the epidermis.

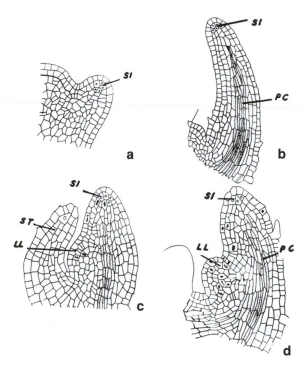

Fig. 2–37. Early stages of trifoliolate leaf development. (a) Very young stage, before vasculation. (b) Later stage, with procambium in future midrib. (c) Lateral leaflet initiation. (d) Young lateral leaflet. Abbreviations: LL—lateral leaflet primordium, PC—procambium, SI—subapical initials (nucleated cells), and ST—stipule. From Sun (1957a).

Shortly after a leaf primordium is initiated, it is penetrated by a procambial strand developing upward from below the shoot apex. When the primordium reaches 60 to 80 μm in length, stipules are initiated at both sides of its base by cell divisions in the outer tunica layer. The adjacent outer part of the corpus soon contributes additional cells. The stipules grow faster than the leaf primordium until the lateral leaflets are initiated (Fig. 2–37c). When the leaf primordium is 140 to 200 μm in length, two new meristems form on opposite sides toward its base, thus initiating the two lateral leaflets. Table 2–2 and Fig. 2–38 show the relative development of the leaf proper, stipules, and hairs at different stages (Thrower, 1962). Stipules and hairs mature extremely early, before the leaf proper has attained even 1% of its mature size, thereby providing some protection and insulation.

When a leaf primordium is about 400 μm long, the lamina is initiated by increased mitotic activity in cells along the margins. Periclinal and anticlinal divisions in the marginal regions of the blade meristem produce six layers of cells (Decker and Postlethwait, 1960; Sun, 1957a). The top and bottom layers divide only anticlinally and eventually become the epidermis. The four central layers originate from a group of submarginal initial cells.

Sun (1957a) reported that the six organized layers later become seven by periclinal divisions of cells in the third abaxial layer. Decker and Postlethwait (1960)

Table 2–2. Young, expanding soybean leaves. From Thrower (1962).

Expanding leaf no. (see Fig. 2–38)	Leaf length	Leaf form	Length of stipules	Length of leaf hairs	Area of leaf
	mm		mm		% of adult area
i	59.0	Trifoliate	7.0	1.3	31
ii	16.0	Trifoliate	7.0	1.3	<2
iii	9.0	Trifoliate	7.0	1.3	<1
iv	6.0	Trifoliate	7.0	1.3	<1
v	2.3	Transitional	3.4	1.1	<1
vi	0.9	3-lobed	2.2	0.6	<1
vii	0.2	3-lobed	0.7	--†	<1
viii	0.1	Single lobe	0.1	--‡	<1

† Hairs present only on stipule: hairs 0.2-mm long.
‡ No hairs present.

concluded that six layers are present in both immature and mature leaflets of Hawkeye soybean. Figures in Franceschi and Giaquinta (1983b) seem to demonstrate that either six or seven layers may occur (Fig. 2–39).

The adaxial layer becomes the upper epidermis and the second and third layers develop into palisade tissue. A portion of the fourth (middle) layer differentiates into veins but most of it becomes paraveinal mesophyll (Franceschi and Gi-

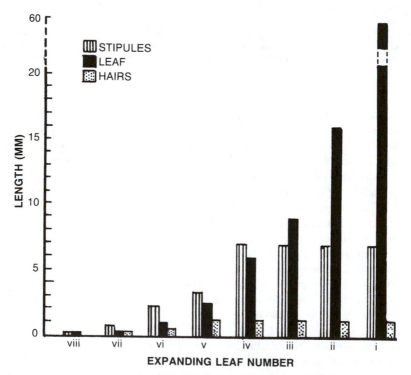

Fig. 2–38. Length of stipules, leaf, and hairs of successive developing soybean leaves. From Thrower (1962).

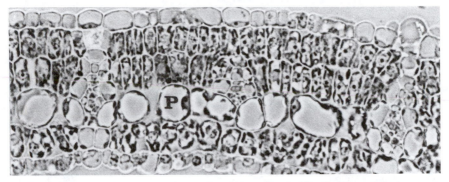

Fig. 2–39. Transection of immature soybean leaflet of 28-mm long leaf. Leaflet is irregularly six to seven cell layers thick. Paraveinal mesophyll layer (P) has differentiated precociously. From Franceschi and Giaquinta (1983b). × 385.

aquinta, 1983b) (Fig. 2–39). The fifth and sixth layers become spongy mesophyll, and the abaxial layer becomes the lower epidermis (Sun, 1957a). Figure 2–12, of a transection of a mature leaflet, shows these layers.

In the midrib of an immature leaflet there is a cambium that forms a limited amount of secondary xylem and phloem. Other veins have only primary vascular tissue. In the mature midrib there is no evidence of a cambium (Decker and Postleth-wait, 1960).

The time elapsed between the initiation of one leaf and another leaf on the opposite side of the stem apex is termed a *plastochron*. Miksche (1961) studied plas-tochronation of the first 10 trifoliolate leaves. Following germination, it took 3 to 3.5 d for the second trifoliolate leaf to be initiated. Subsequent plastochrons took approximately 2 d each. Johnson et al. (1960) found that, by 35 d after planting, when the fifth compound leaf was fully expanded, there were about 19 nodes on the main axis. Since three of these nodes were present in the mature embryo, the average of about 2 d per plastochron reported by Miksche seems to be verified, John-son et al. also concluded that all nodes of the mature plant are already present by the 35th d. The change in the number of nodes on the main axis with reference to days after planting is indicated in Fig. 2–40.

2–5.3 Root Subapical Meristem and Primary Roots

For up to 4 d after germination the primary root tip has distinct stelar initials producing procambium, and another group of initials forming the ground meristem, protoderm, and columella (Miksche, 1961; Sun, 1955) (Fig. 2–40). Sun (1957b) and Anderson and Postlethwait (1960) observed that the apical meristem changes with increasing root age. The apical meristem of the primary root of an 8-d-old seedling has only one group of common initials from which all primary meristems originate (Fig. 2–42 and 2–43). Common initials were also reported by Patel et al. (1975).

Along with the histogens, the size of the quiescent center (an area of almost inert, infrequently dividing cells in the CI zone of Fig. 2–41 and 2–42) also changed in the primary root (Miksche and Greenwood, 1966). The center remained extremely small up to 24 h postgermination, thereafter increasing about 2.5 times to 40 h. After

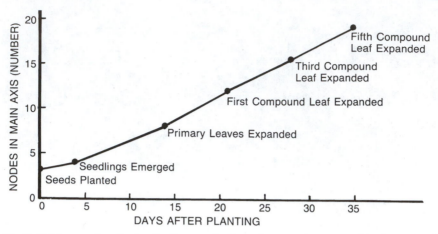

Fig. 2–40. Mean number of nodes of main stem of soybean plants at various stages of development and points in time under simulated normal growing conditions. From Johnson et al. (1960).

that it decreased dramatically to 60 h and shrunk to only about one third its size at 24 h. It stayed at this reduced size until 120 h, when the study ended. The significance of such volume changes is unknown.

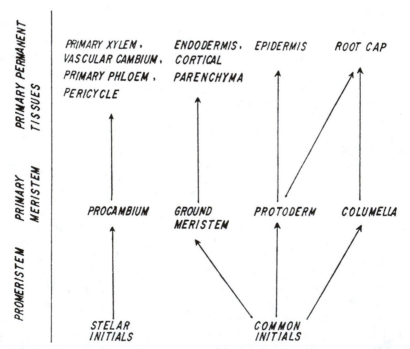

Fig. 2–41. Schematic representation of the pattern of primary tissue differentiation in soybean roots. From Sun (1957b).

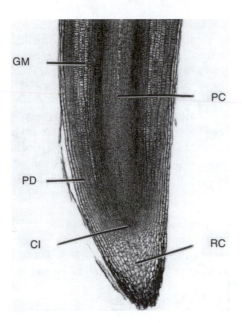

GM

PC

PD

CI

RC

Fig. 2–42. Median longitudinal section of the primary root tip 8 d after planting. Labels as in Fig. 2–43.

The root cap continually wears away, but lost cells are replaced from within by divisions of the columella and protoderm. The life span of individual soybean root cap cells is unknown but in other species they have been shown to exist for 7 to 21 d. Although regular growth patterns occur within the root apex, the shape of the root cap varies greatly (Anderson, 1961).

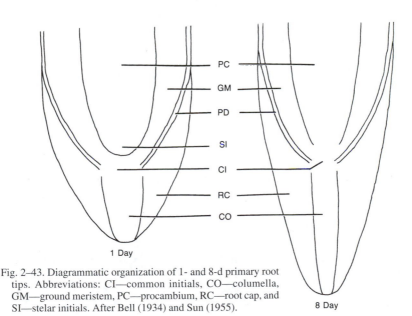

PC
GM
PD
SI
CI
RC
CO

1 Day

Fig. 2–43. Diagrammatic organization of 1- and 8-d primary root tips. Abbreviations: CI—common initials, CO—columella, GM—ground meristem, PC—procambium, RC—root cap, and SI—stelar initials. After Bell (1934) and Sun (1955).

8 Day

All root tip anatomical studies of soybean have used only the radicle and primary root. The assumption that all roots are identical seems warranted, except that
Ransom and Moore (1983) have shown that there are some quantitative differences
in subcellular components between the primary and lateral root of *Phaseolus*. They
tried unsuccessfully to find the anatomical basis for the vertically downward growth
of the primary root as compared to the typically horizontal or obliquely downward
growth of lateral roots. Their work is probably also applicable to soybean roots.

The level at which the various permanent tissues appear and mature is shown
in Fig. 2–44, a composite of data from Sun (1955) and Anderson (1961). The pericycle originates about 160 μm from the stelar initials. Very slightly above this, at
about 200 μm, some vacuolation of the metaxylem elements occurs in the center
of the stele, and at the same level the four primary phloem strands become delimited.

Phloem matures first. The first mature sieve elements are present at about 410
μm from the apex. Slightly behind this, at least one mature sieve element is present for each phloem strand. At least three mature sieve elements are present in each
phloem strand at about 900 μm from the apex. Lateral roots are initiated in the pericycle about 2 mm behind the stelar initials.

Maturation of the protoxylem elements occurs approximately 1 cm from the
apex. In a slightly more mature area, root hairs can be seen. Approximately 2 cm
from the apex, casparian strips are present in endodermal cell walls, and at about
3 cm from the apex the vascular cambium becomes evident. At this level the characteristic tetrarch xylem pattern is visible, with mature protoxylem at the priphery
of the xylem ridges and new elements maturing toward the center. The metaxylem
is evident and some secondary wall thickening is occurring.

The four phloem strands, with mature sieve tube elements and companion
cells, alternate with the protoxylem ridges. Vascular cambium is starting to form
between the phloem strands and the metaxylem. Exarch maturation of the primary
xylem continues until all metaxylem vessels have secondary walls.

The first vascular tissue to differentiate is the phloem. Mature sieve tube members are present 300 to 400 μm from stelar initials in the 3-d-old primary root. Four
protophloem ridges are present in primary roots and normal branch roots (Byrne
et al., 1977a) but triarch and diarch patterns have been reported. The first sieve tube
element matures adjacent to the pericycle, and maturation proceeds centripetally
until about six sieve tube members occur in each protophloem strand. These elements are vacuolate and similar in size, so that companion cells are not distinguishable (Anderson, 1961). The protophloem eventually collapses against the
pericycle because of cambial activity. The cells immediately below crushed protophloem differentiate into primary phloem fibers, and cells located inward from
the fibers mature into metaphloem elements. Metaphloem consists of scattered
parenchyma cells, sieve tube members, and companion cells (Bell, 1934).

Primary xylem is first distinguishable at the same level as phloem, developing acropetally by enlargement and vacuolation of cells and lignification of walls.
As in all rots, transverse maturation within the xylem is exarch, but vacuolation begins first in metaxylem elements at the center of the stele and proceeds centrifu-
gally toward the protoxylem. In contrast, formation of secondary wall and lignification begins first in the protoxylem and then proceeds centripetally until the

largest metaxylem elements differentiate in the center. Protoxylem cells usually have annual or spiral secondary wall thickenings, and they are longer and narrower than metaxylem elements. The first metaxylem elements to mature have a single row of pits on each side, later elements are shorter, larger in diameter, and have densely pitted walls. Some metaxylem parenchyma differentiates into thick-walled con-

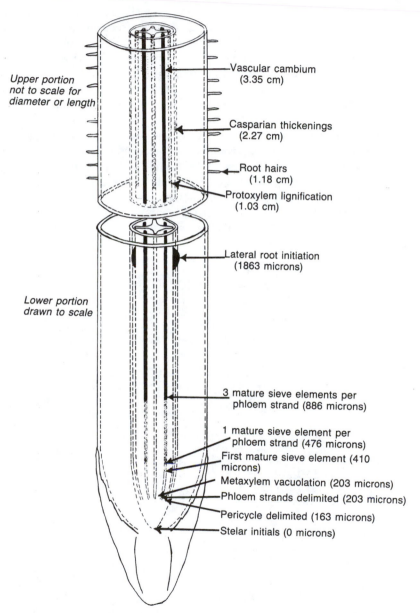

Upper portion
not to scale for
diameter or length

Vascular cambium
(3.35 cm)

Casparian thickenings
(2.27 cm)

Root hairs
(1.18 cm)

Protoxylem lignification
(1.03 cm)

Lateral root initiation
(1863 microns)

Lower portion
drawn to scale

3 mature sieve elements per
phloem strand (886 microns)

1 mature sieve element per
phloem strand (476 microns)

First mature sieve element (410
microns)

Metaxylem vacuolation (203 microns)

Phloem strands delimited (203 microns)

Pericycle delimited (163 microns)

Stelar initials (0 microns)

Fig. 2–44. Diagrammatic scheme of primary tissue differentiation in primary root of soybean. Based on data from Anderson (1961) and Sun (1955).

nective tissue (Bell, 1934). Anderson (1961) found that some xylem cells attain a large diameter but do not become vessel elements. Instead, they have large perforations and bordered pits all over.

2–5.4 Branch Roots

Branch roots arise from cell proliferation in local areas of the pericycle just outside of protoxylem ridges. The xylem pattern, therefore, determines the number of rows of branch roots (Bell, 1934; Byrne et al., 1977a), so tetrarch roots have the potential for four longitudinal rows of branch roots and triarch roots three rows (Fig. 2–51). The first indication of branch root initiation is a radial elongation of pericycle cells followed by tangential and radial divisions (Sun, 1955). This results in a root apex with the same structural organization as the primary root.

Each branch root forces and/or digests its way (Fig. 2–45) through the endodermis, cortex, and epidermis. Flaps of epidermis are pushed aside where roots emerge, and there is a noticeable increase in diameter of the emergent part of the branch root (Fig. 2–46). This may result from the release of pressure by the parent root. This is also where the diarch pattern formed by the branch root's connection to its parent stele ordinarily changes to triarch (Byrne et al., 1977a). The branch root is usually somewhat smaller in diameter than the primary root, and ordinarily it has just one large metaxylem vessel instead of several in the center of the stele. Although

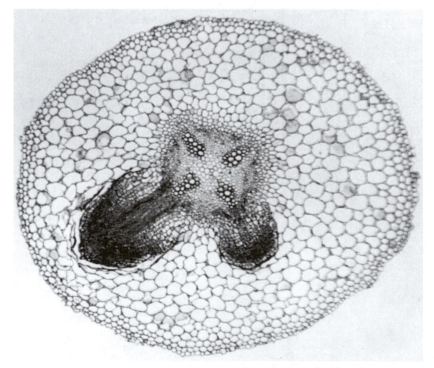

Fig. 2–45. Young branch roots, each opposite a protoxylem ridge, in process of growing through the parent tetrarch root. × 69.

the primary root is tetrarch, secondary branch roots may be triarch. Tertiary, quaternary, and successive orders of smaller branch roots may be either triarch or diarch.

No visible root primorida are ever found closer than 3 cm from the apex of the primary root (Sun, 1955). Branching from the secondary and higher-order roots, however, does occur closer to the apex than in the primary root (Anderson, 1961).

2–6 BACTERIAL ROOT NODULES

Root nodules are conspicuous spheroidal swellings of the root cortex inhabited by *Rhizobium japonicum*, a gram-negative rod-shaped soil bacterium capable of penetrating roots and establishing a N_2-fixing symbiotic relationship. There may be several hundred nodules on a mature plant, distributed at all levels to almost a meter below the surface (Grubinger et al., 1982).

Nodulation begins when rhizobia attach themselves to epidermal cells in a narrow annular zone just behind the root cap. Each bacterium becomes attached only at its tip, a swift recognition reaction occurring in less than 1 min after inoculation in the laboratory (Fig. 2–47). The bacteria remain attached even after repeated washing of the roots (Turgeon and Bauer, 1982).

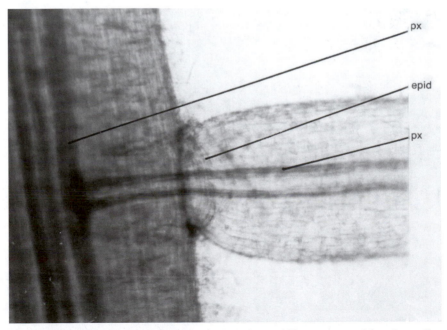

Fig. 2–46. Cleared unstained preparation showing branch root within parent root at left, and extending out to right. Beyond the point of emergence, branch root increases in diameter. Abbreviations: px—protoxylem stands, and epid—flaps of ruptured epidermis. × 55.

Epidermal cells with immature or as yet unformed root hairs are the usual sites for penetration (Bhuvaneswari et al., 1980; Turgeon and Bauer, 1982). This suggests that a certain minimum period of contact between bacteria and host cell is needed, and that only actively growing root hairs can be penetrated. Infected hairs are always shorter than mature intact hairs, and they are markedly curled (Fig. 2–48) (Rao and Keister, 1978; Turgeon and Bauer, 1982; Pueppke, 1983).

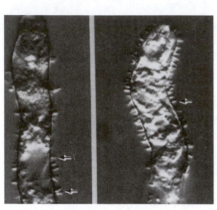

Fig. 2–47. Soybean root hairs after in vitro inoculation by rhizobia, as viewed by differential-interference contrast micrography. *Rhizobia* are attached at one end (*arrows*). (A) root hair with few rhizobia, and (B) root hair with dense rhizobial population. From Turgeon and Bauer (1982). × 1200.

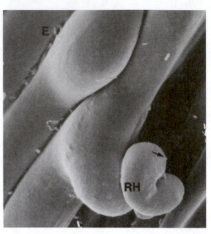

Fig. 2–48. Scanning electron micrograph of small area of root 12 h after inoculation. Curled root hair is the site of bacterial penetration. Arrows indicate bacteria, E—epidermis, and RH—root hair. From Turgeon and Bauer (1982). × 1400.

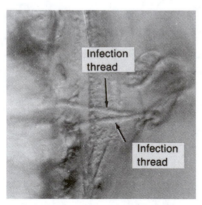

Fig. 2–49. Root hair with two infection threads (*arrows*) viewed by interference-contrast microscopy. From Pueppke (1983).

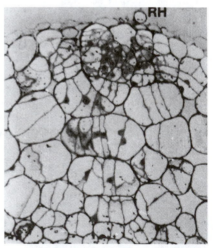

Fig. 2–50. Transection of soybean root at site of rhizobial penetration. Recent cell division is evident in the cortex adjacent to root hair (RH) even before penetration by infection threads. × 750.

At the point of infection, the root hair wall forms a depression that invaginates deeply, forming a tube (infection thread) lined by a continuation of the root hair cell wall and membrane (Goodchild and Bergersen, 1966). Although some root hairs have only one infection thread, two is more common and three or more have been seen (Fig. 2–49) (Rao and Keister, 1978). Infection threads may branch within a root hair (Turgeon and Bauer, 1982; Pueppke, 1983).

It takes about 2 d for the infection thread, with its included dividing bacteria, to grow 60 to 70 μm to the base of the root hair cell (Bieberdorf, 1938; Turgeon and Bauer, 1982). The cortex adjacent to infected root hairs becomes meristematic and produces a wedge-shaped area of dividing cells even before any infection threads enter (Fig. 2–50) (Newcomb et al., 1979; Turgeon and Bauer, 1982). As a thread enters a cortical cell, an invagination of the wall and plasma membrane occurs in the same fashion as in the root hair (Fig. 2–51).

A second wave of mitoses increases cell number in the outer cortical layer, which then becomes the main area of infected cells (Newcomb et al., 1979). The combination of multiple threads and branching of threads in the cortex results in penetration of many, but not all, of these cells (Turgeon and Bauer, 1982). The peripheral uninfected area becomes the nodule cortex, which includes a sclereid layer and several vascular bundles. Numerous cells with calcium oxalate crystals form an irregular reticulum in the outer cortex (Sutherland and Sprent, 1984), a feature in common with other legumes having determinate nodules.

Whether or not the ploidy level of cortical cells increases before or after infection threads penetrate them is a controversial topic (Newcomb, 1981). The only published information about this in soybean is by Kodama (1970), who found only diploid cells in soybean and other Phaseoleae nodules, whereas polyploid cells were the rule in other legumes studies.

At some time during or following mitotic activity, rhizobia are released into cortical cells through small thin areas on the surface of the infection threads. These local areas bulge and release small droplets of thread matrix, with one bacterium per droplet. Each droplet is encased in a peribacteroid membrane contributed by

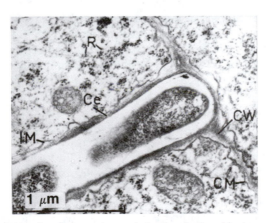

Fig. 2–51. Tip of infection thread penetrating cortical cell wall, viewed by transmission electron microscopy. Abbreviations: Ce—infection thread cellulose, CM—host cell membrane, CW—host cell wall, IM–infection thread membrane, R—ribosome. From Goodchild and Bergersen (1966).

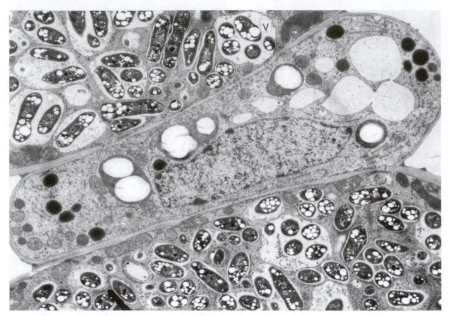

Fig. 2–52. Soybean nodule cells 25 days after inoculation. Middle cell is uninfected but upper and lower cells contain infection vacuoles (V), packets of bacteroids surrounded by a peribacteroid membrane. Transmission electron micrograph from Werner and Mörschel (1978).

the host cell membrane (Verma et al., 1978; Newcomb, 1981). Because of this membrane and the infection thread membrane, rhizobia are never in direct contact with host cell cytoplasm until the nodule senesces.

Rhizobia are called *bacteroids* after their release into the host cell. They continue to divide, and gradually the host cell becomes filled with packets of bacteroids (Werner and Mörschel, 1978). Figure 2–52 illustrates this and also shows that some adjacent cells remain uninfected.

Mitosis in infected cortical cells ceases about 14 d after infection. Subsequent increases in the volume of infected tissue are due entirely to cell enlargement, which often causes infected cells to elongate. As the nodule matures, host cell nuclei become smaller and more regular in shape, and finally disintegrate (Bieberdorf, 1938). During this period oxygen-rich leghemoglobin develops gradually in the host tissue and the nodule becomes pink, remaining so until it begins to senesce. As leghemoglobin forms and bacteria cease dividing, N_2 fixation occurs (Bergersen, 1963).

The increasing volume of the developing nodule causes splitting and sloughing off of the epidermis, thus exposing the outer cortical layer. The second cortical layer becomes meristematic and forms the cortex of the nodule (Fig. 2–53). Division in this layer, which Ikeda (1955) termed the *inner meristem*, permits developing bacteroidal tissue deeper in the nodule to expand. About 8 d after infection, cork cambium forms in the outer layers of the nodule cortex and nonnodulated areas of the root (Bieberdorf, 1938). The cork cambium keeps pace with nodule enlargement and continues to add more corky cells. Lenticels form on the nodule surface and are said to function in gas exchange (Pankhurst and Sprent, 1975).

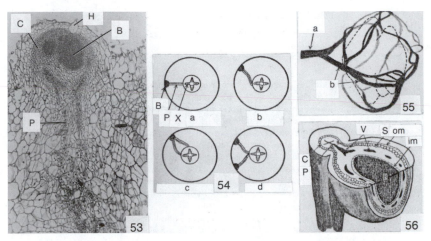

Fig. 2–53 to 2–56. *Fig. 2–53.* A portion of 12-d soybean nodule. Abbreviations: B—bacterial-infected cells, C—meristematic region of nodule cortex, H—hypertrophied root hair, and P—procambium. From Bieberdorf (1938). *Fig. 2–54.* Patterns of procambial strand development connecting nodule to parent stele. Abbreviations: B—developing nodule, P—procambium, and X—protoxylem. From Ikeda (1955). *Fig. 2–55.* Somewhat three-dimensional depiction of vascular network around nodule, a—vascular bundle connecting nodule network with parent root stele, b—one bundle of nodular vascular network. From Ikeda (1955). *Fig. 2–56.* Diagrammatic representation of mature nodule. Abbreviations: B—bacteroidal tissue, C—cork cambium of root, S—sclerenchyma, V—vascular tissue, im—inner meristem, and om–cork cambium of nodule. From Ikeda (1955).

After infection threads have penetrated two or three cortical layers (Bieberdorf, 1938; Ikeda, 1955), parenchyma cells between the nodule primordium and host root stele become meristematic and form a procambial strand that extends toward the protoxylem (Fig. 2–53 and 2–54). This procambium differentiates into pitted tracheids and scalariform vessels surrounded by starch-filled parenchyma cells capable of dividing and adding more xylem cells (Bieberdorf, 1938). Where the nodule vascular bundle meets the vascular system of the parent root, the new xylem from the nodule bundle becomes continuous with the existing secondary xylem. Within the cortex of the nodule, more procambium strands form and differentiate acropetally toward the nodule apex, branching and rebranching, and finally anastomosing into a continuous vascular network (Ikeda, 1955) (Fig. 2–55). Some cells near the xylem differentiate into phloem strands consisting of three to five sieve tubes. As each bundle matures, a well-defined endodermis with a casparian strip encircles it (Fraser, 1942). A mature vascular bundle has a central strand of xylem surrounded by a cylinder of phloem. A sclerenchyma sheath of isodiametric sclereids with simple pits differentiates just outside of the vascular network, maturing about 28 d after infection. It prevents any further increase in the size of the nodule (Bieberdorf, 1938).

A change occurs in vascular tissue of the parent root as rhizobia are being released from infection threads into cortical cells. Wall ingrowths characteristic of transfer cells appear in xylem parenchyma cells adjacent to metaxylem vessels (Newcomb and Peterson, 1979).

Mature nodules at the end of the 4th wk after infection are spheroidal and 3 to 6 mm in diameter (Fig. 2–56). Nodules sometimes become irregular or lobed

Table 2–3. Chronology of nodulation.† Compiled from Bergersen (1958), Bieberdorf (1938), and Ikeda (1955).

Age of nodule	Stage of nodulation
d	
0	Initial invasion of root hair or ordinary epidermal cell by *Rhizobium*.
1–2	Infection thread reaches base of epidermal cell and enters cortex.
3–4	Small mass of infected cells in nodule primordium; procambium strand extends from nodule to stele of root.
5	Very rapid bacterial and host cell division continues about 2 wk
7–9	Nodule visible; procambium of nodule vascular system arises at base of nodule and develops toward nodule apex.
12–18	Continued growth of all nodule tissues; periderm present; some mature cells in sclerenchyma layer; vascular system forms anastomosing network within nodule cortex; bacteroidal tissue is pink at the end of this period and N_2 fixation commences.
23	Most division has ceased in bacterial and host cells; nodule continues growth by cell enlargement for up to 2 more weeks; period of active N_2 fixation.
28–37	Nodule reaches maximum size; vascular tissue and sclerenchyma tissue mature; N_2 fixation continues until nodule degeneration begins.
50–60	Nodule degeneration.

† Absolute times might vary with cultural conditions, time of infection, soybean cultivar, and other factors.

when two or more infected areas develop close together and merge during growth. Soybean nodules are determinate, lacking the apical meristem and extended terminal growth found in the indeterminate nodules of certain other legumes such as *Medicago*, *Melilotus*, and *Trifolium* (Sprent, 1980). The bacteroidal tissue is therefore homogenous in age (Bergersen and Briggs, 1958).

Nodules retain their mature structure until the 6th or 7th wk, then they begin to senesce (Bergersen, 1958). Since reinfection of younger portions of the root system may occur during the growing season, a mature soybean plant may have nodules of several age classes. A chronology of nodulation is presented in Table 2–3.

2–7 MYCORRHIZAL RELATIONS

In addition to bacterial nodules, soybean roots have been shown to form mycorrhizal associations (Bethlenfalvay et al., 1982). In plants inoculated with the vesicular-arbuscular mycorrhizal fungus *Glomus fasciculatus*, hyphae penetrated host cells and formed highly branched endings. These intraradical hyphae, which do not kill the host cell, extended to new roots throughout the life span of the association. Fungal hyphae also extended into the soil (extraradical hyphae) to enhance nutrient absorption. The latter reached maximum development when the pods began rapid maturation and declined thereafter. Knowledge of the anatomical aspects of this association are almost lacking.

REFERENCES

Ali, A., and R.A. Fletcher. 1970. Xylem differentiation in inhibited cotyledonary buds of soybeans. Can. J. Bot. 48:1139–1140.

Anderson, C.E. 1961. The morphogenesis of the root of *Glycine max*. M.S. thesis. Purdue Univ., Lafayette, IN.

Anderson, C.E., and S.N. Postlethwait. 1960. The organization of the root apex of *Glycine max*. Proc. Indiana Acad. Sci. 70:61–65.

Baligar, V.C., V.E. Nash, M.L. Hare, and J.A. Price, Jr. 1975. Soybean root anatomy as influenced by soil bulk density. Agron. J. 67:842–844.

Bán, A.D., L. Muller, B.H. de Souza, T. Strehl, and C.S.A. Martins. 1981. Soybean leaf architecture. (In Portuguese.) Agron. Sulriograndense 17:25–31.

Bán, A.D., T. Strehl, B.H. de Souza, C.S.A. Martins, and L. Muller. 1980. Anatomical study of crystals in soybean leaflets. (In Portuguese.) Agron. Sulriograndense 16:169–179.

Bell, W.H. 1934. Ontogeny of the primary axis of *Soja max*. Bot. Gaz. 95:622–635.

Bergersen, F.J. 1958. The bacterial component of soybean root nodules; changes in respiratory activity, dry cell weight and nucleic acid content with increasing nodule age. J. Gen. Microbiol. 19:312–323.

Bergersen, F.J. 1963. Iron in the developing soybean nodule. Aust. J. Biol. Sci. 16:916–919.

Bergersen, F.J., and M.J. Briggs. 1958. Studies on the bacterial component of soybean root nodules: Cytology and organization in the host tissue. J. Gen. Microbiol. 19:482–490.

Bernard, R.L., and B.B. Singh. 1969. Inheritance of pubescence type in soybeans: Glabrous, curly, dense, sparse, and puberulent. Crop Sci. 9:192–197.

Bethlenfalvay, G.J., M.S. Brown, and R.S. Pacovsky. 1982. Relationships between host and endophyte development in mycorrhizal soybeans. New Phytol. 90:537–543.

Bhuvaneswari, T.V., B.G. Turgeon, and W.D. Bauer. 1980. Early events in the infection of soybean (*Glycine max* L. Merr.) by *Rhizobium japonicum*. I. Localization of infectible root cells. Plant Physiol. 66:1027–1031.

Bieberdorf, F.W. 1938. The cytology and histology of the root nodules of some Leguminosae. J. Am. Soc. Agron. 30:375–389.

Blad, B.L., and D.G. Baker. 1972. Orientation and distribution of leaves within soybean canopies. Agron. J. 64:26–29.

Bohm, W., H. Maduakor, and H.M. Taylor. 1975. Comparison of five methods for characterizing soybean rooting density and development. Agron. J. 69:415–419.

Bostrack, J.M., and B.E. Struckmeyer. 1964. Effects of gibberellic acid on the anatomy of soybeans. Am. J. Bot 51:611–617.

Byrne, J.M., T.C. Pesacreta, and J.A. Fox. 1977a. Development and structure of the vascular connection between the primary and secondary root of *Glycine max* (L.) Merr. Am. J. Bot 64:946–959.

Byrne, J.M., T.C. Pesacreta, and J.A. Fox. 1977b. Vascular pattern change caused by a nematode, *Meloidogyne incognita*, in the lateral roots of *Glycine max* (L.) Merr. Am. J. Bot. 64:960–965.

Ciha, A.J., and W.A. Brun. 1975. Stomatal size and frequency in soybeans. Crop Sci. 15:309–313.

Compton, R.H., 1912. An investigation of seedling structure in the Leguminosae. J. Linn. Soc. Bot. 41:1–119.

Crafts, A.S. 1967. Bidirectional movement of labelled tracers in soybean seedlings. Hilgardia 37:625–638.

Cumbie, B.G. 1960. Anatomical studies in the Leguminosae. Trop. Woods 113:1–47.

Curry, T.M. 1982. Morphological and anatomical comparisons between fasciated and nonfasciated soybeans (*Glycine max* (L.) Merr.). M.S. thesis. Iowa State Univ., Ames.

Datta, P.C., and N. Saha. 1971. Secondary xylem of Phaseoleae. Acta Bot. Acad. Sci. Hung. 17:347–359.

Decker, R.D., and S.N. Postlethwait. 1960. The maturation of the trifoliate leaf of *Glycine max*. Indiana Acad. Sci. Proc. 70:66–73.

Dittmer, H.J. 1940. A quantitative study of the subterranean members of soybean. Soil Conserv. 6(2):33–34.

Dormer, K.J. 1945. An investigation of the taxonomic value of shoot structure in angiosperms with especial reference to Leguminosae. Ann. Bot. 9:141–153.

Dzikowski, B. 1936. Studia nad soja *Glycine hispida* (Moench) Maxim. Cz. 1. Morfologia. Pamietnik Panstwowego Instytutu Naukowego Gospodarstwa Wiejskiego w Pulawach. Tom XVI. zeszyt 2. Rosprawa Nr. 253:Oh 69–100.

Dzikowski, B. 1937. Studia nad soja *Glycine hispida* (Moench) Maxim. Cz. 11. Anatomia. Mem. Inst. Natl. Pol. Econ. Rurale 258:229–265.

Fisher, D.B. 1967. An unusual layer of cells in the mesophyll of the soybean leaf. Bot. Gaz. 128:215–218.

Fisher, D.B. 1975. Structure of functional soybean sieve elements. Plant Physiol. 56:555–569.

Flores, E.M., and A.M. Espinoza. 1977. Epidermis foliar de *Glycine soja* Sieb. y Zucc. Rev. Biol. Trop. 25:263–273.

Franceschi, V.R., and R.T. Giaquinta. 1983a. Glandular trichomes of soybean leaves: Cytological differentiation from initiation through senescence. Bot. Gaz. 144:175–184.

Franceschi, V.R., and R.T. Giaquinta. 1983b. The paraveinal mesophyll of soybean leaves in relation to assimilate transfer and compartmentation. I. Ultrastructure and histochemistry during vegetative development. Planta 157:411–421.

Franceschi, V.R., and R.T. Giaquinta. 1983c. The paraveinal mesophyll of soybean leaves in relation to assimilate transfer and compartmentation. II. Structural metabolic and compartmental changes during reproductive growth. Planta 157:422–431.

Fraer, H.L. 1942. The occurrence of endodermis on leguminous root nodules and its effect upon nodule function. Proc.-R Soc. Edinburgh, Sect. B:Biol. Sci. 61:328–343.

Goodchild, D.J., and F.J. Bergersen. 1966. Electron microscopy of the infection and subsequent development of soybean nodule cells. J. Bacteriol. 92:204–213.

Grandet, J. 1955. Sur le point vegetafif du *Soja hispida* Moench. C.R. Acad. Sci. 240:1003–1005.

Grubinger, V., R. Zobel, J. Vendeland, and P. Cortes. 1982. Nodule distribution on roots of field-grown soybeans in subsurface soil horizons. Crop Sci. 22:153–155.

Gunn, C.R. 1981. Seeds of Leguminosae. p. 913–925. *In* R.M. Polhill and P.H. Raven (ed.) Advances in legume systematics. Ministry of Agriculture. Fisheries and Food. Royal Botanic Garden, Kew, UK.

Ikeda, H. 1955. Histological studies on the root nodules of soybean. (In Japanese.) Kagoshima U. Fac. Agron., Ser. B. 1955(4):54–64.

Johnson, H.W., H.A. Borthwick, and R.C. Leffel. 1960. Effects of photoperiod and time of planting on rates of development of the soybean in various states of the life cycle. Bot. Gaz. 122:77–95.

Kaspar, R.C., C.D. Stanley, and H.M. Taylor. 1978. Soybean root growth during the reproductive stages of development. Agron. J. 70:1105–1107.

Kodama, A. 1970. Cytological and morphological studies on the plant tumors. I. Root nodules of some Leguminosae. J. Sci. Hiroshima Univ., Ser. B., Div. 2 13:223–260.

Kuo, J., J.S. Pate, M. Rainbird, and C.A. Atkins. 1980. Internodes of grain legumes—new location for xylem parenchyma transfer cells. Protoplasma 104:181–185.

Lackey, J.A. 1978. Leaflet anatomy of Phaseoleae (Leguminosae: Papilionoideae) and its relation to taxonomy. Bot. Gaz. 139:436–446.

Lauchli, A., D. Kramer, and R. Stelzer. 1974. Ultrastructure and ion localization in xylem parenchyma cells of roots. p. 363–371. *In* U. Zimmermann and J. Dainty (ed.) Membrane transport in plants. Springer-Verlag New York, New York.

Liu, K-C., A.J. Pappelis, and H.M. Kaplan. 1971. Microbodies of soybean cotyledon mesophyll. Trans. Ill. State Acad. Sci. 64:136–141.

Lugg, D.G., and T.R. Sinclair. 1980. Seasonal changes in morphology and anatomy of field-grown soybean leaves. Crop Sci. 20:191–196.

Miksche, J.P. 1961. Developmental vegetative morphology of *Glycine max*. Agron. J. 53:121–128.

Miksche, J.P., and M. Greenwood. 1966. Quiescent centre of the primary root of *Glycine max*. New Phytol. 65:1–4.

Mitchell, R.L., and W.J. Russell. 1971. Root development and rooting patterns of soybean [*Glycine max* (L.) Merrill] evaluated under field conditions. Agron. J. 63:312–316.

Newcomb, W. 1981. Nodule morphogenesis and differentiation. Int. Rev. Cytol. Suppl. 13:247–298.

Newcomb, W., and R.L. Peterson. 1979. The occurrence and ontogeny of transfer cells associated with root nodules and lateral roots in Leguminosae. Can. J. Bot. 57:2583–2602.

Newcomb, W., D. Sippell, and R.L. Peterson. 1979. The early morphogenesis of *Glycine max* and *Pisum sativum* root nodules. Can. J. Bot. 57:2603–2616.

Nozzolillo, C. 1973. A survey of anthocyanin pigments in seedling legumes. Can. J. Bot. 51:911–915.

Pankhurst, C.E., and J.I. Sprent. 1975. Surface features of soybean root nodules. Protoplasma 85:85–89.

Patel, J.D., J.J. Shah, and K.V. Subbayamma. 1975. Root apical organization in some Indian pulses. Phytomorphology 25:261–270.

Pueppke, S.G. 1983. *Rhizobium* infection thread in root hairs of *Glycine max* (L.) Merr., *Glycine soja* Sieb. & Zucc., and *Vigna unguiculata* (L.) Walp. Can. J. Microbiol. 29:69–76.

Ransom, J.S., and R. Moore. 1983. Geoperception in primary and lateral roots of *Phaseolus vulgaris* (Fabaceae). I. Structure of columella cells. Am. J. Bot. 70:1048–1056.

Rao, R.V., and D.L. Keister. 1978. Infection threads in the root hairs of soybean (*Glycine max*) plants inoculated with *Rhizobium japonicum*. Protoplasma 97:311–316.

Raper, C.D., Jr., and S.A. Barber. 1970. Rooting systems of soybeans. I. Differences in root morphology among varieties. Agron. J. 62:581–584.

Shibles, R., I.C. Anderson, and A.H. Gibson. 1974. Soybean. *In* L.T. Evans (ed.) Crop physiology, some case histories. Cambridge Univ. Press, Cambridge, UK.

Sprent, J.I. 1980. Root nodule anatomy, type of export product and evolutionary origin of some Leguminosae. Plant Cell Environ. 3:35–43.

Struckmeyer, B.E. 1941. Structure of stems in relation to differentiation and abortion of blossom buds. Bot. Gaz. 103:182–191.

Sun, C.N. 1955. Growth and development of primary tissues in aerated and non-aerated roots of soybean. Bull. Torrey Bot. Club 82:491–502.

Sun, C.N. 1957a. Histogenesis of the leaf and structure of the shoot apex in *Glycine max* (L.) Merrill. Bull. Torrey Bot. Club 84:163–174.

Sun, C.N. 1957b. Zonation and organization of root apical meristem of *Glycine max*. Bull. Torrey Bot. Club 84:69–78.

Sutherland, J.M., and J.I. Sprent. 1984. Calcium-oxalate crystals and crystal cells in determinate root nodules of legumes. Planta 161:193–200.

Thrower, S.L. 1962. Translocation of labelled assimilates in the soybean. II. The pattern of translocation in intact and defoliated plants. Aust. J. Biol. Sci. 15:629–649.

Turgeon, B.G., and W.D. Bauer. 1982. Early events in the infection of soybean by *Rhizobium japonicum*. Time course and cytology of the initial infection process. Can. J. Bot. 60:152–161.

Verma, D.P.S., V. Kazazian, V. Zogbi, and A.K. Bal. 1978. Isolation and characterization of the membrane enclosing the bacteroids in soybean root nodules. J. Cell Biol. 78:919–936.

Voroshilova, G.I. 1964. The structure of the vegetative cone and the leaf development in soybeans. (In Russian.) Bot. Zh. (Leningrad) 49(9):1329–1335.

Watari, S. 1934. Anatomical studies on some leguminous leaves with special reference to the vascular system in petioles and rachises. J. Fac. Sci. Imp. Univ. Tokyo, Sect. 3, 4:225–365.

Watson, L., J.S. Pate, and B.E.S. Gunning. 1977. Vascular transfer cells in leaves of Leguminosae–Papilionoideae. Bot. J. Linn. Soc. 74:123–130.

Weaver, H.L. 1960. Vascularization of the root-hypocotyl-cotyledon axis of *Glycine max* (L.) Merrill. Phytomorphology 10:82–86.

Wergin, W.P., and E.H. Newcomb. 1970. Formation and dispersal of crystalline P-protein in sieve elements of soybean (*Glycine max* L.). Protoplasma 71:365–388.

Werner, D., and E. Mörschel. 1978. Differentiation of nodules of *Glycine max*. Planta 141:169–177.

Williams, L.F. 1950. Structure and genetic characteristics of the soybean. p. 111–156. *In* K.S. Markley (ed.) Soybeans and soybean products. Interscience Publ., New York.

Woolley, J.T. 1964. Water relations of soybean leaf hairs. Agron. J. 56:569–571.

3   Reproductive Morphology[1]

JOHN B. CARLSON

NELS R. LERSTEN

Following the period of vegetative growth, which varies depending upon cultivar and environmental conditions such as daylength and temperature, the plant enters the reproductive stage, during which axillary buds develop into flower clusters of 2 to 35 flowers each. There are two types of stem growth habit and floral initiation in soybean (Dzikowski, 1936; Guard, 1931; Williams, 1950).

One type is the indeterminate stem, in which the terminal bud continues vegetative activity during most of the growing season. In this type, the inflorescences are axillary racemes (Fig. 3–1) and the plant at maturity has a sparse and rather even distribution of pods on all branches with a diminishing frequency toward the tip of the stems. The stem may sometimes appear to have a terminal inflorescence but this apparently is a series of small one- or two-flowered axillary inflorescences crowded together by the short internodes at the stem tip.

The second type is the determinate stem, in which the vegetative activity of the terminal bud ceases when it becomes an inflorescence. This type has both axillary racemes and a terminal raceme (Fig. 3–2), and at maturity has pods distributed along the stem as well as a rather dense terminal cluster of pods.

The node of the first flower is related to the development stage of the plant. Since nodes of the cotyledons, the primary leaves, and the first two or three trifoliolate leaves are usually vegetative, the first flowers appear at nodes five or six and sometimes higher. Flowers form progressively toward the tip of the main stem and also toward the tips of the branches. The period of bloom is influenced by the time of planting and may extend from 3 to more than 5 wk (Borthwick and Parker, 1938; Hardman, 1970).

Several investigators have reported that a soybean plant produces many more flowers than can develop into pods. From 20 to 80% of the flowers are reported to abscise for various cultivars (Hansen and Shibles, 1978; Hardman, 1970; Van Schaik and Probst, 1958; Wiebold et al., 1981). Most cultivars with many flowers per node have a higher percentage of flower abscission than those with few flowers per node. Abscission can occur at the time of bud initiation, during the devel-

---

[1] This chapter is reprinted from Chapter 4 of *Soybeans: Improvement, Production, and Uses*, 2nd ed. 1987.

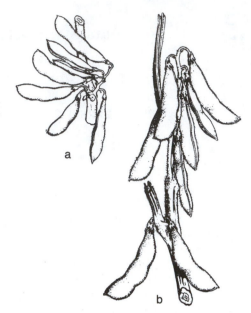

Fig. 3–1. Reduced inflorescence in axil of 'Ottawa-Mandarin' soybean.

Fig. 3–2. Inflorescence types: (a) axillary inflorescence and (b) terminal inflorescence. From Dzikowski (1936).

opment of floral organs, at the time of fertilization, during the early proembryo stage, or at any stage of cotyledon development. Flower or pod abscission occurs most often from 1 to 7 d after flowering (Kato and Sakaguchi, 1954; Kato et al., 1966; Pamplin, 1963; Williams, 1950).

Abernathy et al. (1977) reported that failure of fertilization is insignificant as a cause of floral abscission in soybean. Abscising flowers were mostly all fertilized and usually contained proembryos that had undergone two or three cell divisions. Hansen and Shibles (1978) found that in two indeterminate cultivars, abscission was greatest on the lower stems, whereas pods were retained most often in the middle portions of the plant. In 11 determinate cultivars, in contrast, most harvestable pods were in the top third of the canopy and abscission increased in the lower portions (Weibold et al., 1981). In general, the earliest and latest flowers produced tend to abscise most often.

Individual ovules or entire ovaries may abort. Kato and Sakaguchi (1954) noted that the basal ovule, which is the last one to be fertilized, would frequently abort. Also, the terminal ovule would often abort because of its poorer ability to compete for available water. Thus, normal pods develop with some mature seeds although occasional ovules abort.

When inflorescences are initiated, there are marked changes from the normal vegetative development of the axillary buds. The two opposite prophyll primordia are differentiated as usual, but instead of the typical two-rowed (distichous) arrangement of the trifoliolate leaf primordia, the pattern of initiation changes to a spiral two-fifths phyllotaxy. These primordia do not develop into typical trifoliolate

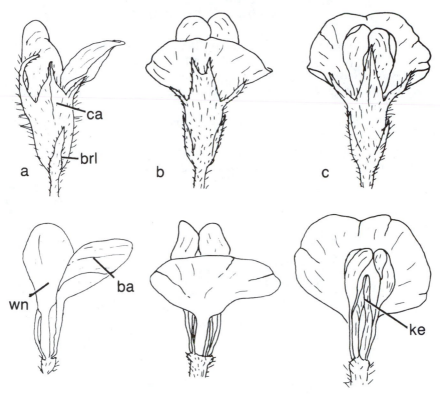

Fig. 3–3. Mature soybean flower. Right-hand column shows flower with calyx removed. (a) Side view showing calyx with bracteoles, banner, and wing petals. (b) Adaxial or top view showing four calyx lobes, both bracteoles, banner, and wing petals. (c) Abaxial or bottom view. Three calyx lobes visible. Abbreviations: ba—banner petal, brl—bracteole, ca—calyx, ke—keel petals, and wn—wing petal. × 6.

leaves with paired stipules, but instead become simple bracts that lack stipules. Growth of both prophylls and bracts is somewhat slower than their vegetative counterparts, but their axillary buds develop much more rapidly, each becoming the primordium of a flower.

In the development of each flower, a similar pair of prophylls is again produced. These prophylls, however, remain small and attached to the lower part of the pedicel of the flower. They become the bracteoles, which are inserted on each side of the calyx of the flower (Borthwick and Parker, 1938; Dzikowski, 1936, 1937; Murneek and Gomez, 1936) (Fig. 3–3).

When a bud in the axil of a trifoliolate leaf develops into an inflorescence, the stalk of that inflorescence remains stem-like, with typical stem anatomy, including epidermis, cortex, endodermis, vascular tissue, and considerable secondary growth from a vascular cambium (Dzikowski, 1937). In the development of an inflorescence, the bract of each flower is homologous to a trifoliolate leaf, and the two bracteoles (Fig. 3–3) are homologous to the prophylls that normally develop at the base of every branch. After forming the primordia of the bracteoles, the apical meristem of the flower gives rise directly to the floral organs.

3–1 FLOWER DEVELOPMENT

Soybean has a typical papilionaceous flower with a tubular calyx of five un-equal sepal lobes, and a five-parted corolla consisting of posterior banner petal, two lateral wing petals, and two anterior keel petals in contact with each other but not fused (Fig. 3–3 and 3–4). The 10 stamens, collectively called the *androecium*, occur in two groups (diadelphous pattern) in which the filaments of nine of the stamens are fused and elevated as a single structure whereas the posterior stamen remains separate (Fig. 3–5). The single pistil is unicarpellate and has one to four campy-lotropous ovules alternating along the posterior suture (Fig. 3–21). The style is about half the length of the ovary and curves backward toward the free posterior stamen, terminating in a capitate stigma (Fig. 3–5). Trichomes occur on the pistil and also cover the outer surfaces of the calyx tube, the bract, and bracteoles. No trichomes are present on the petals or stamens (Dzikowski, 1937; Guard, 1931; Pamplin, 1963).

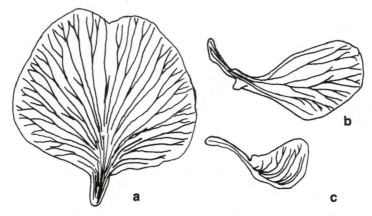

Fig. 3–4. Cleared petals showing venation. (a) Banner petal, (b) Wing petal, and (c) Keel petal. × 7.

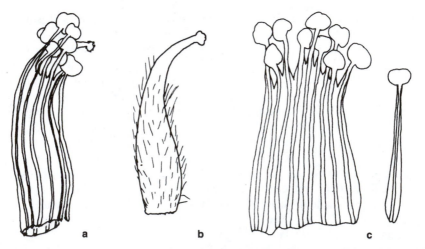

Fig. 3–5. Sexual structures of soybean flower. (a) Diadelphous stamens arranged around pistil, (b) Pis-til, and (c) Nine fused stamens and one separate stamen. × 15.

Guard (1931) gave a detailed description of floral organogeny. The future flower is at first merely a knob-like primordium in the axil of the bract (Fig. 3–6 to 3–8). The sepals are the first whorl of floral organs to be initiated. The anterior, abaxial sepal lobe arises first on the abaxial side of the flower primordium and is followed in rapid succession by the two lateral lobes, and finally, by the two posterior, adaxial lobes. Very early, the bases of these lobes broaden and fuse, and later this becomes the calyx tube (Fig. 3–9 to 3–13).

The petals form next; their primordia alternate with the lobes of the calyx. The keel petals appear first, on the abaxial side, next the two wing petals form laterally and, finally, the banner petal forms on the abaxial side of the flower (Fig. 3–14 and 3–15). Petal primordia develop slowly and are soon surpassed in growth by the stamens. An outer whorl of five stamens appears first, on the anterior side of the flower, just to the inside of the whorl or petal primordia and alternate with them. Initiation of stamen primordia proceeds toward the posterior side of the receptacle. Before the last stamen of the first whorl is visible, a second, inner, whorl of stamens appears, alternating with those of the first, and again starting on the anterior side of the receptacle and progressing toward the posterior, adaxial side. Because of tissue growth below them, these two whorls of stamens quickly align themselves into a single whorl on a staminal tube bearing nine stamens, with the larger and older stamens alternating with the smaller and younger stamens in sequence around the developing pistil. The single free stamen, located between the banner petal and the ventral suture of the pistil, is the last stamen to appear. Although remaining separate, it is a member of the inner whorl of stamens (Fig. 3–16 to 3–19).

The primordium of the pistil appears first as a U-shaped ridge about the same time as the initiation of the last whorl of stamens. The open part of the U, the ventral suture, is on the adaxial side of the flower (Fig. 3–18 and 3–19). All organs of the flower develop rapidly except the petals, which do not elongate much until the anthers have well-developed microsporangia. The staminal tube, the free stamen, and the style elongate at the same pace. Thus, the anthers at maturity are clustered around the stigma (Johns and Palmer, 1982) (Fig. 3–28). At this time the petals grow very rapidly, soon surpassing the calyx, stamens, and pistil to become visible as the flower is in bloom (Fig. 3–20 to 3–27).

Before the margins of the leaf-like pistil fuse, two to four ovule primodia are produced alternately, and develop simultaneously, on the inner surface of the margins, on the placenta (Guard, 1931; Pamplin, 1963). Each ovule becomes campylotropous, with its micropylar end directed upward toward the stigma (Fig. 3–59).

The nectary is visible, about 10 d before anthesis, as a rim of tissue between the base of the pistil and the stamens (Fig. 3–21). At the time of anthesis, the discoidal nectary is a fully formed cup about 0.2 to 0.4 mm in height encircling the base of the staminal sheath (Carlson, 1973; Erickson and Garment, 1979).

The slightly oval nectary stomata are concentrated on each side of the adaxial indentation of the nectary where it contacts the filament of the free ventral stamen. Most of the stomata occur over the rim and ventral interior surface of the nectar cup, occasionally in groups of two or three. On the abaxial side of the cup there are only a few stomata (Erickson and Garment, 1979). Waddle and Lersten (1973) noted that nectaries are vasculated largely by phloem branching from the staminal base. Nectariferous tissue is, therefore, most closely associated with the stamens.

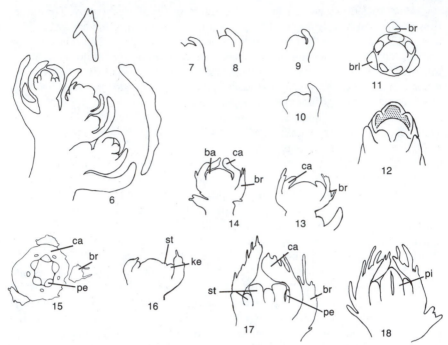

Fig. 3–6 to 3–18. Gross floral development in soybean. Abbreviations: ba—banner, br—bract, brl—bracteole, ca—calyx, ke—keel, ms—microsporangium, ne—nectary, ov—ovule, pe—petal, pi—pistil, and st—stamen. Fig. 3–7 to 3–18. × 60. *Fig. 3–6.* An L-section of raceme with several developing flowers, each in an axil of a bract. × 32. *Fig. 3–7 to 3–8.* Floral primordium in the axil of a bract. *Fig. 3–9.* First sepal primordium on the abaxial side of a flower meristem adjacent to a bract. *Fig. 3–10.* An L-section of a floral primordium showing two sepal primordia. *Fig. 3–11.* Top view of a flower with five sepal primordia. Three abaxial sepal primordia being elevated by growth of basal meristem. Two abaxial sepal primordia not yet being elevated. Abaxial bract and two lateral bracteoles are present. *Fig. 3–12.* Perspective view of same flower shown in Fig. 3–11. *Fig. 3–13.* An L-section through a flower showing bract and elongating calyx tube. Younger flower primordium visible below the upper flower. *Fig. 3–14.* An L-section of older flower through the center of banner petal primordium on adaxial side of flower. All other petal primordia are also present in this stage of floral development. *Fig. 3–15.* Cross section of flower in same stage of development as Fig. 3–14. The section passes through calyx tube, showing five principal veins, alternating with five petal primordia. Stamens and pistil not yet present. *Fig. 3–16.* First stamen primordium appears on the abaxial side between the keel petal primordia. *Fig. 3–17.* Lateral L-section showing outer floral organs. Both whorls of stamens are present, alternating with one another. The older stamens are larger and the younger, inner stamens are smaller. One small petal primordium is present. *Fig. 3–18.* Same flower showing a median L-section passing through the unicarpellate pistil.

3–2 STAMEN DEVELOPMENT, MICROSPOROGENESIS, AND POLLEN MATURATION

The first whorl of five stamen primordia arises shortly after the initiation of the petal primordia, and is quickly followed by the second stamen whorl. The sequence of development is the same for both whorls of stamens, except that it occurs later in the inner whorl. Each stamen primordium contains a more or less ho-

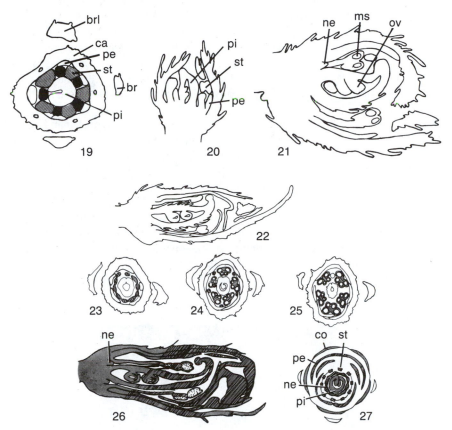

Fig. 3–19 to 3–27. Gross floral development (*continued*). Abbreviations: co—corolla, ne—nectary, pe—petal, pi—pistil, st—stamen. *Fig. 3–19.* Cross section of young flower with all floral organs established. Each whorl of organs is initiated abaxially and alternates with the previously established whorl. The margins of the carpel on adaxial side have not yet fused. × 33. *Fig. 3–20.* Older flower with first indication of ovule primordia in the pistil. The ventral sutures have now fused. × 33. *Fig. 3–21.* An L-section of a flower with ovules at the time of integument initiation. Megasporocytes present in ovule. Anthers have microsporangia with microsporocytes. Nectary is visible encircling the base of the pistil. × 33. *Fig. 3–22.* Older flower in which diadelphous arrangement of the stamens has been established. Ovules have four- or eight-nucleate embryo sacs about 2 d before opening of the flower. × 11. *Fig. 3–23, 3–24, and 3–25.* Cross sections of flower in Fig. 3–22 at different levels. *Fig. 3–23.* Cross section near the base of the flower showing stamen tube and typical diadelphous 9 + 1 arrangement of the stamens. × 11. *Fig. 3–24.* Section through the lower younger anthers. Also visible are the filaments of the upper anthers. × 11. *Fig. 3–25.* Cross section through the upper anthers. × 11. *Fig. 3–26.* Mature flower at the time of anthesis. Calyx, petals, stamen tube, nectary, and pistil with ovules present. × 11. *Fig. 3–27.* Floral diagram showing typical papilionaceous floral organ arrangement. Nectary is about 0.2 mm high at the base of the pistil. × 11.

mogenous mass of cells surrounded by a protoderm layer (Fig. 3–29 and 3–30). As the stamen develops, its apical portion forms a four-lobed anther and a short filament. Each anther lobe consists of a central region of archesporial (primary sporogenous) cells bounded peripherally by four to six layers of cells derived by periclinal divisions of the protoderm. These outer layers mature later into epidermis, endothecium, parietal layers, and tapetum. Toward the center of the anther the arche-

sporium is bound by the centrally located connective tissue, in which the single stamen bundle occurs. The archesporial cells give rise to 25 to 50 microscope mother
cells (MMC) per microsporangium, arranged in two to three columns (Palmer et
al., 1978) (Fig. 3–31 to 3–38).

Each MMC secretes a callose sheath around itself, between plasmalemma and
cell wall. Callose is a nonfibrillar carbohydrate composed of glucose units. The
tapetal cells are reported to be uninucleate (Albertsen and Palmer, 1979; Buss and
Lersten, 1975) or uninucleate and binucleate (Kato et al., 1954; Prakash and Chan,
1976). As meiosis begins in the MMCs, the tapetal cells begin to enlarge and stain
more intensely. Their inner tangential walls, bordering the anther locule, begin to
disorganize. By the end of microsporogenesis, only the outer tangential walls remain. The cytoplasm is then limited only by the plasmalemma.

Osmiophilic orbicules and, later, mucilaginous substances are secreted into
the locule among the young pollen grains (Madjd and Roland-Heydacker, 1978).
The composition of the particles has not been determined. The tapetum becomes
more vacuolate, with diffuse chromatin, and its nuclei are often lobed. The plasmalemma remains intact up to anther dehiscence even though the tapetal cells have
mostly degenerated (Albertsen and Palmer, 1979). Cells of the two innermost pari-

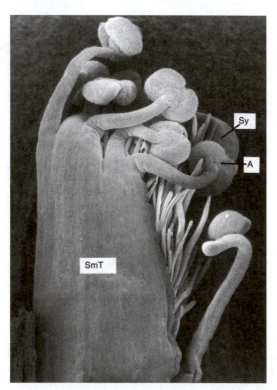

Fig. 3–28. Staminal tube and style growth are synchronized so that the anthers are lifted to the stigma.
Stigma is located behind the labelled anther. Abbreviations: A—anther; SmT—staminal tube, and
Sy—style. From Johns and Palmer (1982). × 27.

etal layers also become somewhat flattened and disorganized. The layer of cells immediately below the epidermis begins to elongate radially and will develop into the endothecium (Fig. 3–38 to 3–46).

During meiosis, the slender threads forming the reticulum within each MMC nucleus become shortened and thickened and, at first metaphase, the haploid number of $n = 20$ chromosomes may be counted. Anaphase and telophase of first meiosis quickly follow. There is then a complete nuclear reorganization resulting in a two-nucleate dyad condition with no intervening cell walls. After a short interphase, a second nuclear division follows, resulting in four microspore nuclei sharing common cytoplasm bounded by the original MMC plasmalemma, the callose layer, and the MMC wall.

Cytokinesis is by simultaneous furrowing, thereby separating the four microspore nuclei into four microspores. The major exine regions are deposited in rudimentary form at this time. The callose disappears later by enzymatic dissolution, the MMC wall breaks, and the individual microspores separate from the tetrad. Each young microspore contains numerous vacuoles, large plastids lacking starch, and a large nucleus appressed to the microspore wall. The wall at this time has the tectum, columellae, and endexine. Three colpi are present (Albertsen and Palmer, 1979).

Mitosis and cytokinesis within the microspore produce the pollen grain, with a large vegetative cell (tube cell) and a small generative cell (Fig. 3–47 and 3–48). The generative cell is at first in contact with the endexine but later is displaced inward by the formation of the intine. Many starch grains accumulate in each plastid of the vegetative cell during intine formation (Fig. 3–49). The mature pollen grain is subtriangular in polar view and spherical to oblate in equatorial view. The exine has both sculptured and smooth areas (Albertsen and Palmer, 1979).

At the time of maturation of the stamens, the anthers are yellow, as is the pollen. The average diameter of the pollen grains varies from 21 to 30 μm (Dzikowski, 1936; Murneek and Gomex, 1936).

During pollen development, the two parietal layers of the microsporangium are crushed. The walls of the endothecium develop U-shaped thickenings on the radial and inner tangential walls (Fig. 3–48 to 3–50). The endothecium is well developed only on the outer walls of the anther and not along the line of dehiscence between the adjacent microsporangia of the anther. As the pollen matures, the septum separating the two microsporangia on either side of the anther ruptures, so that the mature anther has two pollen sacs. The mechanism of dehiscence probably consists of a turning outward of the endothecium as a result of a change in turgor of its cells. The line of dehiscence is along a thin layer of parenchyma tissue, which is easily ruptured under the tension developed by the endothecium.

Microsporogenesis in male sterile (ms_1) and partially male sterile (msp) soybean has been described in detail by Albertsen and Palmer (1979) and Stelly and Palmer (1982). In the ms_1 mutants, microsporogenesis is similar to that of fertile plants from anther ontogeny through telophase II of meiosis, but therefore cytokinesis does not occur; instead, a four-nucleate coenocyte of microspore nuclei forms within the common MMC wall. This coenocyte later develops a pollen wall and accumulates starch and oil reserves as in normal pollen. The coenocyte is round to oval in shape, larger than a fertile pollen grain, and has a variable number of colpi.

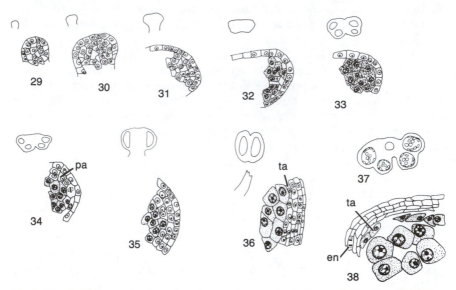

Fig. 3–29 to 3–48. Stamen development and microsporogenesis. Outlines, × 47. Details × 232. Abbreviations: en—endothecium, pa—parietal layer, and ta—tapetum. *Fig. 3–29, 3–30, and 3–31.* Young stamens in outline and detail. Protoderm present but no sporogenous tissue evident. *Fig. 3–32.* Cross section of a young anther. Another becoming four-lobed. Archesporial cells present. *Fig. 3–33.* Anther with four microsporangia. Archesporial cells present in contact with epidermis. *Fig. 3–34.* Periclinal divisions establishing first and second parietal layers. *Fig. 3–35.* Same stage as Fig. 3–34 except in L-section. *Fig. 3–36.* Sporogenous cells are much larger than adjacent parietal and epidermal cells. Two to three parietal layers and future tapetal layer are present. *Fig. 3–37 and 3–38.* Cross section and details of anther. Microsporocytes present. Tapetal cells large. Two parietal layers becoming flattened. Hypodermal cells beginning to elongate radially, initiating the endothecium. *Fig. 3–39 to 3–48 continued on facing page.*

It ultimately collapses but some may first undergo mitosis-like events (Albertsen and Palmer, 1979).

In *msp* mutants, arrest of development and subsequent degeneration of sporogenous tissue may occur at any stage of anther development from sporogenous tissue to almost mature pollen. The onset of abnormalities is most commonly seen near pachytene and at the tetrad and free microspore stages. Cytomixis (movement of chromatin from one cell to another) occurs in sporangia in which development of sporogenous tissue is arrested at pre-meiosis or prophase I. In later stages, fusion of meiocytes results in multinucleate syncytes, which then degenerate. Degenerating sporogenous tissue is usually associated with degenerating tapetal tissue. Although abortion is almost 100% in *msp* anthers, some flowers do contain a few normal pollen grains, thus some successful seed set is possible (Stelly and Palmer, 1982).

3–3 OVULE DEVELOPMENT

Pamplin (1963) studied ovule development of pubescent and glabrous cultivars under both field and greenhouse conditions. He noted that the developmental

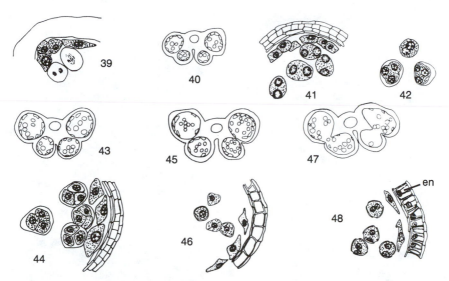

Fig. 3–39 to 3–48. Continued. Fig. 3–39. Meiosis in microsporocyte. Fig. 3–40 and 3–41. Cross section and details of older anther in diad stage. No cell plate forms. Fig. 3–42. Second division has formed tetrad stage. Furrowing of cytoplasm will form four microspores for each microsporocyte. Fig. 3–43 and 3–44. Late stage. Tetrads of microspores. Endothecium continuing to develop. Tapetal cells reach maximum development. Fig. 3–45. Septum between adjacent microsporangia becoming thinner. Fig. 3–46. Microspores have become separated and form thicker walls. Tapetal cells disorganized. Parietal cells are very flattened. Endothecium continuing to differentiate. Fig. 3–47. Almost mature anther. Septum between adjacent microsporangia has ruptured, forming two prominent pollen sacs. Fig. 3–48. Mature pollen grains with three germ pores. Tapetal cells disintegrating. Endothecium has strongly thickened portions along radial and inner tangential walls.

stages were the same under all conditions for the four cultivars he used. George et al. (1979) also concluded that megagametophyte development under greenhouse or field conditions was remarkably similar.

The ovule of soybean has two integuments (bitegmic), and both ovule and embryo sac are bent back on themselves (campylotropous). Megaspores form deep in the nucellus (crassinucellate) (Prakash and Chan, 1976). As many as four ovules first appear as small masses of tissue on the placenta at alternate sides of the posterior suture of the unicarpellate pistil (Fig. 3–51 to 3–53). The cells of an ovule primordium are all about the same size and covered by a single-layered protoderm. Within 1 or 2 d after ovule initiation, several hypodermal archesporial cells are distinguishable (Fig. 3–60). These cells are larger than the neighboring cells and have more densely staining cytoplasm. Soon one of the archesporial cells surpasses the others in size and becomes the functional megasporocyte (Fig. 3–54 and 3–55). The neighboring cells of the archesporium become less prominent and soon resemble the rest of the cells of the young ovule. Periclinal divisions in the hypodermal region produce two parietal layers of nucellus between the elongate megasporocyte and the epidermis of the ovule (Fig. 3–61).

Embryo sac (megagametophyte) development is of the normal or *Polygonum* type, which occurs in more than 75% of angiosperms. Meiosis in the functional megasporocyte results in a linear tetrad of haploid megaspores (Fig. 3–56 and 3–67).

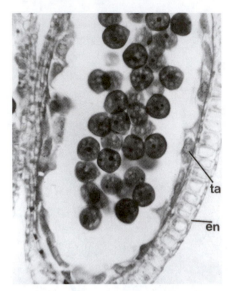

Fig. 3–49. Anther with mature pollen. Pollen grains contain many starch grains. Tapetal cells disintegrating. Endothecium with thickenings in the walls. Abbreviations: en—endothecium, and ta—tapetum. × 350.

Fig. 3–50. Mature anther under polarized light showing crystalline thickenings in endothecial walls and trichome walls. × 94.

Occasionally, a T-shaped tetrad forms (George et al., 1979). The chalazal (i.e., furthest from the micropyle) megaspore continues to enlarge while the three micropylar megaspores become disorganized and soon disintegrate (Fig. 3–57). The first mitotic division of the functional megaspore produces a two-nucleate megagametophyte (Fig. 3–57 and 3–71). These nuclei are displaced to opposite ends by the formation of a large central vacuole. A second mitosis produces the four-nucleate condition (Fig. 3–58 and 3–72). Two successive mitotic divisions result in an eight-nucleate megagametophyte with four nuclei located at the chalazal end and four at the micropylar end of the embryo sac (Fig. 3–62 to 3–73).

Following migration of one nucleus from each end toward the center, and subsequent cell wall formation, the mature megagametophyte consists of seven cells. It is commonly called the *embryo sac*, even though there is no embryo before fertilization. Three chalazal antipodals and an egg and two synergids at the micropylar end are all contained within the large central cell with its two polar nuclei. The egg nucleus is displaced toward the chalazal end of the cell by a large vacuole. Each of the two synergids also has a vacuole, typically at the chalazal end of the cell, which displaces the nucleus toward the micropyle. Tilton et al. (1984a) recently confirmed the presence of a filiform apparatus in each of the synergids. Further maturation of the embryo sac results in the gradual disintegration of the three antipodal cells (Fig. 3–74 and 3–75).

The ovule is supplied by a single vascular bundle, which extends from the posterior bundle of the carpel through the short funiculus. It enters the ovule at the extreme chalazal end just below the hypodermis, where it terminates at the exact center of the chalaza (Pamplin, 1963).

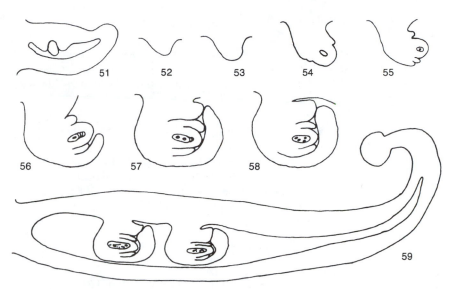

Fig. 3–51 to 3–59. Ovule development. *Fig. 3–52 to 3–58* are × 122. *Fig. 3–51.* Young pistil with three ovule primordia. × 67. *Fig. 3–52 to 3–53.* Ovules appearing as masses of cells alternating on placental margins. *Fig. 3–54.* Integuments being initiated. Ovule bending toward stylar end of pistil. *Fig. 3–55.* Further development of integuments. Megasporocyte present. *Fig. 3–56.* Outer integument has overtaken inner integument. Ovule continues to bend. Four megaspores present. The functional megaspore is larger and at the chalazal end. *Fig. 3–57.* Embryo sac in two-nuclear condition. Ovule now has typical campylotropous form. Outer integument has grown past inner integument and is almost in contact with the placenta. *Fig. 3–58.* Four nucleate embryo sac. Nucellus in contact with outer integument. *Fig. 3–59.* Pistil at time of fertilization. Stylar canal extends almost to capitate stigma. × 53.

The two polar nuclei fuse before fertilization to form a single large diploid secondary nucleus within the large central cell and in close proximity to the egg apparatus. Starch begins to accumulate in the cytoplasm of the central cell, which frequently becomes tightly packed with starch grains. The starch begins to diminish following fertilization and usually has disappeared entirely 1 or 2 d later (Pamplin, 1963). At ovule maturity, antipodals have degenerated and disappeared, the central cell is engorged with starch, and a filiform apparatus (area of conspicuous wall ingrowths) is present in each synergid (Tilton et al., 1984b).

The integuments are initiated from the epidermis of the ovule about the time of the appearance of the megasporocyte. The inner integument arises first, but is quickly followed by the outer integument. Each integument is two cells thick at about the time of division of the functional megaspore (Fig. 3–66), then rapid periclinal divisions and elongation of cells occurs in the outer integument so that it surpasses the inner integument and grows over the apex of the nucellus. The micropyle formed by the outer integument is almost in contact with the placenta of the ventral suture, and has a form described as an inverted Y (Rembert, 1977). The rapid growth of the outer integument results in the apex of the nucellus being in direct contact with the epidermis of the outer integument. The inner integument never forms any part of the micropyle (Fig. 3–54 to 3–59).

During ovule and embryo sac development, the nucellus increased in thickness by cell enlargement and some periclinal divisions. The nucellar cells in contact with the embryo sac become flattened and obliterated at a rate about equal to the production of more nucellar cells. The degeneration of the inner layers of the nucellus is most marked at the micropylar end of the ovule and is first visible when the functional megaspore is elongating (George et al., 1979) (Fig. 3–62).

At the time of fertilization the nucellus still surrounds the embryo sac, but only the epidermis remains intact at the micropylar end, in direct contact with the outer integument (Pamplin, 1963) (Fig. 3–75).

As the seed develops following fertilization, the nucellus ruptures at the micropylar end, exposing the embryo sac so that the suspensor of the embryo is now in direct contact with the epidermis of the outer integument (Fig. 3–84). The chalazal end of the nucellus persists for several days (Fig. 3–87), but continued development of the endosperm finally results in its complete obliteration by 14 d after fertilization (Pamplin, 1963).

3–4 POLLINATION AND DOUBLE FERTILIZATION

By the time of pollination the diadelphous stamens have been elevated so that the anthers form a ring around the stigma. The pollen thus is shed directly on the stigma, resulting in a high percentage of self-fertilization (Williams, 1950) (Fig. 3–28).

Natural crossing varies from <0.5% to about 1%. It has been noted that pollination may occur the day before full opening of the flower; that is, pollination may occur within the bed (Dzikowski, 1936).

We have drawn freely from three recent comprehensive studies by Tilton et al. (1984a, 1984b, 1984c) in the remainder of this section. The wet stigma is overtopped by a proteinaceous film that originates from the cuticle. The film probably prevents dessication of the abundant quantities of lipoidal exudate present at the

---

Fig. 3–60 to 3–75. *Glycine max* L. Merr. From George et al. (1979). *Fig. 3–60.* Several hypodermal archesporial cells divided to form several primary sporogenous and primary parietal cells. *Fig. 3–61.* Megasporocyte just before Meiosis I with the nucleus located in the micropylar third of the cell. *Fig. 3–62.* Abnormally wide megasporocyte; note nucellar degeneration around the megasporocyte. *Fig. 3–63.* Dyad of unequal cells with the chalazal dyad member larger than the micropylar member. *Fig. 3–64.* Meiosis II; nonsynchronous division with the chalazal nucleus in metaphase and the micropylar nucleus in prophase. *Fig. 3–65.* A T-shaped tetrad; no megaspore degeneration. *Fig. 3–66.* Linear tetrad; functional *d* megaspore and degenerating *a, b,* and *c* megaspores. *Fig. 3–67.* Linear tetrad; note vacuoles in the large functional *d* megaspore, the degenerating megaspores and nucellar degeneration. *Fig. 3–68.* One of two linear tetrads in one ovule. *Fig. 3–69.* Isolated nucellus; two-nucleate megagametophyte; note the scattered vacuoles and degenerating megaspores. *Fig. 3–70.* Isolated nucellus; two-nucleate megagametophyte with the nuclei oriented in a line oblique to the long axis of the megagametophyte. *Fig. 3–71.* Isolated nucellus; two-nucleate megagametophyte with nuclei separated by a large central vacuole. *Fig. 3–72.* Isolated nucellus: four-nucleate megagametophyte with the chalazal and micropylar nuclei oriented in a plane almost parallel to the long axis of the megagametophyte. *Fig. 3–73.* Isolated nucellus: eight-nucleate megagametophyte. *Fig. 3–74.* Mature megagametophyte: two polar nuclei and two synergids in front of the egg: note the starch grains in the micropylar nucleus of the egg. *Fig. 3–75.* Mature megagametophyte; degenerating antipodals, two large polar nuclei, a large egg flanked by a synergid; note starch grains in the megagametophyte and the nucellar degeneration. →

distal end of the stigma and confines the exudate to the stigmatic surface. It may also contain recognition factors.

The transmitting tissue of the stigma is made up of papillae with lateral protrusions that anastomose with each other. Papillae of this type occupy the distal end of the stigma and secrete most of the stigmal exudate (Fig. 3–76). Proximal to them are one to three whorls of free papillae lacking protrusions. These are also secretory (Fig. 3–80).

There are numerous exudate-filled channels in the stigma and style. Pollen tubes grow in these channels, which provide nutrition and mechanical guidance. At the base of the stigma, in the transition zone between stigma and style, there is a gradual increase in the amount of exudate between cells except in the center of the style. These cells comprise the stylar-transmitting tissue; they secrete an exudate

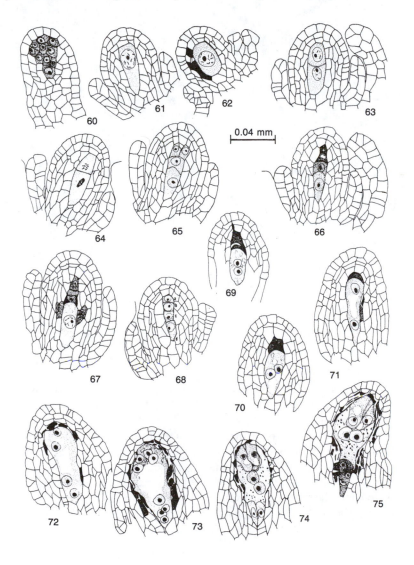

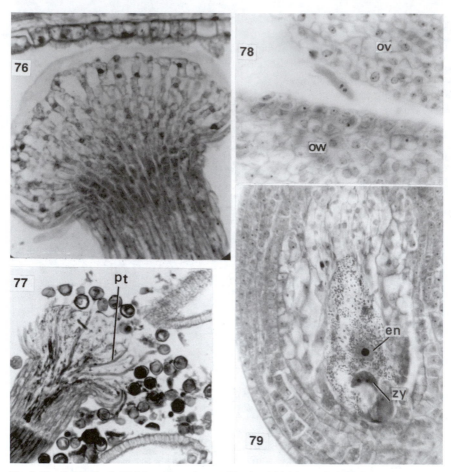

Fig. 3–76 to 3–79. *Fig. 3–76.* Stigma just prior to pollination. Thick layer of stigmatic substance on surface to stigma. × 520. *Fig. 3–77.* Stigma with many pollen tubes on day of flowering. Note broken
pollen sacs in vicinity of stigma. × 236. *Fig. 3–78.* Pollen tube with two male gametes. Ovule to right
and above; ovary wall below. × 380. *Fig. 3–79.* Zygote and primary endosperm nucleus. Abbreviations: en—primary endosperm nucleus, ov—ovule, ow—ovary wall, pt—pollen tube, zy—zygote.
× 375.

similar in appearance to that of the stigma. Stylar-transmitting tissue cells are
mostly free from one another along their axial walls, thus forming a conduit through
which pollen tubes grow.

Pollen usually germinates on the surface of the film overlying the stigmal exudate. Germination can also occur among the lower whorls of papillae but these
tubes then grow into the stigma before entering the style (Fig. 3–77 and 3–81). Although many pollen grains are deposited on the stigma, and most of them germinate and grow into the stigma and upper style, perhaps as many as 90% of the tubes
atrophy and die before reaching the distal end of the ovary. Only a few pollen tubes
reach the locale and compete for ovules to fertilizer.

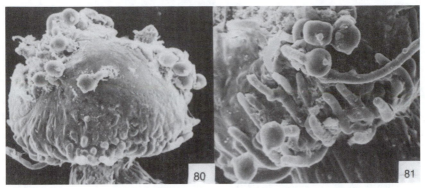

Fig. 3–80. Scanning electron micrograph view of pollen germinating on stigma. From Tilton et al. (1984c). × 230.

Fig. 3–81. Pollen grains lodged between lower-whorl papillae germinate but the pollen tubes grow into the main body of the stigma. From Tilton et al. (1984c). × 335.

Pollen tubes grow between the cells of the stylar-transmitting tissue. The ovarian transmitting tissue forms a secretory obturator on top of which the pollen tubes grow toward the ovules. Its exudate is pectinaceous, which perhaps controls the direction of pollen tube growth chemotactically.

During growth of the pollen tube toward the ovule, the generative cell divides and forms two male gametes, the sperm cells (Fig. 3–78). Finally, the pollen tube grows through the micropyle of the ovule, between nucellar epidermal cells, and enters the filiform apparatus of the degenerate synergid. Here the pollen tube tip bursts and releases the two sperm cells. One sperm fuses with the egg and forms the diploid zygote, the first cell of the embryo, while the other sperm fuses with the secondary nucleus. forming the primary endosperm nucleus (Fig. 3–79). Rustamova (1964) noted that the time from pollination to fertilization varies from about 8 to 10 h. Thus, the day of full opening of the flower is likely the day of fertilization or perhaps is 1 d after fertilization.

Following fertilization, a zone of extensive wall ingrowths called a Wandlabrinthe develops around the micropylar end of the central cell. The Wandlabrinthe encircles the synergids by forming a ridge of cell wall material that projects into the central cell toward the zygote and the synergids at the level of the filiform apparatus (Fig. 3–82). Degenerating nucellar cells are adjacent to, and contiguous with, the Wandlabrinthe, and adjacent cells of both integuments are rich in starch reserves. The Wandlabrinthe may increase the movement of nutrients into the central cell for endosperm and embryo nutrition.

3–5 EMBRYO DEVELOPMENT

The vacuole in the zygote becomes smaller and finally disappears entirely about the time of the first cell division, which occurs about 32 h after pollination (Pamplin, 1963; Rustamova, 1964). Soueges (1949) described soybean embryogeny from the first division of the zygote through the early cotyledon stages (Fig. 3–83).

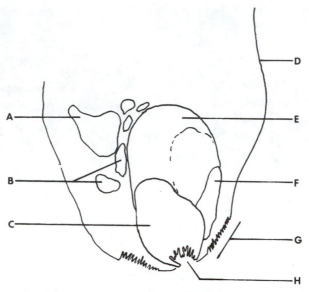

Fig. 3–82. Egg apparatus end of embryo sac, reconstructed in outline from transmission electron micrographs in Tilton et al. (1984). Abbreviations: A—endosperm initial, B—amyloplasts, C—persisting synergid, D—central cell wall, E—Zygote, F—disintegrating synergid, G—Wandlabrinthe, and H—filiform apparatus.

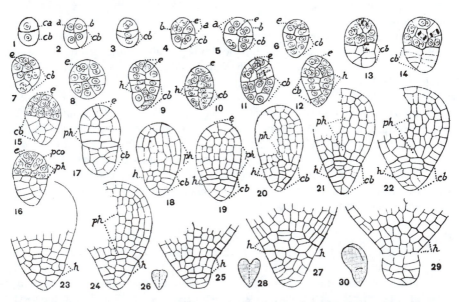

Fig. 3–83. *Glycine soja* Sieb. and Zucc. The principal stages of embryo development. From Soueges (1949). Abbreviations: *a* and *b*—daughter cells of the apical cell *ca*, and *cb*—basal cell, *e*—epiphysis, *h*—hypophysis, *ph*—hypocotyl proper, and *pco*—cotyledon primordium region.

The first division of the zygote is transverse. The apical cell, facing the central cell, will become the embryo. The basal cell, facing the micropyle, forms the suspensor, an ephemeral structure that may aid early embryo growth (Fig. 3–83-1 and 3–84). Continued divisions of the derivatives of the apical cell produce the spherical proembryo at about 3 d (Fig. 3–83-14). The proembryo is approximately the same size as the somewhat conical suspensor (Fig. 3–85). A well-defined protoderm is present in the proembryo by 5 d after fertilization (Fig. 3–83-21 and Fig. 3–90). The suspensor in cv. Hawkeye at 6 d is several cells wide (Fig. 3–90), although suspensors of other cultivars are reported as tiny or rudimentary to small (Lersten, 1983).

About 6 to 7 d after fertilization, localized divisions at opposite sides of the proembryo just below the protoderm initiate the cotyledons. Pamplin (1963) observed that the cotyledon at the chalazal side of the embryo seems to be initiated first but is quickly followed by initiation of the second cotyledon, which grows rapidly and soon is the same size as the first cotyledon. In Fig. 3–91 and 3–92, the chalazal cotyledon appears to be slightly larger than the one toward the anterior surface of the ovule. This, however, may be a consequence of the plane of sectioning rather than an actual difference in size. The cotyledons in this so-called *heart stage* are initiated and developed in a plane that is approximately 90° displaced from their final position in the mature seed. As the cotyledons continue to develop, there is a gradual rotation such that the embryo, with its cotyledons, moves 90° and the cotyle-

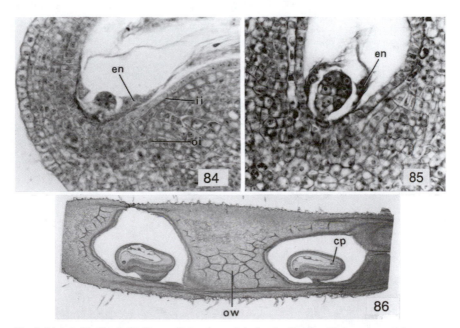

Fig. 3–84 to 3–86. *Fig. 3–84.* Four-celled embryo, 2 d after fertilization. Endosperm nuclei near embryo. × 320. *Fig. 3–85.* Club-shaped embryo, 3 d after fertilization. Endosperm surrounding embryo is acellular. × 320. *Fig. 3–86.* Pod with two ovules, 5 d after fertilization. Abundant endosperm in embryo sac is cellular only near embryo. Both ovules in the same development stage. Abbreviations: cp—chalazal process of endosperm, en—endosperm, ii—inner integument, oi—outer integument, and ow—ovary wall. × 10.

dons assume the position they will have in the mature seed, with their inner surfaces forming a plane to the sides of the ovule (Fig. 3–93 to 3–95).

At this stage the cotyledons appear circular in outline, but rapid growth along the margins, especially toward the chalazal end of the ovule results in a pronounced

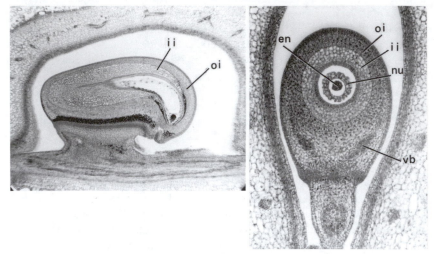

Fig. 3–87. Entire ovule with spherical embryo, 6 d after fertilization. Micropyle visible. Endosperm nuclei line outer surface of embryo sac. Large vacuole present in center of endosperm. × 28.

Fig. 3–88. Cross section of ovule at chalazal end, 6 d after fertilization. Chalazal process of endosperm and nucellus present. Abbreviations: en—chalazal process of endosperm, ii—inner integument, nu–nucellus, oi—outer integument, and vb—vascular bundles. × 73.

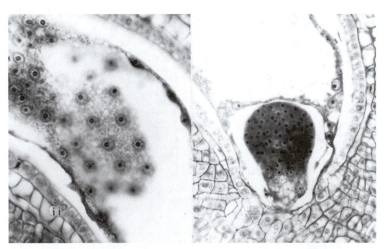

Fig. 3–89. Nuclear, noncellular endosperm showing cytoplasm, symmetrical spacing of nuclei, and large central vacuole of endosperm at right. Near the center of ovule, 6 d after fertilization. × 290.

Fig. 3–90. Embryo at globe stage with lightly stained suspensor, 6 d after fertilization. × 290.

elongation of the cotyledons, giving them their typical reniform shape. Ten to 12 d after fertilization, the tissue systems of the hypocotyl are well blocked out and consist of protoderm, the ground meristem of the cortex, and procambium. The de-

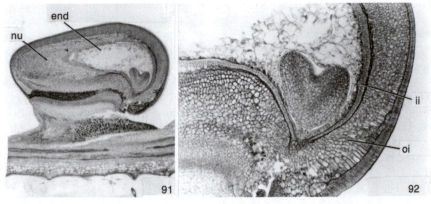

Fig. 3–91. Entire ovule, 8 d after fertilization. Cotyledons starting to develop. All endosperm is cellular. Abbreviations: end—endosperm and nu—nucellus. × 20.

Fig. 3–92. Enlarged portion of Fig. 3–91. Note small suspensor. Palisade epidermal cells of seed coat becoming elongated. Integumentary tapeturm of inner integument prominent. Abbreviations: ii—inner integument and oi—outer integument. × 86.

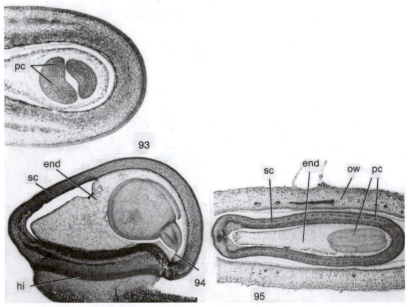

Fig. 3–93 to 3–95. *Fig. 3–93*. Cross section of cotyledons 12 d after fertilization. Cotyledons starting to rotate. × 50. *Fig. 3–94*. Late cotyledon stage, 14 d after fertilization. Abundant cellular endosperm. Suspensor still intact. Double layer of palisade cells visible at hilum. × 19. *Fig. 3–95*. An L-section of 14-d-old seed. Cotyledons have rotated 90° and are now in normal position. Endosperm fills most of the remainder of the seed. Extensive procambium network in seed coat. Abbreviations: end—endosperm, ow—ovary wall, pc—procambium, sc—seed coat, su—suspensor, and hi—hilum. × 19.

rivatives of the hypophysis have formed the initials of the root which, until the time of germination, remain limited to a small area at the end of the hypocotyl just above the point of attachment of the suspensor (Fig. 3–94).

The cotyledons, at about the time of the beginning of rotation of the embryo, already have procambium continuous with the procambium of the hypocotyl. The procambium of the cotyledons continues to develop and forms the finely divided vascular system present in the mature seed (Fig. 3–106).

The epicotyl is initiated simultaneously with the origin of the two cotyledons, as a residual meristem between them. Pamplin (1963) stated that it first appears as an elongated mound of deeply staining cells between the bases of the cotyledons. The outermost cell layer becomes the tunica. About 14 d after fertilization, the epicotyl forms the primordia of the two primary leaves at right angles to the point of attachment of the two cotyledons (Fig. 3–97 and 3–98). The primary leaves continue to enlarge, and by 30 d have reached their maximum dormant embryo size and have assumed the conduplicate vernation characteristic of the plumule of the mature seed (Kato et al., 1954) (Fig. 3–109).

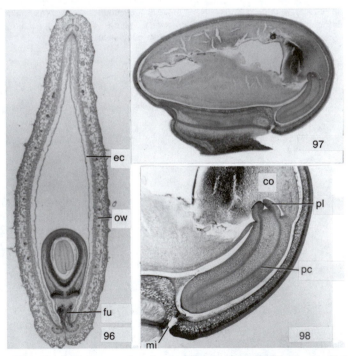

Fig. 3–96 to 3–98. *Fig. 3–96.* Cross section of pod and one seed 14 d after fertilization. Embryo in late cotyledon stage. Vascular bundle of funiculus shown connected to lateral bundle of ventral suture of carpel. × 11. *Fig. 3–97.* Medium L-section of seed 18 d after fertilization. The section was parallel to inner epidermis of both cotyledons and resulted in some tissue breakage. × 9. *Fig. 3–98.* Same seed. Primary leaves are present and oriented at right angles to the point of attachment of the cotyledons. Abbreviations: co—cotyledon, ec—membranous endocarp, fu—funiculus, mi—micropyle, ow—ovary wall, pc—procambium, and pl—primary leaves. × 24.

The first trifoliolate leaf primordium, differentiated about 30 d after fertilization near the base of the two simple leaves, remains reduced in size and does not resume development until the time of germination.

Bils and Howell (1963) described biochemical and cytological changes during the development of soybean cotyledons. They noted that, about 15 to 18 d after flowering, plastids, mitochondria, and some lipid and protein globules were beginning to form. By 26 d, when the cotyledons had reached their maximum size, the cells contained many mitochondria, some lipid granules, and a few protein globules. As fresh weight began dropping during the final stages of maturation, the starch also started to decrease and, by the time of embryo dormancy, starch grains were gone. Lipids comprised 22% of the dry weight of the cotyledons, and proteins about 50%, at 60 d.

3–6 ENDOSPERM DEVELOPMENT

The primary endosperm nucleus divides almost immediately following fertilization. By the time of zygote division, the endosperm already has several free nuclei (Pamplin, 1963; Prakash and Chan, 1976). Divisions of the endosperm nuclei occur as simultaneous cycles for several days following fertilization. The nuclei and common cytoplasm of the endosperm are displaced toward the periphery of the embryo sac by the development of a large vacuole in the center of the mass of endosperm (Fig. 3–87). The free nuclei of the sac-like endosperm are spaced uniformly within the cytoplasm (Fig. 3–89). By 5 d after fertilization, the endosperm begins to become sac and, by 8 d, the heart-shaped embryo is completely embedded in cellular endosperm (Fig. 3–92) (Meng-Yuan, 1963; Prakash and Chan, 1976; Takao, 1962). Endosperm cell walls develop gradually toward the chalazal end of the embryo sac; by 14 d they extend almost to the chalazal end of the ovule (Fig. 3–94).

The chalazal end of the endosperm never becomes cellular but instead forms a rather darkly staining acellular mass, termed the *chalazal process* by Pamplin (1963). This can be seen in the basal ovule in Fig. 3–86 and 3–88. The chalazal process is connected with the degenerated nucellus to form a so-called *chalazal haustorium*. This haustorium adheres to the nucellus, which is connected to the vascular bundles. Takao (1962) concluded that nutrients move from the vascular bundles through the chalazal end of the outer integument, the inner integument, the nucellus, and finally to the fluid cytoplasm in the embryo sac and the endosperm. The acellular chalazal process finally is crushed by the continuous growth of the cellular endosperm and, by 12 or 14 d after fertilization, it has been completely obliterated.

As the ovule continues to enlarge, both embryo and endosperm grow at approximately the same rate so that the relative proportion of endosperm to embryo tissue remains the same until about 14 d after fertilization (Pamplin, 1963). The rapidly growing cotyledons, after they have rotated completely, accumulate food reserves derived from the endosperm. Eighteen or 20 d after fertilization, only remnants of the endosperm remain. It should be emphasized that no evidence exists that cotyledons absorb endosperm directly. In the mature seed, the only evidence of endosperm is a thin aleurone layer and a few crushed endosperm cells (Fig. 3–103). In some cultivars, an aleurone layer is lacking (Prakash and Chan, 1976).

3–7 SEED-COAT DEVELOPMENT

The inner integument consists of two to three cell layers at the time of fertilization. After fertilization, periclinal divisions, especially in the chalazal end of the ovule, result in an increase in thickness of the inner integument to about 10 cell layers (Fig. 3–87 and 3–88). About 10 to 14 d after flowering, the inner layer of the inner integument becomes more densely staining and differentiates as an endothelium or integumentary tapetum, which presumably serves a nutritive function (Fig. 3–89). Thorne (1981) noted the presence of a layer of multicellular tubules, each 50 by 200 µm long, between the seed coat and the cotyledons in soybean seeds nearing maturity. He suggested that these tubules help to transfer nutrients from the endothelium to the embryo. During development of the embryo, there is a gradual crushing of the inner integument, starting at the micropylar end and proceeding toward the chalazal end of the ovule. By 12 to 14 d after fertilization, most of the inner integument has completely disappeared (Kamata, 1952; Pamplin, 1963).

The outer integument at fertilization is two to four cell layers thick except in the region of the micropyle and hilum, where it is considerably thicker (Pamplin, 1963). After fertilization, periclinal divisions occur and the outer integument becomes approximately 12 to 15 cell layers thick (Prakash and Chan, 1976). The epidermis of the outer integument consists of isodiametric cells at the time of fertilization. During growth and maturation of the seed, these cells elongate radially, especially near the hilum. The epidermal cells of the funiculus in the hilum region also elongate radially so that in the hilum there is a double layer of elongate thick-walled epidermal cells (Fig. 3–94, 3–99, and 3–100).

Extending the length of the hilum is a narrow, median strip of epidermal cells that never become thick-walled or elongate, and which separate when dry, leaving

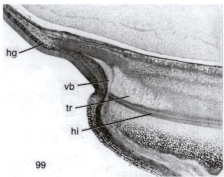

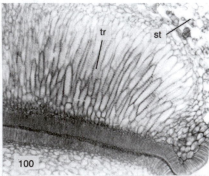

Fig. 3–99. Same seed as in Fig. 3–97 chalazal end of hilum. Single vascular bundle enters at the posterior edge of funiculus. Tracheid bar present adjacent to vascular bundle in the upper part of hilum. × 32.

Fig. 3–100. Same seed. Details of tracheid bar at micropylar end of hilum. Note the retriculate thickenings of tracheid wall. Abbreviations: st—stellate parenchyma, tr—tracheids, and vb—vascular bundle. × 117.

a narrow cleft (hilar groove) in the mature seed (Fig. 3–101). Just below this cleft, some cells of the outer integument differentiate into cells (tracheoids) that resemble tracheids, forming the so-called *tracheid bar* of the mature seed. The tracheoids are lignified, pitted, and oriented with their long axis perpendicular to the hilar groove. Lersten (1982) noted that tracheoids rarely had an intact pit membrane and, in the tribe Phaseoleae, subtribe Glycininae, the pits were warty to vestured. He speculated that the structure of the tracheoid pits enhances the efficiency of the tracheid bar and hilum in gas exchange. The tracheid bar may extend the entire length of the hilum or may be separated into two groups of tracheids, one near the micropyle and the other near the chalazal end of the seed. Although the tracheoids have scalariform pitted walls and resemble true conducted tissue, they seem to have no conducting function (Dzikowski, 1937).

The hypodermal cells of most of the outer integument, by 28 d after fertilization, have differentiated into sclerified cells described in the section on the mature seed (Fig. 3–98 and 3–104).

The entire vascular system of the ovule at the time of fertilization consists of a single median vascular bundle that passes from the vascular bundle of the ventral suture of the pod through the funiculus and enters the ovule at the chalazal end (Fig. 3–96). By 4 or 5 d after fertilization, two lateral procambium strands appear in the outer integument above the hilum and extend most of the length of the ovule near the inner surface of the seed coat (Fig. 3–88). Maturation of these bundles proceeds from the chalazal end, where they are attached to the single median bundle, and progresses forward toward the micropylar end. A number of procambium branches develop from these two lateral veins and, by 12 d, a rather extensive anastomosing retriculate system of veins has formed throughout the entire outer integument (Pamplin, 1963; Thorne, 1981) (Fig. 3–95 to 3–102).

The veins are composed of small, thick-walled sieve tubes surrounded by a bundle shealth of small vascular parenchyma cells. Xylem is absent (Thorne, 1981). As the seed coat continues to mature, the inner parenchyma cells become crushed and flattened, and eventually there is little evidence of a functional vascular system. Presumably the mature seed coat has no functional vascular tissue.

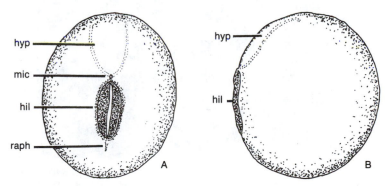

Fig. 3–101. Drawing of 'Chippewa' soybean seed. (A) Top view. (B) Side view, Abbreviations: hyp—hypocotyl-radicle axis, mic–micropyle, hil–hilum with central fissure, and raph—raphe. × 6.

Fig. 3–102. Vascularization of the soybean seed coat. (A)
Sketch of a lateral bundle, illustrating the approximate relationship to the hilum, illustrated in (B). The initial branching of the retriculate venation is also illustrated in (A). (C)
Sketch of a typical transverse section of an entire seed attached to the funiculus, illustrating the approximate location of the lateral vascular bundles in relation to the tracheid
bar, hilum, and cotyledons. Abbreviations: CT—cotyledon, F—funiculus, HL—hilum, P—parenchyma, TB—tracheid bar, and VB—vascular bundle. From Thorne (1981).

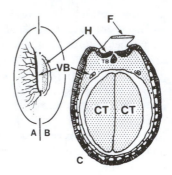

3–8 MATURE SEED

The mature soybean seed, like that of many other legumes, is essentially devoid of endosperm and consists of a seed coat surrounding a large embryo. Seed
shape varies among cultivars from almost spherical to strongly flattened and elongate, but the seeds of most cultivars are oval in outline. The seed coat is marked with
a hilum (seed scar) that varies in shape from linear to oval. At one end of the hilum
is the micropyle, a tiny hole formed by the integuments during seed development
(Fig. 3–97 and 3–98), but covered by a cuticle at maturity. The tip of the hypocotyl-
radicle axis, often visible through the seed coat, is located just below the micropyle
(Fig. 3–101). At the other end of the hilum is the raphe, a small groove extending
to the chalaza, where the integuments were attached to the ovule proper. In most
cultivars, the complete separation of the funiculus from the seed forms a hilum with
a smooth surface except for a narrow fissure running lengthwise down its center.

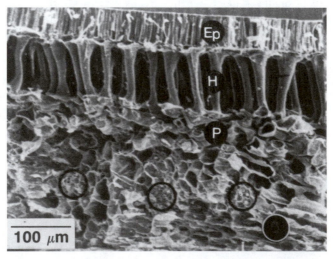

Fig. 3–103. Scanning electron micrograph of a seed-coat portion illustrating the reticulate venation (*cir-
cled*) embedded within the parenchyma layer. The distinctive spindle-shaped cells of the epidermis
and hourglass-shaped cells of the hypodermis provide structural support to the seed. Abbreviations:
Ep—epidermal cells (macrosclereids), H—hypodermis (osteosclereids). From Thorne (1981).

In a few varieties the funiculus remains attached to the hilum by a core of parenchyma (Dzikowski, 1936). When the funiculus finally separates in these cultivars, the hilum is rough and has a wide, white central scar formed by the parenchyma tissue.

The seed coat proper has three distinct layers: (i) epidermis, (ii) hypodermis, and (iii) inner parenchyma layer (Fig. 3–104). The epidermal layer consists of closely packed, thick-walled palisade cells (macrosclereids). These cells, 35 to 70 μm long, are elongated perpendicular to the surface of the seed and have thickened, pitted walls in the outer part of the cell. A cuticle is present on the outer wall of the macrosclereids. As is common in legumes, there is a particularly compact zone present in the walls of the upper part of the macrosclereids that refracts light more strongly than the rest of the wall (Esau, 1965). This characteristic *light line* is visible in seeds of many wild forms of the soybean, but is less prominent in the cultivated forms (Alexandrova and Alexandrova, 1935).

The hypodermis consists of a single layer of sclerified cells variously elongated and separated from each other. These cells range from 30 to 100 μm in length (Patel, 1976). The unevenly thickened cell walls are thin at the ends of the cell and very thick in the central, constricted portion of the cell. These cells thus form a strong supporting layer with considerable intercellular space (Fig. 3–103 and 3–104).

The inner parenchyma tissue consists of six to eight layers of thin-walled, flattened cells that lack contents. This parenchyma tissue is essentially uniform throughout the entire seed coat except at the hilum, where it forms three distinct layers: (i) an outer layer, formed of stellate parenchyma tissue with much intercellular space, in contact with the sclerified hypodermal cells; (ii) a middle parenchyma layer consisting of tiny, flattened cells and containing small bundles of spiral vessels that branch out around the hilum; (iii) an inner layer consisting of more or less typical parenchyma tissue (Dzikowski, 1936).

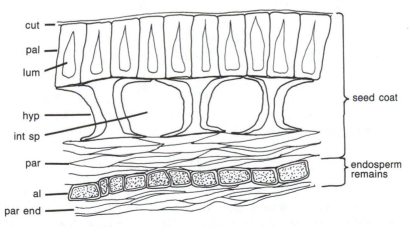

Fig. 3–104. Seed coat of Hawkeye soybean. Abbreviations: al—aleurone cells of endosperm, cut–cuticle, hyp–hourglass cells of hypodermis, int sp–intercellular space, lum—lumen, pal—palisade, par—compressed parenchyma cells, and par end—remains of parenchyma cells of endosperm. × 535.

It is probably the physiological significance that the micropyle and the fissure in the center of the hilum are in direct contact with the loosely packed stellate parenchyma cells and that these are in contact with the extensive intercellular space formed by the sclereid layer. Since the cutinized palisade cells are essentially impermeable to gases, the principal pathway for gas exchange between the embryo and the external environment is through the hilum. Therefore, the structure of the hilum probably has an effect on the metabolism and moisture content of the embryo (Dzikowski, 1936).

The remnants of the endosperm are tightly appressed to the seed coat proper (Fig. 3–104). The outer endosperm layer, called the *aleurone layer*, is composed of small cuboidal cells filled with dense protein. Just to the interior to the aleurone layer are several layers of endosperm cells that have been flattened by growth of the embryo (Dzikowski, 1936; Williams, 1950). In some cultivars there is no distinct aleurone layer (Prakash and Chan, 1976).

Recently, Wolf et al. (1981) and Hill and West (1982) surveyed surface features of soybean seed coats. Many cultivars had abundant pits, up to 277 pits mm^{-2}, while others had few or none. Some "hard" seeds had pits; other seeds lacked them. Pits vary in shape from circular, 15 to 25 µm in diameter, to elongate, 3 × 40 µm. The pits penetrated about 20 to 35% of the thickness of the palisade layer. Oval-shaped cavities were often present below the pit, extending to the sclereid layer. Pits have been shown to provide an entry for fungal hyphae and they may also contribute to water uptake during germination.

The seed coats of some cultivars have a superficial reticulate or honeycomb pattern, which is formed by the residue of the epidermal cell walls of the inner layer of the ovary wall (endocarp), which adheres to the seed coat. Occasionally, small crystals are present as deposits on seed coats (Wolf et al., 1981).

The mature, dormant embryo consists of two large fleshy cotyledons, a plumule with two well-developed primary leaves enclosing one trifoliolate leaf primordium, and a hypocotyl-radicle axis that rests in a shallow depression formed by the cotyledons (Fig. 3–105, 3–108, and 3–109). The tip of the radicle is surrounded by an envelope of tissue formed by the seed coat (Miksche, 1961).

In cross section, each cotyledon is semicircular in outline, bounded by an epidermis of cuboidal cells containing aleurone grains (Fig. 3–105). Stomates are present on both surfaces. The mesophyll of the flat, adaxial side of the cotyledon is made up of one to three layers of palisade tissue that merge with a more spongy type of parenchyma in the central portion of the cotyledon (Fig. 3–106). The abaxial region of the cotyledon consists mainly of elongate parenchyma cells that do not form distinct layers. All cells of the mesophyll are filled with closely packed aleurone grains and oil droplets. Small crystals of calcium oxalate are scattered throughout the cotyledon.

The plasma membrane of parenchyma cells in cotyledons of dry seeds is disorganized, or at least not discernible, in many areas, while elsewhere it has a typical unit membrane structure. In these cells no endoplasmic reticulum can be discerned, and mitochondria are of irregular shape with little evidence of cristae. Nuclei appear round or lobed, bounded by a membrane in which nuclear pores are visible. A 20-min period of imbibition brings about an extensive reorganization of the membranes and organelles (Webster and Leopold, 1977).

Fig. 3–105. Cross section, cotyledons of dormant Hawkeye seed. Abbreviations: mv—midvein, lv—lateral vein. × 16.

A small pit is present in the center of the abaxial surface of the cotyledon above the midvein (Fig. 3–107). It is prominent in some cultivars but barely perceptible in others (Dzikowski, 1937; Miksche, 1961).

The plumule is about 2-mm long and has two opposite simple leaves, each with a pair of stipules at the base. The vascular system of the primary leaves is pin-

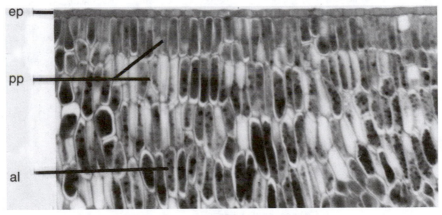

Fig. 3–106. Adaxial side of dormant cotyledon, Hawkeye soybean. Abbreviations: ep—epidermis, pp—palisade parenchyma, and al—aleurone grains. × 130.

nate and consists of protoxylem and metaxylem initials, and some mature protophloem elements (Miksche, 1961). The stem apex has a uniseriate tunica and a
massive corpus. Prior to seed maturation and dormancy, cell divisions in the corpus form the initials of the first trifoliolate leaf.

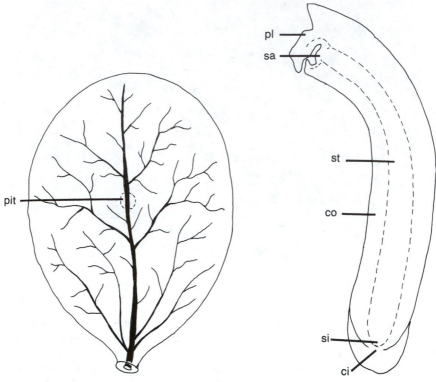

Fig. 3–107. Cleared dormant cotyledon of Hawkeye
soybean, showing netted venation and pit. From
Miksche (1959). × 7.

Fig. 3–108. Longitudinal section of embryo
axis. Most of the primary leaves are removed. From Miksche (1959). Abbreviations: ci—common initials; co—cortex,
pl—primary leaf base, sa—stem apex,
si—stelar initials, and st—stele. × 23.

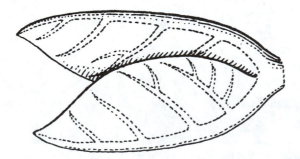

Fig. 3–109. Plumule with two simple leaves showing pinnate venation. From Miksche (1959).

The hypocotyl-radicle axis is about 5-mm long and somewhat flattened both on the outer surface, which is in contact with the seed coat, and on the inner surface, which is tightly appressed to the cotyledons. The radicle, located at the tip of the embryo axis, consists of the stelar initials that produce the stele and a group of common initials that give rise to the root cap, epidermis, and cortex (Miksche, 1961). The transition from root to hypocotyl is not marked by any clear anatomical change in the dormant embryo. The tissue system present in the hypocotyl are the epidermis, cortex, and stele (Fig. 3–108).

Soybean seeds vary in color from yellow, green, brown, to black, and they may be of one color, bicolored, or variegated. Seed-coat pigments, located mainly in the palisade layer, consist of anthocyanin in the vacuole, chlorophyll in the plastids, and various combinations of breakdown products of these pigments. Both the palisade layer and the stellate parenchyma are often pigmented in the hilum, thus giving a more intense coloration to that region (Alexandrova and Alexandrova, 1935).

The cotyledons of the mature embryo may be green, yellow, or chalky yellow, but in most genotypes they are yellow (Williams, 1950). The various combinations of pigments in the seed coat and cotyledons are responsible for the wide range of colors of the seeds in wild and cultivated varieties of soybean.

3–9 POD DEVELOPMENT

The number of pods varies from 2 to more than 20 in a single inflorescence and up to 400 on a plant. A pod may contain from one to five seeds, but in most common cultivars it usually has two or three seeds per pod (Kato et al., 1954; Williams, 1950). Soybean pods are straight or slightly curved and vary in length from <2 cm to up to 7 or more centimeters in some cultivars (Fig. 3–110). Mature pods vary from light yellow to yellow-gray, brown, or black. Pod coloration depends on the presence of carotene and xanthophyll pigments, the color of the trichomes, and the presence or absence of anthocyanin pigments (Dzikowski, 1936).

From the moment of fertilization the ovary starts developing into the fruit, but the style and stigma dry out. The calyx persists during fruit development; remnants of the corolla may also be present when the fruit is mature.

Fig. 3–110. Seventeen-day pods, Ottawa-Mandarin soybean.

The soybean pod is similar to that of other legumes, consisting of two halves of the single carpel joined by dorsal and ventral sutures (Fig. 3–96 and 3–112). The dorsal suture constitutes the main vein of the former carpel, while the ventral suture consists of two principal bundles that correspond to marginal veins of the carpel. On both the dorsal and ventral sutures above the veins, the epidermis of the pod bends inward, forming deep grooves (Dzikowski, 1937) (Fig. 3–112). Extending below the grooves is a vertical layer of parenchyma that separates the conducting tissues into two regions. These layers of parenchyma later help the pods to dehisce.

The wall of a young pod consists of a variously hairy epidermis, a rather wide zone of parenchyma tissue in which the extensive vascular system is embedded, and an inner, very thin layer of parenchyma tissue destined to become the membranous endocarp (Fig. 3–86, 3–95, and 3–96).

As the pod matures, the epidermal cells develop strongly thickened walls covered by a well-developed cuticle. The surface of the epidermis has numerous elevated stomates, each connected by a substomatal chamber to the parenchyma of the inner portion of the pod. The clavate trichomes disappear but the setaceous trichomes develop thick walls and are persistent at maturity (Fig. 3–110) (Dzikowski, 1936, 1937).

Just below the epidermis there is a hypodermal layer consisting of a single layer of short fusiform fibers with very thick walls and numerous small pits (Fig. 3–113 and 3–115). The parenchyma below the hypodermal layer consists of many

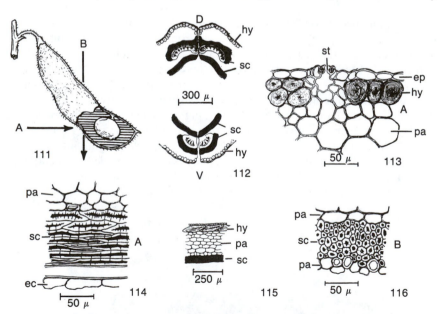

Fig. 3–111 to 3–116. *Fig. 3–111.* Soybean pod showing the direction of the fibers in the fibrous layers. *Fig. 3–112.* Diagram of ventral (V) and dorsal (D) sutures of a pod. *Fig. 3–113.* Portion of a transverse section of the wall of a pod, along the length of fibers (section A, Fig. 3–110). *Fig. 3–114.* The same part of the pod in section A. *Fig. 3–115.* Section of the pod wall, perpendicular to the fibers in the fibrous layer (section B, Fig. 3–110). *Fig. 3–116.* Inner layer of the pod wall. Portion of section B, Fig. 3–110. Abbreviations: ec—endocarp, ep—epidermis, hy—short fibers under the epidermis, pa—parenchyma, sc—sclerenchyma layer, and st—stoma. From Dzikowski (1937).

large, thin-walled isodiametric cells. Within this parenchyma tissue, an extensive vascular system of anastomosing veins interconnects the principle bundles of the dorsal and ventral sutures. Below the parenchyma layer is a rather thick layer of elongate sclerenchyma fibers, which are thick-walled and of small diameter (Fig. 3–116). The orientation of the cells of the hypodermis is almost perpendicular to the orientation of the cells of the inner sclerenchyma layer (Dzikowski, 1936, 1937; Monsi, 1942) (Fig. 3–111). The innermost layer of the pod is a very thin endocarp composed of rather flattened parenchyma cells (Fig. 3–114).

When the pod dehisces, only the inner sclerenchyma layer, termed *motion tissue* by Monsi (1942), seems to be directly involved. A section cut parallel to the long axis of the sclerenchyma cells of the inner layer reveals that there are two distinct layers of cells present. The sclerenchyma cells closest to the central parenchyma tissue are considerably shorter, with blunt ends, and have abundant pits. Monsi (1942) and Dzikowski (1937) determined that these inner cells have an essentially transverse orientation of fibrils in their secondary wall. The sclerenchyma cells of the innermost layers are longer, narrower with more pointed ends, and have fibrils with an essentially longitudinal orientation (Fig. 3–114).

Separation of the two halves of the pod is preceded by the appearance of clefts through the parenchyma of the dorsal and ventral sutures. After separation, the halves twist spirally around the axis, that is, parallel to the direction of the fibers of the inner sclerenchyma layer (Dzikowski, 1937). Dehiscence of the pod must be directly due to differences in tension developed in the cells of the inner sclerenchyma layer

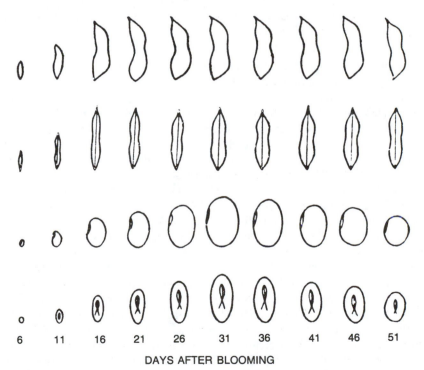

Fig. 3–117. Development of soybean seed and pod. From Suetsugu et al. (1962).

Table 3–1. Chronology of development of flower and ovule soybean.†

Days before flowering	Morphological and anatomical features
25	Initiation of floral primordium in axil of bract.
25	Sepal differentiation.
20–14	Petal, stamen, and carpel initiation.
14–10	Ovule initiation; maturation of megasporocyte; meiosis; four megaspores present.
10–7	Anther initiation; male archesporial cells differentiate; meiosis; microsporogenesis.
7–6	Functional megaspore undergoes first mitotic division.
6–2	Second mitotic division results in four-nucleate embryo sac.
	Third mitotic division results in eight-nucleate embryo sac.
	Cell walls develop around antipodals and egg apparatus forming a seven-celled and eight-nucleate embryo sac.
	Polar nuclei fuse. Antipodal cells begin to degenerate. Nucellus begins to disintegrate at micropylar end and on sides of embryo sac.
	Single vascular bundle in ovule extends from chalaza through funiculus and joins with the carpellary bundle.
1	Embryo sac continues growth; antipodals disorganized and difficult to identify. Synergids with filiform apparatus; one synergid degenerating.
	Tapetum in anthers almost gene. Pollen grains mature; some are germinating.
	Nectary surrounding ovary reached maximum height.
0	Flower opens; usually day of fertilization; resting zygote; primary endosperm nucleus begins dividing.
	Nectary starts collapsing.

† The times are a compilation of data from several soybean cultivars studied by Carlson (1973), Kato et al. (1954), Murneek and Gomez (1936), Pamplin (1961), Prakash and Chan (1976). The sequence of development is essentially the same regardless of cultivar but the absolute times vary with environmental conditions and cultivars.

as a result of loss of moisture. The innermost cells of the sclerenchyma layer, with a parallel orientation of fibrils along the longitudinal axis of the cells, shorten more during the drying process than the upper sclerenchyma cells, which have a transverse orientation of fibrils. The pod thus bends because of differences in changes in cell length in the two layers of the sclerenchyma tissue. Because of the slanted position of these fibers in relation to the axis of the pod, the two halves of the pod will twist. The parenchyma tissue of the pod, the hypodermis, and the epidermal layers, do not seem to have any direct connection with pod dehiscence.

Figure 3–117 is a diagrammatic representation of changes in pod and ovule length, width, and thickness with reference to days after blooming. Although times may vary among individual cultivars and under various environmental conditions, this table illustrates the sequence of changes occurring during pod and seed development. The maximum length of the pod is reached rather early in development, about 20 to 25 d after bloom (Andrews, 1966; Kamata, 1952). At this stage, the seeds have attained an average of 4% of their maximum dry weight (Fraser et al., 1982). The maximum width and thickness of the pod is reached about 30 d after bloom. This corresponds with the time that the seed reaches its maximum size in all dimensions. Maximum fresh seed weight and maximum seed size are reached 5 to 15 d later. As the maturing seed loses moisture, it changes from an elongate reniform shape to the more oval or spherical shape characteristic of the mature seed (Fig. 3–101).

Table 3–2. Chronology of development of seed and pod of soybean.†

Days after flowering	Morphological and anatomical features
0	Resting zygote. Several divisions of primary endosperm nucleus.
1	Two-celled proembryo. Endosperm with about 20 free nuclei.
2	Four- to eight-celled proembryo.
3	Differentiation into proembryo proper and suspensor. Endosperm in peripheral layer with large central vacuole.
4–5	Spherical embryo with protoderm and large suspensor. Endosperm surrounding embryo is cellular but elsewhere it is mostly acellular and vacuolate.
6–7	Initiation of cotyledons. Endosperm mostly cellular.
8–10	Rotation of cotyledons begins. Procambium appears in cotyledons and embryo axis. All tissue systems of hypocotyl present. Root cap present over root initials. Endosperm all cellular.
10–14	Cotyledons have finished rotation and are in normal position with inner surfaces of cotyledons parallel with sides of ovules. Cotyledons elongate toward chalazal end of ovule. Primary leaf primordia present. Endosperm occupies about half of seed cavity. Extensive vascularization of seed coat.
14–20	Continued growth of embryo and seed. Reduction of endosperm tissue by assimilation into cotyledons.
20–30	Primary leaves reach full size. Primordium of first trifoliolate leaf present. Cotyledons reach maximum size. Endosperm almost gone.
30–50	Continued accumulation of dry matter, and loss in fresh weight of seeds and pod. Maturation of pod.
50–80	Various maturity times depending on variety and environmental factors.

† The times are a compilation of data from several soybean cultivars studied by Bils and Howell (1963), Carlson (1973), Fukui and Gotoh (1962), Meng-yuan (1963), Kamata (1952), Kato et al. (1956), Ozaki et al. (1956), Pamplin (1963), Suetsugu et al. (1962). The sequence of development is essentially the same regardless of cultivar but the absolute times vary with environmental conditions and with cultivars.

3–10 CHRONOLOGY

A chronology of flower and ovule development is presented in Table 3–1, and of seed and pod development in Table 3–2. These tables represent a compilation of data from several of the investigations cited in this chapter. The sequence of events remains the same regardless of cultivar or environmental conditions but the absolute times between events may vary by several days as a function of environment and cultivar.

REFERENCES

Abernathy, R.H., R.G. Palmer, R. Shibles, and J.C. Anderson. 1977. Histological observations on abscising and retained soybean flowers. Can. J. Plant Sci. 57:713–716.

Albertsen, M.C., and R.G. Palmer. 1979. A comparative light and electron-microscopic study of microsporogenesis in male sterile (ms$_1$) and male fertile soybeans (*Glycine max* (L.) Merr.). Am. J. Bot. 66:253–265.

Alexandrova, V.G., and O.G. Alexandrova. 1935. The distribution of pigments in the testa of some varieties of soybeans, *Glycine hispida* Maxim. Bull. Appl. Bot. Genet. Plant Breed. 3(4):3–47.

Andrews, C.H. 1966. Some aspects of pod and seed development in Lee soybeans. Diss. Abstr. B 27(5):1347B.

Bils, R.F., and R.W. Howell. 1963. Biochemical and cytological changes in developing soybean cotyledons. Crop Sci. 3:304–308.

Borthwick, H.A., and W.M. Parker. 1938. Influence of photoperiods upon the differentation of meristems and the blossoming of Biloxi soybeans. Bot. Gaz. 99:825–839.

Buss, P.A., and N.R. Lersten. 1975. A survey of tapetal number as a taxonomic character in Leguminosae. Bot. Gaz. 136:388–395.

Carlson, J.B. 1973. Morphology. p. 71–95. *In* B.E. Caldwell (ed.) Soybeans: Improvement, production, and uses. Agron. Monogr. 16. ASA, Madison, WI.

Dzikowski, B. 1936. Studia nad soja *Glycine hispida* (Moench) Maxim. Cz. 1. Morfologia. Mem. Inst. Natl. Pol. Econ. Rurale 254:69–100.

Dzikowski, B. 1937. Studia nad soja *Glycine hispida* (Moench) Maxim. Cz. 11. Anatomia. Mem. Inst. Natl. Pol. Econ. Rurale 258:229–265.

Erickson, E.H., and M.B. Garment. 1979. Soya-bean flowers: Nectary ultrastructure, nectar guides and orientation on the flower by foraging honeybees. J. Agric. Res. 18:3–11.

Esau, K. 1965. Plant anatomy. 2nd ed. John Wiley and Sons, New York.

Fraser, J., D.B. Egli, and J.E. Leggett. 1982. Pod and seed development in soybean cultivars with differences in seed size. Agron. J. 74:81–85.

Fukui, J., and T. Gotoh. 1962. Varietal difference of the effects of day length and temperature on the development of floral organs in the soybean. I. Developmental stages of floral organs of the soybean. Jpn. J. Breed. 12:17–27.

George, G.P., A. George, and J.M. Herr, Jr. 1979. A comparative study of ovule and megagametophyte development in field-grown and greenhouse-grown plants of *Glycine max* and *Phaseolus aureus* (Papilionacea). Am. J. Bot. 66:1033–1043.

Guard, A.T. 1931. Development of floral organs of the soybean. Bot. Gaz. 91:97–102.

Hansen, W., and R. Shibles. 1978. Seasonal log of the flowering and podding activity of field-grown soybeans. Agron. J. 70:47–50.

Hardman, L.L. 1970. The effects of some environmental conditions on flower production and pod set in soybean *Glycine max* (L.) Merrill var. Hark. Diss. Abstr. 31(5):2401-B.

Hill, H. J., and S.H. West. 1982. Fungal penetration of soybean seed through pores. Crop Sci. 22:602–605.

Johns, C.W., and R.G. Palmer. 1982. Floral development of a flower-structure mutant in soybeans, *Glycine max* (L.) Merr. (Leguminosae). Am. J. Bot. 69:829–842.

Kamata, E. 1952. Studies on the development of fruit in soybean 1–2. (In Japanese.) Crop Sci. Soc. Jpn. Proc. 20:296–298.

Kato, I., and S. Sakaguchi. 1954. Studies on the mechanism of occurrence of abortive grains and their prevention on soybeans, *Glycine max*. M. Bull. Div. Plant Breed. Cultiv. Tokai-Kinki Natl. Agric. Exp. Stn. 1:115–132.

Kato, I., and S. Sakaguchi, and Y. Naito. 1954. Development of flower parts and seed in soybean plant, *Glycine max*. M. Bull. Div. Plant Breed. Cultiv. Tokai-Kinki Natl. Agric. Exp. Stn. Bull. 1:96–114.

Kato, I., and S. Sakaguchi, and Y. Naito. 1955. Anatomical observations on fallen buds, flowers, and pods of soybean, *Glycine max*. M. Bull. Div. Plant Breed. Cultiv. Tokai-Kinki Natl. Agric. Exp. Stn. 2:159–168.

Lersten, N.R. 1982. Tracheid bar and vestured pits in legume seeds (Leguminosae: Papilionoideae). Am. J. Bot. 69:98–107.

Lersten, N.R. 1983. Suspensors in Leguminosae. Bot. Rev. 49:233–257.

Linskens, H.F., P.L. Pfahler, and E.L. Knuiman-Stevens. 1977. Identification of soybean cultivars by the surface relief of the seed coat. Theor. Appl. Genet. 50:147–149.

Madjd, A., and F. Roland-Heydacker. 1978. Secretions and senescence of tapetal cells in the anther of *Soja hispida* Moench, Papilionaceae. (In French.) Grana 17:167–174.

Meng-Yuan, H. 1963. Studies on the embryology of soybeans. 1. The development of embryo and endosperm. (In Chinese.) Acta Bot. Sinica. 11:318–328.

Miksche, J.P. 1961. Developmental vegetative morphology of *Glycine max*. Agron. J. 53:121–128.

Moni, M. 1942. Untersuchungen uber den Mechanismus der schleuderbewegung der Sojabohnen-Hulse. Jpn. J. Bot. 12:437–474.

Murneek, A.E., and E.T. Gomez. 1936. Influence of length of day (photoperiod) on development of the soybean plant, *Glycine max* var. Biloxi. Mo. Agric. Exp. Stn. Res. Bull. 242.

Ozaki, K., M. Saito, and K. Nitta. 1956. Studies on the seed development and germination of soybean plants at various ripening stages. (In Japanese.) Res. Bull. Hokkaido Natl. Agric. Exp. Stn. 70:6–14.

Palmer, R.G., M.C. Albertsen, and H. Heer. 1978. Pollen production in soybean with respect to genotype, environment, and stamen position. Euphytica 27:427–434.

Pamplin, R.A. 1963. The anatomical development of the ovule and seed in the soybean. Ph.D. diss. Univ. of Illinois, Urbana. (Diss. Abstr. 63–5128).

Patel, J.D. 1976. Comparative seed coat anatomy of some Indian edible pulses. Phyton 17:287–299.

Prakash, N., and Y.Y. Chan. 1976. Embryology of *Glycine max*. Phytomorphology 26:302–309.

Rembert, D.H., Jr. 1977. Contribution to ovule ontogeny in *Glycine max*. Phytomorphology 27:368–370.

Rustamova, D.M. 1964. Some data on the biology of flowering and embryology of the soybean under conditions prevailing around Tashkent. (In Russian.) Uzb. Biol. Zh. 8(6):49–53.

Soueges, R. 1949. Embryogénie des Papilionacées. Développement de l'embryon chez le *Glycine soja* Sieb. et Zucc. (*Soya hispida* Moench). C.R. Acad. Sci. 229:1183–1185.

Stelly, D.M., and R.G. Palmer. 1982. Variable development in anthers of partially male sterile soybeans. J. Hered. 73:101–108.

Suetsugu, I., I. Anaguchi, K. Saito, and S. Kumano. 1962. Developmental processes of the root and top organs in the soybean varieties. (In Japanese.) p. 89–96. *In* Bull. Hokuriki Agric. Exp. Stn. Takada 3.

Takao, A. 1962. Histochemical studies on the formation of some leguminous seeds. Jpn. J. Bot. 18:55–72.

Thorne, J.H. 1981. Morphology and ultrastructure of maternal seed tissues of soybean in relation to the import of photsynthate. Plant Physiol. 67:1016–1025.

Tilton, V.R., R.G. Palmer, and L.W. Wilcox. 1984a. The female reproductive system in soybeans, *Glycine max* (L.) Merr. (Leguminosae). p. 33–36. *In* Erdelskaan (ed.) Fertilization and embryogenesis in ovulated plants. VEDA, Bratislava, Czechoslovakia.

Tilton, V.R., L.W. Wilcox, and R.G. Palmer. 1984b. Post-fertilization Wandlabrinthe formation and function in the central cell of soybean, *Glycine max* (L.) Merr. (Leguminosae). Bot. Gaz. 145:334–339.

Tilton, V.R., L.W. Wilcox, R.G. Palmer, and M.C. Albertsen. 1984c. Stigma, style, and obturator of soybean, *Glycine max* (L.) Merr. (Leguminosae) and their function in the reproductive process. Am. J. Bot. 71:676–686.

Van Schaik, P.H., and A.H. Probst. 1958. Effects of some environmental factors on flower productive efficiency in soybeans. Agron. J. 50:192–197.

Waddle, R., and N. Lersten. 1973. Morphology of discoid nectaries in Leguminosae, especially tribe Phaseoleae (Papilionoideae). Phytomorphology 23:152–161.

Webster, B.D., and A.C. Leopold. 1977. The ultrastructure of dry and imbibed cotyledons of soybean. Am. J. Bot. 64:1286–1293.

Wiebold, W.J., D.A. Ashley, and H.R. Boerma. 1981. Reproductive abscission levels and patterns for eleven determinate soybean cultivars. Agron. J. 73:43–46.

Williams, L.F. 1950. Structure and genetic characteristics of the soybean. p. 111–134. *In* K.S. Markley (ed.) Soybeans and soybean products. Interscience Publ., New York.

Wolf, W.J., F.L. Baker, and R.L. Bernard. 1981. Soybean seedcoat structural features: Pits, deposits and cracks. Scanning Electron Microsc. 1981(3):531–544.

4   Speciation and Cytogenetics

THEODORE HYMOWITZ

University of Illinois
Urbana, Illinois

Since 1984, there has been an explosion of research activity in the area of speciation and cytogenetics of the genus *Glycine*. This was due to three factors: an increase
in plant exploration and formal taxonomy; an increase in cytogenetic investigations;
and lastly, the application of both biochemical and molecular approaches to phylogenetic studies. For a detailed review of past research especially formal taxonomy and cytogenetics, the reader is directed to Chapter 3 of the Soybean monograph
(Hadley and Hymowitz, 1973) and Chapters 2 and 5 of the Soybean monograph,
2nd edition (Hymowitz and Singh, 1987; Palmer and Kilen, 1987). This chapter covers the literature from January 1984 through January 2002.

4–1 ORIGIN OF THE SOYBEAN AND ITS DISSEMINATION

Linguistic, geographical, and historical evidence suggest that the soybean
[*Glycine max* (L.) Merr.] emerged as a domesticate during the Zhou dynasty in the
eastern half of north China. Domestication is a process of trial and error and not a
time-datable event. In the case of the soybean, this process probably took place during the Shang dynasty (*ca* 1500–1100 B.C.) or perhaps earlier. By the first century
A.D., the soybean probably reached central and south China as well as peninsular
Korea. The movement of soybean germplasm within the primary gene center is associated with the development and consolidation of territories and the degeneration of Chinese dynasties (Ho, 1969; Hymowitz, 1970).

According to Chinese legend, Emperor Shen Nong, the Father of Agriculture
and Medicine, reported the earliest use of the soybean in the herbal, "Ben Cao Gang
Mu". No fewer than six different dates from 2838 B.C. to 2383 B.C. have been ac-
claimed as the publication date for Shen Nong's book (Hymowitz, 1970). However,
the historical analysis of the legitimacy of Emperor Shen Nong by sinologists reveals a completely different story. For example, Hirth (1908) was adamant in his
belief that the value of the works of Shen Nong, who was sometimes represented
as having the body of a man and the head of an ox, were a fabrication of histori-
ans, as perhaps the emperor himself. Statements such as "the soybean is one of the
oldest cultivated crops" or "it has been cultivated for over 5000 years" are thus incorrect.

From about the first century A.D. to the Age of Discovery (15th–16th century), soybeans were introduced into several countries with land races eventually developing in Japan, Indonesia, Philippines, Vietnam, Thailand, Malaysia, Myanmar, Nepal, and north India. These regions comprise a secondary gene center. The movement of the soybean throughout this period was due to the establishment of sea and land trade routes, the migrations of certain tribes from China, and the rapid acceptance of the seeds as a stable food by other cultures (Hymowitz, 1990; Hymowitz and Newell, 1980).

For centuries, the soybean has been the cornerstone of East Asian nutrition. Although many different foods were developed from soybean, the four most important were miso, soy sauce, tempeh, and tofu. These traditional foods have little physical or flavor identity with the original bean. Thus, it is not too surprising that the first Europeans who visited China (e.g., Marco Polo) or Japan did not mention the soybean as a crop in their journals (Hymowitz and Newell, 1981).

Starting in the late 16th century and throughout the 17th century, European visitors to China and Japan noted in their diaries the use of a peculiar bean from which various food products were produced (Carletti, 1964; Satow, 1900). In the 17th century, soy sauce was a common item of trade from the East to the West. For example, in 1679, John Locke noted in his journal that "mango and soy are two sauces brought to England from East India" (King, 1858).

By 1705, European pharmacologists were familiar with the soybean from Japan and its culinary value (Dale, 1705). However, it was not until 1712, when Engelbert Kaempfer, who lived in Japan during 1691 and 1692 as a medical officer of the Dutch East India Company, published his book *Amoenitatum Exoticum...* that the western world fully understood the connection between the cultivation of soybean and the utilization of its seed as a food. Kaempfer's drawing of the soybean is accurate and his detailed description of how to make soy sauce and miso are correct (Kaempfer, 1906).

The soybean must have arrived in the Netherlands before 1737, as Linnaeus described the soybean in the *Hortus Cliffortianus* which was based on plants cultivated in the garden at Hartecamp (Linnaeus, 1737). In 1739, soybean seeds sent by missionaries in China were planted in the Jardin des Plantes, Paris (Paillieux, 1880). In 1790, soybean was planted at the Royal Botanic Garden at Kew, England (Aiton, 1812), and in 1804 were planted near Dubrovnik, Yugoslavia (Buconjic, A. n.d. A Monograph of Dubrovnik 1800–1810. The Church of St. Luka Marunic. Unpubl. Ms. [In Serbo-Croatian]). In the Netherlands, France, and England soybean was grown for taxonomic or display purposes. However, soybean grown in Yugoslavia was harvested, cooked, mixed with cereal grain and then fed to chickens (*Gallus gallus*) for increased egg production.

The soybean was introduced into North America by Samuel Bowen. Henry Yonge, the surveyor General of the Colony of Georgia, planted soybean on his farm at the request of Samuel Bowen in 1765. Mr. Bowen, a former seaman employed by the East India Company, brought soybean to Savannah from China via London. From 1766, Mr. Bowen planted soybean on his plantation "Greenwich" located at Thunderbolt, a few miles east of Savannah. Today the property is a city cemetery. The soybean grown by Bowen were used to manufacture soy sauce and vermicelli (soybean noodles). In addition, he manufactured a sago powder substitute from sweet potato (*Ipomoea batatas* L.). The three products were exported to England. Samuel Bowen received a patent (No. 878) for his manufacturing inventions for producing these products. He was awarded a gold medal from the Society of Arts, Manufactures, and Commerce and received a present of 200 guineas from King George III. In addition, Bowen sent soy sauce and soybean to the American Philosophical Society in Philadelphia and was elected to membership of the society. Unfortunately, when Bowen died in London on 30 Dec.1777, his soybean enterprise in Georgia ended (Hymowitz and Harlan, 1983).

Another early introduction of soybean to North America was by Benjamin Franklin. In 1770, he sent seeds from London to the botanist John Bartram who likely planted them in his garden, which was situated on the west bank of the Schuykill River below Philadephia (Hymowitz and Harlan, 1983).

In 1851, the soybean was introduced first to Illinois and subsequently throughout the Corn Belt. The introduction came through a series of very unusual circumstances. In December 1850, the barque *Auckland* left Hong Kong for San Francisco carrying sugar and other general merchandise. About 830 km (500 mi) off the coast of Japan the ship came across a Japanese junk foundering on the sea. The Japanese crew was removed from the junk and placed on board the Auckland which continued on to San Francisco. In San Francisco, the Japanese fishermen were not permitted to go ashore because of the possibility of spreading diseases. By coincidence, waiting for a passenger ship, to take him back to Alton, IL via the Panama overland route, was Dr. Benjamin Franklin Edwards. Dr. Edwards examined the Japanese fishermen, declared them free of any contagious diseases and received as a gift a packet of soybean that was carried back to Alton. Mr. John H. Lea, an Alton horticulturist, planted the soybean in his garden in the summer of 1851. In 1852, the multiplied soybean were grown in Davenport, IA by Mr. J. R. Jackson and also in Cincinnati, Ohio by Mr. A. H. Ernst. In 1853, Mr. Ernst distributed soybean seeds to the New York State Agricultural Society, the Massachusetts Horticultural Society, and the Commissioner of Patents. The two societies and the Commissioner of Patents sent soybean seeds to dozens of farmers throughout the USA (Hymowitz, 1986).

In mid-1854, Dr. James Morrow, the agriculturist in Commodore Matthew Perry's Expedition to Japan, obtained soybean seed and sent them to the Commissioner of Patents. Subsequently, the seeds were distributed to farmers (Hymowitz, 1986). Thus from 1855 onward, it is difficult to distinguish between soybean seed sources in farmers' reports.

4–2 *GLYCINE* SPECIES

The genus *Glycine* is divided into two subgenera *Glycine* (perennials) and *Soja* (Moench) F.J. Herm. (annuals). The list of species recognized as of 1981, their $2n$ chromosome number and distribution is shown on Table 4–1. The reader is directed to Chapter 2 of the Soybean monograph, Second Edition (Hymowitz and Singh, 1987) for further discussion about each of the seven perennial species described as of 1981. Only new information about each species will be presented in this chapter.

Table 4–1. The genus *Glycine* Willd., subgenera, chromosome number, and distribution (Hymowitz and Newell, 1981).

Species	2n	Distribution
Subgenus *Glycine*		
Glycine clandestina Wendl.	40	Australia
Glycine clandestina var. sericea Benth.	-	Australia
Glycine falcata Benth.	40	Australia
Glycine latifolia (Benth.) Newell and Hymowitz	40	Australia
Glycine latrobeana (Meissn.) Benth.	40	Australia
Glycine canescens F. J. Herm.	40	Australia
Glycine tabacina (Labill.) Benth.	40, 80	Australia, south China, Taiwan, Mariana Island, Ryukyu Island, South Pacific Islands
Glycine tomentella Hayata	38, 40, 78, 80	Australia, south China, Taiwan, Philippines, Papua New Guinea
Subgenus *Soja* (Moench) F. J. Herm.		
Glycine soja Sieb. & Zucc.	40	China, Taiwan, Japan, Korea, Russia
Glycine max (L.) Merr.	40	Cultigen

Since 1981, plant taxonomists have described 15 additional perennial *Glycine* species. This was due primarily to extensive plant exploration activities undertaken by U.S. and Australian scientists (e.g., Anonymous, 1988; Brown et al., 1985; Hymowitz, 1982, 1989, 1998; Newell, 1981). A list of the species in the genus *Glycine* as currently delimited, three-letter code, genomic designation, and distribution is shown on Table 4–2.

The taxonomic description of four of the perennial species occurred while this chapter was being written. These species include *G. aphyonota, peratosa, pullenii*, and *rubiginosa*. Little is known about these species other than what was presented in the initial publications (Brown et al., 2002; Pfeil et al., 2001).

4–2.1 Subgenus *Glycine*

4–2.1.1 *Glycine albicans* Tind. and Craven

Glycine albicans is a nonstoloniferous subshrub that grows up to 0.6 m high. Leaves are digitately three-foliolate. The central leaflets are obovate to obovate-elliptic. Rhizomes are present bearing underground shoots from which new plants develop and produce cleistogamous flowers and fruit. Leaves have white, soft hairs. The corolla is white with a purplish keel. Chasmogamous inflorescences are borne singly in the axils of the leaves. Pods contain two to four black seeds. The species is endemic to the Mitchell Plateau, Western Australia, Australia (Tindale and Craven, 1988). The few accessions studied cytologically are diploid ($2n = 40$). At the University of Illinois, under greenhouse conditions, we have been unsuccessful in the multiplication of seed.

4–2.1.2 *Glycine arenaria* Tind.

Glycine arenaria is a nonstoloniferous herb having a deep woody taproot. The leaves are pinnately three-foliolate. The leaflets are thick, very narrowly lanceolate

Table 4–2. List of species in the genus *Glycine* Willd., three letter code, 2*n*, genome symbol and distribution as of January 2002.

	Code	2*n*	Genome†	Distribution
Subgenus *Glycine*				
1. *G. albicans* Tind. & Craven	ALB	40	II	Australia
2. *G. aphyonota* B. Pfeil	APH	40	?	Australia
3. *G. arenaria* Tind.	ARE	40	HH	Australia
4. *G. argyrea* Tind.	ARG	40	A2A2	Australia
5. *G. canescens* F. J. Herm.	CAN	40	AA	Australia
6. *G. clandestina* Wendl.	CLA	40	A1A1	Australia
7. *G. curvata* Tind.	CUR	40	C1C1	Australia
8. *G. cyrtoloba* Tind.	CYR	40	CC	Australia
9. *G. dolichocarpa* Tateishi and Ohashi	DOL	80	?	(Taiwan)
10. *G. falcata* Benth.	FAL	40	FF	Australia
11. *G. hirticaulis* Tind. & Craven	HIR	40	H1H1	Australia
		80	?	Australia
12. *G. lactovirens* Tind. & Craven	LAC	40	I1I1	Australia
13. *G. latifolia* (Benth.) Newell & Hymowitz	LAT	40	B1B1	Australia
14. *G. latrobeana* (Meissn.) Benth.	LTR	40	A3A3	Australia
15. *G. microphylla* (Benth.) Tind.	MIC	40	BB	Australia
16. *G. peratosa* B. Pfeil & Tind.	PER	40	?	Australia
17. *G. pindanica* Tind. & Craven	PIN	40	H2H2	Australia
18. *G. pullenii* B. Pfeil, Tind. & Craven	PUL	40	?	Australia
19. *G. rubiginosa* Tind. & B. Pfeil	RUB	40	?	Australia
20. *G. stenophita* B. Pfeil & Tind.	STE	40	B3B3	Australia
21. *G. tabacina* (Labill.) Benth.	TAB	40	B2B2	Australia
		80	Complex‡	Australia, West Central and South Pacific Islands
22. *G. tomentella* Hayata	TOM	38	EE	Australia
	TOM	40	DD	Australia, Papua New Guinea
		78	Complex§	Australia, Papua New Guinea
		80	Complex¶	Australia, Papua New Guinea, Indonesia, Philippines, Taiwan
Subgenus *Soja* (Moench) F. J. Herm.				
23. *G. soja* Sieb. & Zucc.	SOJ	40	GG	China, Russia, Taiwan, Japan, Korea (Wild Soybean)
24. *G. max* (L.) Merr.	MAX	40	GG	Cultigen (Soybean)

† Genomically similar species carry the same letter symbols.
‡ Allopolyploids (A and B genomes) and segmental allopolyploids (B genomes).
§ Allopolyploids (D and E, A and E, or any other unknown combination).
¶ Allopolyploids (A and D genomes, or any other unknown combination).

with the terminal leaflet often larger than the lateral leaflets. Chasmogamous inflorescences are borne in the axils of the upper leaves. The corolla is mauve. The style is slightly curved, glabrous having a small capitate stigma. The pods contain two or three black seeds. The species occurs in Western Australia and the Northern Territory, Australia (Tindale, 1986b). The few accessions studied cytologically are diploid (2*n* = 40).

4–2.1.3 *Glycine argyrea* Tind.

Glycine argyrea is a twining herb having slender stems with a woody taproot. The stems are densely covered with downward pointing hairs. The leaves are pinnately trifoliolate and the leaflets are narrowly lanceolate. The inflorescences are axillary. The corolla is light purple. The style is slightly incurved having a terminal capitate stigma. Pods when ripe contain up to nine or more black seeds. The species is endemic to Queensland and New South Wales, Australia (Tindale, 1984). Accessions studied cytologically are diploid ($2n = 40$). This species has both cleistogamous and chasmogamous flowers on the same plant (Schoen and Brown, 1991). Using allozyme polymorphism as measured by starch gel electrophoresis, Brown et al.(1986) determined that natural outcrossing via insect pollination is common in the species.

4–2.1.4 *Glycine canescens* F.J. Herm.

Glycine canescens is restricted to Australia. It is a twining herb with pinnately trifoliolate leaves. The leaflets are elliptic-linear to oblong-lanceolate and pinnately veined. The whole plant processes a hoary, silky-strigose pubescence which produces a silvery appearance in extreme cases. Flowers are pink and pods are linear. Plants are diploid ($2n = 40$) (Hymowitz and Singh, 1987).

4–2.1.5 *Glycine clandestina* Wendl.

Glycine clandestina is a slender twiner restricted to Austrialia. Leaves of *G. clandestina* are digitate; the leaflets range from ovate-lanceolate or obong to linear, and exhibit reticulate venation. Flowers vary in color from pale pink to rose-purple. The accessions studied cytologically are diploid ($2n = 40$) (Hymowitz and Singh, 1987).

4–2.1.6 *Glycine curvata* Tind.

Glycine curvata is a trailing herb having a woody taproot. The leaves are weakly pinnately trifoliolate, and the terminal leaflet is usually slightly larger than the lateral leaflets. The leaflets are very narrowly elliptical to very narrowly lanceolate. The standard is white to cream. First seedling leaves are unifoliolate. The pods are narrowly oblanceolate, curved, and contain five to seven seeds. The seeds are black. The species is endemic to Queensland, Australia (Tindale, 1986a). The plants are diploid ($2n = 40$).

4–2.1.7 *Glycine cyrtoloba* Tind.

Glycine cyrtoloba is a twining herb having a woody taproot. The leaves are pinnately trifoliolate. The leaflets are narrowly lanceolate, ovate, or elliptical. The terminal leaflet is slightly larger than the two lateral leaflets. The standard is pink to purple. The first seedling leaves are trifoliolate. The pods are oblong to oblong-linear, curved, and contain three to nine seeds. The seeds are dark brown to almost black. The plants are diploid ($2n = 40$) (Newell and Hymowitz, 1978). The species is endemic to Queensland and New South Wales, Australia (Tindale, 1984).

4–2.1.8 *Glycine dolichocarpa* Tateishi et Ohashi

Glycine dolichocarpa is a twining or prostrate herb. The hairs on the stems and petioles are deflexed. The leaves are pinnately trifoliate. The leaflets are lanceolate to ovate. The stems and leaves have numerous long tawny hairs. The species has both chasmogamous and cleistoganous flowers. The flowers are pinkish. The pods are linear with constriction between seeds. The five to nine seeds are dark brown or black. The few accessions studied cytologically are polyploid ($2n = 80$).

The species described was collected in Taiwan by Ohashi et al. (1991) and Tateishi and Ohashi (1992). However, it is very likely that the species is endemic to Australia (Hymowitz, 1990) and has been naturalized on Taiwan (Yeh et al., 1997).

4–2.1.9 *Glycine falcata* Benth.

Glycine falcata is restricted to Australia. It has a decumbent or erect growth habit and strigose pubescence. Flower recemes are long and stock with white to pale lilac flowers. Pods are broadly falcate and hirsute-strigose with oblong or ovoid seeds. In addition, the plant contains pods on underground rhizomes. The few accessions studied cytologically are diploid ($2n = 40$) Hymowitz and Singh (1987).

4–2.1.10 *Glycine hirticaulis* Tind. and Craven

Glycine hirticaulis (Tindale and Craven, 1988) is a nonstoloniferous and non-rhizomatous herb. Opportunistic amphicarpy occurs in the species. The leaves are digitately trifoliolate. The leaflets are linear with the middle leaflet longer than the two laterals. Plants have both chasmogamous and cleistogamous seed. The chasmogamous flowers are mauve. The pods contained one to three seeds, not curved. The seeds are purplish black. The few accessions examined are diploid or polyploid ($2n = 40, 80$). The species is endemic to the Northern Territory, Australia.

4–2.1.11 *Glycine lactovirens* Tind. and Craven

Glycine lactovirens is a prostrate nonstoloniferous herb (Tindale and Craven, 1988). The leaves are digitately trifoliolate. The leaflets are light green, narrowly obovate or elliptical. The middle leaflet is larger than the two lateral leaflets. The flower color is milky green. The pods contain two to four black seeds. The few accessions examined cytologically are diploid ($2n = 40$). The species is endemic to Western Australia, Australia. Under greenhouse conditions at the University of Illinois we have not succeeded in the multiplication of seed.

4–2.1.12 *Glycine latifolia* (Benth.) Newell and Hymowitz

Glycine latifolia is a trailing or twining species having long robust stems. It roots freely form stolons. The leaves are pinnately trifoliolate with large rhombic leaflets. The long inflorescence contains flowers from lavender to purple. The species sets seeds from chasmogamous and cleistogamous flowers. The straight pods contain two to four seeds. Accessions studied cytologically are diploid ($2n = 40$). The species is restricted to eastern Australia and shows promise as a pasture legume on clay soils in tropical and subtropical Australia (Newell and Hymowitz, 1980a; Jones et al., 1996).

4–2.13 *Glycine latrobeana* (Meissn.) Benth.

Glycine latrobeana is an uncommon species found in the Australian States of South Australia, Tasmania, and Victoria. The species is considered vulnerable and has been placed on the List of Threatened Australian Flora (ANZECC, 1993). It is a small herb with a compact, decumbent or somewhat twining growth habit and a thickened tap root. The leaves are digitately trifoliolate with obovate or suborbicular leaflets. The flowers are pink to purple. The pods contain three to five cylindrical shaped seeds. Accessions studied cytologically are diploid ($2n = 40$). Under greenhouse conditions at the University of Illinois, seed multiplication has been accomplished, but only rarely and with difficulty (Lynch, 1994).

4–2.1.14 *Glycine microphylla* (Benth.) Tind.

Glycine microphylla (Tindale, 1986b) has prostate stems that can root at the nodes. The stems are almost glabrous. The upper leaves are weakly pinnately trifoliolate. The leaflets vary from obovate, narrowly elliptical to broadly elliptical. The middle leaflet is often larger than the lateral leaflets and has both chasmogamous and cleistogamous inflorescences. The flowers are usually purple. The straight pods have four to six black seeds. Accessions studied cytologically are diploid ($2n = 40$). The species is native to Australia (primarily the eastern states), but also has been collected on Norfolk Island.

4–2.1.15 *Glycine pindanica* Tind. and Craven

Glycine pindanica (Tindale and Craven, 1993) is a climbing or creeping non-rhizomatous and nonstoloniferous species. Opportunist amphicarpy (Tindale and Craven, 1988) occurs in the species. The leaves are trifoliolate. The central leaflet is larger than the lateral leaflets. Chasmogamous and cleistogamous inflorescences are present. The flowers are light purple to dark purple. Pods have two to five seeds. The few accessions studied cytologically are diploid ($2n = 40$). The species appears to be restricted to the Pindan region of Western Australia, Australia.

4–2.1.16 *Glycine stenophita* B. Pfeil and Tind.

Glycine stenophita (Doyle et al., 2000) is a nonstoloniferous scrambling or climbing herb. The leaves are pinnately trifoliolate. The terminal leaflets are usually slightly larger than the lateral leaflets. The stem hairs are white. The flowers are pink to purple. The pods are straight and contain from three to seven seeds. The few accessions studied cytologically are diploid ($2n = 40$). The species is found in southern Queensland and northern New South Wales, Australia.

4–2.1.17 *Glycine tabacina* (Labill.) Benth.

Glycine tabacina is found in Australia, China, West Central and South Pacific Islands (Hymowitz and Singh, 1987; Li et al., 1983). Recently, the species was collected on the Islet of Kinmen (Quemoy) (Yeh et al., 1997; Hymowitz, 1998). The stems trail or twine and bear pinnately trifoliolate leaves. The leaflets have reticulate venation. The deep rose-purple flowers often are fragrant. Pods are stout and linear with oblong or ovoid black, sometimes brown seeds. Diploids ($2n = 40$) and tetraploids ($2n = 80$) have been reported (Hymowitz and Singh, 1987).

4–2.1.18 *Glycine tomentella* Hayata

Glycine tomentella is an extremely variable species distributed in Australia, China (Li et al., 1983), Papua New Guinea, Philippines, and Taiwan. Recently, the species was collected on the Islet of Kinmen (Quemoy)(Hymowitz 1998; Yeh et al., 1997). Four cytotypes have been reported for *G. tomentella* ($2n = 38, 40, 78, 80$). The polyploids form a species complex (Singh et al., 1987b). Hill (1998) conducted an extensive morphological and biochemical analysis of the diploid *G. tomentella*. He concluded that the currently delimited diploid *G. tomentella* ($2n = 38, 40$) is not monophyletic and suggested that the species be divided into four separate species: (i) A central and south Queensland group containing 38 chromosomes; (ii) A northern Queensland-Papua New Guinea group; (iii) A central Queensland big pod group; and (iv) A group of accessions from Western Australia -Northern Territory.

4–2.2 Subgenus *Soja*

The subgenus *Soja* includes the cultivated soybean, *G. max* and *G. soja*, the wild annual soybean (Table 4–2).

4–2.2.1 *Glycine soja* Sieb. and Zucc.

Glycine soja grows wild in the China, Japan, Korea, Russia, and Taiwan. It grows in fields, hedgerows, along roadsides, and riverbanks. The plant is an annual procumbent, or slender twiner having pinnately trifoliolate leaves and often tawny, strigose, or hirsute pubescence. The leaflets may be narrowly lanceolate, ovate, or oblong-elliptic. The purple or very rare white flowers are inserted on short, slender racemes. The pods are short with a strigose to hirsute pubescence and oval-oblong seeds (Hermann, 1962). Plants are diploid ($2n = 40$).

Ohashi (1982) and Ohashi et al. (1984) proposed that the scientific name of the annual wild soybean be changed to *G. max* (L.) Merr. subsp. *soja* (Sieb. and Zucc.) Ohashi. Although the International Legume Database (ILDIS) in the United Kingdom has accepted this proposal in nomenclature (http://www.ildis.org), the U.S. National Plant Germplasm System (USDA-GRIN) has not (http://www.ars.grin.gov).

Glycine max (Merr.) L. subsp. *formosana* (Hosokawa) Tateishi and Ohashi appears to be a form of *G. soja* having abnormally narrow leaflets, and smaller pods (Tateishi and Ohashi, 1992; Thseng et al., 2000).

In this chapter, we will continue to use *G. soja* as the scientific name of the annual wild soybean. The primary reason for conserving the name is that *G. soja* is so well known and hundreds of manuscripts have been published under that name. Secondly, unless there are compelling reasons to make the change, no benefit can be gained for the adaptation of a new name for the wild annual soybean.

4–2.2.2 *Glycine max* (L.) Merr.

Glycine max (L.) Merr., the cultivated soybean, is a true domesticate. In the absence of human intervention, the species would not exist. It is an annual that generally exhibits an erect, sparsely branched, bush-type growth habit with pinnately trifoliolate leaves. The leaflets are broadly ovate, but can be oval to elliptic-lance-olate. The purple, pink, or white flowers are borne on short axillary racemes or re-duced peduncles. The pods are either straight or slightly curved, usually hirsute. The one to three seeds per pod are usually ovoid to subspherical. Seed coats of com-mercial cultivars are yellow. However, the germplasm collections contain soybean with seed coats having olive green, buff, brown, and black to reddish black color-ing. A 100-seed sample of commercial cultivars has a mass of about 10 to 20 g. How-ever, some accessions in the soybean germplasm collection have seeds with a mass of up to 50 g.

The soybean is morphologically extremely variable. This is due primarily to the development of soybean land races in East Asia. Individual farm families have grown soybean containing specific traits generation after generation for use as food, feed, medicinal, religious or ceremonial value. Today, these land races provide the major sources of genetic diversity within soybean germplasm collections. However, these land races or other unusual strains should not to be considered exotic. The term exotic applies only to the wild perennial and annual species (Singh and Hy-mowitz, 1999).

The subgenus *Soja* contains, in addition to *G. max* and *G. soja*, a form known as *G. gracilis*. This semi-cultivated or weedy form known only from Northeast China is somewhat intermediate in morphology between *G. max* and *G. soja* and was pro-posed as a new species of *Glycine* by Skvortzow (1927). However, Hermann (1962) considered *G. gracilis* a variant of *G. max*, a thesis supported by Wang (1976) and Shoemaker et al. (1986). Crossability barriers were not observed in hybrids among *G. soja*, *G. max,* and *G. gracilis* because putative crossed pods matured in the plants, hybrid seed germinated normally, and F_1 plants were totally pollen- and seed-fer-tile (Singh and Hymowitz, 1989). Recently, the taxonomic position of *G. gracilis* was again questioned by Wu et al. (2001). They used SSR markers to evaluate ge-netic diversity among *Glycine* species. They concluded that *G. gracilis* should be separated from *G. max*. However, neither ILDIS nor USDA-GRIN recognizes *G. gracilis* as a species. Thus, the continued use of the gracilis ephithet is not warranted.

4–3 GENERIC RELATIONSHIPS

The genus *Glycine* Willd. is a member of the family Leguminosae, subfam-ily Papilionoideae, and tribe Phaseoleae. The Phaseoleae is the most economically important tribe of the Leguminosae. It contains members that have considerable im-portance as sources of food and feed, such as soybean [*Glycine max* (L.) Merr.]; pigeon pea [*Cajanus cajan* (L.) Millsp.]; sweet-hyacinth bean [*Lablab purpureus* (L.)]; common bean, lima bean, and tepary bean (*Phaseolus* spp.); winged bean [*Psophocarpus tetragonolobus* (L.) DC.]; and cowpea, mung bean, black gram, adzuki bean, and Bambarra groundnut (*Vigna* spp.).

Within the tribe Phaseoleae, Lackey (1981) recognized 16 genera as be-longing to the subtribe Glycininae. These include *Glycine* as well as *Amphicarpaea* Nutt., *Cologania* Kunth., *Dumasia* DC., *Diphyllarium* Gagnep., *Eminia* Taub., *Mastersia* Benth., *Neonotonia* Lackey, *Nogra* Merr., *Pseudeminia* Verdc., *Pseudovi-gna* Verdc., *Pueraria* DC., *Shuteria* W. and A., *Sinodolichos* Verdc., *Teramnus* P.Br., and *Teyleria* Backer.

Table 4–3. The genera within Glycininae, approximate number of species, chromosome numbers, and geographic distribution. (Adapted from Kumar and Hymowitz, 1981; Lackey, 1981; Lee and Hymowitz, 2001.)

General	No. of species	2n	Geographic distribution
Amphicarpaea	3	20, 22, 40	Asia, Africa, and North America
Calopogonium	8	36	South and Central America
Cologania	10	44	Mexico, Central and South America
Dumasia	8	--	Asia and Africa
Diphyllarium	1	20	Indochina
Eminia	5	22	Tropical Africa
Glycine	24	38, 40, 78, 80	Asia and Australia
Mastersia	2	22, 44	Indo-Malaya
Neonotonia	2	22	Africa to Asia
Nogra	3	22	Asia
Pachyrhizus	6	22	Neotropics
Pseudeminia	4	22	Tropical Africa
Pseudovigna	1	22	Tropical Africa
Pueraria	20	22	Asia
Shuteria	5	22	Indo-Malaya
Sinodolichos	2	--	Asia
Teramnus	8	28	Pantropical
Teryleria	1	44	Asia

Viviani et al. (1991) conducted a phenetic analysis of the tribe Phaseoleae. They scored 126 morphological characters and included 11 taxa from eight genera of Glycininae sensu Polhill (1994). Their results revealed that Glycininae was not clustered into a single group because several taxa of Clitoriinae, Diocleinae, and Kennediinae were included within the Glycininae clade. They suggested that the most similar genus to *Glycine* was *Pueraria*, a genus that was considered primitive among the Glycininae by Lackey (1977).

Phylogenetic relationships were investigated within the tribe Phaseoleae based on chloroplast DNA (cpDNA) restriction site mapping of the inverted repeat regions (Doyle and Doyle, 1993; Bruneau, et al., 1994). Glycininae was represented by 12 genera from Lackey's (1977) subtribe Glycininae. The cp DNA restriction study suggested that Glycininae sensu Lackey (1981) are not monophyletic with *Calopogonium* and *Pachyrhizus* of the subtribe Doicleinae arising within the Glycininae clade and with *Shuteria* placed outside of Glycininae. However, the sister genus or genera to *Glycine* within the Glycininae was not clearly resolved. On the basis of these restriction site-based phylogenies, Polhill (1994) transferred *Calopogonium* and *Pachyrhizus* from the subtribe Diocleinae sensu Lackey (1977, 1980) to Glycininae and reorganized 18 genera within Glycininae (Table 4–3).

Lee and Hymowitz (2001) conducted a vigorous analysis of phylogenetic relationships within the subtribe Glycininae inferred from cpDNA *rps16* intron sequence variation. Samples from 15 genera were included in the study. Missing were samples from *Diphyllarium*, *Mastersia*, and *Sinodolichos*. Phylogenies estimated using parisomony, neighbor-joining and maximum likelihood methods clearly revealed that (i) Glycininae is monophyletic if *Pachyrhizus* and *Calopogonium* are included within Glycininae; (ii) the genus *Teramnus* is closely related to *Glycine*, and *Amphicarpaea* showed sister relationship to the clade comprising *Teramnus* and *Glycine*; and (iii) the genus *Pueraria* regarded as a closely related genus to *Glycine* is not monophyletic and needs to be divided into at least four genera (Fig. 4–1).

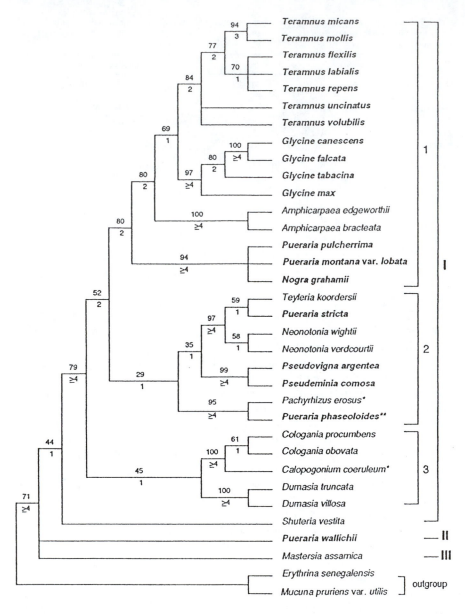

Fig. 4–1. Strict consensus 402 step trees derived from equally weighted parsimony analysis of 34 un-
ambiguously aligned chloroplast DNA *rps*16 intron sequences (consistency indices with and without uninformative characters = 0.769 and 0.627; retention index = 0.751). The numbers above the
branches indicate the number of times a monophyletic group occurred in 100 bootstrap replicates.
The numbers below the branches indicate the additional number of steps over 402 (the total number
of steps in the shortest tree) needed to collapse that branch (decay values). Decay analysis with tree
lengths >4 steps longer than the most parsimonious trees could not be performed owing to the computational constraints. *Glycine* groups of Lackey (1981) are in bold and asterisks indicate species included within Glycininae by Polhill (1994), but not by Lackey (1977a, 1981). Double asterisks indicate *Pueraria phaseoloides* var. *phaseoloides*. (Adapted from Lee and Hymowitz, 2001.)

4–4 ORIGIN OF THE GENUS *GLYCINE*

According to Goldblatt (1981), "The base number for Phaseoleae is almost certainly $x = 11$, which is also probably basic in all tribes." Goldblatt also pointed out that aneuploid reduction ($x = 10$) is prevalent throughout the Papilionoideae. Previously, Darlington and Wylie (1955) proposed a $x = 10$ basic chromosome number for the cultivated soybean. Based upon the above views and on recent taxonomic, cytological, and molecular systematics research on the genus *Glycine* and allied genera, we hypothesize that a putative ancestor of the genus *Glycine* with $2n = 20$ arose in Southeast Asia (Kumar and Hymowitz, 1989; Singh and Hymowitz, 1999; Lee and Hymowitz, 2001; Singh et al., 2001). However, such a progenitor is either extinct or yet to be collected and identified in Cambodia, Laos, or Vietnam (Fig. 4–2).

Tetraploidization ($2n = 2x = 40$) through auto- or allopolyploidy of the progenitor species occurred either prior to or after dissemination from the ancestral region. The progenitor of the wild perennial species radiated out southward, adapting to ecological niches in the Australian continent. These species were not domesticated (Fig. 4–2).

The path of migration northward (Fig. 4–2) from the ancestral region to China from a commmon progenitor is assumed by Singh et al. (2001) as: wild perennial ($2n = 4x = 40$, unknown or extinct) → wild annual ($2n = 4x = 40$; *G. soja*) → soybean ($2n = 4x = 40$; *G. max*, cultigen). All currently described species of the genus

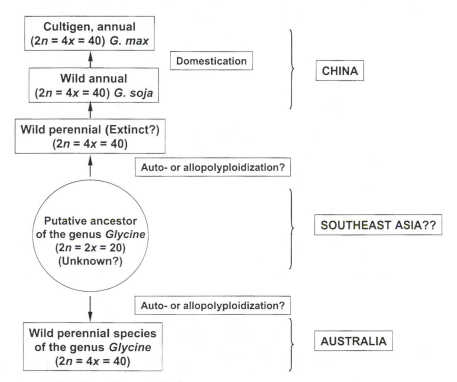

Fig. 4–2. The origin of the genus *Glycine*.

Glycine exhibit diploid-like meiosis, are primarily inbreeders and produce cleis-togamous seed (Singh and Hymowitz, 1985a).

Allopolyploidization (interspecific hybridization followed by chromosome doubling) via unreduced gametes probably played a major role in the speciation of the genus *Glycine*. This assumption infers that the 40-chromosome *Glycine* species and the 80-chromosome *G. tabacina* and *G. tomentella* are tetraploid and octoploid, respectively. The expression of four rDNA loci in *G. curvata* and *G. cyrtoloba* (Singh et al., 2001) strongly supports a hypothesis of allotetraploid origin that was origi-nally proposed on the basis of cytogenetic evidence (Singh and Hymowitz, 1985a, 1985b; Xu et al., 2000a) and molecular studies (Lee and Verma, 1984; Shoemaker et al., 1996).

Hymowitz et al. (1990), based upon cytogenetic studies, hypothesized that the disjunct allopolyploid distribution of *G. tabacina* and *G. tomentella* between Australia and the islands of the west-central Pacific region was due to long-distance dispersal by migrating shore birds. That hypothesis was verified by Doyle et al. (1990a, 1990b), who examined chloroplast DNA and histone H3-D polymorphism patterns within the *G. tabacina* polyploid complex.

Long-distance dispersal of seeds by birds has been well documented (Car-lquist, 1974). Currently, all *G. tabacina* and *G. tomentella* accessions collected on the West-Central Pacific Islands are polyploids. However, all diploids as well as the polyploids are found on the Papua New Guinea-Australian tectonic plate (Table 4–2). Hence, the question must be asked, "Do migrating shore birds selectively ingest seed containing polyploid *Glycine* cytotypes?" (Hymowitz et al., 1990). There is no ev-idence that diploid *Glycine* seeds are not carried by birds. However, the allopoly-ploid forms of *G. tabacina* and *G. tomentella*, unlike their diploid counterparts, are aggressive colonizing species (Singh and Hymowitz, 1985b). Thus the inbreeding tetraploids are able to compete successfully and establish themselves in the West-Central Pacific, where the diploids are unsuccessful. The dispersal process is a chance, but continuous annual event that, apparently, has occurred over many thou-sands of years. Today, the *G. tabacina* populations on the islands off the coast of Taiwan appear to be morphologically similar. However, the populations vary con-siderably as measured by random amplification of polymorphic DNA (RAPD) analysis (Thseng et al., 1997).

4–5 GENOMIC RELATIONSHIPS AMONG DIPLOID SPECIES

Genomic relationships (Fig. 4–3) among diploid species of the subgenus *Glycine* have been established by (i) cytogenetic analyses (Palmer and Hadley, 1968; Putievsky and Broué, 1979; Newell and Hymowitz, 1983; Grant et al., 1984a, 1986; Singh and Hymowitz, 1985b, 1985d; Singh et al., 1988, 1992b, 1997; Kol-lipara et al., 1993), (ii) biochemical techniques (Mies and Hymowitz, 1973; Broué et al., 1977; Vaughan and Hymowitz, 1984; Doyle and Brown, 1985; Doyle et al., 1986; Menancio and Hymowitz 1989; Brown, 1990; Domagalski et al., 1992; Kol-lipara et al., 1995), and (iii) molecular methods (Doyle and Beachy, 1985; Doyle and Brown, 1989; Doyle et al., 1990a, 1990b, 1990c, 1996, 1999a, 1999b; Menan-cio et al., 1990; Zhu et al., 1995b; Kollipara et al., 1997).

The concept of genome designation of the *Glycine* species was first proposed by Singh and Hymowitz (1985b). Based on hybridization success, hybrid seed viability, fertility of F_1 plants in intra- and interspecific hybrids and degree of meiotic chromosome pairing, the authors proposed the following genome symbols for six diploid ($2n = 40$) wild perennial species (Table 4–2, Fig. 4–3): *G. canescens* = AA; *G. clandestina* - Intermediate pod (Ip) = A_1A_1; *G. clandestina* - short pod (Sp) = BB; *G. latifolia* = B_1B_1; *G. tabacina* = B_2B_2; *G. cyrtoloba* = CC; *G. tomentella* = DD). Due to its small pods and leaflets, *G. clandestina* - Sp was removed from *G. clandestina* and taxonomically renamed *G. microphylla* (Tindale, 1986b). In addition, all B- genome species (*G. microphylla*, *G. latifolia*, and *G. tabacina*) have adventitious roots, while the other species lack adventitious roots (Costanza and Hymowitz, 1987).

Genomically similar *Glycine* species are expected to hybridize, produce viable, vigorous, and fertile F_1 plants, and exhibit normal meiotic chromosome pairing at metaphase I. Still, certain hybrid combinations can show a chromatin bridge and an acentric fragment (paracentric inversion) at anaphase I. By contrast, in ge-

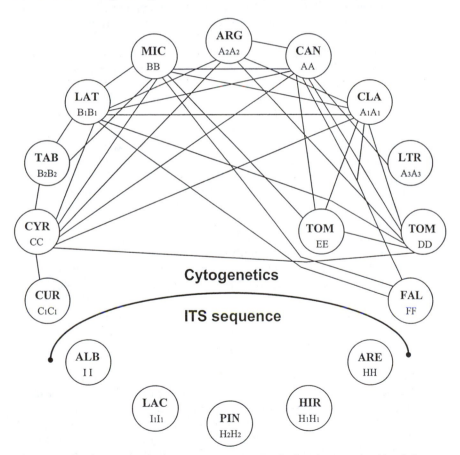

Fig. 4–3. Genomes of the wild perennial *Glycine* species. For the three-letter species abbreviations see Table 4–2.

nomically dissimilar *Glycine* species the interspecific crossability is extremely low, pod abortion is common, and 19- to 21-d old immature seeds must be germinated aseptically in vitro. F_1 hybrids are slow growing, morphologically weak, and completely sterile. The sterility is due to poor chromosome pairing. Furthermore, species distantly related usually produce inviable F_1 seeds, or premature death of germinating seedlings, and seedling and vegetative lethality (Singh et al., 1988, 1992b; Kollipara et al., 1993).

Interspecific crosses within the A-genome species (*G. canescens* = AA; *G. clandestina* = A_1A_1; *G. argyrea* = A_2A_2) or within the B-genome species (*G. microphylla* = BB; *G. latifolia* = B_1B_1; *G. tabacina* = B_2B_2) produced mature pods with normal seed set. F_1 plants were vigorous and completely fertile. Genome A displays partial genome affinities with B- and D-genome species, but a stronger genome homology with D than with B. Genome B shows no genomic affinity with D genome as F_1 hybrids were seedling lethal (Singh et al., 1988). *Glycine cyrtoloba* (C genome) shows slightly stronger (though possibably significant) affinity with B (B × C = 26.2 I + 6.9 II; B_1 × C = 29.9 I + 5.05 II) than with A (A_2 × C = 30.8 I + 4.6 II). Furthermore, A ×C and A_1 × C F_1 plants die prematurely (Singh, 1993). Grant et al. (1984b) reported a higher percentage of meiotic pairing (5.25 I + 17.3 II) in the F_1 hybrid between *G. canescens* and *G. latrobeana*. These results together with seed protein-banding profile similarities within *G. latrobeana, G. canescens, G. clandestina,* and *G. argyrea* prompted the assignment of genome symbol A_3A_3 to *G. latrobeana* (Singh et al., 1992b).

Successful production of hybrid seed is ordinarily a reliable indicator of genomic affinity between species. However, this was not true for *G. cyrtoloba* and *G. curvata*. These two species have curved pods, are morphologically very similar (Tindale, 1984, 1986a), and have similar banding patterns for chloroplast DNA (Doyle et al., 1990a, 1990b) and seed protein (Singh et al., 1992b). Pairwise divergence in the internal transcribed spacer (ITS) sequence of the nuclear ribosomal DNA between *G. cytroloba* and *G. curvata* was 1.3% which was significantly smaller than the mean percent distance between either one and any of other *Glycine* species (Kollipara et al., 1997). Despite close morphological and genetic affinity, the two species did not hybridize even though large number of flowers (748) were pollinated. This suggests that the crossability barrier between *G. curvata* and *G. cyrtoloba* probably are likely physiological than genetic (Singh et al., 1992b). Investigations on cytogenetics (Singh et al., 1988), rDNA loci (Singh et al., 2001), seed protein-banding profiles (Singh et al., 1992b), chloroplast DNA variation (Doyle et al., 1990a, 1990b), and nucleotide sequence variation in the ITS region of nuclear ribosomal DNA (Kollipara et al., 1997) suggest that *G. cyrtoloba* and *G. curvata* contain similar genomes and differ genomically from the other species of the genus *Glycine* (Fig. 4–3).

Intraspecific F_1 hybrids among *Glycine* species usually show normal meiosis and are completely fertile. However, *G. tomentella* ($2n = 38, 40$) is an exception. Morphologically, the 38- and 40-chromosome accessions of *G. tomentella* are indistinguishable. Based on isozyme differences, Doyle and Brown (1985) proposed separating 38-chromosome *G. tomentella* into two groups D1 and D2 (D = diploid), and that 40-chromosome *G. tomentella* be divided into six isozyme groups D3A, D3B, D3C, D4, D5, and D6. Cytogenetic and biochemical results have shown that isozyme groups D1 and D2, and D3A, D3B, and D3C are genomically similar (Kollipara et al., 1993). Isozyme D4 group is much closer genomically to A-genome species than to D1, D2, D3, and D5 groups of *G. tomentella* (Singh et al., 1988; Kollipara et al., 1993). Tindale (1986b) split off the D6 group of *G. tomentella* from the Eastern Kimberley District of Western Australia and described it as a new species, *G. arenaria* (Fig. 4–3).

Based on cytogenetic information, Singh et al. (1988) assigned genome symbol EE to *G. tomentella* accessions with $2n = 38$ chromosomes because in the D3 × E cross at metaphase I, the average chromosome association (range) was 26.6 I (19–33) + 6.2 II (3–10). In these crosses, bivalents are usually rod shaped and are loosely associated at metaphase I. Further cytogenetic and molecular studies (Kollipara et al., 1993, 1997) suggested that the 40-chromosome *G. tomentella* is a species complex. Despite morphological similarity among diploid *G. tomentella* accessions, isozyme D3 group was designated as the true *G. tomentella*. The D4 group accessions are variants close to A-genome species, while the D5 group includes highly heterogeneous accessions from the Western Australia (Singh et al., 1998a). These are tentative desgnations, since classical taxonomy does not split the 38- and 40-chromosome *G. tomentella* into two species.

Of the 22 wild perennial species of the subgenus *Glycine*, *G. falcata* is unique. It differs from the other species in several morphological traits (Hermann, 1962; Hymowitz and Newell, 1975; Newell and Hymowitz, 1978), seed protein composition (Mies and Hymowitz, 1973; Singh et al., 1992b), presence and absence of leaf flavonoids and isoflavonoids (Vaughan and Hymowitz, 1984), seed oil and fatty acid content (Chaven et al., 1982), 5S ribosomal RNA (Doyle and Beachy, 1985), phytoalexin production (Keen et al., 1986), and sequences from the ITS region of nuclear rDNA (Kollipara et al., 1997). Chloroplast DNA data suggests that *G. falcata* should be grouped with species containing the A-chloroplast (plastome) genome (Doyle et al., 1990a, 1990b). However, cytogenetic results do not support this conclusion. *Glycine falcata* showed negligible chromosome homology with the A- and B-genome species and could not be hybridized with the other species (Putievsky and Broué, 1979; Singh et al., 1988). This information led to the assignment of genome symbol FF to *G. falcata* (Singh et al., 1988) (Fig. 4–3).

Genomic relationships of five described diploid species, *G. albicans*, *G. arenaria*, *G. hirticaulis*, *G. lactovirens*, and *G. pindanica* (Tindale, 1986a; Tindale and Craven, 1988, 1993) were not established by cytogenetics. These species are narrowly distributed in Western Australia and do not grow well under greenhouse conditions at Urbana, IL. Kollipara et al. (1997) sequenced the ITS region of the rDNA of 16 wild perennial species of the subgenus *Glycine* and the two annual species of the subgenus *Soja*. Phylogenetic analysis of the ITS region clearly resolved all the genomic groups and verified the cytogenetic results. They assigned new genome symbols HH to *G. arenaria*, $H_1 H_1$ to *G. hirticaulis*, $H_2 H_2$ to *G. pindanica*, II to *G. albicans*, and $I_1 I_1$ to *G. lactovirens* (Fig. 4–3).

Doyle (1991) eloquently reviewed his collaborative research activities utilizing chloroplast DNA (cpDNA) for phylogenetic studies in the genus *Glycine*. Of the 12 perennial *Glycine* species studied, he identified distinct groups of plastomes. The B and C plastomes were congruent with the BB and CC nuclear genome groups of Hymowitz et al. (1998). For contrast, the A plastome group included the AA, DD,

EE, and FF nuclear groups of Singh and Hymowitz (1985b) resulting in apparent incongruence between the cpDNA phylogenetic studies and classical morphological and cytogenetic based taxonomic groupings (Doyle et al., 1999a, 1990b). However, in all cases, the morphological, cytological, and molecular variation seen in the sequenced ITS region of the rDNA were concordant (Kollipara et al., 1997). In this chapter, the genome designations for all the species in the genus *Glycine* primarily are those assigned by the authors of this chapter, and are based upon classical taxonomy and cytogenetics (Table 4–2, Fig. 4–3).

The genome symbol assigned to the recently name of species, *G. stenophita* is B3B3 (A.J. Lee, personal communication, 2001). This was determined similarly to the procedure that Kollipara et al. (1997) reported. The genomic relationship of 80-chromosome *G. dolichocarpa* is unknown; however, plant morphology suggests this species is related to 80-chromosome *G. tomentella*.

4–6 POLYPLOID COMPLEXES

Based on meiotic chromosome pairing in intra- and interspecific F_1 hybrids, Singh and Hymowitz (1985b) proposed that the tetraploid *G. tabacina* and *G. tomentella* are polyploid species complexes and have probably originated through allopolyploidization. These species behave like diploids, that is, chromosomes pair as bivalents (Singh and Hymowitz, 1985a). In contrast to their diploid ($2n = 38$ or 40) counterparts, tetraploid tabacinas and tomentellas are morphologically diverse, have wide geographical distributions, possess aggressive growth habits, and carry immense heterogeneity. These are characteristic features of allopolyploid species complexes (Singh et al., 1987b).

4–6.1 *Glycine tabacina* ($2n = 80$)

Accessions of tetraploid *G. tabacina* in the *Glycine* collection include at least two distinct morphological groups: one with adventitious roots and one without (Costanza and Hymowitz, 1987). Accessions without adventitious roots have longer and narrower leaves than those accessions with adventitious roots. Intraspecific F_1 hybrids within a group were highly fertile. All F_1 plants between groups showed no adventitious roots, suggesting that adventitious rooting is a recessive trait. F_1 hybrids were weak, slow growing, and sterile. This sterility was attributed to disturbed meiotic chromosome association. The mean number of bivalents in the eight hybrid combinations ranged from 14.5 to 19.6 (Singh et al., 1992a). Numerous univalents lagged at the equatorial plate during anaphase I, resulting in an unbalanced chromosome number in male and female spores, which caused sterility.

Singh et al. (1987b) studied meiotic pairing in F_1 hybrids between synthesized allopolyploids (BBB$_2$B$_2$, AABB) and *G. tabacina* ($2n = 80$) accessions with and without adventitious roots. Meiotic pairing of F_1s between the synthesized allopolyploid (BBB$_2$B$_2$) and *G. tabacina* accessions without adventitious roots revealed 40 bivalents at metaphase I and normal seed fertility. Thus, accessions with the adventitious roots are segmental allopolyploids. The genomes may be in any possible combinations (BBB$_1$B$_1$, BBB$_2$B$_2$, B$_1$B$_1$B$_2$B$_2$,) involving only B-genome diploid species. On the other hand, *G. tabacina*, lacking adventitious roots, are true allotetraploids and may constitute any combination of A- and B-genome species (Singh et al., 1992b). The 38-chromosome *G. tomentella* has no genomic affinity with *G. tabacina*, and likewise, *G. latifolia* did not donate its genome to 80-chromosome *G. tomentella* (Singh et al., 1987b).

Menancio and Hymowitz (1989) used isozyme variation to differentiate diploid and tetraploid tabacinas. The number of bands in tetraploids was always greater than in the diploids. Similarly, by using the soybean Bowman-Birk inhibitor as a genome marker, Kollipara et al. (1995) separated tetraploid tabacinas with and without adventitious roots (Costanza and Hymowitz, 1987). Blackhall et al. (1991) and Hammatt et al. (1991) used flow cytometry to measure DNA content in *G. tabacina*. They were able to differentiate diploid from tetraploid tabacinas. With this technique, chromosome counts may not be needed to separate the cytotypes of *G. tabacina*.

4–6.2 *Glycine tomentella* (2n = 78, 80)

Based on cytogenetics, Singh and Hymowitz (1985b, 1985d) proposed that the aneutetraploid (2n = 78) and tetraploid (2n = 80) *G. tomentella* are polyploid species complexes, a proposal supported by additional cytogenetic (Singh et al.,1987b, 1989; Kollipara et al., 1994) and molecular data (Grant et al., 1984a; Doyle and Brown, 1989; Doyle et al., 1990c, 1990d; Kollipara et al., 1994). Isozyme banding patterns indicate three groups (T1, T5, and T6) in aneutetraploid, and three groups (T2, T3, and T4) in tetraploid, *G. tomentella* (Doyle and Brown, 1985; Doyle et al., 1986). Kollipara et al. (1994) tentatively assigned Indonesian accessions to the T7 group based on cytogenetic, biochemical, and molecular studies.

Cytogenetic results, total seed protein profiles, trypsin and chymotrypsin inhibitor migration patterns, anti-KTI, anti-BBI, and anti-SBL immunocrossreactive protein profiles, and RFLP analyses strongly supported three distinct groups in aneutetraploid and four groups in tetraploid *G. tomentella* (Kollipara et al., 1994, 1995). Meiotic chromosome pairing between the synthesized amphidiploid and the aneutetraploid or the tetraploid *tomentella* accessions revealed the genomic constitutions of T1(D3D3EE), T5(AAEE), and T2 (AAD3D3).

The origin of allopolyploidy in 78- and 80-chromosome *G. tomentella* was described by Singh et al. (1987b, 1989). Phylogenetic analyses clearly demonstrated that both 78- and 80-chromosome *G. tomentella* had a common genome. Hybrids among various diploid species were synthesized to determine which diploid ancestors gave rise to polyploid *G. tomentella*. The synthesized amphidiploids (Singh et al., 1987b) were crossed to the T1 through T7 group *G. tomentella* accessions. Morphology, cytology, and seed protein profiles of F_1s showed that the T1 group (2n = 78) originated through allopolyploidization of D3 and E genomes of *G. tomentella,* and the T5 group (2n = 78) was derived from A- and E-genome species. The parents of 80-chromosome *G. tomentella* from Queensland, Australia (T2 group) were determined to be *G. canescens* (AA) and *G. tomentella* (D3 D3). These results suggested that the isozyme groups in both aneutetraploid (T1, T5, and T6) and tetraploid accessions (T2, T3, T4, and T7) probably originated by multiple independent events. Reconstruction of hypothetical ancestors through production of synthetic hybrids provide a technique to analyze this complex (Singh et al., 1989; Kollipara et al., 1994).

Blackhall et al. (1991) and Hammatt et al. (1991) used flow cytometry to measure DNA content in four *G. tomentella* cytotypes. They were able to distinguish diploids from tetraploids. However, they were not able to distinguish between the diploid ($2n = 38$, 40) nor the polyploid ($2n = 78$, 80) cytotypes.

Several investigators (Wang and Zhuang, 1994; Hui et al., 1997; Taylor-Grant and Soliman, 1999; Hsieh et al., 2001; Wu et al., 2001) have conducted phylogenetic or species variation experiments within the genus *Glycine*. The results of those studies were difficult to interpret because diploids (having one genome) were compared to polyploids (having two genomes) or because the experiment lacked standards for comparison.

4–7 GENE POOLS

Harlan and deWet (1971) proposed that the concept of three gene pools, primary (GP-1), secondary (GP-2), and tertiary (GP-3), be based on the success rate of hybridization among species. This system, as applied to *Glycine* (Fig. 4–4) is described as follows:

1. Primary Gene Pool (GP-1)
   GP-1 consists of biological species, and crossing within the gene pool is easy. The hybrids are vigorous, exhibit normal meiotic chromosome pairing, and possess total seed fertility; gene segregation is normal and gene

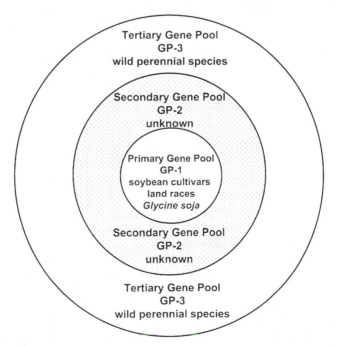

Fig. 4–4. The gene pools of the soybean.

exchange generally is easy. For the soybean, GP-1 includes soybean cultivars, land races, and *Glycine soja* (Fig. 4–4).

2. Secondary Gene Pool (GP-2)
    As defined by Harlan and deWet (1971), GP-2 consists of species that can be crossed with GP-1 and having some fertility in the F1. By this definition, *Glycine* does not have a GP-2.

3. Tertiary Gene Pool (GP-3)
    The GP-3 is the extreme outer limit of potential genetic resources. Hybrids between GP-1 and GP-3 are anomalous, lethal or completely sterile, and gene transfer is not possible or requires rescue techniques (Harlan and deWet, 1971). Based upon this definition, GP-3 includes the 22 wild perennial *Glycine* species of the subgenus *Glycine*. These species are indigenous to Australia and are geographically isolated from *G. max* and *G. soja*. Intersubgeneric hybrids and fertile-derived lines have been produced (Singh et al., 1990, 1993; Singh and Hymowitz, 1999).

4–8 INTERSUBGENERIC HYBRIDS

Ideally, soybean breeders would like to use the genetic diversity found in the wild perennial *Glycine* species to improve soybean yields. Before this can be accomplished, a method must be developed whereby the soybean can be hybridized with a wild perennial species and fertile lines can be derived from the cross. Several research teams have attempted to overcome species hybridization problems and genomic and fertility barriers (Table 4–4).

All of the intersubgeneric hybrids generated to date were obtained via embryo rescue techniques. The wild perennial *Glycine* parent(s) of subgenus *Glycine*, when used in successful hybrids with the soybean subgenus *Soja*, carried the AA, DD, and/or EE genomes (Hymowitz et al., 1998). It is unknown whether the intersubgeneric hybrid failures with AA, DD, and/or EE genomes might be due to lack of crossing effort rather than due to genomic incompatibility (Table 4–4). Chung and Kim (1991) reported a hybrid in a cross between *G. max* and *G. latifolia*. The F_1 carrying the genomes G and B_1 was sterile.

In general, all the F_1 intersubgeneric hybrid plants were vegetatively vigorous and exihibits a growth habit resembling that of the perennial parent. Meiotic analysis of sporophytic cells revealed that univalents predominated. The cells contained a few rod-shaped bivalents each with a terminalized chiasma. The occurrence of several loosely paired rod bivalents suggested the possibility of allosyndetic pairing (pairing between chromosomes of different genomes). All reported F_1 intersubgeneric hybrids were sterile.

When F_1 intersubgeneric hybrids were treated with colchicine (Cheng and Hadley, 1983; Kollipara et al., 1998), fertility was partially restored in the synthetic amphiploids (Newell et al., 1987; Singh et al., 1990). Bodanse-Zanettini et al. (1996) did not mention restored fertility in their $2n = 118$ synthetic amphiploid plant.

Thus far, only Singh et al. (1990, 1993) have reported successfully backcross-derived fertile progeny from the soybean and a wild perennial relative, *G. tomentella*. The soybean cv. Clark 63 was used as the recurrent parent in the backcrossing scheme. Procedures for embryo-rescue were used to obtain F_1, amphiploid,

Table 4-4. Published intersubgeneric hybrids in the genus *Glycine* (Hymowitz et al. 1998).

Intersubgeneric Hybrids	Authors
1. [*G. tomentella* ($2n = 38$, EE) × *G. canescens* ($2n = 40$, AA)] × *G. max* ($2n = 40$, GG) = F_1 ($2n = 59$, EAG), sterile	Broué et al. (1982)
2. *G. max* ($2n = 40$, GG) × *G. tomentella* ($2n = 78$, DDEE) = F_1 ($2n = 59$, GDE), sterile *G. max* ($2n = 40$, GG) × *G. tomentella* ($2n = 80$, AADD) = F_1 ($2n = 60$, GAD) sterile	Newell and Hymowitz (1982)
3. *G. tomentella* ($2n = 78$, DDEE) × *G. max* ($2n = 40$, GG) = F_1 ($2n = 59$, DEG), sterile	Singh and Hymowitz (1985c)
4. *G. max* ($2n = 40$, GG) × *G. tomentella* ($2n = 80$, AADD) = Hybrid embryo cell, $2n = 64$, no plant	Sakai and Kaizuma (1985)
5. *G. argyrea* ($2n = 40$, A_3A_3) × *G. canescens* ($2n = 40$, AA) × *G. max* ($2n = 40$, GG) $F_1 \rightarrow$ CT $\rightarrow$ amphiploid ($2n = 80$), sterile	Grant et al. (1986)
6. *G. max* ($2n = 40$, GG) × *G. clandestina* ($2n = 40$, A_1A_1) = F_1 ($2n = 40$, GA_1) sterile	Singh et al. (1987a)
7. *G. max* ($2n = 40$, GG) × *G. tomentella* ($2n = 78$, DDEE) = F_1 ($2n = 59$, GDE) $\rightarrow$ CT $\rightarrow$ amphiploid ($2n = 118$, GG DDEE), partially fertile *G. canescens* ($2n = 40$, AA) × *G. max* ($2n = 40$, GG) = F_1 ($2n = 40$, AG) $\rightarrow$ CT $\rightarrow$ amphiploid ($2n = 80$, AAGG), sterile *G. tomentella* ($2n = 78$, DDEE) × *G. max* ($2n = 40$, GG) = F_1 ($2n = 59$, DEG) $\rightarrow$ CT $\rightarrow$ amphiploid ($2n = 118$, DDEEGG), sterile	Newell et al. (1987)
8. *G. max* ($2n = 40$, GG) × *G. tomentella* ($2n = 78$, DDEE) = F_1 ($2n = 59$, GDE), sterile	Coble and Schapaugh Jr. (1990) Shen and Davis (1992)
9. *G. max* ($2n = 40$, GG) × *G. tomentella* ($2n =$ AADD) = F_1, ($2n = 60$, GAD), sterile	Chung and Kim (1990) Kwon and Chang (1991)
10. *G. max* ($2n = 40$, GG) × *G. latifolia* ($2n = 40$, B_1B_1) = F_1, cytology not studied, sterile	Chung and Kim (1991)
11. *G. max* ($2n = 40$, GG) × *G. tomentella* ($2n = 78$, DDEE) = F_1 ($2n = 59$, GDE) $\rightarrow$ CT $\rightarrow$ amphiploid ($2n = 118$, GGDDEE), no further information	Bodanse-Zanettini et al. (1996)
12. *G. max* ($2n = 40$, GG) × *G. tomentella* ($2n = 78$, DDEE) = F_1 ($2n = 59$, GDE) $\rightarrow$ CT $\rightarrow$ amphiploid ($2n = 118$, GGDDEE) × *G. max* $\rightarrow$ BC_1 ($2n = 76$, expected = 79) $\rightarrow$ × *G. max* BC_2 ($2n = 58, 56, 55$; GG+D, E) × *G. max* $\rightarrow$ BC_3 - BC_6 ($2n = 40 + 1, 2, 3$; GG + 1D or 1E or 2E...), fertile lines (monosomic alien addition lines)	Singh et al. (1990, 1993)

BC_1, and BC_2 plants (Singh et al., 1987). The BC_1 plant obtained by backcrossing *G. max* to the synthetic amphiploid was male sterile and carried $2n = 76$ (expected $2n = 79$). Continued backcrossing resulted in three sterile BC_2 plants ($2n = 58, 56, 55$). The range of chromosome numbers among backcrossed BC_3 plants was $2n = 41$ to $2n = 52$ and among backcrossed BC_4 plants was $2n = 40$ to 64 (Fig. 4–5).

Because of extremely low crossability and early pod abortion, wild perennial *Glycine* species have not been exploited in soybean-breeding programs (Ladizinsky et al. 1979; Singh and Hymowitz, 1987). The success of Singh et al. (1990, 1993) vs. other scientists was simply due to two factors: (i) persistence in making crosses and (ii) solving the early pod abortion problem. A hormonal mixture was sprayed on pollinated gynoecia 24 h after pollination. The solution consisted of 100 mg GA, 25 mg NAA, and 5 mg kinetin per liter of distilled water. The solution was stored at 4°C (Singh et al., 1990).

Singh et al. (1998b) reported that they isolated 22 individual monosomic alien addition lines (MAALs, $2n = 41$) from fertile lines derived from the cross of the soybean × *G. tomentella* (Fig 4–5). These alien addition lines are excellent sources of economic traits, for example, resistance to pests and pathogens, for broadening the extremely narrow genetic base of the soybean.

Two papers regarding intersubgeneric hybridization have been omitted from Table 4–4. The first is an abstract and a paper by Hood and Allen (1980, 1987) who attempted to cross *G. max* and a wild perennial relative, *G. falcata*. The abstract has been cited by others as an example of successful hybridization, but no hybrid plants were obtained. The other paper is a report by Shoemaker et al. (1990) that vegetative cuttings from a synthetic amphiploid ($2n = 118$) obtained from *G. max* ($2n = 40$) × *G. tomentella* ($2n = 78$) (Newell et al., 1987), were transferred from the Monsanto Company to Iowa State University during October 1987. The origin of this reputed hybrid was examined in a M.S. thesis by Heath (1989). The plants were maintained in the greenhouse for 1 yr. One plant was reported to have 41 pods with more than 100 F_2 seeds. After further cytogenetic evaluation, it was concluded, "that the *G. tomentella* chromosome complement had been eliminated after genetic exchange and/or genetic modification has taken place between the two genomes." (Heath,1989, p. 2). No other evidence for hybridity of this plant was presented. The purported hybridization could not be reproduced by the authors or other researchers, and thus may have been artefactual.

4–9 CYTOGENETIC TECHNIQUES

Since the 1980s cytological studies on the genus *Glycine* have drastically increased. The literature cited in the next section is extensive and thereby offers the opportunity for the reader to examine publications of interest. Palmer and Kilen (1987) reviewed the cytological literature up to 1984.

4–9.1 Mitotic Chromosomes

Singh (1993) provided the details of past protocols developed to count soybean mitotic chromosomes, for example, Palmer and Heer (1973). Xu et al. (1998a)

detailed a procedure for mitotic chromosome counts currently used in our laboratory. The procedure yields numerous cells with metaphase chromosomes that are
well spread for precise counts.

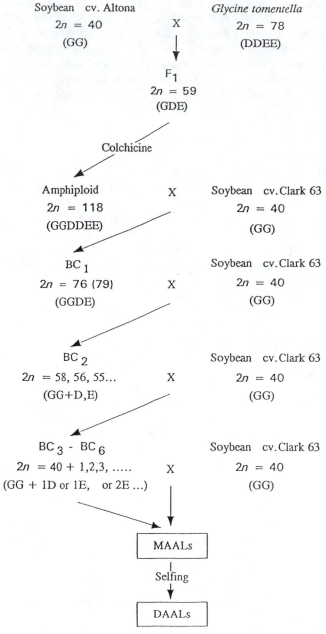

Fig. 4–5. The route to successful backcrossed-derived fertile progeny from the soybean and a wild perennial relative, *G. tomentella*.

The soybean seeds were germinated in a sand bench in the greenhouse. Root tips from actively growing 7- to 10 d-old seedlings were collected in 1.5 mL microcentrifuge tubes containing double-distilled water. Root tips were pretreated with 0.05% 8-hydroxyquinoline for 4 to 5 h at 16°C in a micro-cooler. Pretreated root tips were fixed in a 3:1 (v/v) mixture of 95% ethanol and propionic acid for 24 h. The fixative was removed from the microcentrifuge tubes, and the root-tips were washed once with double-distilled water. Root tips were hydrolyzed in 1 M HCL for 11 to 15 min at 60°C. After hydrolysis, the root tips were rinsed in double distilled water and stained in Schiff's reagent (Fuchsin-sulfate reagent) for 2 to 4 h at room temperature in the dark. The Feulgen stain was removed and the root tops were rinsed with cold double-distilled water and stained with Carbol fuchsin stain overnight at 0 to 4°C in a refrigerator. After Carbol fuchsin staining, the root tips were washed three to four times with cold double-distilled water to remove phenol and stored in cold double-distilled water in a refrigerator. Root tips were squashed in 45% acetic acid on a clean glass slide.

4–9.2 Meiotic Chromosomes

Except for minor changes (Bione et al., 2000) the procedure reported by Singh and Hymowitz (1985a) appears to be generally used. Flower buds undergoing meiosis were fixed in a freshly prepared mixture of 3:1 absolute ethanol: propionic acid. Ferric chloride (1 g 100 mL^{-1} fixative) was added to the fixative to intensify staining of chromosomes. Buds were transferred to 70% ethanol after 48 h of fixation and stored under refrigeration. Anthers with meiotic stages were stained in 0.7% aceto-carmine for 7 d under refrigeration and squashes were made in 45% acetic acid.

4–9.3 Flow Cytometry

The DNA content of nuclei in *Glycine* species has been correlated to genome size and chromosome number as measured by a flow cytometer. Baranyi and Greilhuber (1996) have published the protocols for ethidium bromide flow cytometry.

Hammatt et al. (1991) isolated nuclei from cotyledons of accessions from 14 species of *Glycine*. The 4 C amounts for diploid *Glycine* ranged from 3.80 to 6.59 pg. The accessions were divided into two groups. The first group contained amounts ranging from 3.80 to 5.16 pg and included species from the A, B, D, E, G, and H genomes. The second group had DNA contents ranging from 5.27 to 6.59 pg and consisted of species from the C and F genomes. The polyploid species, *G. tabacina* ($2n = 80$) *and G. tomentella* ($2n = 78, 80$) contained amounts approximating to the sums of the respective parental diploid species. Thus flow cytometry easily distinguishes between diploid and tetraploid forms of either *G. tabacina* or *G. tomentella*.

Graham et al. (1994) reported a significant correlation between maturity and the genome size of 20 soybean cultivars. In addition, they found a 15% difference in genome size among the cultivars. On the other hand, Greilhuber and Obermayer (1997) observed no reproducible genome size difference among cultivars and a significant correlation with maturity group was not confirmed. Rayburn et al. (1997) reported a 12% range in variation in nuclear DNA content among 90 soybean lines from China. Obermayer and Greilhuber (1999) observed no significant differences in the high-ranking DNA content vs. the low-ranking DNA content within soybean groups. Furthermore, no evidence was obtained for a difference in DNA content between Chinese and American soybean. Chung et al. (1998) reported a significant correlation between genome size and leaf and seed size in soybean. This intriguing finding has not, however, been independently confirmed.

For comparative measurements in flow cytometry, the growing conditions of experimental materials must be the same and identical conditions must be present during isolation and processing. The internal standards must be carefully controlled.

4–9.4 Fluorescent In-situ Hybridization

Fluorescent in-situ hybridization (FISH) is a physical mapping approach to detect specific DNA sequences in interphase nuclei or on condensed chromosomes of plants. Zhu et al. (1995a) published a detailed protocol for soybean. Thus far, the use of the procedure has been very limited within the soybean research community. Skorupska et al. (1989) located 18 S and 25 S rDNA sites in interphase cells. The in-situ hybridization results revealed that, for ribosomal RNA genes, G. max behaves as a diploid. Griffor et al. (1991) located rDNA sites in metaphase chromosomes of the soybean. In addition, the rDNA probe was detected in plants that were trisomics for the nucleolar-organizing region (NOR) containing chromosomes. Singh et al. (2001) examined the distribution of rDNA loci in 16 Glycine species by FISH using the internal transcribed spacer (ITS) region of nuclear ribosomal DNA as a probe. All species contained one rDNA site except G. curvata, G. cytoloba, G. tabacina ($2n = 80$) and G. tomentella ($2n = 80$) which had two rDNA sites. The study demonstrated that the distribution of the rDNA gene in the 16 Glycine species studied is highly conserved within the genus.

Shi et al. (1996) reported on the development of FISH, PCR-primed in-situ labeling (PCR-PRINS) procedures, and molecular probes for the cytological identification and physical mapping of soybean somatic chromosomes. They called the method soybean chromosome painting. The researchers demonstrated the utility of in-situ chromosome analysis for soybean by presenting examples of different DNA probes representative of particular sequence categories.

4–9.5 Sterility Systems

Various sterility systems have been reported in soybean. They have been classified as synaptic, structural, partial-male sterile or partial-female sterile, and completely male-sterile, female-fertile. The genetic control may be either nuclear or nuclear-cytoplasmic. Refer to Chapter 5 (Palmer et al., 2004, this publication) for information about the inheritance and gene symbols for specific male-sterile lines.

4–9.6 Diploid-like Meiosis

Chromosome pairing in wheat (Triticum aestivum L.) is controlled by a major gene Ph1 located on the long arm of chromosome 5 and also by minor genes located on other chromosome of wheat (Sears,1976). When Singh and Hymowitz (1985a) examined the chromosome pairing of six amphiploids (2n = 80, 120, or 160), they observed that all the amphiploids showed diploid-like meiosis in the majority of sporocytes and did not exhibit multivalent associations, a classical example of allopolyploidy. The strong preferential pairing between homologous chromosome is likely to be under genetic control.

4–10 KARYOTYPE ANALYSIS

Halvankar and Patil (1990) summarized the literature concerning the karyotype analysis of soybean mitotic chromosomes, and noted that the information about chromosome morphology of *Glycine* species was meager. Therefore, they attempted to provide new information on karyotypes of six *Glycine* species: *G. canescens, G. clandestina, G. latifolia,* and *G. max* were 2n = 40 while *G. tabacina* and *G. tomentella* had 80 chromosomes. The camera lucida drawings revealed that the *Glycine* species had either median or submedian centromeres. The results were identical with those of all previous studies, for example, Ahmad et al. (1984). Because of the high chromosome number, similar chromosome sizes, median to submedian centromere position, and the lack of morphological landmarks, the individual somatic chromosomes of the soybean have not been distinguishable (Zhong et al., 1997).

Singh and Hymowitz (1988) conducted a study to determine the genomic relationship between the cultivated soybean and the wild soybean, G. *soja*. Pachytene analysis of F_1 hybrids aided in the construction of chromosome maps based upon chromosome length, and euchromatic and heterochromatic distribution (Fig. 4–6). The chromosomes were numbered in descending order of size from 1 to 20 (Fig. 4–7). The largest chromosome is about four times the size of the smallest. The NOR is contained on chromosome 13. The smallest arms of six chromosomes—5, 7, 10, 18, 19, and 20—were totally heterochromatic. About 35.8% of the soybean genome was heterochromatic. Except for heteromorphic regions in chromosomes 6 and 11, the pachytene complement of G. *max* and G. *soja* were similar. Thus, the genome symbol GG assigned to G. *max* and G. *soja* by Singh and Hymowitz (1985b) and Singh et al. (1988) was valid. The paper by Singh and Hymowitz (1988) is an important landmark in soybean cytology, genetics, and molecular biology.

4–11 CHROMOSOME ABERRATIONS—NUMERICAL CHANGES

4–11.1 Polyploidy

In general, natural polyploidization in soybean beyond tetraploidy has not been identified. However, numerous synthesized allotetraploids (2n = 4x = 80) have been created to study the progenitors of 80-chromosome G. *tabacina* having or lacking adventitious roots, 78-, 80-chromosome G. *tomentella*, or to elucidate the mechanism of diploid-like meiosis (Singh and Hymowitz, 1985a). See section 4–6 of this chapter for a discussion of the polyploid complexes of *G. tabacina* and *G. tomentella*.

Singh and Hymowitz (1991a) synthesized an allopentaploid ($2n = 5x = 100$). The origin of the pentaploid plant was as follows: *G. clandestina*, $2n = 2x = 40$, A1A1 × *G. canescens*, $2n = 2x = 40$, AA → F_1 ($2n = 2x = 40$, AA1) × *G. tomentella*

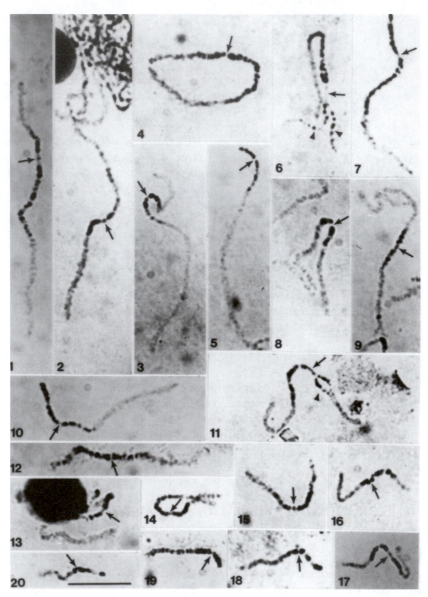

Fig. 4–6. Photomicrographs of the pachytene chromosome complement of *G. max* × *G. soja* F_1 hybrids. Each figure shows a different chromosome. Arrows indicate centromere locations. Arrow heads in chromosomes 6 and 11 show heteromorphic regions. Bar represents 10μ. (Adapted from Singh and Hymowitz, 1988.)

$(2n = 4x = 80, \text{A*A*DD}) \rightarrow F_1(2n = 3x = 60, \text{AA*D}) \rightarrow 0.1\%$ colchicine treatment $\rightarrow 2n = 6x = 120$ (AAA*A*DD) × G. *tomentella* $(2n = 4x = 80, \text{A*A*DD}) \rightarrow BC_1$, $2n = 5x = 100$ (AA*A*DD).

Morphologically, the pentaploid plant very closely resembled the tetraploid G. *tomentella* parent. The pentaploid did not breed true, as chromosomes in the 14 examined progeny ranged from $2n = 86$ to 97. Progeny of a plant with $2n = 90$ segregated for plants with $2n = 81–86$ chromosomes. This suggests that the preferential elimination of G. *canescens* (A genome) chromosomes is rapid and that eventually A*A*DD-genome chromosomes will prevail.

In the process of obtaining backcrossed-derived fertile plants from G. *max* and G. *tomentella* hybrids, an amphiploid containing $2n = 118$ chromosomes was synthesized. For the origin of the amphiploid, refer to section 4–8 of this chapter. The amphiploid plant produced four small black seed that bred true (Singh et al., 1990, 1993).

Singh and Hymowitz (1985a) synthesized four allohexaploids $(2n = 6x = 120)$. The origin of the allohexaploids was as follows: (i) G. *clandestina* $(2n = 40)$ × G. *tabacina* $(2n = 80) = F_1$ $(2n = 3x = 60) \rightarrow 0.1\%$ colchicine treatment $\rightarrow 2n = 6x = 120$, (ii) G. *tabacina* $(2n = 80)$ × G. *canescens* $(2n = 40) = F_1$ $(2n = 3x = 60) \rightarrow 0.1\%$ colchicine treatment $\rightarrow 2n = 6x = 120$, (iii) G. *tomentella* $(2n = 80)$ × G. *canescens* $(2n = 40) = F1$ $(2n = 3x = 60) \rightarrow 0.1\%$ colchicine treatment $\rightarrow 2n = 6x = 120$, and (iv) G. *canescens* $(2n = 40)$ × G. *tomentella* $(2n = 80) = F_1$ $(2n = 3x = 60) \rightarrow 0.1\%$ colchicine treatment $\rightarrow 2n = 6x = 120$.

Sterile F_1 interspecific triploid *Glycine* hybrids set seed only after colchicine treatment (i.e., chromosome doubling), establishing that the mechanism of unreduced gamete formation is a rare event in *Glycine*. The only publication reporting

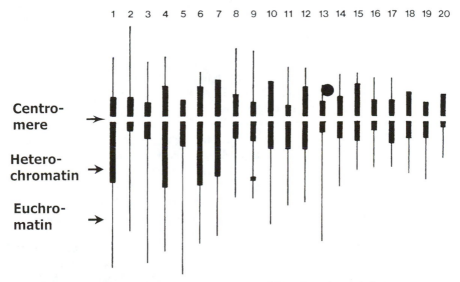

Fig. 4–7. Proposed idiogram of the pachytene chromosomes of the soybean. Arrows indicate centromeres, heterochromatin, and euchromatin. Chromosome 13 carries the nucleolus-organizing region (NOR). (Adapted from Singh and Hymowitz, 1988.)

unreduced gamete formation in *Glycine* was by Newell and Hymowitz (1980b). In a natural population of tetraploid *G. tabacina* they isolated a hexaploid *G. tabacina* ($2n = 6x = 120$) plant. It appears that an unreduced female gamete ($2n = 80$) was fertilized by a reduced ($n = 40$) male gamete. The plant was an autohexapoloid because some cells exhibited up to 16 trivalents.

A synthesized allooctoploid ($2n = 8x = 160$) was produced by Singh and Hymowitz (1985a). This is the highest reported chromosome number in the genus *Glycine*. The origin of the allooctoploid plant was as follows: *G. tomentella* ($2n = 4x = 80$) $\times$ *G. tabacina* ($2n = 4x = 80$) $\rightarrow$ F_1 ($2n = 4x = 80$) $\rightarrow$ 0.1% colchicine treatment $\rightarrow$ $2n = 8x = 160$. The average chromosome association at diakinesis was 10.4 I (4–18) + 74.8 II (71–78). The high frequency of univalents resulted in laggards at anaphase I. This meiotic irregularity in the allooctoploids is believed to be the cause of complete seed sterility in the C1 generation.

4–11.2 Triploidy

Chen and Palmer (1985) obtained autotriploids ($2n = 3x = 60$) from genetic male-sterile (*ms1ms1*) soybean plants. Meiosis in fertile and sterile autotriploids revealed no distinguishable difference in chromosome associations. Chromosomes of the true autotriploids are expected to associate in a trivalent configuration during diakinesis and metaphase I; however, multivalents, bivalents, and univalents also were observed. Pollen fertility in male-fertile triploid plants (*Ms1ms1ms1*) varied from 57 to 82% with an average of about 71%. Chromosome numbers of progenies obtained from these fertile triploids varied from $2n = 40$ to 71. Zhang and Palmer (1990) could not obtain triploids by crossing diploids *ms1ms1* soybean plants with tetraploid plants.

A hypertriploid plant ($2n = 64$) was isolated in BC_4 plants from a cross between the soybean and G. *tomentella*. Singh et al. (1993) believe that an unreduced egg ($n = 44$) was fertilized by a normal haploid ($n = 20$) male spore. The plant contained three doses of the soybean genome (GGG) plus four extra chromosomes of *G. tomentella*. Although vigorous in growth the plant was sterile.

A spontaneous hypertriploid ($2n = 3x + 1 = 61$) in soybean was isolated for the first time from the progeny of the cross between soybean line T31 and a primary trisomic line T190-47-3 (Xu et al., 2000c). Crosses with cultivar Clark 63 and selfed populations were used to evaluate breeding behavior, which revealed that the hypertriploid expressed more vigorous vegetative growth than the corresponding disomic siblings. The hypertriploid showed 63% pollen fertility and produced 98 seeds from self-pollination. Chromosome numbers in the self-pollination of the hypertriploid ranged from $2n = 50$ to $2n = 69$. The chromosome number in F_1 plants from the cross of hypertriploid $\times$ Clark 63 ranged from $2n = 44$ to $2n = 48$ with an exception of one plant that contained $2n = 56$ chromosomes. Several plants with $2n = 44$ to $2n = 47$ had high seed fertility. These data suggest that soybean male and female spores tolerate a higher number of extra chromosmes than most true diploid plant species, corroborating the hypothesis that the soybean is a diploidized polyploid species.

4–12 CHROMOSOME ABERRATIONS—STRUCTURAL CHANGES

4–12.1 Translocations

Translocations are sometimes called interchanges. They are created when there is a reciprocal exchange of terminal segments of nonhomologous chromosomes (Singh, 1993). Translocations have been used very rarely in chromosome mapping studies in the soybean (Sacks and Sadanaga, 1984; Sadanaga and Grindeland, 1984).

A total of 56 G. *soja* accessions from the People's Republic of China and Russia were evaluated for translocations (Palmer et al., 1987). Forty-six accessions had translocations with the same chromosome involved and likely they had the same chromosome structure.

Six different translocation lines were identified genetically and cytologically in soybean. Cytological analysis of F_1 progeny suggested that six of the 20 chromosomes were involved in reciprocal translocations (Mahama et al., 1999).

4–12.2 Inversions

In the genus *Glycine*, paracentric inversions appear to have played a major role in speciation. A dicentric bridge and an acentric fragment at anaphase I is a common configuration for a paracentric inversion. Thus far, inversions have not been used in soybean for mapping studies.

Palmer et al. (2000) conducted a study to determine if the paracentric inversions in two Chinese soybean landraces that were identical to each other and a paracentric inversion identified in *Glycine soja*. The results indicate that the soybean paracentric inversions were the same, but the G. *soja* accession had a chromosome structure differing from that of the two soybean accessions. Thus, the previous reports by Ahmad et al. (1977, 1979) were confirmed.

Extensive hybridization studies among diploid wild perennial *Glycine* species have led to morphologically distinct species being assigned the same genome letter designation but different number subscripts. The latter distinction appeared to be primarily due to paracentric inversions (Hymowitz et al., 1991; Hymowitz et al., 1998).

4–13 CHROMOSOME ABERRATIONS—ANEUPLOIDY

4–13.1 Monosomics

An individual lacking one chromosome is called monosomic and is designated as $2n-1$. Monosomics are useful for locating genes to specific chromosomes and in the assignment of linkage groups. Monosomics are rare in diploid crops. Two monosomic plants were identified among progenies of Triplo 3 (BC3) and Triplo 6 (BC4) trisomic plants, backcrossed to Clark 63 as the recurrent parent. The two monosomics were designated as Mono-3 and Mono-6 (Xu et al., 2000a). Morphologically, Mono-6 was similar to the disomic, while Mono-3 was smaller with reduced vigor. Both monosomics showed 19II +1I chromosome association at Metaphase I. Pollen fertility in Mono-3 was 8.8% and in Mono-6 was 20.0%.

Among the progeny of partially male-sterile plants, a deficiency aneuploid 39-chromosome plant was identified. Of the 130 seeds produced, two plants had 39 chromosomes and the rest were diploids. The low-transmission of n-1 gametes and high abortion rate make these plants unreliable as source of monosomics by sexual reproduction (Skorupska and Palmer, 1987).

4–13.2 Primary Trisomics

An individual with a normal chromosome complement plus an extra complete chromosome ($2n = 2x + 1$) is designated as a simple primary trisomic, and is called a "Triplo". For the soybean, a triplo is $2n = 40 + 1$. Primary trisomics are superb marker stocks for determining the gene-chromosome-linkage group relationships in soybean (Singh, 1993).

Palmer (1974, 1976) made an early attempt to develop primary trisomics of soybean. Five primary trisomics ($2n = 41$) were characterized and arbitrarily designated TRI A, TRI B, TRI C, TRI D, and TRI S (Palmer, 1976; Sadanaga and Grindeland, 1984; Gwyn et al., 1985; Gwyn and Palmer, 1989). The five primary trisomics were not identified karyotypically. Honeycutt et al. (1990) demonstrated that the $v2$ locus was located on the extra chromosome of TRI A. Hedges and Palmer (1991) reported that $Dia1$ was located on the extra chromosome of TRI D.

All 20 primary trisomics of the soybean were tentatively identified by pachytene analysis (Singh and Hymowitz, 1991b; Ahmad et al., 1992; Ahmad and Hymowitz, 1994; Xu et al., 2000b) and were designated as Triplo 1 through Triplo 20. The previously reported primary trisomics TRI A, C, D, and S were identified as Triplo 5, 1, 4, and 13, respectively (Xu et al., 2000b). The sources of primary trisomics in soybean were aneuploid lines ($2n = 41, 42, 43$) derived from asynaptic and desynaptic mutants (Palmer, 1974; Palmer and Heer, 1976), male sterile (ms) lines (Chen and Palmer, 1985; Sadanaga and Grindeland, 1981; Zhang and Palmer, 1990), neutron irradiated plants (Sadanaga and Grindeland, 1981), tissue culture induced sterile mutants (Graybosch et al., 1987; Palmer and Skorupska, 1994), and crosses between 'Funman' sterile plants with Clark 63 (Xu et al., 2000b).

4–13.3 Tetrasomics

An individual with a normal chromosome complement plus two extra similar chromosomes ($2n = 2x + 2$) is designated a tetrasomic and that individual is called "Tetra". For the soybean a tetra is $2n = 40 + 2$. A double trisomic is an individual with a normal chromosome complement plus two extra dissimilar chromosomes ($2n = 2x + 1 + 1$). For the soybean, a double trisomic is $2n = 40 + 1 + 1$.

Gwyn and Palmer (1989) compared 5 tetrasomics, 10 double trisomics and their related disomics for 25 morphological traits. Their analysis suggested that tetrasomics and double trisomics can be accurately distinguished on the basis of phenotype from disomics and from each other. The five tetrasomics were not identified karyotypically.

Tetrasomics have been identified from selfed progenies of primary trisomics by counting somatic chromosomes (Xu et al., 1997, 1998b). A.R.J. Singh (personal communication, 2001) believes that he has isolated and individually identified 15 different tetrasomics.

4–14 CONCLUSIONS

Within the past two decades, due to extensive plant exploration activities and with increased cytological, classical taxonomic and biosytematic investigations, the number of species in the genus *Glycine* has increased from 9 to 24. The genomic relationships among species has been established. As plant exploration activity continues, especially in remote areas of Australia, the number of new species assigned to the genus *Glycine* is expected to increase. The wild perennial species in *Glycine* offers great promise in expanding the narrow germplasm base of cultigen soybean by the introgression of useful traits from the wild species via wide hybridization.

Although soybean is one of the world's major crops, basic cytological and cytogenetic studies have lagged behind crops such as wheat, tomato (*Lycopersicum esculentum* Mill.), maize (*Zea mays* L.), and rice (*Oryza sativa* L.). The soybean chromosomes ($2n = 40$) are smaller than most other crop plants and individual chromosome identification at mitotic metaphase is not possible with current technologies. Within the past two decades, cytological techniques have greatly improved. All 20 chromosome pairs now have been differentiated based upon pachytene chromosome analysis. Cytological and marker stocks have been identified, for example, primary trisomics, tetrasomics, monosomics, translocations, inversions, and monosomic alien addition lines. All of these newly developed stocks have established a firm foundation for conducting genetic studies on the soybean as well as providing information for improving the soybean as a commercial crop

ACKNOWLEDGMENT

I am grateful for the assistance of Dr. Jijun Zou in the preparation of this chapter.

REFERENCES

Ahmad, Q.N., E.J. Britton, and D.E. Byth. 1977. Inversion bridges and meiotic behavior in species hybrids of soybeans. J. Hered. 68:360–364.

Ahmad, Q.N., E.J. Britton, and D.E. Byth. 1979. Inversion heterozygosity in the hybrid soybean × *Glycine soja*. J. Hered. 70:358–364.

Ahmad, Q.N., E.J. Britton, and D.E. Byth. 1984. The karyotype of *Glycine soja* and its relationship to that of the soybean. Cytologia (Tokyo) 49:645–658.

Ahmad, F., and T. Hymowitz. 1994. Identification of five new primary simple trisomics in soybean based upon pachytene chromosome analysis. Genome 37:133–136.

Ahmad, F., R.J. Singh, and T. Hymowitz. 1992. Cytological evidence for four new primary trisomics in soybean. J. Hered. 83:221–224.

Aiton, W.T. 1812. Hortus Kewensis. Vol. 4. p. 295. Longman, Hurst, Rees, Orme, and Brown, London.

Anonymous. 1988. Final report on the Australian *Glycine* and *Gossypium* exploration mission in 1987. Plant Genetic Resources Newsl. 73/74: Rep. 88/9. p.55.

Australian and New Zealand Environment and Conservation Council. 1993. Proposed list of the threatened Australian flora. Report for the ANZECC, ANZECC Endangered Flora Network, Canberra.

Baranyi, M., and J. Greilhuber. 1996. Flow cytometric and Feulgen densitometric analysis of genome size variation in *Pisum*. Theor. Appl. Genet. 92:297–307.

Bione, N.C.P., M.S. Pagliarini, and J.F.F. deToledo. 2000. Meiotic behavior of several Brazilian soybean varieties. Genet. Mol. Biol. 23:623–631.

Blackhall, N.N., N. Hammatt, and M.R. Davey. 1991. Analysis of variation in the DNA content of *Glycine* species: A flow cytometric study. Soybean Genet. Newsl. 18:194–200.

Bodanese-Zanettini, M.H., M.S. Lauxen, S.N.C. Richter, S. Cavalli-Molina, C.E. Lange, P.J. Wang, and C.Y. Hu.1996. Wide hybridization between Brazilian soybean cultivars and wild perennial relatives. Theor. Appl. Genet. 93:703–709.

Broué, P., J. Douglass, J.P. Grace, and D.R. Marshall. 1982. Interspecific hybridization of soybeans and perennial *Glycine* species indigenous to Australia via embryo culture. Euphytica 31:715–724.

Broué, P., D.R. Marshall, and W.J. Müller. 1977. Biosystematics of subgenus *Glycine* (Verdc.): Isoenzymatic data. Aust. J. Bot. 25:555–566.

Brown, A.H.D. 1990. The role of isozyme studies in molecular systematics. Aust. Syst. Bot. 3:39–46.

Brown, A.H.D., J.L. Doyle, J.P. Grace, and J.J. Doyle. 2002. Molecular phylogenetic relationships within and among diploid races of *Glycine tomentella* (Leguminosae). Aust. Syst. Bot. 15:37–47.

Brown, A.H.D., J.E. Grant, J.J. Burdon, J.P. Grace, and R. Pullen. 1985. Collection and utilization of wild perennial *Glycine*. p. 345–352 *In* R. Shibles (ed.) World Soybean Res. Conf., Iowa State Univ., Ames. 12–17 Aug. 1984. Westview Press, Boulder, CO.

Brown, A.H.D., J.E. Grant, and R. Pullen. 1986. Outcrossing and paternity in *Glycine argyrea* by paired fruit analysis. Biol. J. Linn. Soc. 29:283–294.

Bruneau, A., J.J. Doyle, and J.L. Doyle. 1994. Phylogenetic relationships in Phaseoleae: Evidence from choloroplast DNA restriction site characters. p. 309–330. *In* M. Crisp and J.J. Doyle (ed.) Advances in legume systematics. Royal Botanic Gardens, Kew, England.

Carletti, F. 1964. My voyage around the world. Trans. by H. Weinstock. Pantheon Books, NY.

Carlquist, S. 1974. Island biology. Columbia Univ. Press, New York.

Chaven, C., T. Hymowitz, and C.A. Newell. 1982. Chromosome number, oil and fatty acid content of species in genus *Glycine* subgenus *Glycine*. J. Am. Oil Chem. Soc. 59:23–25.

Chen, L.F., and R.G. Palmer. 1985. Cytological studies of triploids and their progeny from male-sterile (*ms1*) soybean (*Glycine max*). Theor. Appl. Genet. 71:400–407.

Cheng, S.H., and H.H. Hadley. 1983. Studies in polyploidy in soybeans: A simple and effective colchicine technique of chromosome doubling for soybean (*Glycine max* (L.) Merr.) and its wild relatives. Soybean Genet. Newsl. 10:23–24.

Chung, G.H., and J.H. Kim. 1990. Production of interspecific hybrids between *Glycine max* and *G. tomentella* through embryo culture. Euphytica 48:97–101.

Chung, G.H., and K.S. Kim. 1991. Obtaining intersubgeneric hybridization between *Glycine max* and *Glycine latifolia* through embryo culture. Korean J. Plant Tissue Cult. 18:39–45.

Chung, J., J.H. Lee, K. Arumuganathan, G.L. Graef, and J.E. Specht. 1998. Relationships between nuclear DNA content and seed and leaf size in soybean. Theor. Appl. Genet. 96:1064–1068.

Coble, C.J., and W.T. Schapaugh, Jr. 1990. Nutrient culture medium components affecting plant recovery from immature embryos of three *Glycine* genotypes and an interspecific hybrid grown *in vitro*. Euphytica 50:127–133.

Costanza, S.H., and T. Hymowitz. 1987. Adventitious roots in *Glycine* subg. *Glycine* (Leguminosae): Morphological and taxonomic indicators of the B genome. Plant Syst. Evol. 158:37–46.

Dale, S. 1705. Pharmacologiae. p. 184. S. Smith and B. Walford, Londoni (Latin).

Darlington, C.D., and A.P. Wylie. 1955. Chromosome atlas of flowering plants. George Allen & Unwin Ltd., London.

Domagalski, J.M., K.P. Kollipara, A.H. Bates, D.L. Brandon, M. Friedman, and T. Hymowitz. 1992. Nulls for the major soybean Bowman-Birk protease inhibitor in the genus *Glycine*. Crop Sci. 32:1502–1505.

Doyle, J.J. 1991. The pros and cons of DNA systematic data: Studies of the wild perennial relatives of soybean. Evol. Trends Plants 5:99–104.

Doyle, J.J., and R.N. Beachy. 1985. Ribosomal gene variation in soybean (*Glycine*) and its wild relatives. Theor. Appl. Genet. 70:369–376.

Doyle, J.J., and A.H.D. Brown. 1989. 5S nuclear ribosomal gene variation in the *Glycine tomentella* polyploid complex (Leguminosae). Syst. Bot. 14:398–407.

Doyle, J.J., and J.L. Doyle. 1993. Chloroplast DNA phylogy of the Papilionoid legume tribe Phaseoleae. Syst. Bot. 18:309–327.

Doyle, J.J., J.L. Doyle, and A.H.D. Brown. 1990a. Analysis of polyploid complex in *Glycine* with chloroplast and nuclear DNA. Aust. Syst. Bot. 3:125–136.

Doyle, J.J., J.L. Doyle, and A.H.D. Brown. 1990b. A chloroplast-DNA phylogeny of the wild perennial relatives of soybean (*Glycine* subgenus *Glycine*): Congruence with morphological and crossing groups. Evolution 44:371–389.

Doyle, J.J., J.L. Doyle, and A.H.D. Brown. 1990c. Chloroplast DNA polymorphism and phylogeny in the B genome of *Glycine* subgenus *Glycine* (Leguminosae). Am. J. Bot. 77:772–782.

Doyle, J.J., J.L. Doyle, and A.H.D. Brown. 1999a. Incongruence in the diploid B-genome species complex of *Glycine* (Leguminosae) revisited: Histone H3-D alleles versus chloroplast haplotypes. Mol. Biol. Evol. 16:354–362.

Doyle, J.J., J.L. Doyle, and A.H.D. Brown. 1999b. Origins, colonization, and lineage recombination in a widespread perennial soybean polyploid complex. Proc. Natl. Acad. Sci. USA 96:10741–10745.

Doyle, J.J., J.L. Doyle, A.H.D. Brown, and J.P. Grace. 1990d. Multiple origins of polyploids in the *Glycine tabacina* complex conferred from chloroplast DNA polymorphism. Proc. Natl. Acad. Sci. USA 87:714–717.

Doyle, J.J., J.L. Doyle, A.H.D. Brown, and B.E. Pfeil. 2000. Confirmation of shared and divergent genomes in the *Glycine tabacina* polyploid complex (Leguminosae) using histone H3-D sequences. Syst. Bot. 25:437–448.

Doyle, J.J., V. Kanazin, and R.C. Shoemaker. 1996. Phylogenetic utility of histone H3 intron sequences in the perennial relatives of soybean (*Glycine*: Leguminosae). Mol. Phylog. Evol. 6:438–447.

Doyle, M.J., and A.H.D. Brown. 1985. Numerical analysis of isozyme variation in *Glycine tomentella*. Biochem. Syst. Ecol. 13:413–419.

Doyle, M.J., J.E. Grant, and A.H.D. Brown. 1986. Reproductive isolation between isozyme groups of *Glycine tomentella* (Leguminosae), and spontaneous doubling in their hybrids. Aust. J. Bot. 34:523–535.

Goldblatt, P. 1981. Cytology and the Phylogeny of Leguminosae. p.427–463 *In* R.M. Polhill and P.H. Raven (ed.) Advances in legume systematics. Part 2. Royal Botanic Garden, Kew, England.

Graham, M.J., D.C. Nickell, and A.L. Rayburn. 1994. Relationship between genome size and maturity group in soybean. Theor. Appl. Genet. 88:429–437.

Grant, J.E., A.H.D. Brown, and J.P. Grace. 1984a. Cytological and isozyme diversity in *Glycine tomentella*. Hayata (Leguminosae). Aust. J. Bot. 32:665–677.

Grant, J.E., J.P. Grace, A.H.D. Brown, and E. Putievsky. 1984b. Interspecific hybridization in *Glycine* Willd. subgenus *Glycine* (Leguminosae). Aust. J. Bot. 32:655–663.

Grant, J.E., R. Pullen, A.H.D. Brown, J.P. Grace, and P.M. Gresshoff. 1986. Cytogenetic affinity between the new species *Glycine argyrea* and its congeners. J. Hered. 77:423–426.

Graybosch, R.A., M.E. Edge, and X. Delannay. 1987. Somaclonal variation in soybean plants regenerated from the cotyledonary node tissue culture system. Crop Sci. 27:803–806.

Greilhuber, J., and R. Obermayer. 1997. Genome size and maturity group in *Glycine max* (soybean). Heredity 78:547–551.

Griffor, M.C., L.O. Vodkin, R.J. Singh, and T. Hymowitz. 1991. Fluorescent *in situ* hybridization to soybean metaphase chromosomes. Plant Mol. Biol. 17:101–109.

Gwyn, J.J., and R.G. Palmer. 1989. Morphological discrimination among some aneuploids of soybean (*Glycine max* [L.] Merr.): 2. Double trisomics, tetrasomics. Can. J. Genet. Cytol. 27:608–613.

Gwyn, J.J., R.G. Palmer, and K. Sadanaga. 1985. Morphological discrimination among some aneuploids in soybean [*Glycine max* [L.] Merr.]: 1. Trisomics. Can. J. Genet. Cytol. 27:608–613.

Hadley, H.H., and T. Hymowitz. 1973. Speciation and cytogenetics. p. 97–116. *In* B.E. Caldwell (ed.) Soybeans: Improvement, production and uses. Agron. Monogr. 16. ASA, Madison, WI.

Halvankar, G.B., and V.P. Patil. 1990. Inter-specific variation in the karyotypes of the genus *Glycine* Willd. Cytologia (Tokyo) 55:273–279.

Hammatt, N., N.W. Blackhall, and M.R. Davey. 1991. Variation in the DNA content of *Glycine* species. J. Exp. Bot. 42:659–665.

Harlan, J.R., and J.M.J. deWet. 1971. Toward a rational classification of cultivated plants. Taxon 20:509–517.

Heath, M.S. 1989. Analysis of hybrid plants and progeny of a cross between *Glycine max* (L.) Merr. and *Glycine tomentella* Hayata. M.S. thesis. Iowa State Univ., Ames.

Hedges, B.R., and R.G. Palmer. 1991. Tests of linkage of isozyme loci with five primary trisomics in soybean. *Glycine max* (L.) Merr. J. Hered. 82:494–496.

Hermann, F.J. 1962. A revision of the genus *Glycine* and its immediate allies. USDA Tech. Bull. 1268. USDA, Washington, DC.

Hill, J.L. 1998. Morphological and biochemical analysis of variation in diploid *Glycine tomentella* Hayata (2n = 38, 40). Ph.D. diss. Univ. of Illinois, Urbana (Diss. Abstr. ADG9921694).

Hirth, F. 1908. The ancient history of China to the end of the Chou Dynasty. Columbia Univ. Press, NY.

Ho, P.T. 1969. The loess and the origin of Chinese agriculture. Am. Hist. Rev. 75:1–36.

Honeycutt, R.J., K.E. Newhouse, and R.G. Palmer. 1990. Inheritance and linkage studies of a variegated leaf mutant in soybean. J. Hered. 81:123–126.

Hood, M.J., and F.L. Allen. 1980. Interspecific hybridization studies between cultivated soybean, *Glycine max* and a perennial wild relative, *G. falcata*. p. 58 *In* Agronomy abstracts. ASA Madison, WI.

Hood, M.J., and F.L. Allen. 1987. Crossing soybeans with a wild perennial relative. Tenn. Farm Home Sci. Issue 144:25–30.

Hsieh, J.S., K.L. Hsieh, Y.C. Tsai, and Y.I. Hsing. 2001. Each species of *Glycine* collected in Taiwan has a unique seed protein pattern. Euphytica 118:67–73.

Hui, D., S.Y. Chen, and B.C. Zhuang. 1997. Phylogeny of 12 species of genus *Glycine* Wild. reconstructed with internal transcribed region in nuclear ribosomal DNA. Sci. China (Ser. C) 40:137–144.

Hymowitz, T. 1970. On the domestication of the soybean. Econ. Bot. 24:408–421.

Hymowitz, T.1982. Exploration for a wild relative of the soybean on Vanuatu. Naika 7:1–4.

Hymowitz, T. 1986. Introduction of the soybean to Illinois. Econ. Bot. 41:28–32.

Hymowitz, T. 1989. Exploration for wild perennial *Glycine* species on Taiwan. Soybean Genet. Newsl. 16:92–93.

Hymowitz, T. 1990. Soybean: The success story. p.159–163. *In* J. Janick and J.E. Simon (ed). Advances in new crops. Timber Press, Portland, OR.

Hymowitz, H. 1998. Plant exploration trip to Taiwan and islet of Kinmen (Quemoy). Soybean Genet. Newsl. 25:127.

Hymowitz, T., and J.R. Harlan. 1983. The introduction of the soybean to North America by Samuel Bowen in 1765. Econ. Bot. 37:371–379.

Hymowitz, T., and C.A. Newell. 1975. A wild relative of the soybean. Illinois Res. 17(4):18–19.

Hymowitz, T., and C.A. Newell. 1980. Taxonomy, speciation, domestication, dissemination, germplasm resources, and variation in the genus *Glycine* p. 251–264. *In* R.J. Summerfield and A.H. Bunting (ed.) Advances in legume science. Royal Botanic Garden, Kew, England.

Hymowitz, T., and C.A. Newell. 1981. Taxonomy of the genus *Glycine*, domestication and uses of soybean. Econ. Bot. 35:272–288.

Hymowitz, T., R.G. Palmer, and R.J.Singh.1991.Cytogenetics of the genus *Glycine*. p. 53–63. *In* T. Tsuchiya and P.K.Gupta (ed.) Chromosome engineering in plants: Genetics, breeding, evolution. Part B. Elsevier, NY.

Hymowitz, T., and R.J. Singh. 1987. Taxonomy and speciation. p. 23–48. *In* J.R. Wilcox (ed.) Soybeans: Improvement, production, and uses. 2nd ed. Agron. Monogr. 16. ASA, CSSA, and SSSA, Madison, WI.

Hymowitz, T., R.J. Singh, and K.P. Kollipara. 1998. The genomes of the *Glycine*. Plant Breed. Rev. 16:289–317.

Hymowitz, T., R.J. Singh, and R.P. Larkin. 1990. Long distance dispersal. The case for the allopolyploids *Glycine tabacina* Benth. and *G. tomentella* Hayata in the west-central pacific. Micronesica 23:5–13.

Jones, R.M., A.H.D. Brown, and J.N. Coote 1996. Variation in growth and forage quality of *Glycine latifolia* (Benth.) Newell and Hymowitz. CSIRO Trop. Agric., Genet. Resources Com. 26:1–11.

Kaempfer, E. 1906. The history of Japan together with a description of the Kingdom of Siam 1690–1692. Translated by J. G. Scheuchzer from the original edition of 1727, Vol I. p. 188. James Maclehose, Glasgow.

Keen, N.L., R.L. Lyne, and T. Hymowitz. 1986. Phytoalexin production as a chemosystematic parameter with *Glycine* spp. Biochem. Syst. Ecol. 14:481–486.

King, P. 1858. The life and letters of John Locke. p.134. Henry G. Bohn, London.

Kollipara, K.P., R.J. Singh, and T. Hymowitz. 1993. Genomic diversity in aneudiploid (2n = 38) and diploid (2n = 40) *Glycine tomentella* revealed by cytogenetic and biochemical methods. Genome 36:391–396.

Kollipara, K.P., R.J. Singh, and T. Hymowitz. 1994. Genomic diversity and multiple origins of the tetraploid (2n = 78, 80) *Glycine tomentella*. Genome 37:448–459.

Kollipara, K.P., R.J. Singh, and T. Hymowitz. 1995. Genomic relationships in the genus *Glycine* (Fabaceae:Phaseoleae): Use of a monoclonal antibody to the soybean Bowman-Birk inhibitor as a genome marker. Am. J. Bot. 82:1104–1111.

Kollipara, K.P., R.J.Singh, and T.Hymowitz.1997. Phylogenetic and genomic relationships in genus *Glycine* Willd. based on sequences from the ITS region of nuclear rDNA. Genome 40:57–68.

Kollipara, K.P., R.J. Singh, and T. Hymowitz. 1998. Induction of amphidiploids through colchicine treatment in the genus *Glycine*. Soybean Genet. Newsl. 16:94–96.

Kumar, P.S., and T. Hymowitz. 1989. Where are the diploid ($2n = 2x = 20$) genome donors of *Glycine* Willd. (Leguminosae, Papilionoideae)? Euphytica 40:221–226.

Kwon, C.S., and K.Y. Chang. 1991. *In vitro* germination of young hybrid embryos between *Glycine max* and *G. tomentella*. Korean J. Breed. 22:379–383.

Lackey, J.A. 1977. A revised classification of the tribe Phaseoleae (Leguminosae, Papilionoideae) and its relation to canavanine distribution. J. Linn. Soc. Bot. 74:163–178.

Lackey, J.A. 1980. Chromosome numbers in the Phaseoleae (Fabaceae; Faboideae) and their relation to taxonomy. Am. J. Bot. 67:595–602.

Lackey, J.A. 1981. Tribe 10 Phaseoleae DC. (1825) p.301–307. *In* R.M. Polhill and P.H. Raven (ed.) Advances in legume systematics. Royal Botanic Gardens, Kew, England.

Ladizinsky, G., C.A. Newell, and T. Hymowitz. 1979. Wide crosses in soybeans: Prospects and limitations. Euphytica 28:421–423.

Lee, J., and T. Hymowitz. 2001. A molecular phylogenetic study of the subtribe Glycininae (Leguminosae) derived from the chloroplast DNA *rps*16 intron sequences. Am. J. Bot. 88:2064–2073.

Lee, J.S., and D.P.S. Verma. 1984. Structure and chromosomal arrangement of leghemoglobin genes in kidney bean suggest divergence in soybean leghemoglobin gene loci following tetraploidization. EMBO J. 3:2745–2752.

Li, F., R. Chang, and S. Shu. 1983. The plants of the genus *Glycine* in China. (In Chinese.) Soybean Sci. 2:109–116.

Linnaeus, C. 1737. Hortus Cliffortianus, Historiae Naturalis Classica. p. 499. *In* J. Cramer and H.K. Swann (ed.) Vol. 63. Reprint 1968. Stechert-Hafner Service Agency, NY.

Lynch, A.J.J. 1994. The identification and distribution of *Glycine latrobeana* (Meissn.) Benth. in Tasmania. Papers and Proc. R. Soc. Tasmania 128:17–20.

Mahama, A.A., L.M. Deaderick, K. Sadanaga, K.E. Newhouse, and R.G. Palmer. 1999. Cytogenetic analysis of translocations in soybean. J. Hered. 90:648–653.

Menancio, D.I., and T. Hymowitz. 1989. Isozyme variation between diploid and tetraploid cytotypes of *Glycine tabacina* (Labill) Benth. Euphytica 42:79–87.

Menancio, D.I., A.G. Hepburn, and T. Hymowitz. 1990. Restriction fragment length polymorphism (RFLP) of wild perennial relatives of soybean. Theor. Appl. Genet. 79:235–240.

Mies, D.W., and T. Hymowitz. 1973. Comparative electrophoretic studies of trypsin inhibitors in seed of the genus *Glycine*. Bot. Gaz. 34:121–125.

Newell, C.A. 1981. Distribution of *Glycine tabacina* (Labill.) Benth. in the West-Central Pacific. Micronesica 17:59–65.

Newell, C.A., X. Delannay, and M.E. Edge. 1987. Interspecific hybrids between the soybean and wild perennial relatives. J. Hered. 78:301–306.

Newell, C.A., and T. Hymowitz. 1978. A reappraisal of the subgenus *Glycine*. Am. J. Bot. 65:168–179.

Newell, C.A., and T. Hymowitz. 1980a. A taxonomic revision in the genus *Glycine* subgenus *Glycine* (Leguminosae). Brittonia 32:63–69.

Newell, C.A., and T. Hymowitz. 1980b. Cytology of *Glycine tabaccina* a wild hexaploid relative of the soybean. J. Hered. 17:175–178.

Newell, C.A., and T. Hymowitz. 1982. Successful wide hybridization between the soybean and a wild perennial relative, G. *tomentella* Hayata. Crop Sci. 22:1062–1065.

Newell, C.A., and T. Hymowitz. 1983. Hybridization in the genus *Glycine* subgenus *Glycine* Willd. (Leguminosae, Papilionoideae). Am. J. Bot. 70:334–348.

Obermayer, R., and J. Greilhuber. 1999. Genome size in Chinese soybean accessions-stable or variable? Ann. Bot. (London) 84:259–262.

Ohashi, H. 1982. *Glycine max* (L.) Merr. subsp. *soja* (Sieb and Zucc.) Ohashi, *comb. nov.* (In Japanese.) J. Jpn. Bot. 57:30

Ohashi, H., Y. Tateishi, T.C. Huang, and T.T. Chen. 1984. Leguminosae of Taiwan. Sci. Rep. Tohoku Univ. 4th Ser. (Biology) 38:315.

Ohashi, H., Y. Tateishi, T. Nemoto, and H. Hoshi. 1991. Taxonomic studies on the Leguminosae of Taiwan IV. Sci. Rep. Tohoku Univ., 4th Ser. (Biology) 40:23–27.

Paillieux, A. 1880. (In French.) Bull. Soc. Natl. Acclim. Fr. Ser. 2. Ann. 71:414.

Palmer, R.G. 1974. Aneuploids in the soybean, *Glycine max*. Can. J. Genet. Cytol. 16:441–447.

Palmer, R.G. 1976. Chromosome transmission and morphology of three primary trisomics in soybeans (*Glycine max*). Can. J. Genet. Cytol. 18:131–140.

Palmer, R.G., and H.H. Hadley. 1968. Interspecific hybridization in *Glycine*, Subgenus *Leptocyamus*. Crop Sci. 8:557–563.

Palmer, R.G., and H. Heer. 1973. A root tip squash technique for soybean chromosome. Crop Sci. 13:389–391.

Palmer, R.G., and T.C. Kilen. 1987. Qualitative genetics and cytogenetics. p. 135–209. *In* J.R. Wilcox (ed.) Soybeans: Improvement, production, and uses. 2nd ed. Agron. Monogr. 16. ASA, CSSA, and SSSA, Madison, WI.

Palmer, R.G., K.E. Newhouse, R.A. Graybosch, and X. Delannay. 1987. Chromosome structure of wild soybean accessions from China and the Soviet Union of *Glycine soja* Sieb. & Zucc. [sic], Chromosome structure of wild soybean (*Glycine soja* Sieb. and Zucc.). Accessions from China and the Soviet Union. J. Hered. 78:243–247.

Palmer, R.G., T.W. Pfeiffer, G.R. Buss, and T.C. Kilen. 2004. Qualitative genetics. p. 137–234. *In* Soybeans: Improvement, production, and uses. 3rd ed. ASA, CSSA, and SSSA, Madison, WI.

Palmer, R.G., and H.T. Skorupska. 1994. Aneuploids from a male sterile mutant from tissue culture. Soybean Genet. Newsl. 21:228–229.

Palmer, R.G., H. Sun, and L.M. Zhao. 2000. Genetics and cytology of chromosome inversions in soybean germplasm. Crop Sci. 40:683–687.

Pfeil, B.E., M.D. Tindale, and L.A. Craven. 2001. A review of the *Glycine clandestina* species complex (Fabaceae, Phaseolae) reveals two new species. Aust. Syst. Bot. 14:891–900.

Polhill, R.M. 1994. Classification of the Leguminosae. p. XXXV–XLVII. *In* F.A. Bisby et al. (ed.) Phytochemical dictionary of the Leguminosae. Chapman and Hall, New York.

Putievsky, E., and P. Broué. 1979. Cytogenetics of hybrids among perennial species of *Glycine* subgenus *Glycine*. Aust. J. Bot. 27:213–223.

Rayburn, A.L., D.P. Birdar, D.G. Bullock, R.L. Nelson, C. Gourmet, and J.B. Wetzel. 1997. Nuclear DNA content diversity in Chinese soybean introductions. Ann. Bot. (London) 80:321–325.

Sacks, J.M., and K. Sadanaga. 1984. Linkage between the male-sterility gene (*ms1*) and a translocation breakpoint in soybean *Glycine max*. Can. J. Genet. Cytol. 26:401–404.

Sadanaga, K., and R.L. Grindeland. 1984. Locating the *w*1 locus on the satellite chromosome in soybean. Crop Sci. 24:147–151.

Sakai, T., and N. Kaizuma. 1985. Hybrid embryo formation in an intersubgeneric cross of soybean (*Glycine max* Merrill) with a wild relative (*G. tomentella* Hayata). Jpn. J. Breed. 35:363–374.

Satow, E.M. 1900. The voyage of Captain John Saris to Japan, 1613. Vol. 5. Ser. 2. p.126. Hakluyt Soc., London.

Schoen, D.J., and A.H.D. Brown. 1991. Whole and part flower self-pollination in *Glycine clandestina* and *G. argyrea* and the evolution of autogamy. Evolution 45:1651–1664.

Shen, B., and L.C. Davis. 1992. Nodulation and nodulin gene expression in an interspecific hybrid between *Glycine max* and *Glycine tomentella*. Austr. J. Plant Physiol. 19:693–707.

Shi, L., T. Zhu, M. Morgante, J.A. Rafalski, and P. Keim. 1996. Soybean chromosome painting: A strategy for somatic cytogenetics. J. Hered. 87:308–313.

Shoemaker, R.C., P.M. Hatfield, R.G. Palmer, and A.G. Atherly. 1986. Chloroplast DNA variation in the genus *Glycine* subgenus *Soja*. J. Hered. 77:26–30.

Shoemaker, R.C., M.S. Heath, H. Skorupska, X. Delannay, M. Edge, and C.A. Newell. 1990. Fertile progeny of a hybridization between soybean [*Glycine max* (L.) Merr.] and *G. tomentella* Hayata. Theor. Appl. Genet. 80:17–23.

Shoemaker, R.C., K. Polzin, J. Labate, J. Specht, E.C. Brummer, T. Olsen, N. Young, V. Concibido, J. Wilcox, J.P. Tamulonis et al. 1996.Genome duplication in soybean (*Glycine* subgenus *soja*). Genetics 144:329–338.

Singh, R.J. 1993. Plant cytogenetics. CRC Press, Boca Raton, FL.

Singh, R.J., and T. Hymowitz. 1985a. Diploid-like meiotic behavior in synthesized amphiploids of the genus *Glycine* Willd. subgenus *Glycine*. Can. J. Genet. Cytol. 27:655–660.

Singh, R.J., and T. Hymowitz. 1985b. The genomic relationships among six wild perennial species of the genus *Glycine* subgenus *Glycine* Willd. Theor. Appl. Genet. 71:221–230.

Singh, R.J., and T. Hymowitz. 1985c. An intersubgeneric hybrid between *Glycine tomentella* Hayata and the soybean, *G. max* (L.) Merr. Euphytica 34:187–192.

Singh, R.J., and T. Hymowitz. 1985d. Intra-and interspecific hybridization in the genus *Glycine*, Subgenus *Glycine* Willd.: Chromosome pairing and genome relationships. Z. Pflanzenzuecht. 95:289–310.

Singh, R.J., and T. Hymowitz. 1987. Intersubgeneric crossability in the genus *Glycine* Willd. Plant Breed. 98:171–173.

Singh, R.J., and T. Hymowitz. 1988. The genomic relationships between *Glycine max* (L.) Merr. and *G. soja* Sieb. and Zucc. as revealed by pachytene chromosome analysis. Theor. Appl. Genet. 76:705–711.

Singh, R.J., and T. Hymowitz. 1989. The genomic relationships among *Glycine soja* Sieb. and Zucc., *G. max* (L.) Merr. and '*G. gracilis*' Skvortz. Plant Breed. 103:171–173.

Singh, R.J., and T. Hymowitz. 1991a. Cytogenetics of a synthesized allopentaploid ($2n = 5x = 100$) in the genus *Glycine* subgenus *Glycine*. Genome 34:751–756.

Singh, R.J., and T. Hymowitz. 1991b. Identification of five primary trisomics of soybean by pachytene chromosome analysis. J. Hered. 82:75–77.

Singh, R.J., and T. Hymowitz. 1999. Soybean genetic resources and crop improvement. Genome 42:605–616.

Singh, R.J., H.H. Kim, and T. Hymowitz. 2001. Distribution of rDNA loci in the genus *Glycine* Willd. Theor. Appl. Genet. 103:212–218.

Singh, R.J., K.P. Kollipara, F. Ahmad, and T. Hymowitz. 1992. Putative diploid ancestors of 80-chromosome *Glycine tabacina*. Genome 35:140–146.

Singh, R.J., K.P. Kollipara, and T. Hymowitz.1987a. Intersubgeneric hybridization of soybeans with a wild perennial species, *Glycine clandestina* Wendl. Theor. Appl. Genet. 74:391–396.

Singh, R.J., K.P. Kollipara, and T. Hymowitz. 1987b. Polyploid complexes of *Glycine tabacina* (Labill.) Benth. and *G. tomentella* Hayata revealed by cytogenetic analysis. Genome 29:490–497.

Singh, R.J., K.P. Kollipara, and T. Hymowitz. 1988. Further data on the genomic relationships among wild perennial species ($2n = 40$) of the genus *Glycine* Willd. Genome 30:166–176.

Singh, R.J., K.P. Kollipara, and T. Hymowitz. 1989. Ancestors of 80- and 78-chromosome *Glycine tomentella* Hayata (Leguminosae). Genome 32:796–801.

Singh, R.J., K.P. Kollipara, and T. Hymowitz. 1990. Backcross-derived progeny from soybean and *Glycine tomentella* Hayata intersubgeneric hybrids. Crop Sci. 30:871–874.

Singh, R.J., K.P. Kollipara, and T. Hymowitz. 1992a. Genomic relationships among diploid wild perennial species of the genus *Glycine* Willd. subgenus *Glycine* revealed by crossability, meiotic chromosome pairing and seed protein electrophoresis. Theor. Appl. Genet. 85:276–282.

Singh, R.J., K.P. Kollipara, and T. Hymowitz. 1993. Backcross (BC$_2$ - BC$_4$) -derived fertile plants from *Glycine max* and *G. tomentella* intersubgeneric hybrids. Crop Sci. 33:1002–1007.

Singh, R.J., K.P. Kollipara, and T. Hymowitz. 1997. Intergenomic relationships among wild perennial *Glycine* species. p. 40–43. *In* B. Napompeth.(ed.) Soybean feeds the world. World Soybean Res. Conf. V Proc., Chiang Mai, Thailand. 21–27 Feb. 1994. Kasetsart Univ. Press, Bangkok, Thailand.

Singh, R.J., K.P. Kollipara, and T. Hymowitz. 1998a. The genomes of *Glycine canescens* F. J. Herm. and *G. tomentella* Hayata of Western Australia and their phylogenetic relationships in the genus *Glycine*. Genome 41:669–679.

Singh, R.J., K.P. Kollipara, and T. Hymowitz. 1998b. Monosomic alien addition lines derived from *Glycine max* (L.) Merr. and *G. tomentella* Hayata.: Production, characterization and breeding behavior. Crop Sci. 38:1483–1489.

Skorupska, H., M.C. Albersten, K.D. Langholz, and R.G. Palmer. 1989. Detection of ribosomal RNA genes in soybean. *Glycine max* (L.) Merr., by in situ hybridization. Genome 32:1091–1095.

Skorupska, H., and R.G. Palmer. 1987. Monosomics from synaptic KS mutant. Soybean Genet. Newsl. 14:174–178.

Skvortzow, B.V. 1927. The soybean-wild and cultivated in Eastern Asia. Proc. Manchurian Res. Soc. Pub. Ser. A. Nat. History Sec. 22:1–8.

Tateishi, Y., and H. Ohashi. 1992. Taxonomic studies on *Glycine* of Taiwan. J. Jpn. Bot. 67:127–147.

Taylor-Grant, N., and K.M. Soliman. 1999. Detection of polymorphic DNA and taxonomic relationships among 10 wild perennial soybean species using specific and arbitary nucleotide primers. Biol. Plant. 42:25–37.

Thseng, F.S., T.K. Lin, and W.S. Tu. 2000. The relations of genus *Glycine* subgenus *soja* and *Glycine formosana* Hosok. collected from Taiwan revealed by RAPD analysis. J. Jpn. Bot. 75:270–279.

Thseng, F.S., H.T. Tsui, and S.T. Wu. 1997. Population variation of wild soybean in Taiwan IV. Intraspecific variation of *Glycine tabacina* (labill.) Benth. in Ponghu Islets off Taiwan: Leaflet and pod morphology and DNA polymorphism. (In Chinese.) J. Agric. Assoc. China No. 178:52–62.

Tindale, M.1984. Two new eastern Australian species of *Glycine* Willd. (Fabaceae). Brunonia 7:207–213.

Tindale, M.D. 1986a. A new north Queensland species of *Glycine* Willd. (Fabaceae). Brunonia 9:99–103.

Tindale, M.1986b. Taxonomic notes on three Australian and Norfolk Island species of *Glycine* Willd. (Fabaceae: Phaseolae) including the choice of a neotype for *G. clandestina* Wendl. Brunonia 9:179–191.

Tindale, M.D., and L.A. Craven. 1988. Three new species of *Glycine* (Fabaceae: Phaseolae) from North-Western Australia, with notes on amphicarpy in the genus. Aust. Syst. Bot. 1:399–410.

Tindale, M.D., and L.A. Craven. 1993. *Glycine pindanica* (Fabaceae, Phaseolae), a new species from West Kimberley, Western Australia. Aust. Syst. Bot. 6:371–376.

Vaughan, D.A., and T. Hymowitz. 1984. Leaf flavonoids of *Glycine* subgenus *Glycine* in relation to systematics. Biochem. Syst. Ecol. 12:189–192.

Viviani, T., L. Conte, G. Christofolini, and M. Speranza. 1991. Sero-systematic and taximetric studies on the Phaseoleae (Fabaceae) and related tribes. Bot. J. Linn. Soc. 105:113–136.

Wang, C.L. 1976. Review on the classification of soybeans. (In Chinese.) Acta Phytotaxon. Sin. 14:22–30.

Wang, Y.M., and B.C. Zhuang. 1994. Study on the biology of the genus *Glycine* III. Analysis of SOD zymogram pattern of different subgenera and species. (In Chinese.) Soybean Sci. 13:340–344.

Wu, X.L., C.Y. He, S.Y. Chen, B.C. Zhuang, K.J. Wang, and X.C. Wang. 2001. Phylogenetic analysis of interspecies in the genus *Glycine* through SSR markers. (In Chinese.) Acta Genet. Sin. 28:359–366.

Xu, S.J., R.J. Singh, and T. Hymowitz. 1997. Establishment of a cytogenetic map of soybean: Progress and prospective. Soybean Genet. Newsl. 24:121–122.

Xu, S.J., R.J. Singh, and T. Hymowitz. 1998a. A modified procedure for mitotic chromosome counts in soybean. Soybean Genet. Newsl. 25:107.

Xu, S.J., R.J. Singh, and T. Hymowitz. 1998b.Establishment of a cytogenetic map of soybean: Current status. Soybean Genet. Newsl. 25:120–122.

Xu, S. J., R. J. Singh, and T. Hymowitz. 2000a. Monosomics in soybean: Origin, identification, cytology and breeding behavior. Crop Sci. 40:985–989.

Xu, S.J., R.J. Singh, K.P. Kollipara, and T. Hymowitz. 2000b. Primary trisomics in soybean: Origin, identification, breeding behavior, and use in linkage mapping. Crop Sci. 40:1543–1551.

Xu, S.J., R.J. Singh, K.P. Kollipara, and T. Hymowitz. 2000c. Hypertriploid in soybean: Origin, identification, cytology, and breeding behavior. Crop Sci. 40:72–77.

Yeh, M.S., C.D. Liu, and S.H. Cheng. 1997. Collection and variation of wild soybean germplasm in Taiwan and Kinmen areas. (In Chinese.) J. Agric. Assoc. China No 177:11–27.

Zhang, F., and R.G. Palmer. 1990. The *ms1* mutation in soybean: involvement of gametes in crosses with tetraploid soybean. Theor. Appl. Genet. 80:172–176.

Zhong, Z.P., D.J. Liu, and Q.F. Chen. 1997. Study on the karyotypes of soybean species. p.12–14. *In* B. Napompeth (ed.) World Soybean Res. Conf. V, Chiang Mai, Thailand. 21–27 Feb. 1994. Kasetsart Univ. Press, Bangkok, Thailand.

Zhu, T., L. Shi, J.J. Doyle, and P. Keim. 1995b. A single nuclear locus phylogeny of soybean based on DNA sequence. Theor. Appl. Genet. 90:991–999.

Zhu, T., L. Shi, and P. Keim. 1995a. Detection by in-situ fluorescence of short, single copy sequence of chromosomal DNA. Plant Mol. Biol. Rep.13:270–277.

5   Qualitative Genetics

REID G. PALMER

USDA-ARS
Iowa State University
Ames, Iowa

TODD W. PFEIFFER

University of Kentucky
Lexington, Kentucky

GLENN R. BUSS

Virginia Polytechnic Institute and State University
Blacksburg, Virginia

THOMAS C. KILEN

USDA-ARS
Crop Genetics and Production Research
Stoneville, Mississippi

In the first edition of *Soybeans: Improvement, Production, and Uses*, a separate chapter was devoted to qualitative genetics (Bernard and Weiss, 1973). The second edition combined qualitative genetics and cytogenetics (Palmer and Kilen, 1987). Recent advances in qualitative genetics and cytogenetics necessitated separate chapters for the third edition. This chapter summarizes existing information and highlights new information on soybean qualitative genetics with an emphasis on publications and data reported since 1986. The tables are inclusive extending from the data provided in the first edition of the monograph.

5–1 SOYBEAN GENETICS COMMITTEE

The Soybean Genetics Committee, established in 1955, carries out the following functions: (i) establishes guidelines and rules for assigning gene symbols and (ii) acts as a review committee for manuscripts concerning qualitative genetic interpretation and gene symbols in the genus *Glycine*. Soybean scientists are encouraged to submit manuscripts to the committee for a gene symbol assignment and to add seeds of the genetic line to the Soybean Genetic Type Collection. This procedure helps to ensure that an orderly nomenclature is followed to symbolize genes and that the line is preserved. Six elected members, the curator of the U.S. Department of Agriculture (USDA) Soybean Germplasm Collection, and the editor of the *Soybean Genetics Newsletter* constitute the Soybean Genetics Committee. The report of the committee and the rules for genetic symbols are published annually in the *Soybean Genetics Newsletter*.

5–2 SOYBEAN GENETICS NEWSLETTER

In 1974, the *Soybean Genetics Newsletter* was established as a means of communication at the international level on topics related to genetics and breeding of the soybean and immediate relatives. Information in the *Soybean Genetics Newsletter* is intended to stimulate thought and exchange ideas. Newsletter articles can be preliminary and speculative. The Soybean Genetics Committee reviews articles concerning qualitative genetic interpretation and gene symbols. Genetic symbols reported in the Newsletter have the same status as those published in refereed journals. The Newsletter is currently available at http://www.soygenetics.org

5–3 USDA SOYBEAN GENETIC TYPE COLLECTION

The Soybean Genetic Type Collection is a part of the USDA Soybean Germplasm Collection. It is comprised of strains with qualitative genetic traits that are not included in any other portion of the Collection (Table 5–1). These strains are assigned T-numbers in the chronological order in which seed are submitted.. The Soybean Genetic Type Collection currently is managed by Dr. R. L. Nelson, USDA-ARS, Univ. of Illinois, Dep. of Crop Sciences, National Soybean Research Center, 110 West Peabody Drive, Urbana, IL, USA 61801.

5–4 SOYBEAN GERMPLASM COLLECTIONS

Collections of the cultivated, wild annual, and wild perennial species are described in Chapter 8 (Carter, Jr. et al., 2004, this publication).

5–5 QUALITATIVE GENETICS

5–5.1 Diseases

Several loci controlling reaction to soybean diseases have been identified. The development of commercial cultivars with multiple-disease resistance has had a major impact in reducing economic losses. Only soybean diseases for which there are published gene symbols for resistance are discussed in this section.

5–5.1.1 Bacterial Blight

A dominant allele (*Rpg1*) was reported by Mukherjee et al. (1966) to control resistance to race 1 of *Pseudomonas syringae* pv. *glycinea* (Coerper) Young, Dye,

Table 5–1. Soybean Genetic Type Collection.

Strain†	Genotype†	Phenotype	Parental origin	When and where found
T16	--	Brown hilum on black seed	Ebony	Before 1930 at Urbana, IL
T31	p2	Puberulent	Soysota × Ogemaw	By F. Wentz, Ames, IA, 1926
T41	ln [d1 d2]‡	Narrow leaflet	Unknown	Before 1930 at Urbana, IL
T43 (Progeny 435B)	P1 [cyt-G1]	Glabrous	Medium Green X "glabrous"	Before 1927 at Urbana, IL
T48	--‡	Spread hilum	Manchu × Ebony	Before 1930 at Urbana, IL
T54	dt1‡	Determinate stem	Manchu	Before 1927 at Urbana, IL
T93	v1 [D1 d2 or d1 D2]	Variegated leaves	Hybrid population	Before 1931 at Urbana, IL
T93A	v1 [d1 d2]	Variegated leaves	T93	At Urbana, IL
T102	y4 le	Greenish-yellow leaves, weak plant; seed lectin absent	Wilson-Five	Before 1932 at Urbana, IL
T104	d1 d2 G cyt-G1	Green seed embryo, green seed coat	T42 (green cotyledon from H. Terao) X "Chromium green"	Before 1932 at Urbana, IL
T116H	y5	Greenish-yellow leaves, very weak plant	Radium-treated PI 65388	Before 1934 at Urbana, IL
T117 (L34-602)	Dt2 lw1 Lw2‡	Semideterminate stem, nonwavy leaf	AK114 X PI 65394	Before 1934 at Urbana, IL
T122	lo [d1 d2]	Oval leaflet, few-seeded pods	Unknown	Before 1934 at Urbana, IL
T134	y5	Greenish-yellow leaves	Illini X Peking	1937 at Urbana, IL
T135	y9	Bright greenish-yellow leaves	Illini	1938 at Urbana, IL
T136	y6 [ln dt1]	Pale green leaves	PI 88351 X Rokusun	1937 at Urbana, IL
T138 (L35-1156)	y7 y8	Yellow growth in cool weather	Unknown	Before 1935 at Urbana, IL
T139	g y3‡	Yellow seed coat, leaves turn yellow prematurely	Illini	About 1936 by Brunson in Kansas
T143	Lf1 g y3 y7 y8‡	5-foliolate, leaves turn yellow prematurely, and in cool weather	T138 X T137 (T137 is y3 from a cross in PI 81029)	By 1935 at Urbana, IL
T144	d1 d2 v1 y7 y8‡	Green seed embryo, variegated leaves, yellow growth in cool weather	LX431 (T93A X T138)	At Urbana, IL
T145 (9-776)	P1‡	Glabrous	Unknown	At Urbana, IL
T146	r-m‡	Brown seed with black stripes	LX286 (PI 82235 X PI 91073)	At Urbana, IL
T152	i	Self dark seed coat	Lincoln	By 1938 at Urbana, IL
T153	k1	Dark saddle on seed coat	Lincoln	By 1938 at Urbana, IL
T157	i	Self dark seed coat	Richland	By 1938 at Urbana, IL
T160	--	Pale green leaves	Hahto (Michigan)	By 1938 at Urbana, IL

(continued on next page)

Table 5–1. Continued.

Strain[†]	Genotype[†]	Phenotype	Parental origin	When and where found
T161	yl0	Greenish-yellow seedling	L36-5 (Mandarin X Mansoy)	1940 at Urbana, IL
T162	yl7	Light yellowish-green leaves	Mandarin	1940 at Urbana, IL
T164	--	Slightly variegated leaves	Morse	1941 at Urbana, IL
T171	--[‡]	Long peduncle	Unknown	At Urbana, IL
T173	f [ln][‡]	Fasciated stem	Keitomame (f) X PI 88351 (ln)	At Urbana, IL
T175	El t[‡]	Late maturity, gray pubescence	Unknown	At Urbana, IL
T176	lw1 lw2 [Dt2][‡]	Wavy leaf	Unknown	At Urbana, IL
T180 (L46-1741-2)	Rjl[‡]	Nodulating	Same F$_3$ plant as T181	At Urbana, IL
T181 (L46-1743-2)	rjl[‡]	Nonnodulating	Lincoln[2] X Richland	At Urbana, IL
T201	rjl[‡]	Nonnodulating	LX1277 (L46-1743 X L46-1741)	At Ames, IA
T202	Rjl[‡]	Nodulating	Sib of T201	At Ames, IA
T204 (L48-101)	ln lo[‡]	Narrow-oval leaflet	T136 X T122	At Urbana, IL
T205 (L48-163)	lw1 lw2[‡]	Wavy leaf	Dunfield X Manchuria 13177	At Urbana, IL
T208 (Ind. Acc. 2300-2)	Se[‡]	Pedunculate inflorescence, small seeds	PI 196176§ ('Yu tae' from Korea)	
T209 (L50-155)	--[‡]	Determinate stem termination	Lincoln X "wild dwarf"	At Urbana, IL
T210 (L49-738)	df2	Dwarf	Colchicine-treated Lincoln	At Urbana, IL
T211H (CX3941-844-2-5)	pm	Dwarf, crinkled leaves, sterile	Kingwa X T161	At Lafayette, IN
T216 (L46-266)	--[‡]	Reddish-black seeds	PI 86038 X PI 88351	1946 at Urbana, IL
T218M	Yl8-m (Urbana)	Chlorophyll chimera, (resembles T225M)	Illini	1952 at Urbana, IL
T218H	yl8 (Urbana)	Nearly lethal, yellow leaves	T218M	By R.G. Palmer, Ames, IA, 1987
T219H (A691-1)	yl1	Lethal yellow, (heterozygote has greenish-yellow leaves)	Richland X Linman 533	1941 at Ames, IA
T220 (L46-431)	--	Greenish-yellow leaves	Lincoln	At Urbana, IL
T221 (L46-426)	--	Yellowish-green leaves	Peking	At Urbana, IL
T223 (L46-429)	--	Yellowish-green leaves	Richland	At Urbana, IL
T224 (L46-428)	--	Greenish-yellow leaves	Richland	At Urbana, IL
T225M	Yl8-m	Unstable gene resulting in chlorophyll chimera	Lincoln	Before 1955 in Iowa
T225H	yl8 (Ames 1)	Nearly lethal, yellow leaves	T225M	By R.G. Palmer, Ames, IA, 1975
T226	--	Greenish-yellow leaves	Lincoln	1943 at Ames, IA
T227	--	Greenish-yellow leaves, becoming green	Illini	1943 at Kanawha, IA

T229	y14	Light green leaves	F$_4$ (Richland X Linman 533)	1943 at Ames, IA
T230 (A43K-643-1)	y13	Whitish-green seedling, greenish-yellow leaves	Mandell X Mandarin (Ottawa)	1944 at Kanawha, IA
T231 (A49-8414)	--	Greenish-yellow leaves, weak plant	AX3015-55 (Richland X Linman 533)	1943 at Ames, IA
T232	--	Yellowish-green leaves	Hawkeye	1950 at Ames, IA
T233	y12	Whitish primary leaves, yellowish-green leaves	Hawkeye	1950 in field N2100 at Ames, IA
T234	y20 (Ames 23) Mdh1-n (Ames 21) (formerly y15)	Yellowish-green leaves, malate dehydrogenase 1 null	L46-2132 (Clark progenitor)	1952 and 1998 at Ames, IA
T235 (L58-274)	wm	Magenta flower	Harosoy	1957 at Urbana, IL
T236 (L46-232)	[Lf1 ln y6]‡	Red-buff seed	T143 X "y6 ln pc dt1 w1"	1945 at Urbana, IL
T238 (S57-3416)	k3	Dark saddle on seed coat	X-rayed Clark	1956 at Columbia, MO
T239 (L63-365)	k2 (Urbana 1)	Tan saddle	Harosoy	1961 at Urbana, IL
T241H	st2	Asynaptic sterile	S54-1714 (from same cross as Wayne)	About 1956 at Columbia, MO
T242H	st3	Asynaptic sterile	AX54-118-2-8 (Blackhawk X Harosoy)	At Lafayette, IN
T243	df2	Dwarf	Colchicine-treated Lincoln	At Ames, IA
T244 (Adams 77-2)	df3	Dwarf	Neutron-irradiated Adams	At Ames, IA
T249H (L67-4408A)	[PI1]	Whitish-yellow seedling, lethal	F$_3$ (Clark^6 X PI 84987) X (Clark^6 X T145)	1964 at Urbana, IL
T250H (L67-4439)	--	Lethal seedling	F$_2$ [Harosoy^5 X (Clark^6 X Chief)]	1964 at Urbana, IL
T251H (L67-4440A)	mn	Miniature plant	F$_2$ (Harosoy^5 X T139)	1961 at Urbana, IL
T252 (L64-2612)	--	Pale green leaves	F$_3$ (Harosoy^6 X T139)	1963 at Urbana, IL
T253 (L67-4415A)	k2 (Urbana 1) y20 (Urbana 1) Mdh1-n (Urbana 1)	Tan saddle, yellowish-green leaves malate dehydrogenase 1 null	T239	1963 at Urbana, IL and y20 Mdh1-n designated in 1996 at Ames, IA
T254 (L67-4412A)	--	Greenish-yellow leaves	F$_2$ (Clark^6 X T176)	1964 at Urbana, IL
T255	lf2	7-foliolate	Hawkeye	1966 at Ames, IA
T256	df4	Dwarf	Hark	1966 at Ames, IA
T257H	y16	Nearly lethal white	C1128^8 X Mukden (C1128 is from Wabash X Hawkeye)	At Lafayette, IN
T258H (A72-1103-6)	st4	Desynaptic sterile	Hark	1968 at Ames, IA
T259H (L71L-06-4)	ms2 (Eldorado)	Male sterile	F$_3$ of SL11 (Wayne-r Rpm Rps1) X L66L-177 [Wayne X (Hawkeye X Lee)]	By R.L. Bernard, Eldorado, IL, 1971

(continued on next page)

Table 5–1. Continued.

Strain†	Genotype†	Phenotype	Parental origin	When and where found
T260H (N69-2774)	ms1 (North Carolina)	Male sterile	Unknown	1966 in a farmer's field in North Carolina
T261 (S56-26)	k2 (Columbia 1) Mdh1-n	Tan saddle, malate dehydrogenase 1 null	Mandarin (Ottawa)	Before 1956 at Columbia, MO, (Columbia 1) and Mdh1-n designated in 1996 at Ames, IA
T262	--	"Double pod"	SRF 200 (Hark-ln)	About 1971 at SRF, Mason City, IL
T263 (A76-2)	df5	Dwarf	Harosoy 63 X PI 257435	1968 at Iowa State Univ. nursery, Hawaii
T264 (L58-2749)	Pd2	Dense pubescence	Neutron-irradiated Blackhawk in the M_2 generation	1956 at Urbana, IL
T265H (L75-0324)	yl9	Delayed albino	F_2 (Williams[6] X T259)	1974 to1975 greenhouse at Urbana, IL
T266H	ms1 (Urbana)	Male sterile, (higher female fertility than T260, T267, and T268)	F_3 row of L67-533 [(Clark[6] X Higan) X SRF 300]	1971 at Urbana, IL
T267H (L56-292)	ms1 (Tonica)	Male sterile	Semisterile plant found in Harosoy	By F.M. Burgess, Tonica, IL, 1955
T268H (A73g-21)	ms1 (Ames 1)	Male sterile	Semisterile plant found in T258H	1970 at Ames, IA
T269H (L70-8654)	fs1 fs2	Structural sterile (T269H is from Fs1 fs1 fs2 fs2 plants)	Flower structure mutant segregating in a plant-progeny row from PI 339868 (Yuwoltae)	1970 at Urbana, IL
T270H (A78-286)	y22	Greenish-yellow leaves, very weak plant	Segregating in an F_2-plant-progeny row from an outcross in T271H	By R.G. Palmer, Ames, IA, 1977
T271H	msp	Partial male sterile	40-parent bulk population [AP6(S1)C1]	By R.G. Palmer, Ames, IA, 1975
T272H (A71-44-13)	st5	Desynaptic sterile	W66-4108 (Merit X W49-1982-32) (W49-1982-32 is from Hawkeye X Manchu[3])	By R.G. Palmer, Ames, IA, 1970
T273H (A72-1711)	ms3 (Washington)	Male sterile	Semisterile plant in an F_3-plant-progeny row from Calland X Cutler	1971 at Washington, IA
T274H (A74-4646)	ms4 (Ames)	Male sterile	Semisterile plant in Rampage	By R.G. Palmer, Ames, IA, 1973

T275 (A77-K150)	cyt-Y2	Yellowish leaves becoming greenish-yellow	Chimeric F$_2$ plant A75-1165-117 from T268H X [PI 101404B (G. soja) X Clark^6]	By R.G. Palmer, Ames, IA, 1975
T276	nr1	Constitutive nitrate reductase absent	M$_2$ generation of Williams treated with EMS, nitroso-guanidine, and X-rays	1979 at Urbana, IL
T277H	ms5	Male sterile	Semisterile plant in the M$_3$ generation of neutron-irradiated Essex	1976 at Blacksburg, VA
T278M	cyt-Y3	Yellow leaves, very weak plant (mutable plants are chlorophyll chimeras)	Chimeric plant of unknown source	By R.G. Palmer, Ames, IA, 1972
T279 (D76-1609)	lps‡	Short petiole	F$_3$ (Forrest^2 X PI 229358) X D71-6234 (D71-6234 is from high-protein Lee type X PI 95960)	1976 at Stoneville, MS
T280 (C1640)	fan	Low linolenic acid	EMS-treated Century (M$_2$ plant)	1981 at West Lafayette, IN
T281 (L58-617)	df7 df8	Dwarf plant, rugose leaf	Off-type in PI 232992	1955 at Urbana, IL
T282 (L81-5482)	--	Curled leaf	Mutant or segregant in F$_3$ of Williams X PI 82278	1980 at Urbana, IL
T283 (A77-86)	--	Chlorophyll deficient	F$_7$ plants of PI 101404B (G. soja) X Clark^6	By R.G. Palmer, Ames, IA, 1977
T284H	ms3 (Flanagan)	Male sterile	Wabash outcross	By H. Chaudhari and W. Davis, Flanagan, IL, 1973
T285 (IL3-1)	fr5	Nonfluorescent seedling	Williams treated with gamma-rays	1981 to 1984 at Ames, IA
T286 (MS2060)	df6	Dwarf	EMS-treated C1421 (Adelphia^8 X Mukden)	Early 1980s at West Lafayette, IN
T287H (S85-62-11)	ms1 (Ames 2)	Male sterile	Segregating in S$_{4,5}$ progeny from AP6(S1)C1	By R. Secrist, Ames, IA, 1984
T288 (Williams 80-7)	y23	Leaves becoming yellow-white and necrotic, viable plant	Williams	By A. Williams, Williams, IN, 1980
T289 (Hardee 2)	Got-c	Glutamate oxaloacetic transaminase variant	Hardee	By Y.T. Kiang, Durham, NH, 1983
T290H	ms1 (Danbury)	Male sterile	Beeson outcross	By M.C. Albertsen, Danbury, IA, before 1988
T291H	ms3 (Plainview)	Male sterile	F$_2$ (Viking X Classic II) X (Mitchell X Columbus), Viking is from Merit X Amsoy	By W. Davis, Plainview, TX, before 1988

(continued on next page)

Table 5–1. Continued.

Strain†	Genotype†	Phenotype	Parental origin	When and where found
T292H	ms4 (Fisher)	Male sterile	Corsoy	By W. Davis, Fisher, AR, before 1988
T293	sp1 (Altona-sp1)	β-amylase null	Altona	By M. Gorman and Y.T. Kiang, Durham, NH, before 1978
T294 (G81-6299)	g3	Green seed coat	F_6 line from Duocrop X G76-57 (G76-57 is from Bragg X Kent)	By H. R. Boerma, Athens, GA, 1982
T295H	ms6 (Ames 1)	Male sterile	A78-245014 from A74-204034 X C1520	By W. R. Fehr and R. G. Palmer, Ames, IA, 1978
T296	--	5 to 7 foliolate	Williams × BB 13 No. 9, which was a mutant in "SJ2"	BB 13 No. 9 found by A. Waranyuwat, Northeast Agric. Center, Khonaen, Thailand, 1976
T297H (L88-3785)	--	Semisterile	F_7 row from Clark X PI 317334B	By R.L. Bernard, Urbana, IL, 1984
T298H (L88-3809)	--	Nearly sterile	F_3 row from L64-2887(Clark-i)[6] X Sooty	By R.L. Bernard, Urbana, IL, 1982
T299H (L88-3834)	-(eu)	Sterile	F_3 row from Williams[6] X PI 229324	By R.L. Bernard, Urbana, IL, 1985
T300H (L88-3854)	--	Sterile	F_7 row from L78-375 (Williams-$Rsv1$) X PI 86740	By R.L. Bernard, Eldorado, IL, 1985
T301H (L88-3886)	--	Semisterile	F_4 row from Beeson X Prize	By R.L. Bernard, Urbana, IL, 1981
T302H (L88-3962)	--	Nearly sterile	Beeson	By R.L. Bernard, Urbana, IL, 1973
T303H (L88-3966)	--	Nearly sterile	L73-5446 from L67-1250 (Harosoy-$Dt2$) X L62-1251(Clark-$Dt2$)	By R.L. Bernard, Urbana, IL, 1975
T304H (L88-3981)	--	Nearly sterile	L73-5741 from Corsoy X Amsoy 71	By R.L. Bernard, Urbana, IL, 1975
T305H (L88-4064)	--	Semisterile	L75-12103 from Wells X Williams	By R.L. Bernard, Urbana, IL, 1975
T306H (L88-4106)	--	Sterile	PI 506669 (Fujihime)	By G. Juvik, Urbana, IL, 1987
T307 (A5)	fan (A5)	Low linolenic acid	EMS mutant of FA 9252 (PI 80476 X PI 85671)	By W.R. Fehr and E. Hammond, Ames, IA, 1980

T308 (C1726)	*fap1*	Low palmitic acid	EMS mutant of Century	By J. Wilcox, West Lafayette, IN, 1982
T309 (C1727)	*fap2*	High palmitic acid	EMS mutant of Century	By J. Wilcox, West Lafayette, IN, 1982
T310 (L81-4148)	--	Weak stem, buff seed coat, and wavy leaflet margin	Harosoy X L67-3391 (the original mutant in Harosoy)	By R.L. Bernard, Urbana, IL, 1965
T311	*shr*	Shriveled seed	F_6 (AP2 X P2180)	By C. Jennings and R. Freestone, Waterloo, IA, 1982
T312	*v2*	Variegated leaves	Clark	By K. Newhouse and R.G. Palmer, Ames, IA, 1978
T313	*lnr*	Narrow, rugose leaflet	EMS-treated C1421	By J. Wilcox, West Lafayette, IN, 1975
T314	*cyt-Y4*	Yellow leaves	Variant in the F_3 generation of a chimeric plant from Clark X T251	By R.G. Palmer, Ames, IA, 1981
T315	*cyt-Y5*	Yellow leaves	Variant in Williams	By A. Williams, Williams, IN, 1981
T316	*cyt-Y6*	Yellow leaves	Variant in F_6 [(Corsoy X Rampage) X Franklin]	By R. Freestone, Waterloo, IA, 1982
T317	*y20* (Ames 1) *Mdh1-n* (Ames 1)	Yellowish-green leaves, malate dehydrogenase 1 null	Somaclonal mutant in Jilin 3 (PI 427099)	By L. Amberger and R.G. Palmer, Ames, IA, 1988
T318	*Aco2-bn*	Aconitase 2 null	Somaclonal mutant in BSR 101	By L. Amberger and R.G. Palmer, Ames, IA, 1988
T319	*cyt-Y7*	Cytoplasmic yellow	Chimeric plant in AX2950 (Hack X PI 407298)	By G. Graef and R.G. Palmer, Ames, IA, 1985
T320	*cyt-Y8*	Cytoplasmic yellow	F_9 chimeric plant from Williams X Essex	By E. Roberts, St. Joseph, IL, 1986
T321	*w4-dp*	Pale purple throat flower	F_{11} plant from T322	By R. Groose and R.G. Palmer, Ames, IA, 1986
T322	*w4-m*	Mutable purple flower	F_7 plant from X1878 X X2717 (X1878 from Corsoy X Essex) (X2717 from Amsoy 71 X AG52109) (AG52109 from Bavender Special X Hark)	By H.D. Weigelt, Stonington, IL, 1983
T323	*y20* (Ames 2) *Mdh1-n* (Ames 2)	Yellowish-green leaves, malate dehydrogenase 1 null	Mutant in T322	By R. Groose and R.G. Palmer, Ames, IA, 1986

Table 5–1. Continued.

Strain†	Genotype†	Phenotype	Parental origin	When and where found
T324	y20 (Ames 3) Mdh1-n (Ames 3)	Yellowish-green leaves, malate dehydrogenase 1 null	Mutant in T322	By R. Groose and R.G. Palmer, Ames, IA, 1986
T325	y20 (Ames 4) Mdh1-n (Ames 4)	Yellowish-green leaves, malate dehydrogenase 1 null	Mutant in T322	By R. Groose and R.G. Palmer, Ames, IA, 1986
T326 (LG87-2116)	br1 br2‡	Few branches only from lower nodes	F₅ row from PI 391583 (Jilin No. 10) X PI 189916 (Ta Ching Mi Hwang Tau Tsa)	By R. Nelson, Urbana, IL, 1987
T327 (LG87-2118)	Br1 Br2‡	Branches originating from upper as well as lower nodes	F₅ row from PI 391583 X PI 189916	By R. Nelson, Urbana, IL, 1987
T328H (NR-1)	rn1 (Ames 1)	Necrotic root	Mutant in T322	By R.G. Palmer, Ames, IA, 1987
T329H (NR-2)	rn1 (Ames 2)	Necrotic root	Mutant in T322	By R.G. Palmer, Ames, IA, 1987
T330H (NR-3)	rn1 (Ames 3)	Necrotic root	Mutant in T322	By R.G. Palmer, Ames, IA, 1987
T331H (Calland TC)	St6 st6 st7 st7	Male sterile, female sterile	Somaclonal mutant in Calland EMS75-4 X AgriPro 1776	By R.G. Palmer, Ames, IA, 1985
T332H	–	Necrotic root	Mutant in T322	By R.G. Palmer, Ames, IA, 1994
T333H (A95-1453)	–	Necrotic root	Mutant in T322	By R.G. Palmer, Ames, IA, 1995
T334 (X-197)	k2 (Urbana 1) y20 (Ames 5) Mdh1-n (Ames 7)	Tan saddle, yellowish-green leaves, malate dehydrogenase 1 null	A1937 X T239	By R.G. Palmer and X. Chen, Ames, IA, 1992
T335 (X-203)	k2 (Urbana 1) y20 (Ames 6) Mdh1-n (Ames 8)	Tan saddle, yellowish-green leaves, malate dehydrogenase 1 null	X2717 X T239	By R.G. Palmer and X. Chen, Ames, IA, 1992
T336 (X-217)	k2 (Urbana 1) y20 (Ames 7) Mdh1-n (Ames 9)	Tan saddle, yellowish-green leaves, malate dehydrogenase 1 null	T239 X X2717	By R.G. Palmer and X. Chen, Ames, IA, 1992
T337 (X-219)	k2 (Urbana 1) y20 (Ames 8) Mdh1-n (Ames 10)	Tan saddle, yellowish-green leaves, malate dehydrogenase 1 null	T239 X Lincoln	By R.G. Palmer and X. Chen, Ames, IA, 1992
T338 (X-241)	k2 (Urbana 1) y20 (Ames 9) Mdh1-n (Ames 11)	Tan saddle, yellowish-green leaves, malate dehydrogenase 1 null	A1937 X T239	By R.G. Palmer and X. Chen, Ames, IA, 1992
T339 (X-451)	k2 (Urbana 1) y20 (Ames 10) Mdh1-n (Ames 12)	Tan saddle, yellowish-green leaves, malate dehydrogenase 1 null	A1937 X T239	By R.G. Palmer and X. Chen, Ames, IA, 1992

T340 (M-7-2)	k2 (Columbia 1) y20 (Ames 11) Mdh1-n (Columbia 1)	Tan saddle, yellowish-green leaves, malate dehydrogenase 1 null	T261 X X2937	By R.G. Palmer and X. Chen, Ames, IA, 1992
T341 (M-11-4)	k2 (Urbana 1) y20 (Ames 12) Mdh1-n (Ames 13)	Tan saddle, yellowish-green leaves, malate dehydrogenase 1 null	T239 X X1878	By R.G. Palmer and X. Chen, Ames, IA, 1992
T342 (M-11-7)	k2 (Urbana 1) y20 (Ames 13) Mdh1-n (Ames 14)	Tan saddle, yellowish-green leaves, malate dehydrogenase 1 null	T239 X X1878	By R.G. Palmer and X. Chen, Ames, IA, 1992
T343 (M-14-23)	k2 (Urbana 1) y20 (Ames 14) Mdh1-n (Ames 15)	Tan saddle, yellowish-green leaves, malate dehydrogenase 1 null	X2937 X T239	By R.G. Palmer and X. Chen, Ames, IA, 1992
T344 (M-19-3)	k2 (Urbana 1) y20 (Ames 15) Mdh1-n (Ames 16)	Tan saddle, yellowish-green leaves, malate dehydrogenase 1 null	X2717 X T239	By R.G. Palmer and X. Chen, Ames, IA, 1992
T345 (M-20-11)	k2 (Urbana 1) y20 (Ames 16) Mdh1-n (Ames 17)	Tan saddle, yellowish-green leaves, malate dehydrogenase 1 null	T239 X X2717	By R.G. Palmer and X. Chen, Ames, IA, 1992
T346 (CD-9)	y20 (Ames 17) Mdh1-n (Ames 19)	Yellowish-green leaves, malate dehydrogenase 1 null	Mutant in T322	By R.G. Palmer and X. Chen, Ames, IA, 1992
T347 (X-193)	k2 (Urbana 1) y20 (Ames 18) Mdh1-n (Ames 6)	Tan saddle, yellowish-green leaves, malate dehydrogenase 1 null	T239 X PI 567630A	By R.G. Palmer and X. Chen, Ames, IA, 1992
T348 (X-194)	k2 (Urbana 1) y20 (Ames 19) Mdh1-n (Ames 6)	Tan saddle, yellowish-green leaves, malate dehydrogenase 1 null	T239 X PI 567630A	By R.G. Palmer and X. Chen, Ames, IA, 1992
T349 (PR-95-649)	k2 (Urbana 1) y20 (Ames 20) Mdh1-n (Ames 6)	Tan saddle, yellowish-green leaves, malate dehydrogenase 1 null	T239 X PI 567630A	By R.G. Palmer and X. Chen, Ames, IA, 1992
T350 (PR-95-650)	k2 (Urbana 1) y20 (Ames 21) Mdh1-n (Ames 6)	Tan saddle, yellowish-green leaves, malate dehydrogenase 1 null	T239 X PI 567630A	By R.G. Palmer and X. Chen, Ames, IA, 1992
T351 (A95-K55)	k2 (Urbana 1) y20 (Ames 22) Mdh1-n (Ames 20)	Tan saddle, yellowish-green leaves, malate dehydrogenase 1 null	T239 X T319	By R.G. Palmer and X. Chen, Ames, IA, 1992

(continued on next page)

Table 5–1. Continued.

Strain[†]	Genotype[†]	Phenotype	Parental origin	When and where found
T352H (A97-1652)	st8	Male sterile, female sterile	Mutant in T322	By R.G. Palmer, Ames, IA, 1989
T353 (E420BC3)	--	Supernodulation	Mutant from EMS-treated Elgin 87	By B. Buttery and R. Buzzell, Harrow, ON, 1990
T354H (A94-JB-124)	ms6 (Ames 2)	Male sterile, female fertile	From F$_2$ plant progeny of A92-JB-13 X BSR 101	By R.G. Palmer, Ames, IA, 1994
T355	--	Wrinkled leaflets (maternal inheritance)	Somaclonal variant from a protoplast culture of Clark 63	By J. Widholm and M. Nguyen, Urbana, IL, 1994
T356 (A95-B66-1)	--	Yellow foliage type derived from chlorophyll-deficient chimeric plant (maternal inheritance)	T323 X L67-3483	By R.G. Palmer, Ames, IA, 1994
T357 (MSM-1)	ms7 [k2 (Columbia 1) Mdh1-n]	Male sterile, female fertile	T261 X T323	By R.G. Palmer, Ames, IA, 1994
T358 (MSM-2)	ms8 [k2 (Urbana 1) Mdh1-n (Ames 1)]	Male sterile, female fertile	T317 X T239	By R.G. Palmer, Ames, IA, 1994
T359 (MSM-3)	ms9 [Mdh1-n (Ames 4)]	Male sterile, female fertile	T325 X L67-3483	By R.G. Palmer, Ames, IA, 1994
T360 (MSM-4)	ms2 (Ames) [k2 (Columbia 1) Mdh1-n]	Male sterile, female fertile	T261 X PI 567630A	By R.G. Palmer, Ames, IA, 1994
T361	y20 (Ames 24) Mdh1-n (Ames 22)	Yellowish-green leaves, malate dehydrogenase 1 null	Somaclonal mutant in Jilin 3 (PI 427099)	By J. Burzlaff and R.G. Palmer, Ames, IA, 1988
T362H	y18 (Ames 2)	Nearly lethal, yellow leaves	Somaclonal mutant in Jilin 3 (PI 427099)	By J. Burzlaff and R.G. Palmer, Ames, IA, 1988
T363	dlm	Disease lesion mimic, necrotic spots with chlorotic halo	Gamma irradiation-induced mutant in Hobbit	By J.E. Specht, Lincoln, NE, 1992

† Where previous strain designations have been used, they are given in parentheses under the T-number. Most of the T-strains are mutants. Those that are segregants from crosses or presumed outcrosses are marked with an ‡ after the genotype. For T-strains with an H suffix (e.g., T211H), the allele is carried as the heterozygote, because the homozygote is lethal, sterile, or very weak. For T-strains with an M suffix (e.g., T225M), the trait is maintained by selecting the mutable genotype. Cytoplasmically inherited traits are prefixed by cyt-. Genes for secondary traits of interest are listed in brackets. Numerical superscripts are used to indicate backcrosses; for example, cv. Lincoln² × cv. Richland means Lincoln × (Lincoln × Richland).

§ Not in USDA Soybean Germplasm Collection, considered a duplicate of PI 196177.

& Wilkie, the causal agent of bacterial blight. Keen and Buzzell (1991) reported that soybean cv. Flambeau and cv. Merit differed in their resistance to *P. syringae* pv. *glycinea* race 4 which carried each of four different avirulence genes. These genes had been cloned from the pathogen and then individually introduced into race 4 by conjugation. Segregation data for F_2 and F_3 progeny of Flambeau × Merit crosses indicated that single dominant and nonallelic genes accounted for resistance to race 4, carrying bacterial avirulence genes *avrA*, *avrB*, *avrC*, or *avrD*. Their results indicated that the previously described resistance allele, *Rpg1*, complemented *avrB*. They proposed the designations *Rpg2*, *Rpg3*, and *Rpg4* for the soybean resistance alleles that complement bacterial genes *avrA*, *avrC*, and *avrD*, respectively. The alleles for susceptibility are *rpg1*, *rpg2*, *rpg3*, and *rpg4*.

5–5.1.2 Bacterial Pustule

Bernard and Weiss (1973) summarized studies showing that resistance to bacterial pustule, caused by *Xanthomonas campestris* pv. *glycines* (Nakano) Dye, found in the cv. CNS, was controlled by a single recessive allele (*rxp*). This allele also controls resistance to the disease wildfire, caused by *Pseudomonas syringae* pv. *tabaci* (Wolf & Foster), Young, Dye, & Wilkie. This allele for resistance from CNS has been transferred to almost all southern U.S. cultivars and many northern ones as well. No additional loci for resistance have been reported.

5–5.1.3 Brown Stem Rot

Brown stem rot (BSR), caused by *Phialophora gregata* (Allington and Chamberlain) W. Gams, is widespread in Canada, and the Midwest and some southern states of the USA (Gray and Grau, 1999). Hanson et al. (1988) reported that the BSR resistance in L78-4094 (donated by PI 84946-2, an off type found in the Korean cv. Kandokon) is conditioned by a dominant allele, *Rbs1*. They also reported that the BSR resistance in PI 437833 (Chinese cv. Curo sengocu) was governed by a dominant allele different from *Rbs1*, designated *Rbs2*. Willmot and Nickell (1989) determined that the BSR resistance found in PI 437970 (VIR 1460 from the Russian collection) was conditioned by a single dominant gene. Crosses of PI 437970 with L78-4094 (*Rbs1*) or with PI 437833 (*Rbs2*) each resulted in segregation data consistent with duplicate dominant epistasis, thus revealing a third dominant resistance allele, designated *Rbs3*.

5–5.1.4 Frogeye Leaf Spot

Bernard and Weiss (1973) summarized studies showing that one dominant allele (*Rcs1*) controlled resistance to race 1 of *Cercospora sojina* Hara, and that at another locus, dominant allele *Rcs2* controlled resistance to race 2. Boerma and Phillips (1983) reported that the cv. Davis had a single dominant allele for resistance to races 2 and 5. Crosses between cv. Kent (*Rcs2*) and Davis showed that the *Rcs* allele in Davis was at a locus different from *Rcs2*. The authors stated that the relationship with the *Rcs1* allele could not be resolved because no race 1 culture of *C. sojina* was available. The allele in Davis was assigned the gene symbol *Rcs3*.

5–5.1.5 Downy Mildew

Bernard and Cremeens (1972) reported that all observed races of *Peronospora manshurica* (Naum.) Syd. ex Gaum were controlled by a single dominant allele (*Rpm*) found in the cv. Kanrich. A newly identified race of *P. manshurica*, virulent on the cv. Union, that carries the allele *Rpm* (transferred from Kanrich) for resistance to all previously known races of downy mildew, was observed in Illinois in 1981 (Lim et al., 1984). Lim (1989) reported that the cv. Fayette and PI 88788 were resistant to races 2 and 33, whereas Union was resistant to race 2 but susceptible to race 33. The allele for resistance in Fayette and PI 88788 was assigned the symbol *Rpm2* because it segregated independently from *Rpm1* in Union.

5–5.1.6 Powdery Mildew

Powdery mildew, caused by *Microsphaera diffusa* Cke. & Pk., occurs frequently on soybean in greenhouses and occasionally in the field. Because the disease occurs only sporadically on field-grown plants, little effort has been devoted to a search for resistance. Grau and Lawrence (1975) reported that resistance was controlled by a single dominant allele found in cv. Chippewa 64, but did not propose a gene symbol. This was perhaps the same allele later designated *Rmd* by Buzzell and Haas (1978), for adult-plant resistance in cv. Blackhawk. The recessive allele (*rmd*) results in susceptibility at all stages of plant development. Lohnes and Bernard (1992) reported that resistance to powdery mildew at both juvenile and adult-plant stages in the line L76-1988 was controlled by a single dominant allele found at the same locus as *Rmd*. L76-1988 was a selection from cv. Williams × (cv. Harosoy × D54-2437). The cv. CNS was the original donor of the resistance allele in D54-2437 and L76-1988. This allele was assigned the symbol *Rmd-c*, which is allelic to *Rmd*. Lohnes et al. (1993) reported linkage between *Rmd, Rj2,* and *Rps2*.

5–5.1.7 Phytophthora Root Rot

Phytophthora root rot (PRR), caused by *Phytophthora sojae* (Kaufmann & Gerdemann), occurs in many of the soybean-producing areas of the USA and Canada. At least 55 physiologic races of *P. sojae* have been identified (Leitz et al., 2000). Palmer and Kilen (1987) summarized studies documenting the identification and assignment of symbols to the following alleles controlling resistance to PRR: *Rps1-a, Rps1-b, Rps1-c, Rps1-k, Rps2, Rps3, Rps4, Rps5,* and *Rps6*. Ploper et al. (1985) identified a second resistance allele at the *Rps3* locus in PI 172901, designated *Rps3-b*. A third resistance allele at the same locus was identified in PI 340046 by Athow et al. (1986), and was assigned the symbol *Rps3-c*. The resistance allele *Rps1-d* was identified by Buzzell and Anderson (1992) while they were studying the inheritance of PRR resistance in a selected breeding line, OX642, that had Chinese cv. Wu An (PI 103091) as its source of resistance. Anderson and Buzzell (1992) reported a new locus (*Rps7*) for resistance to PRR that they found in the cv. Harosoy. In additional crosses, using cv. Harosoy 63, they determined that *Rps7* was linked with *Rps1*. The identified resistance alleles at seven loci are being used either singly or in combination in numerous breeding programs to develop cultivars having resistance to multiple races of *P. sojae*.

5–5.1.8 Stem Canker

Stem canker, caused by *Diaporthe phaseolorum* (Cke. & Ell.) Sacc. var. *caulivora* Athow & Caldwell, was a serious disease of soybean in the midwestern and north-central USA and in Canada in the 1940s and early 1950s. Stem canker became a major disease problem in the southeastern USA in the 1970s and 1980s. Differences in growth and virulence were reported between stem canker isolates from the northern and southern USA (Keeling, 1988). Several schemes for characterizing northern and southern isolates of the pathogen have been proposed, including forma speciales (Morgan-Jones, 1989). Using this system, the southern isolate was named *D. phaseolorum* f. sp. *meridionalis*. Kilen et al. (1985) discovered that stem canker resistance in cv. Tracy-M was controlled by two dominant alleles. Kilen and Hartwig (1987) intercrossed several F_3 lines having one of the alleles and developed populations giving segregation data demonstrating that some of the F_3 lines were homozygous for different dominant resistance alleles. The symbols *Rdc1* and *Rdc2* were assigned. Bowers et al. (1993) crossed the stem canker resistant cvs. Tracy-M, Crockett, and Dowling with each other and with two susceptible cultivars to develop populations of F_1, F_2, F_3, and backcrosses for inoculating with the pathogen. They reported segregation ratios demonstrating the presence of an additional dominant allele for resistance to stem canker in Crockett (*Rdc3*) and another in Dowling (*Rdc4*). Southern isolates of the pathogen were used in both of these studies.

5–5.1.9 Sudden Death Syndrome

Sudden death syndrome is a relatively newly identified disease of soybean caused by blue-pigmented, highly pathogenic strains of *Fusarium solani* (Mart.) Appel & Wollenweb. emend. W. C. Snyder & H. N. Hans. that have now been designated *F. solani* f. sp. *glycines* (Roy et al., 1997). Stephens et al. (1993a) determined that the ability of the cv. Ripley to resist disease symptoms caused by *F. solani* is conditioned by a single dominant nuclear allele (*Rfs*).

5–5.1.10 Soybean Rust

Soybean rust, caused by *Phakopsora pachyrhizi* Syd. & P. Syd., is an economically important fungal disease in many areas of the world, but there are no definitive reports of soybean rust in North America or Europe. McLean and Byth (1980) reported that the cv. Komata (PI 200492) has a single dominant allele (*Rpp*) giving resistance to an Australian rust isolate. Singh and Thapliyal (1977) reported that the cv. Ankur (PI 462312) also has a single dominant allele for resistance. Bromfield and Hartwig (1980) inoculated an F_2 population from the cross cv. Centennial (susceptible) × PI 230970 (introduced from Japan in 1956 without a name) with four rust isolates. The segregation ratios suggested a single dominant allele for resistance. Hartwig and Bromfield (1983) determined the relationships between the three sources of rust resistance known at that time by intercrossing them and inoculating the F_2 and F_3 populations with two rust isolates. Their results demonstrated that each dominant resistance allele in PI 200492, PI 230970, and PI 462312 was at a different locus. They suggested that the *Rpp* symbol for the allele in PI 200492 be changed to *Rpp1*, and assigned the symbols *Rpp2* and *Rpp3* to the alleles in PI 230970 and PI 462312, respectively. Hartwig (1986) reported on studies with the rust resistant cv. Bing Nan (PI 459025) from Fujian Province, China. Crosses were made between cv. Bing Nan and cv. Centennial and the other three known sources of rust resistance. The F_2 and F_3 populations were inoculated with three rust isolates. Data obtained confirmed that PI 459025 had a single dominant resistance allele (*Rpp4*) at a locus different from the other three alleles.

5–5.1.11 Soybean Mosaic Virus

Soybean mosaic virus (SMV) is distributed worldwide wherever soybean is grown and is an important disease in many areas (Hill, 1999). Soybean mosaic virus has been known to cause nearly total crop loss (Kwon and Oh, 1980). There are three basic reactions of soybean when exposed to SMV. The resistant reaction exhibits no response. The typical susceptible reaction is a light and dark green mosaic with later blistering and downward cupping of the leaflets. Infected plants usually produce seeds with mottled seedcoats. The third reaction involves the development of necrotic lesions on the leaflets, petioles, and stems. In its most extreme form, the growing point of young seedlings can be killed, and plants either die or become nonproductive. The necrotic reaction has been classified as susceptible by some workers, because it can be more destructive than the typical mosaic, but necrosis generally has been shown to occur only when alleles are present that exhibit resistance to some SMV strains (Chen et al., 1994). Also, Chen et al. (1991) were not able to produce infection of susceptible checks when inoculated from tissue of necrotic plants, showing that virus titer is very low in necrotic plants. Some alleles exhibit complete resistance to a given SMV strain when in the homozygous state, but show varying degrees of necrosis in the heterozygous condition, while some alleles condition necrosis even when they are homozygous (Kiihl and Hartwig, 1979; Shigemori, 1988; Bowers et al., 1992; Chen et al., 1994).

Cho and Goodman (1979, 1982) developed a system of classification for SMV strains, based on the reactions they produce on differential cultivars. The strain groups were designated G1 to G7, with the lower-numbered strains being able to infect the least number of cultivars, and the higher-numbered strains being able to infect the greatest number of cultivars.

The first report on the inheritance of resistance in soybean to SMV was by Koshimizu and Iizuka (1963). They reported that resistance was due to a single dominant gene in two populations, but appeared to be recessive in a third population, based on F_2 data. Kiihl and Hartwig (1979) assigned the first gene symbol, *Rsv*, to a dominant resistance gene found in PI 96983, an introduction from Korea. The symbol *rsv-t* was proposed for the resistance gene in cv. Ogden, which was allelic to *Rsv* and showed resistance to the common SMV-1 strain, but gave a necrotic reaction to strain SMV-1-B. The recessive symbol was used because the Ogden gene was recessive to *Rsv*, when tested with SMV-1-B. However, it was later shown to be dominant to the cv. York allele and was redesignated *Rsv1-t* by Chen et al. (1991).

In addition to *Rsv1* and *Rsv1-t*, seven other resistance alleles have been identified at the *Rsv1* locus. *Rsv1-y*, *Rsv1-m*, and *Rsv1-k* have been identified in cvs. York, Marshall, and Kwanggyo (PI 406710), respectively (Chen et al., 1991), while

Table 5–2. Reaction of soybean genotypes to seven soybean mosaic virus (SMV) strain groups.

Source	Gene	Soybean mosaic virus strain groups							Reference
		G1	G2	G3	G4	G5	G6	G7	
PI 96983	*Rsv1*	R†	R	R	R	R	R	N	Kiihl and Hartwig (1979)
Ogden	*Rsv1-t*	R	R	N	R	R	R	N	Chen et al. (1991)
York	*Rsv1-y*	R	R	R	N	S	S	S	Chen et al. (1991)
Marshall	*Rsv1-m*	R	N	N	R	R	N	R	Chen et al. (1991)
Kwanggyo	*Rsv1-k*	R	R	R	R	N	N	N	Chen et al. (1991)
PI 507389	*Rsv1-n*	N	N	S	S	N	N	S	Ma et al. (1994)
LR1‡	*Rsv1-s*	R	R	R	R	N	N	R	Ma et al. (1995)
Raiden	*Rsv1-r*	R	R	R	R	N	N	R	Chen et al. (2001)
Suweon 97	*Rsv1-sk*	R	R	R	R	R	R	R	Chen et al. (2002)
OX686‡	*Rsv3*	N	NA	NA	N	NA	NA	R	Buzzell and Tu (1989)
L29‡	*Rsv3-?*	S	S	S	S	R	R	R	Buss et al. (1999)
LR2‡	*Rsv4*	R	R	R	R	R	R	R	Ma et al. (1995)
Peking	*Rsv4*	R	R	R	R	R	R	R	Gunduz (2000)

† R = resistant symptomless; N = necrotic; S = susceptible (mosaic); NA = not available.

‡ LR1 and LR2 derived from cv. Essex × cv. SS74185 (PI 486355); OX686 derived from cv. Columbia × cv. Harosoy; L29 is a cv. Williams BC$_5$ isoline with SMV resistance derived from cv. Hardee.

Rsv1-s and *Rsv1-sk* are respectively present in PI 486355 (SS74185) and PI 483084 (Suweon 97) both developed at the Crop Experiment Station, Suweon, South Korea (Ma et al., 1995; Chen et al., 2002). The allele present in cv. Raiden (PI 360844) is *Rsv1-r* (Chen et al., 2001). An allele at the *Rsv1* locus in PI 507389 (cv. Tousan 50 from Japan) has been tentatively designated *Rsv1-n*. This allele exhibits a severe necrotic reaction with SMV-G1 and is susceptible to the other strain groups (Ma et al., 1994). Cultivars Hill, Essex, and Lee 68, as well as numerous other cultivars, contain the susceptible allele *rsv1*. Table 5–2 shows the reaction of each allele to the SMV strain groups.

Buzzell and Tu (1984) reported that the resistance in strain OX670, presumably derived from cv. Raiden, was at a different locus and designated it as *Rsv2*. However, subsequent studies on Raiden indicated that its resistance gene was at the *Rsv1* locus (Chen et al., 2001). Further analysis of the inheritance of resistance in OX670 by Gunduz et al. (2001) revealed that it actually contained two resistance genes, *Rsv1* and *Rsv3*. The *Rsv3* gene was derived from cv. Harosoy, but Buzzell and Tu (1984) were likely unaware of its existence, since it is susceptible to several strain groups. Thus *Rsv2* is not listed in Table 5–2.

Rsv3 was proposed by Buzzell and Tu (1989) as the symbol for a dominant gene that they identified in cv. Columbia and that was at a previously unreported locus. This gene exhibits an extremely necrotic, hypersensitive-type reaction to strain SMV-G1. Another allele at the *Rsv3* locus, originally from cv. Hardee, was susceptible to the strain groups SMV-G1 to SMV-G4, but was highly resistant to SMV-G5 to SMV-G7 (Buss et al., 1999). It seems that the *Rsv3* alleles in Columbia and Hardee are not identical since they respond differently to SMV strain groups G1 and G4.

Genetic studies by Ma et al. (1995) revealed that PI 486355 contained two different SMV resistance genes. One of the genes was an allele at the *Rsv1* locus, but allelism tests for the other locus were not conducted. Gunduz (2000) subsequently found that the single dominant resistance gene in PI 88788 was allelic to the non-*Rsv1* gene in PI 486355 (Buss et al., 1997) and that both were nonallelic with *Rsv3* and *Rsv1*. The gene symbol *Rsv4* was tentatively assigned to the new locus. The SMV resistance gene in cv. Peking also was demonstrated to be at the *Rsv4* locus. Since all of the sources of *Rsv4* are resistant to SMV-G1 through SMV-G7, no conclusions can be drawn regarding their genetic similarity.

Shigemori (1988) reported that cvs. Hourei, Tousan 140, and Tousan 122 each had single dominant genes for resistance, and they appeared to be at separate loci. Gunduz et al. (2002) conducted allelism tests of cvs. Hourei and Tousan 140 with sources of *Rsv1*, *Rsv3,* and *Rsv4* and found that Hourei and Tousan 140 each had alleles at the *Rsv1* and *Rsv3* loci. The discrepancy with the Shigemori report was probably due to differences in the SMV strains that were used. Shigemori used only a single strain, which probably produced reciprocal reactions on the *Rsv1* and *Rsv3* alleles in Hourei and Tousan 140.

Wang et al. (1998) studied the inheritance of SMV resistance in four Chinese cultivars. Each cultivar was shown to contain a single dominant gene for resistance. Only one, Feng shou haung (PI 458507), had a resistance allele at the *Rsv1* locus. Ke feng No. 1 (PI 556949) had a resistance gene at the *Rsv4* locus. The remaining two cultivars, Da bai ma (PI 556948) and Xu dou No. 1 (PI 556950) had genes that were not at the *Rsv1* locus and probably were not at the *Rsv4* locus, but their actual identity was not established.

Bowers et al. (1992) found that the single dominant resistance genes in cvs. Buffalo and HLS were at separate loci. Gai et al. (1989) concluded from intercrosses among several resistant Chinese cultivars that four different resistance genes were present and they were all in the same linkage group. Since neither of these papers described allelism tests with sources of *Rsv1, Rsv3,* or *Rsv4,* the genetic relationship of these genes to each other and to reported genes is not known. One exception is the gene in HLS, a rogue from cv. Hardee, which was backcrossed into cv. Williams background (Bernard et al., 1991) and has been shown to be an *Rsv3* allele (Buss et al., 1999).

Rsv1, Rsv3, and *Rsv4* have been mapped to the USDA-ARS-ISU soybean molecular linkage groups F (Yu et al., 1994), B2 (Jeong et al., 2002), and D1b (Hayes et al., 2000), respectively.

5–5.1.12 Peanut Mottle Virus

Boerma and Kuhn (1976) found that cvs. Dorman and CNS each had a single dominant gene for resistance to peanut mottle virus (PMV) and assigned the gene symbol *Rpv*, assuming that both cultivars had the same gene. Shipe et al. (1979) reported that cv. Peking had a recessive gene for PMV resistance and labeled it *rpv2*. In a study on the inheritance of resistance to both SMV and PMV in cv. York, Roane et al. (1983) discovered that single dominant genes conditioned resistance to both viruses, and the genes were linked with 3.7 cM between them. It was assumed that cv. York inherited the *Rpv* gene from its parent, cv. Dorman. Later, Buss et al. (1985) studied the inheritance of resistance in cvs. York, Dorman, Shore, and CNS. The conclusion was that York, Dorman, and Shore each contained the *Rpv* gene, but the resistance in CNS was at a different locus. No gene symbol was assigned, because allelism tests with Peking were not conducted.

5–5.1.13 Cowpea Chlorotic Mottle Virus

Boerma et al. (1975) made resistant × susceptible crosses using cvs. Lee, Bragg, and Hill as susceptible parents and cvs. Davis, Jackson, and Hood as resistant parents. The F_2 data indicated that resistance to cowpea chlorotic mottle virus was due to a single dominant gene in each cross. A single gene symbol, Rcv, was assigned, although resistant × resistant crosses were not made to verify allelism among the resistant parents. Goodrick et al. (1991) concluded from a combination of F_2 and F_3 data that resistance in PI 346304 (PLSO-41 from India) was conditioned by two recessive genes. Allelism tests with cv. Bragg were conducted, but the results were inconclusive.

5–5.2 Nematodes

Caviness and Riggs (1976) reported that as many as 50 species of plant-parasitic nematodes feed on soybean. The inheritance of resistance has been reported for the following three species: soybean cyst nematode (*Heterodera glycines* Ichinohe), root-knot nematode (*Meloidogyne* spp.), and reniform nematode (*Rotylenchulus reniformis* Linford & Oliveira).

Caldwell et al. (1960) reported that a combination of three recessive and independent alleles (*rhg1*, *rhg2*, and *rhg3*), controlled resistance to *H. glycines* Ichinohe. Matson and Williams (1965) reported that an additionally independent dominant allele (*Rhg4*) in cv. Peking and the three recessive alleles were required for resistance to *H. glycines* Ichinohe. Rao-Arelli et al. (1992) reported an additional dominant allele in PI 88788 that is at a locus other than those previously reported. Subsequently, this dominant allele was given the gene symbol *Rhg5* (Rao-Arelli, 1994).

Williams et al. (1981) studied the inheritance of reaction to reniform nematode in the cross cv. Forrest (resistant) × cv. Ransom (susceptible). They concluded that resistance was controlled by a single major recessive allele, designated *rrn*, but that one or more minor alleles may contribute to the reaction, giving intermediate infection classes differing from the parents and F_1 plants.

Boquet et al. (1975) reported that resistance to a specific race of root-knot nematode was controlled by one major allele with at least one modifying allele. Luzzi et al. (1994) determined the inheritance of resistance to the southern root-knot nematode, *M. incognita* (Kofoid & White) Chitwood, for the cross cv. Bossier (susceptible) × cv. Forrest (resistant) and its reciprocal. The mean number of galls on the roots of F_1 plants derived from reciprocal crosses were similar to each other and the midparent mean, indicating that inheritance was not maternal and that neither resistance nor susceptibility was dominant. From the segregation of F_2 and F_3 populations, the authors concluded that Forrest contains a single additive gene for resistance to galling, designated *Rmi1*.

5–5.3 Foliar-Feeding Insects

Although breeding for resistance to a few insect species has been conducted at several locations since the 1970s, the genetic basis for resistance has not been clearly established, nor have gene symbols been assigned. Sisson et al. (1976) reported that the resistance of Miyako White (PI 227687) and Sodendaizu (PI 229358) to Mexican bean beetle (*Epilachna varivestis* Mulsant) was quantitatively inherited, but suggested that it resulted primarily from additive gene action of two or three major genes. Studies by Kilen et al. (1977) suggested partial dominance for susceptibility with the action of only a few major genes controlling resistance to soybean looper, *Pseudoplusia includens* (Walker). Kenty et al. (1996) reported that the results of their studies showed a trend toward quantitative inheritance of resistance to soybean looper. Because heritability for resistance was estimated to be 63%, they concluded that progress should be possible from early generation selection.

To determine the possible effects on foliar feeding by lepidopterous insects, the traits dense pubescence(*Pd1*) or glabrous (*P1*) were transferred by backcrossing into cv. Davis (susceptible), cv. Tracy-M (moderately resistant), and breeding line D75-10169 (resistant) (Kilen and Lambert, 1993). Pubescence, in comparison with plants without pubescence, functions as a resistance mechanism to the larval stage of lepidopterous defoliators but enhanced adult oviposition (Lambert et al., 1992).

Genes controlling reaction to bacteria, fungi, viruses, and nematodes in soybean are summarized in Table 5–3.

Table 5–3. Genes affecting pest reaction in soybean.

Gene	Phenotype†	Strain‡	Reference
		1. Bacterial blight	
Rpg1	Resistant, race 1	Norchief, Harosoy	Mukherjee et al. (1966)
rpg1	Susceptible, race 1	Flambeau	
Rpg2	Resistant, race 4 *avrA*	Merit	Keen and Buzzell (1991)
rpg2	Susceptible, race 4 *avrA*	Flambeau	
Rpg3	Resistant, race 4 *avrC*	Flambeau	Keen and Buzzell (1991)
rpg3	Susceptible, race 4 *avrC*	Merit	
Rpg4	Resistant, race 4 *avrD*	Flambeau	Keen and Buzzell (1991)
rpg4	Susceptible, race 4 *avrD*	Merit	
		2. Bacterial pustule	
Rxp	Susceptible	Lincoln, Ralsoy	Hartwig and Lehman (1951), Feaster (1951), Bernard and Weiss (1973)
rxp	Resistant	CNS	
		3. Brown stem rot	
Rbs1	Resistant	L78-4094 (from PI 84946-2)	Hanson et al. (1988)
rbs1	Susceptible	LN78-2714, Century	
Rbs2	Resistant	PI 437833 (Curo sengocu)	Hanson et al. (1988), Nelson et al. (1988)
rbs2	Susceptible	Century	
Rbs3	Resistant	PI 437970 (VIR 1460)	Willmot and Nickell (1989)
rbs3	Susceptible	Pioneer 9271	
		4. Frogeye leaf spot	
Rcs1	Resistant, race 1	Lincoln, Wabash	Athow and Probst (1952, as *Cs*), symbol by Probst et al. (1965)

(continued on next page)

Table 5–3. Continued.

Gene	Phenotype†	Strain‡	Reference
rcs1	Susceptible, race 1	Gibson, Patoka, Hawkeye	
Rcs2	Resistant, race 2	Kent	Probst et al. (1965)
rcs2	Susceptible, race 2	C1043 (PI 70237 X Lincoln), C1270 [Mandarin (Ottawa) X Clark]	
Rcs3	Resistant, races 2 and 5	Davis	Boerma and Phillips (1983)
rcs3	Susceptible races 2 and 5	Blackhawk	

5. Downy mildew

Gene	Phenotype†	Strain‡	Reference
Rpm1	Resistant, race 2	Kanrich	Bernard and Cremeens (1972)
rpm1	Susceptible, race 2	Clark, Chippewa	
Rpm2	Resistant, races 2 and 33	Fayette (from PI 88788)	Lim (1989)
rpm2	Susceptible, race 33	Union	

6. Powdery mildew

Gene	Phenotype†	Strain‡	Reference
Rmd	Resistant (adult plant)	Blackhawk	Buzzell and Haas (1978)
rmd	Susceptible (all stages)	Harosoy 63, PI 65388	
Rmd-c	Resistant (all stages)	L76-1988, CNS	Lohnes and Bernard (1992)
rmd	Susceptible	Harosoy, L82-2024	

7. Phytophthora root rot

Gene	Phenotype†	Strain‡	Reference
Rps1-a	Resistant, races 1, 2, 10, 11, 13-20, 24, 26, 27	Mukden	Bernard et al. (1957, as *Ps*), Lam-Sanchez et al. (1968), Moots et al. (1983), Schmitthenner et al. (1994)
rps1	Susceptible	Lincoln	
Rps1-b	Resistant, races 1, 3-9, 13, 15, 18, 21, 22	D60-9647 (from FC 31745), Sanga, PI 84637	Hartwig et al.(1968, as *rps2*), Mueller et al. (1978), Laviolette and Athow (1983), Schmitthenner et al. (1994)
rps1	Susceptible	Hood	
Rps1-c	Resistant, races 1-3, 6-11, 13-15, 17, 21, 23, 24, 26	PI 54615-1, Arksoy	Mueller et al. (1978), Laviolette and Athow (1983), Schmitthenner et al. (1994)
rps1	Susceptible	Harosoy	
Rps1-d	Resistant, races 1-7, 9-11, 13-16, 18, 21, 22, 24, 25	PI 103091 (Wu An)	Buzzell and Anderson (1992)
rps1	Susceptible	Harosoy	
Rps1-k	Resistant, races 1-11, 13-15, 17, 18, 21-24, 26	Kingwa	Bernard and Cremeens (1981), Laviolette and Athow (1983), Schmitthenner et al. (1994)
rps1	Susceptible	Williams	
Rps2	Resistant, races 1, 2, 10, 12	D54-2437 (from CNS)	Kilen et al. (1974)
rps2	Susceptible	D55-1492	
Rps3-a	Resistant, races 1-5, 8, 9, 11, 13, 14, 16, 18, 23, 25	PI 86972-1, PI 171442	Mueller et al. (1978), Laviolette and Athow (1983)
rps3	Susceptible	Harosoy	
Rps3-b	Resistant, races 1-5, 7, 9-12, 16	PI 172901	Ploper et al. (1985)
rps3	Susceptible	Harosoy	
Rps3-c	Resistant, races 1-4, 12, 13	PI 340046	Athow et al. (1986)
rps3	Susceptible	Harosoy	

(continued on next page)

Table 5–3. Continued.

Gene	Phenotype†	Strain‡	Reference
Rps4	Resistant, races 1-4, 10, 12-16	PI 86050	Athow et al. (1980)
rps4	Susceptible	Harosoy	
Rps5	Resistant, races 1-5, 8, 9, 11, 13, 14, 16	PI 91160	Buzzell and Anderson (1981)
rps5	Susceptible	Harosoy	
Rps6	Resistant, races 1-4, 10, 12, 14-16, 18-21	Altona	Athow and Laviolette (1982), Laviolette and Athow (1983)
rps6	Susceptible	Harosoy	
Rps7	Resistant, races 12, 16, 18, 19	Harosoy	Anderson and Buzzell (1992)
rps7	Susceptible	Williams	

8. Stem canker

Gene	Phenotype	Strain	Reference
Rdc1	Resistant	Tracy-M	Kilen and Hartwig (1987)
rdc1	Susceptible	J77-339	
Rdc2	Resistant	Tracy-M	Kilen and Hartwig (1987)
rdc2	Susceptible	J77-339	
Rdc3	Resistant	Crockett	Bowers et al. (1993)
rdc3	Susceptible	Coker 338, Johnston	
Rdc4	Resistant	Dowling	Bowers et al. (1993)
rdc4	Susceptible	Coker 338, Johnston	

9. Sudden death syndrome

Gene	Phenotype	Strain	Reference
Rfs	Resistant	Ripley	Stephens et al. (1993a)
rfs	Susceptible	Spencer	

10. Soybean rust

Gene	Phenotype	Strain	Reference
Rpp1	Resistant	Komata (PI 200492)	McLean and Byth (1980)
rpp1	Susceptible	Will, Davis	
Rpp2	Resistant	PI 230970	Hartwig and Bromfield (1983)
rpp2	Susceptible	c	
Rpp3	Resistant	Ankur (PI 462312)	Hartwig and Bromfield (1983)
rpp3	Susceptible	c	
Rpp4	Resistant	PI 459025	Hartwig (1986)
rpp4	Susceptible	c	

11. Soybean mosaic virus

Gene	Phenotype	Strain	Reference
Rsv1	Resistant, SMV-1, SMV-1-B, G1 through G6	PI 96983	Kiihl and Hartwig (1979)
rsv1	Susceptible	Hill	Kiihl and Hartwig (1979)
Rsv1-t	Resistant, SMV-1; Susceptible, SMV-1-B, G1, G2, G4, G5, G6	Tokyo, Ogden	Chen et al. (1991)
Rsv1-y	Resistant, G1, G2, G3	York	Chen et al. (1991)
Rsv1-m	Resistant, G1, G4, G5, G7	Marshall	Chen et al. (1991)
Rsv1-k	Resistant, G1, G2, G3, G4	Kwanggyo (PI 406710)	Chen et al. (1991)
Rsv1-n	Necrotic, G1	PI 507389	Ma et al. (1994)
Rsv1-s	Resistant, G1, G2, G3, G4, G7	LR1§, PI 486355	Ma et al. (1995)
Rsv1-r	Resistant, G1, G2, G3, G4, G7	Raiden (PI 360844)	Chen et al. (2001)
Rsv1-sk	Resistant, G1 through G7	Suweon 97 (PI 483084)	Chen et al. (2002)
Rsv3	Resistant, G5, G6, G7	OX686§	Buzzell and Tu (1989)

(continued on next page)

Table 5–3. Continued.

Gene	Phenotype[†]	Strain[‡]	Reference
Rsv3-?	Resistant, G5, G6, G7	L29§ (from Hardee)	Buss et al. (1999)
rsv3	Susceptible	Lee 68	Buss et al. (1999)
Rsv4	Resistant, G1 through G7	LR2§, Peking	Ma et al. (1995), Gunduz (2000)
rsv4	Susceptible	Lee 68	Gunduz (2000)
	12. Peanut mottle virus		
Rpv1	Resistant	Dorman	Boerma and Kuhn (1976)
rpv1	Susceptible	Ransom, Bragg	
rpv2	Resistant	Peking	Shipe et al. (1979)
Rpv2	Susceptible	PI 229315	
	13. Cowpea chlorotic mottle virus		
Rcv	Resistant	Lee, Bragg	Boerma et al. (1975)
rcv	Susceptible	Davis, Hood	
	14. Cyst nematode		
rhg1 with *rhg2 rhg3*	Resistant	Peking	Caldwell et al. (1960)
Rhg1, Rhg2, or *Rhg3*	Susceptible	Lee, Hill	
Rhg4 with *rhg1 rhg2 rhg3*	Resistant	Peking	Matson and Williams (1965)
rhg4	Susceptible	Scott	
Rhg5	Resistant	PI 88788	Rao-Arelli et al. (1992), Rao-Arelli (1994)
rhg5	Susceptible	Essex	
	15. Reniform nematode		
rrn	Resistant	Forrest	Williams et al. (1981)
Rrn	Susceptible	Ransom	
	16. Root-knot nematode		
Rmi1	Resistant	Forrest	Luzzi et al. (1994)
rmi1	Susceptible	Bossier	

† A susceptible phenotype, when specific races are not identified, indicates that the strain was suscep-tible to races used to identify the resistance allele at that locus by authors of the first reference.
‡ Names are released soybean cultivars; c = Indicates that the gene occurs in many cultivars.
§ LR1 and LR2 derived from cv. Essex X cv. SS74185 (PI 486355); OX686 derived from cv. Colum-bia X cv. Harosoy; L29 is a cv. Williams BC5 isoline with SMV resistance derived from cv. Hardee.

5–5.4 Herbicide Reaction

Soybean cultivars, introductions, and breeding lines frequently show differ-ences in the degree of injury to herbicides. Sensitivity to bentazon and to metribuzin is controlled by the single recessive alleles *hb* (Bernard and Wax, 1975) and *hm* (Ed-wards et al., 1976), respectively. Kilen and He (1992) determined that tolerance to metribuzin in the wild annual soybean *Glycine soja* was controlled by alleles at the same locus as the *Hm* gene in cv. Tracy-M.

Cultivar Williams seed was treated with N-ethyl-N-nitrosourea, and four soybean mutants with increased tolerance to sulfonylurea herbicides were recov-ered (Sebastian and Chaleff, 1987). Enhanced tolerance was controlled by a single

Table 5–4. Genes affecting herbicide reaction in soybean.

Gene	Phenotype	Strain	Reference
Als1	Semidominant for resistance to sulfonylurea herbicides	Williams	Sebastian et al. (1989)
als1	Sensitive	Williams 20	
Hb	Tolerant to bentazon	Clark 63	Bernard and Wax (1975)
hb	Sensitive to bentazon	Nookishirohana (PI 229342)	
Hm	Tolerant to metribuzin	Hood, Tracy-M, PI 163453 (*G. soja*), PI 245331 (*G. soja*)	Edwards et al. (1976), Hartwig et al. (1980), Kilen and He (1992)
hm	Sensitive to metribuzin	Semmes, Tracy	
Hs1	Sensitive to sulfonylurea herbicides	Williams	Sebastian and Chaleff (1987)
hs1	Enhanced tolerance	Williams 1-183A, Williams 1-184A	
Hs2	Sensitive to sulfonylurea herbicides	Williams	Sebastian and Chaleff (1987)
hs2	Enhanced tolerance	Williams 1-16A	
Hs3	Sensitive to sulfonylurea herbicides	Williams	Sebastian and Chaleff (1987)
hs3	Enhanced tolerance	Williams 1-126A	

recessive gene in all four mutants. Allelism tests aligned two independent mutations with the *Hs1* locus, but placed two other mutational events at the *Hs2* and *Hs3* loci (Sebastian and Chaleff, 1987). Biochemical studies showed that the four mutants do not contain an altered form of acetolactate synthase, the site of action of the sulfonylurea herbicides.

Seed mutagenesis of cv. Williams with N-methyl-N-nitrosourea, followed by selection for resistance to the herbicide chlorosulfuron, identified a mutant with an increased tolerance to both post- and pre-emergence applications of the sulfonylurea herbicides. Resistance was semidominant and designated *Als1* (Sebastian et al., 1989). Biochemical studies indicated that the mechanism of resistance was reduced sensitivity of acetolactate synthase to sulfonylurea herbicides.

Chlorimuron-tolerant cv. Elgin was crossed with sensitive cv. BSR 101 and breeding line M74-462. Sensitivity was conferred by a single recessive gene (Pomeranke and Nickell, 1988). A gene symbol was not given. Molecular mapping identified one major locus and one minor locus for chlorimuron sensitivity (Mian et al., 1997).

A glyphosate (N-phosphonomethyl-glycine)-tolerant soybean line, 40-3-2, was obtained after biolistic-gun mediated transformation (Padgette et al., 1995). Glyphosate is the active ingredient of Roundup (Monsanto Co., St. Louis, MO) herbicide. Inheritance studies showed that the transgene behaved as a single dominant gene, was stable over several generations (Padgette et al., 1995) and located on the USDA-ARS-ISU soybean molecular map on linkage group D1b (Monsanto Co., unpublished data, 2002). Extensive yield testing of glyphosate-tolerant line 40-3-2 and its derivatives at 58 diverse locations showed no significant yield reduction (Delannay et al., 1995). The development of glyphosate-tolerant soybean (Roundup Ready, Monsanto Co., St. Louis, MO) has provided the farmer with a new weed control system that has gained wide acceptance.

Genes affecting herbicide reaction in soybean are summarized in Table 5–4.

5–5.5 Nodulation Response

Several gene loci control nodulation with the N_2-fixing microsymbionts of soybean. Soybean is nodulated by microsymbionts in at least three diverse genera. These include the slow-growing bacteria *Bradyrhizobium japonicum, B. elkanii,* and *B. liaoningense*, the fast-growing bacterium *Sinorhizobium fredii* and the interme-diate growth form *Mesorhizobium tianshanense* (Kuykendall, 2003, Kuykendall et al., 2000).

The recessive allele *rj1* apparently arose as a spontaneous mutation (Williams and Lynch, 1954). It conditions the nonnodulating response with virtually all *Bradyrhizobia* and *Sinorhizobia*. Strains of *B. elkanii* have a limited ability to nodulate the *rj1 rj1* genotype (Devine and Kuykendall, 1996). The *rj1* locus has been mapped to the *f* locus and the *Idh1* locus in classical linkage group 11 (Devine et al., 1983; Hedges et al., 1990).

Improved bacterial strains that are highly efficient N-fixers usually are not competitive with indigenous strains in the rhizosphere. Studies were conducted to determine whether *S. fredii* USDA 205 could overcome the nodulation block in *rj1 rj1* soybean plants. Rasooly and Isleib (1993) concluded that the soybean genotype *rj1 rj1* cannot be used with *S. fredii* USDA 205 to overcome the problem of bacte-rial competition for nodule occupancy.

The dominant allele *Rj2* occurs frequently in soybean accessions from the lower Yangzi region in the adjacent provinces of Jiangsu and Zhejiang, China (Devine and Breithaupt, 1981; Devine, 1987). The dominant allele conditions an ineffective nodulation response with *B. japonicum* strains in several serogroups (Devine and Kuykendall, 1996). The *Rj2* locus was mapped to classical linkage group 19 with the aconitase locus *Aco2* (Devine et al., 1991b) and with a close link-age to the phytophthora resistance locus *Rps2* (Devine et al., 1991a).

The dominant allele *Rj4* occurs in high frequency in soybean accessions from southeast Asia (Devine and Breithaupt, 1981; Devine, 1987). The dominant allele conditions an ineffective nodulation response particularly with strains of *B. elka-nii* (Devine et al., 1990). *Bradyrhizobium elkanii* strains are often chlorosis-inducing strains and occur in high frequency in the southeastern USA (Devine and Kuyk-endall, 1996; Weber et al., 1989). The *Rj4* gene was found to segregate independ-ently of 12 other loci controlling morphological traits (Devine, 1992). The *Rj4* locus was mapped with molecular markers (Ude et al., 1999).

Genetic allelism tests have established that the *rj1, rj2,* and *rj4* alleles reside at three distinct loci (Devine and O'Neill, 1989, 1993). Near-isogenic lines are avail-able for many of the nodulation response mutants. For example, *rj1 rj1* in cvs. Harosoy and Clark (Bernard et al., 1991) and in cv. Maple Presto (OT89-13) (Vold-eng and Saindon, 1991a), *rj2 rj2* in cv. Clark 63 (BARC-5) (Devine and O'Neill, 1987), and *rj3 rj3* in Clark 63 (BARC-3) (Devine and O'Neill, 1986).

Harper (1989) reported indentification of a nonnodulating mutant, NN5, from N-methyl-N-nitrosourea-treated cv. Williams seed. Pracht et al. (1993a) con-cluded that one of the genes responsible for nonnodulation in NN5 was allelic to the previously identified, but unnamed single recessive gene found in nod139, a nonnodulating mutant from mutagenized cv. Bragg seed (Mathews et al., 1989). The gene symbol *rj6* was assigned to the allele conditioning nonnodulation in NN5 and nod 139 (Pracht et al., 1993a). A second recessive gene controlling nonnodulation, designated *rj5* was identified in NN5 when crosses were made to cv. Harosoy 63 (Pracht et al., 1993a). The NN5 nonnodulating mutant was devoid of nodules unlike the original *rj1* mutant (T201) and the Bragg nod139 mutant which showed occasional nodules when challenged with high doses of inoculum of specific strains.

Carroll et al. (1985) initially isolated several supernodulated and hypernodulated mutants from mutagenized (ethyl methylsulfonate) cv. Bragg and designated them as *nts* mutants. Delves et al. (1988) and Carroll et al. (1988) concluded that all *nts* mutants were inherited as single recessive genes. Gremaud and Harper (1989) isolated and initially characterized hypernodulating plants from mutagenized (N-methyl-N-nitrosourea) cv. Williams seed and designated them as NOD mutants. Initial genetic analysis of hypernodulated NOD mutants identified a single recessive gene, tentatively designated *rjh* (Pracht et al., 1993b), and Harper and Nickell (1995) conducted further genetic analysis of NOD mutants and named the gene *rj7*. Another recessive gene (*rj8*) was tentatively identified in NOD2-4 (Vuong et al., 1996), but this was subsequently shown to be incorrect and all hypernod (NOD) mutants are under control of the *rj7* recessive gene (Vuong and Harper, 2000). It appears that the En6500 and *nts*382 supernoduating mutants are also controlled by the same *rj7* recessive gene (Vuong and Harper, 2000).

The dominant allele *Rfg1* in the cv. Kent conditions an ineffective nodulation response with the fast-growing strain USDA 205 of *Sinorhizobium fredii* (Devine, 1985; Devine and Kuykendall, 1994). Trese (1995) demonstrated that the gene in the cv. McCall conditioning ineffective nodulation with *Sinorhizobium* strain USDA 257 is also the *Rfg1* gene.

Genes affecting *Rhizobium* response in soybean are summarized in Table 5–5.

5–5.6 Roots

Root fluorescence is a phenomenon in which roots of seedlings fluoresce when irradiated with ultraviolet light. Root fluorescence mutants in soybean are defined by five known loci. The single nonfluorescent recessive genes are *fr1, fr2, fr4,* and *fr5,* and the dominant nonfluorescent gene is *Fr3.* The *fr5* gene was identified after ethyl methanesulfonate seed treatment (Sawada and Palmer, 1987) and has not been identified as naturally occurring in germplasm.

In an evaluation of land races from North China (Set 1), 32 nonfluorescent root variants were identified among the 736 accessions evaluated. They included 30 *fr1* variants; 14 from Gansu province, 1 from Shandong province, and 15 from Shanxi province. The two *fr2* variants were from Shanxi province (Torkelson and Palmer, 1997). Among 799 accessions from South China (Set 2), no nonfluorescent root genotypes were identified (Cubukcu et al., 2000). Among an additional 623 accessions from South China (Set 3), one *fr1* genotype was identified (Cubukcu et al., 2000).

The geographical distribution of the four naturally occurring alleles is not random. Perhaps these alleles have a neutral effect on fitness and thus have remained, except for rare outcrosses, in those germplasms where the mutations originally occurred. The chemical basis for root fluorescence/nonfluorescence has not been determined, fluorescence does not seem to be due to isoflavonoid compounds (Grady et al., 1995).

Table 5–5. Genes affecting nodulation response in soybean.

Gene	Phenotype	Strain†	Reference
Rfg1	Ineffective by strain 205	Kent	Devine and Kuykendall (1994)
rfg1	Effective	Peking	
Rj1	Nodulating	T180,T202,c	Williams and Lynch (1954) (as *no*); symbol by Caldwell (1966)
rj1	Nonnodulating	T181,T201	
Rj2	Ineffective by strains b7, b14, and b122	Hardee, CNS	Caldwell (1966)
rj2	Effective	c	
Rj3	Ineffective by strain 33	Hardee	Vest (1970)
rj3	Effective	Clark	
Rj4	Ineffective by strain 61	Hill, Dare, Dunfield	Vest and Caldwell (1972)
rj4	Effective	Lee, Semmes	
Rj5	Nodulating	Williams, Harosoy 63	Pracht et al. (1993a)
rj5	Nonnodulating	Williams NN5, Bragg nod139	
Rj6	Nodulating	Williams	Pracht et al. (1993a)
rj6	Nonnodulating	Williams NN5, Bragg nod139	
Rj7	Nodulating	Williams, Enrei	Kokubun and Akao (1994), Harper and Nickell (1995), Vuong et al. (1996), Vuong et al. (2000)
rj7	Hypernodulating	Williams NOD mutants (NOD1-3, NOD2-4, NOD3-7, NOD4) En6500	

† c = Indicates that the gene occurs in many cultivars.

Both the *Fr1* (Jin et al., 1999) and the *Fr2* (Devine et al., 1993) loci are located on the molecular map. The *Fr1* locus is located in a "segregation distortion region" on molecular linkage group K.

Three necrotic root mutants (*Rn* loci) were recovered in a gene-tagging study with the *w4-m* (T322) mutant (Palmer et al., 1989). Each mutant was the result of a separate genetic event. The phenotype is a progressive necrosis of the root system that can be visualized about 5 to 7 d after germination (Kosslak et al., 1996). This appearance is similar to infection by *Phytophthora sojae* (Kaufmann & Gerdemann). These necrotic root mutants are also known as disease lesion mimics. Allelism tests confirmed that the three mutations are allelic.

Light microscopy studies showed that in the necrotic root mutants browning first occurred in differentiated inner cortical cells adjacent to the stele. This was preceded by a "wave" of autofluorescence that emanated from cortical cells opposite the xylem poles and spread across the cortex (Kosslak et al., 1997). Fragmented DNA was detected before any visible changes in autofluorescence or browning were observed. Electron microscopy and DNA studies suggested that the necrotic root mutants had features commonly observed in animal cell apoptosis and were programmed cell mutants (Kosslak et al., 1997).

When inoculated via the hypocotyl, necrotic root seedlings were susceptible to hyphal infection by a compatible race of the fungal pathogen, *P. sojae*. In contrast, the roots showed partial resistance to infection by *P. sojae* zoospores. This was correlated with the accumulation of isoflavonoid phytoalexins and group 2 peroxidases under axenic conditions (Kosslak et al., 1996).

Table 5–6. Genes affecting root response in soybean.

Gene	Phenotype	Strain†	Reference
Fr1	Fluorescent in UV light	c	Fehr and Giese (1971)
fr1	Nonfluorescent	Minsoy (PI 27890)	
Fr2	Fluorescent in UV light	c	Delannay and Palmer (1982b)
fr2	Nonfluorescent	Noir 1 (PI 290136)	
Fr3	Nonfluorescent	PI 424078	Delannay and Palmer (1982b)
fr3	Fluorescent in UV light	c	
Fr4	Fluorescent in UV light	c	Delannay and Palmer (1982b)
fr4	Nonfluorescent	Dun-cuan (PI 404165)	
Fr5	Fluorescent in UV light	c	Sawada and Palmer (1987)
fr5	Nonfluorescent	T285	
Rn (Ames 1)	Normal	c	Kosslak et al. (1996)
rn (Ames 1)	Necrotic root	Mutant in T322	
Rn (Ames 2)	Normal	c	Kosslak et al. (1996)
rn (Ames 2)	Necrotic root	Mutant in T322	
Rn (Ames 3)	Normal	c	Kosslak et al. (1996)
rn (Ames 3)	Necrotic root	Mutant in T322	

† c = Indicates that the gene occurs in many cultivars.

Two additional independent necrotic root mutations have been characterized. One was identified in a separate gene-tagging study with *w4-m* (Andersen and Palmer, 1997). The other was identified after ethyl methanesulfonate seed treatment of the cv. AgriPro 1776 (Palmer and Wubben, 1998). These two mutants were allelic to the three necrotic root mutants reported by Kosslak et al. (1996), but have not been assigned gene symbols. Linkage studies have failed to locate the trait on the soybean classical linkage map (Kosslak et al., 1996; Wubben and Palmer, 1998).

Genes affecting root response in soybean are summarized in Table 5–6.

5–5.7 Growth and Morphology

5–5.7.1 Flowering and Maturity

Four major gene pairs (*E1 e1–E4 e4*) that affect the time of flowering and maturity have received much attention from soybean researchers. Three additional loci have been identified. A gene for late flowering and maturity, designated *E5*, was identified in L64-4830 (McBlain and Bernard, 1987). Genetic control of flowering and maturity in soybean under short-day conditions differs from that for long days (Hartwig and Kiihl, 1979). Hartwig and Kiihl (1979) used the descriptive phrase "delayed flowering under short-day conditions" to describe the flowering response of PI 159925 (cv. Glycine H. from Peru). The term "long-juvenile" is now used to describe this phenomenon (Sinclair and Hinson, 1992). The long-juvenile trait is controlled by the single recessive gene *j* (Ray et al., 1995). The long-juvenile trait could be important in studying the process of photoperiodic induction, in addition to its use in cultivar development (Ray et al., 1995). Bonato and Vello (1999) designated *E6* for the allele determining earliness in cv. Paranà and *e6* for the allele determining late flowering and maturity in cvs. Paranagoiana and SS-1. The *E7* gene affects maturity and photoperiod sensitivity (Cober and Voldeng, 2001a). The dominant allele results in later flowering and maturity, and sensitivity to incandescent light. The *E7* and *e7* near-isogenic lines were grown in various photoperiods with four red:far-red (R:FR) light qualities. The *E7 E7* lines were sensitive to light quality. The *E7* and *e7* lines were similar under long days of high R:FR light quality, but under natural daylight and under low R:FR light quality, *E7* delayed flowering and *e7* did not (Cober and Voldeng, 2001b).

In controlled-photoperiod and planting-date field experiments at Urbana, IL, flowering and maturity were delayed by *E1, E2,* or *E3* (McBlain et al., 1987). The *E3 e3* gene pair is the locus primarily conferring long-daylength insensitivity in soybean, but both the *E3* and *E4* loci have to be considered when breeding for insensitivity to long daylength using incandescent long days (Saindon et al., 1989).

Seven genetic stocks homozygous for *e3* and *e4*, which are insensitive to incandescent long daylength, were developed (Voldeng and Saindon, 1991b). Photoperiod-sensitivity loci may be useful for short-season cultivar development. Cober et al. (1996a) investigated the photoperiod response of early-maturing cv. Harosoy near-isogenic lines with indeterminate or determinate stems. The *E3* allele exhibited the largest delay in photoperiod response when 20-h photoperiods were compared with 12-h photoperiods. Flowering was delayed by 24 d and maturity by 84 d (Cober et al., 1996a).

Light quality studies, as measured by the biochromatic ratio of R:FR quanta, showed that the *E1* allele was most sensitive, the *E3* allele was least sensitive, and the *E4* allele showed intermediate sensitivity to days to flowering (Cober et al., 1996b).

Six cv. Harosoy near-isogenic lines with different combinations of alleles at the *E1, E3, E4,* and *E7* loci were grown in growth cabinets with 10-, 12-, 14-, 16-, and 20-h photoperiods and either 18 or 28°C constant temperature (Cober et al., 2001). The late-flowering lines flowered earlier under cooler than under warmer temperatures. A Growing Photothermal Day model was developed. Cober et al. (2001) suggested that the model would be useful to predict desirable combinations of alleles for new agronomic production systems.

In a genetic and linkage analysis of cleistogamy in soybean, Takahashi et al. (2001) noticed that one of the genes for cleistogamy was linked to one of the recessive genes for insensitivity to incandescent long daylength. The E_3, E_4, and $e(f)$ loci are known to be involved in the initiation of flowering under incandescent long daylength (Abe et al., 1998), but Takahashi et al. (2001) did not identify the locus.

Molecular markers were used to identify chromosomal regions that control traits for flowering time, maturity, and photoperiod insensitivity in soybean. Tasma et al. (2001) reported that flowering time, maturity, and photoperiod insensitivity may be controlled by the same gene(s) or by tightly clustered genes in the same chromosomal region, but previous studies found no linkages between *E1* to *E5*.

A summary of genes affecting time of flowering and maturity in soybean is given in Table 5–7.

5–5.7.2 Growth of Stem, Petiole, and Inflorescence

Differences in branching pattern in soybean are due to plant spacing, photoperiod, nutrition, pests, genetics, and genotype × environment interactions. Two independent loci controlling differences in branching patterns were observed in the

Table 5–7. Genes affecting growth and morphology in soybean.

Gene	Phenotype	Strain†	Reference
		1. Time of flowering and maturity	
E1	Late	T175	Owen (1927b), Bernard (1971)
e1	Early	Clark	
E2	Late	Clark	Bernard (1971)
e2	Early	PI 86024	
E3	Late and sensitive to to fluorescent light	Harosoy 63	Buzzell (1971), Kilen and Hartwig (1971)
e3	Early and insensitive to fluorescent light	Blackhawk	
E4	Late and sensitive to long daylength	Harcor	Buzzell and Voldeng (1980)
e4	Early and insensitive to long daylength	Urozsajnaja (PI 297550)	
E5	Late	L64-4830	McBlain and Bernard (1987)
e5	Early	Harosoy	
E6	Early	Paranà	Bonato and Vello (1999)
e6	Late	Paranagoiana, SS-1	
E7	Late	Harosoy	Cober and Voldeng (2001a)
e7	Early	PI 196529	
J	Normal	c	Ray et al. (1995)
j	Long juvenile trait	PI 159925	
		2. Growth of stem, petiole, and inflorescence	
Br1 Br2	Branches originating from upper as well as lower nodes	T327	Nelson (1996)
br1 br2	Few branches only from lower nodes	T326	Nelson (1996)
Dt1	Indeterminate stem	Manchu, Clark	Woodworth (1932, 1933), Bernard (1972)
dt1-t	Tall determinate stem	Peking, Soysota	Thompson et al. (1997)
dt1	Determinate stem	Ebony, PI 86024	
Dt2	Semideterminate stem	T117	Bernard (1972)
dt2	Indeterminate stem	Clark	
F	Normal stem	c	Nagai (1926), Takagi (1929), symbol by Woodworth (1932, 1933), Matsuura (1933), Albertsen et al. (1983)
f	Fasciated stem	T173, PI 83945-4, Shakujo (PI 243541)	
Lps1	Normal petiole	Lee 68	Kilen (1983)
lps1	Short petiole	T279	
Lps2	Normal petiole	NJ90L-2	You et al. (1998)
lps2	Short petiole, abnormal pulvinus	NJ90L-1SP	
S	Short, internode length decreased	Higan	Bernard (1975a)
s	Normal	Harosoy	
s-t	Tall, internode length increased	Chief	
Se	Pedunculate inflorescence	T208	VanSchaik and Probst (1958)
se	Subsessile inflorescence	PI 84631	

(continued on next page)

Table 5–7. Continued.

Gene	Phenotype	Strain†	Reference
		3. Dwarfness	
Df2	Normal	c	Porter and Weiss (1948), symbol by Byth and Weber (1969)
df2	Dwarf	T210,T243	
Df3	Normal	c	Byth and Weber (1969)
df3	Dwarf	T244	
Df4	Normal	c	Fehr (1972a)
df4	Dwarf	T256	
Df5	Normal	c	Palmer (1984a)
df5	Dwarf	T263	
Df6	Normal	c	Werner et al. (1987)
df6	Dwarf	T286	
Df7 or *Df8*	Normal	c	Soybean Genetics Committee (1995)
df7 df8	Dwarf	T281	
Mn	Normal	c	Delannay and Palmer (1984)
mn	Miniature plant	T251	
Pm	Normal	c	Probst (1950)
pm	Dwarf, crinkled leaves, sterile	T211	
Sb1 or *Sb2*	Normal	Davis	Kilen and Hartwig (1975), Kilen (1977), Boerma and Jones(1978)
sb1 sb2	Brachytic stem	Ya Hagi (PI 227224)	
		4. Leaf form	
Ab	Abscission at maturity	T161, c	Probst (1950)
ab	Delayed abscission	Kingwa	
Dlm	Normal	Hobbit 87	Chung et al. (1998)
dlm	Necrotic spots with chlorotic halo	T363	
Lf1	5-foliolate	PI 86024	Takahashi and Fukuyama (1919), symbol by Fehr (1972b)
lf1	3-foliolate	c	
Lf2	3-foliolate	c	Fehr (1972b)
lf2	7-foliolate	T255	
Lmn	Normal	c	Yu and Kiang (1993a)
lmn	Leaf margin necrosis	PI 562545 (*G. soja*)	
Ln	Ovate leaflet	c	Takahashi and Fukuyama (1919), Woodworth (1932, 1933), Takahashi (1934), Domingo (1945), symbol by Bernard and Weiss (1973)
ln	Narrow leaflet, 4-seeded pods	T41, PI 84631	
Lnr	Normal	c	Wilcox and Abney (1991)
lnr	Narrow rugose leaf	T313	
Lo	Ovate leaflet	c	Domingo (1945)
lo	Oval leaflet, few-seeded pods	T122	
Lw1 Lw2	Nonwavy leaf	--	Rode and Bernard (1975b)
Lw1 lw2	Nonwavy leaf	Harosoy, Clark	
lw1 Lw2	Nonwavy leaf	T117	
lw1 lw2	Wavy leaf	T176, T205	
Lb1 Lb2	Nonbullate leaf	Harosoy	Rode and Bernard (1975c)

(continued on next page)

Table 5–7. Continued.

Gene	Phenotype	Strain†	Reference
Lb1 lb2	Nonbullate leaf	Clark	
lb1 Lb2	Nonbullate leaf	PI 196166	
lb1 lb2	Bullate leaf	L65-701 (Clark[6] X PI 196166)	

5. Pubescence type

Gene	Phenotype	Strain†	Reference
Pa1 Pa2	Erect	Harosoy, Clark	Karasawa (1936), Ting (1946), symbol by Bernard (1975d)
Pa1 pa2	Erect	L70-4119 (Harosoy[6] X Higan)	
pa1 Pa2	Semiappressed	Scott, Custer, Oksoy	
pa1 pa2	Appressed	Higan	
P1	Glabrous	T145	Nagai and Saito (1923)
p1	Pubescent	c	
P2	Normal	c	Stewart and Wentz (1926)
p2	Puberulent	T31	
Pb	Sharp hair tip	PI 163453 (*G. soja*), Kingwa	Ting (1946)
pb	Blunt hair tip	Clark	
Pc	Normal	Clark, c	Bernard and Singh (1969)
pc	Curly (deciduous)	PI 84987	
Pd1	Dense	PI 80837	Bernard and Singh (1969)
pd1	Normal	Clark, c	
Pd2	Dense	T264	R.L. Bernard (unpublished data, 1975)
pd2	Normal		
Pd1 Pd2	Extra-dense	L79-1815	Gunashinghe et al. (1988)
Ps	Sparse	PI 91160	Bernard and Singh (1969), Bernard (1975c)
Ps-s	Semisparse	Higan	
ps	Normal	c	

6. Seed-coat structure

Gene	Phenotype	Strain†	Reference
B1 B2 B3	Bloom on seed coat	Sooty	Woodworth (1932, 1933), Tang and Tai (1962)
b1, b2, or *b3*	No bloom	c	
N	Normal hilum abscission	c	Owen (1928)
n	Lack of abscission layer	Soysota	

† c = Indicates that the gene occurs in many cultivars.

cross of PI 391583 by PI 189916 (Nelson, 1996). Genetic data supported the hypothesis that two dominant alleles at independent loci were necessary to produce the high-branching phenotype. The gene symbols *Br1* and *Br2* were assigned to the two loci, and the parent line was designated Genetic Type Collection number T327. The low-branching phenotype had fewer branches, which were restricted to the lower portion of the main stem. The gene symbols *br1* and *br2* were assigned, and the parent line was designated T326.

The *Dt1* and *Dt2* loci affect stem termination in soybean. The *dt1-t* allele was identified in cvs. Peking and Soysota (Thompson et al., 1997) when cv. Clark near-isogenic lines were being developed. The *dt1-t* allele has similar effects on plant phenotype as the *dt1* allele, but the *dt1-t* allele significantly delays the expression of stem termination (Thompson et al., 1997). The *Dt1* and *Dt2* near-isogenic lines provide genotypes to study the biochemical-physiological mechanisms for stem termination in soybean.

The *Dt2* (semideterminate stem type) and *S* (short internode stem type) alleles have been shown to reduce excessive vegetative growth and lodging in narrow-row soybean production. Lewers et al. (1998) used the *Dt2* and *S* alleles to test if vegetative heterosis in F_1 hybrid plants could be reduced. The *Dt2* and *S* alleles affected agronomic traits of F_1 hybrids and cultivars (inbreds) similarly. These two alleles hastened maturity, decreased plant height and lodging, and increased harvest index values. The *Dt2* allele decreased seed weight and protein content.

Fasciated soybean plants exhibit a broadened and flattened stem (Albertsen et al., 1983). The effect of the fasciation mutation (*f* allele) was examined in apical meristems in near-isogenic lines and in three mutants of independent origin. Development patterns were similar (Tang and Skorupska, 1997).

In the development of near-isogenic lines for fasciation, selection for seed yield apparently decreased the penetrance and expressivity of the fasciated gene (Leffel et al., 1993). Fasciated genotypes were developed that were comparable to check cultivars in seed yields in lower-yielding but not in higher-yielding environments (Leffel et al., 1993). The fasciated lines were more susceptible to lodging. BARC-10 was selected as an agronomic line from the cross of cv. Hobbit × cv. Shakujo (PI 243541) (Leffel, 1994a). Six pairs of near-isogenic lines, BARC-11-1 to BARC-11-6, with the fasciated gene from PI 83945-4 or PI 243541 were developed and selected for agronomic performance (Leffel, 1994b).

The short-petiole trait in T279 was inherited as a single recessive gene, *lps1*, (Kilen, 1983). A second short-petiole trait with abnormal pulvinus, *lps2*, was identified (You et al., 1998). In certain cross combinations *lps1* and *lps2* behaved as two duplicate recessive loci (You et al., 1998). It is possible that these two genes may control different steps of the same physiological pathway toward the development of the petiole and pulvinus.

A spontaneous mutation was identified that altered flower development and produced apetalous male-sterile flowers. The apetalous male-sterile mutant exhibited floral abnormalities similar to several homeotic gene mutations in *Antirrhinum* and *Arabidopsis*. Male sterility was attributed to tapetum malfunction (Skorupska et al., 1993). The endogenous level of indol-3-acetic acid (IAA) and abscisic acid (ABA) in the wild-type and apetalous mutant were quantified and compared. In normal plants at anthesis, the hormonal balance (IAA/ABA ratio) was favorable to IAA, but at pod-fill stage shifted to ABA. The opposite trend was observed in apetalous plants (Skorupska et al., 1994). The apetalous mutant would be classified as a structural male sterile, which is sporophytically mediated (Johns et al., 1981). The apetalous trait was inherited as a single recessive gene; no gene symbol or Genetic Type Collection number was given.

A summary of genes affecting growth of stem, petiole, and inflorescence in soybean is given in Table 5–7.

5–5.7.3 Dwarfness

A total of six dwarf mutations, *df2* to *df7 df8*, are known in soybean. A dwarf soybean mutant was recovered in cv. C1421 after ethyl methanesulfonate seed treatment. The mutant was inherited as a single recessive gene, designated *df6*, and is maintained in the Genetic Type Collection as T286 (Werner et al., 1987). Leaf epidermal cells of T286 were 76% smaller than C1421 cells, while leaves of T286 were 80% smaller than C1421 leaves. The differences in leaf size were primarily due to differences in cell size (Werner et al., 1987).

A duplicate-factor trait for dwarf plants was designated *df7 df8* and assigned Genetic Type Collection number T281 (Soybean Genetics Committee, 1995).

An ethyleneimine-induced dwarf mutant was identified in cv. Hyuga. The trait was inherited as a single recessive gene (Umezaki et al., 1988). Allelism tests with known dwarf mutants were not reported.

The brachytic stem trait, found in cv. Ya Hagi from Japan, is inherited as a duplicate factor trait, *sb1 sb2*, (Boerma and Jones, 1978). The mainstem growth pattern is characterized by a geniculate internode arrangement with markedly shortened internodes (Kilen, 1977). Light and photoperiod are involved in the phenotypic expression of brachytic stem (Huang et al., 1993). Under low light intensity, expression of the brachytic trait is reduced. In greenhouse experiments, increasing the photoperiod from 14 to 16 h decreased plant height and enhanced brachytic stem development (Huang et al., 1993).

A summary of genes involved in dwarfness in soybean is given in Table 5–7.

5–5.7.4 Leaf Form

The leaf margin necrosis phenotype appears on PI 562545 (*G. soja*) plants about 3 mo after planting, mostly in older leaves. The trait was inherited as a single recessive gene and assigned gene symbol *lmn* (Yu and Kiang, 1993a).

A disease-lesion mimic mutant in soybean was recovered from gamma ray-treated cv. Hobbit 87. Leaves of the mutant became more necrotic and chlorotic as they aged, which resulted in premature leaf senescence. The mutant was inherited as a single recessive gene, *dlm*, and assigned Genetic Type Collection number T363 (Chung et al., 1998). This mutant will be useful in programmed cell death research.

Another possible disease-lesion leaf soybean mutant was noticed after ethyl methanesulfonate seed treatment of cv. C1421 (Wilcox and Abney, 1991). The narrow, rugose-leaf phenotype was not due to soybean mosaic virus, tobacco ring spot virus, tomato ring spot virus, or tobacco streak virus. The mutant trait was inherited as a single recessive gene, *lnr*, and a line with the trait was designated Genetic Type Collection number T313 (Wilcox and Abney, 1991).

A summary of genes involved in leaf form in soybean is given in Table 5–7.

5–5.7.5 Pubescence Type

Bernard et al. (1991) reported the release of cv. Harosoy and Clark near isogenic lines that had trichome density from one-fourth normal to 4 X normal. Gunashinghe et al. (1988) used the cv. Clark isolines to determine the influence of trichome density on the spread of nonpersistently transmitted plant viruses by aphid vectors. Field spread of soybean mosaic virus was negatively correlated with density of pubescence (Gunashinghe et al., 1988).

Three dense (*Pd1 Pd1*) pubescent germplasm lines that quadruple the number of trichomes per unit of epidermal surface were released (Specht and Graef,

1992a). Because an increase in pubescence density can amplify leaf surface reflectivity, which in turn can reduce transpiration, these lines have utility in agronomic and physiology studies.

Zhang et al. (1992) evaluated four dense and four normal pubescent genotypes from each of 20 populations for agronomic traits. Seed yields were somewhat larger for dense-gray pubescent lines when compared with the other combinations and may be of some merit in terms of yield improvement.

A summary of genes involved in pubescence type in soybean is presented in Table 5–7.

5–5.7.6 Seed-Coat Structure

A summary of genes involved in seed-coat structure in soybean is given in Table 5–7.

5–5.8 Fertility-Sterility

Sterility systems in soybean have been classified as synaptic, structural, male-partial sterile or female-partial sterile, and completely male sterile, female fertile. The genetic control may be nuclear or nuclear-cytoplasmic.

Synaptic mutants include mutations that affect chromosome pairing (synapsis) and disjunction. These mutants are characterized as being highly male and female sterile. The soybean Genetic Type Collection contains seven synaptic mutants, $st2$ through $st8$ ($st1$ has been lost). The few seed that occur on these synaptic plants, as a result of self- or cross-pollination, often produce aneuploid or polyploid plants (see Chapter 4, Hymowitz, 2004, this publication).

Genetic Type T331 is the only soybean synaptic mutant that is inherited as a duplicate factor recessive trait ($st6$ $st7$). The mutation was observed in tissue-culture-derived cv. Calland (Illarslan et al., 1997). A single recessive gene mutation, resulting in complete male and female sterility, was recorded among progeny from a gene-tagging study of the $w4$-m line (T322) (Palmer et al., 1989). This synaptic mutant was assigned gene symbol $st8$ $st8$ and the mutant line was designated Genetic Type Collection number T352 (Palmer and Horner, 2000).

Four male-fertile, female-partial-sterile mutations were noticed in a gene-tagging study of the $w4$-m $w4$-m line (Palmer et al., 1989). Embryological studies were done using confocal scanning laser microscopy (Pereira et al., 1996). The partial-sterile (PS-1) mutant was inherited as a single recessive gene and can be maintained as the mutant line. PS-1 is nonallelic to partial steriles 2, 3, and 4 (PS-2, PS-3, and PS-4) and was not linked to the $W4$ locus (Pereira et al., 1997b). Confocal scanning laser microscopy was used to determine that early embryo abortion in PS-1 was due indirectly to abnormal migration of the fused polar nucleus, which prevented the polar nucleus from being fertilized. Subsequent absence of endosperm development led directly to abortion of the proembryo. PS-2, PS-3, and PS-4 are maintained as heterozygotes and upon self-pollination segregate about 1 fertile:1 partial sterile (Pereira et al., 1997a). Genetic tests showed that the gene(s) controlling partial-female sterility in these three mutants was (were) not transmitted through the female when the plants were used as female parents. The three mutants are male fertile. Megagametogenesis studies, using confocal scanning laser microscopy, indicated that the ovules from these three female-partial-sterile plants had normal embryo sac development. Ovule abortion was due to failure of fertilization (Pereira et al., 1997a). The four partial-female-sterile mutants have not been assigned gene symbols and have not been deposited in the soybean Genetic Type Collection.

Additional male-sterile, female-fertile mutants have been identified and added to the soybean Genetic Type Collection. Two *ms1* mutants, T287 and T290, (Skorupska and Palmer, 1990), bring the total to six independent mutations at the *ms1* locus that are included in the Type Collection. Based upon cytological observations, a spontaneous mutant observed in an F_4 population was considered to be a mutation at the *ms1* locus (Mariani et al., 1991). A male-sterile, female-fertile mutant found by Rubaihayo and Gumisiriza (1978) was similar in cytological abnormalities to T260, and may be a mutation at the *ms1* locus.

Although *ms1* does not inhibit female function to the extreme extent it does male function, various chromosome abnormalities have been noted (see Chapter 4, Hymowitz, 2004, this publication). The *ms1* mutation has been used to study embryo-endosperm relationships (Zhang and Palmer, 1990). It also has been proposed to test for apomixis, scale down the ploidy level, and produce deficiency aneuploids at the diploid chromosome level (Palmer et al., 1992).

T360 (*ms2*) originated among progeny of a germinal revertant in a gene-tagging study (Palmer, 2000). Two additional *ms3* mutants, T284 and T291, occurred spontaneously in cultivars (Chaudhari and Davis, 1977; Graybosch and Palmer, 1987; Skorupska and Palmer, 1990). T292 (*ms4*) was recovered in cv. Corsoy (Skorupska and Palmer, 1990).

Nuclear male-sterile-facilitated random mating populations using *ms2* were developed and released by Specht and Graef (1992b). Graef and Specht (1999) developed and released a second population using *ms2* that incorporated the large seed size trait.

Two independent mutations at the *ms6* locus, T295 (Palmer and Skorupska, 1990; Skorupska and Palmer, 1989) and T354 (Ilarslan et al., 1999), have been described. Thirty-four pairs of germplasm lines were developed by backcrossing the *ms6-w1* linked combination to 34 recurrent parents. The lines were developed with either the *ms6-w1* donor cytoplasm or the recurrent parent cytoplasms to form 34 near-isogenic pairs (68 lines total) (Palmer and Lewers, 1998).

The *Ms6* and *W1* loci are tightly linked with F_2 recombination estimates in coupling phase of between 2.48 + 0.1 and 3.18 + 0.1 (Skorupska and Palmer, 1989). In separate experiments, the recombination value from testcross data in the *ms6-w1* donor cytoplasm was 3.14 + 0.80, and in 24 recurrent parent cytoplasms the recombination value averaged 3.62 + 0.89 (Palmer et al., 1998). From F_2 family data, the recombination value in the *ms6-w1* donor cytoplasm was 3.06 + 0.35, but in 24 recurrent parent cytoplasms, the recombination value averaged 4.90 + 0.35 (Palmer et al., 1998).

Lewers et al. (1996) used the *ms6-w1* linkage relationship to develop the Cosegregation Method to produce hybrid soybean seed. The Cosegregation Method was compared with the Traditional Method and the Dilution Method to produce large quantities of hybrid seed for experimental purposes. Lewers et al. (1996) used insect-facilitated cross-pollination in the comparison study and concluded that the Cosegregation Method gave higher seed yield, better efficiency, and equal or better seed quality than the other two methods. Cooper and Tew (2001) used the linkage between *ms5* and one of the two genes for green cotyledon, *d1* or *d2*, to produce seed homozygous for male sterility.

Lewers and Palmer (1997) outlined marker-assisted linkage of a male-fertility (sterility) locus and a marker locus affecting hypocotyl color of the seedling, flower color at anthesis, and hilum color at maturity. The production and evaluation of hybrid soybean was reviewed by Palmer et al. (2001).

The Midwest Oilseed male-sterile mutant was inherited as a single recessive gene, but no gene symbol or Genetic Type Collection number was given (Jin et al., 1997). This mutant is nonallelic to *ms1* to *ms6* (Jin et al., 1997), to *ms7* to *ms9* (Palmer, 2000), and is located on linkage group D1b of the USDA-ARS-ISU soybean molecular genetic map (Jin et al., 1998). For additional information on this mutant see Davis (2000).

Several nuclear-cytoplasmic systems have been reported in China (Palmer et al., 2001). The cytoplasmic male-sterile (CMS) system reported by Sun et al. (1997) was stable under all temperature/photoperiod conditions tested (Smith et al., 2001). As is common with many CMS systems, the Sun et al. (1997) system showed an abnormal development and/or premature degeneration of the tapetum after meiosis II (Smith et al., 2002).

Genetic Type T31, the puberulent pubescent mutant (*p2*), has 10% natural outcrossing in Illinois (Bernard and Jaycox, 1969) and 11 to 58% in India (Singh, 1972). Poor dehiscence of anthers was reasoned to be primarily responsible for poor self-pollination and the high frequency of cross-pollination. Culbertson and Hymowitz (1990) could not detect any abnormalities in number of pollen grains per anther, rate of pollen germination, or self-incompatibility. The *p2* mutant would be classified a structural sterile (Johns et al., 1981).

A summary of genes affecting fertility-sterility in soybean is given in Table 5–8.

5–5.9 Physiology

5–5.9.1 Reaction to Nutritional Factors

Iron (Fe)-deficiency chlorosis is a common problem in soybean grown on calcareous soil. Cultivars differ in their genetic ability to utilize the available Fe. A major problem in determining the genetic control of Fe uptake and utilization is the substantial genotype × environment interaction. Genetic control of Fe-deficiency chlorosis has been reported as a single recessive gene (Weiss, 1943), a major gene with modifying genes (Cianzio and Fehr, 1980), and quantitative inheritance with additive gene action (Cianzio and Fehr, 1982). Lin et al. (1998, 2000a) concluded that the nutrient solution assays and field tests identified similar genetic mechanisms of Fe uptake and/or utilization. Furthermore, two separate genetic mechanisms controlled Fe deficiency in soybean (Lin et al., 1997, 2000b). Gene symbols were not assigned.

Tolerance to certain nutritional factors such as zinc (Zn) and manganese have received some attention. Crosses were made between two breeding lines that were different in Zn absorption. The distribution of F_3 lines for Zn concentration sug-

Table 5–8. Genes affecting fertility-sterility in soybean.

Gene	Phenotype	Strain †	Reference
Fs1 or *Fs2*	Fertile	c	Johns and Palmer (1982)
fs1 fs2	Structural sterile	T269	
Ft	Fertile	c	Singh and Jha (1978)
ft	Structural sterile	Gamma ray-induced mutant	
Ms1	Fertile	c	
ms1 (North Carolina)	Male sterile	T260	Brim and Young (1971)
ms1 (Urbana)	Male sterile	T266	Boerma and Cooper (1978)
ms1 (Tonica)	Male sterile	T267	Palmer et al. (1978b)
ms1 (Ames 1)	Male sterile	T268	Palmer et al. (1978b)
ms1 (Ames 2)	Male sterile	T287	Skorupska and Palmer (1990)
ms1 (Danbury)	Male sterile	T290	Skorupska and Palmer (1990)
Ms2	Fertile	c	
ms2 (Eldorado)	Male sterile	T259	Bernard and Cremeens (1975), Graybosch et al. (1984)
ms2 (Ames)	Male sterile	T360	Palmer (2000)
Ms3	Fertile	c	
ms3 (Washington)	Male sterile	T273	Palmer et al. (1980) Graybosch and Palmer (1987)
ms3 (Flanagan)	Male sterile	T284	Chaudhari and Davis (1977), Graybosch and Palmer (1987)
ms3 (Plainview)	Male sterile	T291	Skorupska and Palmer (1990)
Ms4	Fertile	c	
ms4 (Ames)	Male sterile	T274	Delannay and Palmer (1982a)
ms4 (Fisher)	Male sterile	T292	Skorupska and Palmer (1990)
Ms5	Fertile	c	Buss (1983)
ms5	Male sterile	T277	
Ms6	Fertile	c	
ms6 (Ames 1)	Male sterile	T295	Palmer and Skorupska (1990), Skorupska and Palmer (1989)
ms6 (Ames 2)	Male sterile	T354	Ilarslan et al. (1999)
Ms7	Fertile	c	Palmer (2000)
ms7	Male sterile	T357	
Ms8	Fertile	c	Palmer (2000)
ms8	Male sterile	T358	
Ms9	Fertile	c	Palmer (2000)
ms9	Male sterile	T359	
Msp	Fertile	c	Stelly and Palmer (1980a, 1980b)
msp	Partial male sterile	T271	
St2	Fertile	c	Hadley and Starnes (1964)
st2	Asynaptic sterile	T241	
St3	Fertile	c	Hadley and Starnes (1964)
st3	Asynaptic sterile	T242	
St4	Fertile	c	Palmer (1974)
st4	Desynaptic sterile	T258	
St5	Fertile	c	Palmer and Kaul (1983)
st5	Desynaptic sterile	T272	
St6 or *St7*	Fertile	Calland	Ilarslan et al. (1997)
st6 st7	Male sterile, female sterile	T331	
St8	Fertile	c	Palmer and Horner (2000)
st8	Desynaptic sterile	T352	

†c = Indicates that the gene occurs in many cultivars.

gested that only a few genes controlled Zn-absorption efficiency or inefficiency (Hartwig et al., 1991). Manganese (Mn) deficiency can be a common micronutrient problem with soybean grown on high pH soils. Crosses were made between a tolerant genotype and a genotype intolerant to Mn deficiency. The F_2 and $F_{2:3}$ progeny response to Mn deficiency suggested digenic inheritance, but no gene symbols were given (Graham et al., 1995).

Nutritional quality of soybean may be modified by a reduction in seed phytic acid phosphorous (P). Phytic acid is the storage form of P, and it accounts for about 70% of total soybean seed P. Raboy and Dickinson (1993) reported that essentially all soybean seed P variation is found as variation in phytic acid P. The genetic control is not known, but based upon field studies, numerous genes are involved (Raboy et al., 1984). Wilcox et al. (2000) identified soybean mutants, after seed mutagen treatment, that reduced seed phytic acid P and increased seed inorganic P. These mutants should increase the nutritional value of soy meal and reduce excess P in livestock manure (Wilcox et al., 2000). The genetic control of these low phytic acid mutants was not presented.

A summary of genes affecting nutrition in soybean is given in Table 5–9.

5–5.9.2 Flavonols of Leaves

Four flavonol glycoside genes, *Fg1*, *Fg2*, *Fg3*, and *Fg4,* have been identified in soybean (Buzzell and Buttery, 1973, 1974). The inheritance of a third allele at the *Fg2* locus was determined. The *Fg2-b* allele gave different flavonol classes in combination with *Fg1*, *Fg3*, and *Fg4* than did *Fg2-a* in combination with these alleles (Buzzell and Buttery, 1992). The *Fg* loci in soybean have been used to verify hybrid plants, in cultivar identification, and in chemotaxonomic studies.

A summary of genes affecting flavonols of leaves in soybean is given in Table 5–9.

5–5.9.3 Seed

A shriveled seed mutant was found among progenies of a cross between breeding lines AP2 (Fehr and Clark, 1973) and P2180 (PI 556640). The mutant was inherited as a single recessive gene, designated *shr*, and is maintained in the Genetic Type Collection as T311 (Honeycutt et al., 1989a). The expressivity and penetrance of the recessive allele was influenced by environment, particularly night temperature (Honeycutt et al., 1989a). Seed from *shr shr* plants showed a reduction in the level of the β-subunit of the 7S seed storage proteins (Honeycutt et al., 1989b). Changes in development of protein bodies, accumulation of carbohydrates, and greater water loss in T311 seeds were associated with the ontogeny of shriveled seeds (Chen et al., 1998).

A summary of genes affecting seed in soybean is given in Table 5–9.

5–5.9.4 Chlorophyll Deficiency–Nuclear

In *v1 v1* (T93) plants, variegation usually does not occur on the first trifoliolates produced or on young trifoliolates, but does appear after the leaves have expanded (Woodworth, 1932, 1933). Four variegated plants were observed in a bulk harvest of cv. Clark. Genetic studies confirmed that the trait was inherited as a sin-

Table 5–9. Genes affecting physiology in soybean.

Gene	Phenotype	Strain†	Reference
	1. Reaction to nutritional factors		
Fe	Efficient Fe utilization	c	Weiss (1943)
Fe	Inefficient	PI 54619	
Np	Phosphorus tolerant	Chief	Bernard and Howell (1964)
np	Sensitive to high P level	Lincoln	
Ncl	Chloride excluding	Lee	Abel (1969)
ncl	Chloride accumulating	Jackson	
Nr	Constitutive NO_3^- reductase present	Williams	Ryan et al. (1983a, 1983b)
nr	Constitutive NO_3^- reductase absent	T276	
	2. Flavonol glycosides of leaves		
T	Quercetin and kaempferol present	c	Buttery and Buzzell (1973), (also see Table 5–12)
t	Quercetin absent, kaempferol present	c	
Wm	Glycosides present	c	Buzzell et al. (1977), (also see Table 5–12)
wm	Glycosides absent	T235	
Fg1	β(1-6)-glucoside present	T31, c	Buttery and Buzzell (1975)
fg1	β(1-6)-glucoside absent	Chippewa 64, c	
Fg2-a	Normal kaempferol rutinoside	OX724	Buzzell and Buttery (1992)
Fg2-b	Less kaempferol rutinoside	OX730	
fg2	α(1-6)-rhamnoside absent	Chippewa 64, c	Buttery and Buzzell (1975)
Fg3	β(1-2)-glucoside present	T31, c	Buttery and Buzzell (1975)
fg3	β(1-2)-glucoside absent	Chippewa 64, c	
Fg4	α(1-2)-rhamnoside present	T31, c	Buttery and Buzzell (1975)
fg4	α(1-2)-rhamnoside absent	AK(FC 30761)	
	3. Seed		
Shr	Normal	c	Honeycutt et al. (1989a)
shr	Shriveled	T311	

† c = Indicates that the gene occurs in many cultivars.

gle recessive gene, and that it was nonallelic to *v1* (Honeycutt et al., 1990). The pattern of variegation found on the unifoliolate leaves generally persisted among the trifoliolates of the plant; however, variegation patterns changed from one type to another on individual plants. The mutant was assigned gene symbol *v2* and designated Genetic Type Collection number T312 (Honeycutt et al., 1990). The mutant is true breeding. No green or yellow plants were recovered among 75 000 seedlings descended from self-pollination of *v2 v2* plants.

The nuclear genes affecting chlorophyll deficiency or retention in soybean include two mutable phenotypes, Genetic Types T218M and T225M. Genetic stocks have been constructed that segregate 3 green:1 yellow lethal for T218, designated T218H (Palmer, 1987), and for T225, designated T225H (Sheridan and Palmer, 1975). Genetic type T362H, which originated from a tissue-culture-derived chimeric plant, also segregated 3 green:1 yellow lethal, upon self-pollination (Palmer et al., 2000). The mutable plant, from which T362H was derived, was not added to the Genetic Type Collection.

Genetic Types T218H and T225H are allelic (Hatfield and Palmer, 1986) as well as T225H and T362H. (Palmer et al., 2000). In certain cross-combinations,

T218H, T225H, and T362H gave 15 green:1 yellow lethal (R.G. Palmer, unpublished data, 1985, 2000).

Genetic analyses of the *Y18-m* (T225M) locus suggested that the instability was the result of a transposable element system (Peterson and Weber, 1969; Sheridan and Palmer, 1977). Chandlee and Vodkin (1989b) indicated, however, that the rearrangements of the *Tgm1* transposable element system (Rhodes and Vodkin, 1988) were not associated with the *Y18-m* instability.

The mutation of the unstable allele *Y18-m* (T225M) can be restricted to each of the individual cell layers within the leaf that contain chloroplasts. Biochemical studies have shown that several polypeptides of the thylakoid membranes are missing, and many, including the major light-harvesting complex polypeptides, are reduced (Chandlee and Vodkin 1989a). Additional experiments indicated that a polypeptide with a molecular weight of ~18 kD was absent in the products translated and processed from the mRNA from yellow tissue (Cheng and Chandlee, 1993). Chloroplast gene expression studies revealed a reduction in the levels of 16S rRNA in the yellow tissues (Cheng and Chandlee, 1993).

Several chlorophyll-deficient mutants were recovered among the progeny of independent germinal revertants of *w4-m* (T322) in a gene-tagging study (Palmer et al., 1989). Hedges and Palmer (1992) determined that three viable true-breeding yellow-green mutants were allelic, even though they had distinct phenotypes. Electrophoretic analyses indicated that each mutant lacked two of three mitochondrial malate dehydrogenase (MDH) isozymes (Hedges and Palmer, 1992). The MDH phenotype and the yellow-green phenotype each were inherited as single recessive alleles. Allelism tests showed that these three mutants were allelic to T253 for chlorophyll deficiency (*y20*) and the unique MDH banding pattern (*Mdh1-n*) (Hedges and Palmer, 1992). No recombination between the two traits was observed, suggesting either pleiotropy or chromosomal rearrangements. Molecular studies indicated that the MDH mutations were deletions (Imsande et al., 2001). The phenotypic differences among the near-isogenic chlorophyll-deficient lines suggested that the deletions might have different endpoints. The three mutants were assigned gene symbols and added to the Genetic Type Collection as T323, T324, and T325.

In a separate gene-tagging study with T322, a yellow-green, MDH-null phenotype was identified. Allelism tests showed that the mutant gene was allelic to the *y20 Mdh1-n* genotype, and the mutant strain was designated T317 (Chen et al., 1999).

A total of 25 *y20* mutants have been reported. Of the 25 mutants, 18 are associated with *k2* (tan saddle seed coat), but all 25 are associated with *Mdh1-n*. Chen and Palmer (1998a) have proposed that the instability at the *y20 Mdh1-n k2* chromosomal region was due to transposon activity that could generate chromosomal rearrangements, such as deletions. The percentage recombination between each two loci (*y20-k2*), (*y20-Mdh1-n*) or (*k2-Mdh1-n*) depends upon (i) the class of mutation, for example, spontaneous, tissue-culture-derived, germinal revertants of gene tagging, or progeny of "instability" cross-pollinations; and (ii) whether the alleles are in coupling (cis) or repulsion (trans) phase (Chen and Palmer, 1996, 1998a, 1998b).

Genetic Type T270H (*Y22 y22*) is a conditional lethal, usually viable when grown in the greenhouse, but lethal when grown under field conditions (Palmer et al., 1990b). Genetic Type T288 (*y23 y23*) is a weak plant with leaves that change from green to yellow-white, which eventually become necrotic. In certain genetic backgrounds and environments, the *y23 y23* genotype is a seedling lethal. Both T270 and T288 mutants are inherited as single recessive genes (Palmer et al., 1990b).

Inheritance and allelism tests revealed that cv. Ogden has a dominant nuclear gene for green seed coat, which was designated *G2* (Reese and Boerma, 1989). A breeding line with green seed coat was identified and tested for inheritance and allelism. Genetic data indicated that a single recessive gene at a locus different from *G1* and *G2* was responsible for green seed coat (Reese and Boerma, 1989). The gene symbol *g3* and Genetic Type Collection number T294 were assigned (Reese and Boerma, 1989).

A summary of nuclear genes affecting chlorophyll deficiency or retention in soybean is given in Table 5–10.

Table 5–10. Nuclear genes affecting chlorophyll deficiency or retention in soybean.

Genes	Phenotype	Strain†	Reference
	1. Chlorophyll deficiency		
V1	Normal	c	Woodworth (1932, 1933)
v1	Variegated leaves	T93	
V2	Normal	c	Honeycutt et al. (1990)
v2	Variegated leaves	T312	
Y3	Normal (*y3 G1* is also normal)	c	Nagai (1926), Takagi (1929, 1930), Terao and Nakatomi (1929), symbol by Morse and Cartter (1937)
y3 (with *g1*)	Green seedling, becoming yellow	Kura, T139	
Y4	Normal	c	Symbol by Morse and Cartter (1937), Woodworth and Williams (1938) (as y5 by error)
y4	Greenish-yellow leaves, weak plant	T102	
Y5	Normal	c	Symbol by Morse and Cartter (1937), Woodworth and Williams (1938) (as y4 by error)
y5	Greenish-yellow leaves	T116, T134	
Y6	Normal	c	Symbol by Morse and Cartter (1937), Woodworth and Williams (1938)
y6	Pale green leaves	T136	
Y7 or *Y8*	Normal	c	Morse and Cartter (1937), Probst (1950) (as y8), Williams (1950)
y7 y8	Yellow growth in cool weather	T138	
Y9	Normal	c	Probst (1950)
y9	Bright greenish-yellow leaves	T135	
Y10	Normal	c	Probst (1950)
y10	Greenish-yellow seedling	T161	
Y11	Normal	c	Weber and Weiss (1959)
Y11 y11	Bright greenish-yellow leaves	T219H	
y11	Lethal yellow	T219	
Y12	Normal	c	Weiss (1970a)

(continued on next page)

Table 5–10. Continued.

Genes	Phenotype	Strain†	Reference
y12	Whitish primary leaves, yellowish-green leaves	T233	
Y13	Normal	c	Weiss (1970b)
y13	Whitish-green seedling, greenish-yellow leaves	T230	
Y14	Normal	c	Nissly et al. (1976)
y14	Light green leaves	T229	
Y15‡	Normal	c	Nissly et al. (1976),
y15‡	Yellowish-green leaves	T234	Chen et al. (1999)
Y16	Normal	c	Wilcox and Probst (1969)
y16	Nearly white lethal	T257	
Y17	Normal	c	Nissly et al. (1981)
y17	Light yellowish-green leaves	T162	
Y18	Normal	c	
Y18-m	Unstable allele resulting in chlorophyll chimera	T218M	Palmer (1987)
y18 (Urbana)	Near-lethal yellow	T218H	Palmer (1987)
Y18-m	Unstable allele resulting in chlorophyll chimera	T225M	Peterson and Weber (1969)
y18 (Ames 1)	Near-lethal yellow	T225H	Sheridan and Palmer (1975)
y18 (Ames 2)	Near-lethal yellow	T362H	Palmer et al. (2000)
Y19	Normal	c	Palmer et al. (1990b)
y19	Delayed albino	T265H	
Y20 [*Mdh1 K2*]§	Normal	c	Palmer (1984b)
y20 (Ames 23) [*Mdh1-n* (Ames 21)]	Yellowish-green leaves	T234	Chen et al. (1999)
y20 (Urbana 1) [*Mdh1-n* (Urbana 1) *k2* (Urbana 1)]	Yellowish-green leaves	T253	Palmer (1984b)
y20 (Ames 1) [*Mdh1-n* (Ames 1)]	Yellowish-green leaves	T317	Amberger et al. (1992)
y20 (Ames 2) [*Mdh1-n* (Ames 2)]	Yellowish-green leaves	T323	Hedges and Palmer (1992)
y20 (Ames 3) [*Mdh1-n* (Ames 3)]	Yellowish-green leaves	T324	Hedges and Palmer (1992)
y20 (Ames 4) [*Mdh1-n* (Ames 4)]	Yellowish-green leaves	T325	Hedges and Palmer (1992)
y20 (Ames 5) [*Mdh1-n* (Ames 7) *k2* (Urbana 1)]	Yellowish-green leaves	T334	Chen and Palmer (1998a)
y20 (Ames 6) [*Mdh1-n* (Ames 8) *k2* (Urbana 1)]	Yellowish-green leaves	T335	Chen and Palmer (1998a)
y20 (Ames 7) [*Mdh1-n* (Ames 9) *k2* (Urbana 1)]	Yellowish-green leaves	T336	Chen and Palmer (1998a)

(continued on next page)

Table 5–10. Continued.

Genes	Phenotype	Strain†	Reference
y20 (Ames 8) [Mdh1-n (Ames 10) k2 (Urbana 1)]	Yellowish-green leaves	T337	Chen and Palmer (1998a)
y20 (Ames 9) [Mdh1-n (Ames 11) k2 (Urbana 1)]	Yellowish-green leaves	T338	Chen and Palmer (1998a)
y20 (Ames 10) [Mdh1-n (Ames 12) k2 (Urbana 1)]	Yellowish-green leaves	T339	Chen and Palmer (1998a)
y20 (Ames 11) [Mdh1-n (Columbia 1) k2 (Columbia 1)]	Yellowish-green leaves	T340	Chen and Palmer (1998a)
y20 (Ames 12) [Mdh1-n (Ames 13) k2 (Urbana 1)]	Yellowish-green leaves	T341	Chen and Palmer (1998a)
y20 (Ames 13) [Mdh1-n (Ames 14) k2 (Urbana 1)]	Yellowish-green leaves	T342	Chen and Palmer (1998a)
y20 (Ames 14) [Mdh1-n (Ames 15) k2 (Urbana 1)]	Yellowish-green leaves	T343	Chen and Palmer (1998a)
y20 (Ames 15) [Mdh1-n (Ames 16) k2 (Urbana 1)]	Yellowish-green leaves	T344	Chen and Palmer (1998a)
y20 (Ames 16) [Mdh1-n (Ames 17) k2 (Urbana 1)]	Yellowish-green leaves	T345	Chen and Palmer (1998a)
y20 (Ames 17) [Mdh1-n (Ames 19)]	Yellowish-green leaves	T346	Chen et al. (1999)
y20 (Ames 18) [Mdh1-n (Ames 6) k2 (Urbana 1)]	Yellowish-green leaves	T347	Chen et al. (1999)
y20 (Ames 19) [Mdh1-n (Ames 6) k2 (Urbana 1)]	Yellowish-green leaves	T348	Chen et al. (1999)
y20 (Ames 20) [Mdh1-n (Ames 6) k2 (Urbana 1)]	Yellowish-green leaves	T349	Chen et al. (1999)
y20 (Ames 21) [Mdh1-n (Ames 6) k2 (Urbana 1)]	Yellowish-green leaves	T350	Chen et al. (1999)

(continued on next page)

Table 5–10. Continued.

Genes	Phenotype	Strain†	Reference
y20 (Ames 22) [Mdh1-n (Ames 20) k2 (Urbana 1)]	Yellowish-green leaves	T351	Chen et al. (1999)
y20 (Ames 24) [Mdh1-n (Ames 22)]	Yellowish-green leaves	T361	Palmer et al. (2000)
Y21	Normal	c	Yee et al. (1986)
y21	Lethal yellow	Nennong 2015	
Y22	Normal	c	Palmer et al. (1990b)
y22	Greenish-yellow leaves	T270H	
Y23	Normal	c	Palmer et al. (1990b)
y23	Leaves becoming yellow-white and necrotic	T288	
2. Chlorophyll retention			
D1 or D2	Yellow seed embryo	c	Woodworth (1921), Owen (1927a), Veatch and Woodworth (1930)
d1 d2	Green seed embryo	Columbia, T104	
G1	Green seed coat	Kura	Terao (1918), Takahashi and Fukuyama (1919), Nagai (1921), Woodworth (1921)
g1	Yellow seed coat	c	
G2	Green seed coat	Ogden	Reese and Boerma (1989)
g2	Yellow seed coat	c	
G3	Yellow seed coat	c	Reese and Boerma (1989)
g3	Green seed coat	T294	

† c = Indicates that the gene occurs in many cultivars.
‡ T234 is allelic to T325 (y20 y20). The gene symbol y15 has been deleted. T234 is now y20 (Ames 23) [Mdh1-n (Ames 21)] (Chen et al., 1999).
§ Genes for secondary traits of interest are listed in brackets.

5–5.9.5 Chlorophyll Deficiency and Morphology–Cytoplasmic

Uniparentally-inherited mutants that cause either chlorophyll deficiency in the leaves, chlorophyll retention in the cotyledons, or morphological abnormalities are known in soybean. Cianzio and Palmer (1992) reported the genetics of five yellow foliar mutants. The phenotype was inherited uniparentally in each case through the maternal parent. The mutants were assigned gene symbols cyt-Y4, cyt-Y5, cyt-Y6, cyt-Y7, and cyt-Y8, and added to the Genetic Type Collection as, T314, T315, T316, T319, and T320, respectively. Genetic Type T356 is a cytoplasmically inherited yellow foliage mutant, but a symbol was not assigned (Palmer and Chen, 1996).

A nuclear-cytoplasmic interaction was noticed when cytoplasmic yellows cyt-Y2 through cyt-Y8 were crossed as female parents with either T235 or T323. The F_2 plants that were cyt-yellow y20 y20 Mdh1-n Mdh1-n (K2 K2 or k2 k2) were lethal (Palmer and Cianzio, 1985; Shoemaker et al., 1985; Palmer, 1992; Palmer and Minor, 1994). No lethality was observed in control crosses. The critical cross-pollinations could not be made because the nuclear yellow genotype, y20 y20, does

Table 5–11. Cytoplasmic factors affecting chlorophyll deficiency, retention, or morphology in soybean.

Gene	Phenotype	Strain†	Reference
	1. Chlorophyll deficiency		
cyt-G2	Normal	c	Palmer and Mascia (1980)
cyt-Y2	Yellow leaves, becoming yellowish-green	T275	
cyt-G3	Normal	c	Shoemaker et al. (1985)
cyt-Y3	Yellow leaves, very weak plant, (mutable plants are chlorophyll chimeras)	T278M	
cyt-G4	Normal	c	Cianzio and Palmer (1992)
cyt-Y4	Yellow leaves	T314	
cyt-G5	Normal	c	Cianzio and Palmer (1992)
cyt-Y5	Green-yellow leaves	T315	
cyt-G6	Normal	c	Cianzio and Palmer (1992)
cyt-Y6	Yellow leaves, vigorous	T316	
cyt-G7	Normal	c	Cianzio and Palmer (1992)
cyt-Y7	Yellow leaves, weak	T319	
cyt-G8	Normal	c	Cianzio and Palmer (1992)
cyt-Y8	Green-yellow leaves	T320	
	2. Chlorophyll retention		
cyt-G1	Green seed embryo	T104, Medium Green	Terao (1918), Veatch and Woodworth (1930)
cyt-Y1	Yellow seed embryo	c	
	3. Morphology		
cyt-W1	Wrinkled leaves	Somaclonal variant in Asgrow A3127	Stephens et al. (1991)

† c = Indicates that the gene occurs in many cultivars.

not exist by itself. The *y20 y20* genotype has only been found with *Mdh1-n Mdh1-n* with *K2 K2* or with *k2 k2*.

Two maternally-inherited wrinkled-leaf mutations have been identified in tissue-culture-derived soybean. The cytoplasmic-inherited mutant in cv. Asgrow A3127 was assigned *cyt-W1*, but was not given a Genetic Type Collection number (Stephens et al., 1991). The cytoplasmic-inherited mutant in cv. Clark 63 was not given a gene symbol, but is maintained as T355 in the Genetic Type Collection (Nguyen and Widholm, 1997).

A summary of cytoplasmic factors affecting chlorophyll deficiency, retention, or morphology in soybean is given in Table 5–11.

5–5.10 Pigmentation

5–5.10.1 Flower

Most soybean introductions or cultivars have either purple (*W1*) or white (*w1*) flowers, but tremendous variation is evident among flower color of wild perennial *Glycine* species. There is a pleiotropic effect of the *W1* allele, in that purple pigmentation is evident both in the flower and seedling hypocotyl. The seedling hypocotyls with the *w1* allele are green and the plants have white flowers. The gene action of loci that condition anthocyanin pigmentation in soybean flowers and hypocotyls was evaluated in cvs. Clark and Harosoy. A dosage effect was detected for the *W1* alleles but not for the *W4* alleles (Groose and Palmer, 1991).

Mutable plants that produced both entirely near-white and entirely purple flowers, as well as flowers of a mutable phenotype with purple sectors on near-white petals, were identified (Groose et al., 1988). This line carries an unstable recessive (mutable) allele at the *w4* locus that reverts at high frequency from the recessive form to a stable dominant form (Groose et al., 1990). The mutant gene was designated *w4-m* (*w4*-mutable) and the mutant line was assigned Genetic Type Collection number T322 (Palmer et al., 1990a).

A variant plant of the *w4*-mutable line (T322) had flowers that were phenotypically similar to the dilute-purple phenotype described by Hartwig and Hinson (1962). Genetic analyses established that the mutation was conditioned by a new allele at the *w4* locus, which was assigned the gene symbol *w4-dp,* and this line was designated Genetic Type Collection number T321 (Palmer and Groose, 1993).

Mapping studies with the *w3* locus and anthocyanin genes revealed an absolute cosegregation with dihydroflavonal 4-reductase (Fasoula et al., 1995). It was reasoned that either the *W3* locus encodes dihydroflavonal 4-reductase or that there is a tight coupling phase linkage of dihydroflavonal 4-reductase with the *W3* locus (Fasoula et al., 1995).

Stephens and Nickell (1992) reported a pink-flower mutant that was inherited as a single recessive gene. The mutant was assigned gene symbol *wp*, but was not added to the soybean Genetic Type Collection. The *wp* locus was located on the USDA-ARS-ISU molecular map on linkage group D1b + W (Hegstad et al., 2000b).

A novel variegated flower phenotype was derived from the *wp* line. Flowers were pink and purple sectors or all pink and all purple flowers on the same plant (Johnson et al., 1998). Some of the all pink- and all purple-flowered plants were true breeding, whereas other plants were unstable (mutable) and switched back to other phenotypes when allowed to self-pollinate. During outcrossing, however, the mutability of *wp-m* was not transmissible and stabilized to the recessive pink phenotype in the F_2 plants (Johnson et al., 1998). These data suggested that epigenetic silencing of the mutability was occurring in different genetic backgrounds.

Agronomic traits were measured in stable flower color lines that were derived from *wp-m* (Hegstad et al., 2000a). The majority of purple-flowered revertant lines, derived from a pink-flowered source, were lower in seed protein content than stable pink sister lines. Several pink-flowered lines, derived from a purple-flowered source, however, were significantly higher in seed protein content than stable purple sister lines (Stephens et al. 1993b; Hegstad et al., 2000a). The *wp* allele apparently exerts an influence on both seed protein accumulation and the anthocyanin pathway.

5–5.10.2 Pubescence

The *T* allele has a pleiotropic effect upon seedling hypocotyl color when plants are grown in continuous light. The *T* allele is necessary to produce bronze pigment on seedling hypocotyls in *w1 w1* genotypes, but the *t* allele has minimal effect (Palmer and Payne, 1979). Peters et al. (1984) showed that the *T* allele partially substitutes for the *W1* allele by producing the same pigment, but in lower amounts, in bronze hypocotyls as in purple hypocotyls. The *T* allele does not substitute for the *W4* allele (Groose and Palmer, 1991).

Genetic Type T236 has red-buff seed. In certain cross-combinations with T236 in the F_2 generation, the phenotypic seed coat/hilum colors were not consistent with expectations. Seo et al. (1993) showed that the red-buff seed coat of T236 was due to an allele of the *T* locus, and was assigned gene symbol *t-r*. The red-buff phenotype of *t-r* occurs only in a *R-w1 w1* genetic background. A general scheme outlining the genes involved, and the seed coat, saddle, and hilum color expected, is given by Palmer and Kilen (1987).

The locus is presumed to encode flavonoid 3`-hydroxylase (Toda et al., 2002). Molecular studies of near-isogenic lines, *T T* (tawny pubescence) and *t t* (gray pubescence), showed a single base deletion in the coding region that resulted in a truncated polypeptide (Toda et al., 2002). These authors showed in an F_2 population, cosegregation of the deletion with the *T* locus. Toda et al. (2002) concluded that gray pubescence was the result of a deletion at the *T* locus.

5–5.10.3 Seed

Hilum color and seed-coat color serve as useful marker genes for distinguishing between hybrid and self-pollinated progeny when making cross-pollinations. Williams (1952), Specht and Williams (1978), and Palmer and Stelly (1979) summarized the phenotypes as follows:

Genes	Seed coat and hilum self-color	Saddle and hilum color	Hilum color	Hilum color
	i	*i-k*	*i-i*	*I*
T R	Black	Black	Black	Gray
T r O	Brown	Brown	Brown	Yellow†
T r o	Red brown	Red brown	Red brown	Yellow†
t R W 1	Imperf. black	Imperf. black	Imperf. black	Gray
t R w1	Buff	Buff	Buff	Yellow
t r	Buff	Buff	Buff	Yellow

† Sometimes called imperfect yellow (Cober et al., 1998).

Commercially grown soybean cultivars have seed coats that are either completely yellow or with pigmentation restricted to the hila. Lines with completely pigmented seed coats are usually recessive variants at the *I* locus. Wilcox (1988) reported agronomic tests with near-isogenic lines differing in seed coat pigmentation. There were few differences in agronomic performance among pure-stand comparisons. Interestingly, in three of six blends a significantly higher reproductive rate was noticed for the *i-i* allele vs. the *i* allele.

Very little is known about the biochemical genetics of seed coat color in soybean. Buzzell et al. (1987) reported that the *T* allele is involved in the dihydroxylation of the flavonol B-ring in the formation of cyanidin-3-glucoside, and this was confirmed by Toda (2002). The *W1* allele is involved in trihydroxylation in the formation of delphiniden-3-glucoside. The production of delphiniden is enhanced in the presence of *T*. The *T* locus is thus pleiotropic for hydroxylation of anthocyanidins and flavonols, whereas *W1* is not (Buzzell et al., 1987).

The *I* locus (inhibitor) controls the presence or absence as well as the spatial distribution (*i-i, i-k,* and *i* alleles) (Palmer and Kilen, 1987) of anthocyanins and proanthocyanidins in the epidermal layer of the soybean seed coat (Todd and Vodkin, 1993). Analyses of spontaneous mutations at the *I* locus showed that this locus comprises a region of chalcone synthase duplications (Todd and Vodkin, 1996). Furthermore, the nature of the *I* alleles, and their unusual effect on chalcone synthase expression, indicated that this locus may be a naturally occurring example of homology-dependent gene silencing (Todd and Vodkin, 1996).

Soybean cultivars with the combination of tawny pubescence (*TT*) and yellow hilum (*II*) may exhibit seed discoloration (Takahashi and Asanuma, 1996). Temperatures below 15°C during seed development can result in discolored hila and seed coats (Srinivasan and Arihara, 1994; Takahashi and Abe, 1994; Takahashi and Asanuma, 1996). Also, soybean cultivars with the *TT II* genotype may have very light brown hila. Cober et al. (1998) suggested that hila in cultivars with the *TT II rr* genotype be called imperfect yellow. This designation was approved by the Soybean Genetics Committee.

The *R* locus is involved in the control of black-and-brown anthocyanin pigments in the seed coat. The *r-m* (mutable) allele conditions black spots or stripes on a brown seed coat. Entire branches of seed with brown, black, or mixed are the result of early mutational events. Late mutational events affect single pods or seeds within pods resulting in mixed phenotypes (Chandlee and Vodkin, 1989c). Germinal revertants resulted in all black-seeded plants with brown hila (R*) or all brown-seeded plants (r*). Subsequent generations from revertant lines showed continued instability. During continued self-pollinations, these derivatives of the *r-m* allele switched back and forth in a random fashion and generated all three phenotypes (Chandlee and Vodkin, 1989c).

A summary of genes affecting pigmentation in soybean is given in Table 5–12.

5–5.11 Isoenzymes and Proteins

The 1987 second edition of *Soybeans: Improvement, Production, and Uses* provided a summary of named alleles of genes encoding seed proteins and isozymes. This section updates the information found in Table 5–14 of the 2nd edition. Genes and alleles confirmed by genetic tests and gene symbols approved by the Soybean Genetics Committee are presented in Table 5–13 while reports of variants that have not received official gene symbols are mentioned in this section (5–5.11).

Isozymes are analyzed in starch, polyacrylamide, or an acrylamide/starch combination gel electrophoresis. Extraction, electrophoresis, and staining procedures have been provided in detail (Cardy and Beversdorf, 1984a; Bult et al., 1989a; Rennie et al., 1989; Hedges et al., 1992). Isozymes are detected as either mobility variants or as null variants. Because isozymes exist as monomeric enzymes or as dimeric or higher-order enzyme complexes, banding patterns can be complex and thus difficult to interpret genetically. Also, in many multiple isozyme zymograms, some isozymes are monomorphic. Isozymes can result from multiple alleles of one gene or from multiple genes encoding enzymes with similar functions.

Table 5–12. Genes affecting pigmentation in soybean.

Gene	Phenotype	Strain†	Reference
		1. Flower	
W1	Purple	c	Takahashi and Fukuyama (1919), Woodworth (1923)
w1	White	c	
W3 w4	Dilute purple	L70-4422 [L66 X (Laredo X Harosoy)], Laredo	Hartwig and Hinson (1962)
w3 W4	Purple	c	
w3 w4	Near white	L68-1774 [L66 X (Laredo X Harosoy)]	
w4-dp	Dilute purple	T321	Palmer and Groose (1993)
w4-m	Mutable flower	T322	Palmer et al. (1990a)
Wm	Purple (glycosides present)	c	Buzzell et al. (1977)
wm	Magenta (glycosides absent)	T235	
Wp	Purple	c	Stephens and Nickell (1992)
wp	Pink	(Sherman X Asgrow A2943) X Elgin 87	
wp-m	Mutable flower	(Sherman X Asgrow A2943) X Elgin 87	Johnson et al. (1998)
		2. Pubescence	
T	Tawny (brown); quercetin and kaempferol present	c	Piper and Morse (1910), Nagai (1921) (as C c)
t	Gray; quercetin absent, kaempferol present	c	Woodworth (1921), Williams (1950), Buttery and Buzzell (1973)
t-r	Red-buff seed coat	T236	Seo et al. (1993)
Td	Tawny (brown); flavonol present	Clark	Buttery and Buzzell (1973), Bernard (1975b)
td	Light tawny (near-gray); flavonol absent	Grant, Sooty	
		3. Seed	
I	Light hilum	Mandarin, c	Nagai (1921), Nagai and Saito (1923), Owen (1928), Woodworth (1932, 1933), Mahnud and Probst (1953)
i-i	Dark hilum	Manchu, c	
i-k	Saddle pattern	Black Eyebrow	
i	Self dark seed coat	Soysota, T157	
Im	Nonmottled seed	Merit	Cooper (1966)
im	Dark mottled seed (with SMV infection)	Harosoy	
K1	Nonsaddle	c	Takagi (1929, 1930), Williams (1958)
k1	Dark saddle on seed coat	Kura, T153, Agate	
K2	Yellow seed coat	c	Rode and Bernard (1975a), Palmer (1984b)

(continued on next page)

Table 5–12. Continued.

Gene	Phenotype	Strain†	Reference
k2 (Urbana 1)	Tan saddle on seed coat	T239	
k2 (Columbia 2)	Tan saddle on seed coat	Clark (L67-3483)	Rode and Bernard (1975a)
k2 (Urbana 1) [*Mdh1-n* (Urbana 1) *y20* (Urbana 1)]	Tan saddle on seed coat	T253	Chen and Palmer (1996)
k2 (Columbia 1) [*Mdh1-n* (Columbia 1)]‡	Tan saddle on seed coat	T261	Chen and Palmer (1996)
k2 (Urbana 1) [*Mdh1-n* (Ames 7) *y20* (Ames 5)]	Tan saddle on seed coat	T334	Chen and Palmer (1998a)
k2 (Urbana 1) [*Mdh1-n* (Ames 8) *y20* (Ames 6)]	Tan saddle on seed coat	T335	Chen and Palmer (1998a)
k2 (Urbana 1) [*Mdh1-n* (Ames 9) *y20* (Ames 7)]	Tan saddle on seed coat	T336	Chen and Palmer (1998a)
k2 (Urbana 1) [*Mdh1-n* (Ames 10) *y20* (Ames 8)]	Tan saddle on seed coat	T337	Chen and Palmer (1998a)
k2 (Urbana 1) [*Mdh1-n* (Ames 11) *y20* (Ames 9)]	Tan saddle on seed coat	T338	Chen and Palmer (1998a)
k2 (Urbana 1) [*Mdh1-n* (Ames 12) *y20* (Ames 10)]	Tan saddle on seed coat	T339	Chen and Palmer (1998a)
k2 (Columbia 1) [*Mdh1-n* (Columbia 1) *y20* (Ames 11)]	Tan saddle on seed coat	T340	Chen and Palmer (1998a)
k2 (Urbana 1) [*Mdh1-n* (Ames 13) *y20* (Ames 12)]	Tan saddle on seed coat	T341	Chen and Palmer (1998a)
k2 (Urbana 1) [*Mdh1-n* (Ames 14) *y20* (Ames 13)]	Tan saddle on seed coat	T342	Chen and Palmer (1998a)
k2 (Urbana 1) (Ames 15) *y20* (Ames 14)]	Tan saddle on seed coat	T343	Chen and Palmer (1998a)
k2 (Urbana 1) [*Mdh1-n* (Ames 16) *y20* (Ames 15)]	Tan saddle on seed coat	T344	Chen and Palmer (1998a)
k2 (Urbana 1) [*Mdh1-n* (Ames 17) *y20* (Ames 16)]	Tan saddle on seed coat	T345	Chen and Palmer (1998a)
k2 (Urbana 1) [*Mdh1-n* (Ames 6) *y20* (Ames 18)]	Tan saddle on seed coat	T347	Chen et al. (1999)
k2 (Urbana 1) [*Mdh1-n* (Ames 6) *y20* (Ames 19)]	Tan saddle on seed coat	T348	Chen et al. (1999)

(continued on next page)

Table 5–12. Continued.

Gene	Phenotype	Strain†	Reference
k2 (Urbana 1) [Mdh1-n (Ames 6) y20 (Ames 20)]	Tan saddle on seed coat	T349	Chen et al. (1999)
k2 (Urbana 1) [Mdh1-n (Ames 6) y20 (Ames 21)]	Tan saddle on seed coat	T350	Chen et al. (1999)
k2 (Urbana 1) [Mdh1-n (Ames 20) y20 (Ames 22)]	Tan saddle on seed coat	T351	Chen et al. (1999)
K3	Nonsaddle	c	Bernard and Weiss (1973)
k3	Dark saddle on seed coat	T238	
O	Brown seed coat	Soysota, c	Nagai (1921), Weiss (1970b)
o	Reddish-brown seed coat	Ogemaw	
R	Black seed coat	c	Nagai (1921), Woodworth (1921), Stewart (1930), Williams (1952)
r-m	Black stripes on brown seed	PI 91073, T146	Nagai and Saito (1923), Weiss (1970b)
r	Brown seed coat	c	
		4. Pod	
L1 L2	Black pod	Seneca	Bernard (1967)
L1 l2	Black pod	PI 85505	
l1 L2	Brown pod	Clark, c	
l1 l2	Tan pod	Dunfield, c	

† c = Indicates that the gene occurs in many cultivars.
‡ Genes for secondary traits of interest are listed in brackets.

Isozymes were the initial molecular markers. As such they have been used in soybean cultivar identification (Cardy and Beversdorf, 1984b; Doong and Kiang, 1987c), for verification of F_1 hybrid seeds (Miroslav and Jiri, 1996), in diversity analyses (Zakharova et al., 1989; Kwon et al., 1990; Perry and McIntosh, 1991; Perry et al., 1991; Chung et al., 1995; Griffin and Palmer, 1995; Cerna et al., 1997; Park and Yoon, 1997; Hirata et al., 1999), and to test for association with agronomic traits (Graef et al., 1989; Suárez et al., 1991). While there was a flurry of reports on isozyme inheritance and use prior to 1993, the advent of DNA marker polymorphisms has led to lessened efforts on isozyme discovery.

Where appropriate, the isozymes described in this section are accompanied by Enzyme Commission (EC) numbers, as suggested by the International Union of Biochemists. The numbers serve to identify more precisely the enzyme activity under discussion. A summary of genes involved in isoenzymes and proteins in soybean is given in Table 5–13.

5–5.11.1 Aconitase (EC 4.2.1.3)

Homozygous lines generally display a five-band (zones of activity) isozyme pattern. The observed banding patterns in heterozygous individuals indicate that the aconitase isozymes are monomers. The five zones of activity result from five genes

Table 5–13. Genes controlling inheritance of isoenzymes and protein variants in soybean

Gene	Phenotype	Strain†	Reference
Ap-a	Acid phosphatase mobility variant	Ebony	Gorman and Kiang (1977), Hildebrand et al. (1980)
Ap-b	Acid phosphatase mobility variant	Amsoy 71	
Ap-c	Acid phosphatase mobility variant	Earlyana, Manchu	
Aco1-a	Aconitase mobility variant	Evans	Griffin and Palmer (1987a), Kiang and Bult (1991)
Aco1-b	Aconitase mobility variant	PI 257430	
aco1-n	Aconitase null	Semmes	
Aco2-a	Aconitase mobility variant	PI 464918	Doong and Kiang (1987b), Rennie et al. (1987a)
Aco2-b	Aconitase mobility variant	Evans	
Aco2-bn	Aconitase null	T318	Amberger et al. (1992)
Aco2-c	Aconitase mobility variant	PI 407223	Kiang and Bult (1991)
Aco3-a	Aconitase mobility variant	Evans	Griffin and Palmer (1987a)
Aco3-b	Aconitase mobility variant	PI 437728	
Aco4-a	Aconitase mobility variant	Williams	Griffin and Palmer (1987a)
Aco4-b	Aconitase mobility variant	Evans	
Aco4-c	Aconitase mobility variant	Peking	
Aco4-d	Aconitase mobility variant	PI 407223	
Aco5-a	Aconitase mobility variant	PI 486220	Kiang and Bult (1991)
Aco5-b	Aconitase mobility variant	PI 487341	
aco5-n	Aconitase null	PI 546104	
Adh1	Alcohol dehydrogenase present	Altona, Wilson	Gorman and Kiang (1978), Kiang and Gorman (1983)
adh1	Alcohol dehydrogenase absent	A-100, Lindarin	
Adh2	Alcohol dehydrogenase present	Amsoy, Beeson	Yu and Kiang (1993b)
adh2	Alcohol dehydrogenase absent	Cayuga, Grant	
Adh3	Alcohol dehydrogenase present	PI 564547	Yu and Kiang (1993b)
adh3	Alcohol dehydrogenase absent	PI 562537	
Amy1	α-amylase band 1 present	Harosoy, Clark	Gorman and Kiang (1977, 1978), Kiang (1981)
amy1	α-amylase band 1 absent	Altona, PI 132201	
Amy2	α-amylase band 2 present	Harosoy, Clark	
amy2	α-amylase band 2 absent	Altona, PI 132201	
Sp1-a	β-amylase mobility variant	Amsoy, Evans	Larsen (1967), Larsen and Caldwell (1968), Orf and Hymowitz (1976), Gorman and Kiang (1977, 1978), Hymowitz et al. (1979), Hildebrand and Hymowitz (1980a, 1980b), Kiang (1981)
Sp1-b	β-amylase mobility variant	Williams, Century	Orf and Hymowitz (1976)
Sp1-c	β-amylase mobility variant	PI 464914	Gorman and Kiang (1977, 1978) Hymowitz et al. (1979)
Sp1-an	Seed protein band present, β-amylase activity weak or absent	Chestnut	
sp1	Seed protein band absent, β-amylase activity weak or absent	Altona, PI 132201, T293	Griffin and Palmer (1986)
Cgy1	β-conglycinin subunit α′ present	c	Kitamura et al. (1984)

(continued on next page)

Table 5–13. Continued.

Gene	Phenotype	Strain†	Reference
cgy1	β-conglycinin subunit α' absent	Keburi	
Cgy2-a	α subunit of β-conglycinin produced	Raiden	Davies et al. (1985)
Cgy2-b	α subunit mobility variant	PI 54608-1	
Cgy3	β' subunit of β-conglycinin produced	PI 81041-1	
cgy3	β' subunit of β-conglycinin absent	PI 253651	
Dia1-a	Diaphorase mobility variant	Evans, Elton	Gorman et al. (1983), Kiang and Gorman (1983)
Dia1-b	Diaphorase mobility variant	Cayuga, Kingston	
Dia2-a	Diaphorase mobility variant	BSR 101	Liao and Palmer (1997a)
Dia2-b	Diaphorase mobility variant	Wilson	
dia2-n	Diaphorase band absent	PI 567565	
Dia3	Diaphorase band present	Kingston	
dia3	Diaphorase band absent	Elton	
Enp-a	Endopeptidase mobility variant	Evans	Doong and Kiang (1987a), Griffin and Palmer (1987a), Rennie et al. (1987b)
Enp-b	Endopeptidase mobility variant	PI 257430	
Enp-c	Endopeptidase mobility variant	PI 504287	
Ep	High peroxidase activity	Harosoy 63	Buzzell and Buttery (1969)
ep	Low peroxidase activity	Blackhawk	
Est1-a	Esterase mobility variant	Hardee 2 (T289)	Bult and Kiang (1989)
Est1-b	Esterase mobility variant	AV68	
Eu1-a	Urease-embryo-specific mobility variant	Prize	Buttery and Buzzell (1971)
Eu1-b	Urease-embryo-specific mobility variant	Williams	Kloth and Hymowitz (1985), Holland et al. (1987)
eu1-sun	Embryo-specific urease absent	PI 229324	Kloth et al. (1987)
eu1-n4	Urease null—no mRNA	Williams (NMU mutant)	
eu1-n6	Urease—mRNA present, 5% of normal protein	Williams (NMU mutant)	Meyer-Bothling et al. (1987)
eu1-n7	Urease null—no mRNA	Williams (NMU mutant)	Polacco et al. (1989)
eu1-n8	Urease—mRNA present, 0.5% of normal protein	Williams (NMU mutant)	
Eu2	Urease-normal levels		
eu2	No ubiquitous urease, 0.6% embryo-specific urease	Williams (EMS mutant)	
Eu3	Urease-normal levels		
eu3-e1	Lacks both urease types	Williams (EMS mutant)	
Eu3-e3	Reduced levels of both urease types	Williams (EMS mutant)	
Eu4	Urease-normal levels		
eu4	Normal embryo urease, no ubiquitous urease	Williams (EMS mutant)	
Fle	Fluorescent esterase present	AK (FC 30671)	Doong and Kiang (1988)
fle	Fluorescent esterase absent	Emerald	
Got-a	Glutamate oxaloacetate transaminase mobility variant	PI 407194	Kiang et al. (1987)

(continued on next page)

Table 5–13. Continued.

Gene	Phenotype	Strain†	Reference
Got-b	Glutamate oxaloacetate transaminase mobility variant	PI 407221	
Got-c	Glutamate oxaloacetate transaminase mobility variant	PI 407207	
Gpd	Glucose-6-phosphate dehydrogenase present	Amsoy, Evans	Gorman et al. (1983), Kiang and Gorman (1983)
gpd	Glucose-6-phosphate dehydrogenase (weak)	Chestnut, Cayuga	
Gy1	Glycinin subunit G1 produced	Dare	Nielson et al. (1989)
Gy2	Glycinin subunit G2 produced	Dare	
Gy3	Glycinin subunit G3 produced	Dare	
Gy4-a	G4 subunit of glycinin present	c	Kitamura et al. (1984)
Gy4-b	Mobility variant of glycinin G4 subunit	PI 468916 [G. soja (Sieb. and Zucc.)]	Diers et al. (1994)
gy4	G4 subunit of glycinin absent	Raiden	Kitamura et al. (1984)
Gy5	Glycinin subunit G5 produced	Forrest	Nielsen et al. (1989)
Idh1-a	Isocitrate dehydrogenase mobility variant	Amsoy, Cayuga	Yong et al. (1981, 1982), Gorman et al. (1983), Kiang and Gorman (1983, 1985)
Idh1-b	Isocitrate dehydrogenase mobility variant	Wilson, Evans	
Idh2-a	Isocitrate dehydrogenase mobility variant	Amsoy, Cayuga	
Idh2-b	Isocitrate dehydrogenase mobility variant	Wilson, Evans	
Idh3-a	Isocitrate dehydrogenase mobility variant	Elton, Amsoy	
Idh3-b	Isocitrate dehydrogenase mobility variant	Agate, Wilson	
Lap1-a	Leucine aminopeptidase mobility variant	Norredo, Wilson	Gorman et al. (1982a, 1982b, 1983)
Lap1-b	Leucine aminopeptidase mobility variant	Lindarin	
Lap2	Leucine aminopeptidase present	Amsoy	Kiang et al. (1984)
lap2	Leucine aminopeptidase absent	Jefferson	
Le	Seed lectin present	Harosoy	Orf et al. (1978), Pull et al. (1978), Stahlhut and Hymowitz (1980)
le	Seed lectin absent	T102	
Lx1-a	Lipoxygenase 1 pI 5.85	Century	Hildebrand and Hymowitz (1981, 1982)
Lx1-b	Lipoxygenase 1 pI 5.79	L$_2$-3, PI 86023	Pfeiffer et al. (1993)
lx1	Lipoxygenase 1 absent	L$_1$-5, PI 408251	
Lx2	Lipoxygenase 2 present	c	Davies and Nielsen (1986, 1987)
lx2	Lipoxygenase 2 absent	Suzuyutaka (PI 86023)	
Lx3	Lipoxygenase 3 present	c	Kitamura et al. (1983)
lx3	Lipoxygenase 3 absent	Wase Natsu (PI 417458)	
Mdh1-a	Malate dehydrogenase present	Harosoy	Amberger et al. (1992)

(continued on next page)

Table 5–13. Continued.

Gene	Phenotype	Strain†	Reference
Mdh1-n (Columbia 1) [*k2*(Columbia 1)]‡	Malate dehydrogenase absent	T261	Chen and Palmer (1996)
Mdh1-n (Urbana 1) [*y20*(Urbana 1) *k2*(Urbana 1)]	Malate dehydrogenase absent	T253	Chen and Palmer (1996)
Mdh1-n (Ames 1) [*y20*(Ames 1)]	Malate dehydrogenase absent	T317	Amberger et al. (1992)
Mdh1-n (Ames 2) [*y20*(Ames 2)]	Malate dehydrogenase absent	T323	Hedges and Palmer (1992)
Mdh1-n (Ames 3) [*y20*(Ames 3)]	Malate dehydrogenase absent	T324	Hedges and Palmer (1992)
Mdh1-n (Ames 4) [*y20*(Ames 4)]	Malate dehydrogenase absent	T325	Hedges and Palmer (1992)
Mdh1-n (Ames 5)	Malate dehydrogenase absent	PI 567391	Chen and Palmer (1996))
Mdh1-n (Ames 6)	Malate dehydrogenase absent	PI 567630A	Chen and Palmer (1996)
Mdh1-n (Ames 7) [*y20*(Ames 5) *k2*(Urbana 1)]	Malate dehydrogenase absent	T334	Chen and Palmer (1998a)
Mdh1-n (Ames 8) [*y20*(Ames 6) *k2*(Urbana 1)]	Malate dehydrogenase absent	T335	Chen and Palmer (1998a)
Mdh1-n (Ames 9) [*y20*(Ames 7) *k2*(Urbana 1)]	Malate dehydrogenase absent	T336	Chen and Palmer (1998a)
Mdh1-n (Ames 10) [*y20*(Ames 8) *k2*(Urbana 1)]	Malate dehydrogenase absent	T337	Chen and Palmer (1998a)
Mdh1-n (Ames 11) [*y20*(Ames 9) *k2*(Urbana 1)]	Malate dehydrogenase absent	T338	Chen and Palmer (1998a)
Mdh1-n (Ames 12) [*y20*(Ames 10) *k2*(Urbana 1)]	Malate dehydrogenase absent	T339	Chen and Palmer (1998a)
Mdh1-n (Columbia 1) [*y20*(Ames 11) *k2*(Columbia 1)]	Malate dehydrogenase absent	T340	Chen and Palmer (1998a)
Mdh1-n (Ames 13) [*y20*(Ames 12) *k2*(Urbana 1)]	Malate dehydrogenase absent	T341	Chen and Palmer (1998a)
Mdh1-n (Ames 14) [*y20*(Ames 13) *k2*(Urbana 1)]	Malate dehydrogenase absent	T342	Chen and Palmer (1998a)
Mdh1-n (Ames 15) [*y20*(Ames 14) *k2*(Urbana 1)]	Malate dehydrogenase absent	T343	Chen and Palmer (1998a)

(continued on next page)

Table 5–13. Continued.

Gene	Phenotype	Strain†	Reference
Mdh1-n (Ames 16) [*y20*(Ames 15) *k2*(Urbana 1)]	Malate dehydrogenase absent	T344	Chen and Palmer (1998a)
Mdh1-n (Ames 17) [*y20*(Ames 16) *k2*(Urbana 1)]	Malate dehydrogenase absent	T345	Chen and Palmer (1998a)
Mdh1-n (Ames 18)	Malate dehydrogenase absent	Mandell	Chen et al. (1999)
Mdh1-n (Ames 19) [*y20*(Ames 17)]	Malate dehydrogenase absent	T346	Chen et al. (1999)
Mdh1-n (Ames 6) [*y20*(Ames 18) *k2*(Urbana 1)]	Malate dehydrogenase absent	T347	Chen et al. (1999)
Mdh1-n (Ames 6) [*y20*(Ames 19) *k2*(Urbana 1)]	Malate dehydrogenase absent	T348	Chen et al. (1999)
Mdh1-n (Ames 6) [*y20*(Ames 20) *k2*(Urbana 1)]	Malate dehydrogenase absent	T349	Chen et al. (1999)
Mdh1-n (Ames 6) [*y20*(Ames 21) *k2*(Urbana 1)]	Malate dehydrogenase absent	T350	Chen et al. (1999)
Mdh1-n (Ames 20) [*y20*(Ames 22) *k2*(Urbana 1)]	Malate dehydrogenase absent	T351	Chen et al. (1999)
Mdh1-n (Ames 21) [*y20*(Ames 23)]	Malate dehydrogenase absent	T234	Chen et al. (1999)
Mdh1-n (Ames 22) [*y20*(Ames 24)]	Malate dehydrogenase absent	T361	Palmer et al. (2000)
Mpi-a	Mannose-6-phosphate isomerase mobility variant	Kingwa	Gorman et al.(1983), Kiang and Gorman (1983), Chiang and Kiang (1988)
Mpi-b	Mannose-6-phosphate isomerase mobility variant	Amsoy	
Mpi-c	Mannose-6-phosphate isomerase mobility variant	Disoy	
Mpi-d	Mannose-6-phosphate isomerase mobility variant	PI 407192	
Mpi-e	Mannose-6-phosphate isomerase mobility variant	PI 562547	Yu and Kiang (1993b)
mpi	Mannose-6-phosphate isomerase absent	Earlyana	Chiang and Kiang (1988)
Pgd1-a	Phosphogluconate dehydro-genase mobility variant	Agate	Gorman et al. (1983), Kiang and Gorman (1983), Chiang and Kiang (1987)
Pgd1-b	Phosphogluconate dehydro-genase mobility variant	Elton	

(continued on next page)

Table 5–13. Continued.

Gene	Phenotype	Strain†	Reference
Pgd1-c	Phosphogluconate dehydrogenase mobility variant	PI 407160	
pgd1	Phosphogluconate dehydrogenase absent	Hidaka-1 (PI 406684)	
Pgd2-a	Phosphogluconate dehydrogenase mobility variant	PI 487428	
Pgd2-b	Phosphogluconate dehydrogenase mobility variant	PI 407192	
Pgd2-c	Phosphogluconate dehydrogenase mobility variant	PI 407252	
Pgd3-a	Phosphogluconate dehydrogenase mobility variant	PI 486220	
Pgd3-b	Phosphogluconate dehydrogenase mobility variant	PI 487331	
Pgi1-a	Phosphoglucose isomerase mobility variant	PI 424032	Chiang et al. (1987)
Pgi1-b	Phosphoglucose isomerase mobility variant	Beeson	
pgi1	Phosphoglucose isomerase band absent	PI 157401	
Pgi2	Phosphoglucose isomerase mobility variant	Lindarin	Chiang et al. (1987)
pgi2	Phosphoglucose isomerase band absent	PI 157401	
Pgi3-a	Phosphoglucose isomerase mobility variant	PI 407260	Chiang et al. (1987)
Pgi3-b	Phosphoglucose isomerase mobility variant	AV68	
Pgm1-a	Phosphoglucomutase mobility variant	Chestnut, Wells	Gorman et al.(1983), Kiang and Gorman (1983)
Pgm1-b	Phosphoglucomutase mobility variant	Amsoy, Hark	
Pgm2-a	Phosphoglucomutase mobility variant	PI 423990	Yu and Kiang (1993b)
Pgm2-b	Phosphoglucomutase mobility variant	Amsoy	
Pgm2-c	Phosphoglucomutase mobility variant	PI 562547	Yu and Kiang (1993b)
Pgm2-d	Phosphoglucomutase mobility variant	PI 562550	
Pgm3	Phosphoglucomutase mobility variant	?	
pgm3	Phosphoglucomutase band absent	?	
Pi1	Trypsin inhibitor present	c	Kollipara et al. (1996)
pi1	Trypsin inhibitor absent	PI 440998 [*G. tomentella* (Hayata)]	
Pi2	Trypsin inhibitor present	c	Kollipara et al. (1996)
pi2	Trypsin inhibitor absent	PI 373987 [*G. tomentella* (Hayata)]	
Pi3	Bowman-Birk inhibitor band BBI′ present	c	Kollipara et al. (1996)

(continued on next page)

Table 5–13. Continued.

Gene	Phenotype	Strain†	Reference
pi3	Bowman-Birk inhibitor band BBI′ absent	PI 440998 [G. tomentella (Hayata)]	
Sdh-a	Shikimate dehydrogenase mobility variant	PI 562533	Yu and Kiang (1993b)
Sdh-b	Shikimate dehydrogenase mobility variant	PI 562554	
Sod1	Superoxide dismutase bands 4 and 5 present	c	Gorman and Kiang (1978), Gorman et al. (1982b, 1984), Griffin and Palmer (1984, 1989)
sod1	Superoxide dismutase bands 4 and 5 absent	Evans	
Sod2-a	Superoxide dismutase mobility variant	Polysoy	Griffin and Palmer (1989)
Sod2-b	Superoxide dismutase mobility variant	Evans	
Ti-a	Kunitz trypsin inhibitor mobility variant	Harosoy	Singh et al. (1969), Hymowitz and Hadley (1972), Orf and Hymowitz (1977, 1979)
Ti-b	Kunitz trypsin inhibitor mobility variant	Aoda	
Ti-c	Kunitz trypsin inhibitor mobility variant	PI 86084	
Ti-x	Kunitz trypsin inhibitor mobility variant	?	Zhao et al. (1995b)
ti	Kunitz trypsin inhibitor absent	PI 157440 (Kin-du)	

† c indicates that the gene occurs in many cultivars.
‡ Genes for secondary traits of interest are listed in brackets.

Aco1, Aco2, Aco3, Aco4, and *Aco5* with *Aco1* controlling the least mobile band and *Aco5* controlling the most mobile band. Mobility variants of the five aconitase isozymes are conditioned by independent loci with codominant alleles [*Aco1-(a, b)*; *Aco2-(a, b, c)*; *Aco3-(a, b)*; *Aco4-(a, b, c, d)*; *Aco5-(a, b)*] (Doong and Kiang, 1987b; Griffin and Palmer, 1987a; Rennie et al., 1987a; Kiang and Bult, 1991). Recessive null alleles, *aco1-n* and *aco5-n*, each eliminating a specific isozyme band, have been identified in soybean accessions (Kiang and Bult, 1991), and *aco2-bn*, in plants regenerated via somatic embryogenesis (Amberger et al., 1992).

5–5.11.2 Alcohol Dehydrogenase (EC 1.1.1.1)

Alcohol dehydrogenase loci *Adh1* and *Adh2* were described in the previous soybean monograph, each having a dominant allele for the presence of specific isozyme band(s) and a recessive null allele that reduces the number of isozymes.

A third alcohol dehydrogenase locus was described from the wild annual species *G. soja* (Sieb. and Zucc.) (Yu and Kiang, 1993b) that also has a dominant functional allele (*Adh3*) and a recessive null allele (*adh3*). Five isozymes were present, and anodal band four was lacking in seeds homozygous for *adh3*. All six bands were present with *Adh3*.

5–5.11.3 Diaphorase (EC 1.6.4.3)

Three diaphorase loci have been characterized in soybean (Gorman et al., 1983; Kiang and Gorman, 1983) with two codominant alleles at *Dia1* and *Dia2* and a recessive null allele *dia3* described in the previous soybean monograph. *Dia1* produces a mitochondrial diaphorase, while *Dia2* and *Dia3* encode cytoplasmic enzymes.

The diaphorase pattern of accession PI 567565, characterized as a *Dia2* null mutant, lacks the two bands of intermediate mobility seen in the *Dia2-a* homozygous genotype. A single gene encodes the two bands, and seeds homozygous for the recessive allele *dia2-n* (proposed gene symbol) lack both bands (Liao and Palmer, 1997a). Abe et al. (1992) proposed a third codominant allele for *Dia1* (*Dia1-c*) and a fourth diaphorase locus *Dia4* with a null allele.

5–5.11.4 Endopeptidase (EC 3.4.9.9)

Endopeptidases are proteolytic enzymes that break polypeptides into smaller fragments, defined in these reports as an enzyme capable of hydrolyzing α-N-benzoyl-DL-arginine-β-naphthylamide HCl. Four homozygous endopeptidase zymograms were observed (Doong and Kiang, 1987a), but only three were analyzed genetically. Heterozygotes showed two parental bands with no intermediate band, indicating that the enzyme exists as a monomer. Three bands were controlled by a single locus with three codominant alleles, *Enp-a*, *Enp-b*, and *Enp-c* (Doong and Kiang, 1987a; Griffin and Palmer, 1987a). *Enp-a* has the slowest mobility while *Enp-c* has the fastest mobility. The *Enp-c* allele has only been found in *G. soja*. A fourth codominant allele at the *Enp* locus was suspected to control the fourth zymogram band. Liao et al. (1996) reported an endopeptidase null in PI 567365 from China that was inherited as a single gene with the null allele recessive to *Enp-a*. An official gene symbol, however, has not been assigned.

5–5.11.5 Esterase (EC 3.1.1.1 and EC 3.1.1.2)

An eight-band (seven anodal bands and one cathodal band) zymogram was detected in homozygous seeds (Bult and Kiang, 1989). Both carboxylesterases and arylesterases contributed to this pattern. Variation in mobility was seen in the one cathodal zone. The two cathodal mobility variants were controlled by two codominant alleles of one gene, *Est1-a* for the slower-migrating band, and *Est1-b* for the faster-migrating band. Abe et al. (1992) observed a null form for this cathodal isozyme, but a genetic test was not reported.

5–5.11.6 Peroxidase

The peroxidase test for soybean is a standard assay used in the identification of cultivars. Germplasm can be divided into two groups, based upon the presence of high (*Ep*) or low (*ep*) seed coat peroxidase activity (Buzzell and Buttery, 1969). High or low activity is based upon the observed color change associated with oxidized guaiacol (Buttery and Buzzell, 1968). Seed coat peroxidase activity also can be measured using the soybean peroxidase capture assay (SPCA), which uses a monoclonal antibody to isolate the enzyme (Vierling and Wilcox, 1996). Vierling et al. (1998) reported that testing of soybean seed coat peroxidase is more accurately determined using the SPCA than the guaiacol method. Molecular analyses indicated that the low level of peroxidase accumulation in *ep ep* seed coats is due to a mutation of the structural gene that reduces transcript abundance (Gijzen, 1997).

5–5.11.7 Urease (EC 3.5.1.5)

Buttery and Buzzell (1971) reported that soybean seed urease isozymes were inherited as a dominant (*Eu*, fast migration) and recessive (*eu*, slow migration) allele pair. Kloth and Hymowitz (1985) showed the inheritance to be codominant and renamed the gene symbols *Eu1-a* (slow form, hexamer) and *Eu1-b* (fast form, trimer). The different conclusions resulted from the extraction buffer used; the hexameric form in heterozygotes is unstable when extracted with water and converted to the trimeric form.

Kloth et al. (1987) found four Japanese soybean cultivars lacking urease. The locus was originally defined as *sun*, linked 1 cM from *Eu1*. The data, however, were also consistent with *Eu1* and *sun* being separate parts of one locus. Soybean mutants were studied to clarify this. In the process, two distinct ureases were discovered, an embryo (seed)-specific urease produced by *Eu1* and a ubiquitous (leaf) urease. Multiple alleles at the *Eu1* locus indicated *Eu1* and *sun* were at the same locus (Meyer-Bothling and Polacco, 1987). Recessive allele *eu2* eliminates the ubiquitous urease, and the embryo-specific urease is greatly reduced. Three alleles have been identified at the *Eu3* locus (Meyer-Bothling et al., 1987). The dominant allele *Eu3* produces normal embryo-specific and ubiquitous urease levels compared with recessive allele *eu3-e1* that lacks both urease activities. Allele *Eu3-e3* produces 0.1% ubiquitous and embryo-specific urease activity when homozygous, but when heterozygous with *Eu3*, 5 to 10% of the normal urease activity results. Therefore, *Eu3-e3* is considered a partial dominant allele. Recessive allele *eu4* eliminates the ubiquitous urease; the embryo-specific urease is unaffected (Polacco et al., 1989).

5–5.11.8 Fluorescent Esterase (EC 3.1.1.2)

Two zymogram types were reported for a fluorescent esterase (Doong and Kiang, 1988). Type 1 had five bands, while type 2 had four bands, the least mobile band was lacking. The lack of band 1 is inherited as a single gene recessive (*fle*) trait. Heterozygotes showed five anodal bands, however, the intensity of the first band was weaker than in an individual homozygous for *Fle*.

5–5.11.9 Glutamate Oxaloacetic Transaminase (EC 2.6.1.1)

A four-band zymogram was detected (Kiang et al., 1987). Three bands produced by the cytosolic enzyme were constant. The fastest-migrating band consisted of the plastid-associated enzyme and had three mobility variants. These three variants were produced by three codominant alleles of one gene: *Got-a*, the slowest-migrating band; *Got-b* and *Got-c*, the fastest-migrating band. *Got-b* is the most common allele. *Got-a* and *Got-c* were identified in South Korean accessions of *G. soja* with a frequency of 0.02.

5–5.11.10 Lipoxygenase (EC 1.13.11.12)

Soybean seeds contain three lipoxygenase isozymes: lipoxygenase-1, lipoxygenase-2, and lipoxygenase-3. Each enzyme can be eliminated in an individual homozygous for a recessive null allele (*lx1, lx2,* or *lx3*), reviewed in the previous soybean monograph. A second allele for a functional lipoxygenase-1 was reported (Pfeiffer et al., 1993). Allele *Lx1-b* is codominant with *Lx1-a* (renamed from *Lx1*), and the *Lx1-b* isozyme has a more acidic isoelectric point than the *Lx1-a* isozyme. The *Lx1-b* allele was first identified tightly linked to the lipoxygenase-2 null allele, *lx2.* It was later also identified in cv. McCall, tightly linked to the *Lx2* allele.

5–5.11.11 NAD-Dependent Malate Dehydrogenase (EC 1.1.1.37)

Malate dehydrogenase exists as a dimeric enzyme. The interaction of two duplicate loci produces three isozymes, two homodimers, and a heterodimer with different electrophoretic mobilities. Null variants in one gene, *Mdh1*, have been described. Plants homozygous for a null allele, *Mdh1-n*, produce only one isozyme instead of three. The active allele is designated *Mdh1-a* (Hedges and Palmer, 1992).

Numerous *Mdh1* null alleles have been reported, as *Mdh1* is located in an unstable chromosomal region that also contains the *k2* and *y20* genes. The 24 described *Mdh1-n* alleles, each a unique and separate occurrence, are recessive and allelic. Mutations from *Mdh1-a* to *Mdh1-n* have occurred both independently and concurrently with mutations at *k2* and *y20*. The *Mdh1-n* allele has been found in germplasm accessions [*Mdh1-n*(Ames 18) (cv. Mandell), *Mdh1-n*(Ames 5) (PI 567391), *Mdh1-n*(Ames 6) (PI 567630A)], as spontaneous mutations [*Mdh1-n*(Urbana 1) (T253), *Mdh1-n*(Columbia 1) (T261)], induced in tissue culture [*Mdh1-n*(Ames 1) (T317), *Mdh1-n*(Ames 22)], and induced in genetic backgrounds that activate transposable elements [*Mdh1-n*(Ames 2) (T323), *Mdh1-n*(Ames 3) (T324), *Mdh1-n*(Ames 4) (T325), *Mdh1-n*(Ames 19) (T346)] (Amberger et al., 1992; Hedges and Palmer, 1992; Chen and Palmer, 1996; Chen et al., 1999). It has been proposed that the high frequency of mutation at this chromosomal region is due to transposable elements generating chromosome deletions of varying sizes (Chen and Palmer, 1998a,1998b).

A mobility MDH variant was identified, but no gene symbol or Genetic Type Collection number was assigned. A percentage recombination value of 15.2 + 3.8 was calculated for the MDH variant and the *Rxp* (bacterial pustule) locus (Palmer et al., 1992).

5–5.11.12 Mannose-6-Phosphate Isomerase (EC 5.3.1.8)

Six alleles have been described at the *Mpi* locus. Five alleles [*Mpi-(a, b, c, d, e)*] are codominant alleles that alter isozyme mobility (Chiang and Kiang, 1988). Each homozygous genotype exhibits a two-band zymogram, and both bands show altered mobility. Heterozygotes exhibit four bands unless the migration distances of two of the bands overlap. The order of migration distance for the four originally described codominant alleles is *Mpi-a* with the slowest migration to *Mpi-d* with the fastest migration. Allele *Mpi-e* (Yu and Kiang, 1993b) was found in *G. soja* and produces a mobility variant migrating between the *Mpi-b-* and *Mpi-c-*encoded isozymes. The recessive null allele *mpi* generally showed no enzyme activity but in a few seeds showed a very faint two-band zymogram.

5–5.11.13 6-Phosphogluconate Dehydrogenase (EC 1.1.1.44)

There are four 6-phosphogluconate dehydrogenase bands that are controlled by three genes (Chiang and Kiang, 1987). Bands 1 and 3 are homodimers, and band 2 is the interlocus heterodimer. Band 1 is the product of *Pgd1* with three codominant alleles [*Pgd1-(a, b, c)*], and band 3 is the product of *Pgd2* with three codominant alleles [*Pgd2-(a, b, c)*]. Each of the alleles specifies an isozyme band of differing mobility. Seeds homozygous for the null allele *pgd1* (not confirmed by genetic tests) lack bands 1 and 2 (Chiang and Kiang, 1987). Two monomeric mobility variants of band 4 have been described, produced by two codominant alleles, *Pgd3-a* and *Pgd3-b*.

A second recessive *pgd1* null allele from PI 567399 was reported but not tested for allelism (Liao and Palmer, 1997b).

5–5.11.14 Phosphoglucose Isomerase (EC 5.3.1.9)

Soybean phosphoglucose isomerase isozymes from homozygous individuals produce bands in as many as six zones. Both *Pgi1* and *Pgi3* enzymes are dimeric. Each of the two loci has two codominant alleles, and each allele specifies a single isozyme monomer. Interlocus heterodimers are formed. In heterozygotes, intralocus heterodimers are also formed that migrate to intermediate positions between the two corresponding homodimers. The recessive allele *pgi1* produces no homodimer band 6 and no interlocus heterodimer band 5. Allele *Pgi2* produces a band at the same location as *Pgi3-a*; it is only interpretable in *Pgi3-b* homozygous types. The recessive null allele *pgi2* produces no homodimer band 1. *Pgi2* produces interlocus heterodimers with *Pgi3* but not with *Pgi1* (Chiang et al., 1987).

Liao and Palmer (1997c) reported a new phosphoglucose isomerase isozyme variant in the Chinese accession PI 567365. The pattern was inherited codominantly and thought to be controlled by a fourth locus; a gene symbol has not been assigned.

5–5.11.15 Phosphoglucomutase (EC 2.7.5.1)

Three phosphoglucomutase isozymes controlled by three genes have been reported. The *Pgm1-a*, *Pgm1-b*, *Pgm2-a*, *Pgm2-b*, and *Pgm2-c* alleles were described in the previous soybean monograph. A null allele, *pgm3*, was identified for the fastest-migrating isozyme in seeds that were also *Pgm2-b/Pgm2-b*. The *Pgm3* genotype cannot be determined in *Pgm2-c* plants, as the bands of isozymes 2 and 3 are located in the same gel position (Yu and Kiang, 1993b). A fourth variant of isozyme 2 that migrates farther than isozyme 3 is controlled by a fourth codominant allele, *Pgm2-d* (Yu and Kiang, 1993b).

5–5.11.16 Protease Inhibitors

Each gene of three seed protein protease inhibitors in the wild perennial species *G. tomentella* (2*n* = 38) had a functional dominant allele and a recessive null allele. Alleles *Pi1*, *pi1*, and *Pi2*, *pi2* encode the presence and absence of two trypsin inhibitors of differing mobility. *Pi3* codes for an anti-soybean Bowman-Birk inhibitor immunocross-reactive band BBI', while *pi3* does not produce this band (Kollipara et al., 1996).

A fourth mobility variant of the Kunitz trypsin inhibitor (Zhao et al., 1995a) migrated slower than isozymes encoded by *Ti-a*, *Ti-b*, and *Ti-c*. The slowest band was produced by an allele *Ti-x*, codominant with the other functional alleles and dominant to the null allele *ti* (Zhao et al., 1995b). The *Ti-x* allele was found at a frequency of 1 in 16 344 Chinese cultivars (Zhao et al., 1993).

Zhao and Wang (1992) reported a mobility variant of the Kunitz trypsin inhibitor, tentatively designated *Ti-d* by Wang et al. (2001), which showed slower mobility than the *Ti-b* type, but its inheritance was not clearly defined. A total of 11 081 Chinese germplasms were examined. Two additional variants were identified in a sample of 173 cultivars of *G. max* and 890 accessions of *G. soja* (Wang et al., 1996). One is a variant (previously designated *Ti-a-s*), which is now tentatively designated *Ti-e*, which shows a slightly slower electrophoretic mobility than the *Ti-a* type (Wang et al., 2001). The other variant (tentatively designated *Ti-b-f*) has a slightly faster mobility than the *Ti-b* allele (Wang et al., 2001).

5–5.11.17 Seed Storage Proteins

Five genes encode glycinin subunits denoted by gene symbols *Gy1* to *Gy5* (Cho et al., 1989). A third allele for gene *Gy4* was identified in *G. soja* germplasm. The allele produces a higher molecular weight variant. Allele *Gy4* (Scallon et al., 1987) was renamed *Gy4-a* and considered the common allele, while gene symbol *Gy4-b* was assigned to the variant allele. The two alleles are codominant, and both are dominant to the *gy4* null allele (Kitamura et al., 1984).

5–5.11.18 Shikimate Dehydrogenase (EC 1.1.1.25)

The shikimate dehydrogenase zymogram has three isozymes. A codominant allele in *G. soja* increases the mobility of all three isozymes. Allele *Sdh-a* encodes the slow variant, and allele *Sdh-b* encodes the fast variant (Yu and Kiang, 1993b).

5–5.11.19 Superoxide Dismutase (EC 1.15.1.1)

Nine superoxide dismutase bands occur in three distinct groups. In zone 2 the *Sod1* gene encodes the presence (*Sod1*) or absence (*sod1*) of bands 4 and 5 (described in the previous soybean monograph). Two different homozygous patterns occur in zone 3. In the fast pattern, the mobility of two bands (8 and 9) increases compared with the slow pattern. Codominant alleles (*Sod2-a*, slow pattern and *Sod2-b*, fast pattern) of one gene encode the different variants (Griffin and Palmer, 1989).

5–5.11.20 β-Amylase (EC 3.2.1.2)

The four electrophoretic forms of β-amylase controlled by *Sp1-a*, *Sp1-b*, *Sp1-an*, and *sp1* were described in the previous soybean monograph. Accession PI 464918 has a beta-amylase with mobility intermediate to the isozymes encoded by alleles *Sp1-a* and *Sp1-b*. In a survey of 1500 accessions of *G. max* and *G. soja*, PI 464914 (cv. Dan Dou 3 from Liaoning Province, China) was the only accession found with the codominant *Sp1-c* allele (Griffin and Palmer, 1986).

A summary of genes involved in isoenzymes and proteins in soybean is given in Table 5–13.

5–5.12 Seed Fatty Acids

The identification of mutations and a few naturally occurring variants affecting soybean seed fatty acid synthesis has had a major impact on both the human and industrial use of soybean oil. Saturated fatty acids that have no double bonds in their carbon chains include palmitate (16:0) and stearate (18:0). The fatty acids oleate (18:1), linoleate (18:2), and linolenate (18:3) are unsaturated, with a respective one, two, or three double bonds. The fatty acid content of conventional soybean cultivars is approximately 11% palmitate, 4% stearate, 24% oleate, 54% linoleate, and 7% linolenate (Fehr, 1991).

5–5.12.1 Palmitate

Six mutant alleles have been identified that elevate palmitate content. These include the *fap2* allele (T309) (Erickson et al., 1988), the *fap2-b* allele (Fehr et al., 1991b; Schnebly et al., 1994), the *fap4* allele (Fehr et al., 1991b; Schnebly et al., 1994), the *fap5* allele (Stoltzfus et al., 2000a), the *fap6* allele (Narvel et al., 2000), and the *fap7* allele (Stoltzfus et al., 2000b).

Eight mutant alleles have been identified that reduce palmitate content: the *fap1* allele (Erickson et al., 1988; Wilcox and Cavins, 1990), the *fap2* allele (J10) (Rahman et al., 1999), the *fap3* allele (Fehr, 1991a; Schnebly et al., 1994), the *fap3-nc* allele (Wilson et al., 2001), the *fapx* allele (ELLP-2) (Stojšin et al., 1998a), the *fapx* allele (KK7) (Rahman et al., 1999), the *fap?* allele (J3) (Takagi et al., 1995), and the *fap?* allele (ELHP-1) (Primomo et al., 2002b).

Lines N79-2077-12 and N87-2122-4 (Burton et al., 1994) and N90-2013 (Wilcox et al., 1994) with reduced palmitate content were developed through recurrent selection programs. This new *fap* allele is allelic to *fap3*, and was designated *fap3-nc*. Wilson et al. (2001) have determined by Southern hybridizations that the *fap3-nc* allele is a deletion mutant in the Fat B gene. Primomo et al. (2002b) developed three lines by combinations of loci, and these include CLP-1 (*fap1* with *fan*), RG1 [*fap1, fapx* (ELLP-2) with *fan*], and RG3 [*fap1, fapx* (ELLP-2)].

The effect of environment on individual palmitate alleles and combinations of alleles has been investigated. Modifying genes affecting some, but not all, fatty acid genes have been reported (Horejsi et al., 1994; Hartmann et al., 1996; Rebetzke et al., 1998a, 1998b; Primomo et al., 2002b). Seed yields with altered palmitate levels may not be significantly changed (Horejsi et al., 1994) or may be significantly reduced (Rebetzke et al., 1998a; Primomo et al., 2002b). In addition, seed oil content may be reduced (Horejsi et al., 1994) or may be elevated (Rebetzke et al., 1998a). Molecular mapping with N87-2122-4 with simple sequence repeat markers accounted for 51% of total phenotypic variation for palmitic acid content in the F_2 generation (Li et al., 2002).

5–5.12.2 Stearate

Five mutant alleles have been identified that elevate stearate content. Graef et al. (1985) reported on the *fas*, *fas-a*, and *fas-b* alleles that were identified after seed treatment with ethyl methanesulfonate. Rahman et al. (1997) used X-ray seed irradiation and identified the *st1* and *st2* alleles, which are at different loci. The *st1 st1 st2 st2* genotype gave a stearic acid content greater than 30%, but the seedlings failed to develop into plants (Rahman et al., 1997).

Lundeen et al. (1987) and Hartmann et al. (1997) evaluated lines with normal and elevated stearate levels for agronomic traits. In crosses with A9 and A10 (Lundeen et al. 1987), there were no significant differences in average seed yield between the high-and low-stearate crosses. However, in crosses with A6 (Hammond and Fehr, 1983b), the low stearate lines had significantly greater seed yield than the high stearate lines (Lundeen et al., 1987).

Hartmann et al. (1997) reported that lines with elevated stearate content had significantly lower seed yield than normal stearate lines. The lines with elevated stearate content had a significant reduction in palmitate, oleate, and linoleate and a significant increase in linolenate when compared with lines having normal stearate.

5–5.12.3 Oleate

Brossman and Wilcox (1984) observed a distribution of oleate content skewed towards elevated levels following ethyl methanesulfonate seed treatment. Rahman et al. (1996b) and Takagi and Rahman (1996) reported on two X-ray irradiation-induced oleate mutants. Genetic data indicated that each mutant had a different allele, *ol* (M23), and *ol-a* (M11), at the same locus for oleate content. Because of the complete inverse relationship between oleate and linoleate in both mutants, Rahman et al. (1996b) proposed that the two mutant alleles may control linoleic acid content by blocking the synthesis of linoleic acid at the step of oleic acid desaturation.

Transgenic approaches to modify soybean oil have been successful. Yadav (1995) used an antisense oleate desaturase DNA and increased oleate content from 20% to about 80%.

5–5.12.4 Linoleate

No gene symbols have been assigned for linoleate content mutants. Brossman and Wilcox (1984) noted that distributions were skewed toward reduced levels of linoleate content following ethyl methanesulfonate seed treatment. Transgenic approaches to modify soybean oil quality have been successful. An antisense oleate desaturase DNA was used to reduce the linoleate content from 65 to 3% (Yadav, 1995).

5–5.12.5 Linolenate

Linolenate is the main cause of off-flavor and odor in soybean oil because it oxidizes more rapidly than other unsaturated fatty esters. Hydrogenation reduces the linolenate content, but trans fatty acids are produced during hydrogenation, which are believed to increase the risk of coronary heart disease. Thus, a major effort in soybean breeding has been to reduce the amount of linolenate in seed oil.

Initial efforts to reduce the linolenate content were recurrent selection programs (Carver et al., 1986). This breeding strategy did not result in the desired level of reduction of linolenate content. Mutagenesis was used to produce mutations in the linolenate biosynthetic pathway. Most mutations caused a reduction in linolenate content, but Takagi et al. (1989) reported a high linolenate mutant induced by X-ray radiation. Accessions of the wild annual species *G. soja* were surveyed for linolenate content, and PI 424031 was identified with 15% linolenate (Pantalone et al., 1997). Hymowitz et al. (1972) reported a high of 21.8% linolenate in the wild perennial species *G. tabacina* (PI 319697).

Five alleles have been identified at the *Fan1* locus with reduced linolenate. *Fan2* and *Fan3* loci each have one mutant allele. Allelism tests with *fanx* and *fanx-a* alleles have not been completed. The *Fan1* locus is located on classical linkage group 17 (Rennie et al., 1988; Rennie and Tanner, 1989a), which corresponds to molecular linkage group B2 (Brummer et al., 1995).

Molecular analysis of the linolenate content of *fan1* (A5 soybean genotype) has shown that the reduced level of linolenate content was the result of a deletion of the microsomal omega-3 fatty acid desaturase gene (Byrum et al., 1997). Yadav (1995) used an antisense linoleate desaturase DNA to reduce linolenate content to <2%, and the transgene behaved as a single dominant trait.

Germplasm with reduced linolenate has been released (Burton et al., 1989; Leffel, 1994c). Most efforts have been directed toward combining different genes (alleles) with reduced linolenate, and the identification of transgressive segregates with acceptable agronomic performance (Wilcox et al., 1993; Rahman et al., 1998; Walker et al., 1998; Ross et al., 2000; Primomo et al., 2002a, 2002b).

A summary of genes involved in seed fatty acid composition in soybean is given in Table 5–14.

Table 15–14. Genes affecting seed fatty acid composition in soybean.

Genes	Phenotype	Strain†	Reference
		1. Palmitate	
Fap1	Normal palmitic acid level	c	Erickson et al. (1988), Wilcox and Cavins (1990)
fap1	Reduced palmitic acid level	C1726 (T308)	
Fap2	Normal palmitic acid level	c	Erickson et al. (1988), Wilcox and Cavins (1990)
fap2	Elevated palmitic acid level	C1727(T309)	
fap2	Reduced palmitic acid level	J10	Rahman et al. (1999)
fap2-b	Elevated palmitic acid level	A21	Fehr et al. (1991b), Schnebly et al. (1994)

(continued on next page)

Table 15–14. Continued.

Genes	Phenotype	Strain	Reference
Fap3	Normal palmitic acid level	c	Fehr et al. (1991a), Schnebly et al. (1994)
fap3	Reduced palmitic acid level	A22	
fap3-nc	Reduced palmitic acid level	N79-2077-12	Burton et al. (1994), Wilson et al. (2001)
Fap4	Normal palmitic acid level	c	Fehr et al. (1991b), Schnebly et al. (1994)
fap4	Elevated palmitic acid level	A24	
Fap5	Normal palmitic acid level	c	Stoltzfus et al. (2000a)
fap5	Elevated palmitic acid level	A27	
Fap6	Normal palmitic acid level	c	Narvel et al. (2000)
fap6	Elevated palmitic acid level	A25	
Fap7	Normal palmitic acid level	c	Stoltzfus et al. (2000b)
fap7	Elevated palmitic acid level	A30	
fapx	Reduced palmitic acid level	ELLP-2	Stojšin et al. (1998b)
fapx	Reduced palmitic acid level	KK7	Rahman et al. (1999)
fap?	Reduced palmitic acid level	J3	Takagi et al. (1995)
fap?	Reduced palmitic acid level	ELHP-1	Primomo et al. (2002b)

2. Stearate

Fas	Normal stearic acid level	c	Graef et al. (1985), Hammond and Fehr (1983b)
fas	Elevated stearic acid level	A9	
fas-a	Elevated stearic acid level	A6	
fas-b	Elevated stearic acid level	A10	
St1	Normal stearic acid level	c	Rahman et al. (1997)
st1	Elevated stearic acid level	KK-2	
St2	Normal stearic acid level	c	Rahman et al. (1997)
st2	Elevated stearic acid level	M-25	

3. Oleate

Ol	Normal oleic acid level	c	Rahman et al. (1996b), Takagi and Rahman (1996)
ol	Elevated oleic acid level	M-23	
ol-a	Elevated oleic acid level	M-11	

4. Linolenate

Fan1	Normal linolenic acid level	c	
fan1	Reduced linolenic acid level	PI 123440 (from Burma)	Rennie and Tanner (1989b)
fan1	Reduced linolenic acid level	A5, T307	Hammond and Fehr (1983a)
fan1	Reduced linolenic acid level	C1640, T280	Wilcox and Cavins (1985, 1987)
fan1	Reduced linolenic acid level	PI 361088B (Mira Ungara)	Rennie et al. (1988)
fan1	Reduced linolenic acid level	M-5	Rahman et al. (1996a)
fan1-b	Reduced linolenic acid level	RG10	Stojšin et al., (1998b)
Fan2	Normal linolenic acid level	c	Fehr et al. (1992), Fehr and Hammond (1996)
fan2	Reduced linolenic acid level	A23	
Fan3	Normal linolenic acid level	c	Ross (1999), Ross et al. (2000)
fan3	Reduced linolenic acid level	A26	
fanx	Reduced linolenic acid level	KL-8	Rahman et al. (1996a), Rahman and Takagi (1997)
fanx-a	Reduced linolenic acid level	M-24	Rahman et al. (1998)

† c = Indicates that the gene occurs in many cultivars.

5–6 LINKAGE GROUPS

Twenty linkage groups have been identified that are designated the Classical Genetic Linkage Map (Table 5–15). Several linkage groups have only two loci, and two linkage groups have nine loci.

Inversions have not been used in linkage studies in soybean to date. Chromosome interchanges (translocations) have been used to place gene order for mutants of linkage groups 6 and 8. In fact, these translocations indicated that linkage groups 6 and 8 were the same linkage group (same chromosome) (Mahama and Palmer, 2003). Thus linkage group 6 has been merged with linkage group 8, and is no longer considered a separate linkage group.

Two or more of our current linkage groups may in fact be the same linkage group. Studies with primary trisomics are being used to assign linkage groups to their respective chromosomes (see Chapter 4, Hymowitz, 2004, this publication).

Table 5–15. Genetic linkage groups in soybean.

Linkage group	Linked genes	Linkage intensity map†	References
1	Aco3 — Aconitase	Aco3 —12.4 ± 0.8— Sp1 —16.3 ± 1.5— y12 —22.0 ± 1.5— t	Buzzell (1974, 1977, 1979)
	df5 — Dwarf	Sp1 —30.2 ± 1.5— t	Buzzell and Palmer (1985)
	E1 — Late flowering and maturity		Cober and Voldeng (2001a)
	E7 — Late flowering and maturity	Aco3 —33.6 ± 3.7— y12	Griffin and Palmer (1987b)
	fg3 — Flavonol glycoside absent	Aco3 —39.2 ± 2.1— y12	Hanson (1961)
	fg4 — Flavonol glycoside absent		Kiang and Bult (1991)
	Sp1 — β-amylase	y12 —20.2 ± 1.1— E1 —2.4 ± 0.4— t —3.9— E7	Kiang and Chiang (1988)
	t — Gray pubescence	y12 —21.8 ± 1.1— t	Palmer (1977, 1984a)
	y12 — Chlorophyll deficient	fg3 —12.0 ± 1.8— fg4 —3.9 ± 0.9— t	Weiss (1970a)
		fg3 —13.5 ± 5.9— t	
		df5 —15.4 ± 1.0— t	

Linkage group		Linked genes	Linkage intensity map†	References
2	$P1$	Glabrous	$P1$ ⊢— 20.9 ± 2.4 —⊣ r	Weiss (1970b)
	r	Brown seed		
3	G	Green seed coat	G ⊢— 4.2 ± 0.6 —⊣ $d1$	Weiss (1970b)
	$d1$	Green seed embryo		
4	Enp	Endopeptidase	$v1$ ⊢— 35.6 ± 0.9 —ln— 26.4 ± 1.4 —⊣ $p2$	Abe et al. (1997)
	$Est1$	Esterase	$Est1$ ⊢— 20.4 ± 2.2 —ln— 17.6 ± 1.8 —⊣ Enp	Muehlbauer et al. (1989)
	ln	Narrow-leaflet, four-seeded pods	$Est1$ ⊢— 32.3 ± 2.1 —⊣ Enp	Weiss (1970c)
	$p2$	Puberulent		
	vl	Variegated leaves		

(continued on next page)

Table 5–15. Continued.

Linkage group		Linked genes	Linkage intensity map†	References
5	dt1	Determinate stem	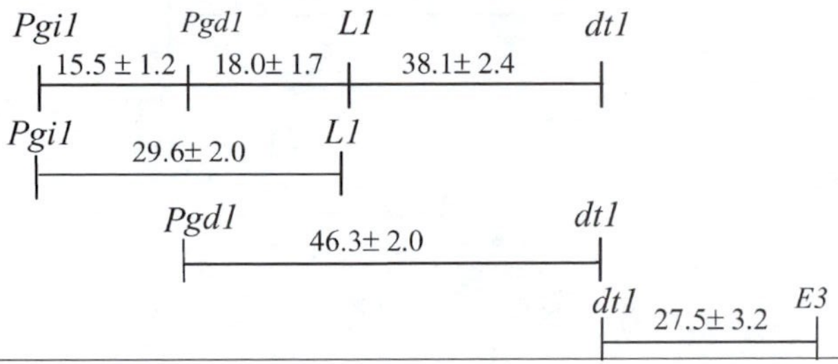	Cober and Voldeng (1996)
	E3	Late and sensitive to fluorescent light		Kiang (1990a)
	L1	Black pod		Weiss (1970d)
	Pgd1	Phosphogluconate dehydrogenase		
	Pgi1	Phosphoglucose isomerase		
7	i	Self dark seed	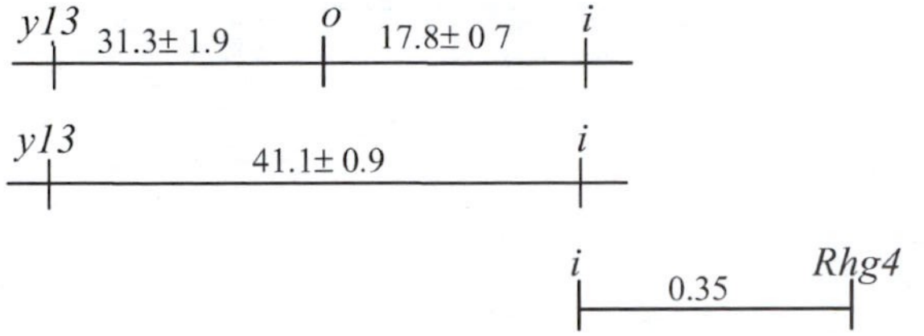	Matson and Williams (1965)
	o	Reddish brown seed		Weiss (1970e)
	Rhg4	Cyst nematode resistance (with rhg1, rhg2, and rhg3)		
	y13	Chlorophyll deficient		

Linkage intensity map† (group 5):

Pgi1 —15.5±1.2— Pgd1 —18.0±1.7— L1 —38.1±2.4— dt1

Pgi1 —29.6±2.0— L1

Pgd1 —46.3±2.0— dt1

dt1 —27.5±3.2— E3

Linkage intensity map† (group 7):

y13 —31.3±1.9— o —17.8±07— i

y13 —41.1±0.9— i

i —0.35— Rhg4

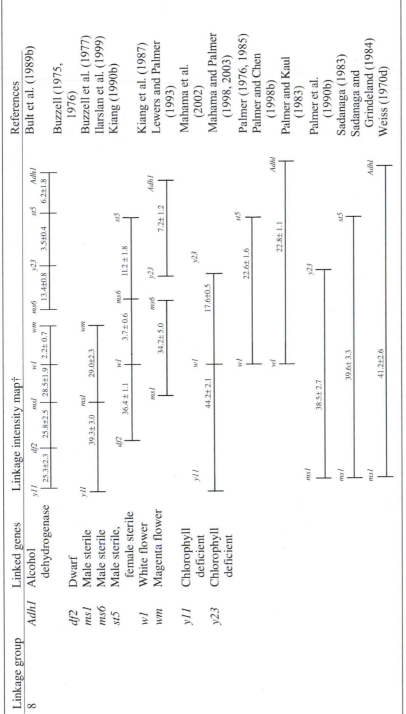

Linkage group	Linked genes		Linkage intensity map†	References
8	Adh1	Alcohol dehydrogenase		Bult et al. (1989b)
				Buzzell (1975, 1976)
	df2	Dwarf		Buzzell et al. (1977)
	ms1	Male sterile		Ilarslan et al. (1999)
	ms6	Male sterile		Kiang (1990b)
	st5	Male sterile, female sterile		Kiang et al. (1987)
	w1	White flower		Lewers and Palmer (1993)
	wm	Magenta flower		Mahama et al. (2002)
	y11	Chlorophyll deficient		Mahama and Palmer (1998, 2003)
	y23	Chlorophyll deficient		Palmer (1976, 1985)
				Palmer and Chen (1998b)
				Palmer and Kaul (1983)
				Palmer et al. (1990b)
				Sadanaga (1983)
				Sadanaga and Grindeland (1984)
				Weiss (1970d)

Linkage intensity map values:

y11 — 25.3±2.3 — df2 — 25.8±2.5 — ms1 — 28.5±1.9 — w1 — 2.2±0.7 — wm / ms6 — 13.4±0.8 — y23 — 3.5±0.4 — st5 — 6.2±1.8 — Adh1

y11 — 39.3±3.0 — ms1 — 29.0±2.3 — wm

df2 — 36.4±1.1 — w1 — 3.7±0.6 — ms6 — 11.2±1.8 — y23 — st5

ms1 — 34.2±5.0 — ms6 — 7.2±1.2 — y23 — Adh1

y11 — 44.2±2.1 — w1 — 17.6±0.5 — y23

w1 — 22.6±1.6 — st5

w1 — 22.8±1.1 — Adh1

ms1 — 38.5±2.7 — y23

ms1 — 39.6±3.3 — st5

ms1 — 41.2±2.6 — Adh1

(continued on next page)

Table 5–15. Continued.

Linkage group	Linked genes		Linkage intensity map†	References
9	Ap	Acid phosphatase	Fr3 —7.4± 0.7— Ap —9.5± 0.8— Ti —10.9± 1.0— Lap1 —20.6± 1.0— Pgd2	Chiang and Kiang (1987)
	Fr3	Nonfluorescent root in UV light	Fr3 —11.9± 3— Ti	Hildebrand et al. (1980)
	Lap1	Leucine aminopeptidase		Kiang et al. (1985)
	Pgd2	Phosphogluconate dehydrogenase	Fr3 —29.3± 1.4— Ap —22.3± 0.9— Lap1	Palmer and Chen (1998a)
	Ti	Kunitz trypsin inhibitor	Ap —39.8± 1.0— Lap1 ... Pgd2	
10	hm	Metribuzin sensitive	Rps7 —12.5 ± 2.7— Rps1	Anderson and Buzzell (1992)
	L2	Pod color		Kilen and Barrentine (1983)
	Rps1	Resistant to phytophthora root rot	Rps1 —7.0 ± 1.2— hm	Kilen and Tyler (1993)
	Rps7	Resistant to phytophthora root rot	Rps1 —27.5 ± 4.1— L2	Wang et al. (2001)

Linkage group	Linked genes		Linkage intensity map†	References
11	f	Fasciated stem	rj1 — 26.9 ± 3.4 — Idh1 — 24.7 ± 1.9 — f	Devine et al. (1983)
	Idh1	Isocitrate dehydrogenase	rj1 — 40.0 ± 2.2 — f	Hedges et al. (1990)
	rj1	Nonnodulating		
12	ep	Low seedcoat peroxidase level	fr1 — 42.6 ± 1.8 — ep	Griffin et al. (1989)
	fr1	Nonfluorescent root in UV light		
13	Rpv1	Resistant to peanut mottle virus	Rsv1 — 3.7 ± 0.8 — Rpv1	Roane et al. (1983)
	Rsv1	Resistant to soybean mosaic virus		
14	Pb	Sharp pubescent tip	y9 — 27.3 ± 1.1 — Pb	Devine (1998)
	y9	Chlorophyll deficient	y17 — 27.0 ± 4.0 — Pb	Thorson et al. (1989)
	y17	Chlorophyll deficient		

(continued on next page)

Table 5–15. Continued.

Linkage group		Linked genes	Linkage intensity map†	References
15	ms2 Pgm1	Male sterile Phosphoglucomutase	Pgm1 —18.7 ± 2.4— ms2	Sneller et al. (1992)
16	lf2 Pd2	Seven foliolate Dense pubescence	Pd2 —12.1 ± 2.2— lf2	Devine (2003)
17	Idh2 Fan Fas	Isocitrate dehydrogenase Seed linolenic acid Seed stearic acid	Idh2 —27.2 ± 2.0— Fan —21.6 ± 1.7— Fas Idh2 —37.0 ± 2.7— Fas	Rennie and Tanner (1989a) Rennie et al. (1988)
18	Dt2 Mpi	Semideterminate stem Mannose-6-phosphate isomerase	Mpi —16.1 ± 6.4— Dt2	Muehlbauer et al. (1989)

Linkage group		Linked genes	Linkage intensity map†	References
19	Aco2	Aconitase	Aco2 —— 44.7 ± 2.0 —— Rj2	Devine et al. (1991a, 1991b)
	Rj2	Ineffective nodulation		Lohnes et al. (1993)
	Rmd	Resistant to powdery mildew	Rj2 —1.9 ± 0.6— Rmd —2.3 ± 0.7— Rps2	
	Rps2	Resistant to phytophthora root rot	Rj2 —— 3.6 ± 0.9 —— Rps2	
20	MDH	Malate dehydrogenase	Rxp —— 15.2 ± 3.8 —— MDH	Palmer et al. (1992)
	Rxp	Bacterial pustule resistance		
21	Dia2	Diaphorase	Fle —— 9.8 ± 1.3 —— Dia2	Yu and Kiang (1993b)
	Fle	Fluorescent esterase		

† Linkage intensity map given as percentage recombination with standard error.

REFERENCES

Abe, J., T. Hirata, and Y. Shimamoto. 1997. Assignment of *Est1* locus to soybean linkage group 4. J. Hered. 88:557–559.

Abe, J., K. Komatsu, and Y. Shimamoto. 1998. A new gene for insensitivity of flowering to incandescent long day-length (ILD). Soybean Genet. Newsl. 25:92.

Abe, J., M. Ohara, and Y. Shimamoto. 1992. New electrophoretic mobility variants observed in wild soybean (*Glycine soja*) distributed in Japan and Korea. Soybean Genet. Newsl. 19:63–72.

Abel, G.H. 1969. Inheritance of the capacity for chloride inclusion and chloride exclusion by soybeans. Crop Sci. 9:697–698.

Albertsen, M.C., T.M. Curry, R.G. Palmer, and C.E. LaMotte. 1983. Genetics and comparative growth morphology of fasciation in soybeans (*Glycine max*) (L.) Merr.). Bot. Gaz. 144:263–275.

Amberger, L.A., R.C. Shoemaker, and R.G. Palmer. 1992. Inheritance of two independent isozyme variants in soybean plants derived from tissue culture. Theor. Appl. Genet. 84:600–607.

Andersen, J.J., and R.G. Palmer. 1997. Allelism of a necrotic root mutant in *Glycine max*. Soybean Genet. Newsl. 24:145–146.

Anderson, T.R., and R.I. Buzzell. 1992. Inheritance and linkage of the *Rps7* gene for resistance to *Phytophthora* rot of soybean. Plant Dis. 76:958–959.

Athow, K.L., and F.A. Laviolette. 1982. *Rps6*, a major gene for resistance to *Phytophthora megasperma* f. sp. *glycinea* in soybean. Phytopathology 72:1564–1567.

Athow, K.L., F.A. Laviolette, A.C. Layton Hahn, and L.D. Ploper. 1986. Genes for resistance to *Phytophthora megasperma* f. sp. *glycinea* in PI 273483D, PI 64747, PI 274212, PI 82312N, and PI 340046. Soybean Genet. Newsl. 13:119–131.

Athow, K.L., F.A. Laviolette, E.H. Mueller, and J.R. Wilcox. 1980. A new major gene for resistance to *Phytophthora megasperma* var. *sojae* in soybean. Phytopathology 70:977–980.

Athow, K.L., and A.H. Probst. 1952. The inheritance of resistance to frogeye leaf spot of soybeans. Phytopathology 42:660–662.

Bernard, R.L. 1967. The inheritance of pod color in soybeans. J. Hered. 58:165–168.

Bernard, R.L. 1971. Two major genes for time of flowering and maturity in soybeans. Crop Sci. 11:242–244.

Bernard, R.L. 1972. Two genes affecting stem termination in soybeans. Crop Sci. 12:235–239.

Bernard, R.L. 1975a. An allelic series affecting stem length. Soybean Genet. Newsl. 2:28–30.

Bernard, R.L. 1975b. The inheritance of near-gray pubescence color. Soybean Genet. Newsl. 2:31–33.

Bernard, R.L. 1975c. The inheritance of semi-sparse pubescence. Soybean Genet. Newsl. 2:33–34.

Bernard, R.L. 1975d. The inheritance of appressed pubescence. Soybean Genet. Newsl. 2:34–36.

Bernard, R.L, and C.R. Cremeens. 1972. A gene for general resistance to downy mildew of soybean. J. Hered. 62:359–362.

Bernard, R.L, and C.R. Cremeens. 1975. Inheritance of the Eldorado male-sterile trait. Soybean Genet. Newsl. 2:37–39.

Bernard, R.L, and C.R. Cremeens. 1981. An allele at the *rps* locus from the variety 'Kingwa'. Soybean Genet. Newsl. 8:40–42.

Bernard, R.L, and R.W. Howell. 1964. Inheritance of phosphorus sensitivity in soybeans. Crop Sci. 4:298–299.

Bernard, R.L, and E.R. Jaycox. 1969. A gene for increased natural crossing in soybean. p. 3. In 1969 Agronomy abstracts. ASA, Madison, WI.

Bernard, R.L, R.L. Nelson, and C.R. Cremeens. 1991. USDA soybean genetic collection: Isoline collection. Soybean Genet. Newsl. 18:27–57.

Bernard, R.L, and B.B. Singh. 1969. Inheritance of pubescence type in soybeans: Glabrous, curly, dense, sparse, and puberulent. Crop Sci. 9:192–197.

Bernard, R.L, P.E. Smith, M.J. Kaufmann, and A.F. Schmitthenner. 1957. Inheritance of resistance to phytophthora root and stem rot in the soybean. Agron. J. 49:391.

Bernard, R.L, and L.M. Wax. 1975. Inheritance of a sensitive reaction to bentazon herbicide. Soybean Genet. Newsl. 2:46–47.

Bernard, R.L, and M.G. Weiss. 1973. Qualitative genetics. p. 117–154. In B.E. Caldwell (ed.) Soybeans: Improvement, production, and uses. Agron. Monogr. 16. ASA, Madison, WI.

Boerma, H.R., and R.L. Cooper. 1978. Increased female fertility associated with the *ms1* locus in soybeans. Crop Sci. 18:344–346.

Boerma, H.R., and B.G. Jones. 1978. Inheritance of a second gene for brachytic stem in soybeans. Crop Sci. 18:559–560.

Boerma, H.R., and C.W. Kuhn. 1976. Inheritance of resistance to peanut mottle virus in soybeans. Crop Sci. 16:533–534.

Boerma, H.R., C.W. Kuhn, and H.B. Harris. 1975. Inheritance of resistance to cowpea chlorotic mottle virus (soybean strain) in soybeans. Crop Sci. 15:849–850.

Boerma, H.R., and D.V. Phillips. 1983. Genetic implications of the susceptibility of Kent soybean to *Cercospora sojina*. Phytopathology 73:1666–1668.

Bonato, E.R., and N.A. Vello. 1999. *E6*, a dominant gene conditioning early flowering and maturity in soybeans. Genet. Molec. Biol. 22:229–232.

Boquet, D., C. Williams, and W. Birchfield. 1975. Inheritance of resistance to the Wartelle race of root-knot nematode in soybeans. Crop Sci. 16:783–785.

Bowers, G.R., Jr., K. Ngeleka, and O.D. Smith. 1993. Inheritance of stem canker resistance in soybean cultivars Crockett and Dowling. Crop Sci. 33:67–70.

Bowers, G.R., E.H. Paschal, R.L. Bernard, and R.M. Goodman. 1992. Inheritance of resistance to soybean mosaic virus in 'Buffalo' and HLS soybean. Crop Sci. 32:67–72.

Brim, C.A., and M.F. Young. 1971. Inheritance of a male-sterile character in soybeans. Crop Sci. 11:564–566.

Bromfield, K.R., and E.E. Hartwig. 1980. Resistance to soybean rust and mode of inheritance. Crop Sci. 20:254–255.

Brossman, G.D., and J.R. Wilcox. 1984. Induction of genetic variation for oil properties and agronomic characteristics of soybean. Crop Sci. 24:783–787.

Brummer, E.C., A.D. Nickell, J.R. Wilcox, and R.C. Shoemaker. 1995. Mapping the *Fan* locus controlling linolenic acid content in soybean oil. J. Hered. 86:245–247.

Bult, C.J., and Y.T. Kiang. 1989. Inheritance and genetic linkage tests of an esterase locus in the cultivated soybean, *Glycine max*. J. Hered. 80:82–85.

Bult, C.J., Y.T. Kiang, Y.C. Chiang, J.Y.H. Doong, and M.B. Gorman. 1989a. Electrophoretic methods for soybean genetics studies. Soybean Genet. Newsl. 16:175–187.

Bult, C.J., Y.T. Kiang, T.E. Devine, J.J. O'Neill, and J.Y.H. Doong. 1989b. Testing for genetic linkage of morphological and electrophoretic loci in the cultivated soybean. Soybean Genet. Newsl. 16:168–174.

Burton, J.W., R.F. Wilson, and C.A. Brim. 1994. Registration of N79-2077-12 and N87-2122-4, two soybean germplasm lines with reduced palmitic acid in seed oil. Crop Sci. 34:313.

Burton, J.W., R.F. Wilson, C.A. Brim, and R.W. Rinne. 1989. Registration of soybean lines with modified fatty acid composition of seed oil. Crop Sci. 29:1583.

Buss, G.R. 1983. Inheritance of a male-sterile mutant from irradiated Essex soybeans. Soybean Genet. Newsl. 10:104–108.

Buss, G.R., G. Ma, P. Chen, and S.A. Tolin. 1997. Registration of V94-5152 soybean germplasm resistant to soybean mosaic potyvirus. Crop Sci. 37:1987–1988.

Buss, G.R., G. Ma, S. Kristipati, P. Chen, and S.A. Tolin. 1999. A new allele at the *Rsv3* locus for resistance to soybean mosaic virus. p. 490. *In* H.E. Kauffman (ed.) Proc. World Soybean Res. Conf., Chicago. 4–7 Aug. 1999. VI. Superior Printing, Champaign, IL.

Buss, G.R., C.W. Roane, S.A. Tolin, and T.A. Vinardi. 1985. A second dominant gene for resistance to peanut mottle virus in soybeans. Crop Sci. 25:314–316.

Buttery, B.R., and R.I. Buzzell. 1968. Peroxidase activity in seeds of soybean varieties. Crop Sci. 8:722–725.

Buttery, B.R., and R.I. Buzzell. 1971. Properties and inheritance of urease isoenzymes in soybean seeds. Can. J. Bot. 49:1101–1105.

Buttery, B.R., and R.I. Buzzell. 1973. Varietal differences in leaf flavonoids of soybeans. Crop Sci. 13:103–106.

Buttery, B.R., and R.I. Buzzell. 1975. Soybean flavonol glycosides: Identification and biochemical genetics. Can. J. Bot. 53:219–224.

Buzzell, R.I. 1971. Inheritance of a soybean flowering response to fluorescent-daylength conditions. Can. J. Genet. Cytol. 13:703–707.

Buzzell, R.I. 1974. Soybean linkage tests. Soybean Genet. Newsl. 1:11–14.

Buzzell, R.I. 1975. Soybean linkage tests. Soybean Genet. Newsl. 2:10–11.

Buzzell, R.I. 1976. Soybean linkage and allelism tests. Soybean Genet. Newsl. 3:11–14.

Buzzell, R.I. 1977. Soybean linkage tests. Soybean Genet. Newsl. 4:12–13.

Buzzell, R.I. 1979. Soybean linkage tests. Soybean Genet. Newsl. 6:15–16.

Buzzell, R.I, and T. R. Anderson. 1981. Another major gene for resistance to *Phytophthora megasperma* var. *sojae* in soybeans. Soybean Genet. Newsl. 8:30–33.

Buzzell, R.I., and T.R. Anderson. 1992. Inheritance and race reaction of a new *Rps1* allele. Plant Dis. 76:600–601.

Buzzell, R.I., and B.R. Buttery. 1969. Inheritance of peroxidase activity in soybean seed coats. Crop Sci. 9:387–388.

Buzzell, R.I., and B.R. Buttery. 1973. Inheritance of flavonol glycosides in soybeans. Can. J. Genet. Cytol. 15:865–867.

Buzzell, R.I., and B.R. Buttery. 1974. Flavonol glycoside genes in soybeans. Can. J. Genet. Cytol. 16:897–899.

Buzzell, R.I., and B.R. Buttery. 1992. Inheritance of an anomalous flavonoid glycoside gene in soybean. Genome 35:636–638.

Buzzell, R.I., B.R. Buttery, and R.L. Bernard. 1977. Inheritance and linkage of a magenta flower gene in soybeans. Can. J. Genet. Cytol. 19:749–751.

Buzzell, R.I., B.R. Buttery, and D.C. MacTavish. 1987. Biochemical genetics of black pigmentation of soybean seed. J. Hered. 78:53–54.

Buzzell, R.I., and J.H. Haas. 1978. Inheritance of adult plant resistance to powdery mildew in soybeans. Can. J. Genet. Cytol. 20:151–153.

Buzzell, R.I., and R.G. Palmer. 1985. Soybean linkage group 1 tests. Soybean Genet. Newsl. 12:32–33.

Buzzell, R.I., and J.C. Tu. 1984. Inheritance of soybean resistance to soybean mosaic virus. J. Hered. 75:82.

Buzzell, R.I., and J.C. Tu. 1989. Inheritance of a soybean stem-tip necrosis reaction to soybean mosaic virus. J. Hered. 80:400–401.

Buzzell, R.I., and H.D. Voldeng. 1980. Inheritance of insensitivity to long daylength. Soybean Genet. Newsl. 7:26–29.

Byrum, J.R., A.J. Kinney, K.L. Stecca, D.J. Grace, and B.W. Diers. 1997. Alteration of the omega-3 fatty acid desaturase gene is associated with reduced linolenic acid in the A5 soybean genotype. Theor. Appl. Genet. 94:356–359.

Byth, D.E., and C.R. Weber. 1969. Two mutant genes causing dwarfness in soybeans. J. Hered. 60:278–280.

Caldwell, B.E. 1966. Inheritance of a strain-specific ineffective nodulation in soybeans. Crop Sci. 6:427–428.

Caldwell, B.E., C.A. Brim, and J.P. Ross. 1960. Inheritance of resistance of soybeans to cyst nematode, *Heterodera glycines*. Agron. J. 52:635–636.

Cardy, B.J., and W.D. Beversdorf. 1984a. A procedure for the starch gel electrophoretic detection of isozymes of soybean. Tech. Bull. 119/8401. Univ. of Guelph, Guelph, ON, Canada.

Cardy, B.J., and W.D. Beversdorf. 1984b. Identification of soybean cultivars using isoenzyme electrophoresis. Seed Sci. Technol. 12:943–954.

Carroll, B.J., P.M. Gresshoff, and A.C. Delves. 1988. Inheritance of supernodulation in soybean and estimation of the genetically effective cell number. Theor. Appl. Genet. 76:54–58.

Carroll, B.J., D.L. McNeil, and P.M. Gresshoff. 1985. Isolation and properties of soybean (*Glycine max* L. Merr.) mutants that nodulate in the presence of high nitrate concentrations. Proc. Natl. Acad. Sci. USA. 82:4162–4166.

Carter, T.E., Jr., R.L. Nelson, C.H. Sneller, and Z. Cui. 2004. Genetic diversity in soybean. p. 303–416. *In* H.R. Boerma and J.E. Specht (ed.) Soybeans: Improvement, production, and uses. 3rd ed. Agron. Monogr. 16. ASA, CSSA, and SSSA, Madison, WI.

Carver, B.F., J.W. Burton, R.F. Wilson, and T.E. Carter, Jr. 1986. Cumulative response to various recurrent schemes in soybean: Oil quality and correlated agronomic traits. Crop Sci. 26:853–858.

Caviness, C.E., and R.D. Riggs. 1976. Breeding for nematode resistance. p. 594–601. *In* L.D. Hill (ed.) Proc. World Soybean Research Conf. I, Champaign, IL. 3–8 Aug. 1975. The Interstate Printers and Publ., Danville, IL.

Cerna, F.J., S.R. Cianzio, A. Rafalski, S. Tingey, and D. Dyer. 1997. Relationship between seed yield heterosis and molecular heterozygosity in soybean. Theor. Appl. Genet. 95:460–467.

Chandlee, J.M., and L.O. Vodkin. 1989a. Unstable genes affecting chloroplast development in soybean. Dev. Genet. 10:532–541.

Chandlee, J.M., and L.O. Vodkin. 1989b. Analysis of the *Y18* gene of soybean and its associated alleles. Soybean Genet. Newsl. 16:199–201.

Chandlee, J.M., and L.O. Vodkin. 1989c. Unstable expression of a soybean gene during seed coat development. Theor. Appl. Genet. 77:587–594.

Chaudhari, H.K., and W.H. Davis. 1977. A new male-sterile strain in Wabash soybeans. J. Hered. 68:266–267.

Chen, P., G.R. Buss, C.W. Roane, and S.A. Tolin. 1991. Allelism among genes for resistance to soybean mosaic virus in strain differential soybean cultivars. Crop Sci. 31:305–309.

Chen, P., G.R. Buss, C.W. Roane, and S.A. Tolin. 1994. Inheritance in soybean of resistant and necrotic reactions to soybean mosaic virus strains. Crop Sci. 31:414–422.

Chen, P., G. Ma, G.R. Buss, I. Gunduz, C.W. Roane, and S.A. Tolin. 2001. Inheritance and allelism tests of Raiden soybean for resistance to soybean mosaic virus. J. Hered. 92:51–55.

Chen, P., G. Ma, S.A. Tolin, I. Gunduz, and M. Cicek. 2002 Identification of a valuable gene in Suweon 97 for resistance to all strains of soybean mosaic virus. Crop Sci. 42:333–337.

Chen, X.F., J. Imsande, and R.G. Palmer. 1999. Eight new mutants at the *k2 Mdh1-n y20* chromosomal region in soybean. J. Hered. 90:399–403.

Chen, X.F., and R.G. Palmer. 1996. Inheritance and linkage with the *k2* and *Mdh1-n* loci in soybean. J. Hered. 87:433–437.

Chen, X.F., and R.G. Palmer. 1998a. Instability at the *k2 Mdh1-n y20* chromosomal region in soybean. Mol. Gen. Genet. 260:309–318.

Chen, X.F., and R.G. Palmer. 1998b. Recombination and linkage estimation between the *k2* and *Mdh1-n y20* loci in soybean. J. Hered. 89:488–494.

Chen, Z., H. Ilarslan, R.G. Palmer, and R.C. Shoemaker. 1998. Development of protein bodies and accumulation of carbohydrates in a soybean (Leguminosae) shriveled seed mutant. Am. J. Bot. 85:492–499.

Cheng, T.-S., and J. M. Chandlee. 1993. The recessive mutation (*y18*) at the *Y18* locus results in the loss of a component of the oxygen evolving complex in photosystem II. Soybean Genet. Newsl. 20:160–163.

Chiang, Y.C., M.B. Gorman, and Y.T. Kiang. 1987. Inheritance and linkage analysis of phosphoglucose isomerase isozymes in soybeans. Biochem. Genet. 25:893–900.

Chiang, Y.C., and Y.T. Kiang. 1987. Inheritance and linkage relationships of 6-phosphogluconate dehydrogenase isozymes in soybean. Genome 29:786–792.

Chiang, Y.C., and Y.T. Kiang. 1988. Genetic analysis of mannose-6-phosphate isomerase in soybeans. Genome 30:808–811.

Cho, E.K., and R.M. Goodman. 1979. Strains of soybean mosaic virus: Classification based on virulence in resistant soybean cultivars. Phytopathology 69:467–470.

Cho, E.K., and R.M. Goodman. 1982. Evaluation of resistance in soybeans to soybean mosaic virus strains. Crop Sci. 22:1133–1136.

Cho, T.-J., C.S. Davies, and N.C. Nielsen. 1989. Inheritance and organization of glycinin genes in soybean. Plant Cell 1:329–337.

Chung, J., P.E. Staswick, G.L. Graef, D.S. Wysong, and J. E. Specht. 1998. Inheritance of a disease lesion mimic mutant in soybean. J. Hered. 89:363–365.

Chung, S.D., H.W. Huh, and M.G. Chung. 1995. Genetic diversity in Korean populations of *Glycine soja* (Fabaceae). J. Plant Biol. 38:39–45.

Cianzio, S.R., and W.R. Fehr. 1980. Genetic control of iron deficiency chlorosis in soybean. Iowa State J. Res. 54:367–375.

Cianzio, S.R., and W.R. Fehr. 1982. Variation in the inheritance of resistance to iron deficiency chlorosis in soybeans. Crop Sci. 22:433–434.

Cianzio, S.R., and R.G. Palmer. 1992. Genetics of five cytoplasmically inherited yellow foliar mutants in soybean. J. Hered. 83:70–73.

Cober, E.R., G.R. Ablett, R.I. Buzzell, B.M. Luzzi, V. Poysa, A.S. Sahota, and H.D. Voldeng. 1998. Imperfect yellow hilum color in soybean is conditioned by *Il rr TT*. Crop Sci. 38:940–941.

Cober, E.R., D.W. Stewart, and H.D. Voldeng. 2001. Photoperiod and temperature responses in early-maturing near-isogenic soybean lines. Crop Sci. 41:721–727.

Cober, E.R., J.W. Tanner, and H.D. Voldeng. 1996a. Genetic control of photoperiod response in early-maturing, near-isogenic soybean lines. Crop Sci. 36:601–605.

Cober, E.R., J.W. Tanner, and H.D. Voldeng. 1996b. Soybean photoperiod-sensitivity loci respond differentially to light quality. Crop Sci. 36:606–610.

Cober, E.R., and H.D. Voldeng. 1996. *E3* and *Dt1* linkage. Soybean Genet. Newsl. 23:56–57.

Cober, E.R., and H.D. Voldeng. 2001a. A new soybean maturity and photoperiod-sensitivity locus linked to *E1* and *T*. Crop Sci. 41:698–701.

Cober, E.R., and H.D. Voldeng. 2001b. Low R:FR light quality delays flowering of *E7 E7* soybean lines. Crop Sci. 41:1823–1826.

Cooper, R.L. 1966. A major gene for resistance to seed coat mottling in soybean. Crop Sci. 6:290–292.

Cooper, R.L., and J. Tew. 2001. Use of leaf cutter bees in the production of hybrid soybean seed. *In* Annual meeting abstracts. [CD-ROM.] ASA, CSSA, and SSSA, Madison, WI.

Cubukcu, P., S.S. Kang, and R.G. Palmer. 2000. Genetic analysis of a root fluorescence mutant from Yunnan Province China. Soybean Genet. Newsl. 27 [Online journal]. Available at http://www.soy-genetics.org/articles/sgn2000-015.htm (verified 10 July 2000).

Culbertson, R.D.R., and T. Hymowitz. 1990. The cause of high natural cross pollination rates in T31 soybean, *Glycine max* (L.) Merr. Leg. Res. 13:160–168.

Davies, C.S., J.B. Coates, and N.C. Nielsen. 1985. Inheritance and biochemical analysis of 4 elec-trophoretic variants of beta-conglycinin from soybean. Theor. Appl. Genet. 71:351–358.

Davies, C.S., and N.C. Nielsen. 1986. Genetic analysis of a null-allele for lipoxygenase-2 in soybean. Crop Sci. 26:460–463.

Davies, C.S., and N.C. Nielsen. 1987. Registration of soybean germplasm that lacks lipoxygenase isozymes. Crop Sci. 27:370–371.

Davis, W. H. 2000. Mutant male-sterile gene of soybean. U.S. Patent 6 046 385. Date issued: 4 Apr. 2000.

Delannay, X., T.T. Bauman, D.H. Beighley, M.J. Buettner, H.D. Coble, M.S. Defelice, C.W. Derting, T.J. Diedrick, J.L. Giffin et al. 1995. Yield evaluation of a glyphosate-tolerant soybean line after treatment with glyphosate. Crop Sci. 35:1461–1467.

Delannay, X., and R.G. Palmer. 1982a. Genetics and cytology of the *ms4* male-sterile soybean. J. Hered. 73:219–223.

Delannay, X., and R.G. Palmer. 1982b. Four genes controlling root fluorescence in soybean. Crop Sci. 22:278–281.

Delannay, X., and R.G. Palmer. 1984. Inheritance of a miniature mutant in soybean. Soybean Genet. Newsl. 11:92–93.

Delves, A.C., B.J. Carroll, and P.M. Gresshoff. 1988. Genetic analysis and complementation studies on a number of mutant supernodulating soybeans. J. Genet. 67:1–8.

Devine, T.E. 1985. Nodulation of soybean (*Glycine max* L. Merr.) plant introduction lines with the fast-growing rhizobial strain USDA 205. Crop Sci. 25:354–356.

Devine, T.E. 1987. A comparison of rhizobial strain compatibilities of *Glycine max* and its progenitor species *Glycine soja*. Crop Sci. 27:635–639.

Devine, T.E. 1992. Genetic linkage tests for the *Rj4* gene in soybean. Crop Sci. 32:961–964.

Devine, T.E. 1998. Assignment of the *Y17* locus to classical soybean linkage group 14. Crop Sci. 38:696–697.

Devine, T.E. 2003. Genetic linkage of the soybean genes *Pd2* and *Lf2*. Crop Sci. 43:(in press).

Devine, T.E., and B.H. Breithaupt. 1981. Frequencies of nodulation response alleles, *Rj2* and *Rj4*, in soybean plant introductions and breeding lines. Tech. Bull.1628. U.S. Gov. Print. Office, Wash-ington, DC.

Devine, T.E., T.C. Kilen, and J.J. O'Neill. 1991a. Genetic linkage of the phytophthora resistance gene *Rps2* and the nodulation response gene *Rj2* in soybean. Crop Sci. 31:713–715.

Devine, T.E., and L.D. Kuykendall. 1994. *Rfg1*, a soybean gene controlling nodulation with fast grow-ing *Rhizobium fredii* strain 205. Plant Soil 158:47–51.

Devine, T.E., and L.D. Kuykendall. 1996. Host genetic control of symbiosis in soybean (*Glycine max* L.). Plant Soil 186:173–187.

Devine, T.E., L.D. Kuykendall, and J.J. O'Neil. 1990. The *Rj4* allele in soybean represses nodulation by chlorosis-inducing *Bradyrhizobia* classified as DNA homology group II by antibiotic resist-ance profiles. Theor. Appl. Genet. 80:33–37.

Devine, T.E., and J.J. O'Neil. 1986. Registration of BARC-2 (*Rj4*) and BARC-3 (*rj4*) soybean germplasm. Crop Sci. 26:1263–1264.

Devine, T.E., and J.J. O'Neil. 1987. Registration of BARC-4 (*Rj2*) and BARC-5 (*rj2*) soybean germplasm. Crop Sci. 27:1322–1323.

Devine, T.E., and J.J. O'Neil. 1989. Genetic allelism of nodulation response genes *Rj1*, *Rj2*, and *Rj4* in soybean. Crop Sci. 29:1347–1350.

Devine, T.E., and J.J. O'Neil. 1993. Genetic independence of the nodulation response gene loci -*Rj1*, *Rj2*, and *Rj4* -in soybean. J. Hered. 84:140–142.

Devine, T.E., J.J. O'Neil, Y.T. Kiang, and C.J. Bult. 1991b. Genetic linkage of the *Rj2* gene in soybean. Crop Sci. 31:665–668.

Devine, T.E., R.G. Palmer, and R.I. Buzzell. 1983. Analysis of genetic linkage in soybean. J. Hered. 74:457–460.

Devine, T.E., J.M. Weisemann, and B.F. Matthews. 1993. Linkage of the *Fr2* locus controlling soybean root fluorescence and four loci detected by RFLP markers. Theor. Appl. Genet. 85:921–925.

Diers, B.W., V. Berlinson, N.C. Nielsen, and R.C. Shoemaker. 1994. Genetic mapping of the *Gy4* and *Gy5* glycinin genes in soybean and the analysis of a variant *Gy4*. Theor. Appl. Genet. 89:297–304.

Domingo, W.E. 1945. Inheritance of number of seeds per pod and leaflet shape in the soybean. J. Agric. Res. 70:251–268.

Doong, J.Y.H., and Y.T. Kiang. 1987a. Inheritance of soybean endopeptidase. Biochem. Genet. 25:847–853.

Doong, J.Y.H., and Y.T. Kiang. 1987b. Inheritance of aconitase isozymes in soybean. Genome 29:713–717.

Doong, J.Y.H., and Y.T. Kiang. 1987c. Cultivar identification by isozyme analysis. Soybean Genet. Newsl. 14:189–226.

Doong, J.Y.H., and Y.T. Kiang. 1988. Inheritance study on a soybean fluorescent esterase. J. Hered. 79:399–400.

Edwards, C.J., Jr., W.L. Barrentine, and T.C. Kilen. 1976. Inheritance of sensitivity to metribuzin in soybeans. Crop Sci. 16:119–120.

Erickson, E.A., J.R. Wilcox, and J.F. Cavins. 1988. Inheritance of palmitic acid percentages in two soybean mutants. J. Hered. 79:465–468.

Fasoula, D.A., P.A. Stephens, C.D. Nickell, and L.O. Vodkin. 1995. Cosegregation of purple-throat flower color with dihydroflavonol reductase polymorphism in soybean. Crop Sci. 35:1028–1031.

Feaster, C.V. 1951. Bacterial pustule disease in soybeans: Artificial inoculation, varietal resistance, and inheritance of resistance. Bull. 487. Mo. Agric. Exp. Res. Stn., Columbia.

Fehr, W.R. 1972a. Inheritance of a mutation for dwarfness in soybeans. Crop Sci. 12:212–213.

Fehr, W.R. 1972b. Genetic control of leaflet number in soybeans. Crop Sci. 12:221–224.

Fehr, W.R. 1991. A plant breeder's response to changing needs in soybean oil quality. p. 79–87. In D. Wilkinson (ed.) Proc. XXI Soybean Seed Res. Conf., Chicago, IL. 10–11 Dec. 1991. Am. Seed Trade Assoc., Washington, DC.

Fehr, W.R., and R.C. Clark. 1973. Registration of five soybean germplasm populations. Crop Sci. 13:778.

Fehr, W.R., and J.H. Giese. 1971. Genetic control of root fluorescence. Crop Sci. 11:771.

Fehr, W.R., and E.G. Hammond. 1996. Soybean having low linolenic acid content and method of production. U.S. Patent 5 534 425. Date issued: 9 July 1996.

Fehr, W.R., G.A. Welke, E.G. Hammond, D.N. Duvick, and S.R. Cianzio. 1991a. Inheritance of reduced palmitic acid content in seed oil of soybeans. Crop Sci. 31:88–89.

Fehr, W.R., G.A. Welke, E.G. Hammond, D.N. Duvick, and S.R. Cianzio. 1991b. Inheritance of elevated palmitic acid content in seed oil of soybeans. Crop Sci. 31:1522–1524.

Fehr, W.R.,G.A. Welke, E.G. Hammond, D.N. Duvick, and S.R. Cianzio. 1992. Inheritance of reduced linolenic acid content in soybean genotypes A16 and A17. Crop Sci. 32:903–906.

Gai, J., Y.Z. Hu, Y.D. Zhang, Y.D. Xiang, and R.H. Ma. 1989. Inheritance of resistance of soybeans to four local strains of soybean mosaic virus. p. 1182–1187. In A.J. Pascale (ed.) Proc. World Soybean Res. Conf. IV, Buenos Aires, Argentina. 5–9 Mar. 1989. Orientacion Grafica Editora S.R.L., Buenos Aires, Argentina.

Gijzen, M. 1997. A deletion mutation at the *ep* locus causes low seed coat peroxidase activity in soybean. Plant J. 12:991–998.

Goodrick, B.J., C.W. Kuhn, and H.R. Boerma. 1991. Inheritance of nonnecrotic resistance to cowpea chlorotic mottle virus in soybean. J. Hered. 82:512–514.

Gorman, M.B., and Y.T. Kiang. 1977. Variety specific electrophoretic variants of four soybean enzymes. Crop Sci. 17:963–965.

Gorman, M.B., and Y.T. Kiang. 1978. Models for the inheritance of several variant soybean electrophoretic zymograms. J. Hered. 69:255–258.

Gorman, M.B., Y.T. Kiang, and Y.C. Chiang. 1984. Electrophoretic classification of selected *G. max* plant introductions and named cultivars in the late maturity groups. Soybean Genet. Newsl. 11:135–140.

Gorman, M.B., Y.T. Kiang, Y.C. Chang, and R.G. Palmer. 1982a. Preliminary electrophoretic observations from several soybean enzymes. Soybean Genet. Newsl. 9:140–143.

Gorman, M.B., Y.T. Kiang, Y.C. Chang, and R.G. Palmer. 1982b. Electrophoretic classification of the early maturity groups of named soybean cultivars. Soybean Genet. Newsl. 9:143–156.

Gorman, M.B., Y.T. Kiang, R.G. Palmer, and Y.C. Chiang. 1983. Inheritance of soybean electrophoretic variants. Soybean Genet. Newsl. 10:67–84.

Grady, H., R.G. Palmer, and J. Imsande. 1995. Isoflavonoids in root and hypocotyl of soybean seedlings (*Glycine max*, Fabaceae). Am. J. Bot. 82:964–968.

Graef, G.L, W.R. Fehr, and S.R. Cianzio. 1989. Relation of isozyme genotypes to quantitative characters in soybean. Crop Sci. 29:683–688.

Graef, G.L., W.R. Fehr, and E.G. Hammond. 1985. Inheritance of three stearic acid mutants of soybean. Crop Sci. 25:1076–1079.

Graef, G.L., and J.E. Specht. 1999. Registration of the SG1LS soybean population with large seed size and *ms2* nuclear male sterility. Crop Sci. 39:1261–1262.

Graham, M.J., C.D. Nickell, and R.G. Hoeft. 1995. Inheritance of tolerance to manganese deficiency in soybean. Crop Sci. 35:1007–1010.

Grau, C.R., and J.A. Lawrence. 1975. Observations on resistance and heritability of resistance to powdery mildew of soybean. Plant Dis. Rep. 59:458–460.

Gray, L.E., and C.R. Grau. 1999. Brown stem rot. p. 28–29. *In* G.L. Hartman et al. (ed.) Compendium of soybean diseases. 4th ed. The Am. Phytopathol. Soc., St. Paul, MN.

Graybosch, R.A., R.L. Bernard, C.R. Cremeens, and R.G. Palmer. 1984. Genetic and cytological studies on a male-sterile, female-fertile soybean mutant. J. Hered. 75:383–388.

Graybosch, R.A., and R G. Palmer. 1987. Analysis of a male-sterile character in soybeans. J. Hered. 78:66–70.

Gremaud, M.F., and J.E. Harper. 1989. Selection and initial characterization of partially nitrate tolerant nodulation mutants of soybean. Plant Physiol. 89:169–173.

Griffin, J.D., S.L. Broich, X. Delannay, and R.G. Palmer. 1989. The loci *Fr1* and *Ep* define soybean linkage group 12. Crop Sci. 29:80–82.

Griffin, J.D., and R.G. Palmer. 1984. Superoxide dismutase (SOD) isoenzymes in soybean. Soybean Genet. Newsl. 11:91–92.

Griffin, J.D., and R.G. Palmer. 1986. An additional beta-amylase mobility variant conditioned by the *Sp1* locus. Soybean Genet. Newsl. 13:150–151.

Griffin, J.D., and R.G. Palmer. 1987a. Inheritance and linkage studies of five isozyme loci in soybean. Crop Sci. 27:885–893.

Griffin, J.D., and R.G. Palmer. 1987b. Locating the *Sp1* locus on soybean linkage group 1. J. Hered. 78:122–123.

Griffin, J.D., and R.G. Palmer. 1989. Genetic studies with two superoxide dismutase loci in soybean. Crop Sci. 29:968–971.

Griffin, J.D., and R.G. Palmer. 1995. Variability of thirteen isozyme loci in the USDA soybean germplasm collections. Crop Sci. 35:897–904.

Groose, R.W., and R.G. Palmer. 1991. Gene action for anthocyanin pigmentation in soybean. J. Hered. 82:498–501.

Groose, R.W., S.M. Schulte, and R.G. Palmer. 1990. Germinal reversion of an unstable mutation for anthocyanin pigmentation in soybean. Theor. Appl. Genet. 79:161–167.

Groose, R.W., H.D. Weigelt, and R.G. Palmer. 1988. Somatic analysis of an unstable mutation for anthocyanin pigmentation in soybean. J. Hered. 79:263–267.

Gunashinghe, U.B., M.E. Irwin, and G.E. Kampmeir. 1988. Soybean leaf pubescence affects aphid vector transmission and field spread of soybean mosaic virus. Ann. Appl. Biol. 112:259–272.

Gunduz, I.. 2000. Genetic analysis of soybean mosaic virus resistance in soybean. Ph.D. Diss. Virginia Polytechnic Inst. and State Univ., Blacksburg.

Gunduz, I., G.R. Buss, P. Chen, and S.A. Tolin. 2002. Characterization of SMV resistance genes in Tousan 140 and Hourei soybean. Crop Sci. 42:90–95.

Gunduz, I., G.R. Buss, G. Ma, P. Chen, and S.A. Tolin. 2001. Genetic analysis of resistance to soybean mosaic virus in OX670 and Harosoy soybean. Crop Sci. 41:1785–1791.

Hadley, H.H., and W.J. Starnes. 1964. Sterility in soybeans caused by asynapsis. Crop Sci. 4:421–424.

Hammond, E.G., and W.R. Fehr. 1983a. Registration of A5 germplasm line of soybean. Crop Sci. 23:192.

Hammond, E.G., and W.R. Fehr. 1983b. Registration of A6 germplasm line of soybean (Reg. No. GP45). Crop Sci. 23:192–193.

Hanson, P.M., C.D. Nickell, L.E. Gray, and S.A. Sebastian. 1988. Identification of two dominant genes conditioning brown stem rot resistance in soybean. Crop Sci. 28:41–43.

Hanson, W.D. 1961. Effect of calcium and phosphorus nutrition on genetic recombination in the soybean. Crop Sci. 1:384.

Harper, J.E. 1989. Nitrogen metabolism mutants of soybean. p. 212–216. *In* A.J. Pascale (ed.) Proc. World Soybean Res. Conf. IV. Orientacion Grafica Editora S.R.L., Buenos Aires, Argentina.

Harper, J.E., and C.D. Nickell. 1995. Genetic analysis of nonnodulating soybean mutants in a hypernodulated background. Soybean Genet. Newsl. 22:185–190.

Hartmann, R.B., W.R. Fehr, G.A. Welke, E.G. Hammond, D.N. Duvick, and S.R. Cianzio. 1996. Association of elevated palmitate content with agronomic and seed traits of soybean. Crop Sci. 36:1466–1470.

Hartmann, R.B., W.R. Fehr, G.A. Welke, E.G. Hammond, D.N. Duvick, and S.R. Cianzio. 1997. Association of elevated stearate with agronomic and seed traits of soybean. Crop Sci. 37:124–127.

Hartwig, E.E. 1986. Identification of a fourth major gene conferring resistance to soybean rust. Crop Sci. 26:1135–1136.

Hartwig, E.E., W.L. Barrentine, and C.J. Edwards, Jr. 1980. Registration of Tracy-M soybeans. Crop Sci. 20:825.

Hartwig, E.E., and K.R. Bromfield. 1983. Relationships among three genes conferring specific resistance to rust in soybeans. Crop Sci. 23:237–239.

Hartwig, E.E., and K. Hinson. 1962. Inheritance of flower color of soybean. Crop Sci. 2:152–153.

Hartwig, E.E., W.F. Jones, and T.C. Kilen. 1991. Identification and inheritance of inefficient zinc absorption in soybean. Crop Sci. 31:61–63.

Hartwig, E.E., B.L. Keeling, and C.J. Edwards, Jr. 1968. Inheritance of reaction to phytophthora rot in the soybean. Crop Sci. 8:634–635.

Hartwig, E.E., and R.A.S. Kiihl. 1979. Identification and utilization of a delayed flowering character in soybeans for short-day conditions. Field Crops Res. 2:145–151.

Hartwig, E.E., and S.G. Lehman. 1951. Inheritance of resistance to the bacterial pustule disease in soybeans. Agron. J. 43:226–229.

Hatfield, P.M., and R.G. Palmer. 1986. Allelism test of T218H and T225H. Soybean Genet. Newsl. 13:147–149.

Hayes, A.J., G. Ma, G.R. Buss, and M.A. Saghai-Maroof. 2000. Molecular marker mapping of *Rsv4*, a gene conferring resistance to all known strains of soybean mosaic virus. Crop Sci. 40:1434–1437.

Hedges, B.R., and R.G. Palmer. 1992. Inheritance of malate dehydrogenase nulls in soybean. Biochem. Genet. 30:491–502.

Hedges, B.R., R.G. Palmer, and L.A. Amberger. 1992. Electrophoretic analysis of soybean seed protein. p. 143–158. *In* H.F. Linskens and J.F. Jackson (ed.) Modern methods of plant analysis. Springer-Verlag, Berlin.

Hedges, B.R., J.M. Sellner, T.E. Devine, and R.G. Palmer. 1990. Assigning isocitrate dehydrogenase to linkage group 11 in soybean. Crop Sci. 30:940–942.

Hegstad, J.M., L.O. Vodkin, and C.D. Nickell. 2000a. Genetic and agronomic evaluation of *wp-m* in soybean. Crop Sci. 40:346–351.

Hegstad, J.M., J.A. Tarter, L.O. Vodkin, and C.D. Nickell. 2000b. Positioning the *wp* flower color locus on the soybean genome map. Crop Sci. 40:534–537.

Hildebrand, D.F., and T. Hymowitz. 1980a. The *Sp1* locus in soybean codes for β-amylase. Crop Sci. 20:165–168.

Hildebrand, D.F., and T. Hymowitz. 1980b. Inheritance of β-amylase nulls in soybean seeds. Crop Sci. 20:727–730.

Hildebrand, D.F., and T. Hymowitz. 1981. Two soybean genotypes lacking lipoxygenase-1. J. Am. Oil Chem. Soc. 58:583–586.

Hildebrand, D.F., and T. Hymowitz. 1982. Inheritance of lipoxygenase –1 activity in soybean seeds. Crop Sci. 22:851–853.

Hildebrand, D.F., J.H. Orf, and T. Hymowitz. 1980. Inheritance of an acid phosphatase and its linkage with the Kunitz trypsin inhibitor in seed protein of soybeans. Crop Sci. 20:83–85.

Hill, J.H. 1999. Soybean mosaic. p. 70–71. *In* G.L. Hartman et al. (ed.) Compendium of soybean diseases. 4th ed. The Am. Phytopathol. Soc., St. Paul, MN.

Hirata, T., J. Abe, and Y. Shimamoto. 1999. Genetic structure of the Japanese soybean population. Genet. Res. Crop Evol. 46:441–453.

Holland, M.A., J.D. Griffin, L.E. Meyer-Bothling, and J.C. Polacco. 1987. Developmental genetics of soybean urease isozymes. Dev. Genet. 8:375–387.

Honeycutt, R.J., J.W. Burton, R.C. Shoemaker, and R.G. Palmer. 1989a. Expression and inheritance of a shriveled seed mutant in soybean. Crop Sci. 29:704–707.

Honeycutt, R.J., J.W. Burton, R.G. Palmer, and R.C. Shoemaker. 1989b. Association of major seed components with a shriveled seed trait in soybean. Crop Sci. 29:804–809.

Honeycutt, R.J., K.E. Newhouse, and R.G. Palmer. 1990. Inheritance and linkage studies of a variegated leaf mutant in soybean. J. Hered. 81:123–126.

Horejsi, T.F., W.R. Fehr, G.A. Welke, D.N. Duvick, E.G. Hammond, and S.R. Cianzio. 1994. Genetic control of reduced palmitate content in soybean. Crop Sci. 34:331–334.

Huang, S., D.A. Ashley, and H.R. Boerma. 1993. Light intensity, row spacing, and photoperiod effects on expression of brachytic stem in soybean. Crop Sci. 33:29–37.

Hymowitz, T. 2004. Speciation and cytogenetics. p. 97–136. *In* H.R. Boerma and J.E. Specht (ed.) Soybeans: Improvement, production, and uses. 3 rd ed. Agron. Monogr. 16. ASA, CSSA, and SSSA, Madison, WI.

Hymowitz, T., and H.H. Hadley. 1972. Inheritance of a trypsin inhibitor variant in seed protein of soybeans. Crop Sci. 12:197–198.

Hymowitz, T., N. Kaizuma, J.H. Orf, and H. Skorupska. 1979. Screening the USDA soybean germplasm collection for *Sp1* variants. Soybean Genet. Newsl. 6:30–32.

Hymowitz, T., R.G. Palmer, and H.R. Hadley. 1972. Seed weight, protein, oil, and fatty acid relationships within the genus *Glycine*. Tropic. Agric. 49:245–250.

Ilarslan, H., H.T. Horner, and R.G. Palmer. 1999. Genetics and cytology of a new male-sterile, female-fertile soybean [*Glycine max* (L.) Merr.] mutant. Crop Sci. 39:58–64.

Ilarslan, H., H.T. Skorupska, H.T. Horner, and R.G. Palmer. 1997. Genetics and cytology of a tissue-culture derived soybean genic male-sterile female-sterile. J. Hered. 88:129–138.

Imsande, J., J. Pittig, R.G. Palmer, C. Wimmer, and C. Gietl. 2001. Independent spontaneous mitochondrial malate dehydrogenase null mutants in soybean are the result of deletions. J. Hered. 92:333–338.

Jeong, S.C., S. Kristipati, A.J. Hayes, P.J. Maughan, S.L. Noffsinger, I. Gunduz, G.R. Buss, and M.A. Saghai Maroof. 2002. Genetic and sequence analysis of markers tightly linked to the soybean mosaic virus resistance gene, *Rsv3*. Crop Sci. 42:265–270.

Jin, W., H.T. Horner, and R.G. Palmer. 1997. Genetics and cytology of a new male-sterile soybean [*Glycine max* (L.) Merr.]. J. Sex. Plant Reprod. 10:13–21.

Jin, W., R.G. Palmer, H.T. Horner, and R.C. Shoemaker. 1998. Molecular mapping of a male-sterile gene in soybean. Crop Sci. 38:1681–1685.

Jin, W., R.G. Palmer, H.T. Horner, and R.C. Shoemaker. 1999. *Fr1* (root fluorescence) locus is located in a segregation distortion region on linkage group K of soybean genetic map. J. Hered. 90:553–556.

Johns, C.W., X. Delannay, and R.G. Palmer. 1981. Structural sterility controlled by nuclear mutations in angiosperms. The Nucleus 24:97–105.

Johns, C.W., and R.G. Palmer. 1982. Floral development of a flower-structure mutant in soybeans, *Glycine max* (L.) Merr. (Leguminosae). Am. J. Bot. 69:829-842.

Johnson, E.O.C., P.A. Stephens, D.A. Fasoula, C.D. Nickell, and L.O. Vodkin. 1998. Instability of a novel multicolored flower trait in inbred and outcrossed soybean lines. J. Hered. 89:508–515.

Karasawa, K. 1936. Crossing experiments with *Glycine soja* and *G. ussuriensis*. Jpn. J. Bot. 8:113–118.

Keeling, B.L. 1988. Influence of temperature on growth and pathogenicity of geographic isolates of *Diaporthe phaseolorum* var. *caulivora*. Plant Dis. 72:220–222.

Keen, N.T., and R.I. Buzzell. 1991. New disease resistance genes in soybean against *Pseudomonas syringae* pv.*glycinea*: Evidence that one of them interacts with a bacterial elicitor. Theor. Appl. Genet. 81:133–138.

Kenty, M.M., K. Hinson, K.H. Quesenberry, and D.S. Wofford. 1996. Inheritance of resistance to the soybean looper in soybean. Crop Sci. 36:1532–1537.

Kiang, Y.T. 1981. Inheritance and variation of amylase in cultivated and wild soybeans and their wild relatives. J. Hered. 72:382–386.

Kiang, Y.T. 1987. Mapping three protein loci on a soybean chromosome. Crop Sci. 27:44–46.

Kiang, Y.T. 1990a. Linkage analysis of *Pgd1*, *Pgi1*, pod color (*L1*), and determinate stem (*dt1*) loci on soybean linkage group 5. J. Hered. 81:402–404.

Kiang, Y.T. 1990b. Mapping the alcohol dehydrogenase locus (*Adh1*) in soybean Linkage Group 8. J. Hered. 81:488–489.

Kiang, Y.T., and C.J. Bult. 1991. Genetic and linkage analysis of aconitate hydratase variants in soybean. Crop Sci. 31:322–325.

Kiang, Y.T., and Y.C. Chiang. 1987. Genetic linkage of *Adh1* and *W1* loci in soybean. Genome 29:582–583.

Kiang, Y.T., and Y.C. Chiang. 1988. Mapping the beta amylase locus (*Am3*) on soybean linkage group 1 chromosome. J. Hered. 79:107–114.

Kiang, Y.T., Y.C. Chiang, and C.J. Bult. 1987. Genetic study of glutamate oxaloacetic transaminase in soybean. Genome 29:370–373.

Kiang, Y.T., Y.C. Chiang, and M.B. Gorman. 1984. Inheritance of a second leucine aminopeptidase locus and its linkage with other loci. Soybean Genet. Newsl. 11:143–145.

Kiang, Y.T., and M.B. Gorman. 1983. Soybean. p. 295–328. *In* S.D. Tanksley and T.J. Orton (ed.) Isoenzymes in plant genetics and breeding. Part B. Elsevier Publ. Co., New York.

Kiang, Y.T., and M.B. Gorman. 1985. Inheritance of NADP-active isocitrate dehydrogenase isozymes in soybeans. J. Hered. 76:279–284.

Kiang, Y.T., M.B. Gorman, and Y.C. Chiang. 1985. Genetic and linkage analysis of a leucine aminopeptidase in wild and cultivated soybeans. Crop Sci. 25:319–321.

Kiihl, R.A.S., and E.E. Hartwig. 1979. Inheritance of reaction to soybean mosaic virus in soybeans. Crop Sci. 19:372–375.

Kilen, T.C. 1977. Inheritance of a brachytic character in soybeans. Crop Sci. 17:853–854.

Kilen, T.C. 1983. Inheritance of a short petiole trait in soybean. Crop Sci. 23:1208–1210.

Kilen, T.C., and W.L. Barrentine. 1983. Linkage relationships in soybean between genes controlling reactions to phytophthora rot and metribuzin. Crop Sci. 23:894–896.

Kilen, T.C., and E.E. Hartwig. 1971. Inheritance of a light-quality sensitive character in soybeans. Crop Sci. 11:559–561.

Kilen, T.C., and E.E. Hartwig. 1975. Short internode character in soybeans and its inheritance. Crop Sci. 15:878.

Kilen, T.C., and E.E. Hartwig. 1987. Identification of single genes controlling resistance to stem canker in soybean. Crop Sci. 27:863–864.

Kilen, T.C., E.E. Hartwig, and B. L. Keeling. 1974. Inheritance of a second major gene for resistance to phytophthora rot in soybeans. Crop Sci. 14:260–262.

Kilen, T.C., B.L. Keeling, and E.E. Hartwig. 1985. Inheritance of reaction to stem canker in soybean. Crop Sci. 25:50–51.

Kilen, T.C., J.H. Hatchett, and E.E. Hartwig. 1977. Evaluation of early generation soybeans for resistance to soybean looper. Crop Sci. 17:397–398.

Kilen, T.C., and G.H. He. 1992. Identification and inheritance of metribuzin tolerance in wild soybean. Crop Sci. 32:684–685.

Kilen, T.C., and L. Lambert. 1993. Registration of three glabrous and three dense pubescent soybean germplasm lines susceptible (D88-5320, D88-5295), moderately resistant (D88-5328, D88-5272), or resistant (D90-9216, D90-9220) to foliar-feeding insects. Crop Sci. 33:215.

Kilen, T.C., and J.M. Tyler. 1993. Genetic linkage of the *Rps1* and *L2* loci in soybean. Crop Sci. 33:437–438.

Kitamura, K., C.S. Davies, N. Kaizuma, and N.C. Nielsen. 1983. Genetic analysis of a null-allele for lipoxygenase-3 in soybean seeds. Crop Sci. 23:924–927.

Kitamura, K., C.S. Davies, and N.C. Nielsen. 1984. Inheritance of alleles for *Cgy1* and *Gy4* storage protein genes in soybean. Theor. Appl. Genet. 68:253–257.

Kloth, R.H., and T. Hymowitz. 1985. Re-evaluation of the inheritance of urease in soybean seed. Crop Sci. 25:352–354.

Kloth.R.H., J.C. Polacco, and T. Hymowitz. 1987. The inheritance of a urease-null trait in soybeans. Theor. Appl. Genet. 73:410–418.

Kokubun, M., and S. Akao. 1994. Inheritance of supernodulation in soybean mutants EN6500. Soil Sci. Plant Nutr. 40:715–718.

Kollipara, K.P., R.J. Singh, and T. Hymowitz. 1996. Inheritance of protease inhibitors in *Glycine tomentella* Hayata (2n = 38), a perennial relative of soybean. J. Hered. 87:461–463.

Koshimizu, S., and T. Iizuka. 1963. Studies on soybean virus diseases in Japan. Bull. 27. Tohoku Natl. Agric. Exp. Stn., Tohoku, Japan.

Kosslak, R.M., J.R. Dieter, R.L. Ruff, M.A. Chamberlin, B.A. Bowen, and R.G. Palmer. 1996. Partial resistance to root-borne infection by *Phytophthora sojae* in three allelic necrotic root mutants in soybean. J. Hered. 87:415–422.

Kosslak, R.M., M.A. Chamberlin, R.G. Palmer, and B.A. Bowen. 1997. Programmed cell death in the root cortex of soybean *root necrosis* mutants. The Plant J. 11:729–745.

Kuykendall, L.D. 2003. Genus *Bradyrhizobium*, family *Bradyrhizobiaceae*. Bergey's manual of systematic bacteriology. (In press.)

Kuykendall, L.D., F.M. Hashem, R.B. Dadson, and G.H. Elkan. 2000. Nitrogen fixation. p. 492–505. *In* J. Lederberg and M. Alexander (ed.) Encyclopedia of microbiology. Vol. 1. Academic Press, New York.

Kwon, S.H., M.R. Chae, K.S. Park, and H.S. Song. 1990. Trypsin inhibitor variants in Korean land races and wild soybeans. Korean J. Crop Sci. 35:171–175.

Kwon, S.H., and J.H. Oh. 1980. Resistance to a necrotic strain of soybean mosaic virus in soybeans. Crop Sci. 20:403–404.

Lam-Sanchez, A., A.H. Probst, F.A. Laviolette, J.F. Schafer, and K.L. Athow. 1968. Sources and inheritance of resistance to *Phytophthora megasperma* var. *sojae* in soybeans. Crop Sci. 8:329–330.

Lambert, L., R.M. Beach, T.C. Kilen, and J.W. Todd. 1992. Soybean pubescence and its influence on larval development and oviposition preference of lepidopterous insects. Crop Sci. 32:463–466.

Larsen, A.L. 1967. Electrophoretic differences in seed proteins among varieties of soybean. Crop Sci. 7:311–313.

Larsen, A.L., and B.E. Caldwell. 1968. Inheritance of certain proteins in soybean seed. Crop Sci. 8:474–476.

Laviolette, F.A., and K.L. Athow. 1983. Two new physiologic races of *Phytophthora megasperma* f. sp. *glycinea*. Plant Dis. 67:497–498.

Leffel, R.C. 1994a. Registration of fasciated soybean germplasm line BARC-10. Crop Sci. 34:318.

Leffel, R.C. 1994b. Registration of six pairs of BARC-11 soybean near-isogenic lines, fasciated vs. normal. Crop Sci. 34:321.

Leffel, R.C. 1994c. Registration of BARC-12, a low linolenic acid soybean germplasm line. Crop Sci. 34:1426–1427.

Leffel, R.C., R.L. Bernard, and J.O. Yocum. 1993. Agronomic performance of fasciated soybean genotypes and their isogenic lines. Crop Sci. 33:427–432.

Leitz, R.A., G.L. Hartman, W.L. Pedersen, and C.D. Nickell. 2000. Races of *Phytophthora sojae* in Illinois. Plant Dis. 84:487.

Lewers, K.S., and R.G. Palmer. 1993. Genetic linkage in soybean: Linkage group 8. Soybean Genet. Newsl. 20:118–124.

Lewers, K.S., and R.G. Palmer. 1997. Recurrent selection in soybean. Plant Breed. Rev. 16:275–313.

Lewers, K.S., S.K. St. Martin, B.R. Hedges, and R.G. Palmer. 1998. Effects of the *Dt2* and *S* alleles on agronomic traits of F_1 hybrid soybean. Crop Sci. 38:1137–1142.

Lewers, K.S., S.K. St. Martin, B.R. Hedges, M.P. Widrlechner, and R.G. Palmer. 1996. Hybrid soybean seed production: A comparison of three methods. Crop Sci. 36:1560–1567.

Li, Z., R.F. Wilson, W.E. Rayford, and H.R. Boerma. 2002. Molecular mapping genes conditioning reduced palmitic acid content in N87-2122-4 soybean. Crop Sci. 42:373–378.

Liao, W., X. Chen, and R.G. Palmer. 1996. Inheritance of an endopeptidase null mutant. Soybean Genet. Newsl. 23:130–133.

Liao, W., and R.G. Palmer. 1997a. Genetic study of a diaphorase-2 null mutant. Soybean Genet. Newsl. 24:157–159.

Liao, W., and R.G. Palmer. 1997b. Inheritance and linkage studies of 6-phosphogluconate dehydrogenase in soybean. Soybean Genet. Newsl. 24:164–167.

Liao, W., and R.G. Palmer. 1997c. A new variant of phosphoglucose isomerase. Soybean Genet. Newsl. 24:179–181.

Lim, S. M. 1989. Inheritance of resistance to *Peronospora manshurica* races 2 and 33 in soybean. Phytopathology 79:877–879.

Lim, S., R.L. Bernard, C.D. Nickell, and L.E. Gray. 1984. New physiological race of *Peronospora manshurica* virulent to the gene *Rpm* in soybeans. Plant Dis. 68:71–72.

Lin, S.F., J.S. Baumer, D. Ivers, S.R. Cianzio, and R.C. Shoemaker. 1998. Field and nutrient solution tests measure similar mechanisms controlling deficiency chlorosis in soybean. Crop Sci. 38:254–259.

Lin, S.F., J.S. Baumer, D. Ivers, S.R. Cianzio, and R.C. Shoemaker. 2000a. Nutrient solution screening of Fe chlorosis resistance in soybean evaluated by molecular characterization. J. Plant Nutr. 23:1915–1928.

Lin, S.F., S.R. Cianzio, and R.C. Shoemaker. 1997. Mapping genetic loci for iron deficiency chlorosis in soybean. Mol. Breed. 3:219–229.

Lin, S.F., D. Grant, S.R. Cianzio, and R.C. Shoemaker. 2000b. Molecular characterization of iron deficiency chlorosis in soybean. J. Plant Nutr. 23:1929–1939.

Lohnes, D.G., and R.L. Bernard. 1992. Inheritance of resistance to powdery mildew in soybeans. Plant Dis. 76:964–965.

Lohnes, D.G., R.E. Wagner, and R.L. Bernard. 1993. Soybean genes *Rj2, Rmd,* and *Rps2* in Linkage Group 19. J. Hered. 84:109–111.

Lundeen, P.O., W.R. Fehr, E.G. Hammond, and S.R. Cianzio. 1987. Association of alleles for high stearic acid with agronomic characters of soybean. Crop Sci. 27:1102–1105.

Luzzi, B.M., H.R. Boerma, and R.S. Hussey. 1994. A gene for resistance to the southern root-knot nematode in soybean. J. Hered. 85:484–486.

Ma, G., G.R. Buss, and S.A. Tolin. 1994. Inheritance of lethal necrosis to soybean mosaic virus in PI 507389 soybean. p.106. *In* Agronomy abstracts. ASA, Madison, WI.

Ma, G., P. Chen, G.R. Buss, and S.A. Tolin. 1995. Genetic characteristics of two genes for resistance to soybean mosaic virus in PI 486355 soybean. Theor. Appl. Genet. 91:907–914.

Mahama, A.A., and R.G. Palmer. 1998. Genetic linkage in soybean: Classical linkage groups 6 and 8, and 'Clark' translocation. Soybean Genet. Newsl. 25:139–140.

Mahama, A.A., and R.G. Palmer. 2003. Translocation breakpoints in soybean classical genetic linkage groups 6 and 8. Crop Sci. 43:1602–1609.

Mahama, A.A., K.S. Lewers, and R.G. Palmer. 2002. Genetic linkage in soybean: Classical genetic linkage groups 6 and 8. Crop Sci. 42:1459–1464.

Mahmud, I., and A.H. Probst. 1953. Inheritance of gray hilum color in soybeans. Agron. J. 45:59–61.

Mariani, P., M. Lucchin, F. Guzzo, S. Varotto, and P. Parrini. 1991. Cytological evidence of a new male-sterile mutant in soybean. [*Glycine max* (L.) Merr.]. J. Sex. Plant Reprod. 4:197–202.

Mathews, A., B.J. Carroll, and P.M. Gresshoff. 1989. A new nonnodulation gene in soybean. J. Hered. 80:357–360.

Matson, A.L., and L.F. Williams. 1965. Evidence of a fourth gene for resistance to the soybean cyst nematode. Crop Sci. 5:477.

Matsuura, H. 1933. *Glycine soja*. p. 100–110. *In* A bibliographical monograph on plant genetics. 2nd ed. Hokkaido Imperial Univ., Tokyo.

McBlain, B.A., and R.L. Bernard. 1987. A new gene affecting the time of maturity in soybeans. J. Hered. 78:160–162.

McBlain, B.A., J.D. Hesketh, and R.L. Bernard. 1987. Genetic effects on reproductive phenology in soybean isolines differing in maturity genes. Can. J. Plant Sci. 67:105–116.

McLean, R.J., and D.E. Byth. 1980. Inheritance of resistance to rust *Phakopsora pachyrhizi* in soybeans. Aust. J. Agric. Res. 31:951–956.

Meyer-Bothling, L.E., and J.C. Polacco. 1987. Mutational analysis of the embryo-specific urease locus of soybean. Mol. Gen. Genet. 209:439–444.

Meyer-Bothling, L.E., J.C. Polacco, and S.R. Cianzio. 1987. Pleiotropic soybean mutants defective in both urease isozymes. Mol. Gen. Genet. 209:432–438.

Mian, M.A.R., E.R. Shipe, J. Alvernaz, J.D. Mueller, D.A. Ashley, and H.R. Boerma. 1997. RFLP analysis of chlorimuron ethyl sensitivity in soybean. J. Hered. 88:38–41.

Miroslav, K., and L. Jiri. 1996. Identification of soybean cultivars through isozymes. Soybean Genet. Newsl. 23:89–91.

Moots, C.K., C.D. Nickell, L.E. Gray, and S.M. Lim. 1983. Reaction of soybean cultivars to 14 races of *Phytophthora megasperma* f. sp. *glycinea*. Plant Dis. 67:764–767.

Morgan-Jones, G. 1989. The *Diaporthe/Phomopsis* complex: Taxonomic considerations. p. 1699–1706. *In* A.J. Pascale (ed.) Proc. World Soybean Res. Conf. IV, Buenos Aires, Argentina. 5–9 Mar. 1989. Orientacion Grafica Editora S.R.L., Buenos Aires, Argentina.

Morse, W.J., and J.L. Cartter. 1937. Improvement in soybeans. p. 1154–1189. *In* Yearbook agriculture. USDA. U.S. Gov. Print. Office, Washington, DC.

Muehlbauer, G.J., J.E. Specht, P.E. Staswick, G.L. Graef, and M.A. Thomas-Compton. 1989. Application of the near-isogenic line gene mapping technique to isozyme markers. Crop Sci. 29:1548–1553.

Mueller, E.H., K.L. Athow, and F.A. Laviolette. 1978. Inheritance to four physiologic races of *Phytophthora megasperma* var. *sojae*. Phytopathology 68:1318–1322.

Mukherjee, D., J.W. Lambert, R.L. Cooper, and B.W. Kennedy. 1966. Inheritance of resistance to bacterial blight in soybeans. Crop Sci. 6:324–326.

Nagai, I. 1921. A genetico-physiological study on the formation of anthocyanin and brown pigments in plants. Tokyo Univ. Coll. Agric. J. 8:1–92.

Nagai, I. 1926. Inheritance in the soybean. (In Japanese.) Nogyo Oyobi Engei 1:14,107–108.

Nagai, I., and S. Saito. 1923. Linked factors in soybeans. Jpn. J. Bot. 1:121–136.

Narvel, J.M., W.R. Fehr, J. Ininda, G.A. Welke, E.G. Hammond, D.N. Duvick, and S.R. Cianzio. 2000. Inheritance of elevated palmitate in soybean seed oil. Crop Sci. 40:635–639.

Nelson, R.L. 1996. The inheritance of a branching type in soybean. Crop Sci. 36:1150–1152.

Nelson, R.L., P.J. Amdor, J.H. Orf, and J.F. Cavins. 1988. Evaluation of the USDA Soybean Germplasm Collection: Maturity groups 000 to IV (PI 427136 to PI 445845). Tech. Bull. 1726. U.S. Dep. of Agric.-Agric. Res. Serv., Washington, DC.

Nguyen, M.V., and J.M. Widholm. 1997. A second tissue culture-induced wrinkled-leaf mutation is also maternally inherited. Soybean Genet. Newsl. 24:142.

Nielsen, N.C., C.D. Dickinson, T.-J. Cho, V.H. Thanh, B.J. Scallon, R.L. Fischer, T.L. Sims, G.N. Drews, and R.B. Goldberg. 1989. Characterization of the glycinin gene family in soybean. Plant Cell 1:313–328.

Nissly, C.R., R.L. Bernard, and C.N. Hittle. 1976. Inheritance in chlorophyll-deficient mutants. Soybean Genet. Newsl. 3:31–34.

Nissly, C.R., R.L. Bernard, and C.N. Hittle. 1981. Inheritance of two chlorophyll-deficient mutants in soybeans. J. Hered. 72:141–142.

Orf, J.H., and T. Hymowitz. 1976. The gene symbols *Sp1-a* and *Sp1-b* assigned to Larsen and Caldwell's seed protein bands A and B. Soybean Genet. Newsl. 7:64–66.

Orf, J.H., and T. Hymowitz. 1977. Inheritance of a second trypsin inhibitor variant in seed protein of soybeans. Crop Sci. 17:811–813.

Orf, J.H., and T. Hymowitz. 1979. Inheritance of the absence of the Kunitz trypsin inhibitor in seed protein of soybeans. Crop Sci. 19:107–109.

Orf, J.H., T. Hymowitz. S.P. Pull, and S.G. Pueppke. 1978. Inheritance of a soybean seed lectin. Crop Sci. 18:899–900.

Owen, F.V. 1927a. Inheritance studies in soybeans. I. Cotyledon color. Genetics 12:441–448.

Owen, F.V. 1927b. Inheritance studies in soybeans. II. Glabrousness, color of pubescence, time of maturity, and linkage relations. Genetics 12:519–529.

Owen, F.V. 1928. Inheritance studies in soybeans. III. Seed coat color and summary of all other mendelian characters thus far reported. Genetics 13:50–79.

Padgette, S.R., K.H. Kolacz, X. Delannay, D.B. Re, B. J. LaVallee, C.N. Tinius, W.K. Rhodes, Y.I. Otero, G.F. Barry et al. 1995. Development, identification, and characterization of a glyphosate-tolerant soybean line. Crop Sci. 35:1451–1461.

Palmer, R.G. 1974. A desynaptic mutant in the soybean. J. Hered. 65:280-286.

Palmer, R.G. 1976. Cytogenetics in soybean improvement. p. 56–66. In H.D. Loden and D. Wilkinson (ed.) Proc. 6th Soybean Seed Res. Conf., Chicago, IL. 9–10 Dec. 1976. Publ. 6. Am. Seed Trade Assoc., Washington, DC.

Palmer, R.G. 1977. Soybean linkage tests. Soybean Genet. Newsl. 4:40–42.

Palmer, R.G. 1984a. Genetic studies with T263. Soybean Genet. Newsl. 11:94–97.

Palmer, R.G. 1984b. Pleiotropy or close linkage of two mutants in soybean. J. Hered. 75:457–462.

Palmer, R.G. 1985. Soybean cytogenetics. p. 337–344. In R. Shibles (ed.) Proc. World Soybean Res. Conf. III, Ames, IA. 12–17 Aug. 1984. Westview Press, Boulder, CO.

Palmer, R.G. 1987. Inheritance and derivation of T218H. Soybean Genet. Newsl. 14:183–185.

Palmer, R.G. 1992. Conditional lethality involving a cytoplasmic mutant and chlorophyll-deficient malate dehydrogenase mutants in soybean. Theor. Appl. Genet. 85:389–393.

Palmer, R.G. 2000. Genetics of four male-sterile, female-fertile soybean mutants. Crop Sci. 40:78–83.

Palmer, R.G., M.C. Albertsen, H.T. Horner, H. Skorupska. 1992. Male sterility in soybean and maize: Developmental comparisons. The Nucleus 35:1–18.

Palmer, R.G., J.D. Burzlaff, and R.C. Shoemaker. 2000. Genetic analyses of two independent chlorophyll-deficient mutants identified among the progeny of a single chimeric foliage soybean plant. J. Hered. 91:297–303.

Palmer, R.G., and X.F. Chen. 1996. A chimeric soybean plant. Soybean Genet. Newsl. 23:123–125.

Palmer, R.G., and X.F. Chen. 1998a. Assignment of the Fr3 locus to soybean linkage group 9. J. Hered. 89:181–184.

Palmer, R.G., and X.F. Chen. 1998b. Genetic linkage in soybean: Classical linkage groups 6 and 8. Soybean Genet. Newsl. 25:138.

Palmer, R.G., and S.R. Cianzio. 1985. Conditional lethality involving nuclear and cytoplasmic chlorophyll mutants in soybeans. Theor. Appl. Genet. 70:349–354.

Palmer, R.G., J. Gai, H. Sun, and J.W. Burton. 2001. Production and evaluation of hybrid soybean. Plant Breed. Rev. 21:263–307.

Palmer, R.G., and R.W. Groose. 1993. A new allele at the w4 locus derived from the w4-m mutable allele in soybean. J. Hered. 84:297–300.

Palmer, R.G., R.W. Groose, H.D. Weigelt, and J.E. Miller. 1990a. Registration of a genetic stock (w4-m w4-m) for unstable anthocyanin pigmentation in soybean. Crop Sci. 30:1376–1377.

Palmer, R.G., B.R. Hedges, R.S. Benavente, and R.W. Groose. 1989. The w4-mutable line in soybean. Devel. Genet. 10:542–551.

Palmer, R.G., J.B. Holland, and K.S. Lewers. 1998. Recombination values for the Ms6 – W1 chromosome region in different genetic backgrounds in soybean. Crop Sci. 38:293–296.

Palmer, R.G., and H.T. Horner. 2000. Genetics and cytology of a genic male-sterile, female-sterile mutant from a transposon containing soybean population. J. Hered. 91:378–383.

Palmer, R.G., and M.L.H. Kaul. 1983. Genetics, cytology, and linkage studies of a desynaptic soybean mutant. J. Hered. 74:260–264.

Palmer, R.G., and T.C. Kilen. 1987. Qualitative genetics and cytogenetics. p.135–209. In J.R. Wilcox (ed.) Soybeans: Improvement, production, and uses. Agron. Monogr. 16. ASA, CSSA, and SSSA, Madison, WI.

Palmer, R.G., and K.S. Lewers. 1998. Registration of 68 soybean germplasm lines segregating for male sterility. Crop Sci. 38:560–562.

Palmer, R.G., S.M. Lim, and B.R. Hedges. 1992. Testing for linkage between the *Rxp* locus and nine isozyme loci in soybean. Crop Sci. 32:681–683.

Palmer, R.G., and P.N. Mascia. 1980. Genetics and ultrastructure of a cytoplasmically inherited yellow mutant in soybeans. Genetics 95:985–1000.

Palmer, R.G., and V.C.M. Minor. 1994. Nuclear-cytoplasmic interaction in chlorophyll-deficient soybean, *Glycine max* (Fabaceae). Am. J. Bot. 81:997–1003.

Palmer, R.G., R.L. Nelson, R.L. Bernard, and D.M. Stelly. 1990b. Linkage and inheritance of three chlorophyll-deficient mutants in soybean. J. Hered. 81:404–406.

Palmer, R.G., and R.C. Payne. 1979. Genetic control of hypocotyl pigmentation among white-flowered soybeans grown in continuous light. Crop Sci. 19:124–126.

Palmer, R.G., and H. Skorupska. 1990. Registration of a male-sterile line (T295H) of soybean. Crop Sci. 30:244.

Palmer, R.G., and D.M. Stelly. 1979. Reference diagrams of seed coat colors and patterns for use as genetic markers in crosses. Soybean Genet. Newsl. 6:55–57.

Palmer, R.G., C.L. Winger, and M.C. Albertsen. 1978b. Four independent mutations at the *msl* locus in soybeans. Crop Sci. 18:727–729.

Palmer, R.G., C.L. Winger, and P.S. Muir. 1980. Genetics and cytology of the *ms3* male-sterile soybean. J. Hered. 71:343–348.

Palmer, R.G., and M. Wubben. 1998. Inheritance and allelism of an EMS-generated necrotic root mutant. Soybean Genet. Newsl. 25:142–143.

Pantalone, V.R., G.J. Rebetzke, J.W. Burton, and R.F. Wilson. 1997. Genetic regulation of linolenic acid concentration in wild soybean *Glycine soja* accessions. J. Am. Oil. Chem. Soc. 74:159–163.

Park, K.S., and M.S. Yoon. 1997. Variation of leucine aminopeptidase isozyme in Korean landraces and wild soybeans. Korean J. Crop Sci. 42:129–133.

Pereira, T.N.S., H. Ilarslan, and R.G. Palmer. 1996. Embryological study of a female partial-sterile soybean mutant by using confocal scanning laser microscopy. Braz. J. Genet. 19:435–440.

Pereira, T.N.S., H. Ilarslan, and R.G. Palmer. 1997a. Genetic and cytological analyses of three lethal ovule mutants in soybean (*Glycine max*; Leguminosae). Genome 40:273–285.

Pereira, T.N.S., N.R. Lersten, and R.G. Palmer. 1997b. Genetic and cytological analyses of a partial-sterile mutant (PS-1) in soybean (*Glycine max*; Leguminosae). Am. J. Bot. 84:781–791.

Perry, M.C., and M.S. McIntosh. 1991. Geographical patterns of variation in the USDA soybean germplasm collection: I. Morphological traits. Crop Sci. 31:1350–1355.

Perry, M.C., M.S. McIntosh, and A.K. Stoner. 1991. Geographical patterns of variation in the USDA soybean germplasm collection: II. Allozyme frequencies. Crop Sci. 31:1356–1360.

Peters, D.W., J.R. Wilcox, J.J. Vorst, and N.C. Nielsen. 1984. Hypocotyl pigments in soybeans. Crop Sci. 24:237–239.

Peterson, P.A., and C.R. Weber. 1969. An unstable locus in soybeans. Theor. Appl. Genet. 39:156–162.

Pfeiffer, T.W., D.F. Hildebrand, and J.H. Orf. 1993. Inheritance of a lipoxygenase—1 allozyme in soybean. Crop Sci. 33:691–693.

Piper, C.G., and W.J. Morse. 1910. The soybean: History, varieties, and field studies. USDA Bureau of Plant Industry Bull. 197. U.S. Gov. Print. Office, Washington, DC.

Ploper, L.D., K.L. Athow, and F.A. Laviolette. 1985. A new allele at the *Rps3* locus for resistance to *Phytophthora megasperma* f. sp. *glycinea* in soybean. Phytopathology 75:690–694.

Polacco, J.C., A.K. Judd, J.K. Dybing, and S.R. Cianzio. 1989. A new mutant class of soybean lacks urease in leaves but not in leaf-derived callus or in roots. Mol. Gen. Genet. 217:257–262.

Pomeranke, G.J., and C.D. Nickell. 1988. Inheritance of chlorimuron ethyl sensitivity in the soybean strains BSR 101 and M74-462. Crop Sci. 28:59–60.

Porter, K.B., and M.G. Weiss. 1948. The effect of polyploidy on soybeans. J. Am. Soc. Agron. 40:710–724.

Pracht, J.E., C.D. Nickell, and J.E. Harper. 1993a. Genes controlling nodulation in soybean: *Rj5* and *Rj6*. Crop Sci. 33:711–713.

Pracht, J.E., C.D. Nickell, and J.E. Harper.. 1993b. Genetic analysis of a hypernodulating mutant of soybean. Soybean Genet. Newsl. 20:107–111.

Primomo, V.S., D.E. Falk, G.R. Ablett, J.W. Tanner, and I. Rajcan. 2002a. Inheritance and interaction of low palmitic and low linolenic soybean. Crop Sci. 42:31–36.

Primomo, V.S., D.E. Falk, G.R. Ablett, J.W. Tanner, and I. Rajcan. 2002b. Genotype X environment interactions, stability, and agronomic performance of soybean with altered fatty acid profiles. Crop Sci. 42:37–44.

Probst, A.H. 1950. The inheritance of leaf abscission and other characters in soybeans. Agron. J. 42:35–45.

Probst, A.H., K.L. Athow, and F.A. Laviolette. 1965. Inheritance of resistance to race 2 of *Cercospora sojina* in soybeans. Crop Sci. 5:332.

Pull, S.P., S.G. Pueppke, T. Hymowitz, and J.H. Orf. 1978. Soybean lines lacking the 120,000 dalton seed lectin. Science (Washington DC) 200:1277–1279.

Raboy V., and D.B. Dickinson. 1993. Phytic acid levels in seeds of *Glycine max* and *G. soja* as influenced by phosphorus status. Crop Sci. 33:1300–1305.

Raboy V., D.B. Dickinson, and F.E. Below. 1984. Variation in seed total phosphorus, phytic acid, zinc, calcium, magnesium, and protein among lines of *Glycine max* and *G. soja.* Crop Sci. 24:431–434.

Rahman, S.M., T. Kinoshita, T. Anai, S. Arima, and Y. Takagi. 1998. Genetic relationships of soybean mutants for different linolenic acid contents. Crop Sci. 38:702–706.

Rahman, S.M., T. Kinoshita, T. Anai, and Y. Takagi. 1999. Genetic relationship between loci for palmitate contents in soybean mutants. J. Hered. 90:423–428.

Rahman, S.M., and Y. Takagi. 1997. Inheritance of reduced linolenic acid content in soybean seed oil. Theor. Appl. Genet. 94:299–302.

Rahman, S.M., Y.Takagi, and T. Kumamaru. 1996a. Low linolenate sources at the *Fan* locus in soybean lines M-5 and IL-8. Breed. Sci. 46:155–158.

Rahman, S.M., Y.Takagi, and T. Kinoshita. 1996b. Genetic control of high oleic acid content in the seed oil of two soybean mutants. Crop Sci. 36:1125–1128.

Rahman, S.M., Y.Takagi, and T. Kinoshita. 1997. Genetic control of high stearic acid content in seed oil of two soybean mutants. Theor. Appl. Genet. 95:772–776.

Rao-Arelli, A.P. 1994. Inheritance of resistance to *Heterodera glycines* race 3 in soybean accessions. Plant Dis. 78:898–900.

Rao-Arelli, A.P., S.C. Anand, and J.A. Wrather. 1992. Soybean resistance to soybean cyst nematode race 3 is conditioned by an additional dominant gene. Crop Sci. 32:862–864.

Rasooly, A., and T.G. Isleib. 1993. Epistasis of *rj1* nonnodulation of soybean to nodulation by *Sinorhizobium fredii*. Crop Sci. 33:329–331.

Ray, J.D., K. Hinson, J.E.B. Mankono, and M.F. Malo. 1995. Genetic control of a long-juvenile trait in soybean. Crop Sci. 35:1001–1006.

Rebetzke, G.J., J.W. Burton, T.E. Carter, Jr., and R.F. Wilson. 1998a. Changes in agronomic and seed characteristics with selection for reduced palmitic acid content in soybean. Crop Sci. 38:297–302.

Rebetzke, G.J., J.W. Burton, T.E. Carter, Jr., and R.F. Wilson. 1998b. Genetic variation for modifiers controlling reduced saturated fatty acid content in soybean. Crop Sci. 38:303–308.

Reese, P.F. Jr., and H.R. Boerma. 1989. Additional genes for green seed coat in soybean. J. Hered. 80:86–88.

Rennie, B.D., W.D. Beversdorf, and R.I. Buzzell. 1987a. Genetic and linkage analysis of an aconitate hydratase variant in soybean. J. Hered. 78:323–326.

Rennie, B.D., W.D. Beversdorf, and R.I. Buzzell. 1987b. Inheritance and linkage analysis of two endopeptidase variants in soybeans. J. Hered. 78:327–328.

Rennie, B.D., and J.W. Tanner. 1989a. Mapping a second fatty acid locus to soybean linkage group 17. Crop Sci. 29: 1081–1083.

Rennie, B.D., and J.W. Tanner. 1989b. Genetic analysis of low linolenic acid levels in the line PI 123440. Soybean Genet. Newsl. 16:25–26.

Rennie, B.D., M.L. Thorpe, and W.D. Beversdorf. 1989. A procedural manual for the detection and identification of soybean [*Glycine max* (L.) Merr.] isozymes using the starch gel electrophoretic system. Dep. Crop Sci. Tech. Bull. 1889. Univ. of Guelph, ON, Canada.

Rennie, B.D., J. Zilka, M.M. Cramer, and W.D. Beversdorf. 1988. Genetic analysis of low linolenic acid levels in the soybean line PI 361088B. Crop Sci. 28:655–657.

Rhodes, P.R., and L.O. Vodkin. 1988. Organization of the *Tgm* family of transposable elements. Genetics 120:597–604.

Roane, C.W., S.A. Tolin, and G.R. Buss. 1983. Inheritance of reaction to two viruses in the soybean cross 'York' x 'Lee 68'. J. Hered. 74:289–291.

Rode, M.W., and R.L. Bernard. 1975a. Inheritance of a tan saddle mutant. Soybean Genet. Newsl. 2:39–42.

Rode, M.W., and R.L. Bernard. 1975b. Inheritance of wavy leaf. Soybean Genet. Newsl. 2:42–44.

Rode, M.W., and R.L. Bernard. 1975c. Inheritance of bullate leaf. Soybean Genet. Newsl. 2:44–46.

Ross, A.J. 1999. Inheritance of reduced-linolenate soybean oil and its influence on agronomic and seed traits. M.S. thesis. Iowa State Univ., Ames.

Ross, A.J., W.R. Fehr, G.A. Welke, E.G. Hammond, and S.R. Cianzio. 2000. Agronomic and seed traits of 1%-linolenate soybean genotypes. Crop Sci. 40:383–386.

Roy, K.W., J.C. Rupe, D.E. Hershman, and T.S. Abney. 1997. Sudden death syndrome of soybean. Plant Dis. 81:1100–1110.

Rubaihayo, P.R., and G. Gumisiriza. 1978. The causes of genetic male sterility in three soybean lines. Theor. Appl. Genet. 53:257–260.

Ryan, S.A., R.S. Nelson, and J.E. Harper. 1983a. Selection and inheritance of nitrate reductase mutants in soybeans. Soybean Genet. Newsl. 10:33–35.

Ryan, S.A., R.S. Nelson, and J.E. Harper. 1983b. Soybean mutants lacking constitutive nitrate reductase activity. II. Nitrogen assimilation, chlorate resistance, and inheritance. Plant Physiol. 72:510–514.

Sadanaga, K. 1983. Locating *wm* on linkage group 8. Soybean Genet. Newsl. 10:39–41.

Sadanaga, K., and R.L. Grindeland. 1984. Locating the *w1* locus on the satellite chromosome in soybean. Crop Sci. 24:147–151.

Saindon, G., H.D. Voldeng, W.D. Beversdorf, and R.I. Buzzell. 1989. Genetic control of long daylength response in soybean. Crop Sci. 29:1436–1439.

Sawada, S., and R.G. Palmer. 1987. Genetic analysis of nonfluorescent root mutants induced by mutagenesis in soybean. Crop Sci. 27:62–65.

Scallon, B.J., C.D. Dickinson, and N.C. Nielson. 1987. Characterization of a null-allele for the *Gy4* glycinin gene from soybean. Mol. Gen. Genet. 208:107–113.

Schmitthenner, A. F., M. Hobe, and R.G. Bhat. 1994. *Phytophthora sojae* races in Ohio over a 10-year interval. Plant Dis. 78:269–276.

Schnebly, S.R., W.R. Fehr, G.A. Welke, E.G. Hammond, and D.N. Duvick. 1994. Inheritance of reduced and elevated palmitate in mutant lines of soybean. Crop Sci. 34:829–833.

Sebastian, S.A., and R.S. Chaleff. 1987. Soybean mutants with increased tolerance for sulfonylurea herbicides. Crop Sci. 27:948–952.

Sebastian, S.A., G.M. Fader, J.F. Ulrich, D.R. Forney, and R.S. Chaleff. 1989. Semidominant soybean mutation for resistance to sulfonylurea herbicides. Crop Sci. 29:1403–1408.

Seo, Y.W., J.E. Specht, G.L. Graef, and R.A. Graybosch. 1993. Inheritance of red-buff seed coat in soybean. Crop Sci. 33:754–758.

Sheridan, M.A., and R.G. Palmer. 1975. Inheritance and derivation of T225H, *Y18 y18*. Soybean Genet. Newsl. 2:18–19.

Sheridan, M.A., and R.G. Palmer. 1977. The effect of temperature on an unstable gene in soybeans. J. Hered. 68:17–22.

Shigemori, I. 1988. Inheritance of resistance to soybean mosaic virus (SMV) C-strain in soybeans. Jpn. J. Breed. 38:346–356.

Shipe, E.R., G.R. Buss, and S.A. Tolin. 1979. A second gene for resistance to peanut mottle virus in soybeans. Crop Sci. 19:656–658.

Shoemaker, R.C., A.M. Cody, and R.G. Palmer. 1985. Characterization of a cytoplasmically inherited yellow foliar mutant (*cyt-Y3*) in soybean. Theor. Appl. Genet. 69:279–284.

Sinclair, T.R., and K. Hinson. 1992. Soybean flowering in response to the long-juvenile trait. Crop Sci. 32:1242–1248.

Singh, B.B. 1972. High frequency of natural cross-pollination in a mutant strain of soybean. Curr. Sci. 41:832–833.

Singh, B.B., and A.N. Jha. 1978. Abnormal differentiation of floral parts in a mutant strain of soybean. J. Hered. 69:143–144.

Singh, B.B., and P.N. Thapliyal. 1977. Breeding for resistance to soybean rust in India. p. 62–65. *In* R.E. Ford and J.B. Sinclair (ed.) Rust of soybeans, the problem and research needs. Int. Agric. Publ. INTSOY Ser. 12. Univ. of Illinois, Urbana.

Singh, L., C.M. Wilson, and H.H. Hadley. 1969. Genetic differences in soybean trypsin inhibitors separated by disc electrophoresis. Crop Sci. 9:489–491.

Sisson, V.A., P.A. Miller, W.V. Campbell, and J.W. Van Duyn. 1976. Evidence of inheritance of resistance to the Mexican bean beetle in soybeans. Crop Sci. 16:835–837.

Skorupska, H., N.V. Desamero, and R.G. Palmer. 1993. Possible homeosis in flower development in cultivated soybean [*Glycine max* (L.) Merr.]: morphoanatomy and genetics. J. Hered. 84:97–104.

Skorupska, H., N.V. Desamero, and R.G. Palmer. 1994. Developmental hormonal expression of apetalous male-sterile mutations in soybean, *Glycine max* (L.) Merr. Ann. Biol. 10:152–164.

Skorupska, H., and R.G. Palmer. 1989. Genetics and cytology of the *ms6* male-sterile soybean. J. Hered. 80:304-310.

Skorupska, H., and R.G. Palmer. 1990. Additional sterile mutations in soybean [*Glycine max* (L.) Merr.]. J. Hered. 81:296–300.

Smith, M.B., H.T. Horner, and R.G. Palmer. 2001. Temperature and photoperiod effects on sterility in a cytoplasmic male-sterile soybean. Crop Sci. 41:702–704.

Smith, M.B., R.G. Palmer, and H.T. Horner. 2002. Microscopy study of a cytoplasmic male-sterile soybean from an interspecific cross between *Glycine max* and *G. soja*. Am. J. Bot. 89:417–426.

Sneller, C.H., T.G. Isleib, and T.E. Carter, Jr. 1992. Isozyme screening of near-isogenic male-sterile soybean lines to uncover potential linkages: Linkage of the *Pgm1* and *ms2* loci. J. Hered. 83:457–459.

Soybean Genetics Committee. 1995. Soybean genetics committee report. Soybean Genet. Newsl. 22: 11–14.

Specht, J.E., and G.L. Graef. 1992a. Registration of soybean germplasm lines possessing a dense pubescence (*Pd1Pd1*) phenotype. Crop Sci. 32:501.

Specht, J.E., and G.L. Graef. 1992b. Registration of soybean germplasm SG1E6. Crop Sci. 32:1080–1082.

Specht, J.E., and J.H. Williams. 1978. Hilum color as a genetic marker in soybean crosses. Soybean Genet. Newsl. 5:70–73.

Srinivasan, A., and J. Arihara. 1994. Soybean seed discoloration and cracking in response to low temperatures during early reproductive growth. Crop Sci. 34:1611–1617.

Stahlhut, R.W., and T. Hymowitz. 1980. Screening the USDA soybean germplasm collection for lines lacking the 120,000 dalton seed lectin. Soybean Genet. Newsl. 7:41–43.

Stelly, D.M., and R.G. Palmer. 1980a. A partially male-sterile mutant line of soybeans, *Glycine max* (L.) Merr.: Inheritance. Euphytica 29:295–303.

Stelly, D.M., and R.G. Palmer. 1980b. A partially male-sterile mutant line of soybeans *Glycine max* (L.) Merr.: Characterization of *msp* phenotype variability. Euphytica 29:539-546.

Stephens, P.A., U.B. Barwale-Zehr, C.D. Nickell, and J.M. Widholm. 1991. A cytoplasmically inherited, wrinkled-leaf mutant in soybean. J. Hered. 82:71–73.

Stephens, P.A., and C.D. Nickell. 1992. Inheritance of pink flower in soybean. Crop Sci. 32:1131–1132.

Stephens, P.A., C.D. Nickell, and F. L. Kolb. 1993a. Genetic analysis of resistance to *Fusarium solani* in soybean. Crop Sci. 33:929–930.

Stephens, P.A., C.D. Nickell, and L.O. Vodkin. 1993b. Pink flower color associated with increased protein and seed size in soybean. Crop Sci. 33:1135–1137.

Stewart, R.T. 1930. Inheritance of certain seed-coat colors in soybeans. J. Agric. Res. 40:829–854.

Stewart, R.T., and J.B. Wentz. 1926. A recessive glabrous character in soybeans. J. Am. Soc. Agron. 18:997–1009.

Stojšin, D., G.R. Ablett, B.M. Luzzi, and J.W. Tanner. 1998a. Use of gene substitution values to quantify partial dominance in low palmitic acid soybean. Crop Sci. 38:1437–1441.

Stojšin, D., B.M. Luzzi, G.R. Ablett, and J.W. Tanner. 1998b. Inheritance of low linolenic acid level in the soybean line RG10. Crop Sci. 38:1441–1444.

Stoltzfus, D.L., W.R. Fehr, G.A. Welke, E.G. Hammond, and S.R. Cianzio. 2000a. A *fap5* allele for elevated palmitate in soybean. Crop Sci. 40:647–650.

Stoltzfus, D.L., W.R. Fehr, G.A. Welke, E.G. Hammond, and S.R. Cianzio. 2000b. A *fap7* allele for elevated palmitate in soybean. Crop Sci. 40:1538–1542.

Suárez, J.C., G.L. Graef, W.R Fehr, and S.R. Cianzio. 1991. Association of isozyme genotypes with agronomic and seed composition traits in soybean. Euphytica 52:137–146.

Sun, H., L. Zhao, and M. Huang. 1997. Cytoplasmic-nuclear male-sterile soybean line from interspecific crosses between *G. max* and *G. soja*. p. 99–102. *In* C. Chainuvat and N. Sarobol (ed.) World Soybean Res. Conf. V, Chiang Mai, Thailand. 21–27 Feb. 1994. Kasetsart Univ. Press, Bangkok, Thailand.

Takagi, F. 1929. On the inheritance of some characters in *Glycine soja*, Bentham (soybean). (In Japanese.) Sci. Rep. Tohoku Univ., Ser. 4:577–589.

Takagi, F. 1930. On the inheritance of some characters in *Glycine soja*, Bentham (soybean). (In Japanese.) Jpn. J. Genet. 5:177–189.

Takagi, Y., A.B.M.M. Hossain, T. Yanagita, and S. Kusaba. 1989. High linolenic acid mutant in soybean induced by X-ray irradiation. Jpn. J. Breed. 39:403–409.

Takagi, Y., and S.M. Rahman. 1996. Inheritance of high oleic acid content in the seed oil of soybean mutant M23. Theor. Appl. Genet. 92:179–182.

Takagi, Y., S.M. Rahman, H. Joo, and T. Kawakita. 1995. Reduced and elevated palmitic acid mutants in soybean developed by X-ray irradiation. Biosci. Biochem. 59:1778–1779.

Takahashi, N. 1934. Linkage relation between the genes for the forms of leaves and the number of seeds per pod of soybeans. (In Japanese, English summary.) Jpn. J. Genet. 9:208–225.

Takahashi, R., and J. Abe. 1994. Genetic and linkage analysis of low temperature-induced browning in soybean seed coats. J. Hered. 85:447–450.

Takahashi, R., and S. Asanuma. 1996. Association of *T* gene with chilling tolerance in soybean. Crop Sci. 36:559–562.

Takahashi, R., H. Kurosaki, S. Yumoto, O.K. Han, and J. Abe. 2001. Genetic and linkage analysis of cleistogamy in soybean. J. Hered. 92:89–92.

Takahashi, Y., and J. Fukuyama. 1919. Morphological and genetic studies on the soybean. (In Japanese.). Hokkaido Agric. Exp. Stn. Rep. 10.

Tang, W.T., and G. Tai. 1962. Studies on the qualitative and quantitative inheritance of an interspecific cross of soybean, *Glycine max* X *G. formosana*. Bot. Bull. Acad. Sin. 3:39–60.

Tang, YuHong, and H.T. Skorupska. 1997. Expression of fasciation mutation in apical meristems of soybean, *Glycine max* (Leguminosae). Am. J. Bot. 34:328–335.

Tasma, I.M., L.L. Lorenzen, D.E. Green, and R.C. Shoemaker. 2001. Mapping genetic loci for flowering time, maturity, and photoperiod insensitivity in soybean. Mol. Breed. 8:25–35.

Terao, H. 1918. Maternal inheritance in the soybean. Am. Nat. 52:51–56.

Terao, H., and S. Nakatomi. 1929. On the inheritance of chlorophyll colorations of cotyledons and seedcoats in the soybean. (In Japanese, English summary.) Jpn. J. Genet. 4:64–80.

Thompson, J.A., R.L. Bernard, and R.L. Nelson. 1997. A third allele at the soybean *dt1* locus. Crop Sci. 37:757–762.

Thorson, P.R., B.R. Hedges, and R.G. Palmer. 1989. Genetic linkage in soybean: Linkage group 14. Crop Sci. 29:698–700.

Ting, C.L. 1946. Genetic studies on the wild and cultivated soybeans. J. Am. Soc. Agron. 38:381–393.

Toda, K., D. Yang, N. Yamanaka, S. Watanabe, K. Harada, and R. Takahashi. 2002. A single-base deletion in soybean flavonoid 3-hydroxylase gene is associated with gray pubescence color. Plant Mol. Biol. 50:187–196.

Todd, J.J., and L.O. Vodkin. 1993. Pigmented soybean (*Glycine max*) seed coats accumulate proanthocyanins during development. Plant Physiol. 102:663–670.

Todd, J.J., and L.O. Vodkin. 1996. Duplications that suppress and deletions that restore expression from a chalcone synthase multigene family. The Plant Cell 8:687–699.

Torkelson, J., and R.G. Palmer. 1997. Genetic analysis of root fluorescence of Central China germplasm. Soybean Genet. Newsl. 24:147–149.

Trese, A.T. 1995. A single dominant gene in McCall soybean prevents effective nodulation with *Rhizobium fredii* USDA 257. Euphytica 81:279–282.

Ude, G.N., T.E. Devine, L.D. Kuykendall, B.F. Matthews. J.A. Saunders, W. Kenworthy, and J.J. Lin. 1999. Molecular mapping of the soybean nodulation gene, *Rj4*. Symbiosis 26:101–110.

Umezaki, T., S. Matsumoto, and I. Shimano. 1988. Studies on dwarf lines in soybean. I. Growth habits and inheritance of Hyuga-dwarf line. (In Japanese, English summary.) Jpn. J. Crop Sci. 57:512–521.

VanSchaik, P.H., and A.H. Probst. 1958. The inheritance of inflorescence type, peduncle length, flowers per node, and percent flower shedding in soybean. Agron. J. 50:98–102.

Veatch, C., and C.M. Woodworth. 1930. Genetic relations of cotyledon color types of soybeans. J. Am. Soc. Agron. 22:700–702.

Vest, G. 1970. *Rj3* -a gene conditioning ineffective nodulation in soybean. Crop Sci. 10:34–35.

Vest, G., and B.E. Caldwell. 1972. *Rj4* -a gene conditioning ineffective nodulation in soybean. Crop Sci. 12:692–693.

Vierling, R.A., R.G. Palmer, and J.R. Wilcox. 1998. Non-peroxidase oxidation of guaiacol. Seed Sci. Technol. 20:91–93.

Vierling, R.A., and J.R. Wilcox. 1996. Microplate assay for soybean seed coat peroxidase activity. Seed Sci. Technol. 24:485–494.

Voldeng, H.D., and G. Saindon. 1991a. Registration of four pairs of 'Maple Presto' derived soybean genetic stocks. Crop Sci. 31:1398–1399.

Voldeng, H.D., and G. Saindon. 1991b. Registration of seven long-daylength insensitive soybean genetic stocks. Crop Sci. 31:1399.

Vuong, T.D., and J.E. Harper. 2000. Inheritance and allelism analysis of hypernodulating genes in the NOD3-7 and NOD2-4 soybean mutants. Crop Sci. 40:700–703.

Vuong, T.D., C.D. Nickell, and J.E. Harper. 1996. Genetic and allelism analyses of hypernodulation soybean mutants from two genetic backgrounds. Crop Sci. 36:1153–1158.

Walker, J.B., W.R. Fehr, G.A. Welke, E.G. Hammond, D.N. Duvick, and S.R. Cianzio. 1998. Reduced-linolenate content associations with agronomic and seed traits of soybean. Crop Sci. 38:352–355.

Wang, K.-J., N. Kaizuma, Y. Takahata, and S. Hatakeyama. 1996. Detection of two new variants of soybean Kunitz trypsin inhibitor through electrophoresis. Breed. Sci. 46:39–44.

Wang, K.-J., Y. Takahata, K. Ito, Y.-P. Zhao, K. Tsutsumi, and N. Kaizuma. 2001. Genetic characterization of a novel soybean Kunitz trypsin inhibitor. Breed. Sci. 51:185–190.

Wang, Y., R.L. Nelson, and Y. Hu. 1998. Genetic analysis of resistance to soybean mosaic virus in four soybean cultivars from China. Crop Sci. 38:922–925.

Weber, C.R., and M.G. Weiss. 1959. Chlorophyll mutant in soybeans provides teaching aid. J. Hered. 50:53–54.

Weber, D.F., H.H. Keyser, and S.L. Uratsu. 1989. Serological distribution of *Bradyrhizobium japonicum* from US soybean production areas. Agron. J. 81:786–789.

Weiss, M.G. 1943. Inheritance and physiology of efficiency in iron utilization in soybeans. Genetics 28:253–268.

Weiss, M.G. 1970a. Genetic linkage in soybeans: Linkage group I. Crop Sci. 10:69–72.

Weiss, M.G. 1970b. Genetic linkage in soybeans. Linkage groups II and III. Crop Sci. 10:300–303.

Weiss, M.G. 1970c. Genetic linkage in soybeans. Linkage group IV. Crop Sci. 10:368–370.

Weiss, M.G. 1970d. Genetic linkage in soybeans. Linkage groups V and VI. Crop Sci. 10:469–470.

Weiss, M.G. 1970e. Genetic linkage in soybeans: Linkage group VII. Crop Sci. 10:627–629.

Weng, C., K Yu, T.R. Anderson, and V. Poysa. 2001. Mapping genes conferring resistance to *Phytophthora* root rot of soybean, *Rps1a* and *Rps7*. J. Hered. 92:442–446.

Werner, B.K., J.R. Wilcox, and T.L. Housley. 1987. Inheritance of an ethyl methanesulfonate-induced dwarf in soybean and analysis of leaf cell size. Crop Sci. 27:665–668.

Wilcox, J.R. 1988. Performance and use of seedcoat mutants in soybean. Crop Sci. 28:30–32.

Wilcox, J.R., and T.S. Abney. 1991. Inheritance of a narrow, rugose-leaf mutant in *Glycine max*. J. Hered. 82:421–423.

Wilcox, J.R., J.W. Burton, G.L. Rebetzke, and R.F. Wilson. 1994.Transgressive segregation for palmitic acid in seed oil of soybean. Crop Sci. 34:1248–1250.

Wilcox, J.R., and J.F. Cavins. 1985. Inheritance of low linolenic acid content of the seed oil of a mutant in *Glycine max*. Theor. Appl. Genet. 71:74–78.

Wilcox, J.R., and J.F. Cavins. 1987. Gene symbol assigned for linolenic acid mutant in the soybean. J. Hered. 78:410.

Wilcox, J.R., and J.F. Cavins. 1990. Registration of C1726 and C1727 soybean germplasm with altered levels of palmitic acid. Crop Sci. 30:240.

Wilcox, J.R., A.D. Nickell, and J.F. Cavins. 1993. Relationships between the *fan* allele and agronomic traits in soybean. Crop Sci. 33:87–89.

Wilcox, J.R., G.S. Premachandra, K.A. Young, and V. Raboy. 2000. Isolation of high seed inorganic P, low-phytate soybean mutants. Crop Sci. 40:1601–1605.

Wilcox, J.R., and A.H. Probst. 1969. Inheritance of a chlorophyll-deficient character in soybeans. J. Hered. 60:115–116.

Williams, C., D.F. Gilman, D.S. Fontenot, and W. Birchfield. 1981. Inheritance of reaction to the reniform nematode in soybean. Crop Sci. 21:93–94.

Williams, L.F. 1950. Structure and genetic characteristics of the soybean. p. 111–134. *In* K.S. Markley (ed.). Soybean and soybean products. Vol. 1. Interscience Publ., New York.

Willliams, L.F. 1952. The inheritance of certain black and brown pigments in the soybean. Genetics 37:208–215.

Willliams, L.F. 1958. Alteration of dominance and apparent change in direction of gene action by a mutation at another locus affecting the pigmentation of the seedcoat of the soybean. Proc. Int. Congr. Genet., 10th 2:315–316 (Abstr.).

Willliams, L.F, and D.L. Lynch. 1954. Inheritance of a non-nodulating character in the soybean. Agron. J. 46:28–29.

Willmot, D.B., and C.D. Nickell. 1989. Genetic analysis of brown stem rot resistance in soybean. Crop Sci. 29:672–674.

Wilson, R.F., J.W. Burton, W.P. Novitzky, and R.E. Dewey. 2001. Current and future innovations in soybean (*Glycine max* L. Merr.) oil composition. J. Oleo Sci. 50:353–358.

Woodworth, C.M. 1921. Inheritance of cotyledon, seed-coat, hilum, and pubescence colors in soy-beans. Genetics 6:487–553.

Woodworth, C.M. 1923. Inheritance of growth habit, pod color, and flower color in soybeans. J. Am. Soc. Agron. 15:481–495.

Woodworth, C.M. 1932. Genetics and breeding in the improvement of the soybean. Bull. Agric. Exp. Stn. (Ill.) 384:297–404.

Woodworth, C.M. 1933. Genetics of the soybean. J. Am. Soc. Agron. 25:36–51.

Woodworth, C.M., and L.F. Williams. 1938. Recent studies on the genetics of the soybean. J. Am. Soc. Agron. 30:125–129.

Wubben, M., and R.G. Palmer. 1998. Linkage studies with necrotic root mutants. Soybean Genet. Newsl. 25:145.

Yadav, N.S. 1995. Genetic modification of soybean oil quality. p. 165–188. *In* D.P.S. Verma and R.C. Shoemaker (ed.) Soybean biotechnology. CAB Int., London.

Yee, C.C., J. Li, and Z.G. Yu. 1986. Genetic studies with Shennong 2015, a lethal yellow mutant (*y21*) in soybean. Hereditas (Beijing) 8:13–16.

Yong, H.S., K.L. Chan, C. Mak, and S.S. Dhaliwal. 1981. Isocitrate dehydrogenase gene duplication and fixed heterophenotype in the cultivated soybean *Glycine max*. Experientia 37:130–131.

Yong, H.S., C. Mak, K.L. Chan, and S.S. Dhaliwal. 1982. Inheritance of isocitrate dehydrogenase in the cultivated soybean. Malay. Nat. J. 35:225–228.

You, M., T. Zhao, J. Gai, and Y. Yen. 1998. Genetic analysis of short petiole and abnormal pulvinus in soybean. Euphytica 102:329–333.

Yu, H., and Y.T. Kiang. 1993a. Genetic characterization of a leaf margin necrosis mutant in wild annual soybean (*Glycine soja*). Genetica 90:31–33.

Yu, H., and Y.T. Kiang. 1993b. Inheritance and genetic linkage studies of isozymes in soybean. J. Hered. 84:489–492.

Yu, Y.G., M.A. Saghai Maroof, G.R. Buss, P.J. Maughan, and S.A. Tolin. 1994. RFLP and microsatellite mapping of a gene for soybean mosaic virus resistance. Phytopathology 84:60–64.

Zakharova, E.S., S.M. Epishin, and Yu.P. Vinetski. 1989. An attempt to elucidate the origin of cultivated soybean via comparison of nucleotide sequences encoding glycinin B4 polypeptide of cultivated soybean, *Glycine max*, and its presumed wild progenitor, *Glycine soja*. Theor. Appl. Genet. 78:852–856.

Zhang F., and R.G. Palmer. 1990. The *ms1* mutation in soybean: Involvement of gametes in crosses with tetraploid soybean. Theor. Appl. Genet. 80:172–176.

Zhang, J., J.E. Specht, G.L. Graef, and B.E. Johnson. 1992. Pubescence density effects on soybean seed yield and other agronomic traits. Crop Sci. 32:641–648.

Zhao, S., G. Qimin, and W. Hai. 1995a. The zymogram pattern of the new type of SBTi-A2 seed protein and the crossing with other alleles. Soybean Genet. Newsl. 22:83–84.

Zhao, S., G. Qimin, and W. Hai. 1995b. Inheritance of a new variant of SBTi-A2 in seed protein of soybean (*Glycine max*) in China. Soybean Genet. Newsl. 22:85–88.

Zhao, S., L. Xingynan, and L. Guoqing. 1993. Gene frequency of SBTi-A2 allele in soybean (*G. max*) seed storage protein in China. Soybean Genet. Newsl. 20:45–47.

Zhao, S., and H. Wang. 1992. A new electrophoretic variant of SBTi-A2 in soybean seed protein. Soybean Genet. Newsl.19:22–24.

6   Soybean Genomics

RANDY C. SHOEMAKER

Iowa State University
Ames, Iowa

PERRY B. CREGAN

Beltsville Agriculture Research Center
Beltsville, Maryland

LILA O. VODKIN

University of Illinois
Urbana, Illinois

6–1 THE SOYBEAN GENOME

Soybean [*Glycine max* (L.) Merr.] has emerged as a model crop system because of its densely saturated genetic map (Cregan et al., 1999), a well-developed genetic transformation system (Clemente et al., 2000; Xing et al., 2000; Zhang et al., 1999), and the growing number of genetic tools applicable to this biological system (reviewed in Shoemaker, 1999). It is also the number one oilseed crop in the world and a multibillion-dollar crop of the USA (Riley, 1999; SoyStats, 1997).

The soybean genome is of average size compared to that of many other plants. It is comprised of about 1.1 Mbp/C (Arumuganathan and Earle, 1991). This makes it about seven and one-half times larger than the genome of Arabidopsis and two and one-half times larger than rice (*Oryza sativa* L.). Still the soybean genome is less than half the size of the corn (*Zea mays* L.) genome and more than 14 times smaller than the genome of bread wheat (*Triticum aestivum* L.) (Arumuganathan and Earle, 1991). Approximately 40 to 60% of the soybean-genome sequence can be defined as repetitive (Gurley et al., 1979; Goldberg, 1978). One family of repetitive sequence, STR120, is comprised of an approximate 120 bp monomer (Morgante et al., 1997). This family is estimated to consist of 5000 to 10 000 copies. Some repetitive sequence families may be species-specific (Morgante et al., 1997).

Bacterial artificial chromosome (BAC) libraries (Marek and Shoemaker, 1997; Danesh et al., 1998; Tomkins et al., 1999; Salimath and Bhattacharyya, 1999; D. Lightfoot, personal communication, 2002) also have been produced, which together cover the soybean genome many times over. Detailed physical contigs have already been developed and reported using some of these libraries (Marek and Shoemaker, 1997). These libraries have been made from different genotypes and with a variety of enzymes and most have been made available to the public. Yeast Artificial Chromosomes have also been created for the purpose of chromosome walking and in situ hybridization (Zhu et al., 1996).

The degree to which soybean chromosomes constrict during meiosis has made it difficult to conduct cytogenetic analyses. However, a complete karyotype has now been reported (Singh and Hymowitz, 1988) based upon pachytene analysis. Analysis of pachytene chromosomes has shown that more than 35% of the genome is made up of heterochromatin, with the short arm of six of the 20 bivalents being completely heterochromatic (Singh and Hymowitz, 1988).

Primary trisomics (genomic complements containing one additional chromosome; $2n = 41$) are useful for quickly locating genes onto a specific chromosome and for associating linkage groups with specific chromosomes. By using aneuploid lines that Dr. R. Palmer (USDA-ARS) supplied, and some generated at the Soybean Cytogenetics Lab at Urbana-Champaign, IL a complete set of 20 trisomics, in which each chromosome exists in an extra copy, is now completed (Xu et al., 1998). This work will undoubtedly be useful for integrating classical and molecular genetics, much as similar cytogenetic collections have been important for maize, rice, barley (*Hordeum vulgare* L.), and tomato (*Lycopersicum esculentum* Mill.).

Soybean has a diploid chromosome number of $2n = 40$. However, most genera in the *Phaseolae* have a genome complement of $2n = 22$ (Hymowitz et al., 1998). This led Lackey (1980) to suggest that *Glycine* was probably derived from a diploid ancestor ($n = 11$) which underwent aneuploid loss to $n = 10$ and subsequent polyploidization to yield the present $2n = 40$.

Some type of polyploidization event has very likely occurred in the soybean's distant past. However, in spite of being a polyploid, the genome, for the most part, acts like a diploid. The 'diploidization' of polyploids is a well-known process and is caused by additions, deletions, mutations, and rearrangements that rapidly inhibit nonhomologous pairing of linkage groups (Ohno, 1970). Examples of divergence of duplicated genes have been reported for soybean receptor-like protein kinases (Yamamoto and Knap, 2001) and CLV1-like genes (Yamamoto et al., 2000). However, this may be a relatively slow process and there remain many exceptions reminding us that soybean is a tetraploid.

Many examples of duplicate factor genes (two independent genes controlling the same trait) can be found in the soybean germplasm collection (Palmer and Kilen, 1987). An analysis of the average number of fragments detected by hybridization to restriction-digested soybean-genomic DNA by each of 280 randomly chosen *Pst*I genomic clones emphasized the abundant duplications found in the genome (Shoemaker et al., 1996). More than 90% of the probes detected more than two fragments and nearly 60% detected three or more fragments. This suggests that <10% of the genome may be single copy sequence and that large amounts of the genome may have undergone genome duplication in addition to the presumed tetraploidization event.

Another analysis of hypomethylated sequences using methylation sensitive restriction enzymes indicated that slightly more than 15% of the hypomethylated genome remains as single-copy DNA (Zhu et al., 1994). The remainder of the hypomethylated genome appeared to be duplicated or middle-repetitive sequence. No evidence of silencing of the duplicated regions was observed by methylation.

Hybridization-based mapping has resolved many duplicated regions of the genome (Cho et al., 1989). These homoeologous regions reflect segmental and whole-genome duplication events and can provide much information about the evolution of the genome (Fig. 6–1). Sequences are often duplicated in the soybean genome in a manner not easily explained by a tetraploidization event. For example, most linkage groups contain markers that can also be found on other linkage groups but examples of this can sometimes be extreme. For any given linkage group, duplicate markers may be present on more than one other linkage group. The average linkage group contains markers that can be found on eight other linkage groups (Shoemaker et al., 1996).

Mapping of duplicated genes controlling pubescence morphology (appressed and non-appressed) provided interesting insight into the evolution of the genome. Lee et al. (1999) mapped *Pa*1 and *Pa*2 to LG-B1/S and LG-F, respectively. It was expected that these genes would map to homoeologous segments of the linkage groups. However, other than the genes, no markers appeared in common between these regions. Unexpectedly, the gene regions were implicated as paralogs through an intermediate linkage group, LG-H, which connected to LG-B1/S and LG-F regions through multiple markers. This suggested that regions of LG-B1/S and LG-F (as well as LG-H) were evolutionarily related and that perhaps a third pubescence gene existed and remained undetected.

Except for many disease-resistance genes, most agronomically important traits are controlled by several to many genes acting in concert. The genetic locations of the quantitative gene(s) are known as quantitative trait loci (QTL). Because of the genetic by environment interactions on most quantitative traits, breeding for them requires replicated field trials conducted over 2 or more years in a variety of locations. This is obviously time consuming and expensive. The ability to select for an

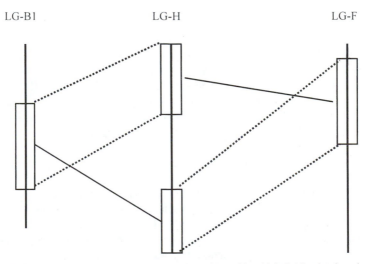

Fig. 6–1. Examples of homeologous regions in soybean detectable with hybridization-based mapping techniques. In this example LG-H has homeologs on both LG-B1 and LG-F, while segments of LG-B1 and LG-F are connected through a single marker with homology to LG-H. From Lee et al. (2001).

easily identifiable marker that is a good predictor of the presence or absence of a QTL trait can save time and money in a breeding program. Discovery and tagging of QTL is a prerequisite of this type of Marker Assisted Selection (MAS).

The first reported mapping of QTL in soybean was in 1990 (Keim et al., 1990). Since then literally dozens of reports have flowed out of research laboratories with numerous quantitative traits, with perhaps the most critical dealing with seed composition (Diers et al., 1992a), other traits have included reproductive characters, seed characters, maturation traits, vegetative/morphological traits, disease resistance, nutritional efficiency, and more. A thorough coverage of QTL mapping in soybean is too extensive for this chapter. Detailed summarizations of these studies and others can be found in SoyBase, the USDA-sponsored genomic database for soybean, at URL: http://soybase.agron.iastate.edu.

Comparative mapping among grasses has identified highly conserved genome structures and has led to the suggestion that grasses can be considered to have a single genome. This has important implications in our ability to transfer genomic information obtained in one grass species to that of another grass species (Bennetzen and Freeling, 1993). Comparative mapping among legumes has not been as simple. The substantial rearrangements that have occurred within the soybean genome, probably as part of the process of diploidization, make it difficult to identify lengthy stretches of syntenic chromosome segments between soybean and related legumes (Boutin et al., 1995). Although mung bean [*Vigna radiata* (L.) Wilczek. var. radiata] ($2n = 11$) and common bean (*Phaseolus vulgaris* L.) ($2n = 11$) (both belonging to the subtribe Phaseolinae) exhibit a high degree of linkage conservation and preservation of marker order, the situation is substantially different when comparing either one to soybean ($2n = 20$) (subtribe Glycininae). While most linkage groups of mung bean consist of only one or a few linkage blocks from common bean (and vice versa) linkage groups of mung bean and common bean are comprised generally of mosaics of short soybean linkage blocks (Boutin et al., 1995).

A more detailed analysis of homologous segments of soybean, common bean, and mung bean genomes was supportive of the hypothesis that homoeologous chromosome blocks within soybean arose through ancient whole-chromosome duplications (Lee et al., 2001). By focusing on soybean genomic regions containing known duplicated appressed pubescence genes (*Pa*1, *Pa*2), these authors showed that homoeologous segments of soybean linkage groups showed a high degree of synteny with large portions of single chromosomes of *Phaseolus* and *Vigna*. These authors further showed that each of the duplicated and homologous regions among the legumes were homologous with duplicated regions of Arabidopsis.

The first example of inter-family synteny was shown by Grant et al. (2000). Taking advantage of the vast amount of DNA sequence data generated by the Arabidopsis Genome Initiative, Grant et al. (2000) was able to demonstrate synteny between Arabidopsis and soybean. These findings were surprising given the millions of years since the divergence of their lineages. These authors were also able to show that only a limited number of chromosomal events was required to explain the structural differences between soybean and Arabidopsis chromosomes.

For a molecular genetic map to be fully exploited it is necessary to incorporate positions of genes or QTLs in an unambiguous manner. Often, this can be accomplished only through painstaking analysis of segregation data obtained one trait at a time. However, Shoemaker and Specht (1995) integrated 18 genes into the map in a single experiment. This was done through careful 'stacking' of mutations into parents to be used in constructing the mapping population. Because many of these genes had been put into the classical genetic map, this study resulted in half of the classical genetic linkage groups being integrated into the molecular genetic map in a single experiment. Today, more than 60 loci for qualitative traits have been placed onto molecular maps.

Using an integrated map containing more than 800 markers and combining data from nine different populations, extensive homoeologous relationships were detected using RFLP hybridization techniques (Shoemaker et al., 1996). The average size of these internal duplications is approximately 45 cM, with some duplicated segments covering more than 100 cM. These authors also observed 'nested' duplications that suggested at least one of the original genomes of soybean may have undergone an additional round of tetraploidization in the far distant past (Shoemaker et al., 1996).

6–2 DNA MARKERS AND MOLECULAR GENETIC LINKAGE MAPS

The development of molecular genetic maps based upon DNA sequence polymorphisms was initiated by the suggestion that restriction fragment length polymorphisms (RFLP) could serve as an approach for the development of numerous DNA markers (Botstein et al., 1980). The application of RFLP technology to numerous animal and plant species began shortly thereafter. Subsequently, the availability of the polymerase chain reaction (PCR) (Mullis et al., 1986) as a tool to detect sequence polymorphism led to the development of numerous additional classes of DNA markers. These included (i) microsatellite or simple sequence repeat (SSR) markers (Litt and Luty, 1989; Weber and May, 1989), (ii) random amplified polymorphic DNA (RAPD) (Williams et al., 1990) or arbitrary primer PCR (AP-PCR) markers (Welsh and McClelland, 1990), (iii) DNA amplification fingerprinting (DAF) markers (Caetano-Anolles et al., 1992), and (iv) amplification fragment length polymorphism (AFLP) markers (Vos et al., 1995).

6–2.1 Restriction Fragment Length Polymorphisms-Based
Genetic Linkage Maps

The first demonstrations of RFLP in soybean were by Apuya et al. (1988) and Keim et al. (1989) and in 1990 the first RFLP-based map of the soybean genome was published (Keim et al., 1990). To maximize molecular diversity, Keim et al. (1990) constructed their map using a mapping population derived from a cross of cultivated × wild soybean (Table 6–1). This map was developed jointly by the USDA-ARS and Iowa State University with support from the American Soybean Association and saw further expansion during the 1990s with the addition of more than 350 RFLP loci (Shoemaker and Olson, 1993) (Table 6–1). Concurrently, the DuPont corporation (Rafalski and Tingey, 1993) developed an extensive RFLP map with more than 600 loci. Like the USDA-ARS/Iowa State map, Rafalski and Tingey

Table 6–1. Population size and type, linkage group number, map length, and numbers of restriction fragment length polymorphisms (RFLP), simple sequence repeat (SSR), random amplified polymorphic DNA (RAPD), amplification fragment length polymorphism (AFLP), isozyme, and classical genetic markers positioned in published soybean mapping populations.

Mapping population and literature citation	Population size and type	Linkage groups	Total map length	Marker type						
				Total	RFLP	RAPD	SSR	AFLP	Isozyme	Classical
		no.	-cM-	no.						
USDA/Iowa St.										
(A81-356022 × G. soja PI 468.916)	59 F2 lines									
Keim et al. (1990)		26	1200	153	150					3
Diers et al. (1992a)		31	2147	246	238				5	3
Shoemaker and Olson (1993)		23	3371	371	355	10			3	3
Shoemaker and Specht (1995)		25	2473	375	358	10			4	3
Cregan et al. (1999)		23	3003	1004	501	10	486	11	4	3
E. I. DuPont Bonus × G. soja PI 81762	68 F2 lines									
Rafalski and Tingey (1993)		21	2678	600+	600+					
Univ. of Utah (Minsoy × Noir 1)										
Lark et al. (1993)	69 F3 families	31	1551	140	132				3	5
Mansur et al. (1996)	240 RILs	35	1981	266	224		41		1	6
Cregan et al. (1999)	240 RILs	22	2413	633	209		412		2	10
Univ. of Nebraska (Clark × Harosoy isolines)	57 F2 lines									
Shoemaker and Specht (1995)		26	1004	99	80	3			4	12
Akkaya et al. (1995)		29	1486	123	80	4	34		5	12
Cregan et al. (1999)		28	2787	535	95	69	339	11	7	14
PI 437654 × BSR301	42 RILs									
Keim et al. (1997)		28	3441	840	165	25		650		
Ferreira et al. (2000)		35	3275	356	250	106				
Changnong 4 × Xinmin 6	88 RILs									
Liu et al. (2000)		22	3713	240	100	62	33	42		2
Misuzudaizu × Moshidou Gong 503	190 F2 plants									
Yamanaka et al. (2001)		20	2909	503	401	1	96			5
Kefeng 1 × Nannong 1138-2	201 RILs									
Wu et al. (2001)		24	2321	792						
Noir 1 × BARC-2	149 F2 plants									
Matthews et al. (2001)		35	1400	207	39	17	25	105		4

(1993) relied upon a cultivated × wild soybean cross to increase the relatively low level of RFLP present in cultivated soybean. However, a large proportion of the loci on these two maps would not be expected to segregate in crosses among cultivated soybean genotypes. For example, Shoemaker and Specht (1995) used 358 RFLP markers from the *G. max* × *G. soja* USDA/Iowa State map to genotype progeny derived from a cross of isolines of the cv. Clark and Harosoy. A total of 118 (33%) were polymorphic in the Clark × Harosoy population. In a previous report, Keim et al. (1992) analyzed 38 diverse soybean genotypes with 132 RFLP probes and found 31% to be monomorphic and further, that more than two alleles were detected at only three RFLP loci. In addition to the relatively low level of polymorphism, another complicating factor with the use of RFLP in soybean is the duplicated nature of the soybean genome, to which RFLP probes will hybridize on an average of 2.55 times (Shoemaker et al., 1996). The duplicated nature of the genome results in multiple banding patterns with most RFLP probes. One hybridizing fragment may be mapped in one population and a different or an additional band in another. This requires that an RFLP locus be defined not only by the probe and restriction enzyme being used, but also by the molecular weight of the segregating band(s). Despite these complications, numerous successful analyses designed to discover QTL and characterize genetic variation in soybean germplasm were conducted and reported using RFLP probes from the USDA/Iowa State RFLP map. The details of this and other similar maps and large amounts of related information can be accessed on the World Wide Web in SoyBase, the USDA-ARS Soybean Genome Database (http://soybase.agron.iastate.edu/)

6–2.2 Simple Sequence Repeat Markers

The desire for soybean DNA markers with greater polymorphism was stimulated by the discovery of high levels of allelic variation associated with microsatellite or SSR markers in human (Litt and Luty, 1989; Weber and May, 1989). The fact that SSR markers are PCR based rather than hybridization based was another attractive feature of this DNA marker system. In the early 1990s, two research groups published similar reports demonstrating the high levels of polymorphism, co-dominant inheritance, and the locus specificity of SSR markers in soybean (Akkaya et al., 1992; Morgante and Olivieri, 1993). Akkaya et al. (1992) found as many as eight SSR alleles at one locus in a set of 38 *G. max* and five *G. soja* genotypes. Subsequent reports of SSR allelic variation in cultivated and wild soybean (Cregan et al.,1994; Maughan et al., 1995; Morgante et al., 1994; Rongwen et al., 1995) detected very high levels of allelic variation, including one locus with 26 alleles among a group of 91 cultivated and five wild soybean genotypes. In addition, data analyses suggested little evidence of the clustering of SSR loci in the soybean genome (Akkaya et al., 1995). Because of their high levels of polymorphism, single locus nature, and random distribution in the genome, it was concluded that SSR markers would provide an excellent complement to RFLP markers for use in soybean molecular biology, genetics, and plant-breeding research. The major drawback to SSR markers is the high cost of development, which requires firstly the discovery of SSR motifs and secondly, knowledge of the flanking sequence to permit the design of locus-specific PCR primers. An additional technical difficulty associated with SSR technology is the frequent need to distinguish alleles that vary by only one or a few repeat units in size.

6–2.3 Restriction Fragment Length Polymorphisms and DNA Amplification Fingerprinting Markers

In contrast to SSRs, RAPD or AP-PCR markers require no prior knowledge of DNA sequence and as a dominant marker, alternative alleles are detected simply as the presence or absence of a PCR product. Thus, genotypes can be readily determined using agarose gel electrophoresis without the need for more sophisticated systems to detect allelic variation. Nonetheless, RAPDs have not been widely used in soybean genetic map development. The exception is the RFLP/RAPD map constructed by Ferreira et al. (2000), which incorporated 106 RAPD markers into a framework of 250 existing RFLP loci using a subset of RILs from the PI 437654 × BSR 101 population (Table 6–1). Like RAPD markers, DNA amplification fingerprint or DAF markers are amplified using a single arbitrary primer (Caetano-Anolles et al., 1992). The differences between RAPD and DAF technology are the shorter arbitrary primer in DAF vs. RAPD (generally eight nucleotides), and the use of polyacrylamide gel electrophoresis with silver staining in the case of DAF markers vs. agarose gels for RAPDs. Predigestion of genomic DNA with a restriction enzyme before PCR amplification is sometimes used to optimize DAF amplification products. A limited number of DAF-generated polymorphisms were mapped in the Univ. of Utah, Minsoy × Noir 1 RIL population (Prabhu and Gresshoff, 1994). No genetic maps were developed in soybean using DAF markers.

6–2.4 Amplification Fragment Length Polymorphism Markers

Like RAPD markers, the generation of AFLP requires no prior knowledge of DNA sequence and as a result, numerous marker loci can be rapidly developed. The AFLP markers are generated based on restriction fragment length polymorphisms. The DNA adaptors are ligated to the ends of restriction fragments and PCR primers homologous to the adaptors are used to amplify selected subpopulations of the pool of fragments. Selectivity results from the addition of two or three arbitrary nucleotides to the 3′ ends of the PCR primers. One of the largest available AFLP maps of any plant species was developed in soybean (Keim et al., 1997). These loci were mapped in a subset of 42 RILs from the 330 RIL PI 437654 × BSR 101 population. This map has a total of 840 loci, 650 of which are AFLPs, whereas the USDA/Iowa State map has 1004 loci, most of which are RFLPs and SSRs (Table 6–1). The authors noted significant clustering with AFLP markers that were generated using *Eco*RI/*Mse*I restriction enzymes (Young et al., 1999). Thirty-four percent of the loci displayed dense clustering. In contrast, *Pst*I/*Mse*I-generated AFLP loci did not cluster. *Pst*I is known to be sensitive to cytosine methylation and in relation to *Eco*RI/*Mse*I-generated AFLPs, those produced using *Pst*I/*Mse*I appeared to eliminate marker clustering. Keim et al. (1997) also noted that despite numerous marker loci and the framework of 165 RFLP loci, 11 of 28 linkage groups could not be aligned with a homologous linkage group on the USDA/Iowa State map. This result is indicative of one shortcoming of AFLP markers, which is the difficulty of comparing AFLP loci across populations. As a result, it is generally necessary to create a new linkage map for every new population, rather than simply using a set of informative marker loci with previously defined positions in the genome. However, because of the large amount of marker data that can be obtained with AFLP without the need for previous knowledge of DNA sequence, AFLPs have proven useful for saturating specific genomic regions using bulked segregant analysis (Michelmore et al., 1991) or for the comparison of near-isogenic lines for a trait of interest (Muehlbauer et al., 1988, 1991). The loci so identified can then be associated with an anchor marker to determine a definitive position on a linkage map.

6–2.5 A Simple Sequence Repeat-Based Soybean Genome Map

The development and mapping of a large set of soybean SSR markers was initiated in 1995 with the support of the United Soybean Board. As a result of this effort more than 600 SSR loci were developed. These markers were derived almost exclusively from genomic libraries. Clearly, the large collection of soybean EST data is another source of sequence data that could be exploited for the development of additional SSR markers. The virtue of EST-derived SSRs is the close association of the resulting SSR markers with expressed genes. The 600 SSR markers developed to date were mapped in one, two, or when possible, three different mapping populations (Cregan et al., 1999). One of these was the USDA/Iowa State population and the second was the 240 RIL University of Utah population developed from a cross of the cultivated soybean genotypes Minsoy and Noir 1 (Table 6–1). The third was the University of Nebraska Clark × Harosoy isoline population consisting of 57 F2-derived lines. In every case, SSR markers mapped to a single locus in the genome with a map order that was essentially identical in all three populations. A total of 187 loci were mapped in each of the three populations, making it a simple matter to align homologous linkage groups. An example of this alignment is presented in Fig 6–2, which pictures one of the 20 homologous soybean linkage groups. In this case, 12 loci were mapped in all three populations and an additional 20 in two populations. Thus, linkage groups U13a and U13b on the Univ. of Utah map could be joined based upon markers in common with linkage groups F-ISU and F-CH21 from the USDA/Iowa State Univ. and the Univ. of Nebraska maps, respectively. As a result, the 20+ linkage groups derived from each of the three populations (Table 6–1) were aligned into a consensus set of 20 homologous groups presumed to correspond to the 20 pairs of soybean chromosomes. Likewise, classical linkage group 8 (*Ms1, W1, Ms6, Y23, St5*, and *Adh1* loci) was associated with linkage group F by in situ segregation of *W1* as were classical loci *Rpg1* and *B1* that had not previously been linked to any other classical loci. Reports in the literature indicated that both *Rps3* (Diers et al., 1992b) and *Rsv1* (Yu et al., 1994) mapped to linkage group F, thereby associating classical linkage group 13 with linkage group F. Based upon in situ segregation or linkage reports in the literature, all but one of the 20 classical linkage groups (Palmer and Shoemaker, 1998) were assigned to a corresponding molecular linkage group. Work is being completed to position approximately 100 additional classical loci on the integrated map using bulked hybrid segregation (Rector et al., 1999). Information relating to the sequence of PCR primers to the 600+ SSR loci reported in Cregan et al. (1999) and a standard protocol for their amplification can be obtained on the SoyBase website. Additional information relating to SSR allele sizes in a set of 10 diverse soybean genotypes, as well as gel images of the alleles produced with the same 10 genotypes, is available on SoyBase. Data relating to the mapping of more than 600 SSR in the University of Utah Minsoy × Noir 1, Minsoy × Archer, and Archer × Noir recombinant inbred line populations can be obtained on the Lark Lab, Univ. of Utah website (http://www.larklab.4biz.net/)

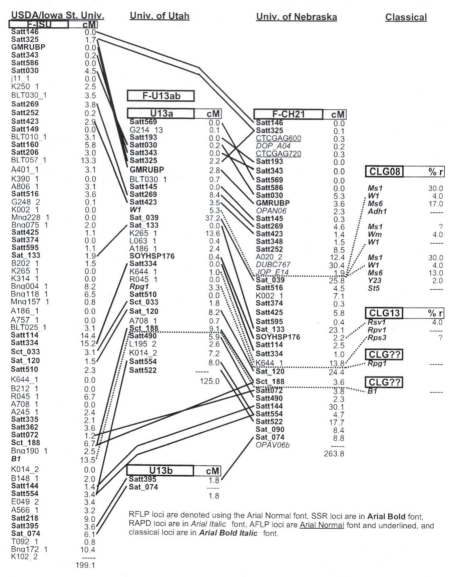

Fig. 6–2. Consensus soybean molecular linkage group F defined using three mapping populations: The USDA/Iowa State Univ., *Glycine max* × *G. soja* population; the Univ. of Utah, Minsoy × Noir 1 population; and the Univ. of Nebraska, Clark × Harosoy population. From Cregan et al. (1999).

The definition of 20 consensus linkage groups with an average of 30 locus-specific SSR markers per group (Cregan et al., 1999) provided a resource that has facilitated the rapid alignment of linkage groups in existing or newly created linkage maps with the consensus linkage groups in the SSR-based genome map. In four instances, a relatively small number of SSR markers ranging from an average of one to as many as five per linkage group was used to associate linkage groups with corresponding linkage groups on the SSR-based soybean genome map. These included two fairly extensive maps, one with 792 markers (Wu et al., 2001) and another with more than 500 markers (Yamanaka et al., 2001) (Table 6–1). Details of the Misuzudaizu × Moshidou Gong 503 map (Yamanaka et al., 2001) are available on the World Wide Web (http://dna-res.kazusa.or.jp/8/2/02/HTMLA/). Matthews et al. (2001) successfully positioned cDNA and genomic clones on consensus linkage groups in the SSR-based soybean genome map using a small number of SSR loci to align corresponding linkage groups (Table 6–1). Liu et al. (2000) provided the fourth example of the use of SSRs to align linkage groups with the consensus map.

6–2.6 Single Nucleotide Polymorphism Markers

Single DNA base changes between homologous DNA fragments plus small insertions and deletions, collectively referred to as single nucleotide polymorphisms (SNPs), are by far the most abundant source of DNA polymorphisms in humans (Collins et al., 1998; Kruglyak 1997; Kwok et al., 1996) and mice (*Mus musculus*) (Lindblad-Toh et al., 2000). In humans, these variations are estimated to occur at a frequency of about one per 1000 bp when any two homologous DNA segments are compared (Cooper et al., 1985; Kwok et al., 1996). In plants, relatively limited data on the frequency of SNPs are available. Cho et al. (1999) compared the DNA sequence of more than 500 kbp of the two *Arabidopsis thaliana* genotypes Columbia and Landsberg erecta and detected one SNP every 1034 bp. However, most other reports have indicated much higher levels of sequence variation in *Arabidopsis* (Kawabe et al., 2000; Kawabe and Miyashita, 1999; Kuittinen and Aguade, 2000; Purugganan and Suddith, 1999). In maize (*Z. mays* ssp. *mays* L.), Tenaillon et al. (2001) sequenced more than 14 kb of coding and noncoding DNA from 21 loci on chromosome 1 in each of 25 genotypes and discovered a mean of 9.6 SNPs per kbp between any two randomly selected genotypes. In soybean, SNP DNA markers are already in use in industrial-breeding programs (Cahill, 2000) using allele-specific hybridization (ASH) for SNP detection similar to the procedure described by Coryell et al. (1999). It is thus apparent that SNP markers are likely to have an important role in the future of soybean genome analysis and manipulation.

Until recently, the comparison of variation in DNA sequence among soybean genotypes has been confined to the assay of single genes or DNA fragments, generally with the purpose of defining gene structure or function or evolutionary relationships. For example, Scallon et al. (1987) compared 3543 bp of the Gy_4 glycinin gene plus flanking DNA in two genotypes and found three SNPs. Zakharova et al. (1989) compared 789 bp of cDNA sequence encoding the A_3B_4 glycinin subunit of the soybean cv. Mandarin, Rannaya-10, and Mukden and found two single nucleotide polymorphisms. Xue et al. (1992) discovered 20 single base changes and four indels (insertion-deletions) in a comparison of 2942 bp of the Gy_4 gene and flanking DNA in the soybean genotypes Forrest, Raiden, and Dare. Zhu et al. (1995) sequenced 400 bp of RFLP probe A-199a in the cv. BSR-101 and A81-356022 and the *G. max* germplasm line PI437654 and found a total of nine SNPs. To permit the comparison of SNP frequency among loci of varying length and between populations that vary in size, measures of nucleotide diversity such as π (Tajima, 1983) and Watterson's theta (θ_w) (Watterson, 1975) have been devised that are standardized for length and adjusted for sample size. Nucleotide diversity from the four aforementioned studies range from $\theta_w = 0.85$ SNPs/kbp (Scallon et al., 1987) to $\theta_w = 15$ SNPs/kbp (Zhu et al., 1995). The wide diversity of values suggested that a systematic study of SNP frequency in soybean was needed.

In recently completed work to assess the SNP frequency in soybean, a group of 25 soybean genotypes that represented 18 ancestral varieties from which North American soybean plants are derived (Gizlice et al., 1994) as well as seven parents of RIL mapping populations was analyzed (Zhu et al., 2003). A total of more than 28.5 kbp of coding sequence and 37.9 kb of noncoding (introns, 3' and 5' UTR, and flanking genomic sequence) from 116 genes was sequenced in each of the 25 genotypes. The SNP frequency in coding and noncoding DNA was 1.98 kbp $^{-1}$ and 4.19 kbp^{-1}, respectively. Nucleotide diversity was $\theta_w = 0.53$ and 1.11 in coding and noncoding sequence, respectively. The mean $\theta_w = 0.97$ was similar to reports of SNP frequency in humans (Wang et al., 1998; Cargill et al., 1999; Halushka et al., 1999) and 5- to 10–fold lower than reports in maize (Remington et al., 2001; Tenaillon et al., 2001). Despite the relatively low frequency, SNPs were discovered in or around 74 of the 116 genes for which sequence data were obtained. These data suggested that SNP discovery focused on noncoding sequence, where greater sequence polymorphism is present, will permit successful SNP discovery in soybean. One obvious target for SNP discovery is 3' UTRs of cDNAs. Discovery and mapping of SNPs in 3' UTRs will not only create useful genetic markers but will position the corresponding expressed gene on the genetic map. The resulting transcript map will provide a powerful tool to associate QTL with candidate genes. An alternative approach to SNP discovery is the sequence analysis of polymorphic AFLP bands or adjacent polymorphic sites as suggested by Meksem et al. (2001). This approach was successful in discovering numerous indels and SNPs in soybean.

6–3 TECHNOLOGIES FOR DNA MARKER ANALYSIS

6–3.1 Restriction Fragment Length Polymorphisms and Random Amplified Polymorphic DNA

The electrophoretic separation of DNA fragments, transfer and immobilization on a membrane, and detection of specific sequences was outlined by Southern (1975) and is the basis for the detection of RFLP. Numerous descriptions of the procedure are available (Sambrook et al., 1989; Grant and Shoemaker, 1997) and as a result there is little need for detailed description here. Likewise, there is little need to describe the analysis of RAPD markers. As indicated earlier, two of the important virtues of RAPD markers are ease of use and broad applicability. Almost without exception, RAPD fragments are analyzed using agarose gel electrophoresis (Rafalski, 1997). High resolution agaroses such as Metaphor agarose (FMC Bioproducts), Agarose 3:1 (Amresco, Solon, OH), and Synergel (Diversified Biotech, Inc., Boston, MA) are also used for RAPD fragment analysis.

6–3.2 Amplification Fragment Length Polymorphism

The procedures that Vos et al. (1995) outlined for the detection of AFLP used standard sequencing gels for the analysis of ^{32}P end-labeled AFLP fragments. In developing the soybean AFLP-based map, Keim et al. (1997) used a slightly modified AFLP protocol as described by Travis et al. (1996). Techniques for chemiluminescent detection of AFLP fragments are available (Lin et al., 1999). The AFLP technologies have been modified to function with small slab gels, a discontinuous buffer system, and silver staining (Mano et al., 2001). The AFLP fragment sizing has also been adapted to numerous semi-automated analysis platforms including the ABI PRISM 377 DNA and the Licor Global IR^2 DNA Analyzer.

6–3.3 Simple Sequence Repeat

Because SSR alleles are defined by the number of repeat units present in the SSR, the basis of allele sizing is the determination of PCR product size. As a result, numerous methods can be used for allele discrimination. With the exception of time-of-flight mass spectrometry, these procedures depend upon analyses using either slab gel or capillary electrophoresis. A necessarily brief overview of available options for SSR allele sizing follows.

6–3.4 Agarose Gel Electrophoresis

In those instances when allele sizes vary by six to eight or more basepairs, high resolution agarose gels are frequently quite adequate for purposes of mapping and marker assisted selection. Appropriate high-resolution agaroses are listed in the preceding paragraph.

6–3.5 Polyacrylamide Gel Electrophoresis

Initially, most SSR allele sizing was performed on high-resolution denaturing polyacrylamide sequencing gels to distinguish fragments that might differ in length by only one or two bases. This standard procedure was described by Kraft et al. (1988) and is used to separate products that are either internally labeled or end-labeled with ^{32}P or ^{33}P. Frequently, numerous "shadow bands" or "stutter bands" are associated with an SSR-containing fragment, making accurate size determination difficult, especially in the case of dinucleotide repeats. Some reduction in stutter bands can be achieved by using sequencing gels containing formamide in addition to urea as suggested by Litt et al. (1993). Silver staining is also frequently used to visualize SSR alleles as applied by Mansur et al. (1996). To resolve SSR alleles that differ by three or more bases, Cregan and Quigley (1997) suggested SSR allele sizing on 1.5-mm thick denaturing formamide/urea gels followed by staining with the DNA-specific stains SYBR-green or SYBR-gold (FMC Bioproducts) and detection on a UV transilluminator. Numerous other gel and capillary electrophoresis systems are available for high throughput automated or semi-automated SSR allele sizing. These include gel and capillary electrophoresis systems from Applied Biosystems (Foster City, CA), LI-COR, Inc. (Lincoln, NE), Beckman Coulter (Fullerton, CA) and Amersham Biosciences (Piscataway, NJ). Advantages of these systems over standard sequencing gels for sizing SSR-containing PCR products are (i) single-base resolution over a wide size range from 75 to 500 bases, (ii) automated sizing, (iii) automated data output, and (iv) elimination of radioactivity. Numerous soybean researchers have reported the use of the ABI PRISM 377 DNA for SSR allele sizing in soybean (Diwan and Cregan, 1997; Mian et al., 1999; Song et al., 1999; Narvel et al., 2000a, 2000b).

6–3.6 Capillary Electrophoresis

The same fluorescent chemistry employed in the Perkin-Elmer DNA sequencers described above is used in capillary electrophoresis systems with 1, 16, or 96 capillary capacity available from Perkin-Elmer Applied Biosystems. Beckman Coulter, Inc. manufactures an eight capillary machine and Amersham Biosciences has three systems with 48, 96, or 384 capillaries that can be used for multiplex SSR allele sizing.

6–3.7 Mass Spectrometry

Braun et al. (1997) proposed the use of matrix-assisted laser desorption/ionization time-of-flight mass spectrometry (MALDI-TOF) for SSR allele sizing. This system requires the annealing of a single detection primer within a few bases of the 3′-end of the SSR. A DNA polymerase extends this primer through the SSR by primer-directed DNA synthesis. A dideoxynucleotide triphosphate (ddNTP) is included to terminate the reaction at a point past the 5′-end of the SSR. Extension reactions from different SSR alleles yield products that differ in length by the number of bases in the alleles. These products are resolved using MALDI-TOF mass spectrometry.

6–3.8 Single Nucleotide Polymorphisms

The promise of SNP markers is the efficiencies in cost per data point and speed of data acquisition that will result from the technological innovations likely to be forthcoming from intensive research that has and continues to be focused on SNP detection. Numerous reviews of these technologies are available (Brookes, 1999; Gut, 2001; Kwok, 2000; Kwok and Chen, 1998; Shi, 2001; Syvanen, 2001). There are four basic approaches to SNP detection. These include (i) allele-specific hybridization (ASH) or allele-specific oligonucleotide hybridization, (ii) single-base extension (SBE) or minisequencing, (iii) the oligonucleotide ligation assay (OLA), and (iv) allele-specific cleavage of a "flap probe". These approaches have been combined with numerous different detection technologies. A number of these are summarized in Table 6–2. In many cases a SNP-containing PCR product is the neces-

Table 6–2. A partial listing of technologies used in the detection of single nucleotide polymorphisms (SNPs).

Allele specific hybridization	
[32]P-labelled probe hybridized to a SNP-containing fragment immobilized on a membrane	Coryell et al. (1999)
5′ nuclease assay	Lee et al. (1993)
Molecular beacons	Tyagi et al. (1998)
Electronic dot blot on semiconductor microchips	Gilles et al. (1999)
Electric field denaturation	Sosnowski et al. (1997)
Affymetrix oligo chip	Sapolsky et al. (1999)
Masscodes cleaved from allele specific oligos detected via mass spectrometry	Kokoris et al. (2000)
Randomly ordered fiber-optic gene arrays	Steemers et al. (2000)
Flow cytometry	
eSensor™	Yu et al. (2001)
Dynamic allele-specific hybridization (DASH)	Prince et al. (2001)
Single-base extension or minisequencing	
Single base extension-Tag array on glass slides (SBE-TAGS)	Hirschhorn et al. (2000)
Matrix-assisted laser desorption ionization-Time of flight (MALDI-TOF) mass spectrometry	Little et al. (1997); Ross et al. (1998)
Fluorescent dideoxynucleotide triphosphates (ddNTPs)	Lindblad-Toh et al. (2000)
Flow cytometry	Chen et al. (2000)
Pyrosequencing (multiple base extension)	Alderborn et al. (2000)
Denaturing high performance liquid chromatography	Hoogendoorn et al. (1999)
Fluorescence polarization	Chen et al. (1999)
Oligonucleotide ligation assay	
Rolling circle amplification	Qi et al. (2001)
Flow cytometry	Iannone et al. (2000)
Allele-specific cleavage of a "flap probe"	Fors et al. (2000)

sary target for detection; however, in some instances genomic DNA is the target and a PCR step is not required. As of December 2002, nearly 5 million putative human SNPs had been submitted to dbSNP (http://www.ncbi.nlm.nih.gov/SNP/index.html), the National Center for Biotechnology Information, National Institutes of Health, SNP database. Clearly, the human genetics community is focusing on SNPs as a major research tool for drug discovery, diagnostics, gene discovery, population genetics, and other applications. The costs of SNP detection are likely to decrease while the ease and speed of detection improve. The plant genetics community stands to be a beneficiary of the investments in technology being made by human geneticists.

6–4 GENE DISCOVERY

6–4.1 Expressed Sequence Tags

For nearly two decades random sequencing of gene transcripts has been recognized as a simple and efficient method of identification of many of the expressed genes in an organism (Putney et al., 1983). These sequences, known as Expressed Sequence Tags (ESTs), have become a valuable and efficient method for gene discovery (Sterky et al., 1998; Hillier et al., 1996; Marra et al., 1999). When sampling is random, the frequency of appearance of any given EST permits the identification of differential patterns of gene expression (Manger et al., 1998; Tanabe et al., 1999, Ewing et al., 1999). The ESTs also provide an opportunity to study gene evolution, to make comparative analyses between genera, and coupled with genetic mapping can identify candidates for important biological processes (Hatey et al., 1998).

Global, multi-tissue EST projects have been reported for *Arabidopsis* (Delseny et al., 1997) and rice (Ewing et al., 1999). More specialized, tissue-specific EST projects have been reported from root-hair-enriched *Medicago truncatula* tissue (Covitz et al., 1998), flower buds of Chinese cabbage [*Brassica rapa* (Pekinensis Group)](Lim et al., 1996), and wood-forming tissues of poplar (*Populus* spp.) (Sterky et al., 1998).

Shoemaker et al. (2002) reported on a global, multi-tissue EST analysis for soybean. More than 120 000 ESTs were generated from more than 50 cDNA libraries representing a wide range of organs, developmental stages, genotypes, and environmental conditions. This study was able to demonstrate correlated patterns of gene expression across cDNA libraries. As a result, gene expression profiles could be evaluated across libraries. cDNA libraries with similar EST composition and genes with similar expression patterns and potentially similar functions were grouped. These studies provide a large resource of publicly available genes and gene sequences and provide valuable insight into structure, function, and evolution of a model crop legume.

6–4.2 Genome Sequencing

Genome sequencing is the cornerstone of functional analyses and is fundamental to understanding the genetic composition of an organism. Whole-genome sequencing is currently underway for rice and is complete for all five of the *Arabidopsis* chromosomes, with gap-filling and annotation in progress (Theologis et al., 2000; Salanoubat et al., 2000; Tabata et al., 2000; Mayer et al., 1999; Lin et al., 1999). Because of the size and complexity of the soybean genome, it is unlikely that, given the current technology, the entire genome will be sequenced in the near future.

'Genomic Sampling' of nearly 2700 DNA sequences from more than 600 mapped loci has provided a glimpse of the composition and general structure of the soybean genome (Marek et al., 2001). For example, approximately one-third of all sequences sampled near SSR markers corresponded to repetitive DNA, while a little more than 18% of the sequences sampled near RFLP markers (putative hypomethylated regions) corresponded to repetitive DNA. Surprisingly, about 7 to 10% of the sequences sampled had significant similarity to an existing soybean EST sequence (Marek et al., 2001). Additionally, the clustering of BAC-end sequences around an anchored locus provided the opportunity to look for micro-synteny between soybean and other species. Few examples of microsynteny were observed between soybean and *Arabidopsis*, while about 33% of the sequence clusters detected microsynteny between soybean and *Medicago truncatula*, another legume (Marek et al., 2001).

Other 'sampling' approaches such as sequencing of hypomethylated regions or selection for open-reading frames (ORFs) may provide a wealth of information about gene-rich regions without the expense of whole-genome sequencing. Approaches that select for genes whose expression varies by organ and tissues, and during various stages of development or under different environmental and biological stresses can also provide a wealth of information. These combinations of data provide essential information about the regulation of genes and about metabolic regulation of the organism.

6–5 FUNCTIONAL GENOMICS

Breakthrough technologies stimulated by the human genome project are fundamentally changing the way in which biology is conducted in the genomics and postgenomics era. The approach has shifted from experimental analysis of "one gene at a time" to "thousands at a time" (with microarrays and chips), from "one protein at the time" to "thousands at a time" (proteomics), and to whole metabolite profiling (metabolomics) of cells and tissues (Brent, 2000; Lander and Weinberg, 2000).

In the broadest view, functional genomics is defined as the process of generating, integrating, and using information from genomics (sequencing), gene expression profiling (microarrays and chips), proteomics, metabolic profiling, and large-scale genotyping and trait analysis to understand the function of genes. Bioinformatics, statistical sciences, and computational sciences are increasingly indispensable for functional genomics research as the field becomes more driven by large-scale data and information processing (Heiter and Boguski, 1997).

The new Plant Genome Program, which received $85 million of funding during 1998 from the National Science Foundation (NSF), stimulated the development of structural and functional genomics resources for plants other than the model plant *Arabidopsis*. The program funded collaborative, multidisciplinary research in maize, soybean, tomato, cotton (*Gossypium hirsutum* L.), and other plants of agronomic importance to speed the development of tools that would be publicly available for gene expression analysis, gene tagging, and mapping (Walbot, 1999).

6–5.1 Creating a Soybean "Unigene" Set

One goal of the NSF-sponsored program for "Soybean Functional Genomics" is to develop a set of 30 000 unique genes from soybean. To accomplish this, the EST data from the commodity-sponsored "Public EST Project" (Shoemaker et al., 2002) is used as the raw material. The mRNAs that are more abundant in various tissues will be more highly represented in the EST collections. The ESTs are compared by computer programs such as PHRAP (Green, 2001) to assemble them into overlapping clones that have sequence similarity known as contiguous segments (contigs). In this way, longer sequences representing expressed genes are assembled and identical sequences that represent redundant clones are recognized. For example, contigs representing the storage protein Kunitz trypsin inhibitor consist of 118 ESTs that form a contig from among 2600 sequences in a cDNA library from the developing cotyledons of soybean whereas the contig representing soybean lipoxygenase-2 consists of six member sequences. The number of sequences in the contigs in a non-normalized cDNA library is a rough approximation of the relative abundance of the mRNAs within that tissue.

To develop a "unigene" set for soybean, the EST representing the most 5' sequence read in each contig is selected to represent that particular gene as it will likely represent the longest clone. In addition, the many sequences that occur only once in the EST collection (singletons) are also selected. Thus, the combined number of singletons and independent contigs that result from a computer assembly of ESTs becomes an estimate of the number of unique genes in the organism. This process is continuous because as the number of ESTs grows, the number of unique genes in the organism will continue to be refined. To date, this number for soybean is exceeding 40 000 (Shoemaker et al., 2002; Vodkin et al., 2003) from a collection of more than 208 000 ESTs. The *Arabidopsis* genome sequence has revealed approximately 26 000 genes (Arabidopsis Genome Initiative, 2000; Martienssen and McCombie, 2001). The number of human genes has been estimated at approximately 30 000 sequences from the complete human genome sequence (International Human Genome Sequencing Consortium, 2001); however, that number is still under debate (Hegenesch, et al., 2001). Alternative splicing appears to result in a much larger number of proteins in the human transcriptome than previously thought. The degree of alternative splicing in soybean or other higher plants has not been assessed. However, many higher plants, including soybean, have extensive duplications of sequences. Many functionally equivalent proteins are encoded by small gene families. One of the surprises to come from the knowledge of the complete *Arabidopsis* genome sequence was the extent to which the small genome of this plant contained duplicated sequences (Marteinssen and McCombie, 2001).

6–5.2 Global Gene Expression Analysis

The genomes of higher eukaryotes contain a very large number of genes. Depending upon the organism, this predicted number ranges from 20 000 to >60 000 unique genes. All of these genes are expressed in a coordinated fashion across tissues, across developmental time, and in response to particular physiological conditions. This regulation may be tightly linked for suites of genes (e.g., a particular biochemical pathway) and some transcripts will alter patterns of expression for other genes. Traditionally, single gene expression patterns have been studied to understand how different temporal, developmental, and physiological processes affected gene expression. This has resulted in our current models for gene regulation but has limited our ability to understand complex regulatory relationships among genes. With recent advances in genomics, very large numbers of genes can now be simultaneously analyzed for their expression levels in a comparative fashion between two biological states using microarray or biochip technology.

Several techniques for high throughput or global analysis of gene expression have been described as an outgrowth of the human genome project (Velculescu et al., 1995; Schena et al., 1995, 1996; DeRisi et al., 1997; Marshall and Hodgson, 1998). These include (i) high density expression arrays of cDNAs on conventional nylon filters with radioactive probing, (ii) microarrays or "chips" using fluorescent probes, and (iii) serial analysis of gene expression (SAGE).

6–5.2.1 High Density cDNA Arrays with Radioactive Probing

High density arrays are one method for assessing gene expression. The cDNA clones are arrayed in 384-well plates and spotted by robots onto nylon membranes. Either the bacterial colony, plasmid DNAs, or PCR products can be spotted. As many as 18 000 cDNAs can be arrayed on a filter of about 20 by 20 cm using robotic technology. The hybridization method is analogous to that used for conventional DNA or RNA blots on nitrocellulose membranes. The membranes are generally probed with ^{33}P-cDNA label produced by reverse transcription of mRNA. After hybridization, the membrane is imaged using a phosphorimager and the highly complex pattern must be read by image analysis software. This technology can be used to select specific cDNAs that are very weakly expressed in the library for further analysis. This "filter normalization" method is a tool for gene discovery of more weakly expressed genes. High density expression arrays have been used to increase gene discovery in soybean (Vodkin et al., 2000). In addition, one can compare the expression within a single library of numerous genes. If a "unigene" collection is used, then the relative expression of various genes within a single sample is easily obtained from high density membranes. The disadvantage is that without dual labeling, one cannot easily compare between two different mRNA samples. Dual labeling is one of the main advantages of microarrays using fluorescent detection.

6–5.2.2 DNA Microarrays or Chips to Analyze Global Gene Expression Patterns

An alternative method to the high density filters is microarray technology using fluorescent probes (DiRisi et al., 1997; Schena et al., 1995). In this method, the inserts from cDNA clones are amplified by PCR with vector primers and the amplified DNAs are arrayed onto glass microscope slides by a computer-controlled printing device. After fixation to the slide, the DNAs on the array are probed with fluorescently labeled cDNAs made from total mRNA of a particular tissue. Two different fluorescently labeled probes can be used simultaneously on the same slide, that is, one in the red range and one in the green range. The fluorescent images are captured with a scanning laser microscope and the intensity of each spot can be compared to standards of known concentrations to give quantitative data on gene expression. An arrayed set of 2375 unique cDNA from *Arabidopsis* has been used to examine changes in gene expression during pathogen challenge (Schenk et al., 2000) and will be used to address important problems in many other plant systems (Bouchez and Hofte, 1998; Mazur et al., 1999; DellaPenna, 1999; Somerville and Somerville, 1999; Somerville, 2000).

Microarrays are most effective when all genes within an organism are represented. Current technology for cDNA arrays are typically in the range of 5000 to 10 000 genes on glass slides. Thus, it is important to reduce the redundancy of the cDNA libraries before they are spotted to maximize the number of genes on the array by printing from the "unigene" set instead of from non-normalized libraries. Currently, arrays of 27 000 cDNAs (9k per array) for soybean have been printed containing low redundancy cDNA from many of the 80 cDNA libraries of the EST project. Experiments to examine differential expression during the process of induction of somatic embryos from tissue culture have been conducted (Thibaud-Nissen et al., 2003). The applications of microarray technology to soybean are enormous. A few include profiling the genes that respond to challenges by various pathogens and environmental stresses such as drought, heat, cold, flooding, and herbicide application. In addition, expression profiling of isoline genotypes that differ in protein or oil content and other quantitative traits will yield significant clues to the genes involved in those pathways and traits.

The future of functional genomics research will include oligonucleotide-based glass arrays that will allow distinguishing gene family members. Full-length sequencing of cDNA clones will include information at the 5' and 3' untranslated regions. The 3' region has more sequence variability and can be used to design oligonucleotides that distinguish gene family members. Assembly of the full-length cDNAs will also allow prediction of the entire ORF. Affymetrix chip technology in which approximately 20 nucleotides are synthesized directly using a photolithographic mask (Marshall and Hodgson, 1998) is very expensive and also requires full-length sequence of the expressed soybean genes. Full-length cDNA information is also critical for large-scale proteomics research on soybean so that peptide masses or partial protein sequences can be matched to the predicted ORFs. Although gene knockout and transposon tagging are not easy to obtain in soybean on a global scale, the application of microarrays for gene expression profiling and proteomics to investigate natural genetic variation in soybean promises to yield substantial information over the next decade that will aid in determining the function of soybean genes.

6–5.2.3 Serial Analysis of Gene Expression

Serial analysis of gene expression represents both a qualitative and quantitative method to characterize gene expression and to compare these patterns across tissues (Velculescu et al., 1995). SAGE captures short 10 to 20 nucleotide "tags" near the 3' end of individual mRNA molecules. It has been shown that 10 nucleotides uniquely identify >95% of human EST sequences. The tags are ligated into concatamers and are then cloned into plasmid vectors for sequencing (~40 tags per clone). Hundreds or even thousands of clones are sequenced to identify SAGE tags and to determine frequency in the plasmid library. The frequency of appearance of tags in the library has been shown to accurately estimate expression levels in the mRNA source tissue. Messages at a very low expression level (one per cell) can be quantitatively detected by sequencing numerous plasmid clones. More highly and moderately expressed genes are easily detected with only a few hundred sequenced clones. SAGE analysis requires an extensive 3' sequence data base. A disadvantage is that low abundance tags are not detected without extensive sequencing of the SAGE tag library from that source, which can be expensive. Initial analysis from 20 SAGE libraries in soybean has resulted in 132 992 SAGE tags of which 40 121 are unique (J. Schupp and P. Keim, personal communication, 2003).

6–5.3 Functional Consequences of Gene Duplication

Soybean has many gene family members that consist of 2 to 10 members that are very similar in sequence (Graham et al., 2000). In addition to gene duplications that arose because of the ancient tetraploid nature of soybean, there are more recent duplications that give rise to gene family members. For example, the chalcone synthase (CHS) family in soybean consists of six gene family members that are >97% similar in the protein-coding region (Akada and Dube, 1995). Five of these reside on the same BAC clone and one is unlinked. Gene family members contain more variation in the 5′ and 3′ untranslated regions that allow for gene-specific primers that can be used for experiments to detect differential expression (Todd and Vodkin, 1996).

In addition to providing the raw material for evolution, gene duplications sometimes lead to gene silencing or cosuppression. Cosuppression is a phenomenon, first described in plant systems in the early 1990s, whereby additional copies of a CHS gene in transgenic petunia (*Petunia hybrida* Vilm.) leads to suppression of both the endogenous and transgenic CHS transcripts (reviewed in Jorgensen, 1995). Homology-dependent gene silencing is now recognized widely in eukaryotic systems (Wolffe and Matzke, 1999). In soybean, the *I* (inhibitor) locus is a classical dominant genetic marker that results in yellow seed coats. The dominant form of the locus is present in most commercially used soybean varieties but spontaneous mutations to the recessive *i* allele occur and result in pigmented seed coats. Total levels of the CHS mRNAs and enzyme activity are reduced in yellow seeded soybean with the dominant *I* allele as compared to pigmented varieties with the recessive *i* allele (Wang et al., 1994). Thus, the biochemical block in pigmentation is via reduction of CHS mRNA and activity. Polymorphisms in CHS genes were found and cosegregated with the *I* locus. Paradoxically, CHS gene duplications suppress and deletions of promoter regions restore expression of the other CHS gene family members (Todd and Vodkin, 1996). These data showed that the *I* locus is a cluster of CHS genes and cosuppression results from naturally occurring duplications leading to lack of pigmentation.

6–6 SUMMARY

Soybean research benefits greatly from a "check-off" program in which the soybean producers have voluntarily chosen to contribute a percentage of their crop proceeds each year to research and marketing. As part of this commitment, the soybean commodity boards have funded an expressed sequence tag (EST) project that will result in approximately 300 000 gene sequence tags being generated and deposited into public databases over the next several years. This will provide the raw material for many gene discovery, evolution, and expression projects including some from the new National Science Foundation Plant Genome Program created in 1998. One of the major objectives of that program is to provide genomic tools for public and private research on economic crop species. The increased rate of discovery from structural and functional genomics research in plants will lead to new products from soybean and production of varieties with improved nutritional and agronomic characteristics through breeding and genetic engineering approaches.

REFERENCES

Akada, S., and S.K. Dube. 1995. Organization of soybean chalcone synthase gene clusters and characterization of a new member of the family. Plant Mol. Biol. 29:189–199.

Akkaya, M.S., A.A. Bhagwat, and P.B. Cregan. 1992. Length polymorphism of simple sequence repeat DNA in soybean. Genetics 132:1131–1139.

Akkaya, M.S., R.C. Shoemaker, J.E. Specht, A.A. Bhagwat, and P.B. Cregan. 1995. Integration of simple sequence repeat DNA markers into soybean linkage map. Crop Sci. 35:1439–1445.

Alderborn, A., A. Kristofferson, and U. Hammerling. 2000. Determination of single-nucleotide polymorphisms by real-time pyrophosphate DNA sequencing. Genome Res. 10:1249–1258.

Apuya, N., B.L. Frazier, P. Keim., E.J. Roth, and K.G. Lark. 1988. Restriction fragment length polymorphisms as genetic markers in soybean, *Glycine max* (L.) Merr. Theor. Appl. Genet. 75:889–901.

Arabidopsis Genome Initiative 2000. Nature (London) 408:796–815.

Arumuganathan, K., and E.D. Earle. 1991. Nuclear DNA content of some important plant species. Plant Mol. Biol. Rep. 9:208–219.

Bennetzen, J., and M. Freeling. 1993. Grasses as a single genetic system: Genome composition, collinearity and compatibility. Trends Genet. 9:259–261.

Botstein, D., R.L White, M. Skolnick, and R.W. Davis. 1980. Construction of a genetic linkage map in man using restriction fragment length polymorphisms. Am. J. Hum. Genet. 32:314–331.

Bouchez, D., and H. Hofte. 1998. Functional genomics in plants. Plant Physiol. 118:725–732.

Boutin, S.R.Y., N.D. Young, T.C. Olson, Z-H. Yu, R.C. Shoemaker, and C.E. Vallejos. 1995. Genome conservation among three legume genera detected with DNA markers. Genome 38:928–937.

Braun, A., D.P. Little, D. Reuter, B. Muller-Mysok, and H. Koster. 1997. Improved analysis of microsatellites using mass spectrometry. Genomics 46:18–23.

Brent, R. 2000. Genomic biology. Cell 100:169–183.

Brookes, A.J. 1999. The essence of SNPs. Gene 234:177–186.

Caetano-Anolles, G., B.J. Bassam, and P.M. Gresshoff. 1992. Primer-template interactions during DNA amplification fingerprinting with single arbitrary oligonucleotides. Mol. Gen. Genet. 235:157–165.

Cahill, D. 2000. High throughput marker assisted selection. p. A02. *In* G.B. Collins et al. (ed.) Proc. 8th Biennial Conf. on the Cellular and Molecular Biology of the Soybean, Lexington, KY. 13–16 Aug. 2000. Univ. of Kentucky, Lexington.

Cargill, M., D. Altshuler, J. Ireland, P. Sklar, K. Ardlie, N. Patil, N. Shaw, C.R. Lane, E.P. Lim, N. Kalyanaraman, J. Nemesh, L. Ziaugra, L. Friedland, A. Rolfe, J. Warrington, R. Lipshutz, G.Q. Daley, and E.S. Lander. 1999. Characterization of single-nucleotide polymorphisms in coding regions of human genes. Nat Genet. 22:231–238.

Chen, J., M.A. Iannone, M.S. Li, J.D. Taylor, P. Rivers, A.J. Nelsen, K.A. Slentz-Kesler, A. Roses, and M.P. Weiner. 2000. A microsphere-based assay for multiplexed single nucleotide polymorphism analysis using single base chain extension. Genome Res. 10:549–557.

Chen, X., L. Levine, and P.Y. Kwok. 1999. Fluorescence polarization in homogeneous nucleic acid analysis. Genome Res. 9:492–498.

Cho, R.J., M Mindrinos, D.R. Richards, R.J. Sapolsky, M. Anderson, E. Drenkard, J. Dewdney, T.L. Reuber, M. Stammers, N. Federspiel, A. Theologis, W.H. Yang, E. Hubbell, M. Au, E.Y Chung, D. Lashkari, B. Lemieux, C. Dean, R.J. Lipshutz, F.M. Ausubel, R.W. Davis, and R.J. Oefner. 1999. Genome-wide mapping with biallelic markers in *Arabidopsis thaliana*. Nat. Genet. 23:203–207.

Cho, T-J., C.S. Davies, and N.C. Nielsen. 1989. Inheritance and organization of glycinin genes in soybean. Plant Cell 1:329–337.

Clemente, T., B.J. LaValle, A.R. Howe, D.C. Ward, R.J. Rozman, P.E. Hunter, D.L. Broyles, D.S. Kasten, and M.A. Hinchee. 2000. Progeny analysis of glyphosate selected transgenic soybeans derived from Agrobacterium-mediated transformation. Crop Sci. 40:797–803.

Collins, F.S., L.D. Brooks, and A. Charkravarti. 1998. A DNA polymorphism discovery resource for research on human genetic variation. Genome Res. 8:1229–1231.

Cooper, D.N., B.A. Smith, H.J. Cooke, S. Niemann, and J. Schmidtke. 1985. An estimate of unique DNA sequence heterozygosity in the human genome. Hum. Genet. 69:201–205.

Coryell, V.H., H. Jessen, J.M. Schupp, D. Webb, and P. Keim. 1999. Allele-specific hybridization markers for soybean. Theor. Appl. Genet. 98:690–696.

Covitz, P.A., L.S. Smith, and S.R. Long. 1998. Expressed sequence tags from a root-hair-enriched medicago truncatula cDNA library. Plant Physiol. 117(4):1325–1332.

Cregan, P.B., A.A. Bhagwat, M.S. Akkaya, and J. Rongwen. 1994. Microsatellite fingerprinting and mapping of soybean. Methods Cell. Mol. Biol. 5:49–61.

Cregan, P.B., and C.V. Quigley. 1997. Simple sequence repeat DNA marker analysis. p. 173–185. *In* G. Caetano-Anolles and P.M. Gresshoff (ed.) DNA markers: Protocols, applications and overviews. John Wiley & Sons, New York.

Cregan P.B., T. Jarvik, A.L. Bush, R.C. Shoemaker, K.G. Lark, A.L. Kahler, N. Kaya, T.T. VanToai, D.G. Lohnes, J. Chung, and J.E. Specht. 1999. An integrated genetic linkage map of the soybean genome. Crop Sci. 39:1464–1490.

Danesh, D., S. Penuela, J. Mudge, R.L. Denny, H. Nordstrom, J.P. Martinez, and N.D. Young. 1998. A bacterial artificial chromosome library for soybean and identification of clones near a major cyst nematode resistance gene. Theor. Appl. Genet. 96:196–202.

DellaPenna, D. 1999. Nutritional genomics: Manipulating plant micronutrients to improve human health. Science (Washington DC) 285:375–379.

Delseny, M., R. Cooke, M. Raynal, and F. Grellet. 1997. The arabidopsis thaliana cDNA sequencing projects. FEBS Lett. 405:129–132.

Diers, B.W., P. Keim, W.R. Fehr, and R.C. Shoemaker. 1992a. RFLP analysis of soybean seed protein and oil content. Theor. Appl. Genet. 83:608–612.

Diers, B.W., L. Mansur, J. Imsande, and R.C. Shoemaker. 1992b. Mapping of Phytophtora resistance loci in soybean with restriction fragment length polymorphism markers. Crop Sci. 32:377–383.

DeRisi, J.L, V.R. Iyer, and P.O. Brown. 1997. Exploring the metabolic and genetic control of gene expression on a genomic scale. Science (Washington DC) 278:680–686.

Diwan, N., and P.B. Cregan. 1997. Automated sizing of fluorescent labeled simple sequence repeat markers to assay genetic variation in soybean. Theor. Appl. Genet. 95:723–733.

Ewing, R.M., A.B. Kahoa, O. Poirot, F. Lopez, S. Audic, and J.M. Claverie. 1999. Large-scale statistical analyses of rice ESTs reveal correlated patterns of gene expression. Genome Res. 9:950–959.

Ferreira, A.R., K.R. Foutz, and P. Keim. 2000. Soybean genetic map of RAPD markers assigned to an existing scaffold RFLP map. J. Hered. 91:392–396.

Fors, L., K.W. Lieder, S.H. Vavra, and R.W. Kwiatkowski. 2000. Large-scale SNP scoring from unamplified genomic DNA. Pharmacogenomics 1:219–229.

Gilles, P.N., D.J. Wu, C.B. Foster, P.J. Dillon, and S.J. Chanock. 1999. Single nucleotide polymorphic discrimination by an electronic dot blot assay on semiconductor microchips. Nature Biotechnol. 17:365–370.

Gizlice, Z., T.E. Carter, and J.W. Burton. 1994. Genetic base for North American public soybean cultivars released between 1947 and 1988. Crop Sci. 34:1143–1151.

Goldberg, R.B. 1978. DNA sequence organization in the soybean plant. Biochem. Genet. 16:45–68.

Graham, M.A., L.F. Marek, D. Lohnes, P. Cregan, and R.C. Shoemaker. 2000. Expression and genome organization of resistance gene analogs in soybean. Genome 43:86–93.

Grant, D., P. Cregan, and R.C. Shoemaker. 2000. Genome organization in dicots: Genome duplication in Arabidopsis and synteny between soybean and Arabidopsis. Proc. Natl. Acad. Sci. USA 97(8):4168–4173.

Grant, D., and R. Shoemaker. 1997. Molecular hybridization.p. 15–26. *In* G. Caetano-Anolles and P.M. Gresshoff (ed.) DNA markers: Protocols, applications and overviews. John Wiley & Sons, New York.

Green, P. 2001. Documentation for Phrap. http://bozeman.mbt.washington.edu, Genome Center, Univ. of Washington, Seattle.

Gurley, W.B., A.G. Hepburn, and J.L. Key. 1979. Sequence organization of the soybean genome. Biochem. Biophys. Acta 561:167–183.

Gut, I.G. 2001. Automation in genotyping of single nucleotide polymorphisms. Hum. Mutat. 17:475–492.

Halushka, M.K., J.B. Fan, K. Bentley, L. Hsie, N. Shen, A. Weder, R. Cooper, R. Lipshutz, and A. Chakravarti. 1999. Patterns of single-nucleotide polymorphisms in candidate genes for blood-pressure homeostasis. Nat. Genet. 22:239–247.

Hatey, F., G. Tosser-Klopp, C. Clouscard-Martinato, P. Mulsant, and F. Gasser. 1998. Expressed sequence tags for genes: A review. Genet. Select. Evol. 30:521–541.

Hegenesch, J.B., K.A. Ching, S. Batalov, A.I. Su, J.R. Walker, Y. Zhou, S.A. Kay, R.G. Schultz, and M.P. Cooke. 2001. A comparison of the Celera and Ensemble predicted gene sets reveals little overlap in novel genes. Cell 106:413–415.

Heiter P., and M. Boguski. 1997. Functional genomics: It's all how you read it. Science (Washington DC) 278:601–602.

Hillier, L.D., G. Lennon, M. Becker, M.F. Bonaldo, B. Chiapelli, S. Chissoe, N. Dietrich, T. DuBuque, A. Favello, et al. 1996. Generation and analysis of 280,000 human expressed sequence tags. Genome Res. 6:807–828.

Hirschhorn, J.N., P. Sklar, K. Lindblad-Toh, Y.M. Lim, M. Ruiz-Gutierrez, S. Bolk, B. Langhorst, S. Schaffner, E. Winchester, and E.S. Lander. 2000. SBE-TAGS: An array-based method for efficient single-nucleotide polymorphism genotyping. Proc. Natl. Acad. Sci. USA 97:12164–12169.

Hoogendoorn, B., M.J. Owen, P.J. Oefner, N. Williams, J. Austin, and M.C. O'Donovan. 1999. Geno-
    typing single nucleotide polymorphisms by primer extension and high performance liquid chro-
    matography. Hum. Genet. 104:89–93.

Hymowitz, T., R.J. Singh, and K.P. Kollipara. 1998. The genomes of the Glycine. Plant Breed. Rev.
    16:289–317.

Iannone, M.A., J.D. Taylor, J. Chen, M.S. Li, P. Rivers, K.A. Slentz-Kesler, and M.P. Weiner. 2000. Mul-
    tiplexed single nucleotide polymorphism genotyping by oligonucleotide ligation and flow cy-
    tometry. Cytometry 39:131–140.

International Human Genome Sequencing Consortium. 2001. Initial sequencing and analysis of the
    human genome. Nature (London) 409:860–921.

Jorgensen, R. 1995. Cosuppression, flower color patterns, and metastable gene expression states. Sci-
    ence (Washington DC) 268:686–691.

Kawabe, A., and N.T. Miyashita. 1999. DNA variation in the basic chitinase locus (*ChiB*) region of the
    wild plant *Arabidopsis thaliana*. Genetics 153:1445–1453.

Kawabe A., K. Yamane, and N.T. Miyashita. 2000. DNA polymorphism at the cytosolic phosphoglu-
    cose isomerase (*PgiC*) locus of the wild plant *Arabidopsis thaliana*. Genetics 156:1339–1347.

Keim, P., W. Beavis, J. Schupp, and R. Freestone. 1992. Evaluation of soybean RFLP marker diversity
    in adapted germplasm. Theor. Appl. Genet. 85:205–212.

Keim, P., B.W. Diers, T.C. Olson, and R.C. Shoemaker. 1990. RFLP mapping in soybean: Association
    between marker loci and variation in quantitative traits. Genetics 126:735–742.

Keim, P., J.M. Schupp, S.E. Travis, K. Clayton, T. Zhu, L. Shi, A. Ferreira, and D.M. Webb. 1997. A
    high-density soybean genetic map based on AFLP markers. Crop Sci. 37:537–543.

Keim, P., R.C. Shoemaker, and R.G. Palmer. 1989. Restriction fragment length polymorphism diver-
    sity in soybean. Theor. Appl. Genet. 77:786–792.

Kokoris, M., K. Dix, K. Moynihan, J. Mathis, B. Erwin, P. Grass, B. Hines, and A. Duesterhoeft. 2000.
    High-throughput SNP genotyping with the Masscode system. Mol. Diagn.5:329–340.

Kraft, R., J. Tardiff, K.S. Krauter, and L.A. Leinwand. 1988. Using mini-prep plasmid DNA for se-
    quencing double stranded templates with Sequenase. Biotechniques. 6:544–549.

Kruglyak, L. 1997 The use of a genetic map of biallelic markers in linkage studies. Nat. Genet. 17: 21–24.

Kuittinen, H., and M. Aguade. 2000. Nucleotide variation at the *CHALCONE ISOMERASE* locus in *Ara-
    bidopsis thaliana*. Genetics 155:863–872.

Kwok, P.Y. 2000. High-throughput genotyping assay approaches. Pharmacogenomics 1:95–100.

Kwok P.Y. and X. Chen. 1998. Detection of single nucleotide variations. p. 125–134. *In* J.K. Setlow (ed.)
    Genetic engineering. Vol. 20. Plenum Press, New York.

Kwok, P.Y., Q. Deng, H. Zakeri, and D.A. Nickerson. 1996. Increasing the information content of STS-
    based genome maps: Identifying polymorphisms in mapped STSs. Genomics 31:123–126.

Lackey, J. 1980. Chromosome numbers in the Phaseoleae (Fabaceae:Faboideae) and their relation to
    taxonomy. Am. J. Bot. 67:595–602.

Lander, E.S., and R.A. Weinberg. 2000. Genomics: Journey to the center of biology. Science (Wash-
    ington, DC) 287:1777–1782.

Lark, K.G., J.M. Weisemann, B.F. Matthews, R. Palmer, K. Chase, and T. Macalma. 1993. A genetic
    map of soybean (*Glycine max* L.) using an intraspecific cross of two cultivars: 'Minsoy' and 'Noir
    1'. Theor. Appl. Genet. 86:901–906.

Lee, J.M., D. Grant, C.E. Vallejos, and R.C. Shoemaker. 2001. Genome organization in dicots. II. Arabi-
    dopsis as a 'bridging species' to resolve genome evolution events among legumes. Theor. Appl.
    Genet. 103:765–773.

Lee, J.M., A. Bush, J.E. Specht, and R.C. Shoemaker. 1999. Mapping of duplicate genes in soybean.
    Genome 42:829–836.

Lim, C.O., H.Y. Kim, M.G. Kim, S.I. Lee, W.S. Chung, S.H. Park, I. Hwang, and M. J. Cho. 1996. Ex-
    pressed sequence tags of chinese cabbage flower bud cDNA. Plant Physiol. 111:577–588.

Lin, J.J., J. Ma, and J. Kuo. 1999. Chemiluminescent detection of AFLP markers. Biotechniques
    26:344–348.

Lindblad-Toh, K., E. Winchester, M.J. Daly, D.G. Wang, J.N. Hirschhorn, J.P. Laviolette, K. Ardlie, D.E.
    Reich, E. Robinson, P. Sklar, N. Shah, D. Thomas, J.B. Fan, T. Gingeras, J. Warrington, N. Patil,
    T.J. Hudson, and E.S. Lander. 2000. Large-scale discovery and genotyping of single-nucleotide
    polymorphisms in the mouse. Nature Genet. 24:381–386.

Litt, M., X. Hauge, and V. Sharma. 1993. Shadow bands seen when typing polymorphic dinucleotide
    repeats: Some causes and cures. Biotechniques 15:280–284.

Litt, M., and J.A. Luty. 1989. A hypervariable microsatellite revealed by in vitro amplification of a din-
    ucleotide repeat within the cardiac muscle actin gene. Am. J. Hum. Genet. 44:397-401.

Little, D.P., T.J. Cornish, M.J. O'Donnell, A. Braun, R.J. Cotter, and H. Kster. 1997. MALDI on a chip: Analysis of arrays of low-femtomole to subfemtomole quantities of synthetic oligonucleotides and DNA diagnostic products dispensed by a piezoelectric pipet. Anal. Chem. 69:4540–4546.

Liu, F., B.C. Zhuang, J.S. Zhang, and S.Y. Chen. 2000. Construction and analysis of soybean genetic map. Yi Chuan Xue Bao 27:1018–1026.

Manger, I.D., A. Hehl, S. Parmley, L. D. Sibley, M. Marra, L. Hillier, R. Waterston, and J.C. Boothroyd. 1998. Expressed sequence tag analysis of the bradyzoite stage of Toxoplasma gondii: Identification of developmentally regulated genes. Infect. Immun. 66(4):1632–1637.

Mano, Y., S. Kawasaki, F. Takaiwa, and T. Komatsuda. 2001. Construction of a genetic map of barley (*Hordeum vulgare* L.) cross 'Azumamugi' x 'Kanto Nakate Gold' using a simple and efficient amplified fragment-length polymorphism system. Genome 44:284–292.

Mansur, L.M., J.H. Orf, K. Chase, T. Jarvik, P.B. Cregan, and K.G. Lark. 1996. Genetic mapping of agronomic traits using recombinant inbred lines of soybean [*Glycine max* (L.) Merr.]. Crop Sci. 36:1327–1336.

Marek, L.F., J. Mudge, L. Darnielle, D. Grant, N. Hanson, M. Paz, Y. Huihuang, R. Denny, K. Larson, D. Foster-Hartnett et al. 2001. Soybean genomic survey: BAC-end sequences near RFLP and SSR markers. Genome 44:572–581.

Marek, L.F., and R.C. Shoemaker. 1997. BAC contig development by fingerprint analysis in soybean. Genome 40:420–427.

Marra, M., L. Hillier, T. Kucaba, M. Allen, R. Barstead, C. Beck, A. Blistain, M. Bonaldo, Y. Bowers, L. Bowles et al. 1999. An encyclopedia of mouse genes. Nature Genet. 21(2):1991–1994.

Marshall, A., and J. Hodgson. 1998. DNA chips: An array of possibilities. Nature Biotechnol. 16:27–31.

Martienssen, R., and W.R. McCombie. 2001. The first plant genome. Cell 105:571–574.

Matthews, B.F., T.E. Devine, J.M. Weisemann, H.S. Beard, K.S. Lewers, M.H. MacDonald, Y.B. Park, R. Maiti, J.J. Lin, et al. 2001. Incorporation of sequenced cDNA and genomic markers into the soybean genetic map. Crop Sci. 41:516–521.

Maughan, P.J., M.A. Saghi Maroof, and G.R. Buss. 1995. Microsatellite and amplified sequence length polymorphisms in cultivated and wild soybean. Genome 38:715–723.

Mayer, K., C. Schuller, R. Wambutt, G. Murphy, G. Volckaert, T. Pohl, A. Dusterhoft, W. Stiekema, K. D. Entian, N. Terryn et al. 1999. Sequence and analysis of chromosome 4 of the plant Arabidopsis thaliana. Nature (London) 402(6763):769–777.

Mazur, B., E. Krebbers, and S. Tingey. 1999. Gene discovery and product development for grain quality traits. Science (Washington DC) 285:372–375.

Meksem, K., E. Ruben, D. Hyten, K. Triwitayakorn, and D.A. Lightfoot. 2001. Conversion of AFLP bands into high-throughput DNA markers. Mol. Genet. Genomics 265:207–214.

Mian, M.A.R., T. Wang, D.V. Phillips, J. Alvernaz, and H.R. Boerma. 1999. Molecular mapping of the Rcs3 gene for resistance to frogeye leaf spot in soybean. Crop Sci. 39:1687–1691.

Michelmore, R.W., I. Paran, and R.V. Kesseli. 1991. Identification of markers linked to disease resistance genes by bulked segregant analysis: A rapid method to detect markers in specific genomic regions by using segregating populations. Proc. Natl. Acad. Sci. USA 88:9828–9832.

Morgante, M., I. Jurman, L. Shi, T. Zhu, P. Keim, and J.A. Rafalski. 1997. The STR120 satellite DNA of soybean: Organization, evolution and chromosomal specificity. Chromosome Res. 5:363–373.

Morgante, M.,and A.M. Olivieri. 1993. PCR-amplified microsatellites as markers in plant genetics. Plant J. 3:175–182.

Morgante, M., J.A. Rafalski, P. Biddle, S. Tingey, and A.M. Olivieri. 1994. Genetic mapping and variability of seven soybean simple sequence repeat loci. Genome 37:763–769.

Muehlbauer, G.J., J.E. Specht, M.A. Thomas-Comton, P.E. Staswick, and R.L. Bernard. 1988. Near isogenic lines—A potential source in the integration of conventional and molecular marker linkage maps. Crop Sci. 28:729–735.

Muehlbauer, G.J., P.E. Staswick, J.E. Specht, G.L. Graef, R.C. Shoemaker, and P. Keim. 1991. RFLP mapping using near-isogenic lines in the soybean [*Glycine max* (L.) Merr.]. Theor. Appl. Genet. 81:189–198.

Mullis, K., F. Faloona, S. Scharf, R. Saiki, G. Horn, and H. Erlich. 1986. Specific enzymatic amplification of DNA in vitro: The polymerase chain reaction. Cold Spring Harbor Symp. Quant. Biol. 51:263-273.

Narvel, J.M., W.-C. Chu, W.R. Fehr, P.B. Cregan, and R.C. Shoemaker. 2000a. Development of multiplex sets of simple sequence repeat DNA markers covering the soybean genome. Mol. Breed. 6:175–183.

Narvel, J.M., W.R. Fehr, W.-C. Chu, D. Grant, and R.C. Shoemaker. 2000b. Simple sequence repeat diversity among soybean plant introductions and elite genotypes. Crop Sci. 40:1452–1458.

Ohno, S. 1970. Evolution by gene duplication. Springer Verlag, New York.

Palmer, R.G., and T.C. Kilen. 1987. Qualitative genetics and cytogenetics. p. 135–209. *In* J.R. Wilcox (ed.) Soybeans: Improvement, production and uses. 2nd ed. ASA, CSSA, and SSSA, Madison, WI.

Palmer, R.G., and R.C. Shoemaker. 1998. Soybean genetics. p. 45–82. *In* M. Vidic and D. Jockovic (ed.) Soja. Soybean Inst. of Field and Vegetable Crops, Novi Sad, Yugoslavia.

Prabhu, R.R., and P.M.Gresshoff. 1994. Inheritance of polymorphic markers generated by DNA amplification fingerprinting and their use as genetic markers in soybean. Plant. Mol. Biol. 26:105–116.

Prince, J.A., L. Feuk, W.M. Howell, M. Jobs, T. Emahazion, K. Blennow, and A.J. Brookes. 2001. Robust and accurate single nucleotide polymorphism genotyping by dynamic allele-specific hybridization (DASH): Design criteria and assay validation. Genome Res. 11:152–162.

Purugganan, M.D., and J.I. Suddith. 1999. Molecular population genetics of floral homeotic loci: Departures from the equilibrium-neutral model at the *APETALA3* and *PISTILLATA* genes of *Arabidopsis thaliana*. Genetics 151:839–848.

Putney, S.D., W.C. Herlihy, and P. Schimmel. 1983. A new troponin T and cDNA clones for 13 different muscle proteins, found by shotgun sequencing. Nature (London) 302(5910):718–721.

Qi, X., S. Bakht, K.M. Devos, M.D. Gale, and A. Osbourn. 2001. L-RCA (ligation-rolling circle amplification): A general method for genotyping of single nucleotide polymorphisms (SNPs). Nucleic Acids Res. 29:E116.

Rafalski, A. 1997. Randomly amplified polymorphic DNA (RAPD) analysis. p. 75–83. *In* G. Caetano-Anolles and P.M. Gresshoff (ed.) DNA markers: Protocols, applications and overviews. J. Wiley & Sons, New York.

Rafalski, A., and S. Tingey. 1993. RFLP map of soybean (*Glycine max*). p. 6.149–6.156. *In* S.J. O'Brien (ed.) Genetic maps: Locus maps of complex genomes. Cold Spring Harbor Lab. Press, New York.

Rector, B.G., A. Demirbas, M.J. Livingston, H.L. Olsen, R.A. Ritchie, G.L. Graef, and J.E. Specht. 1999. Integration of the soybean microsatellite and classical marker maps. P. 136. *In* Proc. Plant and Animal Genome VII. 17–21 Jan. 1999. Scherago Int., New York.

Remington D.L., J.M. Thornsberry, Y. Matsuoka, L.M. Wilson, S.R Whitt, J. Doebley, S. Kresovich, M.M. Goodman, and E.S. Buckler. 2001. Structure of linkage disequilibrium and phenotypic associations in the maize. genome. Proc. Natl. Acad. Sci. USA. 98:11479–11484.

Riley, P.A. 1999. USDA expects record soybean supply. Inform 10(5):503–506.

Rongwen, J., M.S. Akkaya, U. Lavi, and P.B. Cregan. 1995. The use of microsatellite DNA markers for soybean genotype identification. Theor. Appl. Genet. 90:43–48.

Ross, P., L. Hall, I. Smirnov, and L. Haff. 1998. High level multiplex genotyping by MALDI-TOF mass spectrometry. Nature Biotech. 16:1347–1351.

Salanoubat, M., K. Lemcke, M. Rieger, W. Ansorge, M. Unseld, B. Fartmann, G. Valle, H. Blocker, M. Perez-Alonso, B. Obermaier et al. 2000. Sequence and analysis of chromosome 3 of the plant Arabidopsis thaliana. Nature (London) 408(6814):820–822.

Salimath, S., and M.K. Bhattacharyya. 1999. Generation of a soybean BAC library, and identification of DNA sequences tightly linked to the Rps1-k disease resistance gene. Theor. Appl. Genet. 98:712–720.

Sambrook, J., E.F. Fritsch, and T. Maniatis. 1989. Molecular cloning: A laboratory manual. 2nd ed. Cold Spring Harbor Lab., Cold Spring Harbor, New York.

Sapolsky, R., L. Hsie, A. Berno, G. Ghandour, M. Mittman, and J.B. Fan. 1999. High-throughput polymorphism screening and genotyping with high-density oligonucleotide arrays. Genet. Anal. Biomolecular Eng. 14:187–192.

Scallon, B.J., C.D. Dickinson, and N.C. Nielsen. 1987. Characterization of a null-allele for the Gy_4 glycinin gene from soybean. Mol. Gen. Genet. 208:107–113.

Schena, M., D. Shalon, R.W. Davis, and P.O. Brown. 1995. Quantitative monitoring of gene expression patterns with a complementary DNA microarray. Science (Washington DC) 270:467–470.

Schena, M., D. Shalon, R. Heller, A. Chai, P.O. Brown, and R.W. Davis. 1996. Parallel human genome analysis: Microarray-based expression monitoring of 1000 genes. Proc. Natl. Acad. Sci. USA 93:10614–10619.

Schenk, R.M., K. Kazan, I. Wilson, J.P. Anderson, T. Richmond, S.C. Somerville, J.M. Manners. 2000. Coordinated plant defense responses in *Arabidopsis* revealed by microarray analysis. Proc. Natl. Acad. Sci. 97:11655–11660.

Shi, M.M. 2001. Enabling large-scale pharmacogenetic studies by high-throughput mutation detection and genotyping technologies. Clin. Chem. 47:164–172.

Shoemaker, R., P. Keim, L. Vodkin, E. Retzel, S.W. Clifton, R. Waterston, D. Smoller, V. Coryell, A. Khanna, J. Erpelding, X. Gai, V. Brendel, C. Raph-Schmidt, E.G. Shoop, C.J. Vielweber, M.

Schmatz, D. Pape, Y. Bowers, B. Theising, J. Martin, M. Dante, T. Wylie, and C. Granger. 2002. A compilation of soybean ESTs: generation and analysis. Genome 45:329–338.

Shoemaker, R.C. 1999. Soyabean genomics from 1985–2002. AgBiotechNet 1:1–4.

Shoemaker, R.C., and T.C. Olson. 1993. Molecular linkage map of soybean (*Glycine max* L. Merr.). p. 6.131–6.138. *In* S.J. O'Brien (ed.) Genetic maps: Locus maps of complex genomes. Cold Spring Harbor Lab. Press, New York.

Shoemaker, R.C., K. Polzin, J. Labate, J. Specht, E.C. Brummer, T. Olson, N. Young, V. Concibido, J. Wilcox, J.P. Tamulonis, G. Kochert, and H.R. Boerma. 1996. Genome duplication in soybean (*Glycine* subgenus *soja*). Genetics 144:329–338.

Shoemaker, R.C., and J.E. Specht. 1995. Integration of the soybean molecular and classical genetic linkage groups. Crop Sci. 35:436–446.

Singh, R.J., and T. Hymowitz. 1988. The genomic relationship between *Glycine max* (L.) Merr. and *G. soja* Sieb. and Zucc. as revealed by pachytene chromosomal analysis. Theor. Appl. Genet. 76:705–711.

Somerville, C. 2000. The twentieth century trajectory of plant biology. Cell 100:13–25.

Somerville, C., and S. Somerville. 1999. Plant functional genomics. Science (Washington DC) 285:380–383.

Song, Q., C.V. Quigley, T.E. Carter, R.L. Nelson, H.R. Boerma, J. Strachan, and P.B. Cregan. 1999. A selected set of trinucleotide simple sequence repeat markers for soybean variety identification. Plant Varieties Seeds 12:207–220.

Sosnowski, R.G., E. Tu, W.F. Butler, J.P. O'Connell, and M.J. Heller. 1997. Rapid determination of single base mismatch mutations in DNA hybrids by direct electric field control. Proc. Natl. Acad. Sci. USA. 94:1119–1123.

Southern, E.M. 1975. Detection of specific sequences among DNA fragments separated by gel electrophoresis. J. Mol. Biol. 98:503–517.

SoyStats: A reference guide to important soybean facts and figures. 1997. Am. Soybean Assoc., St. Louis, MO.

Steemers, F.J., J.A. Ferguson, and D.R. Walt. 2000. Screening unlabeled DNA targets with randomly ordered fiber-optic gene arrays. Nat. Biotechnol. 18:91–94.

Sterky, F., S. Regan, J. Karlsson, M. Hertzberg, A. Rohde, A. Holmberg, B. Amini, R. Bhalerao, M. Larsson, R. Villarroel, M. Van Montagu, G. Sandberg, O. Olsson, T.T. Teeri, W. Boerjan, P. Gustafsson, M. Uhlen, B. Sundberg, and J. Lundeberg. 1998. Gene discovery in the wood-forming tissues of poplar: Analysis of 5, 692 expressed sequence tags. Proc. Natl. Acad. Sci. USA 95(22):13330–13335.

Syvanen, A.C. 2001. Accessing genetic variation: Genotyping single nucleotide polymorphisms. Nat. Rev. Genet. 2:930–942.

Tabata S., T. Kaneko, Y. Nakamura, H. Kotani, T. Kato, E. Asamizu, N. Miyajima, S. Sasamoto, T. Kimura, T. Hosouchi et al. 2000. Sequence and analysis of chromosome 5 of the plant Arabidopsis thaliana. Nature (London) 408(6814):823–826.

Tajima, F. 1983. Evolutionary relationship of DNA sequences in finite populations. Genetics 105:437–460.

Tanabe, K., S. Nakagomi, S. Kiryu-Seo, K. Namikawa, Y. Imai, T. Ochi, M. Tohyama, and H. Kiyama. 1999. Expressed-sequence-tag approach to identify differentially expressed genes following peripheral nerve axotomy. Molecular Brain Res. 64:34–40.

Tenaillon, M.I., M.C. Sawkins, A.D. Long, R.L. Gaut, J.F. Doebley, and B.S. Gaut. 2001. Patterns of DNA sequence polymorphism along chromosome 1 of maize (*Zea mays* ssp. *mays* L.) Proc. Natl. Acad. Sci. USA 98:9161–9166.

Theologis, A., J.R. Ecker, C.J. Palm, N.A. Federspiel, S. Kaul, O. White, J. Alonso, H. Altafi, R. Araujo, C.L. Bowman et al. 2000. Sequence and analysis of chromosome 1 of the plant Arabidopsis thaliana. Nature (London) 408(6814):816–820.

Thibaud-Nissen, F., R.T. Shealy, A. Khanna, and L.O. Vodkin. 2003. Clustering of microarray data reveals transcript patterns associated with somatic embryogenesis in soybean. Plant Physiol. 132. (In press.)

Todd, J.J., and Vodkin, L.O. 1996. Duplications that suppress and deletions that restore expression from a chalcone synthase multigene family. Plant Cell 8:687–699.

Tomkins, J.P, R. Mahalingam, H. Smith, J.L. Goicoechea, H.T. Knap, and R.A. Wing. 1999. A bacterial artificial chromosome library for soybean PI 437654 and identification of clones associated with cyst nematode resistance. Plant Mol. Biol. 41(1):25–32.

Travis, S.E., J. Maschinski, and P. Keim. 1996. An analysis of genetic variation in *Astragalus cremnophylax* var. *cremnophylax*, a critically endangered plant, using AFLP markers. Mol. Ecol. 5:735–745.

Tyagi, S., D.P. Bratu, and F.R. Kramer. 1998. Multicolor molecular beacons for allele discrimination. Nature Biotech. 16:49–53.

Velculescu, V.E., Ahang, L., Vogelstein, B., Kinzler, K.W. 1995. Serial analysis of gene expression. Science (Washington DC) 270:484–487.

Vodkin, L.O., A. Khanna, S. Clough, R. Shealy, R. Philip, J. Erplending, M. Paz, R. Shoemaker, V. Coryell, J. Schupp, P. Keim, A. Rodriquez-Huete, P. Zeng, J. Polacco, J. Mudge, R. Denny, N. Young, C. Raph, L. Shoop, E. Retzel. 2003. Structural and functional genomics projects in soybean. Plant Mol. Biol. Rep. Supplement 18:2, pS1.

Vos, P., R. Hogers, M. Bleeker, M. Rijans, T. van der Lee, M. Hornes, A. Frijters, J. Pot, J. Peleman, M. Kuiper, and M. Zabeau. 1995. AFLP: A new technique for DNA fingerprinting. Nucleic Acids Res. 23:4407–4414.

Walbot, V. 1999. Genes, genomes, genomics. What can plant biologists expect from the 1998 National Science Foundation Plant Genome Research Program. Plant Physiol. 119:1151–1155.

Wang, C. Todd, J.J., and Vodkin, L.O. 1994. Chalcone synthase mRNA and activity are reduced in yellow soybean seed coats with dominant I alleles. Plant Physiol. 105:739–748.

Wang, D.G., J.B. Fan, C.J. Siao, A. Berno, P. Young, R. Sapolsky, G. Ghandour, N. Perkins, E. Winchester, J. Spencer et al. 1998. Large-scale identification, mapping, and gentyping of single-nucleotide polymorphisms in the human genome. Science (Washington DC) 280:1077–1082.

Watterson, G.A. 1975. On the number of segregating sites in genetical models without recombination. Theor. Pop. Biol. 7:256–276.

Weber, J.L., and P.E. May. 1989. Abundant class of human DNA polymorphisms which can be typed using the polymerase chain reaction. Am. J. Hum. Genet. 44:388–396.

Welsh, J., and M. McClelland. 1990. Fingerprinting genomes using PCR with arbitrary primers. Nucleic Acids Res. 18:7213–7218.

Williams, J.K.G., A.R. Kubelik, K.J. Livak, J.A. Rafalski, and S.V. Tingey. 1990. DNA polymorphisms amplified by arbitrary primers are useful as genetic markers. Nucleic Acids Res. 18:6531–6535.

Wolffe, A.P., and M.A. Matzke. 1999. Epigenetics: Regulation through repression. Science (Washington DC) 286:481–486.

Wu, X.L., C.Y. He, Y.J. Wang, Z.Y. Zhang, Y. Dongfang, J.S. Zhang, S.Y. Chen, and J.Y. Gai. 2001. Construction and analysis of a genetic linkage map of soybean. Yi Chuan Xue Bao 28:1051–1061.

Xing, L., C. Ge, R. Zeltser, G. Maskevitch, B. J. Mayer, and K. Alexandropoulos. 2000. c-Src signaling induced by the adapters Sin and Cas is mediated by Rap1 GTPase. Mol. Cell Biol. 20(19):7363–7377.

Xu, S., R. Singh, and T. Hymowitz. 1998. Establishment of a cytogenetic map of soybean: Current status. Soybean Genet. Newsl. 25:120–122.

Xue, Z.T., M.L. Xu, W. Shen, N.L. Zhuang, W.M. Hu, and S.C. Shen. 1992. Characterization of a Gy_4 glycinin gene from soybean *Glycine max* cv. Forrest. Plant Mol. Biol. 18:897–908.

Yamamoto, E., H.C. Karakaya, and H.T. Knap. 2000. Molecular characterization of two soybean homologs of Arabidopsis thaliana CLAVATA1 from the wild type and fasciation mutant. Biochim. Biophys. Acta 1491:333–340.

Yamamoto, E., and H.T. Knap. 2001. Soybean receptor-like protein kinase genes: Paralogous divergence of a gene family. Mol. Biol. Evol. 18(8):1522–1531.

Yamanaka, N., S. Ninomiya, M. Hoshi, Y. Tsubokura, M. Yano, Y. Nagamura, T. Sasaki, and K. Harada. 2001. An informative linkage map of soybean reveals QTLs for flowering time, leaflet morphology and regions of segregation distortion. DNA Res. 8:61–72.

Young, W.P., J.M. Schupp, and P. Keim. 1999. DNA methylation and AFLP marker distribution in the soybean genome. Theor. Appl. Genet. 99:785–792.

Yu, C.J., Y. Wan, H. Yowanto, J. Li, C. Tao, M.D. James, C.L. Tan, G.F. Blackburn, and T.J. Meade. 2001. Electronic detection of single-base mismatches in DNA with ferrocene-modified probes. J. Am. Chem. Soc. 123:11155–11161.

Yu, Y.G., M.A. Saghai Maroof, G.R. Buss, P.J. Maughan, and S.A. Tolin. 1994. RFLP and microsatellite mapping of a gene for soybean mosaic virus resistance. Phytopathology 84:60–64.

Zakharova, E.S., S.M. Epishin, and Y.P. Vinetski. 1989. An attempt to elucidate the origin of cultivated soybean via comparison of nucleotide sequences encoding glycinin B_4 polypeptide of cultivated soybean, *Glycine max*, and its presumed wild progenitor, *Glycine soja*. Theor. Appl. Genet. 78:852–856.

Zhang, H., L. Yu, N. Mao, Q. Fu, Q. Tu, J. Gao, and S. Zhao. 1999. Cloning, characterization, and chromosome mapping of RPS6KC1, a novel putative member of the ribosome protein S6 kinase family, to chromosome 12q12–q13.1. Genomics 61(3):314–318.

Zhu, T., J.M. Schupp, A. Oliphant, and P. Keim. 1994. Hypomethylated sequences: Characterization of the duplicate soybean genome. Mol. Gen. Genet. 244:638–645.

Zhu, T., L. Shi, J.J. Doyle, and P. Keim. 1995. A single nuclear locus phylogeny of soybean based on DNA sequence. Theor. Appl. Genet. 90:991–999.

Zhu, T., I. Shi, P. Gresshoff, and P. Keim. 1996. Characterization and application of soybean YACs to molecular cytogenetics. Mol. Gen. Genet. 252:483–488.

Zhu, Y.-L, Q.-J. Song, D.L. Hyten, C.P. Van Tassell, L.K. Matukumalli, D.R. Grimm, S.M. Hyatt, E.W. Fickus, N.D. Young, and P.B. Cregan. 2003. Single nucleotide polymorphisms (SNPs) in soybean. Genetics 163:1123–1134.

7

Transgenic Soybean

WAYNE A. PARROTT

The University of Georgia
Athens, Georgia

THOMAS E. CLEMENTE

University of Nebraska
Lincoln, Nebraska

Genetic transformation of soybean [*Glycine max* (L.) Merr.] has become firmly established as a viable breeding and research tool. The use of transgenic soybean for product development is best exemplified by the Roundup Ready (RR) soybean (Monsanto Company, St. Louis, MO). More than any other genetically engineered plant, the RR soybean has demonstrated the advantages of genetically engineered crops and the willingness of producers to accept them. It has been suggested that the RR bean is the most rapidly adopted agricultural technology in history.

The ability to develop the RR soybean and other transgenic soybean plants is the result of technical advances in cell and molecular biology achieved since the 1980s. Nevertheless, implementation of transgenic technology and the rapid adoption of the RR soybean has raised several issues centering on regulatory approval, IP protection, international trade, and the detection of engineered DNA and/or protein in the food supply.

This chapter begins with an overview of the development of the RR and the high oleic (HO) soybean and addresses the issues which have come to the forefront with the introduction of transgenic crops. This chapter also covers the methodology and technical advances made in soybean transformation, and the potential that transgenic soybean has for the future development of novel cultivars and for furthering biological studies of soybean by complementing functional genomics programs.

7–1 TRANSGENIC SOYBEAN AS A PRODUCT

The RR soybean, developed by Monsanto, St. Louis is the most recognized transgenic soybean product on the market. A second product in commercial production is a HO soybean developed by DuPont, Wilmington, DE, which was obtained by inserting an extra copy of the endogenous Δ-12 desaturase gene from soybean. Addition of the extra gene results in overexpression, a phenomenon that in turn triggers gene silencing (Meyer and Saedler, 1996). As a consequence, oleic acid accumulates because it is not converted to linoleic acid. Lack of approval for the European market has limited the hectarage of this soybean.

In addition, two different Aventis' Liberty Link (LL) soybean plants (Aventis, Strasbourge, France), engineered with a phosphinothricin acetyl transferase, one from *Streptomyces hygroscopicus* Crandell and Hamill and the other from *S. viridochromogenes* (Krainsky) Waksman and Henrici, have received regulatory approval for commercialization in the USA. Both are resistant to the herbicide, glufosinate. Lack of regulatory approval in Europe, which stopped approving the sale of additional genetically engineered crops in October 1998, has kept these soybean products off the market.

7–1.1 The Roundup Ready Soybean

7–1.1.1 History and Development

The RR soybean was among the first transgenic crops to reach market. First commercialized in 1996, RR soybean was grown on 54% of U.S. soybean hectarage by 2000 (Carpenter, 2001a). Hectarage for 2002 was expected to reach 74% (Agricultural Statistics Board, 2002).

Glyphosate, the active ingredient in the commercial herbicide, Roundup (Monsanto Company, St. Louis, MO), is a nonselective, broad-spectrum herbicide with favorable environmental attributes, including strong adsorption to soil and negligible animal toxicity. Roundup is one of the most widely used postemergent herbicides, but its nonselective property prevents application to crop plants. Glyphosate specifically acts on the plastid-localized enzyme 5-enolpyruvyl shikimic acid-3-phosphate (EPSP) synthase (Steinrücken and Amrhein, 1980). The inhibition of EPSP synthase by glyphosate blocks the shikimic acid pathway, which in turn prevents aromatic amino acid production. The EPSP synthase genes are ubiquitous in the environment, as they are present in all bacterial and green plants, and therefore have always been part of the human diet.

Two general schemes have been pursued to impart glyphosate tolerance to crop plants. The first is based on metabolic detoxification of the molecule. Glyphosate is readily detoxified by numerous microbes via two metabolic processes, one that generates sarcosine and inorganic phosphate, and the other that produces aminomethylphosphonic acid (Barry et al., 1992). The latter pathway occurs most frequently in soils conducive for glyphosate breakdown (Nomura and Hilton, 1977; Rueppel et al., 1977). A glyphosate oxidoreductase gene was isolated from *Achromobacter* sp. strain LBAA, and expression of a codon-optimized version in plants resulted in glyphosate tolerance (Barry et al., 1992).

The second strategy is based on the identification of a glyphosate-tolerant EPSP synthase. Several such tolerant genes have been identified (Stalker et al., 1985; Klee et al., 1987; Barry et al., 1992). One of these, the *aroA* gene from *Agrobacterium* sp. strain CP4, imparts a high level of tolerance to Roundup when targeted to the chloroplasts of higher plants (Barry et al., 1992) and was used to develop RR soybean, as described below.

For transformation, pPV-GMGT04 (Padgette et al., 1995) was assembled with two CP4 EPSP synthase constructs, one under the control of the enhanced 35S cauliflower [*Brassica oleracea* (Botrytis Group)] mosaic virus (CaMV) promoter (Kay et al.,1987) and the other fused to the figwort mosaic virus (FMV) 35S promoter (Gowda et al., 1989). In both CP4 constructs, the gene was fused to the petunia (*Petunia hybrida* Vilm.) EPSP synthase transit peptide. Plasmid pPV-GMGT04 also carries the ß-glucuronidase gene (Jefferson et al., 1986), under the control of the *Agrobacterium tumefaciens* (Smith & Townsend) Conn. mannopine synthase (MAS) promoter (Velten et al., 1984), and also the *nptII* gene for bacterial selection (Fig. 7–1). The pPV-GMGT04 was delivered into Asgrow cv. A5403 via microprojectile bombardment as McCabe et al. (1988) described. The β-glucuronidase (GUS) expression patterns in the primary transformants were used to predict germline transformation events (Christou and McCabe, 1992). T_1 seed derived by self-pollination from more than 300 transformed lines were screened under greenhouse conditions for tolerance to Roundup (Padgette et al., 1995). One line, designated 40-3, was identified that displayed a high degree of tolerance. Four lines were subsequently generated by selfing line 40-3, of which line 40-3-2 exhibited a consistent high Roundup tolerance under field conditions (Padgette et al., 1995). The pedigrees of all subsequent RR germplasm have 40-3-2 as the parental source of the CP4 EPSPS gene.

Molecular characterization of line 40-3-2 indicates it contains a single functional insert of the 35S CaMV CP4 cassette, an additional 254-bp fragment of the CP4 gene adjacent to the 3′ untranslated region of the 35S CaMV CP4 cassette

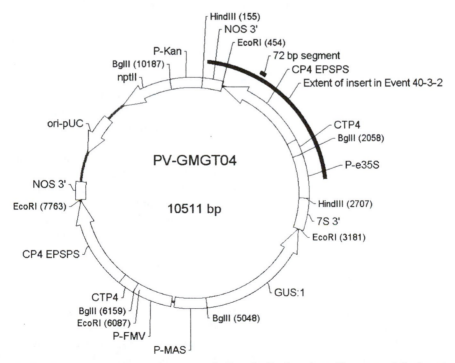

Fig. 7–1. Map of the construct used to generate the Roundup Ready soybean. The commercialized event only contains a copy of the CP4 EPSPS with part of the enhanced CaMV 35S promoter and the nos terminator, as indicated by the heavy line. A 72-bp fragment is also found 3′ to the main insert.

(Windels et al., 2001), and a 72-bp segment of the CP4 gene that co-segregates with the functional CP4 cassette. This transgenic locus is inherited as a single, dominant gene. Neither the GUS gene nor the FMV/EPSP cassette was retained in the final product, having been integrated into a separate locus that segregated out during the early generations.

The RR trait has been incorporated into the majority of soybean-breeding programs and more than 1000 RR cultivars. Initial reports indicated no negative impact on agronomic performance in soybean cultivars which express the RR phenotype (Delannay et al., 1995). Nevertheless, whether the RR gene is truly devoid of negative consequences remains controversial. One reported effect of the transgene is over-deposition of lignin, leading to stem splitting under hot, dry conditions (Gertz et al., 1999). Another reported effect is yield loss (Benbrook, 1999), which has been attributed to both the transgene itself (Elmore et al., 2001), or to simple yield drag that is being eliminated with additional breeding efforts (Carpenter, 2001b). The lack of true near-isogenic lines with and without the RR gene has made it impossible to determine whether observed effects are due to linkage drag or pleiotropic effects from the transgene.

Finally, the level of phytoestrogens has been reported to be lower in RR soybean (Lappé et al., 1998). However, the American Soybean Association (www.bio.org/food&ag/soy/asa.html, accessed January 2002) has pointed out that Lappé et al. (1998) did not consider the impact of environmental effects on isoflavone levels, and that the reported variability is well within the variability known to exist for isoflavone levels in soybean.

7–1.1.2 Regulatory Approval

Prior to release, the RR soybean, like all products derived from genetic engineering, was subjected to a rigorous safety assessment. Information was gathered on the DNA sequence of the junction fragments flanking the foreign locus, expression level of CP4 in vegetative and seed tissue, nutritional composition, and toxicology of the CP4 protein.

Initial safety assessment studies on the CP4 protein compared the physical characteristics of the protein derived from either bacterial expression (*Escherichia coli*) and from in planta expression. Data were compared for molecular weight, immuno-reactivity, N-terminal sequence, glycosylation and functional activity from the two sources. Subsequent studies investigated the potential of the CP4 EPSP synthase to elicit an allergenic response. The ability to withstand degradation in the gastrointestinal tract is but one characteristic of a potential food allergen (Astwood et al., 1996). Accordingly, in vitro digestibility studies were conducted on the protein using simulated gastric intestinal fluids. The protein was degraded within 15 s in simulated stomach fluid and within 10 min in simulated intestinal fluid. Moreover, all known food allergens are present in the ingested product at levels ranging from 1 to 80% (Taylor, 1992), whereas CP4 EPSPS is present in whole soybean seed at approximately 0.04% (Padgette et al., 1995), and levels are significantly lowered during processing. Collectively, these data provide convincing evidence that products derived from RR soybean are as safe as those derived from conventional soybean.

In accordance with standard regulatory requirements currently in place for transgenic crops, the nutritional and compositional characteristics of the RR soybean were compared with those of commodity soybean. The data have been published (Harrison et al., 1996) and made available to the public on www.agbios.com. No significant differences were observed for any characteristic, including total protein, total oil, and amino acid and fatty acid composition (Taylor et al., 1999). A series of feeding studies on a variety of animals including broiler chickens *(Gallus domesticus)*, catfish *(Ictalurus punctatus)*, and dairy cows *(Bos taurus)* were conducted to compare growth and feeding efficiency between RR and conventional soybean-derived meal. The data revealed no significant difference in any of the characteristics measured (Hammond et al., 1996). Together, these data show that feed derived from RR soybean is nutritionally equivalent to that derived from standard commodity soybean. A complete review on the topic has been provided by Yoshida (2000).

It is worth noting here that FDA regulations mandate testing for allergenicity, particularly when the transgene is derived from a known allergen source (FDA, 1992). Accordingly, soybean engineered by Pioneer HiBred International to produce a high-methionine protein from a known allergen, the Brazil-nut *(Bertholletia excelsa* Humb. and Bonpl.), was found to be allergenic when the required testing was performed (Nordlee et al., 1996). Development of this transgenic soybean was ultimately suspended due to these findings, and no soybean with the Brazil-nut protein was ever commercialized or released to the public. This case demonstrates the ability of the current regulatory system to detect possible problems at an early stage.

Finally, the use of glyphosate on soybean inevitably leads to the presence of glyphosate residues in the soybean plant and seed. Accordingly, the Environmental Protection Agency (EPA) (2000) established acceptable glyphosate residue levels of 20 mg kg^{-1} for the soybean seed itself, 100 mg kg^{-1} for the soybean hulls, 50 mg kg^{-1} for aspirated grain fractions, 100 mg kg^{-1} for soybean forage, and 200 mg kg^{-1} for soybean hay.

7–1.1.3 Economic Impact

The RR soybean has rapidly been adopted by soybean producers. Prior to the introduction of RR soybean, producers relied on a herbicide regime that sometimes damaged the developing soybean and/or left residues in the soil that injured the subsequent crop (Gianessi and Carpenter, 2000). The planting of RR soybean has decreased herbicide application costs, and led to competition-induced decreases in the prices of other herbicides. Furthermore, the price of Roundup itself has decreased following expiration of the original patent.

Some of these savings are lost, as a technology fee surcharge is paid in addition to the price of seed. It has been estimated that for 1999, when approximately 48% of the soybean acreage was transgenic, farmers saved approximately $216 million in weed control costs relative to 1995, the last year before the advent of transgenic soybean, with a reduction in 19 million herbicide applications per year, in spite of an 18% increase in the soybean acreage from 1995 to 1999 (Carpenter and Gianessi, 2001). Nevertheless, critics of the technology claim that the economic benefit derived from RR soybean will be diminished as repeated use of glyphosate selects for more glyphosate-tolerant weeds, thus requiring the use of additional herbicides (Benbrook, 1999).

7–1.1.4 Environmental Benefits

Adoption of RR soybean results in benefits which are difficult to quantify in economic terms. Because RR soybean facilitates no-till practices, an estimated 37 million tonnes of soil will be spared from erosion by 2020. Furthermore, reduced tillage will save 81.5 L of fuel per hectare, preventing the release of 400 000 tonnes of CO_2 into the atmosphere (Barnes, 2000). Nevertheless, there is a concern that lower cost of production will make it economically feasible to expand soybean production into agronomically marginal but ecologically sensitive areas, particularly in South America (Pengue, 2000), and cause their degradation.

7–1.2 High-Oleic Soybean

Agriculture biotechnology is implementing the tool of genetic engineering to modify the lipid profile of oilseed crops. This area has received much attention because oils derived from oilseed crops such as soybean can be used in industrial and food applications (Mazur et al., 1999). Soybean oil is the largest source of vegetable oil in the USA and a prime resource for bio-based fuels. The average production of soybean oil from 1993 to 1995 was approximately 450 000 kg (14.9 million lb) (www.nass.usda.gov). Soybean seed is approximately 18% oil, and standard commodity soybean oil is primarily composed of palmitic acid (11%), stearic acid (4%), oleic acid (22%), linoleic acid (53%), and linolenic acid (8%).

Soybean producers have targeted improved soybean oil traits as a priority area of research to enhance the market share of U.S. soybean. Oils high in monounsaturated fatty acids (i.e., oleic acid) possess increased oxidative stability, which in turn negates the need for hydrogenation and thus eliminate production of trans-fatty acids, which have been attributed with negative health aspects (Mazur et al., 1999). Moreover, oil with high oleic acid and low-saturated fatty acid content will have improved performance in biodiesel blends (Duffield et al., 1998). A HO soybean has been generated via mutational breeding (Burton et al., 1983; Takagi and Rahman, 1996) and also with genetic engineering (Mazur et al., 1999). The HO phenotype derived from the former approach is subject to large environmental effects (Martin et al., 1986), while the latter has resulted in a more stable phenotype. In addition, oleic acid content in the seed storage lipids ranges from 30 to 65% in soybean developed via mutagenesis, while the use of genetic engineering resulted in oleic acid content of more than 80%.

In soybean seeds, oleic acid is converted to linoleic acid in a single step catalyzed by a Δ12 desaturase encoded by the *FAD2-1* gene (Heppard et al., 1996). Down-regulation of this gene results in accumulation of up to 80% oleic acid in the seed storage lipids (Mazur et al., 1999; Buhr et al., 2002). The *FatB* gene encodes palmitoyl-thioesterase (Kinney, 1997). Down-regulation of this gene reduces the accumulation of palmitic acid in soybean. The simultaneous down-regulation of *FatB* and *FAD2-1*, in a seed-specific fashion, results in oleic acid levels in soybean seed reaching 90%, with saturated fatty acids below 6% (Buhr et al., 2002).

A HO acid soybean line developed by DuPont Company was first marketed in 1998 and is currently available through DuPont Protein Technologies, St. Louis, MO. The details of the development and characterizations of the HO line were described by Kinney and Knowlton (1998). The following is a summation of the Kinney and Knowlton publication (1998) and includes information found in the Australia New Zealand Food Authority (ANZFA) Draft Risk Analysis Report Application A387, titled, "Food derived from high oleic acid soybean lines G94-1, G94-19 and G168" (www.anzfa.gov.au).

The HO line was derived from microprojectile bombardment as described by McCabe et al. (1988). Two constructs, designated pBS43 and pML102, were delivered into the shoot apex of the Asgrow cv. A2396. The pBS43 harbored two gene cassettes: (i) the *FAD2-1* gene under the control seed specific *a1*-subunit of ß-conglycinin and terminated by the 3' UTR of phaseolin from common bean (*Phaseolus vulgaris* L.) and (ii) ß-glucuronidase gene (*gusA*) from *E. coli* under the control of the 35S CaMV promoter and terminated with the 3' region of the *A. tumefaciens* nopaline synthase gene (*nos*). The second construct, pML102, carried the *Corynebacterium dapA* gene that encodes a lysine-insensitive version of dihydrodipicolinic acid synthase (DHDPS) (Falco et al., 1995) coupled with the transit peptide from the soybean small subunit of ribulose 1,5-bisphosphate carboxylase/oxygenase (rubisco). The 3' region of the Kunitz trypsin inhibitor gene 3 from soybean terminates the genetic element. Vector pBS43 was designed to induce the post-transcriptional gene-silencing phenomenon (Matzke et al., 2001; Marx, 2000) through the use of a plus-sense cassette of the target gene, *FAD2-1*, resulting in what is referred to as cosuppression in plant systems (Matzke and Matzke, 1995). A transgenic line derived from the cotransformations with pBS43 and pML102 was designated 260-05. Progeny derived from selfing this line were categorized into four groups, seeds with >80% oleic acid and wild-type lysine levels, seeds with mid-oleic (70%) and enhanced lysine content, seeds with low oleic (4%) and high lysine content, and seeds with wild-type levels of both oleic acid and lysine content. Molecular characterization on the plants from the respective categories revealed that the primary transformant contained two loci, one that harbored only an element from plasmid pBS43 and the other that possessed inserts from both plasmids. The data demonstrated that the locus with only the pBS43 element was actually the silencing allele. Families derived from plants that contained only the silencing allele were subsequently designated G168, G94-1, and G94-19.

Molecular characterization of the insert harbored in the three lines, G168, G94-1, and G94-19, revealed two copies of the pBS43 cassette at the transgenic locus (Fig. 7–2). The first copy is inverted relative to the second, forming an inverted repeat. A fragment of pML102 that contains only the *bla* gene (ß-lactamase) for streptomycin resistance and a region from the origin of replication (*ori*) is located at the 5' end of the locus just upstream of the first copy of pBS43. The first copy of pBS43 just proximal to this fragment resides starts at the *bla* gene. Hence the *FAD2-1* and *gusA* cassettes are delineated by fragments of the *bla* gene and the *ori* resides at the very 3' end. Just downstream of the first copy is a truncated *gusA* cassette missing the 3' nos region and portion of the open-reading frame. The *FAD2-1* cassette, the entire *bla* gene, and *ori* element of the plasmid are proximal to the truncated *gusA*.

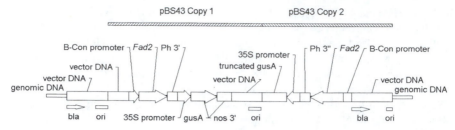

Fig. 7-2. The transgenic locus of DuPont's High Oleic soybean. The locus contains two copies of the Fad2 gene arranged as an inverted repeat, and flanked by the bla gene for bacterial streptomycin resistance and the bacterial origin of replication. One intact copy and one truncated copy of the gusA gene is located between the two copies of Fad2.

The regulatory package for the HO soybean mirrored that developed for the RR soybean. A comprehensive compositional analysis included detailed northern and protein analyses, which revealed that no new proteins are expressed in the HO soybean. However, the protein profile of the seed was found to be different from the parental line. Reduced concentrations were found of the alpha and alpha[1] subunits of β-conglycinin, along with a concomitant increase in A and B subunits of glycinin, along with A2B1A glycinin precursor. Levels of all other storage proteins were similar to those of the parental line. The alterations in the protein profile were subsequently attributed to cosuppression (Kinney et al., 2001).

Additional studies were conducted to determine if the alterations in the protein profile modulated the allergenicity relative to the parental line through radioallergosorbent reactivity, using sera from 31 subjects with known reactions to soybean extracts or who possessed food allergies, along with control sera from tolerant individuals. The data revealed no alteration in allergenicity relative to the parental line.

To evaluate the nutritional composition, HO soybean was compared to conventional soybean for various parameters including proximate analysis, amino acid content, total oil and protein. The data demonstrated that the HO soybean is substantially equivalent to conventional soybean, except for the enhanced oil profile and alteration in the protein profile as mentioned above. Feeding trials were also conducted on pigs (*Sus scrofa*) and chickens. The data showed similar growth performance in the pig-feeding study when the meal was processed above 80 to 85C, and no significant differences in daily weight gain or feed conversion in the chicken-feeding trials were observed between the HO-derived meal and conventional meal. These data indicate that soybean meal derived from the value-enhanced line are equivalent to their commercial counterparts with respect to its ability to support development of both pigs and chickens.

7–1.3 Potential Future Products

In addition to RR and HO soybeans, numerous other traits derived from genetic engineering of soybean have been reported in the literature, although none are currently commercially available (Table 7–1). The engineered traits can be cate-

Table 7–1. Reports of transgenic soybean with the potential for commercialization.

Trait	Reference
α-eleosteric acid	Cahoon et al. (1999)
α-parinaric acid	
Bacillus thuringiensis -endotoxin	Parrott et al. (1994); Stewart, Jr. et al. (1996); Su et al. (1999)
Bean pod mottle virus resistance	Reddy et al. (2001)
Bovine casein	Maughan et al. (1999)
Δ^5-eicosenoic acid	Cahoon et al. (2000)
High oleic acid	Buhr et al. (2002)
Lysine levels elevated	Falco et al. (1995)
Monoclonal antibody	Zeitlin et al. (1998)
P34 suppression (allergenicity reduction)	Herman et al., reported by Suszkiw (2002)
Oxalate oxidase for white mold resistance	Cober et al. (2003)
Phytase	Denbow et al. (1998)
15 kD Zein	Dinkins et al. (2001)

gorized as (i) agronomic, (ii) protein and oil quality, (iii) specialty oils, (iv) the removal of allergenic proteins, and (v) pharmaceuticals. The list in Table 7–1 is growing rapidly, and transgenic soybean producing specialty oils, targeted for contract production on limited hectarage instead of incorporation into commodity soybean, might be the next transgenic soybean to be commercialized. The use of soybean for specialty protein production also remains an option. Generally speaking, seeds with a high protein content appear able to accumulate much higher levels of transgenic proteins than do vegetative tissues (Giddings et al., 2000), and the seed may allow for stable storage of the desired protein (Kusnadi et al., 1997). There might be more than one target tissue in a seed. A company in Indiana, Producers' Natural Processing Inc, Brookston, claims technology to express transgenic proteins in the seed coat of the soybean seed (www.pnpi.com, accessed 4 Jan. 2003).

In the end, to be commercially viable, new transgenic cultivars must have sufficient value added to recover the costs of research and development, compensate all owners of the intellectual property (IP) used to make the cultivar, recover all regulatory costs not to mention the agronomic production costs. The latter two topics will be covered later in this chapter. At present, there are few genes that clearly meet these criteria, but with the advent of functional genomics, the number of genes available for transformation and which can provide sufficient added value will increase exponentially.

7–2 TRANSGENIC SOYBEAN AS A RESEARCH TOOL

Significant public and private investments have been made to support research in soybean genomics. Soybean programs targeting soybean genomics tool development, namely DNA microarrays, serial analysis of gene expression, and the assembly of a physical map that will be ultimately integrated to the genetic map, have been ongoing for a number of years. These efforts have already identified numerous gene sequences whose functional characterization promises to greatly enhance our understanding of soybean biology and thus lead to means of potentially improving this crop plant. A variety of methodologies are being developed and applied to functional analysis of soybean gene sequences. These include comparative analysis of the sequences across genotypes and taxa (Hartnett-Foster et al., 2002, genome mapping; Cregan et al., 1999), and profiling of mRNA expression patterns (Shoemaker et al., 2002). Each of these tools rapidly generates data that permit the formulation of a hypothesis regarding gene function on a genomic scale, but each is limited by our ability to experimentally test such hypothesis. Thus, additional tools are needed to complement existing soybean functional genomics programs.

The ability to specifically modulate gene expression represents an extremely powerful approach to directly characterize gene function. Numerous strategies have been employed in plant systems for gene down-regulation. RNA antisense (Ecker and Davis, 1986), whereby the gene of interest, or portion thereof, has a reverse orientation relative to the promoter, but still retains an appropriate 3′ terminal sequence, has been effective in numerous plant systems to specifically down-regulate gene expression. Staswick et al. (2001) used this approach to down-regulate the soybean vegetative storage proteins. Transgenic soybean harboring an antisense *VspA* construct exhibited a 50-fold reduction in total vegetative storage proteins (Staswick et al., 2001). Attenuation of endogenous soybean gene expression by introduction of a plus sense construct, so called cosuppression (Matzke and Matzke, 1995), effectively down-regulated the Δ-12 fatty acid desaturase gene, *FAD2-1*, in an embryo-specific fashion resulting in a HO acid phenotype (Mazur et al., 1999). The same approach has been used to confirm the role of proline accumulation during drought stress in soybean (de Ronde et al., 2001) and to study the effects of suppressing the alpha subunits of β-conglycin on the formation of protein storage bodies (Kinney et al., 2001).

In animal cells, nuclear localization of antisense transcripts has been shown to be an effective strategy to down-regulate targeted gene expression (Liu and Carmichael, 1994). A transcript can be retained in the nucleus by replacing the 3′ untranslated region with a self-cleaving ribozyme. Buhr et al. (2002) used this approach to specifically down-regulate two soybean fatty acid biosynthesis genes *FAD2-1* and *FatB*. The latter gene is a palmitoyl-preferred thioesterase, which, when suppressed, reduces the percentage of saturated fatty acids. A dual cassette harboring the *FatB* and *Fad2-1* genes, both in sense orientation and under the control of a single promoter, effectively down-regulated expression of both genes, generating a HO acid phenotype, coupled with a level of saturated fatty acid content below 6% in the seed storage lipids (Buhr et al., 2002). Down-regulation of gene expression in soybean using double-stranded RNA molecules (Wang and Waterhouse, 2000), referred to as RNAi in animals (Sharp and Zamore, 2000), has not been tested to our knowledge. However, it has been demonstrated to be a highly efficient strategy for targeted gene attenuation in *Arabidopsis thaliana* (L.) Heynh. (Levin et al., 2000), and rice (*Oryza sativa* L.) (Wang and Waterhouse, 2000) and undoubtedly can be a powerful tool for understanding gene functionality in soybean.

The strategy described above can provide an effective tool that can complement soybean functional genomics programs. However, they all suffer from similar drawbacks, namely multiple transgenic events that are generally required to identify a down-regulated line with a suitable phenotype. One problem is that entire gene families may be simultaneously down-regulated, thereby complicating hypothesis testing. Therefore, additional gene knockout and/or gain of function approaches would be useful to augment the gene down-regulation strategies available for soybean.

7–3 TRANSFORMATION TECHNOLOGY FOR SOYBEAN

The advent of transgenic soybean in the marketplace and as a basic research tool is possible due to the development of several soybean transformation protocols. All successful systems for genetic transformation have at least three features in common. First, there must be a way to deliver DNA into a cell. Second, the cell must be competent for integration of the foreign DNA. Third, the lineage of that cell must participate in the formation of the germline. It follows from this that transformation must occur in a single cell; yet most eukaryotes are multicellular organisms. Hence, there must be a mechanism present whereby a whole plant can be obtained from the transgenic cell. The latter component generally requires an in vitro regeneration scheme. The exception is the in planta *Agrobacterium*-mediated transformation protocol developed by Bechtold et al (1993), in which the integration event occurs in the egg cell (Desfeux et al., 2000). This protocol has thus far only been successful in *Arabidopsis*, and attempts to replicate it in soybean have not been successful. In addition, there must be a way to selectively distinguish and ultimately separate transgenic cells from nontransgenic cells.

7–3.1 Soybean Regeneration Schemes

As just mentioned, the first requisite for transformation is the ability to regenerate fertile plants from cultured cells or tissues. In vitro regeneration of plants can be classified as happening by one of two general pathways, organogenesis and somatic embryogenesis. Each pathway can be further characterized as being indirect or direct, depending upon whether the cell or cells of interest form a dedifferentiated tissue (callus) or immediately differentiate into an organized structure. Depending on the source tissue and the growth regulator regime subjected, soybean explants are capable of undergoing both somatic embryogenic and organogenic routes for in vitro regeneration. These are illustrated in Fig. 7–3.

7–3.1.1 Organogenesis

Organogenesis is the formation of new organs, in this case, shoots, from cultured cells or tissues. As with any other regeneration system, a successful organogenic protocol requires responsive genotypes, regenerable tissues within those genotypes, and appropriate culture protocols. Cheng et al. (1980) first reported on organogenesis from cotyledonary explants derived from soybean seedlings. In this early work, soybean seeds were directly subjected to elevated levels of benzyl aminopurine (BAP) during germination. These high levels ($>2~\mu M$) apparently were sufficient to overcome apical dominance in the developing seedling and led to multiple buds emanating from the axillary meristem (commonly referred to as the cotyledonary node). Subculturing of the excised bud-induced nodal region onto medium

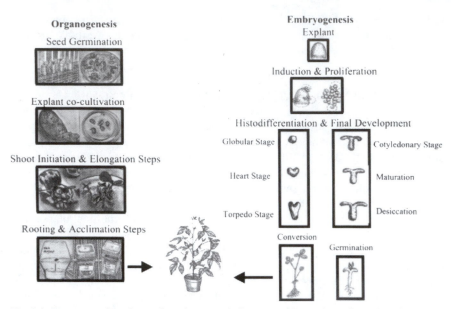

Fig. 7-3. Organogenesis and somatic embryogenesis for regeneration and transformation of soybean. Organogenesis. Mature seeds are allowed to germinate, after which the cotyledons are separated, wounded and inoculated with *Agrobacterium*. Explants are placed on a shoot induction medium with a selection agent to foster the production of transgenic shoots. These are excised, allowed to root, and transferred to soil. For somatic embryogenesis, immature cotyledons are placed on induction medium to induce the formation of globular stage embryos. These are subjected to microprojectile bombardment, and placed on medium with a selection agent to obtain transgenic embryos. These are then transferred to an auxin-free medium, whereby they undergo histodifferentiation. Fully differentiated embryos reach physiological maturity, dry-down, and then germinate into plants.

supplemented with BAP resulted in further development of shoots and eventual recovery of whole, in vitro-derived soybean plants (Cheng et al., 1980). The in vitro culture technique using the cotyledonary node explant was modified by Wright et al. (1986), who observed maximum in vitro response on a reduced-salt medium supplemented with 5 μM BAP from germination through successive subcultures. Immature leaves (Kim et al., 1990; Wright et al., 1987), hypocotyl sections (Dan and Reichert, 1998), embryonic axes (McCabe et al., 1988), and cotyledon explants (Cheng et al., 1980) of soybean have been reported to undergo organogenesis. Soybean embryonic axes as a target tissue for in vitro regeneration have the advantages of requiring a minimal time in culture, thus reducing the probability of a tissue-culture-induced mutation, and provide a readily available explant source that is easily maintained.

7–3.1.2 Somatic Embryogenesis

Somatic embryogenesis is a process by which a cell, other than a zygote, divides to form an embryo that is capable of functioning as a synthetic seed, albeit one obtained without sexual recombination. Somatic embryos are useful for the large-scale propagation of elite genotypes of long-lived species or valuable genotypes, such as used by the forestry industry. Somatic embryos have also been extraordinarily useful for the genetic transformation of a wide range of crop species, when the transformed cells eventually lead to the formation of somatic embryos. Briefly, an embryogenic state is induced by the presence of an auxin. Once the embryogenic state is induced, embryogenesis (i.e, the development of embryos) is suppressed by the continued presence of the inducing auxin, so it is necessary to lower the levels of exogenous auxins to a point that embryogenesis can proceed. The somatic embryos then undergo histodifferentiation, passing through globular, heart, torpedo, and cotyledonary stages of development. The embryos must then undergo a period of maturation, until they reach physiological maturity, at which point they require a period of desiccation. Upon imbibition, the desiccated embryos germinate and convert into plants (Parrott, 2000). Somatic embryogenesis is thus a multistep process. Genotypes that are particularly amenable to induction of embryogenesis may not necessarily perform well for the other stages of the embryogenic process (Bailey et al., 1993; Meurer et al., 2001).

As with organogenesis, a successful somatic embryogenic protocol also requires the identification of responsive genotypes, regenerable tissues within those genotypes, and the development of appropriate culture protocols. The most embryogenic tissue identified to date is the immature cotyledon (Lippmann and Lippmann, 1984). However, embryonic axes (Christianson et al., 1983), callus (Phillips and Collins, 1981; Gamborg et al., 1983; Yang et al., 1991), microspores (Hu et al., 1996; Kaltchuk-Santos et al., 1997; Zhao et al., 1998), and embryonic leaves (Rajasekaran and Pellow, 1997) also have been reported to be embryogenic.

Soybean somatic embryos were first reported in suspension cultures obtained from hypocotyl-derived callus (Phillips and Collins, 1981; Gamborg et al., 1983). The first successful recovery of a plant from a soybean somatic embryo may have been a haploid plant derived from microspores cultured on the auxin, 2,4-dichlorophenoxyacetic acid (2,4-D) (Yin et al., 1982). Rooted shoots were first obtained from somatic embryos forming in liquid culture derived from hard tissue derived from embryonic axes (Christianson et al., 1983). The identification of immature cotyledons as a highly embryogenic tissue (Lippmann and Lippmann, 1984) was a major breakthrough. Nevertheless, the use of 2,4-D to induce somatic embryos frequently resulted in abnormal embryos that could not be converted into plants. The use of naphthaleneacetic acid (NAA) (Barwale et al., 1986; Lazzeri et al., 1985) or indolebutyric acid (IBA) (Hammatt and Davey, 1987) as an inducing auxin permitted the recovery of somatic embryos that were morphologically normal, and in turn facilitated the identification of conditions which facilitated the conversion of somatic embryos into plants. Among these were the realization that exposure to exogenous auxins was detrimental to normal ontogeny of somatic embryos, that the use of exogenous plant growth regulators was counterproductive for embryo maturation (Parrott et al., 1988), and that a desiccation phase between maturation and germination greatly enhanced germination (Hammatt and Davey, 1987; Parrott et al., 1988). Plant recovery was facilitated using continuous illumination (Ghazi et al., 1986), which is unusual in tissue culture protocols, but explained by the fact that soybean is sensitive to photoperiod, and it is essential that the photoperiod is long enough to ensure plants remain vegetative during embryo germination. Other factors found to be important were genotype (Komatsuda and Ohyama,

1988; Parrott et al., 1989b), the orientation of the cotyledon explant on the medium, which should be abaxial (Hartweck et al., 1988), and the use of pH 7.0 for induction, which is a very unusual pH for tissue culture (Komatsuda and Ko, 1990). Wounding and the use of gellan gum rather than agar as the medium solidifying agent also increases embryogenesis (Santarem et al., 1997).

Somatic embryos induced on 2,4-D have a decided advantage in that they are able to proliferate continuously whenever the levels of 2,4-D in the culture medium are at or above 10 mg L^{-1} (Ranch et al., 1985), in a process known as repetitive or proliferative embryogenesis. Proliferating somatic embryos of soybean are particularly amenable to transformation. In the presence of sufficiently high levels of 2,4-D, somatic embryos usually fail to proceed past a globular stage of development, and instead, give rise to another somatic embryo from the apical region (Finer, 1988). The process continues indefinitely, creating large masses of proliferating embryos, as long as the level of 2,4-D remains high. Current protocols have settled on 40 mg L^{-1} for induction of somatic embryos (Ranch et al., 1986; Finer, 1988) and of 20 mg L^{-1} for maintenance (Ranch et al., 1986; Wright et al., 1991). The initial problem of 2,4-D giving rise to abnormal somatic embryos was overcome simply by adding 0.5% activated charcoal to the histodifferentiation medium (Buchheim et al., 1989). Presumably, the charcoal adsorbs any 2,4-D that may otherwise interfere with establishment of the proper polarity of the developing embryo (Cooke et al., 1993; Schmidt et al., 1994).

Altering the 1:1 ratio of ammonium (NH_4^+) to nitrate (NO_3^-) ratio in Murashige and Skoog (1962) basal salts to 1:4 in liquid medium (named Finer and Nagasawa medium) further facilitated the growth of proliferative embryo cultures (Finer and Nagasawa, 1988). Further reductions in total N to 35 mM and the sucrose to 1% led to further growth increases (Samoylov et al., 1998b). Reducing the N content alone facilitated the histodifferentiation of somatic embryos, permitting the recovery of about 10 cotyledon-stage embryos per milligram of tissue in just 21 d (Samoylov et al., 1998a). Nevertheless, such embryos suffered from a low conversion frequency, which was partly solved by the addition of 3% sorbitol to the histodifferentiation and maturation medium (Walker and Parrott, 2001). Further gains were obtained by addition 30 mM filter-sterilized glutamine and 1 mM methionine to the histodifferentiation medium, allowing the recovery of plants from up to 60% of all somatic embryos (W.A. Parrott, unpublished data, 2003). Overall, such medium modifications allow for the regeneration of large numbers of plants from very little tissue, resulting in savings of time, labor, and supplies.

7–3.1.3 Shoot Apices

De novo regeneration may not be technically necessary for the recovery of transgenic plants. Shoot apices, excised from the seed and grown in vitro using shoot-elongation media from organogenic protocols, have led to the recovery of transgenic plants, and is the way the RR soybean was developed (McCabe et al., 1988). More details will be provided subsequently in this chapter. Overall, the transformation ability of shoot apices is too low for this to be a popular transformation target.

7–3.2 DNA Delivery Strategies

Earlier, it was stated that a method to deliver DNA into regenerable tissues is a prerequisite for a transformation system. Today, there are two major DNA delivery strategies that are used for the vast majority of crops, and soybean is no exception. The first centers on the use of the soil bacterium, *Agrobacterium tumefaciens*, which has the natural ability to transfer part of its DNA into plant cells (Stafford, 2000). The second technique uses DNA-coated microprojectiles which are mechanically propelled into plant cells (Klein et al., 1987). In the case of soybean, cotyledonary nodes, and to a limited extent, somatic embryos (Parrott et al., 1989a; Yan et al., 2000), have been good targets for *Agrobacterium*, while shoot apices and somatic embryos are used as targets for microprojectile bombardment.

7–3.2.1 *Agrobacterium*-Mediated Transformation

Soybean was originally not considered to be a susceptible host for infection by *Agrobacterium* (De Cleene and De Ley, 1976). Nevertheless, it has become possible to combine *Agrobacterium*-mediated transformation with organogenesis to develop successful soybean-engineering protocols. The two most widely used tissues for organogenesis-based transformation are the embryonic axis and the cotyledonary node, which undergo direct organogenis from the apical meristem or cotyledonary meristems, respectively. The initial preparation of the axis for transformation and subsequent in vitro development involves removal of the embryonic leaves, which exposes the target meristematic cells (McCabe et al., 1988). More recently the embryonic axis has been successfully used in an *Agrobacterium*-mediated protocol (Williams et al., 2000), similar to an earlier approach using germinating soybean seeds (Chee et al., 1989).

Hinchee et al. (1988) were the first to exploit the cotyledonary node in a soybean genetic engineering protocol. Cotyledonary-node explants were prepared from soybean seeds pregerminated for 4 to 10 d on water agar, and cultured on Gamborg's B5 medium supplemented with 5 μM BAP. A nopaline strain of *A. tumefaciens* was used in the transformations, coupled with the *nptII* gene as the selectable marker. The cotyledonary-node explants were subsequently placed on medium containing kanamycin levels ranging from 200 to 300 mg L^{-1} for selection of transgenic shoots. Following the shoot induction period on 5 μM BAP, the induced explants were subcultured to medium supplemented with reduced cytokinin levels for shoot elongation. Transformation frequencies ranged from 0.3 to 2.2% on a transformant per explant basis (Hinchee et al., 1988).

Di et al. (1996) implemented numerous modifications to the Hinchee et al. (1988) procedure, as well as modifications described by Townsend et al. (1996) to the cotyledonary-node transformation system to introduce the bean pod mottle virus coat protein gene for viral resistance. The modifications included the addition of 100 μM of the *Agrobacterium* virulence-inducing agent acetosyringone (Stachel et al., 1985) during inoculation and co-cultivation, a reduced seed germination period (3 d) and wounding of the explant in the nodal region prior to inoculation. The wounding of the meristematic cells of the node aids in preventing the axillary shoot from emerging, which in turn impedes apical dominance and thus enhances multiple shoot induction from the nodal region (T.E. Clemente, unpublished data, 2003). Proper wounding of the nodal region will influence the in vitro regeneration potential of the explant. The modified explant preparation involves separating the cotyledons from the germinated seedlings by making a cut through the hypocotyl approximately 5 mm below the cotyledon. A vertical slice initiated between the cotyledons and continued through the center of the hypocotyl segment then separates the cotyledons. Using a scalpel blade the embryonic axis is removed and 7 to 12 slices parallel with the axis are conducted to wound the meristematic cells of the node.

Following the co-cultivation period the explants were cultured in liquid shoot induction medium composed of Gamborg's salts, 5 μM BAP, supplemented with 200 mg L^{-1} kanamycin and an antibiotic regime consisting of vancomycin, carbenicillin, and cefotaxime, to eliminate A. tumefaciens cells. The liquid culture period was conducted for 3 d followed by a selection period on solid medium for up to 8 wk. Cotyledonary nodes with developing shoots were subsequently transferred to elongation medium composed of Gamborg's basal medium supplemented with 0.6 mg L^{-1} GA$_3$ and a reduced kanamycin selection pressure of 50 mg L $^{-1}$. The elongated shoots were rooted and established in soil.

Further changes to the soybean cotyledonary-node transformation system focused on modifications during the co-cultivation and elongation steps, and incorporation of herbicide selection (Zhang et al., 1999; Clemente et al., 2000). The co-cultivation modifications include pH reduction to 5.4 (Godwin et al., 1991), level of BAP raised to 7.5 μM, and the addition of 0.25 mg L^{-1} GA$_3$ (Zhang et al., 1999; Clemente et al., 2000). In addition the co-cultivation plates were overlaid with filter paper (Mullins et al., 1990). The co-cultivation step was conducted at a reduced temperature 24C (Fullner et al., 1996). Moreover, the incorporation of the herbicide marker genes *CP4* or *bar* provides for efficient selection of transgenic soybean when glyphosate is used at levels of 0.05 to 0.15 mM with the *CP4* gene (Clemente et al., 2000) or glufosinate is used at 3 mg to 5 mg L^{-1} with the *bar* gene (Zhang et al., 1999).

Modifications in the elongation and rooting steps of the protocol targeted alterations in the medium composition. Following the 4-wk shoot initiation period on Gamborg's basal medium supplemented with 7.5 μM BAP, shoots were subcultured to elongation medium composed of MS salts, Gamborg's vitamins, supplemented with 1 mg L^{-1} zeatin riboside, 0.5 mg L^{-1} GA$_3$, and 0.1 mg L^{-1} IAA. Elongated shoots were rooted without the herbicide selection agent or incorporation of the antibiotics to eliminate A. tumefaciens cells. This is necessary because a dramatic negative impact on root induction is observed when the antibiotics are incorporated into the rooting medium (T.E. Clemente, unpublished data, 2003). Total removal of the antibiotic regime during the rooting step enhances the rooting efficiency without A. tumefaciens cell overgrowth being observed, suggesting that the antibiotic during the initial phases was sufficient to prevent A. tumefaciens colonization in the elongated shoots. Transformation frequencies observed as a result of implementing the respective modifications ranged from 3 to 5% on the basis of number of soil-acclimatized transformants per explant.

Agrobacterium tumefaciens triggers a browning response in soybean cotyledonary explants about the target region, axillary meristem. This necrotic response was presumed to negatively impact T-DNA transfer to the target cells (Olhoft and Somers, 2001). To address whether blocking the necrotic response induced by *A. tumefaciens* would enhance soybean transformation frequencies, L-cysteine supplements were evaluated during the co-cultivation step of the cotyledonary-node transformation protocol (Olhoft and Somers, 2001). The cysteine supplements enhanced transient expression of GUS approximately 2.5 fold in cotyledons. More importantly this increase in transient expression is correlated with improved recovery of stable soybean transformants (Olhoft and Somers, 2001). The use of thiol compounds appears to have the same beneficial effect (Olhoft et al., 2001).

This advancement, coupled with the other modifications implemented to the original cotyledonary-node transformation protocol described by Hinchee et al. (1988) and described above, has led to the adoption of this technique in numerous public sector laboratories, simply because important elite soybean cultivars are amenable to genetic transformation with this system, although success is still strongly genotype-dependent. Most recently, the use of the antibiotic, hygromycin, for selection of transgenic tissues, combined with the use of L-cysteine and other thiol compounds, has led to a transformation rate of 16.4% (Olhoft et al., 2003), the highest transformation frequency ever reported for soybean.

7–3.2.2 Microprojectile Bombardment

Two types of soybean tissue have been shown to be particularly amenable to microprojectile bombardment-mediated transformation: somatic embryos and shoot apices. Of these two, somatic embryos are the more popular target.

7–3.2.2.1 Somatic Embryos. As discussed earlier, somatic embryo cultures undergoing repetitive proliferation are particularly amenable to transformation via microprojectile bombardment using standard bombardment protocols, as described in Trick et al. (1997). Other than the inclusion of bombardment and selection stages, transformation protocols differ very little from the protocol already described in this chapter for somatic embryogenesis.

In early transformation attempts with embryogenic tissue, cell lines would not become amenable to transformation until they were at least 4 to 6 mo old, by which time the resulting plants were sterile (Hazel et al., 1998). This limitation was removed with the reformulation of the culture medium discussed on the section on somatic embryogenesis. The reformulated medium, called Finer and Nagasawa Lite (FNL) (Samoylov et al., 1998b) medium, permits the transformation of cell lines as young as 4 to 6 wk post-induction (W.A. Parrott, unpublished data). Cell lines undergo a burst of mitosis about 4 d after transfer to fresh medium, and appear to be most transformable when bombarded 2 d prior to this mitotic burst (Hazel et al., 1998).

One week after bombardment, the cultures are subjected to selection by the antibiotic, hygromycin at 20 mg L^{-1}. The medium is replaced at biweekly intervals, by pipetting out the old medium and replacing it with fresh. The tissue itself remains in the original container. Transgenic cell lines, which are green in color, appear after 6 to 8 wk in selection.

Although no work has been published, selection protocols theoretically could be devised for use with other antibiotics or herbicides. Selection protocols are fairly flexible, as selection is possible on both solidified medium (Stewart et al., 1996; Santarem and Finer, 1999) and in liquid medium (Finer and McMullen, 1991). Selection on solid medium requires low labor inputs, while selection in liquid medium is faster. Transgenic cell lines appear in 6 to 8 wk in liquid selection, and in about 3 mo on solidified medium. Plants are recovered using the protocols described previously in this chapter for plant recovery from somatic embryos.

This transformation method will dependably produce at least one stable transgenic line per bombardment, and currently may be the most widely used transformation protocol, having been adopted numerous private and public sector laboratories. Although it remains strongly genotype-dependent, several elite cultivars are amenable to this protocol.

7–3.2.2.2 Shoot Apices. The shoot apex from embryonic axes (McCabe et al.,1988), in which the embryonic axis is targeted for microprojectile bombardment and subsequently cultured on a growth medium supplemented with BAP and NAA, also has been a suitable target, and is the target tissue used to obtain the RR soybean. Explants are cultured for 14 d followed by a brief culture period on a high BAP to NAA ratio to induce multiple shoot formation about the apical and axillary meristems.

The apex is a multicellular target, so only one component cell might be transformed. The resulting plants will be chimeric, having transgenic and non-transgenic sectors. Only the L2 layer gives rise to gametes, so a transgenic L2 layer is a prerequisite for the transgene to be transmitted to the progeny. Not all L2 layers within transgenic sectors will themselves be transgenic. Initial efforts centered on the use of *gus*A to sample various tissues of regenerated plants to identify those plants with L2 transgenic sectors (Christou et al., 1989, 1990; Christou, 1990a, 1990b; Christou and McCabe, 1992; Sato et al., 1993). This protocol was later adapted to include glyphosate as a selection agent, thus significantly reducing the labor requirement for the germline prediction assay (Martinell et al., 1999). The *pat* and *bar* genes for resistance to phosphinothricin also have been used for this purpose (Russell et al., 1991), as has the *ahas* gene for resistance to imazapyr (Aragão et al., 2000).

7–3.2.3 Microprojectile vs. *Agrobacterium*-Mediated Transformation

At the present time, both are reasonably efficient for the production of transgenic soybean, both methods have been shown to be portable to various different laboratories, and elite genotypes exist that are amenable to one protocol or the other.

One limitation commonly cited to microprojectile-mediated transformation is that the number of transgene copies engineered into the plant is too high, though this reflects perception more than reality. Single-copy transgenics can be obtained from microprojectile bombardment (e.g., Stewart et al., 1996). Alternatively, the use of *Agrobacterium* does not guarantee single- or even low-copy transgenics, as is clearly documented in the Southern blots published by Olhoft and Somers (2001) and Olhoft et al. (2002). Reducing the amount of DNA used with microprojectile bombardment makes it possible to control the number of transgenes in the resulting plants (W.A. Parrott, unpublished data).

In addition, since the presence of flanking sequences and selectable markers increases difficulties for the regulatory approval of transgenic plants, current protocols will continue to be modified to ensure delivery of only the gene cassette. In the case of *Agrobacterium*, two-vector strategies (Xing et al., 2000) will become increasingly important, while isolated cassettes, in the form of linear DNA, will be used increasingly with microprojectile bombardment.

In the end, the choice of transformation system may not depend on biology as much as on issues related to IP and freedom to operate (FTO) that is, whether the owners of the technology are willing to license it to other users. These will be discussed fully later in this chapter.

7–3.2.4. Miscellaneous Techniques

A variety of different transformation techniques have been applied to soybean, but all suffer from irreproducibility or lack of sufficient evidence to support the claim of stable transformation. One of the first reports of transformation not requiring tissue culture relied on a whole-plant approach. Inoculation of 4000 germinating seeds with *Agrobacterium* resulted in one germline transformant 1996. Recovery of transgenic soybean has also been reported following electroporation of axillary buds, but the transgene frequently was absent in the progeny (Chowrira et al., 1995, 1996).

The *psb*A gene from the chloroplast of black nightshade (*Solanum nigrum* L.) was cloned and microinjected into soybean ovules at the zygote stage to obtain transplastomic soybean. Resulting plants were initially screened by spraying progeny leaves with atrazine. Two plants, obtained from 1220 seeds recovered from 507 flowers, were identified as being atrazine-resistant. Plastid DNA was extracted, and dot blots showed the presence of the vector sequences (Liu et al., 1989, 1990). Transmission of the resistance trait was verified for three generations, though its transmission was not dependable (Yue et al., 1990), suggesting heteroplasty, that is, not all plastids were transgenic, and transgenic and nontransgenic plastids underwent vegetative sorting into sectors containing one plastid or the other. Alternatively, the plants may not have been transgenic at all, with the positive signals coming from DNA maintained by endophytes within the tissue, as has been reported for wheat (*Triticum aestivum* L.) (Chen et al., 1994).

Pollen tubes reportedly also have been used to introduce DNA into recently fertilized ovules. DNA from *Glycine gracilis* Skvortz was isolated and introduced into soybean by first removing the upper stigma after pollination, and adding DNA in solution to the cut stigma. Phenotypic variation among the resulting progeny has been taken as proof of transformation (Lei et al., 1989). A high-yielding, high protein cv. Heisheng 101 was developed from these transgenic progeny (Hu and Wang, 1999). In later work, using DNA isolated from *G. soja* Sieb. & Zucc., random amplified polymorphic DNA (RAPD) in the progeny were used to detect the transgenic soybean (Xie et al., 1995). In addition, a partially sterile soybean was recovered following application of *Cicer arietinum* L. DNA. The sterility was attributed to the wide hybrid nature of the plants (Zhao et al., 1995). Supporting molecular evidence is lacking in all cases, and a biological explanation for how such transformation works and effects these changes is not immediately obvious.

7–3.2.5 Special-Purpose Procedures

The recovery of stable transformants of soybean may not always be necessary. Transient transgene expression can suffice in many cases. As an example, soybean has been inoculated with alfalfa (*Medicago sativa* L.) mosaic virus engineered to produce a rabies antigen. Upon inoculation with the engineered virus, the virus infects the plant, and the soybean plant was able to produce the antigen, useful for a rabies vaccine, in its leaves and seed coats (Fleysh et al., 2001).

Alternatively, recovery of a transgenic organ, rather than a whole plant, can prove useful. For example, hairy roots are particularly useful, as they will grow indefinitely in culture. Hairy roots are obtained using a close relative of *A. tumefaciens* called *A. rhizogenes* (Riker) Conn. In the case of soybean, hairy root cultures have been used to propagate the soybean cyst nematode, *Heterodera glycines* Ichinohe (Cho et al., 2000). In principle, such hairy roots could be used to study root-specific gene function without the need to recover whole, transgenic plants.

7–3.3 Selectable Markers

Generally speaking, it is easier to manipulate tissues than isolated cells. Genetic transformation is usually a low frequency event, so only a few cells within a tissue are transformed. Several strategies have been devised to separate transgenic from nontransgenic cells. Initial strategies centered on the use of selectable markers, that is, genes which make the transgenic cell resistant to a chemical that kills nontransgenic cells. Originally these were genes for antibiotic resistance or for herbicide resistance (Parrott et al., 1991). In the case of soybean, the antibiotic resistance gene most commonly used is the hygromycin phosphotransferase (*hph*) gene from *E. coli* (Finer and McMullen, 1991; Stewart et al., 1996; Maughan et al., 1999). The use of neomycin phosphotransferase II (*nptII*), also from *E. coli*, has been less successful (Hinchee et al., 1988; Sato et al., 1993; Townsend and Thomas, 1993; Di et al., 1996). Genes for herbicide resistance are also used. These usually consist of orthologues of genes already in the plant, but which are insensitive to the herbicide. Alternatively, they can be a gene that detoxifies the herbicide. The first herbicide resistance gene to be successfully used as a selectable marker in soybean was the EPSPS from *Agrobacterium* sp. strain CP4 (Clemente et al., 2000). The use of *pat* for selection with glufosinate has also been successful (Simmonds and Donaldson, 2000; Zhang et al., 1999), as has the use of acetolactate synthase (*ahas*) from *Arabidopsis thaliana* for selection with imazapyr (Aragão et al., 2000).

It is also possible to recover transgenic plants using a visual marker or reporter gene rather than a selectable marker. Such a marker is useful for making transgenic cells visually distinct, but does not affect the viability of transgenic tissue. The most frequently used visual marker is *gusA*, which codes for β-glucuronidase. Transgenic cells can be treated to obtain a distinctive color (Jefferson, 1989). In the case of soybean, it has been possible to submit tissues to the transformation treatment, then assay large numbers of resulting plants for the presence of sectors exhibiting GUS (Christou et al., 1989, 1990; Christou, 1990b). Depending on the tissue types involved, some transgenic sectors are more likely to give rise to transgenic progeny than others. The use of a visual marker thus facilitates the selection of transgenic tissues likely to be true-breeding (Christou, 1990a; Christou and McCabe, 1992).

The use of GUS has traditionally suffered from the fact that tissue must be killed before it can be assayed. An ideal reporter gene is one whose presence can be detected nondestructively. A new GUS, called *gus*PLUS, derived from *Salmonella,* is now available and can be assayed nondestructively (www.cambia.org). However, the most popular nondestructive marker is currently the green fluorescent protein (GFP) gene from the jellyfish (*Aequorea victoria*). The use of nondestructive reporter genes offers the possibility of manually selecting transgenic tissues prior to plant recovery, a strategy which has been successful in oat (*Avena sativa* L.) (Kaeppler et al., 2000). As GFP can be expressed in soybean (Ponappa et al., 1999), the possibility remains that such a strategy will be useful in soybean.

7–3.4 Remaining Hurdles

7–3.4.1 Biological

Remaining hurdles fall into two categories: biological and legal. Biological hurdles center on the need to recover greater numbers of transgenic events per unit of labor, while legal hurdles center on the need to address IP and keep up with rapidly changing regulations on a global level. A case in point is the current European directive against the use of antibiotic genes as selectable markers in plants. Whereas the use of antibiotic resistance in plants has been widespread and very successful in many plant species, there is a concern that such genes could be transferred to pathogens, thus rendering ineffective the use of those antibiotics. In its Guidance to Industry, the Food and Drug Administration (FDA) dismissed such concerns because the antibiotic resistance genes used in transgenic plants are for antibiotics that are not of major use for human therapy, or the resistance genes which are ubiquitous in the environment (FDA, 1998). Nevertheless, all genetically engineered crops containing antibiotic resistance genes must be removed from the European Union market by the end of 2004 (European Parliament legislative resolution on the joint text approved by the Conciliation Committee for a European Parliament and Council directive on the deliberate release into the environment of genetically modified organisms and repealing Directive 90/220/EEC (C5-0685/2000–1998/0072(COD)).

The use of genes for herbicide resistance carries the risk that cross-pollination of the transgenic crop with wild relatives will create herbicide-resistant weeds that are more difficult to control. Environmental groups also allege that the presence of herbicide resistance will also encourage the use of greater amounts of herbicides.

To mitigate this concern, all current and future transformation strategies will need to focus on the development of transgenic soybean developed without the use of antibiotic or herbicide resistance transgenes, or from which the selectable marker has been removed. One possible strategy is the placement of the selectable marker and the gene of interest on separate plasmids. This two-vector strategy can permit the selectable marker to be segregated away if it integrates into a genetic locus distinct from that of the gene of interest (Xing et al., 2000). Other approaches have been developed for other crops, and could conceivably be applied to soybean. One strategy involves enzymatic removal of the selectable marker, which must be flanked by *cre/lox* recombination sites (Dale and Ow, 1991). Another strategy permits transgenic cells to grow on mannose, xylose, or other C source, which are not metabolizable by nontransgenic cells (Bojsen et al., 1999).

Some proposed directives, if enacted, may also insist that all transgenes be present only as a single copy. The efficient recovery of transgenic plants with a single transgene is difficult to achieve, because multiple copies of the transgene are usually integrated, regardless of the method of transformation (Yin and Wang, 2000; Pawlowski and Somers, 1998). A strategy based on site-specific recombination (Srivastava et al., 1999) may be necessary to solve this dilemma. Alternatively, if large numbers of transgenic plants can be produced, then it would be possible to screen for those events with single integration events.

Although soybean transformation may not be as efficient as that of crops such as tobacco (*Nicotiana tabacum* L.), the greatest limitation is no longer the time spent on recovering transgenic plants, but rather the time spent analyzing transgenics to ensure stable transgene expression, quantify the level of transgene expression, verify the desired phenotype is present, ensure the transgene is transmitted to the progeny, establish copy number, etc. It follows that improvements in analytical techniques can make as great a contribution to transgenic technology as improvements in cell culture and transformation protocols. For example, the use of real-time polymerase chain reaction (PCR) can identify cell lines at an early stage that have low copy number of DNA inserts. These lines can be regenerated and the rest discarded before time and supplies are spent in their regeneration and subsequent analysis (Schmidt and Parrott, 2001). Real-time PCR also can be performed on cDNA to provide a rapid quantification of gene expression, thus facilitating the selection of desirable cell lines (Schmidt and Parrott, unpublished data, 2002). A new technique, competitive PCR (Honda et al., 2002) can replace Southern blots to determine gene copy number. Though this technique has yet to be tried out in soybean, adapting it to soybean should be straightforward.

Finally, the number of traits that can be manipulated via single-gene transformation is limited. Transformation with multiple genes will be necessary to modify metabolic pathways and alter polygenic traits. Because of the time necessary to recover an engineered plant, it is desirable to engineer multiple genes simultaneously, rather than sequentially. Microprojectile bombardment has been effective for the introduction of multiple plasmids into soybean (Hadi et al., 1996). However, the use of multiple plasmids makes it difficult to control the coordinated integration of each gene. An alternative approach places all the desired genes into one plasmid. Assembly of multiple genes into one plasmid has traditionally been difficult, as the same enzyme sites used to introduce new genes into the plasmid tend to occur at random in the genes already inserted. The net result is that the first genes are destroyed as new genes are added to the vector. This limitation has been overcome through the use of intron- and intein-encoded endonucleases, whose long recognition sequences make it extremely unlikely that they will cut a previously assembled string of genes. An artificial gene cluster, assembled with this type of enzymes, has been introduced into soybean (Thomson et al., 2002). Further refinements in this technology may greatly facilitate engineering with multiple genes.

7–3.4.2 Intellectual Property

7–3.4.2.1 **Background Information.** The tools of biotechnology have brought innovations to agriculture by drawing upon the wealth of information produced from both public and private sector research efforts. Historically, agricultural advances have been primarily released to the public domain. The Plant Patent Act (PPA) of 1930 was the first attempt to provide protection of IP developed in the agricultural sciences. Patents per se were not issued for biological organisms due to the perception, at that time, that living cells were the products of nature (Baenziger et al., 2000). The PPA provided protection for novel asexually propagated plant varieties, effectively recognizing that human intervention was able to modulate biological organisms to generate novel genetic variation. The Plant Variety Protection Act (PVP) of 1970 is analogous to the PPA of 1930 in that it protects the IP of novel plant varieties developed through traditional breeding efforts that possess a stable distinct phenotype, but it extends this protection to sexually propagated species. A PVP certificate still permits both research and seed saving by farmers (Greengrass, 1993; Jondle, 1989).

Utility patents have been issued since 1985 for plant methodologies and processes if the invention is novel and not apparent, based on information found in the public domain (Jondle, 1989). A utility patent differs from PVP in that a full disclosure of the invention must be submitted in the patent application, which permits anyone "practiced in the art" to reproduce the invention. A utility patent precludes farmers from saving seed. A utility patent may be issued on a plant variety, in which the full disclosure aspect can be met by making the seed available to the public. A typical utility patent will include an introduction to the subject, review of the literature in this area (referred to as prior art), the benefits the invention provides, a description on how the invention solves a problem, how the invention can be used, examples of the invention, and finally the claims the inventor is making for exclusive rights (Jong and Cypess, 1998). One objective of a patent is to teach someone skilled in the art how to perform the invention covering each of the respected claims presented within the document, and perhaps improve upon it. A utility patent can be issued on inventions derived from agricultural biotechnology, including genes, seeds, plant components, in-vitro regeneration schemes and genetic engineering techniques.

Numerous utility patents have been issued that impact the freedom to operate for potential products derived from the implementation of soybean transformation techniques. Nevertheless, until now, a patent per se has not precluded a researcher in the public domain from using a patented technology. Research using the patented material could be conducted, but a commercial product could not be developed and marketed without a license, within the life of the patent. To gain access to a patented biological, the recipient could agree to the provisions outlined within a Material Transfer Agreement (MTA). An MTA is essentially a contract between a supplier and a user, and can be drawn up regardless of the patent status of a biological (Jong and Cypess, 1998). All biological materials have some potential for future discoveries that may lead to a commercial products, hence it is considered prudent to draw up an MTA that will provide the framework of the legal rights of the provider and recipient.

As mentioned above, until now, a patent governing an aspect of a soybean transformation protocol would not block a researcher in a public institution from using the patented technology for research purposes that are not directed towards a commercial product. The FTO issues only came into play if an individual and/or institution attempted to market a product utilizing the patented invention. Researchers at public institutions enjoyed this privilege because the U.S. Supreme Court had recognized that states and state agencies enjoy sovereign immunity, under the 11th Amendment of the Constitution, from suits in federal court for patent infringement (Florida Prepaid Postsecondary Education Expense Board v College Savings Bank, 119th Supreme Court 2199, 1999). In addition, work within the experimental use exemption could be used as an affirmative defense to a charge of patent infringement. Hence, the critical issue was the intended use of the patented material in a research program, that is, purely for research or commercially directed. A researcher could use a patented soybean transformation protocol to gain insight on functionality of a gene and file for a patent on the utility of the gene under study. However, if a researcher were to introduce the novel gene into soybean using patented technologies and attempt to market the germplasm derived from the transformations, then a patent infringement suit would be justified, in addition to an injunction against marketing of the product.

The ability of scientists in public institutions to use IP for their basic research has been stopped by the Federal Circuit Court of Appeals, in its ruling on Madey v. Duke University No 01-1567 on 3 Oct. 2002. Ultimately, a ruling from the Supreme Court will be required to definitively settle the issue.

7–3.4.2.2 Intellectual Property for Soybean Transformation. Utility patents have been issued for all methods of introducing DNA into a plant cell. Chief among these are patents for microprojectile bombardment (Sanford et al., 1993a, 1993b; Fitzpatrick-McElligott et al., 1995) and for *Agrobacterium*-mediated (Barton et al., 2000) transformation. The IP covering the use of microprojectile bombardment has broad claims that include any method, regardless of the delivery apparatus, that introduces DNA into a cell biological material by adsorbing it onto a particle (Sanford et al., 1993a; Fitzpatrick-McElligott et al., 1995).

Intellectual property issues covering *Agrobacterium*-mediated transformation protocols are still being disputed, and undoubtedly litigation will continue for a number of years. The patent recently issued to Barton et al. (2000) includes claims that cover the integration of a disarmed T-DNA element in a dicotyledonous plant, the regenerated plant, and subsequent progeny derived from the transformant that carries the T-DNA element (Barton et al., 2000). Due to its broad nature, this patent will impact the FTO on all genetically engineered dicotyledonous plant species that are commercialized if these are derived from *Agrobacterium*-mediated gene transfer protocols.

Furthermore, utility patents have been issued for numerous plant in vitro regeneration methodologies. Among these are patents for the in vitro regeneration of soybean via somatic embryogenesis (Ranch and Buchheim, 1991; Hemphill and

Warshaw, 1989; Collins et al, 1991) and organogenesis from the cotyledonary-node (Wright, 1991). In some patents, claims have been granted for modification of techniques that have been clearly established in the public domain. Such patents represent judgment calls by the patent examiner, who must determine if the modification meets the criterion of not being obvious based on information available to the public. For example, somatic embryogenic protocols for the regeneration of soybean plants were first reported on in the 1980s (Christianson et al., 1983; Barwale et al., 1986; Ghazi et al., 1986; Lazzeri et al., 1985: Parrott et al., 1988). Hemphill et al. (1989), Ranch and Buchheim (1991), and Collins et al. (1991) were later issued patents covering various aspects of the somatic regeneration scheme, despite the fact that the Hemphill et al. (1989) claims were filed approximately a year earlier than the Collins et al. (1991) claims, and 2 yr prior to the Ranch and Buchheim (1991) claims. A similar sequence of events occurred with respect to organogenesis initiated from the axillary meristem of the soybean cotyledons. Cheng et al. (1980) were the first report on the in vitro response from the so-called cotyledonary-node explant of soybean. Wright et al. (1986) later modified the protocol and was issued claims acquiring IP on the method in 1991 from a patent application originally filed in 1983, for which a continuation was re-filed in 1984.

A wealth of claims has been issued to a variety of soybean transformation protocols implementing both *Agrobacterium*-mediated and microprojectile bombardment strategies combined with either organogenic or somatic embryogenic regeneration schemes. Chee et al. (1989) described a method for the genetic engineering of soybean via a nontissue culture approach by directly infecting germinating seeds. This strategy has the advantage of eliminating the need for tissue culture, but it is severely handicapped by its lack of reproducibility and low efficiency (Chee and Slightom, 1995). Nonetheless, IP was granted to various aspects of this methodology in 1994 (Chee et al., 1994), including methods that use *A. tumefaciens* as a vehicle to transfer transgenes to soybean seed germinated during a 1- to 2-d period.

In a related strategy, McCabe et al. (1988) described the apical meristem of the embryonic axis, obtained from the embryonic axis of imbibed seed, as a target for transgenes delivered via microprojectiles. Using a visual marker, they were able to reliably predict germline transformation events based on the patterns of expression in the primary regenerant (Christou and McCabe, 1992). A patent was subsequently granted on this method (Christou et al., 1991), covering the various steps of the protocol and derived soybean seed. Additional patents were later issued on this method that claim improvements to the original protocol (Christou and McCabe, 1998) and incorporate glyphosate as a selection agent to reduce the labor input (Martinell et al., 1999). Thus, whenever the soybean embryonic axis is used as the target for transformation, additional FTO would have to consider the Chee et al. (1994) protocol and the use of a disarmed *Agrobacterium* strain (Barton et al., 2000). The use of microprojectiles in the Christou et al. (1992) method would violate claims within the Sanford et al. (1993a) patent.

The soybean cotyledonary-node explant was first used as a target for *Agrobacterium*-mediated transformation in the mid-1980s (Hinchee et al., 1988), and has since been employed successfully by numerous independent laboratories (Di et al., 1996; Zhang et al., 1999; Olhoft et al., 2001) to recover transgenic soybean plants. The IP has been acquired covering aspects of this protocol (Hinchee and Conner-

Ward, 1998). Claims granted include the preparation of the explant with and without a wounding step, the derived soybean plants, and seed derived from a transformed soybean plant generated from the nonwounded method.

Townsend and Thomas (1996) also were granted IP on *Agrobacterium*-mediated transformation of soybean using the cotyledonary-node explant. The invention described in this patent was a medium composition that inventors claim optimizes induction of virulence gene expression in *A. tumefaciens*, and thus enhances transformation frequency. Additional claims cover modifications to the soybean tissue culture media for the in vitro differentiation of soybean cells into a whole plant. The FTO issues underlying soybean genetic engineering via the cotyledonary-node coupled with a disarmed *A. tumefaciens* strain are complicated again, since potential products derived from this method may infringe upon at least three issued utility patents, namely Barton et al. (2000), Hinchee et al. (1998), and Townsend and Thomas (1996).

Soybean somatic embryos have been used successfully to introduce transgenes into soybean via both *Agrobacterium*-mediated (Yan et al., 2000) and microprojectile-derived (Parrott et al, 1996; Reddy et al., 2001) strategies. However, there are at least three key utility patents issued, granting IP on methodologies for in vitro regeneration of soybean via a somatic embryo regime (Collins et al., 1991; Ranch and Buchheim, 1991; Hemphill and Warshaw, 1989). The IP from the Hemphill and Warshaw patent (1989) and Ranch and Buchheim (1991) only covers the regeneration method, while the Collins et al. (1991) patent granted expanded claims on transformation of this target tissue via *A. tumefaciens*, and to the regenerated transgenic soybean plant. Transgenic soybean derived from a microprojectile-mediated transformation of somatic embryos may infringe upon at least four patents, Hemphill and Warshaw (1989), Ranch and Buchheim (1991), and Collins et al. (1991) for the regeneration component and Sanford et al. (1993a) for the transformation component. More likely, current protocols have been sufficiently changed as to no longer infringe on the Hemphill and Warshaw (1989) and the Collins et al. (1991) patents. With the pertinent IP resting on only two patents, microprojectile-mediated transformation of somatic embryos might have greater FTO than any other method for soybean transformation, though it may require a trial to settle the issue.

The FTO issues underlying soybean genetic engineering methods can very rapidly become an impediment, particularly if the engineered soybean lacks sufficient added value to meet the remuneration demands of all the parties holding IP. The FTO is further complicated by the multitude of utility patents issued on the various selectable marker genes, promoters, and the genes of interest. Given this level of complexity, it is difficult for a soybean, derived from genetic engineering, to be commercialized without a licensing agreement from at least one institution. A prime example is RR soybean. Monsanto Company owns the IP on the Roundup-tolerant EPSP synthase gene, CP4 (Eichholtz et al., 2001), and the CaMV35S promoter element (Fraley et al., 1994), but DuPont Corporation owns the microprojectile bombardment (Sanford et al., 1993a; Fitzpatrick-McElligott et al., 1995) that was used to introduce the 35S CaMV-CP4 cassette into the soybean genome (Padgette et al., 1995). Therefore, Monsanto Company must possess a licensing agreement with DuPont Corporation to legally market the product. The FTO may be more elusive for public institutions, which lack the necessary resources to pursue development and marketing of a transgenic soybean. Hence, any potential soybean product derived at a public institution would require partnering with other public and/or private sector organizations to commercialize a transgenic soybean.

The IP issues arising in agriculture as the result of the introduction of biotechnology tools have altered the way researchers communicate research results. Patent disclosures must be submitted prior to public disclosure of data, so researchers may be hesitant to discuss ideas in a conference setting for fear of negating potential IP, leading to a delay in release of information. Biological materials are rarely distributed without an attached MTA.

Nevertheless, the obstacles that IP imposes on FTO are minor when weighed against the benefits. In a setting without IP, there would be a drastic redistribution of monies earmarked by the private sector for agriculture research that undoubtedly would have a negative impact on the rate of advances in enhancement of crop germplasm. Moreover, access without IP to a product of biotechnology may limit access of the product for future research. For example, without IP on the product, the inventors may pursue alternative avenues to protect the product, such as trade secret protection (Janis and Kesan, 2001).

7–4 ACCEPTANCE CONSIDERATIONS

7–4.1 Regulatory Considerations

Once FTO issues are settled, regulatory issues pose the final hurdle to the commercialization of transgenic crops. The regulations in place for genetically engineered crops are rapidly evolving, and will likely continue to do so for the foreseeable future. Due to the global economy, regulatory approval from the country of origin is not sufficient. Regulatory approval is necessary wherever the product will be marketed. Yet, internationally accepted regulatory standards are lacking altogether, so regulatory approval must face a patchwork of rapidly changing standards around the world. As an example, the LL soybean received regulatory approval for commercialization in the USA, but not in the European Union (EU). Since there is no way to segregate LL soybean from other soybean produced in the USA, there is no way to ensure that LL would not be mixed in with soybean destined for Europe. Consequently, the LL soybean has never been marketed.

The impact of current and pending regulations on the development of transgenic products is already obvious. Given the current regulatory framework, the cost of regulatory approval may greatly exceed the cost of developing a transgenic cultivar. There are no figures publicly available to substantiate the cost of regulatory approval, but this cost is generally acknowledged to be in the 10s of millions of dollars. The types of transgenes used, and deployment strategies of these transgenes, are being affected by the need for regulatory approval. A lot of emphasis was placed initially on the ability to engineer multiple elite soybean cultivars with a gene of interest, but obtaining regulatory approval for each of multiple engineering events makes such an approach cost prohibitive. Instead, once a transgenic individual has been identified with all the necessary traits for commercialization, that transgene, once deregulated, will be backcrossed or bred into additional varieties. Since only one transgene will be deployed, the need for transformation protocols that can engineer elite genotypes is greatly diminished. However, the need for marker-assisted breeding programs that can rapidly and efficiently deploy transgenes into elite cultivars will be essential.

7–4.2 International Trade Considerations

International trade in living genetically engineered organisms, such as seed, will be governed by the Cartagena Protocol on Biosafety to the Convention on Biological Security (www.biodiv.org/biosafety/), once the treaty has been ratified by the required 50 countries. Although the USA is not a party to the convention, all of its trading partners are signatories. The main purpose of the Cartagena Protocol was to achieve uniform standards for world trade in living genetically modified organisms, while acknowledging the right of each country to protect its biodiversity from potential damage from a living genetically engineered organism, as provided by the Convention on Biological Diversity Articles 8(g) and 19.3. Article 10, Paragraph 6 and Article 11, Paragraph 10 of the Cartagena protocol contain the controversial text that has become known as the Precautionary Principle: "Lack of scientific certainty due to insufficient relevant scientific information and knowledge regarding the extent of the potential adverse effects of a living modified organism on the conservation and sustainable use of biological diversity in the Party of import, taking also into account risks to human health, shall not prevent that Party from making a decision, as appropriate, with regard to the import of the living modified organism in question ... in order to avoid or minimize such potential adverse effects."

A commonly used rewording of the Precautionary Principle comes from the Wingspread Conference, January 1998 (www.sehn.org/precaution.html), "When an activity raises threats of harm to the environment or human health, precautionary measures should be taken even if some cause and effect relationships are not fully established scientifically." As applied to the Cartagena Protocol, it means that a country can ban imports of genetically engineered seed unless provided with proof that the engineered seed will not cause any harm. The exporting party is then faced with the impossible task of proving a negative, for every new concern raised, with no limit on the latter.

For trade in foods made with genetically engineered ingredients, the World Trade Organization (WTO) designated the *Codex Alimentarius*, which governs international food standards, as the official body setting the international standards for transgenic food products. Nevertheless, the *Codex* currently lacks standards for transgenic foods, and the negotiations to write such standards are expected to take years. Progress on these international standards may be monitored at www.codexalimentarius.net/. The only thing that is certain at the moment is that rules for the international trade of genetically modified organisms are in a state of flux, and will not be settled for many years to come.

7–4.3 Detection and Labeling

Labeling regulations are rapidly evolving, and differ greatly from one country to another. Labeling requirements fall into one of three categories. First, when the genetically engineered product is considered to be substantially equivalent to the non-engineered version, labeling is not required. Second, labeling is required when the amount of transgenic DNA or resulting protein exceeds a certain threshold. Third, labeling can be required when the product was derived from a transgenic plant, whether or not transgenic DNA or protein is present.

"Substantial equivalence" is primarily used in the USA and Canada. Under this doctrine, the levels of both nutritional and anti-nutritional compounds must fall within the range of those of nontransgenic products. The transformation process must not result in the production of toxins or allergens. Furthermore, the transformation process must not affect the functional quality or use of the product (FDA, 1992). Using this criterion, products derived from the RR soybean do not require a label. However, oil derived from HO soybean must be labeled as HO, but not as genetically engineered. This labeling requirement also applies to oil from HO soybean developed using conventional technology. Hence, it is the product that determines the need for a label, not the process by which the product was obtained.

In contrast, labeling regulations based on presence of transgenic DNA or its resulting protein are process-based, rather than product-based. In the examples above, products derived from RR or transgenic HO soybean would both require labeling as genetically engineered if enough DNA or protein is present to make the determination and meet an arbitrary threshold. Currently the threshold level to require a label varies from country to country, and ranges from 0.9% currently proposed in the EU (Regulation on Genetically Modified Food and Feed, proposed 28 Nov. 2002) to 5% in Japan. Products without DNA or protein, such as starch, are thus exempt from labeling.

In addition, the EU Regulation on Genetically Modified Food and Feed will mandate labeling of all food and feed products derived from transgenic crops that exceed the 0.9% threshold, even if transgenic DNA or protein is not present. The difficulty posed by this requirement is that it is impossible to determine the origin of those products that lack the protein or DNA necessary to make the identification, such as oil, sugar, or starch. To overcome this limitation, the EU is implementing a traceability requirement, whereby the identity of every crop must be preserved from the time it is harvested in the field, until the time it is used in consumer products.

Labeling regulations cannot be enforced without the ability to detect transgenic DNA in food products. Although immunological detection methods are available and useful at the elevator level (Fagan et al., 2001), detection technology in Europe has centered on detection of transgenes, as DNA is less subject than protein to denaturation during food processing (Hübner et al., 1999). Initial work used the PCR to detect the presence of transgenes (Jankiewicz et al., 1999; Lipp et al., 1999). However, the limitation of PCR is that it is not quantitative. Hence, quantitative techniques have been proposed using competitive PCR (Pietsch et al., 1999), real-time PCR (Hagen and Beneke, 2000), and PCR-ELISA (Brunnert et al., 2001). Based on trends from the recent literature, detection methodology is gravitating towards real-time PCR (Terry and Harris, 2001; Berdal and Holst-Jensen, 2001; Hubner et al., 2001; Waiblinger et al., 2001; Alray et al., 2002; Terry et al., 2002). It is clear that accurate detection and quantification will become an ever increasing challenge as additional transgenic crops are permitted on the market, which is one reason the food industry may be reluctant to accept new engineered crops.

7–5 CONCLUSIONS

Today, soybean is among the easier crops to genetically engineer. Yet, even 5 yr ago, the soybean was considered to be one of the most difficult crops to engineer. Research funded by grower check-off funds provided the necessary resources to develop the protocols available today. The protocols for soybean engineering continue to be modified and refined, and additional advances in engineering technology can be expected.

The legal and cultural frameworks that regulate the commercialization and consumer acceptance of transgenic crops are still being formulated. Currently, transgenic crops face regulatory uncertainty, and their acceptance by consumers is far from guaranteed. Hence, the greatest use of transgenic soybean in the foreseeable future may be to address issues of basic biology.

In the long run, transgenic soybean cultivars will prevail, as the advent of effective transformation protocols for soybean has provided a tool of unprecedented power to develop cultivars with traits that facilitate its production, provide added value, and lead to new uses for soybean.

ACKNOWLEDGMENTS

The authors wish to express their gratitude to Anthony Kinney for helping provide information on DuPont's High Oleic soybean, to Peter LaFayette for the diagrams of the genes used for the RR and HO soybeans, to Karina Nersesova for the diagram on soybean regeneration, and to Carolyn Hightower for assistance with preparation of the typescript.

REFERENCES

Agricultural Statistics Board. 2000. Prospective plantings. Natl. Agric. Statistics Serv., USDA, Washington, DC.

Alary, R., A. Serin, D. Maury, H.B. Jouria, J.-P. Sirven, M.-F. Guatier, and P. Joudrier. 2002. Comparison of simplex and duplex real-time PCR for the quantification of GMO soybean. Food Contr. 13:235–244.

Aragão, F.J.L., L. Sarokin, G.R. Vianna, and E.L. Rech. 2000. Selection of transgenic meristematic cells utilizing a herbicidal molecule results in recovery of fertile transgenic soybean [*Glycine max* (L.) Merril] plants at a high frequency. Theor. Appl. Genet. 101:1–6.

Astwood, J.D., J.N. Leach, and R.L. Fuchs. 1996. Stability of food allergens to digestion in vitro. Nature Biotechnol. 14:1269–1273.

Baenziger, P.S., A Mitra, and I. B. Edwards. 2000. Protecting the value in value added crops: Intellectual property rights. p. 239–248. *In* C.F. Murphy and D.M. Peterson (ed.) Designing crops for added value. ASA, Madison, WI.

Bailey, M.A., H.R. Boerma, and W.A. Parrott. 1993. Genotype effects on proliferative embryogenesis and plant regeneration of soybean. In Vitro Cell. Dev. Biol. 29P:102–108.

Barnes,R.L. 2000. Why the American Soybean Association supports transgenic soybeans. Pest Manag. Sci. 56:580–583.

Barry, G., G. Kishore, S. Padgette, M. Taylor, K. Kolacz, M. Weldon, D. Re, D. Eichholtz, K. Fincher, and L. Hallas. 1992. Inhibitors of amino acid biosynthesis: Strategies for imparting glyphosate tolerance to crop plants. p.139–145. *In* B.K. Singh et al. (ed) Biosynthesis and molecular regulation of amino acids in plants. Am. Soc. of Plant Physiol., Rockville, MD.

Barton, K.A., A.N. Binns, M-D Chilton, and A.J.M. Matzke. 2000. Regeneration of plants containing genetically engineered T-DNA. U.S. Patent 6 051 757. Date issued: 18 Apr. 2000.

Barwale, U.B., H.R. Kerns, and J.M. Widholm. 1986. Plant regeneration from callus cultures of several soybean genotypes via embryogenesis and organogenesis. Planta 167:473B481.

Bechtold, N., J. Ellis, and G. Pelletier. 1993. *In planta Agrobacterium* mediated gene transfer by infiltration of adult *Arabidopsis thaliana* plants. C.R. Acad. Sci. Paris Life Sci. 316:1194-1199.

Benbrook, C. 1999. Evidence of the magnitude and consequences of the Roundup Ready soybean yield drag from university-based varietal trials in 1998. AgBioTech InfoNet Tech. Paper No. 1.

Berdal, K.G., and A. Holst-Jensen. 2001. Roundup Ready soybean event-specific real-time quantitative PCR assay and estimation of the practical detection and quantification limits in GMO analyses. Eur. Food Res. Technol. 213:432–438.

Bojsen, K., I. Donaldson, A. Haldrup, M. Joersbo, J.D. Kreiberg, D., J. Nielsen, F.T. Okkels, S.G. Petersen, and R.J. Whenham. 2000. Positive selection. Novartis AG. U.S. Patent 5 994 629. Date issued: 18 Apr. 2000.

Brunnert, H.J., F. Spener, and T. Börchers. 2001. PCR-ELISA for the CaMV-35S promoter as a screening method for genetically modified Roundup Ready soybeans. Eur. Food Res. Technol. 213:366–371.

Buchheim, J.A., S.M. Colburn, and J.P. Ranch. 1989. Maturation of soybean somatic embryos and the transition to plantlet growth. Plant Physiol. 89:768–775.

Buhr, T., S. Sato, F. Ebrahim, A.Q. Xing, Y. Zhou, M. Mathiesen, B. Schweiger, A. Kinney, P. Staswick, and T. Clemente. 2002. Nuclear localization of RNA transcripts down-regulate seed fatty acid genes in transgenic soybean. Plant J. 30:155–163.

Burton, J.W., R.F. Wilson, and C.A. Brim. 1983. Recurrent selection in soybeans. IV. Selection for increased oleic acid percentage in seed oil. Crop Sci. 23:744–747.

Cahoon, E.B., T.J. Carlson, K.G. Ripp, B.J. Schweiger, G.A. Cook, S.E. Hall, and A.J. Kinney. 1999. Biosynthetic origin of conjugated double bonds: Production of fatty acid components of high-value drying oils in transgenic soybean embryos. Proc. Natl. Acad. Sci. USA 96:12935–12940.

Cahoon, E.B., E.-F. Marillia, K.L. Stecca, S.E. Hall, D.C. Taylor, and A.J. Kinney. 2000. Production of fatty acid components of meadowfoam oil in somatic soybean embryos. Plant Physiol. 124:243B251.

Carpenter, J.E. 2001a. Case studies in benefits and risks of agricultural biotechnology: Roundup Ready soybeans and Bt field corn. Natl. Center for Food and Agric. Policy, Washington, DC.

Carpenter, J.E. 2001b. Comparing Roundup Ready yields and conventional soybean yields 1999. Natl. Ctr. for Food and Agric. Policy, Washington, DC.

Carpenter, J.E, and L.P. Gianessi. 2001. Agricultural biotechnology: Updated benefit estimates. Natl. Ctr. for Food and Agric. Policy, Washington, DC.

Chee, P.P., K.A. Fober, and J. L. Slightom. 1989. Transformation of soybean (*Glycine max*) by infecting germinating seeds with *Agrobacterium tumefaciens*. Plant Physiol. 91:1212–1218.

Chee, P.P., S. L. Goldman, A.C.F. Graves, and J. L. Slightom. 1994. *Agrobacterium* mediated transformation of germinating plant seeds. U.S. Patent 5 376 543. Date issued: 27 Dec. 1994.

Chee, P.P., and J.L. Slightom. 1995. Transformation of soybean (*Glycine max*) via *Agrobacterium tumefaciens* and analysis of transformed plants. p. 101–119. *In* K.M.A. Gartl and M.R. Davey (ed.) Methods in molecular biology. Vol. 44. *Agrobacterium* protocols. Humana Press, Inc., Totowa, NJ.

Chen, D.F., P.J. Dale, J.S. Heslop-Harrison, J.W. Snape, W. Harwood, S. Bean, and P.M. Mullineaux. 1994. Stability of transgenes and presence of N^6 methyladenine DNA in transformed wheat cells. Plant J. 5:429-436.

Cheng, T-Y., H. Saka, and T.H. Voqui-Dinh. 1980. Plant regeneration from soybean cotyledonary node segments in culture. Plant Sci. Lett. 19:91–99.

Cho, H.-J., S.K. Farrand, G.R. Noel, and J.M. Widholm. 2000. High-efficiency induction of hairy roots and propagation of the soybean cyst nematode. Planta 210:195–204.

Chowrira, G.M., V. Akella, P.E. Fuerst, and P.F. Lurquin. 1996. Transgenic grain legumes obtained by *in planta* electroporation-mediated gene transfer. Mol. Biotechnol. 5:85–96.

Chowrira, G.M., V. Akella, and P.F. Lurquin. 1995. Electroporation-mediated gene transfer into intact nodal meristems *in planta*-generating transgenic plants without *in vitro* tissue-culture. Mol. Biotechnol. 3:17–23.

Christianson, M.L., D.A. Warnick, and P.S. Carlson. 1983. A morphogenetically competent soybean suspension culture. Science (Washington DC) 222:632–634.

Christou, P. 1990a. Morphological description of transgenic soybean chimeras created by the delivery, integration and expression of foreign DNA using electric discharge particle acceleration. Ann. Bot. 66:379–386.

Christou, P. 1990b. Soybean transformation by electric discharge particle acceleration. Physiol. Plant. 79:210–212.

Christou, P., and D.E. McCabe. 1992. Prediction of germ-line transformation events in chimeric R_0 trans-genic soybean plantlets using tissue-specific expression patterns. Plant J. 2:283B290.

Christou, P., and D. McCabe. 1998. Plant transformation process with early identification of germ line transformation events. U.S. Patent 5 830 728. Date issued: 3 Nov. 1998.

Christou, P., D.E. McCabe, B.J. Martinell, and W.F. Swain. 1990. Soybean genetic engineering: Com-mercial production of transgenic plants. Tibtech 8:145–151.

Christou, P., D. McCabe, W. Swain, and K.A. Barton. 1991. Particle-mediated transformation of soy-bean plants and lines. U.S. Patent 5 015 580. Date issued: 14 May 1991.

Christou, P., W.F. Swain, N.S. Yang, and D.E. McCabe. 1989. Inheritance and expression of foreign genes in transgenic soybean plants. Proc. Natl. Acad. Sci. USA 86:7500–7504.

Clemente, T., B.J. LaValle, A.R. Howe, D.C. Ward, R.J. Rozman, P.E. Hunter, D.L. Broyles, D.S. Kas-ten, M.A. Hinchee. 2000. Progeny analysis of glyphosate selected transgenic soybeans derived from *Agrobacterium*-mediated transformation. Crop Sci. 40:797-803.

Cober, E.R., S. Rioux, I. Rajcan, P.A. Donaldson, and D.H. Simmonds. 2003. Partial resistance to white mold in a transgenic soybean line. Crop Sci. 43:92–95.

Collins, G.B., D.F. Hilderbrand, P.A. Lazzeri, T.R. Adams, W.A. Parrott, and L.M. Hartweck. 1991. Trans-formation, somatic embryogenesis and whole plant regeneration method for *Glycine* species. U.S. Patent 5 024 944. Date issued: 18 June 1991.

Cooke, T.J., R.H. Racusen, and J.D. Cohen. 1993. The role of auxin in plant embryogenesis. Plant Cell 5:1494–1495.

Cregan, P.B., T. Jarvik, A.L. Bush, R.C. Shoemaker, K.G. Lark, A.L. Kahler, N. Kaya, T.T. VanToai, D.G. Lohnes, J. Chung, and J.E. Specht. 1999. An integrated genetic linkage map of the soy-bean genome. Crop Sci. 39:1464–1490.

Dale, E.C., and D.W. Ow. 1991. Gene transfer with subsequent removal of the selection gene from the host genome. Proc. Natl. Acad. Sci. USA 88:10558–10562.

Dan, Y., and N.A. Reichert. 1998. Organogenic regeneration of soybean from hypocotyl explants. In Vitro Cell. Dev. Biol. Plant 34:12-21.

De Cleene, M., and J. De Ley. 1976. The host range of crown gall. Bot. Gaz. 42:389–466.

Delannay, X., T.T. Bauman, D.H. Beighley, M.J. Buettner, H.D. Coble, M.S. DeFelice, C.W. Derting, T.J. Diedrick, J.L. Griffin, E.S. Hagood, et al. 1995. Yield evaluation of a glyphosate-tolerant soybean line after treatment with glyphosate. Crop Sci. 35:1461–1467.

Denbow, D.M., E.A. Grabau, G.H. Lacy, E.T. Kornegay, D.R. Russell, and P.F. Umbeck. 1998. Soy-beans transformed with a fungal phytase gene improve phosphorus availability for broilers. Poul. Sci. 77:878–881.

de Ronde, J.A., M.H. Spreeth, and W.A. Cress. 2001. Effect of antisense L-D^1-pyrroline-5-carboxylate reductase transgenic soybean plants subjected to osmotic and drought stress. Plant Growth Reg. 32:13–26.

Desfeux, C., S.J. Clough, and A.F. Bent. 2000. Female reproductive tissues are the primary target of *Agrobacterium*-mediated transformation by *Arabidopsis* floral-dip method. Plant Physiol. 123:895-904.

Di, R., V. Purcell, G.B. Collins, and S.A. Ghabrial. 1996. Production of transgenic soybean lines ex-pressing the bean pod mottle virus coat protein precursor gene. Plant Cell Rep. 15:746–750.

Dinkins, R.D., M.S.S. Reddy, C.A. Meurer, B. Yan, H. Trick, F. Thibaud-Nissen, J.J. Finer, W.A. Par-rott, and G.B. Collins. 2001. Increased sulfur amino acids in soybean plants overexpressing the maize 15 kDa zein protein. In Vitro Cell. Dev. Biol.-Plant 37:742–747.

Duffield, H., H.S. Shapouri, M. Graboski, R. McCormick, R. Wilson. 1998. U.S. biodiesel development: New markets for conventional and genetically modified agricultural products. USDA Econ. Res. Serv. Rep. 770.

Ecker, J.R., and R.W. Davis. 1986. Inhibition of gene expression in plant cells by expression of anti-sense RNA. Proc. Natl. Acad. Sci. USA 83:5372–5376.

Eichholtz, D.A., C.S. Gasser, and G.M. Kishore. 2001. Modified gene encoding glyphosate-tolerant-5-enolpruvyl-3-phosphoshikimate synthase. U.S. Patent 6 225 114. Date issues: 1 May 2001.

Elmore, R.W., F.W. Roeth, L.A. Nelson, C.A. Shapiro, R.N. Klein, S.Z. Knezevic, and A. Martin. 2001. Glyphosate-resistant soybean cultivar yields compared with sister lines. Agron. J. 93:408–412.

Environmental Protection Agency. 2000. 40 CFR part 80. Glyphosate; pesticide residues. Fed. Reg. 65:52660–52667.

Fagan, J., B. Schoel, A. Haegert, J. Moore, and J. Beeby. 2001. Performance assessment under field con-ditions of a rapid immunological test for transgenic soybeans. Int. J. Food Sci. Technol. 36:357–367.

Falco, S.C., T. Guida, M. Locke, J. Mauvais, C. Sanders, R.T. Ward, and P. Webber. 1995. Transgenic canola and soybean seeds with increased lysine. Bio/Technol. 13:577–582.

Finer, J.J. 1988. Apical proliferation of embryogenic tissue of soybean [*Glycine max* (L.) Merrill]. Plant Cell Rep. 7:238–241.

Finer, J.J., and M.D. McMullen. 1991. Transformation of soybean via particle bombardment of embryogenic suspension culture tissue. In Vitro Cell. Dev. Biol. 27P:175–182.

Finer, J.J., and A. Nagasawa. 1988. Development of an embryogenic suspension culture of soybean (*Glycine max* Merrill). Plant Cell Tiss. Organ Cult. 15:125–136.

Fitzpatrick-McElligott, S.G., J.G. Lavine, G.F. Rivard, and S. Subramoney. 1995. Method for introducing a biological substance into a target. U.S. Patent 5 466 587. Date issued: 14 Nov. 1995.

Fleysh, N., D. Deka, M. Drath, H. Koprowski, and V.Yusibov. 2001. Pathogenesis of alfalfa mosaic virus in soybean (*Glycine max*) and expression of chimeric rabies peptide in virus-infected soybean plants. Phytopathology 91:941–947.

Food and Drug Administration. 1992. Statement of policy: Foods derived from new plant varieties. Fed. Reg. 57:22984–23001.

Food and Drug Administration. 1998. Guidance for industry: Use of antibiotic resistance marker genes in transgenic plants. U.S. FDA, Center for Food Safety and Applied Nutrition, Office of Premarket Approval. Available at http://vm.cfsan.fda.gov/~dms/opa-armg.html.

Fraley, R.T., R.B. Horsch, and S.G. Rogers. 1994. Chimeric genes for transforming plant cells using viral promoters. U.S. Patent 5 352 605. Date issued: 4 Oct. 1994.

Fullner, J.K., and E.W. Nester. 1996. Temperature affects the T-DNA transfer machinery of *Agrobacterium tumefaciens*. J. Bacteriol. 178:1498–1504.

Gamborg, O.L., B.P. Davis, and R.W. Stahlquist. 1983. Somatic embryogenesis in cell cultures of *Glycine* species. Plant Cell Rep. 2:209–212.

Gertz, Jr. J.M., W.K. Vencill, and N.S. Hill. 1999. Tolerance of transgenic soybean (*Glycine max*) to heat stress. p. 835–840. *In* The 1999 Brighton Conf., Weeds: Proc. of an Int. Conf., Brighton Centre, Brighton, UK. 15–18 Nov. 1999. Organized by the British Crop Protection Council. The Council, Farnham, Surrey, UK.

Ghazi, T.D., H.V. Cheema, and M.W. Nabors. 1986. Somatic embryogenesis and plant regeneration from embryogenic callus of soybean, *Glycine max* L. Plant Cell Rep. 5:452–456.

Gianessi, L.P., and J.E. Carpenter. 2000. Agriculture biotechnology: Benefits of transgenic soybeans. Natl. Ctr. for Food and Agric. Policy, Washington, DC.

Giddings, G., G. Allison, D. Brooks, and A. Carter. 2000. Transgenic plants as factories for biopharmaceuticals. Nat. Biotechnol. 18:1151–1155.

Godwin, I., T. Gordon, B. Ford-Lloyd, and H.J. Newbury. 1991. The effects of acetosyringone and pH on *Agrobacterium*-mediated transformation vary according to plant species. Plant Cell Rep. 9:671-675

Gowda, S., F.C. Wu, and R.J. Shepard. 1989. Identification of promoter sequences for the major RNA transcripts of figwort mosaic and peanut chlorotic streak viruses (Caulimovirus group). J. Cell. Biochem. 13D (suppl.):301.

Greengrass, B. 1993. Non-U.S. protection procedures and practices-implications for U.S. innovators? p. 41–59. *In* P.S. Baenziger et al. (ed.) Intellectual property rights: Protection of plant materials. ASA, Madison, WI.

Hadi, M.Z., M.D. McMullen, and J.J. Finer. 1996. Transformation of 12 different plasmids into soybean via particle bombardment. Plant Cell Rep. 15:500-505.

Hagen, M., and B. Beneke. 2000. Nachweisbarkeit der gentechnischen Veränderung von Soja (Roundup-Ready Soja) in weiterverarbeiteten Produckten. Berl. Münch. Tierärztl. Wschr. 113:454–458.

Hammatt, N., and M.R. Davey. 1987. Somatic embryogenesis and plant regeneration from cultured zygotic embryos of soybean (*Glycine max* L. Merr.). J. Plant Physiol. 128:219–226.

Hammond, B.G., J.L. Vicini, G.F. Hartnell, M.W. Naylor, C.D. Knight, E.H. Robinson, R.L. Fuchs, and S.R. Padgette. 1996. The feeding value of soybeans fed to rats, chickens, catfish, and dairy cattle is not altered by genetic incorporation of glyphosate tolerance. J. Nutr. 126:717–727.

Harrison, L.A., M.R. Bailey, M.W. Naylor, J.E. Ream, B.G. Hammond, D.L. Nida, B.L. Burnette, T.E. Nickson, T.A. Mitsky, M.L. Taylor, R.L. Fuchs, and S.R. Padgette. 1996. The expressed protein in glyphosate-tolerant soybean, 5-enolypyruvylshikimate-3-phosphate synthase from *Agrobacterium* sp. strain CP4, is rapidly digested in vitro and is not toxic to acutely gavaged mice. J. Nutr. 126:728-740.

Hartnett-Foster, D., J. Mudge, D. Larsen, D. Danesh, H. Yan, R. Denny, S. Penuela, and N.D. Young. 2002. Comparative genomic analysis of sequences sampled from a small region on soybean (*Glycine max*) molecular linkage group G. Genome 45:634-645.

Hartweck, L.M., P.A. Lazzeri, D. Cui, G.B .Collins, and E.G.Williams. 1988. Auxin-orientation effects on somatic embryogenesis from immature soybean cotyledons. In Vitro Cell. Dev. Biol. 24:821–828.

Hazel, C.B., T.M. Klein, M. Anis, H.D. Wilde, and W.A. Parrott. 1998. Growth characteristics and transformability of soybean embryogenic cultures. Plant Cell Rep. 17:765-772.

Hemphill, J.K., and C. A. Warshaw. 1989. Process for regenerating soybean. U.S. Patent 4 837 152. Date issued: 6 June 1989.

Hinchee, M.A., and D.-C. Ward. 1998. Method for soybean transformation and regeneration. U.S. Patent 5 824 877. Date issued: 20 Oct. 1998.

Hinchee, M.A., D.-C. Ward, C.A. Newell, R.E. McDonnell, S.J. Sato, C.S. Gasser, D.A. Fischhoff, D.R. Re, R.T. Fraley, and R.B. Horsch. 1988. Production of transgenic soybean plants using *Agrobacterium*-mediated DNA transfer. Bio/Technol. 6:915–922.

Honda, M, Y. Muramoto, T. Kuzuguchi, S. Sawano, M. Machida, and H. Koyama. 2002. Determination of gene copy number and genotype of transgenic *Arabidopsis thaliana* by competitive PCR. J. Exp. Bot. 53:1515–1520.

Hu, C.-Y., and L. Wang. 1999. *In planta* soybean transformation technologies developed in China: Procedure, confirmation and field performance. In Vitro Cell Dev. Biol.-Plant 35:417–420.

Hu, C.-Y., G.-C. Yin, and M. H. Bodanese Zanettini. 1996. Haploids of soybeanCA review article. p. 377–395. *In* S.M. Jain et al. (ed.) *In Vitro* haploid production in higher plants. Vol 2. Kluwer Academic Publ., Dordrecht, the Netherlands.

Hübner, P., E. Studer, and J. Lüthy. 1999. Quantitation of genetically modified organisms in food. Nat. Biotechnol. 17:1137–1139.

Hübner, P., H.U. Waiblinger, K. Pietsch, and P. Brodmann. 2001. Validation of PCR methods for quantitation of genetically modified plants in food. J. AOAC Int. 84:1855–1864.

Janis, M.D., and J. P. Kesan. 2001. Designing an optimal intellectual property system for plants: A US Supreme Court debate. Nat. Biotechnol. 19:981–983.

Jankiewicz, A., H. Broll, and J. Zagon. 1999. The official method for the detection of genetically modified soybeans (German Food Act LMBG section 35): A semi-quantitative study of sensitivity limits with glyphosate-tolerant soybeans (Roundup Ready) and insect-resistant Bt maize (Maximizer). Eur. Food Res. Technol. 209:77–82.

Jefferson, R.A. 1989. The GUS reporter gene system. Nature (London) 342:837–838.

Jefferson, R.A., S.M Burgess, and D. Hirsch. 1986. ß-glucuronidase from *Escherichia coli* as a gene-fusion marker. Proc. Natl. Acad. Sci. USA 83:8447–8451.

Jondle, R.J. 1989. Overview and status of plant proprietary rights. p. 5–15. *In* J.H. Barton et al. (ed.) Intellectual property rights associated with plants. ASA, Madison, WI.

Jong, S.C., and R.H Cypess. 1998. Managing genetic material to protect intellectual property rights. J. Ind. Microbiol. Biotechnol. 20:95–100.

Kaeppler, H.F., G.K. Menon, R.W. Kadsen, A.M. Nuutila, and A.R. Carlson. 2000. Transgenic oat plants via visual selection of cells expressing green fluorescent protein. Plant Cell Rep. 19:661–666.

Kaltchuk-Santos, E., J.E. Mariath, E. Mundstock, C.-Y. Hu, and M.H. Bodanese-Zanettini. 1997. Cytological analysis of early microspore divisions and embryo formation in cultured soybean anthers. Plant Cell Tiss. Organ Cult. 49:107–115.

Kay, R., A. Chan, M. Daly, and J. McPherson. 1987. Duplication of the CaMV 35S promoter sequences creates a strong enhancer for plant genes. Science (Washington DC) 236:1299–1302.

Kim, J., C.E. LaMotte, and E. Hack. 1990. Plant regeneration *in vitro* from primary leaf nodes of soybean *Glycine max* seedlings. J. Plant Physiol. 136:664-669.

Kinney, A.J. 1997. Genetic engineering of oilseeds for desired traits. p. 149-166. *In* J.K. Setlow (ed.) Genetic engineering. Vol 19. Plenum Press, New York.

Kinney, A.J., R. Jung, and E.M. Herman. 2001. Cosuppression of the alpha subunits of β-conglycinin in transgenic soybean seeds induces the formation of endoplasmic reticulum-derived protein bodies. Plant Cell 13:1165–1178.

Kinney, A.J., and S. Knowlton. 1998. Designer oils: The high oleic soybean. p. 193–213. *In* S. Roller and S. Harlander (ed.) Genetic modification in the food industry. Blackie, London.

Klee, H.J., Y.M. Muskopf, and C.S. Gasser. 1987. Cloning of an *Arabidopsis thaliana* gene encoding 5-enolpyruvylshikimate-3-phosphate synthase: Sequence analysis and manipulation to obtain glyphosate-tolerant plants. Mol. Gen. Genet. 210:437–442.

Klein, T.M., E.D. Wolf, R. Wu, and J.C.Sanford. 1987. High-velocity microprojectiles for delivering nucleic acids into living cells. Nature (London) 327:70–73.

Komatsuda, T., and S.-W.Ko. 1990. Screening of soybean [*Glycine max* (L.) Merrill] genotypes for somatic embryo production from immature embryo. Jpn. J. Breed. 40:249–251.

Komatsuda, T., and K. Ohyama. 1988. Genotypes of high competence for somatic embryogenesis and plant regeneration in soybean *Glycine max*. Theor. Appl. Genet. 75:695–700.

Kusnadi, A.R., Z.L. Nikolov, and J.A. Howard. 1997. Production of recombinant proteins in transgenic plants: Practical considerations. Biotech. Bioeng. 56:473B484.

Lappé, M.A., E.B. Bailey, C.C hildress, and K.D.R. Setchell. 1998. Alterations in clinically important phytoestrogens in genetically modified, herbicide-tolerant soybeans. J. Med. Food 1:241–245.

Lazzeri, P.A., D.F. Hildebrand, and G.B. Collins. 1985. A procedure for plant regeneration from immature cotyledon tissue of soybean. Plant Mol. Biol. Rep. 3:160–167.

Lei, B., G. Yin, S. Wang, C. Lu, H. Qian, S. Zhou, K. Zhang, D. Sui, and Z. Lin. 1989. Variation occurring by introduction of wild soybean DNA into cultivar soybean. Oil Crops China 3:11–14.

Levin, J.Z., A.J. deFramond, A. Tuttle, M.W. Bauer, and P.B. Heifetz. 2000. Methods of double-stranded RNA-mediated gene inactivation in *Arabidopsis* and their use to define an essential gene in methionine biosynthesis. Plant Mol. Biol. 44:759–775.

Lipp, M., P. Brodmann, K. Pietsch, J. Pauwels, and E. Anklam. 1999. IUAPC collaborative trial study of a method to detect genetically modified soybeans and maize in dried powder. J. AOAC Int. 82:923–928.

Lippmann, B., and G. Lippmann. 1984. Induction of somatic embryos in cotyledonary tissue of soybean, *Glycine max* L. Merr. Plant Cell Rep. 3:215–218.

Liu, B.-L., S.-X.Yue, N.-B. Hu, X.-B. Li, W.-X. Zhai, N. Li, R.-H. Zhu, L.-H. Zhu, D.-Z. Mao, and P.-Z. Zhou. 1989. Transfer of atrazine resistant gene from nightshade to soybean chloroplast genome and its expression in transgenic plants. Ch. Sci. Bull. 34:1670–1672.

Liu, B.-L., S.-X.Yue, N.-B. Hu, X.-B. Li, W.-X. Zhai, N. Li, R.-H. Zhu, L.-H. Zhu, D.-Z. Mao, and P.-Z. Zhou. 1990. Transfer of the atrazine-resistant gene of black nightshade to soybean chloroplast genome and its expression in transgenic plants. Sci. Chi. Ser. B. 33:444–452.

Liu, Z., and G.G. Carmichael. 1994. Nuclear antisense RNA. Mol. Biotechnol. 2:107–117.

Martin, B.A., R.F. Wilson, and R.W. Rinne. 1986. Temperature effects upon the expression of a high oleic-acid trait in soybean *Glycine max*. J. Am. Oil Chem. Soc. 63:346-352.

Martinell, B., L.A. Julson, M.A.W. Hinchee, D.-C. Ward, D. McCabe, and C. Emler. 1999. Efficiency soybean transformation protocol. U.S. Patent 5 914 451. Date issued: 22 June 1999.

Marx, J. 2000. Interfering with gene expression. Science (Washington DC) 288:1370-1372.

Matzke, M.A., and J.M. Matzke. 1995. How and why do plants inactivate homolgous (trans) genes? Plant Physiol. 107:679–685.

Matzke, M.A., A.J.M. Matzke, G.J. Pruss, and V.B. Vance. 2001. RNA-based silencing strategies in plants. Curr. Opin. Genet. Dev. 11:221-227.

Maughan, P.J., R. Philip, M.-J. Cho, J.M. Widholm, and L.O. Vodkin. 1999. Biolistic transformation, expression, and inheritance of bovine β-casein in soybean (*Glycine max*). In Vitro Cell Dev. Biol.-Plant 35:334–349.

Mazur, B., E. Krebbers, and S. Tingey. 1999. Gene discovery and product development for grain quality traits. Science (Washington DC) 285:372–375.

McCabe, D.E., W.F. Swain, B.J. Martinell, and P. Christou. 1988. Stable transformation of soybean (*Glycine max*) by particle acceleration. Bio/Technol. 6:923B926.

Meurer, C.A., R.D. Dinkins, C.T. Redmond, K.P. McAllister, D.M. Tucker, D.R. Walker, W.A. Parrott, H.N. Trick, J.S. Essig, H.M.Franz, J.J. Finer, and G.B. Collins. 2001. Embryogenic response of multiple soybean [*Glycine max* (L.) Merrill] cultivars across three locations. In Vitro Cell Dev. Biol.-Plant 37:62–67.

Meyer, P., and H. Saedler. 1996. Homology-dependent gene silencing in plants. Ann. Rev. Plant Physiol. Plant Mol. Biol. 47:23B48.

Mullins, M.G., F.C.A. Tang, and D. Facciotti. 1990. *Agrobacterium*-mediated genetic transformation of grapevines: Transgenic plants of *Vitis rupestris* Scheele and buds of *Vitis vinifera* L. Bio/Technol. 8:1041-1045.

Murashige, T., and F. Skoog. 1962. A revised medium for rapid growth and bio assays with tobacco tissue cultures. Physiol. Plant. 15:473B497.

Nomura, N.S., and H.W. Hilton. 1977. The adsorption and degradation of glyphosate in five Hawaiian sugarcane soils. Weed Res. 17:113B121.

Nordlee, J.A., S.L. Taylor, J.A. Townsend, L.A. Thomas, and R.K. Bush. 1996. Identification of a Brazil-nut allergen in transgenic soybeans. New Engl. J. Med. 334:688–692.

Olhoft, P.M, L.E. Flagel, C.M. Donovan, and D.A. Somers. 2003. Efficient soybean transformation using hygromycin B selection in the cotyledonary method. Planta 216:723–735. DOI 10.1007/s00425-002-0922-2.

Olhoft, P.M., K. Lin, J. Galbraith, N.C. Nielsen, and D.A. Somers. 2001. The role of thiol compounds in increasing Agrobacterium-mediated transformation of soybean cotyledonary-node cells. Plant Cell Rep. 20:731–737.

Olhoft, P.M., and D.A. Somers. 2001. L-cysteine increases *Agrobacterium*-mediated T-DNA delivery into soybean cotyledonary-node cells. Plant Cell Rep. 20:706–711.

Padgette, S.R., K.H. Kolacz, X. Delannay, D.B. Re, B.J. LaVallee, C.N. Tinius, W.K. Rhodes, Y.I. Otero, G.F. Barry, D.A. Eichholtz, V.M. Peschke, D.L. Nida, N.B. Taylor, and G.M. Kishore. 1995. Development, identification and characterization of a glyphosate-tolerant soybean line. Crop Sci. 35:1451–1461.

Parrott, W.A. 2000. Embryogenesis in angiosperms, somatic. p. 661–666. *In* R.E. Spier (ed.) The encyclopedia of cell biology. John Wiley & Sons, New York.

Parrott, W.A., M.J. Adang, M.A. Bailey, J.N. All, H.R. Boerma, and N.S. Stewart. 1994. Recovery and evaluation of soybean plants transgenic for a *Bacillus thuringiensis* var. kurstaki insecticidal gene. In Vitro-Plant. 30P:144-149.

Parrott, W.A., J.N. All, M.J. Adang, M.A. Bailey, H.R. Boerma, and C.N .Stewart, Jr. 1994. Recovery and evaluation of soybean (*Glycine max* [L.] Merr.) plants transgenic for a *Bacillus thuringiensis* var. *kurstaki* insecticidal gene. In Vitro Cell. Dev. Biol. 30P:144–149.

Parrott, W.A., G. Dryden, S. Vogt, D.F. Hildebrand, G.B. Collins, and E.G. Williams. 1988. Optimization of somatic embryogenesis and embryo germination in soybean. In Vitro Cell. Dev. Biol. 24:817–820.

Parrott, W.A., L.M. Hoffman, D.F. Hildebrand, E.G. Williams, and G.B. Collins. 1989a. Recovery of primary transformants of soybean. Plant Cell Rep. 7:615–617.

Parrott, W.A., S.A. Merkle, and E.G. Williams. 1991. Somatic embryogenesis: Potential for use in propagation and gene transfer systems. p. 158–200. *In* D.R. Murray (ed.) Advanced methods in plant breeding and biotechnology. CAB Int., Wallingford.

Parrott, W.A., E.G. Williams, D.F. Hildebrand, and G.B.Collins. 1989b. Effect of genotype on somatic embryogenesis from immature cotyledons of soybean. Plant Cell Tissue Organ Cult. 16:15–21.

Pawlowski, W.P., and D.A. Somers. 1998. Transgenic DNA integrated into the oat genome is frequently interspersed by host DNA. Proc. Natl. Acad. Sci. USA 95:12106–12110.

Pengue, W.A. 2000. Herbicide tolerant soybean: Just another step in a technology treadmill? Biotechnol. Dev. Monitor 43:11–14.

Phillips, G.C., and G.B. Collins. 1981. Induction and development of somatic embryos from cell suspension cultures of soybean. Plant Cell Tiss. Organ Cult. 1:123–129.

Pietsch, K., A. Bluth, A. Wurz, and H.U. Waiblinger. 1999. Kompetitive PCR zur Quantifizierung konventioneller und transgener Lebensmittelbestandteile. Deutsche Lebensmittel-Rundschau 95:57–59.

Ponappa, T., A.E. Brzozowski, and J.J. Finer. 1999. Transient expression and stable transformation of soybean using the jellyfish green fluorescent protein. Plant Cell Rep. 19:6–12.

Rajasekaran, K., and J.W. Pellow. 1997. Somatic embryogenesis from cultured epicotyls and primary leaves of soybean [*Glycine max* (L.) Merrill]. In Vitro Cell Dev. Biol.-Plant 33:88–91.

Ranch, J.P., and J.A. Buchheim. 1991. Propagating multiple whole fertile plants from immature leguminous. U.S. Patent 5 008 200. Date issued: 16 Apr. 1991.

Ranch, J.P., L .Oglesby, and A.C. Zielinski. 1985. Plant regeneration from embryo-derived tissue cultures of soybean by somatic embryogenesis. In Vitro Cell. Dev. Biol. 21:653B657.

Ranch, J.P., L. Oglesby, and A.C. Zielinski. 1986. Plant regeneration from tissue cultures of soybean by somatic embryogenesis. p. 97–110. *In* I.K. Vasil (ed.) Cell culture and somatic cell genetics of plants. Academic Press, New York.

Reddy, M.S.S., S.A. Ghabrial, C.T. Redmond, R.D. Dinkins, and G.B. Collins. 2001. Resistance to bean pod mottle virus in transgenic soybean lines expressing the capsid polyprotein. Phytopathology 91:831–838.

Rueppel, M.L., B.B. Brightwell, J. Schaefer, and J.T. Marvel. 1977. Metabolism and degradation of glyphosate in soil and water. J. Agric. Food Chem. 25:517–528.

Russell, D.R., W.F. Swain, and J.T. Fuller. 1991. Soybean plants resistant to glutamine synthetase inhibitors. European Patent 90312498.0(430 511 A1).

Samoylov, V.M., D.M. Tucker, and W.A. Parrott. 1998a. A liquid medium-based protocol for rapid regeneration from embryogenic soybean cultures. Plant Cell Rep. 18:49–54.

Samoylov, V.M., D.M. Tucker, and W.A. Parrott. 1998b. Soybean [*Glycine max* (L.) Merrill] embryogenic cultures: The role of sucrose and total nitrogen content on proliferation. In Vitro Cell Dev. Biol.-Plant 34:8–13.

Sanford, J.C., M.J. Devit, R.F. Bruner, and S.A. Johnston. 1993a. Method and apparatus for introducing biological substances into living cells. U.S. Patent 5 204 253. Date issued: 20 Apr. 1993.

Sanford, J.C., E.D. Wolf, and N.K. Allen. 1993b. Biolistic apparatus for delivering substances into cells and tissues in a non-lethal manner. U.S. Patent 5 179 022. Date issued: 12 Jan. 1993.

Santarem, E.R., and J.J. Finer. 1999. Transformation of soybean [*Glycine max* (L.) Merrill] using proliferative embryogenic tissue maintained on semi-solid medium. In Vitro Cell Dev. Biol.-Plant 35:451–455.

Santarem, E.R., B. Pelissier, and J.J. Finer. 1997. Effect of explant orientation, pH, solidifying agent and wounding on initiation of soybean somatic embryos. In Vitro Cell Dev. Biol.-Plant 33:13–19.

Sato, S., C. Newell, K. Kolacz, L. Tredo, J. Finer, and M. Hinchee. 1993. Stable transformation via particle bombardment in two different soybean regeneration systems. Plant Cell Rep. 12:408–413.

Schmidt, E.D.L., A.J. De Jong, and S.C. de Vries. 1994. Signal molecules involved in plant embryogenesis. Plant Mol. Biol. 26:1305–1313.

Schmidt, M.A., and W.A. Parrott. 2001. Quantitative detection of transgenes in soybean [*Glycine max* (L.) Merrill] and peanut (*Arachis hypogaea* L.) by real-time polymerase chain reaction. Plant Cell Rep. 20:422–428.

Sharp, P.A., and P.D. Zamore. 2000. RNA interference. Science (Washington DC) 287:2431–2433.

Shoemaker, R., P. Keim, L. Vodkin, E. Retzel, S.W. Clifton, R. Waterston, D. Smoller, V. Coryell, A. Khanna, J. Erpelding, X. Gai, V. Brendel, C.R. Schmidt, E.G. Shoop, C.J. Vielweber, M. Schmatz, D. Pape, Y. Bowers, B. Theising, J. Martin, M. Dante, T. Wylie, and C. Granger. 2002. A compilation of soybean ESTs: Generation and analysis. Genome 45:329-338.

Simmonds, D.H., and P.A. Donaldson. 2000. Genotype screening for proliferative embryogenesis and biolistic transformation of short-season soybean genotypes. Plant Cell Rep. 19:485–490.

Srivastava, V., O.D. Anderson, and D.W. Ow. 1999. Single-copy transgenic wheat generated through the resolution of complex integration patterns. Proc. Natl. Acad. Sci. USA 96:11117–11121.

Stachel, S.E., E. Messens, M. Van Montagu, and P. Zambryski. 1985. Identification of the signal molecules produced by wounded plant cells that activate T-DNA transfer in *Agrobacterium tumefaciens*. Nature (London) 318:624-629.

Stafford, H.A. 2000. Crown gall disease and *Agrobacterium tumefaciens*: A study of the history, present knowledge, missing information, and impact of molecular genetics. Bot. Rev. 66:99–118.

Stalker, D.M., W.R. Hiatt, and L. Comai. 1985. A single amino acid substitution in the enzyme 5-enolpyruvylshikimate-3-phosphate synthase confers resistance to glyphosate. J. Biol. Chem. 260:4724–4728.

Staswick, P.E., Z. Zhang, T.E. Clemente, and J.E. Specht. 2001. Efficient down-regulation of the major vegetative storage protein genes in transgenic soybean does not compromise plant productivity. Plant Physiol. 127:1819–1826.

Steinrücken, H.C., and N. Amrhein. 1980. The herbicide glyphosate is a potent inhibitor of 5-enolpyruvylshikimate-3-phosphate synthase. Biochem. Biophys. Res. Commun. 94:1207–1212.

Stewart Jr., C.N., M.J. Adang, J.N. All, H.R. Boerma, G. Cardineau, D. Tucker, and W.A. Parrott. 1996. Genetic transformation, recovery, and characterization of fertile soybean (*Glycine max* (L.) Merrill) transgenic for a synthetic *Bacillus thuringiensis CRYIA(c)* gene. Plant Physiol. 112:121–129.

Su, Y.-H., H.-L. Wang, M.-M. Yu, D.-Y. Lu, and S.-D. Guo. 1999. Studies on transfer of Bt gene into *Glycine max*. Acta Bot. Sin. 41:1046–1051.

Suszkiw, J. 2002. Researchers develop first hypoallergenic soybeans. Agric. Res. 50:16–17.

Takagi, Y., and S.M. Rahman. 1996. Inheritance of high oleic acid content in the seed oil of soybean mutant M23. Theor. Appl. Genet. 92:179–182.

Taylor, N.B., R.L. Fuchs, J. MacDonald, A.R. Shariff, and S.R. Padgette. 1999. Compositional analysis of glyphosate-tolerant soybeans treated with glyphosate. J. Agric. Food Chem. 47:4469–4473.

Taylor, S.L. 1992. Chemistry and detection of food allergens. Food Technol. 46:146–152.

Terry, C.F., and N. Harris. 2001. Event-specific detection of Roundup Ready Soya using two different real time PCR detection chemistries. Eur. Food Res. Technol. 213:425–431.

Terry, C.F., D.J. Shanahan, L.D. Ballam, N. Harris, D.G. McDowell, and H.C. Parkes. 2002. Real-time detection of genetically modified soya using Lightcylcer and ABI 7700 plaforms with TaqMan, Scorpion, and SYBR Green I chemistries. J. AOAC Int. 85:938–944.

Thomson, J.M., P.R. Lafayette, M.A. Schmidt, and W.A. Parrott. 2002. Artificial gene-clusters engineered into plants using a vector system based on intron- and intein-encoded endonucleases. In Vitro Cell. Dev.-Plant 38:537–542.

Townsend, J.A., and L.A. Thomas. 1996. Method of *Agrobacterium*-mediated transformation of cultured soybean cells. U.S. Patent 5 563 055. Date issued: 8 Oct. 1996.

Trick, H.N., R.D. Dinkins, E.R. Santarem, R. Di, V. Samoylov, C. Meurer, D. Walker, W.A. Parrott, J.J. Finer, and G.B. Collins. 1997. Recent advances in soybean transformation. Plant Tiss. Cult. Biotechnol. 3:9-26.

Velten, J., L. Velten, R, Hain, and J. Schell. 1984. Isolation of a dual plant promoter fragment from the Ti plasmid of *Agrobacterium tumefaciens*. EMBO J. 3:2723–2730.

Waiblinger, H.U., M. Gutmann, J. Hadrich, and K. Pietsch. 2001. Validation of real-time PCR for the quantification of genetically modified soybeans. Deutsche Lebensmittel-Rundschau 97:121–125.

Walker, D.R., and W.A. Parrott. 2001. Effect of polyethylene glycol and sugar alcohols on soybean somatic embryo germination and conversion. Plant Cell Tiss. Organ Cult. 64:55–62.

Wang, M.B., and P.M. Waterhouse. 2000. High efficiency silencing of a beta-glucuronidase gene in rice is correlated with repetitive transgene structure but is independent of DNA methylation. Plant Mol. Biol. 43:67–82.

Williams, E., C.A. Emler, L.S. Julson, B. Martinell, D. E. McCabe, and Y. Huang. 2000. Soybean transformation method. Int. Nat. Patent Application WO 00/42207.

Windels, P., I. Taverniers, A. Depicker, E. Van Bockstaele, and M. De Loose. 2001. Characterisation of the Roundup Ready soybean insert. Eur. Food Res. Technol. 213:107–112.

Wright, M. 1991. Method of regenerating soybeans from cultured soybean cotyledonary nodes. U.S. Patent 4 992 375. Date issued: 12 Feb. 1991.

Wright, M.S., K.L. Launis, R. Novitzky, J.H. Duesing, and C.T. Harms. 1991. A simple method for the recovery of multiple fertile plants from individual somatic embryos of soybean [*Glycine max* (L.) Merrill]. In Vitro Cell. Dev. Biol. 27P:153–157.

Wright, M.S., D.V. Ward, M.A. Hinchee, M.G. Carnes, and R.J. Kaufman. 1987. Regeneration of soybean *Glycine max* L. Merr. from cultured primary leaf tissue. Plant Cell Rep. 6:83-89.

Xie, W., B. Wang, L. Bojun, X. Li, C. Lu, H. Qian, and S. Zhou. 1995. Introduction of wild soybean DNA into cultivated soybean and RAPD molecular verification. Sci. Chin. Ser. B. 38:1195–1201.

Xing, A.Q., Z.G. Zhang, S. Sato, P. Staswick, and T. Clemente. 2000. The use of the two T-DNA binary system to derive marker-free transgenic soybeans. In Vitro Cell Dev. Biol.-Plant 36:456–463.

Yan, B., M.S. Srinivasa Reddy, G.B. Collins, and R.D. Dinkins. 2000. *Agrobacterium* t*umefaciens*-mediated transformation of soybean (*Glycine max* (L.) Merrill.) using immature zygotic cotyledon explants. Plant Cell Rep. 19:1090–1097.

Yang, Y.-S., K. Wada, M. Goto, and Y .Futsuhara. 1991. *In vitro* formation of nodular calli in soybean (*Glycine max* L.) induced by cocultivated *Pseudomonas maltophilia*. Jpn. J. Breed. 41:595–604.

Yin, G.-C., Z.-Y .Zhu, Z.Xu, L. Chen, X.-Z .Li, and F.-Y. Bi. 1982. Studies on induction of pollen plant and their androgenesis in *Glycine max* (L.) Merr. Soybean Sci. 1:69–76.

Yin , Z., and G.-L. Wang. 2000. Evidence of multiple complex patterns of T-DNA integration into the rice genome. Theor. Appl. Genet. 100:461–470.

Yoshida, S.H. 2000. The safety of genetically modified soybeans: Evidence and regulation. Food Drug Law J. 55:193–208.

Yue, S.-X., B.-L. Liu, D.-Z. Mao, X.-B. Li, N.-B. Hu, J.-H. Fu, L.-C. Li, and L.-B. Zhu. 1990. A genetic analysis of the progenies of atrazine-resistant transgenic soybean plant. Acta Bot. Sin. 32:343–349.

Zeitlin, L., S.S. Olmsted, T.R. Moench, M.S. Co, B.J. Martinell, V.M. Paradkar, D.R. Russell, C. Queen, R.A. Cone, and K.J. Whaley. 1998. A humanized monoclonal antibody produced in transgenic plants for immunoprotection of the vagina against genital herpes. Nat. Biotechnol. 16:1361–1364.

Zhang, Z., A. Xing, P. Staswick, and T.E. Clemente. 1999. The use of glufosinate as a selective agent in *Agrobacterium*-mediated transformation of soybean. Plant Cell Tiss. Organ Cult. 56:37–46.

Zhao, L.M., D.P. Liu, H.A. Sun, Y.Yuan, and M. Huang. 1995. A sterile material of soybean gained by introducing exogenous DNA. Soybean Sci. 14:83–87.

Zhao, G., Y. Liu, A. Yin, and J. Li. 1998. Germination of embryo in soybean anther culture. Chin. Sci. Bull. 43:1991–1995.

8   Genetic Diversity in Soybean

THOMAS E. CARTER, JR.

USDA-ARS
Raleigh, North Carolina

RANDALL L. NELSON

USDA-ARS
Urbana, Illinois

CLAY H. SNELLER

The Ohio State University
Wooster,Ohio

ZHANGLIN CUI

North Carolina State University
Raleigh, North Carolina

8–1 A CONCEPTUAL FRAMEWORK FOR GENETIC DIVERSITY

Before written history, Chinese farmers used genetic diversity found in the wild to produce the crop species that we now know as domesticated soybean [*Glycine max* (L.) Merr.]. Continued farmer use of genetic diversity since domestication has improved the soybean crop to a remarkable degree, raising the yield of soybean over that of the original wild soybean by at least an order of magnitude, enhancing its resistance to an array of important diseases, and adapting the crop to grow in extreme climates. This transformation of soybean from wild plant to modern crop is one of the more remarkable accomplishments in agriculture, past or present.

Not surprisingly, the technical details of this achievement are lost to antiquity. However, historical records reveal the important fact that farmers developed many phenotypically diverse landraces in the process (landraces defined in this chapter as cultivars that predate scientific breeding). By the early 20th century, Chinese farmers were growing 20 000 to 45 000 distinct types. These landraces are irreplaceable agricultural heirlooms because they are the sum of genetic diversity that was amassed by farmers during the 3000-yr conversion of soybean from wild plant to modern crop. Many of these landraces were collected by plant explorers, germplasm curators, and breeders and are preserved today in extensive germplasm collections. Globally, hundreds of breeders use this diversity as the basis for virtu-ally all genetic improvement in the soybean crop.

In considering the larger purpose of genetic improvement of soybean, it is im-portant to appreciate that genetic crop improvement is essential to long-term food security. New cultivars provide the flexibility to meet societal food demands related to population increase and protection of the food supply from disease epidemics and climate change that may devastate the crop. Genetic improvement also serves more immediate needs of society through lowering prices for the consumer (by in-creasing supply), improving profitability for the food supplier (i.e., the farmer, through increased production efficiency), and improving the nutritional quality of the human diet. Development of new cultivars, the means of genetic improvement of the crop, has received considerable attention in agriculture for these reasons. The premise for the present review of genetic diversity in soybean is that society's re-quirement for new cultivars can be best met through comprehensive and system-atic use of the reservoir of genetic diversity that has been handed down by farmers through the millennia.

There are four main aspects of genetic diversity: formation, collection, eval-uation, and utilization. Each is covered in this review. Because the benefit of ge-netic diversity can be realized only by providing usable products to society, this re-view is oriented to utilization. Effective utilization of diversity requires both germplasm collections and applied breeding programs and, for that reason, the au-thors review diversity in each. It is a hope that the emphasis upon utilization will facilitate significant new research, and promote the best use of diversity in soybean for humankind.

It is said that the legacy of genetic diversity from the past is best applied to the future when germplasm collections and breeding programs operate in concert. Many successful examples of cooperation between breeding programs and germplasm collections are documented in this chapter. However, global germplasm collections have not been used to the full extent possible in soybean breeding, and this is a critically important concept when considering genetic diversity. The *prima facia* evidence for the under-use of germplasm collections in soybean breeding is that, although more than 170 000 soybean accessions are maintained in germplasm collections (perhaps as many as 45 000 are unique), fewer than 1000 have been used in applied breeding. Important secondary evidence is that, although extensive col-lections of the wild progenitor of soybean (*G. soja* Seib. et Zucc.) exist and can be crossed easily with soybean, no commodity and only a few specialty cultivars have *G. soja* in the pedigree.

If soybean collections have indeed been under-utilized, then one must ask why. There are five key factors that should be considered. The most important and straightforward factor is that germplasm accessions are usually inferior, agronom-ically, compared to commercial cultivars. This agronomic inferiority hinders the use of germplasm accessions as parents in breeding, simply because it increases the ef-forts required to produce an improved cultivar. For this reason, many breeders are reluctant to tap into the genetic diversity found in germplasm collections unless there is a clear and present need, such as a disease outbreak.

The second factor is that the extent and nature of favorable alleles in soybean germplasm collections are not well known, nor is the relation of such alleles to those

already used in breeding. Although essential for effective use of germplasm collections, the characterization of economically important alleles is currently under-
researched. One reason for the dearth of information on allelism in germplasm collections is that allele discovery is the primary assignment of only a few geneticists
working with soybean. Breeders do *not* have this assignment per se because their
charge is to deliver a commercial product. Collection curators do *not* have this assignment because their charge is to preserve, characterize, and distribute germplasm.
These charges usually consume most of the scientists' available resources. Systematic allele discovery in collections has not received high priority because collections were under construction for most of the past 20th century. As such, they
were not sufficiently complete to justify the devotion of scientific careers to the discovery and understanding of allelic diversity. Recently, the situation has changed
for the better, and germplasm collections now approximate the true global reservoir of soybean diversity. Their completeness establishes a compelling need for research positions in the area of allele discovery. Recent molecular advances have produced appropriate genetic tools for this work.

The third factor hindering germplasm utilization in breeding programs is that
it is a formidable task (despite molecular advances) to assess the value of alleles
for yield and other complexly inherited traits. Alleles that increase yield potential
(hereafter termed yield alleles), arguably the most important aspect of diversity for
breeders, have been difficult to identify in germplasm collections or breeding programs. The existence of yield alleles can be inferred from breeding progress (Specht
and Williams, 1984; Wilcox, 2001) and most breeders accept that landraces probably contain unique yield alleles not present in applied populations. However,
without knowledge as to which accessions hold unique yield alleles, and which genetic backgrounds might benefit from these alleles, it is a challenge to utilize exotic germplasm in yield improvement schemes. Many breeders are reluctant to com-
mit resources to the large field programs which are needed to capture yield alleles
from exotic accessions, unless they first have some evidence that unique yield alleles are actually present in those accessions. Unfortunately, extensive field testing
is required to produce the needed evidence, even with the aid of molecular tools.
Thus, more extensive field programs are needed with the explicit goal of identifying unique beneficial alleles from exotic germplasm.

The fourth factor is the need for a stronger research link between public programs and commercial companies to identify and transfer genes from germplasm
collections to cultivars. Most successful commercial cultivars in North and South
America are being developed by commercial companies [although there are notable
exceptions], while the public programs are most active in the assessment of genetic
diversity. Cooperation and interaction between the public and commercial programs
will be essential to build on the strengths of each and fully capture the value in the
germplasm collections. Such collaboration might offer a new avenue for the flow
of diverse materials to commercial breeding and also counteract the effects of current legal restrictions on germplasm exchange between commercial companies
(Carter et al., 2000). In this regard, the introduction of utility patents into North
American breeding and the logistical complexities of licensing genetic materials
have constrained almost completely what used to be free germplasm exchange
among commercial seed companies. This trend is likely to extend to public breed-

ers as they release cultivars with transgenic or other protected traits. Limited introgression of external germplasm may curtail breeding progress by decreasing effective population size and increasing relatedness of breeding materials over time (Falconer and Mackay, 1996; St. Martin, 1982; Sneller, 2003). The actual impact of reduced germplasm exchange on breeding progress remains to be seen, but is a potential problem.

A fifth factor is the need to fully exploit the concepts of genetic diversity in breeding theory and practice. The theory of selection efficiency is extensive and its application to breeding is a key to success. Plot combines, the single-seed-descent breeding method, winter nurseries, computer analysis of data, and marker assisted selection have all been well accepted in the breeding community, in part, because of their easily measured impact on selection efficiency. Although these important innovations clearly improve breeding efficiency, they all address breeding efficiency *after* hybridization of parents. Breeding efficiency as a concept needs to be extended to address breeding efficiency *before* hybridization: the selection of parents. Genetic diversity measures may provide the tools for this advance.

In a broad sense, selection of parents may range from commercial cultivars to germplasm accessions. Selection of parents, especially to improve complex traits such as yield, is often regarded as part of the 'art of breeding', because genetic theory to assess the selection of parents has been slow to advance and its hypotheses are often difficult to test. Nonetheless, recent advances in parental assessment may enable breeders to better manage diversity in their programs, better gauge the need to use exotic diversity, and, when needed, more efficiently apply exotic diversity to their programs. For example, multivariate analysis software programs and extensive data bases are now available for detailed pedigree studies of genetic diversity. Breeder-accessible molecular marker systems now allow scientists to estimate genetic diversity of germplasm. Through linkage analysis of this germplasm, the value of individual chromosome segments can be determined, thus, greatly increasing our ability to glean useful alleles (and possibly yield alleles) from exotic sources. Also, a great number of modern cultivars (products of scientific plant breeding) are now in collections from genetically near-independent breeding programs from around the world. These cultivars have been developed through decades of selection for yield and represent potential contrasting reservoirs of parents for yield alleles. These advances are providing opportunities to increase breeding efficiency through efficient selection of parents.

8–2 FORMATION OF THE GLOBAL RESERVOIR OF GENETIC DIVERSITY IN SOYBEAN

8–2.1 Domestication of Soybean

It is commonly believed that soybean was domesticated in ancient China perhaps 3000 to 5000 yr ago from wild soybean (*Glycine soja* Seib. et Zucc.) (Gai, 1997; Gai and Guo, 2001; Hymowitz and Newell, 1980, 1981). This estimate is derived in part from references to soybean that appeared in Chinese literature almost as soon as written characters were developed during the period of the Shang dynasty between 1700 to 1100 BCE. (Qiu et al., 1999). Proverbs and other oral traditions recorded during that time period clearly describe the integration of soybean into daily life and actually suggest a much older association of soybean with Chinese culture. Poems from 600 to 300 BCE mention the first specific uses of soybean as food: soup made from soybean leaves and stew made from seed. Soybean and chicken (*Gallus domesticus*) were described as the major daily food for an emperor from this period (Gai and Guo, 2001). The use of soybean as tofu and green vegetable were first documented around 1000 and 1550 CE, respectively (Gai and Guo, 2001), although tofu was probably in use by 0 CE.

Current theories suggest that soybean could have been domesticated in either southern China (Wang et al., 1973), the middle and lower Yellow River Valley of central China (Hymowitz and Newell, 1980, 1981; Xu, 1986; Chang, 1989; Guo, 1993; Zhou et al. 1998b), northeastern China (Fukuda, 1933), or simultaneously at multiple centers (Lu, 1977). No broad consensus has been reached in this matter, because wild soybean grows commonly in eastern China (from 24–53°N lat), Japan, Korea, and the eastern extremes of Russia. Theoretically, domestication of soybean could have taken place in almost any region of China. However, many researchers are comfortable with the hypothesis that soybean was domesticated in the Yellow River or Yangzi River valleys of central or southern China. Two recent studies illustrate the arguments for these areas as centers of origin.

In the first study, DNA marker data were employed to compare landraces and wild soybean accessions in order to infer domestication patterns using germplasm from the southern, central, and northeastern areas of China. The premise of the study was that *G. max* would be most closely related to the *G. soja* pool from which it was domesticated. Gai et al. (1999) evaluated 200 *G. max* and 200 *G. soja* accessions for chloroplast and mitochondrial restriction fragment length polymorphism (RFLP) markers. These organellar genomes were selected rather than the nuclear genome because their cytoplasmic nature had the presumed advantage of being uncompromised by potential cross-pollination between *G. soja* and *G. max* over millenia. Assuming that the geographical distribution of present-day *G. soja* has not changed appreciably from that of ancient times, the marker data of Gai et al. (1999, 2000) indicated that significant differentiation of wild soybean may have occurred long before domestication of the cultivated soybean. *Glycine max* accessions from all regions of China were more closely related to southern accessions of *G. soja* from the Yangzi River valley than to *G. soja* accessions from any other region. These data support the idea that *G. max* was domesticated from southern *G. soja* types and then spread to other regions (Gai et al., 1999, 2000). The marker data also indicated that the initial spread of domesticated soybean to central and northeastern China from the south may have preceded a further significant differentiation of soybean for maturity within these geographic regions which adapted the crop to spring, summer, and fall planting systems. The rationale is that DNA marker diversity among the three regions of China was much larger than that among maturity types within a region (Gai et al., 1999, 2000).

In the second study, Zhou et al. (1999) employed Vavilov's concept that the greatest genetic diversity for a species should be at the center of domestication. They evaluated 22 695 *G. max* accessions from China for 15 morphological and biochemical traits and concluded that the center of diversity for soybean resided within a corridor from southwest to northeast China which included Sichuan, Shaanxi, Shanxi, Hebei, and Shandong provinces. This corridor connected two centers of early agriculture in China, the Yellow River Valley and Yangzi River Valley. These two ancient agriculture centers have a long history of soybean and millet [*Panicum miliaceum* L. and *Setaria italica* (L.) Beauv.] cultivation, the two major food crops in ancient China, and agricultural exchange between these areas was extensive (Gai and Guo, 2001).

8–2.2 The Rise of Genetic Diversity in Soybean

8–2.2.1 Farmer Improvement of Soybean

Chinese farmers have practiced genetic selection since the domestication of soybean, by saving seed from desirable plants to sow in the following year (Gai, 1997; Qiu et al., 1999). Because soybean is extremely photoperiod sensitive, a single genotype can be grown successfully only over a narrow range in latitude (100 km, typically). Thus farmer selection of maturity variants was required to enable the spread of the crop north and south. Domesticated germplasm likely spread from its center of origin to more northern and southern regimes via trade routes, with farmers practicing selection for early or late-maturing types along the way, adapting soybean to their particular latitude. In addition, selection was practiced to adapt the introduced soybean to new cropping systems and food uses. Over the millennia, farmer-saved selections led to a wide array of genes for disease resistance, seed characters, morphological, and biochemical traits (Fig. 8–1). Farmer selection thus established a global reservoir of diversity in soybean, much of which is available to modern soybean breeders.

Distinct landraces were documented by at least 1116 CE, when Chinese authors listed soybean types that had green, brown, and black seed coats, and large and small seed (Gai and Guo, 2001). By the beginning of the 20th century, at least 20 000 landraces may have been in existence in China (Chang et al., 1999) and perhaps more than 40 000 based on current inventories of germplasm collections. When systematic soybean germplasm collection efforts began early in the last century, many unique landraces could be found within relatively small areas of China. The use of DNA markers has also demonstrated strong distinctions among landraces from diverse geographic origins within China (Li and Nelson, 2001). The preservation of local and regional diversity patterns to the present indicates that landrace exchange between neighbors and neighboring regions may have been limited in early Chinese agriculture. Once soybean was spread and became an established crop, such restricted exchange may have helped to preserve the genetically diverse types which are found today. Outcrossing between adjacently grown landraces, although infrequent, may have provided the needed genetic recombination for continued on-farm selection and plant improvement.

Currently, agronomic and morphological genetic diversity in germplasm collections is greater for *G. max* than for *G. soja*, although molecular marker diversity is greater for *G. soja*. The explanation for greater agronomic diversity in *G. max* is not obvious, given that the initial domestication of *G. max* probably caused a genetic bottleneck (defined as a drastic reduction in population size, Hartl and Clark,

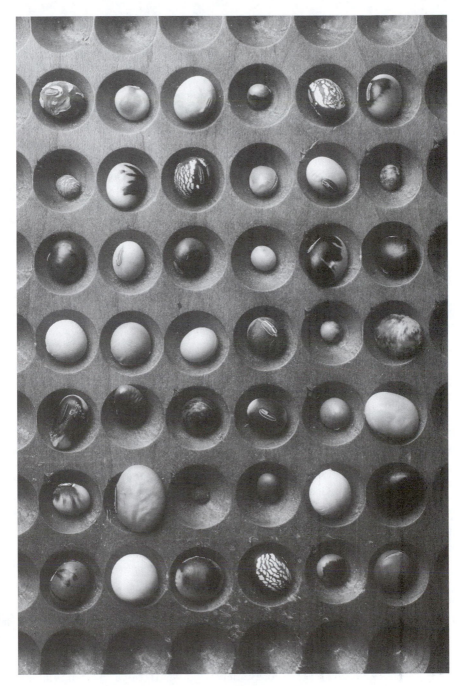

Fig. 8–1. Counting board containing seed of *Glycine max* genotypes that vary in color and size. Reprinted with permission of the University of Illinois.

1997) for diversity in *G. max* relative to *G. soja* (consistent with DNA marker results). However, farmer selection is presumed to play a key role. One possibility is that the number of domesticated soybean plants growing in Asia may have been much greater than for *G. soja,* once soybean became a common part of the Asian diet. This larger domesticated population may have provided a greater opportunity for natural mutation events than in *G. soja.* Coupled with the ability of farmers to save agronomic and morphological mutants (mutants which would probably have been lost had they occurred in *G. soja*), soybean production may have led naturally to increased agronomic diversity over that observed in *G. soja.* Farmers probably selected and saved mutations not only visually, but, perhaps, unconsciously as well through normal harvest of the crop. The rationale is that agronomically useful mutants would have had an advantage in the field and, therefore, by producing more seed than typical plants, would have had the opportunity to proliferate. In contrast to agronomically useful mutations, those with neutral selective advantage, such as many DNA markers, may have been impacted less by humankind's husbandry of the soybean, which could account for the reduced variability in DNA markers in *G. max* in comparison to *G. soja* (again associated with the genetic bottleneck from domestication). A second alternate possibility in explaining the greater agronomic and morphological diversity in *G. max* relative to *G. soja* is that alleles with the potential to affect agronomic productivity resided in *G. soja*, but had no opportunity to express, phenotypically, because of the diminutive *G. soja* plant type (i.e., epistatic interactions among genes limited their effects). After domestication, agronomic productivity in *G. max* may have increased through farmer selection to the point where these alleles had an opportunity to express and affect agronomic growth. Research to test these alternate hypotheses is lacking. Agronomic diversity in maize [*Zea mays* (L.)] is also larger than its wild relatives (Brown and Goodman, 1977; Wilkes, 1967; Sanchez and Ordaz, 1987).

8–2.2.2 Modern Breeding

Modern soybean breeding based on manual cross-pollination began about 1927 in China at the Gongzhuling Agricultural Experiment Station (now Jilin Academy of Agricultural Sciences). Two landraces were hybridized to develop Man Cang Jin, which was released in 1941 (Cui et al., 1999). In North America, the earliest breeding efforts are not well documented. Breeders in the Midwest and Canada were clearly interested in cross-pollination as a breeding technique as early as 1902. By 1922, Ogemaw and Mamloxi were released as products from presumed natural pollination events in the southern USA (Bernard et al., 1988a). Pagoda, released in 1939 in Ontario, Canada may have been the first modern cultivar developed from a documented manual cross-pollination. C.M. Woodworth, a USDA researcher located at the Univ. of Illinois, was the first to describe the technique for manual cross-pollination of soybean (Woodworth, 1932) and he developed the cv. Chief and Lincoln (released in 1940 and 1943, respectively) using this technique. Modern breeding efforts began in Japan shortly after World War II (Miyazaki et al., 1995a). By 2000, breeders had released approximately 800 public cultivars in China, 100 public cultivars in Japan, and more than 600 public and 2000 proprietary cultivars in Canada and the USA (Cui et al, 1999; Zhou et al., 2002; Carter et al., 1993).

Breeders in Brazil, Australia, Korea, India, and Argentina have released more than 400 cultivars in total. Collectively, these cultivars trace to about 700 of the approximately 40 000 estimated unique accessions that are preserved today in the germplasm collections in the USA, China, and Japan (Miyazaki et al., 1995a; Palmer et al., 1996; Xu et al., 1999b; Cui et al. 1998, 1999).

8–3 STATUS OF GLOBAL SOYBEAN GERMPLASM COLLECTIONS

Germplasm is the sum total of living organisms. For a plant breeder, germplasm refers to the genetic resources that are available to improve a cultivated or potentially cultivated crop. This usually means the genetic diversity within a particular crop species plus a set of related species. Because species are based on subjective criteria, they suffer from taxonomic change over time and do not always provide the best germplasm classification for breeders. Harlan and de Wet (1971) recognized this problem and proposed a gene pool system for cultivated plants that defines three levels of genetic resources based on breeding difficulty. The primary gene pool is similar to the biological definition of a species and includes all types that can be crossed easily and produce fertile progeny. For soybean, this includes the two annual species, *Glycine max* (Table 8–1, Fig. 8–1) and its progenitor *Glycine soja* (Table 8–2 and Fig. 8–2). These two species have the same chromosome number ($2n = 40$) and represent a continuum of phenotypes. Accessions at the boundary between *G. max* and *G. soja* can be difficult to classify, and as a result a third species (*Glycine gracilis*) was once defined to encompass this boundary area (Skvortzow, 1927). Hermann (1962) did not consider it to be a separate species, but the *Glycine gracilis* designation is still used in some literature. These intermediate forms are labeled as semi-wild types in some germplasm collections. The primary gene pool *G. max* and *G. soja* includes well over 95% of the ex situ

Table 8–1. The major *Glycine max* germplasm collections. The data in this table were gathered from the database maintained by the International Plant Genetic Resources Institute (http://www.ipgri.org/). Some numbers were updated (at press time) via direct contact with the holding institutions.

Institution	Country	Accessions
		no.
Institute of Crop Germplasm Resources, CAAS#	China	23 578
USDA Soybean Germplasm Collection	USA	18 076
Asian Vegetable Research and Development Centre (AVRDC)	Taiwan	12 508
Soybean Research Institute, Nanjing Agricultural University	China	10 000
Department of Genetic Resources I, National Institute of Agrobiological Resoures	Japan	8 630
Institute of Agroecology and Biotechnology	Ukraine	7 000
N.I. Vavilov Research Institute of Plant Industry	Russia	6 126
Centro Nacional de Pesquisa de Recursos Genéticos e Biotec. (CENARGEN)	Brazil	4 693
Soybean Research Institute, Jilin Academy of Agricultural Science	China	4 200
All India Coordinated Research Project on Soybean, Govind Bal. Plant University	India	4 015
Centro Nacional de Pesquisa de Soja (CNPSO) Embrapa††	Brazil	4 000

(continued on next page)

Table 8–1. Continued.

Institution	Country	Accessions
		no.
Crop Experiment Stn. Upland Crops Research Div.	Korea, Republic of	3 678
Australian Tropical Crops Genetic Research Centre	Australia	3 144
Genebank, Institute for Plant Genetics and Crop Plant Res. (IPK)	Germany	3 063
Regional Station National Bureau of Plant Genetic Resources (NBPGR)	India	2 808
Taiwan Agricultural Research Institute (TARI)	Taiwan	2 699
National Research Centre for Soybean	India	2 500
Crop Breeding Institute DR & SS	Zimbabwe	2 236
Sukamandi Research Institue for Food Crops (SURIF)	Indonesia	2 194
Instituto Agronômico de Campinas (I.A.C.)	Brazil	2 000
National Plant Genetic Resources Lab. IPB/UPLB	Philippines	1 764
CSIRO, Division of Tropical Crops and Pastures‡‡	Australia	1 600
Genetic Resources Department—Research Institute for Cereals and Ind. Crops	Romania	1 600
G.I.E. Amelioration Fourragere	France	1 582
Soyabean Research Institute, Heilongjiang Academy of Agricultural Science	China	1 558
Institute of Oil Crops Research, CAAS	China	1 529
Institute of Plant Breeding, College of Agriculture, UPLB	Philippines	1 508
Instituto Nacional de Investigaciones Agricolas, Station de Iguala	Mexico	1 500
Stat. De Genetique et Amelioration des Plantes INRA C.R. Montpellier	France	1 404
Kariwano Lab., Tohoku National Agricultural Experiment Station	Japan	1 400
Hokkaido Agricultural Experiment Station	Japan	1 383
International Institute of Tropical Agriculture	Nigeria	1 358
Centro de Investigación La Selva (CORPOICA)	Colombia	1 219
Institute of Crop Breeding and Cultivation CAAS#	China	1 200
Institute for Field and Vegetable Crops	Yugoslavia	1 200
Institute of Industrial Crops Jiangsu Academy of Agric. Sci.	China	1 199
Corporacion Colombiana de Investigacion Agropecuaria (CORPOICA)	Colombia	1 170
Genebank Cereal & Oil Crops Inst. Hebei Academy of Agricultural Sciences	China	1 154
Instituto Nacional de Investigaciones Forestales Agrícolas y Pecuarias (INIFAP)	Mexico	1 124
Maharashtra Association for the Cultivation of Science	India	1 081
Total		156 849

† UPLB = University of the Philippines at Los Baños.
‡ IPB/UPLB = Institute of Plant Breeding/University of the Philippines at Los Baños.
§ DR & SS = Department of Research and Specialist Services.
¶ INRA = Institut National de la Recherche Agronomique.
# CAAS = Chinese Academy of Agricultural Science.
††Embrapa = Empresa Brasileira de Pesquisa Agropecuaria.
‡‡ CSIRO = Commonwealth Scientific and Industrial Research Organization.

Glycine germplasm in the world. Harlan and deWit (1971) defined the secondary gene pool as other species that will cross with the crop species, but with great difficulty. For example, the secondary gene pool for maize includes the *Tripsacum* species, and for wheat [*Triticum aestivum* (L.)] includes the *Aegilops* species. There is no secondary gene pool for soybean. The tertiary gene pool includes species which cross to the crop species, but the resulting hybrids are either not viable or infertile. Only through extraordinary procedures, such as embryo culture,

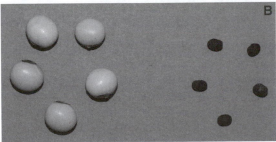

Fig. 8–2. A: *Glycine max* and staked plants of *G. soja* growing at Clayton, NC in August of 1999. B: Mature seed representing *G. max* and *G. soja*.

is gene transfer possible. The tertiary gene pool for soybean includes the perennial *Glycine* species (Table 8–3). One of these distantly related species (*G. tomentella*) has been successfully crossed to soybean (Singh et al., 1990, 1993). With the advent of techniques for creating transgenic plants, the tertiary gene pool can now include any organism, limited only by technology and societal acceptance. However, the traditional definition of the tertiary gene pool continues to have importance in the management of germplasm collections. In this section, genetic diversity is reviewed for the primary and tertiary gene pools of soybean.

8–3.1 Global *Glycine max* Germplasm Collections

According to data collected by the International Plant Genetic Resources Institute (IPGRI) (2001) and augmented with updated data, more than 170 000 *Glycine max* accessions are maintained by more than 160 institutions in nearly 70

Table 8–2. The major *Glycine soja* germplasm collections. The data in this table were gathered from the database maintained by the International Plant Genetic Resources Institute (http://www.ipgri.org/). Some numbers were updated (at press time) via direct contacts with the holding institutions.

Institution	Country	Accessions
		no.
Institute of Crop Germplasm Resources, CAAS†	China	6 172
USDA Soybean Germplasm Collection	United States	1 114
Soybean Research Institute, Nanjing Agricultural Univ.	China	1 000
Soybean Research Institute, Jilin Academy of Agric. Sci.	China	600
Soyabean Research Institute, Heilongjiang Academy of Agricultural Sciences	China	400
Crop Experiment Station Upland Crops Research Division	Korea, Republic of	342
Asian Vegetable Research and Development Centre (AVRDC)	Taiwan	339
N.I. Vavilov Research Institute of Plant Industry	Russia	310
Breeding Laboratory Faculty of Agriculture, Iwate University	Japan	151
CSIRO, Division of Tropical Crops and Pastures‡	Australia	60
Taiwan Agricultural Research Institute (TARI)	Taiwan	46
Hunan Academy of Agriculture Sciences	China	45
Tieling District Agricultural Research Institute	China	29
Department of Agronomy National Chung Hsing University	Taiwan	20
Eastern Cereal & Oilseed Research, Centre Sasketoon Research Centre	Canada	18
Soybean Breeding Laboratory, Tokac. Agricultural Experiment Station	Japan	15
Instituto Nacional de Investigaciones Forestales Agrícolas y Pecuarias (INIFAP)	Mexico	9
All India Coordinated Res. Project on Soybean Govind Bal. Plant University	India	7
Maharashtra Association for the Cultivation of Science	India	6
Sukamandi Research Institute for Food Crops (SURIF)	Indonesia	4
Research Institute for Food Crops Biotechnology (RIFCB)	Indonesia	4
Kariwano Laboratory Tohoku National Agricultural Experiment Station	Japan	3
Genebank Institute for Plant Genetics and Crop Plant Research (IPK)	Germany	2
S.K. University of Agricultural and Technology	India	1
Total		10 697

† CAAS = Chinese Academy of Agricultural Science.
‡ CSIRO = Commonwealth Scientific and Industrial Research Organization.

countries (Table 8–1). China has the largest soybean germplasm collection in the world with nearly 26 000 accessions of *G. max* (Wang, 1982; Chang and Sun, 1991; Chang et al., 1996b, 1999) housed at the Institute of Crop Germplasm Resources of the Chinese Academy of Agricultural Science in Beijing (Table 8–4). Approximately 2000 of the lines in this collection originated from outside of China and most of those are modern cultivars or genetic stocks from the USA. The USDA Soybean Germplasm Collection is the second largest collection and is described in a later section. The Asian Vegetable Research and Development Centre (AVRDC) has the third largest collection with more than 14 000 accessions. The AVRDC collection has been assembled from many countries with approximately 40% obtained from the USDA collection. Nanjing Agricultural University maintains more than 10 000 soybean accessions collected from southern China, of which 70% are common with the larger collection in Beijing. The Institute of Agroecology and Biotechnology

Table 8–3. The major perennial *Glycine* collections. The data in this table were gathered from the database maintained by the International Plant Genetic Resources Institute (http://www.ipgri.org/). Some numbers were updated (at press time) *via* direct contacts with the holding institutions. Number of accessions are reported by country and in some cases there may be more than one collection per country.

Species	Country							
	Australia	USA	South Africa	Taiwan	Russia	Japan	United Kingdom	Total
G. albicans	5							5
G. aphyonota	1							1
G. arenaria	6	3						9
G. argyrea	16	12						28
G. canescens	222	119	1	2	3		1	348
G. clandestina	411	116	7	3	6	5		548
G. curvata	10	6			1			17
G. cyrtoloba	51	44			1			96
G. falcate	54	26		2		1		83
G. hirticaulis	8							8
G. lactovirens	10							10
G. latifolia	120	43	21		2			186
G. latrobeana	19	6			1			26
G. microphylla	211	34	50					295
G. peratosa	1							1
G. pindanica	5	1						6
G. pullenii	4							4
G. rubiginosa	53							53
G. stenophita	42							42
G. tabacina	303	229	111	4	13	15		675
G. tomentella	493	279	113	5	4	6		900
Glycine spp.	139	1	1	53				194
Total	2184	919	304	69	31	27	1	3535

Table 8–4. Geographical origin of soybean accessions collected and maintained in the National Gene Bank in Beijing, China.

Province of origin	*Glycine max*	*Glycine soja*	Province of origin	*Glycine max*	*Glycine soja*
	no.			no.	
Anhui	1 084	117	Liaoning	1 348	1 100
Beijing	97	5	Neimenggu (Inner Mongolia)	310	85
Fujian	591	370	Ningxia	107	24
Gansu	350	90	Shandong	1 134	120
Guangdong	345	17	Shanxi	2 282	424
Guangxi	590	90	Shaanxi	1 035	400
Guizhou	2 068	86	Shanghai	90	
Hebei	1 249	44	Sichuan	2 069	35
Henan	626	305	Taiwan	12	
Heilongjiang	960	739	Xinjiang	42	
Hubei	1 679	70	Xizang (Tibet)	20	11
Hunan	554	65	Yunnan	582	2
Jilin	1232	878	Zhejiang	927	166
Jiangsu	1 782	110			
Jiangxi	422	64			
			Total	23 587	5 417

Source: Chang et al. (1999).

(Ukraine), the N.I. Vavilov Research Institute of Plant Industry (Russia), the Department of Genetic Resources in the National Institute of Agrobiology Resources (Japan), the Crop Experiment Station Upland Crops Research Division (Korea), the Australian Tropical Crops Genetic Research Center, the National Bureau of Plant Genetic Resources (India), and the National Research Centre for Soybean (India) have somewhat smaller but genetically important collections. Some of the South American collections are also large, but were established rather recently. Records indicate that South American accessions were derived almost exclusively from other collections. The European collections are small but may be genetically important because they contain unique landraces or their derivatives that were introduced from Asia more than 100 yr ago.

Accessions that predate scientific plant breeding are probably the most diverse genetic resources in *G. max* collections and are a good measure of the effective size of a collection. Not all *G. max* germplasm currently preserved in east Asian countries predates scientific plant breeding, but estimates based upon the IPGRI database indicate that approximately 40 000 of the approximately 93 000 soybean accessions in Asia may fit this category. Accessions that fit this description originated in east Asia from an area bounded roughly by India on the west, Japan on the east, Indonesia on the south, and Russia on the north. At the center of this region is China, the most important source of accessions.

Thousands of *G. max* germplasm accessions are in collections maintained outside of Asia and many of these soybean accessions were introduced to Europe and North America from China, Japan, Korea, and Russia in the 19th and early 20th century, before formal collections were established in the originating countries. It is likely that some of these accessions are not in the collections maintained in Asian countries. Two DNA marker studies lend support to this theory. In the first, Li et al. (2001b) determined that the genetic base of modern Chinese cultivars is quite distinct from that of North American cultivars, even though most of the ancestral lines of both countries originated from China. In the second, a DNA marker analysis of accessions used in a U.S. breeding revealed that some Asian accessions obtained from European and African collections were genetically distinct from any that were obtained directly from Asia (Thompson et al., 1998; Brown-Guedira et al., 2000). Guided by these results, we used records from the USDA Soybean Germplasm Collection and IPGRI to estimate that as many as 7000 of the accessions currently held outside of Asia are traditional Asian cultivars (or their direct descendents) not currently preserved within Asia. Using this estimate, the total *G. max* germplasm collection of unique Asian landraces worldwide is approximately 47 000 accessions. This is *only* an approximation, but it suggests that perhaps more than two-thirds of the total current world holdings are duplicated accessions. Duplication is not bad because it provides security against loss and makes the soybean germplasm more readily available to those who can use it. However, recognizing where duplication exists is critical for effective management and use of germplasm.

There are a few remote places in Asia where primitive *G. max* cultivars can still be collected in farmers' fields. However, such cultivars in China and elsewhere have largely been replaced by modern cultivars produced through scientific plant breeding. Chinese scientists have systematically and extensively collected *G. max* germplasm throughout their country in the past 50 yr. Recent research has documented the existence of large inter- and intra-provincial variation within China (Li and Nelson, 2001), so landraces not in current collections should be collected and preserved. Outside of China, genetic diversity is generally less than, but often distinct from, that observed in China (Li and Nelson, 2001). Collecting in southern Asia from India to Vietnam may yield diverse germplasm not already preserved, but opportunities will likely be available for only a short time because modern cultivars are becoming more common in these areas. Future collecting efforts might be expected to add thousands but not likely 10s of thousands of new unique accessions. Thus, the current global collection of *G. max* germplasm will probably not increase substantially in the future. Germplasm exchange among existing collections will remain important, but intellectual property rights issues have the potential to make the exchange process more cumbersome.

8–3.2 Global *Glycine soja* Germplasm Collections

According to the IPGRI database, there are fewer than 10 000 *G. soja* accessions currently within ex situ collections and perhaps no more than 8500 unique accessions (Table 8–2). As with the *G. max* collection, the Institute of Crop Germplasm Resources in China holds the largest *G. soja* collection with approximately 6200 accessions (Li, 1990, 1994). All other collections are relatively small. The USDA Soybean Germplasm Collection contains more than 1100 accessions. Nanjing Agricultural University maintains about 1000 accessions collected from southern China, but about 70% are in common with the Beijing collection. The Crop Experiment Station Upland Crops Research Division, (Korea), the Asian Vegetable Research and Development Center (Taiwan), and the N.I. Vavilov Research Institute of Plant Industry (Russia) each have more than 300 accessions.

Glycine soja collections are considerably smaller than *G. max* collections. However, larger collections of *G. soja* may be critical in the future to the long-term genetic improvement of the soybean. As genetic technology improves, it will be easier to identify and isolate economically important genes from *G. soja*. Additional collection of *G. soja* in Russia, China, Korea, and Japan is warranted and could preserve significant new diversity.

8–3.3 Global Perennial *Glycine* Germplasm Collections

Currently, 23 perennial *Glycine* species have been documented and more than 3500 accessions are maintained world-wide in nine collections (Table 8–3). The seven species [*G. clandestina* J.C. Wendl., *G. falcata* Benth., *G. latifolia* (Benth.) C. Newell & Hymowitz, *G. latrobeana* (Meissner) Benth., *G. canescens* F.J. Herm., *G. tabacina* (Labill.) Benth., and *G. tomentella* Hayata] available when the previous monograph was published are summarized by Newell and Hymowitz (1980). *Glycine argyrea* Tindale and *G. cyrtoloba* Tindale were named in 1984 (Tindale, 1984); *G. arenaria* Tindale, *G. microphylla* (Benth.) Tindale, and *G. curvata* Tindale were named in 1986 (Tindale, 1986a, 1986b); *G. albicans* Tindale & Craven, *G. lactovirens* Tindale & Craven, *G. hirticaulis* Tindale & Craven were named in 1988 (Tindale and Craven, 1988), *G. dolichocarpa* Tateishi & Ohashi was named in 1992 (Tateishi and Ohashi, 1992); *G. pindanica* Tindale & Craven was named in 1993 (Tindale and Craven, 1993); *G. stenophita* B. Pfeil & Tindale was named in 2000 (Doyle et al., 2000); *G. peratosa* B. Pfeil & Tindale and *G. rubiginosa* Tindale & B. Pfeil were named in 2001 (Pfeil et al., 2001); and *G. aphyonota* B. Pfeil and *G. pullenii* B. Pfeil, Tindale & Craven were named in 2002 (Pfeil and Craven, 2002). Some accessions previously thought to be *G. tabacina* have been reclassified as *G. pescadrensis* Hayata (Hayata, 1919) by taxonomists in Australia. For more details on *Glycine* speciation see Chapter 4 (Hymowitz and Singh, 2004, this publication). All of the species are native to Australia, and Australia has the largest collection of perennial *Glycine* with more than 2100 accessions at the CSIRO Division of Plant Industry. The USDA Soybean Germplasm Collection has approximately 1000 accessions but more than 90% are duplicates of the Australian collection. There are smaller collections in South Africa, Japan, Taiwan, Russia, and the United Kingdom.

8–4 THE USDA SOYBEAN GERMPLASM COLLECTION

8–4.1 History

Nearly 8300 documented soybean introductions entered the USA between 1898, when records were first kept, and 1949 when the USDA Soybean Germplasm Collection was established. More than 5000 of these lines came from the plant collecting expeditions of P. H. Dorsett and W. J. Morse in China, Korea, and Japan between 1924 and 1932 (Dorsett, 1927; Bernard et al., 1988a; Piper and Morse, 1910, 1923, and 1925). Only 1677 of these soybean accessions remained in individual U.S. research programs in 1949 and they (along with a few soybean genotypes introduced prior to 1898) became the initial USDA collection. The explanation for the loss of so many accessions is that during most of the first half of the 20th century, soybean was a minor crop in the USA and introduced germplasm was not regarded as a reservoir of diversity to be saved, but rather as applied breeding material. Accessions which appeared to have value were selected and the rest were discarded. The oldest known introduction extant in the current collection is Mammoth Yellow which was named sometime before 1895 (Bernard et al., 1988a) after being selected from seed that were probably introduced from Japan prior to 1880.

After 1949, the number of accessions in the collection increased at a moderate but steady rate and more than doubled in size in the next 25 yr (Fig. 8–3). The southern leaf blight epidemic (caused by *Helminthosporium maydis,* Nisikado & Miyake) (Sprague, 1971) in maize in 1970 focused national attention on the genetic base of U.S. crops. Two National Academy of Science reports resulted in the elevation of the curation of genetic resources to a higher priority (Committee on Genetic Vulnerability, National Academy of Science, 1972; Conservation of Germplasm Resources, An Imperative, 1978). This national focus, accompanied by the establishment of working relationships with soybean scientists in Asia, led to a 4.5-fold increase in the Collection numbers during the next quarter century, from just more than 4000 *G. max* accessions in 1974 to more than 18 000 in 2001 (Fig.8–3). All of the accessions within the USDA Soybean Germplasm Collection are freely available to scientists worldwide. Some modern cultivars in the collec-

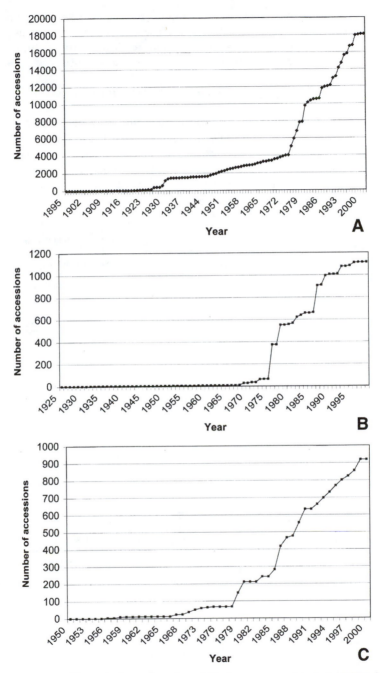

Fig. 8–3. Panel 1A: Change in the total number of *Glycine max* accessions in the USDA Soybean Germplasm Collection from 1895 to 2000. Panel 1B: Change in the total number of *G. soja* accessions in the USDA soybean germplasm collection from 1925 to 2000. Panel 1C: Change in the total number of perennial *Glycine* accessions in the USDA soybean germplasm collection from 1950 to 2000.

tion carry plant variety protection and cannot be commercialized without permission of the developer. However, these lines can be used for research and as parents in crosses. Orders for germplasm can be placed through the National Plant Germplasm System (http://www.ars-grin.gov/npgs/orders.html).

8–4.2 Organization of the USDA Collection

Introduced *Glycine max* is the largest, but is only one of the five *Glycine* subcollections maintained by the USDA (Tables 8–5 and 8–6). The other four are *Glycine soja*, perennial *Glycine* species, genetic stocks, and domestic cultivars. Only six G. *soja* accessions were available when the USDA collection was established. A rapid expansion that began in 1975 increased the G. *soja* collection from 65 to

Table 8–5. Numbers of *Glycine max* accessions, by maturity group, that are maintained in the USDA-ARS Soybean Germplasm Collection. Small seed quantities are freely available upon request from the National Plant Germplasm System (http://www.ars-grin.gov/npgs/orders.html).

Maturity Group	Glycine max	Glycine soja
000	140	109
00	475	50
0	1 113	53
I	1 502	61
II	1 744	95
III	1 674	51
IV	3 618	85
V	2 618	360
VI	1 308	165
VII	750	78
VIII	796	1
IX	717	3
X	111	3
Total	16 566	1 114

Source: R. Nelson (personal communication, 2002).

Table 8–6. Geographical origin of *Glycine max* and *Glycine soja* accessions maintained in the USDA-ARS Soybean Germplasm Collection. Small seed quantities are freely available upon request from the National Plant Germplasm System (http://www.ars-grin.gov/npgs/orders.html).

Country	Glycine max	Glycine soja
China	6 226	198
Japan	2 979	294
Korea	3 614	350
Russia	644	272
Europe	1 087	-
Africa	157	-
Americas	1 596	-
Australia	10	-
Other Asian countries	1 648	-
Unknown	153	-
Total	18 114	1 114

Source: R. Nelson (personal communication, 2002).

more than 1100 accessions curently. The USDA perennial *Glycine* collection grew from six species in 1975 to a current 13 species and approximately 1000 accessions in 2001 (Hymowitz and Newell, 1980). Most of the growth in the subcollections of introduced *G. soja and G. max* came from exchanges of germplasm with other collections, whereas growth in the perennial collection came from exploration in Australia, Taiwan, and South Pacific islands (Hymowitz, 2001). Additional perennial species are likely to be added to the collection in the future.

All accessions of the introduced annual species in the USDA Soybean Germplasm Collection are maintained as pure lines. Each accession in the collection descended from a single seed from the original seed lot and multiple accessions are preserved from introduced heterogeneous samples. The only genetic variation that can exist within an accession is the result of heterozygosity in the original progenitor seed. Thus, genetic drift and natural selection are not factors in the maintenance of the collection. Both *G. max* and *G. soja* are self-pollinated, so accessions within the collection can be assumed to be homozygous and homogeneous. Having genetically uniform accessions in the collection greatly facilitates evaluation. It is also critical in maintaining integrity of the accessions. Extensive descriptive data on morphology and pigmentation allow most contaminants to be easily detected and removed.

8–4.3 Domestic Cultivar Collection

The domestic cultivar collection is divided into three sections: old (named or released prior to 1945), modern (released after 1945), and proprietary. There are 207 named North American cultivars in the old collection, including Asian landraces and introduced and domestic cultivars of uncertain ancestry. There are currently 474 public North American cultivars in the modern collection and 31 cultivars in the proprietary subcollection. The proprietary subcollection was established in 1990 to preserve selected, proprietary cultivars that have unique pedigrees, were widely used in the USA, or were the basis of published research. Proprietary cultivars that were part of the domestic cultivar collection prior to 1990 were included without regard to the three criteria for preservation.

8–4.4 Genetic Stock Collection

The genetic stock collection consists of the type collection, the isoline collection, and germplasm releases registered in Crop Science. The current purpose of the type collection is to preserve mutations not available in other germplasm accessions. Some of the older lines in this subcollection do not meet the present criteria but have been preserved for their historical value. There are 178 lines in this collection and more than 75% of the mutations have been genetically characterized (Palmer et al., 2004, this publication). Chlorophyll mutations are the most common with 67 entries, followed by mutations affecting fertility (36), seed pigmentation (30), and leaflet morphology (12). Mutants preserved in the type collection are critical for allelism tests, which are necessary in the identification of new loci and are used in other research to determine the effects of these unusual phenotypes.

8–4.5 Isoline Collection

The isoline collection contains nearly 600 near-isogenic lines produced by backcross introgression of more than 80 genes into one or more of 11 recurrent parents (Table 8–7). The present collection was developed almost entirely by R. L. Bernard, the former curator of the USDA Soybean Germplasm Collection and research geneticist with USDA-ARS. The most common recurrent parents are Clark (294), Harosoy (139), and Williams (100). This subcollection has been widely distributed and used by many soybean scientists, and continues to be an important research tool. Applications of isolines to research include the following. Isolines differing in maturity genes have been used to determine the effect of maturity on morphological and phenotypic traits (Curtis et al., 2000; Hartung et al., 1981), intergenotypic competition (Wilcox and Schapaugh, 1978), the relationships between reproductive development and N metabolism (Nelson-Schreiber and Schweitzer, 1986), maturity effects arising from E (maturity) alleles present in cultivars (McBlain et al., 1987; Cober et al., 1996), and the relationships between maturity and date of planting on yield and vegetative growth (Pfeiffer and Harris, 1990). The effect of low temperature on seed coat discoloration and cracking has been ascertained using near-isogenic lines differing in hilum color, pubescence color, and maturity (Takahashi, 1997; Morrison et al., 1998; Takahashi and Abe, 1999; Tang et al., 2002). Isolines differing in stem termination and leaf morphology have been examined for effects on yield (Hartung et al., 1980), nodal pod distribution (Hartung et al., 1981), microclimate in the crop canopy (Baldocchi et al., 1985), and to test for traits that may improve adaptation to planting date or row spacing (Raymer and Bernard, 1988; Robinson and Wilcox, 1998). Non-nodulating isolines have been used to compare differential responses to sources of N (Ham et al., 1975; Israel et., 1985), to estimate N loss from nodulated roots (Burton et al, 1983), nodulating patterns of different *Bradyrhizobium* strains (Payne and Pueppke, 1985), to estimate N_2 fixation rates (Vasilas et al., 1990), and to identify biochemical factors that may affect the nodulation process (Suganuma and Satoh, 1991). Isolines that differ in

Table 8–7. A list of the most common traits in the near-isogenic line collection, a component of the USDA soybean germplasm collection. Novel traits, as they were discovered, were backcrossed into at least one and as many as seven common cultivar backgrounds to facilitate trait study. These near-isogenic lines have also been used extensively as diagnostic research tools. Small seed quantities are freely available upon request from the National Plant Germplasm System (http://www.ars-grin.gov/npgs/orders.html).

Trait	No. of near-isogenic lines
Leaf morphology	24
Maturity	30
Chlorophyll	33
Stem growth	38
Pubescence	48
Disease resistance	65
Pigmentation	125
Gene combinations	189
Total	552

Source: R. Nelson (personal communication, 2002).

pubescence density and type have been used to determine pubescence effects on plant growth (Clawson et al., 1986a) and yield (Specht et al., 1985; Zhang et al., 1992), estimate effects on water-use efficiency and canopy radiation balance (Clawson et al., 1986b; Specht et al., 1986; Nielson et al., 1984; Baldocchi et al., 1983a, 1993b), measure the presence and impact of potato leafhopper (*Homoptera*: *Cicadellidae*) (Hammond and Jeffers, 1990), determine the influence on oviposition preference by insects (Lambert and Kilen, 1989; Lambert et al., 1992), measure the effects on Mexican bean beetle (*Epilachna varivestis* (Mulsant)) feeding (Gannon and Bach, 1996), and determine the impact on soybean mosaic virus (SMV) spread (Ren et al., 2000). Genetic male-sterile isolines have been used to manipulate the source-sink ratio to determine its effect on N_2 fixation (Riggle et al., 1984) and produce research quantities of hybrid soybean seed (Nelson and Bernard, 1984). Isolines have been used to identify biochemical compounds or enzymes that may be associated with specific phenotypes (Birnberg et al., 1987; Fasoula et al., 1995). Isolines that differ in disease resistance alleles have been used to quantify disease damage (Wilcox and St. Martin, 1998). Near isogenic lines have also been used to associate molecular markers with phenotypic and biochemical traits (Muehlbauer et al., 1988, 1989, 1991; Maughan et al., 1996).

8–4.6 Registered Germplasm Releases

All germplasm releases registered in the journal *Crop Science* are preserved, with approximately 140 such lines or populations currently being maintained. Some of these releases are isolines, but most are improved experimental lines that have economically or scientifically important genes or combination of genes generally transferred from exotic germplasm. These lines are generally not agronomically acceptable as cultivars, but are valuable as parental lines. New genes for disease or insect resistance or improved seed composition are common reasons for registering germplasm and making it available to the scientific community. Germplasm releases often do not have economic value over long periods of time because new lines with better agronomic characteristics and/or improved function are developed and released, but preserving these lines is important for potential future research.

8–5 GENETIC DIVERSITY IN GERMPLASM COLLECTIONS

8–5.1 Phenotypic Diversity of *Glycine max*

The number of accessions has been a traditional measure of the importance of germplasm collections, but as collections become larger and duplication increases, it is not necessarily the best measure. The goal of germplasm preservation is to maintain maximum genetic diversity, thus the degree of genetic diversity is the ultimate measure of the value of a collection. For decades, genetic diversity has been estimated by phenotypic diversity, and this is still an important aspect of germplasm management and evaluation. Phenotypic diversity in soybean is extensive and is under both qualitative and quantitative genetic control.

8–5.1.1 Qualitative Phenotypic Traits

Qualitative traits provide convenient markers for identifying accessions and classifying differences. For example, more than 20 different categories describe seed coat color and 11 categories are used to describe pubescence form and density (Nelson et al., 1987, 1988; Coble et al., 1992; Bernard et al., 1998; Hill et al., 2001). The range and distribution of qualitative traits is described on the National Plant Germplasm System web site [http://www.ars-grin.gov/npgs/], in numerous USDA technical bulletins and reviews (Nelson et al., 1987, 1988; Juvik et al., 1989b; Coble et al., 1992; Bernard et al., 1998a; Hill et al., 2001; Palmer and Kilen, 1987; Bernard and Weiss, 1973). For more on this topic see Chapter 5 (Palmer et al., 2004, this publication). Qualitative diversity, for disease resistance, is of great economic importance and is reviewed in later sections of this chapter.

8–5.1.2 Quantitative Phenotypic Traits

Most economically important seed traits and plant characteristics are quantitatively inherited and have a continuous range in phenotype. For example, seed oil concentration ranges from 8% in *G. soja* to approximately 25% in *G. max* and seed protein concentration ranges from <35% to >50%. *Glycine max* seeds can weigh as much as 45 cg seed^{-1} and *G. soja* as little as 1 cg seed $^{-1}$. Photoperiod response is generally regarded as continuous and quantitative in nature although several qualitative genes for photoperiod sensitivity have been identified (McBlain et al., 1987). Annual *Glycine* accessions have an extreme range in photoperiod response such that one accession may mature as another begins to flower, when planted on the same day in adjacent plots.

Photoperiod response is important in the evaluation of almost all quantitative traits, because the photoperiodic response has large pleiotropic effects, especially for plant height and seed yield. In North America, maturity also exerts an indirect effect on quantitative traits through temperature. Soybean types which are later maturing tend to ripen under cooler temperatures, which can affect seed composition and other traits. Thus, estimates of the genetic range for many metric traits in soybean are most meaningful when comparing genotypes of similar maturity in common environments.

Data collected on soybean germplasm accessions reveal a large range of values for most traits and wide adaptation to many environments; however, the extremes that are identified in gemplasm are sometimes not as great as those generated through plant breeding where gene recombinations can create more extreme phenotypes than exist in germplasm collections. Extremes in protein concentration (Leffel, 1992), fatty acid concentration in soybean oil (Rebetzke et al., 1997), and yield (Wilcox, 2001) are notable examples.

8–5.2 Molecular Diversity of *Glycine max*

Molecular characterization at the protein, enzyme, and DNA level is a useful approach for assessing genetic diversity. Isozymes were first used in diversity research (Gorman, 1984; Chen et al., 1989; Perry et al., 1991; Cox et al., 1985a, 1985b). Based on data from 13 enzyme systems, germplasm accessions from several major regions of Asia were categorized into four groups: (i) Korea and Japan, (ii) China and eastern Russia, (iii) southeast Asia, and (iv) south central Asia (Griffin and Palmer, 1995). Restriction fragment length polymorphism markers were the first DNA markers employed to document genetic relationships among lines (Grabau et al., 1989, 1992; Chen et al., 1993). Diversity patterns based on isoenzymes and RFLP markers were similar (Clikeman et al., 1998), but the number of polymorphic DNA markers far exceeded that available for enzymes or proteins. With advances in technology, DNA markers quickly became the primary method for assessing diversity. Exotic germplasm accessions identified as having agronomic potential were shown to be genetically distinct from the ancestors of North American cultivars using RFLPs (Kisha et al., 1998) and random amplified polymorphic DNA (RAPDs) (Thompson et al., 1998; Brown-Guedira et al., 2000). By specifically selecting diverse, primitive germplasm from multiple locations, a strong geographical component to genotypic differences was found (Li and Nelson, 2001). Cluster analyses of DNA marker data for primitive germplasm from China, Japan, and the Republic of Korea showed that Chinese accessions generally grouped into clusters based on province of origin within China. Accessions from Japan and the Republic of Korea were clearly distinct from Chinese accessions, but not from each other, and were less diverse than the accessions from China (Li and Nelson, 2001). Using both simple sequence repeat (SSR) and amplified fragment length polymorphisms (AFLP) markers, Choi et al. (2000) grouped 108 Korean soybean accessions into 12 clusters. These results agreed with Li and Nelson (2001) in that the clusters did not represent the geographical origins of the lines, but significant differences were found between some clusters for agronomic traits.

8–5.3 Diversity of *Glycine soja*

8–5.3.1 Molecular Variation in *Glycine soja*

Molecular research has consistently shown that *G. soja* is much more diverse than *G. max* based on assays with isoenzymes (Griffin and Palmer, 1995), RFLPs (Apuya et al., 1988), AFLPs (Maughan et al., 1996), and RAPDs (Li and Nelson, 2002; Dae et al., 1995). This finding is consistent with the theory that *G. soja* is the progenitor of *G. max*. Regional distinctions among *G. soja* populations have also been reported. Analysis of the phenotypic diversity for the large *G. soja* collection in China indicates three major centers of diversity in China: the northeast, the Yellow River valley, and the southeast coast (Dong et al., 1999). These are also the geographical origins for most of the accessions in the Chinese collection. Simple sequence repeat (SSR) marker analysis of cytoplasmic DNA also indicates regional distinctions in *G. soja* within China (northeast, central, and south) (Gai et al., 1999). *Glycine soja* lines from China could also be separated by province of origin using RAPD markers with several subgroups identified within each province (Li and Nelson, 2002). Yu and Kiang (1993) found very high levels of isozyme variation among six natural populations in Korea.

8–5.3.2 Morphological Variation in *Glycine soja*

Kiang and Chiang (1990) analyzed 12 *G. soja* populations from Japan and Korea and found that genetic distance estimated by isozyme variation was correlated with distance estimated by phenological and agronomic traits. However, in another study of seven natural *G. soja* populations from central Japan, Bult and Kiang (1992) reported uniformity of isozymes but highly variable phenotypes within populations. Significant differences for mean seed size (1.9 vs. 2.8 g per 100 seeds) were found among populations in southwestern Hokkaido (Ohara et al., 1989). Morphological differences among *G. soja* populations may be related to adaptation to specific environments. Two distinct types, twining and branching, were found in contrasting ecological niches on the Saru River in Hokkaido (Ohara and Shimamoto, 1994). Chen (2002) found significant variation for leaflet shape and size related to geographical origin among *G. soja* accessions from Russia, China, Japan, and Korea.

8–5.3.3 Intrapopulation Variation and Sampling Strategies for Collecting *Glycine soja*

Variation *within* natural *G. soja* populations is substantial. Highly divergent genotypes have been found in samples collected within the same field (Chen, 2002; Li and Nelson, 2002). Kiang et al. (1992) showed that within-population diversity of four natural populations in Iwate province in Japan was greater than the diversity found between populations. One explanation for large within-population diversity could be higher levels of outcrossing in *G. soja* than in *G. max*. Rates of outcrossing in the two species were found to be low and similar in two studies (Kiang et al., 1992; Abe et al., 2000), but in other studies, outcrossing rates in *G. soja* of 9 and 19% were detected (Fujita et al., 1997) and levels of heterozygosity were nine times higher than *G. max* (Gorman, 1984). The existence of large intra-population diversity has a major impact on devising a collecting strategy for *G. soja* accessions, and this is an active area of research. Based on isoenyzme variation, Shung et al. (1995) found that South Korean *G. soja* populations have more than 30% polymorphic loci and recommended that multiple representatives of populations be sampled when collecting. Using RAPD markers, Vaughan et al. (1995) found a high degree of local variation in *G. soja* populations. The variation was distributed in a mosaic rather than clinal pattern which further complicates sampling strategies for collecting. It would appear that *G. soja* populations can vary from highly structured to more randomly variable (Abe et al., 2000).

8–5.3.4 Unique Alleles in *Glycine soja*

Glycine soja may have important phenotypic characters or specific alleles not present in *G. max*. Hajika et al. (1996) discovered a wild soybean line in the Amakusa Islands (southwestern Japan) that either lacked or had extremely low levels of all three of the 7S subunits, a phenotype that was controlled by a single dominant gene (Hajika et al., 1998). Wang et al. (2001) identified resistance to soybean cyst nematode (SCN, *Heterodera glycines* Ichinohe) in *G. soja* PI 468916. Results from quantitative trait loci (QTL) mapping indicate that this resistance is at loci not previously identified in *G. max*. Both Li and Nelson (2002) and Chen (2002) have reported RAPD fragments in *G. soja* lines that have not been found in *G. max*. Chen (2002) also found extremes in leaflet shape and rapid early growth not documented in *G. max*. There is evidence that *G. soja* and *G. max* have different desaturase alleles that determine oil composition (Pantalone et al., 1997b). Similarly, Kwanyuen et al. (1997) speculate that the large amount of phenotypic diversity in *G. soja* protein concentration and composition may be due to a different complement of genes that control expression of 11S and 7S proteins than in *G. max.* A QTL allele in *G. soja* is associated with increased protein (Sebolt et al., 2000). The frequency of saponin types is reported to be different between *G. max* and *G. soja* but all types are found in both species (Tsukamoto et al., 1993). Some common traits in *G. soja* that may be associated with unique alleles relative to *G. max* are small seed size, lower oil and oleic acid concentrations, and higher linolenic acid concentrations in the seed (Juvik et al., 1989a; Rebetzke et al., 1997).

8–5.3.5 *Glycine soja* in Applied Breeding

Despite the appeal of *G. soja,* as a source of diversity and its ability to cross freely with *G. max,* the only breeding success in North America with *G. soja,* at present, has been the transfer of small seed size to specialty cultivars for the Japanese natto market (LeRoy et al., 1991). This accomplishment has been the basis for much of the natto soybean breeding effort in North America and has resulted in the release of 17 and 12 cultivars with small seed from Iowa State University and Canada, respectively. Three *G. soja* accessions from the USDA collection (PI 81762, PI 135624, and/or PI 101404) contributed from 1/8 to 1/32 of the parentage of these cultivars. All of these specialty cultivars are substantially lower yielding than commodity cultivars. In China, two *G. soja* accessions (Hong Ye-1 and GD 50477) have been used to develop two cultivars with small seed (Hong Feng Xiao 1 Hao and Ji Lin Xiao Li 1 Hao) for the natto export market. No Japanese cultivars have been developed from *G. soja.* Breeding success has been limited with *G. soja* because it has many undesirable phenotypic traits (e.g., extreme vining, shattering, and lodging). Polygenic inheritance of such traits makes it extremely difficult to recover a *G. max* phenotype in the F_2 generation following a *G. max* × *G. soja* mating, even when thousands of F_2 plants are grown.

Despite the difficulty of using *G. soja* in practical breeding, it is theorized that *G. soja* accessions might be important for yield improvement of cultivars in the future. The greater genetic diversity within *G. soja* for DNA markers suggests that *G. soja* may also have greater genetic diversity for alleles controlling metabolic function. These alleles could be beneficial if found and transferred to *G. max.* Early efforts to investigate this possibility relied solely on phenotypic evaluations of progeny from *G. max* × *G. soja* crosses (Carpenter and Fehr, 1986; Ertl and Fehr, 1985). Carpenter and Fehr (1986) reported that at least two backcrosses to *G. max* were required to recover any lines that were agronomically acceptable, and that three backcrosses were needed to create sufficiently large populations of such lines for effective selection in a cultivar development program. No backcross-derived line (BC_1–BC_5) yielded better than the *G. max* recurrent parent (Ertl and Fehr, 1985), when two different populations of *G. soja* × *G. max* matings were tested. Population sizes were small in these studies (nine lines from each backcross generation and cross) and it is likely that larger populations would be required to detect positive transgressive segregants, were they to occur, because of the poor agronomic value of the *G. soja* parent. Graef et al. (1989) reported that selection based on only five isozymes was effective in helping to recover the phenotype of recurrent parent in a population of BC_2F_4 lines derived from two *G. max* × *G. soja* crosses.

Recent advances in molecular marker technology and theory have generated renewed interest in using *G. soja* accessions to improve the productivity and value of soybean. For example, using a combination of markers, backcrossing and phenotypic testing, the existence of beneficial genes in diverse sources can be determined without the need of finding rare transgressive segregants (de Vicente and Tanksley, 1993). One of the first DNA marker mapping populations in soybean was derived from a *G. max* × *G. soja* hybridization (Diers et al., 1992a, 1992b). Analysis of seed protein concentration in that population revealed the presence of *G. soja* marker alleles positively associated with the trait (Diers et al., 1992a). One of the *G. soja* QTL alleles for increased seed protein has been backcrossed into several soybean cultivars (Sebolt et al., 2000). The allele increased protein in moderate protein genetic backgrounds but did not increase protein when introgressed into a soybean that already had high seed protein. Several U.S. research groups have recently initiated marker-based studies of yield in *G. soja* using this approach, but few results have been published (Concibido et al., 2002).

At present, the true value of *G. soja* in cultivar yield improvement remains to be determined. An important factor limiting the discovery and introgression of beneficial yield alleles from *G. soja* is that agronomic assessment of *G. soja* accessions is not very helpful in selecting parents. All *G. soja* accessions yield only a small fraction of commercial cultivars, so that phenotypic proof of yield superiority of one *G. soja* accession over another is not possible. In breeding for disease resistance, by contrast, a *G. soja* accession is likely to be phenotypically superior to other *G. max* lines for the trait in question, and there is an expectation that the *G. soja* allele controlling the phenotype will represent an extreme that does not occur within the commercially-used gene pool. Because direct assay of yield is not useful for *G. soja* breeding efforts, physiological and genomic evaluation of *G. soja* accessions may assist in selecting *G. soja* accessions for studies related to yield improvement. However, few *G. soja* accessions have been evaluated extensively.

8–5.5 Genetic Diversity of Perennial *Glycine* Species

The 23 perennial *Glycine* species represent a potentially very useful, but elusive, source of genetic diversity for soybean. Crosses between perennial and annual *Glycine* are difficult and, at best, result in sterile progeny (Broue et al., 1982). Doubling the chromosome number of a *G. max* × *G. tomentella* hybrid produced seed (Newell et al., 1987) but subsequent backcrossing to create fully fertile progeny has been reported only by one group (Singh et al., 1990, 1993). One of the first *core* collections for any species was that established with the perennial *Glycine* species (Brown et al., 1987, 2000; Brown, 1995,1989a, 1989b), primarily to identify a broad range of diversity for these species. An evaluation of 28 accessions from seven species using RFLPs indicated the greatest within species variation existed for *G. tabacina* and *G. clandestina* (Menancio et al., 1990). Additional cytological and genetic research found contrasting races within the *G. tabacina* polyploid complex (Brown et al., 2000). Morphological and biochemical analysis has also revealed substantial variation with diploid *G. tomentella* (Hill, 1999).

Perennial species are resistant to pathogens that cause sclerotinia stem rot (SCSR, caused by the fungus *Sclerotinia sclerotiorum*), sudden death syndrome [SDS, *Fusarium solani* (Mart.) Sacc. F. sp. *glycines)*, (Hartman et al., 2000), and soybean rust (*Phakopsora pachyrhizi*) (Hartman et al., 1992). Resistance has also been observed to SCN (Riggs et al., 1998), high chloride levels (Pantalone et al., 1997a), glyphosate (Loux et al., 1987), and 2,4-D (Hart et al., 1991).

8–6 GENETIC DIVERSITY IN SOYBEAN BREEDING —AN OVERVIEW

More than 3500 cultivars have been released globally since 1920. These constitute an important genetic resource that has been the subject of many genetic diversity studies. Diversity among cultivars from the USA and Canada (hereafter referred to as North America), China, and Japan, is reviewed in this section. The decision to include only specific countries in this review was based on the availability of summarized data in the literature. Although not reviewed here, it is important to note that Brazil, Argentina, Mexico, Australia, and India have all used U.S. and Canadian cultivars extensively as founding stock in cultivar development (Abdelnoor et al., 1995).

Two major goals in most breeding programs are to genetically raise the yield potential of cultivars and to protect these yield gains with pest resistance. The relation between breeding for pest resistance and genetic diversity is discussed later in this chapter. Cultivar diversity is examined here as it relates to breeding for yield improvement. Three questions underlie much of the interest in genetic diversity pertaining to yield: (i) Does sufficient genetic diversity exist in applied breeding to sustain long term genetic progress for improved yield?, (ii) Can parental matings be selected more wisely to avoid unproductive crosses and thereby improve breeding efficiency?, and (iii) Can new alleles from exotic germplasm accelerate yield improvement when introduced into commercially used gene pools? These questions remain largely unanswered in soybean and many other crops. Knowledge of the exact number, chromosomal location, and effect of all yield alleles would provide the most straight forward route to the answers for these questions. Unfortunately, due to the expense and difficulty of identifying yield alleles, it is unlikely that we will know the chromosomal location of yield alleles except for those with the very largest effects (Beavis, 1998).

Given that complete information on yield alleles will remain elusive, one must ask what other approaches are available to help the breeder capture value from the diversity present in germplasm for yield improvement. A growing body of literature is beginning to point to ways in which genetic diversity estimates can be employed for this purpose.

8–7 TECHNIQUES FOR QUANTITATIVE MEASUREMENT OF GENETIC DIVERSITY IN APPLIED SOYBEAN BREEDING

Several techniques and concepts have been employed as quantitative measures of genetic diversity and are reviewed under separate headings in this section.

This section introduces these topics. Genetic diversity measures are usually based on DNA markers, pedigree lineages, or morphological traits and they have value in estimating how closely or distantly related genotypes and gene pools may be. Genetic base, coefficient of parentage (CP), and SSR-based genetic similarity are common examples. When considering genetic diversity measures, it is important to note that although the various measures of genetic diversity often have some general agreement, they may not all capture the same pattern or extent of diversity. Results are influenced by the number and type of DNA markers, completeness and structure of pedigree data, and/or the statistical analysis employed. Thus, their interpretation is relative and made stronger when more than one diversity measure is employed in germplasm evaluation.

The relation of genetic diversity measures to economically important traits is a topic of current research. However, results are sufficiently positive to suggest that diversity measures have application in practical breeding. The rationale for their use is that genetic diversity will identify contrasting lines, that contrasting lines are likely to contrast for agronomic alleles, and that, consequently, diversity information may aid in the selection of parents for breeding. In terms of yield more distantly related parents (within a species or breeding pool) might be expected to contrast for more yield alleles and, thus, improve the odds that a breeder will select a superior recombinant among progeny which exhibits high yield.

Because of the current interest in yield improvement using gene tagging (used here to mean the linking of DNA markers to yield QTL), it is important to provide a clear rationale for use of genetic diversity measures as an alternative to gene tagging. Both concepts have application to yield improvement. However, the distinction is that diversity measures emphasize *individuals in their entirety* rather than *individual yield alleles*. The notion of considering genotypes *in their entirety* rests on the assumption that yield is a quantitative trait controlled by numerous genes that cannot be identified easily by gene tagging (Burton, 1987). Suppose, for example, that yield in North American soybean breeding was controlled by as many as 50 genes. (The long-term incremental yield advances noted in North America [Specht and Williams, 1984] are consistent with this supposition). The number of populations and the degree of field testing required to locate 50 positive alleles for yield using markers would fall far beyond the means of most breeding programs. In addition to the cost factor, only a portion of the positive alleles could be expected to be identified and mapped, based on theoretical considerations related to precision of testing (Beavis, 1998). Also, many QTLs are subject to QTL × environment interaction. Thus, extensive field testing would be required to identify stable yield alleles vs. those that are environmentally specific.

When definitive data are lacking regarding the location and number of yield alleles, (a likely case based on the considerations outlined above) one can turn to probability-based 'whole genotype' measures of diversity to identify desirable parental matings. The concept of genetic closeness or distance of parents, embodied in genetic diversity measures, addresses this notion. It is important to note that diversity measures and gene tagging approaches can be used in conjunction for yield improvement, where diversity measures are used to identify parents for subsequent gene tagging studies.

8–7.1 Genetic Base

The success of plant breeding is dependent, in part, upon the genetic diversity inherent in its founding stock. To characterize this genetic diversity, plant breeders have introduced the concept of genetic base, which can be defined as the complete set of ancestors that contributed to a breeding population. A breeding population, for example, might be all cultivars released from a particular country, region, or period of time. Two basic components in the description of a genetic base are the number of ancestors in the genetic base and percentage contribution of each ancestor to the total genetic complement of the base. These components are usually estimated by pedigree analysis. The percentage contribution derived from each ancestor provides a quantitative assessment, whereas the number of ancestors provides a less useful qualitative assessment, of the genetic base. An example of the qualitative and quantitative contribution of an ancestor to a genetic base is provided by the soybean plant introduction, CNS, an ancestor of North American soybean breeding. CNS is just one of 39 ancestors of the *southern* soybean cultivars released during 1953 to 1988 in the USA. However, CNS contributes approximately 25% of the parentage of these cultivars (Gizlice et al., 1994) and is therefore considered a major ancestor. The distinction of *all* ancestors vs. *major* ancestors is made possible by the CP analysis of ancestral contribution and is roughly analogous to the distinction of population size vs. effective population size in population genetics. The average CP of an ancestral line with all cultivars in a population is the estimate of the proportion of the genetic base derived from that ancestor (Gizlice et al., 1994).

The narrowness of the genetic base of major crops has been a continuing concern for those who study genetic diversity (Committee on Genetic Vulnerability, National Academy of Science, 1972; Rodgers et al., 1983; Smith, 1988; Smith et al., 1992). There is no widely accepted basis for concluding that a genetic base is too narrow, unless the total number of major ancestors is less than perhaps five (Falconer and Mackay, 1996). Unfortunately, neither is it easy to determine if a genetic base is sufficiently large so that it poses no near-term limitation on selection. Selection limits, effective population size, and other population genetics concepts which may be used to address this question are generally applicable only when the number of generations of intermating (or breeding cycles in the context of this chapter) becomes much higher than the six to eight cycles of breeding which have taken place thus far in Asian and North American soybean breeding programs (Cui et al., 1999; Carter et al., 1993). Thus, although a narrow genetic base may be important, there are no powerful tools for assessing this importance. Perhaps the only meaningful way to gauge the narrowness of a genetic base for a soybean breeding program is to compare it with that of other crops or other soybean breeding programs and then relate the genetic base to historical rates of yield progress. Despite the relative and imperfect nature of such comparisons, they are nevertheless extremely useful in genetic diversity studies. If, for example, two breeding programs are found to have genetic bases with similar genetic diversity and one program experiences a yield plateau, one might assume that the genetic base of the other is insufficient and that it too could develop a similar problem with yield advancement in the future. A third hypothetical program with a much larger genetic base may be perceived as under no threat of genetic exhaustion despite such concerns for the other two. The genetic bases of Chinese, Japanese, North American, and Brazilian (Hiromoto and Vello, 1986) soybean breeding have been characterized and can be compared as outlined above.

8–7.1.1 Using the Genetic Base as a Screening Tool

Although not related to yield improvement per se, the concept of genetic base has another use as is relates to qualitative diversity. Screening of the ancestral base is a low cost method to identify adapted descendent cultivars with a desirable qualitative trait. The approach is to assay the genetic base for the qualitative trait or allele of interest. If it is present in the base, then pedigree analysis can be used to predict which descendent cultivars are most likely to carry the trait, thus circumventing the need to screen the entire set of released cultivars derived from the base. Because the number of ancestors is usually much smaller than the number of cultivars derived from them, analysis of the genetic base identifies a relatively small number of ancestors and old cultivars for screening. For traits that are expensive or difficult to measure, such as ozone tolerance, salt tolerance, protein functionality, or characteristics affecting sensory preference for soyfoods, this approach can be economically important. The genetic base of North American soybean has been screened to detect SSR marker polymorphisms; variation in human allergens; unusually high rates of electron transport in chloroplasts; and resistances to the herbicide sulfentrazone, ultraviolet radiation, and SDS (Gizlice et. al., 1994; Burkey et al., 1996; Diwan and Cregan, 1997; Reed et al., 1991; Song et al., 1999, Mueller et al., 2003, Hulting et al, 2001; Yaklich et al., 1999).

8–7.2 Coefficient of Parentage

Pedigree analysis of cultivars provides a means for estimating genetic diversity of cultivars based on ancestral lineages. Pedigree analyses encompass all ancestors and breeding lines used to develop cultivars as well as the cultivars themselves. Thus, they offer valuable historical insight into, and quantitative assessment of, breeder decisions that shaped the current genetic structure of breeding programs. They can also be used to assess the potential relationship between genetic diversity and yield improvement. The most common applications of pedigree analysis are CP between cultivars, and CP between ancestors and descendent cultivars which can be used to quantify the genetic base of cultivars.

Coefficient of parentage has been used to examine genetic diversity of soybean breeding in China, Japan, Brazil, and North America. In CP, familial relationships are used to calculate the probability that two cultivars each have an allele derived from a common ancestor, i.e. that they have alleles that are identical by descent (Malecot, 1948). Stated another way, CP is an estimate of the proportion of the genome for which two cultivars share alleles that are identical by descent (Falconer and Mackay, 1996). The CP calculation assumes that pedigrees are accurate (which is generally regarded as so in soybean breeding programs), that each parent contributes approximately equally to progeny (regarded as true) (St Martin, 1982; Lorenzen et al., 1995), and that breeder selection, while altering the frequency of certain alleles, does not greatly alter the approximately equal genomic contribution of each parent to the progeny. A CP value of 0 indicates no known pedigree relationship between cultivars (i.e., no ancestors in common), whereas values of 0.25, 0.50, and 1.0 indicate half-sib, full-sib, and identical-twin relationships, respectively, when cultivars are highly inbred and true-breeding, which is the case in soybean but not for most cross-pollinated crops.

Analysis of CP has been used to determine genetic structure of populations, show loss of diversity from genetic drift during genetic bottlenecks (reflected as increases in cultivar relatedness over time), and devise parent selection strategies that avoid rapid depletion of genetic diversity. By contrast pedigree analyses are not useful in directly analyzing the diversity between populations that have little pedigree in common, such as that between modern Asian and North American cultivars. However, separate CP analysis of near independent groups such as North American and Chinese breeding pools can be very important when considering the intercrossing of these pools, in that it may aid in the selection of appropriate representative members from each pool as specific parents.

A potentially important yield-breeding application of CP analysis is in the prioritization of pairs of productive parents for mating. In this context, it is important to note that a CP value of 1.0 between parents indicates an absolutely poor choice for mating, whereas a CP relationship of 0 between two parents may simply indicate that little or no pedigree information is available. The inference is that CP can be used to identify poor parental mating choices but is less useful in identifying the best mating choices. Thus, the potential importance of CP analysis is in risk reduction by eliminating poor crosses, that is, those parental matings that may not produce sufficient variation for effective selection. Unfortunately, there is no exact 'cutoff CP value' above which parental matings should be avoided. In the context of risk avoidance, there is little disagreement about the interpretation of extreme CP values for pairs of parents. Values >0.50 clearly indicate that two parents are highly related and, if mated, would be unlikely to produce much genetic variation for yield improvement (unless the parents differ for a major qualitative gene, such as that for disease resistance). Values <0.25 generally indicate an acceptably low relationship between parents, in the absence of more definitive information, and that desirable progeny might be derived from such matings (Manjarrez-Sandoval et al., 1997a; Cui et al., 2000b; Gizlice et al., 1996). However, the interpretation is less clear for a CP value from 0.26 to 0.49 between parents. Using adapted breeding populations from the southern USA, Manjarrez-Sandoval et al. (1997a) found reduced genetic variance for yield in populations derived from parents whose CP was 0.25 or greater, suggesting that one can expect reduced breeding progress from mating half sibs or more closely related relatives. In Midwestern cultivars, there was generally less genetic variance in populations derived from parental matings with a high CP than from parental matings with a low CP (Kisha et al., 1997; Helms et al., 1997). No studies have shown an absolute association between CP estimates of diversity and genetic variance, when CP values are <0.5. It is interesting to note that all studies showed that many matings among recently developed lines produce negligible genetic variance for yield, suggesting that a lack of diversity is common in applied breeding. Resources devoted to unproductive crosses are wasted and reduce overall gain from selection. It should be noted that these results contrast with those of Rasmusson and Phillips (1997) who found that parents with a very high CP could produce progeny with useful variation for yield in barley *(Hordeum vulgare* L.). The authors theorized that de novo appearance of yield genes and epistasis were responsible for the yield advances in barley. Coefficient of parentage has also been used to predict heterosis in soybean (Gizlice et al., 1993b, Manjarrez-Sandoval, 1997b).

8–7.2.1 Adjusting Coefficient of Parentage Values with DNA Marker or Phenotypic Data

Because the pedigree relationships among ancestors in a CP analysis are unknown, relationships of 0 are ordinarily assigned between all pairs of ancestors. However, phenotypic and DNA marker analyses have shown that genetic similarity among ancestors can vary significantly and that clear ancestral patterns of diversity exist (Diwan and Cregan, 1997; Gizlice et al., 1993a; Keim et al., 1992; Kisha et al., 1998; Skorupska et al., 1993; Brown-Guedira et al., 2000; Li et al., 2001b). To augment CP analysis, such information has been incorporated into the initial phase of CP analysis to better reflect the underlying true genomic relationship between cultivars (Gizlice et al., 1993a; Kisha et al., 1998). This adjustment has the potential to improve the use of CP for the prioritization of pairwise matings among a set of productive parents, if the similarity estimates between pairs of important ancestors in the genetic base deviate greatly from the mean relatedness of all ancestors. The approach in the calculation is to simply replace the assumed zero CP between ancestor pairs with the nonzero estimates of similarity derived from phenotypic or DNA-marker analysis. Note that all CP values between the descendent cultivars increase accordingly, but respectively more or less for those descendents whose ancestors are more or less similar than the average ancestral pair. A result of adjustments is that the range in adjusted CP values is much smaller than for unadjusted CP values. Only a few studies have related unadjusted CP values to genetic variation for yield and none have attempted to relate *adjusted* CP values (calculated as outlined above) to yield variances. Thus, a breeding interpretation of adjusted CP values is difficult. For North American soybean cultivars derived from several generations of breeding, adjusted and unadjusted CP values tend to correlate highly ($r > 0.90$), indicating that such adjustments had little impact on the relative ranking of diversity for pairs of current cultivars and that adjusted and unadjusted CP values would be equally predictive of genetic variation for yield (Gizlice et al., 1993a; Kisha et al., 1998).

8–7.3 Genetic Diversity Based on DNA Markers

DNA markers have been used to estimate the genetic diversity of North American cultivars and, to a lesser extent, those from China and Japan. There are two basic types of parameters used to estimate genetic diversity. One is the genetic similarity between two genotypes. DNA-based measures of genetic similarity estimate the degree to which two genotypes share alleles that appear identical (i.e., alleles that are identical in state but not necessarily identical by descent). Genetic similarity is generally calculated as the proportion of tested markers for which two genotypes appear similar. This differs from CP, an estimate of identity by descent, but the two approaches are analogous otherwise. Indeed, positive relationships between genetic similarity using DNA markers and CP have been shown for North American soybean cultivars (Kisha et. al., 1998; Manjarrez-Sandoval et al., 1997a) and in other crops (Souza and Sorrels, 1991; Hellend and Holland, 2003). Estimates of genetic similarity generally range from 0 (no alleles in common) to 1 (all alleles the same). Genetic distance, a related term, is the complement of genetic similarity (defined as 1 – genetic similarity) and, thus, estimates the degree to which two genotypes have non-identical alleles. Molecular estimates of genetic similiarity are almost always greater than CP estimates, and extremely low values of genetic similarity are rare. Generally, only a fraction of the theoretical range (zero to one) is observed for DNA-based diversity measures. There are three reasons for this. First, many marker loci in a given parental pair are monomorphic. Secondly, for polymorphic loci with only two or three alleles, the frequency of matches of alleles between pairs of genotypes must be at least one-half or one-third, respectively, forcing estimates of genetic similarity upward. Third, markers typically reveal considerable genetic similarity between ancestral pairs, and this limits the range of genetic similarity that is possible among their progeny. The degree to which these three factors influence results varies by marker system. Thus, estimates of genetic similarity with different marker systems are not directly comparable and must be interpreted in a relative way. To address the problems of narrow range in marker-based diversity estimates and differential ranges encountered from marker system to the next, monomorphic loci could be excluded from calculations to more nearly standardize estimates of genetic relatedness on a 0 to 1 scale. Adoption of this practice varies.

Another commonly used measure of diversity is the polymorphism information content (PIC), which is also referred to as gene diversity (Weir, 1996). A PIC value is used to describe the diversity of a population for a given DNA marker, while genetic similarity and distance are estimated for a pair of genotypes. A PIC value estimates the probability that two alleles sampled at random from a locus in a population will be different. The PIC values can range from 0 (monomorphic locus) to a maximum value that is dependent on the number of alleles present at a locus in a population. For example, the maximum PIC value for a two-allele locus would be 0.5 while the maximum would be 0.67 for a three-allele locus. The PIC values can be calculated for individual markers, or averaged over all markers. As with estimates of genetic similarity, PIC values are relative. Estimates of genetic similarity and PIC values and their utility depend on the number and type of markers employed, the degree of marker polymorphism (within and among loci) present in the germplasm, the genomic distribution of the marker loci, and the total number of DNA fragments or alleles used in the calculations of similarity or PIC.

Unlike pedigree similarity analyses, which are based only on recorded pedigree information, DNA marker-based estimates require the physical existence of the genotype for DNA extraction and typing. When ancestors are the subject of study, this can present a special problem, because some ancestors have not been preserved. In such cases, progeny from the ancestor, rather than the ancestor itself, are used in the marker analysis. For example, PI 54610, an ancestor of southern cultivars in North America has been lost. However, most genes which were transferred from this ancestor to modern cultivars were passed through two old cultivars, Jackson and Ogden (Gizlice et al, 1994) (Majos was a third, but substantially less important, conduit). Thus, Jackson and Ogden, although not true ancestors themselves, have been included in molecular analyses of the genetic base of North American soybean as surrogates for PI 54610.

Although marker-based estimates of genetic similarity (and distance) of cultivars have been used to suggest which pairwise matings offer the most potential, empirical validation has been scant. As with CP, there is no consensus among breeders as to a threshold value for parental pairwise DNA similarity beyond which a mating should not be made. The best use of marker-based estimates of genetic similarity in applied breeding, at present, may be simply the prioritization of matings among a set of parents. Kisha et al. (1998), Manjarrez-Sandoval et al. (1997a), and Helms et al. (1997) detected a trend toward increased genetic variance in progeny derived from crosses of parents with lower than average DNA-based genetic similarity.

Another use of genetic distance measures in breeding has been suggested by Cornelious and Sneller (2002). In this approach, progeny are developed from crosses of exotic accessions by current domestic cultivars and then yield tested. The genetic distance of the progeny from the domestic parent is determined with markers. Progeny yield is regressed on genetic distance. Progeny with large positive deviations from the regression are likely to have beneficial yield alleles from the exotic parent.

8–7.4 Genetic Similarity Based on Phenotypic Data

Phenotypic (i.e., morphological and biochemical) data from controlled growth chambers have been used to estimate genetic diversity among U.S. and Chinese cultivars (Cui et al., 2001). The rationale for the use of phenotypic data to quantify diversity is that phenotypic data could be more closely related to breeding behavior than pedigree or DNA-marker based data. Many molecular markers are not part of the DNA coding region of the genome whereas genes controlling phenotype are. As the number of phenotypic traits increases in a comparison, the number of genes involved in the control of phenotypic variation should increase accordingly and, thereby, improve the utility of diversity estimates. Hundreds of genes may underlie phenotypic diversity estimates when many traits are included in the analysis. Phenotypic similarity is calculated by performing a principal component analysis of the metric data (qualitative data should be avoided) and then using the results to develop a similarity estimate between all genotypes on a scale from 0 to near 1 (Goodman, 1972; Gizlice et al., 1993a). Phenotypic diversity has been employed in clustering soybean accessions in the USDA Collection (Perry and McIntosh, 1991).

8–7.4.1 Maturity Impact on Phenotypic Data

One difficulty in the assessment of phenotypic diversity is that differential photoperiod sensitivity (maturity) among soybean genotypes exerts overwhelming effects on plant growth and compromises estimates of diversity. To minimize this source of bias in phenotypic diversity estimates, morphological and biochemical data are usually collected in controlled growth chambers where the genotypic range in photoperiod response can be muted (e.g., using 12 h or shorter photoperiods to reduce asynchrony of flowering). When grown under short photoperiod, correlation of a trait with photoperiod sensitivity is less common, and maturity effects can be avoided in most cases (Gizlice et al., 1993a; Cui et al., 2001). A constant temperature regime in growth chambers during pod filling makes genotypic comparison of seed composition possible even for genotypes that differ in maturity. When maturity is taken into account, as described above, and a large number of traits are measured, then phenotypic diversity has the potential to be a very good estimate of genetic diversity. The concept of phenotypic diversity has been used successfully in several other crops (Grafius et al., 1976; Grafius, 1978; van Beuningen and Busch, 1997; Johns et al., 1997; Autrique et al., 1996; Tatineni et al., 1996).

8–8 STATUS OF DIVERSITY IN NORTH AMERICAN SOYBEAN BREEDING

Although soybean breeding commenced in North America before 1900, consistent and organized programs began in the early 1930s at state agricultural experiment stations and with the USDA (Bernard et al., 1988a). The primary foci of these early programs were high yield, broad adaptation, and, to a lesser extent, increased seed oil content. Disease resistance was added as a breeding objective in the late 1940s through early 1950s as production problems became apparent from bacterial pustule (*Xanthomonas axonopodis* pv. *glycinea*), SCN, and phytophythora (*Phytophthora sojae*). Germplasm was shared extensively between these early programs, especially between those working with breeding materials of similar maturity. Commercial companies invested heavily in soybean breeding programs in the early 1970s with the advent of the Plant Variety Protection Act (Fig.8–4). Propri-

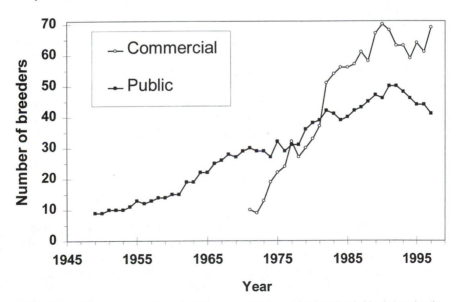

Fig. 8–4. Change in the total number of public and proprietary soybean breeders in North America (i.e., USA and Canada) from 1945 to 2000.

etary programs were founded primarily upon publicly developed cultivars and germplasm. Public and proprietary programs frequently exchanged advanced germplasm up to the late 1980s. Since then, corporate restrictions on proprietary germplasm beyond the Plant Variety Protection act, and more recently utility patent protection, have suppressed germplasm exchange in the commercial sector. Because of the relatively free exchange of breeding materials among soybean breeders for most of the past century, and the cyclic long-term nature of breeding progress, it can be argued that North American breeding efforts have been one large informal recurrent selection program (St. Martin, 1982; Luedders, 1977; Specht and Williams, 1984), although the working populations therein were stratified as result of maturity constraints. This description of soybean breeding is consistent with the view that yield is under the control of numerous genes and related to later discussions in this chapter regarding the use of genetic diversity to improve yield.

8–8.1 North American Genetic Base

8–8.1.1 Overview

Early studies of diversity in North American soybean cultivars reported a limited genetic base (Committee on Genetic Vulnerability, 1972; Specht and Williams, 1984). Analysis of 258 public cultivars released from 1947 to 1988 showed that 80 ancestors accounted for 99% of the parentage in cultivars (Gizlice et al., 1994). While 80 ancestors seems to be a relatively large number, just 26 ancestors accounted for nearly 90% of the total ancestry, with the remaining ancestors each contributing <1% of the total ancestry (Table 8–8). For the northern and southern North America breeding pools, 19 ancestors (17 in common) contributed 85% of the genes to each region (Gizlice et al., 1994). Northern and southern breeding pools are quite distinct in terms of major ancestors and cultivar pedigrees. This topic is discussed in detail in section 8–13.

The most important ancestors of the current North American genetic base were used from the inception of North American soybean breeding (Delannay et al., 1983). Approximately 85% of the current North American genetic base was complete by 1954, and 95% by 1970 (Thompson and Nelson, 1998b). Only a few new ancestors have made large contributions to a single cultivar or the overall genetic base over the past 25 yr (Delanney et al., 1983; Gizlice et el., 1994; Sneller, 1994, 2003), even though many plant introductions have been used as sources of resistance to diseases, especially nematodes. The explanation for this apparent contradiction is that most of the recently added ancestors were often employed as donor parents in backcrossing schemes and, theoretically, contributed very little parentage to breeding other than the genes targeted for introgression. Thus, the composition of the base has remained relatively constant. Among commodity-type cultivars released since 1970, only five have a pedigree in which a plant introduction contributed at least 25% of the pedigree. These PIs are PI 91110-1 (antecedent of cv. IVR 1120), PI 257435 (antecedent of cv. Northrup King S1346), PI 75106 (antecedent of cv. Hutcheson and Ripley), and PI 416937 (parent of N7001) (Sneller, 1994, 2003; Carter et al., 2001).

Table 8–8. Major ancestors of North American soybean cultivars and their percentage contribution to the genetic base. Estimates of contributions are based on pedigree analysis of all public cultivars released from 1947 to 1988 and both public and private commodity cultivars released from 1999 to 2001.

Ancestor	All		South		North	
	1947–1988†	1999–2001‡	1947–1988	1999–2001	1947–1988	1999–2001
Lincoln††	17.9	16.8	2.9	8.4	24.1	23.8
CNS	9.4	14.1	24.7	21.6	3.0	7.7
S-100#	7.5	10.7	21.3	18.1	1.8	4.6
Mandarin (Ottawa)	12.2	6.8	0.0	1.3	17.2	11.3
Richland	8.2	6.5	0.8	2.7	11.3	9.7
Dunfield	3.6	4.5	3.9	4.5	3.5	4.5
Roanoke	2.1	3.8	6.5	5.3	0.2	2.6
AK (Harrow) #	4.9	3.5	0.0	0.6	6.9	6.0
Illini#	2.2	3.0	0.0	1.5	3.1	4.2
Mukden	3.5	2.2	0.0	0.4	4.9	3.7
Patoka	1.0	2.1	1.0	2.7	1.0	1.7
PI 81041	0.0	2.1	0.0	2.7	0.0	1.7
Tokyo	3.8	2.0	7.7	4.1	2.2	0.3
PI 54610	3.7	1.7	7.3	3.5	2.2	0.2
PI 88788	0.5	1.3	0.7	1.4	0.4	1.2
No. 171	1.7	1.3	0.0	0.2	2.4	2.3
PI 257345	0.0	1.2	0.0	0.1	0.0	2.1
Palmetto	0.8	1.1	2.7	2.3	0.0	0.2
Haberlandt	0.8	1.1	2.5	2.1	0.1	0.3
PI 71506	0.2	0.9	0.3	2.0	0.1	0.0
Ralsoy	0.6	0.6	1.9	1.3	0.1	0.0
Total	84.6	87.3	84.2	86.8	84.5	88.1

† From an analysis of 258 public lines released during 1947 to 1988 (Gizlice et al., 1994).
‡ From analysis of 312 public and proprietary commodity cultivars released during 1999 to 2001 (Sneller, 2002).
§ North, refers to cultivars released that are Maturity Group mid-V and earlier.
¶ South, refers to cultivars released that are later than Maturity Group mid-V.
# AK (Harrow) and Illini are indistiguishable phenotypically and are selections from a single plant introduction from China. S-100 was derived from a late maturing off-type plant in Illini. DNA marker analysis is consistent with the theory that S-100 arose as an natural outcross between Illini and an unknown pollen donor source.
†† The identity of the parents of Lincoln has been lost, but they are presumed to be Chinese landraces.

8–8.1.2 History of the Formation of the Genetic Base

At the beginning of the 20th century there was only exotic germplasm in North America, and introduced cultivars provided the basis for production until the 1940s. The Uniform Soybean Tests, which provided coordinated regional testing for breeders under the auspices of the USDA, began in the northern USA in 1939 and in the southern USA in 1943. To better understand how the genetic base was established, we examined the pedigrees of the lines tested through 1955 in the Uniform Tests. During this period approximately 900 experimental lines and cultivars were tested. Nearly 20% of these lines were introductions from other countries, or selections from these introductions, whereas the remainder were derived from natural or manual cross-pollinations among these existing stock. Approximately 150 plant introductions were tested and/or used in crosses during the first 15 yr of cooperative regional testing. Eighty six of these 150 were actually used as parents to develop lines that appeared in USDA tests and other plant introductions were evaluated directly as entries. Over the course of continuous testing and crossing, 37 of the 86 lines eventually became part of the more than 80 members of the genetic base of North American soybean. These 37 lines included 17 of the 36 ancestral lines that contribute 95% of the genes to the current U.S. genetic base. Most other members of the genetic base were selected from the collection as sources of disease resistance (Gizlice et al., 1994).

Most PIs used by breeders as parents prior to 1955 were undoubtedly chosen based on performance in yield trials. It is not clear, however, why some PIs, but not others, became part of the U.S. genetic base. Some of the PIs that were unsuccessfully used to develop cultivars prior to 1955 appeared in the pedigree of only a few entries and may not have been adequately evaluated as parents. Other PIs were extensively tested as parents but failed to produce any exceptional progeny. Notable examples of PIs used extensively but relatively unsuccessfully as parents are: PI 70.218-2-6-7 (selected from the same introduction as Patoka) found in the pedigree of 29 lines, Mandarin in 50 lines, Earlyana in 21 lines, Manchu in 31 lines, other Manchu derivatives in 29 lines, and Macoupin in 18 lines. In contrast, Lincoln, derived from unknown parents (but presumed Chinese landraces) and released in 1943, became the single most important contributor to the U.S. genetic base. In the Group III tests from 1942 to 1945, Lincoln yielded 7% more and was 8 d earlier than the best domestically developed cultivar, Chief. Lincoln yielded 11% more and was 10 d earlier than the best introduced cultivar, Patoka, and Lincoln yielded 16% more than the plant introduction Dunfield, which was of similar maturity. The yield advantage of Lincoln made it a very attractive parent.

The first lines with Lincoln as a parent were entered in the Uniform Tests in 1948. Between 1948 and 1955, 65% of all of the lines tested in Maturity Groups (MGs) 0 to IV had Lincoln as a parent. It was common at that time to share partially inbred segregating populations among states. As a result, many of the lines tested from different states during this period had the same pedigree. Seven states had Uniform Test entries in MG 0 to IV derived from the Lincoln (2) × Richland cross. The release of the southern cv. Lee in 1953 had a similar effect on breeding in the South. Lee and its sister lines from the cross of S-100 × CNS dominated the pedigrees of the Southern Uniform Tests for a decade after the release of Lee. The cv. Lincoln and Lee now constitute major pillars in the foundation of the current genetic base of U.S. cultivars.

In assessing the formation of the genetic base in North America, it is useful to speculate upon other breeding factors that may have caused breeders to expand the genetic base relatively slowly after the releases of Lee and Lincoln. In the early years of North American breeding, soybean was not well adapted to harvest procedures, which were already mechanical. Shattering and lodging were common undesirable traits in North American soybean breeding populations. In addition, most southern cultivars were susceptible to bacterial pustule (BP, caused by *Xanthomonas axonopodis* pv. *glycinea*). Breeding in North America through the mid 1950s produced cultivars with improved yields and resistance to shattering, bacterial pustule, and lodging. Thus, the new cultivars were substantially better adapted than most exotic germplasm available as parental stock at that time. It is theorized that this early success in adapting the crop to North America may have made breeders reluctant to continue with the extensive use of new exotic germplasm in breeding, because it would reintroduce a considerable number of undesirable genes (e.g., for shattering or disease susceptibility or colored seed) into their programs that had just been eliminated or reduced. In other words the negative consequences of using exotic germplasm were obvious, in terms of a large nursery selection effort to eliminate deleterious traits, while the benefits in terms of improved end product were not. An example of this effect may be apparent in the use of the southern maturity accession Tokyo, introduced to North America in 1902. Tokyo yielded about as well as the landmark cv. Lee, which was released in 1953 (Gizlice et al., 1993b). Nevertheless, Tokyo was not used extensively by breeders and now contributes only 8% of the genes to southern cultivars whereas the parents of Lee contribute 40% (Gizlice et al., 1994; Sneller, 2003). Despite its good yielding ability, Tokyo was underutilized in southern breeding because it is susceptible to shattering and bacterial pustule, and has a green seed coat which is undesirable for oil processing (E.E. Hartwig, USDA, personal communication, 1991). Selection against such traits is frustrating and progeny escapes occur commonly, because expression of these traits can vary over years. Such experiences may have discouraged the use of Tokyo and perhaps other PIs, given that breeding programs were making progress using the already formed genetic base.

Another factor affecting the use of PIs after the releases of Lincoln and Lee is that many of the first North American soybean breeders came out of the discipline of plant pathology and were well trained and interested in pathology (Campbell et al., 1999). Diseases were quickly recognized as problems requiring breeding attention in the 1950s. Examples were phytophthora root rot (PRR, caused by *Phytophthora sojae*) in the Midwest and SCN in the South. The early North American breeders recognized that cultivars from Asia were important sources of qualitative genes for disease resistance, but probably viewed the Asian germplasm as undesirable for other field-related traits. To avoid the poor genome in which valuable disease-resistance genes resided, breeders commonly resorted to backcross breeding schemes to extract desirable genes from Asian germplasm. This method required small breeding populations and relatively little yield testing to recover the phenotype of the adapted cultivar plus the added trait. This was very important in an era where plots were hand threshed. Backcrossing allowed rather limited numbers of yield plots to be reserved for testing the premeir agronomic biparental crosses from their breeding programs that were expected to produce the most desirable progeny. Thus, North American breeders very likely intentionally minimized the introduction of new exotic germplasm into the genetic base as they backcrossed in new qualitative traits.

The cooperative nature of soybean breeding in the USA from the 1940s through the 1990s may also have been a key factor in limiting the rapid expansion of the genetic base. Most breeders jointly participated in the USDA cooperative regional tests of breeding lines and used the experimental lines in them freely for crossing. The easy availability of breeding material may have deflected breeding attention away from exotic introductions.

8–8.1.3 Diversity of the North American Genetic Base

Although small in number, the North American ancestral base apparently possessed enough diversity to fuel continual yield improvement over the past 70 yr. Evidence is accumulating that indicates it to be quite diverse for DNA markers as well (Diwan and Cregan, 1997; Keim et al., 1992; Kisha et al., 1998; Skorupska et al., 1993; Thompson and Nelson, 1998a). Marker analyses show that the diversity of major North American ancestors is no less than that of a relatively large set of PIs that has been used in recent population development and selection for yield improvement, nor less than that of the 32 major ancestors of modern Chinese cultivars (Brown-Guedira et al., 2000; Kisha et al., 1998; Li et al., 2001b). The major ancestors of North American soybean also display clear variation for morphological and biochemical traits (Gizlice et al., 1993a). They also possess genes for resistance to some of the major diseases affecting North American soybean production (see section 8–14).

Marker analyses indicate that there are patterns of diversity among ancestral lines (Brown-Guedira et al., 2000; Kisha et al., 1998; Thompson et al., 1998). Four major groupings of ancestors were indicated by these studies: (i) Lincoln, S-100, A.K. (Harrow), Illini, and Dunfield; (ii) Arksoy, Ralsoy, and Haberlandt; (iii) Mandarin (Ottawa), Richland, and Capital; and (iv) Jackson, Ogden, Roanoke, and Mukden. Of the remaining ancestral lines that contributed at least 1% of genes to the current gene pool, Korean, Perry and CNS were diverse from each other and the four other groups. These four major ancestral groupings were generally consistent with what is known about pedigrees, MG, and geographical origin of the ancestors. For example, in the first group, AK (Harrow) and Illini trace to a single introduction (A.K.) from China, and S-100 was selected from Illini (Bernard et al., 1988a). All three members of the second group have their putative origin in Korea, and Ralsoy was selected from Arksoy (Bernard et al., 1988a). In the fourth group, Tokyo, the only major plant introduction from Japan used in North American breeding, is in the pedigree of both Jackson and Ogden. Ancestors within a group tended to be somewhat similar in maturity. The only exception was Mukden, which was grouped with three much later maturing ancestors. Capital has been recorded as derived from the cross of the ancestor Strain 171 (now lost) × A.K. (Harrow), but DNA marker analysis does not support a relationship between the latter parent and Capital. Based on CP analysis, the first group of ancestors above contributed more than 35% of the parentage to recent U.S. cultivars, the second contributed 4%, the third group 22%, and the fourth group 14%. (Table 8–8).

8–8.2 Loss of Diversity as a Consequence of Breeding

While diversity in the North American ancestral base seems noteworthy, CP and molecular marker studies indicate that we have lost some of that diversity during cultivar development. Polymorphism information content values of RFLP markers indicated that recent northern cultivars were only 93% as polymorphic as the ancestral pool while southern cultivars were 82% as polymorphic as the ancestral pool (Table 8–9) (Keim et al., 1989; Kisha et al., 1998). These results agree well with CP analysis that indicated that public cultivars released from 1983 through 1988

Table 8–9. Mean coefficient of parentage (CP), genetic similarity (GS), and polymorphic information content (PIC) of North American soybean cultivars and their ancestors. Values for CP and GS estimates range from 0 (no similarity) to 1 (identical). Reported genetic distance (GD) measures ranged from 0 to 1 and were converted to GS values (GS = 1-GD).

Regional designation of cultivars or ancestors	Release era	Measure of similarity†	Mean value	Reference
Northern plus southern‡	1954–1958	CP	0.10	Gizlice et al. (1993a)
Northern plus southern‡	1969–1973	CP	0.11	Gizlice et al. (1993a)
Northern plus southern‡	1984–1988	CP	0.15	Gizlice et al. (1993a)
Northern plus southern	1989–1991	CP	0.17	Sneller (1994)
Northern plus southern	1999–2001	CP	0.17	Sneller (2003)
Northern vs. southern ‡#	1954–1958	CP	0.00	Gizlice et al. (1993a)
Northern vs. southern ‡#	1969–1973	CP	0.02	Gizlice et al. (1993a)
Northern vs. southern ‡#	1984–1988	CP	0.07	Gizlice et al. (1993a)
Northern vs. southern #	1989–1991	CP	0.10	Sneller (1994)
Northern vs. southern #	1999–2001	CP	0.12	Sneller (2003)
Northern only ‡	1954–1958	CP	0.18	Gizlice et al. (1993a)
Northern only ‡	1969–1973	CP	0.16	Gizlice et al. (1993a)
Northern only	1976–1980	CP	0.25	St. Martin (1982)
Northern only ‡	1984–1988	CP	0.21	Gizlice et al. (1993a)
Northern only	1989–1991	CP	0.23	Sneller (1994)
Northern only	1999–2001	CP	0.21	Sneller (2003)
Southern only ‡	1954–1958	CP	0.24	Gizlice et al. (1993a)
Southern only ‡	1969–1973	CP	0.25	Gizlice et al. (1993a)
Southern only ‡	1984–1988	CP	0.26	Gizlice et al. (1993a)
Southern only	1989–1991	CP	0.26	Sneller (1994)
Southern only	1999–2001	CP	0.22	Sneller (2003)
Northern plus southern		GS_{RFLP}	0.70	Keim et al. (1992)
Northern vs. southern[#]		GS_{RFLP}	0.62	Kisha et al. (1998)
Northern only		GS_{RFLP}	0.66	Kisha et al. (1998)
Northern only		GS_{RFLP}	0.84	Keim et al. (1989)
Southern only		GS_{RFLP}	0.70	Kisha et al. (1998)
Southern only		GS_{RFLP}	0.73	Sneller et al. (1997)
Ancestral lines		GS_{RFLP}	0.74	Keim et al. (1989)
Ancestral lines		GS_{RFLP}	0.62	Kisha et al. (1998)
Ancestral lines		GS_{RAPD}	0.58	Li et al. (2001), Thompson et al. (1998)
Northern plus southern		GS_{SSR}	0.36–0.42	Diwan and Cregan (1997)
Northern only		GS_{SSR}	0.50	Narvel et al. (2000)
Northern plus southern		PIC_{SSR}	0.69	Diwan and Cregan (1997)
Ancestral lines		PIC_{SSR}	0.80	Diwan and Cregan (1997)
Northern, southern, and ancestors		PIC_{SSR}	0.74	Rongwen et al. (1995)
Northern plus southern		PIC_{RFLP}	0.30	Keim et al. (1992)
Northern only		PIC_{SSR}	0.50	Narvel et al. (2000)
Ancestors		PIC_{RFLP}	0.39	Kisha et al. (1998)
Northern only		PIC_{RFLP}	0.36	Kisha et al. (1998)
Southern only		PIC_{RFLP}	0.32	Kisha et al. (1998)

† The RFLP, RAPD, and SSR subscripts are acronyms for restriction fragment length polymorphism, randomly amplified polymorphic DNA, and simple sequence repeat markers, respectively.

‡ Both commodity and specialty cultivars included in the analysis.

§ Northern refers to cultivars released that are Maturity Group mid-V and earlier.

¶ Southern refers to cultivars released that are later than Maturity Group mid-V.

The designation 'vs.' indicates that the measure of similarity is that between cultivars grown in two contrasting regions.

have twice as many genes in common as cultivars released prior to 1954 (Gizlice et al., 1993a). The average CP among commodity cultivars increased from the 1950s until the late 1980s, but has since remained relatively constant at 0.17 to 0.19 within the northern and southern regions of North America (Table 8–9).

8–8.3 Diversity in the North vs. South

Cultivars from the northern and southern regions of North America are not equally diverse (Table 8–9). Comparing cultivars within succeeding decades of release from 1960 through the 1990s, CP analyses indicated less diversity among southern lines than northern lines for most decades (Gizlice et al., 1993a, 1996; Sneller, 1994), a finding that is supported by molecular data (Kisha et al., 1998). This trend is particularly pronounced when diversity among MG VI or later cultivars is compared to diversity among MG V and earlier maturing cultivars (Gizlice et al., 1993a, 1996; Sneller, 1994, 2003). This is not surprising because nearly 40% of southern parentage derives from two ancestors, CNS and S-100 (Delannay et al., 1983; Gizlice et al., 1994; Sneller, 1994, 2003). These ancestors were used primarily as parents to produce a F_2 plant from which the cv. Lee and three sibs were derived (Carter et al., 1993). This single F_2 plant was the primary conduit for 37% of the parentage in southern soybean cultivars.

Another major factor that led to reduced genetic diversity in the southern region was the emphasis on breeding for SCN resistance in the South. The SCN was discovered in the South about 1953. The first resistant southern cultivar, Pickett (a backcross derivative of Lee), was released in 1965 (Brim and Ross, 1966). The SCN is a more recent pest in the northern growing area although the first northern SCN-resistant cultivar, Custer, was released in 1967 (Luedders et al., 1968), and the second, Franklin, was released in 1977 (Bernard and Shannon, 1980). At present, the majority of SCN-resistant cultivar releases have been in the South.

The CP relationships among southern SCN-resistant cultivars released in the 1980s were twice the average CP among SCN susceptible cultivars (0.39 vs. 0.18), and in the 1990s they were 1.6 times more related (0.29 vs. 0.18) (Carter et al., 2001). The high CP among SCN resistant cultivars resulted from two main factors. First, screening for SCN resistance is expensive and most lines from resistant × susceptible crosses are susceptible (because resistance is multigenic and largely recessive). For this reason, only a few adapted (but susceptible) cultivars were selected as recurrent parents to receive the SCN resistance, in the early days to SCN resistance breeding. In later years, to minimize further screening, resistant × resistant crosses were often used to develop new SCN-resistant cultivars. In the South, 54 and 42% (before and after 1988) of the publicly released SCN-resistant cultivars were derived from resistant × resistant crosses. In the North, this tendency has been much less (25 and 13% before and after 1988). Second, as new genes for SCN resistance were discovered over the decades, the five main exotic sources were all backcrossed into cultivars and breeding lines that were highly related to Lee prior to their further use in breeding. As breeders worked to pyramid these new SCN-resistance genes into cultivars, the genetics of an aging Lee were continually reincorporated into their programs.

Until recently, new SCN-resistant cultivars have been lower yielding than the best SCN-susceptible cultivars, under SCN-free conditions (Koenning, 2000). Many cultivars would not have been released except for their SCN resistance. In the South, where most SCN resistant cultivars have been bred, the low yield of SCN-resistant cultivars may be due to the genetic bottleneck for yield resulting from extensive parentage from Lee in SCN-resistant cultivars as described above (Hartl and Clark, 1997). The continuing trend of close relationships among SCN-resistant cultivars in the 1990s (both in the North and the South) suggests that breeders may want to take steps to incorporate SCN-resistance genes into less related lineages (McBroom, 1999; Carter et al., 2001).

In addition to the potential genetic bottleneck, the yield disparity between SCN-susceptible and -resistant cultivars may also be explained partially by the negative association of at least one SCN-resistance gene with yield in the absence of the pest (Kopisch-Obuch et al., 2002; Mudge et al; 1996). However, this linkage may have been broken in some midwestern cultivars. An additional factor related to low yields in SCN-resistant cultivars is that exotic sources of SCN genes have very low yields. Some of this poor genetic background may have been unintentionally incorporated into the SCN-resistant cultivars.

8–8.4 Other Patterns of Diversity in North American Breeding

In addition to the clear separation of northern and southern cultivars in North America, there are diversity patterns among cultivars within each region. These patterns can be explained by the predominant use of a few high-yielding cultivars as parents in cultivar development. Multivariate analysis of CP data led to the categorizing of public cultivars released from 1947 to 1988 into nine clusters (Gizlice et al., 1996); cultivars from the late 1980s were placed in four northern clusters and four southern clusters (Sneller, 1994). Molecular markers were also used to place cultivars in six clusters (Kisha et al., 1998). For most clusters, there is an obvious founder effect related to the extensive use of a historically important cultivar. Prominent successful cultivars that have been used extensively and thus form the nucleus of a cluster are Lee, Williams, A3127, Corsoy, Evans, S1346, Essex, Forrest, and Bedford (Gizlice et al., 1996; Kisha et al., 1998; Sneller, 1994). The lineages of North American clusters are not mutually exclusive, although a single ancestor is usually a dominant contributor to only one or two clusters (Gizlice et al., 1996; Sneller, 1994). Similar cluster patterns have been found for the Chinese gene pool (Cui et al, 2000b).

Recognizable diversity patterns among cultivars can also be related to breeding for local adaptation. The rationale for selection of parents, though it varies by breeding program, is largely a function of local adaptation. Maturity group is a key factor in adaptation, and breeders often hybridize parents of roughly similar maturity to develop new cultivars. Over time, this practice has resulted in higher CP relationships for cultivars within a MG than for cultivars as a whole (Gizlice et al, 1996; Sneller, 1994). Maturity group accounted for about 20% of the variation in CP among current cultivars released from 1999 to 2001 and 32% for public cultivars released before 1988 (Gizlice et al, 1996; Sneller, 2003). Another important trend in local adaptation is that North American breeders tend to favor the use of their own recently developed adapted genetic materials, so most crosses produced in a breeding program will have at least one parent developed in the local program. This practice has resulted in generally higher CP relationships for cultivars within a given North American breeding program than for cultivars as a whole (Gizlice et al, 1996; Sneller, 2003). This trend is sufficiently strong that multivaritate analysis of CP for breeding programs can be used to construct a genetic graph of breeding program relatedness that is remarkably consistent with a geographical map (Fig. 8–5) (Gizlice et al., 1996). Breeding programs accounted for 14 to 19% of the variation in CP values within northern and southern regions and 42% overall (Gizlice et al, 1996; Sneller, 2003). The originating breeding program and MG of cultivars in combination accounted for about 48% of the total variation in CP values among cultivars (Gizlice et al., 1996; Sneller, 2003).

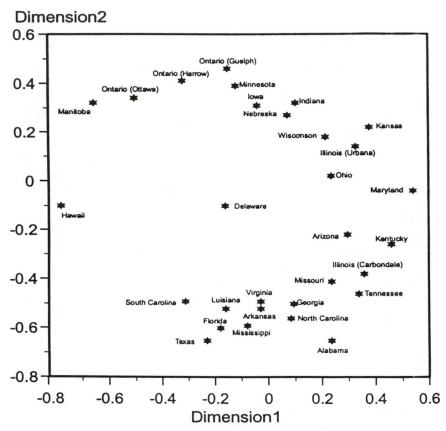

Fig. 8–5. Two-dimensional plot of 30 public breeding programs (BR) that released 258 cultivars together between 1947 to 1988. The complement of the linear distances (1-Distance) between any two cultivars estimates the coefficient of parentage (CP) between them. Distances ≥1 indicate no relationships. Multivariate analysis of CP values provided the results needed to produce a graph with clear geographical interpretation. From Gizlice et al. (1996).

The mean CP among current lines within some proprietary programs is substantially higher than the mean CP among cultivars as a whole (Sneller, 2003). This current limited diversity combined with recent trends toward limited germplasm exchange could be significant for the future of North American soybean breeding. Continued research to expand the genetic diversity of breeding programs by the public sector could be crucial in ameliorating this effect.

Lines from proprietary programs dominate the soybean seed market in most regions of North America. Thus far, there is very little divergence between recently released cultivars from public and proprietary sources in terms of their genetic base or in the mean CP averaged over all breeding programs (Sneller, 1994, 2003). Only 3% of the total variation in CP values among 312 MG 00 to VII cultivars from 1999 to 2001, could be accounted for by the contrast of public vs. proprietary lines when sampled from the same MG (Sneller, 2003). Proprietary lines appear to have more ancestry from Lincoln, and less from Mandarin (Ottawa), CNS, and S-100, than public lines, probably due to more extensive use of A3127 and its progeny by proprietary programs than public programs (Sneller, 1994, 2003).

Since 1996, many North American soybean cultivar releases have been Roundup Ready (Monsanto Co., St. Louis, MO), meaning they possess a transgene that confers tolerance to the herbicide glyphosate. The incorporation of this gene into proprietary cultivars has not had an impact on diversity among cultivars (Sneller, 2003). Recently released Roundup Ready and conventional cultivars tend to have the same mean CP relationship within and between the two groups. Two main factors have tended to minimize the impact of Roundup Ready breeding on diversity. First, the gene conferring tolerance to glyphosate was made available to many proprietary programs and was initially backcrossed at least once into 30 recurrent parents that effectively represented the commercial gene pool. This practice avoided potential bottleneck effects (Sneller, 2003). The final derivatives from these backcrossing programs were then used as parents with other recently released cultivars to generate the Roundup Ready cultivars available in 1999 to 2001. The second factor is a strong tendency for soybean breeders to mate glyphosate susceptible and resistant parents. Glyphosate can be sprayed as a screening tool to select true hybrid F_1's during vegetative growth and eliminate the contamination from selfing that commonly occurs. This susceptible × resistant crossing continues to further dilute possible founder effects of the original 30 recurrent parents.

8–8.5 Comparison of Diversity in North American Soybean and Other Crops

From the literature, it would appear that both pedigree and molecular analyses support the contention that the North American soybean cultivars are less diverse than U.S. sorghum [*Sorghum bicolor* (L.) Moench], U.S. maize (*Zea mays* L.), European maize, and Argentinean wheat [*Triticum aestivum* (L.)] populations, but more diverse than European barley, and Australian wheat populations (Tables 8–9 and 8–10). Based on pedigree analyses only, the North American soybean population is less diverse than U.S. oat (*Avena sativa* L.), and equally diverse as recent U.S. cotton lines (*Gossypium hirsutum* L.), U.S. rice (*Oryza sativa* L.), and U.S.

Table 8–10. Estimates of mean coefficient of parentage (CP), genetic similarity (GS), and polymorphic information content (PIC) from the analysis of elite populations in other crop species. Values for CP and GS measures range from 0 (no similarity) to 1 (identical). Reported genetic distance (GD) measures ranged from 0 to 1 and were converted to GS values (GS = 1-GD).

Description of population†	Measure of similarity ‡	Value	Reference
Barley, Canada, spring (2 and 6 row)	CP	0.08	Tinker et al. (1993)
Barley, Europe, spring	CP	0.13	Schut et al. (1997)
Barley, Europe, spring and winter	CP	0.26	Graner et al. (1994)
Barley, NA, spring, after 1970§	CP	0.12–0.19	Martin et al. (1991)
Barley, NA, spring, before 1971§	CP	0.10–0.11	Martin et al. (1991)
Cotton, USA, 1970–1990	CP	0.07	Bowman et al. (1996)
Cotton, USA, 1995	CP	0.20	Van Esbroeck et al. (1998)
Maize, Europe, flint and dent	CP	0.14	Messmer et al. (1993)
Oats, NA, 1951–1960§	CP	0.09	Souza and Sorrells (1989)
Oats, NA, 1961–1980§	CP	0.13–0.28	Rodgers et al. (1983)
Oats, NA, 1976–1985§	CP	0.08	Souza and Sorrells (1989)
Peanuts, after 1969	CP	0.21	Knauft and Gorbet (1989)
Rice, USA, Arkansas releases	CP	0.25–0.38	Dilday (1990)
Rice, USA, California releases	CP	0.37–0.50	Dilday (1990)
Rice, USA, Louisiana releases	CP	0.13–0.28	Dilday (1990)
Rice, USA, medium and long grain	CP	0.25	Cao and Oard (1997)
Rice, USA, Texas releases	CP	0.28–0.38	Dilday (1990)
Sorghum, USA	CP	0.07–0.08	Ahnert et al. (1996)
Sunflower, North America	CP	0.10–0.31	Cheres and Knapp (1998)
Wheat, Argentina	CP	0.12	Manifesto et al. (2001)
Wheat, Australia spring	CP	0.30	Souza et al. (1998)
Wheat, Canada, hard red spring, 1981–1990	CP	0.18	Mercado et al. (1996)
Wheat, CIMMYT, red spring	CP	0.19	van Beuningen and Busch (1997)
Wheat, durum, ICARDA, CIMMYT	CP	0.21	Autrique et al. (1996)
Wheat, Eastern USA, soft red winter	CP	0.15	Kim and Ward (1997)
Wheat, Eastern USA, soft white winter	CP	0.51	Kim and Ward (1997)
Wheat, Eastern USA, soft winter	CP	0.21	Kim and Ward (1997)
Wheat, Europe	CP	0.29	Plaschke et al. (1995)
Wheat, Germany and Austria	CP	0.05	Bohn et al. (1999)
Wheat, North American, spring	CP	0.18	Souza et al. (1998)
Wheat, Pacific NW, many classes	CP	0.04	Barrett et al. (1998)
Wheat, USA, hard red winter	CP	0.24	Cox et al. (1985b)
Wheat, USA, hard red winter	CP	0.26	Murphy et al. (1986)
Wheat, USA, hard red winter, 1984	CP	0.22	Cox et al. (1986)
Wheat, USA, soft red winter	CP	0.19	Murphy et al. (1986)
Wheat, USA, soft red winter, 1984	CP	0.15	Cox et al. (1986)
Wheat, USA, hard red spring, 1981–1990	CP	0.14	Mercado et al. (1996)
Wheat, USA-Canada, red spring	CP	0.16	van Beuningen and Busch (1997)
Wheat, USA-Canada, white spring	CP	0.25	van Beuningen and Busch (1997)
Barley, Europe, spring	GS_{AFLP}	0.80	Schut et al. (1997)
Canola, Europe, spring and winter	GS_{AFLP}	0.35–0.48	Lombard et al. (2000)
Rice, japonica	GS_{AFLP}	0.78	Mackill (1995)
Wheat, Argentina	GS_{AFLP}	0.55	Manifesto et al. (2001)
Wheat, Germany and Austria	GS_{AFLP}	0.61	Bohn et al. (1999)
Wheat, Pacific NW, many classes	GS_{AFLP}	0.46–0.42	Barrett et al. (1998)
Wheat, UK, 1990s	GS_{AFLP}	0.27	Donini et al. (2000)
Barley, Canada, spring (2 and 6 row)	GS_{RAPD}	0.68	Tinker et al. (1993)

(continued on next page)

Table 8–10. Continued.

Description of population†	Measure of similarity ‡	Value	Reference
Rice, japonica	GS_{RAPD}	0.75	Mackill et al. (1995)
Rice, USA, medium and long grain	GS_{RAPD}	0.73	Cao and Oard (1997)
Sorghum, diverse set of enhanced germplasm	GS_{RAPD}	0.87	Tao et al. (1993)
Barley, Europe, spring	GS_{RFLP}	0.85	Melchinger et al. (1994)
Barley, Europe, winter	GS_{RFLP}	0.84	Melchinger et al. (1994)
Maize, European	GS_{RFLP}	0.50	Dubreuil et al. (1996)
Maize, European flint and dent	GS_{RFLP}	0.34–0.43	Messmer et al. (1992)
Maize, European flint and dent	GS_{RFLP}	0.41–0.48	Messmer et al. (1993)
Maize, NA§	GS_{RFLP}	0.40	Dubreuil et al. (1996)
Maize, USA	GS_{RFLP}	0.37–0.42	Melchinger et al. (1990)
Maize, USA inbreds	GS_{RFLP}	0.40–0.57	Melchinger et al. (1991)
Oats, USA	GS_{RFLP}	0.87	Moser and Lee (1994)
Sorghum	GS_{RFLP}	0.67–0.76	Ahnert et al. (1996)
Sorghum, diverse set of enhanced germplasm	GS_{RFLP}	0.85	Tao et al. (1993)
Wheat, Australia	GS_{RFLP}	0.82	Paull et al. (1998)
Wheat, durum, ICARDA, CIMMYT	GS_{RFLP}	0.79	Autrique et al. (1996)
Wheat, Eastern USA, soft red winter	GS_{RFLP}	0.92	Kim and Ward (1997)
Wheat, Eastern USA, soft white winter	GS_{RFLP}	0.98	Kim and Ward (1997)
Wheat, Europe, spring	GS_{RFLP}	0.89	Siedler et al. (1994)
Wheat, Europe, winter	GS_{RFLP}	0.92	Siedler et al. (1994)
Wheat, Germany and Austria	GS_{RFLP}	0.65	Bohn et al. (1999)
Maize, Modern USA inbreds	GS_{SSR}	0.35–0.38	Lu and Bernardo (2001)
Rice, japonica	GS_{SSR}	0.64	Mackill et al. (1995)
Wheat, Argentina	GS_{SSR}	0.29	Manifesto et al. (2001)
Wheat, Europe	GS_{SSR}	0.44	Plaschke et al. (1995)
Wheat, Germany and Austria	GS_{SSR}	0.57	Bohn et al. (1999)
Wheat, UK, 1990s	GS_{SSR}	0.47	Donini et al. (2000)
Wheat, Argentina	PIC_{AFLP}	0.30	Manifesto et al. (2001)
Wheat, Germany and Austria	PIC_{AFLP}	0.32	Bohn et al. (1999)
Maize, USA, inbreds	PIC_{RFLP}	0.62	Smith et al. (1997)
Sorghum	PIC_{RFLP}	0.62	Smith et al. (2000)
Wheat, Eastern USA, soft red winter	PIC_{RFLP}	0.16–0.24	Kim and Ward (1997)
Wheat, Eastern USA, soft white winter	PIC_{RFLP}	0.07–0.09	Kim and Ward (1997)
Wheat, Eastern USA, soft winter	PIC_{RFLP}	0.15–0.20	Kim and Ward (1997)
Wheat, Germany and Austria	PIC_{RFLP}	0.33	Bohn et al. (1999)
Maize, USA, inbreds	PIC_{SSR}	0.59	Senior et al. (1998)
Maize, USA, inbreds	PIC_{SSR}	0.62	Smith et al. (1997)
Sorghum	PIC_{SSR}	0.58	Smith et al. (2000)
Wheat, Germany and Austria	PIC_{SSR}	0.30	Bohn et al. (1999)

† Barley (*Hordeum vulgare* L.), cotton (*Gossypium hirsutum* L.), maize (*Zea mays* L.), oat (*Avena sativa* L.), peanut (*Arachis hypogea* L.), Rice (*Oryza sativa* L.), Sorghum [*Sorghum bicolor* (L.) Moench],Wheat [*Triticum aestivum* (L.)].

‡ The AFLP, RFLP, RAPD, and SSR subscripts are acronyms for amplified fragment length polymorphism, restriction fragment length polymorphism, randomly amplified polymorphic DNA, and simple sequence repeat markers, respectively.

§ NA, North American.

barley. Comparison of soybean and wheat diversity is sometimes dependent on the populations and techniques employed in the comparison. Pedigree data suggest that the North American soybean population is less diverse than some U.S. and European wheat populations, while molecular data indicate the opposite.

8–9 STATUS OF DIVERSITY IN CHINESE SOYBEAN BREEDING

Soybean breeding emerged in China as early as 1913 with the establishment of the first soybean breeding institution at the Gongzhuling Agricultural Experiment Station in Jilin in the Northeast (Chang et al., 1993). Chinese breeding began with evaluation and selection of local landraces. Professor Shou Wang from the University of Jinling at Nanjing in southern China was the first recorded soybean breeder in China. He released the first improved soybean cv. Jin Da 332 for the lower Changjiang valley in 1923 as a selection from a landrace. Manual cross-pollination was first employed in 1927 at the Gongzhuling Agricultural Experiment Station. The first cultivar from hybridization, 'Man Cang Jin', was developed in 1935 and released in 1941. Artificial cross-pollination was slow to be adopted in most provincial breeding programs. Approximately 70% of the soybean cultivars released before 1960 in China originated as selections from existing soybean strains. From 1961 to 1980, 70% of the cultivars were developed using cross-pollination. Since 1981, hybridization has been the main method of generating variation, and selection within landraces is rarely used. Mutation breeding is less important than hybridization, but has been employed by several institutions. Although few North American cultivars have been developed through mutation breeding, 8% of the cultivars in China were developed using either mutagens alone (radiation or chemical), or hybridization combined with radiation or chemical mutagens applied to the F_1 (Cui et al., 1999). Prior to 1980, various forms of pedigree and mass selection were used after hybridization to develop inbred lines for testing. In the past 10 yr, single-seed descent has been used widely. As in North America, almost all cultivars are derived from plants in the F_4 or more highly inbred generations. A few breeders practice recurrent selection as well. Backcrossing and mating of full sibs are almost absent in Chinese breeding. No marker-assisted selection has been reported in China. Germplasm exchange and joint release of cultivars among breeding programs has generally been less common in China than in public programs in North America.

About 70 soybean breeding programs released cultivars before 1980. The primary breeding objectives were increased yield and lodging resistance (Cui et al., 1998). Sixty of them continued to develop cultivars after 1980. Today, the number of breeding programs has increased to 112 in China and more than 400 scientists are involved with soybean breeding research in 22 of the 28 provinces that grow soybean (Cui et al., 1998,1999; Chang et al., 1996b, 1999, Chang and Sun, 1991). Most of the individual Chinese breeding programs are small in comparison with those in the USA. A typical Chinese breeding program encompasses about 1 to 3 ha of plots as compared with at least 15 ha for a typical U.S. public program (Junyi Gai, Nanjing Agric. Univ., personal communication, 2001). Larger Chinese programs may have 5 to10 ha of field plots per year. Plot combines and mechanical planters are rare in China, but uniformity of field sites for yield testing, and statistical data analysis are comparable to that in North America. Soybean breeders in China earn much more academic credit for cultivar release than for publication of research. This reward system and a lack of extensive proprietary efforts has helped to keep an applied focus in public soybean breeding programs. Private breeding did

Table 8–11. Number of cultivars released in major soybean-producing provinces of China during 1923 to 1995.

Province of origin	Cultivars released	Province of origin	Cultivars released
	no.		no.
Anhui	22	Jiangxi	5
Beijing	23	Liaoning	55
Fujian	12	Neimenggu	7
Guangdong	4	Ningxia	2
Guangxi	2	Shandong	49
Guizhou	7	Shanxi	31
Hebei	23	Shaanxi	8
Henan	32	Sichuan	17
Heilongjiang	162	Tianjin	2
Hubei	11	Xinjiang	3
Hunan	15	Yunnan	2
Jilin	103	Zhejiang	9
Jiangsu	45		
		Total	651

Source: Cui et al. (1999).

not emerge until the late 1990s and currently plays a minor role in soybean improvement.

The National Soybean Breeding Program is equivalent roughly to a combination of the USDA-ARS soybean breeding effort, plus a grants program to affiliated institutions. The National Soybean Breeding Program in China has been the focus of an effort to strengthen soybean breeding and has been coordinated by the Soybean Research Institute at Nanjing Agricultural University (NAU) from 1986 to 1998. In 1998, the central government established the National Center for Soybean Improvement at NAU which currently coordinates national soybean research in China. The purpose of the center is to conduct and coordinate applied and basic research on new germplasm, new technology, and cultivar development, while the older provincial programs emphasize only cultivar development. The advent of regional testing programs for advanced breeding lines and the national coordination of soybean research has helped to promote germplasm exchange among breeding programs.

By 2000, more than 800 soybean cultivars had been released from Chinese public institutions (Table 8–11). Although improved soybean yield has always been a priority in China, it is estimated that yield gains for Chinese soybean breeding programs were low prior to 1970. Yield gains after 1980 have averaged approximately 1% per year for provincial programs and 1.5 to 2% per year for the National Soybean Breeding Program, which was initiated in the early 1980s (Cui et al., 1998).

8–9.1 Genetic Base of Chinese Cultivars

The genetic base of soybean breeding in China is quite large with 339 ancestors identified in the pedigrees of 651 Chinese soybean cultivars released from 1923 to 1995 (Cui et al., 2000a). Nearly 45% of these ancestors were added to the

Chinese base from the 1960s to 1995, demonstrating a continual expansion of the genetic base over time. Ancestors originating from China contributed 88% of the parentage to the Chinese genetic base. Pedigree analysis showed that 35 ancestors contributed 50% and 190 contributed 80% of the parentage to Chinese soybean cultivars. Unlike the North American genetic base, there are remarkably few dominant ancestors in the Chinese base (Table 8–12). The genetic bases of the three major soybean growing regions of China, Northeastern, Northern, and Southern were quite distinct and constituted almost independent gene pools (Cui et al., 2000a). A RAPD DNA-marker analysis of Chinese ancestors also indicated substantial differences among the three regional gene pools (Li et al, 2001b).

The continual expansion of the genetic base in China is explained by three main factors. First, most breeding institutions have maintained a local germplasm collection since the mid 1950s, allowing each breeding program easy access to landraces and instilling familiarity with the value of accessions. Thus, breeders continually identified promising landraces from the local collections and used them as parents. These local collections remained in place until the early 1990s, when a modern germplasm preservation facility, the National Gene Bank in Beijing became the repository for most collections. A few large breeding institutions still maintain germplasm collections today. A second factor fueling the expansion of the genetic base was the emergence of the National Soybean Breeding Program in the early 1980s. Development and utilization of new germplasm has been a major objective of the program. A third factor was the implementation of a research initiative, entitled "Broadening and Improvement of Genetic Base of Soybean in Northeast China" in the late 1980s. This project was heavily funded by the China National Natural Science Foundation for 5 yr in response to breeder concerns that soybean cultivars in Northeast had become closely related. The close relationship of cultivars was exemplified by the relatively high average CP values between all pairs of cultivars released from Heilongjiang province in the 1960s and 1970s (0.26 and 0.27, respectively), which were similar to the mean CP values between all pairs of Midwestern cultivars in North America breeding programs at the same period. The project involved breeders from all regions of China and reduced the mean CP among cultivars released from the Northeast after 1990 to <0.10 (Wang, 1994). Twelve recurrent selection populations were synthesized combining genetically distinct, agronomically superior parents from wide geographic areas. These recurrent selection populations have been employed in cultivar development.

Foreign cultivars have been used to develop Chinese soybean cultivars with high seed yield, lodging resistance, high protein and oil content, and resistance to soybean cyst nematode and frogeye leaf spot (FELS, caused by the fungus *Cercospora sojina* Hara) and early maturity (Cui et al., 1998, 1999). For example, Si Dou 11 was selected from the cross of Si Dou 2 Hao × Williams. It was released in 1987 in Jiangsu province and is still among the highest yielding cultivars in the province. Ji Dou 7 Hao was developed from the cross Williams × Cheng Dou 1 Hao and released in 1992 in Hebei province (Cui et al., 1998). A total of 54 exotic cultivars were used as parents in Chinese soybean breeding: 24 from the USA, 13 from Japan, the rest from Canada, England, Russia, and Sweden. Mamotan, Amsoy, Clark 63, and Beeson from the USA, and Shi Sheng Chang Ye and Ye Qi 1 Hao from Japan appear more frequently in pedigrees of Chinese soybean cultivars than do other ex-

Table 8–12. Genetic contribution of 35 ancestors that contribute 50% of the pedigree to 651 modern Chinese soybean cultivars in three growing regions of China: northeastern China (NEC), northern China (NC), and southern (SC) China. Contribution is based on coefficient of parentage analysis. Ancestors are ranked in terms of overall genetic contribution to Chinese cultivars. The column marked 'Cultivars derived' sums to a much greater number than the number of released cultivars in China because multiple ancestors usually appear in the pedigree of a single cultivar.

				Genetic contribution of ancestor to			
			All				
Code†	Ancestor name	Country of origin	Chinese cultivars	Cumu-lative	NEC cultivars	NC cultivars	SC cultivars	Cultivars derived
					%			no.
A142	Jin Yuan	China	6.583	6.58	11.947	1.469	0.310	244
A203	Si Li Huang	China	5.038	11.62	9.369	0.795	0.188	219
A019	Bai Mei	China	3.389	15.01	6.598	0.139	0.000	132
A071	Du Lu Dou	China	3.209	18.22	5.020	1.701	0.676	93
A219	Tie Jia Si Li Huang	China	2.537	20.76	4.189	1.104	0.338	89
A309	Shi Sheng Chang Ye	Japan	2.047	22.80	3.924	0.179	0.000	52
A288	Unnamed	China	1.958	24.76	0.032	5.574	0.845	60
A129	Ji Mo You Dou	China	1.584	26.34	0.000	4.489	0.798	55
A149	Ke Shan Si Li Jia	China	1.565	27.91	3.002	0.134	0.000	57
A327	Mamotan	USA	1.468	29.38	0.038	4.045	0.845	61
A033	Bin Hai Da Bai Hua	China	1.451	30.83	0.000	3.609	1.680	62
A081	Feng Xian Sui Dao Huang	China	1.301	32.13	0.000	0.357	6.957	20
A224	Tong Shan Tian E Dan	China	1.223	33.35	0.032	3.371	0.704	61
A253	Yong Feng Dou Huang	China	1.152	34.50	1.970	0.238	0.451	20
A247	Yi Du Ping Ding	China	1.151	35.65	0.000	3.358	0.394	53
A234	Xiao Jin Huang	China	1.133	36.79	2.150	0.134	0.000	27
A240	Xiong Yue Xiao Huang Dou	China	1.041	37.83	1.607	0.580	0.225	58
A220	Tie Jia Zi	China	1.005	38.83	1.616	0.575	0.000	30
A002	51–83	China	0.922	39.75	0.000	0.447	4.562	19
A221	Tie Jiao Huang	China	0.843	40.60	0.000	2.406	0.394	49
A284	Unnamed	China	0.841	41.44	0.000	0.000	4.932	11
A235	Xiao Jin Huang	China	0.819	42.26	1.236	0.536	0.113	29
A128	Hun Chun Dou	China	0.759	43.02	1.307	0.179	0.225	14
A202	Si Li Huang	China	0.752	43.77	1.484	0.000	0.000	19
A317	Amsoy	USA	0.749	44.52	1.099	0.595	0.000	19
A319	Clark 63	USA	0.749	45.27	0.455	1.369	0.451	13
A318	Beeson	USA	0.730	46.00	0.379	0.833	1.577	12
A204	Si Li Huang	China	0.638	46.63	1.146	0.179	0.000	21
A209	Suo Yi Ling	China	0.595	47.23	1.174	0.000	0.000	19
A340	Williams	USA	0.538	47.77	0.000	1.429	0.451	9
A038	Da Bai Ma	China	0.528	48.29	0.000	1.637	0.000	7
A194	Shang Hai Liu Yue Bai	China	0.528	48.82	0.000	0.000	3.097	9
A118	Hua Xian Da Lu Dou	China	0.518	49.34	0.000	1.607	0.000	12
A127	Hui Nan Qing Pi Dou	China	0.510	49.85	0.974	0.050	0.000	20
A104	Hai Lun Jin Yuan	China	0.490	50.34	0.597	0.580	0.000	21
Total			50.344	50.34	61.345	43.698	30.213	

† Code numbers with the A prefix (abbreviation for ancestor) are consistent with those of Cui et al. (1999). Consult this reference for a more detailed description of origin, phenotype, and use in breeding.

otic cultivars. The cv. Mamotan and Shi Sheng Chang Ye contributed genes to more than 50 Chinese soybean cultivars. Interestingly, Mamotan is not a part of the genetic base for modern U.S. cultivars. Of the 651 Chinese soybean cultivars released between 1923 to 1995, 222 contain genes from exotic cultivars (Cui et al., 2000a).

Of these 222, 79 have a foreign cultivar as a direct parent, 86 have an foreign grand-parent, and three were developed by selection within foreign cultivars. Although foreign cultivars have been used as parents to develop successful Chinese cultivars, the nature of the agronomic alleles derived from the foreign cultivars has not been determined and awaits study. Cultivars from the USA are perceived by Chinese breeders as more lodging resistant than most Chinese cultivars (Junyi Gai, Nanjing Agric. Univ., personal communication, 2001). This observation may relate to the use of U.S. cultivars in Chinese breeding programs.

8–9.2 Coefficient of Parentage Among Chinese Cultivars

The large number of ancestors employed in Chinese soybean breeding, plus the virtual absence of half-sib, full-sib and backcross matings, have led to very low mean CP among Chinese cultivars. The mean CP among Chinese soybean culti-vars released from 1923 through 1995 was only 0.02, indicating the potential for a high level of genetic diversity in Chinese soybean breeding (Cui et al., 2000b). The CP relationships for cultivars between and within major regions, provinces, crop-ping systems, and release eras were also very low. Use of exotic germplasm (i.e., cultivars from other countries), continued use of landraces in breeding, and initia-tives by the Chinese National Soybean Breeding Program to increase diversity in applied breeding were apparently responsible for a decline in mean CP relationships from 0.07 to 0.02 for cultivars released in the 1960s vs. the 1990s.

Although the mean CP among cultivars was low, cluster analysis based on CP analysis detected pronounced patterns of diversity. Multivariate analysis of CP for provinces produced a graph with clear geographical interpretation (Fig. 8–6). Of 651 cultivars, 270 were assigned to 20 clusters, each with a mean CP >0.25. These clusters explained 41% of the total variability in CP. Parent-offspring and grand-parent-grandchild relationships accounted for a large portion of the pedigree rela-tionships among cluster members. The close relationships among lines within clus-ters also tended to result from long pedigrees involving the mating of cousins. Most or all members of a cluster came from a single province or breeding program. Each well-defined cluster represented a historical path of breeding success in China and had a clear breeding interpretation that could be seen in the well-defined and con-trasting genetic base for each cluster. Although there are few dominant ancestors in the Chinese genetic base overall, the majority of the cultivars within a single clus-ter were derived primarily from a few important ancestors (Cui et al., 2000b). Sim-ilar positive clustering results have been found in the North American gene pool (Gizlice et al., 1996; Kisha et al., 1998; Sneller, 1994). Clustering of a subset of these same Chinese cultivars using phenotypic similarity estimates produced roughly similar results compared with CP-based clustering (Cui et al., 2001).

8–10 STATUS OF DIVERSITY IN JAPANESE SOYBEAN BREEDING

Soybean arrived in Japan from Korea or China approximately 2000 yr ago (Kihara, 1969). Soybean accessions from Japan and Korea have been shown to be genetically distinct from those in China (Li and Nelson, 2001), and Japan is an important source of germplasm with very large seed, extremely early maturity, insect resistance, and slow wilting under drought stress (Voldeng et al., 1997; Sloane et al., 1990; Van Duyn et al., 1971). Currently, there are 11 public breeding programs releasing soybean cultivars in Japan.

8–10.1 Genetic Base of Japanese Cultivars

The genetic base of 86 public Japanese cultivars registered during 1950 to 1988 consisted of 74 ancestors (Zhou et al., 2000). Eighteen and 53 ancestors contributed 50 and 80% of the genetic base, respectively (Table 8–13). Ancestors originating from Japan contributed 91% of the parentage to the Japanese genetic base,

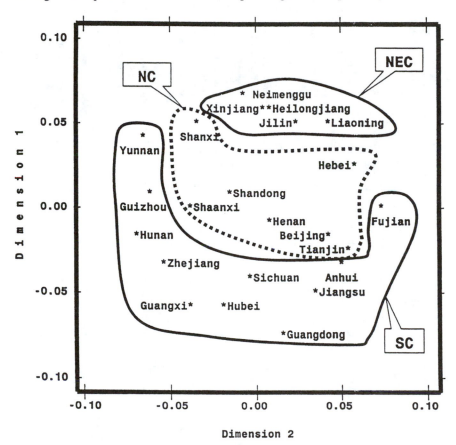

Fig. 8–6. Graph of coefficient of parentage (CP) relationship among 23 of 25 Chinese provinces that released a total of 651 soybean cultivars during 1923 to 1995. Coordinates for a province were obtained from a multidimensional scaling procedure employing two dimensions based on the CP relations. The 23 provinces were superimposed upon the graph to clarify geographical interpretation of the analysis (northeastern China, NEC; northern China, NC; and southern China, SC). The CP between any two provinces can be estimated as 0.0985 - linear distance between them, where 0.0985 is the maximum off-diagonal CP value in Table 8–3. Distances ≥0.0985 indicate no relationship. Ningxia and Jiangxi provinces were deleted from the graph because cultivars developed and released in these provinces were not related to any other provinces. From Cui et al. (2000b).

Table 8–13. Genetic contribution of 35 major ancestors to 86 modern public Japanese soybean cultivars and their first progeny in three regions of Japan: Central Japan (CJ), Northern Japan (NJ), and Southern Japan (SJ). Genetic contribution is based on coefficient of parentage analysis. Ancestors are ranked in terms of overall contribution to Japanese cultivars. First progeny cultivars were defined as those cultivars directly descended from one or more of the 35 ancestors. In some cases, a cultivar may be a first progeny for more than one ancestor. The column marked 'Total cultivars derived' sums to a much greater number than the number of released cultivars in China because multiple ancestors usually appear in the pedigree of a single cultivar.

Code†	Ancestor name	Genetic contribution of ancestor to					Total cultivars derived	First progeny ‡
		Japanese cultivars	Cumulative	CJ cultivars	NJ cultivars	SJ cultivars		
		%					no.	
A014	Geden Shirazu	6.00	6.00	9.17	3.91	0.00	22	Fuku Shirome, Hourai, Nema Shirazu, Toyosuzu
A004	Ani §	5.37	11.37	9.83	0.00	0.59	16	Bon Minori, Enrei, Fuji Otome, Fujimijiro, Hourei, Misuzu Daizu, Nasu Shirome, Oku Mejiro, Shiromeyutaka, Tachi Suzunari, Tachinagaha, Tanrei, Ugo Daizu
A044	Ooyachi	4.52	15.89	2.09	12.21	0.00	24	Fukunagaha, Hokkai Hadaka, Hourai, Hourei, Karikachi, Kitahomare, Kitamusume, Kogane Jiro, Mutsu Mejiro, Nagaha Jiro, Nanbu Shirome, Oshima Shirome, Shinsei, Tachinagaha, Tokachi Shiro, Wase Kogane, Wase Shirome
A013	Daizu Hon 326	4.43	20.32	0.61	14.71	0.00	22	Fukunagaha, Hokkai Hadaka, Hourai, Hourei, Karikachi, Kitahomare, Kitamusume, Kogane Jiro, Nagaha Jiro, Nanbu Shirome, Oshima Shirome, Shinsei, Tachinagaha, Tokachi Shiro, Wase Kogane
A068	Tsuru No Ko§	3.16	23.48	1.36	8.72	0.00	9	Fukunagaha, Himeyutaka, Kitahomare, Kitakomachi, Mutsu Shiratama, Toyomusume, Tsurukogane, Yuuzuru
A055	Shiroge	2.89	26.37	5.20	0.00	0.59	11	Fuji Otome, Fujimijiro, Shiromeyutaka, Tachinagaha, Wase Shiroge
A003	Akasaya§	2.62	28.99	1.63	0.00	9.38	6	Aki Sengoku, Gogaku, Hyuuga, Shiro Sennari
A037	Nezumi Saya	2.62	31.61	4.89	0.00	0.00	7	Fukumejiro, Kokeshi Jiro, Shin Mejiro
A020	Kamishunbetsu Zairai	2.54	34.15	0.00	9.11	0.00	8	Karikachi, Shinsei
A019	Iyo	2.33	36.48	0.00	0.00	12.50	5	Aki Sengoku, Aso Musume, Hougyoku
A035	Nangun Takedate	2.03	38.51	3.80	0.00	0.00	5	Oku Shirome
A016	Hanayome Ibaraki 1	1.74	40.25	3.26	0.00	0.00	5	Fukumejiro, Shin Mejiro
A032	Matsuura	1.74	41.99	0.00	0.00	9.38	3	Fuji Musume, Kogane Daizu, Sayohime
A036	Nattou Kotsubu	1.74	43.73	2.17	2.08	0.00	2	Kosuzu, Suzumaru
A060	Si Li Huang‖	1.74	45.47	0.00	6.25	0.00	5	Nagaha Jiro, Oshima Shirome, Tokachi Shiro, Tsurukogane

A063	Tamanishiki§	1.74	47.21	0.00	0.00	9.38	4	Aso Musume, Hougyoku
A057	Shirosaya§	1.60	48.81	1.90	0.00	3.13	5	Higo Musume, Orihime, Tamamusume
A062	Takiya	1.60	50.41	0.27	0.00	7.81	6	Fuji Musume, Higo Musume, Kogane Daizu, Orihime, Sayohime, Tachiyutaka
A066	Toshi Dai 7910	1.60	52.01	0.00	5.73	0.00	4	Toyosuzu
A011	Chuu Teppou	1.45	53.46	2.17	0.00	1.56	3	Shirotae, Tamahikari, Toyoshirome
A051	Shiro Daizu 3	1.45	54.91	0.00	0.00	7.81	3	Akiyoshi, Fukuyutaka
A061	Souga Zairai	1.45	56.36	2.17	0.00	1.56	3	Shirotae, Tamahikari, Toyoshirome
A029	Kuro Daizu	1.38	57.74	2.58	0.00	0.00	6	Bon Minori, Enrei, Tachi Suzunari, Tanrei
A018	Houjaku	1.16	58.90	2.17	0.00	0.00	2	Miyagi Oojiro, Nakasennari
A033	Misao	1.16	60.06	1.09	0.00	3.13	2	Aso Aogari, Himeshirazu
A034	Miyagi Shirome	1.16	61.22	2.17	0.00	0.00	3	Miyagi Oojiro, Tachinagaha, Tachiyutaka
A058	Shirosota #	1.16	62.38	2.17	0.00	0.00	2	Mutsu Mejiro, Wase Shirome
A015	Hakubi‖	1.09	63.47	0.00	3.91	0.00	5	Kogane Jiro, Wase Kogane
A028	Kuma	1.02	64.49	0.00	0.00	5.47	3	Aso Masari
A042	Ooita Aki Daizu 2	1.02	65.51	0.00	0.00	5.47	3	Aso Masari
A052	Shiro Hachikoku	0.90	66.41	1.68	0.00	0.00	4	Hourei, Misuzu Daizu, Nasu Shirome, Oku Mejiro
A023	Kimusume	0.76	67.17	1.43	0.00	0.00	3	Tamamusume
A031	Mandarin (Ottawa) ††	0.76	67.93	1.43	0.00	0.00	3	Dewa Musume, Suzuyutaka, Tachiyutaka
A008	Chougetsu	0.73	68.66	1.36	0.00	0.00	3	Daruma Masari, Hatsukari, Mutsu Shiratama
A009	Choutan	0.73	69.39	0.00	0.00	3.91	5	Fuji Musume, Higo Musume, Kogane Daizu, Orihime, Sayohime
A022	Kariha Takiya	0.73	70.12	0.00	0.00	3.91	5	Fuji Musume, Higo Musume, Kogane Daizu, Orihime, Sayohime
Total		70.1	70.1	66.6	66.6	85.9		

† Code numbers with the A prefix (abbreviation for ancestor) are those of Zhou et al. (2000).

‡ Just five Japanese cultivars were released during 1950 to 1988 that were not first progeny: Raiden, Raikou, Suzukari, Tachikogane, and Wasesuzunari.

§ The ancestor is present as multiple accessions, but with differing phenotypes in the Japanese collection.

¶ These ancestors were introduced from China.

Ancestors introduced from Korea. PI 84751 was introduced to Japan via USA.

†† These ancestors introduced from the USA: CNS, S-100, Tokyo, and PI 54610 were used to develop NC1-2-2, which appears in the pedigree of Hourei; CNS and S-100 were used to develop Lee, which appears in the pedigree of Tomahomare; AK (Harrow) and Mandarin (Ottawa) were used to develop Harosoy, which appears in the pedigree of Dewa Musume. CNS, PI 54610, S-100, AK (Harrow) and Mandarin (Ottawa) were off-type selections found in field evaluation plots of plant introductions brought to the USA from China before 1945. Tokyo was introduced to the USA from Japan in 1902, prior to the application of manual hybridization to soybean breeding.

while ancestors from the North America, China, and Korea contributed only 2, 5, and 2%, respectively. Cultivars from the northern, central, and southern growing regions of Japan have very distinct genetic bases with at least 50% of ancestral composition of each region unique to that region. Many of these ancestors are preserved in the Japanese soybean germplasm collection in Tskuba. However, multiple contrasting landraces bearing the name of an ancestor are also present, so that the authentic ancestor can be difficult to determine. Passport data for landraces have been used to suggest which accessions are most likely the true ancestors of Japanese breeding (Miyazaki et al.,1995b).

Twelve Japanese cultivars released from 1950 through 1988 derived at least 25% of their parentage from improved U.S. or Chinese breeding materials. Japanese cultivars derived from North American germplasm were all released after 1977, whereas, most cultivars derived as direct progeny from Chinese parents were released before 1960, indicating a shift in the source of diverse germplasm used by breeders over time. If one assumes that a cross was made 10 yr before release of a cultivar and 8 yr before release (or use) of a breeding line from that cross, then almost all Chinese ancestors used as parents were employed before 1955 and all other exotic ancestors, except one, were used after 1965. This switch coincided with historical events. From 1931 to 1945, Japan occupied northeast China, providing Japanese breeders with access to Chinese germplasm. Commercial trade and presumably germplasm exchange continued between Japan and China until the Korean conflict in the early 1950s, but was minimal thereafter for some years (Schaller, 1985).

In contrast, the Japanese began to show much interest in U.S. germplasm only until around 1958 (Creech, 2000). The main impetus for Japanese breeders to use North American germplasm was the liberalization of Japanese soybean import policies in 1961 (S. Miyazaki, National Inst. Agrobiological Resources, personal communication, 1999). As a result of the changes, domestic soybean production lost competitiveness on the world market and decreased dramatically. Therefore, Japanese soybean breeders experienced increased pressure to breed new and more profitable cultivars with superior food processing quality, yield, and better disease resistances. The availability and extensive characterization of U.S. breeding materials for traits such as SCN resistance, the more difficult access to Chinese germplasm after the Korean conflict of the 1950s, and the incomplete characterization of germplasm in the Chinese collection at that time, induced Japanese soybean breeders to turn their attention to North American germplasm.

8–10.2 Coefficient of Parentage Analysis of Japanese Cultivars

The mean CP for the 86 cultivars released from 1950 to 1988 was low (0.04), indicating a potentially high degree of diversity in Japanese breeding (Zhou et al., 2002; Miyazaki et al., 1995a). Eighty percent of all pairs of cultivars were completely unrelated by pedigree. The low mean CP was attributed to continual incorporation of unique Japanese landraces into the genetic base over time, the introduction of foreign germplasm from China and North America, and minimal exchange of germplasm among Japanese breeding programs in different regions. Despite the low overall CP relation for the Japanese cultivars, multivariate analysis clearly identified meaningful clusters with geographical interpretation (Fig. 8–7). Fifty-four of the 86 cultivars were placed in six clusters that had a mean CP of 0.25 or greater. The remaining cultivars could not be assigned effectively to a cluster. The clusters explained 57% of the CP variation and had few ancestors in common. Cultivars within a cluster were derived primarily from a single or few breeding programs. Clusters represented primarily family relationships such as parent-offspring, full-sibs, and half-sibs. Backcrossing and matings of half- and full-sibs were absent in Japanese pedigrees. Despite the low CP relationships among growing regions, release eras, MGs, and developing institutions, these factors did not explain much of the pedigree diversity among the Japanese cultivars. This result reflects the fact that CP relationships within these factors were usually as low as the CP relationships among them so that, in a statistical sense, they were not an important discriminator of patterns of diversity. Although lines within a cluster were generally derived

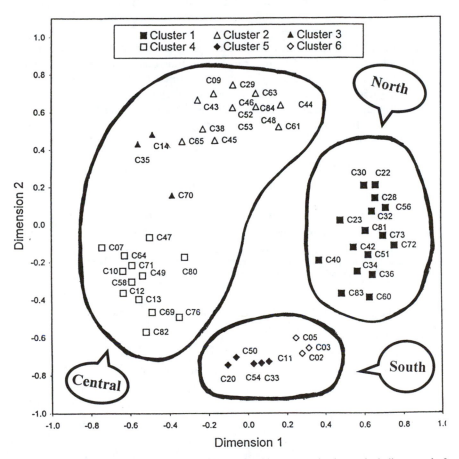

Fig. 8–7. Graph of coefficient of parentage (CP) relationships among six clusters including a total of 54 Japanese soybean cultivars released from 1950 to 1988. Coordinates for the cultivars were obtained from a multidimensional scaling analysis employing two dimensions based on CP relations. The complement of the linear distances (1-Distance) between any two cultivars estimates the CP between them. Distances ≥1 indicate no relationship. From Zhou et al. (2002).

from a single or few breeding programs, a single program may have contributed to more than one cluster.

8–11 COMPARISON OF CHINESE, JAPANESE, AND NORTH AMERICAN BREEDING

8–11.1 Limited Exchange Among Breeding Programs

The breeding programs of Japan, China, and North America have all produced a large number of cultivars over the past 60 yr (Table 8–14) and have achieved their successes in relative isolation from each other. This isolation resulted primarily from the language barrier between Asia and North America and from political barriers which inhibited germplasm exchange with China after WW II until the decline of the the the cultural revolution in the mid-1970s. North Americans have not used modern cultivars from Asia in developing commodity cultivars.

A common feature of diversity in each program is extensive regional independence of breeding programs, in terms of pedigree, that coincides with latitude (i.e.. the northern and southern regions of North America, and the northern, central, and southern regions of China and Japan). The regional isolation within a country or continent arises primarily from the sensitivity of soybean to photoperiod, which makes it difficult to assess and use germplasm outside its adapted latitude, thus preserving regional distinctions in pedigrees.

8–11.2 Relationships Among Genetic Bases of North America, China, and Japan

In North America, the Asian names of introduced soybean genotypes were corrupted or simply replaced by English names at the time of entry. Thus, names

Table 8–14. Number of publicly released cultivars released in China, Japan, and North America (USA and Canada). Both specialty and commodity cultivars are included.

Country	Region	1950–1959	1960–1969	1970–1979	1980–1989	1990–2000	Total
Japan	Northern	2	9	4	9†		24
	Central	7	17	11	11†		46
	Southern	4	9	1	2†		16
	Total	13	35	16	22		86
China	Northeastern	26	41	72	119	57‡	315
	Northern	4	15	54	86	49‡	208
	Southern	2	5	13	50	40‡	110
	Total	32	61	139	255	146	633
USA and	Northern	19	39	53	107	237	455
Canada§	Southern	9	10	28	39	63	149
	Total	28	49	81	146	300	604

† Period spans only 1980 to 1988.
‡ Period spans 1990 to 1995.
§ More than 2000 private cultivars have now been released in North America (data not shown).

of North American ancestors are not helpful in determining their genetic relationship to ancestors of Japanese and Chinese breeding. Passport data for these ancestors indicated that most were collected in China. There are a few ancestor names in common between the Chinese and Japanese genetic bases (Cui et al., 2000a; Zhou et al., 2000); however, the genetic similarity of ancestors with common names is unknown.

Pedigree analysis has shown that the genetic base of China is considerably larger than that of Japan or North America. Eighty percent of the genetic base in China, Japan, and North America could be accounted for by 190, 53, and 13 ancestors, respectively (Table 8–15). Each of the three major Chinese growing regions had more ancestors and a more evenly distributed ancestral contribution compared to that of the total North American genetic base. Although North America and Japanese genetic bases had a similar total number of ancestors, the contribution of Japanese ancestors was more evenly distributed than that of North American ancestors (Table 8–15). Soybean cultivars in Japan, China, and North America are derived from a combined total of only about 700 distinctly named ancestors (Table 8–15). Because only a few of many existing landraces became ancestors of modern breeding in the respective countries, it is improbable that the ancestors for the North American and Asian breeding programs overlapped greatly. Indeed, a survey of RAPD markers for ancestors indicated that the genetic bases of Chinese and North American breeding are quite distinct and that there was as much marker diversity among the major North American ancestors as among 32 major Chinese ancestors (Li et al., 2001b). Regional distinctness of ancestors within China was greater than for North America, however.

8–11.3 Genetic Relatedness of Cultivars between and within China, Japan, and North America

Pedigree analysis of cultivars revealed that cultivar CP relationships were much lower in China and Japan than in the North America (i.e., 0.03, 0.09, and 0.21, when averaged within regions for each, respectively). This result reflects the smaller genetic base employed, the greater tendency to share genetic materials across breeding programs, and the more dominating effect of a 'breakthrough cultivar' in North America than in Asia. Coefficient of parentage reveals that although breeding programs are quite distinct in China, Japan, and North America, there has been more isolation between breeding programs within China and Japan, historically, than in North America. The greater diversity of Asian vs. North American soybean and the genetic improvement in yield in Asia suggests that Asian breeding populations have sufficient genetic diversity to fuel breeding progress for decades. Although North America has adequate genetic diversity to make progress in applied breeding, it is possible that North America may experience a yield plateau from limited diversity well before Asia.

Marker analyses (SSR, AFLP, RAPD, and RFLP) all indicate that Japanese cultivars are strikingly different from North American and Chinese cultivars. There is some overlap between Chinese and North American cultivars, but they also form fairly distinct populations (Carter et al., 2000). (Fig. 8–8, 8–9, 8–10, and 8–11). A RAPD marker analysis of landraces from China and Japan showed that Chinese

Table 8–15. A summary comparison of the Chinese, Japanese, and North American (NA) soybean cultivars developed through modern breeding procedures.

	Chinese	Japanese	NA (by 1988)	NA † (by 2000)
Cultivars released, no.	651	86	258	572
Ancestors in genetic base, no.	339	74	80	152 ‡
Ancestors originating from abroad, no.	47	16	80	--
Ancestors which contributed 50% of the genes to modern cultivars, no.	35	18	5	5
Ancestors which contributed 80% of the genes to modern cultivars, no.	190	53	13	15
Ratio of ancestors in genetic base to cultivars released§	0.53 (China) 0.7 (NEC) 1.0 (NC) 1.4 (NE)	0.86 -- -- --	0.32 (NA) 0.40 (N) 0.50 (S)	0.27 (NA) 0.28 (N) 0.47 (S)
Mean ancestors per cultivar 1951–1960, no.	1.78	2.62	2.9	--
Mean ancestors per cultivar 1961–1970, no.	2.26	2.71	5.3	--
Mean ancestors per cultivar 1971–1980, no.	2.58	2.69	7.0	--
Mean ancestors per cultivar 1981–1990, no.	4.21	4.68	10.0	--
Mean ancestors per cultivar 1951–1990, no.	3.79	3.20	6.7	--
Average coefficient of parentage for all cultivars	0.02	0.04	0.13	0.13
Average coefficient of parentage for cultivars within a defined region§	0.06 (NEC) 0.04 (NC) 0.02 (SC)	0.13 (NJ) 0.07 (CJ) 0.07 (SJ)	0.18 (N) 0.23 (S)	0.17 (N) 0.22 (S)
Average coefficient of parentage of cultivars between two defined regions§	0.007 (NEC-NC) 0.002 (NEC-SC) 0.007 (NC-SC)	0.012 (NJ-CJ) 0.000 (NJ-SJ) 0.006 (CJ-SJ)	0.04 -- --	0.06 -- --
Average coefficient of parentage within era of release				
1923–1950	0.041	--		0.108
1951–1960	0.048	0.021		0.088
1961–1970	0.072	0.047		0.111
1971–1980	0.023	0.034		0.137
1981–1990	0.017	0.037		0.127
1991–2000	0.016	--		0.138

Source: Carter et al. (1993 and 2001), Gizlice et al. (1994 and 1996), Zhou et al. (2000), Cui et al. (2000a, 2000b).

† Data from analysis of 572 public cultivars, including 97 specialty cultivars, released from 1947 to 2000 in North America.

‡ Ninety-four ancestors from crosses for commodity cultivars, 34 ancestors unique to specialty cultivars, and 23 ancestors included in those random mating populations from which one or more cultivars were released.

§ Regional acronyms are NEC, northeast China; NC, north China; SC, south China; NJ, northern Japan; CJ, central Japan; SJ, southern Japan; N, northern North America; and S, southern North America.

germplasm were much more diverse (Li and Nelson, 2001). Modern Japanese cultivars also tended to be less diverse for DNA markers than cultivars from the USA, indicating the possibility that Japanese soybean may have experienced a genetic bottleneck that is not reflected in pedigree analyses when soybean was introduced from the mainland of Asia.

Phenotypic diversity of modern Chinese and North American cultivars has been compared directly. Forty-seven Chinese and 25 North American cultivars were

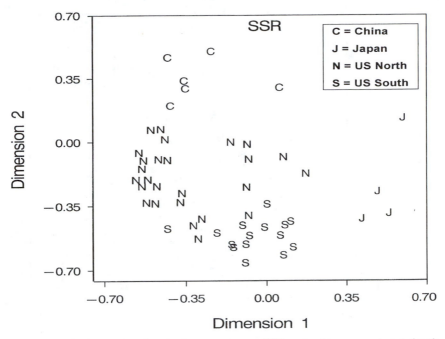

Fig. 8–8. Genetic distances, based on simple sequence repeat (SSR) marker data, among selected of northern and southern U.S. cultivars to those Project SAVE (Soybean Asian Variety Evaluation) Asian cultivars which yielded at least 80% of U.S. cultivars in U.S.-grown yield trials. Plots were obtained from
multidimensional scaling analysis and retain genetic distance relations on a scale from 0 to1. Distances >1 indicate no relationships. Distances <1 estimate the percentage of marker loci for which
two genotypes differ. From Carter et al. (2000).

evaluated for 25 phenotypic traits in growth chambers (Cui et al., 2001). Both sets
showed diversity, though the Chinese cultivars were more diverse than North American cultivars for 24 of the 25 traits. Multivariate analysis of these phenotypic traits
indicated that, with a few exceptions, phenotypic diversity in U.S. cultivars was a
subset of the diversity found in Chinese cultivars (Fig. 8–12). Cultivars from the
southern region of China were the most unique having, among other traits, the highest levels of seed protein content, thinnest leaves, and lowest leaf N and chlorophyll
contents (Cui et al., 2001).

8–12 THE ROLE OF EXOTIC GERMPLASM IN BREEDING FOR
INCREASED YIELD POTENTIAL IN NORTH AMERICA

Yield improvement is the most sought after objective in soybean breeding.
An important question for soybean breeders is whether or not diversity for seed yield
within the commercial gene pool is sufficient to sustain desirable rates of gain in
the future. It is theoretically possible that yield plateaus in breeding progress could
develop if current diversity is inadequate. When yield plateaus are not evident, inadequate diversity may lead to rates of gain from selection that are less than max-

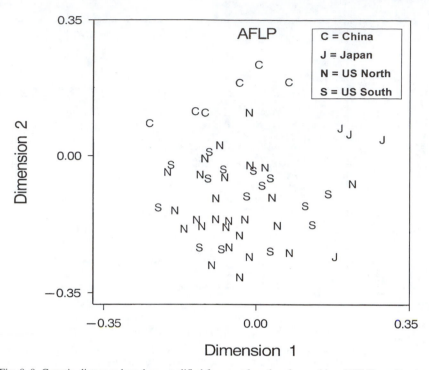

Fig. 8–9. Genetic distances based on amplified fragment length polymorphism (AFLP) marker data, among selected of northern and southern U.S. cultivars to those Project SAVE (Soybean Asian Variety Evaluation) Asian cultivars which yielded at least 80% of U.S. cultivars in U.S.-grown yield trials. Plots were obtained from multidimensional scaling analysis and retain genetic distance relations on a scale from 0 to1. Distances >1 indicate no relationships. Distances <1 estimate the percentage of marker loci for which two genotypes differ. From Carter et al. (2000).

imum. Despite enormous advancements in genetic science and technology, however, very little is known about the genetics of seed yield. Thus, a definitive answer to this question is currently not available. If diversity is not adequate, however, the challenge is to identify the best source of alternative germplasm and most efficient strategy for extracting the useful diversity. In this section, we will use results from more than 50 yr of breeding and research in North America to present what is known and define the critical research issues regarding the role of exotic germplasm in breeding for increased yield potential.

8–12.1 Breeding Progress for Yield Improvement in North America

Specht and Williams (1984) documented steady improvement of yield to approximately 1975 through breeding. Recent evidence indicates that breeding progress continued in the northern USA and was greater in the 1990s than in the 1980s (25 vs. 8 kg ha^{-1} yr^{-1}) (Zhou et al., 1998a), a result consistent with the conclusions of others that genetic gain for yield may be increasing (Specht et al., 1999; Wilcox, 2001). This increase in gain is directly proportional to the increase in the

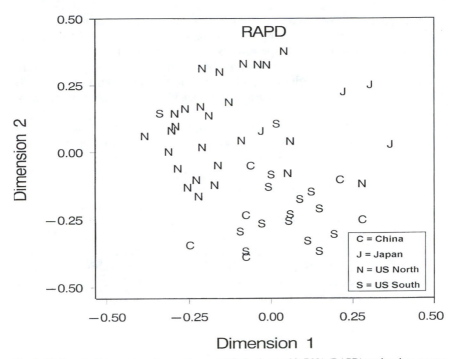

Fig. 8–10. Genetic distances based on random amplified polymorphic DNA (RAPD) marker data, among selected of northern and southern U.S. cultivars to those Project SAVE (Soybean Asian Variety Evaluation) Asian cultivars which yielded at least 80% of U.S. cultivars in U.S.-grown yield trials. Plots were obtained from multidimensional scaling analysis and retain genetic distance relations on a scale from 0 to1. Distances >1 indicate no relationships. Distances <1 estimate the percentage of marker loci for which two genotypes differ. From Carter et al. (2000).

number and size breeding programs in the midwestern USA (Zhou et al., 1998a). There is evidence that the rate of yield improvement in Canada is also increasing (Voldeng et al., 1997). The improved rate of gain observed in Canada has been attributed to an infusion of cold tolerance from exotic sources (Voldeng et al., 1997), while increases in gain in the USA are derived from a relatively unchanged genetic base. Based on the effective population size of the current gene pool in the Midwest, it is estimated that the yield of our best current cultivars are not close to a theoretical limit to yield (St. Martin, 2001).

    In the USA, there is less diversity among southern cultivars than among northern cultivars (Gizlice et al., 1993a, 1996; Sneller, 1994, 2003; Kisha et al., 1998), especially in MG VI or later. The major ancestors of southern cultivars were less diverse based on marker analysis (Kisha et al., 1998), and effective population size for southern breeding was more restricted than for the North (i.e., there were fewer southern breeders, they developed fewer lines and, thus, there were fewer desirable lines overall for use as parents). While gain from selection appears to be increasing in the North, research indicates it decreased in the South from 17 to 2 kg ha^{-1} yr^{-1} from the 1980s to the 1990s (Zhou et al., 1998a). The higher rate of progress in the 1980s in the South was related in part to the high-yielding southern cv. Hutcheson released in 1987 (Ustun et al., 2001). This cultivar continued to out-perform most publicly developed breeding lines and cultivars through the 1990s. Exotic germplasm contributes 25% of the parentage of Hutcheson.

While overall yield improvement is obvious in the northern regions of North America, there is evidence that a lack of diversity may limit gain from selection for specific bi-parental crosses. Most breeders work with populations derived from bi-parental crosses. Populations derived from bi-parental crosses of genetically distant parents are more likely to have high genetic variance for yield than are populations derived from crosses of parents that are more related (Kisha et al., 1998; Helms et al., 1997; Manjarrez-Sandoval et al., 1997a). See section 8–7 for more details. These studies indicate that sufficient genetic variation for yield can be obtained from the crossing of the highest yielding materials available, but that limited genetic diversity for yield between parents renders many such crosses useless. Overall genetic gain for yield in North America would be greater than at present if breeders could accurately identify and dedicate resources to crosses with the most potential for yield advancement and avoid crosses that generate little variation.

Collectively, results to date suggest that there is still sufficient diversity within the commercial gene pool to sustain yield progress. Results do not address whether added diversity from outside the current genetic base would further enhance

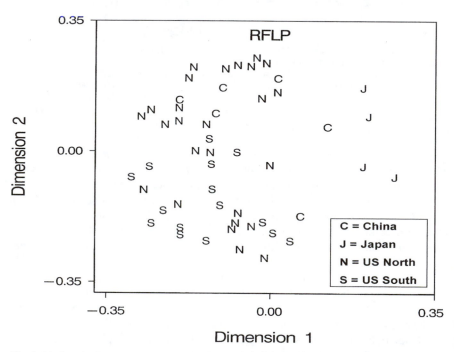

Fig. 8–11. Genetic distances based on restriction fragment length polymorphism (RFLP) marker data, among selected of northern and southern U.S. cultivars to those Project SAVE (Soybean Asian Variety Evaluation) Asian cultivars which yielded at least 80% of U.S. cultivars in U.S.-grown yield trials. Plots were obtained from multidimensional scaling analysis and retain genetic distance relations on a scale from 0 to1. Distances >1 indicate no relationships. Distances <1 estimate the percentage of marker loci for which two genotypes differ. From Carter et al. (2000).

breeding progress. It is well known that the initial North American genetic base did not include many important disease resistance alleles and that they were added later from the germplasm collection. By analogy, it seems highly likely that the initial North American genetic base did not include all of the favorable alleles for seed yield which may be present in *G. max* globally. Analysis of the genetic base indicates that the North American commercially used gene pool includes <1% of the

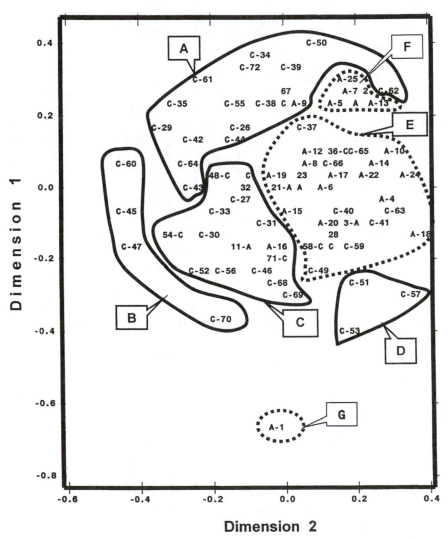

Fig. 8–12. Two-dimensional representation of genetic relations among 47 modern Chinese and 25 North American (NA) soybean cultivars derived from a two-dimensional multidimensional scaling (MDS) analysis based on phenotypic similarity (PS) estimates. These PS estimates were derived from multivariate analysis of phenotypic traits for plants grown in temperature- and photoperiod-controlled growth chambers. A = NA cultivar, C = Chinese cultivar, numbers suffixed to the A or C abbreviation correspond to those in Table 1 of Cui et al. (2001). Seven clusters identified using Ward's minimum variance cluster analysis are superimposed on the MDS graph. From Cui et al. (2001).

G. *max* landraces or derivatives now in collections. A recent survey of soybean breeders by the United Soybean Board indicated that expansion of diversity in applied breeding was an important long-term objective (Kent Van Amburg, personal communication, 2002). The full impact of such an expansion, if it occurred, may not be realized until two or more cycles of breeding are completed with new materials. The discovery of individual yield genes from exotic sources and their introgression into current breeding lines could have a more immediate impact.

8–12.2 Techniques for Selection of Appropriate Exotic Germplasm for Use in North American Yield Improvement Programs

Selecting exotic parents for a yield improvement program is just as critical as selecting adapted parents; however, selecting exotic parents is more difficult. Exotic parental lines will almost always be lower yielding than domestic cultivars, primarily because they are not as well adapted to specific environmental conditions, are susceptible to disease complexes (Hartwig and Lehman, 1951; Wilcox and St. Martin, 1998), and/or have lower genetic potential for yield. The assumption in using exotic parents in yield improvement is that, despite their low yield, they might contain specific alleles that increase yield of productive cultivars. Exotic parents can be selected based on their phenotype per se or in testcrosses (Kenworthy, 1980). To judge the useful of testcrosses, F_5 and F_6 populations (Thorne and Fehr, 1970), F_3 bulk populations (St. Martin and Aslam, 1986), F_2 bulk populations (Reese et al., 1988), F_2-derived lines (Sweeney and St. Martin, 1989), and F_1 hybrids (Lewers et al., 1998) have been evaluated. All indicated that a per se evaluation was as effective as a testcross evaluation and required fewer resources. Measuring heterosis in the F_1 generation may have some utility in identifying low yielding parents with genetic potential, but the choice of tester is obviously important and producing sufficient quantities of hybrid seeds limits the usefulness of this technique in soybean (Lewers et al., 1998).

Typically, the exotic parents used in yield enhancement programs have been selected based on their agronomic performance (Fehr and Cianzio, 1981; Thompson et al., 1998). This practice does not appear to reduce the genetic differences between the exotic germplasm and the domestic gene pool. Comparisons between the more agronomically desirable exotic parents and domestic cultivars using RFLPs (Sneller et al., 1997), RAPDs (Thompson et al., 1998), or SSRs (Narvel et al., 2000) have consistently found distinct DNA marker differences between the exotic and domestic groups. Within the exotic groups, marker-based genetic distance of the exotic accessions from domestic cultivars was unrelated to agronomic performance of the exotic type (Sneller et al., 1997). Plant introductions have also been selected successfully as parents for yield improvement based on their tolerance to drought (T. Carter, personal communication, 2002).

8–12.3 Breeding With Exotic Germplasm In Yield Improvement Programs: Case Studies

Efforts to incorporate new genetic diversity into the base of U.S. soybean breeding have been a small part of U.S. soybean breeding since the 1960s. Plant introductions used for such breeding efforts have been selected for agronomic performance, although most of these exotic lines yielded considerably less than the commercial cultivars (Nelson et al., 1987; Sneller et al., 1997; Thompson and Nelson, 1998b).

Thorne and Fehr (1970) evaluated two-way (domestic × PI) and three-way crosses ([domestic × PI] × domestic) and found that the crosses with 75% domestic pedigree were higher yielding and had a larger genetic variance than the crosses with only 50% domestic parentage. Khalaf et al. (1984) found that progenies from three-parent crosses ([exotic × domestic] × domestic) were more variable and had higher frequencies of superior lines than two- or four-parent crosses with equal numbers of exotic and domestic parents. Wilcox et al. (1984) recommended maintaining the identity of different F_1 families derived from ([exotic × domestic] × domestic) crosses because of the significant differences that exist among these families. Hartwig (1972) crossed 12 exotic lines, obtained from three countries, with the cv. Hill. The best progeny yielded only 80% of the yield of Hill. Schoener and Fehr (1979) evaluated the relationship between the percentage of exotic parentage in a population (100, 75, 50, 25, and 0%) and its agronomic performance. In general, average yield increased as the percentage of exotic parentage decreased (Fehr and Clark, 1973). While all populations possessed some high-yielding lines, the population with 0% exotic parentage had the most lines that were more than one standard deviation above the mean of the test. The authors concluded that using exotic germplasm for short-term yield improvement may not be warranted; however, a long-term selection program using populations with 50% exotic parentage might be successful. Follow-up research was conducted and, after three cycles of recurrent selection, the greatest mean yield and gain from selection continued to come from the population derived only from domestic parentage (Ininda et al., 1996). Vello et al. (1984) evaluated genetic variability in populations developed from 40 exotic lines and 40 domestic cultivars (Fehr and Cianzio, 1981). They also found that as the percentage of exotic parentage increased, the rate of yield advance decreased and also concluded that exotic germplasm cannot enhance the short-term yield improvement of soybean cultivars.

Research has failed to identify breeding procedures that can systematically produce new superior cultivars derived from exotic germplasm and maintain a high level of genetic variation for yield. However, several high-yielding lines have been identified from crosses involving exotic germplasm. The cv. S1346 was developed by the Northrup-King Company from a cross with PI 257435 as one of the parents. This cultivar has been widely used as a parent and PI 257435 now is a key ancestor in explaining clustering of current northern cultivars using pedigree data (Table 8–8) (Sneller, 1994, 2003). The cv. IVR 1120 was derived from a three-way cross involving PI 91110-1. Through IVR 1120, this PI is now in the ancestry of cv. IA2038. PI 71506 is a grandparent of Ripley, a northern cultivar, and of Hutcheson, one the most successful publicly developed cultivars of the 1990s. Fifty percent of the parentage of N7001, released in 2000, is from PI 416937. This PI was selected as a parent because of its drought tolerance (Sloane et al., 1990). Selected breeding lines derived from N7001 (MG VII and VIII) have consistently yielded higher than the standard reference cultivars and almost all other entries in recent USDA regional uniform tests (Paris and Bell, 2002). Thompson et al. (1999) and

Brown-Guedira et al. (2003) released six high-yielding experimental lines with 10 exotic lines in their pedigrees that had not been previously used in U.S. soybean breeding. These introductions represent at least five genetic groups, as defined by RAPD markers, that are distinct from the major North American ancestral lines (Brown-Guedira et al., 2000). In 2001 and 2002, the USDA and the University of Illinois released seven additional high-yielding experimental lines derived from 11 different exotic lines. The highest yielding line, released in 2001, was derived from four introductions and no domestic germplasm. This line yielded 95% of the best cultivar in the USDA regional test (Nowling, 2000). An experimental line with 25% exotic parentage released in 2002 was the highest yielding entry in the 2001 Preliminary IIB Regional Test (Nowling, 2001).

It has been demonstrated that introgressing genes from exotic germplasm can increase the yield of specific commercial cultivars. Lines derived from one back-cross (BC_1) that yielded significantly more than the recurrent parent (Beeson 80) were developed using PI 407720 and PI 68658 as donor parents (Thompson and Nelson, 1998b). Using Elgin as a recurrent parent, significantly higher yielding BC_1 lines were developed using PI 436684, PI 253665D, and PI 283331 as donor parents and yields of some BC_2 lines exceeded that of Beeson 80 using PI 297544 and PI 407720 as donor parents (Procupiuk et al., 2001). Five of the six donor parents were tested in 11 environments over 2 yr (Carter et al., 2000) and averaged only 65 to 75% of the yield of the mean of the commercial cultivars in the test.

Knowledge of the genetics of yield would help evaluate the need and facilitate the utilization of exotic germplasm to improve yield. Recent mapping studies are beginning to provide this information. In a large recombinant inbred line (RIL) population derived from the cross of Minsoy (PI 548389) and Noir 1 (PI 290136) (both introductions from France), three QTLs for yield were identified. Segregation at the QTLs explained 12, 7, and 6% of the genetic variation for yield, respectively (Mansur et al., 1996). These chromosome segments were also associated with height, stem termination type, time of flowering, and maturity (Mansur et al., 1996; Specht et al., 2001), traits that are well known to affect yield and already in use by breeders. A less important QTL for yield was also detected and the allele for higher yield was negatively associated with chlorosis. In a population with 50% exotic germplasm pedigree, SSR markers were used to find nine putative QTLs with the positive yield allele coming from the exotic parent that based on means from six environments (Kabelka et al., 2003). All were independent of known genes or QTL affecting maturity, height, and other plant traits and at least two of these yield QTL were previously identified in two other populations (Orf et al., 1999; Specht et al., 2001; Yuan et al., 2002). At least one putative QTL with a positive allele for seed yield was found in both exotic germplasm (Kabelka et al., 2003) and in a U.S. cultivar (Yuan et al., 2002). The value of these positive yield alleles in backgrounds other than the indicated mapping populations is not known. Three yield QTLs were identified in a RIL population developed from the mating of Noir 1 to Archer, a northern cultivar (Orf et al., 1999). At one locus, the positive yield allele came from Noir 1 and was not associated with height or maturity. At the two other loci, the positive yield allele came from the adapted cultivar Archer. When these loci were introgressed into southern cultivars, there was no increase in yield (Reyna and Sneller, 2001). While there are reports of favorable yield alleles which are stable over environments, there are no reports of favorable yield alleles that have been successfully introduced into multiple, contrasting, adapted genetic backgrounds (Concibido et al., 2002; Reyna and Sneller, 2001).

The above studies make it clear that yield-enhancing alleles do exist in exotic germplasm. They do not indicate that the yield genes in the exotic sources are actually novel or superior to those in the existing breeding pool. Putatively beneficial alleles should be confirmed or validated in additional high-yielding backgrounds to assess whether the alleles have widespread or limited value in the gene pool. This is an active area of study, but answers regarding uniqueness of yield alleles will be slow in forthcoming because of extensive resources required for adequate yield testing. Increased use of exotic germplasm in applied breeding will probably be driven as much by the availability of high-yielding cultivars with exotic pedigrees as by detailed studies which exhaustively prove that a QTL allele from a PI is unique for yield. Making significant changes in the genetic composition of a national soybean breeding program as large as the one in the USA is a slow process. The cultivars and experimental lines mentioned above are being used as parents in breeding programs. The ultimate genetic contribution of the exotic parents to the North American gene pool will depend on the extent to which their progeny are used in future breeding efforts.

Given the theoretical and practical concerns of using traditional QTL approaches to identify yield alleles, alternatives approaches should be evaluated. It has been postulated that QTL discovery could be more relevant and efficient if conducted in complex breeding populations commonly used by variety development programs (Beavis, 1998). For example, instead of employing a single biparental population, an alternative could be to use testcross or factorial crossing programs to discover QTL and simultaneously assess their value over multiple genetic backgrounds. Adapted parents could be selected to represent the main ancestors of the commercial pool. This approach would involve many populations, probably too many to use traditional mapping within each. Thus, one may need to use marker-association or marker-sib analyses on small selected set of lines from each population to discover putative QTLs that are stable over genetic backgrounds (Bink et al., 2002; Fulker and Cardon, 1994).

8–12.4 Modern Asian Cultivars as Sources of Genetic Diversity for Yield Improvement

The relative independence of the large national breeding programs of Japan, China, and North America, and the apparent diversity between these pools (see previous sections), has prompted breeders interested in genetic diversity to consider these programs as independent reservoirs of genetic diversity. Four different marker systems have shown that cultivars developed in the late 20th century in each country are genetically quite distinct (Nelson et al., 1998; Carter et al., 2000) (Fig. 8–8, 8–9, 8–10, and 8–11). In addition, RAPD analysis shows that the major ancestors of Chinese and North American soybean cultivars are genetically different (Li and Nelson, 2001). Chinese and U.S. cultivars have also been shown to differ markedly from each other phenotypically (Cui et al., 2001). It is possible that unique yield genes could have been accumulated into cultivars during 60 yr of breeding in each

Table 8–16. Promising Asian cultivars from modern breeding expressed as a percentage of adapted public and private that yielded from 80 to 88% of elite U.S. check cultivars of comparable maturity. Results were compiled from a minimum of five locations per test in each of 2 yr. None of these Asian cultivars have parents in common with the North Amercan cultivars, except Ji Dou 7 Hao and Tamahomare (see footnotes). These modern Asian cultivars are among 108 modern that were tested in a series of experiments.

Asian cv.	USDA accession	MG†	Origin
Hei Nong 37	PI 592921	I	China
Nakasennari	PI 561388	V	China
Tamahomare‡	PI 507327	VI	Japan
Nen Feng 9 Hao	PI 511866	0	China
Ji Dou 7 Hao§	PI 592936	II	China
Misuzudaizu	PI 423912	V	Japan
Hyuuga	PI 506764	VIII	Japan
Tong Nong 8 Hao	PI 592926	I	China
Ken Nong 2 Hao	PI 592923	0	China
Tsurokogame	PI 594304B	II	Japan
Suzumaru	PI 593972	I	Japan

Source: Carter et al. (2000).
† Maturity group.
‡ Fifty percent Lee in the pedigree (parent).
§ Fifty percent Williams in the pedigree (parent).

national program and these accrued differences could be exploited in intercountry crosses. Pedigree analysis was used to identify 108 diverse modern cultivars from Japan and China for extensive multi-year yield trials. Yield results indicated that ten yield much better than typical landraces from Asia and only 12 to 20% less than adapted U.S. cultivars (Table 8–16, Carter et al., 2000; Manjarrez-Sandoval et al., 1998). Population development using these modern Asian cultivars is underway and some success has been obtained (T.E. Carter, Jr. and R.L. Nelson, USDA, personal communication, 2002).

8–13 NORTHERN AND SOUTHERN NORTH AMERICA AS MUTUALLY IMPORTANT RESERVOIRS OF DIVERSITY FOR YIELD IMPROVEMENT

8–13.1 Regional Distinctions in North America

The gene pool of North American soybean breeding has been characterized for genetic diversity using pedigree, DNA marker, and phenotypic data. All studies of diversity in North American soybean clearly separate cultivars from the northern and southern growing regions. More than 20% of the variation of CP values among North American public commodity cultivars can be explained by the north-south classification (Gizlice et al., 1996). This separation is also readily detected by genetic markers (Diers et al., 1997; Keim et al., 1992; Kisha et al., 1998; Nelson et al., 1998; Sneller et al., 1997). Historical analyses indicated that this separation has existed from the inception of North American soybean breeding due to differential use of ancestors adapted to northern and southern regions (Table 8–8; see also Delannay et al., 1983; Sneller, 1994, 2003; Gizlice et al., 1994). All publicly released cultivars which are earlier than Group II maturity have very little pedigree in common with cultivars which are MG V and later (Gizlice et al., 1996). Group II and III cultivars, by virtue of their median maturity and the large number of breeding programs in these maturities, have a modest pedigree association with cultivars of other MGs.

In North America the most important ancestors of northern cultivars are from northern China, whereas the most important ancestors of southern cultivars for North America are from southern China. Phenotypically, there is a trend for modern cultivars from northern North America and northeastern China to be lower in seed protein while most modern cultivars from southern North America and southern China tend to be higher in protein (Cui et al., 2001; Burton, 1989). This trend can also be seen in the ancestors of modern U.S. cultivars (Gizlice et., 1994). The northern and southern regions of North America also tend to be very distinct in terms of stem termination (mostly determinates in the South and indeterminates in the North). In China, there is also a numerical trend toward separation of stem-termination types from Northeast to South, although both indeterminates and determinates are found in most provinces, as well as semi-determinates (Cui et al., 1999). Modern Chinese cultivars released through 1995 have a distribution of indeterminant, semi-determinant, and determinant stem types as follows: Northeast China (171, 100, 59), North China (58, 47, 103), and South China (1, 18, 91). This distinction may be less drastic among the landraces, however. The underlying basis for the less distinct separation of stem termination types by region in China compared to North America is not clear.

8–13.2 Cultivar Development from North × South Crosses

Breeders have exploited the contrasting diversity in northern and southern North America to some extent through intercrossing. This breeding approach is an area of active research. Some North × South crosses have been motivated by the desire to transfer SCN resistance from southern cultivars to midwestern cultivars (see the disease resistance section of this chapter). Other crosses were the result of ideotype breeding, which introduced determinate growth habit (conferred by the $dt1$ gene) from the South into lines adapted to the northern USA (Cooper, 1981). The $dt1$ gene greatly reduces plant height in the short-season cultivars adapted to the northern USA and Canada, thereby reducing lodging that can limit yield in highly productive environments (Cooper, 1985). The first determinate cultivars adapted to the Midwest (Elf, Gnome, Hobbit, and Sprite) were all derived from the cross of Williams × Ransom. Incorporation of other genes in addition to $dt1$ from southern cultivars into northern cultivars probably contributed to the high yield of the determinate cultivars in productive environments; a line near isogenic to Williams possessing the $dt1$ allele for determinancy does not yield as much as Hobbit, which is similar in maturity. Other crosses between parents adapted to the northern and southern regions have been made primarily to capitalize upon yield advances in contrasting regions. Asgrow A3127, one of the most successful cultivars and parents in North American soybean breeding history (Sneller, 1994), derives from a cross of the cv. Essex, adapted to the southern USA, with Williams, adapted to the North. These parents have almost no pedigree in common. Interestingly, at least 16 other cultivars have been derived from the same cross. Other cultivars (e.g., Graham, Burlison, Delsoy 4900, Duocrop, and Egyptian) have also been derived from crosses between cultivars of northern and southern maturity.

The release of cultivars derived from parents adapted to different regions is anecdotal evidence that the genes of cultivars of one region are diverse and complementary to those in the other region. However, it is important to note that many such crosses do not lead to cultivar releases. Few studies have attempted to determine the underlying basis for successful choices of northern × southern crosses or the nature of yield advantages that can occur from northern × southern matings. It is impossible to directly compare the yield potential of northern and southern cultivars because of the effect of photoperiod. Indirect evaluation of the yield potential of northern and southern cultivars can be achieved by crossing northern and southern parents and evaluating selected progeny with uniform maturity. In such tests, several cultivars adapted to the northern USA appeared to produce southern maturity progeny with yield potential equal to current southern cultivars (Sneller et al., 1997). In several crosses of northern and southern cultivars, breeding lines that yield better than the southern parent have been derived, with some northern parents appearing better than others (Cornelious and Sneller, 2002). Feng (2001) mated a series of old and modern northern cultivars to recently released southern cultivars and developed southern maturity breeding lines from them. Yield trials at four sites indicated that populations derived from the recent northern cultivars were higher yielding than populations derived from older northern cultivars. Because the modern northern cultivars are higher yielding than the older cultivars when tested in the Midwest, these results indicate that breeding progress for yield in the North can be capitalized upon in the South.

Although the uniqueness of yield genes from midwestern and southern breeding programs has not been established, high-yielding cultivars from each region appear to have mutual utility as parents. Given that breeding effort and breeding progress is greater in the Midwest than in the South, interregional crosses may have increased merit in the South in the future. It is important to note that successful exploitation of North × South crosses requires nursery populations that are 5 to 10 times larger than for intra-regional crosses, so that selection may be practiced successfully to eliminate undesirable traits such as shattering, disease susceptibility, and unadapted maturity prior to yield evaluations.

8–13.3 Impact of North × South Breeding on Cultivar Relatedness

Development of cultivars through the hybridization of northern and southern North American cultivars has increased the percentage of parentage from major southern (CNS, S-100) and northern (Lincoln) ancestors in the genetic base of the complementing region (Sneller, 2003). This appears to be due in part to the extensive use of cv. A3127, itself derived from a cross between a northern and southern parent, and its progeny in both northern and southern breeding programs. Not all historically important lineages have been transferred between regions via interregional crossing. For example, Mandarin (Ottawa), a major ancestor of northern cultivars that matures much earlier than Lincoln or A3127, continues to make almost no contribution to the southern gene pool (Table 8–8). There has been a slight decrease in average CP within the northern and southern regions in recent years for proprietary but not public commodity cultivars (Table 8–9) (Sneller, 2003; Carter et al., 2001). The small decline in CP for proprietary cultivars has been attributed to hybridization of parents derived from inter-regional crosses, particularly in the South, and the significant contribution of two plant introductions (PI 257435 and PI 75106) (Sneller, 2003). Consequently, commercial cultivars from the two regions are becoming somewhat more related to one another over time (Table 8–9).

8–14 BREEDING FOR IMPROVED PEST RESISTANCE USING EXOTIC GERMPLASM IN NORTH AMERICA

One of the most successful uses of soybean diversity is in the improvement of pest resistance, particularly in North American breeding. Perhaps the foremost reason for success is the fact that the benefit of exotic alleles for disease resistance can often be readily determined by phenotypic evaluation. Phenotypic distinction facilitates the introgression of exotic resistance alleles into adapted material and allows breeders to readily capture value from exotic disease-resistant germplasm using backcrossing and other breeding schemes that do not require extensive resources. The following sections describe examples of successes and indicate areas where there is a need to further exploit genetic diversity. Readers are referred to Chapters 4 (Hymowitz and Singh, 2004, this publication) and 16 (Niblack et al., 2004, this publication) for detailed discussion of soybean diseases and insects. North American and Chinese breeding is the reference point for much of this discussion, primarily because of availability of literature.

There are several lessons that breeders, pathologists, and entomologists can learn from the history of using diversity for pest resistance in soybean. Perhaps the most important is the value of extensive screening of germplasm. For several soybean diseases (SCN, PRR, root-knot nematodes, and SMV), the best sources of resistance were discovered long after initial sources were identified and used in breeding. Most scientists realize the importance of germplasm screening, though professional recognition and funding for such endeavors is usually lacking. Germplasm screening efforts have been difficult to publish in refereed journals and, thus, professional credit was difficult to garner. Recently, the journal *Crop Science* has devoted a section to plant genetic resources and germplasm screening, facilitating the professional publication of germplasm screening efforts. The United Soybean Board has funded screening programs extensively over the past 6 yr and this funding has resulted in the identification of many new genetic resources with disease resistance. These developments, plus new screening techniques, should encourage the screening of more accessions in the future.

A second lesson gleaned from the history of using diversity for pest resistance is that breeders must pay attention to diversity for other traits while using new sources of resistance. Extensive use of new resistance by breeders can lead to genetic bottlenecks for other traits if the genes for resistance are not incorporated into a diverse array of genetic backgrounds. An example is described in the section on SCN resistance breeding. Another example is found in Brazilian soybean production (Yorinori, 1999). In 1987 to 1988, nearly 60% of the production area was planted to two adapted cultivars that were susceptible to FELS, leading to severe yield losses.

These two cultivars were subsequently replaced primarily with a single cv. Cristalina which was resistant to FELS, but unfortunately very susceptible to stem canker [SC, caused by the fungus *Diaporthe phaseolorum* (Cooke & Ellis) Sacc. F.sp. *meridionalis* Morgan-Jones]. Large yield losses occurred in the mid-1990s when SC became prevalent and devastated Cristalina. Also extensive use of some lineages to improve yield may also lead to disease susceptibility. Extensive use of parentage from A3127 and Williams has been associated with susceptibility to SCSR (Kim et al., 1999).

8–14.1 Genetic Diversity for Disease Resistance in North America

Resistance to 10 major diseases and four major nematode species are discussed in this section. For the purpose of this section, if resistance was first noted in improved North American cultivars, it was considered derived from an adapted source; resistance derived from other germplasm was considered exotic. Although most ancestors of North American breeding were identified and used for reasons other than disease resistance (CNS is the one exception), some degree of resistance was found in the adapted North American cultivars for 10 of 13 major diseases and nematodes discussed in this section. Discovery of resistance in adapted lines has an obvious benefit for breeders and growers: quick release of resistant cultivars. Such fortuitous and timely discoveries illustrate the inherent worth of diversity within a genetic base.

Despite the economically useful range of disease resistance for many major diseases in North American cultivars, it is important to note that resistance genes from exotic sources have been needed to increase or broaden protection for nearly all diseases discussed. New genes for PRR resistance from exotic sources have replaced the first PRR-resistance genes found in adapted cultivars. All major genes for resistance to SCN, BP, and brown stem rot [BSR, caused by the fungus *Phialophora gregata* (Allington & Chamberlain) W. Gams] were derived from exotic sources. The literature shows that the level of resistance in the cultivar pool is inadequate under heavy pressure for some diseases such as root-knot nematodes, SCSR, and SDS. Pathogen variability may reduce the effectiveness of current resistance genes for other diseases, such as SCN, PRR, FELS, SC, and SMV. New resistance genes will likely be required as new pests arise and thus diversity will remain crucial in protecting the North American soybean crop from pests.

The remaining portion of this section discusses the history and use of diversity in breeding for resistance to several major diseases. Each disease has a unique breeding history that was influenced by the level of resistance in the current cultivar population, the variability in the pathogen population, and the potential for the pathogen to develop new strains or races. Some sources of resistances have been partial and others complete, and conditioned by as few as one gene or as many as 13. Pyramiding of multiple genes, each conferring only partial or incomplete resistance, has been attempted in soybean for some diseases. Much germplasm screening in soybean was motivated by the desire to a find simply inherited and complete resistance to replace partial resistance controlled by multiple genes. These factors are discussed for each disease in the context of further improving resistance.

8–14.1.1 Breeding for Resistance to Several Major Diseases

8–14.1.1.1 Bacterial Pustule (BP, caused by *Xanthomonas axonopodis* pv. *glycinea*) and **Bacterial blight** (BB, caused by *Pseudomonas savastanoi* pv. *Glycinea*). The first example of breeding success for disease resistance in North America was the discovery of resistance to BP and incorporating this resistance into the cv. Lee, released in 1953. Bacterial pustule was a common disease problem in the southern USA when soybean was first introduced to the region, causing an average 10% yield reduction (Hartwig and Johnson, 1953). A very late maturing and agronomically poor introduction from China, Clemson Non-Shatter (CNS) was resistant to BP and shattering (hence the name). Clemson Non-Shatter was crossed to earlier maturing S-100 to produce the agronomic, BP- and shattering-resistant cv. Lee and several important sibling breeding lines (Hartwig and Lehman, 1951). The resistance gene, *rxp*, is still effective today and no races of BP have been noted. Due to the durability and simplicity of using the *rxp* gene, there have not been extensive searches for additional alleles or genes to use in North American breeding.

Bacterial blight occurs in the Midwest more than in the South, and many midwestern cultivars are resistant, apparently deriving resistance from the ancestral base of North American soybean. Breeding for this disease generally consists of selecting against the susceptible types that occur in the progeny of resistant × susceptible matings. In the South, the cv. Hutcheson is susceptible to bacterial blight (BB, caused by *Pseudomonas savastanoi* pv. *Glycinea*.) and has been used heavily as a parent in southern breeding. Thus, there is a possibility that BB may become important in the South in the future. Interestingly, Hutcheson is one of the few recently released cultivars that has a significant amount of parentage from a new ancestor. It is possible that this novel ancestry may be the source of the BB susceptibility. Bacterial pustule and BB are important diseases to consider when working with exotic germplasm because many accessions from Asia appear very susceptible to one or both of these diseases.

8–14.1.1.2 Phytophthora Root Rot (PRR, caused by the oomycete *Phytophthora sojae*). Phytophthora root rot is a widespread disease in poorly drained soils of North and South America and is found in some regions of Asia. The pathogen has differentiated into races with susceptible reaction arising from a combination of specific genes for virulence in the pathogen and genes for resisitance in the host. Thus, PRR is one of the few major soybean diseases with a clear race structure that currently influences breeding for resistance. The disease was first described in Indiana in 1948 and later in Ohio in 1951 (Laviolette and Athow, 1977). Most cultivars at that time were susceptible to PRR, though screening identified several resistant cultivars (Blackhawk, Illini, and Monroe) that had a single resistance gene (*Rps*1a) from the ancestor Mukden, introduced from northeast China. The *Rps*2 gene was later identified in a breeding line derived from the ancestor CNS. Dominant resistance genes were also discovered in other adapted cultivars and lines (Altona, *Rps*6; D60-9647, *Rps*1b; Harosoy, *Rps*7). Exotic sources have also been important sources of *Rps* genes, including those most widely used today (*Rps* 1-c, *Rps* 1-k, and *Rps* 3-a). A recently discovered gene (*Rps*8) from a Korean accession (PI 399073) appears to confer resistance to a broad array of *P. sojae* biotypes (Burnham et al., 2003). There are currently 14 major *Rps* alleles at eight loci for resistance to PRR, with eight discovered in exotic sources (see Table 4–1 in Chapter 4). These single dominant genes have been effective in managing the disease. The economic impact of resistance genes has been examined using near-isogenic lines (Wilcox and St. Martin, 1998).

There are currently 44 races of *P. sojae* (Grau, 1999). None of the current *Rps* genes confer resistance to all races and multiple races can be found within a single field (Abney et al., 1997; Grau, 1999). Thus, combinations of non-allelic *Rps* genes will often be required to obtain adequate resistance in many fields. Races that overcome currently employed resistance genes are becoming more frequent including races that overcome the widely used *Rps* 1-c, 1-k, and 3-a genes (Ryley et al., 1998; Schmitthenner et al., 1994). In addition, new races are being discovered including some that are virulent on widely used *Rps* genes (Abney et al., 1997). Schmitthenner (1985) estimated that an *Rps* gene may remain effective for 8 to 15 yr. These factors indicate that pathologists and breeders need to continue to identify new resistance genes from diverse sources to maintain adequate protection of the crop.

There have been several recent assessments of PRR resistance among accessions in the USDA germplasm collection (Dorrance and Schmitthenner, 2000; Lohnes et al., 1996; Nelson et al., 1987; Rennie et al., 1992). Many accessions have resistance that appears to be similar to the previously described *Rps* genes. These efforts have also been successful in identifying several accessions that appear to have unique and potentially useful *Rps* genes. Interestingly, many resistant Chinese accessions with a range of *Rps* genes have been found, even though the disease was not reported there until 1989 (Tan et al., 1999). Other root rot diseases are common in China, and it is possible that genes which condition PRR resistance are also involved in resistance to other root rots.

Field resistance (also referred to as partial resistance, tolerance, and slow rotting) to PRR has also been reported in soybean (Schmitthenner and Walker, 1979; Tooley and Grau, 1982). In field resistance, the plant becomes infected (as opposed to the hypersensitive reaction conditioned by most *Rps* genes), but damage is limited such that root growth and yield is not reduced as much as in susceptible cultivars. No alleles for field resistance have been identified, although it appears to be controlled by multiple genes and mapping studies are underway. The degree of field resistance can be influenced by the specific PRR race encountered. It is not clear whether the current level of field resistance derived from U.S. cultivars is adequate for field protection in the absence of *Rps* genes. Added diversity may be useful to improve field resistance. Many South Korean plant introductions appear to have good levels of field resistance (Dorrance and Schmitthenner, 2000).

8–14.1.1.3 Sclerotinia Stem Rot (SCSR, caused by the fungus *Sclerotinia sclerotiorum*). Sclerotinia stem rot (also commonly called white mold) is an emerging disease problem in North America, but has been a major disease in South America since the 1970s (Ploper, 1999). Sclerotinia stem rot has also been noted in Asia (Tan et al., 1999). Resistance to SCSR has become a major objective of soybean breeding in the midwestern region of North America. In the field, soybean lines may have low disease incidence or low severity due either to avoidance (arising from an open canopy that allows drying), physiological resistance, or a combination of both mechanisms. Breeders have used both mechanisms to reduce disease losses, though much of the current focus is on physiological resistance.

At present, resistance appears to be partial and under multigenic control (Arahana et al., 2001; Kim and Diers, 2000). Several partially resistant, adapted cultivars have been identified through field and greenhouse screening (Boland and Hall, 1986, 1987; Grau et al., 1982; Kim et al., 1999; Kim et al., 2000; Nelson et al., 1991). Northrup King S19-90, a proprietary cultivar, is perhaps the best source of resistance. It has been noted that susceptibility to SCSR is associated with parentage from Asgrow A3127 or Williams (a parent of A3127) two of the most prominent parents in North American soybean breeding (Kim et al., 1999).

While the level of resistance in current cultivars is useful and stable (races of *Sclerotinia sclerotiorum* have not yet been identified), it is not strong enough to prevent yield losses in moderate or severe infestations. It may be possible to increase the level of resistance by combining partial resistance genes from different sources. This hypothesis has not been tested, but it is the driving force behind several germplasm screening efforts. More than 6400 accessions from the USDA Soybean Germplasm Collection were screened for resistance in the field (Hoffman et al., 2002), and several accessions were identified that had resistance superior to that in S19-90. The confounding effect of avoidance makes it difficult to discern physiological resistance in the field when screening unadapted germplasm, but an overhead misting system can reduce the effect of escape mechanisms such as short plant height and reduced canopy closure by helping to maintain high humidity within the canopy (Hoffman et al., 2002). Greenhouse techniques have been used to screen all accessions of perennial *Glycine* species for reaction to SCSR (Hartman et al., 2000). No immunity was reported, though several accessions of *G. tabicina* appeared to be as resistant as S19-90. The genetic mechanisms underlying resistance in the new sources are not known. Partial resistance has been noted in some South American cultivars (Ploper, 1999). The combination of multigenic control, partial resistance, and only moderate levels of resistance among recently released cultivars indicate that additional diversity will be needed in future SCSR breeding efforts.

8–14.1.1.4 Soybean Cyst Nematode Resistance (SCN, *Heterodera glycines* Ichinohe). Although SCN has been a production problem in soybean in China and other Asian countries for centuries (Riggs and Schmitt, 1989), it was first detected in the USA in 1954 (Winstead et al., 1955) and may have entered the USA undetected before 1900 (Noel and Liu, 1999). Since 1960, SCN has been one of the most important pests of soybean across the USA, causing losses of more than $420 million in 1979 and $760 million in the late 1980s, and 1990s (Brewer, 1981; Noel, 1992; Sciumbato, 1993). Until recently, SCN damage was much more serious in the South than in the North. Soybean cyst nematode damage is now even greater in the North, and SCN damage has been reported as far north as Minnesota (Wrather et al., 2001).

Nematicides have not been used extensively to control SCN on the farm and, thus, genetic resistance has been the primary agricultural remedy. No SCN resistance was present in modern North American cultivars, and, thus, the USDA Soybean Germplasm Collection was screened to identify resistance. In the late 1950s, Ilsoy, Peking, PI 79693, PI 84751, PI 90763, PI 209332, and PI 88788 were iden-

Table 8–17. Sources of soybean cyst nematode (SCN) resistance and their pedigree-based percentage contribution to 103 SCN-resistant public cultivars released by North American (USA and Canada) soybean breeding programs. Estimates are based on all SCN-resistant public cultivars released from 1947 to 2000.

Source of resistance†	Cultivars derived	Genetic contribution
	no.	%
PI 88788	62	4.95
Peking	79	1.95
PI 209332	2	0.50
PI 437654	5	0.38
PI 90763	2	0.13
Total	103	7.92

†Source: Z. Cui (personal communication, 2001)

tified as SCN resistant (Ross and Brim, 1957; Ross, 1962). More than 20 yr later in an expanded screening effort, PI 437654 was discovered to be resistant to all known SCN races at that time (Anand and Gallo, 1984; Anand et al., 1985, 1988; Rao-Arelli et al., 1992). Currently, more than 50 resistant sources have been identified in the USDA Soybean Germplasm Collection (Anand and Rao-Arelli, 1989; Dong et al., 1997; Arelli et al., 2000;). Five resistant sources (Peking, PI 88788, PI 90763, PI 209332, and PI 437654) have contributed SCN resistance genes to the 103 cultivars and constitute almost 8% of their genetic base (Table 8–17). The resistance in all but nine cultivars is derived solely from Peking and PI 88788.

The economic return from the use of resistant cultivars has been substantial. During a 6-yr period, planting Forrest soybean, resistant to race 3, increased farmers' income in infested fields by more than $400 million (Young, 1999). By 2000, a total of 103 public cultivars with different levels of resistance to SCN were released from public institutions in the USA (64 from the South and 39 from the North).

Early inheritance studies revealed three recessive genes and one dominant gene (*rhg1, rhg2, rhg3,* and *Rhg4*) in Peking conferring resistance to cyst nematode (Caldwell et al., 1960; Matson and Williams, 1965; Hartwig and Epps, 1970). Additional major genes (e.g., *Rhg5*), multiple alleles at a single locus, and linkage of genes conditioning resistance were reported later (Myers et al. 1988; Anand and Rao-Arelli, 1989; Hancock et al., 1987; Hartwig, 1985). Although the exact number is unclear, there may be as many as eight genes from exotic germplasm that condition SCN resistance. Soybean genome mapping has recently located several genes involved in SCN resistance (Boehlke et al., 1999; Concibido et al., 1996, 1997; Bell-Johnson et al., 1998; Cregan et al., 1999; Corbin, 1999; McBroom, 1999; Meksem et al., 1999; Webb et al., 1995; Mudge et al., 1997).

Soon after resistant sources were found in the late 1950s, breeding for SCN resistance was initiated using the plant introduction Peking, a low yield, shattering susceptible, black seed introduction from China (Caviness, 1992). Resistance genes from Peking were backcrossed into Lee, Scott, and R54-168 to develop resistant lines NC55, Custer, and RA63-19-2, respectively. These three resistant lines were used to develop Pickett (released in 1965), Pickett 71, Mack, Dyer, Forrest, Centennial, Gordon, Bedford, and Nathan. Almost all subsequently developed southern resistant cultivars were derived from these early cultivars. As SCN resistant cultivars were released and grown widely in the south, pathologists quickly realized that multiple SCN races were present and race profiles were shifting in response to the growing number of resistant cultivars. Race indices were developed, and it was determined that cultivars derived from Peking were resistant to Race 3, the dominant race in the southern USA when SCN was first detected. However, these cultivars were susceptible to races 6, 9, and 14 and these quickly became common as cultivar resistance exerted pressure against race 3. PI 88788 was then employed to develop new resistant cultivars. PI 88788 was initially crossed with D68-18, a southern breeding line with SCN resistance derived from Peking. Lines from the cross were further mated or backcrossed with Forrest, D70-3045 (a sib of Centennial), and R72-2647 to develop the SCN resistant lines Bedford, Nathan, HX37-3-16, R75-12L, and a series of USDA breeding lines similar to J74-39. These genetic materials were the source of all of SCN resistance in southern cultivars tracing to PI 88788. In recent years, race 2 has emerged as perhaps the most damaging race in the South as a result of growing cultivars resistant to race 14 for several years. PI 437654 identified through further screening of the USDA soybean collection in the 1980s was used to develop race 2 resistant cultivars such as Hartwig, Delsoy 5710, Anand, and Fowler (Young, 1999; Anand, 1992).

In the North, SCN resistance traces primarily to PI 88788. This PI has better agronomic value than Peking, and resistance from PI 88788 was initially backcrossed only once into the cv. Williams and Union (a BC4 line derived from Williams) to develop SCN-resistant lines L77-994 and L77-1233 and then the cv. Fayette (Bernard et al., 1988b) (a reselection from L77-994 released in 1981), and Cartter (Bernard et al., 1988c). These four genotypes passed resistance from PI 88788 to almost all resistant cultivars in the North. Of the 16 SCN-resistant cultivars released by public sector in the Midwest during the 1990s, 14 had SCN resistance from PI 88788 (Diers et al., 1999). PI 88788 is also the predominant source of resistance in commercial sector (Corbin, 1999; McBroom, 1999). Most midwestern soybean breeders are actively developing SCN-resistant cultivars. The heavy dependence on resistance genes from PI 88788 and the fact that many of these genes may also be present in other resistant sources used by breeders (e.g., PI 90763, PI 207332, Peking, and PI 437654) suggests that new SCN-resistant genes should be identified for future breeding (McBroom 1999; Diers et al., 1999). The USDA collection of *G. soja* and perennial species may carry additional resistance genes (Diers et al., 1999; Riggs et al., 1998).

8–14.1.1.5 Root-Knot Nematodes (MI, *Meloidogyne incognita* (Kofoid and White); **MA,** *Meloidogyne arenaria* (Neal) Chitwood; **MJ,** *Meloidogyne javanica* (Treub) Chitwood]. Root-knot nematode species (MI, MA, MJ) cause more damage than any other nematode group globally. The MI, MA, and MJ species are the most common root-knot nematodes that attack soybean in the southeastern USA. A heavy infestation of MI can reduce yield of susceptible cultivars by as much as 85% (Kinloch, 1974). *Meloidogyne incognita* (Kofoid and White) has been detected in 70% of soybean fields in Florida, and MA has been detected in about 5% of the fields tested in South Carolina and Florida (Garcia and Rich, 1985). *Meloidogyne* *javanica* (Treub) was found less frequently. Sciumbato (1993) estimated that U.S. soybean growers lose $25 to 30 million annually to root-knot nematode. Currently available resistant cultivars are not immune to root-knot species and that 30 to 50% yield loss can occur on resistant cultivars when root-knot nematodes are abundant (Kinloch, 1974; Kinloch et al.,1990). Resistance to all three species is present in the genetic base of North American soybean. In a survey of 139 southern maturity cultivars, 28% were resistant to MI, and 6 and 3% were resistant to MA and MJ, respectively (Hussey et al., 1991; Riggs et al., 1988). The publicly released cv. Bedford, Forrest, Braxton, Gordon, Jackson, and Kirby appeared resistant to all three species. A single gene for resistance to MI was detected in the cv. Forrest and was designated *Rmi*1 (Luzzi et al., 1994). Tracking of SSR marker alleles indicates that the resistant allele traces to the southern ancestor, Palmetto (H.R. Boerma, personal communication, 2002).

Because resistant cultivars can be damaged by root-knot species, about 3000 accessions of the USDA collection have been surveyed to find additional sources of resistance (Luzzi et al., 1987; Harris et al., 2003). PI 96354, PI 200538, and PI 230977 were identified as having the highest levels of resistance to MI, MA, and MJ, respectively, and they were more resistant than current resistant cultivars. PI 230977 was also resistant to MA. Greenhouse studies indicated that resistance in these PIs (and others) was quantitatively inherited and that single genes could not be identified (Luzzi et al., 1994, 1995a, 1995b). Crosses of resistant PIs with resistant cultivars produced segregating progeny, indicating that the PIs carried unique resistance alleles not found in the U.S. cultivars (Luzzi et al., 1994, 1995a, 1995b). More recently, QTL analysis has been applied to PI 96354, PI 200538, and PI 230977 to further elucidate the genetic control of their resistance to MI, MA, and MJ, respectively (Tamulonis et al., 1997a,1997b,1997c). In each case, two QTL explained most of the genetic variation for resistance. One of the QTL for resistance in PI 96543 was associated with previously identified allele *Rmi*1. The results of the QTL analysis for PI 96543 were confirmed in a second study (Li et al, 2001a). Three breeding lines have been developed and released using these three PIs (Luzzi et al., 1996a, 1996b, 1997). These breeding lines have resistance levels similar to the three PIs above, but with higher yields. The most recent screening of the USDA collection for resistance has identified additional PIs with levels of resistance to MI and MA similar to that found in PI 96354 and PI 200538, respectively (Harris, 2003). The pattern of egg production and galling on these recently identified PIs suggests, however, that they may have contrasting mechanisms of resistance, and that, consequently, they may carry new alleles for resistance.

8–14.1.1.6 Sudden Death Syndrome (SDS, caused by the fungus *Fusarium solani* (Mart.) Sacc. F. sp. glycines). Sudden death syndrome is another recent disease that plagues many growing region in North and South America and is receiving considerable attention from breeders. Resistance to SDS is partial and no races of the pathogen have yet been reported. There are different manifestations of resistance including resistance to leaf scorching (Chang et al., 1996a) and resistance to root colonization (Njiti et al., 1997). Resistance appears to be under multigenic control (Chang et al., 1996a, 1997; Hnetkovsky et al., 1996; Njiti et al., 1996, 1997) with continuous distributions of disease scores in field trials. There is one report of single gene inheritance based on greenhouse screening (Stevens et al., 1993). Useful levels of resistance have been noted in adapted North American cultivars and are the primary sources of resistance for current North American breeding. The current level of resistance will not prevent significant yield losses when conditions are favorable for disease development and there is current research to identify better sources of resistance. Some novel sources of partial resistance have been identified in exotic soybean germplasm (Hartman et al., 1997) and within the USDA collection of perennial *Glycine* species (Hartman et al., 2000). Resistant reactions approaching immunity have also been noted in some South American cultivars (Ploper, 1999).

8–14.1.1.7 Brown Stem Rot (BSR, caused by the fungus *Phialophora gregata* (Allington & Chamberlain) W. Gams]. Brown stem rot is a major disease in northern regions of North America and has been noted in South America. There are two different pathotypes of the fungus, but races have not been differentiated. Type I causes necrosis, defoliation, and vascular discoloration while Type II causes only vascular discoloration (Gray, 1971). A variable DNA region in the intergenic spacer of the nuclear rDNA was identified by Chen et al. (2000) that can be used to distinguish between Type I and II strains. Brown stem rot was first reported in 1944 in Illinois and little resistance was noted in adapted cultivars at that time. By screening the germplasm collection, PI 84.946-2 was identified as resistant and is the most widely deployed source of resistance in breeding (Chamberlain and Bernard, 1968). PI 84.946-2 has two dominant alleles for resistance, *Rbs1* and *Rbs3* (*Rbs3* was also identified in PI 437.970) (Eathington et al., 1995; Hanson et al., 1988; Wilmot and Nickell, 1989). The *Rbs2* gene was later identified in PI 437.833 (Hanson et al., 1988), which has been the source of resistance for some recent cultivar releases (Eathington et al., 1995). There is a report of quantitative type resistance derived from an adapted cultivar (Asgrow A3733) that is unrelated to any of the sources of the single genes (Waller et al., 1991).

Current sources of resistance are not immune to *P. gregata* and can be overcome under some field conditions (Bachman et al. 1997) and by using specific strains in growth chamber experiments (Chen et al., 2001b). Continual use of the same genes may promote selection for virulence in this variable pathogen (Bachman and Nickell, 2000). Potential sources of new resistance genes have been identified among accessions from China (Bachman et al., 1997; Bachman and Nickell, 2000), and the USDA germplasm collection (Nelson et al., 1989). Nearly 30% of the accessions from central and southern China were resistant (Bachman and Nickell, 2000). To date, none of the screenings have discovered immunity or resistance that appears better than the current sources. The genetics of the new resistance is unknown.

8–14.1.1.8 Frogeye Leaf Spot (FELS, caused by the fungus *Cercospora sojina* Hara). Frogeye leaf spot is primarily a foliar disease that can be prevalent in warm humid environments and is common in the southern regions of North America, South America, and in parts of Asia. Frogeye leaf spot was the first major disease of soybean in Brazil (Yorinori, 1999) and was first reported in the USA in 1924. There are multiple races of the *C. sojina* (Phillips and Boerma, 1981; Yorinori, 1992) with more than 40 races identified using 38 host differentials (D.V. Phillips, unpublished data, cited in Yang and Weaver, 2001). Three major resistance genes have been recognized: *Rcs1* (from the cv. Lincoln), *Rcs2* (from the cv. Kent), and *Rcs3* (from the cv. Davis). The *Rcs3* gene was later identified in PI 54610 (Baker et al., 1999), though the germplasm used in this study is probably not the PI 54610 used as an ancestor of Davis. Thus, all resistance to date is conferred by major genes found in adapted cultivars. The *Rcs3* gene confers resistance to all U.S. and Brazilian races of *C. sojina*. Other non-*Rcs3* and non-*Rcs2* single resistance genes have been noted in other adapted cultivars ('Lee', 'Ransom', 'Stonewall') and in 'Peking' the original source of cyst nematode resistance in the USA, suggesting that additional genes may be present in the cultivar gene pool (Pace et al., 1993; Baker et al., 1999). It will be difficult to determine the allelic relationship among putatively novel *Rcs* genes, especially as compared to *Rcs1*, because races 1 and 2 of *C. sojina* are no longer available (Pace et al., 1993).

While the current *Rcs* genes, especially *Rcs3*, appear to confer adequate and durable resistance to date, the pathogen may evolve virulence to this limited array of *Rcs* genes. The existence of 44 *C. sojina* races shows that this pathogen is quite adaptive. Additional genes may be needed, such as the *Rcs* gene from Peking that confers resistance to 39 of 44 races (Baker et al., 1999). All MG VI to VII accessions of the USDA germplasm collection have been screened for reaction to *C. sojina* race 4, and 12 accessions with immune reactions were identified (Yang and Weaver, 2001). The genetic mechanisms underlying the resistance in these new sources are not known. Resistance is common among Chinese cultivars (Tan et al., 1999).

8–14.1.1.9 Soybean Mosaic Virus (SMV) and Bean Pod Mottle Virus (BPMV). Soybean mosaic virus is reported wherever soybean is grown and considered the most important viral disease of soybean. Based on observations of symptoms, there are at least seven SMV strains (Cho and Goodman, 1979). Additional strains have been noted in China and Japan in recent years, although their relationship to the previously established strains of Cho and Goodman (1979) is not clear. Soybean genotypes are classified as resistant (symptomless), necrotic, or susceptible (mosaic) to different SMV strains. Single dominant genes usually confer resistant or necrotic reactions, and most genes do not confer resistance to all strains (Hayes et al., 2000). The first resistance gene (*Rsv1*) was identified in PI 96983 (Kiihl and Hartwig, 1979) and additional *Rsv1* resistance alleles have been identified in adapted cultivars Ogden (*Rvs1-t*), York (*Rsv1-y*), and Marshall (*Rsv1-m*) (Chen et al., 1991). Resistance at the *Rvs1* locus has also been identified in 'Raiden', a Japanese cultivar (*Rsv1-r*, Chen et al., 2001a); PI 406710 (*Rsv1-k*, Chen et al., 1991); PI 486355 (*Rsv1-s*, Ma et al., 1995a); PI 458507 (Wang et al., 1998); and Tousan 140 (Gunduz et al., 2002). *Rsv1* is the most commonly deployed *Rsv* allele (Hayes et al., 2000). The *Rsv3* allele was identified in Columbia (Buzzel and Tu, 1989) and additional *Rsv3* alleles have been found in the adapted cv. Hardee (Buss et al., 1999) and Harosoy (Gunduz et al., 2001) and in Tousan 140 (Gunduz et al., 2002). Recently, the *Rsv4* allele that confers resistance to all seven groups of SMV was identified in PI 486355 (Hayes et al., 2000; Ma et al., 1995a). Other non-*Rsv1* resistance genes have been noted in accessions from China (Wang et al., 1998). Soybean genome mapping has recently led to the identification of several genes involved in

SMV resistance (Yu et al., 1994; Hayes et al., 2000, Chen et al., 2001a; Jeong et al., 2002)

While resistance was found in the domestic cultivars, the discovery of the *Rsv4* gene in PI 486355 conferring resistance to all characterized SMV strains illustrates the importance of diversity in protecting soybean cultivars. It is possible that pyramiding *Rsv* genes will broaden the effectiveness of SMV resistance. Combinations of *Rsv* genes that provide broader resistance than individual genes have been found in soybean germplasm. This suggests that breeders may have unintentionally selected for gene combinations (G.R. Buss, Univ. of Virginia, personal communication, 2001) and that breeding may benefit from discovering new genes to further exploit this strategy. Large scale screening of germplasm in North America has not been reported. More than 17 000 Chinese accessions have been screened and 140 SMV-resistant genotypes have been identified (Tan et al., 1999), including 16 highly resistant sources. Soybean mosaic virus incidence may increase in North America because the soybean aphid (*Aphis glycines* Mats.) is a known SMV vector and is increasing in North America.

Bean pod mottle virus (BPMV) warrants discussion here because it is widespread in the southern and eastern USA and is becoming more common in the midwestern USA. It acts synergistically with SMV to reduce yield (Ross, 1968). No commercial soybean cultivar or *G. max* accession has been found to be resistant to BPMV. To date, limited germplasm has been screened, but that work has been recently restarted. Bean pod mottle virus resistance has been noted in an accession of *G. tomentella* (R. Gergerich, personal communication, 2001).

8–14.1.1.10 Stem Canker [SC, caused by the fungus *Diaporthe phaseolorum* (Cooke & Ellis) Sacc. F. sp. *meridionalis* Morgan-Jones]. Stem canker was first noted in the northern USA in the late 1940s after the release of two susceptible cultivars (Hawkeye and Blackhawk) (Kilen and Hartwig, 1987) and in the southern USA in 1975. The disease became a major problem in the mid-south region in the late 1980s when considerable acreage was planted to a high-yielding susceptible cultivar, A5980, (L. Ashlock, Univ. of Arkansas, personal communication, 2001). The northern and southern strains of *D. phaseolorum* are considered to be different *forma speciales* (two forms of the same species) and most breeding efforts have been focused on the southern strain, *f.sp. meridionalis*. Stem canker (f.sp. *meridionalis*) became a significant disease in South America in the 1980s (Ploper, 1999; Yorinori, 1999).

Resistance is common in the North American gene pool, though the allelic nature of the many sources is not clear. Single genes were first identified in adapted cultivars and confer a high level of resistance. Two genes (*Rdc1*, *Rdc2*), each conferring complete resistance were identified in Tracy-M (Kilen et al., 1985; Kilen and Hartwig, 1987). The gene *Rdc3* was later identified in Crockett and *Rdc4* in Dowling (Bowers et al., 1993) and Hutcheson (Tyler, 1996). Moderate resistance has also been noted among other cultivars, though the underlying genetic mechanism is not known.

There is little published research on screening exotic germplasm for SC resistance. In part, this is because there appears to be little immediate need for additional genes. Current resistance in cultivars is durable and nearly complete, and races of *D. phaseolorum* have not been identified. There is some evidence of pathogen variability with some strains killing more than 50% of the plants with *Rdc2* (Pioli et al., 1999). This suggests that the pathogen may be adapting to the wide spread deployment of the current *Rdc* genes. Some accessions from the USDA germplasm collection have been screened for resistance (Keeling, unpublished data,1989, reported by Tyler, 1995) and several resistant accessions were identified. Genetic analysis of two accessions (PI 230976, PI 398469) indicated that they have novel *Rdc* genes (Tyler, 1995), and that they could be used to increase diversity of *Rdc* genes. High levels of resistance have also been noted among South American cultivars (Ploper, 1999), though this gene pool is highly related to the southern U.S. breeding pool.

8–14.1.1 Other Potentially Important Diseases

Soybean production history is punctuated with the occurrence of new diseases that require host resistance for effective control. Some diseases warrant special mention and speculation on the need for diversity to adequately protect North American soybean. Red crown rot [caused by *Cylindrocladium crotalariae* (Loos) Bell and Sobers] has been detected in China and southern North America in recent years (Gai et al., 1992a, 1992b; Berner et al., 1986) Resistance has been noted, but North American germplasm has not been extensively screened for resistance to this disease. Rhizoctonia foliar blight (caused by *Rhizoctonia solani* Kuhn anastomosis group1 IA and IB) has become a problem in the southern USA in recent years and there appears to be little useful resistance among North American cultivars (Harville et al., 1996). Breeders have recently become concerned about resistance to soybean rust (caused by the fungus *Phakopsora pachyrhizi* Sydw.) although this disease has not been detected in North America (Hartwig and Bromfield, 1983; Hartwig, 1988). This disease is common in southern China and other parts of Asia, and there are several Chinese sources of resistance to this highly differentiated pathogen. In addition to new diseases, there are several endemic diseases such as Charcoal rot (*Macrophomina phaseolina* (Tassi) Goid), Phomopsis seed decay [*Phomopsis longicolla* (Hobbs)], and seedling diseases (*Fusarium* and *Pythium* spp.) that cause economic loss but have not been addressed extensively by breeders. This is due in part to a lack of strong resistance in adapted germplasm. A heritable form of resistance to purple stain (*Cercospera kikuchii*) was reported to exist in PI 80837, but resistance appears to be influenced greatly by environment (Wilcox et al., 1975). An allele for resisitance to Phomopsis seed decay was found in PI 417479 (Berger and Minor, 1999)

8–14.2 Genetic Diversity for Insect Resistance in North America

Insect pests of soybean cause significant losses in many regions of the world, including North America. Breeding for insect resistance has not been a high priority of most programs, perhaps due to the effectiveness of chemical control, difficulty in screening for resistance, influence of environment on insect presence, impact of soybean MG on expression of resistance, lack of consistent insect damage over environments, and competing breeding objectives such as yield improvement and disease resistance.

There are many insect pests of soybean in North America. The most damaging appear to be the lepidopterous insects, primarily corn earworm (CE, *Heliothos zea*), bollworm [BW, *Helicoverpa zea* (Boddie)], tobacco budworm [TB, *Heliothos virescens* (F.)], green cloverworm [GC, *Plathypena scabra* (Fabricius)], beet armyworm [BA, *Spodoptera exigua* (Hübner)], fall armyworm [FA, *Spodoptera frugiperda* (J.E. Smith)], cabbage looper [CL, *Trichoplusia ni* (Hubner)], velvet bean caterpillar [VBC, *Anticarsia gemmatalis* (Hübner)], soybean looper [SL, *Pseudoplusia includens* (Walker)]. Coleopterous insects (larval and adult forms) are also serious pests, for example Mexican bean beetle [MBB, *Epilachna varivestis* (Mulsant)], striped blister beetle [SBB, *Epicauta vittatum* (F.)], bean leaf beetle [BLB, *Cerotoma trifurcata* (Frster)], and stink bugs [SB, *Nezara viridula* (L.), *Acrosternum hilare* (Say), and *Euschistus* spp.], as well as sucking insects such as aphids (*Aphis glycines Mats.*), potato leafhopper [PL, *Empoasca fabae* (Harris)], and white fly (*Trialeurodes vaporariorum, Bemisia argentifolii*, and *Trialeurodes abutilonia*).

Some genetic variation for resistance to insect pests has been noted among cultivars (Kraemer et al., 1994; Lambert and Kilen, 1984a; Rector et al., 2000; Van Duyn et al., 1971), though exhaustive screenings and high levels of resistance have not been reported. Three plant introductions from Japan (PIs 171451, 229358, and 227687) were identified as having good resistance to MBB in the early 1970s (Van Duyn et al., 1971). These PIs are also resistant to BLB, BW, CE, TB, CL, SBB, SL, and VBC (All et al., 1989; Clark et al., 1972; Hatchett et al., 1976; Luedders and Dickerson; 1977; Kilen et al., 1977; Lambert and Kilen, 1984a, 1984b; Talekar et al., 1988) and express both antibiosis (detrimental effect on insect development) and antixenosis (nonpreference) resistance mechanisms (All et el., 1989; Clark et al., 1972). These three PIs have been noted to have some resistance to SB as well (Gilman et al., 1982; Turnipseed and Sullivan, 1976).

The resistance from the PIs has a low heritability (Kenty et al., 1996; Kilen and Lambert, 1998; Luedders and Dickerson, 1977; Rufener et al., 1989; Sisson et al., 1976), though it appears to be controlled by relatively few genes (Kenty et al., 1996; Rufener et al., 1989; Rector et al., 1998, 1999, 2000). Some resistance genes appear common to more than one PI, while other genes appear to be unique (Kilen and Lambert, 1986; Rector et al., 1998, 1999, 2000). Genetic analyses show that genes controlling antibiosis and antixenosis are independent. Thus, it should be possible to combine the genes from the three exotic sources into one cultivar (Rector et al., 1998, 1999, 2000). Evidence suggests that genes from two or more of these exotic sources may be required to obtain resistance to multiple-pest complexes that are often encountered in a field. In tests with multiple insect species, each PI showed a unique profile of resistance to defoliating insect species (Talekar et al., 1988). Researchers have noted that resistance to some insects (e.g., MBB and CW) may be independent among progeny derived from the same source (Smith and Brim, 1979), while others have suggested that selection for resistance to one insect may increase resistance to others (All et al., 1989; Lambert and Kilen, 1984a). More research is needed to sort out the genetics of resistance to multiple pests. The durability and economic importance of the resistance from the three key PIs has not been reported. Some of the insect resistance noted in these PIs is related to trichome shape or sharpness (R. Boerma, Univ. of Georgia, personal communication, 2001).

The three PIs have been used extensively to develop germplasm and four cultivars ('Shore', 'Crocket', 'Lamar', and 'Lyon') with resistance to defoliating insects (Hammond and Cooper, 1998; Kilen and Lambert, 1986). To date, no cultivar has been developed that carries all identified resistance genes found in the PIs (Narvel et al., 2001). Insect resistant cultivars have not been grown widely due to low yields (Rector et al., 2000). The low yields may result from the poor agronomic qualities of these PIs, that were transferred unintentionally to the cultivars. Collectively, the PIs lodge severely, shatter, are late in maturity (late Group VIII), and susceptible to several diseases. Additional exotic soybean germplasm lines have been screened for resistance to defoliating insects (Hammond et al., 1998; Kraemer et al., 1988, 1990) including some soyfoods types (Kraemer et al., 1994). Extreme variants for pubescence density, derived from exotic sources, have also been examined for insect resistance (Lambert and Kilen, 1989; Lambert et al., 1992; Gannon and Bach, 1996; Ren et al., 2000). Resistant accesions were identified although none appeared to have greater resistance than the three aforementioned Japanese PIs. One of these more recent sources, PI 417061, has been shown to have major gene(s) not found in PIs 171451 and 227687, but these genes are probably in common with those of PI 229358 (Kilen and Lambert, 1998). Beach et al. (1985) noted that PI 423968 had moderate resistance to SL. The genetics of resistance from these sources and their relationship to previously used sources is mostly unknown. The PI 416937, a slow wilting accession from Japan, is also highly resistant to MBB; the genetic relationship between MBB and drought tolerance has not been established (Kraemer et al., 1988; Sloane et al., 1990). Levels of insect resistance found in accessions recently introduced from China are lower than those in the widely used Japanese lines but the genetics of resistance in the much earlier Chinese lines may be different (Hammond et al., 1998).

8–15 GENETIC DIVERSITY FOR PEST RESISTANCE IN CHINA

8–15.1 Genetic Diversity for Disease Resistance in China

Science-based breeding for disease resistance is a recent endeavor in China. The reasons for this are not clear. Soybean is an ancient crop in China, so pathogens are likely to have developed and presented production problem for centuries. It is possible that the diverse nature of the Chinese germplasm allowed early farmers and subsequently modern breeders to readily select for resistance without the rigors of science by simply selecting for yield and healthy appearance. Today, improved cultivars have largely replaced farmer-developed soybean cultivars so that genetic diversity on the farm in China is greatly reduced compared to previous centuries and no longer offers the natural degree of protection against diseases that it once did. Demands for increased production also preclude a casual approach to improving disease resistance.

Formal breeding for disease resistance began in 1986 with the National Soybean Breeding Program. The breeding objective was resistance to four major diseases, SMV, SCN, FELS, and soybean rust. The SMV is the most important disease problem in China. The SCN damage has become increasingly severe in northeastern and northern China and is spreading to southern China. Frogeye leaf spot is common only in northeastern China while SR is found only in southern China, usually only 1 yr out of 3 or 4. Large-scale screening of Chinese germplasm for disease resistance began in China after 1980, and highly resistant sources derived from these evaluations have been used by breeders since the mid 1980s (Table 8–18). To date, most disease resistance used in Chinese breeding programs is derived from Chinese accessions. However, there are important exceptions for SCN and FELS. For example, Chinese breeders introduced FELS resistance genes from Amsoy, Clark 63, Corsoy, and an introduction designated "Ohio" into local parents and developed cultivars with high levels resistance to the disease (He Feng 27, He Feng 28, He Feng 29, Hei Nong 33, Hei Nong 36, Bao Feng 2 Hao, Sui Nong 8 Hao, and Mu Feng 6 Hao). These cultivars have been well accepted by farmers in northeastern China since 1986. By incorporating SCN-resistance genes from Franklin, two cultivars, Kang Xian Chong 1 Hao and Kang Xian Chong 2 Hao, were developed in Heilongjiang province in northeastern China after 1992.

China has released 173 disease-resistant cultivars from 1986 to 1995, including 107 resistant to SMV, 20 to SCN, 33 to FELS, and more than 30 resistant to rust and downy mildew (Gai et al., 1997). Because systematic breeding for disease resistance commenced only in the 1980s and is only now beginning to have impact, most of the resistance in released cultivars was the by-product of breeding aimed at yield improvement. Disease resistance was identified in the latter stages of testing for breeding lines prior to release.

8–15.2 Genetic Diversity for Insect Resistance in China

Agromyzid bean fly (*Melanagromyza sojae* Zehntner), and several leaf-feeding insects [*Ascotis selenarria* (Schiffemuller et Denis), *Prodenia litura* Fabricius, *Hedylepta indicata* (Fabricius), etc.] in the south, and northern pod borer (*Leguminivora glycinivorella* Mats.) in the North are the most important insects affecting soybean in China. The bean fly, not known to occur in North America, bores into, and lives within, the stem and petiole as opposed to attacking the leaf or pod. Several insect-resistant cultivars have been released in China: Xiao Jing Huang 1 Hao in the 1950s, Jilin 3 Hao and Jilin 4 Hao in 1960s, and Sui Nong 7 Hao and Jiu Nong 17, in 1988 and 1990, respectively (Cui et al., 1998, 1999; Zhang, 1985). These cultivars were identified as resistant to the northern pod borer prior to release. However, they were not derived from crosses designed to produce insect-resistant cultivars. Rather, insect-resistance genes were already present in the adapted breeding materials. Extensive germplasm screening for insect resistance began about 1983 in the Northeast and South, and accessions were identified with resistance to the northern pod borer, the bean fly, and leaf-feeding insects (Table 8–18) (Cui et al., 1995a, 1995b, 1996; Guo et al., 1986; Xu et al., 1999a; Gai et al., 1989a, 1989b, 1991). Hybridization to develop cultivars resistant to these insects did not begin until after 1990, and no cultivars have been released from this research effort yet. Improved cultivars appear to be a primary source of resistance to the northern pod borer, whereas land races appear to provide the primary resistance to the other insects. Thus, sources from outside of China have not been important for insect resistance in China. The three Japanese land races that are the basis for insect-resistance breed-

Table 8–18. Screening for pest resistance in China.

Diseases and insects†	Reported representative resistant sources	No. of resistant accessions found	No. of accessions evaluated	Citation
SMV	Da Ba Yue Huang	46	16 969	Xu et al. (1999)
		16	17 497	Tan et al. (1999)
	Lu Dou 4 Hao‡, Da Bai Ma, Yang Huang, You Bian 30‡, Ji Lin 1 Hao‡, Ji Lin 17‡, Dong Nong 35‡	15	856	Zhong et al. (1986) Chang (1997)
	Wilkin‡, He Jiao 83-590‡	10	236	Li et al. (1986)
	Ke Feng 1 Hao‡, Shan Dong 7222‡, Li Shui Zhong Zi Huang Dou	28	6 000	Gai et al. (1989) Hu et al. (1994)
	Xi Cao Huang, Yang Huang 1 Hao‡, Qi Huang 23‡, Qi Huang 22‡, Da Bai Ma	10	116	Pu and Cao (1983) Chang (1997)
SCN Race 1		15	13 350	Xu et al. (1999)
Race 1	Chang Hei Dou, Xiao Li Hei Dou (Liao 678), Xiao Li Hei (Liao 1119), Lian Mao Hui Hei Dou (Liao 683)	4	1 170	Zhang (1985b) Chang (1997)
Race 1	Mu Shi Hei Dou, Xiao Li Hei Dou, Xiao Li Hei, Bei Piao Da Bai Qi			Liu (1985) Chang (1997)
Race 2		26	3 205	Xu et al. (1999)
Race 3		30	10 167	Xu et al. (1999)
Race 3		34	10 000+	Ma et al. (1991, 1995)
Race 3	Ying Xian Xiao Hei Dou, Ha Er Bing Xiao Hei Dou		809	Wu (1982) Chang (1997)
Race 3	Ha Er Bing Xiao Hei Dou, Lian Mao Hui Hei Dou, Xiao Li Hei, Mu Shi Dou, Xiao Li Hei Dou			Liu (1985) Chang (1997)
Race 4		8	13 039	Xu et al. (1999)
Race 4	Ying Xian Xiao Hei Dou, Xing Xian Hui Pi Zhi Hei Dou	11	1 920	Li (1987b) Chang (1997)
Race 4	Hui Bu Zhi Hei Dou, Wu Zhai Chi Bu Liu Hei Dou	11	13 000+	Li et al. (1991)
Race 5		27 G. max and 3 G. soja	3 391 G. max and 169 G. soja	Zhang et al. (1998)
Frogeye leaf spot		46	108	Zhang et al. (1997)
		14	961	Xu et al. (1999); (N. Qi, personal communication, 1986)
	Qing Pi Dou (GD3125), Da Jing Huang (Liao Bian 46), Wu Ming 13 (Feng Bian 237), Hei Mo Shi Dou (GD3284), Jiao He Tian E Dan (GD2980), Hui Tie Jia (Liao Bian 53)		1 193	(X.M. Zhu, personal communication, 1986) Chang (1997)

(continued on next page)

Table 8–18. Continued.

Diseases and insects[†]	Reported representative resistant sources	No. of resistant accessions found	No. of accessions evaluated	Citation
	Hu Lin 1 Hao, Long Quan 1 Hao (He Qi), Ba Wang Bian, Xiao Qing Dou	14	961	(N. Qi, Personal communication, 1986); Chang (1997)
Soybean rust	Da Jiang Se Dou, Guo Tian Ling Hei Mao Dou, Tian Deng Hei Dou, Jiu Yue Huang, Yu Shan Qing Pi Dou, Zhong Dou 19‡, Zhang You 84-87‡, Zhong You R-34‡	74	8 711	Tan et al. (1997, 1999)
		69	8 654	Xu et al. (1999)
Soybean Downy		319	1 300	Li (1992)
Mildew	Mu Zhuan 1 Hao‡, Sui Nong 6 Hao‡, He Feng 25‡, Dong Nong 36‡, Hei Nong 21‡, Sui Hua Kang Mei 3 Hao‡, Ba Yan Ping Ding Xiang, Si Li Huang, Wu Ding Zhu		1 300	Li (1987a) Chang (1997)
	Si Ping Tou, Si Li Huang Dou, He Qi Feng Di Huang, Yu Shu Si Li Huang, Huai De Lan Qi, Hun Chun Da Dou	98	1 321	(J.C. Hu, Personal communication, 1986) Chang (1997)
Agromyzid bean fly	Wu Xi Chang Qi Guang, Jiang Ning Ci Wen Dou, Hai Men Xi Feng Qing, Yi Wu Da Dou, Shi Yue Bai	10	4 582	Gai et al. (1989)
	Sui Dao Huang, Mei 2, 7205-18, Huai 253, Huai 258			Zhang et al. (1984); Chang (1997)
	Xu Dou 1 Hao, Zheng 76046-1, Zhou 7327-3, Yang Chun Qing Dou	7	1 181	Wang (1985) Chang (1997)
Leaf feeding insects	N2549-2, N1178-2-2, N5454-3, N3687, N3039, N4908	6	6 724	Cui et al. (1995, 1996)
Northern pod borer	ERI1001, ERI1003, Ji Lin 1 Hao‡, Ji Lin 3 Hao‡, Ji Lin 13‡, Ji Lin 16‡	18	3 109	Guo et al. (1986); Chang (1997)
	Ji Lin 1 Hao‡, Ji Lin 3 Hao‡, Ji Lin 4 Hao‡, Ji Lin 13‡, Hei He 3 Hao‡, Tie Jia Si Li Huang, Hun Chun Da Dou, Tie Jia Qing, Guo Yu 98-4, Tie Jia Dou		3 000+	Guo (1983); Chang (1997)
Soybean aphid	Guo Yu 98-2, Guo Yu 98-4, Guo Yu 100-4, Zhong Sheng Luo, Dan Dong Fu Shou, Sun Wu Xiao Bai Mei, Xiong Yu Xiao Li Huang	15		Feng (1984) Chang (1997)

† Agromyzid bean fly, *Melanagromyza sojae* Zehntner; Frog eye leaf spot, *Cercospora sojina* Hara.; Leaf-feeding insects: *Ascotis selenarria* (Schiffemuller et Denis), *Prodenia litura* Fabricius, *Hedylepta indicata* (Fabricius), etc.; Downy mildew, *Peronospora manschurica* (Naoum.) Sydow. Northern pod borer, *Leguminivora glycinivorella* Mats. Soybean aphid (*Aphis glycines* Mats.) Soybean cyst nematode (SCN), *Heterodera glycines* Ichinohe Soybean mosaic virus (SMV); Soybean rust, *Phakopsora pachyrhizi* Sydow.

‡ Improved cultivar or breeding line, others are landraces.

ing in the USA appeared only moderately resistant to leaf-feeding insects in China and several Chinese land races were more resistant (Cui et al., 1995a, 1995b, 1996) (Table 8–18). This finding may have importance in breeding in the USA.

8–15.2.1 Insect Resistance Reported from Taiwan and Egypt

More than 5000 accessions have been screened for resistance to lima bean podborer [or southern pod borer] (*Etiella zinckenelle* (Treitschke)] at the Asian Vegetable Research and Development Center (AVRDC) in Taiwan, resulting in the identification of seven resistant accessions (Talekar and Chen, 1983). Diverse germplasm has also been screened at AVRDC for resistance to beanfly (*Melanagromyza sojae* Zehtner) (Chiang and Talekar, 1980). Germplasm screening for stink bug reaction has identified useful sources of resistance (Gilman et al., 1982; Jackai et al., 1988; Turnipseed and Sullivan, 1976). In Egypt, the dense-pubescence U.S. cultivar Celest is resistant to the cotton leaf worm (*Spodoptera Littoralis Boisd*) and has been used in practical breeding as a source of resistance (Awadallah et al, 1990) (M. Hassan, personal communication, 1998).

8–16 BREEDING FOR VALUE ADDED TRAITS USING EXOTIC GERMPLASM IN NORTH AMERICA

Public breeders have been very active in the development of specialty cultivars in North America. Since the release of the first specialty-use cultivar, Kanrich (a soyfoods type), in 1956, a total of 131 public specialty cultivars have been released (97 with complete pedigree information) through 2001 (Table 8–19). Seventy percent of the specialty cultivars were released after 1990, accounting for one

Table 8–19. Release of 131 specialty-use public cultivars over time and by region in North America (USA and Canada) from 1947 to 2001.

Primary specialty trait (s)†	Release era					Region		Total
	1950s	1960s	1970s	1980s	1990s	North	South	
Large seed	1	2	3	4	27	36	1	37
Small seed				11	32	37	6	43
High protein		6	5	6		16	1	17
High protein, large seed		1	1	1	5	8		8
Lipoxygenase free					9	9		9
High protein, low lipoxygenase					5	5		5
Low linolenic acid oil					1		1	1
Low palmitic, low linolenic acid oil ‡							1	1
Yellow hila, high yield					1	1		1
Null kunitz trypsin inhibitor				1		1		1
Forage					3		3	3
Long juvenile			2	2	1		5	5
Total	1	9	6	24	90	113	18	131

† Name of specialty trait as noted in release or registration report.
‡ Satelite, a cultivar with a low concentration of both palmitic and low linolenic acid was released in 2001.

third of the total public cultivar releases in that same time period. The majority of specialty North American cultivars were developed in the northern USA and Canada. Private companies also develop specialty type cultivars, but no summarized data are available.

Introduction of specialty traits into breeding has often been achieved through the mating of exotic germplasm with adapted breeding stock. The strategy has produced a genetic base for specialty cultivars that is substantially different from that of commodity cultivars. Thus, specialty cultivars are a potential genetic reservoir for the breeding of commodity cultivars in North America. At present, specialty cultivars that diverge most from commodity cultivars (in terms of pedigree) tend to be low yielding (most yield <90% of commodity types) and for this reason have rarely been used as parents in breeding for commodity cultivars. However, continuing selection for improved yield has produced recent specialty types which yield only slightly lower than commodity cultivars. For example, N7103 and N6201, large and small seeded cultivars, respectively, yield only about 5 and 8% below commodity cultivars (Carter et al., 2003a, 2003b). Thus, specialty cultivars may play a more important role in commodity breeding in the future.

8–16.1 Types of Specialty Soybean Cultivars

Specialty-use cultivars are those with special traits aside from high seed yield or pest resistance. Specialty cultivars have been developed with altered seed appearance (i.e., unusually large or small seed, and yellow hila), altered seed composition (i.e., low linolenic or palmitic acid content, high seed protein content, low lipoxygenase activity, or kunitz trypsin inhibitor null), or increased vegetative growth (i.e., forage type that may express the long juvenile trait) (Table 8–19). Seed appearance is often associated with a particular food use. For example, tofu and natto are generally produced from large and small seeded soybean (100-seed-weight > 20 or < 11 g, respectively). Green vegetable or immature soybean is usually produced from cultivars that have a 100 seed weight >25 g at maturity. Altered seed composition may be associated with a niche market use, but is also important in industry breeding efforts to increase the competitiveness of soybean in the world trade. For example, soybean oil that is low in both saturated and polyunsaturated fats promotes human health, eliminates the need for hydrogenation of soybean cooking oil, and is competitive with canola oil in the market place. Cultivars with increased vegetative growth are associated with niche production of soybean as a forage or in double cropping after small grains. In most cases, increased vegetative growth is associated with delayed flowering (Tompkins and Shipe, 1977; Sheaffer et al., 2001).

8–16.2 Genetic Diversity of Specialty Soybean Cultivars

Twenty-nine ancestors were employed in the development of specialty cultivars that do not appear (in a substantial way) in the genetic base of publicly released commodity soybean cultivars in North America (Table 8–20). Eight new ancestors of public cultivars (PI 153293, PI 159925, PI 189880, PI 257435, PI 261475, PI 90406, PI 92567 and T215) had high seed protein content (>44% on a dry weight

Table 8–20. Genetic contribution of 34 ancestors to the 97 specialty-use cultivars developed in North America (USA and Canada) from 1956 to 2000 for the soyfoods market.

Ancestor	Trait	Genetic contribution to specialtybase	Genetic contribution to commodity base‡
		%	
Kanro	Large seed	3.645	0.025
Jogun	Large seed	3.614	0.024
Unknown male parent of Vance	Small seed	2.062	0.000
Bansei	High protein content	2.062	0.001
PI 101404, *G. soja*	Small seed	1.740	0.000
H-24	Small seed	1.546	0.000
PI 240664	Long juvenile	1.289	0.000
Jizuka	Small seed	1.031	0.000
PI 196176	Small seed	1.031	0.000
Aoda	Large seed	1.031	0.000
PI 437267	Small seed	0.773	0.000
PI 86023	Null liproxygenase-lx_2	0.644	0.000
Nakasennari	Large seed	0.515	0.000
PI 81762, *G. soja*	Small seed	0.515	0.000
PI 408016B	Small seed	0.515	0.000
Unknown male parent of Danatto	Small seed?	0.515	0.000
PI 135624, *G. soja*	Small seed	0.451	0.000
PI 261475	High protein content	0.451	0.000
PI 189880	High protein content	0.387	0.000
DSR 252	High yield?	0.322	0.000
PI 19183	Forage	0.290	0.000
Pridesoy II	High yield?	0.258	0.000
PI 153293	High protein content	0.258	0.000
T215	High protein content	0.258	0.000
PI 437296	Small seed	0.258	0.000
PI 65338 †	Low protein content	0.258	0.001
Hahto	Green seed coat, high protein content	0.258	0.001
PI 159925	Long juvenile, high protein content	0.161	0.000
JA42	Small seed	0.161	0.000
PI 123440	Low linolenic acid content	0.129	0.000
PI 189950	Small seed	0.129	0.000
PI 92567	High protein and high oleic acid content	0.032	0.000
PI 90406	High protein and high oleic acid content	0.032	0.000
PI 157440	Null kunitz inhibitor	0.016	0.000
Total genetic contribution		26.637	0.053

† PI 65338 is a low protein content accession that appears in the pedigree of the high protein content cv. Protana.

‡ Kanro, Jogun, Bansei, Hahto, and PI 65338 contributed some parentage to commodity cultivars but were otherwise used predominantly to develop specialty-use cultivars. The other 29 ancestors contributed exclusively to specialty cultivars.

basis). (PI 257435 was used to develop a proprietary commodity cultivar.) Thirteen new ancestors (H-24, JA42, Jizuka, PI 189950, PI 196176, PI 408016B, PI 437267, PI 437296, the two unknown small-seeded parents of Vance and Danatto, PI 101404, PI 135624, and PI 81762) were used in small-seeded cultivar development. The latter three are accessions of the small-seeded *G. soja*. Other important specialty traits from the new ancestors included the lx_2 null lipoxygenase gene (from

PI 86023), the null kunitz inhibitor (from PI 157440), low linolenic acid (from PI 123440), large seed (from Aoda andNakasennari), the long juvenile or delayed flowering trait (from PI 159925, PI 240664), and forage production (from PI 19183).

In general, specialty cultivars have received about one fourth of their pedigree from ancestors that were not involved in commodity breeding. The ancestors showed a remarkably uniform distribution, indicating that no single ancestor dominated in the specialty use areas. This reflects the many purposes and varying MGs for which specialty cultivars are bred. The mean CP for specialty cultivars was smaller than that for commodity cultivars in both northern and southern North America (0.15 vs. 0.18 in the North and 0.18 vs. 0.24 in the South) indicating substantial genetic diversity in specialty cultivars (Cui, Z., North Carolina State Univ., personal communication, 2002). The mean CP value between specialty and commodity cultivars was also small (0.12 in the North and 0.17 in the South) suggesting that specialty and commodity cultivars are potential reservoirs of diversity for each other.

8–17 SUMMARY

There are four main aspects of genetic diversity: formation, collection, evaluation, and utilization. Each is covered in this chapter. Before written history, Chinese farmers used genetic diversity to develop the species that we now know as a *Glycine max*. Following this ancient, and perhaps diffuse, domestication, soybean spread to diverse areas, a dispersion that required selection for regional adaptation. The results of this dispersion and selection is still evident in our current collections.

Soybean germplasm collections exist in many countries. In total, these collections are quite extensive and anticipated future growth is likely to be modest because modern cultivars have already replaced landraces in most places. The collections of other *Glycine* species are relatively small, and there are significant opportunities for expansion. Efforts should be made to collect accessions of these wild relatives and employ molecular genetic techniques in their characterization.

Evaluation of genetic diversity is of paramount importance to its utilization. Evaluation of diversity is always a formidable task, especially given the immense sizes of the *G. max* collections. Easily scored traits such as maturity, plant height, plant and seed pigments, and seed size are recorded for most accessions. Meaningful data for more economically important traits such as seed composition, disease resistance, and yield require substantially more resources, and fewer accessions are characterized for these traits. The advent of new genetic technology based on gene expression patterns, marker polymorphisms, or biochemical traits may facilitate screening the collections for useful alleles.

Genetic diversity has no impact unless it is utilized. The extent of utilization in soybean breeding varies widely by region and trait. A key factor driving utilization of exotic germplasm is potential benefit. Benefit can be quite apparent for characteristics such as disease resistance or specialty traits, but vague for yield or abiotic stress resistance. This has limited the use of exotic germplasm to improve complex traits in the North America. Even with QTL analyses and gene mapping, it will be a difficult task to find beneficial exotic alleles that complement current commercial breeding efforts. New procedures and technology will be required to effectively utilize the diversity available.

Successful utilization requires the interaction of the germplasm collections and breeding programs. In the USA, this will require new models of cooperation between the commercial companies who release most new cultivars and public institutions that do most germplasm evaluation and development. Identifying mutual interests, agreeing on needs, and overcoming legal barriers to access and ownership of germplasm will be key issues to address in forming effective partnerships.

Despite the key role of genetic diversity in the history of science and agriculture, scientists have recognized only in recent decades the significance of systematic collection and preservation of germplasm, and the importance of a comprehensive understanding of genetic diversity. The research tools to find and manipulate genes continue to be more powerful and less expensive. The confluence of technology, genetic resources, and human need may make the efficient and effective utilization of genetic diversity one of the notable accomplishments of the 21st century.

REFERENCES

Abdelnoor, R.V., E.G. de Barros, and M.A. Moreira. 1995. Determination of genetic diversity within Brazilian soybean germplasm using random amplified polymorphic DNA techniques and comparative analysis with pedigree data. Braz. J. Genet. 18(2): 265–273.

Abe, J., K. Oono, D. Vaughan, N. Tomooka, A. Kaga, and S. Miyazaki, 2000. The genetic structure of natural populations of wild soybeans revealed by isozymes and RFLPs of mitochondrial DNAs: Possible influence of seed dispersal, cross-pollination and demography. p.143–158. In D. Vaughan et al. (ed.) The Ministry of Agriculture, Forestry and Fisheries (MAFF), Japan. Int. Workshop on Genetic Resources. Part 1. Wild legumes, Ibaraki, Japan. 13–15 Oct. 1999. Natl. Inst. of Agrobiological Resources (NIAR), Tsukuba, Japan.

Abney, T.S., J.C. Melgar, T.L. Richards, H. Scott, J. Grogan, and J. Young. 1997. New races of *Phytophthora sojae* with *Rps*1-d virulence. Plant Dis. 81:653–655

Ahnert, D., M. Lee, D.F. Austin, C. Livini, W.L. Woodman, S.J. Openshaw, J.S.C Smith, K. Porter, and G. Dalton. 1996. Genetic diversity among elite sorghum inbred lines assessed with DNA markers and pedigree information. Crop Sci. 36:1385–1392.

All, J.N, H.R. Boerma, and J.W. Todd. 1989. Screening soybean genotypes in the greenhouse for resistance to insects. Crop Sci.29:1156–1159.

Anand, S.C. 1992. Registration of Hartwig soybean. Crop Sci. 32:1069–1070.

Anand, S.C., and K.M. Gallo. 1984. Identification of additional soybean germplasm with resistance to race 3 of the soybean cyst nematode. Plant Dis. 68: 593–595.

Anand, S.C., K.M. Gallo, I.A. Baker, and E.E. Hartwig. 1988. Soybean plant introduction with resistance to races 4 or 5 of soybean cyst nematode. Crop Sci. 28:563–564.

Anand, S.C., and A.P. Rao-Arelli. 1989. Genetic analysis of soybean genotypes resistant to soybean cyst nematode race 5. Crop Sci. 29: 1181–1184.

Anand, S.C., J.A. Wrather, and C.R. Shumway. 1985. Soybean genotypes with resistance to races of soybean cyst nematode. Crop Sci. 25:1073–1075.

Apuya, N., B.L. Frazier, P. Keim, E. Jill Roth, and K.G. Lark. 1988. Restriction length polymorphisms as genetic markers in soybean, *Glycine max* (L.) Merrill. Theor. Appl. Genet. 75:889–901.

Arahana, V.S., G.L. Graef, J.E. Specht, J.R. Stedman, and K.M. Eskridge. 2001. Identification of QTLs for resistance to *Sclerotinia sclerotiorum* in soybean. Crop Sci. 41:180–188.

Arelli, P.R., D.A. Sleper, P. Yue, and J.A. Wilcox. 2000. Soybean reaction to races 1 and 2 of *Heterodera glycines*. Crop Sci 40:824–826.

Autrique, E., M.M. Nachit, P. Monneveux, S.D. Tanksley, and M.E. Sorrells. 1996. Genetic diversity in durum wheat based on RFLPs, morphophysiological traits, and coefficient of parentage. Crop Sci. 36:735–742.

Awadallah, W.H., M.F. Lutfallah, Eglal M.A. El-Moneim, M.F. El-Metwally, and M.Z. Hassan. 1990. Evaluation of some soybean genotypes for their resistance to the cotton leaf worm, *Spodoptera Littoralis Boisd*. Agric. Res. Rev. 68:121–126.

Bachman, M.S., and C.D. Nickell. 2000. High frequency of brown stem rot resistance in soybean germplasm from central and southern China. Plant Dis. 84:694–699.

Bachman, M.S., C.D. Nickell, P.A. Stevens, and A.D. Nickell. 1997. Brown stem rot resistance in soybean germplasm from central China. Plant Dis. 81:953–956.

Baker, W.A., D.B. Weaver, J. Qiu, and P.F. Pace. 1999. Genetic analysis of frogeye leaf spot resistance in PI54160 and Peking soybean. Crop Sci. 39:1021–1025.

Baldocchi, D.D., S.B. Verma, N.J. Rosenberg, B.L. Blad, A. Garay, and J.E. Specht. 1983a. Influence of water stress on diurnal exchange of mass and energy between the atmosphere and a soybean (*Glycine max* (L.) Merrill) canopy. Agron. J. 75:537–543.

Baldocchi, D.D., S.B. Verma, N.J. Rosenberg, B.L. Blad, A. Garay, and J.E. Specht. 1983b. Leaf pubescence effects on mass and energy exchange between soybean (*Glycine max* (L.)Merrill) canopies and the atmosphere. Agron. J. 75:543–548.

Baldocchi, D.D., S.B. Verma, N.J. Rosenberg, B.L. Blad, and J.E. Specht. 1985. Microclimate-plant architectural interactions: Influence of leaf width on the mass and energy exchange of a soybean canopy. Agric. Forest. Meteor. 35:1–10.

Barrett, B.A., K.K. Kidwell, and P.N. Fox. 1998. Comparison of AFLP and pedigree-based genetic diversity assessment methods using wheat cultivars from the pacific northwest. Crop Sci. 38:1271–1278.

Beach, R.M., J.W. Todd, and S.H. Baker. 1985. Antibiosis of four insect-resistant soybean genotypes to soybean looper (Lepidoptera: Noctuidae). Environ. Entomol. 14:531–534.

Beavis, W.D. 1998. QTL analyses: Power, precision, and accuracy. p. 145–162. *In* A.H. Patterson (ed.) Molecular dissection of complex traits. CRC Press, Boca Raton, FL.

Bell-Johnson, B., G. Garvey, J. Johnson, D. Lightfoot, K. Meksem. 1998. Biotechnology approaches to improving resistance to SCN and SDS: Methods for high throughout marker assisted selection. Soybean Genet. Newsl. 25:115–117.

Berger, G.U., and H.C. Minor. 1999. A RFLP marker associated with resistance to phomopsis seed decay in soybean PI 417479. Crop Sci. 39:800–805.

Bernard, R.L., C.R. Cremeens, R.L. Cooper, F.L. Collins, O.A. Krober, K.L. Athow, F.A. Laviolette, C.J. Coble, and R.L. Nelson. 1998. Evaluation of the USDA Soybean Germplasm Collection: Maturity Groups 000 to IV (FC 01.547-PI 266.807). USDA, Tech. Bull. 1844. U.S. Gov. Print. Office, Washington, DC.

Bernard, R.L., G.A. Juvik, E.E. Hartwig, and C.J. Edwards, Jr. 1988a. Origins and pedigrees of public soybean varieties in the United States and Canada. USDA Tech. Bull. 1746. U.S. Gov. Print. Office, Washington, DC.

Bernard, R.L., G.R. Noel, S.C. Anand, and J.G. Shannon. 1988b. Registration of 'Fayette' soybean. Crop Sci. 28:1028–1029.

Bernard, R.L., G.R. Noel, S.C. Anand, and J.G. Shannon. 1988c. Registration of 'Cartter' soybean. Crop Sci. 28: 1029–1030.

Bernard, R.L., and J.G. Shannon. 1980. Registration of Franklin soybean. Crop Sci. 20:825.

Bernard, R.L., and M.G. Weiss. 1973. Qualitative genetics. p. 117–154. *In* B.E. Caldwell (ed.) Soybean: Improvement, production, and uses. Agron. Monogr. 16. ASA, Madison, WI.

Berner, D.K., G.T. Berger, M.E. White, J.S. Gerahey, J.A. Freedman, and J.P. Snow. 1986. Red crown rot: Now a major disease of soybean. La. Agric. 29:4–5.

Bink, M.C.A.M., P. Uimari, M.J. Sillanpaa, L.L.G. Janss, and R.C. Jansen. 2002. Multiple QTL mapping in related plant populations via a pedigree-analysis approach. Theor. Appl. Genet. 104:751–762.

Birnberg, P.R., R.F Cordero, and M.L. Brenner. 1987. Characterization of vegetative growth of dwarf soybean genotypes including a gibberellin-insensitive genotype with impaired cell division. Am. J. Bot. 74:868–876.

Boehlke, E.G., J. Mudge, R.L. Denny, J.H. Orf, P.B. Cregan, and N.D. Young. 1999. Comparison between marker assisted selection for *rhg1* and greenhouse screening. p. 45. *In* Natl. Soybean Cyst Nematode Conf. Proc., Orlando, FL. 7–8 Jan. 1999.

Bohn, M., H.F. Utz, and E. Melchinger. 1999. Genetic similarities among winter wheat cultivars determined on the basis of RFLPs, AFLPs, and SSRs and their use for predicting progeny variance. Crop Sci. 39:228–237.

Boland, G.J., and R. Hall. 1986. Growth room evaluation of soybean cultivars for resistance to *Sclerotinia sclerotiorum*. Can. J. Plant Sci. 66:559-564.

Boland, G.J., and R. Hall. 1987. Evaluating soybean cultivars for resistance to *Sclerotinia sclerotiorum* under field conditions. Plant Dis. 71:934–936.

Bowers, G.R. Jr., K. Ngeleka, and O.D. Smith. 1993. Inheritance of stem canker resistance in soybean cultivars Crokett and Dowling. Crop Sci. 33:67–70.

Bowman, D.T. O.L. May, and D.S. Calhoun. 1996. Genetic base of upland cotton cultivars released between 1970 and 1990. Crop Sci. 36:577–581.

Brewer, F.L. 1981. Special assessment of the soybean cyst nematode problem. Joint Planning and Evaluation Staff Paper, USDA. Washington, DC.

Brim, C.A., and J.P. Ross. 1966. Registration of Pickett soybean. Crop Sci. 6:305.

Broue, P, J. Douglass, J.P. Grace, and D.R. Marshall. 1982. Interspecific hybridisation of soybeans and perennial Glycine species indigenous to Australia via embryo culture. Euphytica 31:715–724

Brown, A.H.D. 1995. The core collection at crossroads. p. 3–19. In T. Hodgkin. (ed.) Core collection of plant resources. John Wiley and Sons, Chichester, UK.

Brown, A.H.D. 1989a. The case of core collections. p. 135–156. In A.H.D. Brown et al. (ed.) The use of plant resources. Cambridge Univ. Press, Cambridge, UK.

Brown, A.H.D. 1989b. Core collections: A practical approach to genetic resources management. Genome 31: 818–824.

Brown A.H.D., C.L. Brubaker, D.J. Coates, S.D. Hopper, and S.L. Farrer 2000. Genetics and the conservation and use of Australian wild relatives of crops. Aust. J. Bot. 48:297–303.

Brown, A.H.D., J.P. Grace, and S.S. Speer. 1987. Designation of a core collection of perennial Glycine. Soybean Genet. Newsl. 14: 59–70.

Brown-Guedira, G.L., J.A. Thompson, R.L. Nelson, and M.L. Warburton. 2000. Evaluation of genetic diversity of soybean introductions and North American ancestors using RAPD and SSR markers. Crop Sci. 40:815–823.

Brown-Guedira, G.L., M.L. Warburton, and R.L. Nelson. 2003. Registration of LG92-1255, LG93-7054, LG93-7654, and LG93-7792 soybean germplasm. Crop Sci. (In press.)

Brown, W.L., and M.M. Goodman. 1977. Races of corn. p. 49–88. In G.F. Sprague (ed.) Corn and corn improvement. CSSA, Madison, WI.

Bult, C.J., and Y.T. Kiang. 1992. Electrophoretic and morphological variation within and among natural populations of the wild soybean, *Glycine soja* Sieb. & Zucc. Bot. Bull. Acad. Sinica 33:111–122.

Burkey, K.O., Gizlice, Z., and Carter, T.E., Jr. 1996.Genetic variation in photosynthetic electron transport capacity is related to plastocyanin concentration in the chloroplast. Photosyn. Res. 49:141–149.

Burnham, K.D., A.E. Dorrance, D.M. Francis, R.J. Fioritto, and S.K. St. Martin. 2003. Rps8, a new locus in soybean for resistance to *Phytophthora soja*e. Crop Sci. 43:101–105.

Burton, J.W. 1987. Quantitative genetics: results relevant to soybean breeding. p. 211–247. In J.R. Wilcox (ed.) Soybeans: Improvement, production, and uses. 2nd ed. Agron. Monogr. 16. ASA, CSSA, and SSSA, Madison, WI.

Burton, J.W. 1989. Breeding soybean cultivars for increased soybean percentage. p1079–1085. In A.J. Pascale (ed.) Proc. World Soybean Res. Conf. IV, Buenos Aires, Argentina. 5–9 Mar. 1989. Asociacion Argentina de la Soja, Buenos Aires.

Burton, J.W., C.A. Brim, and J.O. Rawlings. 1983. Perfomance of non-nodulating and nodulating soybean isolines in mixed culture with nodulating cultivars. Crop Sci. 23:469–473.

Buss, G.R., G. Ma, S. Krstipati, P. Chen, and S.A. Tolin. 1999. A new allele at the *Rsv3* locus for resistance to soybean mosaic virus. p. 490. In H. E. Kauffman (ed.) Proc. World Soybean Res. Conf. VI, Chicago, IL. 4–7 Aug.1999. Superior Print., Champaign, IL.

Buzzel, R.I., and J.C. Tu. 1989. Inheritance of soybean resistance to soybean mosaic virus. J. Hered. 75:82.

Caldwell, B.E., C.A. Brim, and J.P. Ross. 1960. Inheritance of resistance to soybean cyst nematode, *Heterodera glycines* Ichinohe. Agron. J. 52:635–636.

Campbell, C.L., P.D. Peterson, and C.S. Griffith. 1999. The formative years of plant pathology in the United States. APS Press, St. Paul, MN.

Cao, D., and J.H. Oard. 1997. Pedigree and RAPD-based DNA analysis of commercial U.S. rice cultivars. Crop Sci. 37:1630–1635.

Carpenter, J.A. and W.R. Fehr. 1986. Genetic variability for desirable agronomic traits in populations containing *Glycine soja* germplasm. Crop Sci. 26:681–686.

Carter, T.E. Jr., J.W. Burton, Z. Cui, X. Zhou, M.R. Villagarcia, M.O. Fountain, and A.S. Niewoehner. 2003a. Registration of 'N6201' soybean. Crop Sci. 43:1125–1126.

Carter, T.E. Jr., J.W. Burton, M.R. Villagarcia, Z. Cui, X. Zhou, M.O. Fountain, D.T. Bowman, and A.S. Niewoehner. 2003b. Registration of 'N7103' soybean. Crop Sci. 43:1128.

Carter T.E., Jr., Z. Cui, M. R. Villagarcia, X. Zhou, and J.W. Burton. 2001. Recent changes in genetic diversity patterns for publicly released North American soybean cultivars. In Annual meetings abstracts. [CD-ROM.] ASA, CSSA, and SSSA, Madison, WI.

Carter, T.E. Jr., Z. Gizlice, and J.W. Burton. 1993. Coefficient of parentage and genetic similarity estimates for 258 North American soybean cultivars released by public agencies during 1954-88. USDA Tech. Bull.1814. U.S. Gov. Print. Office, Washington, DC.

Carter, T. E., Jr., R. L. Nelson, P. B. Cregan, H. R. Boerma, P. Manjarrez-Sandoval, X. Zhou, W. J. Kenworthy, and G. N. Ude. 2000. Project SAVE (Soybean Asian Variety Evaluation) —Potential new sources of yield genes with no strings from USB, public, and private cooperative research. p. 68–83. In B. Park (ed.) Proc. of the 28th Soybean Seed Res. Conf.1998. Am. Seed Trade Assoc., Washington DC.

Caviness, C. E. 1992. Breeding for resistance to soybean cyst nematode. p. 143–156. In R.D. Riggs and J.A. Wrather (ed.) Biology and management of soybean cyst nematode. APS Press, St. Paul, MN.

Chamberlain, D.W., and R.L. Bernard. 1968. Resistance to brown stem rot in soybeans. Crop Sci. 8:728–729.

Chang, R.Z. 1989. Study on the origin of cultivated soybean. (In Chinese.) Oil Crops of China (1):1–6.

Chang, R.Z. 1997. Collection, conservation and evaluation of soybean genetic resources in China. (In Chinese.) p. 55–63. In Chinese agricultural sciences. China Agric. SciTech Press, Beijing, China.

Chang, S.J.C., T.W. Doubler, V. Kilo, R. Suttner, J. Klein, M.E. Schmidt, P.T. Gibson, and D.A. Lightfoot. 1996a. Two additional loci underlying durable field resistance to soybean sudden death syndrome (SDS). Crop Sci. 36:1684–1688.

Chang, S.J.C., T.W. Doubler, V.Y. Kilo, J. Abu-Thredeih, R. Prabhu, V. Freire, R. Suttner, J. Klein, M.E. Schmidt, P.T. Gibson, and D.A. Lightfoot. 1997. Association of loci underlying field resistance to soybean sudden death syndrome (SDS) and cyst nematode (SCN) race 3. Crop Sci. 37:965–971.

Chang, R., L. Qiu, J. Sun, Y. Chen, X. Li, and Z. Xu. 1999. Collection and conservation of soybean Germplasm in China. p. 172–176. In H.E. Kauffman (ed.) Proc. World Soybean Res. Conf. VI, Chicago, IL. 4–7 Aug.1999. Superior Print., Champaign, IL.

Chang, R., and J. Sun. (ed.) 1991. The catalogue of soybean germplasm in China (Suppl. 1). (In Chinese.) Agric. Publ. House, Beijing.

Chang, R.Z., J.Y. Sun, and L.J. Qiu. 1993. Evolution and development of soybean varieties and research plans of soybean germplasm. (In Chinese.) Soybean Bull. 2(3):35–36.

Chang, R.Z., J.Y. Sun. L.J. Qiu, and Y. Chen. 1996b. Chinese Soybean Collection Catalog (continued 2). (In Chinese.) China Agric. Press, Beijing.

Chen, Y. 2002. Evaluation of diversity in Glycine soja and genetic relationships within the subgenus Soja. Ph.D. diss. Univ. of Illinois, Urbana-Champaign (Diss. Abstr. Int. 63-02B:0596).

Chen, P., G.R. Buss, I. Gunduz, C.W. Roane, and S.A. Tolin. 2001a. Inheritance and allelism tests of Raiden soybean for resistance to soybean mosaic virus. J. Hered. 92:51–55.

Chen, P., G.R. Buss, C.W. Roane, and S.A. Tolin. 1991. Allelism among genes for resistance to soybean mosaic virus in strain-differential soybean cultivars. Crop Sci. 31:305–309.

Chen, L.F.O., G.C.Chen, S.F. Lin, and S.C.G. Chen 1993. Polymorphic differentiation and genetic variation of soybean by RFLP analysis. Bot. Bull. Acad. Sin. 34:249–259.

Chen, W., B. W. Diers, and R. L. Nelson. 2001b. Phialophora gregata strains pathogenic to resistance allele sources Rbs1 and Rbs3. Phytopathology 91(Suppl.):S16.

Chen, W., C. R. Grau, E.A. Adee, and X. Meng. 2000. A molecular marker identifying subspecific populations of the soybean brown stem rot pathogen, Phialophora gregata. Phytopathology 90:875–883.

Chen, L.F.O.,W.C. Hu, and S.C.G. Chen. 1989. Analysis of zymogram variations on cultivated soybean (Glycine max (L.) Merr.) of Taiwan. Bot. Bull. Acad. Sin. (Taipei) 30(3):179–190.

Cheres, M.T., and S.J. Knapp. 1998. Ancestral origins and genetic diversity of cultivated sunflower: Coancestry analysis of public germplasm. Crop Sci. 38(6):1476–1482.

Chiang, H.S., and N.S. Talekar. 1980. Identification of sources of resistance to the beanfly and two other agromyzid flies in soybean and mungbean. J. Econ. Entomol. 73:197–199.

Cho, E.K., and R.M. Goodman. 1979. Strains of soybean mosaic virus: classification based on virulence in resistant soybean cultivars. Phytopathology 69:467–470.

Choi, I.Y., S.H. Lim, D.W. Kim, Y.S. Choi, Y.B. Shin, and N.S. Kim. 2000. Classification of diverse soybean germplasm with morphological characters and molecular markers. Korean J. Genet. 22: 87–100.

Clark, W.J., F.A. Harris, F.G. Maxwell, and E.E. Hartwig. 1972 Resistance of certain soybean cultivars to bean leaf beetle, striped blister beetle, and bollworm. J. Econ. Entomol. 65:1669–1672.

Clawson, K.L., J.E. Specht, and B.L.Blad. 1986a. Growth analysis of soybean isolines differing in pubescence density. Agron. J. 78:164–172.

Clawson, K.L., J.E. Specht, B.L. Blad, and A.F. Garay. 1986b. Water use efficiency in soybean pubescence density isolines—A calculation procedure for estimating daily values. Agron. J. 78:483–487.

Clikeman, A.D., Palmer, R.G. and R.C. Shoemaker. 1998. The effect of pre-selection on diversity detection in exotic germplasm. Soybean Genet. Newsl. 25:149.

Cober, E.R., J.W. Tanner, and H.D. Voldeng. 1996. Soybean Photoperiod-sensitivity loci respond differentially to light quality. Crop Sci. 36:606–610.

Coble, C. J., G. L. Sprau, R. L. Nelson, J. L. Orf, D. I. Thomas, and J. F. Cavins. 1992. Evaluation of the USDA Soybean Germplasm Collection: Maturity Groups 000 to IV (PI 490.765 to PI 507.573). USDA Tech. Bull. 1802. U.S. Gov. Print. Office, Washington, DC.

Committee on Genetic Vulnerability of Major Crops. 1972. Genetic vulnerability of major crops Natl. Acad. of Sci. Washington, DC.

Concibido, V.C., R.L. Danny, D.A. Lange, J.H. Orf, and N.D. Young. 1996. RFLP mapping and marker-assisted selection of soybean cyst nematode resistance in PI 209332. Crop Sci. 36:1643–1650.

Concibido, V.C., D.A. Lange, R.L. Denny, J.H. Orf, and N.D. Young. 1997. Genome mapping of soybean cyst nematode resistance genes in 'Peking', PI 90763, and PI 88788 using DNA markers. Crop Sci. 37:258–264.

Concibido, V.C., B. LaVallee, J. Meyer, P. McLaird, L. Hummel, K. Wu, and X. Delannay.2002. Mining for yield genes in wild soybean. Abstract no. 201. 9th Biennial Conf. of the Cellular and Molecular Biology of the Soybean. 11–14 Aug. 2002. Univ. of Illinois, Urbana-Champaign.

Conservation of Germplasm Resources, An Imperative. 1978. Nat. Acad. of Sci. Washington, DC.

Cooper, R.L. 1981. Development of short statured soybean cultivars. Crop Sci. 21:127–131.

Cooper R.L. 1985. Breeding semidwarf soybeans. p. 289–309. In J. Janick (ed) Plant breeding reviews. Vol. III. ABI Publ. Co., Inc., Westport, CT.

Corbin, T. 1999. Developing soybean cyst nematode resistant cultivars—One company's strategy. p. 8. In Natl. Soybean Cyst Nematode Conf. Proc., Orlando, FL. 7–8 Jan. 1999.

Cornelious, B.K., and C.H. Sneller. 2002. Yield and molecular diversity of soybean lines derived from crosses of northern and southern elite parents. Crop Sci. 42:642–647.

Cox, T.S., Y.T. Kiang, M.B. Gorman, and D.M. Rodgers. 1985a. Relationship between coefficient of parentage and genetic similarity indices in the soybean. Crop Sci. 25:529–532.

Cox, T.S., G.L. Lockhart, D.E. Walker, L.G. Harrel, L.D. Albers, and D.M. Rodgers. 1985b. Genetic relationship among hard red winter wheat cultivars as evaluated by pedigree analysis and gliadin polyacrylamide gel electrophoretic patterns. Crop Sci. 25:1058–1063.

Cox, T.S., J.P. Murphy, and D.M. Rodgers, 1985c. Coefficients of parentage for 400 winter wheat cultivars. Agron. Dep., Kansas State Univ., Manhattan.

Cox, T.S., J.P. Murphy, and D.M. Rodgers. 1986. Changes in genetic diversity in the red winter wheat regions of the United States. Proc. Natl. Acad. Sci. 83:5583–5586.

Creech, J.L. 2000. The diplomacy of genetic resources: The key role of plant introduction in U.S./Japanese relations before and after World War II. Diversity 15:18–21.

Cregan, P.B., J. Mudge, E.W. Fickus, D. Danesh, R. Denny, and N.D. Young. 1999. Two simple sequence repeat markers to select for soybean cyst nematode resistance conditioned by the rhg1 locus. Theor. Appl. Genet. 99:811–818.

Cui, Z., T.E. Carter, Jr., and J.W. Burton. 2000a. Genetic base of 651 Chinese soybean cultivars released during 1923 to 1995. Crop Sci. 40:1470–1481.

Cui, Z., T.E. Carter, Jr., and J.W. Burton. 2000b. Genetic diversity patterns in Chinese soybean cultivars based on coefficient of parentage. Crop Sci. 40:1780–1793.

Cui, Z., T.E. Carter, Jr., J.W. Burton, and R. Wells. 2001. Phenotypic diversity of modern Chinese and North American soybean cultivars. Crop Sci. 41:1954–1967.

Cui, Z., T.E. Carter, Jr., J. Gai, J. Qiu, and R.L. Nelson. 1999. Origin, description, and pedigree of Chinese soybean cultivars released from 1923 to 1995. USDA Tech. Bull. 1871. U.S. Gov. Print. Office, Washington, DC.

Cui, Z., J. Gai, T.E. Carter, Jr., J. Qiu, and T. Zhao. 1998. The released soybean cultivars and their pedigree analyses (1923–1995). (In Chinese.) China Agric. Press, Beijing.

Cui, Z., J. Gai, D. Ji, Z. Ren, and Z. Sun. 1995a. Screening for resistant sources of soybeans to leaf-feeding insects. Soybean Genet. Newsl. 22:49–55.

Cui, Z., J. Gai, D. Ji, Z. Ren, and T. Zhao. 1996. Evaluation of soybean germplasm for resistance to leaf-feeding insects in Nanjing. (In Chinese.) Sci. Agric. Sin. 29(4):95–96.

Cui, Z., J. Gai, F. Pu, D. Qian, Q. Wang, J. Mao, D. Cheng, and Y. Wu. 1995b. Survey of leaf-feeding insects on soybeans in Nanjing. Soybean Genet. Newsl. 22:43–48.

Curtis, D.F., J.W. Tanner, B.M. Luzzi, and D.J. Hume. 2000. Agronomic and phenological differences of soybean isolines differing in maturity and growth habit. Crop Sci. 40:1624–1629.

Dae, H.P., K.M. Shim, Y.S. Lee, W.S. Ahn, J.H. Kang, and N.S. Kim. 1995. Evaluation of genetic diversity among the Glycine species using isozymes and RAPD. Korean J. Genet. 17:157–168.

Delannay, X., D.M. Rodgers, and R.G. Palmer. 1983. Relative contribution among ancestral lines to North American soybean cultivars. Crop Sci. 23:944–949.

de Vicente, M.C., and S.D. Tanksley. 1993. QTL analyses of transgressive segregation in an interspecific tomato cross. Genetics 134:585–596.

Diers, B. W., P. Arelli, and S.R. Cianzio. 1999. Management of SCN through conventional breeding for resistance—Midwest perspective. p. 5. In Natl. Soybean Cyst Nematode Conf. Proc., Orlando, FL. 7–8 Jan. 1999.

Diers, B.W., P. Keim, W.R. Fehr, and R.C. Shoemaker. 1992a. RFLP analysis of soybean seed protein and oil content. Theor. Appl. Genet. 83:608–612.

Diers, B.W., L. Mansur, J. Imsande, and R.C. Shoemaker. 1992b. Mapping phytophthora resistance loci in soybean with resistance fragment length polymorphism markers. Crop Sci. 32:377–383.

Diers, B.W., H.T. Skorupska, A.P. Rao-Arelli, and S.R. Cianzio. 1997. Genetic relationships among soybean plant introductions with resistance to soybean cyst nematodes. Crop Sci. 37:1966–1972.

Dilday, R.H. 1990. Contribution of ancestral lines in the development of new cultivars of rice. Crop Sci. 30:905-911.

Diwan, N., and P.B. Cregan. 1997. Automated sizing of fluorescent-labeled simple sequence repeat (SSR) markers to assay genetic variation in soybean. Theor. Appl. Genet. 95:723–733.

Dong, Y.S. et al. 1999. The genetic diversity in annual wild soybean. p. 147–155. In H.E. Kauffman (ed.) Proc. World Soybean Res. Conf. VI, Chicago, IL. 4–7 Aug.1999. Superior Print., Champaign, IL.

Dong, K., K.R. Barker, and C.H. Opperman. 1997. Genetics of soybean-Heterodera glycines interaction. J. Nematol. 29:509–522.

Donini, P., J.R. Law, R.M.D. Koebner, J.C. Reeves, and R.J. Cooke. 2000. Temporal trends in the diversity of UK wheat. Theor. Appl. Genet. 100:912–917.

Dorrance, A.E., and A.F. Schmitthenner. 2000. New sources of resistance to *Phytophthora sojae* in the soybean plant introductions. Plant Dis. 84:1303–1308.

Dorsett P.H. 1927. Soybeans in Manchuria. Publ. 1928. Proc. Am. Soybean Assoc. 1:173–176.

Doyle, J., J.L. Doyle, A.H.D. Brown, and B.E. Pfeil. 2000. Confirmation of shared and divergent genomes in the *Glycine tabacina* polyploid complex (Leguminosae) using histone H3-D sequences. Syst. Bot. 25:437–448.

Dubreuil, P., P. Dofour, E. Krejci, M. Causse, D. de Vienne, A.Gallais, and A. Charcosset. 1996. Organization of RFLP diversity among inbred lines of maize representing the most significant heterotic groups. Crop Sci. 36:790–799.

Eathington, S.R., C.D. Nickell, and L.E. Gray. 1995. Inheritance of brown stem rot resistance from soybean cultivar BSR 101. J. Hered. 86:55–60.

Ertl, D.S. and W.R. Fehr. 1985. Agronomic performance of soybean genotypes from *Glycine max* x *Glycine soja* crosses. Crop Sci. 25:589-592.

Falconer, D.S., and T.F.C. Mackay. 1996. Introduction to quantitative genetics. Longman, Essex, England.

Fasoula, D.A., P.A. Stephens, C.D. Nickell, and L.O.Vodkin. 1995. Cosegregation of purple-throat flower color with dihydroflavonol reductase polymorphism in soybean. Crop Sci. 35:1028–1031.

Fehr, W.R., and S.R. Cianzio. 1981. Registration of soybean germplasm populations AP10 to AP14. Crop Sci. 21:477–478.

Fehr, W.R., and R.C. Clark. 1973. Registration of five soybean germplasm populations. Crop Sci. 13:778.

Feng, L. 2001. Genetic analysis of populations derived from matings of southern and northern U.S. southern soybean cultivars, and analysis of isoflavone concentration in soybean seeds. Ph.D. diss. North Carolina State Univ., Raleigh (Diss. Abstr. AAI3019226).

Fujita, R., M. Ohara, K. Okazaki, and Y. Shimamoto. 1997. The extent of natural cross-pollination wild soybean (*Glycine soja*). J. Hered. 88:124–128.

Fukuda, Y. 1933. Cytogenetical studies on wild and cultivated Manchurian soybean (Glycine L.). Jpn. J. Bot. 6:489–506.

Fulker, D.W., and L.R. Cardon. 1994. A sib-approach to interval mapping of quantitative trait loci. Am. J. Hum. Genet. 54:1092–1103.

Gai, J. 1997. Soybean breeding. In J. Gai (ed.) Plant breeding: Crop species. (In Chinese.) China Agric. Press, Beijing.

Gai, J.-Y., D.-H. Xu, Z. Gao, Y. Shimamoto, J. Abe, H. Fukushi, and S. Kitajima. 2000. Studies on the evolutionary relationship among eco-types of G. max and G. soja in China. Acta Agron. Sin. 26(5):513–520.

Gai, J., Z. Cui, and M. Lin. 1992a. Black root rot of soybeans in Jiangsu, China. Soybean Genet. Newsl. 19:30–32.

Gai, J., Z. Cui, and M. Lin. 1992b. A report on black root rot of soybeans in Jiangsu, China. (In Chinese.) Soybean Sci. 11:113–119.

Gai, J., and W. Guo. 2001. History of Maodou production in China. p. 41–47.*In* T.A. Lumpkin and S. Shanmugasundaram (ed.) Proc. of the 2nd Int. Vegetable Soybean Conf. (Edamame/Maodou), Tacoma WA. 10–11 Aug. 2001. Washington State Univ., Pullman.

Gai, J., Y. Hu, Z. Cui, H. Zhi, W. Hu, and Z. Ren. 1989a. An evaluation of resistance of soybean germplasm to strains of Soybean Mosaic Virus. (In Chinese.) Soybean Sci. 8:323–330.

Gai, J., F. Liu, Z. Cui, J. Xia, and Y. Ma. 1991. Inheritance of resistance to the beanfly (*Melanagromyza sojae* Zehntner) in soybeans. Soybean Genet. Newsl. 18:104–106.

Gai, J., J. Xia, Z. Cui, Z. Ren, F. Pu, and D. Ji. 1989b. A study on resistance of soybeans from southern China to soybean agromyzid fly (*Melanagromyza sojae* Zehntner). p. 1240–1245. *In* A.J. Pascale (ed.) Proc. World Soybean Res. Conf. IV, Buenos Aires, Argentina. 5–9 Mar. 1989. Asoc. Argentina de la Soja, Buenos Aires.

Gai, J., D. Xu, Z. Gao, T. Zhao, Y. Shimamoto, J. Abe, H. Fukushi, and S. Kitajima. 1999. Genetic diversity of annual species of soybeans and their evolutionary relationship in China. p. 515. *In* H.E. Kauffman (ed.) Proc. World Soybean Res. Conf. VI, Chicago, IL. 4–7 Aug.1999. Superior Print., Champaign, IL.

Gai, J., T. Zhao, and J. Qiu. 1997. A review on the advances of soybean breeding since 1981 in China. p. 168–174. *In* Seed industry and agricultural development, CAASS. China Agric. Press, Beijing.

Gannon, A.J., and C.E. Bach. 1996. Effects of soybean trichome density on Mexican bean beetle (*Coleoptera: Coccinellidae*) development and feeding preference. Environ. Entomol. 25:1077–1082.

Garcia, M.R., and J.R. Rich. 1985. Root-knot nematodes in north-central Florida soybean fields. Nematropica 15:43–48

Gilman, D.F., R.M. McPherson, L.D. Newsom, D.C. Herzog, and C. Williams. 1982. Resistance in soybeans to the Southern green stink bug. Crop Sci. 22:573–576.

Gizlice, Z., T.E. Carter, Jr., and J.W. Burton. 1993a. Genetic diversity in North American soybean: I. Multivariate analysis of founding stock and relation to coefficient of parentage. Crop Sci. 33:614–620.

Gizlice, Z., T.E. Carter Jr., and J.W. Burton. 1993b. Genetic diversity in North American soybean: II. Prediction of heterosis in F2 populations of southern founding stock using genetic similarity measures. Crop Sci. 33:620–626.

Gizlice, Z., T. E. Carter, Jr., and J. W. Burton. 1994. Genetic base for North American public soybean cultivars released between 1947 and 1988. Crop Sci. 34:1143–1151.

Gizlice, Z., T.E. Carter, Jr., T.M. Gerig, and J.W. Burton. 1996. Genetic diversity patterns in North American public soybean cultivars based on coefficient of parentage. Crop Sci. 36:753–765.

Goodman, M.M. 1972. Distance analysis in biology. Syst. Zool. 21:174–186.

Gorman, M.B. 1984. An electrophoretic analysis of the genetic variation in the wild and cultivated soybean germplasm. Ph.D. diss. Univ. of New Hampshire, Durham (Diss. Abstr. Int. 44B:3300).

Grabau, E. A., W. H. Davis, and B. G. Gengenbach. 1989. Restriction fragment length polymorphism in a subclass of the 'Mandarin' soybean cytoplasm. Crop Sci. 29:1554–1559.

Grabau, E.A., W. H. Davis, N.D. Phelps, and B.G. Gengenbach. 1992. Classification of soybean cultivars based on mitochondrial DNA restriction fragment length polymorphism. Crop Sci. 32:271–274.

Graef, G.L., W.R. Fehr, and S.R. Cianzio. 1989. Relation of isozyme genotypes to quantitative characters in soybean. Crop Sci. 29:683–688.

Grafius, J.E. 1978. Multiple characters and correlated response. Crop Sci. 18:931–934.

Grafius, J.E., R.L. Thomas, and J. Bernard. 1976. Effect of parental component complementation on yield and components of yield in Barley. Crop Sci. 16:673–677.

Graner, A., W.F. Ludwig, and A.E. Melchinger. 1994. Relationships among European barley germplasm: 11. Comparison of RFLP and pedigree data. Crop Sci. 34:1199–1205.

Grau, C.R., V.L. Radke, and F.L. Gillespie. 1982. Resistance of soybean cultivars to *Sclerotinia sclerotiorum*. Plant Dis. 66:506–508

Grau, G. 1999. Management of economically important soybean diseases in the United States. p. 264–268. *In* H.E. Kauffman (ed.) Proc. World Soybean Res. Conf. VI, Chicago, IL. 4–7 Aug.1999. Superior Print., Champaign, IL.

Gray, L.E. 1971. Variation in pathogenicity of *Cephalosporium gregatum* isolates. Phytopathology 61:1410–1411.

Griffin, J.D., and R.G. Palmer. 1995. Variability of thirteen isozyme loci in the USDA soybean germplasm collections. Crop Sci. 35:897–904.

Gunduz, I., G.R. Buss, G. Ma, P. Chen, and S.A. Tolin. 2001. Genetic analysis of resistance to soybean mosaic virus in OX670 and Harosoy soybean. Crop Sci. 41:1785–1791.

Gunduz, I., G.R. Buss, G. Ma, P. Chen, and S.A. Tolin. 2002. Characterization of SMV resistance genes in Tousan 140 and Hourei soybean. Crop Sci. 42:90–95.

Guo, W. 1993. The history of soybean cultivation in China. (In Chinese.) Hehai Univ. Press, Nanjing, Jiangsu, China.

Guo, S.G. 1983. Preliminary study on resistance of soybean varieties to northern pod borer. (In Chinese.) Soybean Sci. 2(3):200–206.

Guo, S.G., D.R. Yue, and J.L. Lu. 1986. Study on resistance of soybean varieties to northern pod borer. II. Evaluation of soybean germplasm for resistance to northern pod borer. (In Chinese.) Soybean Sci. 5(3):233–238.

Hajika M., M. Takahashi, S. Sakai, and K. Igita. 1996. A new genotype of 7 S globulin (beta-conglycinin) detected in wild soybean. Breed. Sci. 46:385–386.

Hajika, M, M. Takahashi, S. Sakai, and R.Matsunaga.1998. Dominant inheritance of a trait lacking beta-conglycinin detected in a wild soybean line. Breed. Sci. 48:383–386.

Ham, G.E., I.E. Liener, S.D. Evans, R.D. Frazier, and W.W. Nelson. 1975. Yield and composition of soybean seed as affected by N and S fertilization. Agron. J. 67:293–297.

Hammond, R.B., and R.L. Cooper. 1998. Antibiosis of released soybean germplasm to Mexican bean beetle (Coleoptera: Coccinellidae). J. Entomol. Sci. 34:183–190.

Hammond, R.B., C.G. Helms, and R. Nelson. 1998. Introduced soybean lines from China: Screening for insect resistance. J. Econ. Entomol. 91:546–551.

Hammond, R.B., and D.L. Jeffers. 1990. Potato leafhopper (Homoptera-Cicadellidae) populations on soybean relay intercropped into winter-wheat. Environ. Entomol. 19 (6):1810–1819.

Hancock, J.A., F.G. Hancock, C.E. Caviness, and R.D. Riggs. 1987. Genetics of resistance in soybean to "Race X" of soybean cyst nematode. Crop Sci. 27:704–707.

Hanson, P.M., C.D. Nickell, L.E. Gray, and S.A. Sebastian. 1988. Identification of two dominant genes conditioning brown stem rot resistance in soybean. Crop Sci. 28:41–43.

Harlan, J.R., and J.M.J. deWet. 1971. Toward a rational classification of cultivated plants. Taxon. 20:509–517.

Harris, D.K., H. R. Boerma, R.S.Hussey, and S.L. Finnerty. 2003. Additional sources of soybean germplasm with resistance to two species of root-knot nematode. Crop Sci. 43:1848–1851.

Hart, S.E., S. Glenn, and W.W. Kenworthy. 1991. Tolerance and the basis for selectivity to 2,4-D in peren-nial Glycine species. Weed Sci. 39:535–539.

Hartl, D.L., and A.G. Clark. 1997. Principles of population genetics.3rd ed. Sinauer Assoc., Sunder-land, MA.

Hartman, G.L., Gardner, M.E., T. Hymowitz, and G.C. Naidoo. 2000. Evaluation of perennial *Glycine* species for resistance to soybean fungal pathogens that caused Sclerotinia stem rot and sudden death syndrome. Crop Sci. 40:545–549.

Hartman, G.L., Y.H. Huang, R.L. Nelson, and G.R. Noel. 1997. Germplasm evaluation of *Glycine max* for resistance to *Fusarium solani*, the causal agent of sudden death syndrome. Plant Dis. 81:515–518.

Hartman, G.L., T.C. Wang, and T. Hymowitz. 1992. Sources of resistance to soybean rust in perennial Glycine species. Plant Dis. 76:396–399.

Hartung, R.C., J.E. Specht, and J.H. Williams. 1980. Agronomic performance of selected soybean mor-phological variants in irrigation culture with two row spacings. Crop Sci. 20:604–609.

Hartung, R.C., J.E. Specht, and J.H. Williams. 1981. Modification of plant architecture by genes for stem growth habit and maturity. Crop Sci. 21:51–56

Hartwig, E.E. 1972. Utilization of soybean germplasm strains in a soybean improvement program. Crop Sci. 12:856–859.

Hartwig, E.E. 1985. Breeding productive soybeans with resistance to soybean cyst nematode. P. 394–399. *In* R.A. Shibles (ed.) Proc. World Soybean Res. Conf. III, Iowa State Univ., Ames. 12–17 Aug. 1984. Westview Press, Boulder, CO.

Hartwig, E.E. 1988. Registration of soybean germplasm line D86-8286 resistant to rust. Crop Sci 28:1038.

Hartwig, E.E., and K.R. Bromfield. 1983. Relationships among 3 genes conferring specific resistance to rust in soybeans. Crop Sci. 23:237–239

Hartwig, E.E., and J.M. Epps. 1970. An additional gene for resistance to the soybean cyst nematode, *Heterodera glycines*. Phytopathology 60:584.

Hartwig, E.E. and H.W. Johnson. 1953. Effect of the bacterial pustule on yield and chemical composition of soybeans. Agron. J. 45:22-23.

Hartwig, E.E., and S.G. Lehman. 1951. Inheritance of resistance to the bacterial pustule disease in soybean. Agron. J. 43:226–229.

Harville, B.G., J.S. Russin, and R.J. Habetz. 1996. Rhizoctonia foliar blight reactions and seed yields in soybean. Crop Sci. 36:563–566.

Hatchett, J.H., G.L. Beland, and E.E. Hartwig. 1976. Leaf-feeding resistance to bollworm and tobacco budworm in three soybean plant introductions. Crop Sci. 16:277–280.

Hayata, B. 1919.Incones Plantarum Formosanarum 9:26–27.Bureau of Productive Industry, Formosa.

Hayes, A.J., G. Ma, G.R. Buss, and M.A.S. Maroof. 2000. Molecular marker mapping of Rsv4, a gene conferring resistance to all known strains of soybean mosaic virus. Crop Sci. 40:1434–1437.

Hellend, S.J., and J.B. Holland. 2003. Genome-wide genetic diversity among components does not cause cultivar blend responses. Crop Sci. 43:1618–1627.

Helms, T., J. Orf, G. Vallad, and P. McClean. 1997. Genetic variance, coefficient of parentage, and genetic distance of six soybean populations. Theor. Appl. Genet. 94:20–26.

Hermann, F.J. 1962. A revision of genus Glycine and its immediate allies. USDA Tech. Bull. 1268. U.S. Gov. Print. Office, Washington, DC.

Hill, J.L. 1999. Morphological and biochemical analysis of variation in diploid *Glycine tomentella* Hayata (2n = 38, 40) (denatured proteins, genomic DNA).Ph.D. diss. Univ. of Illinois, Urbana-Champaign. (Diss. Abstr. Int. 60B: 0881).

Hill, J.L., K. Peregrine, G.L. Sprau, C.R. Cremeens, R.L. Nelson, M.M. Kenty, T.C. Kilen, and D.A. Thomas. 2001. Evaluation of the USDA Soybean Germplasm Collection: Maturity Groups VI to VIII (FC 03.659 to PI 567.235B. USDA Tech. Bull. 1894. U.S. Gov. Print. Office, Washington, DC.

Hiromoto, D.M., and N.A. Vello. 1986. The genetic base of Brazilian soybean cultivars. Braz. J. Genet. 9:295–306.

Hnetkovsky, N., S.J.C. Chang, T.W. Doubler, P.T. Gibson, and D.A. Lightfoot. 1996. Genetic mapping of loci underlying field resistance to soybean sudden death syndrome (SDS). Crop Sci. 36:393–400.

Hoffman, D.D., B.W. Diers, G.L. Hartman,C.D. Nickell, R.L. Nelson, W.L. Pedersen, E.R. Cober, G.L. Graef, J.R. Steadman, C.R. Grau, B.D. Nelson, L.E. del Rio, T. Anderson, V. Poysa, I. Rajcan, T. Helms, and W.C. Stienstra. 2002. Selected soybean plant introductions with partial resistance to *Sclerotinia sclerotiorum*. Plant Dis. 86:971–980.

Hu, Y., H. Zhi, Z. Cui, W. Hue, and J. Gai. 1994. Utilization and evaluation of resistance of soybean germplasm to soybean mosaic virus (SMV). (In Chinese.) p.174–181. *In* W. Jinling et al. (ed.) Broadening and improvement of genetic background of soybeans in northeast China. Heilongjiang Sci. and Technol. Publ. House, Harbin.

Hulting, A.G., L.M. Wax, R.L. Nelson, and F.W. Simmons. 2001. Soybean (*Glycine max* (L.) Merr.) cultivar tolerance to sulfentrazone. Crop Prot. 20:679–683.

Hussey, R.S., H.R. Boerma, P.L. Raymer, and B.M. Luzzi.1991. Resistance in soybean cultivars from maturity groups V–VIII to soybean cyst and root knot nematodes. J. Nematol. 24(4S):576–583.

Hymowitz, T. 2001. Plant exploration for wild perennial Glycine. *In* Annual meetings abstracts. [CD-ROM.] ASA, CSSA, and SSSA, Madison, WI.

Hymowitz, T. 2004. Speciation and cytogenetics. p. 97–136. *In* H.R. Boerma and J.E. Specht (ed.) Soybeans: Improvement, production, and uses. 3rd ed. Agron. Monogr. 16. ASA, CSSA, and SSSA, Madison, WI.

Hymowitz, T., and C.A. Newell. 1980. Taxonomy, speciation, domestication, dissemination, germplasm resources and variation in the genus Glycine. p. 251–264. *In* R.J. Summerfield and A.H. Bunting (ed.) Advances in legume science. Royal Botanic Gardens. Kew, Richmond, Surrey, UK.

Hymowitz, T, and C.A. Newell. 1981. Taxonomy of the genus Glycine, domestication in China, and uses, of soybeans as human food or animal feed. Econ. Bot. 35:272–288.

Ininda, J., W.R. Fehr, S. Cianzio, and S. Schnebly. 1996 Genetic gain in soybean populations with different percentages of plant introduction parentage. Crop Sci. 36:1470–1472.

International Plant Genetic Resources Institute. 2001. Directory of Germplasm Collections [Online] Available at http://www.ipgri.org/system/page.asp?theme=1 (verified 25 Nov. 2002).

Israel, D.W., J.W.Burton, and R.F. Wilson. 1985. Studies on genetic male sterile soybeans. IV. Effect of male sterility and source of nitrogen nutrition on accumulation, partitioning, and transport of nitrogen. Plant Physiol. 78:762–767.

Jackai, L.E.N., K.E. Dashiell, and L.L. Bello. 1988. Evaluation of soybean genotypes for field resistance to stink bugs in Nigeria. Crop Prot. 7:48–54.

Jeong, S.C., S. Kristapati, A.J. Hayes, P.J. Maughan, S.L. Noffsinger, I. Gunduz, G.R. Buss, and M.A. Saghai Maroof. 2002. Genetic and sequence analysis of markers tightly linked to the soybean mosaic virus resistance gene, *Rsv3*. Crop Sci. 42:265–270.

Johns, M.A., P.W. Skroch, J. Nienhuis, P. Hinrichsen, G. Bascur, and C. Munoz-Schick. 1997. Gene pool classification of common bean landraces from Chile based on RAPD and morphological data. Crop Sci. 37:605–613.

Juvik, G., R.L. Bernard, R.Z. Chang, and J.F. Cavins. 1989a. Evaluation of the USDA Wild Soybean Germplasm Collection: Maturity Group 000 to IV (PI 65.549 to PI 483.464). USDA Tech. Bull. 1761. U.S. Gov. Print. Office, Washington, DC.

Juvik, G.A., R.L. Bernard, J.H. Orf, J.F. Cavins, and D.I. Thomas.1989b. Evaluation of the USDA Soybean Germplasm Collection: Maturity Groups 000 to IV (PI 446.983 to PI 486.355). USDA Tech. Bull. 1760. U.S. Gov. Print. Office, Washington, DC.

Kabelka, E.A., B.W. Diers, W.R. Fehr, A.R. LeRoy, I. Baianu, T. You, D.J. Neece, and R.L. Nelson. 2003. Identification of putative yield enhancing quantitative trait loci from exotic soybean germplasm. Crop Sci. 43:(in press).

Keim, P., W. Beavis, J. Schupp, and R. Freestone. 1992. Evaluation of soybean RFLP marker diversity in adapted germplasm. Theor. Appl. Genet. 85:205–212.

Keim, P., R.C. Shoemaker, and R.G. Palmer. 1989. Restriction fragment length polymorphism diversity in soybean. Theor. Appl. Genet. 80:786–791.

Kenty, M.M., K. Hinson, K.H. Quisenberry, and D.S. Wofford. 1996. Inheritance of resistance to the soybean looper. Crop Sci. 36:1532–1537.

Kenworthy, W.J. 1980. Strategies for introgressing exotic germplasm in breeding programs. p. 217–223. *In* F.T. Corbin (ed.) *In* R.A. Shibles (ed.) World Soybean Res. Conf., 2nd, Raleigh, NC. 26–29 Mar. 1979. Westview Press, Boulder, CO.

Khalaf, A.G.M., G.D. Brossman, and J.R. Wilcox. 1984. Use of diverse populations in soybean breeding. Crop Sci. 24:358–360.

Kiang, Y.T., and Y.C. Chiang. 1990. Comparing differentiation of wild soybean (*Glycine soja* Sieb. & Zucc.) populations based on isozymes and quantitative traits. Bot. Bull. Acad. Sin. 31:129–142.

Kiang, Y.T., Y.C. Chiang, and N. Kaizuma.1992. Genetic diversity in natural populations of wild soybean in Iwate prefecture, Japan. J. Hered. 83:325–329.

Kihara, H. 1969. History of biology and other sciences in Japan in retrospect. p. 49–70. *In* C. Oshima (ed.) Proc. XII Int. Congr. Genetics, Tokyo. 19–28 Aug. 1968. Vol. 3. Sci. Council of Japan, Tokyo.

Kiihl, R.A.S., and E.E. Hartwig. 1979. Inheritance of reaction to soybean mosaic virus in soybeans. Crop Sci. 19:372–375.

Kilen, T.C., and E.E. Hartwig. 1987. Identification of single genes controlling resistance to stem canker in soybean. Crop Sci. 27:863–864.

Kilen, T.C., J.H. Hatchett, and E.E. Hartwig. 1977. Evaluation of early generation soybeans for resistance to soybean loopers. Crop Sci. 17:397–398.

Kilen, T.C., B.L. Keeling, and E.E. Hartwig. 1985. Inheritance of reaction to stem canker in soybean. Crop Sci. 25:50–51.

Kilen, T.C., and L. Lambert. 1986. Evidence for different genes controlling insect resistance in three soybean genotypes. Crop Sci. 26:869–871.

Kilen, T.C., and L. Lambert. 1998. Genetic control of insect resistance in soybean germplasm PI 417061. Crop Sci. 38:652–654.

Kim, H.S., and B.W. Diers. 2000. Inheritance of partial resistance to sclerotinia stem rot in soybean. Crop Sci. 40:55–61.

Kim, H.S., G.L. Hartman, J.B. Manandhar, G.L. Graf, J.R. Stedman, and B.W. Diers. 2000. Reaction of soybean cultivars to sclerotinia stem rot in field, greenhouse, and laboratory evaluations. Crop Sci. 40:665–669.

Kim, H.S., C.H. Sneller, and B.W. Diers. 1999. Evaluation of soybean cultivars for resistance to sclerotinia stem rot in field environments. Crop Sci. 39:64–68.

Kim, H.S., and R.W. Ward. 1997. Genetic diversity in eastern U.S. soft winter wheat (*Triticum aestivum* L. em. Thell.) based on RFLPs and coefficients of parentage. Theor. Appl. Genet. 94:472–479

Kinloch, R.A. 1974. Response of soybean cultivars to nematicidal treatments of soil infested with *Meloidogyne incognita*. J. Nematol. 6:7–11.

Kinloch, R.A., C.K. Hiebsch, and H.A. Peacock. 1990. Evaluation of soybean cultivars of soybean cultivars for production in *Meloidogyne arenaria* race-2 infested soil. J. Nematol. 22:740–744.

Kisha, T. J., B.W. Diers, J.M. Hoyt, and C.H. Sneller. 1998. Genetic diversity among soybean plant introductions and North American germplasm. Crop Sci. 38:1669–1680.

Kisha, T., C.H. Sneller, and B.W. Diers. 1997. The relation of genetic distance and genetic variance in populations of soybean. Crop Sci. 37:1317–1325.

Knauft, D.A., and D.W. Gorbet. 1989. Genetic diversity among peanut cultivars. Crop Sci. 29:1417–1422.

Koenning, S.R. 2000. Density dependent yield of *Heterodera glycines*-resistant and -susceptible cultivars. J. Nematol. 32 (4S)(Suppl.):502–507.

Kopisch-Obuch, F.J.A., R.L. McBroom, and B.W. Diers. 2002. Association between SCN resistance QTL and yield in soybean. *In* Annual meetings abstracts. [CD-ROM.] ASA, CSSA, and SSSA, Madison, WI.

Kraemer, M.E., T. Mebrahtu, and M. Rangappa. 1994. Evaluation of vegetable soybean genotypes for resistance to Mexican bean beetle (Coleoptera: Coccinellidae). J. Econ. Entomol. 87:252–257.

Kraemer, M.E., M. Rangappa, P.S. Benepal, and T. Mebrahtu. 1988. Field evaluation of soybeans for Mexican bean beetle resistance. I. Maturity groups VI, VII, and VIII. Crop Sci. 28:497–499.

Kraemer, M.E., M. Rangappa, T. Mebrahtu, and P.S. Benepal. 1990. Field evaluation of soybean for Mexican bean beetle resistance. Crop Sci. 30:374–377.

Kwanyuen, P., V.R.Pantalone, J.W.Burton, and R.F.Wilson.1997.A new approach to genetic alteration of soybean protein composition and quality. JAOCS 74:983–987.

Lambert, L., R.M. Beach, T.C. Kilen, and J.W. Todd. 1992. Soybean pubescence and its influence on larval development and oviposition preference of lepidopterous insects. Crop Sci. 32:463–466.

Lambert, L., and T.C. Kilen. 1984a. Multiple insect resistance in several soybean genotypes. Crop Sci. 24:887–890.

Lambert, L., and T.C. Kilen. 1984b. Influence of three soybean plant genotypes and their F_1 intercross on the development of five insect species. J. Econ. Entomol. 77:622–625.

Lambert, L., and T.C. Kilen. 1989. Influence and performance of soybean lines isogenic for pubescence type on oviposition preference and egg distribution by corn earworm (*Lepidoptera: Noctuidae*). J. Entomol. Sci. 24:309–316.

Laviolette, F.A., and K.L. Athow, 1977. Three new physiological races of *Phytophthora megasperma* var *sojae*. Phytopathology 67:267–268.

Leffel, R.C. 1992. Registration of high-protein soybean germplasm lines BARC-6, BARC-7, BARC-8, and BARC-9. Crop Sci. 32:502.

LeRoy, A.R., W.R. Fehr, and S.R. Cianzio. 1991. Introgression of genes for small seed size from *Glycine soja* into *G. max*. Crop Sci. 31:693–697

Lewers, K.S, S.K. St.Martin, B.R. Hedges, and R.G. Palmer. 1998. Testcross evaluation of soybean germplasm. Crop Sci. 38:1143–1149.

Li, F.S. 1990. Chinese *G. soja* Collection Catalog. (In Chinese.) China Agric. Press, Beijing.

Li, F.S. 1994. Chinese *G. soja* Collection Catalog (Continued). (In Chinese.) China Agric. Press, Beijing.

Li, M. 1987a. Study on screening of resistance resources to downy mildew of soybean. (In Chinese.) Soybean Sci. 6:71–77.

Li, M.R., and Y.C. Geng. 1986. Evaluation of soybean varieties for low SMV seed-transmission frequency. (In Chinese.) Soybean Sci. 3:245–248.

Li, Y. 1987b. Evaluation and use of new source of resistance to soybean cyst nematode race 4. (In Chinese.) Soybean Sci. 5(4):291–297.

Li, Y., Y.P. Li, and X.Y. Zhang. 1991. Evaluation of Chinese soybean germplasm for resistance to SCN race 4. (In Chinese.) Sci. Agric. Sin. 24:64–69.

Li, Z, L. Jakkula, R.S. Hussey, J.P. Tamulonis, and H.R. Boerma. 2001a. SSR mapping and confirmation of the QTL from PI 96354 conditioning soybean resistance to southern root-knot nematode. Theor. Appl. Genet. 103:1167–1173.

Li, Z., and R.L. Nelson. 2001. Genetic diversity among soybean accessions from three countries measured by RAPDs. Crop Sci. 41:1337–1347.

Li, Z., and R.L. Nelson. 2002. RAPD marker diversity among cultivated and wild soybean accessions from four Chinese provinces. Crop Sci. 42:1737–1744.

Li, Z., L. Qiu, J.A. Thompson, M.M. Welsh, and R.L. Nelson. 2001b. Molecular genetic analysis of U.S. and Chinese soybean ancestral lines. Crop Sci. 41:1330–1336.

Liu, W.Z. 1985. Evaluation of Liaoning soybean germplasm for resistance to SCN race 3. (In Chinese.) Sci. Agric. Sin. (4):25–29.

Lohnes, D., C.D. Nickell, and A.F. Schmitthenner. 1996. Origin of soybean alleles for *Phytophthora* resistance in China. Crop Sci. 36:1689–1692.

Lombard, V., C.P. Baril, P. Dubreuil, F. Blouet, and D. Zhang. 2000. Genetic relationships and finger-printing of rapeseed cultivars by RFLP: Consequences for varietal registration. Crop Sci. 40:1417–1425.

Lorenzen, L. L., S. Boutin, N. Young, J.E. Specht, and R.C. Shoemaker. 1995. Soybean pedigree analysis using map-based molecular markers: I. Tracking RFLP markers in cultivars. Crop Sci. 35:1326–1336.

Loux. M.M. R.A. Liebl, and T. Hymowitz. 1987.Examination of wild perennial Glycine species for glyphosate tolerance. Soybean Genet. Newsl. 14:268–271.

Lu, H., and R. Bernardo. 2001. Molecular marker diversity among current and historical maize inbreds. Theor. Appl. Genet. 103:613–617.

Lu, S.L. 1977. The origin of cultivated soybean (*G. max*). (In Chinese.) *In* J.L.Wang (ed.) Soybean. Shanxi People's Press, Shanxi, China.

Luedders, V. D. 1977. Genetic improvement in yield of soybeans. Crop Sci. 17:971–972.

Luedders, V.D., and W.A. Dickerson. 1977. Resistance of selected soybean genotypes and segregating populations to cabbage looper feeding. Crop Sci. 17:395–397.

Luedders, V.E., L.F. Williams, and A. Matson. 1968. Registration of Custer Soybeans. Crop Sci. 8:402

Luzzi, B.M., H.R. Boerma, and R.S. Hussey. 1987. Resistance to three species of root knot nematode in soybean. Crop Sci. 27:258–262.

Luzzi, B.M., H.R. Boerma, and R.S. Hussey. 1994. A gene for resistance to the southern root knot nematode in soybean. J. Hered. 85:484–486.

Luzzi, B.M., H.R. Boerma, and R.S. Hussey. 1995a. Inheritance of resistance to the peanut root-knot nematode in soybean. Crop Sci. 35:50–53.

Luzzi, B.M., H.R. Boerma, R.S. Hussey, D.V. Phillips, J.P. Tamulonis, S.L. Finnerty, and E.D. Wood. 1996a. Registration of southern root-knot nematode resistant soybean germplasm line G93-9009. Crop Sci. 36:823.

Luzzi, B.M., H.R. Boerma, R.S. Hussey, D.V. Phillips, J.P. Tamulonis, S.L. Finnerty, and E.D. Wood. 1996b. Registration of soybean germplasm line G93-9106 resistant to peanut root-knot nematode. Crop Sci. 36:1423–1424.

Luzzi, B.M., H.R. Boerma, R.S. Hussey, D.V. Phillips, J.P. Tamulonis, S.L. Finnerty, and E.D. Wood. 1997. Registration of root-knot nematode resistant soybean germplasm line G93-9223. Crop Sci. 37:1035–1036.

Luzzi, B.M., J.P. Tamulonis, R.S. Hussey, and H.R. Boerma. 1995b. Inheritance of resistance to the Javanese root-knot nematode in soybean. Crop Sci. 35:1372–1375.

Ma, S.J. 1991. Evaluation of soybean germplasm for resistance to SCN race 3. (In Chinese.) Soybean Sci. 10(3):165–171.

Ma, G., P. Chen, G.R. Buss, and S.A. Tolin. 1995a. Genetic characteristics of two genes for resistance to soybean mosaic virus in PI486355 soybean. Theor. Appl. Genet. 91:907–914.

Ma, S.J., Y.H. Zhang, and Q.X Xue. 1995b. Evaluation of Chinese soybean germplasm for resistance to SCN race 3. (In Chinese.) Soybean Sci. 15(2):97–102.

Mackill, D.J. 1995. Classifying japonica rice cultivars with RAPD markers. Crop Sci. 35:889–894.

Malecot, G. 1948. Les mathematiquea de l'heredite. Masson, Paris. (English translation. The mathematics of heredity. 1969.) W.H. Freeman and Co., San Francisco, CA.

Manifesto, M.M., A.R. Schlatter, H.E. Hopp, E.Y. Suárez, and J. Dubcovsky. 2001. Quantitative evaluation of genetic diversity in wheat germplasm using molecular markers. Crop Sci. 41:682–690.

Manjarrez-Sandoval, P., T.E. Carter, Jr., R.L. Nelson, R.E. Freestone, K.W. Matson, and B.R. McCollum. 1998. Soybean Asian variety evaluation (SAVE): Agronomic performance of modern Asian cultivars in the U.S. 1997. USDA-ARS, Raleigh, NC.

Manjarrez-Sandoval, P., T.E. Carter, Jr., D.M. Webb, and J.W. Burton. 1997a. RFLP genetic similarity estimates and coefficient of parentage as genetic variance predictors for soybean yield. Crop Sci. 37:698–703.

Manjarrez-Sandoval, P., T.E. Carter Jr., D.M. Webb, and J.W. Burton. 1997b. Heterosis in soybean and its prediction by genetic similarity measures. Crop Sci. 37:1443–1452.

Mansur, L.M., J.H. Orf, K. Chase, T. Jarvik, P.B. Cregan, and K.G. Lark. 1996. Genetic mapping of agronomic traits using recombinant inbred lines of soybean. Crop Sci. 36:1327–1336.

Martin, J.M, T.K. Blake, and E.A Hockett. 1991. Diversity among North American spring barley cultivars based on coefficient of parentage. Crop Sci. 31:1131–1137.

Matson, A.L., and L.F. Williams. 1965. Evidence of four genes for resistance to the soybean cyst nematode. Crop Sci. 22:588–590.

Maughan, P.J., M.A. Saghai-Maroof, G.R. Buss, and G.M. Huestis. 1996. Amplified fragment length polymrphism (AFLP) in soybean: Species diversity, inheritance, and near-isogenic line analysis. Theor. Appl. Genet. 93:392–401.

McBlain, B.A., R.L. Bernard, C.R. Cremeens, and J.F. Korczak. 1987. A procedure to identify genes affecting maturity using soybean isoline testers. Crop Sci. 27:1127–1132.

McBroom, R.L. 1999. Breeding for soybean cyst nematode resistance at Novartis Seeds, Inc. p. 7 (Abstr.). *In* National Soybean Cyst Nematode Conf. Proc., Orlando, FL. 7–8 Jan. 1999.

Meksem K., D. Hyten, K. Chancharoenchai, B. Ruben, V.N. Njiti, and D.A. Lightfoot. 1999. Automated marker assisted selection for dual resistance: The soybean cyst nematode and *Fusarium solani*. p. 42 (Abstr.) *In* W.R. Fehr (ed.) Natl. Soybean Cyst Nematode Conf. Proc., Orlando, FL. 7–8 Jan. 1999. Agron. Dep., Iowa State Univ., Ames.

Melchinger, A.E., A. Graner, M. Singh, and M.M. Messmer. 1994. Relationships among European barley germplasm: 1. Genetic diversity among winter and spring cultivars revealed by RFLPs. Crop Sci. 34:1191–1199.

Melchinger, A.B., M. Lee, K.R. Lamkey, and W.L. Woodman. 1990. Genetic diversity for restriction fragment length polymorphisms: Relation to estimated genetic effects in maize inbreds. Crop Sci. 30:1033–1040.

Melchinger, A.E., M.M. Messmer, M. Lee, W.L. Woodman, and K.R. Lamkey. 1991. Diversity and relationships among U.S. maize inbreds revealed by restriction fragment length polymorphisms. Crop. Sci. 31:669–678.

Menancio, D.I., A.G. Hepburn, and T. Hymowitz. 1990. Restriction fragment length polymorphism (RFLP) of wild perennial relatives of soybean. Theor. Appl. Genet. 79:235–240.

Mercado, L.A., E. Souza, and K.D. Kephart. 1996. Origin and diversity of North American hard spring wheats. Theor. Appl. Genet. 93:593–599.

Messmer, M.M., A.E. Melchinger, J. Boppenmaier, E. Brunklaus-Jung, and R.G. Herrmann. 1992. Relationships among early European maize inbreds: I. Genetic diversity among flint and dent lines revealed by RFLPs. Crop Sci. 32:1301–1309.

Messmer, M.M., A.E. Melchinger, R.G. Herrmann, and J. Boppenmaier. 1993. Relationships among early European maize inbreds: II. Comparison of pedigree and RFLP data. Crop Sci. 33:944–950.

Miyazaki, S., T. E. Carter, Jr., S. Hattori, H. Nemoto, T. Shina, E. Yamaguchi, S. Miyashita, and Y. Kunihiro. 1995a. Identification of representative accessions of Japanese soybean varieties registered by Ministry of Agriculture, Forestry and Fisheries, Based on passport data analysis. Misc. Publ. No. 8. Natl. Inst. of Agrobiol. Resources, Tsukuba, Ibaraki, Japan.

Miyazaki, S., T. E. Carter, Jr., S. Hattori, T. Shina, T. Chibana, S. Miyashita, and Y. Kunihiro. 1995b. Identification of representative accessions of old cultivars that contribute to the pedigree of modern Japanese soybean varieties, Based on passport data analysis. Misc. Publ. No. 8. Natl. Inst. of Agrobiol. Resources, Tsukuba, Ibaraki, Japan.

Morrison, M.J., L.N. Pietrzak, and H.D. Voldeng. 1998. Soybean seed coat discoloration in cool-season climates. Agron. J. 90:471–474.

Moser, H., and M. Lee. 1994. RFLP variation and genealogical distance, multivariate distance, heterosis, and genetic variance in oats. Theor. Appl. Genet. 87:947–956.

Mudge, J., V. Concibido, R. Denny, N. Young, and J. Orf. 1996. Genetic mapping of a yield depression locus near a major gene for cyst nematode resistance. Soybean Genet. Newsl. 23:175–178.

Mudge, J., P.B. Cregan, J.P. Kenworthy, W.J. Kenworthy, J.H. Orf, and N.D. Young. 1997. Two microsatellite markers that flank the major soybean cyst nematode resistance locus. Crop Sci. 37:1611–1615.

Muehlbauer, G.J., J.E. Specht, P.E. Staswick, G.L. Graef, and M.A. Thomas-Compton. 1989. Application of the near-isogenic line gene mapping technique to isozyme markers. Crop Sci. 29:1548–1553.

Muehlbauer, G.J., J.W. Specht, M.A. Thomas-Compton, P.E. Staswick, and R.L. Bernard. 1988. Near-isogenic lines—A potential resource in the integration of conventional and molecular marker linkage maps. Crop Sci. 28:729–735.

Muehlbauer, G.J., P.E. Staswick, J.E. Specht, G.L. Graef, R.C. Shoemaker, and P. Keim. 1991. RFLP mapping in the soybean (Glycine max (L.) Merr.) using near-isogenic lines. Theor. Appl. Genet. 81:189–198.

Mueller, D.S, G. L. Hartman, R. L. Nelson, and W. L. Pedersen. 2003. Response of U.S. soybean cultivars and ancestral soybean lines to *Fusarium solani* f. sp. *Glycines*. Plant Dis. 87:827–831.

Murphy, J.P., T.S. Cox, and D.M. Rodgers. 1986. Cluster analysis of red wheat winter cultivars based upon coefficients-of-parentage. Crop Sci. 26:672–676.

Myers, G.O., S.C. Anand, and A.P. Rao-Arelli. 1988. Inheritance of resistance to race 5 of *Heterodera glycines* in soybeans. Agron. J. 80:90.

Narvel, J.M., W.R. Fehr, W. Chu, D. Grant, and R.C. Shoemaker. 2000. Simple sequence repeat diversity among soybean plant introductions and elite genotypes. Crop Sci. 40:1452–1458.

Narvel, J.M., L.R. Jakkula,D.V. Phillips, T. Wang, S.K. Lee, and H.R. Boerma. 2001. Molecular mapping of *Rpx* conditioning reaction to bacterial pustule in soybean. J. Hered. 92:267–270.

Nelson, R.L., P.J. Amdor, J.H. Orf, and J.F. Cavins. 1988. Evaluation of the USDA Soybean Germplasm Collection: Maturity Groups 000 to IV (PI 427.136 to PI 445.845). USDA Tech. Bull. 1726. U.S. Gov. Print. Office, Washington, DC.

Nelson, R. L., P. J. Amdor, J. H. Orf, J. W. Lambert, J. F. Cavins, R. Kleiman, F. A. Laviolette and K. A. Athow. 1987. Evaluation of the USDA Soybean Germplasm Collection: Maturity Groups 000 to IV (PI 273.483 to PI 427.107). USDA Tech. Bull. 1718. U.S. Gov. Print. Office, Washington, DC.

Nelson, R.L. and R.L. Bernard. 1984. Production and performance of hybrid soybeans. Crop Sci. 24:549–553.

Nelson, R.L., P.B. Cregan, H.R. Borma, T.E. Carter, Jr., C.V. Quigley, M.M Welsh, J. Alvernaz, and R. Main. 1998. DNA marker diversity among modern North American, Chinese and Japanese soybean cultivars. p. 164. *In* 1998 Agronomy abstracts. ASA, Madison, WI.

Nelson, B.D., T.C. Helms, and M.A. Olson. 1991. Comparison of laboratory and field evaluations of resistance in soybean to *Sclerotinia sclerotiorum*. Plant Dis. 75: 662–665.

Nelson, R.L, C.D. Nickell, J.H. Orf, H. Tachibana, E.T. Gritton, C.R. Grau, and B.W. Kennedy. 1989. Evaluating soybean germplasm for brown stem rot resistance. Plant Dis. 73:110–114

Nelson-Schreiber, B.M., and L.E. Schweitzer. 1986. Limitations of leaf nitrate reductase activity during flowering and podfill in soybean. Plant Physiol. 80:454–458.

Newell, C.A., X. Delannay, and M.E. Edge. 1987. Interspecific hybrids between the soybean and wild perennial relatives. J. Hered. 78:301–306.

Newell C.A., and T. Hymowitz. 1980. A taxonomic revision in the genus *Glycine* subgenus *Glycine (Leguminosae)*. Brittonia 32:63–69.

Niblack, T.L., G.L. Tylka, and R.D. Riggs. 2004. Nematode pathogens of soybean. p. 821–852. *In* H.R. Boerma and J.E. Specht. (ed.) Soybeans: Improvement, production, and uses. 3rd ed. Agron. Monogr. 16. ASA, CSSA, and SSSA, Madison, WI.

Nielson, D.C., B.L. Blad, S.B. Verma, N.J. Rosenberg, and J.E. Specht. 1984. Influence of soybean pubescence type on radiation balance. Agron. J. 76:924–929.

Njiti, V.N., L. Gray, and D.A. Lightfoot. 1997. Rate reducing resistance to *Fusarium solani* f.sp. *glycines* underlies field resistance to soybean Sudden Death Syndrome (SDS). Crop Sci. 37:132–138.

Njiti, V.N., M.A. Shenaut, R.J. Suttner, M.E. Schmidt, and P.T. Gibson. 1996. Soybean response to sudden death syndrome: Inheritance influenced by cyst nematode resistance in Pyramid x Douglas progenies. Crop Sci. 36:1165–1170.

Noel, G.R. 1992. History, distribution and economics. p. 1–13. *In* R.D. Riggs and J.A. Wrather (ed.) Biology and management of the soybean cyst nematode. The Am. Phytopathol. Soc. Press, St. Paul, MN.

Noel, G., and Z. Liu. 1999. Chinese origin of soybean cyst nematode. Proc. of the Natl. Soybean Cyst Nematode Conf. p. 10 (Abstr.) *In* Natl. Soybean Cyst Nematode Conf. Proc., Orlando, FL. 7–8 Jan. 1999.

Nowling, G. L. 2000. The uniform soybean tests, northern region 2000. Dep. of Agron., USDA-ARS, Purdue Univ., West Lafayette, IN.

Nowling, G.L. 2001. The uniform soybean tests, northern region 2001. Dep. of Agron., USDA-ARS, Purdue Univ., West Lafayette, IN.

Ohara, M and Y. Shimamoto. 1994. Some ecological and demographic characteristics of two growth forms of wild soybean (*Glycine soja*) Can. J. Bot. 72:486–492.

Ohara, M, Y. Shimamoto, and T. Sanbuichi. 1989. Distribution and ecological features of wild soybeans (*Glycine soja*) in Hokkaido. J. Fac. Agric., Hokkaido Univ. 64:43–50.

Orf, J.H., K. Chase, T. Jarvik, L.M. Mansur, P.B. Cregan, F.R. Adler, and K.G. Lark. 1999. Genetics of soybean agronomic traits: I. Comparison of three related recombinant inbred populations. Crop Sci. 39:1642–1651.

Pace, P.F., D.B. Weaver, and L.D. Ploper. 1993. Additional genes for resistance to Frogeye leaf spot races 5 in soybean. Crop Sci. 33:1144–1145.

Palmer, R.G., T. Hymowitz, and R.L. Nelson. 1996. Germplasm diversity within soybean. p. 1–35. *In* D.P.S. Verma, and R.C. Shoemaker (ed.) Soybean: Genetics, molecular biology and biotechnology. CAB Int., Wallingford, UK.

Palmer, R.G., and T.C. Kilen. 1987. Qualitative genetics and cytogenetics. p. 135–209. *In* J.R. Wilcox (ed.) Soybeans: Improvement, production, and uses. 2nd ed. Agron. Monogr. 16. ASA, CSSA, and SSSA, Madison, WI.

Palmer, R.G., T.W. Pfeiffer, G.R. Buss, and T.C. Kilen. 2004. Qualitative genetics. p.137–234. *In* H.R. Boerma and J.E. Specht (ed.) Soybeans: Improvement, production, and uses. 3rd ed. Agron. Monogr. 16. ASA, CSSA, and SSSA, Madison, WI.

Pantalone, V.R., W.J. Kenworthy, L.H. Slaughter, and B.R. James. 1997a. Chloride tolerance in soybean and perennial Glycine accessions. Euphytica 97:235–239.

Pantalone, V.R., G.J. Rebetzke, J.W.Burton, and R.F. Wilson.1997b. Genetic regulation of linolenic acid concentration in wild soybean *Glycine soja* accessions. JAOCS. 74:159–163.

Paris, R.L., and P.P. Bell (ed.) 2002. Uniform soybean tests—Southern states 2001. USDA-ARS, Stoneville, MS.

Paull, J.G., K.J. Chalmers, A. Karakousis, J.M. Kretschmer, S. Manning, and P. Langridge. 1998. Genetic diversity in Australian wheat varieties and breeding material based on RFLP data. Theor. Appl. Genet. 96:435–446.

Payne, J.H., and S.G. Pueppke. 1985. *Rj1* and *rj1* soybean isolines: Adsorption of rhizobia to roots and distribution of primary root nodules. Symbiosis 1:163–176.

Perry, M.C, and M.S. McIntosh. 1991. Geographical patterns of variation in the USDA soybean germplasm collection: I. Morphological traits. Crop Sci. 31:1350–1355.

Perry, M.C, M.S. McIntosh, and A.K. Stoner. 1991. Geographical patterns of variation in the USDA soybean germplasm collection: II. Allozyme frequencies. Crop Sci. 31:1356–1360.

Pfeiffer T.W., and L.C. Harris. 1990. Soybean yield in delayed plantings as affected by alleles increasing vegetative weight. Field Crops Res. 23:93–101.

Pfeil, B.E., and L.A. Craven. 2002. New taxa in *Glycine (Fabaceae:Phaseolae)* from north-western Australia. Aust. Syst. Bot. 15:565–573.

Pfeil, B.E., M.D. Tindale, and L.A. Craven. 2001. A review of the *Glycine clandestina* species complex *(Fabaceae:Phaseolae)* reveals two new species. Aust. Syst. Bot. 14:891–900.

Phillips, D.V., and H.R. Boerma. 1981. *Cercospora sojina* race 5: A threat to soybeans in southeastern United States. Phytopathology 71:334–336.

Pioli, R.N., E.N. Morandi, C.O. Gosparini, and A.L. Borghi. 1999. First report of pathogenic variability of different isolates of *Diaporthe phaseolorum* var *meridionales* on soybean in Argentina. Plant Dis. 83:1071.

Piper, C.V., and W.J. Morse. 1910. The soybean: History, varieties, and field studies. USDA Bur. Plant Ind. Bull. 197.

Piper, C.V., and W.J. Morse. 1923. The soybean. McGraw Hill, New York.

Piper, C.V., and W.J. Morse. 1925. How we got our soybeans. Publ. 1928. Proc. Am. Soybean Assoc. 1:58.

Plaschke, J., M. Ganal, and M.S. Roder. 1995. Detection of genetic diversity in closely related bread wheat using micorsatellite markers. Theor. Appl. Genet. 91:1001–1007

Ploper, L.D. 1999. Management of economically important diseases of soybean in Argentina. p. 269–280. *In* H.E. Kauffman (ed.) Proc. World Soybean Res. Conf. VI, Chicago, IL. 4–7 Aug.1999. Superior Print., Champaign, IL.

Procupiuk, A.M., R.L. Nelson, and B.W. Diers. 2001. Increasing yield of soybean cultivars through backcrossing with exotic germplasm. *In* 2001 Agronomy abstracts. ASA. Madison, WI.

Pu Z.Q., and Q. Cao. 1983. Resistance of soybean varieties to six races of soybean mosaic virus. (In Chinese.) J. Nanjing Agric. College (3):41–45.

Qiu, L., R. Chang, J. Sun, X. Li, Z. Cui, and Z. Li. 1999. The history and use of primitive varieties in Chinese soybean breeding. p.165–172. *In* H.E. Kauffman (ed.) Proc. World Soybean Res. Conf. VI, Chicago, IL. 4–7 Aug.1999. Superior Print., Champaign, IL.

Rao-Arelli, A.P., S.C. Anand, and J.A. Wrather. 1992. Soybean resistance to soybean cyst nematode race 3 is conditioned by an additional dominant gene. Crop Sci. 32:862–864.

Rasmusson, D.C., and R.L. Phillips. 1997. Plant breeding progress and genetic diversity from *de novo* variation and elevated epistasis. Crop Sci. 37:303–310

Raymer, P.L., and R.L. Bernard. 1988. Effects of some qualitative genes on soybean performance in lateplanted environments. Crop Sci. 28:765–769.

Rebetzke, G.J., V.R. Pantalone, J.W. Burton, T.E. Carter, Jr., and R.F. Wilson. 1997. Genotypic variation for fatty acid content in selected *Glycine max* x *Glycine soja* populations. Crop Sci. 37:1636–1640.

Rector, B.G., J.N. All, W.A. Parrott, and H.R. Boerma. 1998. Identification of molecular markers linked to quantitative trait loci for soybean resistance to corn earworm. Theor. Appl. Genet. 96:786–790.

Rector, B.G., J.N. All, W.A. Parrott, and H.R. Boerma. 1999. Quantitative trait loci for antixenosis resistance to corn earworm in soybean. Crop Sci. 39:531–538.

Rector, B.G., J.N. All, W.A. Parrott, and H.R. Boerma. 2000. Quantitative trait loci for antibiosis resistance to corn earworm in soybean. Crop Sci. 40:233–238.

Reed, E.H., A.H. Teramura, and W.J. Kenworthy. 1991. Ancestral U.S. soybean cultivars characterized for tolerance to ultraviolet-B radiation. Crop Sci. 32:1214–1219.

Reese, P.F. Jr., W.J. Kenworthy, P.B. Cregan, and J.O. Yocum. 1988. Comparison of selection systems for the identification of exotic soybean lines for use in germplasm development. Crop Sci. 28:237–241.

Ren, Q., T.W. Pfeiffer, and S.A. Ghabrial. 2000. Relationship between soybean pubescence density and soybean mosaic virus field spread. Euphytica 111:191–198.

Rennie, B.D., R.I. Buzzell, T.R. Anderson, and W.D. Beversdorf. 1992. Evaluation of four Japanese cultivars for *Rps* alleles conferring resistance to *Phytophthora megasperma* f. sp. *glycinea*. Can. J. Plant. Sci. 72:217–220.

Reyna, N., and C.H. Sneller. 2001. Evaluation of marker-assisted introgression of yield QTL alleles into adapted soybean. Crop Sci.: 41:1317

Riggle, B.D., W.J. Wiebold, and W.J. Kenworthy. 1984. Effect of photosynthate source-sink manipulation on dinitrogen fixation of male-fertile and male-sterile soybean isolines. Crop Sci. 24:5–8.

Riggs, R.D., M.L.Hamblen, and L.Rakes. 1988. Resistance of commercial soybean cultivars to six races of *Heterodera glycines* and to *Meloidogyne incognita*. Ann. Appl. Nematol. 2:70–76.

Riggs, R.D., and D.P. Schmitt. 1989. Soybean cyst nematode. p. 1448–1453. *In* A.J. Pascale (ed.) Proc. World Soybean Res. Conf. IV, Buenos Aires, Argentina. 5–9 Mar. 1989. Asociacion Argentina de la Soja, Buenos Aires.

Riggs, RD, S. Wang, R.J. Singh, and T. Hymowitz. 1998. Possible transfer of resistance to *Heterodera glycines* from *Glycine tomentella* to *Glycine max*. J. Nematol. (Suppl.) 30:4, 547–552.

Robinson, S.L., and J.R. Wilcox. 1998. Comparison of determinate and indeterminate soybean near-isolines and their response to row spacing and planting date. Crop Sci. 38:1554–1557.

Rodgers, D.M., J.P. Murphy, and K.J. Frey. 1983. Impact of plant breeding on the grain yield and genetic diversity of spring oats. Crop Sci. 23:737–740.

Rongwen, J., M.S. Akkaya, A.A. Bhagwat, U. Lavi, and P.B. Cregan. 1995. The use of microsatellite DNA markers for soybean genotype identification. Theor. Appl. Genet. 90:43–48.

Ross, J.P. 1962. Physiological strains of *Heterodera glycines*. Plant Dis. Rep. 46:766–769.

Ross, J.P. 1968. Effect of single and double infection of soybean mosaic and bean pod mottle viruses on soybean yield and seed characteristics. Plan Dis. Rep. 52:344–348.

Ross, J.P, and C.A. Brim. 1957. Resistance of soybean to the soybean cyst nematode as determined by a double-row method. Plant Dis. Rep. 41:923–924.

Rufener, G.K., S.K. St.Martin, R.L. Cooper, and R.B. Hammond. 1989. Genetics of antibiosis resistance to Mexican bean beetle. Crop Sci. 29:618–622.

Ryley, M.J., N.R. Obst, J. Irwin, and A. Drenth. 1998. Changes in racial composition of *Phytophthora sojae* in Australia between 1979–1996. Plant Dis. 82:1048–1054.

Sanchez G., J.J., and L. Ordaz S. 1987. Systematic and ecogeographic studies on crop genepools II. El Teocintle en Mexico. Int. Plant Genetic Resources Inst., Rome.

Schaller, M. 1985. The American occupation of Japan: The origins of the cold war in Asia. Oxford Univ. Press, New York.

Schmitthenner, A.F. 1985. Problems and progress in control of Phytophthora root rot of soybean. Plant Dis. 69:362–368.

Schmitthenner, A.F., M. Hobe, and R.G. Bhat. 1994 *Phytophthora sojae* races found in Ohio over a 10-year interval. Plant Dis. 78:269–276.

Schmitthenner, A.F., and A.K. Walker. 1979. Tolerance versus resistance for control of Phytophthora rot in soybeans. p. 35–44. *In* H.D. Loden and D. Wolkenson (ed.) Proc. Soybean Res. Conf. 9th, Chicago, IL. 13–14 Dec. 1979. Am. Seed Trade Assoc., Washington, DC.

Schoener, C.S., and W.R. Fehr. 1979. Utilization of plant introductions in soybean breeding populations. Crop Sci.19:185–188.

Schut, J.W., X. Qi, and P. Stam. 1997. Association between relationship measures based on AFLP markers, pedigree data and morphological traits in barley. Theor. Appl. Genet. 95:1161–1168.

Sciumbato, G.L. 1993. Soybean disease estimates for the southern United States during 1988–1990. Plant Dis. 77:954–956.

Sebolt, A.M., R.C. Shoemaker, and B.W. Diers. 2000. Analysis of a quantitative trait locus allele from wild soybean that increases seed protein concentration in soybean. Crop Sci. 40:1438–1444.

Seidler, H., M.M. Messmer, G.M. Schachermayr, H. Winzeler, M. Winzeler, and B. Keller. 1994. Genetic diversity in European wheat and spelt breeding material based on RFLP data. Theor. Appl. Genet. 88:994–1003.

Senior, M.L., J.P. Murphy, M.M. Goodman, and C.W. Stuber. 1998. Utility of SSRs for determining genetic similarities and relationships in maize using an agarose gel system. Crop Sci. 38:1088–1098.

Sheaffer, C., J.H. Orf, T.E. Devine, and J.J. Grimsbo. 2001. Yield and quality of forage soybean. Agron. J. 93:99–106.

Shung, S.D., H.W. Huh, and M.G. Chung.1995. Genetic diversity in Korean populations of Glycine *soja*. J. Plant Biol. 38:39–45.

Singh, R.J., K.P. Kollipara, and T. Hymowitz. 1990. Backcross-derived progeny from soybean and *Glycine tomentella* Hayata. Crop Sci. 30:871–874

Singh, R.J., K.P. Kollipara, and T. Hymowitz. 1993. Backcross (BC2-BC4)-derived fertile plants from *Glycine max* and *G. tomentella* intersubgeneric hybrids. Crop Sci. 33:1002–1007

Sisson, V.A., P.A. Miller, W.V. Campbell, and J.W. Van Duyn. 1976. Evidence of inheritance of resistance to the Mexican bean beetle. Crop Sci. 16:835–837.

Skorupska, H.T., R.C. Shoemaker, A. Warner, E.R. Shipe, and W.C. Bridges. 1993. Restriction fragment length polymorphism in soybean germplasm of the southern USA. Crop Sci. 33:1169–1176.

Skvortzow, B.V. 1927. The soybean-wild and cultivated in Eastern Asia. Proc. Manchurian Res. Soc. Publ. Ser. A. Nat. History Sec. 22:1–8.

Sloane, R.J., R.P. Patterson, and T.E. Carter, Jr. 1990. Field drought tolerance of a soybean plant introduction. Crop Sci. 30:118–123.

Smith, C.M., and C.A. Brim. 1979. Resistance to Mexican bean beetle and corn earworm in soybean genotypes derived from PI 227687. Crop Sci. 29:313–314.

Smith, J.S.C. 1988. Diversity of United States hybrid maize germplasm; isozymic and chromatographic evidence. Crop Sci. 28:63–69.

Smith, J.S.C., E.C.L. Chin, H. Shu, O.S. Smith, S.J. Wall, M.L. Senior, S.E. Mitchell, S. Kresovich, and J. Ziegle. 1997. An evaluation of the utility of SSR loci as molecular markers in maize (*Zea mays* L.): Comparisons with data from RFLPS and pedigree. Theor. Appl. Genet. 95:163–173.

Smith, J.S.C., S. Kresovich, M.S. Hopkins, S.E. Mitchell, R.E. Dean, W.L. Woodman, M. Lee, and K. Porter. 2000. Genetic diversity among elite sorghum inbred lines assessed with simple sequence repeats. Crop Sci. 40:226–232.

Smith, J.S.C., O.S. Smith, S. Wright, S.J. Wall, and M. Walton. 1992. Diversity of U.S. hybrid maize germplasm as revealed by restriction fragment length polymorphisms. Crop Sci. 32:598–604.

Sneller, C.H. 1994. Pedigree analysis of elite soybean lines. Crop Sci. 34:1515–1522.

Sneller, C.H. 2003. Impact of transgenic genotypes and subdivision on diversity within elite North American soybean. Crop Sci. 43:409–414.

Sneller, C.H., J. Miles, and J.M. Hoyt. 1997. Agronomic performance of soybean plant introduction and their genetic similarity to elite lines. Crop Sci. 37:1595–1600

Song, Q.J., C.V. Quigley, R.L. Nelson, T.E. Carter, H.R. Boerma, J.L. Strachan, and P.B. Cregan. 1999. A selected set of trinucleotide simple sequence repeat markers for soybean cultivar identification. Plant Var. Seeds 12:207–220.

Souza, E., P.N. Fox, and B. Skovmand. 1998. Parentage analysis of international spring wheat yield nurseries 17 to 27. Crop Sci. 38:337–341.

Souza, E., and M.E. Sorrells. 1989. Pedigree analysis of North American oat cultivars released from 1951 to 1985. Crop Sci. 29:595–601.

Souza, E., and M.E. Sorrells.1991. Relationships among 70 North American oat germplasms. II. Cluster analysis suing qualitative characters. Crop Sci. 31:605–612.

Specht, J.E., K. Chase, M. Macrander, G.L. Graef, J. Chung, J.P. Markwell, M. Germann, J.H. Orf, and K.G. Lark. 2001. Soybean response to water: A QTL analysis of drought tolerance. Crop Sci. 41:493–509.

Specht, J.E., D.J. Hume, and S.V. Kumudini. 1999. Soybean yield potential—A genetic and physiological perspective. Crop Sci. 39:1560–1570.

Specht, J.E., and J.H. Williams. 1984. Contribution of genetic technology to soybean productivity-retrospect and prospect. p. 49–74. *In* W.H. Fehr (ed.) Genetic contributions to grain yields of five major crop plants. CSSA Spec. Publ. 7. CSSA and ASA Madison, WI.

Specht, J.E., J.H. Williams, and D.R. Pearson. 1985. Near-isogenic analysis of soybean pubescence genes. Crop Sci. 25:92–96.

Specht, J.E., J.H. Williams, and C. J. Weidenbenner. 1986. Differential responses of soybean genotypes subjected to a seasonal soil water gradient. Crop Sci. 26:922–934.

Sprague, G.F. 1971. Genetic vulnerability in corn and sorghum. Proc. Annual Corn and Sorghum Res. Conf. 26:96–104.

St. Martin, S.K. 1982. Effective population size for the soybean improvement program in maturity groups 00 to IV. Crop Sci. 22:151–152.

St. Martin. 2001. Selection limits—How close are we? Soybean Genetics Newsletter [Online] Available at http://www.soygenetics.org/articles/sgn2001-003.htm (verified 25 Nov. 2002)

St. Martin, S.K., and M. Aslam. 1986. Performance of progeny of adapted and plant introduction soybean lines. Crop Sci. 26: 753–756.

Stevens, P.A., C.D. Nickell, and F.L. Kolb. 1993. Genetic analysis of resistance to *Fusarium solani* in soybean. Crop Sci. 33:929–930.

Suganuma, N., and S. Satoh. 1991. Contents of isoflavones and effect of isoflavones on root hair curling in non-nodulating (*rj1rj1*) soybean plant. Soil Sci. Plant Nutr. 37:163–166.

Sweeney, P.M., and S. St. Martin. 1989. Testcross evaluation of exotic soybean germplasm of different origins. Crop Sci. 29:289–293.

Takahashi, R. 1997. Association of soybean genes I and T with low-temperature induced seed coat deterioration. Crop Sci. 37:1755–1759.

Takahashi, R, and J. Abe. 1999. Soybean maturity genes associated with seed coat pigmentation and cracking in response to low temperatures. Crop Sci. 39:1657–1662.

Talekar, N.S., and B.S. Chen. 1983. Identification of sources of resistance to limabean podborer (*Lepidotera: Pyralidae*). J. Econ. Entomol. 76:38–39.

Talekar, N.S., H.R. Lee, and Suharsono. 1988. Resistance of soybean to four defoliator species in Taiwan. J. Econ. Entomol. 81:1469–1473.

Tamulonis, T.P., B.M. Luzzi, R.S. Hussey, W.A. Parrott, and H.R. Boerma. 1997a. DNA markers associated with resistance to Javanese root-knot nematode in soybean. Crop Sci. 37:783–788.

Tamulonis, T.P., B.M. Luzzi, R.S. Hussey, W.A. Parrott, and H.R. Boerma. 1997b. RFLP mapping of resistance to southern root-knot nematode in soybean. Crop Sci. 37:1903–1909.

Tamulonis, T.P., B.M. Luzzi, R.S. Hussey, W.A. Parrott, and H.R. Boerma. 1997c. DNA marker analysis of loci conferring resistance to peanut root-knot nematode in soybean. Theor. Appl. Genet. 95:664–670.

Tan, Y.1991. Evaluation of soybean germplasm of China for resistance to soybean rust. (In Chinese.) Soybean Sci. 16(3):205–209.

Tan, Y., Z. Yu, and Y. Peng. 1999. Management of economically important soybean diseases in China. p.281–289. *In* H.E. Kauffman (ed.) Proc. World Soybean Res. Conf. VI, Chicago, IL. 4–7 Aug.1999. Superior Print., Champaign, IL.

Tang, D., T. Nakamura, N. Ohtsubo, K. Takahashi, J. Abe, and R. Takahashi. 2002. Seed coat cracking in soybean isolines for pubescence color and maturity. Crop Sci. 42:71–75.

Tao, Y., J.M. Manners, M.M. Ludlow, and R.G. Henzell. 1993. DNA polymorphism in grain sorghum (*Sorghum bicolor* (L.) Moench). Theor. Appl. Genet. 86:679–688.

Tateishi, Y., and H. Ohashi. 1992. Taxonomic studies on *Glycine* of Taiwan. J. Jpn. Bot. 67:127–147.

Tatineni, V., R.G. Cantrell, and D.D. Davis. 1996. Genetic diversity in elite cotton germplasm determined by morphological characteristics and RAPDs. Crop Sci. 36:186–192.

Thompson, J.A., and R.L. Nelson. 1998a. Core set of primers to evaluate genetic diversity in soybean. Crop Sci. 38:1356–1362.

Thompson, J.A., and R.L. Nelson. 1998b. Utilization of diverse germplasm for soybean yield improvement. Crop Sci. 38:1362–1368.

Thompson, J.A., R.L. Nelson, and L.O. Vodkin. 1998. Identification of diverse soybean germplasm using RAPD markers. Crop Sci. 38:1348–1355.

Thompson, J.A., P.J. Amdor, and R.L. Nelson. 1999. Registration of LG90-2550 and LG91-7350R Soybean Germplasm. Crop Sci. 39:302–303.

Thorne, J.C., and W.R. Fehr. 1970. Exotic germplasm for yield improvement in 2-way and 3-way soybean crosses. Crop Sci. 10:677–678.

Tindale, M.D. 1984. Two new eastern Australian species of Glycine Willd. (Fabaceae). Brunonia 7:207–213.

Tindale, M.D. 1986a. A new north Queensland species of Glycine Willd. (Fabaceae). Brunonia 9: 99–103.

Tindale, M.D. 1986b. Taxonomic notes on three Australian and Norfolk Island species of *Glycine* Willd. (Fabaceae: Phaseolae) including the choice of a neotype for *G. clandestina* Wendl. Brunonia 9:179–191.

Tindale, M.D., and L.A. Craven 1988. Three new species of Glycine Willd. (*Fabaceae:Phaseolae*) from north western Australia, with notes on amphicarpy in the genus. Austral. Syst. Bot. 1:399–410.

Tindale, M.D, and L.A. Craven 1993. Glycine pindanica (Fabaceae:Phaseolae) a new species from West Kimberley, western Australia. Austral. Syst. Bot. 6:371–376.

Tinker, N.A., M.G. Fortin, and D.E. Mather. 1993. Random amplified polymorphic DNA and pedigree relationships in spring barley. Theor. Appl. Genet. 85:976–984.

Tompkins, J.P., and E.R.Shipe. 1977. Environmental adaptation of long juvenile soybean cultivars and strains. Agron. J. 89:257–262.

Tooley, P.W., and C.R. Grau. 1982. Identification and quantitative characterization of rate-reducing resistance in *Phytophthora megasperma* f. sp. *glycinae* in soybean seedlings. Phytopathology 72:727–733.

Tsukamoto, C., A. Kikuchi, K. Harada, K. Kitamura, and K. Okubo. 1993. Genetic and chemical polymorphisms of saponins in soybean seed. Phytochemistry 34:1351–1356.

Turnipseed, S.G., and M.J. Sullivan. 1976. Plant resistance in soybean-insect management. p.549–560. *In* L.D. Hill (ed.) World soybean research. Proc. World Soybean Res. Conf., 1st, Champaign, IL. 3–8 Aug. 1975. Interstate Publ., Danville, IL.

Tyler, J.M. 1995. Additional sources of stem canker resistance in soybean plant introduction. Crop Sci. 35:376–377.

Tyler, J.M. 1996. Characterization of stem canker resistance in 'Hutcheson' soybean. Crop Sci. 36:591–593.

Ustun, A., F.L. Allen, and B.C. English. 2001. Genetic progress in soybean of the U.S. midsouth. Crop Sci. 41:993–998

van Beuningen, L.T., and R.H. Busch. 1997. Genetic diversity among North American spring wheat cultivars: I. Analysis of the coefficient of parentage matrix. Crop Sci. 37:570–579.

Van Duyn, J.W., S.G. Turnipseed, and J.D. Maxwell. 1971. Resistance in soybeans to the Mexican bean beetle. I. Sources of resistance. Crop Sci. 11:572–573.

Van Esbroeck, G.A. D.T. Bowman, D.S. Calhoun, and O.L. May. 1998. Changes in the genetic diversity of cotton in the USA from 1970 to 1995. Crop Sci. 38:33–37.

Vasilas, B.L., R.L. Nelson, and R.M. Vanden Heuvel. 1990. Effect of method and plant sample on nitrogen fixation estimates in soybean. Agron. J. 82: 673–676.

Vaughan D.A., S. Fukuoka, and K. Okuno.1995. Population genetic diversity of wild soybeans on the Kanto plain, Japan. Soybean Genet. Newsl. 22:145–146.

Vello, N.A., W.R. Fehr, and J.B. Bahrenfus. 1984. Genetic variability and agronomic performance of soybean populations developed from plant introductions. Crop Sci. 24:511–514.

Voldeng, H.D., E.R. Cober, D.J. Hume, C. Gilard, and M.J. Morrison. 1997. Fifty-eight years of genetic improvement of short-season soybean cultivars in Canada. Crop Sci. 37:428–431.

Waller, R.S., C.D. Nickell, D.L. Drzyycimski, and J.E. Miller. 1991. Genetic analysis of the inheritance of brown stem rot resistance in the soybean cultivar Asgrow A3733. J. Hered. 82:412–417.

Wang, G.X. 1982. Soybean collection catalog. (In Chinese.) China Agric. Press, Beijing.

Wang, J. 1994. Broadening and improvement of genetic base of soybeans in northeast China. (In Chinese.) Heilongjiang Sci. and Technol. Press, Harbin.

Wang, D., P.R. Arelli, R.C. Shoemaker, and B.W. Diers. 2001. Loci underlying resistance to Race 3 of soybean cyst nematode in *Glycine soja* plant introduction 468916. Theor. Appl. Gen. 103:561–566.

Wang, Y., R.L. Nelson, and Y. Hu. 1998. Genetic analysis of resistance to soybean mosaic virus in four soybean cultivars from China. Crop Sci. 38:922–925.

Wang, J.L., Y.X. Wu, H.L. Wu, and S.C. Sun. 1973. Analysis of the photoperiod ecotypes of soybeans from northern and southern China (In Chinese). J. Agric. 7(2):169–180.

Webb, D.M., B.M. Baltazar, A.P. Rao-Arelli, J. Schupp, K. Clayton, P. Keim, and W.D. Beavis. 1995. Genetic mapping of soybean cyst nematode race-3 resistance loci in the soybean PI 437.654. Theor. Appl. Genet. 91:574–581.

Weir, B.S. 1996. Genetic data analysis II. Sinauer Assoc., Inc. Publ., Sunderland, MA.

Wilcox, J.R. 2001. Sixty years of improvement in publicly developed elite soybean lines. Crop Sci. 41:1711–1716.

Wilcox, J.R., A.G.M. Khalaf, and G.D. Brossman. 1984. Variability within and among F1 families from diverse three-parent soybean crosses. Crop Sci. 24:1055–1058.

Wilcox, J.R., F.A. Laviolette, and R.J. Martin. 1975. Heritability of purple seed stain resistance in soybean. Crop Sci. 15:525–526.

Wilcox, J.R., and W.T. Schapaugh, Jr. 1978. Competition between two soybean isolines in hill plots. Crop Sci.18:346–348.

Wilcox, J.R., and S.K. St. Martin. 1998. Soybean genotypes resistant to *Phytophthora sojae* and compensation for yield losses of susceptible isolines. Plant Dis. 82:303–306.

Wilkes, H.G. 1967. Teosinte: the closest relative of maize. The Bussey Inst., Harvard Univ., Cambridge, MA.

Wilmot, D.B., and C.D. Nickell. 1989. Genetic analysis of brown stem rot resistance in soybean. 1989. Crop Sci. 29:672–674.

Winstead, N.N., C.B. Skotland, and J.N. Sasser. 1955. Soybean cyst nematodes in North Carolina. Plant Dis. Rep. 39:9–11.

Woodworth, C.M. 1932. Genetics and breeding in the improvement of the soybean. Bull. 384. Univ. of Ill. Agric. Exp. Stn., Champaign.

Wrather, J.A., W. Stientra, and S.R. Koenning. 2001. Soybean disease loss estimates for the United States from 1996 to 1998. Can. J. Plant Pathol. 23:122–131.

Wu, H.L. 1982. Screening for source of resistance to soybean cyst nematode (In Chinese.) Sci. Agricultia Agric. Sin. (6):19–24.

Xu, B. 1986. New evidence about the geographic origin of soybean. (In Chinese.) Soybean Sci. 5(2):123–130.

Xu, Z., R. Chang, L. Qiu, J. Sun, and X. Li. 1999a. Evaluation of soybean germplasm in China. p. 156–165. In H.E. Kauffman (ed.) Proc. World Soybean Res. Conf. VI, Chicago, IL. 4–7 Aug.1999. Superior Print., Champaign, IL.

Xu, Z., L. Qiu, R. Chang, X. Li, P. Guo, and C. Zheng. 1999b. Using SSR markers to evaluate genetic diversity of soybean in China. p. 510–511. In H.E. Kauffman (ed.) Proc. World Soybean Res. Conf. VI, Chicago, IL. 4–7 Aug.1999. Superior Print., Champaign, IL.

Yaklich, R.W., R.M. Helm, G. Cockrell, and E.M. Herman. 1999. Analysis of the distribution of major soybean seed allergens in a core collection of G. max accessions. Crop Sci. 39:1444–1447.

Yang, W., and D.B. Weaver. 2001. Resistance to Frogeye leaf spot in maturity groups VI and VII of soybean germplasm. Crop Sci. 41:549–552.

Yorinori, J.T. 1992. Management of foliar diseases in soybean in Brazil. p. 185–195. In L.G. Copping et al. (ed.) Pest management in soybean. Elsevier Appl. Sci., New York.

Yorinori, J.T. 1999. Management of economically important diseases in Brazil. p. 290. In H.E. Kauffman (ed.) Proc. World Soybean Res. Conf. VI, Chicago, IL. 4–7 Aug.1999. Superior Print., Champaign, IL.

Young, L.D. 1999. Management of SCN through conventional breeding for resistance—Southern perspective. Proc. of the Natl. Soybean Cyst Nematode Conf. p. 6 (Abstr.) In Natl. Soybean Cyst Nematode Conf. Proc., Orlando, FL. 7–8 Jan. 1999.

Yu H.G., and Y.T. Kiang. 1993.Genetic variation in South Korean natural populations of wild soybean (Glycine soja). Euphytica 68:213–221.

Yu, Y.G., M.A. Saghai Maroof, G.R. Buss, P.J. Maughan, and S.A. Tolin. 1994. RFLP and microsatellite mapping of a gene for soybean mosaic virus resistance. Phytopathology. 84:60–64.

Yuan, J., N.J. Meksem, M.J. Iqbal, K. Triwitayakorn, M.A. Kassem, G.T. Davis, M.E. Schmidt, and D.A. Lightfoot. 2002. Quantitative trail loci in two soybean recombinant inbred line populations segregating for yield and disease resistance. Crop Sci. 42:271–277.

Zhang F.N., Q.H. Feng, and D. Zhang. 1984. Studies on the resistance of soybean to soybean agrimizid fly (Malanagromyza sojae) III. A screen test of Huai Bei local summer soybean varieties resistant to agromyzid fly. (In Chinese.) J. Nanjing Agric. College 1984(2):26–30.

Zhang, J., J.E. Specht, G.L. Graef, and B.L. Johnson. 1992. Pubescence density effects on soybean (Glycine max (L.) Merrill) seed yield and other agronomic characteristics. Crop Sci. 32:641–648.

Zhang, L., O.H. Dai, and J.M. Liu. 1998. Evaluation of Chinese soybean germplasm for resistance to SCN race 5. (In Chinese.) Soybean Sci. 17(2):172–175.

Zhang, L.J., and Q.K. Yang. 1997. Screening for soybean germplasm for resistance to multiple races of frog eye leaf spot. (In Chinese.) Soybean Sci. 16(1):38–41.

Zhang, Z. 1985a. Annals of soybean cultivars in China. (In Chinese.) Agriculture Publ. House, Beijing.

Zhang, Z. 1985b. Evaluation and screening of source of varieties for resistance to soybean cyst nematode. (In Chinese.) Soybean Sci. 4(2):137–140.

Zhong, Z.X., Z.P. Wu, F.L. Gao, and B.X. Lin. 1986. Preliminary study on screening varieties resistant to SMV. (In Chinese.) Soybean Sci. 3:239–242.

Zhou, X. T.E. Carter, Z. Cui, S. Miyazaki, and J.W. Burton. 2000. Genetic base of Japanese soybean cultivars released during 1950 to 1988. Crop Sc. 40:1794–1802.

Zhou, X. T. E. Carter, Jr., Z. Cui, S. Miyazaki, and J. W. Burton. 2002. Genetic diversity patterns in Japanese soybean cultivars based on coefficient of parentage. Crop Sci. 42:1331–1342.

Zhou, X., T.E. Carter, Jr., R.L. Nelson, H.R. Boerma, and B.R. McCollum. 1998a. Breeding progress vs. breeding effort: the road ahead in soybean variety development. p. 81. In 1998 Agronomy abstracts. ASA, Madison, WI.

Zhou, X.A., Y.H. Peng, G.X. Wang, and R.Z. Chang. 1998b. The genetic diversity and center of origin of Chinese cultivated soybean. (In Chinese.) Sci. Agric. Sin. 31:37–43.

Zhou, X., Y. Peng, G. Wang, and R. Chang. 1999. Study on the center of genetic diversity and origin of cultivated soybean. p. 510. *In* H.E. Kauffman (ed.) Proc. World Soybean Res. Conf. VI, Chicago, IL. 4–7 Aug.1999. Superior Print., Champaign, IL.

9

Genetic Improvement: Conventional and Molecular-Based Strategies

JAMES H. ORF

University of Minnesota,
St. Paul, Minnesota

BRIAN W. DIERS

University of Illinois
Urbana, Illinois

H. ROGER BOERMA

University of Georgia
Athens, Georgia

Genetic improvement in soybean continues. Recent estimates indicate soybean yields are improving at a rate of about 23 kg ha^{-1} yr^{-1} due to improved genetics, production practices, and higher atmospheric CO_2 levels (Specht et al.,1999). The genetic improvements have been accomplished mainly through the use of conventional breeding methods; however molecular- based plant-breeding techniques are assuming an increasingly more important role in genetic improvement. Since conventional breeding techniques will continue to play an important role, a review of the recent literature on methods for improving yield, disease and pest resistance, and quality will be discussed first. Subsequently, molecular-based plant-breeding strategies that may enhance the rate of gain will be presented. The emphasis in this chapter will be on selected information reported since the 1987 edition of the Soybean Monograph as it relates methods and strategies for the genetic improvement of soybean.

9–1 CONVENTIONAL BREEDING METHODS

Conventional breeding strategies have been very successful in improving soybean productivity and quality. Yield has been and remains the trait of greatest emphasis by breeders, as it is the trait with the greatest effect on a producer's net income. Wilcox (2001) estimated that public breeders in the northern soybean production areas have increased seed yield about 60% over the past 60 yr. Estimates of progress in the public sector of a similar magnitude have been made in other areas of North America (Salado-Navarro et al.,1993; Specht et al.,1999; Voldeng et al.,1997). Similar progress has undoubtedly been made by breeding efforts in the private sector (Specht et al.,1999), although significant breeding efforts have occurred there only in the last two or three decades. Besides yield, progress has also been made in selecting for resistance to pathogens, insects and nematodes, tolerance to other production hazards, improvement in seed protein, oil, and other quality traits, as well as other agronomic characteristics, such as standability.

Each cycle of genetic improvement begins with the breeder making choices as to the parents to be used to create segregating populations. Those populations are then advanced toward homozygosity with or without selection to produce relatively homozygous lines that are then subject to yield and other trait evaluation. A given cycle ends when its best lines are released as improved pure-line cultivars. Since a cycle is initiated each year, genetic improvement is a continuous process.

The pure-line cultivar is what is grown by the farmer. Although there has been discussion about the development of single-cross F_1 hybrids for use by farmers, acceptance by producers requires a yield advantage of hybrids (over pure-line cultivars) sufficient to consistently return more net revenue. The difficulty of producing large quantities of hybrid seed needed for commercial planting has precluded the use of hybrids in soybean. However, recent reports suggest that hybrid soybean may become commercially available in the near future (Anonymous,1999).

9–1.1 Selection of Parents

Selection of parents is extremely important to success in genetic improvement. Parents used to create segregating populations can be either adapted elite material or exotic germplasm. Generally elite parents of diverse origin are more likely to produce progeny superior to either parent (*and* superior to existing releases) than parents of similar ancestry (Burton, 1997). The parental selection method used depends on many factors, including the trait(s) of interest, the purpose of the cross, the relative importance of characters other than yield, the ancestry of the lines, and the resources and time available. Parents can be selected on the basis of comparative evaluation per se, or by test-cross evaluation. In many cases, per se evaluation data are readily available in the form of breeder-directed or fee-based yield performance tests. However, if the objective is to identify parental germplasm with favorable alleles not present in existing cultivars, test-cross evaluations may be a more mechanistic approach. A method for soybean was suggested by Kenworthy (1980). The details of the procedure are discussed in Chapter 8 of this monograph (Carter et al., 2004, this publication). St. Martin et al. (1996) also developed a test-cross procedure for identifying germplasm lines with the potential to contribute favorable alleles for improving pure-line cultivars. The procedure requires producing a set of all possible hybrids between the cultivars or lines and the germplasm lines and evaluating them along with the parents. Test-cross statistics T_{ij} and T_i are calculated and can be used to measure the potential for favorable alleles from the germplasm lines to improve the cultivars. Two examples using previously published soybean data indicated the method may be useful in parental selection. Lewers et al. (1998) reported on the use of the above test-cross procedure for selecting germplasm to be used in an existing cultivar development program. The results indicated that the testcross statistic T_i was better than F_1 mid-parent heterosis for identifying lines that might be used to construct hybrids. However, the F_1 mid-parent heterosis was better for identifying lines that required additional breeding work before being used directly in an existing cultivar development program.

The selection of parents to improve yield using best linear unbiased predictions (i.e., the use of a mixed linear model; Henderson, 1975) was investigated by Panter and Allen (1995a). The objectives of their research were to compare the efficiency of two methods of selecting parents. These methods involved calculating mid-parent value or the best linear unbiased prediction for identifying superior cross combinations. They explored situations when there were equal or unequal amounts of yield data available on potential parents, or when unequal amounts of data were available on some parents but no data were available on other parents. They used the results of 24 crosses to evaluate the methods. They found that in every case best linear unbiased predictions identified a higher percentage of superior crosses (with higher rank correlations and lower standard errors) than calculating mid-parent values.

Selection of parents will continue to be very challenging but extremely important for success in genetic improvement. In many programs where resources are quite limited, use of existing comparative data using the best linear unbiased predictions appears useful. If more resources and time are available, test-cross evaluations can be used to identify parental germplasm with favorable alleles not present in current cultivars. As molecular and genomic data for potential parents become available, this information could assist in the selection of parents.

9–1.2 Selection

Selection can be practiced among plants during early generations before yield tests are initiated or later among lines during yield testing. The amount and effectiveness of selection depends on the heritability of the trait or character and the environment where the population or lines are grown. Visual selection is mainly carried out in early generations, while selection based on data from unreplicated or replicated plots is carried out in later generations.

Breeding populations for cultivar development are typically developed from 2- or 3- parent hybridizations involving cultivars, breeding lines, or other germplasm. The populations are then advanced through generations of selfed inbreeding. Nearly homozygous lines are then created from individually harvested inbred plants. These lines are then extensively evaluated to identify those with a performance superior to existing cultivars. Procedures for advancing crosses to homozygosity include: pedigree selection, which involves visual selection of the best- appearing families in each generation, followed by within-family selection of one or more plants to advance to the next generation; single-seed-descent or modified single-seed-descent (pod-bulk method), which involves advancing a single seed or pod from each plant to the next generation, to develop nearly homozygous lines yet still preserve most of the original genetic variation in a population; and bulk breeding where the population is advanced in bulk with no artificial selection until later generations when nearly homozygous lines are selected for yield testing. Other procedures used in soybean cultivar development include early generation testing where yield testing is initiated on the F_2 or F_3 generation; backcrossing, in which the recurrent parent is essentially recreated but with a trait transferred from a donor parent; and population improvement via recurrent selection, which may involve a genetic male sterility system.

9–1.2.1 Comparison of Breeding Methods

A comparison of selection methods using the family method (where selection is among and within F_2 and later generation families) and the line method (where selection is among F_3- derived lines without regard to F_2 family identity) for elevated palmitate in soybean was carried out by Bravo et al. (1999). They compared the methods in four populations segregating for the major genes *fap2-b* and *fap4* and modifier genes. On the basis of selection for palmitate content from one environment compared to the mean palmitate in three other test environments, the family method identified 63% of the best lines and the line method 60% of the best lines. Thus the authors concluded that the family method, which requires more record keeping and labor, is no more effective or efficient than the line method in breeding for elevated palmitate.

Byron and Orf (1991) compared three selection procedures (pedigree, single seed descent, and single seed descent with early maturity selection) for identifying early maturity lines originating from crosses of parents differing significantly in maturity. Four populations of crosses of Maturity Group 0 or I with Maturity Groups II and III were used. The results indicated there were no consistent differences among the procedures for maturity, yield, height, lodging, seed weight, or length of reproductive periods in the four populations. Since single seed descent coupled with concurrent selection for early maturity required the fewest resources, the authors concluded that this procedure was the most efficient for development of early maturing cultivars.

Degago and Caviness (1987) evaluated four soybean bulks that had been grown for 10 to 18 yr at two Arkansas locations for yield at those sites and at a third site. The bulks developed at a site that was significantly and consistently plagued with phytophthora root and stem rot infection (caused by *Phytophthora sojae* Kauf. and Gerd.) produced higher yields at that site than the bulks developed at the site with little phytophthora. The only exception was for a bulk where both parents were resistant to the disease. The bulk method of breeding is apparently effective in improving yield at those locations where there is consistent natural selection pressure from a disease.

Numerous cultivars have been developed using early generation evaluation procedures. The basis for early generation testing is the early identification of heterogenous populations or lines from which there is an above-average probability of isolating superior pure-line cultivars. Cooper (1990) reported a modified early generation testing procedure that reduces the number of yield test plots required for testing. Single location, single replication data are used for selection in $F_{2:3}$ through $F_{2:4:6}$. This procedure reduces by 10-fold the number of yield test plots compared to a previously described procedure (Boerma and Cooper, 1975). With fewer plots, the breeder can evaluate the same number of crosses at a reduced cost, or use the same number of plots, but increase the number of crosses evaluated each year.

Cober and Voldeng (2000) evaluated the use of rapid backcrossing and single crosses to develop lines with both high seed yield and high protein content. They developed 886 single cross lines and 800 backcross lines for evaluation. The results showed the average seed yield and protein content of the backcross-derived lines and single-cross lines were not significantly different, although both methods produced lines with higher seed protein content than the recurrent parent. Backcrossing to the high-yielding parent did not result in higher yielding lines. Hintz et al. (1987) also compared backcross and single cross lines. In their study, they were selecting for high yield with iron-deficiency chlorosis resistance. The single cross involved a high-yielding line crossed to a line highly resistant to chlorosis. The backcross was made to the high-yielding parent. The backcross populations had a higher mean yield but lower iron- deficiency chlorosis resistance than the single cross populations. However, when breeding for high-yielding cultivars with acceptable levels of chlorosis resistance, both methods were effective. Thus, the best method to use will depend on other objectives or traits of the parents and potential new cultivars.

The use of backcrossing to introgress genes for small seed size from *G. soja* into *G. max* was studied by LeRoy et al. (1991b). They compared selecting for seed size between backcross generations to no selection during backcrossing. The backcrossed progeny, where selection was practiced, resulted in lines with smaller seed size. Cianzio and Voss (1994) compared three strategies (single cross, three-way cross, and backcross) for developing high-yielding cultivars with improved iron efficiency. The populations developed by backcrossing had the highest yield followed by the three-way crosses and then the single-cross populations. The single-cross populations had the highest iron efficiency. Considering both traits and making comparisons to a check cultivar with high yield and moderate iron efficiency, 15% of the backcross lines, 9% of the three-way lines and 5% of the single cross lines were superior in yield and iron efficiency to check cultivars. The transfer of high seed protein to high yielding soybean cultivars was investigated by Wehrmann et al. (1987). They crossed 'Pando' (high protein line) to three high- yielding lines. Selection for high protein was practiced during two generations of backcrossing. No BC_2F_2 derived lines had protein content as high as the donor parent, but 72% of the lines were higher in protein than the recurrent parents. About 19% of the lines were not significantly lower in yield from the recurrent parents.

9–1.2.2 Recurrent Selection

Recurrent selection is a cyclic method of population improvement. The basic steps in a cycle are intermating, evaluation, and selection. Since recurrent selection is designed to improve the frequency of favorable alleles in a population for quantative traits, additional effort is needed to develop a finished cultivar from a recurrent selection population. Lewers and Palmer (1997) published an excellent review of recurrent selection in soybean. They addressed the standard recurrent selection techniques that have been used in soybean. They also listed studies that were done, the selection method used, the intermating method, and the trait or traits used for selection. They also discuss marker-assisted recurrent selection techniques. It is evident from the literature they cite in their review that recurrent selection is effective in soybean.

Table 9–1. Selected recurrent selection studies.

Objective or trait	Reference
Increased tolerance to phytophthora	Walker and Schmitthenner (1984)
Creation of an ms_2-based population of genetically diverse ancestral material	Specht et al. (1985)
Oil quality	Carver et al. (1986b)
Yield	Piper and Fehr (1987)
Yield	Guimaraes and Fehr (1989)
Yield and protein level	Holbrook et al. (1989)
Iron efficiency	Beeghly and Fehr (1989)
Yield	Burton et al. (1990)
Yield	Werner and Wilcox (1990)
Seed size	Tinius et al. (1991)
Flowering duration, seed filling duration, reproductive period	Hanson (1992)
Yield	Rose et al. (1992)
Recurrent annual introgression of elite material into an ms_2-based population	Specht and Graef (1992)
Early maturity and protein level	Xu and Wilcox (1992)
Seed size and seed growth rate	Tinius et al. (1993)
Seed composition and seed size	Tinius et al. (1992)
Cytoplasmic diversity	Lee et al. (1994)
Yield	Uphoff et al. (1997)
Seed protein level	Wilcox (1998)

The main challenge in using recurrent selection in soybean has been the difficulty of the intermating step. Accomplishing the number of random matings needed in a large-scale population improvement scheme can be a taxing effort if such matings must be made manually. An efficient method for producing hybrid soybean seed was developed by Lewers and Palmer (1997) that takes advantage of the closely linked loci Ms_6 and W_1. In this method, the fertile $Ms_6\_ W_1\_$ plants have purple flowers and hypocotyls while the male-sterile $ms_6ms_6 w_1w_1$ plants have white flowers and green hypocotyls. The W_1/W_1 locus pleiotropically governs hypocotyl color of the seedling, flower color at anthesis, and hilum color at maturity. The male-sterility or fertility of an individual plant in a recurrent selection program is thus inferable at the seedling stage, flowering, and maturity. In addition to facilitating the production of hybrid seed this marker system allows for the elimination of almost all male-sterile plants, shortly after emergence (by hypocotyl color) from evaluation plots. The authors also describe how this method can be used for half-sib recurrent selection methods in soybean. A comparison of three methods of hybrid soybean seed production for use in recurrent selection or other research areas was conducted by Lewers et al. (1996). They compared the traditional method (using ms_2ms_2), the dilution method (also using ms_2ms_2), and the cosegregation method (using linkage between Ms_6 and W_1). They concluded the cosegregation method produced F_1 seed of greater quantity, of equal or better quality, and was more efficient than the other two methods.

Recurrent selection in soybean was discussed by Burton (1987) in the last soybean monograph. Since that time there have been recurrent selection studies published on numerous traits. Table 9–1 presents examples of recurrent selection studies published since 1984. Yield, as expected, has been studied in a number of different populations. Seed traits have also received considerable attention. Other traits include tolerance to phytophthora, adoption of ancestral material, and reproductive traits. As researchers take advantage of better systems for intermating to produce hybrid seeds, more recurrent selection populations and studies will likely be developed and reported and more traits will undergo recurrent selection.

Fatmi et al. (1992) compared four intermating methods for synthesizing base populations for recurrent selection with regard to maintenance of alleles, the amount of recombination, and genetic variability. The four methods were chain cross, convergent cross, diallel cross following pair crossing, and pair crosses following a diallel. In evaluating the results, the authors reported the allele frequencies were in the expected ranges and no differences were found for recombination or genetic variability. Thus, the choice of intermating method for developing recurrent selection populations depends on resources available.

9–1.2.3 Other Studies on Methods of Selection

Smith and Nelson (1986) conducted divergent selection for length of the reproductive period (R1–R7; as defined by Fehr et al. [1971]) in the F_2. In the F_4 and F_5 they conducted divergent selection for seed-filling period (R5–R7). They evaluated lines in the F_6 and F_7 for seed-filling period. They concluded there was no relationship between the reproductive period in the F_2 and seed-filling period in later generations; however, there was a positive association between the classification for seed-filling period in earlier generations (F_4 and F_5) and later generations (F_6 and F_7). In another study, Smith and Nelson (1987) selected lines for differences in reproductive growth period and evaluated the response in seed yield. Genotypic correlations between yield and reproductive growth periods were small. The predicted gain in seed yield by selecting for increased reproductive growth periods was small. They concluded that selection in early generations for reproductive growth periods was not effective in identifying high-yielding lines. Nelson (1988) used pedigree selection to identify lines from a cross that had similar maturity but differed in flowering date. Lines with similar maturities differed by as much as 24 d in time of flowering. The length of the seed-filling period did not vary nearly as much as the reproductive period.

Plantevin et al. (1987) reported the effect of geographic location on selection for earliness of a single seed descent population. A segregating population was advanced by single seed descent in nine European countries to the F_4. The F_5 generation was evaluated at a single location. The results indicated there were significant differences in earliness in the populations advanced in the different countries, with those selected at more northerly latitudes maturing earlier.

Carver et al. (1986a) investigated the response of lines selected for altered unsaturated fatty acids (high oleic and linoleic and low linolenic) to environmental variation. They used lines (classified into two maturity groups) from the original population and from the third and sixth cycles of selection for high oleic acid content and grew them in eight environments and measured seed fatty acid composition. Regression analysis showed significant genotype × environment interactions for each unsaturated fatty acid. Lines selected for high oleic and high linoleic acid were more sensitive to environmental variation than unselected lines. In contrast, low linolenic lines were less sensitive to environmental variation.

The F_3 generation means and variances among F_3 families were used by Troller and deToledo (1996) to predict the genetic potential of crosses. They used yield and plant height to determine the crosses with the best genetic potential. They noted that if G × E was present, data from two environments improved the yield predictions.

9–1.3 Plot Type

Selection among lines that are near homozygosity is another step in the cultivar development process. A commonly used procedure is to select individual plants in the F_4 to F_6 generation, grow out progeny rows and visually select desirable lines, and then conduct replicated yield tests. Recently, many programs have conducted yield tests on progeny rows that may or may not be replicated.

Pfeiffer and Pilcher (1989) compared hill and row plots for predicting yield in full-season and late-planted soybean tests. They concluded that hill plots were as good as row plots for predicting yields (based on mean productivity for both planting times) in both full-season and late planted tests. However, hill plots were not adequate for predicting yields if only late-planted data were used. St. Martin et al. (1990) compared hill, short row, and bordered four-row plots for selection of yield, maturity, height, and lodging among F_2-derived soybean lines. The results indicated that direct selection for maturity, height, and lodging was more effective in four-row plots than smaller plots. The response to direct selection for yield was about 30% greater in the short-row plots compared to the hill plots. They concluded short-row plots would be an effective way for yield testing single-plant progeny. The authors recommended short-row plots for yield be conducted as a single replication at two or three locations. The use of single-row, early generation plant row yield trials to predict yield in multiple-environment advanced yield trials was further investigated by Hegstad et al. (1999). Five populations from crosses between elite cultivars were used with the goal of identifying elite lines that could become new cultivars. They identified the highest, average, and lowest 10 yielding lines from each population from plant-row yield trials for further testing. The yield and yield rank correlations between the plant-row yield tests and advanced tests were highest for the populations, 'Jack' × 'Resnik' and 'Asgrow A3733' × 'Burlington'. About five lines from each population could be identified that were either highest or lowest yielding in both the plant-row and advanced yield tests. The authors concluded that early generation plant-row yield testing was useful in identifying elite lines with high yield potential.

Pazdernik et al. (1996) compared selection for protein and oil content in row and hill plots. They compared short one-row plots at one location and long, four-row plots at two locations and sampling at the end of a row plot vs. the center of the row plot. They reported protein and oil content was not affected by plot type. Sampling position in the row had a large effect with higher protein content and lower oil content in seed collected at the end of the row. Selection for protein and oil content in short, single-row plots at one location appeared to be equivalent to longer multiple-row plots at multiple locations; however, year effects were present. They suggested that either short single-row plots or hill plots could be used to select for protein and oil content. Helms and Orf (1998) evaluated selection for increased protein content using nonreplicated plots to determine the correlated response for yield, oil content, gross value per hectare, and direct response for protein content. Thirty lines from 10 populations were evaluated in seven environments. Selection among lines for increased protein led to increases in protein content but reduction in yield and oil content and gross value per hectare. When selection for protein content was based on population means, the results were similar to selection based on individual lines.

9–1.4 Selection Procedures

The use of best linear unbiased predictions to select superior crosses from a limited number of yield trials was studied by Panter and Allen (1995b). They compared best linear unbiased predictions (with and without using historical parental data) with least square means in 24 soybean crosses. They concluded that either form of best linear unbiased predictions was superior to least square means in ranking the 24 crosses for yielding ability. They noted that the 24 crosses were related, thus making difficult a broader assessment of the usefulness of historical parental data in breeding programs. Helms et al. (1995) investigated the use of nearest-neighbor-adjusted means as a way to select genotypes for yield compared to using unadjusted means. Three hundred experimental lines were evaluated in six environments, with each environment considered a selection environment and actual yield advance measured in the other five environments. In 11 out of 12 cases, the lines selected by nearest neighbor adjusted means were equal in yield to using unadjusted means. Thus, in the populations used, nearest-neighbor-adjusted means did not offer an advantage for selection for yield.

Direct and indirect selection for small seed in temperate (Iowa) and tropical (Puerto Rico) environments was compared by LeRoy et al. (1991a) in three interspecific crosses. Indirect selection was based on pod width. Selection on a plant, plot, or entry-mean basis was also evaluated. The results indicate selection for seed size or pod width on a plant basis was almost as effective as selection on an entry-mean basis. Also, both direct and indirect selection for small seed size were effective. Selection was also effective in both temperate and tropical environments.

9–1.5 Genetic Gain

A procedure for estimating genetic gains for each stage in a multi-stage testing program was described by St. Martin and McBlain (1991). In the procedure, a pairing is made between a set of strains evaluated at one-stage and the selections obtained from that set at the next stage. Realized gains were estimated by comparing common checks in the two tests. By regressing genetic gain on the selection differential, estimates of the proportion of the phenotypic superiority that was realized as genetic gain were obtained. They applied this procedure to regional soybean test data (St. Martin and McBlain, 1991), and to data from the Ohio soybean breeding program (St. Martin and Futi, 2000). The results from the Uniform Soybean Tests, Northern States showed gains were small for maturity, lodging, and seed weight. For yield, the genetic gains and regression coefficients of gain on selection differential varied by maturity group. The estimates of gain were near zero (slightly negative) for Maturity Group 00, but positive for other maturity groups. In the Ohio breeding program, the mean selection differential and genetic gain were determined at the F_3, F_4, and F_6 generations for 1985 to 1997. Maturity changed little at each stage. Lodging improved at the F_4 and F_6 stages. Genetic gain for yield averaged −1.4% in the F_3, +3.7% in the F_4, and +9.1% in the F_6 using the mean of common checks. The negative yield gains in the F_3 may have resulted from significant selection pressure for early maturity and the use of unreplicated yield plots. Since gain at one stage affects gain at another stage the authors concluded nearly equal selection intensity at each stage would be the most beneficial for the breeding program.

Genetic gains for seed yield in recurrent selection programs have been examined by Piper and Fehr (1987), Guimaraes and Fehr (1989), and Uphoff et al. (1997). The first two studies compared genetic gain for yield when 100 F_4-derived lines from 45 single crosses were compared to 10 F_4-derived lines from each of 10 crosses. They found that after three cycles of selection, genetic gain was greater when more lines were evaluated from fewer populations. The third study, Uphoff et al. (1997), added a strategy of three intermating generations to the 100 F_4- derived lines from 45 crosses. The results showed that the genetic gain per cycle was greater when only one generation of intermating was used as opposed to three generations of intermating. After six cycles of selection, the genetic gain per cycle was greater when the 100 F_4 lines were from 45 crosses than from 10 crosses. This indicated that the greatest gains in yield improvement using recurrent selection would be obtained by selecting from the largest number of single cross matings possible.

9–1.6 Male-Sterile-Facilitated Cyclic Breeding

Soybean breeders continue to develop breeding methods to improve the effectiveness and efficiency of their cultivar development programs. Specht and Graef (1990) described a breeding method called the male-sterile-facilitated cyclic breeding (MSFCB) method for cultivar development (Fig. 9–1). The authors describe in detail the theoretical aspects behind the method and suggest that this method combines the best aspects of conventional breeding and diallele selective mating as described by Jensen (1970). They used the original SG1 population (Specht et al. (1985) with the $ms_2 ms_2$ form of soybean nuclear genetic male sterility to create a population of more elite germplasm for use in a variety development program. As of 2002, nine high-yielding cultivars have been released using the MSFCB scheme.

In the MSFCB method (see Fig. 9–1 and 9–2) the authors noted that annually chosen elite parents (cultivars, breeding lines, or elite introduced material) can be placed in a checkerboard row pattern in an isolation nursery containing rows of male-sterile parents (the male-fertile sibs are rogued at flowering). Insects (an appropriate species of bees) transfer the pollen from the elite parents to the male-sterile plants. At least one F_1 seed (or a F_1 seed of a half-sib family) is then harvested from each male-sterile plant. The F_1 seeds can then be grown in a winter nursery and plants threshed in bulk to provide F_2 seed (Fig. 9–2). The majority of the F_2 seed is advanced in the breeding program by selfing (single seed descent of male-fertile plants), followed by selecting plants, evaluating lines, and identifying superior lines for cultivar release. Because male-sterile plants segregate during gener-

A Male-Sterile Facilitated Cyclic Breeding Scheme

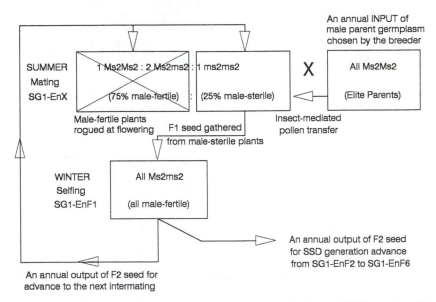

Fig. 9–1. An illustration of how a soybean population segregating for the ms_2 allele for male-sterility (SG1) can be used to facilitate the intermating needed to annually introgress a set of elite (E) germplasm the breeder chooses each year (n). To ensure an intermating each summer, the F_1 to F_2 generation advance must occur in a winter nursery. See Specht et al. (1985), Specht and Graef (1990, 1992) for additional details.

ation advance, the number of plants in each generation will need to be adjusted upward (i.e., more nursery space). The cyclic portion of the method is continued by using a small portion of the F_2 seed for the next year's isolation nursery (Fig. 9–2). These F_2 seed are the source of the male-sterile plants that will be pollinated (using insects) by a new set of elite parents. Since the F_2 plants that will serve as the male steriles segregate three male-fertile to one male-sterile, the male-fertile plants need to be rogued as soon as they produce a flower that can be examined (the male-fertile plants produce normal anthers with considerable pollen while the male-sterile plants produce shrunken anthers with no pollen). The F_1 seed is again harvested from male-sterile plants and the cycle is repeated. The procedure works well, but would be more convenient to implement if male-fertile sibs in the intermating nursery could be identified and destroyed before flowering, and if the heterozygous male-fertile plants could be identified and destroyed in the early stages of generation advance.

9–1.7 Hybrid Soybean Cultivars

Soybean is considered a self-pollinating species. Soybean geneticists and breeders typically make manual cross-pollinations to produce hybrid seeds as part of breeding programs. However, the production of large quantities of hybrid seed, even for experimental purposes, is difficult and very time consuming. The grow-

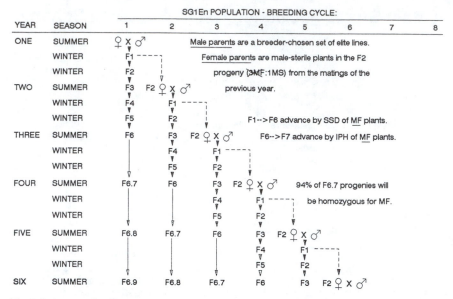

Fig. 9–2. An example of how cultivar development can be coupled to a random-mated population (SG1) undergoing annual introgression of breeder-chosen elite (E) germplasm. The portion of F_2 seed not used for intermating the following year undergoes generation advance. This can be to the F_6 plant generation (as shown) or to an earlier or later generation according to breeder preference. The percentage of homozygous male-fertile plants rises during generation advance (33% in F_2, 60% in F_3, 78% in F_4, 88% in F_5, 94% in F_6).

ing of F_1 hybrids on a commercial scale has not been possible because of the difficulty of producing large quantities of hybrid seed economically. Some reports suggest, however, that hybrid soybean may become commercially available in the near future (Anonymous, 1999).

The production of hybrid soybean seed became easier with the identification of mutations that affect male reproductive function but do not affect female reproductive function. Palmer et al. (2001) listed the nuclear male-sterile mutants, temperature-sensitive nuclear male-sterile mutants, photoperiod-sensitive nuclear male-sterile mutants, and the cytoplasmic-nuclear male-sterile mutants (including maintainer and restorer lines) found in soybean. They provide some details about the inheritance and functioning of the mutants. From a plant-breeding perspective, male sterility can be used to (i) facilitate random mating in a recurrent selection program (Burton, 1987), (ii) facilitate crossing in a combined population improvement and cultivar development program (Specht and Graef, 1990), (iii) aid in introgressing exotic germplasm into elite material (Lewers et al., 1996), and (iv) produce F_1 hybrid seeds for research or commercial purposes.

In their article about hybrid soybean, Palmer et al. (2001) listed five components that are critical for developing commercial hybrid soybean. They are: (i) parental combinations that produce heterosis levels superior to the best pure line cultivars, (ii) a stable male-sterile, female-fertile sterility system, (iii) a selection system to obtain 100% female (pod parent) plants that set seed normally and can be harvested mechanically, (iv) an efficient pollen transfer mechanism from pollen parent to pod parent, and (v) an economical level of seed increase for the seedsman and growers that ultimately benefits the consumer. Palmer et al. (2001) discussed results from research plots that show some hybrids exhibit 10 to 20% high-parent heterosis, which the authors viewed was enough of a yield increase to make hybrids attractive. The cytoplasmic-nuclear male sterility systems developed in China may have the characteristics needed for some of the other necessary components. The economic aspects of hybrid seed will be more fully known once the components of the system can be tested on a commercial scale. Thus, although there appears to be some problems to be overcome, hybrid soybean may be grown on a commercial scale soon.

9–1.8 Impact of Transgenes on Breeding

From a cultivar development perspective, the development of a glyphosate-tolerant soybean line (Padgette et al., 1995) has had a major impact on the soybean industry since the late 1990s. Initially the glyphosate-tolerance gene was transferred to cultivars via backcrossing, but more recently many programs have incorporated the gene into the majority of the breeding material in their programs. Details of an example of the development of a glyphosate-tolerant soybean cultivar are discussed later in this chapter.

9–1.9 New Tools—Field Equipment

The development of better techniques and equipment for plot management, especially plots designed for yield evaluation continues on a regular basis. Major changes have occurred over time that have increased the efficiency of planting, managing, and harvesting plots. Significant changes have also occurred in the preparation of seed for planting, and the weighing and analysis of seed after harvest as well as data manipulation and analysis. These changes have resulted in larger numbers of plots handled per person in a breeding program.

Most soybean-breeding programs use commercial or internally developed computer programs to make entry and plot designations, randomize plots, prepare field plans and field books, print labels, and organize tests. In many cases seed packets or trays may be filled using seed counters while others are filled by hand. Continued improvements are being made on planters used for yield plot testing. Newly designed planters have better seed placement, more accurate plot length, and use air systems for more uniform placement of seed.

Even with improved planting equipment, many breeding programs trim plots to a uniform length at some point during the growing season. Fehr (1987) discussed the aspects related to end-trimming and concluded most breeders end-trim plots at maturity for lines in the advanced stages of yield testing. The increased use of hill plots or short row plots for yield evaluations requires that plots be as uniformly spaced and aligned as possible.

Harvesting of plots for yield is now performed using self-propelled combines. Specialized combines have been developed for the various plot sizes from hill plots to large demonstration plots. In fact, some combines have been developed that are separated down the middle, such that two different plots can be harvested at the same time. Most combines have electronic systems for determining the weight and moisture content of seed harvested from plots. Electronic systems for on-the-go measurements of seed quality (protein and oil) are currently being tested. Combines are also available that will automatically bag and label seed from plots, making the harvesting process a one-person operation. Data are generally collected on a combine computer and then downloaded to office or laboratory computers for analysis.

9–2 MOLECULAR MARKERS

Molecular marker technologies have facilitated the construction of genetic maps of many organisms including important crop species. An analysis of the association of phenotypic trait and genetic marker data from a population of lines segregating for genes conditioning a quantitative trait provides the unique ability to dissect the quantitative trait into a set of discrete quantitative trait loci or QTL (Tanksley et al., 1989). Genetic marker analysis can measure the effect of an individual QTL at the allelic level (i.e., the contribution of an allele from each parent at a particular QTL can be determined separately from the other QTL affecting the trait). This important attribute of genetic markers allows pyramiding of the desirable alleles for a complex or otherwise intractable trait into an adapted genetic background. In addition, genetic markers are potentially useful in plant-breeding programs for parental selection, donor trait and recurrent parent selection in backcross breeding, trait selection in segregating populations, and identification of new positive alleles in exotic germplasm.

Researchers have developed robust sets of DNA markers that span the soybean genome. Soybean restriction fragment length polymorphism (RFLP) markers were introduced in the late-1980s (Apuya et al., 1988; Keim et al., 1989). The arbitrarily primed marker systems, random amplified polymorphic DNA (RAPD) and DNA amplification fingerprint (DAF), were developed in the early 1990s (Caetano-Anolles et al., 1991; Williams et al., 1990). In the mid-1990s, soybean short (simple) sequence repeat (SSR) markers were introduced (Akkaya et al., 1995; Morgante et al., 1994; Rongwen et al., 1995). Soybean SSR markers are highly polymorphic in elite breeding populations, and amendable to automation. In the past decade, several moderate to high density genetic maps of soybean have become available (Cregan et al., 1999a; *SoyBase*, http://soybase.ncgr.org/cgi-bin/ace/generic/search/soybase).

9–2.1 Mapped Genes and Quantitative Trait Loci

Molecular markers have been used to map the genomic location of both major genes and QTL for many agronomic, physiological, pest resistance, and seed composition traits in soybean (Table 9–2). These data provide insight into the successful application of DNA marker technologies by soybean scientists and the progress that has been accomplished. This summary was compiled by a review of available information in *SoyBase* and published literature. Based on these sources of information, we applied the following criteria in the development of the data in Table 9–2: (i) when a QTL for a given trait was reported in several studies, but seems to be a common QTL based on genomic map position, list it only once, (ii) when a QTL for a given trait seems to be conditioned by a known major gene(s) (i.e., qualitative trait) do not list it, and (iii) identify independent QTL that explained 10% or more of the phenotypic variation for each trait. The rationale for identifying QTL conditioning 10% or more of the phenotypic variation was to provide the relative magnitude of the effects of the reported QTL for the various traits. As an example, Mian et al. (1996a, 1998a) reported a total of five QTL conditioning water-use efficiency in two soybean populations (Table 9–2). Two of these QTL conditioned >10% of the phenotypic variation in water-use efficiency.

Our review indicates there are at least 319 QTL reported for the various quantitative traits (Table 9–2). This is approximately one-half the number of QTL reported in *SoyBase*. This difference is largely due to the accounting methods used. In *SoyBase*, a QTL is indicated at any marker or interval significantly associated with the trait in the original study regardless of degree of linkage of the markers. The Table 9–2 summary only reports a single QTL from a group of linked markers that were significantly associated with the QTL. In addition, we have identified the number (total of 162) of unique QTL reported to condition 10% or more of the phenotypic variation in the trait. Compared to the total number of QTL listed for a trait, these QTL have a higher probability of being confirmed in the original population or in another population in which the parents contain these alleles at the QTL. This assumption is based on the population sizes and marker saturation of the current studies (Beavis, 1998).

Across the various categories of traits (i.e., pest resistance, physiological, etc.), 51% of the total reported QTL conditioned 10% or more of the phenotypic variation in the population in which they were identified (Table 9–2). Sixty-one percent of the seed composition QTL conditioned 10% or more of the phenotypic variation, while 40% of the soybean sprout QTL and 47% of the physiological trait QTL were of this magnitude. It is expected that the high percent of QTL conditioning 10% of more of the phenotypic variation in these traits is a result of the small population sizes used in most of the original mapping studies. The use of small populations in QTL mapping studies limits the precision of QTL detection and often results in overestimation of QTL effects (Beavis, 1998).

Given the high heritability of resistance to many soybean diseases and pests, the overall relatively low percent of these QTL (49%) conditioning 10% or more of the phenotypic resistance was somewhat unexpected. When the reported QTL for resistance to sclerotinia stem rot (also known as white mold) (caused by *Sclerotinia sclerotiorum* (Lib.) deBary) are removed from this group, 73% of the 51 remaining pest resistance QTL conditioned 10% or more of the phenotypic variation. Given the current difficultly in experimentally assessing soybean resistance to sclerotinia stem rot, the fact that only 2 of 29 QTL conditioning 10% or more of the resistance to this disease is not totally unexpected (Arahana et al., 2001; Kim and Diers, 2000).

Another interesting observation from the summary of QTL data presented in Table 9–2 is that soybean scientists have primarily applied DNA marker analysis

Table 9–2. A summary of reported soybean quantitative trait loci (QTL).

Traits	Popula-tions(s)	Reported QTL	References
	——— no.———		
	A. Pest resistance traits		
Insect			
Corn earworm	5	16[†] (12)[‡]	Rector et al. (1998a,1998b, 1999, 2000); Terry et al. (1999, 2000)
Root-knot nematodes			
Southern	1	2 (2)	Tamulonis et al. (1997c)
Peanut	1	2 (2)	Tamulonis et al. (1997b)
Javanese	1	2 (2)	Tamulonis et al. (1997a)
Soybean cyst nematode	12	20 (11)	Concibido et al. (1994, 1996a, 1996b, 1997); Heer et al. (1998); Mahalingam and Skorupska (1995); Qiu et al. (1999); Vierling et al. (1996); Wang et al. (2001); Webb et al. (1995); Yue et al. (2001a, 2001b)
Sudden death syndrome	4	7 (7)	Chang et al. (1996,1997); Hnetkovsky et al., 1996; Iqbal et al. (2001)
Brown stem rot	1	2 (1)	Bachman et al. (2001); Lewers et al. (1999)
Sclerotinia stem rot	6	29 (2)	Arahana et al. (2001); Kim and Diers (2000)
	B. Physiological traits		
Water-use efficiency	2	5 (2)	Mian et al. (1996a, 1998a)
Carbon isotope discrimination and beta	1	1 (0)	Specht et al. (2001)
Aluminum tolerance	1	6 (1)	Bianchi-Hall et al. (2000)
Soil waterlogging	2	1 (1)	Van Toai et al. (2001)
Specific leaf weight	1	6 (4)	Mian et al. (1998c)
Leaf area	3	6 (3)	Mansur et al. (1993b, 1996); Mian et al. (1998c); Orf et al. (1999b)
Leaf ash	1	5 (2)	Mian et al. (1996a)
Early plant vigor	1	5 (3)	Mian et al. (1998b)
Leaf length	3	6 (4)	Keim et al. (1990a); Mansur et al. (1996); Orf et al. (1999b)
Leaf width	4	10 (4)	Keim et al. (1990a); Mansur et al. (1996); Orf et al. (1999b)
	C. Seed composition traits		
Protein content	14	32 (18)	Brummer et al. (1997); Csanadi et al. (2001); Diers et al. (1992b); Lee et al. (1996b); Mansur et al. (1993b,1996); Orf et al., (1999b); Qiu et al. (1999)
Oil content	14	24 (13)	Brummer et al. (1997); Csanadi et al. (2001); Diers et al. (1992b); Lee et al. (1996b); Mansur et al. (1993b, 1996); Orf et al. (1999b); Qiu et al. (1999)
Linolenic acid	2	3 (3)	Brummer et al. (1995); Diers and Shoemaker (1992)
Linoleic acid	1	3 (3)	Diers and Shoemaker (1992)
Palmitic acid	1	3 (3)	Diers and Shoemaker (1992)
Oleic acid	1	3 (3)	Diers and Shoemaker (1992)
Sucrose content	1	7 (3)	Maughan et al. (2000)

(continued on next page)

Table 9–2. Continued.

Traits	Popula-tions(s)	Reported QTL	References
	— no. —		
	D. Agronomic traits		
Plant height	5	14 (4)	Keim et al. (1990a); Lee et al. (1996a, 1996c); Mansur et al. (1993b, 1996); Orf et al. (1999b); Specht et al. (2001)
Maturity	6	10 (7)	Keim et al. (1990a); Lee et al. (1996a, 1996c); Mansur et al. (1993b, 1996); Orf et al. (1999b); Specht et al. (2001)
Lodging	7	10 (5)	Lee et al. (1996a, 1996c); Mansur et al. (1993b, 1996); Orf et al. (1999b); Specht et al. (2001)
Pod dehiscence	2	5 (3)	Bailey et al. (1997); Saxe et al. (1996)
Seed coat hardness	1	5 (5)	Keim et al. (1990b)
Flowering date	5	8 (4)	Keim et al. (1990a); Mansur et al. (1993b, 1996); Orf et al. (1999b)
Reproductive period	3	8 (5)	Mansur et al. (1993b,1996); Orf et al. (1999b)
Seed weight	6	23 (6)	Csanadi et al. (2001); Mansur et al. (1993b, 1996); Maughan et al. (1996); Mian et al. (1996b); Orf et al. (1999b); Specht et al. (2001)
Seed yield	2	4 (2)	Mansur et al. (1996); Orf et al. (1999b); Specht et al. (2001)
Canopy height	2	3 (3)	Keim et al. (1990a); Mansur et al. (1996)
Chlorimuron ethyl sensitivity	1	3 (1)	Mian et al. (1997)
Iron deficiency chlorosis	3	7 (1)	Diers et al. (1992a); Lin et al. (1997, 1998)
Stem diameter	1	3 (3)	Keim et al. (1990a)
	E. Soybean sprouts		
Sprout yield	1	4 (2)	Lee et al. (2001)
Hypocotyl length	1	3 (2)	Lee et al. (2001)
Abnormal seedlings	1	3 (0)	Lee et al. (2001)
TOTAL		319 (162)	

† Total number of different QTL identified in the references listed to the right.
‡ Number of QTL that explained 10% or more of the phenotypic variation are shown in parentheses.

to the discovery of QTL for traits of relatively high heritability and high economic importance to soybean growers. Twelve populations have been evaluated for QTL-conditioning resistance to soybean cyst nematode (SCN, *Heterodera glycines* Ichinohe) and 14 populations for seed protein and oil content. Although seed yield is considered the trait of highest priority in most soybean-breeding programs, its relatively low heritability, requirement for extensive phenotypic data collection across environments, and the expectant large number of QTL, with the majority conditioning small effects, has most likely limited the number of QTL identified and the number of QTL mapping studies undertaken for this trait.

9–3 APPLICATION OF GENETIC MARKERS
IN BREEDING PROGRAMS

Soybean breeders and geneticists have been successful in applying genetic markers in mapping genes controlling numerous economically important traits. A greater challenge will be the development of methods for the widespread application of marker-based techniques in breeding programs. Markers will be widely applied by breeders only after the markers have been shown to be both effective and economical. Most of the time when markers will be applied in breeding, the markers will complement traditional breeding methods that are refined and proven successful. In a review of DNA markers in plant breeding, Lee (1995) states "The challenges loom large for marker-assisted selection (MAS); in many crops, conventional selection has had several decades to evolve into a very effective technology. Plant-breeding programs in most crops conduct simultaneous selection for several (>10) traits, in many populations across diverse environments." The experience so far of incorporating MAS in soybean has been mixed. Currently, MAS might be best viewed as a supplement to conventional breeding methods.

9–3.1 Parental Selection Based on Genetic Markers

A potential use of genetic marker information in applied breeding programs is aiding breeders in the selection of those parental combinations with the greatest potential in leading to new, high-yielding cultivars. Cox et al. (1985) stated that optimal breeding populations should have high means for traits of interest and a high genetic diversity. It has been shown that the means of parents of soybean populations are predictive of the means of populations developed from crosses (Panter and Allen, 1995b). However, predicting genetic variance in populations is difficult in typical breeding programs where related, elite parents are crossed. According to quantitative genetic theory, the genetic variance of populations for a metric trait, such as yield, will increase as the parents differ for more genes that affect the trait (Falconer, 1981). It is generally assumed that genetic diversity between parents is inversely related to coefficient of parentage (CP). It may be possible to estimate the number of genes differing between parents with either CP or genetic markers.

The usefulness of genetic distance estimates between parents in predicting genetic variance of agronomic traits in soybean populations has been evaluated in three published studies. In one study, Helms et al. (1997) studied the value of genetic distance between parents based on RAPD markers and CP in predicting genetic variance of six populations. They studied genetic variance for yield, lodging, physiological maturity, and plant height. Three of their populations were developed from crosses between North American developed cultivars and three populations were developed from crosses between North American cultivars and foreign plant introductions. In their set of populations, they found CP was not correlated with genetic distance based on markers. Moreover, they found no relationship between genetic variance for agronomic traits and either CP or genetic distance based on markers.

In a second study, Manjarrez-Sandoval et al. (1997) evaluated five populations to study the correlation between genetic variance within populations and genetic distance among parents based on RFLP markers and CP. In contrast to the results of Helms et al. (1997), they were able to predict the population with the greatest genetic variance for yield using both CP and genetic distance. Across populations, CP was significantly ($P < 0.05$) associated with yield genetic variance, although marker distance was not significantly associated with yield genetic variance.

In a third study, Kisha et al. (1997) evaluated the association between genetic variance for agronomic traits and genetic distance estimates among parents based on both CP and RFLP markers. A total of 53 populations divided into three sets were used in this study. Across these populations, they were unsuccessful in predicting genetic variance for yield with either CP or marker genetic distance. In some sets, they did identify a significant ($P < 0.05$) association between CP or marker genetic distance estimates and genetic variance for plant height and maturity. When populations in each set were divided into those that were the most and least genetically distant based on CP or markers, they observed a trend where the more distant populations had greater genetic variability for yield, maturity, and height. This trend was significant ($P < 0.05$) based on RFLP distance in one set.

There are numerous explanations for why genetic marker distance among parents has not been entirely successful in predicting genetic variability for agronomic traits in soybean populations. The first is that obtaining precise estimates of genetic variability in populations is difficult. These estimates are expensive to obtain because they require the testing of a large number of lines in replicated field tests (Kisha et al., 1997). Even after this testing, there will be high standard errors for the estimates of genetic variation. Additionally, CP calculations require certain assumptions, such as ancestral lines being unrelated, which may or may not be correct. The distance estimates based on genetic markers are also subject to errors depending on the type of marker used. These factors make it difficult to detect an association between genetic variance and CP or marker distance even if it is present. An even more intractable problem is that the association between QTL that control traits like yield and genetic markers used in these studies is unknown. The utility of markers for identifying populations with high genetic variability will most certainly improve if QTL controlling these traits are identified and only markers within or very tightly linked to these QTL are used in the prediction.

9–3.2 Use of Markers in Backcrossing

Backcrossing is a widely used method in soybean breeding and genetics research. Soybean breeders have employed backcrossing to add new genes, especially resistance genes into cultivars (Fehr, 1987) and to study the genetic basis of traits. The use of markers can speed the development of backcross lines through the selection for, or against, genetic regions during backcrossing. This includes selecting for one or more genetic regions from a donor parent that contains gene(s) of interest as well as selecting for the recurrent parent's genome to speed its recovery. The use of markers in backcrossing is especially applicable in transferring, into elite breeding material, novel genes that were obtained through genetic transformation or from exotic germplasm (Frisch et al., 1999; McElroy, 1999; Ribaut and Hoisington, 1998).

The use of markers in selecting for the gene being backcrossed can be especially helpful when the phenotype of the gene is not readily assayable in backcross F_1 individuals. This would occur when the gene(s) being backcrossed are recessive, a destructive test is needed for phenotyping, the phenotype is only expressed after pollination, or the gene contributes to a quantitative trait that can only be evaluated in replicated tests. In all four of these cases, if markers are not available, backcrossing would be slowed because progeny testing would be required during each backcross generation. Marker-assisted selection for the gene being backcrossed is especially useful in soybean, where manual pollinations are very tedious, many pollinations are unsuccessful, and successful pollinations rarely produce more than two seeds. Labor can be focused on the few plants that, based on marker analysis, are presumed to carry the gene(s) of interest.

Hospital et al. (1992) studied the optimal use of markers to increase the recovery rate of the recurrent parent genome during backcrossing. Through the use of mathematical models and computer simulations, they determined that markers could save breeders approximately two backcross generations. When <10% of the individuals are selected based on markers each generation, the amount of the recurrent parent genome recovered after three backcross generations is about equal to or greater than expected without selection at the fifth backcross generation. For example, they found that if 5% of the BCF_1s were selected based on markers in each generation of backcrossing, the proportion of recurrent parent genome in selected $BC2F_1$s would be 0.952 compared to 0.875 without selection. These calculations are valid only for the chromosomes that are not carrying gene(s) being backcrossed. The calculations also assume an organism with 20 chromosome pairs that are 100 cM each and that background selection was done based on two markers on each chromosome that are 20 cM from the telomeric ends.

There are pitfalls in the application of marker-assisted selection during soybean backcrossing. The first pitfall is the difficulty in generating enough backcross F_1 seed to produce a population of sufficient size to practice selection. In contrast to maize (*Zea mays* L.), producing several hundred backcross F_1 seed to select among is difficult in soybean. This is especially problematic when MAS is coupled with rapid cycling of generations in off-season nurseries or greenhouses. Soybean breeders can grow three generations a year through the employment of these facilities (Fehr, 1987). However, unless the breeding organization has an established site in an off-season location, the production of large numbers of hybrid seed is difficult and expensive. For example, when a breeder is backcrossing two genes into a recurrent parent, only one-fourth of the gametes from a backcross F_1 plant will carry both genes. If the goal is to select for the genes being backcrossed, and for the recurrent parent genome, breeders would need to generate 459 backcross plants to have a 95% chance of obtaining 100 plants to select among that carry both genes (Fehr, 1987; Sedcole, 1977).

An additional pitfall faced in backcrossing QTL is the imprecision of locating QTL. Immediately after an important QTL has been mapped, researchers may want to backcross this gene into an elite genetic background for confirmation testing. Unfortunately, even in well- designed experiments, the QTL may not be localized to an interval smaller that 20 cM (Stuber et al., 1999). To make certain the gene is not lost during backcrossing, markers flanking the region need to be used in selection. In many regions of the soybean genome, there is poor marker saturation, which results in the need to incorporate a large region from the donor parent to ensure the QTL is not lost during incorporation.

9–3.3 Recovery of Recurrent Parent

An example of the use of genetic markers for the recovery of a recurrent parent genome is the development of a backcross-derived, glyphosate-tolerant cv. Benning at the Univ. of Georgia. The original cultivar had been released in 1996 (Boerma et al., 1997). The transformation of soybean with a gene from *Agrobacterium* sp. strain CP4 by the Monsanto Company resulted in the development of the glyphosate-tolerant (GT) soybean line 40-3-2 (Padgette et al., 1995). This line provided the opportunity to breed elite cultivars that were insensitive to the application of herbicidal rates of glyphosate without loss of soybean yield (Delannay et al., 1995). The transgene conditioning GT is inherited as a single dominant gene (Padgette et al., 1995). The Univ. of Georgia Research Foundation received access to a GT line for use in cultivar development from the Monsanto Company in May 1996. The backcross breeding strategy is outlined in Table 9–3. When SSR primers became publicly available in December of 1997, they were employed to speed the

Table 9–3. Development of glyphosate-tolerant Benning soybean.

Month/Year	Activity
May 1996	Received a glyphosate tolerant (GT) line from Monsanto Co.
May to Oct. 1996	Cross: Benning × GT line
Nov. 1996 to Apr. 1997	Treated (Benning × GT line) F_1 plants with glyphosate
	1st backcross: Benning × tolerant (Benning × GT line) F_1
May to Oct. 1997	Treated [Benning(2) × GT line] F_1 plants with glyphosate
	2nd backcross: Benning × tolerant [Benning(2) × GT line] F_1
Nov. 1997 to Apr. 1998	Treated [Benning(3) × GT line] F_1 plants with glyphosate
	Used 60 SSR to fingerprint 30 tolerant [Benning(3) × GT line] F_1s
	Expected = 87% Benning Actual = 82% to 91% Benning
	3rd backcross: Benning × tolerant [91% Benning(3) × GT line] F_1
May to Oct. 1998	Treated [Benning(4) × GT line] F_1 plants with glyphosate
	SSR markers to select tolerant [Benning(4) × GT line] F_1
	Expected = 93% Benning Actual = 95 to 97% Benning
	Produced F_2 seed from tolerant [97% Benning(4) × GT line] F_1
Nov. 1998 to Feb. 1999	Treated [Benning(4) × GT line] F_2 plants with glyphosate
	SSRs to select tolerant [Benning(4) × GT line] F_2 plants
	Selected 54 tolerant [99% Benning(4) × GT line] F_2 plants
Feb. to Apr. 1999	Single row increase of 54 (99% Benning-BC3) $F_{2:3}$ lines in unlighted
	Puerto Rican Nursery
	Treated BC3$F_{2:3}$ lines with glyphosate; Selected homozygous
	GT-tolerant rows
May to Oct. 1999	Yield test of 19 BC3$F_{2:4}$ tolerant lines at Athens and Plains, GA
	Obtained maximum seed increase of all 19 lines
Nov. 1999 to Jan. 2000	Grew 0.2 ha of each of the three 99% Benning tolerant lines
	Composited seed of the three lines to create G99-G3438
Feb. 2000 to Apr. 2000	Grew a 6-ha increase of G99-G3438
May 2000 to Oct. 2000	Yield test of G99-G3438 at 15 locations in AL, GA, and SC
	Obtained maximum increase of G99-G3438
Nov. 2000 to Mar. 2001	Official Variety Release of G99-G3438 as H7242 RR
Apr. 2001	Commercial production of ~ 4000 ha of H7242 RR

recovery of Benning's genome. Three evenly spaced SSR markers per linkage group (total of 60 SSRs) were selected and used to fingerprint 30 GT, $BC2F_1$ plants. The variation among $BC2F_1$, $BC3F_1$, and $BC3F_2$ plants in the percentage of Benning genome (estimated by SSR fingerprinting) is shown in Table 9–3. The initial yield tests of 19 GT $BC3F_{2:4}$ lines were conducted at two locations in 1999. The overall means of the 19 GT lines for seed yield, maturity, and plant height were similar to those of Benning. Seed of three GT $BC3F_{2:5}$ lines were increased in a winter nursery. In February 2000 seed of the three lines was composited to create G99-G3438. Seed of G99-G3438 was increased in the winter nursery from February 2000 to June 2000. In the summer of 2000, G99-G3438 was evaluated in 15 environments across the southeastern USA. The mean in seed yield, maturity, and plant height of G99-G3438 were similar to those of Benning across the 15 environments. In addition, a seed increase for commercial production was grown in Georgia. In spring of 2001, G99-G3438 was released as 'H7242 RR' and sufficient seed of this cultivar was available to plant approximately 4000 ha for commercial production.

The use of SSR markers in this backcross program required collection of DNA from a total of 202 plants. In the $BC2F_1$ generation, data on 60 SSR markers were collected from 30 GT plants. Since the initial generation of selection allowed identification of plants homozygous for the Benning alleles at 52 of the 60 markers, data on only 8 SSR markers were assayed in the $BC3F_1$. The DNA marker cost (including labor, prorated equipment, and reagents) for the development of H7242 RR was estimated at approximately $2500.

9–3.4 Marker-Assisted Selection in Segregating Populations

Another application of MAS in soybean breeding is the selection among plants or experimental lines within segregating populations. Compared to traditional breeding approaches that rely on phenotypic selection, MAS could potentially increase genetic gains by reducing costs, increasing the total number of plants or lines available for selection, increasing the effectiveness of selection, and reducing the time required for a breeding cycle. However, there are many questions that need to be resolved before the full potential of MAS in breeding programs will be realized.

Numerous researchers have evaluated MAS for quantitative traits using modeling and simulation methods (Lande and Thompson, 1990; Zhang and Smith, 1993; Gimelfarb and Lande, 1994a, 1994b, 1995; Wittaker et al., 1995). The overall conclusions of these studies are that MAS could be more efficient than phenotypic selection when populations are large and traits have a low heritability. Knapp (1998) developed a theory for estimating the probability of selecting one or more superior genotypes through MAS. He found that MAS could substantially decrease the resources needed in selection of a trait with low to moderate heritability when both the selection goal and intensity are high. In addition, he found that the most efficient MAS strategy would include only those markers associated with highly significant additive effects for the selected trait. The returns of MAS are diminished by including additional markers in an index.

The first trait in soybean that has been widely selected with markers is resistance to SCN. Because of the importance of SCN, breeding for resistance to this pathogen is currently a major objective for both public- and private-sector breeders worldwide. Phenotypic selection for SCN resistance is done by growing plants in SCN-infested soil in the field or greenhouse and the number of nematode females on the roots are counted. Although the heritability for SCN resistance can be high, up to 0.97, the phenotypic evaluation of plants for resistance is expensive and tedious (Webb et al., 1995).

Although SCN resistance in populations is quantitative (Mansur et al., 1993a), a few major QTL control a large proportion of the resistance. For example, *rhg1*, which has been mapped to linkage group (LG) G (Cregan et al., 1999a), has been shown to control 36 to 86% of the variability for SCN resistance in populations segregating for the PI 88788 source of resistance (Concibido et al., 1997; Diers and Arelli, 1999). This gene has also been found to be an important source of resistance in other genetic backgrounds including PI 437654 (Webb et al., 1995), PI 209332, 'Peking', and PI 90763 (Concibido et al., 1996a, 1997). Using two SSR markers that flank *rhg1*, Mudge et al. (1997) identified 98% of the lines that had a female index (FI) below 30% in a population segregating for PI 209332 resistance. Female index is defined as the percent females on the roots of a test genotype compared to an SCN susceptible check. Two additional SSR markers that map within 1 cM of *rhg1* were more recently identified and these could be used to further improve MAS efficiency for *rhg1* (Cregan et al., 1999b). Meksem et al. (2001a) found that in Forrest, a cultivar that derives its SCN resistance from Peking, resistance can be selected using markers linked to *rhg1* and additional markers linked to *Rhg4*, a major SCN resistance QTL on LG A2 (Cregan et al., 1999a). In a population developed by crossing Forrest with Essex, a susceptible cultivar, lines predicted to be homozygous for both resistance alleles based on linked markers had a mean FI of 5%. Diers and Arelli (unpublished data, 2002) studied MAS in a population developed by crossing 'Bell', which derived its SCN resistance from PI 88788, with a susceptible cultivar. Using a marker linked to *rhg1*, they found that 79% of the lines homozygous for the marker allele from Bell had a FI below 10%.

Early attempts at MAS for SCN resistance were done with RFLP markers (Concibido et al., 1996a). Although these markers could be used to successfully identify resistant genotypes, they were not readily adaptable to the high throughput requirements of MAS. Simple sequence repeat markers are much more suitable for high throughput genotyping and are currently being used in MAS. Additional genotyping methods such as allele-specific hybridizations (Coryell et al., 1999) and Taqman (Roche Molecular Systems, Pleasanton, CA) probes (Meksem et al., 2001b) also have been developed for automated assays of markers of breeder interest. Breeders in both the private and public sectors are currently conducting MAS for SCN resistance using SSR markers and other methods.

Resistance to sudden death syndrome [SDS, *Fusarium solani* (Mort.) Sacc. f. sp. *glycines* Roy] is an additional trait that would benefit from MAS. Like SCN, SDS resistance testing is time-consuming and costly (Njiti et al., 2002). Numerous SDS resistance QTL have been identified (Hnetkovsky et al.,1996; Chang et al., 1996; Iqbal et al., 2001; Njiti et al., 2002) and the pyramiding of resistance alleles at these QTL might provide a high level of SDS resistance. Markers have been identified that are linked to resistance to numerous other pathogens and pests including brown stem rot [BSR, *Phialophora gregata* (Allington and Chamberlain) W. Gams f. sp. *sojae* Kobayasi, Yamamoto, Negishi et Ogosh] (Lewers et al., 1999;

Bachman et al., 2001), phytophthora stem rot (Diers et al., 1992c; Demirbas et al., 2001), frog eye leaf spot (*Cercospora sojina* K. Hara) (Mian et al., 1999), root-knot nematodes (*Meloidogyne* spp.) (Tamulonis et al., 1997a, 1997b, 1997c), sclerotinia stem rot (Kim and Diers, 2000; Arahana et al., 2001), and corn earworm [*Helicoverpa* zea (Boddie)] (Rector et al.,1998a, 1998b, 1999, 2000). As MAS systems are further developed, they will likely be employed in the selection for resistance to at least some of these diseases.

Breeders would benefit from the ability to select for seed yield using MAS because yield improvement is the highest priority objective of most soybean breeders and it has a relatively low heritability (Burton, 1987). The results of mapping yield QTL have been reported in the literature (Mansur et al., 1996; Orf et al., 1999a, 1999b). However, the effect of these mapped yield QTL are population and/or environmentally specific (Orf et al., 1999a; Reyna and Sneller, 2001) or are simply pleiotropic effects of genes that govern disease resistance, plant morphology, and/or maturity (Specht et al., 2001). In addition, an important block in the use of MAS for yield is the difficulty in predicting what yield QTL alleles are present in each parent of elite populations. There will likely be insufficient linkage disequilibrium to make these predictions based on all but very tightly linked markers or markers within the yield genes. These and other difficulties will likely limit the use of MAS for yield in elite populations.

9–3.5 Mining New Alleles from the Germplasm Collections with Markers

An important use of markers in the coming decade will likely be the mining of new genetic diversity from soybean germplasm collections and collections of related species. It is well known that the genetic base of elite soybean germplasm in North America is narrow (Gizlice et al., 1994; Sneller, 1994) and that there may be favorable alleles in the accessions banked in the germplasm collections that could improve elite material. The challenge is to identify which accessions in the collections might have these favorable alleles, given that, for traits like yield, the phenotypes of those accessions provide little or no clue as to their identity.

The number of genes controlling the trait being improved and the level of performance of elite germplasm for the trait are important factors in determining the difficulty in improving the trait through introgression from germplasm collections. An example is yield, which has a complex inheritance and is bred to a high level in elite germplasm. Although there are likely yield-improving alleles in the collection, plant introductions (PIs) that carry these new genes have been difficult to identify through traditional breeding methods. This is because the PIs and progeny derived from crossing a PI with an elite soybean line generally fail to exceed the performance of elite germplasm. This poor performance of progeny is caused mainly by the PI bringing more deleterious than favorable alleles to the mating.

Conversely, many disease-resistance traits, such as resistance to phytophthora or brown stem rot, have been effectively improved using germplasm collections because the elite germplasm started at a low level (susceptible) and the disease-resistance traits tend to have a relatively simple inheritance. When resistance to a new pathogen was needed and this resistance was not available in elite germplasm, breeders could simply screen germplasm, identify lines that were resistant, and incorporate the resistance genes into elite germplasm. The incorporation of resistance does become more difficult when the inheritance of the resistance is complex. Still, if each resistance allele is flanked with markers, improvement of elite lines via backcross MAS is possible. Examples of this more complex situation are discussed by Narvel et al. (2001) for insect resistance and Concibido et al. (1996b) for SCN resistance.

The mining of new genes from germplasm collections can be facilitated through the use of genetic markers. This would require the identification of PIs that have a high probability of having new, favorable alleles for the trait of interest, and the identification of markers flanking those alleles.

Genetic similarities among genotypes based on markers has been tested as a method for finding PIs with a high probability of having new, favorable alleles. Genetic diversity studies have shown that genetic markers are informative in describing diversity in soybean (Narvel et al., 2000; Li et al., 2001; Kisha et al., 1998). These surveys have shown that genetic distance estimates based on markers correlate with genetic distances estimated through other means such as pedigree analysis (Kisha et al., 1998). However, there has been little success in predicting allelic diversity per se for specific traits. An example that was mentioned previously is that genetic distance estimates based on markers were found to be a relatively poor predictor of genetic variance for yield and other agronomic traits in populations (Helms et al., 1997; Manjarrez-Sandoval et al., 1997; Kisha et al., 1997). In addition, Diers et al. (1997) surveyed sources of SCN resistance and identified sources that were genetically distant from those being used by breeders. When the genetics of the resistance from three of the most distant sources were studied in segregating populations, they found that markers were unsuccessful in identifying PIs with novel SCN-resistance genes. These distant sources were found to carry many of the same major resistance genes as current resistance sources (Diers and Arelli, unpublished). Our ability to use marker surveys to identify PIs with favorable diversity may only be successful when markers very tightly linked or within the genes of interest are used to predict genetic diversity.

Markers will likely be employed in gene mining through the direct mapping of favorable alleles from exotic germplasm. This can be especially useful in situations where the PI has a poor performance for the trait of interest. In these situations, if a favorable allele from a PI is identified in a population developed from crossing a PI and an elite line, this would be direct evidence that the PI made a positive contribution in the population. Markers linked to the gene can then be used in MAS to incorporate the allele into elite germplasm and test the effect of the allele in diverse genetic backgrounds.

A method used to identify favorable alleles from exotic germplasm in numerous species is the advanced backcross method, which was proposed by Tanksley and Nelson (1996). In the advanced backcross method, backcross populations are developed with an elite recurrent parent and an unadapted donor parent. These populations are evaluated both for genetic markers and phenotypically for the traits of interest. Because backcross populations are evaluated, this method is especially useful when the exotic parent has poor performance for the traits of interest in addition to allowing the evaluation of the trait in a relatively homogeneous background. The advanced backcross method has been used successfully to identify favorable alleles from exotic germplasm that improve yield and fruit quality in tomato (*Lycopersicon esculentum* L.) (de Vicente and Tanksley, 1993; Tanksley et al., 1996; Fulton et al., 1997; Bernacchi et al., 1998) and to map QTL alleles from wild relatives of rice (*Oryza sativa* L.) that enhance rice yield (Xiao et al., 1996, 1998; Moncada et al., 2001). This method was recently employed by Concibido et al. (2002) to map a QTL allele from *G. soja* that had a positive effect on seed yield in soybean. They were able to confirm this QTL in a second population in the background of the soybean recurrent parent. However, when they tested this gene in additional soybean backgrounds, they found that the allele had a positive effect in some genetic backgrounds but not in others.

An example of marker-based introgression of an allele from exotic germplasm is described in Sebolt et al. (2000). Two protein QTL were originally mapped by Diers et al. (1992b) in a population developed by crossing a *G. soja* PI with a *G. max* experimental line. Alleles from the *G. soja* parent for markers linked to both QTL were associated with greater protein concentration. The *G. soja* genomic segment on LG I, which had the greatest effect, was backcrossed into the background of the experimental line with the assistance of one marker linked to the gene. The BC3 F_4-derived lines that were homozygous for the *G. soja* QTL allele had a 20 g kg^{-1} greater seed protein concentration than lines homozygous for the soybean allele. They also crossed the *G. soja* QTL into the backgrounds of two high-yielding cultivars and one high protein experimental line. They found a significant ($P <$ 0.0001) increase in protein concentration associated with the *G. soja* QTL in the former two populations but not the population derived from the high protein experimental line. The lack of a significant effect in the high protein concentration background suggests that an allele with an effect similar to the one from *G. soja* was already in the background of the experimental line. These experiments demonstrate the usefulness of markers in introgressing genes from exotic sources but also the fact that "mined" alleles may already be present in elite germplasm.

9–4 CONCLUSIONS

Soybean scientists are positioned to increase the rate of genetic gain in seed yield, pest resistance, and soybean seed composition by application of new conventional and molecular breeding tools. The increased mechanization of field plot equipment, use of progeny row yield evaluations, new computer software programs to manage all aspects of a breeding program, along with the potential for development of hybrid cultivars should contribute to this progress. The successful incorporation of MAS for SCN and other disease resistance traits into the breeding program has already been accomplished in several private-sector breeding organizations and MAS has already been used to develop three backcross derived cultivars in a public-sector breeding program. In the next decade we should see many additional applications of MAS in soybean breeding. The development of single nucleotide polymorphisms (SNP) markers linked to genes of interest should greatly accelerate the utilization of MAS for these applications and more complex soybean traits.

The challenge for soybean breeders will be the efficient and effective introgression of the new conventional and molecular technologies into their existing cultivar development programs. Based on the availability of these new conventional and molecular technologies along with the availability of new technologies already under development, we predict a substantial increase in the rate of genetic gain for economically important soybean traits in the next decade.

REFERENCES

Akkaya, M.S., R.C. Shoemaker, J.E. Specht, A.A. Bhagwat, and P.B. Cregan. 1995. Integration of simple sequence repeat DNA markers into a soybean linkage map. Crop Sci. 35:1439–1445.

Anonymous. 1999. Developing hybrid soybean seed. Plant breeding news. ed. 103. 28 July. FAO, Rome.

Apuya, N.R., B.L. Frazier, P. Keim, E.J. Roth, and K.G. Lark. 1988. Restriction fragment length polymorphisms as genetic markers in soybean, *Glycine max* (L.) Merrill. Theor. Appl. Genet. 75:889–901.

Arahana, V.S., G.L. Graef, J.E. Specht, J.R. Steadman, and K.M. Eskridge. 2001. Identification of QTLs for resistance to *Sclerotinia sclerotiorum* in soybean. Crop Sci. 41:180–188.

Bachman, M.S., J.P. Tamulonis, C.D. Nickell, and A.F. Bent. 2001. Molecular markers linked to brown stem rot resistance genes, *Rbs1* and *Rbs2*, in soybean. Crop Sci. 41:527–535.

Bailey, M.A., M.A.R. Mian, T.E. Carter, Jr., D.A. Ashley, and H.R. Boerma. 1997. Pod dehiscence of soybean: Identification of quantitative trait loci. J. Hered. 88:152–154.

Beavis, W.D. 1998. QTL analyses: Power, precision and accuracy. *In* A.H. Paterson (ed.) Molecular dissection of complex traits. CRC Press, Boca Raton, FL.

Beeghly, H.H., and W.R. Fehr. 1989. Indirect effects of recurrent selection for Fe efficiency in soybean. Crop Sci. 29:640–643.

Bernacchi, D., T. Beck Bunn, Y. Eshed, J. Lopez, V. Petiard, J. Uhlig, D. Zamir, and S. Tanksley. 1998. Advanced backcross QTL analysis in tomato. I. Identification of QTLs for traits of agronomic importance from *Lycopersicon hirsutum*. Theor. Appl. Genet. 97:381–397.

Bianchi-Hall, C.M., T.E. Carter, Jr., M.A. Bailey, M.A.R. Mian, T.W. Rufty, D.A. Ashley, H.R. Boerma, C. Arellano, R.S. Hussey, and W.A. Parrott. 2000. Aluminum tolerance associated with quantitative trait loci derived from soybean PI 416937. Crop Sci. 40:538–545.

Boerma, H.R., and R.L. Cooper. 1975. Effectiveness of early generation yield selection of heterogeneous lines in soybeans. Crop Sci. 15:313–315.

Boerma, H.R., R.S. Hussey, D.V. Phillips, E.D. Wood, G.B. Rowan, and S.L Finnerty. 1997. Registration of 'Benning' soybean. Crop Sci. 37:1982.

Bravo, J.J., W.R. Fehr, G.D. Welke, E.G. Hammond, and S.R. Cianzio. 1999. Family and line selection for elevated palmitate of soybean. Crop Sci. 39:679–682.

Brummer, E.C., G.L. Graef, J. Orf, J.R. Wilcox, and R.C. Shoemaker. 1997. Mapping QTL for seed protein and oil content in eight soybean populations. Crop Sci. 37:370–378.

Brummer, E.C., A.D. Nickell, J.R. Wilcox, and R.C. Shoemaker. 1995. Mapping the *Fan* locus controlling linolenic acid content in soybean oil. J. Hered. 86(3):245–247.

Burton, J.W. 1987. Quantative genetics: Results relevant to soybean breeding. p. 211–247. *In* J.R. Wilcox (ed.) Soybean: Improvement, production, and uses. 2nd ed. Agron. Monogr. 16. ASA, CSSA, and SSSA, Madison, WI.

Burton, J.W. 1997. Soyabean (*Glycine max* (L.) Merr.) Field Crops Res. 53:171–186.

Burton, J.W., E.M.K. Koinange, and C.A. Brim. 1990. Recurrent selfed progeny selection for yield in soybean using genetic male sterility. Crop Sci. 30:1222–1226.

Byron, D.F., and J.H. Orf. 1991. Comparison of three selection procedures for development of early-maturing soybean lines. Crop Sci. 31:656–660.

Caetano-Anolles, G., B.J. Bassam, and P.M. Gresshoff. 1991. DNA amplification fingerprinting: A strategy for genome analysis. Plant Mol. Biol. Rep. 9:294–307.

Carter, T.E., Jr., R.L. Nelson, C.H. Sneller, Z. Cui. 2004. Genetic diversity in soybean. p. 303–416. *In* H.R. Boerma and J.E. Specht (ed.) Soybeans: Improvement, production, and uses. 3rd ed. Agron. Monogr. 16. ASA, CSSA, and SSSA, Madison, WI.

Carver, B.F., J.W. Burton, T.E. Carter, Jr., and R.F. Wilson. 1986a. Response to environmental variation of soybean lines selected for altered unsaturated fatty acid composition. Crop Sci. 26:1176–1181.

Carver, B.F., J.W. Burton, R.F. Wilson, and T.E. Carter, Jr. 1986b. Cumulative response to various recurrent selection schemes in soybean: Oil quality and correlated agronomic traits. Crop Sci. 26:853–858.

Chang, S.J.C., T.W. Doubler, V.Y. Kilo, J. Abu-Thredeith, R. Prabau, V. Freire, R. Suttner, J. Klein, M.E. Schmidt, P.T. Gibson, and D.A. Lightfoot. 1997. Association of loci underlying field resistance to soybean sudden death syndrome (SDS) and cyst nematode (SCN) race 3. Crop Sci. 37:965–971.

Chang, S.J.C., T.W. Doubler, V.Y. Kilo, R. Suttner, J. Klein, M.E. Schmidt, P.T. Gibson, and D.A. Light-foot. 1996. Two additional loci underlying durable field resistance to soybean sudden death syndrome (SDS). Crop Sci. 36:1684–1688.

Cianzio, S.R., and B.K. Voss. 1994. Three strategies for population development in breeding high yielding soybean cultivars with improved iron efficiency. Crop Sci. 34:355–359.

Cober, E.R., and H.D. Voldeng. 2000. Developing high-protein, high yield soybean populations and lines. Crop Sci. 40:39–42.

Concibido, V.C., R.L. Denny, S.R. Boutin, R. Houtea, J.H. Orf, and N.D. Young. 1994. DNA marker analysis of loci underlying resistance to soybean cyst nematode (*Heterodera glycines* Ichinohe). Crop Sci. 34:240–246.

Concibido, V.C., R.L. Denny, D.A. Lange, J.H. Orf, and N.D. Young. 1996a. RFLP mapping and marker-assisted selection of soybean cyst nematode resistance in PI 209332. Crop Sci. 36:1643–1650.

Concibido, V.C., D.A. Lange, R.L. Denny. J.H. Orf, and N.D. Young. 1997. Genome mapping of soybean cyst nematode resistance genes in Peking, PI90763, and PI88788 using DNA markers. Crop Sci. 37:258–264.

Concibido, V.C., B.L. Vallee, P. Mclaird, N. Pineda, J. Meyer, L. Hummel, J. Yang, K. Wu, and X. De-lannay. 2002. Introgression of a quantitative trait locus for yield from *Glycine soja* into commercial soybean cultivars. Theor. Appl. Genet. 106:575–582.

Concibido, V.C., N.D. Young, D.A. Lange, R.L. Denny, D. Danesh, and J.H. Orf. 1996b. Targeted comparative genome analysis and qualitative mapping of a major partial-resistance gene to the soybean cyst nematode. Theor. Appl. Genet. 93:234–241.

Cooper, R.L. 1990. Modified early generation testing procedure for yield selection in soybean. Crop Sci. 30:417–419.

Coryell, V.H., H. Jessen, J.M. Schupp, D. Webb, and P. Keim. 1999. Allele-specific hybridization markers for soybean. Theor. Appl. Genet. 98:690–696.

Cox, T.S., Y.T. Kiang, M.B. Gorman, and D.M. Rodgers. 1985. Relationship between coefficient of parentage and genetic similarity indicies in the soybean. Crop Sci. 25:529–532.

Cregan, P.B.,T. Jarvik, A.L. Bush, K.G. Lark, R.C. Shoemaker, A.L. Kahler, T.T. vanToai, D.G. Lohnes, J. Chung, and J.E. Specht. 1999a. An integrated genetic linkage map of the soybean. Crop Sci. 39:1464–1490.

Cregan, P.B., J. Mudge, E.W. Fickus, D. Danesh, R. Denny, and N.D. Young. 1999b. Two simple sequence repeat markers to select for soybean cyst nematode resistance conditioned by the *rhg1* locus. Theor. Appl. Genet. 99:811–818.

Csanadi, G., J. Vollmann, G. Stift, and T. Lelley. 2001. Seed quality QTLs identified in a molecular map of early maturing soybean. Theor. Appl. Genet. 103:912–919.

Degago, Y., and C.E. Caviness. 1987. Seed yield of soybean bulk populations grown for 10 to 18 years in two environments. Crop Sci. 27:207–210.

Delannay, X., T.T. Bauman, D.H. Beighly, M.J. Buettner, H.D. Coble, M.S. DeFelice, C.W. Dertling, T.J. Diedrick, J.C. Griffin, G.S. Hagood et al. 1995. Yield evaluation of glyphosate-tolerant soybean line after treatment with glyphosate. Crop Sci. 35:1461–1467.

Demirbas, A., B.G. Rector, D.G. Lohnes, R.J. Fioritto, G.L. Graef, P.B. Cregan, R.C. Shoemaker, and J.E. Specht. 2001. Simple sequence repeat markers linked to the soybean *Rps* genes for phytophthora resistance. Crop Sci. 41:1220–1227.

de Vicente, M.C., and S.D. Tanksley. 1993. QTL analysis of transgressive segregation in an interspecific tomato cross. Genetics 134:585–596.

Diers, B.W., and P.R. Arelli. 1999. Management of parasitic nematodes of soybean through genetic resistance. p. 330–306. *In* H.E. Kauffman (ed.) World Soybean Res. Conf. VI. Proc. World Soybean Res. Conf., 6th, Chicago, IL. 4–7 Aug. 1999. Superior Printing, Champaign, IL.

Diers, B.W., S.R. Cianzio, and R.C. Shoemaker. 1992a. Possible identification of quantitative trait loci affecting iron efficiency in soybean. J. Plant Nutr.. 15:2127–2136.

Diers, B.W., P. Keim, W.R. Fehr, and R.C. Shoemaker. 1992b. RFLP analysis of soybean seed protein and oil content. Theor. Appl. Genet. 83:608–612.

Diers, B.W., L. Mansur, J. Imsande, and R.C. Shoemaker. 1992c. Mapping phytophthora resistance loci in soybean with restriction fragment length polymorphism markers. Crop Sci. 32:377–383.

Diers, B.W., and R.C. Shoemaker. 1992. Restriction fragment length polymorphism analysis of soybean fatty acid content. JAOCS 69:1242–1244.

Diers, B.W., H.T. Skorupska, A.P. Rao-Arelli, and S.R. Cianzio. 1997. Genetic relationships among soybean plant introductions with resistance to soybean cyst nematodes. Crop Sci. 37:1966–1972.

Falconer, D.S. 1981. Introduction to quantitative genetics. 2nd ed. Longman Inc., New York.

Fatmi, A., D.B. Wagner, and T.W. Pfeiffer. 1992. Intermating schemes used to synthesize population are equal in genetic consequences. Crop Sci. 32:89–93.

Fehr, W.R. 1987. Breeding methods for cultivar development. p. 249–293. *In* J.R. Wilcox (ed.) Soybean: Improvement, production, and uses. 2nd ed. Agron. Monogr. 16. ASA, CSSA, and SSSA, Madison, WI.

Fehr, W.R., C.E. Caviness, D.T. Burmood, and J.S. Pennington. 1971. Stage of development descriptions for soybeans, *Glycine max* (L.) Merrill. Crop Sci. 11:929–931.

Frisch, M., M. Bohn, and A.E. Melchinger. 1999. Minimum sample size and optimal positioning of flanking markers in marker-assisted backcrossing for transfer of a target gene. Crop Sci. 39:967–975.

Fulton, T.M., T. Beck Bunn, D. Emmatty, Y. Eshed, J. Lopez, V. Petiard, J. Uhlig, D. Zamir, and S.D. Tanksley. 1997. QTL analysis of an advanced backcross of *Lycopersicon peruvianum* to the cultivated tomato and comparisons with QTLs found in other wild species. Theor. Appl. Genet. 95:881–894.

Gimelfarb, A., and R. Lande. 1994a. Simulation of marker-assisted selection in hybrid populations. Genet. Res. 63:39–47.

Gimelfarb, A., and R. Lande. 1994b. Simulation of marker-assisted selection for non-additive traits. Genet. Res. 64:127–136.

Gimelfarb, A., and R. Lande. 1995. Marker-assisted selection and marker-QTL associations in hybrid populations. Theor. Appl. Genet. 91:522–528.

Gizlice, Z., T.E. Carter, Jr., and J.W. Burton. 1994. Genetic base for North American public soybean cultivars released between 1947 and 1988. Crop Sci. 34:1143–1147.

Guimaraes, E.P., and W.R. Fehr. 1989. Alternative strategies of recurrent selection for seed yield of soybean. Euphyticia 40:111–120.

Hanson, W.D. 1992. Phenotypic recurrent selection for modified reproductive period in soybean. Crop Sci. 32:968–972.

Heer, J.A., H.T. Knap, R. Mahalingam, E.R. Shipe, P.R. Arelli, and B.F. Matthews. 1998. Molecular markers for resistance to *Heterodera glycines* in advanced soybean germplasm. Mol. Breed. 4:359–367.

Hegstad, J.M., G. Bollero, and C.D. Nickell. 1999. Potential of using plant row yield trials to predict soybean yield. Crop Sci. 39:1671–1675.

Helms, T.C., and J.H. Orf. 1998. Protein, oil and yield of soybean lines selected for increased protein. Crop Sci. 38:707–711.

Helms, T.C., J.H. Orf, and R.A. Scott. 1995. Nearest neighbor-adjusted means as a selection criterion within two soybean populations. Can. J. Plant Sci. 75:857–863.

Helms, T., J. Orf, G. Vallad, and P. McClean. 1997. Genetic variance, coefficient of parentage, and genetic distance of six soybean populations. Theor. Appl. Genet. 94:20–26.

Henderson, C.R. 1975. Best linear unbiased estimation and prediction under a selection model. Biometrics 31:423–477.

Hintz, R.W., W.R. Fehr, and S.R. Cianzio. 1987. Population development for selection of high-yielding soybean cultivars with resistance to iron-deficiency chlorosis. Crop Sci. 22:433–434.

Hnetkovsky, N., S.J.C. Chang, T.W. Doubler, P.T. Gibson, and D.A. Lightfoot. 1996. Genetic mapping of loci underlying field-resistance to soybean sudden death syndrome (SDS). Crop Sci. 36:393–400.

Holbrook, C.C., J.W. Burton, and T.E. Carter, Jr. 1989. Evaluation of recurrent restricted index selection for increasing yield while holding seed protein constant in soybean. Crop Sci. 29:324–329.

Hospital, F., C. Chevalet, and P. Mulsant. 1992. Using markers in gene introgression breeding programs. Genetics 132:1199–1210.

Iqbal, M.J., K. Meksem, V.N. Njiti, M.A. Kassem, and D.A. Lightfoot. 2001. Microsatellite markers identify three additional quantitative trait loci for resistance to soybean sudden-death syndrome (SDS) in Essex x Forrest RILs. Theor. Appl. Genet. 102:187–192.

Jensen, N.F. 1970. A diallele selective mating system for cereal breeding. Crop Sci. 10:629–635.

Keim, P., B.W. Diers, T.C. Olson, and R.C. Shoemaker. 1990a. RFLP mapping in soybean: Association between marker loci and variation in quantitative traits. Genetics 126:735–742.

Keim, P., B.W. Diers, and R.C. Shoemaker. 1990b. Genetic analysis of soybean hard seededness with molecular markers. Theor. Appl. Genet. 79:465–469.

Keim, P., R.C. Shoemaker, and R.G. Palmer. 1989. Restriction fragment length polymorphism diversity in soybean. Theor. Appl. Genet. 77:786–792.

Kenworthy, W.J. 1980. Strategies for introgressing exotic germplasm in breeding programs. p. 217–233. *In* F.T. Corbin (ed.) Proc. World Soybean Res. Conf. II, Raleigh, NC. Westview Press, Boulder, CO.

Kim, H.S., and B.W. Diers. 2000. Inheritance of partial resistance to sclerotinia stem rot in soybean. Crop Sci. 40:55–61.

Kisha, T.J., B.W. Diers, J.M. Hoyt, and C.H. Sneller. 1998. Genetic diversity among soybean plant introductions and North American germplasm. Crop Sci. 38:1669–1680.

Kisha, T.J., C.H. Sneller, and B.W. Diers. 1997. Relationship between genetic distance among parents and genetic variance in populations of soybean. Crop Sci. 37:1317–1325.

Knapp, S.J. 1998. Marker-assisted selection as a strategy for increasing the probability of selecting superior genotypes. Crop Sci. 38:1164–1174.

Lande, R., and R. Thompson. 1990. Efficiency of marker-assisted selection in the improvement of quantitative traits. Genetics 124:743–756.

Lee, D.J., C.A. Caha, J.E. Specht, and G.L. Graef. 1994. Analysis of cytoplasmic diversity in an outcrossing population of soybean. Crop Sci. 34:46–50.

Lee, M. 1995. DNA markers and plant breeding programs. Adv. Agron. 35:265–344.

Lee, S.H., M.A. Bailey, M.A.R. Mian, T.E. Carter, Jr., D.A. Ashley, R.S. Hussey, W.A. Parrott, and H.R. Boerma. 1996a. Molecular markers associated with soybean plant height, lodging, and maturity across locations. Crop Sci. 36:728–735.

Lee, S.H., M.A. Bailey, M.A.R. Mian, T.E. Carter, Jr., E.R. Shipe, D.A. Ashley, W.A. Parrott, R.S. Hussey, and H.R. Boerma.1996b. RFLP loci associated with soybean protein and oil content across populations and locations. Theor. Appl. Genet. 93:649–657.

Lee, S.H., M.A. Bailey, M.A.R. Mian, E.R. Shipe, D.A. Ashley, W.A. Parrott, R.S. Hussey, and H.R. Boerma. 1996c. Identification of quantitative trait loci for plant height, lodging, and maturity in a soybean population segregating for growth habit. Theor. Appl. Genet. 92:516–523.

Lee, S.H., K.Y. Park, H.S. Lee, E.H. Park, and H.R. Boerma. 2001. Genetic mapping of QTLs conditioning sprout yield and quality. Theor. Appl. Genet. 103:702–709.

LeRoy, A.R., S.R. Cianzio, and W.R. Fehr. 1991a. Direct and indirect selection for small seed of soybean in temperature and tropical environments. Crop Sci. 31:697–699.

LeRoy, A.R., W.R. Fehr, and S.R. Cianzio. 1991b. Introgression of genes for small seed size from *Glycine soja* into *G. max*. Crop Sci. 31:693–697.

Lewers, K.S., E.H. Crane, C.R. Bronson, J.M. Schupp, P. Keim, and R.C. Shoemaker. 1999. Detection of linked QTL for soybean brown stem rot resistance in 'BSR 101' as expressed in a growth chamber environment. Mol. Breed. 5:33–42.

Lewers, K.S., and R.G. Palmer. 1997. Recurrent selection in soybean. Plant Breed. Rev. 15:275–313.

Lewers, K.S., S.K. St. Martin, B.R. Hedges, and R.G. Palmer. 1998. Testcross evaluation of soybean germplasm. Crop Sci. 38:1143–1149.

Lewers, K.S., S.K. St. Martin, B.R. Hedges, M.P. Widrlechner, and R.G. Palmer. 1996. Hybrid soybean seed production: Comparison of three methods. Crop Sci. 36:1560–1567.

Li, Z., L. Qiu, J.A. Thompson, M.M. Welsh, and R.L. Nelson. 2001. Molecular genetic analysis of U.S. and Chinese soybean ancestral lines. Crop Sci. 41:1330–1336.

Lin, S., J.S. Baumer, D. Ivers, S.R. Cianzio, and R.C. Shoemaker. 1998. Field and nutrient solution tests measure similar mechanisms controlling iron deficiency chlorosis in soybean. Crop Sci. 38:254–259.

Lin, S., S. Cianzio, and R.C. Shoemaker. 1997. Mapping genetic loci for iron deficiency chlorosis in soybean. Mol. Breed. 3:219–229.

Mahalingam, R., and H.T. Skorupska. 1995. DNA markers for resistance to *Heterodera glycines* I. Race 3 in soybean cultivar Peking. Breed. Sci. 45:435–443.

Manjarrez-Sandoval, P., T.E. Carter, Jr., D.M. Webb, and J.W. Burton. 1997. RFLP genetic similarity estimates and coefficient of parentage as genetic variance predictors for soybean yield. Crop Sci. 37:698–703.

Mansur, L.M., A. Carriquiry, and A.P. Rao-Arelli. 1993a. Generation mean analysis of resistance to race 3 of soybean cyst nematode. Crop Sci. 33:1249–1253.

Mansur, L.M., K.G. Lark, H. Kross, and A. Oliveira. 1993b. Interval mapping of quantitative trait loci for reproductive, morphological, and seed traits of soybean (*Glycine max* L.). Theor. Appl. Genet. 86:907–913.

Mansur, L.M., J.H. Orf, K. Chase, T. Jarvik, P.B. Cregan, and K.G. Lark. 1996. Genetic mapping of agronomic traits using recombinant inbred lines of soybean. Crop Sci. 36:1327–1336.

Maughan, P.J., M.A. Saghai Maroof, and G.R. Buss. 1996. Molecular-marker analysis of seed weight: Genomic locations, gene action, and evidence for orthologous evolution among three legume species. Theor. Appl. Genet. 93:574–579.

Maughan, P.J., M.A. Saghai Maroof, and G.R. Buss. 2000. Identification of quantitative trait loci controlling sucrose content in soybean (*Glycine max*). Mol. Breed. 6:105–111.

McElroy, D. 1999. Moving agbiotech downstream. Nature Biotechnol. 17:1071–1074.

Meksem, K., P. Pantazopoulos, V.N. Njiti, L.D. Hyten, P.R. Arelli, and D.A. Lightfoot. 2001a. 'Forrest' resistance to the soybean cyst nematode is bigenic: Saturation mapping of the *Rhg1* and *Rhg4* loci. Theor. Appl Genet. 103:710–717.

Meksem, K., E. Ruben, D.L. Hyten, M.E. Schmidt, and D.A. Lightfoot. 2001b. High-throughput genotyping for a polymorphism linked to soybean cyst nematode resistance gene *Rhg4* by using Taqman probes. Mol. Breed. 7:63–71.

Mian, M.A.R, D.A. Ashley, and H.R. Boerma. 1998a. Additional QTL for water use efficiency in soybean. Crop Sci. 38: 390–393.

Mian, M.A.R., D.A. Ashley, W.K. Vencill, and H.R. Boerma. 1998b. QTLs conditioning early growth in a soybean population segregating for growth habit. Theor. Appl. Genet. 97:1210–1216.

Mian, M.A.R., M.A. Bailey, D.A. Ashley, R. Wells, T.E. Carter, Jr., W.A. Parrott, and H.R. Boerma. 1996a. Molecular markers associated with water use efficiency and leaf ash in soybean. Crop Sci. 36:1252–1257.

Mian, M.A.R., M.A. Bailey, J.P. Tamulonis, E.R. Shipe, T.E. Carter, Jr., W.A. Parrott, D.A. Ashley, R.S. Hussey, and H.R. Boerma. 1996b. Molecular markers associated with seed weight in two soybean populations. Theor. Appl. Genet. 93:1011–1016.

Mian, M.A.R., E.R. Shipe, J. Alvernaz, J.D. Mueller, D.A. Ashley, and H.R. Boerma. 1997. RFLP analysis of chlorimuron ethyl sensitivity in soybean. J. Hered.. 88:38–41.

Mian, M.A.R., T. Wang, D.V. Phillips, J. Alvernaz, and H.R. Boerma. 1999. Molecular mapping of the *Rcs3* gene for resistance to frogeye leaf spot in soybean. Crop Sci. 39:1687–1691.

Mian, M.A.R., R. Wells, T.E. Carter, Jr., D.A. Ashley, and H.R. Boerma. 1998c. RFLP tagging of QTLs conditioning specific leaf weight and leaf size in soybean. Theor. Appl. Genet. 96:354–360.

Moncada, P., C.P. Martinez, J. Borrero, M. Chatel, H. Gauch, Jr., E. Guimaraes, J. Tohme, and S.R. McCouch. 2001. Quantitative trait loci for yield and yield components in an *Oryza sativa* x *Oryza rufipogon* BC2F2 population evaluated in an upland environment. Theor. Appl. Genet. 102:41–52.

Morgante, M., J. Rafalski, P. Biddle, S. Tingey, and A.M. Olivieri. 1994. Genetic mapping and variability of seven soybean simple sequence repeat loci. Genome 37:763–769.

Mudge, J., P.B. Cregan, J.P. Kenworthy, W.J. Kenworthy, J.H. Orf, and N.D. Young. 1997. Two microsatellite markers that flank the major soybean cyst nematode resistance locus. Crop. Sci. 37:1611–1615.

Narvel, J.M., W.R. Fehr, W. Chu, D. Grant, and R.C. Shoemaker. 2000. Simple sequence repeat diversity among soybean plant introductions and elite genotypes. Crop Sci. 40:1452–1458.

Narvel, J.M., D.R. Walker, B.C. Rector, J.N. All, W.A. Parrott, and H.R. Boerma. 2001. A retrospective DNA marker assessment of the development of insect resistant soybean. Crop Sci. 41:1931–1939.

Nelson, R.L. 1988. Response in selection for time of flowering in soybeans. Crop Sci. 28:623–626.

Njiti, V.N., K. Meksem, J.J. Iqbal, J.E. Johnson, K.F. Zobrist, V.Y. Kilo, and D.A. Lightfoot. 2002. Common loci underlie field resistance to soybean sudden death symdrome in Forrest, Pyramid, Essex, and Douglas. Theor. Appl. Genet. 104:294–300.

Orf, J.H., K. Chase, F.R. Adler, L.M. Mansur, and K.G. Lark. 1999a. Genetics of soybean agronomic traits: II. Interactions between yield quantitative trait loci in soybean. Crop Sci. 39:1652–1657.

Orf, J.H., T. Jarvik, L.M. Mansur, P.B. Cregan, F.R. Adler, and K.G. Lark. 1999b. Genetics of soybean agronomic traits: I. Comparison of three related recombinant inbred populations. Crop Sci. 39:1642–1651.

Padgette, S.R., K.H. Kolacz, X. Delannay, D.B. Re, B.J. LaVallee, C.N. Tinius, W.K. Rhodes, Y.I. Otero, G.E. Barry, D.A. Eichholtz, V.M. Peschke, D.L. Nida, N.B. Taylor, and G.M. Kishore. 1995. Development, identification and characterization of a glyphosate-tolerant soybean line. Crop Sci. 35:1451–1461.

Palmer, R.G., J. Gai, H. Sun, and J.W. Burton. 2001. Production and evaluation of hybrid soybean. Plant Breed. Rev. 21:263–307.

Panter, D.M., and F.L. Allen. 1995a. Using best linear unbiased predictions to enhance breeding for yield in soybean. I. Choosing parents. Crop Sci. 35:397–405.

Panter, D.M., and F.L. Allen. 1995b. Using best linear unbiased predictions to enhance breeding for yield in soybean. II. Selection of superior crosses from a limited number of yield trials. Crop Sci. 35:405–410.

Pazdernik, D.L., L.L. Hardman, J.H. Orf, and F. Clotaire. 1996. Comparison of field methods for selection of protein and oil content in soybean. Can. J. Plant Sci. 76:721–725.

Pfeiffer, T.W., and D.L. Pilcher. 1989. Hill vs. row plots in predicting full season and late planted soybean yield. Crop Sci. 29:286–288.

Piper, T.E., and W.R. Fehr. 1987. Yield improvement in a soybean population by utilizing alternative strategies of recurrent selection. Crop Sci. 27:172–178.

Plantevin, A., J. Dayde, and R. Ecochard. 1987. Effect of geographic location on earliness of an SSD population of soybean. Eurosoya 5:57–63.

Qiu, B.X., P.R. Arelli, and D.A. Sleper. 1999. RFLP markers associated with soybean cyst nematode resistance and seed composition in a 'Peking' x 'Essex' population. Theor. Appl. Genet. 98:356–364.

Rector, B.G., J.N. All, W.A. Parrott, and H.R. Boerma. 1998a. Identification of molecular markers linked to quantitative trait loci for soybean resistance to corn earworm. Theor. Appl. Genet. 96:786–790.

Rector, B.G., J.N. All, W.A. Parrott, and H.R. Boerma. 1998b. Inheritance of insect resistance in three wild soybean genotypes. Theor. Appl. Genet. 96:790–794.

Rector, B.G., J.N. All, W.A. Parrott, and H.R. Boerma. 1999. Quantitative trait loci for antixenosis resistance to corn earworm in soybean. Crop Sci. 39:531–538.

Rector, B.G., J.N. All, W.A. Parrott, and H.R. Boerma. 2000. Quantitative trait loci for antibiosis resistance to corn earworm in soybean. Crop Sci. 40:233–238.

Reyna, N., and C.H. Sneller. 2001. Evaluation of marker-assisted introgression of yield QTL alleles into adapted soybean. Crop Sci. 41:1317–1321.

Ribaut, J., and D. Hoisington. 1998. Marker-assisted selection: New tools and strategies. Trends Plant Sci. 3:236–239.

Rongwen, J., M.S. Akkaya, A.A. Bhagwat, U. Lavi, and P.B. Cregan. 1995. The use of microsatellite DNA markers for soybean genotype identification. Theor. Appl. Genet. 90:43–48.

Rose, J.L., D.G. Butler, and M.J. Ryley. 1992. Yield improvement in soybeans using recurrent selection. Aust. J. Agric. Res. 43:135–144.

Salado-Navarro, L.R., T.R. Sinclair, and K. Hinson. 1993. Changes in yield and seed growth traits in soybean cultivars released in the southern USA from 1945 to 1983. Crop Sci. 33:1204–1209.

Saxe, L.A., C. Clark, S.F. Lin, and T.A. Lumpkin. 1996. Mapping the pod-shattering trait in soybean. Soybean Genetic Newsl. 23:250–253.

Sebolt, A.M., R.C. Shoemaker, and B.W. Diers. 2000. Analysis of a quantitative trait locus allele from wild soybean that increases seed protein concentration in soybean. Crop Sci. 40:1438–1444.

Sedcole, J.R. 1977. Number of plants necessary to recover a trait. Crop Sci. 17:667–668.

Smith, J.R., and R.L. Nelson. 1986. Selection for seed filling period in soybean. Crop Sci. 26:466–469.

Smith, J.R., and R.L. Nelson. 1987. Predicting yield from early generation estimates of reproductive growth periods in soybean. Crop Sci. 27:471–474.

Sneller, C.H. 1994. Pedigree analysis of elite soybean lines. Crop Sci. 34:1515–1522.

Specht, J.E., K. Chase, M. Macrander, G.L. Graef, J. Chung, J.P. Maxwell, M. German, J.H. Orf, and K.G. Lark. 2001. Soybean response to water: A QTL analysis of drought tolerance. Crop Sci. 41:493–509.

Specht, J.E., and G.L. Graef. 1990. Breeding methodologies for chickpea: New avenues to greater productivity. p. 217–223. In B.J. Walby and S.D. Hall (ed.) Chickpea in the nineties. Proc. 2nd Int. Workshop on Chickpea Improvement. ICRISAT Center, India.

Specht, J.E., and G.L. Graef. 1992. Registration of soybean germplasm SG1E6. Crop Sci. 32:1080–1082.

Specht, J.E., D.J. Hume, and S.V. Kumundini. 1999. Soybean yield potential—A genetic and physiological perspective. Crop Sci. 39:1560–1570.

Specht, J.E., J.H. Williams, W.J. Kenworthy, J.H. Orf, D.G. Helsel, and S.K. St. Martin. 1985. Registration of SG1 germplasm. Crop Sci. 25:717–718.

St. Martin, S.K., B.W. Dye, and B.A. McBlain. 1990. Use of hill and short row plots for selection of soybean genotypes. Crop Sci. 30:74–79.

St. Martin, S.K., and X. Futi. 2000. Genetic gain in early stages of a soybean breeding program. Crop Sci. 40:1559–1564.

St. Martin, S.K., K.S. Lewers, R.G. Palmer, and B.R. Hedges. 1996. A testcross procedure for selecting exotic strains to improve pure-line cultivars in predominantly self-fertilizing species. Theor. Appl. Genet. 92:78–82.

St. Martin, S.K., and B.A. McBlain. 1991. Procedure to estimate genetic gain by stages in multistage testing programs. Crop Sci. 31:1367–1369.

Stuber, C.W., M. Polacco, and M.L. Senior. 1999. Synergy of empirical breeding, marker-assisted selection, and genomics to increase crop yield potential. Crop Sci. 39:1571–1583.

Tamulonis, J.P., B.M. Luzzi, R.S. Hussey, W.A. Parrott, and H.R. Boerma. 1997a. DNA markers associated with Javanese root-knot resistance in soybean. Crop Sci. 37:783–788.

Tamulonis, J.P., B.M. Luzzi, R.S. Hussey, W.A. Parrott, and H.R. Boerma. 1997b. DNA marker analysis of loci conferring resistance to peanut root-knot nematode in soybean. Theor. Appl. Genet. 95:664–670.

Tamulonis, J.P., B.M. Luzzi, R.S. Hussey, W.A. Parrott, and H.R. Boerma. 1997c. RFLP mapping of resistance to southern root-knot nematode in soybean. Crop Sci. 37:1903–1909.

Tanksley, S.D., S. Grandillo, T.M. Fulton, D. Zamir, Y. Eshed, V. Petiard, J. Lopez, and T. Beck Bunn. 1996. Advanced backcross QTL analysis in a cross between elite processing line of tomato and its wild relative *L. pimpinellifolium*. Theor. Appl. Genet. 92:213–224.

Tanksley, S.D., and J.C. Nelson. 1996. Advanced backcross QTL analysis: A method for the simultaneous discovery and transfer of valuable QTLs from unadapted germplasm into elite breeding lines. Theor. Appl. Genet. 92:191–203.

Tanksley, S.D., N.D. Young, A.H. Paterson, and M.W. Bonierbale. 1989. RFLP mapping in plant breeding: New tools for an old science. Biotechnology 7:257–264.

Terry, L.I., K. Chase, T. Jarvik, J. Orf, L. Mansur, and K.G. Lark. 2000. Soybean quantitative trait loci for resistance to insects. Crop Sci. 40:375–382.

Terry, L.I., K Chase, J. Orf, T. Jarvik, L. Mansur, and K.G. Lark. 1999. Insect resistance in recombinant inbred soybean lines derived from non-resistant parents. Entomol. Exp. Appl. 91:465–476.

Tinius, C.N., J.W. Burton, and T.E. Carter, Jr. 1991. Recurrent selection for seed size in soybeans. I. Response to selection in replicate populations. Crop Sci. 31:1137–1141.

Tinius, C.N., J.W. Burton, and T.E. Carter, Jr. 1992. Recurrent selection for seed size in soybean. II. Indirect effects on seed growth rate. Crop Sci. 32:1480–1483.

Tinius, C.N., J.W. Burton, and T.E. Carter, Jr. 1993. Recurrent selection for seed size in soybean. III. Indirect effects on seed composition. Crop Sci. 33:959–962.

Troller, C., and J.F.F. de Toledo. 1996. Using F_3 generation for predicting the breeding potential of soybean crosses. Braz. J. Genet. 19:289–294.

Uphoff, M.D., W.R. Fehr, and S.R. Cianzio. 1997. Genetic gain for soybean seed yield by three recurrent selection methods. Crop Sci. 37:1155–1158.

Van Toai, T.T., S.K. St. Martin, K. Chase, G. Boru, V. Schnipke, A.F. Schmitthenner, and K.G. Lark. 2001. Identification of a QTL associated with tolerance of soybean to soil waterlogging. Crop Sci. 41:1247–1252.

Vierling, R.A., J. Faghihi, V.R. Ferris, and J.M. Ferris. 1996. Association of RFLP markers with loci conferring broad-based resistance to soybean cyst nematode (*Heterodera glycines*). Theor. Appl. Genet. 92:83–86.

Voldeng, H.D., E.R. Cober, D.J. Hume, C. Gillard, and M.J. Morrison. 1997. Fifty-eight years of genetic improvement of short-season soybean cultivars. Crop Sci. 37:428–431.

Walker, A.K., and A.F. Schmitthenner. 1984. Recurrent selection for tolerance to phytophthora root rot in soybean. Crop Sci. 24:495–497.

Wang, D., Arelli, P.R., R.C. Shoemaker, and B.W. Diers. 2001. Loci underlying resistance to race 3 of soybean cyst nematode in *Glycine soja* plant introduction 468916. Theor. Appl. Genet. 103:561–566.

Webb, D.M., B.M. Baltazar, A.P. Rao-Arelli, J. Schupp, K. Clayton, P. Keim, and W.D. Beavis. 1995. Genetic mapping of soybean cyst nematode race-3 resistance loci in soybean PI 437654. Theor. Appl. Genet. 91:574–581.

Wehrmann, V.K., W.R. Fehr, S.R. Cianzio, and J.F. Cavins. 1987. Transfer of high seed protein to high-yielding soybean cultivars. Crop Sci. 27:927–931.

Werner, B.K., and J.R. Wilcox. 1990. Recurrent selection for yield in *Glycine max* using genetic male sterility. Euphytica 50:19–26.

Wilcox, J.R. 1998. Increasing seed protein in soybean with eight cycles of recurrent selection. Crop Sci. 38:1536–1540.

Wilcox, J.R. 2001. Sixty years of improvement in publicly developed elite soybean lines. Crop Sci. 41:1711–1716.

Williams, J.G.K., A.R. Kubelik, J.K. Livak, J.A. Rafalski, and S.V. Tingey. 1990. DNA polymorphisms amplified by arbitrary primers are useful as genetic markers. Mol. Gen. Genet. 244:638–645.

Wittaker, J.C., R.N. Curnow, C.S. Haley, and R. Thompson. 1995. Using marker-maps in marker-assisted selection. Genet. Res. 66:255–265.

Xiao, J., S. Grandillo, S.N. Ahn, S.R. McCouch, S.D. Tanksley, J. Li, L. Yuan, J.H. Xiao, J.M. Li, and L.P. Yuan. 1996. Genes from wild rice improve yield. Nature (London) 384:223–224.

Xiao, J., J. Li, S. Grandillo, S. Ahn, L. Yuan, S.D. Tanksley, and S.R. McCouch 1998. Identification of trait-improving quantitative trait loci alleles from a wild rice relative, *Oryza rufipogon*. Genetics 150:899–909.

Xu, H., and J.R. Wilcox. 1992. Recurrent selection for maturity and percent seed protein in *Glycine max* based on 50 evaluations. Euphyticia 62:51–57.

Yue, P., P.R. Arelli, and D.A. Sleper. 2001a. Molecular characterization of resistance to *Heterodera glycines* in soybean PI 438489B. Theor. Appl. Genet. 102:921–928.

Yue, P., D.A. Sleper, and P.R. Arelli. 2001b. Mapping resistance to multiple races of *Heterodera glycines* in soybean PI 89772. Crop Sci. 41:1589–1595.

Zhang, W., and C. Smith. 1993. Simulation of marker-assisted selection utilizing linkage disequilibrium: The effects of several additional factors. Theor. Appl. Genet. 86:492–496.

10 Managing Inputs for Peak Production

LARRY G. HEATHERLY

USDA-ARS
Crop Genetics and Production Research Unit
Stoneville, MS

ROGER W. ELMORE

University of Nebraska
Lincoln, Nebraska

Successful soybean [*Glycine max* (L.) Merr.] production requires the integration of inputs into a system that contains only those items necessary to optimize the amount of a quality product or net return. Inputs such as seed, pesticides, fertilizers, labor, machinery, and fuel that are basic to any management system, plus costs associated with financing, land (rent or ownership), and irrigation where used, must all be considered and manipulated to provide the optimum opportunity for profit. The selection of a management system for a given farm or production entity must be based on local conditions such as weather, soil properties, land cost, markets, land-use restrictions, and environmental constraints.

This chapter presents discussion based on crop management research results reported since the soybean monograph chapters by Johnson (1987) and Van Doren and Reicosky (1987). Our goal is to synopsize recent research on and recommendations for specific areas of soybean management to maximize production. References that present information about a particular subject are used to suggest components of a production plan for both the southern and northern USA. Such references include recently published management guides (Honeycutt, 1996; Heatherly and Hodges, 1999; Hoeft et al., 2000). Sections on cultivar selection, tillage, soil fertility, planting practices, cropping systems, and post-planting management and harvesting are included. For reader convenience, both SI and English units of measure are often presented (rounding may cause slight disagreement between values). Where only SI units are shown, an equation for conversion to English units is given. Designations for both vegetative [fully developed unifoliolate leaves (V1) to fully developed last leaf (Vn)] and reproductive [beginning bloom (R1) to full maturity (R8)] developmental stages are used as defined by Fehr and Caviness (1977).

Soybean management often differs between the northern and southern USA. For purposes of presentation of information in this chapter, southern USA refers to the states of Alabama, Arkansas, Florida, Georgia, Louisiana, Maryland, Mississippi, North Carolina, South Carolina, Virginia, Tennessee, and Texas. The remaining soybean-producing states are referred to as the northern USA (or the Midwest). Most of the U.S. soybean production is concentrated in the northern or midwestern states; between 1995 and 2000, only 14.5 to 18% of the U.S. hectarage was in the southern states, while production in the northern states ranged from 85.2 to 89.9% of total U.S. production. Average yield in the southern USA was only 74% of the average yield for the entire USA during the same period.

10–1 CULTIVAR SELECTION

Cultivar selection is the first step to successful soybean management. New and improved cultivars are continually released to producers by both public and private soybean breeders. These cultivars are evaluated in different production environments to determine their yield potential and supplemental traits such as resistance to diseases, nematodes, and insects, as well as tolerance to commonly used herbicides. Producers should always consider annually released new cultivars given that genetic improvement in yield is occurring at the rate of about 30 kg ha^{-1} yr^{-1} (about 0.45 bu acre^{-1} yr^{-1}) (Specht et al., 1999). Results from cultivar trials conducted by both public (Table 10–1) and private institutions should be consulted each year to determine if a new cultivar offers higher yield potential than one currently being grown by the producer. Maturity and seed cost also should be considered when selecting a cultivar. With hundreds of soybean cultivars available in the USA, selection can be based on very narrow criteria to ensure that the chosen cultivar matches as many of the requirements as possible for a particular production scheme. When selecting a cultivar for special environments, it is important to ensure that unnecessary risks are not assumed. For instance, a cultivar with genetic resistance to soybean cyst nematode (SCN; *Heterodera glycines* Ichinohe) is not necessary for maximum yield potential on the fine-textured clay soils in the lower Mississippi River alluvial flood plain because populations of SCN are not maintained in these soils (Heatherly and Young, 1991).

The choice of cultivars should be based on desired plant properties or growing conditions. However, seed of public cultivars generally are less expensive than those of private nontransgenic cultivars, which in turn are cheaper than those of most transgenic cultivars. Thus, if cultivars are determined to be equal in desired traits, the choice can be based solely on seed cost.

10–1.1 Maturity Classification

Soybean cultivars are classified by maturity group (MG). The thirteen MGs are ordinarily expressed as Roman numerals (MG 000 being the earliest and MG X being the latest) that are used for identifying the region of adaptation for soybean. In this chapter, Arabic numbers will be used to accommodate fractional MG designations. Maturity Group zones represent defined areas where a cultivar is best adapted; however, this does not imply that cultivars of a specific MG can be grown only in that particular region. Cultivars of two to three MGs are often grown suc-

Table 18-1. Internet addresses of soybean cultivar trial results for U.S. states and Ontario, Canada, and parameters evaluated or measured.

State	Site address†	Yield	Pro-tein	Oil	Lodging	Mature height	Relative maturity	Seed wt.	Disease/Herbicide reaction‡	Plant traits§
Alabama	www.aces.edu/department/cotton/soybean.html	X			X	X	X		PRO; CHLORIDE	X
Arkansas	www.aragriculture.org/cropsoilwtr/soybeans/varietytests/default.asp	X			X	X	X			X
Georgia	www.griffin.peachnet.edu/swvt	X			X	X	X	X		X
Illinois	http://vt.cropsci.uiuc.edu/soybean.html	X	X	X	X	X	X		SCN; SDS; WM	X
Indiana	www.agry.purdue.edu/ext/variety.htm	X	X	X	X	X	X			
Iowa	www.agron.iastate.edu/icia/YieldTesting3.html	X	X		X	X	X		BSR; CHL; PRR; SCN; WM	X
Kansas	www.ksu.edu/kscpt/	X			X	X	X		SCN; PRR	
Kentucky	http://www.uky.edu/Ag/GrainCrops/varietytesting.htm (verified 31 July 2003)	X			X	X	X		BSR; FLS; PRR; SC; SCN; SDS; VIR	
Louisiana	www.agctr.lsu.edu/Subjects/soybean	X	X	X	X	X	X		SCN	X
Maryland	www.nrsl.umd.edu/extension/crops/soybeans	X	X	X	X	X	X	X	PRR; WM	
Michigan	www.css.msu.edu/varietytrials/	X	X	X	X	X	X			
Minnesota	www.extension.umn.edu/farm/	X	X	X	X		X	X	BSR; CHL; PRR; SCN; WM	X
Mississippi	http://msucares.com/pubs/infobulletins/ib384.pdf	X	X	X	X	X	X	X	BB; BP; BS; CLS; DM; FLS; MET; PRR; SC; SCN; SDS; VIR	X
Missouri	www.agebb.missouri.edu/index.htm	X	X		X	X	X		PRR; SCN	X
New Jersey	www.rce.rutgers.edu/pubs/pdfs/e041n.pdf	X	X	X	X	X	X			
Nebraska	http://varietytest.unl.edu/soytst/2002/index.htm	X	X	X	X	X	X	X	CHL; PRR	X
N. Carolina	www.cropsci.ncsu.edu/ovt/cotton_soy/2002/toc.htm	X	X		X	X	X			
Ohio	http://www.oardc.ohio-state.edu/soy2002/	X	X		X	X	X	X	PRR; WM	
Oklahoma	http://www.agr.okstate.edu/soybeans/variety results2002.html	X			X	X	X	X		X
Ontario	www.gov.on.ca/OMAFR/english/crops/index.html	X	X		X	X	X	X	PRR; WM	X
Pennsylvania	www.agronomy.psu.edu/Extension/Extension.html	X			X	X	X	X		
S. Carolina	cropweb.clemson.edu/	X			X	X	X			
S. Dakota	www.sdstate.edu/~wpls/http/var/vartrial.html	X	X	X	X	X	X		SCN; CHL; PRR	X
Tennessee	web.utk.edu/~taescomm/research/variety.html	X	X		X	X	X			
Texas	www.tamu-commerce.edu/coas/agscience/jjh.html	X				X	X		PSS	
Virginia	www.vaes.vt.edu/tidewater/soybean/variety.html	X			X		X			
Wisconsin	http://soybean.agronomy.wisc.edu/soyvar.htm	X			X		X		BSR; PRR; WM	X

† Websites verified on 27 Nov. 2002.
‡ BB = bacterial blight; BS = brown spot; BSR = brown stem rot; BP = bacterial pustule; CHL = chlorosis; CLS = cercospora leafspot; DM = downy mildew; FLS = frogeye leafspot; MET = metribuzin; PRR = phytophthora root rot; PRO = propanil; PSS = purple seed stain; SC = stem canker; SCN = soybean cyst nematode; SDS = sudden death syndrome; VIR = virus (SMV/BPMV); WM = white mold.
§ Includes bloom/flower/pubescence/hilum/pod wall color, seed size, shatter rating, and height of lowest pod, but not necessarily all.

cessfully at a specific site within a MG zone. The MG(s) adapted to a particular area can be determined from those that are tested in each state's cultivar trial (Table 10–1).

Plant development, from germination through the onset of flowering and on to maturity, is controlled by photoperiod and temperature (Major et al., 1975). How cultivars respond to these abiotic factors determines which MG they fall into. Soybean is a short-day plant species, because floral induction in apical and axillary meristems occurs only when days are shorter than some critical length. After floral induction occurs, temperature determines the time required for the appearance of flowers. Floral induction in southern U.S. cultivars is delayed by long days, making these cultivars too late to be grown in the northern USA. Conversely, northern U.S. cultivars flower and mature too early when grown in shorter daylengths of the southern USA. Later, we discuss reasons cultivars of a particular MG might be grown outside their region of photoperiodic adaptation.

Cultivars are often arbitrarily designated as early, mid-, or full-season (Johnson, 1987). These terms describe the relative maturity of cultivars based on the length of growing season in a given region. The early, mid-, and full-season classification is thus location-specific, since a cultivar classified as full-season in one location would be considered early season in a more southerly location. For instance, in east-central Nebraska, MG 2.0 cultivars are considered early season, MG 2.5 to 3.0 cultivars are mid-season, and MG 3.5 cultivars are considered full-season. In the southern USA, MG 3 and 4 cultivars are considered early season and are used in the early soybean production system (ESPS; approximate 4.5-mo growing season from late March/early April through mid-August), whereas cultivars in MGs 5, 6, and 7 are considered mid- to full-season and are used for plantings that encompass the previous normal-length season of 5.5 mo from early May to mid-October. Thus, it is important to have a specific growing-season length or period in mind when selecting cultivars for any region.

Soybean breeders assign a cultivar to a MG based on its adaptation to the conventional planting practices used in the region. The ESPS in the southern USA is an example of using cultivars outside their assigned MG region of adaptability. In this system, indeterminate cultivars in MGs 3 and 4 are planted in late March and April in the zones ordinarily assigned to MGs 5, 6, and 7. These cultivars begin blooming (R1) in May, start setting pods in late May to early June, and reach full seed (R6) in mid-July to early August. The reason for using this system and its requisite early-maturing cultivars is to avoid drought that can adversely affect the later-maturing, full-season cultivars that are normally assigned to the region. The later-maturing cultivars are in reproductive phases during July and August when conditions that favor drought stress are common. Conversely, northern growers may use late-maturing cultivars for forage production. However, in Minnesota, late cultivars (MGs 5, 6, and 7) did not reach R6 before frost when planted from early to late May, produced forage yields that were similar to those from adapted cultivars (MGs 1 and 2), and had lower forage quality because of the low percentage of grain (Sheaffer et al., 2001).

10–1.2 Stem Growth Habit

Cultivars in MGs 000 through 4 generally are classified as having indeterminate growth habit, whereas those in MGs 5 through 10 are classified as having determinate growth habit. Stem growth habit is governed primarily by a genetic locus (Dt_1/dt_1) whose contrasting genes give rise to an apical meristem that is (dt_1) or is not (Dt_1) florally induced when daylengths are shorter than a critical daylength (Bernard, 1972). In determinate (dt_1dt_1) plants, stem apices and axillary meristems on main-stem nodes are converted immediately from vegetative to floral meristems. Later near-simultaneous appearance of flowers occurs at all nodes plus the stem tip. The uppermost leaf is usually as large as the leaf below, but can be smaller if its vegetative development was arrested during its primordial stage by the floral induction process. In the southern USA, indeterminate cultivars generally produce lateral branches only at the lowermost nodes (Pattern 12 in Lersten and Carlson, 1987), while determinate cultivars produce lateral branches from nodes over the entire length of the main stem (Pattern 22 in Lersten and Carlson, 1987). Thus, plants of determinate cultivars grown in the southern USA usually have a bushier canopy than those of indeterminate cultivars. Hartung et al. (1981) found that the primary effect of the determinate gene (dt_1) in isolines with similar maturity when grown in Nebraska was a severe shortening of stem length that resulted in pod distribution being compressed into few nodes along the main stem. This resulted in more pods at the upper and lower portions of the shorter stems of the determinate isolines.

Determinate plants will usually have a distinct cluster of pods borne on a pronounced apical raceme. In contrast, Dt_1Dt_1 indeterminate plants bear apical meristems that are resistant to floral induction, whereas axillary meristems undergo floral induction, though not simultaneously. In general, the first flower appears at the V6 node, with floral appearance spreading downward and upward from there. The stem apices almost invariably remain vegetative, producing more leaves. However, as reproductive development proceeds, photosynthate is preferentially allocated to the developing pods, thus leading to ever smaller leaves at the stem apex. Eventually, apical stem growth slows and then ceases vegetative activity before floral induction and flower appearance. In optimum environments, the pod clusters on indeterminate plant stem tips may appear to be determinate, but the clustering is almost always due to short internodes at the stem tip.

Cooper (1981) suggested that lodging was a yield-reducing factor in traditional indeterminate cultivars. Subsequently, semi-dwarf (actually dt_1dt_1 determinate) cultivars adapted for northern U.S. latitudes have been developed and released. These cultivars achieve 80% of their main stem height by R1 and 92 to 93% of their final height within 7 d after R1 (Lin and Nelson, 1988). They generally are shorter, have fewer nodes (generally half as many), have lower pod heights on the main stem, and lodge less than indeterminate types. Stem growth habit can be modified to semideterminacy via a second genetic locus (Dt_2/dt_2). Semideterminate ($Dt_1Dt_1Dt_2Dt_2$) plants have stem tips that are less responsive to floral induction than are determinate ($dt_1dt_1\_\_$) plants, allowing more vegetaive stem growth (i.e., more nodes) because of a less abrupt conversion of the stem apex from its vegetative to reproductive state. Semideterminate cultivars lodge less than indeterminate cultivars, but have more nodes than semidwarf cultivars. A highly productive semideterminate cultivar ('NE 3001') recently was released in Nebraska (George Graef, personal communication, 2002).

Determinate cultivars have played an important role in the northern production area since the late 1970s. In high-yield environments, determinate cultivars in MGs 2 and 3 yielded better than indeterminate cultivars and yielded similar following stress that occurred during the late vegetative through reproductive phases (Elmore et al., 1987). Determinate cultivars in MGs 0 to 2 grown in Minnesota were found useful in improving lodging resistance and yield (Foley et al., 1986). Work with MG 1 and MG 2 cultivars in Ontario, Canada shows that breeding for high yield and yield stability in determinate and semideterminate cultivars is possible (Ablett et al., 1989, 1994). Foley et al. (1986) and Cober and Tanner (1995) found that the reproductive period in indeterminate cultivars of the early MGs was longer than that of the determinate cultivars. This resulted in the suggestion that indeterminate cultivars would better adjust to the effects of short-term stresses. However, others have found that cultivars of the two types have equal-length reproductive periods (Wilcox and Frankenberger, 1987; Ablett et al., 1989), or that determinate cultivars actually have a longer reproductive period (Ablett et al., 1989; Saindon et al., 1990). Thus, the effect of length of the reproductive period on performance of indeterminate vs. determinate cultivars in the early-maturing MGs is not clearcut.

Robinson and Wilcox (1998) found no association between determinate and indeterminate isolines for seed yield, suggesting that neither growth habit nor plant type per se affected seed yield. They found an absence of any interaction of determinate and indeterminate isolines with row spacings of either 0.2 m (8 in) or 0.6 m (24 in), which indicates that high-yielding lines of both determinate and indeterminate types can be identified in either row spacing. Their data indicate that genetic loci contributing to high seed yield are expressed in both plant types.

A concern among northern U.S. producers when growing determinate cultivars is the typically high positive correlation between plant height and distance from ground to lowest pod (Johnson, 1987). Beaver and Johnson (1981) indicate that pods that are within 10 cm (4 in) of the ground are subject to loss during harvest. Saindon et al. (1990) successfully isolated nondwarf (tall) MG 0 determinate soybean lines adapted to Ontario, Canada. The determinate lines were shorter than their MG 0 indeterminate sister lines, but the lowest pods of the determinate cultivar were 2.5 cm (1 in) higher off the ground. The MG 0 indeterminate and determinate lines had similar lodging and similar seed quality, but the determinate line produced the greater yield (Cober and Tanner, 1995). In a more recent comparison of indeterminate cultivars and tall determinate lines, height-to-lowest pod was greater in the determinate line. Height of plants of the two types was similar, but lodging of the determinate lines was greater and yield was lower (Cober et al., 2000). This work supports the possibility of developing early-maturing determinate genotypes with acceptable height-to-lowest pod. In a survey of Kentucky producers, average combine cutting height was 10.7 cm (4.2 in) aboveground (Grabau and Pfeiffer, 1990). This resulted in an average yield loss of about 1.4% with a range of 0 to 3.8%. Seventy percent of the fields in the survey had stubble heights of between 7.5 and 12.5 cm (3–4.9 in) following harvest. Iowa work reported by Hoeft et al. (2000) indicates yield losses of 5.4, 9.4, and 12.2% for cutting heights of 8.9, 12.7, and 16.5 cm (3.5, 5.0, and 6.5 in), respectively. It is obvious from these two cases that even small increases in height-to-lowest pod are important.

Determinate and indeterminate cultivars are known to have similar grain protein contents, and grain protein content increases from the lowest to the highest nodes (Escalante and Wilcox, 1993). It appears that the normal negative correlation between seed yield and grain protein is true for indeterminate but not necessarily for determinate lines (Wilcox and Zhang, 1997). Thus, determinacy may be needed to develop cultivars with both high yield and high grain protein. Thomison et al. (1990) reported that seed of determinate isolines compared to indeterminate isolines were more susceptible to infection by *Phomopsis longicolla* Hobbs, and germination of seed from the determinate isolines was also lower. They concluded that seed of early-maturing determinate cultivars may be more susceptible to this disease than are seed from early-maturing indeterminate cultivars if weather conditions during seed development and after maturation are conducive to its development.

Generally, cultivars in MGs 5 through 9 are grown in the southern USA. These determinate cultivars have a more uniform podset up the stalk, and generally branch more profusely up the stem than do indeterminate cultivars. Contrary to previous information (Fehr and Caviness, 1977; Johnson, 1987), determinate cultivars in the southern USA generally increase considerably in height after flowering begins. There are three or more unextended internodes and unexpanded leaves in the tissue cluster at the main stem terminal of determinate cultivars when flowering begins, and prevailing weather conditions after R1 dictate whether or not this unexpanded tissue will reach full size and increase height after R1 (L.G. Heatherly, personal communication, 2002). Appearance of lateral branches in determinate cultivars continues well after the onset of flowering; therefore, canopy development in determinate cultivars continues both vertically and laterally after beginning flowering.

Kilgore-Norquest and Sneller (2000) used near-isogenic pairs that contrasted in stem type to assess effect of stem type on performance in Arkansas environments. The indeterminate lines were taller and had greater lodging in all environments. Regression techniques determined that indeterminate growth habit is likely to confer a yield advantage over determinate growth habit in southern U.S. environments with limited growth and yield potential. Panter and Allen (1989) in Tennessee reported that determinate lines had greater yield than indeterminate lines from late plantings (early to mid-June) compared to early plantings (late April to mid-May). Parvez et al. (1989) in Florida reported the opposite response; that is, a determinate cultivar outyielded an indeterminate cultivar in mid-May plantings, while there was little difference in yield between the two in early July plantings. Ouattara and Weaver (1994), using near-isogenic lines planted in late June to early July in Alabama, found that the reproductive period for indeterminates was only 2 d (2.5%) longer than for determinates. However, indeterminates averaged 41% greater height and 21% more mainstem nodes per plant than determinates. They also found that determinate lines had better yield than indeterminate lines in a higher-yield environment, while indeterminate lines produced greater yield than determinate lines in a lower-yield environment. Kilgore-Norquest and Sneller (2000) suggested that indeterminate cultivars may be useful to fully exploit the yield potential of low-yield environments in the southern USA.

Early and late cultivars in a particular region should be managed differently because of differences in calendar days to R1. Thus, management inputs that are aligned with R1 will need to address this. For instance, guidelines for initiation of irrigation of soybean in the midsouthern USA have centered on R1. Beginning bloom or R1 for MGs 2 through 4 indeterminate cultivars used in the ESPS in the southern USA occurs sooner after planting than for MG 5 and later determinate cultivars. Thus, other criteria for irrigation initiation such as soil moisture content or tension, or cumulative soil water loss, should be used instead of the plant criterion of R1. This makes the use of irrigation scheduling models more appropriate.

10–1.3 Soil Type Effects

Soil texture affects soybean growth and development by affecting availability of water to the plant (Heatherly and Russell, 1979), and thus affects the amount of water that is in the plant to promote cell expansion and subsequent growth. On soils such as clays in the midsouthern USA that have a relatively low available water-holding capacity and low hydraulic conductivity, this may result in plants that are too short at maturity when early-maturing cultivars are planted early. On soils such as deep sandy and silt loams that have a relatively high available water-holding capacity, rapid growth of cultivars or cultivars with a long juvenile period may have increased lodging. Soils with fine texture, such as the clay soils in the alluvial flood plain of the Mississippi River and fine-textured, low lying soils in the northern USA, provide soil-water environments that are more favorable for seedling diseases such as *Phytophthora* spp. and *Pythium* spp. This results in unique stand establishment problems for susceptible cultivars or for early plantings such as those of the ESPS (Bowers and Russin, 1999). Also, low-lying soils that are subject to extended periods of saturation are a poor environment for those cultivars with poor tolerance of these conditions (Heatherly and Pringle, 1991). On the other hand, coarse-textured soils provide a more favorable environment for soybean cyst nematode (SCN) development (Young and Heatherly, 1990). Thus, selection of cultivars for these soils must consider resistance to SCN. The effect of soils with differing fertility and pH levels on cultivar selection is discussed later.

10–1.4 Cultivar Trials

Soybean cultivars are assessed in field trials conducted each year by agronomists in the soybean-producing states and regions of the USA and Canada. These trials provide yield information for cultivars grown at multiple locations within a state, plus many other details about each cultivar. The amount of information provided varies among the states. Many of the states also provide multi-year yield averages for cultivars. Results from cultivar trials conducted in the USA and Ontario, Canada are published annually, and all are available on the worldwide web. Internet addresses and a list of information provided by each state in its cultivar trial publication are shown in Table 10–1.

10–1.5 Pest Resistance Considerations

Cultivars should be selected for a particular set of production conditions, to include resistance to or tolerance of prevalent pests. This information should be available from the originator of a cultivar, as well as from state cultivar trial results. Soybean cultivars are available that are resistant to soybean cyst nematode, root knot nematode, diseases, and insects.

The information in Table 10–2 is for the most prominent diseases that occur in the southern (Bowers and Russin, 1999) and the northern USA (Hoeft et al., 2000). According to Bowers and Russin (1999), disease development is controlled by the interaction between host plant, the pathogen, and the environment. The basis for this is genetic, and genetic resistance is, in most instances, the best for disease-management strategy. However, there is no resistance to some prominent diseases in soybean; therefore, prophylactic measures must be taken against these diseases if successful culture of soybean is to occur.

Resistance to a disease refers to the ability of the host to interfere with the normal growth and/or development of the pathogen organism. Resistance does not mean that a particular disease has no effect on the host. Rather, a resistant plant may support some disease development but to a lesser degree than that exhibited by a susceptible plant. Symptoms of a particular disease on resistant plants generally are localized (affect only a small area on the plant) within the growing season and there is no additional spread of the pathogen organism within the plant. Tolerance to a disease is the ability of the host plant to perform effectively even though it exhibits the symptoms of a susceptible host plant. The performance of an infected tolerant plant is ordinarily expected to be similar to that of a plant without infection. Tolerance of a cultivar to a particular disease organism generally is not known; thus, this information is not presented in cultivar trial information. Avoidance or escape of susceptible plants from infection by a particular disease results from chance occurrences related to pattern of pest progression and environmental conditions that allow these plants to remain uninfected even in the presence of the causal organism, and management practices. For example, manipulating the planting date of a susceptible cultivar may allow it to avoid pest pressures and environmental conditions that promote the aggressive development of a particular disease. Disease resistance/tolerance should be balanced against other desirable traits of a cultivar. Resistance to disease(s) that is/are not prevalent where a cultivar will be grown is not a concern. Yield-performance data reflect reaction to diseases present at test locations, and at least indirectly reflect reaction to those diseases. Realistically, the best-yielding cultivars have adequate resistance or tolerance to diseases common to the test site, or they would not have yielded well in those environments.

Soybean is attacked by numerous insect pests, but only a few pose a serious economic threat in North America (Funderburk et al., 1999). Table 10–3 provides information about the most prominent or damaging insect pests that affect soybean in the USA. The tabled information and the following narrative summarizes insect management information adapted from Higley and Boethel (1994), Higgins (1997), and Funderburk et al. (1999).

Injury from insects can occur during any soybean growth stage, but the greatest threat of economic loss occurs from infestations during reproductive development. Most insect pests of soybean are detected by scouting during periods of greatest potential loss. Soybean developmental stage and number and developmental stage of individual insect species should be documented for determination of effective control measures. Management decisions for control of insect infestations

Table 10–2. Common pathogens that affect soybean in the northern and southern USA, indication of soybean resistance, and management/control measures. Adapted from Bowers and Russin (1999), Hartman et al. (1999), Hoeft et al. (2000), and Loren Giesler (personal communication, 2001).

Pathogen		Cultivar resistance	Management/Control
Common name	Causal organism(s)		
Anthracnose	*Colletotrichum truncatum* (Schw.) Andrus & W.D. Moore	No	Plant disease-free seed, treat seed with fungicide†, apply foliar fungicide during reproductive development, plow under crop residue, rotate with nonlegume crops
Bacterial blight	*Pseudomonas savastanoi* pv. *glycinea*	Yes	Do not save seed from infected fields, plant high-quality seed, clean till to destroy infected residue, rotate
Bacterial pustule	*Xanthomonas axonopodis* pv. *glycines*	Yes	Do not save seed from infected fields, plant high-quality seed, clean till to destroy infected residue, rotate
Bean pod mottle	Bean pod mottle virus	No	Transmitted by insects feeding in other legumes like alfalfa and clovers, destroy alternative broadleaf weed hosts
Brown spot	*Septoria glycines* Hemmi	No	Plant disease-free seed, rotate with nonlegume crop, apply foliar fungicide during reproductive development
Brown stem rot	*Phialophora gregata*	Yes	Clean till, plant late, rotate
Cercospora leaf blight, purple seed stain	*Cercospora kikuchii* (T. Matsu. & Tomoyasu) Gardner	Yes	Plant seed of resistant cultivars, plant disease-free seed, late planting, apply foliar fungicide during reproductive development, rotate with nonlegume crop
Charcoal rot	*Macrophomina phaseolina* (Tassi) Goid	No	Rotate with nonsusceptible (cereals and cotton) crops, plant tolerant cultivars, avoid excessive seeding rates, minimize plant stresses, conservation tillage
Downy mildew	*Peronospora manshurica*	Yes	Plant in wide rows, clean till
Frogeye leaf spot	*Cercospora sojina* Hara	Yes	Plant seed of resistant cultivars, plant disease-free seed, rotation with nonlegume crop, apply foliar fungicide during reproductive development
Fusarium root rot and seedling blight	*Fusarium solani* and *F. oxysporum*	Yes	Plant resistant cultivars, plant late, treat seed with fungicide†, plant high-quality seed, clean till
Phytophthora rot	*Phytophthora megasperma* Drechs. f. sp. *glycinea* (Hildeb.) Kuan and Erwin	Yes	Plant seed of race-resistant cultivars, treat seed with fungicide†, maintain good surface drainage, use conventional tillage

Disease	Pathogen	Fungicide	Management
Pod and stem blight and Phomopsis seed decay	*Diaporthe phaseolorum* (Cke. & Ell.) Sacc. f. sp. *sojae* (Lehman) Wehm., *Phomopsis sojae* Lehman, and *Phomopsis longicolla* Hobbs	Yes	Plant disease-free seed, treat seed with fungicide†, plant early (north), apply a foliar fungicide during reproductive development, harvest promptly at maturity, plow under crop residue
Powdery mildew	*Microsphaera diffusa*	Yes	Plant resistant cultivars, apply foliar fungicides
Red crown rot	*Calonectria ilicicola* Boedijn and Reitsma	No	Delay planting, plant on coarse-textured soils, plant tolerant cultivars
Rhizoctonia root and stem decay (damping off)	*Rhizoctonia solani*	No	Plant late, plant high-quality seed, treat seed with fungicide†, minimize stresses, clean till
Rhizoctonia foliar blight (aerial blight)	*Rhizoctonia solani* anastomosis group (AG) 1	No	No complete control; plant tolerant cultivars, apply foliar fungicide during reproductive period, avoid excessive irrigation
Sclerotinia stem rot	*Sclerotinia sclerotiorum*	Yes	Use row spacings >90 cm (36 in), plant late, plant sclerotia-free seed, treat seed with fungicide†, rotate, clean till
Sclerotinia blight (southern blight, white mold)	*Sclerotinia rolfsii* Sacc.	No	Rotate with nonhost crop, avoid post-plant cultivation, bury residue 15 to 25 cm (6–10 in) deep
Seed rots and seedling diseases	*Pythium, Phytophthora, Fusarium, Rhizoctonia, Sclerotinia,* and *Phomopsis* spp.	Yes	Treat seed with a fungicide†, plant disease-free seed, delay planting until soil temperatures are warm
Soybean mosaic virus	Soybean mosaic virus	Yes	Plant early, plant virus-free seed, plant seed of resistant cultivars
Soybean cyst nematode	*Heterodera glycines*	Yes	Plant seed of race-resistant cultivars, control weeds, balance fertility, rotate sources of resistance and with non-host crop
Stem canker	*Diaporthe phaseolorum* var. *meridionalis* (south) and var. *caulivora* (north)	Yes	Plant seed of resistant cultivars, plant disease-free seed, rotate with other crops, plow under crop residue
Sting nematode	*Belonolaimus* spp.	No	Rotate
Sudden death syndrome	*Fusarium solani* (Mart.) Sacc. f. sp. *glycines*	No	Use resistant or moderately resistant cultivars, control soybean cyst nematode, clean till, plant early (south), plant late (north)

† See Table 10–6 for proper fungicide.

Table 10–3. Major insect pests that affect soybean in the USA, plant parts injured, and important management considerations. Adapted from Funderburk et al. (1999), Higley and Boethel (1994), and Higgins (1997).

| Insect pest | | Injurious insect stage: | |
Common name	Scientific name	Plant parts injured	Management considerations[†]
Bean leaf beetle	*Cerotoma trifurcata* (Forster)	Adult: leaves, stems, blooms, pods. Larva: roots and underground stem	Infestation predominates at seedling stage and during flowering and pod-forming through seed-filling period; adult feeding results in greatest injury; may transmit viruses; encouraged by reduced tillage systems
Beet armyworm	*Spodoptera exigua* (Hübner)	Leaf blades	Late-season infestation
Blister beetle	*Epicauta* spp.	Adult: leaves and flowers.	Mid- to late-summer infestation; may cause complete defoliation; scout after mowing of nearby fields
Corn earworm	*Helicoverpa zea* (Boddie)	Adults: leaf blades, pods, seed	July and August infestations associated with hot and dry conditions
Grasshopper	*Melanoplus femurrubrum* (DeGeer) *M. differentialis* (Thomas)	Nymph and adult: leaves, pods, seed in pods	Early summer through harvest infestations; Monitor field edges close to grassy areas; encouraged by reduced tillage
Green cloverworm	*Plathypena scabra* (Fabricius)	Larva: Leaf blades in upper canopy	Early to mid-season infestation
Japanese beetle	*Popillia japonica* (Newman)	Adult: leaves will be skeletonized	Full summer infestation; infrequent pest in eastern portion of midwest; manage in association with other defoliators
Lesser cornstalk borer	*Elasmopalpus lignosellus* (Zeller)	Larva: Lower stems	Late-season infestations associated with hot and dry conditions; seedling damage most injurious
Mexican bean beetle	*Epilachna varivestis* Mulsant	Adult and larva: leaf blades between veins	Early season infestation; greater threat under moderate weather conditions of coastal areas
Potato leafhopper	*Empoasca fabae* (Harris)	Nymph and adult: Leaf blades and veins	Full-summer infestation; Mainly in southern USA, but migrates north; dense leaf pubescence provides mechanical barrier

Common name	Scientific name	Location/feeding	Comments
Saltmarsh caterpillar	*Estigmene acrea* (Drury)	Larva: Leaves in upper canopy	Similar to woollybear caterpillar
Seed corn maggot	*Delia platura* (Meigen)	Larva: Underground cotyledons	May reduce emergence from cool, wet soils with recent organic matter incorporation; delay planting after residue incorporation or use chemical seed treatments
Soybean aphid	*Aphis glycines* (Matsumura)	Leaves	Feeding may cause stunted plants with distorted leaves; peak populations during V2 to R2; overwinters on *Rhamnus* spp. (buckthorn); no economic thresholds; late plantings possibly at greater risk.
Soybean looper	*Pseudoplusia includens* (Walker)	Larva: Leaf blades	Mid- to late-season infestation; worse in soybean–cotton regions
Stink bugs: Southern green, Green, Brown	*Nezara viridula* (L.), *Acrosternum hilare* (Say), *Euschistus servus* (Say)	Adult: Pods, seeds	Mid-season infestation; most injurious damage to seed during early seed formation; treatment of field borders may be sufficient
Thistle caterpillar	*Vanessa cardui* (Linnaeus)	Larva: Leaves webbed, skeletonized	Full-summer infestation period; may require treatment after large migrations
Threecornered alfalfa hopper	*Spissistilus festinus* (Say)	Nymph and adult: Lower stems, petioles	Early season infestation that may go unnoticed until lodging occurs
Two-spotted spider mite	*Tetranychus urticae* (Koch)	Larva, nymph, adult: Leaf sucking, leaf yellowing, dead lower leaves	Full-summer infestation period; populations can increase rapidly during hot, rain-free periods, and may require immediate treatment
Velvetbean caterpillar	*Anticarsia gemmatalis* Hübner	Larva: Leaf blades	Late-season infestation
Wireworms	*Melanotus* spp.	Larva: Seed, roots, underground stem	Spring infestation period; may reduce germination in fields with grass prior to soybean.
Woollybear caterpillar	*Spilosoma virginica* (Fabricius)	Larva: Leaves in upper canopy	Rare late summer outbreaks may require treatment

† Best pest management involves identifying species, sampling to estimate numbers of each species, and consulting economic threshold values provided by Cooperative Extension Service personnel, university entomologists and specialists, and/or crop consultants.

are based on predetermined economic injury level, which is the lowest population density of each pest that is likely to cause economic damage. The economic injury threshold usually changes during the growing season, and is affected by soybean developmental stage, changes in the growing season environment, and crop market value.

Significant yield losses occur when lepidopterous defoliators remove >35% of leaf area before R1, and >15 to 20% after R1. Injury from these insects can be avoided in ESPS plantings in the southern USA, which results in leaves maturing during July and early August before major infestation peaks occur (Baur et al., 2000). Dry soil limits larval weights and larval development, especially on the clay soils that are dominant in the midsouthern USA (Lambert and Heatherly, 1991, 1995). Thus, yield reduction resulting from lepidopterous defoliators is greater for irrigated than for rainfed plants. In fact, dryland producers may find that control measures for soybean plants growing under drought stress can be delayed or not applied because of retarded insect development and lower yield and profit potential from rainfed production.

Host-plant resistance (both chemically-derived and morphological) to insects has been identified for pest species in Coleoptera, Hemiptera, Homoptera, and Lepidoptera insect families (Todd et al., 1994). These are available in plant introductions and in crosses derived from them. However, few insect-resistant cultivars are currently available. Still, the potential for developing insect-resistant cultivars to fit into Integrated Pest Management systems is significant (Todd et al., 1994).

Insect-resistant cultivars have been developed (Bowers, 1990; Hartwig et al., 1990) and offer some resistance to foliage feeders. However, these cultivars are not planted for production in the southern USA because level of resistance is insufficient for effective control. There is a difference in cultivar preference among some insect species, which means that some cultivars may be defoliated sooner than others or to the exclusion of others. This information should be available from the originator of a particular cultivar. The best method for determining the need for insect control in soybean is scouting during periods of risk to determine the species and population density of insects, and selecting a curative measure based on economic injury levels.

With ESPS plantings in the southern USA, injury from bean leaf beetle [*Cerotoma trifurcata* (Forster)], three-cornered alfalfa hopper [*Tetranychus urticae* (Koch)], and stink bug [*Nezara viridula* (L.), *Acrosternum hilare* (Say), and *Euschistus servus* (Say)] infestations may be more pronounced and infestations may require more intense management than normal for conventional May or later plantings. This is because ESPS plantings are earlier than conventional plantings, and thus provide an immediate host at the time of insect emergence.

10–1.6 Nematode Considerations

The SCN is the most serious nematode pest of soybean in the USA (Lawrence and McLean, 1999). In areas with severe infestations, soybean production without control measures is not economically feasible. Soil texture affects movement of SCN in the soil (Young and Heatherly, 1990; Heatherly and Young, 1991) and also may affect their reproduction and development. Basically, major damage to soybean by

SCN infestation occurs when the crop is grown on medium- and coarse-textured soils. Apparently, populations of SCN are not sustainable in soils series classified as clay (Heatherly and Young, 1991).

Determination of the presence, race, and density of SCN is important to prevent losses. A cultivar with resistance to a specific population of a race of SCN should not be planted year after year. Continuous planting of a cultivar could lead to the development of a different SCN race that damages the crop, making that cultivar useless for SCN control (Young, 1994).

Crop rotation is an effective tool for managing SCN (Wrather et al., 1992; Young, 1994). Nonhost crops such as corn (*Zea mays* L.), cotton (*Gossypium hirsutum* L.), and grain sorghum [*Sorghum bicolor* (L.) Moench] successfully reduce SCN populations. Young (1998a) determined that rotation of resistant and susceptible soybean cultivars with a nonhost crop produced greater long-term soybean yields and slowed the shift toward new SCN races in the field. It is important to determine the race of SCN in a field and the race specificity of the resistance gene of a previously planted soybean cultivar when planning to use a new resistant cultivar in a crop-rotation system for SCN management. The originator of a soybean cultivar should furnish information about the race-specific resistance of that cultivar. Cultivars with resistance to SCN are available in all MGs (Young, 1998b).

Early planting of soybean may benefit soybean production in fields that are infested with SCN. Wrather et al. (1992) showed that SCN populations were lower at harvest on early-maturing cultivars compared with those maturing later. Wang et al. (1999) determined that resistant cultivars of earlier MGs appeared to be more effective in reducing nematode numbers than were those of later MGs in several environments of 10 north-central states in the USA. With susceptible cultivars, this was not the case. In contrast, Todd (1993) found only a small influence of MG on SCN reproduction in MG 3 and MG 4 cultivars grown in Kansas. Wang et al. (1999) concluded that planting susceptible cultivars of early MGs is not effective in reducing nematode population densities and resulted in lower yields than from similar later-maturing cultivars.

Nematicides can be effective in controlling SCN populations in infested fields, but their use should be based on expected yield and subsequent income, given that lessened yield loss in low-yield environments may not be sufficient to offset nematicide cost. Heatherly et al. (1992b) determined that soybean irrigation did not affect cultivar response to infection by SCN, the capability of SCN to maintain cysts on any cultivar, or the yield-limiting effect of SCN on susceptible cultivars. Irrigation may increase yield of susceptible cultivars grown on SCN-infested fields, but often yields will be less than those from irrigated susceptible cultivars grown on non-infested fields as well as those from irrigated resistant cultivars grown on infested fields. Thus, irrigation of SCN-susceptible cultivars grown on infested fields should not be considered since irrigation efficiency (amount of yield increase unit^{-1} of applied water) will be low and subsequent yields will be unprofitable.

Root-knot nematodes (*Meloidogyne incognita, M. arenaria,* and *M. javanica*) and reniform nematode (*Rotylenchulus reniformis*) are significant pests of soybean grown in the southeastern USA, especially in the drought-sensitive soils of the southeastern Coastal Plain (Kinloch, 1992; Riggs, 1992). The use of resistant cultivars is the most effective tool for management of the root-knot nematode. Resistance to *M. incognita* is more prevalent in MG 6 through MG 8 cultivars than in MG 5 and earlier cultivars. Recent adoption of the use of MG 4 and earlier cultivars in the southern USA points to the need for *M. incognita* resistance in earlier-maturing cultivars. Continuous use of cultivars with resistance to *M. incognita* could lead to the prevalence of the other two species for which there is no cultivar with resistance. Management of root-knot nematode by crop rotation is complicated by the wide range of hosts for the three root-knot species. Cultivars resistant to *M. arenaria* and *M. javanica* have not been widely or adequately developed; therefore, rotation of soybean with other crops may be the only means of nematode management. Use of resistant cultivars is effective in the management of the reniform nematode. Breakdown of resistance to *R. reniformis* has not been reported. Rotation to grasses, which are poor hosts for *R. reniformis*, is an effective management tactic. Nematicides are not an economical control practice for either root-knot or reniform nematodes.

Fields in the southern USA are often infested with both SCN and root-knot nematodes. Cultivars with resistance to both SCN and root-knot nematodes are more common in the later maturity groups. Today, resistant cultivars are frequently the most productive ones when grown in both infested and noninfested fields (Young, 1998b). Still, producers can test this thesis by conducting an on-farm strip test of a susceptible and resistant pair of cultivars that are best adapted to their area.

10–1.7 Herbicide Resistance/Tolerance

Traditionally, herbicides were designed largely for crops rather than cultivars designed to tolerate a specific herbicide. During the past decade, advances in biotechnology coupled with plant breeding have resulted in the development of herbicide-resistant soybean cultivars. Currently, glyphosate-resistant (GR), glufosinate-resistant, and sulfonylurea-resistant soybean cultivars are available for use in soybean production systems (Reddy et al., 1999). Well over half of the U.S. soybean area is planted to GR soybean cultivars, with some states having more than three-fourths of their soybean area in GR soybean. Glyphosate does not seem to alter the chemical composition of harvested soybean seed (Taylor et al., 1999), and GR genotypes are equivalent in seed composition to parental lines and other soybean cultivars (Padgette et al., 1996).

Reddy et al. (1999) and Reddy (2001a) summarized the current situation pertaining to the use of GR soybean cultivars. Glyphosate has low mammalian toxicity and is considered environmentally safe. After more than two decades of use on cropland, weed resistance to glyphosate has occurred, but genetic shifts in weed populations have not yet been documented. Glyphosate is a nonselective herbicide that kills most annual and perennial grass and broadleaf weeds. Thus, there is no sequence-of-application concern as there is with herbicides that kill either grass weeds or broadleaf weeds, but not both. Control of weeds of the same species that differ in size can be attained simply by increasing the rate of glyphosate. Thus, herbicide application timing for adequate weed control is of less concern than when using nonglyphosate herbicides. Since glyphosate has no soil persistence, a glyphosate-only weed management program can be used with no concern for choice of following crops from a herbicide carryover standpoint.

Table 10–4. Comparisons of costs using preemergent (PRE) and/or postemergent (POST) weed management in conventional (CONV) and glyphosate-resistant (GR) soybean cultivars grown in narrow row culture (no mechanical weed management) in the southern USA using 2001 prices. Adapted from Reddy et al. (1999).

Cultivar/weed management	Inputs	Rate	Cost $ ha⁻¹ ($ acre⁻¹)
			$ ha⁻¹ ($ acre⁻¹)
CONV public/PRE + POST	Seed†	296 000 seed ha⁻¹ (120 000 acre⁻¹)	19.77 (8.00)
	PRE metribuzin + chlorimuron premix	420 g a.i. ha⁻¹ (0.375 lb a.i. acre⁻¹)	53.97 (21.84)
	POST sethoxydim‡	210 g a.i. ha⁻¹ (0.19 lb a.i. acre⁻¹)	29.45 (11.92)
	POST 2,4-DB + linuron§ tankmix	224 g a.i. + 560 g a.i. ha⁻¹ (0.2 lb + 0.5 lb. a.i. acre⁻¹)	33.28 (13.47)
	Total		136.47 (55.23)
CONV public/POST	Seed†	296 000 seed ha⁻¹ (120 000 acre⁻¹)	19.77 (8.00)
	POST bentazon + acifluorfen‖ premix	840 g a.i. ha⁻¹ (0.75 lb. a.i. acre⁻¹)	39.29 (15.90)
	POST sethoxydim‡	210 g a.i. ha⁻¹ (0.19 lb a.i. acre⁻¹)	29.45 (11.92)
	POST 2,4-DB + linuron§ tankmix	224 g a.i. + 560 g a.i. ha⁻¹ (0.2 lb + 0.5 lb. a.i. acre⁻¹)	33.28 (13.47)
	Total		121.80 (49.29)
CONV private/PRE + POST	Seed†	296 000 seed ha⁻¹ (120 000 acre⁻¹)	32.62 (13.20)
	PRE metribuzin + chlorimuron premix	420 g a.i. ha⁻¹ (0.375 lb a.i. acre⁻¹)	53.97 (21.84)
	POST sethoxydim‡	210 g a.i. ha⁻¹ (0.19 lb a.i. acre⁻¹)	29.45 (11.92)
	POST 2,4-DB + linuron§ tankmix	224 g a.i. + 560 g a.i. ha⁻¹ (0.2 lb + 0.5 lb. a.i. acre⁻¹)	33.28 (13.47)
	Total		149.32 (60.43)
CONV private/POST	Seed†	296 000 seed ha⁻¹ (120 000 acre⁻¹)	32.62 (13.20)
	POST bentazon + acifluorfen‖ premix	840 g a.i. ha⁻¹ (0.75 lb. a.i. acre⁻¹)	39.29 (15.90)
	POST sethoxydim‡	210 g a.i. ha⁻¹ (0.19 lb a.i. acre⁻¹)	29.45 (11.92)
	POST 2,4-DB + linuron§ tankmix	224 g a.i. + 560 g a.i. ha⁻¹ (0.2 lb + 0.5 lb. a.i. acre⁻¹)	33.28 (13.47)
	Total		134.64 (54.49)
GR/POST	Seed#	296 000 seed ha⁻¹ (120 000 acre⁻¹)	49.42 (20.00)
	POST glyphosate	1120 g a.i. ha⁻¹ (1 lb a.i. acre⁻¹)	25.45 (10.30)
	POST glyphosate	1120 g a.i. ha⁻¹ (1 lb a.i. acre⁻¹)	25.45 (10.30)
	Total		100.32 (40.60)

† Based on planting CONV public cultivar at $0.44 kg⁻¹ seed ($10.00 per 50 lb bag) or a CONV private cultivar at $0.73 kg⁻¹ seed ($16.50 per 50 lb bag), both with 6.6 seed g⁻¹ (3000 seed lb⁻¹).

‡ Includes herbicide at $22.98 ha⁻¹ ($9.30 acre⁻¹) + crop oil at $6.47 ha⁻¹ ($2.62 acre⁻¹).

§ Includes herbicides at $31.06 ha⁻¹ ($12.57 acre⁻¹) + surfactant at $2.22 ha⁻¹ ($0.90 acre⁻¹).

‖ Includes herbicides at $36.08 ha⁻¹ ($14.60 acre⁻¹) + crop oil at $3.21 ha⁻¹ ($1.30 acre⁻¹).

Based on planting GR cultivar at $1.10 kg⁻¹ seed ($25.00 per 50 lb. bag) with 6.6 seed g⁻¹ (3000 seed lb⁻¹).

Glyphosate-resistant cultivars offer the flexibility to control a broad spectrum of weeds in soybean with no concern for crop safety (Reddy, 2001a). Cost of weed control should be less, even with the higher cost for seed of most GR cultivars (Table 10–4). This could translate to increased profits if yields from GR cultivars are equal or nearly equal to those from non-GR cultivars. Use of GR cultivars should preempt the use of tillage and preemergent herbicides for weed management. Research has shown that nonglyphosate herbicides applied to continuously cropped GR soybean or soybean grown in rotation with corn do not adversely affect GR soybean (Bennett et al., 1998; Hofer et al., 1998; Nelson and Renner, 1999; Webster et al., 1999). This increases options and flexibility for weed control when GR cultivars are used. If weeds are present that are difficult to control with nonglyphosate herbicides, use of GR cultivars may result in greater profit, especially in low-yield environments where costs must be minimized. The advantages of GR cultivars should translate to a reduction in management decisions for producers related to weed control in soybean.

Early nonpublished Monsanto research with six pairs of iso-populations with and without the GR gene indicated that no yield suppression was associated with the GR gene (X. Delannay, personal communication, 1999). Glyphosate has no negative effect on GR cultivar growth, development, and yield (Nelson and Renner, 1999; Elmore et al., 2001a). However, comparisons in side-by-side cultivar performance trials indicated that a yield suppression may exist with GR soybean relative to non-GR soybean (Nelson et al., 1997, 1998, 1999; Oplinger et al., 1998a; Minor, 1998; H.C. Minor, personal communication, 1999; Nielsen, 2000). Yield suppressions may result from either cultivar genetic differentials or the GR gene/gene-insertion process. The GR gene, CP4 EPSPS from breeding line 40-3-2 (Delannay et al., 1995), remains the source for resistance in current GR cultivars (X. Delannay, personal communication, 1999). In another study, five backcross-derived pairs of GR and non-GR soybean sister lines were compared along with three high-yielding, nonherbicide-resistant cultivars and five other herbicide-resistant cultivars. In contrast to the unpublished Monsanto report (X. Delannay, personal communication, 1999), GR sister lines yielded 5% (200 kg ha^{-1}; 3 bu acre^{-1}) less than the non-GR sister lines (Table 10–5) (Elmore et al., 2001b). High-yielding, nonherbicide-resistant cultivars included for comparison yielded 5% more than the non-GR sister lines and 10% more than the GR sister lines. The potential for a 5 to 10% yield advantage for non-GR cultivars vs. GR cultivars should be considered in the evaluation of profit opportunity of the two systems, especially in high-yield environments such as those with irrigation (Heatherly et al., 2002a). This is an area of evolving technology, and unpublished results from recently completed research indicate that the yield relationship between GR and non-GR cultivars is not clearcut in favor of either GR or non-GR cultivars.

There are disadvantages to the GR soybean weed management system. Lengthy periods of windy conditions in the spring may limit preplant spraying opportunities because of drift concerns. The GR system is most advantageous when used in a total postemergent weed management program; thus, lack of weed management with traditional residual herbicides will necessicate multiple sprayings with glyphosate. Lengthy periods of wet soil likewise may cause delays in applications of glyphosate by ground equipment, thus allowing weed competition with the crop

Table 10–5. Growth, development and yield of nonglyphosate-resistant soybean sister lines (non-GR sisters) and GR sister lines averaged over Nebraska locations for 2 yr. Adapted from Elmore et al. (2001b).

Sister-line groups	Plant density	R1 bloom	R7 maturity	R8 maturity	R7 plant height	R7 Lodging†	Seed wt.	Yield	Grain moisture
	plants ha⁻¹ (acre⁻¹) × 1000	———d from 31 May———			cm (in)	1 to 5	mg seed⁻¹	Mg ha⁻¹ (bu acre⁻¹)	%
Non-GR sisters	266 (108) a‡	43.6 a	111.9 a	120.4 a	86 (34) b	1.6 a	147 a	3.68 (54.8) a	10.0 a
GR sisters	267 (108) a	43.7 a	112.7 a	121.7 a	88 (35) b	1.4 a	141 b	3.48 (51.9) b	10.0 a
No. locations: 1998/1999	4/4	2/4	3/4	3/1	4/4	4/4	0/3	4/4	4/4

† 1 to 5 scale with 1 = erect and 5 = prostrate.
‡ Means followed by the same letter within a column are not significantly different at $P \leq 0.05$. Means were separated with single-degree-of-freedom comparisons.

in the early season to become yield limiting to soybean before spraying is possible. Although timing of glyphosate application is not as critical as with non-GR postemergent herbicides, some weeds such as morningglories (*Ipomoea* spp.) are more easily killed by glyphosate when they are small. Thus, significant delays in glyphosate application result in using more expensive higher rates. The higher cost for seed of most GR cultivars increases the importance of using seeding rates that are within the optimal range.

Cultivar trials conducted by the entities listed in Table 10–1 typically include assessment of GR genotypes, usually in separate trials. Because these are separate trials, caution should be used in comparing GR and non-GR cultivar performance. The usual criteria for selection of cultivars should be used when selecting herbicide-resistant cultivars. Choice of GR and non-GR cultivars should be based on (i) previous weed pressure and success of control measures in specific fields, (ii) availability and cost of herbicides, (iii) availability and cost of GR cultivars, and (iv) yield potential of a specific field. The herbicide-resistance component of a cultivar's genetics should be viewed as a weed management option rather than a cultivar selection criteria.

10–1.8 Seed Quality and Germination

Experts in some southern states recommend the application of a foliar fungicide at beginning pod (R3) and beginning seed (R5) growth stages in seed production fields to prevent seed diseases that reduce germination quality of seed (Moore, 1996; Bowers and Russin, 1999). However, these applications may be too early if environmental conditions that favor seed decay occur near soybean maturity. Control of stink bugs is especially critical for soybean seed crops because of their association with seed injury and their serving as a vector for the transmitting of seed diseases. Results from research in the southern USA indicate that soybean grown for seed should be irrigated during reproductive development to ensure the highest germination (Heatherly, 1999b). Seed should be harvested as soon as they reach 14% moisture content to ensure the least possible damage from weathering and the least amount of seed damage during the threshing process.

Seed diseases often affect seed germination. Seed-treatment fungicides can reduce germination problems associated with these seed diseases (Table 10–6), but they are no substitute for high-quality, disease-free seed. There are two classes or types of seed-treatment fungicides: contact or protectant fungicides that are active against pathogenic organisms that are present on the planted seed, and systemic fungicides that are active against pathogenic organisms that are soil- or residue-borne and that attack planted seed if conditions are conducive for disease development. It is generally a good practice to treat seed with a product that contains a combination of the two classes of fungicides when planting in cool, wet soils that provide a favorable environment for seedling disease development. Information in Table 10–6 gives common pathogens of planted soybean seed, along with the fungicides that provide control of these diseases.

Seed quality (germination, discoloration, shriveling, etc.) of harvested seed is of paramount importance in soybean production. Mayhew and Caviness (1994) grew four MG 3 and four MG 4 April-planted soybean cultivars under nonirrigated conditions in 1989 and 1990 in Arkansas. Average seed germination for MG 3 and

Table 10–6. Seed-treatment fungicides for control of soybean seed and seedling diseases, type of control, and organisms controlled or suppressed by each fungicide. Adapted from Anonymous (2001), L. Geisler (personal communication, 2001), Keith and Delouche (1999), and Mueller et al. (1999).

Fungicide			
Trade name	Common name	Type†	Pathogen‡ controlled
Apron XL	Mefenoxam	S	PRR, PYT
ApronMaxx	Mefenoxam + fludioxonil	C,S	FUS, PHO§, PRR, PYT, RHI, SCL
Maxim	Fludioxonil	C	FUS, RHI, SCL
Mertect	Thiabendazole	S	FUS, RHI, SCL
Rival	Captan¶ + PCNB¶ + thiabendazole	C,S	FUS, PHO, RHI, SCL
Stilleto	Carboxin + Thiram + metalaxyl	C,S	ANT, FUS, PHO, PYT, RHI, SCL
Terraclor	PCNB¶	C	RHI
Vitavax CT	Carboxin + thiram	C,S	ANT, FUS, PHO, RHI, SCL

† C = contact (protectant); S = systemic.
‡ ANT = Anthracnose; FUS = *Fusarium* spp.; PHO = *Phomopsis* spp.; PRR = *Phytophthora* root rot; PYT = *Pythium* spp. seedling rot; RHI = *Rhizoctonia solani* root rot; SCL = *Sclerotinia sclerotiorum*.
§ Suppression.
¶ Captan and PCNB have an adverse effect on *Bradyrhizobia japonicum* inoculant (Curley and Burton, 1975). Avoid these materials if seed is directly inoculated or use an in-furrow application of the inoculant if captan is used. Check product label for compatibility with *B. japonicum* inoculant when using any seed treatment fungicide.

MG 4 cultivars was 28 and 42%, respectively. Germination percentage was significantly and negatively correlated ($r = -0.72$) with infection with *Phomopsis longicolla*. They did not grow MG 5 or later cultivars in this study, so it is impossible to say that only seed of early plantings of early-maturing cultivars are susceptible to low germination. They concluded that cultivars resistant to *Phomopsis* seed decay are necessary if production of planting seed for early-planted, short-season soybean is to be viable in the southern USA. Elmore et al. (1998) reported that soybean lines resistant to *Phomopsis* seed decay can provide effective control without fungicide application. Heatherly (1993, 1996) measured significantly higher germination of seed harvested from irrigated vs. nonirrigated MG 4 and MG 5 cultivars at Stoneville, MS, but this improvement was not always sufficient to impart acceptable levels of germination. Heatherly (1996) also determined that planting MG 4 and MG 5 cultivars in May and later in the midsouthern USA and irrigating almost always ensured seed with adequate germination, although lower yields were obtained. The conclusion drawn from current knowledge is that production of seed from early-maturing cultivars in the southern USA results in a product with germinability that is unpredictable. Seed for ESPS plantings should be obtained from reputable sources whose seed production was conducted in locations with environments known to produce quality, germinable seeds. For soybean production where seed will be for uses other than planting seed, *Phomopsis* seed decay is not usually a concern. However, quality of harvested seed from early planted, early-maturing cultivars in the southern portion of the midsouthern USA in 2001 was so adversely affected by seed decay that a large portion of the crop was significantly reduced in yield and value (Dorris, 2001). Thus, genetic resistance to seed decay is now important. In both the southern and midwestern USA, good quality seed of early-maturing cultivars is usually obtained from the northernmost region of their adaptation. This is because seed quality is greater when seed mature under the cooler temperatures of early fall rather than the hotter temperatures of late summer.

10–1.9 Specialty Cultivars

High protein soybean cultivars may become important to U.S. livestock producers. Unfortunately, the strong inverse relationship between seed protein and grain yield has limited progress. Correlations for this inverse relationship have typically ranged from $r = -0.023$ to -0.86. New developments have decreased this, suggesting the absence of physiological barriers between high seed yield and high seed protein (Wilcox and Cavins, 1995). For example, new populations derived from 'Maple Glen' show low association between seed yield and protein, with r values ranging from -0.06 to -0.21 (Cober and Voldeng, 2000). Additionally, determinate cultivars apparently are a better source of selections that combine high seed yield and high protein than are indeterminate cultivars for the northern production areas of the USA (Wilcox and Zhang, 1997). High-protein MG 2 through MG 5 lines are now available that produce more than 46% seed protein (dry matter basis) and more than 50% meal protein (Table 10–7). New high-yielding soybean cultivars with higher protein content are under development.

Markets are available for specialty soybean with different seed sizes. Natto is a Japanese food product made from small, mature soybean seed that are cooked and fermented. Cultivars that produce small seed also are used for sprouting. Specialty soybean cultivars that produce large seed are used as edamame or vegetable soybean, and these seed are harvested before maturity (at R6) when the seed have filled 80 to 90% of the pod width. The pods are boiled to assist shelling, and the seeds are eaten as a vegetable in a variety of ways, especially in East Asia. Other products like tofu and miso also call for mature, large soybean seed from specialty cultivars. Tofu consumption is growing rapidly in the USA (Rao et al., 2002). Cultivars used for natto and sprouts produce small seed that weigh ≤ 80 mg seed^{-1} (>5700 seed lb^{-1}) when mature. The large seed produced by cultivars that are used for edamame, tofu, and miso weigh more than 220 mg seed^{-1} (<2000 seed lb^{-1}). For comparison, seed of conventional cultivars weigh from 120 to 180 mg seed^{-1} (3800–2500 seed lb^{-1}).

Table 10–7. Summary of north-central U.S. regional high protein soybean test, 2000. Adapted from George Graef (personal communication, 2001).

Maturity Group	No. of locations	Trait	Range among standard cultivars (checks)	Range among high protein strains
II	11	Seed protein %†	33.9–35.7	35.5–42.3
		Meal protein %	48.3–49.4	49.1–54.8
		Yield (% of checks)		70.7–107.0
III	9	Seed protein %	35.1–36.6	38.6–41.4
		Meal protein %	48.8–50.3	50.3–54.7
		Yield (% of checks)		82.7–96.9
IV	6	Seed protein %	36.4–37.3	40.3–43.4
		Meal protein %	50.4–50.5	53.9–55.8
		Yield (% of checks)		78.4–94.3
V	6	Seed protein %	36.5–36.9	38.0–46.7
		Meal protein %	50.1–51.3	51.2–59.0
		Yield (% of checks)		80.2–112.4

† Dry matter basis.

In addition to seed size, breeding for specific characteristics is important within each of the specialty types mentioned. Seed used for natto should be round and have easily permeable seed coats for rapid water uptake during soaking and rapid water loss after steaming. Hilum color is not critical. Total protein and oil content of seed is correlated with natto quality or desired sugar content (Geater et al., 2000), and can be used as a criterion by breeders to select for cultivars for the natto industry. Seed for tofu uses should have a clear hilum and high protein (Wang et al., 1983). It is possible to develop broadly adapted short-season natto cultivars for production in areas as far north as eastern Canada (Cober et al., 1997).

The ideal ideotype for edamame soybean includes 40 to 50 pods plant^{-1}, unblemished dark green pods that have dimensions of > 4.5 cm by 1.3 cm (1.8 by 0.5 in), 2.5 to 3 g fresh weight and 2 to 3 seeds pod^{-1}, round seeds with a clear hilum, and gray (white) pubescence (Konovsky et al., 1994; Cober et al., 1997; Nguyen, 1998). Two aspects of edamame soybean complicate breeding efforts. First, both greater number of pods plant^{-1} and greater pod weights are desired; however, these two traits are negatively correlated (Mebrahtu et al., 1991). Second, a tendency to shatter at maturity is common among edamame cultivars, and this a negative factor for seed production and breeding progress. Once harvested, however, release of seed from pods is a positive trait for vegetable soybean because consumers desire pods that open more easily.

Seed yields from specialty cultivars with both large and small seed are less than yields from cultivars with normal seed size. Cultivars with large seed yielded 82% of check cultivars, while cultivars with small seed yielded 72% of check cultivars in a 4-yr study in Nebraska. Seed weights were not affected greatly by either row spacing or seeding rate (Hoffmeister and Elmore, 1999). Seed weight can, however, be altered significantly by irrigation timing. Irrigation during flowering (R1–R3) almost invariably increases number of seed plant^{-1}, and may result in a large number of smaller seed if irrigation is discontinued before seedfill. Inversely, not irrigating soybean during flowering and early pod development (R1–R4) followed by adequate irrigation thereafter can increase weight of the fewer seed that are produced (Korte et al., 1983a, 1983b; Kadhem et al., 1985a, 1985b). Irrigation timing can thus be used as a management factor to exert control over the final weight of seed of specialty soybean cultivars developed for either the small- or large-seed markets. For production of specialty cultivars to be profitable and attractive to producers, any underperformance in yield must be offset by premiums paid for the seed.

10–2 TILLAGE

Tillage in soybean management systems is used to prepare a seedbed, remedy compaction, incorporate fertilizers and herbicides, and control weeds. Hoeft et al. (2000) have given definitions for common tillage terms: those definitions with modifications are used in this chapter. Clean tillage (synonymous with conventional tillage and often associated with moldboard plowing and disk harrowing) is a term used to describe a production system that uses tillage for any purpose at any time. Clean-till systems employ any implement or implements that leave <10% of the soil surface covered with residue. Reduced-till systems refer to those practices that leave between 10 and 30% residue on the soil surface, and are often a compromise choice between clean-till and conservation tillage systems. When used, reduced-till systems may increase dependence on both pre- and post-planting chemical weed control. Conservation tillage refers to any tillage system that maintains at least 30% of the soil surface covered with residue up to planting. The intention when using conservation tillage is to conserve soil and water, and reduce fuel, labor, and equipment inputs. Conservation tillage systems may include reduced-till, mulch-till, ecofallow, strip-till, ridge-till, and no-till.

No-till refers to a system where tillage is essentially eliminated during both the growing season and the off-season. However, some tillage is conducted in the process of creating a seed trench or strip with a coulter or disk-opener during planting (Jasa et al., 1991). Use of no-till and narrow rows places total dependence on herbicides for both pre- and post-plant weed management. A no-till system does not allow for any correction of soil surface or subsurface problems with tillage. Most soybean management systems combine components of conventional, reduced, and no-till approaches over a period of years. The rigid use of any one tillage approach can lead to production problems resulting from either too much tillage or too little tillage in situations where an appropriate tillage operation may offer the only solution to a particular problem.

In the midwestern USA, tillage systems used for soybean are varied (Jasa et al., 1991). In rotation systems involving soybean, a commonly used scheme is no-till planting. In this case, it is common for one or two disk harrowings followed by a field cultivation (shallow tillage with an implement having spring-tooth tines or sweeps) or shallow chisel plowing to be done following corn, grain sorghum, or wheat (*Triticum aestivum* L. emend. Thell.) harvest preceding the soybean crop in the rotation, with no tillage following the soybean crop. These operations in combination with no-till planting will leave at least 30% residue cover following both growing seasons, and provide the most erosion control while still allowing for some tillage of the less fragile nonsoybean residue. In ridge systems, all crops are planted into ridges formed during cultivation of the previous row crop. Soil is undisturbed between harvest and planting of the next crop, and residue cover during this period is maximum. Post-plant tillage (cultivation) is used to maintain the ridges at least 15 to 20 cm (6–8 in) tall. In a ridge-plant system, no soil disturbance occurs prior to planting. A row cleaning device on the planter may be used to push a small amount of debris off the top of the ridge. In a ridge-till system, some tillage prior to planting may be done, but it is shallow and disturbs only the ridge tops without destroying them. This tillage may be necessary to flatten or smooth peak-shaped ridges so the planter will stay on the row, and/or remove excess residue. Ridge systems are well-suited for level or gently sloping fields (especially those having soils with poor internal drainage) and for fields that will be furrow-irrigated. The use of a ridge system will dictate that row spacing will be the same for all crops in a rotation, and that the rows be wide enough to accommodate effective post-plant cultivation. This likely will require rows that are at least 75 cm (30 in) apart.

Deep tillage (sometimes referred to as "subsoiling" or "deep ripping") refers to operations that affect soil 15 cm (6 in) or deeper (Hoeft et al., 2000). These operations are used to fracture or loosen deep soil barriers, improve rainfall infiltration, and mix residue and nutrients deep into the profile. Deep tillage can be part of a conservation tillage system if it minimally disturbs the soil surface. Shallow tillage, or secondary tillage, refers to operations that affect soil to depths up to 15 cm. These operations are used to kill weeds, incorporate herbicides and fertilizers, level soil, and prepare a smooth seedbed. Shallow tillage operations usually result in a clean-till environment (low soil surface residue) since they significantly disturb the soil surface and residue cover.

Bedding describes ridging soil so that the seedbed is raised on poorly drained soils. This concept is useful where early planting occurs on soils such as flat alluvial clays of the lower Mississippi River flood plain. If a disk hipper is used for bed forming, row spacing should be 76 cm (30 in) or wider because individual row beds cannot be effectively formed, maintained, and planted in more narrow rows. If narrow-row planting is desired on beds in these environments, a wide bed capable of supporting several rows per bed must be constructed. Recent equipment developments (Ginn et al., 1998) allow this to be done, and a management system of narrow rows planted on beds is possible.

Cultivar rankings do not vary among tillage systems (Elmore, 1987, 1990; Guy and Oplinger, 1989). Cultivar performance trials conducted in conventional tillage systems can therefore be used for selecting cultivars for conservation tillage systems, and vice versa. In addition, tillage system seldom interacts with planting date or seeding rate (Elmore, 1990, 1991). Thus, similar management practices are optimum for various tillage systems. Weed control in conservation tillage systems is now simplified because of herbicide-resistant cultivars. In fact, GR soybean cultivars are well matched to reduced tillage systems because weed-management expenses associated with their use should be no higher than when they are used with conventional tillage systems.

10–2.1 Deep Tillage

Deep tillage (DT) is conducted postharvest in the fall and is used to disrupt the soil profile below 15 cm with implements such as "subsoilers", rippers, and chisel plows that have curved shanks or standards spaced 50 to 100 cm (20–40 in) apart. Operation of these implements is intended to lift and shatter the soil profile to the depth of operation. Correct operation of these implements should minimally disturb the soil surface. Heatherly (1981) measured almost identical yields among treatments in studies on Sharkey clay (very-fine, smectitic, thermic Chromic Epiaquert) where DT performed in late winter or early spring when soil was wet was compared to shallow, disk-harrow spring tillage preceding soybean planting. Popp et al. (2001) found that DT of wet clay soil in late winter or early spring in Arkansas resulted in net returns that were similar to those resulting from conventional shallow tillage. Thus, DT of wet soils was not effective in increasing net return. In some cases, DT to a depth of 40 to 45 cm in the fall following harvest is used to disrupt soil barriers and increase water held in the soil profile. Koskinen and McWhorter (1986) reported increased perennial and biennial weeds with no-tillage systems; thus, deep tillage of dry soil in the fall could be considered for suppressing problem perennial weeds such as redvine [*Brunnichia ovata* (Walt.) Shinners], which is deep-rooted. Wesley and Smith (1991) performed DT on a Tunica silty clay (clayey over loamy, smectitic, nonacid, thermic, Vertic Haplaquept) in the fall in Mississippi following soybean harvest when the soil profile was dry. They measured large, significant yield increases from soybean planted in May in years when drought occurred during the growing season, and determined that net return was greatly increased from this practice (Wesley et al., 2000). The increased production was associated with increased moisture content in the soil, presumably because of greater infiltration and storage resulting from DT. Wesley et al. (2001) concluded that fall deep tillage should be performed once every 3 yr on a Tunica silty clay. This work has been used to promote DT of all dry clay soils in the fall in the midsouthern USA.

Studies on Sharkey clay in Arkansas (Popp et al., 2001) and Mississippi (Wesley et al., 2001) showed average increases in yield of 580 kg ha^{-1} (8.6 bu acre^{-1}) and 365 kg ha^{-1} (5.4 bu acre^{-1}), respectively, and average increases in net return of $96 and $71 ha^{-1} ($39 and $29 acre^{-1}), respectively, from fall DT (Table 10–8). In the Arkansas study, yields following fall DT were significantly greater than those from conventional tillage even though drought was not severe. The Mississippi study used estimated DT costs that were $17 to $20 ha^{-1} ($7–$8 acre^{-1}) more than those for a treatment that received only secondary tillage [≤10 cm (4 in)]. Heatherly and Spurlock (2001) and Heatherly et al. (2002c) determined that profits from producing soybean following DT of Sharkey clay were significantly greater than those from conventional tillage only when plantings were made in April vs. May and later (Table 10–8). In their study, costs associated with DT were $29 to $42 ha^{-1} ($12–$17 acre^{-1}) greater than those for a conventional shallow tillage system (fall tillage with a disk

Table 10–8. Yield and net return for soybean grown in tillage studies in the southern USA.

State (reference)	Soil series/texture	Tillage treatment†	Yield	Net return
			kg ha^{-1} (bu acre^{-1})	$ ha^{-1} ($ acre^{-1})
Arkansas (Popp et al., 2001)	Sharkey clay‡	Conventional	2715 (40.4)	398 (161)
		DT	3235 (48.1)	480 (194)
Mississippi	Tunica silty clay‡	Conventional	2435 (36.2)	220 (89)
(Wesley et al., 2000)		DT1	3450 (51.3)	436 (176)
		DT2	3260 (48.5)	413 (167)
		DT3	3255 (48.4)	417 (169)
		DT4	3160 (47.0)	395 (160)
		DT5	2840 (42.3)	321 (130)
Mississippi	Sharkey clay‡	Conventional	1860 (27.7)	166 (67)
(Wesley et al., 2001)		DT	2225 (33.1)	237 (96)
Mississippi (Heatherly and	Sharkey clay‡	Conventional	2050 (30.5)	240 (97)
Spurlock, 2001)		DT	2465 (36.7)	305 (123)
	Sharkey clay§	Conventional	1650 (24.6)	110 (44)
		DT	1785 (26.6)	105 (42)
Mississippi	Tunica‡	Conventional	2000 (29.8)	156 (63)
(Heatherly et al., 2002c)		DT	3165 (47.1)	370 (150)
South Carolina	Eunola loamy sand§	No-till	2160 (32.1)	NA
(Frederick et al., 2001)		DT	2415 (35.9)	NA

† Conventional = shallow tillage (≤10 cm) with chisel plow, disk harrow, or spring-tooth cultivator; DT = deep-tilled to 38 to 46 cm (15–18 in) depth; DT1 = deep-tilled annually, DT2 = deep-tilled every other year, DT3 = deep-tilled every 3rd yr, DT4 = deep-tilled every fourth year, and DT5 = deep-tilled every 5th yr.

‡ April-planted.

§ May and later-planted (South Carolina study followed wheat harvest).

harrow and/or a spring-tooth harrow). In extremely dry years (yield levels <1000 kg ha^{-1} or 15 bu acre^{-1}), or in production systems where irrigation was applied, deep tillage provided no yield or economic benefit (Heatherly et al., 2002c). On a Coastal Plain loamy sand soil in South Carolina, Frederick et al. (2001) measured a 12% yield increase from DT compared to no DT (2415 vs. 2160 kg ha^{-1}) just prior to May planting of soybean that was not irrigated (Table 10–8). They also measured a 50% yield increase when irrigation was applied following no DT compared to no DT and no irrigation (3201 kg ha^{-1} vs. 2160 kg ha^{-1}). Thus, both the Frederick et al. (2001) and Heatherly et al. (2002c) studies indicate that DT with irrigation is not necessary. In another South Carolina study using late May/early June plantings of determinate soybean following wheat on a loamy sand, DT combined with no surface tillage compared to only surface tillage prior to planting of soybean resulted in significantly greater yields in 19-cm-wide (7.5-in) but not in 76-cm-wide (30-in) rows (Frederick et al., 1998). Highest yields in the Frederick et al. (1998) study were achieved when both fall and spring deep tillage were conducted on the sandy soil.

In the northern USA, DT that is performed is done in the fall, followed by one or more secondary tillage operations in the spring. Chisel plows typically operate between 15 (6 in) and 30 cm (12 in) deep. Subsoilers or rippers operate at 46 cm (18 in) or deeper. The subsoiler shanks are usually spaced 50 to 100 cm (20–40 in) apart. They are designed to create deep slots in the soil profile to open a channel for water infiltration and root penetration in soils with natural hard pans. Subsoilers are also operated at <30 cm (12 in) deep to break up surface compaction. Rolling coulters may be placed in front of the shanks to improve performance in heavy crop residues. Descriptions and pictures of various tillage tools can be found in Hoeft et al. (2000).

Costs of tillage operations play a major role in the selection of a tillage system for soybean production. Yields of dryland soybean following deep tillage of the clay soils in Table 10–8 did not approach the high yield and net return levels obtained from irrigated plantings of soybean at this location (Heatherly and Spurlock, 1999). These yield and net return responses are marginal when measured against high fuel prices and low commodity prices. Using a $0.184 kg^{-1} seed ($5.00 bu$^{-1)}$ price for soybean, a 160 to 230 kg ha^{-1} (2.4–3.4 bu acre^{-1}) yield increase would be required to break even using the $29 to $42 ha^{-1} higher tillage cost associated with DT in the Heatherly et al. (2002c) studies. Thus, with low commodity prices, significant profitability from DT of these clay soils in the fall will require consistent yield increases such as those obtained in the cited studies. The use of DT on the clay soils should be based on anticipated early April planting and expected commodity price since significant economical yield increases are not consistently achieved.

10–2.2 Preplant (Secondary) Tillage

Preplant tillage is conducted to remedy soil surface problems such as rutting that were created during harvest, to destroy weed vegetation so that the crop is planted in a clean seedbed, to disrupt restrictive layers in the soil profile that may interfere with root penetration and soil moisture extraction, and to promote soil

Table 10–9. Annual soil loss from plots with 5% slope in the brown loam soil region of Mississippi (Triplett and Dabney, 1999).

	Conventional tillage		No-till		
	Soil loss yr^{-1}		Soil loss yr^{-1}		
Crop	C Factor[†]	Mg ha^{-1} (ton acre^{-1})	C factor[†]	Mg ha^{-1} (ton acre^{-1})	Reference
Sorghum	0.04	9.4 (4.2)	0.005	1.3 (0.6)	McGregor and Mutchler (1992)
Corn (grain)	0.09	16.1 (7.2)	0.005	0.9 (0.4)	McGregor and Mutchler (1983)
Corn (silage)	0.14	25.1 (11.2)	0.003	0.7 (0.3)	McGregor and Mutchler (1983)
Soybean	0.12	47.3 (21.1)	0.006	2.7 (1.2)	McGregor (1978)
Soybean	0.10	43.9 (19.6)	0.008	3.1 (1.4)	Mutchler and Greer (1984)
Cotton	0.31	69.9 (31.2)	0.053	12.1 (5.4)	Mutchler et al. (1985)

† Factor used in the Universal Soil Loss Equation to reflect influence of soil management and cropping methods on water erosion. Kind and time of tillage, implements used, time of planting, crops planted, postemergence cultivation, crop sequence, residue cover on the soil surface, and changes in soil organic matter all affect C factor.

warming prior to planting in the northernmost soybean-growing regions. The old adage that tillage is needed for seedbed preparation is not valid. Significant soybean hectarage in the USA is now planted in environments with no preplant tillage. Tillage systems for corn and soybean production in the midwestern USA were nearly identical in the late 1980s and into 1991. In 1992, dramatic changes in tillage systems for full-season soybean occurred. No-till production rose from 8.3 to 23.8% of the total production area between 1991 and 1992 (Conservation Technology Information Center, 2002). No-till corn increased from 9 to 16.8% between the same years. Since 1992, no-till soybean area has risen gradually to 33.4% in 2002, and no-till corn production has remained around 17%. Average corn area managed with conventional tillage (37%) and total corn area (25 400 000 ha or 63 000 000 acres) stayed about the same from 1989 until 2002. However, proportional area of conventional-tilled soybean declined from 42.7 to 21.1% as area increased from 17 900 000 ha (44 300 000 acre) to 24 400 000 ha (60 400 000 acre).

In the midsouthern USA, no tillage between harvest and planting of the subsequent soybean crop resulted in yields and net returns that were similar to or greater than those resulting from soybean being planted following fall or spring tillage for seedbed preparation on clay soil (Heatherly et al., 1990, 1993). Greater number of soybean *Bradyrhizobial* cells and *Bradyrhizobial* diversity have been measured in no-till compared to conventional-tillage systems in Brazil (Ferreira et al., 2000).

All tillage operations affect the erosion potential of any soil used for soybean production. Moldboard plowing buries almost all residue, whereas chisel plowing loosens the soil but leaves considerable residue on the soil surface (Erbach, 1982). However, multiple passes with a chisel plow, disk harrow, or field cultivator used in conservation-tillage production systems can result in residue cover being reduced to <5% at planting and lead to increased soil loss (Triplett and Dabney, 1999). Table 10–9 shows annual soil loss from conventional and no-till production systems with various crops in Mississippi. Similar results from field tests in Nebraska are shown in Table 10–10. These data indicate that no-till management can reduce soil loss from all crops, especially soybean. This was also the case for a corn–soybean rotation system using conservation tillage measures on a large watershed in Ohio (Ed-

Table 10–10. Measured surface cover and soil loss for various tillage systems used for corn and soybean production in Nebraska. Adapted from Dickey et al. (1986).

Residue type/tillage system	Residue cover	Erosion	Erosion reduction from moldboard plow
	%	Mg ha^{-1} (ton acre^{-1})	%
	Corn residue†		
Moldboard plow, disk 2X, plant	7	17.5 (7.8)	--
Chisel plow, disk, plant	35	4.7 (2.1)	74
Disk 2X, plant	21	4.9 (2.2)	72
Rotary-till, plant	27	4.3 (1.9)	76
Till-plant	34	2.5 (1.1)	86
No-till, plant	39	1.6 (0.7)	92
	Soybean residue‡		
Moldboard plow, disk 2X, plant	2	32.0 (14.3)	--
Chisel plow, disk, plant	7	21.5 (9.6)	32
Disk, plant	8	23.8 (10.6)	26
Field cultivate, plant	18	17.0 (7.6)	46
No-till, plant	27	11.4 (5.1)	64

† Nebraska tests after tillage and planting on a silt loam soil having a 10% slope and 5 cm (2 in) water applied in 45 min.
‡ Nebraska tests after tillage and planting on a silty clay loam soil having 5% slope and 5 cm (2 in) water applied in 45 min.

wards et al., 1993). However, use of no-till can lead to increased runoff of applied herbicides in some cropping systems (Shipitalo et al., 1997). Influence of soil management and cropping methods on water erosion for selected soybean management systems in Mississippi are given by Triplett and Dabney (1999).

Soybean residue needs special consideration when preparing soil for subsequent crops (Erbach, 1982). First, residue levels following soybean may be sufficient to meet requirements to reduce erosion for highly erodible land, but any fall or spring tillage and even the planting operation will easily destroy the residue because of its fragileness (Erbach, 1982). Both soybean and corn produce a 1:1 ratio of residue to grain. Since soybean yields about 33% as much grain as corn, it follows that soybean residue is only about 33% that of corn. Second, soybean residue degrades quickly because of its high N content. These two factors—a small amount of residue following soybean and the fragileness of that residue—may lead to increased soil erosion following a soybean crop. Erosion following soybean is about 50% greater than that from areas where corn is grown when the same tillage system is used (Fig. 10–1). Use of conservation tillage in a soybean production system thus becomes an important consideration. No-till systems may be the only ones that consistently leave at least 20% residue cover following soybean (Erbach, 1982; Dickey et al., 1986). Use of production systems with a tillage rotation on a given field will allow some tillage to control problem weeds, bury shallow-germinating weed seeds, and incorporate P and K fertilizers (Johnson, 1987).

Any tillage practice may leave the soil prone to erosion and result in some degree of soil moisture loss. These moisture losses could reach the equivalent of 1.5 cm (0.6 in) of rainfall per tillage operation and affect soybean stand in droughty soil environments (Paul Jasa, personal communication, 2001). Thus, on well-

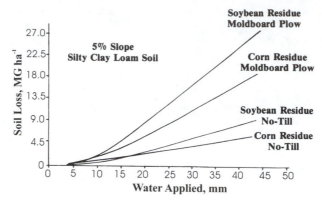

Fig. 10–1. Soil loss associated with moldboard plow and no-till systems with either corn or soybean residue at the Univ. of Nebraska, Lincoln. Water was applied at 63.5 mm hr^{-1}. Adapted from Dickey et al. (1986).

drained, coarse-textured, or drought-prone soils, conservation-tillage systems often result in greater yield than do clean-till systems (Dick et al., 1991). On moderately to poorly drained, fine-textured soils, the opposite is often true. If soil moisture is excessive to the point where denitrification, N leaching, or plant diseases increase, yield probably will decrease with conservation-tillage systems (Hoeft et al., 2000). Thus, it is advisable to avoid using no-till systems on poorly drained soils. Conservation-tillage systems seldom show a benefit when soils are not subject to early season moisture-deficit stress (Hoeft et al., 2000). Cool soil temperatures can result in variable stands, and slower crop development compared to clean-till systems.

Yiridoe et al. (2000) found that net returns from a corn–soybean rotation grown using conventional tillage and reduced tillage systems on clay soils in Ontario, Canada were similar. No-till systems generated lower net returns compared with conventional and reduced tillage systems because of lower yields and higher no-till machinery-related costs. They also found that tillage systems that use a common set of machinery for different crops in rotation production systems resulted in a savings on annual average machinery costs. The tillage system that results in the highest yields and/or the lowest management risk likely will result in the most profit (Hoeft et al., 2000). For example, soybean in pivot-irrigated conventional tillage and ridge-till systems in Nebraska have similar costs of materials, services, and field operations (Roger Selley, personal communication, 2001). Yield and risk are the important deciding factors. On the other hand, costs of rain-dependent, conventional-tillage systems are about 8% less than those for no-till systems because labor and machinery savings in no-till systems do not offset increased costs for herbicides, sprayers, and planters. These factors should be considered when selecting a tillage system.

Soybean yields on well-drained soils are often similar regardless of tillage system (Bharati et al., 1986; Elmore, 1987, 1990, 1991). However, many factors such as poor drainage, poor weed control, soybean following soybean, herbicide injury, nematodes, and diseases have been reported to reduce yields of soybean grown without preplant tillage relative to yields of soybean grown with preplant tillage (Burnside et al., 1980; Webber et al., 1987; Edwards et al., 1988; Vasilas et al., 1988). In some situations, soybean yields with no tillage are less than yields with tillage for unknown reasons (Guy and Oplinger, 1989; Philbrook and Oplinger, 1989). Van Doren and Reicosky (1987) have a detailed section on the effects of soil type and tillage on soybean yields.

Secondary tillage conducted near planting time in the midsouthern USA can delay planting of soybean on clay soils. On these poorly drained clay soils, that delay frequently becomes extended to weeks because of inconveniently-timed spring rains, and results in reduced yield and net return (Heatherly, 1999a). A stale seedbed planting system (Heatherly and Elmore, 1983; Heatherly, 1999c) has been adopted on a large hectarage of the alluvial soils of the lower Mississippi River Valley that are normally saturated or nearly saturated in the spring. The stale seedbed is described as "a seedbed that has received no seedbed preparation tillage just prior to planting. It may or may not have been tilled since harvest of the preceding crop. Any tillage conducted in the fall, winter, or early spring will have occurred sufficiently ahead of intended planting time to allow the seedbed to settle or become stale. A crop is planted in the unprepared seedbed, and weeds present before or at planting are killed with herbicides" (Heatherly, 1999c). The stale seedbed planting system does not preclude tillage; rather, it is a minimum or reduced-tillage concept where tillage is relegated to those times that will not result in delayed planting.

Weed seedlings that emerge after harvest of the preceding crop or since the last tillage operation must be dead or killed at planting in a stale seedbed (Elmore and Heatherly, 1988; Bruff and Shaw, 1992a, 1992b; Heatherly et al., 1994; Lanie et al., 1994a; Hydrick and Shaw, 1995). This can be accomplished with a preplant, foliar-applied herbicide and the crop can be planted into the stale seedbed with the dead weed residue remaining on the soil surface. If existing weeds are not killed at planting, yields (Heatherly et al., 1994; Lanie et al., 1994a; Hydrick and Shaw, 1995) and net returns (Heatherly et al., 1994) are reduced. The use rate of burndown herbicides is critical for achieving complete weed kill (Lanie et al., 1993; Hydrick and Shaw, 1994; Lanie et al., 1994a) and subsequent maximum yield potential (Lanie et al., 1993). Herbicides with either soil activity or soil and foliar activity can be applied at or after planting to manage weeds postemergence in the stale seedbed planting system (Heatherly et al.,1992a; Lanie et al., 1994b). Use of pre- and postemergent herbicides in addition to a preplant foliar-applied herbicide results in increased yield (Heatherly et al., 1993; Hydrick and Shaw, 1995) and net return (Heatherly et al., 1993) when highly competitive weeds appear after crop emergence. The effectiveness of pre- and postemergent herbicides following application of preplant, foliar-applied herbicides in stale seedbed soybean plantings depends on the rate of burndown herbicide used and weed size at burndown application (Lanie et al., 1993). If existing weeds are not killed with burndown herbicides at planting, then application of pre- and postemergent herbicides will not be effective (Oliver et al., 1993) in this system.

Annually generated state budgets can be consulted for guidance in choosing among the various tillage management systems for soybean. An example of such budgets for soybean enterprises in the southern USA (Mississippi) can be obtained from Spurlock (2002). An example of such budgets for soybean enterprises in the northern USA (Nebraska) can be obtained from Selley et al. (2001). These budgets provide cost information for equipment, fuel consumption, and labor for various implements and different sized tractors used in the various tillage systems.

10–2.3 Postplant Tillage

Rotary hoeing is effective as a weed-management tool when used shortly after soybean emergence. It is especially effective in controlling small-seeded broadleaf weeds that germinate <5 cm (2 in) from the soil surface, but is relatively ineffective on large-seeded weeds that germinate deeper in the soil, on no-till fields, and on fields with > 20 to 30% residue cover (Gunsolus, 1990). Stand loss of up to 10% during the operation may not lower yields if initial stands are as intended (Gunsolus, 1990). Rotary hoes also improve soybean emergence from crusted soils. Both rotary hoeing and interrow cultivation are best performed on dry soils with weather conditions appropriate for rapid desiccation of disrupted weeds.

There is general agreement that interrow cultivation after soybean emergence is needed only for weed management. Interrow cultivation can contribute to excessive soil loss in conventional tillage cropping systems (Edwards et al., 1993). Soybean plantings in the USA are made in rows ranging in width from about 20 to 102 cm (8–40 in). Post-plant tillage for soybean planted in wide rows (>50 cm or 20 in) may involve one to three passes with a row crop cultivator as needed for weed control. The ability to use postplant cultivation for weed management is one of the few reasons to plant soybean in wide rows. Postplant cultivation is most cost-effective when herbicides are applied on a narrow band over the row.

Banded herbicide application combined with interrow cultivation in wide rows can be used to effectively manage weeds (Buhler et al., 1992; Poston et al., 1992; Heatherly et al., 2001a, 2001b), reduce weed control costs (Buhler et al., 1997; Heatherly et al., 2001a, 2001b), and reduce amount of herbicide introduced into the environment (Poston et al., 1992; Swanton et al., 1998). Use of combinations of preemergent (PRE) and postemergent (POST) herbicides with POST cultivation is common in wide-row production systems in the midsouthern USA (Heatherly and Elmore, 1991; Poston et al., 1992; Heatherly et al., 1993, 1994; Oliver et al., 1993; Hydrick and Shaw, 1995; Askew et al., 1998). Herbicides banded over the crop row and cultivation of interrow areas can provide complementary weed control (Griffin et al., 1993; Newson and Shaw, 1996), and may result in lower weed management costs than for broadcast applications of herbicides (Krausz et al., 1995; Heatherly et al, 2001a) in any row spacing. Interrow cultivation alone will not control weeds over time, and will result in lower yield and net returns (Buhler et al., 1997) than when supplemented with herbicide weed control. Narrow-row systems preclude POST cultivation normally used in wide rows (Newsom and Shaw, 1996; Buhler et al., 1997; Hooker et al., 1997; Swanton et al., 1998). In narrow-row soybean plantings, effective weed management systems almost exclusively involve herbicides (Oliver et al., 1993; Johnson et al., 1997, 1998a). This can lead to improved weed control in narrow-row systems that result in greater yield and net returns compared with wide-row systems (Mickelson and Renner, 1997; Swanton et al., 1998; Heatherly et al., 2001a, 2001b). However, increased net returns are dependent on both economical weed management for, and increased yield from, narrow-row systems. Both of these requirements may not occur, and if not, can lead to lower net returns. The use of narrow rows and post-emergence weed management with herbicides has replaced between-row cultivation on a large portion of U.S. soybean plantings.

10–3 SOIL FERTILITY

10–3.1 Nitrogen

From 25 to 75% of N in mature soybean plants is from symbiotic dinitrogen (N_2) fixation by *Bradyrhizobia japonicum*; the remaining is from soil N supply (Varco, 1999). Physiological analysis of energy requirements indicates that N assimilation via N_2 fixation requires more photosynthate than does NO_3 uptake and reduction. Still, both sources of N are essential for maximum yield. High soil NO_3 inhibits symbiotic fixation. Considerable conflicting research surrounds the question of soybean responses to both preplant and post-plant N fertilizer application.

Soybean grown on most soils does not respond to preplant N fertilization (Johnson, 1987; Varco, 1999; Hoeft et al., 2000). The exceptions cited by Johnson (1987) were applications made to soils that were somewhat poorly drained, were low in organic matter, and/or were strongly acid. Ferguson et al. (2000) summarized work from Nebraska that showed positive responses to preplant N applications about half the time, but determined that it was not possible to predict soybean response to N fertilizer based on soil properties. The situations with positive responses often either had very low residual N, low N mineralization capability, or soil pH so low that it inhibited nodulation and N_2 fixation. In these cases, 56 to 112 kg N ha^{-1} (50–100 lb N acre^{-1}) increased yields. Kansas scientists found that soybean planted into large amounts of wheat residue responded to 11 to 22 kg N ha^{-1} (10–20 lb N acre^{-1}) of starter N because inorganic N is temporarily immobilized by soil microorganisms decomposing the straw. They also found that soybean planted on recently leveled soils may respond to 33 to 45 kg N ha^{-1} (30–40 lb N acre^{-1}) because of low soil N (Whitney, 1997).

Soybean N uptake reaches a maximum rate of up to 4.5 kg N ha^{-1} d^{-1} (4 lb N acre^{-1} d^{-1}) between the R3 and full pod (R4) growth stages. Because of this, several researchers have attempted to increase yields by applying N during late vegetative and early reproductive growth stages. Conflicting results are reported; however, most show no positive response. Nevertheless, a recent report from Kansas found that N applications at R3 significantly increased yields and net returns at six of eight irrigated sites (Wesley et al., 1998). Generally, rates of 22 and 45 kg N ha^{-1} (20 and 40 lb N acre^{-1}) provided similar increases when compared to a 0 N rate. Responsive soils generally had low organic matter, low soil profile N, and were relatively high yielding (>3700 kg ha^{-1} or >55 bu acre^{-1}). These Kansas scientists concluded that applications of additional N during reproductive development should be considered for irrigated soybean with high-yield potential. This small amount of N could be applied through a center pivot irrigation system, but Ferguson et al. (2000) suggests that it be considered on an experimental basis until more consistently positive results are reported. In most cases, N fertilization of soybean is an unnecessary expenditure on nonproblem soils (Varco, 1999; Hoeft et al., 2000). In addition, adding starter N fertilizers to soybean delays or impedes nodulation, and thus can reduce N₂ fixation.

10–3.2 Lime and Soil pH

Liming is an important prerequisite for profitable soybean production on acid soils (Johnson, 1987). Soil pH values of 6 to 6.5 are suitable to optimize yield and performance in corn–soybean rotations. Slightly higher pH values of 6.5 to 7 are needed if alfalfa (*Medicago sativa* L.) or clover (*Trifolium* spp. L.) are included in the rotation (Hoeft et al., 2000). Liming acid soils to achieve these pH levels improves the ability of a plant to take up nutrients; reduces concentrations of potentially toxic elements such as H, Al, and Mn; increases the availability of Ca, Mg, and Mo; and improves N₂ fixation by *B. japonicum* (Mengel et al., 1987). Figure 10–2 graphically shows the importance of pH maintenance for optimum availability of essential nutrients. In addition, liming acid soils enhances microbial breakdown of crop residues.

Lime sources vary in their neutralizing capability and fineness of grind. These factors, plus the soil pH and the depth to which neutralization is necessary, dictate the amount of lime required. Variation in soil pH occurs naturally among and within soil series. It is possible to improve soil pH and more accurately predict lime requirements on a site-specific basis with site-specific lime applications based on spatial variability. This may improve soybean yields on a whole-field basis as well (Pierce and Warncke, 2000).

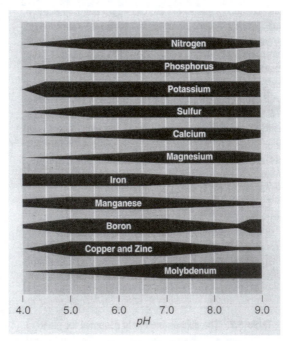

Fig. 10–2. Soil pH effects the availability of plant nutrients. The thicker the bar, the more of the nutrient is available. The best overall balance is between pH 6.0 and 7.0. From Hoeft et al. (2000).

Alkaline soils present problems for soybean production. Availability of Fe, Mn, Cu, B, Zn, and P all decrease with increasing pH (Fig. 10–2). Iron chlorosis is common on calcareous soils with a high pH. Damaging effects from using some soil-applied herbicides (e.g., metribuzin) and carryover of triazine herbicides is more likely on alkaline soils, and can result in loss of plants in an emerging stand. Because lowering soil pH is not practical for soybean production, management practices for alkaline soils include using tolerant cultivars and increasing seeding rates to about 40 m^{-1} (12 ft^{-1}) of row length to ensure plant adjacency since soybean (even an intolerant cultivar) tolerates alkaline soils better with close intrarow spacing (Ferguson et al., 2000). This precludes using narrow rows (<50 cm or <20 in) since seeding rates would exceed 775 000 seed ha^{-1} (314 000 acre^{-1}) and be prohibitively expensive. In the most difficult situations, iron chelate applied with the seed at planting may improve soybean performance (Penas and Wiese, 1989; Ferguson et al., 2000).

10–3.3 Phosphorus and Potassium

Soybean is less responsive to P applications than are corn, wheat, alfalfa, and clover (Ferguson et al., 2000; Hoeft et al., 2000). Although soybean P$_2$O$_5$ requirements are considerably less than those of either N or K (Table 10–11), all three are equally important for plant growth and productivity. Soil test P levels of 22.5 to 45 kg P ha^{-1} (20–40 lb P acre^{-1}) are considered adequate for maximum yield (Varco, 1999). Broadcast application of P fertilizer is better than banded application at planting unless P values are low. Soybean grown in rotation with well-fertilized crops such as corn and wheat requires minimal fertilizer P to optimize yields (Varco, 1999). Phosphorus-deficiency symptoms are most obvious on small plants; young plants need higher P content in tissues than do older plants. This is accentuated by the fact that P is less available for uptake in cool soils typical of early season growing conditions. More than 90% of the alluvial and coastal plain soils in the midsouthern USA are in the medium to high P category, and thus require no P fertilizer for optimum soybean yield (Varco, 1999).

Soybean requires large amounts of K (Table 10–11). In contrast to P, nearly all northern soils except sands (K readily leaches from sandy soils) have substantial amounts of K within the rooting zone (Hoeft et al., 2000). Only a small portion of the K in soils is available for plant growth, yet K is rarely required in northern states like Nebraska. In contrast to P, seedling demands for K are relatively small. Potassium deficiencies generally appear between late flowering and early seedfill.

Table 10–11. Nutrient content of soybean with a 3400 kg ha^{-1} seed yield (50 bu acre^{-1}). Adapted from Ferguson et al. (2000).

Nutrient	Grain	Plant	Total	Grain	Plant	Total
	kg ha^{-1}			lb acre^{-1}		
Nitrogen	211	142	353	188	127	315
Phosphorus (P$_2$O$_5$)	49	34	83	44	30	74
Potassium (K$_2$O)	74	646	720	66	576	642
Sulfur	6	17	23	5	15	20
Zinc	0.06	0.34	0.40	0.05	0.3	0.35

Table 10–12. Soil test P and K categories used by the Mississippi State University Soil Testing Laboratory, and recommended P and K fertilization rates for soybean as recommended by Louisiana State University (Funderburg, 1996) and Mississippi State University (Varco, 1999).†

Soil test category	Soil test P	Recommended P rate§	Cation exchange capacity‡				Recommended K rate¶
			<7	7 to 14	15 to 25	>25	
			Soil test K				
			kg ha^{-1}				
Very low	0–20	39#, 58††	0–56	0–67	0–78	0–90	75#, 112††
Low	21–40	29#,††	57–123	68–157	79–179	91–202	56#,††
Medium	41–81	15#,††	124–179	158–213	180–235	203–269	28#, 56††
High	82–161	0#,††	180–314	214–376	236–415	270–471	0#,††
Very high	161+	0#,††	314+	376+	415+	471+	0#,††

† Multiply all tabled values by 0.893 to convert to lb acre^{-1}.
‡ Increasing values indicate increased capacity for adsorbing cations such as K. As the CEC increases, a given amount of exchangeable K will equilibrate with less K^+ in solution. Fine-textured soils (clays) require a higher level of exchangeable K to produce the same available K^+ that coarse-textured soils (sands) do (Foth and Ellis, 1997).
§ Multiply values by 2.29 to convert to P_2O_5 fertilizer rates.
¶ Multiply values by 1.2 to convert to K_2O fertilizer rates.
# Recommended by Louisiana State University.
†† Recommended by Mississippi State University.

More than 85% of the alluvial and coastal plain soils in the midsouthern U.S. test in the medium to high category, and require no K fertilizer for optimum soybean yield (Varco, 1999).

Recommendations in the midsouthern USA for the addition of P and K to soils are based on soil test values (Varco, 1999). Recommended P and K additions based on soil test categories used by the Louisiana State University (Funderburg, 1996) and Mississippi State University (Varco, 1999) Soil Testing Laboratories are shown in Table 10–12. The recommended rate of P at the medium soil test level is essentially a maintenance fertilization rate with a low probability of a yield response. The underlying philosophy in the K categorization is that greater soil test K levels are required with increasing cation exchange capacity (Foth and Ellis, 1997).

Two philosophical approaches to P and K fertilization are followed in the soybean production area of the northern states (Frank, 2000; Sander and Penas, 2000). The more western states in the North (Kansas, Nebraska, and South Dakota) use a deficiency correction approach (Whitney, 1997; Ferguson et al., 2000), whereas the eastern states use a modified crop removal or maintenance approach (Hoeft et al., 2000; Vitosh et al., 2001). In the deficiency correction approach, both P and K are applied for crops on soils where yield increases are expected. This approach requires accurate soil testing and analysis. In the crop-removal or maintenance approach, nutrient removal amounts of the previous crops are replaced once the nutrient levels of the soils are increased to a specific maintenance range. Once a maintenance level is achieved, soil sampling may or may not be necessary with this approach. The two approaches result in different P applicaton rates on soils with the same P levels. For example, P application in Nebraska is not triggered until the soil test level is ≤10 ppm P (Bray-1) using the deficiency correction approach. In states where the crop removal/maintenance approach is used, fertilizer recommendations are equal to crop removal on soils testing 15 to 30 ppm (Bray-1 P). For soils testing <15 ppm, additional P is recommended to build soil levels. Reduced P rates are suggested for soils >30 ppm P. Soils in the eastern and southeastern USA as well as some in Wisconsin and Minnesota do not have the capacity to quickly release K to rapidly growing plants. In those areas, K recommendations are inversely correlated with cation exchange capacity (CEC) of the soil (Vitosh et al., 2001). Soils in the western states are relatively unweathered and release K almost as rapidly as plants need it.

Band application of P is more efficient than broadcast application if soil P values are low. However, bands at least 2.5 cm (1 in) from the seed are necessary to prevent seedling injury. A broadcast-incorporated application of K before planting is efficient. In conservation tillage and no-till systems, nutrient application should occur with some amount of tillage to incorporate the relatively immobile P and K. Since approaches to soil testing vary with soils and states, P and K recommendations provided by soil-testing laboratories in the state or region where the crop will be grown should be followed.

10–3.4 Secondary/Micro Nutrients

Micronutrient deficiencies are the exception rather than the rule in soybean-producing areas of the USA. In many cases, simply maintaining a proper pH level prevents many problems. Two perspectives exist on the use of secondary (Ca, Mg, and S) and micronutrients (Cu, Fe, Mn. Mo, Zn). The first is preventative application, and the second is deficiency correction. There is often a narrow range between deficiency and toxicity; thus, application techniques and rates are critical. Hoeft et al. (2000) has suggested seven practices in dealing with micronutrients: (i) know deficiency symptoms and then watch for them in the first 1 to 2 mo after emergence; (ii) observe for deficiency symptoms on more sensitive crops to provide advanced warning of problems that may develop on soybean (Table 10–13); (iii) know those soil situations where deficiencies are likely to develop (Table 10–13); (iv) test soils for micronutrients, but use more reliable plant analyses to determine if adequate nutrients are being supplied to the crop; (v) avoid crop injury by applying the proper form of a micronutrient in the proper place; (vi) control soil pH by liming acid soil; and (vii) consult experts and reputable testing laboratories for recommendations after determining a problem does indeed exist.

10–4 PLANTING PRACTICES

10–4.1 Planting Date

Soybean tolerates a relatively wide range of planting dates in both the northern and southern soybean regions of the USA. Optimum planting dates for most of the northern states range from early to mid-May. This was the previous paradigm for soybean planting date in the southern region, but has been replaced by earlier planting from late March through late April in the midsouthern USA. Planting date affects the plant size attained before floral induction. Yields in most cases decline rapidly with June and later planting in both the north and south.

Table 10–13. Likely soil conditions for secondary- and micro-nutrient deficiencies observed in soybean. Adapted from Hoeft et al. (2000), Johnson (1987), and Vitosh et al. (2001).

Nutrient	Soil conditions likely to create deficiency	Most sensitive crop	Relative ratings of soil test procedures†
Calcium	pH < 5.0	Alfalfa, clovers	40
Copper	High organic matter (mucks, peat soils with pH < 5.3); highly weathered, sandy soils	Peanut	Organic soils = 20; Mineral soils = 5
Iron	pH > 7.3; wet soils; poorly aerated soil; cool temperature	Soybean, navy bean, millet, grain sorghum	pH > 7.5 = 30; pH < 7.5 = 10
Magnesium	Acid soils; sandy soils; high K levels	Corn	40
Manganese	pH > 7.3 (Mn deficiency); mucks, peat soils with pH > 5.8; black sands and lake-bed depression soils with pH > 6.2; (Note: pH < 5.2 = Mn toxicity)	Soybean, navy bean, oat	pH > 7.5 = 40; pH <7.5 = 10
Molybdenum	pH < 5.0; strongly weathered soils; soils mostly east of Miss. River with moderate to heavy rainfall	Soybean, alfalfa, pea (affects primarily nodulation and N_2 fixation)	
Zinc	Exposed subsoil; areas leveled for irrigation; peat and muck soils and mineral soils with pH < 6.5; soils, especially sandy, with low organic matter; high pH, very high P soils; cool, wet soils	Corn	45

† Adapted from Hoeft et al. (2000). Other relative ratings are water pH = 100; P = 85; K = 70; organic matter = 75.

In some regions of the northern USA, indeterminate soybean cultivars planted earlier and later than the recommended planting date range of 1 May to mid-June often are shorter and have fewer nodes (Beaver and Johnson, 1981; Wilcox and Frankenberger, 1987; Hoeft et al., 2000). Determinate cultivars planted from mid-May through mid-June often have similar or greater height and number of nodes as those planted earlier. Beaver and Johnson (1981) determined that node numbers of indeterminate cultivars steadily declined as planting date was delayed from mid-May through early July, whereas node numbers of determinate cultivars remained fairly constant over this range of planting dates. Planting dates of 10 to 20 May are considered optimum for achieving adequate vegetative growth and maximum yield potential.

Planting after 1 June generally results in lower yields due to a reduction in size of plants. Research in Ohio with both determinate and indeterminate cultivars determined that yield declines about 22 kg ha^{-1} (0.33 bu acre^{-1}) d^{-1} of planting date delay after the first of May (Beuerlein, 1988). In an Illinois study, seed yields of indeterminate cultivars declined linearly and averaged 33% as date of planting was delayed from early May to early July. Seed yields of determinate cultivars did not begin to decline appreciably until planting dates were delayed past early June. In plantings after early June, they declined at a greater rate than did yields of indeterminate cultivars (Beaver and Johnson, 1981). Determinate cultivars in both Nebraska and Indiana differed in response to planting date compared to indeterminate cultivars (Wilcox and Frankenberger, 1987; Elmore, 1990). Determinate cultivar yields were best with late May to early June planting, while indeterminate cultivar yields were best with early to late May planting. However, seeding rates and cultivar growth habit are often confounded in many northern U.S. studies (Elmore, 1990) because earlier work showed that determinate cultivars should be planted at higher rates than indeterminate cultivars (Cooper, 1981). More recent work has shown this is not necessary (see section 10–4.3). Protein levels generally increase with delayed planting, but these increases do not compensate for the associated reductions in oil content and yield (Helms et al., 1990). Along with yield reductions, delayed planting can reduce severity of brown stem rot [*Phialophora gregata* (Allington and S.W. Chamberlain) W. Gams] (Grau et al., 1994) and sudden death syndrome [*Fusarium solani* (Mort.) Sacc. f. sp. *glycines*] in susceptible cultivars.

In the northern USA, planting the latest adapted cultivars early in the growing season followed by planting early to mid-season adapted cultivars during mid-May through early June has some merit. Planting early-maturing cultivars early (before 1 May) could result in flowering in late May/early June and subsequent short stature, and the occurrrence of critical reproductive stages during the moisture-deficit periods of July and August. Planting late-adapted cultivars at the above-mentioned very early planting dates avoids these potential problems. Mid-season adapted cultivars are advised for later planting dates, including doublecrop systems. These cultivars will grow taller and have more nodes than will shorter-season cultivars when planted late, and will have less risk of late-season frost injury compared with full-season cultivars. Frost injury to soybean after beginning maturity (R7) will not reduce yield, but frost before this stage can reduce yield and seed quality. This is important since half of the potential seed dry matter accumulation in soybean occurs after R6.5 [pod cavites filled (Whiting et al., 1988)].

In traditional northern corn–-soybean rotation systems, producers usually plant corn before planting soybean to realize maximum yield from corn. However, this may result in not having adequate rainfall or soil moisture to sustain soybean podfill during August. In addition, an early fall frost can reduce late-planted soybean yields. Since it appears that soybean yield is relatively stable over a wide range of planting dates, some producers are planting soybean before corn to alleviate machinery management constraints. Information from March- or April-planted soybean in the northern USA is limited, but available data indicate that yields from April-planted soybean can be about 1000 kg ha^{-1} (15 bu acre^{-1}) greater than those from June plantings (Paul Jasa, personal communication, 2001). These data are from trials planted in no-till seedbeds and using a seed-applied fungicide. Seed germination and growth were slower with the early plantings. In preliminary studies in Illinois, yields from early April soybean plantings have been about 17% less than those from late-April plantings (E.D. Nafziger, personal communication, 2002).

Concerns with early (March and April) plantings in the northern USA include early season frost injury and insect feeding. Frost injury to newly emerged plants with unfolding cotyledonary leaves in early plantings can significantly reduce stands (E.D. Nafziger, personal communication, 2002). There is some evidence that soybean in the early vegetative growth stages is more frost tolerant than at later growth stages. If the terminal growing point of soybean is killed, regrowth can occur

Table 10–14. Summary of average temperature and rainfall and pan evaporation for growing season months, Stoneville, MS, 1964 to 1993 (Boykin et al., 1995), and Sikeston, MO, 1961 to 1990 (temperature and rain) and 1985 to 1997 (pan evap.) (Owenby and Ezell, 1992; J. Henggeler, unpublished data, 1998).†

	Stoneville					Sikeston				
	Air temp.		Rain	Pan evap.	Diff.	Air temp.		Rain	Pan evap.	Diff.
Month	Max	Min				Max	Min			
	°C		cm			°C		cm		
Apr.	23.3	11.7	13.7	15.5	−1.8	20.6	8.3	11.7	16.3	−4.6
May	27.8	16.7	12.7	19.6	−6.9	26.1	13.3	13.2	19.8	−6.6
June	32.2	20.6	9.4	21.6	−12.2	30.6	18.3	9.4	23.1	−13.7
July	32.8	22.2	9.4	20.8	−11.4	32.8	20.6	9.6	24.1	−14.5
Aug.	32.2	21.1	5.8	18.5	−12.7	31.1	18.9	8.4	20.3	−11.9
Sept.	29.4	17.2	8.6	14.7	−6.1	27.8	15.0	9.9	15.0	−5.1

† Multiply temperature values by 1.8 and add 32 to convert to °F; multiply rain and evaporation values by 0.394 to convert to inches.

from the cotyledonary node or the lower nodes if the lateral buds were not injured. Regrowth from the cotyledonary nodes results in an abnormal plant with two equally dominant stems. The effect of this abnormal plant on later development, lodging and stem breakage, and yield is not well-documented. Soybean stands from plantings made before mid-May in the northern USA attract adult bean leaf beetles and offer an ideal environment for egg laying. Even though mid-May and later planting may minimize the initial colonization by beetles (Hunt et al., 1994), the insect often migrates into these later-planted fields from surrounding areas that were planted earlier. Soybean cultivars are not resistant to bean leaf beetle feeding.

In the midsouthern USA, the ESPS is the new paradigm for soybean production (Bowers, 1995; Boquet, 1998; Heatherly, 1999a; Heatherly and Spurlock, 1999). The ESPS may use both indeterminate (MG 2 through 4) and determinate (MG 5) cultivars (Heatherly and Spurlock, 1999; Bowers et al., 2000). This system replaces the conventional soybean production system (CSPS) which includes May and June planting of later-maturing cultivars. Choice of row spacing in the ESPS depends on whether indeterminate or determinate cultivars are used (Heatherly and Bowers, 1998; Bowers et al., 2000; Heatherly et al., 2002b). Indeterminate cultivars should be planted in narrow (<50 cm or 20 in) rows, while determinate cultivars can be planted in either wide or narrow rows. The purpose of using this earlier planting system is to avoid much of the drought stress that is associated with the high temperatures and moisture deficits that result from decreasing rainfall and increasing evaporative demand in July, August, and September, as verified by long-term weather records for Stoneville, MS and Sikeston, MO (Table 10–14). Increasing drought stress during the growing season is detrimental especially to yield of MG V and later soybean cultivars that are planted in May and later because they are setting pods and filling seed during this period. Use of the ESPS also lowers production risks (Boquet, 1998). A detailed outline of this system has been presented by Heatherly and Bowers (1998) and Heatherly (1999a).

The data in Table 10–15 show nonirrigated (NI) and irrigated (I) soybean yields from research at Stoneville, MS for the 1979 through 1990 period. These data

Table 10–15. Yield of nonirrigated (NI) and irrigated (I) soybean cultivars grown in a conventional soybean production system at Stoneville, MS, 1979 to 1990. From Heatherly (1999a). Adapted from Heatherly (1983, 1988), Heatherly and Elmore (1986), Heatherly and Pringle (1991), Heatherly and Spurlock (1993), and Heatherly et al. (1994).

Year	Date of planting	Cultivar (MG)	Irrigation level			
			NI		I	
			kg ha^{-1}	bu acre^{-1}	kg ha^{-1}	bu acre^{-1}
1979	13 June	Bedford (5)	2748	40.9	2668	39.7
		Tracy (6)	3367	50.1	3373	50.2
		Bragg (7)	3165	47.1	3588	53.4
1980	8 May	Bedford	732	10.9	1996	29.7
		Tracy	1149	17.1	2809	41.8
		Bragg	1317	19.6	3555	52.9
1981	13 May	Bedford	981	14.6	2775	41.3
		Bragg	1028	15.3	3273	48.7
1982	12 May	Bedford	974	14.5	2244	33.4
		Braxton (7)	1008	15.0	2715	40.4
1984	14 May	Braxton	1357	20.2	3494	52.0
1985	2 May	Braxton	1599	23.8	2876	42.8
1986	15 May	Braxton	101	1.5	2594	38.6
1986	3 June	Sharkey (6)	376	5.6	2950	43.9
1987	11 May	Sharkey	706	10.5	2688	40.0
1988	16 May	Sharkey	2278	33.9	2675	39.8
1987	8 June	A 5980 (5)	914	13.6	2614	38.9
1987	6 May	Leflore (6)	1102	16.4	2903	43.2
1988	25 May	A 5980	2641	39.3	3649	54.3
		Leflore	2211	32.9	3084	45.9
1989	8 May	A 5980	2675	39.8	2769	41.2
		Leflore	1781	26.5	2150	32.0
1990	2 May	A 5980	1277	19.0	2977	44.3
		Leflore	1068	15.9	3326	49.5

show that planting cultivars in MGs 5, 6, and 7 in May and June and not irrigating was a high-risk enterprise during this period. In many years, NI yields were below 1345 kg ha^{-1} (20 bu acre^{-1}) and only infrequently exceeded 1680 kg ha^{-1} (25 bu acre^{-1}). There was usually large response to irrigation in dry years, but even this large response to irrigation resulted in only modest yields [2850—3150 kg ha^{-1} (mid-40s bu acre^{-1})] of I soybean. Irrigated yields of May-planted soybean ranged from 2000 kg ha^{-1} (29.7 bu acre^{-1}) to 3650 kg ha^{-1} (54.3 bu acre^{-1}), but the frequency of I yields exceeding 3365 kg ha^{-1} (50 bu acre^{-1}) was low. Bowers (1995) conducted 3 yr (1986–1988) of NI studies at two northeast Texas locations (Blossom and Hooks—Table 10–16). Two facts are obvious from this report: (i) early-maturing cultivars planted in April yielded more than later-maturing cultivars planted in May, and (ii) early-maturing cultivars planted in May yielded as much as or more than later-maturing cultivars planted in May. Heatherly and Spurlock (1999) conducted NI and I studies at Stoneville on Sharkey clay in 1992 and 1994 through 1997 (Table 10–17). The following conclusion can be drawn from those data: In most years, cultivars in MGs 4 and 5 that are planted in April and grown with or without irrigation produced greater yields and net returns compared to conventional I and NI May and later plantings.

Stink bug management in ESPS plantings in the southern portions of the mid-southern USA is as critical as for conventional plantings (Baur et al., 2000). Early

Table 10–16. Yield of MG III through VII soybean cultivars planted in April and May at Blossom and Hooks, Texas in 1986, 1987, and 1988. Adapted from Bowers (1995).

Planting date†	Cultivar (MG)	Year 1986	Year 1987	Year 1988
		kg ha^{-1} (bu acre^{-1})		
	Blossom			
Apr.	Williams 82 (3)	2956 (44.0)	2842 (42.3)	1518 (22.6)
	Crawford (4)	1720 (25.6)	1787 (26.6)	2318 (34.5)
	Forrest (5)	531 (7.9)	1176 (17.5)	2526 (37.6)
	Leflore (6)	289 (4.3)	524 (7.8)	1566 (23.3)
	Bragg (7)	215 (3.2)	356 (5.3)	1082 (16.1)
May	Williams 82	1008 (15.0)	927 (13.8)	--
	Crawford	961 (14.3)	947 (14.1)	--
	Forrest	867 (12.9)	860 (12.8)	--
	Leflore	719 (10.7)	289 (4.3)	--
	Bragg	255 (3.8)	148 (2.2)	--
	Hooks			
Apr.	Williams 82	3675 (54.7)	1612 (24.0)	2392 (35.6)
	Crawford	3238 (48.2)	2116 (31.5)	3218 (47.9)
	Forrest	2473 (36.8)	759 (11.3)	3144 (46.8)
	Leflore	2862 (42.6)	443 (6.6)	2728 (40.5)
	Bragg	1848 (27.5)	752 (11.2)	2553 (38.0)
May	Williams 82	2419 (36.0)	1693 (25.2)	--
	Crawford	2150 (32.0)	726 (10.8)	--
	Forrest	2943 (43.8)	544 (8.1)	--
	Leflore	2452 (36.5)	1068 (15.9)	--
	Bragg	1915 (28.5)	1384 (20.6)	--

† Blossom: 16 Apr. and 15 May 1986; 17 Apr. and 12 May 1987; 22 Apr. and 6 May 1988. Hooks: 17 Apr. and 14 May 1986; 15 Apr. and 11 May 1987; 21 Apr. and 7 May 1988.

Table 10–17. Average seed yields and net returns from irrigated and nonirrigated April and May plantings of Maturity Group (MG) 4 and 5 soybean cultivars at Stoneville, MS, 1992 and 1994 through 1997. Adapted from Heatherly and Spurlock (1999).

MG	Seed yield — Planting date† April	May	Avg.	Net return — Planting date April	May	Avg.
	kg ha^{-1} (bu acre^{-1})			$ ha^{-1} ($ acre^{-1})		
	Irrigated					
4	3770 (56.2)	3350 (49.9)	3560 (53.0)	395 (160)	283 (114)	339 (137)
5	3890 (57.9)	3430 (51.1)	3660 (54.5)	418 (169)	301 (122)	359 (145)
Avg.	3830 (57.0)	3390 (50.5)		406 (164)	292 (118)	
	Nonirrigated					
4	2245 (33.4)	1905 (28.4)	2075 (30.9)	205 (83)	109 (44)	157 (63)
5	2630 (39.2)	2210 (32.9)	2420 (36.0)	285 (115)	186 (75)	235 (95)
Avg.	2440 (36.3)	2060 (30.7)		245 (99)	148 (60)	

† 15 Apr. and 27 May 1992; 21 Apr. and 13 May 1994; 18 Apr. and 9 May 1995; 30 Apr. and 15 May 1996; 9 Apr. and 12 May 1997.

planting of early-maturing cultivars results in more early-season insect predators and in a lower likelihood of economic injury from lepidopterous and coleopterous defoliators that occur late in the growing season (Baur et al., 2000). Either ESPS alone or in combination with CSPS (depending on availability of seasonal labor) in eastern Kansas offers farmers in that region a diversification strategy for greater farm net returns (Casey et al., 1998). The use of ESPS allowed the distribution of labor and machinery field time requirements over more time and resulted in greater farm income even though soybean seed costs in the ESPS were arbitrarily $64 ha^{-1} ($26 acre^{-1}) higher. In the more northern regions of the southern USA (Tennessee and Kentucky), or the transition zone between southern and northern production areas, use of the ESPS may not be advantageous (Pfeiffer et al., 1995; Kane et al., 1997; Logan et al., 1998). In a Kentucky study where early-maturing cultivars (MGs 1 through 3) were planted late to simulate the planting date of soybean double-cropped with wheat, there was no alleviation of the yield penalty associated with the late planting of the usual MG 3 and MG 4 cultivars (Egli and Bruening, 2000). On the other hand, these results do indicate that early-maturing cultivars can be used in late plantings for a particular region. This shortened growing season for late plantings may mean lower management costs (fewer inputs) and lower risk since the early-maturing cultivars will be in the field for a shorter time.

10–4.2 Row Spacing

Certain tenets pertaining to row spacing for soybean have become accepted. Soybean grown in narrow rows (<50 cm [20 in] in the southern USA and <38 cm [15 in] in the northern USA) canopies sooner, and thus intercepts radiation that would have been expended on the soil surface in a wide-row environment. Soybean grown in narrow rows uses more soil water or depletes soil water more rapidly during vegetative development (Taylor, 1980; Van Doren and Reicosky, 1987). This enhanced early season water use is usually beneficial; however, it may be detrimental in rainfed environments where stored soil water from early season rainfall is not sufficient to compensate for low rainfall during reproductive development. Soybean grown in narrow rows results in less weed presence than when grown in wide-row systems due to suppression of weed seed germination in soil surfaces shaded by a closed canopy. Soybean grown in narrow rows precludes postemergent cultivation in most cases, thus requiring weed management by herbicides. This may lead to greater weed management expense in narrow- vs. wide-row soybean.

In the northern soybean-growing region of North America, soybean grown in narrow rows generally outyields soybean grown in wide rows (Devlin et al., 1995; Mickelson and Renner, 1997; Elmore, 1998; Swanton et al., 1998; Nelson and Renner, 1999). Reasons for this narrow-row advantage may be related to better weed control in narrow rows (Mickelson and Renner, 1997; Nelson and Renner, 1999), drought-free growing seasons (Devlin et al., 1995), and less weed resurgence following early season weed management in narrow rows (Yelverton and Coble, 1991).

When the only factor limiting productivity is light, equidistant plant spacings result in maximum crop yields (Johnson, 1987). Many recent research reports from northern states like Illinois, Iowa, Indiana, Minnesota, Missouri, and Ohio indicate that soybean grown in rows 25 cm (10 in) or less in width have greater yields relative to wider rows, which is similar to what Johnson (1987) reported. However, there is an increasing amount of research from western (Kansas and Nebraska) and northern (Michigan and Wisconsin) states that indicates that soybean in wider rows (>50 cm or 20 in) may yield more than those in rows that are 25 cm (10 in) or less in width (Devlin et al., 1995 [Kansas]; Graterol et al., 1996 [Nebraska]; Elmore, 1998 [Nebraska]; Nelson and Renner, 1999 [Michigan]; Bertram and Oplinger, 2000 [Wisconsin]). These reports of low yields from narrow rows are often related to situations with poor early season growing conditions or poor environments. Kansas data showed that yields were greater in 20-cm-(8-in-) wide rows than in 76-cm- (30-in-) wide rows in "high yield" environments, with the reverse occurring in "low yield" environments (Devlin et al., 1995). They classified "high yield" environments as those that produced yields greater than about 3400 kg ha^{-1} (50 bu acre^{-1}) and "low yield" environments as those that produced yields less than about 2700 kg ha^{-1} (40 bu acre^{-1}). A report from Michigan showed that soybean in 19-cm- (7.5-in-) wide rows with good weed control from either hand weeding or glyphosate yielded the same as soybean in 76-cm- (30-in-) wide rows (Nelson and Renner, 1999). Narrow rows provided better weed control with all other herbicides tested, and thus yields were better in narrow rows than in wide rows in nonglyphosate treatments. Even with early season stress that limits yield responses in narrow rows, canopy closure rates are faster with narrow rows than with wide rows. For double-cropped soybean in the northern USA which are planted from early to late June, Beuerlein (2001b) states that narrow rows (18 cm or 7 in) are required for maximum yield. Indeterminate and determinate cultivars often respond the same to row width if early season stress is absent.

Sclerotinia stem rot [*Sclerotinia sclerotiorum* (Lib.) de Bary] is a greater problem in narrow rows because of canopy microclimate and more interrow shading in narrow rows (25–38 cm or 10–15 in) vs. wide rows (76 cm or 30 in) (Grau and Radke, 1984). However, narrow rows (17 vs. 76 cm or 7 vs. 30 in) do not appear to affect brown stem rot [*Phialophora gregata* (Allington and D.W. Chamberlain) W. Gams] severity (Grau et al., 1994). If a drill is used to plant narrow rows, seeding rates should be increased by 10 to 15% to improve plant emergence and subsequent stands. Also, avoid using large seed (<5300 seed kg^{-1} or <2400 seed lb^{-1}) in drills since they may be damaged by the seed-metering device (Beuerlein, 1995).

Soybean production using row widths of < 25 cm (10 in) is giving way to production in mid-width row spacings of 38 to 50 cm (15–20 in) in the major northern soybean-producing states. Studies that have included a mid-range of row widths often show that soybean yields are optimized in row spacings of 38 to 50 cm (15–20 in) (Elmore, 1998; Bertram and Oplinger, 2000), which is consistent with Johnson (1987). Bullock et al. (1998) found that yields of an indeterminate cultivar were increased as row widths were reduced from 114 (45 in) and 76 cm (30 in) to 38 cm (15 in) as a result of increased pods plant^{-1}, plant height, and harvest index. They suggested that these responses were due to the beneficial effects of narrow rows prior to the main grain-fill period, which is similar to results reported by Duncan (1986). In contrast, Singer (2001) found no differences in yield, pods plant^{-1}, branches, or harvest index between 18- to 20-cm- (7- to 8-in-) wide rows and 76-cm- (30-in-) wide rows at relatively high yield levels. In a series of narrow-row, no-till, multi-state studies, soybean grown in rows spaced 18 to 25 cm (7–10 in) apart yielded more than soybean grown in rows spaced 76 to 91 cm (30–36 in) apart at 6 of 21 sites. At one site, soybean in wide rows outyielded soybean in narrow rows (Oplinger et al., 1998b). Narrow rows yielded the same as intermediate rows (37–56 cm or 15–22 in).

Reasons for planting soybean in mid-width row spacings (38–50 cm or 15–20 in) in the northern states are given by Hoeft et al. (2000). They are: (i) white mold is becoming more of a problem with drilled soybean; (ii) the higher cost for seed of transgenic cultivars makes the typically higher seeding rates required for drilled plantings less attractive; (iii) producers are recognizing that row widths typical for grain drills (17–25 cm or 7–10 in) are not necessary to maximize yield; (iv) some corn and sugar beet (*Beta vulgaris* L.) producers are shifting to mid-width row spacings which means that a single planter can be used for all crops; and (v) variability in seed-to-seed distance is greater the narrower the row spacing (drill seed-metering imprecision), which makes achieving true equidistant plant spacing difficult.

In the southern USA, recent results indicate that use of narrow rows (50 cm or 20 in) in ESPS plantings results in taller plants and better weed control in both nonirrigated and irrigated environments (Heatherly et al., 2002b). Others have reported varying degrees of enhanced weed control in narrow rows vs. wide rows (Mickelson and Renner, 1997; Nelson and Renner, 1999). However, costs for weed management in narrow rows is greater. Choice of row width for MG 5 cultivars in ESPS plantings that are not irrigated appears arbitrary, but MG 4 cultivars in nonirrigated ESPS plantings have done best in narrow rows. In irrigated environments, both MG 4 and MG 5 cultivars have higher yields and greater net returns when grown in narrow rows. Bowers et al. (2000) determined that yields of MG 3 and MG 4 indeterminate cultivars grown in narrow rows were greater than yields from wide rows at 50% of the sites in a regional study (Arkansas, Louisiana, and Texas). However, both narrow- and wide-row treatments were kept weed-free in these studies, with no comparison of the costs for this factor. The economic value of the yield advantage of narrow over wide rows might have been nil if the additional revenue was offset by greater weed control costs in narrow- vs. wide-row systems.

In the southern USA, conventional plantings (May and later) of soybean grown in narrow rows ($\leq$50 cm or 20 in wide) generally produce higher yields than soybean grown in wide rows (Heatherly, 1988; Ethredge et al., 1989; Boquet, 1990; Oriade et al., 1997). However, the yield advantage of narrow rows is inconsistent over years and relatively small without irrigation (Heatherly, 1988). Thus, choice of row spacing should not be based solely on the presumption that narrow-row soybean systems will yield more than wide-row systems. A yield advantage for narrow rows should be measured against the economics of each system. Use of narrow-row systems is important when the ESPS is used because indeterminate cultivars planted in this system have only upright branching from the lower stem and are short-statured. Thus, they will not form a canopy in wide rows. In double-cropped systems (May and June planting of soybean) in the southern USA, soybean grown in narrow rows results in greater yields (Frederick et al., 1998; Ball et al., 2000). Wide-row systems should be used only where special circumstances are present, such as rotations with crops such as cotton where wide rows are considered necessary, or the need to replace broadcast herbicide applications with banded applications in conjunction with use of mechanical weed control. If a wide-row system of production is used in the southern USA, determinate cultivars should be used because of their bushier canopy structure which is more likely to result in a closed canopy (Heatherly et al., 2001b).

The preponderance of research results indicates that soybean in all regions of the USA should be grown in intermediate or narrow-row systems [50 cm (20 in) or less row width]. The review and results given by Bullock et al. (1998) support the hypothesis that yield increases from growing soybean in narrow vs. wide rows result from more vigorous early season growth and development that occurs before about R5. Soybean grown in narrow-row production systems enhances weed management by forming a quicker canopy, and produces a higher net return.

10–4.3 Seeding Rate/Plant Density

Results over the years from numerous seeding rate experiments across the northern U.S. soybean production area have shown the same thing: seeding from 300 000 to 370 000 viable seeds ha^{-1} (120 000–150 000 seed acre^{-1}) optimizes yield in wide rows when conventional tillage and indeterminate cultivars are used. Fig. 10–3 shows data from one of these studies. Seeding rates in this range result in 250 000 mature, harvestable plants ha^{-1} (100 000 acre^{-1}) if normal plant losses during emergence and the remaining growing season occur. Soybean responses to seeding rates are the same in both rainfed and irrigated systems, and low- and high-yield environments. Plants in fields with low population densities are often short, thick-stemmed, heavily branched at the lower nodes, and will have more pods close to the ground. Weed control is more difficult because of an incomplete canopy. Plants in productive fields resulting from seeding rates above 370 000 seed ha^{-1} (150 000 acre^{-1}) and following good emergence are tall, spindly, and more susceptible to lodging. Lodging disrupts the canopy structure, and if it occurs at R3, will limit pod set, seed development, and thus yield, as well as reduce harvest efficiency. Determinate cultivars generally follow the same response as indeterminate cultivars to seeding

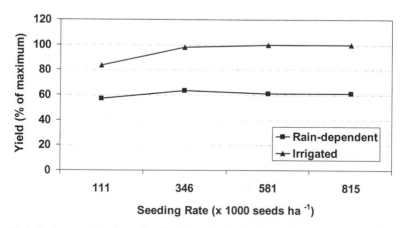

Fig. 10–3. Soybean seeding rate effect in irrigated and rain-dependent environments in Nebraska. Adapted from Elmore (1998).

rates. However, higher seeding rates for the ordinarily short determinate cultivars will result in taller plants and pods higher off the ground, which often improves harvest efficiency.

The general recommendation of planting 300 000 to 370 000 viable seed ha^{-1} (120 00–150 000 acre^{-1}) is based on wide-row, conventional-tillage systems. Special circumstances that may involve tillage system, planting date, and row spacing will require modification of this recommendation to achieve the desired goal of 250 000 plants ha^{-1} (100 000 plants acre^{-1}) at harvest, and these are given in Table 10–18. For example, fewer seedlings survive when no-till or minimum-till planting systems are used. Studies conducted over several northern states have shown that seeding rates of around 550 000 seed ha^{-1} (225 000 seed acre^{-1}) may be necessary to achieve maximum yields from no-till environments (Oplinger et al., 1998b). Yields increased 111 kg ha^{-1} for every 100 000 increase in planted seed ha^{-1} (1.65 bu acre^{-1} for every 40 000 seed acre^{-1}).

In the southern USA, the preponderance of research results and information indicate that a population of 200 000 to 300 000 plants ha^{-1} (80 000–120 000 plants acre^{-1}) provides optimum yield opportunity. Information in Table 10–19 can be used to determine the seeding rate to achieve a desired plant population in selected row spacings, as well as cost associated with the different seeding rates for cultivars differing in seed size and price. A website calculator is available to determine this information (Anonymous, 2002). Soil moisture conditions and seed germination quality should be determined to select a seeding rate that will likely produce these populations.

Seeding rates should be based on seed per unit area rather than on weight per unit area. Seed of cultivars grown under optimum conditions differ greatly in size, and this is under genetic control. The size of any seed lot typically is stamped on the originator's bag. Generally, seed sizes range from 5300 to 7950 seed kg^{-1} (2400–3600 seed lb^{-1}), but sizes of common cultivars and specialty cultivars can be outside this range. This variation in seed size requires that planters are calibrated

Table 10–18. Special conditions that warrant seeding rate deviation from the recommended 300 000 to 370 000 viable seed ha^{-1} (120 000–150 000 acre^{-1}) rate in the northern USA. Adapted from Beuerlein (1995), Hoeft et al. (2000), and Oplinger et al. (1998b).

Condition	Recommendation and reason†
Row width <25 cm (10 in) or drill-planted	Increase seeding rate up to one-third because of imprecision of seed metering system.
Poor seedbed (cloddy, high-residue)	Increase seeding rate 10% because of poor seed-soil contact.
Early-maturing cultivar	Increase seeding rate 10% if planting seed produced in the same region.
Reduced tillage system	Increase seeding rate up to 50% because of more obstacles to germination; that is, cool soil, poor seed-soil contact, less precise planting depth, possible drying of seed drill resulting from residue.
Planting before or after optimum date	Increase seeding rate 20% because of cooler soil (before optimum date) and shorter plants (before and after optimum date).
High-cost seed	Do not seed more than 300 000 seed ha^{-1}

† Do not accumulate seeding rate increases. If more than one special condition exists, use the highest recommended increase.

Table 10–19. Number of seed per 30 cm or 1 ft of row and expense for soybean seed of varied size and cost planted at different seeding rates in five row spacings. Adapted from Heatherly et al. (1999).

Seed size	Seeding rate	Row spacing—cm (in)					Cost per 22.7 kg (50 lb) of seed			
		18 (7)	38 (15)	51 (20)	76 (30)	102 (40)	$10	$15	$20	$25
No. kg⁻¹ (lb⁻¹)	ha⁻¹ (acre⁻¹) × 1000	No. seed per 30 cm or 1 ft of row					$ per 0.4 ha (1.0 acre) cost			
5300 (2400)	198 (80)	1.1	2.3	3.1	4.6	6.1	6.67	10.00	13.33	16.67
	247 (100)	1.3	2.9	3.8	5.7	7.7	8.33	12.50	16.67	20.83
	296 (120)	1.6	3.4	4.6	6.9	9.2	10.00	15.00	20.00	25.00
	346 (140)	1.9	4.0	5.4	8.0	10.7	11.67	17.50	23.33	29.17
	395 (160)	2.1	4.6	6.1	9.2	12.2	13.33	20.00	26.67	33.33
	445 (180)	2.4	5.2	6.9	10.3	13.8	15.00	22.50	30.00	37.50
	494 (200)	2.7	5.7	7.6	11.5	15.3	16.67	25.00	33.33	41.67
	544 (220)	2.9	6.3	8.4	12.6	16.8	18.33	27.50	36.67	45.83
5750 (2600)	198 (80)	1.1	2.3	3.1	4.6	6.1	6.15	9.23	12.31	15.38
	247 (100)	1.3	2.9	3.8	5.7	7.7	7.69	11.54	15.38	19.23
	296 (120)	1.6	3.4	4.6	6.9	9.2	9.23	13.85	18.46	23.08
	346 (140)	1.9	4.0	5.4	8.0	10.7	10.77	16.15	21.54	26.92
	395 (160)	2.1	4.6	6.1	9.2	12.2	12.31	18.46	24.62	30.77
	445 (180)	2.4	5.2	6.9	10.3	13.8	13.85	20.77	27.69	34.62
	494 (200)	2.7	5.7	7.6	11.5	15.3	15.38	23.08	30.77	38.46
	544 (220)	2.9	6.3	8.4	12.6	16.8	16.92	25.38	33.85	42.31
6150 (2800)	198 (80)	1.1	2.3	3.1	4.6	6.1	5.71	8.57	11.43	14.29
	247 (100)	1.3	2.9	3.8	5.7	7.7	7.14	10.71	14.29	17.86
	296 (120)	1.6	3.4	4.6	6.9	9.2	8.57	12.86	17.14	21.43
	346 (140)	1.9	4.0	5.4	8.0	10.7	10.00	15.00	20.00	25.00
	395 (160)	2.1	4.6	6.1	9.2	12.2	11.43	17.14	22.86	28.57
	445 (180)	2.4	5.2	6.9	10.3	13.8	12.86	19.29	25.71	32.14
	494 (200)	2.7	5.7	7.6	11.5	15.3	14.29	21.43	28.57	35.71
	544 (220)	2.9	6.3	8.4	12.6	16.8	15.71	23.57	31.43	39.29

6600 (3000)	198 (80)	1.1	2.3	3.1	4.6	6.1	5.33	8.00	10.67	13.33
	247 (100)	1.3	2.9	3.8	5.7	7.7	6.67	10.00	13.33	16.67
	296 (120)	1.6	3.4	4.6	6.9	9.2	8.00	12.00	16.00	20.00
	346 (140)	1.9	4.0	5.4	8.0	10.7	9.33	14.00	18.67	23.33
	395 (160)	2.1	4.6	6.1	9.2	12.2	10.67	16.00	21.33	26.67
	445 (180)	2.4	5.2	6.9	10.3	13.8	12.00	18.00	24.00	30.00
	494 (200)	2.7	5.7	7.6	11.5	15.3	13.33	20.00	26.67	33.33
	544 (220)	2.9	6.3	8.4	12.6	16.8	14.67	22.00	29.33	36.67
7050 (3200)	198 (80)	1.1	2.3	3.1	4.6	6.1	5.00	7.50	10.00	12.50
	247 (100)	1.3	2.9	3.8	5.7	7.7	6.25	9.38	12.50	15.63
	296 (120)	1.6	3.4	4.6	6.9	9.2	7.50	11.25	15.00	18.75
	346 (140)	1.9	4.0	5.4	8.0	10.7	8.75	13.13	17.50	21.88
	395 (160)	2.1	4.6	6.1	9.2	12.2	10.00	15.00	20.00	25.00
	445 (180)	2.4	5.2	6.9	10.3	13.8	11.25	16.88	22.50	28.13
	494 (200)	2.7	5.7	7.6	11.5	15.3	12.50	18.75	25.00	31.25
	544 (220)	2.9	6.3	8.4	12.6	16.8	13.75	20.62	27.50	34.38
7500 (3400)	198 (80)	1.1	2.3	3.1	4.6	6.1	4.71	7.06	9.41	11.76
	247 (100)	1.3	2.9	3.8	5.7	7.7	5.88	8.82	11.76	14.71
	296 (120)	1.6	3.4	4.6	6.9	9.2	7.06	10.59	14.12	17.65
	346 (140)	1.9	4.0	5.4	8.0	10.7	8.24	12.35	16.47	20.59
	395 (160)	2.1	4.6	6.1	9.2	12.2	9.41	14.12	18.82	23.53
	445 (180)	2.4	5.2	6.9	10.3	13.8	10.59	15.88	21.18	26.47
	494 (200)	2.7	5.7	7.6	11.5	15.3	11.76	17.65	23.53	29.41
	544 (220)	2.9	6.3	8.4	12.6	16.8	12.94	19.41	25.88	32.35
7950 (3600)	198 (80)	1.1	2.3	3.1	4.6	6.1	4.44	6.67	8.89	11.11
	247 (100)	1.3	2.9	3.8	5.7	7.7	5.56	8.33	11.11	13.89
	296 (120)	1.6	3.4	4.6	6.9	9.2	6.67	10.00	13.33	16.67
	346 (140)	1.9	4.0	5.4	8.0	10.7	7.78	11.67	15.56	19.44
	395 (160)	2.1	4.6	6.1	9.2	12.2	8.89	13.33	17.78	22.22
	445 (180)	2.4	5.2	6.9	10.3	13.8	10.00	15.00	20.00	25.00
	494 (200)	2.7	5.7	7.6	11.5	15.3	11.11	16.67	22.22	27.78
	544 (220)	2.9	6.3	8.4	12.6	16.8	12.22	18.33	24.44	30.56

to accommodate this variation when different cultivars are used. This is easily accomplished by counting the number of seed dropped over a 6-m (20-ft) distance and then referring to Table 10–19 to estimate seeding rate. Using the same planter settings for a 7950 seed kg^{-1} (3600 seed lb^{-1}) cultivar as for a 5300 seed kg^{-1} (2400 seed lb^{-1}) cultivar results in overseeding, a population density that is too high, and extra expense since cost of both cultivars is based on weight. From Table 10–19, this unnecessary extra cost for overseeding of the small-seeded cultivar at a rate of 296 000 seed ha^{-1} (120 000 seed acre^{-1}) will be about \$20.80 ha^{-1} (\$8.32 acre^{-1}) when using the \$25 cost for a bag of seed.

10–4.4 Inoculation with *Bradyrhizobia japonicum*

The relationship between *Bradyrhizobia* spp. bacteria and plants is unique to legumes. When infected by *B. japonicum*, the soybean and bacteria form special structures called nodules. The plant provides carbohydrates and mineral nutrients to the bacteria which in turn provides N to the host soybean plant. This relationship is symbiotic (beneficial to both). The *B. japonicum* organism that "infects" soybean is not native to the USA and acts in symbiosis only with soybean.

Soybean can obtain up to 75% of its N requirements from the air when N-fixing *B. japonicum* bacteria are present in the soil, have infected the roots of soybean, and functioning nodules are present on those roots. Establishing *B. japonicum* (inoculation) in a field where soybean has never been grown is necessary to ensure N_2 fixation. There is inconsistency in results from inoculation of fields with a previous history of soybean culture. For example, *B. japonicum* numbers were similar among treatments in a long-term crop rotation and tillage study even though some plots had not had soybean for more than 5 yr (Triplett et al., 1993). The currently established *B. japonicum* strains were introduced early in the last century and are typically less efficient at N_2 fixation than modern strains. Unfortunately, modern strains often do not compete well with established strains and may not overwinter (Jim Beuerlein, personal communication, 2001). In these cases, reinoculation with modern strains may increase yields even in fields with a recent history of soybean culture. Nguyen (1998) states that edamame (vegetable) soybean seed should be inoculated with *B. japonicum* strain CB1809.

Aggressive new strains of *B. japonicum* from public and commercial laboratories are introduced periodically and incorporated into inoculant products. Commercial firms typically rotate strains or use a blend of strains in their products. Products currently available are dry peat- and clay-based products for planter box treatment, liquid products for planter box and in-furrow treatment, and granular products for in-furrow treatment. Two cautions are important when in-furrow products are used: (i) they are not economical for drilled plantings and (ii) they must be placed within 13 mm (0.5 in.) of soybean seed (Jim Beuerlein, personal communication, 2001). The first caution may be ignored when new labels are approved to allow a lower inoculant rate per linear row length of drilled plantings.

Most inoculant products contain more than 2×10^9 *B. japonicum* cells gm^{-1} and deliver more than the previously recommended minimum of 10^5 cells seed^{-1}. However, Hume and Blair (1992) found that increasing *B. japonicum* cells to 10^6 bacterial cells seed^{-1} increased soybean yields. Not all products they tested provided

Table 10–20. Conditions that relate to *B. japonicum* inoculation of soybean and author recommendations.

Condition	Recommendation
Fields with no soybean history or poor nodulaton history	Inoculate seed with 10^5 to 10^6 bacteria cells seed^{-1}
Fields with nodulated soybean in previous 5 yr	Inoculation not necessary except in northern states with cool soils at planting
Optimum N_2 fixation	Maintain soil pH in the 6 to 7 range
Inoculation/planting time interval	Plant seed within 4 h of inoculation
Fungicide-treated seed	Inoculate with *B. japonicum* only after fungicide is dry
Fungicide/inoculant compatibility	Check with inoculant manufacturer; if in doubt, use in-furrow inoculation
Flooded soils or sandy soils (northern USA)	Always inoculate
Acid soils (pH < 6.0)	Add lime or add seed treatment with molybdenum
Well-nodulated soybean plant	5 to 7 nodules on primary root 2 wk after emergence, or 5 nodules cm^{-1} of tap root at flowering

that many cells. New products that may improve nodule development in cool soils early in the growing season are being evaluated (Beuerlein, 1999). Also, new seed treatment processes may make pre-inoculated seed a viable option for producers.

Early season soil temperature differences apparently are responsible for differences in *B. japonicum* inoculation recommendations in the northern USA. Inoculation resulted in yield increases of 8.6% with an associated 13% reduction in cost of production in several northern states (Michigan, Minnesota, South Dakota, and Wisconsin), whereas inoculated soybean in warmer, more southern states (Illinois, Indiana, and Ohio) in the study performed the same as uninoculated controls (Oplinger et al., 1998b). Inoculation of previously inoculated fields in Nebraska has not improved soybean yields. Other data from Ohio indicate that responses to inoculation are frequent and profitable (Beuerlein, 2001a).

Glyphosate inhibits the enzyme 5-enolpyruvylshikimate-3-phosphate synthase (EPSPS) and thus blocks aromatic amino acid synthesis. While GR soybean cultivars contain resistant EPSPS, *B. japonicum* does not contain a resistant enzyme. Thus, glyphosate applied to GR soybean may interfere with the symbiotic relationship (King et al., 2001). In greenhouse and growth chamber studies, early applications of glyphosate generally delayed N_2 fixation and decreased biomass and N accumulation. However, plants had recovered by 40 d after emergence. In growth chamber studies, N_2 fixation was more sensitive to water deficits in GR plants treated with glyphosate.

Conditions that may justify *B. japonicum* inoculation of soybean, and author recommendations for its use are given in Table 10–20. Inoculation failures are infrequent but do occur, especially on soils that never have had well-nodulated soybean. Most failures are probably due to heat- or desiccation-induced killing of *B. japonicum* prior to, during, or following the inoculation process, or an incompatible inoculant–seed fungicide treatment. Post-plant inoculant applications should not be considered in failure cases. If an inoculation failure is discovered early in the growing season (< 30 d after planting), apply N fertilizer. A "rule of thumb" can be used to determine the amount of N to apply to overcome an inoculation failure. First, assume that 25% of the N contained in harvested soybean seed comes from the soil; inversely, 75% comes from N_2 fixation in a well-nodulated crop. If the goal is to harvest 3350 kg ha^{-1} (50 bu acre^{-1}) of seed, then about 160 kg N ha^{-1} or about 140 lb N acre^{-1} (projected yield × 0.0625 N in seed) will be required to offset the lack of N$_2$ fixation.

10–4.5 Fungicide Treatment of Seed

Fungicide treatment of seed can help control seedling damping off and seed rot problems caused by fungi. Several materials are available for commercial and/or on-farm use (Table 10–6). *Pythium* spp., *Phytophthora sojae*, *Rhizoctonia* spp., and *Fusarium* spp. are the most common pathogens associated with reduced soybean germination and emergence and subsequent stand failures (Table 10–2). Conditions which favor responses to fungicide seed treatments are early plantings in cool, wet soils with anticipated slow seedling emergence and growth, minimum-till or no-till systems, high amounts of surface residue, deep planting, fields that have continuous or frequently grown soybean, and fields with a previous history of seedling diseases. Soybean replanted into a failed stand situation is especially prone to fungal disease infection since the soil likely will have a high level of fungal activity. In this case, a fungicide seed treatment should be used to maximize plant stands from replanting. In cases where replanting is sufficiently later so that soils are warmer and drier, use of fungicide seed treatment is optional. Treatment of seed for ESPS plantings in the midsouthern USA is recommended since a stand failure results in the cost of replanting plus the lost benefit of early planting.

The primary benefit of fungicide seed treatment is to improve crop stands rather than to improve seed viability. Because soybean will yield well under a wide range of plant densities, yield responses to fungal seed treatments are not always observed. Oplinger et al. (1998b) found that fungicides improved plant stands 2% in 2 of 3 yr in no-till tests across several states. However, there was no yield response in five of eight Iowa trials although plant stands were increased by 20 000 ha^{-1} (8000 acre^{-1}). In trials conducted in Wisconsin, fungicides increased stands by 19% and yields by 11%. The difference between the Iowa and Wisconsin results was likely due to cooler soil temperatures and high crop residues in Wisconsin.

Three cautions are worthy of consideration when using fungicide seed treatments. First, some seed treatment fungicides are incompatible with *B. japonicum*. See Table 10–20 for ways to handle this situation. Second, feeding or selling treated seed is prohibited by federal law; therefore, treat only those seed that will be planted. Third, fungicide seed treatments will not improve the quality or viability of a seed lot. Therefore, plant high-quality seed even if a seed treatment is considered necessary.

10–5 CROPPING SYSTEMS

10–5.1 Crop Rotation

Crop rotation is a term used to describe the temporal pattern of occurrence of two or more crop species in the production history of a given field. Soybean is commonly rotated with corn, wheat, cotton, rice (*Oryza sativa* L.), or grain sorghum.

Table 10–21. Corn and soybean yields when grown continuously and in rotation with each other. All data but those from NE were summarized from several sources by Hoeft et al. (2000); NE data are from an irrigated trial (Roger Selley, personal communication, 2000).

State	Site-yr no.	Yield of corn following:			Yield of soybean following:		
		Corn	Soybean	Advantage†	Soybean	Corn	Advantage†
		— kg ha⁻¹ (bu acre⁻¹) —		%	— kg ha⁻¹ (bu acre⁻¹) —		%
IL	17	9 030 (144)	10 660 (170)	18			
IN	20	10 410 (166)	11 230 (179)	8	3 070 (45.7)	3 420 (50.9)	11
IA	8	8 030 (128)	9 090 (145)	13	2 140 (31.9)	2 410 (35.8)	12
MN	20	7 650 (122)	8 530 (136)	12	2 420 (36.0)	2 740 (40.8)	13
NE	8	10 280 (164)	10 910 (174)	6			
NY	12	7 960 (127)	8 720 (139)	9			
WI	9	8 220 (131)	9 530 (152)	16	3 510 (52.2)	3 700 (55.0)	5

† Advantage to rotation.

The growing of soybean rotated with wheat and other small grains within a 12-mo period is referred to as doublecropping or intercropping (crops grown in sequential seasons of the same year) and are discussed later. Johnson (1987) and Wesley (1999b) have presented extensive reviews of crop-rotation research. A summary and update of their presentations follow.

Reasons for growing soybean in rotation rather than continuously are: (i) higher yields of one or both crops; (ii) a decreased need for N fertilizer on the crop following soybean; (iii) increased residue cover; (iv) mitigation of pest and weed cycles; and (v) distribution of labor and machine requirements over a larger portion of the growing season. Studies cited by Johnson (1987) and Wesley (1999b) show that rotated soybean generally yields more than continuous soybean and that other crops benefit from rotations that include soybean. Ferreira et al. (2000) measured greater *Bradyrhizobia* diversity and higher rates of N_2 fixation in cropping systems where soybean was rotated with wheat or corn in Brazil.

Continuous soybean is not a common cropping practice in the northern USA. Yields of both corn and soybean are increased when planted in rotation (Table 10–21). Some evidence indicates that soybean responds more to crop rotation than does corn (Table 10–22). Perhaps the main reason for this is that soybean may be

Table 10–22. Effect of number of years on yield of corn and soybean averaged over two locations in Minnesota and one location in Wisconsin. Based on Porter et al. (1997); adapted from Hoeft et al. (2000).

Years of continuous cropping following 5 yr of other crop	Corn yield	Soybean yield
	— kg ha⁻¹ (bu acre⁻¹) —	
1	9000 (143) a†	3260 (49) a
2	8040 (128) b	2990 (45) b
3	7900 (126) b	2840 (42) c
4	7900 (126) b	2820 (42) cd
5	7880 (126) b	2800 (42) cd
Continuous	7810 (124) b	2770 (41) d
Rotated corn/soybean	8830 (141) a	3050 (45) b

† Means in individual columns followed by the same letter are not significantly different at $P \leq 0.05$.

more affected by soil-borne diseases than is corn. Not all situations favor short-term rotations with soybean. For example, severity of brown stem rot increased and soybean yield and seed weight decreased as soybean frequency in rotation with corn increased (Adee et al., 1994). Soil organic C and N are greater with continuous corn than with a corn–soybean rotation (Omay et al., 1997). These differences were related to the amount of crop residues returned to the soil. Soil microbial biomass and potentially mineralizable N were not affected by rotation with soybean. Nevertheless, economic and agronomic incentives favor a 2-yr corn–soybean rotation in the northern states.

Crops that are rotated with soybean produce more dry matter and subsequent residue than does soybean. In an Iowa study, residue cover after planting soybean no-till following corn exceeded 50%, whereas residue cover after planting corn no-till following soybean was only 37% (Erbach, 1982). The same relative differences in after-planting residue cover following corn compared to following soybean were measured in various tilled systems as well. This increased residue resulting from rotation of soybean with other crops may lead to improved water infiltration, soil tilth, and organic matter.

Rotation of soybean with a crop that is not a host to SCN can be used effectively to help alleviate damage to soybean by the pest in addition to delaying or preventing buildup of new SCN races (Dabney et al., 1988). Soybean in a rotation with corn may mitigate the need for pesticides to control pests of corn such as corn rootworm (*Diabrotica* spp.). Longer crop rotation cycles between soybean crops can break pest cycles and thus require less expenditure for control of insects and diseases (Adee et al., 1994; Hoeft et al., 2000). The continuous growing of either crop maximizes the opportunities for those weed species best adapted to compete with the crop to increase. Crop rotation, on the other hand, limits the potential for establishment of weed species that are most competitive with a given crop species (Gunsolus, 1990). Rotation of corn and soybean also allows the rotation of herbicides, which may limit occurrence of resistant weed species. In New York, Katsvairo and Cox (2000) found that a soybean–corn rotation resulted in greater net returns and reduced fertilizer, herbicide, and pesticide use compared to a continuous corn system.

A full economic analysis to include the different equipment complements necessary for culture of the different crops should be used to determine economic feasibility of any cropping system. Yield response alone may not be an adequate guide for determining whether or not to adopt a rotational system using soybean. The presence or absence of irrigation plays an integral part in response of soybean to rotation in the midsouthern USA, and this should be a key factor to consider. Long-term commodity price prospects should be used to project the potential net returns of varying cropping systems that may involve rotation. Machinery costs that are crop specific increase production costs and therefore reduce net returns from rotational systems (Yiridoe et al., 2000). The decision to rotate soybean with other crops thus should be evaluated from both agronomic and economic perspectives.

In a study conducted in the midsouthern USA using systems of continuous soybean and soybean rotated with corn or grain sorghum, plus doublecrop systems of wheat–soybean alone and in rotation with corn and grain sorghum (Wesley et al. 1994, 1995), the analysis of net returns to eight cropping systems over an 8-yr

Table 10–23. Average crop yield and net return from eight nonirrigated and irrigated cropping systems on Tunica clay near Stoneville, MS (1984 to 1991). Adapted from Wesley et al. (1994, 1995).

Cropping system†	Crop	Nonirrigated		Irrigated	
		Yield	Net return‡	Yield	Net return§
		kg ha⁻¹ (bu acre⁻¹)	$ ha⁻¹ ($ acre⁻¹)	kg ha⁻¹ (bu acre⁻¹)	$ ha⁻¹ ($ acre⁻¹)
1	Corn	4085 (60.8)	−22 (−9) e	7850 (116.8)	195 (79) c
2	Soybean	1445 (21.5)	62 (25) cd	2760 (41.1)	131 (53) d
3	Grain sorghum	5370 (79.9)	148 (60) ab	6300 (93.8)	47 (19) e
4	Corn	3615 (53.8)	104 (42) bc	8520 (126.8)	272 (110) ab
	Soybean	2710 (40.3)		3345 (49.8)	
5	Grain sorghum	4020 (59.8)	188 (76) a	7100 (105.7)	208 (84) c
	Soybean	2955 (44.0)		3465 (51.6)	
6	Wheat/	2565 (38.2)	119 (48) abc	2950 (43.9)	304 (123) a
	Soybean	725 (10.8)		2190 (32.6)	
7	Corn	2710 (40.3)	40 (16) de	8405 (125.1)	336 (136) a
	Wheat/	2895 (43.1)		3280 (48.8)	
	Soybean	1335 (19.9)		2565 (38.2)	
8	Sorghum	3910 (58.2)	158 (64) ab	6940 (103.3)	235 (95) bc
	Wheat/	3030 (45.1)		2970 (44.2)	
	Soybean	1480 (22.0)		2505 (37.3)	

† Cropping systems were: 1 = continuous corn; 2 = continuous soybean; 3 = continuous sorghum; 4 = biennial rotation of corn—soybean; 5 = biennial rotation of sorghum—soybean; 6 = continuous wheat—soybean doublecrop; 7 = biennial rotation of corn and wheat—soybean doublecrop; 8 = biennial rotation of sorghum and wheat—soybean doublecrop.
‡ Values in individual columns followed by the same letter are not significantly different at $P \leq 0.05$.

period (Table 10–23) provided the following conclusions: (i) without irrigation, grain sorghum was the more desirable component crop for rotation with soybean and in rotation with a wheat–soybean doublecropping sequence; (ii) with irrigation, net returns to cropping systems that included corn rotated with soybean and in rotation with a wheat--soybean doublecropping sequence were greater than those from continuous single-crop systems or the other rotations; and (iii) in both nonirrigated and irrigated systems, rotated crop sequences provided greater net returns than a continuous soybean system. With the advent of the higher-yielding ESPS in the mid-southern USA, these findings need updating. In an 8-yr study at Stoneville, MS, Kurtz et al. (1993) reported yields of 1235 and 1860 kg ha⁻¹ (18.4 and 27.7 bu acre⁻¹) from nonirrigated soybean that was grown continuously and in rotation with rice, respectively. Respective net returns were $20 and $161 ha⁻¹ ($8 and $65 acre⁻¹). Rice yields and net returns also were increased by rotation with soybean, and 8-yr average net returns from rice–soybean rotations exceeded those from both continuous nonirrigated soybean and continuous rice. This same result from nonirrigated soybean following rice was also acheived in later work at this location (Wesley, 1999b). Where soybean was irrigated (which will be the case in a soybean/rice rotation), soybean that was cropped in a 1:1 rotation with rice produced yields and net returns that were similar to those from continuous soybean (Wesley, 1999b). Since irrigated soybean yields following rice do not appear to be enhanced by the rotation with rice, the advantages of rotating soybean with rice where both are irrigated must accrue from benefits such as enhanced rice yields and disruption of pest and weed cycles rather than a yield benefit to the soybean.

Crop rotation can be used to decrease erosion potential. As shown in Table 10–9, culture of some crops results in more of an erosion hazard than others. Soils planted to soybean or cotton may have as much as 10 to 100% greater soil loss potential than do soils planted to corn or grain sorghum (Triplett and Dabney, 1999). Reasons for this are: (i) neither soybean nor cotton produce a large volume of residue that covers the soil during the off-season; and (ii) soybean residue decomposes more rapidly than the stalks and leaves of nonleguminous crops. Rotation of corn and soybean with soybean planted no-till allows corn residue cover to persist into the soybean-growing season, thus reducing erosion potential during the soybean-growing season. Small grain straw also provides extensive, persistent cover, making a soybean–small grain doublecrop system effective in controlling soil loss. For this system to function well, the small grain straw should not be burned and soybean should be planted no-till.

10–5.2 Doublecropping

Doublecropping refers to the practice of growing two crops in 1 yr. The potential advantages of doublecropping are: (i) increased cash flow that results from having income from two crops in one 12-mo period; (ii) reduced soil and water losses by having the soil covered with a plant canopy most of the year; (iii) more intensive use of land, machinery, labor, and capital investments; and (iv) harvesting more of the solar radiation available in a given year by deploying two crop canopies. Doublecropping is practiced in the southern portion of the soybean-growing region of the USA, and the majority of the doublecropped hectarage in this region involves soybean and soft red winter wheat. Dabney et al. (1988) found that doublecropped soybean planted from early June through early July yielded significantly less than full-season soybean that was planted in early to mid-May. Wesley et al. (1994, 1995) confirmed this, and determined that a wheat–soybean doublecrop system should be used only with irrigation for profitable production on the clayey soils in the midsouthern USA. With the increased use of early planting in the ESPS and subsequent higher yields from continuous soybean in the midsouthern USA, producers should compare the economics of continuous soybean using the ESPS vs. doublecropping when determining which system to use.

Using results from dryland wheat–soybean doublecropping and continuous wheat research in Mississippi, Spurlock et al. (1997) determined that doublecropping was less profitable than continuous wheat if the soybean price is less than $0.184 kg^{-1} ($5 bu^{-1}) and the wheat price is greater than $0.101 kg^{-1} ($2.75 bu^{-1}). At a soybean price of $0.220 kg^{-1} ($6 bu^{-1}) and a wheat price of $0.101 kg^{-1} ($2.75 bu^{-1}), doublecropping was slightly more profitable. A wheat price of $0.165 kg^{-1} ($4.50 bu^{-1}) and a soybean price of $0.220 kg^{-1} ($6 bu^{-1}) is required to exceed net returns from continuous dryland soybean resulting from use of the ESPS [2350 kg ha^{-1} (35 bu acre^{-1}) yield, $0.196 kg^{-1} ($5.35 bu^{-1}) price] in the Mississippi Delta region.

Most farmers in the southern USA generally decide to plant wheat following soybean based on the expected price of wheat, and government programs in force at that time. The decision to plant soybean following wheat is influenced by both agronomic and economic factors. Agronomic factors include harvest date of the wheat crop (which dictates soybean planting date), soil moisture status for soybean planting and emergence, and availability of seed of desired cultivars. Economic factors that influence planting soybean following wheat are the return realized from the wheat crop, expected soybean price, and the expected soybean yield compared to the known cost of production.

Wesley (1999a) has compiled a detailed listing of management practices for doublecropping wheat and soybean in the midsouthern USA. The following information is summarized from that publication, plus additional sources. For wheat, use shallow tillage to prepare a seedbed (number of seedbed preparation tillage trips depends on preceding crop and rutting from harvest). Plant in 15- to 25-cm-wide (6- to 10-in-wide) rows using a seeding rate of 100 to 135 kg seed ha^{-1} (90–120 lb acre^{-1}). Apply 22.5 to 34 kg N ha^{-1} (20–30 lb acre^{-1}) if wheat follows a summer grass crop or fallow. Use a fungicide seed treatment if planting on low-lying soils that are subject to submergence or prolonged saturation. If ryegrass (*Lolium multiflorum* Lam.) infestations are present after wheat emergence, make fall applications of herbicide before ryegrass reaches the five-leaf stage. Apply appropriate herbicides in late winter to control winter weeds such as wild garlic (*Allium vineale* L.), curly dock (*Rumex crispus* L.), and annual broadleaf weeds, if needed. Apply 100 to 135 kg N ha^{-1} (90–120 lb acre^{-1}) in late February/early March, using split applications on soils with poor internal drainage. Harvest wheat with a combine having a straw shredder/spreader. If soybean is to be planted no-till, cut wheat 15 to 30 cm (6–12 in) aboveground. If wheat stubble is to be burned prior to soybean planting, ensure that conditions are conducive for the complete burning of the wheat stubble. Results from previous research in the midsouthern USA indicate that burning of wheat straw is a management practice that enhances net returns if soybean is planted no-till following burning (Wesley and Cooke, 1988). Kelley and Sweeney (1998) found that burning wheat straw over a 12-yr cycle of doublecropped wheat–soybean followed by continuous full-season soybean had no long-term effect on soil properties.

Soybean cultivars selected for superior performance in conventional environments (early planting) can be expected to be among the superior cultivars in doublecrop or later-planted environments (Panter and Allen, 1989). Soybean that is doublecropped should be planted in narrow rows (<50 cm or 20 in) as soon as possible after wheat harvest. The least planting delay occurs when soybean is planted into standing or burned wheat stubble. Conventional recommendations in the midsouthern USA promote a seeding rate to achieve a final stand that is about 10 to 30% higher than that for conventional earlier plantings (Boquet, 1996; Wesley, 1999a). However, recent results from Arkansas research show that there is no irrefutable evidence to support a higher seeding rate for doublecropped later plantings (Ball et al., 2000). As shown in Table 10–19, using a higher seeding rate results in additional cost for soybean seed in doublecrop plantings. Application of a preplant nonselective herbicide to kill weeds in standing wheat stubble is recommended at the time of soybean planting. Planting soybean in a prepared seedbed is recommended if problem weeds are present or adverse weather conditions resulted in rutting during wheat harvest. Use broadcast-applied postemergent herbicides to ensure the least cost associated with weed management. Irrigation where available will ensure maximum emergence, growth, and yield on droughty soils. Production of soybean after wheat on clayey soils in the midsouthern USA without irrigation generally is not profitable due to the effects of normal summer drought.

In the midwestern USA, doublecropping of soybean and wheat can be practiced in the more southerly portions of the region such as southern Ohio. Beuerlein (2001a) has listed management considerations for soybean following wheat in that area. Add required P and K fertilizer at the time of wheat planting. Plant an early-maturing wheat cultivar to ensure early wheat harvest and the earliest possible soybean planting time of early to mid-June. Cut wheat to leave ≈30 cm (12 in) of stubble to provide mulch cover for the soybean. Wheat straw passing through the combine should be shredded; otherwise, bale and remove. Plant soybean as early as possible after wheat harvest, but no later than 10 July. If soil is quite dry at the time of wheat harvest and will not be irrigated, do not plant soybean. Plant soybean no-till at 13 seed m^{-1} (4 seed ft^{-1}) in narrow rows (≈18 cm or 7 in), and use a mid-season maturity (MG 3.4–MG 3.8) cultivar. Plant no-till, and kill existing weeds with a nonselective herbicide. Use GR cultivars and glyphosate for the most satisfactory and economical weed management in doublecropped soybean.

10–5.3 Intercropping

Relay intercropping has been used to extend multiple cropping further northward in the USA or to improve yields compared to those from doublecropping. In this system, soybean is planted into growing small grains, which means that both crops occupy the same area until small grain harvest. Usually, small grain rows are widened to accommodate soybean-planting equipment. Thus, small grain yields may be reduced in accordance with the area lost by the widened rows of the small grain crop, or by injury to small grain plants during soybean planting (Reinbott et al., 1987). Soybean emergence and seedling growth may be adversely affected by the removal of soil moisture by the small grain crop from the soybean seeding zone (Duncan and Schapaugh, 1997). Soybean planting date should be selected with regard to the growth stage of the small grain or the anticipated small grain harvest date to ensure that soybean plants are sufficiently short to be unaffected by small grain harvest (Reinbott et al., 1987). Wallace et al. (1992) determined that the overlap between soybean planting and wheat harvest must be relatively short (2–3 wk in the southern USA) to prevent negative effects of relay intercropping on soybean yield.

Jacques et al. (1997) compared net returns from continuous soybean, doublecropped soybean and wheat, and relay intercropped soybean and wheat in Arkansas. Doublecropping produced a higher net return than relay intercropping, which produced a higher net return than continuous soybean. However, yields from continuous soybean were below 2000 kg ha^{-1} because of a late May planting date. Yield from soybean that was intercropped was below that from both continuous and doublecropped soybean. Duncan and Schapaugh (1997) conducted research in Kansas and concluded that supplemental irrigation must be available for soybean that is intercropped into standing wheat on soils that are droughty or that have low moisture-holding capacity. They also determined that the height of wheat cultivars (determines degree of shading of emerging soybean seedlings) was instrumental in early season soybean seedling survival until wheat harvest. Net returns from intercropping were less than returns from continuous irrigated soybean. Reinbott et al. (1987) in a Missouri study measured a 3-yr average intercropped soybean yield that was 25% less than yield from a continuous soybean treatment, but 52% greater than yield from late June/early July-planted doublecropped soybean.

Strip intercropping is a variation of intercropping. In this system, soybean is grown simultaneously with other crops such as corn or grain sorghum in contiguous alternating strips. When soybean and corn have been grown in strip intercrop systems, increased corn yields have usually been offset by reduced soybean yields from rows that bordered the corn strips (Crookston and Hill, 1979; Iragavarapu and Randall, 1996). The use of wheat as a strip crop between corn and soybean resulted in the determination by Iragavarapu and Randall (1996) that alternate three-crop strips of wheat–soybean–corn could be planted in a north-south row direction to optimize production in Minnesota. Lesoing and Francis (1999a) determined that a corn–soybean strip intercropping system in Nebraska has more potential under irrigation than under rainfed conditions. In fact, their irrigated intercropping system was economically competitive with continuous single-crop systems. Lesoing and Francis (1999b) determined that grain sorghum and soybean yields from intercropped strips in a 2-yr rotation were similar to single-crop yields in the same rotation under both rainfed and irrigation conditions.

10–5.4 Cover Crops

Growing vegetation is often used as a source of soil cover during the winter and spring seasons when summer annual crops are not present. This vegetation can consist of annual weeds or a crop planted specifically to provide cover. Where a cover crop is used, wheat or some other small grain is ideal to precede soybean. Cover crops usually are destroyed prior to planting of the summer row crop with no short-term economic gain from their use. Thus, the potential long-term benefits for integrating cereal cover crops into a soybean production system would have to arise from erosion control, increasing soil organic matter, and increasing soil productivity. Cover crops can play a role in decreasing erosion and increasing soil organic matter following soybean, since soybean typically leaves the soil more prone to wind and water erosion than either corn or grain sorghum. This is especially valued on low organic matter soils and rolling loess soils (Wilson et al., 1993). Shipitalo at al. (1997) proposed using a rye (*Secale cereale* L.) cover crop following soybean harvest in a corn–soybean rotation as an acceptable management practice for reducing total herbicide loss in runoff in the northern Appalachian region of the USA.

Cover crops are not used widely in northern U.S. soybean production systems. When used in conventional soybean production systems (i.e., not organic production systems), cover crops are grown between harvest and planting of row crops rather than in place of them (Hoeft et al., 2000). They typically are planted at or slightly before the fall harvest of the summer crop and are allowed to grow during the fall, winter, and early spring. Moore et al. (1994) found that soybean yields were comparable with or without cover crops in an Ontario, Canada study. Thus, net returns in their study effectively would have been reduced because of the expense incurred for seeding the cover crops. In Nebraska, a planting of winter rye following soybean harvest resulted in as much post-planting residue cover as a crop of corn (Kessavalou and Walters, 1997). Rye was destroyed in the spring by disking prior to corn planting. In 2 of 3 yr, the rye produced enough residue to limit erosion and did not affect corn yields. In the third year, rye development was delayed in the spring and corn plant density and yields were decreased, possibly because of allelopathic effects of the rye. In addition to this concern, the large immobilization of nitrate (NO_3)–N and fertilizer N by cover crop residues may result in N deficiency in corn unless corrected with N application. In Iowa, Karlen and Doran (1991) found that a winter cover crop resulted in a 10% lower corn yield, which they attributed to the depletion of soil NO_3 levels by cover crop decomposition that was not overcome by postemergence broadcast application of N. In the mid-Atlantic states, Lu et al. (1999) found that net returns from a 2-yr rotation of corn–wheat–soybean grown in a no-till system with and without winter cover crops were essentially identical over a 4-yr period ($233 ha^{-1} vs. $238 ha^{-1}). However, the no-till system without cover crops was determined to have the lowest economic risk (CV = 1.14 vs. 1.24; lower limit = $53 ha^{-1} vs. $39 ha^{-1}).

In the upper midwestern USA, there is concern about the small amount of time for cover crop growth and development when planted after harvest of the summer row crop. To address this, Johnson et al. (1998b) overseeded oat (*Avena sativa* L.) and rye as monocultures (sole crops) and mixtures into standing soybean in August. This practice allowed more dry matter production from the cover crop than was obtained from postharvest plantings of cover crops in the fall. Oat was more advantageous than rye because it was winter-killed and thus required no herbicide treatment in the spring. Soybean yield in the overseeded treatments was slightly below (but not statistically so) that from a continuous soybean control treatment with no winter cover crop. A rye cover crop was associated with reduced corn yields the following year, but yields of corn following oat were not affected. In Nebraska, aerial seeding of winter rye at beginning leaf drop of soybean is possible on sand and sandy loam soils; light irrigations are sometimes necessary to promote rye germination and development (Wilson et al., 1993).

Cover crops also may have a role when planted prior to soybean in a rotation. They have suppressed early season weeds for the first 3 to 5 wk after soybean planting (Williams et al., 1998). Efficacy of other integrated weed management tactics possibly is enhanced when used in conjunction with cover crops (Williams et al., 1998). Cover crops planted in the fall and killed with herbicides before soybean planting the following spring favored soybean emergence and growth over that of weeds (Williams et al., 1998). In 1 yr of their study, however, abnormally heavy cover crop residues (6.3–7.2 Mg ha^{-1} or 5 625–6 430 lb acre^{-1}) of both winter rye and wheat interfered with soybean seed placement and reduced soybean plant densities.

The limited use of cover crops in northern U.S. soybean production systems probably is due to the shorter and cooler growing season available for cover crop growth. Development in the fall and early spring is slow, and cover crops impede soil warming (Hoeft et al., 2000). In addition, cover crop residue with unfavorable spring weather is difficult to handle and makes the practice unattractive. In drier areas, allowing the cover crop to grow too long in the spring can use soil moisture that will be needed for soybean germination and emergence.

Table 10–24. Effect of preplant tillage, wheat cover crop, and planting date (PD) on average yields and net returns from irrigated 'Leflore' soybean planted in a stale seedbed at Stoneville, MS (1986 and 1987). Adapted from Heatherly (1999c).

Preplant tillage	Planting date†			Planting date		
	Early	Late	Avg.	Early	Late	Avg.
	kg ha⁻¹ (bu acre⁻¹)			$ ha⁻¹ ($ acre⁻¹)		
Fall disk	3225 (48)	2555 (38)	2890 (43) a‡	232 (94)	89 (36)	161 (65) b
Spring disk	3225 (48)	2620 (39)	2890 (43) a	257 (104)	121 (49)	188 (76) ab
Prepared seedbed	3360 (50)	2620 (39)	2955 (44) a	294 (119)	143 (58)	220 (89) a
None after harvest	3290 (49)	2690 (40)	2955 (44) a	267 (108)	136 (55)	203 (82) a
Fall disk + wheat	3225 (48)	2690 (40)	2955 (44) a	163 (66)	64 (26)	114 (46) c
Avg.	3225 (48) a	2620 (39) b		242 (98) a	111 (45) b	

† Early = 6 May 1986 and 5 May 1987; Late = 16 June 1986 and 28 May 1987.
‡ Average values for yield or net returns that are followed by the same letter are not significantly different at $P \leq 0.05$.

In the southern USA, cover crops have numerous benefits in row crop production, and can be used effectively in the stale seedbed system (Griffin and Dabney, 1990; Elmore et al., 1992; Heatherly et al., 1993; Reddy, 2001b). The cover crop can be killed at the appropriate time with foliar-applied herbicides (Griffin and Dabney, 1990; Elmore et al., 1992). However, using wheat or other cereals as a winter cover crop may result in lower net returns (Table 10–24; Reddy, 2001b, 2003) due to the expense incurred in establishing the small grain cover with no resulting soybean yield increase (Elmore et al., 1992; Heatherly et al., 1993), whereas doublecropping soybean and a small grain provides both soil cover and the potential for extra income. Thus, the potential long-term benefits of erosion control, increases in soil organic matter, and increases in soil productivity resulting from use of cover crops in soybean production systems provide the only advantages for their use.

Where soil protection provided by a cover crop is needed to reduce erosion potential on slopes, drainage usually is adequate for wheat to be well-adapted. Under these conditions, doublecrop soybean following wheat has been more profitable than either wheat or soybean grown as a sole crop. On a poorly drained Tunica clay, however, a wheat cover crop killed before planting did not increase soybean yield, reduced net returns, and increased the percentage winter ground cover only in years when fall tillage reduced the populations of volunteer winter annual weeds (Elmore et al., 1992).

Winter cover crops offer the potential to overcome weed problems that may be otherwise unmanageable in the winter and spring (Reddy et al., 1999). In the southern USA, using volunteer winter weeds themselves as a cover crop in a soybean production system has merit. There is no expense associated with their establishment and they can be killed in the spring with preplant, foliar-applied herbicides. Successful use of winter weeds as cover crops may depend on amount and time of fall tillage, since some of the winter annual species emerge in late summer/early fall and tillage after this time may jeopardize volunteer stands of weeds. Recently imposed label restrictions on the latest date for late winter/early spring aerial application of some preplant, foliar-applied herbicides may reduce the value of winter weeds as cover crops if they are killed too far ahead of planting soybean.

10–6 POST-PLANTING MANAGEMENT

10–6.1 Replanting Decisions

Replanting as a result of poor emergence should be considered only when plant densities are below the desired ranges given previously. A decision to replant also should be evaluated on an economic basis because of the cost of seed and replanting. Cost for replanting can be significant and minimally will include cost of seed (Table 10–19), seed treatment (if used), and planting [>\$27 ha^{-1} (\$11 acre^{-1}); Spurlock, 2000]. Also, even a small delay in planting, or in this case, a later planting date resulting from replanting, can result in much lower seed yield in some regions if weather conditions during the growing season are less favorable for development of the later planting (Gaska, 2000; Heatherly and Spurlock, 2002a).

Farmers faced with poor stands have three choices: keep the stand, replant affected areas of a field, or plant an additional row alongside each existing but poorly emerged row. In the northern USA, Vasilas et al. (1990) investigated the three options in Illinois and found that even a 66% reduction from an original stand of 16 to 19 plants m^{-1} (4.8–5.8 ft^{-1}) of 76-cm- (30-in-) wide rows did not justify replanting. An offset row [20 cm (8 in) to the side and parallel to each row] increased yield only in the cases where >66% of the stand was lost in randomly placed gaps. Soybean is extremely tolerant of poor stands, assuming they are uniformly spaced. Yields from stands as low as 120 000 plants ha^{-1} (50 000 acre^{-1}) are often the same as those from higher populations. When stands are less than this level, replanting should be considered a viable option. Mid-season adapted cultivars are advised for replantings at later-than-optimum dates. These cultivars provide greater height and node numbers than shorter-season cultivars when planted late, and will have reduced risk of late-season frost injury.

Some stands may be adequate in terms of number of plants, but crusting-induced cotyledon shearing at emergence or hail or insect damage soon after emergence can greatly slow vegetative growth of surviving plants and irreparably lower their yield potential. In these cases, a determination must be made that the conditions causing plant injury can be remedied or likely will not be repeated for a replanting. It also must be determined that the yield potential of and projected net return from the replanted crop will exceed yield potential of and net return from the damaged crop. Objective assessment of these criteria is difficult. A situation that likely justifies remedying is that of a marginal stand of damaged plants. If a stand is at the lower end of an acceptable population range, but a significant number of emerged plants is damaged, then replanting is recommended.

Assessment of hail-damaged stands requires an estimation of total plants (stand count) and damaged plants (defoliation and stem breakage). Soybean plants can recover from stem damage if the stem is not severed below the cotyledonary (seed leaf) node. This is the first node on the seedling; the two fleshy, dark-green cotyledons are attached on opposite sides of this node. Buds on each side of this node can grow new branches if they are still present and undamaged. A plant that is broken below this node will not survive. Hail stones often damage stem tissue at the base of plants which may lead to stem lesions at or near the soil surface. These injuries may contribute to lodging susceptibility later in the season as the plant canopy reaches maximum size and weight of filling pods becomes too great for the injured stem to support. Shapiro et al. (1985) provide a detailed guide for estimating soybean yield loss due to hail damage.

In general, determining plant density on an area basis is preferred. To determine if a soybean stand is adequate, a systematic sampling of plant density and adjacency should be used (Willers et al., 1999). This involves using line-intercept sampling (LIS) in a management unit (i.e., a 20- to 40-ha [50- to 100-acre] field). The first step in using LIS is to divide the field into subunits where crop phenology and soil type are similar. Within each subunit, locations for transect lines (a string or rod that lies perpendicular to row direction) are chosen randomly. Generally, each transect line should be the width of one planter pass, with longer lines encompassing multiples of whole planter passes. Plant counts are taken on segments (typically 0.3–1 m [1–3.3 ft]) of each of the rows emanating from one side of the transect line. The shorter length sample is sufficient where stands are dense and uniform, whereas the longer length sample should be used where the stand is sparse and/or nonuniform. The length of row sampled should be the same for all rows within and among transect lines within a subunit. Use of such a system allows an objective determination of the number of plants present on an area basis after final emergence and the uniformity of their distribution.

Key points to consider before replanting have been adapted from Martin (2001). (i) Use an objective method (such as that given by Willers et al. [1999] and described above) for assessing stand and compare the sample numbers to the values in Table 10–19 that are needed for a given population range. (ii) Assess the health and vigor of plants in stands that are at the lower end of the acceptable population range. Compare estimated yield and net return from damaged stands with estimated yield from a replanted crop. This should weigh the cost of replanting and estimated effect of later planting on yield against the cost of protecting/recovering plants in damaged stands. This also assumes that the factor(s) (such as hail, insects, or soil crusting) causing damage to the affected stand will not be repeated. Gaska (2000) provides estimated yields based on replanting dates in Wisconsin. (iii) Consider replanting only if the cause of a stand failure can be determined and corrected with the replanting. (iv) Determine availability of seed of preferred cultivars since replanting with substandard cultivars is discouraged.

10–6.2 Weed/Pest Management

Inputs used for weed management in soybean represent a significant cost (Heatherly et al., 1994; Buhler et al., 1997; Johnson et al., 1997), and must be managed early (PRE) or on an as-needed basis (POST). In narrow-row soybean plantings, effective weed management systems almost exclusively involve herbicides (Oliver et al., 1993; Johnson et al., 1997, 1998a) because of the inability to effectively conduct interrow cultivation. However, this can lead to improved weed control in narrow-row systems and result in greater yield and net returns than from wide-row systems (Mickelson and Renner, 1997; Swanton et al., 1998). Use of combinations of PRE and POST herbicides with POST cultivation for broadleaf and grass weed control is common in wide-row soybean production systems in the mid-southern USA (Poston et al., 1992; Heatherly et al., 1993, 1994; Oliver et al., 1993;

Hydrick and Shaw, 1995; Askew et al., 1998), while a weed management system that is totally dependent on herbicides is used in narrow-row systems.

A comprehensive summary of weed management for soybean grown in the southern USA is presented by Reddy et al. (1999). Comprehensive summaries of disease, insect, and nematode management for soybean grown in the southern USA are presented by Bowers and Russin (1999), Funderburk et al. (1999), and Lawrence and McLean (1999), respectively. Chapter 18 (Buhler and Hartzler, 2004, this publication) contains an up-to-date summary of weed management issues for the soybean producing region of the USA. Summaries of management practices for diseases and nematodes are presented in Chapters 14 (Grau et al., 2004, this publication), 15 (Tolin and Lacy, 2004, this publication), and 16 (Niblack et al., 2004, this publication). Management of insect pests is addressed in Chapter 17 (Boethel, 2004, this publication).

10–6.3 Irrigation

It is widely thought that crops adapt to drought stress and become capable of withstanding drought. There is no evidence to support this view when it is considered on the basis of producing an economic yield. The limited adaptation that does occur only increases the plant's ability to survive during drought. This may be a valuable mechanism for a desert shrub, but it is of little value where production of a profitable seed yield from a crop such as soybean is important. The moisture status of plants is a function of soil water supply, evaporative demand of the atmosphere, and the capacity of the soil to release water. In the field, significant water deficits develop on hot sunny days even in well-watered plants. As water is transpired from the leaves, the moisture tensions that develop increase the rate of water uptake from the soil. If roots cannot absorb water rapidly enough, plant water tension increases. These tensions often become growth-limiting. See Chapter 12 (Purcell and Specht, 2004, this publication) for extensive details on the impact of water-deficit stress on soybean.

Irrigation that is properly managed or applied is important for soybean production in several areas of the USA, especially if consistent profits are expected. The lower Mississippi River alluvial flood plain and eastern Nebraska have high concentrations of irrigated soybean. Irrigated soybean yields in Nebraska have increased 40% faster than those of rainfed soybean [35.1 vs. 24.9 kg ha^{-1} yr^{-1} (0.52 vs. 0.37 bu acre^{-1} yr^{-1}); Specht et al., 1999], probably because of better management of irrigation. Inadequate water supply to soybean limits absolute crop yield and appears as an obstacle to yield improvements (Specht et al., 1999). In the midsouthern USA, soybean irrigation is required to make a profit on a consistent basis (Heatherly, 1999b).

Because of the importance of irrigation for optimizing yield and maximizing efficient use of inputs, and because of restrictions on water use imposed by regulatory agencies in the central Great Plains, many researchers have proposed soybean irrigation-scheduling strategies to optimize productivity and/or irrigation water-use efficiency. Water use (evaporation plus transpiration, or ET) of a fully irrigated crop of full-season soybean ranges from about 47 to 61 cm (18.5–24 in) per growing season (Benham et al., 1998). About 75% of this is used during reproductive development (Fig. 10–4). Table 10–25 lists irrigation water requirements for soy-

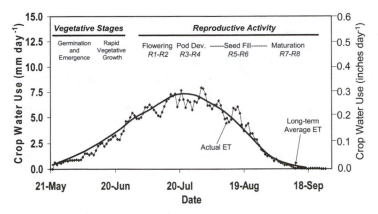

Fig. 10–4. Soybean crop water use (evapotranspiration) and growth stages. Data are averages from 1987 to 2002 at Clay Center, NE. Adapted from Benham et al. (1998).

bean during the reproductive stages when grown in Nebraska assuming the soil water reservoir is at or near capacity to a 1.5-m (5-ft) depth at planting. This usually is the case in the eastern half of Nebraska if the soils were irrigated the previous season and if there was sufficient off-season rain to recharge the profile. Peak water use is about 0.76 cm d^{-1} (0.3 in. d^{-1}), which occurs from R2 to R3. Yield responses to factorially and serially timed irrigation during reproductive development indicate that irrigations coinciding with the pod elongation (R3–R4) and seedfill (R5–R6) periods are the most effective (Korte et al., 1983a, 1983b; Kadhem et al., 1985a, 1985b). This of course assumes that plants had adequate water up to this time, which may not be the case for soybean grown on sandy soils or during years with a dry early season. In these cases, irrigation during vegetative and early reproductive development may be necessary to ensure optimum growth and development of plants, with careful attention to avoiding irrigation-induced excessive vegetative growth which will result in lodging (Benham et al., 1998). The use of determinate or semi-determinate cultivars can mitigate the lodging problem. Irrigation initiated before or during flowering must be followed with adequate water for the remainder of the season to ensure maximum number of seed and seed weight.

Soybean does best on soils with good internal and surface drainage. Although soybean roots may reach depths of 2 m (6.5 ft), irrigation management should concentrate on the top meter (3.3 ft) of the soil profile since most roots proliferate there. Soil type, irrigation system, and system capacity are important considerations for

Table 10–25. Irrigation water requirements in Nebraska for soybean during reproductive growth stages when grown on deep medium- and fine-textured soils. This assumes the soil water reservoir is at or near field capacity to 1.5-m (5-ft) depth. Adapted from Benham et al. (1998).

Growth stage	Reproduction stage irrigation water requirement
	cm (in)
Full flower (R2–R3)	7.6 (3)
Pod elongation (R3–R4)	7.6 (3)
Seedfill (R5–R6)	11.4 (4.5)
Total irrigation water required	26.7 (10.5)

irrigation management. Soil type determines available water-holding capacity and infiltration, irrigation system determines how water is delivered and affects irrigation efficiency, and irrigation system capacity determines the amount of time required to deliver an amount of irrigation water.

Irrigation scheduling is a means of accurately forecasting the times and amounts of water application to ensure that irrigation-mediated yield enhancement is economical. Factors that affect irrigation amount and frequency are determined by the amount of water applied by the previous irrigation (minus runoff), effective rainfall (amount that entered the soil), and estimated water use by the soybean crop since the previous irrigation and/or rain. The sensitivity of the developmental stage to water-deficit stress must also be considered. Rainfall measurements at the field site can be made easily, and well capacities or irrigation system outputs and efficiencies can be measured and/or calculated. Crop water use can be estimated by using pan evaporation numbers from the nearest weather station since actual evapotranspiration during the R1 to R6 period closely resembles pan evaporation (Reicosky and Heatherly, 1990). Estimates of water use based on pan evaporation can be combined with estimates of water supplied by irrigation and rainfall to predict the soil water deficit in the effective rooting zone.

The following guidelines for soybean irrigation management are adapted from recommendations for Nebraska by Benham et al. (1998). For coarse-textured or sandy soils with less than 12.5 cm m^{-1} (1.5 in ft^{-1}) water-holding capacity, allow no more than 50% water depletion in the top 0.6 m (2 ft) of soil during flowering (R1–R2). Allow no more than 50% depletion in the top 0.9 m (3 ft) of soil during the pod elongation to seedfill period (R3–R6). For deep medium- and fine-textured soils (silt loams, silty clay loams, and silty clays) with more than 12.5 cm m^{-1} (1.5 in ft^{-1}) water-holding capacity, allow no more than 50% water depletion in the top 1 m (3.3 ft) of soil during the R1 to R6 period. Producers with the latter soils often use an irrigation trigger criterion of 25% during the water-deficit-stress-sensitive pod elongation (R3–R4) and seed enlargement (R5–R6) periods, but 50 to 60% for other periods. Paraphrasing a comment made in the prior monograph (Van Doren and Reicosky, 1987), the sensitivity of the plant to water-deficit stress should signal when to irrigate, with the soil water status determining how much to irrigate.

Soil water levels at specified depths can be determined by soil sampling and drying, by instruments such as tensiometers, gypsum blocks, and neutron probes, or by ET estimates. Scheduling based solely on reproductive stage sensitivity or according to stage of development (Specht et al., 1989), or solely on soil and weather parameters, can be used with equal effectiveness for irrigation management of soybean. However, application amounts applied using the different methods may be different. For example, in a series of Nebraska studies (Klocke et al., 1989), irrigation initiated at growth stage R3 to R4 with soil water content not considered resulted in 40 kg ha^{-1} of seed produced per centimeter of irrigation water (1.52 bu acre^{-1} in^{-1}). Irrigation scheduled using the 50% soil water depletion parameter resulted in 25 kg ha^{-1} of seed produced per centimeter of irrigation water (0.95 bu acre^{-1} in^{-1}). The growth stage technique is simple but requires the soil profile over the entire potential rooting zone to be at or near field capacity at planting. This is usually the situation in eastern Nebraska and the midsouthern USA, assuming normal off-season and preplant rainfall patterns.

In the midsouthern USA, moisture deficits become more negative from April through August, as indicated from weather data collected at Stoneville, MS (Table 10–14). This leads to serious drought stress during reproductive development of soybean nearly every growing season. Since pod and seed growth, which are quite sensitive to plant water deficits, occur later in the season when soil moisture and rainfall are at the lowest seasonal levels, the potential for significant reductions in their growth and development and subsequent yield is great. Drought stress conditions can also result in greater infection of soybean roots by *Macrophomina phaseolina*, the causal organism of the yield-reducing disease charcoal rot (Kendig et al., 2000).

The advantages of irrigating soybean in the southern USA are well-documented (Reicosky and Heatherly, 1990; Heatherly, 1999b). Irrigation of soybean significantly increases yields by overcoming drought. The effectiveness of irrigation in alleviating the effects of drought on soybean in the southern USA is accepted, and is typically profitable (Heatherly, 1999b). If an irrigation system is in place, then it should be used since the fixed costs associated with the equipment exist regardless of whether or not the system is used. The question, then, is not whether to irrigate soybean for significant yield enhancement, but how to do it properly for maximum profit.

Weather data and measurements of the amount of water applied at each irrigation at Stoneville, MS, and the recent summary of irrigation research results (Heatherly, 1999b) have resulted in the following practical approach to scheduling irrigation for soybean in the midsouthern USA. Pan evaporation in the region ranges from 6.4 to 7.1 mm d^{-1} (0.25–0.28 in d^{-1}) during the months of June, July, and August (Boykin et al., 1995). Water use by MG V irrigated soybean that is in reproductive development during this period is about 7.7 mm d^{-1} (0.3 in d^{-1}) (Heatherly, 1986). Thus, in the absence of rain, about 7.5 cm (3 in) of water (net applied to soil) is needed about every 10 to 12 d. This is the amount typically supplied by a normal furrow or flood irrigation to cracking clay soils (Heatherly, 1999b). Therefore, furrow or flood irrigation should be planned every 10 to 12 d in the absence of rain to match the normal water deficit that occurs in the period since the last irrigation. To modify this approach, results from studies conducted in Arkansas (Tacker et al., 1997) should be used. These results show that irrigation scheduled to replace a 5 cm (2 in) soil water loss since the last irrigation resulted in significantly greater soybean yield than irrigation scheduled to replace a 7.5 cm (3 in) deficit. This simple approach results in a successful strategy for irrigating soybean in most situations, and ensures that clay soils are irrigated before noticeable cracking occurs. For sprinkler irrigation and assuming no runoff, an overhead irrigation system that applies 3 gross cm (1.2 in) should be scheduled to irrigate about every 3 to 4 d (assuming 80–85% efficiency).

Crusting soils with a low capacity for water infiltration, or shallow soils that have a relatively low total water-holding capacity, can experience runoff if large amounts of water are applied in a short period of time. In these situations, less water must be applied at each irrigation, but irrigation should be more frequent. On a silt loam site at Stoneville, MS, runoff of irrigation water applied through an overhead system occurred when an application exceeded 2 cm ha^{-1} $event^{-1}$ or 0.8 in $acre^{-1}$ $event^{-1}$ (Heatherly et al., 1992b). Thus, frequency of irrigation on this site was greater than on sites discussed above.

In Arkansas (Tacker et al., 1994), inadequate irrigation and/or improper timing of irrigations are the major reasons for lower-than-expected soybean yield responses from irrigation. They conclude that a water-balance approach has the most potential for properly irrigating soybean. They use two irrigation scheduling methods that are based on soil moisture accounting procedures. The Arkansas Checkbook Method uses a daily water-use chart and a computation table for updating soil moisture content (Tacker, 1993). The University of Arkansas Irrigation Scheduling Program operates basically the same, but uses a computer program to perform the computations (Tacker et al., 1997). The computer program requires the emergence date, the soil moisture deficit at planting, and a predetermined allowable soil moisture deficit of 5, 7.5, or 10 cm (2, 3, or 4 in). The daily information required to use either of these methods is maximum air temperature, rainfall, and irrigation amounts.

Numerous studies in the southeastern USA have investigated yield response of determinate soybean cultivars grown continuously to both full-season irrigation (water applied as needed during both the vegetative and reproductive phases of development) vs. irrigation during reproductive development only (water applied as needed from R1–R6). Reicosky and Heatherly (1990) summarized the results of these studies. The conclusions from these many studies, plus the additional information supplied by Heatherly (1999b), follow: (i) irrigation before R1 produced no appreciable yield advantage above that realized from irrigation applied only during reproductive development; and (ii) irrigation efficiency, defined here as the increase in seed yield ha^{-1} cm^{-1} of water applied, was usually higher for the reproductive phase irrigation. Thus, irrigation of monoculture soybean prior to R1 appears to be of little benefit, even though atmospheric demand for water increases through R1. In some years, significant drought during vegetative development may justify irrigation prior to bloom to ensure adequate vegetative framework to support a yield response to reproductively timed irrigations. Most soil types, assuming periodic rainfall, can supply the water necessary to meet atmospheric demands and support adequate growth during the vegetative phase. Exceptions to this are those soils that have a shallow rooting depth (Griffin et al., 1985) or low available water-holding capacity, or doublecropped soybean that is planted in dry soil.

Delaying initiation of irrigation until R4 or R5 in years when rainfall is limited during early reproductive stages results in seed yields that are lower than those realized from irrigation started at or about R1 (Elmore et al., 1988; Reicosky and Heatherly, 1990). Number of pods and seeds is increased if irrigation occurs during early reproductive development, but only the weight of seeds is increased if irrigation is delayed until later stages. Where drought stress is severe but alleviated by irrigation during early reproductive development, the biggest percentage yield increase comes from increased number of seeds. If irrigation is applied only after pods are set and seeds are filling, increase in weight of individual seeds is the major contributor to increased yields. Numerous research reports support the conclusion that the major effect of drought stress on seed yield is a reduced number of seeds (Reicosky and Heatherly, 1990). Frederick et al. (2001) found that increased yield resulting from irrigation of determinate cultivars grown on a Coastal Plain soil in South Carolina came from increased branch seed yield vs. main stem seed yield.

Irrigation that is started during early reproductive development must be continued into the seedfill stage (Griffin et al., 1985; Reicosky and Heatherly, 1990; Heatherly and Spurlock, 1993) so that soil moisture is readily available through the full seed stage. This ensures that yield potential is realized, and prevents increased infestations by the charcoal rot fungus (Kendig et al., 2000). Stress that occurs during seedfill results in smaller seeds, but will not reduce the total number of seeds below the number produced by plants that are irrigated during all stages of reproductive development (Reicosky and Heatherly, 1990). Thus, the number of seeds that are set is maintained during any drought stress that occurs after seed formation (except in extremely severe drought conditions), but maximum weight of individual seeds is not realized if drought occurs during seedfill. Irrigation during the full reproductive period is required to maximize both number of seeds (established by early alleviation of drought stress) and weight of seeds (maximized by later irrigations).

There may be cases where only a limited amount of irrigation water is available, and it is not enough for full reproductive phase irrigation. It can be allocated for use during early reproductive development to establish a maximum number of seeds, or to the latter stages of reproductive development to maximize weight of seeds. However, neither of these practices produces the maximum yield that may be required to maximize net returns unless adequate rainfall is received during the times of no irrigation (Heatherly, 1983; Elmore et al., 1988; Heatherly and Spurlock, 1993). The use of limited irrigation early in the reproductive phase can be advantageous if rains are received during the latter stages of reproductive development. Late-occurring rain has the greatest effect if relatively large numbers of seed are set as a result of irrigation during early reproductive development. However, the probability of late summer (August and September) rain in the midsouthern USA is low. The use of limited irrigation during the seedfill period can be advantageous for ensuring maximum weight of seeds. In cases of limited irrigation water, irrigation during the seedfill period appears to provide the greatest probability for maximizing yield. This appears less risky than using it earlier and depending on late-season rain to enlarge seeds that were set as a result of irrigation during early reproductive development. Unfortunately, producers using surface-water rights of lower priority may have those water rights halted before that occurs. Moreover, this premise assumes that a reasonably high number of seeds were set in the absence of irrigation during early reproductive development.

Irrigation of soybean interacts with other management practices such as cultivar, planting date, and row spacing. In reality, response of cultivars to irrigation probably is related more to time of reproductive development in relation to planting date than to cultivar per se. Early planted, early-maturing cultivars in the midsouthern USA sometimes require less irrigation and often produce greater irrigated yields than later-maturing cultivars planted during May and June because the irrigation period (R1–R6) of the later-planted, later-maturing cultivars is closely aligned with the period experiencing the greatest moisture deficit (Heatherly, 1999a, 1999b). In the midwestern USA, irrigation invariably mitigates the hastening of maturity induced by water-deficit stress. Indeed, in irrigated production, full-season cultivars recommended for rainfed culture may actually mature as much as 7 to 14 d later, which may not be the best adaptation. Lodging can result in reduced response to irrigation, especially when overhead irrigation is used and irrigation is applied before beginning bloom of determinate cultivars in the southern USA (Boquet, 1989). When both irrigation and row spacing are considered, proper irrigation is more important; that is, much greater yield responses can be achieved with irrigation of any row spacing than can be achieved by changing row spacing in the absence of adequate water (Elmore et al., 1988; Heatherly, 1999b).

From 1980 through 1997, various experiments that have involved irrigation of soybean planted in April, May, and June were conducted at Stoneville, MS (Table 10–26). The yield data led to several general but unmistakable conclusions. First, irrigation of soybean cultivars planted in April or early May almost always resulted in greater yields than did irrigation of the same cultivars planted later. Prior to 1992, when the earliest plantings were in early to mid-May and the late plantings were in late May to late June, more irrigations of the early plantings were required to achieve these higher yields. From 1992 to 1997, when the early plantings were in April and the late plantings were in early to late May, the earlier plantings required the same or fewer irrigations. Second, in the absence of irrigation, planting dates ranging from early May to late June (1980 through 1986) had little effect on soybean yield. April plantings of soybean that were not irrigated yielded more than later plantings (1992 through 1997). These first two conclusions led to a third. For fields that are to be irrigated, plant at the earliest acceptable time for the given cultivar and location to provide opportunity for the maximum seed yield response from MG 4 and MG 5 cultivars with the least irrigation.

Soybean grown on the flat alluvial flood plain of the lower Mississippi River Valley in the midsouthern USA is often irrigated using flood irrigation. The following summary regarding management of flood irrigation for soybean is condensed from information presented by Heatherly (1999b). Flood irrigation results in an inundation of a field with water amounts that result in standing water on some portion of an enclosed area. The flow rate of the water source and the size of the enclosed area being irrigated determine the time required to complete the flood. During the time of flooding, an increasingly larger area is covered with water until the entire area within the levees or borders is finally inundated. Numerous studies (Griffin et al., 1988; Scott et al., 1989; Heatherly and Pringle, 1991; Heatherly and Spurlock, 2000) have been conducted in the midsouthern USA to investigate the response of soybeans to flood irrigation. Results from these studies show the following: (i) soybean exposed to longer than 2 d of standing water was more tolerant of flooding during the vegetative period (V4) than during the reproductive period (R2); (ii) the damaging effect of prolonged flooding (more than 2 d) was more severe for soybean grown on a clay vs. a silt loam soil; (iii) differences exist in cultivar sensitivity to conditions resulting from flood irrigation; (iv) flood irrigation duration of 3 d resulted in less than maximum yield increase from irrigation, while that of 2 d or less ensured the greatest yield increase when using flood irrigation; and (v) properly timed and managed flood irrigation resulted in yields of soybean that were comparable to those resulting from proper furrow irrigation (Heatherly and Spurlock, 2000). Thus, for highest yield response from flood irrigation, it should be managed so that all area within a set of levees or borders will have the process started and finished within 2 d. Longer flood irrigation periods will lessen the expected yield response to irrigation due to root oxygen deprivation.

Determining the need and timing of a last irrigation application to soybean is important. Irrigation that exceeds that amount necessary to maximize net return is a waste of a valuable resource, increases labor costs and fuel consumption, and may result in reduced yield if lodging occurs. Drought still may occur late in the season, but the soybean plant's ability to use added water for additional increases

Table 10–26. Irrigation and planting date effects on seed yield and number of irrigations (No.) for soybean grown on Sharkey clay at Stoneville, MS. Adapted from Heatherly (1999b).

Year	Cultivar (MG)	Planting date	Seed yield†			No.
			I	NI	I -NI	
			kg ha^{-1} (bu acre^{-1})			
1980	Bedford (4)	12 May	2730 (40.6)	990 (14.7)	1740 (25.9)	7
		3 June	3145 (46.8))	1155 (17.2)	1990 (29.6)	5
	Bragg (7)	12 May	3520 (52.4)	1330 (19.8)	2190 (32.6)	7
		3 June	2975 (44.3)	1515 (22.6)	1460 (21.7)	5
1981	Bedford (5)	13 May	2775 (41.3)	980 (14.6)	1795 (26.7)	3
		4 June	2375 (35.3)	1050 (15.6)	1325 (19.7)	2
	Braxton (7)	13 May	3275 (48.7)	1030 (15.3)	2245 (33.4)	4
		4 June	2935 (43.7)	1695 (25.2)	1245 (18.5)	3
1982	Bedford (5)	12 May	2245 (33.4)	975 (14.5)	1270 (18.9)	3
		28 May	1665 (24.8)	880 (13.1)	785 (11.7)	3
	Braxton	12 May	2715 (40.4)	1010 (15.0)	1705 (25.4)	4
		28 May	2345 (34.9)	1195 (17.8)	1150 (17.1)	3
1984	Braxton	14 May	3570 (53.1)	1405 (20.9)	2165 (32.2)	5
		25 June	3110 (46.3)	1580 (23.5)	1530 (22.8)	4
1985	Braxton	2 May	2955 (44.0)	1860 (27.7)	1095 (16.3)	6
		24 June	1895 (28.2)	1655 (24.6)	240 (3.6)	3
1986	Braxton	15 May	2690 (40.0)	110 (1.6)	2580 (38.4)	7
		24 June	1425 (21.2)	260 (3.9)	1165 (17.3)	4
1986	Leflore (6)	6 May	3595 (53.5)	--		7
		16 June	2735 (40.7)	--		5
1987	Leflore	5 May	2895 (43.1)	--		7
		28 May	2515 (37.4)	--		7
1992	RA 452 (4)	15 Apr	4180 (62.2)	2840 (42.3)	1340 (19.9)	2
		27 May	3035 (45.2)	2175 (32.4)	860 (12.8)	2
	A 5979 (5)	15 Apr	4315 (64.2)	3555 (52.9)	760 (11.3)	2
		27 May	2935 (43.7)	2230 (33.2)	705 (10.5)	2
1994	RA 452	21 April	3360 (50.0)	2645 (39.4)	715 (10.6)	4
		13 May	3245 (48.3)	2155 (32.1)	1090 (16.2)	4
	A 5979	21 April	3440 (51.2)	2595 (38.6)	845 (12.6)	4
		13 May	3365 (50.1)	2265 (33.7)	1100 (16.4)	4
1995	DP 3478 (4)	18 April	4440 (66.1)	2905 (43.2)	1535 (22.9)	3
		9 May	3620 (53.9)	2035 (30.3)	1585 (23.6)	3
	A 5979	18 April	3845 (57.2)	1740 (25.9)	2105 (31.3)	4
		9 May	3890 (57.9)	1405 (20.9)	2485 (37.0)	4
1996	DP 3478	30 April	3835 (57.1)	2170 (32.3)	1665 (24.8)	4
		15 May	3515 (52.3)	1950 (29.0)	1565 (23.3)	5
	Hutcheson (5)	30 April	4200 (62.5)	3035 (45.2)	1165 (17.3)	5
		15 May	4110 (61.2)	3035 (45.2)	1075 (16.0)	5
1997‡	DP 3478	9 April	4205 (62.6)	2015 (30.0)	2190 (32.6)	4
		12 May	4150 (61.8)	2045 (30.4)	2105 (31.4)	6
	Hutcheson	9 April	3620 (53.9)	2420 (36.0)	1200 (17.9)	5
		12 May	4240 (63.1)	2235 (33.3)	2005 (29.8)	7

† NI = nonirrigated; I = irrigated; I - NI = irrigated minus nonirrigated yield.
‡ 1997 irrigation scheduled more frequently.

Table 10-27. Components and variables for equation to determine last irrigation of soybean in Nebraska. Adapted from Klocke et al., 1991.

Component	Inputs for determining value for factor
Water requirement based on crop stage and water use to R7 stage	23, 16.5, and 9 cm (9.0, 6.5, and 3.5 in) are required from the R4, R5, and R6 stages, respectively.
Available water content (AWC)	60% × AWC × 1.2 m (3.9 ft); 60% of the AWC in the top 1.2 m (3.9 ft) of the root zone can be depleted at maturity and not reduce yield.
Current soil moisture in 1.2-m- (3.9-ft) deep root zone	Determined by gravimetric sampling, hand-feel method, crop water-use scheduling method, soil moisture blocks, tensiometers, etc.

in seed dry matter is limited at some point by the physiological processes of the maturing plant system. On the other hand, one additional irrigation may mean optimizing yield via further increase in seed size. An optimal date for the last seasonal irrigation requires consideration of two conflicting goals: (i) supply enough root-zone water for the crop to produce maximum yield and (ii) terminate soon enough to allow for the depletion of soil profile water so that off-season precipitation storage will be maximized (Klocke et al., 1991). Field information on crop growth stage, expected water use to R7, remaining useable soil profile moisture, and probability of precipitation are factors that should be considered in determining need for a final irrigation. For the northern USA, an equation adapted from Klocke et al. (1991) for use in Nebraska may be useful: Remaining irrigation amount required = Water requirement – available water capacity (AWC) – current soil moisture status. Components for calculations are presented in Table 10–27. The principal difficulty is forecasting the calendar date when R7 will occur. Physiological maturity of a pod is achieved when dry matter accumulation ceases. Some researchers use the degree to which the pod membrane clings to the pod wall instead of the seed as a key visual criterion of physiological maturity. This normally occurs in the pod about 3 to 7 d before that pod achieves a mature pod color (R7 criterion) (James Specht, personal communication, 2002).

In the midsouthern USA, an effective surface irrigation at stage R5.5 supplied enough soil moisture to finish filling seeds of MG 5 and 6 cultivars that were planted in May. Irrigation later than R5.5 did not increase yield or net returns. Termination of irrigation at an earlier stage resulted in lower yields and net returns. Irrigation terminated during early bloom (R1–R2), full bloom (R2–R3), podset (R3–R4), and full pod/beginning seedfill (R4–R5) periods resulted in negative or only slightly positive net returns, even though yields were increased significantly above nonirrigated yields by all of these early-terminated treatments (Heatherly and Spurlock, 1993). These results point out the importance of starting irrigation at early reproductive development of soybean and continuing well into the seedfill period when using surface application methods. When irrigation is supplied by overhead systems that may apply less water per event than is applied by surface methods, the last irrigation should be later since an overhead irrigation applied at stage R5.2 to R5.5 may not provide enough water to finish filling seeds. In southeast Missouri, the recommendation is that irrigation should continue past R6 on all soils that have a coarser texture than silt loam, whereas flood irrigation can be terminated at about mid-R5 on all soils but coarse sand (Henggeler, 2002). For ESPS plantings in the midsouthern USA, indications from the extreme drought periods in the late summer of 1999 and 2000 are that surface irrigation should be continued to R6. A fully charged soil profile at R6 is needed because the R6 to R7 period occurs during the hottest, driest part of the growing season and water use is still occurring.

10–6.4 Harvesting

Physiological maturity of soybean occurs when maximum dry matter has accumulated. The easiest visual indicator of this is the presence of one normal pod on the plant that has reached the mature pod color (R7) (However, see previous section). Seed moisture at this time may range between 40 and 60%. An autumn frost before R7 can result in premature plant death, with subsequent reduced yield and green pods and seeds. Leaves of frosted plants do not drop normally and field dry-down of seed is slower. Only 1% green seed are allowable for U.S. no. 1 soybean and 2% for U.S. no. 2 grade. Oil produced from green soybean has a green tint and requires additional refining. Thus, a frost which results in green seed may reduce market value of the seed. Seed from plants frosted within 14 d of maturity, or seed from plants only partially frosted, may lose their green color if allowed to field dry (Hurburgh et al., 2001). Seed that has a light brown coloration will lose that coloration after a few months in storage. Protein content of frosted soybean seed is similar to that of nonfrosted soybean seed; however, oil content and germination are both reduced. Germination of seed from a yellow pod (55% seed moisture content) is decreased when exposed to 8 h at -7°C, while germination of seed from brown pods (35% seed moisture content) is decreased when exposed to 8 h at -12°C (Johnson, 1987). Frost-damaged or green soybean seed can be marketed as livestock feed. For swine (*Sus* spp.), extruded frost-damaged soybean may totally replace soybean meal. For sheep (*Ovis* spp.) and cattle (*Bos* spp.), frost-damaged soybean should be limited to 14% of total dry matter intake by the animals (Loy and Holden, 1993). Because of variability in composition, frost-damaged soybean seed should be analyzed before use in formulating rations.

Purposeful desiccation prior to soybean harvest is an option in some instances, such as ESPS plantings in the midsouthern USA where an earlier open canopy allows weed resurgence. The following thoughts are presented based on discussion by Ellis et al. (1998), Heatherly (1999a), and Reddy et. al. (1999). A preharvest desiccant will be needed if weed densities are high enough to lead to increases in soybean seed moisture and damaged soybean seed, more foreign material in the seed, and/or decreased combine speed and subsequent harvest efficiency. A preharvest desiccant will not be needed if: (i) weeds present at maturity emerged late in the growing season and their size will not interfere with harvest; (ii) the weeds present have not produced mature seeds that will contaminate the grain; and (iii) the desiccant cannot be applied sufficiently ahead of harvest so as to ensure that the weeds are dry at harvest, or that time interval restrictions in the harvest aid label can be met. This may be the case in the high temperature, low humidity conditions common to August when the time between maturity of soybean, or 95% mature pod color, and harvest maturity may be as little as 5 to 7 d and the required interval between desiccant application and harvest is 7 to 15 d, depending on the desiccant. The budgets for ESPS soybean in Mississippi (Spurlock, 2000) do not include a desiccation input since conditions rarely justify the need for or allow the effective use of desiccants.

The maturity stage of soybean seed best for harvest is when seed moisture falls to 15% or lower (Keith and Delouche, 1999). Harvest of soybean seed production fields should commence as soon as seed moisture content reaches 14% (Moore, 1996). Susceptibility to harvest-mediated mechanical damage (i.e., cracked seed coat and split seed) occurs when seed moisture content falls below about 13% (Keith and Delouche, 1999). This lowers seed quality and may result in dockage at the delivery point or reduced germination of seed beans. Combine settings should be checked over the course of each day's harvest to allow for changes in seed moisture content that occurs from the morning through the afternoon and evening of the same day.

The harvest environment is dictated by soil conditions, weather, presence of weeds, and the condition of the seed to be harvested. Harvesting is a critical phase since the condition of the final saleable product is determined at this time. Weathering or field deterioration of seeds may be associated with harvest delays beyond the harvest maturity stage. Therefore, it is critical to commence and continue harvest whenever proper conditions are present. Highest harvest efficiency occurs when the harvesting machine travels on dry soil. This allows combine ground speed and speeds of the machine components to be matched for greatest effect.

An important consideration when choosing cultivars at the beginning of the growing season in the midsouthern USA is their maturity date in relation to optimum soil conditions for harvesting on clay soils. Harvesting efficiency is reduced on soils of this texture when they are wet because of combine wheel spinning and slipping. Fields are rutted by the harvesting machine, and remediation of these ruts requires otherwise unnecessary tillage. Rutted fields must be repaired in the fall following harvest; otherwise, their remediation the following spring may cause planting delays that result in lowered yield potential caused by the delayed planting. Thus, consideration of optimum soil conditions for harvest in this region is an important component of a soybean production system.

Seed quality can be affected by the harvesting operation. Improper combine settings can result in excessive splitting of seed and/or the failure of the machine to remove seed-contaminating foreign matter. These problems can result in lowered seed quality and dockage at the delivery point.

10–7 SUMMARY

This chapter is intended to supplement those of Johnson (1987) and Van Doren and Reicosky (1987), which centered on midwestern U.S. soybean production. Much of the material in those chapters is still pertinent to today's production systems that involve soybean. The information in this chapter in many cases is an addendum to information in the aforementioned chapters. Most of the information in this chapter pertains to subjects that have received considerable research attention since the previous chapters, or to information about new technologies that have emerged since the writing of the previous chapters.

In the last 15 yr, new technologies and problems have become significant contributing factors to soybean production in the USA. This chapter presents details for solving those problems as well as details of new technologies that should be considered in soybean management systems. First, glyphosate-resistant soybean is a significant new technology that affects cultivar development and selection, as well as systems of management in the U.S. soybean production zone. The incorporation of this new technology into current management systems is requiring a reassessment of weed management for soybean and rotational systems using soybean. Second, new paradigms regarding planting date/MG combinations are being accepted as methods of avoiding pest and weather stresses. The use of cultivars from early-maturing MGs in regions outside their perceived area of adaptation is being used as a tool to manage biotic and abiotic stresses of soybean. Third, SCN incursion into the midwestern/northern U.S. soybean-producing areas has added a new dimension to management systems in this region. This has placed more emphasis on pest resistant cultivars and rotational schemes as management tools. Fourth, irrigation has become an increasingly important component in soybean management systems where summer drought conditions are prominent. The need to reduce seasonal irrigation requirement for soybean and avoid major water-deficit periods of the growing season has placed increased emphasis on management practices that enhance drought avoidance and increase water use efficiency to produce the highest possible economic yield. Fifth, the need to lower producer risk (reduce costs, decrease management time, and stablize higher yields) in the face of stiffer global competition requires that all components of soybean management be assessed in relation to each other to ensure the most efficient combination of production variables.

Competition for soybean market share has become pronounced in the global marketplace. The producer with the most efficient and economical production system will be the most competitive. Therefore, it is paramount that proven new technology and new paradigms for soybean production be adopted quickly. This requires that new and emerging educational tools be used to place new and improved management concepts into the hands of the producer as quickly as possible. This chapter attempts to accomplish this through presentation of information on new soybean production/management technology and where to find it.

ACKNOWLEDGMENTS

The authors sincerely appreciate the reviews provided by Drs. Seth Naeve, C. Wayne Smith, Don Boquet, Jeff Ray, and Emerson Nafziger.

REFERENCES

Ablett, G.R., W.D. Beversdorf, and V.A. Dirks. 1989. Performance and stability of indeterminate and determinate soybean in short-season environments. Crop Sci. 29:1428–1433.

Ablett, G.R., R.I. Buzzell, W.D. Beversdorf, and O.B. Allen. 1994. Comparative stability of 40 indeterminate and semideterminate soybean lines. Crop Sci. 34:347–351.

Adee, E.A., E.S. Oplinger, and C.R. Grau. 1994. Tillage, rotation sequence, and cultivar influences on brown stem rot and soybean yield. J. Prod. Agric. 7:341–347.

Anonymous. 2001. Crop protection guide. C&P Press, New York.

Anonymous. 2002. Seed drop rate calculator. *In* Illinois agronomy handbook. Dep. of Crop Sciences, Univ. of Illinois, Urbana-Champaign, IL [Online]. Available at http://web.aces.uiuc.edu/aim/IAH/drop.html (verified 26 Nov. 2002).

Askew, S.D., J.E. Street, and D.R. Shaw. 1998. Herbicide programs for red rice (*Oryza sativa*) control in soybean (*Glycine max*). Weed Technol. 12:103–107.

Ball, R.A., L.C. Purcell, and E.D. Vories. 2000. Optimizing soybean plant population for a short-season production system in the southern USA. Crop Sci. 40:757–764.

Baur, M.E., D.J. Boethel, M.L. Boyd, G.R. Bowers, M.O. Way, L.G. Heatherly, J. Rabb, and L. Ashlock. 2000. Arthropod populations in early soybean production systems in the mid-south. Environ. Entomol. 29:312–328.

Beaver, J.S., and R.R. Johnson. 1981. Response of determinate and indeterminate soybeans to varying cultural practices. Agron. J. 73:833–838.

Benham, B.L., J.P. Schneekloth, R.W. Elmore, D.E. Eisenhauer, and J.E. Specht. 1998. Irrigating soybean. NebGuide G98-1367-A. Univ. of Nebraska Coop. Ext. Serv., Lincoln. Available online at http://www.ianr.unl.edu/pubs/fieldcrops/g1367.htm (verified 25 Nov. 2002).

Bennett, A.C., D.R. Shaw, and S.M. Schraer. 1998. Effect of conventional herbicide programs and irrigation on glyphosate-tolerant soybean yield. p. 270–271. *In* J.A. Dusky (ed.) Southern Weed Sci. Soc. Proc. Southern Weed Sci. Soc. Mtg., Birmingham, AL. 26–28 Jan. 1998. Weed Sci, Champaign, IL.

Bernard, R.L. 1972. Two genes affecting stem termination in soybeans. Crop Sci. 12:235–239.

Bertram, M.G., and E.S. Oplinger. 2000. Agronomy advice. Dep. of Agronomy, Univ. of Wisconsin. Madison, WI [Online]. Available at http://soybean.agronomy.wisc.edu (verified 29 Nov. 2002).

Beuerlein, J.E. 1988. Yield of indeterminate and determinate semidwarf soybean for several planting dates, row spacings, and seeding rates. J. Prod. Agric. 1:300–303.

Beuerlein, J.E. 1995. Adjusting a grain drill for planting soybeans. Ohio State Univ. Ext. Serv. FactSheet AGF-114-95, Columbus, OH. Available online at http://ohioline.osu.edu/agf-fact/0114.html (verified 25 Nov. 2002).

Beuerlein, J.E. 1999. Soybean inoculation and nitrogen nutrition. Ohio State Univ. Ext. Serv. FactSheet AGF-137-99, Columbus, OH. Available online at http://ohioline.osu.edu/agf-fact/0137.html (verified 25 Nov. 2002).

Beuerlein, J.E. 2001a. 2000 Ohio soybean inoculation trials. Ohio State Univ. Ext. Serv. FactSheet AGF-137-01, Columbus, OH. Available online at http://ohioline.osu.edu/agf-fact/0137.html (verified 25 Nov. 2002).

Beuerlein, J.E. 2001b. Doublecropping soybean following wheat. Ohio State Univ. Ext. Serv. FactSheet AGF-103-01, Columbus, OH. Available online at http://ohioline.osu.edu/agf-fact/0103.html (verified 25 Nov. 2002).

Bharati, M.P., D.K. Whigham, and R.D. Voss. 1986. Soybean response to tillage and nitrogen, phosphorus, and potassium fertilization. Agron. J. 78:947–950.

Boethel, D.J. 2004. Integrated management of soybean insects. p. 853–882. *In* H.R. Boerma and J.E. Specht (ed.) Soybeans: Improvement, production, and uses. 3rd ed. Agron. Monogr. 16. ASA, CSSA, and SSSA, Madison, WI.

Boquet, D.J. 1989. Sprinkler irrigation effects on determinate soybean yield and lodging on a clay soil. Agron. J. 81:793–797.

Boquet, D.J. 1990. Plant population density and row spacing effects on soybean at post-optimal planting dates. Agron. J. 82:59–64.

Boquet, D.J. 1996. Row spacings and plant population density. p. 90–92. *In* J. Honeycutt (ed.) Louisiana soybean handbook. Publ. 2624. Louisiana State Univ., Baton Rouge.

Boquet, D.J. 1998. Yield and risk utilizing short-season soybean production in the mid-southern USA. Crop Sci. 38:1004–1011.

Bowers, G.R., Jr. 1990. Registration of 'Crockett' soybean. Crop Sci. 30:427.

Bowers, G.R. 1995. An early soybean production system for drought avoidance. J. Prod. Agric. 8:112–119.

Bowers, G.R., J.L. Rabb, L.O. Ashlock, and J.B. Santini. 2000. Row spacing in the early soybean production system. Agron. J. 92:524–531.

Bowers, G.R., and J.S. Russin. 1999. Soybean disease management. p. 231–270. *In* L.G. Heatherly and H.F. Hodges (ed.) Soybean production in the mid-south. CRC Press, Boca Raton, FL.

Boykin, D.L., R.R. Carle, C.D. Ranney, and R. Shanklin. 1995. Weather data summary for 1964–1993, Stoneville, Mississippi. Tech. Bull. 201. Mississippi Agric. and For. Exp. Stn., Mississippi State.

Bruff, S.A., and D.R. Shaw. 1992a. Early season herbicide applications for weed control in stale seedbed soybean (*Glycine max*). Weed Technol. 6:36–44.

Bruff, S.A., and D.R. Shaw. 1992b. Tank-mix combinations for weed control in stale seedbed soybean. Weed Technol. 6:45–51.

Buhler, D.D., J.L. Gunsolus, and D.F. Ralston. 1992. Integrated weed management techniques to reduce herbicide inputs in soybean. Agron. J. 84:973–978.

Buhler, D.D., and R.G. Hartzler. 2004. Weed biology and management. p. 883–918. *In* H.R. Boerma and J.E. Specht (ed.) Soybeans: Improvement, production, and uses. 3rd ed. Agron. Monogr. 16. ASA, CSSA, and SSSA, Madison, WI.

Buhler, D.D., R.P. King, S.M. Swinton, J.L. Gunsolus, and F. Forcella. 1997. Field evaluation of a bioeconomic model for weed management in soybean (*Glycine max*). Weed Sci. 45:158–165.

Bullock, D., S. Khan, and A. Rayburn. 1998. Soybean yield response to narrow rows is largely due to enhanced early growth. Crop Sci. 38:1011–1016.

Burnside, O.C., G.A. Wicks, and D.R. Carlson. 1980. Control of weeds in an oat (*Avena sativa*)-soybean (*Glycine max*) ecofarming rotation. Weed Sci. 28:46–50.

Casey, W.P., T.J. Dumler, R.O. Burton, D.W. Sweeney, A.M. Featherstone, and G.V. Granade. 1998. A whole-farm economic analysis of early-maturing and traditional soybean. J. Prod. Agric. 11:240–246.

Cober, E.R., J. Madill, and H.D. Voldeng. 2000. Early tall determinate soybean genotypes E1E1e3e3e4e4dt1dt1 sets high bottom pods. Can. J. Plant Sci. 80:527–531.

Cober, E.R., and J.W. Tanner. 1995. Performance of related indeterminate and tall determinate soybean lines in short-season areas. Crop Sci. 35:361–364.

Cober, E.R., and H.D. Voldeng. 2000. Developing high-protein, high-yield soybean populations and lines. Crop Sci. 40:39–42.

Cober, E.R., H.D. Voldeng, and J.A. Frègeau-Reid. 1997. Heritability of seed shape and seed size in soybean. Crop Sci. 37:1767–1769.

Conservation Technology Information Center. 2002. National crop residue management survey, 2002 [Online]. Available at http://www.ctic.purdue.edu/Core4/CT/CT.html (verified 20 Nov. 2002).

Cooper, R.L. 1981. Development of short-statured soybean cultivars. Crop Sci. 21:127–131.

Crookston, R.K., and D.S. Hill. 1979. Grain yields and land equivalent ratios from intercropping corn and soybeans in Minnesota. Agron. J. 71:41–44.

Curley, R.L., and J.C. Burton. 1975. Compatibility of *Rhizobium japonicum* with chemical seed protectants. Agron. J. 67:807–808.

Dabney, S.M., E.C. McGawley, D.J. Boethel, and D.A. Berger. 1988. Short-term crop rotation systems for soybean production. Agron. J. 80:197–204.

Delannay, X., T.T. Bauman, D.H. Beighley, M.J. Buettner, H.D. Coble, M.S. DeFelice, C.W. Derting, T.J. Diedrick, J.L. Griffin et al.1995. Yield evaluation of a glyphosate-tolerant soybean line after treatment with glyphosate. Crop Sci. 35:1461–1467.

Devlin, D.J., D.L. Fjell, J.P. Shroyer, W.B. Gordon, B.H. Marsh, L.D. Maddux, V.L. Martin, and S.R. Duncan. 1995. Row spacing and seeding rates for soybean in low and high yielding environments. J. Prod. Agric. 8:215–222.

Dick, W.A., E.L. McCoy, W.M. Edwards, and L.R. Lal. 1991. Continuous application of no-tillage to Ohio soils. Agron. J. 83:65–73.

Dickey, E.C., D.P. Shelton, and P.J. Jasa. 1986. Residue management for soil erosion control. NebGuide G81-544. Univ. of Nebraska Coop. Ext. Serv., Lincoln. Available online at http://www.ianr.unl.edu/pubs/fieldcrops/g544.htm (verified 25 Nov. 2002).

Dorris, E.A. 2001. Keeping the faith. Miss. Farmer 11(12):12–14.

Duncan, W.G. 1986. Planting patterns and soybean yield. Crop Sci. 26:584–588.

Duncan, S.R., and W.T. Schapaugh, Jr. 1997. Relay-intercropped soybean in different water regimes, planting patterns, and winter wheat cultivars. J. Prod. Agric. 10:123–129.

Edwards, J.H., D.L. Thurlow, and J.T. Eason. 1988. Influence of tillage and crop rotation on yields of corn, soybean, and wheat. Agron. J. 80:76–90.

Edwards, W.M., G.B. Triplett, D.M. Van Doren, L.B. Owens, C.E. Redmond, and W.A. Dick. 1993. Tillage studies with a corn-soybean rotation: Hydrology and sediment loss. Soil Sci. Soc. Am. J. 57:1051–1055.

Egli, D.B., and W.P. Bruening. 2000. Potential of early-maturing soybean cultivars in late plantings. Agron. J. 92:532–537.

Ellis, J.M., D.R. Shaw, and W.L. Barrentine. 1998. Soybean seed quality and harvesting efficiency as affected by low weed densities. Weed Technol. 12:166–173.

Elmore, C. D., and L. G. Heatherly. 1988. Planting system and weed control effects on soybean grown on clay soil. Agron. J. 80:818–821.

Elmore, C.D., R.A. Wesley, and L.G. Heatherly. 1992. Stale seedbed production of soybeans with a wheat cover crop. J. Soil Water Conserv. 74:187–190.

Elmore, R.W. 1987. Soybean cultivar response to tillage systems. Agron. J. 79:114–119.

Elmore, R.W. 1990. Soybean cultivar response to tillage systems and planting date. Agron. J. 82:69–73.

Elmore, R.W. 1991. Soybean cultivar response to planting rate and tillage. Agron. J. 83:829–832.

Elmore, R.W. 1998. Soybean cultivar responses to row spacing and seeding rates in rainfed and irrigated environments. J. Prod. Agric. 11:326–331.

Elmore, R.W., D.E. Eisenhauer, J.E. Specht, and J.H. Williams. 1988. Soybean yield and yield component response to limited capacity sprinkler irrigation systems. J. Prod. Agric. 1:196–201.

Elmore, R.W., M.D. MacNeil, and R.F. Mumm. 1987. Determinate and indeterminate soybeans in low-yield and high-yield environments. Appl. Agric. Res. 22:74–80.

Elmore, R.W., H.C. Minor, and B.L. Doupnik, Jr. 1998. Soybean genetic resistance and benomyl for *Phomopsis* seed decay control. Seed Technol. 20:23–31.

Elmore, R.W., F.W. Roeth, R. Klein, S.Z. Knezevic, A. Martin, L. Nelson, and C.A. Shapiro. 2001a. Glyphosate-resistant soybean cultivar response to glyphosate. Agron. J. 93:404–407.

Elmore, R.W., F.W. Roeth, L.A. Nelson, C.A. Shapiro, R.N. Klein, S.Z. Knezevic, and A. Martin. 2001b. Glyphosate-resistant soybean cultivar yields compared with sister lines. Agron. J. 93:408–412.

Erbach, D.C. 1982. Tillage for continuous corn and corn-soybean rotation. Trans. ASAE 25:906–911, 918.

Escalante, R.B., and J.R. Wilcox. 1993. Variation in seed protein among nodes of determinate and indeterminate soybean near-isolines. Crop Sci. 33:1166–1168.

Ethredge, W.J., Jr., D.A. Ashley, and J.M. Woodruff. 1989. Row spacing and plant population effects on yield components of soybean. Agron. J. 81:947–951.

Fehr, W.R., and C.E. Caviness. 1977. Stages of soybean development. Spec. Rep. 80. Ia. Agric. Exp. Stn., Ames.

Ferguson, R.B., E.J. Penas, and W.B. Stevens. 2000. Soybean. p. 121–125. *In* R.B. Ferguson and K.M. DeGroot (ed.) Nutrient management for agronomic crops in Nebraska. EC-01-155. Univ. of Nebraska Coop. Ext. Serv., Lincoln.

Ferreira, M.C., D. de S. Andrade, L.M. de O. Chueire, S.M. Takemura, and M Hungria. 2000. Tillage method and crop rotation effects on the population sizes and diversity of bradyrhizobia nodulating soybean. Soil Biol. Biochem. 32:627–637.

Foley, T.C., J.H. Orf, and J.W. Lambert. 1986. Performance of related determinate and indeterminate soybean lines. Crop Sci. 26:5–8.

Foth, H.D., and B.G. Ellis. 1997. Soil fertility. 2nd ed. Lewis Publ., Boca Raton, FL.

Frank, K.D. 2000. Potassium. p. 23–31. *In* R.B. Ferguson and K.M. DeGroot (ed.) Nutrient management for agronomic crops in Nebraska. EC-01-155. Univ. of Nebraska Coop. Ext. Serv., Lincoln.

Frederick, J.R., P.J. Bauer, W.J. Busscher, and G.S. McCutcheon. 1998. Tillage management for doublecropped soybean grown in narrow and wide row width culture. Crop. Sci. 38:755–762.

Frederick, J.R., C.R. Camp, and P.J. Bauer. 2001. Drought-stress effects on branch and mainstem seed yield and yield components of determinate soybean. Crop Sci. 41:759–763.

Funderburg, E.R. 1996. Fertilization and liming. p. 74–77. *In* J. Honeycutt (ed.). Louisiana Soybean Handb. Publ. 2624. Louisiana State Univ., Baton Rouge.

Funderburk, J., R. McPherson, and D. Buntin. 1999. Soybean insect management. p. 273–290. *In* L.G. Heatherly and H.F. Hodges (ed.) Soybean production in the mid-south. CRC Press, Boca Raton, FL.

Gaska, J. 2000. Soybean replant and late plant issues. Wisconsin Crop Manager 7(12):75 [Online]. Available at http:/soybean.agronomy.wisc.edu/publications/wcm/00wcm_soybean_replant.htm (verified 29 Nov. 2002).

Geater, C.W., W.R. Fehr, and L.A. Wilson. 2000. Association of soybean seed traits with physical properties of natto. Crop Sci. 40:1529–1534.

Ginn, L.H., L.G. Heatherly, E.R. Adams, and R.A. Wesley. 1998. A modified implement for constructing wide beds for crop production. Bull. 1072. Miss. Agric. and For. Exp. Stn., Mississippi State.

Grabau, L.J., and T.W. Pfeiffer. 1990. Assessment of soybean stubble losses in different cropping systems. Appl. Agric. Res. 5:96–101.

Graterol, Y.E., R.W. Elmore, and D.E. Eisenhauer. 1996. Narrow-row planting systems for furrow-irrigated soybean. J. Prod. Agric. 9:546–553.

Grau, C.R., A.E. Dorrance, J. Bond, and J.S. Russin. 2004. Fungal diseases. p. 679–764. *In* H.R. Boerma and J.E. Specht (ed.) Soybeans: Improvement, production, and uses. 3rd ed. Agron. Monogr. 16. ASA, CSSA, and SSSA, Madison, WI.

Grau, C.R., E.S. Oplinger, E.A. Adee, E.A. Hinkens, and M.J. Martinka. 1994. Planting date and row width effect on severity of brown stem rot and soybean productivity. J. Prod. Agric. 7:347–351.

Grau, C.R., and V.L. Radke. 1984. Effects of cultivars and cultural practices on sclerotinia stem rot of soybean. Plant Dis. 68:56–58.

Griffin, J.L., and S.M. Dabney. 1990. Preplant-postemergence herbicides for legume cover-crop control in minimum tillage systems. Weed Technol. 4:332–336.

Griffin, J.L., R.J. Habetz, and R.P. Regan. 1988. Flood irrigation of soybeans in Southwest Louisiana. Louisiana Agric. Exp. Stn. Bull 795. Louisana State Univ., Baton Rouge.

Griffin, J.L., D.B. Reynolds, P.R. Vidrine, and S.A. Bruff. 1993. Soybean (*Glycine max*) tolerance and sicklepod (*Cassia obtusifolia*) control with AC 263,222. Weed Technol. 7:331–336.

Griffin, J.L., R.W. Taylor, R.J. Habetz, and R.P. Regan. 1985. Response of solid-seeded soybeans to flood irrigation. I. Application timing. Agron. J. 77:551–554.

Gunsolus, J.L. 1990. Mechanical and cultural weed control in corn and soybeans. Am. J. Alternative Agric. 5:114–119.

Guy, S.O., and E.S. Oplinger. 1989. Soybean cultivar performance as influenced by tillage system and seed treatment. J. Prod. Agric. 2:57–62.

Hartman, G.L., J.B. Sinclair, and J.C. Rupe (ed.) 1999. Compendium of soybean diseases. 4th ed. Am. Phytopathol. Soc., St. Paul, MN.

Hartung, R.C., J.E. Specht, and J.H. Williams. 1981. Modification of soybean plant architecture by genes for stem growth habit and maturity. Crop Sci. 21:51–56.

Hartwig, E.E., L. Lambert, and T.C. Kilen. 1990. Registration of 'Lamar' soybean. Crop Sci. 30:231.

Heatherly, L.G. 1981. Soybean response to tillage of Sharkey clay soil. Bull. 892. Miss. Agric. and For. Exp. Stn., Mississippi State.

Heatherly, L.G. 1983. Response of soybean cultivars to irrigation of a clay soil. Agron. J. 75:859–864.

Heatherly, L.G. 1986. Water use by soybeans grown on clay soil. p. 113–121. *In* Proc. Delta Irrig. Workshop, Greenwood, MS. 28 Feb.1986. Miss. Coop. Ext. Serv., Starkville.

Heatherly, L.G. 1988. Planting date, row spacing, and irrigation effects on soybean grown on clay soil. Agron. J. 80:227–231.

Heatherly, L.G. 1993. Drought stress and irrigation effects on germination of harvested soybean seed. Crop Sci. 33:777–781.

Heatherly, L.G. 1996. Yield and germination of harvested seed from irrigated and nonirrigated early and late planted MG IV and V soybean. Crop Sci. 36:1000–1006.

Heatherly, L.G. 1999a. Early soybean production system (ESPS). p. 103–118. *In* L.G. Heatherly and H.F. Hodges (ed.) Soybean production in the mid-south. CRC Press, Boca Raton, FL.

Heatherly, L.G. 1999b. Soybean irrigation. p. 119–142. *In* L.G. Heatherly and H.F. Hodges (ed.) Soybean production in the mid-south. CRC Press, Boca Raton, FL.

Heatherly, L.G. 1999c. The stale seedbed planting system. p. 93–102. *In* L.G. Heatherly and H.F. Hodges (ed.) Soybean production in the mid-south. CRC Press, Boca Raton, FL.

Heatherly, L.G., A. Blaine, H. Hodges, and R.A. Wesley. 1999. Cultivar selection, planting date, row spacing, and seeding rate. p. 41–52. *In* L.G. Heatherly and H.F. Hodges (ed.) Soybean production in the mid-south. CRC Press, Boca Raton, FL.

Heatherly, L.G., and G.R. Bowers (ed.) 1998. Early soybean production system handbook. USB 6009-091998-11000. United Soybean Board, St. Louis, MO.

Heatherly, L.G., and C.D. Elmore. 1983. Response of soybeans (*Glycine max*) to planting in untilled, weedy seedbed on clay soil. Weed Sci. 31:93–99.

Heatherly, L.G., and C.D. Elmore. 1986. Irrigation and planting date effects on soybeans grown on clay soil. Agron. J. 78:576–580.

Heatherly, L.G., and C.D. Elmore. 1991. Grass weed control for soybean (*Glycine max*) on clay soil. Weed Technol. 5:103–107.

Heatherly, L. G., C. D. Elmore, and S. R. Spurlock. 1994. Effect of irrigation and weed control treatment on yield and net return from soybean (*Glycine max*). Weed Technol. 8:69–76.

Heatherly, L.G., C.D. Elmore, and S.R. Spurlock. 2001a. Row width and weed management systems for conventional soybean plantings in the midsouthern USA. Agron. J. 93:1210–1220.

Heatherly, L.G., C.D. Elmore, and S.R. Spurlock. 2002a. Weed management systems for conventional and glyphosate-resistant soybean with and without irrigation. Agron. J. 94:1419–1428.

Heatherly, L.G., C.D. Elmore, S.R. Spurlock, and R.A. Wesley. 2001b. Row spacing and weed management systems for nonirrigated early soybean production system plantings in the midsouthern USA. Crop Sci. 41:784–791.

Heatherly, L. G., C. D. Elmore, and R. A. Wesley. 1990. Weed control and soybean response to preplant tillage and planting time. Soil Tillage Res. 17:199–210.

Heatherly, L. G., C. D. Elmore, and R. A. Wesley. 1992a. Weed control for soybean (*Glycine max*) planted in a stale or undisturbed seedbed on clay soil. Weed Technol. 6:119–124.

Heatherly, L.G., and H.F. Hodges (ed.) 1999. Soybean production in the midsouth. CRC Press, Boca Raton, FL.

Heatherly, L.G., and H.C. Pringle, III. 1991. Soybean cultivars' response to flood irrigation of clay soil. Agron. J. 83:231–237.

Heatherly, L.G,, H.G. Pringle, III, G.L. Scuimbato, L.D. Young, M.W. Ebelhar, R.A. Wesley, and G.R. Tupper. 1992b. Irrigation of soybean cultivars susceptible and resistant to soybean cyst nematode. Crop Sci. 32:802–806.

Heatherly, L.G., and W.J. Russell. 1979. Vegetative development of soybeans grown on different soil types. Field Crops Res. 2:135–143.

Heatherly, L.G., and S.R. Spurlock. 1993. Timing of furrow irrigation termination for determinate soybean on clay soil. Agron. J. 85:1103–1108.

Heatherly, L.G., and S.R. Spurlock. 1999. Yield and economics of traditional and early soybean production system (ESPS) seedings in the midsouthern USA. Field Crops Res. 63:35–45.

Heatherly, L.G., and S.R. Spurlock. 2000. Furrow- and flood-irrigated ESPS MG IV and V soybean rotated with rice. Agron. J. 92(4):785–791.

Heatherly, L.G., and S.R. Spurlock. 2001. Economics of fall tillage for early and conventional soybean plantings in the midsouthern USA. Agron. J. 93:511–516.

Heatherly, L.G., and S.R. Spurlock. 2002a. Small differences in planting dates affect soybean performance. Res. Rep. 23(4). Miss. Agric. and For. Exp. Stn., Mississippi State.

Heatherly, L.G., S.R. Spurlock, J.G. Black, and R.A. Wesley. 2002c. Fall tillage for soybean grown on Delta clay soils. Bull. 1117. Miss. Agric. and For. Exp. Stn., Mississippi State, MS.

Heatherly, L.G., S.R. Spurlock, and C.D. Elmore. 2002b. Row width and weed management systems for early soybean production system plantings in the midsouthern USA. Agron. J. 94:1172–1180.

Heatherly, L.G., R.A. Wesley, C.D. Elmore, and S.R. Spurlock. 1993. Net returns from stale seedbed plantings of soybean (*Glycine max*) on clay soil. Weed Technol. 7:972–980.

Heatherly, L.G., and L.D. Young. 1991. Soybean and soybean cyst nematode response to soil water content in loam and clay soils. Crop Sci. 31:191–196.

Helms, T.C., C.R. Hurburgh, Jr., R.L. Lussenden, and D.A. Whited. 1990. Economic analysis of increased protein and decreased yield due to delayed planting of soybean. J. Prod. Agric. 3:367–371.

Henggeler, J. 2002. When should the last irrigation of soybeans occurr? [Online]. Available at http://agebb.missouri.edu/irrigate/tips/lastsoy.htm (verified 26 Nov. 2002).

Higgins, R.A. 1997. Soybean insects. p. 19–23. *In* Soybean production handbook. C-49. Kansas State Univ., Manhattan.

Higley, L.G., and D.J. Boethel (ed.) 1994. Handbook of soybean insect pests. Entomol. Soc. of Am., Lanham, MD.

Hoeft, R.G., E.D. Nafziger, R.R. Johnson, and S.R. Aldrich. 2000. Modern corn and soybean production. 1st ed. MCSP Publ., Champaign, IL.

Hofer, J.M., D.E. Peterson, W.B. Gordon, S.A. Staggenborg, and D.L. Fjell. 1998. Yield potential and response of glyphosate-resistant soybean varieties to imidazolinone herbicides. p. 25–26. *In* Proc. North Central Weed Sci. Soc., North Central Weed Sci. Soc., Champaign, IL.

Hoffmeister, G.F., Jr., and R.W. Elmore. 1999. Row spacing and seeding rates for small- and large-seeded soybean. p. 559–560. *In* Proc. World Soybean Res. Conf., 6th, Chicago, IL. 4–7 Aug. 1999. Superior Printing, Champagne, IL.

Honeycutt, J. 1996. Louisiana soybean production. Publ. 2624. Louisiana State Univ., Baton Rouge.

Hooker, D.C., T.J. Vyn, and C.J. Swanton. 1997. Effectiveness of soil-applied herbicides with mechanical weed control for conservation tillage systems in soybean. Agron. J. 89:579–587.

Hume, D.J., and D.H. Blair. 1992. Effect of numbers of *Bradyrhizobium japonicum* applied in commercial inoculants on soybean seed yield in Ontario. Can. J. Microbiol. 38:588–593.

Hunt, T., J.F. Witkowski, R. Wright, and K. Jarvi. 1994. The bean leaf beetle in soybeans. Univ. of Nebraska Coop. Ext. Serv. NebGuide G90-974 (revised 9/94). Lincoln, NE. Available online at http://www.ianr.unl.edu/pubs/insects/g974.htm (verified 25 Nov. 2002).

Hurburgh, C.R., D.E. Farnham, and K. Whigham. 2001. Frost damage to corn and soybeans [Online]. Available at http://www.exnet.iastate.edu/Pages/grain/publications/grprod/010927frostdam.pdf (verified 27 Nov. 2002).

Hydrick, D.E., and D.R. Shaw. 1994. Sequential herbicide applications in stale seedbed soybean (*Glycine max*). Weed Technol. 8:684–688.

Hydrick, D.E., and D.R. Shaw. 1995. Non-selective and selective herbicide combinations in stale seedbed soybean (*Glycine max*). Weed Technol. 9:158–165.

Iragavarapu, T.K., and G.W. Randall. 1996. Border effects on yields in a strip-intercropped soybean, corn, and wheat production system. J. Prod. Agric. 9:101–107.

Jacques, S., R.K. Bacon, and L.D. Parsch. 1997. Comparison of single cropping, relay cropping, and doublecropping of soyabeans with wheat using cultivar blends. Exp. Agric. 33:477–486.

Jasa, P.J., D.P. Shelton, A.J. Jones, and E.C. Dickey. 1991. Conservation tillage and planting systems. NebGuide G91-1046. Univ. of Nebraska Coop. Ext. Serv., Lincoln. Available online at http://www.ianr.unl.edu/pubs/fieldcrops/g1046.htm (verified 25 Nov. 2002).

Johnson, R.R. 1987. Crop management. p. 355–390. In J.R. Wilcox (ed.) Soybeans: Improvement, production, and uses. 2nd ed. Agron. Monogr. 16. ASA, CSSA, and SSSA, Madison, WI.

Johnson, W.G., J.S. Dilbeck, M.S. DeFelice, and J.A. Kendig. 1998a. Weed control with reduced rates of chlorimuron plus metribuzin and imazethapyr in no-till narrow-row soybean (Glycine max). Weed Technol. 12:32–36.

Johnson, T.J., T.C. Kaspar, K.A. Kohler, S.J. Corak, and S.D. Logsdon. 1998b. Oat and rye overseeded into soybean as fall cover crops in the upper Midwest. J. Soil Water Conserv. 53:276–279.

Johnson, W.G., J.A. Kendig, R.E. Massey, M.S. DeFelice, and C.D. Becker. 1997. Weed control and economic returns with postemergence herbicides in narrow-row soybeans. Weed Technol. 11:453–459.

Kadhem, F.A., J.E. Specht, and J.H. Williams. 1985a. Soybean irrigation serially timed during stages R1 to R6. I. Agronomic responses. Agron. J. 77:291–298.

Kadhem, F.A., J.E. Specht, and J.H. Williams. 1985b. Soybean irrigation serially timed during stages R1 to R6. II. Yield component responses. Agron. J. 77:299–304.

Kane, M.V., C.C. Steele, and L.J. Grabau. 1997. Early-maturing soybean cropping system: I. Yield responses to planting date. Agron. J. 89:454–458.

Karlen, D.L., and J.W. Doran. 1991. Cover crop management effects on soybean and corn growth and nitrogen dynamics in an on-farm study. Am. J. Sustainable Agric. 6:71–82.

Katsvairo, T.W., and W.J. Cox. 2000. Economics of cropping systems featuring different rotations, tillage, and management. Agron. J. 92:485–493.

Keith, B.C., and J.C. Delouche. 1999. Seed quality, production, and treatment. p. 197–230. In L.G. Heatherly and H.F. Hodges (ed.) Soybean production in the mid-south. CRC Press, Boca Raton, FL.

Kelley, K.W., and D.W. Sweeney. 1998. Effects of wheat residue management on doublecropped soybean and subsequent crops. J. Prod. Agric. 11:452–456.

Kendig, S.R., J.C. Rupe, and H.D. Scott. 2000. Effect of irrigation and soil water stress on densities of Macrophomina phaseoline in soil and roots of two soybean cultivars. Plant Dis. 84:895–900.

Kessavalou, A., and D.T. Walters. 1997. Winter rye as a cover crop following soybean under conservation tillage. Agron. J. 89:68–74.

Kilgore-Norquest, L., and C.H. Sneller. 2000. Effect of stem termination on soybean traits in southern US production systems. Crop Sci. 40:83–90.

King, C.A., L.C. Purcell, and E.D. Vories. 2001. Plant growth and nitrogenase activity of glyphosate-tolerant soybean in response to foliar glyphosate applications. Agron. J. 93:179–186.

Kinloch, R. 1992. Management of root-knot nematodes in soybean. p. 147–154. In L.G. Copping et al. (ed.) Pest management in soybean. Elsevier Sci. Publ. LTD, London.

Klocke, N.L., D.E. Eisenhauer, and T.L. Bockstadter. 1991. Predicting the last irrigation for corn, grain sorghum, and soybean. NebGuide G82-602. Univ. of Nebr. Coop. Ext. Serv., Lincoln. (Available online at http://www.ianr.unl.edu/pubs/irrigation/g602.htm.) (verified 25 Nov. 2002).

Klocke, N.L., D.E. Eisenhauer, J.E. Specht, R.W. Elmore, and G.W. Hergert. 1989. Irrigate soybeans by growth stages in Nebraska. Appl. Eng. Agric. 5:361–366.

Konovsky, J., T.A. Lumpkin, and D. McClary. 1994. Edamame: The vegetable soybean. p. 173–181. In A.D. O'Rourke (ed.) Understanding the Japanese food and agrimarket: A multifaceted opportunity. Haworth Press, Binghampton, NY.

Korte, L.L., J.E. Specht, J.H. Williams, and R.C. Sorensen. 1983b. Irrigation of soybean genotypes during reproductive ontogeny. II. Yield component responses. Crop Sci. 23:528–533.

Korte, L.L., J.H. Williams, J.E. Specht, and R.C. Sorensen. 1983a. Irrigation of soybean genotypes during reproductive ontogeny. I. Agronomic responses. Crop Sci. 23:521–527.

Koskinen, W.C., and C.G. McWhorter. 1986. Weed control in conservation tillage. J. Soil Water Conserv. 41:365–370.

Krausz, R.F., G. Kapusta, and J.L. Matthews. 1995. Evaluation of band vs. broadcast herbicide applications in corn and soybean. J. Prod. Agric. 8:380–384.

Kurtz, M.E., C.E. Snipes, J.E. Street, and F.T. Cooke, Jr. 1993. Soybean yield increases in Mississippi due to rotations with rice. Bull. 994. Miss. Agric. and For. Exp. Stn., Mississippi State.

Lambert, L., and L.G. Heatherly. 1991. Soil water potential: Effects on soybean looper feeding on soybean leaves. Crop Sci. 31:1625–1628.

Lambert, L., and L.G. Heatherly. 1995. Influence of irrigation on susceptibility of selected soybean genotypes to soybean looper. Crop Sci. 35:1657–1660.

Lanie, A.J., J.L. Griffin, D.B. Reynolds, and P.R. Vidrine. 1993. Influence of residual herbicides on rate of paraquat and glyphosate in stale seedbed soybean. (*Glycine max*). Weed Technol. 7:960–965.

Lanie, A.J., J.L. Griffin, P.R. Vidrine, and D.B. Reynolds. 1994a. Weed control with non-selective herbicides in soybean (*Glycine max*) stale seedbed culture. Weed Technol. 8:159–164.

Lanie, A.J., J.L. Griffin, P.R. Vidrine, and D.B. Reynolds. 1994b. Herbicide combinations for soybean (*Glycine max*) planted in stale seedbed. Weed Technol. 8:17–22.

Lawrence, G.W., and K.S. McLean. 1999. Plant-parasitic nematode pests of soybean. p. 291–310. *In* L.G. Heatherly and H.F. Hodges (ed.). Soybean production in the mid-south. CRC Press, Boca Raton, FL.

Lersten, N.R., and J.B. Carlson. 1987. Vegetative morphology. p. 49–94. *In* J.R. Wilcox (ed.) Soybeans: Improvement, production, and uses. 2nd ed. Agron. Monogr. 16. ASA, CSSA, and SSSA, Madison, WI.

Lesoing, G.W., and C.A. Francis. 1999a. Strip intercropping of corn-soybean in irrigated and rainfed environments. J. Prod. Agric. 12:187–192.

Lesoing, G.W., and C.A. Francis. 1999b. Strip intercropping of grain sorghum/soybean in irrigated and rainfed environments. J. Prod. Agric. 12:601–606.

Lin, M.S., and R.L. Nelson. 1988. Relationship between plant height and flowering date in determinate soybean. Crop Sci. 28:27–30.

Logan, J., M.A. Mueller, and C.R. Graves. 1998. A comparison of early and recommended soybean production systems in Tennessee. J. Prod. Agric. 11:319–325.

Loy, D., and P. Holden. 1993. Using frost-damaged soybeans in livestock rations. Dep. of Animal Sci. Iowa State Univ., Ames. Available at http://www.extension.iastate.edu/Publications/DR28.pdf (verified 10 Aug. 2003).

Lu, Yao-Chi, B. Watkins, and J. Teasdale. 1999. Economic analysis of sustainable agricultural cropping systems for mid-Atlantic states. J. Sustainable Agric. 15:77–93.

Major, D.J., D.R. Johnson, J.W. Tanner, and I.C. Anderson. 1975. Effects of daylength and temperature on soybean development. Crop Sci. 15:174–179.

Martin, A. (ed.) 2001. Nebraska soybean field guide. EC-01-146. Univ. of Nebr. Coop. Ext. Serv., Lincoln.

Mayhew, W.L., and C.E. Caviness. 1994. Seed quality and yield of early-planted, short-season soybean genotypes. Agron. J. 86:16–19.

McGregor, K.C. 1978. C factors for no-till and conventional-till soybean from plot data. Trans. ASAE 21:1119–1122.

McGregor, K.C., and C.K. Mutchler. 1983. C factors for no-till and reduced-till corn. Trans. ASAE 26:785–788, 794.

McGregor, K.C., and C.K. Mutchler. 1992. Soil loss from conservation tillage for sorghum. Trans. ASAE 35:1841–1845.

Mebrahtu, T., A. Mohamed, and W. Mersie. 1991. Green pod yield and architectural traits of selected vegetable soybean genotypes. J. Prod. Agric. 4:395–399.

Mengel, D.B., W. Segars, and G.W. Rehm. 1987. Soil fertility and liming. p. 461–496. *In* J.R. Wilcox (ed.) Soybeans: Improvement, production, and uses. 2nd ed. Agron. Monogr. 16. ASA, CSSA, and SSSA, Madison, WI.

Mickelson, J.A., and K.A. Renner. 1997. Weed control using reduced rates of postemergence herbicides in narrow and wide row soybean. J. Prod. Agric. 10:431–437.

Minor, H. 1998. Performance of GMOs vs. traditional varieties: a southern perspective. *In* Proc. 53rd Corn and Sorghum Res. Conf., Chicago, IL. December 1998. Am. Seed Trade Assoc., Washington, DC.

Moore, M.J., T.J. Gillespie, and C.J. Swanton. 1994. Effect of cover crop mulches on weed emergence, weed biomass, and soybean (*Glycine max*) development. Weed Technol. 8:512–518.

Moore, S.H. 1996. Soybean seed quality. p. 16–23. *In* J. Honeycutt (ed.) Louisiana soybean handbook. Publ. 2624. Louisiana State Univ., Baton Rouge.

Mueller, D.S., G.L. Hartman, and W.L. Pedersen. 1999. Development of sclerotia and apothecia of *Sclerotinia sclerotiorum* from infected soybean seed and its control by fungicide seed treatment. Plant Dis. 83:1113–1115.

Mutchler, C.K., and J.D. Greer. 1984. Reduced tillage for soybeans. Trans. ASAE 27:1364–1369.

Mutchler, C.K., L.L. McDowell, and J.D. Greer. 1985. Soil loss from cotton with conservation tillage. Trans. ASAE 28:160–163, 168.

Nelson, K.A., and K.A. Renner. 1999. Weed management in wide- and narrow-row glyphosate resistant soybean. J. Prod. Agric. 12:460–465.

Nelson, L.A., R.W. Elmore, R.N. Klein, and C. Shapiro. 1997. Nebraska soybean cultivar tests-1997. E.C. 97-104-A. Nebr. Coop. Ext., Lincoln.

Nelson, L.A., R.W. Elmore, R.N. Klein, and C. Shapiro. 1998. Nebraska soybean cultivar tests-1998. E.C. 98-104-A. Nebr. Coop. Ext., Lincoln.

Nelson, L.A., R.W. Elmore, R.N. Klein, and C. Shapiro. 1999. Nebraska soybean cultivar tests-1999. E.C. 99-104-A. Nebr. Coop. Ext., Lincoln.

Newsom, L.J., and D.R. Shaw. 1996. Cultivation enhances weed control in soybean (*Glycine max*) with AC 263,222. Weed Technol. 10:502–507.

Nguyen, V.Q. 1998. Edamame (vegetable green soybean). *In* K. Hyde (ed.). The new rural industries: A handbook for farmers and investors. Australian Rural Industries Res. and Development Corp. [Online]. Available at http:\\www.rirdc.gov.au/pub/handbook/edamame.html (Verified on 27 Nov. 2002).

Niblack, T.L., G.L. Tylka, R.D. Riggs. 2004. Nematode pathogens of soybean. p. 821–852. *In* H.R. Boerma and J.E. Specht (ed.) Soybeans: Improvement, production, and uses. Agron. Monogr. 16. ASA, CSSA, and SSSA, Madison, WI.

Nielsen, R.L. 2000. Transgenic crops in Indiana: Short-term issues for farmers. Agron. Dep., Purdue Univ., West Lafayette, IN. [Online]. Available at http://www.agry.purdue.edu/ext/corn/news/articles.00/GMO_Issues_000203.html (verified 27 Nov. 2002).

Oliver, L.R., T.E. Klingaman, M. McClelland, and R.C. Bozsa. 1993. Herbicide systems in stale seedbed soybean (*Glycine max*) production. Weed Technol. 7:816–823.

Omay, A.B., C.W. Rice, L.D. Maddux, and W.B. Gordon. 1997. Changes in soil microbial and chemical properties under long-term crop rotation and fertilization. Soil Sci. 61:1672–1678.

Oplinger, E.S., M.J. Martinka, and K.A. Schmitz. 1998a. Performance of transgenic soybeans—Northern United States. p. 10–14. *In* Proc. 28th Soybean Seed Res. Conf., Chicago, IL. December 1998. Am. Seed Trade Assoc., Washington, DC.

Oplinger, E.S., K. Whigham, and J. Beuerlein. 1998b. No-till soybean practices for the midwest. NTSP-1. North Central Soybean Res. Program, Madison, WI.

Oriade, C.A., C.R. Dillon, E.D. Vories, and M.E. Bohanan. 1997. An economic analysis of alternative cropping and row spacing systems for soybean production. J. Prod. Agric. 10:619–624.

Ouattara, S., and D.B. Weaver. 1994. Effect of growth habit on yield and agronomic characteristics of late-planted soybean. Crop Sci. 34:870–873.

Owenby, J.R., and D.S. Ezell. 1992. Monthly station normals of temperature, precipitation, and heating and cooling degree days, 1961–1990. Missouri. Climatography of the U.S. no. 81. NOAA, Natl. Climatic Data Center, Asheville, NC.

Padgette, S.R., N.B. Taylor, D.L. Nida, M.R. Bailey, J. MacDonald, L.R. Holden, and R.L. Fuchs. 1996. The composition of glyphosate-tolerant soybean seeds is equivalent to that of conventional soybeans. J. Nutr. 126:702–716.

Panter, D.M., and F.L. Allen. 1989. Simulated selection for superior yielding soybean lines in conventional vs. doublecrop nursery environments. Crop Sci. 29:1341–1347.

Parvez, A.Q., F.P. Gardner, and K.J. Boote. 1989. Determinate- and indeterminate-type soybean cultivar responses to pattern, density, and planting date. Crop Sci. 29:150–157.

Penas, E.J., and R.A. Wiese. 1989. Soybean chlorosis management. NebGuide G89-953-A. Univ. of Nebr. Coop. Ext. Serv., Lincoln. Available online at http://www.ianr.unl.edu/pubs/fieldcrops/g953.htm (verified 25 Nov. 2002).

Pfeiffer, T.W., L.J. Grabau, and J.H. Orf. 1995. Early maturity soybean production system: Genotype x environment interaction between regions of adaptation. Crop Sci. 35:108–112.

Philbrook, B.D., and E.S. Oplinger. 1989. Soybean seeding rates for reduced tillage. p. 144–148. *In* B. Jensen (ed.) Proc. of the 1989 Integrated Crop and Pest Management Workshop, Madison, WI. 14–16 Feb. 1989. Univ. of Wisconsin Coop. Ext. Serv., Madison.

Pierce, F.J., and D.D. Warncke. 2000. Soil and crop response to variable-rate liming for two Michigan fields. Soil Sci. 64:774–780.

Popp, M.P. T.C. Keisling, C.R. Dillon, and P.M. Manning. 2001. Economic and agronomic assessment of deep tillage in soybean production on Mississippi River valley soils. Agron. J. 93:164–169.

Porter, P.M., J.G. Lauer, W.E. Lueschen, J.H. Ford, T.R. Hoverstad, E.S. Oplinger, and R.K. Crookston. 1997. Environment affects the corn and soybean rotation effect. Agron. J. 89:442–448.

Poston, D.H., E.C. Murdock, and J.E. Toler. 1992. Cost-efficient weed control in soybean (*Glycine max*) with cultivation and banded herbicide application. Weed Technol. 6:990–995.

Purcell, L.C., and J.E. Specht. 2004. Physiological traits for ameliorating drought stress. p. 569–620. *In* H.R. Boerma and J.E. Specht (ed.) Soybeans: Improvement, production, and uses. 3rd ed. Agron. Monogr. 16. ASA, CSSA, and SSSA, Madison, WI.

Rao, M.S.S., B.G. Mullinix, M. Rangappa, E. Cebert, A.S. Bhagsari, V.T. Sapra, J.M. Joshi, and R.B. Dadson. 2002. Genotype x environment interactions and yield stability of food-grade soybean genotypes. Agron. J. 94:72–80.

Reddy, K.R. 2001a. Glyphosate-resistant soybean as a weed management tool: Opportunities and challenges. Weed Biol. Manage. 1:193–202.

Reddy, K.R. 2001b. Effects of cereal and legume cover crop residues on weeds, yield, and net return in soybean (*Glycine max*). Weed Technol. 15:660–668.

Reddy, K.R. 2003. Impact of rye cover crop and herbicides on weeds, yield, and net return in narrow-row transgenic and conventional soybean (*Glycine max*). Weed Technol. 17:28–35.

Reddy, K.R., L.G. Heatherly, and A. Blaine. 1999. Weed management. p. 171–195. *In* L.G. Heatherly and H.F. Hodges (ed.) Soybean production in the mid-south. CRC Press, Boca Raton, FL.

Reicosky, D.A., and L.G. Heatherly. 1990. Soybean. p. 639–674. *In* B.A. Stewart and D.A. Nielsen (ed.) Irrigation of agricultural crops. Agron. Monogr. 30. ASA, Madison, WI.

Reinbott, T.M., Z.R. Helsel, D.G. Helsel, M.R. Gebhardt, and H.C. Minor. 1987. Intercropping soybean into standing green wheat. Agron. J. 79:886–891.

Riggs, R.D. 1992. Management of nematode problems on soybean in the United States of America. p. 128–136. *In* L.G. Copping et al. (ed.) Pest management in soybean. Elsevier Sci. Publ. LTD, London.

Robinson, S.L., and J.R. Wilcox. 1998. Comparison of determinate and indeterminate soybean near-isolines and their response to row spacing and planting date. Crop Sci. 38:1554–1557.

Saindon, G., H.D. Voldeng, and W.D. Beversdorf. 1990. Adjusting the phenology of determinate soybean segregants grown at high latitude. Crop Sci. 30:516–521.

Sander, D.H., and E.J. Penas. 2000. Phosphorus. p. 17–21. *In* R.B. Ferguson and K.M. DeGroot (ed.) Nutrient management for agronomic crops in Nebraska. EC-01-155. Univ. of Nebr. Coop. Ext. Serv., Lincoln.

Scott, H.D., J. DeAngulo, M.B. Daniels, and L.S. Wood. 1989. Flood duration effects on soybean growth and yield. Agron. J. 81:631–636.

Selley, R.A., T. Barrett, R.T. Clark, R.N. Klein, and S. Melvin. 2001. Nebraska crop budgets. EC01-872-S. Univ. of Nebr. Coop. Ext. Serv., Lincoln. Available online at http://www.ianr.unl.edu/pubs/farmmgt/ec872/procedures.htm (verified 27 Nov. 2002).

Shapiro, C.A., T.A. Peterson, and A.D. Flowerday. 1985. Soybean yield loss due to hail damage. NebGuide G85-762-A. Univ. of Nebr. Coop. Ext. Serv., Lincoln. Available online at http://www.ianr.unl.edu/pubs/fieldcrops/g762.htm (verified 22 Oct. 2002).

Sheaffer, C., J.H. Orf, T.E. Devine, and J.G. Jewett. 2001. Yield and quality of forage soybean. Agron. J. 93:99–106.

Shipitalo, M.J., W.M. Edwards, and L.B. Owens. 1997. Herbicide losses in runoff from conservation-tilled watersheds in a corn-soybean rotation. Soil Sci. Soc. Am. J. 61:267–272.

Singer, J. 2001. Soybean light interception and yield response to row spacing and biomass removal. Crop Sci. 41:424–429.

Specht, J.E., R.W. Elmore, D.E. Eisenhauer, and N.W. Klocke. 1989. Growth stage scheduling criteria for soybeans. Irrig. Sci. 10:99–111.

Specht, J.E., D.J. Hume, and S.V. Kumudini. 1999. Soybean yield potential—A genetic and physiological perspective. Crop Sci. 39:1560–1570.

Spurlock, S.R. 2000. Soybeans: 2001 planning budgets. Agric. Econ. Rep. 117. Mississippi State Univ., Mississippi State.

Spurlock, S.R. 2002. Soybeans: 2002 planning budgets [Online]. Available at http://www.agecon.msstate.edu/Research/budgets.php (verified 26 Nov. 2002).

Spurlock, S.R., J.G. Black, L.G. Heatherly, C.D. Elmore, and R.A. Wesley. 1997. Economics of monocrop winter wheat on clay soils in the Delta area of Mississippi. Miss. Agric. and For. Exp. Stn. Res. Rep. 22, no. 1. Mississippi State, MS.

Swanton, C. J., T. J. Vyn, K. Chandler, and A. Shrestha. 1998. Weed management strategies for no-till soybean (*Glycine max*) grown on clay soils. Weed Technol. 12:660–669.

Tacker, P., L. Ashlock, E. Vories, L. Earnest, R Cingolani, D. Beaty, and C. Hayden. 1997. Field demonstration of Arkansas Irrigation Scheduling Program. p. 974–979. *In* C.R. Camp et al. (ed.) Evaporation and Irrigation Scheduling Conf., San Antonio, TX. 3–6 Nov. 1996. Am. Soc. Agric. Eng, St. Joseph, MO. Available online at http://www.aragriculture.org/agengineering/irrigation/default.asp (verified 25 Nov. 2002).

Tacker, P.L. 1993. Irrigation scheduling—Arkansas checkbook method user's guide. Univ. of Arkansas Coop. Ext. Serv., Little Rock.

Tacker, P.L., E.D. Vories, and L.O. Ashlock. 1994. Drainage and irrigation. *In* L.O. Ashlock (ed.) Technology for optimum production of soybeans. Publ. AG411-12-94. Univ. of Arkansas Coop. Ext. Serv. Little Rock.

Taylor, H.M. 1980. Soybean growth and yield as affected by row spacing and by seasonal water supply. Agron. J. 72:543–547.

Taylor, N.B., R.L. Fuchs, J. MacDonald, A.R. Shariff, and S.R. Padgette. 1999. Compositional analysis of glyphosate-tolerant soybeans treated with glyphosate. J. Agric. Food Chem. 47:4469–4473.

Thomison, P.R., W.J. Kenworthy, and M.S. McIntosh. 1990. Phomopsis seed decay in soybean isolines differing in stem termination, time of flowering, and maturity. Crop Sci. 30:183–188.

Todd, J.W., R.M. McPherson, and D.J. Boethel. 1994. Management tactics for soybean insects. *In* L.G. Higley and D.J. Boethel (ed.). Handbook of soybean insect pests. Entomol. Soc. of Am., Lanham, MD.

Todd, T.C. 1993. Soybean planting date and maturity effects on *Heterodera glycines* and *Macrophomina phaseolina* in southeastern Kansas. J. Nematol. 25:731–737.

Tolin, S.A., and G.H. Lacy. 2004. Viral, bacterial, and phytoplasmal diseases. p. 765–820. *In* H.R. Boerma and J.E. Specht (ed.) Soybeans: Improvement, production, and uses. 3rd ed. Agron. Monogr. 16. ASA, CSSA, and SSSA, Madison, WI.

Triplett, E.W., K.A. Albrecht, and E.S. Oplinger. 1993. Crop rotation effects on populations of *Bradyrhizobium japonicum* and *Rhizobium meliloti*. Soil Biol. Biochem. 25:781–784.

Triplett, G.B., and S.M. Dabney. 1999. Soil erosion and soybean production. p. 19–39. *In* L.G. Heatherly and H.F. Hodges (ed.) Soybean production in the mid-south. CRC Press, Boca Raton, FL.

Van Doren, D.M., Jr., and D.C. Reicosky. 1987. Tillage and irrigation. p. 391–428. *In* J.R. Wilcox (ed.) Soybeans: Improvement, production, and uses. 2nd ed. Agron. Monogr. 16. ASA, CSSA, and SSSA, Madison, WI.

Varco, J.J. 1999. Nutrition and fertility requirements. p. 53–70. *In* L.G. Heatherly and H.F. Hodges (ed.) Soybean production in the mid-south. CRC Press, Boca Raton, FL.

Vasilas, B.L., R.W. Esgar, W.M. Walker, R.H. Beck, and M.J. Mainz. 1988. Soybean response to potassium fertility under four tillage systems. Agron. J. 80:5–8.

Vasilas, B.L., G.E. Pepper, and M.A. Jacob. 1990. Stand reductions, replanting, and offset row effects on soybean yield. J. Prod. Agric. 3:120–123.

Vitosh, M.L., J.W. Hohnson, D.B. Mengel (ed.) 2001. Tri-state fertilizer recommendations for corn, soybeans, wheat, and alfalfa. Bull. E-2567. Ohio State Univ. Ext. Serv., Columbus. Available at http://www.ag.ohio-state.edu/~ohioline/e2567/index.html (verified 27 Nov. 2002).

Wallace, S.U., T. Whitwell, J.H. Palmer, C.E. Hood, and S.A. Hull. 1992. Growth of relay intercropped soybean. Agron. J. 84:968–973.

Wang, H.L., E.W. Swain, W.F. Kwolek, and W.R. Fehr. 1983. Effect of soybean varieties on the yield and quality of tofu. Am. Assoc. Cereal Chem. 60:245–248.

Wang, J., P.A. Donald, T.L. Niblack, G.W. Bird, J. Faghigi, J.M. Ferris, D.J. Jardine, P.E. Lipps, A.E. MacGuidwin et al. 1999. Soybean cyst nematode reproduction in the north central United States. Plant Dis. 84:77–82.

Webber, C.L., III, M.R. Gebhardt, and H.D. Kerr. 1987. Effect of tillage on soybean growth and seed production. Agron. J. 79:952–956.

Webster, E.P., K.J. Bryant, and L.D. Earnest. 1999. Weed control economics in nontransgenic and glyphosate-resistant soybean. Weed Technol. 13:586–593.

Wesley, R.A. 1999a. Double cropping wheat and soybeans. p. 143–156. *In* L.G. Heatherly and H.F. Hodges (ed.) Soybean production in the mid-south. CRC Press, Boca Raton, FL.

Wesley, R.A. 1999b. Crop rotation systems for soybean. p. 157–170. *In* L.G. Heatherly and H.F. Hodges (ed.) Soybean production in the mid-south. CRC Press, Boca Raton, FL.

Wesley, R.A., and F.T. Cooke. 1988. Wheat-soybean doublecrop systems on clay soil in the Mississippi Valley area. J. Prod. Agric. 1:166–171.

Wesley, R.A., L.G. Heatherly, C.D. Elmore, and S.R. Spurlock. 1994. Net returns from eight irrigated cropping systems on clay soil. J. Prod. Agric. 7:109–115.

Wesley, R.A., L.G. Heatherly, C.D. Elmore, and S.R. Spurlock. 1995. Net returns from eight nonirrigated cropping systems on clay soil. J. Prod. Agric. 8:514–520.

Wesley, R.A., and L.A. Smith. 1991. Response of soybean to deep tillage with controlled traffic on clay soil. Trans. Am. Soc. Agric. Eng. 34:113–119.

Wesley, R.A., L.A. Smith, and S.R. Spurlock. 2000. Residual effects of fall deep tillage on soybean yields and net returns on Tunica clay soil. Agron. J. 92:941–947.

Wesley, R.A., L.A. Smith, and S.R. Spurlock. 2001. Fall deep tillage of Tunica and Sharkey clay: Residual effects on soybean yield and net return. Bull. 1102. Mississippi Agric. and For. Exp. Stn., Mississippi State.

Wesley, T.L., R.E. Lamond, V.L. Martin, and S.R. Duncan. 1998. Effects of late-season nitrogen fertilizer on irrigated soybean yield and composition. J. Prod. Agric. 11:331–336.

Whitney, D.A. 1997. Fertilization. p. 11–13. *In* Soybean production handbook. C-449. Kansas State Univ. Agric. Exp. Stn. and Coop. Ext. Serv., Manhattan, KS.

Whiting, K.R., R.K. Crookston, and W.A. Brun. 1988. An indicator of the R6.5 stage of development for indeterminate soybean. Crop Sci. 28:866–867.

Wilcox, J.R., and J.F. Cavins. 1995. Backcrossing high seed protein to a soybean cultivar. Crop Sci. 35:1036–1041.

Wilcox, J.R., and E.M. Frankenberger. 1987. Indeterminate and determinate soybean responses to planting date. Agron. J. 79:1074–1078.

Wilcox, J.R., and G. Zhang. 1997. Relationships between seed yield and seed protein in determinate and indeterminate soybean populations. Crop Sci. 37:361–364.

Willers, J.L., G.W. Hergert, and P.D. Gerard. 1999. Sampling tips and analytical techniques for soybean production. p. 311–353. *In* L.G. Heatherly and H.F. Hodges (ed.). Soybean production in the midsouth. CRC Press, Boca Raton, FL.

Williams, M.M. II, D.A. Mortensen, and J.W. Doran. 1998. Assessment of weed and crop fitness in cover crop residues for integrated weed management. Weed Sci. 46:595–603.

Wilson, R., J. Smith, and R. Moomaw. 1993. Cover crop use in crop production systems. NebGuide G93-1146-A. Univ. of Nebr. Coop. Ext. Serv., Lincoln. Available online at http://www.ianr.unl.edu/pubs/fieldcrops/g953.htm (verified 25 Nov. 2002).

Wrather, J.A., S.C. Anand, and S.R. Koenning. 1992. Management by cultural practices. p. 125–131. *In* R.D. Riggs and J.A. Wrather (ed.) Biology and management of the soybean cyst nematode. Am. Phytopathol. Soc., St. Paul, MN.

Yelverton, F.H., and H.D. Coble. 1991. Narrow row spacing and canopy formation reduces weed resurgence in soybeans (*Glycine max*). Weed Technol. 5:169–174.

Yiridoe, E.K., A. Weersink, D.C. Hooker, T.J. Vyn, and C. Swanton. 2000. Income risk analysis of alternative tillage systems for corn and soybean production on clay soils. Can. J. Agric. Econ. 48:161–174.

Young, L.D. 1994. Changes in the *Heterodera glycines* female index as affected by ten-year cropping sequencess. J. Nematol. 26:505–510.

Young, L.D. 1998a. Influence of soybean cropping sequences on seed yield and female index of the soybean cyst nematode. Plant Dis. 83:615–619.

Young, L.D. 1998b. Breeding for nematode resistance and tolerance. p. 187–207. *In* K.R. Barker et al. (ed.) Plant and nematode interactions. Agronomy 36:187–207.

Young, L.D., and L.G. Heatherly. 1990. *Heterodera glycines* invasion and reproduction on soybean grown in clay and silt loam soils. J. Nematol. 22:618–619.

11 Improved Carbon and Nitrogen Assimilation for Increased Yield[1]

THOMAS R. SINCLAIR

USDA-ARS
University of Florida
Gainesville, Florida

Soybean [*Glycine max* (L.) Merr.] monographs in the past have had individual chapters on carbon (C) assimilation and nitrogen (N) metabolism. While allocating two chapters to these critical processes of soybean growth has resulted in excellent, comprehensive reviews of these topics, such a segregated presentation diminishes an integrated perspective on the high degree of interaction that occurs in C and N accumulation. Further, examining C and N assimilation separately may foster questions about which process is most constraining in increasing soybean yields.

To broaden the perspective of C and N accumulation in this monograph, it was decided to consider both topics in the same chapter and give emphasis to their interaction. Consequently, this chapter has the dual objectives of reviewing the critical advances in the last 15 yr in regards to C and N accumulation by soybean, and examining insights gained about the interaction between the processes that can influence rate and amount of accumulation. Still, to structure the discussion the first parts of this chapter have been divided into general sections on C and N accumulation. In each section, attention will be given to those reports that have addressed the influence of one process on the other. The final section of this chapter considers directly the consequences of variation in C and N accumulation on soybean yield. The combined constraints of C and N accumulation on soybean plant and seed growth will be considered to help to understand realistic limits on soybean yield.

11–1 CARBON ASSIMILATION

11–1.1 Leaf Photosynthesis

11–1.1.1 Photosynthesis Rates and Nitrogen

In the previous Soybean monograph, Shibles et al. (1987) concluded that leaf photosynthetic rate is "determined by the amount of photosynthetic apparatus per unit leaf area". Research since that time has only strengthened this conclusion. One

---

[1] Mention of a trademark or proprietary product does not constitute a guarantee or warranty of the product by the U.S. Department of Agriculture and does not imply approval or the exclusion of other products that may also be suitable.

of the simplest correlations in support of this view has been between leaf photosynthesis rates per unit leaf area and leaf mass per unit leaf area (specific leaf weight). For example, a high, nonlinear correlation ($R^2 = 0.96$) was reported between leaf photosynthetic rates and specific leaf weight for a single cultivar subjected to differing temperature treatments (Bunce, 1991). Specific leaf weight, however, may not necessarily directly reflect photosynthetic capacity across cultivars. In a comparison of leaf photosynthesis rates among 16 soybean lines, there was no consistent correlation with specific leaf weight across years (Thompson et al., 1995).

Instead of using specific leaf weight as an index of "photosynthetic apparatus," much higher and consistent correlations are obtained between photosynthesis rates and N per unit leaf area. Such results have been reported in field studies (Lugg and Sinclair, 1981b) and growth chamber studies with plants grown on differing N treatments (Tolley-Henry and Raper, 1986; Tolley-Henry et al., 1992). The results from one experiment of Tolley-Henry et al. (1992), plotted in Fig. 11–1, illustrate the high correlation ($R^2 = 0.85$) that is observed between leaf photosynthesis rate and reduced N content. This high correlation can be anticipated because roughly half of the N in soybean leaves is in rubisco (ribulose-1,5-bisphosphate carboxylase/oxygenase) (Wittenbach et al., 1980), which is the rate-limiting enzyme in carbon dioxide (CO_2) fixation.

The next advance in improving the definition of the amount of photosynthetic apparatus has been to measure correlations between leaf photosynthetic rates and soluble leaf protein or rubisco levels per unit leaf area. Crafts-Brandner and Egli (1987) found that declines in photosynthesis rates following the initiation of flowering were correlated with decreasing rubisco levels per unit leaf area in three cultivars. Similarly, Ford and Shibles (1988) observed that losses in leaf photosynthesis activity of six cultivars corresponded to a decrease in the amount of leaf soluble protein, however, no change was observed in the number of chloroplasts per unit leaf area during the senescence period. Comparison of photosynthetic rates following flowering in Clark isolines with normal and lancelot leaflets (Sung and Chen, 1989) also showed a parallel decrease in soluble protein and rubisco activity with the decrease in leaf photosynthetic rate. Jiang et al. (1993) found that losses in photosynthetic activity and rubisco content were coordinated with decreases in the levels of mRNA of the small and large subunits of rubisco, indicating the possibility of a slower transcription rate contributing to the decline in rubisco content.

Since leaf photosynthesis rate is closely associated with the protein levels in leaves, it is not surprising that numerous studies have shown that leaf photosynthesis rates are quite sensitive to N supply to soybean. In the early stage of growth, increased nitrate (NO_3^-) supply to the plants resulted in increased leaf photosynthesis rates, sucrose phosphate synthase activity (E.C. 2.4.2.14), sucrose concentration, and assimilate export rates from leaves (Kerr et al., 1986). In a similar study involving the growth of soybean on nutrient solutions of high or low NO_3^- concentrations, Robinson (1996, 1997) found that high NO_3^- resulted in leaves in increased photosynthetic rate, higher soluble protein, decreased starch levels, and greatly decreased levels of inorganic phosphorus (P). The decreased photosynthetic rates with low NO_3^- treatment was concluded to be associated with decreased leaf soluble protein (Robinson, 1996). Robinson and Burkey (1997) also compared the response of two NO_3^- treatments, although the low concentration (6 mM NO_3^-) was not as low as in the previous experiment (3 mM NO$_3^-$). In this later comparison, leaf photosynthesis rates and leaf sucrose concentrations were generally not different between the two NO$_3^-$ treatments. Strikingly, leaf starch concentration was much greater in the moderate NO$_3^-$ treatment as compared to the high NO$_3^-$ treatment.

Tolley-Henry et al. (1992) found that switching plants from a nutrient solution with adequate N (1 mM NO$_3^-$) at 36 d to a solution containing no NO$_3^-$ resulted

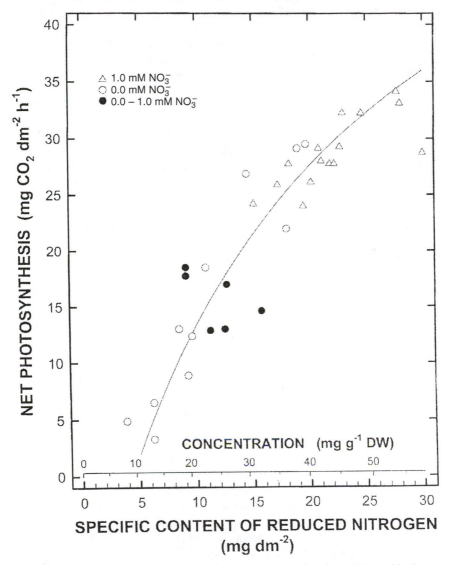

Fig. 11–1. Net leaf photosynthetic rate measured on soybean grown on nutrient solutions with adequate nitrate (NO$_3^-$) and after 36 d transferred to various NO$_3^-$ treatments. Each datum is the mean of three leaves measured on various dates and plotted against the mean reduced N content of the leaves. An approximate conversion to N concentration on a dry weight basis is shown by the inside scale of the abscissa. From Tolley-Henry et al. (1992).

in a steady decline in photosynthesis rate during the subsequent 25 d. There was no decline in photosynthesis rates for those plants maintained on the 1 mM NO_3^- solution. In a study of the short-term responses to shifting soybean seedlings to a nutrient solution without NO_3^-, Rufty et al. (1988) found that photosynthetic rates decreased after 3 d without NO_3^- and those rates continued to decrease to the end of the study at 6 d. Leaf starch levels, however, were substantially increased within 24 h for those plants shifted to the nutrient solution without NO_3^-. The starch levels continued to increase gradually in the leaves for the remainder of the experiment. While there was an increase in leaf sucrose concentrations with the zero NO_3^- treatment, the increase occurred in the first few hours, with no further increase thereafter (Rufty et al., 1988). In general, cessation or decreasing N supply results in rapid increases in leaf carbohydrate levels, but decreases in photosynthetic rates are delayed and then decrease slowly. These results indicate that leaf starch and sucrose concentrations do not result in immediate or direct limitations on leaf photosynthetic rates.

The type of N source seems to have little influence on leaf photosynthesis, other than the differences in the total N levels in leaves. de Veau et al. (1990, 1992) studied photosynthate metabolism in leaves of soybean plants relying on dinitrogen (N_2) fixation only, NO_3^- only, or a combination of the two. There was no differences in leaf photosynthesis rates observed among the three N treatments. Those plants relying on N_2 fixation, however, had much higher foliar carbohydrate levels, particularly starch, than those that were supplied with NO_3^-. The specific activities of enzymes in the starch synthesis pathway were substantially greater in the N_2-fixing plants when expressed on a soluble protein basis. These results further support the conclusion of some level of independence between leaf carbohydrate concentration and photosynthetic rates.

There is a single report, however, that soybean plants dependent on N_2 fixation as their N source had greater photochemical efficiency than plants dependent on NO_3^- (Maury et al., 1993). Measurements made at stage R5 showed a higher opening of oxidized PSII centers and/or a higher initial rate of energization of the thylakoid membrane. It was suggested that the high demand for energy as a result of N_2 fixation stimulated the increase in quantum yield. That study, however, was done with plants grown in growth chambers with only 350 μmol m^{-2} s^{-1} PAR so it is not clear that these results are relevant to the field environment.

11–1.1.2 Response to Photosynthate Supply and Demand

A critical topic of continued study has been the alteration of photosynthetic rates as a result of overall changes in photosynthate supply and demand within the plant. In general, there is little evidence that carbohydrate accumulation in the leaves causes a direct limitation on photosynthesis rates. As noted previously, starch accumulation in soybean leaves in response to low or terminated N supply was not associated with decreases in photosynthetic rates (Rufty et al., 1988; de Veau et al., 1990, 1992; Robinson and Burkey, 1997).

Heat girdling of petioles to prevent the export of photosynthate from leaves has been used to examine the response on photosynthesis rate. This treatment resulted in both a substantial increase in leaf starch content and a decrease in photosynthesis rate 7 d after the petiole girdling treatment (Goldschmidt and Huber, 1992).

However, caution concerning the interpretation of these results was suggested based on a parallel study with a starchless mutant of tobacco (*Nicotiana sylvestris* L.) (Goldschmidt and Huber, 1992). Photosynthesis rate was also decreased by the girdling treatment in this mutant even though no starch was accumulated in the leaves. Therefore, starch accumulation in the leaves was not necessary to cause the loss in photosynthetic activity.

Numerous experiments have been reported in which the overall supply or demand of photosynthate in the plants were altered. In many cases, these treatments have involved removal of leaf area and/or removal of developing flowers and pods. The results of these experiments are variable and seem to be especially dependent on the growth conditions and severity of the treatments. Lauer and Shibles (1987) found in a field study that partial depodding had no influence on leaf photosynthesis rate. Similarly, Saravitz et al. (1994) found, in a depodding experiment performed in a high-light growth chamber, no decrease in leaf photosynthetic rates except in the top leaves at the very end of the normal seed-fill period.

Suwignyo et al. (1995) initiated defoliation and depodding treatments at full flowering on soybean plants grown in a net house. Leaf photosynthetic rates, measured 3 wk after the treatment, were increased with defoliation and decreased with depodding. Leaf photosynthetic rates of the treated plants were positively correlated with sucrose phosphate synthase activity and negatively correlated with sucrose synthase activity. The authors concluded that, in the long-term, sucrose phosphate synthase was coupled to leaf photosynthetic activity.

In an experiment in which the tops of the shoots were trimmed from vegetative plants, Morcuende et al. (1996) reported the rather surprising result that 6 to 9 d after partial removal of the tops of vegetative plants the photosynthetic rates of the remaining leaves were increased. In this case, the removal of the tops resulted in large increases in leaf protein content and rubisco activity of the leaves remaining on the plants so that stimulated photosynthetic activity seemed to be a response to increased photosynthetic apparatus in the leaves left on the plant.

Sawada et al. (1992) used a rooted-leaf culture system to study the regulation of photosynthesis in a situation where there was virtually no demand for photosynthate. They concluded that the decrease in photosynthetic activity of this system was associated with a deactivation of rubisco resulting from a decrease in free inorganic P in the chloroplast stroma. This experiment was followed by one using intact plants in which all the flower buds were removed as they appeared (Sawada et al., 1995). Initially, there was little influence of flower bud removal on leaf photosynthesis rate, but a significant decrease was eventually observed 16 d after the initial removal of buds. While the total activity of rubisco was not influenced by the treatment, rubisco activation was decreased substantially by the treatment. The authors hypothesized that under conditions in which sucrose accumulated in leaves, sucrose synthesis was inhibited as a result of P being bound in phosphorylated intermediates of C metabolism. Sawada et al. (1995) hypothesized that it was the decreased levels of free P that resulted in decreased rubisco activity and photosynthetic rates. This hypothesis, and the consequences on short-term regulation of photosynthetic rates, were recently reviewed in detail by Paul and Foyer (2001).

There is abundant evidence that P deficiency is associated with lower rates of leaf photosynthesis in soybean (Lauer et al., 1989; Fredeen et al., 1989; Qiu and

Israel, 1992). The loss in photosynthesis activity is associated with decreased soluble protein per unit leaf area (Lauer et al., 1989) and increased starch concentrations in leaves (Lauer et al., 1989; Fredeen et al., 1989; Qiu and Israel, 1992). It appears that low concentrations of inorganic P relative to 3-phosphoglyceric acid activates ADP-glucose pyrophosphorylase in soybean leaves resulting in starch synthesis (Qiu and Israel, 1992). Moreover, exports of carbohydrates from the leaves decrease under low P treatment (Qiu and Israel, 1992). Consequently, P deficiency seems to cause a decrease in the export rate of photosynthate from the leaves, an accumulation of carbohydrates, and a decrease in P availability in the chloroplast.

11–1.1.3 Conclusions

The regulation of leaf photosynthesis appears to result from two major mechanisms. The first, and it seems most prevalent under field conditions, is a change in the soluble protein per unit leaf area, which also defines a change in the amount of the photosynthetic apparatus. There is a close linkage between leaf photosynthesis rate and the level of leaf protein (Fig. 11–1). In particular, those treatments that allow more N to be accumulated in the photosynthetic leaves are likely to increase photosynthetic rates. Nitrogen accumulation by plants and leaves is critical to obtaining and sustaining high photosynthetic activity.

A second possible mechanism influencing leaf photosynthesis rates is a feedback system associated with the accumulation of photosynthate in the leaves. It is clear that starch accumulation per se does not play a direct role, but rather, the feedback seems to be associated with an inability to continue the synthesis and export of sucrose. The hypothesis of Sawada et al. (1995) that sucrose synthesis can explain the availability of inorganic P to activate rubisco is an intriguing possibility. However, it should be remembered that photosynthate-feedback inhibition of photosynthesis rates is not commonly observed under natural, field conditions (Lauer and Shibles, 1987).

11–1.2 Canopy Carbon Accumulation

11–1.2.1 Influence of Canopy Structure

Carbon accumulation by a leaf canopy depends on the photosynthetic capability of the individual leaves and the distribution of the interception of photosynthetically active radiation (PAR) among the leaves. In the development of a soybean canopy, young leaves emerge and develop at the top of the canopy in a high radiation environment, and then as new leaves develop they become shaded for a progressively greater fraction of the time. This progression from a high radiation environment to a shaded environment is commonly reflected in the photosynthetic capacity of the leaves. In continuous measurements of leaf photosynthetic rates of soybean leaves, Lugg and Sinclair (1981a) tracked the decrease over time in the photosynthetic activity of leaves as the developing canopy caused them to be subjected to greater shading. Burkey and Wells (1991) also showed the decline in photosynthetic activity of individual leaves with canopy shading and that the decline could be arrested by thinning canopies and exposing leaves to increased radiation levels.

SOYBEANS:
Improvement, Production, and Uses

Third Edition

© 2004 ASA-CSSA-SSSA

Chapter 1—World Distribution and Trade of Soybean
James R. Wilcox

Plates 1–1, 1–2, and 1–3

Chapter 14—Fungal Diseases
Craig R. Grau, Anne E. Dorrance, Jason Bond, and John S. Russin

Plates 14–1, 14–2, 14–3, 14–4, 14–5, 14–6, 14–7, 14–8, 14–9, 14–10, 14–11, 14–12, 14–13, 14–14, 14–15, 14–16, 14–17, 14–18, 14–19, and 14–20

Chapter 16—Nematode Pathogens of Soybean
Terry L. Niblack, Gregory L. Tylka, Robert D. Riggs

Plates 16–1, 16–2, 16–3, 16–4, 16–5, 16–6, 16–7, 16–8, 16–9, 16–10, 16–11, 16–12, 16–13, 16–14, 16–15, 16–16, and 16–17

Chapter 17—Integrated Management of Soybean Insects
David J. Boethel

Plates 17–1, 17–2, 17–3, 17–4, 17–5, 17–6, 17–7, 17–8, 17–9, 17–10, 17–11, 17–12, and 17–13

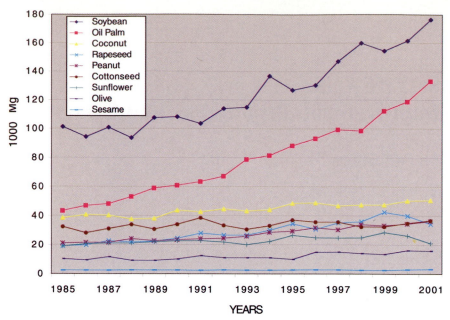

Plate 1–1. World production of major oilseeds, 1985 to 2001.

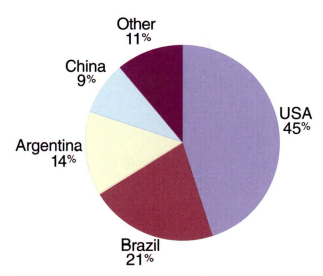

Plate 1–2. World soybean production by major countries, 1999/2000.

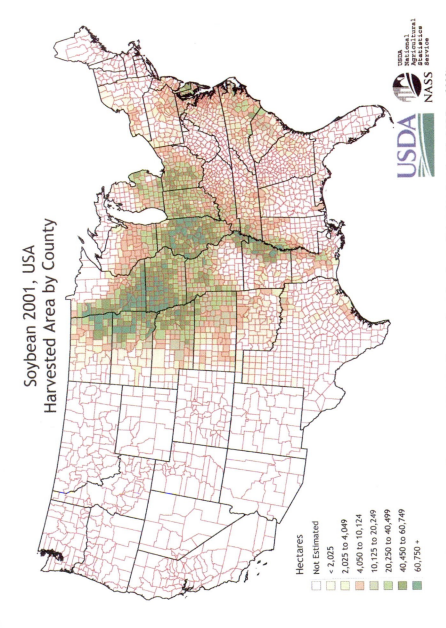

Soybean 2001, USA
Harvested Area by County

Hectares

Not Estimated
< 2,025
2,025 to 4,049
4,050 to 10,124
10,125 to 20,249
20,250 to 40,499
40,450 to 60,749
60,750 +

Plate 1–3. United States soybean area harvested, by counties, 2001. http://www.usda.gov/nass/aggraphs/graphics.htm (verified 17 Dec. 2002).

Plate 14–1. Cankered lesions on soybean hypocotyls caused by *Rhizoctonia solani*.

Plate 14–2. Comparison of soybean seedling populations resulting from treatment of seed with a fungicide (left) to sparse seedling populations resulting from no fungicide seed treatment.

Plate 14–3. Charcoal rot of soybean. (Photo courtesy of the American Phytopathological Society.)

Plate 14-5. Plant mortality caused by *Phytophthora sojae* at mid-vegetative growth stages.

Plate 14-6. Field scene depicting a common distribution pattern of soybean plants killed by *Phytophthora sojae*.

Plate 14-4 Charcoal rot of soybean. (Photo courtesy of the American Phytopathological Society.)

Plate 14–8. Soybean cultivars and lines express varying degrees of resistance to *Phytophthora sojae* after inoculation in controlled environments. Reaction phenotypes range from complete race specific resistance on left, to partial resistance, to moderate susceptibility and complete susceptibility.

Plate 14–10. Pattern of chlorosis and interveinal necrosis symptomatic of sudden death syndrome.

Plate 14–9. Soybean plants expressing symptoms of Rhizoctonia root rot.

Plate 14–7. Soybean plant expressing symptoms of wilt and characteristic lower stem lesion caused by *Phytophthora sojae*.

Plate 14–11. Symptoms of defoliation and petiole retention associated with sudden death syndrome.

Plate 14–12. Anthracnose of soybean. (Photo courtesy of the American Phytopathological Society.)

Plate 14–13. Internal stem symptoms of brown stem rot.

Plate 14–14. Foliar symptoms of brown stem rot.

Plate 14–15. Symptoms of Sclerotinia stem rot at canopy level.

Plate 14–16. Characteristic symptoms and signs of Sclerotinia stem rot.

Plate 14–17. Stem canker of soybean.

Plate 14–18. Brown spot of soybean.

Plate 14–19. Cercospora leaf blight. (Photo courtesy of the American Phytopathological Society.)

Plate 14–20. Downy mildew of soybean.

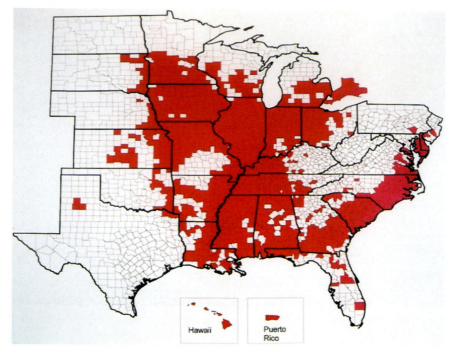

Plate 16–1. Distribution of *Heterodera glycines* in the USA in 2002.

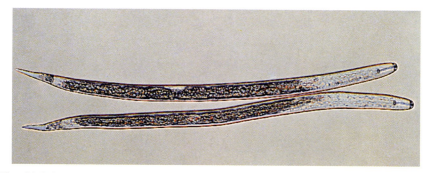

Plate 16–2. Second-stage juveniles of *Heterodera glycines*.

Plate 16–3. Virgin female *Heterodera glycines* with gelatinous matrix, on the surface of a soybean root.

Plate 16–4. Brown cyst of *Heterodera glycines* broken open to reveal eggs retained within the cyst wall. (Photo courtesy of E. Sikora).

Plate 16–5. Stunting and chlorosis caused by severe infestation with *Heterodera glycines*.

Plate 16–6. Aerial photo of severe damage to soybean caused by *Heterodera glycines* in Iowa.

Plate 16–7. Healthy-looking soybean plants in a field infested with yield-reducing levels of *Heterodera glycines*.

Plate 16–8. Resistant (right) and susceptible (left) soybean plants in a field infested with *Heterodera glycines* causing a 20% yield loss in the susceptible.

Plate 16–9. Soybean plants stressed by *Heterodera glycines* expressing potassium deficiency symptom.

Plate 16–10. Early-season injury due to *Heterodera glycines* infection.

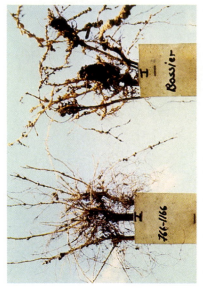

Plate 16–12. Root symptoms of root-knot nematode on resistant (left) and susceptible (right) soybean plants (By permission of the Society of Nematologists).

Plate 16–14. Soybean root with *Rotylenchulus reniformis* egg masses. Note soil clinging to the egg mass.

Plate 16–11. *Heterodera glycines* females visible on soybean roots.

Plate 16–13. Females of *Rotylenchulus reniformis* (stained pink) on the surface of a soybean root (By permission of the Society of Nematologists).

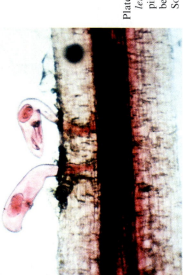

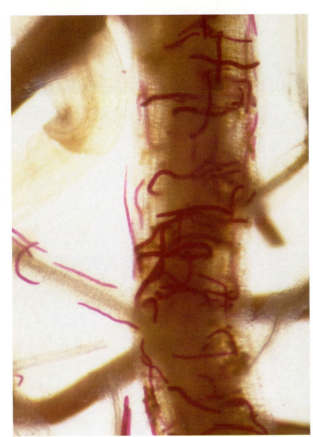

Plate 16–16. Lance nematodes (stained pink) within a soybean root.

Plate 16–15. Head of a lance nematode, *Hoplolaimus* sp. (Photo courtesy of E. C. Bernard).

Plate16–17. Damage to soybean roots caused by sting nematode, *Belonolaimus* sp. (By permission of the Society of Nematologists).

Plate 17-3. Bean leaf beetle adult. Reprinted with permission of the Entomological Society of America (L.G. Higley, photographer).

Plate 17-1. Velvetbean caterpillar, larva (light form). Reprinted with permission of the Entomological Society of America (M. Shepard, photographer).

Plate 17-2. Soybean looper larva. Reprinted with permission of the Entomological Society of America (M. Shepard, photographer).

Plate 17–5. Southern green stink bug nymph (dark form) (D. Boethel, photographer).

Plate 17–4. Southern green stink bug nymph (light form). Reprinted with permission of the Entomological Society of America (D. Boethel, photographer).

Plate 17–8. Green cloverworm larva (J. Lenhard, photographer).

Plate 17–6. Green stink bug adult. Reprinted with permission of the Entomological Society of America (G.R. Carner, photographer).

Plate 17–7. Green stink bug nymph. Reprinted with permission of the Entomological Society of America (D. Boethel, photographer).

Plate 17–10. Threecornered alfalfa hopper adult. Reprinted with permission of the Entomological Society of America (J. Lenhard, photographer).

Plate 17–11. Threecornered alfalfa hopper nymph. Reprinted with permission of the Entomological Society of America (J. Lenhard, photographer).

Plate 17–9. Brown stink bug adult and nymph. Reprinted with permission of the Entomological Society of America (D. Boethel, photographer).

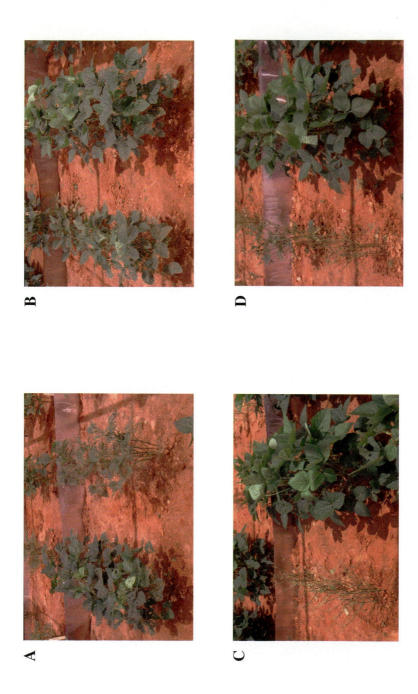

Plate 17–12. Defoliation of entries planted in adjacent rows in 1997, shown 19 d after initial infestation with larvae of H. zea or A. gemmatalis. The top photos show H. zea damage to (A) Jack-Bt (left) and Jack (right), and to (B) IR 81-296 (left) and Jack-Bt (right). The bottom photos show A. gemmatalis damage to (C) Jack (left) and Jack-Bt (right), and to (D) IR 81-296 (left) and Jack-Bt (right). Reprinted with permission of the Entomological Society of America.

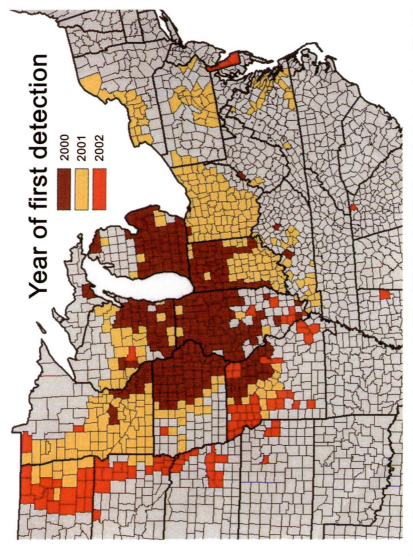

Plate 17–13. Distribution of the soybean aphid since its discovery in the USA in 2000 (November 2002). Reprinted with the permission of David Ragsdale, Univ. of Minnesota.

Light level on individual leaf surfaces in the soybean canopy, however, is highly dynamic because of reorientation of leaves during the daily cycle (Kao and Forseth, 1992) and wind-driven leaf movements (Pearcy et al., 1990). Angles of soybean leaves relative to the angle of incidence of direct beam radiation change to some extent during the day, in that individual leaves orient to increase light interception in the morning and late afternoon, and decrease light interception at midday (Kao and Forseth, 1992). Plants subjected to deficient N maintained the light-avoidance leaf angle for a longer period at midday resulting in less interception of excessively high solar radiation.

Wind-driven movements of leaves high in the canopy are especially important to leaves within the canopy. Pearcy et al. (1990) reported that a large fraction of radiation received by within-canopy soybean leaves was via sunflecks. Sunfleck durations were typically <30 s and for 80% of the locations in the leaf canopy the durations were <10 s. There was a larger range in the photosynthetic induction state of leaves in the soybean canopy as determined by the photosynthetic rates after 1 min in full sunlight than there was after 20 min (Pearcy and Seemann, 1990). Activation of fructose-1,6-bisphosphatase appears to be especially limiting for leaves at the bottom of the canopy that are intermittently subjected to low light conditions of <70 μmol m^{-2} s^{-1} (Sassenrath-Cole and Pearcy, 1994). Both rubisco and fructose-1,6-bisphosphatase required up to 10 min to be fully activated when leaves are transiently exposed to high light following low light, and these enzymes were not fully activated until light levels reached 400 μmol m^{-2} s^{-1} (Sassenrath-Cole and Pearcy, 1992, 1994).

Sometimes it is suggested that erect leaves might be helpful to increase the photosynthetic capability of a canopy. Unfortunately, the intuitive attraction of this hypothesis is not supported by theoretical analyses of integrated assimilation over the full day. The original analysis of canopy assimilation by deWit (1965) showed only a small influence of leaf angle. Duncan et al. (1967) also examined the issue of leaf angle and showed that a leaf area index of at least five was needed to have even a small advantage for more erect leaves. In this case, leaf angle has to be large compared to the more normal angle of 40°. In fact, a leaf angle of 40° was advantageous at a leaf area index of four or less, which is the range of leaf area index for much of the life cycle of soybean. No experimental measurements have been reported for soybean on variation in canopy assimilation in response to leaf angle.

Sinclair and Sheehy (1999) hypothesized that erect leaves were advantageous in increasing crop yield, however, this advantage was not related to increased rate of canopy photosynthesis. They suggested that a major limitation to obtaining high yields was the supply of N required by developing seeds during reproductive growth. This is especially true for soybean with seeds of high protein content. To store N in vegetative tissue for later transfer to developing seeds, it would be advantageous to retain a large leaf area as storage tissue for N. Erect leaves allow a small amount of light penetration to the bottom leaves so that these leaves might be retained on the plant rather than senesce before seed fill. Consequently, erect leaves were hypothesized to allow leaf canopies to maintain greater leaf area indices, leading to greater N storage during vegetative development for use in generating higher seed yields. This hypothesis has yet to be tested experimentally.

11–1.2.2 Canopy Assimilation Rates

Measurement of CO_2 exchange by soybean canopies is very challenging, and consequently, only a limited number of reports for soybean have been added to the literature. Nevertheless, these data are quite useful in documenting the C accumulation capacity of the entire leaf canopy. Boerma and Ashley (1988) placed a mylar chamber on a 1-m segment of row and measured the CO_2 depletion over a 1- to 2-min period. Measurements of canopy CO_2 exchange rates of 20 soybean genotypes showed that rates reached a maximum between full bloom (R2) and the beginning of seed fill (R5). Thereafter, CO_2 exchange rates decreased during seed fill. A mean CO_2 exchange rate was calculated from measurements on five fixed dates during reproductive development. Seed yield was positively correlated with duration of seed fill and CO_2 exchange rates. Similar results were obtained by Ashley and Boerma (1989) in an experiment that included cv. Tracy and Davis and 38 F_4 and F_6 lines derived from a Tracy × Davis mating.

Wells (1991) made frequent measurements of canopy CO_2 exchange rates through two seasons on a single soybean cultivar. In 1 yr, canopy CO_2 exchange rates peaked at flowering (R1) and in the second year at beginning seed fill (R5). The difference in the period when maximum CO_2 exchange rates occurred between years reflected differences in the time when maximum radiation interception occurred. In both years, however, there was a slow decline in CO_2 exchange rates until maturity following the period of maximum rates. Canopy CO_2 exchange rates up to the time of maximum rate were linearly correlated to the fraction of radiation intercepted. The slope of CO_2 exchange as a function of intercepted radiation was essentially equal between the two seasons.

The relationship between canopy CO_2 exchange and radiation was also reported by Campbell et al. (1990) for soybean plants grown in naturally lit growth chambers. Data were presented for the period when the leaf area index was 4.5 to 5.7 and nearly all of the incident radiation was intercepted by the leaf canopy. A curvilinear relationship between CO_2 exchange rate and incident solar radiation was obtained from observations made throughout a single day. Exclusion of the observations obtained at low radiation level, which represented a small fraction of the daily cycle, resulted in a nearly linear relationship between CO_2 exchange rates and incident solar radiation for much of the data.

The most intense measurements of canopy CO_2 exchange rates were reported by Rochette et al. (1995). Canopy net CO_2 exchange rates for a large soybean field were measured at hourly intervals throughout the growing season using an eddy correlation approach. The CO_2 exchange rates were plotted against measured intercepted radiation. Using data from all situations during the season when leaf area index was >2, the regression slope of the relationship was 0.708 mg CO_2 μmol^{-1} PAR ($r^2 = 0.82$). This result supported the view that CO_2 assimilation was linked linearly to the amount of radiation intercepted by the soybean crop.

11–1.2.3 Radiation-Use Efficiency

A near-linear relationship between canopy CO_2 exchange and intercepted radiation (Wells, 1991; Campbell et al., 1990; Rochette et al., 1995) lends support to the idea that the growth in soybean can be expressed by a radiation-use efficiency

Table 11–1. Radiation-use efficiency (RUE) obtained for selected intervals during the growing season at Ottawa, Canada for the cv. Maple Glen (Rochette et al., 1995). The RUE were calculated from mass accumulation data measured weekly and radiation interception measured continuously.

Period	Leaf area index	Dry matter	RUE
Day of yr	$m^2\,m^{-2}$	$g\,m^{-2}$	$g\,MJ^{-1}$
161–195	LAI < 1	57.31	0.36
196–203	$1 \leq LAI < 2$	88.74	0.68
204–208	$2 \leq LAI < 3$	102.78	1.00
209–214	$3 \leq LAI < 4$	113.08	0.92
215–227	$4 \leq LAI < 5$	173.61	0.84
228–252	$4 \leq LAI < 5$	333.89	1.02

(RUE), which is defined here as the ratio of accumulated aboveground crop mass ($g\,m^{-2}$) to cumulative intercepted total solar radiation ($MJ\,m^{-2}$). Since RUE can be useful in understanding and describing crop growth, the field experiments in which estimates of RUE for soybean were presented are reviewed in detail. For the purpose of this review, values of RUE given on a PAR basis in the literature were converted approximately to a total solar radiation basis by multiplying by 0.5 (Sinclair and Muchow, 1999).

The experiment on CO_2 exchange by Rochette et al. (1995) also included data collected to calculate RUE. This experiment was carried out in Ottawa, Canada on the cv. Maple Glen. Twenty plants (plant density of 42 plants m^{-2}) were harvested weekly to determine crop mass accumulation and the interception of PAR was measured continuously. The mean RUE for intervals during the season when leaf area index was >2, was 0.93 g MJ^{-1}. However, the value of RUE was lower early in the season when leaf area index was <2 (Table 11–1).

The first attempt to measure RUE for soybean was apparently done by Nakaseko and Gotoh (1983) in Sapporo, Japan using cv. Tokachinagaha and Harosoy. Dry matter accumulation was measured five times during the growing season. Radiation interception was calculated from the leaf area index and an extinction coefficient determined three times during the growing season. The RUE estimates for the two cultivars were essentially equal and calculated to have a value of 1.36 g MJ^{-1}.

Daughtry et al. (1992) determined RUE from a study conducted at two locations, West Lafayette, IN (cv. Century) and Beltsville, MD (cv. Williams 82). Weekly measurements were made of plant mass by harvesting 0.5-m lengths of row during the season from sowing to mid-grain fill. The intercepted radiation was calculated at the West Lafayette site by spot measurements at midday on clear days and was measured at the Beltsville site continuously with a line quantum sensor at the soil surface. The RUE at West Lafayette was 0.68 g MJ^{-1} ($r^2 = 0.98$) and at Beltsville was 0.99 g MJ^{-1} ($r^2 = 0.91$).

The RUE was determined by Muchow et al. (1993a) for two seasons at Katherine, Australia (cv. Buchanan) and one season at Lawes, Australia (cv. Davis). Plants were harvested every 7 to 10 d from a 1.73 m^2 area at Katherine and a 2.0 m^2 at Lawes. Tube solarimeters were placed on the soil surface to measure continuously intercepted solar radiation. Fitted step-wise linear regressions resulted in estimates of RUE of 0.89 g MJ^{-1} ($r^2 = 0.98$) and 0.81 g MJ^{-1} ($r^2 = 0.99$) at Katherine and 0.88 g MJ^{-1} ($r^2 = 0.96$) at Lawes.

Most recently, RUE was measured for cv. A4922 and Manokin grown over a wide range of plant densities at Keiser, AR by Purcell et al. (2002). At approximately 2-wk intervals, light interception was measured using a line-quantum sensor at midday and plant mass was measured from 1 m^2 harvest areas. The influence of plant population on RUE was not significant in 1 yr, but in a second year a statistically significant, but small, decrease in RUE was detected with increased population. In the first year the RUE was 0.66 g MJ^{-1}, and in the second year at a plant population of 40 plants m^{-2} RUE was interpolated to be 0.67 g MJ^{-1}.

Finally, Sinclair and Shiraiwa (1993) published estimates for RUE obtained for six cultivars grown in Gainesville, FL and four cultivars grown in Azuchi, Japan. Plants were harvested from 0.56 m^2 areas at approximately 2-wk intervals following canopy closure at Gainesville and from 1.0 m^2 areas at weekly intervals at Azuchi. Since the data from Gainesville were only for the period when leaf area index was >5 and there was complete canopy closure, intercepted radiation was set equal to incident radiation. Intercepted radiation was measured weekly in Japan. The values of RUE ranged from 0.41 to 0.66 g MJ^{-1} at Gainesville and 0.84 to 1.15 g MJ^{-1} in Japan (Fig. 11–2). Differences in RUE were attributed to differences in the fraction of diffuse radiation in the incident radiation between the two locations, and in the N content of the leaves among cultivars (Sinclair and Shiraiwa, 1993).

The sensitivity of RUE to leaf N observed by Sinclair and Shiraiwa (1993) is not surprising considering the fact that leaf photosynthesis rate is sensitive to leaf N (Fig. 11–2). Sinclair and Horie (1989) derived a relationship between RUE and leaf N for various crop species based on the responses of leaf photosynthesis to specific leaf N. The relationship they derived for soybean (Fig. 11–2) illustrates a nearly

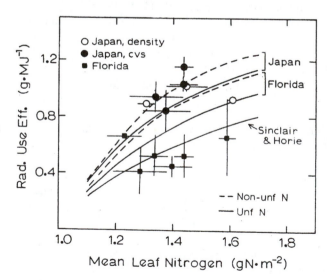

Fig. 11–2. Radiation-use efficiency (RUE) plotted against mean canopy N per unit area. The individual data are from experiments performed in Shiga, Japan and Gainesville, FL. The theoretical relationship derived by Sinclair and Horie (1989) is shown by the bottom solid line. The upper lines include modifications for differences in the fraction of diffuse radiation at each location. Also, calculations were performed at both locations assuming either a uniform or a nonuniform distribution of N with depth of leaf area index (LAI) of the canopy. From Sinclair and Shiraiwa (1993).

linear increase in RUE with increasing leaf N. At a specific leaf N of 2.5 g m^{-2}, the theoretical soybean RUE would be about 1.1 g MJ^{-1}.

A nonuniform distribution of leaf N has the possibility of increasing RUE if more of the N is present in top leaves and less in the bottom leaves. The advantage of the nonuniform N distribution results because the top leaves intercept much more of the incident solar radiation than the bottom leaves, and therefore, the top leaves are responsible for much of the photosynthetic activity. Indeed, theoretical analyses by Sinclair and Shiraiwa (1993) (Fig.11–2) and Hammer and Wright (1994) demonstrated that this is the case. The advantage of a nonuniform N distribution, however, was calculated to be greatest when the overall N availability to the leaves is low. At high N availability, the top leaves are already approaching near maximum photosynthetic rates and shifting more N into these leaves would have little positive effect on their photosynthesis rate.

11–1.2.4 Conclusions

Canopy structure is a useful consideration in understanding crop photosynthetic capacity. In most situations in the field when row spacing is not wide and closed canopies develop, canopy structure seems to have a fairly minor influence on CO_2 exchange. The interception of the incident radiation and N levels of the leaves appear to have the major influences on canopy CO_2 exchange. Consequently, maximizing radiation interception over the growing season and sustaining leaf photosynthetic rates seem to be critical factors in obtaining high C accumulation by soybean. The persistence of high CO_2 exchange during the latter stages of reproductive growth appear to be particularly critical in obtaining increased yields.

Radiation-use efficiency has developed as a useful approach to summarizing the photosynthetic capacity of crops. The observed range in RUE for soybean appears to be quite large with reported values ranging from 0.41 to 1.36 g MJ^{-1}. The two reports of highest RUE were both from Japan, where enhanced diffuse radiation likely contributes to an especially high RUE. Many of the observations place RUE for soybean in the range of 0.65 to 1.15 g MJ^{-1} with 0.9 g MJ^{-1} as an approximate median. The upper end of this range is similar to the proposed asymptotic level for soybean RUE (Sinclair and Horie, 1989). Clearly, most of the measured values of RUE are below the theoretical estimate of about 1.1 g MJ^{-1}. The challenge in achieving high soybean yield will be in obtaining, and more importantly, sustaining high RUE through the season. This requires high leaf photosynthetic rates, that is, high leaf N content per unit area, and the interception of large fractions of available solar radiation.

11–2 NITROGEN ACCUMULATION

11–2.1 Symbiotic Dinitrogen Fixation

11–2.1.1 Photosynthate Supply and Nodule Activity

The reduction of atmospheric N_2 to ammonia is energetically an expensive process, and this has led to the intuitive hypothesis that photosynthate demands in the nodule may frequently impose an immediate limitation on the rate of N_2 fixation. Considerable research in recent years has helped to clarify the role of photosynthate availability in limiting N_2 fixation activity. Photosynthate availability does not seem to change rapidly in nodules, nor cause rapid changes in N_2 fixation activity. This is evidenced by the fact that during the diurnal cycle there is no decrease in N_2 fixation rates during the night period associated with inadequate photosynthate. In studies of field-grown soybean plants, no night-time decreases in N_2 fixation rates other than those associated with decreases in soil temperature have been reported (Denison and Sinclair, 1985; Weisz and Sinclair, 1988). Millhollon and Williams (1986) measured N_2 fixation rates and found no decrease in rates until after 14 h of subjecting plants to darkness.

Similarly, attempts to stimulate N_2 fixation rates per unit nodule weight by increasing rates of photosynthesis have offered little evidence of a photosynthate limitation. Vidal et al. (1996) increased the light exposure of plants from 400 to 1200 $\mu mol \ m^{-2} \ s^{-1}$ PAR, but this resulted in no marked change in the N_2 fixation rate. Further, an increase in CO_2 concentration around the shoots, which substantially increased photosynthesis rate, resulted in no change in N_2 fixation rate. Serraj et al. (1998) compared the N_2 fixation rates of soybean plants exposed to 360 or 700 $\mu mol \ CO_2 \ mol^{-1}$ for 17 d and found no difference in N_2 fixation rates per plant.

An intriguing variant of experiments employing CO_2 treatments was a treatment where the plants were subjected to their CO_2 compensation point for photosynthesis, which was approximately 60 $\mu mol \ CO_2 \ mol^{-1}$ (Vidal et al., 1995). Dinitrogen fixation rate declined markedly between 6 and 16 h after initiating the low CO_2 treatment. The rapid response to the CO_2 decrease indicated N_2 fixation inhibition in advance of a photosynthate-limited decline. Indeed, exposure of the nodules subjected to low CO_2 to elevated oxygen concentrations resulted in N_2 fixation rates equal to or greater than the control. These data were interpreted by Vidal et al. (1995) to demonstrate that photosynthate was not limiting, but that the diffusion of oxygen into the nodule had been decreased. That is, there was a sequence of events in the plant triggered by low CO_2 concentration that caused permeability of the oxygen diffusion barrier in the nodules to decrease so as to result in a decrease in N_2 fixation rate.

Walsh et al. (1998) measured ^{11}C-photosynthate transport to nodules and concluded that there was no direct link between photoassimilate import to the nodule and the status of the oxygen diffusion barrier in the nodule. Nevertheless, phloem flow to the nodule has been shown to be essential to sustain nodule activity because N_2 fixation rates decrease severely about 30 min following stem girdling (Vessey et al., 1988), stem chilling, and leaf removal (Walsh et al., 1987).

11–2.1.2 Nitrogen Feedback and Nodule Activity

An alternate hypothesis to a direct limitation of photosynthate on N_2 fixation has focused on the possibility of a feedback based on the transport in the phloem of nitrogenous compounds to nodules. It has been hypothesized by several investigators (Streeter, 1993; Parsons et al., 1993; Hartwig et al., 1994; Serraj et al., 1999a) that excess N compound(s) in the shoot feedback to decrease N_2 fixation rates. The importance of the shoot in regulating N_2 fixation rates was shown in an experiment with reciprocal grafts in which N_2 fixation activity was closely linked to the shoot and not to the root (Fujita et al., 1991). Further, Zhu et al. (1991) blocked the accumulation of nitrogenous compounds by eliminating the source of N_2 by growing nodulated plants in an $Ar:O_2$ (79:21) atmosphere. They measured a three-fold greater nitrogenase activity for the $Ar:O_2$ treated plants compared to the controls that were exposed to $N_2:O_2$.

Bacanamwo and Harper (1997) suggested that asparagine might be the N signal compound from the shoot because of dramatic increases in asparagine concentration in the shoot after subjecting plants to a NO_3^- treatment. Since there was no significant change in nodule concentration of asparagine, it was concluded that the response did not require an accumulation of asparagine in the nodule. Vadez et al. (2000a) added asparagine to the hydroponic solution on which soybean plants were grown and caused an increase in asparagine concentration in the nodules, particularly. The N_2 fixation rates of the plants supplied with asparagine were observed to be decreased 24 h after initiating the asparagine treatment.

Neo and Layzell (1997) exposed shoots of soybean to ammonia and found large increases in nearly all nitrogenous compounds in the phloem at the base of the stem. The increase in nitrogenous compounds was associated with a decrease in nodule N_2 fixation rates and in the permeability of the oxygen diffusion barrier. They suggested that glutamine was a likely signal compound because its concentration change in the phloem sap was relatively greater than any of the other nitrogenous compounds.

Another group of compounds that seem intimately involved in feedback regulation of N_2 fixation are the ureides, allantoin and allantoic acid. Ureides are the main product of N_2 fixation in soybean nodules and are transported from the nodules to the shoots. Ureides are very efficient compounds for transport because these molecules require only one C per N. There are, therefore, two specific advantages to soybean resulting from ureide synthesis and transport: (i) it is more economical per unit N in the use of imported C to synthesize ureides instead of amino acids (Pate, 1985), and (ii) the amount of C tied up in N transport to the shoots is considerably less than that in transporting amino acids.

The feeding of ureides to soybean plants grown on hydroponic solution resulted in large increases in ureide levels in both shoots and nodules and decreases in N_2 fixation rates (Serraj et al., 1999b; Vadez et al., 2000a). A decrease in N_2 fixation rates occurred 2 d after beginning ureide feeding, or about 1 d later than the initial decrease observed following asparagine feeding. Nodule oxygen permeability decreased in a pattern similar to that of N_2 fixation (Serraj et al., 1999b).

Ureide degradation rates in the leaves seems to be a critical factor in the feedback regulation of N_2 fixation. Circumstances that favor ureide catabolism in the leaves seem to minimize feedback inhibition by ureides. Increased CO_2 concentrations around soybean shoots eliminated the increase in leaf and nodule ureide concentration that was observed when plants were fed 10 mM allantoic acid (Serraj and Sinclair, 2003). The lack of increase in ureide concentrations under exposure to high CO_2 concentrations was associated with a much smaller influence of the allantoic acid treatment on N_2 fixation rates for those plants grown under 700 μmol CO_2 mol^{-1} as compared to those grown in 360 μmol CO_2 mol^{-1} treatment. Vadez et al. (2000b) found that increasing the concentration of manganese (Mn) in the nutrient solution on which plants were grown influences the rate of ureide catabolism in the leaves. Increased Mn levels resulted in less sensitivity of N_2 fixation of the nodules to the feeding of ureides. Manganese is a cofactor in allantoic acid degradation by the enzyme allantoate amidinohydrolase (E.C. 3.5.3.9) (Winkler et al., 1987).

The decrease in specific and total N_2 fixation activity associated with P deficiency may also involve N feedback (Sa and Israel, 1991). Ureides accumulated in nodules of plants subjected to a P deficiency, but the ureide concentration in the xylem sap was actually decreased (Sa and Israel, 1995). Energy charge in the bacteroids was little changed by the P deficiency but the plant cell cytoplasm had decreased energy charge and ATP concentration. Sa and Israel (1995) concluded that a decrease in oxidative phosphorylation might have impaired ureide transport from the nodule to the xylem causing ureide accumulation in the nodule.

11–2.1.3 Water and Nodule Activity

It is now well recognized that N_2 fixation in soybean is highly sensitive to soil drying, and appears to be more sensitive to water-deficit stress than essentially all other physiological processes in the plant (Serraj et al., 1999a). In particular, the greater sensitivity of N_2 fixation as compared to leaf gas exchange and C accumulation has been documented both in controlled environment studies (Kuo and Boersma, 1971; Sinclair, 1986; Durand et al., 1987; Djekoun and Planchon, 1991) and in field studies (Sinclair et al., 1987). Indeed, Purcell and King (1996) found in a field experiment comparing yield between irrigated and rainfed treatments that fertilization with 33.6 g NO_3^- m^{-2} resulted in equivalent yields between watering regimes, but dependence on N_2 fixation for N resulted in a significant yield decrease in the rainfed treatment as compared to the irrigated treatment. Pot experiments included in the study by Purcell and King (1996) confirmed a much greater decrease in response to soil drying by N_2 fixation than N uptake from the soil amended with NO_3^-.

Figure 11–3 presents N_2 fixation rates plotted against transpiration rates from dry-down experiments reported by Serraj and Sinclair (1996), where in each case the rates during the dry-down are normalized to the rates of well-watered control plants. Much of the data in this plot were below the 1:1 line, denoting a greater decrease in N_2 fixation rates than transpiration rates at all stages of soil drying. Subsequent to these initial studies, the problem of decreased N_2 fixation activity in soybean during soil drying has been a topic of considerable study.

Sinclair and Serraj (1995) compared the sensitivity of N_2 fixation in nine grain legumes and found that N_2 fixation in soybean and cowpea (*Vigna unguiculata* L.) was substantially more sensitive to soil drying than the other species. They suggested that transport of N from the nodules predominately as ureides in soybean and cowpea, in contrast to transport of amides in other legume species, might be a critical aspect of the water-deficit sensitivity. Since then, several studies have shown dramatic increases in the level of ureides in leaves and petioles of soybean in response to drying soil in controlled environments (de Silva et al., 1996; Serraj and Sinclair, 1996; King and Purcell, 2001), and in the field (Serraj and Sinclair, 1997; Purcell et al., 1998). Ureide concentrations of nodules also increased greatly under devel-

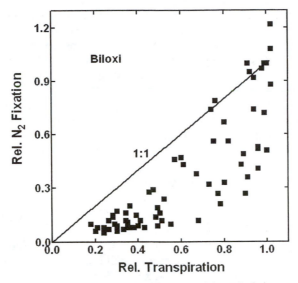

Fig. 11–3. Comparison of relative dinitrogen (N_2) fixation activity and relative transpiration rates on individual days of a water-deficit cycle imposed on the cv. Biloxi. From Serraj and Sinclair (1996).

oping water deficit (Purcell and Sinclair, 1995; Gordon et al., 1997; Serraj et al., 1998).

The accumulation of ureides under water-deficit conditions is consistent with changes in the water economy of nodules and the plant under these conditions (Serraj et al., 1999a). Figure 11–4 is a schematic showing that water flow in the

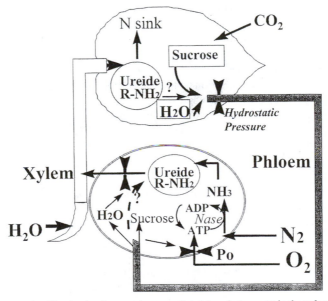

Fig. 11–4. Schematic of interaction between leaves and nodules via transport in the xylem and phloem of water, sucrose, and ureides. From Serraj et al. (1999a).

phloem from the shoot to nodules and flow in the xylem from nodules to shoots is a key feature of the N_2 fixing system. The importance of this flow scheme is the fact that the nodule is essentially a closed system and dependent almost exclusively on phloem flow for water supply to the nodule. Phloem supply of water received by the nodules defines the volumetric water flux into the nodule, and therefore, the volumetric flux out of the nodule into the xylem. Hence, phloem flow determines both the delivery rate of materials to the nodule, as well as the export rate of materials from the nodules (Walsh, 1995). A steady flow from nodules is especially required because of the very low solubility of ureides in water (Sprent, 1980). Raven et al. (1989) warned that the balance between phloem flux and xylem flux is precarious and sustaining xylem flow from the nodule may require special mechanisms in dealing with any water shortfall. An advantage for soybean in transporting ureides, however, is that N in the form of ureides instead of amides generates less osmoticum, which in sufficient concentrations could influence water potential gradients that drive water flux from the nodules.

A slowing of phloem flow from nodules to the leaves has the potential to slow the flow of ureides from the nodule, causing a build-up of N compounds and hence an immediate feedback inhibition of N_2 fixation (Streeter, 1993). The dramatic decreases in N_2 fixation rates when phloem flow is disrupted (Walsh et al., 1987) are consistent with a decreased water flow into and out of the nodules causing accumulation of ureides in the nodule and a rapid inhibition of N_2 fixation (Walsh et al., 1989; Walsh, 1995). Streeter and Salminen (1993) also noted shoot decapitation influenced specifically the nodule apoplasm such that ureide concentration was increased and the fraction of the total ureides and amino N in the apoplast was increased.

Another factor involved in the response of nodules to water deficit may be the inability, upon stress, to catabolize sucrose in the nodules. Sucrose, along with other compounds, accumulates in nodules when subjected to water-deficit stress (Fellows et al., 1987; González et al., 1995). Sucrose synthase (E.C. 2.4.2.13), the key sucrose hydrolytic enzyme, had decreased activity by the second day of the stress treatment, and its mRNA had already declined on day one. However, the addition of abscisic acid to a nutrient solution fed to soybean plants, which induced a response that mimicked water deficit, resulted in no change in sucrose synthase activity (González et al., 2001). The dramatic decrease in transpiration rate in their experiment may have resulted in a change in the nodule water flux and observed decrease in N_2 fixation rate.

11–2.1.4 Conclusions

Photosynthate levels in the nodules do not seem to have a primary role in regulating N_2 fixation rates. Rather, phloem flow and the feedback of a nitrogenous compound(s) have been identified as a key feature of changes in nodule activity. It does not appear that the bulk accumulation of the signal compound in the nodule is necessarily a key feature of the regulation. One possibility is that localized increases in the concentration of a signal compound in nodules might be sufficient to trigger a decrease in N_2 fixation rate. For example, changes only in nodule cortical cells associated either with xylem loading or the oxygen diffusion barrier could ultimately have consequences on the overall rate of N_2 fixation.

The sensitivity of N_2 fixation in soybean to water deficit (Fig. 11–3) appears to be a key constraint on N accumulation, and in the absence of soil sources of N, ultimately yield. This sensitivity may result from the water balance of the nodule, which depends nearly exclusively on phloem flow as a water source. Changes in phloem flow will immediately influence the water status and/or export rate of materials from the nodules. Changes in water flow, along with changes in the flux of a signal compound from the shoot, could well explain the sensitivity of N_2 fixation in soybean to drying soil (Fig.11– 4).

11–2.2 Nitrate and Dinitrogen Fixation

The negative interaction between NO_3^- uptake and N_2 fixation has been a topic that has received considerable research attention in the past (Harper, 1987). The benefit of having a soybean crop with a dual capacity for NO_3^- uptake and symbiotic N_2 fixation would be increased cropping flexibility and increased yield. An attractive possibility for soybean is to have active uptake of soil NO_3^- during the first part of the cropping season without inhibition of the nodule development and the initiation of N_2 fixation. Therefore, later in the cropping season when the soluble soil NO_3^- has been depleted, N_2 fixation would be able to immediately provide N to the plants. Further, there appears to be no fundamental incompatibility of selecting cultivars that can exhibit both high N_2 fixation activity and high NO_3^- uptake capacity. Zhang et al. (1997) measured N accumulation in six cultivars under conditions favorable to N_2 fixation and NO_3^- uptake and found no identifiable relationship between plant performance on the two N sources. The fundamental problem is an inability to develop and sustain N_2 fixation activity in the presence soil NO_3^- at any greater than small concentrations of a 'starter' fertilizer (Wiersma and Orf, 1992; Yinbo et al., 1997).

11–2.2.1 Nitrate Uptake

Soybean plants readily accumulate NO_3^- during the vegetative growth phase but there are reports that soybean is unable to sustain high utilization rates of NO_3^- during seed fill (Imsande, 1986; Imsande and Edwards, 1988). In contrast, studies that assured sustained utilization of NO_3^- by plants offered evidence for continued NO_3^- uptake capacity until very late in the growth cycle. Indeed, Vessey et al. (1990) sustained NO_3^- uptake until near maturity when plants were spaced to maximize light interception and C accumulation. Zapata et al. (1987) also found no decrease in N uptake from soil through stage R7 for field-grown plants.

Nitrate uptake by the roots appears to be the crucial step in the use of rhizosphere NO_3^-. Similar to N_2 fixation, phloem flow is crucial in sustaining NO_3^- uptake. Nitrate uptake differs, however, in that it appears that photosynthate translocation from the shoot to the root is critical to support NO_3^- uptake process (Touraine et al., 1992; Delhon et al., 1996). Stem-girdling treatments in each of these studies resulted in a rapid loss in the capacity of soybean roots to uptake NO_3^-. Delhon et al. (1995) identified photosynthate transport as necessary for NO_3^- uptake because uptake by soybean roots decreased during the dark portion of the daily cycle, and the uptake rate was essentially independent of transpiration rate and the accumulation of NO_3^- and asparagine in the roots. Touraine et al. (1992) identified malate availability in the roots as critical in sustaining NO_3^- uptake. They supplied K-malate to the external solution of one half of the roots in a split-root experiment and found in the other half of the roots an increased phloem malate concentration and an increased NO_3^- uptake rate.

11–2.2.2 Inhibition of Nitrogen Fixation by Nitrate

Application of NO_3^- to soybean plants that are dependent on N_2 fixation results in dramatic declines in N_2 fixation rates within 1 to 2 d (Streeter, 1985; Arrese-Igor et al., 1997). Removal of NO_3^- after 7 d resulted in full recovery of N_2 fixation activity within about another 7 d (Streeter, 1985). The recovery is consistent with the results of Arrese-Igor et al. (1997) showing that there was no change in components 1 and 2 of the nitrogenase complex for 9 d after treating plants with NO_3^-. Consequently, NO_3^- uptake does not appear to be directly damaging to the potential activity of N_2 fixation machinery.

The fact that N_2 fixation rate is severely decreased in the presence of NO_3^- has resulted in considerable attention to understanding the physiological basis for this response in the hope that such knowledge can be used to overcome this limitation. Streeter (1988) published a comprehensive review on the status of various hypotheses explaining the inhibition of nodule formation and N_2 fixation by NO_3^-. His conclusion at that time was that neither carbohydrate deprivation nor NO_3^- toxicity, the two prevalent hypotheses for N_2 fixation inhibition, were supported by convincing evidence. Vessey and Waterer (1992) updated the analysis of N_2 fixation inhibition and offered additional evidence in support of the conclusions of Streeter (1988). They also offered a new hypothesis, specifically that NO_3^- may result in a decrease in the permeability of the oxygen diffusion barrier of the nodule, and as a result the nodule becomes oxygen limited causing a restriction on nitrogenase activity. Consistent with this hypothesis, Arrese-Igor et al. (1997) found a large decrease in nodule oxygen permeability 4 d after a NO_3^- treatment, but N_2 fixation had already decreased dramatically after only 2 d. Further, there is little response in N_2 fixation activity to a treatment of increasing oxygen concentration around nodules in an attempt to alleviate the oxygen limitation (Heckmann et al., 1989; Serraj et al., 1992).

Another hypothesis to explain NO_3^- inhibition of N_2 fixation is based on the depletion of reductant at the bacteroids (Heckmann et al., 1989). Their suggestion is that reductant is used preferentially by NO_3^- reductase in the cytosol in the presence of NO_2 so that reductant for N_2 fixation is greatly diminished. Unfortunately, only data on whole nodule levels of energy charge and ATP/ADP ratios were presented and the decreases in levels associated with NO_3^- treatment were modest. Nevertheless, a negative correlation ($r^2 = 0.90$) has been reported between N_2 fixation activity in the presence of NO_3^- and NO_3^- reductase activity extracted from the cytosol of nodules (Serraj et al., 1992).

Feedback of a nitrogenous compound from the shoot might be crucial in the inhibition of N_2 fixation in the presence of NO_3^-. Bacanamwo and Harper (1996) altered C and N levels in soybean plants by adjusting light and CO_2 levels around the shoots, and measuring the extent of the N_2 fixation decrease 24 h following a treatment with 15 mM NO_3^-. The inhibition of N_2 fixation rate was positively correlated with shoot N concentration, but negatively correlated with shoot total nonstructural carbohydrate and with C:N ratio in the shoot. These results for the three shoot variables contrasted with those same variables in the nodules. The C and N variables in the nodules were not significantly correlated to N_2 fixation inhibition. The conclusion was that N_2 fixation depression occurring as a result of NO_3^- treatment resulted from a feedback based on C and N levels in the shoot, and not those in the nodules.

In conclusion, the inhibition of N_2 fixation by NO_3^- is rapid and dramatic. Despite intensive study for a number of years, there is as of yet no clear, consistent hypothesis to explain the physiological basis of why NO_3^- uptake suppresses N_2 fixation. This is a critical issue that will need resolution in an effort to develop soybean plants without a NO_3^- induced suppression of N_2 fixation.

11–2.2.3 Genetic Variation in Nitrate Response

Considerable variability appears to exist among soybean cultivars in their ability for sustained N_2 fixation activity when subjected to NO_3^-. Several cultivars have been identified with N_2 fixation that is less sensitive to the presence of NO_3^- including 'Dunadja' (Hardarson et al., 1984), 'Elf' (Gibson and Harper, 1985), and 'Tielingbaime' (Serraj et al., 1992). In addition, a comparison of the N_2 fixation of 47 plant introduction lines grown on low NO_3^- and high NO_3^- field sites showed relative differences in total N_2 fixed between the two locations (Neuhausen et al., 1988). Since yield was highly correlated to the amount of N_2 fixation in each location, breeding for increased N_2 fixation on high N soils seemingly would be advantageous.

Betts and Herridge (1987) undertook a major effort to screen for NO_3^- tolerance of N_2 fixation across a diversity of germplasm. Starting with an initial population of 489 genotypes, they eventually identified 32 lines as being NO_3^- tolerant. Nine of the identified genotypes came from an original population of genotypes originating from Korea. Crosses using four of the NO_3^- –tolerant Korean lines showed that NO_3^- tolerance was heritable, but unfortunately none of the crosses resulted in yields equaling that of a commercial parent (Herridge and Rose, 1994).

In contrast to the experience of Herridge and Rose (1994), Raffin et al. (1995) found more positive results in tests of progeny from the NO_3^-–tolerant cv. Tielingbaime (Serraj et al., 1992). Nineteen F_8 progeny of Tielingbaime × Kingsoy were grown on moderate and high NO_3^- soil treatments. In comparing lines based on the fraction of seed N derived from fixation on the high NO_3^- soil, those genotypes that had high N_2 fixation on the high NO_3^- soil also produced high vegetative mass and seed yield. It was concluded that breeding for N_2 fixation tolerance to soil NO_3^- could be a positive selection criterion (Raffin et al., 1995).

In an alternate approach to using existing genetic variation, Carroll et al. (1985) employed mutagenesis to develop soybean lines with differing tolerance of N_2 fixation to NO_3^-. Using the cv. Bragg, mutant lines were identified that nodulated well when subjected to high NO_3^- conditions. These lines also produced very large numbers of nodules when NO_3^- levels were low, and hence they were identified as 'supernodulating' (Carroll et al., 1985). When challenged with NO_3^-, mutant *nts* 382 had decreased N_2 fixation activity on a plant basis but the rate was greater than that of normal plants (Day et al., 1987). The fact that the supernodulating lines still had decreased N_2 fixation when treated with NO_3^- was taken as an indication that the NO_3^- response mechanism in these lines was unchanged, and that the mutants simply had a higher base number of nodules so more nodules remained when the plants were treated with NO_3^- (Eskew et al., 1989; Hansen et al., 1992). Nitrogen accumulation and yield in field tests have not been enhanced by the supernodulating trait (Maloney and Oplinger, 1997).

Gremaud and Harper (1989) also developed mutants for high nodulation from the cv. Williams. When subjected to NO_3^-, these lines also had decreased nodule number but were able to sustain some N_2 fixation activity. The ability of these lines to sustain N_2 fixation activity under high NO_3^- also existed under field conditions, but these lines had lower seed yields than the parent cv. Williams (Wu and Harper, 1991; Pracht et al., 1994).

11–2.2.4 Conclusions

There is an obvious attraction to eliminating the incompatibility between N_2 fixation and soil NO_3^- uptake, if doing so would increase crop yields and improve the economy of N use in cropping systems that include soybean. Despite considerable research on this topic, the negative influence of NO_3^- on N_2 fixation remains an enigma. Exposure of soybean plants to NO_3^- clearly causes a severe decrease in N_2 fixation rates. Even though several hypotheses have been explored to explain this behavior, no definitive explanation has emerged. The most likely hypothesis for the NO_3^- inhibition of N_2 fixation seems to involve a NO_3^- induced or derived nitrogenous compound causing a feedback inhibition of nodule activity.

The practical challenge of developing cultivars with sustained N_2 fixation in the presence of NO_3^- has also received considerable attention. The efforts based on mutants with large numbers of nodules, which retain some nodules with N_2 fixing capability in the presence of NO_3^-, has not led to genotypes with yields equivalent to parental lines. Similarly, efforts to use natural genotypic variation for N_2 fixation in the presence of NO_3^- have not shown consistent yield improvement. One practical limitation in developing cultivars will be the commitment of scientists, time, and resources required to screen, breed, and verify lines that have NO_3^- tolerance trait. Nevertheless, the potential benefits of having N_2 fixation compatible with soil NO_3^- would seemingly justify such an effort.

11–3 YIELD INCREASE

1–3.1 Increasing Photosynthesis

Increasing soybean yield through the selection of genotypes with superior photosynthetic activity was the focus of considerable research activity from roughly 1970 to 1985. In the previous Soybean monograph, Shibles et al. (1987) discussed in detail the various investigations aimed at increasing photosynthesis and consequently yield. Photosynthesis rates were found to be heritable, but relationships between photosynthesis and yield, when found, were obtained based on photosynthesis measurements made during reproductive development. Shibles et al. (1987) concluded that "Although single-leaf and canopy photosynthesis can be enhanced by standard breeding techniques, the utility of such an endeavor (to achieve yield improvement) remains to be proved". That conclusion has not been altered by recent evidence. For example, canopy photosynthesis rates for 20 selected genotypes showed that values integrated over five measurement dates during reproductive development did not correlate with yield (Boerma and Ashley, 1988). Removal of maturity effects did, however, results in a significant partial correlation ($r^2 = 0.63$). A similar result was obtained in comparing cv. Tracy, Davis, and 38 progeny lines (Ashley and Boerma, 1989). Integration of canopy photosynthesis measurements on three dates during reproductive development resulted in significant partial correlations with yield ($r^2 = 0.53$) after removing the effect of maturity and year of test.

Thompson et al. (1995) returned to the idea of using specific leaf weight as a surrogate for leaf photosynthesis to select for increased leaf photosynthesis rate and yield. Crosses were made among eight plant introduction lines with extreme specific leaf weight, and 16 progeny lines were eventually selected for study. In the four environments in which the genotypes were tested, only one showed a correlation ($r^2 = 0.54$) between leaf photosynthesis and seed yield.

Selection for high photosynthesis rates as an independent variable no longer seems to be a high priority option. Efforts to increase yields based on photosynthetic measurements must seemingly be done in the context of the many factors that influence crop growth. Among these factors, the dynamics of N accumulation and partitioning to the leaves and the retention of N in the leaves are crucial contributions to C accumulation and yield increase.

11–3.2 Increasing Nitrogen Accumulation

The genetic approach to increasing yield and N_2 fixation has resulted in some positive results. Ronis et al. (1985) created two populations and measured yield and N_2 fixation in two environments of 170 individual plants. Nitrogen fixation was estimated by ^{15}N dilution of the same plants on which yield was measured. The heritability of the amount of N in the seed resulting from fixation was high. Further, the correlation between seed yield and the amount of N in the seed was extremely high (0.99) indicating a very close coupling between N and C accumulation. Because of this high correlation, Ronis et al. (1985) concluded that there was no need to select specifically for N_2 fixation beyond a selection for seed yield.

The correlation between N_2 fixation and seed yield obtained in a study of 210 plants from three crosses by Burias and Planchon (1990) was not as large as that reported by Ronis et al. (1985). In this more recent study, N_2 fixation was measured on F_2 plants grown in growth chambers on a nutrient solution and yield was measured on derived F_4 plants in the field. Nevertheless, after combining results from all crosses, the correlation between seed yield and N_2 fixation was highly significant ($r^2 = 0.29$). In a continuation of this investigation, F_6 filial progeny were studied for N_2 fixation activity and seed yield (Burias and Planchon, 1992). A high heritability was found for N_2 fixation. The correlation between yield of the F_6 plants and N_2 fixation measured for the F_2 plants was highly significant ($r^2 = 0.30$).

Similar results were reported by Pazdernik et al. (1997) from a study of 20 lines on a field site where N_2 fixation was the main source of accumulated N. There was a significant correlation between yield and total N accumulation up to 60 d after sowing ($r^2 = 0.26$) and between 60 d after sowing and the R6 stage ($r^2 = 0.30$). Further, there was a correlation between field ranking for yield and N accumulation with early nodulation observed in a growth chamber study (Pazdernik et al., 1996).

In a 2-yr field study with 63 lines, Shibles and Sundberg (1998) found significant correlations ($r^2 = 0.37$ and 0.31) between yield and total leaf N at stage R5 across all the lines. The difficulty, however, was that the rankings of genotypes were different between years leading Shibles and Sundberg to conclude that conventional breeding for yield increase based on N accumulation would be difficult. This conclusion is supported by an early comparison among 315 lines in two single-cross populations for yield and plant N accumulation at stage R5 (Zeinali-khanghah et al., 1993). Selection for plant N resulted in a very small yield increase that was concluded not to be of biological importance.

As discussed previously, attempts to increase N accumulation by sustained N_2 fixation activity in the presence of NO_3^- have yet to result in cultivars with increased yields. As an alternative, Herridge and Brockwell (1988) undertook a field test to explore options to manipulate the relative contribution of N_2 fixation and NO_3^- availability in an effort to maximize yield. All combinations of four levels of NO_3^- fertility and four levels of bacteria inoculation were applied. The highest amounts of crop N accumulation during reproductive development were with inoculation and no NO_3^- application, with very high levels of inoculation at 100 and 200 kg NO_3^- ha^{-1}, and with 300 kg NO_3^- ha^{-1} application. These same treatments also generally resulted in the highest yield. These results demonstrated that either source of N can result equally in maximizing yield.

Similar to photosynthesis, it seems that selection for improved N_2 fixation rates to increase yield has not given consistent positive responses. Interaction between N_2 fixation and other physiological processes in the plant make it difficult to simply select for this single trait. The challenge again appears to be to increase N_2 fixation within the framework of other physiological processes within the plant and cropping environment.

11–3.3 Limits to Yield Increase

11–3.3.1 Maximum Yield Limits

Assuming that either C accumulation or N accumulation could be increased independently, what are the maximum yield increases that can be realistically expected by altering either process? This question can be answered by calculating upper limits to yield using relatively simple assumptions. This approach calculates yield limits based on resource input (Sinclair, 1998). In this case, the maximum yield based on C input can be calculated from a dependence on the intercepted solar radiation resource and the maximum yield based on the N input can be calculated from the use of the N resource in producing seed mass.

Soybean mass accumulation is dependent directly on the quantity of intercepted solar radiation (I) multiplied by RUE. As discussed previously, the maximum RUE for soybean appears to be approximately 1.1 g MJ^{-1}. Seed yield is esti-mated by multiplying the mass accumulation by an estimate of harvest index (HI). For this analysis, it will be assumed that 0.55 approaches a realistic maximum HI for high-yielding soybean. Therefore, maximum soybean yield (Y_{max}) can be cal-culated from the following simple equation.

$$Y_{max} = HI \times RUE \times I = 0.55 \times 1.1 \times I \qquad [1]$$

Figure 11–5 is the linear plot of Y_{max} as a function of I as given in Eq. [1]. Assuming solar radiation intercepted over the growing season is 1200 MJ m^{-2}, Y_{max} is calcu-lated to be 726 g m^{-2}, or 7.3 t ha^{-1} (about 110 bu acre^{-1} in English units). This es-timate of Y_{max} is obviously a very high yield, but it is not greatly in excess of yields sometimes reported for experimental situations. Such calculations show that soy-bean yields are not unbounded, because the amount of solar radiation received and intercepted by a soybean crop through the season defines a rigid ceiling for yield increases.

Maximum yield based on N accumulation (N_a) can be calculated based on the amount of N that must be partitioned into developing soybean seeds. The par-titioning of N to the seeds can be expressed as the nitrogen harvest index (NHI), which is defined as the ratio of the amount of N in the seeds to the total N accu-mulated in the aboveground crop mass. For a highly productive soybean crop, NHI are likely to be no >0.8 at maturity. Soybean seeds commonly have a N content of about 65 mg N g^{-1} (G_N). Given the variables NHI and G_N, the following simple equa-tion gives an estimate of Y_{max} for soybean as a function of N_a.

$$Y_{max} = N_a \ NHI \ / \ G_N = N_a \ 0.8 \ / \ 0.065 \qquad [2]$$

Figure 11–6 is the linear plot of Y_{max} as a function of N_a as given in Eq. [2]. The very large amount of N that must be accumulated to support seed production in soy-

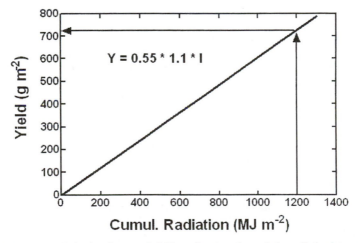

Fig. 11–5. Theoretical limit of soybean seed yield as a function of cumulative radiation intercepted dur-ing the growing season. This result was obtained using Eq. [1] in the text and assuming harvest index equal to 0.55 and radiation-use efficiency (RUE) equal to 1.1 g MJ^{-1}.

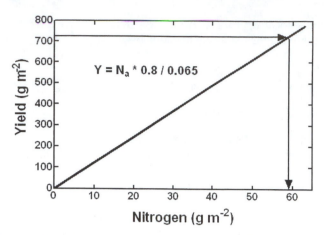

Fig. 11–6. Theoretical limit of soybean seed yield as a function of total N accumulation during the growing season. This result was obtained using Eq. [2] in the text and assuming nitrogen harvest index (NHI) equals 0.8 and soybean seed N concentration equals 65 mg N g^{-1} seed.

bean is clearly illustrated in this figure. For example, a highly productive cereal crop is likely to accumulate no more than 25 g N m^{-2} from the soil. If soybean accumulated only this amount from the soil or by N$_2$ fixation, then Y_{max} would be limited to 308 g m^{-2}, or 3.1 t ha^{-1}. This is a reasonable yield but certainly well below what is now produced by many growers. To achieve the previous Y_{max} estimate of 7.3 t ha^{-1} calculated for the radiation limitation example, nearly 59 g N m^{-2}, or 590 kg N ha^{-1}, would need to be accumulated by this soybean crop, whether by NO$_3^-$ uptake or by N$_2$ fixation. There is clearly a huge challenge in accumulating N from various sources to achieve sufficiently large values of N$_a$ required in the production of large Y_{max}.

11–3.3.2 Realistic Yield Limits

Equations [1] and [2] are helpful to set the upper boundaries of soybean yields but they are, of course, quite simplistic. No interaction is assumed in such simple analyses, which is in direct contrast to the conclusion of much of the reviewed research that C and N accumulation are mutually dependent. That is, C accumulation depends directly on the amount of N in the leaves, and N accumulation depends both on C supply to the roots for assimilation and on C in the shoots for metabolism of N products. The separation of C and N accumulation is artificial and it seems virtually impossible to identify one process as limiting to increased yield independent of the other. Sinclair (1989) previewed this conclusion when he argued, based on evidence available at that time, that C and N accumulation *simultaneously* limit increases in soybean yield.

In addition to the mutual dependence of C and N that has been implicit in much of the research already reviewed, several studies have explicitly demonstrated this interaction. For example, Imsande (1988) grew soybean plants in a chamber on nutrient solutions with differing N levels and obtained a close correspondence between N and mass accumulation among treatments. Similar results have been reported from field tests. In a comparison of lines with differing seed-filling periods, Smith et al. (1988) found that the high-yielding lines accumulated significantly more N and dry matter during seed filling than the low-yielding lines.

Muchow et al. (1993b) found across environments and harvest dates during the season that the accumulation of N was linearly correlated ($r^2 = 0.99$) with accumulation of aboveground mass. Hayati et al. (1995) removed shading from plants that had been previously shaded and found an increase in N accumulation corresponding to the time of shade removal. Vasilas and Fuhrmann (1993) reported that increased N_2 fixation rates resulting from the use of different bacteria inoculants were associated with both increased plant mass and seed yield. Further, in a comparison of N accumulation and use by eight soybean lines, Vasilas et al. (1995) found a consistent correlation during seed fill between seed yield and dry matter accumulation and N_2 fixation.

To consider conceptually the limitations of C and N accumulation on soybean yield, it is necessary to set aside the 'limiting-factor' concept originally proposed in 1840 by Liebig (Sinclair and Park, 1993). That is, yield must be considered within a systems text where interactive and mutual influence among processes are taken into consideration. Of course, this is a much more complicated and less intuitive perspective than the Liebig idea of looking for 'the shortest stave of the barrel'. To understand the contributions of various factors to yield increase, it is necessary to consider the complexity of C and N interactions.

Soybean models have been constructed to attempt quantifications of the interactions within the soybean system. This approach has been used effectively, for example, to examine the environmental conditions causing soybean yield variations in high-yielding experiments in Japan (Spaeth et al., 1987). Indeed, the simulation analysis of Spaeth et al. (1987) reproduced the yield variation across a number of years for soybean growth on both an upland field and a converted rice (*Oryza sativa* L.) paddy. The basis of the yield differences was attributed to variations in temperature and solar radiation to which the crops were exposed in various years. The simulations also mimicked the maximum yield of 649 g m^{-2} that was achieved 1 yr on a converted rice paddy. Based on the model analysis, Spaeth et al. (1987) concluded that high temperature early in the growing season and high seasonal levels of solar radiation resulted in the highest yields.

Boote et al. (2001) examined yield responses to changes in a number of 'genetic coefficients' that are incorporated into the relatively complex CROPGRO model. The feedback interactions between traits in this model resulted in very modest yield gains as a result of adjustments in any single trait. Increasing maximum leaf photosynthesis rate was found to have very little influence on yield and it was concluded that this trait is not "particularly important" for increasing yield. Nitrogen accumulation in the model is limited by various feedback relationships, and again large increases in yield were not simulated in this model by changing N levels in the plant. Their analysis clearly indicated the need to optimize simultaneously a number of traits in the soybean to achieve substantial yield gain.

Recognition of the close interaction between C and N accumulation only heightens the challenge in altering these processes to increase yield. Can research be integrated to consider the processes of C and N accumulation simultaneously in anything other than the well-tested, empirical approach of yield trials? Can the specific control points for the various processes be traced to specific events, or even identifiable proteins and genes? Even if this approach is successful, however, the challenge remains in integrating the change in the activity of a single protein so that it has the desired influence on a particular process, and even more challenging to achieve the expected response in the overall integrated process of yield formation.

11–4 SUMMARY

Carbon and N are clearly the fundamental building blocks required to form soybean yield. Each of these resources have to be accumulated in the plant over the growing season and finally partitioned to the seed to form yield. While research in recent years has resulted in considerable progress in understanding the processes influencing C and N accumulation, this knowledge has not yet led directly to yield increases. A possible limitation in achieving yield progress may have been a conceptual framework that has tended to consider C and N accumulation as independent processes.

Review of recent progress on C and N accumulation by soybean demonstrates a large mutual dependence between these two processes. Leaf photosynthetic rate and canopy radiation-use efficiency are both very closely linked to the amount of N in the leaves. The amount of photosynthetic apparatus and the regulation of the process are both highly dependent on the amount of N in the leaves. Substantial increases in C accumulation over either a few hours or a number of days seem possible only by increasing the amount of N available in the leaves and plant.

Similarly, N accumulation by either N_2 fixation or soil N uptake are closely linked to C accumulation rate. Nitrate uptake is directly dependent on the availability of photosynthate in the roots. Dinitrogen fixation appears to be closely tied to the accumulation of nitrogenous compounds in the shoot of the plant. Those treatments that are associated with an accumulation of leaf N result in decreases in both N_2 fixation rates and soil N uptake. More knowledge on the precise regulatory mechanisms of these processes may help sustain N accumulation under conditions that result in decreased N_2 fixation rates, in particular.

The critical challenge in the coming years will be understanding the mutual regulation, and ultimately, the yield limit imposed by C and N accumulation. Are there regulatory approaches in the plant that may allow decoupling of the overall processes of C and N accumulation so that these resources can be stored in the plant and eventually used in seed production? As a result of the high integration of C and N assimilation, it may be difficult to identify decoupling between the processes that may be advantageous. On the other hand, differences in sensitivity between the processes under stress conditions such as temperature or water deficit may offer an opportunity to bring the processes into balance under these conditions and increase yield under field conditions where these stresses are frequently encountered.

Resource availability imposes firm limits on achievable soybean yield. The amount of light available to fuel CO_2, N_2 fixation, NO_3^- uptake, and the metabolism of the C and N products into plant components ultimately constrains yield. Model analyses of soybean development and growth illustrate the dependence of the crop on environmental conditions and the difficulty in modifying only a single trait to increase yield. While possible approaches for additional yield increases are indicated in recent advances, a repeat of earlier quantum increases in soybean yield do not seem likely based on close analysis of the processes influencing the mutual interaction of C and N accumulation leading to seed production.

REFERENCES

Arrese-Igor, C., F.R. Minchin, A.J. Gordon, and A.K. Nath. 1997. Possible causes of the physiological decline in soybean nitrogen fixation in the presence of nitrate. J. Exp. Bot. 48:905–913.

Ashley, D.A., and H.R. Boerma. 1989. Canopy photosynthesis and its association with seed yield in advanced generations of a soybean cross. Crop Sci. 29:1042–1045.

Bacanamwo, M., and J.E. Harper. 1996. Regulation of nitrogenase activity in *Bradyrhizobium japonicum*/soybean symbiosis by plant N status as determined by shoot C:N ratio. Physiol. Plant. 98:529–538.

Bacanamwo, M., and J.E. Harper. 1997. The feedback mechanism of nitrate inhibition of nitrogenase activity in soybean may involve asparagine and/or products of its metabolism. Physiol. Plant. 100:371–377.

Betts, J.H., and D.F. Herridge. 1987. Isolation of soybean lines capable of nodulation and nitrogen fixation under high levels of nitrate supply. Crop Sci. 27:1156–1161.

Boerma, H.R., and D.A. Ashley. 1988. Canopy photosynthesis and seed-fill duration in recently developed soybean cultivars and selected plant introductions. Crop Sci. 28:137–140.

Boote, K.J., M.J. Kropff, and P.S. Bindraban. 2001. Physiology and modelling of traits in crop plants: Implications for genetic improvement. Agric. Syst. 70:395–420.

Bunce, J.A. 1991. Control of the acclimation of photosynthesis to light and temperature in relation to partitioning of photosynthate in developing soybean leaves. J. Exp. Bot. 240:853–859.

Burias, N., and C. Planchon. 1990. Increasing soybean productivity through selection for nitrogen fixation. Agron. J. 82:1031–1034.

Burias, N., and C. Planchon. 1992. Divergent selection for dinitrogen fixation and yield in soybean. Theor. Appl. Genet. 83:543–548.

Burkey, K.O., and R. Wells. 1991. Response of soybean photosynthesis and chloroplast membrane function to canopy development and mutual shading. Plant Physiol. 97:245–252.

Campbell, W.J., L.H. Allen, Jr., and G. Bowes. 1990. Response of soybean canopy photosynthesis to CO_2 concentration, light, and temperature. J. Exp. Bot. 41:427–433.

Carroll, B.J., D.L. McNeil, and P.M. Gresshoff. 1985. A supernodulation and nitrate-tolerant symbiotic (*nts*) soybean mutant. Plant Physiol. 78:34–40.

Crafts-Brandner, S.J., and D.B. Egli. 1987. Sink removal and leaf senescence in soybean. Plant Physiol. 85:662–666.

Daughtry, C.S.T., K.P. Gallo, S.N. Goward, S.D. Prince, and W.P. Kustas. 1992. Spectral estimates of absorbed radiation and phytomass production in corn and soybean canopies. Remote Sens. Environ. 39:141–152.

Day, D.A., G.D. Price, K.A. Schuller, and P.M. Gresshoff. 1987. Nodule physiology of a supernodulating soybean (*Glycine max*) mutant. Aust. J. Plant Physiol. 14:527–538.

Delhon, P., A. Gojon, P. Tillard, and L. Passama. 1995. Diurnal regulation of NO_3^- uptake in soybean plants. II. Relationship with accumulation of NO_3^- and asparagine in the roots. J. Exp. Bot. 46:1595–1602.

Delhon, P., A. Gojon, P. Tillard, and L. Passama. 1996. Diurnal regulation of NO_3^- uptake in soybean plants. IV. Dependence on current photosynthesis and sugar availability to the roots. J. Exp. Bot. 47:893–900.

Denison, R.F., and T.R. Sinclair. 1985. Diurnal and seasonal variation in dinitrogen fixation (acetylene reduction) rates by field-grown soybeans. Agron. J. 77:679–684.

de Silva, M., L.C. Purcell, and C.A. King. 1996. Soybean petiole ureide response to water deficits and decreased transpiration. Crop Sci. 36:611–616.

de Veau, E.J.I., J.M. Robinson, R.D. Warmbrodt, and D.F. Kremer. 1992. Photosynthate metabolism in the source leaves of N_2-fixing soybean plants. Plant Physiol. 99:1105–1117.

de Veau, E., J.M. Robinson, R.D. Warmbrodt, and P. van Berkum. 1990. Photosynthesis and photosynthate partitioning in N_2-fixing soybeans. Plant Physiol. 94:259–267.

deWit, C.T. 1965. Photosynthesis of leaf canopies. Agric. Res. Rep. 663. Inst. for Biological and Chemical Res. on Field Crops and Herbage, Wageningen, the Netherlands.

Djekoun, A., and C. Planchon. 1991. Water status effect on dinitrogen fixation and photosynthesis in soybean. Agron. J. 83:316–322.

Duncan, W.G., R.S. Loomis, W.A. Williams, and R. Hanau. 1967. A model for simulating photosynthesis in plant communities. Hilgardia 38:181–205.

Durand, J.L., J.E. Sheehy, and F.R. Minchin. 1987. Nitrogenase activity, photosynthesis and nodule water potential in soybean plants experiencing water deprivations. J. Exp. Bot. 38:311–321.

Eskew, D.L., J. Kapuya, and S.K.A. Danso. 1989. Nitrate inhibition of nodulation and nitrogen fixation by supernodulating nitrate-tolerant symbiosis mutants of soybean. Crop Sci. 29:1491–1496.

Fellows, R.J., R.P. Patterson, C.D. Raper, Jr., and D. Harris. 1987. Nodule activity and allocation of photosynthate of soybean during recovery from water stress. Plant Physiol. 84:456–460.

Ford, D.M., and R. Shibles. 1988. Photosynthesis and other traits in relation to chloroplast number during soybean leaf senescence. Plant Physiol. 86:108–111.

Fredeen, A.L., I.M. Rao, and N. Terry. 1989. Influence of phosphorus nutrition on growth and carbon partitioning in *Glycine max*. Plant Physiol. 89:225–230.

Fujita, K., T. Masuda, and S. Ogata. 1991. Analysis of factors controlling dinitrogen fixation in wild and cultivated soybean (*Glycine max*) plants by reciprocal grafting. Soil Sci. Plant Nutr. 37:233–240.

Gibson, A.H., and J.E. Harper. 1985. Nitrate effect on nodulation of soybean by *Bradyrhizobium japonicum*. Crop Sci. 25:497–501.

Goldschmidt, E.E., and S.C. Huber. 1992. Regulation of photosynthesis by end-product accumulation in leaves of plants storing starch, sucrose, and hexose sugars. Plant Physiol. 99:1443–1448.

González, E.M., L. Gálvez, and C. Arrese-Igor. 2001. Abscisic acid induces a decline in nitrogen fixation that involves leghaemoglobin, but is independent of sucrose synthase activity. J. Exp. Bot. 52:285–293.

González, E.M., A.J. Gordon, C.L. James, and C. Arrese-Igor. 1995. The role of sucrose synthase in the response of soybean nodules to drought. J. Exp. Bot. 46:1515–1523.

Gordon, A.J., F.R. Minchin, L. Skøt, and C.L. James. 1997. Stress-induced declines in soybean N_2 fixation are related to nodule sucrose synthase activity. Plant Physiol. 114:937–946.

Gremaud, M.F., and J.E. Harper. 1989. Selection and initial characterization of partially nitrate tolerant nodulation mutants of soybean. Plant Physiol. 89:169–173.

Hammer, G.L., and G.C. Wright. 1994. A theoretical analysis of nitrogen and radiation effects on radiation use efficiency in peanut. Aust. J. Agric. Res. 45:575–589.

Hansen, A.P., T. Yoneyama, and H. Kouchi. 1992. Short-term nitrate effects on hydroponically-grown soybean cv. Bragg and its supernodulating mutant. J. Exp. Bot. 43:1–7.

Hardarson, G., F. Zapata, and S.K.A. Danso. 1984. Effect of plant genotype and nitrogen fertilizer on symbiotic nitrogen fixation by soybean cultivars. Plant Soil 82:397–405.

Harper, J.E. 1987. Nitrogen metabolism. p. 497–533. *In* Soybeans: Improvement, production, and uses. 2nd ed. Agronomy Monogr.16. ASA, CSSA, and SSSA, Madison, WI.

Hartwig, U.A., I. Heim, A. Loscher, and J. Nosberger. 1994. The nitrogen sink is involved in the regulation of nitrogenase activity in white clover after defoliation. Physiol. Plant. 92:375–382.

Hayati, R., D.B. Egli, and S.J. Crafts-Brandner. 1995. Carbon and nitrogen supply during seed filling and leaf senescence in soybean. Crop Sci. 35:1063–1069.

Heckmann, M.-O., J.-J. Drevon, P. Saglio, and L. Salsac. 1989. Effect of oxygen and malate on NO_3^- inhibition of nitrogenase in soybean nodules. Plant Physiol. 90:224–229.

Herridge, D.F., and J. Brockwell. 1988. Contributions of fixed nitrogen and soil nitrate to the nitrogen economy of irrigated soybean. Soil Biol. Biochem. 20:711–717.

Herridge, D.F., and I.A. Rose. 1994. Heritability and repeatability of enhanced N_2 fixation in early and late inbreeding generations of soybean. Crop Sci. 34:360–367.

Imsande, J. 1986. Ineffective utilization of nitrate by soybean during pod fill. Physiol. Plant. 68:689–694.

Imsande, J. 1988. Enhanced nitrogen fixation increases net photosynthetic output and seed yield of hydroponically grown soybean. J. Exp. Bot. 39:1313–1321.

Imsande, J., and D.G. Edwards. 1988. Decreased rates of nitrate uptake during pod fill by cowpea, green gram, and soybean. Agron. J. 80:789–793.

Jiang, C.-Z., S.R. Rodermel, and R.M. Shibles. 1993. Photosynthesis, Rubisco activity and amount, and their regulation by transcription in senescing soybean leaves. Plant Physiol. 101:105–112.

Kao, W.-Y., and I.N. Forseth. 1992. Diurnal leaf movement, chlorophyll fluorescence and carbon assimilation in soybean grown under different nitrogen and water availabilities. Plant Cell Environ. 15:703–710.

Kerr, P.S., D.W. Israel, S.C. Huber, and T.W. Rufty, Jr. 1986. Effect of supplemental NO_3^- on plant growth and components of photosynthetic carbon metabolism in soybean (*Glycine max*). Can. J. Bot. 64:2020–2027.

King, C. A., and L.C. Purcell. 2001. Soybean nodule size and relationship to nitrogen fixation response to water deficit. Crop Sci. 41:1099–1107.

Kuo, T., and L. Boersma. 1971. Soil water suction and root temperature effects on nitrogen fixation in soybeans. Agron. J. 63:901–904.

Lauer, M.J., S.G. Pallardy, D.G. Blevins, and D.D. Randall. 1989. Whole leaf carbon exchange characteristics of phosphate deficient soybeans (*Glycine max* L.). Plant Physiol. 91:848–854.

Lauer, M.J., and R. Shibles. 1987. Soybean leaf photosynthetic response to changing sink demand. Crop Sci. 27:1197–1201.

Lugg, D.G., and T.R. Sinclair. 1981a. Seasonal changes in photosynthesis of field-grown soybean leaflets. 1. Relation to leaflet dimensions. Photosynthetica 15:129–137.

Lugg, D.G., and T.R. Sinclair. 1981b. Seasonal changes in photosynthesis of field-grown soybean leaflets. 2. Relation to nitrogen content. Photosynthetica 15:138–144.

Maloney, T.S., and E.S. Oplinger. 1997. Yield and nitrogen recovery from field-grown supernodulating soybean. J. Prod. Agric. 10:418–424.

Maury, P., S. Suc, M. Berger, and C. Planchon. 1993. Response of photochemical processes of photosynthesis to dinitrogen fixation in soybean. Plant Physiol. 101:493–497.

Millhollon, E.P., and L.E. Williams. 1986. Carbohydrate partitioning and the capacity of apparent nitrogen fixation of soybean plants grown outdoors. Plant Physiol. 81:280–284.

Morcuende, R., P. Pérez, R. Martínez-Carrasco, I.M. del Molino, and L.S. de la Puente. 1996. Long- and short-term responses of leaf carbohydrate levels and photosynthesis to decreased sink demand in soybean. Plant Cell Environ. 19:976–982.

Muchow, R.C., M.J. Roberston, and B.C. Pengelly. 1993a. Radiation-use efficiency of soybean, mungbean and cowpea under different environmental conditions. Field Crops Res. 32:1–16.

Muchow, R.C., M.J. Roberston, and B.C. Pengelly. 1993b. Accumulation and partitioning of biomass and nitrogen by soybean, mungbean and cowpea under contrasting environmental conditions. Field Crops Res. 33:13–36.

Nakaseko, K., and K. Gotoh. 1983. Comparative studies on dry matter production, plant type and productivity in soybean, azuki bean and kidney bean. VII. An analysis of the productivity among the three crops on the basis of radiation absorption and its efficiency for dry matter accumulation. Jpn. J. Crop Sci. 52:49–58.

Neo, H.H., and D.B. Layzell. 1997. Phloem glutamine and the regulation of O_2 diffusion in legume nodules. Plant Physiol. 113:259–267.

Neuhausen, S.L., P.H. Graham, and J.H. Orf. 1988. Genetic variation for dinitrogen fixation in soybean of Maturity Group 00 and 0. Crop Sci. 28:769–772.

Parsons, R., A. Stanforth, J.A. Raven, and J.I. Sprent. 1993. Nodule growth and activity may be regulated by a feedback mechanism involving phloem nitrogen. Plant Cell Environ. 16:125–136.

Pate, J.S. 1985. Partitioning of carbon and nitrogen in N_2-fixing grain legumes. p. 715–727. *In* R. Shibles (ed.) World Soybean Res. Conf. III: Proc. Westview Press, Boulder, CO.

Paul, M.J., and C.H. Foyer. 2001. Sink regulation of photosynthesis. J. Exp. Bot. 52:1383–1400.

Pazdernik, D.L., P.H. Graham, and J.H. Orf. 1997. Variation in the pattern of nitrogen accumulation and distribution in soybean. Crop Sci. 37:1482–1486.

Pazdernik, D.L., P.H. Graham, C.P. Vance, and J.H. Orf. 1996. Host variation in traits affecting early nodulation and dinitrogen fixation in soybean. Crop Sci. 36:1102–1107.

Pearcy, R.W., J.S. Roden, and J.A. Gamon. 1990. Sunfleck dynamics in relation to canopy structure in a soybean (*Glycine max* (L.) Merr.) canopy. Agric. For. Meteorol. 52:359–372.

Pearcy, R.W., and J.R. Seemann. 1990. Photosynthetic induction state of leaves in a soybean canopy in relation to light regulation of ribulose-1-5-bisphosphate carboxylase and stomatal conductance. Plant Physiol. 94:628–633.

Pracht, J.E., C.D. Nickell, J.E. Harper, and D.G. Bullock. 1994. Agronomic evaluation of non-nodulating and hypernodulating mutants of soybean. Crop Sci. 34:738–740.

Purcell, L.C., R.A. Ball, J.D. Reaper, III, and E.D. Vories. 2002. Radiation use efficiency and biomass production in soybean at different plant population densities. Crop Sci. 42:172–177..

Purcell, L.C., and C.A. King. 1996. Drought and nitrogen source effects on nitrogen nutrition, seedgrowth, and yield in soybean. J. Plant Nutr. 19:969–993.

Purcell, L.C., R. Serraj, M. de Silva, T.R. Sinclair, and S. Bona. 1998. Ureide concentration of fieldgrown soybean in response to drought and the relationship to nitrogen fixation. J. Plant Nutr. 21:949–966.

Purcell, L.C., and T.R. Sinclair. 1995. Nodule gas exchange and water potential response to rapid imposition of water deficit. Plant Cell Environ. 18:179–187.

Qiu, J., and D.W. Israel. 1992. Diurnal starch accumulation and utilization in phosphorus-deficient soybean plants. Plant Physiol. 98:316–323.

Raffin, A., P. Roumet, and M. Obaton. 1995. Tolerance of nitrogen fixation to nitrate in soybean: A progeny (tolerant × non-tolerant) evaluation. Eur. J. Agron. 4:143–149.

Raven, J.A., J.I. Sprent, S.G. McInroy, and G.T. Hay. 1989. Water balance of N_2-fixing root nodules: Can phloem and xylem transport explain it? Plant Cell Environ. 12:683–688.

Robinson, J.M. 1996. Leaflet photosynthesis rate and carbon metabolite accumulation patterns in nitrogen-limited, vegetative soybean plants. Photosynth. Res. 50:133–148.

Robinson, J.M. 1997. Influence of daily photosynthetic photon flux density on foliar carbon metabolite levels in nitrogen-limited soybean plants. Int. J. Plant Sci. 158:32–43.

Robinson, J.M., and K.O. Burkey. 1997. Foliar CO_2 photoassimilation and chloroplast linear electron transport rates in nitrogen-sufficient and nitrogen-limited soybean plants. Photosynth. Res. 54:209–217.

Rochette, P., R.L. Desjardins, E. Pattey, and R. Lessard. 1995. Crop net carbon dioxide exchange rate and radiation use efficiency in soybean. Agron. J. 87:22–28.

Ronis, D.H., D.J. Sammons, W.J. Kenworthy, and J.J. Meisinger. 1985. Heritability of total and fixed N content of the seed in two soybean populations. Crop Sci. 25:1–4.

Rufty, T.W., Jr., S.C. Huber, and R.J. Volk. 1988. Alterations in leaf carbohydrate metabolism in response to nitrogen stress. Plant Physiol. 88:725–730.

Sa, T-M., and D.W. Israel. 1991. Energy status and functioning of phosphorus-deficient soybean nodules. Plant Physiol. 97:928–935.

Sa, T-M., and D.W. Israel. 1995. Nitrogen assimilation in nitrogen-fixing soybean plants during phosphorus deficiency. Crop Sci. 35:814–820.

Saravitz, C., J.W. Rideout, and C.D. Raper, Jr. 1994. Nitrogen uptake and partitioning in response to reproductive sink size of soybean. Int. J. Plant Sci. 155:730–737.

Sassenrath-Cole, G.F., and R.W. Pearcy. 1992. The role of ribulose-1,5-bisphosphate regeneration in the induction requirement of photosynthetic CO_2 exchange under transient light conditions. Plant Physiol. 99:227–234.

Sassenrath-Cole, G.F., and R.W. Pearcy. 1994. Regulation of photosynthetic induction state by the magnitude and duration of low light exposure. Plant Physiol. 105:1115–1123.

Sawada, S., S. Enomoto, T. Tozu, and M. Kasai. 1995. Regulation of the activity of ribulose-1,5-bisphosphate carboxylase in response to changes in the photosynthetic source-sink balance in intact soybean plants. Plant Cell Physiol. 36:551–556.

Sawada, S., H. Usuda, and T. Tsukui. 1992. Participation of inorganic orthophosphate in regulation of the ribulose-1,5-bisphosphate carboxylase activity in response to changes in the photosynthetic source-sink balance. Plant Cell Physiol. 33:943–949.

Serraj, R., J.-J. Drevon, M. Obaton, and A. Vidal. 1992. Variation in nitrate tolerance of nitrogen fixation in soybean (*Glycine max*) -*Bradyrhizobium* symbiosis. J. Plant. Physiol. 140:366–371.

Serraj, R., and T.R. Sinclair. 1996. Processes contributing to N_2-fixation insensitivity to drought in the soybean cultivar Jackson. Crop Sci. 36:961–968.

Serraj, R., and T.R. Sinclair. 1997. Variation among soybean cultivars in dinitrogen fixation response to drought. Agron. J. 89:963–969.

Serraj, R., and T.R. Sinclair. 2003. Evidence that carbon dioxide enrichment alleviates ureide-induced decline of nodule nitrogenase activity. Ann. Bot. 91:85–89..

Serraj, R., T.R. Sinclair, and L.H. Allen. 1998. Soybean nodulation and N_2 fixation response to drought under carbon dioxide enrichment. Plant Cell Environ. 21:491–500.

Serraj, R., T.R. Sinclair, and L.C. Purcell. 1999a. Symbiotic N_2 fixation response to drought. J. Exp. Bot. 50:143–155.

Serraj, R., V. Vadez, R.F. Denison, and T.R. Sinclair. 1999b. Involvement of ureides in nitrogen fixation inhibition in soybean. Plant Physiol. 119:289–296.

Shibles, R., J. Secor, and D.M. Ford. 1987. Carbon assimilation and metabolism. p. 535–588. *In* Soybeans: Improvement, production, and uses. 2nd ed. Agronomy Monogr.16. ASA, CSSA, and SSSA, Madison, WI.

Shibles, R., and D.N. Sundberg. 1998. Relation of leaf nitrogen content and other traits with seed yield in soybean. Plant Prod. Sci. 1:3–7.

Sinclair, T.R. 1986. Water and nitrogen limitations in soybean grain production. I. Model development. Field Crops Res. 15:125–141.

Sinclair, T.R. 1989. Simultaneous limitation to soybean yield increase by carbon and nitrogen. p. 183–188. *In* A.J. Pascale (ed.) World Soybean Res. Conf., Buenos Ares, Argentina. 5–9 Mar. 1989. Asociaion Argentina de la Soja, Buenos Aires, Argentina.

Sinclair, T.R. 1998. Options for sustaining and increasing the limiting yield-plateaus of grain crops. Jpn. J. Crop Sci. 67:65–75.

Sinclair, T.R., and T. Horie. 1989. Leaf nitrogen, photosynthesis, and crop radiation use efficiency: A review. Crop Sci. 29:90–98.

Sinclair, T.R., and R.C. Muchow. 1999. Radiation use efficiency. Adv. Agron. 65:215–266.

Sinclair, T.R., R.C. Muchow, J.M. Bennett, and L.C. Hammond. 1987. Relative sensitivity of nitrogen and biomass accumulation to drought in field-grown soybean. Agron. J. 79:986–991.

Sinclair, T.R., and W.I. Park. 1993. Inadequacy of the Liebig limiting-factor paradigm for explaining varying crop yields. Agron. J. 85:742–746.

Sinclair, T.R., and R. Serraj. 1995. Dinitrogen fixation sensitivity to drought among grain legume species. Nature (London) 378:344.

Sinclair, T.R., and J.E. Sheehy. 1999. Erect leaves and photosynthesis in rice. Science (Washington DC) 283:1456–1457.

Sinclair, T.R., and T. Shiraiwa. 1993. Soybean radiation-use efficiency as influenced by nonuniform specific leaf nitrogen distribution and diffuse radiation. Crop Sci. 33:808–812.

Smith, J.R., R.L. Nelson, and B.L. Vasilas. 1988. Variation among soybean breeding lines in relation to yield and seed-fill duration. Agron. J. 80:825–829.

Spaeth, S.C., T.R. Sinclair, T. Ohnuma, and S. Konno. 1987. Temperature, radiation, and duration dependence of high soybean yields: Measurement and simulation. Field Crops. Res. 16:297–307.

Streeter, J. 1988. Inhibition of legume nodule formation and N_2 fixation by nitrate. Crit. Rev. Plant Sci. 7:1–23.

Sprent, J.I. 1980. Root nodule anatomy, type of export product and evolutionary origin in some *Leguminosae*. Plant Cell Environ. 3:35–43.

Streeter, J.G. 1985. Nitrate inhibition of legume nodule growth and activity. II. Short term studies with high nitrate supply. Plant Physiol. 77:325–328.

Streeter, J.G. 1993. Translocation—A key factor limiting the efficiency of nitrogen fixation in legume nodules. Physiol. Plant. 87:616–623.

Streeter, O.G., and S.O. Salminen. 1993. Alterations in apoplastic and total solute concentrations in soybean nodules resulting from treatments known to affect gas diffusion. J. Exp. Bot. 44:821–828.

Sung, F.J.M., and J.J. Chen. 1989. Changes in photosynthesis and other chloroplast traits in lanceolate leaflet isoline of soybean. Plant Physiol. 90:773–777.

Suwignyo, R.A., A. Nose, Y. Kawamitsu, M. Tsuchiya, and K. Wasano. 1995. Effects of manipulations of source and sink on the carbon exchange rate and some enzymes of sucrose metabolism in leaves of soybean [*Glycine max* (L.) Merr.]. Plant Cell Physiol. 36:1439–1446.

Thompson, J.A., R.L. Nelson, and L.E. Schweitzer. 1995. Relationships among specific leaf weight, photosynthetic rate, and seed yield in soybean. Crop Sci. 35:1575–1581.

Tolley-Henry, L., and C.D. Raper, Jr. 1986. Expansion and photosynthetic rate of leaves of soybean plants during onset of and recovery from nitrogen stress. Bot. Gaz. 147:400–406.

Tolley-Henry, L., C.D. Raper, Jr., and J.W. Rideout. 1992. Onset of and recovery from nitrogen stress during reproductive growth of soybean. Int. J. Plant Sci. 153:178–185.

Touraine, B., B. Muller, and C. Grignon. 1992. Effect of phloem-translocated malate on NO_3^- uptake by roots of intact soybean plants. Plant Physiol. 99:1118–1123.

Vadez, V., T.R. Sinclair, and R. Serraj. 2000a. Asparagine and ureide accumulation in nodules and shoots as feedback inhibitors of N_2 fixation in soybean. Physiol. Plant. 110:215–223.

Vadez, V., T.R. Sinclair, R. Serraj, and L.C. Purcell. 2000b. Manganese application alleviates the water deficit-induced decline of N_2 fixation. Plant Cell Environ. 23:497–505.

Vasilas, B.L., and J.L. Fuhrmann. 1993. Field response of soybean to increased dinitrogen fixation. Crop Sci. 33:785–787.

Vasilas, B.L., R.L. Nelson, J.J. Fuhrmann, and T.A. Evans. 1995. Relationship of nitrogen utilization patterns with soybean yield and seed-fill period. Crop Sci. 35:809–813.

Vessey, J.K., C.D. Raper, Jr., and L. Tolley-Henry. 1990. Cyclic variations in nitrogen uptake rate in soybean plants: Uptake during reproductive growth. J. Exp. Bot. 41:1579–1584.

Vessey, J.K., K.B. Walsh, and D.B. Layzell. 1988. Oxygen limitation of N_2 fixation in stem-girdled and nitrate-treated soybean. Physiol. Plant. 73:113–121.

Vessey, J.K., and J. Waterer. 1992. In search of the mechanism of nitrate inhibition of nitrogenase activity in legume nodules: Recent developments. Physiol. Plant. 84:171–176.

Vidal, R., A. Gerbaud, and D. Vidal. 1996. Short-term effects of high light intensities on soybean nodule activity and photosynthesis. Environ. Exp. Bot. 36:349–357.

Vidal, R., A. Gerbaud, D. Vidal, and J.-J. Drevon. 1995. A short-term decrease in nitrogenase activity
(C_2H_2 reduction) is induced by exposure of soybean shoots to their CO_2 compensation point.
Plant Physiol. 108:1455–1460.

Walsh, K.B. 1995. Physiology of the legume nodule and its response to stress. Soil Biol. Biochem.
27:637–655.

Walsh, K.B., M.J. Canny, and D.B. Layzell. 1989. Vascular transport and soybean nodule function: II.
A role for phloem supply in product export. Plant Cell Environ. 12:713–723.

Walsh, K.B., M.R. Thorpe, and P.E.H. Minchin. 1998. Photoassimilate partitioning in nodulated soybean. II. The effect of changes in photoassimilate availability shows that nodule permeability
to gases is not linked to the supply of solutes or water. J. Exp. Bot. 49:1817–1825.

Walsh, K.B., J.K. Vessey, and D.B. Layzell. 1987. Carbohydrate supply and N_2 fixation in soybean. Plant
Physiol. 85:137–144.

Weisz, P.R., and T.R. Sinclair. 1988. Soybean nodule gas permeability, nitrogen fixation and diurnal cycles in soil temperature. Plant Soil 109:227–234.

Wells, R. 1991. Soybean growth response to plant density: Relationships among canopy photosynthesis, leaf area, and light interception. Crop Sci. 31:755–761.

Wiersma, J.V., and J.H. Orf. 1992. Early maturing soybean nodulation and performance with selected
Bradyrhizobium japonicum strains. Agron. J. 84:449–458.

Winkler, R.G., D.G. Blevins, J.C. Polacco, and D.D. Randall. 1987. Ureide catabolism in soybeans. II.
Pathway of catabolism in intact leaf tissue. Plant Physiol. 83:585–591.

Wittenbach, V.A., R.C. Ackerson, R.T. Giaquinta, and R.R. Hebert. 1980. Changes in photosynthesis,
ribulose bisphosphate carboxylase, proteolytic activity, and ultrastructure of soybean leaves during senescence. Crop Sci. 20:225–231.

Wu, S., and J.E. Harper. 1991. Dinitrogen fixation potential and yield of hypernodulating soybean mutants: A field evaluation. Crop Sci. 31:1233–1240.

Yinbo, G., M.B. Peoples, and B. Rerkasem. 1997. The effect of N fertilizer strategy on N_2 fixation, growth
and yield of vegetable soybean. Field Crops Res. 51:221–229.

Zapata, F., S.K.A. Danso, G. Hardarson, and M. Fried. 1987. Time course of nitrogen fixation in field-
grown soybean using nitrogen-15 methodology. Agron. J. 79:172–176.

Zeinali-khanghah, H., D.E. Green, and R.M. Shibles. 1993. Use of morphological, developmental, and
plant nitrogen traits in a selection scheme in soybean. Crop Sci. 33:1121–1127.

Zhang, F., M. Dijak, D.L. Smith, J. Lin, K. Walsh, H. Voldeng, F. Macdowell, and D.B. Layzell. 1997.
Nitrogen fixation and nitrate metabolism for growth of six diverse soybean [*Glycine max* (L.)
Merr.] genotypes under low temperature stress. Environ. Exp. Bot. 38:49–60.

Zhu, Y., K.R. Schubert, and D.H. Kohl. 1991. Physiological responses of soybean plants grown in a ni-
trogen-free or energy limited environment. Plant Physiol. 96:305–309.

12 Physiological Traits for Ameliorating Drought Stress

LARRY C. PURCELL

University of Arkansas
Fayetteville, Arkansas

JAMES E. SPECHT

University of Nebraska
Lincoln, Nebraska

In Chapter 15 (entitled Stress Physiology) of the second edition of the Soybean monograph, Raper and Kramer (1987) provided soybean [*Glycine max* L. (Merr.)] researchers with an excellent review of the physiological responses of soybean to abiotic stresses of temperature, water, light, carbon dioxide (CO_2)concentration, and metal toxicity. That same year, Hale and Orcutt (1987) published a comprehensive review of plant stress physiology that covered all plant species. In the 15 yr since then, research in this subject area has continued, leading to new information, concepts, and theories. Thus, a revisit of the topic of soybean stress physiology was deemed appropriate for this, the third edition of the Soybean monograph.

Nilsen and Orcutt (1996) reviewed the physiology of plants experiencing the stresses of radiation deficit or excess, drought or flooding, and freezing, chilling, or high temperatures. These authors, in a second volume, extended that review to the physiology of plants subjected to abiotic stresses associated with soil processes (e.g., nutrient deficiencies or toxicities and salinity), biotic factors, and anthropogenic pollutants in the soil and atmosphere (Orcutt and Nilsen, 2000). Research on the molecular biology underlying plant response to stress was discussed in a chapter (Bray et al., 2000) of the recently published tome on plant biochemistry and molecular biology. In the preparation and writing of this chapter, it was not our goal to reiterate the information presented in the aforementioned treatises. Instead, we elected to focus on physiological mechanisms and traits involved in the *yield response* of a soybean crop subjected to stress. Our perspective in this review was thus *agro*-physiological (i.e., plant biomass and grain yield), rather than *eco*-physiological (i.e., plant survival or competitiveness). Given this perspective, we consider a stress-related trait to be of value only if its manipulation (via genetic and/or agronomic means) is possible, and then only if such manipulation can lessen the reduction in soybean yield caused by the stress. Our goals are that this review will be of broad interest to scientists in the public and private sectors and to crop consultants. Furthermore, we hope that the perspective presented will stimulate discussion about the agronomic value of traits and provide a useful framework for their critical evaluation.

12–1 SOME POINTS TO CONSIDER

A few other comments are in order before commencing this review. First, drought is a meteorological phenomenon that often results in injury to crops grown in field environments. We have distinguished drought stress from water-deficit stress, which is an imposed treatment usually in a greenhouse of growth chamber. Although water-deficit treatments in controlled greenhouse or growth chamber experiments may provide meaningful insights to the drought response of plants grown in field environments, the results should be interpreted within their domain of relevance.

Second, it is critical to distinguish between an injury (e.g., *symptom*) induced in the plant by the stress, and a phenotypic response (e.g., *acclimation*[1]) induced in the plant to mitigate the injury or avoid the stress. The distinction is not always obvious, but it is certainly not trivial. Consider, for example, stress-induced leaf abscission. Is this induced trait *symptomatic* or *acclimative*? If the crop response could be altered, would one select for decreased symptomology (i.e., lessened, slower, or no leaf abscission upon exposure to stress) or for increased acclimative capacity (i.e., genotypes with a greater and/or faster leaf abscission upon exposure to stress)? Clearly, a fundamental understanding of the complexities of a stress-related trait is necessary in assessing the utility of that trait in selecting for stress tolerance.

Third, it is well known that specific traits, when altered, can indeed mitigate a yield reduction resulting from a specific stress (Blum, 1988, 1996). However, in many instances, such an alteration may prevent or lessen a yield increase under favorable conditions (Richards, 1996). A common example is the selective use of early-maturity genes to create shorter season cultivars that are less susceptible to yield reduction caused by terminal drought. Muchow and Sinclair (1986) showed that when water supply was limited to the extent that the yields of soybean cultivars of normal maturity were reduced by 40%, the yield reduction for early-maturing cultivars was less. However, early-maturing cultivars typically yield less in high rainfall years, because normal-maturing cultivars have more developmental time to exploit the abundance of water in those years. Such a trade-off is not unusual in the search for drought tolerant traits. For example, the most common plant response to an increasing soil water deficit is a reduction in growth, more so in the shoot than in the root, which leads to an increasing root/shoot dry matter ratio (Blum, 1996). Thus, a plant species or genotype that *constitutively* allocates more of its accumulating dry matter to roots (at the expense of shoots) will yield better under drought, assuming that this leads to a deeper and/or denser root system that can gather more soil water (Taylor and Klepper, 1978). On the other hand, the natural response of

---

[1] Acclimation and adaptation are *not* synonymous terms (Jones et al., 1989; Nilsen and Orcutt, 1996). Acclimation is a phenotypic change (e.g., hardening) arising from a stress-mediated induction of the regulatory genes governing signal transduction pathways specific for that stress. Adaptation is a change in gene frequency, mediated by evolution or breeding, leading to a population (or individual) expressing phenotypes offering a better (adaptive) fit to an evolving (or targeted) environment. The *capacity* for acclimation, of course, can be altered by changes in gene frequency.

most crop plants to soil water abundance is a proportionately greater allocation of dry matter growth to the shoots (Monneveux and Belhassen, 1996), a response that is clearly incompatible with any constitutive forcing of more dry matter into roots. Given that soybean seed dry matter closely tracks total aboveground dry matter over a range of seasonal soil water amounts (Spaeth et al., 1984), forcing more dry matter into roots could lead to less aboveground dry matter in high rainfall years, thereby compromising yield potential in those seasons.

The two foregoing examples reflect the fact that some traits represent a *conservative* means of dealing with varied seasonal water amounts (i.e., *stable yield*), while other traits represent an *opportunistic* means of dealing with that same variation (i.e., *potential yield*). Ludlow and Muchow (1990) and Evans and Fischer (1999) have noted that yield-stable cultivars are of benefit to risk-adverse low-input producers whose economic future is uncertain should their low crop yields be made even lower by stress. However, much of the world's soybean hectarage—perhaps nearly all of that in the USA—falls into the category of *intensified* (modern), rather than *subsistence* (low yield, hand harvest, etc.) agriculture.

Would U.S. soybean producers freely adopt a genetic (or agronomic) alteration purported to improve stress tolerance if, in years more favorable than normal, it renders a *yield penalty* that is *significantly greater* than the *yield premium* it renders in years more unfavorable than normal? Consider the 1983 yields in water-scarce and water-abundant 1983 Nebraska environments reported by Specht et al. (1986) for the drought-insensitive cv. Williams 82 (i.e., 1460 and 3300 kg ha^{-1}, respectively) and the drought-sensitive cv. Asgrow A3127 (970 and 4130 kg ha^{-1}, respectively). The yield *advantage* (i.e., 490 kg ha^{-1}) of Williams 82 in the drier environment translates into an economic *premium* (at a \$180 Mg^{-1} soybean price) of \$90.16 ha^{-1}. However, its yield *disadvantage* (i.e., 830 kg ha^{-1}) in the wetter environment translates into an economic *penalty* of \$151.90 ha^{-1}. If the contrasting seasonal water environments were equally probable (but still not predictable), then choosing to grow the yield-stable Williams 82 entails an "upside" cost (opportunity risk) that on average would be \$62 ha^{-1} more than the "downside" cost (stability risk) of choosing to grow Asgrow A3127. Clearly, any physiological manipulation purported to be a possible means of improving stress tolerance must always be critically examined to determine its impact, absent the given stress, on yield potential.

Our fourth and final point is that large intraseason and interannual variability in rainfall amount and distribution is a ubiquitous feature of nearly every soybean production area. Of all of the abiotic factors whose seasonal variability impacts crop productivity, insufficiency in the amount of water available for crop transpiration is fundamentally the most recurrent, if not annually prevalent, reducer of crop yield (Boyer, 1982; Waggoner, 1994). In that the uptake of atmospheric CO_2 in terrestrial plants allows the escape of water vapor from a humid leaf interior, the processes of photosynthesis and transpiration are inseparably coupled. Therefore, crop biomass accumulation (and grain yield), in any given season, is directly related to crop transpiration (Sinclair, 1994, 1998). If transpiration is impeded because of inadequate water, crop biomass accumulation invariably will be lessened. For that reason, much of the emphasis in this chapter is directed towards the physiology of soybean response to drought stress and plant mechanisms and traits that might be genetically modified to ameliorate the yield-reducing effects of that stress. The reader is advised to refer to the literature mentioned above relative to the physiology of other kinds of stresses.

12–2 CROP YIELD POTENTIAL—A REFERENCE POINT

To fully comprehend the impact of water stress on crop productivity, it is necessary to examine what crop yield might be in the absence of any stress. The *maximum* yield potential of a given crop species is set by the total amount of solar radiation a crop can maximally intercept during a growing season and the maximum efficiency with which it uses that solar energy to acquire, fix, and reduce C and N (Sinclair, 1994). While maximum yield potential is useful in crop species comparisons, of more practical value is a crop's *attainable* yield potential. Evans and Fischer (1999) defined it as *the yield of a cultivar when grown in an environment in which it is adapted, with water and nutrients nonlimiting, and pests, diseases, weeds, lodging, and other stresses effectively controlled.* Implicit in the cultivar, environment, and adaptation terms used in this definition is the notion that long-term selection by breeders has generated cultivars with reasonably adaptive phenological fits to the photo- and thermo-periodic patterns of the target environment.

Sinclair theorized that the *maximum* grain yield potential of soybean was probably no greater than 7300 (Sinclair, 2004, this publication) to 8000 (1998) kg ha^{-1}. Specht et al. (1999), using an alternate line of reasoning, came to the same conclusion, but noted that the *attainable* grain yield potential of soybean was probably 6660 kg ha^{-1}—the highest verifiable yield reported to date in the irrigated production category of Nebraska yield contests. The attainable grain-yield estimate of Specht et al. (1999) is similar to the value of 7000 kg ha^{-1} suggested in an earlier publication by Sinclair (1998). His estimate originated from an assumption that 500 mm of seasonal water was probably the maximum available to a crop freely transpiring into an atmosphere of a temperate region where the daily vapor pressure deficit (vpd) might average about 2.0 kPa.

During the last 5 yr of the 20th century, annual U.S. soybean yields averaged only 2550 kg ha^{-1}. Thus, less than two-fifths (38%) of the *attainable* soybean yield potential is currently being *realized* on the average U.S. farm. It is of interest to estimate just how much of the unrealized yield differential might be due to drought.

12–3 DROUGHT—ITS IMPACT ON SEED YIELD

Some 20 yr ago, Boyer (1982) observed that average on-farm U.S. crop yield was just 20% or so of the (then) attainable crop yield potential. He attributed much (69%) of the total unrealized yield differential (80%) to abiotic stresses arising from unfavorable climatic (meteorological) and edaphic (soil-related) conditions, with the remainder (11%) attributable to biotic stresses (pathogens, insects, and weeds). He also noted that soils with low water availability (based on soil type and/or region) constituted 45% of the U.S. land surface, and that from 1939 to 1978, drought had accounted for 41% of the insurance indemnities for crop losses.

Repetitive occurrences of severe drought have continued to plague U.S. crop production. Ross and Lott (2000) noted that in 18 of the 20 yr between 1980 to 1999, the USA experienced 46 weather-related disasters. A disaster was defined as causing a loss of a billion dollars or more (normalized to 1998 dollars). Hurricanes and tropical storms were the most frequent (14 of the 46), but these disasters accounted for only 30% of the 20-yr total of economic damages. In contrast, *43% of the total damages* arising from the 46 USA weather-related disasters were attributable to eight major droughts that occurred during the 20-yr period. After the Ross and Lott report, another drought occurred in 2000, making it nine major U.S. droughts in 22 yr. Given the frequency and vast economic impact of drought, it is not surprising that the National Climate Data Center now offers a web site that monitors the development of U.S. droughts on a weekly basis (http://lwf.ncdc.noaa.gov/oa/climate/research/dm/weekly-DM-animations.html).

The impact of drought on soybean crop production obviously depends upon the given soybean production region and reference time span. Specht et al. (1999) noted that the average difference between the yield trend lines for irrigated and rainfed soybean production in Nebraska from 1972 to 1997 was about 800 kg ha^{-1}. However, this differential cannot be entirely water-related, because in high rainfall years the differential did not go to zero (Fig. 12–1). Instead, it narrowed no closer than about 270 kg ha^{-1}—a differential likely attributable to the more intense management of irrigated production systems. Another noteworthy feature of the data presented in Fig. 12–1 is that the annual variability of irrigated yields ($R^2 = 0.73$) was just one-third of that of rainfed yields ($R^2 = 0.22$). Specht et al. (1999) determined that the magnitude of the annual differential between irrigated and rainfed

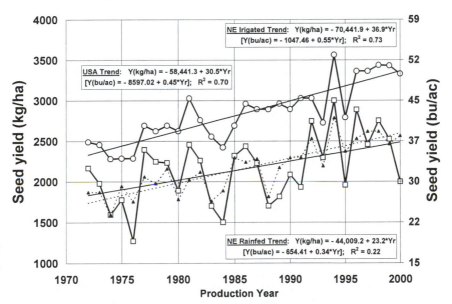

Fig. 12–1. Average annual seed yield for U.S. soybean production (solid triangles) and for irrigated (open circles) and rainfed (open squares) Nebraska soybean production from 1972 to 2000. Modified (with permission) from panel A of Fig. 2 in Specht et al. (1999), using updated yield data obtained from a web site operated by the National Agricultural Statistics Service (http://www.nass.usda.gov:81/ipedb/).

yields was negatively associated with the annual rainfed yield (i.e., $R^2 = 0.36$; $b = -0.39$), but detected no relationship of the annual irrigated-rainfed yield differential with annual irrigated yield (i.e., $R^2 = 0$). These data indicate that, in Nebraska at least, insufficient and variable seasonal rainfall has a substantial impact on rainfed soybean yields.

12–4 ASSESSING TRAIT UTILITY IN WATER-LIMITED ENVIRONMENTS

A now well-accepted framework for evaluating the value of plant traits and physiological mechanisms in water-limited environments is that first proposed by Passioura (1977, 1994, 1996) and subsequently used in the reviews by Ludlow and Muchow (1990) and Turner et al. (2001). This framework treats water as a resource available for crop production, thus making grain yield (Y) in water-limited environments a function of three largely independent entities, amount of water transpired (T), water-use efficiency (WUE), and harvest index (HI):

$$Y = T \times WUE \times HI \qquad [1]$$

In Eq. [1], Y is grain yield (g grain m^{-2}) and T refers to the total quantity of water (g water m^{-2}) transpired over the course of crop progression from emergence to physiological maturity. The WUE refers to the relative amount of shoot biomass produced per unit of water transpired (g biomass g water^{-1}), and HI is defined as the ratio of grain mass to total shoot mass. It should be noted that biomass composition affects WUE. For example, production of grain with a high oil concentration requires more photosynthate (and hence water) than production of grain composed primarily of starch, resulting in WUE that is low when oil concentration is high (Tanner and Sinclair, 1983).

The relationship described by Eq. [1] is graphically displayed in Fig.12–2. The graph makes it readily apparent that WUE is a response coefficient (i.e., slope) reflecting the degree of change in biomass (BM) that occurs per unit change in transpiration (T). Thus, for any given WUE, if the amount of water transpired by the crop can be increased, the amount of biomass produced by the crop will also be increased. Biomass production is also improved if the WUE coefficient itself can be increased, and in fact, some plant species have accomplished just that by evolving from a C3 to a C4 pathway of photosynthetic CO_2 acquisition, a topic that will be considered later. So long as HI (whatever its value) remains constant vis-à-vis variable amounts of T, the response of grain mass to transpiration will be as linear as that of the biomass to transpiration relationship. Note, however, that an increased HI steepens the grain yield response to transpiration.

Before proceeding, it is important to point out that researchers can (and have) defined the BM, T, and WUE terms differently than just stated. In most instances, biomass will ordinarily consist only of shoots, but may include roots (especially in pot studies). In grain legumes, it may not include abscised leaves (if these are not collected prior to abscission). If crop water consumption is measured without partitioning out evaporative (E) water loss from the soil surface, then an ET term

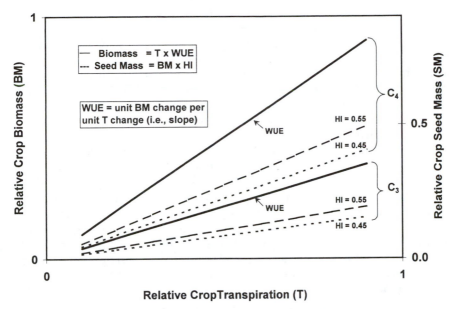

Fig. 12–2. Displayed in this figure is the well-known linear relationship (solid lines) between crop bio-mass (BM) and crop transpiration (T) in water-limited production environments. The slope of that relationship is water-use efficiency (WUE), defined as the change in BM accruing per unit change in T. A C4 photosynthetic species (e.g., maize) will typically exhibit a WUE more than twice that of a C3 photosynthetic species (e.g., soybean). The ratio of crop seed mass to crop biomass is defined as the harvest index (HI). In the graph, HI is assumed to be constant, which makes crop seed mass a lin-ear function of crop transpiration amount. We have arbitrarily chosen to graph two HI values to il-lustrate how a change in HI would impact the relationship between seed mass and crop transpiration in the two species. See Eq. [1] in the text and the associated discussion.

replaces the T term in Eq. [1]. Crop water input (I), in the form of combined pre-cipitation and irrigation amount (after adjustments for surface runoff and deep per-colation), is sometimes used in lieu of T or ET measurements. Relative to the der-ivation of WUE, the biomass and T amounts collected in treatments that consist of genotypes or cultural practices must be assessed on some common time scale, whether this is seasonal, monthly, weekly, daily, hourly, etc. When those measure-ments are more or less instantaneous, it is CO_2 flux to, and H_2O flux from, the leaf that is measured. Thus CO_2 assimilation (A) replaces the BM component of Eq. [1], and the WUE derived from such measurements is more generally known as tran-spiration efficiency (TE). The energy costs of producing biomass of different com-positions (i.e., different proportions of oil, starch, and protein) also affect WUE. The relevant equations for TE are discussed later in this review. The papers of Tanner and Sinclair (1983), Sinclair et al. (1984), and Sinclair (1994) can be consulted if more detail is desired as to the alternative formulations of Eq. [1].

As noted by Passioura (1994), traits that increase any component in Eq. [1] could potentially increase grain yield in water-limited environments. Conversely, traits that do not affect any of these components positively are unlikely to have a benefit in water-limited environments. In the discussion that follows, we shall use this conceptual framework to critically examine soybean traits that have the potential to lessen the yield reduction arising from seasonal drought.

Soybean seed is unique among other legume and cereal grains in that it is high in both protein (i.e., N) and oil (i.e., C) (Sinclair and deWitt, 1975), and, therefore, the nutritional demands for grain filling require consideration. In that the responses of N nutrition and dinitrogen (N_2) fixation were not included in the conceptual framework of Ludlow and Muchow (1990), we have included one section of our review to examine the effects of water deficit on N nutrition and N_2 fixation.

One final point deserves discussion before proceeding. Some researchers have argued that the narrow genetic base in elite soybean germplasm imposes a significant limit on the degree of genetic advance possible in drought tolerance (Carter et al., 1999; 2004, this publication). There is validity to this argument in that if a specific trait is postulated to offer adaptive advantages in drought conditions, then genetic variability in that trait is obviously critical. Ceccarelli and Grando (1996) note that achieving a higher economic return in a crop routinely subjected to drought is often an elusive goal from a genetic point of view. Still, the probability of finding genes to advance that goal is most likely the greatest in germplasm such as land races that have been exposed to thousands of years of farmer selection in drought-prone areas.

12–5 TRAITS INFLUENCING THE AMOUNT OF WATER AVAILABLE FOR TRANSPIRATION

It is quite clear from examining Eq. [1] and Fig. 12–2 that, if WUE and HI remain constant, *greater* (not less) T is required to enhance grain yield. The quantity of water transpired by a crop depends upon management, environmental conditions, genetic factors, and their interactions. Examples of how management may increase the quantity of water transpired include increasing the water infiltration rate and the water-storage capacity by deep tillage in fall (Wesley, 1999) and residue management (Hatfield et al., 2001). The mass balance of soil water is affected by precipitation and evaporative demand, and both are manageable to some extent. In the mid-southern USA, rainfall distribution is generally adequate for crop needs from April through June (Heatherly, 1999), and cooler temperatures during this period decrease evaporative demand. These observations have led to specific crop management systems to synchronize crop phenology with predictable periods of precipitation sufficient to support crop growth. Early sowing dates and the switch from Maturity Group (MG) VI and VII cultivars to MG III and IV cultivars in Mississippi and Louisiana have stabilized yield and resulted in higher non-irrigated yields in most years (Heatherly, 1999). Other management strategies to increase soil-water storage are discussed in Chapter 10 (Heatherly and Elmore, 2004, this publication).

Crop traits that influence the T component of Eq. [1] were reviewed by Ludlow and Muchow (1990), and those specific to grain legumes were recently reviewed by Turner et al. (2001). Ludlow and Muchow (1990), in Table I of their paper, listed 16 plant traits or physiological mechanisms that they postulated as having an impact on crop yield in water-limited environments. Twelve of the original 16 traits

Table 12–1. Traits that in theory or practice impact soybean yield (Y) via transpiration (T), water-use efficiency (WUE), or harvest index (HI)—components of the conceptual framework of Y = T × WUE × HI proposed by Passioura (1977, 1996). Modified from Table 1 of Ludlow and Muchow (1990) to show trait importance in intensified soybean systems.

Trait or physiological mechanism	Increase[†] crop transpiration?	Increase[†] water-use efficiency?	Increase[†] harvest index?[‡]
1. Matching phenology to water supply	Yes	No	Yes
2. Photoperiod sensitivity	Yes	No	Yes
3. Developmental plasticity	Yes	No	Yes
4. Leaf area maintenance	Yes	No	Yes
5. Heat tolerance	Yes?	No	No
6. Osmotic adjustment	Yes	No	Yes
7. Early vigor	Yes	No	No
8. Rooting depth and density	Yes	No	Yes
9. Transpiration efficiency	No	Yes	No
10. Leaf reflectance	No	Yes	No
11. High axial root resistance	No	No	Yes
12. Mobilization of pre-anthesis dry matter	No	No	Yes
13. Drought tolerant N_2 fixation§	No	No	Yes

† The general assertions listed in this table come from research summarized in Ludlow and Muchow (1990) and from research reports since then (discussed in this chapter).

‡ Soybean harvest index is affected primarily by the available amount of soil water supply after R4, when the crop begins dry matter deposition into seeds. See text.

§ Drought tolerant N_2 fixation was not included in the putative list of mechanisms conferring drought tolerance that was proposed by Ludlow and Muchow (1990).

proposed by Ludlow and Muchow (1990) are discussed in this review, and we have added to this list the trait for prolonged N_2 fixation during drought (Table 12–1). The genetic manipulation of eight of these traits in Table 12–1 could increase or possibly maintain crop T during drought, and they are examined here.

Three of the traits listed in Table 12–1 are analogous to what conceptually are known as drought escape mechanisms (Turner et al., 2001). These include: (Trait 1) genetic alteration of developmental ontogeny to achieve a better "fit" with the "historically normal" pattern of seasonal rainfall distribution, (Trait 2) genetic manipulation of cultivar photoperiod response to ensure crop flowering just before a relatively predictable meteorological event (i.e., transition from a rainy to dry season), and (Trait 3) genetic deployment of developmental plasticity (e.g., indeterminate, not determinate flowering) as an ontogenetic hedge to mitigate the uncertainty of rainfall timing. These phenological adjustments *enhance* transpiration in the sense that a cultivar with an optimized adaptive fit to the production environment should (on average) be able to access more water for transpiratory purposes than a cultivar lacking such a fit. In any event, selection for adaptation is an intrinsic element of just about every empirically driven breeding program, and because those in the USA have been ongoing for many years, today's cultivars are likely to already have near-optimized adaptive fits to their (targeted) production areas (Sinclair, 1994). Still, breeding for adaptation must come into play whenever new cultivars need to be developed for new production areas, or for new cultural practices, such as the early-sowing–early-maturing production system now being used in the mid-southern USA that was mentioned earlier.

Decreasing harmful consequences resulting from water deficits is a key component in three other traits listed in Table 12–1. These include: (Trait 4) genetic factors that help maintain a leaf area index (LAI) of at least 3.3 when drought induces leaf abscission, (Trait 5) genetically elevating temperature tolerance with respect to heat-induced stomata closure, and (Trait 6) genetically enhancing the *capacity* for osmotic adjustment. These traits are postulated to ensure continuation of radiation interception, stomatal conductance, and turgor-driven growth processes, thereby minimizing the reduction in transpiration induced by drought (and by the high temperatures that often accompany drought). If so, biomass accumulation can also be maintained (as per Eq. [1]). Ludlow and Muchow (1990) noted that leaf area maintenance was risky strategy in that continuing transpiration could quickly exhaust a limited water supply. They also noted that this trait should not be confused with the so-called "stay-green" trait found in some sorghum [*Sorghum bicolor* (L.) Moench] hybrids (Borrell et al., 2000). While leaf tolerance to high temperature is, in theory, a useful trait, a causal relationship with biomass grain and/or grain yield has not been satisfactorily documented.

Ludlow and Muchow (1990) and Turner et al. (2001) considered osmotic adjustment (Trait 6) a useful trait because it provides a means for maintaining cell turgor when tissue water potential declines. Osmotic adjustment is defined as the *active* accumulation of solutes in plant tissues that occurs in response to an increasing external and internal water deficit. Osmotic adjustment is distinguished from the *passive* accumulation of solutes that occurs during water deficit since the latter is simply a concentration effect arising from decreased cellular volume. Osmotic adjustment is illustrated in Fig. 12–3. Osmotic adjustment in soybean ranges from 0.3 to 1.0 MPa, which is not as great as the ranges observed in some grain legumes (Muchow, 1985; Cortes and Sinclair, 1986; Turner et al., 2001).

It is usually assumed that osmotic adjustment increases drought tolerance because a decrease in plant water potential would increase the water-potential gradient from the soil to the plant and, therefore, increase water uptake. In theory, this assumption is correct, but in practical terms osmotic adjustment has little effect on the ability to increase soil-water extraction (Munns, 1988; Serraj and Sinclair, 2002). The reason for this is that plants can normally extract water from soil until the matric potential is about −1.5 MPa, which is traditionally considered the permanent wilting point (Kramer and Boyer, 1995). After osmotic adjustment of 0.5 MPa, a plant may be able to extract water from soil having a matric potential of −2.0 MPa, but the quantity of water in soil at matric potentials <−1.5 MPa is insignificant with the exception of clay soils. The water release properties of clay soils are discussed subsequently with specific reference to Fig. 12–4.

If the purpose of osmotic adjustment is to maintain turgor and continue expansive growth and photosynthesis during drought, then osmotic adjustment may increase transpiration and the severity of water deficit (Munns, 1988; Serraj and Sinclair, 2002). If, in fact, osmotic adjustment was successful in delaying the decrease in transpiration, Sinclair and Muchow (2001) concluded that it was unlikely to increase grain yield, at least for a maize (*Zea mays* L.) crop. A sensitivity analysis of putative traits affecting drought tolerance in maize indicated that osmotic adjustment was yield-neutral in most years. In several years, however, simulated yield decreased because prolonged transpiration exhausted soil water supply (Sinclair and

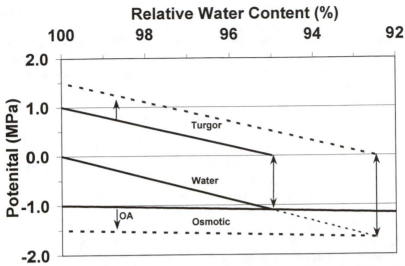

Fig. 12–3. Displayed in this figure is a generalized description of the nature of osmotic adjustment (OA). For simplicity, we have let leaf *water* potential (ψ) equal the sum of the *osmotic* potential (s) and the *pressure* (*turgor*) potential (p), such that ψ = s + p. The solid lines indicate values *before* OA; the dashed lines *after* OA. When leaf water potential declines, turgor also declines to the point where ψ = s (to the first double-arrowed line), where p = 0 and thus the cell plasmolyzes (i.e., dies). With OA, cells *actively* accumulate solutes upon exposure to water deficits, and these additional solutes make s more negative (downward single arrow labeled OA). The cell's turgor can now be maintained at a relatively higher level (upward arrow) at the same leaf water potential as before, and can also be maintained to a lower leaf water potential (dashed line extension from solid line) and to a lower relative water content (RWC). See Jones (1992) and Nilsen and Orcutt (1996) for details.

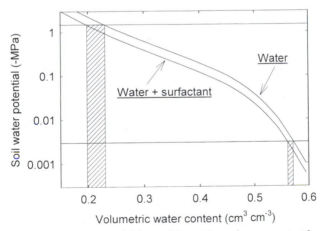

Fig. 12–4. Hypothetical soil water potential (log scale) vs. volumetric water content for a clay soil that has been moistened with water or water plus a surfactant. Calculations were based on surface tension of water at 20°C (7.28×10^{-8} MPa m) and for water plus surfactant (4.4×10^{-8} MPa m). The upper horizontal line in the graph denotes a soil water potential of −1.5 Mpa (traditionally considered the permanent wilting point), and the lower horizontal line represents a soil water potential of −0.003 MPa (field capacity). The hatched areas denote the differential in soil volumetric water content between release curves at field capacity (0.01 cm^3 cm^{-3}) and the permanent wilting point (0.035 cm^3 cm^{-3}) for water alone and water plus a surfactant. Water-release curve from clay soil for pure water was based upon Brady and Weil (2002).

Muchow, 2001). The exception where osmotic adjustment may have an important benefit during drought is when it is directed to roots where turgor maintenance might lead to an increased rooting depth (Serraj and Sinclair, 2002). In this case, osmotic adjustment would increase the amount of water available to the crop, and increased rooting depth is considered likely to increase grain yield in many environments (Ludlow and Muchow, 1990; Turner et al., 2001; Sinclair and Muchow, 2001).

Two additional traits listed in Table 12–1 actually have the potential to *enhance* the amount of crop transpiration and biomass produced (as per Eq. [1]). These include: (Trait 7) genetic improvement of early season vegetative vigor, and (Trait 8) genetically designing a root system capable of exploiting more of the annually recharged, plant-available water in the soil zones that can be reached by roots. By extension of the preceding discussion, *osmotic adjustment of roots*, which results in deeper rooting, would also fall into this category of enhancing the amount of water for transpiration. During the initial phases of crop production, much soil water is lost via evaporation from moist soil surfaces, especially in production areas where spring rainfall constitutes a large fraction of the total summer precipitation. Early season plant vigor can decrease the time required for canopy closure. This would result in less solar radiation being intercepted by the soil surface, where it simply provides the energy for soil water evaporation, and more interception by the crop for use in photosynthesis and transpiration. Moreover, if early season shoot vigor is accompanied by vigorous rooting in the topmost soil zone, a greater fraction of soil water just below the soil surface might be transpired prior to its possible loss via evaporation. Earlier canopy closure thus lessens the soil E fraction by elevating the crop T fraction of the crop ET measure of consumptive crop water use. Genetic improvement in early season soybean vigor is possible and has probably been improving over time as breeders select and release cultivars with ever-greater yield potential. Marker-assisted selection for this trait may also be possible, given the recent report of Mian et al. (1998b) documenting quantitative trait loci (QTL) for plant height and canopy width during the early vegetative stages of soybean growth.

For environments where severe or terminal drought frequently occurs near the beginning of grain filling, rapid canopy establishment may deplete soil water that otherwise could be used during grain filling. Decreasing soil evaporation by means other than canopy shading (e.g., decreased tillage and residue management) could conserve water under these conditions and be more effective in increasing rainfed yields.

12–5.1 Extracting More Soil Water for Crop Transpiratory Purposes

Many researchers have hypothesized that the best means for increasing yields in water-limited environments is genetic manipulation of Trait 8, that is, a rooting system capable of extracting water from inadequately explored soil volumes, presumably that in the deeper soil zones. However, implicit in this hypothesis are three assumptions: (i) that water is actually present and available in the inadequately explored, or deeper soil volumes, (ii) that this water is routinely replenished before each cropping cycle, and (iii) that the cost in photosynthetic assimilate diverted from the shoot to the root system to tap those soil volumes is not appreciable (Ludlow and Muchow 1990; Sinclair, 1994). Passioura (1983) has questioned the value of deep roots on the basis that production of the extra dry matter needed for deeper roots might *offset* the amount of water actually recovered by those roots and made available to the crop. The solution to this dilemma is not clear, although he believes that drought-mediated regulation of root growth, whether in a relative sense (i.e., shoot/root ratio) or in an absolute sense, is probably mediated by hormones like abscisic acid (ABA) (Passioura, 1994). Genetic variation in the soybean root system phenotype relative to drought stress seems to exist (Boyer et al., 1980; Goldman et al., 1989). Moreover, soybean root depth seems to be positively correlated with grain yield (Cortes and Sinclair, 1986). Because making more soil water available for transpiration is a logical means of improving yield in water-limited soybean production environments, we have elected to provide the reader with a detailed review of crop rooting and soil water uptake.

Taylor and Klepper (1978) provided a detailed framework for evaluating rooting and rooting characteristics important for water uptake from soils. Their analysis is brought forth here, with the exception that thermodynamic considerations have been minimized. For more details on the thermodynamics of water uptake by roots, the reader is referred to several reviews (Taylor and Klepper, 1978; Rendig and Taylor, 1989; Klepper and Rickman, 1990).

If the total rooting volume of a soil is divided into discrete zonal depths (d; from 1 to n), the maximum amount of crop-available water (CAW_d, cm^3 water cm^{-3} soil) in a given soil-layer can be expressed as the difference between the soil volumetric water content at the upper-drained limit (θmax_d, cm^3 cm^{-3}) and soil volumetric water content after water has been exhausted by the plant (θmin_d, cm^3 cm^{-3}), multiplied by the root water extraction (RWE_d). The RWE_d may be considered, for this discussion, as a coefficient ranging from 0 to 1 and which is composite of a complex relationship between root length density (RLD_d, cm root cm^{-3} soil) and the effective root extraction zone ($EREZ_d$, cm^3 soil cm^{-1} root):

$$CAW_d = (\theta max_d - \theta min_d) \times RWE_d \qquad [2]$$

The relationship between RLD and EREZ and their relationship to RWE are discussed in detail in a subsequent section. The total amount of CAW (CAW_t) is obtained by summing CAW_d present in soil zones whose depths (d) extend from the soil surface zone (d = 1) to the zone (d = n) corresponding to the maximum rooting depth (MRD, cm):

$$CAW_t = \sum_{d=1}^{n} (\theta max_d - \theta min_d) \times RWE_d \qquad [3]$$

The terms in Eq. [3], and implicitly MRD, depend upon soil and crop factors. Each term will now be examined relative to its potential for increasing CAW_t, whether by genotypic or agronomic improvement.

12–5.1.1 Volumetric Water Content

The volumetric water content at the upper-drained limit (θmax) is dictated primarily by soil texture (Rendig and Taylor, 1989) and organic matter (Hudson, 1994). As soil-particle size decreases, the amount of water that can be held against gravity in a soil increases. Because θmax is a soil characteristic and depends upon physical properties of the soil, it is not easily changed or manipulated. Strategies to increase soil-water storage include optimizing conditions whereby θ approaches θmax such as increasing water infiltration to lessen rainfall run-off and decreasing soil-surface evaporative losses to make more water available for crop transpiration.

The importance of soil-particle size to the matric potential of the soil (P, MPa) is described as a function of the surface tension of water (γ, MPa m), the contact angle (α) between the water and soil, and the pore radius (r, m) between two soil particles (Nobel, 1991):

$$P = -(2\gamma \times \cos \alpha) \times r^{-1} \qquad [4]$$

In soil, γ is generally considered to be equivalent to that of pure water with a value of approximately 7.28×10^{-8} MPa m at 20°C (Nobel, 1991), and soil particles are considered to be hydrophilic with α approaching 0 degrees. For complex biological systems in the rhizosphere, however, the assumed pure-water value for γ may not be entirely correct, since minute quantities of surfactant-like materials found in organic matter may lower γ to values as low as 4.4×10^{-8} MPa m (Chen and Schnitzer, 1978). In any event, decreasing γ at any given θ makes P less negative, presumably enhancing the amount of water available to plants (Passioura, 1988). As noted previously in the discussion of osmotic adjustment, there is little water present in most soils at water potential <-1.5 MPa. The exception to this is for fine-textured clay soils that have large amounts of water when $P \leq -1.5$ MPa.

For a clay soil, a plot of soil matric potential vs. the volumetric water content (generates a water release curve similar to that illustrated in Fig. 12–4. The upper line in this figure is the water release of a clay soil assuming that the soil water has a surface tension equivalent to that of pure water (i.e., 7.28×10^{-8} MPa m at 20°C). The lower line in the figure represents the water release of the same soil assuming the surface tension is decreased (i.e., 4.4×10^{-8} MPa m) due to the presence of surfactant compounds, similar to that found in soil organic matter (Chen and Schnitzer, 1978). The total amount of water available to a crop may be approximated by the difference between θmax, when the soil matric potential is at field capacity (–0.003 MPa), and θmin, when the matric potential is –1.5 MPa (Kramer and Boyer, 1995). Figure 12–4 indicates that for a clay soil, a decreased surface tension decreases the amount of water held against gravity (θmax) by approximately 0.01 cm³ cm⁻³. At field capacity, a decreased surface tension would, therefore, lessen the amount of water held against gravity and the amount available to plants (stippled area on the right side of Fig. 12– 4). This could benefit a soybean crop by increasing soil aeration, which can be a limitation in clay soils following extended rainfall (VanToai et al., 1994; 2001). As a soil dries, the matric potential at any given θ remains greater (less negative) for soil water with a decreased surface tension than that for pure soil water. Figure 12–4 indicates that at a matric potential of –1.5 MPa, a decreased surface tension of soil water would result in an increase in crop-available water of approximately 0.035 cm³ cm⁻³ (stippled area on the left side of Fig. 12–4). For a crop with a 600-mm rooting depth, this corresponds to an additional 21 mm of available water. Roots, microorganisms in the rhizosphere, and soil organic matter may function to decrease γ (Nimmo and Miller, 1986; Passioura, 1988). Engineering or selecting soybean cultivars whose roots would somehow make P (cf., Eq. [4]) less negative in the surrounding soil has not been attempted, but may provide opportunities for increasing water extraction by crops.

Soil organic matter has long been recognized as beneficial to soils and soil structure, but it was generally considered to be innocuous in its effect on the quantity of water available to plants (Hudson, 1994). Early experiments with peat-moss amendments to soil indicated that there was little effect of organic matter in soils on plant-water availability (Feustal and Byers, 1936 as cited by Hudson, 1994). This conclusion was generally the consensus view for the next 50 to 60 yr despite evidence to the contrary (see Hudson, 1994). For sand, silt loam, and silty-clay loams, Hudson (1994) found that θmax increased linearly as percent organic matter increased whereas the permanent wilting point (i.e., a soil matric potential of -1.5 MPa) was only slightly affected by organic matter (Fig. 12–5). As soil organic matter increased from 1 to 3% by weight, the volumetric water content that was available to plants increased 73% for sands, 59% for silt loams, and 61% for silty-clay loams.

Soil organic matter can be increased over time by maximizing crop residues while minimizing the oxidative loss of those crop residues. Assuming that future genetic improvement in soybean grain yield comes about by higher biomass yield instead of greater harvest index, those future cultivars will produce greater amounts of crop residue. No-till and minimum tillage practices conserve mulch on the soil

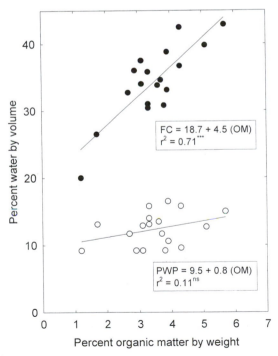

Fig. 12–5. Volumetric water content vs. percent organic matter in a silt-loam soil at field capacity (FC) and the permanent wilting point (PWP, -1.5 MPa) Redrawn (with permission) from Fig. 2 of Hudson (1994).

surface, thereby lessening soil moisture evaporation, increasing infiltration, and decreasing soil erosion loss (reviewed by Hatfield et al., 2001). Sowing soybean in a stale seed-bed system also has advantages in conserving soil moisture (Heatherly, 1999). Spring cultivation in Iowa resulted in a 10 to 12 mm evaporative water loss over a 3-d period following each cultivation compared with <2 mm evaporative loss from no-tilled fields (Hatfield et al., 2001). An estimated 20 to 30 mm of soil moisture may be lost from the seed zone from traditional cultivation in the spring (Hatfield et al., 2001). Diversion of some of that E loss into T would obviously be of value, since it would enhance the amount of early spring crop biomass, particularly the leaf area needed for earlier canopy closure.

Although there are great differences among soils in the maximum amount of water that may be held (θmax), the fraction of water extractable by roots among soils, on a volumetric basis, is remarkably similar (Ratliff et al., 1983; Sinclair et al., 1998). Ratliff et al. (1983) surveyed 401 soils from across the USA that ranged in textural classes from sands to clays. For all the soils that they examined except sands, they found that the difference in the volumetric water content of a soil between the well-drained upper limit (θmax) and the lower limit (when the soil was extremely dry and plants were dead or dormant, θmin) averaged 13.2%. Sandy soils differed significantly from the other textural classes, and the mean difference between θmax and θmin for sands was 8%.

Ritchie (1981) defined the fraction of the total extractable water (FTEW) in the root zone at any given θ as:

$$FTEW = (\theta - \theta min) \times (\theta max - \theta min)^{-1} \qquad [5]$$

Ritchie (1981) noted that many physiological processes were unaffected as a soil dried until a threshold value of the extractable water had been reached. Photosynthesis and transpiration, for example, were unaffected until FTEW was approximately 0.30. As soil moisture decreased below this threshold value, leaf gas exchange decreased linearly to 0 at a FTEW = 0 (Fig. 12–6).

The approach of using FTEW as a measure of physiological responses to water deficit is attractive for several reasons. First, the volumetric water content available to plants is essentially a constant 13% (Ratliff et al., 1983), except for sandy soils (Ratliff et al., 1983; Sinclair et al., 1998), in which the sand content exceeds 55% (Sinclair et al., 1998). Organic matter also may increase the amount of water available to plants, as discussed previously (Hudson, 1994). A second advantage of using FTEW is that its derivation does not require measurement of soil water potential. In that regard, it is directly related to the physiological activity of the crop, whereas soil-water potential has no direct relationship with plant response or function (Sinclair and Ludlow, 1986; Passioura, 1988; Sinclair et al., 1998). Third, using FTEW to characterize plant response to water deficits allows comparisons of sensitivity for different physiological responses to water deficit (Ritchie, 1981; Sinclair and Ludlow, 1986). For example, Ritchie (1981) found that leaf expansion decreased at an FTEW threshold that was higher than for leaf gas exchange. Sinclair and Ludlow (1986), in an evaluation of physiological responses of FTEW, discovered that N_2 fixation in soybean was considerably more sensitive to water deficit than was transpiration. A fourth advantage of this technique is that threshold values of

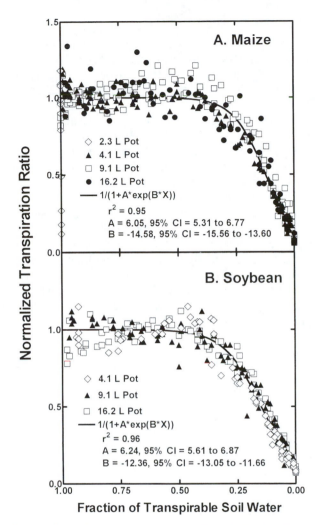

Fig. 12–6. Relative transpiration for maize (A.) and soybean (B.) vs the fraction of total extractable water for pot volumes of 4.1, 9.1, and 16.2 L. Redrawn (with permission) from Fig. 3 of Ray and Sinclair (1998).

FTEW at which leaf gas exchange begins to decrease are apparently independent of RLD (Ray and Sinclair, 1998; Fig. 12–6). For soybean and maize grown in pots ranging from 4 to 16 L, plant size doubled with increasing pot size but the FTEW threshold at which transpiration began to decrease was constant for the different container sizes. This observation is particularly important in that it suggests that physiological responses in container-grown plants can be evaluated for different genotypes or treatments as a function of FTEW even though there may be some degree of root restriction.

12–5.1.2 Root Water Extraction

As discussed previously, RWE is a coefficient (ranging from 0–1) that depends upon complex relationship between root length density (RLD) and the effective root extraction zone (EREZ). Intuitively, the EREZ can be thought of as the halfway distance in the soil between two roots through which water must move to reach the root surface. This relationship indicates that RLD and EREZ are inversely related. A second factor affecting EREZ is the activity of individual roots for water uptake. Certainly, not all roots have the same ability to extract water from soil, and this ability changes with transpiration demand, root age, and with radial and axial resistances to water movement in the root. The relationships of RLD and EREZ to RWE are considered in this section of the review.

Since the publication of the last monograph, there has been great interest in genetic differences in RLD among soybean genotypes (e.g., review by Carter et al., 1999). At a given soil depth, water uptake for soybean cultivars differing in RLD would seemingly be proportional to those RLD differences, provided that concomitant changes EREZ were strictly associated with a decreased distance that water would be required to move to the root surface in the soil as RLD increased. However, changes in EREZ occur independently of changes of RLD due to changes in crop evaporative demand and efficiency of individual roots for water uptake (Taylor and Klepper, 1978). The efficiency of roots in water uptake is considered in a subsequent section.

A simplification of soil rooting that has been helpful in understanding requirements of RLD and water movement to the root is the single-root model, originally proposed by Gardner (1960). Passioura (1985, 1988) discussed the strengths and weaknesses of this model. Roots in a given volume of soil are envisioned as evenly distributed and consisting of cylinders of radius **a**, with each root surrounded by a soil cylinder of radius **b**, from which the root can draw soil water. The dimension of radius **b** that the root has sole access to soil water is related to RLD by the relationship:

$$\mathbf{b} = (\pi\text{RLD})^{-0.5} \qquad [6]$$

Water flux to the root surface (Q) depends upon the changing gradient in θ from the outer rim of the soil cylinder (θ_b) to the surface of the root cylinder (θ_a), and changes in the soil-water diffusivity, D, from the outer edge of b to the root surface. Between FTEW values of 0 and 0.25, D is very low and approximately constant at values near 2 cm^2 d^{-1} (Passioura, 1985). With this simplification, we may write:

$$Q = D\,(\theta_b - \theta_a) \times (2b^2)^{-1} \qquad [7]$$

In dry soil, θ_b is approximately the same as of the bulk soil but measurement of θ_a is difficult because of rapid changes induced by changes in transpirational demand and the activity of individual roots for water uptake. Nevertheless, if θ_a is as low as possible for a particular root (and constant over time), then we can write an expression for the time constant of water movement through the soil. This time constant, t, represents the time required in days to decrease θ_b to θ_a by 50% when the limitation to water movement through the soil is strictly soil related:

$$t = (2\,b^2) \times D^{-1} \qquad [8]$$

If we substitute **b** in this equation with its equivalent, $(\pi RLD)^{-0.5}$, (Eq. [6]), an expression relating the time required for water extraction to RLD is given by:

$$t = 2 \times (\pi RLD \times D)^{-1} \qquad [9]$$

The relationship of t and RLD, as expressed in Eq. [9], is shown in Fig. 12–7 for a dry soil in which $D = 2\ cm^2\ d^{-1}$. This ideal approximation indicates that as RLD increases from 0.1 to 0.5 cm cm^{-3} the time required for water to move to the root surface from the bulk soil decreases rapidly. However, given this asymptotic response, when RLD exceeds 1 to 1.5 cm cm^{-3}, there is little further decrease in the time required for water to move to the root surface, indicating little advantage within a given soil layer for RLD to exceed 1 to 1.5 cm^{-2} (Gardner, 1964).

The second factor interacting with RLD to affect RWE is the effective root extraction zone (EREZ). The EREZ can be thought of as the halfway point between two roots in a soil, and, therefore, EREZ decreases as RLD increases, as described by Eq. [6]. Equations [7] through [9] in the preceding section examined the special case of water movement when the soil was extremely dry which makes water uptake limited primarily by D. As the volumetric water content of a soil declines from θmax to θmin, water diffusivity in the soil (D, Eq. [7]) decreases by approximately four orders of magnitude (Rendig and Taylor, 1989).

In a moist soil (FTEW > 0.5), D is large and generally does not limit water uptake by roots. Water moves into roots in response to a decrease in energy or, more conventionally, a decrease in pressure (energy per unit volume). Equations describing the movement of water from the soil to the plant as a function of pressure potential are discussed elsewhere (e.g., Taylor and Klepper, 1978; Passioura, 1985; Steudle and Peterson, 1998). There are two key factors in addition to RLD that govern the EREZ in a moist soil. One is radial root resistance, which results in a difference in pressure potential between the soil and root xylem. The second is axial

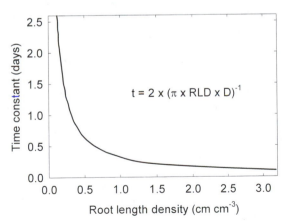

Fig. 12–7. Time required for a 50% decrease in soil water content from the outer edge of the root extraction zone to the root surface over a range of root length densities. Calculations assume that the soil is dry with a hydraulic conductivity of 2 cm^2 d^{-1} and that the soil moisture content next to the root is very low and constant.

root resistance, which causes a decrease in the pressure potential between the xylem elements in the leaf and xylem elements in the root. Although both of these resistances may limit water uptake, radial resistance is considered to be of greater magnitude and importance than is axial resistance (Steudle and Peterson, 1998).

The pressure in the xylem of a root depends upon the evaporative demand at the leaf surface, which is transmitted through the continuous xylem-water column to the root. Not all of this pressure, however, is necessarily transmitted to the root. As the axial resistance in the xylem pathway increases, it effectively decreases the pressure difference between the xylem in the root and the soil matric pressure (potential). Because dicots are capable of secondary growth in roots, the number of xylem elements increases over time per unit root length, which would effectively decrease axial resistance (Passioura, 1983). This appears to be the case in cotton (*Gossypium hirsutum* L.) (Taylor and Kleeper, 1978), but axial resistance in soybean is substantial (Willatt and Taylor, 1978), and increases with soil rooting depth. Decreasing axial resistance would increase the ability of roots to extract water deep in the soil profile, which is an effective means of exploiting the soil moisture residing at lower soil horizons, provided that this soil moisture is recharged before the next cropping season (Passioura, 1983).

For conditions where a crop is dependent upon soil-stored moisture for the later portion of a growing season (when rainfall does not occur or irrigation is not possible), an increased axial resistance may be beneficial (Passioura, 1983). A high axial resistance restricts water use during vegetative growth, thus allowing soil moisture deep in the soil profile to be conserved for later use during grain filling. Ludlow and Muchow (1990) also noted that increased root hydraulic resistance could be beneficial for conditions where a crop was grown primarily on stored soil moisture (Trait 11, Table 12–1). A breeding program to increase axial resistance of wheat (*Triticum aestivum* L.) by selecting for narrow xylem elements in seminal roots was successful in conserving soil moisture for grain filling (Richards and Passioura, 1989). In low yield, drought-stressed environments, wheat lines selected for narrow xylem vessels had yields 3 to 11% greater than that of the unselected controls, whereas in wetter environments there was no difference between selected and unselected lines. The selected lines, in addition to having greater yield in these low-yield water-limited environments, also had higher HI.

Radial resistance of water movement from the soil into the xylem decreases the effective "suction" that roots apply to water at the root-soil interface, which decreases the water uptake rate and the EREZ. The radial resistance consists of several individual resistances that are in series and in parallel (Steudle and Peterson, 1998). The largest resistance in maize seedlings is not associated with the Casparian strip, as is commonly assumed, but is evenly distributed among cells of the endodermis (Steudle and Peterson, 1998). Shrinkage of roots during water-deficit stress may decrease root-soil contact, and decrease water uptake because of an increased radial resistance (Huck et al., 1970). In cotton (Huck et al., 1970), roots that adhered to the glass wall of a rhizotron shrunk up to 40% in diameter during water deficit, but the importance of this phenomena in undisturbed soil is less clear (Passioura, 1988).

The contribution of the endodermal cells to the axial resistance is probably associated with water having to cross at least two membranes before reaching the xylem elements in roots. Of particular interest is whether aquaporin activity modulates the axial resistance of roots (Maurel, 1997). Aquaporins are protein channels found in membranes that selectively allow the passive transport of water. It has been suggested that the large and rapid changes in hydraulic conductivity of roots that occur in response to environmental factors may be associated with abundance and activity of membrane aquaporins (Maurel, 1997). Decreasing axial root resistance by increased expression of aquaporins in roots may allow for increased EREZ and water uptake.

There is also evidence that root hydraulic resistance changes diurnally, perhaps in response to evaporative demand and water flux through the roots (Kramer and Boyer, 1995; Zwieniecki et al., 2001). At low flow rates, root hydraulic resistance is relatively high, but as flow rates increase, resistance typically decreases. Although the exact reason for changes in resistance is not known with certainty, Zwieniecki et al. (2001) found that hydraulic resistance in stem segments of *Laurus nobilis* increased as the KCl concentration in a perfusing solution increased. This response was completely reversible and occurred only when the pathway crossed xylem to xylem connections. The authors proposed that as salt concentration in the xylem increases, as would be expected under conditions of low flux, hydrogel swelling in the bordered pits of xylem connections is enhanced, increasing axial resistance and decreasing flow. Under conditions of high volumetric flow in the xylem, the salt concentration in the xylem would be expected to decrease, leading to the shrinkage of hydrogels in bordered pits, and consequently decreased resistance and increased flow.

Roots differ greatly in their ability to extract water from the soil, which suggests that the resistance to water transport also varies. For example, it has long been recognized that certain portions of roots are more active than others in water uptake (Kramer and Boyer, 1995), and that the most active region for water uptake is generally about 10 to 100 mm behind the root tip (Kramer and Boyer, 1995). Older roots tend to undergo suberization and are generally less efficient in water uptake than are nonsuberized roots (Kramer and Boyer, 1995). Moreover, roots growing in dry soil tend to undergo more suberization, which increases their axial resistance (Passioura, 1988).

12–5.1.3 Maximum Rooting Depth

In many soils, soybean roots are confined to a relatively shallow depth, although a significant amount of water may be present at depths below the rooting zone. This unused soil moisture represents an untapped resource for providing transpirational water during drought. The inability of roots to grow deep into the profile may be due to one or a combination of several factors: physical restrictions in the soil structure, chemical restrictions associated with toxic compounds in the soil, and insufficient time during the cropping season for roots to reach available soil moisture. Overcoming these restrictions to deep rooting is an important breeding and management goal. Management techniques that might increase rooting depth, including deep tillage, are discussed in Chapter 10 (Heatherly and Elmore, 2004, this publication).

Genetic differences among cultivars in root penetration of compacted soil have been noted for several crops including cotton (Kasperbauer and Busscher, 1991), common bean (*Phaseolus vulgaris* L.) (Asady et al., 1985), rice (*Oryza sativa* L.) (Ray et al., 1996), maize, and soybean (Bushamuka and Zobel, 1998a). In soybean, Bushamuka and Zobel (1998a) compared the ability of taproots and basal roots of different cultivars to penetrate a compacted soil layer. They found large genotypic differences among cultivars in both taproot length and basal-root length. Taproot length below the compacted soil layer relative to that above the compacted soil layer ranged from approximately 97% for PI416937 to 30% for 'Weber'. Basal-root length below the compacted soil layer relative to that above the restriction layer was reversed for these genotypes compared to those of the taproots. Basal-root length below the compacted soil layer was approximately 12% for PI416937 and 72% for Weber. If the observed negative correlation between taproot penetration and basal-root penetration is not an allometric one (i.e., amount of root dry matter is fixed, but more is allocated to the root type that penetrates first), breeders might be able to specifically combine these traits. In rice, QTL have been identified that are associated with root penetration capability (Ray et al., 1996). Given the existence of genetic variation in soybean root penetration (Bushamuka and Zobel, 1998a), it is possible that molecular marker analysis may enable soybean breeding programs to select for this trait.

12–5.2 Aluminum Toxicity—Impact on Roots

Restriction of soybean rooting in lower soil horizons by toxic concentrations of subsoil aluminum (Al^{3+}) affects wide geographic regions of production, primarily those high-rainfall areas in the lower latitudes. It is estimated that low soil pH and high soil Al affect approximately 40% of the arable land worldwide (Foy et al., 1978). Whereas the high Al^{3+} concentration in the upper 15 cm may be ameliorated by liming, this option is not considered practical for decreasing subsoil Al^{3+} (Hammel et al., 1985). Therefore, there has been a considerable effort in identifying and selecting soybean lines tolerant of high Al^{3+} (Carter et al., 1999).

Genotypic differences in the tolerance to Al^{3+} in soybean were first reported in 1968 (Armiger et al., 1968). Armiger et al. (1968) grew 48 genotypes, ranging in maturity from MG 00 to VIII, in pots containing Al-toxic subsoil (pH 4.8). Shoot mass was decreased 19% for the most tolerant genotypes (Biloxi, Norchief, and Wayne) to 66% for the most sensitive line (PI 85666). The authors noted the importance of Al tolerance for drought avoidance and the role of plant breeding in improving Al tolerance. Nevertheless, there was little progress in screening and selecting for Al and drought tolerance until the early 1980s (Carter et al., 1999). At that time T.E. Carter, Jr. and J.J. Cappy evaluated approximately 220 nonprostrate plant introductions from the USDA Soybean Germplasm Collection on a sandy soil in North Carolina. They noted that during a prolonged drought, leaf wilting in PI416937 was delayed for approximately 2 wk beyond the date of wilting in the other genotypes (Sloane et al., 1990). Further work established that the ability of PI416937 to maintain leaf turgor during prolonged drought was at least partially related to tolerance to soil Al^{3+} (Goldman et al., 1989; Carter and Rufty, 1993).

Not only are there genotypic differences in tolerance of soybean roots to Al, but different root types within a cultivar express differential tolerance to Al (Bushamuka and Zobel, 1998b). Bushamuka and Zobel (1998b) found that taproots and lateral roots of PI416937 were tolerant of soil Al, but basal roots were relatively sensitive to soil Al. Conversely, taproots and lateral roots of 'Essex' were sensitive to soil Al, but basal roots of Essex were extremely tolerant of soil Al. These results were similar to the differences found in the ability of roots to penetrate compacted layers (Bushamuka and Zobel, 1998a) and indicate that a breeding program aimed at increasing Al tolerance should consider combining tolerance in tap, lateral, and basal roots.

The mechanism by which plants exhibit tolerance to Al has generally been attributed to exclusion of Al from the root tip, or to a detoxification of Al once it enters the cell (Kochian, 1995). Silva et al. (2000) used the Al-specific stain lumogallion in conjunction with confocal laser scanning microscopy to determine the cellular and subcellular locations of Al following exposure of soybean roots to 1.45 μM Al^{3+}. Aluminum was found in much greater concentrations in root meristems after 30 min of exposure to Al^{3+} for the Al-sensitive cv. Young compared with the Al-tolerant line PI416937, and Al was bound to both the cell walls and cell nuclei. The accumulation of Al in the nucleus of roots may be responsible for decreased mitotic activity and growth. That Al concentration in cell walls and nuclei was considerably less in PI416937 than in Young, may indicate that tolerance in this line was associated with exclusion mechanisms, limiting the entry of Al into root tissues.

The exclusion of Al from plant roots is thought to be due to the release (by roots) of Al-chelating compounds (de la Fuente et al., 1997) or due to an alkalization of the root rhizosphere, although the latter mechanism has not been documented (Kochian, 1995). Root-released organic acids are believed responsible for Al tolerance in snapbean (*Phaseolus vulgaris* L.) cultivars that are Al tolerant (Miyasaka et al., 1991). The Al-tolerant snapbean cv. Dade exuded 70 times the concentration of citric acid into the growth media in the presence of Al as in its absence, and Dade exuded 10 times the concentration into the medium in the presence of Al as did an Al-sensitive cultivar, Romano. Similar responses have been found for high levels of malate excretion in roots of Al-tolerant isolines of wheat (Delhaize et al., 1993) and for citrate release from Al-tolerant maize lines (Pellet et al., 1995).

In soybean, citrate and malate production in roots also confer Al, or at least the ability of roots to elongate in the presence of solutions containing Al. Citrate and malate efflux increased in all of the eight genotypes evaluated during the first 6 h following exposure of roots to Al, but only Al-tolerant lines continued organic acid efflux for extended periods (Silva et al., 2001c). For eight lines selected for contrasting tolerance to Al, the ability of primary roots to elongate in the presence of 1.4 μM Al^{3+} for 72 h was highly dependent ($r^2 = 0.75$) upon citrate concentration in root tips. Citrate concentrate ranged from approximately 100 nmol g^{-1} fw for the most Al-sensitive genotype ('Dunfield') to 900 (PI416937) and 1350 (Biloxi) nmol g^{-1} fw for the most Al-tolerant genotypes. The addition of 50 μM Mg to the hydroponic solution containing 4.6 μM Al^{3+} stimulated the citrate efflux of roots six- to ninefold, and increased citrate concentration of root tips three- to fivefold, relative to roots exposed to 4.6μM Al^{3+} in the absence of magnesium (Silva et al., 2001a). Furthermore, magnesium increased root elongation in the presence of Al, particularly for Al-sensitive lines, and diminished genotypic differences among lines for Al tolerance (Silva et al., 2001b).

The production of citrate is governed by citrate synthase activity, and this gene from *Pseudomonas aeruginosa* (CSb) has been transferred behind a 35S promoter to the genomic DNA of tobacco (*Nicotiana tabacum* L.) and papaya (*Carica papaya* L.) (de la Fuente et al., 1997). High expression of citrate synthase in transformed tobacco plants greatly increased the tolerance of root growth to Al. For example, root growth of nontransformed control plants was inhibited approximately 80% by 200 µ*M* Al whereas root growth of the transformed line CSb-18 was inhibited approximately 30% by 200 µ*M* Al. In subsequent experiments, however, Delhaize et al. (2001) compared the amount of organic acids released from roots of the tobacco CSb-transformed lines developed by de la Fuente et al. (1997), plus independent CSb-transformants, and wild-type plants. They found no difference in organic acid release by roots among these plants despite the fact that roots of transformed plants had up to twice the amount of the citrate synthase protein as wild-type plants. Certainly, further work is required to establish the utility of expressing CSb in plant roots for citrate extrusion and Al tolerance.

Although there is tolerance to Al in naturally occurring soybean germplasm (Bianchi-Hall et al., 1998, 2000), QTL analysis indicates that the trait is multigenic with each gene exerting only a modest effect (Bianchi-Hall et al., 2000). Pyramiding the five tolerance alleles identified in QTL analysis along with other desired traits may be considerably more difficult than selecting a single dominant gene derived from high expression in engineered plants, such as that reported for engineered tobacco plants expressing a constructed citrate synthase gene (de la Fuente et al., 1997). It remains to be determined if expression of citrate synthase in roots of genetically engineered soybean would lead to citrate extrusion and confer Al tolerance in field environments, but this strategy appears attractive. Regulatory hurtles for transgenic crops and consumer preference for nontransgenic crops, however, increases the importance of a traditional breeding approach, augmented by marker-assisted selection, for pyramiding Al tolerance alleles in soybean.

12–5.3 Rates of Root Growth

As noted earlier, early seedling vigor results in rapid establishment of the leaf cover needed to minimize solar-mediated evaporation of water from the soil surface. Likewise, the quick establishment of a deep root system may also be an important means of avoiding drought during a growing season. In a greenhouse experiment, growth rates of soybean taproots during vegetative development ranged from 5.3 to 3.0 cm d^{-1} (Kaspar et al., 1984). Cultivars with high rates of root elongation had roots that were approximately 9- to 10-cm deeper at 46 and 56 d after sowing than cultivars that had low rates of root elongation. The cultivars from the faster elongating group also had three times the root density at 150-cm depth than did the slower elongating group.

Equally important as roots growing fast during early developmental stages is the continued growth of roots during reproductive growth. For the seven cultivars evaluated by Kaspar et al. (1978), the maximum rate of downward root growth occurred during flowering with rates ranging from 8.1 to 3.3 cm d^{-1}. Similar rates of downward root growth were measured until R5, at which time roots of most of the cultivars had reached the bottom of the rhizotron. During vegetative growth, rates of root growth ranged from 1.1 to 1.8 cm d^{-1} and were significantly less for all cultivars than growth rates during flowering.

The rate of root growth is also responsive to soil temperature (Bland, 1993), which is particularly important for the early-sowing management practice used throughout the soybean production region. As soil temperature increased from the soil surface to lower soil horizons, the rate of downward root growth increased from 0.9 cm d^{-1} in a slowly warming soil to 1.2 cm d^{-1} in a rapidly warming soil, which contrasted with a rate of root growth of 2.6 cm d^{-1} in an isothermal soil (Bland, 1993). Stone and Taylor (1983) found that soybean taproot extension rates increased linearly from 0 cm d^{-1} at 13°C to 6.5 cm d^{-1} at 26°C. Genotypic variation for root growth in cool soil has not been reported but could be important in improving soil water extraction for young plants in the spring and plants approaching maturity in the fall.

12–6 TRAITS INFLUENCING WATER-USE EFFICIENCY

To produce biomass via photosynthesis, atmospheric CO_2 must diffuse to carboxylation sites in the chloroplasts of the leaf mesophyll cells, which requires open stomates in the epidermis. However, water evaporating from cell walls inside the leaf can then escape to the atmosphere by diffusing out of those same stomates. Condon et al. (2002) stated the problem succinctly: "Water is the required unit of exchange for the acquisition of CO_2 by plants." The degree to which this tight coupling of photosynthesis and transpiration can be altered is the subject of this section of our review.

As noted before, WUE is a response coefficient reflecting the unit change in biomass per unit change in transpiration (i.e., $\Delta BM/\Delta T$). An enhancement in WUE would steepen the slope of the biomass-transpiration relationship shown in Fig. 12–2 and, given a constant HI, would also steepen the slope of the seed mass-transpiration relationship. To enhance WUE, one must either increase the increment of biomass (i.e., ΔBM) generated per incremental unit of transpiration (i.e., ΔT), or inversely, decrease the transpiratory increment (i.e., ΔT) associated with a given incremental unit of biomass (i.e., ΔBM). It is important for the reader to recognize the distinction between the component of the WUE term and the T term in Eq. [1], since decreasing the "transpiration" component (i.e., ΔT) of WUE is *not incompatible* with our prior statement of the desirability of increasing transpiration (T) term in Eq. [1].

Of the traits listed by Ludlow and Muchow (1990), only two—(Trait 9) greater leaf transpiration efficiency (TE), and (Trait 10) increased leaf reflectivity—were expected to possibly improve WUE (Table 12–1). Transpiration efficiency (TE) is defined as the mass or moles of C or CO_2 fixed per unit of water lost from the *leaf*, and it will be discussed first.

Before proceeding, some theoretical considerations important for understanding WUE are briefly discussed. In an exceptionally thorough and insightful paper reviewing this subject, Tanner and Sinclair (1983) noted that evaporative demand was driven by the leaf-to-air vpd, which is dependent upon site- and year-specific meteorological conditions. They partitioned WUE into a biological (crop) and a physical (meteorological) component:

$$\mathrm{WUE} = k \times vpd^{-1} \qquad\qquad [10]$$

Tanner and Sinclair (1983) defined k as a species-specific coefficient of crop seasonal water use, whereas vpd was defined as a seasonal integration of the average daily atmospheric vpd experienced by the crop during transpiration.

The k coefficient in the analysis of Tanner and Sinclair (1983) was estimated to be 4 Pa for the C3 soybean vs. 12 Pa for C4 maize. The mechanistic factors the authors coalesced into the k coefficient were described in detail in their classic paper, and later succinctly restated by Sinclair (1994). One of the key components included in k was the term c, defined as $c = 1 - C_i/C_a$, where C_i denotes the mole fraction of the internal (leaf) CO_2, and C_a denotes the atmospheric CO_2. Tanner and Sinclair (1983) noted that the maximal value for this c term was about 0.3 for many C3 species, and about 0.7 for many C4 species. Sinclair (1994) suggested that c, after many years of empirical breeding, had probably been optimized to a near-constant value for many C3 crop species ($c = 0.3$) and for many C4 crop species ($c = 0.7$). He noted that while genetic variation in WUE had been reported for a given C3 or C4 species, the upper range of that reported variability rarely exceeded those given c values. Because of that, he speculated that, within a species, there was probably little opportunity to make more improvement in WUE.

Other scientists have, however, disputed that notion. Let us consider their view in light of the definition of TE. As noted previously, TE reflects the change in CO_2 uptake per unit change in H_2O evolution. Given that TE is governed by the concentration gradients of CO_2 and H_2O between the inside of the leaf and the atmosphere, it can be described mathematically as:

$$\mathrm{TE} = [C_a \times (1 - C_i/C_a)] \times [1.6 \times (e_i - e_a)]^{-1} \qquad\qquad [11]$$

where C_i and C_a were defined as above, and e_i and e_a refer to the water vapor concentrations inside the leaf and in the surrounding atmosphere, respectively. The 1.6 constant in the denominator of Eq. [11] accounts for the diffusivity difference between H_2O and (the heavier) CO_2 molecule. Equation [11] indicates that increases in TE can be accomplished via decreases in the internal CO_2 concentration, increases in the external CO_2 concentration, or decreases in the water vapor concentration gradient from the leaf to the atmosphere. A decrease in C_i is an effective means of improving TE if other terms in the equation do not change. Indeed, because C_i is considerably lower in C4 leaves (~100 μL L^{-1}) compared to C3 leaves (~200 μL L^{-1}), C4 plants have considerably higher TE than C3 plants (Fig. 12–2). The higher TE allows C4 plants to operate at lower values of leaf conductance for gas exchange (Brown, 1994). On the other hand, the TE for soybean and other C3 species will likely increase relatively more rapidly than the TE in C4 species with respect to the rising level of atmospheric CO_2. However, such increased TE may be mitigated by unfavorable seasonal temperature changes that may be induced by the "greenhouse" effects of greater CO_2 (Allen, 1994).

Turner et al. (2001) disagree with the notion that intraspecies variation in WUE is small or nonexistent. Their argument is based on a technique that provides a retrospective (integrated) measure of crop TE up to the day a tissue sample is collected. This technique takes advantage of the fact that the heavier ^{13}C isotope of C is discriminated against relative to ^{12}C (most notably in C3 species) during the processes of CO_2 diffusion and photosynthetic carboxylation. The amount of discrimination (Δ, ‰) that occurs in C3 photosynthesis can be described by this simple relationship (Farquhar et al., 1982):

$$\Delta = a + (b - a) \times (C_i/C_a) \qquad [12]$$

where a represents the diffusitivity difference between $^{13}CO_2$ and $^{12}CO_2$ (~4.4‰), and b is the degree of isotopic discrimination effected by the enzyme ribulose 1,5-bisphosphate (RuBP) carboxylase (~27 ‰). The terms in the C_i/C_a ratio were defined above. TE, as defined by Eq. [11], is related to Δ as defined in Eq. [12], by their common dependence upon C_i/C_a. A close examination of the two equations reveals that TE increases linearly with declining values of Δ. Measurement of Δ thus provides an indication of TE in comparative leaf tissues that will be integrative over leaf developmental ontogeny up to the time those tissues were sampled for measurements. The negative relationship between Δ and TE in the field implies that low Δ genotypes would have greater biomass (and thus potentially greater seed yield), assuming of course that both the low and high Δ genotypes use the same amount of water for transpiration (Richards, 1996).

The reader interested in the genetic applications of this technique needs to be aware of the fact that, in some papers, carbon isotopic composition (δ), which has negative values, is often reported when δ is not converted to carbon isotopic discrimination (Δ), and the latter has positive values. This awareness is important since TE correlates negatively with Δ, but positively with δ.

Evidence of genotypic differences in TE within a crop species, based on Δ or δ measurements, or based on other forms of WUE measurement, has been reported in several C3 species. These include wheat (Farquhar and Richards, 1984; Sayre et al., 1995), peanut (*Arachis hypogaea* L.) (Hubick et al., 1986; Wright et al., 1994), cowpea [*Vigna unguiculata* (L.) Walp.] (Hall et al., 1992; Ismail and Hall, 1992, 1993), lentil (*Lens culinaris* Medik) (Matus et al., 1995), and soybean (Mian et al., 1996, 1998a; Purcell et al., 1997a; Specht et al. 1986, 2001).

Mian et al. (1996), who measured WUE in soybean by metering water to the potted plants grown in a greenhouse, reported that Young had considerably higher WUE (4.4 g dw L^{-1}) than did PI416937 (3.7 g dw L^{-1}). The 120 F4-derived families from a Young × PI416937 mating showed continuous variation in WUE, which is typical of a quantitative trait. Four independent restriction fragment length polymorphism (RFLP) markers were associated with WUE, which accounted for 38% of the phenotypic variation. In the same study, leaf ash was mapped, and six independent markers accounted for 53% of the variation. Leaf ash is negatively associated with WUE (Masle et al., 1992), and Mian et al. (1996) found a −0.40 correlation between WUE and leaf ash. The physiological basis of this correlation has not, however, been established (Turner et al., 2001). Mian et al. (1996) also determined that two QTLs were associated with both WUE and leaf ash, and that the alleles conditioning higher WUE at these two loci also conditioned lower leaf ash. In a subsequent experiment with an F2-mapping population from 'S100' × 'Tokyo' (Mian et al., 1998a), two independent RFLP markers were identified as being associated with WUE in this population. One marker was also associated with WUE in the Young × PI416937 population, and the other marker was novel, accounting for 14% of the variation in WUE in the S100 × Tokyo population. Specht et al. (2001) noted that in well-watered pot experiments, the amount of transpired water should be adjusted for any pot-to-pot differences in transpiring leaf area. The four WUE QTLs in the Young × PI416937 population reported by Mian et al. (1996) did have a different map position than the three QTLs for leaf size (cm^2 $leaf^{-1}$) at the 8th and 9th nodes that Mian et al. (1998c) identified in a later phenotyping of that population. The authors noted that the total leaf area of a soybean plant is a function of the size of individual leaves multiplied by the total number of leaves, but did not quantify that function in this mapping population.

Purcell et al. (1997a) compared the WUE of 'Jackson' with PI416937 and found in greenhouse experiments that under water-deficit conditions, Jackson accumulated more biomass and total N than did PI416937 with similar transpirational losses. Additionally, Δ was significantly lower for Jackson than for PI416937, as would be expected for a line with superior WUE. Jackson was noted previously as having the ability to continue N_2 fixation at soil-moisture contents considerably lower than that of other lines (Sall and Sinclair, 1991; Serraj and Sinclair, 1996b). The authors concluded that the superiority of N_2 fixation under water-deficits for Jackson was apparently due to the continued ability to sustain biomass production during water deficit and to allocate greater amounts of photosynthate to nodules prolonging nitrogenase activity. Differences among soybean genotypes in their response of N_2 fixation to water deficit are discussed in more detail in a subsequent section.

Specht et al. (2001) conducted a QTL analysis of soybean response to differing water levels generated by a sprinkler-gradient system that replenished 100, 80, 60, 40, 20, and 0% of weekly crop ET loss. The genotypic material was a population of 236 recombinant inbred lines from a 'Minsoy' × 'Noir 1' mating. Because of the fortuitous occurrence of a severe drought during one of the years, tissue samples from the genotypes grown in the 100% ET (water freely available for transpiration) and in the 0% ET (water supply severely limited) treatments were collected for carbon isotopic composition (δ) analyses. Plots of seed yield vs. crop water input in the 0, 20, 40, 60, 80, and 100% ET treatments were quite linear, and for each genotype a WUE coefficient was derived (cf., Fig. 12–2). The authors preferred to use the term *beta* for that coefficient, since it was actually a seed mass WUE, not a biomass WUE (total biomass was not determined in this field study). Significant genetic differences were observed for *beta*, which is an empirical, seed yield-based estimation of genotypic WUE, and for δ, which is purported to be a mechanistic, leaf tissue-based indication of TE. The genotypic correlation of *beta* and δ was positive confirming that the latter is indicative of WUE; however the correlation was low ($r = 0.26$). The major QTL identified as affecting these two traits turned out to be coincident with two major genes for maturity and one major gene for stem growth habit. Such genes have pleiotropic impacts on many traits. While some minor QTL were detected for *beta* and δ, their effects were too small to be of use in a marker-assisted selection program. The strong positive relationship ($r = 0.71$) between genotypic *beta* and genotypic mean yield indicated that selection for a high yield potential would produce cultivars with a high *beta*, and because of the latter's positive correlation with δ probably a higher TE as well. The authors believed that genotypic *beta*, a seed-based measure of WUE, was probably very reflective of biomass-based genotypic WUE, but HI data would be needed to confirm that contention.

The results of Specht et al. (2001) confirm in soybean the close association of seed yield and transpiration (i.e., compare Fig. 12–3 in Specht et al. with Fig. 12–2 in this chapter). Moreover, the close association between genotypic *beta* (a field-measured proxy for WUE) and genotypic mean yield provides some credence that soybean WUE and thus drought tolerance can be improved by simply selecting for genotypic mean yield. Although the Minsoy and Noir 1 parents of the mapping population used by Specht et al. (2001) differed significantly in both *beta* and mean yield, and produced transgressive segregation, Minsoy was not a priori known to be drought tolerant, nor was Noir 1 a priori known to be drought intolerant. The goal of identifying alleles conditioning even greater enhancements WUE than was seen in Specht et al. (2001) will likely require the evaluation of populations derived from matings in which one parent has an exceptionally high WUE (T.E. Carter, Jr., personal communication, 2002). Still, the caveat of Sinclair (1994)—that the upper limit of the reported genetic variation in WUE for most C3 crop species may already be near its genetic maximum—should not be taken lightly, particularly given the effort required to make WUE measurements.

Although there are numerous reports documenting differences in Δ within any given crop species, these differences do not routinely translate into differences into greater grain yield in field studies (Ludlow and Muchow, 1990; Turner, 1993; Turner et al. 2001). One possible reason for this discrepancy is that TE would be expected to increase if stomatal conductance is decreased while holding photosynthesis constant. However, with less transpirational cooling, leaf temperature rises, thereby *increasing* the vpd (vpd in Eq. [10]) between the leaf and atmosphere. Thus, an increased leaf temperature and a greater vpd would likely negate any increase in TE arising from decreased stomatal conductance. Moreover, higher leaf temperatures may also increase respiration, which would decrease net crop biomass gain, thus depressing the numerator portion of TE. It would be much better to increase the numerator of TE, given that the largest crop-specific difference in TE (i.e., WUE) is that existing between C4 and C3 plant species (Fig. 12–2), and it arose from a genetic improvement in the carboxylation numerator of TE. Improving soybean TE would thus necessitate its conversion from a C3 to a C4 species. This is not entirely impossible, given transgenic progress towards that goal in C3 rice reported by Ku et al. (1999). Small improvements in TE may be of little practical use, since a marginally higher TE is more beneficial when the amount of seasonal transpiration is high (right side of Fig. 12–2) than when it is low (left side of Fig.12– 2) as a result of a water deficit.

Nevertheless, an intensive Australian program for increasing drought tolerance of wheat has recently demonstrated the value of low Δ (i.e., a proxy for high TE) for yield improvement, and not just in low-yield production environments (Condon et al., 2002; Rebetzke et al., 2002; Richards et al., 2002). Rebetzke et al. (2002) backcrossed low Δ into an elite cultivar, resulting in 30 lines ($BC_2F_{4:6}$) with low Δ and 30 lines with high Δ. Averaged over eight environments, the mean yield of the low-Δ group was significantly greater (230 kg ha^{-1}). Surprisingly, the yield advantage of the low-Δ lines over high-Δ lines was not only larger (ca. 11% greater) in the environment lowest in yield (i.e., 1000 kg ha^{-1}), it persisted in environments of much higher yield (ca. 3% greater in 6000 kg ha^{-1} environments). Using trend line analysis, the yield advantage did not zero out until extrapolated to (predicted, not actual) environments of relatively high yield (i.e., 7700 kg ha^{-1}), indicating that yield potential had not been compromised in the low-Δ lines. Other advantages (i.e., greater biomass, enhanced HI) were also reported for wheat lines selected to have a low Δ and thus higher TE (the latter by implication, since it was not directly confirmed in these lines). Richards et al. (2002) noted that a new wheat cultivar, selected for low Δ was to be released in 2001 in the northern wheat area of Australia. Although a comparable effort has not been undertaken in soybean to increase TE, these results offer a promising case study of how drought tolerance was achievable via a multidisciplinary, long-term research approach and a targeted objective.

Greater leaf reflectivity to incoming solar radiation should lower leaf temperature and thus the *vpd* component of WUE (Trait 10, Table 12–1), assuming no offsetting decrease in biomass accumulation occurs as a result of the nonabsorption of the solar radiation that is reflected. These conditions are most likely to occur if the canopy was light saturated. If daily ΔBM is not appreciably reduced by greater leaf reflectivity, then its new value, now inextricably coupled with a fractionally smaller daily ΔT would be an improvement in the denominator part of WUE. This might be of particular value in a terminal stress environment, assuming that the unused water (resulting from a lessening of daily ΔT per unit of daily ΔB) can be retained and made available for crop use later in crop development. The value of such conserved water in an intermittent stress environment would depend on the duration of each intermittent stress. Soybean leaf reflectivity can be genetically manipulated (via a dominant gene) to achieve a denser population of trichomes on the leaf surface (Clawson et al., 1986). This does seem to improve WUE (Fig. 12–8). However, the denser leaf pubescence must be gray, not tawny, in color, and isogenic analysis of normal vs. dense pubescence in multiple cultivar backgrounds has revealed that densely pubescent lines do not always have higher grain yield than their nondense counterparts (Zhang et al., 1992).

More realistic increases in WUE on a field basis may be realized by focusing on the physical component of its denominator of Eq. [10]. If more of the crop's life cycle coincides with periods of cooler temperatures (i.e., lower evaporative demand) and more frequent rainfall, the crop will have an inherently higher WUE (Sinclair et al., 1984). In cropping systems where water deficit is primarily associated with late-summer drought, early sowing of early-maturing cultivars at high population densities minimizes evaporative water loss from the soil and increases WUE due to cooler temperatures and a decreased vpd. Often the crop can use the conserved water later in its reproductive development. The Early Soybean Production System, now widely adopted in the southern Mississippi Delta, utilizes this strategy (Bowers, 1995; Heatherly, 1999).

12–7 TRAITS INFLUENCING HARVEST INDEX

Yield under water-limited conditions, as defined by Eq. [1], is the product of the amount of transpired water, WUE, and HI. The product of the first two terms determines the total amount of aboveground biomass at harvest maturity, but it is

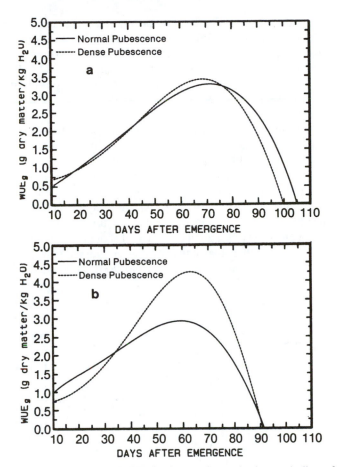

Fig. 12–8. Daily water-use efficiency (WUE) for dense and normal pubescent isolines of the soybean cultivars (a) Clark (tawny pubescence) and (b) Harosoy (gray pubescence). Crop evapotranspiration (ET) was estimated using a soil water budget computed from weekly measurements of soil water content and precipitation, and converted into daily values of ET loss per unit land area. Crop biomass (excluding roots and naturally abscised leaves) was collected twice per week, and converted into daily values of Crop Growth Rate (CGR), which was assumed to be analogous to net photosynthesis per unit land area. Daily WUE was then computed as the CGR/ET ratio. Redrawn (with permission) from Fig. 3 of Clawson et al. (1986).

HI that determines what proportion of that biomass is or will be allocated to grain (Fig. 12–2). Therefore, HI depends upon both the growth rate of seeds and the duration of seed growth, and these concepts are reviewed elsewhere (Egli and Crafts-Brandner, 1996).

Soybean grain yield is the functional product of the mean number of plants per unit area, pods per plant, seeds per pod, and the average mass of the individual seed. Each of these yield components is sequentially fixed (in the order just stated) during specific timeframes of soybean ontogenetic development (Shaw and Laing, 1965). Water scarcity, if confined to a specific developmental phase, reduces the yield component that is established and fixed in that particular phase. Conversely, water abundance during a specific developmental phase enhances that particular component. Shaw and Laing (1965) were the first to demonstrate this by applying serially timed stress periods from the beginning of flowering to the end of seed-fill. Their work can be summarized by stating that (i) stress early in R1 to R6 (Fehr and Caviness, 1977) ontogeny reduces seed number, but upon cessation of the stress, that reduction is offset by enhanced individual seed mass (i.e., 100-seed weight), and (ii) stress later in R1 to R6 greatly depresses seed size, the last yield component to be fixed during reproductive ontogeny. Those observations have since been verified by others (Sionit and Kramer, 1977; Constable and Hearn, 1978; Martin et al., 1979; Momen et al., 1979), and are illustrated again in the data presented in Table 12–2. Other workers (Kadhem et al,1985a, 1985b) have used timed peaks of water abundance (i.e., irrigation) during R1 to R6 to evaluate yield component response. Those studies show that as those peaks are delayed after R1, the response of seed number to water declines linearly (Fig. 12–9, top), whereas seed size is actually depressed below that of a rainfed control until about R4, but is enhanced thereafter to a level exceeding the control that was never stressed (Fig. 12–9, bottom). Korte et al. (1983a, 1983b) and Specht et al. (1986, 2001) documented that drought

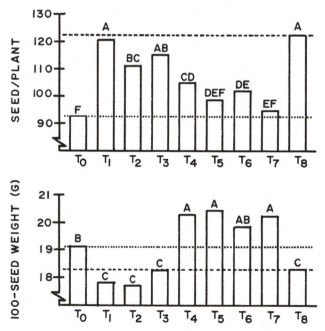

Fig. 12–9. Response of the soybean seed number (per plant) and seed mass (100-seed weight) yield to *peaks of water abundance* (i.e., T_1 to T_7; each a single irrigation) that were scheduled at 10-d intervals over a 60-d span of reproductive ontogeny (R1–R6.5). The controls included a rainfed (non-irrigated) treatment (T_0), and a well-watered treatment (T_8) irrigated every 10 d (i.e., each coincident with one of the T1–T7 single irrigation treatments). The response data are 2-yr, 16-cultivar, 4-replicate means. The *reversal* in seed mass response between T3 (seed mass smaller than T0) and T4 (seed mass larger than T8) occurred between stages R3.7 and R4.7, respectively, near the end of pod elongation but before the beginning of seed enlargement. Redrawn (with permission) from Fig. 3 of Kadhem et al. (1985).

accelerated crop senescence, and noted that the yield component bearing the brunt of this earlier crop maturity was almost always 100-seed weight. Both effects impact HI, as Specht et al. (1986) observed.

The traits in Table 12–1 that we have previously mentioned, with the exception of Traits 5 (heat tolerance), 7 (early vigor), 9 (transpiration efficiency), and 10 (leaf reflectance), were also expected to impact HI. Two additional traits affecting HI include: Trait 11 (high axial root resistance, discussed previously as a means of limiting water use during the vegetative phase so that it can be conserved for use during the reproductive use), and Trait 12 (remobilization of pre-anthesis dry matter into the grain).

The assimilate required for seed growth is derived from a combination of two sources (Ludlow and Muchow, 1990; Turner et al., 2001): (i) the production of current assimilate that is partitioned for grain filling during reproductive growth, and (ii) the redistribution of assimilate from sources produced earlier in the season, primarily during vegetative growth, to seeds. Drought, as discussed in previous sections, directly affects the production of biomass. Prior to anthesis, drought results in a smaller pool of vegetative mass from which remobilization can occur. After anthesis, drought decreases the production of assimilate used to support concurrent grain growth, and this increases the reliance upon stored assimilate. The depletion of C and N reserves from vegetative sources during grain filling is associated with premature senescence and a decreased length of the seed-fill period (Sinclair and deWitt, 1975; Cure et al., 1983; Korte et al., 1983a; Specht et al., 1986, 2001; Salado-Navarro et al., 1985; De Souza et al., 1997). Soybean exposed to water-deficit stress and short photoperiods during seed fill (late R5–R6) were unable to recover photosynthetic capacity following stress removal (Cure et al., 1983). In contrast, plants with a decreased demand for nutrient remobilization due to a decreased seed-filling rate (Thomas and Raper, 1976), were able to recover photosynthetic activity following stress removal (Cure et al., 1983). These results are similar to the observation that drought occurring late during reproductive development shortened the seed-fill period compared to well-watered treatments (Meckel et al., 1984; De Souza et al., 1997).

If remobilization is important in yield formation, then increasing vegetative mass prior to seed filling could serve as a buffer for stresses that occur during reproductive development. Under well-watered conditions in Kentucky, MG V cultivars had approximately twice the vegetative mass as MG I cultivars at the beginning of R6, and at maturity, grain yield was about 6% greater for MG V cultivars than for MG I cultivars (Egli, 1997). Nevertheless, imposing a 63% shade treatment from R6 to maturity decreased yield similarly (23–26%) for early- and late-maturing cultivars. Given that this shading treatment had no effect on crop senescence or maturity, which is typical of late-season drought-stress effects (Cure et al., 1983; De Souza et al., 1997), it is difficult to extend these results to drought conditions. The speculation that increased vegetative mass at beginning of R6 ameliorates yield losses due to late-season drought thus needs confirmation.

Water deficit in a greenhouse experiment (Table 12–2) from R1 to R5 had virtually no effect on HI, whereas water deficit from R5 to maturity decreased HI from 0.61 in well-watered plants to 0.48. Decreased HI due to water-deficit stress in field experiments also has been shown to be a major determinant of yield loss. Specht

Table 12–2. Response of seed yield, individual seed mass, seed number, and harvest index to water deficit in greenhouse-grown soybean. Values were averaged over the Maturity Group 00 cv. McCall and Glacier (cultivar and cultivar × treatment were not significant). Unpublished data of Reaper and Purcell, 1999.

Water treatment†	Seed yield‡	Seed number	Individual seed mass	Harvest index
	g pot^{-1}	no. pot^{-1}	mg seed^{-1}	
Control	29.3 a*	240 a	120 b	0.61 a
R1–R5	24.1 b	129 c	190 a	0.62 a
R5–R7	16.9 c	196 b	90 c	0.48 b

* Means in a column followed by different letters are significantly different as determined by an LSD ($P = 0.05$).

† Plants of the control treatment were watered to 70% of pot capacity weight daily, plants of the R1–R5 treatment were watered to 35% of pot capacity weights daily from R1 to R5, and plants of the R5–R7 treatment were watered to 35% of the pot capacity weight daily from R5 to R7.

‡ Each pot had six plants, and each pot was considered an experimental unit.

et al. (1986) used a line-source irrigation system in their evaluation of 16 soybean cultivars during the drought years of 1983 and 1984 in Nebraska. After threshing, the grain and haulm (i.e., stem and pod wall) fractions were weighed to generate an *apparent* HI (Schapaugh and Wilcox, 1980). The 1983 data are presented in Fig. 12–10. Note that the grain response to water (0.070 Mg ha^{-1} cm^{-1}) was greater than that of the haulm (0.051 Mg ha^{-1} cm^{-1}). The ratio of the grain mass and biomass regression coefficients (0.070/0.121) suggested that, for each incremental amount of biomass generated per added centimeter of crop water use, a constant 58% of that increment was grain. However, the response of apparent HI to water was asymptotic, going from 0.25 in the nonwatered (drought-stressed) plots to a probable near-plateau value of 0.41 in the well-watered plots. In 1984, the crop was planted 2 wk later than in 1983 and grain yields were lower, but a similar asymptotic response between 0.35 and 0.47 was observed (Specht et al., 1986). In both 1983 and 1984, rainfall abruptly ended at the end of June–early July (just before R1), and remained scarce until a drought-terminating rainfall event occurred in each year about 20 August (about R6 in 1983; about R5.5 in 1984). Drought hastened senescence by 13 d in 1983, and 9 d in 1984. The HI data in Fig. 12–10 and in Table 12–2 indicate that, when severe drought spans the seed enlargement phase *and* hastens maturity, HI may not be completely independent of T (as hypothesized in Eq. [1] and illustrated in Fig. 12–2).

Selection for greater HI in drought conditions might help lessen the asymptotic HI response. A mild drought during grain filling in Arkansas resulted in a 20 to 30% decrease in yield for two different cultivars (Purcell et al., 1997b). Approximately 90% of the yield decrease was attributable to decreased HI, with the remainder attributable to decreased crop biomass at maturity. At earlier developmental stages, during vegetative, flowering, and pod set stages, HI within a given cultivar was relatively stable in response to both drought (Spaeth et al., 1984) and defoliation treatments (Board and Tan, 1995), and yield decreases due to drought were associated with decreased biomass.

Selection for high HI during drought has not been a priority for soybean breeders because of the uncertainty of its validity as a selection criterion, the additional

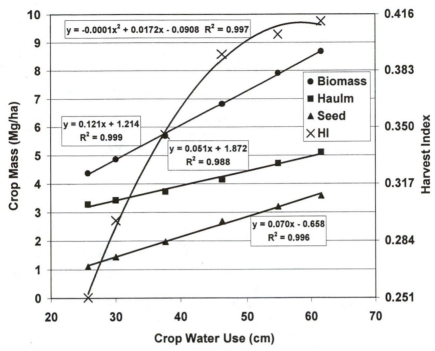

Fig. 12–10. Response of soybean mature standing biomass, haulm (mature stems and pod walls), seed mass, and (apparent) harvest index (HI) to six seasonal water amounts during a severe drought (no rainfall) spanning 50 d after R1. Data are the means of 16 cultivars and four reps of a 1983 experiment. The boxes containing the regression data correspond to the trend lines that the box corner touches. Redrawn (with permission) from data presented in the last four panels of Fig. 3 in Specht et al. (1986).

effort in measuring HI, and because HI is affected by cultivar, environment, and developmental stage when the crop is exposed to drought. Direct selection for yield would appear to be a more direct means of increasing yield in drought environment, but this assumes that HI does not have a large genotype × environment interaction due to drought. For example, suppose two cultivars ('A' and 'B') have similar maturity, yield potential (e.g., 4000 kg ha^{-1}), and HI (0.55) under well-watered conditions. If under drought, both also have similar yield (1500 kg ha^{-1}), but the HI of 'A' remains constant at 0.55 while that of 'B' is decreased to 0.40, then 'B' actually produced approximately 1000 kg ha^{-1} more biomass than did 'A'. If only the grain yield of these two cultivars was measured under drought, the drought-tolerant biomass production of cv. 'B' would likely go unrecognized. Assuming that the drought-stable HI of cv. 'A' and the drought-tolerant biomass production of 'B' are genetically independent characters, then the combination of both traits into a single cultivar would be expected to lead to better performance in drought environments. However, Schapaugh and Wilcox (1980) reported that HI, measured either with or without the collection of abscised leaves, was negatively correlated with total biomass ($r = -0.73$), vegetative biomass ($r = -0.86$), mature (i.e., no leaves and petioles) biomass ($r = -0.67$), and haulm weight ($r = -0.85$). Likely to be of most concern to breeders selecting for greater HI was the highly negative correlation of HI with cultivar maturity ($r = -0.75$), with early-maturing cultivars exhibiting high HI (ca. 0.45–0.50) and late-maturing cultivars exhibiting low HI (ca. 0.30–0.35). Studies will need to be conducted to determine if HI QTLs exist, and then whether the alleles at those HI QTLs can significantly enhance HI and do so without pleiotropically depressing total biomass.

12–8 DROUGHT STRESS AND PLANT NITROGEN NUTRITION

12–8.1 Nitrogen Redistribution—Impact of Drought

The high protein concentration of soybean grain requires a tremendous amount of N and much of this N is derived from vegetative sources. Sinclair and deWitt (1975, 1976) proposed that, because of the high protein concentration of soybean grain, N needs could not be fully met by concurrent uptake of soil N or from N_2 fixation. Consequently, a remobilization of N from vegetative tissues was required to supplement the N supply, and they suggested that when N demand still exceeded N supply, the result was a loss of photosynthetic capacity and premature senescence. Environmental conditions that increased the plant's N requirement or decreased N supply would exacerbate those effects. By and large, experimental data are consistent with this hypothesis.

Under well-watered conditions, the nitrogen HI (NHI) of a high protein soybean genotype was 0.90 compared to a standard cultivar with a NHI of 0.80 (Salado-Navarro et al., 1985). In contrast, HI of these lines, based upon grain mass and dry matter at maturity, was identical at 0.48. This indicates that the mobilization of N from vegetative tissues to the seed tissue is actually more intensive than the mobilization of C, which agrees with other data on N (Zeiher et al, 1982) and ^{14}C remobilization during seed fill (Hume and Criswell, 1973). In that the length of the seed-fill period was negatively correlated with seed-protein concentration ($r = -0.58$), Salado-Navarro et al. (1985) concluded that the rapid remobilization of N from vegetative tissues to seed in high protein lines accelerated senescence and decreased the duration of seed-fill. Kumudini et al. (2001, 2002) found that the yield advantage of newer, early-maturing cultivars over older cultivars of similar maturity was associated with a greater capacity to continue N accumulation during the seed-filling period and with a longer leaf area duration. In contrast to these results, Zeiher et al. (1982) concluded that although a large proportion of seed N was derived from vegetative tissues, the inherent demand of N by the seed and that redistribution of N did not appear to determine the length of the seed-fill period. An alternative hypothesis for senescence in soybean (Egli and Crafts-Brandner, 1996) is that photosynthate utilization by developing seeds decreases photosynthate availability in leaves below a threshold value that is required for maintenance of leaf physiological activity, ultimately resulting in leaf senescence.

Certainly, redistribution of N from vegetative sources to seed appears to be important in contributing to the high protein concentration of soybean seed. Under water-deficit conditions the remobilization of N from vegetative tissues may be particularly important in supplying carbohydrate and N to seed. In chickpea (Cicer arietinum L.) under well-watered conditions, 9% of the C and 67% of the N in the seed, in a desi-type cultivar, were remobilized from vegetative sources, but under water-deficit stress, 13 and 88% of the C and N, respectively, were derived from vegetative tissues (Davies et al., 2000). Decreased photosynthesis and N_2 fixation during pod-filling increase the importance of remobilized C and N, respectively, in meeting the needs of the seed. In soybean, N_2 fixation is more sensitive to water-deficit stress than are many other processes (Sinclair, 1986; Sinclair et al., 1987; Sall and Sinclair, 1991; Sinclair and Serraj, 1995; Serraj et al., 1999a), which may exacerbate N remobilization and thereby accelerate plant senescence (Cure et al., 1983; Salado-Navarro et al., 1985; De Souza et al., 1997). The importance of N_2 fixation to continued physiological activity is discussed in the subsequent section.

12–8.2 Drought Stress—Impact on Dinitrogen Fixation

Dinitrogen fixation in soybean is more sensitive to drought stress than is photosynthesis (Durand et al. 1987), biomass accumulation (Sinclair et al., 1987; Serraj et al., 1997), transpiration (Sall and Sinclair, 1991), and accumulation and assimilation of N from the soil (Purcell and King, 1996). Similarly, under root hypoxia and flooding stresses, N accumulation in plants entirely dependent upon N_2 fixation was affected more than that in plants supplemented with nitrate (NO_3^-) (Bacanamwo and Purcell, 1999). In that a large proportion of the N in soybean seed is derived from N_2 fixation (Matheny and Hunt, 1983), the sensitivity of N_2 fixation to drought probably imposes more severe limitations to grain production than those processes that are less sensitive to drought (for a review, see Serraj et al., 1999a). Prolonging N_2 fixation during late-season drought may be one means of preventing the premature senescence associated with late-season drought (Cure et al., 1983; Korte et al., 1983a, 1983b; Specht et al., 2001; Salado-Navarro et al., 1985; De Souza et al., 1997). From this standpoint, increased drought tolerance of N_2 fixation may be one means of maintaining HI during late-season drought (Table 12–1).

The importance to grain yield of a continued N supply during drought is indicated by nonirrigated field studies in which grain yield increased from 2373 kg ha^{-1} in the absence of N fertilizer to 2798 kg ha^{-1} with the addition of 336 kg ha^{-1} of N (Purcell and King, 1996). In the irrigated treatments, N fertilizer did not affect yield. During a moderate drought, high rates of NH_4NO_3 addition increased the rates of biomass and N accumulation to levels similar to those of irrigated treatments without N addition. Subsequent field experiments indicated that N application had no significant effect on HI and that the increased yield was due to greater biomass (De, 1997).

Other experiments evaluating the response of yield to N fertilizer have generally not included irrigation and drought as experimental treatments. Literature indicates that yield responses to fertilizer are either neutral or positive to N fertilizer when soybean is irrigated or grown with an adequate water supply but that yield responses are more common when crop-water supply is limiting. Weber (1966) grew nodulating and nonnodulating near isolines with different amounts of N fertilizer for 5 yr under rainfed conditions in Iowa. He found that the yield of the nonnodulating line increased as the amount of fertilizer N increased but that there was no increase in yield of the nodulating line with fertilizer application as high as 672 kg

N ha^{-1}. In contrast, Brevedan et al. (1978) found that N fertilization of 168 kg N ha^{-1} under well-watered conditions increased yield from 2500 kg ha^{-1} (0 N ha^{-1}) to 3300 kg ha^{-1} when applied at the beginning of flowering and to 3326 kg ha^{-1} when applied at the end of flowering. Lower amounts of N fertilizer (50–100 kg N ha^{-1}) increased pod yield up to 64% for vegetable soybean grown during the rainy season in Thailand (Yinbo et al., 1997) and increased seed yield by 453 kg ha^{-1} for irrigated soybean that was foliarly fertilized (Wesley et al., 1998). Sorensen and Penas (1978) found that N fertilization increased yield in 9 out of 13 on-farm trials. The four sites that did not respond to N fertilizer had the highest yield (2740–3650 kg ha^{-1}) and those that had the greatest response to N fertilizer had the lowest yields (1420 and 1830 kg ha^{-1}) and most severe drought. Lyons and Earley (1952) also found that N fertilization at two sites in Illinois resulted in large yield increases in a year that was hot and dry but that there was no yield response to N fertilizer in years in which rainfall was adequate and temperature was moderate. Although increasing the drought tolerance of soybean by addition of costly N fertilizer is not a viable economic option, these data provide strong support for the view that increased drought tolerance of N$_2$ fixation would result in yield increases.

There are differences among soybean genotypes in their sensitivity of N$_2$ fixation to drought. The cv. Jackson was noted for its ability to continue accumulating N at higher rates during drought than eight other cultivars (Sall and Sinclair, 1991). The superiority of Jackson in continuing N$_2$ fixation during drought was confirmed in subsequent greenhouse (Sall and Sinclair, 1991; Serraj and Sinclair, 1996b; Purcell et al., 1997a, 2000) and field experiments (Serraj et al., 1997).

Drought tolerance that is found in Jackson is apparently heritable and is realized through the genetic contributions of its maternal parent, 'Volstate', instead of its paternal parent, 'Palmetto' (Serraj and Sinclair, 1996b). Analysis of segregating F4:5 families of a cross between Jackson and a drought-sensitive cultivar, KS4895, indicates that drought-tolerant N$_2$ fixation is controlled by only a few genes with a narrow-sense, heritability of about 0.36, when progeny performance is expressed as the mean of four observations (Sneller et al., 1997). Furthermore, upon field evaluation of approximately 20 lines derived from this cross, several lines had yields greater than checks under drought stress but not under irrigated conditions (Sneller et al., 2000). Additional yield evaluations will be required to assess the ultimate value of drought-tolerant N$_2$ fixation in increasing non-irrigated soybean yield.

Understanding the mechanism of drought tolerant N$_2$ fixation may also lead to more efficient selection tools, such as marker-assisted selection, that can be used to move drought tolerance into elite cultivars. Physiological explanations for decreased N$_2$ fixation in response to water-deficit stress have generally focused on one of three different mechanisms (reviewed by Serraj et al., 1999a): nodule oxygen supply, nodule carbohydrate supply, and regulation of N$_2$ fixation by accumulation of N-containing compounds. Serraj et al. (1999a) point out that each of these mechanisms are commonly linked to the volumetric flow of water to nodules, which is delivered almost exclusively through the phloem (Walsh et al., 1989). A decreased volumetric flow of water in the phloem to nodules may constitute the fundamental elicitor of the observed changes in these three putative mechanisms regulating N$_2$

fixation under water deficit. Differences have been noted between drought-tolerant and drought-sensitive lines for each of these three mechanisms.

Nodule oxygen supply is closely regulated by alterations in the permeability of nodules to oxygen (Hunt and Layzell, 1993). During the early stages of water-deficit stress, decreased nodule permeability to oxygen limits nodule respiration and the energy equivalents required for N_2 fixation (Weisz et al., 1985; Serraj and Sinclair, 1996a, 1996b; King and Purcell, 2001). Serraj and Sinclair (1996a) found that the response of Jackson to oxygen was different from that of Biloxi, a drought sensitive cultivar. Under mild moisture deficit, nitrogenase activity at 21 kPa oxygen of both Jackson and Biloxi was approximately 50% of the well-watered plants, but nitrogenase activity was completely restored when the oxygen partial pressure was increased to 40 kPa (Serraj and Sinclair, 1996b), indicating an oxygen limitation to nodule activity. Under well-watered conditions, a 40 kPa rhizosphere oxygen treatment increased nitrogenase activity of Jackson approximately 50% compared to the rate at 21 kPa oxygen, whereas there was no response of oxygen enrichment to nodule activity in well-watered Biloxi plants. Although nodule permeability was not measured directly in this report, the authors speculated that permeability may have been lower in Jackson than in drought-sensitive lines. In that the specific activity of nodules for acetylene reduction in well-watered plants was considerably less for Jackson than for Biloxi (110 vs. 212 mol g^{-1} DW h^{-1}), Serraj and Sinclair (1996b) speculated that drought tolerance of Jackson might be associated with low nodule permeability and low rates of specific nodule activity. Subsequently, direct measurements of nodule permeability determined that Jackson had considerably higher nodule permeability to oxygen than did a drought-sensitive line, KS4895 (King and Purcell, 2001). Furthermore, under well-watered conditions, nodule specific activity for acetylene reduction was not significantly different between Jackson (372 mol g^{-1} DW h^{-1}) and KS4895 (264 mol g^{-1} DW h^{-1}), but the nodule activity for Jackson under well-watered conditions was approximately 3.3-fold higher than those reported by Serraj and Sinclair (1996b).

A higher oxygen flux to nodules during water deficit would be expected to increase nodule respiration and N_2 fixation, provided that an adequate carbohydrate supply was also available. Differences among soybean genotypes for drought tolerant N_2 fixation may be associated with photosynthate and water supply to nodules. Several lines of evidence have led to this conclusion. First, in comparisons of ^{14}C movement from leaves to nodules, Jackson accumulated considerably more ^{14}C than did drought-sensitive KS4895 (Purcell et al., 1997a; King and Purcell, 2001). The greater ^{14}C concentration in Jackson may indicate a greater allocation of carbohydrate and water to nodules. Jackson has been noted as having larger individual nodules than drought-sensitive lines (Purcell et al., 1997a; King and Purcell, 2001). Serraj and Sinclair (1998) found that drought-tolerant N_2 fixation for several genotypes was positively correlated with the ratio of individual nodule mass of plants from a drought treatment to plants from a control treatment ($r = 0.84$). Large nodules may be advantageous during drought because the proportion of N_2-fixing tissue in large nodules increases from approximately 0% for a nodule with a radius of 1 mm to 70% for a nodule with a 2-mm radius (based on equations found in Weisz and Sinclair, 1988). The greater volume of N_2-fixing tissue would create a greater sink demand for photosynthate and result in greater supply of both carbohydrates and water, which are delivered to nodules in the phloem (Walsh et al., 1989). In agreement with this proposal, it was found that nodule relative water content increased, for water-deficit treatments, as nodule diameter increased from <2 mm to >4 mm (King and Purcell, 2001).

A third mechanism that appears important in conferring drought tolerant N_2 fixation is associated with potential feedback inhibition of N_2 fixation by accumulation of N compounds in soybean during drought (DeSilva et al., 1996; Serraj and Sinclair, 1996b; Serraj et al., 1999b; Purcell et al., 1998, 2000; King and Purcell, 2001). During the initial stages of water-deficit stress, the products of N_2 fixation, the ureides, accumulate in shoot tissues in controlled environments (DeSilva et al., 1996; Serraj et al., 1996b; Purcell et al., 2000; King and Purcell, 2001) and in field experiments (Serraj et al., 1997; Purcell et al., 1998, 2000). The accumulation of ureides in shoots during water deficit, however, is considerably greater in drought-sensitive lines than in drought-tolerant lines (Serraj et al., 1996b, 1997; Purcell et al., 1998, 2000; King and Purcell, 2001).

Not only do ureides accumulate in shoot tissues during water deficit, but they also increase in nodules (Purcell and Sinclair, 1995; Serraj et al., 1999b; Vadez et al., 2000; King and Purcell, 2001). This accumulation of ureides in nodules apparently results from an insufficient supply of phloem-derived water that is required for nodule ureide export (Purcell and Sinclair, 1995). In contrast to shoot ureides, drought-tolerant Jackson accumulates substantially more ureides in nodules than does drought-sensitive KS4895 (King and Purcell, 2001). The greater concentration of ureides in nodules of Jackson during drought likely reflects continued N_2 fixation relative to drought-sensitive lines, whereas tolerant and sensitive lines may both have restricted export from nodules. The higher concentration of ureides in nodules of drought-tolerant lines relative to drought-sensitive lines precludes the direct involvement of ureides as a signal to decrease N_2 fixation (Serraj et al., 1999b; King and Purcell, 2001). Other N compounds closely associated with N_2 fixation may directly serve as signals between shoots and nodules to decrease N_2 fixation during drought (Serraj et al., 1999a, 1999b; Vadez et al., 2000).

Differences in shoot ureide concentrations between drought-tolerant and drought-sensitive soybean have been utilized in identification of new sources of drought-tolerant germplasm. Approximately 3000 soybean lines were evaluated for petiole ureide concentration as the first tier of a screening program (Sinclair et al., 2000). There was a wide range of ureide concentration among lines, but the distribution was skewed towards the low concentration ranges (Fig. 12–11). Approximately 200 selections for the second phase of the screening were made from the 10th percentile of the natural-log-normalized distribution of ureide concentrations (Fig. 12–11). These 200 lines were evaluated for N accumulation during drought in a field environment with very low soil N. Based on this second tier screening data, 18 lines were advanced to the final tier of the screen. By following nodule activity under a progressive water deficit in a controlled environment, eight lines of the original 3000 were identified as drought tolerant for N_2 fixation. A key question that remains to be determined is whether tolerance among these lines is due to a common mechanism and, if not, whether the underlying genes in each line can be combined into one genotype to further increase N_2 fixation drought tolerance.

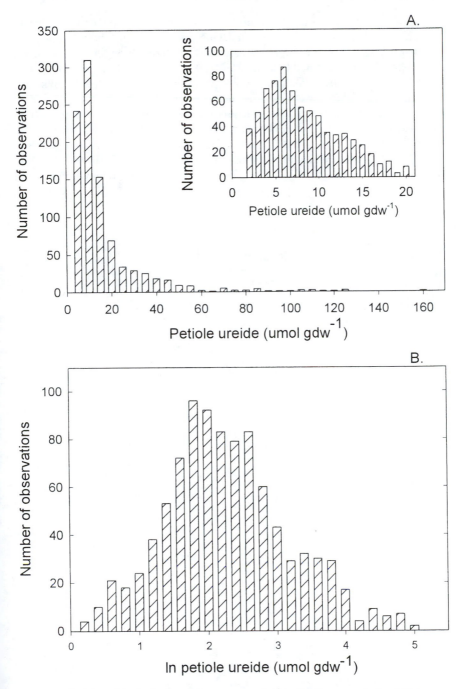

Fig. 12–11. Petiole ureide distribution (A.) among plant introduction lines as part of an initial screen for drought-tolerant N₂ fixation. The petiole ureide data were transformed by the natural logarithm and the lowest 10th percentile were advanced to the second stage of a screening program. Redrawn (with permission) from Fig. 1 of Sinclair et al. (2000).

The reason for the increased ureide concentration in shoot tissues during drought for soybean is not well understood. One possibility is that the catabolic pathway that is responsible for ureide breakdown is inhibited during drought (Purcell et al., 1998, 2000; Vadez and Sinclair, 2000, 2001; Vadez et al., 2000). The possible pathways for ureide breakdown are shown in Fig. 12–12. The initial step in ureide catabolism is the conversion of allantoin to allantoate by allantoinase (E.C. 3.5.2.5). The catabolism of allantoate to ureidoglycolate may follow one of two different routes, both of which have been reported in soybean. In the first route, Shelp and Ireland (1985), using 'Maple Arrow', found that allantoate amidinohydrolase (E.C. 3.5.3.4) converted allantoate to urea and ureidoglycolate. In the second route, Winkler et al. (1987), using 'Williams', found that allantoate amidohydrolase (E.C.

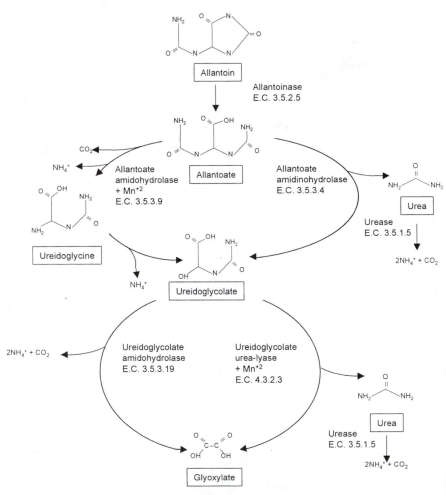

Fig. 12–12. Possible pathways for ureide (allantoin and allantoate) catabolism in soybean. Allantoate amidohydrolase is believed to have catalytic activity on both allantoate and ureidoglycine. There is a cofactor requirement for Mn^{+2} for both allantoate amidohydrolase and ureidoglycolate urea-lyase.

3.5.3.9) converted allantoate to ureidoglycine and released CO_2 and NH_4^+. The ureidoglycine was subsequently converted to ureidoglycolate by ureidoglycine aminohydrolase (Winkler et al., 1987). The catabolism of ureidoglycolate is also branched (Fig. 12–12); ureidoglycolate aminohydrolase (E.C. 3.5.3.19) releases two NH_4^+ molecules and glyoxylate. Ureidoglycolate urea-lyase (E.C. 4.3.2.3) is a Mn-dependent enzyme producing urea, two NH_4^+ molecules, CO_2, and glyoxylate. Ureidoglycolate urea-lyase is responsible for ureidoglycolate breakdown in chickpea (Munoz et al., 2001), but this pathway has not yet been characterized in soybean. Both the allantoate amidinohydrolase and the ureidoglycolate urea-lyase pathways (diagramed in Fig. 12–12) generate urea, but urea does not accumulate during N_2 fixation in urease-null plants from a Williams genetic background. (Stebbins and Polacco, 1995). This provides evidence that Williams soybean plants catabolize ureidoglycolate via the ureidoglycolate amidohydrolase route.

There is evidence that not all soybean lines use the same pathway for ureide breakdown as does Williams, and that this, in part, confers drought tolerance or sensitivity for N_2 fixation (Vadez and Sinclair, 2000, 2001). As mentioned in the previous discussion, 'Maple Arrow' was originally found to catabolize ureides through the allantoate amidinohydrolase route (Shelp and Ireland, 1985), in contrast to Williams which uses the allantoate-amidohydrolase route (Winkler et al., 1987). Vadez and Sinclair (2000) evaluated both pathways of ureide catabolism in these cultivars by loading leaves with allantoate and subsequently evaluated in situ ureide degradation in the presence or absence of asparagine or boric acid. Asparagine and boric acid serve to chelate Mn and would decrease ureide degradation if it occurred by the Mn-dependent allantoate amidohydrolase route (Lukaszewski et al., 1992). In a series of experiments, data were consistent with Mn being required by Williams, but not by Maple Arrow (Vadez and Sinclair, 2000). It would be of interest to determine which route of ureide degradation predominates among the soybean cultivars currently being used by producers in each U.S. maturity group, given the fact that Williams and Maple Arrow are MG III and 00, respectively.

It is important to note that the route of ureide breakdown through allantoate amidohydrolase requires Mn as a cofactor (Lukaszewski et al., 1992). Supplemental Mn leads to decreased tissue ureide concentration during drought and to an increased tolerance of N_2 fixation to water deficit (Purcell et al., 2000; Vadez et al., 2000; Vadez and Sinclair, 2001). On soils prone to Mn deficiency, such as those high in pH, excessively leached, or high in organic matter (Marschner, 1995), Mn fertility may be particularly important in managing non-irrigated soybean crops. Genetic differences exist in tolerance to Mn deficiency for soybean (Graham et al., 1995), which may offer genetic routes for lessening ureide concentration and prolonging N_2 fixation during drought.

12–9 CONCLUSIONS

Soybean grain production is directly related to the crop's ability to accumulate the large amounts of C and N that are needed for the high oil and protein concentrations of the seed. The accumulation of plant mass through photosynthesis is inseparably linked to transpiratory losses of water, and crop growth and yield are, therefore, directly dependent upon the quantity of water that the crop transpires. Frequently, insufficiencies of water limit the acquisition of C and N and the genetic yield potential of the crop.

Our purpose in this review was to focus on those traits that under drought conditions would most likely benefit soybean grain production and that could be manipulated through crop management, breeding and selection, or other genetic modification. In evaluating the potential benefit of specific traits (Table 12–1), we have considered their impacts on any or all of the three entities that determine grain yield in water-limited environments: total quantity of water available for transpiration (T), water-use efficiency (WUE), and harvest index (HI). Additionally, because of the importance of N acquisition for grain production, we considered the potential for increasing the tolerance of N_2 fixation to drought stress.

It is clear, based on the literature reviewed, that an increased rooting depth has much potential for increasing the water available for T, and thereby increasing grain yield. Root-growth restriction, due to soil physical and chemical barriers (Al toxicity), limit the volume of soil from which water is extracted. Incorporating traits for superior root penetration and tolerance to high soil Al may benefit soybean yield in environments where these factors are limitations. Additionally, osmotic adjustment in roots may be one means of achieving increased root growth under drought conditions, but the value of osmotic adjustment in leaves for increasing T is questioned.

Although we have noted reports in the literature of differences among soybean lines in transpiration efficiency, we have been unable to document any cases where superior transpiration efficiency in soybean resulted in superior grain yield or biomass production in field environments. Therefore, the superior transpiration efficiency that generally has been determined on a gas-exchange basis in controlled environments has not resulted in superior WUE in the field. One case where superiorWUE can be realized in field environments is by shifting the period of crop growth to periods when temperatures are cooler and evaporative demand is lessened. Likewise, early season vigor encourages quick canopy establishment and lessens the amount of water that evaporates from the soil. These strategies of conserving early-season soil moisture may be particularly important where late-season drought results in severe water limitations during grain filling. The water conserved by crop growth during periods of lessened evaporative demand may be utilized during grain filling, resulting in increased HI.

The final trait that we have evaluated in this review is that of prolonged N_2 fixation during water deficit. The high protein concentration of soybean grain and the apparent sensitivity of N_2 fixation to water-deficit stress make this a particularly attractive trait for improvement. Dinitrogen fixation tolerance to water deficits has been identified in several plant introductions and in an older cultivar. Incorporation of this trait into improved germplasm may provide a means of lessening the effects of late-season drought on premature crop senescence and continued grain growth.

The improvement of drought tolerance in soybean will likely require the combination of several traits. The specific combination of traits most beneficial for drought tolerance may differ from one geographic region to another and among production systems. Nevertheless, realistic improvements in drought tolerance, using the framework outlined in this chapter, appear to be a worthy and achievable goal. Incorporation of specific multi-genic traits in adapted cultivars will likely require a new paradigm in cultivar improvement. Characterization of these putatively beneficial traits at the physiological and molecular levels, and the combination of these traits into adapted, high-yielding cultivars will require close cooperation among agronomists, crop physiologists, breeders, and molecular biologists.

ACKNOWLEDGMENTS

The authors gratefully acknowledge support for their research programs from the United Soybean Board (L.C. Purcell and J.E. Specht), Arkansas Soybean Promotion Board (L.C. Purcell), and the Nebraska Soybean Development, Utilization, and Marketing Board (J.E. Specht).

REFERENCES

Allen, Jr., H.A. 1994. Carbon dioxide increase: Direct impacts on crops and indirect effects mediated through anticipated climatic changes. p. 425–459. *In* K.J. Boote et al. (ed.) Physiology and determination of crop yield. ASA, Madison, WI.

Armiger, W.H., C.D. Foy, A.L. Fleming, and B.E. Caldwell. 1968. Differential tolerance of soybean varieties to an acid soil high in exchangeable aluminum. Agron. J. 60:67–70.

Asady, G.H., A.J.M. Smucker, and M.W. Adams. 1985. Seedling test for the quantitative measurement of root tolerance to compacted soil. Crop Sci. 25:802–806.

Bacanamwo, M., and L.C. Purcell. 1999. Soybean dry matter and N accumulation responses to flooding stress, N sources, and hypoxia. J. Exp. Bot. 50:689–696.

Bianchi-Hall, C.M., T.E. Carter, Jr., M.A. Bailey, M.A.R. Mian, T.W. Rufty, D.A. Ashley, H.R. Boerma, C. Arellano, R.S. Hussey, and W.A. Parrott. 2000. Aluminum tolerance associated with quantitative trait loci derived from soybean PI 416937 in hydroponics. Crop Sci. 40:538–545.

Bianchi-Hall, C.M., T.E. Carter, Jr., T.W. Rufty, C. Arellano, H.R. Boerma, D.A. Ashley, and J.W. Burton. 1998. Heritability and resource allocation of aluminum tolerance derived from soybean PI 416937. Crop Sci. 38:513–522.

Bland, W.L. 1993. Cotton and soybean root system growth in three soil temperature regimes. Agron. J. 85:906–911.

Blum, A. 1988. Plant breeding for stress environments. CRC Press, Boca Raton, FL.

Blum, A. 1996. Crop responses to drought and the interpretation of adaptation. p. 57–70. *In* E. Belhassen (ed.) Drought tolerance in higher plants: Genetical, physiological, and molecular biological analysis. Kluwer Academic Publ., the Netherlands.

Board, J.E., and Q. Tan. 1995. Assimilatory capacity effects on soybean yield components and pod number. Crop Sci. 35:846–851.

Borrell, A.K., G.L. Hammer, and R.G. Henzell. 2000. Does maintaining green leaf area in sorghum improve yield under drought? II. Dry matter production and yield. Crop Sci. 40:1037–1048.

Bowers, G.R. 1995. An early season production system for drought avoidance. J. Prod. Agric. 8:112–119.

Boyer, J.S. 1982. Plant productivity and environment. Science 218:443–448.

Boyer, J.S., R.R. Johnson, and S.G. Saupe. 1980. Afternoon water deficits and grain yields in old and new soybean cultivars. Agron. J. 72:981–986.

Brady, N.C., and R.R. Weil. 2002. The nature and properties of soils. 13th ed. p. 188–189. Prentice Hall, Upper Saddle River, NJ.

Bray, E.A., J. Bailey-Serres, and E. Weretilnyk. 2000. Responses to abiotic stresses. p. 1158–1203. *In* B.B. Buchanan et al. (ed.) Biochemistry and molecular biology of plants. Am. Soc. of Plant Physiologists, Rockville, MD.

Brevedan, R.E., D.B. Egli, and J.E. Leggett. 1978. Influence of N nutrition on flower and pod abortion and yield of soybeans. Agron. J. 70:81–84.

Brown, R.H. 1994. The conservative nature of crop photosynthesis and the implications of carbon dioxide fixation pathways. p. 211–219. *In* K.J. Boote et al. (ed.) Physiology and determination of crop yield. ASA, Madison, WI.

Bushamuka, V.N., and R.W. Zobel. 1998a. Differential genotypic and root type penetration of compacted soil layers. Crop Sci. 38:776–781.

Bushhamuka, V.N., and R.W. Zobel. 1998b. Maize and soybean tap, basal, and lateral root responses to a stratified acid, aluminum-toxic soil. Crop Sci. 38:416–421.

Carter, T.E., Jr., P.I. DeSouza, and L.C. Purcell. 1999. Recent advances in breeding for drought and aluminum resistance in soybean. p. 106–125. *In* H.E. Kauffman (ed.) Proc. of World Soybean Res. Conf. VI, Chicago. 4–7 Aug. 1999. Superior Print., Champaign, IL.

Carter, T.E., Jr., R.L. Nelson, C.H. Sneller, and Z. Cui. 2004. Genetic diversity in soybean. p. 303–416. *In* H.R. Boerma and J.E. Specht (ed.) Soybeans: Improvement, production, and uses. 3rd ed. Agron. Monogr. 16. ASA, CSSA, and SSSA, Madison, WI.

Carter, T.E., Jr., and T.W. Rufty. 1993. Soybean plant introductions exhibiting drought and aluminum tolerance. p. 331–346. *In* C.C. Kuo (ed.) Adaptation of food crops to temperature and water stress. Proc. of an Int. Symp. Asian Vegetable Res. and Development Center, Publication 93-410. Taipei, Taiwan.

Ceccarelli, S., and S. Grando. 1996. Drought as a challenge for the plant breeder. p. 71–77. *In* E. Belhassen (ed.) Drought tolerance in higher plants: Genetical, physiological, and molecular biological analysis. Kluwer Academic Publ., the Netherlands.

Chen, Y., and M. Schnitzer. 1978. The surface tension of aqueous solutions of soil humic substances. Soil Sci. 125:7–15.

Clawson, K.L., J.E. Specht, B.L. Blad, and A.F. Garay. 1986. Water-use efficiency in soybean pubescence density isolines—A calculation procedure for estimating daily values. Agron. J. 78:483–487.

Condon, A.G., R.A. Richards, G.J. Rebetzke, and G.D. Farquhar. 2002. Improving intrinsic water-use efficiency and crop yield. Crop Sci. 42:122–131.

Constable, G.A., and A.B. Hearn. 1978. Agronomic and physiological responses of soybean and sorghum crops to water deficits. I. Growth, development, and yield. Aust. J. Plant Physiol. 5:159–167.

Cortes, P.M., and T.R. Sinclair. 1986. Water relations of field-grown soybean under drought. Crop Sci. 26:993–998.

Cure, J.D., C.D. Raper, Jr., R.P. Patterson, and W. A. Jackson. 1983. Water stress recovery in soybeans as affected by photoperiod during seed development. Crop Sci. 23:110–115.

Davies, S.L., N.C. Turner, J.A. Palta, K.H.M. Siddique, and J.A. Plummer. 2000. Remobilisation of carbon and nitrogen supports seed filling in chickpea subjected to water deficit. Aust. J. Agric. Res. 51:855–866.

De, A. 1997. Sensitivity of N_2 fixation in soybean under drought stress with N fertilizer application at different growth stages. M.S. thesis. Univ. of Arkansas, Fayetteville.

de la Fuente, J., V. Ramirez-Rodriguez, J.L. Cabrera-Pounce, and L. Herrera-Estrella. 1997. Aluminum tolerance in transgenic plants by alteration of citrate synthesis. Science (Washington DC) 276:1566–1568.

Delhaize, E., S. Craig, C.D. Beaton, R.J. Bennet, V.D. Jagadish, and P.J. Randall. 1993. Aluminum tolerance in wheat (*Triticum aestivum* L.). I. Uptake and distribution of aluminum in root apices. Plant Physiol. 103:685–693.

Delhaize, E., D.M. Hebb, and P.R. Ryan. 2001. Expression of a *Pseudomonas aeruginosa* citrate synthase gene in tobacco is not associated with either enhanced citrate accumulation or efflux. Plant Physiol. 125:2059–2067.

DeSilva, M., L.C. Purcell, and C.A. King. 1996. Soybean petiole ureide response to water deficits and decreased transpiration. Crop Sci. 36:611–616.

De Souza, P.I., D.B. Egli, and W.P. Bruening. 1997. Water stress during seed filling and leaf senescence in soybean. Agron. J. 89:807–812.

Durand, J.L., J.E. Sheehy, and F.R. Minchin. 1987. Nitrogenase activity, photosynthesis and nodule water potential in soyabean plants experiencing water deprivation. J. Exp. Bot. 38:311–321.

Egli, D.B. 1997. Cultivar maturity and response of soybean to shade stress during seed filling. Field Crops Res. 52:1–8.

Egli, D.B., and S.J. Crafts-Brandner. 1996. Soybean. p. 595–623. *In* E. Zamski and A.A. Schaffer (ed.) Photoassimilate distribution in plants and crops: Source-sink relationships. Marcel Dekker, New York.

Evans, L.T., and R.A. Fischer. 1999. Yield potential: Its definition, measurement, and significance. Crop Sci. 39:1544–1551.

Farquhar, G.D., M.H. O'Leary, and J.A. Berr. 1982. On the relationship between carbon isotope discrimination and the intercellular carbon dioxide concentration in leaves. Aust. J. Plant Physiol. 9:121–137.

Farquhar, G.D., and R.A. Richards. 1984. Isotopic composition of plant carbon correlates with water-use efficiency of wheat genotypes. Aust. J. Plant Physiol. 11:539–552.

Fehr, W.R., and C.E. Caviness. 1977. Stages of soybean development. Spec. Rep. 80. Coop. Ext. Serv., Agric. Exp. Stn., Iowa State Univ., Ames.

Foy, C.D., R.L. Chaney, and M.C. White. 1978. The physiology of metal toxicity in plants. Ann. Rev. Plant Physiol. 29:511–566.

Gardner, W.R. 1960. Dynamic aspects of water availability to plants. Soil Sci. 89:63–73.

Gardner, WR. 1964. Relation of root distribution to water uptake and availability. Agron J. 56:41–45.

Goldman, I.L., T.E. Carter, Jr., and R.P. Patterson. 1989. Differential genotypic response to drought stress and subsoil aluminum in soybean. Crop Sci. 29:330–334.

Graham, J.J., C.D. Nickell, and R.G. Hoeft. 1995. Inheritance of tolerance to manganese deficiency in soybean. Crop Sci. 35:1007–1010.

Hale, M.G., and D.M. Orcutt 1987. The physiology of plants under stress. John Wiley & Sons, New York.

Hall, A.E., R.G. Mutters, and G.D. Farquhar. 1992. Genotypic and drought-induced differences in carbon isotope discrimination and gas exchange of cowpea. Crop Sci. 32:1–6.

Hammel, J.E., M.E. Sumner, and H. Shahandeh. 1985. Effect of chemical profile modification on soybean and corn production. Soil Sci. Soc. Am. J. 49:1508–1511.

Hatfield, J.L., T.J. Sauer, and J.H. Prueger. 2001. Managing soils to achieve greater water use efficiency: A review. Agron J. 93:271–280.

Heatherly, L.G. 1999. Early soybean production system (ESPS). p. 103–118. In L.G. Heatherly and H.F. Hodges (ed.) Soybean production in the midsouth. CRC Press, Boca Raton, FL.

Heatherly, L.G., and R.W. Elmore. 2004. Managing inputs for peak production. p. 451–536. In H.R. Boerma and J.E. Specht (ed.) Soybeans: Improvement, production, and uses. 3rd ed. Agron. Monogr. 16. ASA, CSSA, and SSSA, Madison, WI.

Hubick, K.T., G.D. Farquhar, and R. Shorter. 1986. Correlation between water-use efficiency and carbon isotope discrimination in diverse peanut (Arachis) germplasm. Aust. J. Plant Physiol. 13:803–816.

Huck, M.G., B. Klepper, and H.M. Taylor. 1970. Diurnal variations in root diameter. Plant Physiol. 45:529–530.

Hudson, D.B. 1994. Soil organic matter and available water capacity. J. Soil Water Conserv. 49:189–194.

Hume, D.J., and J.G. Criswell. 1973. Distribution and utilization of ^{14}C-labelled assimilates in soybean. Crop Sci. 13:519–524.

Hunt, S., and D.B. Layzell. 1993. Gas exchange of legume nodules and the regulation of nitrogenase activity. Ann. Rev. Plant Physiol. Plant Mol. Biol. 44:483–511.

Ismail, A.M., and A.E. Hall. 1992. Correlation between water-use efficiency and carbon isotope discrimination in diverse cowpea genotypes and isogenic lines. Crop Sci. 32:7–12.

Ismail, A.M., and A.E. Hall. 1993. Carbon isotope discrimination and gas exchange of cowpea accessions and hybrids. Crop Sci. 33:788–793.

Jones, H.G. 1992. Plants and microclimate. 2nd ed. Cambridge University Press.

Jones, H.G., T.J. Flowers, and M.B. Jones. 1989. Plants under stress: Biochemistry, physiology, and ecology, and their application to crop improvement. Cambridge Univ. Press, Cambridge.

Kadhem, F.A., J.E. Specht, and J.H. Williams. 1985a. Soybean irrigation serially timed during stages R1 to R6. II. Yield component responses. Agron. J. 77:299–304.

Kadhem, F.A., J.E. Specht, and J.H. Williams. 1985b. Soybean irrigation serially timed during stages R1 to R6. I. Agronomic responses. Agron. J. 77:291–298.

Kaspar, T.C., C.D. Stanley, and H.M. Taylor. 1978. Soybean root growth during the reproductive stages of development. Crop Sci. 70:1105–1107.

Kaspar, T.C., H.M. Taylor, and R.M. Shibles. 1984. Taproot-elongation rates of soybean cultivars in the glasshouse and their relation to field rooting depth. Crop Sci. 24:916–920.

Kasperbauer, M.J., and W.J. Busscher. 1991. Genotypic differences in cotton root penetration of a compacted subsoil layer. Crop Sci. 31:1376–1378.

King, C.A., and L.C. Purcell. 2001. Soybean nodule size and relationship to nitrogen fixation response to water deficit. Crop Sci. 41:1099–1107.

Klepper, B., and R.W. Rickman. 1990. Modeling crop root growth and function. Adv. Agron. 44:113–132.

Kochian, L.V. 1995. Cellular mechanisms of aluminum toxicity and resistance in plants. Ann. Rev. Plant Physiol. Plant Mol. Biol. 46:237–260.

Korte, L.L., J.E. Specht, J.H. Williams, and R.C. Sorenson. 1983a. Irrigation of soybean genotypes during reproductive ontogeny. II. Yield component responses. Crop Sci. 23:528–533.

Korte, L.L., J.H. Williams, J.E. Specht, and R.C. Sorenson. 1983b. Irrigation of soybean genotypes during reproductive ontogeny. I. Agronomic responses. Crop Sci. 23:521–527.

Kramer, P.J., and J.S. Boyer. 1995. Water relations of plants and soils. Academic Press, New York.

Ku, M.S.B., S. Agarie, M. Nomura, H. Fukayama, H. Tsuchida, K. Ono, S. Toki, M. Miyao, and M. Matsuoka. 1999. High-level expression of maize phosphoenolpyruvate carboxylase in transgenic rice plants. Nat. Biotechnol. 17:76–80.

Kumudini, S., D.J. Hume, and G. Chu. 2001. Genetic improvement in short season soybeans: I. Dry matter accumulation, partitioning and leaf area duration. Crop Sci. 41:391–398.

Kumudini, S., D.J. Hume, and G. Chu. 2002. Genetic improvement in short-season soybeans: II. Nitrogen accumulation, remobilization and partitioning. Crop Sci. 42:141–145.

Ludlow, M.M., and R.C. Muchow. 1990. A critical evaluation of traits for improving crop yields in water-limited environments. Adv. Agron. 43:107–153.

Lukaszewski, K.M., D.G. Blevins, and D.D. Randall. 1992. Asparagine and boric acid cause allantoate accumulation in soybean leaves by inhibiting manganese-dependent allantoate amidohydrolase. Plant Physiol. 99:1670–1676.

Lyons, J.E., and E.B. Earley. 1952. The effect of ammonium nitrate applications to field soils on nodulation, seed yield, and nitrogen and oil content of the seed of soybeans. Soil Sci. Am. Proc. 16:259–263.

Marschner, H. 1995. Mineral nutrition of higher plants. 2nd ed. Academic Press, New York.

Martin, C.A., D.K. Cassel, and E.J. Kamprath. 1979. Irrigation and tillage effects on soybean yield in a coastal plain soil. Agron. J. 71:592–594.

Masle, J., G.D. Farquhar, and S.C. Wong. 1992. Transpiration ratio and plant mineral content are related among genotypes of a range of species. Aust. J. Plant Physiol. 19:709–721.

Matheny, T.A., and P.G. Hunt. 1983. Effects of irrigation on accumulation of soil and symbiotically fixed N by soybean grown on a Norfolk loamy sand. Agron J. 75:719–722.

Matus, A., A.E. Slinkard, and C. Van Kessel. 1995. Carbon isotope discrimination and indirect selection for seed yield in lentil. Crop Sci. 35:679–684.

Maurel, C. 1997. Aquaporins and water permeability of plant membranes. Ann. Rev. Plant Physiol. Plant Mol. Biol. 48:399–429.

Meckel, L., D.B. Egli, R.E. Phillips, D. Radcliffe, and J.E. Leggett. 1984. Effect of moisture stress on seed growth in soybeans. Agron. J. 76:647–650.

Mian, M.A.R., M.A. Bailey, D.A. Ashley, R. Wells, T.E. Carter, Jr., W.A. Parrott, and H.R. Boerma. 1996. Molecular markers associated with water use efficiency and leaf ash in soybean. Crop Sci. 36:1252–1257.

Mian, M.A.R., D.A. Ashley, and H.R. Boerma. 1998a. An additional QTL for water use efficiency in soybean. Crop Sci. 38:390–393.

Mian, M.A.R., D.A. Ashley, W.K. Vencill, and H.R. Boerma. 1998b. QTLs conditioning early growth in a soybean population segregating for growth habit. Theor. Appl. Genet. 97:1210–1216.

Mian, M.A.R., R. Wells, T.E. Carter, Jr., D.A. Ashley, and H.R. Boerma. 1998c. RFLP tagging of QTLs conditioning specific leaf weight and leaf size in soybean. Theor. Appl. Genet. 96:354–360.

Miyasaka, S.C., J.G. Buta, R.K. Howell, and C.D. Foy. 1991. Mechanism of aluminum tolerance in snapbeans: Root exudation of citric acid. Plant Physiol. 96:737–743.

Momen, N.N., R.E. Carlson, R.H. Shaw, and O. Arjmand. 1979. Moisture stress effect on the yield components of two soybean cultivars. Agron. J. 71:86–90.

Monneveux, P., and E. Belhassen. 1996. The diversity of drought adaptation in the wild. p.7–14. In E. Belhassen (ed.) Drought tolerance in higher plants: Genetical, physiological, and molecular biological analysis. Kluwer Academic Publ., the Netherlands.

Muchow, 1985. Stomatal behavior in grain legumes grown under different water regimes in a semi-arid tropical environment. Field Crops Res. 11:291–307.

Muchow, R.C., and T.R. Sinclair. 1986. Water and nitrogen limitations in soybean grain production. II. Field and model analyses. Field Crops Res. 15:143–156.

Munns, R. 1988. Why measure osmotic adjustment? Aust. J. Plant Physiol. 15:717–726.

Munoz, A., P. Piedraw, M. Aguilar, and M. Pineda. 2001. Urea is the product of ureidoglycolate degradation in chickpea. Purification and characterization of the ureidoglycolate urea-lyase. Plant Physiol. 125:828–834.

Nilsen, E.T., and D.M. Orcutt. 1996. Physiology of plants under stress. Abiotic factors. John Wiley & Sons, New York.

Nimmo, J.R., and E.E. Miller. 1986. The temperature dependence of isothermal moisture vs. potential characteristics of soils. Soil Sci. Soc. Am. J. 50:1105–1113.

Nobel, P.S. 1991. Physicochemical and environmental plant physiology. Academic Press, New York.

Orcutt, D.M., and E.T. Nilsen. 2000. Physiology of plants under stress: Soil and biotic factors. John Wiley & Sons, New York.

Orcutt, D.M., and E.T. Nilsen. 2000. Physiology of plants under stress. Soil and biotic factors. John Wiley & Sons, New York.

Passioura, J.B. 1977. Grain yield, harvest index and water use of wheat. J. Aust. Inst. Agric. Sci. 43:117–120.

Passioura, J.B. 1983. Roots and drought resistance. Agric. Water Manage. 7:265–280.

Passioura, J.B. 1985. Roots and water economy of wheat. p. 185–198. In W. Day and R.K. Atkin (ed.) Wheat growth and modeling. Plenum Press, New York.

Passioura, J.B. 1988. Water transport in and to roots. Ann. Rev. Plant Physiol. Plant Mol. Biol. 39:245–265.

Passioura, J.B. 1994. The yield of crops in relation to drought. p. 343–359. In K.J. Boote et al. (ed.) Physiology and determination of crop yield. ASA, Madison, WI.

Passioura, J.B. 1996. Drought and drought tolerance. p. 1–5. In E. Belhassen (ed.) Drought tolerance in higher plants: Genetical, physiological, and molecular biological analysis. Kluwer Academic Publ., the Netherlands.

Pellet, D.M., D.L. Grunes, and L.V. Kochian. 1995. Organic acid exudation as an aluminum-tolerance mechanism in maize. Planta 196:788–795.

Purcell, L.C., M. deSilva, C. Andy King, and W.H. Kim. 1997a. Biomass accumulation and allocation in soybean associated with genotypic differences in tolerance of nitrogen fixation to water deficits. Plant Soil 196:101–113.

Purcell, L.C., and C.A. King. 1996. Drought and nitrogen source effects on nitrogen nutrition, seed growth, and yield in soybean. J. Plant Nutr. 19:969–993.

Purcell, L.C., C.A. King, and R.A. Ball. 2000. Soybean cultivar differences in ureides and the relationship to drought tolerant nitrogen fixation and manganese nutrition. Crop Sci. 40:1062–1070.

Purcell, L.C., R. Serraj, M. DeSilva, T.R. Sinclair, and S. Bona. 1998. Ureide concentration of field-grown soybean in response to drought and the relationship to nitrogen fixation. J. Plant Nutr. 21:949–966.

Purcell L.C., and T.R. Sinclair. 1995. Nodule gas exchange and water potential response to rapid imposition of water deficits. Plant Cell Environ. 18:179–187.

Purcell, L.C., E.D. Vories, P.A. Counce, and C. A. King. 1997b. Soybean growth and yield response to saturated soil culture in a temperate environment. Field Crops Res. 49:205–213.

Raper, Jr.,C.D., and P.J. Kramer. 1987. Stress physiology. p. 589–642. In J.R. Wilcox (ed.) Soybeans: Improvement, production, and uses. 2nd ed. ASA, CSSA, and SSSA, Madison, WI.

Ratliff, L.F., J.T. Ritchie, and D.K. Cassel. 1983. Field-measured limits of soil water availability as related to laboratory-measured properties. Soil Soc. Am. J. 47:770–775.

Ray, J.D., and T.R. Sinclair. 1998. The effect of pot size on growth and transpiration of maize and soybean during water deficit stress. J. Exp. Bot. 49:1381–1386.

Ray, J.D., L.X. Yu, S.R. McCouch, M.C. Champoux, G. Wang, and H.T. Nguyen. 1996. Mapping quantitative trait loci associated with root penetration ability in rice (Oryza sativa L.). Theor. Appl. Genet. 42:627–636.

Rebetzke, G.J., A.G. Condon, R.A. Richards, and G.D. Farquhar. 2002. Selection for reduced carbon isotope discrimination increases aerial biomass and grain yield of rainfed bread wheat. Crop Sci. 42:739–745.

Rendig, V.V., and H.M Taylor. 1989. Principles of soil-plant interrelationships. McGraw-Hill, New York.

Richards, R.A. 1996. Defining selection criteria to improve yield under drought. p. 79–88. In E. Belhassen (ed.) Drought tolerance in higher plants: Genetical, physiological, and molecular biological analysis. Kluwer Academic Publ., the Netherlands.

Richards, R.A., and J.B. Passioura. 1989. A breeding program to reduce the diameter of the major xylem vessel in the seminal roots of wheat and its effect on grain yield in rain-fed environments. Aust. J. Agric. Res. 40:943–950.

Richards, R.A., G.J. Rebetzke, A.G. Condon, and A.F. van Herwaarden. 2002. Breeding opportunities for increasing the efficiency of water use and crop yield in temperate cereals. Crop Sci. 42:111–121.

Ritchie, J.T. 1981. Water dynamics in the soil-plant-atmosphere system. Plant Soil 58:81–96.

Ross, T., and N. Lott. 2000. A climatology of recent extreme weather and climate events. National Climatic Data Center Technical Rep. 2000-2. Updated data available at http://lwf.ncdc.noaa.gov/oa/reports/billionz.html#extremes (verified 1 May 2003).

Salado-Navarro, L.R., K. Hinson, and T.R. Sinclair. 1985. Nitrogen partitioning and dry matter allocation in soybeans with different seed protein concentration. Crop Sci. 25:451–455.

Sall, K., and T.R. Sinclair. 1991. Soybean genotypic differences in sensitivity of symbiotic nitrogen fixation to soil dehydration. Plant Soil 133:31–37.

Sayre, K.D., E. Acevedo, and R.B. Austin. 1995. Carbon isotope discrimination and grain yield of three bread wheat germplasm groups grown at different levels of water stress. Field Crops Res. 41:45–54.

Schapaugh, Jr., W.T., and J.R. Wilcox. 1980. Relationships between harvest indices and other plant characteristics in soybeans. Crop Sci. 20:529–533.

Serraj, R., S. Bono, L.C. Purcell, and T.R. Sinclair. 1997. Nitrogen fixation response to water-deficits in field-grown 'Jackson' soybean. Field Crops Res. 52:109–116.

Serraj, R., and T.R. Sinclair. 1996a. Inhibition of nitrogenase activity and nodule permeability by water deficit. J. Exp. Bot. 47:1067–1073.

Serraj, R., and T.R. Sinclair 1996b. Processes contributing to N_2 fixation insensitivity to drought in the soybean cultivar Jackson. Crop Sci. 36:961–968.

Serraj, R., and T.R. Sinclair. 1998. Soybean cultivar variability for nodule formation and growth under drought. Plant Soil 202:159–166.

Serraj, R., and T.R. Sinclair. 2002. Osmolyte accumulation: Can it really help increase crop yield under drought conditions? Plant Cell Environ. 25:333–341.

Serraj, R., T.R. Sinclair, and L.C. Purcell. 1999a. Symbiotic N_2 fixation response to drought. J. Exp. Bot. 50:143–155.

Serraj, R., V. Vadez, R.F. Denison, and T.R. Sinclair. 1999b. Involvement of ureides in nitrogen fixation inhibition in soybean. Plant Physiol. 119:289–296.

Shaw, R.H., and D.R. Laing. 1965. Moisture stress and plant response. p. 73–92. In W.H. Pierre (ed.) Plant environment and efficient water use. ASA, Madison, WI.

Shelp, B.J., and R.J. Ireland. 1985. Ureide metabolism in leaves of nitrogen-fixing soybean plants. Plant Physiol. 77:779–783.

Silva, I.R., T.J. Smyth, D.F. Moxley, T.E. Carter, N.S. Allen, and T.W. Rufty. 2000. Aluminum accumulation at nuclei of cells in the root tip. Fluorescence detection using lumogallion and confocal laser scanning microscopy. Plant Physiol. 123:543–552.

Silva, I.R., T.J. Smyth, D.W. Israel, C.D. Raper, and T.W. Rufty. 2001a. Magnesium ameliorates aluminum rhizotoxicity in soybean by increasing citric acid production and exudation by roots. Plant Cell Physiol. 42:546–554.

Silva, I.R., T.J. Smyth, D.W. Israel, and T.W. Rufty. 2001b. Altered aluminum inhibition of soybean root elongation in the presence of magnesium. Plant Soil 230:223–230.

Silva, I.R., T.J. Smyth, C.D. Raper, T.E. Carter, and T.W. Rufty. 2001c. Differential aluminum tolerance in soybean: An evaluation of the role of organic acids. Physiol. Plant. 112:200–210.

Sinclair, T.R. 1986. Water and nitrogen limitations in soybean grain production. I. Model development. Field Crops Res. 15:125–141.

Sinclair, T.R. 1994. Limits to crop yield? p. 509–532. In K.J. Boote et al. (ed.) Physiology and determination of yield. ASA., Madison, WI.

Sinclair, T.R. 1998. Limits to crop yield. In Nina V. Fedoroff and Joel E. Cohen (ed.) Plants and populations: Is there time? National Academy of Science Colloquium, Irvine, CA. 5–6 Dec 1998. Available at http://www.lsc.psu.edu/nas/Panelists/Sinclair%20Comments.html (verified 1 May 2003).

Sinclair, T.R. 2004. Improved carbon and nitrogen assimilation for increased yield. p. 537–568. In H.R. Boerma and J.E. Specht (ed.) Soybeans: Improvement, production, and uses. 3rd ed. ASA, CSSA, and SSSA, Madison, WI.

Sinclair, T.R., and C.T. deWitt. 1975. Photosynthate and nitrogen requirements for seed production by various crops. Science (Washington DC) 189:565–567.

Sinclair, T.R., and C.T. deWitt. 1976. Analysis of the carbon and nitrogen limitations to soybean yield. Agron. J. 69:274–278.

Sinclair, T.R., L.C. Hammond, and J. Harrison. 1998. Extractable soil water and transpiration rate of soybean on sandy soils. Agron. J. 90:363–368.

Sinclair T.R., and M.M. Ludlow. 1986. Influence of soil water supply on the plant water balance of four tropical grain legumes. Aust. J. Plant Physiol. 13:329–341.

Sinclair, T.R., and R.C. Muchow. 2001. System analysis of plant traits to increase grain yield on limited water supplies. Agron. J. 93:263–270.

Sinclair, T.R., R.C. Muchow, J.M. Bennett, and L.C. Hammond. 1987. Relative sensitivity of nitrogen and biomass accumulation to drought in field-grown soybean. Agron. J. 79:986–991.

Sinclair, T.R., L.C. Purcell, V. Vadez, R. Serraj, C.A. King, and R. Nelson. 2000. Identification of soybean genotypes with N_2 fixation tolerance to water deficits. Crop Sci. 40:1803–1809.

Sinclair, T.R., and R. Serraj. 1995. Dinitrogen fixation sensitivity to drought among grain legume species. Nature (London) 378:344.

Sinclair, T.R., C.B. Tanner, and J.M. Bennett. 1984. Water-use efficiency in crop production. Bioscience 34:36–40.

Sionit, N., and P.J. Kramer. 1977. Effect of water stress during different stages of growth of soybeans. Agron. J. 74:721–725.

Sloane, R.J., R.P. Patterson, and T.E. Carter, Jr. 1990. Field drought tolerance of a soybean plant introduction. Crop Sci. 30:118–123.

Sorensen, R.C., and E.J. Penas. 1978. Nitrogen fertilization of soybeans. Agron. J. 70:213–216.

Sneller, C.H., L.C. Purcell, C.A. King, T.R. Sinclair, and R. Serraj. 1997. Inheritance of drought tolerant nitrogen fixation in soybean. p. 81. In 1997 Agronomy abstracts. ASA, Madison, WI.

Sneller, C.H., L.C. Purcell, T.R. Sinclair, and T.E. Carter, Jr. 2000. Yield of soybean lines derived from a parent with sustained nitrogen fixation during drought. p. 112. In 2000 Agronomy abstracts. ASA, Madison, WI.

Spaeth, S.C., H.C. Randall, T.R. Sinclair, and J.S. Vendeland. 1984. Stability of soybean harvest index. Agron. J. 76:482–486.

Specht, J.E., K. Chase, M. Macrander, G.L. Graef, J. Chung, J.P. Markwell, M. Germann, J.H. Orf, and K.G. Lark. 2001. Soybean response to water: A QTL analysis of drought tolerance. Crop Sci. 41:493–509.

Specht, J.E., D.J. Hume, and S.V. Kumudini. 1999. Soybean yield potential—A genetic and physiological perspective. Crop Sci. 39:1560–1570.

Specht, J.E., J.H. Williams, and C.J. Weidenbenner. 1986. Differential responses of soybean genotypes subjected to a seasonal soil water gradient. Crop Sci. 26:922–934.

Stebbins, N.E., and J.C. Polacco. 1995. Urease is not essential for ureide degradation in soybean. Plant Physiol. 109:169–175.

Steudle, E., and C.A. Peterson. 1998. How does water get through roots? J. Exp. Bot. 49:775–788.

Stone, J.A., and H.M. Taylor. 1983. Temperatures and the development of the taproot and lateral roots of four indeterminate soybean cultivars. Agron. J. 75:613–618.

Tanner, C.B., and T.R. Sinclair. 1983. Efficient water use in crop production: Research or re-search. p. 1–27. In H.M. Taylor et al. (ed.) Limitations to efficient water use in crop production. ASA, Madison, WI.

Taylor, H.M., and B. Klepper. 1978. The role of rooting characteristics in the supply of water to plants. Adv. Agron. 30:99–128.

Thomas, J.F., and C.D. Raper, Jr. 1976. Photoperiodic control of seed filling for soybeans. Crop Sci. 16:667–672.

Turner, N.C. 1993. Water use efficiency of crop plants: Potential for improvement. p. 75–82. In D.R. Buxton et al. (ed.) International crop science I. CSSA, Madison, WI.

Turner, N.C., G.C. Wright, and K.H.M. Siddique. 2001. Adaptation of grain legumes (pulses) to water-limited environments. Adv. Agron. 71:193–231.

Vadez, V., and T.R. Sinclair. 2000. Ureide degradation pathways in intact soybean leaves. J. Exp. Bot. 51:1459–1465.

Vadez, V., and T.R. Sinclair. 2001. Leaf ureide degradation and N_2 fixation tolerance to water deficit in soybean. J. Exp. Bot. 52:153–159.

Vadez, V., T.R. Sinclair, R. Serraj, and L.C. Purcell. 2000. Manganese application alleviates the water deficit-induced decline in N_2 fixation. Plant Cell Environ. 23:497–505.

VanToai, T.T., J.E. Beuerlein, A.F. Schmitthenner, and S.K. St. Martin. 1994. Genetic variability for flooding tolerance in soybeans. Crop Sci. 34:1112–1115.

VanToai, T.T., S.K. St. Martin, K. Chase, G. Boru, V. Schnipke, A.F. Schmitthenner, and K.G. Lark. 2001. Identification of a QTL associated with tolerance of soybean to soil waterlogging. Crop Sci. 41:1247–1252.

Waggoner, P.E. 1994. How much land can ten billion people spare for nature? Task Force Rep. 121. Feb. 1994. Council for Agric. Sci. and Technol., Ames, IA.

Walsh, K.B., M.J. Canny, and D.B. Layzell. 1989. Vascular transport and soybean nodule function. II. A role for phloem supply in product export. Plant Cell Environ. 12:713–723.

Weber, C.R. 1966. Nodulating and nonnodulating soybean isolines: I. Agronomic and chemical attributes. Agron. J. 58:43–46.

Weisz, P.R., R.F. Denison, and T.R. Sinclair. 1985. Response to drought stress of nitrogen fixation (acetylene reduction) rates by field-grown soybeans. Plant Physiol. 78:525–530.

Weisz, P.R., and T.R. Sinclair. 1988. A rapid non-destructive assay to quantify soybean nodule gas permeability. Plant Soil 105:69–78.

Wesley, R.A. 1999. Tillage systems for soybean production. p. 81–92. *In* L.G. Heatherly and H.F. Hodges
(ed.) Soybean production in the midsouth. CRC Press, Boca Raton, FL.

Wesley, T.L., R.E. Lamond, V.L. Martin, and S.R. Duncan. 1998. Effects of late-season nitrogen fertilizer on irrigated soybean yield and composition. J. Prod. Agric. 11:331–336.

Willatt, S.T., and H.M. Taylor. 1978. Water uptake by soya-bean roots as affected by their depth and by
soil water content. J. Agric. Sci. 90:205–213.

Winkler, R.G., D.G. Blevins, J.C. Polacco, and D.D. Randall. 1987. Ureide catabolism of soybeans. II.
Pathway of catabolism in intact leaf tissue. Plant Physiol. 83:585–591.

Wright, G.C., R.C.N. Rao, and G.D. Farquhar. 1994. Water-use efficiency and carbon isotope discrimination in peanut under water deficit conditions. Crop Sci. 34:92–97.

Yinbo, G., M.B. Peoples, and B. Rerkasem. 1997. The effect of nitrogen fertilizer strategy on N_2 fixation, growth and yield of vegetable soybean. Field Crops Res. 51:221–229.

Zeiher, C., D.B. Egli, J.E. Leggett, and D.A. Reicosky. 1982. Cultivar differences in nitrogen redistribution in soybeans. Agron. J. 74:375–379.

Zhang, J., J.E. Specht, G.L. Graef, and B.L. Johnson. 1992. Pubescence density effects on soybean seed
yield and other agronomic characteristics. Crop Sci. 32:641–648.

Zwieniecki. M.A., P.J. Melcher, and N.M. Holbrook. 2001. Hydrogel control of xylem hydraulic resistance in plants. Science (Washington, DC) 291:1059–1062.

13   Seed Composition

RICHARD F. WILSON

USDA-Agricultural Research Service
Beltsville, Maryland

This chapter provides an overview of recent research accomplishments and understandings that relate to the genetic alteration of soybean [*Glycine max* (L.) Merr.] seed composition. In a relatively short time, these activities have evolved from basic academic interest in the fundamental biological mechanisms that mediate protein and oil synthesis, into emerging practical applications of the science. Greater knowledge, coupled with an ability to regulate seed composition, has fostered a renaissance in the genetic enhancement of this important commodity. For example, prominent innovations in oil composition are embodied in lower saturated fat content and acceptable flavor stability without hydrogenation. Advances in genetic technology also enable increased digestibility and improved functional properties of soyprotein. Collectively, these traits will likely provide the soybean industry with a high degree of flexibility to meet consumer and end-user preferences in a wide range of food and feed products. Thus, there is a clear expectation that an appropriate use of genes that determine beneficial change in seed composition will have significant influence on the quality, utility, and economic worth of soybean. Although market trends, industry priorities, and government policies will influence the course of scientific endeavors to genetically alter traditional soybean composition, enhancing the competitive position of soybean in domestic and global markets will remain the ultimate goal. In effect, the intent is to make a good product even better.

13–1 CURRENT CHALLENGES IN THE OILSEED MARKET

At the beginning of the 21st century, the foremost problem facing the global oilseed industry was the farm price for soybean, which in the USA dropped to a 20-yr low in Year 2001 (USDA, FAS, 2002). Historically, periods of ample supply have led to increased end-stocks and depressed prices (Fig. 13–1). As demand for soybean strengthened, disappearance of end-stocks eventually resulted in higher prices. Typically, when global end-stocks have fallen below 20 million metric tons (MMT), soybean prices have averaged about $6 per bushel. Because of the strong negative correlation between end-stocks and price, it was logical to anticipate a down turn in future soybean prices when global end-stocks began to rise after 1996. However, during that period the amplitude of this economic cycle was distorted by a sig-

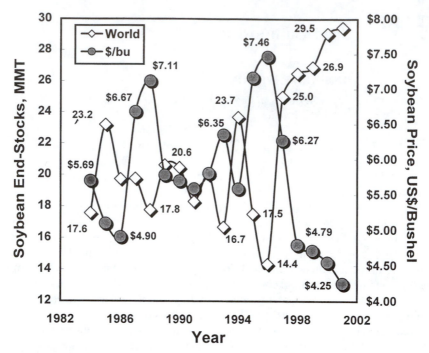

Fig. 13–1. Cyclic relation between annual global end-stocks and the U.S. farm price for soybean.

nificant expansion of soybean production in South America. In recent years, South American soybean production has accounted for about 75% of the surplus in the world soybean market. Hence, soybean prices have languished in a range from $4 to $5 per bushel, even though USA-held end-stocks have averaged about 7 MMT. The near-term outlook for a rapid up-turn in the price of soybean is not optimistic.

This situation is a function of increased global competition not only for commodity soybean, but also in markets for soybean products. The impact on U.S. share of global markets for soy-oil and meal is now obvious (Fig. 13–2). Since 1992, there has been a 5-basis point drop in the U.S. share of world consumption of soybean meal, and a drop of nearly 7-basis points in the U.S. share of world consumption of soybean oil. Although these trends demonstrate a significant shift in the competitive forces affecting a major commodity, changes in consumer preference also impact the entire soybean market (Fig. 13–3). For example, global soybean oil consumption has grown at a rate of about 1 MMT per year since 1994. Until recent years, growth in consumption of low-saturate oils such as canola and sunflower (*Helianthus annuus* L.) kept pace with soybean. However, world consumption of oils with higher oleic acid (18:1) content (including palm-olein, canola, sunflower, olive, and peanut) grew at a faster rate of 1.8 MMT per year. Obviously, the world market for oilseed products continues to change. It is becoming more competitive, and this situation establishes motive to genetically tailor soybean composition in ways that help to expand the utility of this important commodity. Therefore, efforts to enhance the economic value or market share of soybean must address changes in seed composition that merit a competitive advantage.

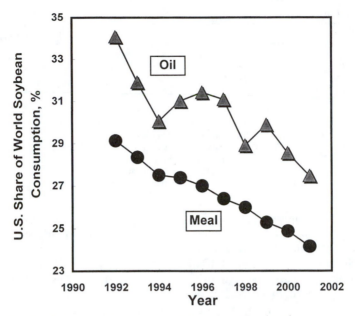

Fig. 13–2. Trends in U.S. share of the global market for soybean oil and meal.

13–1.1 Soybean Oil

The average fatty acid composition of commercial soybean oil is about: 10% palmitic acid (16:0), 4% stearic acid (18:0), 22% oleic acid (18:1), 54% linoleic acid

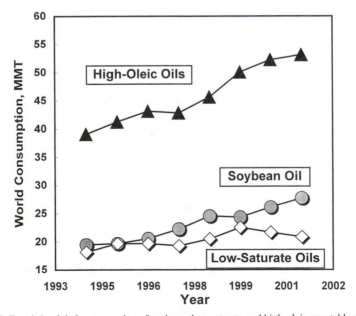

Fig. 13–3. Trends in global consumption of soybean, low-saturate and high-oleic vegetable oils.

Table 13–1. Fatty acid nomenclature.

Abbreviation[†]	Systemic name[‡]	Common name
16:0	Hexadecanoic acid	Palmitic
18:0	Octadecanoic acid	Stearic
18:1	Δ-9c Octadecenoic acid	Oleic
18:2	Δ-9c,12c Octadecenoic acid	Linoleic
18:3	Δ-9c,12c,15c Octadecenoic acid	Linolenic

[†] The first number of the fatty acid abbreviation indicates the number of carbon atoms, and the second number indicates the total number of unsaturated bonds in the molecule.

[‡] A Δ designation in the systematic name indicates that the position of unsaturated carbon-carbon bonds is determined by counting carbon atoms from the carboxyl-end of the molecule (an ω designation would denote the carbon number starting from the terminal methyl-end of the molecule). The 'c' designation indicates a 'cis' unsaturated bond configuration (as opposed to a 'trans' bond).

(18:2), and 10% linolenic acid (18:3). For reference, the first number of these fatty acid abbreviations indicates the number of carbon atoms and the second number represents the number of unsaturated carbon-carbon bonds in the molecule (Table 13–1). Since the 1970s, research has led to a better understanding of how to genetically alter the fatty acid composition of soybean oil. The original impetus for this work emerged during the fuel crisis in the early 1970s when U.S. oilseed processors desired a lower cost alternative to catalytic hydrogenation for producing vegetable oil products with desired functionality and flavor stability. A soybean with naturally lower 18:3 concentration was deemed a reasonable means to attain that goal. A decade later, concern for dietary health issues stimulated research to develop soybean with lower levels of saturated fat (principally 16:0 and 18:0) in U.S. food products. Then, during the 1990s, health concerns resurfaced in regard to the level of '*trans*-isomers' of unsaturated fatty acids (a geometric isomer formed during hydrogenation from the naturally occurring 'cis' unsaturated carbon-carbon bonds in 18:1, 18:2 and 18:3) in some foods (Puri, 1978; Carpenter et al., 1976). In this case, an increased level of 18:1 was suggested as an approach to reduce the need for hydrogenation. Each of these research priorities has led to innovative technologies that have progressed from ideas to accomplishments; and thence toward commercial application.

Interestingly, U.S. soybean producers provided the driving force, through local and national Soybean Check-Off programs, that has enabled scientists to address the priorities of the oilseed industry. The nature of this support also has evolved with time. Originally, state commodity organizations such as the North Carolina Soybean Producers' Association championed the initial research to genetically alter the fatty acid composition of soybean oil. Then, the American Soybean Association escalated this work to a national level. Today, the United Soybean Board (USB), a farmer-led organization with 62 directors representing all soybean-producing states in the USA, sponsors public research and interfaces with private companies in an effort to accelerate the development and marketing of soybean with genetic traits that enhance customer demand for soybean oil and meal. The focal point of this thrust is the 'Better Bean Initiative' (BBI), a program launched in 2000. The BBI mission specifically addresses the improvement of soybean composition for food and feed applications, and facilitates the development of marketing processes to commercialize these products. The main BBI targets for soybean oil are lower saturated fat and lower *trans*-isomers. Soybean oil with lower saturated fat content should compare favorably to oils like sunflower and canola. Soybean oil with higher 18:1 plus lower 18:3 also should help manufacturers produce foods with lower *trans*-isomer content (through reduced need for hydrogenation). Both of these strategic goals formulate corrective measures to bring conventional soybean oil in compliance with current and pending U.S. Food and Drug Administration (FDA) rules.

13–1.1.1 United States Food and Drug Administration Rules on Saturated Fat

A major factor that has influenced the development of soybean oil with lower levels of saturated fatty acids relates to claims regarding the nutrient content of certain conventional foods (e.g., cooking oil, shortening, and margarine). The allowed extent of such claims is defined in FDA regulations pursuant to the Nutrition Labeling and Education Act of 1990 (U.S. Food and Drug Administration, 1999b). The final regulation limited claims such as "low in saturated fat" to foods containing <1 g of saturated fat per serving. This means that the fatty acid composition of vegetable oil ingredients in a food product should have less than 7% saturated fat to warrant such a claim. In practice, the amount of saturated fat present in one serving of many conventional foods containing soybean oil is generally >1 g, thus precluding the inclusion of low in saturated fat nutrient content claim on the product label. But, this claim would be possible if the concentration of saturated fats in soybean oil was reduced to a level equal to or below 7% of total oil. Hence, this innovation should help make soybean oil more attractive as an ingredient to food manufacturers.

13–1.1.2 Oxidative Stability and *Trans*-isomers

Perhaps the most significant factor that has guided the development of crude soybean oil with lower 18:3 and 18:2 concentrations is the need to decrease the use of catalytic hydrogenation as a means to improve the oxidative stability of vegetable oil. Both 18:3 and 18:2 are polyunsaturated fatty acid (PUFA) constituents of glycerolipids in soybean oil. The olefinic (unsaturated) bonds between carbon atoms in these molecules are labile to oxygen, resulting in the formation of free radicals that may induce the autoxidation of other PUFA (Cosgrove et al., 1987). Under controlled conditions, the rate of spontaneous oxidation of 18:3 (ca. 4×10^{-2} $M^{-1/2}$ $s^{-1/2}$) is twice the rate for 18:2. However, these rates become greater under high temperature frying conditions. In general, the various ketone, aldehyde, enal, and dienal products from the degradation of PUFA have significant negative impact on flavor stability and intensity of oxidized fats (Warner and Frankel, 1985; Dixon and Hammond, 1984).

Among other techniques, catalytic hydrogenation of unsaturated lipids (a process of adding molecular hydrogen to olefinic bonds) has been used to help minimize the oxidation of soybean oil (Ilsemann and Mukherjee. 1978). The full measure of this process may sequentially and completely convert 18:3 to 18:2, thence to 18:1 and finally to 18:0. However, for most food applications, vegetable oils are only partially hydrogenated, if at all, to ensure that functional liquid properties are retained at room temperature. Thus, hydrogenated soybean oils typically contain lower levels of PUFA and exhibit enhanced long-term oxidative stability compared to soybean oils that are only refined, deodorized and bleached. Yet the apparent solution of one problem may give rise to another. In this case, the olefinic bonds of unsaturated fatty acids in crude vegetable oils are predominately found in a '*cis*' configuration that introduces a natural bend in the molecule. During hydrogenation, a portion of these bonds may be rearranged in a '*trans*' configuration that effectively straightens the structure of an unsaturated fatty acid in a way that may emulate a saturated fatty acid. The degree to which *trans*-isomers of fatty acids may produce physiological effects in humans similar to saturated fatty acids has been reviewed extensively (Judd et al., 1994; Applewhite, 1997; Hayakawa et al., 2000).

In response to the controversy surrounding the *trans*-isomer issue, FDA proposed a regulation in November 1999 (U.S. Food and Drug Administration, 1999a) that would require listing the amount of *trans*-isomers of fatty acids on the label of foods when measurable quantities are present. In addition, the proposal would define the level of *trans*-fatty acid allowable in foods to make certain nutrient content claims, and could possibly require the amount of *trans*-fatty acid to be included in the total amount of saturated fat. If the latter becomes reality, many margarines and foods containing shortenings made with conventional soybean oil probably would not qualify for a label claiming the product is "low in saturates". In anticipation of a final ruling, which is expected before 2004, food companies have begun to seek alternative oil and fat ingredients that may replace soybean oil to ensure their food products contain lower levels of *trans*-fatty acids (List et al., 1995a). As an example, Fig. 13–4 shows a facsimile of the label of a popular product that claims a zero *trans*-fatty acid content. Previously, this product was made with soybean oil, but now contains cottonseed and rapeseed oil. Consequently, the potential loss of market share has drawn oilseed processing industry attention and support toward the development of soybean cultivars containing oil with a fatty acid composition that requires less or no hydrogenation.

As mentioned, genetic approaches to reduce 18:3 and 18:2 (and/or increase 18:1) have been employed to improve the oxidative stability of soybean oil. The measure of how well that approach fulfills this goal is extensively documented. Comparisons of the performance of refined, bleached, deodorized (RBD) soybean oils (with a range from 2–7% 18:3) have demonstrated a strong negative correlation between 18:3 concentration and both flavor quality and intensity of fried foods (Liu and White, 1992a, 1992b; Warner and Mounts, 1993). In similar trials, soybean oils with 2 to 5.5% 18:3 compared favorably to hydrogenated soybean oils, in terms of reduced room odor and fried food flavor (Tompkins and Perkins, 2000; Mounts et al., 1994). Therefore based on frying applications, crude soybean oil with lower 18:3 concentration was deemed to be an acceptable alternative to hydrogenated soybean oil.

13–1.1.3 Designing Oil Composition for Specific End-products

In addition to transgenic approaches to alter fatty acid levels (Budziszewski et al., 1996; Kinney, 1994), several natural gene mutations have been discovered that enable genetic flexibility in tailoring the fatty acid composition in soybean oil.

0 GRAMS TRANS FAT PER SERVING

Nutriton Facts

Serving Size 2 tbsp (32 g)

Servings Per Container about 16

Amount Per Serving

Calories 190   Calories from Fat 140

	% Daily Value*
Total Fat 17 g	26%
Saturated Fat 3.5 g	18%
Cholesterol 0 mg	0%
Sodium 150 mg	6%
Total Carbohydrate 6 g	2%
Dietary Fiber 2 g	8%
Sugars 3 g	
Protein 8 g	
Iron 2%	Niacin 20%

Not a significant source of vitamin A, vitamin C, and calcium

*Percent Daily Values are based on a 2000 calorie diet

Fig. 13–4. A facsimile of nutritional claims listed on the label of a well-known food product.

This ability not only allows the manipulation of single genes that regulate the activity of an enzyme in the lipid metabolic pathway, but also the melding of functional combinations of genes to produce novel fatty acid profiles. However, for economic reasons, it is impractical to commercially develop all possible oil phenotypes. Thus, developmental research priorities have been set, with guidance from consumers and end-users, to initially target fatty acid profiles that have the highest probability to facilitate expanded use of soybean oil in edible and industrial applications (Wilson, 1998). Such deliberations have focused on three different oil phenotypes (Table 13–2). These phenotypes were chosen with respect to the relative volume of soybean oil used in specific products. In that regard, salad dressings and cooking oil account for about 44% of U.S. soybean oil consumption. About 52% of U.S.

Table 13–2. Goals for redesigning soybean oil composition for specific food and industrial applications.

Fatty acid†	Normal oil	Desired composition for specific use		
		Frying	Baking	Industrial
		% Crude soybean oil		
Saturated	15	7	42	11
Oleic	23	60	19	12
Linoleic	53	31	37	55
Linolenic	9	2	2	22

† Saturated, 16:0 + 18:0; Oleic, 18:1; Linoleic, 18:2; Linolenic, 18:3.

soybean oil is consumed as baking fats including shortening and margarine. Inedible oil products represent the remaining primary usage of soybean oil.

In the first category (frying products), a crude soybean oil having low-16:0, mid to high-18:1 and low-18:3 concentration offers competitive market opportunities in numerous edible and industrial applications. Such oil would be widely used in salad/cooking oil, baking, and frying fats. As an example, the U.S. snack food industry uses more than 453.6 Gg of oil annually. Considering consumer trends toward products that are fried in oils with low-saturated fat content (Massiello, 1978), the industry may favor use of canola oil, which meets FDA guidelines for a "Low in Saturated Fat" product label. However, canola accounts for <0.4% of total U.S. vegetable oil production (USDA, FAS, 2002), and contains relatively high levels of 18:3 (Salunkhe et al., 1992). Hence, there still is need to hydrogenate canola oil. Thus, to abate consumer concern for *trans*-isomers, low-saturated soybean oil with 45 to 60% 18:1 plus 3% (or less) 18:3 should provide a viable domestic source for Low in Saturated Fat foods made with vegetable oil.

Soybean oil with much higher 18:1 concentration also could be used in the manufacture of lubricants and hydraulic oil base stocks, which is an 362.9 Gg annual market (USDA, 1996a). Soy-diesel is another industrial application for this type of oil. The USA consumes 177.9 GL of petroleum diesel annually. Research shows that soy-diesel helps reduce emissions in compliance with federal clean air standards (Graboski et al., 1996). A 20% blend of methyl-soyate in petrol-diesel could capture 2% of this market (0.68 Gg of soybean oil) by Year 2003 (Duffield et al., 1998). However, improved ignition and cold flow properties are needed to accelerate acceptance of soy-diesel. A high 18:1 level (increased cetane index) and a lower 16:0 level (improved cold-flow) will help overcome ignition problems and poor performance in cooler climates (Dunn et al., 1996). In addition, a high 18:1 level should extend the utility of soybean oil in manufacture of pharmaceuticals and cosmetic products including: bath oils, emollient creams, lipstick, make-up bases, aftershave lotions, post-operative nutrients, suppository bases, foam builders, detergents, shampoo, and clear gels (USDA. 1996b).

In the baking fat category (Table 13–2), solid fat content at certain temperatures determines the utility of vegetable oil in the manufacture of baking fats. Solid fat index (SFI) is used as a relative measure of the temperature at which the mixture turns from a solid to a liquid (Segura et al., 1995). This index is directly related to the concentration of saturated fatty acids in the oil. The SFI of refined soybean oil ordinarily is too low to be used directly as a shortening, but it may be increased through processes like hydrogenation to form base stocks with varied levels of *trans*-isomers. Soybean oil with naturally higher SFI could then minimize the degree of hydrogenation needed to form ingredients for products, such as margarine. With inter-esterification, a soybean oil containing 30 to 35% 16:0 + 18:0 can make an acceptable, *trans*-free, margarine (List et al., 1995b, 2000). Thus, soybean oil with inherently greater SFI and possibly improved functional properties, such as the β′ (beta-prime) crystalline structure, should protect and expand the utilization of soybean oil in edible applications for margarine and shortening based products.

The industrial oil category (Table 13–2) relates almost exclusively to inedible applications. Although inedible uses of soybean oil have been limited by oleochemical industry dependence on lower-priced petroleum ingredients, opportunities could be realized if soybean oil contained greater levels of PUFA. As an example, soybean oil with higher PUFA concentration could replace traditional 'drying oils' such as tung and linseed in oil-based paints and coatings, which is a 4.35 Tg annual market (USDA, 2001). Inks made with conventional soybean oil already are a viable industrial application, and account for a 500 million lb domestic market. Yet, even this use could be improved with a low-saturate plus high-PUFA composition. The recent discovery of alternative desaturase genes in *Glycine soja*, the wild ancestor of *Glycine max*, enables breeding soybean with more chemically reactive oils for a variety of bio-based products (Pantalone et al., 1997a). Such genes may mediate soybean oil composition with up to 85% total PUFA to satisfy this emerging market for industrial products.

13–1.2 Soybean Protein

Protein and oil determine soybean value, even at low prices, but world demand for soybean meal (SBM) drives the soybean market primarily because of its use as a source of amino acids in livestock feed. Soybean meal has the highest level of crude protein in comparison to other sources of vegetable protein, and its amino acid balance nearly provides the essential amino acid nutritional requirements of swine and poultry (Smith, 1997). Apparent nutritional deficiencies in essential acids that may limit livestock performance are commonly supplemented with synthetic amino acid additives, such as methionine, cysteine, and lysine. However, as livestock feeding operations become more sophisticated and geneticists continue to enhance livestock germplasm, there is an emerging need for complementary changes in soybean protein quality. Thus, in recognition of the need to enhance the economic value of soybean, a portion of the USB's BBI research program has been directed toward the improvement of soybean protein composition.

The primary goals for enhancing soybean protein quality are: to improve essential amino acid balance, increase digestibility of the meal, and to help reduce the environmental impact of animal production. Several strategies have been outlined to achieve these innovations. Essential amino acid balance may be augmented via regulation of the expression of genes in particular amino acid pathways, or by increasing the concentration of total crude protein in SBM above 48% crude protein (48% is the current standard for high-protein SBM). Digestibility and total metabolizable energy of SBM in livestock rations could be improved by reducing the level of oligosaccharides (raffinose and stachyose) in soybean seed. Environmental impact of organic P in livestock waste may be minimized by genetic regulation of phytic acid accumulation in developing seed. In addition, the food industry would benefit from genetic traits that improve the functional characteristics of soy-protein. These attributes are needed to expand applications for vegetable protein-based products. Also, it is important to ensure that soy-based foods contain an adequate supply of isoflavones, which may convey certain health benefits. Of course, these attributes must be affected in soybean that have very good yielding ability. These are no small tasks, but progress has been made toward these goals.

13–1.2.1 Overcoming the Negative Genetic Correlation between Protein, Oil, and Yield

There is a wide range of genetic variation in protein (Fig. 13–5) and oil (Fig. 13–6) concentration among accessions of the USDA Soybean Germplasm Collection (USDA, ARS, 2001). The reported range of protein concentration is 34.1 to 56.8% of seed dry mass, with a mean of 42.1%. Oil concentration among the accessions in the Collection may range from 8.3 to 27.9%, with a mean of 19.5%. As indicated in a sampling of commercial northern and southern soybean varieties, there generally is a strong negative correlation between protein and oil concentration in soybean (Hurburgh et al., 1990). This means a genetic or environmental influence that causes an increase in protein, often results in a decline in oil. In addition, varieties adapted to the southern USA appear to have a higher mean protein concentration and a wider range in protein and oil levels than those adapted to northern or Midwest states (Fig. 13–7). Thus, it is extremely rare to find germplasm in which the concentration for both protein and oil is relatively high. There also is a negative genetic correlation between protein and yield, but a positive association between oil and yield (Wilcox and Guodong, 1997). The former relation has significantly impeded commercial production of soybean with greater-than-average protein concentration. Depending on comparative yielding ability, a soybean with greater-than-average protein (dry mass) may not be profitable to grow. Considering the composition of selected cultivars shown in Table 13–3, estimates of constituent value-based 1992 prices, which are equivalent to the median oil/meal price ratio from 1984 to 2000 (USDA, FAS, 2002), reveal an economic plateau when seed protein concen-

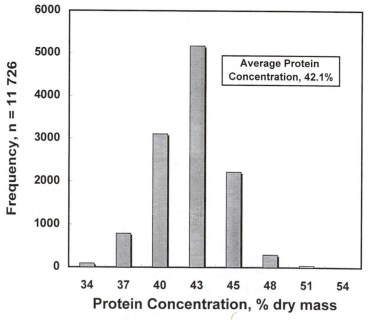

Fig. 13–5. Distribution of protein concentration among 11 726 accessions of the USDA Soybean Germplasm Collection.

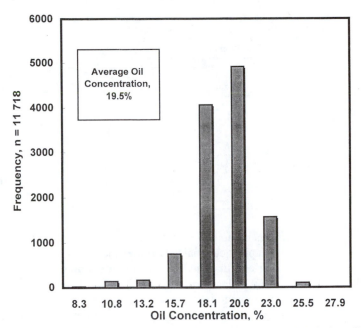

Fig. 13–6. Distribution of oil concentration among 11 718 accessions of the USDA Soybean Germplasm Collection.

tration nears 48% (Fig. 13–8). However, it may be possible to select gene combinations that enable higher than normal protein concentration in germplasm that maintains acceptable levels of oil and yielding ability.

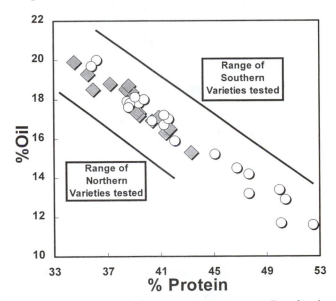

Fig. 13–7. Apparent range of protein and oil concentration among a sampling of modern commercial soybean varieties with Northern and Southern maturity.

Table 13–3. Examples of the range of genetic variation for protein, oil, and carbohydrate concentration in selected soybean germplasm.

Variety	Protein	Oil	Residual
		% seed dry mass	
N88-480	35.6	23.7	40.7
Ransom	39.3	22.9	37.8
Young	42.9	20.5	36.6
Nakasenari	43.5	20.3	36.2
Prolina	44.8	19.5	35.7
N88-438	46.2	18.8	35.0
N88-252	47.7	17.7	34.6
F76-1771	49.8	15.8	34.4
NC-112	51.7	14.2	34.1
$LSD_{0.05}$	3.4	2.1	1.4

If soybean accessions in the USDA collection average about 42.1% protein and 19.5% oil (dry mass), a practical goal to optimize the constituent value of soybean under most market price situations is to achieve a commercial cultivar with about 44 to 45% protein and no less than 18% oil. The fact that such a phenotype is not common in current commercial soybean varieties indicates the complexity of the problem. Protein synthesis is embodied by a multitude of enzymatic reac-

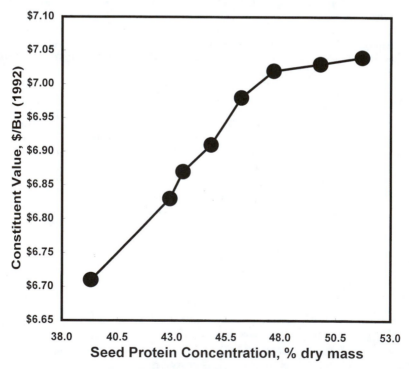

Fig. 13–8. Relation the estimated constituent value of soybean to seed protein concentration. Constituent value was based on oil and meal prices reported by USDA-FAS for 1992. These 1992 values equaled the mean of annual prices for oil and meal from 1984 to 2000.

tions, ranging from amino acid synthesis to the final assembly of storage protein (Shotwell and Larkins, 1989; Herman and Larkins, 1999). Each enzymatic reaction is governed by the expression of a specific gene, and gene expression may be regulated or influenced by environmental factors such as growth temperature or the amount of fertilizer that is applied to a crop. Yet, specialized breeding methods, such as a recurrent selection index, have been used to increase yield in a high-protein population (Kenworthy and Brim, 1979; Wilcox, 1998). With this technique, significant gain in yield may be achieved without substantial reductions in the high-protein trait. Several agronomic high-protein varieties have been developed in this manner. The first commercial cultivar of this type was named 'Prolina' (Burton et al., 1999). Several other lines with similar composition, such as S96-2641 and S97-1688 from the University of Missouri (S.C. Anand, personal communication, 1999) are beginning to emerge. These soybean varieties demonstrate that it is possible to achieve simultaneous gain in protein, oil, and yield. Still, understanding of the causal relations for this accomplishment significantly lags behind the ability to achieve this accomplishment. For example, the high protein alleles at statistically significant quantitative trait loci (QTL) in some populations have pleiotropic effects or linkages to low oil and low yield alleles (Chung et al., 2003).

13–1.2.2 Enhancing Feeding Efficiency and Protein Functionality

There is no question that soybeans with higher protein concentration produce meal with more than 48% crude protein (CP), the current standard for high-protein SBM. Soybean cultivars such as Prolina typically produce meal with up to 54% CP (Bajjalieh, 1996). In poultry diets, ultra-high protein meals contribute greater metabolizable energy than the standard high-protein SBM (Edwards et al., 2000). The impact of more energy is evidenced by at least a 5% weight gain in broilers performance on diets containing SBM with 50+% CP (J.T. Brake, North Carolina State University, personal communication, 2000). This effect may be directly attributed to the greater total supply of essential amino acids provided in an equivalent weight of high-protein meal.

Subtle changes in essential amino acid balance also may be achieved through conventional breeding for increased protein concentration. When the relative amino acid composition of SBM from the high-protein cv. Prolina was compared to standard SBM with 48% CP, positive change is found in the concentration of amino acids such as lysine, arginine, and leucine. The β-conglycinin fraction (see 13–2.1.1 for definition) of the storage protein in this cultivar also contained additional cysteine residues (Kwanyuen et al., 1998). Although these changes may appear to be insignificant, they are important in the manufacture of vegetable-protein based foods. Such shifts in amino acid balance probably occur as a result of macro-molecular changes in protein composition (Utsumi and Kinsella, 1985). In that regard, alteration of the level of storage protein subunits can have significant effect on the functional properties of protein isolates. As an example, isolate from the high-protein line Prolina exhibits increased gel strength due in part to a higher proportion of the 7S or β-conglycinin fraction, and to increased 'hydrogen-bonding' among amino acids of adjacent proteins (Luck et al., 2001). In practice, the addition of isolate with greater gel strength to commuted meat systems improves water retention capacity, compared to normal soy-isolate. This attribute makes it possible to formulate a wider range of food products, from vegetable-protein based sausages to low-fat ice cream.

13–2 GENETIC AND ENVIRONMENTAL INFLUENCE ON SEED COMPOSITION

The previous section briefly has described the predominant physical and regulatory issues that influence or may determine soybean value, use, and market share. That information should complement discussions of the global market situation and the economic aspects of soybean production that are presented in Chapter 1 (Wilcox, 2004, this publication) and Chapter 19 (Sonka et al., 2004, this publication). In addition, numerous recommended changes in seed composition also were outlined that may stimulate the future use of soybean products in foods and feed. However, attainment of these putative genetic changes must be based on a sound understanding of the biological systems and the factors that regulate or are associated with the expression of the respective metabolic pathways in developing soybean seed. Accordingly, this section focuses on the nature of biological mechanisms that may affect change in the intrinsic composition of this important commodity.

13–2.1 Variation and Associations among Seed Constituents

As previously shown for protein and oil concentration, an inspection of available data on the composition of genetically diverse accessions of the USDA Soybean Germplasm Collection (USDA, ARS, 2001) suggests a relatively wide concentration range for other seed constituents, such as carbohydrates, amino acids, and fatty acids. However, these data sets often represent normal distributions where the respective phenotypic traits are fairly well conserved for a majority of the accessions. Since these distributions usually are homogenous, it is difficult to identify obvious and bona fide deviations from the norm that may represent a genetic event that influences component levels of the major or minor constituents. When a natural or induced mutation does occur in a gene that is involved in the synthesis of a seed constituent, its expression may not be easily distinguished by a significant change in phenotype. Because of potential interactions with other genes that mediate metabolic pathways, care also must be taken to ensure that the observed phenotype is not unduly influenced by environmental factors that may effect seed composition. Thus, knowledge or estimates of genotypic differences in the timing and rates of constituent deposition during seed development under varied environmental conditions have become important assets in the discovery and characterization of a suspected gene mutation.

One approach toward improved evaluation of genotypic differences in seed composition is to model the deposition of constituents during reproductive development as a function of time (Vereshchagin, 1991). This is possible when the seed component of interest is an end product of a metabolic process, and when the deposition of that product follows a sigmoidal pattern (Settlage, et al., 1998). In practice, the cumulative deposition (mg seed^{-1}) of applicable compounds may be modeled during seed development by application of the logistic function (Eq. [1]). After model fitting, the first derivative of the logistic function (Eq. [2]) may be used to estimate the incremental rate of product accumulation (mg seed^{-1} d^{-1}). In each case, **W** is the amount of seed product at **T** days after flowering (DAF); **a**, is the amount of constituent at seed maturity; **dW/dT** is the rate of constituent deposition at T; and **k** and **b** are empirically derived constants.

$$W = a/[1 + be^{(-kT)}] \qquad\qquad [1]$$

$$dW/dT = kabe^{-kT} / (1 + be^{-kT})^2 \qquad\qquad [2]$$

Examples presented in this chapter demonstrate how these equations may be used to model temporal expression of major seed constituents (protein, oil, carbohydrate, seed mass) or components of major constituents (such as an individual fatty acid in triacylglycerol or a subunit of storage protein). In general, the pattern of estimated rates for seed constituent accumulation parallel the intensity of complementary gene transcription during seed development (Wilson et al., 2001a). Thus, a well-modeled pattern of seed constituent deposition should indicate the stage of seed development that is best suited for comparison of genotypic differences in metabolic activities associated with a given trait.

13–2.1.1 Major Constituents: Protein, Oil, and Carbohydrate

The USDA Soybean Germplasm Collection contains about 18 800 accessions of cultivated soybean. As mentioned earlier, these accessions exhibit a wide range of protein and oil concentration. In general, soybean storage protein primarily is composed of three fractions defined by sedimentation value as: 2S (α-conglycinin), 7S (β-conglycinin) and 11S (conglycinin). The 2S fraction typically contains proteins such as protease inhibitors (Mies and Hymowitz, 1973; Mityko et al., 1990; Werner and Wemmer, 1991); the 7S fraction is composed of trimers of α, α' and β subunits (Coates et al., 1985; Maruyama et al., 2001); and the 11S fraction is composed of hexamers of various acidic and basic subunits (Miles et al., 1985; Nielsen, 1985; Nielsen et al., 2001). The assembly, structure, and nature of genes that encode these proteins are well documented (Nielsen et al., 1995, 1989; Beachy et al., 1983b; Watanabe and Hirano, 1994). Crude soybean oil contains various glycerolipids (primarily phospholipids, diacylglycerol, and triacylglycerol). Triacylglycerol is the main component of the oil. Phospholipids (such as phosphatidylcholine, phosphatidylethanolamine, and phosphatidylinositol) have structural function in cell membranes and may be metabolically involved in triacylglycerol synthesis (Wilson et al., 1980). Each glycerolipid class is composed of molecular species formed by various combinations of the five fatty acids (16:0, 18:0, 18:1, 18:2, and 18:3) or polar groups that are esterified at the *sn-1, sn-2,* and *sn-3* stereospecific positions (hydroxyl groups) of a glycerol molecule (Carver and Wilson, 1984a, 1984b).

13–2.1.1.1 Accumulation During Seed Development. As shown in Table 13–3, certain soybeans may be classified as having a 'high-protein' (e.g., NC-112) or 'high-oil' trait (e.g., N88-480). Although the genes that determine these traits are unknown, information regarding the nature of trait expression and affected biological mechanisms often can be derived from the patterns of constituent deposition in contrasting lines. For example, temporal patterns and the incremental rates of accumulation of these constituents have been derived with the logistic function for protein, oil, residuals or carbohydrate, and seed dry mass (Fig. 13–9) for the high-

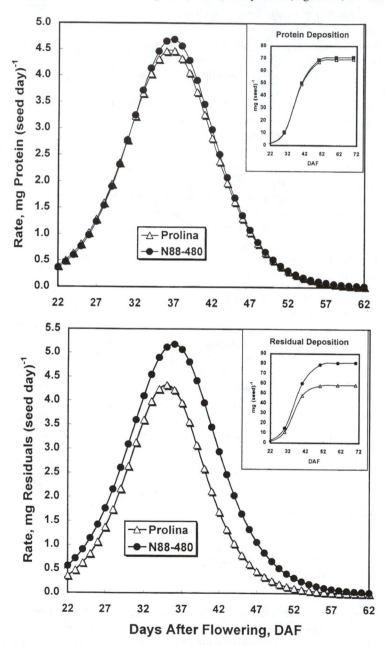

Fig. 13–9. Estimates of the pattern of protein, oil, carbohydrate, and dry mass accumulation (inserts) and the rate of constituent deposition during soybean seed development. Trend lines are derived from a logistic function analysis of data from the high protein cv. Prolina and the high oil cv. N88-480.

protein cv. Prolina (Burton et al., 1999) and for the high-oil germplasm N88-480 (Burton and Wilson, 1994). These data demonstrate that the respective peak rates for total protein, oil, carbohydrate, and dry mass accumulation occur at the R5.5 stage of mid-seed fill (about 35–37 DAF). In general, the near simultaneous tim-

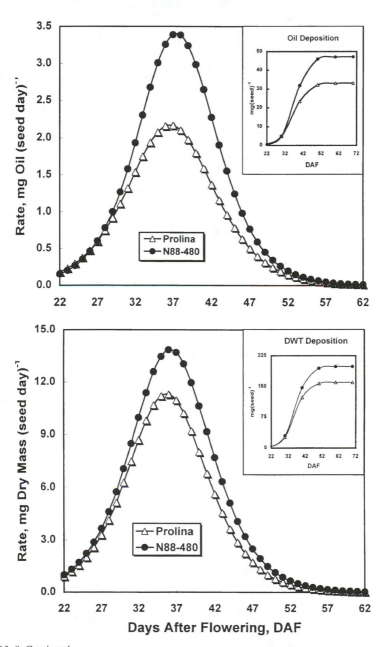

Fig. 13–9. Continued.

ing of the peak rates in the two cultivars suggests that this trend may be representative of that in other soybean genotypes. The mid-seed-fill stage of seed development would then appear to be optimal for the evaluation of metabolic processes underlying differences in seed constituent deposition among soybean cultivars. It is important to note that cultivar differences in seed mass must be considered when directly comparing data derived from logistic function analyses among genotypes. For example, Prolina has less mature seed mass than N88-480, despite only small differences in protein deposition per seed. However, when the peak rates for protein synthesis is considered in light of the peak rate for dry mass accumulation, Prolina seed exhibited a greater capacity to accumulate protein than N88-480. Similarly, the peak rates for oil and carbohydrate deposition (relative to the peak rate of dry mass accumulation) are significantly greater in N88-480 than Prolina. Hence, on a comparable level, the high-protein and high-oil traits in these lines appear to be determined by their respective abilities to make protein and oil, but the question of the causal mechanism remains unanswered. In that regard, we may not be certain that a 'high protein' trait in soybean is due to enhanced ability to synthesize protein, or a diminished ability to synthesize oil, or a permutation of these probabilities.

Although there is little evidence in the published literature to establish a defensible conclusion as to why some soybean has higher protein or oil, some inferences may be drawn from modeling constituent deposition in other high-protein and high-oil genotypes. In a recent investigation (Settlage et al., 1998), a strong positive correlation was found between the estimated peak rate of oil deposition (near mid-seed development) and oil concentration in mature seed. In addition, the activity of EC 2.3.1.20 diacylglycerol acyltransferase (DGAT), the enzyme that catalyzes synthesis of triacylglycerol from acyl-CoA and diacylglycerol, measured in seed at or near the peak rate of oil deposition, was correlated positively with oil phenotype. Although oil concentration in soybean is a quantitative trait, DGAT activity may be an indicator of coordinated genetic expression of gene-products in the entire glycerolipid synthetic pathway for a given oil phenotype. Variation in DGAT activity is possibly a contributing factor underlying the variation in oil content among soybean germplasm, and this association may be documented when DGAT activities are observed to correspond to the rates of oil deposition during seed development.

Carbohydrate accounts for approximately 86% of the residual dry mass of mature soybean seed. The primary constituents are starch, sucrose, and other soluble sugars, and oligosaccharides (raffinose and stachyose). As shown in electron micrographs (Norby et al., 1984) and chemical analyses (Rubel et al., 1972; Yazdi-Samadi et al., 1977), starch is the predominate carbohydrate early in seed development. Starch deposition peaks near mid-seed fill, then declines and is nearly absent in mature seed. In conjunction with starch hydrolysis, soluble sugars (sucrose, fructose, glucose) begin to accumulate prior to mid-pod fill as a function of elevated invertase and sucrose synthase activity (Phillips et al., 1984). Raffinose and stachyose accumulate later in seed development (Obendorf et al., 1998). Typical ranges reported for mature seed are: sucrose (41–67%), raffinose (5–16%) and stachyose (12–35%) of total soluble carbohydrates (USDA, ARS, 2001).

13–2.1.1.2 Influence of Growth Temperature on Composition. In studies of 'trait' inheritance, the interpretation of genotypic differences in seed composition always is based on phenotypic observations, from which a putative gene with at least two alleles is inferred to 'govern' the trait. In practice, numerous factors may impede or impair the attainment of accurate phenotypic estimates of seed compositional trait expression. As an example, the quality of analytical data determines the credibility of conclusions regarding the degree to which a given phenotype deviates from the expected 'norm' for the trait among genetically diverse soybean, or the degree of variation about an expected phenotypic value for a highly inbred line. Greater confidence levels in these estimates may be achieved if analytical procedures are conducted in an accredited manner. Yet, assuming that analytical protocols such as ISO 17025 are followed (ISO/IEC, 1998), deviations from expected values may persist and cast doubt on the 'genetic stability' of the trait. In many cases, barring physical mixtures of seed or the segregation of genes that mediate the trait, the apparent anomaly may be attributed to the response of the gene(s) or gene product(s) to environmental factors that influence trait expression. Perhaps, the most obvious of these factors is the influence of temperature of the environment during reproductive plant growth.

This consideration becomes acute when genotype × environmental (GxE) interaction strongly influences the expression of a potentially 'value-added' change in seed content. It is well known that growth temperature affects the protein and oil concentration of soybean seed (Wolf et al., 1982). Typically, protein and oil vary in an inverse relation with changes in growth temperature (Fig. 13–10). These relations are well documented (Piper and Boote, 1999). Plants subjected to water stress also may produce seed with higher oil and lower protein concentration (Specht et

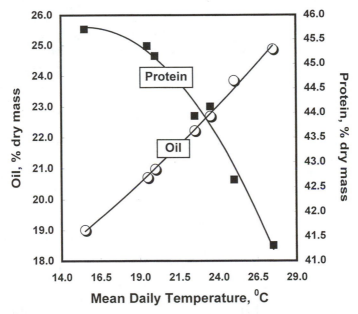

Fig. 13–10. Effect of growth temperature on protein and oil concentration in mature soybean seed. Collected from the cv. Dare grown to maturity under controlled environmental conditions.

Table 13–4. Growth temperature effects on the composition of mature seed of high-protein germplasm.

Growth temperature	NC-111		NC-106	
	Oil	Protein	Oil	Protein
d/night (°C)	——— % dry wt. ———		——— % dry wt. ———	
30/26	19.7	43.8	20.3	52.9
26/22	19.4	45.2	18.4	54.4
22/18	15.8	50.4	16.8	55.9
18/14	14.4	58.4	16.1	56.8
Slope†	0.5	−1.2	0.4	−0.3
R^2	0.91	0.92	0.97	0.99
	——— mg seed^{-1} ———		——— mg seed^{-1} ———	
30/26	35.7	79.5	39.6	103.4
26/22	34.2	79.6	39.7	117.6
22/18	25.7	81.7	39.9	132.4
18/14	19.7	79.7	39.6	139.9
$LSD_{0.05}$	7.1	1.0	0.1	15.3

† Regression of oil or protein concentration on the daily mean growth temperature under 9-h day lengths.

al., 2001), but this was likely caused by warmer temperatures within transpiration-limited canopies in drought-stricken environments. Still, the causal basis for the seed constituent response to temperature is unknown. In fact, one cannot assume that all soybean plants respond to temperature in the same manner. This becomes evident in an examination of growth temperature effects on the protein and oil concentration of mature seed of two high-protein inbred lines from different parentage (Carter et al., 1986). Under controlled environments, the genotypes NC-106 and NC-111 respond in an expected manner, that is, oil concentration trends lower and protein concentration trends higher with lower treatment temperature (Table 13–4). However, the change in protein concentration across treatments is significantly greater in NC-111. Treatment effects on the actual amounts of protein and oil per seed reveals further differences in response of these genotypes to growth temperature. In NC-111, oil content declines but protein content remains essentially constant at progressively lower treatment temperature. A contrasting scenario occurs in NC-106, where protein content increases and oil content remains unaffected by lower temperature. These phenomena seem to be consistent with the observed trends in the apparent energy of activation for protein and oil synthesis among these lines (Moorman, 1990). Therefore, it appears that the efficiency of metabolic mechanisms may mediate G×E effects on protein and oil concentration among soybean.

Although the means and extent to which changes in growth temperature may affect the activity of 'rate-limiting' enzymatic steps in the protein and lipid metabolic pathways is unknown, an emerging notion suggests that reversible protein phosphorylation may play an important role in the regulation of plant metabolism. The activity of several enzymes, including sucrose-phosphate synthase, sucrose synthase, hydroxymethylglutaryl-CoA-reductase, phosphoenolpyruvate carboxylase, and NADH:nitrate reductase is known to be controlled by protein phosphorylation (Winter and Huber, 2000; Toroser and Huber, 2000). In many cases, regulatory phosphorylation sites have been identified on these enzymes, and such sites are the tar-

get of specific groups of protein kinase (Hardie, 1999). A minimal phosphorylation motif has been developed for these protein kinases (-ϕ-x-Arg-x-x-Ser/Thr-)
where 'x' can be any amino acid and 'ϕ' can only be a hydrophobic residue (Halford and Hardie, 1998; Lee et al., 1998). These phosphorylation motifs occur in several enzymes, such as cytosolic pyruvate kinase. Hence, protein phosphorylation
may regulate a wide variety of processes in developing seeds including protein synthesis. Therefore, it is possible that subtle changes in gene domains that encode protein phosphorylation sites in key metabolic enzymes may help explain the range of
genetic variation and the nature of environmental variation in protein and oil composition among soybean cultivars.

13–2.1.2 Predominate Fatty Acids

As mentioned before, 16:0, 18:0, 18:1, 18:2, and 18:3 are the predominate
fatty acids of soybean oil. The acyl-ACP (acyl carrier protein) derivatives of 16:0,
18:0, and 18:1 are the products of fatty acid synthesis in plastids of developing soybean seed. These acyl-ACP moieties are then converted to acyl-CoA derivates and
released into the cell cytoplasm by an unknown mechanism. In the cytoplasm, acyl-
CoA is the preferred substrate for the synthesis of various glycerolipids (Browse
and Somerville, 1991). The classical glycerolipid pathway features the esterification of varying molecular acyl-CoA species to the *sn*-1 and *sn*-2 positions of glyc-
erol-3-PO$_4$ to form phosphatidic acid, which in turn may be converted to diacylglycerol (DG). At this point in the pathway, DG may be catalytically esterified with
a third acyl-CoA from the cytoplasm to form triacylglycerol (TG) via DGAT, or may
be diverted toward the synthesis of phospholipids. In that regard, DG may be used
directly in the synthesis of phosphatidylcholine (PC) or phosphatidylethanolamine;
or DG may be converted to the CDP-DG derivative and become a substrate for phosphatidylinositol synthesis. Phospholipids such as PC, that contain 18:1, are a preferred substrate for the predominant ω-6 and ω-3 desaturases that catalyze the respective conversion of 18:1 to 18:2 and 18:2 to 18:3 in the endoplasmic reticulum
(Ohlrogge and Browse, 1995). The PC also may directly contribute acyl-CoA as a
substrate for the catalytic conversion of diacylglycerol to triacylglycerol. This reaction may be catalyzed by a recently discovered enzyme, the phospholipid:dia-
cylglycerol acyltransferase (Banas et al., 2000; Cases et al., 1998).

13–2.1.2.1 Accumulation During Seed Development. The majority of fatty
acids synthesized during the development of soybean seed accumulate as esters of
TG. A typical time-delineated pattern of individual fatty acid concentration in TG
is shown in Table 13–5 for the conventional cv. Dare. The TG composition in developing seed is altered most noticeably by a substantial increase in 18:2 concentration, and decline in 18:3 concentration as seeds mature. These patterns reflect
the activity of the predominant ω-6 and ω-3 desaturases that catalyze PUFA synthesis. However, direct assays for ω-6 and ω-3 desaturase activity are impeded by
the inability to purify functionally active preparations of these membrane-bound
enzymes. Therefore, a practical means of estimating these activities was developed
(Wilson et al., 1990). Relative estimates of ω-6 desaturase activity may be calculated from the fatty acid composition of the oil at a given stage of seed development with Eq. [3]. Relative ω-3 desaturase activity may be calculated with Eq. [4].

Table 13–5. Changes in fatty acid concentration during the development of soybean seed (cv. Dare).

Days after flowering	Fatty acid class†				
	16:0	18:0	18:1	18:2	18:3
	% Total fatty acid in triacylglycerol				
25	11.8	3.4	25.9	43.6	15.3
30	11.8	3.3	25.4	48.7	10.8
40	11.2	3.7	24.0	51.3	9.7
45	10.5	4.0	22.9	53.1	9.5
50	10.6	4.1	21.4	54.9	9.0
55	10.9	4.1	20.4	55.9	8.7
60	10.3	4.1	20.3	56.8	8.5
65	10.7	4.1	20.3	56.4	8.5
70	10.2	4.0	20.5	56.8	8.5

† Abbreviation represents the number of carbon atoms:number of unsaturated carbon-to-carbon bonds in each fatty acid molecule.

Appling these equations to the data set in Table 13–4 reveals the pattern of expression of these desaturase activities during seed development (Fig. 13–11).

Relative ω-6 desaturation = %[(18:2 + 18:3)/(18:1 + 18:2 + 18:3)]*100   [3]

Relative ω-3 desaturation = %[(18:3)/(18:2 + 18:3)]*100   [4]

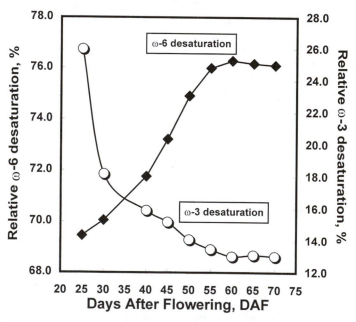

Fig. 13–11. Trend in relative estimates of ω-6 and ω-3 desaturase activity during soybean seed development in the cv. Dare. Relative ω-6 desaturation is determined as: % [(18:2 + 18:3)/(18:1 + 18:2 + 18:3)] × 100. Relative ω-3 desaturation is determined as: % [(18:3)/(18:2 + 18:3)] × 100.

Table 13–6. Effect of growth temperature on fatty acid composition in mature seed of soybean germplasm with normal (cv. Dare) and genetically modified acyl-desaturase activity.

Growth temperature	Fatty acid class					Desaturation	
	16:0	18:0	18:1	18:2	18:3	ω–6	ω–3
d/night (°C)	% dry wt. in the cv. Dare					%	
30/26	11.1	3.2	24.8	52.3	8.3	71.0	13.6
26/22	10.3	3.3	22.1	55.4	8.6	74.4	13.4
22/18	10.5	3.4	17.1	59.7	9.2	80.1	13.3
18/14	10.4	3.1	16.3	60.7	9.4	81.2	13.4
	% dry wt. in the cv. N78-2245						
30/26	10.9	3.6	56.1	25.1	4.3	34.4	14.6
26/22	11.1	3.3	48.8	31.4	5.4	43.0	14.7
22/18	11.2	3.6	37.4	40.4	7.2	56.0	15.1
18/14	10.5	3.5	30.9	47.2	7.9	64.1	14.3
	% dry wt. in the cv. N85-2176						
30/26	11.0	3.1	44.7	38.1	3.0	47.9	7.3
26/22	11.3	3.2	40.1	41.9	3.5	53.1	7.7
22/18	12.5	3.3	35.3	44.5	4.3	58.0	8.8
18/14	12.6	3.4	33.0	45.4	5.6	60.7	11.0

13–2.1.2.2 Influence of Growth Temperature. As with protein and oil concentration, growth temperature also influences fatty acid composition (Table 13–6). Under controlled conditions, temperature has little effect on the concentration of the saturated fatty acids, 16:0 and 18:0. However, 18:1 concentration is negatively correlated with 18:2 and 18:3 as growth temperature declines. This relation is particularly evident in developing seed of the germplasm N78-2245, which carries a natural recessive mutation in the gene that encodes the ω-6 desaturase. The sensitivity of ω-6 desaturase activity to changes in growth temperature is such that 18:1 concentration in N78-2245 may vary from about 30% (at lower temperatures) to more than 50% (in higher temperature treatments). Noticeably, ω-3 desaturase activity in these germplasm is relatively constant over a wide range of growth temperatures. When a recessive gene that encodes the ω-3 desaturase is combined with the recessive gene for the ω-6 desaturase (as in the germplasm N85-2176), the additive effects contribute to more stable expression of lower 18:3 concentrations over a wider range of growth temperatures (Burton et al., 1989). Many theories have been proposed to explain the relative increase in PUFA at lower growth temperatures. For example, the solubility of O_2 in the cell cytoplasm at different temperatures once was thought to mediate desaturase activity, but this thesis has since been discounted (Harris and James, 1969). Alternative theories suggest that this phenomenon may be due to different temperature optima for the desaturase enzymes (Martin et al., 1986) and/or modulation of membrane fluidity in a manner that effects the physical position and exposure of the enzyme active site to the cytosol (Inbar and Shinitzky, 1974). All are quite interesting, but none of these theories have adequately described the biological mechanism(s) that mediate temperature effects on ω-6 desaturase activity. Currently, the most plausible theory suggests that ω-6 desaturase activity may be regulated by signal transduction events involving reversible phosphorylation of the enzyme (Topfer et al., 1995). Indeed, phosphorylation sites have been detected within the primary structure of the primary ω-6 desaturase in soybean seed (S.C. Huber, USDA-ARS, Raleigh NC, personal communication, 2000).

13–2.1.3 Minor Constituents of Oil and Protein

Soybean contains several highly valued minor constituents, such as tocopherol, sterol, and isoflavone. Tocopherols and sterols may be removed as a by-product in deodorizer distillate during oil refining (see Chapter 20, Lusas, 2004, this publication). Soybean is the predominant commercial source of α-tocopherol (natural vitamin E), and the phytosterol, stigmasterol, is widely used for commercial synthesis of steroid hormones and other pharmaceutical products. The isoflavones, principally diadzein and genistein, are physiologically active components of SBM. It is believed that isoflavones possess antioxidant properties, and that these properties are associated with numerous health benefits (see Chapter 21, Birt et al., 2004, this publication).

13–2.1.3.1 Tocopherols. Tocopherols serve as antioxidants that help protect PUFA from oxidative decomposition. These compounds may interrupt the chain reaction of lipid peroxidation by scavenging peroxyl radicals, and also may inhibit triacylglycerol peroxidation at initiation by accepting free radicals (Kamal-Eldin and Appelqvist, 1996). Soybean oil typically contains three primary types of tocopherol: delta (2,8-dimethyl-2-(4,8,12-trimethyltridecyl)-, gamma (2,7,8-dimethyl-2-(4,8,12-trimethyltridecyl)-, and alpha (2,5,7,8-dimethyl-2-(4,8,12-trimethyltridecyl)-tocopherol (Gutfinger and Letan, 1974). The relative effectiveness of these compounds as antioxidants are, in decreasing order: delta-(Δ), gamma-(γ) and alpha-(α) tocopherol (Sherwin, 1976).

Although a considerable amount of tocopherol (ca. 1000–2000 mg kg^{-1}) remains in refined soybean oil, its effect on the oxidative stability of soybean oil must still be augmented by a reduction of the native concentration of 18:2 and 18:3. This is achieved most commonly by hydrogenating the oil, but also may be accomplished through genetic means. Indeed, genetically modified oils have been shown to exhibit improved stability under high-temperature frying conditions, resulting in superior flavor quality than foods fried in hydrogenated soybean oil (Mounts et al., 1994). Mounts et al. (1996) also reported the first estimates of tocopherol content in soybean oils exhibiting genetically modified 18:3 composition. Almonor et al. (1998) later showed that the impact that environmental conditions may have on tocopherol composition in soybean exhibiting improved oil quality. They found that total tocopherol accumulation paralleled oil deposition during soybean seed development, with the most active rate of tocopherol synthesis occurring between 30 and 45 DAF. However, the rate of accumulation and total amount of tocopherol was significantly diminished when plants were grown in lower temperature treatments. Still, at seed maturity, tocopherol content exhibited a strong positive relation with growth temperature (Table 13–7). In addition, the concentration of both Δ- and α-tocopherol tended to increase as growth temperature declined, whereas, γ-tocopherol concentration was positively correlated with growth temperature. When expressed on a g kg^{-1} oil basis, change in amount of γ-tocopherol accounted for 73.4 ± 2.7% of the change in total tocopherol between the highest and lowest temperature treatments over both genotypes. The average amount of Δ-tocopherol (280.4 ± 11.0 g

Table 13–7. Effect of growth temperature on tocopherol composition in mature seed of soybean germplasm with normal (cv. Dare) and genetically modified acyl-desaturase activity.

Growth temperature	Tocopherol class			
	delta	gamma	alpha	Total
d/night (°C)	—— % total tocopherol in the cv. Dare ——			mg kg^{-1}
30/26	19.4	71.2	9.4	1529.2
26/22	20.0	70.7	9.3	1392.0
22/18	20.8	68.9	10.3	1213.2
18/14	21.7	68.0	10.4	1064.2
	— % total tocopherol in the cv. N85-2176—			
30/26	23.7	61.9	14.4	1283.1
26/22	24.3	61.4	14.3	1078.3
22/18	25.6	60.4	13.9	1000.6
18/14	28.8	59.4	13.8	848.3
LSD$_{0.05}$	2.0	3.0	1.4	136.9

kg^{-1} oil) and α-tocopherol (139.0 ± 5.1 g kg^{-1} oil) in the treatment/genotype combinations was not influenced by growth temperature. Therefore, the apparent increase in α- and Δ-tocopherol concentration within genotypes was a function of treatment effects on the amount of γ-tocopherol.

A striking feature of these data was the apparent effect of alleles governing expression of the ω-6 desaturase on the response on tocopherol content to growth temperature. Tocopherol content in each temperature treatment was significantly greater in genotypes carrying homozygous dominant alleles for ω-6 desaturase (in the cv. Dare) than those with homozygous recessive alleles for ω-6 desaturase (in N85-2176). Furthermore, it appeared that genetic regulation of ω-6 desaturase activity not only imposed a considerable influence on tocopherol content, but also on tocopherol composition. In general, there was a positive relation between γ-tocopherol and 18:3 concentration in the oil of mature soybeans (Fig. 13–12). Thus, lower γ-tocopherol concentration may be expected in varieties having genetically reduced levels of 18:3. By the same token, low-18:3 soybean oils exhibited elevated levels of α-tocopherol or vitamin E. The apparent enrichment of total tocopherol in α-tocopherol was a function of loss of γ-tocopherol within genotype × treatment combinations. Therefore, soybean varieties exhibiting an oil with a lower 18:3 concentration should contain more α-tocopherol, especially when grown under warmer commercial production environments. Although the biochemical basis for these observations is unknown, it would appear that γ-tocopherol synthesis may be regulated as a function of endogenous oxidative activity. Moreover, conditions that enrich the amount of extractable vitamin E should provide an additional beneficial aspect of genetic approaches to improve soybean oil quality.

13–2.1.3.2 Sterols. Phytosterols in crude soybean oil occur as various free, acyl, glucoside, and acyl-glucoside esters of three secondary alcohols: stigmasterol, campesterol, and β-sitosterol (in order of lowest to greatest concentration). The accumulation of these compounds (hydrolyzed to the free form for analysis) during seed development does not conform to a sigmoid pattern, and hence is not amenable to modeling with the logistic function. Total sterol content in mature seed is posi-

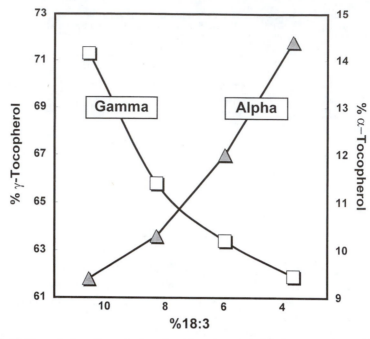

Fig. 13–12. Concentration of γ- and α-tocopherol in mature seed of soybean germplasm with altered linolenic acid concentration. Each data point represents one of the four homozygous classes of segregates for *Fan* and *Fan* alleles from the mating N93-194 × N85-2176.

tively related with growth temperature (Table 13–8), although the variation in the concentration of stigmasterol, campesterol, and β-sitosterol is small (Vlahakis and Hazebroek, 2000). However, as shown for tocopherols, there is an apparent effect of alleles governing expression of the ω-6 desaturase on sterol composition. Stigmasterol concentration was significantly greater in genotypes carrying homozygous

Table 13–8. Effect of growth temperature on the sterol composition of mature seed of soybean germplasm with normal (cv. Dare) and genetically altered acyl-desaturase activity.

Growth temperature	Phytosterol class			
	Campesterol	Stigmasterol	β-Sitosterol	Total
d/night (^{0}C)	———— % total sterol in the cv. Dare ————			g kg^{-1} oil
30/26	24.9	10.9	64.3	1.36
26/22	25.8	11.1	63.1	1.34
22/18	25.8	11.5	62.7	1.30
18/14	26.9	13.2	60.0	1.15
	———— % total sterol in the cv. N85-2176 ————			
30/26	24.2	23.3	52.5	1.17
26/22	24.4	22.8	52.8	1.11
22/18	25.1	22.5	52.4	1.04
18/14	25.7	22.5	51.8	1.01
LSD$_{0.05}$	0.5	3.7	3.4	0.08

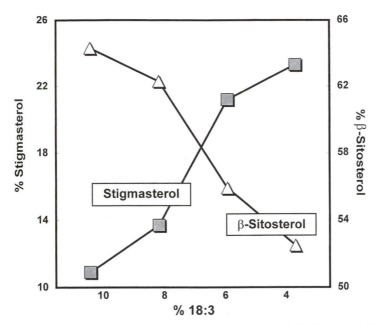

Fig. 13–13. Concentration of β-sitosterol and stigmasterol in mature seed of soybean germplasm with altered linolenic acid concentration. Each data point represents one of the four homozygous classes of segregates for *Fan* and *Fan* alleles from the mating N93-194 × N85-2176.

recessive alleles (in N85-2176) than those with homozygous dominant alleles for ω-6 desaturase (in Dare). In general, it appears that genetic regulation of ω-6 desaturase activity imposes a considerable influence on sterol composition. For example, stigmasterol concentration was negatively correlated with the 18:3 concentration in crude oil of mature soybean (Fig. 13–13). Thus, greater stigmasterol concentration may be expected in soybean varieties having a genetically reduced 18:3 concentration. By the same token, low-18:3 soybean oils exhibited lower levels of β-sitosterol. The apparent enrichment of the stigmasterol fraction of total phytosterol was a function of loss of β-sitosterol within genotype × treatment combinations. Therefore, lower 18:3 soybean varieties should contain more stigmasterol, without significant influence of growth environment. Although the biochemical basis for these observations is unknown, detailed information on the genes and genetic systems that regulate plant sterol biosynthesis has been compiled by Bach and Benveniste (1997).

13–2.1.3.3 Isoflavones. Isoflavones are a class of flavanoids that exist in soybean as free (aglycones) or glycoside derivatives. The fundamental aglycone compounds are: diadzein, genistein, and glycitein (Fig. 13–14). Only the aglycone form of these compounds is believed to contribute the physiological activities that are attributed to the entire complement of isoflavones in soybean seed (Lichtenstein, 1998). The glycosides (diadzin, genistin, and glycitin) may also occur as 6″-O-malonyl- or 6″-O-acetyl-derivatives of the three fundamental aglycones (Table 13–9). Malonyl-esters, the dominant form of isoflavone, are thermally and chemically unstable, and thus are easily converted during soybean processing to the more stable

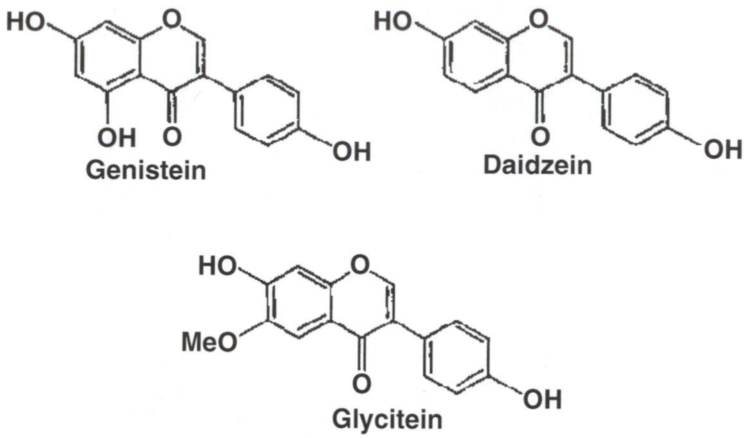

WILSON

Fig. 13–14. Fundamental aglycone structures of isoflavanoids in soybean.

acetyl-esters (Wang et al., 1998). Total isoflavone content in SBM ranges from 300 to >3000 mg kg^{-1} among accessions of the USDA Soybean Germplasm Collection (USDA, ARS, 2001). Although little is known about the genetic regulation of isoflavone synthesis in soybean, several genes in the phenylpropanoid synthetic pathway (Fig. 13–15) have been isolated and cloned (Jung et al., 2000). Isoflavone synthase (IFS) catalyzes the first committed step of the isoflavone branch of this pathway. Two genes have been identified that encode IFS, a type of cytochrome P450 protein, in soybean. Understanding the genetic regulation of this pathway may become necessary because of interest to maintain adequate isoflavone levels in response to certain genetic and environmental influences. For example, total isoflavone content of soybean seed appears to be negatively related with growth temperature (Tsukamoto et al., 1995). In addition, a negative relation may exist between total isoflavone and 18:3 concentration, and more recently with higher protein concentration (Fig. 13–16). Therefore, genetic regulation of isoflavone content and composition may become an important consideration in the development of high-protein soybean with improved oil quality.

Table 13–9. The isoflavone composition of typically processed soybean meal.

Isoflavone class	Composition	
	mg kg^{-1}	% of total
Daidzein	13.9	0.7
Daidzin	141.1	7.1
Acetyl daidzin	91.2	4.6
Malonyl daidzin	477.8	24.2
Glycitin	49.1	2.5
Malonyl glycitin	82.8	4.2
Genistein	16.7	0.8
Genistin	210.4	10.6
Acetyl genistin	28.3	1.4
Malonyl genistin	864.4	43.8
Total	1975.7	100.0

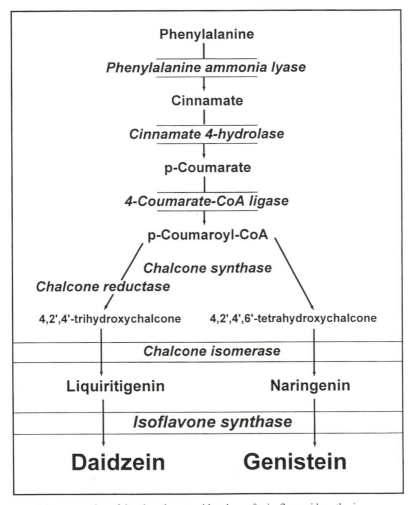

Fig. 13–15. Representation of the phenylpropanoid pathway for isoflavanoid synthesis.

13–2.2 Genetic Regulation of Fatty Acid Composition

Molecular genetic technologies have opened new insight into the biological mechanisms that govern fatty acid composition in soybean. Considerable information (Fig. 13–17) has been gathered from DNA sequences of nearly every gene that encodes an enzyme in the fatty acid synthetic pathway (Ohlrogge and Jaworski, 1997). These advances in knowledge have led to directed genetic modification of soybean oil composition (Hitz et al., 1995) and better understanding of the functional structure of enzymes, such as acyl-desaturases (Shanklin et al., 1997). Significant progress also has been made in the development of molecular genetic markers that facilitate the identification of genotypes in populations segregating for fatty acid traits, and the positioning of these genes on genetic maps of the soybean

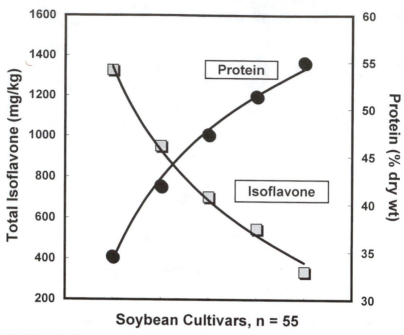

Fig. 13–16. Total isoflavone concentration among 55 commercial soybean cultivars that were classified in five groups based on statistical similarities in protein concentration.

genome (Diers and Shoemaker, 1992; Li et al., 2002). However, the foundation for all of this technology rests upon the discovery or creation of natural mutations in genes that mediate altered oil phenotypes. An arsenal of germplasm carrying such mutations has been amassed. These genetic resources are being used to determine the inheritance of traits and transfer desirable genes to agronomic cultivars. Still, there is need to associate a phenotypic trait with a gene and its functional products. Also, there is a need to determine the nature of the mutation that confers the altered phenotype. These are perhaps the greatest challenges that lie ahead for both science and industry. To help facilitate the process required to meet that challenge, the next section documents the discovery of germplasm resources and reports on research progress aimed at a functional characterization of natural gene mutations that affect the fatty acid composition of soybean oil.

13–2.2.1 Genetic Modification of Saturated Fatty Acid Composition

13–2.2.1.1 Low-16:0 Phenotypes. N79-2077-12 was the first soybean germplasm developed with a lower than normal 16:0 concentration (Wilson et al., 1988; Burton et al., 1994), and is still the only known germplasm that carries a serendipitously discovered natural mutation, designated as the recessive fap_{nc} allele (Table 13–10). Other soybean germplasm exhibiting about half of the 16:0 levels found in normal soybean oil have been induced with chemical mutagens such as ethylmethanesulfonate (EMS). The alleles in these germplasm include homozygous recessive fap_1 (C1726, [Erickson et al., 1988]), fap_3 (A22, [Schnebly et

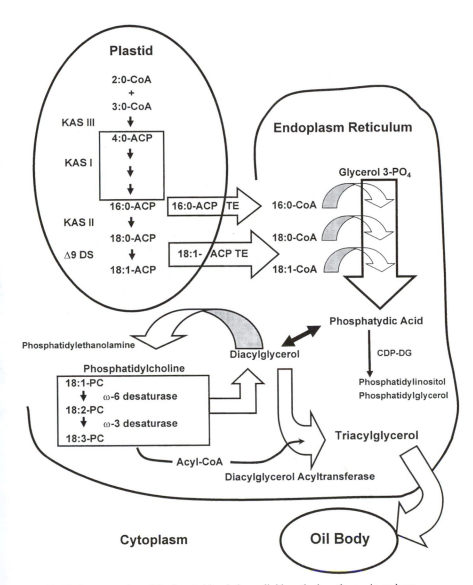

Fig. 13–17. Representation of the fatty acid and glycerolipid synthetic pathways in soybean.

al., 1994]), or the temporarily designated *fap** [ELLP2, (Stojsin et al., 1998)]. Homozygous combinations of *fap$_1$* and *fap$_3$* alleles (Horejsi et al., 1994) or of *fap$_1$* and *fap$_{nc}$* alleles (Wilcox et al., 1994) or of *fap$_1$* and *fap** alleles (Stojsin et al., 1998) that are obtained as progeny from repulsion phased parental mating, reportedly constitute transgressive segregates for 16:0 concentration (i.e., <4.5% 16:0). The inbred lines C1943 (northern U.S. maturity) and N94-2575 (southern U.S. maturity) are examples of selections in which *fap$_1$* and *fap$_{nc}$* have been combined (Burton et al., 1998). Based on this information, it is highly probable that the mutations rep-

Table 13–10. Fatty acid composition of germplasm with homozygous alleles at genomic loci governing a low-palmitic acid trait in soybean oil.

Germplasm	Fatty acid class					Putative genotype
	16:0	18:0	18:1	18:2	18:3	
	% of total					
Normal	12	4	21	54	9	$FapFap$
N79-2077-12	6	4	39	44	7	$fap_{nc}fap_{nc}$
C1726	8	3	26	58	5	fap_1fap_1
A22	4	3	27	58	9	fap_3fap_3
ELLP2	6	3	22	61	8	fap^*fap^*
N94-2575	4	3	23	62	8	$fap_1fap_1 fap_{nc}fap_{nc}$
C1943	4	4	31	55	6	$fap_1fap_1 fap_{nc}fap_{nc}$

resented by the fap_3, fap_{nc}, and fap^* descriptors are different and distinct from fap_1. However, it is not known whether fap_3, fap_{nc}, and fap^* are allelic to each other, or whether they represent three different loci.

Given that fap_{nc} and fap_1 segregate as independent loci, efforts have been made to identify the enzyme(s) they encode. It has been shown with in vivo saturation kinetics of acetate metabolism that both of these alleles affect reduced 16:0-ACP TE activity (Wilson et al., 2001b). Genetic effects on the activity of this enzyme also were apparent at the transcriptional level when mRNA from cotyledons at 35 DAF was hybridized with soybean cDNA probes derived from the 16:0-ACP thioesterase (FAT B) gene sequence. The steady-state levels of FAT B transcripts were directly related to the 16:0 concentration exhibited by Fap_1Fap_1 $Fap_{nc}Fap_{nc}$, fap_1fap_1 $Fap_{nc}Fap_{nc}$, Fap_1Fap_1 $fap_{nc}fap_{nc}$ or $fap_1fap_1 fap_{nc}fap_{nc}$ genotypes (Wilson et al., 2001b). In addition, Southern blots and DNA sequencing showed that the mutation in the fap_{nc} allele was a natural genomic deletion of one isoform of the FAT B gene, whereas fap_1 represented a point mutation in another isoform of the FAT B gene (Wilson et al., unpublished data, 2000). Therefore, major genetic effects of the fap_1 and fap_{nc} alleles are the result of mutations in two different genes that encode the 16:0-ACP TE in soybean. However, the same may not be presumed to be the case for other fap alleles until the mutations and products of those putative genes have been determined.

13–2.2.1.2 High 16:0 Phenotypes. The concentration of 16:0 in soybean also may be genetically elevated to levels above 11% of crude oil by chemical mutagenesis of gene(s) presumed to be at Fap loci. Germplasm with about 150 % or greater levels of 16:0 normally found in soybean oil include lines that are homozygous recessive for fap_2 (C1727, [Erickson et al., 1988]), fap_{2b} (A21, [Schnebly et al., 1994]), fap_4 (A24, [Schnebly et al., 1994]), or fap_5 (A27, [Stoltzus et al., 2000b]) alleles (Table 13–11). Various combinations of these alleles may elevate 16:0 concentration up to 40% of crude oil (Stoltzus et al., 2000b). However, classical Mendelian genetic studies are needed to help determine whether the fap_2, fap_{2b}, fap_4, and fap_5 alleles reside at independent gene loci. The gene products (enzymes) of these alleles are unknown. Recent reports of other putative high 16:0 mutations, such as fap_6 (Narvel et al., 2000) and fap_7 (Stoltzfus et al., 2000a), accentuates the need to determine the enzyme these alleles encode and the exact nature of the mutations.

Table 13–11. Fatty acid composition of germplasm with homozygous alleles at genomic loci governing a high palmitic acid trait in soybean oil.

Germplasm	Fatty acid class					Putative genotype
	16:0	18:0	18:1	18:2	18:3	
	% of total					
Normal	12	4	21	54	9	*FapFap*
C1727	18	3	16	55	8	*fap₂fap₂*
A21	16	4	21	52	7	*fap₂bfap₂b*
A24	15	4	25	48	8	*fap₄fap₄*
A27	17	4	26	46	7	*fap₅fap₅*

Mutations at *Fap* loci could affect the activity of numerous enzymes within the fatty acid synthetic pathway (Ohlrogge and Browse, 1995). The possibilities include at least eight enzymes (3-keto-acyl-ACP synthetase III [KAS-III], 3-keto-acyl-ACP synthetase II [KAS-II], 18:0-ACP desaturase [Δ9DES], 16:0-ACP thioesterase [16:0-ACP TE], 18:1-ACP thioesterase [18:1-ACP TE], glycerol-3-phosphate acyltransferase [G3PAT], lysophosphatidic acid acyltransferase [LPAAT], and DGAT). Any of these enzyme activities might also be influenced by 'modifier' genes (genes of minor effect) that may modulate expression of *fap* loci (Rebetzke et al., 1998).

Only a few investigations have explored the potential metabolic targets for *fap* and related modifier gene action in soybean. In some published reports (Wilson et al., 1988, 2001b), tests were conducted relative to the hypothesis that *fap₂* alleles altered the activity of glycerolipid acyltransferase enzymes. However, based on a comparison of the in vivo saturation kinetics for the synthesis of triacylglycerol (TG) from exogenous acetate and 16:0-CoA, it was apparent that *fap₂* alleles did not effect the activities of G3PAT, LPAAT, DGAT, or any downstream activity leading to acyl-CoA formation. Rather, higher 16:0 concentration appeared to be a function of endogenous synthesis or transport of 16:0-CoA from plastids of developing soybean cotyledons. This finding focused attention on KAS-II, Δ9DES, 16:0-ACP TE, or 18:1-ACP TE as the prime candidates for the *fap₂* gene product. KAS-II catalyzes the elongation of 16:0-ACP to 18:0-ACP, Δ9DES desaturates 18:0-ACP to 18:1-ACP, 16:0-ACP TE initiates the conversion of 16:0-ACP or 18:1-ACP (with little activity on 18:0-ACP) to their respective acyl-CoA derivatives, and 18:1-ACP TE is reported to have a preference for 18:1-ACP (Ohlrogge and Browse, 1995). In cottonseed (Pirtle et al., 1999) and high-16:0 sunflower germplasm (Martinez-Force et al., 1999), 16:0-ACP TE activity may be primarily responsible for genetically enhanced 16:0 content of vegetable oils (albeit lower KAS-II activity also may be involved). In addition, seed of transgenic canola expressing a 16:0-ACP TE gene from *Cuphea* is reported to have 35% 16:0 (Jones et al., 1995), and similar transgenic events appear to increase 16:0 levels in *Arabidopsis* (Leonard et al., 1998). In transgenic soybean embryos (Kinney, 1994), it is observed that increased 16:0-ACP TE activity leads to elevated esterification of glycerolipids with 16:0 and decreased export of 18:1-CoA from plastids. Conversely, antisense constructs or sense suppression of the 16:0-ACP TE gene produces the opposite effect on the levels of 16:0-CoA and 18:1-CoA that are available for glycerolipid synthesis. In comparison, overexpression of the KAS-II or the Δ9DES gene had little effect on the

Table 13–12. Fatty acid composition of germplasm with homozygous alleles at genomic loci governing a high stearic acid trait in soybean oil.

Germplasm	Fatty acid class					Putative genotype
	16:0	18:0	18:1	18:2	18:3	
			% of total			
Normal	12	4	21	54	9	$FasFas$
A6	8	28	20	38	7	$fas^a fas^a$
FA41545	8	15	23	45	9	$fas^b fas^b$
A81-606085	7	19	20	45	9	$fasfas$
FAM94-41	10	9	25	51	5	$fas_{nc} fas_{nc}$
KK-2	11	7	25	50	7	$st_1 st_1$
M-25	9	20	18	45	8	$st_2 st_2$

16:0 concentration of soybean oil. Therefore, it would appear that genetic manipulation of 16:0-ACP TE activity is the most effective way to increase the 16:0 concentration of soybean oil.

However, in developing seed of the high-16:0 soybean germplasm, C1727,
acetate saturation kinetics also showed a significantly lower V_{max} for 18:0 synthesis and a considerably longer half-life ($t_{0.5}$) for 18:0 incorporation into phospholipids (Wilson et al., 2001b). This observation suggested that the $fap_2 fap_2$ genotype
might mediate an increase in 16:0 synthesis through gene action that reduces KAS-
II activity, rather than (or in addition to) increased 16:0-ACP TE activity. Furthermore, pulse-chase experiments with exogenous acetate removed the possibility that
C1727 exhibited a greater rate of 18:0 metabolism or transfer from TPL toward final
deposition in TG. The $t_{0.5}$ for 18:0 (turn-over from pulse-chase experiments) was
not significantly different among genotypes. In contrast, significant genotypic differences were found in the first-order decay rates for 16:0, and a strong negative
correlation was found between the V_{max} for 16:0 synthesis (from acetate saturation
kinetics) and the $t_{0.5}$ for 16:0 turnover (from acetate pulse-chase experiments). These
findings supported previous data showing an apparent lack of major genetic effects
of fap_2 alleles on glycerolipid acyltransferase activities or on any event downstream
of acyl-CoA formation (Wilson et al., 2001b). Thus, the putative mechanism of the
fap_2 allele may represent a mutation in the FAT B or the KAS-II gene, or in both
genes as suggested in *Arabidopsis* germplasm with elevated lauric acid (12:0) concentration (Voelker and Kinney, 2001).

13–2.2.1.3 High-18:0 Phenotypes. The average 18:0 concentration in soybean averages about 3% of crude oil (USDA, ARS, 2001). However, this phenotypic trait may be genetically altered by certain mutations at gene loci designated
Fas. The currently described mutations at *Fas* typically result in elevated 18:0 concentration (Table 13–12). Nearly all of these variants have been induced by chemical or X-ray mutagenesis, with the exception of a newly developed line, FAM94-
41 (Pantalone et al., 2002). FAM94-41 (9% 18:0) is the only known soybean
germplasm that carries a serendipitously identified natural mutation, presumably
at *Fas*, which has been designated as the recessive fas_{nc} allele. Prior to the discovery of fas_{nc}, five germplasm lines were reported to carry homozygous recessive *fas*
alleles {fas^a (A6, [Hammond and Fehr, 1983]), fas^b (FA41545, [Graef et al., 1985]),

fas (A81-606085, [Graef et al., 1985]), *st₁* (KK-2 [Rahman et al., 1997]) or *st₂* (M25, [Rahman et al., 1997])}. It is known that *fasᵃ* (30% 18:0), *fasᵇ* (15% 18:0), and *fas* (19% 18:0) are allelic to each other (Graef et al., 1985), and presumably represent different mutations in the same gene. In an F_3 population segregating for *st₁* and *st₂*, the observation of progeny with 18:0 concentrations above and below the respective parental means (transgressive segregation) provides evidence for two independently inherited genes governing higher 18:0 concentrations in soybean. The proposed *st₁ st₁ st₂ st₂* genotype may elevate 18:0 concentration beyond 35% of crude oil (Rahman et al., 1997). However, it is unknown whether *st₁* or *st₂* are allelic to *fasᵃ*, *fasᵇ* or *fas*. Mendelian genetic studies for uniqueness and allelism among these *fas* and *st* alleles have not been conducted. Still, other evidence supports the hypothesis that genes at different *Fas* loci may contribute to a very high 18:0 phenotype. In the mating of A6 and ST2 (another line derived from chemical mutagenesis), the F_1 progeny mean was greater than either parent (Bubeck et al., 1989). Based on this information, it is probable that mutations in at least two independent genes determine the activity of one or more enzymes involved in 18:0 synthesis in soybean.

It has been reported that *fasᵃ*, *fasᵇ*, or *fas* alleles cause severe reduction in the yielding ability of high 18:0 soybean germplasm (Lundeen et al., 1987; Hartmann et al., 1997). Poor yielding ability attributed to these *fas* alleles have been a significant obstacle to the development of acceptable commercial soybean varieties with higher-18:0 content. However, *fas_{nc}* in FAM94-41 may be an exception because about 80% of its pedigree comes from the high-yielding cv. Brim. Thus, FAM94-41 has agronomic potential to overcome the apparent yield depression intrinsic to material developed with the *fasᵃ*, *fasᵇ*, or *fas* alleles.

A6 and FAM94-41 have been mated to determine whether the inheritance of *fas_{nc}* is independent or allelic to *fasᵃ* (Pantalone et al., 2002). The F_2 progeny segregating for *fas_{nc}* and *fasᵃ* exhibited 6 to 26% 18:0. Within statistical limits of certainty, the variation in 18:0 concentration among all of these progeny fell between the two parents. A lack of transgressive segregates (progeny with 18:0 concentration above the high-parent value or below the low-parent value) in this population removed the possibility that this phenotypic variation arose from the combination of complementary or independently inherited genes. Based on chi-square analyses, *fas_{nc}* and *fasᵃ* were determined to be allelic to each other, and represented different mutations in the same gene locus. An analogous conclusion was reported from the inheritance pattern for *fasᵃ*, *fasᵇ*, and *fas* alleles (Graef et al., 1985) where the *fasᵃ* allele was dominant over the *fas* allele, but not dominant over *fasᵇ*. The genetic analysis of A6 × FAM94-41 progeny suggested that the *fas_{nc}* allele is different and partially dominant over the *fasᵃ* allele in A6. Thus, all five of these *fas* alleles probably contributed major genetic effects upon the activity of the same enzyme in the fatty acid or glycerolipid synthetic pathway.

Although the gene product (enzyme) of these *fas* alleles is unknown, further interpretation of the population segregating for *fas_{nc}* and *fasᵃ* provides useful information regarding the nature of these genetic effects. As an example, a strong negative relation was found between 18:1 and 18:0 concentration (Pantalone et al., 2002). In fact, the decline in 18:1 accounted for 89% of the increase in 18:0 within the F_2 progeny of A6 × FAM94-41. Only a weak negative relation existed between

16:0 and 18:0. Among all enzymes in the fatty acid and glycerolipid pathways, the enzymes: KAS-II, Δ9DES, 16:0-ACP TE, or 18:1-ACP TE have the highest probability of affecting change in 18:0 concentration in soybean oil. In view of the weak relation between 16:0 and 18:0 in this population, it was deemed unlikely that the high 18:0 trait arose from gene action that increased conversion of 16:0 to 18:0, as may be envisioned by greater KAS-II or reduced 16:0-ACP TE activities. However, mutations in fas_{nc} and fas^a could lead to either reduced Δ9DES or perhaps reduced 18:1-ACP TE activity. Notably, the high 18:0 mutation at the fab_2 locus of *Arabidopsis* led to reduced Δ9DES activity (Lightner et al., 1994), and 30% 18:0 in *Brassica* seed oil was accomplished via anti-sense expression of Δ9DES (Knutzon et al., 1992). Yet, there also is at least one report that a FAT A class acyl-ACP thioesterase from mangosteen (*Garcinia mangostana* L.) shows relatively high substrate preference for 18:0-ACP (Hawkins and Kridl, 1998). When expressed in transgenic canola, the gene encoding this enzyme enabled production of 22% 18:0 in seed oil.

13–2.2.2 Genetic Modification of Unsaturated Fatty Acid Composition

13–2.2.2.1 High 18:1 Phenotypes. Soybeans typically contain about 24.2% 18:1 (USDA, ARS, 2001). As mentioned earlier, the germplasm N78-2245 was perhaps the first soybean developed with higher levels (about 42%) of 18:1. This phenotype is attributed to a natural mutation in the FAD2-1 gene that encodes the predominant ω-6 desaturase in soybean seed. When a normal FAD2 gene is expressed in antisense orientation (or by co-suppression) in transgenic soybeans, the seed oil may contain up to 80% 18:1 (Hitz et al., 1995). Therefore, it may be presumed that natural mutations at *Fad* gene loci determine the high-18:1 trait in nontransgenic soybean. Until recently, transgenic events such as the seed specific down-regulation of both the Fat B and FAD2-1 genes (T. Clemente, Univ. of Nebraska, Lincoln, personal communication, 2001) appeared to be the only feasible approach to achieve soybean oil with exceptionally high levels of 18:1. However, through natural gene recombination, J.W. Burton (USDA-ARS at Raleigh, NC) has developed a population with segregates that range from 45 to 70% 18:1. An experimental inbred line (Wilson, 1999), N98-4445, containing about 60% 18:1 has been selected from this population (Table 13–13). It is believed that this line contains mutations that affect the product of two different isoforms of the FAD2-1 gene (R.E. Dewey, North Carolina State Univ., Raleigh, personal communication, 2001). Apparently, this natural mutation confers the high 18:1 trait without the deficiencies in plant germination that are encountered in transgenically derived high-18:1 germplasm (Miquel and Browse, 1994).

13–2.2.2.2 Low-18:3 Phenotypes. About 30 yr ago, the collaborative team of R.W. Howell, R.W. Rinne, and C.A. Brim proposed to develop soybean with lower levels of 18:3 through natural gene recombination (Howell et al., 1972). At the time, virtually nothing was known about the genetic or biochemical regulation of the trait. However, it was suspected that 18:3 synthesis in soybean probably was mediated by two enzymes that catalyzed sequential desaturation of 18:1 to 18:3, and that the inheritance of 18:1 was negatively correlated with PUFA concentration. Even though genetic variation for 18:3 concentration was practically nonexistent in the

Table 13–13. Fatty acid composition of germplasm with homozygous alleles at genomic loci governing a low linolenic acid trait in soybean oil.

Germplasm	16:0	18:0	18:1	18:2	18:3	Putative genotype
	\multicolumn{5}{c}{Fatty acid class}	
	\multicolumn{5}{c}{% of total}	
Normal	12	4	21	54	9	*FadFad FanFan*
N78-2245	10	4	42	39	6	fad₁fad₁ FanFan
PI-123440	10	4	30	51	5	FadFad fanfan
PI-361088B	12	4	19	61	4	FadFad fanfan
C1640	10	4	25	57	4	FadFad fanfan
A5	9	4	47	36	3	FadFad fan₁fan₁
N85-2176	10	3	44	40	3	fad₁fad₁ fanfan
N87-2120-3	6	4	39	48	3	FadFad fanfan
N87-2122-4	5	3	48	39	5	fadfad FanFan
DuPont†	9	3	79	3	6	anti-FAD2-1
N98-4445	8	4	60	26	2	fad₂fad₂ fanfan

† Transgenic.

USDA Soybean Germplasm Collection, Brim devised a recurrent mass and within half-sib family breeding method for selection of germplasm with higher 18:1 to indirectly reduce 18:3 (Wilson et al., 1976, 1981). After several selection cycles, 18:1 was increased from 22 to 42%, and 18:3 was reduced from 9 to 6% of crude oil. These traits were captured in the germplasm line, N78-2245 (Burton et al., 1989), which was the first soybean intentionally bred for lower 18:3 concentration (Table 13–13).

Other breeding programs soon followed with low 18:3 germplasm developed through chemical mutagenesis. Mutagenesis of the cv. Century with EMS, followed by selection led to a line, C1640 that contained about 3.5% 18:3 (Wilcox et al., 1984). Inheritance studies revealed that this trait was controlled by a single recessive allele, designated *fan* (Wilcox and Cavins, 1985). Hawkins et al. (1983) mutagenized the line FA9525 and selected a line, A5, that contained about 4% 18:3. The single recessive allele in A5 was designated, *fan₁*. Subsequently, two additional mutations were described, *fan₂* and *fan₃*, at *Fan* loci. When combined (*fan₁fan₁fan₂fan₂fan₃ fan₃*) in the germplasm line A29, these alleles reportedly produced soybean oil with 1.1% 18:3 (Fehr et al., 1992). In addition, two low-18:3 plant introductions from the USDA Soybean Collection (PI123440, identified by C.A. Brim [Howell et al., 1972]; and PI361088B, identified by Rennie et al. [1988]) contained natural mutations at *Fan* loci that were shown to be either allelic or identical to the original *fan* allele in C1640 (Rennie and Tanner, 1989). All of these respective *fan* alleles represent mutations in different genes or possibly different mutations in the same gene.

Early attempts to associate alleles governing 18:3 concentration with specific genes and altered enzyme activities suggested that the mutation resident in N78-2245 (*fadfad FanFan*) affected ω-6 desaturase activity, and the mutation in PI123440 (*FadFad fanfan*) affected ω-3 desaturase activity. Evidence in support of this theory was obtained from a mating of N78-2245 × PI123440 where the frequency distribution for 18:3 concentration among four phenotypic classes in the F₃ generation fit a model for two independently segregating gene loci (Wilson and Burton,

1986). An inbred line, N85-2176, derived from this population contained just 3.5% 18:3, and currently represents the only known soybean having recessive mutations affecting both ω-6 desaturase and ω-3 desaturase activity. This was confirmed by statistical analysis of the genotypic frequency distributions for ω-6 desaturase and ω-3 desaturase activity among F_4 progenies from N83-375 (*FadFad FanFan*) × N85-2176 (*fadfad fanfan*) (Wilson et al., 1990). Then more recently, soybean cDNA for FAD2-1 (ω-6 desaturase) and FAD3 (ω-3 desaturase) was used to probe mRNA and DNA from inbred lines representing the major phenotypic and predicted genotypic classes from F_4 progenies of N83-375 × N85-2176. These data further document the premise that mutations at *Fad* loci determine desaturation of 18:1 to 18:2, and that mutations at *Fan* loci determine conversion of 18:2 to 18:3. Furthermore, a separate investigation has shown that the *fan₁* allele in A5 germplasm is associated with the FAD3 gene (Byrum et al., 1995).

13–2.2.2.3 High-18:3 Phenotypes. Oil in seed of the wild ancestor of cultivated soybean, *Glycine soja* (Sieb. & Zucc), may contain up to 23% 18:3 (USDA, ARS, 2001). Little is known of the genetic mechanisms that govern 18:3 synthesis in wild soybean, although, overexpression of the Fad2-2 gene has been reported to elevate 18:3 (Hitz et al., 1995). It has been proposed that a different complement of *Fad* and *Fan* alleles may direct expression of the high 18:3 trait. To test that hypothesis, the inheritance of 18:3 was investigated among inbred progeny of four *G. max* × *G. soja* populations (Pantalone et al., 1997a). Interspecific hybridization of N87-2120-3 (*G. max*) × PI 342434 (*G. soja*) or PI 424031 (*G. soja*) revealed a wide range of segregation for estimates of ω-6 desaturation among F_3 progeny of these populations. However, the frequency distribution in the respective interspecific populations segregating for *Fad* alleles was distinct or independent from each other. This observation suggested that these *G. soja* parents contributed different *Fad* alleles to the population. Transgressive segregates in both populations, where ω-6 desaturation exceeded the *G. soja* parents, also indicated that the parents carried different *Fad* alleles. In addition, Chi-square analyses of the progeny class frequencies showed that each population fit the phenotypic distribution expected for a single gene in the F_3 generation. Thus, PI 342434 and PI 424031 may carry 'alternative' *Fad* alleles governing ω-6 desaturase activity.

Interspecific hybridization of N87-2122-4 (*G. max*) with PI 342434 or PI 424031 exhibited a wide range of segregation for ω-3 desaturation in the F_3 generation. Each population gave a normal distribution without transgressive segregates. However, these segregation patterns failed to overlap, as would be expected if the putative *Fan* alleles were identical. Thus, dissimilar distributions between these data sets again indicated that different *Fan* alleles were present. Chi-square analyses of these data confirmed this assumption, by showing that the progeny class frequencies from the respective populations each fit the phenotypic distribution expected for a single gene in the F_3 generation. Thus, PI 342434 and PI 424031 also may carry 'alternative' *Fan* alleles governing ω-3 desaturase activity.

To gain a better understanding of the inheritance of alternative desaturase genes in wild soybean, their mode of gene interaction was evaluated in a *G. soja* × *G. soja* population (Wilson, unpublished data, 1999). The distribution of relative ω-6 desaturase activity among F_2 progeny from PI 342434 × PI 424031 revealed a significant number of transgressive segregates, but the frequency distribution was heavily skewed toward higher 18:1. A 13:3 phenotypic ratio, confirmed by analysis of more than 1000 $F_{2:3}$ progeny, was consistent with a model for epistatic gene interaction. Considering the putative alternative *Fan* alleles, few transgressive segregates were found in ω-3 desaturase activity among F_2 progeny from *PI342434 × PI424031*. This suggested that the putative *Fan* genes from each parent were allelic to each other (i.e., represented different mutations in the same gene). This frequency distribution was heavily skewed toward lower 18:3, also in a 13:3 phenotypic ratio. Statistical tests supported the interpretation that this pattern fit a model for epistatic gene interaction. Overall, these observations suggested that wild soybean contained unique desaturase alleles, which apparently are not expressed in cultivated soybean. However, when recombined with *G. max*, these alleles acted in an additive genetic manner, to affect a higher 18:3 concentration (Rebetzke et al., 1997; Pantalone et al., 1997b). Thus, transfer of the wild soybean desaturase genes to *G. max* germplasm could facilitate the development of high PUFA soybean 'drying' oils.

13–2.2.3 Influence of Multiple Gene Combinations on Oil Composition

The genetic resources documented above represent a positive avenue toward resolving current industry issues relating to improved soybean oil quality. Developing this technology may lead to the creation of specific oil phenotypes for each of the major domestic markets to ensure greatest impact. Such innovations must involve the combination of multiple gene mutations to produce economically viable commercial products. Depending on the goal, research is needed to determine the combining ability of specific mutations (Streit et al., 2001; Rahman et al., 2001). As an example, it apparently is relatively easy to develop experimental oil phenotypes with high or low concentrations of 16:0 and 18:3 (Reske et al., 1997). These phenotypes include soybean oil with: 4% 16:0 plus 3% 18:3, 6% 16:0 plus 18% 18:3, 19% 16:0 plus 3% 18:3, or 15% 16:0 plus 19% 18:3. However, knowledge of the genomic properties and functional products of the mutations to be combined will improve experimental efficiency in attaining more complex goals. In addition, it will be important to ensure that expression of the desired traits does not adversely impact the function of membrane lipids in developing seed (Hazebroek, 2000).

13–2.2.3.1 General-Purpose Applications. Currently, it appears that soybean oil with a lower 16:0 plus lower 18:3 concentration will be acceptable to the oilseed industry. The development of germplasm with that characteristic would be an initial step to strengthen the position of soybean oil in general-purpose applications. To achieve that goal, national and local Soybean Check-Off funds have been invested in public and private research to breed soybean with oils having lower saturated fat content plus improved oxidative stability. This work, coordinated by the USB and USDA-ARS, involves 11 breeding programs in Georgia, Indiana, Iowa, Maryland, Minnesota, Missouri, North Carolina, Ohio, South Carolina, Tennessee, and Virginia. This collaborative group is developing numerous agronomic low 16:0 plus low 18:3 soybean varieties with adaptation to respective areas of the U.S. soybean production region (Maturity Groups [MG] I through VIII). The first of these new varieties to be released is the MG V cv. Satelite (Wilson, 1999). The next improvement in oil quality for general-purpose applications involves transfer of the mid 18:1 trait from germplasm such as N98-4445 to varieties like Satelite. N98-4445 (derived from N97-3363-4 by J.W. Burton at Raleigh NC) represents the only known mid 18:1 soybean in the public sector.

13–2.2.3.2 Low-Trans Isomer Margarine. Although soybean oil with higher saturated fat content may not be the highest priority for the U.S. oilseed industry, such oil may become an important ingredient of the emerging market for *trans*-free margarine. Several studies have shown that experimental high 16:0 high 18:0 soybean oils have good potential for this type of margarine production (List et al., 1995b; Neff et al., 1999). However, commercial viability of varieties containing higher saturated oil will hinge upon the ability to correct agronomic deficiencies in the yielding of these lines, and upon the development of effective and rapid methods that can identify progeny containing homozygous *fap* and *fas* alleles. The latter may not be easy, given what appears to be an epistatic interaction between these two alleles. For example, a population (FAM94-41 × C1727) was created to test the hypothesis that fap_2 and fas_{nc} alleles could be combined to produce a high 16:0 plus high 18:0 oil, presumably via a greater 16:0-ACP TE activity plus a reduced Δ9DES activity (Pantalone et al., 2002). In the F2 generation, the population exhibited transgressive segregation in that the total saturated fatty acid (TS) concentration ranged from 13.8 to 23.2%. Approximately, 12.5 or 25% of the inbred progeny of this population would be expected to be $fap_2 \, fap_2 fas_{nc} \, fas_{nc}$ homozygotes. However, far fewer progeny than that genetic expectation achieved or exceeded the combined parental phenotype (18% 16:0 plus 9% 18:0). Most of these progeny exhibited either high 16:0 plus moderately high 18:0, the converse composition, or moderately high 16:0 and moderately high 18:0 concentration. Nevertheless, those few individuals with higher than normal 18:0 plus higher than normal 16:0 concentration demonstrated that the combination of fas_{nc} and fap_2 with a suitable complement of other parental genes might increase TS. Albeit mutations in these alleles probably targeted adjacent enzymes in the fatty acid synthetic pathway, other mutations may be found that enable the full potential of fas_{nc} or fap_2 recombinants to elevate TS levels in soybean oil.

13–2.2.3.3 Industrial Applications. Although industrial applications, such as inks, resins, and coatings only account for about 4% of the domestic market for soybean oil, some applications could make greater use of a more reactive soybean product. A promising approach toward increasing PUFA levels in soybean oil for that purpose would be the introgression of alternative desaturase genes from wild soybean through interspecific hybridization with *G. max* varieties. Significant genetic gain in 18:3 concentration is observed in such populations. However, even greater PUFA levels might be achieved if the *G. max* parent exhibits a $fap_1 fap_1 fap_{nc} fap_{nc}$ genotype, for lower 16:0 concentration (Fig. 13–18). The 16:0 component of soybean oil often is regarded as a waste material in industrial oils. Hence, a low 16:0 trait not only would enhance total polyunsaturate levels, but also provide an economic advantage. Thus, it would appear that inbred lines could be selected with >75% 18:2 plus 18:3. Current soybean oil typically contains about 58% PUFA.

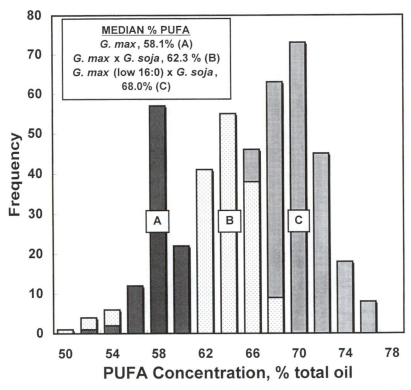

Fig. 13–18. Effect of the combination of alleles governing alternative ω-6 and ω-3 desaturases from *G. soja* and *fap* alleles from *G. max* on polyunsaturate levels in soybean oil.

13–2.3 Enhancement of Soybean Meal Quality

Approximately 85% of U.S. produced SBM is fed to nonruminent animals. Poultry (52%) and swine (29%) are the largest users, followed by beef (7%), dairy (6%), and pet foods or aquaculture (6%). Although the nutritional requirements for livestock vary, the primary criteria for formulating higher energy rations are amino acid balance and protein digestibility. For poultry, amino acid balance is assessed relative to the levels of available lysine in SBM (Mack et al., 1999; O'Quinn et al., 2000; Klemesrud et al., 2000). Compared to the 'ideal' amino acid composition established by the National Research Council (Table 13–14), methionine (*met*) and cysteine (*cys*) are present in commercial SBM at levels that may limit poultry performance. Commercial SBM also may be somewhat limiting in threonine (thr) and tyrosine (tyr). Otherwise, SBM provides nearly a complete balance of amino acids to finished feed. Synthetic amino acids and/or complementary meals such as corn (*Zea mays* L.) may be added to correct the diet for these apparent deficiencies. However, the ability to develop agronomic 'high-protein' soybean varieties provides livestock nutritionists with additional flexibility in designing the most cost-effective diets with enhanced feed-conversion. As an example, the essential amino acid balance

Table 13–14. Relative amounts of 10 essential amino acids in soybean meal from normal and the high-protein cv. Prolina. Data includes a comparison of published and commercial values from conventional soybeans, a high protein cv. Prolina, and ideal ratios for poultry diets.

Essential amino acid	Published value[†]	Commercial SBM	Prolina	Ideal amino acid balance[‡]
	——— Amount of amino acid relative to the amount of lysine ———			
Thr	0.63	0.59	0.91	0.69
Val	0.75	0.75	0.99	0.79
Ile	0.72	0.70	0.82	0.68
Leu	1.26	1.18	1.50	1.09
Tyr	0.66	0.45	0.48	0.53
Phe	0.79	0.77	0.85	0.53
Lys	1.00	1.00	1.00	1.00
Met	0.23	0.24	0.22	0.41
Cys	0.24	0.24	0.09	0.36
Arg	1.18	1.16	1.31	1.09

† National Research Council (1994).
‡ Mack et al. (1999).

of SBM from high-protein cultivars such as Prolina may supply a large proportion of the nutritional requirement for *met* and *cys*, thus obviating dependence on synthetic amino acid additives. Still, it is important to understand the biological mechanisms that contribute to enhanced levels of essential amino acids, and the regulation of other factors (such as oligosaccharides and phytic acid levels) that pose problems in the digestibility of feeds. The pursuit of more knowledge will not only impact the quality of feeds beneficially, but also will effect the physiochemical properties of soybean protein in ways that facilitate use of soybean isolates and concentrates in the manufacture of food products based on vegetable protein.

13–2.3.1 Biological Potential for Altering Protein Composition

Among all vegetable sources, soybean protein may provide the most complete amino acid balance for food and feed. However, as described above, additional improvements in the quality of SBM would benefit the soybean industry. Many seed storage protein genes from soybean have been isolated, sequenced, and expressed in transgenic plans to gain a better understanding of their function and regulation. The potential of genetic engineering to modify soybean protein composition is evident. However, problems relating to the control of gene copy number, the site of transgene insertion, and the effects of amending the native primary structure of polypeptides may impede progress toward improvements in protein quality. With the exception of the introduction of novel proteins from sources such as the Brazil-nut (*Bertholletia excelsa* Humb.& Bonpl.) (Altenbach et al., 1992), molecular genetic manipulation of specific genes that encode these storage proteins has not yielded significant or obvious changes in the concentration of essential amino acids such as *met* and *lys* in total protein (Mandal and Mandal, 2000). Transgenic manipulation of regulatory steps in the synthesis of the amino acids (*met, cys, lys, thr, ile*) derived from aspartic acid may lead to increased accumulation of free *thr* or *lys*, but do not elevate the level of these two amino acids in storage proteins. This result may be attributed to the complexity of the protein synthetic pathway, and to effects of various environmental influences on the constituent enzyme systems. Yet, significant knowledge on the biological mechanisms that regulate protein composition has been gained from these studies, and future progress will be aided by the investigation of natural or induced mutations in the subject storage protein genes.

13–2.3.1.1 Mutations in 7S Storage-Protein Genes. As mentioned before, β-conglycinin (7S protein) is composed of three different subunits, α, α' and β. There are at least 15 members of the gene family that governs 7S protein synthesis. These β-conglycinin genes are clustered in several regions in the soybean genome, and the full-length sequences are highly homologous (Harada et al., 1989). Apparently, β-conglycinin gene expression is subject to both transcriptional and post-translational regulation (Lessard et al., 1993). The gene sequence for the subunit (Tierney et al., 1987) and the α' subunits are known (Beachy et al., 1983a). Although the structure of the gene that encodes the subunit has not been completely determined, it may be composed of six exons that have similar organization to that found in the α' subunit gene (Yoshino et al., 2001). When the α and α' subunits are suppressed by sequence-mediated gene silencing in transgenic soybean seed, no significant depression of total protein content was detected, but 11S protein content increased at the expense of 7S protein (Kinney et al., 2001). Similar elevation in 11S protein content was detected in soybean germplasm (with induced mutations) that lacks the α and β subunits (Phan et al., 1996) or all three subunits (Hayashi et al., 2000). Given that 11S proteins are enriched in sulfur (S)-containing amino acids compared to 7S proteins, the higher 11S/7S ratio in these germplasm would be expected to provide a more favorable amino acid composition. However, the individual concentrations of *met*, *cys*, and *lys* in soybean seed with low β-conglycinin levels was only marginally greater that ordinary varieties (Ogawa et al., 1989). It should be noted that routine methods of measuring amino acid concentration in whole seeds may not have sufficient resolution to identify genotypic differences. For example, comparison of amino acid residues per mole of purified 11S and 7S proteins from the high-protein cv. Prolina and the high-oil cv. Dare revealed a significant increase from 1 to 5 *cys* residues per mole 7S protein in the former line (Kwanyuen et al., 1998; Luck et al., 2001).

13–2.3.1.2 Mutations in 11S Storage Protein Genes. The glycinin gene family encoding 11S subunits of soybean storage protein is composed of at least five (Gy_1 to Gy_5) gene members (Nielsen et al., 1989). The inheritance and organization of glycinin gene has been documented extensively (Cho et al., 1989a; Yagasaki et al., 1996; Staswick and Nielsen, 1983). The products of these major glycinin genes have been classified into two major subunit groups based on their sequence homologies. Group I contains: $A_{1a}B_{1b}$, A_2B_{1a}, and $A_{1b}B_2$ subunits. Group II contains: $A_5A_4B_3$ and A_3B_4 subunits (Cho et al., 1989a). Genes sequences have been reported for Gy_1 (Sims and Goldberg, 1989), Gy_2 and Gy_3 (Kitamura et al., 1990), and Gy_4 (Xue et al., 1992). Several of these genes have been mapped to positions in the soybean genome (Diers et al., 1994; Chen and Shoemaker, 1998). Natural aberrations occur in these genes, such as the recessive Gy_3 allele in the cv. Forrest (Cho et al., 1989b).

Table 13–15. Effect of N nutrition on the protein content of mature seed from normal (NC-107) and high protein (NC-111) soybean germplasm.

Germplasm	Treatment	Peak rate†	Mature seed		
	mM N	mg protein seed^{-1} d^{-1}	mg DWT seed^{-1}	mg protein seed^{-1}	% protein
NC-111	10	5.43	200.8	94.8	47.0
	30	7.39	234.6	129.1	54.7
NC-107	10	3.41	173.2	59.7	34.3
	30	5.85	216.3	102.2	46.8
LSD$_{0.05}$		1.20	19.1	21.0	6.19

†Regression protein concentration vs. peak rate: R^2, 0.904; SE, 0.577; Slope, 0.215.

13–2.3.1.3. Influence of Nutrition on Storage Protein Gene Expression.

Soybean plants grown with varied levels of nutrients, such as nitrate and sulfate, produce seed with significant variation in the amount of *met* and *cys*, particularly in S-rich proteins which likely occur in the 2S protein fraction (Staswick and Nielsen, 1983). The elevated expression of such proteins has a pronounced effect on the normal complement of 7S and 11S proteins. For example, when S is limiting, soybean seed typically contain lower levels of glycinin, and greater amounts of the β subunit of β-conglycinin (Naito et al., 1994). The latter effect is mediated by up-regulation of transcription of the *Cgy$_3$* gene that encodes the β-subunit of 7S protein. Application of nitrate to nitrogen (N) deficient soybean may elicit a similar response in an elevation of mRNA for the β-subunit (Ohtake et al., 2001). Concomitant effects may be observed in the expression of the *Cgy$_2$* gene (α′ subunit), which is genetically linked to the *Cgy$_3$* gene (Davies et al., 1985; Chen et al., 1989). These observations demonstrate that the supply and balance of N and S nutrients exert regulatory effects on the relative abundance of specific soybean storage proteins (Sexton et al., 1998a, 1998b). Indeed, the protein product of the *Cgy$_3$* gene is difficult to detect in non-nodulating soybean grown under normal field conditions (Ohtake et al., 1994), but this protein is induced in these seed by applying supplemental N treatments (Ohtake et al., 1996).

In general, increased supply of N and S nutrients also increases total protein content. This observation has been documented in soybean germplasm with typically normal protein concentration. In addition, a few reports have evaluated the response of high-protein soybean to elevated N and S nutrition. These investigations also show that such lines respond to fertilization treatments under controlled growth conditions (Nakasathien et al., 2000). As shown in Table 13–15, the application of super optimal levels of N has a positive influence on protein content and concentration in mature seed of both normal and high-protein germplasm. This suggests the possibility that the full capacity of soybean (even those selected for higher protein concentration) may not be realized when plants are grown under standard cultural practices. It also is interesting to note the impact of nutrient supply on the patterns of 11S and 7S protein accumulation in developing seed of NC-111 and a genetically related low-protein germplasm line NC-107 that were grown with a 30 mM N treatment (Fig. 13–19). This experiment shows the temporal pattern of 7S and 11S protein deposition in seed produced by each line. The 7S pattern is simi-

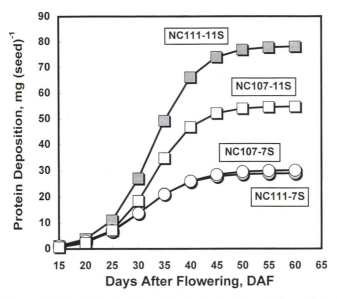

Fig. 13–19. Pattern of 11S and 7S subunit deposition during development of normal and high-protein soybean seed. Trend lines are derived from a logistic function analysis of the 7S and 11S protein fractions collected from the cv. NC-111 (high-protein) and NC-107 (normal protein) grown with 30 m*M* N.

lar for both lines, but more 11S is deposited in NC-111 seed. This result is attributable to nutrient effects on transcriptional regulation of *Gly* and *Cgy* genes. Although these genes were up-regulated by high-N nutrition in both lines, the N induced net gain in 11S protein was relative to the steady-state levels of *Gly* gene mRNA. The apparent advantage of NC-111 in 11S protein synthesis (compared to NC-107) is believed to be a function of the level of transcription of enzymes involved in 11S protein synthesis (D.W. Israel, USDA-ARS, Raleigh, NC, personal communication, 1998).

13–2.3.1.4 Association with Protein Functionality. Manufacture of dietary products from soy-flour, concentrates, and isolates is increasing in the USA because it generally is perceived that plant proteins offer greater nutritional and health benefits than animal proteins. However, functional qualities inherent to plant proteins often limit their utility in soymilk and vegetable-protein food formulations (Morr, 1990). Compared to egg white albumin and casein, protein from commercial soybean varieties has major limitations in solubility, water absorption/binding, and viscosity. These properties are determined by size, flexibility, and the three-dimensional conformation of the protein molecules. An elegant experiment conducted by Yagasaki et al. (2000) has demonstrated the impact of altered 7S and 11S content on the gelation properties of soymilk prepared from a low-β-conglycinin soybean line lacking α and α′ subunits and from a low-glycinin soybean line lacking various 11S subunit groups (*I, IIa, IIb, I + IIa, I + IIb, or IIa + IIb*). The induced genetic mutations in these genes enabled significant variation in the 11S:7S ratio (from 3.8 to 0.1) in the soymilk treatments. Results showed that protein gel strength from low-β-conglycinin soybean (greater 11S protein) was about fourfold greater than that in low-glycinin soybean (greater 7S protein). Thus, there was a strong positive relation between protein gel strength and the 11S:7S ratio.

In addition, it has been shown that protein functionality or its physiochemical properties may be influenced by the number of disulfide bridges between *cys* residues in the 11S and 7S proteins (Kwanyuen et al., 1998). In this case, the cv. Prolina and Dare had the same number of *cys* residues per mole 11S protein, but Prolina exhibited a fivefold increase in *cys* residues per mole-purified 7S protein compared to Dare. As is typical of conventional soybean, 11S and 7S proteins purified from Dare exhibited soft or poor heat-induced gelation properties. Similar results were found for the gelation properties of purified 11S protein from Prolina. The 11S proteins from both Prolina and Dare formed very soft gels that collapsed upon storage overnight, and purified 7S protein from Dare did not form gel. However, 7S purified protein from Prolina became very viscous upon solubilization in buffer and formed a firm gel that was strong enough for shear stress and strain tests. Therefore, the gelation property of 7S protein from Prolina may have been attributed to greater hydrogen bonding among the constituent proteins. Hence, subtle variation in the primary structure of 11S and 7S subunits may be equally effective in enhancing the functional properties of soybean protein.

13–2.3.2 Genetic Regulation of Oligosaccharide Content

As research progress continues to fine-tune soy-meal quality, attention will turn to the complex carbohydrates, raffinose, and stachyose. These constituents, though minor in the seeds of soybean and many other legumes, are not well metabolized by monogastric mammals (Lowell and Kuo, 1989), and the resultant fermentation products have unpleasant odors. Removal of these oligosaccharides also should boost digestibility and total metabolizable energy of soybean meal. The primary enzymes in the oligosaccharide synthetic pathway (Fig. 13–20) that could be targeted for modification are: galactinol synthase, raffinose synthase, and stachyose synthase (Handley et al., 1983; Hoch et al., 1999). Genotypic differences in oligosac-

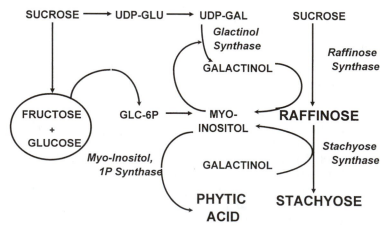

Fig. 13–20. Diagram of the stachyose and phytic acid synthetic pathways in soybean.

charide composition suggest natural allelic variation is present for the genes that encode these synthases (Bianchi et al., 1984; Hymowitz et al., 1972; Jones et al., 1999; Openshaw and Hadley, 1981). Indeed, recessive alleles already have been identified at *Stc-1* loci that presumably encode one or more of the enzymes of the pathway (Kerr and Sebastian, 1998). Two of these natural gene mutations mediate reduced raffinose synthase activity; the third recessive mutation apparently causes lower galactinol synthase activity. The combination of all three recessive alleles eliminates at least 97% of the normal levels of raffinose plus stachyose in soybean seed. (Table 13–16). Unfortunately, these low-stachyose beans have poor seed germination, suggesting that stachyose may play a vital biological role (Hsu et al., 1973; Geater and Fehr, 2000; Main et al., 1983). This problem has impeded the use of these valuable traits in commercial variety development.

13–2.3.3 Genetic Regulation of Phytic Acid Content

Soybean is one of the best agronomic crops for removing available phosphorus (P) from soils. Most of this P is stored as phytic acid. Based on global annual soybean production, it is estimated that harvested seed contains more than 2 MMT of phytic acid (Lott et al., 2000). Only three other crops (wheat [*Triticum aestivum* L.], corn, and rice [*Oryza sativa* L.]) produce more phytic acid than soybean. However, the phytic acid levels in soybean is an issue because of the formation of insoluble phytate-protein (primarily with 7S proteins) complexes during crushing and processing (Honig et al., 1984). These complexes tend to exert negative effects on mineral bioavailability and feed conversion in livestock (Selle et al., 2000). As a result, significant quantities of nondigestible P are returned to soils in the form of livestock waste. Hence, phytate levels in feeds with SBM ingredients contribute to an undesirable impact of animal production on the environment (Nahm, 2000). Reduction of phytate levels in typical corn:soybean poultry diets significantly improves the digestibility of eight amino acids (including *lys, met*, and arginine) and increases P bioavailability two- to threefold (Douglas et al., 2000). Therefore, considerable effort has been invested in strategies that would help reduce phytic acid levels in SBM. A prominent method has involved the amendment of SBM with the addition of microbial phytase enzymes, which may catalyze the hydrolysis of phytate to phosphate (Han, 1989). However, commercial application of this technology may be impeded by economic concerns. Another promising approach for reducing phytic acid is based on the genetic regulation of phytate synthesis. In maize, endogenous phytase activity is determined by two distinct genes, *PHYT-I* and *PHYT-II* (Maugen-

Table 13–16. Effect of mutations in genes that encode enzymes in the oligosaccharide synthetic pathway on the soluble carbohydrate constituents of soybean seed.

Enzyme targeted by genetic modification	Stachyose	Raffinose	Sucrose
	% of total soluble carbohydrate†		
Control	43.5	9.3	47.2
Galactinol synthase	16.9	5.2	77.9
Galactinol + raffinose synthases	6.6	1.3	92.1
Galactinol + myoinositol-1P Synthases	0.0	0.9	99.1

† Total soluble carbohydrate of seed is typically 7 to 12% of seed dry mass.

Table 13–17. Effect of gene mutations on the phytic acid concentration and content of mature soybean seed.

| Cultivar | Phytic acid | Phosphorus concentration | | Total |
		Organic	Inorganic	
	g kg⁻¹	——— % of total ———		g kg⁻¹
Normal	2.7	89.1	10.9	5.5
M153	1.1	34.5	65.5	5.5

est et al., 1999). Overexpression of these genes may help control the accumulation of phytic acid. A low phytic acid phenotype in maize also may be mediated by either of two independently inherited recessive alleles at Lpa loci (Murthy et al., 2000; Raboy, 2000). These alleles are believed to encode defective copies of D-myo-inositol-3-PO_4 synthase, MIPS (Hegeman et al., 2001). The MIPS catalyses the first step in de novo synthesis of myo-inositol, a primary substrate of phytic acid biosynthesis (Fig. 13–20), and the constituent gene has been characterized in soybean (Hegeman and Grabau, 2001).

Although MIPS activity is necessary for oligosaccharide synthesis, recent evidence suggests that the genes for the low stachyose and low phytic acid traits in soybean are not genetically linked (J.R. Wilcox, USDA-Agricultural Research Service, personal communication, 2001). This opinion is based on studies of soybean germplasm carrying EMS-induced mutations that confer a low-phytic acid phenotype without effect on oligosaccharide content (Wilcox et al., 2000). Presumably, these mutations determine MIPS activity. This phenotype is distinguished not only by a threefold reduction in phytic acid concentration, but also by a concomitant increase in inorganic P (Table 13–17), which is metabolized by poultry and swine. This low-phytate gene will play a large role in attaining goals for reducing environmental impacts of livestock production (Raboy, 1997, 2000).

13–3 OVERALL SYNOPSIS

The information presented in the chapter outlined the need to make fundamental changes in the constituent composition of soybean seed. It also appears that much of the technology is available that will be required to attempt this task. Still, it will be difficult to assess the impact of all possible combinations of genes that individually mediate desired changes in protein, oil, and carbohydrate in an agronomic background. Necessarily, pyramiding of available genes for these traits must be approached in a stepwise and orderly manner to better evaluate genic interactions (i.e., epistasis). This will ensure gradual improvement, and will result in the successful conversion of the traditional commodity soybean. The 'Better Bean' ultimately will be distinguished by higher protein and oil, improved amino acid balance, and increased protein functionality. The meal will have lower phytic acid, stable isoflavone content, and possibly reduced oligosaccharides. Not withstanding important alterations in oil, fatty acid, and minor constituent composition, the foremost criterion to be met must be no reduction in yield potential. This goal may seem futuristic, but it is attainable and will be achieved. Together, these innovations in soybean seed composition should stimulate long-term benefits in the market for U.S. produced soybean.

REFERENCES

Almonor, G.O., G.P. Fenner, and R.F. Wilson. 1998. Temperature effects on tocopherol composition in soybeans with genetically improved oil quality. J. Am. Oil Chem. Soc. 75:591–596.

Altenbach, S.B., C.-C. Kuo, L.C. Staraci, K.W. Pearson, C. Wainwright, A. Georgescu, and J. Townshend. 1992. Accumulation of a Brazil nut albumin in seeds of transgenic canola results in enhanced levels of seed protein methionine. Plant Mol. Biol. 18:235–245.

Applewhite, T.H. 1997. Trans-isomers in human health and nutrition (update 97). Am. Soybean Assoc., Brussels, Belgium.

Bach, T.J., and P. Benveniste. 1997. Cloning of cDNAs or genes encoding enzymes of sterol biosynthesis from plants and other eukaryotes: Heterologous expression and complementation analysis of mutations for functional characterization. Prog. Lipid Res. 36:197–226.

Bajjalieh, N.L. 1996. Added-value grains to have expanded value in feed. Feedstuffs 68(10):1, 23–26.

Banas, A., A. Dahlqvist, U. Stahl, M. Lenman, and S. Stymne. 2000. The involvement of phospholipid:diacylglycerol acyltransferases in triacylglycerol production. Biochem. Soc. Trans. 28:703–705.

Beachy, R.N., J. Bryant, J.J. Doyle, K. Kitamura, and B.F. Ladin. 1983a. Molecular characterization of a soybean variety lacking a subunit of the 7S seed storage protein. Plant Mol. Biol. 23:413–422.

Beachy, R.N., J.J. Doyle, B.F. Ladin, and M.A. Schuler. 1983b. Structure and expression of genes encoding the soybean 7S seed storage proteins. Natl. Adv. Sci. Inst. Ser. 63:101–112.

Bianchi, M.de L. P., H.C. Silva, and G.L. Braga. 1984. Oligosaccharide content of ten varieties of darkcoated soybeans. J. Agric. Food Chem. 32:355–357.

Birt, D.F., S. Hendrich, D.L. Alekel, and M. Anthony. 2004. Soybean and the prevention of chronic human disease. p. 1047–1118. In H.R. Boerma and J.E. Specht (ed.) Soybeans: Improvement, production, and uses. 3rd ed. Agron. Monogr. 16. ASA, CSSA, and SSSA, Madison, WI.

Browse, J., and C. Somerville. 1991. Glycerolipid synthesis: Biochemistry and regulation. Annu. Rev. Plant Physiol. Plant Mol. Biol. 42:467–506.

Bubeck, D.M., W.R. Fehr, and E.G. Hammond. 1989. Inheritance of palmitic and stearic acid mutants of soybean. Crop Sci. 29:652–656.

Budziszewski, G.J., K.P.C. Croft, and D.F. Hildebrand. 1996. Uses of biotechnology in modifying plant lipids. Lipids 31:557–569.

Burton, J.W., T.E. Carter, Jr., and R.F. Wilson. 1999. Registration of 'Prolina' soybean. Crop Sci. 39:294–295.

Burton, J.W., J.R. Wilcox, R.F. Wilson, W.P. Novitzky, and G.J. Rebetzke. 1998. Registration of low palmitic acid soybean germplasm lines N94-2575 and C1943. Crop Sci. 38:1407.

Burton, J.W., and R.F. Wilson. 1994. Registration of N88-480, a soybean germplasm line with a high concentration of oil in seeds. Crop Sci. 34:313–314.

Burton, J.W., R.F. Wilson, and C.A. Brim. 1994. Registration of N79-2077-12 and N87-2122-4, two soybean germplasm lines with reduced palmitic acid in seed oil. Crop Sci. 34:313.

Burton, J.W., R.F. Wilson, C.A. Brim, and R.W. Rinne. 1989. Registration of soybean germplasm lines with modified fatty acid composition of seed oil. Crop Sci. 29:1583.

Byrum, J.R., A.J. Kinney, R.C. Shoemaker, and B.W. Diers. 1995. Mapping of the microsomal and plastid omega-3 fatty acid desaturases in soybean [Glycine max (L.) Merr.]. Soybean Genet. Newsl. 22:181–184.

Carpenter, D.L., J. Lehmann, B.S. Mason, and H.T. Slover. 1976. Lipid composition of selected vegetable oils. J. Am. Oil Chem. Soc. 53:713–718.

Carter, Jr., T.E., J.W. Burton, and C.A. Brim. 1986. Registration of NC 101 to NC 112 soybean germplasm lines contrasting in percent seed protein. Crop Sci. 26:841–842.

Carver, B.F., and R.F. Wilson. 1984a. Polar-glycerolipid metabolism in soybean seed with genetically altered unsaturated fatty acid composition. Crop Sci. 24:1023–1026.

Carver, B.F., and R.F. Wilson. 1984b. Triacylglycerol metabolism in soybean seed with genetically altered unsaturated fatty acid composition. Crop Sci. 24:1020–1023.

Cases, S., S.J. Smith, Y.-W. Zheng, H.M. Myers, S.R. Lear, E. Sande, S. Novak, C. Collins, C.B. Welch et al. 1998. Identification of a gene encoding an acyl CoA:diacylglycerol acyltransferase, a key enzyme in triacylglycerol synthesis. Proc. Natl. Acad. Sci. USA 95:13018–13023.

Chen, Z., and R.C. Shoemaker. 1998. Four genes affecting seed traits in soybeans map to linkage group F. J. Hered. 89:211–215.

Chen, Z.-L., S. Naito, I. Nakamura, and R.N. Beachy. 1989. Regulated expression of genes encoding soybean β-conglycinins in transgenic plants. Dev. Genet. 10:112–122.

Cho, T.-J., C.S. Davies, R.L. Fischer, N.E. Turner, R.B. Goldberg, and N.C. Nielsen. 1989b. Molecular characterization of an aberrant allele for the Gy3 glycinin gene: A chromosomal rearrangement. Plant Cell. 1:339–350.

Cho, T.-J., C.S. Davies, and N.C. Nielsen. 1989a. Inheritance and organization of glycinin genes in soybean. Plant Cell. 1:329–337.

Chung, J., J.L. Babka, G.L. Graef, P.E. Staswick, D.J. Lee, P.B. Cregan, R.C. Shoemaker, and J.E. Specht. 2003. The seed protein, oil and yield QTL on soybean linkage group I. Crop Sci. 43:1053–1067.

Coates, J.B., J.S. Medeiros, V.H. Thanh, and N.C. Nielsen. 1985. Characterization of the subunits of β-conglycinin. Arch. Biochem. Biophys. 243:184–194.

Cosgrove, J.P., D.F. Church, and W.A. Pryor. 1987. The kinetics of the autoxidation of polyunsaturated fatty acids. Lipids 22:299–304.

Davies, C.S., J.B. Coates, and N.C. Nielsen. 1985. Inheritance and biochemical analysis of four electrophoretic variants of β-conglycinin from soybean. Theor. Appl. Genet. 71:351–358.

Diers, B.W., V. Beilinson, N.C. Nielsen, and R.C. Shoemaker. 1994. Genetic mapping of the Gy4 and Gy5 glycinin genes in soybean and the analysis of a variant of Gy4. Theor. Appl. Genet. 89:297–304.

Diers, B.W., and R.C. Shoemaker. 1992. Restriction fragment length polymorphism analysis of soybean fatty acid content. J. Am. Oil Chem. Soc. 69:1242–1244.

Dixon, M.D., and E.G. Hammond. 1984. The flavor intensity of some carbonyl compounds important in oxidized fats. J. Am. Oil Chem. Soc. 61:1452–1456.

Douglas, M.W., C.M. Peter, S.D. Boling, C.M. Parsons, and D.H. Baker. 2000. Nutritional evaluation of low phytate and high protein corns. Poultry Sci. 79:1586–1591.

Duffield, J., H. Shapouri, M. Graboski, R. McCormick, and R.F. Wilson. 1998. Biodiesel development: New markets for conventional and genetically modified agricultural products. Agric. Econ. Rep. 770. USDA Econ. Res. Serv., Washington, DC.

Dunn, R.O., M.W. Shockley, and M.O. Bagby. 1996. Improving the low-temperature properties of alternative diesel fuels: Vegetable oil-derived methyl esters. J. Am. Oil Chem. Soc. 73:1719–1728.

Edwards III, H.M., M.W. Douglas, C.M. Parsons, and D.H. Baker. 2000. Protein and energy evaluation of soybean meals processed from genetically modified high-protein soybeans. Poultry Sci. 79:525–527.

Erickson, E.A., J.R. Wilcox, and J.F. Cavins. 1988. Inheritance of altered palmitic acid percentage in two soybean mutants. J. Hered. 465–468.

Fehr, W.R., G.A. Welke, E.G. Hammond, D.N Duvick, and S.R. Cianzio. 1992. Inheritance of reduced linolenic acid content in soybean genotypes A16 and A17. Crop Sci. 32:903–906.

Geater, C.W., and W.R. Fehr. 2000. Association of total sugar content with other seed traits of diverse soybean cultivars. Crop Sci. 40:1552–1555.

Graboski, M.S., R.L. McCormack, and J.D. Ross. 1996. Transient emissions from No.2 diesel and biodiesel blends in a DDC Series 60 engine. Tech. Paper 961166. Soc. Automotive Engine Tech.,Washington, DC.

Graef, G.L., L.A. Miller, W.R. Fehr, and E.G. Hammond. 1985. Fatty acid development in a soybean mutant with high stearic acid. J. Am. Oil Chem. Soc. 62:773–775.

Gutfinger, T., and A. Letan. 1974. Studies of unsaponifiables in several vegetable oils. Lipids 9:658–663.

Halford N.G., and D.G. Hardie. 1998. SNF1-related protein kinases: Global regulators of carbon metabolism in plants? Plant Mol.Biol.37: 735–748.

Hammond, E.G., and W.R. Fehr. 1983. Registration of A6 germplasm line of soybean. Crop Sci. 23:192–193.

Han, Y.W. 1989. Use of microbial phytase in improving the feed quality of soya bean meal. Anim. Feed Sci. Technol. 24:345–350.

Handley, L.W., D.M. Pharr, and R.F. McFeeters. 1983. Relationship between galactinol synthase activity and sugar composition of leaves and seeds of several crop species. J. Am. Soc. Hortic. Sci.. 108:600–605.

Harada, J.J., S.J. Barker, and R.B. Goldberg. 1989. Soybean β-conglycinin genes are clustered in several DNA regions and are regulated by transcriptional and posttranscriptional processes. Plant Cell 1:415–425.

Hardie D.G. 1999. Plant protein serine/threonine kinases: Classification and functions. Annu Rev Plant Physiol. Mol. Biol. 50:97–131.

Harris, P., and A.T. James. 1969. Effect of low temperature on fatty acid biosynthesis in seeds. Biochim. Biophys. Acta 187:13–18.

Hartmann, R.B., W.R. Fehr, G.A. Welke, E.G. Hammond, D.N. Duvick, and S.R. Cianzio. 1997. Association of elevated stearate with agronomic and seed traits of soybean. Crop Sci. 37:124–127.

Hawkins, D.J., and J.C. Kridl. 1998. Characterization of acyl-ACP thioesterases of mangosteen (Garcinia mangostana) seed and high levels of stearate production in transgenic canola. Plant J. 13:743–752.

Hawkins, S.E., W.R. Fehr, and E.G. Hammond. 1983. Resource allocation in breeding for fatty acid composition of soybean oil. Crop Sci. 23:900–904.

Hayakawa, K., Y.-Y. Linko, and P. Linko. 2000. The role of trans fatty acids in human nutrition. Eur. J. Lipid Sci. Technol. 102:419–425.

Hayashi, M., M. Nishioka, K. Kitamura, and K. Harada. 2000. Identification of AFLP markers tightly linked to the gene for deficiency of the 7S globulin in soybean seed and characterization of abnormal phenotypes involved in the mutation. Breed. Sci. 50:123–129.

Hazebroek, J.P. 2000. Analysis of genetically modified oils. Prog. Lipid Res. 39:477–506.

Hegeman, C.E., L.L. Good, and E.A. Grabau. 2001. Expression of D-myo-inositol-3-phosphate synthase in soybean. Implications for phytic acid biosynthesis. Plant Physiol. 125:1941–1948.

Hegeman, C.E., and E.A. Grabau. 2001. A novel phytase with sequence similarity to purple acid phosphatases is expressed in cotyledons of germinating soybean seedlings. Plant Physiol. 126:1598–1608.

Herman, E.M., and B.A. Larkins. 1999. Protein storage bodies and vacuoles. Plant Cell 11:601–613.

Hitz, W.D., N.S. Yadav, R.S. Reiter, C.J. Mauvais, and A.J. Kinney. 1995. Reducing polyunsaturation in oils of transgenic canola and soybean. p. 506–508. In J.-C. Kader et al. (ed.) Plant lipid metabolism. Kluwer Academic Publ., Dordrecht, the Netherlands.

Hoch, G., T. Peterbauer, and A. Richter. 1999. Purification and characterization of stachyose synthase from lentil (Lens culinaris) seeds: Galactopinitol and stachyose synthesis. Arch. Biochem. Biophys. 366:75–81.

Honig, D.H., W.J. Wolf, and J.J. Rackis. 1984. Phytic acid and phosphorus content of various soybean protein fractions. Cereal Chem. 61:523–526.

Horejsi, T.F., W.R. Fehr, G.A. Welke, D.N. Duvick, E.G. Hammond, and S.R. Cianzio. 1994. Genetic control of reduced palmitate content in soybean. Crop Sci. 34:331–334.

Howell, R.W., C.A. Brim, and R.W. Rinne. 1972. The plant geneticist's contribution toward changing lipid and amino acid composition of soybeans. J. Am. Oil Chem. Soc. 49:30–32.

Hsu, S.H., H.H. Hadley, and T. Hymowitz. 1973. Changes in carbohydrate contents of germinating soybean seeds. Crop Sci. 13:407–410.

Hurburgh, Jr., C.R., T.J. Brumm, J.M. Guinn, and R.A. Hartwig. 1990. Protein and oil patterns in U.S. and world soybean markets. J. Am. Oil Chem. Soc. 67:966–973.

Hymowitz, T., W.M. Walker, F.I. Collins, and J. Panczner. 1972. Stability of sugar content in soybean strains. Commun. Soil. Sci. Plant Anal. 3:367–373.

Ilsemann, K., and K.D. Mukherjee. 1978. Continuous hydrogenation of fats and fatty acids at short contact times. J. Am. Oil Chem. Soc. 55:892–896.

Inbar, M., and M. Shinitzky. 1974. Increase of cholesterol level in the surface membrane of lymphoma cells and its inhibitory effect on ascites tumor development. Proc. Natl. Acad. Sci. USA 71:2128–2130.

International Standards Organization/International Evaluations Committee. 1998. ISO/IEC Method no. 17025:1999. International Organization of Standardization, Geneva, Switzerland. Available at http://www.iso.ch/iso/en.

Jones, A., H.M. Davies, and T.A. Voelker. 1995. Palmitoyl-acyl carrier protein (ACP) thioesterase and the evolutionary origin of plant acyl-ACP thioesterases. Plant Cell 7:359–371.

Jones, D.A., M.S. DuPont, M.J. Ambrose, J. Frias, and C.L. Hedley. 1999. The discovery of compositional variation for the raffinose family of oligosaccharides in pea seeds. Seed Sci. Res. 9:305–310.

Judd, J.T., B.A. Clevidence, R.A. Muesing, J. Wittes, M.E. Sunkin, and J.J. Podczasy. 1994. Dietary trans fatty acids: Effects on plasma lipids and lipoproteins of healthy men and women. Am. J. Clin. Nutr. 59:861–868.

Jung, W., O. Yu, S.-M.C. Lau, D.P. O'Keefe, J. Odell, G. Fader, and B. McGonigle. 2000. Identification and expression of isoflavone synthase, the key enzyme for biosynthesis of isoflavones in legumes. Nature Biotechnol. 18:208–212.

Kamal-Eldin, A., and L.-A. Appelqvist. 1996. The chemistry and antioxidant properties of tocopherols and tocotrienols. Lipids 31:671–701.

Kenworthy, W.J., and C.A. Brim. 1979. Recurrent selection in soybeans. I. Seed yield. Crop Sci. 19:315–318.

Kerr, P.S., and S.A. Sebastian. 1998. Soybean products with improved carbohydrate composition and soybean plants. U.S. Patent 5 710 365. Date issued: 20 January.

Kinney, A.J. 1994. Improving soybean seed quality. Curr. Opin. Biotechnol. 5:144–151.

Kinney, A.J., R. Jung, and E.M. Herman. 2001. Cosuppression of the α subunits of β-conglycinin in transgenic soybean seeds induces the formation of endoplasmic reticulum-derived protein bodies. Plant Cell 13:1165–1178.

Kitamura, Y., M. Arahira, Y. Itoh, and C. Fukazawa. 1990. The complete nucleotide sequence of soybean glycinin A2B1a gene spanning to another glycinin gene A1aB1b. Nucleic Acids Res. 18:4245–4246.

Klemesrud, M.J., T.J. Klopfenstein, and A.J. Lewis. 2000. Metabolizable methionine and lysine requirements of growing cattle. J. Anim. Sci. 78:199–206.

Knutzon, D.S., G.A. Thompson, S.E. Radke, W.B. Johnson, V.C. Knauf, and J.C. Kridl. 1992. Modification of Brassica seed oil by antisense expression of a stearoyl-acyl carrier protein desaturase gene. Proc. Natl. Acad. Sci. USA 89:2624–2628.

Kwanyuen, P., R.F. Wilson, and J.W. Burton. 1998. Soybean protein quality. p. 284–289. In S.S. Koseoglu et al. (ed.). Oilseed and edible oils processing. Vol 1. Am. Oil Chem. Soc. Press, Champaign, IL.

Lee J-Y., B-C. Yoo, and A.C. Harmon. 1998. Kinetic and calcium-binding properties of three calcium-dependent protein kinase isoenzymes from soybean. Biochemistry 37:6801–6809.

Leonard, J.M., S.J. Knapp, and M.B. Slabaugh. 1998. A Cuphea b-ketoacyl-ACP synthase shifts the synthesis of fatty acids towards shorter chains in Arabidopsis seeds expressing Cuphea FatB thioesterases. Plant J. 13:621–628.

Lessard, P.A., R.D. Allen, T. Fujiwara, and R.N. Beachy. 1993. Upstream regulatory sequences from two β-conglycinin genes. Plant Mol. Biol. 22:873–885.

Li, Z., R.F. Wilson, W.E. Rayford, and H.R. Boerma. 2002. Molecular mapping of genes controlling reduced palmitic acid content in soybean. Crop Sci. 42:373–378.

Lichtenstein, A.H. 1998. Soy protein, isoflavones and cardiovascular disease risk. J. Nutr. 128:1589–1592.

Lightner, J., J. Wu, and J. Browse. 1994. A mutant of Arabidopsis with increased levels of stearic acid. Plant Physiol. 106:1443–1451.

List, G.R., T.L. Mounts, F. Orthoefer, and W.E. Neff. 1995b. Margarine and shortening oils by interesterification of liquid and trisaturated triglycerides. J. Am. Oil Chem. Soc. 72:379–382.

List, G.R., T. Pelloso, F. Orthoefer, M. Chrysam, and T.L. Mounts. 1995a. Preparation and properties of zero Trans soybean oil margarines. J. Am. Oil Chem. Soc. 72:383–384.

List, G.R., K.R. Steidley, and W.E. Neff. 2000. Commercial spreads formulation, structure and properties. INFORM 11:980–986.

Liu, H.-R., and P.J. White. 1992a. Oxidative stability of soybean oils with altered fatty acid compositions. J. Am. Oil Chem. Soc. 69:528–532.

Liu, H.-R., and P.J. White. 1992b. High-temperature stability of soybean oils with altered fatty acid compositions. J. Am. Oil Chem. Soc. 69:533–537.

Lott, J.N.A., I. Ockenden, V. Raboy, and G.D. Batten. 2000. Phytic acid and phosphorus in crop seeds and fruits: A global estimate. Seed Sci. Res. 10:11–33.

Lowell, C.A., and T.M. Kuo. 1989. Oligosaccharide metabolism and accumulation in developing soybean seeds. Crop Sci. 29:459–465.

Luck, P.J., T.C. Lanier, C.R. Daubert, R.F. Wilson, and P. Kwanyuen. 2001. Functionality and visoelastic behavior of Prolina soybean isolate. p. 197–202. In R.F. Wilson (ed.) Oilseed processing and utilization. AOCS Press, Champaign, IL.

Lundeen, P.O., W.R. Fehr, E.G. Hammond, and S.R. Cianzio. 1987. Association of alleles for high stearic acid with agronomic characters of soybean. Crop Sci. 27:1102–1105.

Lusas, E.W. 2004. Soybean processing and utilization. p. 949–1046. In H.R. Boerma and J.E. Specht (ed.) Soybeans: Improvement, production, and uses. 3rd ed. Agron. Monogr. 16. ASA, CSSA, and SSSA, Madison, WI.

Mack, S., D. Bercovici, G. de Groote, B. Leclerq, M. Lippens, M. Pack, J.B. Schutte, and S. van Cauwenberghe. 1999. Ideal amino acid profile and dietary lysine specification for broiler chickens of 20 to 40 days of age. Br. Poult. Sci. 40:257–265.

Main, E.L., D.M. Pharr, S.C. Huber, and D.E. Moreland. 1983. Control of galactosyl-sugar metabolism in relation to rate of germination. Physiol. Plant. 59:387–392.

Mandal, S., and R.K. Mandal. 2000. Seed storage proteins and approaches for improvement of their nutritional quality by genetic engineering. Curr. Sci. 79:576–589.

Martin, B.A., R.F. Wilson, and R.W. Rinne. 1986. Temperature effects upon the expression of a high oleic acid trait in soybean. J. Am. Oil Chem. Soc. 63:346–352.

Martinez-Force, E., R. Alvarez-Ortega, and R. Garces. 1999. Enzymatic characterization of high-palmitic acid sunflower (*Helianthus annuus* L.) mutants. Planta 207:533–538.

Maruyama, N., M. Adachi, K. Takahashi, K. Yagasaki, M. Kohno, Y. Takenaka, E. Okuda, S. Nakagawa, B. Mikami, and S. Utsumi. 2001. Crystal structures of recombinant and native soybean β-conglycinin β homotrimers. Eur. J. Biochem. 268:3595–3604.

Massiello, F.J. 1978. Changing trends in consumer margarines. J. Am. Oil Chem. Soc. 55:262–265.

Maugenest, S., I. Martinez, B. Godin, P. Perez, and A.-M. Lescure. 1999. Structure of two maize phytase genes and their spatio-temporal expression during seedling development. Plant Mol. Biol. 39:503–514.

Mies, D.W., and T. Hymowitz. 1973. Comparative electrophoretic studies of trypsin inhibitors in seed of the genus glycine. Bot. Gaz. 134:121–125.

Miles, M.J., V.J. Morris, D.J. Wright, and J.R. Bacon. 1985. A study of the quaternary structure of glycinin. Biochim. Biophys. Acta 827:119–126.

Miquel, M.F., and J.A. Browse. 1994. High-oleate oilseeds fail to develop at low temperature. Plant Physiol. 106:421–427.

Mityko, J., J. Batkai, and G. Hodos-Kotvics. 1990. Trypsin inhibitor content in different varieties and mutants of soybean. Acta Agron. Hungarica 39:401–405.

Moorman, D.D. 1990. Evaluation of genotypic differences in protein and oil metabolism in developing soybean seeds. Thesis (M.S.), North Carolina State Univ., Raleigh.

Morr, C.V. 1990. Current status of soy protein functionality in food systems. J. Am. Oil Chem Soc. 67:265–271.

Mounts, T.L., S.L. Abidi, and K.A. Rennick. 1996. Effect of genetic modification on the content and composition of bioactive constituents in soybean oil. J. Am. Oil Chem. Soc. 73:581–586.

Mounts, T.L., K. Warner, G.R. List, W.E. Neff, and R.F. Wilson. 1994. Low-linolenic acid soybean oils-Alternatives to frying oils. J. Am. Oil Chem. Soc. 71:495–499.

Murthy, W.F. Sheridan, and D.S. Ertl. 2000. Origin and seed phenotype of maize low phytic acid 1-1 and low phytic acid 2-1. Plant Physiol. 124:355–368.

Nahm, K.H. 2000. A strategy to solve environmental concerns caused by poultry production. World's Poult. Sci. J. 56:379–388.

Naito, S., M.Y. Hirai, M. Chino, and Y. Komeda. 1994. Expression of a soybean (*Glycine max* [L.] Merr.) seed storage protein gene in transgenic Arabidopsis thaliana and its response to nutritional stress and to abscisic acid mutations. Plant Physiol. 104:497–503.

Nakasathien, S., D.W. Israel, R.F. Wilson, and P. Kwanyuen. 2000. Regulation of seed protein concentration in soybean by supra-optimal nitrogen supply. Crop Sci. 40:1277–1284.

Narvel, J.M., W.R. Fehr, J. Ininda, G.A. Welke, E.G. Hammond, D.N. Duvick, and S.R. Cianzio. 2000. Inheritance of elevated palmitate in soybean seed oil. Crop Sci. 40:635–639.

Neff, W.E., G.R. List, and W.C. Byrdwell. 1999. Effect of triacylglycerol composition on functionality of margarine base stocks. Lebensm.-Wiss. u.-Technol. 32:416–424.

Nielsen, N.C. 1985. The structure and complexity of the 11S polypeptides in soybeans. J. Am. Oil Chem. Soc. 62:1680–1685.

Nielsen, N.C., V. Beilinson, R. Bassuner, and S. Reverdatto. 2001. A Gb-like protein from soybean. Physiol. Plant. 111:75–82.

Nielsen, N.C., C.D. Dickinson, T.-J. Cho, V.H. Thanh, B.J. Scallon, R.L. Fischer, T.L. Sims, G.N. Drews, and R.B. Goldberg. 1989. Characterization of the glycinin gene family in soybean. Plant Cell 1:313–328.

Nielsen, N.C., R. Jung, Y.-W. Nam, T.W. Beaman, L.O. Oliveira, and R. Bassuner. 1995. Synthesis and assembly of 11S globulins. J. Plant Physiol. 145:641–647.

Norby, S.W., C.A. Adams, and R.W. Rinne. 1984. An ultrastructural study of soybean seed development. Dep. of Agron., Univ. of Illinois, Urbana.

Obendorf, R.L., M. Horbowicz, A.M. Dickerman, P. Brenac, and M.E. Smith. 1998. Soluble oligosaccharides and galactosyl cyclitols in maturing soybean seeds in planta and in vitro. Crop Sci. 38:78–84.

Ogawa, T., E. Tayama, K. Kitamura, and N. Kaizuma. 1989. Genetic improvement of seed storage proteins using three variant alleles of 7S globulin subunits in soybean (Glycine max L.). Jpn. J. Breed. 39:137–147.

Ohlrogge, J., and J. Browse. 1995. Lipid biosynthesis. Plant Cell 7:957–970.

Ohlrogge, J., and J.G. Jaworski. 1997. Regulation of fatty acid synthesis. Annu. Rev. Plant Physiol. Plant Mol. Biol. 48:109–136.

Ohtake, N., T. Kawachi, A. Sato, I. Okuyama, H. Fujikake, K. Sueyoshi, and T. Ohyama. 2001. Temporary application of nitrate to nitrogen-deficient soybean plants at the mid- to late-stages of seed development increased the accumulation of the β-subunit of β-conglycinin, a major seed storage protein. Soil Sci. Plant Nutr. 47:195–203.

Ohtake, N., T. Nishiwaki, K. Mizukoshi, T. Chinushi, Y. Takahashi, and T. Ohyama. 1994. Lack of β-subunit of β-conglycinin in non-nodulating isolines of soybean. Soil Sci. Plant Nutr. 40:345–349.

Ohtake, N., M. Suzuki, Y. Takahashi, T. Fujiwara, M. Chino, T. Ikarashi, and T. Ohyama. 1996. Differential expression of β-conglycinin genes in nodulated and non-nodulated isolines of soybeans. Physiol. Plant. 96:101–110.

Openshaw, S.J., and H.H. Hadley. 1981. Selection to modify sugar content of soybean seeds. Crop Sci. 21:805–808.

O'Quinn, P.R., J.L. Nelssen, R.D. Goodband, D.A. Knabe, J.C. Woodworth, M.D. Tokach, and T.T. Lohrmann. 2000. Nutritional value of a genetically improved high-lysine, high-oil corn for young pigs. J. Anim. Sci. 78:2144–2149.

Pantalone, V.R., G.J. Rebetzke, J.W. Burton, and R.F. Wilson. 1997a. Genetic regulation of linolenic acid concentration in wild soybean Glycine soja accessions. J. Am. Oil Chem. Soc. 74:159–163.

Pantalone, V.R., G.J. Rebetzke, R.F. Wilson, and J.W. Burton. 1997b. Relationship between seed mass and linolenic acid in progeny of crosses between cultivated and wild soybean. J. Am. Oil Chem. Soc. 74:563–568.

Pantalone, V.R., R.F. Wilson, W.P. Novitzky, and J.W. Burton. 2002. Genetic regulation of elevated stearic acid concentration in soybean oil. J. Am. Oil Chem. Soc. 79:549–553.

Phan, T.H., N. Kaizuma, H. Odanaka, and Y. Takahata. 1996. Specific inheritance of a mutant gene controlling α, β subunits-null of β-conglycinin in soybean (Glycine max (L.) Merrill) and observation of chloroplast ultrastructure of the mutant. Breed. Sci. 46:53–59.

Phillips, D.V., D.O. Wilson, and D.E. Dougherty. 1984. Soluble carbohydrates in legumes and nodulated nonlegumes. J. Agric. Food Chem. 32:1289–1291.

Piper, E.L., and K.J. Boote. 1999. Temperature and cultivar effects on soybean seed oil and protein concentrations. J. Am. Oil Chem. Soc. 76:1233–1241.

Pirtle, R.M., D.W. Yoder, T.T. Huynh, M. Nampaisansuk, I.L. Pirtle, and K.D. Chapman. 1999. Characterization of a palmitoyl-acyl carrier protein thioesterase (FatB1) in cotton. Plant Cell Physiol. 40:155–163.

Puri, P.S. 1978. Correlations for Trans-isomer formation during partial hydrogenation of oils and fats. J. Am. Oil Chem. Soc. 55:865–869.

Raboy, V. 1997. Low phytic acid mutants and selection thereof. U.S. Patent 5 689 054. Date issued: 18 November.

Raboy, V. 2000. Low phytic acid mutants and selection thereof. U.S. Patent 6 111 168. Date issued: 29 August.

Rahman, S.M., T. Kinoshita, T. Anai, and Y. Takagi. 2001. Combining ability in loci for high oleic and low linolenic acids in soybean. Crop Sci. 41:26–29.

Rahman, S.M., Y. Takagi, and T. Kinoshita. 1997. Genetic control of high stearic acid content in seed oil of two soybean mutants. Theor. Appl. Genet. 95:772–776.

Rebetzke, G.J., J.W. Burton, T.E. Carter, Jr., and R.F. Wilson. 1998. Genetic variation for modifiers controlling reduced saturated fatty acid content in soybean. Crop Sci. 38:303–308.

Rebetzke, G.J., V.R. Pantalone, J.W. Burton, T.E. Carter, Jr., and R.F. Wilson. 1997. Genotype variation for fatty acid content in selected Glycine max × Glycine soja populations. Crop Sci. 37:1636–1640.

Rennie, B.D., and J.W. Tanner. 1989. Genetic analysis of low linolenic acid levels in the line PI 123440. Soybean Genet. Newsl. 16:25–26.

Rennie, B.D., J. Zilka, M.M. Cramer, and W.D. Beversdorf. 1988. Genetic analysis of low linolenic acid levels in the soybean line PI 361088B. Crop Sci. 28:655–657.

Reske, J., J. Siebrecht, and J. Hazebroek. 1997. Triacylglycerol composition and structure in genetically modified sunflower and soybean oils. J. Am. Oil Chem. Soc. 74:989–998.

Rubel, A., R.W. Rinne, and D.T. Canvin. 1972. Protein, oil and fatty acid composition in developing soybean seeds. Crop Sci. 12:739–741.

Salunkhe, D.K., J.K. Chavan, R.N. Adsule, and S.S. Kadam. 1992. Canola. p. 9–58. In World oilseeds: Chemistry, technology and utilization. AVI, New York.

Schnebly, S.R., W.R. Fehr, G.A. Welke, E.G. Hammond, and D.N. Duvick. 1994. Inheritance of reduced and elevated palmitate in mutant lines of soybean. Crop Sci. 34:829–833.

Segura, J.A., M.L. Herrera, and M.C. Anon. 1995. Margarines: A rheological study. J. Am. Oil Chem. Soc. 72:375–378.

Selle, P.H., V. Ravindran, R.A. Caldwell, and W.L. Bryden. 2000. Phytate and phytase: Consequences for protein utilization. Nutr. Res. Rev. 13:255–278.

Settlage, S.B., P. Kwanyuen, and R.F. Wilson. 1998. Relation between diacylglycerol acyltransferase activity and oil concentration in soybean. J. Am. Oil Chem. Soc. 75:775–781.

Sexton, P.J., S.L. Naeve, N.C. Paek, and R. Shibles. 1998b. Sulfur availability, cotyledon nitrogen: Sulfur ratio, and relative abundance of seed storage proteins of soybean. Crop Sci. 38:983–986.

Sexton, P.J., N.C. Paek, and R. Shibles. 1998a. Soybean sulfur and nitrogen balance under varying levels of available sulfur. Crop Sci. 38:975–982.

Shanklin, J., E.B. Cahoon, E. Whittle, Y. Lindqvist, W. Huang, G. Schneider, and H. Schmidt. 1997. Structure-function studies on desaturases and related hydrocarbon hydroxylases. p. 6–10. In J.P. Williams et al. (ed.) Physiology, biochemistry, and molecular biology of plant lipids. Kluwer Academic Publ., Dordrecht, the Netherlands.

Sherwin, E.R., 1976. Antioxidants for vegetable oils. J. Am. Oil Chem. Soc. 53:430–436.

Shotwell, M.A., and B.A. Larkins. 1989. The biochemistry and molecular biology of seed storage proteins. p. 297–345. In A. Moses et al. (ed.) The biochemistry of plants. Vol. 15. Academic Press, New York.

Sims, T.L., and R.B. Goldberg. 1989. The glycinin Gy1 gene from soybean. Nucleic Acids Res. 17:4386.

Smith, K. 1997. Advances in feeding soybean meal [Online]. Keith Smith & Assoc., Farmington, MA. Available at http://www.soymeal.rog/ksmith1.html (verified 23 Apr. 1998).

Sonka, S.T., K.L. Bender, and D.K. Fisher. 2004. Economics and marketing. p. 919–948. In H.R. Boerma and J.E. Specht (ed.) Soybeans: Improvement, production, and uses. 3rd ed. Agron. Monogr. 16. ASA, CSSA, and SSSA, Madison, WI.

Specht, J.E., K. Chase, M. Macrander, G.L. Graef, J. Chung, J.P. Markwell, M. Germann, J.H. Orf, and K.G. Lark. 2001. Soybean response to water: A QTL analysis of drought tolerance. Crop Sci. 41:493–509.

Staswick, P.E., and N.C. Nielsen. 1983. Characterization of a soybean cultivar lacking certain glycinin subunits. Arch. Biochem. Biophys. 223:1–8.

Stojsin, D., G.R. Ablett, B.M. Luzzi, and J.W. Tanner. 1998. Use of gene substitution values to quantify partial dominance in low palmitic acid soybean. Crop Sci. 38:1437–1441.

Stoltzfus, D.L., W.R. Fehr, G.A. Welke, E.G. Hammond, and S.R. Cianzio. 2000a. A fap7 allele for elevated palmitate in soybean. Crop Sci. 40:1538–1542.

Stoltzus, D.L., W.R. Fehr, G.A. Welke, E.G. Hammond, and S.R. Cianzio. 2000b. A fap5 allele for elevated palmitate in soybean. Crop Sci. 40:647–650.

Streit, L.G., W.R. Fehr, G.A. Welke, E.G. Hammond, and S.R. Cianzio. 2001. Family and line selection for reduced palmitate, saturates, and linolenate of soybean. Crop Sci. 41:63–67.

Tierney, M.L., E.A. Bray, R.D. Allen, Y. Ma, R.F. Drong, J. Slightom, and R.N. Beachy. 1987. Isolation and characterization of a genomic clone encoding the β-subunit of β-conglycinin. Planta 172:356–363.

Tompkins, C., and E.G. Perkins. 2000. Frying performance of low-linolenic acid soybean oil. J. Am. Oil Chem. Soc. 77:223–229.

Topfer, R., N. Martini, and J. Schell. 1995. Modification of plant lipid synthesis. Science (Washington DC) 268:681–686.

Toroser, D., and S.C. Huber. 2000. Carbon and nitrogen metabolism and reversible protein phosphorylation.p. 435–458. In M. Kreis and J.C. Walker (ed.) Plant protein kinases. Advances in botanical research including advances in plant pathology. Academic Press, New York.

Tsukamoto, C., S. Shimada, K. Igita, S. Kudou, M. Kokubun, K. Okubo, and K. Kitamura. 1995. Factors affecting isoflavone content in soybean seeds: Changes in isoflavones, saponins, and composition of fatty acids at different temperatures during seed development. J. Agric. Food Chem. 43:1184–1192.

U.S. Department of Agriculture. 1996a. Fats and oils—Production, consumption and stocks: Annual Summaries 1993–1995. USDA, Washington, DC.

U.S. Department of Agriculture. 1996b. Industrial uses of agricultural materials situation and outlook report. IUS-6:22–23.

U.S. Department of Agriculture. 2001. Industrial uses of agricultural materials situation and outlook report. IUS-11:6–9.

U.S. Department of Agriculture, Agricultural Research Service, National Genetic Resources Program. Germplasm Resources Information Network-(GRIN). [Online Database] National Germplasm Resources Lab., Beltsville, MD. Available: http://www.ars-grin.gov/var/apache/cgi-bin/npgs/html/ (verified 1 Oct. 2001).

U.S. Department of Agriculture, Foreign Agricultural Research Service. 2002. Oilseeds: World markets and trade FOP 12:02. USDA, Washington, DC.

U.S. Food and Drug Administration. 1999a. FDA proposes new rules for trans fatty acids in nutrition labeling, nutrient content claims and health claims. Federal Register. 12 Nov. 1999.

U.S. Food and Drug Administration, Center for Food Safety and Applied Nutrition. 1999b. A food labeling guide—Appendix A: Definitions of nutrient content claims. (Available at http://vm.cfsan.fda.gov/~dms/flg-6a.html.) (verified 11 Dec. 2000.)

Utsumi, S., and J.E. Kinsella. 1985. Structure-function relationships in food proteins: Subunit interactions in heat-induced gelation of 7S, 11S, and soy isolate proteins. J. Agric. Food Chem. 33:297–303.

Vereshchagin, A.G. 1991. Comparative kinetic analysis of oil accumulation in maturing seeds. Plant Physiol. Biochem. 29:385–393.

Vlahakis, C., and J. Hazebroek. 2000. Phytosterol accumulation in canola, sunflower and soybean oils: Effects of genetics, planting location and temperature. J. Am. Oil Chem. Soc. 77:49–53.

Voelker, T., and A.J. Kinney. 2001. Variations in the biosynthesis of seed-storage lipids. Annu. Rev. Plant Physiol. plant Mol. Biol. 52:335–361.

Wang, C., Q. Ma, S. Pagadala, M.S. Sherrard, and P.G. Krishnan. 1998. Changes of isoflavones during processing of soy protein isolates. J. Am. Oil Chem. Soc. 75:337–341.

Warner, K., and E.N. Frankel. 1985. Flavor stability of soybean oil based on induction periods for the formation of volatile compounds by gas chromatography. J. Am. Oil Chem. Soc. 62:100–103.

Warner, K., and T.L. Mounts. 1993. Frying stability of soybean and canola oils with modified fatty acid compositions. J. Am. Oil Chem. Soc. 70:983–988.

Watanabe, Y., and H. Hirano. 1994. Nucleotide sequence of the basic 7S globulin gene from soybean. Plant Physiol. 105:1019–1020.

Werner, M.H., and D.E. Wemmer. 1991. H assignments and secondary structure determination of the soybean trypsin/chymotrypsin Bowman-Birk inhibitor. Biochemistry 30:3356–3364.

Wilcox, J.R. 1998. Increasing seed protein in soybean with eight cycles of recurrent selection. Crop Sci. 38:1536–1540.

Wilcox, J.R. 2004. World distribution and trade of soybean. p. 621–678. In H.R. Boerma and J.E. Specht (ed.) Soybeans: Improvement, production, and uses. 3rd ed. Agron. Monogr. 16. ASA, CSSA, and SSSA, Madison, WI.

Wilcox, J.R., J.W. Burton, G.J. Rebetzke, and R.F. Wilson. 1994. Transgressive segregation for palmitic acid in seed oil of soybean. Crop Sci. 34:1248–1250.

Wilcox, J.R., and J.F. Cavins. 1985. Inheritance of low linolenic acid content of the seed oil of a mutant in Glycine max. Theor. Appl. Genet. 71:74–78.

Wilcox, J.R., J.F. Cavins, and N.C. Nielsen. 1984. Genetic alteration of soybean oil composition by a chemical mutagen. J. Am. Oil Chem. Soc. 61:97–100.

Wilcox, J.R., and Z. Guodong. 1997. Relationships between seed yield and seed protein in determinate and indeterminate soybean populations. Crop Sci. 37:361–364.

Wilcox, J.R., G.S. Premachandra, K.A. Young, and V. Raboy. 2000. Isolation of high seed inorganic P, low-phytate soybean mutants. Crop Sci. 40:1601–1605.

Wilson, R.F. 1998. New commodity products from soybean through biotechnology. Oil Mill Gazetteer 104:27–33.

Wilson, R.F. 1999. Alternatives to genetically-modified soybeans: The Better Bean Initiative. Lipid Technol. 11:107–110.

Wilson, R.F., and J.W. Burton. 1986. Regulation of linolenic acid in soybeans and gene transfer to high yielding, high protein germplasm. p. 386–391. In A.R. Baldwin (ed.) Proc. World Conf. on Emerging Technol. in the Fats and Oils Industry, Cannes, France. 3–8 Nov. 1985. AOCS, Champaign, IL.

Wilson, R.F., J.W. Burton, and C.A. Brim. 1981. Progress in the selection for altered fatty acid composition in soybeans. Crop Sci. 21:788–791.

Wilson, R.F., J.W. Burton, and P. Kwanyuen. 1990. Effect of genetic modification of fatty acid composition of soybeans on oil quality. p. 355–359. In D.R. Erickson (ed.) Edible fats and oils processing: Basic principles and modern practices. AOCS Press, Champaign, IL.

Wilson, R.F., P. Kwanyuen, and J.W. Burton. 1988. Biochemical characterization of a genetic trait for low palmitic acid content in soybean. p. 290–293. In T.H. Applewhite (ed.) Proc. World Conf. on Biotechnol. for the Fats and Oils Industry, Hamburg, Germany. 21–24 Oct. 1987. AOCS, Champaign, IL.

Wilson, R.F., T.C. Marquardt, W.P. Novitzky, J.W. Burton, J.R. Wilcox, and R.E. Dewey. 2001a. Effect of alleles governing 16:0 concentration on glycerolipid composition in developing soybeans. J. Am. Oil Chem. Soc. 78:329–334.

Wilson, R.F., T.C. Marquardt, W.P. Novitzky, J.W. Burton, J.R. Wilcox, A.J. Kinney, and R.E. Dewey. 2001b. Metabolic mechanisms associated with alleles governing the 16:0 concentration of soybean oil. J. Am. Oil Chem. Soc. 78:335–340.

Wilson, R.F., R.W. Rinne, and C.A. Brim. 1976. Alteration of soybean oil composition by plant breeding. J. Am. Oil Chem. Soc. 53:595–597.

Wilson, R.F., H.H. Weissinger, J.A. Buck, and G.D. Faulkner. 1980. Involvement of phospholipids in polyunsaturated fatty acid synthesis in developing soybean cotyledons. Plant Phyiol. 66:545–549.

Winter, H., and S.C. Huber. 2000. Regulation of sucrose metabolism in higher plants. Localization and regulation of activity of key enzymes. Crit. Rev. Plant Sci. 19:31–67.

Wolf, R.B., J.F. Cavins, R. Kleiman, and L.T. Black. 1982. Effect of temperature on soybean seed constituents: Oil, protein, moisture, fatty acids, amino acids, and sugars. J. Am. Oil. Chem. Soc. 59:230–232.

Xue, Z.-T., M.-L. Xu, W. Shen, N.-L. Zhuang, W.-M. Hu, and S.C. Shen. 1992. Characterization of a Gy4 glycinin gene from soybean Glycine max cv. Forrest. Plant Mol. Biol. 18:897–908.

Yagasaki, K., N. Kaizuma, and K. Kitamura. 1996. Inheritance of glycinin subunits and characterization of glycinin molecules lacking the subunits in soybean (Glycine max (L.) Merr.). Breed. Sci. 46:11–15.

Yagasaki, K., F. Kousaka, and K. Kitamura. 2000. Potential improvement of soymilk gelation properties by using soybeans with modified protein subunit compositions. Breed. Sci. 50:101–107.

Yazdi-Samadi, B., R.W. Rinne, and R.D. Seif. 1977. Components of developing soybean seeds: Oil, protein, sugars, starch, organic acids, and amino acids. Agron. J. 69:481–486.

Yoshino, M., A. Kanazawa, K.-I. Tsutsumi, I. Nakamura, and Y. Shimamoto. 2001. Structure and characterization of the gene encoding a subunit of soybean β-conglycinin. Genes Genet. Syst. 76:99–105.

14   Fungal Diseases

CRAIG R. GRAU

University of Wisconsin-Madison
Madison, Wisconsin

ANNE E. DORRANCE

The Ohio State University
Wooster, Ohio

JASON BOND AND **JOHN S. RUSSIN**

Southern Illinois University
Carbondale, Illinois

Soybean, *Glycine max* (L.) Merr., plant health is a critical component of profitable soybean production. Plant pathogenic fungi are an important group of organisms that compromise soybean health. Fungal pathogens of soybean vary greatly in importance based on frequency of occurrence and associated yield loss (Doupnik, 1993; Wrather et al., 1997, 2001a, 2001b). Many diseases regarded as unimportant today were at one time important yield-limiting diseases of soybean. Several fungal pathogens have emerged from minor to significant importance, or were recently discovered. More than 40 fungal pathogens are reported to cause significant disease problems worldwide (Hartman et al., 1999). Disease severity and prevalence and agronomic loss are closely associated with environmental conditions (Yang and Feng, 2001), crop management practices, and reaction of soybean cultivars to plant pathogenic fungi. Frequently some soybean diseases are geographically limited, whereas others are distributed widely (Athow, 1987; Bowers and Russin, 1998; Dickson, 1956; Hartman et al., 1999; Ling, 1951; Nyvall, 1989; Sinclair and Backman, 1989; Wyllie and Scott, 1988).

Fungi pathogenic to soybean may be restricted to a specific plant part, or infect several or all parts of the plant. Many soybean pathogens share common survival strategies, whereas others have interesting and unique characteristics that make them successful pathogens. Plant pathogenic fungi are capable of not only reducing yield, but also can modify seed composition. The diseases we have selected for discussion are grouped according to the plant part most typically affected and coverage of specific diseases will differ according to importance and availability of published literature.

In many cases, general management tactics are applicable to several diseases, whereas many are managed by unique or complex management strategies. Cultural practices are frequently employed to reduce the impact of soybean pathogens and directed at removal or destruction of infested residue, proper soil preparation and drainage, optimal plant nutrition, crop rotation, and use of pathogen-free seed. A key factor in management of soybean diseases is breeding cultivars that express varying degrees of resistance to specific pathogens. Yield potential has steadily increased each decade (Wilcox, 2001), but protection of yield potential can be achieved by further genetic modification of soybean germplasm by incorporating resistance genes to important soybean pathogens. Soybean breeders have been responsive to the changing importance of specific fungal pathogens. In some cases host resistance is not a control option for pathogenic fungi, thus cultural practices or fungicides are deployed for their control. Genetic improvement of soybean cultivars is augmented by cultural practices designed to target inoculum reduction or modification of soil and crop canopy environment to disfavor pathogen activity or reduce plant stress due to abiotic causes. Although fungicides are registered to control many fungal pathogens of soybean, the application of these products to soil or foliage is not a common practice due to economic considerations. Application of fungicides to seed, however, has increased in practice in some regions due to improved efficacy of newer active ingredients and proven need. Economics of soybean production mandate prevention rather than "rescue" treatments to manage diseases caused by fungi. Thus, planting site preparation and planting disease resistant cultivars are common options to manage soybean diseases.

14–1 SEEDLING DISEASES

14–1.1 Soilborne Pathogens

14–1.1.1 Pathogens

Oomycetes and many soilborne fungi infect planted soybean seeds and seedlings. Many of these pathogens continue to cause disease of roots and lower stems at later plant growth stages. Several species of *Pythium* and *Fusarium* are pathogenic on soybean as well as *Rhizoctonia solani* Kühn, *Phytophthora sojae* Kaufmann and Gerdemann, *Macrophomina phaseolina* (Tassi) Goidanich, and *Mycoleptodiscus terrestrus* (Gerd.) Ostazeski. All of these pathogens survive in the soil and infect soybean over a wide range of environmental conditions. *Phytophthora sojae* is described in greater detail in Section 14–2.1of this chapter. *Pythium* species that are pathogenic on soybean can be found wherever this crop is grown. *Pythium aphanidermatum* (Edson) Fitzp., *Pythium debaryanum* Auct. non R. Hesse, *Pythium irregulare* Buisman, *Pythium myriotylum* Drechs, *Pythium ultimum* Trow, and most recently *Pythium torulosum* Coker and F. Patterson and *Pythium vexans* de Bary have all been identified either alone or in combination as pathogenic on soybean plants collected from fields (Griffin, 1990; Kirkpatrick, 1999; Rizvi and Yang, 1996; Yang, 1999a; Zhang et al., 1996, 1998). Griffin (1990) described a decline in the soybean cv. Essex in which *P. ultimum* infects and kills seedlings early in the growing season when soil moisture is high and temperatures are low. In a recent survey in Iowa, species of *Pythium* were the predominant pathogens isolated from dying soybean seedlings (Rizvi and Yang, 1996). Under flooded field conditions, *P. ultimum* was the most prevalent pathogen isolated from soybean roots in

Arkansas as well as *P. aphanidermatum, P. irregulare,* and *P. vexans* (Kirkpatrick et al., 1998, 1999; Kirkpatrick, 1999).

Rhizoctonia solani strains that attack soybean can be differentiated based on the teleomorph (sexual stage), number of nuclei within each cell, and on the anastomosis group (AG) of the asexual anamorph. There are currently 11 anastomosis groups within *R. solani* and these are distinguished by examining the cellular morphology in forced matings of isolates. Hyphae of isolates within the same AG group will fuse followed by plasmolysis of the cell contents (Sneh et al., 1991). Different AG groups appear to have different host ranges. Two AG groups, AG-2-2 and AG-4, have been found most often associated with seed and seedling diseases of soybean (Baird et al., 1996; Muyolo et al., 1993; Nelson et al., 1996; Rizvi and Yang, 1996). *Rhizoctonia solani* isolates that were pathogenic on both rice (*Oryza sativa* L.) and soybean in Arkansas were characterized as AG-11 (Carling et al., 1994). In the Red River Valley of Minnesota and North Dakota, AG-5 has also been identified as a serious pathogen of soybean (Nelson et al., 1996). In addition, isolates of *R. solani* vary in the severity of disease they cause (Nelson et al., 1996) and the environmental conditions which favor infection (A.E. Dorrance, unpublished data, 2002; Lewis and Papavizas, 1977; Tachibana, 1968).

Fusarium oxysporum (Schlechet.) emend Snyd.& Hans. and *F. solani* (Mart.) Appel & Wollenweb. emend. W.C. Snyder & H.N. Hans. cause seedling blight and root rot of soybean in several states in the USA (Farias and Griffin, 1989; Killebrew et al., 1993). In a survey of soybean seedling pathogens in Mississippi, Fusaria were isolated most frequently (Killebrew et al., 1993). However, in this study as well as others, not all of the Fusaria were pathogenic on soybean (Killebrew et al., 1993; Rizvi and Yang, 1996). Fusaria can often be secondary invaders of root tissue. *Fusarium solani* form B (non-sudden death syndrome [SDS], blue isolate) was shown to cause both damping-off and root lesions on seedlings in the greenhouse as well as yield losses in field studies (Killebrew et al., 1988). In this same study the authors noted an increase in disease severity when seed quality of planted seed was poor. Nelson et al. (1997) reported a frequent occurrence of *F. solani* associated with both young and older plants in North Dakota and Minnesota.

Macrophomina phaseolina causes charcoal rot in soybean and can be found worldwide. *Macrophomina phaseolina* produces numerous small black microsclerotia in host tissue and in culture. Infected seedlings can die but may also serve as an inoculum source for the mature plant phase of this disease (Meyer et al., 1974).

Mycoleptodiscus terrestrus has been identified as a root and crown rot of many species in the Leguminosae. This fungus has been reported from North Africa, India, and in the USA from Illinois, Ohio, Maryland, and Virginia mainly on red clover (*Trifolium pratense* L.) and alfalfa (*Medicago sativa* L.). *Mycoleptodiscus terrestrus* causes pre-emergence damping-off of red clover and alfalfa but also root and crown rot in soybean (Gerdemann, 1953, 1954b; Gray, 1978; Oztazeski, 1967).

14–1.1.2 Symptoms and Losses

Seedling pathogens of soybean commonly infect plants individually or in combination (Farias and Griffin, 1989; Killebrew et al., 1993; Rizvi and Yang, 1996).

The most common symptom caused by soilborne pathogens is reduced stands. Some soybean seedlings die prior to emergence (pre-emergence damping-off) and others die after emergence (post-emergence damping-off). Symptoms expressed on dying seedlings may include distinct lesions on the roots, hypocotyl or cotyledon, or an overall root decay with hard to define margins ranging from a pale brown to deep brown, black, or brick red. Brick red hypocotyl lesions are typical symptoms caused by some isolates of *R. solani* (Plate 14–1). Soybean plants infected by *R. solani* in the spring may not express symptoms until drought conditions occur or until pods begin to fill.

Pythium spp., *P. sojae*, and *R. solani* were the major pathogens associated with a seedling disease complex in Iowa (Rizvi and Yang, 1996). In contrast, *Fusarium* spp. are the predominant pathogen isolated from soybean seedlings in Mississippi (Killibrew et al., 1993). Seedling diseases were reported to cause estimated losses of 751 455 Mg during 1989, 1990, and 1991 in the north central USA (Doupnik, 1993). In survey of the top 10 soybean-producing countries during 1994 and 1998, losses to soybean seedling diseases caused by *Rhizoctonia, Pythium,* and/or *Fusarium* were estimated to be 289 900 and 924 100 Mg, respectively (Wrather et al., 1997, 2001). In controlled field studies, where seed was inoculated with conidia, two *F. solani* strains reduced yields significantly in two successive years whereas another strain did not (Killebrew et al., 1988). Different strains of *F. solani,* environmental conditions, and seed health appear to interact in the degree of damage caused by this pathogen (Killebrew et al., 1993). In tropical regions, seedling blight caused by *M. phaseolina* resulted in losses up to 77% of seedling stand.

14–1.1.3 Disease Cycle and Epidemiology

Soilborne pathogens produce thick-walled spores that survive in soil for long periods of time. Common structures are oospores for *Pythium* spp. and *P. sojae*, chlamydospores for *Fusarium* spp., and microsclerotia for *R. solani*. When environmental conditions are favorable, the overwintering propagules will germinate and infect soybean plants. Cooler temperatures favor some species of *Pythium* while warmer temperatures favor germination of *P. sojae* and other species of *Pythium*.

Pythium spp. and *P. sojae* will infect seed and roots especially when soils are water saturated. *Fusarium* spp. infects roots and crowns, and *R. solani*, can infect seeds, roots, and hypocotyls depending on the isolate. Soybean plants with hypocotyl lesions from *R. solani* infections in the spring may not express symptoms until either drought conditions occur or as pods begin to fill which result in early-maturing plants. Overwintering structures, oospores for *Pythium* spp. and *P. sojae*, chlamydospores for *Fusarium* spp., and minute sclerotia of *Rhizoctonia* are formed in the infected root tissue. Some mycelia of these root pathogens will also survive in the decaying soybean roots and lower stems.

Lewis and Papavizas (1977) reported that high temperatures (26–32°C) and high soil moisture (>70% moisture holding capacity) favored infection by *R. solani* whereas, the largest documented epidemic of *R. solani* in soybean in Iowa occurred during a dry summer (Tachibana, 1968). Farias and Griffin (1989) reported that *F. oxysporum* infections occurred when soil was maintained at field capacity at 20°C. In the north central USA, *M. phaseolina* is noted for dry-weather wilt, where symptom development occurs under dry environmental conditions (Smith and Wyllie, 1999).

14–1.1.4 Management

Soilborne pathogens are best managed with a combination of improved soil drainage, fungicide seed treatments, planting high quality seed, crop rotation, and tillage. These management tactics reduce inoculum in fields and limit amount of water available for the pathogens to germinate and infect. Host resistance is used effectively for management of *P. sojae* (see Section 14–2.1.1). Moderate levels of resistance have been identified in soybean to *R. solani* (Bradley et al., 2001) and *M. phaseolina* (Smith and Carvil, 1997). The soybean cultivar, Archer, was found to have significantly less Pythium root rot than the cv. Hutcheson (Kirkpatrick, 1999). The resistance in Archer was found to be effective against four different species of *Pythium*, regardless of seed quality (Nanayakkara, 2001).

Crop rotation reduces pathogen populations and prevents rapid build-up of pathogen population density. This is critical for some soilborne fungi such as *Fusarium* spp. and *Rhizoctonia*. However, alternate crops must be found which are not hosts for soybean pathogens. Nelson et al. (1996) proposed that the effectiveness of crop rotation as a management tool may be limited in some cases. For example, *Rhizoctonia* AG groups AG-2-2, 4, and 5 found in the Red River Valley are pathogenic on many of the crops produced in this region; sunflower (*Helianthus annuus* L.), dry bean (*Phaseolus vulgaris* L,), sugarbeet (*Beta vulgaris* L. subsp. *vulgaris*), mustard (*Brassica juncea* L., and *Sinapis alba* L.), and flax (*Linum usitatissimum* L.). Similarly, *P. ultimum* and *P. torulosum* are pathogens of both corn (*Zea mays* L.) and soybean (Zhang et al., 1998; Zhang and Yang, 2000). *Pythium* population levels were highest in the corn-soybean and continuous soybean compared to continuous corn crop rotations. These findings thus limit the effects of this rotation sequence to manage pathogen population density.

Numerous fungicide and biological seed treatment materials are labeled or in experimental phase for management of soilborne seedling pathogens (Dorrance and McClure, 2001; Osburn et al., 1995). These products may be beneficial in fields with a history of disease or that are prone to disease development (Plate 14–2). For example, *P. sojae* and *Pythium* spp. are prevalent in soils with higher clay content and minimal crop rotations. Individual fungicide and biological seed treatment products are efficacious on specific pathogens, but for broad-spectrum control, several active ingredients are needed in one formulated product or multiple products applied to seed.

Conservation tillage prevents water and wind erosion of soil by maintaining plant residue on the soil surface. However, soils in conservation tillage systems are slower to warm in the spring and take longer to dry following rains providing the environmental conditions conducive to seedling diseases. Thus, fungicide treatments are more beneficial in conservation tillage compared to conventional tillage systems. Under conservation tillage systems, much of the inoculum remains near or at the soil surface. For example, there was a greater recovery rate of *P. sojae* from 0 to 7.5 cm in depth from fields with conservation compared to conventional tillage systems in some locations where soil samples were collected (Workneth et al., 1998).

14–2 ROOT AND STEM DISEASES

14–2.1 Charcoal Rot

Charcoal rot is prevalent throughout the world but is most severe between 35° N and 35° S lat (Wyllie, 1976). Symptoms associated with this disease, which affects health of roots and lower stems, are often attributed to hot and dry soil conditions rather than by the charcoal rot pathogen.

14–2.1.1 Pathogen

The fungus *Macrophomina phaseolina*, which causes charcoal rot, has both a broad host range and wide geographic distribution. Known hosts include soybean, alfalfa, corn, cotton (*Gossypium hirsutum* L.), grain sorghum [Sorghum bicolor (L.) Moench], peanut (*Arachis hypogaea* L.), and numerous other cultivated crop species. This fungus exists in soil as microsclerotia, which are black, spherical to oblong, and typically measure 50 to 200 μm in diameter (Smith and Wyllie, 1999). Microsclerotia are produced in host tissues in large numbers, which are released into soil as these tissues decay (Short et al., 1980).

There are differences among isolates of *M. phaseolina* in their ability to grow in the presence of chlorate (Pearson et al., 1987; Cloud and Rupe, 1991). These early reports described a relationship between chlorate utilization and host preference. In a recent study of *M. phaseolina* from different hosts, Su et al. (2001) found that isolates from corn, soybean, grain sorghum, and cotton were genetically distinct from one another. In addition, chlorate-sensitive isolates were distinct from chlorate-resistant isolates within a given host. However host specialization occurred only in isolates from corn (Su et al., 2001).

Variability in aggressiveness among isolates of *M. phaseolina* has long been noted. This variability has been associated with many factors ranging from type of plant tissue to geographic region (Dhingra and Sinclair, 1973a, 1973b). Chitima-Matsiga and Wyllie (1987) observed differences in aggressiveness among isolates derived from single conidia from the same pycnidium. Recently, Pecina et al. (2000) detected ds-RNA in isolates of the fungus from USA and Mexico and showed further that isolates that contained ds-RNA were less virulent than those that were free of ds-RNA.

14–2.1.2 Symptoms and Losses

Charcoal rot symptoms are evident in soybean plants of all ages. Infected seed are either asymptomatic or have microsclerotia, appearing as black spots of variable size, embedded in cracks of the seedcoat (Gangopadhyay et al., 1970). These microsclerotia may be visible in seed coat cracks or over the seed surface. Infected seeds may germinate, but resultant seedlings usually die within a few days (Kunwar et al., 1986). The fungus invades cotyledon and embryo tissues and generally produces microsclerotia within 3 to 5 d (Kunwar et al., 1986; McGee, 1992). Hypocotyl lesions on young diseased seedlings are reddish brown initially but subsequently turn gray to black (Meyer et al., 1974). The incidence of seedling disease is greatest at warm soil temperatures and low soil moistures (Smith and Wyllie, 1999). Surviving seedlings may show little or no external symptoms and thus serve as latent sources of inoculum (Meyer et al., 1974).

Plants infected after the seedling stage generally show no aboveground symptoms until later in the growing season. Symptoms of charcoal rot typically appear after mid-season. Diseased plants initially show non-pecific symptoms such as reduced leaf size and stem height, which indicate loss of vigor. McGee (1992) described occasional superficial lesions that extend from the soil line. Wilting, which is an indicator of root dysfunction, also may be evident (Plate 14–3). Beginning at flowering, a light gray discoloration develops on the epidermal and subepidermal tissues of both tap and secondary roots and lower stems. Black microsclerotia are produced in these tissues and in pith as well. Microsclerotia can vary in density in soybean tissue (Plate 14–4). They can be so abundant that diseased tissue appears grayish black, hence the name charcoal rot, or sufficiently sparse so that they appear as random black specks (McGee, 1992; Smith and Wyllie, 1999). Microscleoria can be particularly noticeable at stem nodes (Wyllie, 1988). The fungus also can cause reddish-brown discoloration in vascular tissues of roots and stems.

Macrophomina phaseolina can reduce plant height, root volume, and root weight by more that 50% (Wyllie, 1976). These deleterious effects on roots are most evident during the pod formation and filling stages, when demand is high for water and nutrient absorption. Because diseased plants have smaller root systems, seeds that develop tend to be fewer and lighter. Diseased plants also will mature several weeks earlier than healthy plants, which further contributes to yield loss. Accurate figures for yield loss are difficult to determine, but Wrather et al. (1995) estimated average yield losses in 16 southern states were nearly 1% annually for 1988 to 1994. This translates into a yearly loss of nearly 2×10^5 Mg during this period. Before 1988, charcoal rot was not recognized as a serious problem across the southern USA (Wrather et al., 1995).

14–2.1.3 Disease Cycle and Epidemiology

The charcoal rot fungus survives in soil and soybean debris as microsclerotia, the primary source of inoculum. Population density of microsclerotia in soil can vary based on cropping history. Short et al. (1980) showed that population density was directly correlated with the number of years that a given field had been cropped to soybean or corn. Redistribution of microsclerotia in soil occurs with normal tillage practices, although pathogen survival generally is not affected by burial at depths down to 20 cm (Short et al., 1980; Olanya and Campbell, 1988). Microsclerotia germinate on the surfaces of or in close proximity to roots, and germination can occur throughout the growing season as long as environmental conditions remain favorable (Wyllie, 1988).

Root infection can occur very early in soybean plant development. Wyllie (1976) reported that *M. phaseolina* could infect up to 100% of soybean plants within 3 to 4 wk after planting. Fungal hyphae contact host roots and penetrate by means of mechanical pressure combined with pectinolytic enzyme activity (Ammon et al., 1974, 1975). Invasion of the root cortex is followed by colonization of vascular tissues. The fungus then grows outward throughout the root and stem tissue, and eventually produces characteristic microsclerotia later in the growing season (Wyllie,

1988). Disease is most evident at this time, however most root damage and yield reduction already has occurred. Affected plants mature prematurely, normal leaf abscission is not initiated, foliage may appear chlorotic and senescent, and pods generally fail to fill completely (Wyllie, 1988).

Numerous environmental factors can influence production and germination of microsclerotia, root colonization, and symptom expression. A temperature range of 30 to 35°C was described by Meyer et al. (1974) to be optimal for seedling disease expression. Microsclerotia are produced abundantly under favorable conditions, but this can increase in response to various stresses. Microsclerotia are more abundant in very acidic (pH 4.3) and alkaline soils (pH 8.0) compared to neutral soils; in low (5°C) and high (40°C) temperature soils than at more moderate temperatures; at low soil moistures; and at low-moderate soil bulk densities (Gangopadhyay et al., 1982; Mukherjee et al., 1983; Olaya and Abawi, 1996). Production of microsclerotia was also inversely correlated with soil depth and O_2 concentration (Wyllie et al., 1984). Microsclerotia germination is a complex process that is affected by certain root exudates, soil moisture, and oxygen levels, organic soil amendments, and inorganic N (Ayanru and Green, 1974; Dhingra and Sinclair, 1975; Filho and Dhingra, 1980a; Wyllie et al., 1984). Severity of charcoal rot generally correlates positively with soil temperature and inversely with soil moisture.

Several studies examined effects of herbicides on *M. phaseolina*. Colonization of soybean roots by this fungus was increased by 2,4-DB and chloramben, but was decreased by trifluralin and alachlor (Canaday et al., 1986). Saprophytic colonization of soybean stems by *M. phaseolina* decreased following application of the desiccant herbicides paraquat, glyphosate, and sodium chlorate: sodium borate (1:1 weight:weight [w:w]) (Cerkauskas et al., 1982). Fewer microsclerotia were recovered from a sandy loam soil treated with several herbicides, including alachlor, compared to untreated soil (Filho and Dhingra, 1980b). Growth of *M. phaseolina* was reduced by ametryn at high concentrations (Anahosur et al., 1984). Atrazine, metolachlor, and alachlor stimulated microsclerotia production, whereas slight (10%) reductions in germination were induced by both metolachlor and alachlor (Russin et al., 1995).

14–2.1.4 Management

Corn, grain sorghum, and cotton are hosts for the pathogen, but generally support lower populations of microsclerotia in soil than does soybean. Wyllie (1988) reported that 15 or fewer microsclerotia per gram of soil is an acceptable population density for production of soybean. Rotation with these less-susceptible crops should continue for sufficient time to allow reduction of soil populations to this level. Once levels >15 microsclerotia per gram of soil are found, rotating away from soybean for 1 yr using any of these crops should reduce soil populations of microsclerotia to acceptable levels (Wyllie, 1988).

Most management recommendations focus on reducing crop stress and maintaining plant vigor. Avoiding excessive seeding rates will minimize competition stress (Bowen and Schapaugh, 1989), and proper fertilization will ensure vigorous, rapidly growing plants. However, Wyllie (1988) reports that there is little direct evidence on the role of plant nutrition or soil fertility on charcoal rot disease. Evidence suggests that the charcoal rot fungus is very efficient at extracting available nutrients from soil, which may give this fungus a competitive advantage under conditions of low soil nutrient levels, especially when soil temperatures are high (Koover, 1954; Wyllie, 1988). Irrigation also is recommended to help alleviate stress, but recent work (Kendig et al., 2000) showed that water management limits, but does not prevent root infection by *M. phaseolina*. *Macrophomina phaseolina* infects soybean plants early and across a wide range of environmental conditions. Infected plants remain asymptomatic unless the plant is compromised by various abiotic and other biotic stress factors.

Host resistance generally is not available for management of charcoal rot because soybean cultivars with high levels of resistance have not been developed. However cultivars react differently to root infection by *M. phaseolina*, and those with the lowest levels of root infection often produce the highest yields (Short et al., 1980; Pearson et al., 1984). Short et al. (1978) suggested that the differences in numbers of propagules in diseased tissues reflect differences in compatibility between selected soybean cultivars and *M. phaseolina*. Recently, Smith and Carvil (1997) reported that counts of microsclerotia in lower stem and taproot tissues at R7 (Fehr et al., 1971) were useful to compare cultivar reactions to this fungus. Pearson et al. (1984) found that cultivars that mature later in the season may escape some of the stress associated with high temperature and low moisture in soil. *Macrophomina phaseolina* and *Heterodera glycines*, the soybean cyst nematode (SCN), commonly occur together. Todd et al. (2000) reported greater root colonization by *M. phaseolina* increased in the presence of high populations of *H. glycines*.

The fungicides benomyl, thiabendazole, thiram, triforine, and PCNB reduced seedling infection and increased emergence in greenhouse evaluations, but not in field tests. Several of these, particularly benomyl, reduced numbers of microsclerotia (McGee, 1992), but their usefulness under field conditions has not been demonstrated. Results from a study in India suggest that biological control of *M. phaseolina* using microbial parasites of microsclerotia may have some potential (Srivastava et al., 1996).

14–2.2 Fusarium Wilt and Root Rot

Fusarium wilt was first reported in 1917 and occurs in most soybean-producing regions of the world. Whereas, Fusarium root rot was first reported in 1961 and occurs throughout the world (Athow, 1987; Nelson, 1999). *Fusarium* species are common to soybean roots and stems, thus making it difficult to determine pathological importance of specific species and variants within species of *Fusarium*.

14–2.2.1 Pathogen

Various species, formae species, and physiological races of *Fusarium* cause diseases of soybean (Nelson, 1999). The cause of sudden death syndrome, caused by a specific form of *F. solani*, will be discussed in Section 14–2.5. Fusarium blight or wilt is caused by race 1 of *Fusarium oxysporum* Schlect. f. sp. *tracheiphilum* Amst. & Amst. (Armstrong and Armstrong, 1950) and race 2 of *F. oxysporum* Schlect. f. sp. *vasinfectum* Amst. & Amst. (Armstrong and Armstrong, 1958).

Fusarium root rot is caused by *Fusarium oxysporum* Schlect. emend. Synd. & Hans. in the Elegans group and *F. solani* (Killebrew et al., 1993; 1988; Nelson, 1999; Nelson et al., 1997).

14–2.2.2 Symptoms and Losses

Symptoms of blight or wilt occur at midseason under hot, dry, stressful conditions. Infected plants exhibit characteristic browning of the vascular tissue in the root and stems. Foliar symptoms include chlorotic leaves that often wither and drop from the plant. Root symptoms when present are minor (Nelson, 1999). As a result of disease damage, pods will not fully develop resulting in yield reductions that can be as high as 59% (Ferrant and Carroll, 1981).

Young plants are at the greatest risk to Fusarium root rot particularly under cool conditions (≤14 °C). When host tissue is colonized, seedling emergence may be hindered and seedlings may be stunted and weak. Cotyledons on infected plants are chlorotic and in time become necrotic. Usually, the lower portion of the tap root and lateral roots are destroyed and result in a shallow, fibrous root system. Consequently, plants wilt when subjected to low soil moisture and high soil temperature conditions. Highly pathogenic isolates of *F. solani* form B were reovered from soybean grown in the Red River Valley Region of Minnesota and North Dakota (Nelson et al., 1997). These isolates caused severe root rot and foliar symptoms of chlorosis, necrosis, and defoliation in greenhouse studies. Plants affected by combinations of biotic and abiotic stress factors will produce seed, however the seed are often reduced in size and quality. Under humid conditions, the fungus may invade pods and progress into seed and as a result, may serve as a potential inoculum source. Yield reductions can be severe and can approach 64% (Nelson, 1999). In 1998 and 1999, estimated annual losses exceeded 81.1 thousand Mg in the USA. In 2000, drought conditions were prevalent and yield reductions attributed to Fusarium root rot were only 40.5 thousand Mg (Wrather et al., 2001b).

14–2.2.3 Disease Cycle and Epidemiology

The inoculum sources for Fusarium wilt and root rot persist in the soil for extended periods. *Fusarium oxysporum, F. oxysporum* f. sp. *tracheiphilum*, and *F. oxysporum* f. sp. *vasinfectum* overwinter as chlamydospores or mycelium in the soil and in various types of plant debris (Nelson, 1999). *Fusarium* species are common in most agricultural soils, but species recovered from soybean are generally different from species recovered from corn and sorghum (Leslie et al., 1990). Dicotyledonous weeds may serve as symptomless hosts to forms of *F. oxysporum* pathogenic to soybean (Helbig and Carroll, 1984).

Fusarium oxysporum gains access into seedlings by directly penetrating the host epidermis, and indirectly through lenticels, stomates, natural wounds, and wounds caused parasites and pathogens. The fungus quickly colonizes the host growing intercellularly in the cortex while leaving the stele unaffected. *Fusarium oxysporum* f. sp. *tracheiphilum* uses similar strategies to invade the plant. Following penetration, the fungus colonizes the xylem vessels, and in later stages of infection, the vessels become filled with mycelium and the parenchyma cells are infected (Datnoff, 1988).

As with many soilborne pathogens, infection and disease development are influenced greatly by environmental conditions. Fusarium infection and disease severity are favored by cool (14–23° C), damp conditions, and saturated soils. Following infection and the resulting damage to the plant, disease severity is exacerbated when soil moisture is limited (Nelson, 1999).

The severity of Fusarium root rot can be accentuated when plant health is compromised by biotic and abiotic factors. Herbicides (Carson et al., 1991) and other abiotic agents can cause injury and predispose the plant to greater disease severity. Other important soybean pathogens such as nematodes (Sumner and Minton, 1987) and *Rhizoctonia solani* (Datnoff and Sinclair, 1988) also render the plant more vulnerable to attack by *Fusarium* spp.

14–2.2.4 Management

Soybean cultivars possess varying degrees of susceptibility to *F. solani* and *F. oxysporum*, but resistance has not been reported (Nelson et al., 1997). Thus, planting less susceptible cultivars is the most economical management strategy. Planting dates that correspond to warmer soils should be employed in problem fields. In addition, planting high-quality seed and/or applying fungicide seed treatments can help reduce disease severity. Additionally, increased drainage and reducing soil compaction via tillage results in greater aeration and conditions that are less conducive for disease development. Although crop rotation can help to reduce inoculum levels in the soil (Nelson, 1999), the pathogens' wide host range and overwintering strategies require that additional management options be employed.

14–2.3 Phytophthora Root and Stem Rot

Phytophthora root and stem rot is now known to occur worldwide. The disease was first noted in Indiana in 1948 and again in Ohio in 1951 but the causal agent was not described until 1958 (Kaufmann and Gerdemann, 1958). Phytophthora root rot has mainly been reported from major soybean production regions in the USA, but also in Australia (Pegg et al., 1980; Ryley et al., 1998), Argentina, and Brazil (Wrather et al., 1997). More recently, the disease was described in Republic of Korea (Jee et al., 1998) and People's Republic of China (Yanchun and Chongyao, 1993).

14–2.3.1 Pathogen

Phytophthora sojae Kaufmann and Gerdemann (1958) causes root and stem rot of soybean when soils are saturated for prolonged periods of time and susceptible cultivars are planted. High levels of soil moisture favor zoospore formation and subsequent disease development. Synonyms include *Phytophthora megasperma* var. *sojae* Hildebrand, *P. megasperma* f. sp. *glycinea* (Kuan and Erwin, 1980), and *P. sojae* f.sp. *glycines* (Faris et al., 1989). *Phytophthora* spp. are members of the Oomycetes, also known as water molds, and are classified mainly on the morphological traits of the sexually produced oospore and asexually produced sporangia. The oospores and sporangia produced by *P. sojae* are almost indistinguishable from *P. megasperma* which has a very wide host range and is also the reason for changes in taxonomic classification. Hansen and Maxwell (1991) redescribed *P. sojae* as well as the other species in the *P. megasperma* complex based on genetic data. *Phytophthora sojae* has a very narrow host range where it is mainly a pathogen of soybean but it also has been reported to infect lupine *Lupinus angustifolius* L., *L. luteus* L., and *L. albus* L. (Jones and Johnson, 1969). *Phytophthora sojae* populations in the USA are comprised of many physiologic races with virulence to many of the *Rps* genes that are currently deployed in Australia, Canada, and USA (see Section 14–2.1.4).

14–2.3.2 Symptoms and Losses

Phytophthora sojae causes seed, seedling pre- and post-emergence damping-off and stem rot. Seed and pre-emergence damping-off result mainly in reduced stands. Post-emergence damping-off occurs when young seedlings are infected at the hypocotyl, cotyledon, or root. Plant tissues will turn tan to brown in color. Phytophthora root rot symptoms are dependent on the resistance level in the cultivar, and can range from no symptoms of stems and leaves, to stunted, chlorotic, and wilting plants, to dead and dying plants (Plates 14–5 and 14–6). Plants with Phytophthora stem rot first wilt, followed by chlorosis that is usually associated with a characteristic chocolate brown discoloration that occurs from the base of the stem and slowly advances up the plant as *P. sojae* slowly girdles its host (Plate 14–7). The stem canker symptom can occur anytime from the first trifoliate to late R9 stage of development on highly susceptible cultivars.

Estimated combined losses due to Phytophthora root and stem rot during 1994 were 597 800 Mg of grain in Argentina, Brazil, Canada, Italy, and the USA (Wrather et al., 1997). From 1989 to 1991 and 1996 to 1998, 929 877 and 2 625 700 Mg of grain were lost to *P. sojae,* respectively, in the North Central region of the USA (Doupnik, 1993; Wrather et al., 2001b). In the southern USA, losses due to *P. sojae* were estimated at 0.2% of a total of 1.5 million Mg of harvested soybean (Pratt and Wrather, 1998). Wilcox and St. Martin (1998) compared near isolines and blends of soybean with *Rps* genes that were resistant and susceptible to the predominant *P. sojae* populations. In field locations where *P. sojae* damping-off and stem rot were evident, susceptible near isolines averaged 65 to 93% of the yield of the resistant near isoline. An earlier study by Tooley and Grau (1984b) identified losses from *P. sojae* were mainly due to reductions in stand but reductions in weight per seed and number of pods per plant also occurred.

14–2.3.3 Disease Cycle and Epidemiology

Large numbers of oospores are formed in the roots and stems of susceptible cultivars and can persist for long periods of time in soil. The oospore serves as the primary inoculum and they do not all germinate at the same time (Schmitthenner, 1985). Saturated soil conditions and temperatures of 25 to 30°C favor oospore germination and production of sporangia. Zoospores that are formed in the sporangia are released and are attracted to soybean root exudates, genistein, daidzein (Morris and Ward, 1992), and other isoflavones (Tyler et al., 1996). Although genistein is reported as an attractant to zoospores, Vedenyapina et al. (1996) report genistein to also be inhibitory to *P. sojae* at very low concentrations. Once zoospores reach the root, they encyst (lose their flagella) begin to germinate, and mycelia penetrate directly between the cell walls of the epidermis (Beagle-Ristaino and Rissler, 1983). *Phytophthora sojae* colonizes soybean plants with and without *Rps* genes (Beagle-Ristaino and Rissler, 1983; Enkerli et al., 1997). In the *Rps* gene-mediated interaction, the hypersensitive response results and the pathogen is contained within numerous necrotic or dead cells and there is no development of haustoria (Enkerli et al., 1997). In the susceptible interaction, mycelium initially grows intracellularly and form many haustoria within root cells without triggering any response from the plant (Enkerli et al., 1997). In susceptible cultivars, the mycelium continues to colonize the root and begins to form oospores. A few oospores have been observed in the *Rps* gene-mediated resistance response (Anderson, 1986; Beagle-Ristaino and Rissler, 1983). Sporangia are readily formed from infected tissues if the soil becomes flooded (Schmitthenner, 1985). However, sporangia are thought to play a very limited role in secondary spread of this disease. It is generally thought that very high levels oospores already exist in the soil and these are the primary source of inoculum throughout the growing season (Tooley and Grau, 1984a).

14–2.3.4 Management

Phytophthora damping-off, root, and stem rot have been successfully managed with host resistance. There are three types of host resistance described in soybean to *P. sojae*: (i) single dominant *Rps* genes detected with the hypocotyl inoculation test; (ii) single dominant gene, *Rps2*, that has an intermediate or partial kill in the hypocotyl inoculation test but a hypersensitive response with root inoculation (Kilen et al., 1974; Thomison et al., 1991); and (iii) partial resistance, which following root inoculation, is expressed as fewer rotted roots, and disease progresses at a much slower rate than highly susceptible cultivars (Schmitthenner, 1985; Tooley and Grau, 1984a). In commercial cultivars, single dominant *Rps* genes have proven to be the most economical and efficient means of control. However, as in many other host-pathogen systems which are governed by a gene-for-gene system (Flor, 1955), the pathogen does adapt to specific *Rps* genes and they need to be replaced with one of the following: (i) novel *Rps* genes for resistance; (ii) novel combinations of *Rps* genes; or (iii) *Rps* genes combined with partial resistance (synonyms include: tolerance [Walker and Schmitthenner, 1984]; general resistance or field resistance). The first *Rps* resistance gene was identified in the early 1950s, subsequently, eight different *Rps* loci have been identified to date with multiple alleles at the *Rps1* and *Rps3* loci (Table 14–1). All of these loci have been placed on the soybean genetic map (Burnham et al., 2003; Cregan et al., 1999; Diers et al., 1992). For marker-assisted selection, markers (RFLP and SSR) have been identified for *Rps1*, *Rps2*, *Rps3*, *Rps4*, and *Rsp7* (Diers et al., 1992; Demirbas et al., 2001; Lohnes and Schmitthenner, 1997; Cregan et al., 1999).

Recently, four surveys for new sources of resistance in the soybean plant introductions from central and southern China as well as South Korea (Dorrance and Schmitthenner, 2000; Kyle et al., 1998; Lohnes et al., 1996; Smith, 2001) have been completed. Accessions from each of these studies were identified that have resistance to multiple physiological races of *P. sojae*. Dorrance and Schmitthenner (2000) used a strategy in which isolates chosen had a virulent interaction with all of the currently identified *Rps* genes, two-*Rps* gene, and all but four 3-*Rps* gene combi-

Table 14–1 Genes and sources for resistance to *Phytophthora sojae*.

Allele	Source (s)	Citation
*Rps*1a	Mukden	Bernard et al. (1957)
*Rps*1b	PI 84637	Mueller et al. (1978);
	D60-9647	Hartwig, et al. (1968)
*Rps*1c	PI 54615-1	Mueller et al. (1978)
*Rps*1d	PI 103091	Buzzell and Anderson (1992)
		Laviolette and Athow (1977)
*Rps*1k	Kingwa	Bernard and Cremeens (1981)
*Rps*2	CNS	Kilen et al. (1974)
*Rps*3a	PI 86972-1	Mueller et al. (1978);
	PI 171442	Kilen and Keeling (1981)
*Rps*3b	PI 172901	Ploper et al. (1985)
*Rps*3c	PI 340046	R. Nelson, personal communication, 2002
*Rps*4	PI 86050†	Athow et al. (1980)
*Rps*5	L62-904	Buzzel and Anderson (1981)
*Rps*6	Altona	Athow and Laviolette (1982)
*Rps*7	Harosoy	Anderson and Buzzell (1992)
*Rps*8	PI 399073	Burnham et al. (2003)

† PI 86050 has two genes, *Rps*1c and *Rps*4.

nations. One of the 32 accessions identified in this study was found to contain a novel *Rps* gene (*Rps*8, Burnham et al., 2003). In 7 of 18 accessions from an earlier survey, Hegstad et al. (1998) identified unique RFLP-banding patterns and hypocotyl inoculations potentially indicating a novel *Rps* gene.

During 1960s much of the north central USA was planted to cultivars containing *Rps*1a, subsequently, new races of *P. sojae* were identified, race 2 in 1965 (Morgan and Hartwig, 1965); race 3 in 1972 (Schmitthenner, 1972); race 4 in 1974 (Schwenk and Sim, 1974); races 5 and 6 in 1976 (Haas and Buzzel, 1976) and races 7, 8, and 9 in 1977 (Laviolette and Athow, 1977) (Table 14–2). As more *Rps* alleles have been identified, changes in the differentials used to characterize *P. sojae* were made. However, it was generally agreed that differentials with *Rps*1a, *Rps*1b, *Rps*1c, *Rps*1d, *Rps*1k, *Rps*3a, *Rps*6, and *Rps*7 were the standards for characterization of the physiological races of *P. sojae* (Abney et al., 1997; Förster et al., 1994; Henry and Kirkpatrick, 1995; Laviolette and Athow, 1981; Schmitthenner et al., 1994; Wagner and Wilkinson, 1992; Ward, 1990; Yang et al., 1996). This is true up to physiologic race 45. For *P. sojae* races 46 to 55, additional differentials were added, but the original isolates of races 1 to 45 were not re-classified on the additional differentials (Ryley et al., 1998; Leitz et al., 2000). Hence, race 48 = race 1 on the standard differential set; 49 = 5; 50 = 13; and 52 = 1. With eight differentials representing alleles, *Rps*1a, *Rps*1b, *Rps*1c, *Rps*1d, *Rps*1k, *Rps*3a, *Rps*6, and *Rps*7, there is the possibility of classifying 256 races of *P. sojae* if all virulence pathotypes were identified. If the differential set was expanded to include all known 13 *Rps* genes, there would be the possibility of classifying 8192 physiologic races. At this time, with the great variability that has been identified in *P. sojae*, including the identification of virulence to *Rps* alleles that have not knowingly been deployed (Abney et al., 1997; Cochran and Abney, 1999; Leitz et al., 2000; Schmitthenner et al., 1994) it is more important that the virulence pathotype be reported for a region than to codify with race classifications (Table 14–2). In addition, it is espe-

Table 14–2. Virulence formulae for reported† physiologic races of *Phytophthora sojae* on soybean differentials with *Rps*1a, *Rps*1b, *Rps*1c, *Rps*1d, *Rps*1k, *Rps*3a, *Rps*6 and *Rps*7.

Race	Virulence pathotype	Race	Virulence pathotype	Race	Virulence pathotype
0	Avir	19	1a,1b,1c,1d,1k,3a	38	1a, 1b, 1c, 1d, 1k, 3a, 6, 7
1	7	20	1a,1b,1c,1k,3a,7	39	1a, 1b, 1c, 1k, 3a, 6, 7
2	1b,7	21	1a,3a,7	40	1a,1c,1d,1k,7
3	1a,7	22	1a,1c,3a,6,7	41	1a,1b,1d,1k,7
4	1a,1c,7	23	1a,1b,6,7	42	1a, 1d, 3a, 7
5	1a,1c,6,7	24	1b,3a,6,7	43	1a, 1c, 1d, 7
6	1a,1d,3a,6,7	25	1a,1b,1c,1k,7	44	1a, 1d, 7
7	1a,3a,6,7	26	1b,1d,3a,6,7	45	1a, 1b, 1c, 1k, 6, 7
8	1a,1d,6,7	27	1b,1c,1k,3a,6,7	46	1a,1c,3a,5‡,7
9	1a,6,7	28	1a,1b,1k,7	47	1a,1b,1c,7
10	1b,3a,7	29	1a,1b,1k,6,7	48=1	5‡,7
11	1b,6,7	30	1a,1b,1k,3a,6,7	49=5	1a,1c,4‡,6,7
12	1a,1b,1c, 1d, 1k,3a	31	1b,1c,1d,1k,6,7	50=13	4‡,6,7
13	6,7	32	1b,1k,6,7	51	1c,5‡,6,7
14	1c,7	33	1a,1b,1c,1d,1k,7	52=1	3b‡,5‡,7
15	3a,7	34	1a,1k,7	53	1a,1b,1c,3a,5,7
16	1b,1c,1k	35	1a, 1b, 1c, 1d, 1k	54	1d,7
17	1b,1d,3a,6,7	36	3a, 6	55	1d,3a,3c‡,4‡,5‡,6,7
18	1c	37	1a, 1c, 3a, 6, 7		

† Abney et al. (1997), Förster et al. (1994), Haas and Buzzell (1976), Henry and Kirkpatrick (1995), Keeling (1982, 1984), Laviolette and Athow (1977), Leitz et al. (2000), Morgan and Hartwig (1965), Ryley et al. (1998), Schmitthenner (1972), Schmitthenner et al. (1994), Schwenk and Sim (1974), Wagner and Wilkinson (1992), Ward (1990), and Yang et al. (1996).

‡ Additional differentials have been incorporated with *Rps*2, *Rps*3b, *Rps*3c, *Rps*4 and *Rps*5 but the original isolates were not reclassified (P134, L3-4).

cially important to have the differentials chosen for a study match the most recently deployed *Rps* genes. For example, Ryley et al. (1998) reported in a survey of *P. sojae* in Australia, that 'Davis' had recently become susceptible to *P. sojae*. The soybean cv. Davis has *Rps*2, but the differentials used in this study did not have *Rps*2. Interestingly, the increase in disease was proposed to be due to the increase in *P. sojae* race 15 (virulence to 3a, 7).

Characterizing virulence pathotypes for *P. sojae* has become especially critical in the North Central region of the USA and Ontario, where there are increasing reports of populations that can cause disease on cultivars with the most recently deployed *Rps* genes, *Rps*1c and *Rps*1k (Abney et al., 1997; Dorrance et al., 2003a; Kaitany et al., 2001; Schmitthenner et al., 1994; X.B. Yang, unpublished data, 2001). As *Rps* genes have been deployed there have been changes in the virulence pathotypes across this region. Prior to 1990 (Fig. 14–1A), the *Rps1a* and *Rps1c* had failed in the North Central regions. More states began reporting failures of *Rps* genes during the mid-1990s (Fig. 14–1B) and by the late 1990s 60, 50, 35, and 1% of the fields surveyed in Ohio, Indiana, Michigan, and Wisconsin, respectively, as well as reports from Iowa and Minnesota indicated hat races of *P. sojae* existed that can cause disease on cultivars with the *Rps*1k gene (Fig. 14–1C). In addition, many of the isolates recovered from these studies are highly complex, in that they can cause disease on five or more of the differentials (T.S. Abney, unpublished data, 2001; Dorrance et al., 2003b; Kaitany et al., 2001).

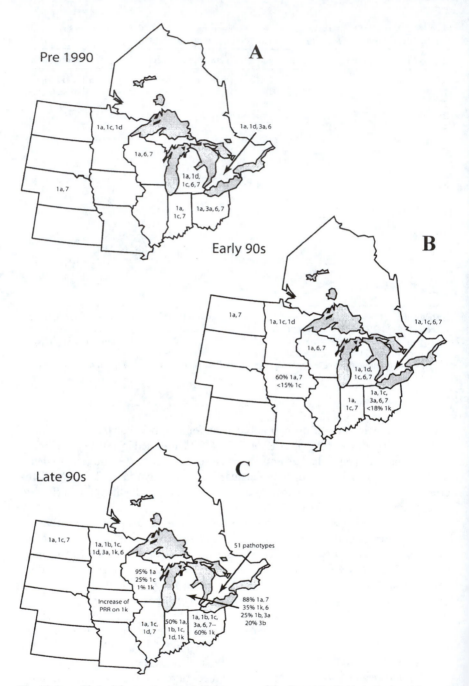

Fig. 14–1a–c. *Phytophthora sojae* has adapted to many of the *Rps* genes that have deployed in commercial cultivars in the north central region. The average "life" of an *Rps* gene is 8 to 20 yr. This graph illustrates the changes in effectiveness of the *Rps* genes over time. In each state, the *Rps* genes which are not effective to the majority of locations are listed during (A) pre-1990s, (B) early 1990s, and (C) early 2000.

With deployment of single dominant *Rps* genes, breeders must continually respond to the shifts in the pathogen *P. sojae* population, but this disease may also be managed with partial resistance. Partial resistance, (also termed field resistance, general resistance, or tolerance), has been shown to be effective against all races of *P. sojae* (Tooley and Grau, 1982, 1986; Schmitthenner, 1985), although Thomison et al. (1988) reported that isolates of races 1, 5, 10, and 24 differentially interacted with soybean genotypes characterized as partially resistant. Buzzell and Anderson (1982) proposed that selection for partial resistance combined with *Rps* genes would in fact provide long-term disease control of *P. sojae* in soybean. This resistance is also thought to be rate-limiting (Tooley and Grau, 1984a, 1984b). Tooley and Grau (1984a) reported that the primary component in rate-reducing resistance to *P. sojae* was the ability to restrict fungal colonization of the plant tissue. From field evaluations, they reported that area under the disease progress curve, simple interest infection rate, and disease incidence at growth stages V7 (late vegetative) and R5 (beginning seed development) were able to differentiate soybean cultivars with varying degrees of rate-reducing resistance. In addition, they found that differences in yield among cultivars in *P. sojae*-infested soil could be attributed to the degree of rate-reducing resistance present in the cultivar (Tooley and Grau, 1984b).

In studies with other soybean lines with partial resistance there is evidence that it is highly heritable and a quantitative trait (Buzzell and Anderson, 1982; Walker and Schmitthenner, 1984). In addition, St. Martin et al. (1994) reported that partial resistance [tolerance] should not negatively impact yield potential of soybean cultivars. This finding was confirmed by Glover and Scott (1998) who concluded that neither yield potential, or maturity should impact the presence of partial resistance to Phytophthora root rot. Field evaluations of cultivars with high levels of partial resistance were compared to cultivars with low levels of partial resistance and cultivars with *Rps* genes in seven environments, including simple and complex pathotypes and low and moderate disease pressure (Dorrance et al., 2003a). Under very low disease pressure, yields were not significantly different for cultivars with high levels of partial resistance and those with an *Rps* gene, with one exception. More importantly, cultivars with high levels of partial resistance combined with an *Rps* gene were most consistent for yield across all of the environments compared to those with moderate levels of partial resistance plus an *Rps* gene regardless of the pathogen population.

Partial resistance is identified by challenging soybean lines with a single compatible race to determine the amount of the plant that is colonized in greenhouse assays (Plate 14–8). In the greenhouse assay, there is little to no colonization of roots by *P. sojae* in soybean lines with high levels of partial resistance. Soybean plant introductions with very high levels of partial resistance may contain other types of resistance that cannot be identified with these methods such as that described earlier for *Rps*2. A characterization of the reaction in soybean lines that have a high degree of partial resistance, to ensure that it is not a single major gene but a quantitatively inherited trait, is necessary to avoid selecting for these single genes. Further field evaluations with a diversity of races of *P. sojae* are warranted.

Other management strategies for *P. sojae* are crop rotation, tillage, tiling, and seed treatment with metalaxyl or mefenoxam. While crop rotation does not allow for reduction of inoculum (Schmitthenner, 1985) it does prevent the immediate build-up of *P. sojae* populations. Anderson (1986) found that disease severity and plant loss was the same for cultivars with partial resistance compared to highly susceptible cultivars when grown in a monoculture system compared to a rotation with a cultivar with an effective *Rps* gene. Oospores are produced in soybean cultivars with partial resistance and under high inoculum pressure this type of resistance can be overwhelmed.

Tillage and tiling both work to promote soil drainage, which limits the infection period. Soils prone to Phytophthora root and stem rot are often tiled at very narrow widths (4.5–12.5 m) to promote quick drainage. Tillage helps to bury oospores deeper in the soil profile while also promoting drainage of the seedbed (Workneh et al., 1998; Kittle and Gray, 1979) reported that decreases in soil porosity increased Phytophthora root rot severity in a greenhouse study. Therefore, soils that are prone to compaction may also have high levels of disease incidence. Workneh et al. (1999b) reported a similar finding that a high incidence of *P. sojae* occurs in conventional tillage systems. In the same study, tillage had very little effect on disease incidence in soils of high clay content.

Seed or in-furrow treatments with metalaxyl have been recommended for management of *P. sojae* in fields where Phytophthora root rot is chronic (Guy et al., 1989; Schmitthenner, 1985, 1988). Metalaxyl applied in-furrow or as a seed treatment has reduced plant emergence loss and increased yields of susceptible soybean cultivars when *P. sojae* was present and the environment was conducive to infection (Anderson and Buzzel, 1982; Guy et al., 1989). However the seed treatments had no effect on yields of partially resistant soybean cultivars in these same studies. Soybean seed treatment with metalaxyl has increased yields in conservation tillage production systems (Guy and Oplinger, 1989). Several studies have shown no benefit of seed treatment with metalaxyl for some cultivars (Anderson and Buzzel, 1982; Guy and Oplinger, 1989; Guy et al., 1989). These conflicting results could be partially explained by a recent study where timing of infection of cultivars with *Rps* genes or partial resistance was compared (Dorrance and McClure, 2001). No disease developed on seed or seedlings of a resistant soybean cultivar with an *Rps* gene following inoculation with an incompatible race of *P. sojae*. However, the soybean cultivar, Conrad, with a high level of partial resistance was highly susceptible to infection by *P. sojae* from the day of planting until 5 d after planting after which only limited disease developed on the roots of the plants.

14–2.4 Rhizoctonia Root Rot

Rhizoctonia root rot is common throughout the world and can be a severe problem in the USA (Athow, 1987; Yang, 1999b). The cosmopolitan distribution of the pathogen greatly enhances the risk yield loss attributed to Rhizoctonia root rot. However, environment and pathogen variation modify its occurrence and agronomic affect on soybean.

14–2.4.1 Pathogen

Rhizoctonia root rot is caused by the soilborne fungus *Rhizoctonia solani* Kühn. Isolates of *R. solani* are grouped into 12 anastomosis groups (AGs) (Carling and Kuninaga, 1990; Carling et al., 1994; Sneh et al., 1991). Isolates that cause root rot of soybean are primarily in AG-4, however some isolates within AG-2-2, AG-5, and AG-7 are capable of causing root rot of soybean (Nelson et al., 1996; Yang, 1999b). *Rhizoctonia solani* has a broad host range and attacks fruit, vegetable, and field crops. Many of the AGs that attack soybean also attack rotational crops such as corn, alfalfa, and cereals (Nelson et al., 1995; Yang, 1999b).

14–2.4.2 Symptoms and Losses

Seedlings are most vulnerable and are attacked usually before the first trifoliate leaf develops (Yang, 1999b). Plants infected by *R. solani* exhibit a reddish-brown lesion on the hypocotyls at or near the soil line (Athow, 1987). The lesions may enlarge and girdle lower stems or root rot will occur later as the plant matures from late vegetative stages to the reproductive stages. Infected plants commonly are stunted and may develop chlorotic foliage followed by wilt especially during hot, dry conditions. Infected plants may not express foliar symptoms if soil moisture is optimal, but shallow rooted plants will succumb if dry soil conditions prevail.

Infected plants may be alone or within groups that are scattered throughout fields (Plate 14–9). Yield reductions are minimal if scattered plants are diseased. However, significant yield loss occurs when plants throughout the entire field are symptomatic. Yield reductions have approached 48% in Iowa (Tachibana et al., 1971) and as high as 80% in Brazil (Yang, 1999b).

14–2.4.3 Disease Cycle and Epidemiology

Rhizoctonia solani is an excellent saprophyte that can survive for long periods in the absence of host plants (Bell and Sumner, 1987; Kamal and Weinhold, 1967), as mycelium or sclerotia on or in plant debris (Papavizas and Davey, 1961). The disease is monocyclic and infection begins soon after planting. Infection takes place near the hypocotyl at or near the soil line. Infected plants may or may not die depending on the virulence of the isolate and the environmental conditions. The pathogen will spread to the root system in seedlings that are not killed by the damping-off phase. As a result, the lateral roots rot leaving only the taproot and secondary roots (Yang, 1999b). Infected plants frequently wilt and die or show wilting during the hot part of the day. Loss of lateral roots may be followed by formation of new roots just below the soil line when soil environment become less favorable for disease development. Plants symptomatic of foliar symptoms may occur in patches, but more often occur as individual plants or smaller groups of plants scattered throughout the field.

Warm, moist soils are conducive for the pathogen and disease development. Isolates vary regarding optimal temperature and moisture requirements. In greenhouse studies, the pathogen and disease development were favored by high temperatures (26–32°C), moisture in the soil at 30 to 60% water-holding capacity, and soil pH > 6.6 (Lewis and Papavizas, 1977). However, loss of oxygen in the soil environment resulting from excess water will reduce or arrest the pathogen and disease development. Once plants are infected, disease is more severe in areas of the field that are deficient in nutrients and/or water.

14–2.4.4 Management

Cultural practices that improve drainage and growing conditions of the plant are important in disease management. Stimulation of new root growth may also help to compensate for roots that are infected or rotted. Injury by nematodes, other pathogens, and herbicides may compromise host defenses and should be minimized (Harikrishnan and Yang, 2002; Yang, 1999b).

Both introduced and naturally occurring biocontrol agents offer some promise of control of Rhizoctonia root rot. A binucleate form of *Rhizoctonia* occurs naturally on the surfaces, and in epidermal cells of hypocotyls, roots, and root hairs. Infection of soybean tissues by *R. solani* is less in the presence of binucleate *Rhizoctonia*. The mode of action is believed not to involve forms of antagonism, but rather induced resistance (Poromarto et al., 1998). *Trichoderma harzianum* is also reported to express biocontrol activity against *R. solani*, but enzymes directed at cell wall components of the pathogen are reported as the putative modes of action (Soglio et al., 1998).

Variation in reaction to *R. solani* has been observed in soybean germplasm. Bradley et al (2001) evaluated 90 ancestral soybean lines and 700 commercial cultivars for reaction to *R. solani* (AG 2-2) based on severity of symptoms on hypocotyls and roots. 'CNS', 'Mandarin (Ottawa)' and 'Jackson' are examples of 21 ancestral cultivars that expressed partial resistance to *R. solani*. Only 20 of 700 commercial cultivars expressed partial resistance.

Numerous seed treatments, both fungicides as well as biological control agents, are now available which increase stands when *R. solani* is active as a pathogen. However, the economics of seed treatment application and prediction on which fields have disease potential to warrant fungicide-treated seed are still needed.

14–2.5 Sudden Death Syndrome

Sudden death syndrome was observed first in Arkansas in 1971 and is now widespread in the major soybean-producing areas of the USA. It recently was reported in Delaware, Maryland, and Pennsylvania (Pennypacker, 1999). In most of these areas, SDS causes chronic problems that vary in intensity from asymptomatic to severe. Although symptoms of SDS were reported in 1971, it was years before the causal agent was identified and characterized.

14–2.5.1 Pathogen

Sudden death syndrome is caused by the soilborne fungus *Fusarium solani* f. sp. *glycines* (Roy, 1997a), which has a narrow host range. In greenhouse studies, soybean and mung bean [*Vigna radiata* (L.)Wilcek] expressed both foliar and root symptoms of SDS following inoculation of nonwounded plants (Melgar and Roy, 1994). Green bean (*Phaseolus vulgaris* L.), lima bean (*Phaseolus lunatus* L.), and cowpea [*Vigna unguiculata* (L.) Walp.] also showed root symptoms of SDS, but only on plants wounded before inoculation. Isolates of *F. solani* f. sp. *glycines* from throughout the USA are reported to have a low level of genetic variation (Achenbach et al., 1997). However, differences in aggressiveness among isolates of this fungus are common (Rupe, 1989; Melgar and Roy, 1994; Achenbach et al., 1996). Data suggest that random amplified polymorphic DNA (RAPD) markers may be useful to distinguish isolates of *F. solani* f. sp. *glycines* from isolates of *F. solani* that do not cause SDS (Achenbach et al., 1996). Rupe et al. (2001) provided additional evidence of limited genetic variability within *F. solani* f. sp. *glycines*, and a distinct genetic separation of the SDS pathogen from other forms of *F. solani*. Rupe et al. (2001) evaluated many isolates of *F. solani* f.sp. *glycines* and characterized all to a single mtDNA random fragment length polymorphism (RFLP) haplotype in contrast to other isolates of *F. solani* that belonged to nine distinctly different mtDNA RFLP haplotypes. Only isolates of *F. solani* f.sp. *phaseoli* were genetically similar to isolates of *F. solani* f.sp. *glycines* in culture. Current data support the concept that isolates of *F. solani* f.sp. *glycines* represent a genetically distinct subgroup within *F. solani,* but share genetic relatedness to *F. solani* f. sp. *phaseoli*.

14–2.5.2 Symptoms and Losses

Symptoms of SDS generally appear during the reproductive stages, but onset varies with geographic location. Symptoms can appear as early as V4 (Fehr et al., 1971) in the southern USA but not until R4 to R6 in the northern part of its range (Rupe and Hartman, 1999). Initial symptoms are interveinal chlorotic spots on leaves that expand into interveinal chlorotic streaks and become necrotic (Plate 14–10). In severely affected foliage, only the leaf veins remain green. There is no obvious wilting of foliage, but symptomatic leaflets can abscise and leave only petioles attached to stems (Plate 14–11). Root systems of diseased plants are smaller and exhibit varying degrees of necrosis. Vascular tissue shows light brown discoloration that can extend several nodes up the stem, but pith tissue remains white. Sporulation by the pathogen often is visible as dark blue to blue-green areas on root surfaces from plants showing severe foliar symptoms and when soil moisture levels are high (Roy, 1997b). In infested fields, symptoms appear in somewhat circular or elongated patches, which can coalesce into large, irregular areas of diseased plants.

Yield losses can be sporadic because SDS is not a severe problem every year or in every part of an affected field. However, yield losses can approach 100% in areas where the disease is severe (Hartman et al., 1995). Sudden death syndrome affects yield by reducing both seed size and seed number (Hershman et al., 1990; Rupe at al., 1993). Smaller seed size results from leaf area reductions due to foliar necrosis and premature defoliation, whereas lower seed number results from flower and pod abortion.

14–2.5.3 Disease Cycle and Epidemiology

Propagules of *F. solani* f. sp. *glycines* are greatest in the upper 15 cm of soil, which is where most soybean roots are located (Rupe et al., 1999). The pathogen penetrates root tissue throughout the life of the plant (Rupe, 1989). Recent studies reported detectable levels of root infection as early as 15 to 30 d after planting (Njiti et al., 1997; Luo et al., 1999). Although the lower portion of the taproot is first to discolor (Scott, 1988), all portions of the root system can show necrosis. Roy et al. (1989) reported development of leaf symptoms 3 to 4 wk after planting in greenhouse tests. In similar tests, the fungus produced red-brown external stem lesions at the soil line beginning 7 d after inoculation (Rupe, 1989). Foliar symptoms developed subsequently and were most severe 20 to 43 d after inoculation.

Foliar symptoms of SDS may be due in part to a phytotoxic polypeptide produced by the fungus in liquid culture (Jin et al., 1996b). Culture filtrates from the fungus caused browning in soybean calli along with foliar symptoms suggestive of those in the field. Positive correlations between calli browning and SDS severity in microplots suggest potential for this approach as a tool to screen for resistance (Jin et al., 1996a, 1996c).

Herbicides, soil moisture, and temperature can affect disease severity, although the exact nature of these relationships remains unclear. Early reports indicated that symptoms appeared following passage of a weather front that brought cooler temperatures and rain (Scott, 1988). Rupe (1989) associated cool, wet weather near the time of flowering with symptom development. Observations suggested that symptoms generally were less severe in dry years and more severe in irrigated fields (Roy et al., 1989; Rupe et al., 1993). A recent study in Illinois (Vick et al., 2001) reported that differences in SDS severity among years correlated strongly with early season (June) rainfall and less so with late season (August) rainfall. Scherm and Yang (1996) found that severity of foliar symptoms increased as soil moisture decreased, but reported no close correlation between root disease and foliar disease severities. Studies in controlled temperature baths showed that disease was more severe at cooler soil temperatures and higher soil moistures, and that each factor affected disease development independent of the other (Vest and Russin, 2001). Acifluorfen, glyposate, and imazethapur are reported to increase severity of SDS (Sanogo et al., 2001).

Several reports addressed the role of soil fertility in development of SDS. In Arkansas, Rupe et al. (1993) found SDS associated with increased levels of available soil P, soluble salts, organic matter, and exchangeable Na, Ca, and Mg. This supported original observations that SDS was more severe in fields with high yield potential. However, results from an Iowa study (Scherm et al., 1998) identified only available K as a possible disease-enhancing factor and did not observe consistent relationships of SDS with any other factor. Applying high rates of Cl reduced SDS incidence and severity in Tennessee (Howard et al., 1999) but increased SDS on selected cultivars in Arkansas (Rupe et al., 2000). High Cl rates also increased populations of SCN. Continued studies by Sanogo and Yang (2001) suggest that potassium chloride decreased SDS severity compared to the control, but different forms of phosphate fertilizers enhanced severity of SDS. Although difficult to interpret, such contrasting results reflect the complex relationship between SDS and soil fertility. Soil compaction also is associated with increase disease severity. In areas where compaction is problematic, subsoiling can increase porosity, decrease water-holding capacity, and reduce disease severity substantially (Vick et al., 2001). Severity of SDS increased as sand content in soil increased in soil, but decreased as soil pH was lowered from 7.7 to 5.5 (Sanogo and Yang, 2001).

The SCN and SDS frequently occur together in the same field (Sanogo et al., 2001). Greenhouse and field studies have shown that foliar symptoms of SDS developed more rapidly and were more severe when SCN also parasitized soybean roots (Mclean and Lawrence, 1992; Rupe and Gbur, 1995). Hershman et al. (1990)

found that cultivars resistant to SCN showed fewer symptoms of SDS than did susceptible cultivars. In pot studies, symptoms were more severe when both pathogens were present than those caused by the fungus alone (Roy et al., 1989). In contrast, results from Iowa found only a weak relationship between SDS severity and populations of SCN in soil (Scherm et al., 1998). An Illinois study found no differences in SCN populations between fields with or without SDS (Hartman et al., 1995). Mclean and Lawrence (1995) reported that lower populations of SCN on roots that also were infected by *F. solani* f. sp. *glycines*. Collectively these results suggest that SCN may increase severity of SDS but is not required for disease development. *Fusarium solani* f. sp. *glycines* and *F. solani* both are reported to colonize cysts of SCN. However, the non-SDS form of *F. solani* was recovered more frequently than was *F. solani* f. sp. *glycines* (Roy et al., 2000). It is common to detect *F. solani* f. sp. *glycines* in SCN cysts from fields of plants lacking symptoms of SDS (Roy et al., 2000).

14–2.5.4 Management

The most effective approach to SDS control is use of resistant cultivars. Inheritance of resistance to SDS is reported as both qualitative and quantitative. Stephens et al. (1993) reported that qualitative resistance to leaf scorch phase of SDS in cv. Ripley resulted from a single, dominant gene (*Rfs*). Bigenic resistance to the leaf scorch was identified in 'P9451' (Ringler and Nickell, 1996). Several studies described quantitative resistance in resistant cultivars, including 'Jack', 'Forrest', and 'Pyramid' (Chang et al., 1996; Njiti et al., 1996; Rupe and Hartman, 1999). Reaction of cultivars in the field is expressed as partial resistance derived from root resistance loci (Njiti et al., 1997, 1998; Prabhu et al., 1999) and leaf scorch loci (Gibson et al., 1994; Meksem et al., 1999). Selection for resistance to SDS has been successful in both greenhouse (Stephens et al., 1993a, 1993b) and field environments (Gibson et al., 1994; Njiti et al., 1997). However cultivar reactions in greenhouse trials have not accurately predicted reactions of soybean cultivars in the field (Torto et al., 1996). More precise control of inoculum density has resulted in improved prediction of field reactions of soybean in greenhouse trials (Njiti et al., 2001).

Because cultivar reactions to SDS can be mitigated by SCN (Rupe et al., 1991), resistance to both pathogens is desirable. The germplasm lines DS83-3349, LS93-0375, and LS94-3207 are resistant to both *F. solani* f. sp. *glycines* and *H. glycines*. Potential new sources of resistance to SDS include recent plant introductions from China (Hartman et al., 1997) and several perennial *Glycine* spp. (Hartman et al., 2000).

Mechanisms of resistance to SDS are not well understood. Luo et al. (1999, 2000) reported that foliar symptoms and yield loss related directly to levels of fungal colonization in tap roots. Njiti et al. (1997) described rate-reducing resistance to SDS in several resistant cultivars and indicated that this resistance may function by extending the disease latent period. Rupe and Gbur (1995) found similar results and concluded that resistance was expressed as a delay in disease onset (i.e., extended latent period) coupled with a reduced rate of disease increase.

Several cultural practices provide reasonable control options. Symptoms often are less severe in delayed plantings and in early-maturing cultivars (Hershman, 1990; Rupe and Gbur, 1995). Both can delay SDS development until later re-productive stages, which may reduce yield loss. Wrather et al. (1995) also observed fewer symptoms following delayed planting, but only in a no-till system. Severity of SDS was greater under no-till than conventional tillage (Von Qualen et al., 1989; Wrather et al., 1990). Control of SDS in various crop rotation studies has been inconsistent. Howard et al. (1999) reported that a corn-soybean rotation increased incidence of SDS, whereas Von Qualen (1989) found less severe disease in that rotation compared to continuous soybean. However, Rupe and Hartman (1999) indicated that cropping sequence had no effect on SDS.

14–3 STEM DISEASES

14–3.1 Anthracnose

Anthracnose of soybean was first reported in Korea and now is a disease of world-wide distribution. The disease is more of a problem in warm and humid production regions. Anthracnose is one of several stem diseases that occur together and at differing degrees of severity.

14–3.1.1 Pathogen

Anthracnose of soybean is caused by several *Colletotrichum* spp. The most frequently associated species is *Colletotrichum truncatum* (Schw.) Andrus & Moore. Other species that cause anthracnose are *C. coccodes*, *C. destructivum* O'-Gara, *C. gloeosporioides* Penz., and *C. gramincola* (Ces.) Wilson. *Colletotrichum truncatum* has a host range of more than 28 plant species (Sinclair, 1988a). These fungi are present throughout temperate and subtropical environments where warm and humid conditions prevail (Manandhar and Hartman, 1999; Sinclair, 1988a).

14–3.1.2 Symptoms and Losses

All soybean growth stages may be affected by anthracnose with symptoms usually appearing during germination, seedling stages, or during pod filling and senescence (Manandhar and Hartman, 1999). On emerging seedlings, reddish-brown, sunken lesions are concentrated on the outer surface of the cotyledons. Cankers gradually extend up toward the epicotyl and down to the root. Rotting of the epicotyl and hypocotyls results in pre-and post-emergence damping off. On seedlings that survive, one or both cotyledons may be destroyed or the fungus may continue to grow into the stem tissues. Symptoms on stems include many small, shallow, elongated, reddish-brown lesions or large, dark-brown lesions that can kill seedlings or young plants. Symptoms may disappear on surviving seedlings, but their development is slowed when compared to that of healthy seedlings. During the vegetative stages, stems, pods, leaves, and roots may be infected, but with no apparent symptoms. Symptom expression usually follows favorable conditions for spore production (Bowers and Russin, 1999).

Leaf symptoms may vary with environmental conditions and severity of infection. Generally, necrotic lesions can be seen on the underside of the leaflet, and in response to severe infections, leaves may cup downward and fall prematurely. On petioles, dark-brown lesions are nearly rectangular in shape (Bowers and Russin, 1999). Defoliation will occur if the lesions girdle the leaf petiole. On stems and pods, irregularly-shaped brown areas develop and coalesce to cover the entire surface (Plate 14–12). Within the diseased areas, numerous acervulli arise with several short, dark spines or setae within each fruiting body (Manandhar and Hartman, 1999).

Seed may not develop or be reduced in size and quantity if pods are subjected to early infection. Subsequent infections cause local lesions on the outer surface of the pod. Seed in diseased pods may be shriveled, dark brown, and moldy, or be asymptomatic. Occasionally the pod cavity can be completely filled with mycelium of the pathogen (Bowers and Russin, 1999).

Yield loss resulting from anthracnose is primarily by damage to the pod and seed. Yield losses range from 16 to 26%, in naturally infected plants (Backman et al., 1982) and 17 to 30% in artificially infected plants (Khan and Sinclair, 1992). From 1996 to 1999, estimated annual losses exceeded 81.1 thousand Mg. In 2000, drought conditions were prevalent throughout the mid-South and yield reductions attributed to anthracnose were only 40.5 thousand Mg (Wrather et al., 2001).

14–3.1.3 Disease Cycle and Epidemiology

Colletotrichum spp. responsible for anthracnose have wide geographic ranges and are good saprophytes of many plant species (Sinclair, 1988a). *Colletotrichum truncatum* overwinters in diseased crop residue left in the field and in infected seed. Infected seed serve not only as the primary inoculum but also as a means for widespread distribution of the pathogen to new areas. Secondary inoculum is produced as the lower leaves die and infections resulting from this inoculum cause the most damage to yield and seed quality (Bowers and Russin, 1999).

Severe disease occurs when there are extended durations of high moisture, humidity, and temperature (Sinclair, 1988a). Spore germination and penetration of the host epidermis requires free water. Spores germinate within 4 to 16 h at 20 to 30°C and fungal growth is sustained once relative humidity exceeds 83%. Infection of the host will occur when free moisture is available for 12 or more hours (Khan and Sinclair, 1991). Weed competition greatly increases the severity of anthracnose creating conducive environmental conditions in the microenvironment. In addition, many weeds are hosts and serve as reservoirs for the pathogen, as well as a means of overwintering (Bowers and Russin, 1999).

14–3.1.4 Management

Resistant cultivars are not currently available for the management of anthracnose (Bowers, 1984; Bowers and Russin, 1999), however the susceptibility of soybean cultivars may vary (Manandhar and Hartman, 1988). Pathogen-free seed can be used to reduce the incidence and severity of anthracnose. Fungicides are useful primarily when used as a seed treatment, but can also be used in foliar applications (Sinclair, 1988a). Cultural practices such as deep plowing of infected crop residue and rotation with nonhost crops are helpful in disease management.

14–3.2 Brown Stem Rot

Brown stem rot was first observed in 1944 in central Illinois (Abel, 1977; Allington and Chamberlain, 1948). This disease of the vascular system has become widespread and prevalent in the North Central states and Canada, but to a lesser degree in the southern USA (Gray and Grau, 1999; Workneh et al., 1999a). Outside of North America, brown stem rot is most common in Japan, but is reported to occur in Egypt (Abel, 1977; Gray and Grau, 1999). Currently brown stem rot is ranked as the fifth most important disease of soybean in the USA (Wrather et al., 2001b), but the disease is less important in other countries (Wrather et al., 2001a).

14–3.2.1 Pathogen

Brown stem rot of soybean is caused by the fungus *Phialophora gregata* (Allington and Chamberlain) W. Gams f.sp. *sojae* Kobayasi, Yamamoto, Negishi et Ogoshi (Gray and Grau. 1999). The causal organism was initially identified as *Cephalosporium gregatum* Allington and Chamberlain (Allington and Chamberlain, 1948), but this species designation was relegated to a synonym after the latin binomial was changed to *Phialophora gregata* (Allington and Chamberlin.) W. Gams (Gams, 1971). Reproduction of *P. gregata* is restricted to mycelium and single-celled hyaline conidia formed on single phialides. The conidia are ovoid to elliptical and 1.7 to 4.3 µm × 3.4 to 9.4 µm in size (Gray and Grau, 1999). A sexual stage of *P. gregata* has not been identified nor have specialized survival structures such as chlamydospores or sclerotia (Allington and Chamberlain, 1948). Colonies of *P. gregata* on potato dextrose agar (PDA) are characteristically slow growing, dense circular colonies with radial folds and vary in color from tan to degrees of gray. Colony margins are typically lobed in appearance (Allington and Chamberlain, 1948). Due to similar morphological and cultural traits and frequent co-isolation from symptomatic soybean stems, there is possible confusion between *P. gregata* and *Plectosporium tabacinum* (van Beyma) M.E. Palm (teleomorph = *Plectosphaerella cucumerina*), a common stem inhabitant of soybean (Chen et al., 1996, 1999; Harrington et al., 2000; Hamilton and Boosalis, 1955; Mengistu and Grau, 1986; Palm et al., 1995). Isolates of *P. tabicinum* sporulate profusely on PDA in contrast to isolates of *P. gregata* that do not sporulate on PDA. RAPD markers are now available to clearly differentiate isolates of *P. gregata* and *P. tabicinum* (Chen et al., 1999). Isolates of *P. tabacinum* generally cause minimal severity of internal stem browning and no significant foliar symptoms (Mengistu and Grau, 1986). *Verticillium dahliae* also infects soybean, but is not commonly isolated nor should be confused with *P. gregata* (Sickinger et al., 1987)

Phialophora gregata is readily isolated from soybean tissue using PDA or a green bean based (GBB) semi-selective medium (Mengistu et al., 1991). The pathogen can also be detected by direct PCR of host tissues (Chen et al., 1999). Isolates of *P. gregata* are variable with respect to phenotypes related to colony growth rate, colony morphology, and sporulation. Radial growth of *P. gregata* is slow, regardless of incubation temperature or culture medium (Allington and Chamberlain, 1948). Isolates of *P. gregata* may be cultured on lima bean agar (LBA), nutrient agar (NA), soybean stem agar (SStA), soybean seed agar (SSA), potato agar (PA), and

PDA (Gray and Grau, 1999). Colony phenotypes are commonly modified by culture medium (Phillips, 1973; Allington and Chamberlain, 1948). Several optimal temperature ranges for colony growth are reported and differ by culture medium (Allington and Chamberlain, 1948; Gray and Grau, 1999; Phillips, 1973; Mengistu and Grau, 1986). Usually 18 to 25°C is optimal for vegetative growth, but growth decreases dramatically at 32°C (Mengistu and Grau, 1986). Allington and Chamberlain (1948) reported sporulation of *P. gregata* occurs on PA, string-bean agar (SBA), rice polish agar (RPA), cucumber agar (CA), and SSA, and green bean broth (GBB) (Mengistu et al., 1991). Sporulation is inhibited if dextrose (2%) is added to PA, SBA, RPA, CA, and SStA (Allington and Chamberlain, 1948; Mengistu and Grau, 1986).

Severity of foliar symptoms is commonly used to characterize variability in aggressiveness among isolates of *P. gregata* (Gray, 1971; Hughes et al., 2002; Phillips, 1973). Gray (1971) proposed isolates could be characterized into two groups, those that caused both foliar and internal stem symptoms, Type I-defoliating, or those that cause only internal stem symptoms, Type II-nondefoliating pathotypes. This phenotypic concept of pathotypes is supported by DNA-based characterization of isolates. Chen et al. (2000) identified a region of variability among isolates of *P. gregata* within the intergenic spacer region (IGS) of rDNA and developed species-specific primers capable of separating isolates by polymerase chain reaction (PCR), into two distinct genotypes, A (1020 bp) and B (830 bp). Results from pathogenicity tests suggest that genotypic characterization of *P. gregata* is strongly predictive of symptoms caused by each genotype of *P. gregata*. Genotype A isolates are capable of causing both foliar symptoms and internal stem symptoms, but B genotype isolates cause only internal stem symptoms (Hughes et al., 2002). The concept of Type I-defoliating and Type II-nondefoliating pathotypes, originally proposed by Gray (1971), is supported by recent phenotypic and genotypic data (Chen et al., 2000; Hughes et al., 2002). Isolates of *P. gregata* derived from stems of brown stem rot resistant cultivars differed in DNA content from isolates recovered from susceptible cultivars (Yeater et al., 2001).

14–3.2.2 Symptoms and Losses

The symptoms associated with brown stem rot are characterized as a browning of the vascular and pith tissues of the stem and root (Plate 14–13), which may be accompanied by a sudden interveinal chlorosis and necrosis, wilting, and defoliation of the leaves (Plate 14–14) (Gray and Grau, 1999). The progression of symptoms is related to plant growth and development (Phillips, 1972) besides soil and climatic conditions and soybean cultivar (Mengistu and Grau, 1987; Mengistu et al., 1987; Tachibana, 1982). Mild internal stem and root symptoms can be observed during the vegetative stages of soybean development as early as V4 stage of development. However, internal stem symptoms intensify as the plant progresses into reproductive stages. There is no external evidence of the disease at this time and symptoms of early infection generally go unnoticed unless the stems are cut open and examined. The onset of foliar symptoms typically does not occur until growth stages R4 and R5 and foliar symptoms peak at R7 (Gray and Grau, 1999; Mengistu and Grau, 1987). The disease is rarely observed until late July and early

August when plants are normally in the R4 to R5 growth stage. Gregatins, a family of compounds produced by *P. gregata*, are believed to have an important role in pathogenesis and possibly the types of symptoms observed (Gray et al., 1999).

Yield loss estimates for brown stem rot have been obtained primarily by comparing yield of susceptible and resistant cultivars grown in field plots infested with *P. gregata*. The loss of yield attributed to brown stem rot is greater if stem and foliar symptoms are present compared to if only stem symptoms are observed (Adee et al., 1995; Mengistu and Grau, 1987). Yield losses of 10 to 30% are common (Bachman et al., 1997a; Gray and Sinclair, 1973; Mengistu and Grau, 1987), but have been reported as high as 44% (Dunleavy and Weber, 1967). The effect of brown stem rot on yield depends greatly upon the environmental conditions (Gray and Grau, 1999) and losses will be greatest in environments conducive for high soybean yield potential (Grau et al., 1994). Greatest yield loss occurs in seasons with a cool period in early August followed by hot, dry weather during late pod-fill (Mengistu and Grau, 1987; Mengistu et al., 1987). Weber et al. (1966) showed that yield loss reached 44% at 100% disease incidence. Two-thirds of the yield loss was due to a reduction in seed number and one-third to a reduction in seed size. Severely diseased plants lodge extensively which results in greater mechanical harvest loss in addition to physiological yield loss. Yield loss associated with brown stem rot is greatest if soybean fields are planted early compared to delayed planting dates (Grau et al., 1994).

14–3.2.3 Disease Cycle and Epidemiology

Phialophora gregata is a soil-borne pathogen that survives in soybean residue previously colonized during the pathogen's parasitic phase (Adee et al., 1995, 1997). By soybean stage V3, the pathogen has invaded the main and lateral roots and progressively colonizes the xylem vessels of the root. Once established as a parasite, *P. gregata* systemically colonizes stem, petiole, and leaf tissues (Gray and Sinclair, 1973; Mengistu and Grau, 1987; Schneider et al., 1972). Since the host range of *P. gregata* is limited to soybean, adzuki bean [*Vigna angularis* (Willd.) Ohwi and H. Oashi], and mung bean (Gray and Grau, 1999; Gray and Pataky, 1994), extended periods of cropping to nonhosts effectively lowers inoculum of *P. gregata*. The rate of inoculum decline is directly related to rate of soybean residue decomposition (Adee et al., 1995, 1997). The incidence and severity of brown stem rot is modified by ambient and soil environments, and crop management systems. Brown stem rot is a disease of cooler temperatures that range between 15 and 27°C (Abel, 1977; Allington and Chamberlain, 1948; Gray, 1974; Mengistu and Grau, 1986; Phillips, 1971). Foliar symptoms are greatly affected by air temperature during reproductive growth stages (Gray and Grau, 1999). High temperatures during these stages, specifically R3 and R4, suppress foliar symptom development. Foliar symptoms are reportedly most severe if optimal soil moisture is present after flowering (Mengistu and Grau, 1987) compared to lower symptom severity if soils are dry during the early stages of pod development following flowering. The timing of soil moisture deficit stress is important to symptom expression and yield loss due to brown stem rot. Greater severity of stem symptoms is reported if moisture deficits occur during the R6 to R8 stages of development (Mengistu et al., 1987). The severity of brown stem rot is greater if soils are low in P and K (Waller et al., 1992) and soil pH is below 6.5 (Kurtzweil et al., 2002; Waller et al. 1992).

Phialophora gregata and the SCN frequently occur together, but information is limited about potential interactions. Sugawara et al. (1997) observed, in split-root experiments, an increase in severity of internal stem symptoms when SCN and *P. gregata* were present together. The colonization of soybean stems by *P. gregata* was 38 to 79%, when both pathogens were present, compared to 13 to 33% in the presence of *P. gregata* alone (Tabor et al., 2001). *Phialophora gregata* and *H. glycines* commonly occur and the significance of this common occurrence should be investigated further (Workneh et al., 1999a, 1999b).

14–3.2.4 Management

Successful control of brown stem rot has been obtained through crop rotation (Adee et al., 1994; Dunleavy and Weber, 1967; Kennedy and Lambert, 1981) and planting brown stem rot resistant cultivars (Bachman et al., 1997a; Mengistu et al., 1986; Tachibana, 1982). The limited host range of *P. gregata* allows crop rotation to be an effective tactic to control brown stem rot (Allington and Chamberlain, 1948). The incidence and severity of brown stem rot is greater in no-till systems vs. tilled systems (Adee et al., 1994; Workneh et al., 1999b). Thus, the importance of longer crop rotations, up to 5 yr, and brown stem rot resistant cultivars is greater if soybean fields are in a no-till system (Adee et al., 1994). The practice of crop rotation normally involves rotating a different crop with soybean, but rotation can also apply to rotation of brown stem rot resistant and susceptible cultivars during the course of a rotation sequnece. Studies in Iowa indicated agronomic performance of susceptible cultivars is improved following brown stem rot resistant cultivars compared to if susceptible cultivars followed other susceptible cultivars (Tachibana et al., 1989). The decline in disease potential has not been experimentally determined. Possible mechanisms are related to less pathogen reproduction in resistant cultivars leading to a decline in pathogen population density, and/or a shift from the more aggressive A genotype to the less aggressive B genotype of *P. gregata* (Adee et al., 1995, 1997; Hughes et al., 2002).

Cultivar selection has become an effective tactic to control brown stem rot. In the past, cultivar selection was based on the concept that early-maturing cultivars escaped severe symptom development, but this tactic is done at the expense of yield potential (Weber et al., 1966). However, resistance to *P. gregata* has been incorporated into elite cultivars. Brown stem rot resistant cultivars are extremely important because of the desired trend for shorter crop rotations and less soil tillage. Complete resistance to brown stem rot has not been reported, but several sources of partial resistance are available to soybean breeders (Table 14–3). Resistance to brown stem rot has been defined as the "delay or lack of symptom expression after an incubation period, which causes an advanced expression of symptoms in susceptible check genotypes" (Sebastian et al., 1985). Since 1968, significant progress was made to identify sources of resistance to brown stem rot, and understand the inheritance of resistance genes (Bachman and Nickell, 2000a; Chamberlain and Bernard, 1968; Eathington et al., 1995; Hanson et al., 1988; Sebastian et al., 1985). Resistance genes were found to agronomically inferior plant introduc-

Table 14–3. Genes and sources for resistance to *Phialophora gregata*.

Gene	Source	Citation
Rbs1	L78-4094	Sebastian and Nickell (1985)
	PI 437833	Hanson et al. (1988)
Rbs2	LN92-12033	Nickell et al. (1997)
	PI 437970	Hanson et al. (1988)
Rbs3		
Rbs1 & *Rbs3*	PI 84946-2	Chamberlain and Bernard (1968)
	BSR 101	Eathington et al. (1995)
Rbs1, *Rbs2* & *Rbs3*	PI 567609	Bachman and Nickell (2000a)

tions (Bachman and Nickell, 2000b; Bachman et al., 1997b; Nelson et al., 1989), but soybean breeders have transferred resistance genes to adapted soybean germplasm (Nickell et al., 1990, 1997). Methods have been established to identify resistance and relate results in controlled environments to field performance (Sebastian et al., 1986). Reported resistance loci are the *Rbs1* locus that traces to PI 84946-2, an accession from Korea, the *Rbs2* locus that traces to PI 437833, and the *Rbs3* locus that traces to both PI 84946-2 and PI 437970 (Bachman and Nickell, 2000a). Resistance to brown stem rot has been achieved by phenotypic selection, but molecular studies suggest that selection methodology can now employ several molecular markers located on Molecular Linkage Group J (Bachman et al., 2001; Lewers et al., 1999). Although brown stem rot is most noted as a stem disease, evidence suggest that resistance to *P. gregata* is expressed in roots and pathogen and host interactions in roots regulates the expression of symptoms of stems and leaves (Bachman and Nickell, 1999).

Soybean cultivars with SCN resistance tracing to the cv. Fayette, derived from PI 88788, are frequently observed to express none to mild symptom severity when infected by *P. gregata* (Hughes et al., 2001; MacGuidwin et al., 1985; Waller et al., 1992). A similar situation is not observed for cultivars derived from other sources of SCN resistance such as PI 209332, PI 89772, PI 437654, 'Cloud' and 'Peking' (Hughes et al., 2001). The cv. Bell, derived from PI 88788, commonly expresses low brown stem rot symptom severity in the field (MacGuidwin et al., 1985). The form of resistance to *P. gregata* expressed by Bell is correlated by a quantitative trait locus (QTL) located on MLG J (B.W. Diers and M. Kirsh, unpublished data, 2003).

Bachman and Nickell (2000a) proposed an alternative model to explain the genetic basis of resistance to *P. gregata*. The model was based on the segregating progeny of crosses between PI 567609, a resistant plant introduction, with 'Century 84', a susceptible parent, and L78-4094 (*Rbs* 1), PI 437833 (*Rbs* 2), and PI 437970 (*Rbs* 3), resistant parents. Resistance to brown stem rot is further hypothesized as a two to four gene system and genes located at independently segregating loci. The loci *Rbs* 1, *Rbs* 2, and *Rbs* 3, are referred to as "genes necessary for resistance to brown stem rot" (Nrb genes). One or more of the newly designated Nrb genes must then interact with a fourth locus or clustered loci, designated *R*, to confer the brown stem rot-resistant interaction phenotype. From this model, genotypes of L78-4094, PI 437833, and PI 437970 would, respectively, be as follows: Nrb 1 nrb 2 nrb 3 R; nrb 1 Nrb 2 nrb 3 R; nrb1 nrb 2 Nrb 3 R. Likewise, the genotype of PI 567609, which was shown to be multigenic and allelic to *Rbs* 1, *Rbs* 2, and *Rbs* 3, would be Nrb 1 Nrb 2 Nrb 3 R. The concept of brown stem rot resistance linked to PI 88788-derived cultivars supports the proposed model of Bachman and Nickell (2000a). A working hypothesis is that a PI 88788-derived cultivar would contain an Nrb gene(s) contributed by 'Williams' or 'Williams 82' and an R gene(s) contributed by PI 88788 (Hughes et al., 2001; B.W. Diers and M. Kirsh, unpublished data, 2003). PI 88788 is susceptible to genotype A isolates of *P. gregata*, but appears to contribute to resistance when crossed with Williams (Hughes et al., 2001).

Isolate by soybean genotype interactions are reported for brown stem rot, but data does not support a concept of distinct physiologic races like reported for other soybean-pathogen systems (Sills et al., 1991; Willmot et al., 1989). This research was conducted without knowledge of the concept that populations of *P. gregata* pathogenic to soybean are composed of two subpopulations (genotypes A and B), and that genotypes are correlated to pathogenic effects on soybean (Chen et al., 2000; Hughes et al., 2001). It is not known whether different sources of brown stem rot resistance differentially interact with isolates within each pathogen genotype.

14–3.3 Pod and Stem Blight and Phomopsis Seed Decay

Pod and stem blight and Phomopsis seed decay are two diseases within the *Diaporthe-Phomopsis* complex of soybean (Sinclair, 1999a). Northern and Southern stem canker fungi are the two remaining components of the complex and are discussed in Section 14–3.6. Pod and stem blight (Kulik and Sinclair, 1999b) and Phomopsis seed decay (Kulik and Sinclair, 1999a) are endemic throughout all the soybean-producing areas in the USA and many other countries (Athow, 1987). Pod and stem blight is a common disease, but there is debate whether the cause of pod and stem blight is an aggressive pathogen or rather an opportunistic fungus that sporulates abundantly on normally maturing plants or plants prematurely killed by other causes. Phomopsis seed decay is generally recognized as a disease of greater agronomic importance. Phomopsis seed decay and pod and stem blight will be discussed together because of the close relationship of the causal pathogens.

14–3.3.1 Pathogen

Diaporthe phaseolorum is a complex species composed of three subspecific forms that cause separate diseases of soybean. Pod and stem blight is caused by *Diaporthe phaseolorum* (Cke. & Ell.) Sacc. var. *sojae* (Lehman) Wehm (anamorph = *Phomopsis phaseoli*; Sinclair, 1999a). *Diaporthe phaseolorum* var. *caulivora* and *D. phaseolorum* var. *meridionalis* cause stem canker and will be discussed in Section 14–3.5. *Phomopsis longicolla* Hobbs, is the primary cause of Phomopsis seed decay, but *P. phaseoli* also causes seed decay (Kulik, 1984; Kulik and Sinclair, 1999a, 1999b). Pathogenicity, host tissues infected, symptoms, and distinct morphological phenotypes are the basis of the multiple taxa system. Random amplified polymorphic DNA markers have confirmed the taxa phenotypes of the Diaporthe-Phomopsis complex (Zhang et al., 1997). Athow (1987) reviews the historical aspects of the nomenclature of these pathogens.

Diaporthe phaseolorum var. *sojae* is the cause of pod and stem blight, however the anamorphic stage, *P. phaseoli*, is most commonly observed on diseased tissues. *Phomopsis phaseoli* forms pycnidia in compact black stromata located beneath the epidermis of petioles, stems, and pods. Conidia are borne on short and unicellular conidiophores within the pycnidium. Alpha-conidia are unicellular, hyaline, fusiform-elliptical, and contain two oil guttulates. Conidia are 5 to 10 μm × 2 to 3 μm in size. Beta-conidia are rarely formed. *Diaporthe*, the telomorphic stage, produce perithecial, but appear less frequently than the pycnidia on host residue. The perithecia are produced singly in black stromatic tissues embedded in cortical tissues. Asci range from 38.0 to 51.2 μm × 3.3 to 5.6 μm in size (Athow, 1987; Kulik and Sinclair, 1999a, 1999b). Ascospores are bicelluar, each cell contains two guttulates, and are similar in size and morphology to alfalfa conidia. Ascospores are 2 to 9 × 9-13 μm and are freed from perithecia in extrusion droplets (Kulik and Sinclair, 1999b).

Diaporthe phaseolorum and *P. longicolla* have wide host ranges including crops frequently planted in rotation with soybean and weed species common to soybean fields. Common bean, pea (*Pisum sativum* L.) and cotton are reported hosts along with velvet leaf (*Abutilon theophrasti* L.) and pigweed (*Amaranthus retroflexus* L.), two weeds common in soybean fields (Sinclair, 1988b, 1991).

14–3.3.2 Symptoms and Losses

Pod and stem blight is recognized more commonly by signs, reproductive structures of the pathogen, than by symptoms. The most common sign is appearance of linear rows of dark, speck-sized pycnidia that cover physiologically mature stems, petioles, and pods. The appearance of pycnidia coincides with senescent plant parts such as abscised petioles, broken branches, and prematurely killed tissues during pod-fill stages and at harvest maturity.

Phomopsis seed decay problems are diagnosed by direct observation of seed for symptoms and signs. However, many infected seed are asymptomatic, thus such seed must be incubated on culture media suitable for growth of *Phomopsis* spp. (McGee, 1986). Ultrasound analysis was recently established as a means to detect asymptomatic soybean seeds infected by *Phomopsis* species (Walcot et al., 1998).

Symptoms of Phomopsis seed decay are cracked and shriveled seedcoats, covered with a chalky white-gray mold, and seeds that are smaller in size and seed weight. Seedlings derived from infected seed frequently express necrosis of hypocotyls tissues. Seedlings derived from infected seed may express brown to red lesions on cotyledons, and frequently the seed coat remains attached to cotyledons by strands of mycelium after emergence of seedlings (Athow, 1987; Kulik and Sinclair, 1999a, 1999b). Infection of seedlings may result in stunting and less productive plants or seedling mortality.

Pod and stem blight is an economically important disease of soybean during harvest seasons that have excessive amounts of rain (McGee, 1988). Wrather et al. (1997, 2001a) reported worldwide soybean yield losses of 1 334 000 Mg in 1998. Total losses in the USA from 1996 to 1998 were estimated at 630 000 Mg (Wrather et al., 2001b). Pod and stem blight, with or without the development of Phomopsis seed decay, reduces yield by a shortened seed-filling period related to premature plant death. Benomyl applied at soybean stages R3 and R5 resulted in increased yield for early and less so for late maturity cultivars in Nebraska studies. Yield effects were related to pod and stem blight but less to Phomopsis seed decay based on seed symptoms in high-yield environments. Benomyl has been frequently used in yield loss studies involving soybean and members of the Diaporthe-Phomopsis complex. Data from such studies should be carefully interpreted because of the growth regulator activity of benomyl (Slater et al., 1991).

Phomopsis spp. are major causes of poor quality soybean seed, especially affecting food-grade quality seed. Reduced seed germination, seedling vigor, and reduced field emergence are common problems associated with seed lots with a high incidence of seed infected by *Phomopsis* spp. (Athow and Laviolette, 1973; Hepperly and Sinclair, 1981; Kulik and Sinclair, 1999a; Wall et al., 1983; Zorrilla et al., 1994). The effect of seed borne *Phomopsis* is greatest when seed is planted in dry soil marginally conducive for seed germination compared with moist soil. The percentage of seed infected by *P. longicolla* is reported to decline as seed is stored. The decline of the pathogen results in improved germination and seedling performance (Wallen and Seaman, 1963). *Phomopsis* can lower seed grade by reducing test weight, increasing the number of split seed, and lower quality of flour and oil (Hepperly and Sinclair, 1978).

14–3.3.3 Disease Cycle and Epidemiology

Diaporthe phaseolorum f. sp. *sojae* and *P. longicolla* readily survive in soybean residue and seed (Athow, 1987; Garzonio and McGee, 1983; Kmetz and Ellett, 1979; Kulik and Sinclair, 1999a, 1999b). Primary inocula are conidia produced in pycnidia embedded in host residue present on the soil surface (Hilbebrand, 1954; Kmetz et al., 1978). Although seedborne inoculum is regarded as important, the incidence of *Phomopsis*-infected seed planted does not relate to incidence of *Phomopsis*-infected seed at harvest (Garzonio and McGee, 1983). *Diaporthe phaseolorum* f. sp. *sojae* and *P. longicolla* are commonly the first fungi to infect soybean stems (Baker et al., 1987; Kmetz et al., 1974). Plants are infected early in the growing season following dissemination of rain-splashed conidia from crop residue or seedlings derived from infected seed. Initially, each pathogen causes a localized infection and remains latent until later reproductive growth stages (Sinclair, 1999a). Hyphae of *D. phaseolorum* f. sp. *sojae* and *P. longicolla* remain localized at points of infections and do not systemically colonize the host (Hill et al., 1981). The extensive appearance of pycndia is likely the result of multiple points of localized points of infection. *Diaporthe phaseolorum* f. sp. *sojae* has been recovered from lower stems and roots (Gerdemann, 1954a), and its role as a root rotting pathogen merits investigation. Internal movement of the pathogen within the plant occurs with tissue senescence or plant death. Thus, signs of these pathogenic fungi are delayed until well into the reproductive stages of growth. Sinclair (1991) addressed the concept and importance of latent infection of soybean plants and seed by *D. phaseolorum* f. sp. *sojae* and *P. longicolla* as suggested by Gerdemann (1954a). The incidence of plants infected by *P. phaseoli* is greatest at V4 if soybean were grown continuously compared to soybean in an alternating corn-soybean rotation (Garzonio and McGee, 1983). Pods are infected between the R5 to R6 stages, but both *P. longicolla* and *P. Phaseoli* remain latent until pods begin to mature (Baker et al., 1987; Kulik, 1984; Kmetz et al., 1978).

Seed infection by *P. phaseoli* and *P. longicolla* occurs by the beginning of stage R7. Inoculum for late-season seed infections originates from latent infections of petioles and stems (Athow and Laviolette, 1973; Kmetz et al., 1978; McGee and Brandt, 1979). Inoculum for seed infection originates from pods rather than a systemic movement of pathogen from stems or other plant structures (Baker et al., 1987; Kmetz and Ellett, 1979). The pathogen progresses from the pod wall into the seed beginning at stage R6 and colonizes seed prior to harvest maturity. Moisture content of tissue greatly influences susceptibility to *P. longicolla* and *P. phaseoli* (Rupe and Ferris, 1986). The ability of each pathogen to invade host tissues decreases as moisture content of tissue declines. Furthermore, colonization of pods decreases dramatically as pods reach stage R7 and water content of seed declines from 55% to about 12 to 14% at harvest maturity. Temperatures in the range of 25°C favor seed infection if seed water content falls within a range of 19 to 35% seed moisture (Rupe and Ferris, 1986). Seed infection will not occur if seed moisture drops below 19% (Balducchi and McGee, 1987). The incidence of seed infection will continue to occur after seed maturation and the pod wall is in a state of deterioration. Periods of warm and wet weather greatly favor seed infection although rainfall is more predictive of seed infection than temperature during pod fill. There is a high positive correlation between disease incidence and rainfall during pod fill (Shortt et al., 1981). Periods of high relative humidity are important for infection of pods to occur, but high air temperatures are conducive for colonization of seed. The duration of optimal weather conditions will greatly influence incidence of seed infected by *P. longicolla* and *P. phaseolii*.

Typically, Phomopsis seed decay is enhanced by delayed harvest provided fall weather conditions are favorable for seed infection (TeKrony et al., 1984). Early-maturity cultivars are at greater risk of Phomopsis seed decay compared to later, full-season maturity group cultivars. Early-maturity cultivars express greater incidence because such cultivars are likely to mature during periods of higher air temperatures and higher relative maturity (Balducchi and McGee, 1987). These conditions favor the movement of *Phomopsis* from the pod wall to the developing seed. Late-maturity cultivars may escape infection. Therefore seed production of a soybean cultivar at the northern most point of its adapted range may reduce the risk of Phompsis seed decay.

Phomopsis spp. are reported to interact with other seedborne fungi and viruses to modify the incidence of Phomopsis seed decay. *Phomopsis* is reported to interact inversely with *Alternaria* (Ross, 1975), and *Cercospora kikuchii*, the cause of purple seed stain (Hepperly and Sinclair, 1981). However, the incidence of Phomopsis seed decay may be greater if plants are infected with *bean pod mottle virus* (Stuckey et al., 1982) or *soybean mosaic virus* (Koning et al., 2001). Furthermore, the incidence of Phomopsis seed decay is increased in pods injured by insect feeding.

14–3.3.4 Management

Management of the *Diaporthe/Phomopsis* complex is primarily focused on reducing the incidence of Phomopsis seed decay rather than the infection of stems. As with other residue-borne pathogens, crop rotation and thorough incorporation of soybean residue by tillage will reduce inoculum of the *Diaporthe/Phomopsis* complex. Management of residue-borne inoculum can be rendered less effective by the introduction of inoculum on seed (Garzonio and McGee, 1983; Jeffers et al., 1984). Phomopsis seed decay was found to be highest in monoculture soybean systems, less in a corn-soybean rotation, and least in soybean following several years of corn. However, *Phomopsis* was still detected in seed harvested from soybean grown in a field that had been continuously cropped to corn for 10 yr (Garzonio and McGee, 1983). Phomopsis seed decay is reported to be greater in early rather than later plantings, but incidence of infected seed was low regardless of planting date (Wrather et al., 1996).

The commercialization of benzimidazole fungicides made available commercial products of increased efficacy against the *Diaporthe/Phomopsis* complex and are reported to effectively reduce the incidence of seed infection by *P. longicolla* and *P. phaseoli* (TeKronly et al., 1985a). Foliar-applied fungicides are most practical for the production of seed rather than cash grain. A single application of fungicide between R3 and R6 can control seed infection, however applications at R7 or later generally are less effective (TeKrony et al., 1985a). Seed yield is frequently not increased by fungicides even though seed quality is improved by the application of a fungicide (Slater et al., 1991). The need for fungicide applications has been a topic of critical importance because of the added expense to seed production. The incidence of infected pods at R6 positively correlates to the incidence of *Phomopsis*-infected seed at harvest (McGee, 1986; TeKrony et al., 1985b). This knowledge was used to develop a method to predict the need for fungicide application (McGee, 1986. TeKrony et al. (1985b) developed a prediction system that employed cropping history, cultivar maturity group, planting date, and existing and predicted rainfall during the R2 to R7 stages. Both prediction methods have most application to soybean seed producers and making the decision whether or not to apply fungicide to reduce the incidence of Phomopsis seed decay.

Cultivars and breeding lines are reported to differ in reaction to *P. longicolla* and *P. phaseoli* based on host physiology and relative maturity (Anderson et al., 1995; Wrather et al., 1996). PI 80837 has low seed infection in the Midwest, but becomes heavily infected when grown to the south of its range of adaptation (Brown et al., 1987). Minor et al. (1993) released a soybean-breeding line resistant to Phomopsis seed decay derived from PI 417479. Resistance derived from PI 417479 is inherited as two complimentary dominant genes (Zimmerman and Minor, 1993). Genetic resistance to Phomopsis seed decay, derived from PI 417479, was superior to appliclations of benomyl fungicide to control seed infection (Elmore et al., 1998). Progress has been slow to incorporate resistance to *Phomopsis* in elite soybean cultivars. Variation in incidence of infected seed among soybean genotypes of identical maturity has been reported (TeKrony et al., 1984). However, information on the relative susceptibility of currently grown cultivars to Phomopsis seed decay is generally unavailable. Several traits related to soybean growth and development have been associated with lower incidence of Phomopsis seed decay. Thomison et al. (1990) compared near isolines and reported that delayed flowering and later maturity was associated with lower incidence of Phomopsis seed decay. Determinate near isolines also expressed greater incidence of Phomopsis seed decay compared to indeterminate near isolines. Soybean genotypes with permeable seed coats are reported as susceptible compared to soybean genotypes with a impermeable seedcoat (Roy et al., 1994). In the future, breeders may consider developing cultivars with a shorter time period between R7 and R8 to escape seed infection compared to genotypes with a relatively longer maturation period between R7 and R8.

A fungicidal seed treatment will increase emergence and reduce the number of diseased seedlings if the planted seed lot has a high incidence of seed infected by *Phomopsis* spp. (Hall and Xue, 1995; Wall et al., 1983; Wallen and Seaman, 1963). Seed lots should be avoided for planting if the incidence of Phomopsis-infected seed exceeds 50% (Wall et al., 1983).

Several studies have associated lower seed germination with K-deficient soils and Phomopsis seed decay (Jeffers et al., 1984). Application of K improved seed germination but did not always reduce the incidence of Phomopsis seed rot. However, seed performance was frequently improved by optimal K nutrition, thus soybean seed growers should strive to monitor soil K to maintain a high quality seed product.

14–3.4 Sclerotinia Stem Rot

Sclerotinia stem rot of soybean, commonly called white mold, is a chronic to epidemic disease of soybean grown throughout the world (Grau and Hartman, 1999). The disease is most destructive in North America and South America (Wrather et al., 2001a, 2001b; Yorinori and Homechin, 1985). Sclerotinia stem rot was reported first in the USA in 1924 (Grau and Hartman, 1999), but it was not until 1948 that the disease was observed in central Illinois, a major soybean production area (Chamberlain, 1951). This disease was a chronic, but locally severe problem in Michigan, Minnesota, and Wisconsin during the 1970s. Sclerotinia stem rot was uncommon in the remainder of the North Central States. Beginning in 1990, Sclerotinia stem rot became widespread in each of the Great Lakes States and by 1992, was prevalent throughout the North Central Region of the USA. Sclerotinia stem rot has progressed from a sporadic disease of localized importance, to an annual threat to soybean production throughout the upper North Central Region (Wrather et al., 2001a). Reasons for the sudden increase of Sclerotinia stem rot are not fully understood, but the unexpected outbreaks of the disease are possibly related to changes in cultural practices, changes in the genetic base of current soybean cultivars, and a high incidence of pathogen-infested seed.

14–3.4.1 Pathogen

Sclerotinia stem rot is caused by *Sclerotinia sclerotiorum* (Lib.) deBary, a long-lived soilborne fungus. The fungus is easily recognized by the presence of the fluffy white mycelium that develops on the surface of stem lesions. Sclerotia range in diameter (2–5 mm) and length (3–40 mm) and are observed internally and externally associated with stems and pods. Sclerotia resume growth as myceliogenic germination (mycelial growth) or by carpogenic germination resulting in the formation of numerous tan apothecia. Apothecia range from 3 to 10 mm in diameter at maturity and release ascospores for several days. *Sclerotinia sclerotiorum* grows on PDA at 1 to 30°C and an optimum at 24°C (Grau and Hartman, 1999). The fungus is highly variable in culture, producing white to chocolate brown aerial mycelium and black sclerotia. Sclerotia will be larger if cultures are incubated at

10 to 20°C and may not form if grown at 30°C or higher. Apothecia and ascospores develop in alternating light and darkness at 12 to 15°C, provided the sclerotia are incubated in a water-saturated medium. *Sclerotinia sclerotiorum* may be isolated from diseased plants by collecting sclerotia from tissue or incubating host tissue on PDA. Isolates of *S. sclerotiorum* may be obtained by incubating sclerotia on carrot (*Daucus carota* var. *sativa*) discs in a moist environment. Mycelium may be removed from carrot disc and transferred to PDA. *Sclerotinia minor*, which typically infects peanut, is reported to infect soybean in the southeastern USA (Phipps and Porter, 1982), has not been reported to infect soybean in the North Central Region of the USA (Grau and Hartman, 1999).

Sclerotinia sclerotiorum infects many broadleaf plants although varying degrees of susceptibility are found among these potential hosts (Boland and Hall, 1994). Common hosts include green bean, cabbage (*Brassica oleacea* L.), sunflower (*Helianthus annuus* L.), peanut, and numerous other cultivated crops. Examples of nonhost crops are corn, small grains, and all forage grasses. Sclerotinia stem rot is frequently most severe when soybean is grown in rotation with other susceptible crops (Phipps and Porter, 1982; Grau et al., 1982).

14–3.4.2 Symptoms and Losses

Symptoms and signs of Sclerotinia stem rot normally do not appear until early reproductive stages (Grau and Hartman, 1999). Wilt and eventual death of upper leaves at growth stages R3 to R4 are the first canopy level symptoms of Sclerotinia stem rot (Plate 14–15). Leaves become grayish green as necrosis begins and eventually turn brown and frequently remain attached to stems even past maturity. Leaf symptoms are secondary in nature resulting from stem lesions that occur 10 to 50 cm above the soil line. Nodes are infected at or shortly after flowering and the pathogen progresses acropetally and basipetally into internode tissues. Lesions are white, form three to four nodes above the soil surface, vary in size, and usually encircle the stem and disrupt vascular transport systems of the plant. Lesions range in size from 1 to 30 cm or more. White fluffy mycelium will cover the lesions, especially during periods of high relative humidity (Plate 14–16). The characteristic black sclerotia are, in time, readily visible on the surface of stem lesions and within the stem. By plant maturity, stem tissues are white and tissues have a shredded appearance. The border of lesions frequently is discolored a reddish to purple color. At harvest, diseased stems are characterized by differing degrees of pod development. Pods have a white appearance, are smaller and form fewer seeds that are frequently moldy and shriveled. Sclerotia are formed in and on pods. Sclerotia that originate in stems and pods are observed with harvested grain (Fig. 14–2). Diseased plants stand out at harvest as "bleached white" stems in contrast to normal plants especially if the soybean cultivar has a tawny pubescence phenotype. The presence of sclerotia inside stems and pods is diagnostic evidence that *S. sclerotiorum* caused premature death of the plant. Occasionally sclerotia are observed mixed with seed, but dead plants are not obvious in a field.

Yield loss caused by *S. sclerotiorum* is dependent on the percent of the plant population killed by the pathogen and how quickly individual plants die in relationship to reproductive development. Field trials indicate that for each 1% incre-

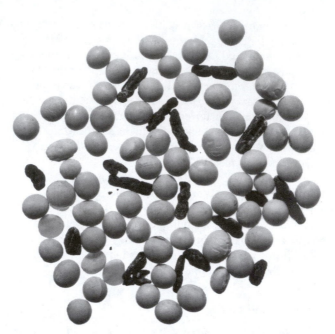

Fig. 14–2. Sclerotia of *Sclerotinia sclerotiorum* are commonly mixed with soybean seed. The presence of sclerotia indicates that Sclerotinia stem rot was present and represents an efficient means to disseminate the pathogen.

ment of plant mortality at the R6 to R7 stages of development, soybean yield is reduced 17 to 49 kg ha^{-1} (Chun, 1987; Grau, 1988; Hoffman et al., 1998; Kurle et al., 2001). For example, a field with 20% plant mortality would lose 340 to 680 kg ha^{-1} of yield potential to Sclerotinia stem rot. However in Wisconsin studies, yield loss was 32.6 kg ha^{-1} for a partially resistant cultivar vs. 48.9 kg ha^{-1} for a susceptible cultivar (Kurle et al., 2001). Sclerotia of *S. sclerotiorum* are not toxic to livestock.

14–3.4.3 Disease Cycle and Epidemiology

Sclerotinia stem rot is governed by a synchronized sequence of events involving host canopy closure, germination of sclerotia, formation of apothecia, release of ascospores, a flowering host, and a canopy microclimate conducive for infection and colonization of host tissue. Sclerotia of *S. sclerotiorum* survive in soil for several years although population density declines with time (Kurle et al., 2001; Mueller et al., 2002b). Sclerotia are replenished in soil by reproduction on susceptible hosts and soybean seed lots contaminated with sclerotia (Grau, 1988; Hartman et al., 1998; Mueller et al., 1999; Yang et al., 1998). Mortality of sclerotia is related to natural depletion of nutrients needed for germination, and by parasitism and predation by soil microflora and fauna. Sclerotia of *S. sclerotiorum* are parasitized by *Coniothyrium minitans* Campbell, *Gliocladium catenulatum* Gilman and Abbott, *Sporidesium sclerotivorum* Uecker, Adams et Ayers, and *Tricoderma viride* Pers. ex Fr (Bardin and Huang, 2001; del Río et al., 2002; Yang and del Río,

2002). Biological control of *S. sclerotiorum* was achieved by del Río et al. (2002) in field-scale studies in Iowa. Conidia of *S. sclerotivorum* were applied to soil and 50 to 100% control of Sclerotinia stem rot was achieved in these studies.

The life cycle of the pathogen begins with the germination of sclerotia positioned within 5.0 cm or less from the soil surface. Sclerotia will be positioned within the soil profile (plow layer) depending on degree and type, or lack, of tillage (Kurle et al., 2001). Prolonged periods of relatively low soil temperatures (5–15°C) and soil matric potentials less than −5 bars for 10 to 14 d are most favorable for sclerotia to germinate and form apothecia. A dense crop canopy is needed to create a shaded soil surface that is needed for ascospore bearing apothecia to develop (Boland and Hall, 1988a,1988b; Sun and Yang, 2000). Filtered light reaching the soil surface at 276 and 319 nm was conducive for normal development and maturation of apothecia. Wavelengths required to stipe formation are not essential for apothecia development indicating there are different photoreceptors for phototropism and photomorphogenesis required for formation of apothecia (Thaning and Nilsson, 2000).

The disease cycle of Sclerotinia stem rot begins when apothecia form and ascospores are forcibly ejected from asci. Ascospores initially are ejected into the crop canopy, but quickly escape the canopy and are wind disseminated up to 100 m from a point source (Wegulo et al., 2000). Ascospores land on senescing flower petals that adhere to emerging pods (Grau, 1988; Sutton and Deverall, 1983). Ascospores germinate and hyphae colonize senescent tissues allowing the pathogen to further advance into pod and eventually node tissues. Canopy temperatures of <28°C and plant surface wetness for 42 to 72 h are needed for disease development. Canopy temperatures above 30°C are extremely suppressive to disease development in spite of other factors, such as adequate soil moisture and relative humidity, being favorable for *S. sclerotiorum* (Boland and Hall, 1988b). Dissemination and germination of ascospores, infection of plants, and disease development are favored by high relative humidity in crop canopy (Wegulo et al., 2000). If adjacent plants come into contact with an infected plant, they may also become infected, but plant-to-plant spread of the pathogen is minimal. Sclerotia are formed from mycelium on the stem surface and inside stems and pods. Sclerotia that are formed externally on stems and pods eventually fall to the soil surface. Sclerotia that form inside stems and pods are released when plants are mechanically harvested and are either deposited on the soil surface or physically mixed with harvested grain. However, many sclerotia are removed from the field with the grain. *Sclerotinia sclerotiorum* may be introduced to a field as sclerotia physically mixed with seed or as mycelium in infected seed (Mueller, et al., 1999).

14–3.4.4 Management

Sclerotinia stem rot is best managed by an integrated approach of selecting soybean cultivars with the highest level of resistance and adjusting cultural practices to minimize environmental factors that favor disease development. This approach requires a coordinated plan that matches the level of resistance in a soybean cultivar to expected disease potential and cropping practices that influence crop canopy closure. No single tactic will completely control Sclerotinia stem rot.

Sclerotinia stem rot is a disease of high yield potential soybean production. Although several factors are believed responsible for the increased occurrence of white mold, none may be more important than management practices or environmental conditions that promote rapid and complete crop canopy closure (Oplinger and Philbrook, 1992). Sclerotinia stem rot is particularly favored by dense soybean canopies created by plantings in narrow row widths, high plant populations, early planting, high soil fertility, or other management practices that promote rapid and complete canopy closure (Grau, 1988; Grau and Radke, 1984). The effect of row width on incidence of Sclerotinia stem rot and subsequent yield varies by year and is strongly controlled by annual climatic conditions (Buzzell et al., 1993; Grau and Radke, 1984). Frequently, the yield advantage of narrowed row widths, compared to wide widths, is expressed even though the incidence of Sclerotinia stem rot may be greater in narrow row systems(Fig. 14–3).

Crop rotations that employ nonhosts result in a reduced incidence of Sclerotinia stem rot, but some nonhosts are better than others (Gracia-Garza et al., 2002;

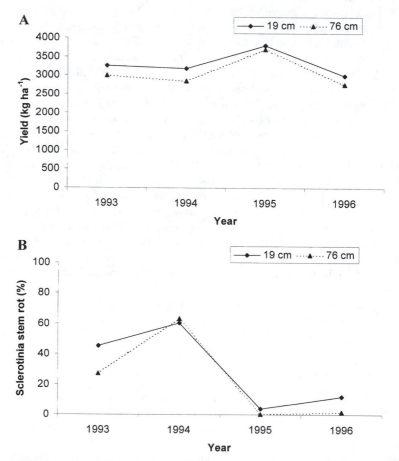

Fig. 14–3. Seed yield (panel A) and incidence of Sclerotinia stem rot (panel B) of soybean plants grown in 19- and 76-cm rows for 4 yr.

Kurle et al., 2001; Mueller et al., 2002b). In Wisconsin studies, short-term crop rotations with corn and oat had a moderate effect on reducing the severity of Sclerotinia stem rot (Kurle et al., 2001). A preceding crop of small grain, in contrast to corn, had a greater impact on reducing the incidence of Sclerotinia stem rot (Kurle et al., 2001). Rotation with nonhosts resulted in fewer apothecia formed under the soybean canopy (Gracia-Garza et al., 2002; Kurle et al., 2001). Tillage rather than crop rotation sequence had the most pronounced effect on the population density of apothecia (Gracia-Garza et al., 2002; Kurle et al., 2001). The population density of apothecia was greatest in moldboard plow systems compared to no-tillage systems (Gracia-Garza et al., 2002; Kurle et al., 2001). Fewer apothecia in no-tillage systems is a partial explanation why lower incidence of Sclerotinia stem rot is observed in no-till fields compared to fields receiving some degree of tillage (Gracia-Garza et al., 2002; Kurle et al., 2001; Workneh and Yang, 2000). The distribution and incidence of Sclerotinia stem rot is correlated with spatial patterns and population density of apothecia (Boland and Hall, 1998a).

Soybean cultivars have been observed for years to differ in reaction to *S. sclerotiorum* (Boland and Hall, 1986, 1987; Chun et al., 1987; Grau et al., 1982; Kim et al., 1999; Nelson et al., 1991b; Yang et al., 1999; Wegulo et al., 1998). A point of continued debate is why do soybean cultivars differ in incidence of plants infected by *S. sclerotiorum*. Several mechanisms are likely involved and are speculated to involve both disease escape and a physiological resistance to *S. sclerotiorum*. Plant architecture, lodging potential, and maturity group are most often cited as canopy modifying phenotypes that that influence cultivar reactions to *S. sclerotiorum*. Ample evidence, however, is reported to support the concept that physiological resistance to *S. sclerotiorum* is operative in soybean. Glyceolin, a putative phytoalexin, is induced when soybean are inoculated with ascospores or mycelium of *S. sclerotiorum* (Nelson et al., 2002; Sutton and Deverall, 1984). Although complete resistance is yet to be identified, partial resistance to *S. sclerotiorum* is functioning in cultivars and plant introductions (Arahana et al., 2001; Kim and Diers, 2000; Kim et al., 2000; Hoffman et al., 2002). A broad-sense heritability estimate of 0.59 has been calculated for partial resistance to *S. sclerotiorum* (Kim and Diers, 2000). Several QTLs are reported that confer physiological resistance to *S. sclerotiorum* (Arahana et al., 2001; Kim and Diers, 2000). However in natural epidemics, the reaction of soybean to *S. sclerotiorum* is likely governed by both physiologically based resistance, and various traits related to disease escape. Two QTLs were associated with traits related to disease escape (Arahana et al., 2001; Kim and Diers, 2000). Although lodging and disease incidence are positively correlated, lodging a result of Sclerotinia stem rot rather than a trait that promotes greater incidence of the disease. Lodging data should be recored during early reproductive growth stages and at harvest marturity to separate cause and affects.

Abiotic factors such as temperature and moisture have a significant impact on the incidence and severity of Sclerotinia stem rot (Nelson et al., 1991a). The reaction soybean cultivars to *S. sclerotiorum* are differentially sensitive to photon flux density of photosynthetically active radiation (Pennypacker and Risius, 1999). Thus, light is an important environmental variable to regulate when soybean germplasm is evaluated for reaction to *S. sclerotiorum* in controlled environments. Mechanisms of resistance have not been conclusively described. Although glyceolin, a putative phytoalexin, is involved in resistance to *S. sclerotiorum* (Nelson et al., 2002; Sutton and Deverall, 1984), partially resistant soybean genotypes may also be tolerant to, or metabolize oxalic acid and polygalacturonase. These compounds are reported pathogenicity factors produced by *S. sclerotiorum* (Cessna et al., 2000; Favaron et al., 1988, 1994). Soybean has been transformed with a gene from wheat that codes for oxalate oxidase, an oxalic acid degrading enzyme, resulting in lower incidence of Sclerotinia stem rot in field trials (Cober et al., 2003).

The population density of sclerotia of *S. sclerotiorum* decline with time and the decline has been related to crop rotation and tillage practices (Kurle et al., 2001; Mueller, 2002). One explanation for the disappearance of sclerotia is that sclerotia are parasitized by several species of fungi (Bardin and Huang, 2001; del Río, 2002). Although sclerotia are parasitized by *G. catenulatum* and *T. viride*, *S. sclerotivorum* and *C. minitans* appear to be most important (de Vrije et al., 2001; Bardin and Huang, 2001; del Río et al., 2002). Both fungi have been investigated as candidates for commercialization as a biological control product. *Coniothyrium minitans* has shown promise as a biological control agent and a potential alternative for chemical fungicides to control diseases caused by *S. sclerotiorum* (de Vrije et al., 2001).

Thiophanate-methyl will reduce the incidence of Sclerotinia stem rot if applied during flowering and the product penetrates into the lower regions of the canopy to protect pod tissues (Mueller et al., 2002a). Thiabendazole, applied to seed, is effective against seedborne inoculum of *S. sclerotiorum* (Mueller et al., 1999). The use of fungicides to control Sclerotinia stem rot is more feasible for seed production rather than grain production because of economic factors.

Common agricultural herbicides have various direct affects on *S. sclerotiorum*, or the development of Sclerotinia stem rot. Lactofen and thifensulfuron, two herbicides applied to emerged soybean plants, are reported to reduce the incidence and severity of Sclerotinia stem rot (Dann et al., 1998, 1999; Nelson et al., 2002). Lactofen applied at the R1 stage of development has a suppressive effect on Sclerotinia stem rot, especially if applied to partially susceptible cultivars. Although the incidence of Sclerotinia stem rot is reduced by lactofen, a positive effect on yield is not always achieved by this practice (Dann et al., 1998, 1999; Nelson et al., 2002). Lactofen reduces the incidence and severity of Sclerotinia stem rot by inducing increased levels of phytoalexins in the host, reduce canopy density, and delay flowering of the host allowing disease escape (Nelson et al., 2002). The ability of thifensulfuron to reduce the incidence of Sclerotinia stem rot is related to its ability to reduce canopy density (Nelson et al., 2002). Trifluralin and metribuzin increase germination of sclerotia and number of apothecia per sclerotium, but atrazine causes malformed apothecia (Casale, and Hart, 1986; Radke and Grau, 1986). The use of atrazine in corn production has declined and may be a contributing factor to the expanded distribution of Sclerotinia stem rot in the North Central Region of the USA. Glyphosate has become a widely used herbicide in most soybean-growing regions. Nelson et al. (2002) reported that glyphosate did not modify host factors commonly accepted to affect the development of Sclerotinia stem rot. However, several glyphosate resistant cultivars of Maturity Group II, treated with glyphosate, expressed greater incidence of Sclerotinia stem rot compared to untreated plants of the same soybean cultivars. Not all soybean cultivars express greater incidence of

Sclerotinia stem rot when treated with glyphosate (Lee et al., 2000; Nelson et al., 2002).

14–3.5 Sclerotium Blight

Sclerotiorum blight, also known as southern blight, occurs primarily in the southern USA. The disease is found in the sandy-soil areas of the South where high temperatures occur. Soybean is only one of many crops affected by this disease (Rupe, 1999).

14–3.5.1 Pathogen

Sclerotium rolfsii Sacc. is the reported cause of Sclerotium blight, also known as southern blight, southern stem rot, and white mold. The teleomorph is *Arthelia rolfsii* (Rupe, 1999). This soilborne pathogen has a host range of more than 500 plant species, including soybean and practically all of the summer legumes adapted to the South (Athow, 1987). The pathogen is found most commonly in, but not restricted to, the warmer regions of the world (Rupe, 1999). *Sclerotium rolfsii* is characterized by broad hyphae (5–9 μm wide), clamp connections at most septa, and small tan sclerotia.

14–3.5.2 Symptoms and Losses

Southern blight generally develops on isolated or small groups of plants scattered throughout the field. Symptoms can develop at all growth stages and include light brown lesions that quickly darken and girdle the hypocotyl or stem. Yellowing and wilting of foliage follows quickly, and necrotic leaves cling to the dead stem. Foliage also may show circular, tan to brown, zonate lesions with dark brown margins. Plants infected as seedlings may develop twin stems. A white mat of fungus mycelium commonly forms on bases of infected stems, leaf debris, and the soil surface. Tan to brown, spherical sclerotia are 1 mm in diameter and form on diseased plant debris and the soil surface (Rupe, 1999; McGee 1992). Sclerotium blight develops as a rot at the base of the stem. Infected plants die prematurely, sometimes before the seed has formed.

Although symptoms can be striking, southern blight usually is a minor disease. Quebral and Pua (1976) reported seedling losses of up to 10%, but reliable estimates of yield loss in mature soybean are lacking. Although plant mortality may reach 20 to 30% incidence in specific fields, the disease is more commonly observed in small, scattered patches that results in minimal yield loss.

14–3.5.3 Disease Cycle and Epidemiology

Sclerotium rolfsii survives as sclerotia throughout the soil profile and as mycelia in colonized plant debris. Sclerotia viability can remain high for 8 to 10 mo (Beute and Rodríguez-Kabana, 1981). Infection of stems occurs at or below the soil surface (McGee, 1992), and the fungus can use decomposing plant material as a food base for attacking healthy plants (Beute and Rodríguez-Kabana 1979). Although symptoms can appear throughout the season, disease severity generally is greatest during the reproductive stages of plant growth.

Most available information about the epidemiology of southern blight is from studies on peanut. Southern blight is more severe in wet years (Aycock, 1966), but the disease can occur in dry years as well. Cool temperatures seem to limit disease development, and both soil moisture and temperature interact to influence survival of sclerotia. Beute and Rodríguez-Kabana (1981) reported that temperature had no effect on sclerotia survival in dry soils, but reduced survival in moist field soil when the temperature was >20°C. Furthermore, mycelia died rapidly in moist field soil but survived at least 6 mo in dry soil (Beute and Rodríguez-Kabana, 1981). In crops other than soybean, addition of Ca fertilizer or ammonium bicarbonate reduced losses to this disease. Although mechanisms for these effects are unclear, they may result from increased levels of Ca in tissues or direct toxicity to the fungus (Punja et al., 1986). In soybean, moderate to high levels of soil chloride increased severity of southern blight (Rupe et al., 2000). Root injury by the root knot nematode, *Meloidogyne arenaria*, as well as stem girdling by threecornered alfalfa hopper [*Spissistilus festinus* (Say)] can predispose plants to infection by *S. rolfsii* (Herzog et al., 1975; Minton et al., 1975).

14–3.5.4 Management

Several crops, including corn, grain sorghum, and pasture grasses, are nonhosts for the fungus. Two to four years of rotation with these crops may be required to significantly reduce levels of inoculum in soil. Resistant cultivars are not available in the USA, but cultivars were classified as susceptible to moderately resistant in trials in India (Agarwal and Kotasthane, 1971). Deep plowing inhibited sclerotia germination and reduced disease incidence in carrot (Punja et al. 1986).

14–3.6 Stem Canker

Stem canker has been widely recognized as an important soybean disease, but recently has been divided into northern stem canker and southern stem canker based on two causal agents (Fernandez et al., 1999). Northern stem canker was first reported in the late 1940s in Iowa, and by the 1950s, the disease was observed throughout the midwestern USA and Ontario, Canada. Southern stem canker was reported in the south in 1973, and by 1984, had been detected in all southern states (Athow, 1987). Southern stem canker has recently been reported in the North Central States (Gravert et al., 2001). Stem canker has been reported in both South America and Europe.

14–3.6.1 Pathogen

Northern stem canker and southern stem canker are caused by *Diaporthe phaseolorum* var. *caulivora* Athow and Caldwell and *Diaporthe phaseolorum* var. *meridionalis* Morgan-Jones, respectively. Each biotype of the pathogen differs with regards to the number of sporocarps that develop in infected tissue and colony phenotype on synthetic culture media. *Diaporthe phaseolorum* var. *caulivora* produces perithecia singly or in groups of 2 to 12 in infected tissue, and on PDA the fungus produces white colonies with cottony tufts of mycelium. *Diaporthe phaseolorum* var. *meridionalis* produces single-borne perithecia in infected tissue and produces white colonies on PDA, but with age, the white buff colonies become light to dark brown (Fernandez et al., 1999).

The host range of both pathogens has not been studied extensively, however, more than 16 weed species are known to harbor *D. phaseolorum* var. *meridionalis* (Roy and Miller, 1983; Smith and Backman, 1988).

14–3.6.2 Symptoms and Losses

Initial expression of symptoms occurs during the early reproductive stages, with the development of a small, reddish-brown superficial stem lesion at the base of branches or petioles. The lesion is first observable in the leaf scar after the petiole has fallen from one of eight nodes. The lesion elongates and becomes dark brown or black, sunken in appearance and often girdles the stem (Plate 14–17). As a result of a phytotoxin produced by the fungus, interveinal chlorosis and necrosis are expressed in the leaves and is soon followed by plant death. Above and below the canker, green tissue is present and the leaves on the dead plant wither, but remain attached. A top dieback can occur and results in a characteristic shepherd's crook curling of the terminal bud (Fernandez et al., 1999).

Yield losses have been reported to be as high as 50 to 80% (Keeling, 1982; Krausz and Fortnum, 1983) in naturally infested fields. Yield reductions resulting from stem canker have increased dramatically over the past 2 yr. Estimated yield losses to stem canker were 89.2 and 59.5 thousand Mg in 1999 and 2000, respectively (Wrather et al., 2001).

Both pathogens overwinter in colonized stems and infected seed (Athow, 1987; Backman et al., 1985; McGee and Biddle, 1987). Long distance dissemination of the pathogens is made possible by the movement of infested soybean residue and to a lesser extent by infected seed. Seed infection by *D. phaseolorum* var. *caulivora* can be as high as 10 to 20%; however, seed transmission of *D. phaseolorum* var. *meridionalis* generally does not exceed 1% (Fernandez et al., 1999). Severe epidemics caused by *D. phaseolorum* var. *meridionalis* have occurred with infected seed being the only conceivable inoculum source, especially in the northern regions of the north central USA.

As epidemiological information regarding *D. phaseolorum* var. *caulivora* is lacking, information on *Diaporthe phaseolorum* var. *meridionalis* will be presented. In late winter, pycnidia begin to develop and conidia are released beginning in late April continuing into June and serve as the primary inoculum. Splashing, blowing rain, and wind disperse spores up to 2 m from the point source to petioles, petiole bases, stems, and leaves (Damicone et al., 1990). The growth stage of the plant at the time of exposure to the inoculum heavily influences the incidence and severity of disease. Exposure to inoculum at V3 corresponds to the highest severity of disease. Disease severity is progressively reduced when first contact is delayed from V3 to V10 growth stages (Bowers and Russin, 1999). Secondary inoculum is released from pycnidia present in stem cankers. Conidia produced at this time could be responsible for late season infections and thereby increase the inoculum potential for the next growing season.

Environmental conditions during the vegetative stages govern disease development. Temperature greatly influences infection, with the highest levels of infection occurring when the air temperature is between 28 and 34°C, with an optimal temperature of 28.5°C (Rupe et al., 1996). The frequency of infection is reduced when the temperature is between 10 and 22°C and is arrested at 36°C (Rupe and Sutton, 1994). Temperature and wetness period are significantly related to infection (Rupe et al., 1996).

Rainfall during plant vegetative growth is critical for the development of stem canker epidemics. Cumulative rainfall (Damicone et al., 1987), not the number of rainy days, was related to higher disease severity, but this is in contrast to other research that indicates the number of rainy days and not the cumulative rainfall results in greater disease (Subbarao et al., 1992).

Other pests can influence the severity of stem canker. A synergistic interaction can occur between *Spissitilus festinus* (threecornered alfalfa hopper) and *D. phaseolorum* var. *caulivora* (Russin et al., 1986). In this case, the length of the cankers is larger on plants that have been girdled by the insect. In contrast, stem canker damage can be reduced when plants are parasitized by *Heterodera glycines* (SCN) or if the plants have been defoliated by *Psuedoplusia includens* Walker (soybean looper) (Russin et al., 1989a, 1989b).

14–3.6.3 Management

Stem canker is effectively managed by the combination of planting resistant cultivars and reducing infested soybean residue on the soil surface (Fernandez et al., 1999). Four resistance genes have been described and are designated *Rdc1*, *Rdc2*, *Rdc3*, and *Rdc4*. Research is lacking regarding the additive effects of these genes. Deep plowing can reduce crop residue prior to planting a soybean crop.

The benefits of crop rotation to reduce stem canker have not been demonstrated in production fields. Also, a double-cropping system of wheat and soybean may result in greater disease than monocropping soybean (Rothrock et al., 1985, 1988). Delayed planting can reduce the incidence and severity of stem canker. However, loss of yield potential that accompanies delayed planting makes this as a questionable control strategy.

Seed that are to be used for planting should not be harvested from fields with a history of stem canker. Fungicides applied to seed greatly reduce stem canker but will not completely eliminate the pathogen (Crawford, 1984). Foliar fungicides can be effective when applied during vegetative stages (Backman et al., 1985). However, results are inconsistent, and in most cases foliar fungicides would not be an economical management strategy.

14–4 LEAF DISEASES

14–4.1 Brown Spot

Brown spot is a common disease of soybean in the USA and throughout the world (Athow, 1987; Sinclair and Hartman, 1999a). This disease was first reported in Japan in 1915 and followed closely by reports in North America (Athow, 1987; Sinclair and Hartman, 1999a). This disease is primarily a leaf spot disease, although stems, pods, and seeds of maturing plants are also infected. The disease is commonly observed during the final stages of pod fill and intensifies as the crop progresses to harvest maturity.

14–4.1.1 Pathogen

Brown spot is caused by *Septoria glycines* Hemmi. Brown pycnidia are imbedded in mature lesions and are more numerous on the upper side of the leaves. Conidia are produced in pycnidia and are hyaline, filiform, and curved in shape, and have two to four cells per conidium. Conidia range in size from 1.4 to 2.1 μm × 21 to 50 μm (Sinclair and Hartman, 1999a).

Septoria glycines is reported to produce a host specific pathotoxin that in vitro causes typical symptoms similar to symptoms caused in vivo by the pathogen itself. The pathotoxin has been characterized as a polysaccharide with a high content of uronic acid and low levels of mannose, galactose, and glucose (Song et al., 1993). Isolates of *S. glycines* have not expressed variability in virulence to different soybean genotypes (Kamicker and Lim, 1985). The host range of *S. glycines* has been expanded to include most species of *Glycine*, several legume species (Lee and Hartman, 1996), and common weeds such as velvetleaf (*Abutilon theophrasti*) (Sinclair and Hartman, 1999a).

14–4.1.2 Symptoms and Losses

Symptoms of brown spot are first observed early in the season as irregular, dark-brown patches on the cotyledons followed by the appearance of reddish brown angular spots, 1 to 5 mm in diameter, on unifoliolate leaves (Plate 14–18). Severely diseased unifoliate leaves become cholorotic and senesce faster than normal. Symptoms of brown spot are frequently absent or extremely mild during the vegetative growth stages. Symptoms begin to reappear after flowering and progress upward from the lower leaves. This gap in seasonal symptom progression is a result of a longer latent period during preflower growth stages (Young and Ross, 1979). Numerous irregular, light brown lesions develop on trifoliolate leaves and gradually darken to chocolate brown and eventually a blackish brown appearance. Adjacent lesions frequently merge resulting in larger and irregular blotches of symptomatic tissue. By late in the growing season, symptoms are observed throughout the canopy and leaves become rusty brown and drop prematurely. Brown spot causes plants to prematurely defoliate, and this event is frequently confused with "normal maturity". Symptoms of brown spot develop on stems and pods of plants approaching maturity. Stem and pod lesions have indefinite margins, are dark in appearance and range in size from flecks to lesions several centimeters in length, but are not distinct enough to be diagnostic. Seed are infected but symptoms are not conspicuous.

Yield loss associated with brown spot varies from region to region in the USA (Wrather et al., 2001b). Greater yield losses are reported to occur in northern compared to southern states (Wrather et al., 1995, 2001b). Yield loss occurs most often in high yield environments and yield loss associated with brown spot is related to timing and rate of defoliation (Cooper, 1989). Most yield loss estimates are in the 8 to 34% range and occur when 25 to 50% of the canopy prematurely defoliates (Backman et al., 1979; Lim, 1980, 1983; Pataky and Lim, 1981a; Williams and Nyvall, 1980; Young and Ross, 1979). Disease severity at the R6 stage is predictive of yield (Lim, 1980). Severe brown spot results in smaller seed size (Young and Ross, 1979).

14–4.1.3 Disease Cycle and Epidemiology

Septoria glycines survives as mycelia and pycnidia embedded in leaf and stem residue (Sinclair and Hartman, 1999a). Sporulation occurs in spring and the pathogen is water-splash dispersed to leaves. Germ tubes of germinated conidia penetrate through stomates and grow intercellularly. Infected seed is another possible source of primary inoculum. Unifoliate leaves are infected early in the season and serve as source of secondary inoclum. *Septoria glycines* is capable of infecting soybean leaves if temperatures are between 15 and 30°C, with an optimum temperature of 25°C (Schuh and Adamowicz, 1993). Conidia of *S. glycines* do not require free water to germinate as do most other plant pathogenic fungi. However, disease severity is reported to increase with increasing duration of leaf wetness (Peterson and Edwards, 1982; Schuh and Adamowicz, 1993). The duration of leaf wetness is critical with infection reported to occur after only 6 h of leaf wetness, but severity is increased if the leaf surface is wet for 36 continuous hours (Ross, 1982). This pathogen invades stomata of pods and progresses into seed directly, or systemically through placental and funicular tissue (MacNeill and Zalasky, 1957).

The incidence and severity of brown spot is greatest in the spring and latter weeks of the growing season as soybean enter full pod development and eventually maturity. Mid-summer suppression of symptoms is likely due to high air temperatures unfavorable to the pathogen. Symptom severity is greater late in the growing season compared to symptom severity in the spring. Inoculum for late season epidemics originates from lesions latent since early reproductive growth stages that progress to lesions capable of supporting sporulation (Young and Ross, 1979).

14–4.1.4 Management

Brown spot is most prevalent in fields frequently planted to soybean and its prevalence is further enhanced if no tillage or minimal soil tillage leave soybean residue on the soil surface. Crop rotation has a limited effect in current management systems because of the popular alternate year rotation of soybean and corn, and the practice of maintaining crop residue on the soil surface (Lim, 1980). Row spacing has a minimal impact on brown spot severity or incidence (Pataky and Lim, 1981b).

Soybean cultivars and lines are reported to differ in reaction to *S. glycines* in the early vegetative stages, but these differences are not readily apparent at late reproductive stages (Lim, 1979). The onset of brown spot symptoms is influenced by the relative maturity of the soybean cultivar and symptoms appear earlier in the season on an early-maturing cultivar. Complete resistance has not been identified in soybean cultivars or lines (Young and Ross, 1978; Lim, 1979, 1983). Partial or rate-reducing resistance varies among cultivars and is of practical importance to reduce yield loss caused by the brown spot pathogen. Variability in reaction to *S. glycines* was observed among R7 generation soybean lines regenerated from calli exposed to host-specific pathotoxin produced by the pathogen (Song et al., 1994). Intermediate levels of resistance had low heritability (23%) for resistance to *S. glycines*. Selected lines with an intermediate reaction to *S. glycines* expressed a lower area-under the disease progress curve (AUDPC), matured later, and produced yields compared to susceptible lines (Lee et al., 1996). Cooper (1989) reported yield reductions due to brown spot varied by soybean genotype, but variation in yield loss was not associated with determinate or indeterminate growth types. A chlorotic and nonchlorotic lesion phenotype was associated with yellow and green seed soybean genotypes, however disease severity and yield were identical in field trials to compared performance of near isolines differing for these traits (Lim, 1983).

Application of fungicides to soybean foliage from bloom to pod fill has effectively reduced the severity of brown spot and subsequently increased seed yield. Thiophanate-methyl is registered in the USA as a fungicide to control brown spot of soybean. Fungicides, if applied at stages R3 and R6, effectively slow the rate of brown spot development (Backman et al., 1979; Cooper, 1989; Pataky and Lim, 1981a).

14–4.2 Cercospora Leaf Blight and Purple Seed Stain

Cercospora leaf blight and purple seed stain are caused by the same pathogen and occur together throughout the world and in most parts of the USA where soybean is grown. Purple seed stain was first detected in the USA in 1924 (Athow, 1987; Schuh, 1999). Purple seed stain is agronomically more important than Cercospora leaf blight.

14–4.2.1 Pathogen

Cercospora leaf blight and purple seed stain are caused by *Cercospora kikuchii* (Matsumoto & Tomoyasu) M.W. Gardner. Although 10 other species of *Cercospora* isolated from other hosts are reported to cause purple discoloration of soybean seed, *Cercospora kikuchii* (Mat. & Tomoy.) is regarded as the primary cause of purple seed stain (Athow, 1987; Jones, 1959; Schuh, 1999). The fungus produces abundant filiform, hyaline conidiophores, 0 to 22 septa and 70 to 164 μm × 4 to 5 μm in size on the diseased seedlings.

14–4.2. 2 Symptoms and Losses

The foliar symptoms are most evident during late pod-filling stages as plants approach maturity. The first symptoms begin as light purple discolorations, pinpoint spots to irregular patches up to 1 cm in diameter, that occur on both upper and lower surfaces of leaves in the upper canopy. Affected leaves appear leathery and develop deep red to purple discoloration (Plate 14–19). Lesions frequently coalesce to form large necrotic areas. On susceptible cultivars, this discoloration may intensify and expand to cover the entire leaf surface. Angular lesions may be evident and can vary in size from pinpoint spots to irregular patches up to 1 cm in diameter. Lesions may coalesce to form large necrotic areas. Veins also can be necrotic. Severe symptom severity may lead to defoliation of the uppermost leaves, which can be confused with early crop maturity. However, green leaves usually remain below the defoliated area within the crop canopy (Schuh, 1999). Upper stems, petioles, and pods often show deep red to purple discoloration similar to that on leaves. Lesions on petioles and stems can develop and are similarly colored and slightly sunken, but eventually may have a dull gray to dark brown appearance and desiccate prematurely.

Infection of petioles increases defoliation, but the petioles remain attached to the plants. Heavily infected stems have a dull gray to dark brown appearance and desiccate 7 to 10 d prematurely. Infected seed can be asymptomatic or show discoloration of seedcoat. This discoloration is the most common and easily distinguished phase of the disease and varies from pink to dark purple. Discolored areas vary in size from small spots to the entire surface of the seed coat, which can be cracked. Cotyledons from symptomatic seeds generally show no symptoms.

Yield losses due to the foliar phase of this disease are believed to occur but are not well documented. Most economic losses result from purple seed stain. Incidence of the seed stain phase can be up to 50% (Roy and Abney, 1976). Economic loss due to purple seed stain relates to lower grain grade. The purple discoloration is not detrimental to soybean for processing because the purple color disappears when the seed is heated. Estimated yield losses from Cercospora leaf blight in southern USA have been as high as 3.5 million Mg in some years (Wrather et al., 2001b). The amount of purple discoloration in seed lots is important from the grading standpoint, and the value of the seed may be lowered by excessive amounts of purpling. The U.S. grading standards do not allow more than 5% purple stain in No. 1 yellow soybean. Several studies found little relationship between seed infection by *C. kikuchii* and germination (Schuh, 1992; TeKrony et al., 1985a). However, other studies found reduced or slower emergence of seedlings derived from stained seed (Wilcox and Abney, 1973).

14–4.2.3 Disease Cycle and Epidemiology

Cercospora kikuchii overwinters in infested residue of leaves, petioles, stems, and seeds. Athow (1987) reported limited movement of the pathogen from infected seed to young seedlings. The incidence of the disease in harvested seed did not differ appreciably whether planted with purple stained or purple stain-free seed (Wilcox and Abney, 1973). However, when diseased seeds are planted, the fungus grows from the seed coat into the cotyledons and from these into the stem of a small percentage of the seedlings. The spores are blown by wind and splashed by rain to other leaves. Most early season infections are latent (Orth and Schuh, 1994) and latent infections, along with colonized crop residue, are sources of secondary inoculum for infection of foliage and pods (Schuh, 1991). Seed infection results directly from pod infection during the pod-filling stages. Flowers are not infection courts for this fungus, but inoculating pods results in seed infection and the production of purple-stained seed (Roy and Abney, 1976, Schuh, 1992).

Weather at inoculation and several days afterward is most critical for development of purple seed stain (Roy and Abney, 1976). Infection occurs between 15 and 32°C, with the optimum at 25°C (Martin and Walters, 1982; Schuh, 1991, 1992). Splashing rain is not required for spore dispersal but sufficient relative humidity and dew are required for spore germination and infection (Schuh, 1991). A leaf wetness period of 18 to 24 h generally is required for leaf and pod infection (Martin and Walters, 1982; Schuh, 1991, 1992). However, infection can occur under shorter (8–10 h) dew periods if relative humidity exceeds 90% (Schuh, 1993).

Leaf spots are caused by primary inoculum followed by the production of secondary inoculum that causes infection of leaves, stems, and pods. Weather conditions during flowering and plant maturity have a pronounced affect on the incidence of purple stained seed. The incidence of purple seed stain varies among years, geographic areas, and soybean cultivar. Roy and Abney (1976) reported that the highest incidence of purple stained seed occurred from controlled inoculations at full-bloom or early pod stages, but also found a high percentage of purple stained seed if inoculations were made at the green bean stage (R5).

Harvest delays do not increase the incidence of purple seed stain (Athow and Laviolette, 1973) compared to other seedborne fungi such as *Phomopsis* (Athow and Laviolette, 1973). The incidence of Phomopsis seed decay is reduced in the presence of *C. kikuchii*. Hepperly and Sinclair (1981) reported higher germination of soybean seed at harvest maturity if purple seed stain was present.

14–4.2.4 Management

Soybean cultivars are reported to differ in reaction to Cercospora leaf blight and purple seed stain (Orth and Schuh, 1994; Walters, 1980, 1985). Separate genetic systems confer resistance to foliar and seed stain phases of the disease (Orth and Schuh, 1994). PI 80837 and 'CNS' are sources of resistance to purple seed stain (Athow, 1987). Resistance to purple seed stain is highly heritable and one to two genes govern the expression of this trait in PI 80837 (Athow, 1987; Wilcox et al., 1975). Unfortunately, developing cultivars with resistance to *C. kikuchii* receives little emphasis in public or private soybean-breeding programs.

Application of foliar fungicides during pod-filling stages reduced seed infection and the incidence of purple seed stain, but had little effect on yield (TeKrony et al., 1985a). Other studies suggest that a significant reduction of purple seed stain will result from an application of benomyl applied at full-bloom to early pod stage (Abney et al., 1975; Tenne and Sinclair, 1978; Miller and Roy, 1982). Benomyl applied in the R5 to R6 stages has been reported to increase yield by reducing foliar infection when conditions were favorable for its development. Benomyl was formerly labeled for management of several soybean pathogens but the registration has since been cancelled (Federal Register Aug. 8, 2001; Vol. 66, No. 123)

14–4.3 Downy Mildew

Downy mildew is a common disease of soybean and found wherever soybean is grown in the world. The disease was first reported in Manchuria and reported on soybean in the USA in 1923 (Athow, 1987; Lehman and Wolf, 1924). The downy mildew pathogen and soybean constitute an interesting host-parasite relationship that is rich in genetic variation for both organisms.

14–4.3.1 Pathogen

Downy mildew is caused by *Peronospora manshurica* (Naumov) Syd. in Gäum (Phillips, 1999a). Like many pathogens of soybean, the nomenclature of *P.*

manshurica has evolved with time (Athow, 1987). The pathogen is a biotrophic Oomycete and cannot be cultured on synthetic culture media. Thus, all reproductive structures are observed in association with its host. Intercellular hyphae are observed in host tissue and are coenocytic and 7 to10 μm wide. Haustoria are specialized intracellular structures derived from intercellular hyphae and allow the pathogen to establish a parasitic relationship with a compatible host genotype. Sporangia are 19 to 24 μm in diameter and are produced on dichotomously branched sporangiophores that commonly appear on the underside of leaves in association with lesions. Oospores are embedded in host tissue or are encrusted on the surface of seed. Oospores are light brown, smooth-walled, globose, and 20 to 23 μm in diameter (Phillips, 1999a).

Geeseman (1950) was the first to report variation in virulence among populations of *P. manshurica* and established the concept of physiologic races of the pathogen. Other investigators discovered additional races as new host differentials were identified (Grabe and Dunleavy, 1959). Thirty-three races are described for *P. manshurica* (Dunleavy, 1971; Lim, 1989) and the number of described races will likely increase as new host resistance genes are used to assess genetic variation in the pathogen population (Lim et al., 1984). Dunleavy (1977) and Lim et al. (1984) have reported evidence that *P. manshurica* is capable of rapid genetic change in response to resistance genes that predominant in commercially used cultivars of the day.

14–4.3.2 Symptoms and Losses

Symptoms of downy mildew of soybean change with plant growth and development (Athow, 1987; Phillips, 1999a). Chlorotic spots and blotches of irregular size and shape are early symptoms on the upper surface of leaves (Plate 14–20). The centers of chlorotic lesions eventually become necrotic and progress from dull to dark brown and bordered by chlorotic margins. Some soybean genotypes express mild leaf distortion that may resemble symptoms caused by common soybean viruses. Leaves of highly susceptible soybean cultivars may express extensive brown necrosis and premature defoliation. Downy mildew is recognized by reproductive structures of the pathogen that develop on the underside of the leaf, and in conjunction with necrotic or chlorotic lesions observed on the upper leaf surface. Signs of the pathogen are fluffy tufts of mycelium ranging from white to light purple in color and are most evident during periods of high relative humidity. A microscopic examination of mycelium allows observation of sporangiophores and sporangia of *P. manshurica*. Pods are infected, but symptoms are not obvious. Most obvious, however, are seeds with white-gray patches of encrusted oospores on the seedcoat that may cover the entire seed. *Peronospora manshurica* is capable of systemic infection, primarily in seedlings derived from seedborne inoculum. Chlorotic tissues are observed on both unifoliate and trifoliolate leaves starting at the leaf base and progressing along veins towards the leaf tip. Systemically infected leaves are generally smaller, curl downward, and support massive amounts of sporulation on the underside of leaves (Hildebran and Koch, 1951; Jones and Torrie, 1946). Symptoms of systemic infection are first expressed by unifoliate leaves and generally cease by the V3 stage. Symptoms of systemic infection may be confused with phytotoxicity caused by herbicides, especially triazine class compounds.

Yield loss caused by downy mildew is usually minimal, but occasional epidemics result in yield losses of 9 to 18% (Dunleavy, 1987). Loss estimates differed by soybean cultivar.

14–4.3.3 Disease Cycle and Epidemiology

This obligate biotrophic pathogen survives as oospores in infested crop residue or associated with soybean seed most typically as oospores encrusted on the seed coat, but less commonly as mycelium infecting cells of the seedcoat. The pathogen infects hypocotyls and advances systemically into leaves. Extended periods of leaf wetness are favorable for sporulation and sporangia are disseminated by rain and wind currents. Period of high relative humidity and air temperatures in the range of 20 to 22°C are favorable for infection by *P. manshurica*. Sporulation occurs between 10 to 25°C, but ceases above 30°C or below 10°C. Immature leaves are more susceptible thus, symptoms frequently are observed in the upper regions of the soybean canopy (Phillips, 1999a). The pathogen can spread quickly from centers of primary infection and infect 100% of the plants by the end of the growing season (Lim, 1978, 1985).

14–4.3.4 Management

Oospores in crop residue and oospores on the surface of seed are common forms of overseasoned inoculum that can be managed to reduce or delay infection of soybean early in the growing season. Crop rotation or deep burial of infested crop residue can reduce primary inoculum, but these practices can be negated by seedborne inoculum (Hildebrand and Koch, 1951). Thus, planting pathogen-free seed is an effective practice.

Numerous sources of resistance to *P. manshurica* are present in soybean germplasm, but only two non-allelic genes, *Rpm* and *Rpm2*, have been characterized by inheritance studies (Bernard and Cremeens, 1971; Lim, 1989). The *Rpm* gene was found in the cv. Union, and the *Rpm2* gene traces to PI 88788, a popular source of resistance to the SCN. These resistance genes, and other uncharacterized genes are reported to interact with host differential genotypes to establish 33 races of *P. manshurica* (Dunleavy, 1970, 1977; Dunleavy and Hartwig, 1970; Lim, 1989). Host genotypes with the *Rpm* gene are resistant to all races except race 33 (Lim, 1989). *Peronospora manshurica* must be monitored closely because of its explosive ability to genetically recombine and generate new races that defeat resistant genes. Although downy mildew is currently not considered an important disease, the status of downy mildew must be carefully monitored as the soybean genome is restructured by new sources of soybean germplasm and transgenic technology.

Foliar-applied fungicide products are available for control of downy mildew, but are seldom recommended or economically feasible for grain production (Dunleavy, 1987; Miller and Roy, 1982). However, an application of metalaxyl has been reported to reduce the frequency of oospore-encrusted seed and may be feasible for seed production (Dunleavy, 1987). Fungicide seed treatment products containing metalaxyl or mefenoxam can be used to manage seedborne inoculum. Frequently, products formulated with these active ingredients are applied to seed to control soil-borne pathogens, thus the ability to control *P. manshurica* is an additional benefit.

14–4.4 Frogeye Leaf Spot

Frogeye leafspot was first reported on soybean in the USA in 1924 (Athow, 1987) and occurs throughout the world (Athow, 1987; Phillips, 1999b). The disease is most common in production areas in the southern USA causing chronic problems on an annual basis and has recently become more severe as far north as central Iowa (Yang et al., 2001) and southern Wisconsin (Mengistu et al., 2002). Reasons for this increase may include warmer winter temperatures and susceptible soybean germplasm coupled with conservation tillage practices that retain pathogen-infested plant debris on the soil surface.

14–4.4.1 Pathogen

Frogeye leaf spot is caused by *Cercospora sojina* K. Hara. Early literature reports *C. daizu* to be the cause of frogeye leaf spot, but *C. sojina* is the currently accepted cause (Athow, 1987; Phillips, 1999b). Hyaline conidia are 5 to 7 μm × 39 to 70 μm, have one to eight cells, produced on leaf and stem residue, or seed, and serve as primary and secondary inoculum.

Isolates of *C. sojina* express physiologic specialization for soybean cultivars (Phillips, 1999b; Phillips and Yorinori, 1989). Twelve races of *C. sojina* are reported from several states in the USA, but more are likely present (Athow, 1987; Ross, 1968; Phillips and Boerma, 1981). Race 2 of *C. sojina* was identified in the 1950s (Athow et al., 1962); races 3 and 4 in the 1960s (Ross, 1968); and race 5 was identified in 1978 (Phillips and Boerma, 1981). Physiologic specialization for host cultivars is also reported for 22 races in Brazil (Yorinori, 1992) and 14 races in China (Ma and Li, 1997). However, different sets of differentials were used in these studies so it is difficult to compare reactions. There is a high degree of probability that more of races of the pathogen will be determined if more differential cultivars are used to characterize isolates of *C. sojina*.

14–4.4.2 Symptoms and Losses

Frogeye leaf spot is primarily a foliar disease, but stems, pods, and seeds also are infected. The most common symptoms are foliar lesions that are angular spots up to 5 mm in diameter. Leaf spots typically have light gray centers with a distinct purple to red-brown margins. Chlorotic halos surrounding lesions are characteristically absent. Leaf spots can be single or coalesce to form larger lesions with irregular margins. Numerous lesions result in premature leaf drop. Symptoms on stems generally appear late in the season when foliar disease is severe. Stem lesions are elongate and can obtain a size sufficiently large to surround half the stem circumference. Young lesions have dark margins with red-brown centers that become light gray as they mature. Stem infections may be latent until the plant approaches maturity. Lesions on pods resemble those on stems, are circular to elongate, are initially red-brown changing to gray as they mature, and have ark borders. Seeds in close proximity to pod lesions can be infected and are discolored light to dark gray or brown. Symptomatic areas on seed vary in size and can cover the entire seed coat in severe cases (Bisht and Sinclair 1985; Phillips, 1999b). Infected seed germinates poorly and the resulting seedlings are generally low in vigor (Sherwin and Kreitlow, 1952). Spores formed on cotyledons of infected seedlings are a source of inoculum and infected seeds are a means of distant dissemination of the fungus.

Significant yield losses of 30 to 73% have been reported for susceptible cultivars grown in the southern regions of the USA. A recent study compared yield of susceptible cultivars to yield of resistant near isolines of these cultivars with the *Rcs3* gene which produced 31% more seed yield than their susceptible counterparts (Mian et al., 1998). Reported yield losses were dependent on the cumulative level of disease that occurred during the growing season. Yield loss in northern states is reported as less. Laviolette et al. (1970) reported yield losses of 21% from frogeye leaf spot in Indiana. Similar studies in Illinois showed 10% yield loss (Bisht and Sinclair, 1985). Laviolette et al. (1970), using closely related susceptible and resistant lines, showed that moderate severity of frogeye symptoms resulted in yield losses of 12 to 15%. Severely infected seeds may fail to germinate, and percent germination in seed lots frequently relates inversely to the percentage of discolored seed (Phillips 1999b). Resultant seedlings from infected seed are often weak and stunted.

14–4.4.3 Disease Cycle and Epidemiology

Frogeye leaf spot is endemic throughout the southern USA, where warm, humid conditions prevail. *Cercospora sojina* survives as mycelia in infected seed and soybean debris. Seed infection is common and lesions on cotyledons are a source of primary inoculum for the leaf phase of the disease (Sherwin and Kreitlow, 1952). Conidia production is abundant in warm, humid weather, and secondary infection of leaves, stems, and pods occurs throughout the season (Laviolette et al., 1970). Specific effects of temperature and moisture on disease development are not well understood. With prolonged adequate moisture, infection of new leaves occurs as leaves develop. This results in symptoms throughout the soybean canopy. However, leaves produced during dry periods may remain relatively symptom-free, so that layered patterns of heavily and lightly diseased leaves can exist on the same plant (Phillips, 1999b). The latent period for infection is up to 2 wk, so leaf spots are generally common on fully expanded leaflets and rare on young expanding leaves. However, infection of young leaf tissue occurs more readily than that of older tissue (Phillips, 1999b). The increase in the acidity of irrigation water was shown to reduce severity of frogeye leaf spot (Walker et al., 1994).

14–4.4.4 Management

The most effective management tactic for frogeye leaf spot is use of resistant cultivars. Epidemics of frogeye were common in Indiana in the 1940s. Much of the problem was closely linked to the cv. Hawkeye and other commonly grown cultivars that expressed high susceptible. In contrast, the cv. Lincoln and Wabash expressed a form of complete resistance to *C. sojina* that was inherited as a single dominant gene (Athow and Probst, 1952). Susceptible cultivars were discontinued and frogeye leaf spot became an insignificant disease until 1959, when resistance failed due to the emergence of a new physiologic race of the pathogen, race 2 (Athow et al., 1962). Many of the cultivars resistant to race 1 were susceptible to race 2.

However, some cultivars expressed resistance to race 1 and race 2, and one culti-var expressed resistance to race 2, but was susceptible to race 1 (Athow et al., 1962). Additional races and resistance genes have been reported since the early findings of Athow et al. (1962). Eventually race 5 emerged and has been the focus of breed-ing efforts and genetic studies (Boerma and Phillips, 1984; Phillips and Boerma, 1982; Mian et al., 1999; Pace et al., 1993). Three major dominant genes in the cul-tivars Lincoln (Rcs_1), Wabash, Kent (Rcs_2), and Davis (Rcs_3) confer resistance to specific races of *C. sojina* (Pace et al., 1993). Although progress has been made by phenotypic selection, soybean breeders have several markers to use in marker-as-sisted selection programs for resistance to *C. sojina*. Using simple sequence repeat (SSR), Mian et al. (1999) have mapped *Rcs3* to MLG-J using an isolate of race 5. Additional sources of resistance in cv. Ransom, Stonewall, and Lee express resist-ance to Race 5 and have genes non-allelic to *Rcs3* and to each other (Pace et al., 1993). *Rcs3* is an important allele because it confers resistance to all races of the pathogen from USA, South America, and China (Phillips, 1999b; Phillips and Yorinori, 1989). The soybean cv. Peking is resistant to most isolates of *C. sojina*. Molecular markers linked to the Peking resistance gene were mapped using am-plified fragment length polymorphism (AFLP) and microsatellite techniques in a F_2 population derived from the cross of Peking (resistant) and Lee (susceptible). One AFLP marker was identified and mapped the Peking resistance gene to mo-lecular linkage group MLG-J of the soybean genome. More recently, 660 soybean plant introductions in Maturity Groups VI and VII were identified as resistant to one isolate of an unknown race of *C. sojina* (Yang and Weaver, 2001).

Cultural practices have shown promise for management of frogeye leaf spot. High quality, pathogen-free seed should be used for planting. Deep plowing of crop residues and 2-yr rotations with crops other than soybean can lower inoculum lev-els and reduce symptom severity. Foliar fungicides, benomyl, triphenyltin hy-droxide, and thiabendazole, applied from bloom to early pod set are reported to ef-fectively reduce severity of frogeye leaf spot and improve seed yield as much as 73% (Akem, 1995; Backman et al., 1979; Horn et al., 1975). Benomyl was formerly labeled for management of several soybean pathogens but the registration has since been cancelled (Federal Register August 8, 2001; Vol. 66, Number 123).

14–4.5 Powdery Mildew

Powdery mildew is reported throughout the world (Sinclair, 1999b), but has been most researched in the USA (Grau, 1985). The disease was reported on soy-bean in North Carolina in 1947 by Lehman (1947) and has since become widespread in the USA (Arny et al., 1975; Dunleavy, 1980; Phillips, 1984). Although initially regarded a problem on plants grown in greenhouses (Paxton and Rogers, 1974), pow-dery mildew became common in soybean fields since 1973 (Arny et al., 1975; Dun-leavy, 1980; Leath and Carroll, 1982; Phillips, 1984).

14–4.5.1 Pathogen

Powdery mildew is caused by *Microsphaera diffusa* Cooke & Peck. Al-though initially identified as *Erysiphe polygoni* D.C. or *E. glycines* Tai (Athhow,

1987), *M. diffusa* is the commonly accepted cause of powdery mildew on soybean (Lehman, 1947; McLaughlin et al., 1977; Paxton and Rogers, 1974). Roane and Roane (1976), however, suggest that both *M. diffusa* and *E. polygoni* are found together on soybean leaves. The fungus forms abundant conidia 2.8 to 5.4 μm long and 1.7 to 2.1 μm wide borne in chains on short, simple conidiophores. Conidia of *M. diffusa* are barrel-shaped with flattened ends. Cleistothecia have characteristic appendages dichotomously branched three to five times at the tip. Cleistothecia are hemispherical, light yellow to tan when immature, turning rusty brown and then black at maturity with 4 to 50 appendages. The pyriform asci are borne on short stalks and contain up to six yellow, ovoid ascospores measuring 9 × 18 μm.

Soybean cultivar resistance to *M. diffusa* in field trials was reported as stable against multiple populations of the pathogen (Phillips, 1984). However, one report suggests that isolates of *M. diffusa* and soybean cultivars differentially interact (Grau and Laurence, 1975). Further research is needed to determine the threat of resistance genes being defeated by specific populations of *M. diffusa*.

14–4.5.2 Symptoms and Losses

White, powdery patches composed of mycelium and conidia develop on cotyledons, stems, pods, and particularly on the upper surface of leaves. Small colonies form initially and enlarge and coalesce until the entire surface of infected plant parts are covered with mycelium and conidia. Symptoms are less common than *signs*, reproductive structures of the pathogen. Symptoms on the leaves are green and yellow islands, interveinal necrosis, necrotic specks, and crinkling of the leaf blade followed by defoliation. These symptoms may be almost absent when mycelial growth is abundant. Chlorotic spots and necrosis of veins are phenotypes typically expressed by resistant cultivars challenged in a controlled environment.

Estimates of yield loss have been achieved by comparing yield of plots treated or not treated with a fungicide or comparing yield of resistant and susceptible cultivars in the presence of powdery mildew epidemics. Measured yield losses up to 26%, with an average loss of 13%, were estimated in Iowa studies from 1976 through 1978 by comparing both resistant and susceptible cultivars treated or not treated with a fungicide (Dunleavy, 1978, 1980). Similar data were reported from Georgia (Phillips, 1984). Host near isolines with resistance genes yielded 18% more than susceptible near isolines in Illinois field studies (Lohnes and Nickell, 1994). Powdery mildew is not a common disease and if it does occur, it often develops too late in the season to cause yield loss.

14–4.5.3 Disease Cycle and Epidemiology

It is assumed that *M. diffusa* survives as cleistothecia associated with host residue and ascospores are primary inoculum early in the growing season. Powdery mildew is most common in years with below normal air temperatures, and especially in soybean fields planted later than normal planting dates (Mignucci et al., 1977). *Microsphaera diffusa* caused reduced photosynthesis and transpiration, which conceivably contributes to lower yield in Illinois studies (Mignucci and Boyer, 1979). Other plant species are susceptible to *M. diffusa*, but it is not known whether the form of *M. diffusa* that infects soybean is a host specific form of the pathogen.

Legume hosts of *M. diffusa* other than soybean are common bean, mung bean, pea, cowpea, and other species of *Glycine* (Sinclair, 1999b).

14–4.5.4 Management

Cultural practices such as crop rotation have limited value as tactics to manage powdery mildew. Conida, secondary inoculum of *M. diffusa*, are readily introduced into fields by wind. Most cultural practices, such as row spacing and plant population, are not known to affect the incidence and severity of powdery mildew. One exception is planting date. Frequently, powdery mildew is more severe in late-planted soybean fields.

Soybean germplasm expresses considerable variation in reaction to *M. diffusa* (Arny et al., 1975; Demski and Phillips, 1974; Dunleavy, 1980; Grau and Laurence, 1975). Two resistant phenotypes are expressed by soybean. Complete resistance to *M. diffusa* is expressed in field and controlled environments. Adult-plant resistance is expressed as a total absence of signs and symptoms in the field, but plants challenged in controlled environments are infected and support mycelium growth on unifoliate leaves and two or three trifoliolate leaves before signs of powdery mildew cease. Furthermore, although infection occurs and mycelium is present on the leaf surface, sporulation is absent or minimal (Demski and Phillips, 1974; Grau and Laurence, 1975; Mignucci and Lim, 1980).

Complete resistance and adult-plant resistance are controlled by single dominant alleles at the *Rmd* locus (Buzzell and Haas, 1978; Grau and Laurence, 1975). *Rmd-c* confers complete resistance at all host growth stages, *Rmd* confers adult-plant resistance, and *rmd* conditions susceptibility (Lohnes and Bernard, 1992; Lohnes and Nickell, 1994). Soybean cultivars that trace to Corsoy are highly susceptible to *M. diffusa*.

Fungicides effectively control powdery mildew (Dunleavy, 1980). The application of fungicides may be restricted to soybean fields for seed production or other value-added traits that commend a higher market value for grain. Powdery mildew is frequently a problem in greenhouses. Elemental S can be placed on a heat source and the resulting volatilized S will provide continuous control and replaces the need for frequent applications of an organic fungicide.

14–4.6 Rhizoctonia Foliar Blight

Rhizoctonia foliar blight, also known as aerial blight or web blight, is characterized by premature defoliation and pod abortion of soybean. First reported from Louisiana in 1951, this disease was epidemic by 1973 (Atkins and Lewis, 1954; O'Neill et al., 1977) and now affects a large portion of the soybean hectarage in Louisiana and other states along the Gulf of Mexico and Mississippi River delta (Russin and Stetina, 1999). Rhizoctonia foliar blight is a problem of increasing importance in several other countries, particularly Brazil and Argentina (Hepperly et al., 1982).

14–4.6.1 Pathogen

Anastomosis group (AG) 1 of *Rhizoctonia solani* causes Rhizoctonia foliar blight. Two distinct types of *R. solani* AG-1 are associated with this disease; intraspecific group A (IA), which causes aerial blight, and intraspecific group B (IB), which causes web blight (Yang et al., 1990a). Large sclerotia (2–4 mm in diameter) are produced by IA isolates, whereas microsclerotia produced by IB isolates resemble sand grains. The IA isolates of *R. solani* AG1 also causes sheath blight in rice (Rush and Lee, 1992), and rice-soybean double cropping may contribute to increased disease incidence on both crops. Black et al. (1996a) reported 14 weed species common to Louisiana soybean fields were hosts for both types of *R. solani* AG-1. These included species of broadleaf and grass weeds as well as sedges.

14–4.6.2 Symptoms and Losses

Both types of *R. solani* AG-1 can infect seeds and seedlings (Yang et al., 1990b). They cause pre-emergence seed decay along with post-emergence lesions on hypocotyls and shoot apices. Lesions on hypocotyls are initially red but later turn brown and dry. Mycelial webbing along with sclerotia or microsclerotia occur on diseased seedlings. These symptoms and signs differ from those described previously for early season damping-off caused by other intraspecific groups of *R. solani* (Yang, 1999b). Foliar blight is expressed as water-soaked lesions on roots and stems and can cause up to 50% reduction in soybean stand.

Foliar symptoms occur throughout the soybean canopy later in the season. Lesions first appear water soaked and grayish green, with mycelium spreading over the leaf surface in advance of the lesion (O'Neill et al., 1977). Mycelium spreads by contact between leaves, which results in mats of webbed, diseased foliage. Defoliation soon follows. Diseased leaves frequently fall onto lower leaves, stems, and pods, which then can be colonized by the pathogen. Yang et al. (1990a) described differences in symptoms caused by isolates of *R. solani* IA and IB. In addition to production of larger sclerotia, infections caused by isolates of *R. solani* IA isolates are typically characterized by extensive blighting of foliage. These symptoms are generally restricted to within the soybean crop canopy. In contrast, infections from IB isolates of *R. solani* are frequently associated with distinct leaf spots. These can result in a shot-hole appearance when weather becomes unfavorable for disease development. Microsclerotia are common on such plants. Leaf spot symptoms occur on more exposed portions of the canopy, which may suggest a role for airborne microsclerotia in disease development. Both IA and IB isolates of *R. solani* can occur in the same field and even on the same plant. Therefore, symptoms tend to be a combination of both types and are frequently difficult to distinguish. Disease severity assessments do not distinguish between these two types but rather combine all symptoms together (Harville et al., 1996). Severely infected leaves collapse, with the petioles remaining attached to the stems. Petioles of infected leaves remain green after the leaflets become necrotic, and dark brown sclerotia may be observed on necrotic petiole and leaf tissues.

Other plant parts also can show symptoms (Russin and Stetina, 1999). Reddish-brown to brown lesions can form on stems and pods after canopy closure. Pod symptoms are observed as irregularly shaped, tan, sunken lesions that develop where dead, blighted leaf tissues come in contact with pods. Severe infection can cause blighting and death of young pods. Seed infection can result from pod infection and infected seeds have a tan discoloration, with mycelium growing on the seed surface.

The pod-blighting phase is believed to be responsible for most yield loss caused by Rhizoctonia foliar blight. Soybean yield loss in Louisiana was estimated at 3% in 1992 (Wrather and Sciumbato, 1995), but losses up to 70% have been reported in localized areas within fields (Russin and Stetina, 1999). However, controlled studies to accurately measure yield loss in soybean are lacking.

14–4.6.3 Disease Cycle and Epidemiology

There are two distinct phases in the epidemiology of Rhizoctonia foliar blight, occurring before and after canopy closure (Yang et al., 1990e). Early in the season sclerotia or microsclerotia in soil and diseased plant debris germinate and infect seedlings directly or are splashed onto stems and foliage during rainfall. Infected seedlings become sources of inoculum and have a significant impact on subsequent disease development (Yang et al., 1990b). Under very wet conditions, mycelia also can reach foliage by growing directly up the outside of stems. Following canopy closure, the pathogens can spread extensively along and across rows. It is at this time that disease foci expand most rapidly (Yang et al., 1990e) and production of sclerotia or microsclerotia is abundant (Yang et al., 1990a). Diseased soybean debris supporting mycelia, sclerotia, or microsclerotia eventually falls to the soil surface. Because it has both soilborne and foliage components, Rhizoctonia foliar blight is a highly clustered disease (Yang et al., 1991).

Black et al. (1996a) showed that *R. solani* AG-1 IA and IB can spread from inoculated common weed plants to adjacent healthy soybean plants, thereby using weeds as "bridge hosts". Sclerotia and microsclerotia occurred on the weeds, suggesting that infected weeds are important for overwintering of inoculum and survival of the pathogens in the absence of soybean.

The primary means of pathogen spread in the soybean canopy is through water splashing of inoculum onto foliage or by growth of mycelia on overlapping leaves. Therefore, disease development is favored by high temperatures and extended periods of high humidity, rainfall, and cloudy weather (Atkins and Lewis, 1954; O'Neill et al., 1977; Yang et al., 1990e). Free moisture on plant surfaces is very important for disease development (Yang et al., 1990c).

Several post-emergence herbicides influenced the severity of Rhizoctonia foliar blight (Black et al., 1996b). Notable among these was paraquat, which reduced disease severity under conditions of both low and high disease pressure.

14–4.6.4 Management

Because Rhizoctonia foliar blight can affect seeds and seedlings, seed treatment may be effective to control this phase of the disease. Unfortunately, no controlled studies have yet been done to demonstrate this. Limited control in experimental plots has been achieved using high rates of several fungicides. Host resistance can be effective for disease control. Harville et al. (1996) screened 91 soybean cultivars and advanced breeding genotypes for reaction to Rhizoctonia foliar blight. Although most cultivars are very susceptible, several expressed low to moderate disease severity. In a field study, Joye et al. (1990) reported reduced disease severity when row spacing was 50 cm compared to narrower row spacings. However, disease severity was not affected by within-row plant populations of 10 to 39 plants per meter of row. Although environmental conditions were not measured, wider row spacing presumably allowed greater air circulation within canopies, which reduced disease development.

14–4.7 Rust

Soybean rust is a devastating disease found in tropical and subtropical regions in the Eastern and Western Hemispheres. In the Eastern Hemisphere, soybean production in Asia, Australia, Japan, Taiwan, and the Philippines is affected by this disease. Soybean rust has been restricted to Africa, Brazil, Paraguay, Central America, and the Caribbean in the Western Hemisphere (Sinclair and Hartman, 1996). Soybean rust was detected in Puerto Rico in 1976 (Vakili and Bromfield, 1976) and in Hawaii in 1987 (Killgore et al., 1994). Although soybean rust has not been detected on the USA mainland, it is a potential threat to soybean production in the USA (Bromfield, 1976, 1984; Sinclair, 1989).

14–4.7.1 Pathogen

Phakospora pachyrhizi Sydow and *Phakospora meibomiae* (Arthur) Arthur are the casual agents of rust, however *P. pachyrhizi* is the most aggressive and predominant species. *Phakospora pachyrhizi* has an extensive host range of more than 75 plant species, including soybean and other legumes (Rytter et al., 1984; Sinclair, 1982; Sinclair and Hartman, 1999b). *Phakospora meibomiae* also has a large host range with 66 potential hosts including soybean (Sinclair and Hartman, 1999b).

14–4.7.2 Symptoms and Losses

Lesions are primarily on the leaves, however they may occur on petioles and stems. Symptoms begin as water-soaked lesions, which increase in size and become chlorotic. The angular lesions quickly expand to a size of 2 to 3 mm and are restricted by the leaf veins. Depending on factors such as the age of the lesion, pathogen virulence, and host genotype, the color of the lesion may be grayish brown, tan to dark brown, or reddish brown (Sinclair and Hartman, 1999b). Erumpent, globuse uredia are found primarily on the undersides of leaves and increase in number as the lesion ages. Teliospores are produced although germination has not been observed (Bromfield, 1976)

Yield reduction results from fewer pods, pods with less seed, and reduced seed weight (Melching et al., 1989). Yield losses can be dramatic and range from 13 to 80% (Sinclair and Hartman, 1996). In severe cases, yield losses have been reported as high as 40% in Japan (Bromfield, 1984) and as high as 80% in Taiwan (Yang et al., 1990d).

14–4.7.3 Disease Cycle and Epidemiology

The overwintering source(s) of primary inoculum for the pathogens is unknown. Epidemics of rust occur when soybean leaves are infected early in the season, however leaves are susceptible at all stages of plant growth (Sinclair and Hartman, 1999b). Free moisture and moderate temperatures are required for germination of urediniospores and host penetration (Melching et al., 1989). Urediniospores directly penetrate the host epidermis and produce hyphae that grow intercellulary and colonize the mesophyll cells. In most cases, uredia (sporocarps) can be observed producing urediniospores 9 to 10 d following penetration. The secondary inoculum quickly moves from the lower to upper leaves and to adjacent plants with dispersal over greater distances facilitated by windblown rain and air currents (Sinclair and Hartman, 1996).

Epidemics of soybean rust are greatly influenced by environmental conditions. Weather conditions at spore dispersal, deposition, and germination greatly influence the success rate of lesion development. Germination of urediniospores and subsequent host penetration and lesion development require 6 to 7 h of continual wetness with an optimum of a 12-h dew period with temperatures of 18 to 26.5°C (Melching et al., 1989). Spores can remain viable in the absence of moisture for durations of no more than 8 d (Kitani and Inoue, 1960; Melching et al., 1989). Spore viability is also influenced by ultraviolet light. Exposure to sunlight reduces viability of spores compared to spores exposed to cloudy conditions (Melching et al., 1989).

14–4.7.4 Control

Phakospora pachyrhizi is a variable pathogen, with multiple races that differ in virulence on a range of soybean genotypes (Burdon and Speer, 1984). Four dominant resistance genes have been identified in PI 200692 (*Rpp1*), PI 230970 (*Rpp2*), PI 462312 (*Rpp3*), and PI 459025 (*Rpp4*) (Sinclair and Hartman, 1996). Resistance genes that have yet to be identified are contained in other plant introductions and among wild *Glycine* spp. (Sinclair and Hartman, 1999b). The threat of soybean rust invading the USA mainland, has initiated research in the USA to identify resistant sources. Resistance has been identified in accessions of perennial *Glycine* spp. (Hartman et al., 1991), and efforts are being made to transfer this resistance to *G. max* (Sinclair and Hartman, 1996). In addition, germplasm lines with partial resistance to rust have been developed (Kilen, 1997).

Foliar-applied fungicides can reduce the severity of rust epidemics; however, frequent applications of the fungicide are required (Sinclair and Hartman, 1996). Under severe disease pressure and when losses exceed 10 to 15%, chemical control would be economically justified. There is little information on cultural practices that may reduce disease severity, and clearly this may be an area of future research.

14–4.8 Target Spot

Target spot was first reported in the USA in 1945 (Athow, 1987) and was believed to be limited to the southern USA. However, target spot has since been reported in northern regions of North America (Raffel et al., 1999; Sinclair, 1999c) and now has a world-wide distribution. Although the common name of target spot implies a leaf disease, the target spot pathogen has been found to infect other plant parts. Evidence is mounting that multiple species may be involved with this disease.

14–4.8.1 Pathogen

Target spot is caused by the fungus *Corynespora cassiicola* (Ber. & M.A. Curtis) C.T. Wei. The fungus has a wide host range (Jones, 1961; Spencer and Walters, 1969). Isolates of *C. cassiicola* have also been characterized into distinct groups based on host range and plant parts infected. Spencer and Walters (1969) differentiated isolates of the pathogen into those that infected cowpea and those that infected soybean. They also found isolates pathogenic to soybean to differentiate into two pathogenic types, those that infect hypocotyls, roots, and stems, and those that infect leaves, pods, and seeds. Isolates of each group differ morphologically from the other and are likely different species. There is not an indication that one group is more specific than the other to soybean.

The fungus is reported to produce chlamydospores in older cultures (Sinclair, 1999c). Isolates produce conidia ranging from 7.7 to 13.0 μm in width and 66 to 326 μm in length (Spencer and Walters, 1969). Spencer and Walters (1969) further report that conidia of isolates from southern states are smaller compared to isolates from the northern USA and Canada. These morphological differences may imply two species of *Corynespora* are infecting soybean in the USA.

14–4.8.2 Symptoms and Losses

Target spot was originally reported as a disease of leaves, but symptoms also develop on petioles, stems, pods, and seeds. Isolates are also reported to cause infection of hypocotyls, roots, and stems. Leaf lesions are somewhat unique from those caused by other leaf-infecting fungi by being larger and reddish brown. Circular to irregular lesions start as specks, but progress to 10 to 15 mm in diameter as lesions mature and frequently concentric rings are readily observed within the lesion, thus the name of "target spot". Lesions are frequently bordered by dull-green or yellowish green halos. When severe, target spot may result in premature defoliation following lesion development.

Isolates of *C. cassiicola* that cause leaf lesions also cause spots on petioles, stems, and pods. Stem lesions are dark brown and range from specks to elongated spindle-shaped lesions. Pod lesions are generally circular, about 2 mm in diameter, with slightly depressed, purple-black centers and brown margins. The pathogen colonizes pods and may penetrate the pod wall and infect seed causing small, blackish brown spots.

Corynespora cassiicola has been reported as a cause of a root and stem rot of soybean in Nebraska (Boosalis and Hamilton, 1957), Canada, (Seaman et al., 1965), and more recently in Wisconsin (Raffel et al., 1999). Symptoms are observed at seedling emergence, late vegetative, and mid-reproductive growth stages. Reddish to dark brown longitudinal lesions develop on hypocotyls, upper taproots, and lower stems. Lateral roots may develop extensive necrosis. Isolates that infect below ground plant tissue are reported not to infect leaves, stems, and pods (Seaman et al., 1965; Spencer and Walters, 1969).

Yield loss associated with the leaf phase of target spot is reported to range from 18 to 32% in the Delta area of Mississippi (Hartwig, 1959). Yield loss estimates are not available for the root and stem phase of the disease.

14–4.8.3 Disease Cycle and Epidemiology

The long-term survival of *C. cassiicola* in soil is likely attributed to chlamydospores embedded in host residue, a wide host range, and the frequent colonization of cysts of the SCN (Carris et al., 1986; Seaman et al., 1965; Sinclair, 1999c). Leaf infection is favored by high relative humidity. Temperature for growth of leaf infecting isolates is reported to be optimal at 28°C and growth ceases at 40°C. Roots and lower stems are infected during seedling emergence (Raffel et al., 1999; Seaman et al., 1965) and favored by temperatures of 15 to 20°C (Seaman et al., 1965). Although *C. cassiicola* is closely associated with the SCN, it is not known whether these root-infecting pathogens interact (Carris et al., 1986).

14–4.8.4 Management

Athow (1987) reports most soybean cultivars adapted to the southern USA express resistance and thus, target spot is not a problem. The root and lower stem phase of the disease in the northern USA and Canada is common and presumably not controlled by common rotation sequences or other management practices. Improved seedling health and yield have been achieved with seed-applied fungicides such as thiabendazole, carboxin, and fludioxynil. It is speculated that control of *C. cassiicola* may be a partial explanation for increased yields associated with applications of these fungicides. Data is not available on reaction of northern soybean germplasm and cultivars to *C. cassiicola*.

14–5 FUNGI ASSOCIATED WITH OTHER DISEASES

Numerous fungi are pathogens of soybean, but are not considered to cause major yield loss, or are of localized importance. However, the history of soybean is full of examples of relatively benign pathogens becoming important because of climatic changes, or human redirection of cultural practices or genetic manipulation of soybean. Presented below are fungi that may be of interest to some readers. The text is a slight modification of a summary previously prepared (Athow, 1987). The third (Sinclair and Backman, 1989) and fourth editions of the *Compendium of Soybean Diseases* (Hartman et al., 1999), and the second edition of *Field Crop Diseases Handbook* (Nyvall, 1989) are excellent resources to obtain further information.

Several species of *Alternaria* cause Alternaria leaf spot, which is uniquely characterized by concentric rings within the lesion. *Alternaria alternata* and *A. tenuissima* are reported to infect seed, especially if bean leaf beetles or stinkbugs have wounded pods. *Alternaria atrans* is a pathogen of leaves, but again is more successful if leaf tissues are mechanically damaged.

Botrytis cinerea is commonly isolated from soybean seeds in the upper regions of the USA. The pathogen has a wide host range and a life cycle similar to *S. sclerotiorum*.

Calonectria pyrochroa causes red crown rot or black root rot of soybean. The pathogen is an important pathogen of peanut, thus soybean grown in rotation with peanut are at greatest risk of developing this disease. *Calonectria pyrochroa* is common in the southern USA and Brazil. Leaf symptoms are similar to symptoms of southern stem canker. However, symptoms of red crown rot is characterized by internal stem tissues that are discolored gray-brown, and red-orange perithecia that form on stems close to the soil line after plant mortality.

Choanephora infundibulifera causes Choanephora leaf blight in Eastern Asia, especially Thailand. This pathogen infects the distal half of leaves, followed by extensive necrosis and defoliation.

Leptosphaerulina trifolii causes Leptosphaerulina leaf spot, and is also common on forage legumes. The pathogen has been reported in India and USA (state of Maryland).

Nematospora coryli is the cause of yeast spot, or sometimes called Nematospora spot. The disease occurs in Africa, Brazil, and the southern USA. The pathogen is not only transmitted by several species of stinkbugs, primarily by *Acrosternum hilare*, but survives in stinkbugs. Stinkbugs actively feed on pods and transmit the fungus to seed. Symptoms range from small and shriveled seed, to absence of seed in injured pods.

Neocosmospora vasinfecta causes Neocosmospora stem rot of soybean. Symptoms of the disease may be confused with those associated with brown stem rot. However, confusion is unlikely because Neocosmospora stem rot occurs at high air temperatures in Japan, Nigeria, and the southeastern USA.

Phyllosticta sojicola causes Phyllosticta leaf spot and is common worldwide and favored by cool, wet weather. The pathogen survives in seed and host residue as pycnidia and mycelium.

Phymatotrichum omnivorum causes Phymatotrichum root rot, sometimes called Texas root rot, in the southwestern USA and northern Mexico. The pathogen has a wide host-range and forms long-lived sclerotia. *Phymatotrichum omnivorum* causes root rot that results in secondary symptoms of wilt and death of plants in patches.

Pyrenochaeta glycines (syn. *Dactuliochaeta glycines*) is reported to cause red leaf blotch and significant yield losses in Africa. Red leaf blotch is a recently described soybean disease. The pathogen forms sclerotia, which serve as water-splashed inoculum.

Sphaceloma glycines is the cause of scab of soybean in Japan. Symptoms first appear on leaves, but pods are eventually infected and pods with numerous lesions are void of seed.

Stemphylium botryosum causes Stemphylium leaf blight and occurs in India. The pathogen has a wide host range.

Thielaviopsis basicola causes Thielaviopsis root rot of soybean also named black rot. The pathogen has a wide host range, thus high incidence of black rot in preceding crops affects the occurrence in soybean.

Verticillium dahliae, V. albo-atrum, and *V. nigrescens* are reported as pathogens of soybean, but have not reached importance on soybean compared to their pathogenic activity on other crops. *Verticillium dahliae* is a severe and common pathogen of velvetleaf, a common weed in soybean fields. However, isolates of *V. dahliae* from velvetleaf are reported to cause symptoms on soybean only in controlled environments and not in the field even though the fungus is readily isolated from soybean stems.

REFERENCES

Abel, G.H. 1977. Brown stem rot of soybean-*Cephalosporium gregatum*. Rev. Plant Pathol. 56:1065–1075.

Abney, T.S. J.C. Melgar, T.L. Richards, D.H. Scott, J. Grogan, and J. Young. 1997. New races of *Phytophthora sojae* with *Rps*1-d virulence. Plant Dis. 81:653–655.

Abney, T.S., T.L. Richards, and K.W. Roy. 1975. Effect of benomyl foliar applications on purple stain of soybeans. Proc. Am. Phytopathol. Soc. 2:82.

Achenbach, L.A., J. Patrick, and L. Gray. 1996. Use of RAPD markers as a diagnostic tool for the identification of *Fusarium solani* isolates that cause soybean sudden death syndrome. Plant Dis. 80:1228–1232.

Achenbach, L.A., J.A. Patrick, and L.E. Gray. 1997. Genetic homogeneity among isolates of *Fusarium solani* that cause soybean sudden death syndrome. Theor. Appl. Genet. 95:474–478.

Adee, E. A., C.R. Grau, and E.S. Oplinger. 1995. Inoculum density of *Phialophora gregata* related to severity of brown stem rot and yield of soybean in microplot studies. Plant Dis. 79:68–73.

Adee, E. A., C.R. Grau, and E.S. Oplinger. 1997. Population dynamics of *Phialophora gregata* in soybean residue. Plant Dis. 81:199–203.

Adee, E.A., E.S. Oplinger, and C.R. Grau. 1994. Tillage, rotation sequence, and cultivar influences on brown stem rot and soybean yield. J. Prod. Agric. 7:341–347.

Agarwal, S.C., and S.R. Kotasthane. 1971. Resistance in some soybean varieties against *Sclerotium rolfsii* Sacc. Indian Phytopathol. 24:401–403.

Akem, C.N. 1995. The effect of timing of fungicide applications on control of frogeye leafspot and grain yield of soybeans. Eur. J. Plant Pathol. 10:183–187.

Allington, W. B., and D.W. Chamberlain. 1948. Brown stem rot of soybean. Phytopathology 38:793–802.

Ammon, V.D., T.D. Wyllie, and M.F. Brown. 1974. An ultrastructural investigation of pathological alterations induced by *Macrophomina phaseolina* Tassi (Goid.) in seedlings of soybean *Glycine max* (L.) Merrill. Physiol. Plant Pathol. 4:1–4.

Ammon, V.D., T.D. Wyllie, and M.F. Brown. 1975. Investigation of the infection process of *Macrophomina phaseolina* on the surface of soybean roots using scanning electron microscopy. Mycopathologica 55:77–81.

Anahosur, K.H., S.H. Patil, and R.K. Hegde. 1984. Effect of herbicides on *Macrophomina phaseolina* (Tassi) Goid. causing charcoal rot of sorghum. Pesticides 18:11–12.

Anderson, T.R. 1986. Plant losses and yield responses to monoculture of soybean cultivars susceptible, tolerant and resistant to *Phytophthora megasperma* f. sp. *glycinea*. Plant Dis. 70:468–471.

Anderson, T.R. and R.I. Buzzell. 1982. Efficacy of metalaxyl in controlling Phytophthora root and stalk rot of soybean cultivars differing in field tolerance. Plant Dis. 66:1144–1145.

Anderson, T.R. and R.I. Buzzell. 1992. Inheritance and linkage of the *Rps*7 gene for resistance to Phytophthora rot of soybean. Plant Dis. 76:958–959.

Anderson, T.R., R.I. Buzzell, and B.R. Buttery. 1995. Incidence of pod and seed infection in two soybean lines differing in resistance to Phomopsis seed decay. Can. J. Plant Sci. 75:543–545.

Arahana, V.S., G.L. Greaef, J.E. Specht, J.R. Steadman, and K.M. Eskridge. 2001. Identification of QTLs for resistance to *Sclerotinia sclerotiorum* in soybean. Crop Sci. 41:180–188.

Armstrong, G.M., and J.K. Armstrong. 1950. Biological races of the Fusarium causing wilt of cowpeas and soybeans. Phytopathology 40:181–193.

Armstrong, J.K., and G.M. Armstrong. 1958. A race of the cotton wilt Fusarium causing wilt of Yelredo soybean and flue-cured tobacco. Plant Dis. Rep. 42:147–151.

Arny, D.C., E.W. Hanson, G.L. Worf, E.S. Oplinger, and W.H. Hughes. 1975. Powdery mildew on soybean in Wisconsin. Plant Dis. Rep. 59:288–290.

Athow, K.L. 1987. Fungal diseases. p. 687–727. *In* J.R. Wilcox (ed) Soybeans: Improvement, production, and uses. 2nd ed. Agron. Monogr. 16. ASA, CSSA, and SSSA, Madison, WI.

Athow, K.L., and F.A. Laviolette. 1973. Pod protection effects on soybean seed germination and infection with *Diaporthe phaseolorum* var. *sojae* and other microorganisms. Phytopathology 63:1021–1023.

Athow, K.L., and F.A. Laviolette. 1982. *Rps*6, a major gene for resistance to *Phytophthora megapserma* f. sp. *glycinea* in soybean. Phytopathology 72:1564–1567.

Athow, K.L., F.A. Laviolette, E.H. Mueller, and J.R. Wilcox. 1980. A new major gene for resistance to *Phytophthora megasperma* var. *sojae* in soybean. Phytopathology 70:977–980.

Athow, K.L., and A.H. Probst. 1952. The inheritance of resistance to frog-eye leaf spot of soybean. Phytopathology 42:660–662.

Athow, K.L., A.H. Probst, C.P. Kurtzman, and F.A. Laviolette. 1962. A newly identified physiological race of *Cercospora sojina* on soybean. Phytopathology 52:712–714.

Atkins, J.G., Jr., and W.D. Lewis. 1954. Rhizoctonia aerial blight of soybeans in Louisiana. Phytopathology 44:215–218.

Ayanru, D.K.G., and R.J. Green, Jr. 1974. Alteration of germination patterns of sclerotia of *Macrophomina phaseolina* on soil surfaces. Phytopathology 64:595–601.

Aycock, R. 1966. Stem rot and other diseases caused by *Sclerotium rolfsii*. Tech. Bull. 174. North Carolina Agric. Exp. Stn., Raleigh.

Bachman, M. S., and C.D. Nickell. 1999. Use of reciprocal grafting to study brown stem rot resistance in soybean. Phytopathology 89:59–63.

Bachman, M. S., and C.D. Nickell. 2000a. Investigating the genetic model for brown stem rot resistance in soybean. J. Hered. 91:316–321.

Bachman, M.S., and C.D. Nickell. 2000b. High frequency of brown stem rot resistance in soybean germ plasm from Central and Southern China. Plant Dis. 84:694–699.

Bachman, M.S., C.D. Nickell, P.A. Stephens, A.D. Nickell, and L.E. Gray. 1997a. The effect of *Rbs2* on yield of soybean. Crop Sci. 37:1148–1151.

Bachman, M. S., C.D. Nickell, P.A. Stephens, and A.D. Nickell. 1997b. Brown stem rot resistance in soybean germplasm from central China. Plant Dis. 81.953–956.

Bachman, M.S., J.P. Tamulonis, C.D. Nickell, and A.F. Bent. 2001. Molecular markers linked to brown stem rot resistance genes, Rbs1 and Rbs2, in soybean. Crop Sci. 41:527–535.

Backman, P.A., R. Rodríquez-Kabana, J.M. Hammond, and D.L. Thurlow. 1979. Cultivar, environment, and fungicide effects on foliar disease losses in soybeans. Phytopathology 69:562–564.

Backman, P.A., D.B. Weaver, and G. Morgan-Jones. 1985. Etiology, epidemiology, and control of stem canker. p. 589–597. *In* R. Shibles (ed.) World Soybean Res. Conf. III: Proc., Ames, IA. 12–17 Aug. 1984.Westview Press, Boulder, CO.

Backman, P.A., J.C. Williams, and M.A. Crawford. 1982. Yield losses in soybeans from anthracnose caused by *Colletotrichum truncatum*. Plant Dis. 66:1032–1034.

Baird, R.E., D.E. Carling, and B.G. Mullinix. 1996. Characterization and comparison of isolates of *Rhizoctonia solani* AG-7 from Arkansas, Indiana, and Japan, and select AG-4 isolates. Plant Dis. 80:1421–1424.

Baker, D.M., H.C. Minor, M.F. Brown, and E.A. Brown. 1987. Infection of immature soybean pods and seed by *Phomopsis longicolla*. Can. J. Microbiol. 33:797–801.

Balducchi, A.J., and D.C. McGee. 1987. Environmental factors influencing infections of soybean seeds by *Phomopsis* and *Diaporthe* species during seed maturation. Plant Dis. 71:209–212.

Bardin, S.D., and H.C. Huang. 2001. Research on biology and control of *Sclerotinia* diseases in Canada. Can. J. Plant Pathol. 23:88–98.

Beagle-Ristaino, J.E., and J.F. Rissler. 1983. Histopathology of susceptible and resistant soybean roots inoculated with zoospores of *Phytophthora megasperma* f. sp. *glycinea*. Phytopathology 73:590–595.

Bell, D.K., and D.R. Sumner. 1987. Survival of *Rhizoctonia solani* and other soilborne basidiomycetes in fallow soil. Plant Dis. 71:911–915.

Bernard, R.L., and C.R. Cremeens. 1971. A gene for general resistance to downy mildew of soybeans. J. Hered. 62:359–362.

Bernard, R.L., and C.R. Cremeens. 1981. An allele at the rps_1 locus from the variety 'Kingwa'. Soybean Genet. Newslett. 8:40–42.

Bernard, R.L., P.E. Smith, M.J. Kaufmann, and A.F. Schmitthenner. 1957. Inheritance of resistance to Phytophthora root and stem rot in soybean. Agron. J. 49:391.

Beute, M.K., and R. Rodríguez-Kabana. 1979. Effect of volatile compounds from remoistened plant tissues on growth and germination of sclerotia of *Sclerotium rolfsii*. Phytopathology 69:802–805.

Beute, M.K., and R. Rodríguez-Kabana. 1981. Effects of soil moisture, temperature, and field environment on survival of *Sclerotium rolfsii* in Alabama and North Carolina. Phytopathology 71:1293–1296.

Bisht, V.S., and J.B. Sinclair. 1985. Effect of *Cercospora sojina* and *Phomopsis sojae* alone or in combination on seed quality and yield of soybeans. Plant Dis. 69:436–439.

Black, B.D., J.L. Griffin, J.S. Russin, and J.P. Snow. 1996a. Weeds as hosts for *Rhizoctonia solani* AG-1, causal agent for Rhizoctonia foliar blight of soybean (*Glycine max*). Weed Technol. 10:865–869.

Black, B.D., J.S. Russin, J.L. Griffin, and J.P. Snow. 1996b. Herbicide effects on *Rhizoctonia solani* AG-1 IA and IB in vitro and on Rhizoctonia foliar blight of soybean (*Glycine max*). Weed Sci. 44:711–716.

Boerma, H.R., and D.V. Phillips. 1984. Genetic implications of the susceptibility of Kent soybean to *Cercospora sojia*. Phytopathology 74:1666–1669.

Boland, G.J., and R. Hall. 1986. Growth room evaluation of soybean cultivars for resistance to *Sclerotinia sclerotiorum*. Can. J. Plant Sci. 66:559–564.

Boland, G.J., and R. Hall. 1987. Evaluating soybean cultivars for resistance to *Sclerotinia sclerotiorum* under field conditions. Plant Dis. 71:934–936.

Boland, G.J. and R. Hall. 1988a. Relationships between the spatial pattern and number of apothecia of *Sclerotinia sclerotiorum* and stem rot of soybean. Plant Pathol. 37:329–336.

Boland, G.J. and R. Hall. 1988b. Epidemiology of Sclerotinia stem rot of soybean in Ontario. Phytopathology 78:1241–1245.

Boland, G.J. and R. Hall. 1994. Index of plant hosts of *Sclerotinia sclerotiorum*. Can. J. Plant Pathol. 16:93–100.

Boosalis, M.G., and R.I. Hamilton. 1957. Root and stem rot of soybean caused by *Corynespora cassiicola* (Berk. & Curt.) Wei. Plant Dis. Rep. 41:696–698.

Bowen, C.R., and W.T Schapaugh. 1989. Relationships among charcoal rot infection, yield, and stability estimates in soybean blends. Crop Sci. 29:42–46.

Bowers, G.R., Jr. 1984. Resistance to anthracnose. Soybean Genet. Newsl. 11:150–151.

Bowers, G.R., and J.S. Russin. 1999. Soybean disease management. p. 231–271. *In* L.G. Heatherly and H.F. Hodges (ed.) Soybean production in the Midsouth. CRS Press, Boca Raton, FL.

Bradley, C.A., G.L. Hartman, R.L. Nelson, D.S. Mueller, and W.L. Pedersen. 2001. Response of ancestral soybean lines and commercial cultivars to Rhizoctonia root and hypocotyl rot. Plant Dis. 85:1091–1095.

Bromfield, K.R. 1976. World soybean rust situation. p. 491–500-26. *In* L.D. Hui (ed.) World Soybean Res. Proc., World Soybean Res. Conf., Champaign, IL. Interstate Printers and Publ., Danville, IL.

Bromfield, K.R. 1984. Soybean rust. Monogr. 11. APS Press, St. Paul, MN.

Brown, E.A., H.C. Minor, and O.H. Calvert. 1987. A soybean genotype resistant to Phomopsis seed decay. Crop Sci. 27:895–898.

Burdon, J.J., and S.S. Speer. 1984. A set of differential hosts for the identification of pathotypes of *Phakopsora pachyrhizi* Syd. Euphytica 33:891–896.

Burnham, K.D., A.E. Dorrance, D.M. Francis, R.J. Fioritto, and S.K. St. Martin. 2003. *Rps*8, a new locus in soybean for resistance to *Phytophthora sojae*. Crop Sci. 43:101–105.

Buzzell, R.I., and T.R. Anderson. 1982. Plant loss response of soybean cultivars to *Phytophthora megasperma* f.sp. *glycinea* under field conditions. Plant Dis. 66:1146–1148.

Buzzell, R.I., and T.R. Anderson. 1992. Inheritance and race reaction of a new soybean *Rps*1 allele. Plant Dis. 76:600–601.

Buzzell, R.I., and J.H. Haas. 1978. Inheritance of adult plant resistance to powdery mildew in soybeans. Can. J. Genet. Cytol. 20:151–153.

Buzzell, R.I., T.W. Welacky, and T.R. Anderson 1993. Soybean cultivar reaction and row width effect on Sclerotinia stem rot. Can. J. Plant. Sci. 73:1169–1175.

Canaday, C.H., D.G. Helsel, and T.D. Wyllie. 1986. Effects of herbicide-induced stress on root colonization of soybeans by *Macrophomina phaseolina*. Plant. Dis. 70:863–866.

Carling, D.E., and S. Kuninaga. 1990. DNA base sequence homology in *Rhizoctonia solani* Kühn: Inter- and intragroup relatedness of anastomosis group-9. Phytopathology 80:1362–1364.

Carling, D.E., C.S. Rothrock, G.C. Macnish, M.W. Sweetingham, K.A. Brainard, and S.W. Winters. 1994. Characterization of anastomosis group 11 (AG-11) of *Rhizoctonia solani*. Phytopathology 84:1387–1393.

Carris, L.M., D.A. Glawe, and L.E. Gray. 1986. Isolation of the soybean pathogens *Cornyespora cassiicola* and *Phialophora gregata* from cysts of *Heterodera glycines* in Illinois. Mycologia 78:503–506.

Carson, M.L., W.E. Arnold, and P.E. Todt. 1991. Predisposition of soybean seedlings to Fusarium root rot with trifluralin. Plant Dis. 75:342–347.

Casale, W.L., and L.P. Hart. 1986. Influence of four herbicides on carpogenic germination and apothecium development of *Sclerotinia sclerotiorum*. Phytopathology 76:980–984.

Cerkauskas, R.F., O.D. Dhingra, and J.B. Sinclair. 1982. Effect of herbicides on competitive saprophytic colonization by *Macrophomina phaseolina* of soybean stems. Trans. Br. Mycol. Soc. 79:201–205.

Cessna, S.G., V.E. Sears, M.B. Dickman, and P.S. Low. 2000. Oxalic acid, a pathogenicity factor for *Sclerotinia sclerotiorum*, suppresses the oxidative burst of the host plant. Plant Cell 12:2191–2199.

Chamberlain, D.W. 1951. Sclerotinia stem rot of soybeans. Plant Dis. Rep. 35:490–491.

Chamberlain, D.W., and R.L. Bernard. 1968. Resistance to brown stem rot in soybeans. Crop Sci. 8:728–729.

Chang, S.J.C., T.W. Doubler, V. Kilo, R. Suttner, J. Klein, M.E. Schmidt, P.T. Gibson, and D.A. Lightfoot. 1996. Two additional loci underlying durable field resistance to soybean sudden death syndrome (SDS). Crop Sci. 36:1684–1688.

Chen, W., C.R. Grau, E.A. Adee, and X. Meng. 2000. A molecular marker identifying subspecific populations of the brown stem rot pathogen, *Phialophora gregata*. Phytopathology 90:875–883.

Chen, W., L.E. Gray, and C.R Grau. 1996. Molecular differentiation of fungi associated with brown stem rot and detection of *Phialophora gregata* in resistant and susceptible soybean cultivars. Phytopathology 86:1140–1148.

Chen, W., L.E. Gray, J.E. Kurle, and C.R Grau. 1999. Specific detection of *Phialophora gregata* and *Plectosporium tabacinum* in infected soybean plants using polymerase chain reaction. Mol. Ecol. 8:871–877.

Chitima-Matsiga R.T., and T.D. Wyllie. 1987. Variability among single conidial isolates of *Macrophomina phaseolina*. Phytopathology 77:1702.

Chun, D., L.B. Kao, J.L. Lockwood, and T.G. Isleib. 1987. Laboratory and field assessment of resistance in soybean to stem rot caused by *Sclerotinia sclerotiorum*. Plant Dis.71:811–815.

Cloud, G.L., and J.C. Rupe. 1991. Morphological instability on a chlorate medium of isolates of *Macrophomina phaseolina* from soybean and sorghum. Phytopathology 81:892–895.

Cober, E.R., S. Rioux, I. Rajcan, P.A. Donaldson, and D.H. Simmonds. 2003. Partial resistance to white mold in a transgenic soybean line. Crop Sci. 43:92–95.

Cochran, A.J., and T.S. Abney. 1999. *Rps* gene combinations needed to control diverse pathotypes of *Phytophthora sojae*. Phytopathology 89:S104.

Cooper, R.L. 1989. Soybean yield response to benomyl fungicide application under maximum yield conditions. Agron. J. 81:847–849.

Crawford, M.A. 1984. Seed treatments and tillage practices as they affect spread and control of stem canker. *In* M.M. Kulick (ed.) Proc. Conf. on *Diaporthe/Phomopsis*. Ft populations of the soybean brown stem rot pathogen, *Phialophora gregata*. Phytopathology 90:875–883.

Cregan, P.B., T. Jarvik, A.L. Bush, R.C. Shoemaker, K.G. Lark, A.L. Kahler, N. Kaya, T.T. VanToai, D.G. Lohnes, and J. Chung. 1999. An integrated genetic linkage map of the soybean genome. Crop Sci. 39:1464–1490.

Damicone, J.P., G.T. Berggren, and J.P. Snow. 1987. Effect of free moisture on soybean stem canker development. Phytopathology 77:1568–1572.

Damicone, J.P., J.P. Snow, and G.T. Berggren. 1990. Spatial and temporal spread of soybean stem canker from an inoculum point source. Phytopathology 80:571–578.

Dann, E.K., B.W. Diers, J. Byrum, and R. Hammerschmidt. 1998. Effect of treating soybean with 2,6-dichloroisonicotinicacid (INA) and benxothiadiazole (BTH) on seed yields and level of disease caused by *Sclerotinia sclerotiorum* in field and greenhouse studies. Eur. J. Plant Pathol. 104:271–278.

Dann, E.K., B.W. Diers, and R. Hammerschmidt. 1999. Suppression of Sclerotinia stem rot of soybean by lactofen herbicide treatment. Phytopathology 89:598–602.

Datnoff, L.E. 1988. Fusarium blight or wilt, root rot, and pod and collar rot. p. 33–35. *In* J.B. Sinclair and P.A. Backman (ed.) Compendium of soybean diseases. 3rd ed. St. Paul, MN.

Datnoff, L.E., and J.B. Sinclair. 1988. The interaction of *Fusarium oxysporum* and *Rhizoctonia solani* in causing root rot of soybeans. Phytopathology 78:771–777.

del Río, L.E., C.A. Martinson, and X.B. Yang. 2002. Biological control of Sclerotinia stem rot of soybean with *Sporidesmium sclerotivorum*. Plant Dis. 86:999–1004.

de Vrije, T., N. Antoine, R.M. Buitelaar, S. Bruckner, M. Dissevelt, A. Durand, M. Gerlagh, E.E. Jones, P. Luth et al. 2001. The fungal biocontrol agent *Coniothyrium minitans*: Production by solid-state fermentation, application and marketing. Appl. Microbiol. Biotechnol. 56:58–68.

Demirbas, A., B.G. Rector, D.G. Lohnes, R.J. Fioritto, G.L. Graef, P.B. Cregan, R.C. Shoemaker, and J.E. Specht. 2001. Simple sequence repeat markers linked to the soybean *Rps* genes for Phytophthora resistance. Crop Sci. 41:1220–1227.

Demski, J.W., and D.V. Phillips. 1974. Reaction of soybean cultivars to powdery mildew. Plant Dis. Rep. 58:723–726.

Dhingra, O.D., and J.B. Sinclair. 1973a. Location of *Macrophomina phaseolina* on soybean plants related to cultural characteristics and virulence. Phytopathology 63:934–936.

Dhingra, O.D., and J.B. Sinclair. 1973b. Variation among isolates of *Macrophomina phaseolina* (*Rhizoctonia bataticola*) from different regions. Phytopathol. Z. 76:200–204.

Dhingra, O.D., and J.B. Sinclair. 1975. Survival of *Macrophomina phaseolina* sclerotia in soil: Effects of soil moisture, carbon: nitrogen ratios, carbon sources, and nitrogen concentrations. Phytopathology 65:236–240.

Dickson, J.G. 1956. Soybean diseases. *In* Diseases of field crops. McGraw-Hill Book Co., New York.

Diers, B.W., L. Mansur, J. Imsande, and R.C. Shoemaker. 1992. Mapping Phytophthora resistance loci in soybean with restriction fragment length polymorphism markers. Crop Sci. 32:377–383.

Dorrance, A.E., and S.A. McClure. 2001. Beneficial effects of fungicide seed treatments for soybean cultivars with partial resistance to *Phytophthora sojae*. Plant Dis. 85:1063–1068.

Dorrance, A.E., S.A. McClure, and A. de Silva. 2003a. Pathogenic diversity of *Phytophthora sojae* in Ohio soybean fields. Plant Dis. 87:139–146.

Dorrance, A.E., S.A. McClure, and S.K. St. Martin. 2003b. Effect of partial resistance on Phytophthora stem rot incidence and yield of soybean in Ohio. Plant Dis. 87:308–312.

Dorrance, A.E., and A.F. Schmitthenner. 2000. New sources of resistance to *Phytophthora sojae* in the soybean plant introductions. Plant Dis. 84:1303–1308.

Doupnik Jr., B. 1993. Soybean production and disease loss estimates for north central United States from 1989 to 1991. Plant Dis. 77:1170–1171.

Dunleavy, J.M. 1970. Sources of immunity and susceptibility to downy mildew of soybean. Crop Sci. 10:507–509.

Dunleavy, J.M. 1971. Races of *Peronospora manshurica* in the United States. Am. J. Bot. 58:209–211.

Dunleavy, J.M. 1977. Nine new races of *Peronospora manshurica* found on soybeans in the Midwest. Plant Dis. Rep. 61:661–663.

Dunleavy, J.M. 1978. Soybean seed yield losses caused by powdery mildew. Crop Sci. 18:337–339.

Dunleavy, J.M. 1980. Yield loses in soybeans included by powdery mildew. Plant Dis. 64:291–292.

Dunleavy, J.M. 1987. Yield reduction in soybeans caused by downy mildew. Plant Dis. 71:1112–1114.

Dunleavy, J.M., and E.E.Hartwig. 1970. Sources of immunity from and resistant to nine races of the soybean downy mildew fungus. Plant Dis. Rep. 54:901–902.

Dunleavy, J.M., J.W. Keck, K.S. Gobelman-Werner, and M.M. Thompson. 1984. Prevalence of soybean downy mildew in Iowa. Plant Dis. 68:778–779.

Dunleavy, J. M., and C.R. Weber. 1967. Control of brown stem rot of soybeans with corn-soybean rotations. Phytopathology 57:114–117.

Eathington, S. R., C.D. Nickell, and L.E. Gray. 1995. Inheritance of brown stem rot resistance in soybean cultivar BSR 101. J. Hered. 86:55–60.

Elmore, R.W, H.C. Minor, and B.L. Doupnik, Jr. 1998. Soybean genetic resistance and benomyl for phomopsis seed decay control. Seed Technol. 20:23–31.

Enkerli, K., M.G. Hahn, and C.W. Mims. 1997. Ultrastructure of compatible and incompatible interactions of soybean roots infected with the plant pathogenic oomycete *Phytophthora sojae*. Can J. Bot. 75:1493–1508.

Farias, G.M., and G.J. Griffin. 1989. Roles of *Fusarium oxysporum* and *F. solani* in Essex disease of soybean in Virginia. Plant Dis. 73:38–42.

Faris, M.A., F.E. Sabo, D.J.S. Barr, and C.S. Lin. 1989. The systematics of *Phytophthora sojae* and *P. megasperma*. Can J. Bot. 67:1442–1447.

Favaron, F., R. D'Ovidio, E. Porceddu, and P. Alghisi. 1994. Purification and molecular characterization of a soybean polygalacturonase-inhibiting protein. Planta 195:80–87.

Favaron, F., P. Marcino, and P. Magro. 1988. Polygalcturonase isoenzymes and oxalic acid produced by *Sclerotinia sclerotiorum* in soybean hypocotyls as elicitors of glyceolin. Physiol. Mol. Plant Pathol. 33:385–395.

Fehr, W.R., C.E. Caviness, D.T. Burmood, and J.S. Pennington. 1971. Stage of development descriptions for soybeans, *Glycine max* (L.) Merrill. Crop Sci. 11:929–931.

Fernandez, F.A., D.V. Phillips, J.S. Russin, and J. C. Rupe. 1999. Stem canker. p. 33–35 *In* G.L. Hartman et al. (ed). Compendium of soybean diseases. 4th ed. APS Press, St. Paul, MN.

Ferrant, N.P., and R.B. Carroll. 1981. Fusarium wilt of soybean in Delaware. Plant Dis.65:596–599.

Filho, E.S., and O.D. Dhingra. 1980a. Survival of *Macrophomina phaseolina* sclerotia in nitrogen amended soils. Phytopathol. Z. 97:136–143.

Filho, E.S., and O.D. Dhingra. 1980b. Effect of herbicides on survival of *Macrophomina phaseolina* in soil. Trans. Br. Mycol. Soc. 74:61–64.

Flor, H.H. 1955. Host parasite interaction in flax rust-its genetics and other implications. Phytopathology 45:680–685.

Förster, H., B.M. Tyler, and M.D. Coffey. 1994. *Phytophthora sojae* races have arisen by clonal evolution and by rare outcrosses. MPMI 7:780–791.

Gams, W. 1971. *Cephalosporium*-artige Schimmelpilze (Hyphomycetes). Gustav Fisher Verlag, Stuttgart.

Gangopadhyay, S., T.D. Wyllie, and V.D. Luedders. 1970. Charcoal rot disease of soybeans transmitted by seeds. Plant Dis. Rep. 54:1088–1091.

Gangopadhyay, S., T.D. Wyllie, and W.R. Teague. 1982. Effect of bulk density and moisture content of soil on the survival of *Macrophomina phaseolina*. Plant Soil 68:241–247.

Garzonio, D.M., and D.C. McGee. 1983. Comparison of seeds and crop residue as sources of inoculum for pod and stem blight of soybeans. Plant Dis. 67:1374–1376.

Geeseman, G.E. 1950. Physiologic races of *Peronospora manshurica* on soybeans. Agron. J. 42:257–258.

Gerdemann, J.W. 1953. An undescribed fungus causing a root rot of red clover and other Leguminosae. Mycologia 45:548–553.

Gerdemann, J.W. 1954a. The association of *Diaporthe phaseolorum* var. *sojae* with root and basal root rot of soybean. Plant Dis. Rep. 38:742–743.

Gerdemann, J.W. 1954b. Pathogenicity of *Leptodiscus terrestris* on red clover and other Leguminosae. Phytopathology 44:451–455.

Gibson, P.T., M.A. Shenaut, R.J. Suttner, V.N. Njiti, and O. Myers, Jr. 1994. Soybean varietal response to soybean death syndrome. p. 20–40 *In* D. Wilkinson (ed.) Proc. 24th Soybean Seed Res. Conf., Chicago, IL. 9 Dec. 1994. Am. Seed Trade Assoc., Washington, DC.

Glover, K.D., and R.A. Scott. 1998. Heritability and phenotypic variation of tolerance to Phytophthora root rot of soybean. Crop Sci. 38:1495–1500.

Grabe, D.F., and J.M. Dunleavy. 1959. Physiologic specialization in *Peronospora manshurica*. Phytopathology 49:791–793.

Gracia-Garza, J.A., S. Neumann, T.J. Vyn, and G.J. Boland. 2002. Influence of crop rotation and tillage on production of apothecia by *Sclerotinia sclerotiorum*. Can. J. Plant Pathol. 24:137–143.

Grau, C.R. 1985. Powdery mildew, a sporadic but damaging disease of soybean. World Soybean Res. Conf. III, Ames, IA. 12–17 Aug. 1984. p. 658–574. *In* R. Shibles (ed.) Proceedings. Westview Press, Boulder, CO.

Grau, C.R. 1988. Sclerotinia stem rot of soybean. p. 56–66. *In* T.D. Wyllie and D.H. Scott (ed.) Soybean diseases of the north central region. APS Press, St. Paul, MN.

Grau, C.R., and G.L. Hartman. 1999. Sclerotinia stem rot. p. 46–48. *In* G.L. Hartman et al. (ed.) Compendium of soybean diseases. 4th ed. APS Press, St. Paul, MN.

Grau, C.R., and J.A. Laurence. 1975. Observations on resistance and heritability of resistance to powdery mildew of soybean. Plant Dis. Rep. 59:458–460.

Grau, C.R., E.S. Oplinger, E.A. Adee, E.A. Hinkens, and M.J. Martinka. 1994. Plant date and row width effect on severity of brown stem rot and soybean productivity. J. Prod. Agric. 7:347–351.

Grau, C.R., and V.L. Radke. 1984. Effects of cultivars and cultural practices on Sclerotinia stem rot of soybean. Plant Dis. 68:56–58.

Grau, C.R., V.L. Radke, and F.L. Gillespie. 1982. Resistance of soybean cultivars to *Sclerotinia sclerotiorum*. Plant Dis. 66:506–508.

Gravert, C.E, S. Li, and G.L. Hartman. 2001. Inoculation studies and resistance screening for stem canker on soybean. Phytopathology 91:S32.

Gray, L.E. 1971. Variation in pathogenicity of *Cephalosporium gregatum* isolates. Phytopathology 61:1410–1411.

Gray, L.E. 1974. Role of temperature, plant age, and fungus isolate in the development of brown stem rot in soybeans. Phytopathology 64:94–96.

Gray, L.E. 1978. *Mycoleptodiscus terrestris* root rot of soybeans. Plant Dis. Rep. 62:72–73.

Gray, L.E., H.W. Gardner, D. Weisleder, and M. Leib. 1999. Production and toxicity of 2,3-dihydro-5-hydroxy-2-methyl-4H-1-benzopyran-4-one by *Phialophora gregata*. Phytochemistry 50:1337–1340.

Gray, L.E., and C.R. Grau. 1999. Brown stem rot. p. 28–29. *In* G.L. Hartman et al. (ed.) Compendium of soybean diseases. 4th ed. APS Press, St. Paul, MN.

Gray, L.E., and J.K. Pataky. 1994. Reaction of mung bean plants to infection by isolates of *Phialophora gregata*. Plant Dis. 78:782–785.

Gray, L.E., and J.B. Sinclair. 1973. The incidence, development, and yield effects of *Cephalosporium gregatum* on soybeans in Illinois. Plant Dis. Rep. 57:853–854.

Griffin, G.J. 1990. Importance of *Pythium ultimum* in a disease syndrome of cv. Essex soybean. Canadian J. Plant Pathol. 12:135–140.

Guy, S.O., and E.S. Oplinger. 1989. Soybean cultivar performance as influenced by tillage system and seed treatment. J. Prod. Agric. 2:57–62.

Guy, S.O., E.S. Oplinger, and C.R. Grau. 1989. Soybean cultivar response to metalaxyl applied in furrow and as a seed treatment. Agron. J. 81:529–532.

Haas, J.H., and R.I. Buzzell. 1976. New races 5 and 6 of *Phytophthora megasperma* var. *sojae* and differential reactions of soybean cultivars for races 1 and 6. Phytopathology 66:1361–1362.

Hall, R., and A.G. Xue. 1995. Effectiveness of fungicidal seed treatments applied to smooth or shrivelled soybean seeds contaminated by *Diaporthe phaseolorum*. Phytoprotection 76:47–56.

Hamilton, R.I., and M.G. Boosalis. 1955. Asexual reproduction in *Cephalosporium gregatum*. Phytopathology 45:293–294.

Hansen, E.M., and D.P. Maxwell. 1991. Species of the *Phytophthora megasperma* complex. Mycologia 83:376–381.

Hanson, P. M., C.D. Nickell, L.E. Gray, and S.A. Sebastian. 1988. Identification of two dominant genes conditioning brown stem rot resistance in soybean. Crop Sci. 28:41–43.

Harikrishnan, R., and X.B. Yang. 2002. Effects of herbicides on root rot and damping-off caused by *Rhizoctonia solani* in glyphosate-tolerant soybean. Plant Dis. 86:1369–1373.

Harrington, T.C., J. Steimel, F. Workneh, and X.B. Yang. 2000. Molecular identification of fungi associated with vascular discoloration of soybean in the north central United States. Plant Dis. 84:83–89.

Hartman, G.L., M.E. Gardner, T. Hymowitz, and G.C. Naidoo. 2000. Evaluation of perennial *Glycine* species for resistance to soybean fungal pathogens that cause Sclerotinia stem rot and sudden death syndrome. Crop Sci. 40:545–549.

Hartman, G.L., Y.H. Huang, R.L. Nelson, and G.R. Noel. 1997. Germplasm evaluation of *Glycine max* for resistance to *Fusarium solani*, the causal organism of sudden death syndrome. Plant Dis. 81:515–518.

Hartman, G.L., L. Kull, and Y.H. Huang. 1998. Occurrence of *Sclerotinia sclerotiorum* in soybean fields in East-Central Illinois and enumeration of inocula in soybean seed lots. Plant Dis. 82:560–564.

Hartman, G.L., G.R. Noel, and L.E. Gray. 1995. Occurrence of soybean sudden death syndrome in east-central Illinois and associated yield losses. Plant Dis. 79:314–318.

Hartman, G.L., J.B. Sinclair, and J.C. Rupe (ed.) 1999. Compendium of soybean diseases. 4th ed. APS Press, St. Paul, MN.

Hartman, G.L., T.C. Wang, and T. Hymowitz. 1991. Sources of resistance to soybean rust in perennial *Glycine* species. Plant Dis. 76:396–399.

Hartwig, E.E. 1959. Effect of target spot on yield of soybeans. Plant Dis. Rep. 43:504–505.

Harville, B.G., J.S. Russin, and R.J. Habetz. 1996. Rhizoctonia foliar blight reactions and seed yields in soybean. Crop Sci. 36:563–566.

Hegstad, J.M., C.D. Nickell, and L.O. Vodkin. 1998. Identifying resistance to *Phytophthora sojae* in selected soybean accessions using RFLP techniques. Crop Sci. 38:50–55.

Helbig, J.B., and R.B. Carroll. 1984. Dicotyledonous weeds as a source of *Fusarium oxysporum* pathogenic on soybean. Plant Dis. 68:694–696.

Henry, R.N., and T.L. Kirkpatrick. 1995. Two new races of *Phytophthora sojae*, causal agent of Phytophthora root and stem rot of soybean, identified from Arkansas soybean fields. Plant Dis. 79:1074.

Hepperly, P.R., J.S. Mignucci, J.B. Sinclair, R.S.Smith, and W.H. Judy. 1982. Rhizoctonia web blight of soybean in Puerto Rico. Plant Dis. 66:256–257.

Hepperly, P.R., and J.B. Sinclair. 1978. Quality losses in Phomopsis-infected soybean seeds. Phytopathology 68:1684–1687.

Hepperly, P.R., and J.B. Sinclair. 1981. Relationships among *Cercosopra kikuchii*, other seed mycoflora, and germination of soybeans in Puerto Rico and Illinois. Plant Dis. 65:130–132.

Hershman, D.E., J.W. Hendrix, R.E. Stuckey, P.R. Bachi, and G. Henson. 1990. Influence of planting date and cultivar on soybean sudden death syndrome in Kentucky. Plant Dis. 74:761–766.

Herzog, D.C., J.W. Thomas, R.L. Jensen, and L.D. Newsom. 1975. Association of sclerotial blight with *Spissistilus festinus* gridling injury on soybean. Environ. Entomol. 4:986–988.

Hildebrand, A.A. 1954. Observation on the occurrence of the stem canker and pod and stem blight fungus on mature stems of soybeans. Plant Dis. Rep. 38:640–646.

Hildebrand, A.A., and L.W. Koch. 1951. A study of the systemic infection by downy mildew of soybean with special reference to symptomatology, economic significance and control. Sci. Agric. 31:505–518.

Hill, H.C., N.L. Horn, and W.L. Steffan. 1981. Mycelial development and control of *Phomopsis sojae* in artificially inoculated soybean stems. Plant Dis. 65:132–134.

Hoffman, D.D., B.W. Diers, G.L. Hartman, C.D. Nickell, R.L. Nelson, W.L. Pedersen, E.R. Cober, G.L. Graef, J.R. Steadman et al. 2002. Selected soybean plant introductions with partial resistance to *Sclerotinia sclerotiorum*. Plant Dis. 86:971–980.

Hoffman, D.D., G.L. Hartman, D.S. Mueller, R.A. Leits, C.D. Nickell, and W.L. Pedersen. 1998. Yield and seed quality of soybean cultivars infected with *Sclerotinia sclerotiorum*. Plant Dis. 82:826–829.

Horn, N.L., F.N. Lee, and R.B. Carver. 1975. Effects of fungicides and pathogens on yields of soybeans. Plant Dis. Rep. 59:724–728.

Howard, D.D., A.Y. Chambers, and M.A. Newman. 1999. Reducing sudden death syndrome in soybean by amending the soil with chloride. Commun. Soil Sci. Plant Anal. 30:545–555.

Hughes, T.J., W. Chen, and C.R Grau. 2002. Pathogenic characterization of genotypes A and B of *Phialophora gregata* f.sp. *sojae*. Plant Dis. 86:729–735.

Hughes T.J, N.C. Kurtzweil, and C. R. Grau. 2001. Reaction of soybean germplasm with SCN resistance to brown stem rot development. Phytopathology (Suppl.): 91:S178.

Jee, H., W. Kim, and W. Cho. 1998. Occurrence of Phytophthora root rot on soybean (*Glycine max*) and identification of the causal fungus. Crop Prot. 40:16–22.

Jeffers, D.L. A.F. Schmitthenner, and M.E. Kroetz. 1982. Potassium fertilization effects on Phomopsis seed infection, seed quality, and yield of soybean. Agron. J. 74:886–890.

Jin, H., G.L. Hartman, Y.H. Huang, C.D. Nickell, and J.M. Widholm. 1996a. Regeneration of soybean plants from embryogenic suspension cultures treated with toxic culture filtrates of *Fusarium solani* and screening of regenerants for resistance. Phytopathology 86:714–718.

Jin, H., G.L. Hartman, C.D. Nickell, and J.M. Widholm. 1996b. Characterization and purification of a phytotoxin produced by *Fusarium solani*, the causal agent of soybean sudden death syndrome. Phytopathology 86:277–282.

Jin, H., G.L. Hartman, C.D. Nickell, and J.M. Widholm. 1996c. Phytotoxicity of culture filtrate from *Fusarium solani*, the causal agent of sudden death syndrome of soybean. Plant Dis. 80:922–927.

Jones, F.R., and J.H. Torrie. 1946. Systemic infection of downy mildew in soybean and alfalfa. Phytopathology 36:1057–1059.

Jones, J.P. 1959. Purple stain of soybean seeds incited by several *Cercospora* species. Phytopathology 49:430–432.

Jones, J.P. 1961. A leaf spot on cotton caused by *Corynespora cassiicola*. Phytopathology 51:305–308.

Jones, J.P., and H.W. Johnson. 1969. Lupine, a new host for *Phytophthora megasperma* var. *sojae*. Phytopathology 59:504–507.

Joye, G.F., G.T. Berggren, and D.K. Berner. 1990. Effects of row spacing and within-row population on Rhizoctonia aerial blight of soybean and soybean yield. Plant Dis. 74:158–160.

Kaitany, R.C., L.P. Hart, and G.R. Safir. 2001. Virulence composition of *Phytophthora sojae* in Michigan. Plant Dis. 85:1103–1106.

Kamal, M., and A.R. Weinhold. 1967. Virulence of *Rhizoctonia solani* as influenced by age of inoculum in soil. Can. J. of Bot. 45:1761–1765.

Kamicker, T.A., and S.M. Lim. 1985. Field evaluation of pathogenic variability in isolates of *Septoria glycines*. Plant Dis. 69:744–746.

Kaufmann, M.J., and J.W. Gerdemann. 1958. Root and stem rot of soybean caused by *Phytophthora sojae* n. sp. Phytopathology 48:201–208.

Keeling, B.L. 1982. A seedling test for resistance to soybean stem canker caused by *Diaporthe phaseolorum* var. *caulivora*. Phytopathology 72:807–809.

Keeling, B.L. 1982. Four new physiologic races of *Phytophthora megasperma* f. sp. *glycinea*. Plant Dis. 66:334–335.

Keeling, B.L. 1984. A new physiologic race of *Phytophthora megasperma* f. sp. *glycinea*. Plant Dis. 68:626–627.

Kendig, S.R., J.C. Rupe, and H.D. Scott. 2000. Effect of irrigation and soil water stress on densities of *Macrophomina phaseolina* in soil and roots of two soybean cultivars. Plant Dis. 84:895–900.

Kennedy, B.W., and J.W. Lambert. 1981. Influences of brown stem rot and cropping history on soybean performance. Plant Dis. 65:896–897.

Khan, M., and J.B. Sinclair. 1991. Effect of soil temperature on infection of soybean roots by sclerotia-forming isolates of *Colletotrichum truncatum*. Plant Dis. 75:1282–1285.

Khan, M., and J.B. Sinclair. 1992. Pathogenicity of sclerotia- and nonsclerotia-forming isolates of *Colletotrichum truncatum* on soybean plants and roots. Phytopathology 82:314–319.

Kilen, T.C. 1997. Identification of a soybean breeding line resistant to rust in the Philippines. Soybean Genet. Newsl. 24:199–200.

Kilen, T.C., E.E. Hartwig, and B.L. Keeling. 1974. Inheritance of a second major gene for resistance to Phytophthora root rot in soybeans. Crop Sci. 14:260–262.

Kilen, T.C., and B.L. Keeling. 1981. Genetics of resistance to Phytophthora rot in soybean cultivar PI 171442. Crop Sci. 21:873–875.

Killebrew, J.F., K.W. Roy, and T.S. Abney. 1993. Fusaria and other fungi on soybean seedlings and roots of older plants and interrelationships among fungi, symptoms, ans soil characteristics. J. Plant Pathol. 15:139–146.

Killebrew, J.F., K.W. Roy, G.W. Lawrence, K.S. McLean, and H.H. Hodges. 1988. Greenhouse and field evaluation of pathogenicity to soybean seedlings. Plant Dis. 72:1067–1070.

Killgore, E., R. Heu, and D.E. Gardner. 1994. First report of soybean rust in Hawaii. Plant Dis. 78:1216.

Kim, H.S., and B.W. Diers. 2000. Inheritance of partial resistance to Sclerotinia stem rot in soybean. Crop Sci. 40:55–61.

Kim, H.S., G.L. Hartman, J.B. Manadhar, G.L. Graef, J.R. Steadman, and B.W. Diers. 2000. Reaction of soybean cultivars to Sclerotinia stem rot in field, greenhouse and laboratory evaluations. Crop Sci. 40:665–669.

Kim, H.S., C.H. Sneller, and D.W. Diers. 1999. Evaluation of soybean cultivars for resistance to Sclerotinia stem rot in field environments. Crop Sci. 39:64–68.

Kirkpatrick, M.T. 1999. The effect of *Pythium* spp. on soybeans subjected to periodic soil flooding. M.S. thesis. Univ. of Arkansas, Fayetteville.

Kirkpatrick, M.T., J.C. Rupe, and C.S. Rothrock. 1998. Association of *Pythium* spp. with roots of soybean subjected to soil flooding. Phytopathology (Suppl.) 88:S49.

Kirkpatrick, M.T., J.C. Rupe, and C.S. Rothrock. 1999. Effect of *Pythium ultimum* on soybean subjected to soil flooding. Phytopathology (Suppl.) 89:S95.

Kitani, K., and Y. Inoue. 1960. Studies on the soybean rust and its control measure. Part I. Studies on the soybean rust. Bull. Shik. Agric. Exp. Stn. (Zentsuji, Japan) 5:319–342.

Kittle, D.R., and L.E. Gray. 1979. The influence of soil temperature, moisture, porosity, and bulk density on the pathogenicity of *Phytophthora megasperma* var. *sojae*. Plant-Dis. Rep. 63:231–234.

Kmetz, K., and C.W. Ellett. 1979. Soybean seed decay: Source of inoculum and nature of infection. Phytopathology 69:798–801.

Kmetz, K., C.W. Ellett, and A.F. Schmitthenner. 1974. Isolation of seedborne *Diaporthe phaseolorum* and *Phomopsis* from immature soybean plants. Plant Dis. Rep. 58:978–982.

Kmetz, K., A.F. Schmitthenner, and C.W. Ellett. 1978. Soybean seed decay: Prevalence of infection and symptom expression caused by *Phomopis* sp., *Diaporthe phaseolorum* var. *sojae,* and *D. phaseolorum* var. *caulivora*. Phytopathology 68:838–840.

Koning, G., D.M. TeKrony, T.W. Pfeiffer, and S. A. Ghabrial. 2001. Infection of soybean with *Soybean mosaic virus* increases susceptibility to *Phomopsis* spp. seed infection. Crop Sci. 41:1850–1856.

Koover, A.T.A. 1954. Some factors affecting the growth of *Rhizoctonia bataticola* in soil. J. Madras Univ. 24:47–52.

Krausz, J.P., and B.A. Fortnum. 1983. An epiphytotic of Diaporthe stem canker of soybean in South Carolina. Plant Dis. 67:1128–1129.

Kuan, T.L., and D.C. Erwin. 1980. *Formae speciales* differentiation of *Phytophthora megasperma* isolates from soybean and alfalfa. Phytopathology 70:333–338.

Kulik, M.M. 1984. Symptomless infection, persistence, production of *Phomopsis phaseoli* and *Phomopsis sojae*, and the taxonomic implications. Mycologia 76:274–291.

Kulik, M.M., and J.B. Sinclair. 1999a. Phomopsis seed decay. p. 31–32. *In* G.L.Hartman et al. (ed) Compendium of soybean diseases. 4th ed. APS Press, St. Paul, MN.

Kulik, M.M., and J.B. Sinclair. 1999b. Pod and stem blight. p. 32–33. *In* G.L. Hartman et al. (ed) Compendium of soybean diseases. 4th ed. APS Press, St. Paul, MN.

Kunwar, I.K., T. Singh, C.C. Machado, and J.B. Sinclair. 1986. Histopathology of soybean seed and seedling infection by *Macrophomina phaseolina*. Phytopathology 76:532–535.

Kurle, J.E., C.R. Grau, and E.S. Oplinger. 2001. Tillage, crop sequence, and cultivar effects on Sclerotinia stem rot incidence and yield in soybean. Agron. J. 93:973–982.

Kurtzweil, N. C.,.E.A. Kinziger, and C.R. Grau, C. R. 2002. Effect of soil pH on symptom development and pathogen reproduction of *Phialophora gregata* in soybean. Phytopathology (Suppl.) 92:S43.

Kyle, D.E., C.D. Nickell, R.L. Nelson, and W.L. Pederson. 1998. Response of soybean accessions from provinces in southern China to *Phytophthora sojae*. Plant Dis. 82:555–559.

Laviolette, F.A., and K.L. Athow. 1977. Three new physiologic races of *Phytophthora megasperma* var. *sojae*. Phytopathology 67:267–268.

Laviolette, F.A., and K.L. Athow. 1981. Physiologic races of *Phytophthora megasperma* f. sp. *glycinea* in Indiana, 1973–1979. Plant Dis. 65:884–885.

Laviolette, F.A., K.L. Athow, A.H. Probst, J.R. Wilcox, and T.S. Abney. 1970. Effect of bacterial pustule and frogeye leafspot on yield of Clark soybean. Crop Sci. 10:418–419.

Leath, S., and R.B. Carroll. 1982. Powdery mildew on soybean in Delaware. Plant Dis. 66:70–71.

Lee, C.D., D. Penner, and R. Hammerschmidt. 2000. Influence of formulated glyphosate and activator adjuvants on *Sclerotinia sclerotiorum* in glyphosate-resistant and susceptible *Glycine max*. Weed Sci. 48:710–715.

Lee, G.B., and G.L. Hartman. 1996. Reactions of *Glycines* species and other legumes to *Septoria glycines*. Plant Dis. 80:90–94.

Lee, G.B., G.L. Hartman, and S.M. Lim. 1996. Brown spot severity and yield of soybeans regenerated from calli resistant to a host-specific pathotoxin produced by *Septoria glycines*. Plant Dis. 80:408–413.

Lehman, S.G. 1947. Powdery mildew of soybean. Phytopathology 37:434.

Lehman, S.G., and F.A. Wolf. 1924. A new downy mildew on soybean. J. Elisha Mitchell Sci. Soc. 39:164–169.

Leitz, R.A., G.L. Hartman, W.L. Pedersen, and C.D. Nickell. 2000. Races of *Phytophthora sojae* on soybean in Illinois. Plant Dis. 84:487.

Leslie, J.F., C.A.S. Pearson, P.E. Nelson, and T.A. Toussoun. 1990. Fusarium spp. From corn, sorghum, and soybean fields in the central and eastern United States. Phytopathology 80:343–350.

Lewers, K. S., E.H. Crane, C.R. Bronson, J.M. Schupp, P. Keim, and R.C. Shoemaker. 1999. Detection of linked QTL for soybean brown stem rot resistance in 'BSR 101' as expressed in a growth chamber environment. Mol. Breed. 5:33–42.

Lewis, J.A., and G.C. Papavizas. 1977. Factors affecting *Rhizoctonia solani* infection of soybean in the greenhouse. Plant Dis. Rep. 61:196–200.

Lim, S.M. 1978. Disease severity gradient of soybean downy mildew from a small focus of infection. Phytopathology 68:1774–1778.

Lim, S.M. 1979. Evaluation of soybean for resistance to Septoria brown spot. Plant Dis. Rep. 63:242–245.

Lim, S.M. 1980. Brown spot severity and yield reduction in soybean. Phytopathology 70:974–977.

Lim, S.M. 1983. Response to *Septoria glycines* of soybean nearly isogenic except for seed color. Phytopathology 73:719–722.

Lim, S.M. 1985. Epidemiology of soybean downy mildew. p. 555–561 *In* R. Shibles (ed.) World Soybean Res. Conf. III Proc. Westview Press, Boulder, CO.

Lim, S.M. 1989. Inheritance of resistance to *Peronospora manshurica* races 2 and 33 in soybean. Phytopathology 79:877–879.

Lim, S.M., R.L. Bernard, C.D. Nickell, and L.E. Gray. 1984. New physiological race of *Peronospora manshurica* virulent to the gene *Rpm* in soybeans. Plant Dis. 68:71–72.

Ling, L. 1951. Bibliography of soybean diseases. Plant Dis. Rep. Suppl. 244:109–173.

Lohnes, D.G., and R.L. Bernard. 1992. Inheritance of resistance to powdery mildew in soybean. Plant Dis. 76:964–965.

Lohnes, D.G, and C.D. Nickell. 1994. Effects of powdery mildew alleles *Rmd-c*, *Rmd*, and *rmd* on yield and other characteristics in soybean. Plant Dis. 78:299–301.

Lohnes, D.G., C.D. Nickell, and A.F. Schmitthenner. 1996. Origin of soybean alleles for Phytophthora resistance in China. Crop Sci. 36:1689–1692.

Lohnes, D.G., and A.F. Schmitthenner. 1997. Position of the Phytophthora resistance gene *Rps7* on the soybean molecular map. Crop Sci. 37:555–556.

Luo, Y., K. Hildebrand, S.K. Chong, O. Myers, and J.S. Russin. 2000. Soybean yield loss to sudden death syndrome in relation to symptom expression and root colonization by *Fusarium solani* f. sp. *glycines*. Plant Dis. 84:914–920.

Luo, Y., O. Myers, D.A. Lightfoot, and M.E. Schmidt. 1999. Root colonization of soybean cultivars in the field by *Fusarium solani* f. sp. *glycines*. Plant Dis. 83:1155–1159.

Ma, S.M., and B.Y. Li. 1997. Primary report on the identification for physiological races of *Cercospora sojina* Hara in Northeast China. Acta Phytopathol. Sin. 27:180.

MacGuidwin, A.E., C.R. Grau, and E.S. Oplinger. 1985. Impact of planting 'Bell', a soybean cultivar resistant to *Heterodera glycines*, in Wisconsin. J. Nematol. 27:78–85.

MacNeill, B.H., and H. Zalasky. 1957. Historical study of host-parasite relationships between *Septoria glycines* Hemmi and soybean leaves and pods. Can.J. Bot. 35:501–505.

Manandhar, J.B., and G.L. Hartman. 1999. Anthracnose. p. 13–14 *In* G.L. Hartman et al. (ed) Compendium of soybean diseases. 4th ed. APS Press, St. Paul, MN.

Manandhar, J.B., G.L. Hartman, and J.B. Sinclair. 1988. Soybean germplasm evaluation for resistance to *Colletotrichum truncatum*. Plant Dis. 72:56–59.

Martin, K.F., and H.J. Walters. 1982. Infection of soybean by *Cercospora kikuchii* as affected by dew temperature and duration of dew periods. Phytopathology 72:974.

McGee, D.C. 1986. Prediction of Phomopsis seed decay by measuring soybean pod infection. Plant Dis. 70:329–333.

McGee, D.C. 1988. Evaluation of current predictive methods for control of Phomopsis seed decay of soybeans. p. 22–25. *In* T.D.D. Wyllie and D.H. Scott (ed). Soybean diseases of the North Central Region. St. Paul, MN.

McGee, D.C. 1992. Soybean diseases: A reference source for seed technologists. APS Press, St. Paul, MN.

McGee, D.C., and J.A. Biddle. 1987. Seedborne *Diaporthe phaseolorum* var. *caulivora* in Iowa and its relationship to soybean stem canker in the southern United States. Plant Dis. 71:620–622.

McGee, D.C., and C.L. Brandt. 1979. Effect of foliar application of benomyl on infection of soybean seeds by *Phomopsis* in relation to time of inoculation. Plant Dis. Rep. 63:675–677.

McLaughlin, M.R., Mignucci, J.S., and G.M. Milbrath. 1977. *Microsphaera diffusa*, the perfect stage of the soybean powdery mildew pathogen. Phytopathology 67:726–729.

Mclean, K.S., and G.W. Lawrence. 1992. Interrelationships of *Heterodera glycines* and *Fusarium solani* in sudden death syndrome of soybean. J. Nematol. 25:434–439.

Mclean, K.S., and G.W. Lawrence. 1995. Development of *Heterodera glycines* as affected by *Fusarium solani*, the causal agent of sudden death syndrome of soybean. J. Nematol. 27:70–77.

Melching, J.S., W.M. Dowler, D.L. Koogle, and M.H. Royer. 1989. Effects of duration, frequency, and temperature of leaf wetness periods on soybean rust. Plant Dis. 73:117–122.

Melgar, J., and K.W. Roy. 1994. Soybean sudden death syndrome: Cultivar reactions to inoculation in a controlled environment and host range and virulence of causal agent. Plant Dis. 78:265–268.

Mengistu, A., and C.R. Grau. 1986. Variation in morphological, cultural, and pathological characteristics of *Phialophora gregata* and *Acremonium* sp. recovered from soybean in Wisconsin. Plant Dis. 70:1005–1009.

Mengistu, A., and C.R. Grau. 1987. Seasonal progress of brown stem rot and its impact on soybean productivity. Phytopathology 77:1521–1529.

Mengistu, A., C.R. Grau, and E.T. Gritton. 1986. Comparison of soybean genotypes for resistance to and agronomic performance in the presence of brown stem rot. Plant Dis. 70:1095–1098.

Mengistu, A., N.C. Kurtzweil, and C.R. Grau. 2002. First report of frogeye leaf spot (*Cercospora sojina*) in Wisconsin. Plant Dis. 86:1272.

Mengistu, A., H. Tachibana, A.H. Epstein, K.G. Bidne, and J.D. Hatfield. 1987. Use of leaf temperature to measure the effect of brown stem rot and soil moisture stress and its relation to yields of soybeans. Plant Dis. 71:632–634.

Mengistu, A., H. Tachibana, and C.R. Grau. 1991. Selective medium for isolation and enumeration of *Phialophora gregata* from soybean straw and soil. Plant Dis. 75:196–199.

Meyer, W.A., J.B. Sinclair, and M.N. Khare. 1974. Factors affecting charcoal rot of soybean seedlings. Phytopathology 64:845–849.

Mian, M.A.R., H.R. Boerma, D.V. Phillips, M.M. Kenty, G. Shannon, E.R. Shipe, A.R. Soffes Blount, and D.B. Weaver. 1998. Performance of frogeye leafspot-resistant and susceptible near-isolines of soybean. Plant Dis. 82:1017–1021.

Mian, M.A., T. Wang, D.V. Phillips, J. Alvernaz, and H.R. Boerma. 1999. Molecular mapping of the *Rcs3* gene for resistance to frogeye leaf spot in soybean. Crop Sci. 39:1687–1691.

Mignucci, J.S., and J.S. Boyer. 1979. Inhibition of phytosynthesis and transpiration in soybean infected by *Microsphaera diffusa*. Phytopathology 69:227–230.

Mignucci, J.S., and S.M. Lim. 1980. Powdery mildew development on soybeans with adult-plant resistance. Phytopathology 70:919–921.

Mignucci, J.S., S.M. Lim., and P.R. Hepperly. 1977. Effects of temperature on reactions of soybean seedlings to powdery mildew. Plant Dis. Rep. 61:122–124.

Miller, W.A., and K.W. Roy. 1982. Effect of benomyl on the colonization of soybean leaves, pods, and seeds by fungi. Plant Dis. 66:918–920.

Minor, H.C., E.A. Brown, B. Doupnik, Jr., R.W. Elmore, and M.S. Zimmerman. 1993. Registration of Phomopsis seed decay resistant soybean germplasm MO/PSD-0259. Crop Sci. 33:1105.

Minton, N.A., M.B. Parker, and R.A. Flowers. 1975. Response of soybean cultivars to *Meloidogyne incognita* and to the combined effects of *M. arenaria* and *Sclerotium rolfsii*. Plant Dis. Rep. 59:920–923.

Morgan, F.L., and E.E. Hartwig. 1965. Physiologic specialization in *Phytophthora megasperma* var. *sojae*. Phytopathology 55:1277–1279.

Morris, P.F., and E.W.B. Ward. 1992. Chemoattraction of zoospores of the soybean pathogen, *Phytophthora sojae*, by isoflavones. Physiol. Molec. Plant Pathol. 40:17–22.

Mueller, D.S., A.E. Dorrance, R.C. Derksen, E. Ozkan, J.E. Kurle, C.R. Grau, J.M. Gaska, G.L. Hartman, C.A. Bradley, and W.L. Pedersen. 2002a. Efficacy of fungicides on *Sclerotinia sclerotiorum* and potential control of Sclerotinia stem rot on soybean. Plant Dis. 86:26–31.

Mueller, D.S., G.L. Hartman, and W.L. Pedersen. 1999. Development of sclerotia and apothecia of *Sclerotinia sclerotiorum* from infected soybean seed and its control by fungicide seed treatment. Plant Dis. 83:1113–1115.

Mueller, D. S., G.L. Hartman, and W.L. Pedersen. 2002b. Effect of crop rotation and tillage system on Sclerotinia stem rot on soybean. Can. J. Plant Pathol. 24:450–456.

Mueller, E.H., K.L. Athow, and F.A. Laviolette. 1978. Inheritance of resistance to four physiologic races of *Phytophthora megasperma* var. *sojae*. Phytopathology 68:1318–1322.

Mukherjee, B., S. Banerjee, and C. Sen. 1983. Influence of soil pH, temperature, and moisture on the ability of mycelia of *Macrophomina phaseolina* to produce sclerotia in soil. Indian Phytopathol. 36:158–160.

Muyolo, N.G., P.E. Lipps, and A.F. Schmitthenner. 1993. Anastomosis grouping and variation in virulence among isolates of *Rhizoctonia* associated with dry bean and soybean in Ohio and Zaire. Phytopathology 83:438–444.

Nanayakkara, R. 2001. Influence of soybean cultivar, seed quality, and temperature on seed exudations and *Pythium* disease development. Ph.D diss. Univ. of Arkansas, Fayetteville (Diss. Abstr. AAT3025499).

Nelson, B.D. 1999. Fusarium blight or wilt, root rot, and pod and collar rot. p. 35–36. *In* G.L. Hartman et al. (ed.) Compendium of soybean diseases. 4th ed. APS Press, St. Paul, MN.

Nelson, B.D., J.M. Hansen, C.E. Windels, and T.C. Helms. 1997. Reaction of soybean cultivars to isolates of *Fusarium solani* from the Red River Valley. Plant Dis. 81:664–668.

Nelson, B.D., T. Helms, T. Christianson, and I. Kural. 1996. Characterization and pathogenicity of *Rhizoctonia* from soybean. Plant Dis. 80:74–80.

Nelson, B.D., T.C. Helms, and I. Kural. 1991a. Effect of temperature and pathogen isolate on laboratory screening of soybean resistance to *Sclerotinia sclerotiorum*. Can. J. Plant Sci. 71:347–352.

Nelson, B.D., T.C. Helms, and M.A. Olson. 1991b. Comparison of laboratory and field evaluations of resistance in soybean to *Sclerotinia sclerotiorum*. Plant Dis. 75:662–665.

Nelson, K.A., K.A. Renner, and R. Hammerschmidt. 2002. Cultivar and herbicide selection affects soybean development and the incidence of sclerotinia stem rot. Agron. J. 94:1270–1281.

Nelson, R. L., C.D. Nickell, J.H. Orf, H. Tachibana, E.T. Gritton, C.R. Grau, and B.W. Kennedy. 1989. Evaluating soybean germplasm for brown stem rot resistance. Plant Dis. 73:110–114.

Nickell, C. D., M.S. Bachman, P.A. Stephens, A.D. Nickell, T.R. Cary, and D.J. Thomas. 1997. Registration of LN92-12033 and LN92-12054 soybean germplasm lines near-isogenic for brown stem rot resistance gene *Rbs* 2. Crop Sci. 37:1978.

Nickell, C. D., P.M. Hanson, L.E. Gray, D.J. Thomas, and D.B. Willmot. 1990. Registration of soybean germplasm lines LN86-1595 and LN86-1947 resistant to brown stem rot. Crop Sci. 30:241.

Njiti, V.N., J.E. Johnson, T.A. Torto, L.E. Gray, and D.A. Lightfoot. 2001. Inoculum rate influences selection for field resistance to soybean sudden death syndrome in the greenhouse. Crop Sci. 41:1726–1731.

Njiti, V.N., M.A. Shenaut, R.J. Suttner, M.E. Schmidt, and P.T. Gibson. 1996. Soybean response to sudden death syndrome: inheritance influenced by cyst nematode resistance in Pyramid x Douglas progenies. Crop Sci. 36:1165–1170.

Njiti, V.N., M.A. Shenaut, R.J. Suttner, M.E. Schmidt, and P.T. Gibson. 1998. Relationship between sudden death syndrome disease measures and yield components in F6-derived lines. Crop Sci. 38:673–678.

Njiti, V.N., R.J. Suttner, L.E. Gray, P.T. Gibson, and D.A. Lightfoot. 1997. Rate-reducing resistance to *Fusarium solani* f. sp. *phaseoli* underlies field resistance to soybean sudden death syndrome. Crop Sci. 37:132–138.

Nyvall, R.F. 1989. Field crop diseases handbook. 2nd ed. Van Nostrand Reinhold, New York.

O'Neill, N.R., M.C. Rush, N.L. Horn, and R.B. Carver. 1977. Aerial blight of soybean caused by *Rhizoctonia solani*. Plant Dis. Rep. 61:713–717.

Olanya, O.M., and C.L. Campbell. 1988. Effects of tillage on the spatial pattern of microsclerotia of *Macrophomina phaseolina*. Phytopathology 78:217–221.

Olaya, G., and G.S. Abawi. 1996. Effect of water potential on mycelial growth and on production and germination of sclerotia of *Macrophomina phaseolina*. Plant Dis. 80:1347–1350.

Oplinger, E.S., and B.D. Philbrook. 1992. Soybean planting date, row width, and seeding rate response in three tillage systems. J. Prod. Agric. 5:94–99.

Orth, C.E., and W. Schuh. 1994. Resistance of 17 soybean cultivars to foliar, latent, and seed infection by *Cercospora kikuchii*. Plant Dis. 78:661–664.

Osburn, R.M., J.L. Milner, E.S. Oplinger, R.S. Smith, and J. Handelsman. 1995. Effect of *Bacillus cereus* UW85 on the yield of soybean at two field sites in Wisconsin. Plant Dis. 79:551–556.

Ostazeski, S.A. 1967. An undescribed fungus associated with a root and crown rot of birdsfoot trefoil (*Lotus corniculatus*). Mycologia 59:970–975.

Pace, P.F., D.B. Weaver, and L.D. Ploper. 1993. Additional genes for resistance to frogeye leaf spot race 5 in soybean. Crop Sci. 33:1144–1145.

Palm, M.E., W. Gams, and H.I. Nirenberg. 1995. *Plectosporium*, a new genus for *Fusarium tabacinum*, the anamorph of *Plectosphaeerella cucumerina*. Mycologia 87:397–406.

Papavizas, G.C., and C.B. Davey. 1961. Saprophytic behavior of *Rhizoctonia* in soil. Phytopathology 51:693–699.

Pataky, J.K., and S.J. Lim. 1981a. Efficacy of benomyl for controlling septoria brown spot of soybeans. Phytopathology 71:438–442.

Pataky, J.K., and S.J. Lim. 1981b. Effects of row width and plant growth habit on *Septoria glycines* brown spot development and soybean yield. Phytopathology 71:1051–1056.

Paxton, J.D., and D.P. Rogers. 1974. Powdery mildew of soybeans. Mycologia 66:894–896.

Pearson, C.A.S., J.F. Leslie, and F.W. Schwenk. 1987. Host preference correlated with chlorate resistance in *Macrophomina phaseolina*. Plant Dis. 71:828–831.

Pearson, C.A.S., F.W. Schwenk, F.J. Crowe, and K. Kelly. 1984. Colonization of soybean roots by *Macrophomina phaseolina*. Plant Dis. 68:1086–1088.

Pecina, V., M.J. Alvarado, H.W. Alanis, R.T. Almaraz, and G.J. Vandemark. 2000. Detection of double-stranded RNA in *Macrophomina phaseolina*. Mycologia 92:900–907.

Pegg, K.G., J.K. Kochman, and N.T. Vock. 1980. Root and stem rot of soybean caused by *Phytophthora megasperma* var. *sojae*. Australas. Plant Pathol. 9:15.

Pennypacker, B.W. 1999. First report of sudden death syndrome caused by *Fusarium solani* f. sp. *glycines* on soybean in Pennsylvania. Plant Dis. 83:879.

Pennypacker, B.W., and M.L. Risius. 1999. Environmental sensitivity of soybean cultivar response to *Sclerotinia sclerotiorum*. 89:618–622.

Peterson, D.J., and H.H. Edwards. 1982. Effects of temperature and leaf wetness period on brown spot disease of soybeans. Plant Dis. 66:995–998.

Phillips, D.V. 1971. Influence of air temperature on brown stem rot of soybean. Phytopathology 61:1205–1208.

Phillips, D.V. 1972. Influence of photoperiod, plant age, and stage of development on brown stem rot of soybean. Phytopathology 62:1334–1337.

Phillips, D.V. 1973. Variation in *Phialophora gregata*. Plant Dis. Rep. 57:1063–1065.

Phillips, D.V. 1984. Stability of *Microsphaera diffusa* and the effect of powdery mildew on yield of soybean. Plant Dis. 68:953–956.

Phillips, D.V. 1999a. Downy mildew. p. 18–19. *In* G.L. Hartman et al. (ed.) Compendium of soybean diseases. 4th ed. APS Press, St. Paul, MN.

Phillips, D.V. 1999b. Frogeye leaf spot. p. 20–21. *In* G.L. Hartman et al. (ed) Compendium of soybean diseases. 4th ed. APS Press, St. Paul, MN.

Phillips, D.V., and H.R. Boerma. 1981. *Cercospora sojina* race 5: A threat to soybeans in the southeastern United States. Phytopathology 71:334–336.

Phillips, D.V., and H. R. Boerma. 1982. Two genes for resistance to race 5 of *Cercospora sojina* in soybeans. Phytopathology 72:764–766.

Phillips, D.V, and J.T. Yorinori. 1989. Frogeye leaf spot. p. 19–21 *In* J.B. Sinclair, and P.A. Backman (ed.) Compendium of soybean diseases. 3rd ed. APS Press, St. Paul, MN.

Phipps, P.M., and D.M. Porter. 1982. Sclerotinia blight of soybean caused by *Sclerotinia minor* and *Sclerotinia sclerotiorum*. Plant Dis. 66:163–165.

Ploper, L.D., K.L. Athow, and F.A. Laviolette. 1985. A new allele at the *Rps*3 locus for resistance to *Phytophthora megasperma* f. sp. *glycinea* in soybean. Phytopathology 75:690–694.

Poromarto, S.H., D.D. Nelson, and T.P. Freeman. 1998. Association of binucleate Rhizoctonia with soybean and mechanism of biocontrol of *Rhizoctonia solani*. Phytopathology 88:1056–1067.

Pratt, P.W., and J.A. Wrather. 1998. Soybean disease loss estimates for the southern United States, 1994 to 1996. Plant Dis. 82:114–120.

Punja, Z.K., J.D. Carter, G.M. Campbell, and E.L. Rossell. 1986. Effects of calcium and nitrogen fertilizers, fungicides, and tillage practices on incidence of *Sclerotium rolfsii* on processing carrots. Plant Dis. 70:819–824.

Quebral, F.C., and D.R. Pua. 1976. Screening of soybeans against *Sclerotium rolfsii*. Trop. Grain Leg. Bull. 6:22–23.

Radke, V.L., and C.R. Grau. 1986. Effects of herbicides on carpogenic germination of *Sclerotinia sclerotiorum*. Plant Dis. 70:19–23.

Raffel, S.J., E.R. Kazmar, R. Winberg, E.S. Oplinger, J. Handelsman, R. M. Goodman, and C.R. Grau. 1999. First report of root rot of soybeans caused by *Corynespora cassiicola* in Wisconsin. Plant Dis. 83:696.

Ringler, G.A., and C.D. Nickell. 1996. Genetic resistance to *Fusarium solani* in Pioneer Brand 9451 soybean. Soybean Genet. 23:144–148.

Rizvi, S.S.A., and X.B. Yang. 1996. Fungi associated with soybean seedling disease in Iowa. Plant Dis. 80:57–60.

Roane, C.W., and M.K. Roane. 1976. *Erysiphe* and *Microsphaera* as dual causes of powdery mildew of soybeans. Plant Dis. Rep. 60:611–612.

Ross, J.P. 1968. Additional physiologic races of *Cercospora sojina* on soybeans in North Carolina. Phytopathology 58:708–709.

Ross, J.P. 1975. Effect of overhead irrigation and benomyl sprays on late season foliar diseases, seed infection, and yields of soybean. Plant Dis. Rep. 59:809–813.

Ross, J.P. 1982. Effect of simulated dew and postinoculation moist periods on infection of soybean by *Septoria glycines* brown spot. Phytopathology 72:236–238.

Rothrock, C.S., T.W. Hobbs, and D.V. Phillips. 1985. Effects of tillage and cropping system on incidence and severity of southern stem canker of soybean. Phytopathology 75:1156–1159.

Rothrock, C.S., D.V. Phillips, and T.W. Hobbs. 1988. Effects of cultivar, tillage, and cropping system on infection of soybean by *Diaporthe phaseolorum* var. *caulivora* and southern stem canker symptom development. Phytopathology 78:266–270.

Roy, K.W. 1997a. *Fusarium solani* on soybean roots: nomenclature of the causal agent of sudden death syndrome and identity and relevance of *F. solani* form B. Plant Dis. 81:259–266.

Roy, K.W. 1997b. Sporulation of *Fusarium solani* f. sp. *glycines*, causal agent of sudden death syndrome, on soybeans in the midwestern and southern United States. Plant Dis. 81:566–569.

Roy, K.W., and T.S. Abney. 1976. Purple seed stain of soybeans. Phytopathology 66:1045–1049.

Roy, K.W., and T.S. Abney. 1988. Colonization of pods and infection of seeds by *Phomopsis longicolla* in susceptible and resistant lines inoculated in the greenhouse. Can.J. Plant Pathol. 10:317–320.

Roy, K.W., B.C. Keith, and C.H. Andrews. 1994. Resistance of hardseeded soybean lines to seed infection by *Phomopsis*, other fungi and soybean mosaic virus. Can. J. Plant Pathol. 12:43–47.

Roy, K.W., G.W. Lawrence, H.H. Hodges, K.S. McLean, and J.F. Killebrew. 1989. Sudden death syndrome of soybean: *Fusarium solani* as incitant and relation of *Heterodera glycines* to disease severity. Phytopathology 79:191–197.

Roy, K.W., and W.A. Miller. 1983. Soybean stem canker incited by isolates of *Diaporthe* and *Phomopsis* spp. from cotton in Mississippi. Plant Dis. 67:135–137.

Roy, K.W., M.V. Patel, and R.E. Baird. 2000. Colonization of *Heterodera glycines* cysts by *Fusarium solani* form A, the cause of sudden death syndrome, and other *Fusaria* in soybean fields in the midwestern and southern United States. Phytoprotection 81:57–67.

Rupe, J.C. 1989. Frequency and pathogenicity of *Fusarium solani* recovered from soybeans with sudden death syndrome. Plant Dis. 73:581–584.

Rupe, J.C., C.M. Becton, K.J. Williams, and P. Yount. 1996. Isolation, identification, and evaluation of fungi for the control of sudden death syndrome of soybean. Can. J. Plant. Pathol. 18:1–6.

Rupe, J.C., J.C. Correll, J.C. Guerber, C.M. Becton, E.E. Gbur, M.S. Cummings, and P.A. Yount. 2001. Differentiation of the sudden death syndrome pathogen of soybean, *Fusarium solani* f.sp *glycines*, from other isolates of *F. solani* based on cultural morphology, pathogenicity, and mitochondrial DNA restriction fragment length polymorphisms. Can. J. Bot. 79:829–835.

Rupe, J.C., and R.S. Ferris. 1986. Effects of pod moisture on soybean seed infection by *Phomopsis longicolla*. Phytopathology 76:273–277.

Rupe, J.C., and E.E. Gbur, Jr. 1995. Effect of plant age, maturity group, and the environment on disease progress of sudden death syndrome of soybean. Plant Dis. 79:139–143.

Rupe, J.C., E.E. Gbur, and D.M. Marx. 1991. Cultivar responses to sudden death syndrome of soybean. Plant Dis. 75:47–50.

Rupe, J.C. and G.L. Hartman. 1999. Sudden death syndrome. p. 37–39 *In* G.L. Hartman et al. (ed.) Compendium of soybean diseases. 4th ed. APS Press, St. Paul, MN.

Rupe, J.C., R.T. Robbins, C.M. Becton, W.A. Sabbe, and E.E. Gbur, Jr. 1999. Vertical and temporal distribution of *Fusarium solani* and *Heterodera glycines* in fields with sudden death syndrome of soybean. Soil Biol. Biochem. 31:245–251.

Rupe, J.C., W.E. Sabbe, R.T. Robbins, and E.E. Gbur, Jr. 1993. Soil and plant factors associated with sudden death syndrome of soybean. J. Prod. Agric. 6:218–221.

Rupe, J.C., and E.A. Sutton. 1994. Effect of temperature on infection of soybeans by the southern biotype of the stem canker pathogen, *Diaporthe phaseolorum* var. *caulivora*. Phytopathology 84:1120.

Rupe, J.C., E.A. Sutton, C.M. Becton, and E.E. Gbur, Jr. 1996. Effect of temperature and wetness period on recover of the southern biotype of *Diaporthe phaseolorum* var. *caulivora* from soybean. Plant Dis. 80:155–257.

Rupe, J.C., J.D. Widick, W.E. Sabbe, R.T. Robbins, and C.B. Becton. 2000. Effect of chloride and soybean cultivar on yield and the development of sudden death syndrome, soybean cyst nematode, and southern blight. Plant Dis. 84:669–674.

Rupe, J.R. 1999. Sclerotium blight. p. 48. *In* G.L. Hartman et al. (ed.) Compendium of soybean diseases. 4th ed. APS Press, St. Paul, MN.

Rush, M.C., and F.N. Lee. 1992. Sheath blight. p. 22–23. *In* R.K. Webster and P.S. Gunnell (ed.) Compendium of rice diseases. APS Press, St. Paul, MN.

Russin, J.S., D.J. Boethel, G.T. Berggren, and J.P. Snow. 1986. Effects of girdling by three-cornered alfalfa hopper on symptom expression of soybean stem canker and associated soybean yields. Plant Dis. 70:759–761.

Russin, J.S., C.H. Carter, and J.L. Griffin. 1995. Effects of preemergence herbicides for grain sorghum (*Sorghum bicolor*) on the charcoal rot fungus. Weed Technol. 9:343–351.

Russin, J.S., M.B. Layton, D.J. Boethel, E.C. McGawley, J.P. Snow, and G.T. Berggren. 1989a. Severity of soybean stem canker disease affected by insect-induced defoliation. Plant Dis. 73:144–147.

Russin, J.S., M.B. Layton, D.J. Boethel, E.C. McGawley, J.P. Snow, and G.T. Berggren. 1989b. Development of *Heterodera glycines* on soybean damaged by soybean looper and stem canker. J. Nematol. 21:108–114.

Russin, J.S., and S.R. Stetina. 1999. Rhizoctonia foliar blight. p. 24–25. *In* J.B. Sinclair et al. (ed) Compendium of soybean diseases. 4th ed. APS Press, St. Paul, MN.

Ryley, M.J., N.R., Obst, J.A.G., Irwin, and A. Drenth. 1998. Changes in the racial composition of *Phytophthora sojae* in Australia between 1979 and 1996. Plant Dis. 82:1048–1054.

Rytter, J.L., W.M. Dowler, and K.R. Bromfield. 1984. Additional alternative hosts of *Phakospora pachyrhizi*, causal agent of soybean rust. Plant Dis. 68:818–819.

Sanogo, S., and X.B. Yang. 2001. Relation of sand content, pH, and potassium and phosphorus nutrition to the development of sudden death syndrome in soybean. Can. J. Plant Pathol. 23:174–180.

Sanogo, S., X.B. Yang, and P. Lundeen. 2001. Field response of glyphosate-tolerant soybean to herbicides and sudden death syndrome. Plant Dis. 85:773–779.

Scherm, H., and X.B. Yang. 1996. Development of sudden death syndrome of soybean in relation to soil temperature and soil matric water potential. Phytopathology 86:642–649.

Scherm, H., X.B. Yang, and P. Lundeen. 1998. Soil variables associated with sudden death syndrome in soybean fields in Iowa. Plant Dis. 82:1152–1157.

Schmitthenner, A.F. 1972. Evidence for a new race of *Phytophthora megasperma* var. *sojae* pathogenic to soybean. Plant Dis. Rep. 56:536–539.

Schmitthenner, A.F. 1985. Problems and progress in control of Phytophthora root rot of soybean. Plant Dis. 69:362–368.

Schmitthenner, A.F. 1988. Phytophthora rot of soybean. p. 71–80. *In* T.D. Wyllie and D.H. Scott (ed.) Soybean diseases of the north central region. APS Press, St. Paul, MN.

Schmitthenner, A.F., M. Hobe, and R.G. Bhat. 1994. *Phytophthora sojae* races in Ohio over a 10-year interval. Plant Dis. 78:269–276.

Schneider, R.W., J.B. Sinclair, and L.E. Gray. 1972. Etiology *of Cephalosporium gregatum* in soybean. Phytopathology 62:345–349.

Schuh, W. 1991. Influence of temperature and leaf wetness period on conidial germination in vitro and infection of *Cercospora kikuchii* on soybean. Phytopathology 81:1315–1318

Schuh, W. 1992. Effect of pod development stage, temperature, and pod wetness duration on the incidence of purple seed stain of soybeans. Phytopathology 82:446–451.

Schuh, W. 1993. Influence of interrupted dew periods, relative humidity, and light on disease severity and latent infections caused by *Cercospora kikuchii* on soybean. Phytopathology 83:109–113.

Schuh, W. 1999. Cercospora blight, leaf spot, and purple seed stain. p. 17–18 *In* G.L. Hartman et al. (ed.) Compendium of soybean diseases. 4th ed. APS Press, St. Paul, MN.

Schuh, W., and A. Adamowicz. 1993. Influence of assessment time and modeling approach on the relationship between temperature-leaf wetness periods and disease parameters of *Septoria glycines* on soybeans. Phytopathology 83:941–948.

Schwenk, F.W., and T. Sim. 1974. Race 4 of *Phytophthora megasperma* var. *sojae* from soybeans proposed. Plant Dis. Rep. 58:353–354.

Scott, D.H. 1988. Soybean sudden death syndrome. p. 67–70 *In* T.D. Wyllie and D.H. Scott (ed.) Soybean diseases of the North Central Region. APS Press, St. Paul, MN.

Seaman, W.L., R.A. Shoemaker, and E.A. Peterson. 1965. Pathogenicity of *Corynespora cassiicola* on soybean. Can. J. Bot. 43:1461–1469.

Sebastian, S. A., and C.D. Nickell. 1985. Inheritance of brown stem rot resistance in soybeans. J. Hered. 76:194–198.

Sebastian, S. A., C.D. Nickell, and L.E. Gray. 1985. Efficient selection for brown stem rot resistance in soybean under greenhouse screening conditions. Crop Sci. 25:753–757.

Sebastian, S. A., C.D. Nickell, and L.E. Gray. 1986. Relationship between greenhouse and field ratings for brown stem rot reaction in soybean. Crop Sci. 26:665–667.

Sherwin, H.S., and K.W. Kreitlow. 1952. Discoloration of soybean seeds by the frogeye fungus, *Cercospora sojina*. Phytopathology 42:568–572.

Short, G.E., T.D. Wyllie, and V.D. Ammon. 1978. Quantitative enumeration of *Macrophomina phaseolina* in soybean tissue. Phytopathology 68:736–741.

Short, G.E., T.D. Wyllie, and P.R. Bristow. 1980. Survival of *Macrophomina phaseolina* in soil and in residue of soybean. Phytopathology 70:13–17.

Shortt, B.J., A.P. Grybauskas, F.D. Tenne, and J.B. Sinclair. 1981. Epidemiology of Phomopsis seed decay of soybean in Illinois. Plant Dis. 65:62–64.

Sickinger, S. M., Grau, C. R., and Harvey, R. G. 1987. Verticillium wilt of velvetleaf (*Abutilon theophrasti* Medic.). Plant Dis. 71:415–418.

Sills, G. R., E.T. Gritton, and C.R. Grau. 1991. Differential reaction of soybean genotypes to isolates of *Phialophora gregata*. Plant Dis. 75:687–690.

Sinclair, J.B. (ed.) 1982. Compendium of soybean diseases. 2nd ed. APS Press, St. Paul, MN.

Sinclair, J.B. 1988a. Anthracnose of soybeans. p. 92–95. *In* T.D. Wyllie and D.H. Scott (ed.) Soybean diseases of the North Central Region. APS Press, St. Paul, MN.

Sinclair, J.B. 1988b. Diaporthe/Phomopsis complex of soybeans. p. 96–101. *In* T.D. Wyllie and D.H. Scott (ed.) Soybean diseases of the north central region. APS Press, St. Paul, MN.

Sinclair, J.B. 1989. Threats to soybean production in the tropics: Red leaf blotch and leaf rust. Plant Dis. 73:604–606.

Sinclair, J.B. 1991. Latent infection of soybean plants and seeds by fungi. Plant Dis. 75:220–224

Sinclair, J.B. 1999a. Diaporthe-Phomopsis complex. p. 31. *In* G.L. Hartman et al. (ed.) Compendium of soybean diseases. 4th ed. APS Press, St. Paul, MN.

Sinclair, J.B. 1999b. Powdery mildew. p. 22–23. *In* G.L. Hartman et al. (ed) Compendium of soybean diseases. 4th ed. APS Press, St. Paul, MN.

Sinclair, J.B. 1999c. Target spot. p. 27. *In* G.L. Hartman et al. (ed.) Compendium of soybean diseases. 4th ed. APS Press, St. Paul, MN.

Sinclair, J.B., and P.A Backman (ed.). 1989. Compendium of soybean diseases. 3rd ed. APS Press, St. Paul, MN.

Sinclair, J.B., and G.L. Hartman (ed.) 1996. Proceedings of the Soybean Rust Workshop, Urbana, IL. Natl. Soybean Res. Lab. Publ. 1. Urbana-Champaign, IL.

Sinclair, J.B., and G.L. Hartman. 1999a. Brown spot. *In* G.L. Hartman et al. (ed.) Compendium of soybean diseases. 4th ed. APS Press, St. Paul, MN.

Sinclair, J.B., and G.L. Hartman. 1999b. Soybean rust. p. 25–26. *In* G.L. Hartman et al. (ed.) Compendium of soybean diseases. 4th ed. APS Press, St. Paul, MN.

Slater, G.P., R.W. Elmore, B.L. Doupnik, Jr., and R.B. Ferguson. 1991. Soybean cultivar yield response to benomyl, nitrogen, phosphorus, and irrigation levels. Agron. J. 83:804–809.

Smith, E. F., and P.A. Backman. 1988. Soybean stem canker: An overview. p. 47–55. *In* T.D. Wyllie and D.H. Scott (ed.) Soybean diseases of the north central region. APS Press, St. Paul, MN.

Smith, G.S., and O.N. Carvil. 1997. Field screening of commercial and experimental soybean cultivars for their reaction to *Macrophomina phaseolina*. Plant Dis. 81:363–368.

Smith, G.S., and T.D. Wyllie. 1999. Charcoal rot. p. 29–31. *In* G.L. Hartman et al. (ed.) Compendium of soybean diseases. 4th ed. APS Press, St. Paul, MN.

Smith, R.L. 2001 Soybean Phytophthora RS [Online database]. Available at http://www.ars-grin.gov/cgi-bin/npgs/html/eval.pl/491592 (verified 25 Oct. 2002) Natl. Germplasm Resources Lab., Beltsville, MD.

Sneh, B., L. Burpee, and A. Ogoshi. 1991. Identification of *Rhizoctonia* Species. APS Press, St.Paul, MN.

Soglio, F.K., B.L. Bertagnolli, J.B. Sinclair, G.Y. Yu, and D.M. Eastburn. 1998. Production of chitinolytic enzymes and endoglucanase in the soybean rhizosphere in the presence of *Trichoderma harzianum* and *Rhizoctonia solani*. Biocontrol 12:111–117.

Song, H.S., S.M. Lim, and J.M. Clark, Jr. 1993. Purification and partial characterization of a host-specific pathotoxin from culture filtrates of *Septoria glycines*. Phytopathology 83:659–661.

Song, H.S., S.M. Lim, and J.M. Widholm. 1994. Selection and regeneration of soybeans resistant to the pathogenic culture filtrates of *Septoria glycines*. Phytopathology 84:948–951.

Spencer, J.A., and H.J. Walters. 1969. Variations in certain isolates of *Corynespora cassiicola*. Phytopathology 59:58–60.

Srivastava, A.K., D.K. Arora, S. Gupta, R.R. Pandey, and M.W. Lee. 1996. Diversity of potential microbial parasites colonizing sclerotia of *Macrophomina phaseolina* in soil. Biol. Fertil. Soils 22:136–140.

St. Martin, S.K., D.R. Scott, A.F. Schmitthenner, and B.A. McBlain. 1994. Relationship between tolerance to Phytophthora rot and soybean yield. Plant Breed. 113:331–334.

Stephens, P.A., C.D. Nickell, and F.L. Kolb. 1993. Genetic analysis of resistance to *Fusarium solani* in soybean. Crop Sci. 33:929–930.

Stuckey, R.E., S.A. Ghabrial, and D.A. Reicosky. 1982. Increased incidence of *Phomopsis* sp. in seed from soybeans infected with bean pod mottle virus. Plant Dis. 66:826–829.

Su, G., S.-O. Suh, R.W. Schneider, and J.S. Russin. 2001. Host specialization in the charcoal rot fungus, *Macrophomina phaseolina*. Phytopathology 91:120–126.

Subbarao, K.V., J.P. Snow, G.T. Berggren, J.P. Damicone, and G.B. Padgett. 1992. Analysis of stem canker epidemics in irrigated and nonirrigated conditions on differentially susceptible soybean cultivars. Phytopathology 82:1251–1256.

Sugawara, K., K. Kobayashi, and A. Ogoshi. 1997. Influence of the soybean cyst nematode (*Heterodera glycines*) on the incidence of brown stem rot in soybean and adzuki bean. Soil Biol. Biochem. 29:1491–1498.

Sumner, D.R., and N.A. Minton, 1987. Interaction of Fusarium wilt and nematodes in Cobb soybean. Plant Dis. 71:20–23.

Sun, P, and X.B. Yang. 2000. Light, temperature, and moisture effects on apothecium production of *Sclerotinia sclerotiorum*. Plant Dis. 84:1287–1293.

Sutton, D.C., and B.J. Deverall. 1983. Studies on infection of bean (*Phaseolus vulgaris*) and soybean (*Glycine max*) by ascospores of *Sclerotinia sclerotiorum*. Plant Pathol. 32:251–261.

Sutton, D.C., and B.J. Deverall. 1984. Phytoalexin accumulation during infection of bean and soybean by ascospores and mycelium of *Sclerotinia sclerotiorum*. Plant Pathol. 33:377–383.

Tabor, G. M., G.L. Tylka, and C.R. Bronson. 2001. *Heterodera glycines* increases susceptibility of soybeans to *Phialophora gregata*. Phytopathology (Abstr.) 91:S87.

Tachibana, H. 1968. Rhizoctonia solani root rot epidemic of soybeans in central Iowa 1967. Plant Dis. Rep. 52:613–614.

Tachibana, H. 1982. Prescribed resistant cultivars for controlling brown stem rot of soybeans and managing resistance genes. Plant Dis. 66:271–274.

Tachibana, H., A.H. Epstein, and B.J. Havlovic. 1989. Effects of four years of continuous cropping of maturity group II soybeans resistant to brown stem rot on brown stem rot and yield. Plant Dis. 73:846–849.

Tachibana, H., D. Jowett, and W.R. Fehr. 1971. Determination of losses in soybeans caused by *Rhizoctonia solani*. Phytopathology 61:1444–1446.

TeKrony, D.M., D.B. Egli, J. Balles, L. Tomes, and R.E. Stucky. 1984. Effect of date of harvest maturity on soybean seed quality and *Phomopsis* sp. seed infection. Crop Sci. 24:189–193.

TeKrony, D.M., D.B. Egli, R.E. Stucky, and T.M. Loeffler. 1985a. Effect of benomyl applications on soybean seedborne fungi, seed germination, and yield. Plant Dis. 69:763–765.

TeKrony, D.M., R.E. Stucky, D.B. Egli, and L. Tomes. 1985b. Effectiveness of a point system for scheduling foliar fungicides in soybean seed fields. Plant Dis. 69:962–965.

Tenne, F.D., and J.B. Sinclair. 1978. Control of internally seedborne microorganisms of soybean with foliar fungicides in Puerto Rico. Plant Dis. Rep. 62:459–463.

Thaning, C., and H.E. Nilsson. 2000. A narrow range of wavelengths active in regulating apothecial development in *Sclerotinia sclerotiorum*. J. Phytopathol. 148:627–631.

Thomison, P.R., W.J. Kenworthy, and M.S. McIntosh. 1990. Phomopsis seed decay in soybean isolines differing in stem termination, time of flowering and maturity. Crop Sci. 30:183–188.

Thomison, P.R., C.A. Thomas, and W.J. Kenworthy. 1991. Tolerant and root-resistant soybean cultivars: Reactions to Phytophthora rot in inoculum-layer tests. Crop Sci. 31:73–75.

Thomison, P.R., C.A. Thomas, W.J. Kenworthy, and M.S. McIntosh. 1988. Evidence of pathogen specificity in tolerance of soybean cultivars to phytophthora rot. Crop Sci. 28:714–715.

Todd, T.C, J.H. Long, Jr., and T.R. Oakley. 2000. Effects of maturity and determinacy in soybean on host-parasite relationships of *Heterodera glycines*. J. Nematol. 32:584–590.

Tooley, P.W., and C.R. Grau. 1982. Identification and quantitative characterization of rate-reducing resistance to *Phytophthora megasperma* f. sp. *glycines* in soybean seedlings. Phytopathology 72:727–733.

Tooley, P.W., and C.R. Grau. 1984a. Field characterization of rate-reducing resistance to *Phytophthora megasperma* f. sp. *glycinea* in soybean. Phytopathology 74:1201–1208.

Tooley, P.W., and C.R. Grau. 1984b. The relationship between rate-reducing resistance to *Phytophthora megasperma* f.sp. *glycinea* and yield of soybean. Phytopathology 74:1209–1216.

Tooley, P.W., and C.R. Grau. 1986. Microplot comparison of rate-reducing and race-specific resistance to *Phytophthora megasperma* f. sp. *glycinea* in soybean. Phytopathology 76:554–557.

Torto, T.A., V. Njiti, and D.A. Lightfoot. 1996. Loci underlying resistance to sudden death syndrome and *Fusarium solani* in field and greenhouse assays do not correspond. Soybean Genet. Newsl. 23:163–166.

Tyler, B.M., M. Wu, J. Wang, W. Cheung, and P.F. Morris. 1996. Chemotactic preferences and strain variation in response of *Phytophthora sojae* zoospores to host isoflavones. Appl. Environ. Microbiol. 62:2811–2817.

Vakili, N.G., and K.R. Bromfield. 1976. Phakospora rust on soybean and other legumes in Puerto Rico. Plant Dis. Rep. 60:995–999.

Vedenyapina, E.G., G.R. Safir, B.A. Niemira, and T.E. Chase. 1996. Low concentrations of the isoflavone genistein influence in vitro asexual reproduction and growth of *Phytophthora sojae*. Phytopathology 86:144–148.

Vest, J.D., and J.S. Russin. 2001. Effects of soil temperature, moisture, and planting date on soybean sudden death syndrome root and foliar disease. Phytopathology 91:S91–92.

Vick, C.M, S.K. Chong, J.P Bond, and J.S. Russin. 2001. Sudden death syndrome and the soil physical environment: a two-year study. Phytopathology 91:S92.

Von Qualen, R.H., T.S. Abney, D.M. Huber, and M.M. Schreiber. 1989. Effects of rotation, tillage, and fumigation on premature dying of soybeans. Plant Dis. 73:740–744.

Wagner, R.E., and H.T. Wilkinson. 1992. A new physiological race of *Phytophthora sojae* on soybean. Plant Dis. 76:212.

Walcot, R.R., D.C. McGee, and M.K. Misra. 1998. Detection of asymptomatic fungal infections of soybean seeds by ultrasound analysis. Plant Dis. 82:584–589.

Walker, A.K., and A.F. Schmitthenner. 1984. Heritability of tolerance to Phytophthora rot in soybean. Crop Sci. 24:490–491.

Walker, J.T., D.V. Philips, J. Melin, and D. Spradlin. 1994. Effects of intermittent acidic irrigations on soybean yields and frogeye leaf spot. Environ. Exp. Bot. 34:311–318.

Wall, M.T., D.C., McGee, and J.S. Burris. 1983. Emergence and yield of fungicide-treated soybean seed differing in quality. Agron. J. 75:969–973.

Wallen, V.R., and W.L. Seaman. 1963. Seed infection of soybean by *Diaporthe phaseolorum* and its influence on host development. Can. J. Bot. 41:13–21.

Waller, R. S., C.D. Nickell, and L.E. Gray. 1992. Environmental effects on the development of brown stem rot in soybean. Plant Dis. 76:454–457.

Walters, H.J. 1980. Soybean leaf blight caused by *Cercospora kikuchii*. Plant Dis. 64:961–962.

Walters, H.J. 1985. Purple seed stain and Cercospora leaf blight. *In* R. Shibles (ed.) World Soybean Res. Conf. III: Proc., Ames, IA. 12–17 Aug. 1984. Westview Press, Boulder, CO.

Ward, E.W.B. 1990. The interaction of soya beans with *Phytophthora megasperma* f.sp. *glycinea*: Pathogenicity. p. 311–327. *In* D. Hornby (ed.) Biological control of soil-borne plant pathogens. CAB International, Wallingford, England.

Weber, C.R., J.M. Dunleavy, and W.R. Fehr. 1966. Influence of brown stem rot on agronomic performance of soybeans. Agron. J. 258:519–520.

Wegulo, S.N., C.A. Martinson, and X.B. Yang. 2000. Spread of Sclerotinia stem rot of soybean from area and point sources of apothecial inoculum. Can. J. Plant Sci. 80:389–402.

Wegulo, S.N., X.B. Yang, and C.A. Martinson. 1998. Soybean cultivar responses to *Sclerotinia sclerotiorum* in field and controlled environment studies. Plant Dis. 82:1264–1270.

Wilcox, J.R. 2001. Sixty years of improvement in publicly developed elite soybean lines. Crop Sci. 49:1711–1716.

Wilcox, J.R., and T.S. Abney. 1973. Effects of *Cercospora kikuchii* on soybeans. Phytopathology 63:796–797.

Wilcox, J.R., F.A. Laviolette, and R.J. Martin. 1975. Heritability of purple seed stain resistance in soybean. Crop Sci. 15:525–526.

Wilcox, J.R. and S.K. St. Martin. 1998. Soybean genotypes resistant to *Phytophthora sojae* and compensation for yield losses of susceptible isolines. Plant Dis. 82:303–306.

Williams, D.J., and R.F. Nyvall. 1980. Leaf infection and yield losses caused by brown spot and bacterial blight diseases of soybean. Phytopathology 70:900–902.

Willmot, D. B., C.D. Nickell, and L.E. GrayE. 1989. Physiological specialization of *Phialophora gregata* on soybean. Plant Dis. 73:290–294.

Workneh, F., G.L. Tylka, X.B. Yang, J. Faghihi, and J.M. Ferris. 1999a. Regional assessment of soybean brown stem rot, *Phytophthora sojae*, and *Heterodera glycines* using area-frame sampling: Prevalence and effects of tillage. Phytopathology 89:204–211.

Workneh, F., and X.B. Yang. 2000. Prevalence of Sclerotinia stem rot of soybeans in the north-central United States in relation to tillage, climate and latitudinal positions. Phytopathology 90:1375–1382.

Workneh, F., X.B. Yang, and G.L. Tylka. 1998. Effect of tillage practices on vertical distribution of *Phytophthora sojae*. Plant Dis. 82:1258–1263.

Workneh, F., X.B. Yang, and G.L. Tylka. 1999b. Soybean brown stem rot, *Phytophthora sojae*, and *Heterodera glycines* affected by soil texture and tillage relations. Phytopathology 89:844–850.

Wrather, J.A., T.R. Anderson, D.M. Arsyad, Y. Tan, L.D. Ploper, A. Porta-puglia, H.H. Ram, and J.T. Yorinori. 2001a. Soybean disease loss estimates for the top ten soybean-producing countries in 1998. Can. J. Plant Pathol. 23:115–121.

Wrather, J.A., A.Y. Chambers, J.A. Fox, W.F. Moore, and G.L. Sciumbato. 1995. Soybean disease loss estimates for the southern United States, 1974 to 1994. Plant Dis. 79:1076–1079.

Wrather, J.A., S.R. Kendig, S.C. Anand, T.L. Niblack, and G.S. Smith. 1995. Effects of tillage, cultivar, and planting date on percentage of soybean leaves with symptoms of sudden death syndrome. Plant Dis. 79:560–562.

Wrather, J.A., S.R. Kendig, W.J. Wiebold, and R.D. Riggs. 1996. Cultivar and planting date effects on soybean stand, yield and *Phomopsis* sp. seed infection. Plant Dis. 80:622–624.

Wrather, J.A., and G.L. Sciumbato. 1995. Soybean disease loss estimates for the southern United States during 1992 and 1993. Plant Dis. 79:84–85.

Wrather, J.A., W.C. Stienstra, and S.R. Koenning. 2001b. Soybean disease loss estimates for the United States from 1996 to 1998. Can. J. Plant Pathol. 23:122–131.

Wyllie, T.D. 1976. *Macrophomina phaseolina*—Charcoal rot. p. 482–484. *In* L.D. Hill (ed.) World Soybean Research. Proc. of the World Soybean Res. Conf., Champaign, IL. Interstate, Danville, IL.

Wyllie, T.D. 1988. Charcoal rot of soybeans—Current status. p. 106–113. *In* T.D. Wyllie and D.H. Scott (ed.) Soybean diseases of the north central region. APS Press, St. Paul, MN.

Wyllie, T.D., S. Gangopadhyay, W.R. Taegue, and R.W. Blanchar. 1984. Germination and production of *Macrophomina phaseolina* microsclerotia as affected by oxygen and carbon dioxide concentration. Plant Soil 81:195–201.

Wyllie, T.D., and D.H. Scott (ed.). 1988. Soybean diseases of the North Central Region. APS Press, St. Paul, MN.

Yanchun, S., and S. Chongyao. 1993. The discovery and biological characteristics studies of *Phytophthora megasperma* f.sp. *glycinea* on soybean in China. ACTA Phytopathol. Sin. 23:341–347.

Yang, W., and D.B. Weaver. 2001. Resistance to frogeye leaf spot in maturity groups VI and VII of soybean germplasm. Crop Sci. 41:549–552.

Yang, W., D.B. Weaver, B.L. Nielsen, and J. Qiu. 2001. Molecular mapping of a new gene for resistance to frogeye leaf spot of soya bean in 'Peking'. Plant Breed. 120:73–78.

Yang, X.B. 1999a. Pythium damping-off and root rot. p. 42–44 *In* G.L. Hartman et al. (ed.) Compendium of soybean diseases. 4th ed. APS Press, St.Paul, MN.

Yang, X.B. 1999b. Rhizoctonia damping-off and root rot. p. 45–46. *In* G.L. Hartman et al. (ed.) Compendium of soybean diseases. 4th ed. APS Press, St. Paul, MN.

Yang, X.B., G.T. Berggren, and J.P. Snow. 1990a. Types of Rhizoctonia foliar blight on soybean in Louisiana. Plant Dis. 74:501–504.

Yang, X.B., G.T. Berggren, and J.P. Snow. 1990b. Seedling infection of soybean by isolates of *Rhizoctonia solani* AG-1, causal agent of aerial blight and web blight of soybean. Plant Dis. 74:485–488.

Yang, X.B., G.T. Berggren, and J.P. Snow. 1990c. Effects of free moisture and soybean growth stage on focus expansion of Rhizoctonia foliar blight. Phytopathology 80:497–503.

Yang, X.B., and L.E. del Río. 2002. Implementation of biological control of plant diseases in Integrated Pest Management systems. p. 339–354. *In* E. Gnanamanickam (ed.) Biological control of crop diseases. Marcel Dekker, New York.

Yang, X.B, and F. Feng. 2001. Ranges and diversity of soybean fungal diseases in North America. Phytopathology 91:769–775.

Yang, X.B., P. Lundeen, and M.D. Uphoff. 1999. Soybean varietial response and yield loss caused by *Sclerotinia sclerotiorum*. Plant Dis. 83:456–461.

Yang, X.B., M.H. Royer, A.T. Tschantz, and B.Y. Tsia. 1990d. Analysis and quantification of soybean rust epidemics from seventy-three sequential planting experiments. Phytopathology 80:1421–1427.

Yang, X. B., R.L. Ruff, X.Q. Meng, and F. Workneh. 1996. Races of *Phytophthora sojae* in Iowa soybean fields. Plant Dis. 80:1418–1420.

Yang, X.B., J.P. Snow, and G.T. Berggren. 1990e. Analysis of epidemics of Rhizoctonia aerial blight of soybean in Louisiana. Phytopathology 80:386–392.

Yang, X.B., J.P. Snow, and G.T. Berggren. 1991. Patterns of Rhizoctonia foliar blight on soybean and effect of aggregation on disease development. Phytopathology 81:287–293.

Yang, X.B, M.D. Uphoff, and S. Sanogo. 2001. Outbreaks of soybean frogeye leaf spot in Iowa. Plant Dis. 85:443.

Yang, X.B., F. Workneh, and P. Lundeen. 1998. First report of sclerotium production by *Sclerotinia sclerotiorum* in soil on infected soybean seed. Plant Dis. 82:264.

Yeater, K.M., C.R. Grau, and A.L. Rayburn. 2002. Flow cytometric analysis of *Phialophora gregata* isolated from soybean plants resistant and susceptible to brown stem rot. J. Phytopathol.150:258–262.

Yorinori, J.T. 1992. Management of foliar fungal diseases in Brazil. p.185–193. *In* L.G. Copping et al. (ed.) Pest management in soybean. Elsevier Appl. Sci., London.

Yorinori, J.T., and M. Homechin. 1985. Sclerotinia stem rot of soybeans, its importance and research in Brazil. p. 582–588. *In* R. Shibles (ed.) World Soybean Res. Conf. III, Ames, IA. 12–17 Aug. 1984. Westview Press, Boulder, CO.

Young, L.D., and J.P. Ross. 1978. Resistance evaluation and inheritance of a nonchlorotic response to brown spot of soybean. Crop Sci. 18:1075–1077.

Young, L. D., and J.P. Ross. 1979. Brown spot development and yield response of soybean inoculated with *Septoria glycines* at various growth stages. Phytopathology 69:8–11.

Zhang, B.Q., W.E. Chen, and X.B. Yang. 1998. Occurrence of *Pythium* species in long-term maize and soybean monoculture and maize/soybean rotations. Mycol. Res. 102:1450–1452.

Zhang, A.W., G.L. Hartman, L. Riccioni, W.D. Chen, R.Z. Ma, and W.L. Pedersen. 1997. Using PCR to distinguish *Diaporthe phaseolorum* and *Phomopsis longicolla* from other soybean fungal pathogens and to detect them in soybean tissues. Plant Dis. 81:1143–1149.

Zhang, B.Q., and X.B. Yang. 2000. Pathogenicity of *Pythium* populations from corn-soybean rotation fields. Plant Dis. 84:94–99.

Zhang, B.Q, X.B. Yang, and W.D. Chen. 1996. First report of the pathogenicity of *Pythium torulosum* on soybean. Plant Dis. 80:710.

Zimmerman, M.S., and H.C. Minor. 1993. Inheritance of Phomopsis seed decay resistance in soybean PI 417479. Crop Sci. 33:96–100.

Zorrilla, G., A.D. Knapp, and D.C. McGee. 1994. Severity of phomopsis seed decay, seed quality evaluation, and field performance of soybean. Crop Sci. 34:172–177.

15   Viral, Bacterial, and Phytoplasmal Diseases of Soybean

SUE A. TOLIN AND GEORGE H. LACY

Virginia Polytechnic Institute and State University
Blacksburg, Virginia

In this chapter, the characteristics of the diseases of soybean [*Glycine max* (L.) Merr.] caused by viruses, bacteria, and phytoplasmas are summarized, including information about the pathogens, their distribution, epidemiology, effects on soybean, and management approaches. At least 46 viruses belonging to 27 distinct taxonomic groups, eight bacteria, and five or more phytoplasmas have been reported from soybean. Aphids (Aphidadae: Hemiptera) transmit soybean viruses in either a stylet-borne, nonpersistent (15 viruses) or circulative (4 viruses) manner, by beetles (Chrysomelidae: Coleoptera) (5 viruses), thrips (Thripidae: Thysanoptera) (4 viruses), whiteflies (Aleyrodidae: Hemiptera) (7 viruses), or nematodes (Dorylaimida: Nematoda) (3 viruses). There are also viruses for which no biological vector has been reported. Transmission of the pathogen through soybean seed has been demonstrated for eight viruses, four bacteria, and one phytoplasma. Sources of the pathogens other than soybean seed may be other crops, weeds, or vectors. Control can be by elimination of pathogen source, vector management, crop management to escape pathogens, or use of resistant cultivars, which are available for only a few viruses and bacteria. Several genes and alleles for resistance to viruses or bacteria have been mapped onto the molecular soybean genetic linkage map, but none as yet have been cloned. Pathogen-mediated, genetically engineered resistance has been demonstrated for two viruses.

15–1 VIRAL PATHOGENS OF SOYBEAN

Soybean diseases caused by viruses are found in all soybean-growing areas worldwide. The emphasis in this chapter will be on viruses that have been described to naturally occur in soybean and cause agronomic losses to soybean production in the USA or some soybean production area of the world. Table 15–1 lists 67 names and abbreviations (acronyms) of viruses, and includes naturally-occurring viruses as well as those capable of replicating in soybean following artificial inoculation and having the potential to be introduced into new areas and emerge or re-emerge as disease threats. Information about the genome of each virus is included. Table 15–2 lists 46 viruses found to occur in soybean in nature, giving the transmissibility of the virus mechanically through rubbing extracted sap onto leaves, vertically

Table 15–1. Classification, nomenclature, and genome characteristics of viruses that infect, or have the potential to infect, soybean.

| Genome\taxon† | Virus species‡ | Acronym§ | Segments | Genome‖ | | Soybean as a host# | |
				Size in nucleotides	Entries	Natural	Inoculated
dsDNA-RT							
Caulimoviridae							
'SbCMV-like'	Peanut chlorotic streak	PCSV	1	8174	9		+
'SbCMV-like'	Soybean chlorotic mottle	SbCMV	1	8175	4	+	+
ssDNA							
Geminiviridae							
Begomovirus	Abutilon mosaic	AbMV	2	2581 + 2629	18	+	
Begomovirus	African soybean dwarf			--	0	+	
Begomovirus	Bean calico mosaic	BCaMV	2	2572 + 2603	9	+	+
Begomovirus	Bean golden mosaic	BGMV	2	2585 + 2646	29	+	
Begomovirus	Horsegram yellow mosaic	HgYMV		--	0		
Begomovirus	Mung bean yellow mosaic	MYMV	2	2675+2723	19	+	
Begomovirus	Soybean yellow mosaic	SYMV	2 ?			+	
Begomovirus	Soybean crinkle leaf	SCLV	1	2737	1	+	
Uncharacterized	Soybean geminivirus	'SbGV'		--	0	+	
Unassigned							
Nanovirus	Milk vetch dwarf	MDV	10	~1000 each	10	+	
NssRNA							
Bunyaviridae							
Tospovirus	Groundnut (peanut) bud necrosis	PBNV	3	3057 + 4801 + 8911	3	+	
Tospovirus	Groundnut (peanut) yellow spot	GYSV		--	0		
Tospovirus	Tomato spotted wilt	TSWV	3	2916 + 4851 + 8897	86	+	+
ssRNA							
Comoviridae							
Comovirus	Bean pod mottle	BPMV	2	3662 + 5995	10	+	
Comovirus	Bean rugose mosaic	BRMV		--	0	+	
Comovirus	Cowpea mosaic	CPMV	2	3481 + 5889	13		+
Comovirus	Cowpea severe mosaic	CPSMV	2	3732 + 5957	3	+	+
Comovirus	Glycine mosaic	GMV		--	0		
Comovirus	Pea mild mosaic	PMiMV		--	0		
Comovirus	Quail pea mosaic	QPMV		--	0	+	+ +

Genus / Family	Common name	Acronym	n	Genome	No.	Markers
Comovirus	*Red clover mottle*	RCMV	2	3543 + 6033	5	+ +
Fabavirus	*Broad bean wilt*	BBWV	2	3667 + 5951	9	+ +
Nepovirus	*Tobacco ringspot*	TRSV	2	7514	7	+ +
Nepovirus	*Tomato ringspot*	ToRSV	2	7273 + 8214	14	+ + +
Potyviridae						
Potyvirus	*Bean common mosaic*	BCMV	1	9992	56	+ +
Potyvirus	*Azuki bean mosaic*	'BCMV-AzB'	1	Incomplete	2	+
Potyvirus	*Peanut stripe*	'BCMV-PSt'	1	10 056	33	+
Potyvirus	*Bean yellow mosaic*	BYMV	1	9532	36	+
Potyvirus	*Clover yellow vein*	CYVV	1	9584	22	+
Potyvirus	*Passionfruit woodiness*	PWV	1	Incomplete	5	+
Potyvirus	*Peanut mottle*	PeMoV	1	9709	8	+
Potyvirus	*Soybean mosaic*	SMV	1	9588	84	+ + +
Potyvirus	*Watermelon mosaic*	WMV	1	Incomplete	9	+ + +
Luteoviridae						
Unassigned	*Bean leaf roll*	BLRV	1	5964	4	+
Luteovirus	*Beet western yellows*	BWYV	1	5642	64	+ +
Unassigned	*Soybean dwarf*	SbDV	1	5708	9	+ +
Unassigned	*Indonesian soybean dwarf*	ISbDV	1	--	0	+
Unassigned	*Subterranean clover red leaf*	'SbDV-SCRL'	1	--	0	+
Tombusviridae						
Carmovirus	*Bean mild mosaic*	BMMV	1	--	0	
Carmovirus	*Blackgram mottle*	BMoV	1	--	0	+
Carmovirus	*Cowpea mottle*	CPMoV	1	4029	3	+
Dianthovirus	*Red clover necrotic mosaic*	RCNMV	2	1448 + 3890	4	+ + +
Necrovirus	*Tobacco necrosis*	TNV	1	3684	11	+ + +
Bromoviridae						
Alfamovirus	*Alfalfa mosaic*	AMV	3	2037 + 2593 + 3644	70	+ +
Bromovirus	*Cowpea chlorotic mottle*	CCMV	3	2170 + 2774 + 3171	11	+ + +
Bromovirus	*Spring beauty latent*	SBLV	3	--	0	
Cucumovirus	*Cucumber mosaic*-soybean stunt	'CMV-SS'	3	2216 + 3050 + 3357	416	+ + +
Cucumovirus	*Peanut stunt*	PSV	3	2188 + 2947 + 3357	21	+ + +
Ilarvirus	*Tobacco streak*	TSV	3	2205 + 2926 + 3491	10	+ + +

(continued on next page)

Table 15–1. Continued.

Genome\taxon†	Virus species‡	Acronym§	Segments	Genome‖ Size in nucleotides	Entries	Soybean as a host# Natural	Inoculated
Unassigned Genera							
Carlavirus	*Cowpea mild mottle*	CPMMV	1	--	0	+	
Carlavirus	*Pea streak*	PeSV	1	Incomplete	2		+
Carlavirus	*Poplar mosaic*	PopMV	1	Incomplete	4		+
Pecluvirus	*Peanut clump*	PCV	2	4504 + 5897	12		+
Tobravirus	*Pea early browning*	PEBV	2	3374 + 7073	9		+
Tobravirus	*Tobacco rattle* -soybean fleck	'TRV-SF'	2	Variable	97	+	
Tobamovirus	*Tobacco mosaic* -soybean	'TMV-S'	2	Variable	247	+	+
Potexvirus	*Asparagus 3*	AV-3	1	--	0		
Potexvirus	*Clover yellow mosaic*	CYMV	1	7015	7	+	
Potexvirus	*Commelina X*	ComXV	1	--	0		+
Potexvirus	*White clover mosaic*	WCMV	1	5846	3		+
Sobemovirus	*Southern bean mosaic*	SBMV	1	4194	8	+	
Sobemovirus	*Subterranean clover mottle*	SCMoV	1	--	0		
Umbravirus	*Groundnut rosette*	GRV	1	4019	28		+
Unclassified	Soybean yellow vein	SbYVV		--	0	+	
Unclassified	Soybean mild mosaic	'SbMMV'		--	0	+	
Uncharacterized	Soybean rhabdovirus	'SbRhV'		--	0	+	

† NA refers to type of nucleic acid comprising the genome. dsDNA-RT = double-stranded DNA utilizing reverse transcription to RNA in its replication cycle; ssDNA = single-stranded DNA; NssRNA = negative sense single-stranded RNA; ssRNA = single-stranded RNA. Taxon refers to Family (ending in –viridae) or Genus name (ending in –virus), if italicized. "Unassigned" means the genus has not been assigned to a Family and is in the category of "Floating Genera", or the virus species has not yet been assigned to a recognized genus.

‡ Species names accepted by the International Committee for the Taxonomy of Viruses (ICTV) (van Regenmortel et al., 2000) are italicized. Non-italicized species are described isolates or strains that have characteristics of the taxon into which they are placed, or have been insufficiently characterized to name as a distinct virus.

§ Abbreviations are those accepted by the ICTV and are from Fauquet and Mayo (1999) and Hull (2001). Abbreviations in single quotes are not yet accepted, and are assigned for this chapter only.

‖ Number of unique nucleic acid segments comprising the complete viral genome. Size of each segment in number of nucleotide base pairs (SbCMV) or bases, from sequence data. Information is from Brunt et al. (1996) and GenBank (www.ncbi.nlm.nih.gov). "No data" indicates that the complete genome of the virus has not been sequenced. "Entries" refers to the number of nucleotide sequence entries in GenBank/Entrez for the virus. "Incomplete" indicates only partial sequences are recorded. Variable indicates several strains, other than those from soybean, have been sequenced and genome sizes vary with strain.

Data are from Brunt et al. (1996). Only selected viruses infecting soybean by artificial inoculation are included.

Table 15–2. Transmission and biological properties of viruses naturally infecting soybean.†

Virus taxon	Viral species‡	Seed-borne in soybean§	Mechanical transmission	Vector¶
'SbCMV-like'	SbCMV	–	+	Unreported
Begomovirus	AbMV, BGMV, HgYMV, MYMV, SCLV, SbGV	–	–	Whitefly
Nanovirus	MDV	–	–	Aphid (CP)
Tospovirus	PBNV, TSWV	–	+	Thrips (PP)
Comovirus	BPMV	+/–	+	Beetle
	BRMV, CPMV, CPSMV, QPMV	–	+	Beetle
Fabavirus	BBWV	–	+	Aphid (NP)
Nepovirus	TRSV	+	+	Nematode, thrips
	ToRSV	+	+	Nematode
Potyvirus	BCMV-AzB, BCMV, BYMV, CYVV, PWV, PeMoV	–	+	Aphid (NP)
	BCMV-PStV	0.3–3%	+	Aphid (NP)
	SMV	2–75%	+	Aphid (NP)
Luteoviridae	BLRV, BWYV, SbDV, ISbDV, SbDV-SCRL	–	–	Aphid (CP)
Carmovirus	BMoV	–	+	Beetle
Alfamovirus	AMV	+	+	Aphid (NP)
Bromovirus	CCMV	–	+	Beetle
Cucumovirus	CMV-SS	4–100%	+	Aphid (NP)
	PSV	–	+	Aphid (NP)
Ilarvirus	TSV	30%	+	Thrips
Carlavirus	CPMMV	?	+	Whitefly
Tobravirus	TRV-SF	?	+	Nematode
Tobamovirus	TMV-S	–	+	Unknown
Potexvirus	CYMV	?	+	Unknown
Sobemovirus	SBMV	?	+	Beetle
Unclassified	SbMMV	–	+	Aphid (NP)
Unclassified	SbYVV	?	+	Unknown
Rhabdoviridae	SbRhV	–	–	Unknown

† Data compiled from Brunt (1996) and Hartman et al. (1999).
‡ See Table 15–1 for explanation of virus abbreviations.
§ (+) = seed transmission demonstrated; (-) = no seed transmission in tests; (+/-) = report of low percentage and of zero seed transmission; (?) = no reports of tests conducted.
¶ Relationship of virus with the vector:NP = nonpersistent or stylet-borne; CP = circulative persistent; PP = propagative persistent.

through seed to emerging seedlings from seed grown on infected plants, and horizontally by insect vectors. Table 15–3 lists 27 viruses on which the greatest number of studies have been conducted worldwide, and which either are, or have the potential to be, important viral diseases of soybean worldwide. For each virus, information is given on the geographical distribution, distinguishing symptom on leaves and stems, and whether resistance in soybean has been recognized.

15–1.1 Naming and Classifying Viruses

The classification and nomenclature of viruses has undergone extensive revision since the last Soybean Monograph (Ross, 1987), and the new system is reflected in the text and tables of this chapter. Periodic revisions by the International Committee for the Taxonomy of Viruses (ICTV) are expected to continue (van Re-

Table 15–3. Distribution, symptoms, and resistance of selected viral diseases of soybean.†

Family, genus, or virus‡	Virus distribution	Distinguishing foliar and plant symptoms	Resistance in soybean§
Caulimoviridae			
SbCMV	Japan	Chlorotic mottle and leaf curl	NR
Begomovirus			
SCLV	Thailand	Crinkle leaf, vein yellow netting	NR
MYMV	SE Asia	Yellow mosaic	Observed
SbGV	Argentina	Severe stunting, intense chlorotic mottling, leaf distortion	Some observed
Tospovirus			
GBNV	India, Taiwan	Bud blight	Observed
TSWV	USA-WI, Iran, Brazil	Systemic chlorosis and necrosis	NR
Comoviridae			
BPMV	USA, South America	Chorotic mottle, mild to severe strains; green stems after pod maturity	None found
BRMV	Argentina, Brazil, Central America	Puffy areas along veins, chlorotic blotches	NR
CPSMV	Brazil, Puerto Rico, Trinidad and Tobago USA—AR, LA, IL	Severe necrosis and bud blight	NR
TRSV	USA, People's Republic of China (PRC)	Bud blight:curved terminal bud; lateral bud proliferation; dwarf bronze leaflets	Observed
Potyvirus			
BYMV	Worldwide	Yellow mottling, rusty necrotic spotting, often not until late in season	Observed
PeMoV	USA, PRC, Thailand, Argentina	Mosaic with green islands; line pattern	*Rpv*1; *Rpv*1, *rpv*1
BCMV-PSt	USA, PRC, Taiwan	Mild mottle	Dominant gene
SMV	Worldwide	Green to yellow mosaic, leaf curl, rugosity; pods lacking hairs. Systemic necrosis with some strains on cultivars with some R genes	*Rsv*1, *Rsv*3, *Rsv*4
Luteoviridae			
ISbDV	Indonesia, Thailand	Leaf rugosity, greening, brittle; shortened petioles and internodes.	NR
SbDV	Japan, Indonesia, some of Africa, Australia, NZ	Shortened petioles and internodes. D strains: Leaf curl, greening, brittle. Y strains: Interveinal chlorosis, brittle.	NR
Bromoviridae			
AMV	USA; prob. worldwide	Bright yellow mosaic and mottle	NR
CCMV	USA—AR, GA, SC	Stunting, chlorosis, yellow stipple	Single dominant gene
PSV	USA—IL,KY,VA Japan	Mild mottle and leaf crinkle	Observed
CMV-SS	Japan	Mottle, crinkle, and stunting	Observed
TSV	USA—IA,OH,OK, WI,VA	Bud blight:axillary branches, stunting	NR
Others			
BMoV	Thailand	Mild mottle	NR
CPMMV	SE Asia, Brazil, Africa	Vein clearing, leaf roll; vein necrosis	NR, reaction varies
SbRhV	South Africa		NR
SbYVV	Thailand	Yellow veins	NR
SBMV	Africa, NA, SA, CA	Vein clearing and mottle	NR
TMV-S	USA, Eastern Europe	Very mild vein clearing, chlorotic mosaic	NR

† Data from Hartman et al. (1999) and Brunt et al. (1996) and other cited literature.
‡ Refer to Table 15–1 for key to virus abbreviations.
§ NR = no report of a search for resistance.

genmortel et al., 2000). The traditional biological characters used in virus characterization, such as host range, transmission characteristics, and symptoms, are now much less important in classification than are characteristics of the viral nucleic acid and genome. The system for assigning viruses to genera and families relies primarily on viral genomics—the nucleotide sequence and genome organization, expression, and replication (van Regenmortel et al., 2000). The order in which the viruses are listed in the tables corresponds to that used by the ICTV. Web-based searchable databases make extensive information about each virus readily available (Brunt et al., 1996).

There are new rules of virus nomenclature regarding the writing of names of viruses, or orthography. The name of the virus (equivalent to species) is italicized in a title and the first time of use in a paper, with the first letter of the first word of the name capitalized. The acronymn or abbreviation is given in parentheses following the virus name the first time it is used in the text, followed by the Genus and Family (if assigned) of the virus. For example: *Bean pod mottle virus* (BPMV; Genus:*Comovirus*; Family:*Comoviridae*). There is now a standardized system for the naming viruses and designating acronyms (Fauquet and Mayo, 1999).

15–1.2 Methods for Detecting and Identifying Viruses

Approaches to detecting and identifying soybean viruses discussed in this section include those based on pathogenicity to a range of plant species and cultivars, the detection of viral coat protein by serological methods, detection and analysis of specific nucleotide sequences, and the morphological and physical properties of the virus particles or its gene products. Additionally, a discussion is included of importance for diagnosis purposes of knowledge of transmissibility of the virus by seed, a biological vector, or mechanically, as well as disease management. Identification of a new virus or characterization of new isolates will often have data from all of the above approaches. At some point, the causality of a particular symptom with, and only with, a distinct virus must be confirmed biologically. Rapid detection of viruses can now be conducted by rapid or high-throughput serological or nucleic acid-based methods. However, such tests will give positive results only for those viruses for which immunological or nucleic acid probes are used. The method of choice should be based on the desired efficiency, sensitivity, specificity, and convenience.

15–1.2.1 Biological Methods

If a virus can be transmitted mechanically, the symptoms developing on a range of plant species can be used to determine the identity of a specific virus by its pathogenicity to a range of host plant species. Results of such transmission tests can also be invaluable in determining whether a field sample contains a single or multiple viruses, and in separating individual viruses from a mixed infection. For example, inoculating a SMV-resistant soybean cultivar should permit recovery of a BPMV culture free of SMV from a mixed infection. Disadvantages of biological methods include the length of time to perform the assay (weeks to months), the variability in host response under various environmental conditions, and the lack of readily available, genetically uniform seed sources. Dually infected plants may give confusing results, and will require careful testing by back-inoculation to soybean and parallel testing by other methods to support conclusions.

Host plant assays were used to verify identification of BPMV by symptoms (Pitre et al., 1979). Host range comparisons helped to distinguish the comoviruses BRMV, BPMV, and CPSMV in Brazil (Anjos et al., 1999; Bertacini et al., 1998; Iizuka and Charchar, 1991). Isolates of BCMV-PSt from Taiwan were characterized by Vetten et al. (1992) relative to each other and other potyviruses (AzMV, BlCMV, and SMV) from soybean by symptoms on species useful in diagnosing legume viruses including cultivars of bean (*Phaseolus vulgaris* L.) known to differentiate strains of BCMV. In this study, SMV caused necrotic local lesions on 'Topcrop' and 'Black Turtle-1' bean but did not infect *Chenopodium amaranticolor*. In contrast, BlCMV and BCMV-PSt caused chlorotic local lesions on *C. amaranticolor* but did not infect either bean cultivar (Vetten et al., 1992). Strains of SMV are identified by the differential responses of cultivars of a single species, soybean (Cho and Goodman, 1979; see also Section 15–2.3.2).

15–1.2.2 Serological Methods

For most soybean viruses, specific polyclonal antisera or monoclonal antibodies have been developed and used for reliable, rapid, and accurate virus identification and detection. Antibodies can often be obtained from fellow researchers. Antisera that can be purchased from the American Type Culture Collection (http://www.atcc.org) include AMV, BCMV-PSt, BPMV, BWYV, BYMV, CCMV, CPMV, CPSMV, CYMV, PeMoV, PSV, SbDV, SMV, TMV-S, ToRSV, TRSV, TSV, TSWV, and comparable culture collections or national or international agricultural research organizations may have other antisera. Commercial testing services are available from Agdia (www.agdia.com) as their "Bean Screen" tests for 14 viruses:AMV, BPMV, CPMV, CMV, PSV, BCMV as POTY group, SBMV, SMV, TMV, TRSV, TSV, ToRSV, TSWV, and WMV2. A Tospovirus, *Impatiens necrotic spot virus,* is included, even though it and others on the list have not been reported on soybean. Test kits for most of these viruses can also be purchased from Agdia, mostly in the double-antibody sandwich enzyme-linked immunosorbent assay (DAS-ELISA) format. Tests for TSWV, CMV, and TMV are also available as immunostrips that require 15 min or less for completion. Specific serological assay methods used for soybean viruses include gel diffusion (Hopkins and Mueller, 1984a; Iizuka and Charchar, 1991), ELISA using plate-trapped antigen, and versions of DAS and triple-antibody sandwich (TAS) ELISA; and dot and tissue immunobinding assays using membranes. Polyclonal and monoclonal antibodies were used to distinguish SMV-G5 from other SMV strains (Hill et al., 1984, 1994) and conduct within-field spread studies to detect movement of inoculated SMV-G5 without detecting indigenous non-G5 strains (Nutter et al., 1998). Immunoassays on nitrocellulose membranes (Lin et al., 1990) have also proved useful for field detection of SMV and BPMV (S.A. Tolin, unpublished data, 2000). Microtiter plate-based ELISA with both polyclonal antisera and monoclonal antibodies has been used extensively with SDV and other luteoviruses (Damsteegt et al., 1999; D'Arcy and Hewings, 1986; D'Arcy et al., 1989). The presence of BPMV particles was detected in bean leaf beetle vectors by ELISA (Ghabrial and Schultz, 1983).

India, the tospovirus PBNV was associated with bud blight symptoms by direct antigen coating ELISA and dot immunobinding assays (Thakur et al., 1996). The DAS-ELISA successfully detected BBWV in soybean, as well as other crops including pea (*Pisum sativum* L.), cowpea [*Vigna unguiculata* (L.) Walp.], pepper (*Capsicum annuum* L.), and eggplant (*Solanum melongena* L.) in China (Zhou et al., 1996).

15–1.2.3 Nucleic Acid-Based Methods

15–1.2.3.1 Sequences and Genomes. Nearly all viruses have a similar, basic minimal complement of genes, namely those involved in replication of the genome, in movement of the genome or particle within the host cell and plant, and in encapsidation of the genomes. Depending upon the virus, certain parts of the genome determine specificity of interaction between the virus and its host or vector. Knowledge of the genome of viruses enables use of the very sensitive nucleic acid-based methods for detection and identification. Table 15–1 gives those soybean viruses for which sequence data is currently in GenBank, searchable at Entrez Nucleotide (www.ncbi.nlm.nih.gov).

Sequences known for viruses with a DNA genome include: SbCMV from Japan (Iwaki et al., 1984); the monopartite *Begomovirus* SCLV from Thailand (Samretwanich et al., 2001); the bipartite begomoviruses MYMV from India (Pant et al., 2001) and SYMV (Girish and Usha, 2001); and three New World begomoviruses AbMV, BCaMV, and BGMV. The *Nanovirus* MDV is an unusual and newly characterized virus having a genome of 10 single-stranded DNAs each of about 1000 nucleotides (nt) (Sano et al., 1998).

Sequencing the ambisense genome of *Tospovirus* members has enabled separation of this genus into several distinct viruses, and confirmed prior serological groupings. TSWV is reported infrequently on soybean (Golnaraghi et al., 2001), and sequences of such isolates have not yet been reported. In India, PBNV is common on soybean, and sequencing of its three genomic RNAs has clearly shown that PBNV is distinct from TSWV and is most closely related to *Watermelon silver mottle virus*, a virus widespread in Taiwan (Gowda et al., 1998; Jain et al., 1998; Satyanarayana et al., 1996a, 1996b).

Among members of the *Comoviridae*, CPSMV was one of the first single-stranded RNA viruses for which sequence of its bipartite genome (RNA1 = 5957nt; RNA2 = 3732nt) was determined (Chen and Bruening, 1992). Complete genome sequences are also known for CPMV, RCMV from the United Kingdom, and BPMV from the USA (Di et al., 1999). Because of the increasing incidence of BPMV in the USA, sequence of a segment of 653 nt of the coat protein region has recently been determined for several diverse isolates of BPMV (Gu et al., 2002). Undoubtedly, genome analysis will be used to monitor the genetic diversity and track the dissemination of BPMV in the USA (Giesler et al., 2002). Of the nepoviruses, sequences are known for several strains of TRSV including a bud blight strain from soybean in the USA. Several isolates of ToRSV have been sequenced but most of them have originated in fruit crops.

Sequence data have clarified the phylogeny of legume-infecting *Potyviruses*. Comparisons of the complete genomic sequence for two strains of SMV (Jayaram et al., 1992), and partial sequences for a large number of strains and isolates have shown the diversity of SMV as well as its close affinity to the BCMV-subgroup (Berger et al., 1997; Ghabrial et al., 1990b; Sano et al., 1997). However, a detailed phylogenetic analysis of the large number of partial sequences of SMV deposited in GenBank has not been published. Assigned to the BCMV subgroup viruses of serotype B are the viruses originally named *Azuki bean mosaic virus* and *Peanut stripe virus* (PStV) (Berger et al., 1997; Collmer et al., 1996). Monoclonal antibody analysis of 29 strains of this subgroup confirmed that serological distinction was related to nucleotide sequence, but not to symptomatology, host range, or pathotype (Mink et al., 1999). The genome of BCMV-PSt is larger than BCMV, and isolates infecting bean and soybean could be distinguished from bean-infecting BCMV through sequence and host range (Higgins et al., 1998). Analysis of the PeMoV genome confirmed its status as a distinct potyvirus (Flasinski et al., 1997).

Several strains of SDV have been sequenced totally or in part, showing that it is a distinct member in the family *Luteoviridae* as it contains five major open reading frames (Rathjen et al., 1994; Smith et al., 1993, 1998; Terauchi et al., 2001). This genome organization is unlike other *Luteoviridae* genera, giving SDV its status as not yet assigned to a genus. From Japan, the SDV strains DS and DP (each 5708 nt) are smaller than SDV strains YS (5853 nt) and YP (5841 nt) (Terauchi et al., 2001).

Among the *Bromoviridae*, which includes AMV, CCMV, CMV-SS, PSV, and TSV strains known to replicate in soybean, sequences are known for many viruses, but none of the sequenced isolates are from soybean (Cook et al., 1999; Hajimorad et al., 1999). A comparison of the sequence or RNA1 of the type virus for the *Ilarvirus* genus, TSV, suggests that AMV and TSV are closely related and should perhaps be in the same genus (Scott et al., 1998). These data are consistent with results demonstrating the functional equivalence of AMV and TSV coat protein in early, but not late, functions in the replication cycle by addition of purified protein (van Vloten-Doting, 1975) or construction of a chimeric infectious clone of AMV containing TSV coat protein (Reusken et al., 1995). In these studies, other members of the *Bromoviridae* which do not require coat protein for genome activation, CMV and *Brome mosaic virus*, had no interaction with either AMV or TSV.

Sequences of other soybean viruses that are deposited in GenBank include PCV, PEBV, CYMV, WCMV, SBMV, and GRV. The isolates are unlikely to have come from soybean, but nonetheless sequences are probably closely aligned with those infecting soybean. Sequences of two Arkansas isolates of SBMV have seven nucleotide differences between the wild-type strain and one that overcomes resistance in cowpea (Lee and Anderson, 1998). For other soybean viruses, including CPMMV, TRV-SF, TMV-S, SbYVV, SbMMV, and SbRhV, sequence data are not currently available.

15–1.2.3.2 Polymerase Chain Reaction-based Methods. Increasingly, the reverse transcription-polymerase chain reaction (RT-PCR) method is used for high-throughput, sensitive detection of viruses in soybean and other crops. Primers specific for viral genomes can be designed from the available sequences for nearly all soybean viruses, but there are as yet few reports of its use. The first use of RT-PCR for SMV was to detect and differentiate between SMV G2 and G7 using primers in the center of the cylindrical inclusion (CI) region of the genome to generate fragments of 380 and 280 base pairs (bp), respectively (Omunyin et al., 1996). Upon restriction endonuclease digestion of the product with *Hae*III, strains G3, G4, G5, and G7a yielded 155 and 122 bp fragments, but G1 was not digested. In a later study, primers farther upstream in the CI the region of SMV amplified a 1385 bp DNA fragment conserved in several strains (Kim et al., 1999). A universal primer for a conserved sequence in the NIb region of the potyvirus genome was used to amplify 3' terminal genome regions of 21 different viruses in the family, including SMV (Chen et al., 2001). Primers within a highly conserved region of the coat protein gene of SMV were also used to detect a 469 bp fragment by RT-PCR from single individuals of the aphid, *Aphis glycines,* following brief or overnight times on infected plants (Wang and Ghabrial, 2002).

A protocol for RT-PCR based on a highly conserved region of the capsid polyprotein coding sequences effectively detected all isolates of BPMV, regardless of strain classification (Gu et al., 2002). The sensitivity was three orders of magnitude greater than ELISA. However, the use of certain primers may give false positive results because of the somewhat unexpected finding of near complete sequence BPMV RNA-2 in the genome of non-infected soybean (Sundararaman et al., 2000). In forage legumes, multiplex PCR was five orders of magnitude more sensitive than ELISA for AMV, BYMV, CYVV, CMV, and SCMoV in subterranean clover (*Trifolium subterreanean* L.) seeds (Bariana et al., 1994).

Specific primers are also used frequently for those viruses occurring in low concentration in tissue and not mechanically transmissible, especially SDV and other luteoviruses (Robertson et al., 1991). The method of choice for confirming a virus as a geminivirus often involves PCR, and certain sequences are highly conserved.

15–1.2.3.3 Hybridization. Labeled cloned probes have been used in northern hybridization experiments with SMV and other potyviruses (Frenkel et al., 1992), BPMV strains (Gu et al., 2002), and SDV (Martin and D'Arcy, 1990). The cloned coat protein gene of the Sa strain of SMV was labeled by nick translation and used to detect as little as 0.4 *ng* of purified virus or virus in soybean leaves by spot hybridization (Zhou et al., 1994). RNA dot blot hybridization, as well as ELISA and Western blotting, were used to monitor concentration of SMV, CPMV, and BPMV in soybean (Anjos et al., 1992).

15–1.2.4 Morphological and Physical Characters of Viruses

Most soybean-infecting viruses have been purified and the particles characterized as to sedimentation properties and dimensions in transmission electron microscopy (Brunt et al., 1996). Members of the *Comovirus, Fabavirus, Nepovirus, Luteoviridae, Carmovirus, Bromovirus, Cucumovirus,* and *Sobemovirus* all have particles 25 to 30 nm in diameter. *Ilarvirus* particles are similar in size but capsids are more irregular and less rigid, a property shared by a strain of the *Alphamovirus* AMV (Roosien and van Vloten-Doting, 1983). Typically AMV has a series of small bacilliform particles 18 nm in diameter and 30 to 57 nm in length, each encapsidating a different RNA species. The *Caulimoviridae* have larger icosahedral particles (50 nm) and *Nanovirus* particles are smaller (20 nm). Flexuous rod particles 10 to 12 nm in diameter are found in the taxa *Potyvirus* (680–900 nm), *Carlavirus*

(650 nm), and *Potexvirus* (540 nm). Viruses of the *Tobravirus* and *Tobamovirus* genera and SbYVV have rigid rods with a visible central core with dimensions of 20 × 215 nm, 15 × 300 nm, and 20 × 550 nm, respectively. The geminiviruses have 18 × 30 nm particles of icosahedral symmetry that usually occur in pairs, leading to the name suggesting 'twin' particles. The rhabdovirus SbRhV has 80 × 290 to 440 nm bacilliform particles of a complex structure with a lipid membrane. The *Tospovirus* members have a membrane surrounding an icosahedral core, giving an irregularly-shaped 90-nm virion that is rather unstable in vitro.

Relative molecular mass (M_r) of protein(s) making up the viral capsid, as determined by gel electrophoresis, is often used as a confirming, identifying character to genus or family. Minipurification of virus particles directly from field samples will generally concentrate viral particles and capsid proteins such that M_r values and estimates of sedimentation can be readily determined (see http://plantpath.unl.edu/llane/). Size of the major protein for each virus particle is in the range of 20 to 60 kDa, and can be found on line (Brunt et al., 1996). Viral capsid proteins digested enzymatically showed similarities and differences among strains of SMV, and strengthened the argument for not calling SMV a strain of WMV-2 (Hill et al., 1989; Jain et al., 1992).

Electron microscopic examination of ultrathin sections showing viruses and inclusion bodies aided in distinguishing isolates of BCMV-PSt collected in Taiwan (Vetten et al., 1992). In tissues of bean and soybean with bean angular mosaic virus, the presence of falcate bundles verified that the virus was the same as CPMMV and was a *Carlavirus* (Gaspar and Costa, 1993). Particle size and protein M_r were also used to confirm these conclusions, and purified virus reproduced the symptoms of this disease.

Determination of M_r and detection of protein bands by Western blotting was used to confirm identification of a comovirus in Brazil as BPMV (Anjos et al., 1999). The first identification of BPMV in Peru from soybean plants showing chlorotic mottle symptoms was by production of necrotic lesions on bean cultivars, detection by SDS gel electrophoresis of two protein bands of 21 600 and 41 200 daltons (characteristic of comoviruses) 30-nm particles, and positive reactions against BPMV antiserum in ELISA and nitrocellulose membrane tests (Fribourg and Perez, 1994).

15–1.2.5 Vector, Seed, and Mechanical Transmission

Knowledge of the transmissibility of viruses can help in identifying viruses at least to genera, as well as in developing disease management strategies (see Section 15–2.3).

15–1.2.5.1 Vectors. Categories of vectors known to transmit 46 viruses are given in Table 15–2. Names of specific vectors are not given as they are discussed in Section 15–2 and can also be accessed elsewhere (Brunt et al., 1996; Hull, 2001). Vectors of the greatest numbers of soybean viruses are aphids, which transmit members of the *Fabavirus, Potyvirus, Cucumovirus,* and *Alfamovirus* genera, and SbMMV in a nonpersistent or stylet-borne manner. The short acquisition and inoculation times required for transmission of these viruses permits virus transmission by transient aphids (Halbert et al., 1981). Although there is usually little vector specificity with this type of transmission, certain potyviruses are known to be nontransmissible due to a lack of specific conformational receptors in the coat protein-helper component-protease complex (Jayaram et al., 1998). Naturally occurring SMV strains vary in their transmissibility (Lucas and Hill, 1980). *Nanovirus* and *Luteoviridae* members are circulative and persistent in their aphid vectors, requiring transit through membrane systems into salivary glands (Gildow et al., 2000). Because of this unique biological interaction, specific strains of SDV are transmitted only by specific aphid species (Damsteegt and Hewings, 1986; Terauchi et al., 2001).

Vectors of *Comovirus, Carmovirus, Bromovirus,* and *Sobemovirus* members are leaf-feeding beetles, flea beetles, and certain blister beetles (Fulton et al., 1987). Whiteflies (vectors of *Begomovirus*, CPMMV:*Carlavirus*) and thrips (*Tospovirus*) are becoming increasingly important vectors of soybean viruses in some parts of the world, but not yet in the USA. TRSV (*Nepovirus*) and TSV (*Ilarvirus*) have also been shown to have a thrips vector, but the mechanism of transmission is suggested to be via pollen-carried virus introduced by mechanical abrasion during feeding (Hull, 2001). Soil-borne vectors include nematodes of *Xiphinema* spp. (TRSV and ToRSV:*Nepovirus*) and *Paratrichodorus* spp. and *Trichodorus* spp. (TRV:*Tobravirus*), but these vectors appear to be of little economic significance except in the sandy soils of Delaware where an apparent strain of TRSV caused a serious severe stunting disease of soybean (Weldekidan et al., 1992; Evans, 1999). Interestingly, leafhoppers are known to be major vectors of many plant viruses, but none have been demonstrated to be the vector of a soybean virus. Demonstration that leafhoppers can transmit phytoplasmas is evidence, however, that these insects can establish a feeding relationship with soybean that could result in transmission of a virus (see Section 15–5).

15–1.2.5.2 Seed. Seed transmissibility is a property of relatively few soybean viruses, and this characteristic is usually shared with members of the virus taxa in at least some host (Hull, 2001). Those viruses demonstrated to be transmitted through soybean seed to progeny plants, and not just detected in seed, include the nepoviruses TRSV and to a lesser incidence ToRSV, the potyviruses SMV and BCMV-PSt, and the *Bromoviridae* (AMV, the soybean stunt strain of CMV, and TSV) (Table 15–2). The comovirus BPMV has a reported low incidence of seed transmission that varies with cultivar (Lin and Hill, 1983; Ross, 1986b; Giesler et al, 2002). The significance of seed transmissibility to disease epidemiology is discussed in Section 15–2. Taxons not generally known to be seed-borne in any host are the caulimoviruses, begomo-, nano-, and tospoviruses, and the *Luteoviridae*. Soybean viruses from these taxa have all been tested and found to be not seed-borne. For several other viruses, however, definitive data are not readily available.

15–1.2.5.3 Mechanical. Mechanical or sap transmission is a property of most soybean viruses and is a valuable tool for propagation of viruses or testing for host range and disease resistance. There has been speculation that mechanical transfer from leaf to leaf by contact, perhaps aided by mechanized operations such as spraying, may occur in the field with BPMV (Ross, 1987). The high incidence of virus that occurs in breeders' crossing blocks might also be the result of inoculation during the mechanical pollination process. The source of virus could either be floral parts or contamination of workers' hands or tools from virus in leaves.

Lack of mechanical or sap transmissibility of a causal agent of a virus-like disease from soybean is an indication that the agent, if biotic, is probably a member of the *Luteoviridae, Geminiviridae,* or *Rhabdoviridae* families, the *Nanovirus* genus, or the prokaryotic phytoplasma (Section 15–5).

15–1.3 Virus Disease Symptomatology

15–1.3.1 Symptoms on Foliage and Whole Plants

Most typical symptoms resulting from virus replication in soybean are on leaves and displayed as mosaic, mottle, and veinal chlorosis (Table 15–3). Plates showing foliar symptoms for most soybean viruses can be found in the *Compendium of Soybean Diseases* (Hartman et al., 1999) and on the webpages of many individual scientists in soybean-growing areas. Leaf distortion, described as puckering, blistering, or rugosity, and accompanied by leaf cupping, rolling, and downward curling are associated with SMV. Other potyviruses such as the BCMV strains AzMV and PStV, and PeMoV usually cause less leaf distortion. Symptom severity is affected by host genotype, SMV strain, growth stage of the plant when infected, and environmental conditions. High temperature has been attributed to causing a reduction in symptom severity late in the season, particularly in the North Central Region of the USA (Hill, 1999), but few definitive studies in the literature verify this attribution. Late-season symptoms in southern regions of the USA persist, in spite of high temperatures, suggesting that cultivar difference is a likely factor as well. Conversely, attenuated isolates of SMV were derived by maintaining diseased plants at lower temperatures (Kosaka and Fukunishi, 1993).

Foliar symptoms associated with comoviruses are similar, but with generally less leaf blade distortion with most strains in comparison to that caused by SMV. BRMV and CPSMV symptoms are more severe than those of BPMV. Bright yellow chlorotic mosaic and mottle symptoms are common with AMV, CCMV, and sometimes BYMV. Leaves on plants infected with TRSV or TSV may be severely stunted or dwarfed, with chlorosis and bronzing.

15–1.3.2 Symptoms on Stems and Pods

Stem and pod symptoms, if severe, are usually called "bud blight". This name was apparently first used to describe TRSV on soybean, and usually refers to terminal necrosis starting with a downward curling of the apical meristem, and proliferation of axillary buds. Few pods form, and those that do may have a single, large seed (Ross, 1987). Both leaves and stems of plants remain green late in the season because maturation is delayed when pod and seed formation is reduced. Symptomatology of ToRSV is essentially the same, but it has seldom been observed.

It is now evident that TRSV is not the only cause of bud blight. Several other viruses or other pathogens (see Section 15–5) are associated with the syndrome. BPMV was consistently detected serologically from soybean stems that remained green after pods matured and to which petioles were still attached and green but leaves had fallen (Schwenk and Nickell, 1980). Virus was detected from pulvini and pith, and from extracts from seeds, but not from seedlings that developed from the seeds. Plants inoculated as very young seedlings, with cotyledons still folded but green, were stunted, some having terminal and/or axillary necrosis, and some remaining green for 200 d.

Bud blight symptoms in Brazil have been associated with six different viruses, each of which varied in the reaction induced on different cultivars (Costa, 1988). For example, TSV was described as 'Brazilian bud blight' (Costa and Carvalho, 1961; Hartman et al., 1999). More recently, CPSMV was isolated from Brazilian soybean plants with bud blight symptoms (Bertacini et al., 1998). In India, the incidence of bud blight disease of soybean is consistent with the presence of the *Tospovirus* PBNV (Thakur et al., 1998). The titer of infectious PBNV was higher in first trifoliolate leaflets, compared to second and third, in which none was detected. Bud blight symptoms were most severe when plants were infected at flowering and pod initiation, and much less if infection was at pre-flowering or podding.

Pods on plants with SMV are often smaller and in some cultivars show a distinct decrease in pubescence, giving a glabrous or shiny appearance of immature pods. Upon mixed infection with SMV and BPMV, pods may have necrotic streaks and show distortion (Hartman et al., 1999). Pods on plants infected with TRSV may also be distorted and have necrotic blotches, and often have poor seed set (Ross, 1987).

15–1.3.3 Symptoms on Seeds

Seeds produced on virus-diseased plants are generally reduced in size, except those in single-seeded pods that may be larger. Most strikingly, viruses often cause patterns of discoloration on the seed coat, commonly called seed mottling or seed coat mottling.

Early studies described in the last monograph concluded that mottling of seeds was not a reliable indication that the mother plant had SMV, nor is lack of mottling a clear indication that it did not (Ross, 1987). Extent of mottling in seeds of plants with SMV increased as temperatures decreased (Almeida et al., 1995a; Morrison et al., 1998; Ross, 1970; Tu, 1989, 1992). Delayed planting significantly reduced SMV foliar symptoms, percentage of mottled seeds, and mottling index, perhaps because of higher temperatures (Sun et al., 1992).

In Brazil, the symptom caused by SMV is known as "mancha-café" or coffee-spot (Lourencao and de Miranda, 1996). The color of the seed coat mottling is usually like that of the hilum, which is under complex genetic control and ranges from clear to black. Streaks of color may emanate from the seed hilum, often called "bleeding hilum". However, bleeding hilum can occur in the absence of virus in some cultivars. Other viruses, including the severe stunt strain of CMV, cause seed coat mottling and interact with SMV (Koshimizu and Iizuka, 1963).

Strain specificity of SMV in relation to causing seed coat mottling has been demonstrated (Bowers and Goodman, 1991). The genetic complement of the host plant greatly influences both the susceptibility to SMV and seed coat mottling (Almeida et al., 1995a; Sun et al., 1992; Tu, 1989). A study of 'Clark' and 'Harosoy' near-isolines with various combinations of genes for maturity, stem termination, SMV resistance, and seed coat pigment control loci indicated that seed coat mottling was reduced by $tE1$, Im, and $Rsv1$, increased by $E1T$, i^k, and iR, and not affected by R and $Dt1$ (Sun et al., 1992).

The mechanism by which SMV, BPMV, and other viruses cause seed coat mottling remains obscure and has seldom been investigated with current molecular approaches used to examine the pathways involved. The pigmentation of soybean seed coats is black or brown in all wild soybean lines because of homozygous recessive i allele. In the dominant condition I inhibits synthesis and spatial array of anthocyanin pigments in the epidermal layer of the seed coat resulting in a yellow seed coat at maturity. The R and T loci control the specific color of the seed coat, if it is not inhibited by I. Black seed coats were found to have a much reduced level of PRP1, a developmentally-regulated 35 kDa proline-rich protein also expressed in cell walls of soybean hypocotyls (Lindstrom and Vodkin, 1991). In the presence of I, the mRNA for chalcone synthase, a key enzyme required for the synthesis of anthocyanin pigments, is greatly reduced (Wang et al., 1994). Thus, this gene reduces the expression of a flavonoid pathway gene required for color formation. Perhaps the mechanism by which viruses cause seed coat mottling involves the silencing of one or more alleles of the I gene permitting expression of genes on the seed coat resulting in spatially-arrayed mottling.

With the increased prevalence of BPMV has come an increased effort on its effect on seed quality, not only in mottling but also by reducing the sheen of the seed coat. In an earlier report, an increased incidence of the fungus *Phomopsis* occurred in plants infected with BPMV infection (Stuckey et al., 1982). It was later observed that BPMV delayed harvest maturity (Abney and Ploper, 1994). Pod colonization by the fungus *Phomopsis* was consistently increased by BPMV, whereas seed colonization by the fungus was increased only if BPMV delayed the rate of seed maturation.

The SMV has also been associated with increased incidence of species of *Phomopsis* (Hepperly et al., 1979) and *Colletotrichum* (Porta Puglia et al., 1990). Experiments by Koning et al. (2001) demonstrated infection of SMV-susceptible lines, 'Clark' and 'Williams', prior to the R2 developmental stage increased susceptibility to *Phomopsis* spp. resulting in poor seed quality and low vigor. Lack of infection with SMV, not the resistance gene itself, was deemed to be the controlling factor (Koning et al., 2002).

Controlled inoculation experiments of 'Williams' near-isolines with and without resistance to SMV confirmed that both BPMV and SMV, alone or in combination, caused seed coat mottling (Hobbs et al., 2001). Cultivar differences in response to BPMV were observed in the percentage of seed mottling, from 59 to 92%, and in seed quality rating (Ziems et al., 2001a).

15–1.4 Distribution of and Diversity among Major Soybean Viruses

The worldwide distribution of 27 major soybean viruses is given in Table 15–3. In the USA, only 15 of these viruses are reported to be isolated from soybean. These include members of the following genera: *Potyvirus*: SMV, BYMV, PeMoV, BCMV-PSt; *Comovirus*: BPMV, CPSMV; *Nepovirus*: TRSV, ToRSV; *Alfamovirus*: AMV; *Bromovirus*: CCMV; *Cucumovirus*: PSV; *Ilarvirus*:TSV; *Sobemovirus*:SBMV; *Tobamovirus*:TMV-S; and *Tospovirus*: TSWV. The most prevalent and widespread are BPMV and SMV, whereas most other viruses occur sporadically and/or in isolated areas.

There are several soybean viruses not yet reported in the USA. Among these viruses, the geminiviruses and luteoviruses cause severe foliar symptoms of extensive stunting, crinkling, and yellowing, whereas bud blight and necrosis may be caused by CPSMV and PBNV, and may pose a threat to U.S. soybean production (Table 15–3).

15–1.4.1 Potyvirus

Strains of SMV are recognized by a pattern of virulence on sets of soybean cultivars to SMV, although the possibility of using genomic analysis is increasing (see Section 15.1–3). Differential soybean cultivars with varying degrees of resistance were initially used to classify SMV isolates from the U.S. germplasm collection into nine strain groups (G1-G7, G7A, C14) based on a systemic, necrotic, or near immune response (Cho and Goodman, 1979). This system has become the basis for strain identification in other countries as well, including Canada (Tu, 1989), Korea (Lee et al., 1991), South Africa (Pietersen and Garnett, 1992), and Brazil (Almeida et al., 1995a, 1995b). In Japan, scientists have used a different system to distinguish five strains (A–E). Two additional differential cultivar systems have been used by scientists to distinguish three strain groups (S1–S3) in northern China and eight strains (SA–SH) in the Jiangsu Province of central China. Other sets of differential cultivars are used in other regions of China (Shang et al., 1999).

Considerable diversity was exhibited within a collection of eight isolates of PeMoV, including their pattern of infecting soybean and overcoming resistance in soybean genotypes such as 'York' and PI 96983 that carry a gene for resistance to PeMoV (Bijaisoradat et al., 1988).

15–1.4.2 Comoviridae

Viruses of the *Comoviridae* are most prevalent in the New World. The diversity of different viruses of the *Comovirus* genus appears to be greatest in South America. The different viruses are readily distinguished by serology and host range, with some differences in vector specificity. The relationship among comoviruses has not yet been addressed molecularly, and may reveal some interesting aspects regarding viral genome diversification.

The comoviruses BRMV (Iizuka and Charchar, 1991), CPSMV (Bertacini et al., 1998), and BPMV (Anjos et al., 1999) have been described from soybean in Brazil, and BPMV has been identified from soybean in a Peruvian jungle (Fribourg and Perez, 1994). The Brazilian BPMV was identified as strain C and transmitted by *Cerotoma arcuata* Oliver at an average 66.7% efficiency (Anjos et al., 1999). A new strain of CPMV transmitted by *C. ruficornis* Oliver and *Diabrotica balteata* Leconte was reported in Cuba (Fernandez Suarez and Lastres Gonzalez, 1984). It is likely that this was CPSMV, which was identified, together with bean yellow stipple virus (CCMV), by inoculation and serological tests (Crespo Romero et al., 1990). In the USA, CPSMV has appeared sporadically in soybean in Arkansas and Illinois, but it is not recognized as a widely prevalent virus. *Glycine mottle virus* (GMV) was described as a new member of the *Comovirus* genus, isolated from two *Glycine* species indigenous to Australia and the western Pacific, *G. clandestina* Wendl. and *G. tabacina* (Labill.) Benth. (Bowyer et al., 1980). *Glycine mottle virus* was not found naturally in soybean, but replicated and caused disease in all 21 *G. max* cultivars inoculated.

The geographical distribution of BPMV in soybean in the USA has increased in recent years since its discovery in Arkansas (Walters, 1958) and recognition in southeastern and southern states (Ross, 1987). Walters and Lee (1969) commented that the virus was found in 1 to 100% of the plants in every field checked in Arkansas in the previous 3 yr. A detailed review by Giesler et al. (2002) documents the expansion of BPMV and its vector into the north central and Northern Great Plains of the USA. Recent reports of new incidence include South Dakota (Langham et al., 1999), Ohio (Dorrance et al., 2001), Wisconsin (Lee et al., 2001b), Nebraska (Ziems et al., 2001b), Indiana (Clark and Perry, 2002; S.A. Tolin, unpublished data, 2001), and Canada (Michelutti et al., 2002). Analysis of a collection of several BPMV isolates from Kentucky, Virginia, Mississippi, Arkansas, and elsewhere has demonstrated that BPMV exists as two genetically distinct subgroups, I and II, detectable by nucleic acid cross-hybridization (Gu et al., 2002; Giesler et al., 2002). Some BPMV isolates were reassortants between the subgroups, and these isolates often caused more severe disease symptoms in soybean.

The nepoviruses TRSV and ToRSV have had no recent reports relative to the diversity of isolates from soybean. TRSV is identified frequently but at a low incidence or geographical distribution (Evans, 1999), whereas reports of natural infection of soybean with ToRSV are rare.

15–1.4.3 Luteoviridae

Soybean dwarf virus diversity has recently been re-examined in Japan where its incidence is the greatest (Terauchi et al., 2001). Genomic analyses were combined with symptomatology on soybean and vector specificity to recognize four distinct strains, whereas previously only two were recognized in Japan and distinguished in studies conducted under quarantine conditions in the USA (Damsteegt et al., 1990). Both SDV and closely related viruses have now been recognized in the southeastern USA in forage legumes, and are readily transferred to soybean in vector inoculation experiments (Damsteegt et al., 1995, 1999).

15–1.4.4 Bromoviridae

Seven strains of CCMV were distinguished by their response on six soybean genotypes (Paguio et al., 1988). The most severe strain, capable of breaking resistance of 'Bragg', was the bean yellow stipple (BYS) strain. Strains D and N were derived by passage of BYS through Bragg. Genomic analyses of these strains have not apparently been conducted. Although there are no reports on the diversity of soybean isolates of AMV, CMV, PSV, and TSV, these viruses are inherently varied and strains are quite likely to exist.

15–1.4.5 Tospovirus

Viruses in the genus *Tospovirus* have been more intensely studied in crops other than soybean. As discussed in Section 15–1.2, genomic and serological analyses have been used to both identify viruses and address their diversity, including those occurring in soybean. TSWV has been identified recently in the middle East on soybean by ELISA and by the reaction of 26 test plant species (Shahraeen et al., 2001).

15–1.5 Impact of Soybean Virus Diseases

Assessing the impact of viruses on soybean productivity is a multi-faceted task, yet the percentage yield loss that a soybean virus causes is the only impact commonly reported. The primary cause of seed coat mottling is virus infection, yet this impact of viruses is seldom considered. Additionally, the contribution of viruses to enhanced seed-borne fungi and reduced seedling vigor has seldom been included in impact assessments.

Most research has reported yield loss data from plots inoculated to achieve 100% disease incidence at a defined growth stage, which does not represent a pattern of field spread that varies with natural vector populations. On an individual plant basis, individual seed weight and number of seed per pod and pods per plant are decreased, and seed viability and seedling vigor are reduced. On a field basis, however, lower yields from a BPMV- or SMV-diseased plant may be compensated by adjacent healthy plants (Quiniones et al., 1971; Windham and Ross, 1985). For example, uninfected plants adjacent to BPMV-infected plants yielded 50% more than uninfected plants in an uninfected stand (Windham and Ross, 1985).

Two principal factors to determining yield loss due to BPMV are judged to be the stage of plant development when infection occurs and the incidence of virus-infected plants in the crop (Ross, 1986a). Yield reduction relationship with phenological age of the crop rather than to the calendar date when infected was confirmed by Ren et al. (1997a). The R1 stage of development (first flower) was the critical stage for infection of soybean with SMV in order for yield loss to be significant. The effect of SMV on seed quality traits, such as seed coat mottling and reduced seed germination, also depended upon disease incidence at plant growth stages, and was greater if infection occurred before, rather than after, flowering (Ren et al., 1997a). Mixed cropping of soybean with sorghum [*Sorghum bicolor* (L.) Moench] reduced SMV-caused seed coat mottling in rows outside an inoculated area, presumably by delaying virus spread by aphids (Bottenberg and Irwin, 1992).

Data are also reported from experiments comparing yields of resistant and susceptible cultivars either inoculated or naturally infected. These experiments may be small or large replicated plots or simply replicated hill plots of 6 to10 plants (Ren et al., 1997b). Infection of plants by more than one virus results in a disease synergy in many cases. In particular, SMV and BPMV interact to drastically reduce yield and seed quality (Anjos et al., 1992; Calvert and Ghabrial, 1983; Ross, 1968).

The challenge of interpreting these data and projecting an impact of viruses on agronomic performance of soybean will remain. The potential for extremely high impact will occur under a given set of circumstances resulting in early, uniform infection of susceptible cultivars. A model for SMV in soybean predicts six times greater virus incidence if timing and all factors are optimal (Irwin et al., 2000). The impact on seed quality was not included in the model, however.

15–2 ECOLOGY, EPIDEMIOLOGY, AND MANAGEMENT OF SOYBEAN VIRUSES

15–2.1 Ecological Factors Affecting Survival and Persistence of Viruses

When soybean is introduced into a new cropping area, viruses that are indigenous to that area in either crops or wild plants may be disseminated to soybean and cause disease. Indigenous viruses will be those that persist in perennial wild plants or weeds, or are seed transmitted in annual plants or crops grown in the area, and have a means of horizontal transmission to new hosts.

Alternatively, soybean seeds distributed in world trade and by plant breeders and seedsmen may carry viruses and thus introduce a virus into a new ecosystem. Major factors in whether a particular virus survives the introduction, becomes established, and causes a disease of economic importance in the new agroecosystem include the quantity and diversity of virus initially introduced, and the time and extent of spread into and within the crop.

A vector-borne virus or phytoplasma is likely to become established in a new area if a suitable vector and host for the virus are present. The host could be susceptible soybean cultivars, other susceptible crops, or wild plants. Viruses and phytoplasmas nearly always persist in association with a living host or vector, thus the possible reservoir sources are limited to seed, other plants—wild flora, weeds, nearby crops—or the vector itself. Soybean viruses are transmitted by aphids, beetles, whiteflies, nematodes, or thrips (Table 15–2). Each invertebrate vector transmits only a particular virus or set of viruses. Species or biotypes of vectors often transmit virus strains with differing efficiencies. Phytoplasmas, but not soybean viruses, are reported to be transmitted by leafhoppers (see Section 15–5).

15–2.2 Epidemiology

A high incidence, severe disease in soybean is favored by conditions that result in high vector populations having access to a source of pathogen when soybean plants are in a sensitive, early vegetative phase of growth. Because of these complex interactive requirements, outbreaks of viruses and phytoplasmas in soybean are often sporadic in some areas and chronically severe in others.

15–2.2.1 Seeds as Sources of Soybean Pathogens

Viruses are often transmitted through the seeds of large-seeded legumes (Hull, 2001). Soybean seeds are known to serve as a pathogen source for eight viruses (Table 15–2). In the classic work by Iizuki (1973), SMV was found to be transmitted to 34% of the progeny seedlings, which is the true test of seed transmissibility. Virus was detected in embryos and seed coats. Rates of transmission varied with cultivar and were higher if plants were infected prior to flowering. Many other studies have been done since then, utilizing ELISA, electron microscopy, and other approaches to detect SMV in embryos and seed coats, and comparing detection rates to seed coat mottling incidence and rate of seed transmissibility (Tu, 1975; Lister, 1978). The rate of seed transmission is affected by cultivars, SMV strains, and environmental conditions (Almeida et al., 1995a; Bowers and Goodman, 1991; Tu, 1992).

New strains and incidences of SMV are usually attributed to seed transmission (Almeida et al., 1995b; Zhang et al., 1986). Based on the finding that 30% or more of the seeds on a diseased plant contained SMV, Taraku and Juretic (1990) concluded that seeds were the primary inoculum source in the Province of Kosovo. In addressing the question of long-distance movement of SMV through exotic soybean germplasm in relation to quarantine restrictions, Khetarpal et al. (1992) tested 31 accessions, finding 19 testa positive in ELISA for SMV, 10 embryos positive, and four of these giving rise to diseased seedlings. A second paper from India reported detecting SMV in 5 of 163 accessions from Taiwan and in 13 of 74 accessions from USA, and suggested testing procedures to minimize the entry of virulent exotic SMV strains (Parakh et al., 1994).

BPMV has been reported to have an incidence of 0.1% (7 of 6976) seed transmission (Lin and Hill, 1983), 0.0013% in beetle-inoculated 'Centennial' (1 of 10 377) and 'Forrest' (1 of 4400) plants (Ross, 1986a), and 0.037% in one industry-developed cultivar but none in a second (Giesler et al., 2002). As BPMV is quite stable and easily transmitted by contact, and can be detected in seed coats (Schwenk and Nickell, 1980), Giesler et al. (2002) suggested that the low levels of transmission could be the result of virus transfer through mechanical injury of seedlings by the seed coat, rather than by true embryo transmission. Regardless of the mechanism, the low number of infected plants was recognized as being a significant source because of the high vector populations (Giesler et al, 2002). Circumstantially, the sudden appearance of BPMV in a Peruvian jungle was attributed to seed transmission and subsequent dissemination in the field by beetles (Fribourg and Perez, 1994). In Virginia, BPMV has been detected primarily in USDA-ARS Uniform Regional cultivar trials in plants grown from seed harvested in areas of high incidence of the virus, suggesting that seed may have been the source (S.A. Tolin, unpublished data, 2000).

The seed transmissibility of other soybean viruses does not seem to be a major factor in these diseases. Because of symptom severity and bud blight, poor seed development occurs on plants with TRSV, ToRSV, and TSV. The vivid symptoms induced by most AMV strains would be noticed by most seed production programs and plants eliminated. The CMV severe stunt strain, though transmitted at a high rate, is not reported outside of Japan where it is apparently a minor disease. Interestingly, all of these viruses have wide host ranges and likely persist in native flora. Conversely, SMV has a limited host range and is dependent upon seed transmission in soybean for its persistence.

Elimination of virus-infected plants from germplasm collections, breeding nurseries, and foundation seed is an obvious approach to greatly reducing the incidence and worldwide dissemination of soybean viruses. Roguing symptomatic plants is a strategy to eliminate those viruses causing obvious symptoms. However, some viruses, particularly SMV, may show very mild or no symptoms and would not be eliminated without indexing by a detection method such as serology.

15–2.2.2 Dissemination of Viruses by Vectors

15–2.2.2.1 Aphids. The transmission of SMV within soybean fields by aphids has been studied extensively by researchers at the Univ. of Illinois, and has served as a model system for many epidemiological predictive systems (Irwin and Goodman, 1981; Irwin et al., 2000). In this system, the important vector species were those having seasonal abundance patterns coinciding with the growth stage of soybean vulnerable to SMV infection. By trapping live winged aphids, Halbert et al. (1981) found that *Aphis craccivora* Koch, *Macrosiphum euphorbiae* (Thos.), *Myzus persicae* (Schulz.), *Rhopalosiphum maidis* (Fitch), and *R. padi* (L.) accounted for more than 93% of the SMV transmissions, even though at least 31 species of aphids have been reported to transmit SMV. Leaves of diseased soybean plants were lighter yellow and attracted *M. persicae* and *R. maidis* more than green leaves of healthy soybean plants (Fereres et al., 1999). Aphids also tend to leave diseased plants soon after probing, increasing the probability of making an infective probe before the inoculation potential is lost, and thus spreading virus rapidly.

An extensive study was also done in China where the important aphid species varied with the spring and summer soybean crop. Even though *Aphis glycines* Matsumura, the soybean aphid, was present and colonized the summer soybean crop, it was not the most abundant species at the time infection occurred and thus was not considered the most important vector (Halbert et al., 1986). Vectors of SMV identified in India include *M. persicae, A. craccivora, A. gossypii,* and *A. nerii* Boyer de Fonscolombe, having transmission efficiencies of 90, 45, 35, and 25%, respectively (Naik and Venkatesh, 1997). In Korea, the predominant aphid species on soybean also varied with seeding date, and were *Acyrthosiphon solani* Kaltenbach at early seeding and *Aphis glycines* at late seeding (Kim et al., 2000).

A report from northern Italy attempted to identify aphids responsible for the extensive spread of SMV in 1988 and 1989 (Coceano et al., 1998). Suction trapping of live winged forms yielded 64 different aphid species, whereas by direct collection from soybean plants only nine species were found and identified. All successful transmissions of SMV were with *A. craccivora*, but since the efficiency was much less than had been obtained by Halbert et al. (1981), Coceano et al. (1998) concluded that this species could not explain the extensive spread of virus. However, high population density of alate aphids may have overcome the low vector efficiency. An extensive study in the Philippines of population density of *A. glycines* on soybean, and the extent to which it and other aphid species transmitted SMV to soybean, demonstrated that *A. glycines* and *A. gossypii* were the most important vectors (Quimio and Calilung, 1993). The seasonal phenology and abundance of this aphid, which has been observed to increase in the vegetative plant stage and decline rapidly afterwards, was shown to be affected by predation in Indonesia (van den Berg et al., 1997).

Aphis glycines, which has recently been introduced into the USA, has the potential to become an important vector of SMV and other viruses (Hartman et al., 2001; Hill et al., 2001). The efficiency of the soybean aphid as a vector of SMV was shown by Wang and Ghabrial (2002) to be greater following a brief, 1 min acquisition probe than after overnight feeding on SMV-infected plants. It was shown to be a vector of newly collected field isolates of SMV and AMV from Indiana, but not of BYMV, PSV, TSV, or BPMV (Clark and Perry, 2002). One isolate identified as TRSV was transmitted after an extended feeding by *A. glycines*, but not after brief probes. An aerial vector of TRSV has long been suspected (Bergeson et al., 1964), but aphids have seldom been tested.

15–2.2.2.2 Beetles. Many factors related to beetle transmission of plant viruses were discussed in the reviews by Fulton et al. (1987), Gergerich and Scott (1996), and Gergerich (1999). Viruses are acquired and transmitted during the feeding process by a mechanism in which virus particles circulate through the beetle, but there is no evidence of viral propagation (Wang et al., 1992). The best-studied beetle-host system is that of the bean leaf beetle, *Cerotoma trifurcata* (Forst.), and BPMV on soybean in the USA. Studies relating the spatial and temporal distribution with beetle population density peaks suggested the second beetle generation was the most important as a vector (Hopkins and Mueller, 1983). The detection of BPMV serologically in bean leaf beetles was used as a predictor of the incidence of BPMV in soybean fields (Ghabrial et al., 1990a). The puzzling aspect of this system is that although the BPMV detected in overwintering beetles was presumed to provide the initial virus inoculum, the virus is rendered nontransmissible by protease degradation of the capsid protein in beetle regurgitant (Langham et al., 1990). Recent studies in the USA are examining the relative importance of beetles and seed as inoculum sources for this increasingly important virus (Daniels et al., 2001; Giesler et al., 2002).

15–2.2.2.3 Thrips. In Brazil, 10 species of thrips were identified in soybean fields. The most common, *Haplothrips robusta* and *Frankliniella schultzei* Trybom, were associated with transmission and incidence of TSV (Almeida et al., 1994). Regarding the transmission of tospoviruses in Brazil, the more efficient vector for TSWV was *F. occidentalis* Pergande, and for groundnut ringspot virus, *F. shultzei*. Both viruses are capable of infecting soybean, but only TSWV has been detected in diseased plants (Williams et al., 2001). The primary vector of PBNV in India is the melon thrips, *Thrips palmi* Karny (Reddy et al., 1992).

15–2.3 Virus Disease Management

An understanding of pathogen sources and vectors allows design of management practices to reduce the initial inoculum source, vector populations, or transmission efficiency, and the rate of spread into and within the soybean crop. Application of the practices should reduce the disease progress curve, resulting in decision-making systems and decreased yield losses (Guo et al., 1991; Nutter et al., 1998). Intercropping, for example, can cut virus incidence and economic losses to one-third (Irwin et al., 2000).

15–2.3.1 Reduction in Disease Incidence and Early Infection

15–2.3.1.1 Seed-Borne Pathogen Inoculum. When the initial inoculum source is from seed, it seems logical that efforts would be made to reduce the level of virus-carrying seed. However, this is not always the case. In the USA there are few programs to assure that certified seeds are pathogen-free. Seedlings in State and

USDA Uniform Regional cultivar trials, as well as in germplasm collections and breeders' crossing blocks, often have a high incidence of SMV (S.A. Tolin, unpublished data, 1998–2001). Seed coat mottling is often evident in such pre-commercial seed lines, but this alone is no evidence that virus is present. Conversely, absence of mottling does not assure that no virus is in the seed (Hill, 1999; Giesler, et al., 2002). The threshold level for the amount of seed transmission that should be tolerated in a seed lot is related to the aphid population size in the field (Irwin and Goodman, 1981), which is unpredictable. The recent invasion into the USA of a soybean-colonizing aphid capable of transmitting viruses may reduce the acceptable threshold level in the future (Hill et al., 2001; Wang and Ghabrial, 2002). The same principle applies to the low seed transmissibility of BPMV and the increased populations of its bean leaf beetle vector (Giesler et al., 2002).

Elimination of diseased plants in seed production fields through inspection, testing, and roguing can greatly reduce the potential for seed-transmitted viruses. Visual examination for discolored, symptomatic soybean seeds has become increasingly automated (Ahmad et al., 1994, 1999). Heat treatment was tested as a method of freeing soybean seed of SMV and was effective as hot water treatment at 70°C , but seed germination was greatly reduced (Haque et al., 1993).

Quarantine measures have undoubtedly had a role in reducing the introduction of viruses in seeds. Although the U.S. plant pest rules do not require examination of imported seeds for viruses, other countries frequently perform tests and in some cases report the results. For example, India reported detection of BCMV-PSt in soybean seeds imported from China (Prasada Rao et al., 1997), as well as SMV in soybean seed from USA and Taiwan.

15–2.3.1.2 Pathogen Inoculum Associated with Plant Reservoirs. Viruses that are not seed-transmitted usually have another host that will serve as a source of virus for its spread into soybean and a reservoir during noncrop periods. Many bean and peanut (*Arachis hypogaea* L.) viruses, seed-borne in these hosts, also replicate in soybean, and growing these crops nearby provides a source of inoculum for soybean. Interestingly, SMV has a very limited host range and is not known to occur outside of Asia in any crop other than soybean. In Japan, tsurumame (*G. soja* Siebold & Zucc.) is recognized as a weed host of SMV (Kosaka and Fukunishi, 1994). In the USA, the only geographic area in which PeMoV is found in soybean is in those areas of the southern states that also grow peanut, as it is seed-transmitted in peanut (Gillaspie and Hopkins, 1991). Forage legumes in the southeastern USA including alfalfa (*Medicago sativa* L.), alsike clover (*Trifolium hybridum* L.), arrowleaf clover (*T. vesiculosum* L.), crimson clover (*T. incarnatum* L.), red clover (*T. pratense* L.), and white clover (*T. repens* L.) are known to be sources of AMV, BYMV, CYVV, and PSV, all of which have aphid vectors (McLaughlin and Boykin, 1988). SDV, a persistently transmitted virus also harbored by clover, may pose an increasing threat to the U.S. soybean crop with the presence of a potential aphid vector, *A. glycines* (Damsteegt et al., 1999). Harvesting these crops by mowing has the potential to disturb large numbers of aphid vectors, which may uniformly inoculate soybean plants as they seek another host.

Beetle-transmitted viruses are found in perennial weed hosts. The leguminous weeds *Desmodium canadense* (L.) DC. and *D. paniculatum* (L.) DC. are natural reservoirs of BPMV and fed on by beetle vectors such as *C. trifurcata*, which transmit the virus to soybean (Moore et al., 1969; Walters and Lee, 1969). CPSMV in *D. canascens* (L.) DC. was transmitted to cowpea but not to soybean by *C. trifurcata* (McLaughlin et al., 1978). The simultaneous cultivation of soybean and bean provided a source of BRMV, which interacted synergistically with SMV to cause severe losses (Martins et al., 1994).

In India, a soybean disease caused by PBMV occurs in fields near peanut (Thakur et al., 1998). The geminivirus SYMV was found in 13 weed and wild hosts, and *Corchorus olitorius* L. and *C. capsularis* L. were the most important bridge hosts for the virus to infect soybean plants within the first 2 wk of planting (Keshwal et al., 1999).

15–2.3.1.3 Management to Reduce or Avoid Vector Populations. Attempting to control aphids on soybean to reduce SMV incidence has not been a practice in the USA since there were no colonizing aphids. Also, a high incidence of noncolonizing aphids occurs later in the season (Irwin and Goodman, 1981), whereas infections causing losses in yield occur early. With the introduction of the colonizing soybean aphid (*A. glycines*), this situation has changed but reports of trials are just becoming available. In countries having this aphid, there have been reports of decreasing virus disease incidence after insecticide treatments for aphid vectors. For example in Korea, foliar sprays of soybeans at V4/V6 stage of development with acephate WP (*O,S*-dimethyl acetylphosphoramidothioate) were effective in reducing colonizing *A. solani* and *A. glycines,* and in reducing SMV incidence (Kim et al., 2000). Imidacloprid WP (1-[(6-chloro-3-pyridinyl)methyl]-*N*-nitro-2-imidazolidinimine) or imidacloprid G applied in furrow and benfuracarb EC (ethyl *N*-[[[[(2,3-dihydro-2,2-dimethyl-7-benzofuranyl)oxy]carbonyl] methylamino]thio]-*N*-(1-methylethyl)-β-alaninate) were also effective in aphid control (Kim et al., 2000).

Imidacloprid was also found effective as a seed treatment (1.4 g a.i. kg^{-1} seed) in reducing the incidence and spread of two legume viruses (*Bean leaf roll virus*: family *Luteoviridae* and *Faba bean necrotic yellows virus*: genus *Nanovirus*), persistently transmitted by the aphid *Acyrthosiphon pisum*, in faba bean (*Vicia faba* L.) and lentil (*Lens culinaris* Medik.) in Syria (Makkouk and Kumari, 2001). Against SDV, imidacloprid was effective under glasshouse conditions but not in the field, presumably because the aphid vector transmits more efficiently and quickly in the field.

The management of BPMV by controlling bean leaf beetle populations is reviewed by Giesler et al. (2002). This strategy is considered potentially effective because control of spring colonizing beetles should reduce early spread of BPMV. Early application of foliar insecticides with residual activity would be expected to protect soybean from spring colonizing beetles and thus from BPMV. Preventative treatments being investigated are soil-applied insecticides and systemic seed treatment insecticides (Giesler et al., 2002).

Adjustment of date of seeding is a common practice to avoid peaks in vector populations, and has been suggested for management of BPMV (Giesler, 2002). Recent agronomic practice of earlier planting of soybean in the North Central states favors increased bean leaf beetle densities and, preliminarily, higher BPMV incidence relative to later seeding dates. Inoculation at earlier growth stages is known to result in greater yield reduction by BPMV (Hopkins and Mueller, 1984b; Ross, 1986a). Incidence of bud blight caused by TSV in Brazil was reduced from 80 to 20% when seeding was delayed until accumulated rainfall over 300 mm had reduced the thrips vector population (Corso and Almeida, 1991).

15–2.2.1.4 Management of the Cropping System. Few approaches falling into the category of agronomic practices, apart from delaying time of infection by vectors, have been shown to affect the incidence of virus, the disease progress curve, and impact of viruses on the plant. The extensive changes in the management of soybean cropping systems in recent years, such as double cropping, no-till, and new weed management strategies enabled by herbicide-tolerant cultivars all have the potential to change the exposure of soybean plants to viruses through changes in surrounding weeds and other vegetation at the time of seedling emergence. The same practices could also lead to changes in seasonal abundance and diversity of vectors.

Experiments by Pacumbaba et al. (1997) demonstrated complex interactions between fertilizers and rates of application on incidence of several diseases, including SMV. Plants getting complete fertilizer at a 20.0, 8.7, and 16.6 kg ha^{-1} at a 100 kg ha^{-1} rate of N, P, and K, respectively, had lowest incidence of SMV (about 20%) and highest yield, whereas those with high potash and phosphorus fertilizers had nearly twice the incidence of SMV and lower yield. Incidence of SMV decreased with doubled rates of nitrate (NO_3^-), but increased with the other fertilizers (Pacumbaba et al., 1997). There are few other reports of the influence of abiotic factors, such as temperature or moisture stress, on virus incidence.

Several other factors that can be manipulated to influence virus impact are discussed theoretically by Irwin et al. (2000). The SMV pathosystem simulation model is developed to illustrate the effect of varying factors in a pliant environment to predict whether the factor will reduce percentages of SMV infection at 50 d after planting, yield loss, and seed transmission of virus into progeny plants. The factors include; virus and vector manipulations, crop manipulations such as seeding rate, row spacing, planting date, and mixtures of susceptible and resistant cultivars, and cropping system manipulations such as intercropping.

15–2.3.2 Reduction in Host Susceptibility

15–2.3.2.1 Host Genes for Resistance to Virus. Deploying soybean cultivars with resistance to viruses that occur in the ecosystem or which may be introduced through seed or vectors is usually considered the best means of disease management. Specific genes conditioning resistance have been identified for several potyviruses including SMV (Chen et al., 1991, 1994), PeMoV (Boerma and Kuhn, 1976; Shipe et al., 1979), BCMV-PSt (Choi et al., 1989), and CCMV:*Bromovirus* (Boerma et al., 1975). Details of the genetics of resistance are given for SMV, PeMoV, and CCMV in Chapter 5. The genes for resistance to the three potyviruses, SMV, PeMoV, BCMV-PSt, map to a cluster of resistance genes at soybean molecular linkage group (MLG) F (Yu et al., 1994). Two additional genes for resistance to SMV, *Rsv3* and *Rsv4* have been mapped to soybean MLG B2 (Jeong et al., 2002) and D1b (Hayes et al., 2000), respectively. Apparently the location of *Rcv* for resistance to CCMV is not yet known.

VIRAL, BACTERIAL, AND PHYTOPLASMAL DISEASES

Screening cultivars for resistance to virus is an ongoing process worldwide, and nearly every announcement of release of a new cultivar in several countries includes a description of the viruses, if any, to which it is resistant. Screening a wider range of cultivars for BCMV-PSt resistance in Taiwan may have identified some new sources of resistance (Green and Lee, 1989). In Korea, screening has identified several good sources of SMV resistance, and has linked this to SMV pathotypes (Kim et al., 2001). Several cultivars and germplasm releases of Maturity Group V and later in the USA now have resistance to SMV, much of which is derived from PI 96983 (Kiihl and Hartwig, 1979), 'York' (Roane et al., 1983), or 'Hutcheson', which has both York and 'Essex' in its pedigree and was selected for resistance to SMV (Buss et al., 1988). The infusion of SMV resistance into Midsouth germplasm is thought to have decreased yield loss due to viruses by avoiding the synergistic effect of infection by both SMV and BPMV (Ross, 1968a; Calbert and Ghabrial, 1983). However, isolates of SMV able to overcome the $Rsv1y$ allele in Hutcheson are now being detected in Virginia (S.A. Tolin, unpublished data, 2001).

There has been an active search for BPMV resistance in soybean germplasm, but there are currently no commercial cultivars with resistance to BPMV. No resistance was found in early testing of 239 lines and cultivars of soybean and 123 *G. max* plant introductions, but introductions of *G. falcata* Benth., *G. javanica* Thunb., *G. tomentella* Hayata, and some *G. wightii* (Wight and Arn.) Verdc. were not infected by BPMV (Scott et al., 1974). Although severity of symptoms varies among entries in cultivar trials (Ziems et al., 2001a) and BPMV isolate (Gu et al., 2002; S.A. Tolin, unpublished data, 2001), there has been no report of heritable resistance since that of Ross (1986a), which was based on absence of symptoms. Based on analogy to other comovirus-host systems, Giesler et al. (2002) predicted that a single gene for resistance to BPMV should exist in soybean. Cultivars with resistance to CPSMV (Fernandez et al., 1989) or SDV (Damsteegt et al., 1990) were not found. In southern regions of Brazil, genetic resistance to the comovirus BRMV was identified in three cultivars, and its inheritance suggested a single recessive gene (Martins et al., 1995).

Among the other viruses listed in Table 15–3, resistance has been observed for certain begomoviruses (Siddiqui and Trimohan, 1999), PBNV:*Tospovirus*, TRSV:*Nepovirus* (Gopal, 1996), BYMV:*Potyvirus*, and the cucumoviruses CMV and PSV (Xu et al., 1988). For the 15 remaining viruses listed in Table 15–3, no report of a search for resistance was noted in the literature.

15–2.3.2.2 Transgenic Resistance to Virus. Successful development of resistance through insertion of specific genes into soybean has been slow in coming, with the first reports of stable transformation events appearing quite recently for SMV (Wang et al., 2001) and BPMV (Di et al., 1996; Reddy et al., 2001). A line homozygous for BPMV coat protein with systemic resistance to BPMV was produced that is expected to be useful in generating commercial cultivars resistant to BPMV by classical breeding. SMV coat protein transformed lines had a significantly lower infection rate, less seed coat mottling, and lower final disease incidence that did control plants without the transgene (Steinlage et al., 2002). Even though resistance to SMV was not absolute, a rate-limiting type of resistance might be effective and could place less selection pressure on the virus to develop resistance-breaking strains.

15–2.2.2.3 Resistance Induced by Cross-Protection. Classical cross-protection occurs when infection with a mild strain of a virus protects the plant from super-infection with a more severe strain of the same virus (Tolin, 2001). In what is claimed to be the first use of cross-protection in soybean, chronic damage of SMV to a black-seeded soybean, 'Tambaguro', was effectively reduced by prior inoculation with strains derived by attenuation at 15°C for 2 wk or more (Kosaka and Fukunishi, 1993). An attenuated SMV strain, which was efficiently aphid-transmitted but poorly seed-transmitted, was shown to protect plants when introduced by transplanting inoculated seedlings into growers' fields. Although in-field spread of the attenuated SMV strain by aphids reduced infection by more virulent strains, it was not entirely prevented (Kosaka and Fukunishi, 1994). The authors commented that development of a seed-transmitted, attenuated strain would provide a more efficient means of delivery.

Symptoms caused by diverse isolates of BPMV vary from very severe to very mild (Gu et al., 2002), and it was speculated that a mild strain might be a candidate for management of BPMV by cross-protection. Ghabrial (S.A. Ghabrial, personal communication, 2000) inoculated soybean plants with a mild BPMV strain, then challenged them with severe strains and was able to demonstrate cross-protection. Beetles disseminate this virus rapidly and widely, suggesting that beetle transmission from bait plants may provide a delivery means, at least experimentally. However, since there are risks of recombination of the protecting strain with native strains to yield even more virulent isolates, widespread application of cross-protection is limited to a few examples (Tolin, 2001).

15–2.3.2.4 Resistance to Inoculation by Vectors. Any resistance to inoculation by vectors has the potential to decrease incidence of diseased plants and delay time of infection in relation to soybean developmental stage, both important factors affecting yield loss magnitudes (Ren et al., 1997a). For example, trichome density may influence the spread of nonpersistently transmitted virus by aphids by affecting the ability of aphids to probe on soybean leaves and acquire virus, thus reducing transmission efficiency. Earlier work showed that denser leaf pubescence decreased probing activity of the aphid vectors *M. persicae*, *R. maidis*, and *Aphis citricola* Van der Goot, and also retarded development of field epidemics of SMV (Gunasinghe et al., 1988). A later study confirmed the results using some of the same 'Clark' near-isolines, showing a decrease in the SMV disease progress curves of dense, and extra dense pubescence lines compared to glabrous and normal lines, which was interpreted as a delay in time of inoculation by aphids (Ren et al., 2000).

Host resistance to the aphid *A. glycines* has the potential to reduce virus transmission. In China, resistance was found in 3 of 1000 *G. soja* genotypes, and was inherited as two independent recessive genes (Sun et al., 1990). However, reports of incorporation of resistance into soybean cultivars and demonstration of a reduction in SMV were not located. In theory, the spread of a persistently-transmitted virus like SDV is more likely to be reduced by using resistant cultivars, as long-term feeding needed for acquisition and inoculation of the virus should not occur. The report by Wang and Ghabrial (2002) in which SMV was more readily transmitted by *A. glycines* allowed short probes than by those feeding overnight would suggest that resistance to the aphid vector may actually increase spread of virus since aphid movement might be greater as a suitable host is sought.

15–3 BACTERIAL PATHOGENS OF SOYBEAN

Bacterial pathogens are widespread on soybean, but do not appear in general to seriously limit production in the USA. Gram-negative bacteria cause two major diseases of soybean, bacterial blight, caused by *Pseudomonas syringae* pv. *glycinea*, and bacterial pustule, caused by *Xanthomonas campestris* pv. *glycines*, and minor diseases including wildfire, caused by *P. syringae* pv. *tabaci*, and bacterial wilt caused by *Ralstonia solanacearum*. Gram-positive actinobacteria *Curtobacterium flaccumfaciens* pv. *flaccumfaciens* causes tan spot and *Rhodococcus facians* causing fasciation.

15–3.1 Naming Bacteria

The names and current classifications of bacteria causing disease on soybean are listed in Table 15–4. Naming bacteria, especially pseudomonads and xanthomonads, is complicated by the current state of the taxonomy of these organisms. In 1976, the Approved Lists of Bacterial Names was created to provide a new start for bacteriological nomenclature (Lapage et al., 1975). Species epithets were included only if a published description existed including valid name and adequate descriptive information to differentiate the species from other species. Although most plant pathogenic bacterial species can be differentiated from other species by host range assays, they are not clearly separable from other species by commonly used bacteriological methods. Therefore, the names of more than 200 species of plant pathogens were excluded from the list and after 1980, had "no further standing…" and were "available for reuse in the naming of new taxa" (Lapage et al., 1975). To prevent confusion in the literature, Dye et al. (1980) proposed to conserve the epithets of the excluded species of pseudomonads and xanthomonads as "pathovars", artificial infrasubspecific taxa, of *Pseudomonas syringae* or *Xanthomonas campestris*. Therefore, pathovars of these two species may not be as closely related phylogenetically as their inclusion with these species seems to suggest.

Several extensive studies have failed to include all pseudomonad and xanthomonad pathovars (reviewed in Kersters et al., 1996; Vauterin et al., 1995) and some of the recommendations, especially the grouping of large numbers of xanthomonad pathovars in *Xanthomonas axonopodis*, remain controversial (Schaad et al., 2000; Young et al., 2001).

15–3.2 Identifying Bacteria

Methods available for identifying bacteria known to cause disease on soybean in the field are summarized in Table 15–5. The subject of identification is reviewed by group in depth by Schaad et al. (2001a); agrobacteria (Moore et al., 2001), cur-

Table 15–4. Classification of bacteria and phytoplasmas and the diseases they cause on soybean.

Prokaryotae (Kingdom)
  Proteobacteria (Division)
    Alphaproteobacteria (Subdivision)
      Rhizobiaceae (Family)
        Agrobacterium tumefaciens : Crown gall†
        Agrobacterium rhizogenes: Hairy root†
        Bradyrhizobium japonicum: Rhizobitoxin damage under conditions in which high
            N is supplied to the host plant
    Betaproteobacteria
      Ralstoniaceae
        Ralstonia solanacearum: Bacterial wilt
      Enterobacteriaceae
        Erwinia amylovora: Fire blight†
        Erwinia (Pectobacterium) carotovora subsp. *atroseptica*:soft rot†
    Gammaproteobacteria
      Pseudomonadaceae
        Pseudomonas syringae pv. *glycinea*: Bacterial blight
        Pseudomonas syringae pv. *phaseolicola*: Halo blight†
        Pseudomonas syringae pv. *syringae*:brown spot†
        Pseudomonas syringae pv. *tabaci*: Wildfire
      Xanthomonadaceae
        Xanthomonas campestris pv. *cannabis*: Leaf spot†
        Xanthomonas campestris pv. *glycines* : Bacterial pustule
  Firmicutes
      Bacillus subtilus: Reduced germination
  Mollicutes
    Acholoplasmataceae
      Phytoplasma spp.: Machismo, bud proliferation, witches'-broom, phyllody,
        and stem-greening
  Actinobacteria
    Micrococcaceae
      Curtobacterium flaccumfaciens pv. *flaccumfaciens*: Bacterial tan spot
    Nocardiaceae
      Rhodococcus fasians: fasciation

† Pathogens that are only known from laboratory or greenhouse inoculation experiments (controlled conditions).

tobacteria (Davis and Vidaver, 2001), erwiniae (DeBoer and Kelman, 2001; Jones and Geider, 2001), pseudomonads (Braun-Wiewnick and Sands, 2001), rhodococci (Davis and Vidaver, 2001), and xanthomonads (Schaad et al., 2001b). Generally, rapid identification methods include polymerase chain reaction (PCR) amplification of diagnostic DNA products, fatty acid methyl ester (FAME) analyses of lipids, carbon utilization profiles, and serology. Semiselective media are useful for primary isolation of the presumptive pathogen from diseased plant tissues, but should not be considered as definitive in identification.

15–3.3 Descriptions of Diseases Caused by Bacteria

Many aspects of bacteria-caused diseases are similar enough that control measures are often also similar. Because of these commonalities, the strategies for ecological, epidemiological, and disease management are presented in a separate section.

15–3.3.1 Bacterial Blight

Bacterial blight, caused by *Pseudomonas syringae* pv. *glycinea* (Coerper) Young, Dye, and Wilke (formerly *Pseudomonas glycinea*), is the most common bacterial disease of soybean worldwide. Bacterial blight occurs during periods (about 5 d duration) of wet, cool weather and decreases with warm, dry weather. Yield reduction estimates range from 4 to 40% (Hartman et al., 1999; Hwang and Lim, 1992). However, this disease rarely devastates a crop.

15–3.3.1.1 Symptoms. Bacterial blight causes small (<1mm), translucent, water soaked angular areas (delimited by veinlets) on leaves. Seedlings may die if apical meristems are affected (Hartman et al., 1999). Young leaves are most susceptible and may be deformed. Lesions appear as yellow to light brown spots on leaves. Necrotic centers of lesions dry and become reddish to very dark brown surrounded by halos of chlorotic tissue. Necrotic areas may coalesce with dead tissue tearing away to give a ragged appearance to the leaf, and defoliation may occur. Cotyledons, petioles, and pods are also affected. Pod lesions often involve contamination or colonization of seed. Infested seeds may appear normal or shriveled.

15–3.3.1.2 The Pathogen. The bacterium is a Gram-negative rod, with one to several polar flagella. In culture under iron deprivation, the pathogen produces water soluble, pyoverdine siderophores visible by weak yellow-green fluorescence under ultraviolet light. Like other *P. syringae* pathovars, the pathogen induces a hypersensitive reaction in tobacco, nucleates ice formation, produces levan, does not produce oxidase or arginine dihydrolase, and utilizes sorbitol, sucrose, and *meso*-tartrate (Schaad et al., 2001). Toxins such as coronatine and pectinases aid pathogenicity.

Eight races of the pathogen have been distinguished in the USA by reactions on differential cultivars (Cross et al., 1966; Leary et al., 1984). However, five additional strains have been described that could not be characterized within any of the known races (Prom and Venette, 1997). Races of the pathogen contain *avr* genes eliciting resistant reactions in soybean cultivars carrying specific resistance genes (Nickell et al., 1994). The bacterial blight pathogen may also cause disease in cowpea and field bean, lima bean (*Phaseolus lunatus* L.), and tepary beans (*Phaseolus acutifolius* A). It is related closely to *P. syringae* pvs. *syringae* and *phaseolicola*, which cause disease in soybean but only upon artificial inoculation (Nikitina and Korsakov, 1978; Stall and Kucharek, 1982).

15–3.3.2 Bacterial Pustule

Bacterial pustule occurs in soybean crops where warm, moist conditions occur. Crop defoliation and reduction in individual seed weight affect yields. Measured yield reductions ranged from 4 to 35% (Hwang and Lim, 1992; Prathuangwong, 1985). Like bacterial blight, this disease is not likely to devastate a crop.

15–3.3.2.1 Symptoms. The earliest symptoms are minute (<1mm), light-green spots with slightly raised centers that form on the lower surface of the leaf, but some form on the upper surface of the leaf. Pustules form by cellular hypertrophy and, occasionally, hyperplasia, and the raised centers become quite large

Table 15–5. Bacterial disease and pathogen identification systems.

Disease (pathogen)	Identification aid (reference)

Crown gall (*Agrobacterium tumefaciens*)
    Semiselective media:1A (Brisbane and Kerr, 1983)
    Serological: None available
    Carbon utilization: GN2 MicroPlate system - Biolog, Inc. †
    Fatty acid methyl esters (FAME): Microbial Identification System, Microbial ID, Inc.
        (MIDI)‡
    PCR primers: Available for detection of *virC* gene in oncogenic strains of all species
        and biotypes of agrobacteria (Haas et al., 1995)
Hairy root (*Agrobacterium rhizogenes*)
    Semiselective media: 2E (Brisbane and Kerr, 1983)
    Serological: None available.
    Carbon utilization: GN2 MicroPlate system—Biolog, Inc. †
    FAME: MIDI‡
    PCR primers: Available for detection of *virC* gene in oncogenic strains of all species
        and biotypes of agrobacteria (Haas et al., 1995)
Fire blight (*Erwinia amylovora*)
    Semiselective media: MS (Miller and Schroth, 1972)
    Serological: No reliable method available.
    Carbon utilization: GN2 MicroPlate system—Biolog, Inc. †
    FAME: MIDI‡
    PCR primers: Based on detection of universal plasmid pEA29 (Bereswill et al., 1992)
Soft rot [*Erwinia (Pectobacterium) carotovora* subsp. *atroseptica*]
    Semiselective media: CVP (Cuppels and Kelman, 1974)
    Serological: Multiple serological groups make identification difficult (DeBoer and Kelman,
        2001).
    Carbon utilization: GN2 MicroPlate system—Biolog, Inc. †
    FAME: MIDI‡
    PCR primers: Several primer sets are available (DeBoer and Ward, 1995; Frechon et al.,
        1995; Smid et al., 1995)
Bacterial blight (*Pseudomonas syringae* pv. *glycinea*)
    Semiselective media: M-71 (Leben, 1972) and BANQ (Fieldhouse and Sasser, 1982)
    Serological: ELISA kit from ADGEN§
    Carbon source utilization: GN2 MicroPlate system—Biolog, Inc. †
    FAME: MIDI‡
    PCR primers: Available for coronatine toxin genes (Bereswill et al., 1994)
Halo blight (*Pseudomonas syringae* pv. *phaseolicola*)
    Semiselective media: MSP (Mahan and Schaad, 1987)
    Serological: ELISA kits from Agdia¶, ADGEN§, D-Genos#
    Carbon source utilization: GN2 MicroPlate system—Biolog, Inc. †
    FAME: MIDI‡
    PCR primers: Available for detection of phaseolotoxin gene region (Prosen et al., 1993)
Brown spot (*Pseudomonas syringae* pv. *syringae*)
    Semiselective media: MSP (Mohan and Schaad, 1987) and KBC (Schaad et al., 2001a)
    Serological: None available.
    Carbon source utilization: GN2 MicroPlate system—Biolog, Inc. †
    FAME: MIDI‡
    PCR primers: Available for lipodepsinonopeptide (syringotoxin) toxin genes (Sorensen et al.,
        1998)
Wildfire (*Pseudomonas syringae* pv. *tabaci*)
    Semiselective media: None available
    Serological: None available
    Carbon source utilization: GN2 MicroPlate system—Biolog, Inc. †
    FAME: MIDI‡
    PCR primers: None available
    Other: Southern probes for tabtoxin genes (Kinscherf et al., 1991)

(continued on next page)

Table 15–5. Continued.

Disease (pathogen)	Identification aid (reference)

Leaf spot (*Xanthomonas campestris* pv. *cannabis*)
    Semiselective media: None available
    Serological: None available
    Carbon source utilization: GN2 MicroPlate system—Biolog, Inc. †
    FAME: MIDI‡
    PCR primers: None available
Bacterial pustule (*Xanthomonas campestris* pv. *glycines*)
    Semiselective media: None available
    Serological: None available
    Carbon source utilization: GN2 MicroPlate system—Biolog, Inc. †
    FAME: MIDI‡
    PCR primers: None available
Bacterial tan spot (*Curtobacterium flaccumfaciens* pv. *flaccumfaciens*)
    Semiselective media: None available
    Serological: None available
    Carbon source utilization: GP2 MicroPlate system—Biolog, Inc. †
    FAME: MIDI‡
    PCR primers: None available
    Other: Immunodiagnosis from mung bean seeds (Diatloff et al., 1993)
Bacterial wilt (*Ralstonia solanacearum*)
    Semiselective media: SMSA (French et al., 1995)
    Serological: Not currently available (ELISA kits may soon be available from Agdia¶)
    Carbon source utilization: GP2 MicroPlate system—Biolog, Inc. †
    FAME: MIDI‡
    PCR primers: Species specific primers (Gillings et al., 1993; Hartung et al., 1998; Seal et al., 1992)
Bacterial fasciation (*Rhodococcus fascians*)
    Semiselective media: None available
    Serological: None available
    Carbon source utilization: GP2 MicroPlate system—Biolog, Inc. †
    FAME: MIDI‡
    PCR primers: Species specific primers (Stange et al., 1996)

† BIOLOG, Hayward, Calif. 3938 Trust Way, Hayward CA 94545 USA; http://www.Biolog.com; info@biolog.com. Commercial organizations are listed here as a resource to the reader. No endorsement of their products or services is intended or implied.
‡ MIDI, 125 Sandy Dr., Newark, DE 19713, USA:http://www.microbialid.com; ServiceLab@microbialid.com
§ ADGEN, Nellies Gate, Auchincruive, Ayr. KA6 5HW, UK; http://adgen.co.uk; info@agden.co.uk
¶ Agdia, Inc., 30380 Cty Rd 6, Elkhart, IN 46514 USA; http://www.agdia.com/; info@agdia.com
# D-Genos, Parc Technologique des Capuchins, BP 50750-49007, Angers Cedex 01, France; http://www.d-genos.com; a.barthelaix@d-genos.com.

(Hokawat and Rudolf, 1993). This form may be confused with soybean rust, *Phakospora pachyrhizi* Sydow, just prior to spore eruption. However bacterial pustules are surrounded by distinct chlorotic yellow halos, while rust lesions form tan or brown lesions with a central pore for spore release. In severe bacterial pustule disease, premature defoliation may occur. Coalescing lesions cause brown dry tissue necrosis and death. If these dead areas tear away, the leaf may appear ragged. Small, reddish-brown, slightly raised spots form on petioles and pods. On resistant cultivars either no symptoms occur or symptoms are less intense (i.e., small chlorotic spots without raised centers may form or pustules may be less numerous).

15–3.3.2.2 The Pathogen. Bacterial pustule was first observed in the USA by Smith (1905) and the causal organism was described by Clinton (1916). Nakano (1919), in Japan, published the first adequate description of the pathogen as *Pseudomonas glycines* which was more accurately reclassified as *Xanthomonas campestris* pv. *glycines* (Nakano) Dye. In the USA and without knowledge of the Japanese work, Hedges (1922, 1924) named it *Bacterium phaseoli* var. *sojense*. Later, the American bacterium was shown to be identical to *X. campestris* pv. *glycines*. *X. campestris* pv. *glycines* (Nakano) Dye is synonymous with *X. axonopodis* pv. *glycines*, *X. phaseoli* var. *sojensis* and *X. campestris* pv. *phaseoli* var. *sojensis*.

The bacterium is a Gram-negative rod with one polar flagellum. On common bacteriological media, colonies are yellow due to xanthomonadin pigments in the bacterial cell walls. Like most *X. campestris* strains, the pathogen grows at 35C, hydrolyzes starch, produces acid from arabinose, and grows on the selective medium, SX (Schaad et al, 2001a). Little is known of how the bacterium causes pathogenesis. However, the pathogen produces auxins, extracellular polysaccharides, and may have biological races (Hartman et al., 1999).

15–3.3.3 Wildfire

Wildfire is a minor disease of soybean caused by *Pseudomonas syringae* pv. *tabaci* (Wolf and Forster) Young, Dye and Wilkie (formerly *Pseudomonas syringae*), which also causes wildfire of tobacco (Wolf and Foster, 1917). In soybean, the pathogen enters stomates, lenticels, and wounds by wind-driven rain. Small (1–2 mm) dark brown lesions are surrounded by broad chlorotic halos that are produced by the action of tabtoxin. Lesions may coalesce, dead tissue is torn away, and the leaf appears ragged. Premature defoliation may occur.

The pathogen is very similar to *P. syringae* pv. *glycinea*, but pv. *tabaci* produces β-glucosidase, liquefies gelatin, and utilizes erythritol, L(+)-tartrate, and DL-glycerol while pv. *glycinea* does not (Schaad et al., 2001a). Races of the pathogen may occur because soybean strains are equally damaging to tobacco (*Nicotiana tabacum* L.), while tobacco strains are less damaging to soybean (Hartman et al., 1999).

15–3.3.4 Bacterial Wilt

Bacterial wilt is considered to be a minor disease of soybean in southeastern USA, but it can cause considerable damage in the Ukraine (Hartman et al., 1999). The causal bacterium *Ralstonia solanacearum* (formerly *Burkholderia solanacearum, Pseudomonas solanacearum*) causes Granville wilt or bacterial wilt in tobacco and, occasionally causes wilt in peanut (Denny and Hayward, 2001; Hayward, 1991). *Ralstonia solanacearum* may also cause foliar and/or wilt symptoms in soybean, especially in regions where bacterial wilt is endemic in tobacco and/or peanut.

Wilt symptoms in soybean may be mild showing either a progressive withering or mild stunting. Other plants may wilt rapidly and severely. Polysaccharides are important in wilting (Orgambide et al., 1991). The bacterial wilt pathogen may also affect only the leaves of plants. These leaves become slightly chlorotic and develop small dark necrotic lesions. The lesions develop dark borders, elongate, and the dried tissue falls off causing the leaves to appear torn.

Ralstonia solanacearum is a Gram-negative rod with a polar flagellum. It is an obligate aerobe, fails to produce yellow nonwater soluble or fluorescent green or blue water-soluble pigments, accumulates poly-β-hydroxybutyrate granules, grows at 37C but not at 40C, does not utilize arginine or betaine, and induces a hypersensitive response in tobacco (Denny and Hayward, 2001). Five biovars of the pathogen are known and may be separated by their ability to utilize or oxidize various substrates (Denny and Hayward, 2001).

15–3.3.5 Tan Spot

Curtobacterium flaccumfaciens pv. *flaccumfaciens* (formerly *Corynebacterium flaccumfaciens*) as been implicated as a causal agent of bacterial wilt of beans (Hedges, 1926). Dunleavy (1963) described the same pathogen on soybean causing tan spot and wilt. The disease is episodic in nature (Dunleavy et al., 1983) and seed yield losses as high as 13% have been measured (Dunleavy, 1984).

15–3.3.5.1 Symptoms. Small (1–2 mm), translucent to yellowish lesions form on leaves. Oval or elongated chlorotic regions, usually beginning near leaf margins, proceed from the leaf margin toward the midrib. Affected tissues die, dry, and may be blown away causing the leaf to appear torn. Young leaves or seedlings may be misshapen and have fused leaflets.

15–3.3.5.2 The Pathogen. Young cultures contain Gram-positive short, irregular rods and pairs are often arranged in a "V" shape. In older cultures, rods segment into small cocci. On common media, aerial mycelia do not form and colonies are yellow, smooth, slightly convex, and circular. Curtobacteria are nonmotile or motile by lateral flagella. They are obligate aerobes, hydrolyze esculin, utilize acetate and formate, and produce acid from ribose (Davis and Vidaver, 2001). Diversity occurs among pathogenic strains—soybean strains cause wilt on bean, but bean wilt strains do not cause soybean tan spot (Hartman et al., 1999). The pathogen also causes disease in mung bean [*Vigna radiata* (L.) R. wilczek] and cowpea (Wood and Easdown, 1990).

15–3.3.6 Bacterial Seed Rot

Bacterial seed rot occurs when seeds are stored in hot moist conditions (25–35C) followed by planting at temperatures above 30C (Cubeta et al., 1985). In the field, seed yield losses as high as 100% have been reported (Hartman et al., 1999). The pathogen, *Bacillus subtilis*, is a common soil microbe often associated with leaves and stems of plants as casual contaminants resulting from rain splash and wind-dispersed soil. The bacterium is a Gram-positive, large rod with peritrichous flagella that forms endospores and produces copious enzymes capable of breaking down seed tissues. The disease can be managed by avoiding hot, moist conditions for both extended seed storage and planting. Antibiotic treatment may be effective (Hartman et al., 1999), and biological control using strains of the same organism are promising (Sheng and Blackman, 1996).

15–3.4 Other Bacterial Associations with Soybean

15–3.4.1 Bacteria That May be Pathogens

Other Gram-negative pathogens of soybean are of minor or local importance. Those pathogens, indicated as pathogens under "artificial conditions" in Table 15–4, are only known to be pathogenic from inoculation studies (Garcia and Zak, 1983; Nikitina and Korsakov, 1978; Severin, 1978; Stall and Kucharek, 1982). These include *Agrobacterium* spp. causing crown gall and hairy root on many hosts, *Erwinia amylovora* causal agent of fire blight on roseaceous hosts, *Erwinia* (*Pectobacterium*) *carotovora* subsp. *atroseptica* causal agent of soft rot and blackleg in many crops, and *Xanthomonas campestris* pv. *cannabis* causal agent of leaf spot on marijuana (*Cannabis sativa* L.). It is not known if these pathogens are or will be important on soybean in the field. They are listed for the information of the reader and may be of occasional importance in localized regions.

15–3.4.2 Ineffective Dinitrogen Fixation

Conversion of nitrogen (N_2) from the atmosphere to ammonium nitrogen (NH_4^{1-}) by bacterial symbionts provides 40 to 85% of the total N requirement for soybean (Hartman et al., 1999). Nodules, or gall-like knots of tissue on the primary and secondary roots, are the sites of N_2 fixation. Leghemaglobin in nodules provides the reducing environment for fixation. Bradyrhizobia infect and colonize soybean roots to cause gall formation, induce leghemaglobin production, and produce the nitrogenase necessary for reducing gaseous N. Commercial inoculants contain several to many strains of the bacterium *Bradyrhizobium japonicum* (formerly *Rhizobium japonicum*) and its subspecies (1 and 2) because they vary in their ability to effectively nodulate different soybean genotypes under different soil conditions. Ineffective N_2 fixation is the failure of the symbiotic relationship of soybean with the bacterium and result in poor seed yield from chlorotic plants. Symbiosis may fail for several reasons. *Bradyrhizobium japonicum* strains in the rhizosphere may not produce effective nodules on the cultivars planted. Soils containing toxic levels of aluminum (southeastern USA) are detrimental to *B. japonicum* survival. Flooded soils reduce or eliminate nodulation (Hartman et al., 1999). Soybean plants may become chlorotic for other reasons. Under reduced tillage conditions, organic N in crop debris may be released too slowly to provide inorganic N to supplement N_2 fixation for crop growth. Soils amended with high levels of inorganic N or organic materials including sewage sludge or manure may prevent nodulation. Under high N fertilization and on certain soybean genotypes (e.g., 'Hood'), some strains of *B. japonicum* produce phytotoxic rhizobitoxin that induces chlorosis and, occasionally, tissue death at apical growing points (Johnson et al., 1959; Xie et al., 1996). Rhizobitoxin effects are most prevalent in new leaves in plants grown under conditions of high N. This suggests a source sink affect with the newly expanding leaves acting as a sink for rhizobitoxin. The chlorosis seems to be caused by inhibition of cystathionine beta-lyase through covalent binding with rhizobitoxin (Owens et al., 1968). Rhizobitoxin has also been shown inhibit 1-aminocyclopropane-1-carboxylic acid synthase, an enzyme critical to ethylene biosynthesis (Owens et al., 1971). On legumes, inhibition of ethylene biosynthesis improves, nodule formation, numbers, position, and the morphology. On roots, rhizobitoxin reverses the inhibitory effect of exogenous ethylene on nodule formation and enhances nodule formation. Rhizobitoxin-less bradyrhizobia mutants no longer inhibited 1-aminocyclopropane-1-carboxcylic acid synthase resulting in a decrease in effective nodulation (Duodu et al., 1999). This suggests that rhizobitoxin may have a beneficial role in the mutalistic symbiosis leading to N_2 fixation. However, under conditions of high N, rhizobitoxin production may become unbalanced converting *B. japonicum*, into an antagonistic symbiont or pathogen of soybean.

Management of ineffective nodulation includes balancing inorganic N availability in reduced tillage with added organic N (manure) with strains of bradyrhizobia effective for nodulating the cultivars planted.

15–4 ECOLOGY, EPIDEMIOLOGY, AND MANAGEMENT OF BACTERIAL DISEASES

Knowledge of the ecology and epidemiology is important in controlling diseases caused by bacteria. This is because: (i) bacteria have extremely short generation times allowing very rapid development of high populations; and (ii) disease spread and disease development usually depend upon the presence of free-water for bacterial growth. The conjunction of short generation times, the presence of free-water, and rapid disease development result in disease problems that are best managed prophylactically before symptoms are evident rather than after symptoms develop. This is consistent with soybean production strategies that emphasize low inputs and management with little or no pesticides for disease control after planting.

15–4.1 Ecology

15–4.1.1 Overseasoning Modes

Management of diseases starts with knowing how the pathogen overseasons. In temperate areas, soybean pathogens often overseason in seeds and crop debris. These pathogens include those causing bacterial blight (*P. syringae* pv. *glycinea*), bacterial pustule (*X. campestris* pv. *glycines*), bacterial wilt (*R. solanacearum*), wildfire (*P. syringae* pv. *tabaci*), and bacterial seed rot (*B. subtilis*).

15–4.1.2 Phyllosphere Residence Phase

Once germination occurs, pathogens such as *P. syringae* pv. *glycinea* and *X. campestris* pv. *glycines,* often persist and replicate slowly as apparently harmless residents in the phyllosphere (aerial plant surfaces). This resident phase is extremely important for bacterial pathogens because, as a group, they lack resistant spores, such as fungi have, for survival until conditions are right for penetration into the host and disease development.

15–4.1.3 Rhizosphere Residence Phase

In addition to plant debris, the bacterial wilt pathogen, *R. solanacearum*, in the rhizosphere (soil associated with roots) of many plants including disease hosts (tobacco, banana [*Musa paradisiaca* L.], peanut, potato [*Solanum tuberosum* L.], and tomato [*Lycopersicon esculentum* Mill.]) and nondisease hosts (pea, corn [*Zea mays* L.]) (Michiko et al., 1991). The wildfire pathogen, *P. syringae* pv. *tabaci*, also overseasons in the rhizosphere of many crop plants.

15–4.1.4 Weed Hosts

Some pathogens extend their epiphytic resident phases to weed hosts. *Xanthomonas campestris* pv. *glycines* is found in wheat (*Triticum aestivum* L.) rhizospheres (associated with soil on roots) and in the phyllosphere of weeds such as redvine (*Brunnichia ovata* [Walter] Shinners) and horse-gram (*Macrotyloma uniforum* [Lam.] Verdc. [*ex Dolichos biflorus*]).

15–4.2 Epidemiology

15–4.2.1 Wind Driven Rain

During rain, wind aids pathogen penetration into stomates, lenticels, and wounds (including wind-abrasion wounds). Pathogens multiply in intercellular spaces producing copious water-retaining extracellular polysaccharides that contribute to the water-soaked appearance of lesions. Pathogens associated with bacterial blight (*P. syringae* pv. *glycinea*) , bacterial pustule (*X. campestris* pv. *glycines*), bacterial wilt (*R. solanacearum*), and tan spot (*C. flaccumfaciens* pv. *flaccumfaciens*) are commonly dispersed in this manner. Overhead irrigation has many of the features of wind-driven rain that are important in pathogen dissemination and should be avoided. In arid regions, furrow irrigation is preferred for producing pathogen-free seed (Civerolo, 1982).

15–4.2.2 Wounding

The bacterial wilt pathogen, *R. solanacearum*, gains entry to roots via wounds caused by tillage, nematodes, insects, or lateral and adventitious root emergence. Other pathogens, including *P. syringae* pv. *glycinea* and *X. campestris* pv. *glycines*, may gain entry through wind abrasion wounds on leaves.

15–4.3 Management

15–4.3.1 Seeds

Pathogen-free seed should be used for planting. Seed should be stored under dry and cool conditions to prevent bacterial seed rot caused by *B. subtilis* and to reduce the opportunity for early replication of pathogens causing bacterial blight (*P. syringae* pv. *glycinea*), bacterial pustule (*X. campestris* pv. *glycines*), or tan spot (*C. flaccumfaciens* pv. *flaccumfaciens*). Antibiotic treatment may be effective in reducing bacterial seed rot (Hartman et al., 1999).

15–4.3.2 Planting Conditions

Planting time and conditions may be manipulated for disease management. To avoid tan spot, caused by *C. flaccumfaciens* pv. *flaccumfaciens,* planting early before temperatures reach 25C reduces systemic dispersal of the pathogen (Dunleavy, 1985). Likewise, to avoid bacterial seed rot, planting above 30C should be avoided (Cubeta et al., 1985).

15–4.3.3 Rotation

Crop rotation may be manipulated for disease control. Rotating out of soybean and other legumes, including field beans, is helpful in reducing bacterial blight caused by *P. syringae* pv. *glycinea*. The same strategy is important in controlling bacterial pustule, caused by *X. campestris* pv. *glycines,* because this pathogen also may cause disease in peanut, cowpea, field bean, lablab bean [*Lablab purpureus* (L.) Sweet], lima bean, and yam (or jicama) bean [*Pachyrhizus erosus* (L.) Urb.] (Hokawat and Rudolph, 1993). The bacterial wilt pathogen, *R. solanacearum,* causes special concern because it not only survives and multiplies in the rhizosphere of many plants including disease host plants (tobacco, banana, peanut, potato, and tomato), but it also is harbored on the roots of nondisease hosts (pea, corn, sorghum [*Sorghum bicolor* (L.) Moench.], and bean). For disease control, it is best to avoid planting soybean in areas where bacterial wilt has occurred or is endemic (Civerolo, 1982; Michiko et al., 1991).

15–4.3.4 Cultivation

Poorly-timed cultivation may contribute to mechanical spread of the pathogen. Because of the need for wounds and free-water, cultivation in fields should be avoided when the foliage is wet especially during cool periods for bacterial blight and during warm wet periods for bacterial pustule.

15–4.3.5 Tillage

Tillage reduces the amount of crop debris and soybean straw in the field. However, conservation or reduced-tillage strategies to protect crop land from erosion increases the amount of plant debris in the field. Leaving debris in the field results in lower temperatures at the soil level, higher humidity, and traps moisture—all conditions favoring bacterial pathogenesis. This increases the possibility that pathogen-infested plant debris will contribute to epiphytotics, particularly of bacterial blight, bacterial pustule, and wildfire.

15–4.3.6 Resistance

With reduced tillage becoming more important in the cultivation of soybean, host resistance emerges as increasingly important in disease management. Unfortunately, resistance to bacterial diseases has not been a priority in soybean breeding programs and options are rather limited. Resistance of soybean to particular races of the bacterial blight pathogen, *P. syringae pv. glycinea*, has been found (Cross et al., 1966; Leary et al., 1984) and multiple resistance genes have been combined into some soybean lines (Nickell et al., 1994). *Pseudomonas syringae* pathovars, including the blight pathogen pv. *glycinea*, has become a model for understanding interactions between plants and their pathogens especially concerning the gene-for-gene hypothesis for host resistance (Staskawicz et al., 2001). Numerous reviews of the secretion of avirulence gene products with host-resistance genes to induce the apoptotic hypersensitive resistance reaction (Li et al., 2002; Mackey et al., 2002; Schneider, 2002; Yuan et al., 2000). A draft genome sequence of *P. syringae* is now available for analyses and annotation will spur further development in our understanding (ORNL, 2002).

Resistance is available to bacterial pustule caused by *X. campestris* pv. *glycines* (Hartwig and Johnson, 1953; Prathuangwong, 1985; Weber et al., 1966). Some cultivars are less susceptible than others to tan spot, caused by *C. flaccumfaciens* pv. *flaccumfaciens* (Dunleavy, 1984, 1985).

15–4.3.7 Biological Control

Biological control of soybean diseases caused by bacteria has not yet matured to the point of application. Studies with suppressive soils indicate some progress in biological control of *R. solanacearum*-, the causal agent of bacterial wilt (Michiko et al., 1991). Seed treatments with some strains of *B. subtilis* also suggest that seed rot caused by other strains of *B. subtilis* may be controlled by biological means (Sheng and Blackman, 1996).

15–5 PHYTOPLASMAS ASSOCIATED WITH SOYBEAN DISEASES

Phytoplasmas cause several problems including the better-studied diseases machismo and bud proliferation. Yield losses as high as 80% have been reported (Sinclair, 1999). Usually, the incidence, severity, and yield losses are much lower than this. Less well-described problems, which may also have a phytoplasmal etiology, include witches' broom, phyllody, and a green-stem syndrome. These symptoms are also associated with some viruses.

15–5.1 Classification

Phytoplasmas (formerly mycoplasmalike-organisms [MLOs]) are pleomorphic Mollicutes (prokaryotes lacking cell walls) that are most closely related to Firmicutes such as bacilli and clostridia. These organisms are obligate parasites that have alternate insect hosts, usually leafhoppers, and many and diverse plant hosts.

Phytoplasmas are currently classified phylogenetically into 14 16S rRNA groups determined by sequence analyses (Lee et al., 2001a). Phytoplasmas pathogenic on soybean have been found in three groups; 16SrI (aster yellows group phytoplasmas), 16SrIII (X-disease group of phytoplasmas), and 16SrVI (clover proliferation group of phytoplasmas) (Grau et al., 1999; Jomantiene et al., 1999, 2002; Lee et al., 2000). Recent studies using 16S rRNA methods to identify the causal agent of the green-stem syndrome failed to establish a connection with phytoplasmas (Lee et al., 2002). This study did, however, detect two phytoplasmas, 16SR groups I-A and I-O, that have not yet been associated with specific symptoms in soybean.

15–5.2 Identification

Identification of phytoplasmas is complicated by their absolute requirement for living hosts. Phytoplasmal pathogens of soybean have been identified by

Table 15–6. Symptoms and other characteristics of soybean phytoplasmal diseases.

Phytoplasma-incited disease	Symptoms and other features
Machismo (Granada, 1979a)	Yield losses to 80% (Sinclair, 1999) Symptoms begin soon after flowering. Flower sepals may be longer, petals remain closed, pod-like structure may protrude from closed flower. Pods may be upright, flat, thin, and lack beans or they may have a leaf-like structure. Buds proliferate in axils. Transmission by *Scaphytopius fuliginosus* (Osborn) Not seed transmitted. Other hosts affected: *Cajanus cajan* DC., *Catharanthus roseus* (L.) G. Don, *Crotolaria* spp., *Desmodium* spp., *Galactia glaucescens* Kunth, *Phaseolus vulgaris, Rhynchosia minima* (L.) DC. 16SrRNA sequences have not been determined.
Machismo-like disease (Fletcher et al., 1984)	Probably machismo (Sinclair, 1999). 16SrRNA sequences have not been determined.
Phyllody and witches' broom (Dhingra and Chenulu, 1983)	Symptoms resemble machismo, but in addition leaflets are reduced in size. Transmitted by leafhoppers, *Orosius argentatus* Evans, *O. orientalis* (Matsumura), and *Orosius* sp. 16SrRNA sequences have not been determined.
Soybean bud proliferation (Derrick and Newsom, 1984)	Symptoms resemble machismo. Transmitted through soybean seed and by the leafhopper, *Scaphytopius acutus* Say. 16SrRNA sequences have not been determined.
Soybean veinal necrosis (Jomantiene et al., 1999, 2002)	Symptoms: Necrotic veins. The soybean veinal necrosis phytoplasma has been identified by 16S rRNA sequencing to belong in the X-disease (16SrIII) group. Leafhopper vectors have not been determined.
Un-named disease (Grau et al., 1999; Lee et al., 2000).	Symptoms: Stunted plants with dark or bluish foliage. Bud proliferation causes bushy appearance. Stems marked with purple patches or streaks; stems remain green after pods have matured. Pods reduced in number. Seed coats dull. Associated with two unnamed phytoplasmas, identified by 16S rRNA sequencing as members of the aster yellows (16SrI) and clover proliferation (16SrVI) groups. Some plants are colonized by both phytoplasmas. Leafhopper vectors have not been determined.

leafhopper transmission (Table 15–6) and comparison of symptoms produced on a range of host plants. Comparison of phytoplasma-caused diseases from different geographical regions is especially difficult because these studies require transport of living hosts.

Polymerase chain reaction (PCR) amplification from plant tissue extracts using primers specific for phytoplasmas allows identification by proxy via sequence analyses of the ribosomal small subunit RNA (16S rRNA) (Gundersen et al., 1994; Lee et al., 1993). In this method, ribosomal RNAs (rRNAs) are considered to be phytoplasmal in origin if they are amplified from diseased tissue but not from apparently healthy tissue. Similarly, amplification of specific DNAs from leafhoppers or dodder (*Cuscuta* spp.) that have been exposed to diseased plant tissue but not from those exposed to apparently healthy tissue is also presumptive evidence for isolation of phytoplasmal rRNAs.

The PCR primers are available for detecting and identifying phytoplasmal 16S rRNAs. For the three groups of phytoplasmas known to contain soybean pathogens, Lee et al. (2001a) list the sequences of 11 primer pairs for the 16SrI group and subgroups, three for the 16SrIII group, and two for the 16SrVI group and one subgroup. For detection and identification of other soybean pathogens, nine primer pairs effective for detecting all phytoplasmas are listed or available from other sources (Gundersen et al., 1996; Lee et al., 1993, 1997).

15–5.3 Soybean Diseases Associated with Phytoplasmas

Table 15–6 lists soybean diseases caused by phytoplasmas and provides details on their symptoms. Because phytoplasmas are difficult to study separately from their hosts, mechanisms for host damage are essentially unknown. However, some symptoms apparently are linked to mechanical plugging at phloem sieve plates, others seem to be related to phytohormone interference with plant development, and some seem to be due to effects of toxins.

The best-studied disease is machismo, which has been reported from Colombia (Granada, 1979a) . Because direct host transmission tests and 16S rRNA assays have not been done, several diseases in other geographic regions with symptoms similar to machismo may actually be caused by the machismo pathogen. These diseases include bud proliferation, phyllody, witches'-broom, and machismo-like diseases (Table 15–6). However, great care should be exercised in making this assumption because similar symptoms in an unnamed soybean disease symptomatically distinct from machismo were found to be caused, separately and together, by two very different phytoplasmas belonging to 16SrI and 16SrVI groups (Grau et al., 1999; Lee et al., 2000). Soybean veinal necrosis disease, caused by a phytoplasma belonging to group 16SIII, has been reported recently from eastern Europe (Jomantiene et al., 1999, 2002).

15–5.4 Diagnostic Approaches

Phytoplasmal-caused diseases are difficult to diagnose and differentiate from virus-caused diseases. In soybean, phytoplasmal-diseases may be suspected when common symptoms are observed usually including phyllody (atypical leaf-like development) of flowers and pods and witches'-broom (proliferation of lateral buds). Other symptoms, such as veinal chlorosis, are difficult or impossible to differentiate from symptoms caused by viruses.

Tetracycline remission of symptoms is commonly used to separate phytoplasma-caused diseases from diseases caused by viruses (as well as diseases caused by fungi). Because the protein synthesis machinery of prokaryotic phytoplasmas (70S ribosomes) is inhibited by tetracycline, but eukaryotic protein synthesis (80S ribosomes) is not, phytoplasma-caused symptoms remit with treatment while viral symptoms do not (Granada, 1979b).

15–5.5 Epidemiology and Transmission

15–5.5.1 Life Cycle

Phytoplasmas replicate and overseason in the phloem sieve tubes of plants. Phytoplasmas do not overseason in leafhoppers as they are not passed to the next generation through eggs. For dissemination from one plant to another, phytoplasmas depend upon phloem-feeding leafhoppers, which ingest phloem sap contaminated with phytoplasmas. Phytoplasmas pass from the insect lumen to hemolymph and are carried to the salivary glands where they replicate. Phytoplasma-contaminated saliva inoculates phloem during insect feeding to complete the disease cycle. Phytoplasmas often cause disease in both plant and insect hosts. In other instances, no apparent disease may occur in one or both hosts.

15–5.5.2 Transmission

For experimental transmission of the pathogen, in addition to using leafhoppers, phytoplasmas may be horizontally transmitted among plants by grafting or via parasitic dodder (Lee et al., 2001a). These methods succeed because they establish the intimate phloem-to-phloem contact necessary for the transfer of the pathogen, and have often been used for viruses. Phytoplasmas are usually passed horizontally from plant to insect and back to insect without vertical transmission, or passage through sexual reproductive structures such as insect eggs or plant seeds. However, there is an interesting report that soybean bud proliferation may be passed vertically from one generation of plants to the next through seed (Derrick and Newsom, 1984; Sinclair, 1999). Confirmation of this report may be important to disease control strategies.

15–5.6 Management

Soybean diseases caused by phytoplasmas are episodic, making their control difficult. Epidemics are related to the amount of inoculum available in the host or reservoir hosts and the number of vectors. Tetracycline, while useful for diagnosis, provides only partial and/or transient control and is not approved for use on field crops. Prophylactic removal of weed hosts and roguing diseased plants prevents build-up of inoculum levels. However, there is currently very little information on the nature of the weed hosts for phytoplasmas pathogenic on soybean. Obviously, leafhopper control is important in control of phytoplasmal spread, but as yet no trials to test the feasibility of this approach have been published.

REFERENCES

Abney, T.S., and L.D. Ploper. 1994. Effects of bean pod mottle virus on soybean seed maturation and seedborne *Phomopsis* spp. Plant Dis. 78:33–37.

Ahmad, I.S., J.F. Reid, and M.R. Paulsen. 1994. Morphological feature extraction by machine vision of damaged soybean seeds. p. 1–10. Presented at the 19–22 June 1994 meeting. Paper 94-3014. Am. Soc. of Agric. Eng., St. Joseph, MI.

Ahmad, I.S., J.F. Reid, M.R. Paulsen, and J.B. Sinclair. 1999. Color classifier for symptomatic soybean seeds using image processing. Plant Dis. 83:320–327.

Almeida, A.M.R., L.A. Almeida, M.C.N. Oliveira, F.A. Paiva, and W.A. Moreira. 1995a. Strains of soybean mosaic virus identified in Brazil and their influence on seed mottling and rate of seed transmission. Fitopatol. Bras. 20:227–232.

Almeida, A.M.R., L.A. de Almeida, R.A. d. S. Kiihl, and L.A. Domit. 1995b. Identification of a new strain of soybean mosaic virus in Brazil. Arq. Biologia Tecnol. 38:1095–1100.

Almeida, A.M.R., S. Nakahara, and D.R. Sosa Gomez. 1994. Thrips species identified in soyabean fields in Brazil. An. Soc. Entomol. Bras. 23:363–365.

Anjos, J. R., U. Jarlfors, and S. A. Ghabrial. 1992. Soybean mosaic potyvirus enhances the titer of two comoviruses in dually infected soybean plants. Phytopathology 82:1022–1027.

Anjos, J.R.N., P.S.T. Brioso, and M.J.A. Charchar. 1999. Partial characterization of bean pod mottle virus in soyabeans in Brazil. (In Portuguese: Caracterizacao parcial de um isolado do "bean pod mottle virus" (BPMV) no Brasil.) Fitopatol. Bras. 24:85–87.

Bariana, H.S., A.L. Shannon, P.W.G. Chu, and P.M. Waterhouse. 1994. Detection of five seedborne legume viruses in one sensitive multiplex polymerase chain reaction test. Phytopathology 84:1201–1205.

Bereswill, S., P.Bugert, B. Völksch, M.Ullrich, C. L. Bender, and K. Geider. 1994. Identification and relatedness of coronatine-producing Pseudomonas syringae pathovars by PCR analysis and sequence determination of the amplification products. Appl. Environ. Microbiol. 57:2924–2930.

Bereswill, S., A. Pahl, P. Bellemann, W. Zeller, and K. Geider. 1992. Sensitive and species-specific detection of Erwinia amylovora by polymerase chain reaction analysis. Appl. Environ. Microbiol. 58:3522–3526.

Berger, P.H., S.D. Wyatt, P.J. Shiel, M.J. Silbernagel, K. Druffel, and G.I. Mink. 1997. Phylogenetic analysis of the Potyviridae with emphasis on legume-infecting potyviruses. Arch. Virol. 142:1979–1999.

Bergeson, G.B., K.L. Athow, F.A. Laviolette, and M. Thomasine. 1964. Transmission, movement and vector relationships of tobacco ringspot virus in soybean. Phytopathology 54:723–728.

Bertacini, P.V., A.M.R. Almeida, J.A.A. Lima, and C.M. Chagas. 1998. Biological and physicochemical properties of cowpea severe mosaic comovirus isolated from soybean in the State of Parana. Brazil. Arch. Biol. Technol. 41:409–416.

Bijaisoradat, M., C.W. Kuhn, and C.P. Benner. 1988. Disease reactions, resistance, and viral antigen content in six legume species infected with eight isolates of peanut mottle virus. Plant Dis. 72:1042–1046.

Boerma, H.R., and C.W. Kuhn. 1976. Inheritance of resistance to peanut mottle virus in soybeans. Crop Sci. 16:533–534.

Boerma, H.R., C.W. Kuhn, and H.B. Harris. 1975. Inheritance of resistance to cowpea chlorotic mottle virus (soybean strain) in soybeans. Crop Sci. 15:849–850.

Bottenberg, J., and M.I. Irwin. 1992. Using mixed cropping to limit seed mottling induced by soybean mosaic virus. Plant Dis. 76:304–306.

Bowers, G.R., Jr., and R.M. Goodman. 1991. Strain specificity of soybean mosaic virus seed transmission in soybean. Crop Sci. 31:1171–1174.

Bowyer, J.W., J.L. Dale, and G.M. Behncken. 1980. Glycine mosaic virus:A comovirus from Australian native Glycine species. Ann. Appl. Biol. 95:385–390.

Braun-Kiewnick, A., and D.C. Sands. 2001. Pseudomonas. p. 84–120 In N.W. Schaad et al. (ed.) Laboratory guide for identification of plant pathogenic bacteria. 3rd ed. APS Press, St. Paul, MN.

Brisbane, P.G., and A. Kerr. 1983. Selective media for three biovars of Agrobacterium. J. Appl. Bacteriol. 54:425–431.

Brunt, A.A., K. Crabtree, M.J. Dallwitz, A.J. Gibbs, L. Watson, and E.J. Zurcher. 1996. Plant viruses online:Descriptions and lists from the VIDE database. Version:20 Aug. 1996. Mirror URL: http://image.fs.uidaho.edu/vide/ (verified 13 May 2003).

Buss, G.R., H.M. Camper, Jr., and C.W. Roane. 1988. Registration of 'Hutcheson' soybean. Crop Sci. 28:1024–1025.

Calvert, L.A., and S.A. Ghabrial. 1983. Enhancement by soybean mosaic virus of bean pod mottle virus titre in doubly infected soybeans. Phytopathology 73:992–997.

Chen, J., J. Chen, and M.J. Adams. 2001. A universal PCR primer to detect members of the Potyviridae and its use to examine the taxonomic status of several members of the family. Arch. Virol. 146:757–766.

Chen, P., G.R. Buss, C.W. Roane, and S.A. Tolin. 1991. Allelism among genes for resistance to soybean mosaic virus in strain differential soybean cultivars. Crop Sci. 31:305–309.

Chen, P., G.R. Buss, C.W. Roane, and S.A. Tolin. 1994. Inheritance in soybean of resistant and necrotic reactions to soybean mosaic virus strains. Crop Sci. 34:414–422.

Chen, X., and G. Bruening. 1992. Nucleotide sequence and genetic map of cowpea severe mosaic virus RNA2 and comparisons with RNA2 of other comoviruses. Virology 187:682–692.

Cho, E.K., and R.M. Goodman. 1979. Strains of soybean mosaic virus:Classification based on virulence in resistant soybean cultivars. Phytopathology 69:467–470.

Choi, S.H., S.K. Green, and D.R. Lee. 1989. Linkage relationship between two genes conferring resistance to peanut stripe virus and soybean mosaic. Euphytica 44:163–169.

Civerolo, E.L. 1982. Disease management by cultural practices and environmental control. p. 344–351. In M.S. Mount and G.H. Lacy (ed.), Phytopathogenic prokaryotes. Vol. 2. Academic Press, New York.

Clark, A.J., and K.L. Perry. 2002. Transmissibility of field isolates of soybean viruses by *Aphis glycines*. Plant Dis. 86:1219–1222.

Clinton, G.P. 1916. Notes on plant diseases in Connecticut. Connecticut Agric. Exp. Stn. Rep. 1915:444–446.

Coceano, P.G., S. Peressini, and G.L. Bianchi. 1998. The role of winged aphid species in the natural transmission of soybean mosaic potyvirus to soybean in North-east Italy. Phytopathol. Mediterr. 37:111–118.

Collmer, C.W., M.F. Marson, S.M. Albert, S. Bajaj, H.A. Maville, S.E. Ruuska, E.J. Vesely, and M.M. Kyle. 1996. The nucleotide sequence of the coat protein gene and the 3' untranslated region of Azuki bean mosaic potyvirus, a member of the bean common mosaic virus subgroup. Mol. Plant-Microbe Interact. 9:758–761.

Cook, G., J.R. De Miranda, M.J. Roossinck, and G. Pietersen. 1999. Tobacco streak ilarvirus detected on groundnut in South Africa. Afr. Plant Protect. 5:13–19.

Corso, I.C. and A.M.R. Almeida. 1991. Effect of sowing time on the incidence of bud blight in soybean (*Glycine max* L. Merr.). J. Phytopathol. 132:251–257.

Costa, A.S. 1988. Bud blight, symptoms produced on soyabean by various virus-cultivar combinations. Fitopatol. Brasil. 13:180–182.

Costa, A.S., and A.M.B. Carvalho. 1961. Studies on Brazilian tobacco streak. Phytopathol. Z. 42:113–138.

Crespo Romero, J., C. Garcia Guerra, and T. Fernandez. 1990. The presence of viral diseases in soyabeans in the provinces of Granma and Holguin, Cuba. Ciencas Agric. 40:167.

Cross, J.E., B.W. Kennedy, J.W. Lambert, and R.L. Cooper. 1966. Pathogenic races of the bacterial blight pathogen of soybeans, *Pseudomonas glycinea*. Plant Dis. Rep. 50:557–560.

Cubeta, M.A., G.L. Hartman, J.B. Sinclair. 1985. Interaction between *Bacillus subtilis* and fungi associated with soybean seeds. Plant Dis. 69:506–506.

Cuppels, D., and A. Kelman. 1974. Evaluation of selective media for isolation of soft-rot bacteria from soil and plant tissue. Phytopathology 64:468–475.

Damsteegt, V.D., and A.D. Hewings. 1986. Comparative transmission of soybean dwarf virus by three geographically diverse populations of *Aulacorthum* (=*Acyrthosiphon*) *solani*. Ann. Appl. Biol. 109:453–463.

Damsteegt, V.D., A.D. Hewings, and A.B. Sindermann. 1990. Soybean dwarf virus:Experimental host range, soybean germplasm reactions, and assessment of potential threat to U.S. soybean production. Plant Dis. 74:992–995.

Damsteegt, V.D., A.L. Stone, and A.D. Hewings. 1995. Soybean dwarf, bean leaf roll, and beet western yellows luteoviruses in southeastern U.S. white clover. Plant Dis. 79:48–50.

Damsteegt, V.D., A.L. Stone, A.J. Russo, D.G. Luster, F.E. Gildow, and O.P. Smith. 1999. Identification, characterization, and relatedness of luteovirus isolates from forage legumes. Phytopathology 89:374–379.

Daniels, J.L., G.P. Munkvold, and D.C. McGee. 2001. Comparison of infected soybean seed and bean leaf beetles as inoculum sources for *Bean pod mottle virus*. Phytopathology 91:S20.

D'Arcy, C.J., and A.D. Hewings. 1986. Enzyme-linked immunosorbent assay for study of serological relationships and detection of three luteoviruses. Plant Pathol. 35:288–293.

D'Arcy, C.J., L. Torrance, and R.R. Martin. 1989. Discrimination among luteoviruses and their strains by monoclonal antibodies and identification of common epitopes. Phytopathology 79:869–873.

Davis, M.J., and A.K. Vidaver. 2001. Coryneform plant pathogens. p. 218–235. In N.W. Schaad et al. (ed.) Laboratory guide for the identification of plant pathogenic bacteria. 3rd ed. APS Press, St. Paul, MN.

DeBoer, S.H., and A. Kelman. 2001. *Erwinia* soft rot group. p. 56–72. In N. W. Schaad et al. (ed.) Laboratory guide for the identification of plant pathogenic bacteria. 3rd ed. APS Press, St. Paul, MN.

DeBoer, S.H., and L.J. Ward. 1995. PCR detection of *Erwinia carotovora* subsp. *atroseptica* associated with potato tissue. Phytopathology 85:854–858.

Denny, T.P., and A.C. Hayward. 2001. *Ralstonia*. p. 151–174. *In* N. W. Schaad et al. (ed.) Laboratory guide for the identification of plant pathogenic bacteria. 3rd ed. APS Press, St. Paul, MN.

Derrick, K.S., and L.D. Newsom. 1984. Occurrence of a leafhopper-transmitted disease of soybeans in Louisiana. Plant Dis. 68:343–344.

Dhingra, K.L., and V. Chenulu. 1983. Symptomatology and transmission of witches' broom disease of soybean in India. Curr. Sci. 52:603–604.

Di, R., C.-C. Hu, and S.A. Ghabrial. 1999. Complete nucleotide sequence of bean pod mottle virus RNA1:Sequence comparisons and evolutionary relationships to other comoviruses. Virus Genes 18:129–137.

Di, R., V. Purcell, G.B. Collins, and S.A. Ghabrial. 1996. Production of transgenic soybean lines expressing the bean pod mottle virus coat protein precursor gene. Plant Cell Rep.15:746–750.

Diatloff, A., W.C. Wong, and B.A. Wood. 1993. Non-destructive methods for detecting *Curtobacterium flaccumfaciens* pv. *flaccumfaciens* in mung bean seed. Let. Appl. Microbiol. 16:269–273.

Dorrance, A.E., D.T. Gordon, and A.F. Schmitthenner. 2001. First report of *Bean pod mottle virus* in soybean in Ohio. Plant Dis. 85:1029. (On line D-2001-0702-01N, 2001.) Available at http://www.apsnet.org/pd (verified 28 May 2003).

Dunleavy, J.B. 1963. A vascular disease of soybeans caused by *Corynebacterium* sp. Plant Dis. Rep. 47:612–613.

Dunleavy, J.B. 1984. Yield losses in soybeans caused by bacterial tan spot. Plant Dis. 68:774–776.

Dunleavy, J.B. 1985. Spread of bacterial tan spot of soybean in the field. Plant Dis. 68:1036–1039.

Dunleavy, J.M., J.W. Keck, J.W. Gobelman, K.S. Reddy, and M.M. Thompson. 1983. Prevalence of *Corynebacterium flaccumfaciens* as an incitant of bacterial tan spot in Iowa. Plant Dis. 67:1277–1279.

Duodu, S., T.V. Bhuvaneswari, T.J.W. Stokkermans, and N.K. Peters. 1999. A positive role for rhizobitoxine in *Rhizobium*-legume symbiosis. Mol. Plant-Microbe Interac. 12:1082–1089.

Dye, D.W., J.F. Bradbury, M. Goto, A.C. Hayward, R.A. Lelliott, and M.N. Schroth. 1980. International standards for naming pathovars of phytopathogenic bacteria and a list of pathovar names and pathotype strains. Rev. Plant Pathol. 59:153–168.

Evans, T.A. 1999. Soybean severe stunt. p. 65–66. *In* G.L. Hartman et al. (ed.) Compendium of soybean diseases. 4th ed. APS Press. St. Paul MN.

Fauquet, M.C., and M.A. Mayo. 1999. Abbreviations of plant virus names—1999. Arch. Virol. 144:1250–1273.

Fereres, A., G.E. Kampmeier, and M.E. Irwin. 1999. Aphid attraction and preference for soybean and pepper plants infected with *Potyviridae*. Ann. Entomol. Soc. Am. 92:542–548.

Fernandez Suarez, R., and N. Lastres Gonzalez. 1984. Characteristics of a strain of cowpea mosaic virus on soyabean. Ciencas Agric.19:7–12.

Fernandez, T., J. Crespo Romero, and G. Croche Alfonso. 1989. Response of soyabean varieties to cowpea severe mosaic virus (CpSMV) under experimental conditions. Ciencas Agric. 37–38:5–9.

Fieldhouse, D.J., and M. Sasser. 1982. A medium highly selective for *Pseudomonas syringae* pv. *glycinea*. Phytopathology 72:706 (abstr.).

Flasinski, S., R.A. Gonzales, and B.G. Cassidy. 1997. The complete nucleotide sequence of peanut mottle virus strain M. GenBank Accession NC_002600.

Fletcher, J., M.E. Irwin, O.E. Bradfute, and G.A. Granada. 1984. A machismo-like disease of soybeans in Mexico. Phytopathology 74:857 (abstr).

French, E.R., L. Gutarra, P. Aley, and J. Elphinstone. 1995. Culture media for *Pseudomonas solanacearum* isolation, identification, and maintenance. Fitopatologia 30:126–130.

Frechon, D., P. Exbrayat, O. Gallet, V. Le Clerc, N. Payet, and Y. Bertheau. 1995. Sequences nuceotidiques pour la detection des *Erwinia carotovora* subsp. *atrospectica*. Inst. Natl. Re. Agron., Sanofi Diagnostics Pasteur. Brevet 95:12–805.

Frenkel, M.J., J.M. Jilka, D.D. Shukla, and C.W. Ward. 1992. Differentiation of potyviruses and their strains by hybridization with the 3' non-coding region of the viral genome. J. Virol. Meth. 36:51–62.

Fribourg, C.E., and W. Perez. 1994. Bean pod mottle virus (BPMV) affecting *Glycine max* (L.) Merr. in the Peruvian jungle. Fitopatologia 29:207–210.

Fulton, J.P., R.C. Gergerich, and H.A. Scott. 1987. Beetle transmission of plant viruses. Ann. Rev. Phytopathol. 25:111–123.

Garcia, J.R., and L. Zak. 1983. Effect of various bacteria on the germination and seedling emergence of soybean. Agrocienc. (Spanish) Mex. 51:69–80.

Gaspar, J.O., and A.S. Costa. 1993. Virus do mosaico angular di feijoeiro:Purificacao e ultraestrutura dos tecidos infectados. (Bean angular mosaic virus:Purification and ultrastructure of infected tissues). Fitopatol. Bras. 18:534–540.

Gergerich, R. 1999. Bean pod mottle virus. p. 61–62. *In* G.L. Hartman et al. (ed.) Compendium of soybean diseases. 4th ed. APS Press, St. Paul MN.

Gergerich, R.C., and H.A. Scott. 1996. Comoviruses:Transmission, epidemiology, and control. p. 77–98. *In* B.D. Harrison and A.F. Murant (ed.) The plant viruses 5:Polyhedral virions and bipartite RNA genomes. Plenum Press, New York.

Giesler, L.J., S.A. Ghabrial, T.E. Hunt, and J.H. Hill. 2002. *Bean pod mottle virus*:A threat to U.S. soybean production. Plant Dis. 86:1280–1289.

Ghabrial, S.A., D.E. Hershman, D.W. Johnson, and D. Yan. 1990a. Distribution of bean pod mottle virus in soybeans in Kentucky. Plant Dis. 74:132–134.

Ghabrial, S.A., and F.J. Schultz. 1983. Serological detection of bean pod mottle virus in bean leaf beetles. Phytopathology 73:480–483.

Ghabrial, S.A., H.A. Smith, T.D. Parks, and W.G. Dougherty. 1990b. Molecular genetic analyses of the soybean mosaic virus NIa proteinase. J. Gen. Virol. 71:1921–1927.

Gildow, F.E., V.D. Damsteegt, A.L. Stone, O.P. Smith, and S.M. Gray. 2000. Virus-vector cell interactions regulating transmission specificity of soybean dwarf luteoviruses. J. Phytopathol. 148:333–342.

Gillaspie, A.G., Jr., and M.S. Hopkins. 1991. Spread of peanut stripe virus from peanut to soybean and yield effects on soybean. Plant Dis. 75:1157–1159.

Gillings, M., P. Fahy, and C. Davies. 1993. Restriction analysis of an amplified polygalacturonase gene fragment differentiates strains of the phytopathogenic bacterium *Pseudomonas solanacearum*. Lett. Appl. Microbiol. 17:44–48.

Girish, K.R., and R. Usha. 2001. Molecular characterization of *Soybean yellow mosaic virus* from India. GenBank Accession AJ421642.

Golnaraghi, A.R., N. Shahraeen, R. Pourrahim, S. Ghorbani, and S. Farzadhar. 2001. First report of *Tomato spotted wilt virus* on soybean in Iran. Plant Dis. 85:1290. (On line D-2001-1016-01N, 2001.) Available at http://www.apsnet.org/pd (verified 28 May 2003).

Gopal, K. 1996. Screening soybean for budblight resistance caused by tobacco ring spot virus. Plant Dis. Res. 11:89–90.

Gowda, S., T. Satyanarayana, R.A. Naidu, A. Mushegian, W.O. Dawson, and D.V.R. Reddy. 1998. Characterization of the large (L) RNA of peanut bud necrosis *tospovirus*. Arch. Virol. 143:2381–2390.

Granada, G.A. 1979a. Machismo disease of soybeans. I. Symptomatology and transmission. Plant Dis. Rep. 63:47–50.

Granada, G.A. 1979b. Machismo disease of soybeans. II. Suppressive effects of tetracycline on symptom development. Plant Dis. Rep. 63:309–312.

Grau, C.R., I.-M. Lee, M.E. Lee, M.F. Heimann, and L.A. Lukaesko. 1999. Update on Phytoplasmas in the United States. p. 11. *In* G.L. Hartman et al. (ed.) Compendium of soybean diseases. 4th ed. APS Press, St. Paul, MN.

Green, S.K., and D.R. Lee. 1989. Occurrence of peanut stripe virus (PStV) on soybean in Taiwan—Effect on yield and screening for resistance. Tropical Pest Manage. 35:123–126.

Gu, H., A.J. Clark, P. de Sa, T.W. Pfeiffer, S.A. Tolin, and S.A. Ghabrial. 2002. Diversity among isolates of the comovirus *Bean pod mottle virus*. Phytopathology 92:446–452.

Gunasinghe, U.B., M.E. Irwin, and G.E. Kampmeier. 1988. Soybean leaf pubescence affects aphid vector transmission and field spread of soybean mosaic virus. Ann. Appl. Biol. 112:259–272.

Gundersen, D.E., I.-M. Lee, S.A. Rehner, R.E. Davis, and D.T. Kingsbury. 1994. Phylogeny of mycoplasmalike organisms (phytoplasmas):A basis for their classification. J. Bacteriol. 176:5244–5254.

Gundersen, D.E., I.-M. Lee, D.A. Schaff, N.A. Harrison, C.J. Chang, R.E. Davis, and D.T. Kingsbury. 1996. Genomic diversity and differentiation among phytoplasma strains in 16S rRNA groups I (aster yellows and related phytoplasmas) and III (X-disease and related phytoplasmas). Int. J. Syst. Bacteriol. 46:64–75.

Guo, J.Q., M.H. Zhang, and H.C. Dai. 1991. SIMSMV, a simulation model for soybean mosaic virus epidemiology and management decision systems. (In Chinese, with English abstract.) Sci. Agric. Sin. 24:34–42.

Haas, J.H., L.W. Moore, W. Ream, and S. Manulis. 1995. Universal primers for detection of pathogenic *Agrobacterium* strains. Appl. Environ. Microbiol. 6:2879–2884.

Hajimorad, M.R., C.-C. Hu, and S.A. Ghabrial. 1999. Molecular characterization of an atypical Old World strain of *Peanut stunt virus*. Arch. Virol. 144:1587–1600.

Halbert, S.E., M.E. Irwin, and R.M. Goodman. 1981. Alate aphid (*Homoptera:Aphididae*) species and their relative importance as field vectors of soybean mosaic virus. Ann. Appl. Biol. 97:1–9.

Halbert, S.E., G.X. Zhang, and Z.Q. Pu. 1986. Comparison of sampling methods for alate aphids and observations on epidemiology of soybean mosaic virus in Nanjing, China. Ann. Appl. Biol. 109:473–483.

Haque, G.U., S. Hassan, A. Ali, and M. Arif. 1993. Control of soybean mosaic virus through thermotherapy. Sarhad J. Agric. 9:177–180.

Hartman, G.L., L.L. Domier, L.M. Wax, C.G. Helm, D.W. Onstad, J.T. Shaw, L.F. Solter, D. J. Voegtlin, C.J. D'Arcy, M.E. Gray et al. 2001. Occurrence and distribution of *Aphis glycines* in Illinois in 2000 and its potential control. Plant Health Progress doi:10.1094/PHP-2001-0205-01-HN. Available at http://www.plantmanagementnetwork.org/php (verified 13 May 2003).

Hartman, G.L., J.B. Sinclair, and J.C. Rupe (ed.) 1999. Compendium of soybean diseases. 4th ed. APS Press, St. Paul, MN.

Hartung, F., R. Werner, H.P. Muhlbach, and C. Buttner. 1998. Highly specific PCR-diagnosis to determine *Pseudomonas solanacearum* strains of different geographic origins. Theor. Appl. Genet. 96:797–802.

Hartwig, E.E., and H.W. Johnson. 1953. Effect of bacterial pustule disease on yield and chemical composition of soybeans. Agron. J. 45:22–23.

Hayes, A.J., G. Ma, G.R. Buss, and M.A. Saghai-Maroof. 2000. Molecular marker mapping of *Rsv4*, a gene conferring resistance to all known strains of soybean mosaic virus. Crop Sci. 40:1434–1437.

Hayward, A.C. 1991. Biology and epidemiology of bacterial wilt caused by *Pseudomonas solanacearum*. Annu. Rev. Phytopathol. 29:65–87.

Hedges, F. 1922. A bacterial wilt of the bean caused by *Bacillus flaccumfaciens* nov. sp. Science (Washington DC) 55:433–434.

Hedges, F. 1924. Bean wilt (*Bacillus flaccumfaciens* Hedges) including comparisons with *Bacterium phaseoli*. Phytopathology 16:1–21.

Hedges, F. 1926. Bacterial wilt of beans (*Bacterium flaccumfaciens* Hedges), including comparisons with *Bacterium phaseoli*. Phytopathology 16:1–22.

Hepperly, P.R., G.R. Bowers, Jr., J.B. Sinclair, and R.M. Goodman. 1979. Predisposition to seed infection by *Phomopsis sojae* in soybean plants infected with soybean mosaic virus. Phytopathology 69:846–848.

Higgins, C.M., B.G. Cassidy, P.-Y. Teycheney, S. Wongkaew, and R.G. Dietzgen. 1998. Sequences of the coat protein gene of five peanut stripe virus (PStV) strains from Thailand and their evolutionary relationship with other bean common mosaic virus sequences. Arch. Virol. 143:1655–1667.

Hill, E.K., J.H. Hill, and D.P. Durand. 1984. Production of monoclonal antibodies to viruses in the potyvirus group:Use in radioimmunoassay. J. Gen. Virol. 65:525–532.

Hill, J.H. 1999. Soybean mosaic. p. 70–71. *In* G. L. Hartman et al. (ed.) Compendium of soybean diseases. 4th ed. APS Press, St. Paul, MN.

Hill, J.H., R. Alleman, D.B. Hogg, and C.R. Grau. 2001. First report of transmission of *Soybean mosaic virus* and *Alfalfa mosaic virus* by *Aphis glycines* in the New World. Plant Dis. 85:561.

Hill, J.H., H.I. Benner, T.A. Permar, T.B. Bailey, R.E. Andrews, Jr., D.P. Durand, and R.A. van Deusen. 1989. Differentiation of soybean mosaic virus isolates by one-dimensional trypsin peptide maps immunoblotted with monoclonal antibodies. Phytopathology 79:1261–1265.

Hill, J.H., H.I. Benner, and R.A. Van Deusen. 1994. Rapid differentiation of soybean mosaic virus isolates by antigenic signature analysis. J. Phytopathol. 142:152–162.

Hobbs, H.A., G.L. Hartman, R.L. Bernard, L.L. Domier, W.L. Pedersen, and D.M. Eastburn. 2001. Effects of *Bean pod mottle virus* and *Soybean mosaic virus* alone and in combination, on soybean seed coat mottling. Phytopathology 91:S39 (abstr.).

Hokawat, S., and K. Rudolph. 1993. *Xanthomonas campestris* pv. *glycines*: Cause of bacterial pustule of soybean. p. 44–48. *In* J. G. Swings and E. L. Civerolo (ed.) *Xanthomonas*. Chapman and Hall, New York.

Hopkins, J.D., and A.J. Mueller. 1983. Distribution of bean pod mottle virus in Arkansas soybean as related to the bean leaf beetle, *Cerotoma trifurcata*, (*Coleoptera:Chrysomelidae*) population. Environ. Entomol. 12:1564–1567.

Hopkins, J.D., and A.J. Mueller. 1984a. Efficiency of the Ouchterlony gel diffusion test in detecting bean pod mottle virus infection in soybean. Environ. Entomol. 13:1135–1137.

Hopkins, J.D. and A.J. Mueller. 1984b. Effect of bean pod mottle virus on soybean yield. J. Econ. Entomol. 77:943–947.

Hull, R. 2001. Matthews' plant virology. 4th ed. Academic Press, Orlando, FL.

Hwang, I., and S.M. Lim. 1992. Effects of individual and multiple infections with three bacterial pathogens in disease severity and yield of soybeans. Plant Dis. 76:195–198.

Iizuka, N., and M. A. Charchar. 1991. Occurrence of a strain of bean rugose mosaic virus on soybean (*Glycine max*) in Central Brazil. Japan Agric.Res. Quart. 24:260–264.

Irwin, M.E., and R.M. Goodman. 1981. Ecology and control of soybean mosaic virus. p. 181–200. *In* K. Maramorosch and K. F. Harris (ed.) Plant diseases and their vectors:Ecology and epidemiology. Academic Press, New York.

Irwin, M.E., W.G. Ruesink, S.A. Isard, and G.E. Kampmeier. 2000. Mitigating epidemics caused by nonpersistently transmitted aphid-borne viruses:The role of the pliant environment. Virus Res. 71:185–211.

Iwaki, M., Y. Isogawa, H. Tsuzuki, and Y. Honda. 1984. Soybean chlorotic mottle, a new caulimovirus on soybean. Plant Dis. 68:1009–1011.

Jain, R.K., N.M. McKern, S.A. Tolin, J.H. Hill, O.W. Barnett, M. Tosic, R.E. Ford, R.N. Beachy, M.H. Yu, C.W. Ward, and D.D. Shukla. 1992. Confirmation that fourteen potyvirus isolates from soybean are strains of one virus by comparing coat protein peptide profiles. Phytopathology 82:294–299.

Jain, R.K., H.R. Pappu, S.S. Pappu, M. Krishna Reddy, and A. Vani. 1998. Watermelon bud necrosis tospovirus is a distinct virus species belonging to serogroup IV. Arch. Virol. 143:1637–1644.

Jayaram, C., J.H. Hill, and W.A. Miller. 1992. Complete nucleotide sequences of two soybean mosaic virus strains differentiated by response of soybean containing the Rsv resistance gene. J. Gen. Virol. 73:2067–2077.

Jayaram, C., R.A. van Deusen, A.L. Eggenberger, A.W. Schwabacher, and J.H. Hill. 1998. Characterization of a monoclonal antibody recognizing a DAG-containing epitope conserved in aphid transmissible potyviruses:Evidence that the DAG motif is in a defined conformation. Virus Res. 58:1–11.

Jeong, S.C., S. Kristipati, A.J. Hayes, P.J. Maughan, S.L. Noffsinger, I. Gunduz, G.R. Buss, and M.A. Saghai Maroof. 2002. Genetic and sequence analysis of markers tightly linked to the soybean mosaic virus resistance gene, *Rsv3*. Crop Sci. 42:265–271.

Johnson, H.W., U.M. Means, and F.E. Clark. 1959. Responses of seedlings to extracts of soybean nodules bearing selected strains of *Rhizobium japonicum*. Nature (London) 183:308–309.

Jomantiene, R., R.E. Davis, A. Alminaite, J. Staniulis, and D. Valiunas. 2002. Ribosomal RNA interoperon sequence heterogeneity in new phytoplasma lineages infecting oak, campion, thistle, and dandelion. Publ. P-2003-0010-PTA of the 2002 Potomac Div. Abstr. of the Am. Phytopathol. Soc. Available at http://www.apsnet.org/meetings/div/po02abs.asp (verified 13 May 2003).

Jomantiene, R., R.E. Davis, L. Antoniuk, and J. Staniulis. 1999. First report of phytoplasmas in soybean, alfalfa, and *Lupinus* sp. in Lithuania:Soybean veinal necrosis phytoplasma. GenBank accession AF177383.

Jones, A.L.,and K. Geider. 2001. Erwinia amylovora group. p. 40–55 *In* N. W. Schaad et al. (ed.) Laboratory guide for the identification of plant pathogenic bacteria. 3rd ed. APS Press, St. Paul, MN.

Kersters, K., W. Ludwig, M. Vancanneyt, P. De Vos, M.Gillis, and K.H. Schleifer. 1996. Recent changes in the classification of pseudomonads: An overview. Syst. Appl. Microbiol. 19:465–477.

Keshwal, R.L., K.N. Gupta, and R.K. Khatri. 1999. Appearance and sources of soybean yellow mosaic virus in Jabalpur (M.P.). Indian J. Virol. 15:97–99.

Khetarpal, R.K., D.B. Parakh, S. Shamsher, and N. Ram. 1992. ELISA detection of soybean mosaic virus in testas, embryos and seedlings from mottled and unmottled seeds of imported soybean germplasm. Indian J. Virol. 8:106–110.

Kiihl, R.A.S., and E.E. Hartwig. 1979. Inheritance of reaction to soybean mosaic virus in soybeans. Crop Sci. 19:372–376.

Kim, Y.-H., O.-S. Kim, B.-C. Lee, J.-H. Roh, M.-K. Kim, D.-J. Im, I.-B. Hur, and S.-C. Lee. 1999. Detection of soybean mosaic virus using RT-PCR. Korean J. Crop Sci. 44:253–255.

Kim, Y.-H., O.-S. Kim, J.-K. Moon, J.-H. Roh, D.-J. Im, I.-B. Hur, and S.-C. Lee. 2001. Evaluation of Korean recommended soybean cultivars of resistance to soybean mosaic virus. Korean J. Crop Sci. 46:17–21.

Kim, Y.-H., J.-H. Roh, M.-K. Kim, D.-J. Im, and I.-B. Hur. 2000. Seasonal occurrence of aphids and selection of insecticides for controlling aphids transmitting soybean mosaic virus. (In Korean, with English summary.) Korean J. Crop Sci. 45:353–355.

Kinscherf, T.G., R.H. Coleman, T.M. Barta, and D.K. Willis. 1991. Cloning and expression of the tabtoxin biosynthetic region from *Pseudomonas syringae*. J. Bacteriol. 173:4124–4132.

Koning, G., E.M. TeKrony, S.A. Ghabrial, and T.W. Pfeiffer. 2002. *Soybean mosaic virus* (SMV) and the SMV resistance gene (*Rsv₁*): Influence on *Phomopsis* spp. Seed infection in an aphid free environment. Crop Sci. 42:178–185.

Koning, G., D.E. TeKrony, T.W. Pfeiffer, and S.A. Ghabrial. 2001. Infection of soybean with *Soybean mosaic virus* increases susceptibility to *Phomopsis* spp. seed infection. Crop Sci. 41:1850–1856.

Kosaka, Y., and T. Fukunishi. 1993. Attenuated isolates of soybean mosaic virus derived at a low temperature. Plant Dis. 77:882–886.

Kosaka, Y., and T. Fukunishi. 1994. Application of cross-protection to the control of black soybean mosaic disease. Plant Dis. 78:339–341.

Koshimizu, Y., and N. Iizuka. 1963. Studies on soybean virus diseases in Japan. Bull. Tohoku Natl. Agric. Exp. Stn. 27:1–103.

Langham, M.A.C., D.C. Doxtader, J.D. Smolik, and R.A. Scott. 1999. Outbreak of a viral disease affecting soybeans (*Glycine max* (L.) Merrill) in South Dakota. Phytopathology 89:S43 (abstr.).

Langham, M.A.C., R.C. Gergerich, and H.A. Scott. 1990. Conversion of comovirus electrophoretic forms by leaf-feeding beetles. Phytopathology 80:900–906.

Lapage, S.P., P.H.A. Sneath, R.A. Lelliot, V.B.D. Skerman, H.P.R. Seeliger, and W.A. Clark. (ed.) 1975. International code of nomenclature of bacteria (1976 Revision). Bacteriological code. Am. Soc. for Microbiol., Washington, DC.

Leary, J.V., M.D. Thomas, and E. Allingham. 1984. Conjugal transfer of *E. coli* Flac from *Erwinia chrysanthemi* to *Pseudomonas syringae* pv. *glycinea* and the apparent stable incorporation of the plasmid into the pv. *glycinea* chromosome. Mol. Gen. Genet. 198:125–127.

Leben, C. 1972. The development of a selective medium for *Pseudomonas glycinea*. Phytopathology 62:674–676.

Lee, I.-M., R.E. Davis, and J. Fletcher. 2001a. Spiroplasmas and phytoplasmas. p. 283–320. *In* N.W. Schaad, J.B. Jones, and W. Chun (ed.) Laboratory guide for the identification of plant pathogenic bacteria. 3rd ed. APS Press, St. Paul, MN.

Lee, I.-M., R. W. Hammond, R.E. Davis, and D.E. Gundersen. 1993. Universal amplification and analysis of pathogen 16S rDNA for classification and identification of mycoplasmalike organisms. Phytopathology 83:834–842.

Lee, I.-M., M. Pastore, M. Vibio, A. Danielli, S. Attathom, R.E. Davis, and A. Bertaccini. 1997. Detection and characterization of the phytoplasma associated with annual blue grass (*Poa annua*) white leaf in southern Italy. Eur. J. Plant Pathol. 103:251–254.

Lee, L., and E.J. Anderson. 1998. Nucleotide sequence of a resistance breaking mutant of southern bean mosaic virus. Arch. Virol. 143:2189–2201.

Lee, M.E., R.J. Alleman, L.A. Lukaesko, I.-M. Lee, and C.R. Grau. 2000. Molecular detection and comparison of aster yellows phytoplasmas in soybeans and carrots in Wisconsin. GenBank accession AF268403-5.

Lee, M.E., C.R. Grau, L.A. Lukaesko, and I.-M. Lee. 2002. Identification of aster yellows phytoplasmas in soybean in Wisconsin based on RFLP analysis of PCR-amplified products (16S rDNAs). Can. J. Plant Pathol. 24:125–130.

Lee, M.E., N.C. Kurtzweil, and C.R. Grau. 2001b. Prevalence and agronomic effects of viruses in Wisconsin soybeans. Phytopathology 92:S139 (abstr.) Publ. P-2002-0015-NCA. Available at http://www.apsnet.org/meetings/div/ (verified 13 May 2003).

Lee, Y.C., J.J. Kim, and E.K. Cho. 1991. Classification of seed-borne SMV strains and resistance to SMV in leading soybean cultivars. Korean J. Breed. 23:53–58.

Li, C.-M., I. Brown, J. Mansfield, C. Stevens, T. Boureau, M. Romantschuk, and S. Taira. 2002. The Hrp pilus of *Pseudomonas syringae* elongates from its tip and acts as a conduit for translocation of the effector protein HrpZ. Europ. Molec. Biol. Org. J. (EMBO) 21:1909–1915.

Lin, M.T., and J.H. Hill. 1983. Bean pod mottle virus: Occurrence in Nebraska and seed transmission in soybeans. Plant Dis. 67:230–233.

Lin, N.S., Y.H. Hsu, and H.T. Hsu. 1990. Immunological detection of plant viruses and a mycoplasmalike organism by direct tissue blotting on nitrocellulose membranes. Phytopathology 80:824–828.

Lindstrom, J.T., and L.O. Vodkin. 1991. A soybean cell wall protein is affected by seed color genotype. Plant Cell 3:561–571.

Lister, R.M. 1978. Application of the enzyme-linked immunosorbent assay for detecting viruses in soybean seed and plants. Phytopathology 68:1393–1400.

Lourencao, A.L., and M.A.C. de Miranda. 1996. "Mancha cafe" em soja:Selecao para resistencia e interacao entre genotipos e epocas de inoculacao. (Seedcoat mottling in soyabeans:Selection for resistance and interaction between genotypes and inoculation date). Bragantia 55:105–116.

Lucas, B.S., and J.H. Hill. 1980. Characteristics of the transmission of three soybean mosaic virus isolates by *Myzus persicae* and *Rhopalosiphum maidis*. Phytopathol. Z. 99:47–53.

Mackey, D., B.F. Holt III, A. Wiig, and J.L. Dangl. 2002. RIN4 interacts with *Pseudomonas syringae* type III effector molecules and is required for RPM1-mediated resistance in *Arabidopsis*. Cell 108:743–754.

Makkouk, K.M., and S.G. Kumari. 2001. Reduction of incidence of three persistently transmitted aphid-borne viruses affecting legume crops by seed-treatment with the insecticide imidacloprid (Gaucho®). Crop Prot. 20:433–437.

Mahan, S.K., and N.W. Schaad. 1987. Semiselective agar medium for isolating *Pseudomonas syringae* pv. *syringae* and pv. *phaseolicola* from bean seed. Phytopathology 77:1390–1395.

Martin, R.R., and C.J. D'Arcy. 1990. Relationships among luteoviruses based on nucleic acid hybridization and serological studies. Intervirology 31:23–30.

Martins, T.R., L.A. Almeida, A.M.R. Almeida, and R.A.S. Kiihl. 1995. Inheritance of soybean resistance to bean "mosaic-em-desenho" virus. Fitopatol. Bras. 20:241–243.

Martins, T.R., A.M.R. Almeida, L.A. Almeida, A. Nepomuceno, C.M. Chagas, and J.F.F. de Toledo. 1994. Sinergismo observado em plantas de soja infectadas pelos virus do mosaico comum da soja e do mosaico em desenho do feijoeiro. [Synergism between soybean mosaic virus and bean rugose mosaic virus in soybeans]. Fitopatol. Bras. 19:430–436.

McLaughlin, M.R. and D.L. Boykin. 1988. Virus diseases of seven species of forage legumes in the southeastern United States. Plant Dis. 72:539–542.

McLaughlin, M.R., P. Thongmeearkom, R.M. Goodman, G.M. Milbrath, S.M. Ries, and D.J. Royse. 1978. Isolation and beetle transmission of cowpea mosaic virus (severe subgroup) from *Desmodium canascens* and soybeans in Illinois. Plant Dis. Rep. 62:1069–1073.

Michelutti, R., J. C. Tu, D.W.A. Hunt, D. Gagnier, T.R. Anderson, and T.W. Welacky. 2002. First report of *Bean pod mottle virus* in soybean in Canada. Plant Dis. 86:330. (On line D-2002-0109-03N, 2002). Available at http://www.apsnet.org/pd (verified 13 May 2003).

Michiko, A., U. Kyoko, and K. Koshi. 1991. An antimicrobial substance produced by *Pseudomonas cepacia* B5 against the bacterial wilt pathogen, *Pseudomonas solonacearum*. Agric. Biol. Chem. 55:715–722.

Miller, T.D., and M.D. Schroth. 1972. Monitoring the epiphytic population of *Erwinia amylovora* on pear with a selective medium. Phytopathology 62:1175–1182.

Mink, G.I., H.J. Vetten, S.D. Wyatt, P.H. Berger, and M.J. Silbernagel. 1999. Three epitopes located on the coat protein amino terminus of viruses in the bean common mosaic potyvirus subgroup. Arch. Virol. 144:1173–1189.

Mohan, S.K., and N.W. Schaad. 1987. Semiselective medium for isolating *Pseudomonas syringae* pv. *syringae* and pv. *phaseolicola* from bean seed. Phytopathology 77:1390–1395.

Moore, B.J., H.A. Scott, and H. J. Walter. 1969. *Desmodium paniculatum*, a perennial host of bean pod mottle virus in nature. Plant Dis. Rep. 53:154–155.

Moore, L.W., H. Bouzar, and T. Burr. 2001. *Agrobacterium*. p. 17–35. *In* N. W. Schaad et al. (ed.) Laboratory guide for the identification of plant pathogenic bacteria. 3rd ed. APS Press, St. Paul, MN.

Morrison, M.J., L.N. Pietrzak, and H.D. Voldeng. 1998. Soybean seed coat discoloration in cool-season climates. Agron. J. 90:471–474.

Naik, R.G., and Venkatesh. 1997. *Myzus persicae* Sulz. is an efficient vector of soyabean mosaic virus (SMV). Insect Environ. 3:78.

Nakano, K. 1919. Bacterial leaf-blight of soybean. J. Plant Prot. (Tokyo) 6:217–221.

Nickell, C.D., S.D. Lim, and S. Eathington. 1994. Registration of soybean germplasm line LL 89-605, resistant to brown stem rot and bacterial blight. Crop Sci. 34:1134.

Nikitina, K.V, and N. I. Korsakov. 1978. Bacterial diseases of soybean in the Soviet Far East and in southern regions of the USSA. (In Russian, with English abstract.) Tr. Pril. Bot. Genet. Sel. 62:13–18.

Nutter, F.W., Jr., P.M. Schultz, and J.H. Hill. 1998. Quantification of within-field spread of soybean mosaic virus in soybean using strain-specific monoclonal antibodies. Phytopathology 88:895–901.

Omunyin, M.E., J.H. Hill, and W.A. Miller. 1996. Use of unique RNA sequence-specific oligonucleotide primers for RT-PCR to detect and differentiate soybean mosaic virus strains. Plant Dis. 80:1170–1174.

Orgambide, G., H. Montrozier, P. Servin, J. Roussel, D. Trigalet-Demery, and A. Trigalet. 1991. High heterogeneity of the exopolysaccharides of *Pseudomonas solanacearum* strain GMI 1000 and the complete structure of the major polysaccharide. J. Biol. Chem. 266:8312–8321.

Oak Ridge National Laboratory. 2002. Computational Biology Program of the Life Sciences Div. of Oak Ridge Natl. Lab., Oak Ridge, TN. Available at http://genome.ornl.gov/microbial/psyr/ (verified 13 May 2003).

Owens, L.D., S. Guggenheim, and J.L. Hilton. 1968. *Rhizobium*-synthesized phytotoxin:An inhibitor of Beta-cystathionase in *Salmonella typhimurium*. Biochim. Biophys. Acta 158:219–225.

Owens, L.D., M. Lieberman, and A. Kunishi. 1971. Inhibition of ethylene production by rhizobitoxine. Plant Physiol. 48:1–4.

Pacumbaba, R.P., G.F. Brown, and R.O. Pacumbaba, Jr. 1997. Effect of fertilizers and rates of application on incidence of soybean diseases in Northern Alabama. Plant Dis. 81:1459–1460.

Paguio, O.R., C.W. Kuhn, and H.R. Boerma. 1988. Resistance-breaking variants of cowpea chlorotic mottle virus in soybean. Plant Dis. 72:768–770.

Pant, V., D. Gupta, N.R. Choudhury, V.G. Malathi, A. Varma, and S.K. Mukherjee. 2001. Molecular characterization of the Rep protein of the blackgram isolate of Indian mungbean yellow mosaic virus. J. Gen. Virol. 82:2559–2567.

Parakh, D.B., R.K. Khetarpal, S. Shamsher, and N. Ram. 1994. Post entry quarantine detection of soybean mosaic virus by ELISA in soybean germplasm. Indian J. Virol. 10:17–21.

Pietersen, G., and H.M. Garnett. 1992. Properties and strain identity of soybean mosaic virus isolates from the Transvaal, South Africa. Phytophylactica 24:279–283.

Pitre, H.N., V.C. Patel, and B.L. Keeling. 1979. Distribution of bean pod mottle disease on soybeans in Mississippi. Plant Dis. Rep. 63:419–423.

Porta Puglia, A., S. Peressini, G. Conca, and G.L. Bianchi. 1990. *Colletotrichum truncatum* associated with SMV on soyabean seeds in Italy. Informat. Fitopatolog. 40:55–57.

Prasada Rao, R., S.K. Chakrabarty, and A.S. Reddy. 1997. Interception of peanut stripe virus in soybean seeds imported from China. Indian J. Plant Protect. 25:81.

Prathuangwong, S. 1985. Soybean bacterial pustule research in Thailand. p. 40–41. *In* G.J. Persley (ed.) Tropical legume improvement. Proc.7 of Thai/ACIAR Planning and Coordination Worksh., Bangkok, Thailand. 10–12 Oct. 1983. ACIAR Proc. Ser. 8. CSIRO Publ., Melbourne, Australia.

Prom, L.K., and J.R. Venette. 1997. Races of *Pseudomonas syringae* pv. *glycinea* on commercial soybean in eastern North Dakota. Plant Dis. 81:541–544.

Prosen, D., E. Hatziloukas, N.W. Schaad, and N.J. Panopoulos. 1993. Specific detection of *Pseudomonas syringae* pv. *phaseolicola* in bean seed by PCR-based amplification of a phaseolotoxin gene region. Phytopathology 83:965–970.

Quimio, G.M., and V.J. Calilung. 1993. Survey of flying viruliferous aphid species and population build-up of *Aphis glycines* Matsumura in soybean fields. Philipp. Entomol. 9:52–100.

Quiniones, S.S., J.M. Dunlevy, and J.W. Fisher. 1971. Performance of three soybean varieties inoculated with soybean mosaic virus and bean pod mottle virus. Crop Sci. 11:662–664.

Rathjen, J.P., L.E. Karageorgos, N. Habili, P.M. Waterhouse, and R.H. Symons. 1994. Soybean dwarf luteovirus contains the third variant genome type in the luteovirus group. Virology 198:671–679.

Reddy, D.V.R., A.S. Ratna, M.R. Sudarshana, F. Poul, and I.K. Kumar. 1992. Serological relationships and purification of bud necrosis virus, a tospovirus occurring in peanut (*Arachis hypogaea* L.) in India. Ann. Appl. Biol. 120:279–286.

Reddy, M.S.S., S.A. Ghabrial, C.T. Redmond, R.D. Dinkins, and G.B. Collins. 2001. Resistance to *Bean pod mottle virus* in transgenic soybean lines expressing the capsid polyprotein. Phytopathology 91:831–838.

Ren, Q., T.W. Pfeiffer, and S.A. Ghabrial. 1997a. Soybean mosaic virus incidence level and infection time: Interaction effects on soybean. Crop Sci. 37:1706–1711.

Ren, Q., T.W. Pfeiffer, and S.A. Ghabrial. 1997b. Soybean mosaic virus resistance improves productivity of double-cropped soybean. Crop Sci. 37:1712–1719.

Ren, Q., T.W. Pfeiffer, and S.A. Ghabrial. 2000. Relationship between soybean pubescence density and soybean mosaic virus field spread. Euphytica 111:191–198.

Roane, C.W., S.A. Tolin, and G.R. Buss. 1983. Inheritance of reaction to two viruses in the soybean cross 'York' x 'Lee 68'. J. Heredity 74:289–291.

Robertson, N.L., R. French, and S.M. Gray. 1991. Use of group-specific primers and the polymerase chain reaction for the detection and identification of luteoviruses. J. Gen. Virol. 72:1473–1477.

Roosien, J. and L. van Vloten-Doting. 1983. A mutant of alfalfa mosaic virus with an unusual structure. Virology 126:155–167.

Ross, J.P. 1968. Effect of single and double infections of soybean mosaic and bean pod mottle viruses on soybean yield and seed characters. Plant Dis. Rep. 52:344–348.

Ross, J.P. 1970. Effect of temperature on mottling of soybean seed caused by soybean mosaic virus. Phytopathology 60:1798–1800.

Ross, J.P. 1986a. Registration of four soybean germplasm lines resistant to bean pod mottle virus. Crop Sci. 26:210.

Ross, J.P. 1986b. Response of early-and late-planted soybeans to natural infection by bean pod mottle virus. Plant Dis. 70:222–224.

Ross, J.P. 1987. Viral and bacterial diseases. p. 729–755. *In* J.R. Wilcox (ed.) Soybeans: Improvement, production, and uses. ASA, CSSA, and SSSA, Madison, WI.

Samretwanich, K., K. Kittipakorn, P. Chiemsombat, and M. Ikegami. 2001. Complete nucleotide sequence and genome organization of soybean crinkle leaf virus. J. Phytopathol. 149:333–336.

Sano, Y., M. Tanahashi, M. Kawata, and M. Kojima. 1997. Comparative studies on soybean mosaic virus strains B and C: Nucleotide sequences of the capsid protein genes and virulence in soybean cultivars. Ann. Phytopathol. Soc. Jpn. 63:381–384.

Sano, Y., M. Wada, Y. Hashimoto, T. Matsumoto, and M. Kojima. 1998. Sequences of ten circular ssDNA components associated with the milk vetch dwarf virus genome. J. Gen. Virol. 79:3111–3118.

Satyanarayana, T., S.E. Mitchell, D.V. Reddy, S. Kresovich, R. Jarret, R.A. Naidu, S. Gowda, and J.W. Demski. 1996a. The complete nucleotide sequence and genome organization of the M RNA segment of peanut bud necrosis tospovirus and comparison with other tospoviruses. J. Gen. Virol. 77:2347–2352.

Satyanarayana, T., S.E. Mitchell, D.V.R. Reddy, S. Brown, S. Dresovich, R. Jarret, R.A. Naidu, and J.W. Demski. 1996b. Peanut bud necrosis tospovirus S RNA: Complete nucleotide sequence, genome organization, and homology to other tospoviruses. Arch. Virol. 141:85–98.

Schaad, N.W., J.B. Jones, and W. Chun. (ed.) 2001a. Laboratory guide for the identification of plant pathogenic bacteria. 3rd ed. APS Press, St. Paul, MN.

Schaad, N.W., J.B. Jones, and G.H. Lacy. 2001b. Xanthomonas. p. 175–200. In N.W. Schaad et al. (ed.) Laboratory guide for the identification of plant pathogenic bacteria. 3rd ed. APS Press, St. Paul, MN.

Schaad, N.W., A.K. Vidaver, G.H. Lacy, K. Rudolph, and J.B. Jones. 2000. Evaluation of proposed amended names of several pseudomonads and xanthomonads and recommendations. Phytopathology 90:208–213.

Schneider, D.D. 2002. Plant immunity and film noir: What gumshoe detectives can tell us about plant-pathogen interactions. Cell 109:537–540.

Schwenk, F.W., and C.D. Nickell. 1980. Soybean green stem caused by bean pod mottle virus. Plant Dis. 64:863–865.

Scott, H.A., J.V. van Scyoc, and C.E. van Scyoc. 1974. Reactions of Glycine spp. to bean pod mottle virus. Plant Dis. Rep. 58:191–192.

Scott, S.W., M.T. Zimmerman, and X. Ge. 1998. The sequence of RNA 1 and RNA 2 of tobacco streak virus:additional evidence for the inclusion of alfalfa mosaic virus in the genus Ilarvirus. Arch. Virol.143:1187–1198.

Seal, S.E., L.A. Jackson, and M.J. Daniels. 1992. Isolation of a Pseudomonas solanacearum-specific DNA probe by subtraction hybridization and construction of species-specific oligonucleotide primers for sensitive detection by the polymerase chain reaction. Appl. Environ. Microbiol. 58:3751–3758.

Severin, V. 1978. A new pathogenic bacterium on hemp— Xanthomonas campestris pv. cannabis (German). Arch. Phytopathol. Pflanzenschutz 14:7–15.

Shahraeen, N., R. Pourrahim, S. Farzadfar, and S. Ghorbani. 2001. First report of tomato spotted wilt virus on soybean in Iran. Plant Dis. 85:1290.

Shang, Y., J. Zhao, C. Yang, C. Li, X. Lu, X. Xin, and R. Luo. 1999. Classification and distribution of strains of soybean mosaic virus in the Huang-Huai area of China. Acta Phytopathol. Sin. 29:115–119.

Sheng, H.-Y. and P.A. Blackman. 1996. Effect of combination of Bacillus subtilis biocontrol strains GB03 and GB07 on soybean rhizosphere colonization. Phytopathology 86:S36 (abstr.).

Shipe, E.R., G.R. Buss, and S.A. Tolin. 1979. A second gene for resistance to peanut mottle virus in soybeans. Crop Sci. 19:656–658.

Siddiqui, K.H., and Trimohan. 1999. Resistant sources amongst soybean genotypes against yellow mosaic disease transmitted by whitefly. Shashpa 6:153–166.

Sinclair, J.B. 1999. Diseases caused by mollicutes. p. 10–11 . In G.L. Hartman et al. (ed.) Compendium of soybean diseases. 4th ed. APS Press, St. Paul, MN.

Smid, E.J., A.H.J. Jansen, and L.G.M. Gorris. 1995. Detection of Erwinia carotovora subsp. carotovora and Erwinia chrysanthemi in potato tubers using polymerase chain reaction. Plant Pathol. 44:1058–1069.

Smith E.F. 1905. Bacteria in relation to plant disease. Vol. 1. Carnegie Inst., Washington, DC.

Smith, O.P., V.D. Damsteegt, K.F. Harris, and R.V. Haar. 1993. Nucleotide sequence and E. coli expression of the coat protein gene of the yellowing strain of soybean dwarf luteovirus. Arch. Virol. 133:223–231.

Smith, O.P., S.A. Durkin, D.G. Luster, L.L. McDaniel, A.J. Russo, and V.D. Damsteegt. 1998. Sequence and expression in Escherichia coli of the coat protein gene of the dwarfing strain of soybean dwarf luteovirus. Virus Genes 17:207–211.

Sorensen, K.N., K.H. Kim, and J.K. Takemoto. 1998. PCR detection of cyclic lipodepsinonapeptide-producing Pseudomonas syringae pv. syringae and the similarity of strains. Appl. Environ. Microbiol. 64:226–230.

Stall, R.E., and T.A. Kucharek. 1982. A new bacterial disease of soybean in Florida. Phytopathology 72:990 (abstr.).

Stange, R.R., Jr., D. Jeffares, C. Young, D.B. Scott, J.R. Eason, and P.E. Jameson. 1996. PCR amplification of the fas-1 gene for detection of virulent strains of *Rhododocossus fascians*. Plant Pathol. 45:407–417.

Staskawicz, B.J., and M.B. Mudgett, J.L. Dangl, and J.E. Galan. 2001. Common and contrasting themes of plant and animal diseases. Science (Washington DC) 292:2285–2289.

Steinlage, T.A., J.H. Hill, and F.W. Nutter, Jr. 2002. Temporal and spatial spread of *Soybean mosaic virus* (SMV) in soybeans transformed with the coat protein gene of SMV. Phytopathology 92:478–486.

Stuckey, R.E., S.A. Ghabrial, and D.A. Reicosky. 1982. Increased incidence of *Phomopsis* sp. in seeds from soybeans infected with bean pod mottle virus. Plant Dis. 66:826–829.

Sun, Z., J.E. Specht, and G. Graef. 1992. Effects of planting dates, maturity and some qualitative genes on the formation of mottled seeds in soyabeans. (In Chinese, with English abstract.) Soybean Sci. 11:204–213.

Sun, Z.Q., P.Z. Tian, and J. Wang. 1990. Study on the uses of aphid-resistant character in wild soybean. I. Aphid-resistance performance of F2 generation from crosses between cultivated and wild soybeans. Soybean Genet. Newsl. 17:43–48.

Sundararaman, V.P., M.V. Stromvik, and L.O. Vodkin. 2000. A putative defective interfering RNA from *Bean pod mottle virus*. Plant Dis. 84:1309–1313.

Taraku, N., and N. Juretic. 1990. Distribution of soybean mosaic virus within soybean embryo. Acta Bot. Croat. 49:13–17.

Terauchi, H., S. Kanematsu, K. Honda, Y. Mikoshiba, K. Ishiguro, and S. Hidaka. 2001. Comparison of complete nucleotide sequences of genomic RNAs of four *Soybean dwarf virus* strains that differ in their vector specificity and symptom production. Arch. Virol. 146:1885–1898.

Thakur, M.P., D.V.R. Reddy, A.S. Reddy, A.S. Ratna, M. Al Nasiri, and K.C. Agrawal. 1996. Identification of bud blight of soybean (*Glycine max* (L.) Merr.) through ELISA and infectivity assay. Indian J. Virol. 12:79–82.

Thakur, M.P., K.P. Verma, and K.C. Agrawal. 1998. Characterization and management of bud blight disease of soybean in India. Int. J. Pest Manage. 44:87–92.

Tolin, S.A. 2001. Cross protection. p. 263–264. *In* O.C. Maloy and T.D. Murray (ed.) Encyclopedia of plant pathology. John Wiley and Sons, New York.

Tu, J.C. 1975. Localization of infectious soybean mosaic virus in mottled soybean seeds. Microbios 14:151–156.

Tu, J.C. 1989. Effect of different strains of soybean mosaic virus on growth, maturity, yield, seed mottling and seed transmission in several soybean cultivars. J. Phytopathol. 126:231–236.

Tu, J.C. 1992. Symptom severity, yield, seed mottling and seed transmission of soybean mosaic virus in susceptible and resistant soybean: The influence of infection stage and growth temperature. J. Phytopathol. 135:28–36.

van den Berg, H., D. Ankasah, A. Muhammad, R. Rusli, H.A. Widayanto, H.B. Wirasto, and H. Yully. 1997. Evaluating the role of predation in population fluctuations of the soybean aphid *Aphis glycines* in farmers' fields in Indonesia. J. Appl. Ecol. 34:971–984.

van Regenmortel, M.H.V., C.M. Fauquet, D.H.L. Bishop, E.B. Carstens, M.K. Estes, S.M. Lemon, J. Manjiloff, M.A. Mayo, D.J. McGeoch, C.R. Pringle, and R.B. Wickner. 2000. Virus taxonomy: The classification and nomenclature of viruses. The seventh report of the international committee on taxonomy of viruses. Academic Press, San Diego.

van Vloten-Doting, L. 1975. Coat protein is required for infectivity of tobacco streak virus: Biological equivalence of the coat proteins of tobacco streak and alfalfa mosaic viruses. Virology 65:215–225.

Vauterin, L., B. Hoste, K. Kersters, and J. Swings. 1995. Reclassification of *Xanthomonas*. Int. J. Syst. Bacteriol. 45:472–489.

Vetten, H.J., S.K. Green, and D.E. Lesemann. 1992. Characterization of peanut stripe virus isolates from soybean in Taiwan. J. Phytopathol. 135:107–124.

Walters, H.J. 1958. A virus disease complex in soybeans in Arkansas. Phytopathology 48:346.

Walters, H.J. and F.N. Lee. 1969. Transmission of bean pod mottle virus from *Desmodium paniculatum* to soybean by the bean leaf beetle. Plant Dis. Rep. 53:411.

Wang, C.S., J.J. Todd, and L.O. Vodkin. 1994. Chalcone synthase mRNA and activity are reduced in yellow soybean seed coats with dominant *I* alleles. Plant Physiol. 105:739–748.

Wang, R.Y., R.C. Gergerich, and K.S. Kim. 1992. Noncirculative transmission of plant viruses by leaf-feeding beetles. Phytopathology 82:946–950.

Wang, R.Y., and S.A. Ghabrial. 2002. Effect of aphid behaviour on efficiency of transmission of *Soybean mosaic virus* by the soybean-colonizing aphid, *Aphis glycines*. Plant Dis. 86:1260–1264.

Wang, X., A.L. Eggenberger, F.W. Nutter, Jr., and J.H. Hill. 2001. Pathogen-derived transgenic resistance to soybean mosaic virus in soybean. Mol. Breed. 8:119–127.

Weber, C.R., J.M. Dunleavy, and W.R. Rehr. 1966. Effect of bacterial pustule on closely related soybean lines. Agron. J. 58:544–545.

Weldekidan, T., T.A. Evans, R.B. Carroll, and R.P. Mulrooney. 1992. Etiology of soybean severe stunt and some properties of the causal virus. Plant Dis. 76:747–750.

Williams, L.V., P.M. Lopez Lambertini, K. Shohara, and E.B. Biderbost. 2001. Occurrence and geographical distribution of tospovirus species infecting tomato crops in Argentina. Plant Dis. 85:1227–1229.

Windham, M.T., and J.P. Ross. 1985. Transmission of bean pod mottle virus in soybeans and effects of irregular distribution of infected plants on plant yield. Phytopathology 75:310–313.

Wolf, F.A., and A.C. Foster. 1917. Bacterial leafspot of tobacco. Science (Washington DC) 46:361–362.

Wood, B.A., and W. J. Easdown. 1990. A new bacterial disease of mung bean and cowpea for Australia. Australas. Plant Pathol. 19:16–21.

Xie, Z.-P., C. Staehelin, A. Weinmken, and T. Boller. 1996. Ethylene responsiveness of soybean cultivars characterized by leaf senescence, chitinase induction, and nodulation. J. Plant Physiol. 149:690–694.

Xu, Z.Y., Z.Y. Zhang, J.X. Chen, and X.H. Deng. 1988. Tests of soybean varieties and germplasm lines for resistance to peanut mild mottle virus and peanut stunt virus. Soybean Sci. 7:53–60.

Young, J.M., C.T. Bull, S.H. DeBoer, G. Firrao, L. Gardan, G.E. Saddler, D.E. Stead, and Y. Takikawa. 2001. Classification, nomenclature, and plant pathogenic bacteria—A clarification. Phytopathology 91:617–620.

Yu, Y.G., M.A. Saghai Maroof, G.R. Buss, P.J. Maughan, and S.A. Tolin. 1994. RFLP and microsatellite mapping of a gene for soybean mosaic virus resistance. Phytopathology 84:60–64.

Yuan, J., W. Wei, S. Gopalan, W. Hu, Q. Jin, A. Plovanich-Jones, L. Muncie, and S.Y. He. 2000. Hrp genes of *Pseudomonas syringae*. p.1–20. *In* G. Stacey and N.T. Keen (ed.) Plant-microbe interactions. Vol.5. APS Press, St. Paul, MN.

Zhang, M.H., W. Q. Lu, Z.X. Zhong, R.Y. Wang, and Y.H. Li. 1986. The importance of the diseased seedlings from SMV infected seeds and the vector of the virus in the epidemic. Acta Phytopathol. Sin. 16:151–158.

Zhou, X., Z. Pu, Z. Fang, R. Chu, and Z. Chen. 1994. Construction and use of a cloned soybean mosaic virus (SMV) coat protein gene as probe for the detection of SMV. (In Chinese, with English abstract.) J. Virol. 10:81–85.

Zhou, X., Y. Yu, Y. Qi, Z. Chen, D. Li, X.P. Zhou, Y.J. Yu, Y.J. Qi, Z.X. Chen, and D.B. Li. 1996. Detection of broad bean wilt virus by double antibody sandwich ELISA. Acta Phytopathol. Sin. 26:347–352.

Ziems, A.D., L.J. Giesler, G.L. Graef, and L.C. Lane. 2001a. Effect of bean pod mottle virus on soybean seed quality. Phytopathology 91:S100 (abstr.).

Ziems, A.D., L.J. Giesler, and L.C. Lane. 2001b. Incidence of *Bean pod mottle virus* and *Soybean mosaic virus* in Nebraska. Phytopathology 92:S141 (abstr.) Publ. P-2002-0030-NCA. Of the 2001 North Central Div. Abstr. of the Am. Phytopathol. Soc. Available at http://www.apsnet.org/meetings/div/ (verified 13 May 2003).

16 Nematode Pathogens of Soybean

TERRY L. NIBLACK

University of Illinois
Urbana, Illinois

GREGORY L. TYLKA

Iowa State University
Ames, Iowa

ROBERT D. RIGGS

University of Arkansas
Fayetteville, Arkansas

16–1 NEMATODE PARASITES AND DAMAGE ESTIMATES

Plant-parasitic nematodes are found in all soybean, *Glycine max* (L.) Merr., fields. Although over 100 species have been reported in association with soybean (Schmitt and Noel, 1984), fortunately only a relative few have been shown to be soybean pathogens, i.e., those capable of causing consistently measurable yield loss.

Unlike other microscopic soybean pathogens, nematodes are animals— roundworms, or eelworms, to be more specific. All the known soybean-pathogenic nematodes are soil-dwelling or soil-borne, and live in or on roots. This characteristic of their life histories makes them easy to overlook as the source of soybean yield reductions, since it generally requires technical expertise to diagnose them as disease agents. The aboveground damage, stunting and chlorosis, they can cause is usually nonspecific and quite similar to that caused by numerous other biotic and abiotic agents. Nematodes can damage soybean plants directly, as a result of their migration or feeding activities, or indirectly as partners in disease complexes with other organisms.

16–1.1 Nematodes Pathogenic to Soybean

The most important soybean-pathogenic nematodes, on a worldwide basis, are the soybean cyst nematode, *Heterodera glycines*, and some of the root-knot nematodes, *Meloidogyne* spp. (Koenning et al., 1999; Wrather et al., 2001). These nematodes will receive the emphasis in this chapter. Other species, of more local or regional importance, are also covered briefly. These include *Hoplolaimus* spp. (lance nematodes), *Pratylenchus* spp. (lesion or root-lesion), and *Rotylenchulus reni-*

formis (reniform). Additional information on these and other nematode parasites and pathogens of soybean may be found in Riggs and Niblack (1993), Schmitt and Noel (1984), and Sikora and Greco (1990).

16–1.2 Worldwide Economic Loss Estimates

Due to the very nature of damage caused by root-parasitic nematodes, estimates of losses incurred by soybean producers are very hard to confirm. Scientifically rigorous nematode distribution surveys are very few, due in large part to the reluctance of granting agencies to fund such research (Koenning et al., 1999). In addition, damage thresholds used to calculate losses are affected by innumerable factors. Nonetheless, several individuals and groups have made commendable efforts to keep us apprised of the economic importance of nematodes in soybean production.

In 1987, a committee of the Society of Nematologists (SON Crop Loss Committee, 1987) estimated soybean yield losses due to plant-parasitic nematodes in the USA to be 8% of total production, 5.8% of which was due to *H. glycines* alone. A subsequent group (Koenning et al., 1999) found that nematode-induced losses in 1994 were higher than in 1987, despite the considerable success of soybean breeding programs in producing resistant and tolerant cultivars (Young, 1998a).

J. A. Wrather (Univ. of Missouri Delta Center), alone and with various colleagues, has been compiling loss estimates due to the major yield-reducing diseases of soybean in the southern USA since the mid-1970s (e.g., Wrather et al., 1995; Wrather and Sciumbato, 1995), in all soybean-producing states (e.g., Wrather et al., 2000) since Doupnik (1993) published the first estimate for the North Central states, and in the top 10 soybean-producing countries since the mid-1990s (Wrather et al., 1997, 2001). The most damaging pathogen of soybean in the world remains *H. glycines*, which caused an estimated $2 billion in lost soybean yields in 1998. The highest losses were reported from the USA, primarily from the North Central states Iowa, Illinois, Minnesota, Indiana, Ohio, Missouri, Nebraska, and South Dakota. Root-knot and other nematodes were reported as the cause of soybean yield reductions valued at $117 million worldwide in 1998, only 24% of which were from the USA.

16–2 SOYBEAN CYST NEMATODE

16–2.1 Distribution

Soybean cyst nematode has infested fields in Asia for hundreds or even thousands of years according to Chinese scientists. These infestations occur in Japan, Korea, People's Republic of China, and Russia (Noel, 1992). In the last half of the 20th century, this nematode has been discovered in Argentina, Brazil, Canada, Colombia, Indonesia, Taiwan, and the USA (Noel, 1992). There also is an unsubstantiated report of its occurrence in Egypt (Diab, 1968). Little is known about the actual distribution within Argentina, Colombia, Taiwan, Indonesia, Japan, Korea, and People's Republic of China. Information from a recent survey of soybean

Table 16–1. Distribution of races of *Heterodera glycines* in Brazil.

State	Races found
Goias	3,4,6,9,14
Mato Grosso do Sul	3,4,6,9,10,14
Mato Grosso	1,2,3,4,5,6,9,10,14, 14+†
Minas Gerais	3
Sao Paulo	3
Parana	3
Rio Grande do Sul	3,6

† 14 + = Race 14 that reproduces on the soybean cv. Hartwig.

fields in Brazil shows the widespread occurrence of *H. glycines* and of numerous races there (Table 16–1).

Following the discovery of *H. glycines* in North Carolina in 1954 (Winstead et al., 1955) it soon was found in several other states including Arkansas, Illinois, Mississippi, Missouri, Kentucky, South Carolina, Tennessee, and Virginia (Noel, 1992). Generally, the areas of infestation were initially small. During the next 47 yr, the nematode has been found in all other states that produce an appreciable hectarage of soybean, except North Dakota, and in Ontario in Canada (Noel, 1992). In addition, the infested area in each state has expanded (Plate 16–1). This spread was the result of many agents. Spread can be attributed to wind currents, water movement, foot and machinery traffic, soil peds in seed stocks, and birds (the cysts can be ingested and pass through the digestive system) (Epps, 1971; Smith et al., 1992).

16–2.2 Life Cycle and Disease Cycle

The life cycle of *H. glycines* is typical of that for a cyst nematode in the genus *Heterodera*. Vermiform, second-stage juveniles (J2) hatch from the eggs and are the only life stage of the nematode capable of infecting a root and establishing a feeding relationship with the soybean plant (Plate 2–2). Upon penetrating the root, the juvenile migrates intracellularly through the cortex tissue to the periphery of the vascular cylinder. There, the second-stage juvenile begins probing one cortical parenchyma, endodermal, or pericycle cell with its pointed, protrusible, sclerotized stylet. This host plant cell, referred to as the initial syncytial cell, undergoes dramatic changes in structure and function and eventually begins fusing with adjacent cells (Endo, 1991). The cellular changes that occur in the host are presumed to be in response to secretions from the juvenile nematode's esophageal glands (Davis et al., 2000). Fusion of host cells occurs repeatedly until a large, multinucleate feeding site, called a syncytium, forms. The exact location of the syncytium can be affected by the amount of water available at the time of nematode infection. Johnson et al. (1993) discovered that *H. glycines* formed syncytia in the root cortex in irrigated research plots and in the vascular cylinder (stele) in non-irrigated plots.

The life cycle of *H. glycines* takes 21 d to complete on soybean root explants growing at 25°C in the laboratory (Lauritis et al., 1983). Following establishment of the syncytia, both male and female juvenile nematodes swell and develop to third- and then fourth-stage juvenile stages before they become reproductive adults (Ichinohe, 1955). Throughout their development, females continue to increase in size and eventually become so large that they rupture out from the root and are exposed on the surface (Plate 16–3). The bodies of male nematodes also swell in the third and fourth juvenile stages, but then revert back to a vermiform shape after the fourth juvenile stage. Adult *H. glycines* males emigrate from the root and re-enter the soil environment. Mating occurs with males in the soil and females exposed on the surface of the root. *Heterodera glycines* females lay eggs in a gelatinous egg mass on the posterior surface of the female body and also retain eggs within the body cavity. Eventually the *H. glycines* female dies, and the nematode body wall forms a durable, protective covering for the eggs that were deposited within (Plate 16– 4). *Heterodera glycines* eggs begin hatching when soil temperature and moisture conditions become favorable and the life cycle begins again (Schmitt and Riggs, 1989).

The life cycle of the nematode is well adapted to that of the host crop. *Heterodera glycines* hatching decreases sharply beginning mid season as the soybean crop matures and remains at a low level until the next soybean crop begins to grow (Yen et al., 1995). Also, the eggs contained within the *H. glycines* cyst can survive for several years in the absence of a host crop (Inagaki and Tsutsumi, 1971). Egg-mass eggs hatch more rapidly and to a greater extent in deionized water than encysted eggs (Thompson and Tylka, 1997).

16–2.3 Interactions with Other Pathogens

Heterodera glycines interacts with numerous other soybean pathogens. These interactions may exacerbate the other disease. Conversely, nematode reproduction and development may be affected by the interaction as well. Foliar symptoms of Fusarium wilt, caused by *Fusarium oxysporum*, were greater for plants of 'Jackson' soybean grown in greenhouse soil infested with *F. oxysporum* and *H. glycines* than in soil infested with the fungus alone (Ross, 1965). Similarly, severity of *Phytophthora sojae* infection of 'Corsoy' and 'Dyer' soybean was greater in *H. glycines*-infested than in *H. glycines*-free greenhouse soil (Adeniji et al., 1975). In field microplot experiments, greater numbers of *H. glycines* females were produced on Fusarium-infected than uninfected 'Lee' soybean (Ross, 1965). Conversely, fewer *H. glycines* females developed on *P. sojae*-infected than on *P. sojae*-free Corsoy grown in a greenhouse experiment (Adeniji et al., 1975). *Phytophthora sojae* resistance of 'Harosoy 63' soybean in a greenhouse experiment was not affected by *H. glycines* infection, nor was *H. glycines* resistance in Dyer soybean affected by *P. sojae* infection (Adeniji et al., 1975). Colonization of *H. glycines*-susceptible soybean plants by the charcoal rot fungus, *Macrophomina phaseolina*, was significantly greater than that of *H. glycines*-resistant plants in soil infested with the soybean cyst nematode (Todd et al., 1987; Winkler et al., 1994).

Heterodera glycines also interacts with the fungal pathogen *Fusarium solani* f. sp. *glycines* (Roy, 1997), the causal agent of soybean sudden death syndrome (SDS). Hershman et al. (1990) and Rupe et al. (1991) detected greater severity of SDS foliar symptoms for *H. glycines*-susceptible soybean cultivars than for *H. glycines*-resistant cultivars when grown in fields infested with both pathogens. The *H. glycines*-susceptible 'Coker 156' soybean had greater SDS incidence and severity in field microplots infested with both *H. glycines* and *F. solani* than when grown in microplots infested with the fungus alone (McLean and Lawrence, 1993).

Additionally, fewer *H. glycines* eggs were produced on *F. solani*-infected than on *F. solani*-free Coker 156.

Researchers noticed increased symptoms of brown stem rot (BSR), causal agent *Phialophora gregata*, in both BSR-susceptible and BSR-resistant soybeans grown in *H. glycines*-infested fields, and results of preliminary experiments based on these observations also suggested increased symptoms of the fungal disease when soybean plants also were infected with the nematode (Tubajika et al., 1994). In 1997, Sugawara et al. (1997) reported that *H. glycines* increased internal stem discoloration caused by BSR in a soybean cultivar susceptible to both pathogens, but not in two other cultivars, one resistant to the fungus and another resistant to *H. glycines*. Most recently, Tabor et al. (2001) have shown that *H. glycines* significantly increases internal stem discoloration as well as colonization by *P. gregata* when plants also are infected with *H. glycines* relative to plants not infected with the nematode.

16–2.3.1 Symptoms

Heterodera glycines can cause soybean yield loss in the complete absence of symptoms (Mendes and Dickson, 1993; Wang et al., 2002; Young, 1996). The "typical" symptoms reported for the disease caused by *H. glycines* are plant stunting and leaf chlorosis (Plate 16–5), in oval patches elongate in the direction of tillage (Plate 16–6). In high yield environments, such symptoms may not be readily visible (Plate 16–7) until yield reduction is high, 30% or more, in the affected areas. Infestations that produce lower yield reductions may not even be obvious when resistant and susceptible cultivars are planted side-by-side (Plate 16–8).

Intensive studies of the effect of *H. glycines* on soybean growth and development in the Midwest showed that the primary effect of the nematode on the plant was to reduce the number of pods and seeds (Wang et al., 2002). Comparisons of the growth parameters of resistant and susceptible cultivars showed that yield loss could be substantial (15% or more) in the absence of visible symptoms. For example, in a field with an infestation level high enough to cause a 54% reduction in soybean yield on a susceptible cultivar, there was no difference in plant height between the susceptible and resistant cultivars (Fig. 16–1).

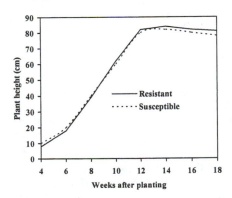
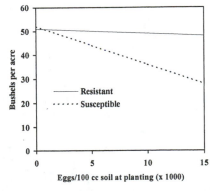

Fig. 16–1. Effects of *Heterodera glycines* on (left) plant height and (right) soybean yield of resistant and susceptible cultivars in two locations over four growing seasons in Iowa and Missouri.

Heterodera glycines infection stresses soybean plants. In plants already under stress from some additional factor, such as a nutrient deficiency, the symptom expressed may be that of the additional stress factor. For example, infected plants in K-deficient soils are likely to express the K-deficiency symptom (Plate 16–9). Adjusting K fertility may alleviate the symptom, but mislead the grower into thinking there is no additional problem if the soil is not also checked for the presence or population density of the nematode. Symptoms caused by the nematode may also be mistaken for those caused by Phytophthora root rot, charcoal rot, herbicide injury, and iron deficiency chlorosis, among others (Plate 16–10). The latter situation, chlorosis caused by iron deficiency in soils with pH > 7.5 in parts of the Midwest, is complicated by the direct relationship between *H. glycines* population densities and elevated soil pH in the Midwest (Tylka et al., 1998).

Because the nematode is so widely distributed in soybean production areas, and because it can cause substantial yield loss even in the absence of obvious symptoms, it would be prudent to sample soybean fields periodically for the presence of the nematode (if the field is thought to be uninfested) or the infestation level (if the field if known to be infested).

16–2.4 Diagnosis

16–2.4.1 Sampling and Confirmation

Because of the inconsistent nature of the appearance of aboveground symptoms of damage caused by *H. glycines*, it is essential that all fields be tested for the presence of the nematode before soybean is grown in areas where the nematode is known to exist. There are two ways to accurately determine if a field is infested with *H. glycines*. An inexpensive and relatively quick way to check for the presence of *H. glycines* in a field is to carefully observe the roots of susceptible soybean plants for the presence of *H. glycines* females, which can be seen with the unaided eye (Plate 16–11). This must be done during the growing season. It will take several weeks for the first *H. glycines* females to become readily observable on the roots, but then small, white females should be apparent on the roots for several months. The *H. glycines* females on the roots will not be readily observable once the crop has matured.

The other accurate way to determine if a field is infested with *H. glycines* is to collect a soil sample and have it tested for the presence of the nematode. Typically, it is recommended to collect 10 to 20 2-cm-diameter, 15- to 20-cm-deep soil cores from a designated sampling area using a systematic zigzag or M-shaped pattern. Alternatively, one could collect soil cores from areas of a field that are likely to be infested with the nematode, such as areas that may have had soil introduced from surrounding fields. Such areas include near field entryways where soil from equipment may be introduced, along fence lines where windblown soil may accumulate, and in low spots and previously flooded areas of fields where soil may have been introduced from surrounding fields by runoff of surface water. To increase the accuracy of the sample results, one can increase the number of cores collected (Francl, 1986) and/or decrease the area from which the sample is collected.

Depending on the specific purpose of the sampling, it may be desirable to not only extract and enumerate cysts from the soil, but also to extract and count the eggs within the recovered cysts. Both cyst and egg population densities are more reliable measures of the nematode population than densities of second-stage juveniles (Barker et al., 1987). For advisory purposes and certain research objectives, it may be better to use cyst population densities than egg population densities because of the lower variability of cyst population densities. For example, Gavassoni et al. (2001) found greater spatial dependence using cyst population densities than egg population densities. But cyst population densities are not as precise an estimate of *H. glycines* populations as are egg population densities. Consequently, better results usually will be obtained with egg rather than cyst population densities when assessing relationships between *H. glycines* population densities and other factors.

Cysts of *H. glycines* can be recovered from soil collected at anytime during the year. There are several techniques for extracting *H. glycines* cysts from soil. Cysts can be recovered on 250-μm-pore sieves with a semi-automatic elutriator (Byrd et al., 1976). This specialized apparatus uses flowing water and air to separate cysts and similarly sized materials from heavier materials in the soil. Unwanted debris collect in a sieve nested above the 250-μm-pore sieve, and materials smaller than *H. glycines* cysts pass through the 250-μm-pore sieve. Another commonly used technique for extracting *H. glycines* cysts from soil is an adaptation of a wet-sieving and decanting technique originally developed for recovery of spores of mycorrhizal fungi (Gerdemann, 1955). In this technique, soil is suspended in water, then allowed to settle for a short time before the suspension is passed through nested sieves. As with the elutriator, unwanted debris collect in a sieve nested above the 250-μm-pore sieve, and materials smaller than *H. glycines* cysts pass through the 250-μm-pore sieve in this technique.

Numerous techniques are used to recover *H. glycines* eggs from cysts. Acedo and Dropkin (1982) reported extracting eggs from *H. glycines* cysts by rupturing the cysts through a 150-μm-pore sieve manually using a rubber stopper, then subsequently separating eggs from unwanted debris by centrifugal flotation. Faghihi and Ferris (2000) modified this technique somewhat by attaching the stopper to a motor and creating a 250-μm-pore sieve on which the cysts would be ruptured, then recovering the eggs on a 38-μm-pore sieve situated below the 250-μm-pore sieve. Eggs can be chemically extracted from *H. glycines* cysts by dissolving the cyst walls in a dilute solution of sodium hypochlorite and subsequently recovering the eggs on a 38-μm-pore sieve (Faghihi et al., 1986). Alternatively, a stainless steel pestle can be used to rupture *H. glycines* cysts, followed by recovery of the eggs on a 25-μm-pore sieve (Boerma and Hussey, 1984, Niblack et al., 1993).

Several different cyst nematode species can occur in soybean production fields, including the carrot cyst nematode, *Heterodera carotae*, the sugar beet cyst nematode, *Heterodera schachtii*, the clover cyst nematode, *Heterodera trifolii*, and the smartweed cyst nematode, *Cactodera weissi*. Some of these other cyst nematode species also can parasitize soybean. Consequently, it is necessary at times to determine the species identity of cysts recovered from the soil to confirm diagnosis of a *H. glycines* infestation. The cysts of *H. glycines* are lemon-shaped, 500 to 900 μm long, and 200 to 700 μm wide (Taylor, 1974). Mulvey and Golden (1983) published an illustrated key to species of cyst nematode.

16–2.5 Genetic Variability: Races

For everyone involved in management of *H. glycines*, one of the most difficult issues is that of races, or more precisely, variation among populations in their ability to develop and reproduce on resistant soybean lines. There are no known soybean lines with complete resistance to all populations of *H. glycines*. Even highly resistant lines allow a few females to develop and reproduce, which results eventually in selection of populations adapted to resistant lines.

For the past 30 yr, variation among populations has been measured by means of a race test. This is a test of the ability of a population to develop on a susceptible cultivar, Lee, and four soybean lines, called differentials, representing different genetic sources of resistance to the nematode (Golden et al., 1970; Riggs and Schmitt, 1988; Table 16–2). The test is a bioassay conducted in a greenhouse usually for a period of 30 d. The test begins with infestation of soil in containers planted separately to each of the differentials and Lee, and replicated several times. After 30 d, the plants are removed and the female nematodes (cysts) are washed from the roots and counted. For each differential, a female index (FI) is calculated as follows: (mean number of females on the differential) (mean number of females on Lee) × 100. If the mean FI for a differential is less than 10, the score for that differential is "–", and the score for differentials with FI of 10 or more is "+". The pattern of "–" and "+" scores for the four differentials is compared to the standards (Table 16–2), and the matching pattern reveals the race designation for that population.

When genetic variation among *H. glycines* populations was first noted, almost all were what is now known as race 3, with relatively few that were identified as races 1 or 2 (Golden et al., 1970). As the nematode spread, resistant cultivars were planted. In many fields, resistant cultivars were planted repeatedly until the selec

Table 16–2. Races of the soybean cyst nematode, *Heterodera glycines*, according to the race determination schemes of Golden et al. (1970) and Riggs and Schmitt (1988).

Race†	Pickett	Peking	PI 88788	PI 90763
1	–	–	+	–
2	+	+	+	–
3	–	–	–	–
4	+	+	+	+
5	+	–	+	–
6	+	–	–	–
7	–	–	+	+
8	–	–	–	+
9	+	+	–	–
10	+	–	–	+
11	–	+	+	–
12	–	+	–	+
13	–	+	–	–
14	+	+	–	+
15	+	–	+	+
16	–	+	+	+

† Race determination is made on the basis of the pattern of "+" and "–" ratings for each race. A "+" rating is given if the number of females produced by an *H. glycines* population on each soybean differential is equal to or greater than 10% of the number produced on the standard susceptible cv. Lee. If the number of females is <10%, a "–" rating is given.

tion pressure resulted in nematode populations that were virulent on most resistant cultivars, such as races 2, 4, and 5. Race 3 is almost never found in some areas now.

Originally, race 1 was found only in North Carolina and race 2 only in Virginia. Race 3 was found in all infested areas. Resistance to races 1 and 3 was obtained from Peking and was used extensively in southern U.S. soybean fields. In midwestern U.S. soybean fields, resistance primarily came from Plant Introduction (PI) 88788. As a result, races 1, 2, and 5 were more common after resistance had been used for a few years. In the South, cultivars with resistance from the Peking were used at first, then PI 88788 was added and a broad range of races were subsequently noted, including races 1, 2, 4, 5, 6, 9, and 14.

An accurate distribution of the races of *H. glycines* by state is not available. However, reports from the various states show a great variation in the number of races in any given state (Table 16–3). Florida reports that soybean is planted on very few acres now. During the last several years, damage caused by *H. glycines* in the southern states has been less than in earlier years. Many growers began to feel that *H. glycines* was not a problem any longer. However, in 2001 damage from *H. glycines* in Arkansas became much greater again. The reason for the decline in damage is not known but is probably related to a higher percentage of growers rotating to manage the nematode, buildup of natural enemies that reduced populations, and (or) the number of cultivars with various levels of resistance. The damage in 2001 may be the result of stopping rotation programs or continuous planting of resistant cultivars that resulted in the selection of virulent populations against which we have little resistance in cultivars that are being planted.

16–2.6 Alternative Means of Assessing Variability

There are numerous conceptual problems with defining genetic variability according to the race scheme (Niblack, 1992). Foremost is the mistaken assumption that a race designation reveals a genotype, and the resulting assumption that there can be "mixtures of races" within a field, or that two populations with the same race designation are genotypically equivalent. When used properly, race testing and similar bioassays (e.g., Young, 1989, 1998a) can reveal much about within- and between-population variability, but it has rarely been used in this manner. Labeling of resistant cultivars as "resistant to race 3" or "resistant to races 3 and 14", for example, is biologically incorrect and is based on the assumption that all populations designated race 3 or 14 are equivalent to the population used initially to classify the cultivars as resistant. To see many examples, inspect the data generated by cultivar testing programs in such states as Georgia, Iowa, Illinois, and Missouri. Note also that many cultivars in these screens that are not highly "resistant," when this is defined by a FI < 10, will outperform susceptible cultivars in yield trials in infested fields.

Because the race test was established before there were many known sources of resistance to *H. glycines*, it does not allow us to monitor variation among *H. glycines* populations that are virulent on other sources of resistance. For this and other reasons, a new test has been developed (Niblack et al., 2002) by a group of scientists, including nematologists, soybean breeders, and geneticists, who agreed that the race test should be changed, but only in the most conservative way possi-

Table 16–3. Distribution of races of *Heterodera glycines* according to nematologists or recent surveys in the various states or provinces.

State†	Races												
	1	2	3	4	5	6	7	8	9	10	11	13	14
Arkansas	x‡	x	x	x_p§	x	x_p	x	x	x_p	x	x		x
Florida		3%	x										x
Illinois	27%		64%	3%	3%	x		x			x		
Indiana	x		x_p		x	x							
Iowa	26%		58%		6%	3%		3%				3%	
Kansas	37%	6%	12%	6%		12%			6%				12%
Kentucky	x		x_p	x		x			x				x
Louisiana	21%	11%	5%	5%		31%			11%	5%			11%
Maryland	x							x					
Minnesota	18%		62%	2%	2%	13%							2%
Mississippi		9%		9%	2%	55%			18%				9%
Missouri	28%	22%	33%	5%					5%	1%			1%
Nebraska	33%		67%										
North Carolina	24%	38%	4%	17%	4%	10%	1%		1%	4%			4%
Ohio		40%	60%										
South Carolina				x									
South Dakota				x			x						
Tennessee			x_p 14%	7%	57%	29%			7%				
Virginia			x_p 31%			x							
Ontario	15%	15%			15%	15%				8%			7%

†Little is known about the races that occur in Alabama, Delaware, Georgia, Hawaii, Michigan, New Jersey, Oklahoma, and Wisconsin.
‡ x = race present but percentage of total is not known.
§ x_p = a predominant race in the area.

Table 16–4. List of indicator lines for the HG Type classification scheme for genetically diverse populations of *Heterodera glycines*.

Chronological order	Indicator line	Registration in *Crop Science*
1	PI 548402 (Peking)	Brim and Ross (1966)
2	PI 88788	Hartwig and Epps (1978)
3	PI 90763	Hartwig and Young (1990)
4	PI 437654	Anand (1992a)
5	PI 209332	Anand (1992b)
6	PI 89772	Nickell et al. (1994b)
7	PI 548316 (Cloud)	Nickell et al. (1994a)

ble, until such time as we can do direct genetic testing on *H. glycines* for the genes for avirulence that interact with soybean resistance genes.

For the new test, the term "race" has been dropped to avoid the implication that the test is for clearly identifiable genotypes. The name is now "HG Type"; "HG" for *H. glycines*, and "type" because it has no specific genetic implications. The HG Type test differs from the race test in only four significant ways.

First, the cv. Pickett was dropped from the list of differentials, now called "indicator lines," because discriminant function analysis has shown that Pickett and Peking do not discriminate *H. glycines* populations on the basis of FI (Tourjee and Niblack, 2000).

Second, four new indicator lines were added. The criterion for addition of a line to the HG Type test is its registration in *Crop Science* as a source of resistance to *H. glycines* in a germplasm or cultivar (Table 16–4). If the same model for naming races had been used for naming HG Types, that is, a specific pattern of – and + phenotypes, then we would now have a chart containing 128 possible combinations.

Therefore, rather than promulgate the confusion that would ensue with 128 HG Types, *H. glycines* populations will be classified based on the indicator lines for which the FI of a population is 10 or greater (Table 16–5). A population with

Table 16–5. Examples of HG Type classifications for field and greenhouse populations of *Heterodera glycines*.

Population[†]	Source[‡]	Females on Lee 74 (no.)	Female indices[§] on indicator lines							HG Type
			1	2	3	4	5	6	7	
Wind1	Field	188	2	1	3	0	0	0	4	0
Farr16	Field	299	1	5	0	0	16	0	40	5.7
MacG3	Field	232	1	26	0	0	22	0	44	2.5.7
TN10	GH	232	12	1	3	0	1	9	0	1
TN20	GH	112	70	57	79	78	80	67	88	1–7

† Population names are given arbitrarily for field populations, and according to published rules for strains (Bird and Riddle, 1994).

‡ The source of the sample from which the nematodes were extracted for the HG Type test. GH = greenhouse.

§ Female index = (mean no. females on test line/mean no. females on Lee 74) × 100.

no FI of 10 or greater on any indicator line is an HG Type 0 (zero). Note that FI are calculated relative to 'Lee 74' rather than Lee.

Finally, for any report of an HG Type, the FI must accompany the HG Type designation. For example, for *H. glycines* population MacG3 in Table 16–5, the HG Type should be reported as "HG Type 2.5.7, with FI of 26, 22, and 44, respectively." This requirement will help avoid the implication that two HG Types 2.5.7 are identical.

Additional reasoning behind the HG Type test has been summarized (Niblack et al., 2002). In the more distant future, it is to be hoped that sufficient information about the genes for resistance to *H. glycines* in the soybean, and genes for avirulence in the nematode, will be developed into informative genetic tests that will tell us more than do bioassays such as the race test or HG Type test.

16–2.7 Management

For all practical purposes, *H. glycines* cannot be eliminated from a field once it is infested. However, there are things that can be done to manage the nematode to maximize soybean yields and minimize reproduction of the nematode. Effective management involves an integrated approach comprising scouting for early detection of infestations (examining roots or collecting soil samples as described above) followed by proper use of resistant soybean cultivars in rotation with nonhost crops in infested fields. Chemical control through use of nematicides also is an option.

Nematicides once were a cost-effective and widely used option for management of soybean cyst nematode. However, most effective and relatively inexpensive nematicides have been removed from the market due to environmental and human health concerns. Nematicides available for management of nematodes on soybean in recent years include aldicarb, carbofuran, 1,3 dichloropropene, and isothiocyanate-generating compounds such as metam sodium. Growers and pesticide applicators should carefully check the labels of nematicides prior to use because pesticide regulations change annually and some compounds are labeled for use only with some crops in some states. Management of soybean cyst nematode with nematicides can be uneconomical or unpopular to some individuals because of human and environmental toxicity, the high cost of the available compounds, the relatively low per hectare value of soybean, and the fact that nematicides usually do not result in season-long control. Nonetheless, soybean productivity can be significantly increased in *H. glycines*-infested soils treated with nematicides (Smith et al., 1991).

Being an obligate parasite, *H. glycines* must feed on living host roots to mature and multiply and cannot reproduce on nonhost crops. Consequently, population densities will decline in soils during any season that nonhost crops are grown (Francl and Dropkin, 1986; Koenning et al., 1993; Schmitt and Riggs, 1991). However, the magnitude of decline throughout a single growing season may vary greatly among geographical locations. Population densities generally decrease similarly regardless of the nonhost crop species grown (Francl and Dropkin, 1986). Short-term use of nonhost crops will not eliminate the nematode from the soil because some eggs are capable of surviving for years in the absence of a host crop (Inagaki and Tsutsumi, 1971). Nevertheless, use of nonhost crops is the most effective means to decrease population densities of *H. glycines* in infested soils. The use of nonhost crops for nematode management is limited by economic constraints of the agricultural production systems in a particular region. Soybean producers are unable to grow nonhost crops for the sole purpose of nematode management; the nonhost crop must have some market value. Unfortunately, multiple successive years of nonhost crops often are needed to reduce *H. glycines* population densities to levels at which soybean may be profitably grown again.

Soybean cultivars with genetic resistance to *H. glycines* are now widely available in maturity groups grown in the USA. Resistant soybean cultivars limit nematode development, resulting in no increase, or at times a decrease, in population densities of the nematode (Chen et al., 2001; Noel and Sikora, 1990; Riggs et al., 1995; Todd et al., 1995; Wang et al., 2000; Wheeler et al., 1997; Young and Hartwig, 1988). Additionally, resistant soybean cultivars also result in greater yields in fields infested with the nematode than yields produced by susceptible cultivars (Chen et al., 2001; Noel and Sikora, 1990; Todd et al., 1995; Wheeler et al., 1997; Young and Hartwig, 1988). The resistance or susceptibility of a soybean cultivar is assessed by determining the percentage of adult females that develop on its roots relative to the number formed on a known, susceptible soybean cultivar. Resistant soybean plants allow 10% or fewer of the number of females that form on the standard susceptible soybean cultivar (Schmitt and Shannon, 1992). Plants that support ≥10% but <30% of the number of females produced on a susceptible cultivar are designated as moderately resistant (Schmitt and Shannon, 1992).

The soybean accessions PI 88788, Peking, and PI 90763 commonly are used as sources of *H. glycines* resistance in soybean cultivars. Additionally, there are a few cultivars with resistance genes from PI 209332 and PI 437654. All of these sources of resistance allow reproduction by a limited number of individuals in most natural soybean cyst nematode populations. The reproduction of even a few individuals on a resistant soybean cultivar creates the potential for selection and eventual increase of that nematode subpopulation (Luedders and Dropkin, 1983; McCann et al., 1982; Young, 1982). PI 437654 appears to be more durable and effective source of resistance than the aforementioned sources (Anand et al., 1985). Resistance genes from PI 437654 have been incorporated into the public cv. Hartwig (Anand, 1992a), and both PI 437654 and Hartwig are currently used as sources of resistance genes in many public and private breeding programs. Initial indications were that PI 437654 was a source of complete resistance to numerous *H. glycines* races (Anand et al., 1985). However, adult females have been observed on the roots of PI 437654 and Hartwig infected by some soybean cyst nematode populations (Davis et al., 1996; Young 1998b, 1999, 2000).

In addition to resistance, tolerance has been proposed as a strategy for management of *H. glycines*. Host plant tolerance to nematodes is defined as the sensitivity of the plant to, or the amount of yield loss incurred due to nematode parasitism (Cook, 1974). Tolerant soybean cultivars have less yield loss than intolerant cultivars when grown in *H. glycines*-infested fields. Numerous susceptible soybean cultivars have been identified as tolerant to *H. glycines* (Anand and Koenning, 1986; Behm et al., 1996; Boerma and Hussey, 1984). Such tolerant cultivars would be useful in an integrated management program when it is desirable to grow a susceptible soybean cultivar to lessen or offset selection for a nematode population capable of reproducing on a resistant cultivar.

Various cultural practices also have been investigated as possible management strategies for the soybean cyst nematode, including manipulation of planting date and tillage. Effects of early or late planting on *H. glycines* population densities and soybean yields vary. Delay of soybean planting in Missouri resulted in exposure of the soybean crop to significantly lower population densities of the nematode, but end-of-season nematode population densities were as great or greater than with soybean planted at the ideal time (Koenning and Anand, 1991). Late planting did not affect *H. glycines* populations densities in Georgia and the effect of the nematode on soybean yield was increased (Hussey and Boerma, 1983). In Arkansas, overall *H. glycines* reproduction was less on early planted (April) soybean plants than on those planted later (June or July), but soybean yields were not consistently correlated with nematode population densities (Riggs et al., 2000). In North Carolina, reproduction of the nematode was less with late planting of shorter season soybean cultivars than planting full-season soybean cultivars on the recommended date (Hill and Schmitt, 1989; Schmitt, 1991). Late planting of soybean cultivars in Kansas did not consistently reduce season-long reproduction of the nematode (Todd, 1993).

The effect of tillage on soybean cyst nematode population densities and soybean yields also varies. Population densities and reproduction of *H. glycines* were greater in soils that were tilled (Workneh et al., 1999) or disturbed (Young, 1987) than in undisturbed or no-till soils in some research. However, tillage had little or no effect on *H. glycines* numbers or soybean yields in studies in Kentucky and Tennessee (Hershman and Bachi, 1995; Baird and Bernard, 1984). Interestingly, *H. glycines* population densities were less on soybean grown in soils in which residue from a preceding wheat (*Triticum aestivum* L.) crop had been incorporated than in unamended soils in the Kentucky and Tennessee research. In contrast, Koenning and Anand (1991) report that a preceding wheat crop had no direct effect on *H. glycines* in research conducted in Missouri.

Although it is not currently a viable management option, there is potential for biological control of the soybean cyst nematode in the future. Numerous soil fungi have been recovered from cysts of the soybean cyst nematode (Carris and Glawe, 1989; Chen et al., 1994), but not all have been shown to adversely affect the nematode in controlled experiments (Chen et al., 1996). Among the most promising microorganisms for biological control of *H. glycines* are Arkansas Fungus 18 (ARF18) (Kim and Riggs 1991), *Hirsutella minnesotensis* (Chen et al., 2000), *H. rhossiliensis* (Chen et al., 1996), and *Verticillium chlamydosporium* (Chen et al., 1996). Soybean cyst nematode population densities were significantly lowered in soils infested with *H. rhossiliensis* and *V. chlamydosporium* relative to densities in untreated soils (Chen et al., 1996). Likewise, reproduction of the nematode was 70% less in field soil infested with ARF18 than in autoclaved field soil (Kim and Riggs, 1991). Another promising potential biological control organism is the soil bacterium *Pasteuria*. This unidentified species of *Pasteuria* was first isolated from second-stage juveniles and males of *H. glycines* in field experiments in Illinois (Noel and Stanger, 1994). The bacterium is capable of becoming established when introduced into a soil and reducing soybean cyst nematode populations substantially (Atibalentja et al., 1998).

16–3 ROOT-KNOT NEMATODES

16–3.1 Distribution

Root-knot nematodes, *Meloidogyne* spp., comprise the second most destructive nematode group on soybean. Seven species, *Meloidogyne arenaria, M. hapla, M. incognita, M. javanica, M. bauruensis, M. inornata,* and *M. trifoliophila* are known to parasitize soybean, and the first four account for 95% of the root-knot nematodes found in agricultural soils by the International *Meloidogyne* Project (Sasser and Carter, 1982). Only *M. incognita* is worldwide in occurrence. *Meloidogyne arenaria, M. hapla,* and *M. javanica,* though able to cause serious damage, are limited in occurrence because of temperature relationships (Sasser, 1977). *Meloidogyne hapla* does not do well in tropical and subtropical areas, and *M. arenaria* and *M. javanica* do their greatest damage in those areas. *Meloidogyne bauruensis* and *M. inornata* have been reported on soybean only in Brazil (Lordello, 1956a, 1956b), and *M. trifoliophila* is a parasite of white clover (*Trifolium repens* L.) that has been shown to parasitize soybean (Bernard and Jennings, 1997). In addition, most root-knot nematode species are found most often in sandy or sandy loam soil (Riggs and Schmitt, 1987).

Root-knot nematodes were first reported to parasitize soybean in 1882 when it was observed in a greenhouse (Franck, 1882). However, the first report on field-grown soybean was in 1922 (McClintock, 1922) and it was not known as a widespread parasite until the 1940s (Atkinson, 1944).

16–3.2 Life Cycle and Disease Development

Root-knot nematodes overwinter as eggs in egg masses attached to root debris or free in the soil (Kinloch, 1982). Second-stage juveniles (J2) that hatch before the soil becomes cold (below 15° C) overwinter in the soil. The J2 penetrate soybean roots, whether susceptible or resistant (Dropkin and Nelson, 1960; Veech and Endo, 1970). The feeding of the nematodes initiates "giant cells", which provide feeding sites for the females to develop to maturity. These cells and associated hyperplastic cells may interrupt the continuity of the xylem and phloem. Such an interruption leads to water and nutrient deficiencies and poor plant growth. The life cycle of the nematode may take as long as 39 d in soybean and each female may produce 300 or more eggs, usually at the root surface (Gommers and Dropkin, 1977). Males are generally few in number. A growing season of 150 d would accommodate as many as four generations, therefore the populations level could increase greatly in one season. In one field test, the population level on 3 June was an average of 11 J2/250 cm^3 soil (R.D. Riggs, unpublished data, 2001) and on 3 October the levels ranged from 4580 to 22 800/250 cm^3 soil with 12 914 in an untreated check. In northern areas where *M. hapla* is more common, fewer generations would occur and possibly lower population levels in the shorter growing season. In coastal and more tropical areas, where *M. arenaria, M. javanica,* and other species may be more common, the number of life cycles would increase as would the final population level if food supply did not become limiting.

16–3.3 Damage and Symptoms

The damage threshold, number of nematode units (eggs or juveniles divided by volume of soil) necessary to cause damage, varies with many factors, including moisture, temperature, soil type, and cultivar. The first indication of damage is not easily seen, because the root galls are formed under ground, but they may be massive (Plate 16–12). The aboveground symptoms are not definitive, and may be similar to N or other nutrient deficiency that results in chlorosis of the leaves. Other symptoms are less definitive such as stunting of plant growth and reduced yield. Plants may be so stunted that few or no pods develop, and in unusual circumstances plants may be killed. Other organisms may interact with the nematodes to result in greater damage than would occur with the nematodes alone (Goswami and Agrawal, 1978). The leaf chlorosis has not been associated with a reduction in dinitrogen (N_2)-fixing nodules, but galling may be so severe that few nodules can be found (Baldwin et al., 1979). The nematode galls usually are not confused with the nodules because the galls usually include a major portion of the root whereas the nodule is attached to the root by a thin connection.

16–3.4 Diagnosis

As mentioned earlier, aboveground symptoms are not diagnostic of damage by root-knot nematodes. The symptoms can be caused by numerous other agents, biological and physiological. However, the root galling is diagnostic. Galls quite often are associated with swelling of the entire circumference of the root where infection by *M. arenaria, M. incognita,* or *M. javanica* occurs. Galls produced when *M. hapla* juveniles infect soybean, and most other hosts, may be quite small (the size of a green pea or smaller) and may appear as a protrusion on the side of a root. These small galls may be the source of numerous adventitious roots.

A soil sample from a field plot will usually provide adequate evidence of the presence of root-knot nematodes except in late fall and winter. The J2 may be extracted from the soil by the roiling and sieving method followed by flotation-centrifugation in a 456 g L^{-1} sucrose solution (Barker, 1982). After J2 are sieved from the sucrose solution, they may be identified as *Meloidogyne* juveniles with a stereoscopic microscope.

The species may be identified from preparations of female perineal patterns observed with a compound microscope, usually at high magnification (Hartman and Sasser, 1982). This is time-consuming and may require examination of several patterns to make a determination. Another procedure for identification of species is the differential host test (Hartman and Sasser, 1982). This test takes longer because it requires that the nematodes be grown to maturity on a series of hosts, about 60 d. The species (within limits) and race then can be determined by the hosts on which the population reproduced and caused galling.

In addition to the morphological and host methods of identification, molecular methods of identification are available to support the other methods (Noe, 1992; Esbenshade and Triantaphyllou, 1985). Identification by this method requires separation of esterase and malate dehydrogenase isozymes by electrophoresis. The variability of perineal patterns and host tests makes the use of the molecular test almost necessary. A table showing the isozyme patterns for the four major species and races of root-knot nematodes attacking soybean plus 14 other populations (seven species and four undescribed species) has been published (Carneiro et al., 2000).

Assessment of disease severity with root-knot nematodes is based on two factors, gall index and egg counts. The gall index may be rated on a scale such as 0 to 4, where: 0 = no galls on the entire root system; 1 = galls confined to 25% or less of the root system; 2 = galls on 26 to 50% of the root system; 3 = galls on 51 to 75% of the root system; and 4 = galls on > 75% of the root system (Stetina et al., 1997); or on a 0 to 5 scale, where: 0 = 0 galls per root system; 1 = 1 to 2 galls, 2 = 3 to10 galls, 3 = 11 to 30 galls, 4 = 31 to 100 galls and 5 = >100 galls per root system (Ritzinger et al., 1998). The severity of galling could be rated as: 0 = no galls; 1 = galls < 3 mm in diameter, no reduction in number of feeder roots; 2 = galls 3 to 10 mm in diameter, no reduction in number of feeder roots; 3 = galls 11 to 20 mm in diameter with no more than slight reduction in feeder roots; 4 = galls > 20 mm in diameter, moderate reduction in feeder roots; and 5 = galls > 20 mm in diameter, severe reduction in number of feeder roots (Stetina et al, 1997). In some cases the relative number of egg masses is included in the gall rating.

Generally, cultivars and breeding lines are screened for resistance to root-knot nematodes in greenhouse tests. Periodically, cultivar resistance screens are published in the *Supplement to the Journal of Nematology*, but in some cases the results of screens are reproduced for local advising but not published for general use. In the last 10 yr of the *Journal of Nematology*, only one report of cultivar tests against root-knot nematode was found (Davis et al, 1996). On the other hand, the Univ. of Georgia annually publishes the results of tests, conducted by R.S. Hussey, of the prominent new soybean cultivars. These are published in a Georgia Agricultural Experiment Station Research Report (Day et al., 2000). These have been published each year for a number of years. The information in some issues is available at www.griffin.peachnet.edu/swvt.

16–3.5 Management of Root-knot Nematodes

Root-knot nematode parasitism of soybean can be managed and the damage reduced by the use of rotations and resistant cultivars (Acosta and Negron, 1982; Crittenden, 1956; Kim et al., 1982; Kinloch and Henson, 1974; Yoshii, 1977; Minton and Parker, 1987; Weaver et al., 1988). Resistant genotypes used in breeding resistant cultivars include 'Avery', 'Forrest', 'Gordon', 'Jackson', D 83-3349, D 86-3429, G 93-9009, G 93-9106, G 93-9223, PI 80466, PI 96354, PI 200538, PI 230977, and PI 417444. Control of root-knot nematodes is aided by maintaining weed-free fields because so many weeds are good hosts of root-knot nematodes. Weeds may maintain the nematode population levels even when non-hosts or resistant cultivars are planted (Kinloch and Rodriguez-Kabana, 1999). Nematicides that are available for use are expensive and their use is seldom profitable. Rotations with graminaceous plants, other than corn, may result in significant reductions in nematode population levels. Even grain sorghum (*Sorghum bicolor* [L.] Moench), which is considered a host of *M. incognita*, is effective in a rotation program if it is planted for only 1 yr. Biological control would be useful on low-value crops such as soybean, but so far it has not been used effectively as a tool. Observations indicate that population levels are reduced by the bacterium *Pasteuria penetrans* or the fungus *Paecilomyces lilacinus*; the results have not been consistent enough to make a recommendation.

16–4 OTHER NEMATODES

16–4.1 Lesion Nematodes

Numerous species of the lesion nematode, *Pratylenchus*, can parasitize soybean, including *P. agilis*, *P. alleni*, *P. brachyurus*, *P. coffee*, *P. crenatus*, *P. hexincisus*, *P. neglectus*, *P. penetrans*, *P. scribneri*, *P. vulnus*, and *P. zeae* (Acosta and Malek, 1979; Ferris and Bernard, 1962; Golden and Rebois, 1978; Nematode Geographical Distribution Committee, Society of Nematologists, 1984; Schmitt and Noel, 1984; Sydenham and Malek, 1989). The most important species affecting soybean are *P. brachyurus*, *P. hexincisis*, and *P. penetrans*. Distribution of lesion nematode species that parasitize soybean in the USA varies from relatively restricted for *P. agilis* and *P. alleni* to widespread for *P. hexincisus*, *P. neglectus*, *P. penetrans*, *P. scribneri*, and *P. vulnus* (Nematode Geographical Distribution Committee, Society of Nematologists, 1984). Several species are reported to contribute to soybean yield losses in the USA (Koenning et al., 1999).

16–4.1.1 Life Cycle

Lesion nematodes are migratory endoparasites of soybean roots, feeding as they move intracellularly throughout the root cortex (Loof, 1991; Rebois and Huettel, 1986). *Pratylenchus agilis* has been reported to feed ectoparasitically on undifferentiated epidermal and root-hair cells in excised soybean root culture (Rebois and Huettel, 1986). Second-stage juveniles hatch from eggs and develop to third- and then fourth-stage juveniles before reaching adulthood. All stages of the nematode can be found in soil and root tissue, and all stages except unhatched eggs are vermiform and capable of infecting soybean roots. Some *Pratylenchus* species reproduce parthenogenetically, others reproduce through mating of males and females (Roman and Triantaphyllou, 1969). Generation times vary among species, but generally take 30 to 60 d (Loof, 1991; Niblack, 1988).

16–4.1.2 Symptoms and damage/Diagnosis

The foliage of soybean plants damaged by feeding of lesion nematodes becomes yellow (chlorotic) and plants appear stunted. Additionally, often there will be the appearance of discrete, necrotic lesions on infected roots. A single nematode can cause a lesion (Acosta and Malek, 1981). Also, the cortex of areas of infected roots can collapse due to feeding by the nematode, creating sunken areas on the root (Acosta and Malek, 1981; Thomason et al., 1976).

Symptoms of lesion nematode damage are not unique and, consequently, cannot be used to conclusively diagnose a problem. One must extract, identify, and enumerate nematodes from root and soil samples collected from fields suspected to be infested to verify an infestation. Lesion nematodes can be recovered from soil and root tissue using standard nematode-extraction techniques (Barker and Niblack, 1990). Morphologically, lesion nematodes are distinguished from other genera of plant-parasitic nematodes by their short but prominent stylet in the anterior region and a tapering, bluntly rounded tail. Species of lesion nematode are distinguished by lip and tail morphology, stylet size, and the presence of males, among other characters (Café Filho and Huang, 1989; Frederick and Tarjan, 1989). Males are common for some species of *Pratylenchus*, but rare for others. Discovery of lesion nematodes in a field does not necessarily mean that the nematode is causing damage to the soybean crop. The nematode must be present at a population density great enough to cause damage (Acosta, 1982; Koenning et al., 1985b) and this damage threshold varies by species.

16–4.1.3 Management

The host range of several lesion nematode species is broad. For example, population densities of *P. scribneri* increased in a field where corn, sorghum, soybean, and wheat was grown (Todd, 1991). Nonetheless, there may be crop species that can be used as a nonhost crop in rotation with soybean for management of the nematode, depending on the *Pratylenchus* species present in the field. But the host range among *Pratylenchus* species varies, so species identification is essential for developing an effective crop rotation scheme utilizing nonhost crops for management of the nematode. Population densities of *P. hexincisus* decrease on tomato (*Lycopersicon esculentum* L.) and white clover whereas those of *P. scribneri* increase on the same crop species (Sydenham and Malek, 1989). Thus, tomato and white clover could serve as nonhosts in rotation with soybean to manage *P. scribneri*, but not *P. hexincisis*.

Reproduction of *Pratylenchus* species (resistance/susceptibility) and plant response to the nematode parasitism (tolerance/intolerance) varies among soybean cultivars and *Pratylenchus* species (Cook, 1974). Also, the extent of reproduction of several *Pratylenchus* species varies on different soybean cultivars (Koenning and Schmitt, 1987; Lindsey and Cairns, 1971; Schmitt and Barker, 1981; Zirakparvar, 1982). Resistance currently is not widely available for management of lesion nematodes. But increased soybean yields from tolerant soybean cultivars relative to intolerant or "sensitive" soybean cultivars has been documented (Koenning and Schmitt, 1987; Schmitt and Barker, 1981), and tolerance is used today as part of an integrated management program for this nematode pest.

The lesion nematode also can be managed by soil-applied nematicides, including aldicarb, carbofuran, and fenamiphos. All three nematicides have been reported to reduce population densities of lesion nematodes and increase soybean yields (Koenning and Schmitt, 1987; Lawn and Noel, 1986; Minton, 1994), although Niblack (1992) saw no effect of aldicarb on lesion nematode population densities or soybean yield.

Some cultural practices affect lesion nematode population densities. For example, *P. brachyurus* population dynamics and soybean yield loss is affected by planting date of the crop. Nematode population densities were greater and soybean yields less with early-planted soybean relative to soybean planted later (Koenning et al., 1985b). Also, delayed planting decreases survival of *P. brachyurus* in the ab-

sence of a host crop (Koenning et al., 1985a). Interestingly, population densities of
P. brachyurus decrease more in the presence of a winter wheat crop than in a winter fallow (Koenning et al., 1985a). Population densities of *P. penetrans* also are reduced significantly by a cover crop of marigolds, *Tagetes tenuifolia* (Kimpinski et
al., 2000).

16–4.2 Reniform Nematode

The reniform nematode, *Rotylenchulus reniformis* Linford and Oliveira,
1940 (Linford and Oliveira, 1940) has been found in 38 countries, most of which
have tropical or subtropical climates (Heald and Thames, 1982). The first observation of *R. reniformis* on soybean in the USA was in 1965 (Fassuliotis and Rau,
1967). Some have thought that *R. reniformis* would not survive winters in temperate latitudes. It appears to be surviving very well as far north as central Arkansas
and has been found in Mississippi county in northeast Arkansas. However, the population level in the Mississippi County field in September was about 5000/550 cm^3
soil whereas in Monroe County populations were regularly at 50 000 to 60 000/550
cm^3 soil in September (T.L. Kirkpatrick, personal communication, 2001). It was reported in the Bootheel of Missouri (southeastern Missouri) (Wrather and Niblack,
1992), but no subsequent population level information is available. *Rotylenchulus
reniformis* also has been found in three counties in northwest Tennessee (M.E. Newman, personal communication, 2001). They were first found in the early 1990s and
the area of infestation continues to expand. These counties are across the Mississippi River from the Mississippi County, Arkansas, and Missouri Bootheel infestations.

16–4.2.1 Life Cycle and Disease Cycle

Second-stage juveniles emerge from the eggs and mature to males and preadult females in the soil. The preadult females penetrate the root surface but the posterior end remains outside the root tissue. The posterior end enlarges (Plate 16–13)
and becomes enveloped in a gelatinous matrix in which eggs are deposited. Although
R. reniformis can build to very high population levels in a growing season, the numbers of eggs produced by single females on soybean range from 59 to 72 (Rebois,
1973; Lim and Castillo, 1979). The life cycle is completed in 19 d, which facilitates the production of several generations per year and accounts for the large population increases possible in a single growing season (Rebois, 1973). Males are necessary for reproduction in some populations but not in others. When the nematodes
begin to feed on endodermal cells, these cells become syncytial cells (Rebois, 1973;
Rebois et al., 1975). As many as 200 cells may be incorporated into the syncytium.
Invasion is greatest when soil moisture is near the water-holding capacity of the soil
and is less under wet or dry conditions. A temperature of 29°C appears to be optimum for penetration. However, populations vary in their ability to reproduce on certain hosts (McGawley and Overstreet, 1995). When 17 populations from five states
were tested on two cotton (*Gossypium hirsutum* L.) and two soybean cultivars, the
high and low reproductive factor (R = final population level/initial population

level) values were: 'Deltapine 90' cotton, 68.5 and 8.2; 'Deltapine 41' cotton, 29.4 and 9.3; 'Bragg' soybean, 185 and 57.2; and 'Hartz 6686' soybean, 176.3 and 43.9, respectively. Eleven and 14 of the populations caused reductions in root weight of cotton and soybean, respectively.

16–4.2.2 Damage and Symptoms

The nematodes feed in the roots of their host and initiate syncytia. These syncytia are metabolic sinks into which metabolites flow and furnish food for the nematode feeding in the syncytium. The root system becomes necrotic and stunted and, as a result, the aboveground portion of the plant becomes stunted and chlorotic (Singh, 1975). In addition, the reduction in seed yield may be as much as 33% (Rebois, 1971). The nematodes increase the seedling disease caused by *Rhizoctonia solani*.

The most prominent signs that could be used to indicate the presence of nematodes are the egg masses covered with sand particles, attached to the roots. The soil/sand particles are not easily shaken from the roots and they give the root system a characteristic appearance (Plate 16–14). The egg masses have a different appearance than those attached to the posterior ends of root-knot nematodes. When the egg masses are removed the female body is exposed whereas with root-knot nematodes the female body is usually embedded in the root tissue even after the egg mass is removed.

16–4.2.3 Management

As with root-knot nematodes, the use of nematicides to reduce damage by *R. reniformis* is not cost-effective even if effective nematicides were available. Cultivars with resistance to *R. reniformis* are available but the resistance level may not be sufficient to prevent yield reductions. Soybean cultivars that have resistance to soybean cyst nematodes from the cv. Peking or from the accession PI 437654 generally are resistant to *R. reniformis*. For comparison, in a test where the number of nematodes in a pot with a susceptible soybean cultivar ranged from 12 545 to 49 465, the number on Forrest was 6034 and on Hartwig was 5025 and the number in the pots with no plant was 636 (Robbins et al., 1999). Obviously, a moderate population level is maintained on Forrest and Hartwig and some damage would result at that level, but relative to many other cultivars, they are resistant. Among the cultivars tested in Maturity Groups IV to VI, 93 were classed as resistant and 189 were susceptible. Of 45 soybean lines reported to be resistant to soybean cyst nematode, 16 were resistant to *R. reniformis* (R = 2.8–15.3) whereas 29 were susceptible (R = 60.4–265.5) (Robbins and Rakes, 1996). Lee 74, the standard cultivar susceptible to soybean cyst nematode had R = 190.0, and Braxton had R = 308.1. Rotation is a practical and economical method for managing reniform nematode (Kirkpatrick et al., 2000). Crop plants that are poor hosts or nonhosts that could be used in a rotation include grain sorghum, corn, rice, and peanut. Wheat and oat also are good rotation crops, but the greatest benefit from these would be obtained only if the field were fallowed for the remainder of the growing season following grain harvest (McGawley and Overstreet, 1999).

16–4.3 Lance Nematodes

Lance nematodes, *Hoplolaimus* spp., can be serious pests of soybean (Appel and Lewis, 1984; Lewis and Smith, 1976; Lewis et al., 1976) (Plate 16–15). The most pathogenic species of *Hoplolaimus* that infects soybean is *H. columbus* (Fassuliotis, 1974). Soybean also is a host of *H. galeatus*, but damage to the crop by this nematode species is not well documented. *Hoplolaimus columbus* occurs in several states in the southeastern USA; *H. galeatus* is more widely distributed throughout most of the soybean-producing states in the country (Nematode Geographical Distribution Committee, Society of Nematologists, 1984). *Hoplolaimus columbus* has been calculated to cause 30 to 48% losses in soybean yield in Georgia (Noe et al., 1991, 1993).

16–4.3.1 Life Cycle

In general, the lance nematode feeds completely within the soybean root tissue, as a migratory endoparasite, but it can feed ectoparasitically as well (Plate 16–16). Vermiform, second-stage juveniles hatch from the eggs in soil or roots and develop through two subsequent juvenile stages before becoming adults. All juvenile and adult stages of *Hoplolaimus* are vermiform, occur in soil and roots, and are capable of infecting soybean roots. Females of *H. columbus* reproduce parthenogenetically; males occur rarely, and mating is not required for reproduction. At 21 to 27°C, *H. columbus* completes a generation in 45 to 49 d (Schmitt and Noel, 1984). The nematode feeds upon cortex, endodermal, pericycle, and phloem cells within the infected root (Fortuner, 1991).

16–4.3.2 Symptoms and Damage

Lance nematode damage to soybean causes foliage to become yellow (chlorotic) and plants to be stunted. Additionally, parasitism by the nematode reduces pod production of the soybean plant (Lewis et al., 1976). Brown, necrotic lesions and collapse of cortical cells occur in areas of the root fed upon by large numbers of the nematode (Lewis et al., 1976). Secondary root formation also can be reduced. None of the aforementioned symptoms are unique or diagnostic of damage from feeding by the lance nematode.

To conclusively diagnose damage caused by lance nematode parasitism to soybean, the nematode must be extracted, identified, and counted from soil and root samples collected from an affected area. Any of the techniques for extracting nematodes from soil and plant tissue will work in recovering lance nematodes. Lance nematodes are characterized by the overall large, stout appearance of all vermiform stages of the nematode. Additionally, the stylet of *Hoplolaimus* species is very prominent and has unique, tulip-bulb-shaped knobs at the base (Plate 16–15). The mere presence of lance nematodes in a field in which soybean is growing does not necessarily mean that the crop has been damaged by the nematode. The nematode population must exceed a threshold before damage can be measured (Noe et al., 1991).

16–4.3.3 Management

The lance nematode is polyphagus, capable of feeding on numerous crop species (Lewis and Smith, 1976). Consequently, use of other crops as nonhosts for management of lance nematodes is limited. However, good weed control will provide some management benefit because several weed species are hosts for the nematode (Fassuliotis, 1974).

Reproduction of the lance nematode (resistance/susceptibility) and damage caused by the nematode (tolerance/intolerance) on soybean cultivars varies (Lewis et al., 1976; Mueller and Sanders, 1987; Mueller et al., 1988; Nyczepir and Lewis, 1979; Schmitt and Imbriani, 1987). Tolerant cultivars, in particular, have been used in management of the nematode (Lewis et al., 1976; Mueller and Sanders, 1987; Mueller et al., 1988; Nyczepir and Lewis, 1979; Schmitt and Imbriani, 1987).

Nematicides also have worked well in managing lance nematodes. Schmitt and Imbriani (1987) found that 1,3-dichloropropene, aldicarb, carbofuran, and fenamiphos reduced *H. columbus* population densities and increased soybean yields. Mueller and Sanders (1987) had similar results with aldicarb. However, in field research in North Carolina, aldicarb and fenamiphos alone did not affect lance nematode population densities or soybean yields significantly and consistently, although a combination of both nematicides reduced nematode population densities and increased soybean yields (Schmitt and Bailey, 1990).

16–4.4 Sting Nematodes

The designation "sting nematode" has been applied to all species in the genus *Belonolaimus*. The designation may more properly apply to *B. longicaudatus*, the root symptoms of which give the appearance that the root has been "stung" (Plate 16–17). For this discussion we will use the broader concept. In 1991, nine species were listed (Essary, 1991). *Belonolaimus longicaudatus* was synonomized with *B. gracilis* but some still feel they are distinct species (R.T. Robbins, personal communication, 2001) Of the nine species listed only two, *B. gracilis* and *B. nortoni*, have been studied to any extent and only *B. nortoni* has been reported to parasitize soybean (Graham and Holdeman, 1953; Tomerlin and Perry, 1967). In 1961, Riggs (1961) reported the occurrence of *B. longicaudatus* in a single soybean field in Arkansas but this was later identified as *B. nortoni*. Generally, sting nematodes have a broad host range.

Sting nematodes are found only in deep soils that contain about 85% or more sand. They reproduce at temperatures up to 34°C, but not higher. They are ectoparasites and may kill all the secondary roots on a major root leaving a "shoe string" effect. The feeding sites may be large sunken lesions that may eventually girdle the entire root, rendering the distal part of the root useless to the plant and it may become detached leaving a root stub (Shurtleff and Averre, 2000). Shoots show symptoms of nutrient and water deficiency and become stunted. Because the host range is so extensive and the damage so severe, sting nematode would be a major problem were its occurrence not confined to deep sandy soils.

There are no *Belonolaimus*-resistant soybean cultivars. Nematicides and other management tactics, such as weed control, may help alleviate problems. Because of the nematode's large size and sensitivity to soil moisture, irrigation increases the difficulty of managing sting nematodes.

16–5 Issues in Nematode Management

Without question, the major problem in managing soybean-parasitic nematodes is obtaining a proper diagnosis. Nematodes are often overlooked as the cause of yield loss in soybean production. As we have stated many times in the previous text, the aboveground symptoms of nematode injury are nonspecific. Because of this, the symptoms are frequently ascribed to other causes. With certain exceptions, such as severe galling caused by root-knot nematodes, even the root symptoms of nematode injury are not easy to identify. As the causes of additional plant stress, or as direct participants in disease interactions, nematodes may be involved in disease situations that seem to have another obvious cause. Most soybean-parasitic nematodes can be controlled with appropriate management practices, but the identification of such practices hinges on an accurate diagnosis.

Nematodes are ubiquitous in cultivated fields. Soybean is no exception. We have observed that nematodes are often considered the diagnosis of last resort even though the distributions and damage potential of soybean-pathogenic nematodes are familiar to most diagnostic practitioners. For the reasons outlined in the previous paragraph, diagnosis of soybean problems should include sampling for nematodes and professional identification of those present in many more cases than is the case at present.

REFERENCES

Acedo, J.R., and V.H. Dropkin. 1982. Technique for obtaining eggs and juveniles of *Heterodera glycines*. J. Nematol. 14:418–420.

Acosta, N. 1982. Influence of inoculum level and temperature on pathogenicity and population development of lesion nematodes on soybean. Nematropica 12:189–197.

Acosta, N., and R.B. Malek. 1979. Influence of temperature on population development of eight species of *Pratylenchus* on soybean. J. Nematol. 11:229–232.

Acosta, N., and R.B. Malek. 1981. Symptomatology and histopathology of soybean roots infected by *Pratylenchus scribneri* and *P. alleni*. J. Nematol. 13:6–12.

Acosta, N., and J.A. Negron. 1982. Susceptibility of six soybean cultivars to *Meloidogyne incognita* race 4. Nematropica 12:181–187.

Adeniji, M.O., D.I. Edwards, J.B. Sinclair, and R.B. Malek. 1975. Interrelationship of *Heterodera glycines* and *Phytophthora megasperma* var. *sojae* in soybeans. Phytopathology 65:722–725.

Anand, S.C. 1992a. Registration of 'Hartwig' soybean. Crop Sci. 32:1069–1070.

Anand, S.C. 1992b. Registration of 'Delsoy 4710' soybean. Crop Sci. 32:1294

Anand, S.C., and S.R. Koenning. 1986. Tolerance of soybean to *Heterodera glycines*. J. Nematol. 18:195–199.

Anand, S.C., J.A. Wrather, and C.R. Shumway. 1985. Soybean genotypes with resistance to races of soybean cyst nematode. Crop Sci. 25:1073–1075.

Appel, J.A., and S.A. Lewis. 1984. Pathogenicity and reproduction of *Hoplolaimus columbus* and *Meloidogyne incognita* on 'Davis' soybean. J. Nematol. 16:349–355.

Atibalentja, N., G.R. Noel, T.F. Liao, and G.Z. Gertner. 1998. Population changes in *Heterodera glycines* and its bacterial parasite *Pasteuria* sp. in naturally infested soil. J. Nematol. 30:81–92.

Atkinson, R.E. 1944. Diseases of soybeans and peanuts in the Carolinas in 1943. Plant Dis. Rep. Suppl. 148:254–259.

Baird, S.M., and E.C. Bernard. 1984. Nematode population and community dynamics in soybean-wheat cropping and tillage regimes. J. Nematol. 16:379–386.

Baldwin, J.G., K.R. Barker, and L.A. Nelson. 1979. Effects of *Meloidogyne incognita* on soybean in microplots. J. Nematol. 11:156–161.

Barker, K.R.. 1982. Nematode extraction and bioassays. p. 19–35. *In* J.N. Sasser and C.C. Carter (ed.) An advanced treatise on *Meloidogyne*. Vol II. Methodology. North Carolina State Univ. Graphics, Raleigh, NC.

Barker, K.R., and T.L. Niblack. 1990. Soil sampling methods and procedures for field diagnosis. p. 10–19. *In* B.M. Zuckerman et al. (ed.) Plant nematology laboratory manual. The Univ. of Massachusetts Agric. Exp. Stn., Amherst.

Barker, K.R., J.L. Stan, and D.P. Schmitt. 1987. Usefulness of egg assays in nematode population-density determinations. J. Nematol. 19:130–134.

Behm, J.E., G.L. Tylka, and S.R. Cianzio. 1996. Evaluations of soybean genotypes for tolerance to soybean cyst nematode. Phytopathology 86(S):98.

Bernard, E.C., and P.L. Jennings. 1997. Host range and distribution of the clover root-knot nematode, *Meloidogyne trifoliophila*. J. Nematol. Suppl. 29:662–672.

Bird, D.McK., and D.L. Riddle. 1994. A genetic nomenclature for parasitic nematodes. J. Nematol. 26:138–143.

Brim, C.A., and J.P. Ross. 1966. Registration of Pickett soybeans. Crop Sci. 6:305.

Boerma, H.R., and R.S. Hussey. 1984. Tolerance to *Heterodera glycines* in soybean. J. Nematol. 16:289–296.

Byrd, D.W., Jr., K.R. Barker, H. Ferris, C.J. Nusbaum, W.E. Griffin, R.H. Small, and C.A. Stone. 1976. Two semi-automatic elutriators for extracting nematodes and certain fungi from soil. J. Nematol. 8:206–212.

Café Filho, A.C., and C.S. Huang. 1989. Description of *Pratylenchus pseudofallax* n. sp. with a key to species of the genus *Pratylenchus* Filipjev, 1936 (Nematoda: Pratylenchidae). Rev. Nematol. 12:7–15

Carneiro, R.M.D.G., M.R.A. Almeida, and P. Quineharve. 2000. Enzyme phenotypes of *Meloidogyne* spp. populations. Nematology 2:645–654.

Carris, L.M., and D.A. Glawe. 1989. Fungi colonizing cysts of *Heterodera glycines*.Bull. 786. Ill. Agric. Exp. Stn., Urbana-Champaign.

Chen, S.Y., D.W. Dickson, J.W. Kimbrough, R. McSorley, and D.J. Mitchell. 1994. Fungi associated with females and cysts of *Heterodera glycines* in a Florida soybean field. J. Nematol. 26:296–303.

Chen, S.Y., D.W. Dickson, and D.J. Mitchell. 1996. Pathogenicity of fungi to eggs of *Heterodera glycines*. J. Nematol. 28:148–158.

Chen, S.Y., X.Z. Liu, and F.J. Chen. 2000. *Hirsutella minnesotensis* sp. nov., a new pathogen of the soybean cyst nematode. Mycologia 92:819–824.

Chen, S.Y., P.M. Porter, J.H. Orf, C.D. Reese, W.C. Stienstra, N.D. Young, D.D. Walgenbach, P.J. Schaus, T.J. Arlt, and F.R. Breitenbach. 2001. Soybean cyst nematode population development and associated soybean yields of resistant and susceptible cultivars in Minnesota. Plant Dis. 85:760–766.

Cook, R. 1974. Nature and inheritance of nematode resistance in cereals. J. Nematol. 6:165–174.

Crittenden, H.W. 1956. Control of *Meloidogyne incognita acrita* by crop rotations. Plant Dis. Rep. 40:977–980.

Davis, E. L., R.S. Hussey, T.J. Baum, J. Bakker, A. Schots, M.N. Rosso, and P. Abad. 2000. Nematode parasitism genes. Annu. Rev. Phytopathol. 38:365–396.

Davis, E.L., S.R. Koenning, J.W. Burton, and K.R. Barker. 1996. Greenhouse evaluation of selected soybean germplasm for resistance to North Carolina populations of *Heterodera glycines, Rotylenchulus reniformis* and *Meloidogyne* species. J. Nematol. Suppl. 28:590–598.

Day, J.L., A.E. Coy, and P.A. Rose (ed.) 2000. 2000 Soybean, sorghum grain and silage, grain millet, sunflower and summer annual forage performance tests. Res. Rep. 670:1–94. Univ. of Georgia, Athens.

Doupnik, B. 1993. Soybean production and disease loss estimates for north central United States from 1989 to 1991. Plant Dis. 77:1170–1171.

Dropkin, V.H., and P.E. Nelson. 1960. The histopathology of root-knot nematode infection in soybeans. Phytopathology 50:442–447.

Endo, B.Y. 1991. Ultrastructure of initial responses of susceptible and resistant soybean roots to infection by *Heterodera glycines*. Rev. Nematol. 14:73–94.

Esbenshade, P.R., and A.C. Triantaphyllou. 1985. Use of enzyme phenotypes for identification of *Meloidogyne* species. J. Nematol. 17:6–20.

Essary, B.A. 1991. Catalog of the Order Tylenchida (Nematoda). Agriculture Canada, Ottawa, ON.

Faghihi, J., and J.M. Ferris. 2000. An efficient new device to release eggs from *Heterodera glycines*. J. Nematol. 32:411–413.

Faghihi, J., J.M. Ferris, and V.R. Ferris. 1986. *Heterodera glycines* in Indiana: I. Reproduction of geographical isolates on soybean differentials. J. Nematol. 18:169–172.

Fassuliotis, G. 1974. Host range of the Columbia lance nematode, *Hoplolaimus columbus*. Plant Dis. 58:1000–1002.

Fassuliotis, G., and G.J. Rau. 1967. The reniform nematode in South Carolina. Plant Dis. Rep. 51:557.

Ferris, V.R., and R.L. Bernard. 1962. Injury to soybeans caused by *Pratylenchus alleni*. Plant Dis. Rep. 46:181–184.

Fortuner, R. 1991. The Hoplolaiminae. p. 669–719. *In* W.R. Nickle (ed.) Manual of agricultural nematology. Marcel Dekker, Inc., New York.

Franck, A.B. 1882. Gallender *Anguillula radicicola* Greff on *Soja hispida, Medicago sativa, Lactuca sativa, unt Pirus communis*. Ver. Bot. Vereins. Der. Prov. Brandenberg 23:45–55.

Francl, L.J., and V.H. Dropkin. 1986. *Heterodera glycines* population dynamics and relation of initial population to soybean yield. Plant Dis. 70:791–795.

Frederick, J.J., and A.C. Tarjan. 1989. A compendium of the genus *Pratylenchus* Filipjev, 1936 (Nemata: Pratylenchidae). Rev. Nematol. 12:243–256.

Gavassoni, W.L., G.L. Tylka, and G.P. Munkvold. 2001. Relationships between tillage and spatial patterns of *Heterodera glycines*. Phytopathology 91:534–545.

Gerdemann, J.W. 1955. Relation of a large soil-borne spore to phycomycetous mycorrhizal infections. Mycologia 47:619–632.

Golden, A.M., and R.V. Rebois. 1978. Nematodes on soybeans in Maryland. Plant Dis. 62:430–432.

Gommers, F.J., and V.H. Dropkin. 1977. Quantitative histochemistry of nematode-induced transfer cells. Phytopathology 67:869–873.

Goswami, B.K., and D.K. Agrawal. 1978. Interrelationships between species of *Fusarium* and root-knot nematode, *Meloidogyne incognita*, in soybean. Nematologia Mediterranea 6:125–128.

Graham, T.W., and Q.L. Holdeman. 1953. The sting nematode, *Belonolaimus gracilis*, on cotton and other crops in South Carolina. Phytopathology 43:434–439.

Hartman, K.M., and J.N. Sasser. 1982. Identification of *Meloidogyne* species on the basis of differential host test and perineal-pattern morophology.p. 70–77. *In* J.N. Sasser and C.C. Carter (ed.) An advanced treatise on *Meloidogyne*. Vol. II. Methodology. North Carolina State Univ. Graphics, Raleigh, NC.

Hartwig, E.E., and J.M.Epps. 1978. Registration of Bedford soybeans. Crop Sci. 18:915.

Hartwig, E.E., and L.D. Young. 1990. Registration of 'Cordell'soybean. Crop Sci. 30:231.

Heald, C.M., and W.H. Thames. 1982. The reniform nematode, *Rotylenchulus reniformis*. p. 139–143. *In* R.D. Riggs (ed.) Nematology in the Southern Region of the United States. Southern Coop. Ser. Bull. 276. Ark. Agric. Exp. Stn., Fayetteville.

Hershman, D.E., and P.R. Bachi. 1995. Effect of wheat residue and tillage on *Heterodera glycines* and yield of doublecrop soybean in Kentucky. Plant Dis. 79:631–633.

Hershman, D.E., J.W. Hendrix, R.E. Stuckey, P.R. Bachi, and G. Henson. 1990. Influence of planting date and cultivar on soybean sudden death syndrome in Kentucky. Plant Dis. 74:761–766.

Hill, N.S., and D.P. Schmitt. 1989. Influence of temperature and soybean phenology on dormancy induction of *Heterodera glycines*. J. Nematol. 21:361–369.

Hussey, R.S., and H.R. Boerma. 1983. Influence of planting date on damage to soybean caused by *Heterodera glycines*. J. Nematol. 15:253–258.

Ichinohe, M. 1955. Studies on the morphology and ecology of the soybean cyst nematode, *Heterodera glycines*. Report of the Hokkaido Agric. Exp. Stn. 48:1–64.

Inagaki, H., and M. Tsutsumi. 1971. Survival of the soybean cyst nematode, *Heterodera glycines* Ichinohe (Tylenchida: Heteroderidae) under certain storing conditions. Appl. Entomol. Zool. 6:156–162.

Johnson, A.B., K.S. Kim, R.D. Riggs, and H.D. Scott. 1993. Location of *Heterodera glycines*-induced syncytia in soybean as affected by soil water regimes. J. Nematol. 25: 422–426.

Kim, D.G., D.R. Choi, and Y.E. Choi. 1982. Resistance of soybean cultivars to root-knot nematode species in Korea. Korean J. Plant Prot. 21:34–37.

Kim, D.G., and R.D. Riggs. 1991. Characteristics and efficacy of a sterile Hyphomycete (ARF18), a new biocontrol agent for *Heterodera glycines* and other nematodes. J. Nematol. 23:275–282.

Kimpinski, J., W.J. Arsenault, C.E. Gallant, and J.B. Sanderson. 2000. The effect of marigolds (*Tagetes* spp.) and other cover crops on *Pratylenchus penetrans* and on following potato crops. Suppl. J. Nematol. 32:531–536.

Kinloch, R.A. 1982. The relationships between soil population of *Meloidogyne incognita* and yield reduction of soybean in the coastal plain. J. Nematol. 14:162–167.

Kinloch, R.A., and K. Henson. 1974. Comparative resistance in soybean breeding lines to *Meloidogyne javanica*. Nematropica 4:17–18.

Kinloch, R.A., and R. Rodriguez-Kabana. 1999. Root-knot nematodes. p. 55–57. *In* G.L. Hartman et al. (ed.) Compendium of soybean disease. 4th ed. APS Press, St. Paul, MI.

Kirkpatrick, T.L., R.D. Riggs, and R.T. Robbins. 2000. Nematode control. p. 64–69. *In* Soybean Commodity Committee (ed.). Arkansas soybean handb. M P 197. Univ. of Arkansas Coop. Ext. Serv., Little Rock.

Koenning, S.R., and S.C. Anand. 1991. Effects of wheat and soybean planting date on *Heterodera glycines* population dynamics and soybean yield with conventional tillage. Plant Dis. 75:301–304.

Koenning, S.R., C. Overstreet, J.W. Noling, P.A. Donald, J.O. Becker, and B.A. Fortnum. 1999. Survey of crop losses in response to phytoparasitic nematodes in the United States for 1994. Suppl. J. Nematol. 31(4S):587–618.

Koenning, S.R., and D.P. Schmitt. 1987. Control of *Pratylenchus brachyurus* with selected nonfumigant nematicides on a tolerant and a sensitive soybean cultivar. Ann. Appl. Nematol. 1:26–28.

Koenning, S.R., D.P. Schmitt, and K.R. Barker. 1993. Effects of cropping systems on population density of *Heterodera glycines* and soybean yield. Plant Dis. 77:780–786.

Koenning, S.R., D.P. Schmitt, and K.R. Barker. 1985a. Influence of selected cultural practices on winter survival of *Pratylenchus brachyurus* and subsequent effects on soybean yield. J. Nematol. 17:464–469.

Koenning, S.R., D.P. Schmitt, and K.R. Barker. 1985b. Influence of planting date on population dynamics and damage potential of *Pratylenchus brachyurus* on soybean. J. Nematol. 17:428–434.

Lauritis, J.A., R.V. Rebois, and L.S. Graney. 1983. Development of *Heterodera glycines* Ichinohe on soybean, *Glycine max* (L.) Merr., under gnotobiotic conditions. J. Nematol. 15:272–281.

Lawn, D.A., and G.R. Noel. 1986. Field interrelationships among *Heterodera glycines*, *Pratylenchus scribneri*, and three other nematode species associated with soybean. J. Nematol. 18:98–106.

Lewis, S.A., and F.H. Smith. 1976. Host plants, distribution, and ecological associations of *Hoplolaimus columbus*. J. Nematol. 8:264–270.

Lewis, S.A., F.H. Smith, and W.W. Powell. 1976. Host-parasite relationships of *Hoplolaimus columbus* on cotton and soybean. J. Nematol. 8:141–145.

Lim, B.K., and M.B. Castillo. 1979. Screening soybeans for resistance to reniform nematode disease in the Phillipines. J. Nematol. 11:275–282.

Lindsey, D.W., and E.J. Cairns. 1971. Pathogenicity of the lesion nematode, *Pratylenchus brachyurus*, on six soybean cultivars. J. Nematol. 3:220–226.

Linford, M.B., and J.M. Oliveira. 1940. *Rotylenchulus reniformis*, nov. gen., n. sp., a nematode parasite of roots. Proc. Helm. Soc. Wash.7:35–42.

Loof, P.A.A. 1991. The family Pratylenchidae Thorne, 1949. p. 363–421. *In* W.R. Nickle (ed.) Manual of agricultural nematology. Marcel Dekker, Inc., New York.

Lordello, L.G.E. 1956a. Nematoides que parasitam a soja na regaiode Bavru. Bragantia 15 (6):55–64.

Lordello, L.G.E. 1956b. *Meloidogyne inornata* sp. n., a serious pest of soybean in the state of Sao Paulo, Brazil (Nematoda, Heteroderidae). Rev. Brasilian Biol. 16:65–70.

Luedders, V.D., and V.H. Dropkin. 1983. Effect of secondary selection on cyst nematode reproduction on soybeans. Crop Sci. 23:263–264.

McCann, J., V.D. Luedders, and V.H. Dropkin. 1982. Selection and reproduction of soybean cyst nematodes on resistant soybeans. Crop Sci. 22:78–80.

McClintock, J.A. 1922. Resistant plants for root-knot nematode control. Georgia Agric. Exp. Stn. Circ. 77:1

McGawley, E.C., and C. Overstreet. 1995. Reproduction and pathological variation in populations of *Rotylenhulus reniformis*. J. Nematol. 27:508 (abstr).

McGawley, E.C., and C. Overstreet. 1999. Reniform nematode. p. 54–55. *In* G.L. Hartman et al. (ed.) Compendium of soybean diseases. 4th ed. APS Press, St. Paul, MN.

McLean, K.S., and G.W. Lawrence. 1993. Interrelationship of *Heterodera glycines* and *Fusarium solani* in sudden death syndrome of soybean. J. Nematol. 25:434–439.

Mendes, M.L., and D.W. Dickson. 1993. Detection of *Heterodera glycines* on soybean in Brazil. Plant Dis. 77:499–500.

Minton, N.A. 1994. Effects of small grain crops, aldicarb, and *Meloidogyne incognita* resistant soybean on nematode populations and soybean production. Nematropica 24:7–15.

Minton, N.A., and M.B. Parker. 1987. Root-knot nematode management and yield of soybean as affected by winter cover crop, tillage systems, and nematicides. J. Nematol. 19:38–43.

Mueller, J.D., and G.B. Sanders. 1987. Control of *Hoplolaimus columbus* on late-planted soybean with aldicarb. Ann. Appl. Nematol. 1:127–128.

Mueller, J.D., D.P. Schmitt, G.C. Weiser, E.R. Shipe, and H.L. Musen. 1988. Performance of soybean cultivars in *Hoplolaimus columbus*-infested fields. Ann. Appl. Nematol. 2:65–69.

Mulvey, R.H., and A.M. Golden. 1983. An illustrated key to the cyst-forming genera and species of Heteroderidae in the western hemisphere with species morphometrics and distribution. J. Nematol. 15:1–59.

Nematode Geographical Distribution Committee, Society of Nematologists. 1984. Distribution of plant-parasitic nematode species in North America. Soc. of Nematologists, Marceline, MO.

Niblack, T.L. 1988. Soybean nematodes in the north central United States. p. 87–91. In Soybean diseases of the North Central Region. APS Press, St. Paul, MN.

Niblack, T.L. 1992. *Pratylenchus, Paratylenchus, Helicotylenchus*, and other nematodes on soybean in Missouri. Suppl. J. Nematol. 24:738–744.

Niblack, T.L., P.R. Arelli, G.R. Noel, C.H. Opperman, J. Orf, D.P. Schmitt, J.G. Shannon, and G.L. Tylka. 2002. A revised classification scheme for genetically diverse populations of *Heterodera glycines*. J. Nematol. 34:279–288.

Niblack, T.L., R.D. Heinz, G.S. Smith, and P.A. Donald. 1993. Distribution, density, and diversity of *Heterodera glycines* in Missouri. Suppl. J. Nematol. 25:880–886.

Nickell, C.D., G.R Noel, R.L. Bernard, D.J Thomas, and K. Frey. 1994a. Registration of soybean germplasm line 'LN89-5717' resistant to soybean cyst nematode. Crop Sci. 34:1133.

Nickell, C.D., G.R. Noel, R.L. Bernard, D.J Thomas, and J. Pracht. 1994b. Registration of soybean germplasm line 'LN89-5612' moderately resistant to soybean cyst nematode. Crop Sci. 34:1134.

Noe, J.P. 1992. Variability among populations of *Meloidogyne arenaria*. J. Nematol. 24:404–414.

Noe, J.P. 1993. Damage functions and population changes of *Hoplolaimus columbus* on cotton and soybean. J. Nematol. 25:440–445.

Noe, J.P., J.N. Sasser, and J.L. Imbriani. 1991. Maximizing the potential of cropping systems for nematode management. J. Nematol.23:353–361.

Noel, G.R. 1992. History, distribution, and economics. p. 1–14. In R.D. Riggs and J.A. Wrather (ed.) Biology and management of the soybean cyst nematode. APS Press, St. Paul.

Noel, G.R., and E.J. Sikora. 1990. Evaluation of soybeans in maturity groups I – IV for resistance to *Heterodera glycines*. Suppl. J. Nematol. 22:795–799.

Noel, G.R., and B.A. Stanger. 1994. First report of *Pasteuria* sp. attacking *Heterodera glycines* in North America. J. Nematol. 26:612–615.

Nyczepir, A.P., and S.A. Lewis. 1979. Relative tolerance of selected soybean cultivars to *Hoplolaimus columbus* and possible effects of soil temperature. J. Nematol. 11:27–31.

Pratt, P.W., and J.A. Wrather. 1998. Soybean disease loss estimates for the southern United States, 1994-1996. Plant Dis. 82:114–116.

Rebois, R.V. 1973. Effect of soil temperature on infectivity and development of *Rotylenchulus reniformis* on resistant and susceptible soybeans, *Glycine max*. J. Nematol. 5:10–13.

Rebois, R.V., and R.N. Huettel. 1986. Population dynamics, root penetration, and feeding behavior of *Pratylenchus agilis* in monoxenic root cultures of corn, tomato, and soybean. J. Nematol. 18:392–397.

Rebois, R.V., P.A. Madden, and B.J. Eldridge. 1975. Some ultrastructural changes induced in resistant and susceptible soybean roots following infection by *Rotylenchus reniformis*. J. Nematol. 7:122–139.

Riggs, R.D. 1961. Sting nematode in Arkansas. Plant Dis. Rep. 45:392.

Riggs, R.D., and T.L. Niblack. 1993. Nematode pests of oilseed crops and grain legumes. p. 209–258. In K. Evans et al. (ed.) Plant parasitic nematodes in temperate agriculture. CAB Int., Wallingford, UK.

Riggs, R.D., L. Rakes, and D. Dombek. 1995. Responses of soybean cultivars and breeding lines to races of *Heterodera glycines*. Suppl. J. Nematol. 27:592–601.

Riggs, R.D., and D.P. Schmitt. 1987. Nematodes. p. 757–778. In J.R. Wilcox (ed.) Soybeans: Improvement, production, and uses. 2nd ed. ASA, Madison, WI.

Riggs, R.D., and D.P. Schmitt. 1988. Complete characterization of the race scheme for *Heterodera glycines*. J. Nematol. 20:392–395.

Riggs, R.D., J.A. Wrather, A. Mauromoustakos, and L. Rakes. 2000. Planting date and soybean cultivar maturity group affect population dynamics of *Heterodera glycines*, and all affect yield of soybean. J. Nematol. 32:334–342.

Ritzinger, C.H.S.P., R. McSorley, and R.N. Gallaher. 1998. Effect of *Meloidogyne arenaria* and mulch type on okra in microplot experiments. J. Nematol. Suppl. 30:616–623.

Robbins, R.T., L. Rakes, L.E. Jackson, and D.G. Dombeck. 1999. Reniform nematode resistance in selected soybean cultivars. J. Nematol. Suppl. 31:667–677.

Roman, J., and A.C. Triantaphyllou. 1969. Gametogenesis and reproduction of seven species of *Pratylenchus*. J. Nematol. 1:357–362.

Ross, J.P. 1965. Predisposition of soybean to Fusarium wilt by *Heterodera glycines* and *Meloidogyne incognita*. Phytopathology 55:361–364.

Roy, K.W. 1997. *Fusarium solani* on soybean roots: Nomenclature of the causal agent of sudden death syndrome and identity and relevance of *F. solani* form B. Plant Dis. 81:259–266.

Rupe, J.C., E.E. Gbur, and D.M. Marx. 1991. Cultivar responses to sudden death syndrome of soybean. Plant Dis. 75:47–50.

Sasser, J.N. 1977. Worldwide dissemination and importance of the root-knot nematodes, *Meloidogyne* spp. J. Nematol. 9:26–29.

Sasser, J.N., and C.C. Carter. 1982. Overview of the International *Meloidogyne* Project 1975–1984. p. 19–24. *In* J.N. Sasser and C.C. Carter (ed.) An advanced treatise on *Melodogyne*. Vol. 1. Biology and control. North Carolina State Univ. Graphics, Raleigh.

Schmitt, D.P. 1991. Management of *Heterodera glycines* by cropping and cultural practices. J. Nematol. 23:348–352.

Schmitt, D.P., and J.E. Bailey. 1990. Chemical control of *Hoplolaimus columbus* on cotton and soybean. J. Nematol. 22:689–694.

Schmitt, D.P., and K.R. Barker. 1981. Damage and reproductive potentials of *Pratylenchus brachyurus* and *Pratylenchus penetrans* on soybean. J. Nematol. 13:327–332.

Schmitt, D.P., and J.L. Imbriani. 1987. Management of *Hoplolaimus columbus* with tolerant soybean and nematicides. Ann. Appl. Nematol. 1:59–63.

Schmitt, D.P., and G.R. Noel. 1984. Nematode parasites of soybean. p. 13–59. *In* W.R. Nickle (ed.) Plant and insect nematodes. Marcel Dekker, New York.

Schmitt, D.P., and R.D. Riggs. 1991. Influence of selected plant species on hatching of eggs and development of juveniles of *Heterodera glycines*. J. Nematol. 23:1–6.

Schmitt, D.P., and R.D. Riggs. 1989. Population dynamics and management of *Heterodera glycines*. Agric. Zool. Rev. 3:253–269.

Schmitt, D.P., and G. Shannon. 1992. Differentiating soybean responses to *Heterodera glycines* races. Crop Sci. 32:275–277.

Shurtleff, M.C., and C.W. Averre III. 2000. Diagnosing plant diseases caused by nematodes. APS Press, St. Paul. MN.

Sikora, R.A., and N. Greco. 1990. Nematode parasites of food legumes. p. 181–235. *In* Plant parasitic nematodes in subtropical and tropical agriculture. CAB Int., Wallingford, UK.

Smith, G.S., T.L. Niblack, and R.D. Heinz. 1992. Recovery of *Heterodera glycines* from field-collected droppings of lesser snow geese (*Chen caerulescens*) in Missouri. J. Nematol. 24:619. (abstr.).

Smith, G.S., T.L. Niblack, and H.C. Minor. 1991. Response of soybean cultivars to aldicarb in *Heterodera glycines*-infested soils in Missouri. Suppl. J. Nematol. 23:693–698.

Society of Nematologists Crop Loss Assessment Committee. 1987. Bibliography of estimated crop losses in the United States due to plant-parasitic nematodes. Ann. Appl. Nematol. 1:6–12.

Stetina, S.R., E.C. McGawley, and J.S. Russin. 1997. Relationship between *Meloidogyne incognita* and *Rotyenchulus reniformis* as influenced by soybean genotype. J. Nematol. 29:395–403.

Sugawara, K., K. Kobayashi, and A. Ogoshi. 1997. Influence of the soybean cyst nematode (*Heterodera glycines*) on the incidence of brown stem rot in soybean and adzuki bean. Soil Biol. Biochem. 29:1491–1498.

Sydenham, G.M., and R.B. Malek. 1989. Comparative host suitability of selected crop species for *Pratylenchus hexincisus* and *P. scribneri*. J. Nematol. 21:590 (abstr.).

Tabor, G.M., G.L. Tylka, and C.R. Bronson. 2001. *Heterodera glycines* increases susceptibility of soybeans to *Phialophora gregata*. Phytopathology 91 (Suppl.):S87.

Taylor, A.L. 1974. Identification of soybean cyst nematodes for regulatory purposes. Soil Crop Sci. Soc. Florida Proc. 34:201–206.

Thomason, I.J., J.R. Rich, and F.C. O'Melia. 1976. Pathology and histopathology of *Pratylenchus scribneri* infecting snap bean and lima bean. J. Nematol. 8:347–352.

Thompson, J.M., and G.L. Tylka. 1997. Differences in hatching of *Heterodera glycines* egg-mass and encysted eggs in vitro. J. Nematol. 29:315–321.

Todd, T.C. 1991. Effect of cropping regime on populations of *Belonolaimus* sp. and *Pratylenchus scribneri* in sandy soils. Suppl. J. Nematol. 23:646–651.

Todd, T.C. 1993. Soybean planting date and maturity effects on *Heterodera glycines* and *Macrophomina phaseolina* in southeastern Kansas. J. Nematol. 25:731–737.

Todd, T.C., C.A.S. Pearson, and F.W. Schwenk. 1987. Effect of *Heterodera glycines* on charcoal rot severity in soybean cultivars resistant and susceptible to soybean cyst nematode. Ann. Appl. Nematol. 1:35–40.

Todd, T.C., W.T. Schapaugh, Jr., J.H. Long, and B. Holmes. 1995. Field response of soybean in maturity groups III-V to *Heterodera glycines* in Kansas. Suppl. J. Nematol. 27:628–633.

Tomerlin, A.H., and V.G. Perry. 1967. Pathogenicity of *Belonolaimus longicaudatus* to three varieties of soybean. Nematologica 13:154 (abstr.).

Tourjee, K.R., and T. L. Niblack. 2000. *Heterodera glycines* pathogenicity and fecundity: Developing a genetic model of a pathogen's life history traits. J. Nematol. 32:466 (abstr.).

Tubajika, K.M., G.L. Tylka, H. Tachibana, and X.B. Yang, 1994. Incidence of brown stem rot as influenced by soybean cyst nematode. Phytopathology 84:1101.

Tylka, G.L., C. Sanogo, and S.K. Souhrada. 1998. Relationships among *Heterodera glycines* population densities, soybean yields, and soil pH. J. Nematol. 30:519–520.

Veech, J.A., and B.Y. Endo. 1970. Comparative morphology and enzyme histochemistry of root-knot resistant and susceptible soybean. Phytopathology 60:896–902.

Wang, J., P.A. Donald, T.L. Niblack, G.W. Bird, J. Faghihi, J.M. Ferris, D.J. Jardine, C. Grau, P.E. Lipps et al. 2000. Soybean cyst nematode reproduction in the north central United States. Plant Dis. 84:77–82.

Wang, J., T.L. Niblack, J.N. Tremaine, W.J. Wiebold, G.L. Tylka, C.R. Marrett, G.R. Noel, O. Myers, and M.E. Schmidt. 2002. The soybean cyst nematode reduces soybean yield without causing obvious symptoms. Plant Dis. 87:623–628.

Weaver, D.B., R. Rodriguez-Kabana, and E.L. Carden. 1998. Velvetbean and bahiagrass as rotation crops for management of *Meloidogyne* spp. and *Heterodera glycines* in soybean. J. Nematol. Suppl. 30:563–568.

Weaver, D.B., R. Rodriguez-Kabana, G. Robertson, R.L. Akridge, and E.L. Cardere. 1988. Effect of crop rotation on soybean in a field infested with *Meloidogyne arenaria* and *Heterodera glycines*. Ann. Appl. Nematol. 2:106–109.

Wheeler, T.A, P.E. Pierson, C.E. Young, R.M. Riedel, H.R. Wilson, J.B. Eisley, A.F. Schmitthenner, and P.E. Lipps. 1997. Effect of soybean cyst nematode (*Heterodera glycines*) on yield of resistant and susceptible soybean cultivars grown in Ohio. Suppl. J. Nematol. 29:703–709.

Winkler, H.E., B.A.D. Hetrick, and T.C. Todd. 1994. Interactions of *Heterodera glycines*, *Macrophomina phaseolina*, and mycorrhizal fungi on soybean in Kansas. Suppl. J. Nematol. 26:675–682.

Workneh, F., G.L. Tylka, X.B. Yang, J. Faghihi, and J.M. Ferris. 1999. Regional assessment of soybean brown stem rot, *Phytophthora sojae*, and *Heterodera glycines* using area-frame sampling: Prevalence and effects of tillage. Phytopathology 89:204–211.

Wrather, J.A., T.R. Anderson, D.M. Arsyad, J. Gai, L.D. Ploper, A. Porta-Puglia, H.H. Ram, and J.T. Yorinori. 1997. Soybean disease loss estimates for the top 10 soybean producing countries in 1994. Plant Dis. 81:107–110.

Wrather, J.A., T.R. Anderson, D.M. Arsyad, Y. Tan, L.D. Ploper, A. Porta-Puglia, H.H. Ram, and J.T. Yorinori. 2001. Soybean disease loss estimates for the top ten soybean-producing countries in 1998. Can. J. Plant Pathol. 23:115–121.

Wrather, J.A., A.Y. Chambers, J.A. Fox, W.F. Moore, and G.L. Sciumbato. 1995. Soybean disease estimates for the southern United States, 1974 to 1994. Plant Dis. 79:1076–1079.

Wrather, J.A., and T.L. Niblack. 1992. Survey of plant-parasitic nematodes in Missouri cotton fields. Ann. Appl. Nematol. 24:779–782.

Wrather, J.A., and G.L. Sciumbato. 1995. Soybean disease loss estimates for the southern United States during 1992 and 1993. Plant Dis. 79:84–85

Wrather, J.A., W.C. Stienstra, and S.R. Koenning. 2000. Soybean disease loss estimates for the United States from 1996 to 1998. Can. J. Plant Pathol. 23:122–131.

Yen, J.H., T.L. Niblack, and W.J. Wiebold. 1995. Dormancy of *Heterodera glycines* eggs. J. Nematol. 27:153–163.

Yoshii, K. 1977. Reactions of soybean varieties to the root-knot nematode (*Meloidogyne incognita*). Fitopatologia 12:35–38.

Young, L.D. 1982. Reproduction of differentially selected soybean cyst nematode populations on soybeans. Crop Sci. 22:385–388.

Young, L.D. 1989. Use of statistics in race determination tests. J. Nematol. 21:544–546.

Young, L.D. 1987. Effects of soil disturbance on reproduction of *Heterodera glycines*. J. Nematol. 19:141–142.

Young, L.D. 1996. Yield loss in soybean caused by *Heterodera glycines* 1996. Suppl. J. Nematol. 28:604–607.

Young, L.D. 1998a. Breeding for nematode resistance. p. 187–207. *In* K.R. Barker et al. (ed.) Plant and nematode interactions. ASA, CSSA, and SSSA. Madison, WI.

Young, L.D. 1998b. *Heterodera glycines* populations selected for reproduction on Hartwig soybean. J. Nematol. 30:523.

Young, L.D. 1999. Soybeans resistant to *Heterodera glycines* populations attacking Hartwig cultivar. J. Nematol. 31:583.

Young, L.D. 2000. Soybeans resistant to multiple *Heterodera glycines* populations attacking Hartwig cultivar. J. Nematol. 32:472–473.

Young, L.D., and E.E. Hartwig. 1988. Evaluation of soybeans resistant to *Heterodera glycines* race 5 for yield and nematode reproduction. Ann. Appl. Nematol. 2:38–40.

Zirakparvar, M.E. 1982. Susceptibility of soybean cultivars and lines to *Pratylenchus hexincisus*. J. Nematol. 14:217–220.

17   Integrated Management of Soybean Insects

DAVID J. BOETHEL

Louisiana State University Agricultural Center
Baton Rouge, Louisiana

The literature on soybean [*Glycine max* (L.) Merr.] entomology is voluminous and continues to expand. Kogan and Turnipseed (1987) emphasized this fact and conceded that a comprehensive account of the literature could not be provided even after a span of 11 yr since their 1976 review article on the subject. The challenge they faced surfaced again during the formulation of this chapter. Following their lead, new trends and advances in the science and technology of soybean insect management formed the focus of this review. However, because the audience for the Soybean Monograph includes practitioners from other scientific disciplines (i.e., agronomy, physiology, plant pathology, genetics, etc.), this perspective of soybean insects also attempts to provide a historical view of insect problems and the management approaches used to address these problems. The coverage is not comprehensive, but for the interested reader, should serve as an avenue for gathering additional information.

17–1 STATUS OF INSECT PROBLEMS IN THE UNITED STATES

The dramatic increase in soybean production in the Western Hemisphere during the 1960s was accompanied by growing concern about the impact of insect pests as factors in limiting profitable production of the crop (Turnipseed, 1973). These concerns were especially acute in the southern USA, where a subtropical climate characterized by mild winters and long growing seasons harbored a complex of potentially important insect species that attack soybean (Newsom, 1978). Over the last quarter century, the numbers of soybean insect pest species and their densities were found to follow a north-south gradient, and in general, insect pressure has been greatest in the southern states bordering the Gulf of Mexico and the Atlantic Ocean (Way, 1994). Of the eight insect species considered major pests in the USA (Kogan, 1980), at least one or more of six species generally reach population densities that warrant management in some state in the southern soybean production area annually. The lepidopterous defoliators-velvetbean caterpillar, *Anticarsia gemmatalis* Hübner (Plate 17–1) and soybean looper, *Pseudoplusia includens* (Walker) (Plate 17–2); the coleopterous defoliator-bean leaf beetle, *Cerotoma trifurcata* (Forster) (Plate 17–3); the pod feeding stink bug complex-southern green stink bug, *Nezara viridula*

(L.) (Plates 17–4 and 17–5) and green stink bug, *Acrosternum hilare* (Say) (Plates 17–6 and 17–7); and the corn earworm, *Helicoverpa zea* (Boddie), which attacks foliage and pods, frequently require management. In recent years, brown stink bugs in the genus *Euschistus* have become problematic in the region and present interesting challenges related to insecticide management of this guild when occurring simultaneously with the green stink bug complex.

The traditional area of soybean production, the Midwest, has had only sporadic outbreaks of insect pests. However, because of the vast number of hectares in the region, even outbreaks of pests considered minor in other regions, can cause serious economic impact. This was the case in 1988 when drought in the midwestern states resulted in twospotted spider mite, *Tetranychus urticae* Koch, and grasshopper, *Melanoplus* spp., populations that caused yield losses and management costs in the tens of millions of dollars (Way, 1994). Of the eight major pests identified by Kogan (1980), the green cloverworm, *Plathypena scabra* (F.) (Plate 17–8) and Mexican bean beetle, *Epilachna varivestis* (Mulsant) have been more of a problem in the northern production area than in the South. Both regions have suffered damage from the bean leaf beetle, but this pest has become troublesome in the Midwest over the last several years, especially in Iowa and Ohio (Rice et al., 2000). Perhaps more problematic for the region is the invasion of the soybean aphid, *Aphis glycines* Matsumura, a native of the Orient. Since its discovery in mid-July 2000 in Wisconsin, the aphid had been identified in 10 midwestern states by season's end (Marking, 2001a).

Kogan (1981) discussed the development of arthropod communities on soybean after introduction of the crop into the Western Hemisphere. These communities were composed of native fauna that included polyphagous species that added soybean to their diets, oligophagous species that normally were associated with cultivated wild legumes, and oligophagous species that normally fed on other plants but switched to soybean. Twenty years after Kogan's (1981) report and 70 yr after the crop has been widely planted in the New World, the major insect pests have remained rather constant, have emerged from the colonizers described, and have confirmed his observation that few well-adapted soybean feeders occurred in the new growing region. Although the concerns that shifts in composition of arthropod communities to species with greater adaptation to soybean feeding have not materialized, the comments about immigrant species were prophetic. Because the midwestern USA has the same latitude and ecological conditions as the area of the Orient where insect species are highly adapted to soybean, this area is particularly vulnerable to invasions of immigrant species, especially soybean colonizing aphids, pod borers, and root and stem flies (Kogan, 1981). With the discovery of the soybean aphid in the Midwest, and its apparent establishment, this scenario has occurred.

One insect does appear to have undergone a host shift and adapted to soybean. The widespread and long-term use of a corn, *Zea mays* (L.), and soybean crop rotation has been implicated in the selection of a biotype of the western corn rootworm, *Diabrotica virgifera virgifera* LeConte, that prefers soybean environments over corn environments (Sammons et al., 1997; O'Neal et al., 1999). Levine and Oloumi-Sadeghi (1996) found western corn rootworm larval injury to corn in fields that had been planted the prior year to weed-free soybean. The initial explanation was that pyrethroid insecticides used for corn earworm control repelled western corn rootworm female beetles from treated corn to soybean fields where they subsequently oviposited. Sammons et al. (1997) reported increasing rootworm injury in soybean following corn in northwestern Indiana and east central Illinois and identified a variant, the eastern behavioral phenotype, in populations collected in northwestern Indiana that preferred soybean. The ramifications of these findings are that soybean-corn crop rotation, a major strategy for managing western corn rootworm on corn in the Corn Belt, is rapidly diminishing as a viable strategy.

17–2 SOYBEAN INSECT PESTS

Some of the important soybean insect pests in the USA are listed in Table 17–1. Several reviews and books have provided extensive coverage of these and other minor pests. Information on geographic distribution, host range, plant parts, and growth stage of plants attacked, injury, biology of the pests and their natural enemies including pathogens, sampling methodology, management approaches, and diagnostic keys can be found in Kogan and Herzog (1980), Kogan and Turnipseed (1987), Higley and Boethel (1994), and Funderburk et al. (1999).

17–2.1 Pod, Stem, and Seed Feeders

The stink bug complex consisting of the southern green stink bug, green stink bug, and brown stink bug species are annual pests in the southern USA and pose the most serious insect threat to soybean in the region (McPherson and McPherson, 2000). The southern green stink bug is the primary pest in the coastal states with the green stink bug assuming that position in the states of Arkansas, Tennessee, and Kentucky. Although *Euschistus servus* (Say) (Plate 17–9) is the predominant brown stink bug species, *E. tristigmus* (Say) and *E. quadrator* Rolston are frequently encountered. In recent years, brown stink bugs have become more abundant on the crop, especially in the mid-South states of Louisiana, Arkansas, and Mississippi.

Although adults and nymphs feed on several plant parts, the injury caused by the pests is most severe to the seeds which are the preferred feeding site. Feeding injury during early seed formation can result in deformed or aborted seeds (McPherson et al., 1994). This type of injury can directly reduce yield (Yeargan, 1977; McPherson et al., 1979; Miner and Dumas, 1980; Todd, 1989), whereas injury to fully developed seed causes quality losses due to shriveling and discoloration. Heavily damaged seed is frequently discounted when sold or actually rejected for sale (McPherson et al., 1994). The injection of digestive enzymes and the feeding punctures of the pests predispose the seed to invasion by pathogenic and decay organisms. This indirect injury can be substantial, but its impact on quality is dependent on late-season environmental conditions that affect timely harvest (Russin et al., 1988).

In the southern USA, it is not uncommon to observe delayed maturity of soybean wherein entire fields or portions of fields retain their leaves and remain green long after the normal harvest date. Studies in Louisiana demonstrated that southern green stink bugs consistently caused delayed maturity of determinate (Boethel et al., 2000) and indeterminate soybean cultivars (Lingren, 1995). However, this phe-

Table 17–1. Soybean insect pests and selected references.

Common name	Scientific name	Selected references
Pod, stem, and seed feeders		
Southern green stink bug	*Nezara viridula* (L.)	Panizzi and Slansky (1985); Todd (1989)
Green stink bug	*Acrosternum hilare* (Say)	Yeargan (1977); Miner and Dumas (1980)
Brown stink bug	*Euschistus servus* (Say)	McPherson et al. (1979); Jones and Sullivan (1983); McPherson et al. (1994)
	Euschistus spp.	McPherson and McPherson (2000)
Bean leaf beetle[†]	*Ceratoma trifurcata* (Forster)	Kogan et al. (1980); Pedigo et al. (1990); Hunt et. al. (1995)
Corn earworm	*Helicoverpa zea* (Boddie)	Mueller and Engroff (1979); Terry et al. (1987)
Threecornered alfalfa hopper	*Spissistilus festinus* (Say)	Mitchell and Newsom (1984); Sparks (1986)
Lesser cornstalk borer	*Elasmopalpus lignosellus* (Zeller)	Herbert and Mack (1987)
Dectes stem borer	*Dectes texanus texanus* LeConte	Campbell (1980); Lentz (1994)
Seedcorn maggot	*Delia platura* (Meigen)	Hammond (1990); Higley and Pedigo (1991)
Foliage feeders		
Soybean looper	*Pseudoplusia includens* (Walker)	Jensen et al. (1974); Beach and Todd (1986)
Velvetbean caterpillar	*Anticarsia gemmatalis* Hubner	Collins and Johnson (1985); Gregory et al. (1990)
Green cloverworm	*Plathypena scabra* (F.)	Pedigo et al. (1983); Wolf et al. (1987)
Beet armyworm	*Spodoptera exigua* (Hubner)	Huffman and Mueller (1983); Wier and Boethel (1995)
Fall armyworm	*Spodoptera frugiperda* (J. E. Smith)	Pitre and Hogg (1983)
Yellow striped armyworm	*Spodoptera ornithogalli* (Guenee)	Mueller et al. (1979)
Yellow woollybear	*Spilosoma virginica* (F.)	Peterson (1994)
Twospotted spider mite	*Tetranychus urticae* Koch	Mellors et al. (1984); Klubertanz et al. (1990)
Mexican bean beetle	*Epilachna varivestis* Mulsant	Hammond (1984) Edwards et al. (1994)
Potato leafhopper	*Emopasca fabae* (Harris)	Ogunlana and Pedigo (1974); Yeargan (1994a)
Silverleaf whitefly	*Bemisia argentifolii* Bellows and Perring	McPherson and Lambert (1995)
Bandedwinged whitefly	*Trialeurodes abutilonea* (Haldeman)	Hudson et al. (1993)
Grasshopper	*Melanoplus* spp.	Wintersteen et al. (1990); De Gooyer and Browde (1994)
Soybean thrips	*Neohydatothrips variabilis* (Beach)	Mueller and Luttrell (1977); Irwin and Kuhlman (1979)
Root and nodule feeders		
Soybean nodule fly	*Rivellia quadrifasciata* (Macquart)	Eastman and Wuensche (1977); Koethe and Van Duyn (1984)
Banded cucumber beetle	*Diabrotia balteata* LeConte	Waddill et al. (1984)
White grubs	*Phyllophaga* spp.	Lentz (1985)
Grape colaspis	*Colaspis brunnea* (F.)	Lambert (1994)

† Several pests (e.g., bean leaf beetle, Mexican bean beetle, banded cucumber beetle) do not confine their attack only to one part on the soybean plant. For more detail see Kogan and Herzog (1980), Kogan and Turnipseed (1987), Higley and Boethel (1994), and Funderburk et al. (1999).

nomenon occurred only when specific requirements for host phenological stage at time of infestation as well as infestation level were met. Delayed maturity resulted after infestation exclusively during the pod set and seed-filling stages (R3-R5.5; Fehr et al., 1971), and infestations at R3 to R4 and R5 resulted in delayed maturity more consistently than did infestation at R5.5. Data revealed that infestation levels six-fold greater than the economic threshold were required to delay maturity (Boethel et al., 2000). Panizzi et al. (1979) reported similar findings in South America associated with *Piezodorus guildinii* (Westwood), a stink bug species with limited distribution and economic importance in the USA (McPherson et al., 1994).

The decline of soybean hectares in the southern USA has been accompanied by increased hectares of grain crops [corn, grain sorghum, *Sorghum bicolor* (L.), and cotton (*Gossypium hirsutum* (L.), most recently transgenic Bt cotton]. The grain crops serve as early season hosts for stink bugs and along with the wide-spread adoption of Bt cotton and implementation of boll weevil (*Anthonomus grandis grandis* Boheman) eradication, both of which have resulted in reduced insecticide usage on cotton, have contributed to increased populations of stink bugs. Thus, these crops and soybean face increased risk of stink bug attack, with the problem magnified in years when mild winters improve overwintering survival.

In 1980, Kogan (1980) labeled the corn earworm as the number one insect pest of soybean in the USA. Larvae of the corn earworm feed on leaves, stems, flowers, pods, and seeds within pods, with the most severe injury resulting when large larvae occur in synchrony with relatively mature pods. Because feeding at that time is more exclusively on seeds within pods, infestation levels of three large larvae per meter of row have been estimated to reduce yield approximately 134 kg ha^{-1} (Van Duyn et al., 1994). Several studies demonstrated that late instar larvae are the most damaging (Mueller and Engroff, 1979; McWilliams, 1983) and that the soybean flowering period is the most preferred phenological stage for oviposition (Johnson et al., 1975). The latter behavior may explain the sporadic and localized outbreaks of the pest. Lack of synchrony between moth flights from other hosts and soybean in the flowering stage may be the reason the corn earworm is not a more severe pest of soybean. The movement to early maturity group cultivars in the South may have altered this synchrony and surfaced as a causal factor in the decrease in importance of this pest on soybean in the region. In addition, cotton is preferred over soybean as an oviposition host (Hillhouse and Pitre, 1976). The increase in cotton hectares in the mid-Atlantic states following boll weevil eradication and more favorable economics of cotton production relative to soybean in the remainder of the South probably also has resulted in the decline of the corn earworm's pest status on soybean. Kogan (1980) suggested that reassessment 5 to 10 yr hence would be interesting due to the dynamics of the insect pest situation. After 23 yr, the corn earworm can no longer be considered the most important insect pest on the crop.

Among the soybean insect pests, the bean leaf beetle is the most versatile in its injury to the crop. The bean leaf beetle damages soybean by larval feeding on roots and nodules, adults feeding on foliage, blossoms, and pods, and adult transmission of bean pod mottle virus, cowpea mosiac virus, and southern bean mosiac virus. The pest is widespread throughout the soybean growing area of the USA, but in recent years has become a more consistent pest in the North-Central states (Rice et al., 2000).

Little data are available on the effect of bean leaf beetle feeding on soybean roots and nodules. Layton (1983) demonstrated that larval feeding reduced number and total dry mass of nodules per plant and dinitrogen (N_2)-fixing capacity by about 55%. However, yield was not reduced by this level of damage.

Small round holes between the major leaflet veins is characteristic of adult defoliation. Although defoliation can occur throughout the growing season, beetle numbers relative to leaf area rarely are sufficient to warrant management except as it contributes to other insect defoliation (Pedigo, 1994). Foliage feeding during V1 and V2 soybean stages can cause concern but again, management is rarely needed due to the plants' ability to compensate. Hunt et al. (1995) found the economic injury levels (EILs) for bean leaf beetles feeding on seedling soybean were much higher than those currently being used. Pod feeding by adults appears to be the most important type of injury. Lesions on pods expose seed to moisture and pathogen invasion, and seed beneath the lesions become shrunken, discolored, and moldy. Both quantity and quality losses result. This damage along with "pod clipping", a phenomenon wherein feeding on the pod peduncle causes the pods to fall, indicate that the generation of bean leaf beetles that occurs during pod filling should garner increased attention. Smelser and Pedigo (1992) developed economic-injury equations for late-season pod injury based on adult and injured-pod counts for several combinations of pest management costs and crop market values. Their findings revealed that the EIL's developed for seed-loss are exceeded before enough seed damage is incurred to cause discounts at time of sale.

Several studies on the relationship between bean pod mottle virus and bean leaf beetles to include the distribution of the virus and beetle population abundance and occurrence (Myhre et al., 1973; Pitre et al., 1979; Mueller and Haddox, 1980; Hopkins and Mueller, 1983), transmission (Walters, 1964; Patel and Pitre, 1976), impact on N_2 fixation (Hicks et al., 1983), yield (Hopkins and Muller, 1984), and EIL (Horn et al., 1973) have been conducted. However, recent years have seen a paucity of research on the subject. Interest has renewed, however, with the increasing abundance of the pest in the North-Central states and concern about plants demonstrating symptoms of the disease.

The threecornered alfalfa hopper, *Spissistilus festinus* (Say)(Plates 17–10 and 17–11), attacks soybean from plant emergence until harvest maturity. Adults and nymphs feed on phloem and frequently girdle stems and petioles with feeding punctures circumscribing these structures (Mitchell and Newsom, 1984). Main stem girdles near the soil surface frequently cause plants to lodge and die; however, this injury rarely results in economic losses because of the ability of adjacent healthy plants to compensate for those injured (Tugwell et al., 1972). Sparks and Newsom (1984) and Sparks (1986) found no yield reduction associated with main stem girdling, but did report yield loss when large populations occurred after R1.

Feeding on pedicels and peduncles of pods appeared to be the mechanism most responsible for reduced seed weight and pod number. In addition, Hicks et al. (1984) observed that girdling restricted nutrient flow, nodule number and growth, and N_2 fixation. Subsequent studies confirmed yield losses associated with late-season damage and indicated the second generation on soybean as the most damaging (Sparks and Boethel, 1987). Thus, management of the threecornered alfalfa hopper before the onset of flowering is rarely necessary.

Lesser cornstalk borer, *Elasmopalpus lignosellus* (Zeller), outbreaks on soybean are most common during periods of hot, dry weather, in fields with predominately sandy soils, and in double-cropped soybean, especially when residue from the previous crop is burned (Herbert and Mack, 1987; Funderburk and Mack, 1994). The larvae bore into the main stem of the plants just below the soil surface resulting in wilting and death. Heavy infestations can reduce plant stands to levels where economic losses can occur or replanting is necessary. As plants grow beyond the seedling stage, damage greatly diminishes. In fields under conditions of high risk, in-furrow insecticide applications at planting are warranted. Irrigation can reduce damage potential but may be cost prohibitive depending on the price of soybean. Post-planting insecticides must be applied at the first indication of injury to be effective in reducing the negative impact on yield.

The Dectes stem borer, *Dectes texanus texanus* LeConte, is a minor pest of soybean. Infestations occur on field borders near wild hosts such as cocklebur, *Xanthium strumarium* (L.), and giant ragweed, *Ambrosia trifida* (L.) (Campbell, 1980). Newly hatched larvae tunnel in the pith of petioles before moving to the main stem. Feeding in the main stem can cause slight reductions in seed weight, but most yield loss results from lodging associated with late instar girdling of the plant near the soil line (Lentz, 1994). Early-planted, early-maturing cultivars are most vulnerable to infestation. The most effective management tactic is burial of borer-infested stubble to decrease larval survival and adult emergence.

The seedcorn maggot, *Delia platura* (Meigen), is a common pest in the North-Central states, but only severe in fields where a green cover crop has been incorporated prior to planting (Hammond, 1990). These conditions are favorable for oviposition. Larvae feed on germinating seeds, cotyledons, and terminals causing delay in emergence, stand loss, and Y-plants or "snake-heads". Damage is most intense under cool, wet conditions which prolong larval development and delay soybean emergence. Management is rarely required in reduced or no-till fields.

17–2.2 Foliage Feeders

Undoubtedly, the most recognized insect injury to soybean is defoliation caused by lepidopterous pests. Extensive research on these pests has revealed that protection of the crop from flowering (R2) through pod filling (R5) is critical to prevent economic losses. Monitoring of populations and utilization of economic thresholds (ET's) have been the primary mechanisms for making pest management decisions for this group of pests. Although successful, efforts continue to improve the system by focusing on understanding the relationships between defoliation and soybean physiology (e.g., soybean canopy size and similar parameters, as well as pest populations) (Higley, 1992). More definitive information on soybean response to defoliation may be the avenue for development of more advanced decision tools, such as multiple species EIL's (Hutchins et al., 1988). Because soybean defoliators frequently occur as a complex of species, including lepidopterous complexes, this technology would improve management.

Canopy light interception has been shown to be an important factor in determining yield loss associated with insect injury (Higley, 1992; Board et al., 1997; Haile et al., 1998) and is dependent on leaf area index (LAI) (Hunt et al., 1999).

Klubertanz et al., (1996) found that leaf area removed did not predict yield as well as remaining leaf area, but unless remaining leaf area is coupled with data on initial canopy size, pest density, and predictions of future injury, its pest management application is limited. Herbert et al. (1992) had earlier examined the value of using LAI, leaf area removed, and percent defoliation to predict yield and found that LAI was the best predictor of yield. This focus on LAI has spawned studies to find instrumentation to accurately measure LAI (Hunt et al., 1999) and canopy light interception (Board and Boethel, 2001) for incorporation into current insect-monitoring programs. This movement may provide the linkage to precision agriculture technology (GPS, GIS, remote sensing), which also is receiving interest by soybean integrated pest management (IPM) practitioners.

The soybean looper and velvetbean caterpillar are the two most important defoliators of soybean in the southern soybean growing region of the USA. Because both are annual migrants into the area, outbreaks of the pests can be sporadic but are characterized by rapid population surges that require immediate action to prevent crop loss. One or both pests require management on an annual basis in the region.

In general, soybean loopers appear earlier in the season and achieve larger populations in cotton-soybean agroecosystems (Burleigh, 1972; Beach and Todd, 1986). The association with cotton plays a role in insecticide resistance in this species, and for this reason, soybean looper management is challenging and costly (Boethel et al., 1992).

The velvetbean caterpillar arrives later in the season, and severe outbreaks typically occur only in the extreme southern latitudes of the USA. In contrast to soybean loopers, velvetbean caterpillars are easily managed with an array of insecticides and have been the subject of preventive control strategies using insect growth regulators (Willrich et al., 2002). A better understanding of migration of both species might lead to enhanced management in the form of prediction of outbreaks. Progress in this regard has been made for velvetbean caterpillar (Gregory et al., 1990).

The fall armyworm [*Spodoptera frugiperda* (J.E. Smith)], yellow striped armyworm [*Spodoptera ornithogalli* (Guenee)], and beet armyworm [*Spodoptera exigua* (Hübner)] can be found in soybean fields in the South, but rarely require control measures. Beet armyworm problems surface when hot, dry environmental conditions prevail and usually in cotton-growing regions (Wier and Boethel, 1995). McLeod et al. (1978) reported that economic losses could result from pod feeding, but Huffman and Mueller (1983) and Wier and Boethel (1995) observed little pod feeding. Beet armyworms consume less foliage than soybean loopers and velvetbean caterpillars (Wier and Boethel, 1995). Huffman and Mueller (1983) could not demonstrate yield reduction with any larval density examined. Collectively, these studies suggest that the beet armyworm is a relatively minor problem.

On occasion, fall armyworm will attack soybean when sufficient grass hosts are not available (Pitre and Hogg, 1983). Infestations on seedling soybean have caused concern but this situation usually arises following over-the-top herbicide applications for grass control. Removal of grass hosts results in the pest moving to soybean. The yellow striped armyworm can be found in most southern U.S. soybean fields in early season, but densities are extremely low. In Arkansas, seedling soybean subjected to severe larval feeding compensated for damage, making control measures unnecessary (Mueller et al., 1979).

The yellow woollybear, *Spilosoma viriginica* (F.), seldom causes economic injury to soybean (Peterson, 1994). In the South, infestations can be found in early season along field borders, whereas in the North-Central states infestations are most likely to occur in late summer. A related species, the salt-marsh caterpillar, *Estigmene acrea* (Drury), may occur simultaneously. Females oviposite eggs in masses, thus initial infestations are localized in fields facilitating "spot" curative insecticide treatments.

The green cloverworm is commonly found throughout the soybean-growing area of the USA. In the South, the insect rarely requires management unless as part of a complex of late season defoliators. Its presence during the spring on wild legumes, and soybean early in the season, may play a role in maintenance and increase of natural enemies (parasitoids and pathogens) prior to the arrival of migratory lepidopterous pests. Also, it has been suggested that establishment of exotic natural enemies for classical biological control of these migratory pests may depend on the utilization of green cloverworm as a host by the introduced natural enemies in early season (Daigle et al., 1988).

In the Midwest, severe outbreaks of the green cloverworm occur in cycles and are characterized by rapidly increasing larval populations during June and July (Pedigo et al., 1983). In these major outbreak years, millions of hectares of soybean require treatment. In other years, "endemic" green cloverworm populations develop slowly during the season but remain below levels requiring management. Pedigo et al. (1983) hypothesized that the difference in "outbreak" and "endemic" populations was due to the density of immigrating moths. Subsequent studies by Wolf et al. (1987) revealed that a synoptic weather pattern was correlated with the first appearance of the moths each Spring. The number of days that this weather pattern occurred during the Spring was defined as potential migration days. Years in which "outbreak" populations occurred were those with the most migration days. Being able to predict outbreaks of green cloverworm in the Midwest has enormous value due to the pest's sporadic occurrence and the vast number of hectares at risk.

The twospotted spider mite is another sporadic but costly pest in the North-Central states. Prolonged hot, dry conditions appear to precipitate mite problems, and the outbreaks on soybean in the Midwest in 1983 and 1988 fit the historical weather criteria (Way, 1994). Mellors et al. (1984) found that under greenhouse conditions, twospotted spider mites developed larger populations on soybean receiving higher watering rates. Thus, moisture stress may not be a factor in greater spider mite populations, but certainly development is accelerated by higher temperatures. Generally, symptoms of spider mite feeding are detected before the pest is found. Feeding reduces photosynthetic capacity and chlorophyll content of the leaves. Water loss through feeding wounds on the leaves causes plant stress, and severe injury will result in defoliation.

The Mexican bean beetle has been reported to cause economic damage in the mid-Atlantic states, the Carolinas, Georgia, Tennessee, Kentucky, Indiana, and Ohio. In Indiana, the occurrence of Mexican bean beetles increased as the hectares of soybean double-cropped with small grains increased (Edwards et al., 1992). Yield reductions are associated with adult and larval feeding on foliage. Adults consume tissue between the veins while larval feeding skeletonizes the same area (Hammond, 1984). Pod feeding by adults has been observed in late season, and cotyledons and terminals have been attacked in early season. In both instances, damage appears to be inconsequential. Considerable research has been conducted to identify soybean cultivars resistant to the Mexican bean beetle; however, the cultivar released was not widely adopted (Boethel, 1999). Although the Mexican bean beetle received considerable research attention as soybean expansion began in the USA, populations of the insect and its status as a pest have declined in recent years.

The potato leafhopper, *Empoasca fabae* (Harris), overwinters in states along the Gulf Coast and migrates north each year, arriving in northern USA between April and early June (Yeargan, 1994a). Potato leafhopper problems on soybean often are related to large migrating populations attacking vegetative stage plants. Nymphs and adults feed primarily on phloem tissue. Symptoms include distorted leaf veins and curled leaves with yellow necrotic margins. These symptoms are referred to as "hopperburn" and severe stunting can occur. Ogunlana and Pedigo (1974) determined that an infestation level of one leafhopper per plant could cause economic injury to V2 stage soybean. As plants mature, progressively larger populations are required to cause economic damage. Plant pubescence confers resistance to potato leafhopper. Incorporation of this trait into most commercial soybean cultivars has resulted in stable population suppression and virtually relegated the insect to non-pest status on the crop (Boethel, 1999).

Other foliage feeders in Table 17–1 include the silver leaf whitefly, *Bemisia argentifolii* Bellows and Perring; bandedwinged whitefly, *Trialeurodes abutilonea* (Haldeman); soybean thrips, *Neohydatothrips variabilis* (Beach); and grasshoppers, *Melanoplus* spp. They are not considered major pests of soybean but can be found on the crop. Grasshoppers have been problematic on occasion, generally in years of consecutive drought (DeGooyer and Browde, 1994). Infestations typically occur first along field borders. Fields adjacent to grassy sites or small grain fields are most susceptible to infestation. Nymphs and adults feed on leaves and after pod formation, both leaf and pod feeding are common. Depending on population density, defoliation and pod damage, singly or combined, can result in yield loss. The redlegged grasshopper, *M. femurrubrum* (DeGeer), and differential grasshopper, *M. differentialis* (Thomas), are the species most often encountered on soybean.

17–2.3 Root and Nodule Feeders

Turnipseed and Kogan (1976) reported that the economic impact of the soil-inhabiting insect complex on soybean is little understood. Since their assessment, scant progress has been made. Most species are generalists and their occurrence and abundance on soybean is influenced by other hosts near soybean fields and cultural practices.

Larvae of the soybean nodule fly, *Rivella quadrifasciata* (Macquart), were first described attacking nodules of soybean by Eastman and Wuensche (1977). Subsequent studies have shown that injury to soybean nodules can occur, but economic damage has not been confirmed (Koethe and Van Duyn, 1984). Its presence and abundance in Louisiana have declined since the initial report in the state.

The banded cucumber beetle, *Diabrotica balteata* Le Conte, has a wide host range (Waddill et al., 1984), and it is not uncommon to find them in soybean fields in the states along the Gulf Coast. Adults contribute to overall defoliation, and while larvae are known to feed on roots, yield loss has not been attributed to this insect. Troxclair and Boethel (1984) observed larger populations in reduced tillage soybean fields.

White grubs, *Phyllophaga congrua* (Le Conte) and *P. implicita* (Horn), cause damage by larvae feeding on roots and reducing stands (Lentz, 1985). Infestations are not common on soybean. The presence of suitable adult hosts and grasses near fields may enhance infestations.

The grape colaspis, *Colospis brunnea* (F.), and *C. louisianae* Blake are common in soybean fields (Lambert, 1994). Larvae feed on roots causing plant stunting but seldom plant mortality. Larval infestations are frequently localized in small areas of a field, and usually on soils that are poorly drained or high in organic matter. Adults contribute to overall defoliation, but Baur et al. (1999) found that colaspis beetles consume less foliage than bean leaf beetles or banded cucumber beetles.

17–3 INTEGRATED PEST MANAGEMENT TACTICS

As soybean production expanded in the southern USA, entomologists, especially those with experience controlling insects on cotton, warned that the temptation to follow automatic insecticide application practices should be avoided on soybean (Lincoln et al., 1975; Newsom, 1978). The lack of a "key pest", the economics of soybean production, the potential for insecticide resistance, the resurgence of primary pests, the possibility of secondary pest outbreaks, and insecticide contamination of the environment were cited as reasons to develop IPM systems requiring minimum use of conventional insecticides on soybean.

Entomologists with responsibility for developing management strategies for insects on soybean heeded the early warnings. What emerged from a coordinated research effort was a basic IPM system developed primarily for the southern USA where pest pressure was intense. It consisted of systematic scouting during periods of risk to monitor crop growth, damage, pest development, and natural enemy activity; the use of action-decision rules based on EILs; and conservation of natural enemies by the use of minimum rates of insecticides when needed to reduce pests to subeconomic levels (Turnipseed and Kogan, 1994). In the northern USA, where sporadic pest outbreaks are the norm, the IPM system was modified to place emphasis on methods to predict the infrequent outbreaks and reduce reliance on monitoring.

The cornerstone for pest management is the EIL, because it defines how much pest injury can be tolerated (Higley and Pedigo, 1997a). The practical aspect of the EIL is that it forms the basis for the ET, which is generally described as the insect population density that indicates management measures should be initiated to prevent economic loss. More EILs have been developed for soybean than any other crop (Peterson, 1997). In fact, soybean has provided a fertile model for research on both the theory and practice of EILs and ETs (Kogan, 1976; Pedigo et al., 1986, 1989; Hutchins et al., 1986, 1988; Higley and Pedigo, 1997b).

In practice, EILs and ETs have been integral components of soybean IPM systems. Equally important has been the recognition of the need for adequate sampling methodology (Kogan and Herzog, 1980) and the constant refinement of sampling techniques to address the dynamics of soybean IPM (Kogan et al., 1994). These factors undoubtedly have resulted in more judicious use of insecticides, wherein less than half of the soybean hectares in the southern USA and only a small fraction in central and northern USA areas receives an application in an "average" season (Turnipseed and Kogan, 1994).

Chemical insecticides have remained a major management tactic in soybean IPM. Because use has been predicated on population monitoring and ETs, chemical control has been effective and in general, economical and compatible with the conservation of natural enemies. Although organophosphorous (OP) and carbamate compounds are still required for some pests (Fitzpatrick et al., 1999, 2000), for example, brown stink bugs and soybean loopers, respectively, pyrethroids now are commonly used in place of older classes of insecticides. The OP's remain the products of choice in many instances because of their cost-effectiveness for the producer. Newer insecticide chemistries (emamectin benzoate, spinosad, imidacloprid, Bts) being evaluated for use on soybean are less detrimental than the older products or the pyrethroids on generalist predators commonly found in soybean agroecosystems (Boyd and Boethel, 1998a, 1998b). The target specificity characteristic of these products confers protection of beneficials, but they may be cost-prohibitive, if tank mixes of different insecticides are needed to manage pest complexes consisting of pests from different taxa (e.g., stink bugs and lepidopterous defoliators). Currently, the newer products are more expensive than most of the conventional compounds, and unless wide-spread use on other crops results in decreased cost, their adoption on soybean may be limited. The move to the pyrethroids and newer products certainly has resulted in a major reduction of the pesticide load in the environment. Many of the newer insecticides are efficacious at fractions of the amount of active ingredient per hectare than are the OP's and carbamates.

Kogan and Turnipseed (1987) cautioned that extensive use of pyrethroids on cotton and soybean might lead to resistance in the corn earworm. The same year as their warning, pyrethroid resistance did develop but in the soybean looper, not the corn earworm (Leonard et al., 1990). Resistance levels were documented to be greater in areas where soybean was grown near cotton (Boethel et al., 1992), but selection pressure at the point of origin of the migratory populations also has been suggested as a contributing factor in resistance development (Thomas et al., 1994). Other than the soybean looper, insecticide resistance has not been demonstrated in any other soybean insect pest, and the factors responsible have not been associated with insecticide usage on soybean.

Pitre (1983) conducted a comprehensive review and assessment of the value of natural enemies in soybean IPM. Generalist predators and entomopathogens usually are considered the more important natural control agents in maintaining populations of lepidopterous pests at subeconomic levels (Todd et al., 1994). Conservation of predators and parasitoids is at the core of soybean IPM, and programs are designed to avoid action detrimental to these beneficials. The most notable implementation of classical biological control on soybean in the USA has been the annual introduction of *Pediobius foveolatus* (Crawford), a parasitoid native to India, for control of the Mexican bean beetle in the mid-Atlantic states (Edwards et al., 1994). Other attempts at classical biological control have resulted in limited success but with the establishment of the exotic soybean aphid in the Midwest, this management tactic is receiving renewed interest. Foreign exploration to identify potential natural enemies already has begun.

Trap cropping is a cultural management tactic that has been demonstrated to be highly effective for the stink bug complex and the bean leaf beetle (McPherson and Newsom, 1984). By planting the trap crop early (10–14 d) before the main crop, bean leaf beetles can be concentrated in the smaller trap crop and controlled with insecticides (in-furrow at planting or over-the-top applications) before dispersal to the larger main crop planting. If managed properly, bean pod mottle virus can be confined to the trap crop area, also. The limitation to this approach is that when planting conditions are optimum in the Spring, producers are reluctant to delay planting. In the case of stink bugs, planting early or planting an early-maturing cultivar on a small section of a farm will function as a trap crop. The concentration of ovipositing adults and the relatively immobile nymphs in the trap crop allow chemical control to be directed only on that portion of the crop (Todd et al., 1994). The key to trap crop management is to control the colonizers (beetles and stink bugs) before they disperse to the main crop. If not managed in a timely fashion, the impact of the strategy may be lost or the trap crop may serve as a nursery to produce colonizers of the main crop. Trap crops for stink bugs have been more widely accepted because planting delays are not required. With the adoption of Early Season Production Systems (ESPS) in the southern USA, the movement to plant a majority of the hectares to early-planted, early-maturing cultivars may diminish the trap crop as an effective tactic for stink bugs. Other cultural practices such as tillage, planting dates, crop rotation, row spacing, and cultivar selection have been shown to have positive and negative influence on soybean insect abundance and impact (see section 17–4.4 on Preventive Tactics for Insect Management).

After Van Duyn et. al. (1971) screened most of the USDA Germplasm Collection of Maturity Group (MG) VII and VIII soybean for resistance to Mexican bean beetle and found three plant introductions (PIs) (PI 171451, PI 227587, PI 229358) to be highly resistant, one or more of these PIs was included in most germplasm screenings as resistant standards and have been used as donor parents in breeding programs to develop insect-resistant germplasm and cultivars. Although considerable research effort has been expended to develop insect-resistant cultivars for soybean IPM systems, only limited success has been achieved (Boethel, 1999). The four cultivars released with resistance to foliar-feeding insects have not been widely used, primarily because of yield limitations. In a search for new sources of resistance, Hammond et al. (1998) found that a few PIs from China had potentially useable levels of resistance, but none had resistance levels similar to previously identified resistant accessions that have served as donor parents of the released insect-resistant cultivars. Sporadic outbreaks of insect pests, efficient and generally inexpensive insecticides, and an ecologically based IPM system that has been efficacious and cost-effective require that an insect-resistant cultivar exhibit high yield potential for adoption by producers (Boethel, 1999). The obstacles encountered by soybean breeders and entomologists to develop multiple insect-resistant cultivars through traditional breeding methods may have reached a point where focus on transgenic technology to achieve the goal may be imperative.

17–4 CHALLENGES AND OPPORTUNITIES FOR ENHANCED INSECT MANAGEMENT

Over the last decade, advances in soybean cultural practices, insect control strategies, and plant breeding have presented opportunities of enhanced insect management but present challenges that must be dealt with by soybean IPM practitioners. Likewise, the establishment of the soybean aphid, an exotic pest from Asia, may result in increased production costs for growers in areas where insect control is rarely necessary.

17–4.1 Early Soybean Production Systems

Heatherly (1999) reported that the low soybean yields in the midsouthern USA during the 1970s and 1980s could be attributed to drought that occurs in the region from mid-July through mid-September, a period when conventional plantings of MG V, VI, and VII cultivars are in high-water-demanding reproductive stages. This dilemma lead to the move toward drought avoidance by adopting a production practice known as early soybean production system (ESPS) (Heatherly and Bowers, 1998). In ESPS, early-maturing soybean cultivars (MG IV and V; many indeterminate) are planted in early to mid-April, approximately 4 to 6 wk before conventional soybean plantings and are harvested about 4 wk earlier (Baur et al., 2000; McPherson et al., 2001).

Advantages associated with ESPS include higher yields in irrigated and non-irrigated plantings and narrow row culture, compatibility with stale seed-bed planting systems, avoidance of inclement weather during harvest, higher prices associated with early harvest, effective land preparation before fall-planted grain or cover crops due to early harvest, and timely planting of rotation grain crops because of land preparation in the fall following early soybean harvest (Heatherly and Bowers, 1998). These factors, along with the development of early-maturing cultivars adopted for the southern USA, offered producers in the region greater opportunities for profitability.

The adoption of ESPS has been dramatic, especially in the states of Mississippi, Arkansas, Louisiana, and Texas. Studies to assess the impact of ESPS on insect pest abundance and soybean IPM programs accompanied this rapid adoption. Initial studies in Louisiana revealed that early-planted (i.e., April) MG IV cultivars matured early enough to escape damaging populations of soybean loopers and velvetbean caterpillars (Boyd et al., 1997).

Although southern green stink bugs reached populations that required control on both MG IV and V soybean cultivars in southern Louisiana, multiple insecticide applications were necessary to manage the pests on MG IV cultivars. In northern Louisiana, southern green stink bugs only exceeded economic thresholds in the MG IV cultivars suggesting that in some areas ESPS plantings might function as trap crops for this pest. In Georgia, stink bugs, primarily the southern green stink bug, also were found to be more numerous on ESPS soybean in mid-season but moved to conventional planted soybean in late season, reaching densities much greater that those found in ESPS (McPherson et al., 2001). This same phenomenon was noted in other studies in Georgia with stink bugs colonizing ESPS soybean in early season and successively moving to later maturing cultivars as each became attractive due to the presence of developing pods (McPherson, 1996). Baur et al. (2000) also found that conventional soybean production systems (May planted MG V-VIII cultivars) experienced significantly higher populations of late-season lepidopterean defoliators and bean leaf beetles that ESPS soybean. However, ESPS plantings harbored larger populations of stink bugs and threecornered alfalfa hoppers. Generalist predators were more abundant on ESPS soybean in early season but declined in late season in both ESPS and conventional plantings.

Although ESPS offers several agronomic advantages over conventional soybean systems, studies conducted in Georgia and the mid-South states of Texas, Arkansas, Mississippi, and Louisiana where this cultural practice has been widely embraced, have demonstrated positive impact on soybean IPM as well. The escape from defoliation by lepidopterous pests provided by early-planted (i.e., April), early-maturing cultivars is a tremendous advantage of ESPS and should be adapted in areas where soybean looper and velvetbean caterpillar have historically been a problem. With the development of pyrethroid resistance in the soybean looper (Boethel et al., 1992), ESPS offers a cost effective alternative to the limited number of insecticides currently recommended for control of this pest (Baldwin et al., 2001). Costs for insect management for ESPS probably will be comparable to that of conventionally planted soybean primarily because of consistent stink bug pressure in ESPS (Baur et al., 2000). It appears from data collected in Louisiana that the pest pressure during a given year will dictate which planting system is the most economical from an insect management standpoint as shown in Fig. 17–1. In 1996, a year characterized by small stink bug populations and large lepidopterean defoliator populations, control costs were lower in ESPS soybean. In 1997, when the population abundance for these pest complexes reversed, control costs were greater on ESPS. What has emerged from these studies is the realization that the IPM tactics developed for con-

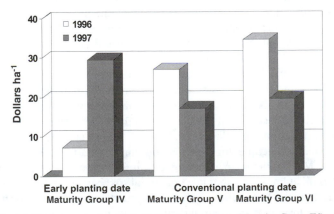

Fig. 17–1. Insecticide cost per hectare for early planted soybeans (Maturity Group IV) and soybeans planted on conventional planting dates (Maturity Groups V and VI).

ventionally planted soybean can be adapted for ESPS plantings, and insect management has not been a limiting factor in adoption of this technology.

17–4.2 Transgenic Herbicide-Tolerant Soybean

The development of the transgenic soybean line, 40-3-2, tolerant to the nonselective herbicide, glyphosate, (Padgette et al., 1995) and the incorporation of this technology (Roundup Ready, Monsanto,Inc., St. Louis, MO) into elite soybean cultivars have changed soybean weed control in the USA and globally. Since the introduction of soybean cultivars with the glyphosate-tolerant gene in 1996 (Bradshaw et al., 1997), the hectares planted to Roundup Ready cultivars has grown substantially. From 1996 to 1998, the hectares planted to transgenic cultivars resistant to Roundup herbicide increased from 11 to 38% in Iowa (Buckelew et al., 2000). In Louisiana, 60% of the hectares in 1999 were planted to glyphosate-resistant cultivars (Anonymous, 2000), and the percentage is expected to increase (Ellis, 2001). This rapid adoption, with its potential for alteration of the soybean agroecosystem, prompted soybean entomologists, under the auspecies of a USDA-CSREES multi-state research project (S-281), to launch studies on the impact of these weed management systems on pest and beneficial insect species on the crop.

Buckelew et al. (2000) reported that although transgenic herbicide-resistant soybean cultivars did not strongly affect insect populations, weed management systems could affect insect populations on the crop. Systems that allowed more weed escapes typically had higher insect population densities, whereas systems with fewer weeds were preferred by potato leafhoppers. In Georgia, higher populations of lepidopteran defoliators (green cloverworm and velvetbean caterpillar) were found in MG VII Roundup Ready soybean cultivars compared to conventional MG VII cultivars (McPherson, 2000), and stink bugs were more numerous on Roundup Ready soybean cultivars in mid-September (McPherson, 1999). However, despite these findings, McPherson (2000) concluded that data collected over 5 yr indicated that herbicide-tolerant transgenic soybean cultivars did not dramatically change the arthropod complex or population densities on the crop. Similar conclusions were drawn in Louisiana (Boethel, 2000), Kentucky (Yeargan, 2000), and Tennessee (Lentz, 2000) after 3 yr of monitoring herbivorous and predaceous arthropods on herbicide-resistant and conventional soybean cultivars. In Tennessee and Louisiana, sulphonyl-urea tolerant (STS) soybean cultivars also did not affect the arthropod community.

Pedigo (2000) reported that spider mite populations were significantly higher (threefold) in Roundup Ready soybean fields than in nontransgenic fields in Ohio and Iowa in 1999. In 2000, five of six Roundup Ready soybean fields sampled required insecticides for control of spider mites. Pedigo (2000) demonstrated that the glyphosate formulation, Roundup Ready-To-Use (Monsanto, Inc., St. Louis, MO), was detrimental to several entomopathogenic fungi, including *Neozygites floridiana* Weiser& Muma, known to attack spider mites. However, six other glyphosate formulations including those used in commercial soybean production and the technical material did not have negative effects on growth of the fungi, *Normuraea rileyi* (Farlow) Samson, frequently found infecting lepidopteran defoliators, was not susceptible to any glyphosate formulation tested.

The bulk of the evidence collected in Georgia, Iowa, Kentucky, Louisiana, Ohio, and Tennessee points to a minimal impact of herbicide-tolerant cultivars on soybean insect management. The findings on spider mites cautions for further study on these pests, however. Research on habitat preference (Alteri and Todd, 1981; Alteri et al.,1981; Shelton and Edwards, 1983) and reduced tillage systems (Troxclair and Boethel, 1984; Lamp et al.,1984; Hammond and Stinner, 1987; Buntin et al., 1995; Lam and Pedigo, 1998) indicate that weed management can influence the diversity and density within the soybean arthropod community. If weed escapes and species shifts occur due to prolonged reliance on herbicide-tolerant technology, weed habitats may emerge that are more suitable to pest and beneficial arthropods.

17–4.3 Transgenic Insect-Resistant Soybean

Soybean is among the economically important plant species that have been successfully transformed with delta-endotoxin genes from *Bacillus thuringiensis* (Bt) (Mazier et al., 1997). Parrott et al. (1994) were the first to report the successful expression of a Bt gene in soybean that involved a truncated native Cry1Ab gene from *B. thuringiensis* subs. *kurstaki* HD-1. These plants demonstrated a level of resistance to velvetbean caterpillar similar to that of GatIR81-296, an insect resistant breeding line developed in Georgia. To obtain greater expression of the Bt toxin, Stewart et al. (1996) developed a soybean line expressing a truncated synthetic Cry1Ac gene. This gene, designed to be highly expressed in plants, is similar to the gene found in Bt transgenic cotton and corn, which were planted on more than 50% of the hectares of these crops in the USA in 1998 (All et al., 1999b). One line that resulted from these transformations of the cv. Jack showed resistance to corn earworm, soybean looper, and velvetbean caterpillar (Stewart et al., 1996). More recently, field studies revealed that a Jack-Bt transgenic line had three-to fivefold and eight-to ninefold less defoliation from corn earworm and velvetbean caterpillar, respectively, when compared with an untransformed Jack (Walker et al., 2000) (Plate 17–12). The transformed line demonstrated significant, but lower resistance to soybean looper, but fourfold greater resistance to lesser cornstalk borer than Jack. The findings relative to lesser cornstalk borer were notable. The sporadic occurrence of this pest and its underground feeding habits make insecticide management difficult. In general, these experiments suggested that Jack-Bt should provide adequate levels of resistance to several lepidopterean pests under field conditions (Walker et al., 2000).

A program to identify soybean genetic markers associated with insect resistance was initiated in the Univ. of Georgia Center for Soybean Improvement in 1993 (All et al.,1999a). Restriction fragment length polymorphism (RFLP) markers have been used in plants resulting from crosses of insect susceptible and resistant genotypes to identify and map major quantitative trait loci (QTL) for insect resistance to facilitate their combined introgression into elite soybean gemoplasm (Boerma and Parrot, 1996; Boerma et al., 1996; Rector et al., 1996, 1999, 2000). Earlier findings are being expanded by mapping the resistance QTLs in insect resistant PIs with PCR-based simple sequence repeat makers (SSR), with the goal of minimizing the linkage drag of undesirable genes linked to insect resistant genes in a backcrossing program (All et al., 1999a; Narvel et al., 2001).

The University of Georgia program is attempting to pyramid Bt genes (Jack-Bt) with native resistance genes from PIs into a single soybean line (Walker et al., 2002). According to All et al. (1999a), this attempt to impart horizontal resistance has three advantages. If successful, the method should help avoid development of resistance to a single resistance gene, enhance resistance by having additive or synergistic interactions between two or more resistance genes, and provide multiple pest resistance with multiple genes that each target a specific pest.

Use of the single Bt toxin, Cry1Ac, in transgenic cotton and in the transgenic soybean lines currently developed has ramifications for both short- and long-term insect management with this technology. On both crops, differential control of noctuid species has been documented (MacIntosh et al., 1990; Walker et al., 2000) requiring other management tactics for some species. Reliance on a single Bt toxin is expected to lead to resistance in some pest species. Mascarenhas et al. (1998) expressed concern that widespread adoption of transgenic Bt cotton could accelerate Bt resistance development in strains of soybean looper in the southern USA, where the pest occurs on both soybean and cotton which are grown in close proximity. Their data from Louisiana indicated that soybean loopers collected from Bt cotton and soybean near Bt cotton had elevated dosage-mortality responses to commercial Bt insecticides.

Bollgard II (Monsanto Co., St. Louis, MO) transgenic cotton will become commercially available in the 2002 growing season. It contains two Cry proteins and has been shown to be more effective against corn earworm and soybean looper than the Bollgard I (Monsanto Co., St. Louis, MO) transgenic lines (Leonard and Torrey, 1999). This pyramiding of multiple Bt toxins has improved efficacy and may delay the development of resistance in pest species. However, the approach being pursued on soybean wherein resistance genes with different modes of action are being incorporated may be a more sustainable strategy for managing the development of resistance to the products of transgenic technology. In addition, this strategy may be the avenue for obtaining regulatory approval for introduction of a Bt soybean. Current Bt resistance management strategies require that a percentage of the crop be planted to non-Bt cultivars to serve as a refugia for the production of susceptible individuals (Gould, 1998). Some have proposed that soybean could serve as a refugia for the pest species that attack transgenic Bt crops. Thus, there may be a reluctance on the part of producers of these crops, regulatory agencies such as the EPA, and commercial companies who market current transgenic technology to introduce a transgenic Bt soybean. These factors along with the determination of the need for a Bt soybean (considering current IPM tactics and infrequent occurrence of the target pests), the cost of a technology fee on a crop with low value per hectare and small insect management costs, and the debate whether insect pests controlled on other Bt cultivars really use soybean enough to consider it refugia surely will surface in the next decade.

Monsanto, the company that has developed many of the transgenic crops grown commercially, has begun to develop transgenic Bt soybean. Several Bt lines have been evaluated in field trials in the USA, Brazil, and Argentina, and some have been shown to be effective against lepidopteran defoliators (Marking, 2001b).

Therefore, it appears that both the private and public sectors are developing another tool to be used in soybean IPM programs, although its adoption may be less essential than it might be for insect management on other crops (Boethel, 1999).

17–4.4 Preventive Tactics for Insect Management

Although current IPM tactics on soybean consisting of pest monitoring, use of economic thresholds, and proper application of pesticides when needed have been successful, the integration of preventive tactics along with these therapeutic tactics has been urged to achieve more sustainable systems (Pedigo, 1992). Some preventive tactics such as biological controls (Pitre, 1983; Shepard and Herzog, 1985; Fuxa, 1994; Yeargan, 1994b), tillage (Troxclair and Boethel, 1984; Hammond and Funderburk, 1985; Hammond and Stinner, 1987; Hammond, 1990; Hammond and Cooper, 1993; Lam and Pedigo, 1998), row spacing (McPherson and Bondari, 1991; Pedigo and Zeiss, 1996), trap crops (McPherson and Newsom, 1984), and cultivar selection (Boethel, 1999) have been intensively studied and implemented with varying degrees of success.

Because of pest resurgence and resistance to pesticides, conventional insecticides have not been considered a sustainable prevention strategy (Pedigo, 1992), but recent findings have challenged this rationale. The insect growth regulator (IGR), diflubenzuron (Dimilin, Uniroyal Chemical Co., Middlebury, CT), has demonstrated effectiveness in the control of velvetbean caterpillar in several studies over a wide geographical range—Georgia, South Carolina, and Brazil (Turnipseed et al., 1974), Louisiana (Layton and Boethel, 1986; Willrich et al., 2002), Florida (Funderburk et al., 1989), and Texas (Way et al., 1995). The extended residual activity of the product and the fact that the pesticide label indicates that use, in the absence of significant insect pressure can result in enhanced yields (Anonymous, 1999), has resulted in interest in the IGR as a preventive management tactic.

In Georgia, a preventive, co-application of diflubenzuron and boron applied at the R2 to R3 growth stage is recommended for yield enhancement (Hudson, 1998). Yield increases of 1075 kg ha^{-1} on loamy soils (McPherson and Gascho, 1999) and 200 to 538 kg ha^{-1} increases on sandy soils (Hudson, 1998) have been demonstrated in the state. In addition, Hudson and Clarke (1997) reported smaller stink bug populations in diflubenzuron-treated plots, with suppression believed to be associated with increased predation and parasitism and not direct mortality from the insecticide.

Studies in other southern states have shown data that contrasts with that found in Georgia. Under irrigated and non-irrigated conditions on both sandy and clay soils in Mississippi, Zhang (2001) reported that applications of diflubenzuron and boron did not significantly improve soybean yield. Willrich et al. (2002) found similar results in trials in Louisiana. Yield enhancement did not materialize in plots oversprayed to remove all insect pressure, thus challenging the claims of yield increases in the absence of insect pests. McPherson and Gascho (1999) attributed the yield increases in their preventively-treated plots in Georgia to control of velvetbean caterpillar. Willrich et al. (2002) found virtually no impact of diflubenzuron on the stink bug complex, confirming the findings seen in previous studies (D.J.

Boethel, 1986. Evaluation of insecticides for control of soybean insect pests. Louisiana Soybean and Grain Research and Promotion Board Annual Report, Baton Rouge, LA.; McPherson and Gascho, 1999). Also, the product did not prove to be effective against soybean loopers, which frequently occur simultaneously with velvetbean caterpillar, and not suitable for ESPS plantings because these soybean plants temporally escape attack from late-season lepidopteran defoliators.

An experimental IGR, methoxyfenozide (Intrepid, Dow Agrosciences, Indianapolis, IN), has exhibited successful control of soybean looper and velvetbean caterpillar in field studies in Louisiana (Fitzpatrick et al.,1999; 2000), including preventive applications (Willrich, 2001). This product, if registered on the crop, potentially has a niche in the cotton-soybean agroecosystems of the mid-South, where soybean loopers are the predominant pest. It also has no effectiveness against stink bugs, and its cost may increase the risk of employing it as a preventive strategy, relative to the low cost of diflubenzuron for preventive treatment of velvetbean caterpillars.

Willrich (2001) discussed the advantages and disadvantages of an alternative, preventive insect management strategy using IGRs. The advantages listed were: (i) insecticide resistance is less likely because velvetbean caterpillar and soybean looper are migratory pests with little evidence of return migration, thus selection is limited to one insecticide application each season; (ii) IGRs are target specific and relatively nontoxic to natural enemies (see Willrich and Boethel, 2001), therefore pest resurgence and secondary pest outbreaks are unlikely; and (iii) less insect monitoring is required, and the problem of inclement weather occurring when late-season insecticide applications are needed is avoided. The disadvantages of the preventive management strategy include: (i) treatments occur before insect population densities are known, and disregard for the economic threshold may result in an unneeded insecticide application; (ii) diflubenzuron is ineffective against soybean looper, thus an additional treatment will be required if this pest reaches the economic threshold; and (iii) neither IGR has activity against the stink bug complex so a class of insecticides capable of controlling this pest complex must be applied when populations reach the economic threshold.

Collectively, the research on IGRs as a preventive management strategy indicate the approach may be practical, but the adoption should be limited to circumstances that meet specific criteria. The IPM for soybean is a balanced strategy employing a multitude of tactics. If preventive application of IGRs is considered an addition to the arsenal of tactics, used when the situation is optimum and not as prophylatic treatment, this management approach may actually have potential as a sustainable prevention strategy.

17–4.5 Soybean Aphid

During late summer and fall of 2000, the soybean aphid, a native of China and Japan, was found in Illinois, Indiana, Iowa, Kentucky, Michigan, Minnesota, Missouri, Ohio, West Virginia, and Wisconsin. This was the first record of an exotic insect pest of soybean in the USA. Based on its broad distribution, it is suspected the soybean aphid had been in North America for more than a year and likely went undetected for 3 or 4 yr. It has spread rapidly. As of November 2002, the soybean aphid had been documented in 20 states, including Georgia and Mississippi in the southern soybean growing area (Plate 17–13).

Little data on the impact of the pest in the USA are available, but it has been reported to cause significant reduction in plant height and up to a 28% yield reduction in experiments conducted in China (DiFonzo and Hines, 2001). Although transmission of viruses by the soybean aphid has not been documented in the USA, the aphid is capable of transmitting several viruses, including alfalfa mosaic, soybean mosaic, bean yellow mosaic, peanut mottle, peanut stunt, and peanut stripe, that naturally infect soybean. Direct feeding injury coupled with virus transmission make the soybean aphid a potential formidable threat. Annual management in the northern USA soybean-growing areas may become imperative and the pest's spread southward may add to the management costs in areas already incurring insect problems. Certainly, it is not unreasonable to predict that the soybean aphid may dominate soybean entomology research over the next decade.

17–5 CLOSING COMMENTS

In the 1970s at a point when national interest in IPM was arguably at its highest level, the soybean IPM program was lauded as a unique, imaginative system designed to prevent unnecessary insecticide usage (Van den Bosch, 1978). Over the years, the system has been durable and adaptable (Turnipseed and Kogan, 1994) and has not experienced serious problems since the basic components were formulated and implemented a quarter century ago. Refinement and improvement of the system remains a goal. Challenges cited in this chapter and those elucidated by others (Kogan and Turnipseed, 1987), including pest complexes, improved pest forecasting, multiple pest-resistant cultivars, and systems integration, will present exciting research opportunities. Accompanying the opportunities, for those that pursue them, will be the responsibility to be stewards of a system that has served the soybean industry well by sustaining profitability and respecting the environment.

REFERENCES

All, J.N., H.R. Boerma, W. Parrott, B. Rector, D. Walker, and C.N. Stewart, Jr. 1999a. New technologies for development of insect resistant soybean. p. 316–318. *In* H.E. Kauffman (ed.) World Soybean Research Conf. VI, Chicago, IL. 4–7 Aug. 1999. Superior Printing, Champaign, IL.

All, J., H.R. Boerma, W. Parrott, C.N. Stewart, Jr., P. Raymer, B. Rector, S. Ramachandran, D. Walker, and M. Tracey. 1999b. Interactions in entomology: Utilization and management of new genetic techniques for insect control in southern field crops. J. Entomol. Sci. 34:2–7.

Alteri, M.A., and J.W. Todd. 1981. Some influences of vegetational diversity on insect communities of Georgia soybean fields. Prot. Ecol. 3:333–338.

Alteri, M.A., J.W. Todd, E.W. Hauser, M. Patterson, G.A. Buschman, and R.H. Walker. 1981. Some effects of weed management and row spacing on insect abundance in soybean fields. Prot. Ecol. 3:339–343.

Anonymous. 1999. Dimilin® 2L insect growth regulator. p. 2179–2181. *In* Crop protection reference -2000. Chemical and Pharmaceutial Press, New York.

Anonymous. 2000. Louisiana Suggested Chemical Weed Control Guide. Publ. 1565. La. Coop. Ext. Serv., Baton Rouge.

Baldwin, J.L., D.J. Boethel, and B.R. Leonard. 2001. Soybean Insect Control. Publ. 2211. La. Coop. Ext. Serv., Baton Rouge.

Baur, M.E., D.J. Boethel, M.L. Boyd, G.R. Bowers, M.O. Way, L.G. Heatherly, J. Rabb, and L. Ashlock. 2000. Arthropod populations in early soybean production systems in the mid-south. Environ. Entomol. 29:312–328.

Baur, M.E., L.M. Hattier, and D.J. Boethel.1999. Comparative feeding by three chrysomelid (Coleoptera:Chrysomelidae) species on eight soybean genotypes. J. Entomol. Sci. 35:283–289.

Beach, R.M., and J.W. Todd. 1986. Comparison of soybean looper (Lepidoptera: Noctuidae) populations in soybean and cotton/soybean agroecosystem. J. Entomol. Sci. 21:21–25

Board, J.E., and D.J. Boethel. 2001. Light interception: A way for soybean farmers to determine when to spray for defoliating insects. Louisiana Agric. 44:8–10.

Board, J.E., A.T. Wier, and D.J. Boethel. 1997. Critical light interception during seed filling for insecticide application and optimum soybean grain yield. Agon. J. 89:374.

Boerma, H.R., J.N. All, and B.G. Rector. 1996. Summary of research by public soybean breeders on soybean looper, velvetbean caterpillar, and corn earworm. Natl. Soybean Breeder/Entomologist Workshop, St. Louis, MO. Dr. E.R. Shipe, Dep. of Agron., Clemson Univ., Clemson, SC.

Boerma, H.R., and W.A. Parrott. 1996. New technologies for the development of pest resistant soybean varieties. Proc. 4th Ann. Southern Soybean Conf., Memphis, TN. 7–9 Feb. 1996.United Soybean Board, St. Louis, MO.

Boethel, D.J. 1999. Assessment of soybean germplasm for multiple insect resistance. p.101–129. In L.L. Clement and S.S. Quisenberry (ed.) Global plant genetic resources for insect-resistant crops. CRC Press, Boca Raton, FL.

Boethel, D.J. 2000. Louisiana annual report. Dynamic soybean insect pest management for emerging agricultural technologies and variable environments. USDA-CSREES Regional Res. Project S-281, Washington, DC.

Boethel, D.J., J.S. Mink, A.T. Wier, J.D. Thomas, B.R. Leonard, and F. Gallardo. 1992. Management of insecticide resistant soybean loopers, *Psuedoplusia includens*, in the southern USA. p. 66–87. In L.G. Copping et al. (ed.) Pest management in soybean. Elsevier, Essex, England.

Boethel, D.J., J.S. Russin, A.T. Wier, M.B. Layton, J.S. Mink and M.L. Boyd. 2000. Delayed maturity associated with southern green stink bug (*Heteroptera:Pentatomidae*) injury at various soybean phenological stages. J. Econ. Entomol. 93:707–712.

Boyd, M.L., and D. J. Boethel. 1998a. Residual toxicity of selected insecticides to Heteropteran predaceous species (Heteroptera: Lygaeidae, Nabidae, Pentatomidae) on soybean. Environ. Entomol. 27:154–160.

Boyd, M.L., and D.J. Boethel. 1998b. Susceptibility of predaceous hemiptearan species to selected insecticides on soybean in Louisiana. J. Econ. Entomol. 91:401–409.

Boyd, M.L., D.J. Boethel B.R. Leonard, R.J. Habetz, L.P. Brown, and W.B. Hallmark. 1997. Seasonal abundance of arthropod populations on selected soybean varieties grown in early season soybean production systems in Louisiana. Bull. 860. La. Agric. Exp. Stn., Baton Rouge.

Bradshaw, L.D., S.R. Padgette, S.L. Kimball, and B. Wells. 1997. Perspectives on glyphosate resistance. Weed Technol.11:189–198.

Buckelew, L.D. L.P. Pedigo, H.M. Mero, M.D.K. Owen, and G.L. Tylka. 2000. Effects of weed management systems on canopy insects in herbicide-resistant soybeans. J. Econ. Entomol. 93:1437–1443.

Buntin, G.D., W.L. Hargrove, and D.L. McCracken. 1995. Populations of foliage-inhabiting arthropods on soybean with reduced tillage and herbicide use. Agron. J. 87:789–794.

Burleigh, J.G. 1972. Population dynamics and biotic control of the soybean looper in Louisiana. Environ. Entomol. 1:290–294.

Campbell, W.V. 1980. Sampling coleopterous stem borers in soybean. p.357–373. In M. Kogan and D.C. Herzog (ed.) Sampling methods in soybean entomology. Springer-Verlag, New York.

Collins, F.L., and S.J. Johnson. 1985. Reproductive response of caged adult velvetbean caterpillar and soybean looper to the presence of weeds. Agric. Ecosyst. Environ. 14:139–149.

Daigle, C.J., D.J. Boethel, and J.R. Fuxa. 1988. Parasitoids and pathogens of green cloverworm (Lepidoptera:Noctuidae) on an uncultivated spring host (Vetch, *Vicia* spp.) and a cultivated summer host (Soybean, *Glycine max*). Environ. Entomol. 17:90–96.

DeGooyer, T.A., and J.A. Browde. 1994. Grasshoppers. p. 57–59. In L.G. Higley and D.J. Boethel (ed.) Handbook of soybean insect pests. Entomol. Soc. Am., Lanham, MD.

DiFonzo, C., and R. Hines. 2001. Soybean aphid in Michigan. Mich. Ext. Bull. E-2748.   Michigan State Univ., East Lansing.

Eastman, C.E., and A.L. Wuensche. 1977. A new insect damaging nodules on soybean: *Rivella quandrifasciata* (Marquart). J. Ga. Entomol. Soc. 12:190–199.

Edwards, C.R., D.A. Herbert, Jr., and J.W. Van Duyn. 1994. Mexican bean beetle. p. 71–92. In L.G. Higley and D.J. Boethel (ed.) Handbook of soybean insect pests. Entomol. Soc. Am., Lanham, MD.

Edwards, C.R., J.R. Obermezer, T.N. Jordan, D.J. Childe, D.H. Scott, J.M. Ferris, R.M. Corigan, and L.W. Bledsoe. 1992. Field crops pest management manual. IPM-1, CS8-59. Purdue Univ. Coop. Ext. Serv., Purdue.

Ellis, J.M. 2001. Glyphosate- and glufosinate-resistant technologies: weed management and off-target crop response. Ph.D. diss. Louisiana State Univ., Baton Rouge (Diss. Abstr. AAI 3016543).

Fehr, W.R., C.E. Caviness, D.T. Burmood, and J.S. Pennington. 1971. Stages of development descriptions for soybean, *Glycine max* (L.) Merrill. Crop Sci. 11:929–931.

Fitzpatrick, B.J., M.E. Baur, and D.J. Boethel. 2000. Evaluation of insecticides against soybean looper and velvetbean caterpillar on soybean. 1999. Arthropod Manage. Tests 25:308–309.

Fitzpatrick, B.J., M.E. Baur, T.S. Hall, D.J. Boethel, and B.R. Leonard.1999. Evaluation of insecticides against soybean looper on soybean in northeast Louisiana. 1998. Arthropod Manage. Tests 24:286–287.

Funderburk, J.E., and T.P. Mack. 1994. Lesser cornstalk borer. p.66–67. *In* L.G. Higley and D.J. Boethel (ed.) Handbook of soybean insect pests. Entomol. Soc. Am., Lanham, MD.

Funderburk, J., J. Maruniak, and D. Boucias. 1989. Velvetbean caterpillar control in soybean. Insecticide Acaracide Tests 16:217.

Funderburk, J., R.M. McPherson, and G.D. Buntin. 1999. Soybean insect management. p. 273–290. *In* L.G. Heatherly and H. F. Hodges (ed.) Soybean production in the mid-South. CRC Press, Boca Raton, FL.

Fuxa, J.R. 1994. Entomopathogens. p. 107–108. *In* L.G. Higley and D.J. Boethel (ed.) Handbook of soybean insect pests. Entomol. Soc. Am., Lanham, MD.

Gould, F. 1998. Sustainability of transgenic insecticidal cultivars: Integrating pest genetics and ecology. Ann. Rev. Entomol. 43:701–726.

Gregory, B.M., Jr., S.J. Johnson, A.N. Lievens, A.M. Hammond, Jr., and A. Delgado-Salinas. 1990. A midlatitude survival model of *Anticarsia gemmatalis* (Lepidoptera: Noctuidae). Environ. Entomol. 19:1017–1023.

Haile, F.J., L.G. Higley, and J.E. Specht. 1998. Soybean cultivars and insect defoliation: Yield loss and economic injury levels. Agron. J. 90:344–352.

Hammond, R.B. 1984. Development and survival of the Mexican bean beetle, *Epilachnia varivestis* Mulsant, on two host plants. J. Kansas Entomol. Soc. 57:695–699.

Hammond, R.B. 1990. Influence of cover crops and tillage on seedcorn maggot (Dipera: Anthomyiidae) populations in soybean. Environ. Entomol. 19:510–514.

Hammond, R.B., and R.L. Cooper. 1993. Interaction of planting times following the investigation of a living green cover crop and control measures on seedcorn maggot populations in soybean. Crop Prot. 12:539–543.

Hammond, R.B., and J.E. Funderburk. 1985. Influence of tillage practices on soil-insect population dynamics in soybeans. *In* R. Shibles (ed.) World Soybean Research. Conf. III, Ames, IA. 12–17 Aug. 1984. Westview Press, Boulder, CO.

Hammond, R.B., C.G. Helm, and R. Nelson. 1998. Introduced soybean lines from China: Screening for insect resistance. J. Econ. Entomol. 91:546–551.

Hammond, R.B., and B.R. Stinner. 1987. Soybean foliage insects in conservation tillage systems: Effects of tillage, previous cropping history, and soil insecticide application. Environ. Entomol. 16:524–531.

Heatherly, L.G. 1999. Early soybean production system. p. 103–118. *In* L.G. Heatherly and H.F. Hodges (ed.) Soybean production in the mid-South. CRC Press, Boca Raton, FL.

Heatherly, L.G., and G.R. Bowers. 1998. Early soybean production system handbook. USB 6009-09 1998-11000. United Soybean Board, St. Louis, MO.

Herbert, D.A., and T.R. Mack. 1987. Identification of lesser cornstalk borer damage to soybeans. Leaflet 103. Ala. Agric. Exp. Stn., Auburn.

Herbert, D.A., Jr., T.P. Mack, P.A. Bachman, and R. Rodriguez-Kabana. 1992. Validation of a model for estimating leaf-feeding by insects in soybean. Crop Prot. 92:27–34.

Hicks, P.M., P.K. Bollick, E.P. Dunigan, B.G. Harville, P. Mitchell, and L.D. Newsom. 1983. Effect of bean pod mottle virus and soybean mosaic virus on nitrogen-fixation in soybean. Proc. La. Acad. Sci. 46:106–108.

Hicks, P.M., P.L. Mitchell, E.P. Dunigan, L.D. Newsom, and P.K. Bollich. 1984. Effect of threecornered alfalfa hopper (Homoptera:Membracidae) feeding on translocation and nitrogen fixation in soybeans. J. Econ. Entomol. 77:1275–1277.

Higley, L.G. 1992. New understandings of soybean defoliation and their implications for pest management. *In* L.G. Copping et al. (ed.) Pest management in soybean. Elsevier Applied Sci., New York.

Higley, L.G., and D.J. Boethel (ed.). 1994. Handbook of soybean insect pests. Entomol. Soc. Am., Lanham, MD.

Higley, L.G., and L.P. Pedigo. 1991. Soybean yield responses and intraspecific competition from simulated seed corn maggot injury. Agron. J. 83:135–139.

Higley, L.G., and L.P. Pedigo. 1997a. The EIL concept. p. 9–21. *In* L.G. Higley and L.P. Pedigo (ed.) Economic thresholds for integrated pest management. Univ. of Nebraska Press, Lincoln.

Higley, L.G., and L.P. Pedigo (ed.). 1997b. Economic thresholds for integrated pest management. Univ. of Nebraska Press, Lincoln.

Hillhouse, T.L., and H.N. Pitre. 1976. Oviposition by *Heliothis* on soybeans and cotton. J. Econ. Entomol. 69:144–146.

Hopkins, J.D., and A.J. Mueller. 1983. Distribution of bean pod mottle virus in Arkansas soybean as related to bean leaf beetle, *Cerotoma trifurcata*, (Coleoptera:Chrysomelidae) population. Environ. Entomol. 12:1564–1567.

Hopkins, J.D., and A.J. Mueller. 1984. Effect of bean pod mottle virus on soybean yield. J. Econ. Entomol. 77:943–947.

Horn, N.L., L.D. Newsom, and R.L. Jensen. 1973. Economic injury thresholds of bean pod mottle and tobacco ringspot virus infection of soybeans. Plant Dis. Rep. 57:811–813.

Hudson, R.D. 1998. Dimilin and boron: Insect control and yield enhancement in soybeans. Publ. Ent 98 RDH (01) July 1998. Univ. of Georgia Coop. Ext. Serv., Tifton.

Hudson, R.D., and J.R. Clarke. 1997. Dimilin and boron for insect control and yield increases in soybeans. p. 1–8. *In* Proc. of the 5th Annual Southern Soybean Conf., Myrtle Beach, SC. 11–13 Feb. 1997. United Soybean Board, St. Louis, MO>

Hudson, R.D., D.C. Jones, and R.M. McPherson. 1993. Soybean insects. p.37–38. *In* R.M. McPherson and G.K. Douce (ed.) Summary of losses from insect damage and costs of control in Georgia. Spec. Publ. 83. Ga. Agric. Exp. Stn., Tifton.

Huffman, F.R., and A.J. Mueller. 1983. Effects of beet armyworm (Lepidoptera: Noctuidae) infestation levels on soybean. J. Econ. Entomol. 76:744–747.

Hunt, T.E., F.J. Haile, W.W. Hoback, and L.G. Higley. 1999. Indirect measurement of insect defoliation. Environ. Entomol. 28:1136–1139.

Hunt, T.E., L.G. Higley, and J. Witkowski. 1995. Bean leaf beetle injury to seedling soybean: Consumption, effects of leaf expansion, and economic injury levels. Agron. J. 87:183–188.

Hutchins, S.H., L.G. Higley, and L.P. Pedigo. 1988. Injury equivalency as a basis for developing multiple-species economic injury levels. J. Econ. Entomol. 81:1–8.

Hutchins, S.H., L.G. Higley, L.P. Pedigo, and P.H. Calkins. 1986. A linear programming model to optimize management decisions in soybean: An integrated soybean pest management example. Bull. Entomol. Soc. Am. 32:96–102.

Irwin, M.E., and D.E. Kuhlman. 1979. Relationships among *Sericothrips variabilis*, systemic insecticides, and soybean yield. J. Ga. Entomol. Soc. 14:148–154.

Jensen, R.L., L.D. Newsom, and J. Gibbens. 1974. The soybean looper: Effects of adult nutrition on oviposition, mating frequency, and longevity. J. Econ Entomol. 67:467–470.

Johnson, M.W., R.E. Stinner, and R.L. Rabb. 1975. Ovipositional response of *Heliothis zea* (Boddie) to its major hosts in North Carolina. Environ. Entomol. 4:291–297.

Jones, W.A., Jr., and M.J. Sullivan. 1983. Seasonal abundance and relative importance of stink bugs in soybean. Bull 1087. South Carolina Agric. Exp. Stn., Clemson.

Klubertanz, T.H., L.P. Pedigo, and R.E. Carlson. 1990. Effects of plant moisture stress and rainfall on population dynamics of the twospotted spider mite (Acari: Tetranychidae). Environ. Entomol. 19:1773–1779.

Klubertanz, T.H., L.P. Pedigo, and R.E. Carlson. 1996. Reliability of yield models of defoliated soybean based on leaf area index versus leaf area removed. J. Econ. Entomol. 89:751–756.

Koethe, R.W., and J.W. Van Duyn. 1984. Aspects of larva/host relations of the soybean nodule fly, *Rivellia quandrifasciata* (Marquart) (Diptera: Platystomatidae). Environ. Entomol. 13:945–947.

Kogan, M. 1976. Evaluation of economic injury levels for soybean insect pests. p. 515–533. *In* L.D. Hill (ed.) World Soybean Research Conf. I, Champaign-Urbana, IL. 3–8 Aug. 1975. Interstate Printers, Danville, IL.

Kogan, M..1980. Insect problems of soybeans in the USA. p. 303–325. *In* F.T. Corbin (ed.) World Res. Conf. II, Raleigh, NC. 26–29 Mar. 1979. Westview Press, Boulder, CO.

Kogan, M.. 1981. Dynamics of insect adaptations to soybeans: Impact of integrated pest management. Environ. Entomol. 10:363–371.

Kogan, M., and D.C. Herzog (ed.) 1980. Sampling methods in soybean entomology. Springer-Verlag, New York.

Kogan, M., and S.G. Turnipseed. 1987. Ecology and management of soybean arthropods. Annu. Rev. Entomol. 32:507–538.

Kogan, M., S.G. Turnipseed, and L.P. Pedigo. 1994. Sampling arthropod populations in soybeans. p. 111–113. *In* L.G. Higley and D.J. Boethel (ed.) Handbook of soybean insect pests. Entomol. Soc. Am., Lanham, MD.

Kogan, M., G.P. Waldbauer, G. Boiteau, and C.E. Eastman. 1980. Sampling bean leaf beetles on soybean. p. 201–236. *In* M. Kogan and D.C. Herzog (ed.) Sampling methods in soybean entomology. Springer-Verlag, New York.

Lam, W.F., and L.P. Pedigo 1998. Response of soybean insect communities to row width under crop-residue management systems. Environ. Entomol. 27:1069–1079.

Lambert, L. 1994. Grape colaspis. p. 56–57. *In* L.G. Higley and D.J. Boethel (ed.). Handbook of soybean insect pests. Entomol. Soc. Am., Lanham, MD.

Lamp, W.O., M..J. Morris, and E.J. Ambrust. 1984. Suitability of common weed species as host plants for the potato leafhopper, *Empoasca fabae*. Entomol. Exp. Appl. 36:586–591.

Layton, M.B. 1983. The effects of feeding by bean leaf beetle larvae, *Cerotoma trifurcata* (Forster), on nodulation and nitrogen fixation of soybeans. M.S. thesis. Louisiana State Univ., Baton Rouge.

Layton, M.B., and D.J. Boethel. 1986. Effects of insecticides on pest and beneficial insects in soybeans, 1985. Insecticide and Acaracide Tests. 11:344–345.

Lentz, G.L. 1985. Occurrence of *Phyllophaga congrua* (LeConte) and *P. implicita* (Horn) (Coleoptera: Scarabaeidae) on soybeans. J. Kans. Entomol. Soc. 58:202–206.

Lentz, G.L. 1994. Dectes stem borer. p. 52–53. *In* L.G. Higley and D.J. Boethel (ed.) Handbook of soybean insect pests. Entomol. Soc. Am., Lanham, MD.

Lentz, G. 2000. Tennessee annual report. Dynamic soybean insect pest management for emerging agricultural technologies and variable environments. USDA-CSREES Regional Res. Project S-281, Washington, DC.

Leonard, B.R., D.J. Boethel, A.N. Sparks, Jr., M. B. Layton, J. S. Mink, A.M. Pavloff, E. Burris, and J. B. Graves. 1990. Variations in the response of soybean looper (Lepidoptera:Noctuidae) to selected insecticides in Louisiana. J. Econ. Entomol. 83:27–34.

Leonard, B.R., and K. Torrey. 1999. Evaluation of selected Bollgard II lines against soybean looper, beet armyworm, bollworm, and tobacco budworm. LAES Res. Summary 120. Louisiana State Univ., Baton Rouge.

Levine, E., and H. Oloumi-Sadeghi. 1996. Western corn rootworm (Coleoptera:Chrysomelidae) larval injury to corn grown for seed production following soybeans grown for seed production. J. Econ. Entomol. 89:1010–1016.

Lincoln, C., W.P. Boger, and F.D. Miner. 1975. The evolution of insect pest management in cotton and soybean: Past experience, present status, and future outlook in Arkansas. Environ. Entomol. 4:1–7.

Lingren, P.S. 1995. Determination of a dynamic economic threshold for the southern green stink bug, *Nezara viridula* (L), (Hemiptera:Pentatomidae) on an indeterminate soybean cultivar. M.S. thesis. Louisiana State Univ., Baton Rouge.

MacIntosh, S.C., T.B. Stone, S.R. Sims, P.L. Hunst, J.T. Greenplate, P.G. Marrone, F.J. Perlak, D.A. Fishhoff, and R.C. Fuchs. 1990. Specificity and efficacy of purified *Bacillus thuringiensis* proteins against agronomically important insects. J. Invertebr. Pathol. 56:258–266.

Marking, S. 2001a. Tiny terrors. Soybean Digest 61:64–65.

Marking, S. 2001b. Next up: Bt soybean? Soybean Digest 61:8–9.

Mascarenhas, R.N., D.J. Boethel, B.R. Leonard, M.L. Boyd, and C.J. Clemens. 1998. Resistance monitoring to *Bacillus thuringiensis* insecticides for soybean loopers collected from soybean and transgenic Bt cotton. J. Econ. Entomol. 91:1044–1050.

Mazier, M., C. Pannetier, J. Tourneur, L. Jouanin, and M. Giband. 1997. The expression of *Bacillus thuringiensis* toxin genes in plant cells. Biotechnol. Annu. Rev. 3:313–347.

McLeod, P.J., S.Y. Young III, and W.C. Yearian. 1978. Effectiveness of microbial and chemical insecticides on beet armyworm larvae on soybeans. J. Ga. Entomol. Sci. 13:266–269.

McPherson, R.M. 1996. Relationship between soybean maturity group and the phenology and abundance of stink bugs (Heteroptera: Pentatomidae): Impact on yield and quality. J. Entomol. Sci. 31:199–208.

McPherson, R.M. 1999. Georgia annual report. Dynamic soybean insect pest management for emerging agricultural technologies and variable environments. USDA-CSREES Regional Res. Project S-281, Washington, DC.

McPherson, R.M. 2000. Georgia annual report. Dynamic soybean insect pest management for emerging agricultural technologies and variable environments. USDA-CSREES Regional Research Project S-281, Washington, DC.

McPherson, R.M., and K. Bondari. 1991. Influence of planting date and row width on abundance of vel-vetbean caterpillars (Lepidoptera:Noctuidae) and southern green stink bugs (Heteroptera:Pen-tatomidae) in soybean. J. Econ. Entomol. 84:311–316.

McPherson, R.M., and G.J. Gascho. 1999. Interactions in entomology: mid-season Dimilin and boron treatment impact on the incidence of arthropod pests and yield enhancement of soybean. J. En-tomol. Sci. 34:17–30.

McPherson, R.M., and A.L. Lambert. 1995. Abundance of two whitefly species (Homoptera: Aleyro-didae) on Georgia soybean. J. Entomol. Sci. 30:527–533.

McPherson, J.E., and R.M. McPherson. (ed.) 2000. Stink bugs of economic importance in America north of Mexico. CRC Press, Boca Raton, FL.

McPherson, R.M., and L.D. Newson. 1984. Trap crops for control of stink bugs in soybean. J. Ga. En-tomol. Soc.19:470–480.

McPherson, R.M., L.D. Newson, and B.F. Farthing. 1979. Evaluation of four stink bug species from there genera affecting soybean yield and quality in Louisiana. J. Econ. Entomol. 72:188–194.

McPherson, R.M., J.W. Todd, and K.V. Yeargan. 1994. Stink bugs. p. 87–90. In L.G. Higley and D.J. Boethel (ed.) Handbook of soybean insect pests. Entomol. Soc. Am., Lanham, MD.

McPherson, R.M., M.L. Wells, and C.S. Bundy. 2001. Impact of early soybean production system on arthropod pest populations in Georgia. J. Econ. Entomol. 30:76–81.

McWilliams, J.M. 1983. Relationship of soybean pod development to bollworm and tobacco budworm damage. J. Econ. Entomol. 76:502–506.

Mellors, W.K.A., A. Allegro, and A.N. Hsu. 1984. Effects of carbofuran and water stress on growth of soybean plants and twospotted spider mite (Acari:Tetranychidae) populations under green-house conditions. Environ. Entomol. 13:561–567.

Miner, F.D., and B.A. Dumas. 1980. Effect of green stink bug damage on soybean seed quality before and after storage. Bull. 844. Ark. Agric. Exp. Stn., Fayetteville.

Mitchell, P.L., and L.D. Newsom. 1984. Seasonal history of the threecornered alfalfa hopper (Ho-moptera:Membracidae) in Louisiana. J. Econ. Entomol. 77:906–914.

Mueller, A.J., and B.N. Engroff. 1979. Effects of infestation levels of Heliothis zea on soybean. J. Econ. Entomol. 73:271–275.

Mueller, A.J., and A.W. Haddox. 1980. Observations on seasonal development of bean leaf beetle, Cero-toma trifurcata (Forster) and incidence of bean pod mottle virus in Arkansas soybean. J. Ga. En-tomol. Soc. 15:398–403.

Mueller, A.J., F.R. Huffman, and B.A. Dumas. 1979. Yellow-striped armyworms on soybeans. Ark. Farm Res. (March–April):4.

Mueller, A.J., and R.G. Luttrell. 1977. Thrips on soybean. Ark.. Farm Res. (July–August):7.

Myhre, D.L., H.N. Pitre, M. Haridasan, and J.D. Hesketh. 1973. Effect of bean pod mottle virus on yield components and morphology of soybeans in relation to soil water regimes:a preliminary study. Plant Dis. Rep. 57:1050–1054.

Narvel, J.M., D.R.Walker, B.E. Rector, J.N. All, W.A. Parrott, and H.R. Boerma. 2001. A retrospective DNA marker assessment of the development of insect resistant soybean. Crop Sci. 41:1931–1939.

Newsom, L.D. 1978. Progress in integrated pest management of soybean pests. p. 157–180. In E.H. Smith and D. Pimental (ed.) Pest control strategies. Academic Press, New York.

Ogunlana, M.O., and L.P. Pedigo. 1974. Economic injury levels of the potato leafhopper on soybeans in Iowa. J. Econ. Entomol. 67:29–32.

O'Neal, M.E., M.E. Gray, and C.A. Smyth. 1999. Population characteristics of a western corn rootworm (Coleoptra: Chrysomelidae) strain in east-central Illinois corn and soybean fields. J. Econ. En-tomol. 92:1301–1310.

Padgette, S.R., K.H. Kolacz, X. Delannay, D.B. Re, B.J. La Vallee, C.N. Tinius, W.K. Rhodes, Y.I. Otero, G.F. Berry, D.A. Eickholtz, V.M. Peschke, D.L. Nida, N.B. Taylor, and G.M. Kishoe. 1995. De-velopment, identification, and characterization of a glyphosate-tolerant soybean line. Crop Sci. 35:1451–1461.

Panizzi, A.R., and F. Slansky. 1985. Review of phytophagous pentatomids (Hemiptera: Pentatomidae) associated with soybean in the Americas. Fla. Entomol. 68:184–214.

Panizzi, A.R., J.G. Smith, L.A.G. Pereira, and J. Yamashita. 1979. Efecitos dos danos de Piezodorous guildinii (Westood 1837) no rendimento é qualidade da soja. An. I Semin. Nac. Perq. Soja 2:59–78.

Parrott, W.A., J.N. All, M.J. Adang, M.A. Bailey, H.R. Boerma, and C.N. Stewart, Jr. 1994. Recovery and evaluation of soybean plants transgenic for a Bacillus thuringiensis var. kurstaki insectici-dal gene. In Vitro Cell. Dev. Biol. Plant 30:144–149.

Patel, V.C., and H.N. Pitre. 1976. Transmission of bean pod mottle virus by bean leaf beetle and me-chanical inoculation to soybeans at different growth stages. J. Ga. Entomol. Soc. 11:289–293.

Pedigo, L.P. 1992. Integrating preventative and therapeutic tactics in soybean insect management. p. 10–19. *In* L.G. Copping et al. (ed.) Pest management in soybean. Elsevier Appl. Sci., New York.

Pedigo, L.P. 1994. Bean leaf beetle. p. 42–44. *In* L.G. Higley and D.J. Boethel (ed.) Handbook of soybean insect pests. Entomol. Soc. Am., Lanham, MD.

Pedigo, L.P. 2000. Iowa annual report. Dynamic soybean insect pest management for emerging agricultural technologies and variable environments. USDA-CREES Regional Research Project S-281, Washington, DC.

Pedigo, L.P., E.J. Bechinski, and R.A. Higgins. 1983. Partial life tables of the green cloverworm (Lepidoptera:Noctuidae) in soybean and a hypothesis of population dynamics in Iowa. Environ. Entomol. 12:186–195.

Pedigo, L.P., L.H. Higley, and P.M. Davis. 1989. Concepts and advances in economic thresholds for soybean entomology. p. 1487–1493. *In* A.J. Pascale (ed.) World Soybean Research Conf. IV, Buenos Aries, Argentina. 5–9 Mar. 1989. Realization Orientation, Grafica Editora, Buenos Aires, Argentina.

Pedigo, L.P., S.H. Hutchins, and L.G. Higley. 1986. Economic injury levels in theory and practice. Annu. Rev. Entomol. 31:341–368.

Pedigo, L.P., and M.R. Zeiss.1996. Effect of soybean planting date on bean leaf beetle (Coleoptera:Chrysomelidae) abundance and pod injury. J. Econ. Entomol. 89:183–188.

Pedigo, L.P., M.R. Zeiss, and M.E. Rice. 1990. Biology and management of bean leaf beetle in soybean. p. 109–117. *In* Proc. of the 1990 Crop Production and Protection Conf. Iowa State Univ. Ext. Serv., Ames.

Peterson, R.K.D. 1994. Yellow woollybear. p. 102. *In* L.G. Higley and D.J. Boethel (ed.) Handbook of soybean insect pests. Entomol. Soc. Am., Lanham, MD.

Peterson, R.K.D. 1997. The status of economic-decision-level development. p. 151–178. *In* L.G. Higley and L.P. Pedigo (ed.) Economic thresholds for integrated pest management. Univ. of Nebraska Press, Lincoln.

Pitre, H.N. (ed.) 1983. Natural enemies of arthropod pests in soybean. Southern Coop. Series Bull. 285. Mississippi State Univ. Mississippi State.

Pitre, H.N., and D.B. Hogg. 1983. Development of fall armyworms on cotton, soybean, and corn. J. Ga. Entomol. Soc. 18:182–187.

Pitre, H.N., V.C. Patel, and B.L. Keeling. 1979. Distribution of bean pod mottle disease on soybeans in Mississippi. Plant Dis. Rep. 63:419–423.

Rector, B.G., J.N. All, and H.R. Boerma. 1996. Identification of molecular markers associated with insect resistance QTLS in soybean. 6th Biennial Conf.-Molecular and Cellular Biology of Soybean, Columbia, MO. 12–14 Aug. 1996. (abstr.) Univ. of Missouri, Columbia, MO.

Rector, B.G., J.N. All, W.A. Parrott, and H.R. Boerma. 1999. Quantitative trait loci for antixenosis resistance to corn earworm in soybean. Crop Sci. 39:531–538.

Rector, B.G., J.N. All, W.A. Parrott, and H.R. Boerma. 2000. Quantitative trait loci for antibiosis resistance to corn earworm in soybean. Crop Sci. 40:233–238.

Rice, M.E., R.K. Krell, W.F. Lam, and L.P. Pedigo. 2000. New thresholds and strategies for management of bean leaf beetles in Iowa soybean. p. 75–84. *In* Proc. of the Integrated Crop Management Conf., Ames, IA. 29–30 Nov. 1990. Iowa State Univ. Ext. Serv., Ames.

Russin, J.S., D.B. Orr, M.B. Layton, and D.J. Boethel. 1988. Incidence of microorganisms in soybean seeds damaged by stink bug feeding. Phytopathology 78:306–310.

Sammons, A.E., C.R. Edwards, L.W. Bledsoe, P.J. Boeve, and J.J. Stuart. 1997. Behavioral and feeding assays reveal a western corn rootworm (Coleoptera:Chrysomelidae) variant that is attracted to soybean. Environ. Entomol. 26:1336–1342.

Shelton, M.D., and C.R. Edwards. 1983. Effects of weeds on the diversity and abundance of insects in soybeans. Environ. Entomol. 12:296–298.

Shepard, M., and D.C. Herzog. 1985. Soybean: Status and current limits to biological control in the southeastern U.S. p. 557–571. *In* M.A. Hoy and D.C. Herzog (ed.) Biological control in agricultural IPM systems. Academic Press, New York.

Smelser, R.B., and L.P. Pedigo. 1992. Soybean seed yield and quality reduction by bean leaf beetle (Coleoptera:Chrysomelidae) pod injury. J. Econ. Entomol. 85:2399–2403.

Sparks, A.N. Jr. 1986. The three cornered alfalfa hopper on soybean: Determination of damaging stages, development of sampling methodology, refinement of the action threshold, and evaluation of insecticidal control. Ph.D. diss. Louisiana State Univ., Baton Rouge (Diss. Abstr. AAI 8625357).

Sparks, A.N., Jr., and D.J. Boethel. 1987. Late-season damage to soybeans by threecornered alfalfa hopper (Homoptera:Membracidae) adults and nymphs. J. Econ. Entomol. 80:471–477.

Sparks, A.N., Jr., and L.D. Newsom. 1984. Evaluation of pest status of the threecornered alfalfa hopper (Homoptera:Membracidae) on soybean in Louisiana. J. Econ. Entomol. 77:1553–1558.

Stewart, C.N., M.J. Adang, J.N. All, H.R. Boerma, G. Cardinceau, and D. Tucker. 1996. Genetic transformation, recovery and characterization of fertile soybean transgenic for a synthetic *Bacillus thuringiensis* crylAc gene. Plant Physiol. 112:121–129.

Terry, I., J.R. Bradley, Jr., and J.W. Van Duyn.1987. Survival and development of *Heliothis zea* (Lepidoptora: Nocutidae) larvae on selected soybean growth stages. Environ. Entomol.16:441–445.

Thomas, J.W., J.S. Mink, D.J. Boethel, A.T. Wier, and B.R. Leonard.1994. Activity of two novel insecticides against permethrin-resistant *Pseudoplusia includens*. Pestic. Sci. 40:239–243.

Todd, J.W. 1989. Ecology and behavior of *Nezara viridula*. Ann. Rev. Entomol. 34:273–292.

Todd, J.W., R. M. McPherson, and D.J. Boethel. 1994. Management tactics for soybean insects. p. 115–117. *In* L.G. Higley and D. J. Boethel (ed.) Handbook of soybean insect pests. Entomol. Soc. Am., Lanham, MD.

Troxclair, N.N., and D.J. Boethel. 1984. Influence of tillage practices and row spacing on soybean insect populations in Louisiana. J. Econ. Entomol. 77:1571–1579.

Tugwell, P., F.D. Miner, and E.E. Davis. 1972. Threecornered alfalfa hopper infestations and soybean yield. J.Econ. Entomol. 65:1731–1733.

Turnipseed, S.G. 1973. Insects. p. 545–572. *In* B.E. Caldwell (ed.) Soybeans: Improvement, production, and uses. ASA, Madison, WI.

Turnipseed, S.G., and M. Kogan. 1976. Soybean entomology. Ann.Rev. Entomol. 21:247–282.

Turnipseed, S.G., and M. Kogan. 1994. Principles and history of soybean pest management. p. 109–110. *In* L.G. Higley and D.J. Boethel (ed.) Handbook of soybean insect pests. Entomol. Soc. Am., Lanham, MD.

Turnipseed, S.G., E.A. Heinrichs, R.F.P. Da Silva, and J.W. Todd. 1974. Response of soybean insects to foliar applications of a chitin synthesis inhibitor TH 6040. J. Econ. Entomol. 67:760–762.

Van den Bosch, R. 1978. Pesticide conspiracy. Univ. of California Press, Berkeley.

Van Duyn, J.W., A.J. Mueller, and C.S. Eckel. 1994. Corn earworm. p. 48–50. *In* L.G. Higley and D.J. Boethel (ed.) Handbook of soybean insect pests. Entomol. Soc. Am., Lanham, MD.

Van Duyn, J.N., S.E. Turnipseed, and J.D. Maxwell. 1971. Resistance in soybeans to the Mexican bean beetle. I. Sources of resistance. Crop Sci. 11:572–573.

Waddill, V., F. Slansky, and J. Strayer.1984. Performance and host preference of adult banded cucumber beetles, *Diabrotica balteata*, when offered several crops. J. Agric. Entomol. 1:330–338.

Walker, D., H.R. Boerma, J. All, and W. Parrott. 2002. Combining crylAc with QTL alleles forum PI 229358 to improve soybean resistance to lepidopteran pests. Mol. Breed. 9:43–51.

Walker, D.R., J.N. All, R.M. McPherson, H.R. Boerma, and W.A. Parrott. 2000. Field evaluation of soybean engineered with a synthetic Cry1Ac transgene for resistance to corn earworm, velvetbean caterpillar (Lepidoptera:Noctuidae) and lesser cornstalk borer (Lepidoptera:Pyralidae). J. Econ. Entomol. 93:613–622.

Walters, H.J. 1964. Transmission of bean pod mottle virus by bean leaf beetles. Phytopathology 54:240.

Way, M.O. 1994. Status of soybean insect pests in the USA. p. 15–16. *In* L.G. Higley and D.J. Boethel (ed.) Handbook of soybean insect pests. Entomol. Soc. Am., Lanham, MD.

Way, M.O., N.G. Whitney, and R.G. Wallace. 1995. Evaluation of Dimilin and Benlate for soybean insect pest and disease control, 1992. Arthropod Manage. Tests. 20:248–249.

Wier, A.T., and D.J. Boethel. 1995. Foliage consumption and larval development of three noctuid pests of soybean and cotton. J. Entomol. Sci. 30: 359–361.

Wintersteen, W.K., J.A. Browde, and M.E. Rice. 1990. Grasshopper management in soybeans. Proc. of the 1990 Crop Production and Protection Conf., Ames, IA. 18–19 Dec. 1990. Iowa State Univ. Ext. Serv., Ames.

Willrich, M.M. 2001. Effects of insect growth regulators on soybeans and soybean insect pests: Insecticidal efficacy, yield enhancement, and preventive use. M.S. thesis. Louisiana State Univ., Baton Rouge.

Willrich, M.M., and D.J. Boethel. 2001. Effects of diflubenzuron on *Pseudoplusia includeus* (Lepidoptera: Noctuidae) and its parasitoid *Copidosoma floridanum* (Hymenoptera: Encrytidae). Environ. Entomol. 30:794–797.

Willrich, M.M., D.J. Boethel, B.R. Leonard, D.C. Blouin, B.J. Fitzpatrick, and R.J. Habetz. 2002. Late-season insect pests of soybean in Louisiana: Preventive management and yield enhancement. Bull. 880. La. Agric. Exp. Stn., Baton Rouge.

Wolf, R.A., L.P. Pedigo, R.H. Shaw, and L.D. Newsom. 1987. Migration/transport of the green cloverworm, *Plathypena vscabra* (F.) (Lepidoptera: Noctuidae), into Iowa as determined by synoptic-scale weather patterns. Environ. Entomol. 16:1169–1174.

Yeargan, K.V. 1977. Effects of green stink bug damage on yield and quality of soybeans. J. Econ. Entomol. 70:619–622.

Yeargan, K.V. 1994a. Potato leafhoppers. p. 75–77. *In* L.G. Higley and D.J. Boethel (ed.) Handbook of soybean insect pests. Entomol. Soc. Am., Lanham, MD.

Yeargan, K.V. 1994b. Predators and parasitoids. p. 105–106. *In* L.G. Higley and D.J. Boethel (ed.) Handbook of soybean insect pests. Entomol. Soc. Am., Lanham, MD.

Yeargan, K. 2000. Kentucky annual report. Dynamic soybean insect pest management for emerging agricultural technologies and variable environments. USDA-CSREES Regional Res. Project S-281, Washington, DC.

Zhang, L. 2001. Effects of foliar application of boron and Dimilin on soybean yield. Mississippi Agric. and For. Exp. Stn. Res. Rep. 22 (16).

18   Weed Biology and Management

DOUGLAS D. BUHLER

Michigan State University
East Lansing, Michigan

ROBERT G. HARTZLER

Iowa State University
Ames, Iowa

18–1 BASIC CONCEPTS AND IMPACTS OF WEEDS

18–1.1 Definitions and Characteristics

Agriculture has been defined as a controversy with weeds, and profitable soybean [*Glycine max* (L.) Merr.] production is indeed dependant upon successfully managing weeds. Many definitions have been used to characterize plants with weedy characteristics, but according to the Weed Science Society of America, a weed is "any plant that is objectionable or interferes with the activities or welfare of man". This definition implies a cost associated with the weed, rather than simply being a plant growing where it is not wanted as described by Bailey and Bailey (p. 778, 1941).

In order for a plant to be a successful weed of soybean and other agronomic crops, the species must have characteristics that allow it to survive and reproduce in spite of repeated attempts to remove it from the field. Baker (1974) developed a list of characteristics that would be favorable for weedy species, and included the following: (i) ability to germinate under many environments, (ii) persistent seed, (iii) rapid growth, (iv) prolonged period of seed production, (v) able to reproduce through both self- and cross-pollination, (vi) cross-pollination not dependent upon specialized vectors, (vii) high seed production under favorable growing conditions, (viii) ability to produce seed under a wide variety of environmental conditions, (ix) adaptations for short- and long-distance dispersal of reproductive units, (x) perennial species that have vegetative reproductive capacity and are not easily removed from the soil, and (xi) increased competitiveness via allelochemicals or other mechanisms. From this list it is clear that reproductive characteristics are very important attributes of weediness. Baker (1974) pointed out that no single species has all of the above characteristics, but that as the number of these characteristics possessed by a species increased, the weediness of a species would also increase.

18–1.2 Monetary Impacts

A survey of Midwest U. S. farmers reported that annual weeds were perceived to be the greatest pest problem of corn (*Zea mays* L.) and soybean production, followed by perennial weeds (Aref and Pike, 1998). Soybean insects and diseases were rated as less problematic among pest complexes. The impact of weeds on soybean production consists of reductions in seed yield and the cost of weed control efforts. Bridges (1994) made a conservative estimate that the annual economic impact of weeds on U.S. agriculture exceeded $15 billion. The impact of weeds on soybean production is poorly documented since damage is field-specific and varies by regions and years. However, in Arkansas during 1997 it was estimated that weeds reduced soybean yields by a value of $37 ha^{-1} on 0.45 million ha for a total loss of more than $16 million (Webster, 1998). An additional cost of $1 million was attributed to increased harvest costs due to effects of weeds on harvesting efficiency.

In addition to direct yield losses, controlling weeds is a significant expense in soybean production. The average cost for herbicides in Iowa soybean production was estimated at $62.20 ha^{-1} in 2000 (Duffy and Smith, 2000). With more than 4.1 million ha planted to soybean in Iowa, the cost for herbicides alone approached $250 million per year in this state. If herbicide costs are assumed to be similar in other soybean-growing regions, this represents a cost of more than $1 billion spent annually for herbicides on the more than 28 million ha of soybean grown in the USA. In addition to herbicides, much of the tillage used during soybean production is a component of the weed control program. Seedbed preparation tillage was used on approximately 70% of the soybean crop in the USA in 2000 (Anonymous, 2000). The cost of seedbed preparation ranges from $10 to $20 ha^{-1}, whereas interrow cultivation costs approximately $7.50 ha^{-1} (Duffy and Smith, 2000). This results in about $500 to $750 million being spent annually for tillage in soybean, with a primary purpose to enhance the effectiveness of weed control programs.

18–1.3 Ecological Concepts

Plant communities in agronomic fields are continually changing in response to the annual disturbances and cultural practices used to control weeds. Plant succession is a change in the composition of species occupying a habitat over time (Radosevich et al., 1997). If an environment is stable, the rate of change slows until the climax vegetation for the habitat is reached. The presence of weeds in a soybean field is a form of annual succession since weeds occupy niches previously vegetated but opened up by tillage and/or herbicide application. Given the repeated disturbance in soybean fields, climax vegetation does not develop.

Grime (1977) stated that there are three types of habitats for which plants have evolved, with plant types having specific differing characteristics that favor their survival in different habitats. Most weeds are classified as ruderals–plants adapted for survival in areas of frequent disturbance and high resource availability. Characteristics of ruderals include an annual or short-lived perennial life cycle, rapid relative growth rate, and a tendency to devote a large proportion of their biomass to reproductive structures. Relatively few species possess the characteristics required to survive the specialized habitat of agricultural fields. Holm et al. (1977, 1997) com-

Table 18–1. Changes in the relative rank of weeds of soybean in the southern USA from 1974 to 1995.†

Species	Relative rank		
	1974	1983	1995
Common cocklebur (*Xanthium strumarium* L.)	1	2	4
Johnsongrass [*Sorghum halepense* (L.) Pers.]	2	3	3
Pigweeds (*Amaranthus* spp.)	3	5	5
Morningglories (*Ipomoea* spp.)	4	1	2
Sicklepod [*Senna obtusifolia* (L.) Irwin and Barneby]	5	4	1
Ragweeds (*Ambrosia* spp.)	6	6	9
Hemp sesbania [*Sesbania exaltata* (Raf.) Rybd.]	7	8	13
Prickly sida (*Sida spinosa* L.)	8	--‡	8
Florida beggarweed [*Desmodium tortuosum* (Sw.) DC.]	9 (tie)	7	6
Nutsedges (*Cyperus* spp.)	9 (tie)	9	7

†Adapted from Webster and Coble (1997).
‡Species was not included in the 1983 survey.

piled a list of 104 species classified as the world's worst weeds. Less than 30 of these species were considered important weeds of soybean.

The composition of a weed community of a specific area of land is influenced by many factors, but tillage and crop rotation usually have a larger effect than specific control tactics (Buhler et al., 2000). The crop rotations practiced in the major soybean-producing regions of the USA are dominated by annual crop species and thus, favor weed communities dominated by a few annual species. Weed communities generally are more diverse when crop rotations are more complex (Aldrich, 1984; Hume, 1982). The diversity of weeds decreased with increasing levels of tillage in an Ohio study (Cardina et al., 1991) and numerous studies have documented significant shifts among weed species as tillage was reduced or eliminated (Buhler, 1995). Changes in herbicide use in agricultural fields generally affect the relative abundance of individual species in a community, rather than determining the absence or presence of species (Hume, 1987; Mahn and Helmecke, 1979). Development of herbicide resistant biotypes is an example of a herbicide-induced shift changing the relative abundance of weed species within a community.

The small number of weeds adapted to soybean fields results in a stable group of dominant weeds. A comprehensive list of important weeds in U.S. soybean production regions was included in an earlier edition of this monograph (Jordan et al., 1987). Surveys of weed scientists in the southern USA found that only one weed species ranked in the top 10 weeds of soybean in 1974 was not included on the similar list compiled in 1995 (Table 18–1) (Webster and Coble, 1997). Changes in the ranking among the weed species were attributed to introductions of new herbicides with greater efficacy on specific species. Occasionally species of little economic importance increase in prevalence due to changes in crop production practices or movement into new regions. For example, adoption of no-tillage systems allows the survival of many winter annual and perennial weeds unable to tolerate intensive soil disturbance (Buhler, 1995). Occasionally, new species will move into an area. Woolly cupgrass [*Eriochloa villosa* (Thunb.) Kunth], a native of China, was first reported in Minnesota in 1965 (Strand and Miller, 1980). During the 1980s and 1990s woolly cupgrass spread throughout much of the northern Corn Belt.

18–2 WEED POPULATION BIOLOGY

Understanding weed biology is critical for the development of efficient weed management systems. As used in this chapter, weed management and weed control are not considered synonymous (see Section 18–5). Weed control refers to actions used to remove an existing weed population at a given point in time. Weed management includes all factors that influence the weed community over a period of years (Buhler, 1996; Zimdahl, 1991). Improved information on weed biology provides the foundation for the development of new strategies and more efficient ways for using existing weed control tools, resulting in more reliable weed management systems that are cost-effective and pose less of a threat to the environment.

Weed control recommendations for soybean production typically provide information on appropriate tillage methods and herbicide selection. The information concerning weed infestations used to base these recommendations typically is not sufficient to optimize the efficiency of these strategies. Information on weed populations can be improved by increasing the time spent scouting or otherwise assessing fields. However, time and cost restrict the amount of scouting on high area, low unit-value crops such as soybean. This problem could be alleviated with an improved understanding of weed population biology, allowing us to predict when best to invest time in scouting and determine the optimum time for weed control operations.

Many biological, cultural, and environmental factors regulate weed population dynamics at various scales (Cardina et al., 1999). Understanding the principles that regulate weed populations at different scales will allow for the anticipation of the effects of weather and management practices on weed behavior in the short- and long-terms and help promote a more holistic approach to weed management. In recent years there has been increasing attention on development of weed management systems that go beyond matching the appropriate herbicide to the primary weeds present in a field. Weed managers are beginning to view weeds as a more complex problem that needs to be treated as part of the cropping system rather than a problem to be solved independently.

Managing weeds at a global scale is not a topic often considered by agriculturalists (Cardina et al., 1999). However, many of our most important weed species are exotic (Holm et al., 1977, 1997) and were accidently or intentionally introduced to distant locations sometime in the past. With global trade in soybean and other grain crops increasing, the potential introduction of weedy species into new habitats will continue to be an issue. Therefore, it is important to understand distribution and adaptation of weeds in all soybean-growing areas and to be aware of the potential of these species moving into new areas.

Understanding weed distributions on a regional basis may also be important in weed management (Burkart and Buhler, 1997). The spatial domains of individual weed species and species assemblages across a region are a reflection of a combination of species adaptation and selection pressure. Changes in weather patterns or production practices may cause significant changes in weed communities over large regions. Understanding these relationships may help predict or forestall shifts in weed populations before they cause yield losses or increase weed control costs.

Of primary interest to most soybean producers and weed managers is how weeds behave in individual fields. The soybean crop and its associated management practices create a matrix of resource conditions, mortality events, and stresses that regulate the ability of various weed species to survive and proliferate (Liebman and Ohno, 1998). Weed population dynamics, crop resource use, and competitive interactions are affected by tillage systems (Buhler, 1995) and seedbed preparation (Mohler, 1993), residues of previous crops and cover crops (Putnam and DeFrank, 1983; Teasdale, 1998), fertility management (Dyck and Liebman, 1994; Tollenaar et al., 1994), manure, composts, and other organic amendments (Bloemhard et al., 1992; Kennedy and Kremer, 1996), soybean density and spatial patterns (Stoller et al., 1987), soybean cultivar (Pester et al., 1999), and rotational sequence of crops (Liebman and Ohno, 1998).

While resources often become limiting at some point in the growing season, resources unused by the crop, especially after crop harvest or soon after crop planting, is a major contributor to the ability of weeds to infest crop land. In annual crops, large areas of bare soil with an ample supply of light, water, and nutrients are usually available to weedy vegetation for a significant portion of the growing season. Since soil devoid of vegetation is not a natural condition (Harper, 1977), nature moves to fill this void, commonly with weeds. Thus, it can be argued that the most basic condition that allows weeds to persist is the availability of resources not used by the crop (Harper, 1977; Radosevich et al., 1997). As such, periods of excess resources are more important in defining weed communities than periods of limiting resources.

18–3 WEED ESTABLISHMENT AND PERSISTENCE

The majority of the weed species of economic importance in soybean production are annual species. The seed bank in the soil is the primary source of new infestations of annual weeds (Cavers, 1983) and weed seed bank characteristics influence both the weed populations that occur in a field and success of weed control practices.

Many factors are involved in the generation and regulation of the weed seed bank in the soil. These factors affect the density, species composition, and spatial distribution of the seed bank and fluctuate with seed introductions and losses (Buhler and Hartzler, 2001; Burnside et al., 1986; Wilson, 1988). Management practices have major impacts on these processes and represent opportunities for regulating seed bank characteristics in crop production systems.

Seed bank composition is influenced by past farming practices and varies from field to field (Fenner, 1985; Robinson, 1949) and within fields (Dieleman et al., 2000; Mortensen et al., 1993). Reports of seed bank size in agricultural land range from near zero to as much as one million seeds m^{-2} (Fenner, 1985). Generally, seed banks are composed of many species, with a few dominant species comprising 70 to 90% of the total seed bank (Wilson, 1988). These species are the primary pests due to the ability to avoid control or their adaptation to the cropping system.

18–3.1 Additions to the Weed Seed Bank

New seeds may enter the seed bank through many sources, but the largest source is plants producing seeds within the field (Cavers, 1983). A characteristic of many weed species is the potential for prolific seed production (Stevens, 1957). However, weeds present in agricultural fields usually produce fewer seeds due to competition from the crop, damage from herbicides, and late emergence. Common cocklebur (*Xanthium strumarium* L.) growing without crop competition produced more than 7000 seeds plant^{-1}, whereas plants growing with soybean produced 1100 seeds plant^{-1} (Senseman and Oliver, 1993). Velvetleaf (*Abutilon theophrasti* Medik.) seed production was reduced up to 82% by competition with soybean (Lindquist et al., 1995). Increased shading, which often occurs when weeds emerge later than the crop, also reduces seed production. For example, 76% shade reduced velvetleaf seed production up to 94% (Bello et al., 1995). Sublethal doses of herbicide reduced seed production by as much as 90% (Biniak and Aldrich, 1986; Salzman, et al., 1988).

Seeds may also enter fields from external sources such as farm equipment, contaminated crop seed, animals, wind, or manure (Cavers, 1983; Mt. Pleasant and Schlather, 1994). The number of seeds introduced into the seed bank by these sources is smaller than those produced by weeds in the field; however, these sources are often important in establishing infestations of new species. Many weeds (e.g., Canada thistle [*Cirsium arvense* (L.) Scop.], horseweed [*Conyza canadensis* (L.) Cronq.], and dandelion (*Taraxacum officinale* Weber in Wiggers) have seeds adapted to wind dispersal. Dandelion and horseweed have become problems in no-tillage systems partially due to the wind transport of their seeds (Buhler, 1995).

While most of the seeds are killed when passing through the digestive tracts of animals, a small percentage typically survives in manure (Harmon and Keim, 1934). A study of 20 New York dairy farms found that, on average, spreading manure introduced 350 weed seeds m^{-2} (Mt. Pleasant and Schlather, 1994). This seemingly high number of seeds was relatively low compared with the number of seeds already in the seed bank. If manure is spread on the fields where the feed was grown, seeds returned to the field will be of little consequence. However, manure can be a source of new weed problems if feed is moved between farms and is contaminated with seed of species not in the field (Eberlein et al., 1992).

Another mechanism of weed seed transport is farm machinery moving between fields. This mechanism has become increasingly important as machinery is moved greater distances due to increasing farm size and contract harvesting. Movement of weed seed by combines and other harvest equipment is of particular concern (Currie and Peeper, 1988; McCanny and Cavers, 1988). Careful management can reduce the risk of spreading weed seed into non-infested fields. Preventive practices include conducting operations in infested fields last or thoroughly cleaning machinery after working in infested fields.

18–3.2 Seed Losses and Seed Persistence

Although seeds of many weed species have the potential for long-term survival in the soil, most seeds have a relatively short life-span (Murdoch and Ellis,

1992). Factors accounting for the loss of weed seeds in the soil include germination, decay, predation, and physical movement. The relative importance of these mechanisms varies with species and environmental conditions.

In weed management we are primarily interested in seeds that germinate and emerge in association with the crop. Sporadic germination in time and space (Forcella et al., 1997; Hartzler et al., 1999) is a characteristic that complicates weed control efforts and allows weeds to survive despite our efforts to remove them from the system. Dormancy is a primary mechanism regulating these variable germination patterns. Several types of seed dormancy exist (Nikolaeva, 1977) and most weed species possess one or more types of dormancy. In a review on exploiting weed seed dormancy through agronomic practices, Dyer (1995) concluded that management practices can influence dormancy. Even slight adjustments in planting date, cultivation timing, harvest method, or residue management may have significant effects on the dynamics of weed seed dormancy.

The percentage of seeds in the seed bank that emerge in a given year is influenced by the species and the environment encountered by the seeds. For common annual species in cultivated soil, approximately 1 to 40% of the seed will generate new seedlings in a given year (Buhler and Hartzler, 2001; Forcella et al., 1992, 1997; Hartzler et al., 1999) with great variation both within and among weed species. Information on weed emergence was collected for 22 site-years in an area ranging from Ohio to Colorado and Missouri to Minnesota (Forcella et al., 1997). Average emergence percentages for some major species were giant foxtail (*Setaria faberi* Herrm.), 31%; velvetleaf, 28%; common ragweed (*Ambrosia artemisiifolia* L.), 15%; pigweed species (*Amaranthus* spp.), 3%; and common lambsquarters (*Chenopodium album* L.), 3%.

Since weed seeds persist in the soil, understanding seed bank dynamics over time is important for devising multi-year weed management strategies. Hartzler (1996) found that 25% of the velvetleaf seeds introduced into the soil produced seedlings over the following four growing seasons with maximum emergence of 11% occurring during the second year. Emergence in the fourth year declined to 2% of the original seed bank. Webster et al. (1998) observed that cumulative emergence of velvetleaf ranged from 6 to 25% over four growing seasons depending on tillage system and burial depth. Hartzler et al. (1999) found that 3-yr cumulative emergence of woolly cupgrass and giant foxtail was greater than velvetleaf and common waterhemp (*Amaranthus rudis* Sauer). In a related study, Buhler and Hartzler (2001) found that seed banks of woolly cupgrass and giant foxtail were depleted within 3 yr, while 12% of the common waterhemp and 5% of the velvetleaf remained viable in the soil after 4 yr of burial.

There are numerous reports of extreme seed longevity when weed seeds were placed in an artificial environment (Burnside et al., 1996 and references therein). The goal of these studies was to determine the maximum potential life of seeds. However, these studies may not be good indicators of the longevity of weed seeds subjected to agricultural practices. For example, tillage increased the depletion rate of itchgrass [*Rottboellia cochinchinensis* (Lour.) W. Clayton] seeds in soil by 32% each year (Bridgemohan et al., 1991). Lueschen et al. (1993) found that after 17 yr with no seed return, soil that had not been tilled still contained 15 to 25% of the original velvetleaf seed compared with 0.8 to 2.5% when the soil was moldboard plowed at least once each year. Buhler (1999) reported that maintaining weed-free conditions through canopy closure for 4 yr greatly reduced weed densities, however, enough weeds remained to reduce soybean yields by 22 to 51% when no herbicide was applied during the fifth year.

18–4 WEED/CROP INTERACTIONS

The primary reason for controlling weeds in soybean is to prevent yield losses. Weeds reduce crop yields primarily by competing for limited resources; however, some weed species may release allelopathic chemicals into the soil that inhibit crop growth (Bhomik and Doll, 1982). The term interference is used to describe the collective interactions among neighboring plants (Harper, 1977).

Competition occurs when a plant reduces the growth of another by acquiring a disproportionate share of a resource or resources potentially available to both (Harper, 1977). Resources for which plants compete include nutrients, light, water, and space. Under typical agronomic conditions, weeds limit soybean productivity primarily through competition for light and/or water (Stoller et al., 1987).

The growth habit of a weed dictates which resources it is most efficient at capturing. Velvetleaf is an efficient competitor for light since it concentrates its leaf area near the top of its canopy and increases light interception through diurnal leaf movements that track the sun (Akey et al., 1990). Soybean yield losses associated with velvetleaf and jimsonweed (*Datura strumarium* L.) were attributed mostly to competition for light, whereas light competition accounted for only 50% of the yield loss from common cocklebur (Stoller and Woolley, 1985). The growth habit of ivyleaf morningglory [*Ipomoea hederacea* (L.) Jacq.] makes it a poor competitor for light since most of its leaves were found in the lower third of the soybean canopy (Cordes and Bauman, 1984).

Whereas available light remains relatively constant during the growing season, water availability varies with soil water-holding capacity and rainfall or irrigation. Thus, competition for soil moisture is more variable than that for light. The relative competitiveness of weeds under different soil moisture conditions varies depending upon the species adaptation for water stress (Pickett and Bazzaz, 1978; Wiese and Vandiver, 1970). Soybean yield losses due to yellow foxtail [*Setaria lutescens* (L.) Beauv.] competition were three times greater when late-season soil moisture was limiting compared to conditions with adequate soil moisture (Staniforth, 1958). Conversely, common cocklebur caused a lower percentage reduction in soybean yield when soil moisture was limiting than with adequate soil moisture (Mortensen and Coble, 1989).

18–4.1 Competition and Yield Losses

The magnitude of interference between soybean and weeds is directly related to both the density of the weed infestation and the length of time that interference occurs. The relationship between weed density and crop yield is best described by a hyperbolic equation in which a linear relationship between weed density and yield loss occurs at low weed densities (Fig. 18–1) (Cousens, 1985). At low densities, each additional weed contributes an equivalent amount of yield loss as the weeds

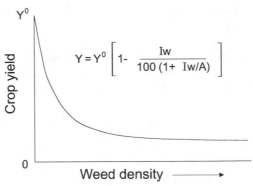

Fig. 18–1. The rectangular hyberbolic yield function for weed-crop interference. Y = actual yield with weed interference; Y^0 = weed-free yield; I = percentage yield loss per weed as weed density approaches zero; w = weed density; A = maximum percentage crop yield loss asymptote as w approaches infinity.

already present. As weed densities increase, each additional weed contributes less to crop yield loss since increasing intraspecific competition among the weeds reduces interspecific competition with soybean.

The length of time that weeds compete with a crop dramatically influences the magnitude of yield loss. The critical period is defined as the maximum length of time weeds can be tolerated without affecting crop yields (Zimdahl, 1987). Two separate critical periods are of interest in weed control. The first involves weeds that emerge at the same time as the crop and compete during the early part of the growing season. This critical period defines how long weeds can be allowed to remain in the crop after planting before yields are affected, and is important when relying on postemergence herbicides. The time required for weeds to reduce soybean yields by 2.5% ranged from 9 to 38 d in southern Ontario (Van Acker et al., 1993). Stoller et al. (1987) concluded that removing weeds within 4 to 6 wk of soybean emergence resulted in minimal (<10%) soybean yield loss.

The second critical period takes into account the effect of weeds that emerge after soybean emergence, and defines how long after crop emergence weeds must be controlled to prevent yield loss. Weeds that emerge after this time are noncompetitive due to the head-start given the crop. The impact of late-emerging weeds on yields decreases rapidly during the first 3 wk after soybean emergence (Stoller et al., 1987). A weed-free period of 8 to 10 wk after soybean emergence was required to prevent yield losses from giant ragweed (*Ambrosia trifida* L.) (Baysinger and Sims, 1991), whereas with Johnsongrass [*Sorghum halepense* (L.) Pers.] the weed-free period ranged from 3 to 6 wk (Williams and Hayes, 1984).

Factors other than weed density and duration of competition influence soybean yield losses from weeds, including soybean planting date (Oliver, 1979), row spacing (Legere and Schreiber, 1989), cultivar (Bussan et al., 1997; Rose et al., 1984), and environment (King and Purcell, 1997; Stahler, 1948). Weeds exposed to sublethal concentrations of herbicides may be less competitive than uninjured weeds (Adcock and Banks, 1991; Holloway and Shaw, 1996). Given all of the potential interacting factors, the ability to accurately predict yield losses based simply on weed densities in a field is very limited (Norris, 1999; O'Donovon, 1996).

18–4.2 Thresholds and Weed Management

Various types of thresholds have been described, most developed with the purpose of assisting farmers in making better weed control decisions (Cousens, 1987). The economic threshold is the most commonly described, and is defined as the weed density at which the cost of control equals the value of the crop that would be lost due to interference if weeds were not treated. At weed densities below the economic threshold it is recommended that weeds be left in the field since net returns would be higher than if they were controlled.

One of the earliest decision tools based on economic thresholds for soybean was developed at North Carolina State Univ. (Wilkerson et al., 1991). This model, called HERB, was based on a system in which weeds were ranked according to their relative competitive ability and yield loss predictions were based on weed densities early in the growing season (Coble, 1986). This decision tool provided recommendations on the control tactics that provided the most effective weed control and the highest economic return. The general concept used in HERB has been adapted by several southern states and is being modified for use in the central USA (Krishnan et al., 2001).

While much effort has been devoted toward developing economic thresholds, their acceptance by both weed scientists and producers has been limited. A survey of Illinois farmers found that only 9% used economic thresholds as a basis for weed control (Czapar et al., 1995). Biological and agronomic limitations of economic thresholds have recently been reviewed (Norris, 1999; O'Donovon, 1996). Norris (1999) concluded that the concept of economic thresholds as initially developed for arthropod control was inappropriately adapted for weed control. He argued that differences in the ecology and population biology between weeds and arthropods limit the transferability of threshold concepts between pest classes and that unique principles and practices need to be developed for weeds.

The primary reasons cited by Illinois farmers for not using economic thresholds were: (i) harvest problems caused by weeds (64%), (ii) landlord concerns (38%), (iii) weed seed production (38%), and (iv) aesthetics (36%) (Czapar et al., 1997). Most farmers base their weed control decisions on the weed infestation the previous year (Czapar et al., 1995). In his discussion of thresholds, Cousens (1987) concluded that it was misleading to specify an exact threshold value to a farmer since it implies a precision that is not possible and that subjectivity is acceptable in decision making.

A primary argument against economic thresholds is the long-term impacts of weed seed production by subthreshold weed densities. Subthreshold densities of velvetleaf resulted in a rapid increase in the weed seed bank and velvetleaf densities (Cardina et al., 1997; Hartzler, 1996; Zanin and Sattin, 1988). Economic optimum thresholds (EOT) differ from economic thresholds in that EOT consider the long-term cost associated with seed production (Cousens, 1987). The EOT for velvetleaf and common sunflower (*Helianthus annuus* L.) in soybean were calculated to be 7.5- and 3.6-fold lower, respectively, than the economic threshold due to the impact of seed production (Bauer and Mortensen, 1992).

The instability of yield losses has also been cited as limitations of economic thresholds. The yield loss attributed to a specific weed infestation can vary by a factor of two or more depending upon the environment and crop production practices (Stoller et al., 1987). Yield losses due to weed interference, expressed as a percentage of weed-free yield, were greater under high yielding environments than low yielding environments in a many studies. However, numerous studies have found the opposite response.

The distribution of weeds within agricultural fields is rarely uniform, but rather weeds typically are found in patches having a high relative density surrounded by areas with a few individual plants (Cardina et al., 1996). Since the spatial pattern of weeds is not regular, the mean density alone is of little value in predicting yield losses associated with a non-uniform weed infestation. Assuming a regular distribution of weeds when predicting yield losses resulted in an overestimation of weed-related yield losses (Wiles et al., 1992). The error associated with patchy weed distributions resulted in control tactics being recommended when losses were actually below the economic threshold level.

18–4.3 Noncompetitive Effects

While preventing soybean yield losses is the main reason for controlling weeds, uncontrolled weeds may also impact soybean production in other ways. Many weed species are prolific seed producers, and a few escaped weeds can produce sufficient seed to infest the field for several years into the future (Baker, 1974; Cardina and Nouquay, 1997; Hartzler, 1996). Because of the importance of seed production, Norris (1999) recommended a no-seed-threshold be established in which the objective of the weed management program would be to allow no weeds to produce seed.

Weeds may reduce soybean harvest efficiency (Ellis et al., 1998; Jordan et al., 1987). Low densities of redroot pigweed (*Amaranthus retroflexus* L.), ivyleaf morningglory, hemp sesbania [*Sesbania exaltata* (Raf.) Rydb. Ex A.W. Hill], and common cocklebur increased foreign material, soybean moisture at harvest, and the number of damaged soybean seeds (Ellis, 1998). The presence of weeds also required slower combine speeds.

Interactions between weeds and other pest complexes (i.e., weeds with insects and diseases) is also an important consideration. In a recent review of insect-weed interactions (Norris and Kogan, 2000), three types of interactions were discussed: (i) trophic relationships in which weeds act as a food source for insect pests or predators, (ii) habitat alterations by the weed that increase or suppress insect infestations, and (iii) control tactics used to mitigate an insect or weed influence nontarget pests. Mexican bean beetle (*Epilachna varivestis* Malsant) densities were found to be highest in weed-free soybean, whereas insect predators were highest in weedy soybean fields (Shelton and Edwards, 1983). In contrast, reproductive efficiencies of velvetbean caterpillar (*Anticarsia gemmatalis* Hubner) and soybean looper (*Pseudoplusia includens* Walker) (Collins and Johnson, 1985) and populations of corn earworm (*Helicoverpa zea* Boddie) (Alston et al., 1991) were greater in weedy than weed-free soybean.

Weeds may serve as inoculum reservoirs and maintain pathogens in fields when soybean is absent. Weeds have been reported to serve as alternate hosts to fungal (Black et al., 1996; McLean and Roy, 1988) and nematode (Riggs and Hamblen, 1966; Smart, 1964) pests that damage soybean.

Herbicides may influence disease development via direct effects on the pathogens (Altman and Campbell, 1977; Rodriguez-Kabana et al., 1966) or by altering soybean resistance to pathogens (Dann et al., 1999; Levene et al., 1998). Increased disease resistance in soybean has been attributed to herbicide-induced glyceollin production (Dann et al., 1999; Levene et al., 1998). In contrast, nonphytotoxic concentrations of glyphosate-reduced glyceollin levels in soybean without glyphosate resistance and suppressed resistance to *Phytophtora sojae* (Keen et al., 1982). The site of action for glyphosate is EPSPS (5-enolpyruvylshikimate-3-phosphate synthase), a component of the shikimate acid pathway that produces precursors of glyceollin (Kishore and Shaw, 1988).

18–5 WEED CONTROL PRACTICES AND WEED MANAGEMENT

18–5.1 Integrated Management

Weed management is more encompassing than the removal of weed populations at a point in time and places greater emphasis on the understanding of the underlying causes of weeds and minimizing the negative impact on the crop (Buhler, 1996; Zimdahl, 1991). Weed management emphasizes integrating techniques and knowledge in a manner that considers the causes of weed problems rather than reacting to problems after they occur. The goal of integrated weed management is to optimize crop production and grower profit through the concerted use of preventive tactics, scientific knowledge, management skills, monitoring procedures, and appropriate use of control practices.

The central challenge of developing effective integrated weed management systems is to create cropping systems unfavorable for weeds and minimize the impact of the weeds that survive. No single practice should be considered as more than part of a total weed management strategy. No single weed management tactic has proven to be the "silver bullet" to eliminate weed problems, and given the nature of weed communities, we should not expect one to be developed in the near future. The best approach may be to integrate cropping system design and available weed control strategies into a comprehensive weed management system that is environmentally and economically viable over the long term.

18–5.2 Control Practices

Control of weeds is a primary function of most crop management practices in soybean. The management inputs that may contribute to the control of weeds include various forms of tillage, crop cultural practices, prevention, biological agents, and herbicides. The relative importance of different practices in controlling weeds depends on production philosophy, tillage systems, rotation options, economics, farm size, and labor availability.

18–5.3 Tillage and Cultivation

Tillage for seed bed preparation can greatly reduce densities of annual weeds, especially if planting is delayed to allow weed seed germination prior to the final tillage operation (Buhler and Gunsolus, 1996; Gunsolus, 1990). In cropping systems that include perennial crops, tillage is typically used to destroy the perennial crop prior to the seeding of an annual crop (Triplett, 1985).

Tillage following soybean planting can also be an effective weed control tool. Rotary hoeing (two passes approximately 7 and 14 d after soybean planting) reduced weed density up to 85% (Buhler and Gunsolus, 1996). However, it must be noted that timing and soil conditions are critical for rotary hoeing (Gunsolus, 1990). If rotary hoeing is conducted too late, weed seedlings will be anchored in the soil and will not be uprooted. Wet soil conditions may also reduce the effectiveness of rotary hoeing as seedlings may reroot in moist soil. Rotary hoeing may also improve soybean stands by breaking the soil crust to allow uniform emergence. On the other hand, soybean should not be rotary hoed when the hypocotyl is exposed at the soil surface during emergence as severe soybean injury may occur.

Interrow cultivation is another form of post-plant tillage that can be used for weed control, especially in conventional tillage, wide-row soybean production. Cultivation between soybean rows can be effective in removing most annual weeds and inhibiting the growth of perennial species. Generally, shallow cultivation is best to minimize soybean root pruning and avoid disturbing weed seeds buried deeper in the soil. In most instances, interrow cultivation alone is not sufficient for complete weed control due to the inability of the cultivator to effectively control weeds within the crop row (Jordan et al., 1987). However, when used in combination with rotary hoeing (Buhler and Gunsolus, 1996; Gunsolus, 1990) or supplemented with broadcast (Gebhardt, 1981; Steckel et al., 1990) or band applications of herbicides (Buhler et al., 1992, 1993), cultivation can provide effective weed control. The role of the supplemental tactics is to create a height differential between the crop and weeds, allowing the cultivator to bury weeds within the crop row without damaging the crop.

18–5.4 Cultural Practices

Crop rotation can be an important component of weed management by reducing weed densities and increasing the diversity of control options. Crop rotations that maximize differences in crop life cycles and competitive characteristics of rotational crops as well as the associated weed control practices generally result in better weed control than continuous production of a single crop. Liebman and Ohno (1998) summarized the results of 25 test crop by rotation combinations and found that in 19 cases, weed density in rotation was less than in monoculture, higher than monoculture in two cases, and equivalent with monoculture in four cases. Yields of test crops were higher in rotation than in monoculture in 9 of 12 cases where the crop yield was reported.

Enhancing the ability of soybean to compete with weeds can be accomplished by providing the best possible environment for soybean growth combined with practices that reduce the density and/or vigor of the weeds. Practices such as narrow row spacing, increased soybean density, appropriate time of planting, and fertility management are capable of shifting the competitive balance to favor soybean over weeds. In an extensive review of the effects of cultural practices on soybean/weed interactions, Stoller et al. (1987) concluded that soybean cultivar selection, row spacing, plant density, planting date, crop rotation, tillage, and herbicides can all be used to maximize the ability of soybean to compete with weeds.

Differential competitive ability has been documented among common cultivars of soybean and several authors have suggested that weed management could be improved through crop breeding (Callaway, 1992; Pester et al., 1999; Rose et al., 1984). Characteristics commonly associated with soybean competitiveness with weeds included rapid germination and root development, rapid early vegetative growth and vigor, rapid canopy closure and high leaf area index, profuse branching, increased leaf duration, and greater plant height (Callaway, 1992; Pester et al., 1999). The most consistent conclusion among many studies has been that vigorous growth characteristics enhance weed competitiveness by reducing light quantity and quality beneath the crop canopy.

18–5.5 Prevention

Using measures to forestall the introduction and spread of weeds is the most basic of all weed control methods. Preventable means by which weeds can be introduced into new areas include contaminated crop seed; transport of plant parts and seeds on planting, tillage, harvest, and processing equipment; livestock; manure and compost; irrigation and drainage water; and forage and feed grains (Walker, 1995). Movement of weeds may be particularly significant for crop seeds and animal feeds that are transported long distances.

Historically, prevention has been an important method of weed control (Walker, 1995). However, the concept of prevention has been deemphasized in recent times due to the availability of effective herbicides and mechanical control measures and increasing farm size. These control tools have led to the impression that weeds can always be controlled after they have become established. This may hold true in many cases, but if new species are tolerant of commonly used herbicides, highly competitive, or have other characteristics that make control difficult, they may increase weed control costs or reduce soybean yields.

While the concept of prevention is quite simple, the success and feasibility are determined by the weed species, the means of dissemination, farm size, and the amount of effort expended. Preventive programs are most successful in situations when humans are the vectors or where they have direct control over the seed source (Walker, 1995). Some programs can be implemented through community action by the enactment and enforcement of laws and regulations (Day, 1972). Seed purity and noxious weed laws are examples of such weed prevention programs.

18–5.6 Biological Control

Much has been written about biological control of weeds with insects and plant pathogens. While biological control of weeds has been successful in systems such as rangelands or water bodies that are infested with one dominant species, it has had minimal impact in soybean production systems. Most biological control agents have a narrow host range and thus will only control a single or a few closely related species (Hasan and Ayres, 1990). One of the major challenges for biological control in crops such as soybean is that weed communities are usually composed of several major species capable of causing economic damage and several other species capable of increasing to damaging levels if not controlled for one or more years (Buhler et al., 2000). Therefore, eliminating an individual species would have minimal impact on the overall weed problem.

Many cultural practices used in soybean production (i.e., tillage and crop rotation) cause disturbances in the environment that may be detrimental to biological control systems (Cardina, 1995). Environment, particularly relative humidity, strongly influences infection rate and effectiveness of many biological control agents, especially fungi (Charudattan, 1991). However, while the prospect for controlling diverse populations of annual weeds with biological control organisms is not promising, control of persistent perennial species such as nutsedges (*Cyperus* spp.), Canada thistle, Johnsongrass, or quackgrass [*Elytrigia repens* (L.) Nevski] may be feasible and would be of potential value in soybean production systems.

18–5.7 Herbicides

Herbicides are the primary means of weed control for most soybean growers. Detailed discussions of the history and development of herbicides in soybean production have been presented in many publications and textbooks, including the previous editions of this monograph (Jordan et al., 1987). Herbicide use patterns vary by region and time, with herbicide labels changing on a frequent basis. Because of the time sensitive nature of this information, readers should refer to the most recent annual weed control recommendations published for their state or region, herbicide product labels, or herbicide reference manuals such as the Weed Science Society of America's *Herbicide Handbook* (Anonymous, 2002) for specific information on herbicide treatments and chemistries.

In recent history, weed control in soybean has relied heavily on herbicides with soil residual activity that were applied before or immediately after crop planting and before weed emergence, with the intention of controlling weeds as they germinate and emerge. These herbicides relied on mechanical incorporation or rainfall to move them into the soil and facilitate weed control. While the relative importance of soil-applied herbicides in soybean production has declined with the introduction of effective postemergence herbicides and herbicide resistant cultivars, these products were still used on at least one-third of the U.S. soybean crop in 1999, often in combination with postemergence herbicides or mechanical control tactics (Table 18–2).

The importance of foliar-applied herbicides has increased since the late 1970s (Hartzler et al., 1997). Foliar-applied herbicides allow the nature of the weed infestation to be determined prior to application and avoid herbicide-soil interactions that contribute to variability in performance of soil-applied herbicides. The most commonly used foliar-applied herbicides are listed in Table 18–3. The introduction of glyphosate [*N*-(phosphonomethyl)glycine]-resistant soybean in the

Table 18–2. Most commonly used soil-applied herbicides on U.S. soybean, 1999.†

Common name	Herbicide family	Mode of action	Area treated
			%
Pendimethalin	Dinitroaniline	Microtubule inhibitor	14
Trifluralin	Dinitroaniline	Microtubule inhibitor	14
Chlorimuron-ethyl	Sulfonylurea	ALS‡ inhibitor	12§
Metribuzin	Triazine	Photosynthesis inhibitor	5
Metolachlor	Acetamide	Unknown	4
Sulfentrazone	Aryl triazolinone	PPO‡ inhibitor	4

† Anonymous. 2000. Agricultural chemical usage. 1999 field crops summary. USDA/NASS. Ag Ch 1(00) a.
‡ ALS, acetolactate synthase; PPO, protoporphyrinogen oxidase.
§ Area treated includes both soil and foliar applied applications of chorimuron-ethyl.

late 1990s resulted in a rapid shift from programs based on acetolactate synthase (ALS)-inhibiting herbicides to glyphosate (Owen, 2000).

Many factors should be considered when determining the appropriate herbicides to use in a specific soybean production situation (Harrison and Loux, 1995). These include product and application costs, crop rotation sequence, tillage system, weed community characteristics, environmental conditions, and availability of other weed control options. Harrison and Loux (1995) present an excellent discussion of the relative advantages and disadvantages of different herbicide application times.

Combinations of soil-applied and postemergence herbicides often provide the most consistent and broadest spectrum of weed control. In most cases, no single herbicide applied annually will selectively control all weeds over an extended period of years. Efficient and consistent weed control with herbicides requires that herbicide selection be based on the weed species present in the area and the recognition that weed populations will adapt to herbicides over time.

Table 18–3. Most commonly used foliar-applied herbicides on U.S. soybean, 1999.†

Common name	Herbicide family	Mode of action	Area treated
			%
Glyphosate	Unclassified	ESPS‡ inhibitor	62§
Imazethapyr	Imidazolinone	ALS‡ inhibitor	16
Sethoxydim, fluazifop, clethodim, quizalofop¶	Aryloxyphenoxy propionate and cyclohexanediones	ACC-ase‡ inhibitor	13
Acifluorfen, fomesafen, lactofen¶	Diphenylether	PPO‡ inhibitor	9
Thifensulfuron	Sulfonylurea	ALS inhibitor	5
Cloransulam-methyl	Sulfonanilide	ALS inhibitor	5
Bentazon	Benzothiadiazole	Photosysthesis inhibitor	4

† Anonymous. 2000. Agricultural chemical usage. 1999 field crops summary. USDA/NASS. Ag Ch 1(00)a.
‡ ACC-ase; acetyl-CoA carboxylase; ALS, acetolactate synthase; ESPS, 5-enolpyruvylshikimate-3-phosphate; PPO, protoporphyrinogen oxidase.
§ Area treated includes both use in no-tillage soybean as burndown treatment and postemergence application in glyphosate-resistant soybean.
¶ Herbicides are grouped together because of similar weed spectrum and use patterns.

Herbicides are often marketed in prepackaged mixtures containing more than one active ingredient. Premixed products are formulated to provide broader-spectrum weed control, prevent excessive persistence, or reduce the potential for crop injury. Similarly, many herbicides can be mixed with other herbicides in a spray mixture just prior to application or applied consecutively in separate applications. As with all herbicide applications, these mixtures must be consistent with product labels.

The use of modern genetic techniques to transfer resistance to nonselective herbicides such as glyphosate and glufosinate [2-amino-4-(hydroxymethylphos-phinyl)butanoic acid] into soybean provides additional herbicide options. Fundamentally, these herbicide resistant cultivars do not change weed control. They are simply an advance in technology that provides herbicides with a broader spectrum of control and more flexibility in application time (Burnside, 1992; Owen, 2001). In the longer term, the ability to alter the crop genome to allow the use of previously nonselective herbicides to the growing crop may continue to generate herbicide options with a broad weed spectrum. On the other hand, the emphasis on genetic techniques to generate new herbicide/crop combinations may reduce or eliminate the development of new herbicide chemistries.

18–6 FACTORS AFFECTING WEED CONTROL EFFECTIVENESS

18–6.1 Tillage Systems

Tillage systems not only influence weed populations, but also impact weed control options available to soybean producers and the effectiveness of those options. Certain herbicides may not be used in conservation tillage systems because of the need for mechanical incorporation into the soil after application. With less tillage and more plant residue on the soil surface, mechanical weed control operations are sometimes less effective (Springman et al., 1989). Rotary hoeing is especially difficult in untilled soil covered with residue of the previous crop. Interrow cultivation remains an option in some conservation tillage systems and is an integral component of ridge-tillage (Buhler, 1992; Forcella and Lindstrom, 1988). Combining interrow cultivation with reduced rates or banded herbicide provided weed control similar to full-rate herbicide treatments in conservation tillage systems (Buhler et al., 1995; Mulder and Doll, 1993).

Plant residue on the soil surface may alter the distribution or behavior of soil-applied herbicides (Banks and Robinson, 1982; Johnson et al., 1989). Depending on percentage surface cover, residue type, and herbicide formulation, up to 60% of the herbicide applied may be intercepted by residue. However, much of this herbicide may be washed off by subsequent rainfall and irrigation. Up to 90% of the herbicide applied to corn residue was washed off with 6.8 cm of simulated rainfall, with most washoff occurring with the first 1.5 cm of rain (Baker and Shiers, 1989).

18–6.2 Cultural Practices and Mechanical Weed Control

Many factors interact to determine the effectiveness of cultural and mechanical weed control practices. For example, narrower row spacing results in more rapid canopy closure, providing soybean an advantage over late-emerging weeds (Yelverton and Coble, 1991). The increased suppression of weeds by the soybean canopy in narrow-rows provides opportunities to reduce herbicide rates with early applications to small weeds (Hartzler, 1999). Cultural practices affect the nature of the weed population and the effectiveness of mechanical weed control treatments (Buhler and Gunsolus, 1996). For example, delaying planting 14 d reduced weed densities in soybean 25 to 90% and greatly increased the effectiveness of mechanical control practices. Timely rotary hoeing reduced weed dry weight in soybean by 72%, but wet soil before or after timely rotary hoeing resulted in only 33% reduction (Lovely et al., 1958).

18–6.3 Soil-Applied Herbicides

In order to be effective, adequate concentrations of soil-applied herbicide must be at the proper depth within the soil profile at the time weed seeds germinate and seedlings develop. Herbicide absorption may occur via roots, seeds, shoots, and vegetative propagules of perennial plants. Soil-applied herbicides are ineffective if they remain on the soil surface, thus they need to be moved to the depth of germinating weed seeds to be effective (Mindreboe, 1970). Rainfall influences herbicide performance through effects on herbicide position within the soil profile and soil moisture content. Mechanical incorporation of herbicides tends to improve performance in the absence of timely rainfall events (Walker and Roberts, 1975). Mechanical incorporation of herbicides was an important component of soybean production systems in the 1970s and 1980s, but the advent of reduced tillage systems and introduction of postemergence herbicides has greatly reduced this practice (Hartzler et al., 1997).

Since soil-applied herbicides are absorbed by plants from the soil solution, adequate soil moisture is required for herbicide activity (Moyer, 1987). When soil moisture approaches the permanent wilting point, herbicide activity is lost because virtually all of the herbicide becomes adsorbed to soil and is unavailable for plant uptake. As soil moisture increases, an equilibrium is reached between the amount of herbicide bound to soil colloids and the amount in soil solution (Bailey and White, 1964). The amount of herbicide available in the soil solution is influenced by both the chemical properties of the herbicide and soil characteristics.

Herbicide adsorption to soil colloids is directly related to the clay and organic matter content. The recommended rates for most soil-applied herbicides are based on the adsorptive capacity of the soil, that is, higher rates are recommended on fine-textured soils with high organic matter than on coarse-textured soils with low organic matter. The organic matter fraction has a larger influence on herbicide adsorption than clay content in most soils (Blumhorst et al., 1990). The adsorptivity of a herbicide generally influences availability more than its water solubility (Bailey and White, 1964). Many soil-applied herbicides are either weak acids or weak bases, and therefore their behavior is dependent upon the pH of the soil and pKa of the herbicide (Harper, 1994).

Decreases in performance of soil-applied herbicides have been observed with increasing weed densities (Dieleman et al., 1999; Hartzler and Roth, 1993; Taylor and Hartzler, 2000). The number of weeds surviving herbicide treatment typically was proportional to the initial weed density until densities became very high. The increased survivorship at high densities was attributed to competition among seedlings for available herbicide, greater likelihood of seedlings emerging from a safe-site, and/or increased diversity in herbicide tolerance within the weed population.

18–6.4 Foliar-Applied Herbicides

Numerous factors influence the performance of postemergence herbicides, including environmental conditions, spray coverage of the target species, growth stage of the weed, type of spray adjuvant included in the carrier, and water quality. Within normal ranges of air temperature and relative humidity, changes in relative humidity have a greater influence than temperature (Ritter and Coble, 1981; Wills, 1984). Herbicide effectiveness generally increases as relative humidity increases due to enhanced herbicide absorption. Weeds under soil moisture stress generally are less susceptible to herbicides than nonstressed plants (Boydston, 1992; Kidder and Behrens, 1988). While changes in herbicide efficacy in response to the environment are well documented, these interactions are too complex to provide most users specific guidelines on adjusting herbicide use in response to weather conditions.

In a review of the effects of spray volume and droplet size on performance of foliar-applied herbicides, Knoche (1994) concluded carrier volume affected herbicide efficacy less consistently than did droplet size. Herbicide efficacy generally increased with decreasing droplet size, and the advantages of small droplets were attributed to increased canopy penetration, increased droplet retention on difficult-to-wet leaf surfaces, decreased deposition variability, and increased herbicide absorption. The optimum spray parameters varied with different herbicides. Since small droplets are subject to drift, a balance between optimizing herbicide efficacy and minimizing drift must be reached. Recently, several new nozzle types have been developed that are designed to reduce drift by reducing the portion of the spray volume found in very small spray droplets (Etheridge et al., 1999). Herbicide efficacy with these nozzles was reported to be equivalent to that achieved with traditional nozzle types (Etheridge et al., 2001; Jensen et al., 2001).

A variety of adjuvants are used with foliar-applied herbicides to modify the characteristics of the spray solution or enhance herbicide efficacy (Hazen, 2000). Activator adjuvants enhance herbicide activity by increasing spray retention and herbicide absorption, and may be a component of the herbicide formulation or added to the spray tank by the end user (Penner, 2000). The specific mechanism by which an adjuvant improves herbicide performance varies among adjuvants, but most are active at the leaf cuticle, the primary barrier to herbicide absorption. Ammonium salts are commonly used with glyphosate and other postemergence herbicides. The NH_4^+ ions reduce the antagonistic effects of calcium and other cations found in hard water due to the formation of NH_4^+-glyphosate (Thelen et al., 1995) and have also been proposed to facilitate movement of weak acid herbicides across the plasmalemma (Wade et al., 1993).

18–6.5 Risk Management

The widespread reliance on chemical weed control in soybean production has been attributed to the ability of herbicides to reduce variability in economic returns (Olson and Eidman, 1992). Gunsolus and Buhler (1999) stated that a producer's perception of risk was the key limiting factor to the adoption of integrated weed management systems. Weed management programs that relied on mechanical tactics or a combination of mechanical and herbicidal tactics were found to be less risky than a program relying solely on herbicides (Gunsolus and Buhler, 1999). This is not surprising because it is well documented that alternative management strategies can provide as effective weed control as herbicide-based systems (Gunsolus, 1990; Hartzler et al., 1993; Peters et al., 1965).

A primary advantage for herbicides over mechanical strategies is that herbicides reduce the time devoted to weed control. A control program based on broadcast soil-applied herbicides required 2.5 h per 40 ha compared with 24 h when the herbicide was applied in a band followed by interrow cultivation (Gunsolus and Buhler, 1999). Total reliance on mechanical weed control (two rotary hoeings plus two interrow cultivations) required 38 h per 40 ha. The importance of time requirements for weed control is magnified by the influence of environmental conditions on field work. The probability of being able to conduct field work on a given day during the early growing season in southwest Minnesota was approximately 60% (Seely, 1995).

The likelihood of being able to implement a control tactic during its peak effectiveness is a critical factor in a producer's decision-making process. Mechanical control tactics generally have a narrower window of opportunity than herbicides (Gunsolus, 1990). The time constraints of specific control tactics become more critical with increasing farm size. The rapid acceptance of glyphosate-resistant soybean is partially due to the wider application window of this product compared to other postemergence herbicide programs (Owen, 2000). Most producers viewed the benefit of increased flexibility in application timing to outweigh a lower yield potential of glyphosate-resistant soybean cultivars compared to conventional varieties (Marking, 1999; Oplinger et al.,1999). In contrast, farmers less reliant on herbicides generally reduce risks through use of more diverse management systems (Liebman and Gallandt, 1997).

18–7 WEED POPULATION SHIFTS AND HERBICIDE RESISTANCE

The development of herbicide resistant weeds (Heap, 2001) and weed species shifts in response to other management practices (Buhler et al., 1997; Holt, 1994) have received considerable attention in recent years. But the ability of weed communities to evolve in response to control practices is not a phenomenon associated only with modern weed control practices. Weeds have the longest history of adapting to control practices among crop pests, largely because they have a great capacity for morphological change. Two of the most striking early examples are the development of common vetch (*Vicia sativa* L.) seeds that mimicked lentil (*Lens culinaris* Moench) seeds in response to winnowing, and the development of rice (*Oryza sativa* L.)-like characteristics by barnyardgrass [*Echinochloa crus-galli* (L.) Beauv.] in response to handweeding (Gould, 1991).

In modern soybean production systems, weed population shifts are most commonly related to tillage systems and crop sequence. Weed species not previously observed have rapidly appeared in fields following elimination of preplant tillage (Buhler, 1995). Species most rapidly and commonly observed are winter annual and biennial species. These weed species are unable to complete their life cycles in association with soybean if the soil is disturbed before planting (Bazzaz, 1990). Schreiber (1992) found that growing corn in a soybean/corn or soybean/wheat (*Triticum aestivum* L.)/corn rotation greatly reduced giant foxtail seed in the soil compared to corn grown continuously, regardless of herbicide use or tillage system. The effects of crop rotation and environmental conditions associated with years and locations were larger than tillage effects on weed species composition and abundance in two studies in Canada (Derkson et al., 1993; Thomas and Frick, 1993). Similarly, Ball (1992) reported that cropping sequence was the most dominant factor influencing weed species composition.

Herbicide resistance has been defined as the evolved capacity of a previously herbicide-susceptible weed population to withstand a herbicide and complete its life cycle when the herbicide is used at its normal rate in an agricultural situation (Heap and LeBaron, 2001). The first herbicide-resistant weed in the USA was found in1968 (Ryan, 1970). By 2001, 249 herbicide-resistant biotypes had been identified worldwide, with approximately 10 new resistant biotypes reported annually since 1980 (Heap, 2001).

A resistant weed biotype develops in response to selection pressure from a single herbicide or class of herbicides (Gressel and Segal, 1978). Selection for change occurs when a small number of plants within a population have a genetic makeup that enables them to survive a particular herbicide. When a herbicide is applied, most of the plants of the susceptible biotypes are killed. The surviving plants mature and produce seed, increasing the proportion of the resistant biotype in the population. If the herbicide is used repeatedly, the resistant biotype dominates the population and creates an agronomic problem.

Several mechanisms have been identified that may confer herbicide resistance in a weed. Many herbicides act by binding or interacting with a protein, therefore interfering with the function of the protein (Preston and Mallory-Smith, 2001). The most common resistance mechanism involves an alteration in the target protein which reduces the affinity of the herbicide for the target site (Guttieri et al., 1995; Hirschberg and McIntosh, 1983). Other mechanisms of resistance involve a reduction in the amount of herbicide reaching the target site due to reductions in absorption or translocation (Preston et al., 1992) or enhanced metabolism of the herbicide (Anderson and Gronwald, 1991).

The frequency at which resistance occurs in the population prior to selection pressure being applied has a large influence on the rate resistance develops in a field (Jasieniuk et al., 1996). Obtaining reliable estimates of the initial frequency of resistance is difficult, but an initial frequency of 1×10^{-6} was assumed in some early herbicide resistance models (Gressel and Segel, 1982; Maxwell et al., 1990). This estimate was based on triazine-resistant biotypes in which the resistance mechanism carried a fitness penalty (Ahrens and Stoller, 1983). The apparent lack of a fitness penalty associated with resistance to certain other herbicide classes may result in higher initial frequencies, and may partly explain the rapid evolution of resistance to several nontriazine herbicides (Jasieniuk et al., 1996). Other biological factors which influence the rate of resistance development in a population include the mode of inheritance of resistance and gene flow within and between populations. Management practices that affect selection of resistance include the effectiveness of the herbicide, the frequency that the herbicide is used, crop rotation, and the use of alternative control practices, either nonchemical or alternative herbicides (Boerboom, 1999).

Herbicide resistance in U.S. soybean producing regions initially was associated with repeated use of triazine herbicides in corn (Bandeen et al., 1982). The first cases of resistance selection in soybean involved ALS-inhibitor herbicides (Horak and Peterson, 1995; Sprague et al., 1997). The ALS-inhibitor resistant weeds were a major problem for soybean producers in the 1990s, and account for 16 of the 22 herbicide resistance biotypes found in soybean (Heap and LeBaron, 2001). Problems in controlling weeds resistant to ALS-inhibitors contributed to the rapid adoption of glyphosate resistant soybean.

The widespread adoption of glyphosate-resistant soybean has led to considerable debate on the likelihood of weeds developing resistance to this herbicide. At the time of introduction of glyphosate-resistant soybean, industry scientists stated that the development of glyphosate-resistant weeds was an unlikely event (Bradshaw et al., 1995). This assumption was based on the long history of glyphosate use resulting in no resistant weeds and the information gained while developing glyphosate-resistant crops. However, Gressel (1996) reviewed the potential mechanisms for resistance and stated that there were few constraints to weeds evolving resistance to glyphosate. Gressel's conclusion was that the discussion should focus on how to institute resistance management strategies rather than debate the potential for glyphosate resistance.

The long-use history of glyphosate suggests that resistance will appear less frequently than with most herbicide classes (Heap and LeBaron, 2001). However, the recent reports of glyphosate resistance reinforced the need to manage glyphosate-resistant crops and glyphosate in a manner that reduces the selection pressure placed on weed populations. The first reported case of selected resistance involved rigid ryegrass (*Lolium rigidum* Gaud.) in Australia (Powles et al., 1998). Glyphosate-resistant goosegrass [*Eleusine indica* (L.) Gaertn.] was identified in Malaysia in 1997 (Lee and Ngim, 2000). Both of these biotypes were found in orchards, rather than in field crops. The first documented case of glyphosate resistance in an agronomic crop involved horseweed in Delaware (VanGessel, 2001).

Few studies have documented the cost of herbicide-resistant weeds in soybean or other grain crops. The greatest cost to a producer probably occurs when resistance first appears in a field. A typical response by a grower is to reapply the same herbicide since it worked effectively in the past (Horak and Peterson, 1995). By the time the grower recognizes resistance is present it may be too late for alternative strategies to be implemented, and thus the weed remains in the field for the entire growing season. Costs associated with the control failure include expense of the additional herbicide treatments, yield losses due to competition, and the increase in the weed seed bank of the resistant biotype.

Due to the large number of herbicides available in soybean, alternative herbicides have been available to control herbicide-resistant weeds (Boerboom, 1999). Soybean producers rated the development of strategies for herbicide resistant weeds 7th of 14 production issues (Stoller et al., 1993). The most common response to herbicide resistance has been a change in herbicides or inclusion of additional herbicides, rather than integration of alternative management strategies into the weed management program (Boerboom, 1999).

18–8 FUTURE OPPORTUNITIES FOR WEED MANAGEMENT IN SOYBEAN

Selective herbicides have been the dominant component of weed management in soybean for the past several decades. Herbicides will continue to play a major role for the foreseeable future, but significant changes in the way herbicides are developed and applied may occur. There is also increasing interest in better methods of cultural, mechanical, and biological weed control that may change how some soybean growers approach weed management. This is especially the case for producers interested in organic and other specialty markets or producers in nations without ready availability of herbicides and herbicide-resistant cultivars.

18–8.1 Herbicides and Herbicide Resistant Crops

Tolerance of crops to herbicides is the basis of selective weed control and has been part of weed management in soybean for about 40 yr. The major change in herbicide development in recent years has been a shift in the way herbicide-resistant crops are developed. In the past, the primary means of developing selective herbicides was through screening of large numbers of organic compounds against major crop and weed species to identify selectivity (Burnside, 1992). With the advent of new genetic technologies (Dunwell, 1996; Mazur and Falco, 1989), genetic resistance to herbicides can be inserted into the crop genome, making the crop resistant to a herbicide that was previously toxic to that species.

The potential advantages and disadvantages of genetically-induced herbicide resistance in soybean and other major crops have been discussed extensively (i.e., Burnside, 1992; Owen, 2001; Young, 2000) and will not be repeated here. The bottom line is that major emphasis has been placed on the development and marketing of herbicide-resistant soybean and it is evident that this technology will likely play a major role in weed management in soybean in the coming years. If used judiciously in weed management systems, this technology provides new tools for producers, but should not be expected to take the place of all other forms of weed management. These herbicides will be subject to population shifts and resistance development just as all the herbicides that have proceeded them.

18–8.2 Intercropping

Intercropping is one of the most widely available and inexpensive methods for increasing crop production per unit area of land (Plucknett and Smith, 1986).

There is considerable evidence that the simultaneous culture of two or more crops on the same piece of land will produce a greater yield than a monoculture of any of the component crops (Barker and Francis, 1986). In regions of the world such as Latin America, Asia, and Africa, intercropping is the dominant cropping method. While most commonly practiced on small farms with minimal mechanization or chemical technology, intercropping need not be restricted to such situations (Jagtap and Adeleye, 1999).

Competitive suppression of weeds takes a different form with intercropping than in crop monocultures. Increasing the complexity of a cropping system by interplanting species of differing growth forms, phenologies, and physiologies can create different patterns of resource availability to weeds (Ballare and Casal, 2000). Because a more diverse crop population can capture resources more efficiently, these resources may be converted to crop yield rather than leaving them available for weed growth. Thus, understanding how intercrops compete with weeds requires a thorough understanding of the growth and development characteristics of the crop and weed species in the intercropping system. Because resource availability is key to weed occurrence (Harper, 1977; Radosevich et al., 1997), increasing resource utilization through intercropping may provide unique opportunities for weed management.

18–8.3 Cover and Smother Crops

Cover crops are included in cropping systems for long- and short-term improvements in soil fertility and crop performance. Long-term benefits are derived from reduced soil erosion, improved soil quality, and increased soil organic matter (Power, 1996). Short-term responses are the result of changes in radiation balance, soil temperature and moisture, nutrient availability, runoff and infiltration, crop establishment, and pest populations. When cover crops are used for weed control, the goal is to replace an unmanageable weed population with a manageable cover crop (Teasdale, 1998). This is accomplished by managing the cover crop to preempt niches previously available to weeds. There are at least two major types of cover crops that can be used for weed control: (i) off-season cover crops and (ii) smother crops (a cover crop grown during part or all of the cropping season). When using off-season cover crops, the goal is to produce sufficient plant residue and/or allelochemicals to create an unfavorable environment for weed germination and establishment. When using a smother crop, the goal is usually to displace weeds from the harvested crop through resource competition.

The effectiveness of cover and smother crops in controlling weeds has been highly variable (Mohler and Teasdale, 1993; Putnam and DeFrank, 1983; Teasdale, 1998). Because of this variability, we need to develop a better understanding of the mechanisms by which cover crops change weed population dynamics, including how the growth of cover crops and the subsequent degradation of their residues change weed/soil interactions in multi-year cropping systems. As Teasdale (1998) concluded, "attention should be focused on defining the impact of cover crops on important rate-defining steps in the life cycle of weeds. This knowledge will help characterize how to use cover crops most effectively to disrupt the succession of important weed species".

18–8.4 Crop Competitiveness

Enhancing the ability of crops to compete with weeds is an attractive approach to improving weed management systems (Pester et al., 1999). Enhancing the ability of a soybean to compete with weeds can be accomplished by providing the best possible environment for crop growth combined with practices that reduce the density and/or vigor of the weeds. Practices such as narrow row spacing, increased plant density, appropriate time of planting, and fertility management are capable of shifting the competitive balance to favor crops over weeds (Buhler and Gunsolus, 1996; Stoller et al., 1987; Teasdale, 1995).

Differential weed competitive ability has been documented among commonly grown cultivars of soybean and several authors have suggested that weed management could be improved through crop breeding (i.e., Callaway, 1992; Pester et al., 1999; Rose et al., 1984). It should be feasible to breed crop cultivars that are genetically superior competitors with weeds through crop tolerance to weeds (maintain yield in presence of weeds) or crop interference with weeds (suppress growth of weeds). Once we understand the genetics of crop tolerance and/or competitiveness, methods used for developing these cultivars will depend on the type of competitive mechanism being selected for and the environment in which they will be grown (Martinez-Ghersa et al., 2000).

18–8.5 Fertility Management

The impact of weeds may be reduced by management strategies that maximize the uptake of nutrients by crops and minimize the availability of nutrients to weeds (DiTomaso, 1995). While we did not find specific research on soybean, changing fertilization practices has been shown to affect the interactions of several other crops with weeds. For example, applying fertilizer 5 cm below the soil surface in every second interrow space reduced weed biomass by 55% and weed density by 10% while increasing barley (*Hordeum vulgare* L.) grain yield by 28% compared with a broadcast fertilizer application (Rasmussen et al., 1996). Weed density, biomass, and N uptake was 20 to 40% less and wheat yield was 12% more where fertilizer was banded beside the crop row compared with broadcast application (Kirkland and Beckie, 1998). Other methods to alter the relative availability of nutrients to crops and weeds include timing of fertilizer applications (Anderson, 1991), altering nutrient sources (DeLuca and DeLuca, 1997), or altering nutrient availability using materials such as nitrification inhibitors (Teyker et al., 1991).

A more fundamental method of manipulating the relative uptake of nutrients by crop and weeds may be through enhancing the mechanisms and kinetics of mineral uptake by crop plants. In almost all cases where nutrient concentrations in crops and associated weeds were compared, accumulation of nutrients in the weeds exceeded the levels in the corresponding crop (DiTomaso, 1995). Management of fertility to benefit the crop may not only increase nutrient uptake by the crop, but likely will improve the competitiveness of the crop for other resources that might otherwise be available for weeds (Anderson et al., 1998; Kirkland and Beckie, 1998).

18–8.6 Organic Amendments/Weed Suppressive Soils

Organic matter amendments such as compost and manure alter temporal patterns of nutrient availability, especially for N and P, compared to pulsed application of synthetic fertilizers (DeLuca and DeLuca, 1997; Gallandt et al., 1998). Because germination and early growth of many weed species are strongly dependent on soil nutrient concentrations (DiTomaso, 1995; Karssen and Hillhorst, 1992), shifts in the timing of nutrient availability may affect weed density, emergence timing, and community composition.

Organic matter amendments are a source of non-nutrient compounds that may affect plant growth. Some of these compounds are growth inhibiting, whereas others are growth promoting (Ozores-Hampton et al., 1999). Organic matter amendments also contain persistent forms of organic C that may affect soil physical properties, such as water-holding capacity (Serra-Wittling et al., 1996), temperature and thermal conductivity (Al Kayassi et al., 1990), and aggregate stability and porosity (Guidi et al., 1981). These changes may affect weeds through changes in moisture and temperature and its related effects on timing of germination and seedling development (Mester and Buhler, 1991; Wiese and Binning, 1987) and aggregation and porosity effects on the abundance of suitable regeneration sites (Gallandt et al., 1999).

Organic matter amendments may also increase soil microbial biomass and activity, and change the incidence and severity of soil-borne diseases. Conklin et al. (1998) reported that wild mustard (*Brassica kaber* L.) seedlings grown in soil amended with compost and red clover (*Trifolium pratense* Sibth.) residue were smaller and had a higher incidence and severity of *Pythium* infection than seedlings grown in soil receiving ammonium nitrate ($NH_4^+NO_3^-$)fertilizer; corn seedlings were unaffected by soil amendment treatments.

Kennedy and Kremer (1996) suggested that it might be possible to develop farming practices that create "weed suppressive soils" in which microbial community composition and activity are altered in ways that would lead to depletion of the weed seed bank, reduced probabilities of weed seedling establishment, and reduced weed growth and competitive ability. They suggested that this may be accomplished by managing residue and microbial activity to establish an area of increased seed decay potential within the residue zone. Muller-Scharer et al. (2000) also addressed this approach by concluding that it should be one of the major routes to developing biological weed control systems for annual weed species.

18–8.7 Site-Specific Management

Site-specific agriculture has been defined as "an information and technology based agricultural management system to identify, analyze, and manage spatial and temporal variability within fields for optimum profitability, sustainability, and protection of the environment" (Robert et al., 1994). This concept has direct application to weed management because of the spatial and temporal heterogeneity of weed populations across agricultural landscapes (Cardina et al., 1996; Dieleman et al., 2000). Significant variation in soil properties may exist both within and among fields in a narrow geographic area (Mulla, 1993) and these differences may be correlated to weed occurrence (Andreasen et al., 1991; Dieleman et al., 2000). While specific soil conditions have been associated with weed infestations, it should also be recognized that these same soil conditions may affect the vigor of the crop, changing its ability to compete with weeds. Therefore, the weeds associated with a specific soil condition may be a secondary effect related to crop vigor rather than a weed response to soil conditions (Buhler et al., 2000).

While weeds are not uniformly distributed across fields, most weed control practices are applied uniformly. This uniform application of inputs over the nonuniform weed population has been identified as an important source of inefficiency in weed management (Cardina et al., 1997). Large portions of crop fields are often below threshold densities when the average field density is above the threshold (Cardina et al., 1995).

One method of dealing with weed patchiness is to develop methods to detect or map weeds and use that information to spatially direct herbicide application. Optical reflectance and image analysis are two approaches that have been used for real-time sensing of weeds to operate "patch sprayers" (Woebbecke et al., 1993). Another approach is to link remote-sensing technology, weed control recommendation models, and herbicide application with global positioning systems for delivering herbicides only to those areas with weed infestations that exceed threshold levels (Medlin et al., 2000). Both of these approaches have shown potential, but do not yet have the capabilities to reliably detect low densities of weeds within crop canopies.

Site-specific management of weeds involves both new concepts of weed biology and new technology. Principles of weed management and biology will need to be applied in a more precise fashion, with as much attention to where control practices are applied as to what is applied and when it is applied.

18–9 CONCLUSION

Weeds are a persistent problem in soybean production and it is the dynamic nature of weed populations that results in the perpetual encounters with this pest complex. The nature of a weed infestation in a field is a product of the production practices (rotation, tillage, fertilization, row spacing, herbicides, etc.) used in recent history. Whereas tillage and rotations generally influence the composition of species found in a field, the specific weed control tactics (i.e., herbicides) used usually determine the relative proportions of individual species.

While new tactics and approaches for weed management need to be developed, we also need to understand the fundamental elements of agroecosystems and the relationship of weeds with edaphic factors. We also need to gather feedback on the effectiveness of control tactics through better monitoring and assessment methods because of the importance of the reproductive output of surviving weeds on future weed populations.

The adoption of herbicide-based weed management systems has allowed the development of relatively simple cropping systems throughout the U.S. soybean production areas and has facilitated earlier soybean planting and increased farm size. With the availability of herbicides, weed management is no longer an integral component of the cropping system. While herbicide use has benefitted the farm community in many ways, the heavy reliance on herbicides creates an environment favorable for weed population shifts and resistence development and creates the potential for environmental contamination. The current challenge for producers is to manage herbicides in a manner that prevents adapted species from reaching troublesome populations and to minimize off-site movement. The challenge for weed scientists is to develop innovative management techniques that can be integrated into soybean production systems that will bring a more diverse and integrative approach to weed management in soybean.

REFERENCES

Adcock, T.D., and P.A. Banks. 1991. Effects of preemergence herbicides on the competitiveness of selected weeds. Weed Sci. 39:54–56.

Ahrens, W.H., and E.W. Stoller. 1983. Competition, growth rate and CO_2 fixation in triazine-susceptible and -resistant smooth pigweed (*Amaranthus hybridus*). Weed Sci. 31:438–444.

Akey, W.C., T.W. Jurik, and J. Dekker. 1990. Competition for light between velvetleaf (*Abutilon theophrasti*) and soybean (*Glycine max*). Weed Res. 30:403–411.

Aldrich, R.J. 1984. Weed-crop ecology: Principles in weed management. Breton, North Scituate, MA.

Al Kayassi, A.W., A.A. al Karaghouli, A.M. Hasson, and S.A. Beker. 1990. Influence of soil moisture content on soil temperature and heat storage under greenhouse conditions. J. Agric. Eng. Res. 45:241–252.

Alston, D.G., J.R. Bradley Jr., D.P. Schmitt, and H.D. Coble. 1991. Response of *Helicoverpa zea* (Lepidoptera: Noctuidae) populations to canopy development in soybean as influenced by *Hederodera glycines* (Nemotoda: Heteroderidae) and annual weed population densities. J. Econ. Entomol. 84:267–276.

Altman, J., and C.L. Campbell. 1977. Effect of herbicides on plant diseases. Ann. Rev. Phytopathol. 15:361–385.

Anderson, M.P., and J.W. Gronwald. 1991. Atrazine resistance in a velvetleaf (*Abutilon theophrasti*) biotype due to enhanced glutatione-S-transferase activity. Plant Physiol. 96:104–109.

Anderson, R.L. 1991. Timing of nitrogen application affects downy brome (*Bromus tectorum*) growth in winter wheat. Weed Technol. 5:582–585.

Anderson, R.L., D.L. Tanaka, A.L. Black, and E.E. Schweizer. 1998. Weed community and species response to crop rotation, tillage, and nitrogen fertility. Weed Technol. 12:531–536.

Andreasen, C., J.C. Streibig, and H. Haas. 1991. Soil properties affecting the distribution of 37 weed species in Danish fields. Weed Res. 31:181–187.

Anonymous. 2002. Herbicide handbook of the Weed Science Society of America. 8th ed. Weed Sci. Soc. Am., Champaign, IL.

Anonymous. 2000. 2000 Crop residue management survey. Conserv. Technol. Info. Ctr., West Lafayette, IN.

Aref, S., and D.R. Pike. 1998. Midwest farmers' perceptions of crop pest infestation. Agron. J. 90:819–825.

Bailey, G.W., and J.L. White. 1964. Review of adsorption and desorption of organic pesticides by soil colloids with implications concerning pesticide bioactivity. J. Agric. Food Chem. 12:324–332.

Bailey, L.H., and E.Z. Bailey. 1941. Hortus the second. Macmillan, New York.

Baker, H.G. 1974. The evolution of weeds. Annu. Rev. Ecol. Syst. 5:1–24.

Baker, J.L., and L.E. Shiers. 1989. Effects of herbicide formulation and application method on washoff from corn residue. Trans. Am. Soc. Agric. Eng. 32:830–833.

Ball, D.A. 1992. Weed seedbank response to tillage, herbicides, and crop rotation sequence. Weed Sci. 40:654–659.

Ballare, C.L., and J.J. Casal. 2000. Light signals perceived by crop and weed plants. Field Crops Res. 67:149–160.

Bandeen, J.D., G.R. Stephenson, and E.R. Cowett. 1982. Discovery and distribution of herbicide-resistant weeds in North America. p. 9–30. *In* H.M. LeBaron and J. Gressel (ed.) Herbicide resistance in plants. John Wiley & Sons, New York.

Banks, P.A., and E.L. Robinson. 1982. The influence of straw mulch on the soil reception and persistence of metribuzin. Weed Sci. 30:164–168.

Barker, T.C., and C.R. Francis. 1986. Agronomy of multiple cropping systems. p. 161–182. *In* C.A. Francis (ed.) Multiple cropping systems. Macmillan, New York.

Bauer, T.A., and D.A. Mortensen. 1992. A comparison of economic and economic optimum thresholds for two annual weeds in soybeans. Weed Technol. 6:228–235.

Baysinger, J.A., and B.D. Sims. 1991. Giant ragweed (*Ambrosia trifida*) interference in soybeans (*Glycine max*). Weed Sci. 39:358–362.

Bazzaz, F.A. 1990. Plant-plant interactions in successional environments. p. 239–263. *In* J.B. Grace and D. Tilman (ed.) Perspectives on plant competition. Academic Press, San Diego, CA.

Bello, I.A., M.D.K. Owen, and H.M. Hatterman-Valenti. 1995. Effect of shade on velvetleaf (*Abutilon theophrasti*) growth, seed production, and dormancy. Weed Technol. 9:452–455.

Bhomik, P.C., and J.D. Doll. 1982. Corn and soybean response to allelopathic effects of weed and crop residues. Agron. J. 74:601–606.

Biniak, B.M., and R.J. Aldrich.1986. Reducing velvetleaf (*Abutilon theophrasti*) and giant foxtail (*Setaria faberi*) seed production with simulated-roller herbicide applications. Weed Sci. 34:256–259.

Black, B.D., G.B. Padgett, J.S. Russin, J.L. Griffin, J.P. Snow, and G.T. Berggren Jr. 1996. Potential weed hosts for *Diaporthe phaseolorum* var. caulivora, causal agent for soybean stem canker. Plant Dis. 80:763–765.

Bloemhard, C.M.J., M.W.M.F. Arts, P.C. Scheepens, and A.G. Elema. 1992. Thermal inactivation of weed seeds and tubers during drying of pig manure. Neth. J. Agric. Sci. 40:11–19.

Blumhorst, M.R., J.B. Weber, and L.R. Swain. 1990. Efficacy of selected herbicides as influenced by soil properties. Weed Technol. 4:279–283.

Boerboom, C.M. 1999. Nonchemical options for delaying weed resistance to herbicides in Midwest cropping systems. Weed Technol. 13:636–642.

Bradshaw, L.D., S.R. Padgette, and B.H. Wells. 1995. Perspectives on the potential of the glyphosate-resistant weeds. Weed Sci. Soc. Am. Abstr. 35:66.

Bridgemohan, P., A.I. Brathwaite, and C.R. McDavid. 1991. Seed survival and patterns of seedling emergence of *Rottboellia cochinchinensis* (Lour.) W. Clayton in cultivated soils. Weed Res. 31:265–272.

Bridges, D.C. 1994. Impact of weeds on human endeavors. Weed Technol. 8:392–399.

Buhler, D.D. 1992. Population dynamics and control of annual weeds in corn (*Zea mays*) as influenced by tillage systems. Weed Sci. 40:241–248.

Buhler, D.D. 1995. Influence of tillage systems on weed population dynamics and management in corn and soybean in the central USA. Crop Sci. 35:1247–1258.

Buhler, D.D. 1996. Development of alternative weed management strategies. J. Prod. Agric. 9:501–505.

Buhler, D.D. 1999. Weed population responses to weed control practices. II. Residual effects on weed populations, control, and soybean yield. Weed Sci. 47:423–426.

Buhler, D.D., J.D. Doll, R.T. Proost, and M.R. Visocky. 1995. Integrating mechanical weeding with reduced herbicide use in conservation tillage corn production systems. Agron. J. 87:507–512.

Buhler, D.D., and J.L. Gunsolus. 1996. Effect of date of preplant tillage and planting on weed populations and mechanical weed control in soybean (*Glycine max*). Weed Sci. 44:373–379.

Buhler, D.D., J.L. Gunsolus, and D.F. Ralston. 1992. Integrated weed management techniques to reduce herbicide inputs in soybean. Agron. J. 84:973–978.

Buhler, D.D., J.L. Gunsolus, and D.F. Ralston. 1993. Common cocklebur (*Xanthium strumarium*) control in soybean (*Glycine max*) with reduced rates of bentazon and cultivation. Weed Sci. 41:447–453.

Buhler, D.D., and R.G. Hartzler. 2001. Emergence and persistence of seed of velvetleaf, common waterhemp, woolly cupgrass, and giant foxtail. Weed Sci. 49: 230–235.

Buhler, D.D., R.G. Hartzler, and F. Forcella. 1997. Implications of weed seed bank dynamics to weed management. Weed Sci. 45:329–336.

Buhler, D.D., M. Liebman, and J.J. Obrycki. 2000. Theoretical and practical challenges to an IPM approach to weed management. Weed Sci. 48:274–280.

Burkart, M.R., and D.D. Buhler. 1997. A regional framework for analyzing weed species and assemblage distributions using a geographic information system. Weed Sci. 45:455–462.

Burnside, O.C. 1992. Rationale for developing herbicide-resistant crops. Weed Technol. 6:621–625.

Burnside, O.C., R.S. Moomaw, F.W. Roeth, G.A. Wicks, and R.G. Wilson.1986. Weed seed demise in soil in weed-free corn (*Zea mays*) production across Nebraska. Weed Sci. 34:248–251.

Burnside, O.C., R.G. Wilson, S. Weisberg, and K.G. Hubbard. 1996. Seed longevity of 41 weed species buried 17 years in eastern and western Nebraska. Weed Sci. 44:74–86.

Bussan, A.J., O.C. Burnside, J.H. Orf, E.A. Ristau, and K.J. Puettmann. 1997. Field evaluation of soybean (*Glycine max*) genotypes for weed competitiveness. Weed Sci. 45:31–37.

Callaway, M.B. 1992. A compendium of crop varietal tolerance to weeds. Am. J. Altern. Agric. 7:169–180.

Cardina, J. 1995. Biological weed management. p. 279–341. *In* A.E. Smith (ed.) Handbook of weed management systems. Marcel Dekker, New York.

Cardina, J., G. A. Johnson, and D.H. Sparrow. 1997. The nature and consequences of weed spatial distribution. Weed Sci. 45:364–373.

Cardina, J., and H.M. Nouquay. 1997. Seed production and seedbank dynamics in subthreshold velvetleaf (*Abutilon theophrasti*) populations. Weed Sci. 45:85–90.

Cardina, J., E. Regnier, and K. Harrison. 1991. Long-term tillage effects of seed banks in three Ohio soils. Weed Sci. 39:186–194.

Cardina, J., D.H. Sparrow, and E.L. McCoy. 1995. Analysis of spatial distribution of common lambsquarters (*Chenopoduim album*) in no-till soybean (*Glycine max*). Weed Sci. 43:258–268.

Cardina, J., D.H. Sparrow, and E.L. McCoy. 1996. Spatial relationships between seedbank and seedling populations of common lambsquarters (*Chenopodium album*) and annual grasses. Weed Sci. 44:298–308.

Cardina, J., T.M. Webster, C.P. Herms, and E.E. Regnier. 1999. Development of weed IPM: Levels of integration for weed management. J. Crop Prod. 2:239–267.

Cavers, P.B. 1983. Seed demography. Can. J. Bot. 61:3678–3690.

Charudattan, R. 1991. The mycoherbicide approach with plant pathogens. p. 24–57. *In* D.O. TeBeest (ed.) Microbial control of weeds. Chapman and Hall, New York.

Coble, H.D. 1986. Development and implementation of economic thresholds for soybean. p. 295–307. *In* R.E. Frisbie and P.L. Adkisson (ed.) CIPM: Integrated pest management on major agricultural systems. Texas A & M Univ., College Station.

Collins, F.L., and S.J. Johnson. 1985. Reproductive response of caged adult velvetbean caterpillar and soybean looper to the presence of weeds. Agric. Ecosyst. Environ. 14:139–149.

Conklin, A.E., M.S. Erich, M. Liebman, and D.H. Lambert. 1998. Disease incidence and growth of wild mustard seedlings in red clover and compost amended soil. p. 279. *In* Agronomy abstracts 90. ASA, Madison, WI.

Cordes, R.C., and T.T. Bauman. 1984. Field competition between ivyleaf morningglory (*Ipomoea hederacea*) and soybeans (*Glycine max*). Weed Sci. 32:364–370.

Cousens, R. 1985. A simple model relating yield loss to weed density. Ann. Appl. Biol. 107:239–252.

Cousens, R. 1987. Theory and reality of weed control thresholds. Plant Prot. Quart. 2:13–20.

Currie, R.S., and T.F. Peeper. 1988. Combine harvesting affects weed seed germination. Weed Technol. 2:499–504.

Czapar, G.F., M.P. Curry, and M.E. Gray. 1995. Survey of integrated pest management practices in central Illinois. J. Prod. Agric. 8:483–486.

Czapar, G.F., M.P. Curry, and L.M. Wax. 1997. Grower acceptance of economic thresholds for weed management in Illinois. Weed Technol.11:828–831.

Dann, E.K., B.W. Diers, and R. Hammerschmidt. 1999. Suppression of sclerotinia stem rot of soybean by lactofen herbicide treatment. Phytopathology 89:598–602.

Day, B.E. 1972. Nonchemical weed control. p. 330–338. *In* Pest control: Strategies for the future. Natl. Acad. of Sci., Washington, DC.

DeLuca, T.H., and D.K. DeLuca. 1997. Composting for feedlot manure management and soil quality. J. Prod. Agric. 10:235–241.

Dieleman, J.A., D.A. Mortensen, D.D. Buhler, C.A. Cambardella, and T.B. Moorman. 2000. Identifying associations among site properties and weed species abundance. I. Multivariate analysis. Weed Sci. 48:567–575.

Dieleman, J.A., D.A. Mortensen, and A.R. Martin. 1999. Influence of velvetleaf (*Abutilon theophrasti*) and common sunflower (*Helianthus annuus*) density variation on weed management outcomes. Weed Sci. 47:81–89.

DiTomaso, J.M. 1995. Approaches for improving crop competitiveness through the manipulation of fertilization strategies. Weed Sci. 43:491–497.

Duffy, M., and D. Smith. 2000. Estimated costs of crop production in Iowa -2001. Bull. Fm-1712. Iowa State Univ. Ext., Ames.

Dunwell, J.M. 1996. Time-scale for transgenic product development. Field Crop Res. 45:135–140.

Dyck, E., and M. Liebman. 1994. Soil fertility management as a factor in weed control: The effect of crimson clover residue, synthetic nitrogen fertilizer, and their interaction on emergence and early growth of lambsquarters and sweet corn. Plant Soil 167:227–237.

Dyer, W.E. 1995. Exploiting weed seed dormancy and germination requirements through agronomic practices. Weed Sci. 43:498–503.

Eberlein, C.V., K. Al-Khatib, M.J. Guttieri, and E.P. Fuerst. 1992. Distribution and characteristics of triazine-resistant Powell amaranth (*Amaranthus powelli*) in Idaho. Weed Sci. 40:507–512.

Ellis, J.M., D.R. Shaw, and W.L. Barrentine. 1998. Soybean (*Glycine max*) seed quality and harvesting efficiency as affected by low weed densities. Weed Technol. 12:166–173.

Etheridge, R.E., W.E. Hart, R.M. Hayes, and T.C. Mueller. 2001. Effect of venturi-type nozzles and application volumes on postemergence herbicide efficacy. Weed Technol. 15:75–80.

Etheridge, R.E., A.R. Womac, and T.C. Mueller. 1999. Characterization of the spray droplet spectra and patterns of four venturi-type drift reduction nozzles. Weed Technol. 13:765–770.

Fenner, M. 1985. Seed ecology. Chapman Hall, New York.

Forcella, F., and M.J. Lindstrom. 1988. Weed seed populations in ridge and conventional tillage. Weed Sci. 36:500–502.

Forcella, F., R.G. Wilson, J. Dekker, R.J. Kremer, J. Cardina, R.L. Anderson, D. Alm, K.A. Renner, R.G. Harvey et al.. 1997. Weed seedbank emergence across the corn belt. Weed Sci. 45:67–76.

Forcella, F., R.G. Wilson, K.A. Renner, J. Dekker, R.G. Harvey, D.A. Alm, D.D. Buhler, and J.A. Cardina. 1992. Weed seedbanks of the U.S. Cornbelt: Magnitude, variation, emergence, and application. Weed Sci. 40:636–644.

Gallandt, E.R., M. Liebman, S. Corson, G.A. Porter, and S.D. Ullrich. 1998. Effects of pest and soil management systems on weed dynamics in potato. Weed Sci. 46:238–248.

Gallandt, E.R., M. Liebman, and D. Huggins. 1999. Improving soil quality: Implications for weed management. J. Crop Prod. 2:95–121.

Gebhardt, M.R. 1981. Preemergence herbicides and cultivation for soybeans (*Glycine max*). Weed Sci. 29:165–168.

Gould, F. 1991. The evolutionary potential of crop pests. Am. Sci. 79:496–507.

Gressel, J. 1996. Fewer constraints than proclaimed to the evolution of glyphosate-resistant weeds. p. 2–5. *In* M. Whalon and R. Hollingworth (ed.) Resistant pest management newsletter. Vol. 8. Pesticide Res. Ctr., Michigan State Univ., East Lansing.

Gressel, J., and L.A. Segel. 1978. The paucity of plants evolving genetic resistance to herbicides: Possible reasons and implications. J. Theor. Biol. 75:349–371.

Gressel, J., and L.A. Segel. 1982. Interrelating factors controlling the rate of appearance of resistance: The outlook for the future. p. 325–347. *In* H.M. LeBaron and J. Gressel (ed.) Herbicide resistance in plants. John Wiley & Sons, New York.

Grime, J.P. 1977. Evidence for the existence of three primary strategies in plants and its relevance to ecological and evolutionary theory. Am. Nat. 111:1169–1194.

Guidi, G., M. Pagliai, and M. Giachetti. 1981. Modification of some physical and chemical soil properties following sludge and compost applications. p. 122–136. *In* G. Catroux et al. (ed.) The influence of sewage sludge application on physical and biological properties of soils. D. Reidel Publ., Dordrecht, the Netherlands.

Gunsolus, J.L. 1990. Mechanical and cultural weed control in corn and soybeans. Am. J. Alt. Agric. 5:114–119.

Gunsolus, J.L., and D.D. Buhler. 1999. A risk management perspective on integrated weed management. J. Crop Prod. 2:167–187.

Guttieri, M.J., C.V. Eberlein, and D.C. Thill. 1995. Diverse mutations in the acetolactate synthase gene confer resistance in kochia (*Kochia scoparia*) biotypes. Weed Sci. 43:175–178.

Harmon, G.W., and F.D. Keim. 1934. The percentage and viability of weed seeds recovered in the feces of farm animals and their longevity when buried in manure. J. Am. Soc. Agron. 26:762–767.

Harper, J.L. 1977. The population biology of plants. Academic Press, London.

Harper, S.S. 1994. Sorption-desorption and herbicide behavior in soil. Rev. Weed Sci. 6:207–225.

Harrison, S.K., and M.M. Loux. 1995. Chemical weed management. p. 101–153. *In* A.E. Smith (ed.) Handbook of weed management systems. Marcel Dekker, New York.

Hartzler, R.G. 1996. Velvetleaf (*Abutilon theophrasti*) population dynamics following a single year's seed rain. Weed Technol. 10:581–586.

Hartzler, R.G. 1999. Reducing herbicide use in field crops. Bull. IPM-36. Iowa State Univ. Ext., Ames.

Hartzler, R.G., D.D. Buhler, and D.E. Stoltenberg. 1999. Emergence characteristics of four annual weed species. Weed Sci. 47:578–584.

Hartzler, R.G., and G.W. Roth. 1993. Effect of prior year's weed control on herbicide effectiveness in corn (*Zea mays*). Weed Technol. 7:611–614.

Hartzler, R.G., B.D. Van Kooten, D.E. Stoltenberg, E.M. Hall, and R.S. Fawcett. 1993. On-farm evaluation of mechanical and chemical weed management practices in corn (*Zea mays*). Weed Technol. 7:1001–1004.

Hartzler, R., W. Wintersteen, and B. Pringnitz. 1997. A survey of pesticides use in Iowa crop production in 1995. Bull. Pm-1718. Iowa State Univ. Ext., Ames.

Hasan, S., and P.G. Ayres. 1990. The control of weeds through fungi. New Phytol. 115:201–222.

Hazen, J.L. 2000. Adjuvants—Terminology, classification, and chemistry. Weed Technol. 14:773–784.

Heap, I.2001. The international survey of herbicide resistant weeds. Online. Internet. Available at www.weedscience.com (verified 15 May 2001).

Heap, I., and H. LeBaron. 2001. Introduction and overview of resistance. p. 1–22. *In* S.B. Powles and D.L. Shaner (ed.) Herbicide resistance and world grains. CRC Press, Boca Raton, FL.

Hirschberg, J., and L. McIntosh. 1983. Molecular basis of herbicide resistance in *Amaranthus hybridus*. Science (Washington DC) 222:1346–1349.

Holloway, J.C., and D.R. Shaw. 1996. Effect of herbicides on ivyleaf morningglory (*Ipomoea hederaceae*) interference in soybean (*Glycine max*). Weed Sci. 44:860–864.

Holm, L., J. Doll, E. Holm, J. Pancho, and J. Herberger.1997. World weeds: Natural histories and distribution. p. 1129. John Wiley & Sons, New York.

Holm, L., D. Plucknett, J. Pancho, and J. Herberger. 1977. The world's worst weeds: Distribution and biology. Univ. Hawaii Press, Honolulu.

Holt, J.S. 1994. Impact of weed control on weeds: New problems and research needs. Weed Technol. 8:400–402.

Horak, M.J., and D.E. Peterson. 1995. Biotypes of Palmer amaranth (*Amaranthus palmeri*) and common waterhemp (*Amaranthus rudis*) are resistant to imazethapyr and thifensulfuron. Weed Technol. 9:192–195.

Hume, L. 1982. The long-term effects of fertilizer application and three rotations on weed communities in wheat. Can. J. Plant Sci. 62:741–750.

Hume, L. 1987. Long-term effects of 2,4-D application on plants. 1. Effects on the weed community in a wheat crop. Can. J. Bot. 65:2530–2536.

Jagtap, S.S., and O. Adeleye. 1999. Land use efficiency of maize and soyabean intercropping and monetary returns. Trop. Sci. 39:50–55.

Jasieniuk, M., A.L. Brule-Babel, and I.N. Morrison. 1996. The evolution and genetics of herbicide resistance in weeds. Weed Sci. 44:176–193.

Jensen, P.K., L.N. Jorgensen, and E. Kirknel. 2001. Biological efficacy of herbicides and fungicides applied with low-drift and twin-fluid nozzles. Crop Prot. 20:57–64.

Johnson, M.D., D.L. Wyse, and W.E. Lueschen. 1989. The influence of herbicide formulation on weed control in four tillage systems. Weed Sci. 37:239–249.

Jordan, T.N., H.D. Coble, and L.M. Wax. 1987. Weed control. p. 429–460. *In* J.R. Wilcox (ed.) Soybeans: Improvement, production, and uses. ASA, CSSA, and SSSA, Madison, WI.

Karssen, C.M., and H.W.M. Hillhorst. 1992. Effect of chemical environment on seed germination. p. 327–348. *In* M. Fenner (ed.) Seeds: The ecology of regeneration in plant communities. CAB Int., Wallingford, UK.

Keen, N.T., M.J. Holliday, and M. Yoshikawa. 1982. Effects of glyphosate on glyceollin production and the expression of resistance to *Phytophthora megasperma* f. sp. glycinea in soybean *Glycine max*. Phytopathology 72:1467–1470.

Kennedy, A.C., and R.J. Kremer. 1996. Microorganisms in weed control strategies. J. Prod. Agric. 9:480–485.

Kidder, D.W., and R. Behrens. 1988. Plant responses to haloxyfop as influenced by drought stress. Weed Sci. 36:305–312.

King, C.A., and L.C. Purcell. 1997. Interference between hemp sesbania (*Sesbania exaltata*) and soybean (*Glycine max*) in response to irrigation and nitrogen. Weed Sci. 45:91–97.

Kirkland, K.J., and H.J. Beckie. 1998. Contribution of nitrogen fertilizer placement to weed management in spring wheat (*Triticum aestivum*). Weed Technol. 12:507–514.

Kishore, G.M., and D.M. Shaw. 1988. Amino acid biosynthesis inhibitors as herbicides. Ann. Rev. Biochem. 57:627–663.

Knoche, M. 1994. Effect of droplet size and carrier volume on performance of foliage-applied herbicides. Crop Prot. 13:163–178.

Krishnan, G., D.A. Mortensen, A.R. Martin, and L.B. Bills. 2001. Regionalizing a locally adapted weed management decision support system. Weed Sci. Soc. Am. Abstr. p. 41. Weed Sci. Soc. Am., Lawrence, KS.

Lee, L.J., and J. Ngim. 2000. A first report of glyphosate-resistant goosegrass (*Eleusine indica* (L.) Gaertn.) in Malaysia. Pest. Manage. Sci. 56:336–339.

Legere, A., and M.M. Schreiber. 1989. Competition and canopy architecture as affected by soybean (*Glycine max*) row width and density of redroot pigweed (*Amaranthus retroflexus*). Weed Sci. 37:84–92.

Levene, B.C., M.D.K. Owen, and G.L. Tylka. 1998. Response of soybean cyst nematodes and soybeans (*Glycine max*) to herbicides. Weed Sci. 46:264–270.

Liebman, M., and E.R. Gallandt.1997. Many little hammers: Ecological management of crop-weed interactions. p. 291–343. *In* L.E. Jackson (ed.) Ecology in agriculture. Academic Press, San Diego, CA.

Liebman, M., and T. Ohno. 1998. Crop rotation and legume residue effects on weed emergence and growth: Applications for weed management. p.181–221. *In* J.L. Hatfield et al. (ed.) Integrated weed and soil management. Ann Arbor Press, Chelsea, MI.

Lindquist, J.L., B.D. Maxwell, D.D. Buhler, and J.L. Gunsolus.1995. Velvetleaf (*Abutilon theophrasti*) recruitment, survival, seed production, and interference in soybean (*Glycine max*). Weed Sci. 43:226–232.

Lovely, W.G., C.R. Weber, and D.W. Staniforth. 1958. Effectiveness of the rotary hoe for weed control in soybeans. Agron. J. 50:621–625.

Lueschen, W.E., R.N. Andersen, T.R. Hoverstad, and B.R. Kanne. 1993. Seventeen years of cropping systems and tillage affect velvetleaf (*Abutilon theophrasti*) seed longevity. Weed Sci. 41:82–86.

Mahn, E.G., and K. Helmecke. 1979. Effects of herbicide treatment on the structure and functioning of agro-ecosystems. II. Structural changes in the plant community after the application of herbicides over several years. Agro-Ecosystems 5:159–179.

Marking, S. 1999. Roundup Ready yields—A summary of university soybean trials reveals slightly lower yield. Soybean Dig. (March):6–7.

Martinez-Ghersa, M.A., C.M. Ghersa, and E.H. Satorre. 2000. Coevolution of agricultural systems and their weed companions: Implications for research. Field Crops Res. 67:181–190.

Maxwell, B.D., M.L. Roush, and S.R. Radosevich. 1990. Predicting the evolution and dynamics of herbicide resistance in weed populations. Weed Technol. 4:2–13.

Mazur, B.J., and S.C. Falco. 1989. The development of herbicide resistant crops. Annu. Rev. Plant Physiol. Plant Mol. Biol. 40:441–480.

McCanny, S.J., and P.B. Cavers. 1988. Spread of proso millet (*Panicum miliaceum* L.) in Ontario, Canada. II. Dispersal by combines. Weed Res. 28:67–72.

McLean, K.S., and K.W. Roy. 1988. Purple seed stain of soybean caused by isolates of *Cercoppora kikuchii* from weeds. Can. J. Plant Pathol. 10:166–171.

Medlin, C.R., D.R. Shaw, P.D. Gerard, and F.E. LaMastrus. 2000. Using remote sensing to detect weed infestations in *Glycine max*. Weed Sci. 48:393–398.

Mester, T.C., and D.D. Buhler. 1991. Effects of soil temperature, seed depth, and cyanazine on giant foxtail (*Setaria faberi*) and velvetleaf (*Abutilon theophrasti*) seedling development. Weed Sci. 39:204–209.

Mindreboe, K.J. 1970. Weed control with preemergence herbicides in relation to rainfall. Weed Res.10:71–74.

Mohler, C.L. 1993. A model of the effects of tillage on emergence of weed seedlings. Ecol. Applic. 3:53–73.

Mohler, C.L., and J.R. Teasdale. 1993. Response of weed emergence to rate of *Vicia villosa* Roth and *Secale cereale* L. residue. Weed Res. 33:487–499.

Mortensen, D.A., and H.D. Coble. 1989. The influence of soil water content on common cocklebur (*Xanthium strumarium*) interference in soybeans. Weed Sci. 37:76–83.

Mortensen, D.A., G.A. Johnson, and L.J. Young.1993. Weed distribution in agricultural fields. p. 113–123. *In* P.C. Robert et al. (ed.) Soil specific crop management. ASA, CSSA, and SSSA, Madison, WI.

Moyer, J.R. 1987. Effect of soil moisture on the efficacy and selectivity of soil-applied herbicides. Rev. Weed Sci. 3:9–32.

Mt. Pleasant, J., and K.J. Schlather. 1994. Incidence of weed seed in cow (*Bos* sp.) manure and its importance as a weed source for cropland. Weed Technol. 8:304–310.

Mulder, T.A., and J.D. Doll. 1993. Integrating reduced herbicide use with mechanical weeding in corn (*Zea mays*). Weed Technol. 7:382–389.

Mulla, D.J. 1993. Mapping and managing spatial patterns in soil fertility and crop yield. p.15–26. *In* P.C. Robert et al. (ed.) Soil specific crop management. ASA, CSSA, and SSSA, Madison, WI.

Muller-Scharer, H., P.C. Scheepens, and M.P. Greaves. 2000. Biological control of weeds in European crops: recent achievements and future work. Weed Res. 40:83–98.

Murdoch, A.J., and R.H. Ellis. 1992. Longevity, viability and dormancy. p. 193–229. *In* M. Fenner (ed.) Seeds: The ecology of regeneration in plant communities. CAB Int., Wallingford, UK.

Nikolaeva, M.G. 1977. Factors controlling the seed dormancy pattern. p. 51–74. *In* A.A. North (ed.) The physiology and biochemistry of seed dormancy and germination. Holland Publ. Co., Khan, Amsterdam.

Norris, R.F. 1999. Ecological implications of using thresholds for weed management. p. 31–58. *In* D.D. Buhler (ed.) Expanding the context of weed management. The Haworth Press, New York.

Norris, R.F., and M. Kogan. 2000. Interactions between weeds, arthropod pests, and their natural enemies in managed ecosystems. Weed Sci. 48:94–158.

O'Donovan, J. T. 1996. Weed economic thresholds: Useful agronomic tool or pipe dream? Phytoprotection 77:13–28.

Oliver, L.R. 1979. Influence of soybean (*Glycine max*) planting date on velvetleaf (*Abutilon theophrasti*) competition. Weed Sci. 27:183–188.

Olson, K.D., and V.R. Eidman. 1992. A farmer's choice of weed control method and the impacts of policy and risk. Rev. Agric. Econ. 14:125–137.

Oplinger, E.S., M.J. Marinka, and K.A. Schmitz. 1999. Performance of transgenic soybeans in the Northern US. Available at www.biotech-info.net/herbicide-tolerance.html#soy (verified 10 Oct. 2001).

Owen, M.D.K. 2001. World maize/soybean and herbicide resistance. p. 101–163. *In* S.B. Powles and D.L. Shaner (ed.) Herbicide resistance and world grains. CRC Press, Boca Raton, FL.

Owen, M.D.K. 2000. Current use of transgenic herbicide-resistant soybean and corn in the USA. Crop Prot. 19:765–771.

Ozores-Hampton, M., P.J. Stoffella, T.A. Bewick, D.J. Cantliffe, and T.A. Obreza. 1999. Effect of age of composted MSW and biosolids on weed seed germination. Compost Sci. Util. 7:51–57.

Penner, D. 2000. Activator adjuvants. Weed Technol. 14:785–791.

Pester, T.A., O.C. Burnside, and J.H. Orf. 1999. Increasing crop competitiveness to weeds through crop breeding. J. Crop Prod. 2:31–58.

Peters, E.J., M.R. Gebhardt, and J.F. Stritzke. 1965. Interrelations of row spacings, cultivations and herbicides for weed control in soybeans. Weeds 13:285–289.

Pickett, S.T.A., and F.A. Bazzaz. 1978. Organization of an assemblage of early successional species on a soil moisture gradient. Ecology 59:1248–1255.

Plucknett, D.L., and N.J.H. Smith. 1986. Historical perspectives on multiple cropping. p. 20–39. *In* C.A. Francis (ed.) Multiple cropping systems. Macmillan, New York.

Power, J.F. 1996. Cover crops. p. 124–126. *In* S.B. Parker (ed.) 1997 McGraw-Hill yearbook of science and technology. McGraw-Hill, New York.

Powles, S., D. Lorraine-Colwill, J.D. Dellow, and C. Preston.1998. Evolved resistance to glyphosate in rigid ryegrass (*L. rigidum*) in Australia. Weed Sci. 46:604–607.

Preston, C., J.A.M. Holtum, and S.B. Powles. 1992. On the mechanism of resistance to paraquat in *Hordeum glaucum* and *H. leporinum*. Plant Physiol.100:630–636.

Preston, C., and C.A. Mallory-Smith. 2001. Biochemical mechanisms, inheritance, and molecular genetics of herbicide resistance in weeds. p. 23–60. *In* S.B. Powles and D.L. Shaner (ed.) Herbicide resistance and world grains. CRC Press, Boca Raton, FL.

Putnam, A.R., and J. DeFrank. 1983. Use of phytotoxic plant residues for selective weed control. Crop Prot. 2:173–181.

Radosevich, S., J. Holt, and C. Ghersa. 1997. Weed ecology: Implications for management. 2nd ed. John Wiley & Sons, New York.

Rasmussen, K., J. Rasmussen, and J. Petersen. 1996. Effects of fertilizer placement on weeds in weed harrowed spring barley. Acta Agric. Scand. 46:192–196.

Riggs, R.D., and M.L. Hamblen. 1966. Additional weed hosts of *Heterodera glycines*. Plant Dis. Rep. 50:15–16.

Ritter, R.L., and H.D. Coble. 1981. Influence of temperature and relative humidity on the activity of acifluorfen. Weed Sci. 29:480–485.

Robert, P.C., R.H. Rust, and W.E. Larson. 1994. Preface. *In* P.C. Robert et al. (ed.) Site-specific management for agricultural systems. ASA, CSSA, and SSSA, Madison, WI.

Robinson, R.G. 1949. Annual weeds, their viable seed populations in the soil and their effects on yields of oats, wheat, and flax. Agron. J. 41:513–518.

Rodriguez-Kabana, R., E.A. Curl, and H.H. Funderburk Jr. 1966. Effect of four herbicides on growth of *Rhizoctonia solani*. Phytopathology 56:1332–1333.

Rose, S.J., O.C. Burnside, J.E. Specht, and B.A. Swisher. 1984. Competition and allelopathy between soybeans and weeds. Agron. J. 76:523–528.

Ryan, G.F. 1970. Resistance of common groundsel to simazine and atrazine. Weed Sci. 18:614–616.

Salzman, F.P., R.J. Smith, and R.E. Talbert. 1988. Suppression of red rice (*Oryza sativa*) seed production with fluazifop and quizalofop. Weed Sci. 36:800–803.

Schreiber, M.M. 1992. Influence of tillage, crop rotation, and weed management on giant foxtail (*Setaria faberi*) population dynamics and corn yield. Weed Sci. 40:645–653.

Seely, M. 1995. Some applications of temporal climate probabilities to site-specific management of agricultural systems. p. 513–530. *In* P.C. Robert et al. (ed.) Site specific management for agricultural systems. ASA, CSSA, and SSSA, Madison, WI.

Senseman, S.A., and Oliver, L.R.1993. Flowering patterns, seed production, and somatic polymorphism of three weed species. Weed Sci. 41:418–425.

Serra-Wittling, C., S. Houot, and E. Barriuso. 1996. Modification of soil water retention and biological properties by municipal solid waste compost. Compost Sci. Util. 4:44–52.

Shelton, M.D., and C.R. Edwards. 1983. Effects of weeds on the diversity and abundance of insects in soybeans. Environ. Entomol. 12:296–298.

Smart, G.C. 1964. Additional hosts of the soybean cyst nematode, *Heterodera glycines*, including hosts in two additional plant families. Plant Dis. Rep. 48:388–390.

Sprague, C.L., E.W. Stoller, and L.M. Wax. 1997. Common cocklebur (*Xanthium strumarium*) resistance to selected ALS-inhibiting herbicides. Weed Technol. 11:241–247.

Springman, R., D. Buhler, R. Schuler, D. Mueller, and J. Doll. 1989. Row crop cultivators for conservation tillage systems. Publ. A3483. Univ. of Wisconsin Ext., Madison.

Stahler, L.M. 1948. Shade and soil moisture as factors in competition between selected crops and field bindweed. Agron. J. 40:490–502.

Staniforth, D.W. 1958. Soybean-foxtail competition under varying soil moisture conditions. Agron. J. 50:13–15.

Steckel, L.E., M.S. DeFelice, and B.D. Sims. 1990. Integrating reduced rates of postemergence herbicides and cultivation for broadleaf weed control in soybeans (*Glycine max*). Weed Sci. 38:541–545.

Stevens, O. A. 1957. Weights of seeds and numbers per plant. Weeds 5:46–55.

Stoller, E.W., S.K. Harrison, L.M. Wax, E.E. Regnier, and E.D. Nafzinger. 1987. Weed interference in soybeans (*Glycine max*). Rev. Weed Sci. 3:155–181.

Stoller, E.W., L.M. Wax, and D.A. Alm. 1993. Survey results on environmental issues and weed science research priorities within the corn belt. Weed Technol. 7:763–770.

Stoller, E.W., and J.T. Woolley. 1985. Competition for light by broadleaf weeds in soybeans (*Glycine max*). Weed Sci. 33:199–202.

Strand, O.E., and G.R. Miller. 1980. Woolly cupgrass—A new weed threat in the Midwest. Weeds Today:16.

Taylor, K.L., and R.G. Hartzler. 2000. Effect of seed bank augmentation on herbicide efficacy. Weed Technol.14:261–267.

Teasdale, J.R. 1998. Cover crops, smother plants, and weed management. p. 247–270. *In* J.L. Hatfield et al. (ed.) Integrated weed and soil management. Ann Arbor Press, Chelsea, MI.

Teasdale, J.R. 1995. Influence of narrow row/high population corn (*Zea mays*) on weed control and light transmission. Weed Technol. 9:113–118.

Teyker, R.H., H.D. Hoelzer, and R.A. Liebl. 1991. Maize and pigweed reponse to nitrogen supply and form. Plant Soil 135:287–292.

Thelen, K.D., E.P. Jackson, and D. Penner. 1995. The basis for the hard-water antagonism of glyphosate antagonism. Weed Sci. 43:541–548.

Thomas, A.G., and B.L. Frick. 1993. Influence of tillage systems on weed abundance in southwestern Ontario. Weed Technol. 7:699–705.

Tollenaar, M., S.P. Nissanka, A. Aguilera, S.F. Weise, and C.J. Swanton. 1994. Effect of weed interference and soil nitrogen on four maize hybrids. Agron. J. 86:596–601.

Triplett, G.B., Jr. 1985. Principles of weed control for reduced-tillage corn production. p. 26–40. *In* A.F. Wiese (ed.) Weed control in limited tillage systems. Weed Sci. Soc. Am., Champaign, IL.

Van Acker, R.C., C.J. Swanton and S.F. Weise. 1993. The critical period of weed control in soybean [*Glycine max* (L.) Merr.]. Weed Sci. 41:194–200.

VanGessel, M.J. 2001. *Conyza canadensis* insensitivity to glyphosate. Abstr. Northeast. Weed Sci. Soc. 55:32.

Wade, B.R., D.E. Riechers, R.A. Liebl, and L.M. Wax. 1993. The plasma membrane as a barrier to herbicide penetration and site for adjuvant action. Pestic. Sci. 37:195–202.

Walker, A., and H.A. Roberts. 1975. Effects of incorporation and rainfall on the activity of some soil-applied herbicides. Weed Res.15:263–269.

Walker, R.H. 1995. Preventative weed management. p. 35–50. In A.E. Smith (ed.) Handbook of weed management systems. Marcel Dekker, New York.

Webster, E.P. 1998. Economic losses due to weeds in southern states. Proc. South.Weed Sci. Soc. 51:314–322.

Webster, T.M., J. Cardina, and H.M. Norquay. 1998. Tillage and seed depth effects on velvetleaf (Abutilon theophrasti) emergence. Weed Sci. 46:76–82.

Webster, T.M., and H.D. Coble. 1997. Changes in the weed species composition of the southern United States: 1974 to 1995. Weed Technol. 11:308–317.

Wiese, A.F., and C.W. Vandiver. 1970. Soil moisture effects on competitive ability of weeds. Weed Sci. 18:518–519.

Wiese, A.M., and L.K. Binning. 1987. Calculating the threshold temperature of development for weeds. Weed Sci. 35:177–179.

Wiles, L.J., G.G. Wilkerson, H.J. Gold, and H.D. Coble. 1992. Modeling weed distribution for improved postemergence control decisions. Weed Sci. 40:546–553.

Wilkerson, G.G., S.A. Modena, and H.D. Coble. 1991. HERB: Decision model for postemergence weed contol in soybean. Agron. J. 83:413–417.

Williams, C.S., and R.M. Hayes. 1984. Johnsongrass (Sorghum halapense) competition in soybeans (Glycine max). Weed Sci. 32:498–501.

Wills, G.D. 1984. Toxicity and translocation of sethoxydim in bermudagrass (Cynodon dactylon) as affected by environment. Weed Sci. 32:20–24.

Wilson, R.G.1988. Biology of weed seeds in the soil. p. 25–39. In M. A. Altieri and M. Liebman (ed.) Weed management in agroecosystems: Ecological approaches. CRC Press, Boca Raton, FL.

Woebbecke, D.M., G.E. Meyer, K. Von Bargen, and D.A. Mortensen. 1993. Plant species identification, size, and enumeration using machine vision techniques on near-binary images. p. 208–219. In J.A. DeShazer (ed.) Proc. of the SPIE Conf. on Optics in Agriculture and Forestry, Boston, MA. Vol. 1836. SPIE, Boston, MA.

Yelverton, F.H., and H.D. Coble. 1991. Narrow row spacing and canopy formation reduces weed resurgence in soybeans (Glycine max). Weed Technol. 5:169–174.

Young, A.L. 2000. Genetically modified crops: the real issues hindering public acceptance. Environ. Sci. Pollut. Res. 7:1–12.

Zanin, G., and M. Sattin. 1988. Threshold level and seed production of velvetleaf (Abutilon theophrasti) in maize. Weed Res. 28:347–352.

Zimdahl, R.L. 1987. The concept and application of the critical weed free period. p.145–155. In M.A. Altieri and M. Liebman (ed.) Weed management in agroecosystems: Ecological approaches. CRC Press, Boca Raton, FL.

Zimdahl, R.L.1991.Weed science—A plea for thought. USDA, Coop. State Res. Serv., Washington, DC.

19   Economics and Marketing

STEVEN T. SONKA
University of Illinois
Urbana, Illinois

KAREN L. BENDER
University of Illinois
Urbana, Illinois

DONNA K. FISHER
Georgia Southern University
Statesboro, Georgia

This chapter will describe today's marketplace for soybean [*Glycine max* (L.) Merr.] and examine key forces likely to shape the soybean marketplace of tomorrow. The description of today's marketplace includes global dimensions of production and utilization as well as the patterns that have emerged to lead to today's setting. Looking to the future, there are many forces that will shape the future evolution of the soybean market. Two of these forces will be considered in detail within this chapter. One force is the future need for protein, in the context of future levels of global population, income, and malnutrition. Another section examines pressures for change in the commodity marketing system. This approach has dominated, and continues to dominate, the soybean sector. However societal desires for more information regarding production and marketing in the food system, in part fueled by concerns about genetic modification of agricultural products, coupled with advances in information technology, combine to make it potentially feasible for alternative market systems to emerge and supplement or supplant the commodity approach. If that change were to occur, it is likely to have ramifications throughout the supply value chain, including the research institutions that traditionally support that chain.

One of the most discussed changes in the soybean marketplace in the 1990s was the introduction and widespread adoption of transgenic soybean in the USA and Argentina. Introduced in the middle of the 1990s, Roundup Ready (Monsanto Co., St. Louis, MO) soybean cultivars increased to more than 71% of soybean production in the USA by 2001 (Monsanto, 2002). Globally, Roundup Ready soybean cultivars accounted for 63% of all transgenic crops grown (Monsanto, 2002). Although producers in general have welcomed these technological innovations, the societal response has been quite negative among some political interest groups. The resulting controversy has intensified the pressure for fundamental change in agricultural commodity systems. These issues will be examined in the section of this chapter that deals with the potential for alternative marketing systems.

19–1 THE SOYBEAN MARKETPLACE TODAY AND HOW WE GOT HERE

This section of the chapter will provide a brief review of the market performance of soybean globally over the last three decades or so. The primary variables of interest will be consumption of soybean meal and oil, production (globally and for key producing nations), trade, value, and prices.

19–1.1 Using the Soybean Crop

Since the 1970s, the global economy performed admirably as people in many parts of the world shared in the increased abundance fueled by globalization (The Economist, 2001). As income levels increased in Asia, Mexico, Latin America, and other nations, consumers chose to enhance their diet by consuming greater quantities of animal protein and fats and oils. The soybean crop, and those farmers around the world who produced it, played a key role by providing the protein for livestock feed and soybean oil for human consumption. In 1999, soybean oil comprised nearly 30% of the global consumption of vegetable oils (Soy Stats, 2002). Even more impressive, soybean meal accounted for almost 70% of the world's protein meal production in that year (Soy Stats, 2002).

Despite considerable interest and expenditures, industrial uses of soybean products, including use as alternative fuels, remain speculative and small. In the USA, for example, soybean meal production exceeded 34 metric tonnes and soybean oil production exceeded 8 metric tonnes in 1999 (Soy Stats, 2002). Industrial uses, in total, were <10 thousand metric tonnes. Industrial and energy uses of soybean products could, of course, dramatically increase in the future. Legislation mandating such use, either for environmental or political reasons, could fuel such a change. But little evidence of such an increase is indicated in marketplace experience to date. Thus, this discussion focuses on food and feed uses for soybean oil and meal, respectively.

As a feedstuff for livestock, soybean meal plays a central role in contributing to protein availability across the world. Direct human consumption has been a far more limited use of soybean, except in certain countries in Asia. For example, data from the United Nations Food and Agriculture Organization (FAO) show that food consumption of soybean amounted to about 8% of the total 1999 soybean crop (FAO of the United Nations, 2002). In Japan, Korea, China, and Indonesia, however, soybean is a dietary staple. Consumption of traditional soybean products such as tofu, tempeh, and miso typically is a key component of the diet in these countries.

In recent years, there has been an explosion of interest in other parts of the world in the human consumption of soy-based products, in addition to those traditional products (United Soybean Board, TalkSoy, U.S. Soy Foods Directory, 2002). This interest has been fueled by several factors, but one widely cited factor is the

Food and Drug Administration (FDA) approval of a health claim for soybean linking soybean consumption to the potential for reducing cholesterol levels (U.S. Soy Foods Directory, 2002). Interest in the use of soy-based products in soymilk, baked goods, cereals, and pastas has greatly intensified. Although growth in the consumption of such products has been sparked in recent years, the total quantity of soybean utilized in these products remains relatively small.

Soybean meal has held and continues to hold a commanding position relative to other sources of meal for protein. Soybean oil, on the other hand, has been subjected to considerable pressure from alternative sources of vegetable oil in recent years. Canola oil and olive oil are two competitors that appeal to some health segments of the market. From a volume perspective, however, the massive increase in palm oil production and use is particularly noteworthy. At nearly 22 metric tonnes in 2000, palm oil production was almost 15 times greater than it was in 1961 (FAO of the United Nations, 2002).

19–1.2 Soybean Production and Exports

Figure 19–1 presents production data for soybean globally and in the USA since 1970. Starting at slightly more than 43 metric tonnes in 1970, global production increased nearly fourfold, first exceeding 160 metric tonnes in 1998 (FAO United Nations, 2002). The 1970s were a period of rapid increase, with global soybean production doubling. The 1980s saw a much slower rate of increase, with average production in the years 1989 to 1991 only 30% higher than at the start of that decade. During the 1990s, however, production leaped upward again. The 160 metric tonnes level now being achieved is 50% greater than the level attained at the start of the 1990s.

In the USA, production in 1970 was nearly 31 metric tonnes, accounting for almost three-fourths of world production. Although production in the USA more than doubled since the 1970s, achieving in excess of 75 metric tonnes in 2000, the USA now accounts for less than half of global soybean production. As was the case globally, production jumped sharply during the 1970s, exceeding 50 metric tonnes for the first time in 1978. Production in the USA also stagnated in the 1980s and

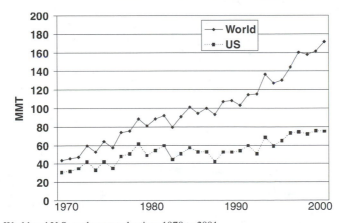

Fig. 19–1. World and U.S. soybean production: 1970 to 2001.

for much of the early 1990s. Since 1994, production increased substantially. Prior to 1994, production in the USA hovered in the 50 to 60 metric tonnes range. Because of good weather, enhanced technology and management, and domestic farm income support policies, production in the USA has jumped to levels well above 70 metric tonnes.

Figure 19–2 provides data on soybean production levels for the top five producing nations. The data for the USA are the same as that shown in Fig. 19–1. China, which started the 1970s as the second largest producing nation, now is fourth in total production. Although production has increased by more than 80% in China, production increases in Brazil and Argentina far outpaced those levels. Production levels in both Brazil and Argentina were inconsequential in 1970 (1.5 and 0.1 metric tonnes, respectively). By the late 1990s, production in those two nations exceeded 50 metric tonnes (32.7 million in Brazil and 20.2 million in Argentina in 2000). India, which produced only a few thousand metric tonnes in the 1970s, attained production levels in excess of 5 metric tonnes by the end of the 1990s.

Fueled by global economic growth, this massive increase in production has been employed to provide soybean meal for livestock feed and soybean oil for consumers. The global increase in exports documents a significant component of this increase in need. Global soybean exports increased over threefold between 1970 and 1999 (Fig. 19–3). Exports from the USA, followed the global upward pattern from 1970 to the early 1980s. Then, however, the pace of increase slowed substantially in both absolute terms and relative to global exports. The early 1990s saw a significant "uptick" in exports from the USA, however, since 1996 export growth has stagnated. A strong U.S. dollar, increased production in other nations, and concerns regarding transgenic soybean are all rationales that could contribute to explaining these trends.

Table 19–1 identifies the top five importing countries for U.S. soybean in 1990 and 2000. The European Union, Japan, Taiwan, and Mexico are on this list for both years, although the quantities imported have slipped for all but Mexico. The introduction of China, at a level exceeding one billion dollars in soybean imports, is a dynamic of considerable interest.

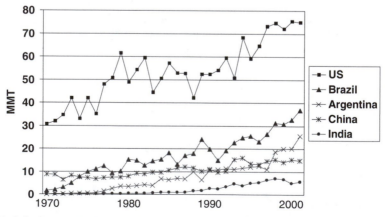

Fig. 19–2. Soybean production in the five largest producing nations: 1970 to 2001.

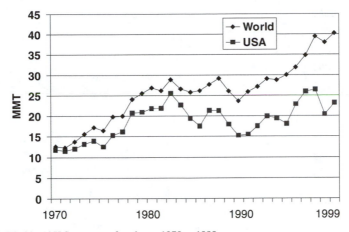

Fig. 19–3. World and U.S. exports of soybean: 1970 to 1999.

Figure 19–4 provides information, not just on the value of soybean exports, but also on the value of soybean meal and oil exports. Although not as widely discussed, the value of soybean meal and oil exports is a substantial component of the international revenue stream for the U.S. soybean industry. For the 1997 to 1999 period, the value of soybean exports exceeded $5.6 billion. During this same 3-yr

Table 19–1. Top five customers for U.S. soybean (1990 vs. 2000).

1990		2000	
	(Million $s)		(Million $s)
European Union	1433	European Union	1143
Japan	821	China	1008
Taiwan	411	Japan	758
Mexico	200	Mexico	678
South Korea	194	Taiwan	385

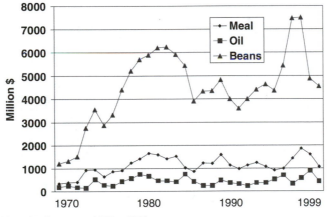

Fig. 19–4. Value of U.S. exports: 1970 to 1999.

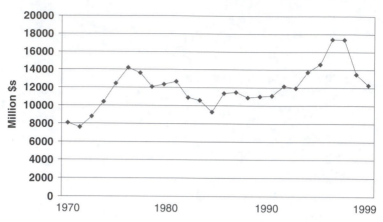

Fig. 19–5. Value of U.S. soybean crop: 1974 to 1999.

period, the combined value of the exports of soybean meal and oil exceeded $2.1 billion dollars.

Figure 19–5 describes the total annual value of the U.S. soybean crop from 1970 to 1999 (Soy Stats, 2002). The rapid drop in crop value since 1996 documents the current economic stress of U.S. soybean farmers. These data do not include the direct financial impact of domestic farm policies which have mitigated, to a large extent, the negative effects of low prices on net income of soybean farmers. Farm program policies, especially the ratio of the soybean to corn (*Zea mays* L.) loan rates in the 1996 legislation, influenced the amount and location of soybean production in the USA. That ratio appears to have fostered increases in soybean hectares relative to other crops. The volatility of crop revenues also is documented in Fig. 19–5. Soybean revenues saw two periods of rapid increase. One was in the first half of the 1970s and the second was in the mid-1990s. Unfortunately both those periods of increased revenue were not sustained.

Figure 19–6 presents data on the farm price received by U.S. farmers for soybean from 1974 to 1999 (Soy Stats, 2002). These are in nominal, not real, terms.

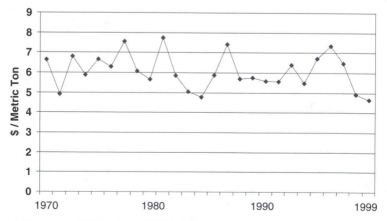

Fig. 19–6. Farm price of U.S. soybean: 1970 to 1999.

Since 1975, soybean prices oscillated around the $6 per bushel level. Recent prices, however, are substantially below that level.

19–1.3 Summing Up

As a major agricultural commodity, the growth in soybean consumption and production has been noteworthy. In general, the last decades of the 20th century saw unprecedented growth in economic well-being. As lower and middle-income consumers experience a gain in income, a traditional response is to upgrade their diet status. Soybean oil and animal protein fed with soybean meal are important vehicles which allow consumers to translate their income gains into enhanced diets.

In the 1970s, production increases were heavily tied to North America. However, in the 1980s and 1990s, production increases in other parts of the world, notably Brazil and Argentina, acted to greatly extend the geographic reach of significant soybean production. The continuing global need for soybean and soybean products is demonstrated by the sharply increasing levels of soybean exports that occurred over the last 30 yr.

19–2 THE DYNAMICS OF TOMORROW'S GLOBAL APPETITE FOR PROTEIN

One of the key issues facing agricultural producers in general, and the soybean industry in particular, is uncertainty regarding the future need for protein and, therefore, for soybean. To address this concern, the National Soybean Research Laboratory (NSRL) developed a system dynamics model to explore the potential appetite for protein in the world food system. The Protein Consumption Dynamics (PCD) Model simulates future global protein appetite scenarios based on population and income growth, highlighting the systematic relationships between population, income, appetite (potential demand), and malnutrition (Fisher, 2000)[1]. It tracks the human appetite for six agricultural commodities (beef, pork, poultry, fish, fats and oils, and vegetable protein), on a global basis, for each year from 2001 to 2025. The model also provides estimates of the extent of malnutrition for the same time period.

19–2.1 Model Structure

Figure 19–7 illustrates the relationships made explicit in the Protein Consumption Dynamics (PCD) Model. In the model, a region's income and population increase each year at specified rates. The effects of cultures and dietary preferences are reflected in the regionally-specific econometric estimates between income and food appetite. The effects of per capita income and cultural influences are combined to develop estimates of appetite (potential demand or consumption) and malnutri-

---

[1] Model development was funded in part by the Illinois Soybean Program Operating Board, the Soybean Industry Chair in Agriculture Strategy, and other supporters of the National Soybean Research Laboratory.

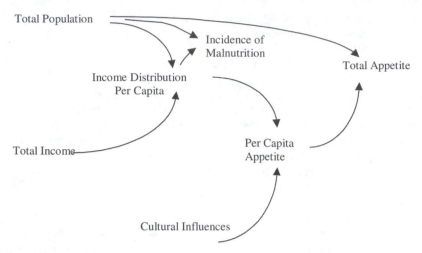

Fig. 19–7. Relationships underlying the protein consumption dynamics model.

tion for each region. These per capita estimates are then multiplied by the appropriate population estimates for each region to compute total potential demand for
the various commodities.

Cultural differences are incorporated through the designation of eight demographic regions. The regions are defined to be relatively homogeneous in terms
of income, income growth levels, food appetite patterns, and cultural characteristics. These regions, which are consistent with those identified by both the World
Bank and the U.N. FAO, are delineated in Table 19–2.

A key reason for the development of this model was to have the capability to
compare the effects of alternative assumptions on the desire and need for protein
across a range of parameter values. To illustrate this capability, two future scenarios are defined.[2] These are:

- Base Case: Employs population growth projections consistent with World
  Bank and U.N. FAO medium-level projections and income growth projections consistent with the actual experience of the last two decades.
- Lower Income Case: Uses the same population projections as the Base Case,
  but income growth rates are 50% smaller than those of the Base Case.

The model framework requires that future income and population growth be
projected for each region. Population and income growth information are based on
secondary data taken from the World Bank and the FAO. The historic consumption
data are taken from the 1997 FAOSTAT Statistical Database. Historic and future
malnutrition data are taken from FAO and Bread for the World (BFW, 1998).[3] It is
difficult to find long-term income growth rate projections from official sources. The

[2] These scenarios are projections as defined by Ferris (1998). For our purposes, the probability of
each scenario occurring is not important. We are more concerned with getting decision makers to consider alternative potential futures, than in predicting the future.
[3] More information on malnutrition can be found at: http://www.fao.org/NEWS/1999/991004-
e.htm.

Table 19–2. Protein consumption dynamics (PCD) regional definitions.

China
East Asia
Transition economies (the countries of the former USSR, Eastern Europe, and Turkey)
Latin America
Middle East and North Africa (MENA)
OECD (The relatively economically well-off nations of Europe, North America, Australia, New Zealand, and Japan)
South Asia
Sub-Saharan Africa

Table 19–3. Annual income growth rates.

Regions	Historic†	Base case	Lower income
		%	
China	10.20	9.00	4.50
East Asia	7.00	6.00	3.00
Transition economies	−0.50	1.60	0.80
Latin America	2.20	2.20	1.10
MENA	−1.40	1.00	0.50
OECD	2.60	2.60	1.30
South Asia	5.60	5.40	2.70
Sub-Saharan Africa	1.50	2.00	1.00
World	3.40	3.70	1.90

† Historic annual income growth rates are averages from 1984 to 1996.

income projections employed here (in Table 19–3) are based upon historic rates of income growth, published literature on future prospects (Coplin and O'Leary, 1994), and expert judgment.

19–2.2 Future Needs and the Role of Income

This subsection describes the modeling results provided by the PCD model. The information provided here is only a small subset of data that the modeling tool can provide. A more complete analysis is available in Fisher (2000). For the Base Case Scenario, Fig. 19–8 and 19–9 present estimates of regional population and per capita income for the Years 2001 through 2025. Although the numbers have profound implications for many measures of human well being, the estimates are likely not too surprising in themselves. Population growth is relatively low in the OECD region (the relatively economically well-off nations of Europe, North America, Australia, New Zealand, and Japan) and the Transition Economies, roughly 5% for both regions. Conversely, Fig. 19–8 shows relatively high population growth in the MENA and Sub-Saharan Africa regions; an increase of more than 50% in population for both.

Based upon these assumptions, very strong per capita income growth would occur in China (585%), East Asia (209%), and South Asia (163%) (Fig. 19–9). Data for the OECD region is not included in Fig. 19–9 because of scaling difficulties. Income growth in the OECD countries is roughly equal to the global average but, of course, starts at a very large initial level. Growth in total gross domestic product (GDP) is weak in the regions with high population growth. The result is a de-

% Change 2001-2025

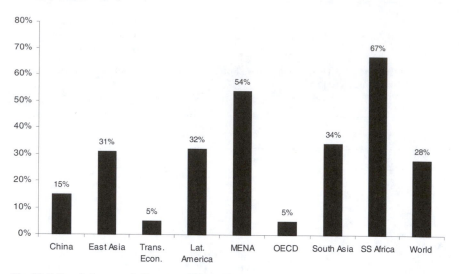

Fig. 19–8. Population growth. Base case 2001 to 2025.

cline in per capita income for the Sub-Saharan Africa and MENA regions, −3% and −17%, respectively.

Table 19–4 indicates how the forces shown in Fig. 19–8 and 19–9 could impact the appetite for animal protein, fish, and vegetable protein. Globally, the appetite for animal protein increases by 81%, fish by 86%, and vegetable protein by

% Change 2001-2025

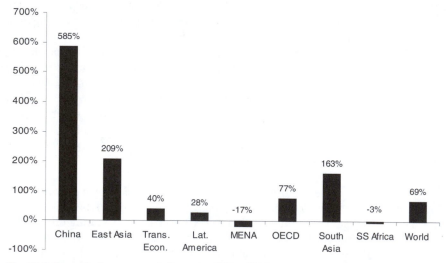

Fig. 19–9. Per capita income growth. Base case 2001 to 2025.

Table 19–4. Base case scenario: Animal protein, fish, and vegetable protein consumption (2001–2025).

Region	Animal protein			Fish			Vegetable protein		
	2001	2025	Change	2001	2025	Change	2001	2025	Change
	million metric tonnes		%	million metric tonnes		%	million metric tonnes		%
China	46.1	113.7	146	18.3	49.8	172	5.9	9.3	57
East Asia	9.9	31.0	213	11.3	23.3	107	5.7	8.2	44
Trans. econ.	21.9	27.3	24	4.7	6.0	28	2.1	2.6	23
Lat. America	24.8	35.7	44	4.8	7.4	54	6.1	7.8	28
MENA	5.1	6.5	27	1.8	2.3	27	3.1	4.9	58
OECD	73.7	79.1	7	29.6	29.2	-1	8.9	7.4	-17
South Asia	5.4	45.7	745	6.6	27.1	310	17.2	24.6	43
SS Africa	6.3	10.4	64	6.7	10.9	63	9.5	15.9	67
World	193.3	349.3	81	83.8	156.0	86	58.6	80.7	38

38%. China, East Asia, and South Asia show large increases in the appetite for animal protein, 146%, 213%, and 745% respectively. Growth in the OECD is relatively modest, 7% for animal protein, −1% for fish, and −17% for vegetable protein. This reflects the fact that people in the OECD countries, in general, will not devote a significant portion of any additional income to expenditures for food. It should be noted, however, that even though the OECD had the smallest level of change in the appetite for animal protein, the region still accounts for nearly 20% of the total appetite (Fisher, 2000).

Over the last three decades, both the number and the proportion of the globe's undernourished population have declined. Table 19–5 indicates that at the global level, continuation of this trend is possible, given the population and income parameters shown in Fig. 19–8 and 19–9. Unfortunately this relatively positive result at the global level masks serious regional problems, especially in the Sub-Saharan Africa and MENA regions, which would experience increases in the number malnourished of 78 and 97%, respectively. By the Year 2025, Sub-Saharan Africa would account for 56% of the world's malnourished.

For the Lower Income Scenario, the annual income growth rates are reduced by 50% from their level in the Base Case Scenario. The result is less income spread across the same number of people as in the Base Case Scenario. Figures 19–10, 19–11, and 19–12 compare the estimated appetite for the various commodities (an-

Table 19–5. Base case scenario: Human malnutrition (2001–2025).

Base case Region	Number undernourished			Proportion undernourished		
	2001	2025	Change	2001	2025	Change
	no. × 10^6		%	no. × 10^6		%
China	182.3	9.7	-95	0.15	0.01	-93
East Asia	78.2	34.9	-55	0.15	0.05	-67
Lat. America	63.4	69.0	9	0.13	0.11	-15
MENA	25.6	50.5	97	0.09	0.11	22
South Asia	275.2	134.2	-51	0.22	0.08	-64
SS Africa	211.6	375.9	78	0.35	0.36	3
World	836.4	674.2	-19	0.14	0.09	-37

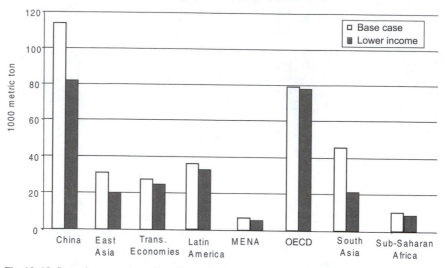

Fig. 19–10. Scenario comparisons for animal protein appetite in the Year 2025.

imal protein, fish, and vegetable protein, respectively) between scenarios. Each figure shows the appetite for a given commodity for the Base Case and the Lower Income scenarios in the Year 2025.

The Transition Economies, Latin America, MENA, and Sub-Saharan Africa regions all show little variation in the appetite for animal protein between the two scenarios. In contrast, the appetite for animal protein in the other regions is more affected by the reduced income. The strong relationship between income and appetite patterns is evidenced by the sharp decline in animal protein appetite for the Asian regions of China, East Asia, and South Asia in the Lower Income Scenario. The appetite for fish exhibits similar patterns, as seen in Fig. 19–11.

Cultural differences influence the appetite for vegetable protein between regions (Fig. 19–12). In the Lower Income Scenario, the appetite for vegetable protein increases slightly in Latin America, MENA, OECD, and Sub-Saharan Africa indicating a substitution of vegetable for animal protein.

Finally we turn to the comparison of malnutrition results between the scenarios. Income and population have a strong influence on malnutrition (Fig. 19–13). The decrease in income associated with the Lower Income Scenario brings about startling increases in malnutrition in all six regions. Sub-Saharan Africa accounts for more than a third of the world's malnutrition, with more than 350 million people underfed, even in the Base Case Scenario. With lower income growth, increases in the number of malnourished are exhibited in all regions. South Asia is a region that is particularly sensitive to income growth, as malnutrition would increase by more than 100 million people in the Lower Income vs. Base Case scenarios. The number of malnourished exceeds 400 million people in Sub-Sahara Africa under the assumptions of the Lower Income Scenario.

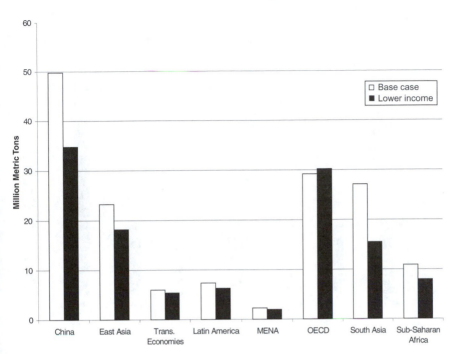

Fig. 19–11. Scenario comparisons for fish appetite in the Year 2025.

19–2.3 Future Agricultural Productivity and Innovation: Who Cares?

Abundant, low cost, safe food supplies are the ultimate product of agricultural productivity and innovation. However, it often is difficult for citizens and public policy makers to see and appreciate the linkage between food in grocery stores and the need for innovations from agricultural research. Indeed, for the last 5 yr citizens of the developed world have benefited from past agricultural research and favorable growing conditions through unusually abundant food supplies. Unfortunately, that abundance reduces the public's willingness to financially support the continual efforts needed for agricultural research innovation. Therefore we undertook to use some recently developed modeling capabilities to explore the notion of agricultural productivity and potential future costs of constraining agricultural productivity.

In the PCD framework, food availability is primarily reflected through the level of food prices. Extending the work just described, we've defined a third scenario, the Food Price Inflation Case:

- Food Price Inflation Case: For the years, 2006 to 2015, commodity food prices follow the annual inflationary pattern that existed during the decade of the 1970s. Before and after those years, prices for food commodities are constant in nominal terms as in the first run.

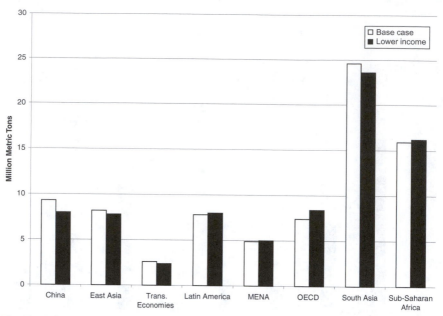

Fig. 19–12. Scenario comparisons for vegetable protein appetite in the Year 2025.

The Base Case scenario, described previously, is the comparison setting. In that situation, future prices for food commodities are held constant in nominal terms (declining in real terms). This stipulation is consistent with the actual experience of the last two decades for numerous agricultural commodities (Fig. 19–6).

Table 19–6 provides key results for selected years. To summarize these effects, estimates for one variable, the number of people malnourished, are shown for the three key regions of China, South Asia, and Sub-Saharan Africa. With constant food prices (and future world economic growth consistent with that of the last two decades of the 1900s and moderate population growth in the future), malnutrition levels would decline dramatically from their current levels in China and South Asia. In Sub-Saharan Africa, however, even constant food prices are not sufficient to keep malnutrition levels from increasing, as population growth would swamp income growth.

The analysis examines the effect of just 10 yr of food inflation, starting in the Year 2006 (but using the same population and economic growth assumptions as previously). If food inflation rates follow those that actually occurred during the decade of the 1970s, malnutrition levels skyrocket. In 2015, the year of peak malnutrition for the three regions, more than 960 million additional people, just in these three regions, would suffer malnutrition because of the lack of agricultural productivity. This analysis assumes that food prices stay constant after 2015. Despite that, the number of malnourished would be more than 700 million greater in the Year 2025 than in the scenario with no food price inflation.

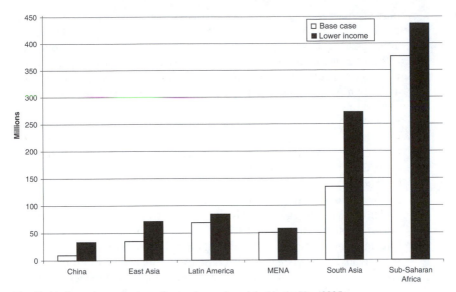

Fig. 19–13. Scenario comparisons for number malnourished in the Year 2025.

As citizens and public policy makers evaluate biotechnology and other agricultural research innovations, it is critically important to understand that restraining agricultural innovation is not a "risk-free" decision. The cost of overly restraining innovation could be hundreds of millions of additional hungry people. The results of this analysis document the stark and dramatic effect of the lack of agricultural research innovation on the well being of the world's poor.

19–2.4 Summing Up

The PCD model tracks annual estimated human appetite (potential demand) from 2001 to 2025, for six agricultural commodities (beef, fish, pork, poultry, fats and oils, and vegetable protein) in eight regions that encompass the world. Four key implications emerge from analysis of the PCD simulation results. These implications form the basis for further strategic discussion.

1. There is an important disconnect between some of the industry's key skills and capabilities and those it will need in the future. The industry's growth since 1975 occurred mainly in the OECD region. Therefore the in-

Table 19–6. Estimates of the number of malnourished humans with constant vs. inflating food prices.

Area	Current	2025 with constant prices	2015 with food price inflation	2025 with food price inflation
		no. of people		
China	182 000 000	11 000 000	432 000 000	201 000 000
South Asia	275 000 000	143 000 000	574 000 000	469 000 000
Sub-Saharan Africa	211 000 000	376 000 000	502 000 000	601 000 000

dustry's marketing and policy expertise, as well as its research direction, are heavily focused on the needs of customers in those regions. However, although likely to still be significant, that region is unlikely to be the source of significant volume growth in the future. With strong income growth, the indicated appetite for animal protein (particularly in the China, East Asia, and South Asia regions) surges. Therefore new skills and capabilities are needed to serve the potential growth markets of the world.

2. Projected growth in appetite is relatively robust with respect to population growth but is more sensitive to income growth. The market need for protein declines significantly between the Base Case and the Lower Income Case. Global economic growth is one of the key strategic issues for the soybean sector.

3. Even with optimistic income growth, malnutrition in the Sub-Saharan Africa and MENA regions persist at frightening levels. With lower income growth, malnutrition intensifies in other areas of the world as well. Therefore, humanitarian need for food is likely to be a fixture of the next 25 yr. Historically, the soybean protein complex has not been a significant component of humanitarian food responses. For moral and business reasons, strategies that heighten the industry's role in humanitarian responses warrant careful consideration.

4. Long-time lags typically exist between investment in research and the resulting gains in agricultural productivity. In times of abundance, such as the world has experienced in the last few years, it is natural for citizens and decision makers to underappreciate the need for continual investment in agricultural research. The analysis of the effects of potential food price inflation (the market's signal of a productivity shortfall) vividly documents that the impact of inadequate productivity falls upon the poorest of the world's population.

19–3 WILL TOMORROW'S MARKETPLACE BE DOMINATED BY TODAY'S COMMODITY MARKET APPROACH?

The second of two forces shaping the future evolution of the soybean market is the pressure for change in the commodity marketing system. Two primary distribution systems exist for soybean in commercial agricultural systems. One distribution system is focused on commodity crops, where the emphasis is on homogeneity. The other distribution system is focused on high-value traits, but has been utilized primarily for very small volumes. The problem with these two primary distribution systems is that neither channel can cost effectively supply the new differentiated value-enhanced crops. It is expected that many of the new value-added crops will be produced in larger volumes, relative to the high-value trait crops. With the growing attention placed on biotechnology and value-added crops, there is a growing need for market channels that will allow distribution of a product that is identity-preserved (Sonka et.al., 2000).

19–3.1 The Structure of Today's Marketplace

Homogeneity is a fundamental attribute that has permeated the traditional soybean supply chain. In the commodity value chain, farmers produce generic soybean crops, all of which are viewed the same, although produced from a choice among hundreds of cultivars. After harvest, farmers deliver their grains to a first-handler or store them on-farm for later delivery. Whether delivered at harvest or from storage, the first handler receiving the crops is not interested in differentiating these grains for different end-uses, but is interested in blending grains to meet physical limits for numerical grades in outbound shipments.

This commodity orientation has important implications. First the capability to coordinate a large and diverse sector such as agriculture with minimal information flow throughout the sector has been a major strength of the sector. Although commodity output meets the general specifications of the customer at the next level, it may not optimally meet the specific needs of any one customer. The associated loss of efficiency at the customer level is offset by the considerable flexibility of supply offered by the commodity system and its low cost. This is one reason that commodity agriculture has been successful. However, one side effect of this structure is that knowledge creation tends to be concentrated within each segment in the chain rather than disseminated throughout the chain.

For domestic purposes, almost all soybean is traded as U.S. no. 1 soybean, while export specification is almost entirely U.S. no. 2 soybean. In general, the soybean is transported to soybean processors who crush (mill) the soybean into two components: soybean oil and soybean meal. Soybean oil is then sold to food and industrial users, while soybean meal is used in feed rations for livestock. More refined protein products can be created for human consumption.

The handlers in the traditional soybean supply chain typically have large volume storage units, and their profit is created by turning over a very large quantity of soybean at very small margins. Because soybean is treated as a homogeneous product, large volume storage units are very efficient and commingling of a multitude of soybean cultivars does not influence the price received for transshipped soybean. Similarly, soybean processors crush large volumes of homogeneous soybean, with an objective of maximizing yield of oil and meal. There are two commercial types of soybean meal that can be produced: low protein meal which contains a minimum of 43.5% protein and high protein meal which contains a minimum of 47.5% protein. Most soybean received at the processing facility can be crushed as they arrive or blended with stored soybean to produce either level of meal, so differentiation before arrival based on protein content is not required.

This traditional supply chain for soybean, emphasizing homogeneity, has been in place since the production of soybean in the USA began in earnest in the 1960s. In this supply chain, much of the domestic and all of the international trade are based on specific tests that determine numerical grade. For soybean, there are four numerical grades established, and each grade is assigned a minimum test weight, and maximum levels of heat damaged kernels, total damaged kernels, foreign materials, splits, and soybean seeds of other colors (U.S. Congress, 1989). The pricing of grains and oilseeds is dependent on these numerical grades. For example, the con-

Table 19–7. Marketing system characteristics.

Commodity	Identify preserved	Societal objectives
Large volumes	Small volumes	Large volumes
Low cost	Higher cost	Low cost
Minimal quality standards	Specific quality standards	Specific quality standards
Purchaser flexibility	Minimal flexibility	Purchaser flexibility

tract specification for soybean futures traded at the Chicago Board of Trade indicates that the deliverable grade is U.S. #2. There are appeal mechanisms in place if there are disagreements about quality delivered, and even if a lower grade is received than contracted for, the characteristics that are part of the grade factors often have little relationship to the grain characteristics that determine value to the end user. The emphasis on homogeneity drives the system toward average quality, and thus limits the opportunity to match the level of specific attributes available in different crop lots to the needs of different buyers.

An alternate supply chain exists in parallel to the homogeneous commodity market and is used for some differentiation of soybean, particularly in markets for tofu and organic soybean. A key differentiating feature of this system is that decision making is administratively coordinated rather than being coordinated primarily through market price signals as is the case within the commodity market. For example, administrative coordination could result in increased consumer demand for a differentiated chicken product for Easter, even though the overall market signals for generic chicken suggest that demand is falling. Historically, it has been much more difficult and expensive to coordinate identity preserved systems, therefore the portion of the crop marketed within this system has been relatively small. Further a value chain optimized to use very specific farm output incurs the risk of restricted flexibility. Because of the biologic variability inherent in agricultural production, output levels can fluctuate both substantially and unpredictably. Substitution in an optimized system (if the optimal farm-level output is not available in sufficient supply) has been costly, further restraining the use of such optimized systems.

The identity-preserved supply chain typically consists of a specialty grain firm contracting cultivar-specific soybean production with farmers, with particular production and management requirements as contract terms. The farmer stores this production on farm, and either delivers it directly for loading onto a container for export shipment, delivers directly to the tofu processor, or delivers for direct loading onto trains for domestic shipment. In any of these cases, the goal is to minimize the number of handlings so as to reduce quality deterioration and to minimize the potential for commingling with nondifferentiated soybean. Generally, the identity-preserved system is focused on small-scale lots of high-valued soybean whose added value (compared to commodity soybean) is greater than the additional costs of production, handling, and segregation.

Table 19–7 lists the relative strengths and weakness of the commodity vs. identity preserved systems. Until relatively recently, the food marketing sectors were comfortable with the need to choose between these two relatively different approaches. However, societal events are pulling and technological changes are pushing our food systems to move to a differing set of alternatives. The general desires of the market are listed in the third column of Table 19–7. It is intriguing that this set of desires is inconsistent with both of the traditional alternatives.

19–3.2 Market and Social Forces for Change

There are many forces, which independently and combined, are putting pressure on the traditional production and marketing practices in agriculture. Increasing consumer sophistication, technological change, competition, environmental concerns, and biotechnology are some of the factors that are influencing today's agricultural marketplace. Whether the confluence of these factors will permanently and significantly alter the structure of U.S. agriculture, or will result only in incremental changes has not yet been determined. How each of these forces may influence commodity agriculture is discussed below.

Consumer sophistication has resulted in interest in foods that go beyond traditional concerns of price and presentation. The enhanced consumer interest primarily involves three aspects of the foods they eat; food safety, health issues, and perceptions associated with particular production practices. Consumers want assurance that the foods they eat are safe. Although the commodity market structure has, in the main, supplied safe foods to consumers, this market structure does not provide consumers information on the production, handling, and processing activities that occurred during the transformation of raw crops into finished consumer goods. Information on production activities on individual grain lots is lost when these lots are commingled with other grain lots. Recent concerns of bioterrorism and agroterrorism potentially could put additional pressure on the commodity market structure, if these concerns lead to emphases on tracking and traceability of raw grains as they are produced and transformed into products.

Consumers also have expressed interest in foods that improve their health, such as products which lower cholesterol or which may reduce their risk of certain cancers. Some consumers are willing to pay for foods with certain production characteristics, such as organic foods, regardless of whether these foods are actually more "safe" than what it generally available. The ability to supply either the attributes that provide the health or nutritional component, or provide the information on production practices, requires an identity-preserved market channel.

Technological innovation also has contributed to the pressure to initiate alternative market channels. New measurement technologies allow for identification of specific attributes in grains that can be rapidly measured at first delivery, including oil and protein in soybean. Improved processing technologies enable the refinement or capture of desirable attributes, such as isoflavones in soybean. In both cases, soybean with that attribute may need to be segregated throughout the market channel to retain their value. Developments in information technology may soon provide the capability to efficiently capture by electronic means the relevant activities related to the production, handling, and processing of identity-preserved crops, and to deliver this information in a timely manner to end-users.

Another force for change is increased competition from South America. The primary U.S. competitor in the global soybean market is Brazil, although soybean production has grown in many South American countries, such as Argentina, Paraguay, and Uruguay. Historically, infrastructure impediments and higher transportation costs have been a competitive disadvantage for Brazilian producers. Efforts to improve road and river transportation, if successful, could enhance Brazil's competitive position. Similar efforts to substantially enhance Brazil's information and coordination infrastructure might allow that system to respond more effectively to pressures to change away from reliance on the traditional commodity approach.

Societal concerns regarding the impact of agricultural practices on the environment also may influence whether alternative market channels will develop for soybean. Desires for reduced use of fertilizer and chemical inputs have led to new production practices based on precision agriculture, where the quantity of inputs applied to a given area of land is matched to the productivity needs of that land (National Research Council, 1997). One result of precision agriculture is that more knowledge is created about each land unit, and information is recorded regarding the timing, quantities, and location of input applications, as well as information on yields during harvest. While the emergence of precision agriculture practices was driven primarily to increase profitability for producers, this data also could be used in identity-preserved systems to verify when and where specific production and harvest activities took place. It might be anticipated that the linkage of this information with value-added soybean might first occur in the high value-added identity preserved system currently in place. However, if the number of producers embracing precision agricultural activities increases, the information collected also might contribute to the development of alternative identity-preserved market channels (Sonka et.al., 1999).

The application of biotechnology to agriculture also may profoundly change the reliance on commodity marketing channels. To date, biotechnology has been focused on developing soybean with transgenic input traits, such as resistance to herbicides or insects. The incorporation of these traits into soybean has no known effect on the end-use characteristics of soybean, such as their protein and oil levels, or processing values. Although it was anticipated that the "first generation" transgenic soybean, that is, soybean modified with input traits, would be marketed using the traditional commodity supply chain, the international trade environment for transgenic soybean has become much more complex. In fact, today the commodity marketing channel is not a sufficient supply chain for all international markets.

The next generation of genetic modification is focused on output traits that are intended to alter end-use characteristics to make soybean products healthier, such as modified oils. Identity-preserved supply chains, which differ from either of the existing primary systems, will be required to maintain the value of these modified soybean cultivars throughout the marketing channel. The next section will present a model of the social construction of alternative market channels.

19–4 THE SOCIAL CONSTRUCTION OF ALTERNATIVE MARKETS

An initial consideration of how rapidly commodity market channels will change to identity-preserved market channels frequently focuses on only the additional costs incurred in segregating and handling value-added goods vs. commodities. If the development of identity-preserved market channels were dependent solely on the sum of these additional costs, then it would be a fairly simple exercise to determine under what scenarios identity-preserved channels would develop. However, although costs are clearly a factor in the development of identity-preserved market channels, reliance on only costs will not help us realistically anticipate the potential for change. This section will introduce and describe a social construction model that incorporates a richer set of parameters necessary to better understand the array of possible outcomes for commodity markets, and the social groups that may influence the final outcome.

The application of biotechnology to soybean was expected to increase the value of the crop to growers and processors, through the development of soybean cultivars with preferred agronomic characteristics and intrinsic product attributes. For example, Roundup Ready soybean was developed to allow the producer to use Roundup herbicide to kill a broad spectrum of weeds effectively and at lower total cost without destroying the soybean plant itself. This type of application is referred to as an input or agronomic trait, since it impacts the producer's use of inputs such as herbicides or pesticides. The second application of biotechnology is the alteration of output traits, where the chemical components of soybean, such as oil, protein or fatty acids, are altered to make the soybean more valuable in various end-uses.

Input trait biotechnologies arrived to market first. This was seen by holders of biotechnology patents, ex ante, as an important element of the rollout of biotechnologies. Transgenic input traits did not affect the use value (in processing or consumption) of the product since its nutritive attributes remained the same. Producers would adopt the transgenic crop for economic reasons, perhaps, with positive environmental impacts. New output traits would follow as the technology developed, after the market—buyers and sellers—had become accustomed to the technology through the use of seeds with the altered input traits.

Traditional supply chains were envisioned for distribution of input trait-transgenic soybean, since these soybean do not affect value to the end-user and therefore are not physically differentiable from nontransgenic cultivars. However, with the recent development of international concerns about the basic technologies of transgenic crops (particularly in Europe, but also extending to Japan and other countries), some or all types of transgenic soybean may need to be segregated from nontransgenic soybean.

The short-term impact of this international resistance to transgenic modification is primarily confusion across the soybean marketplace. In this setting, firms must identify which of their customers require nontransgenic soybean, identify how the soybean can be segregated throughout the supply chain, and evaluate whether existing testing technologies can ascertain if the soybean they receive is transgenic. There are many unresolved issues in the transactions between firms in the production and marketing of soybean. In the long-run, will agronomic trait-transgenic soybean be marketed in the same market as nontransgenic soybean? Will the markets be distinct despite the fact the soybean is homologous in processing? Will there be distinct, separate markets for output trait-transgenic soybean, or will there be some simpler aggregate market where all attribute combinations for soybean will be treated in common? At a practical, immediate level, these uncertainties have profound economic consequences for buyers and sellers in the supply chain. Will nontransgenic soybean command a price premium? What is the liability for delivering contaminated (mixed) loads and can one be insured against this liability? How will technologies for sampling and testing, segregation requirements, and labeling be introduced into the market—as mandatory or voluntary programs?

The longer-term impact from consumer and political resistance to transgenic soybean is by no means apparent today. Whether or not the international resistance to transgenic soybean continues for years to come, is mitigated by some form of bilateral agreements or World Trade Organization (WTO) negotiations, or fades because of diminished interest, currently is not known. Likewise, the U.S. consumer's longer-term reaction to genetic modification in the food supply is uncertain.

19–4.2 The Market Construction Process

The question of whether the dominant marketing system for the heretofore homogeneous soybean can accommodate transgenic soybean, whether it must eventually be replaced by a plethora of smaller markets that bear some resemblance to the small-volume identity-preserved markets that now exist, or whether a continuum of market channels will be developed will eventually be answered in the marketplace. What the answer will be, and the path that leads to that answer, will be "constructed" as we watch. However, as White (1981) warns, "Building a market is a conflict-ridden and erratic process with quite a range of outcomes possible in the form of market schedules" (p. 520).

As White suggests, the construction of a new market such as the one (or several) in which transgenic soybean will change hands is an uncertain process. However, we must remember that all markets are social constructions, even the markets that we now take for granted. These are markets where products are well defined, as are the buyers and sellers; thus, the market boundary is fixed. Today, we routinely speak of the "corn market", the "personal computer market", or the "automobile market". However, at some point in time, each of these markets had to be socially constructed. Drawing upon the literature of social construction of markets allows for informed speculation regarding the dynamics likely to emerge relative to agricultural commodity markets (Bender and Westgren, 2001; K.L. Bender, 2001. Product and exchange of attributes in defining boundaries for value-added corn and soybean markets. Unpublished Ph.D. Diss. Univ. of Illinois at Urbana-Champaign.).

Selected scholars who study organizational behavior within firms and industries examine the process by which markets emerge and develop. Porac et al. (J.F. Porac, A. J. Rosa, and M.S. Saxon. 1997. America's family vehicle: The minivan market as an enacted conceptual system. Paper prepared for the Multidisciplinary International Workshop on Path Creation and Dependence, Copenhagen Business School, August) use the terms *artifacts* and *attributes* in their model of market evolution, and suggest that artifacts can be differentiated based on differing "conceptual clustering of attributes". That is, the artifacts that are socially constructed by buyers and sellers are complex constructs of multiple attributes: products. This conceptual clustering develops over time through social interactions between buyers and sellers who develop shared knowledge of the artifact. Conceptual clustering leads to stable definitions of artifacts, which ultimately allows

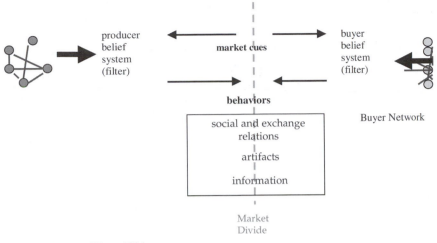

Source: Porac and Rosa, 1996

Fig. 19–14. A model of market construction. Source: Porac and Rosa (1996).

for artifact transactions to extend beyond the initial set of producers and sellers to others diffused in time and space. In their terminology "stable product conceptual systems are an intersection of understood attributes and usage conditions shared across the market divide; a stable product market exists to the extent that there is an equilibrium consensus in core attributes and uses for artifacts considered to be members of the market." In the complex global market for soybean and derivative products, consensus must extend beyond local transactions between farmers and country elevators to include domestic and international merchandisers, processors, consumers, and regulatory agencies.

Markets for artifacts (products) evolve across what Porac and Rosa (1996) call the "market divide," where the social construction process articulates between the collective behavior of buyers and the collective behavior of sellers around product attributes (Fig. 19–14). However, Fig. 19–14 omits an important set of social processes that affect the process. That is, the market is *embedded* in a social context broader than just the market transaction.

> "Actors do not behave or decide as atoms outside a social context, nor do they adhere slavishly to a script written for them by the particular intersection of social categories that they happen to occupy. Their attempts at purposive action are instead embedded in concrete, ongoing systems of social relations." Granovetter (1985, p. 487)

Pinch and Bijker (1987) model the particular case of the social construction of technology. To analyze the social processes around the development, codification, and acceptance of a new technology, one must identify the relevant social groups, problems, and solutions that shape a technological artifact (product). The mapping of the development process of a specific technology provides the opportunity to observe if and when artifacts stabilize, and how this stabilization may differ across social groups. Pinch and Bijker (1987) indicate that closure to problems arising from artifacts may arise in two forms: rhetorical closure and closure by re-

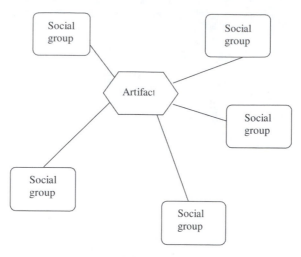

Source: Pinch and Bijker, 1987

Fig. 19–15. Network of an artifact and social groups.

definition of the problem. Rhetorical closure occurs when the relevant social groups feel that the problem has been solved, whether or not the problem has actually been solved. For example, a strict government labeling requirement for products containing transgenic soybean may force rhetorical closure on the discussion of possible health risks. Consumers can opt for soy foods labeled as nontransgenic and avoid the risk, even though the risk may still exist in the transgenic foods. Closure by redefinition of the problem occurs through a shift in focus to another problem. This would occur if the concerns about environment hazards (e.g., genes "escaping" into wild species) are left behind in search for a solution to feed the world's malnourished.

The framework presented by Pinch and Bijker (1987) uses a set of diagrams to show the connectivity among artifacts, social groups, problems, and solutions. In their representation, an artifact is the central feature of the diagram, with links to different social groups (Fig. 19–15).

A schema of the relationship between an artifact, multiple social groups, perceived problems, and solutions is diagrammed in Fig. 19–16. Figure 19–16 includes one refinement of the framework designed by Pinch and Bijker (1987): the inclusion of links among social groups. Given the importance of network ties to social construction, consideration must be made as to whether the social groups are interlinked (Uzzi, 1996). Early in the social construction process, there are many more elements that are being "tested" cognitively by parties on both sides of the market divide than will appear as closure is reached. At closure, path dependence and path destruction will have eliminated some feasible and infeasible solutions. Problems with solutions will become subsumed and no longer be part of the rhetoric, as will problems that are redefined. This should be some comfort to those who see the current state of transgenic-market development as chaotic and intractable.

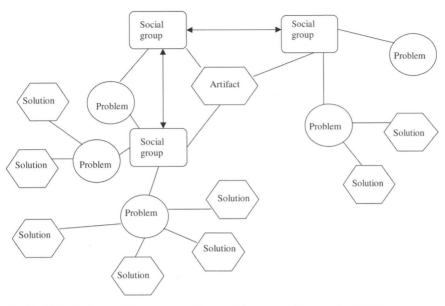

Fig. 19–16. Revised network between an artifact, social groups, problems, and solutions.

19–4.2 Application to Soybean Markets

Some inferences can be drawn about possible pathways into the future for the transgenic problem from reading ex post analyses of other socially constructed markets and technologies. Central among these is that the existing, institutionalized market for homogeneous commodities will be a likely platform for construction of a market for transgenic soybean. The commodity market is well understood and many of the social groups involved in domestic and international commerce in soybean and other commodities have a stake in it as an artifact. Their stake in the commodity market artifact also links these social groups in a network that should facilitate communication, organize stakeholder power, and thereby speed up the cognitive processes on both sides of the market divide.

The pathway towards social construction of the transgenic soybean market(s) will be chosen in a complex process that balances the structural embeddedness of the commodity market among merchants and U.S. producers of soybean with the political and cultural embeddedness in the social discourse of governments, scientists, consumers, and environmentalists. That is, the analysis of the cultural and political landscape of the biotechnology controversy above is the context in which buyers and sellers socially construct the rules of exchange. For example, if substantive equivalence is granted between transgenic and nontransgenic cultivars, then a soybean market driven by homogeneity—the commodity market—can exist. To the extent that some jurisdictions require identification and labeling of transgenics, the commodity market offers a single solution: all soybean must be considered as transgenic. Otherwise, the market channel must develop a system of physical separation,

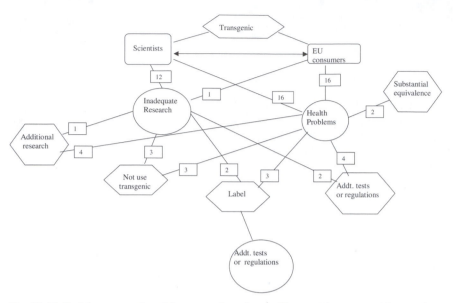

Fig. 19–17. Partial representation of the transgenic soybean artifact, social groups, problems, and solutions.

oversight and testing, and regulation (Fig. 19–17) that creates separate markets for separate socially constructed artifacts: transgenic and nontransgenic soybean.

For some consumers this is the minimal requirement to establish normative legitimacy, which may be augmented by regulation. For other consumers, the separation of transgenic and nontransgenic sources of soybean may be unnecessary and artificial; their tolerance for commingling may be effectively unlimited. Nonetheless, if labeling and segregation are required for transgenics by enough end-users in enough jurisdictions, we likely will see path destruction of the artifacts that tie transgenic soybean to undifferentiated, homogeneous commodities.

Markets are not developing solely on the basis of atomistic behavior. The construction of the transgenic soybean market is influenced by the relationship of many social groups with the artifact of genetic modification, and by the interrelationships among social groups. For example, both labeling and additional testing or regulations are potential solutions that are identified for multiple social groups concerned with a particular problem (adverse health effects of transgenics) as well as solutions for different problems across social groups. The frequency of occurrence of a given problem or solution both within and across social groups may help identify the priority with which these problems or solutions are considered in reaching closure.

As buyers and sellers of soybean and other products that may be transgenic choose solutions across the market divide that meet their particular requirements, they must do so in a way that conforms to the social and political constraints in which these choices are embedded. We are early in the social construction process and the alternative pathways to the eventual socially constructed market(s) are not clearly demarcated. Some may be foreclosed as a result of recent political choices on labeling, but in many jurisdictions the political and cultural discourse is far from complete.

19–4.4 Summing Up

Speculation as to the future course of societal acceptance of transgenic agricultural products is indeed speculation. Events in society and in the laboratory that will happen tomorrow will interact with our current state of acceptance to forge the dynamics that will lead to acceptance or not in the various regions and markets of the world. But it is important to remember that genetic modification of soybean is only one of several factors that are themselves interacting to pressure agricultural commodity markets to evolve to new states. These new states are being socially constructed at this time.

Evaluation of the current situation in the context of social construction yields the following four insights:

- The pressure for increased differentiation of crops at the farm level comes from several interacting sources. Social concern regarding transgenic crops is a key factor in some markets and locales. However, broader forces such as food safety, environmental concerns, and advances in information technology have the potential to interact in complex ways to fuel or retard change.
- The growth in market structures to deliver value-added crops will be determined by the potential value gained, the associated operating costs, and the state of the system capability that exists to track and deliver agricultural products. Traditional identity preserved marketing systems were limited in scale in large part because of the coordination technologies available. Advances in information technology and managerial skills offer the potential to create such systems which are scaleable and, when in place, operate at low cost.
- The demand for public oversight or certification of identity preserved systems is increasing. The current identity preserved market channels tend to be privately developed and implemented. The USDA has moved to certify private testing labs to provide laboratory clientele with assurance of laboratory results. The USDA might pursue certification of private identity preserved processes if there is demand for public involvement in identity preserved systems, or could even choose to develop a U.S. identity-preserved system.
- Internationally the demand for food system traceability is intensifying. The European Union has proposed requirements that would require traceability of soybean imports as to whether they are transgenic or not. Such pressure raises the potential for further decline in the proportion of the world's soybean crop that is marketed as a commodity. Although the controversy regarding transgenic crops is one force encouraging traceability, it is only one among several forces. If traceability requirements are implemented and adopted widely, then significant adaptations will need to be made to agri-

cultural marketing systems. These pressures may continue even if the societal concern regarding genetic modification of soybean were to diminish.

19–5 SUMMARY

This chapter described today's marketplace for soybean and examined key forces likely to shape the soybean marketplace of tomorrow. The growth in soybean production and consumption over the last 30 yr is impressive for an agricultural commodity. The availability of soybean oil and meal allowed consumers to enhance their diets and level of well-being. The future soybean marketplace will be determined by the complex interaction of many forces. Two of those forces will be the effective demand for protein and the nature of the production/marketing system for soybean.

With income growth, the effective global demand for protein should continue to expand. Much of that growth would occur in Asia and other parts of the world where lower income consumers would expend additional income to enhance their diet. Although the USA, Western Europe, and Japan would remain as significant markets, relatively little growth in demand is expected there. Global income growth is not assured, however. Reductions in income growth from levels consistent with those of the latter part of the 20th century would significantly curtail the effective market demand for protein. Although the number of malnourished is negatively correlated with general income growth, significant numbers of malnourished are likely to exist in Africa even with optimistic income growth.

Often, when times are favorable, agricultural productivity and the investment in research needed to maintain productivity diminish as priorities. However, examination of the costs of inadequate productivity (measured in terms of inflation in food prices) shows that inadequate productivity has dire human costs. The world's poor and near-poor bear the primary burden of shortfalls in productivity through decline in diet quality and increase in the number of malnourished.

For many years, the commodity market system has effectively served consumers and the soybean industry. Now, however, numerous forces (including social concerns regarding transgenic crops) suggest that the homogenous nature of the commodity market is no longer adequate. Today we cannot discern how those pressures will eventually effect the soybean production/marketing system, if at all. The soybean market (and numerous similar agricultural markets) is experiencing the type of turbulence that is referred to as social construction of markets.

As the sector moves through this period of turbulence and potential change, scrutiny of four key concepts may offer insights as to the eventual form and character of tomorrow's soybean market. The pressure for differentiated farm output comes from several interacting forces, of which the controversy over transgenic crops is just one. The eventual change in the market will occur as potential benefits are weighed against operational costs and the state of the information systems that underpin the market. The involvement of public oversight or certification can further fuel or retard change. The current move towards demanding traceability in European markets has roots that extend beyond concerns regarding genetic modification and are likely to persist even if those concerns were to dissipate in the future.

REFERENCES

Bender, K.L., and R.E. Westgren, 2001. Social construction of the market(s) for genetically modified and nonmodified crops. Am. Behavioral Sci. 44(8):1350–1370.

Bread for the World Institute. 1998. The changing politics of hunger: Ninth annual report on the state of world hunger, Maryland: BFW Inst., Washington, DC.

Coplin, W.D., and M.K. O'Leary. 1994. The handbook of country and political risk analysis. Political Risk Serv., New York.

Ferris, J.N. 1998. Agricultural prices and commodity market analysis. WCB/McGraw-Hill, Boston.

Fisher, D.K. 2000. Assessing how information affects cognitive maps of strategic issues. Unpublished Ph.D. Diss. Univ. of Illinois.

Food and Agriculture Organization of the United Nations, 2002. Available at http://www.fao.org/.

Granovetter, M. 1985. Economic action and social structure: The problem of embeddedness. Am. J. Sociol. 91(3):481–510.

Monsanto. 2002. Safety assessment of Roundup Ready soybean event 40-3-2. Available at

http://www.monsanto.com/monsanto/content/our_commitments/roundupsoy_product.pdf. (Verified 21 Nov. 21 2002.)

National Research Council. 1997. Precision agriculture in the 21st century: Geospatial and information technologies in crop management. National Academy Press, Washington, DC.

Pinch, T.J., and W.E. Bijker. 1987. The social construction of facts and artifacts: Or how the sociology of science and the sociology of technology might benefit each other. p. 17–50. In W.E. Bijker et al. (ed.) The Social construction of technological systems: New directions in the sociology and history of technology. The MIT Press, Cambridge, MA.

Porac, J., and A.J. Rosa. 1996. Rivalry, industry models, and the cognitive embeddedness of the comparable firm". p. 363–388. In J.A.C. Baum and J.E. Dutton (ed.) Advances in strategic management. Vol. 13. JAI Press, Greenwich, CT.

Sonka, S.T., D.A. Lins, R.C. Schroeder, and S.L. Hofing. 1999. Production agriculture as a knowledge creating system. Int. Food and Agribusiness Manage. Rev. 2:165–178.

Sonka, S.T., R.C. Schroeder, and C. Cunningham, 2000. Transportation, handling, and logistical implications of bioengineered grains and oilseeds: A prospective analysis. USDA Agric. Marketing Serv., Washington, DC.

Soy Stats. 2002. A reference guide to important soybean facts and figures. United Soybean Board. Available at http://www.unitedsoybean.org/soystats2000/index.htm. (verified 21 Nov. 2002.)

Talksoy.com. Promoting improved soy varieties in a growing market. United Soybean Board. 2002. Available at http://www.talksoy.com/Home.htm. (verified 21 Nov. 2002.)

The Economist. 2001. The case for globalization. Survey p. 1–30. 29 Sept.–5 Oct. 2001. The Economist Newspaper Group Ltd., London.

U.S. Congress, Office of Technology Assessment. 1989. Enhancing the quality of U.S. grain for international trade, OTA-F-399. U.S. Gov. Print. Office, Washington, DC.

U.S. Soy Foods Directory. 2002. Available at http://soyfoods.com/. (verified 21 Nov. 2002.)

United Soybean Board. 2002. Available at http://www.unitedsoybean.org/. (verified 21 Nov. 2002.)

Uzzi, B. 1996. The sources and consequences of embeddedness for the economic performance of organizations: The network effect. Am. Sociol. Rev. 61:674–698.

White, H.C. 1981. Where do markets come from? Am. J. Sociol. 87:517–547.

20 Soybean Processing and Utilization

EDMUND W. LUSAS

Ed Lusas, Problem Solvers, Inc.
Bryan, Texas

20–1 DEVELOPMENT OF PRACTICES AND MARKETS

20–1.1 Introduction

The ability of soybean [*Glycine max* (L.) Merr.] to produce more edible protein per hectare of arable land than any other major annual crop or through animal grazing, has been recognized for many years. Among available plant sources, soybean protein is closest to the optimum dietary essential amino acids (EAA) profiles for human and animal nutrition. Currently, soybean meal clearly is the world's major source of feed protein; but the oil palm (*Elaeis guineensis* L.) is competing to become the leading edible oil source. Freshly made and fermented soybean protein products have long been consumed in various forms in China, and Eastern and Southern Asia, and are becoming accepted in other parts of the world. Production of soybean food protein ingredients, and whole seed food uses, also are increasing. Breeders and growers have good reason to take pride in this crop. However, consumer fashions, economic strategies, and politics often have more influence on availability and buying decisions than technical merit.

20–1.1.1 Chapter Objectives

Markets provide the pull for producing soybean and processing it into ingredients for food, feed, and industrial uses. This chapter is written with an industry viewpoint, and differs from reviews in earlier editions. In addition to describing current (2002) practices in soybean processing and utilization, selected research since the last edition and changes in the way business is done also are reviewed. Units are first expressed in the metric system for scientists and for readers in other countries, and also in the English system for convenience of domestic readers. The word "quality" has several everyday meanings including: (i) level of desirability or goodness; and (ii) reliability in producing products consistently at the selected level. The terms "quality level" and "quality reliability," respectively, are used in this chapter when the meaning might not be clear otherwise. Space is limited, and readers desiring more details are directed to the extensive references. Only a few economic and nutrition details, necessary for this chapter, are included. Readers are referred to Chapters 19 (Sonka et al., 2004, this publication) and 21 (Birt et al., 2004, this publication), respectively, and to the literature for reviews of these subjects.

Since the 1987 edition of this monograph, world population has grown by 25%, trade has become global, and products of the Biotechnology Revolution have entered the marketplace. The soybean processing and utilization industries have not operated in a vacuum. Some of the problems are as old as the industries themselves; but, better techniques have been devised for reducing their effects. Today's scientific talent also is directed to modifying facilities, equipment, and processes to comply with new regulations and further business, social, and political objectives that are reshaping today's life. Since the 1980s, many U.S. processors have spent far more on implementing new regulations than on research and development of processes and products. Drivers for change in the industry are identified, and references provided for readers who would like to further explore how today's soybean utilization industry has developed.

More significant changes have occurred in soybean processing and utilization in the last 20 yr than during the crop's prior history in the USA. Several important issues currently faced by the industry are described. Hopefully, they will have been resolved before the next edition, and serve here only as examples of the continuing challenges in soybean utilization.

20–1.1.2 Additional Resources

Increasingly, the Internet is becoming an important source of technical information. Readers should search it and the various computer abstract data bases for technical information beyond this chapter and references. In the USA, the American Oil Chemists' Society (AOCS), Champaign, IL, is the most concentrated information source and publishes the *Journal of the American Oil Chemists' Society*, *INFORM,* and *Lipids.* The AOCS Press publishes books on: oilseeds processing; oils, fats and protein chemistry and applications; soaps and detergents; lipid biochemistry, metabolism and nutrition; and proceedings of AOCS world conferences. The *Oil Mill Gazetteer*, endorsed by the International Oil Mill Superintendents' Association (IOMSA), focuses on oilseeds extraction. The American Association of Cereal Chemists (AACC), St. Paul, MN, publishes *Cereal Chemistry* and *Cereal Foods World*, which occasionally contain articles on fats, oils, and proteins. The latter publication is a major source of information on food and feed extrusion and soybean protein texturization. Research also is reported in the *Journal of Agricultural and Food Chemistry* published by the American Chemical Society (ACS), and in *Food Technology* and *Journal of Food Science* published by the Institute of Food Technologists (IFT). Improved computer programs now enable easy and free searching of patents and applications filed at the U.S. Patent and Trademark Office. The annual *Soya & Oilseed Bluebook* presents a nearly complete listing of oilseed extraction plants and refineries throughout the world, updated USDA statistics on oilseeds, meals and oils production, prices and exports, and lists of suppliers to the industry (Soya and Oilseeds Bluebook, 2003).

Federal agencies that prepare Internet-accessible statistics on soybean and other oilseeds include: the U.S. Department of Agriculture's Foreign Agricultural Service (USDA-FAS), which reports on worldwide production and trade of oilseeds, meals, and oils; the Economic Research Service (USDA-ERS), which reports on domestic oilseeds plantings, harvest, prices, and processing; and the U.S. Bureau of the Census, which reports on domestic oilseeds processing and utilization in end products. The USDA-ERS and USDA-FAS also report on feedstuffs. Federal law prohibits publication of data for domestic industries in which the market consists of four or fewer companies. Such estimates may be found in *Oils and Fats International,* published in Surrey, England, and *Oil World Monthly* published in Hamburg, Germany.

Proposed regulations, their chronology, and final action can be searched in the *Federal Register* on the Internet. Current rules of the various federal agencies are in the *Code of Federal Regulations.* Most of the agencies also provide news on current issues within their jurisdiction. Searching and downloading is free, and has been greatly facilitated by improved retrieval programs at the various sites.

The five volume *Bailey's Industrial Oil & Fat Products,* 5th edition (Hui, 1996) is a significant reference for the oilseeds industry. Other current domestic books which describe soybean processing and utilization, have been prepared by Erickson (1995), Bockish (1998), O'Brien (1998), Liu (1999), and O'Brien et al. (2000). Persons interested in processing and utilization of tropical soybean are referred to Weingartner (1987). A Nutrient Requirements of Domestic Animals series is published by National Research Council of the National Academy of Sciences, and is available from the U.S. Government Printing Office, Washington, DC. Requirements for 12 types of animals are updated as timely. The major analytical methods used in the U.S. oilseeds processing industry are detailed in the *Official Methods and Recommended Practices of the AOCS,* available from the AOCS Press (AOCS Official Methods, 2002). A Spanish edition of selected methods also is available.

20–1.2 Competition

Soybean grown in the USA is exported whole, or "crushed" (solvent extracted or screw pressed) primarily for meal and oil. The oil is refined for food or industrial uses, or is partially degummed to reduce precipitation of phosphatides during storage before refining or export. Meal is fed to domestic animals or exported, with lesser quantities used in industrial applications. Limited amounts of soybean, approximately 4% domestically, are converted into food and feed protein ingredients. In specific areas, locally-produced animal fats and other seed oils compete with soybean for food use.

Some of the more stable processing by-products and feed protein alternatives also are exported. Plantings of crops raised specifically to compete for oil and protein markets are price sensitive, but significant quantities of oil- and protein-bearing by-products are unavoidably produced by allied industries (Lusas, 2000).

Most crops are grown for their primary product, regardless of whether it is the heaviest fraction obtained from processing. Even though oil sells for more per unit weight, meal is the primary product of soybean processing in the USA, and has accounted for the major return per unit weight of seed since the late 1940s (ERS, 1961). The highly-populated, palm-oil-producing Southeast Asia countries can do without soybean oil, but they and the world require soybean protein for their poultry, animal, and aquaculture industries. In contrast, in India, soybean is raised primarily for oil. With a limited animal-feeding industry, approximately 65% of

India's soybean meal is exported, even though the protein might be used to feed the largely vegetarian population (FAS, 2002).

Generally, the prices of primary products are set by supply and demand, except where dictated by governments. Secondary products are less interchangeable and must be moved in the trade to make room for additional by-products from scheduled production of primary products. For example, during 2001 when domestic surpluses of fats and oils existed and fuel oil costs had risen, various vegetable oils and rendered animal fats were burned to generate processing steam and to clear the holding tanks for more scheduled by-product oils.

20–1.2.1 Competing By-products

Production of oilseed meal and by-products always exceeds that of extracted oil, ranging from about 80% of total weight for soybean and seed of cotton (primarily *Gossypium hirsutum* L.), to about 60% for sunflower seed (*Helianthus annus* L.) and rapeseed/canola (*Brassica napus* L., *B. rapa* L.), and slightly more than 50% for peanut (groundnut, *Arachis hypogaea* L.). Essentially all oilseed and grain processing operations produce by-product streams, which often compete with soybean in feed uses.

Cotton is grown for its fiber, with the cottonseed by-product accounting only for about 8 to 12% of total crop returns. Regardless of how much cottonseed oil prices may rise, its supply is relatively inelastic and dictated by cotton plantings in response to world fiber prices. Similarly, increased prices for edible tallow and lard (secondary products) are unlikely to result in significant increases of meat animals.

Competition for cottonseed also affects soybean utilization. Undelinted (fuzzy) cottonseed, contains 23% crude protein, 21% fat, and 24% crude fiber. It increases milk and butterfat production when fed to dairy cattle. Additionally, it contains gossypol, which is toxic to monogastric animals and also limits seed feeding to 6.4 to 8.2 kg (7–9 lb) d^{-1} per milking cow. But, within that limit, cottonseed is a nutritional bargain and can be bought at harvest time for $82.50 to $148.50 t^{-1} (metric ton) or $75 to $135 short ton^{-1}, compared to whole soybean at $183 to $220 t^{-1} ($167–$200 short ton^{-1}). Dairy cattle feeders can outbid oil mills, and take first choice of the crop in years when cottonseed availability is limited. Direct feeding of cottonseed, in fuzzy, extruded, or coated forms (Laird et al., 1997; Bernard, 1999), grew from hardly any in the mid-1970s to more than 60% of the crop in 2001. The trend may lead to the demise of the domestic cottonseed oil industry, but also establishes a strong competitor to soybean and soybean meal in feeding dairy cattle.

Recent expansions in the corn (*Zea mays* L.) wet-milling industry to produce starch, sweeteners, and ethanol, have made large quantities of corn gluten meal and corn germ available. Although statistics are not published, oil extracted from corn germ is the second major edible oil produced domestically. Aflatoxin-contaminated corn can be fermented into alcohol, and the oil from its germ safely refined for food use. Cottonseed and peanut also are aflatoxin-susceptible crops, whose oils may be recovered for food use. For the most part, aflatoxins remain in the meals during extraction, which can only be used for fertilizers, buried, or fed to nonlactating cattle within regulatory guidelines.

The EAA profile of corn gluten meal is not as well balanced nutritionally as soybean protein, and is enhanced with synthetic lysine, or lysine from soybean in formulated feeds. As disposal of food processing wastes becomes more expensive, by-product streams of other crops processing industries will be increasingly inspected for components that can be sold for feed.

Inedible tallows and greases were introduced in high-energy poultry broiler feeds in the 1950s, and offer the advantage of higher caloric density than cereal carbohydrates (9 vs. 4 calories g^{-1}). Their use has broadened into pig, fish, and dairy cattle feeding in the last three decades, and also has encouraged use of full-fat soybean meal (Lesson and Summers, 2001; Lusas and Riaz, 1996a.)

20–1.2.2 Competing Crops

The palm fruit is similar to an olive, but much larger and orange-red in color. The outer fleshy part yields palm oil, and the internal palm kernel contains 12 to 14% oil. Palm oil is used primarily for food; and palm kernel oil is used in soaps, cosmetics, cocoa butter substitutes, and edible toppings in competition with soybean oil. World production of palm and palm kernel oils in 2000/2003 is estimated at 96% of soybean oil. Probably more significant, palm oils account for about 51% of global edible oil exports, compared to 27% for soybean oil which often is partially retained to meet needs of the producing country (FAS, 2002a). Because of ready availability, and prices lower than soybean oil, new fats and oils users are increasingly likely to consider palm oils.

Growing seasons are short in countries with cool summers, including Canada and Northern Europe. Often, rapeseed/canola seed is grown, and exported with oil and meal, to earn foreign exchange while soybean or soybean meal are imported for the local animal feed industry. World rapeseed/canola seed and meal trade is the second largest of the row crop oilseeds after soybean (FAS, 2002a).

Soybean continues to attract rivals as sources of plant protein and oil among row crops. Corn research in recent years has included elusive attempts to develop nitrogen-fixing capabilities in the plant. High lysine content and quality protein maize (QPM) cultivars of corn have been developed and contain as much as 44 to 46 mg lysine 100 g^{-1} protein (Zarkadas et al., 2000). High-oil corn, with ~82% more oil, and slightly more (~6%) protein than traditional cultivars, has been introduced in the animal feed market (Du Pont, 2001).

Low alkaloid-content (nontoxic, "sweet") lupins (*Lupinus* spp.) have shown promise as feed and food protein crops in areas with acidic soils and limited growing seasons. Protein contents of 300 to 450 g kg^{-1}, and more than 100 g kg^{-1} oil, have been demonstrated (Cerletti and Duranti, 1979). Extensive research on the narrow-leaf white lupin (*Lupinus angustifolius)* in Australia has led to shipments of seed to Europe for animal feeding beginning in the early 1990s. Lupin protein contains more fiber, the EAA are not considered as well-balanced as soybean protein and require methionine supplementation, but good results have been reported with dehulled lupin flour in child feeding trials in developing countries.

Research in Italy, France, the USA, and other countries in the 1980s and early 1990s demonstrated that protein fractions, with even better EAA balances than soybean protein, can be extracted from young grass and legume plants (Fantozzi, 1990). But, the processes are energy intensive, and color and flavor problems exist in using the protein concentrates in foods. Development of grass food protein technology has been shelved, at least temporarily. Soybean breeders, producers, and processors must work creatively to keep their crop more valuable than would-be replacements.

20–1.3 Soybean Utilization Technical Developments

20–1.3.1 Recognition of Soybean Protein Quality Level

The comparative value of soybean as a source of effective protein is readily seen by poultry, pig, and dairy and beef cattle feeders. However, it has been less publicly recognized in human nutrition. In the 1970s to early 1980s, protein quality level of ingredients was assessed by relative weight gains of rats during 28 d feeding tests, with casein the control set at a protein efficiency ratio (PER) of 2.5. Animal-source proteins typically surpass this value, but not plant protein sources. Soybean food proteins, in the 1.8 to 2.2 PER range, rank the highest among currently processed crops. For a period, U.S. food labeling law required that the percent of Recommended Daily Allowance (RDA) provided per serving in package labels be calculated on the basis of 45 g protein d^{-1} intake if the source had a PER of 2.5 or higher, but on a 65 g d^{-1} basis for ingredients with lower PERs. This put soybean protein at an economic disadvantage. Through repeated efforts, soybean industry and other plant protein nutritionists convinced the World Health Organization (WHO) and domestic regulators that the PER rat test is inherently biased by the higher sulfur amino acid requirements of a fur-bearing test animal compared to humans. The WHO protein recommendations for different human age groups, first published in 1985, were eventually reviewed in the USA (NRC, 1989) and led to setting a protein requirement of 45 g d^{-1} from all sources for adults in the early 1990s.

Today, the ranking of protein sources for quality level is mainly based on "protein digestibility-corrected amino acid score" (PDCAAS), although some countries still use PER. Current Food and Drug Administration (FDA) labeling laws, published in the Code of Federal Regulations (CFR), require that foods intended for adults and children 4 yr or older have PDCAAS of 20% or greater, and those intended for children between 1 and 4 yr of age must have PDCAAS of 40% or greater. Otherwise: (i) a "not a significant source of protein" label claim; or (ii) a "corrected amount of protein per serving" statement must be made on the label. As an option, PER of the food can be divided by the reference standard of casein, and must exceed 40%. Soybean proteins test well above these limits. Under regulations prevailing on 25 July 2001, 50 g of protein is the current Daily Reference Value for adults and children over 4 yr of age. Special labeling requirements exist for foods intended for infants (under 1 yr age), children 1 to 4 yr age, and pregnant or lactating women (21CFR101.9, 2002). (Note: CFR citations are not listed in the References. The sections can be downloaded directly from the Code of Federal Regulations, and are updated annually to include revisions. New regulations, not yet included in the updated code can be downloaded from the Federal Register.)

Through additional efforts, health claims of coronary heart disease (CHD) reduction by soybean protein have been allowed by the FDA on food labels. Rationale, requirements, and optional details for the claim, are described in 21CFR101.82 (2002) and the following model label claims are suggested:

"(1) 25 grams of soy protein a day, as part of a diet low in saturated fat and choles-terol, may reduce the risk of heart disease. A serving of (name of food) supplies ____ grams of soy protein, or (2) Diets low in saturated fat and cholesterol that include 25 grams of soy protein a day may reduce the risk of heart disease. One serving of (name of food) provides ____ grams of soy protein."

This has opened a new era of developing diets and formulating processed foods to ensure at least 25 g of soy protein is consumed daily.

20–1.3.2 Current Issues

20–1.3.2.1 Marketing New Soybean Types. Although eight classes of wheat are recognized by the USDA Federal Grain Inspection Service (FGIS) and traded internationally, only two classes of soybean ("yellow" and "mixed") are recognized. Limited amounts of specialty (identity preserved, IP) soybean are grown under con-tract in the USA for domestic use and export. But, essentially a multi-purpose soy-bean is traded in world markets. Concerns have been raised about how to market the almost unlimited types of soybean which breeders now can develop. Executives of major soybean processing companies have publicly stated their companies will handle new types when justified by the market (Bastiaens, 2001)—meaning the ad-ditional costs of handling segregated types will be borne by the buyers. Indeed, a market for nontransgenic ("non-GMO") soybean, organized by contract agents, farmer cooperatives, and large processors appeared with about $9.00 to 18.50 t^{-1} ($0.25–$0.50 bushel^{-1}) premiums paid for nontransgenic soybean at elevators in 2001. Segregation expenses in later handling, storage, processing, and distribution of products, further add to price premiums required for the products. Processors have learned: "Don't make it if you can't sell it!" and are not likely to assume the risks of building inventories of new cultivar products, without encouraging estimates of sales. Contract growing and identity preserved (IP) handling for interested users is likely to remain the major market entry route for new soybean types.

20–1.3.2.2 Transgenic Crops and Products. Current controversies about global acceptance of transgenic (GMO, genetically-modified organism; and GE, ge-netically-engineered) crops and foods are not limited to soybean alone, and some federal policies are the result of marketing earlier biotechnology products. A clash of principles is occurring between two cultures—marketers of transgenic technol-ogy products and food additives safety scientists. Both groups have multinational following, but different goals.

Transgenic technology marketers have been very effective in transforming biotechnology discoveries into marketable products, and in persuading regulators in various countries to expedite their approval for sale. Their objective is to gain as much of the market for their companies as rapidly as possible, and is normal busi-ness practice.

Passage of the Delaney Clause in the 1958 Food Additives Amendment (Section 409) to the 1954 Federal Food, Drug, and Cosmetic Act formally recog-nized food ingredients safety as part of the national dialog. The FDA was appre-ciably expanded to implement the amendment, review all food and feed ingredients, and establish procedures for evaluating proposed additions. Principles which emerged from the program include:

1. With the exception of identification of flavor compounds, consumers have the right to know what they are eating through an ingredients list on the product label.
2. Protocols were established for determining whether a proposed ingredient is "generally recognized as safe (GRAS)," or requires further review and additional tests before authorization as a "food additive"—sometimes with specified upper limits on levels of use. .
3. Testing requirements for previously unknown food additive candidates typically included raising generations of laboratory animals, with postmortem examination of critical organs to observe health and growth, and assessment of potential effects on fertility, and teratogenicity in the offspring. Importance of the latter point had been learned earlier in human medical research, where various problems did not appear until maturing of succeeding generations.

Other countries also developed food additives safety programs during the 30+ yr after enactment of the Delaney Clause. Some, including Germany, Japan, and Canada, established requirements even more demanding than the USA., and have capable scientists who advise their governments based on review of data, rather than consensus of selected panels. Informal networks, for sharing experiences and information about methodology, developed globally between scientists working in the field in different countries.

Space limitations do not allow a review of FDA and Environmental Protection Agency (EPA) protocol changes in assessing transgenic product safety, including introduction of the concept of "substantial equivalence," which has been applied to waive animal life-cycle testing. Consumer advocate versions are readily accessible on the Internet, and readers may gain useful insights from the following references: Federal Register (1992), Maryanski (1993), Glickman (1999), IFST (1999), Jacobs (1999), Millstone et al. (1999), Paulson (1999), Herbert (2000), Kerr and Hobbs (2000), North (2000), NRC (2000), Redman (2000), American Soybean Assoc. (2001), Handbook (2001), U.S. Mission (2001), and Vanderveen (2001).

Transgenic crops, like soybean cultivars that can grow closer to the Equator and enlarge productive land area, announced by Brazil in 2001, and soybean cultivars that can grow on previously unusable saline lands, announced by China in 2002, would seem welcome if yields are acceptable. But, suspicion of transgenic products with hormonal, estrogenic, or plant-incorporated protectant (PIP, insecticide) properties, is likely to continue among traditional food additives safety scientists until safety is confirmed by generation trials with appropriate animal species. To the business-oriented processor's mind, obtaining such data simply is the price of giving customers what they want if the potential market is large enough to warrant the cost. To the European Union (EU) regulator's mind, the issue may simply resolve itself in time, as new South American suppliers, anxious to provide the desired type of soybean, develop.

20–1.3.2.3 Looking After Processor Self-Interests. The basic objectives of soybean processors include: (i) satisfying customers, (ii) staying in business by obeying applicable laws, and (iii) and maintaining profitable operations. Processors are not in a position to take sides on controversial issues, unless appreciable ingredient functionality advantages are obvious. Processors must avoid selling products from illegal or contaminated cultivars because of severe potential liabilities under domestic tort law. Although agricultural chemicals manufacturers and seed suppliers, in applying for approval, claim their products are safe and legal when used or grown as directed, they cannot guarantee the growers always will follow instructions, or that crops grown from nontransgenic seed will not become contaminated by pollen drift or volunteers from previous plantings of transgenic seed.

Processors must ensure for themselves that they are receiving what is expected. Service laboratories and kits for identifying various transgenic crops are available, and more receiving dock laboratories or tests are likely in the future. Professional competence and integrity of growers in segregation of crops in the field, and post-harvest handling of transgenic cultivars, is essential. Prior clearance of transgenic cultivars for domestic use and export to other countries, by the supplier before planting seed is released, is critical.

20–1.4 The Regulators

Many groups have become part of the infrastructure that influences soybean-processing facilities and management practices during the last two decades.

20–1.4.1 Occupational Safety and Health Administration

A part of the U.S. Department of Labor, the Occupational Safety and Health Administration (OSHA) has the broad mission to save lives, prevent injuries, and protect the health of U.S. workers under regulations described in the U.S. Code of Federal Regulations (29CFR1900, 2002). Safe practice requirements are too extensive to list here, but affect equipment design, work area conditions, work practices, safety equipment, and instruction programs. Examples include exposure of personnel to dust in soybean handling, solvents in extraction plants, and hazardous materials in laboratories. Periodic testing of personnel whose health might be affected by work conditions, including hearing-loss tests for workers in high-noise areas, is required. The OSHA requirements have forced remodeling of some extraction plants and refineries. The concurrent introduction of small process control and personal computers in the late 1980s to early 1990s led to increased automation, with on-line monitoring of equipment and operation of processing lines moved into remote control rooms. Interestingly, product quality reliability often improved as equipment operators were withdrawn from making fine adjustments on the processing floors.

20–1.4.2 Environmental Protection Agency

The Environmental Protection Agency (EPA) (40CFR1, 2002) is led by an Administrator appointed directly by the U.S. President, and has the mission of protecting human health and safeguarding the natural environment including air, water, and land. This federal agency, in cooperation with OSHA and respective state agencies, has had the most effect in reshaping soybean-processing facilities and practices in the last two decades. EPA requirements include monitoring and controlling:

1. Atmospheric emissions of particulates. The EPA has jurisdiction over dust, smoke, and odors throughout the entire crop collection, processing, and distribution system—from elevators, to processing facilities, to ship loading.
2. Extraction solvent losses. The FDA specifies which solvents can be used in extracting oilseeds. OSHA specifies maximum permissible worker exposure levels (PEL) in workplace air. The EPA is concerned with escape of toxic, flammable, and volatile organic chemicals (VOCs) into soil, water, and the atmosphere. A VOC is defined as any compound of carbon, with specified exceptions, that participates in atmospheric photochemical reactions to form ozone (Wakelyn, 1997). Purchases of makeup solvent after starting an extraction plant are considered proof of loss into the environment and may be subject to penalties. During the last two decades, hexane losses in solvent extraction plants have been reduced from as much as 8.0 L t^{-1} (2 gallons short ton^{-1}) of seed extracted at the worst of extraction plants to an industry norm of about 1 L t^{-1} (0.25 gallon short ton^{-1}), with some plants reported to achieve 0.5 L t^{-1} (0.125 gallon short ton^{-1}) loss.
3. Solvent tank rust-through and leakage. To avoid rust-through and seepage into the soil, solvent tanks now generally are located aboveground, usually under shade, and often in pits where they can be inspected. Tank vents and other equipment are chased with vacuum lines to collect vapors and bring them to a central location for scavenging solvent from the entraining air by mineral oil scrubbers. The concrete-lined pits are not allowed to drain into the soil or water streams. Construction of dikes (nonpermeable basins around tanks storing liquids, large enough to hold tank contents in case of leakage or bursting) has become established in many industries, and includes oil mill tanks holding solvents, crude and refined vegetable oils, and other liquids.
4. Process stream discharges. Aqueous streams that exceed local biological oxidative demand (BOD) or chemical oxidative demand (COD) standards for direct discharge must be sent to a properly operated sewage treatment plant before release into public waters. When local public facilities cannot handle the load, processors install their own facilities for primary treatment (removal of separable solids) and secondary treatment (usually aerobic digestion of solubles).
5. Rain water runoff. Extraction plants have been required to grade their grounds to catch and direct rainwater, including roof and ground washings, into holding ponds where primary settling and initial aerobic digestion can occur. Water that does not percolate into the soil can be spray irrigated over crop land.
6. Surplus gums and soapstock. High disposal costs of gums (phosphatides) and soapstock from alkali neutralization in refineries has led to many com-

panies relocating refineries adjacent to their extraction plants. This enables spreading gums and soap stocks over the extracted soybean before drying, and sale as part of the soybean meal. The number of independent or stand-alone refineries has greatly decreased in the USA.

7. Refined oil wash water. High disposal costs of soap-containing waters from washing alkali-refined oils have led to no-wash ("zero") water discharge processes which absorb soaps directly from the neutralized oil with silica hydrogels (see Section 20–4.4).

20–1.4.3 U.S. Department of Agriculture

Two agencies in the U.S. Department of Agriculture (USDA) affect soybean products utilization. The Federal Safety Inspection Service (FSIS) is responsible for inspecting meat, poultry, and egg products, that enter interstate commerce (9CFR300, 2002). Processed meat or poultry products that use soybean flours or protein concentrates as processing aids, soybean protein concentrates or isolates as restructuring agents or pumping proteins, or canned meat products whose appearance is enhanced by inclusion of texturized soybean proteins, are prepared under FSIS inspection.

Through its Food and Nutrition Service (FNS), USDA oversees the National School Lunch Program (7CFR210, 2002) and other mass feeding programs, and approves their use of soybean food products.

20–1.4.4 Food and Drug Administration

The U. S. Food and Drug Administration (FDA) (21CFR, 2002) is part of the Department of Health and Human Services. Through its various subagencies, including the Center for Veterinary Medicine (CVM), FDA has final responsibility for ensuring safety of the nation's foods, pet foods and feeds, in addition to drugs and medical devices. Safety of foods not inspected by the FSIS is the responsibility of FDA.

Domestic food processors have the legal obligation to offer foods for sale that are safe. In failing to provide safe foods, processors are open to compliance actions by the regulatory agencies, suits by injured consumers under tort laws, and damages to customers under contractual laws. In recent years, the courts have held that FDA no longer has to prove that a specific foreign substance in a food lot is harmful to users in each law suit. Rather, if the substance has not been specifically allowed by an agency (FDA, EPA, FSIS) responsible for ensuring safety of the product, the food automatically is considered "adulterated" and noncompliance action can be brought against the processor. Non-approved substances can include contaminants, pesticides that are not allowed or exceed permissible limits, microbial presence or growth, and others. Foods that have been produced, or stored under conditions where they might become contaminated, for example by rodents or insects, also are considered adulterated even though actual contamination is not found (Johnson, 1996; FDA/USDA, 2000).

20–1.4.5 Good Manufacturing Practices and Hazard Analysis Critical Control Point Programs

The FDA was granted access to inspect all food plants and facilities engaged in interstate commerce by amendments to the Federal Food, Drug, and Cosmetic Act (FFDCA), but funds for inspectors were limited. Thus, it took the route of formalizing a general Good Manufacturing Practices (GMP) umbrella (21CFR110 Good Manufacturing Practice in Manufacturing, Packaging, or Holding Human Food, 2002), which details expectations in (i) buildings and yards design and care; (ii) rodent, pest, and bird control; (iii) ventilation, ceiling condensate, and dust control; (iv) equipment selection, care, and sanitation; (v) process documentation; (vi) raw materials, packaging materials, and product storage; and (vii) restroom facilities and instruction of employee personal hygiene. Detailed GMPs exist for specific operations that make highly sensitive food and feed products.

Specific FDA GMP requirements do not exist for grain and oilseeds ingredients processors, apparently due to (i) higher priorities for still establishing regulations for the more perishable wet-processed foods, (ii) lower inherent spoilage during dry processing, and (iii) placing the primary responsibility for delivering wholesome foods to consumer markets on food assemblers. Food product assemblers periodically visit their ingredients suppliers as part of their corporate quality assurance programs and review GMP practices. Each ingredient processor is likely to be inspected more times a year by its industry customers, than a food product processing plant is visited by an FDA inspector.

The FDA further established industry-wide Hazard Analysis Critical Control Point (HACCP; 21CFR120.8, 2002) programs for food processors which became U.S. law in 1995. HACCP is a problem prevention system based on Critical Control Point Management techniques used earlier in the hard goods industries. HACCP programs require that each food-processing operation develop and implement plans, in which production of each product is reviewed step by step and outlined in a flow sheet. The critical points at which food safety problems might arise are identified and physical conditions (such as temperature), and chemical or microbiological tests, are established for passing or rejecting product at those points. Records must be kept of measurements taken, and of disposition or treatment of products rejected in-process (Johnson, 1996).

A FDA inspector visiting a food plant would expect to find it in compliance with GMP requirements, and would ask to see the HACCP plan and examine the required records. Penalties for noncompliance might be imposed immediately, or recommendations made for changes to be implemented before a revisit. HACCP plans are concerned only with food safety. Commodity processors and food assemblers typically implement additional corporate Critical Control Point programs to ensure that other desired product quality characteristics also are maintained. The USDA-FSIS has established Sanitation Standard Operating Procedures (SSOP; 9CFR416.12, 2002), a pathogenic organism reduction program, and its version of HACCP (9CFR417, 2002).

The FDA has announced that its future priorities include ensuring that food products are adequately labeled concerning allergens. Approximately 1.5% of the U.S. population is estimated to have food allergies.

20–1.4.6 Association of American Feed Control Officials

Feeds and pet foods must be registered in each state where sold, and are regulated by the feed control officials of that state. The officials work with the "Model Bill" of the Association of American Feed Control Officials Incorporated (AAFCO), which is designed to promote establishment of nationwide uniform feed laws. For pet foods, minimum specified nutrient compositions, or defined animal feeding trials, must be satisfied for making specific nutritional claims, and labels must meet guidelines. The ingredients used must be approved by AAFCO, and listed in decreasing order, using names defined by AAFCO or the International Feed Number (IFN) system. The AAFCO typically defers to EPA and FDA for establishing allowable tolerances for residual pesticides and toxic substances (Official Publication, 2002).

20–1.4.7 National Fire Protection Association

Although not a regulatory agency, the National Fire Protection Association, Inc. (NFPA, 2001) plays an important role in oil extraction plant design and operation. Drawing on experienced Industry Advisory Committees, it issues *NFPA 61 - Standard for the Prevention of Fires and Dust Explosions in Agricultural and Food Products Facilities*, and *NFPA 36 - Standard for Solvent Extraction Plants*, which are revised every few years. Oilseed handlers, processors, and solvent extraction plant operators who do not meet the standards, might be denied insurance to operate their facilities, and would be exposed to claims of not following recognized good industry practices in litigations for personal injuries and property damage.

20–1.4.8 International Organization for Standards (ISO 9000)

Agreement on product analytical methods and standards was one of the first problems faced in globalization. The Codex Alimentarius program of the United Nations initially worked to harmonize differences between crops produced in various parts of the world by negotiating broader definitions, composition ranges, and thus variability. But, it became obvious this was inadequate in industries which had already advanced to the point of specifying ingredient compositions and functional properties required for making their products. Also, small importers cannot afford to inspect processing facilities in distant lands.

The problem has been partially solved by establishment of a global nongovernmental International Organization for Standards (ISO) system, which enables volunteer suppliers to obtain training in implementation of internal standards systems and obtain certification that their production and quality control programs meet recognized practices. ISO certification does not guarantee that a supplier's laboratory analyses are correct, but only that the supplier has production and laboratory quality control programs in place, which meet good practice expectations. Processors in the USA and other countries originally obtained ISO certification to do business in Europe, but buyers throughout the world are increasingly accepting ISO certification. Currently, suppliers of nontransgenic crops are seeking ISO 9000 certification to improve their credibility to potential customers. The basic ISO 9000 program was reorganized in 2000, with changes made in some ISO 9000-2000 categories. An ISO 14000 program also exists for environmental management systems. Additional information is readily available on the Internet (ANSI, 2003), and training consultants and accredited certification inspectors are available in a variety of specialties.

20–1.5 Changes in Doing Business

Changes in management techniques have made all domestic and global businesses more competitive in the last two decades. Earlier "production orientation," where the processor decided which products would be made, has been replaced by "market orientation," where ingredients processors first determine what prospective customers want, and develop means to provide the desired products. Food assemblers rely even more heavily on consumer tests in developing and marketing grocery products.

Many companies have found that seeking to be the "best," in converting crops into ingredients, and marketing consumer products is too broad a scope to handle. The food industry has voluntarily divided into two groups: "ingredients producers," who focus on collecting and trading crops, and converting them into food, feed, and industrial ingredients; and "food or feed assemblers" who focus on formulating ingredients into final products, and marketing them to users. Each group has its own specialized equipment, technologies, and marketing challenges. Many grain and oilseed milling companies have divested feed divisions and consumer product lines.

20–1.5.1 Quality

The word "quality" has a specialized definition when producing ingredients, and means "consistency," "reproducibility," or "quality reliability," (Berger et al., 1986; Deming, 1986; Juran, 1986). A food assembler buying ingredients, expects the next shipment will perform the same as the previous one without adjustments to equipment or operating conditions. But, many variations occur in natural products, and the soybean processor is expected to modify crop processing to deliver consistent composition and functionality.

20–1.5.2 Total Quality Management

Quality improvement principles can be applied to all departments within a firm. As used here, total quality management (TQM) means planning, reviewing, and controlling for improved performance—not only in product quality reliability, but also in regulatory compliance, resources utilization, customer service satisfaction, employee job satisfaction, and profitability (owner satisfaction).

20–1.5.2.1 Just-in-Time Delivery. Three to four decades ago, the common practice among food assemblers was to hold received ingredients while laboratory tests were completed to verify usability. Under TQM, corporate finance departments realized that large amounts of capital were tied up in unproductive buildings, storage tanks, and raw materials inventories. The just-in-time (JIT) delivery of components as needed had been developed in the hard goods assembly industries in the

1970s, and was adopted in food assembly operations in the late-1980s and 1990s. Capital tied up in on-site ingredients holding facilities and inventories at food assembly plants was greatly reduced.

20–1.5.2.2 Self-Certification. Food assemblers needed assurance that purchase orders were placed only with suppliers capable of providing ingredients that met sanitation, composition, and functionality requirements. This led to food assemblers seeking out suppliers who could meet their needs, and in some cases teaching them how to make the desired ingredients. Selected processors were invited to "self-certify" (guarantee) they would ship only ingredients that meet the buyer's specifications. In turn, food assemblers switched from buying on the open market to purchasing from a limited number of self-certified suppliers. Earning the position of "Self-Certified Supplier" to a major food assembler is a distinct achievement by an ingredients producer.

20–1.5.2.3 Responsibilities Shifted to Ingredients Suppliers. Just-in-Time Delivery and Self-Certification have not been adopted throughout all countries. But, where implemented, the commitments are serious legal obligations and more is at stake than loss of future business if specifications and contracts are not met. Together, JIT and Self-Certification have shifted responsibility for ingredient reliability control, raw materials warehousing, and delivery functions from the food assembler to the commodity processor. Nowadays, analysis confirmation sheets may accompany shipments, or be faxed or e-mailed while ingredients are in transit, with the buyer randomly testing samples to confirm supplier performance. Generally, ingredients processors have accepted the additional responsibilities in return for more reliable markets. In oils processing, it sometimes has meant installing equipment that has flexibility for producing a variety of smaller batches for more frequent shipments. In granting or renewing self-certification status, ingredients buyers inspect processor facilities for sanitation, GMP and HACCP programs or their equivalent, and quality control practices. Special attention is given to the analytical laboratory, qualifications of its personnel, equipment, internal control checks to validate that assays are performing as expected, and integrity in keeping records (Green, 1996).

20–1.5.2.4 Internal Quality Control-Quality Assurance Programs. Government food safety regulators and inspectors have no interest in a company's profitability or survival. Many processors have programs to collect data to document that the HACCP program is working for inspectors, plus their own Critical Control Point program to ensure that ingredients and products are made to meet customer or the company's own specifications.

20–2 SOYBEAN, MEALS, AND OILS SPECIFICATIONS

Considerable variation can occur in quality levels of soybean, meal, and oil. In the USA, grades for whole soybean are established by the Grain Inspection, Packers, and Stockyard Administration of the USDA, with inspection conducted by the Federal Grain Inspection Service, "FGIS" (7CFR800, 2002). Additionally, purchases are made under a selected set of trading rules which also specify requirements for shipping, method of payment, handling of order shortfalls or overages, sampling and price adjustment procedures, and a referee chemist if assays are needed to settle disputes. Most domestic processors operate at least under the trading rules for soybean meal and soybean oil established by the National Oilseed Processors Association (NOPA, 2000), in addition to customer purchase specifications and their own quality specifications for branded products.

20–2.1 U.S. Standards for Soybeans

(Subpart J of Official United States Standards for Grain)

TERMS DEFINED

7CFR810.1601 Definition of soybeans (2002).

Grain that consists of 50 percent or more of whole or broken soybeans (*Glycine max* (L.) *Merr.*) that will not pass through an 8/64 (inch) round-hole sieve and not more than 10.0 percent of other grains for which standards have been established under the United States Grain Standards Act.

7CFR810.1602 Definition of other terms (2002).

(a) *Classes.* There are two classes for soybeans: Yellow soybeans and Mixed soybeans.
(1) *Yellow soybeans.* Soybeans that have yellow or green seed coats and which in cross section, are yellow or have a yellow tinge, and may include not more than 10.0 percent of soybeans of other colors.
(2) *Mixed soybeans.* Soybeans that do not meet the requirements of the class Yellow soybeans.
(b) *Damaged kernels.* Soybeans and pieces of soybeans that are badly ground-damaged, badly weather-damaged, diseased, frost-damaged, germ-damaged, heat-damaged, insect-bored, mold-damaged, sprout-damaged, stinkbug-stung, or otherwise materially damaged. Stinkbug-stung kernels are considered damaged kernels at the rate of one-fourth of the actual percentage of the stung kernels.
(c) *Foreign material.* All matter that passes through an 8/64 round-hole sieve and all matter other than soybeans remaining in the sieved sample after sieving according to procedures prescribed in FGIS instructions.
(d) *Heat-damaged kernels.* Soybeans and pieces of soybeans that are materially discolored and damaged by heat.
(e) *Purple mottled or stained.* Soybeans that are discolored by the growth of a fungus; or by dirt; or by a dirt-like substance(s) including nontoxic inoculants; or by other nontoxic substances.
(f) *Sieve.* 8/64 round-hole sieve. A metal sieve 0.032 inch thick perforated with round holes 0.125 (8/64) inch in diameter.
(g) *Soybeans of other colors.* Soybeans that have green, black, brown, or bicolored seed coats. Soybeans that have green seed coats will also be green in cross section.

Bicolored soybeans will have seed coats of two colors, one of which is brown or black, and the brown or black color covers 50 percent of the seed coats. The hilum of a soybean is not considered a part of the seed coat for this determination.
(h) *Splits*. Soybeans with more than 1/4 of the bean removed and that are not damaged.

PRINCIPLES GOVERNING THE APPLICATION OF STANDARDS

7CFR810.163 Basis of determination (2002).

Each determination of class, heat-damaged kernels, damaged kernels, splits, and soybeans of other colors is made on the basis of the grain when free from foreign material. Other determinations not specifically provided for under the general provisions are made on the basis of the grain as a whole.

7CFR810.1604 Grades and Grade Requirements for Soybeans (2002)

Grading factors	Grades U.S. Nos.			
	1	2	3	4
Minimum test weight per bushel	56.0	54.0	52.0	49.0
	Maximum percent limits of:			
Damaged kernels:				
Heat (part of total)	0.2	0.5	1.0	3.0
Total	2.0	3.0	5.0	8.0
Foreign material	1.0	2.0	3.0	5.0
Splits	10.0	20.0	30.0	40.0
Soybeans of other color [1]	1.0	2.0	5.0	10.0
	Maximum count limits of:			
Other materials				
Animal filth	9	9	9	9
Castor beans	1	1	1	1
Crotalaria seeds	2	2	2	2
Glass	0	0	0	0
Stones [2]	3	3	3	3
Unknown foreign substance	3	3	3	3
Total [3]	10	10	10	10

U.S. Sample Grade, Soybeans that:
  (a) Do not meet the requirements for U.S. Nos. 1, 2, 3, 4; or
  (b) Have a musty, sour, or commercially objectionable foreign odor (except garlic odor); or
  (c) Are heating or otherwise of distinctly low quality.

[1] Disregard for Mixed soybeans.
[2] In addition to the maximum count limit, stones must exceed 0.1 percent of the sample weight.
[3] Includes any combination of animal filth, castor beans, crotalaria seeds, glass, stones, and unknown foreign substances. The weight of stones is not applicable for total other material.

SPECIAL GRADES AND SPECIAL GRADE REQUIREMENTS

7CFR810.1605 Special grades and special grade requirements, 2002.

(a) *Garlicky soybeans.* Soybeans that contain five or more green garlic bulblets or an equivalent quantity of dry or partly dry bulblets in a 1,000 gram portion.

(b) *Purple mottled or stained.* Soybeans with pink or purple seeds as determined on a portion of approximately 400 grams with the use of an FGIS Interpretive Line Photograph.

(End—U.S. Standards for Soybeans)

Notes. The U.S. regulations mandate that exported soybean be graded and weighed by FGIS inspectors. For trading within the USA, voluntary inspection and weighing is available from authorized state and private agencies who employ personnel trained and licensed by FGIS. Elevators and company receiving stations often use such neutral third parties to minimize arguments with growers about grading of received soybean.

Technologists will recognize that official grades do not include numerical information about moisture, protein and oil contents, free fatty acid (FFA), or peroxide values PV), which affect yields of saleable products. However, the inspectors include a moisture analysis report with the inspection certificate. Persons purchasing soybean are familiar with the types of deviations that can be expected in different grades, for example, lower test weights can indicate smaller and less mature soybean which contain more fiber and yield less oil and protein. Depending on the age of the crop, higher moisture content can indicate higher FFA content in the oil. Increased heat damage can indicate moldiness, and increased FFA and PV values. Higher splits content can mean a seed lot with higher moisture content and FFA, and a soybean more susceptible to absorbing moisture during ocean voyages and holding at high humidity destinations.

Experienced buyers know the general characteristics of soybean drawn from specific growing areas, and possible effects of the current season's weather. They can make additional requests in their order. An example is chlorophyll content, which increases in soybean that have not matured fully in the field, and is detrimental to stability of bottled and frying oils. Independent Certified Samplers and Weighers are available at most shipping ports to draw samples of the lot for local analysis. A Soy Importers' Handbook, prepared by American Soybean Association and United Soybean Board, is available free for the downloading (Am. Soybean Assoc.-United Soybean Board, 2001). An interesting paper entitled "Procuring Beans for Fullfat Soya in Europe" has been prepared by Meyer (1996).

20–2.2 National Oilseed Processors Association Trading Rules (NOPA, 2000)

Soybean Meal Definitions:
- *Soybean Cake or Soybean Chips* - the product after extraction of part of the oil by pressure or solvents from soybeans. A name descriptive of the

process of manufacture, such as expeller, hydraulic or solvent extracted shall be used in the brand name. It shall be designated and sold according to protein content.

- *Soybean Meal* - ground soybean cake, soybean chips, or soybean flakes. A name descriptive of the process of manufacture, such as expeller, hydraulic or solvent extracted shall be used in the brand name. It shall be designated and sold according to protein content.

- *Soybean Mill Feed* - the by-product resulting from manufacture of soybean flour or grits, composed of soybean hulls and offal from the tail of the mill. Typical analysis: 13% crude protein, 32% crude fiber, 13% moisture.

- *Soybean Mill Run* - product resulting from manufacture of dehulled soybean meal, composed of soybean hulls and such bean meats that adhere to the hull in normal milling operations. Typical analysis: 11% crude protein, 35% crude fiber, 13% moisture.

- *Soybean Hulls* - product consisting primarily of outer covering of the soybean. Typical analysis: 13% moisture.

- *Solvent Extracted Soybean Flakes* - product obtained after extracting part of the oil from soybean by use of hexane or homologous hydrocarbon solvents. It shall be designated and sold according to protein content.

- *44% Protein Soybean Flakes or Soybean Meal* - product produced by cracking, heating, flaking, solvent extracting soybean, followed by cooking soybean. Extracted flakes may be marketed as such or ground into meal. Standard specifications: protein, 44.0% min fat, 0.5% min; fiber, 7.0% max; moisture 12.0% max. May contain no more than 0.5% (10 lb/short ton) nonnutritive, inert, agent to reduce caking and improve flowability.

- *High Protein Solvent Extracted Soybean Flakes or Soybean Meal* - product produced by cracking, heating, flaking, solvent extracting dehulled soybean, followed by cooking. Extracted flakes may be marketed as such or ground into meal. Standard specifications: protein, 47.5 to 49.0% min; fat, 0.5% min; fiber, 3.3 to 3.5% max; moisture 12.0% max. May contain no more than 0.5% (10 lb/short ton) non-nutritive, inert, agent to reduce caking and improve flowability.

Soybean Oil Definitions
    NOPA (2000) analytical definitions for Crude Degummed Soybean Oil for Export are:

Analytical Requirements	Maximum	Minimum	AOCS Method
Unsaponifiable Matter	1.5%	---	Ca 6a-40 (97)
Free Fatty Acids, as Oleic	0.75%	---	Ca 5a-40 (97)
Moisture and Volatile Matter			M&V Ca 2d-25 (97)
and Insoluble Impurities	0.3%	---	Ca 3a-46 (97)
Flash Point	---	250°F, 121°C	Cc 9c-95 (97)
Phosphorous	0.02%		Ca 12-55 (97)

The chemical analysis shall also include the qualitative test for fish oil and marine animal oils (AOAC Method No. 28.121) and shall be negative.

NOPA (2000) analytical requirements for Once Refined Soybean Oil for Export are:

- Clear and brilliant in appearance at 70-85°F, 21-29.5°C.
- Free from settlings at 70-85°F, 21-29.5°C.
- No more than 0.10% moisture and volatile materials (AOCS Ca 2d-25 -- 97).
- No more than 0.10% free fatty acids (AOCS Ca 5a-40 -- 97).
- Color when bleached, according to (AOCS Ce 8e-63 -- 97), no darker than 3.5 red and shall not have a predominantly green color.
- Flash point not below 250° F, 121° C (AOCS Cc 9c-95 -- 97).
- Unsaponifiable content less than 1.5% (AOCS Ca 6a-40 -- 97).
- AOAC Method 28.121 for fish and marine oils negative.

NOPA (2000) analytical requirements for Fully Refined Soybean Oil for Export are:

- Flavor shall be bland.
- Color (Lovibond) 20Y/2.0R maximum (AOCS Cc 13b-45 -- 97).
- No more than 0.05% free fatty acids.
- Clear and brilliant in appearance at 70–85°F, 21–29.5°C.
- Cold Test: 5.5 hrs. minimum (AOCS Cc 11-53 -- 97).
- No more than 0.10% moisture and volatile materials.
- Unsaponifiable matter less than 1.5%.
- Peroxide value Meq/kg less than 2.0 (AOCS Cd 8-53 -- 97).
- Stability – AOM: minimum of 8 hrs., 35 Meq/kg. (AOCS Cd 12-57 -- 97).
- Approved preservatives are permitted.
- AOAC Method 28.121 for fish and marine oils negative.

(End—National Oilseed Processors Association Trading Rules)

Notes. Buyers typically include a specification on residual trypsin inhibitor (TI) content in flakes or meals, expressed as urease activity or protein solubility. This is discussed in greater detail in Section 20–8.2.

NOPA (2000) trading rules specify percent discounts applicable to oils which do not meet analytical requirements for the respective grade, in addition to maximum or minimum analysis values for rejection by the buyer. Overseas and domestic buyers may negotiate additional specifications by mutual agreement.

20–3 SOYBEAN OIL EXTRACTION

20–3.1 Drivers for Change

Today's oilseeds extraction and processing industry includes five basic skills, developed by trial and error over more than two millennia:

1. Preservation of seed by natural or artificial drying, and storage to keep it in a dormant state and protected from insects and rodents.

2. Removal of trash and hulls by beating, stamping, threshing, disc hullers, cracking rolls, or other devices, followed by winnowing, sieving, or air aspiration to carry away hulls and chaff.

3. Freeing the oil by pounding the seed with rocks; uses of mortar and pestle grinders (which later were redesigned as ghanis powered by man, beasts, or water wheels), use of edge runners (thin, vertical rolling stone rollers) to crush the seed, and more recently, flaking rolls and expanders.

4. Heating the seed to increase oil recovery by denaturation of seed proteins to make the oil-holding matrix easier to shear, inactivation of enzymes which otherwise would initiate oil degradation processes. Heat and moisture (including steam) may be applied to soften seed tissue before flaking, or heat alone in cooking and drying flakes before pressing.

5. Separation of oil by draining crushed seed; squeezing by lever or wedge presses initially and later by hydraulic presses or continuous screw presses; and now by counter-current solvent extraction.

These skills developed in different sequences throughout the world, resulting in various processes, some still seen in ancient equipment operating in primitive societies. The five skills are simple in concept, and often help explain problems when troubleshooting even in the most modern oil extraction plants. Much of the extraction research conducted today deals with improvements in the five skills, or in better preparing the extracted meal or oil for downstream uses. Reductions in engineering programs at the USDA regional utilization research centers have led to fewer publications on process development research from U.S. public sector laboratories. Publications by engineers and scientists in other countries are seen more frequently now, but some are never translated into English. Most process development now is done by equipment manufacturers and large processors, and kept confidential except for patent disclosures and promotions by vendors.

The major drivers for change in the soybean processing industry during the last two decades include (i) need for food and feed to support the world's growing population, (ii) substantially increased regulations regarding food safety, working conditions, and environmental protection which already have been described, (iii) achieving maximum benefits from inherent nutrients in foods and feeds applications, and (iv) reduction of processing costs to keep firms competitive. Energy now is the second most expensive processing input after cost of the seed. Seemingly small losses can add up to appreciable yearly costs if not stopped. Steps that have been taken to improve energy utilization efficiency include (i) pinpointing sites of heat loss by infrared photography, (ii) insulation of vessels and steam lines where warranted, (iii) minimizing water and steam leaks, (iv) periodic reviews of proper sizing of electric motors, (v) minimization of repeated heating and cooling, as by installation of hot soybean dehulling systems, (vi) changing air circulation patterns in seed dryers for more efficiency, and (vi) installation of heat exchangers in refineries to recapture heat of exiting hot process streams.

Opportunities for economy of scale, including using larger machines, and automation to process larger quantities of seed with the same number or fewer workers are taken when feasible. However, extraction plants, and especially refineries, lose flexibility in responding to special crop problems and market needs as they become larger.

20–3.2 Principles of Oil Chemistry

It is assumed the readers are at least partially familiar with oils and proteins chemistry, and have access to textbooks and other references as needed. This section focuses on control of enzymatic activity and prevention of oil degradation in the seed and during processing. More details will be provided in succeeding sections on oil processing and utilization.

20–3.2.1 Enzymes and Physical Chemistry

A soybean seed is a temporary storehouse of nutrients and enzymes assembled for a relatively short period of dormancy until the next germination and growth cycle of the plant can begin. The seed is protected against physical damage and excessive drying by the hull; the unsaturated fatty acids are protected by natural antioxidants; and, respiration continues at a reduced rate. Reduced water availability supports dormancy, which ends when water becomes plentiful. At that time, life processes accelerate; some enzymes cleave fatty acids from the triglycerides for further metabolism as plant energy sources, while other enzymes focus on developing the new plant. The increased metabolic activity generates heat which further accelerates the biochemical reactions.

Little degradation of oil occurs in soybean as long as the oil-containing spherosomes in the seed tissue are inaccessible to enzymes and water is limited. But, once the structures are sheared and the reactive components blended, oil degradation products accumulate over time, with their production accelerated by heat. Oil degradation processes can occur in seed storage piles and tanks before processing, during preparation of seed for screw pressing or solvent extraction, and continue in the extracted meal unless it has been adequately dried or steam heated to temperatures sufficient to inactivate the enzymes. Enzymes of concern in processing soybean include: lipase, which splits free fatty acids from the triglycerides; lipoxygenase, which degrades fatty acids into oxidation precursors that can cause later flavor reversion; and phospholipase, which splits water-soluble polar compounds from phosphatides, enabling complexing with divalent cations to become nonhydratable and difficult to remove in the refinery. More research is needed to clarify relationships between moisture content in soybean flakes and inactivation temperatures of these three enzymes under processing conditions. However, we know they are active at ambient temperatures and show greatly increased activity at 60 to 65°C (140–150°F), usually with reduction seen at 82°C (185°C) and higher. They typically are inactivated by expanders at 104°C (220°F) with noticeable benefits in reduction of nonhydratable phosphatides (NHP) and p- Anisidine Value (AV; aldehydes, principally 2-alkenals and 2,4-dienals) in the oil. However, much of the FFA rise already will have occurred during seed storage.

The laws of physical chemistry still apply after enzymes have been inactivated. Thermodynamically, oils are destined to degrade, with rates of chemical reactions accelerated by heat, availability of reactants like water and oxygen (air), and the presence of catalysts including copper, other metals, alkalis, residual chlorophyll, and light. Although degradation cannot be arrested, it can be greatly slowed by reducing the accelerants while oil is consumed.

The principles of handling soybean seed to minimize enzymatic activity, in-activate animal antigrowth factors during processing, and slow post-process chem-ical deteriorations, are simple and gradually becoming understood. High quality level refined oils, broiler feed meals, and good tasting soybean food proteins on the mar-ket, generally are made by firms who have mastered the principles and keep end-product standards in mind while processing.

As a practical definition, oils are liquid at ambient temperatures, and fats con-sist of oils and crystals of their triglycerides. Approximately 95% of the lipids (or-ganic compounds that dissolve in nonpolar solvents) in nature are triglycerides. The International Union of Pure and Applied Chemistry (IUPAC) prefers the name "tri-acylglycerols." It is increasingly found in technical writing, but "triglycerides" con-tinues in use throughout industry.

20–3.2.2 Fatty Acids

Fatty acids are the building blocks of triglycerides. Except for fatty acids syn-thesized by microorganisms in fermentation processes and also present in ruminant meat fats and dairy products, the chains have even-number carbon atoms and take the general form:

carboxyl end   methyl end

$$H\text{-}O\text{-}\overset{\overset{\textstyle O}{\|}}{C}\text{-}\ \ldots\ldots\text{-}(CH_2)_n\text{-}\ \ldots\ldots\ \text{-}\underset{\underset{\textstyle H}{|}}{\overset{\overset{\textstyle O}{\|}}{C}}\text{-}H$$

Fats and oils chemists identify the specific carbons by the Systematic System, and count from the carboxyl end, with the carboxyl carbon numbered "1." Biochemists and nutritionists use the Omega System, and count from the methyl end, with the methyl carbon being "1" and designated by a lower case "ω" or "n." Unsaturation occurs when two bonds occur between adjacent carbons, shown as -CH=CH-.

Typical soybean oil consists of two major saturated fatty acids: palmitic (C16:0), 10.7%, and stearic (C18:0), 3.9%; and 3 major unsaturated fatty acids: oleic (C18:1, n-9 or ω-9), 22.8%; linoleic (C18:2, n-6), 50.8%; and linolenic (C18:3, n-3), 6.8% (Perkins, 1995). In these notations, the number immediately after "C" in-dicates the carbon atoms in the chain, and that after the colon indicates the num-ber of unsaturated positions. If the chain contains unsaturated carbons, the number after n (or ω) indicates the carbon after which the first unsaturated bond occurs, counting in the omega system. Fats chemists know that, in systematic numbering, the unsaturated positions typically are after carbon 9 for C18:1, carbons 9 and 12 for C18:2, and carbons 9, 12, and 15 for C18:3. Sometimes, the Greek letter delta (Δ) is used to call attention to the bond after a specific carbon; for example Δ^9 for the bond in C18:1 (n-9). Fatty acids are called "monounsaturated" if they contain one unsaturated position, and "polyunsaturated" if two or more.

Melting points (mp) of individual fatty acids increase with chain length, and decrease as the number of unsaturated positions increase: palmitic, 62.9°C (145.2°F); stearic, 69.6°C (157.3°F); oleic, 16.3°C (61.3°F); linoleic, −6.5°C

(20.3°F); and linolenic, −12.8°C (9°F). The relative reactivity of the C18 fatty acid series increases with the number of unsaturated positions, and often is estimated at 1:12:25 for oxidation, and 1:20:40 for hydrogenation, of oleic, linoleic, and linolenic acids, respectively. The extent of unsaturation is determined by the Iodine Value (IV), a titrametric analysis (AOCS TG 1a-64 for fatty acids, and AOCS Cd 1d-92 for triglycerides). Stearic acid has an IV of 0; oleic acid, 90; linoleic acid, 181; and the C18:3 (n-3) form of linolenic acid, 274. IVs of soybean oils vary in the 125 to 149 range. The major triunsaturated fatty acid form (C18:3, n-3) in soybean oil sometimes is called "α-linolenic acid"; a second, minor, (C18:3, n-6) form, also called "linolenic acid," is present in soybean oil.

20–3.2.3 Fatty Acid Isomerization, cis/trans, CLAs

Two types of fatty acid double bond isomers occur: *cis/trans* and positional. The long-time belief that nonprocessed unsaturated plant fatty acids always are in the *cis* form, has been challenged by recent research. Controlled chemical hydrogenation, and bio-hydrogenation in ruminants, can produce partially-hydrogenated "*trans*" structures whose melting points are increased, but not to the extent of fully saturated structures. This often is depicted in 2-dimension as:

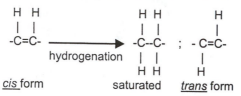

Serum cholesterol levels rise after ingestion of fats with higher melting points. Consumer advocates have brought pressures on the FDA to include the *trans* fat content in the Saturated Fat Total, and to additionally list the amount of *trans* fat in Nutritional Facts Tables on packaged foods. Interested readers can retrieve the initial FDA proposal in the *Federal Register* (1999), and follow responses and revisions to the proposal. A regulation may be enacted in 2003.

In seeking an answer to the question: "If charcoal broiled hamburgers contain carcinogens, where are the patients in our barbecue-enthusiastic society?" Dr. Michael W. Pariza, a cancer research scientist at the Univ. of Wisconsin, reported a mutagenic modulator in fried ground beef in 1983 (Yurawecz et al., 1999). This was welcomed by the beef and dairy industries, long criticized for "unhealthy" food by some nutritionists. The agent was found to be a fatty acid modified through biohydrogenation by rumen microorganisms, which becomes incorporated into beef triglycerides after subsequent digestion of the microorganisms.

Much has been learned about ruminant nutrition in recent years, and may be summarized by the model that digestion and absorbtion in ruminants are basically similar to those in monogastric mammals, but are preceded by fermentative digestion of hemicelluloses and other nonstructural carbohydrates at nearly neutral pH in the rumen section of the stomach. This activity is conducted by bacteria and protozoa, which protect themselves against toxic liquid fats by splitting some triglycerides into fatty acids and glycerin. The glycerin is metabolized in the rumen, and the (even more) toxic fatty acids are biohydrogenated to the *trans* structural isomer.

Also, the di-unsaturated sequences in fatty acids, as between carbon atoms 9 and 13 in linoleic acid, -CH=CH-CH$_2$-CH=CH-, lose the methylene group and become conjugated into a -CH=CH-CH=CH- sequence, although 18 carbons still remain in the chain. Additionally, the conjugated sequences move within the fatty acid chain to form a variety of positional isomers. Having originated from linoleic acid, the isomers are called "conjugated linoleic acids," "CLAs," or "conjugated dienes" (Yurawecz et al., 1999).

Biohydrogenation is accomplished by enzymes in cell walls of the rumen microorganisms, and has the effect of raising the melting point of the fatty acid to slightly above body temperature of the ruminant and preventing entry of the modified fatty acid into the cell. To continue this explanation, it is necessary to adopt CLA terminology, where "c" stands for cis, "t" stands for $trans$, and the Systematic Name of the fatty acid also is used. The melting point of oleic acid (c-9 octadecenoic acid) is 16.3°C; the $trans$ isomer, elaidic acid, (t-9-octadecenoic acid) melts at 43.7°C; and the major octadecenoic (monounsaturated) fatty acid of rumen biohydrogenation, $trans$-vaccenic acid (t-11-octadecenoic acid) melts at 44.0°C.

The CLA which attracted the earliest attention, rumenic acid (c-9,t-11-octadecadienoic acid), has been studied extensively for its ability to inhibit carcinogenesis. Another CLA, t-10,c-12-octadecadienoic acid, has been related to affecting the amount and composition of body fat. Other reported CLA effects include control of Type II diabetes in humans, reduction of atherosclerosis, reduction of fat in humans and pigs, and strengthening of muscle and bone.

Improvements in separation and identification techniques, and recognition of new CLA isomers ranging in positions from t-6,c-8-18:2 to t-12,t-14-18:2, have rapidly obsoleted earlier estimates of total CLA contents of meat and dairy products. In 1998, typical CLA levels in beef fats were estimated at < 1%, and ~3 to 4% in cheese fats with 6% occasionally reached in soft cheeses when cows are on spring pastures. Interestingly, while c-9, t-11-octadecadienoic acid in meat fats is derived from the ruminant's diet, that in milk is believed to be mainly produced endogenously in mammary tissues by Δ^9-desaturase activity on t-11-octadecenoic acid. A website: (http:/www.wisc.edu/fri/clarefs.htm) has been established to maintain a current listing of the scientific literature on CLAs since 1987 (Yurawecz et al., 1999).

Rumen metabolism may seem distant from soybean, but early studies on CLAs coincided with initial feedings of roasted and extruded full-fat soybean to dairy cattle. Indeed, it was found that mono- and polyunsaturated fatty acids in oilseeds elicited responses similar to fats from lush seasonal pastures, but had greater effect because of increased and consistent fat levels in the diet. Since then, biotechnologists have established microbial cultures capable of producing CLAs directly. These are available now as over-the-counter food supplements for humans, and selected forms have been effective in raising leaner pigs.

Conjugation, positional and $cis/trans$ isomers have long been known in the oils hydrogenation literature (Coenen, 1976; Ariaansz, 1993). Yet, as late as 1999, CLA researchers were reporting essentially negligible amounts of $trans$ fats and CLAs in vegetable oils and margarines. Using improved analytical methods, like silver-impregnated high pressure liquid chromatography, gas chromatography of 4,4-dimethyloxazoline (DMOX) derivatives, and mass spectrometers, 20 different CLA isomers have been identified in hydrogenated soybean oil (Mun-Yhung-Jung and

Mi-Ok-Jung et al., 2002.) This group of workers also has studied effects of catalyst types, contents, agitator speed, temperature, and hydrogenation pressure on CLA
production in soybean oil. Selective catalysts have been found more effective than
nonselective types in producing CLAs, with contents as high as 16% attained in soybean oil hydrogenation (Jung and Ha, 1999; Mi-Ok-Jung et al., 2001, 2002).

In addition to intentional production of functional *trans* fats by selective chemical hydrogenation methods, production of *trans* fats have been estimated at 0.5 to
0.7% in high temperature vacuum bleaching processes, 0.75 to 1.75% during deodorization, and up to 15% and more in degraded vegetable frying oils. *Trans* fat
and conjugated diene formation often are included in fatty acids oxidation pathways.
These compounds have been part of the human diet for thousands of years, and recent claims that they are detrimental, but sometimes beneficial, need clarification.
Intensive research activity on production of *trans* fats and CLAs in fats and oils processing, and on roles of these compounds in human and animal nutrition and health
is likely to continue.

20–3.2.4 Triacylglycerols, Triglycerides

Triglycerides are formed by enzymatic esterification of three fatty acids to
glycerin, a trihydroxy alcohol, in a dehydration reaction with one molecule of water
formed for each ester linkage. The water evaporates as the mature seed dries. The
reverse reaction, hydrolysis, is the most frequent form of triglyceride degradation.
Hydrolysis is catalyzed by water, lipase, and alkalis, and gives off heat sufficient
to char soybean in storage or ship holds, and cause cottonseed to break into spontaneous combustion, if not adequately dried before storage. The oil of a mature, well-
handled soybean contains 0.50 to 0.75% non-esterified fatty acids remaining from
earlier synthesis; larger quantities of FFA in seed usually indicate hydrolytic degradation. FFA is determined by AOCS Ac 5-41. Fatty acids retain many of their characteristics even when part of triglycerides.

The saturated fatty acids melt well above human and animal body temperatures, but remain liquid because of their incorporation into triglycerides. Fatty acid
positions on the triglyceride are indicated by two systems. In the R/S (*rectus/sin-
ister*) system, where the specific location of a fatty acid is unknown, the triglyceride
is depicted with the longest chain in the 1 position, the second longest in the 2 position, and the shortest in the 3 position, with preference given for placing unsaturated fatty acids in the 2 position. Knowledge of specific locations of fatty acids is
shown in the *sn* (stereospecific numbering) system. Thus, *sn*-18:0-18:1-16:0, identifies *sn*-glycerol-1-stearate-2-oleate-3-palmitate (Nawar, 1996).

$$
\begin{array}{ll}
\text{H} & \underline{\text{Positions}} \\
| & \\
\text{H-C-O-R}_1 & \text{1 } (\alpha) \\
| & \\
\text{H-C-O-R}_2 & \text{2 } (\beta) \\
| & \\
\text{H-C-O-R}_3 & \text{3 } (\gamma) \\
| & \\
\text{H} &
\end{array}
$$

The positions have very important effects (i) on the melting point of a triglyceride and its use in specific applications, (ii) in metabolism, since mammals possess primarily 1,3 specific lipases, and (iii) in potential effects of the type of fatty acid in the 2 position on atherosclerosis development. These are factors that oil refineries can at least partially control.

Nature takes optimum advantage of the available unsaturated fatty acids by placing them in positions that maximize lowering the melting points of triglycerides. In plants and most animals, the unsaturated fatty acids are placed in the 2 position first, and in the 1 or 3 positions later as available. However, in pork fat, saturated fatty acids are in the 2 position. The melting point of naturally-arranged soybean oil is 8.2°C (14.8°F) lower than for the same oil after chemical randomization (Bockish, 1998). Although soybean oil has only five major fatty acids, 22 to 25 natural configurations of triglycerides have been studied. Oils of the same seed lot that has matured in cooler climates or growing seasons have lower melting points because of increased unsaturation, but their resistance to oxidation also is reduced.

20–3.2.5 Fats and Oils Oxidation

Oxidation is initiated at double-bond sites and is characterized by formation of conjugated dienes, migration of bond position, formation of aldehydes, breakdown into shorter chains, and cross-linking to form polymers. More than 200 different breakdown compounds have been found in degradation of linoleic acid (Sevanian, 1988; Frankel, 1998; Min and Smouse, 1989; Nawar, 1996). Part of the confusion about oxidation reactions results from the several types present. Light-induced photosensitized singlet oxygen oxidation is the fastest (Min et al., 1989). Porphyrins and photosensitizers are detrimental to oil stability. For example (i) great care is taken to adsorb chlorophylls during bleaching to prolong oil shelf life, and (ii) deep frying of blood-containing foods, like chicken giblets, greatly shortens fryer oil life.

The extent of oxidation in a fatty acid or triglyceride is determined as peroxide value (PV), a titrametric method (AOCS Cd 8-53). Readers should be cautioned that peroxides decompose. A PV value may indicate the analyzed oil is in the increasing phase of oxidation, or the major oxidation may have occurred and only the nondecomposed peroxides are determined; age and history of the product should be considered in forming a conclusion. Stability of triglycerides formerly was determined by the Active Oxygen Method (AOM, Cd 12-93), in which heated air was bubbled through a heated liquid sample of the oil or fat, and the number of hours for the sample to reach 100 milliequivalents (meq) of peroxide was determined. The AOM procedure was put in "surplus status" (still official, but not preferred) in 1993 and focus was then placed on the Oil Stability Index (OSI; AOCS Cd 12b-92). In the OSI procedure, heated air is bubbled through heated liquid triglycerides, with the exiting air scrubbed in a bath of deionized water whose conductivity is continuously monitored. Initiation of polar compounds generation is noticed immediately. Whereas the AOM method determined the period for the triglyceride to reach a specific level of oxidation, OSI determines the "induction period," or time required to exhaust antioxidant properties of the oil. As such, OSI values always are lower (less time) than AOM values.

20–3.2.6 Phosphatides, Phospholipids

Phosphatides are a group of natural emulsifiers in soybean oil, sometimes called "lecithins." The typical form is a diacyl glyceride, with fatty acids in positions 1 and 2, which impart a local hydrophobic character, and a polar phosphorous-containing moiety with one of five side chains (choline, ethanolamine, inositol, serine, or an hydrogen proton) in position 3(R').

The proper name of a specific phosphatide molecule includes the two fatty acids and the R' component (for example sn-1-stearoyl-2-linoleoyl-glycero-3-phosphocholine); but specific fatty acids often are ignored, and the fraction would be called "phophatidylcholine." Soybean crude lecithin contains approximately 34.5% phosphatidylcholine (the "purified lecithin" of commerce), 23.2% phosphatidylethanolamine (the "purified cephalin" of commerce), 15.7% phosphatidylinositol, 7.0% phosphatidic acid, and 6.1% phosphatidylserine (Cherry and Kramer, 1989).

Soybean and canola oils are the richest sources of phosphatides among common solvent-extracted vegetable oils; both contain approximately 1.0 to 3.0% phosphatides, and 311 to 940 parts per million phosphorous, calculated from the relationship:

$$\text{(Phosphatide } [\%] \times 10^4)/31.7 = \text{Phosphorous, mg kg}^{-1} \text{ (ppm) (Farr, 2000)}$$

Phosphatides can cause severe problems in edible oil processing, and must be removed. Normally, this is easily done by mixing a small amount of water into the oil to hydrate the phosphatides, and removing the formed gums by centrifugation. However, Phospholipases A1 and A2 (also sometimes known as Phospholipase B) cleave fatty acid ester bonds of phospholipids to produce fatty acids. Phospholipase C hydrolyzes the phosphorous linkage to produce a diglyceride and a base. The main interest is in Phospholipase D, which hydrolyzes the R' chain, and produces phosphatidic acid. The proton (H^+) readily dissociates, leaving two nearly adjacent negative charges that can complex with positive divalent cations (Ca^{2+}, Mg^{2+}, Fe^{2+}, Cu^{2+}, and others) to render the phosphatide nonhydratable by water.

Phospholipase D is readily active in moist, split, cracked or flaked soybean, especially under warm conditions, but its action can be arrested by heating soybean to more than 104°C (220°F) in expanders and extruders. List et al. (1990) have shown that microwave heating is effective for inactivating phospholipase D, but the process has not been applied in commercial operations. They also confirmed that optimum moisture content for phospholipase D activity is 14%, and the enzyme is inactivated at about 110°C.

Some nonhydratable phosphatides always occur in vegetable oils. They can be converted into hydratable forms by one of many acid degumming methods developed during the last two decades. Phosphorous or divalent metal cations, in a water-degummed oil sample, indicate the presence of nonhydratable phosphatides. Various analytical methods have been developed to measure either, with the inductive-coupled plasma spectrograph (ICP) method for metals determination currently the most popular. The instrument can be operated by nonchemists; and midsize refineries (500 t d^{-1}) typically recover investment costs (~$70 000) in under a year through increased yields of saleable oil.

Treatment of nonhydratable phosphatides typically consists of adding sufficient phosphoric or citric acid to the degumming process (if lecithin is not saved), or to the alkali neutralization treatment to chelate the metal cations after water degumming, thus returning the phosphatide to its hydratable form (Segers and van de Sande, 1990; Dijkstra, 1993; Farr, 2000). A process (EnzyMax, registered trademark of Lurgi AG, Frankfurt, Germany has been developed which cleaves the nonhydratable phosphatides by phospholipase A-2 to produce a lysophosphatide that is insoluble in oil and is removed by centrifuging (Dahlke and Eichelsbacher, 1998).

20–3.3 Siting Extraction Plants

A variety of soybean-processing facilities exist, from three 14 000 t (15 428 short ton) d^{-1} solvent extraction plants in Argentina and Brazil, primarily producing meal and oil for export, to 45 t (50 short ton) d^{-1} extruder/screw press plants in the USA which produce full-fat soybean meal for animal feed, and may partially recover oil for sale or pooling in cooperative refineries or biodiesel plants. The major considerations are (i) the market to be served, (ii) sources and availability of soybean, (iii) transportation costs, and (iv) the critical production capacity necessary to sustain a profitable operation. At the U.S. average soybean yield of 2.59 t ha^{-1} (38.5 bushels acre^{-1}), each 1000 t of extraction plant capacity, operating 350 d yr^{-1}, requires the annual soybean output of 134 953 ha or 333 333 acres. New soybean extraction plants currently built in the USA are at least in the 2722 t (3000 short ton) d^{-1} range and require ~ 367 350 ha or 907 500 acres of cropland to supply them. In recent years, much effort has gone into selecting the most profitable domestic sites for obtaining soybean and shipping products, especially feed meals, to customers. Seaports are the obvious sites for major processors of meals and oils for export, or who import whole soybean for local processing and distribution (American Soybean Association-United Soybean Board, 2001).

The following round numbers are sometimes used for crude estimates of yields (Soya and Oilseed Bluebook, 2002). Processing yields: 79.2% soybean meal, 17.8% soybean oil, 3.0% waste; weight yields: 1 t of soybean yields 178.3 kg crude oil, 791.6 kg soybean meal, 649.9 kg soybean flour; 333.3 kg soybean protein concentrate, and 196.7 kg soybean protein isolate (1 short ton of soybean yields 356.7

lb crude oil, 1583.3 lb soybean meal, 1300.0 lb soybean flour, 666.7 lb soybean protein concentrate; 393.3 lb soybean protein isolate).

20–3.4 Soybean Preparation

Soybean entering a modern extraction plant no longer receives the same process regardless of condition. Modern processing recognizes that differences occur in soybean, which require segregated storage, handling, and processing to maximize yields of saleable products. Sensing instruments, data acquisition systems, and computers help keep track of changing conditions of the various lots that might be stored at the extraction plant or other company sites at one time. A typical flow sheet of soybean solvent extraction operations is shown in Fig. 20–1.

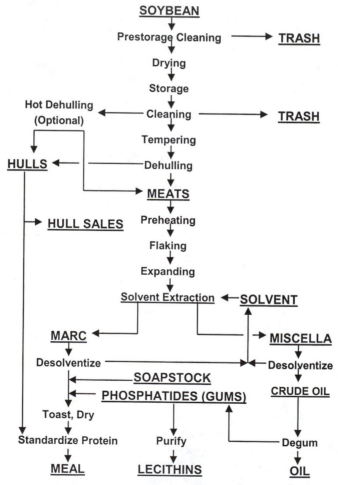

Fig. 20–1. Flow sheet of major soybean oil solvent extraction processes.

20–3.4.1 Seed Precleaning, Drying, and Storage

Soybean is cleaned several times, including on the farm, prestorage, and before preparation for extraction. Magnets are located throughout processing lines to remove ferrous metal pieces from foreign sources, or shed by machinery, and reduce damage to processing equipment. Seed typically is passed through scalping equipment (rotating drums covered with screens of increasing sizes to remove mud, dirt, sticks, plastic, and dust before drying or storing). Destoners may be required for soybean grown in some areas. Removal of stems and leaves is especially important because they can cause fires in dryers and are the first to absorb moisture from the atmosphere and transmit it to soybean during storage (Barger, 1981; Carr, 1993).

The objective in storage is to bring the seed to low physiological activity and keep it in that state as long as possible. Moisture and heat accelerate presprouting changes, including increases in FFA content. Of the two factors, moisture is more important, since physiological activity itself also increases temperature. For minor reductions in moisture content, air drying can be used, provided relative humidity of inlet air is lower than that of air exiting the seed. Fuel drying is expensive, and reasonably efficient commercial dryers require 645 316 to 688 337 J kg^{-1} (1500–1600 btu lb^{-1}) moisture removed (Woerfel, 1995). Seed temperature should not be allowed to exceed 76°C (170°F) during drying.

Silos containing soybean often are equipped with thermocouples to continuously monitor stability of seed during storage by watching for temperature rise. Aeration of stored soybean is necessary, the amount required is positively related to moisture content and temperature of the seed (Barger, 1981).

A headspace relative humidity not exceeding 65 to 70% (~12.5% seed moisture content) is recommended for long-term storage of soybean, although the industry trades at 13.0% moisture content (Gustafson, 1976; Spencer, 1976; Sauer et al., 1992). In humid tropical countries, imported soybean typically is sent directly to extraction as received because of high humidity at seaports. Soybean may be grown during the rainy season at higher elevations, and harvested at as low as 8.5% moisture content when maturing in the early dry season. Typically, care is taken to rush the crop through extraction before soybean has absorbed excessive moisture in the succeeding rainy season.

Soybean damage during handling and shipping is mainly the result of freefall, with ~2 to 3% increase in splits estimated for each transfer (USDA, 1973, 1978). Splits are more susceptible to moisture absorption, mold growth, and enzymatic damage including FFA rise, than whole seed (Hesseltine et al., 1978). Various devices for gently lowering whole soybean into ship holds are ignored currently, but may be brought into use as buyers demand higher grade soybean. Angles of repose and bulk densities are approximately: 27°, 720 to 800 kg m^{-3} (45–60 lb ft^{-3}) for whole soybean; 45°, 93 to 112 kg m^{-3} (6 to 7 lb ft^{-3}) for unground hulls; 35°, 560 to 610 kg/m^3 (35 to 38 lb/ft^3) for solvent-extracted 44% (protein) meal; 32 to 37°, 657 to 673 kg m^{-3} (41–42 lb ft^{-3}) for solvent-extracted 50% meal; and 35°, 575 to 640 kg m^{-3} (36–40 lb ft^{-3}) for screw pressed oil meal (Gustafson, 1976; Barger, 1981; Appel, 1994). Soybean meals are traded at 12% moisture maximum. They

Table 20–1. Chemical compositions of traditional soybean and their components, shown on a dry
weight basis (Perkins, 1995; with permission).

Components	Yield	Protein	Fat	Ash	Carbohydrates
			%		
Whole soybean	100.0	40.3	21.0	4.9	33.9
Cotyledon	90.3	42.8	22.8	5.0	29.4
Hull	7.3	8.8	1.0	4.3	85.9
Hypocotyl	2.4	40.8	11.4	4.4	43.4

will bridge in tanks at more than 13% moisture or if filled too hot. When piled, they
can form steep walls which become cave-in hazards to front-end loader operators.

20–3.4.2 Post-storage Cleaning, Tempering, Dehulling, Cracking, and Flaking

Damage to oil quality before receipt of soybean cannot be reversed. The oil
must be extracted as best as possible, but can be additionally damaged while
preparing soybean for extraction. The general industry practice is to accept process-
able crops as they come, and minimize rejections to avoid alienation of local producers and elevator operators who will be needed in future years. The additional
processing costs of converting lower quality soybean seed into saleable ingredients
are reflected in prices paid for different grades. If soybean has been cleaned by the
processor before drying and storage, it typically is screened again to remove clumps
that may have developed in the interim. Newly purchased soybean usually is
cleaned using at least a two-deck screener to size trash, with a multi-aspirator to
remove light materials and dust (Moore, 1983). A typical composition of traditional
soybean and its components at this point is shown in Table 20–1 (Perkins, 1995).

Cracking, flaking, and extraction of whole soybean, produces a meal containing ~44% protein (on an as is basis). The digestion systems of nonruminants,
like pigs and especially poultry, have limited volumes and ability to handle fiber.
"High protein" content meals (47.5–49.0% minimum), with fiber contents of 3.3
to 3.5% maximum instead of the traditional 7.0% maximum, have been developed
to provide increased nutrient density for these species. Processors have the choice
of "front-end dehulling" before extraction, or "tail-end dehulling" after extraction
(Moore, 1983; Woerfel, 1995). Typically, front-end dehulling is used because removal of hulls enables sending more soybean "meats" through the extractor. The
removed hulls are partially returned after extraction to standardize meals to protein
guarantees.

Conventional front-end dehulling consists of drying 13% moisture soybean
from storage to 10% and holding ("tempering") for 2 to 3 d to equilibrate the moisture and enhance cracking and dehulling. Cracking is conducted by non-inter-
meshing horizontally corrugated rolls, whose ribs have saw-tooth designs cut along
the length of their leading edges. Both rolls turn inward, "sharp to sharp," with the
roll with the larger teeth turning slower and partially holding back the soybean as
the faster, smaller-tooth, roll sweeps by and cuts off part of the seed. Two sets of
cracking rolls, the coarser roll set mounted above the finer set, are used in modern
cracking installations to produce four to eight pieces from each soybean (Moore,
1983; Woerfel, 1995).

The hulls (~8% of the soybean) then are removed by gyratory or aspiration separators. Next, cracked soybean (with or without hulls) is conditioned by steam in vertical stack cookers, or in rotary horizontal cookers equipped with steam pipes and water sprays, to 11% moisture content and ~71°C (161°F), and is flaked by smooth rolls to 0.02 to 0.5 mm (0.008–0.020 in) thickness. Approximately 1.5% moisture is lost during flaking, returning the cotyledons to ~10.0% moisture content (Woerfel, 1995).

20–3.4.3 Hot dehulling

In the last two decades, at least three suppliers have introduced hot dehulling systems for soybean (Fig. 20–2). Cleaned whole soybean seed, at normal storage moisture content, is slowly heated to ~60°C (140°F), over 20 to 30 min in the conditioner to allow moisture to migrate to the surface and make it sweaty. Then, the seed is subjected to hot blasts of 150°C (300°F) dry air in the fluid bed dryer over a short period of time (1–6 min), which causes the hulls to loosen and "pop" away from the meats before leaving the fluid bed at 75 to 85°C (170–190°F). Soybean then passes through a coarse cracking mill, a hulls separator, a finer cracking mill and a flaking mill to produce flakes approximately 0.30-mm (0.012 in) thick. The system also includes a sifter and separator to recover pieces of meats that may adhere to the hulls. Less than 0.9% hulls remain on the meats. Nitrogen Solubility Index (NSI) of soybean is reduced by 5 to 8% (U.V. Keller, Buhler, Inc., Minneapolis, MN, personal communication, 2002).

Systems are available for various-size plants, and use multiples of cracking rolls available in the 900 to 1000 t d^{-1} range and flaking rolls in the 500 t d^{-1} range. Advantages include single-pass conversion of seed into flakes ready for solvent extraction and overall reductions in steam and electricity use (U. Keller, Buhler, Inc., Minneapolis, MN, personal communciation, 2002). Another advantage, claimed significant by some, is the close-coupled design of the system. This reduces the time that seed is subject to enzymatic activity, and accessible by air (including oxygen).

20–3.4.4 Expanders

Expanders are single-screw extruders with a characteristic interrupted-flight crew that conveys the product past homogenizing shearbolts fixed through the barrel wall (Fig. 20–3). The machine initially was developed in 1955 by the Anderson International Company, Cleveland, OH, as a grain cooker (Williams and Baer, 1965). The design was applied to preparing soybean for solvent extraction in Brazil in the late 1960s and early 1970s, and gained worldwide interest in the latter 1980s and early 1990s (Williams, 1995a, 1995b; private correspondence with M.A. Williams, Anderson International Co., Cleveland, OH, 2001). The expander basically homogenizes soybean flakes, and reforms them into collets. In the process, oil-containing spherosomes in seed cells are ruptured, enabling the freed oil to coalesce, foam to the surface of exiting hot collets, and be reabsorbed on cooling. Homogenization changes oilseeds extraction from a diffusion process, where solvent must permeate into individual spherosomes in seed cells, solubilize the oil, and exit, to a leaching process where a solvent wash solubilizes and carries away freed oil. Benefits include:

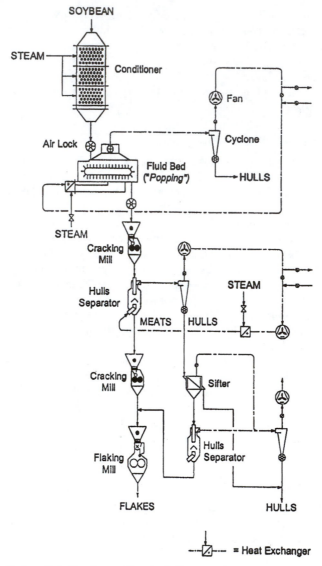

Fig. 20–2. Hot Soybean Dehulling System "Popping." (Courtesy of Buhler, Incorp., Minneapolis, MN.)

1. Reduction of miscella (oil and solvent mixture) holdup in extracted collets, resulting in sufficient savings in desolventizing steam to recover
   costs of installing the expander in less than a year,
2. Opportunity to increase the extractor throughput by 50 to 75% if the accompanying seed preparation and solvent recovery systems also are enlarged.
3. Gross reduction of nonhydratable phosphatides in the oil, provided phospholipase D has not been triggered by earlier seed rupture. Enzymes are

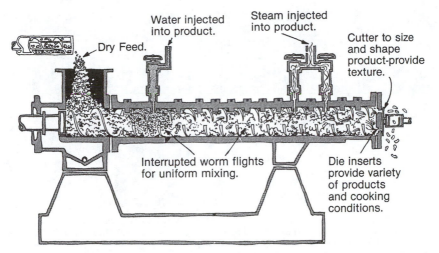

Fig. 20–3. Cut-away drawing of Anderson International Solvex Series Expander (registered trademark, Anderson International Corp., Cleveland, OH); interrupted flight extruder for preparing low-oil-content oilseeds for solvent extraction and for making full-fat soybean meal. (Courtesy of Anderson International Corp., Cleveland, OH).

destroyed almost instantly at ~110°C (230°F) in the expander by heat from injected steam and mechanical shearing of the seed. This has resulted in increased recoveries of saleable oil and improved stability after refining (Watkins et al., 1989).

Installation of expanders proceeded rapidly during the late 1980s and the 1990s, resulting in their broad use throughout the world on a variety of crops. The far majority of expanders, with capacities of up to 1500 t (1700 short ton) d^{-1}, are produced by screw press or extruder manufacturers. Unless specifically requested by the buyer, major oilseeds equipment manufacturers advise larger extractors in new installations, arguing that savings from using expanders are not as significant in well-run plants. However, expanders typically are added in later projects to increase capacities of existing extractors. Plants vary in passing some or all soybean flakes through expanders. Field observations have shown that, although the concept of using expanders is simple, many of the machines are not properly operated for maximum benefits.

Expanders work best with flakes, which can be thicker than sent directly to the solvent extractor. Steam is injected into the expander to raise the temperature, and increases moisture content of the flakes by 2 to 3%; but, much is flashed-off as collets puff when exiting the expander. The collets then are allowed to dry and cool on a conveyor belt under a hood, and must be at least 5°C (9°F) below the boiling point of the solvent before entering the extractor. Many improvements have been made in expander design. Hydraulically-operated cone choke heads (Fig. 20–4) have been developed to replace face-plates and dies, and enable easy adjustment for seeds which vary in extrusion properties (Watkins et al., 1989).

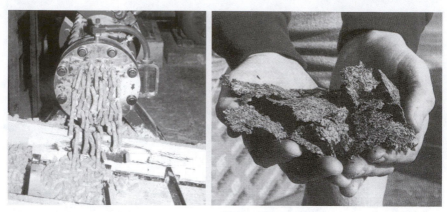

Fig. 20–4. Left: soybean collets exiting die plate head, Solvex Series Expander; strands break into random lengths. Right: puff sheets made by hydraulically positioned cone choke head on expander. (Courtesy of Anderson International Corp., Cleveland, OH.)

20–3.5 Solvent Extraction of Soybean Oil

20–3.5.1 Solvents

Currently, the major solvent used for oilseeds extraction throughout the world is "commercial hexane," with use of "commercial isohexane" slowly being implemented. Both are flammable solvents, and require "explosion-proof" facilities operated to meet NPFA 36 Standard for Solvent Extraction Plants (NFPA, 2001). "Commercial" solvents are not pure compounds, but rather mixtures of compounds distilled between two preset boiling temperatures at the oil refinery. Maximum permissible levels of recognized toxic compounds, for example benzene, have been set in the solvents.

Johnson and Lusas (1983), Lusas et al. (1990), and Johnson (1997) have reviewed the history of extraction solvents. Many of the early extraction solvents would not pass current FDA requirements for food additives. Research on dichloromethane (methylene chloride), the last of the permitted nonflammable halogenated hydrocarbons, showed it to be an effective solvent (Johnson et al., 1986), but it also became a suspected carcinogen. Supercritical extraction, using (nonflammable) liquefied carbon dioxide (CO_2) at 15 bar (5145 psig) or higher, has been proven technically effective for extracting various oils on pilot plant scale and have produced improved oils (Friedrich et al., 1982). However, techniques for continuous loading and unloading of flakes/collets in quantities required in modern oilseed extraction operations, have not been developed. Small bench-top critical CO_2 instruments are sold for analyzing oilseeds and their products for oil content. In 2002, China had eighteen 300 t d^{-1} propane extraction plants in operation.

Ethanol extraction has been reviewed by Hron (1997). In the 1990s, new information was reported for isopropanol (isopropyl alcohol, IPA) by Lusas et al. (1997), Lusas and Hernandez (1997), and Lusas and Gregory (1998). At a concentration of 92% or higher, hot IPA and soybean oil are completely miscible at all ratios and essentially act like hexane-oil mixtures. But, to obtain >92% IPA, distillation-recovered 87.7% IPA was rectified using pervaporation membranes. The technical feasibility of replacing petroleum-based extraction solvents with IPA is promising, but doing so would require extensive retrofitting in existing solvent extraction plants—a change industry is likely to resist while other options remain available.

The focus on extraction solvents in the early 1990s resulted when n-hexane was implicated by EPA as a neurotoxin based on trials in which peripheral nerve damage occurred in rats exposed to high inhalation exposure for several months. The industry responded with rat data showing the problem does not occur with commercial hexane, containing only a portion of n-hexane (Wakelyn, 1997). At the same time, concerns ran high about effects of photochemically active volatile organic compounds (VOCs) and ozone production. The concerns have abated somewhat as the solvent extraction industry reduced hexane losses to <1 L t^{-1} (0.25 gallon short ton^{-1}) of seed processed.

20–3.5.2 Extractors and Extraction

The three major types of extractors currently sold for extracting soybean flakes and collets are shown in Fig. 20–6 to 20–8. Similarities include (i) percolation-type design, in which the solvent is sprayed on top and drains through the bed, (ii) counter-current extraction, in which the oil-ladened flakes or collets are first washed with the miscella richest in oil, and then with progressively more dilute miscellas and fresh solvent, (iii) the flakes/collets are dragged across nonclogging, parallel, wedge-shaped bar, support screens for extraction and drainage, (iv) extractor interiors of all-stainless steel construction, and (v) extractors are operated under slight vacuum to prevent escape of solvent vapors into the atmosphere.

The extractor shown in Fig. 20–5 also is known as "rectangular loop type." Flakes/collets are loaded at the left of the top level and are dragged by paddles suspended from chains in clockwise fashion along a linear drainage screen, while sprayed by increasingly dilute miscellas. The bed turns over as it passes to the lower level, and extraction ends with a final rinse of fresh solvent. After draining, the marc (solvent ladened, extracted flakes/collets) is sent to the desolventizer. Solvent travel is in the counter-current direction. This type of extractor is available in capacities of up to 8000 t d^{-1}. It is considered a "shallow bed" extractor, with product layers up to 1.5 m high on the large machines. It sometimes is used for reextracting "white flakes" with aqueous ethanol in producing soybean protein concentrates.

The machine in Fig. 20–6 is known as a "perforated-belt or a diffusion-type" extractor. The bed is moved as one mass on a belt, consisting of folding sections of linear drainage screens. It is considered to be an intermediate-bed extractor.

The machine in Fig. 20–7 is known as a "basket or circular-type" extractor. The product is held in orange segment-type cells ("baskets"), which are rotated across a supported extraction and drainage screen, and pass under a flake/collet loading device and a series of counter-current miscella sprays, before draining and dropping into a receiving hopper. The rotating basket is the only moving part. Extractors of this type are operating in the 10 000 t d^{-1} range. With product layers up to 3.7 m (12 ft) high, they are known as "deep-bed" extractors.

When flakes are extracted directly, their thickness varies with the type of extractor used; 0.30 to 0.38 mm (0.012–0.015 in) thickness, weighing ~460 kg m,$^{-3}$

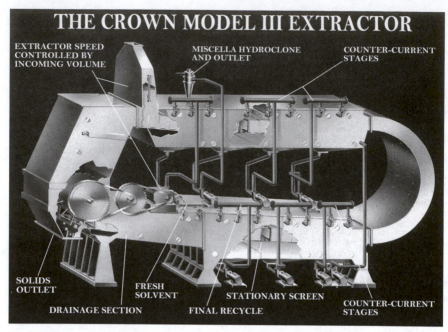

Fig. 20–5. Rectangular loop-type continuous counter-current solvent extractor. (Courtesy of Crown Iron Works Co., Minneapolis, MN.)

(29 lb ft $^{-3}$) is recommended for shallow-bed extractors (under 1.2 m, 48 in) and 0.33 to 0.43 mm (0.013–0.017 in) for deep-bed extractors (G.E. Anderson, Crown Iron Works, Minneapolis, MN, personal communication, 2001). Residence of flakes or collets in continuous extractors typically is 20 to 40 min.

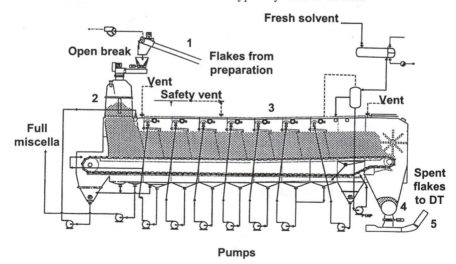

Fig. 20–6. Drawing of DeSmet LM (registered trademark, Extraction De Smet N.V./S.A., Brussels, Belgium) perforated belt diffusion-type extractor. (Courtesy of Extraction De Smet N.V./S.A., Brussels, Belgium.)

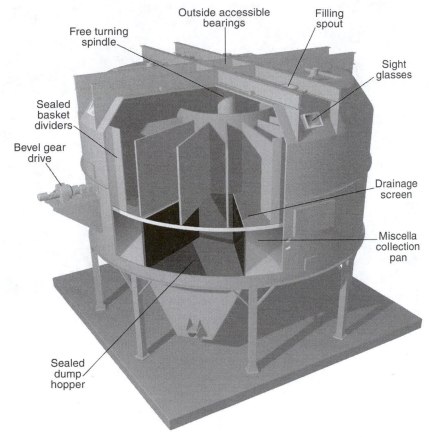

Fig. 20–7. Reflex (registered trademark, Extraction De Smet N.V./S.A., Brussels, Belgium) "basket" or "circular-type" extractor; basket revolves in shell. (Courtesy of Extraction De Smet N.V./S.A., Edegem, Belgium.)

20–3.5.3 Meal Desolventizing, Toasting, and Cooling

After leaving the extractor, soybean marc is desolventized under vacuum with supplemental heat provided by steam injection. In doing so, steam condenses and also provides moisture for "toasting" (cooking) the meal to reduce activities of trypsin inhibitors and other anti-growth factors. Units which desolventize and toast are referred to as "DTs." The moisture content of the meal must be reduced to <12% to prevent spoilage. This typically is done by hot-air drying, followed by cooling the meal in dryer-coolers, known as "DCs." Equipment suppliers provide single units (DTDCs) as one installation, or can separate them into separate DTs and DCs.

A modern DTDC, developed in the last two decades, is shown in Fig. 20–8. The marc is heated in ring cooker-like pans, equipped with sweep arms that mix and move it down through the stacked trays. Steam is sparged into the meal at the bottom of the DT section and raises through hollow bolts into the higher trays, car-

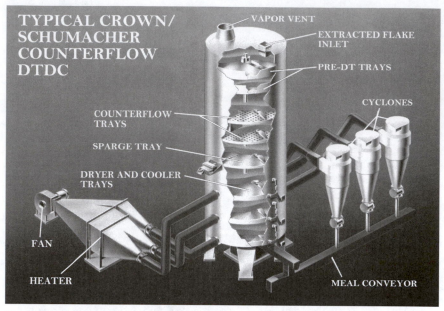

TYPICAL CROWN/
SCHUMACHER
COUNTERFLOW
DTDC

VAPOR VENT

EXTRACTED FLAKE
INLET

PRE-DT TRAYS

CYCLONES

COUNTERFLOW
TRAYS

SPARGE TRAY

DRYER AND COOLER
TRAYS

FAN

HEATER

MEAL CONVEYOR

Fig. 20–8. Schumacher-type desolventizer-toaster-dryer-cooler (DTDC). (Courtesy of Crown Iron Works Company, Minneapolis, MN.)

rying the hexane vapors with it. The moist, cooked, meal then passes into a separated DC section for drying and cooling.

In a typical commercial process, described by Witte (1995), soybean marc leaves the extractor at about 57°C (135°F), and for feed uses is "toasted" by steam at 16 to 24% moisture and 100 to 105°C (212–220°F) for 15 to 30 min. In the DC operation, the meal is dried to <12% and is cooled to <32°C (90°F) or within 6°C (10°F) of ambient temperature, whichever is higher. Additional details of desolventizing flakes for food use, or toasting for specific animal species, are described later.

20–3.5.4 Miscella Desolventizing and Solvent Recovery

Much effort has gone into development of heat recapture systems to conserve energy in oils extraction and processing in the last two decades. The major components of the miscella desolventizing-hexane recovery system consist of:

1. First-stage evaporator. In this unit, sometimes called an "economizer," the miscella (~30% oil content, at slightly less than the operating temperature of the extractor) is pumped upward in a vertical tube-in-shell heat exchanger equipped with a dome. As the oil miscella rises in the tubes, its solvent is vaporized by heat from descending DT vapors in the shell. Energy for evaporation by this recaptured heat is free. The miscella is concentrated to ~65 to 85% oil, settles to the bottom of the evaporator and is sent to the next evaporator. The solvent-water vapors are pulled from the dome to a condenser by vacuum.

2. Second-stage evaporator. The partially desolventized miscella again is pumped upward in a second tube in-shell heat exchanger, but this time the shell is heated by steam. Solvent-water vapors again are flashed off and collected from the dome, with oil content in the miscella raised to ~90 to 95%.

3. Oil stripper. Miscella from the second stage evaporator enters the top of a column that is sparged with steam from the bottom, and removes the rest of the solvent as the oil settles to the bottom. After exiting, the desolventized soybean oil is sent to short-term storage or to degumming.

4. Solvent vapors collection system. Solvent vapors from the first and second stage evaporators and the oil stripper, the extractor, and lines that scavenge vents on storage tanks and other sites where they might escape into the atmosphere, are collected and brought to cooling heat exchangers, where they, and the accompanying water, are condensed.

5. Solvent work tank. The condensed solvent-water mixtures from the economizer and vapor condensers are brought to a tank which allows them to separate by gravity, with hexane rising to the top. Before discarding, the water is sent to a reboiler to reclaim solvent that may have dissolved in it.

6. Mineral oil scrubbers. Some noncondensable gases (mainly air) are produced at each vapor condenser. They are sparged through mineral oil scrubbers to salvage solvent that may have dissolved in them. Periodically, the scrubber oils are heated to distill hexane vapors that may have been collected. These are condensed and returned to the solvent supply system. Modern hexane extraction systems operate at 70°C (158°F) in the DT dome (head space), with the desolventized meal exiting the DT at ~18 to 19% moisture content. The miscella/oil mixture leaves the first-stage evaporator at 85% oil content, and is heated by an in-line heat exchanger for second-stage evaporation 75.5°C (170°F). Modern solvent extraction and recovery uses about 188 kg of steam t^{-1} (375 lb short ton^{-1}) of soybean crushed (DeSmet, 2001).

20–4 SOYBEAN OIL PROCESSING

Technically, "refining" means alkali neutralization of oil. But, over time, all processing which occurs after extraction in production of oil ingredients has become known as refining, and the facility in which it occurs a "refinery." Production of consumer products, like margarine, often is done at a different location or firm. Soybean extraction plants may have simple partial degumming capability for reducing phosphorous content of the oil to <0.02% to prepare it for trading as Crude Degummed Soybean Oil (NOPA, 2000), or an extended refining and processing facility which makes packaged consumer goods, sometimes including soaps and detergents. A flow sheet of the more common operations in fats and oils processing is shown in Fig. 20–9.

20–4.1 Crude Oil Receiving and Handling

Maximizing yields of saleable soybean oil requires even more detailed attention to lot-to-lot differences than for soybean extraction. The first priority after receiving a shipment of oil is to characterize its quality level and determine how to convert it into the most profitable products. If purchased, typical composition checks to determine the price paid typically include:

- Flash point—AOCS Cc 9c-95 (>250°F, to ensure low hexane residues)
- Unsaponifiable Matter—AOCS Ca 6a-40 (<1.5%)
- Free Fatty Acids, as Oleic—AOCS Ca 5a-40 (<0.75%)
- Moisture and Volatile Matter—AOCS Ca 2d-25 and Insoluble Impurities—AOCS Ca 3a-46 (<0.3%)
- Phosphorous - AOCS Ca 12-55 (<0.02%)

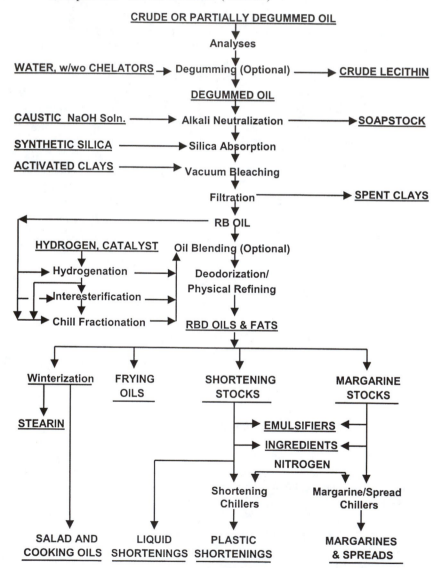

Fig. 20–9. Composite flow sheet of oils and fats refining and processing.

Additionally, the analyst may estimate how much saleable oil can be produced from the lot by Neutral Oil and Loss analysis (AOCS Ca 9f-57). In this procedure, a solvent-diluted sample of the oil is poured over a column packed with activated aluminum oxide. After removal of the solvent, the weight percentage of the oil that passed through the column is considered "neutral" or saleable oil, and the percentage of retained polar compounds is "loss." The analyst may also run a bleach test (AOCS Cc 83-63) if concerns exist about reducing color of the oil to the desired range.

In preparation for processing, a "day tank," large enough to supply the refinery with a uniform oil feedstock for a suitable period, is filled and mixed until uniform. Samples are taken for free fatty acids and phosphorous content of water-degummed oil (or metals analysis if an ICP unit is available), and the several treatments ("treats") are calculated for each day tank before processing begins.

20–4.2 Degumming

The objectives of degumming are to remove the phosphatides (i) for purification and sale as "lecithins," (ii) to reduce viscosity of soapstock and loss of occluded neutral oil in alkali refining, (iii) to prevent their later inactivation of hydrogenation catalysts, and (iv) to prevent darkening of oil during deodorization and in later heated uses like deep fat frying. A surplus of crude soybean lecithin exists in the USA and the major portion of separated gums is sprayed over the meal before drying in the DC of the extraction plant for sale with the soybean meal. The phosphatides also would be removed with the soapstock during alkali neutralization of the oil, but with an increased loss of neutral oil. Thus, most large U.S. soybean oil refiners degum before alkali neutralization. Crude soybean oil contains ~600 to 700 mg kg^{-1} (ppm) phosphorous and traditional screw-pressed oil ~700 to 900 mg kg^{-1} (ppm). Fresh extruder-screw press process oil may contain as much as 1200 mg kg^{-1} (ppm) phosphorous, but phospholipase D is inactivated early in seed extrusion, and the nonhydratable phosphatides content is very low. Refineries that save phosphatides for making lecithins water degum, because addition of acids affect properties of the lecithin (Farr, 2000).

Earlier practices of adding an amount of water equal to the weight of the phosphatides have been made more precise, with

$$\text{Added Water} = (\text{ppm P} \times 3.17 \times 10^{-4})\,0.7$$

now recommended. The amount of phosphoric acid used is

$$H_3PO_4 = [(Ca + Mg)/2] \times 10$$

with all components expressed in mg kg^{-1} or ppm. The use of ICP analysis (see Section 20–3.2.6) enables rapid analysis of Ca^{2+} and Mg^{2+}. The phosphoric acid is added to the crude oil stream through a high-shear mixer as oil and water are pumped to the hydration tank (Farr, 2000).

Details of the process, using an Alfa Laval PX-90 centrifuge rated at 33 000 kg h^{-1} for degumming, are shown in Fig. 20–10. Self-cleaning centrifuges were in-

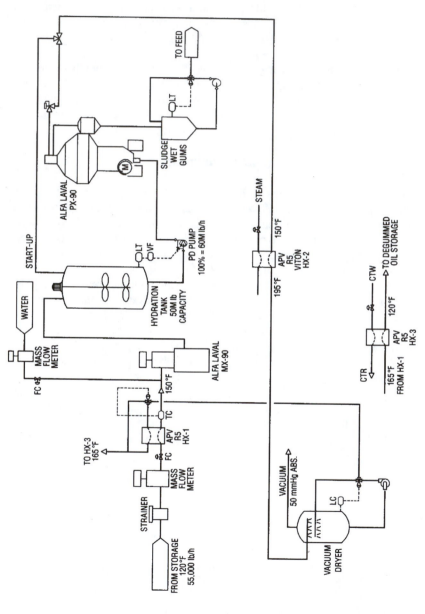

Fig. 20–10. Modern soybean oil phosphatides degumming line (Farr, 2000; with permission).

troduced in the late 1970s, and devices for changing the ratios of the separated streams while the machines are running in the late 1980s and early 1990s. The oil is heated to 65°C (150°F) before acid and water are added. After degumming, the crude oil is vacuum dried to <0.3% moisture and volatiles content, and cooled to 50°C (120°F) for storage or shipment, but this step can be omitted if the oil is refined next (Farr, 2000). Additional information on processing and modifying lecithins is provided by List (1989), Szuhaj (1989), Erickson (1995), and Zieglitz (1995).

20–4.3 Alkali Neutralization

A major breakthrough of the 1990 decade in oils refining is elimination of water washing the neutralized oil, with processes still being optimized. Phosphatides are removed by degumming as already described, and FFA in the crude oil are neutralized with caustic (sodium hydroxide) solution. But, the refinery then has the choice of water-washing the oil after removal of soapstock, or adsorbing the residual soaps onto silica hydrogel before bleaching by the Modified Caustic Refining or Silica Refining Process (registered trademark, W.R. Grace & Co., Baltimore, MD) without creating wash water.

Two major processes exist for alkali neutralization of FFA in row crop oilseed oils—the "long mix" and the "short mix" process. The short mix process evolved in Europe, runs at higher temperature, and reportedly is effective with various oils. The long mix process was developed in the USA, and often is recommended for refining soybean oil. It recognizes that chemical reactions occur more rapidly and are harder to control at higher temperatures. Earlier long mix processes started with crude soybean oil at ambient temperature, and used a low-concentration caustic solution and a mixer retention time of 15 min after which the oil-caustic mixture is heated to 70°C (160°F) to reduce its viscosity before centrifuging. In the short-mix process, crude oil is heated to 90°C (194°F), mixed with high-concentration caustic for 1 min and centrifuged (Erickson, 1995). The throughput per hour of both systems is the same since additional volume for the holding time is built into the long mix line. A flow chart of an updated long-mix neutralization process is shown in Fig. 20–11. The introduction of greatly improved caustic/oil mixers in the mid-1990s now allows a retention mixer time of 6 min for soybean oil. The amount of caustic to be added for neutralization is calculated as:

$$\% \text{ Treat} = ([\{\text{Factor} \times \%\text{FFA}\} + \{\% \text{ Excess}\}]/\% \text{ NaOH}) \times 100,$$

where: Factor = 0.142; % NaOH is determined from the 20° Be of the caustic solution; % Excess is selected from the following ranges based on experience (i) 0.01 to 0.05% for degummed soybean oil, and (ii) 0.15 to 0.25% for nondegummed soybean oil. Crude soybean oil from storage (Fig. 20–11) is heated to 38°C (100°F), strained, mixed with the "treat" in the rapid mixer, held in the retention mixer for 6 min, heated to 60°C (140°F) to reduce viscosity, and passed through the primary (first) centrifuge (Farr, 2000). The soapstock is returned to the extraction plant to be spread, with unsold or acid-degummed phosphatides, over the meal before drying in the DC. The refinery next has the option of traditional water-washing and

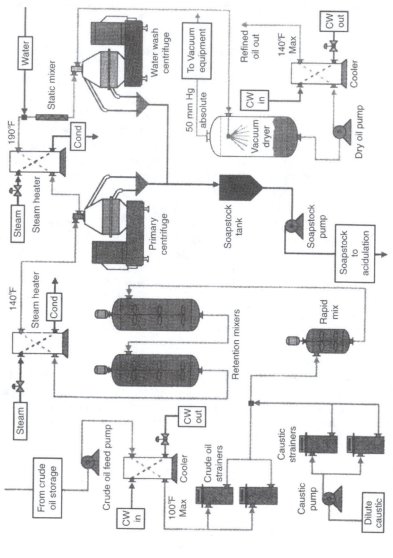

Fig. 20–11. Flow sheet, continuous refining of soybean oil (long mix process) with water wash option (Farr, 2000; with permission).

vacuum drying the oil or passing it on to silica gel adsorption. If a water wash is used, the oil is heated to 88°C (190°F) with 12 to 15% soft water, held in an agitated tank for ~0.5 h, and centrifuged to produce an oil with < 20 mg kg^{-1} (ppm) soap. Soap has been reduced to "0" by addition of a small amount of phosphoric acid in the water-wash retention tank. The oil is next sprayed into a drying tank at 50 mm Hg (Farr, 2000).

20–4.3.1 Sodium Silicate Refining

Sodium silicate has shown promise in simultaneous alkali neutralization and degumming of soybean oil (Hernandez, 2001). Sodium silicate, preferably with a weight ratio of silicon dioxide to sodium oxide between 0.91 to 3.3, is used as a 30% aqueous solution at a preferred temperature between 77 to 93°C (170–200°F) (Hernandez and Rathbone, 2002). The neutralized fatty acids, hydrated gums, and other undesirable components, form an amorphous silica gel floc, which is easily removed with conventional filtration equipment. In addition to saving the costs of purchasing and maintaining centrifuges, which makes establishment of small and medium-size refineries more affordable, less destruction of antioxidants occurs, and longer oil shelf lives and improved frying properties result. The finished soybean oil contains approximately 0.02% FFA and <60 mg kg^{-1} (ppm) soaps. The invention is a neutralization-separation module, which is compatible with most modern upstream and downstream practices, including removal of residual soap.

Various researchers have reported on attempts to remove fatty acids from oils by ultrafiltration membranes. Success generally has been limited by a shortage of membranes that can withstand extraction solvents. However, progress has been made on degumming with membranes (Lin et al., 1998).

20–4.4 Silica Gel Adsorption

Silica hydrogels are very effective in removing phosphatides, residual soaps, and metal ions, thus reducing the amount of bleaching clay required by 50 to 75% and leaving its function primarily to removal of chlorophyll and secondary oxidation products. By eliminating soap wash water, the total water discharge of refineries is reduced by ~50% in volume and has much lower biological or chemical oxidative demands. Additionally, the costs and expenses of a second centrifuge are avoided (Farr, 2000).

The W.R. Grace Company began demonstrating the process in 1986 using Trisyl (registered trademark, W.R. Grace, Baltimore, MD). The method of using silica hydrogel has changed during perfection of the process (Welsh et al., 1990; Bogdanor and Price, 1994; Parker, 1994). Originally, it was mixed with the bleaching clay. Later, time was allowed for its action before exposing the oil to clay. Silica hydrogel can accept oil from the centrifuge at 0.2 to 0.4% water content, and the current recommended process consists of blending the silica hydrogel with oil directly from the soapstock removal centrifuge, with minimal, if any, drying of the oil. Silica hydrogel then is removed by filtration before mixing the oil with the bleaching earth (Farr, 2000). In a variation, the bleaching earth is precoated on the filter to make a "packed bed," and the reacted oil-silica gel mixture is sent directly to it.

20–4.5 Bleaching

"Bleaching" is another misnomer of this industry. It originally was meant to be a process for reducing color in oils. Although this occurs to a limited extent, most reduction of red and yellow colors occurs during the high heat of the deodorization process. The practical functions of bleaching are to remove chlorophyll and oxidation products, and to prepare the oil for hydrogenation by scavenging the remaining soaps, phosphatides, and minerals which might poison the catalyst. Theory and practical aspects of bleaching have been reviewed (Patterson, 1992; Taylor, 1993; Zschau, 2000).

The bleaching earths are made from naturally occurring minerals, including palygorskite—also known as attapulgite, sepiolite, bentonite, montmorillonite—and other minerals belonging to the aluminum silicate family. They may be used as such, but typically are activated by treatment with hydrochloric or sulfuric acids which (i) increases adsorption by increasing the surface several fold, (ii) provides acid centers which have catalytic properties, and (iii) imparts ion-exchange properties to the clay. These properties are important in assisting breakdown of oxidation products, rendering complex organic structures adsorbable, and in adsorbing various undesirable impurities in the oil including pesticides, polycyclic aromatic hydrocarbons, *trans* and conjugated fatty acids, dimers, and polymers (Zschau, 2000).

Bleaching earths also provide an active surface area and are catalysts for breaking down peroxides. Decomposition is an exothermic reaction, with the heat providing the sometimes reported "press effect" of bleaching carotenoids in the filter press. The cation exchange property of the activated earth is credited with removing magnesium from the center of the chlorophyll complex, rendering the pheophytin adsorbable to the clay. Cation exchange also is used for removal of heavy trace metal prooxidants like iron and copper, and for removing trace nickel in post-bleaching of hydrogenated oil (Zschau, 2000). However, not all intermediate oxidation compounds are exhausted on the surface of the bleaching earth. Although peroxides content is reduced, p-anisidine value (AV), determined by AOCS Cs 18-90, often increases. The AV method is believed to estimate the amount of aldehydes (2-alkenals and 2,4-dienals) in animal and vegetable oils with potential for later breakdown, and some chemists favor reporting AV in the final analysis of refined oils. The bleaching clay load (typically 0.1–2.0%) and operating temperature depend on the type and quality level of oil processed (Zschau, 2000). Modern bleaching processes are conducted under vacuum (50 mm Hg) to minimize later oil oxidation, and subsequent nitrogen blanketing is recommended. Close coupling of the refining and bleaching operations is highly recommended, especially when using the Modified Caustic Refining or Silica Refining Process (Farr, 2000). Because of the high degree of oil unsaturation and peroxides content of spent bleaching earth, it is very susceptible to spontaneous combustion unless quenched with water. Obtaining permission for disposal in land fills is becoming increasingly difficult. Research has been conducted on spreading spent bleaching earths over soybean meal for feeding animals. Caution should be taken in implementation because, by absorbing pesticides and mycotoxins, bleaching earth is one of the two safety valves in processing oils; the other is collection of pesticides in the deodorizer distillate.

20–4.6 Hydrogenation

Hydrogenation is the catalytically assisted addition of hydrogen to carbon-carbon double bonds. Its main uses in soybean oil are (i) to produce solids for making fat products, including shortenings, margarines and spreads, and various confectionery and specialty products, and (ii) to increase fryer life of oils and the shelf lives of table oils and bakery products. Hastert (1990), Ariaansz (1993), Kokken (1993), Patterson (1994), Erickson and Erickson (1995), and Hastert (2000) have reviewed the hydrogenation process.

For the food technologist, it would be desirable to minimize the C18:3 content of soybean oil to improve shelf-life. Selective breeding and biotechnology have partially accomplished this, but the crops still are not widely grown. Linoleic (C18:2 n-6) and linolenic (C:18:3 n-3) fatty acids are dietary essential, and should not be completely eliminated. For many applications, the susceptibility of soybean oil to oxidation is stabilized by "brush" (light) hydrogenation selective for the Δ^{15} bond of linolenic acid, but the product must remain liquid at ambient temperatures.

Many applications require fats that soften over a range of temperatures. This is accomplished by preparing four to six "base stocks" of oils hydrogenated to different melting points and blending them with soybean oil and "hard stocks" (nearly completely hydrogenated soybean or cottonseed oils) to achieve the desired melting points. Details of preparing "temperature profiled fats" are discussed in Section 20–6.2.

Hydrogenation is conducted in hardening plants (Fig. 20–12), which consist of a converter (a pressurized reaction tank, equipped with a high speed mixer and baffles, a means to add and remove the oil, a gas distributor, a means to add the catalyst, and heating and cooling coils). The hardening plant additionally has means for premeasuring and heating the oil, drop tank, heat exchangers, and a catalyst fil-

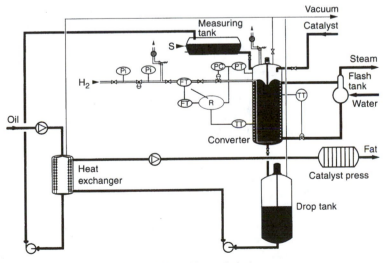

Fig. 20–12. Hydrogenation line (Hastert, 2000; with permission).

ter. Progress in hydrogenation is followed using a Refractive Index curve calibrated to the process or by Mettler Dropping Point (AOCS Cc 18-80).

Purities of the oil and hydrogen affect the life of the catalyst, which often consists of a thin film of nickel on an inert carrier (Hastert, 2000). "Selectivity" has several meanings in hydrogenation catalysts. "Saturate or preferential selectivity" is the tendency to focus on saturating a specific bond, for example the C18:3 bond of linolenic acid to form C18:2 linoleic acid. "*Trans*-isomer selectivity" is less directed and favors formation of *trans* bonds. Catalysts vary in selectivity, and their performance is greatly affected by catalyst dosage, effectiveness of mixing, temperature, and hydrogen pressure in the reactor. Ariaansz and Okonek (1998) have prepared a review on process factors that contribute to *trans* isomer production.

In "brush" hydrogenation of liquid soybean oils (i) a selective catalyst is used, (ii) IV (Iodine Value) is reduced by 15 to 25 to ~115 units, (iii) about 15% *trans* isomers are formed, and (iv) the C18:3 content is reduced to 3% maximum, and C18:0 content increases by ~1%. In margarine bases selective catalysts are used, IV is reduced to ~70, and about 50% *trans* isomers are formed. In shortening bases: selective or nonselective catalysts are used, IV is reduced to ~75; and about 35% *trans* isomers are formed. In producing stearin flakes (hard stock), a high activity catalyst is used and IV is reduced to ~5 to 10; no *trans* isomers remain; the hard stock is kept melted, flaked on ammonia-chilled rolls, or beadlets can be made using a shot tower with chilled air. Maintaining high levels of polyunsaturates is desired in producing coating fats: IV is reduced to ~70; about 65% *trans* isomers are formed with production of saturates; C18:0 production is minimized with only a 2 to 4% increase; and a sulfided nickel catalyst is used (Okonek, 2001).

Factors affecting isomer formation during hydrogenation include: temperature, hydrogen pressure, agitation, and catalyst type and dosage. Low *trans* formation occurs under conditions that favor rapid and complete saturation, including lower temperatures, high hydrogen pressure, fast agitation, high catalyst dosage, and use of nickel catalysts. High *trans* formation is encouraged by high temperature, low hydrogen pressure, slow agitation, low catalyst dosage, and sulfided nickel catalysts (Okonek, 2001).

20–4.7 Interesterification

The major means of obtaining fat solids for food products are by (i) incorporating natural fats with high melting points like tallow or fractionated by-products like palm oil stearins, (ii) hydrogenation, with intentional control of *trans* fats production for different functional properties, and (iii) rearranging the fatty acids on triglycerides within a species or blend.

"Interesterification" (INES) is the exchange of acyl radicals between an ester and an acid, an ester and an alcohol, or an ester and an ester (Nawar, 1996). The same principle can be used to position fatty acids on molecules with hydroxyl sites to produce monoglycerides (emulsifiers), fatty acid methyl esters (FAME) for analytical purposes and fuels use like methyl soyate ("biodiesel"), special dietary fats like medium chain triglycerides (MCT), and noncaloric sugar-ester fat substitutes like olestra (Olean, registered trademark, Procter and Gamble, Co., Cincinatti, OH). Types of interesterification include:

Table 20–2. Triglyceride classes of soybean oil (List et al., 1977; with permission).

Triglyceride class†	Random	Found‡
SSS	0.4	0.07
SUS	2.1	5.2
USS	4.4	0.4
USU	11.2	0.7
UUS	22.4	35.0
UUU	59.7	58.4

† S = Saturated, either palmitic or stearic acids; U = Unsaturated, either oleic, linoleic or linolenic acids.
‡ Determined from lipases hydrolysis data.

1. Acidolysis—position exchange between a triglyceride and a free fatty acid.
2. Alcoholysis—creation of an ester between a fatty acid and an alcohol.
3. Glycerolysis—moving a fatty acid from a triglyceride to glycerin to produce mono- or diglycerides.
4. Sucrosolysis—positioning fatty acids on sugar, as in production of olestra (Olean) or noncaloric fats (Akoh, 1988).
5. Directed randomization—reactions conducted at low temperatures that precipitate low-melting triglycerides as formed; used in rearranging lard.
6. Enzyme-directed—use of site-specific enzymes to position fatty acids at 1,3 or 2 positions on triglycerides.
7. Intraesterification—fatty acid positions are exchanged within one fat species.
8. Randomization—chemical catalyzed rearrangement between all fatty acid positions in the mixture.
9. Special—as in making emulsifiers by esterifying acids to hydroxyl groups of polyols (sorbitol, propylene glycol) or compounds like lactic acid.

The six triglyceride classes in soybean oil are shown in Table 20–2 (List et al., 1977). Their distribution is significantly changed from the natural (found) state by chemical randomization, with an increase of saturated fatty acids in the *sn*-2 position to 15.8% of the total, and ~8.2°C (14.8°F) rise in the melting point (Bockish, 1998). The elements of interesterification to produce *trans*-free fats include:

1. Feedstock. The oil must be degummed, well-refined (<0.05% FFA), and free of moisture and peroxides. Palm oil stearin and completely saturated (5 IV) C18 hydrogenated fats generally are *trans* free and may be used as part of the feed. Reactions are run under vacuum, and oil is best stored under nitrogen in between processes.
2. Catalysts. Currently, sodium methylate and sodium ethylate are popular catalysts because of their efficiency, but others are available. A mechanism for their action has been described by Rozendaal (1993).
3. Equipment. A flow sheet for a semi-continuous interesterification line with post-bleaching is shown in Fig. 20–13. The oil is loaded into the reactor, shown with both an agitator and a pumped circulation-spray loop, and heated under vacuum (110–130°C; 230–266°F) to dry the water and reduce the peroxide content. The oil is cooled to 70 to 90°C (158–194°F) and a catalyst is added as dry powder or suspended in dry oil at 0.05 to

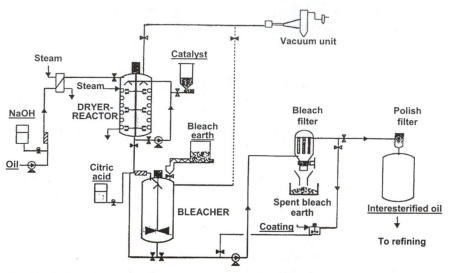

Fig. 20–13. Semi-continuous interesterification line with post-bleaching (Courtesy of Extraction DeSmet, N.V./S.A., Edegem, Belgium.)

0.15%. Randomization requires about 30 min, with an additional 15 to 30 min allowed for completion. Post-esterification refining is necessary to remove soaps, color bodies, and FFA from the rearranged oil. Water washing and or adsorptive bleaching will remove the first two, and deodorization should remove the FFA. Losses from the reaction (as FFA and FAME) are ~10 times the catalyst weight, plus an additional 0.5 to 1.0% for formation of mono- and diglycerides, for a total of 1.5 to 2.0%. Continuous processes also are available. Interesterification progress can be monitored by on-line UV spectrophotometry, and completion of the process by determination of melting point and other measurements. The process is quenched by adding water (Lampert, 2000).

Interesterification is promoted as a means of producing fat solids without creating *trans* fats. "*Trans*-free" margarines already are well established in the European Union (EU) and Canada. But, interesterification may not be the final solution. *Trans* fats are more common in the diet than most people realize. Also, while nature favors placing unsaturated fatty acids in the *sn*-2 triglyceride position, saturated fatty acids appear in this position in proportion to their availability in chemically randomized triglycerides. In mammalian digestion, the pancreatic lipase-cleaved 1, 3 free fatty acids, the *sn*-2 monoglycerides, and biliary and pancreatic secretions form micelles which are absorbed into the small intestine mucosa, where they are resynthesized into triglycerides, formed into chylomicrons, and transported to the lymphatic system (Lyford and Tal Huber, 1988). But, the fatty acid in the *sn*-2 position stays with the monoglyceride and is reconstituted in the same position in the resynthesized triglyceride. Considerable data is accumulating that triglycerides, saturated at the sn-2 position, are atherosclerotic (Kritchevsky, 1998). The preferred type of interesterification may be enzyme-directed rather than chemical. We may be swept into an era focused on reducing intake of measurable *trans* fatty acids, without a holistic understanding of effects of the change on life-cycle health.

20–4.8 Thermal Fractionation

Oil and fat products that contain too much solids for the specific application may be fractionated, which typically consists of chilling to crystallize the higher melting triglycerides and waxes as stearins and separating them from the liquid (oleins) by vacuum and other types of filters, or by centrifugation. Fractionation is common practice with some species, and elaborate systems have been developed. A mild fractionation ("winterization") sometimes is used to remove stearins that have developed during brush hydrogenation of soybean oil to ensure clear table ("salad") oil when stored in household refrigerators. An acceptable oil must pass the (AOCS Cc 11-53) Cold Test and remain clear after 5.5 h at 0°C. Soybean oils used in making mayonnaise and salad-dressings also are winterized to prevent breaking of emulsions by crystallizing when refrigerated (O'Brien et al., 2000). More elaborate equipment, originally developed for the palm oil fractionation industry, is used in thermal fractionation of interesterified oils (Tirtiaux, 1990, 1998; Bockisch, 1998; Kellens and Hendrix, 2000; Widlak et al., 2001).

20–4.9 Deodorization/Physical Refining

Deodorization/physical refining is the final step in producing refined, bleached, deodorized "RBD" soybean oil. The process is called "deodorization" if most of the FFA are removed by alkali neutralization as with row crop ("soft") oils, and "physical refining" if they are left for steam distillation as with palm and coconut oils. Larger yields of neutral oil are obtained from oil species suitable for physical refining, but phosphatides in row crop oils must be removed first. Various technically successful physical refining methods have been developed for soybean oil, but have not been considered sufficiently advantageous to warrant retrofitting current installations.

Basically, deodorization is a steam distillation process, in which the volatile peroxides, other decomposition products, and odoriferous compounds form reduced-boiling point azeotropes with water in the steam at high temperatures (250–260°C, 482–500°F), and very low vacuums (2–4 mbar) created by a series of steam ejectors, usually four stages. The process is conducted above the smoke point of soybean oil, but below the flash point, and oxygen must be excluded. Considerable heat bleaching of yellow-red carotenoids also occurs at these temperatures. *Trans* fatty acid formation is influenced by time and temperature, and generally is negligible below 220°C (428°F); significant between 220 and 240°C (464°F), and exponential above 240°C. The relative isomerization rate is the highest for linolenic acid, and lowest for oleic acid, and has been estimated as: $C_{18:3}(100) \gg C_{18:2}(10) \gg C_{18:1}(6)$. More recently, sublimation (freezing of deodorizer vapors with ammonia heat exchangers) instead of relying on water temperature for condensation, has become popular. It has been estimated that this substitution, with accompanying heat recovery units, can save ~50% of the costs of operating a typical clean water steam ejector system at typical North American costs of steam and electricity (Kellens and De Greyt, 2000).

Typically, the deodorization process requires 20 to 40 min after come-up time, uses 0.5 to 2.0% sparge steam (the higher level if tocopherols are removed), and produces a product with about 0.03 to 0.5% FFA (Kellens and De Greyt, 2000). Historically, the standard U.S. batch deodorizer held one railroad tank car (27 216 kg; 60 000 lb) of oil. Except for refineries that produce one type of refined RBD oil for export, installation of continuous deodorizers slowed with the advent of JIT, self-certification, and customers deciding to buy on the basis of their projected production schedules. This resulted in return to improved batch-continuous systems, but designed to handle many batches of different oil blends per day with minimum cross contamination and delays for temperature and vacuum come-up.

For many years, deodorizers (operating at >270°C, 520°F) were heated by mineral oil-like thermal fluids, which could be provided at 315°C, 600°F, in plumbing and coils at 3.2 bar, 46 psig, by direct-fire furnaces. During the 1970s, user health problems, ascribed to leakage of thermal fluids into the oil during deodorization, arose in Europe. This led to the European market requiring that local and import oil industries shift to using high-pressure steam generators, operating at 80 bar (1160 psig) to provide a temperature of 295°C (560°F). Other importing countries also have adopted European standards. This meant that deodorizer heating coils and jackets of suppliers wanting to sell oil in Europe had to be rebuilt, or replaced by new deodorizers, high pressure steam generators, and supply lines (Zehnder, 1995).

Deodorizers are built in many vertical and horizontal designs (Zehnder, 1995; Kellens and De Greyt, 2000). Figure 20–14 shows a new "soft column" deodorizer system that was introduced by the Alfa Laval Company in 1996. It is claimed to have a highly efficient oil-to-oil heat recovery system of 75%, use less deodorizing steam, and operate at lower temperatures. Although several types of equipment are available, it serves here as a walk-through example of what occurs in a deodorizer. Well-prepared RB oil passes through a heater (Fig. 20–14, No. 6), then through a deaerator (Fig. 20–14, no. 7) to eliminate dissolved air, through the economizer (Fig. 20–14, no. 3) for additional heating, and to the final heater (Fig. 20–14, no. 4). It then passes through the stripping column (Fig. 20–14, no. 1), to the holding section with sparger pipes (Fig. 20–14, no. 2), and the economizer (Fig. 20–14, no. 3) to give up some of the heat. After partial cooling, citric acid (20–50 mg kg^{-1}, ppm) in solution form is added (Fig. 20–14, no. 10) as a sequesterant for iron or copper that may be picked up later by the oil. The oil is still hot and under vacuum, allowing moisture in the citric acid solution to flash off. Next, the oil is cooled and passes through a polishing filter on the way to temporary storage and shipment under nitrogen. The liquid used to "scrub" the vapors (Fig. 20–14, no. 5) is previously condensed deodorizer distillate that is chilled and recycled. It also may entrap vapors of pesticides if they get this far in the process. Deodorizer distillate is rich in tocopherols, some of which have vitamin E activity. With the increasing consumption of Vitamin E supplements in the USA, and strong markets for natural antioxidants, deodorizer distillate is very much in demand and the domestic supply from refineries has long been committed by contracts with Vitamin E proces-

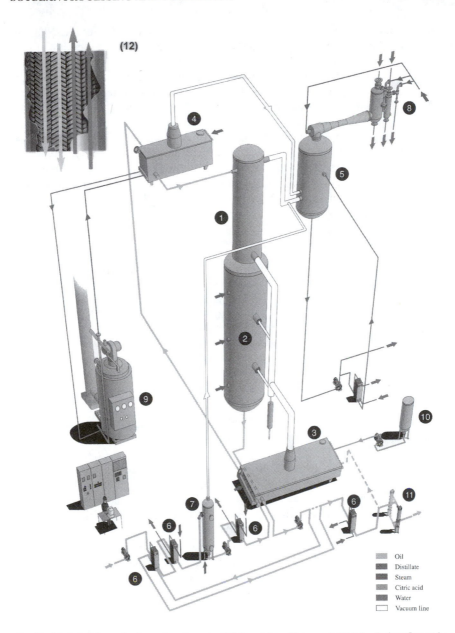

Fig. 20–14. SoftColumn (registered trademark, Alfa Laval Inc., Richmond, VA) deodorizer. Legends: (1) packed column; (2) oil holding section; (3) vacuum heat exchanger; (4) vacuum-sparged final heater; (5) condensate scrubber; (6) plate heat exchangers; (7) deaerator before high temperature oil heating; (8) steam ejection vacuum system; (9) high pressure type steam boiler; (10) citric acid addition system; (11) polishing filters; (12) cut-away of steam stripping section of column 1; preformed plates provide large surface "packing" area; steam rises on the right, oil (from 4) descends on left. Oil route: 6 - 7 - 6 - 3 - 4 - 1 - 2 - 3 - 6 - 11. (Courtesy of Alfa Laval Inc., Richmond, VA.)

sors. As much as 60% of the tocopherols in soybean oil could be extracted at the deodorizer, but in doing so, they no longer are available to stabilize shelf life of soybean oil or provide its full natural nutritional benefits.

At this point in processing, peroxide value in the oil is barely detectable, but will start increasing soon. Thus, margarine and other profiled-temperature fats are blended from base stocks and deodorized just before shipment.

After cooling the oil to appropriate temperatures, processors may add additional oil-soluble ingredients, including emulsifiers, antioxidants, vitamins, colors, and others that customers might have difficulty dispersing in the oil/fat at their processing sites. Unless specifications call for chilling and texturization of the fat into a soft-plastic shortening form, it is kept pumpable at about 10°C above its melting point for shipping and handling. Nitrogen purging of liquid fats during pump transfer, and storage under nitrogen blanket, are common if held for more than several days before use by the processor.

20–5 ALTERNATIVE PROCESSES AND MARKETS

This section focuses on preparation and utilization of soybean food, feed, and industrial ingredients for niche markets which are substantial businesses in their own right.

20–5.1 Screw Pressing of Soybean

20–5.1.1 Hard Pressing

Preparation of soybean for hard pressing is similar to that of solvent extraction, except that the flakes are intercepted and cooked in stack or horizontal cookers at ~12% moisture and 85 to 93.3°C (186–200°F) for 15 to 20 min to denature the protein, then dried to 2.5 to 3.0% moisture before pressing (Williams, 1996). The presence of hulls helps feed screw presses. Hard pressed high-protein meal is seldom made unless tail-end dehulling is used. Generally, protein quality of hard-pressed soybean meal is more heat damaged than meal produced by solvent extraction. Residual oil in the soybean cake varies with the condition of the screw press and is in the 5.0 to 6.5% range. Production of hard-pressed soybean meal had almost disappeared by the early 1990s, with few plants surviving only because of ability to command premium prices for the meal because of its high rumen-bypass properties for cattle, or for uses in fish bait (Johnson et al., 1992b).

20–5.1.2 Extrusion—Screw Pressing

In the latter 1960s, the Triple-F Feed Co., Des Moines, IA, applied dry extruders to inactivating soybean trypsin inhibitors and other anti-growth factors in meals used in poultry, pig, and cattle feeds. These are cast iron, single screw, autogenous (heat-generating) machines that can directly heat 8 to 10% moisture content soybean to 145°C (294°F) by shear (Hanson, 1996; Lusas and Riaz, 1996b). The objective was to make soybean useable for feed at the farm, or at a local cooperative or small feed mill, without first sending it through the commercial grain elevator-solvent plant route. The Insta-Pro (registered trademark, Insta-Pro International, Inc., Des Moines, IA) machines were relatively small, 273 kg (600 lb) h[-1] and 1136 kg (2500 lb) h[-1], made of cast iron, relatively low cost, and did not require prior preparation of seed. The U.S. Agency for International Development (USAID) sponsored a project on food-processing applications of low-cost extruders in developing countries in the 1970s, essentially producing local equivalents of the U.S. PL-480 CS (corn-soya) product (Wilson and Tribelhorn, 1979). Insta-Pro International is the only company from that era that has stayed in business and continues to sell low-entry cost feed and food-processing machines directly in developing countries. The United Nations also uses its extruders in World Food Program nutrition intervention programs.

Nelson et al. (1987) described extrusion of soybean followed by screw pressing to produce both oil and partially-defatted (6–9% fat) meal. The full-fat soybean product, in its normal oil to protein ratio, had contained too much oil for many animals, especially ruminants, and some feeders had been required to purchase solvent-extracted soybean meal to reduce overall fat content. By extrusion-screw pressing, processors could sell both oil and a fat-enriched soybean meal. This process was patented under the name ExPress.(registered trademark, Insta-Pro International, Inc., Des Moines, IA). A later paper described grinding the press cake into flour and passing it through the initial-type extruder to produce texturized soybean protein (TSP) (Lusas and Riaz, 1996b). Insta-Pro International has since modified its product line to include a larger 4.5 t (10 000 lb) h[-1] extruder, several extruder designs (Fig. 20–15), improved screw presses, a low-entry cost oil refining system, and other supporting equipment.

Screw pressing of soybean, which had nearly disappeared, has returned, but for production of full-fat and partially-defatted soybean meals by large feeders, cooperatives, and small processors. Companies/feeders with larger capacity requirements have selected from a variety of oilseeds processing equipment, often available used. When properly processed, phosphatides in the pressed crude soybean oil are substantially higher than in solvent extracted oil, but are readily hydratable because of early destruction of phospholipase D in the extruder. This has led to installation of one of the first soybean oil physical refining facilities to use extruder-screw press-extracted crude soybean oil. Edible dehulled flours, with Protein Dispersibility Index (PDI) as high as 55, have been produced. Although not as versatile as those made by flash desolventization, they can fill a variety of needs and are surprisingly stable against oxidation.

20–5.2 Organic Foods

For many years, a portion of the public in many countries has distrusted "chemicals" in foods. This is separate from vegetarianism, which often is based on religious or personal conviction, although similar foods may be chosen. The number of diet-concerned customers has grown. Successful health food processors, stores, and specialty outlets have been established, and in 2000, more than half of the $7.8 billion of U.S. organic food was purchased in conventional retail outlets (ERS, 2002). Many organic food growers and processors are dedicated to their work; but, objectives and practices in preparing "pure" and "natural" foods had not been defined in the past, and charges of fraud had arisen.

Many U.S. businesses and regulators have wrestled with the situation for years. Regulators sought truth in labeling. Business interests were divided, some feeling that establishment of standards would "legitimize" a previously unrecognized part of the food industry, while others argued the segment was here to stay and had growth potential (Pollan, 2001).

The last revision of the National Organic Foods Production Act of 1990, issued 1 Jan. 2001 (7CFR 205), contains definitions and regulations regarding land use, livestock feed, animal and crop production, and their handling and labeling. The new program is administered by the USDA Agricultural Marketing Service and

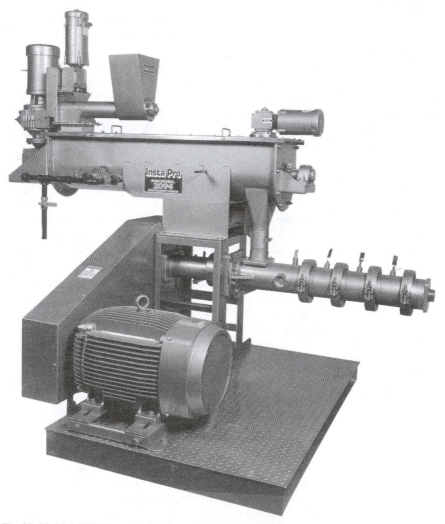

Fig. 20–15. Model 2094 Insta-Pro (registered trademark, Insta-Pro International Co., Des Moines, IA) extruder. Without the shown conditioner or feeder, machine makes full-fat soybean meal for animal feeding or screw pressing to recover oil and partially-defatted meal/flour. With conditioner and feeder, machine converts partially-defatted soybean flour into food grade texturized soybean protein (TSP). (Courtesy of Insta-Pro International Co., Des Moines, IA.)

is considered a handling process or system. Thus, products are subject to existing FDA regulations, labeling policies and practices, but must meet additional requirements. Too detailed to summarize here, they are readily available to Internet browsers under "National Organic Program," and through the Code of Federal Regulations and the Federal Register. Significant points regarding the new National Organic Foods Program include:

1. Individual organic food growers must be certified as detailed in the legislation, except for operations with less than $5000 sales annually.
2. The applicant must file plans on organic food production and land management, or raw materials and processing, that are subject to inspections and periodic reviews.
3. Provisions exist for labeling several types of products:
    a. 100% organic, all organically produced ingredients;
    b. Organic, which must contain not <95% organically produced ingredients, exclusive of water and salt; and
    c. Made with organic, which must contain not <70% organically produced identified ingredients, exclusive of water and salt.

The National List of Allowed and Prohibited Substances is another interesting feature of the new organic food regulations. Uses of numerous natural and synthetic substances are allowed as algaecides, disinfectants, and sanitizers in organic crop production; as soap-based herbicides in farmstead maintenance; and as insecticides, rodenticides, plant disease control agents, and herbicides in crop production. Comparable needs are addressed in animal production, and petitions can be submitted for additional additives for producing crops and animals. An extensive list of organic, nonorganic, and synthetic substances permitted in food processing exists in the initial legislation. Currently, solvent extraction of oil appears excluded. But, soybean oil might be processed by the following mechanical and physical means:

1. Extrusion of good quality level soybean to inactivate phospholipase D.
2. Screw pressing to separate the oil.
3. Water degumming of phosphatides in the oil, with minimum use of citric acid (natural or possibly microbial produced).
4. Bleaching with bentonite is listed, and ashes of crops, for example rice hulls (Proctor et al., 1995; Kalapathy and Proctor, 2000) might be allowed on petition.
5. Physical refining to remove FFA and for lightening red and yellow colors.

20–5.3 Dietary Supplements, Nutraceuticals, and Functional Foods

The Dietary Supplement and Health Education Act of 1994 (DSHEA), amended the Federal Food, Drug, and Cosmetic Act to include several provisions that apply only to dietary supplements. Chief among these was the provision that dietary ingredients used in dietary supplements are no longer subject to premarket safety evaluations required of other new food ingredients, or for new uses of old food ingredients. However, they must meet the requirements of other safety provisions. Before DSHEA, demonstrating the safety of new additives before marketing was the responsibility of the petitioner, with FDA deciding when sufficient convincing safety data had been submitted. After DSHEA, products can be introduced in the markets without prior notification to FDA; yet, FDA still is responsible for seeking out suspected ingredients and products, proving they are not safe, and when necessary ordering their removal from the market. Each situation is a separate case and FDA policing resources are limited. FDA is still organizing procedures for identifying and dealing with potentially hazardous supplements (FDA/CFSAN, 1995, 2001a, 2002). Claims that can be made on the package, and in separate promotional literature, is one of the chief areas of contention (Federal Register, 2000).

Nutraceuticals have been defined as "any substance that is a part of a food that provides medical and/or health benefits, including prevention and treatment of diseases." Functional foods have been described as "foods for which a structure-function claim related to health benefits is made because of the presence of some added active ingredient" (Schmidl and Labuza, 2000). Although nutraceuticals and functional foods are rapidly growing businesses, only three options exist for their sale: drugs, foods, or supplements. Each has its specific protocols regarding obtaining FDA approval, labeling, and types of claims that may be made. The principle that neither foods nor supplements can be marketed as treatments or cures for specific diseases or conditions is clearly established for the three categories. The law permits three types of claims for supplements (FDA/CFSAN, 2001b):

1. Health claims—the relationship between a food substance and a disease or health-related condition; example: "diets high in calcium my reduce the risk of osteoporosis"
2. Structure/function claims—the benefit related to a nutrient deficiency disease, or the role of a nutrient or dietary ingredient intended to affect a structure or function in humans; example: "fiber maintains bowel regularity"
3. Nutrient content claims—terms like "good source," "high," or "free."

In general, structure/functions claims also must include disclaimers like: "This statement has not been evaluated by the Food and Drug Administration." and "This product is not intended to diagnose, treat, cure, or prevent any disease." now found on supplements labels. Labeling regulations are complex, and more information is available at the FDA Center for Food Safety and Applied Nutrition website (www.cfsan.fda.gov/label.html).

Functional foods (not to be confused with food ingredients functionality to be described later), are intended to be eaten for beneficial health reasons. Examples include (i) dietary fiber soft drinks for intestinal regularity, (ii) probiotic yogurts to assist gastrointestinal bacteria, (iii) *omega*-3 fatty acids in fish oils and in eggs from hens on specific diets to reduce risk of cardiovascular disease and improve mental functions, and (iv) stannol-containing spreads intended to reduce blood cholesterol levels. Indeed, effectiveness of some functional foods has been documented by respected research institutes. However, nutraceuticals blur traditional lines between food and medicine and add to public confusion. Additional information on this issue is provided in Chapter 21.

Disclaimers in fine print on packages are a polite way of saying "Proceed at your own risk." Concerns have been expressed by FDA that a significant percentage of middle age and older persons use prescription drugs, and undesirable interactions may occur with phyto remedies. Nevertheless, dietary supplements currently are the fastest growing segment (rate of >20% yr^{-1}) of the food industry.

Several major soybean processors produce isoflavones, including daidzein, genistein, glycitein, and others from the protein fraction of soybean for bulk sales and possibly later pharmaceutical products. Various extraction and purification processes are used, including gross purification by ultrafiltration, concentration by reverse osmosis, adjustments in extracting alcohol concentrations and temperature, and adsorption chromatography. The compounds are of increasing interest in treating female symptoms (hot flashes, osteoporosis, sleep disorders, menopausal symptoms, vaginal dryness, premenstrual syndrome, and other problems related to a reduction of sex hormones and to menstruation), neurological symptoms, and cancer (Gugger and Dueppen, 1997; Gugger and Grabiel, 2000, 2001; Empie and Gugger, 2001, 2002a, 2002b, 2002c). A database on isoflavone technical literature is maintained by the Department of Food Science and Human Nutrition, Iowa State Univ., Ames, IA, at www.nal.usda.gov/fnic/foodcomp/Data/isoflv/isoflav.html.

20–6 EDIBLE USES OF SOYBEAN OIL

In 2001, the USA produced 78.68 Mt of soybean; and 8.5 Mt of soybean oil. Of this, 7.39 Mt was consumed domestically, 45% (3.34 Mt) as salad or cooking oil, 43% (3.17 Mt) as baking and frying fats, 8% (0.59 Mt) as margarines and spreads, <1% (0.06 Mt) as other edible products, and 3% (0.23 Mt) as inedible products. Soybean oil provided ~83% of all fats and oils consumed in the USA (Soy Stats, 2002). Any of the following technologies might be employed. If additives are used, they would be listed on the label.

20–6.1 Shelf Life Extension

Degradation by oxidation usually is the major factor that limits the useful life of fats and oils. Oxidation is delayed by (i) reduction of the more susceptible fatty acids (linolenic and linoleic) by brush hydrogenation, (ii) removal of oxidation catalysts, including chlorophyll and metal ions by bleaching, (iii) removal of peroxides by bleaching and/or deodorization, and (iv) addition of citric acid to sequester copper and iron which may enter the oil later. Packaging in a light-proof container greatly assists in prolonging shelf life, but typically is resisted by brand managers who want customers to see the oil. Natural oil-soluble antioxidants protect the unsaturated bonds by quenching quantums of energy that could attack fatty acids at their unsaturated sites and initiate peroxides formation, but eventually are exhausted (Nawar, 1996). They can be supplemented by natural tocopherols (Gregory, 1996), which are relatively expensive. Synthetic antioxidants like butylated hydroxyanisole (BHA), butylated hydroxytoluene (BHT) or di-*t*-butylhydroquinone (TBHQ), provide greater protection and can be added in combination at up to 0.02% of the fat. Resistance of fats or oils to oxidation is estimated by AOM (AOCS Cd 12-93) or OSI (AOCS Cd 12b-92) methods.

Oxygen also hastens oxidation, and is commonly excluded by vacuum processing, storing oils under nitrogen blanket, and by injection of nitrogen as the aerating agent in plasticized shortenings. As oils are used in frying, their viscosity increases from polymerization of break down products and often results in foaming and retention of oxygen. Foaming can be reduced by addition of 0.5 to 3.0 mg kg^{-1} (ppm) of polydimethylsiloxane (methyl silicone), which greatly extends fryer life of oils (O'Brien, 1998).

20–6.2 Temperature-Profiled Fats and Oils

20–6.2.1 Solid Fat Index and Solid Fat Content

Fats naturally soften and melt as temperature increases. The solids content profile is controlled in many food products. The two major characterization methods are Solid Fat Index (SFI; AOCS Cd 10-57) and Solid Fat Content (SFC; AOCS Cd 16-81). The SFI method uses dilatometry which was developed and still is broadly used in the USA. The SFC method is a pulsed nuclear magnetic resonance procedure developed more recently to accommodate palm and other tropical oils. The SFC method is used throughout much of the world. The SFC reads higher at the lower temperatures (10 and 21.1°C), but both methods have similar readings at higher temperatures (List et al., 2001). The following equations have been developed for converting SFC to SFI:

Commercial spreads: SFI = 1.98 + 0.72*SFC - 0.035*Temp.
Base stocks: SFI = -40.94 + 1.22*SFC + 1.03*Temp.
Blends (base stocks/liquid oil) = −0.94 + 0.82*SFC + 0.02*Temp + 0.02*%.

The SFI curves for stick, soft stick, and tub margarines are shown in Fig. 20–16. The SFI curves for butter vary with the feed cow are on; but are most closely

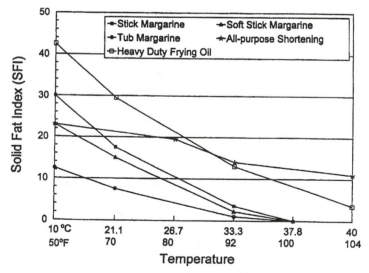

Fig. 20–16. Solid Fat Index (SFI) curves for hard stick, soft stick, and tub margarines, and for all-purpose shortening and heavy duty frying oil (Erickson and Erickson, 1995; with permission).

mimicked by the soft stick margarine curve. Desirable margarine and spread fats are completely melted at body temperature (37.0°C, 98.6°F). Many people find spreads "greasy" when more than 2 to 3% of the fat is in solid form at mouth temperature; the threshold is about 5 to 6% fat solids in cereal and bakery products.

20–6.2.2 Plastic Range

A "base stock" system is often used in preparing temperature-profiled fats, like margarines/spreads and shortenings. Four to six soybean oils bases are hydrogenated to different Iodine Values (Fig. 20–17) and held liquid in tanks. Several of these bases are combined, with soybean oil and a "hard stock" (soybean or cottonseed oil, hydrogenated to <10 IV) to obtain the solids profile desired for the product. The melted mixture then is deodorized, to remove peroxides that may have developed, before shipping to the margarine processor.

Shortenings and baking margarines machine "work" best at a plastic range of 15 to 25% fat solids. Doughs or batters are too firm at higher solids contents, and lose their desirable characteristics because of excessive oil at lower solids. This is demonstrated in Fig. 20–18, where the bakery mixing area must be kept below 13°C (55°F) if 85 IV base is used. But, by adding 12% hardstock, the machining temperature range can be increased up to 39°C (102°F), which is more characteristic

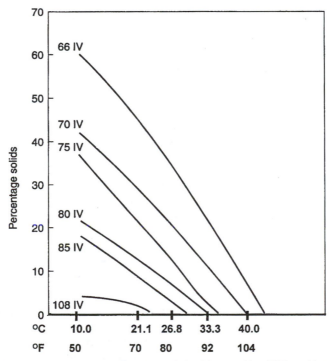

Fig. 20–17. Soybean oil fat solids and Iodine Value relationships over 10 to 40°C; used in preparing base stocks for formulating margarine, spreads, shortenings, frying oils, and other temperature-profiled fats (Hastert, 2000; with permission).

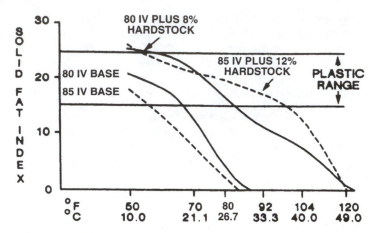

Fig. 20–18. Effects of adding hardstocks on broadening working temperatures in the 15 to 25 SFI plastic range (Hastert, 1990; with permission).

of bakery working conditions (Hastert, 1990). The SFI curves for a variety of fats are shown in the book by O'Brien (1995).

20–6.2.3 Margarines, Spreads, and Plasticized Shortenings

Margarine, made from animal fats, was invented in France in 1869 as a butter substitute and was first produced in the USA in 1873. Partially hydrogenated soybean oil became the lead oil in margarine in 1956—the same year that U.S. per capita consumption of margarine surpassed butter. By law, margarine must contain the same amount of fat as butter, which is 80% in the USA but differs among countries. The other components of margarine are water, salt, and flavorings. Major breakthroughs in technology included invention of the internal scraped surface heat exchanger (SSHE) in 1937, and reduction of emulsifier restrictions in 1992, which permitted development of a wide range of reduced-fat content spreads—some as low as 20% fat content.

Fat polymorphism, which can result in as many as four melting points and six crystal forms in fat solids, is important to achieving smooth textures in margarines/spreads, chocolate, and its extenders, and other applications (Sato and Garti, 1988; Widlak, et al., 2001). Providing a diversity of triglyceride crystallizing species is among the surest means of obtaining small "beta prime" (β') oil-holding crystals, rather than large coarse beta (β) crystals which are the thermodynamic destiny of oils. For example, soybean oil alone is difficult to work into a smooth margarine because it consists mainly of C18 fatty acids. Thus, it was common practice in the past to use cottonseed hardstock, which introduced a substantial amount of palmitic (C16) fatty acid; fractionated palm oil stearins would have the same effect. Each different *trans* triglyceride molecule becomes a nucleating species, which helps explain the effectiveness of selective *trans* hydrogenation in making smooth-textured margarines/spreads. Also, mono- and di-glycerides act as crystallization inhibitors by fouling crystal surfaces for additional lattice building.

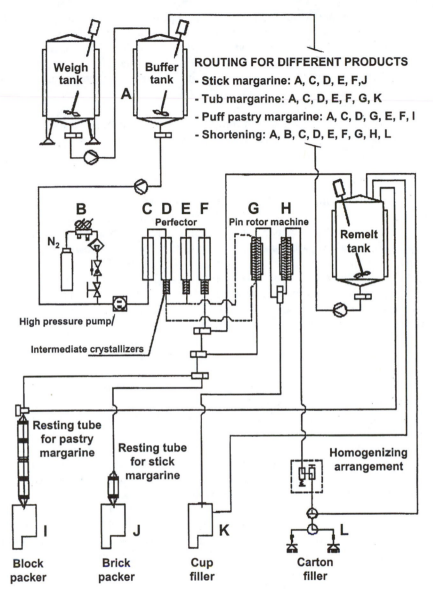

ROUTING FOR DIFFERENT PRODUCTS
- Stick margarine: A, C, D, E, F,J
- Tub margarine: A, C, D, E, F, G, K
- Puff pastry margarine: A, C, D, G, E, F, I
- Shortening: A, B, C, D, E, F, G, H, L

Fig. 20–19. Flexible Perfector Plant (registered trademark, Gerstenber & Agger A.S., Brondby, Denmark) for making stick, tub and pastry margarines, and shortening. (Courtesy of Gerstenberg & Agger A.S., Brondby, Denmark.)

Margarines and spreads are water-in-oil emulsions. Any soluble microbial inhibitors, including salt, are concentrated in the water droplets and microorganisms have difficulties migrating through the oil to inhabit other water droplets. As water content has increased in spreads, microbial inhibitor contents, viscosity-building agents like gums, and the ratio of water:oil soluble flavoring agents in formulas, also have increased.

Margarine/spread preparation includes a repeated series of chilling the emulsion in scrape-surface heat exchangers to initiate crystallization, followed by application of mechanical energy through pin workers to arrest and disperse crystal growth as small β' crystals (Fig. 20–19). High mono-fatty acid oils, including palm and canola, require considerable working to produce smooth-textured products.

Texturized shortenings also can be produced on margarine/spread making equipment (Fig. 20–19), but vary in complexity from single base stocks, nitrogen-plasticized for convenience in handling, to more complex mixtures cake and frosting shortenings.

2–6.2.4 Baking Fats

The term "shortening" basically means an "all fat" product which can be used for baking, frying, and in confections. The historical function of fats in baking is to "shorten" or control development of gluten in wheat to ensure tenderness in the product. Some bakery products ("puff pastries," fillos/phyllos) consist of flaky layers made by rolling the dough thin, covering it with a layer of shortening, folding, rerolling, and repeating the process several times. The gluten layers must stay intact, and the shortening must be flexible but not give off free oil during the machining. One of the more recent innovations is production of puff pastry margarine with a very flat plastic range on margarine equipment.

Numerous emulsifiers are available for dough and cake systems. In dough systems, emulsifiers act as dough conditioners by (i) improving tolerances to variations in flour and other ingredients, (ii) increasing resistance to mixing and mechanical abuse, (iii) providing increased gas retention, shorter proof times, and increased product volume in yeast-leavened systems, (iv) improving uniformity in cells size, finer grain, and stronger cell walls, (v) improving product slicing, and (vi) delaying staling and starch retrogradation to lengthen product freshness (Artz, 1990).

Cakes essentially are emulsified slurries before baking. Emulsifiers have three functions in cake systems (i) improving air incorporation, (ii) dispersing the shortening into smaller particles to maximize the number of air cells, and (iii) improving moisture retention. Fats, emulsifiers, and starches in low-protein flour form complexes which result in smooth, tender, moist cakes (Artz, 1990).

Emulsifiers, made by interesterification of soybean oil, play an important role in formulating and processing bakery products. Broad classes include: lecithin and lecithin derivatives, mono- and di-fatty acid glycerol esters, hydroxycarboxylic acid and fatty acid esters, lactylated fatty acid esters, polyglycerol fatty acid esters, ethylene or propylene glycerol fatty acid esters, ethoxylated derivatives of monoglycerides, and fatty acid esters of sorbitol (Artz, 1990). Emulsifiers sometimes are classified by the hyrophile-lipophile balance (HLB) Index, ranging from 1 to 20. Lower numbers indicate a more lipophylic emulsifier, with HLB 3 to 6 best for water-in-oil emulsions, and 8 to 18 are better for oil-in-water emulsions (Artz, 1990; Walstra, 1996).

20–6.2.5 Frying Fats

Most frying occurs in the 177 to 219°C (350–425°F) range. An excellent review of this subject has been edited by Perkins and Erickson (1996). Oil acts as the

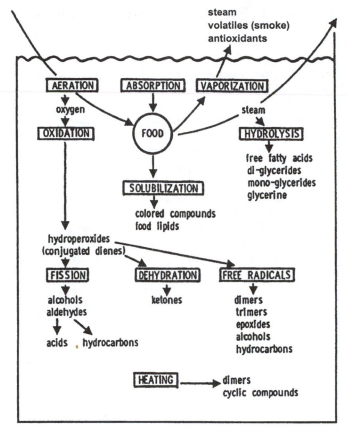

steam
volatiles (smoke)
antioxidants

AERATION   ABSORPTION   VAPORIZATION

oxygen   steam

OXIDATION   FOOD   HYDROLYSIS

free fatty acids
di-glycerides
mono-glycerides
glycerine

SOLUBILIZATION

colored compounds
food lipids

hydroperoxides
(conjugated dienes)

FISSION   DEHYDRATION   FREE RADICALS

alcohols   ketones   dimers
aldehydes   trimers
                   epoxides
                   alcohols
acids , hydrocarbons   hydrocarbons

HEATING ——→ dimers
             cyclic compounds

Fig. 20–20. Changes occurring during deep fat frying (Fritsch, 1981; with permission).

heat transfer agent in a process which rapidly heats, inactivates enzymes, cooks, and may puff and nearly dehydrate the product (Fig. 20–20). The entering food carries some oxygen that temporarily aerates the hot oil, and water that forms steam, but both are rapidly swept out because of low solubility of gases in the hot liquid. Besides enzymes, which are inactivated rapidly, raw materials also carry oxidation catalysts like iron in hemoglobin of fresh meats, chlorophyll, and color pigments. Alkalis, which cleave free fatty acids from triglycerides, may accompany the raw food, especially in chemically-leavened doughnut batters and alkali-treated corn snacks. In short, almost every concern cautioned against thus far in extraction and oil processing occurs almost instantaneously in frying. At the high temperatures of frying, oils begin breaking down into simpler compounds and also form cyclic compounds and polymers. However, some self-cleansing occurs in the system by steam distillation. Studies have shown that carbonyl compounds build up in hot oil pans of restaurant fryers, while inactive between preparation of noon and evening meals, but are partially dissipated when the fryer is returned to use for evening meals.

Most frying is done by commercial operations like snack food processors, producers of convenience fried foods like Chinese egg rolls and frozen meals, and fast food vendors who prepare French fries, fried chicken, and other products. Banks and Lusas (2001) have summarized selection and use of frying oils for these conditions. The operations can be assigned to two groups: those who sell all of the oil with the product, and those who must recondition or dispose of part of it.

Generally, industrial snack food fryers are designed to use all the oil. With clean-out systems and in-line filters, they should operate indefinitely, only adding makeup oil as need. "Clean label" operations carefully inspect and monitor their self-certified suppliers. Antioxidants or silicones defoamers are not permitted in their oils. Oils are received in bulk, typically at <0.05% FFA, <1.0 PV, <4 mg kg^{-1} (ppm) phosphorous and <0.75 mg kg^{-1} (ppm) chlorophyll, and are kept in stainless steel tanks under nitrogen blanket (Banks and Lusas, 2001). Oxidative stability of soybean oil has consistently increased in frying trials with reduction of linolenic acid, whether by breeding, mixing with other oils, or by hydrogenation (Evans et al., 1972; Moulton et al., 1975; Mounts et al., 1994). Less than 2% linolenic acid content has been a long-term industry goal; but, the way (hydrogenation or breeding) by which it is reduced to <2% has not greatly affected fried product stability, with hydrogenation performing slightly better (Tompkins and Perkins, 2000). Commercial processors are reluctant to fry with oils higher than 2.5 PV. Freshness of fried snack foods is dependent on use of moisture and oxygen impermeable packaging, including windowless laminates of aluminized polyvinylidene dichloride films (PVDC; Saran, registered trademark, Dow Chemical Co., Midland, MI) on oriented polypropylene (OPP). The package material is resistant to moisture and oxygen penetration, and blocks 99+% of the light to prevent photo-induced oxidation. Additionally, the pouches are nitrogen flushed before sealing, thus creating pillow packs that protect the product against crushing. The snacks are placed on store shelves or racks by company delivery personnel. The products may have shelf-lives of 6 to 8 wk, and inventories are managed to ensure rapid turnover and fresher products.

Industrial fryers, who cannot sell all the oil in the product, have lengthened its life by 3 to 10 times using polydimethylsiloxane, which is not allowed in some countries. Levels as low as 0.2 to 0.3 mg kg^{-1} (ppm) have been found effective, with commercial usage of 0.5 to 5.0 mg kg^{-1} (ppm) reported. Minimized usage (1.0 to 3.0 mg kg^{-1}, ppm) is advised. Dispersion of polydimethylsiloxane in oil is difficult. The compound operates by suppressing foaming and polymerization, and increases smoke points of oils by up to 13.9°C (25°F). Antioxidants are steam distilled during normal frying, and their initial inclusion essentially protects the oil only until the time of use. For greater effectiveness, antioxidants are best sprayed onto the product in fresh oil after frying, or included in dry seasonings.

Normally, the product's warm surface oil serves as the binder ("tacking agent") for adsorbing salt and dry flavorings to snack foods. The concept of reducing or entirely eliminating oil from snack foods became popular during the early 1990s. Rather than deep fat frying, the products were dried at high temperatures in fluidized bed continuous dryers. A smaller amount of oil was then sprayed on to the dried product for flavor. Where a "fat free" snack food was desired, solutions of edible gums or specialty starches were sprayed onto the snack to serve as flavor binders; the product then required an additional drying step (to <1.5 to 2.0% moisture content to ensure crispiness). Low public acceptance of low-fat, or noncaloric fat, snack foods was a major disappointment of the late 1990s, but some "baked" snack foods have reappeared on grocery store shelves.

Operations that must discard frying oil periodically take extra steps to prolong its life beyond use of polydimethylsiloxane. These include: inline filters, periodic cleaning of fryers to remove settled charred product, neutralization of fatty acids, and refreshing the oil by passing it through adsorbent earth filters continuously or at the end of the work day. Numerous kits and advisory services are available. Large commercial frying operations and fast food franchises that prepare French fries and chicken, at least have provisions and trained personnel to prolong the life of frying oils. The greatest concerns are about small restaurants that do occasional frying during mid-day and evening meals. Several countries have imposed standards on quality levels of used frying oil. Products fried in such oils usually are objectionable in taste to most Americans long before they exceed nonuseable specifications (Firestone, 1996).

20–7 PREPARATION AND USE OF SOYBEAN FOOD PROTEIN INGREDIENTS

Johnson et al. (1992b) and Lusas and Rhee (1995) have reviewed early preparation and uses of soybean food protein ingredients (flours, concentrates, and isolates). Books by Smith and Circle (1972), Wolf and Cowan (1975); Wilcke et al. (1979), and Altschul and Wilcke (1985) still are useful references of technology basics. Proceedings of protein-focused AOCS World Conferences in Munich (JAOCS, 1974), Amsterdam (JAOCS, 1979), Acapulco, Mexico (JAOCS, 1981), and Singapore (Applewhite, 1989) add further depth.

Readers are referred to Chapter 21 for details on human and animal nutrition. Soybean milk and tofu are the fastest growing Asiatic-origin foods in Western markets. The subject is addressed in AOCS World Conference Proceedings, and in Lusas et al. (1989), but technology has advanced rapidly and searching in the latest databases probably better serves the readers. Like earlier vitamin A and D supplementations to make margarines and spreads more nutritionally equivalent to butter, regulators are bringing soybean milks up to the protein content equivalent of cow's milk. Some soybean milks with added calcium are marketed currently. Teaming between disposable milk and juice packaging companies and soybean milk equipment manufacturers has been highly effective in expanding market coverage of soybean milk.

20–7.1 Practical Protein Chemistry

Approximately, 10 to 12% of the proteins of soybean are albumins (soluble in water and dilute salt solutions) and ~90% globulins (sparingly soluble in water but soluble in salt solutions). This explains their compatibility with other proteins, as in processing meat.

Soybean globulins precipitate if pH of their aqueous solution is lowered to the isoelectric point of ~4.2, and will resolubilize with continuing acidification (Fig. 20–21). Addition of alkali (typically NaOH) resolubilizes the protein. Both steps

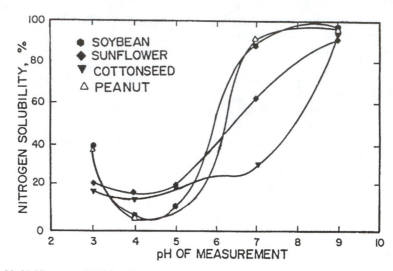

Fig. 20–21. Nitrogen solubilities of soybean, sunflower, cottonseed, and peanut proteins at different pH.

are used in making soybean protein isolates—first, protein solubilization at ~pH 9 to 10, then precipitation to ~pH 4.2 to harvest and wash the protein, followed by reneutralization to ~pH 6.5 to 7.0 to make the preparation more compatible with other ingredients in formulated foods. Also, protein loss, when making concentrates by aqueous extraction, is reduced by acidification to the isoelectric pH. Approximately 20% of total proteins (albumins and short-chain globulins) are lost in commercial extraction of soybean protein isolates unless recovered by ultrafiltration membranes. Although these short chain fractions include trypsin inhibitors and other troublesome compounds, they significantly enhance the EAA profile of soybean protein.

In addition to maximizing recovery of saleable protein, ingredient processors must strike a balance between retaining the protein solubility necessary for desired functional properties in the ingredient, and heat inactivation of anti-growth factors (see Section 20–8.2).

20–7.2 Preparation of Food Protein Ingredients

Three broad types of soybean food protein ingredients are available: (i) full-fat products made from whole soybean, (ii) partially-defatted flours and their texturized forms made by extrusion-screw pressing of soybean, and (iii) flours/grits, protein concentrates, protein isolates, and texturized flours and concentrates, made from hexane-extracted "white flakes" (Kanzamar et al., 1993; Lusas and Rhee, 1995). When making bland food protein ingredients, soybean of the best quality level should be selected, cleaned thoroughly with splits removed, and processed rapidly in close-coupled operations. U.S. No. 1 soybean is expensive, and still contains foreign seeds, splits, or damaged seed. The preferred approach is to have two parallel lines, the smaller one processing food-grade products and the other feed grade; or to at least have silos for holding rejects while the line is processing food ingredients. Separating food quality soybean from U.S. 2 and possibly 3 grade, provided the FFA is low, is common practice, but can mean as much as 25 to 35% rejects going to feed meal extraction. Processors in developing countries sometimes divide flakes leaving the solvent extractor into food and feed directed streams, but typically run into problems from insufficient cleaning as customers become more sophisticated and competitors offer improved products.

20–7.2.1 Full-fat Soybean Flour and Grits

Several types of full-fat soybean flour and grits are available:

1. Enzyme-active full-fat flours. Cracked, dehulled raw soybean is flaked or ground to specification, often with the assistance of an air classifier. Example specification: $42 \pm 1\%$ protein, moisture free basis (mfb); 10.0% max. moisture; 21.0 ± 0.5 fat; and $4.7 \pm 0.2\%$ ash.
   Enzyme-active full-fat flour is used for lipoxygenase bleaching of wheat flours in making white breads.
2. Toasted full-fat soybean flours and grits. Soybean is steamed under light pressure for 20 to 30 min, cooled, dried, cracked, hulls removed by shaker or aspirator, milled with sieving to produce full-fat grits or flour. Specification examples: 20 to 35 PDI; flour ground to pass U.S. No. 100 or 200 mesh screens (150, 73 micron openings, respectively); grit specifications: coarse: through No. 10 screen on No. 20 (2000, 841 microns); medium: through No. 20 on No. 40 (841, 420 microns), fine: through No. 40 on No. 80 (420, 178 microns). Products made from dry toasted soybean also are available.
3. Extruder produced full-fat flours—This is an updated extension of a process pioneered by Mustakas and co-workers at the USDA Northern Regional Research Center, Peoria, IL, in the late 1960s (Mustakas et al., 1970). Product also can be made by heating whole or dehulled soybean on a dry extruder to 138°C (280°F) (Fig. 20–15), cooling, and grinding. This is more than sufficient to inactivate the lipoxygenase that causes beany flavor (Zhu et al., 1996) and has given the best growth in chick studies (Zhang et al., 1993). Trypsin inhibitor activities have been reduced by as much as 98.8% in weaning foods for children.
   Except for bleaching by enzyme-active soybean flours used, full-fat ingredients are primarily used where both caloric and EAA contributions are desired, and where capital for establishing processing facilities is limited.

20–7.2.2 Partially Defatted Extruded and Texturized Flours

Traditionally, texturized soybean protein (TSP) has been made by extrusion of solvent-defatted soybean flour. A process, by-passing solvent extraction, has been described by Lusas and Riaz (1996b). Cleaned soybean (often freshly dehulled and aspirated to ensure thorough cleaning) is passed through a high-shear dry extruder of the Insta-Pro® type, but not using a conditioner, (Fig. 20–15). Temperature is raised to more than 121°C (250°) almost immediately and enzymes, including lipoxygenase which causes beany flavor, are inactivated (Zhu et al., 1996). Next, fat content of the extruded soybean is reduced to 6 to 8% by screw pressing. The cake then is ground into flour which passes a U.S. 80 screen (178 microns). The desired PDI is 40 to 60, but the extruder will texturize flour at 30 PDI or slightly lower. The flour is moistened to about 20%, and passed through the same type of dry extruder again, this time using the preconditioner. The soybean flour starts to "melt" and laminate at approximately 138°C (280°F). The product spews out of the extruder die in expanded irregular form, and is collected and cut into shreds of desirable size by a suitable mill. This process has been optimized and commercial equipment is available for making the product (Wijeratne, 2000). The Insta-Pro extruder has sufficient shear to texturize the product. A Wenger Manufacturing Company (Sabetha, KS) heated-jacket twin-screw extruder, properly configured and operating at 26% moisture, also will texturize the product, but at a lower temperature.

It was noticed early that, even though the flour contains 6 to 9% fat, it has an unexpectedly high resistance to oxidation. Darkening of the product over time, possibly a Maillard reaction because of the high content of reducing sugars, might be reduced by packaging in bags with reduced moisture permeability. Iowa State University researchers have characterized the flours, and concluded their potential usability in many food products (Crowe et al., 2001; Crowe and Johnson, 2001). Insta-Pro International Company researchers have developed various applications for the flours and have greatly improved on extrusion methods.

20–7.2.3 Solvent-Extracted White Flake Products

Ingredients like defatted flours, grits, protein concentrates, and protein isolates are made from "white flakes," whose preparation has previously been described (Johnson, 1989; Fulmer, 1989a; Kanzamar et al., 1993; Witte, 1995). Typically, a smaller, separate, processing line is dedicated to making food grade products, although the main line is "cleaned up" in some installations. When doing so, special attention should be given to cleaning the solvent separation tank and ensuring that clean solvent is used. White flakes are processed in a manner similar to that shown in Fig. 20–1, except they are intercepted as marc after extraction, and most of the solvent evaporated by flash desolventization (FDS) using superheated hexane as the energy source. Temperature of the solvent vapors is in the 135 to 175°C (275–347°F) range, and hexane content of the flakes is reduced to approximately 0.10% (1000 ppm). The flakes next are "stripped" under vacuum, with contact heating and minimum steam to prevent cooking, to attain ~0.0300% (300 ppm) hexane content. More steam is used when lower protein solubilities are desired. Some older installations throughout the world still use the Schnecken system of long closed tubes under vacuum, in which flakes are conveyed by ribbon-flight screw conveyors, and the heat for solvent evaporation is mainly provided by steam jackets. Development of FDS systems is highly active currently, with at least four new proprietary designs in field trials.

Two methods are used for determining protein solubility, Nitrogen Solubility Index (NSI; AOCS Ba 11-65) and Protein Dispersibility Index (PDI; AOCS Ba 10-65), sometimes referred to as the "slow stir" and 'fast stir" methods, respectively. Protein Dispersibility Index is the more complex, requires a specific mixer and calibration of operating conditions, but also is the more reproducible. Their relationship has been estimated by the equation: PDI = 1.07(NSI) + 1. White flake products include the following.

20–7.2.3.1 White Flakes. The Soy Protein Council has classified solvent-extracted, FDS desolventized flakes on the basis of NSI value: "white" (enzyme active): 85+; "cooked," 20 to 60; and "toasted," NSI < 20. White flakes are an item of commerce. One domestic supplier of high-PDI white flakes offers a product with 86 to 88% PDI; 80% lipoxygenase (bleaching activity) with a 2.2 minimum pH rise in urease activity; and the following granulation: 35% on U.S. No. 20 (841 microns); 45% through No. 20 on No. 100 (841, 150 microns); and 15% through No. 100 (150 microns).

20–7.2.3.2 Defatted Flour and Grits. These products typically contain 52 to 54% protein on an as is basis, and are produced by grinding white flakes, into various granulations to meet supplier needs. Domestic manufacturers generally offer soybean flours with PDIs of 90 (enzyme active), 70, 65 and 20; and granulations of U.S. 100, 200, and 300 and 400 mesh on special order (150, 73, 47, and 37 microns).

20–7.2.3.3. Refatted or Lecithinated Soybean Flours. From 1 to 15% refined oil sometimes is mixed back with the flour to reduce dustiness and provide fat for product formulations. Lecithinated flours are offered at 3, 6, and 15% added lecithin to assist dispersion in beverages and batters.

20–7.2.3.4. Soybean Protein Concentrates. These products typically contain 65 to 72% protein on a moisture-free basis. They are made by reextracting white-flakes in shallow-bed, loop-type solvent extractors with aqueous 60 to 70% ethanol, or by hydrochloric acid acidified aqueous leaching of white flake flour in tanks at pH 4.2 (Campbell et al., 1985; Beery, 1989). Either process removes the oligosaccharides, a major cause of soybean flatulence (Rackis, 1981b). Ethanol extraction is the more popular in the USA, but requires explosion-proof facilities because of concentrated solvent flammability. A more bland product, and denaturation of trypsin inhibitors and allergens is claimed, making the concentrate more suitable for use in baby foods and calf and piglet starters. When extracting with water, the dispersed solution is acidified to the isoelectric point (pH 4.2) to minimize protein solubility and concentrated by centrifugation—often using decanters. If flours are extracted, they are sprayed dried from slurries containing about 18 to 20% solids. Acidic water-extracted concentrates may be dried at the isoelectric point, or raised to near neutrality (with sodium or calcium hydroxides) to produce soybean concentrate proteinates. The original functional properties of FDS flour remain, except for the additional heat damage of processing.

Ethanol-extracted soybean protein concentrates reportedly have NSI values of 5 to 10. While they still exhibit valuable functional properties, they are "refunctionalized" to increase solubility by "refolding" the protein by mechanical shearing and working (Howard et al., 1980; Campbell et al., 1985; Beery, 1989; Chajuss, 1993, 2001).

20–7.2.3.5 Soybean Protein Isolates. The protein content of soybean protein isolates typically is 90 to 92% on a moisture-free basis. A flow sheet of the gen-

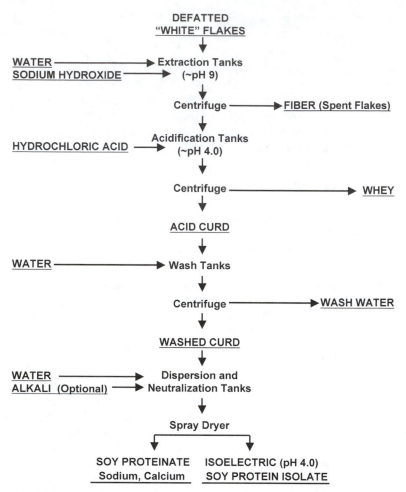

DEFATTED
"WHITE" FLAKES

WATER ──────────────▶ Extraction Tanks
SODIUM HYDROXIDE ──────▶ (~pH 9)

Centrifuge ──────────▶ FIBER (Spent Flakes)

HYDROCHLORIC ACID ─────▶ Acidification Tanks
(~pH 4.0)

Centrifuge ──────────────────────▶ WHEY

ACID CURD

WATER ────────────────▶ Wash Tanks

Centrifuge ──────────────▶ WASH WATER

WASHED CURD

WATER ────────────────▶ Dispersion and
ALKALI (Optional) ──────▶ Neutralization Tanks

Spray Dryer

SOY PROTEINATE   ISOELECTRIC (pH 4.0)
Sodium, Calcium   SOY PROTEIN ISOLATE

Fig. 20–22. Flow diagram for commercial preparation of soybean protein isolate.

eral process is shown in Fig. 20–22. White flakes, or their flours are extracted with sodium hydroxide at ~pH 9, and the fiber separated by centrifugation. The solution next is acidified to ~pH 4.0 to precipitate the protein ("curd"), which is separated by centrifuge. In commercial operations, a second washing often is made of the fiber to recover more protein. The centrifuged curd, which is in slurry form, can then be sprayed to produce an isoelectric soybean protein. Alternatively, the curd may be neutralized to near neutrality with sodium or calcium hydroxides and spray dried.

Ultrafiltration of protein solutions for separation by molecular size, and removal of soluble salts by reverse osmosis, are appropriate tools for producing soybean protein isolates and reducing quantities of waste water that must be discarded. Lawhon and Lusas (1984) reported preparation of a very bland product by separating a soybean protein solution with a 100 000 molecular weight cut off ultrafiltration membrane. The off-flavor compounds either directly permeated the membrane or were adsorbed to the small proteins that passed through (Lawhon,

Table 20–3. Composition of soybean food protein products (Endres, 2001; with permission).

Constituent	Defatted flours and grits		Concentrates		Isolates	
	As is	mfb†	As is	mfb	As is	mfb
			%			
Protein, (N × 6.25)	52–54	56–59	62–69	65–72	86–87	90–92
Fat (petroleum ether)	0.5–1.0	0.5–1.1	0.5–1.0	0.5–1.0	0.5–1.0	0.5–1.0
Crude Fiber	2.5–3.5	2.7–3.8	3.4–4.8	3.5–5.0	0.1–0.2	0.1–0.2
Soluble fiber	2	2.1–2.2	2–5	2.1–5.9	<0.2	<0.2
Insoluble fiber	16	17.0–17.6	13–18	13.5–20.2	<0.2	<0.2
Ash	5.0–6.0	5.4–6.5	3.8–6.2	4.0–6.5	3.8–4.8	4.0–5.0
Moisture	6–8	0	4–6	0	4–6	--
Carbohydrates (by difference)	30–32	32–34	19–21	20–22	3–4	3–4

† mfb = moisture-free basis.

1983). Many techniques are trade secrets, and the extent of membrane utilization in soybean proteins processing today, especially in combination with enzymes, isn't publicly known. Enzymatic modification also would occur before drying of soybean protein concentrates and isolates. It already plays a major role in adapting ingredients to specific applications, but processes are proprietary (Uhlig, 1998).

20–7.2.3.6 Fractionation of Soybean Protein Isolates. Further separation of soybean protein isolates into 7S (β conglycinin) and 11S (glycinin) globulin fractions has long been researched for commercial applications. New product opportunities for the 11S fraction, which forms the matrix in tofu, are anticipated. Cottonseed protein is easily fractionated because of ~2.5 pH difference between the isoelectric points of its 11S "storage protein" (pH 6.5) and 7S "nonstorage protein" (pH 4.0). After cottonseed protein is solubilized at pH 9 and separated from fiber by centrifugation, it is acidified to pH 6.5 to precipitate the storage protein recovered by centrifugation, then further acidified to pH 4.0 to precipitate and recover the nonstorage fraction. The (11S) fraction also is highly soluble in acidic beverages like orange juice (Lusas and Jividen, 1987). Unfortunately, the isoelectric points of 11S and 7S soybean proteins differ by <1 pH, making them more difficult to fractionate by pH precipitation. Earlier attempts were made by Kinsella (1979) and Gibson and Yackel (1989). Recently, researchers at Iowa State University (Wu et al., 1999, 2000) separated soybean 7S and 11S fractions in pilot plant quantities and reported purities of 90.4 and 72.7%, respectively. Enzymes have come into broad use in preparing soybean food and feed ingredients and final food products. Information is proprietary but available from enzyme suppliers to potential customers.

20–7.3 Use of Soybean Food Proteins

Compositions of the major soybean protein food protein ingredients are shown in Table 20–3 (Endres, 2001). Functional properties of soybean ingredients in food formulas include solubility, water absorption and binding, increasing viscosity, firming-gelation, cohesion-adhesion, elasticity, emulsification, fat absorption, flavor-binding, and aeration-foaming in whipped products (Kinsella, 1979).

Table 20–4. Relationship between Protein Dispersibility Index (PDI) and applications of defatted soybean flour in foods (Fulmer, 1989a; with permission).

PDI†	Application
90+	White bread bleaching (enzymatically active) Fermentation Soybean protein isolates, fibers
60–75	Controlled fat and water absorption Doughnut mixes Bakery mixes Pastas Baby foods Meat products Breakfast cereals Soybean protein concentrates
30–45	Meat products Bakery mix Nutrition, fat and water absorption, emulsification
10–25	Baby foods Protein beverages Comminuted meat products Soups, sauces, gravies Hydrolyzed vegetable proteins
Soybean grits	Nutrition, meat extender Patties, meatballs and loaves, chili, sloppy joes Soups, sauces, and gravies

† Protein Dispersibility Index, AOCS Ba 10-67 (97).

Protein Dispersibility Index requirements for various applications of defatted soybean flour/flakes are shown in Table 20–4 (Fulmer, 1989a). Flours with 90+ PDI are "enzymatically active" and can be used at up to 0.5% in the USA for bleaching yellow colors in wheat flour. They are used at 2 to 3% in Chinese noodle and pasta formulas to strengthen wheat gluten in low-protein flours, reportedly through oxidation by lipoxygenase. Generally, water-holding capabilities of soybean proteins decrease, and fat-holding capacities increase, as PDI deceases. But, this does not always happen, and bears checking out in the test kitchen. By selecting the proper protein, formula costs often can be appreciably reduced by retaining more water in baked products, and saleable yields of precooked meats increased by reducing shrinkage from syneresis of fat and juices.

The quality level of soybean food proteins generally has moved up a notch in the last two decades. Whereas food formulators once specified soybean concentrates to avoid the beany flavor of soybean flour, because of better lipoxygenase control, this defect is less common among the large soybean processors. A high-sucrose (low stachyose) soybean, whose flours were anticipated to reduce flatulence problems, has been developed, but is not in broad distribution. Several processes for its use in preparing food ingredients, including soybean flours and protein concentrates, have been patented (Kerr and Sebastian, 1998, 2000; Johnson, 1999). Functional ("refolded") soybean protein concentrates are more bland than flours, and have been applied in meat pumping, which formerly used soybean protein isolates. However, world production of soybean protein isolates still remains appreciably larger than concentrates.

Enzymes are employed in processing or utilization of various soybean ingredients. Suppliers are very willing to recommend types of enzymes and extent of hydrolysis to inactivate allergens or to avoid bitterness in modified proteins to customers. A patent was recently awarded for extracting completely soluble soybean protein isolate. Defatted flour is first extracted with water (plus minimum salts where appropriate). The insoluble protein then is hydrolyzed by peptidases until it also becomes extractable (Jiang et al., 2002).

Various other processes for using soybean food proteins are available, but their capabilities are not widely appreciated. For example, hydrothermal cooking (also known as "steam jet cooking") can convert soybean flours to soybean milk, or modify functionalities of defatted protein flours, concentrates, and isolates in 30 s (Wang and Johnson, 2001).

20–7.3.1 Extrusion Processing

Extruders are continuous cooking devices which mix and compress particulate formulas under high pressure and shear to achieve flowable plastic characteristics. Temperatures typically are well above the boiling point of water. The products are cut after passing through shaping dies and expand ("puff") because pressurized steam is released. Water is not added in dry extrusion, and heat is generated solely by product shear as the motor turns the extruder screw. Dry extruders typically operate with an upper moisture range of ~20 to 22% (Fig. 20–15).

Cooking extruders can accommodate water, steam, and other liquid additives injected into barrel with the dry mix. A significant amount of heat usually is generated by mechanical shear, but the barrel jacket may additionally be heated by high pressure steam, thermal fluid, or electricity. Theoretically, there is no upper limit to moisture content of the formula, although the practical maximum is about 35 to 40% for producing traditional breakfast cereals, pet foods, and texturized soybean flour or concentrate proteins (TSP) that resemble meat. A modern cooking extruder consists of (i) a reliable system for supplying a particulate mix, (ii) an adjustable speed feeder to introduce the mix at a constant rate, (iii) a preconditioner to lengthen equilibration time of dry ingredients and added moisture, (iv) a cooking extruder barrel and screw, (v) a die for shaping the extrudate, and (vi) a knife to cut cross-sections of the extrudate (Fig. 20–23). Cooking extruders usually are followed by a dryer and sometimes a flavor or fat applicator. The twin-screw, self-wiping, co-rotating extruder became popular in the late 1980s and early 1990s. The positive feed characteristics provide a nonpulsating discharge which enables improved consistency of product length (thickness). Twin-screw extruders can handle materials that are very sticky, or contain 6 to 9% higher fat content than single-screw machines. It was believed that PDIs over 60 were required to texturize soybean flours or concentrates when only single-screw extruders were available. Soybean flours with PDIs in the 20s, and ethanol-extracted concentrates at 10 PDI can be texturized with twin-screw machines (Kearns et al., 1989). Currently, the far majority of extrusion sys-

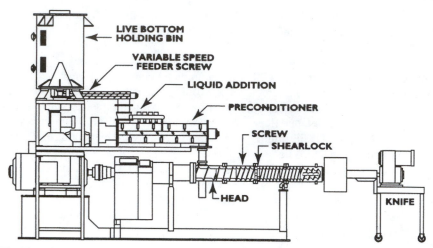

Fig. 20–23. Cooking and texturized soybean protein extruder. (Courtesy of Wenger Manufacturing Co., Sabetha, KS.)

tems are shipped with computer controls that can start-up, adjust extruder feed rates and conditions to ensure consistent products, and shut the machine down without jamming.

Examples of dry meat-like TSP products, that might be used in stews, pizza toppings, and extended hamburgers after hydration, are shown in Fig. 20–24. "Meatless" (vegetarian) prepared foods are available. But, significant amounts of TSP are added to improve appearance of products which already meet legal requirements for meat content. Figure 20–24 also shows a hydrated structured meat analog-type product, which has been marketed for nearly 25 yr, and two recently developed forms (i) a fine fiber-like ingredient that can be used for restructuring finely-ground meat, poultry, or seafood pastes and (ii) a flaky bar-like form that is extruded at 70% moisture content and may have potential as a delicatessen product.

20–7.3.2 Bakery Products

Applications of soybean food proteins in bakery goods is shown in Table 20–5 (Fulmer, 1989b). Their typical rates of use are ~3 to 4%. Competition for water with other ingredients in the formula is a common problem with soybean food proteins. Solutions have included use of emulsifiers, like sodium-strearyl-lactylate, pre-toasting the protein ingredient to reduce its solubility, or ordering a concentrate with a lower PDI.

20–7.3.3 Meat Products

Considerable innovation has been shown in using soybean food proteins in meat products. (Bonkowski, 1989). Soybean flours and concentrates bind up to three times their weight in water, whereas nonfat dry milk solids bind only an equal weight

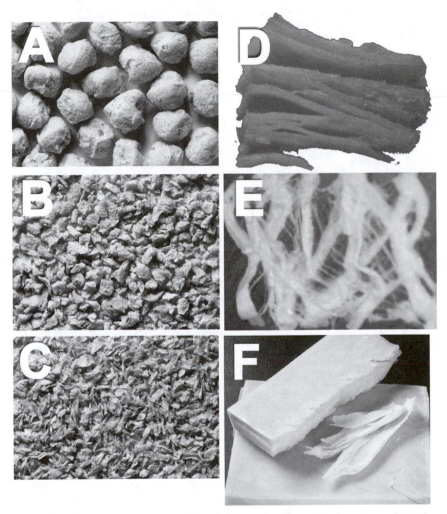

Fig. 20–24. Textured soybean proteins (TSP) made by extrusion. Dry meat replacements and extenders: (A) chunk style; (B) minced style (Urschel cut); (C) flaked (hammer milled). Rehydrated products: (D) structured meat analog (SMA) showing layers; (E) fiber soybean protein (FSP); (F) high moisture meat analog (HMMA) extruded at ~ 70% moisture. (Courtesy of Wenger Manufacturing Co., Sabetha, KS.)

of water. The USDA Food Safety Inspection Service allows uses of soybean proteins (i) as processing aids in manufacture of frankfurters, sausages, and comminuted meats, (ii) in marinades and tumbling solutions for restructuring of meats, (iii) injection pumping of protein solutions to increase the weight of intact muscles and cuts, and (iv) rehydration of extruder-texturized flours and concentrates that can be used at up 30% in hamburgers. Regulations allow:

1. Use of up to 3.5% soybean flour or concentrate, or up to 2% soybean isolate as processing aids in sausages and frankfurters.

Table 20–5. Bakery food applications of soybean food protein ingredients (Fulmer, 1989b; with permission).

Ingredient	White bread and rolls	Specialty bread and rolls	Cakes	Cake doughnuts	Yeast-raised doughnuts	Sweet goods	Cookies
Defatted soybean flour	X	X	X	X	X	X	X
Enzyme-active flour	X						
Low-fat soybean flour			X	X		X	
High-fat soybean flour			X	X		X	
Full-fat soybean flour			X	X		X	
Lecithinated soybean flour		X	X			X	
Defatted soybean grits		X					
Soybean protein concentrates		X					
Soybean protein isolates	X	X	X	X			
Soybean fiber		X					

2. Restructuring of meats by flaking/chunking poultry or red meats into small pieces; tumbling or mixing in marinades (consisting of salt, polyphosphates, soybean flour/concentrates/isolates, flavorings and water); filling into shaping casings or loaf pans; and heat setting at 68°C (154°F). After cooling and firming, the products can be sliced as luncheon meats.

3. Pumping of meats using brines consisting or water, salt, polyphosphates, and soybean protein concentrates or isolates into muscle cuts using stitch pumps. Hams and corned beef can be pumped to achieve cooked yields of 130%, but the product must contain a minimum of 17% protein (Rakes, 1993).

4. Procedures have been developed for preparing a soybean protein isolate, polyphosphate, poultry skin and fat emulsion, into which lower quality meats can later be incorporated in countries with less rigid laws (Bonkowski, 1989).

5. Soybean flours or protein concentrates may be texturized by extrusion; rehydrated to 18% protein content (60–65% moisture content); and used at up to 30% of total weight in ground meat blends and hamburgers. In practice, about 20% extension has been used because of texture and (bland) flavor problems at higher levels.

6. Incorporation of TSP particles into processed meats is common practice in some countries. Some sausage makers have developed procedures for making soybean granules in the silent cutter ("bowl chop") from dry ingredients as needed (Payne and Egbert, 1997, 1999).

20–7.3.4 Dairy and Dairy Showcase Products

Soybean protein concentrates and isolates have long been used in dairy case dips (Pedersen, 1993), hypo-allergenic infant formulas, frozen-tofu desserts, and as part of imitation cheeses made of sodium caseinate. Much research is in progress on soybean protein-extended dairy milks and milk-soybean protein cheeses (Borders et al., 2002a, 2002b; Gottemoller and True, 2002a, 2002b). Such combinations would have been strongly opposed by the dairy industry 20 yr ago.

20–8 SOYBEAN AS ANIMAL FEED

20–8.1 Solvent-extracted Soybean Meals in Feeding

In 2001/2002, soybean provided 68.3% of the world's feed protein meals, including fish meal. The U.S. feed consumption of soybean meal was ~27.6 Mt (30.4 million short tons). Use by species was: poultry (broilers), 44%; pigs, 24%; beef cattle, 13%; poultry (eggs), 7%; dairy cattle, 6%; pet foods, 3%; aquaculture, 1%; and others, 2.5%. Technology-based feed and food protein applications show the most promise for the future of soybean. There simply are no other major competing concentrated protein sources on the horizon currently.

No group of consumers is more knowledgeable about availability and relative values of alternative ingredients than professional feeders, who convert protein and calorie sources into animal products. They rely on buying advice from nutritionists, equipped with personal computers and least cost linear programs, who are less susceptible to promotional programs than the general public in the grocery products industry. By far, the poultry broiler industry, is the largest soybean meal customer, and this market is not saturated yet. Nutrition knowledge and industrialization of pork production have advanced rapidly in the Western countries and China. Many persons are not aware of gains made in improving feed conversion ratios, disease control, and establishment of large sow operations (some with 1000–3000 animals with each sow weaning 10 or more piglets at least twice a year). Aquaculture already is a significant soybean protein user; but only about a dozen and a half-species are cultivated currently, with a potentially 200 or more available. Each new species requires establishing techniques for reproduction in captivity, and nutritional requirements for larval, infantile, growth, and finishing stages diets. Researchers in Israel have reported production of tilapia (*Oreochromis niloticus* L) with soybean meal as the only protein supplement, without any fish meal used during the life cycle (Viola et al., 1988; Digani et al., 1997).

Waste disposal problems and environmental regulations haunt intensive animal production industries, and range from large mounds of dairy cattle manure piled in California, to impoundments of wash-down water from large pig farms nationally, to biofilters and settling lagoons in coastal areas preparing disposal waters from aquaculture operations. Reportedly, soils and waters are becoming overburdened with nitrogen and phosphorous from feed lots run-off and surface applications of wastes. Least-cost feeding formulas do not spare on these two components yet. Major soybean processors have assigned personnel to developing more environmentally friendly feed formulas. Phytase is being fed to poultry to enable utilization of organic phosphorous in soybean meals, and reduce inclusion of inorganic phosphates in feed formulations. Soybean cultivars with reduced phytate content also have been developed.

20–8.2 Protein Solubility and Control of Antinutrients

Protein solubility in soybean meal is positively related to its digestibility and the presence of active anti-nutritional factors. The objective of (steam) toasting, immediately after solvent extraction-desolventizing, is to inactivate trypsin inhibitors, lectins (phytohemagglutinins), and other toxic compounds or anti-nutrients. Sixty percent of growth inhibition in rats fed soybean is estimated due to trypsin inhibitors, with the specific effect dependent on the animal species and amount of inhibitors in the diet (Rackis, 1972, 1981a; Liener, 1979, 1994). Most oilseeds contain anti-growth inhibitors, and animals have varying abilities of detoxification. Thus, 100% destruction of trypsin inhibitors is not necessary and may not solve specific anti-nutrient problems. Research has shown that feeding low-Kunitz inhibitor soybean, or lectin-free soybean is not as effective as toasting seed of traditional cultivars for animal growth (Anderson-Hafferman et al., 1992; Douglas et al., 1999, respectively). Rumen microorganisms are able to inactivate antinutrients, and cattle can tolerate appreciably higher levels of trypsin inhibitors than monogastric animals; unfortunately, they also degrade soybean protein and replace it with lower quality microbial protein.

The problems of optimizing soybean meal for monogastric animals include (i) determining how much moist heat to apply during meal toasting, but (ii) avoiding overheating, which destroys EAA, especially lysine and histidine, reduces metabolizable energy (ME), and increases synthetic lysine supplementation costs for the formula. In ruminants, a means for soybean to "escape" ("by-pass") microbial interception before passing to the abomasum also is desired. Finding an easy and meaningful test that relates the degree of soybean protein "cook" to its nutritional value for the specific animal specie has not been easy.

It would be most desirable to measure residual trypsin inhibitor directly (TI; AOCS Ba 12-75), but the test is considered too complex and lengthy for practical process control use in extraction plants. The heat inactivation curve of urease resembles that of trypsin inhibitors, and the broiler industry has adopted the residual Urease Activity (UA) test (AOCS Ba 9-58), even though urease itself is not important to the bird. Urease is assessed in terms of change in pH during the assay, with broadly accepted values of 0.05 to 0.2 Δ pH for toasted meals, although some growers have reported success with meals as high as 0.4 to 0.6 Δ pH. High UA values indicate undercooked meal; low values indicate trypsin inhibitors have been destroyed, but not the degree of overcooking (Leeson and Summers, 2001).

A 0.2% KOH Protein Solubility Test (PS) was developed to solve this problem, but found extremely sensitive to meal particle size and methods of sample preparation (Araba and Dale, 1990a, 1990b). Some researchers were concerned that pH of the solution at 0.2% KOH is ~12.6, a condition where KOH is known to be a powerful protein dissociator and far above what would be encountered in the digestive tract. Various researchers have reported the test has limited use for detecting under processed soybean meal.

Batal et al. (2000) compared autoclaving time, broiler weight gain, and gain:feed ratio to PS, UA, PDI (AOCS Ba 10-65) and TI, and found the PDI test most indicative of trypsin inhibitor activity in broilers, with PDI values of 45% or lower indicating adequate processing (Table 20–6).

The poultry industry's acceptance of PDI would have advantages of (i) expanding the scope of an established method, with laboratory equipment and 50 yr experience in controlling vegetable food protein solubility already in place and (ii) putting food protein solubility and poultry feed protein solubility on the same scale, which still could include very low-solubility rumen escape protein (often in the 10–12 PDI range).

Table 20–6. Effect of autoclaving soybean flakes on chick performance, protein solubility, urease index, and protein dispensability index (Batal et al., 2000; with permission).

Autoclaving time	Weight gain	Gain:feed ratio	Protein solubility	Urease index	Protein dispersibility index	Trypsin inhibitor
min	g	g:g	%	units of pH change	PDI	mg g^{-1}
0	178	0.578	97	2.40	76	44.2
6	180	0.557	93	2.20	63	31.0
12	189	0.599	93	2.10	63	26.8
18	204	0.671	94	1.80	47	12.3
24	207	0.685	81	0.20	30	3.4
30	205	0.678	81	0.30	32	4.5
36	210	0.682	78	0.10	24	2.6

20–8.3 Direct Feeding of Extruded/Screw Pressed and Roasted Soybean

Whole soybean roasters and low-cost extruders were developed to inactivate soybean anti-nutritional factors in the late 1960s for local feeding, and continue to be improved. Readers have already been introduced to Insta-Pro type extruders and their products (see Sections 20–5.1.2, 20–7.2.2 and Fig. 20–15). Full-fat and partially-defatted soybean meals, made by this equipment and similar processes, have been well accepted in poultry, pig, and cattle feeding operations (American Soybean Association, 1996; American Soybean Association-United Soybean Board, 1996). These products generally are in the 15 to 30 PDI range. Equipment operating conditions could be adjusted to more closely control products within the 12 to 40 PDI range if the feeds industry used PDI as the method of analysis.

Oil in full-fat and partially defatted soybean meal increases its caloric density, which is important in feeding monogastric animals and high-producing dairy cows. An estimated 50% of Wisconsin's soybean crop was home fed in the mid-1990s. Official statistics are not available, but domestic direct feed use of full-fat and partially-defatted soybean has been estimated at ~12%. The portion of exported soybean fed to animals without extraction also is increasing, although industry protection laws in some countries allow only established oilseed extraction plants to also process full-fat soybean meal.

Dry-roasted soybean is used in the dairy industry. Some feeders believe reduction of protein solubility by dry roasting improves rumen escape, and also increases butterfat production (Satter et al., 1991; Bates, 1994). Rather than grind roasted soybean, some feeders prefer to crack the hulls, and feed the seed as splits or no smaller than four pieces per seed. These products are in the 12 to 20 PDI range.

20–9 INDUSTRIAL USES OF SOYBEAN OIL AND MEAL

With near saturation of domestic oil and protein needs for food and feed by soybean, and a significant export market in hand, U.S. research for additional markets has shifted to potential industrial ("nonfood nonfeed") applications. The history of industrial utilization of soybean has been reviewed previously (Johnson et al., 1992a, 1992b; Meyers, 1993; Johnson and Meyers, 1995; Shurtleff and Aoyagi, 1995). As with other oilseed crops, soybean oil was first used in industrial applications, including oil for illuminating, lubrication, coatings, boat caulking, and making soaps and glycerin. Records of meal used in early glues and plastics go back to the early 1900s.

20–9.1 Chemurgy Revisited

The Great Depression of the 1930s demonstrated that soybean oil and meal could serve many industrial needs, once the energies of U.S. industry, federal laboratory and university researchers were focused on utilization problems. Major applications from the 1920s, and later, which continued in use for many years, include (i) moldable plastics from formaldehyde-treated soybean meal, (ii) hot air- and steam-blown soybean oils to increase oxidation and polymerization for coatings use, (iii) soybean oil-meal glue for the plywood industry, (iv) soybean oil alkyd resins used in paints, (v) soybean protein isolate sizings for coating paper and textiles, (vi) improved soybean protein adhesives, (vii) soybean protein isolate textile fibers (which are being researched again in China), and (viii) lime-hydrolyzed soybean protein isolate foams for fire fighting. Post World War II developments include(ix) epoxy plasticizers from soybean oil for use in plastics (1950s), (x) methyl- and ethyl-fatty acid esters for diesel fuel, "Soy Biodiesel," (late 1970s), (xi) soybean oil-based inks (early 1980s), (xi) use of soybean oils for aerial pesticides applications (mid-1980s), and (xii) use of soybean and other edible oil sprays as dust suppressants to reduce explosions in grain elevators, and improve animal health in rearing facilities (1980s) (Johnson et al., 1992a, 1992b; Meyers, 1993; Johnson and Meyers, 1995).

Many of these applications became outdated by availability of lower-cost petroleum carbon sources. Researchers today realize that some of the earlier soybean industrial utilization problems resulted from natural degradability of the raw materials. Many can be solved with the additional knowledge that has accumulated. Also, the deficiencies can be turned to advantages in applications where biodegradability is desired.

Business prospects for chemurgic products are different now than in 1937, especially with the USA committed to an open global trade policy. They are based on three factors: (i) building ever-larger companies and production facilities to maximize economy of scale also reduces their flexibility to respond to smaller projects, (ii) the world's population explosion, and (iii) sensitization of the public to environmental issues. More people, with greater diversity of wants, means markets for a greater diversity of products. Large companies are less flexible in responding to smaller product markets, but still can provide the partially processed ingredients for smaller businesses who pursue smaller markets. We now are entering the "Niche Market Age." While the niches may individually seem small as fractions of a multi-national company's total operations, their aggregate needs can still be a significant and profitable industrial ingredients businesses. Niche opportunities exist in growing special types of crops, and in producing specialized products. For the latter, the processor does not have to go back to purchasing and extracting soybean, but can buy partially processed streams for final modification. The concept of suppliers of food and feed ingredients works equally well for suppliers of industrial ingredients Successes are being found in the niche area in the current chemurgic movement.

20–9.2 Industrial Uses of Soybean Oil

An estimated 288 000 t of soybean oil was used in domestic industrial applications in 1999/2000. Of this, approximately 58% was for fatty acids, soap, and feed uses; 18% for resins and plastics; 17% for inks; 6% for paints and varnishes; 1% for biodiesel, and <1% for solvents. The situation has changed substantially with numerous biodiesel plants built, and more methyl-soyate being shipped 3 yr later. The chemistry of fatty acids processing and uses has been described by Johnson and Fritz (1989), and animal feed uses reviewed by Lusas and Riaz (1996a).

20–9.2.1 Fatty Acid Methyl Esters

Pathways for converting oils and fats into various oleochemicals (Zoebelein, 1992; Johnson and Meyers, 1995) are shown in Fig. 20–25. Fatty acid methyl esters (FAME) are the gateway to many products. The interesterification process is conducted by alcoholysis, and also yields glycerol, a by-product much in demand.

20–9.2.2.1 Biodiesel. This product has been called "methyl soyate" if made from soybean oil. Tests during the latter 1970s to early 1980s showed that diesel engines can initially run on plant triglycerides, or in mixtures with diesel fuel. But, despite various additives to the fuel and engine oil, problems eventually were encountered with fuel injection valve clogging, cylinder head carbon deposits, and engine oil fouling. These were solved by making methyl- or ethyl-fatty acid esters by alcoholysis type of interesterification. Esters do not easily break down to FFA, thus

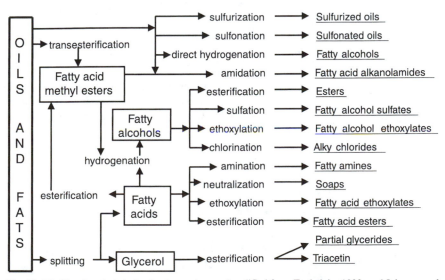

Fig. 20–25. Oleochemical derivatization pathways (modified from Zoebelein, 1992, and Johnson and Meyers, 1995; with permission).

minimizing corrosion problems during storage or use. They can be made in various viscosities by prior hydrogenation of the oil if needed. As equipment operating temperatures become cooler, they will thicken to the point where heat is needed to keep them fluid. Processes and principles described earlier in this chapter helped reduce this problem. Winterization of methyl soyate reduced crystallization temperature by 7.1°C (Lee et al., 1996). Mixing in unlike molecular structures, specifically addition of isopropyl and 2-butyl (branched alcohol) esters lowered crystallization temperature of soybean methyl esters by 7 to 11°C and 12 to 14°C, respectively (Lee et al., 1995).

20–9.2.1.2 Solvent Uses. Methyl esters are effective solvents for grease and resin-type stains. Current uses include graffiti, stains, and sticky deposits cleaners; light lubricants; degreasing baths, inclusion in penetrating oils; asphalt and concrete mold release agents; and adjuvants in various applications. Considerable investments in expanding manufacturing facilities have been made in the past year in expectation of rapid growth of this industry.

20–9.2.2 Other Soybean Oil Industrial Uses

Besides methyl fatty acid esters, current areas of soybean oil industrial research are plastics, coatings, lubricants, and hydraulic fluids. Potential applications are only limited by imagination, economics, and business skills of the respective entrepreneur. The United Soybean Board maintains a website (www.unitedsoybean.org) of current soybean oil-based industrial products manufacturers, listed under the categories of adjuvants, alternative fuels and fuel additives, building and construction, cleaners, concrete, dust suppressants, engine oils, hydraulic fluids, ingredients, metal working fluids, printing, and miscellaneous. The categories have as many as several hundred products and suppliers each. In many cases, suppliers list their web sites for interested persons to learn more about uses and specifications of products.

20–9.3 Industrial Soybean Protein Applications

Gradually, processes and products are being developed to win back nonfood, nonfeed markets formerly held by soybean, and expand into new areas.
Currently progress in industrial soybean protein applications is being made in:
1. Adhesives for wood products
2. Structural and moldable composites
3. Paper coatings
4. Sizings for textiles
5. Texturizing and embodying agents for paints, caulking compounds, etc.
6. Drilling muds

Many applications are announced, then little is heard about them while patents are being pursued and contacts negotiated. As shown in Table 20–7, many diverse technologies are involved. Periodic browsing of the United Soybean Board and American Soybean Association web sites is one of the better means for stay-

Table 20–7. Examples of industrial uses of soybean and soybean products. (Source: various publications and Internet pages, American Soybean Association and United Soybean Board, St. Louis, MO.)

Soybean oils	Soybean oils	Soybean lecithin	Flours, concentrates isolates
Anti-corrosion agents	Linoleum backing	Anti-foaming agents:	Adhesives
Anti-static agents	Lubricants	Alcohol	Antibiotics
Candles	Metal casting/working oils	Yeast	Asphalt emulsions
Caulking compounds	Methyl esters in:	Dispersing agents:	Calf starters
Composite materials	Soybean diesel fuel	Inks	Composite building materials
Concrete release agents	Solvents	Magnetic tapes	Fermentation aids/nutrients
Core oils	Oiled fabrics	Paint	Fibers
Crayons	Paints and finishes	Paper	Packaging films
Dust control agents	Pesticide carriers	Pesticides	Fire fighting foams
Electrical insulation	Plasticizers	Pharmaceuticals	Fire resistant coatings
Epoxy resins	Plastics	Synthetic rubber	Inks
Fatty acids	Printing inks	Softening & curing leather	Leather substitutes
Fatty alcohols	Protective coatings	Viscosity modification,	Particles boards
Form release agents	Putty	concrete, drilling muds	Paper coatings
Fungicides	Soaps/shampoos/detergents		Pesticide/fungicide carriers
Glycerin in	Solvents		Pharmaceuticals
Explosives	Vinyl plastics		Plastics
Industrial uses	Wallboard		Polyesters
Pharmaceuticals	Waterproof cement		Textile sizing
Hydraulic fluids			Water-based paints

ing abreast of research funding and developments in industrial applications of soybean oils and proteins.

REFERENCES

Akoh, C.C. 1998. Reduced-calorie fats and fat replacements. p. 223–234. *In* S.S. Koseoglu et al. (ed.) Current practices, quality control, technology transfer, and environmental issues. AOCS Press, Champaign, IL.

Altschul, A.M., and H.L. Wilcke (ed.) 1985. New protein foods. Vol. 5. Seed storage products. Academic Press, New York.

American National Standards Institute. 2003. The American National Standards Institute, 25 West 43 rd St., 4th Floor, New York, NY. www.ansi.org (verified 26 June 2003).

American Soybean Association. 1996. Fullfat soya handbook. Am. Soybean Assoc., St. Louis, MO.

American Oil Chemists' Society Official Methods. 2002. Official methods and recommended practices of the AOCS. AOCS Press, Champaign, IL.

American Soybean Association-United Soybean Board. 1996. Second International Fullfat Soya Conf.: Processing, Quality Control, Utilization, Budapest, Hungary. 21–24 Aug. 1996. USDA Foreign Agric. Serv., Washington, DC.

American Soybean Associaton. 2001. Genetically modified organisms: A summary of regulatory aspects. Online. Available at www.asa-europe.org/pdf/biotech4.pdf (downloaded 19 Aug. 2001; verified 26 June 2003).

American Soybean Association-United Soybean Board. 2001. Soy importers' handbook. Online. Produced by American Soybean Association, United Soybean Board, and Missouri Soybean Merchandising Council. Available at www.asasea.com/technical/soy.html (downloaded 9 July 2001; verified 26 June 2003).

Anderson-Haffermann, J.S., Y. Zhang, C.M. Parsons, and T. Hymowitz. 1992. Effect of heating on nutritional quality of conventional and Kunitz trypsin inhibitor-free soybean. Poult. Sci. 71:1700–1709.

Appel, W.B. 1994. Appendix E: Physical properties of feed ingredients. p. 347–417. *In* R.R. McEllhiney (ed.) Am. Feed Industry Assoc., Arlington, VA.

Applewhite, T.H. (ed.). 1989. Proceedings of the World Congress: Vegetable Protein Utilization. *In* Human foods and animal feedstuffs, Singapore, October 1988. AOCS Press, Champaign, IL.

Araba, M., and N.M. Dale. 1990a. Evaluation of protein solubility as an indicator of overprocessing soybean meal. Poult. Sci. 69:76–83.

Araba, M., and N.M. Dale. 1990b. Evaluation of protein solubility as an indicator of underprocessing soybean meal. Poult. Sci. 69:1749–1752.

Ariaansz, R.F. 1993. Hydrogenation theory. p. 166–172. *In* T.H. Applewhite (ed.) Proc. of the World Conf. on Oilseed Technology and Utilization, Budapest, Hungary. 1992. AOCS Press, Champaign, IL.

Ariaansz, R.F., and D.V. Okonek. 1998. Isomer control during edible oil processing. p. 77–91. *In* S.S. Koseoglu et al. (ed.) Emerging technologies, current practices, quality control, technology transfer, and environmental issues. AOCS Press, Champaign, IL.

Artz, W.E. 1990. Emulsifiers. p. 347–417. *In* A.L. Branen et al. (ed.) Food additives. Marcel Dekker, New York.

Banks, D.E., and E.W. Lusas. 2001. Oils and industrial frying. p. 137–104. *In* E.W. Lusas and L.W. Rooney (ed.) Snack foods processing. Technomic Press-CRC Publ. Co., Boca Raton, FL.

Barger, W.M. 1981. Handling, transportation and preparation of soybean. J. Am. Oil Chem. Soc. 58:154–156.

Bastiaens, G. 2001. Key factors shaping the edible oil world. INFORM. 12:666–671.

Batal, A.B., M.W. Douglas, A.E. Engram, and C.M. Parsons. 2000. Protein dispersibility index as an indicator of adequately processed soybean meal. Poult. Sci. 79:1592–1596.

Bates, L.S. 1994. Dry heat processing of full-fat soy and other ingredients. Online. Am. Soybean Assoc. Tech. Bull. Available at www.asasea.com/technical/ft11-1994.html. Downloaded 16 Sept. 2001.

Beery, K.E. 1989. Preparation of soy protein concentrate products and their application in food systems. p. 62–65. *In* T.H. Applewhite (ed.) Proc. of the World Congress on Vegetable Protein Utilization in Human Foods and Animal Feedstuffs. Am. Oil Chemists' Soc., Champaign, IL.

Berger, R.W., D.L. Shores, and M.C. Thompson (ed.) 1986. Quality circles: Selected readings. Marcel Dekker, New York.

Bernard, J.K. 1999. Performance of lactating dairy cows fed whole cottonseed coated with gelatinized cornstarch. J. Dairy Sci. 82:1305–1309.

Birt, D.F., S. Hendrich, D.L. Alekel, and M. Anthony. 2004. Soybean and the prevention of chronic human disease. p. 1047–1118. *In* H.R. Boerma and J.E. Specht (ed.) Soybeans: Improvement, production, and uses. 3rd ed. Agron. Monogr. 16. ASA, CSSA, and SSSA, Madison, WI.

Bockish, M. 1998. Fats and oils handbook. AOCS Press, Champaign, IL.

Bogdanor, J.M., and A.L. Price. 1994. Novel synthetic silica adsorbents for refining of edible oils. Oil Mill Gazet. 99(12):32–34.

Bonkowski, A.T. 1989. The utilization of soy proteins from hot dogs to haramaki. p. 430–438. *In* T.H. Applewhite (ed.) Proc. of the World Congress on Vegetable Protein Utilization in Human Foods and Animal Feedstuffs, Singapore. 1988. Am. Oil Chemists' Soc., Champaign, IL.

Borders, C., V. Lobo, W.R. Egbert, L. True, and T. Gottemoller. 2002a. Use of isolated soy protein for making fresh, unripened cheese analogs. U.S. Patent 6 413 569. Date issued: 2 July. Assigned to Archer Daniels Midland Co., Decatur, IL.

Borders, C., V. Lobo, W.R. Egbert, L. True, and T. Gottemoller. 2002b. Method for use of isolated soy protein in the production of fresh, unripened cheese analogs and cheese analogs. U.S. Patent 6 495 187. Date issued: 17 December. Assigned to Archer Daniels Midland Co., Decatur, IL.

Campbell, M.F., C.W. Kraut, W.C. Yackel, and H.S. Yang. 1985. Soy protein concentrate. p. 259–299. *In* A.M. Altschul and H.L. Wilcke (ed.) New food proteins. Vol. 5. Seed storage proteins. Academic Press, New York

Carr, R.A. 1993. Oilseed harvesting, storage and transportation.p. 118–125. *In* T.H. Applewhite (ed.) Proc. of the World Conf. on Oilseed Technology and Utilization. AOCS Press, Champaign, IL.

Cerletti, P., and M. Duranti. 1979. Development of lupine proteins. J. Am. Oil Chem. Soc. 56:460–463.

Chajuss, D. 1993. Process for enhancing some functional properties of proteinaceous material. U.S. Patent 5 210 184. Date issued: 11 May.

Chajuss, D. 2001. Soy protein concentrate: Processing, properties and prospects. INFORM 12(12):1176–1180.

Cherry, J.P., and W.H. Kramer. 1989. Plant sources of lecithin p. 16–31. *In* B.F. Szuhaj (ed.) Lecithins: Sources, manufacture & uses. Am. Oil Chemists' Soc., Champaign, IL.

Coenen, J.W.E. 1976. Hydrogenation of edible oils. J. Am. Oil Chem. Soc. 53(6):382–386.

Crowe, T.W., and L.A. Johnson. 2001. Twin-screw texturization of extruded-expelled soybean flour. J. Am. Oil Chem. Soc. 78:781–786.

Crowe, T.W., L.A. Johnson, and T. Wang. 2001. Characterization of extruded-expelled soybean flours. J. Am. Oil Chem. Soc. 78:775–779.

Dahlke, K., and M. Eichelsbacher. 1998. EnzyMax® and Alcon® - Lurgi's route to physical refining. p. 53–59. *In* S.S. Koseoglu et al. (ed.) Emerging technologies, current practices, quality control, technology transfer, and environmental issues. AOCS Press, Champaign, IL.

Deming, W.E. 1986. What happened in Japan? p. 18–28. *In* R.W. Berger et al. (ed.) Quality circles: Selected readings. Marcel Dekker, New York.

DeSmet. 2001. Raising fuel costs—Extraction energy savers ease the pain. DeSmet Process & Technology News. June issue. DeSmet Process & Technology, Marietta, GA.

Digani, G., S. Viola, and Y. Yehuda. 1997. Apparent digestibility of protein and carbohydrate in feed ingredients for adult tilapia (*Oreochromis aureus* x *O. niloticus*). Israeli J. Aquaculture-Bamidgeh 49(3):113–123.

Dijkstra, A.J. 1993. Degumming, refining, washing and drying fats and oil. p. 138–151. *In* T.H. Applewhite (ed.) Proc. of the World Conf. on Oilseed Technology and Utilization. AOCS Press, Champaign, IL.

Douglas, M.W., C.M. Parsons, and T. Hymowitz. 1999. Nutritional evaluation of lectin-free soybeans for poultry. Poult. Sci. 78:91–95.

Economic Research Service. 1961. FOS-210, Fats and oils situation. November. ERS, USDA, Washington, DC.

Economic Research Service. 2002. Recent growth patterns in the U.S. organic foods market. AIB-777. September. ERS, USDA, Washington, DC.

Empie, M., and E. Gugger. 2001. Method of preparing and using isoflavones. U.S. Patent 6 261 565. Date issued: 17 July. Assigned to Archer Daniels Midland Co., Decatur, IL.

Empie, M., and E. Gugger. 2002a. Method of preparing and using isoflavones for the treatment of female symptoms. U.S. Patent 6 391 309. Assigned to Archer Daniels Midland Co., Decatur, IL.

Empie, M., and E. Gugger. 2002b. Method of preparing and using isoflavones for the treatment of neurological symptoms. U.S. Patent 6 391 310. Date issued: 21 May. Assigned to Archer Daniels Midland Co., Decatur, IL.

Empie, M., and E. Gugger. 2002c. Method of preparing and using isoflavones for the treatment of cancer. U.S. Patent 6 395 279. Date issued: 28 May. Assigned to Archer Daniels Midland Co., Decatur, IL.

Endres, J.G. 2001. Soy protein products: Characteristics, nutritional aspects and utilization. Revised and expanded ed. AOCS Press, Champaign, IL.

Erickson, D.R. (ed.) 1995. Practical handbook of soybean processing and utilization. AOCS Press, Champaign, IL.

Erickson, D.R., and M.D Erickson. 1995. Hydrogenation and base stock formulation. p. 218–238. *In* D.R. Erickson (ed.) Practical handbook of soybean processing and utilization. AOCS Press, Champaign, IL.

Evans, C.D., K. Warner, G.R. List, and J.C. Cowan. 1972. Room evaluation of oils and cooking fats. J. Am. Oil Chem. Soc. 49:578–582.

Fantozzi, P.E. (ed.) 1990. Proc. of the Third Int. Conf. on Leaf Protein Res. Special issue. Italian J. Food Sci. Chiriotti Editori, Pinerolo, Italy.

Farr, W.E. 2000. Refining of fats and oils. p. 136–157. *In* R.D. O'Brien et al. (ed.) Introduction to fats and oils technology. 2nd ed. AOCS Press, Champaign, IL.

Federal Register. 1992. Statement of policy: Foods derived from new plant varieties. Food and Drug Administration. Federal Register 57(104):22984–23005. 29 May.

Federal Register. 1999. Food labeling: *Trans* fatty acids in nutrition labeling, nutrient content claims, and health claims; proposed rule. Federal Register 64(221):62745–62825. 17 November.

Federal Register. 2000. 21 CFR Part 101. Regulations on statements made for dietary supplements concerning the effect of the product on the structure or function of the body; Final Rule. 6 January. p. 999–1050.

Firestone, D. 1996. Regulation of frying fat and oil. p. 323–334. *In* E.G. Perkins and M.D.

Food Agricultural Service. 2002. FOP 07-02, Oilseeds: Markets and trade. July. FAS, USDA, Washington, DC.

Food and Drug Administration/Center for Food Safety and Applied Nutrition. 1995. Dietary Supplement Health and Education Act of 1994. On line. U.S. FDA, CFSAN, Washington, DC. Available at www.cfsan.fda.gov/~dms/dietsupp.html (downloaded 1 Dec. 2002; posted 1 Dec. 1995; verified 26 June 2003).

Food and Drug Administration/Center for Food Safety and Applied Nutrition. 2001a. Overview of dietary supplements. On line. CFSAN. U.S. FDA. Available at www.cfsan.fda.gov/~dms/dsoview.html (downloaded 26 Dec. 2002; posted 3 Jan. 2001; verified 26 June 2003).

Food and Drug Administration/Center for Food Safety and Applied Nutrition. 2001b. Claims that can be made for conventional foods and dietary supplements. On line. CFSAN, U.S. FDA, Revised October 2001. Available at www.cfsan.fda.gov/~dms/hclaims.html (downloaded 26 Dec. 2002; posted October 2001; verified 26 June 2003).

Food and Drug Administration/Center for Food Safety and Applied Nutrition. 2002. Announcing CAERS—The CFSAN adverse event reporting system. On line. Office of Scientific Analysis and Support, CFSAN, U.S. FDA. Available at www.cfsan.fda.gov/~dms/caersltr.html.(downloaded 26 Dec. 2002; posted 29 Aug. 2002; verified 26 June 2003).

Food and Drug Administration/U.S.Department of Agriculture. 2000. United States food safety system. Precaution in US food safety decision making: Annex II to The United States' National food safety system paper. 3 March. FDA and USDA, Washington, DC.

Erickson (ed.) Deep frying: Chemistry, nutrition, and practical applications. AOCS Press, Champaign, Illinois.

Frankel, E.N. 1998. Lipid oxidation. The Oily Press, Dundee, Scotland.

Fulmer, R.W. 1989a. The preparation and properties of defatted soy flour and their products. p. 55–61. *In* T.H. Applewhite (ed.) Proc. of the World Congress on Vegetable Protein Utilization in Human Foods and Animal Feedstuffs, Singapore. 1988. Am. Oil Chemists' Soc., Champaign, IL.

Fulmer, R.W. 1989b. Uses of soy proteins in bakery and cereal production. p. 424–429. *In* T.H. Applewhite (ed.) Proc. of the World Congress on Vegetable Protein Utilization in Human Foods and Animal Feedstuffs, Singapore. 1988. Am. Oil Chemists' Soc., Champaign, IL.

Gibson, P.W., and W.C. Yackel. 1989. Soy protein fractional and applications. p. 507–509. *In* T.H. Applewhite (ed.) Proc. of the World Congress on Vegetable Protein Utilization in Human Foods and Animal Feedstuffs, Singapore. 1988. Am. Oil Chemists' Soc., Champaign, IL.

Glickman, D. 1999. New crops, new century, new challenges: How will scientists, farmers and consumers learn to love biotechnology and what happens if they don't?, On line. Available at www.useu.be/issues/glick0715.html (downloaded 19 Nov. 2001; posted13 July 1999; verified 26 June 2003).

Gottemoller, T., and L. True. 2002a. Soy extended cheese. U.S. Patent 6 383 531. Date issued: 7 May. Assigned to Archer Daniels Midland Co., Decatur, IL.

Gottemoller, T., and L. True. 2002b. Use of soy isolated protein for making fresh cheese. U.S. Patent 6 399 135. Date issued: 4 June. Assigned to Archer Daniels Midland Co., Decatur, IL.

Green, J. 1996. A practical guide to analytical method validation. Anal. Chem. 68:305A–309A

Gregory, J.F., III. 1996. Vitamins. p. 531–617. *In* O. R. Fennema (ed.) Food chemistry. 3rd ed. Marcel Dekker, New York,.

Gugger, E.T., and D.G. Dueppen. 1997. Production of isoflavone-enriched fractions from soy protein extracts. U.S. Patent 5 702 752. Date issued: 30 December. Assigned to Archer Daniels Midland Co., Decatur, IL.

Gugger, E.T., and R.D. Grabiel. 2000. Process for production of isoflavone fractions from soy. U.S. Patent 6 033 714. Date issued: 7 March. Assigned to Archer Daniels Midland Co., Decatur, IL.

Gugger, E., and R.D. Grabiel. 2001. Production of isoflavone enriched fractions from soy protein extracts. U.S. Patent 6 171 638. Date issued: 9 January. Assigned to Archer Daniels Midland Co., Decatur, IL.

Gustafson, E.H. 1976. Loading, unloading, storage, drying and cleaning of vegetable oil-bearing materials. J. Am. Oil Chem. Soc. 53:248–250.

Handbook. 2001. The First Amendment Handbook. On line. The Reporters Committee for Freedom of the Press, Arlington, VA. Available at www.rcfp.org/handbook/viewpage.cgi (downloaded 29 Nov. 2001; verified 26 June 2003).

Hanson, L.J. 1996. Expected animal response to the quality of full fat soya. p. 83–88. *In* Proc. of Second Int. Fullfat Soya Conf., Budapest, Hungary. 21–24 Aug. 1996. Am. Soybean Assoc., St. Louis, MO.

Hastert, R.C. 1990. Cost/quality/health: The three pillars of hydrogenation. p. 142–152. *In* D.R. Erickson (ed.) Edible fats and oils processing: Basic principles and modern practices. Am. Oil Chemists' Soc., Champaign, IL.

Hastert, R.C. 2000. Hydrogenation. p. 136–157. *In* R.D. O'Brien et al. (ed.) Introduction to fats and oils technology. 2nd ed. AOCS Press, Champaign, IL.

Herbert, M.R. 2000. Feasting on the unknown: Being exposed to one of the largest uncontrolled experiments in history. 3 September. Chicago Tribune, Chicago, IL.

Hernandez, E. 2001. Latest developments in refining and filtration of specialty oils with sodium silicates. Paper presented at Am. Oil Chemists' Society Annual Meet., Minneapolis, MN.13–17 May. Food Protein Res. and Develop. Center, Texas A&M Univ., College Station.

Hernandez, E., and S.J. Rathbone. 2002. Refining of glyceride oils by treatment with silicate solutions and filtration. U.S. Patent 6 448 423 B1. 10 September.

Hesseltine, C.W., R. F. Rogers, and R J. Bothas. 1978. Microbiological study of exported soybeans. Cereal Chem. 55:332–340.

Howard, P.A., M.F. Campbell, and D.T. Zollinger. 1980. Water-soluble vegetable protein aggregates. U.S. Patent 4 234 620. Date issued: 18 November.

Hron, R.J., Sr. 1997. Ethanol. p. 192–198. *In* P.J. Wan and P.J. Wakelyn (ed.) Technology and solvents for extracting oilseeds and nonpetroleum oils. AOCS Press, Champaign, IL.

Hui, Y. H. (ed.) 1996. Bailey's industrial oil and fat products. 5th ed. Vol. I–V. John Wiley & Sons, New York.

Institute of Food Science and Technology. 1999. Bovine Somatotropin (BST). On line. Available at www.ifst.org/hottop8a.htm. Downloaded 2 Nov. 2001 (verified. 26 Dec. 2002).

Jacobs, P. 1999. US, Europe lock horns in beef hormone debate. (Available through archives service.) Los Angeles Times, Los Angeles, CA. 9 April. Downloaded 2 Nov. 2001.

Jiang, H., and F. Hongwei. 2002. Method for extracting soybean proteins using an enzyme. U.S. Patent 6 335 043. Date issued: 1January .

Johnson, D.R. 1996. HACCP & US food safety guide. The Food Inst., Fair Lawn, NJ.

Johnson, D.W. 1989. General uses of soybeans. p. 12–38. *In* E.W. Lusas et al. (ed.) Food uses of whole oil and protein seeds. Am. Oil Chemists' Soc., Champaign, Illinois.

Johnson, L.A. 1997. Theoretical, comparative, and historical analyses of alternative technologies for oilseeds extraction. p. 4–47. *In* P.J. Wan and P.J. Wakelyn (ed.) Technology and solvents for extracting oilseeds and nonpetroleum oils. AOCS Press, Champaign, IL.

Johnson, L.A. 1999. Process for producing soy protein concentrate from genetically-modified soybeans. U.S. Patent 5 936 069. Date issued: 10 August.

Johnson, L.A., J.T. Farnsworth, N.Z. Sadek, N. Chamkasem, E.W. Lusas, and B.L. Reid. 1986. Pilot plant studies on extracting cottonseed with methylene chloride. J. Am. Oil Chem. Soc. 63(5): 647–652.

Johnson, R.W., and E. Fritz (ed.) 1989. Fatty acids in industry. Marcel Dekker, New York.

Johnson, L.A., and E.W. Lusas. 1983. Comparison of alternative solvents for oils extraction. J. Am. Oil Chem. Soc. 60:181A–242A.

Johnson, L.A., D.J. Meyers, and D.J. Burden. 1992a. Early uses of soy protein in Far East. U.S. INFORM 3:282–288, 290.

Johnson, L.A., D.J. Meyers, and D.J. Burden. 1992b. Soy protein's history, prospects in food, feed. IN-FORM 3:429–430, 432, 434, 437–438, 440,442–444.

Johnson, L.A., and D.J. Meyers. 1995. Industrial uses for soybeans. p. 380–427. In D.R. Erickson (ed.) Practical handbook of soybean processing and utilization. AOCS Press, Champaign, IL.

Journal of the American Oil Chemistry Society. 1974. Proceedings of World Soy Protein Conf., Munich.11–14 Nov. 1973. J. Am. Oil Chem. Soc. 51(1):47A–207A.

Journal of the American Oil Chemistry Society. 1979. Proceedings of World Conf. on Vegetable Food Proteins, Amsterdam. 29 Oct. – 3 Nov. 1978. J. Am. Oil Chem. Soc. 56(3):99–483.

Journal of the American Oil Chemistry Society. 1981. Proceedings of the World Conf. on Soya Processing and Utilization, Acapulco, Mexico. 9–14 Nov. 1980. J. Am. Oil Chem. Soc. 58(3):121–540.

Jung-M.Y., and Ha-Y.L. 1999. Conjugated linoleic acid isomers in partially hydrogenated soybean oil obtained during nonselective and selective hydrogenation processes. J. Agric. Food Chem. 47(2):204–708.

Juran, J.M. 1986. The quality circle phenomenon. p. 3–17. In R.W. Berger et al. (ed.) Quality circles: Selected readings. Marcel Dekker, New York.

Kalapathy, U., and A. Proctor. 2000. A new method for free fatty acid reduction in frying oil using silicate films produced from rice hull ash. J. Am. Oil Chem. Soc. 77:593–598.

Kanzamar, G.J., S.J. Predlin, D.A. Oreg, and Z.M. Csehak. 1993. Processing of soy flours/grits and texturized soy flours. p. 226–240. In T.H. Applewhite (ed.) Proc. of the World Conf. on Oilseed Technology and Utilization. AOCS Press, Champaign, IL.

Kearns, J.P., G.J. Rokey, and G.R. Huber. 1989. Extrusion of texturized proteins. p. 353–362. In T.H. Applewhite (ed.). Proc. of the World Conf. on Oilseed Technology and Utilization. AOCS Press, Champaign, IL.

Kellens, M., and W. De Greyt. 2000. Deodorization. p. 235–268. In R.D. O'Brien et al. (ed.) Introduction to fats and oils technology. 2nd ed. AOCS Press, Champaign, IL.

Kellens, M., and M. Hendrix. 2000. Fractionation. p. 194–207. In R.D. O'Brien et al. (ed.) Introduction to fats and oils technology. 2nd ed. AOCS Press, Champaign, IL.

Kerr, P.S., and S.A. Sebastian. 1998. Soybean products with improved carbohydrate composition and soybean plants. U.S. Patent 5 710 365. Date issued: 20 January.

Kerr, P.S., and S.A. Sebastian. 2000. Soybean products with improved carbohydrate composition and soybean plants. U.S. Patent 6 147 193. Date issued: 14 November.

Kerr, W.A., and J.E. Hobbs. 2000. The WTO and the dispute over beef produced using growth hormones. CATRN PAPER 2000-07. Canadian Agrifood Trade Research Network. Available at www.eru.ulaval.ca/catrn (downloaded 15 Aug. 2001; posted September 2000; verified 26 June 2003).

Kinsella, J.E. 1979. Functional properties of soy proteins. J. Am. Oil Chem. Soc. 56:242–258.

Kokken, M.J. 1993. Hydrogenation in practice. p. 173–179. In T.H. Applewhite (ed.) Proc. of the World Conf. on Oilseed Technology and Utilization. AOCS Press, Champaign, IL.

Kritchevsky, D. 1998. Triglyceride structure and atherosclerosis. p. 183–188. In A.B. Christophe (ed). Structural modified fats: Synthesis, biochemistry and use. AOCS Press, Champaign, IL.

Laird, W., T.C. Wedegaertner, T.D. Valco, and R.D. Baker. 1997. Engineering factors for coating and drying cottonseed to create a flowable product. Presented at the10–14 Aug. 1997 Am. Soc. of Agric. Eng. Annual Int. Meet. ASAE Paper 979-1015. ASAE, St. Joseph, MI.

Lampert, D. 2000. Processes and products of interesterification. p. 208–234. In R.D. O'Brien et al. (eds.) Introduction to fats and oils technology. 2nd ed. AOCS Press, Champaign, IL.

Lawhon, J.T. 1983. Method for processing protein from nonbinding oilseed by ultrafiltration and solubilization. U.S. Patent 4 420 425. Date issued: 13 December.

Lawhon, J.T., and E.W. Lusas. 1984. New techniques in membrane processing of oilseeds. Food Technol. 38(12):97–106.

Lee, I., L.A. Johnson, and E.G. Hammond. 1995. Use of branched-chain esters to reduce the crystalization temperature of biodiesel. J. Am. Oil Chem. Soc. 72:1155–1160.

Lee, I., L. A. Johnson, and E.G. Hammond. 1996. Reducing the crystallization temperature of biodiesel by winterizing methyl soyate. J. Am. Oil Chem. Soc. 73:631–636.

Leeson, S., and L.S. Summers. 2001. Scott's nutrition of the chicken. 4th ed. Univ. Books, Guelph, Ontario, Canada.

Liener, I. 1979. Significance for humans of biologically active factors in soybeans and other food legumes. J. Am. Oil Chem. Soc. 56:121–129.

Liener, I.E. 1994. Implications of antinutritional components in soybean foods. Crit. Rev. Food Sci. Nutr. 34(1):31–67.

Lin, L., K.C. Rhee, and S.S. Koseoglu. 1998. Recent progress in membrane degumming of crude vegetable oils on a pilot-plant scale. p. 76–82. *In* S.S. Koseoglu et al. (ed.) Advances in oils and fats, antioxidants, and oilseed by-products. AOCS Press, Champaign, IL.

List, G.R. 1989. Commercial manufacture of lecithin. p. 145–161. *In* B.F. Szuhaj (ed.)Lecithins: Sources, manufacture and uses. Am. Oil Chemists' Soc., Champaign, IL.

List, G.R., E.A. Emken, W.F. Kwolek, T.D. Simpson, and H.J. Dutton. 1977. "Zero *trans*" margarines: Preparation, structure, and properties of interesterified soybean oil-soy trisaturated blends. J. Am. Oil Chem. Soc. 54:408–413.

List, G.R., T.L. Mounts, A.C. Lanser, and R.K. Holloway. 1990. Effect of moisture, microwave heating, and live steam treatment on phospholipase activity in soybeans and soybean flakes. J. Am. Oil Chem. Soc. 67:867–871.

List, G.R., K.. Steidley, D. Palmquist, and R. O. Adlof. 2001. Solid fat index vs. solid fat content: A comparison of dilatometry and pulsed nuclear magnetic resonance for solids in hydrogenated soybean oil. p. 146–152. *In* N. Widlak et al. (ed.) Crystallization and solidification properties of lipids. AOCS Press, Champaign, IL.

Liu, K-S. 1999. Soybeans: Chemistry, technology and utilization. Aspen Publ.-Chapman & Hall, Gaithersburg, MS.

Lusas, E.W. 2000. Oilseeds and oil-bearing materials. p. 297–362. *In* K. Kulp and J.G. Ponte, Jr., (ed.) Handbook of cereal science and technology. 2nd ed. rev. Marcel Dekker, New York.

Lusas, E.W., D.R. Erickson, and W-K. Nip (ed.) 1989. Food uses of whole oil and protein seeds. Am. Oil Chemists' Soc., Champaign, IL.

Lusas, E.W., and S.R. Gregory. 1998. New solvents and extractors. p. 204–219. *In* S.S.Koseoglu et al. (ed.) Emerging technologies, current practices, quality control, technology transfer, and environmental issues. AOCS Press, Champaign, IL.

Lusas, E.W., and E. Hernandez. 1997. Isopropyl alcohol. p. 199–266. *In* P.J. Wan and P.J. Wakelyn (ed.) Technology and solvents for extracting oilseeds and nonpetroleum oils. AOCS Press, Champaign, IL.

Lusas, E.W., and G. M. Jividen. 1987. Glandless cottonseed: A review of the first twenty-five years processing and utilization research. J. Am. Oil Chem. Soc. 64:839–854.

Lusas, E.W., and K.C. Rhee. 1995. Soy protein and utilization. p. 117–160. *In* D.R. Erickson (ed.) Practical handbook of soybean processing and utilization. AOCS Press, Champaign, IL.

Lusas, E.W., and M.N. Riaz. 1996a. Fats in feedstuffs and pet foods. p. 255–239. *In* Y.H. Hui (ed.) Bailey's industrial oil & fat products. 5th ed. Vol. 3. Edible oil & fat products: Products and application technology. John Wiley & Sons, New York.

Lusas, E.W., and M.N. Riaz. 1996b. Texturized food proteins from fullfat soybeans at low cost. Extrusion Communiqué 9 (5):15–18.

Lusas, E.W., L.R. Watkins, S.S. Koseoglu, K.C. Rhee, E. Hernandez, M.N. Riaz, W.H. Johnson, Jr., and S.C. Doty. 1997. Final report: IPA as an extraction solvent. INFORM8:290–291, 294–306.

Lusas, E.W., L.R. Watkins, and K.C. Rhee. 1990. Separation of fats and oils by solvent extraction: Nontraditional methods. p. 56–78. *In* D.R. Erickson (ed.) Edible fats and oils processing: Basic principles and modern practices. Am. Oil Chemists' Soc., Champaign, IL.

Lyford, S.J., Jr., and J. Tal Huber. 1988. Digestion, metabolism and nutrient needs in preruminants. p. 401–420. *In* D.C. Church (ed.) The ruminant animal digestive physiology and nutrition. Waveland Press, Prospect Heights, IL.

Maryanski, J. 1993. Genetically engineered foods: Fears and facts. On line. FDA Consumer. January–February 1993. Available at www.fda.gov/bbs/topics/CONSUMER/CON00191.html. Downloaded 4 Dec. 2001.

Meyer, C-H. 1996. Procuring beans for full fat soya in Europe. p. 509–546. *In* Proc. of Second Int. Fullfat Soya Conf., Budapest, Hungary. 21–24 Aug. Am. Soybean Assoc. and United Soybean Board, St. Louis, MO.

Meyers, D.J. 1993. Past, present and potential uses of soy proteins in nonfood Industrial applications. p. 278–285. *In* T.H. Applewhite (ed.) Proc. of the World Conf. on Oilseed Technology and Utilization. AOCS Press, Champaign, IL.

Millstone, E., E. Brunner, and S. Mayer. 1999. Beyond "substantial equivalence." Nature (London) 401:525–526.

Min, D.B., S.H. Lee, and E.C. Lee. 1989. Singlet oxygen oxidation of vegetable oils. p. 57–97. *In* D.B. Min and T.H. Smouse (ed.) Flavor chemistry of lipid foods. Am. Oil Chemists' Soc., Champaign, IL.

Min, D.B., and T.H. Smouse (ed.) 1989. Flavor chemistry of lipid foods. Am. Oil Chemists' Soc., Champaign, IL.

Moore, N.H. 1983. Oilseed handling and preparation prior to solvent extraction. J. Am. Oil Chem. Soc. 60:141A–192A.

Moulton, K.J., R.E. Beal, K. Warner, and B.K. Bount. 1975. Flavor evaluation of copper-nickel hydrogenated soybean oil and blends with unhydrogenated oil. J. Am. Oil Chem. Soc. 52:469–472.

Mounts, T.L., K. Warner, and G. List. 1994. Performance evaluation of hexane-extracted oils from genetically-modified soybeans. J. Am. Oil Chem. Soc. 71:16–161.

Mi-Ok-Jung, Suk-Ho-Yoon, and Mun-Yhung-Jung. 2001. Effects of temperature and agitation rate on the formation of conjugated linoleic acids in soybean oil during hydrogenation process. J. Agric. Food. Chem. 49(6):3010–3016.

Mi-Ok-Jung, Jin-Woo-Ju, Dong-Seong-Choi, Suk-Ho-Yoon, and Mun-Yhung-Jung. 2002. CLA formation in oils during hydrogenation processes as affected by catalyst types, catalyst contents, hydrogenation pressure, and oil species. J. Am. Oil Chem. Soc. 79(5):501–510.

Mun-Yhung-Jung and Mi-Ok-Jung. 2002. Identification of conjugated linoleic acids in hydrogenated soybean oil by silver ion-impregnated HPLC and gas chromatograph-ion impacted mass spectrometry of their 4,4-dimethyloxazoline derivatives. J. Agric. Food Chem. 50:6188–6193.

Mustakas, G.C., W.J. Albrecht, G.N. Bookwalter, J.E. McGhee, W.F. Kwolek, and E.L. Griffin, Jr. 1970. Extruder-processing to improve nutritional quality, flavor, and keeping quality of full-fat soy flour. Food Technol. 24(11):11290–1296.

Nawar, W. 1996. Lipids. p. 225–319. *In* O.R. Fennema (ed.) Food chemistry. 3rd ed. Marcel Dekker, New York.

National Fire Protection Association. 2001. NFPA 36 Standard for solvent extraction plants, 2001 ed. NFPA, Quincy, MA.

National Oilseed Processors Association. 2000. Yearbook & trading rules 2000–2001. NOPA,Washington, DC.

National Research Council. 1989. Recommended Dietary Allowances 10th ed. NRC. Natl. Academy Press, Washington, DC.

Nelson, A.I., W.B. Wijeratne, S.W. Yeh, T.M. Wei, and L.S. Wei. 1987. Dry extrusion as an aid to mechanical expelling of oil from soybeans. J. Am. Oil Chem. Soc. 64:1341–1347.

North, R.D. 2000. The GMO battle: Stories from the troubled beginning of the biological century. On line. Available at www.iea.org.uk/wpapers/northgmo2.htm. Downloaded 6 July 2001.

O'Brien, R.D. 1995. Soybean oil products utilization: Shortenings. p. 363–379. *In* D.R. Erickson (ed.). Practical handbook of soybean processing and utilization. AOCS Press, Champaign, IL.

O'Brien, R.D. 1998. Fats and oils: Formulating and processing for applications. Technomic Press-CRC Publ. Co., Boca Raton, FL.

O'Brien, R.D., W.E. Farr, and P.J. Wan (ed.) 2000. Introduction to fats and oils technology. 2nd ed. AOCS Press, Champaign, IL.

Official Publication. 2002. Association of American Feed Control Officials Incorporated, Available from Sharon Senesac, Assistant Secretary-Treasurer, P. B. Box 478. Oxford, IN.

Okonek, D. 2001. Hydrogenated oil end products. *In* E. Hernandez et al. (ed.) Vegetable oils practical short course processing manual. Texas A&M Univ., College Station.

Parker, P.M. 1994. Innovations in refining technology: Silica adsorbents for edible oil. Oils & Fats Int. 10(6):24, 26–27.

Patterson, H.B.W. 1992. Bleaching and purifying fats and oils: Theory and practice. Am. Oil Chemists' Soc., Champaign, IL.

Patterson, H.B.W. 1994. Hydrogenation of fats and Oils. Am. Oil Chemists' Soc., Champaign, IL.

Paulson M. 1999. WTO case file: The beef hormone case. On line, available through archive service. Seattle Post-Intelligencer, 22 November. Available at seattlep-.nwsource.com/national/case22.shtml. Downloaded 2 Nov. 2001.

Payne, T., and R. Egbert. 1997. Process for making vegetable-based meat extenders. U.S. Patent 5 626 899. Date issued: 6 May. Assigned to Archer Daniels Midland Co., Decatur, IL.

Payne, T., and R. Egbert. 1999. Process for making extenders for lower fat meat systems. U.S. Patent 5 858 442. Date issued: 12 January. Assigned to Archer Daniels Midland Co., Decatur, IL.

Pedersen, H.E. 1993. Nonmeat applications of soy protein concentrates. p. 320–326. *In* T.H. Applewhite (ed.). Proc. of the World Conf. on Oilseed Technology and Utilization. AOCS Press, Champaign, IL.

Perkins, E. 1995. Composition of soybeans and soybean products. p. 9–28. *In* D.R. Erickson (ed.) Practical handbook of soybean processing and utilization. AOCS Press, Champaign, IL.

Perkins, E.G., and M.D. Erickson (ed.) 1996. Deep frying: Chemistry, nutrition, and practical applications. AOCS Press, Champaign, IL.

Pollan, M. 2001. How organic is corporate/industrial organic? On line. Available through archives service. New York Times Magazine.Downloaded copy from http://www.purefood.org/Organic/industrialorganic.cfm (downloaded 26 July 2001; posted 13 May 2001; verified 26 June 2003).

Proctor, A., P.K. Clark, and C A. Parker. 1995. Rice hull ash adsorbent performance under commercial soy oil bleaching conditions. J. Am. Oil Chem. Soc. 72:459–462.

Rackis, J.J. 1972. Biologically active components. p. 158–202. *In* A.K. Smith and S.J. Circle(ed.) Soybeans: Chemistry and technology. Vol. 1. Proteins. Avi Publ. Co.,Westport, CT.

Rackis, J.J. 1981a. Significance of soya trypsin inhibitors in nutrition. J. Am. Oil Chem. Soc. 58(3):495–501.

Rackis, J.J. 1981b. Flatulence caused by soya and its control through processing. J. Am. Oil. Chem. Soc. 58(3):503–509.

Rakes, G.A. 1993. Meat applications of soy protein concentrate. p. 311–319. *In* T.H. Applewhite (ed.). Proc. of the World Conf. on Oilseed Technology and Utilization. AOCS Press, Champaign, IL.

Redman, N.E. 2000. Food safety: A reference handbook. ABC-CLIO, Santa Barbara, CA.

Rozendaal, A. 1993. Interesterification and fractionation. p. 180–185. *In* T.H. Applewhite(ed.) Proc. of the World Conf. on Oilseed Technology and Utilization. AOCS Press, Champaign, IL.

Satter, L.D., M.A. Faldet, and M. Socha. 1991. *In* Symposia Proceedings: Alternative Feeds for Dairy and Beef Cattle. Natl. Invitation Symp. USDA Ext. Serv., in cooperation with Univ. Ext. Conf. Office. Univ. of Missouri, Columbia.

Sauer, D.B., R.A. Meronuck, and C.M. Christensen. 1992. Microflora. p. 313–340. *In* D.B.Sauer (ed.) Storage of cereal grains and their products. 4th ed. Am. Assoc. of Cereal Chemists, St. Paul, MN.

Schmidl, M.K., and T.P. Labuza. 2000. Essentials of functional foods. Aspen Publ., Gaithersburg, MD.

Segers, J.C., and R.L.K.M. van de Sande. 1990. Degumming—Theory and practice. p. 88–93. *In* D.R. Erickson (ed.) World Conf. Proc., Edible Fats and Oils Processing: Basic principles and modern practice. Am. Oil Chemists' Soc., Champaign, IL.

Sevanian, A. (ed.) 1988. Lipid peroxidation in biological systems. Am. Oil Chemists' Soc., Champaign, IL.

Shurtleff, A., and A. Aoyagi. 1995. Industrial utilization of soybeans (non-food, non-feed)—Biography and sourcebook. A.D. 980 to 1994. Soyfoods Center, Lafayette, CA.

Smith, A.K., and S.J. Circle (ed.) 1972. Soybeans: Chemistry and technology. Vol. 1. Proteins. Avi Publ. Co., Westport, CT.

Sonka, S.T., K.L. Bender, and D.K. Fisher. 2004. Economics and marketing. p. 919–948. *In* H.R. Boerma and J.E. Specht (ed.) Soybeans: Improvement, production, and uses. 3rd ed. Agron. Monogr. 16. ASA, CSSA, and SSSA, Madison, WI.

Soya & Oilseed Bluebook. 2003. (issued annually). Soyatech, Inc., Bar Harbor, ME.

Soy Stats. 2002. On line. United Soybean Board. Available at www/soystats.com/2002/Downloaded 14 Dec. 2002.

Spencer, M.R. 1976. Effect of shipping on quality of seeds, meals, fats and oils. J. Am. Oil Chem. Soc. 53:238–240.

Szuhaj, B. F. (ed.) 1989. Lecithins: Sources, manufacture and uses. Am. Oil Chemists' Soc., Champaign, IL.

Taylor, D.R. 1993. Adsorptive purification. p. 152–165. *In* T.H. Applewhite (ed.) Proc. of the World Conf. on Oilseed Technology and Utilization. AOCS Press, Champaign, IL.

Tirtiaux, A. 1990. Dry fractionation: A technique and an art. p. 136–141. *In* D.R. Erickson (ed.) World Conf. Proc., Edible Fats and Oils Processing: Basic Principles and Modern Practices. Am. Oil Chemists' Soc., Champaign, IL.

Tirtiaux, A. 1998. Dry fractionation: The boost goes on. p. 92–98. *In* S. S. Koseoglu et al.(ed.) Emerging technologies, current practices, quality control, technology transfer, and environmental issues. AOCS Press, Champaign, IL.

Tompkins, C., and E.G. Perkins. 2000. Frying performance of low-linolenic soybean oil. J. Am. Oil Chem. Soc. 72:223–229.

Uhlig, H. 1998. Industrial enzymes and their applications. John Wiley & Sons, New York.

U.S. Department of Agriculture. 1973. Marketing research report 968. USDA, Washington, DC.

U.S. Department of Agriculture. 1978. Marketing research report 1078. USDA, Washington, DC.

U.S. Mission. 2001. The US-EU dispute on EU hormone ban. U.S. Mission to the European Union, Brussels, Belgium. On line. Available at www.useu.be/issues/hormonedossier.html (downloaded 2 Nov. 2001; posted 6 Nov. 2000; verified 26 June 2003).

Vanderveen, J. E. 2001. Mandatory special labeling: Consumer right to know vs. material fact. Presented at Biotech Food. 4–5 June 2001. Am. Conf. Inst., Chicago, IL.

Viola, S., Y. Arieli, and G. Zohar. 1988. Animal-protein-free feeds for hybrid tilapia (*Oreochromis niloticus* x *O. aureus*) in intensive culture. Aquaculture 75(1–2):114–125.

Wakelyn, P.J. 1997. Regulatory considerations for extraction solvents for oilseeds and other nonpetroleum oils. p. 48–75. *In* P.J. Wan and P.J. Wakelyn (ed.) Technology and solvents for extracting oilseeds and nonpetroleum oils. AOCS Press, Champaign, IL.

Walstra, P. 1996. Dispersed systems: Basic considerations. p. 95–156. *In* O. R. Fennema (ed.) Food chemistry. 3rd ed. Marcel Dekker, New York.

Wang, C., and L.A. Johnson. 2001. Functional properties of hydrothermally cooked soy protein products. J. Am. Oil Chem. Soc. 78:189–195.

Watkins, L.R., W.H. Johnson, Jr., and S.C. Doty. 1989. Extrusion-expansion of oilseeds for enhancement of extraction, energy reduction and improved oil quality. p. 41–46. *In* T.H. Applewhite (ed.) Proc. of the World Congress on Vegetable Protein Utilization in Human Foods and Animal Feedstuffs. Am. Oil Chemists' Soc., Champaign, IL.

Weingartner, K.E. 1987. Processing, nutrition and utilization of soybeans. p. 149–178. *In* S.R. Singh et al. (ed.) Soybeans for the tropics: Research, production and utilization. John Wiley & Sons, New York.

Welsh, W.A., J.M Bogdonor and G.J. Toeneboen. 1990. Silica refining of oils and fats. p. 189–202. *In* D.R. Erickson (ed.) Edible fats and oils processing: Basic principles and modern practices. Am. Oil Chemists' Soc., Champaign, IL.

Widlak, N., R. Hartel, and S. Narine (ed.) 2001. Crystallization and solidification propertiesof lipids. AOCS Press, Champaign, IL.

Wijeratne, W.B. 2000. Alternative techniques for soybean processing. p. 371–374. *In* Proc. Int. Conf. on Soybean Processing and Utilization, Tsukuba, Japan. Japanese Soc. of Food Sci. & Technol., Organizing Committee for ISPUC-III. Japanese Soc. Of Food Sci. & Technol., Ibarakim, Japan.

Wilcke, H.L., D.T. Hopkins, and D. Waggle (ed.) 1979. Soy protein & human nutrition. Academic Press, New York.

Williams, M.A. 1995a. Extrusion preparation for oil extraction. INFORM 6:289–293.

Williams, M.A. 1995b. Advancement of processing hardware. Oil Mill Gazet. 101(1):20–24.

Williams, M.A. 1996. Obtaining oils and fats from source materials. p. 61–155. *In* Y.H. Hui (ed.) Bailey's industrial oil & fats products. Vol. 4. Edible oil & fat products: Processing technology. John Wiley & Sons, New York

Williams, M.A., and S. Baer. 1965. The expansion and extraction of rice bran. J. Am. Oil. Chem. Soc. 42:151–155.

Wilson, D.E., and R.E. Tribelhorn. 1979. Low-cost extrusion cookers: Second Int. Workshop Proc., Dar es Salaam, Tanzania.15–18 July. Colorado State Univ., Ft. Collins, CO.

Witte, N H. 1995. Soybean meal processing. p. 93–116. *In* D.R. Erickson (ed.) Practical handbook of soybean processing and utilization. AOCS Press, Champaign, IL.

Woerfel, J.B. 1995. Extraction. p. 65–92. *In* D.R. Erickson (ed.) Practical handbook of soybean processing and utilization. AOCS Press, Champaign, IL.

Wolf, W.J., and J.C. Cowan. 1975. Soybeans as a food source. CRC Press, Cleveland,OH.

Wu, S., P.A. Murphy, L.A. Johnson, A.R. Fratzke, and M.A. Reuber. 1999. Pilot-plant fractionation of Soybean Glycinin and β-conglycinin. J. Am. Oil Chem. Soc. 76:285–293.

Wu, S., P.A. Murphy, L.A. Johnson, A.R. Fratzke, and M.A. Reuber. 2000. Simplified pilot-plant process for soybean glycinin and ß-conglycinin fractionation. J. Agric. Food Chem. 48:2702–2708.

Yurawecz, M.P., M.M. Mossoba, J.K.G. Kramer, M.W. Pariza, and G.J. Nelson (ed.) 1999. Advances in conjugated linoleic acid research. Vol. 1. AOCS Press, Champaign, IL.

Zarkadas, C.G., R.I. Hamilton, C. Yu Ziran, V.K. Choi, S. Khanizadeh, N.G.W. Rose, and P.L. Pattison. 2000. Assessment of the protein quality of 15 new northern adapted cultivars of quality protein maize using amino analysis. J. Agric. Food Chem. 48(11): 5351–5361.

Zehnder, C.T. 1995. Deodorization. p. 239–257. *In* D.R. Erickson (ed.) Practical handbook of soybean processing and utilization. AOCS Press, Champaign, IL.

Zhang, Y., C.M. Parsons, K.E. Weingartner, and W.B. Wijerantne. 1993. Effects of extrusion and expelling on the nutritional quality of conventional and Kunitz trypsin inhibitor-free soybeans. Poult. Sci. 72:2299–2308.

Zhu, S., M. Riaz, and E W. Lusas. 1996. Effect of different extrusion temperatures and moisture content on lipoxygenase inactivation and protein solubility in soybeans. J. Agric. Food Chem. 44: 3315–3318.

Zoebelein, H. 1992. Renewable resources for the chemical industry. INFORM 3:721–725.

Zschau, W. 2000. Bleaching. p. 158–178. *In* R.D. O'Brien et al. (ed.) Introduction to fats and oils technology. 2nd ed. AOCS Press, Champaign, IL.

21

Soybean and the Prevention of Chronic Human Disease[1]

DIANE F. BIRT, SUZANNE HENDRICH, AND **D. LEE ALEKEL**

Iowa State University
Ames, Iowa

MARY ANTHONY

Wake Forest University School of Medicine
Winston-Salem, North Carolina

Consumption of foods containing soybean, *Glycine max* (L.) Merr., and soybean constituents is associated with improved heart disease risk factors, bone sparing, reduced cancer, and in a limited number of studies, reduced diabetes. In general soybean-based food (soy food) consumption by humans was suggested to be protective against several chronic diseases of western society that are thought to depend partly on diet in their etiology. But, the diets of western and eastern societies differ in numerous other factors than soy food consumption. For example, it may be the green tea (*Camilia sinensis*) consumption of some Asian societies, the high fat and caloric intake and/or positive energy balance due to a lack of physical activity of many people in western societies that cause the difference in disease risk.

The data showing improved risk factors for heart disease with the consumption of soybean, however, were strong enough to warrant the development of a health claim as is discussed in Section 21–1. Furthermore, ever-increasing amounts of data have associated soy food consumption with bone sparing and/or reduced rates of osteoporosis as is discussed in section 21–2. The relationship between consumption of soy food or soybean constituents and cancer prevention is, unfortunately, much more uncertain as is described in Section 21–3. Fortunately, a growing body of information is being developed by nutritional scientists and pharmacologists on the bioavailability, metabolism, and distribution of the soybean constituents as described below in Section 21–4 and these data will provide a foundation for understanding the interactions of soy food constituents with each other and with other dietary and environmental factors. Clearly, these interactions may have a major impact on the bioactivity and health impact of soy foods, while understanding the interactions will be essential for using dietary soybean in human disease prevention.

Soybean has long been viewed as a highly nutritious food particularly because of the high protein and lipid content of this legume. Unfortunately, sensory and cul-

[1] This work was supported in part by the Center for Designing Foods to Improve Nutrition, Iowa State Univ., Ames, IA, USDA Special Grant 96-341152835.

tural practices have limited the use of soybean in western diets. Soybean cultivars used for human foods were developed through standard plant breeding and, more recently, through genetic engineering. These cultivars were selected and developed to provide improved functional properties in foods and improved environmental, insect, and herbicide tolerance. However, the gross compositional characteristics of dry soybean for food use include > 40% protein, 19% oil, 5% ash, and 30 to 40% carbohydrate of which 5 to15% is soluble. Soybean protein is comprised of amino acids that are well balanced for human growth and development. Indeed, soybean has provided the protein source for human milk substitutes used for infant feeding. Soybean protein, while adequate for human growth and development in the amino acid methionine, is lower than red meat in this amino acid. Further, some scientists attribute some of the health benefit of soy foods to the lower methionine content. Soybean protein abundance and the particular proteins in soy foods differ considerably among soy foods that are widely consumed (Wilson et al., 1991). For example, soybean protein isolates contain about 90% protein and tempeh contains about 18% protein. Soybean proteins include the major proteins, glycinin, and β-conglycinin that comprise 65 to 80% of the protein and other minor proteins, including Kunitz trypsin inhibitor and Bowman-Birk trypsin inhibitor (BBI) (Hammond et al., 1993). The trypsin inhibitors in raw soybean are concentrated enough to cause the raw beans to be inedible (Liener, 1995). However, moist heat processing destroys most (90%) of the trypsin inhibitor activity and the resultant soybean is highly nutritious (Liener, 1995).

Soybean lipids are noted for their high polyunsaturated fatty acid composition and it is possible that this lipid composition may contribute to the vascular health benefits attributed to soy foods. Soybean oil is one of the most commonly used vegetable oils for home use and hydrogenated soybean oil is commonly used in processed foods (e. g., margarines, salad dressings, baked goods).

Some of the minor constituents of soybean include isoflavones, saponins, and phytic acid. The isoflavones are ethanol-extractable soybean constituents that were studied extensively for their role in health and disease, and for their content in soy foods and soy food ingredients. Considerable research has been conducted on the isoflavone forms and isoflavone content of soy foods and food ingredients (Murphy et al., 1999; Song et al., 1998). They found that soybean cultivar and processing were important factors in determining the isoflavone content. Indeed, soybean cultivars varied in isoflavone content by fivefold. Heat treatment decreased the content of malonylglucosides and water exposure, as is used in making tempeh or tofu, increased the concentration of B-glucosides and aglycones. Furthermore, extraction with ethanol removed some or all of the isoflavones. Clearly, processing can impact isoflavone content and fermented soybean foods contained higher concentrations of isoflavone aglycones, while nonfermented foods were richer in glycosides (Song et al., 1998;Wang and Murphy, 1996). This topic is discussed further in Section 21–4. Murphy and colleagues established an extensive database (USDA/ISU Isoflavone Database (1999). Unfortunately, less is known regarding other phytochemicals in soy foods. This chapter on the role of soybean consumption and chronic human disease will review recent studies on soybean and vascular disease, bone health, and cancer prevention. We then discuss aspects of the analysis and bioavailability and toxicity of soybean isoflavones constituents.

21–1 SOYBEAN AND VASCULAR HEALTH

21–1.1 Introduction and Epidemiology of Cardiovascular Disease

Cardiovascular disease (CVD) is the leading cause of death for both men and women in western countries. In the USA in 1998, coronary heart disease (CHD) caused nearly 460 000 deaths (one in every five deaths) and stroke was the cause of about 158 400 deaths (1 in every 15 deaths) (AHA, 2000). Coronary heart disease is the leading cause of premature, permanent disability. About 6 million men and 6.3 million women that are alive currently, have a history of CHD. Coronary heart disease and stroke are also a primary reason for rising health care costs with $10.8 billion paid to Medicare beneficiaries for CHD and $3.8 billion for stroke in 1997 (AHA, 2000). There is great potential to reduce the chronic disease burden by nutritional education and dietary changes. Increasing the consumption of soy foods is one dietary change that could have an important impact on improving cardiovascular health and this recommendation is included in the recent report of the Nutrition Committee of the American Heart Association (Krauss et al., 2000). There is substantial data to support this recommendation.

Atherosclerosis is an underlying process that manifests in clinical events of heart attack and stroke. In addition to the accumulation of lesions that can occlude the lumen of arteries, the atherosclerotic process is associated with endothelial and vascular dysfunction that can result in vasospasm and plaques that become unstable and rupture, resulting in emboli or thrombi. Some of the primary modifiable risk factors for atherosclerosis and cardiovascular disease include dyslipidemia (e.g., elevated low density lipoprotein (LDL) cholesterol, elevated plasma triglycerides, reduced high density lipoprotein (HDL) cholesterol concentrations), elevated blood pressure, and obesity. Much is known regarding effects of soybean on CVD risk factors.

21–1.2 Components of Soybean that Might Affect Cardiovascular Disease

While the atheroprotective effects of soybean protein are well accepted, a lot of research focused on the components in soybean responsible for cardiovascular protection. During the 1970s and early 1980s, the amino acid composition was investigated for its effect on plasma lipid/lipoprotein metabolism and its role in atherosclerosis inhibition. Research has also focused on other protein components, including specific fractions and globulins; saponins in soybean; and more recently, the isoflavones.

Huff and colleagues (1977) did a study that generally reflects the findings of the amino acid composition studies. In that experiment, groups of rabbits (n = 6–10 per group) were fed diets that contained either casein or a mixture of amino acids that duplicated the amino acid composition of casein. The plasma cholesterol concentrations of the two groups were found to be essentially the same. They did the same thing with soybean protein, feeding one group the intact protein and another a mixture of amino acids that was identical to the composition of intact soybean protein. The soybean protein amino acid mixture was not as hypocholesterolemic as the intact protein, providing evidence that there were components of the intact soybean protein other than the amino acids that favorably affected lipoprotein metabolism.

Other protein components including the 7S globulin, β-conglycinin, (Lovati et al., 2000) and the high molecular weight fraction of soybean (Sugano, 1998) were investigated for their effects on plasma lipoprotein metabolism. Lovati et al. (1998) reported that the 7S soybean globulin (β-conglycinin) and in particular the α' subunit can increase LDL receptor expression in a human hepatoma cell line (Hep G2 cells). More recently, these investigators reported that specific low-molecular-weight peptides in soybean appear to be responsible for the regulation of LDL receptor expression and LDL receptor-mediated uptake of LDL in vitro (Lovati et al., 2000). Since it is unlikely that large peptides of soybean reach the hepatocytes, the applicability of these observations must await in vivo confirmation. Sugano et al. (1990) described the hypocholesterolemic effects of a high-molecular-weight fraction of soybean protein. Compared to whole soybean protein, the high-molecular-weight fraction increased neutral and acidic sterol excretion in rats (Sugano et al., 1990) and lowered LDL cholesterol concentrations in women (Wang et al., 1995).

Accumulating evidence suggests that the components of soybean protein responsible for a large part of its hypocholesterolemic and atheroprotective effects are alcohol-extractable (Adams et al., 2002; Anthony et al., 1996, 1997; Balmir et al., 1996; Clarkson et al., 2001; Crouse et al., 1999; Gardner et al., 2001; Kirk et al., 1998; Merz-Demlow et al., 2000; Peluso et al., 2000; Sugano and Koba, 1993; Tovar-Palacio et al., 1998; Wangen et al., 2001; Yamakoshi et al., 2000). The isoflavones (nonsteroidal phytoestrogens) are largely removed from soybean protein when it is washed with alcohol and these molecules are being studied intensively for their role in cardiovascular disease protection. Other alcohol-extractable components that might affect plasma lipid concentrations are the saponins and phytosterols. Alfalfa saponins clearly inhibit cholesterol absorption, reduce plasma cholesterol concentrations, and result in reduced atherosclerosis (Malinow et al., 1983). However, soya saponins are of a different chemical structure and may not have the same effects (Malinow, 1984). The bulk of evidence regarding saponins suggests they exert no hypocholesterolemic effect in the presence of soybean protein (Potter et al., 1993). The approaches used in many of these earlier studies of soybean saponins was to ethanol wash the soybean (which would have extracted both saponins and isoflavones) (Calvert et al., 1981) or to add saponins from other plants to soybean protein (Potter et al., 1993). Neither of these approaches would give a definitive answer to the potential role of soybean saponins in cholesterol metabolism. Therefore, a role for soybean saponins in cholesterol metabolism cannot be ruled out. Most soybean protein isolates have low amounts of phytosterols because of delipidation and extensive processing in making soybean protein isolates (Anderson and Wolf, 1995); therefore, phytosterols as the active components seems unlikely.

21–1.3 Soybean and/or Isoflavones and Cardiovascular Disease

Studies citing the protective effects of soybean for cardiovascular disease, risk factors, and inhibition of atherosclerosis have been in the literature since the 1940s.

Many of the studies focused on the effects of soybean or their constituents on plasma lipid and lipoprotein concentrations. However, more recently, other cardiovascular disease risk factors have been evaluated.

21–1.3.1 Plasma Lipids and Lipoproteins

The atheroprotective effect of soybean seems to be mediated in part by effects on plasma lipoprotein concentrations. A meta-analysis that included 38 clinical trials in humans reported that soybean consumption resulted in 13% lower LDL cholesterol concentrations, 10% lower plasma triglycerides, and slightly, but not significantly, higher HDL cholesterol (about 2% higher) (Fig. 21–1, panel a) (Anderson et al., 1995). There was also evidence from this study that those subjects with higher baseline plasma cholesterol concentrations had a larger decrease in LDL cholesterol with soybean consumption than those with normal cholesterol concentrations at baseline (Fig. 21–1, panel b). These beneficial effects of soybean protein on plasma lipoprotein concentrations have recently culminated in the U.S. Food and Drug Administration approving a health claim that "25 grams of soybean protein a day, as part of a diet low in saturated fat and cholesterol, may reduce the risk of heart disease." A study published more recently suggests that as little as 20 grams of soybean protein per day can reduce concentrations of total cholesterol, "LDL" cholesterol (i.e., non-HDL cholesterol), and apolipoprotein B concentrations with no adverse effects on HDL cholesterol and apolipoprotein A-I concentrations, in moderately hypercholesterolemic individuals (Teixeira et al., 2000).

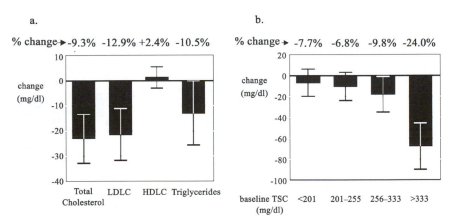

Fig. 21–1. Results of a meta-analysis of 38 clinical studies that evaluated the effect of soybean protein on plasma lipid and lipoprotein concentrations. Panel a: Effects of soybean-containing diets compared to control diet on total cholesterol, low density lipoprotein cholesterol (LDLC), high density lipoprotein cholesterol (HDLC), and triglyceride concentrations. Data are expressed as change from the control diet (mg dL^{-1}) ± standard error, percent change is indicated at the top of each bar. Panel b: Effects of soybean-containing diets compared to a control diet on LDLC concentrations in groups stratified by the baseline total serum cholesterol (TSC) concentrations. Data are expressed as change from the control diet (mg dL^{-1}) ± standard error, percent change is indicated at the top of each bar. Adapted from Anderson et al. (1995). Printed with permission.

Several studies in animal models found that alcohol washing of soybean protein (a process that largely removes the isoflavones in soybean) results in a soybean protein that is less effective at improving plasma lipoprotein concentrations (i.e., lowering LDL cholesterol, increasing HDL cholesterol) than isoflavone-intact soybean protein (Fig. 21–2, panels a and b) (Anthony et al., 1996, 1997; Clarkson et al., 2001; Kirk et al., 1998). Four recent studies in humans suggest that soybean protein with higher levels of isoflavones have more robust effects on lowering LDL cholesterol concentrations than soybean protein with lower isoflavone amounts (Crouse et al., 1999; Gardner et al., 2001; Merz-Demlow et al., 2000; Wangen et al., 2001). In the study by Crouse et al. (1999), there was an increasing reduction in LDL cholesterol concentrations with increasing isoflavone content (from 3–62 mg) in 25 g of soybean protein (Fig. 21–3, panel a) and slightly higher (not statistically significant) HDL cholesterol concentrations (Fig. 21–3, panel b). In a study in normocholesterolemic premenopausal women, soybean protein (53 g d^{-1}) with the highest isoflavone content (129 mg) had a more robust effect than the same amount of soybean protein with about half the isoflavone content (65 mg) on lowering LDL cholesterol concentrations (Merz-Demlow et al., 2000). In postmenopausal women consumption of soybean protein (63 g d^{-1}) with 132 mg isoflavones lowered LDL cholesterol more than the same amount of soybean protein with about 65 mg isoflavones (Wangen et al., 2001). In a second study of post menopausal women, 42 g of soybean protein containing 80 mg isoflavones lowered LDL cholesterol concentration compared to an alcohol-washed soybean protein isolate with 3 mg isoflavones (Gardner, et al., 2001).

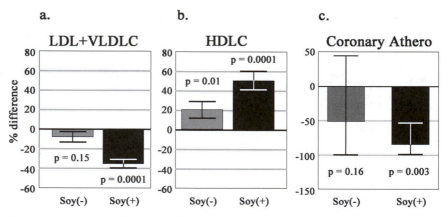

Fig. 21–2. Effects of alcohol washed soybean protein [Soy(−)] or intact soy protein [Soy(+)] on low density lipoprotein + very low density lipoprotein cholesterol (LDL + VLDLC) concentrations ($n = 160$), high density lipoprotein cholesterol (HDLC) concentrations ($n = 160$), and coronary artery atherosclerosis ($n = 33$) in nonhuman primates. Panel a: Percent difference (mean ± SEM) in LDL + VLDLC between the Soy(−) or Soy(+) groups and the casein/lactalbumin-fed control group. Panel b: Percent difference (mean ± SEM) in HDLC between the Soy(−) or Soy(+) groups and the casein/lactalbumin-fed control group. Panel c: Percent difference (mean ± SEM) in coronary artery atherosclerosis (athero) plaque size between the Soy(−) or Soy(+) groups and the casein/lactalbumin-fed control group. Adapted from Anthony et al. (1998). Printed with permission.

On the other hand, there is little evidence that purified soybean isoflavones (such as found in isoflavone pills) can improve plasma lipid concentrations either in humans (Hodgson et al., 1999a; Hsu et al., 2001; Nestel et al., 1997; Samman et al., 1999; Simons et al., 2000), or nonhuman primates (Greaves et al., 1999), rabbits (Yamakoshi et al., 2000) or rats (Peluso et al., 2000). There is only one report in the literature that found improved total and LDL cholesterol concentrations with an isoflavone extract (Han et al., 2002). Thus, it seems important that both soybean protein and the isoflavones be present for maximum effect on improving plasma lipoprotein concentrations.

21–1.3.2 Other Cardiovascular Disease Risk Factors

There is also evidence that soybean protein and some of its components might modify cardiovascular disease independent of effects on plasma lipoprotein concentrations (Anthony, 2000). Some of these potential mechanisms that are independent of plasma lipid concentrations are: blood pressure, vascular function, platelet aggregation and serotonin storage, LDL oxidation, and direct effects on the cells of the artery wall.

21–1.3.2.1 Vascular Function. The effects of soybean protein isolate (SPI) with isoflavones on vascular function is unclear. In a study in female nonhuman primates (n = 11) (Honoré et al., 1997), consumption of SPI with the isoflavones for 6 mo, inhibited coronary artery vascular constriction in response to acetylcholine (an endothelium-dependent vascular response) by about 12%, compared to a group fed alcohol-washed (isoflavone-devoid) SPI. These same results were seen in postmenopausal women (R. DuBroff and P. Decker, 1999, Soy phytoestrogens improve endothelial dysfunction in postmenopausal women. p. 53. *In* North American Menopause Society Meeting Abstracts, New York. 23–25 Sept. 1999. Abstr. no. 99.085). Women with abnormal endothelium dependent flow-mediated dilation (n = 18), assessed by ultrasound of the brachial artery after tourniquet release, were

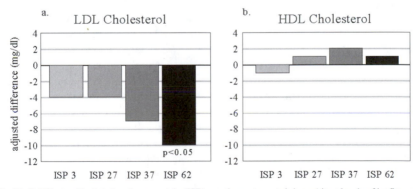

Fig. 21–3. Effects of isolated soybean protein (ISP) supplements containing various levels of isoflavones: 3, 27, 37, and 62 mg in moderately hypercholesterolemic men and women (*n* = 156). Panel a: Low density lipoprotein (LDL) cholesterol in the groups taking ISP supplements shown as the difference from the casein supplement group (mg dL^{-1}). Panel b: High density lipoprotein (HDL) cholesterol in the groups taking ISP supplements shown as the difference from the casein supplement group (mg dL^{-1}). Adapted from Crouse et al. (1999). Printed with permission.

given a beverage with 40 g of soybean protein containing 80 mg of isoflavones daily for 1 mo and then reassessed. Flow-mediated dilation was significantly improved by 5.3% with soybean consumption and the response returned to baseline after a 1-mo washout. More recently, Teede et al. (2001) found no effect on flow-mediated dilation with 3-mo treatment with 40 g of SPI with 118 mg isoflavones in postmenopausal women (n = 83) and an adverse effect (~3.7% vascular constriction) in men (n = 96), compared to a casein placebo. In another study in postmenopausal women (Simons et al., 2000), 80 mg isoflavone pills did not significantly change flow-mediated dilation (n = 20). Thus, more studies are required to understand whether SPI has beneficial effects on flow-mediated dilation, and if so, at what level of isoflavones and in which subjects.

Other indicators of vascular function are vascular elasticity and stiffness. In a placebo-controlled, randomized, cross-over study with peri- and postmenopausal women (n = 21); treatment for 5 wk with 80 mg d^{-1} of soybean isoflavone pills (i.e., no soybean protein) improved systemic arterial compliance, an indicator of vascular elasticity, by 26% (Nestel et al., 1997). In the study by Teede et al (2001) there was a slight nonsignificant improvement in systemic arterial compliance (5.7%) with SPI containing 118 mg isoflavones compared to casein placebo. Finally, in both the study by Teede et al (2001) and an observational study in Dutch postmenopausal women (van der Schouw et al., 2002), soybean-containing isoflavones (or foods containing isoflavones in Dutch women) resulted in improved pulse wave velocity, an indicator of arterial stiffness. At present there is more evidence that soybean and/or isoflavone consumption might improve vascular elasticity than endothelium-mediated vascular function.

21–1.3.3 Platelet Aggregation and Serotonin Storage

Another important mechanism by which soybean with isoflavones might improve cardiovascular disease is effects on platelets. In a study with female nonhuman primates (n = 12), a group fed SPI without the isoflavones had a 26% greater reduction in blood flow after collagen-induced platelet activation compared to a group fed SPI with the isoflavones (Williams and Clarkson, 1998). While the precise mechanism for this protection against reduction in blood flow by the isoflavones could not be determined in this study, there are several possible explanations. When platelets are activated, they release their vasoactive substances, including serotonin, which is a potent vasoconstrictor. These investigators (Williams and Clarkson, 1998) found that in vitro platelet aggregation in response to thrombin and serotonin was reduced in platelets collected from animals fed SPI with the isoflavones compared to platelets from animals fed the alcohol-washed SPI. Likewise, Schoene and Guidry (1999) found that platelets from rats fed isoflavone-intact SPI had apparent volumes that were significantly smaller than platelets from rats fed isoflavone-devoid SPI, suggesting these smaller platelets were in a more quiescent state. Other indicators of platelet activation in that study also suggested that the isoflavones could inhibit platelet activation. Peluso et al. (2000) found that in a study with rats, both isoflavone-depleted and isoflavone-intact soybean protein resulted in a 36% lower serotonin release from unstimulated platelets, compared to a group fed casein. In that same study, the group fed isoflavone-intact soybean protein had a 13% lower thrombin-stimulated serotonin release from platelets, compared to the casein-fed group and the group fed isoflavone-depleted soybean protein had an intermediate response. Helmeste and Tang (1995) found that genistein inhibited serotonin uptake in platelets. Thus, the isoflavones might inhibit platelet activation and aggregation and reduce the amount of serotonin in the platelets, all of which could contribute to a reduction in coronary vasospasm and thrombosis.

21–1.3.4 Lipoprotein Oxidation

Oxidized LDL particles are thought to play an important role in exacerbating atherogenesis. Four recent studies have evaluated the effects of soybean protein with isoflavones on LDL oxidation in humans (Jenkins et al., 2000a, 2000b; Tikkanen et al., 1998; Wiseman et al., 2000). In two randomized cross-over studies by Jenkins et al. (2000a, 2000b), consumption of soy foods containing 86 mg or 118 mg isoflavones resulted in 5 to 9% lower concentrations of conjugated dienes in the LDL fraction compared to a control diet period (n = 25–31 men and women in each study). Wiseman et al. (2000) evaluated both F_2-isoprotanes and LDL resistance to copper-induced conjugated diene formation in 19 premenopausal women and 5 men who consumed textured soybean protein that was either high in isoflavones (56 mg d^{-1}) or low (1.9 mg d^{-1}) for 17-d periods. They reported that the high isoflavone soybean resulted in 19.5% lower concentrations of 8-*epi*-prostaglandin $F_{2\alpha}$, a biomarker of in vivo lipid peroxidation, and a 9% longer lag time for copper-induced LDL oxidation. Tikkanen et al. (1998) found that consumption of soybean protein containing 60 mg isoflavones prolonged copper-induced LDL oxidation lag time by about 20 min after 2 wk of soybean consumption compared to measures done at baseline and after a 2-wk washout in a group of healthy volunteers (n = 6).

Two studies have evaluated the effects of isoflavone pills on LDL oxidation, both with negative results (Hodgson et al., 1999b; Samman et al., 1999). Hodgson et al. (1999b) found no effect on urinary F_2-isoprostane concentrations after 8 wk of treatment with 55 mg of isoflavones in 46 men and 13 women. Samman et al. (1999) reported that in 14 premenopausal women, 86 mg of isoflavones given for two menstrual cycles did not affect lag time for copper-induced LDL oxidation, oxidation rate, or maximum oxidation. Whether these differences in study results are because of the form of the supplements (soybean with isoflavones vs. isoflavone pills), cannot be ascertained; but deserves follow-up.

21–1.3.5 Blood Pressure

In four published trials, soybean protein isolate (SPI) with the isoflavones lowered diastolic blood pressure and in some cases systolic blood pressure (Washburn et al., 1999; Crouse et al., 1999; Teede et al., 2001; Burke et al., 2001). In the first, a cross-over study in perimenopausal women (n = 51), 20 g SPI (containing 34 mg isoflavones) given in a split dose, significantly reduced diastolic blood pressure by 5 mm Hg (Washburn et al., 1999). In a study by Crouse et al. (1999) in which moderately hypercholesterolemic men (n = 94) and women (n = 62) were given supplements containing 25 g of casein or 25 g SPI with different concentrations of isoflavones (3–62 mg isoflavones per 25 g of protein); there was a significant trend for lower diastolic blood pressure with increasing isoflavone dose in the women. There was no effect of the supplements on blood pressure in men. In the study by Teede et al (2001), treatment with 40 g SPI containing 118 mg isoflavones compared to casein placebo resulted in lower diastolic blood pressure by 2.4 mm Hg and systolic blood pressure by 3.9 mm Hg in postmenopausal women (n = 83) and men (n = 96). Finally, in a study by Burke et al (2001), soybean protein containing 23 mg isoflavones, compared to isocaloric carbohydrate supplement, resulted in 5.9 mm Hg lower 24-h systolic blood pressure and 2.6 mm Hg lower diastolic blood pressure in treated hypertensive men and women. Three studies that evaluated the effects of isoflavone pills on blood pressure found no effects (Han et al., 2002; Nestel et al., 1997; Simons et al., 2000). Thus, soybean containing isoflavones has beneficial effects on blood pressure; however, isoflavone pills do not appear to confer these same benefits.

21–1.3.6 Direct Effects on the Artery Wall

There are also several studies that suggest that the isoflavones might have direct effects on the artery wall, including inhibition of the migration and proliferation of smooth muscle cells, which might inhibit the promotion and progression of atherosclerosis (Fujio et al., 1993; Mäkelä et al., 1999; Shimokado et al., 1994, 1995). In a study of arterial injury-induced atherogenesis in rats, Mäkelä et al. (1999) found that the injured arteries markedly increased the expression of estrogen receptor β (ERβ) and that genistein (which binds with high affinity to ERβ) inhibited neointima formation (i.e., reduced the proliferation of cells in the artery wall that form atherosclerotic lesions). These studies suggest that the isoflavones might have direct effects on the artery wall to inhibit atherosclerosis formation and maintain cardiovascular health.

21–1.4 Atherosclerosis

Finally, there are studies in animal models suggesting that soybean protein with the isoflavones can reduce atherosclerosis. In two studies in nonhuman primates, soybean protein with the isoflavones [Soy(+)] inhibited atherosclerosis formation relative to alcohol-washed (isoflavone-devoid) soybean protein [Soy(−)]. In the first study, young males fed casein/lactalbumin (C/L) had the most atherosclerosis, those fed Soy(+) had the least amount of atherosclerosis (90% smaller lesions than the C/L group) and those fed Soy(−) had an intermediate amount (50% smaller lesions than the C/L group) (Fig. 21–2, panel c) (Anthony et al., 1997). In a study in surgically postmenopausal females (Clarkson et al., 2001), compared to the Soy(−) group, the Soy(+) group had significantly less atherosclerosis in the arteries of the head and neck (common carotid: 34% smaller, internal carotid: 61% smaller). In the coronary arteries the Soy(+) group had 25% smaller atherosclerotic lesions than the Soy(−) group, however this did not reach statistical significance.

Kirk et al. (1998) reported an inhibition of atherosclerosis in LDL receptor-intact mice fed Soy(+) relative to a group fed Soy(−). However, there was no difference in atherosclerosis between these diet groups in a strain of mouse without LDL receptors, suggesting that at least a portion of the atheroprotective effect of

Soy(+) is mediated by LDL receptors. Ni et al. (1998) found that in another strain of mouse (Apolipoprotein E-deficient), soybean protein feeding, relative to casein, reduced atherosclerosis in the absence of effects on total plasma cholesterol concentrations. Adams et al (2002) found that Soy (+) reduced atherosclerosis in LDL-receptor null and apo E-null mice compared to casein and Soy(−), an effect that was independent of effects on plasma lipoprotein concentrations. There are also data in animal models to suggest that soybean and/or isoflavones can reduce arterial lipid oxidation (Wagner et al., 1997; Yamakoshi et al., 2000).

There is only one published study that evaluated the effect of an isoflavone extract on atherosclerosis. In rabbits (Yamakoshi et al., 2000), an isoflavone aglycone-rich extract, without soybean protein, attenuated atherosclerosis. However, since purified isoflavones appear to have minimal effects on many mediators of atherosclerosis (plasma lipid concentrations, LDL oxidation, vascular funtion), then it seems reasonable to speculate that purified isoflavones will not have as potent an effect as soybean protein with the isoflavones on atherosclerosis.

Thus, the beneficial effects of soybean on risk factors, appear to translate to reductions in atherosclerosis extent. Alcohol-washed soybean protein does not have the same benefits for inhibiting atherosclerosis that the intact soybean protein does. While at least a portion of the atherosclerosis inhibition is likely mediated by effects on plasma lipid concentrations, there are also suggestions that effects on LDL oxidation and other nonlipid mechanisms might be involved.

21–1.5 Summary of Soybean and Cardiovascular Disease

There are many mechanisms by which soybean protein and/or the isoflavones might decrease atherosclerosis and cardiovascular disease. There are the well-recognized improvements in plasma lipid and lipoprotein concentrations; that is, lower LDL cholesterol, lower triglycerides, and possibly higher HDL cholesterol. There is also evidence that soybean protein and/or isoflavones can have beneficial effects on blood pressure; vascular and endothelial cell function; platelet activation, aggregation, and serotonin storage; LDL oxidation; smooth muscle cell proliferation and migration; and possibly direct effects on the artery wall. There is also data in established animal models showing inhibition of atherosclerosis with soybean consumption. Thus soybean foods containing isoflavones could have an important impact on the reduction of cardiovascular disease burden. However, neither isoflavone-devoid soybean protein nor isoflavone pills appear to have all the same benefits for improving cardiovascular health. Therefore, including foods containing intact soybean with isoflavones in a "heart healthy" diet seems the best recommendation for cardiovascular health.

21–2 SOYBEAN AND BONE HEALTH

Historically, nutritional interest in soybean has focused on their high-quality protein and "anti"-nutritional factors. During the past decade, three international symposia have been convened to advance the study of soybean in disease prevention. Soybean and their constituents have been extensively investigated for their role in preventing chronic disease, with particular focus on cardiovascular health and cancer protection. At this junction, the role of soybean in bone health deserves further consideration. The observations suggesting that soybean may contribute to bone health include the low rates of hip fractures in Asians originating from the Pacific Rim (Ho et al., 1993; Ross et al., 1991), the effectiveness of the isoflavone-derivative ipriflavone to prevent and treat postmenopausal osteoporosis (Fujita et al., 1986; Agnusdei et al., 1992), the in vitro (Markiewicz et al., 1993) and in vivo (Song et al., 1999) estrogenic activity of soybean isoflavones, and the lower urinary calcium losses in soybean vs. animal protein diets (Breslau et al., 1988). Although the evidence suggesting a bone-protective effect of isoflavone-containing soybean is intriguing and encouraging, it is nonetheless speculative. This bone section is a summary of this research, relying primarily upon peer-reviewed papers published in English.

21–2.1 Introduction and Epidemiology of Bone Health

21–2.1.1 Osteoporosis

Osteoporosis is a silent epidemic afflicting almost 35 million humans, accounting for 1.5 million new fractures each year in the USA alone and countless millions worldwide (Melton et al., 1992). It is the most prevalent metabolic bone disease in developed countries including the USA (Wasnich, 1996). Osteoporosis afflicts almost twice as many women as men (Melton et al., 1992), who typically develop fractures 5 yr later than women (De Laet et al., 1997). The longer life expectancy of women amplifies their disease burden. It is estimated that osteoporosis will cost our society $60 billion by the Year 2020 (Tucci, 1998). Many areas of the world are experiencing increases in hip fracture incidence (Gullberg et al., 1997), although it has stabilized in some countries (Melton et al., 1998; Rogmark et al., 1999). The projected rise in the number of older adults could cause the number of hip fractures worldwide to increase from an estimated 1.7 million in 1990 to a projected 6.3 million in 2050 (Cooper et al., 1992). At present, the majority of hip fractures occur in Europe and North America, but enormous increases in the number of elderly in South America, Africa, and Asia will shift this burden of disease from the developed to developing world (Genant et al., 1999). Effective prevention strategies will need to be designed and disseminated in these parts of the world to prevent the expected increase in hip fractures.

Osteoporosis is defined as a "disease characterized by low bone mass and microarchitectural deterioration of bone tissue leading to enhanced bone fragility and a consequent increase in fracture incidence" (Melton and Riggs, 1983). The World Health Organization (WHO) has developed an operational definition of osteoporosis based on bone mineral density (BMD) of young adult Caucasian women (Melton, 2000). Unfortunately, because of insufficient data on the relationship between BMD and fracture risk in men or nonCaucasian women, the WHO did not offer a definition of osteoporosis for groups other than Caucasian women (Kanis et al., 1994). The WHO defines osteoporosis as a BMD less than 2.5 standard deviations (SD) below the mean for young women. The WHO defines osteopenia as a BMD between 1 and 2.5 SD (also referred to as a z-score) below the mean for young women. A z-score (SD) of -1 below the mean or greater indicates normal BMD. Based on these cut-offs, it is estimated that 54% of postmenopausal Caucasian women in the USA are osteopenic and 30% are osteoporotic (Melton 1995). For each SD below (-1 z-score) the mean ("peak bone mass"), a woman's risk of fracture doubles. However, a limitation of using a cut-off is that fracture risk varies directly and continuously with BMD (Kanis et al., 1994), with many risk factors being independent of BMD (Cummings et al., 1995).

21–2.1.2 Caucasian vs. Asian Populations: Bone Density and Fractures

It is known that there are ethnic and genetic differences in bone that may make some groups more susceptible than others to osteoporotic fractures (Anderson and Pollitzer, 1994). For example, Caucasian women are at greater risk than African and Mexican Americans (Looker et al., 1997), who have lower fracture rates (Silverman and Madison, 1988). Vertebral fracture incidence among Taiwanese (Tsai, 1997) is comparable (18%) to Caucasian women, whereas that of hip fracture among elderly Taiwanese (Tsai, 1997) and those from mainland China (Xu et al., 1996) is lower. Despite 10 to 15% lower femoral BMD than Caucasians, Taiwanese have lower hip fracture rates. Researchers have examined determinants of peak bone mass in Chinese women in China (Ho et al., 1997), risk factors for hip fracture in Asian men and women (Lau et al., 2001), and BMD in elderly Chinese vegetarians vs. omnivores (Lau et al., 2001). We have also learned about the contribution of anthropometric and lifestyle factors to peak bone mass in a multiethnic population (Davis et al., 1999), hip fracture prevalence in different racial groups (Parker et al., 1992), as well as hip axis length in premenopausal Chinese women living in Australia (Chin et al., 1997) and women originating from the Indian subcontinent (Alekel et al., 1999). Some differences in osteoporotic risk among ethnic groups are inexplicable, but may be largely due to frame size differences that lead to size-related artifacts in BMD measurements (Prentice et al., 1994; Ross et al., 1996) and to differences in hip axis length (Cummings et al., 1994; Nakamura et al., 1994). Thus, when comparing bone mass across ethnic groups, it is important to correct for frame size to accurately interpret spinal BMD values (Alekel et al., 2002) and to consider hip geometry to accurately assess hip fracture risk (Nakamura et al., 1994). Differences in osteoporotic risk may also be related to culture-specific dietary and exercise-related factors, which are beyond the scope of this review.

21–2.1.3 Soybean Intake, Bone Density, and Fractures

The low hip fracture rate among Asians has been credited to the beneficial effect of isoflavone-containing soybean on bone health (Ho et al., 1993; Ross et al., 1991). However, two human studies found that isoflavone-rich soybean protein (40 g d^{-1}) intake was associated with favorable effects on spinal (Alekel et al., 2000; Potter et al., 1998) but not on femoral (hip) bone. Also, the amount of isoflavones (in aglycone form) consumed by subjects in the high isoflavone groups in these two studies (80 or 90 mg d^{-1}) was greater than what is typically consumed by either Chinese (39 mg d^{-1}), (Chen et al., 1999) or Japanese (23 mg d^{-1}), (Kimira et al., 1998) women or by women from a multiethnic population in Hawaii (ranged from 5 mg

d^{-1} in Filipino to 38.2 mg d^{-1} in Chinese) (Maskarinec et al., 1998). Nevertheless, it is possible that lesser amounts of soybean isoflavones consumed over the course of many years could have significant bone-sparing effects. Still, differences are not apparent in the spinal fracture rate (Ross et al., 1995; Tsai et al., 1996) or in lumbar spine BMD (Ross et al., 1996; Tsai et al., 1991) of Asian compared with Caucasian women. In contrast, higher spine and hip BMD values have been reported in U.S.-born vs. Japan-born Japanese women (Kin et al., 1993). Many factors may contribute to the lower hip fracture rate in Asians, notably the shorter hip axis length of Asians originating from the Pacific Rim (Cummings et al., 1994; Nakamura et al., 1994) and of Asians originating from the Indian subcontinent (Alekel et al., 1999). Other protective factors include the lesser propensity of Asians to fall (Davis et al., 1997) and the shorter stature of Asians (Lau et al., 2001), although these unmodifiable factors have little practical importance in preventing hip fractures.

Few observational data have been published from epidemiologic studies on the relationship between soybean intake and BMD or fracture risk, although there are some recent reports. Somekawa et al. (2001) examined the relationship between soybean isoflavone intake, menopausal symptoms, lipid profiles, and spinal BMD measured by dual-energy x-ray absorptiometry (DXA) in 478 postmenopausal Japanese women. They reported that once BMD was adjusted for weight and years since menopause, BMD values were significantly different among four isoflavone intake levels, ranging from 35 to 65 mg d^{-1}, in both the early and late postmenopausal groups. The women who consumed more soybean isoflavones had higher BMD values. Differences in other characteristics (i.e., height, weight, years since menopause, lipid or lipoprotein concentrations) across isoflavone intakes were not significant. In a study by Horiuchi et al. (2000) in postmenopausal Japanese women (n = 85), dietary intake of soybean protein was positively associated with lumbar spine BMD after controlling for energy, protein, and calcium intake. Based on their lumbar BMD measurements, approximately 60% of the women were osteopenic or osteoporotic. In accord with these findings, Kritz-Silverstein et al. (2002) reported that postmenopausal women with the highest level of isoflavone intake had the greatest lumbar spine (but not hip) BMD values, but also had 18% lower N-Tx values (bone resorption marker). These results suggest that usual intake of dietary isoflavones may be protective against bone loss in postmenopausal women.

Another study in mid-life (40–49 yr) Japanese women (N = 995) examined the relationship of various dietary factors (including soybean intake) to metacarpal BMD as measured by computed x-ray densitometry (Tsuchida et al., 1999). Women who consumed soybean at least twice per week had greater BMD than those who consume soybean once or less per week, with this tendency remaining after controlling for age, height, weight, and weekly calcium intake. In the Netherlands, Kardinaal et al. (1998) tested the hypothesis that the rate of postmenopausal radial bone loss measured by single photon absorptiometry is inversely related to urinary excretion of phytoestrogens as a marker of long-term dietary intake. Contrary to their hypothesis, these researchers reported that women with a relatively high rate (1.91 ± 0.08%) of yearly bone loss had significantly higher urinary excretion of enterolactone (median = 838 vs. 1108 µg g^{-1}, respectively) than women with a low rate (0.27 ± 0.08%) of loss. Colonic bacteria synthesize enterolactone from precursors found in grains, legumes, seeds, and vegetables (Thompson et al., 1991). However, urinary concentrations of genistein, daidzein, and equol did not differ between the two groups of bone losers. It should be noted that the Dutch typically consume very low amounts of dietary phytoestrogens, with this group of women having similarly low intakes. These published studies differ with respect to the type and site of bone measured, as well as the amount of dietary phytoestrogens habitually consumed. Nonetheless, the evidence for an effect of soybean-derived isoflavones on bone appears to be stronger for trabecular (i.e., spinal) than cortical (i.e., radial, metacarpal) bone and is likely dependent upon habitual intakes.

21–2.2 Selective Estrogen Receptor Modulators

Selective estrogen receptor modulators (SERM) is a class of drugs, comprised of a group of chemically diverse nonsteroidal compounds, that bind to and interact with the estrogen receptor. These estrogen-like compounds are designed to have tissue selectivity, such that a given SERM may act as an estrogen agonist in certain tissues and an estrogen antagonist in others (Bryant and Dere, 1998). Pharmacologically and structurally similar to soybean isoflavones, synthetic SERMS (such as ipriflavone, tamoxifen, and raloxifene) are effective in preventing or reducing bone loss. Ipriflavone, an isoflavone derivative of plant origin (Havsteen, 1983), has been used to prevent and treat postmenopausal osteoporosis (Fujita et al., 1986; Agnusdei et al., 1989, 1992) and in several models of experimental osteoporosis (Yamazaki et al., 1986a; Yamazaki, 1986b). Ipriflavone also improves the therapeutic bone response when combined with estrogen, above that of either alone (Agnusdei et al., 1995). Conversely, a muti-center trial (Alexandersen et al., 2001) indicated that ipriflavone did not prevent bone loss or affect bone turnover, casting doubt on its effectiveness. Tamoxifen, widely used in treating breast cancer, has weak estrogenic effects on bone remodeling (Wright et al., 1994). A randomized trial indicated that tamoxifen increased spine BMD 0.6%, while the placebo group declined 1% after 1 yr (Love et al., 1992). Other studies have borne out these beneficial effects of tamoxifen on bone, but its major drawback is its endometrial stimulatory effects (Fornander et al., 1989). Raloxifene is an alternative to hormone therapy for bone, but unlike tamoxifen, does not stimulate the endometrium (Delmas et al., 1997). The FDA has approved raloxifene to prevent postmenopausal osteoporosis because of its favorable effect on bone (Delmas et al., 1997) and remodeling (Heaney and Draper, 1997). Yet, a drawback of raloxifene is that it may increase hot flushes in some women (Draper et al., 1996; Walsh et al., 1998), thus limiting its use to those who are well beyond the menopausal transition. In contrast, we found that isoflavone-rich soybean had no adverse or favorable effects on vasomotor symptoms in perimenopausal women (St.Germain et al., 2001). The purported mechanism of action for SERMs on bone is to decrease bone resorption, unlike the maintenance of bone turnover previously reported (Alekel et al., 2000). Although these potential beneficial effects of SERMs make them very attractive in preventing and treating osteoporosis, naturally occurring soybean isoflavones may be more acceptable to many postmenopausal women than the synthetic analogues.

21–2.2.1 Soybean and Isoflavones and Osteoporosis

The most promising effect of isoflavones in menopausal women may be that of bone-sparing. Soybean protein isolate with isoflavones has been shown to prevent femoral and lumbar bone loss in ovariectomized (ovx) rats (Arjmandi et al., 1996) and lumbar spine bone loss in humans during the short-term (Alekel et al., 2000). Animal research provides valuable information on potential mechanisms of action, but clinical trials ultimately must be conducted to confirm the long-term effects of soybean isoflavones in humans. Before examining the effects of isoflavone-containing soybean on bone, we must consider its endogenous estrogenic activity, other potential mechanisms of action such as its affect on calcium metabolism, and what is currently known about the safety profile of soybean and/or its isoflavones.

21–2.2.2 Estrogenic Effect of Isoflavones

Phytoestrogens or plant estrogens include isoflavones, coumestans, and lignans and have been identified in whole grains, fruits, and vegetables. Soybean isoflavones (genistein, daidzein, glycitein), structurally similar to 17β-estradiol, are hypothesized to protect against chronic diseases like osteoporosis, breast cancer, and cardiovascular disease (Kurzer and Xu, 1997). Isoflavones exert an estrogenic effect on the central nervous system, induce estrus, stimulate growth of the genital tract in female mice, and bind to estrogen receptors (Liberman, 1996). Shutt and Cox (1972) determined that phytoestrogens bind to estrogen receptors, but compared with 17 β-estradiol, relative binding affinity of daidzein and genistein is weak. Genistein has a particular binding affinity for estrogen receptor-β, but removing one hydroxyl group (daidzein) leads to great loss in this affinity (Kuiper et al., 1998). Further, the stronger affinity of isoflavones for estrogen receptor-β may be particularly important because this receptor has been identified in bone tissue (Vidal et al., 1999). Dietary isoflavones are weakly estrogenic, particularly when there is a lack of endogenous estrogen, but preferentially bind to estrogen receptor-β (Setchell, 1995), implying their actions are distinct from those of classical steroidal estrogens that bind preferentially to estrogen receptor-α. Isoflavones indeed may exert tissue selective effects, since some tissues contain predominately estrogen receptor-α or estrogen receptor-β. Interestingly, the in vivo estrogenic activity of glycitein on uterine weight in mice, despite its lower binding affinity for estrogen receptor-α, is three times higher than that of genistein (Song et al., 1999). Thus, the binding affinity of isoflavones to estrogen receptors and their end-organ effects may differ, but one hypothesized mechanism by which these estrogenic compounds act on bone is through estrogen receptor-β (Burke, 2000).

Consuming soybean protein foods rich in isoflavones leads to a significant change in the hormonal characteristics of the menstrual cycle in premenopausal women (Kelly et al., 1995; Cassidy et al., 1994). These changes include a longer menstrual cycle as a result of a longer follicular phase, with marked suppression in mid-cycle surges of luteinizing hormone and follicle-stimulating hormone. Wu et al. (2000) examined the effects of soybean foods on ovarian function in premenopausal women and found a 17.4% reduction in luteal phase estradiol among Asian, but not in non-Asian subjects, with marginally higher urinary isoflavone excretion among Asians (29.2 µmol d^{-1}) than non-Asians (17.1 µmol d^{-1}). A 2-yr-long study on the effects of soybean isoflavones (0, 65, or 130 mg d^{-1}) on estrogen and isoflavonoid metabolism in postmenopausal (n = 17) women indicated that those on the higher dose had a decrease in estrone–SO_4, an increase in sex hormone-binding globulin, and a trend toward decreased 17 β-estradiol and estrone (Duncan et al., 1999b). However, there were no effects of isoflavone intake on vaginal cytology or endometrial biopsy data.

Initially, soybean isoflavones were thought to act purely as hormones. Yet, the actions of isoflavones are quite diverse, with no single action explaining all their in vivo and in vitro effects (Setchell and Adlercreutz, 1988; Adlercreutz, 1995). Isoflavones are thought to inhibit protein tyrosine kinase, suppress angiogenesis, have antioxidant effects (Ruiz-Larrea et al., 1997), and arrest cell growth by interfering with signal transduction (Higashi and Ogawara, 1994). There is some evidence that isoflavones in vitro may inhibit aromatase (Adlercreutz et al., 1993), the rate-limiting enzyme in estrogen synthesis in humans. In addition, genistein also increases in vitro concentrations of transforming growth factor-β (Kim et al., 1998), an important skeletal growth factor (Centrella et al., 1991). Thus, these data suggest that one underlying mechanism by which isoflavones act is hormonal, but this is not likely the sole mechanism.

21–2.2.3 Isoflavones, Protein Intake, Mineral Metabolism, and Calcium Absorption

A soybean protein diet may protect against bone loss indirectly by mechanisms independent of its estrogenic effects on bone. Similar to the estrogen-enhancing effects on calcium uptake in vitro (Arjmandi et al., 1993), isoflavones may improve calcium absorption. Yet, a recent human study reported no evidence that either soybean protein or isoflavones affect fractional calcium absorption or net calcium retention (Spence et al., 2001). Still, a high soybean diet may decrease urinary calcium loss compared to animal-based diets (Spence et al., 2001). Legumes, including soybean, are somewhat lower in sulfur amino acids than meat (Pennington, 1998).

Animal protein has in fact been shown to be more hypercalciuric than soybean protein in human studies (Anderson et al., 1987; Breslau et al., 1988; Pie and Paik, 1986), perhaps due to its lower sulfur-containing amino acid content. Intakes of milk whey (28 mg methionine g^{-1}) and soybean protein (13 mg methionine g^{-1}) were compared acutely over a 24-h period (Anderson et al., 1987). Four hours after ingestion, the urinary calcium:creatinine ratio increased by 45% with the intake of milk whey, but increased by only 3% with a similar amount of soybean as the primary source of protein. After 24 h, the calcium:creatinine ratio was 56% higher than baseline in the whey compared with 27% higher in the soybean group. A longer-term 2-wk feeding study (Watkins et al., 1985) was performed in subjects (N = 9) aged 22 to 69 yr who were fed ~80 g of protein derived primarily from either soybean or chicken, but similar amounts of Ca, P, Mg, and S. They demonstrated that in comparison to baseline values, urinary total titratable acid increased only 4% on the soybean diet but by 46% on the meat diet. Average daily urinary calcium excretion was 169 mg on the soybean vs. 203 mg on the meat diet, demonstrating that soybean was less hypercalciuric than meat protein.

Similarly, Breslau et al. (1988) examined Ca metabolism in 15 subjects 23 to 46 yr of age who consumed in random order (crossover) for 12 d each of three diets: soybean protein (vegetarian), soybean and egg protein (ovo-vegetarian), or animal (beef, chicken, fish, cheese) protein. The diets were kept constant in protein (75 g d^{-1}), Ca (400), P (1000 mg), Na (400 mg), and fluid (3 L), providing sufficient energy for weight maintenance. They reported no difference in fractional ^{47}Ca absorption among the diets, but 24-h urinary Ca excretion increased from 103 ± 15 mg d^{-1} on the vegetarian to 150 ± 13 mg d^{-1} on the animal protein diet. Pie and Paik (1986) fed young Korean women (N = 6) a meat-based (71 g protein d^{-1}) followed by a soybean-based (83 g protein d^{-1}) diet for 5 d each. Despite similar dietary Ca (525 mg d^{-1}) contents, urinary and fecal Ca excretions, respectively, were higher while subjects were on the meat-based (127 and 467 mg d^{-1}) compared with the soybean-based (88 and 284 mg d^{-1}) diet. Consequently, overall Ca balance was more negative on the meat-based (-65.4 mg d^{-1}) vs. soybean-based (155.3 mg d^{-1}) diet.

A related question is whether the Ca recommendation (age 50+yr, 1200 mg d^{-1}) for Caucasians should apply to Asians with a smaller skeletal size who typically consume ~500 mg d^{-1} (Pun et al., 1990). Nonosteoporotic postmenopausal Chinese women had Ca absorption rates equal to 58% with a 600 mg supplement, 60% during the unmodified period, and 71% during Ca deprivation (<300 mg d^{-1}) (Kung et al., 1998). Might this twofold higher calcium absorption rate (Heaney et al., 1989) be related to the high vegetable and soybean intakes of Chinese (Weaver, 1998), which provide 41% of their Ca in contrast to <10% of intake in the USA (Lau, 1995)? Further study is required to determine more precisely how soybean-vs. animal-based diets affect bone and Ca homeostasis. The beneficial effect of soy foods on Ca excretion may be clinically relevant if individuals consume two or three servings per day (20 or more g d^{-1}). By substituting soybean for animal protein over the long-term, the balance could be tipped in favor of Ca retention.

21–2.3 Animal Models of Osteoporosis

21–2.3.1 Isoflavone-Rich Soybean Protein and Bone

The FDA guidelines for evaluation of agents used to treat or prevent osteoporosis indicate that the ovarectomized rat model mimics postmenopausal cancellous bone loss when examined over short periods (i.e., <12 mo) of time (Thompson et al., 1995). Numerous studies have been carried out in the ovx rat model and will be reviewed in this next section. Kalu et al. (1988) reported that both soybean-based and food-restricted diets, compared with casein, delayed the onset of femoral bone loss by 3 to 4 mo in old male Fischer rats (*Rattus norvegicus*). In a shorter-term study, Arjmandi et al. (1996) found that ovx Sprague-Dawley rats fed soybean protein isolate (SPI) for 30 d had significantly higher mean bone densities of the right femur and fourth lumbar vertebra than ovx casein-fed rats (Fig. 21–4). The bone density of the fourth lumbar vertebra in the SPI-fed rats was similar to that of rats given 17 β-estradiol, but significantly higher than the sham-operated controls. Serum total alkaline phosphatase (bone formation marker) and tartrate-resistant acid phosphatase (bone resorption marker) were higher in the SPI-fed and ovx casein-fed rats than in the sham-operated and ovx 17 β-estradiol-treated rats. In an analogously designed study, Harrison et al. (1998) found that sham-operated casein-fed,

Bone Density

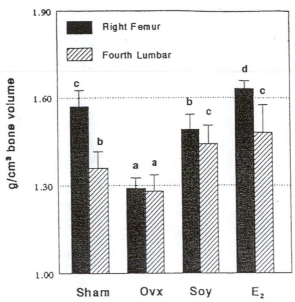

Fig. 21–4. Right femur and fourth lumbar vertebra bone densities of sham-operated (sham), ovariectomized (ovx), ovx + soybean (soy) and ovx + 17 β-estradiol (E_2) rats. Values are means + SD, n = 8 per group. For each bone, densities with different letters are significantly different ($P < 0.05$). Adapted from Arjmandi et al. (1996). Printed with permission.

ovx soybean protein-fed, and ovx casein-fed estrogen-treated Sprague-Dawley rats had higher femur and tibia ash content compared with those in the ovx casein-fed control group. Serum total and bone-specific alkaline phosphatase were both lower in the estrogen-treated and sham-operated groups compared with the soybean- and casein-fed groups. Unlike estrogen, soybean protein did not exert any uterotrophic effect and did not decrease bone formation markers in either of these studies (Arjmandi et al., 1996; Harrison et al., 1998), suggesting a different mechanism of action than that of estrogen. Results from these two latter studies suggest that bone formation stimulated by isoflavone-containing soybean protein exceeds the resorption induced by ovariectomy. Omi et al. (1994) reported a stimulatory effect of soybean milk vs. casein feeding for 28 d on tibial (proximal metaphysis and diaphysis) and vertebral BMD, mechanical bone strength (greater breaking force), and Ca absorption in ovx Sprague-Dawley rats. They suggested that the enhanced Ca absorption was responsible for the bone-related effects.

The studies in the above section were not designed to distinguish between the effects of soybean protein and isoflavones on bone. Employing a similar design as described above, Arjmandi et al. (1998a) reported that the ovx SPI+ and sham-operated casein-fed groups had greater femoral BMD than the ovx casein-fed rats, whereas the soybean protein with reduced (extracted) isoflavones (SPI–) group was similar to the ovx casein-fed rats (Fig. 21–5). The ovx-induced gains in bone formation indicated by histomorphometry and bone turnover reflected by biochemi-

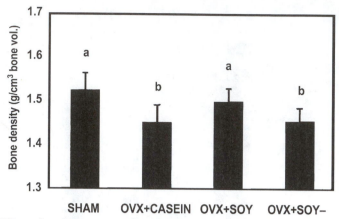

Fig. 21–5. Effects of ovariectomy and soybean treatments on right femur density. Bars represent mean
+ SD. Bars with different letters are significantly different ($P < 0.05$); $n = 12$ per treatment group.
SHAM, sham-operated rats fed a casein-based diet; OVX+CASEIN, ovariectomized rats fed a ca-
sein-based diet; OVX+SOY, ovariectomized rats fed soy protein with normal isoflavone content;
OVX+SOY–, ovariectomized rats fed soy protein with reduced isoflavone content. Adapted from Ar-
jmandi et al. (1998a). Printed with permission.

cal markers were not affected by SPI+ or SPI–. These results suggest that isoflavones
are the bone-active components in soybean. This study was performed in 95-d-old
rats, whereas another study was carried out in ovx rats with established bone loss
(Arjmandi et al., 1998b). With a delay in SPI+ or SPI– feeding after surgery (35
d), the fourth lumbar vertebral BMD was higher in the sham-operated casein than
in the SPI+, SPI–, or ovx casein groups. The femoral BMD was higher in the sham-
operated vs. ovx casein group, but there were no significant differences among the
SPI+, SPI–, and sham-operated casein groups. Our results suggest that a delay in
feeding reduced the efficacy of SPI to reduce trabecular (vertebral) bone loss, but
perhaps permitted some lessening of cortical (femoral) bone loss. These results confirm the idea that the most effective approach is to prevent rather than treat bone
loss once it has occurred (Kanis, 1996).

In contrast to the bone-sparing effects shown in rats, one published study in
ovx cynomologus macaques reported no protective effect of SPI+ fed for 7 mo (Lees
and Ginn, 1998). In comparison, 17 β-estradiol suppressed ovx-induced increases
in bone formation rates, regardless of the dietary protein source. In another study
(Jayo et al. 1996), neither SPI+ nor SPI– slowed bone loss over the course of 23
mo of intervention, whereas the ovx monkeys given estrogen gained whole body
and spinal bone mass. Taken together, these animal studies may illustrate species-
specific differences in response to isoflavone-rich soybean.

21–2.3.2 Soybean and/or Isoflavones and Bone

The effect of isolated (extracted) isoflavones on BMD has been examined in
the ovx rat model, in ovx mice, and on bone cells in vitro, with most reporting beneficial results. Blair et al. (1996) examined the effect of genistein or genistin (44
mmol d^{-1}) for 30 d on the bone mass of ovx Sprague-Dawley rats. They also com-

pared the effects of genistein and daidzein with other tyrosine kinase inhibitors on avian osteoclasts in vitro. Dry femoral mass in the genistein-treated group was 12% greater than controls, whereas genistin did not have a significant effect. Genistein inhibited osteoclastic activity and protein synthesis at pharmacologically attainable concentrations. Ishida et al. (1998) reported that daidzin or genistin at 50 mg kg^{-1} d^{-1} or estrone (7.5 µg kg^{-1} d^{-1}) largely prevented a loss in femoral bone density and bone strength (tested via yield force), a decline in ash weight, and decreases in Ca and P contents in ovx Sprague-Dawley rats fed Ca-deficient diets. Daidzin or estrone, but not genistin, also prevented the ovx-induced rise in urinary bone resorption markers, indicating a suppressive effect of daidzin but not of genistin on bone turnover. In ovx, lactating rats on a low Ca diet, Anderson et al. (1998) investigated the effects of genistein at three doses (0.5, 1.6, or 5.0 mg d^{-1}) compared with conjugated estrogen (16 µg d^{-1}) for 2 wk. Administered in feed, low dose genistein was comparable to estrogen in maintaining cancellous bone (reflected by the number and density of trabeculae and femoral ash weight), whereas genistein was less effective at the higher doses. These findings of a biphasic response to genistein suggest that at the low dose, this isoflavone may have an estrogen agonist effect, whereas at higher doses it is less effective and may in fact have adverse effects on bone cells.

Fanti et al. (1998) also used three doses of genistein (1, 5, or 25 µg kg^{-1} d^{-1}), but they injected ovx rats daily for 21 d following surgery. In contrast to results from Anderson et al. (1998), Fanti et al. (1998) found no effect of low dose (1 µg kg^{-1} d^{-1}) genistein on bone, but significant bone-sparing effects on tibial (trabecular and cortical) bone at higher doses, with the intermediate (5 µg kg^{-1} d^{-1}) being slightly, albeit nonsignificant, better than the high (25 µg kg^{-1} d^{-1}) dose. Genistein did not prevent uterine atrophy except marginally at the highest dose. Genistein was associated with a higher bone formation rate per tissue volume, but did not significantly affect bone resorption, suggesting a different effect of genistein than that of estrogen on bone. Also, genistein blocked the production of cellular tumor necrosis factor-α, suggesting a modulatory effect on this proinflammatory cytokine. Using ovx mice, Ishimi et al. (1999) examined the effects of administering genistein (0.1–0.7 mg d^{-1}) compared with 17 β-estradiol (0.01–0.1 µg d^{-1}) for 2 to 4 wk on bone and bone marrow. The ovx-induced increase in B-lymphopoiesis was completely restored by either genistein or 17 β-estradiol, whereas genistein did not reverse the uterine atrophy. Genistein (0.7 mg d^{-1}) or 17 β-estradiol (0.01 µg d^{-1}) restored trabecular bone volume of the femoral distal metaphysis.

Picherit et al. (2000) investigated the effects of orally-administered genistein or daidzein at similar doses (10 µg kg^{-1} d^{-1}) on ovx-induced bone loss in adult Wistar rats. Daidzein- or 17 α-ethinlyestradiol-treated animals did not lose vertebral, total femur, femur metaphyseal (trabecular-rich), or femur diaphyseal (cortical-rich) bone, whereas genistein only maintained the diaphyseal region. Genistein, daidzein-, or 17 α-ethinlyestradiol-suppressed bone resorption. These results indicate that genistein prevented loss of cortical bone, whereas daidzein prevented loss of both trabecular and cortical bone. Picherit et al. (2001a) examined the effects of various doses (0, 20, 40, or 80 mg kg^{-1} d^{-1}) of soybean isoflavones (similar amounts of genistein/genistin and daidzein/daidzin, less glycitein/glycitin) to reverse established osteopenia in ovx Wistar rats. Despite dose-dependent effects in reducing bone resorption (deoxypyridinoline) and formation (osteocalcin), isoflavones did not reverse established osteopenia, similar to that reported from an earlier study by Arjmandi and colleagues (1998a). In a similar study design but in rats without established osteopenia, Picherit et al. (2001b) found that total femoral and disphyseal BMD and femoral failure (strength) were similar in isoflavone-treated and sham-operated rats. Metaphyseal BMD values were similar in ovx rats and those receiving the lowest isoflavone dose, but those on higher doses were similar to sham-operated rats. Urinary deoxypyridinoline at Day 45 was higher in all ovx vs. sham-operated animals. At Day 91, deoxypyridinoline was comparable between the higher isoflavone dosed and sham-operated rats, whereas plasma osteocalcin was greater in the isoflavone-treated rats. This study indicated that isoflavones prevented ovx-induced bone loss by maintaining bone formation and inhibiting bone resorption (but not initially). The optimal dose used was 40 mg kg^{-1} d^{-1}, since the highest dose exerted a weak uterotrophic effect, whereas both 40 and 80 mg/kg/d preserved cancellous and cortical bone.

Finally, Uesugi et al. (2001) determined the effects of the β-glycosides, genistin (50 or 100 mg kg^{-1} d^{-1}), daidzin (25 or 50 mg kg^{-1} d^{-1}), and glycitin (25 or 50 mg kg^{-1} d^{-1}), given orally to young ovx rats. Isoflavone glycosides at the intermediate dose, similar to estrone (7.5 μg kg^{-1} d^{-1}), prevented femoral bone loss. At this dose, daidzin and glycitin also prevented uterine atrophy, gain in abdominal fat and body weight, and increases in bone resorption markers (pyridinoline and deoxypyridinoline). In contrast, genistin prevented ovx-induced uterine atrophy only at the highest dose. Thus, daidzin and glycitin may prevent bone loss due to suppression of bone resorption, similar to estrogen, whereas genistin may exert its effect via a different mechanism.

These studies suggest that the route of administration, dose and form of isoflavone (i.e., genistein/daidzein/glycitein as the aglycone or genistin/daidzin/glycitin as the β-glycoside), age of the animal, degree of osteopenia, and type (i.e., trabecular or cortical) of bone affected are key in interpreting results. Evidence suggests that once osteopenia is established, isoflavones are not effective in reversing the loss. There is likely an optimal dose, but not yet defined, conceivably dependent upon the size and species of the animal, degree of bone loss, route of administration, and the composition of the isoflavone treatment. These studies suggest that genistein, daidzein, and glycitein are all bone-active compounds, perhaps exerting selective effects on osteoblasts and osteoclasts. It may be that daidzein/daidzin and glycitein/glycitin are more estrogenic than genistein/genistin and hence mixtures with various proportions of isoflavones may exert differential skeletal effects. The mechanism by which isoflavones attenuate bone loss is uncertain, but may involve osteoblast stimulation and thus maintenance of bone formation, not simply by inhibiting osteoclasts, as is the purported mechanism for estrogen. Further study is needed to determine which isoflavones inhibit bone resorption and which stimulate formation.

The next studies reviewed were designed to examine the cellular effects of genistein and daidzein and their β-glycosides genistin and daidzin on bone. Gao and Yamaguchi (1999a, 1999b, 1999c, 2000) have performed numerous studies to examine potential mechanisms of action of isoflavones. The first (Gao and Yamaguchi, 1999a) compared the effect of daidzein with genistein on femoral (cortical) bone in vitro in tissue derived from elderly rats. Either treatment increased alkaline phosphatase activity and the contents of Ca and DNA in the femoral-diaphyseal tissue, similar to that of 17 β-estradiol or zinc sulfate ($ZnSO_4$). In the presence of zinc sulfate, but not with the addition of 17 β-estradiol, the effect of daidzein or genistein on alkaline phosphatase activity was enhanced synergistically. These results suggest that daidzein and genistein exert an osteoblast-stimulating effect. In contrast, the second (Gao and Yamaguchi, 1999b) study reported a potent inhibitory effect of genistein on osteoclast-like cell formation in mouse marrow cultures, perhaps involving cAMP signaling. The inhibitory effect of genistein was equal to that of 17 β-estradiol, calcitonin, or zinc sulfate. The third (Gao and Yamaguchi, 1999c) and fourth (Gao and Yamaguchi, 2000) study examined the effect of genistein on osteoclast-like multinucleated cells from rat femoral tissue. Genistein, daidzein, or calcitonin decreased the number of osteoclasts, which was abolished by inhibitors of Ca-dependent protein kinases (Gao and Yamaguchi, 1999c). The authors suggest that the isoflavones may have inhibited osteoclasts by inducing apoptosis. In the companion study (Gao and Yamaguchi, 2000), genistein decreased the number of osteoclasts, partially through activating protein tyrosine phosphatase and inhibiting protein kinase in osteoclasts. Finally, Yamaguchi and Sugimoto (2000) reported that the effect of genistein or daidzein on osteoblastic MC3T3-E1 cells was to stimulate protein synthesis, activating aminoacyl-tRNA synthetase, with a maximal effect at 48 h. The authors suggest that isoflavones, particularly genistein, have a stimulatory effect on the translational process of protein synthesis in osteoblastic cells. Hence, these in vitro studies indicate that isoflavones both suppress osteoclastic and enhance osteoblastic function.

21–2.4 Human Studies

21–2.4.1 Soybean and/or Isoflavone and Bone Density

Six intervention/prospective studies thus far have been published (Dalais et al., 1998; Potter et al., 1998; Alekel et al., 2000; Hsu et al., 2001; Clifton-Bligh et al., 2001; Ho et al., 2001) that include a measurement of bone mass as an outcome in response to isoflavone intake. Only two of these studies (Alekel et al., 2000; Ho et al., 2001) were designed specifically to examine bone as the primary outcome. Four studies used either soybean protein isolate (Alekel et al., 2000; Potter et al., 1998) or a soy food (Dalais et al., 1998) or usual soy food intake among Asians (Ho et al., 2001) as the source of isoflavones; two used extracted isoflavones (Hsu et al., 2001; Clifton-Bligh et al., 2001).

Dalais et al. (1998) supplied daily 45 g soybean grits (flour) containing 53 mg d^{-1} isoflavones, 45 g linseed (flaxseed with mammalian lignan precursors), or 45 g wheat (*Triticum vulgare*) kibble (control) to 44 postmenopausal women for 12 wk using a crossover design. They found that total body bone mineral content (BMC) increased 5.2% in the soybean, with nonsignificant increases in the linseed and wheat groups, whereas there was no change in BMD. The magnitude of this increase is remarkable, but in addition, there were increases in BMC in the other two groups, one of which was control. Thus, these results should be interpreted with caution. The next published study (Potter et al., 1998), designed to examine the lipid-related effects of soybean protein, randomly assigned 66 postmenopausal women to one of three treatments: (i) casein + nonfat dry milk protein, (ii) soybean protein (40 g d^{-1}) isolate (SPI) with 56 mg d^{-1} of isoflavones, or 3(iii) SPI with 90 mg d^{-1} of isoflavones. The hypercholesterolemic women were heterogeneous with respect to time since menopause and age (49–83 yr). After 6 mo of treatment, women in the high isoflavone group experienced an increase (~2%; $P < 0.05$), whereas those in the casein + milk-based protein had slight decreases, in lumbar spine BMD and BMC. However, women in the high isoflavone group began the study with lower BMD and BMC than the other two groups, but baseline values were not taken into account. The effect of treatment on bone is typically greater in those with lower bone mass (Pines et al., 1999) and hence baseline BMD should be considered.

Nevertheless, the next published study (Alekel et al., 2000) in 69 peri-menopausal women is in general agreement with the previously described work. Subjects were randomized (double-blind) to treatment, with dose expressed as aglycone units: isoflavone-rich soybean (SPI+, 80.4 mg d^{-1}; $n = 24$), isoflavone-poor soybean (SPI–, 4.4 mg d^{-1}; $n = 24$), or whey (control; $n = 21$) protein. The authors found no change in lumbar spine BMD or BMC (Fig. 21–6), respectively, in the SPI+ (−0.2%, +0.6%,) or SPI– (−0.7%, −0.6%,) groups, but losses of 1.3% BMD and 1.7% BMC occurred in the control group. Baseline BMD and BMC negatively affected percentage change in these outcomes, but baseline values were taken into account in the analysis of covariance (ANCOVA) and regression analysis. Results of ANCOVA indicated that treatment had a significant effect on percentage change in BMC, but not on percentage change in BMD. Contrast coding using AN-COVA with BMD or BMC as the outcome revealed that isoflavones, not soybean protein, exerted a positive effect. Taking various contributing factors into account using multiple regression analysis, SPI+ had a significant positive treatment effect on the percentage change in both BMD (5.6%) and BMC (10.1%). Body weight at baseline rather than weight gain or final weight was related to percentage change

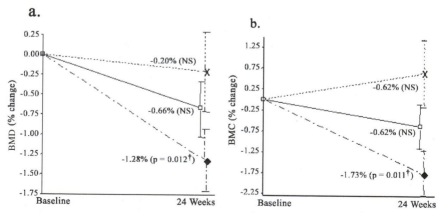

Fig. 21–6. Mean (+SEM) percentage change in lumbar spine bone mineral density (BMD) and bone mineral content (BMC) from baseline to posttreatment in three treatment groups of perimenopausal women: isoflavone-rich soy (SPI+: X; $n = 24$), isoflavone-poor soy (SPI–:□; $n = 24$), and whey (control: ◆; $n = 21$) protein. Lumbar spine bone mass was determined by using dual-energy X-ray absorptiometry. †Significantly different from baseline (paired t-tests with Bonferroni adjustment; P values noted). Adapted from Alekel et al. (2000). Printed with permission.

in BMD, suggesting that weight gain did not confound the effect of SPI+ on bone. Contrary to the hypothesis, no effect of reproductive hormones or estrogen status on bone loss was found. The other treatments had no effect on the spine and treatment in general had no effect on bone sites other than the spine. The previous two studies support the idea that isoflavones are the bioactive component of soybean with respect to bone.

A recent study examined habitual soybean intake and bone mass in premenopausal Chinese women 30 to 40 yr of age living in Hong Kong (Ho et al., 2001). Researchers reported a positive effect of soybean isoflavones on spinal BMD, while adjusting for age and body size (height, weight, and bone area), after an average follow-up time of 38.1 mo. Average percent decline in spinal BMD in 116 women was greater in the lowest (−3.5%) vs. highest (−1.1%) quartile of soybean isoflavone intake. Soybean isoflavone intake (along with lean body mass, physical activity, energy adjusted Ca intake, and follow-up time) accounted for 24% of the variance in spinal BMD in these women. This 3-yr study indicated that soybean isoflavone intake had a positive effect on maintaining spinal BMD in premenopausal women 30 to 40 yr of age. In contrast, the effect of supplementing with soybean isoflavones (150 mg d^{-1}) in 37 postmenopausal women did not produce a significant change in calcaneous BMD after 6 mo of treatment (Hsu et al., 2001). This study had no control group, making interpretation difficult. It may be that because the calcaneous (heel) is weight-bearing and has greater trabecular content (Davis et al., 1999) than vertebral bone, it responds differently to isoflavone treatment than the lumbar spine. An unusual finding in this study was that plasma estradiol increased significantly after 6 mo of treatment. The other published human study also used isoflavones, but they were extracted from red clover (*Trifolium pratense* L.) and not soybean (Clifton-Bligh et al., 2001). The red clover preparation contained genistein, daidzein, formononetin, and biochanin and was administered (doubleblind) to 46 postmenopausal women for 6 mo. Subjects were randomized to one of three treatments: 28.5, 57, or 85.5 mg d^{-1} of isoflavones. Women at the 57 mg dose experienced a 4.1% increase and those at the 85.5 mg dose a 3.0% increase of the proximal radius and ulna BMD, whereas the 2.9% increase at the 28.5 mg dose was not significant. There was no significant change in the distal radius and ulna (predominantly trabecular bone) BMD in relation to isoflavone dose, nor did endometrial thickness increase with treatment. This study had no control group, complicating the interpretation. Yet, results of this study suggest that isoflavones may indeed have a significant effect on cortical (proximal radius and ulna) bone, or that appendicular bone responds differently than the axial (i.e., spine) skeleton. Alternatively, formononetin and biochanin may exert effects on cortical bone, whereas soybean isoflavones may not have an effect on appendicular bone.

Taken together, results of these human studies suggest that isoflavones may attenuate bone loss from the lumbar spine in estrogen-deficient-women, who may otherwise be expected to lose 2 to 3% yearly. This attenuation of loss, particularly if continued throughout the postmenopausal period, could translate into a decrease in lifetime risk of osteoporosis. Since a bone-remodeling cycle ranges from 30 to 80 wk (Heaney, 1994), such short-term preliminary studies cannot answer the question of whether these bone-sparing effects would be sustained over a longer period. From these results, we cannot determine whether the reported bone-sparing effect is due to treatment or is an artifact of the bone-remodeling transient (Heaney, 1994). A study of longer duration is necessary to determine whether soybean isoflavones will affect the remodeling balance, tipping it in favor of bone formation rather than resorption.

21–2.4.2 Bone Turnover

Few studies have been published examining the response of biochemical markers of bone turnover to isoflavone-rich soybean protein or extracted isoflavones in humans. This first section will review studies using soybean protein as the treatment and the next section will cover trials using extracted isoflavones, illustrating the somewhat inconsistent results in the biochemical markers.

Murkies et al. (1995) was the first group to include a measure of bone turnover as part of a study designed to examine hot flushes. They supplemented the diets of postmenopausal women with either wheat or soybean flour (45 g d^{-1}) for 12 wk. Urinary hydroxyproline, a nonspecific marker of bone resorption, increased over time in the wheat flour but not in the soybean flour group, although the difference between groups was not significant. In a study designed to examine the effects of soybean protein on cardiovascular disease risk factors and menopausal symptoms, Washburn et al. (1999) also reported on alkaline phosphatase activity, a bone formation marker. This was a randomized crossover (double-blind) trial in 51 subjects who consumed isocaloric supplements for 6 wk each: (i) 20 g soybean protein (34 mg isoflavones) in a single dose; (ii) 20 g soybean protein (34 mg isoflavones) split into two doses; or (iii) 20 g complex carbohydrate. Compared with the carbohydrate-supplemented group, alkaline phosphatase activity decreased in women on either soybean diet. Although the authors suggest that this decline may reflect a beneficial effect of soybean, these results are difficult to interpret because they did not measure any bone resorption marker and alkaline phosphatase is not at all specific for bone. An inherent drawback in crossover designs is the potential for carryover or contamination in the outcomes of interest. Another randomized (double-blind) placebo-controlled 12-wk study was designed to examine the effects of soybean protein (40 g d^{-1}) on menopausal symptoms in 24 women (Knight et al., 2001). They found no differences between the isoflavone-(77.4 mg d^{-1} aglycone components) and casein-treated control groups with respect to serum alkaline phosphatase or pyridinoline cross-links (bone resorption marker). In contrast to these studies, the recently completed 24-wk trial in perimenopausal women (N = 69) described above did not report any decline in bone resorption (cross-linked N-telopeptides) during the course of treatment (Alekel et al., 2000). These women entered the study in four waves or cohorts, with approximately equal numbers from each of the three treatments in each cohort. Repeated measures ANCOVA indicated that both time and baseline value were significant, whereas treatment per se had no effect on either cross-linked N-telopeptides or bone-specific alkaline phosphatase (Fig. 21–7). However, cohort had a significant effect on N-telopeptides, but not on bone-specific alkaline phosphatase, suggesting that cohort may reflect a seasonal effect on bone resorption. This study appears to corroborate our findings in the rat model (Arjmandi et al., 1996; 1998b), both of which indicate that soybean with isoflavones does not decrease bone turnover. Wangen et al. (2000) designed a randomized,

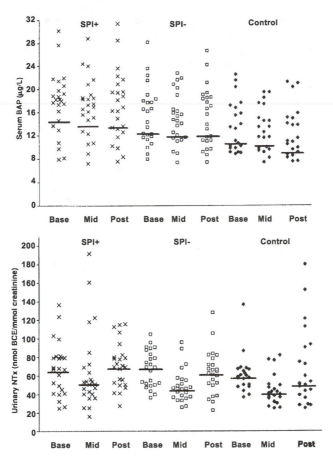

Fig. 21–7. Scatter plot (median values represented by —) of serum bone-specific alkaline phosphatase (BAP) and urinary cross-linked *N*-telopeptide (*N*-Tx) concentrations at baseline (base), midtreatment (mid), and posttreatment (post) in three treatment groups of perimenopausal women: isoflavone-rich soybean (SPI+: X; $n = 24$), isoflavone-poor soybean (SPI–: □; $n = 24$), and whey (control: ◆; $n = 21$) protein. BCE, bone collagen equivalents. Adapted from Alekel et al. (2000). Printed with permission.

crossover 3-mo study to examine the dose-response effect of isoflavones (7+1.1, 65+11, or 132+22 mg d^{-1}, on a per kilogram body weight basis, expressed as agly-cone units) on bone markers in 17 postmenopausal women. The treatment was provided as a 63 g d^{-1} soybean protein isolate. Bone-specific alkaline phosphatase was decreased by the low (65 mg d^{-1}) and high (132 mg d^{-1}) isoflavone diets. There was a trend toward a decrease in osteocalcin and in insulin-like growth factor-I (IGF-I) in the high isoflavone group. All three diets increased bone-specific alkaline phosphatase, osteocalcin, and IGF-I when compared with baseline, although the increases in the latter two markers were not statistically significant at the highest isoflavone dose. The authors submit that although isoflavones modestly affected markers of bone turnover, the changes were small and not likely clinically relevant. Taken together, these results suggest that soybean isoflavones may not decrease bone resorption as does estrogen or bisphosphonate therapy, but indeed may prevent a decline in bone formation. However, this issue is not yet resolved.

In the study described above, Clifton-Bligh et al. (2001), using isoflavones extracted from red clover, reported no significant changes in urinary deoxypyridinoline (bone resorption marker) from baseline to 6 mo and no differences among the doses of isoflavones. Scambia et al. (2000) used a standardized soybean extract (50 mg d^{-1} of isoflavones) or placebo to determine the effects on early climacteric symptoms. Subjects (N = 39) were randomly assigned to soybean or placebo treatment for 6 wk and then given conjugated equine estrogen (0.625 mg d^{-1}) for 4 wk. Soybean-related changes in serum osteocalcin (bone formation marker) were not observed and estrogen-related changes were not modified by soybean extract. Likewise, Upmalis et al. (2000) used a soybean isoflavone extract (50 mg d^{-1} of isoflavones) vs. placebo to determine the effects on climacteric symptoms in postmenopausal women (N = 177) at 15 sites. Although they found a reduction in the number and severity of hot flushes with soybean, they reported no treatment effect or change in either serum osteocalcin or urinary N-telopeptides.

Given the apparent contradictory results regarding biochemical markers of bone in both animal and human models, it is difficult to draw conclusions about the effect of soybean or soybean isoflavones on bone turnover. Some of this discrepancy is likely due to the extreme variability of these markers and the fact that they rise markedly in ovx animals or early menopausal women. In addition, differences in study design do not allow reasonable comparisons.

21–2.4.3 Future Research Directions

At present, there is a paucity of studies in humans, but they do support a modest protective effect of isoflavone-rich soybean on bone. Virtually nothing is known about the effects of soybean in men or elderly women at-risk for osteoporosis. Isoflavone supplements may not perturb eating habits, body weight, or overall dietary intake, thus allowing better long-term compliance and leading to less potential confounding by extraneous factors. Although data suggest that isoflavones are the primary bone-active components of soybean, there may be as yet unknown untoward side effects of long-term use of isolated isoflavones. Moreover, the intake of soybean protein may promote additional health benefits unrelated to bone. Also, we may discover that some non-isoflavone component of soybean contributes to its bone-protective effect. To advance our basic knowledge and pave the way for finding alternatives to steroid hormone therapy for postmenopausal women, a long-term, dose-response study in humans designed to corroborate the findings presented in this review and to examine potential mechanisms is needed.

21–2.5 Summary of Soybean and Bone Health

Overall, the evidence that soybean protein and/or its isoflavones favorably affect BMD is promising but tentative. However, a consensus cannot be reached at this time on whether soybean and/or its isoflavones inhibit bone resorption or stimulate bone formation or some combination of the two. Insufficient evidence is presently available to recommend soy foods as a substitute for estrogen or hormone replacement therapy. However, the public should still be encouraged to include soy foods in their diet because of its nutrient profile and other health benefits. In the future, we may be able to recommend that incorporating isoflavone-containing soybean products into the diets of perimenopausal and early postmenopausal women will serve as an adjunct to treatment for women at-risk of osteoporosis. This may be especially important for women who are poor candidates for or elect not to take steroid hormones. Although few dose-response studies have been conducted, human data suggest that perhaps 60 to 90 mg d^{-1} of isoflavones may be effective. This translates into approximately two to three servings of traditional soy foods, certainly a challenge for many consumers. Perhaps lesser amounts of soybean protein or soybean isoflavones consumed over many years will exert favorable effects on bone. Nonetheless, as the food industry continues to develop new soybean products, this should allow consumers to increase soy-food intake.

21–3 SOYBEAN AND CANCER PREVENTION

Cancer is the second leading cause of death in the USA. It is anticipated that 552 200 people, more than 1500 people a day, died of cancer in 2000. Indeed, one of four deaths in the USA is from cancer. Furthermore, about 1 220 100 new cancer cases were diagnosed in 2000. The leading sites for cancer deaths in males in 2000 were (in order of age adjusted cancer deaths), lung and bronchus, prostate, colon and rectum, pancreas, stomach, and liver. In women the sites were lung and bronchus, breast, colon and rectum, ovary, pancreas, uterus, and stomach (American Cancer Society, 2000). This section will first provide an overview of human, then animal investigations into the role of soy foods and soybean constituents in cancer prevention. We then review the constituents in soy foods that may contribute to cancer prevention, and their potential mechanisms of action.

21–3.1 Human Investigations

The potential role of soybean in the prevention of human cancer was recently reviewed (Birt, 2001; Birt et al., 2001; Fournier et al., 1998; Messina et al., 1994; Messina, 1999). A cancer protective role for soy foods was first proposed to explain correlational studies that noted distinct differences in cancer rates between the western and Asian cultures. For example, the studies by Hirayama (1979) noted that the rate of prostate cancer was lower in Japan than in the USA and, that Japanese who were living in the USA experienced an increase in prostate cancer rates and their rate approached those in U.S. Caucasian (age standardized incidence rates of prostate cancer per 100 000 people: Osaka, Japan, 2.7; Japanese in Hawaii, 24.5; Iowa (primarily Caucasian) 38.6). Other investigations, however, have demonstrated that these associations are not necessarily due to soy foods since other dietary factors appear to contribute. For example, a prospective study of prostate cancer in Hawaiian men of Japanese descent found the lowest rates among men who consumed diets abundant in rice and tofu (modestly protective) and low in seaweed (Severson et al., 1989).

In considering gastric cancer, Chinese cabbage (*Brassica chinensis*) seemed to play an important role in protecting against gastric cancer in a case control study in China (Hu et al., 1988). Furthermore, consumption of fermented and salted soybean paste was positively associated with higher rates of gastric cancer in this study (Hu et al., 1988). In considering lung cancer, a case control study of Chinese women in Hong Kong found an association between fresh fruit and fish consumption and lower cancer rates (Koo, 1988). They observed lower rates of adenocarcinoma or large cell lung tumors in people with higher consumption of leafy green vegetables, carrots, tofu, fresh fruit, and fresh fish (Koo, 1988). A study in Chinese men in the Yunnan Province similarly found an inverse association between lung cancer and the consumption of diets rich in bean curd, meat, eggs, and vegetables (Swanson et al., 1992).

One breast cancer study (Nomura et al., 1978) noted that people who had lower rates of breast cancer ate less meat and butter and more soybean foods, green tea, and seaweed than populations with higher cancer rates. Wu et al. (1998) investigated breast cancer and found considerable challenges in trying to compare studies in Asia with studies from the West because soybean food consumption in Asia is much higher than in the West. In addition, human investigations suggesting that soybean foods may protect against breast cancer have not provided adequate analytical information on the bioactive constituents in the soy foods, the portion size or the other components that may be protective in the diets of people who eat soy foods. Wu et al. (1998) concluded that the data on soybean intake and breast cancer suggest, but not consistently, that soy foods provide some protection against breast cancer.

Many investigations have attempted to use biomarkers of soybean intake as surrogate markers of soy food intake and to thus associate soy food intake with cancer prevention. The most commonly used biomarker is isoflavone concentrations in bodily fluids. This fortuitous use of isoflavones as a marker for soy food intake has lead some scientists and nonscientists to assume that isoflavones are the sole or primary active constituents in soy foods. It is important to note that although isoflavones may be a good marker for consumption of soy food constituents, they are not necessarily the agents responsible for the cancer prevention. Using isoflavone as a biomarker, one recent investigation provided support for soy foods containing isoflavones as protective against cancer. In particular, urinary excretion of total phenols and all individual isoflavonoids, particularly glycitein, was lower in breast cancer patients than in controls in a case control study of breast cancer in Shanghai (Zheng et al., 1999). In comparing women with the highest urinary excretion of phenol and total isoflavonoids with women in the lowest tertile, they observed an adjusted odds ratio for breast cancer of 0.14 (0.02–0.88, 95% confidence interval). However, a more recent Dutch study (Den Tonkelaar et al., 2001) was unable to detect a statistically significant relationship between urinary isoflavones and breast cancer risk in postmenopausal breast cancer. This distinct difference in results from the Shanghai and Dutch studies may certainly be due to the appreciably higher isoflavone excretion (and presumably intakes) in China than in the Netherlands (Table 21–1).

Intriguing results were obtained by Bennink (2001) in a soybean protein (with high isoflavones) intervention in a small number of human subjects. Subjects were

Table 21–1. ORs and 95% CIs for the association of breast cancer with urinary excretion of isoflavonoids.

Urinary excretion (by tertile)†	Cases	Controls	Unadjusted Odds Ratio 95% Confidence Interval	Test for trend
	——— no. ———			
Shanghai‡				
Total isoflavonoids				
<732	35	20	1.00	
732–2446	11	20	0.35 (0.14–0.89)	
>2447	14	20	0.45(0.19–1.10)	$P = 0.04$
Genistein				
<129	32	20	1.00	
130–535	11	20	0.43 (0.19–1.01)	
>536	17	20	0.59 (0.25–1.43)	$P = 0.12$
Glycitein				
<83	35	20	1.00	
83–2.14	13	20	0.43 (0.18–1.04)	
>281	12	20	0.36 (0.14–0.94)	$P = 0.02$
Dutch§				
Enterolactone/creatinine				
7.16-379.0	26	91	1	
379.1–655.9	27	94	1.01 (0.55–1.85)	
656.0–2334.9	34	83	1.43 (0.79–2.59)	$P = 0.25$
Genistein/creatinine				
10.2–67.1	31	86	1	
67.2–112.2	30	92	0.90(0.51–1.62)	
112.3–523.8	27	90	0.83 (0.46–1.51)	$P = 0.60$

† Expressed as $\mu mol\ mol^{-1}$ creatinine for isoflavonoids
‡ Zheng et al. (1999). Printed with permission.
§ Tonkelaar et al. (2001). Printed with permission.

selected to be at high risk for colon cancer and changes in the labeling index and proliferative zone were assessed following a 1 yr intervention with 39 g of isolated soybean protein or casein. Supplementing with soybean protein isolate was associated with a reduction in labeling index and proliferative zone suggesting a reduction in colon cancer risk.

In summary, studies in Asians indicate that soy foods are often associated with reductions in cancer rate, but other dietary constituents appear to contribute to this association and soy foods may not be the primary protective component of the Asian diet. Furthermore, using isoflavones as biomarkers of soy food intake has not consistently demonstrated cancer prevention in association with elevated isoflavone urinary excretion. Several other factors may protect against cancer in Asian populations including high physical activity, high fruit and vegetable intake, and lower fat and energy intakes. It is important to note that few prospective studies have been conducted and no intervention trials in humans with soy foods or soy food constituents were reported with a cancer endpoint.

21–3.2 Animal Investigations

Results of animal investigations have provided evidence for a role of soy foods and soybean ingredients in cancer prevention. Messina et al. (1994) reviewed this data in 1994 and 17 of 26 studies (65%) provided evidence of protection. In 1998 a review by Fournier et al. (1998) reported that 94% of the studies that were published from 1990 to 1998 supported protection against cancer by soy foods or soy food components. Many of the studies reviewed in the more recent evaluation were with models of mammary, colon, and prostate cancer and they used isoflavones as the intervention strategy (Birt et al., 2001; Fournier et al., 1998).

Soybean isolates and soy foods have been studied as cancer preventive agents against radiation and chemically induced mammary cancer in rodents in several studies (Baggott et al., 1990; Barnes et al., 1990; Cohen et al., 2000; Gotoh et al., 1998; Hawrylewicz et al., 1991; Troll et al., 1980). Earlier investigations found striking inhibition of radiation, N-methyl-N-nitrosourea (MNU), and 7,12,Dimethyl-benz[a]anthracene (DMBA)-induced mammary cancer. Mammary cancer was reduced by up to approximately 50% in a dose response manner by dietary soybean protein isolate incorporated in the diet from 2 to 20% in replacement of casein (Barnes et al., 1990; Hawrylewicz et al., 1991). A more recent study found that 10% dietary soybean or miso (Japanese soybean paste) reduced MNU (40 mg kg^{-1} body weight) induced mammary cancer in rats by nearly 50% (Gotoh et al., 1998). However, the diets were fed for only 18 wk following the MNU treatment and the data suggest that the soybean-supplemented groups may have been killed during the phase of rapid tumor development. In another study, miso was compared with a NaCl supplemented diet to control for the salt in miso and a modest inhibition of DMBA (10 mg rat^{-1}) initiated cancer was observed in both groups suggesting that the inhibition in the miso-treated group may have been due to the high salt content of this soybean food (Baggott et al., 1990). Finally, a more recent investigation that had the aim of determining if soybean isoflavones contribute to the potential breast cancer prevention by soybean and soy foods compared soybean based diets with and without isoflavones (Cohen et al., 2000). These investigators compared diets with 10 and 20% soybean protein with diets containing the same amount of isoflavone depleted soybean protein and none of the dietary treatments prevented MNU induced (40mg kg^{-1} body weight) mammary cancer in F-344 rats (Cohen et al., 2000).

Because the soybean isoflavones genistein and daidzein possess estrogen receptor antagonist and agonist activities, numerous experimental investigations with animals and cultured cells have investigated them for anti-breast cancer prevention. Research with cultured human breast cancer cells found that culturing estrogen receptor positive (MCF-7) or estrogen-receptor negative (MDA-MB-468) cells with 30 to 150 mmol L^{-1} genistein-inhibited growth and increased expression of maturation markers (Constantinou et al., 1998). In nude mouse implantation studies, treatment of the above cells with genistein (30 μmol L^{-1}) for 6 d prior to implantation into nude mice decreased the growth of the implanted cells. These studies did not provide support for the hypothesis that the estrogenecity of genistein contributed to the inhibition of human cancer cell growth. This team also assessed the impact of injections of genistein and daidzein (0.8 mg daily for 180 d) against N-methyl-N-nitrosourea-induced mammary tumors in Sprague Dawley rats. These injected isoflavones weakly inhibited breast tumor incidence but, in parallel studies, the isoflavones did not inhibit topoisomerase activity or protein tyrosine kinase activity in tumors. Thus, the reduction of mammary cancer by these isoflavones was not caused by impacts on topoisomerase and/or protein tyrosine kinase activity (Constantinou et al., 1996).

Some of the more conflicting data regarding the impact of isoflavones on mammary carcinogenesis were interpreted to suggest that the time of life and the status of the animal at the time of exposure to genistein may be critical factors in determining whether isoflavones inhibit or enhance breast cancer rates. In particular, when genistein was administered early in life mammary carcinogenesis was inhibited (Barnes, 1997; Lamartiniere et al., 1995). However, when genistein was administered during tumor development, enhanced tumor growth was observed (Hsieh et al., 1998). Studies by Lamartiniere et al. (1995) were designed with an early life exposure to genistein because of the prior observations that neonatal estrogen inhibited both spontaneous and chemically induced breast cancer. Neonatal rats were treated with 5-mg genistein on Day 2, 4, and 6 postpartum and mammary tumors were induced with DMBA on Day 50. They observed a reduction in the number of mammary tumors and a delay in their appearance in the rats that were pretreated with genistein (Fig. 21–8) (Lamartiniere et al., 1995). Furthermore, prepubertal administration of 500 μg g^{-1} body weight genistein with the DMBA protocol inhibited breast cancer development. Furthermore, dose response inhibition of mammary tumors was observed in rats exposed to genistein (0, 25, and 250 mg kg^{-1} diet) from conception to 21 d postpartum prior to treatment with DMBA (50 d postpartum) (Fig. 21–8). In parallel studies, rats treated chronically with genistein had fewer terminal end buds and fewer undifferentiated terminal ductal structures at 21 and 50 d of age (Fritz et al., 1998). Although neonatal administration of

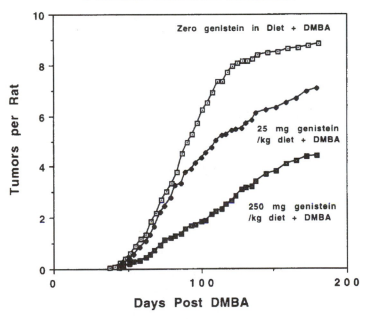

Fig. 21–8. Ontogeny of palpable mammary tumors in female Sprague-Dawley CD rats exposed perinatally to genistein in the diet from conception until 21-d post-partum. After weaning, the offspring were fed AIN-76A diet only. On Day 50 post-partum all animals were treated with 80-mg DMBA kg^{-1} body wt. Adapted from Fritz et al. (1998). Reprinted with permission from C.A. Lamartiniere.

5-mg genistein pup^{-1} was adverse to normal ovarian follicular development, prepubertal genistein (500 μg g^{-1} body weight) did not appear to be toxic (Lamartiniere et al., 1998a, 1998b).

However, as previously observed with estrogen, administration of genistein (750 mg kg^{-1} in the diet) to mice with developing tumors enhanced the growth of estrogen responsive tumors (Hsieh et al., 1998) (Fig. 21–9). Furthermore, both in vitro and in ovariectomized athymic mice, genistein (10 nM–10 μM) enhanced the proliferation of MCF-7 human breast cancer cells. Genistein (1 μM) induced pS2, an estrogen responsive gene, expression demonstrating its action as an estrogen agonist. Thus, while soybean isoflavones show some promise for cancer prevention when administered during the development of the mammary glands, isoflavones may enhance the growth and development of mammary tumors in tumor-bearing animals. Clearly, caution needs to be taken in using isoflavones for cancer prevention in humans. The public should be advised to carefully consider the use of high potency isoflavone preparations that are now available as dietary supplements.

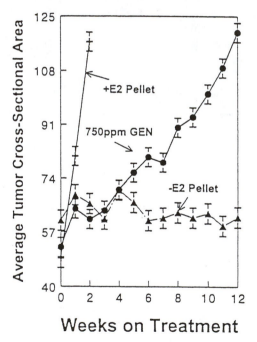

Fig. 21–9. The effect of estrogen pellet (2 mg) and dietary genistein (750 mg kg^{-1}) on MCF-7 tumor growth in athymic nude mice. MCF-7 human breast cancer cells were injected s.c. into four sites on the flanks of mice at 1×10^6 cells/sites. After tumors had formed, the mice were grouped to equalize the tumor area, and dietary treatment was initiated. Experimental groups included negative control AIN-93G (5 mice; n – 15 tumors), positive control implanted with a 2-mg estrogen pellet (5 mice; $n = 17$), and AIN-93G + 750 ppm genistein (5 mice; $n = 17$ tumors). Data are expressed as changes in tumor areas for each week of measurement. The treatment by week interaction is statistically significant ($P < 0.0001$). Treatment means for each week were compared using an LSD test. Adapted from Hsieh et al. (1998). Reprinted with permission from W. G. Helferich.

Since prostate cancer, like breast cancer, has been observed more frequently in Western cultures than in Asia, there has been considerable interest in the ability of soybean, soy foods, and isoflavones to prevent this disease. However, research in this area was complicated by the absence of representative experimental models for prostate cancer. Indeed, while the rodent prostate cancer models with DES, MNU, and DMAB/testosterone propionate have been used as noted below, their relevance to human prostate cancer is not clear. In a rat investigation with soybean flour fed at 33% by weight of the diet an approximate 30 to 40% reduction in the growth of transplanted Dunning R3327 prostatic adenocarcinoma was observed (Zhang et al., 1997). When prostate lesions were induced in male rats by diethylstilbesterol (DES) treatment over 3 d following birth, 7% soybean in the diet reduced the development of severe dysplasia at 9 mo but this diet did not reduce prostatic dysplasia at 12 mo (Makela et al., 1995).

The ability of soybean isoflavones to inhibit prostate cancer has been assessed in several studies. Methylnitrosourea (MNU) induced prostate seminal vesicle adenocarcinomas were inhibited in Lobund-Wistar rats that were fed high-isoflavone $(1.69$ mg g^{-1}) soybean-based diet in comparison with a low isoflavone soybean-based diet before MNU initiation (Pollard and Luckert, 1997). In another study, rats were treated with 3,2'-dimethyl-4-aminobiphenyl (DMAB) and testosterone propionate to induce adenocarcinoma in the prostate and seminal vesicles. Feeding a soybean isoflavone mixture containing 74% genistein and 21% daidzein at total doses of 100 and 400 mg kg^{-1} reduced the incidence of lesions by about 50% in comparison with rats fed control diet (Onozawa et al., 1999). Studies with cultured prostate cancer cell lines (MAT-lylu and the human prostate cancer cell line PC-3) treated with genistein (1 μg mL^{-1} and 100 ng mL^{-1}, respectively) demonstrated that this isoflavone was cytotoxic to cultured cells. In animal studies genistein failed to inhibit the growth of MAT-lylu cells implanted into rats when it was added to the drinking water at doses of 0.07 to 0.285 mg kg^{-1} d^{-1} (Naik et al., 1994).

Contradictory results have been obtained using animal models to assess soybean and soybean isoflavones in the prevention of colon cancer. In studies with genistein, the most extensively studied soybean isoflavone, doses of 75 and 150 mg kg^{-1} were found to inhibit azoxymethane (AOM) induced colonic aberrant crypt foci (Pereira et al., 1994). Furthermore, studies comparing soybean flakes, soybean flour, genistein, and calcium reported that soybean flakes, soybean flour, and genistein each reduced aberrant crypt foci and genistein (0.015%) caused the greatest reduction (Thiagarajan et al., 1998). In contrast, when soybean proteins with and without isoflavones were compared in the prevention of aberrant crypt foci or colon adenocarcinomas in rats treated with AOM (15 mg kg^{-1} body weight, 2X) no inhibition of colon lesions were observed (Davies et al., 1999). In addition, studies with the min mouse that carries a mutant Adenomatosis polyposis coli (APC) gene and develops intestinal tumors without chemical treatment, showed that neither high isoflavone (475 mg kg^{-1} diet) nor low isoflavone (16 mg kg^{-1} diet) soybean protein-based diets reduced in intestinal tumors, although the positive control, sulindac was effective (Sorensen et al., 1998). In a colon biomarker study, Wang and Higuchi (2000) found that 20% dietary soybean protein reduced polyamine levels in rat intestinal mucosa, a biomarker of cellular proliferation and colorectal cancer risk. However, 0.1% dietary soybean isoflavones (genistein and daidzein at 1:1 mixture) did not alter polyamine levels.

Lee et al. (1995) assessed the impact of soybean isoflavone extract on hepatic preneoplastic foci. γ-Glutamyltransferase and placental glutathione S-transferase positive foci were induced by diethylnitrosamine (15 mg kg^{-1} body weight) and promoted with phenobarbital treatment. Soybean isoflavone extract with approximately equal proportions of genistein and daidzein [containing 920 or 1840 μmole (240 or 480 mg) total isoflavones kg^{-1} diet] was fed during initiation and promotion. After 3 mo of feeding isoflavones both doses of isoflavone extract inhibited hepatic foci but feeding the extract for 11 mo revealed a promotion of altered hepatic foci in the absence of phenobarbital. Thus, the high isoflavone dose promoted the development of preneoplastic foci.

Pancreatic cancer models have been the focus of several investigations on soybean protein and cancer because of early studies showing the trophic effect and pancreatic toxicity of raw soybean protein. In studies of acinar pancreatic cancer using the azaserine-induced cancer model Daly et al. (1992) found that acinar cell nodules were induced similarly in chow fed rats and rats fed raw soybean flour for up to 20 wk. However, the raw soybean flour fed rats exhibited elevated labeling index and pancreatic growth rate. In contrast, the dysplastic ductular pancreatic lesions induced by N-nitrosobis(2-oxopropyl)amine (BOP) in the hamster (3 × 10 mg kg^{-1} body weight) were inhibited in animals fed soybean trypsin inhibitor (5% in the diet) for 37 wk following BOP treatment. Furthermore, pancreatic adenocarcinomas followed the same inhibitory tendency, but the differences were not statistically significant (Furukawa et al., 1991). The incidence of dysplastic pancreatic lesions were also reduced in hamsters administered the soybean trypsin inhibitor for 5 wk with simultaneous BOP weekly injections (5 × 10 mg kg^{-1} body weight^{-1})(Furukawa et al., 1992). Although soybean trypsin inhibitors have been implicated in the trophic effect of raw soybean these agents at lower doses have a potentially important role in cancer prevention as is discussed below.

A recent report demonstrated that defatted soybean fed at 20% of the diet synergistically increased the growth of thyroid glands with iodine deficiency (81.7 ± 8.6 mg 100 g^{-1} body weight) in comparison with the control diet (8.4 ± 2.0 mg 100 g^{-1} body weight) or the iodine deficient group (15.5 ± 1.3 mg 100 g^{-1} body weight) (Ikeda et al., 2000). Thyroid-stimulating hormone was elevated in the defatted soybean protein groups. It was not clear if the defatted soybean protein was or was not heat treated and thus it is possible that the soybean trypsin inhibitor content of the soybean may have contributed to the result.

Both soybean protein and soybean isoflavones (genistein and daidzein) (Li et al., 1999) have been assessed for their anti-metastatic potential using the pulmonary metastasis model of B16 BL6 murine melanoma cells injected into C57BL/6 mice (Yan et al., 1997). Diets containing soybean protein isolate (10, 15, or 20%) containing isoflavone or purified genistein and daidzein (113, 225, 450, or 900 μmol kg^{-1} to match the isoflavones in 2.5, 5, 10, or 20% soybean protein diets) were prefed for 2 wk before and after intravenous injection of the melanoma cells (0.75 ×10^{5}cells) and the soybean or isoflavone containing diets inhibited the incidence and number of lung metastases in a dose response manner (Li et al., 1999).

21–3.3 Components of Soybean that Might Affect Cancer Prevention

Soybean isoflavones may certainly contribute to cancer prevention by soy foods. They have been identified to possess numerous bioactivities that may control and prevent cancer including their anti-estrogenic, anti-proliferative, anti-angiogenesis, pro-apoptotic and immune-enhancing properties (Birt et al., 1998, 2001). However, the role of isoflavones in cancer prevention may have been over emphasized since numerous studies have associated soybean isoflavone excretion with reduced cancer rates. It is important to note that soybean is the primary human food that is rich in isoflavonoids, and thus the excretion of these compounds has often used as a biomarker of soy food intake. The abundance of associative data has been potentially misinterpreted as evidence for a decisive role of isoflavones in cancer prevention. Clearly soybean is also low in methionine, an amino acid that may promote the growth of some cancers, and they are a rich source of Bowman-Birk trypsin inhibitor (Kennedy, 1995) and other proteins with health benefits, phosphatidyl inositol, saponins, β-sitosterol, and sphingolipids that have potential health benefits. Each of these soybean properties/constituents has been studied for cancer prevention potential and have been found to possess activity in the prevention of cancer in animal models (Birt, 2001; Fournier et al., 1998).

21–3.3.1 Isoflavones

In considering the bioactivities of isoflavones that may contribute to cancer prevention the anti-estrogenicity of soybean isoflavones has probably received the most attention. Although evidence for anti-estrogenicity has been obtained with cell culture studies and suggested in animal models and humans, estrogenicity has also been observed and it has been suggested that physiologically relevant doses of isoflavone that might be obtained in individuals consuming soybean-based diets would be estrogenic toward breast cancer cells and not antiestrogenic (Birt et al., 1998). Clearly this is a controversial topic and it has been the subject of numerous reviews (Barnes, 1997; Birt et al., 1998). Indeed the results of recent human intervention trials have been mixed with many investigators observing minimal or no evidence of estrogenicity in humans consuming soy foods or isoflavone-rich extracts and other investigators observing striking evidence of estrogenicity in assessing estrogen responses (Birt et al., 1998). One recent study supplemented the daily diets of premenopausal Japanese women with soybean milk containing 109 mg genistein for three consecutive menstrual cycles. They included 31 soybean milk supplemented subjects and 29 control subjects and observed marginal reductions that were not statistically significant in estrone (23% in comparison with basal levels) and estradiol (27%). They concluded that much larger studies will be required if the small impact is to be clearly demonstrated (Nagata et al., 1998).

Isoflavones have also been studied for their ability to alter cell cycle parameters; inhibit cellular proliferation and induce apoptosis. Recent reviews on these activities have appeared (Birt et al., 1998, 2001) and in general the results suggest that it is certainly probable that soybean isoflavones may block cell proliferation through arresting cells in either G1 (gap-1) or G2/M (Gap-2/mitosis). A comprehensive study on the impact of isoflavones (genistein, genistin, daidzein, and biochanin A) on urinary murine and human bladder cancer cell lines demonstrated a G2/M arrest in all cell lines and some lines showed DNA fragmentation consistent with the induction of apoptosis by isoflavones. An increase in the proportion of bladder cancers cells in G2/M was generally observed with 50 μM doses of genistein, but some cell lines had an elevated proportion of cells in G2/M at 10 μM. In general, cell proliferation was reduced with the lower doses of genistein (10 μM) (Zhou et al., 1998). In parallel studies, this laboratory observed a reduction in MB-49 murine bladder cancer volume in C57BL/6 mice that were injected with genistein, fed soybean protein, or soybean phytochemicals in comparison with control mice (Zhou et al., 1998).

Research has suggested that isoflavones may inhibit neovascularization (angiogenesis). Angiogenesis is required for tumor growth and metastasis and many chemotherapeutic agents were designed to prevent angiogenesis and thus block the growth of established tumors. Some studies have suggested that isoflavones inhibit angiogenesis as was reviewed (Birt et al., 1998). For example, genistein inhibited the proliferation of capillary endothelial cells and displayed inhibitory activity in a in vitro angiogenesis assay (Birt et al., 1998). More recently, genistein, soybean protein, and soybean phytochemicals reduced vascular density by up to 50% in MB-49 murine bladder cancers that were growing in mice (Zhou et al., 1998).

Recent reviews report isoflavones, and particularly genistein, have also been shown to enhance the immune response, inhibit the activation of selected chemical carcinogens, serve as antioxidants under some conditions, and inhibit tyrosine kinases and topoisomerases (Birt et al., 1998, 2001). The importance of these isoflavone bioactivities in the prevention of cancer remains to be demonstrated.

21–3.3.2 Methionine Restriction

A well known but often ignored property of soybean protein is the somewhat low content of the amino acid methionine. Interestingly, methionine requirements are relatively high for many tumors and methionine restriction may contribute to cancer prevention by soybean protein in rodent models because of their high methionine requirement. A review by Hawrylewicz et al. (1995) reported on three studies where the prevention of cancer or metastasis was blocked by supplementing the soybean-based diet with methionine up to the amount of methionine in the control, generally casein, based diet. One of these studies used the rat mammary cancer model induced by MNU (40 mg kg^{-1} body weight). Soybean protein isolate was fed alone or supplemented with 0.7% DL-methionine and compared with casein supplemented with 0.3% DL-methionine. Tumor incidence was 80% in the casein group, and 42.3% in the soybean protein isolate group while the addition of methionine to the soybean protein isolate diet resulted in an intermediate incidence of mammary tumors (64%) (Hawrylewicz et al., 1991). A similar approach was used in assessing the contribution of the low methionine content of soybean to the inhibition of metastasis from a rhabdomyosarcoma tumor to the lung. Soybean protein diet inhibited metastasis to the lungs but the inhibition was completely blocked by supplementing the soybean diet with methionine (Hawrylewicz et al., 1995). However, it is not clear if the relatively low methionine content of soy foods would contribute to cancer prevention in humans because the human requirements for methionine is 50% of the rat requirement and soybean protein is adequate in methionine for humans.

21–3.3.3 Trypsin Inhibitor

Specific soybean proteins with trypsin inhibitor activity may also contribute to cancer prevention by soy foods. Excellent and comprehensive reviews on soybean protease inhibitors and cancer prevention have been published (Kennedy, 1998). Two soybean protease inhibitors have been identified and studied for their contribution of cancer prevention by soy foods, Bowman-Birk protease inhibitor and soybean trypsin inhibitor (Kunitz inhibitor). The BBI is effective as an inhibitor of both trypsin and chymotrypsin. It is a polypeptide with seven disulfide bonds and a M_r of 8000 and it is available as the purified protein (PBBI) and as a concentrate from soybean enriched for BBI or BBI concentrate (BBIC). The second protease inhibitor, Kunitz inhibitor, is effective primarily against trypsin. It has a high molecular weight and it is quite heat labile relative to BBI. Studies have been conducted with these inhibitors using cell culture and animal models and human intervention studies are currently underway.

In several cultured cell studies, protease-inhibitor activity blocked carcinogen-induced malignant transformation by radiation, 3-methylcholanthrene, N-methyl-N'-nitroso-guanidine, benzo(a)pyrene, or β-propiolactone (Kennedy, 1998). Extensive animal carcinogenesis studies have been conducted showing that soybean BBI prevented chemically induced mouse lung, liver, and colon, carcinogenesis; rat liver, and esophagus, and hamster oral cancer (Kennedy, 1998). Several other protease inhibitors have been found to block carcinogenesis and isolated protease activity from soybean protein reduced spontaneous mouse liver cancer (Kennedy, 1998). Finally, colon adenomas that developed in mice carrying a truncated adenomotosis polyposis coli (APC) gene were inhibited by BBI (Sorensen et al., 1998).

It is important to note that the cancer prevention potential of the various soybean protease inhibitors correlates strongly with their protease inhibitor activity suggesting that this activity may contribute to the prevention of cancer (Kennedy, 1998). However, although the mechanism of action of protease inhibitors has been extensively studied no clear mechanism has been identified and it is possible that cancer prevention by protease inhibitor may be through several mechanisms.

Anti-proliferative and anti-inflammatory activity of BBI may contribute to cancer prevention. The proto-oncogene c-myc is clearly important in cellular proliferation and exciting results in animals gavaged daily with BBI suggest that inhibition of c-myc expression may contribute to protease inhibitor prevention of cancer. The BBI-treated animals experienced a complete block in carcinogen induced c-myc expression (St.Clair et al., 1990).

A controversial issue related to soybean trypsin inhibitor prevention of cancer has been the bioavailability of these proteins. In reviewing this topic Kennedy (1998) reports that about 50% of the dietary BBI is found in the blood stream and is distributed to tissues in the body. However, not all scientists agree with this evaluation and further data will be needed to resolve the issue (Fournier et al., 1998). Another problem in assessing the role that soybean protease inhibitors play in cancer prevention is assessing exposure to this activity. Clearly heat treatment of soy-based foods reduces soybean trypsin inhibitor activity but quantitating the residual protease inhibitor concentration in soy foods has been problematic.

21–3.3.4 Saponins

Soy foods also provide a dietary source of saponins, amphiphilic glucoside compounds that consist of a hydrophobic steroid or a triterpene moiety bound with sugar hydrophilic sugars. Saponins, like many other plant constituents have been shown to have immunostimulatory, hypocholesterolemic, and anticarcinogenic properties as was previously reviewed (Fournier et al., 1998; Koratkar and Rao, 1997). Dietary soybean saponins (3%) were assessed for their potential in the prevention of aberrant crypt foci (ACF) in mice treated with the colon carcinogen azoxymethane (AOM) (Koratkar and Rao, 1997). The saponin supplement reduced the number of ACF/colon by 67% in comparison with the AOM-treated mice that did not receive saponins. However, clearly the addition of saponins at 3% by weight of the diet was much higher than would be possible with a soybean-based diet since soy foods were reported to contain about 0.3 to 0.4% saponins. However, ongoing studies suggest that saponins are present at equimolar concentrations with isoflavones in soy foods (J. Hu and P. Murphy, unpublished data).

21–3.3.5 Phytic Acid (Inositol hexaphosphate)

Some investigations have provided evidence for potential cancer prevention by inositol phosphate. Inositol phosphate occurs naturally as a salt with monovalent and divalent cations. Inhibition of growth and differentiation of the human colon carcinoma cell line HT-29 was observed when the cells were cultured with inositol phosphate (Yang and Shamsuddin, 1995) and the prevention of colon carcinogenesis was demonstrated in rats given inositol hexaphosphate supplemented in the drinking water (Ullah and Shamsuddin, 1990). Inositol phosphate has been reported to reduce cellular proliferation, increase differentiation, and to be incorporated into and modify the inositol phosphate-signaling pathway in cells (Shamsuddin, 1995; Yang and Shamsuddin, 1995). Further studies will be required to determine if inositol phosphate in soy foods contributes to the cancer prevention potential of diets rich in soybean.

21–3.4 Summary of Soybean and Cancer Prevention

Reduced rates of cancers of the colon, breast, and prostate have been observed in association with the consumption of diets rich in soy foods. However, these reduced cancer rates could certainly be due to other dietary or environmental factors such as the consumption of lower fat and energy, the consumption of more fruits and vegetables, or higher physical activity. Intervention studies will be required to determine if soy foods in general, specific soy-based foods, or soybean constituents contribute to cancer prevention. Unfortunately, at present the data from human and animal systems does not clearly justify the initiation of cancer prevention intervention studies in humans. There are too many inconsistencies in the data and there is some evidence of potential toxicity of isolated soybean constituents. The exception to this may be in the area of soybean BBI studies. Ongoing intervention studies with soybean BBI are focused on cancer therapy rather than on prevention, but if the soybean BBI is highly effective in cancer therapy and toxicity can clearly be avoided or controlled it is possible that cancer prevention trials will eventually be warranted.

Improving our information on the constituents in soybean would strengthen cancer prevention studies with soy foods and soybean constituents. In particular, we have limited information on the level of soybean constituents in soy foods, the impact of processing on these foods, the bioavailability, and interactions of constituents. Furthermore, little is known regarding metabolism, and mechanisms of action of soybean constituents. In particular, research is needed to address the impact of combinations of soybean constituents in disease prevention. The importance of gender and genetic differences between people on bioavailability, metabolism, and bioactivity needs investigation. Furthermore, we need to better understand the potential toxicity of soybean constituents and we need strategies to decrease toxicity. In considering isoflavones we need a better understanding of the impact of phytoestrogens (and estrogens) at different times during life. The emerging story on the role of soy foods and soybean and cancer prevention offers considerable promise. However, research in several critical areas will be required to develop sound strategies for the prevention of human cancer through soybean-based foods or soybean constituents.

21–4 SOYBEAN ISOFLAVONE ANALYSIS, BIOAVAILABILITY, AND TOXICITY

Isoflavones are a signature component of soybean, inasmuch as no other commonly consumed human foods, except for alfalfa (*Medicago sativa* L.) and clover sprouts, contain amounts of isoflavones likely to be significant enough to exert appreciable biological effects (USDA/ISU Isoflavone Database, 1999). The isoflavone content of soybean is about 150 mg 100 g^{-1}, whereas isoflavone contents of the few other foods in which isoflavones may be found (e.g., green tea, chickpea [*Cicer arietinum* L.], kidney [*Phascolus valgaris* L.], and other beans) are generally in the range of 1 mg isoflavones 100 g^{-1} (USDA/ISU Isoflavone Database, 1999). Not all soybean-containing foods are good isoflavone sources, nor are all soybean ingredients. Isoflavones are associated with the protein fraction of soybean. Isoflavones are not easily dissolved in water or highly lipophilic solvents, and they can be removed during some types of soybean food processing. Although isoflavones are heat stable, they are susceptible to microbial fermentation and seeming degradation. For example, soybean oil does not contain isoflavones, and soybean protein isolate derived from ethanol-washed soybean protein and soybean sauce (a fermented product) contain only ~0.1 mg isoflavones g^{-1} (USDA/ISU Isoflavone Database, 1999).

The major soybean isoflavones are daidzin, genistin, and glycitin (Fig. 21–10). In soybean, daidzin and genistin are the major isoflavones, but daidzin and glycitin predominate in soybean hypocotyls (soygerm) (Hendrich and Murphy, 2001). The isoflavones in most soy foods are present mostly as glycosides, malonyl-glycosides, and acetyl-glycosides, with minor proportions of aglycones. The aglycone forms (genistein, daidzein, and glycitein) increase in food products such as

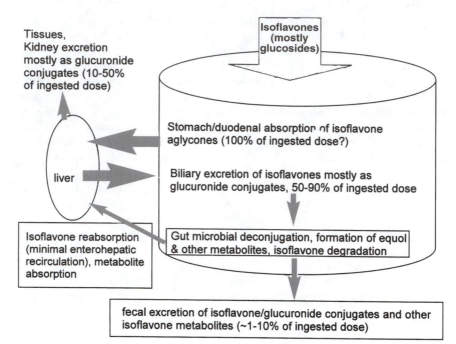

Fig. 21–10. A current hypothetical picture of isoflavone bioavailability. Neither the mechanisms of isoflavone absorption nor of their disappearance (probable degradation) from the gastrointestinal tract are well-characterized. Isoflavone excretion as a proportion of ingested dose and isoflavone metabolism (especially equol formation from daidzein in some individuals, and glucuronidation as a major metabolic fate) are relatively well understood. Adapted from Hendrich and Murphy (2001). Printed with permission.

tempeh, because fermentation introduces microbial β-glucosidases to the isoflavones. The soybean isoflavones differ somewhat in solubility, bioavailability, and in biological effects (glycitein is more estrogenic than genistein in a mouse uterine growth assay (Song et al., 1999), although much work remains to characterize these effects.

The extent to which the active forms of isoflavones reach target tissues is bioavailability. For food components, bioavailability is determined by the interactions of the food component with the gastrointestinal tract, its endogenous digestive enzymes and transport systems, and with gut microorganisms. Once the food component is absorbed, endogenous biotransformation enzymes may act. By biotransformation, the food component is altered in its chemical form and solubility, which influence its distribution, storage, and excretion. This in turn determines the nature and time course of the health effects of the food component. Thus isoflavone bioavailability is of first importance in assessing biological effects of these soybean components. The current overall picture of isoflavone bioavailability is shown in Fig. 21–10.

Although isoflavone bioavailability has been reviewed recently (Kurzer and Xu, 1997; Hendrich et al., 1998, 1999; Hendrich and Murphy, 2001; Murphy and Hendrich, 2002), new information is constantly accumulating. Developing better, more feasible and rapid methods of isoflavone analysis in human body fluids is one such aspect of vitally needed information. Information about isoflavone dynamics is also continuing to develop; their flow and fate in the human body, their mechanisms of absorption, distribution, utilization, and excretion. In particular, the role of gut microorganisms and endogenous biotransformation in isoflavone metabolism and biological efficacy is a current topic of growing interest and complexity. The following is an attempt to comprehensively address the state of knowledge regarding isoflavone bioavailability, and to point out crucial emerging needs in our knowledge base.

21–4.1 Analysis of Isoflavones in Biological Fluids and Tissues

Chromatographic methods for analysis of isoflavones have been reviewed recently (Hendrich, 2002). Isoflavones were first quantified in human urine by gas chromatography (Axelson et al., 1984; Bannwart et al., 1984). Lundh et al. (1988) developed a C_{18} reverse-phase HPLC method for analysis of isoflavones in bovine urine, using UV detection of isoflavone aglycones after glucuronidase/sulfatase treatment. A variety of other HPLC or GC methods, some coupled with mass spectroscopy, and using various internal standards have been developed (Table 21–2). A fluoroimmunoassay has also been recently developed for isoflavone analysis (Uehara et al., 2000). These assays vary in ease of use, expense, and need for extraction methods. The lowest concentrations of isoflavones in tissues, plasma, or urine associated with beneficial or adverse health effects are not known, and highly sensitive isoflavone detection systems (GC or LC coupled with MS, or fluoroimmunoassay) are needed to determine the limits of effective isoflavone doses for various health-related endpoints, especially in tissues. Isoflavone analysis in plasma and urine are most often needed for determining compliance with dietary soybean treatment. With respect to improvement of blood cholesterol status, the most commonly studied health effect of soybean associated with isoflavones, intake of soybean protein containing ~30 mg isoflavones per person per day seemed to be required (Crouse et al., 1999). For studies of soybean and isoflavone health effects (atherosclerosis, osteoporosis, and cancer risk), dietary doses of soybean isoflavones generally are in the range of >20 mg per person per day. The HPLC with UV detection methods seem to be adequate for such studies, and such methods are probably most affordable in terms of equipment, supplies and labor, for a researcher needing to set up isoflavone analysis. HPLC with coulometric detection would be preferable if equol detection were an important endpoint.

21–4.2 Absorption of Isoflavones

Because isoflavones have molecular weights in the range of 240 to 270 g mol^{-1} and isoflavone aglycones are neither very hydrophilic nor lipophilic, the isoflavones may be predicted to be incompletely absorbed through diffusion. There is no evidence for facilitated or active absorption of these compounds, but isoflavone absorption has not been studied yet on a molecular level. Isoflavone glucosides do not seem to be absorbed intact. No reports showing the presence of these isoflavone

Table 21–2. Comparison of major analytical methods for isoflavones in biological samples.

Method	Methodological notes	Internal standard	References
HPLC	Reverse-phase, UV detection, equol not well-detected but detects ~ 50 nmol L^{-1} of other isoflavones	2,4,4'-trihydroxy-deoxybenzoin flavone	Lundh et al. (1988) Zhang et al. (1999a) Franke and Custer (1994) Gamache and Acworth (1998) Nurmi and Adlercreutz (1999)
	Detects O-desmethylangolensin		
	All isoflavones (including equol) detected at ~5 nmol L^{-1}		
	Coulometric electrode array detects isoflavones and lignans (e.g., enterodiol)		
GC	Requires silyl derivatization, detects ~2 nmol L^{-1}		Axelson et al. (1984) Bannwart et al. (1984)
GC/MS	Dihydroisoflavone and other metabolites detected glucuronide	^{14}C-estrone	Adlercreutz et al. (1991) Heinonen et al. (1999)
HPLC/MS	Used to detect biliary isoflavone excretion and isoflavone metabolites such as p-ethylphenol sulfate, no extraction of sample needed		Barnes et al. (1998)
HPLC/electrospray MS	Online solid phase extraction increased speed of analysis compared with HPLC method		Doerge et al. (2000)
Isotope dilution/HPLC/ electrospray MS	Rat tissue isoflavone analysis, ~50 pmol g^{-1}	^{2}H-genistein	Chang et al. (2000)
Time-resolved fluoroimmunoassay	No sample clean-up required, allows rapid screening of large numbers of urine samples, detects ~0.5 nmol L^{-1}	Europium-labeled isoflavones	Uehara et al. (2000) Wang et al. (2000)

forms in plasma or urine, and the analytical systems available readily separate and detect the glucosides as well as the aglycone forms of isoflavones.

In male Wistar rats of unspecified age, isoflavone aglycones seem to be absorbed in the stomach and intestine based on studies of genistein and genistin (genistein glucoside) absorption (Piskula, 2000b). After pyloric ligation that permits absorption in the stomach only, genistein and not genistin was absorbed. Based on detection of isoflavone aglycone after enzymatic hydrolysis of isoflavone glucuronide and sulfate conjugates, genistin was absorbed rapidly as its aglycone, genistein with peak plasma concentrations at about 5 h after dosing. Cleavage of the genistin glucoside occurred in the intestine prior to isoflavone absorption, thus slowing absorption of isoflavone from its genistin form. Given the speed of absorption, mammalian β-glucosidases were likely to be responsible for genistin cleavage. The enzymes involved have not been identified.

A similar pattern of isoflavone absorption was seen in humans. Soy foods or beverages were given as single meals providing ~1 mg isoflavones (mostly as glucosides) kg^{-1} body weight. Apparent peak plasma isoflavone concentrations were achieved at about 6 h after dosing (Watanabe et al., 1998; Xu et al., 1994), suggesting substantial isoflavone glucoside cleavage in the upper small intestine that must be mediated by human β-glucosidases, probably disaccharidases. In four men and four women, 0.11 or 1.7 mmol total isoflavones as glucoside or aglycone were fed as single doses after breakfast. Isoflavone aglycone or glucoside (0.3 mmol d^{-1} divided across three meals) was also given for 4 wk (Izumi et al., 2000). Both daidzein and genistein had threefold greater peak plasma concentrations than did their glucosides after a single dose. This difference remained in plasma samples taken before lunch at 2 and 4 wk during daily isoflavone dosing. Setchell et al. (2001) fed doses of about 0.2 mmol isoflavones as daidzein, genistein, daidzin, or genistin, observing no differences between the aglycone and glucoside forms in AUC (area under curve, representing the total amount of isoflavones in the plasma over time). The time to achieve maximum plasma isoflavone contents was delayed when glucosides were fed compared with aglycones. Because Izumi et al. (2000) only measured plasma from 0 to 6 h after dosing, they may have missed the true peak plasma isoflavone content after glucoside administration. Physiological differences between the subjects in the two studies may have caused the markedly different results. For example, an absence of lactase that might prevent absorption of glucosides, would be more likely in the Japanese than in the American subjects. This remains to be determined.

21–4.3 Gut Microbial Transformation of Isoflavones

Four potential aspects of gut microbial transformation of isoflavones may influence the biological effects of the isoflavones: glucosidases-glucuronidases, other isoflavone-specific metabolism, and isoflavone degradation. The presumed mammalian β-glucosidase cleavage of isoflavone glucosides may not be efficient. Gut microbial mammalian β-glucosidases may also act to free the aglycones for absorption, perhaps enhancing bioavailability of the isoflavones. This is not yet established.

Once isoflavones are initially absorbed, they seem to be immediately metabolized to glucuronide and sulfate conjugates (Sfakianos et al., 1997; Barnes et al., 1998; Coldham et al., 1999). These conjugates are excreted partly in bile, which has been established in rats for genistein glucuronide (Sfakianos et al., 1997). Because of structural similarities, daidzein and glycitein should be converted largely to glucuronide conjugates as well. Gut microbial β-glucuronidases could reconvert these conjugates to their aglycone forms, theoretically permitting enterohepatic recirculation. Enterohepatic recirculation might cause a second peak of isoflavone absorption, but this was not seen in the two most detailed human bioavailability studies (Watanabe et al., 1998; Setchell et al., 2001). Plasma samples were not taken between 12 and 24 h in either study, so the second peak of isoflavone absorption might have been missed, but plasma profiles of an occasional subject showed a second (minor) peak. It may be that further metabolism and degradation of isoflavones in the lower gastrointestinal tract prevented enterohepatic recirculation.

A variety of isoflavone metabolites produced by gut microorganisms have been identified, but the organisms responsible are unknown and the biological effects of these metabolites are generally uncharacterized. Equol is a relatively potent phytoestrogen produced from daidzein in the gut, typically after a few days of eating soy foods, and in about one-third of human subjects (Setchell et al., 1984). One-sixth of male subjects (Lu et al., 1995) and two-thirds of female subjects fed soybean for several weeks were equol producers (Lu et al., 1996). Lampe et al. (1998) showed that about 35% of subjects excreted equol after 3 d of soybean feeding, with no significant gender difference. This study was much larger (n = 60) than any previous study assessing equol production. Equol-producing microorganisms have yet to be identified, and little is known about the physiological determinants of equol production, let alone the health significance of equol production. In one study, after feeding soybean for 3 d, equol production was inversely related to the production of O-desmethylangolensin (ODMA), another gut microbial daidzein metabolite (Kelley et al., 1993). Another study of 19 women and 5 men fed isoflavone-rich texturized protein in vegetarian burgers for 17 d showed that there was no relationship between equol and ODMA production among the one-third of subjects who were equol producers (Rowland et al., 2000). The health effects, if any, of ODMA have not been determined, nor what organisms or other factors drive its production.

Other isoflavone metabolites derived from gut microbial metabolism include dihydrodaidzein, dihydrogenistein, 6-hydroxy-ODMA, cis-4-equol (Heinonen et al., 1999), and p-ethylphenol (Barnes et al., 1998). Gut microbial metabolites of glycitein have not been characterized. In cecal cultures from male and female rats, gut microbial metabolites of genistein were identified including 6-hydroxy-ODMA (Coldham et al., 1999). In Wistar rats dosed with ^{14}C-genistein, urinary genistein metabolites included dihydrogenistein, its glucuronide, and genistein sulfate and glucuronide, as well as a new and major metabolite, 4-hydroxyphenyl-2-propionic acid, but not 6-OH-ODMA which seemed to be rapidly converted to 4-hydroxyphenyl-2-propionic acid. Dihydrogenistein and 4-hydroxyphenyl-2-propionic acid were gut microbial metabolites because rat liver homogenates did not produce these metabolites. 4-ethylphenol represented 42% of a single dose of genistein given to Wistar rats (King, 1998). In the studies by Coldham et al. (1999), this metabolite may not have been identified because the radiolabel of the genistein was at the

C that was removed in transforming 4-hydroxyphenyl-2-propionic acid to 4-ethyl-phenol. Coldham and Sauer (2000) showed that the predominant metabolites of radiolabeled genistein were 4-hydroxyphenyl-2-propionic acid and genistein glucuronide in rat plasma, with about 0.1 to 0.6% of the genistein dose remaining in tissues 168 h after dosing. Reproductive tissues contained significantly greater amounts of genistein and 4-hydroxyphenyl-2-propionic acid than did other tissues except for liver, which contained the greatest amounts of the isoflavone and its metabolite. To our knowledge, there are no quantitative reports of the phenyl metabolites of genistein in human urine or plasma, but given other similarities in rat and human isoflavone metabolism such as production of equol and dihydrogenistein, such metabolites are likely. Their biological effects are unknown, but methyl *p*-hydroxyphenyllactate is a flavonoid metabolite that blocks nuclear estrogen receptor binding and inhibits the growth of MCF-7 breast cancer cells in vitro (Markaverich et al., 1988), suggesting that the monophenolic metabolites of isoflavones are worth investigating for their health-related effects.

When isoflavone "balance" studies have been attempted, generally between 50 to 95% of the isoflavone dose cannot be accounted for in urine and feces, with fecal contents of isoflavones being no more than 1 to 10% of total isoflavone excreted (Xu et al., 1994, 1995). Tissue genistein only accounted for about 0.5% of the total ingested isoflavone dose in two rat studies (Coldham and Sauer, 2000; Chang et al., 2000), suggesting strongly that a significant portion of the isoflavones disappear. In humans, the total urinary daidzein, genistein, equol, and ODMA represented approximately 15% of the ingested dose of isoflavones from roasted soybean fed over 3 d (Franke et al., 1994). Although the theoretically major genistein metabolite, 4-ethylphenol was not included in this total, it seems that a significant proportion of isoflavones disappear from the gastrointestinal tract through gut microbial degradation.

The Hendrich laboratory has studied isoflavone disappearance extensively, characterizing isoflavone disappearance from the human gut in an in vitro model system using anerobic fecal incubations, and characterizing the influence of fecal isoflavone degradation phenotype on apparent isoflavone absorption as reflected in urinary excretion and plasma concentrations of isoflavones. Xu et al. (1995) fed seven women soybean milk during three meals on a single day and noted that two of the subjects excreted 10 to 20-fold greater amounts of isoflavones in feces than did the other subjects. The "high excretors" also had two- to threefold greater urinary and plasma levels of isoflavones. In-vitro fecal isoflavone disappearance half-life studied in one male subject was less for genistein than for daidzein (3.3 vs. 7.5 h). The more rapid disappearance of genistein than of daidzein may account for lesser apparent absorption of genistein than of daidzein. In-vitro human fecal isoflavone disappearance in samples taken from 20 subjects showed three distinct disappearance phenotypes (Hendrich et al., 1998). Disappearance rate constants for genistein of 0.023 h^{-1} (high degrader, n = 5), 0.163 h^{-1} (moderate degrader, n = 10), and 0.299 h^{-1} (low degrader, n = 5) were relatively constant when reexamined 10 mo later with disappearance rate constants of 0.049 h^{-1} (high degrader, n = 5), 0.233 h^{-1} (moderate, n = 4), and 0.400 h^{-1} (low, n = 5). Twelve of 14 subjects who remained in the study after 10 mo maintained their initial isoflavone degradation phenotype. Plasma isoflavone contents was negatively and significantly correlated with degradation rate constant, r = −0.88 for genistein and r = −0.74 for daidzein (Wang, 1997). Perhaps a portion of interindividual variability in apparent absorption, urinary excretion, and plasma contents of isoflavones might be accounted for by variation in gut microbial isoflavone degradation rate. When women of the moderate isoflavone disappearance phenotype were chosen for an isoflavone bioavailability study, the intersubject variation in urinary excretion of isoflavones was five-to eightfold (Zhang et al., 1999b), as compared with 12- to 15-fold variation within treatment in a similar study in which the interindividual variation in isoflavone disappearance was not taken into account (Karr et al., 1997). This suggests that controlling for inter-individual variation in isoflavone disappearance phenotype might make it easier to discern health effects of isoflavones by lessening interindividual variation in response to isoflavones, but this remains to be demonstrated. We have attempted to determine the factors that cause or are closely related to isoflavone disappearance phenotypes. Wang (1997) noted that four of five subjects who had the high isoflavone diappearance phenotype also reported irregular or infrequent bowel movements, which was not noted by subjects of the other disappearance phenotypes. Zheng (2000) studied 35 Asian and 33 Caucasian women, characterizing in-vitro fecal isoflavone disappearance phenotype, physical activity, and dietary habits. Five months later, these characterizations were repeated. Isoflavone excretion and gut transit time after a single dose of soybean beverage were also measured. Significantly more Asian subjects had a high daidzein disappearance phenotype than did Caucasians at both timepoints, (averaging 11 Asians vs. 2 Caucasians of this phenotype). There were no ethnic differences in genistein disappearance phenotype. Although Asians and Caucasians differed markedly in physical activity and dietary habits, there were no diet or activity differences specific to isoflavone disappearance phenotype. Asians were less physically active, had lesser VO_{2max}, ate three times more red meat (0.9 vs. 0.3 servings d^{-1}) and cholesterol (by twofold), and ate less dairy products and dietary fiber (2 vs. 4 g d^{-1}) than did Caucasians. The only difference observed between phenotypes was that the Asians of the high genistein disappearance phenotype had prolonged gut transit time compared with Asians of the low genistein disappearance phenotype (65 vs. 40 h). The low genistein disappearance phenotype was associated with about threefold greater urinary excretion of genistein than occurred in subjects of the high genistein disappearance phenotype, in Asian subjects only. All of the Caucasians, of low and high disappearance phenotypes, had significantly prolonged gut transit time compared with Asian subjects, which may have caused the Caucasians of both phenotypes to have similarly low levels of urinary genistein (about 2% of ingested dose recovered in urine over 24 h after dosing; the same recovery as in Asians of the high genistein disappearance phenotype.) These results imply that significant differences in gut transit time may determine individuals' ability to absorb isoflavones and related compounds such as flavonoids. The differences in gut transit time between Asians and Caucasians may be due to gut microbial differences that cannot be explained by differences in diet and activity habits because none of these habits differed according to isoflavone disappearance phenotype.

In the human gut, isoflavone-degrading microorganisms are not well characterized. Some *Clostridia* strains may cleave the C-ring of flavonoids, including isoflavones (Winter et al., 1989). Genistein, and related compounds with a 5-hydroxyl moiety, seem particularly susceptible to microbial degradation (Griffiths and Smith, 1972). Population of the human gut by *Clostridia* may accompany meat consumption (Mitsuoka, 1982), but based on Zheng (2000), meat consumption was not related to isoflavone disappearance phenotype. Much work remains to characterize gut microbial metabolism of isoflavones and its role in human health through altering bioavailability of these compounds.

21–4.4 Influence of Food Matrix on Apparent Isoflavone Absorption

The effects of other dietary components on isoflavone absorption are not well characterized. Fasted rats given oral genistein in Na_2CO_3, pH 10.6, showed peak plasma genistein at 10 min after administration, whereas fed rats showed a sixfold lesser peak plasma genistein at 4 min after dosing (Piskula, 2000a), presumably due to the much more acidic condition of the stomach in fed rats. Plasma and urinary isoflavones did not differ in eight women eating three daily meals containing soybean milk, when the meals were:(i) provided by the investigators, (ii) self-selected but taken at specified times, or (iii) ad libitum, in a randomized cross-over design. Although protein and fat intakes were greater in subjects eating self-selected or ad libitum diets, neither of these diet components seemed to influence isoflavone absorption (Xu et al., 2000). No significant differences in urinary or plasma isoflavones were noted in women who were fed a single breakfast of isoflavone-containing tofu, tempeh, soybean, or texturized vegetable protein in a randomized cross-over design (Xu et al., 2000). But in men fed soybean pieces or tempeh for 9 d, urinary daidzein and genistein recovery was significantly greater from tempeh than from soybean pieces (Hutchins et al., 1995). Although gender differences in isoflavone bioavailability were noted in some studies (Lu and Anderson, 1998), gender differences in response to background food matrix and isoflavone absorption have not been studied. When women were fed a day's diet containing 40 g dietary fiber, nearly all from wheat, compared with a diet containing 15 g dietary fiber in a randomized crossover design, the bioavailability of genistein (and not daidzein) from a single soybean meal was modestly but significantly decreased (Tew et al., 1996). In rats fed fructooligosaccharides (5% by weight of the diet) for 7 d, a single dose of an isoflavone extract by gavage gave greater genistein absorption as measured by total area under curve of plasma genistein over time, compared with the control group fed AIN-93G diet only (Uehara et al., 2001). Fructooligosaccharides are thought to stimulate growth of bifidobacteria in the gut, suggesting that the presence of bifidobacteria may inhibit genistein-degrading microorganisms. Dietary influences on equol production and excretion have been assessed (Rowland et al., 2000). By 3-d weighed food records, equol excretors consumed significantly less fat (25% of energy vs. 35% of energy) and more carbohydrate (55% of energy vs. 45% of energy than nonproducers of equol. Lampe et al. (1998) also observed that female but not male equol excretors consumed significantly more carbohydrate and dietary fiber than did non-excretors of equol. How these dietary differences alter gut microorganisms to favor equol production is unknown. There seems to be reason to further explore the influence of diet on isoflavone bioavailability, especially with respect to gut microbial metabolism of isoflavones.

21–4.5 Endogenous Biotransformation of Isoflavones

Once isoflavones are absorbed, their phenolic structure is available for mammalian biotransformation by UDPglucuronosyltransferases (UGTs) and sulfotransferases in the intestinal mucosa, liver, and other organs. Isoflavone glucuronides and sulfates are not routinely measured. The more common and inexpensive methods for isoflavone analysis use glucuronidase/sulfatase pretreatment to quantify total isoflavone aglycones (as adapted from Lundh et al., 1988). Standards of isoflavone glucuronide and sulfate metabolites are not readily available. The biological activity of these metabolites is not clear. Peterson et al. (1998) identified sulfate, hydroxylated and methylated metabolites of genistein in breast cancer cell lines. Hydroxylated and methylated genistein metabolites, but not genistein sulfate, correlated with inhibition of cancer cell proliferation. Zhang et al. (1999a) showed that genistein and daidzein glucuronides were about 10-fold less estrogenic than the isoflavone aglycones in ability to bind to mouse uterine cytosolic estrogen receptors. The daidzein and genistein glucuronides activated human natural killer (NK) cells in vitro by about 30% in concentrations of 0.1 to 10 µM, whereas genistein was an effective NK activator at 0.1 to 5.0 µM. Up to 50 µM isoflavone glucuronide was nontoxic to NK cells invitro, whereas 10 µM genistein significantly inhibited NK activity, presumably because genistein is a tyrosine kinase inhibitor (Akiyama and Ogawara, 1991), and tyrosine kinases are necessary for NK activation (Einspahr et al., 1991). Thus, with respect to one biological endpoint, isoflavone glucuronides were as effective as isoflavone aglycones with less toxicity.

Isoflavone glucuronides and/or sulfates have been measured in a few studies in animals and humans. Sfakianos et al. (1997) showed that when genistein was administered by gavage to rats, 90% of the dose was absorbed, 70% of the dose appearing rapidly as glucuronide conjugate in portal blood presumably due to intestinal mucosal UGT. Most of the glucuronide was then excreted in bile, suggesting a capacity for enterohepatic recirculation after gut microbial glucuronidase action, but further metabolism and degradation by gut microflora may interfere with this process.

It is likely that human metabolism of isoflavones would be similar to rats because humans have significant gut mucosal UGT activity. In experiments feeding women single soybean meals or soybean for 6 d, glucuronide conjugates accounted for significantly greater proportions of urinary than plasma isoflavones, presumably because of additional glucuronidation by the kidney after gut mucosal and hepatic biotransformation (Zhang, 2000). The glucuronides were the predominant isoflavone forms in both urine and plasma, with sulfates representing about 20% of total conjugates. Plasma contained more isoflavone aglycones than did urine. In these studies, isoflavone glucuronides were measured directly in comparison with standard glucuronides we prepared, or indirectly after enzymatic cleavage with a pure glucuronidase (not contaminated with sulfatase). Sulfates were measured only indirectly by subtraction of isoflavone aglycone product as measured after treatment with pure glucuronidase from isoflavone aglycones produced after glucuronidase/sulfatase treatment. In six men and six women fed a soybean beverage, about 17% of total genistein and 20% of total daidzein were excreted in urine as sulfate conjugates, with the remaining isoflavone excreted almost entirely as glucuronide conjugates (Shelnutt et al., 2000). In this study, the conjugates were measured after treatment of urine samples with glucuronidase or sulfatase, rather than by direct measurement of the conjugated metabolites. Glucuronide/sulfate diconjugates were not measured in these studies, and might account for a portion of total mammalian biotransformation as well. In male Sprague-Dawley rats, Yasuda et al. (1996) identified genistein 4'-O-sulfate, genistein 7-O-glucuronide, and genistein 4'-O-sulfate 7-O-glucuronide as urinary and biliary metabolites, with the sulfate predominating in urine and the diconjugate the sole biliary metabolite. This result differs markedly from the result of Sfakianos et al. (1997), perhaps because the genistein dose given by Yasuda et al. (1996) was 100 mg kg^{-1}, much greater than the dose given by Sfakianos et al. (1997), which was no more than ~1 mg kg^{-1} body weight infused over 40 min. Large doses are typically metabolized differently from smaller doses, depending on the capacity of the biotransformation enzymes. In Sprague-Dawley rats fed 250 or 1250 mg kg^{-1} genistein in the diet, serum or plasma glucuronides represented nearly all of the isoflavone recovered, using indirect (selective deconjugation) methods of conjugate analysis (Holden et al. 1999). These doses would be estimated to be approximately 15 and 75 mg genistein kg^{-1} body weight, the higher dose being only one-third less than the dose given by Yasuda et al. (1996), but the rats in the experiment by Yasuda et al. (1996) had been fasted for 18 h before the study. Piskula (2000a) showed that the fasted rats showed greater amounts of sulfates and lesser amounts of glucuronides in plasma than did fed rats. Therefore, differences in nutritional status might at least partly account for the different pictures of isoflavone conjugation seen among studies.

Chang et al. (2000) indirectly characterized the extent of genistein glucuronide and sulfate conjugates in rat tissues after feeding varying genistein doses, by comparing the amounts of genistein in tissues before and after glucuronidase/sulfatase treatment, using a highly sensitive LC/ES/MS method. Total amounts of tissue genistein were minute (0.2–2.0 mg g^{-1}, after feeding 500 mg genistein kg^{-1} diet or ~40 mg kg^{-1} body weight). The percentage of genistein aglycone (of total genistein) ranged from 10 to 100% in males (10% in testes, 100% in brain) and 20 to 100% in females (20% in thyroid, 100% in uterus). Tissue differences in isoflavone metabolism might be an important determinant of tissue specificity of isoflavone action, but this remains to be proven.

Many studies of the biological effects of isoflavones examine such effects only in vitro and only using isoflavone aglycones. In light of the complicated patterns of isoflavone biotransformation, most of these in vitro studies are probably quite misleading. Careful attention to modeling the patterns and quantities of isoflavone metabolites in such in vitro studies is absolutely essential in determining mechanisms of the health effects of isoflavones.

21–4.6 Dynamics of Isoflavone Distribution and Excretion

The relevance of many studies of isoflavones to human health effects depends on accounting for the amounts of isoflavones that are likely to occur in the body after dietary exposure to these compounds. Furthermore, the kinetics of isoflavone metabolism may determine the dosing regimens and isoflavone forms that are needed to achieve desirable effects while avoiding adverse effects. Several studies in men or women or both have studied plasma levels and urinary disposition of isoflavones over time. In general, plasma dynamics of daidzein, genistein, and glycitein after a single isoflavone-containing meal are relatively similar with peak plasma concentrations occurring between 6 to 10 h after dosing and elimination half-lives of 4 to 8 h (Setchell et al., 2001; Watanabe et al., 1998; King and Bursill, 1998; Xu et al., 1994; 1995; Lu et al., 1995, 1996). Dose/response studies indicated relatively linear responses in plasma and urinary isoflavone contents to varying isoflavone doses when isoflavones were given in single meals (Xu et al., 1994) or in three meals per day (Xu et al., 1995). Generally a single dose of isoflavones was nearly completely eliminated within 24 h (Xu et al., 1994). End-stage renal patients showed threefold greater plasma isoflavones at about 12 h after isoflavone dosing, and had 10-fold longer elimination half-lives of daidzein and genistein (Fanti et al., 1999). Most studies of isoflavone dynamics showed that when intake of genistein and daidzein were approximately equal, daidzein was about two- to threefold more bioavailable, as reflected in urinary excretion (Xu et al., 1994; 1995; King, 1998). Subjects who were seemingly low degraders of isoflavones (or high excretors), (Xu et al., 1995) showed greater and more prolonged genistein levels in plasma than did the low excretors. Watanabe et al. (1998) also observed greater genistein than daidzein in plasma after similar amounts of both isoflavones were fed, but the relative isoflavone degradation ability of those subjects was not known. Isoflavone bioavailability in infants consuming soybean milk formulas or breast milk after mothers consume soy foods is of interest, because of potential health effects of phytoestrogens. A women who consumed 37 mg isoflavones showed about 0.1 mmol daidzein + genistein L^{-1} in breast milk (Franke et al., 1998). This maternal dose is within the range of what may be consumed daily in regions of the world where soybean intake is common. The infant would be exposed to no more than 25 mg isoflavones per day from such breast milk (~5 mg kg^{-1} body weight), a dose two to three orders of magnitude less than doses showing cholesterol-lowering effects (Crouse et al., 1999). Four infants fed soybean-based formulas were exposed to ~3 mg isoflavones kg^{-1} body weight and excreted 0.5 mg daidzein and 0.2 mg genistein $(kg·d)^{-1}$ when assessed between ~1 and 4 mo of age (Irvine et al., 1998). Relative bioavailability of isoflavones in infants (daidzein > genistein) was similar to adults (Xu et al., 1994). Plasma concentrations of isoflavones averaged 1 mg L^{-1} (4 µM) in 4-mo-old infants fed soybean formula exclusively. These levels are arguably greater than in plasma of most adults consuming soy foods regularly, reflecting the greater dose of isoflavones in the soybean formula than in adult soybean-containing diets. Subjects (N = 248) who were fed soybean milk for the first 4 mo of life were assessed as young adults (ages 20–34) for differences in reproductive function compared with subjects fed cow's milk as infants (N = 563) (Strom et al., 2001). No differences in height, weight, onset of puberty, or fertility were noted between groups. Women fed soybean milk as infants had slightly longer duration of menses, and slightly greater menstrual discomfort than did women fed cow's milk as infants, but neither of these findings is of any clear clinical significance. The isoflavone contents of the soybean formulas fed were not determined, but manufacturing practices are currently similar to those used in the 1960s to 1970s. Thus, infant formula containing isoflavones seems to pose no risk of adverse effects on reproductive health.

21–4.7 Utility of Body Fluid Isoflavones as Biomarkers of Isoflavone Intake and Chronic Disease Risk

The isoflavone contents of body fluids may be a convenient biomarker of the health effects of soy foods. Eventually the measurement of these compounds may be useful predictors of various effects of isoflavone intake. This remains to be established. In 60 case-control pairs from the Shanghai Breast Cancer Study, urinary excretion of isoflavones from breast cancer patients (urine collected before cancer therapy was initiated) was less than in controls by about one-third (Zheng et al., 1999). In an Australian case control study of 18 women recently diagnosed with breast cancer compared with 20 controls, the cases had 14-fold lesser 24-h urinary daidzein than did controls (Murkies et al., 2000). Among dietary factors studied, only β-carotene was significantly less in cases than in controls. Dietary soy food intake was not described in either paper, but both studies suggest that isoflavone intake is protective from breast cancer.

In 30 subjects with blood pressures averaging 130/75 who were given 55 mg isoflavones per day for 8 wk showed no correlation between urinary isoflavone excretion and urinary F2-isoprostanes, a marker of oxidative stress (Hodgson et al., 1999b). Isoflavones were proposed to be antioxidants invivo, but with respect to this marker, they did not seem to be protective. In 39 serum samples taken over the first 6 wk of an intervention in men (average age 65 yr) with elevated prostate specific antigen (PSA) giving ~70 mg isoflavones-1d in a soybean protein beverage, serum cholesterol was negatively correlated with total serum isoflavones $r = -0.55$), but serum PSA was not associated with changes in serum isoflavones during the intervention (Urban et al., 2001). A positive correlation was observed between luteal phase estradiol and daidzein in 40 premenopausal women fed isoflavone tablets (20 and 40 mg total isoflavones d^{-1}) for 1 mo ($r = 0.335$) (Watanabe et al., 2000). This finding is of uncertain health significance, and contrasts with a study by Lu et al. (2000) in which 10 women were given an average of ~150 mg total isoflavones d^{-1} from Day 2 of one menstrual cycle through Day 2 of the subsequent cycle. Decreased follicular and luteal phase estradiol was associated with urinary isoflavone excretion ($R^2 = 0.70, 0.65$, respectively). These changes in estradiol were not associated with isoflavone intake per se, supporting the need for monitoring body fluid isoflavone content because of great interindividual variability in isoflavone bioavailability. Mitchell et al. (2001) gave 15 men 40 mg isoflavones d^{-1} for 2 mo and measured reproductive hormones and semen quality. The isoflavone supplement had no effect on these parameters over this time period. With respect to some biomarkers, body fluid isoflavones may become useful co-predictive markers. Many more biomarkers for various diseases need to be assessed with respect to the role of isoflavones.

Several studies have examined relationships between isoflavone contents of plasma, urine, and feces and dietary isoflavone intake. Dietary questionnaires and interviews assessed soy food intake in 147 subjects (average age 56 yr) in the Singapore Cohort Study. When frequency of overall soybean intake was grouped into low, moderate, and high intakes, daidzein and total isoflavones excreted in urine showed a significant positive relationship with soy food intake (Seow et al., 1998). Isoflavone intake, grouped in three levels of increasing intake, was positively associated with daidzein, genistein, and total isoflavone excretion in spot urine samples from 60 control subjects in the Shanghai Breast Cancer Study (Chen et al., 1999). Twenty-four hour soybean and isoflavone intake recalls were positively correlated with urinary isoflavone contents (r = 0.61–0.62) in 102 Hawaiian women of various ethnicities (Maskarinec et al., 1998). Soy food intake in 49 men and 49 women was positively associated with urinary isoflavone excretion (r = ~0.4 for all isoflavones measured, except that no significant association was observed between equol excretion and soybean intake), when soybean intake was assessed by 5-d food records and food frequency questionnaires (Lampe et al., 1999). In 111 Japanese women, plasma daidzein, and genistein, assayed by time-resolved fluoroimmunoassay, a rapid but expensive method, were correlated with dietary isoflavone intake as assessed by 7-d food record and isoflavone analysis of Japanese soy foods r = 0.29 for genistein, r = 0.31 for daidzein (Uehara et al., 2000). The ability of a modified Block food frequency questionnaire to determine dietary daidzein and genistein intakes was determined in 51 Japanese and 18 Caucasian women (Huang et al., 2000). The food frequency questionnaire was compared with four 48-h dietary recalls per subject, using an isoflavone database to estimate daidzein and genistein. Urinary daidzein and genistein were also measured from four 24-h samples taken at the time of the dietary recalls. Significant but not extremely strong positive correlations were obtained between the questionnaire and dietary recalls of isoflavone intake (r = 0.57 for daidzein, r = 0.51 for genistein). Urinary daidzein was significantly correlated with both food intake methods r = ~0.47), as was genistein, but less strongly so (r = ~0.35), with the food frequency questionnaire being somewhat less strongly correlated with urinary genistein than the food recall data. In 106 middle-aged Japanese women, urinary and plasma genistein and daidzein were correlated with isoflavone intake estimated from 3-d dietary records (r = 0.4). These data support the use of dietary intake data to estimate body isoflavones, in a general sense. For more specific quantitation of disease risk, body fluid isoflavone content seems to be somewhat more useful than dietary estimates of isoflavones. Much work remains to be done to describe the utility of body fluid isoflavone contents as biomarkers.

21–4.8 Toxicity

We must consider the potential of soybean isoflavones to cause toxicity, despite epidemiological evidence in Asian women who consume high soybean diets, but have low rates of breast (Wu et al., 1998) and endometrial (Wynder et al., 1991) cancer. Yet, short-term dietary soybean intake in premenopausal women with benign and malignant breast conditions stimulated breast tissue proliferation (McMichael-Phillips et al., 1998). In response to 14 d of 60 g soybean (45 mg isoflavones) in 84 premenopausal breast cancer patients, breast nipple aspirate levels of apolipoprotein D declined and pS2 levels rose (Hargreaves et al., 1999). This short-term study using dietary soybean intake had a weak estrogenic effect on the breast, but no effect was detected on epithelial cell proliferation, estrogen or progesterone receptor status, apoptosis, mitosis, or Bcl-2 expression. Estrogenic effects in postmenopausal women are even less impressive, with only slight increases in vaginal cell maturation (Baird et al., 1995; Wilcox et al., 1990). Studies have shown no effects of ~130 mg d^{-1} of isoflavones on vaginal cytology or endometrial biopsy data in premenopausal or postmenopausal women (Duncan et al., 1999a, 1999b). Likewise, soybean isoflavone extract did not stimulate the endometrium of postmenopausal women in a multi-center trial (Upmalis et al., 2000) or pilot study (Scambia et al., 2000). Animal studies have shown that, unlike estrogens, soybean isoflavones (Arjmandi et al., 1998a, 1998b) or genistein alone (Murrill et al., 1996) are not uterotropic, except at a high dose (Picherit et al., 2001b). Further, surgically postmenopausal macaques treated with isoflavone-rich soybean showed no signs of induced proliferation in endometrial or mammary tissue (Foth and Cline, 1998). As a basis of comparison, doses of 80 to 120 mg d^{-1} approximate what one might consume with 2 to 4 high isoflavone soy foods d^{-1} (Wang and Murphy, 1994). Yet Asians do not typically consume this high amount of isoflavones. Nonetheless, the long-term safety of isoflavone intake in humans should be further studied. In particular, the response of women with estrogen responsive cancers to dietary isoflavones should be further studied since some studies suggest that isoflavone intake may enhance breast cancer growth as noted in Section 21–3.2.

21–4.9 Chapter Summary

Isoflavone bioavailability varies greatly among individuals, and thus may be a significant influence on the potential of dietary isoflavones to exert health benefits by reducing chronic disease risk. Both gut microbial and endogenous mammalian biotransformation of the isoflavones influence their availability to sites of action. The metabolites of isoflavones may be at least partly responsible for the observed biological effects of these compounds. The utility of isoflavones as biomarkers of disease risk is promising, at least with respect to reduction of blood cholesterol. Understanding isoflavone bioavailability will be a challenging and useful pursuit for the foreseeable future. Data from numerous studies support the safety of isoflavones consumed by healthy populations in the ranges found in Asian diets and in soy foods. However, some data suggest that high intakes of isoflavones should be avoided in women with breast cancer.

21–5 OVERALL CONCLUSIONS

Considerable data have been accumulated suggesting that the consumption of soybean-based foods may contribute to improved health. In particular, improved vascular health, bone health, and cancer prevention have been associated with soy food consumption.

There is convincing evidence that soybean consumption can improve cardiovascular disease risk factors. Clinical trial results show that consumption of soybean protein can improve plasma lipid and lipoprotein concentrations, specifically lower LDL cholesterol concentrations, and in some experiments, higher HDL cholesterol and lower triglyceride concentrations. In addition, some studies have found improvements in diastolic blood pressure and vascular function, particularly in postmenopausal women. Other mechanisms by which soybean consumption might inhibit cardiovascular disease are: (i) reduced oxidative potential for LDL; (ii) lower platelet activation, aggregation, and serotonin storage; and (iii) potentially direct effects on the artery wall. Lower rates of atherosclerosis progression have also been reported with soybean feeding in studies with animal models. These beneficial effects on cardiovascular disease risk factors, suggest that soybean consumption can reduce the risk of coronary heart disease morbidity and mortality; however, there are no studies that have evaluated these outcomes.

Modest protection of dietary soybean or soybean isoflavones against bone loss in perimenopausal and early postmenopausal women has been supported by a limited number of investigations and the data are consistent enough to justify long term studies (2–3 yr). Unfortunately, nothing is known about the effects of soybean against osteoporosis in men or in elderly women. The data suggests that isoflavones or isoflavones in conjunction with soybean protein are responsible for the bone health benefits of dietary soybean but further studies on the role of particular soy food constituents, interactions of constituents, and mechanisms of action are warranted. The data on soy foods and bone health do not justify the use of dietary soy foods or soybean constituents as a substitute for hormone replacement therapy in women, and further investigations are needed to determine the potential role of soy foods in bone health.

The data associating soybean consumption with cancer prevention is intriguing but not yet convincing. While reduced rates of breast, prostate, and colon cancer have been associated with soy food intake, it is possible that other factors, such as reduced fat and/or energy intake, elevated fruit and vegetable intake, or higher rates of physical activity are the true preventive measures in the soy food-consuming populations. Animal studies are supportive of a role for soybean and, potentially for isoflavones, in cancer prevention, but considerable conflicting data make extrapolation impossible. Finally, considerable mechanistic data have been gathered on some soybean constituents suggesting that these constituents may protect against cancer, but intervention studies will be required to definitively demonstrate cancer prevention. Intervention studies are unlikely until safe dietary intakes have been clearly identified and more compelling descriptive data have been gathered.

In designing intervention studies using soybean foods or their components, the bioavailability of the soy components relevant to the particular disease endpoint must be considered. For example, soybean isoflavone bioavailability has been relatively well characterized. This characterization has revealed that substantial interindividual variation in gut microbial metabolism of the isoflavones, either to produce equol, a relatively potent estrogenic metabolite of daidzein, or to cause further breakdown and disappearance of the isoflavones, may be partly controllable. Limiting or at least knowing the variation among individuals or populations in isoflavone metabolism may lead to greater certainty about the health effects caused by these compounds. This could lead to clearer recommendations about effective and safe isoflavone doses to be incorporated into designer foods, disease therapies or food supplies of populations. In addition, clarifying the mechanisms of action of isoflavones in the body depends upon an accurate picture of the active forms of the isoflavones and their sites of action. The biological effects of the major isoflavone metabolites, glucuronide conjugates, are not well understood. But these metabolites are not inert. Researchers need to gain sophistication in their choice of isoflavone compounds for mechanistic studies, if we are to make rapid progress in identifying the soybean components that are of most benefit to human health.

Although it is widely speculated that particular constituents of soy foods and soybean ingredients may be the bioactive ingredients that are responsible for the observed health benefits of soy foods, no one component demonstrates all of the potential health benefits of soy foods. Indeed, isoflavones have been extensively studied for their contribution to all of the purported health benefits of soybean, but it is not clear for any observed health benefit that isoflavones are solely responsible. Indeed, data are being gathered suggesting that components of soy foods may interact to provide health benefits.

There is certainly a need for additional information on the particular constituents and interactions of constituents that may help promote optimal human health. Furthermore, the bioavailability of these constituents and factors that influence bioavailability are unknown and information on metabolism and distribution in the body is unavailable. The compelling data associating soy food consumption with improved health has heightened the interest by food companies, health professionals, and the public. It is now critical that we identify sound science-based dietary strategies for improving health through soy food or soybean constituent consumption.

REFERENCES

Adams, M.R., D.L. Golden, M.S. Anthony, T.C. Register, and J.K. Williams. 2002. The inhibitory effect of soy protein isolate on atherosclerosis in mice does not require the presence of LDL receptors or alteration of plasma lipoproteins. J Nutr 132:43–49.

Adlercreutz, H. 1995. Phytoestrogens: Epidemiology and a possible role in cancer prevention. Environ. Health Perspect 103:103–112.

Adlercreutz, H., C. Bannwart, K. Wähälä, T. Mäkelä, G. Brunow, T. Hase, P.J. Arosemena, J.T. Kellis Jr, and L.E. Vickery. 1993. Inhibition of human aromatase by mammalian lignans and isoflavonoid phytoestrogens. J. Steroid Biochem. Mol. Biol. 44:147–153.

Adlercreutz, H., T. Fotsis, C. Bannwart, K. Wahala, G. Brunow, and T. Hase. 1991. Isotope dilution gas chromatographic-mass spectrometric method for the determination of lignans and isoflavonoids in human urine, including identification of genistein. Clin. Chim.. Acta 199:263–278.

Agnusdei, D., A. Camporeale, F. Zacchei, C. Gennari, M.C. Baroni, D. Costi, M. Biondi, M. Passeri, A. Ciacca et al. 1992. Effects of ipriflavone on bone mass and bone remodeling in patients with established postmenopausal osteoporosis. Curr. Ther. Res. 52:81–92.

Agnusdei, D., C. Gennari, and L. Bufalino. 1995. Prevention of early postmenopausal bone loss using low doses of conjugated estrogens and the non-hormonal, bone-active drug ipriflavone. Osteoporosis Int. 5(6):462–466.

Agnusdei, D., F. Zacchei, S. Bigazzi, C. Cepollaro, P. Nardi, M. Montagnani, and C. Gennari. 1989. Metabolic and clinical effects of ipriflavone in established postmenopausal osteoporosis. Drugs Exp. Clin. Res. 15:97–104.

Akiyama, T., and H. Ogawara. 1991. Use of specificity of genistein as an inhibitor of protein tyrosine kinases. Methods Enzymol. 201:362–370.

Alekel, D.L., E. Mortillaro, E.A. Hussain, B. West, N. Ahmed, C.T. Peterson, R.K. Werner, B.H. Arjmandi, and S.C. Kukreja. 1999. Lifestyle and biologic contributors to proximal femur bone mineral density and hip axis length in two distinct ethnic groups of premenopausal women. Osteoporosis Int 9:327–338.

Alekel, D.L., C.T. Peterson, R.K. Werner, E. Mortillaro, N. Ahmed, and S.C. Kukreja. 2002. Frame size, ethnicity, lifestyle, and biologic contributors to areal and volumetric lumbar spine bone mineral density in Indian/Pakistani and American Caucasian premenopausal women. J. Clin. Densitometry 5:175–186.

Alekel, D.L, A. St.Germain, C.T. Peterson, K.B. Hanson, J.W. Stewart, and T.Toda. 2000. Isoflavone-rich soy protein isolate attenuates bone loss in the lumbar spine of perimenopausal women. Am. J. Clin. Nutr. 72:844–852.

Alexandersen, P., A. Toussaint, C. Christiansen, J-P Devogelaer, C. Roux, J. Fechtenbaum, C. Gennari, and J.Y. Reginster. 2001. For the ipriflavone multicenter European fracture study: Ipriflavone in the treatment of postmenopausal osteoporosis. JAMA 285:1482–1488.

American Cancer Society. 2000. Cancer facts and figures, 2000. Am. Cancer Soc., Atlantia, GA.

American Heart Association. 2000. 2001 Heart and stroke statistical update. Am. Heart Assoc., Dallas, TX.

Anderson, J.J.B., W. W. Ambrose, and S.C. Garner. 1998. Biphasic effect of genistein on bone tissue in the ovariectomoized, lactating rat model. PSEBM 217:345–350.

Anderson, J.J.B., and W.S. Pollitzer. 1994. Ethnic and genetic differences in susceptibility to osteoporotic fractures. p. 129–149. In H.H. Draper (ed.) Advances in nutritional research. Vol. 9. Plenum Press, New York.

Anderson, J.J.B., K. Thomsen, and C. Christiansen. 1987. High protein meals, insular hormones and urinary calcium excretion in human subjects. In C. Christiansen et al. (ed.) Proceedings of the International Symposium on Osteoporosis, Aalborg, Denmark. 27 Sept.–2 Oct. 1987. Nørhaven A/S, Viborg, Denmark.

Anderson, J.W., B.M. Johnstone, and M.E. Cook-Newell. 1995. Meta-analysis of the effects of soy protein intake on serum lipids. New Engl. J. Med. 333:276–282.

Anderson, R.L., and W.J. Wolf. 1995. Compositional changes in trypsin inhibitors, phytic acid, saponins and isoflavones related to soybean processing. J. Nutr 125:581S–588S.

Anthony, M.S. 2000. Soy and cardiovascular disease: Cholesterol lowering and beyond. J. Nutr. 130:662S–663S.

Anthony, M.S., T.B. Clarkson, B.C. Bullock, and J.D. Wagner. 1997. Soy protein versus soy phytoe-strogens in the prevention of diet-induced coronary artery atherosclerosis of male cynomolgus monkeys. Arterioscler. Thromb. Vasc. Biol. 17:2524–2531.

Anthony, M.S., T.B. Clarkson, C.L. Hughes, T.M. Morgan, and G.L. Burke. 1996. Soybean isoflavones improve cardiovascular risk factors without affecting the reproductive system of peripubertal rhe-sus monkeys. J. Nutr 126:43–50.

Arjmandi, B.H., L. Alekel, B.W. Hollis, D. Amin, M. Stacewicz-Sapuntzakis, P. Guo, and S.C. Kukreja. 1996. Dietary soybean protein prevents bone loss in an ovariectomized rat model of osteoporo-sis. J. Nutr. 126:161–167.

Arjmandi, B.H., R. Birnbaum, N.V. Goyal, M.J. Getlinger, S. Juma, L. Alekel, C.M. Hasler, M.L. Drum, B. W. Hollis, and S.C. Kukreja. 1998a. The bone-sparing effect of soy protein in ovarian hor-mone deficient rats is related to its isoflavone content. Am. J. Clin. Nutr. 68(Suppl.):1364S–1368S.

Arjmandi, B.H., M.J. Getlinger, N.V. Goyal, L. Alekel, C.M. Hasler, S. Juma, M.L. Drum, B.W. Hol-lis, and S.C. Kukreja. 1998b. The role of soy protein with normal or reduced isoflavone content in reversing bone loss induced by ovarian hormone deficiency in rats. Am. J. Clin. Nutr. 68(Suppl.):1358S–1363S.

Arjmandi, B.H., M.A. Salih, D.C. Herbert, S.H. Sims, D.N. Kalu. 1993. Evidence for estrogen recep-tor-linked calcium transport in the intestine. Bone Miner. 21:63–74.

Axelson, M., J. Sjovall, B.E. Gustavsson, and K.D.R. Setchell. 1984. Soya-a dietary source of the non-steroidal oestrogen equol in man and animals. J. Endocrinol. 102:49–56.

Baggott, J.E., T. Ha, W.H. Vaughn, M.M. Juliana, J.M. Hardin, and C.J. Grubbs. 1990. Effects of miso (Japanese soybean paste) and NaCl on DMBA-induced mammary tumors. Nutr. Cancer 14:103–109.

Baird, D.D., D.M. Umbach, L. Lansdell, C.L. Hughes, K.D.R. Setchell, C.R. Weinberg, A.F. Haney, A.J. Wilcox, J.A. Mclachlan. 1995. Dietary intervention study to assess estrogenicity of dietary soy among postmenopausal women. J. Clin. Endocrinol. Metab. 80(5):1685–1690.

Balmir, F., R. Staack, E. Jeffrey, M.D. Berber-Jimenez, L. Wang, and S.M. Potter. 1996. An extract of soy flour influences serum cholesterol and thyroid hormones in rats and hamsters. J. Nutr. 126:3046–3053.

Bannwart, C., T. Fotsis, R. Heikkinen, and H. Adlercreutz. 1984. Identification of the isoflavonic phy-toestrogen daidzein in human urine. Clin. Chim. Acta 136:165–172.

Barnes, S. 1997. The chemopreventive properties of soy isoflavonoids in animal models of breast can-cer. Breast Cancer Res. Treatment. 46:169–179.

Barnes, S., L. Coward, M. Kirk, and J. Sfakianos. 1998. HPLC-mass spectrometry analysis of isoflavones. Proc. Soc. Exp. Biol. Med. 217:254–262.

Barnes, S., C. Grubbs, K.D.R. Setchell, and J. Carlson. 1990. Soybean inhibit mammary tumors in mod-els of breast cancer. Prog.Clin.Biol.Res. 347:239–253.

Bennink, M.R. 2001. Dietary soy reduces colon carcinogenesis in humans and rats. p. 11–17. *In* Am. Inst. for Cancer Res. (ed.) Nutrition and cancer prevention: New insights into the role of phytochemicals. Kluwer Academic/Plenum Publ., New York.

Birt, D.F. 2001. Soybean and cancer prevention: A complex food and a complex disease. AICR 492:1–10. Kluwer Academic/Plenum Publishers.

Birt, D.F., S. Hendrich, and W. Wang. 2001. Dietary agents in cancer prevention: Flavonoids and isoflavonoids. Pharmacol. Therapeutics 90:1–21.

Birt, D.F., J.D. Shull, and A.Yaktine. 1998. Chemoprevention of cancer. *In* M.E. Shils et al. (ed.) Modern nutrition in health and disease. Williams & Wilkins, Baltimore, MD.

Blair, H.C., S.E. Jordan, T.G. Peterson, and S. Barnes. 1996. Variable effects of tyrosine kinase inhibitors on avian osteoclastic activity and reduction of bone loss in ovariectomized rats. J. Cell Biochem. 61:629–637.

Breslau, N.A., L. Brinkley, K.D. Hill, and C.Y.C. Pak. 1988. Relationship of animal protein-rich diet to kidney stone formation and calcium metabolism. J. Clin. Endocrinol. Metabol. 66:140–146.

Bryant, H.U., and W. H. Dere. 1998. Selective estrogen receptor modulators: An alternative to hormone replacement therapy. Proc. Soc. Exp. Biol. Med. 217(1):45–52.

Burke, G.L., M.Z. Vitolins, and D. Bland. 2000. Soybean isoflavones as an alternative to traditional hormone replacement therapy: Are we there yet? J. Nutr. 130:664S–665S.

Burke, V., J.M. Hodgson, L.J. Beilin, N. Giangiulioi, P. Rogers, and T.B. Puddey. 2001. Dietary protein and soluble fiber reduce ambulatory blood pressure in treated hypertensives. Hypertension 38:821–826.

Calvert, G.D., L. Blight, R.J. Illman, D.L. Topping, and J.D. Potter. 1981. A trial of the effects of soyabean flour and soya-bean saponins on plasma lipids, faecal bite acids and neutral sterols in hypercholesterolaemic men. Br. J. Nutr. 45:277–281.

Cassidy, A., S. Bingham, K.D.R. Setchell. 1994. Biological effects of a diet of soy protein rich in isoflavones on the menstrual cycle of premenopausal women. Am. J. Clin. Nutr. 60:333–40.

Centrella, M., T.L. McCarthy, and E. Canalis. 1991. Transforming growth factor-beta and remodeling of bone. J. Bone Joint Surg. Am. 73:1418–1428.

Chang, H.C., M.I. Churchwell, K.B. Delclos, R.R. Newbold, and D.D. Doerge. 2000. Mass spectrometric determination of genistein tissue distribution in diet-exposed Sprague-Dawley rats. J. Nutr. 130:1963–1970.

Chen, Z., W. Zheng, L.J. Custer, W. Dai, X-O Shu, F. Jin, and A.A. Franke. 1999. Usual dietary consumption of soy foods and its correlation with the excretion rate of isoflavonoids in overnight urine samples among Chinese women in Shanghai. Nutr. Cancer 33:82–87.

Chin, K., M.C. Evans, J. Cornish, T. Cundy, and I.R. Reid. 1997. Differences in hip axis and femoral neck length in premenopausal women of Polynesian, Asian and European origin. Osteoporosis Int. 7:344–347.

Clarkson, T.B., M.S. Anthony, and T.M. Morgan. 2001. Inhibition of postmenopausal atherosclerosis progression: A comparison of the effects of conjugated equine estrogens and soy phytoestrogens. J. Clin. Endocrinol. Metab. 86:41–47.

Clifton-Bligh P.B., R.J. Baber, G.R. Fulcher, M-L Nery, and T. Moreton. 2001. The effect of isoflavones extracted from red clover (Rimostil) on lipid and bone metabolism. Menopause 8:259–265.

Cohen, L.A., Z. Zhao, B. Pittman, and and J.A. Scimeca. 2000. Effect of intact and isoflavone-depleted soy protein on NMU-induced rat mammary tumorigenesis. Carcinogenesis (New York) 21:929–935.

Coldham, N.G., L.C. Howells, A. Santi, C. Montesissa, C. Langlais, L.J. King, D.D. Macpherson, and M.J. Sauer. 1999. Biotransformation of genistein in the rat: elucidation of metabolite structure by product ion mass fragmentology. J. Steroid Biochem.. Mol. Biol. 70:169–184.

Coldham, N.G., and M.J. Sauer. 2000. Pharmacokinetics of [^{14}C] genistein in the rat: Gender-related differences, potential mechanisms of biological action, and implications for human health. Toxicol. Appl. Pharmacol. 164:206–215.

Constantinou, A.I., A.E. Krygier, and R.R. Mehta. 1998. Genistein induces maturation of cultured human breast cancer cells and prevents tumor growth in nude mice. Am. J. Clin. Nutr. 68:1426S–1430S.

Constantinou, A.I., R.G. Mehta, and A. Vaughan. 1996. Inhibition of N-methyl-N-nitrosourea-induced mammary tumors in rats by the soybean isoflavones. Anticancer Res. 16:3293–3298.

Cooper, C., G. Campion, and L.J. Melton III. 1992. Hip fractures in the elderly: A worldwide projection. Osteoporosis Int. 2:285–289.

Crouse, J.R., T.M. Morgan, J.G. Terry, J. Ellis, M. Vitolins, and G.L. Burke. 1999. A randomized trial comparing the effect of casein with that of soy protein containing varying amounts of isoflavones on plasma concentrations of lipids and lipoproteins. Arch. Intern. Med. 159:2070–2076.

Cummings, S.R., J.A. Cauley, L. Palermo, P.D. Ross, R.D. Wasnich, D. Black, and K.G. Faulkner. 1994. Racial differences in hip axis lengths might explain racial differences in rates of hip fracture. Osteoporosis Int. 4:226–229.

Cummings, S.R., M.C. Nevitt, W.S. Browner, K. Stone, K.M. Fox, K.E. Ensrud, J. Cauley, D. Black, and T.M. Vogt. 1995. Risk factors for hip fracture in white women. Study of Osteoporotic Fractures Research Group. New Engl. J. Med. 332:767–773.

Dalais, F.S., G.E. Rice, M.L. Wahlqvist, M. Grehan, A.L. Murkies, G. Medley, R. Ayton, and B.J.G. Strauss. 1998. Effects of dietary phytoestrogens in postmenopausal women. Climacteric 1:124–129.

Daly, J. M., R.G.H. Morgan, P.S. Oates, G.C.T. Yeoh, and L.B.G. Tee. 1992. Azaserine-induced pancreatic foci: Detection, growth, labelling index and response to raw soya flour. Carcinogenesis (New York) 13:1519–1523.

Davies, M.J., E.A. Bowey, H. Adlercreutz, I.R. Rowland, and P.C. Rumsby. 1999. Effects of soy or rye supplementation of high-fat diets on colon tumour development in azoxymethane-treated rats. Carcinogenesis (New York) 20:927–931.

Davis, J.W., R. Novotny, R.D. Wasnich, and P.D. Ross. 1999. Ethnic, anthropometric, and lifestyle associations with regional variations in peak bone mass. Calcif. Tissue Int. 65:100–105.

Davis, J.W., P.D. Ross, M.C. Nevit, and R.D. Wasnich. 1997. Incidence rates of falls among Japanese men and women living in Hawaii. J. Clin. Epidemiol. 50:589–594.

De Laet, C.E., B.A. van Hout, H. Burger, A. Hofman, and H.A. Pols. 1997. Bone density and risk of hip fracture in men and women: Cross sectional analysis. Br. Med. J. 315:221–225.

Delmas, P.D., N.H. Bjarnason, B.H. Mitlak, A-C Ravoux, A.S. Shah, W.J. Huster, M. Draper, and C. Christiansen. 1997. Effects of raloxifene on bone mineral density, serum cholesterol concentrations, and uterine endometrium in postmenopausal women. New Engl. J. Med. 337:1641–1647.

Den Tonkelaar, I., L. Keinan-Boker, P. Van't Veer, C.J.M. Arts, H. Adlercreutz, J.H.H. Thijssen, and P.H.M. Peeters. 2001. Urinary phytoestrogens and postmenopausal breast cancer risk. Cancer Epidemiol. Biomarkers Prevention 10:223–228.

Draper, M.W., D.E. Flowers, W.J. Huster, J.A. Neild, K.D. Harper, and C. Arnaud. 1996. A controlled clinical trial of raloxifene (LY139481) HCl: Impact on bone turnover and serum lipid profile in healthy postmenopausal women. J. Bone Mineral Res. 11:835–842.

Duncan, A.M., B.E. Merz, X. Xu, T.C. Nagel, W.R. Phipps, and M.S. Kurzer. 1999a. Soy isoflavones exert modest hormonal effects in premenopausal women. J. Clin. Endocrinol. Metab. 84(1):192–197.

Duncan, A.M., K.E.W. Underhill, X. Xu, J. LaValleur, W.R. Phipps, and M.S. Kurzer. 1999b. Modest hormonal effects of soy isoflavones in postmenopausal women. J. Clin. Endocrinol. Metab. 84(10):3479–3484.

Einspahr, K.J., R.T. Abraham, B.A. Binstade, Y. Uehara, and P.J. Liebson. 1991. Tyrosine phosphorylation provides an early and requisite signal for the activation of natural killer cytotoxic function. Proc. Natl. Acad. Sci. USA 88:6279–6283.

Fanti, P., M.C. Monier-Faugere, Z. Geng, J. Schmidt, P.E. Morris, D. Cohen, and H.H. Malluche. 1998. The phytoestrogen genistein reduces bone loss in short-term ovariectomized rats. Osteoporosis Int 8:274–281.

Fanti, P., B.P. Sawaya, L.J. Custer, and A.A. Franke. 1999. Serum levels and metabolic clearance of the isoflavones genistein and daidzein in hemodialysis patients. J. Am. Soc. Nephrol. 10:864–871.

Fornander, T., L.E. Rutquist, B. Cedermark, U. Glas, A. Mattsson, C. Silfversward, L. Skoog, A. Somell, T. Theve, N. Wilking, J. Askergren, and M-L Hjalmar. 1989. Adjuvant tamoxifen in early breast cancer: Occurrence of new primary cancers. Lancet 1:117–120.

Foth, D., and J.M. Cline. 1998. Effects of mammalian and plant estrogens on mammary glands and uteri of macaques. Am. J. Clin. Nutr. 68(Suppl.):1413S–1417S.

Fournier, D.B., J.W. Erdman, Jr., and G.B. Gordon. 1998. Soy, its components, and cancer prevention: A review of the in vitro, animal, and human data. Cancer Epidemiol. Biomarkers Prevention 7:1055–1065.

Franke, A.A., and L.J. Custer. 1994. High-performance liquid chromatographic assay of isoflavonoids and coumestrol from human urine. J. Chromatogr. B Biomed. Appl. 662:47–60.

Franke, A.A., L.J. Custer, C.M. Cerna, and K.K. Narala. 1994. Quantitation of phytoestrogens in legumes by HPLC. J. Agric. Food Chem. 42:1905–1913.

Franke, A.A., L.J. Custer, and Y. Tanaka. 1998. Isoflavones in human breast milk and other biological fluids. Am. J. Clin. Nutr. 68 (Suppl.):1466S–1473S.

Fritz, W.A., L. Coward, J. Wang, and C.A. Lamartiniere. 1998. Dietary genistein: Perinatal mammary cancer prevention, bioavailability and toxicity testing in the rat. Carcinogenesis (New York) 19:2151–2158.

Fujio, Y., Y. Fumiko, K. Takahashi, and N. Shibata. 1993. Responses of smooth muscle cells to platelet-derived growth factor are inhibited by herbimycin-A tyrosine kinase inhibitor+. Biochem. Biophys. Res. Comm. 195:79–83.

Fujita, T., S. Yoshikawa, K. Ono, T. Inoue, and H. Orimo. 1986. Usefulness of TC-80 (ipriflavone) tablets in osteoporosis: Multi-center, double-blind, placebo-controlled study. Igaku Noayumi 138:113–141.

Furukawa, F., K. Imaida, H. Okamiya, K. Shinoda, M. Sato, T. Imazawa, Y. Hayashi, and M. Takahashi. 1991. Inhibitory effects of soybean trypsin inhibitor on induction of pancreatic neoplastic lesions in hamsters by N-nitrosobis(2-oxopropyl)amine. Carcinogenesis (New York) 12:2123–2125.

Furukawa, F., A. Nishikawa, K. Imaida, M. Mitsui, T. Enami, Y. Hayashi, and M. Takahashi. 1992. Inhibitory effects of crude soybean trypsin inhibitor on pancreatic ductal carcinogenesis in hamsters after initiation with N-nitrosobis(2-oxopropyl)amine. Carcinogenesis (New York) 13:2133–2135.

Gamache, P.H., and I.N. Acworth. 1998. Analysis of phytoestrogens and polyphenols in plasma, tissue, and urine using HPLC with coulometric array detection. Proc. Soc. Ex.p Biol. Med. 217:274–280.

Gao, Y.H., and M.Yamaguchi. 1999a. Anabolic effect of daidzein on cortical bone in tissue culture: Comparison with genistein effect. Mol. Cell Biochem. 194:93–98.

Gao, Y.H., and M. Yamaguchi. 1999b. Inhibitory effect of genistein on osteoclast-like cell formation in mouse marrow cultures. Biochem. Pharmacol. 58:767–772.

Gao, Y.H., and M. Yamaguchi. 1999c. Suppressive effect of genistein on rat bone osteoclasts: Apoptosis is induced through Ca^{2+} signaling. Biol. Pharm. Bull. 22:805–809.

Gao, Y.H., and M. Yamaguchi. 2000. Suppressive effect of genistein on rat bone osteoclasts: Involvement of protein kinase inhibition and protein tyrosine phosphatase activation. Int. J. Mol. Med. 5:261–267.

Gardner, C.D., K.A. Newell, R. Cherin, and W.L. Haskell. 2001. The effect of soy protein with or without isoflavones relative to milk protein on plasma lipids in hypercholesterolemic postmenopausal women. Am. J. Clin. Nutr. 73:728–735.

Genant, H.K., C. Cooper, G. Poor, I. Reid, G. Ehrlich, J. Kanis, B.E.C. Nordin, E. Barrett-Connor, D. Black et al. 1999. Interim report and recommendations of the World Health Organization Task-Force for Osteoporosis. Osteoporosis Int. 10:259–264.

Gotoh, T., K. Yamada, H. Yin, A. Ito, T. Kataoka, and K. Dohi. 1998. Chemoprevention of N-nitroso-N-methylurea-induced rat mammary carcinogenesis by soy foods or biochanin A. Japanese J. Cancer Res. (Amsterdam) 89:137–142.

Greaves, K.A., J.S. Parks, J.K. Williams, and J.D. Wagner. 1999. Intact dietary soy protein, but not adding an isoflavone-rich soy extract to casein, improves plasma lipids in ovariectomized cynomolgus monkeys. J. Nutr. 129:1585–1592.

Griffiths, L.A., and G. E. Smith. 1972. Metabolism of apigenin and related compounds in the rat. Biochem.. J. 128:901–911.

Hammond, E. G., L.A. Johnson, and P.A. Murphy. 1993. Properties and analysis.p. 4223–4225. In B. Caballero et al. (ed.) Encyclopedia of food science, food technology and nutrition. Academic Press, New York.

Han, K.K., J. M. Soares, M.A. Haidar, G.R. de Lima, and E.C. Baracat. 2002. Benefits of soy isoflavone therapeutic regimen on menopausal symptoms. Obstet. Gynecol. 99:389–394.

Hargreaves, D.F., C.S. Potten, C. Harding, L.E. Shaw, M.S. Morton, S.A. Roberts, A. Howell, and N.J. Bundred. 1999. Two-week dietary soy supplementation has an estrogenic effect on normal premenopausal breast. J. Clin. Endocrinol. Metab. 84(11):4017–4024.

Harrison, E., A. Adjel, C. Ameho, S. Yamamoto, and S. Kono. 1998. The effect of soybean protein on bone loss in a rat model of postmenopausal osteoporosis. J. Nutr. Sci. Vitaminol. 44:257–268.

Havsteen, B. 1983. Flavonoids, a class of natural products of high pharmacological potency. Biochem. Pharmacol. 32:1141–1148.

Hawrylewicz, E.J., H.H. Huang, and W.H. Blair. 1991. Dietary soybean isolate and methionine supplementation affect mammary tumor progression in rats. J. Nutr. 121:1693–1698.

Hawrylewicz, E.J., J.J. Zapata, and W.H. Blair. 1995. Soy and experimental cancer: Animal studies. J. Nutr. 125 (Suppl.):698S–708S.

Heaney, R.P. 1994. The bone-remodeling transient: Implications for the interpretation of clinical studies of bone mass change. J. Bone Miner. Res. 9:1515–1523.

Heaney, R.P., and M.W. Draper. 1997. Raloxifene and estrogen: Comparative bone-remodeling kinetics. J. Clin. Endocrinol. Metab. 82:3425–3429.

Heaney, R.P., R.R. Recker, M.R. Stegman, and A.J. May. 1989. Calcium absorption in women: Relationships to calcium intake, estrogen status, age. J. Bone Miner. Res. 4:469–475.

Heinonen, S., K. Wahala, and H. Adlercreutz. 1999. Identification of isoflavone metabolites dihydro-daidzein, dihydrogenistein, 6'-OH-O-dma, and cis-4-OH-equol in human urine by gas chromatography-mass spectrometry using authentic reference compounds. Anal. Biochem. 274:211–219.

Helmeste, D.M., and S.W. Tang. 1995. Tyrosine kinase inhibitors regulate serotonin uptake in platelets. Eur. J. Pharmacol. 280:R5–R7.

Hendrich, S. 2002. Bioavailability of isoflavones. J. Chromatog. B 777:203–210.

Hendrich, S., and P. A. Murphy. 2001. Isoflavones: Source and metabolism. p. 55–75. In R.E.C. Wildman (ed.) Handbook of nutraceuticals. CRC Press, Boca Raton, FL.

Hendrich, S., G.-J. Wang, H.-K. Lin, X. Xu, B.-Y. Tew, H.-J. Wang, and P.A. Murphy. 1999. Isoflavone metabolism and bioavailability. p. 211–230. In A. Pappas (ed.) Antioxidant status, diet, nutrition and health. CRC Press, Boca Raton, FL.

Hendrich, S., G.-J. Wang, X. Xu, B.-Y. Tew, H.-J. Wang, and P.A. Murphy. 1998. Human bioavailability of soy bean isoflavones: Influences of diet, dose, time and gut microflora.p. 150–155. In T. Shibamoto (ed.) Functional foods. ACS Monogr. ACS Books, Washington, DC.

Higashi, K., and H. Ogawara. 1994. Daidzein inhibits insulin- or insulin-like growth factor-1-mediated signaling in cell cycle progression of Swiss 3T3 cells. Biochim. Biophys. Acta 1220:29–35.

Hirayama, T. 1979. Epidemiology of prostate cancer with special reference to the role of diet. Natl. Cancer Inst. Monogr. 53:149–155.

Ho, S.C., E. Bacon, T. Harris, A. Looker, and S. Muggi. 1993. Hip fracture rates in Hong Kong and the United States, 1988 through 1989. Am. J. Publ. Health. 83:694–697.

Ho, S.C., S.G. Chan, Q. Yi, E. Wong, and P.C. Leung. 2001. Soy intake and the maintenance of peak bone mass in Hong Kong Chinese women. J. Bone Miner Res. 16:1363–1369.

Ho, S.C., E. Wong, S.G. Chan, J. Lau, C. Chan, and P.C. Leung. 1997. Determinants of peak bone mass in Chinese women aged 21 - 40 years. III. Physical activity and bone mineral density. J. Bone Miner. Res. 12:1262–1271.

Hodgson, J.M., I. B. Puddey, L. J. Beilin, T.A. Mori, and K.D. Croft. 1999a. Supplementation with isoflavonoid phytoestrogens does not alter serum lipid concentrations: a randomized controlled trial in humans. J. Nutr. 128:728–732.

Hodgson, J.M., I.B. Puddey, K.D. Croft, T.A. Mori, J. Rivera, and L.J. Beilin. 1999b. Isoflavonoids do not inhibit in vivo lipid peroxidation in subjects with high-normal blood pressure. Atherosclerosis 145:167–172.

Holder, C.L., M.I. Churchwell, and D.D. Doerge. 1999. Quantification of soy isoflavones, genistein and daidzein, and conjugates in rat blood using LC/ES-MS. J. Agric. Food Chem. 47:3764–3770.

Honoré, E.K., J.K. Williams, M.S. Anthony, and T.B. Clarkson. 1997. Soy isoflavones enhance vascular reactivity in atherosclerotic female macaques. Fertil. Steril. 67:148–154.

Horiuchi, T., T. Onouchi, M. Takahashi, H. Ito, and H. Orimo. 2000. Effect of soy protein on bone metabolism in postmenopausal Japanese women. Osteoporosis Int. 11:721–724.

Hsieh, C.Y., R.C. Santell, S.Z. Haslam, and W.G. Helferich. 1998. Estrogenic effects of genistein on the growth of estrogen receptor-positive human breast cancer (MCF-7) cells in vitro and in vivo. Cancer Res. 58:3833–3838.

Hsu, C-S, W.W. Shen, Y-M Hsueh, and S-L Yeh. 2001. Soy isoflavone supplementation in postmenopausal women. Effects on plasma lipids, antioxidant enzyme activities, and bone density. J. Reprod. Med. 46:221–226.

Hu, J., S. Zhang, E. Jia, Q. Wang, S. Liu, Y. Liu, Y. Wu, and Y. Cheng. 1988. Diet and cancer of the stomach: A case-control study in China. Int. J. Cancer 41:331–335.

Huang, M.H., G.G. Harrison, M.M. Mohamed, J.A. Gornbein, S.M. Henning, V.L.W. Go, and G.A. Greendale. 2000. Assessing the accuracy of a food frequency questionnaire for estimating usual intake of phytoestrogens. Nutr. Ca. 37:145–154.

Huff, M.W., R.M.G. Hamilton, and K.K. Carroll. 1977. Plasma cholesterol levels in rabbits fed low fat, cholesterol-free, semipurified diets: Effects of dietary proteins, protein hydrolysates and amino acid mixtures. Atherosclerosis 28:187–195.

Hutchins, A.M., J.L. Slavin, and J.W. Lampe. 1995. Urinary isoflavonoid phytoestrogen and lignan excretion after consumption of fermented and unfermented soy products. J. Am. Diet. Assoc. 95:545–551.

Ikeda, T., A. Nishikawa, T. Imazawa, S. Kimura, and M. Hirose. 2000, Dramatic synergism between excess soybean intake and iodine deficiency on the development of rat thyroid hyperplasia. Carcinogenesis (New York) 21:707–713.

Irvine, C. H., N. Shand, M. G. Fitzpatrick, and S. L Alexander. 1998. Daily intake and urinary excretion of genistein and daidzein by infants fed soy- or dairy-based infant formulas. Am. J. Clin. Nutr. 68(Suppl.):1462S–1465S.

Ishida, H., T. Uesugi, K. Hirai, T. Toda, H. Nukaya, K. Yokotsuka, and K. Tsuji. 1998. Preventive effects of the plant isoflavones, daidzin and genistin, on bone loss in ovariectomized rats fed a calcium-deficient diet. Biol. Pharm. Bull. 21:62–66.

Ishimi, Y., C. Miyaura, M. Ohmura, Y. Onoe, T. Sato, Y. Uchiyama, M. Ito, X. Wang, T. Suda, and S. Ikegami. 1999. Selective effects of genistein, a soybean isoflavone, on B-lymphopoiesis and bone loss caused by estrogen deficiency. Endocrinology 140:1893–1900.

Izumi, T., M.K. Piskula, S. Osawa, A. Obata, K. Tobe, M. Saito, S. Kataoka, Y. Kubota, and M. Kikuchi. 2000. Soy isoflavone aglycones are absorbed faster and in higher amounts than their glucosides in humans. J. Nutr. 130:1695–1699.

Jayo, M.J., M.S. Anthony, T.C. Register, S.E. Rankin, T. Vest, and T.B. Clarkson. 1996. Dietary soy isoflavones and bone loss: a study in ovariectomized monkeys. J. Bone Mineral Res. 11:S228 (Abstr. no. S555).

Jenkins, D.J.A., C.W.C. Kendall, M. Garsetti, R.S. Rosenberg-Zand, C-J Jackson, S. Agarwal, A.V. Rao, E.P. Diamandis, T. Parker et al. 2000a. Effect of soy protein foods on low-density lipoprotein oxidation and ex vivo sex hormone receptor activity—A controlled crossover trial. Metabolism 49:537–543.

Jenkins, D.J.A., C.W.C. Kendall, E. Vidgen, V. Vuksan, C.J. Jackson, L.S.A. Augustin, B. Lee, M. Garsetti, S. Agarwal et al. 2000b. Effect of soy-based breakfast cereal on blood lipids and oxidized low-density lipoprotein. Metabolism 49:1496–1500.

Kalu, D.N., E.J. Masoro, B.P. Yu, R.R. Hardin, and B.W. Hollis. 1988. Modulation of age-related hyperparathyroidism and senile bone loss in Fischer rats by soy protein and food restriction. Endocrinology 122:1847–1854.

Kanis, J.A. 1996. Estrogens, the menopause, and osteoporosis. Bone 19:185S–190S.

Kanis, J.A., L.J. Melton III, C. Christiansen, C.C. Johnston, and N. Khaltaev. 1994. Perspective: The diagnosis of osteoporosis. J. Bone Miner. Res. 9:1137–1141.

Kardinaal, A.F.M., M.S. Morton, I.E.M. Bruggemann-Rotgans, and E.C.H. van Beresteijn. 1998. Phytooestrogen excretion and rate of bone loss in postmenopausal women. Eur. J. Clin. Nutr. 52:850–855.

Karr, S.C., J.W. Lampe, A.M. Hutchins, and J.L. Slavin. 1997. Urinary isoflavonoid excretion in humans is dose dependent at low to moderate levels of soy-protein consumption. Am. J. Clin. Nutr. 66:46–51.

Kelly, G.E., G.E. Joannou, C. Nelson, A.Y. Reeder, and M.A. Waring. 1995. The variable metabolic responses to dietary isoflavones in human. Proc. Soc. Exp. Biol. Med. 208:40–43.

Kelly, G.E., C. Nelson, M.A. Waring, G.E. Joannou, and A.Y. Reeder. 1993. Metabolites of dietary (soya) isoflavones in human urine. Clin. Chim. Acta 223:9–22.

Kennedy, A.R. 1995. The evidence for soybean products as cancer preventive agents. J. Nutr. 125:733S–743S.

Kennedy, A.R. 1998. Chemopreventive agents: Protease inhibitors. Pharmacol. Therapeutics 78:167–209.

Kim, H., T.G. Peterson, and S. Barnes. 1998. Mechanisms of action of the soy isoflavone genistein: Emerging role for its effect via transforming growth factor b signalling pathways. Am. J. Clin. Nutr. 68:1418S–1425S.

Kimira, M., Y. Arai, K. Shimoi, and S. Watanabe. 1998. Japanese intake of flavonoids and isoflavonoids from foods. J. Epidemiol. 8:168–175.

Kin, K., E. Lee, K. Kushida, D.J. Sartoris, A. Ohmura, P.L. Clopton, and T.Inque. 1993. Bone density and body composition on the Pacific Rim: A comparison between Japan-Born and U.S.-born Japanese American women. J. Bone Miner. Res. 8:861–869.

King, R. 1998. Daidzein conjugates are more bioavailable than genistein conjugates in rats. Am. J. Clin. Nutr. 68 (Suppl.):1496S–1499S.

King, R.A., and D.B. Bursill. 1998. Plasma and urinary kinetics of the isoflavones daidzein and genistein after a single soy meal in humans. Am. J. Clin. Nutr. 67:867–872.

Kirk, E.A., P. Sutherland, S.A. Wang, A. Chait, and R.C. LeBoeuf. 1998. Dietary isoflavones reduce plasma cholesterol and atherosclerosis in C57BL/6 mice but not LDL receptor-deficient mice. J. Nutr. 128:954–959.

Knight, D.C., J.B. Howes, J.A. Eden, and L.G. Howes. 2001. Effects on menopausal symptoms and acceptability of isoflavone-containing soy powder dietary supplementation. Climacteric 4:13–18.

Koo, L.C. 1988. Dietary habits and lung cancer risk among Chinese females in Hong Kong who never smoked. Nutr. Cancer 11:155–172.

Koratkar, R., and A.R. Rao. 1997. Effect of soya bean saponins on azoxymethone-induced preneoplastic lesions in the colon of mice. Nutr. Cancer 27:206–209.

Krauss, R.M., R.H. Eckel, B. Howard, L.J. Appel, S.R. Daniels, R.J. Deckelbaum, J.W. Erdman, P. Kris-Etherton, I.J. Goldberg et al. 2000. AHA dietary guidelines revision 2000: A statement for health-care professionals from the nutrition committee of the American Heart Association. Circulation 102:2284–2299.

Kritz-Silverstein, D., and D.L. Goodman-Gruen. 2002. Usual dietary isoflavone intake, bone mineral density, and bone metabolism in postmenopausal women. J. Womens Health Gend Based Med. 11(1):69–78.

Kuiper, G.G., J.G. Lemmen, B. Carlsson, J.C. Corton, S.H. Safe, P.T. van der Saag, B. van der Burg, and J.A. Gustafsson. 1998. Interaction of estrogenic chemicals and phytoestrogens with estrogen receptor-b. Endocrinology 139:4252–4263.

Kung, A.W.C., K.D.K. Luk, L.W. Chu, and P.K.Y. Chiu. 1998. Age-related osteoporosis in Chinese: An evaluation of the response of intestinal calcium absorption and calcitropic hormones to dietary calcium deprivation. Am. J. Clin. Nutr. 68:1291–1297.

Kurzer, M.S., and X. Xu. 1997. Dietary phytoestrogens. Annu. Rev. Nutr. 17:353–81.

Lamartiniere, C. A., J. Moore, M. Holland, and S. Barnes. 1995. Neonatal genistein chemoprevents mammary cancer. Proceedings Society of Experimental Biology and Medicine, 208:120–123.

Lamartiniere, C. A., W.B. Murrill, P.A. Manzolillo, J.X. Zhang, S. Barnes, X.S. Zhang, H.C. Wei, and N.M. Brown. 1998a. Genistein alters the ontogeny of mammary gland development and protects against chemically-induced mammary cancer in rats. Proc. Soc. Exp. Biol. Med. 217:358–364.

Lamartiniere, C.A., J.X. Zhang, and M.S. Cotroneo. 1998b. Genistein studies in rats: Potential for breast cancer prevention and reproductive and developmental toxicity. Am. J. Clin. Nutr. 68:1400S–1405S.

Lampe, J.W., D.R., Gustafson, A.M. Hutchins, M.C. Martini, S. li, K. Wahala, G.A. Grandits, J.D. Potter, and J.L. Slavin. 1999. Urinary isoflavonoid and lignan excretion on a Western diet: Relation to soy, vegetable, and fruit intake. Cancer Epidem. Biomarkers. Prev. 8:699–707.

Lampe, J.W., S.C. Karr, A.M. Hutchins, and J.L. Slavin. 1998. Urinary equol excretion with a soy challenge: Influence of habitual diet. Proc. Soc. Exp. Biol. Med. 217:335–339.

Lau, E.M.C. 1995. Osteoporosis in Asians—The role of calcium & other nutrients. Challenge Mod. Med. 7:45–54.

Lau, E.M.C., P. Suriwongpaisal, J.K. Lee, S. Das De, M.R. Festin, S.M. Saw, A. Khir, T. Torralba, A. Sham, and P. Sambrook. 2001. Risk factors for hip fracture in Asian men and women: The Asian osteoporosis study. J. Bone Miner. Res. 16:572–580.

Lee, K.W., H.J. Wang, P.A. Murphy, and S. Hendrich. 1995. Soybean isoflavone extract suppresses early but not later promotion of hepatocarcinogenesis by phenobarbital in female rat liver. Nutr. Cancer 24:267–278.

Lees, C-J., and T.A. Ginn. 1998. Soy protein isolate diet does not prevent increased cortical bone turnover in ovariectomized macaques. Calcif. Tissue Int. 62:557–558.

Li, D.H., J.A. Yee, M.H. McGuire, P.A. Murphy, and L. Yan. 1999. Soybean isoflavones reduce experimental metastasis in mice. J. Nutr., 129:1075–1078.

Liberman, S. 1996. Are the differences between estradiol and other estrogens, naturally occurring or synthetic, merely semantical? J. Clin. Endrocinol. Metab. 81:850.

Liener, I.E. 1995. Possible adverse effects of soybean anticarcinogens. Am. Inst. Nutr. 125:744S–750S.

Looker, A.C., E.S. Orwoll, C.C. Johnston Jr, R.L. Lindsay, H.W. Wahner, W.L. Dunn, M.S. Calvo, T.B. Harris, and S.P. Heyse. 1997. Prevalence of low femoral bone density in older U.S. adults from NHANES III. J. Bone Miner. Res. 12:1761–1768.

Lovati, M.R., C. Manzoni, E. Gianazza, A. Arnoldi, E. Kurowska, K.K. Carroll, and C.R. Sirtori. 2000. Soy protein peptides regulate cholesterol homeostasis in Hep G2 cells. J. Nutr. 130:2543–2549.

Lovati, M.R., C. Manzoni, E. Gianazza, and C.R. Sirtori. 1998. Soybean protein products as regulators of liver low-density lipoprotein receptors. I. Identification of active β-conglycinin subunits. J. Agric. Food Chem. 46:2474–2480.

Love, R.R., R.B. Mazess, H.S. Barden, S. Epstein, P.A. Newcomb, V.C. Jordan, P.P. Carbone, and D.L. DeMets. 1992. Effects of tamoxifen on bone mineral density in postmenopausal women with breast cancer. New Engl. J. Med. 326:852–856.

Lu, L.J., J.J. Grady, M.V. Marshall, V.M. Ramanujam, and K.E. Anderson. 1995. Altered time course of urinary daidzein and genistein excretion during chronic soya diet in healthy male subjects. Nutr. Cancer. 24:311–323.

Lu, L.J., S.N. Lin, J.J. Grady, M. Nagamani, and K.E. Anderson. 1996. Altered kinetics and extent of urinary daidzein and genistein excretion in women during chronic soya exposure. Nutr. Cancer 26:289–302.

Lu, L.-J.W., and K.E. Anderson. 1998. Sex and long-term soy diets affect the metabolism and excretion of soy isoflavones in humans. Am. J. Clin. Nutr. 68 (Suppl.):1500S–1504S.

Lu, L.-J.W., K.E. Anderson, J. J. Grady, F. Kohen, and M. Nagamani. 2000. Decreased ovarian hormones during a soya diet: Implications for breast cancer prevention. Cancer Res. 60:4112–4121.

Lundh, T.J., H. Pettersson, and K.H. Kiessling. 1988. Liquid chromatographic determination of the estrogens daidzein, formononetin, coumestrol, and equol in bovine blood plasma and urine. J. Assoc. Off. Anal. Chem. 71:938–41.

Mäkelä, S., H. Savolainen, E. Aavik, M. Myllärniemi, L. Strauss, E. Taskinen, J-A. Gustafsson, and P. Häyry. 1999. Differentiation between vasculoprotective and uterotrophic effects of ligands with different binding affinities to estrogen receptors α and ß. Proc. Natl. Acad. Sci. 96:7077–7082.

Mäkelä, S.I., L. Pylkkanen, S.S.S. Risto, and H. Adlercreutz. 1995. Dietary soybean may be antiestrogenic in male mice. J. Nutr. 125:437–445.

Malinow, M.R. 1984. Saponins and cholesterol metabolism (Letter to the Editors). Atherosclerosis 50:117–119.

Malinow, M.R., P. McLaughlin, C. Stafford, A.L. Livingston, and J.W. Senner. 1983. Effects of alfalfa saponins on alfalfa saponins on regression of atherosclerosis in monkeys. p. 241–254. In W.H. Hauss and Wissler (ed.) Clinical implications of recent research results in arteriosclerosis. Westdeutscler Verlag, Opladen, West Germany.

Markaverich, B.M., R.R. Gregory, M.A. Alejandro, J.H. Clark, G.A. Johnson, and B.S. Middleditch. 1988. Methyl-p-hydroxyphenyllactate—An inhibitor of cell growth and proliferation and an endogenous ligand for nuclear type-II binding sites. J. Biol. Chem. 263:7203–7210.

Markiewicz, L., J. Garey, H. Adlercreutz, and E. Gurpide. 1993. In vitro bioassays of non-steroidal phytoestrogens. J. Steroid Biochem. Molec. Biol. 45:399–405.

Maskarinec, G., S. Singh, L. Meng, and A.A. Franke. 1998. Dietary soy intake and urinary isoflavone excretion among women from a multiethnic population. Cancer Epidemiol. Biomark. Prev. 7:613–619.

McMichael-Phillips, D.F., C. Harding, M. Morton, S.A. Roberts, A. Howell, C.S. Potten, and N.J. Bundred. 1998. Effects of soy protein supplementation on epithelial proliferation in the histologically normal human breast. Am. J. Clin. Nutr. 68(Suppl.):1431S–1436S.

Melton, L.J. III. 1995. How many women have osteoporosis now? J. Bone Miner. Res. 10:175–177.

Melton, L.J. III. 2000. Who has osteoporosis? A conflict between clinical and public health perspectives. J. Bone Miner. Res. 15:2309–2314.

Melton, L.J. III, E.A. Chrischilles, C. Cooper, A.W. Lane, and B.L. Riggs. 1992. Perspective: How many women have osteoporosis? J. Bone Miner. Res. 7:1005–1010.

Melton, L.J. III, and B.L. Riggs. 1983. Epidemiology of age-related fractures. p. 45–72. In A.V. Avioli (ed.) The osteoporotic syndrome. Grune & Stratton, New York.

Melton, L.J. III, T.M. Therneau, and D.R. Larson. 1998. Long-term trends in hip fracture prevalence: The influence of hip fracture incidence and survival. Osteoporosis Int. 8:68–74.

Merz-Demlow, B.E., A.M. Duncan, K.E. Wangen, X. Xu, T.P. Carr, W.R. Phipps, and M.S. Kurzer. 2000. Soy isoflavones improve plasma lipids in normocholesterolemic, premenopausal women. Am. J. Clin. Nutr. 71:1462–1469.

Messina, M.J. 1999. Legumes and soybean: overview of their nutritional profiles and health effects. Am. J. Clin. Nutr. 70:439S–450S.

Messina, M.J., V. Persky, K.D.R. Setchell, and S. Barnes. 1994. Soy intake and cancer risk: A review of the in vitro and in vivo data. Nutr. Cancer 21:113–131.

Mitchell, J.H., E. Cawood, D. Kinniburgh, A. Provan, A.R. Collins, and D.S. Irvine. 2001. Effect of a phytoestrogen food supplement on reproductive health in normal men. Clin. Sci. 100:613–618.

Mitsuoka, T. 1982. Recent trends in research on intestinal flora. Bifidobact. Microflora 3:3–24.

Murkies, A., F.S. Dalais, E.M. Briganti, H.G. Burger, D.L. Healy, M.L. Wahlqvist, and S.R. Davis. 2000. Phytoestrogens and breast cancer in postmenopausal women: A case control study. Menopause 7:289–296.

Murkies, A.L., C. Lombard, B.J.G. Strauss, G. Wilcox, H.G. Burger, and M.S. Morton. 1995. Dietary flour supplementation decreases post-menopausal hot flushes: Effect of soy and wheat. Maturitas 21:189–195.

Murphy, P., and S. Hendrich. 2002. Phytoestrogens in foods. Adv. Food Nutr. Res. 44:195–246.

Murphy, P.A., T. Song, G. Buseman, K. Barua, G.R. Beecher, D. Trainer, and J. Holden. 1999. Isoflavones in retail and institutional soy foods: J. Agric Food Chem. 47:2697–2704.

Murrill, W.B., N.M. Brown, P.A. Manzolillo, J-X. Zhang, S. Barnes, and C.A. Lamartiniere. 1996. Prepubertal genistein exposure suppresses mammary cancer and enhances gland differentiation in rats. Carcinogenesis 17:1451–1457.

Nagata, C., N. Takatsuka, S. Inaba, N. Kawakami, and H. Shimizu. 1998. Effect of soymilk consumption on serum estrogen concentrations in premenopausal Japanese women. J. Natl. Cancer Inst. 90:1830–1835.

Naik, H.R., J.E. Lehr, and K.J. Pienta. 1994. An in vitro and in vivo study of antitumor effects of genistein on hormone refractory prostate cancer. Anticancer Res. 14:2617–2620.

Nakamura, T., C.H. Turner, T. Yoshikawa, C.W. Slemenda, M. Peacock, D.B. Burr, Y. Mizuno, H. Orimo, Y. Ouchi, and C.C. Johnston Jr. 1994. Do variations in hip geometry explain differences in hip fracture risk between Japanese and white Americans? J. Bone Miner. Res. 9:1071–1076.

Nestel, P.J., T. Yamashita, T. Sasahara, S. Pomeroy, A. Dart, P. Komesaroff, A. Owen, and M. Abbey. 1997. Soy isoflavones improve systemic arterial compliance but not plasma lipids in menopausal and perimenopausal women. Arterioscler. Thromb. Vasc. Biol. 17:3392–3398.

Ni, W., Y. Tsuda, M. Sakono, and K. Imaizumi. 1998. Dietary soy protein isolate, compared with casein, reduces atherosclerotic lesion area in apolipoprotein E-deficient mice. J. Nutr. 128:1884–1889.

Nomura, A., B.E. Henderson, and J. Lee. 1978. Breast cancer and diet among the Japanese in Hawaii. Am. J. Clin. Nutr. 31:2020–2025.

Nurmi, T., and H. Adlercreutz. 1999. Sensitive high-performance liquid chromatographic method for profiling phytoestrogens using coulometric electrode array detection: Application to plasma analysis. Anal. Biochem. 274:110–117.

Omi, N., S. Aoi, K. Murata, and I. Ezawa. 1994. Evaluation of the effect of soybean milk and soybean milk peptides on bone metabolism in the rat model with ovariectomized osteoporosis. J. Nutr. Sci. Vitaminol. 40:201–211.

Onozawa, M., T. Kawamori, M. Baba, K. Fukuda, T. Toda, H. Sato, M. Ohtani, H. Akaza, T. Sugimura, and K. Wakabayashi. 1999. Effects of a soybean isoflavone mixture on carcinogenesis in prostate and seminal vesicles of F344 rats. Jpn. J. Cancer Res. (Amsterdam) 90:393–398.

Parker, M., J.K. Anand, J.W. Myles, and R. Lodwick. 1992. Proximal femoral fractures: Prevalence in different racial groups. Eur. J. Epidemiol. 8:730–732.

Peluso, M.R., T.A. Winters, M.F. Shanahan, and W.J. Banz. 2000. A cooperative interaction between soy protein and its isoflavone-enriched fraction lowers hepatic lipids in male obese Zucker rats and reduces blood platelet sensitivity in male Sprague-Dawley rats. J. Nutr. 130:2333–2342.

Pennington, J.A.T. 1998. Bowes and Church's food values of portions commonly used. 17th ed. Harper & Row Publ., New York.

Pereira, M.A., L.H. Barnes, V.L. Rassman, G.V. Kelloff, and V.E. Steele. 1994. Use of azoxymethane-induced foci of aberrant crypts in rat colon to identify potential cancer chemopreventive agents. Carcinogenesis. 15:1049–1054.

Peterson, T.G., G.P.Ji, M. Kirk, L. Coward, C.N. Falany, and S. Barnes. 1998. Metabolism of the isoflavones genistein and biochanin A in human breast cancer cell lines. Am. J. Clin. Nutr. 68(Suppl.):1505S–1511S.

Picherit C, C. Bennetau-Pelissero, B. Chanteranne, P. Lebecque, M-J. Davicco, J-P Barlet, and V. Coxam. 2001a. Soybean isoflavones dose-dependently reduce bone turnover but do not reverse established osteopenia in adult ovariectomized rats. J. Nutr. 131:723–728.

Picherit, C., B. Chanteranne, C. Bennetau-Pelissero, M-J Davicco, P. Lebecque, J-P Barlet, and V. Coxam. 2001b. Dose-dependent bone-sparing effects of dietary isoflavones in the ovariectomized rat. Br. J. Nutr. 85:307–316.

Picherit, C., V. Coxam, C. Bennetau-Pelissero, S. Kati-Coulibaly, M-J Davicco, P. Lebecque, and J-P Barlet. 2000. Daidzein is more efficient than genistein in preventing ovariectomy-induced bone loss in rats. J. Nutr. 130:1675–1681.

Pie, J-E., and H.Y. Paik. 1986. The effect of meat protein and soy protein on calcium metabolism in young adult Korean women. Kor. J. Nutr. 19:32–40.

Pines, A., H. Katchman, Y. Villa, V. Mijatovic, I. Dotan, Y. Levo, and D. Ayalon. 1999. The effect of various hormonal preparations and calcium supplementation on bone mass in early menopause. Is there a predictive value for the initial bone density and body weight? J. Intern. Med. 246:357–61.

Piskula, M. 2000a. Soy isoflavone conjugation differs in fed and food-deprived rats. J. Nutr. 130:1766–1771.

Piskula, M.K. 2000b. Factors affecting flavonoids absorption. Biofactors 12:175–180.

Pollard, M., and P.H. Luckert. 1997. Influence of isoflavones in soy protein isolates on development of induced prostate-related cancers in L-W rats. Nutr. Cancer 28:41–45.

Potter, S.M., J.A. Baum, H. Teng, R.J. Stillman, N.F. Shay, and J.W. Erdman, Jr. 1998. Soy protein and isoflavones: Their effects on blood lipids and bone density in postmenopausal women. Am. J. Clin. Nutr. 68:1375S–1379S.

Potter, S.M., R. Jimenez-Flores, J-A. Pollack, T.A. Lone, and M.D. Berber-Jiminez. 1993. Protein-saponin interaction and its influence on blood lipids. J. Agric. Food Chem. 41:1287–1291.

Prentice, A, T.J. Parsons, and T.J. Cole. 1994.Uncritical use of bone mineral density in absorptiometry may lead to size-related artifacts in the identification of bone mineral determinants. Am. J. Clin. Nutr. 60:837–842.

Pun, K.K., L.W.L. Chan, V. Chung, and F.H.W. Wong. 1990. Calcium and other dietary constituents in Hong Kong Chinese in relation to age and osteoporosis. J. Appl. Nutr. 43:12–17.

Rogmark, C., I. Sernbo, O. Johnell, and J-A. Nilsson. 1999. Incidence of hip fractures Malm`, Sweeden, 1992–1995. Acta Orthop. Scan. 70:19–22.

Ross, P.D., S. Fujiwara, C. Huang, J.W. Davis, R.S. Epstein, R.D. Wasnich, K. Kodama, and J. Melton III. 1995. Vertebral fracture prevalence in women in Hiroshima compared to Caucasians or Japanese in the US. Int. J. Epidemiol. 24:1171–1177.

Ross, P.D., Y-F. He, A.J. Yates, C. Coupland, P. Ravn, M. McClung, D. Thompson, and R.D. Wasnich. 1996. Body size accounts for most differences in bone density between Asian and Caucasian women. Calcif. Tissue Int. 59:339–343.

Ross, P.D., H. Norimatsu, J.W. Davis, K. Yano, R.D. Wasnich, S. Fujiwara, Y. Hosoda, and J. Melton III. 1991. A comparison of hip fracture incidence among native Japanese, Japanese Americans, and American Caucasians. Am. J. Epidemiol. 133:801–809.

Rowland, I.R., H. Wiseman, T.A. Sanders, H. Adlercreutz, and E.A. Bowey. 2000. Interindividual variation in metabolism of soy isoflavones and lignans: Influence of habitual diet on equol production by the gut microflora. Nutr. Cancer. 36:27–32.

Ruiz-Larrea, M.B., A.R. Mohan, G. Paganga, N.J. Miller, G.P. Bolwell, and C.A. Rice-Evans. 1997. Antioxidant activity of phytoestrogenic isoflavones. Free Radic. Res. 26:63–70.

Samman, S., P.M.L. Wall, G.S.M. Chan, S.J. Smith, and P. Petocz. 1999. The effect of supplementation with isoflavones on plasma lipids and oxidisability of low density lipoprotein in premenopausal women. Atherosclerosis 147:277–283.

Scambia, G., D. Mango, P.G. Signorile, R.A. Anselmi-Angeli, C. Palena, D. Gallo, E. Bombardelli, P. Morazzoni, A. Riva, and S. Mancuso. 2000. Clinical effects of a standardized soy extract in postmenopausal women: A pilot study. Menopause 7(2):105–111.

Schoene, N.W., and C.A. Guidry. 1999. Dietary soy isoflavones inhibit activation of rat platelets. J. Nutr. Biochem. 10:421–426.

Seow, A., C.-Y. Shi, A.A. Franke, J.H. Hankin, H.-P. Lee, and M.C. Yu. 1998. Cancer Epidem. Biomarkers Prev. 7:135–140.

Setchell, K.D., S.P. Borriello, P. Hulme, D.N. Kirk, and M. Axelson. 1984. Nonsteroidal estrogens of dietary origin: Possible roles in hormone-dependent disease. Am. J. Clin. Nutr. 40:569–578.

Setchell, K.D.R. 1995. Non-steroidal estrogen of dietary origin: Possible role in health and disease, metabolism and physiological effects. Proc. Nutr. Soc. N. Z. 20:1–21.

Setchell, K.D.R., and H. Adlercreutz. 1988. Mammalian ligands and phyto-oestrogens. Recent studies on their formation, metabolism and biological role in health and disease. p. 315–345. In I.A. Rowland (ed.) The role of gut microflora in toxicity and cancer. Academic Press, New York.

Setchell, K.D.R., N.M. Brown, P. Desai, L. Zimmer-Nechemias, B.E. Wolfe, W.T. Brashear, A.S. Kirschner, A. Cassidy, and J.E. Heubi. 2001. Bioavailability of pure isoflavones in healthy humans and analysis of commercial soy isoflavone supplements. J. Nutr. 131:1362S–1375S.

Severson, R.K., A.M.Y. Nomura, J.S. Grove, and G.N. Stemmermann. 1989. A prospective study of demographics, diet, and prostate cancer among men of Japanese ancestry in Hawaii. Cancer Res. 49:1857–1860.

Sfakianos, J., L. Coward, M. Kirk, and S. Barnes. 1997. Intestinal uptake and biliary excretion of the isoflavone genistein in rats. J. Nutr. 127:1260–1268.

Shamsuddin, A.M. 1995. Inositol phosphates have novel anticancer function. J. Nutr. 125:725S–732S.

Shelnutt, S.R., C.O. Cimino, P.A. Wiggins, and T.M Badger. 2000. Urinary pharmacokinetics of the glucuronide and sulfate conjugates of genistein and daidzein. Cancer Epidemiol. Biomarkers Prev. 9:413–419.

Shimokado, K., K. Umezawa, and J. Ogata. 1995. Tyrosine kinase inhibitors inhibit multiple steps of the cell cycle of vascular smooth muscle cells. Exp. Cell Res. 220:266–273.

Shimokado, K., T. Yokota, K. Umezawa, T. Sasaguri, and J. Ogata. 1994. Protein tyrosine kinase inhibitors inhibit chemotaxis of vascular smooth muscle cells. Arteriosclerosis Thrombosis 14:973–981.

Shutt, D.A., and R.I. Cox. 1972. Steroid and phytoestrogen binding in sheep uterine receptors in vitro. Endocrinology 52:299–310.

Silverman, S.L., and R.E. Madison. 1988. Decreased incidence of hip fracture in Hispanics, Asians, and Blacks: California hospital discharge data. Am. J. Public Health 78:1482–1483.

Simons, L.A., M. von Konigsmark, J. Simons, and D.S. Celermajer. 2000. Phytoestrogens do not influence lipoprotein levels or endothelial function in healthy, postmenopausal women. Am. J. Cardiol. 85:1297–1301.

Somekawa, Y., M. Chiguchi, T. Ishibashi, and T. Aso. 2001. Soy intake related to menopausal symptoms, serum lipids, and bone mineral density in postmenopausal Japanese women. Obstet. Gynecol. 97:109–115.

Song, T., K. Barua, G. Buseman, and P.A. Murphy. 1998. Soy isoflavone analysis: Quality control and new internal standard. Am..J. Clin.Nutr. 68 (Suppl.):1474S–1479S.

Song, T.T., S. Hendrich, and P.A. Murphy. 1999. Estrogenic activity of glycitein, a soy isoflavone. J. Agric. Food Chem. 47:1607–1610.

Sorensen, I.K., E. Kristiansen, A. Mortensen, G.M. Nicolaisen, J.A.H. Wijnandes, H.J. Van Kranen, and C.F. Van Kreijl. 1998. The effect of soy isoflavones on the development of intestinal neoplasia in Apc^{Min} mouse. Cancer Lett. 130:217–225.

Spence, L.A., E.R. Lipscomb, J. Cadogan, B.R. Martin, M. Peacock, and C.M. Weaver. 2001. Effects of soy isoflavones on calcium metabolism in postmenopausal women. FASEB J. 15(5):A728, Abstr. No. 575.4.

St.Germain, A., C.T. Peterson, J. Robinson, and D.L. Alekel. 2001. Isoflavone-rich or isoflavone-poor soy protein does not reduce menopausal symptoms during 24 weeks of treatment. Menopause 8(1):17–26.

Strom, B.L., R. Schinnar, E. Ziegler, K.T. Barnhart, M.D. Sammel, G.A. Macones, V.A. Stallings, J.M. Dralis, S.E. Nelson, and S.A. Hanson. 2001. Exposure to soy-based formula in infancy and endocrinological and reproductive outcomes in young adulthood. J. Am. Med. Assoc. 286:807–814.

Sugano, M., S. Goto, Y. Yamada, K. Yoshida, Y. Hashimoto, T. Matsuo, and M. Kimoto. 1990. Cholesterol-lowering activity of various undigested fractions of soybean protein in rats. J. Nutr. 120:977–985.

Sugano, M., and K. Koba. 1993. Dietary protein and lipid metabolism: A multifunctional effect. Ann. N.Y. Acad. Sci. 676:215–222.

Sugano, M., Y. Yamada, K. Yoshida, Y. Hashimoto, T. Matsuo, and M. Kimoto. 1988. The hypocholesterolemic action of the undigested fraction of soybean protein in rats. Atheroscler 72:115–122.

Swanson, C.A., B.L. Mao, J.Y. Li, J.H. Lubin, S.X. Yao, J.Z. Wang, S.K. Cai, Y. Hou, Q.S. Luo, and W.J. Blot. 1992. Dietary determinants of lung-cancer risk: Results from a case-control study in Yunnan province China. Int. J. Cancer 50:876–880.

Teede, H.J., F.S. Dalais, D. Kotsopoulos, Y-L Liang, S. Davis, and B.P. McGrath. 2001. Dietary soy has both beneficial and potentially adverse cardiovascular effects: A placebo-controlled study in men and postmeopausal women. JCEM. 86:3053–3060.

Teixeira, S.R., S.M. Potter, R. Weigel, S. Hannum, J.W. Erdman, and C.M. Hasler. 2000. Effects of feeding 4 levels of soy protein for 3 and 6 wk on blood lipids and apolipoproteins in moderately hypercholesterolemic men. Am. J. Clin. Nutr. 71:1077–1084.

Tew, B. Y., X. Xu, H.-J. Wang, P. A. Murphy, and S. Hendrich. 1996. A diet high in wheat fiber suppresses the bioavailability of isoflavones in a single meal fed to women. J. Nutr. 126:871–877.

Thamsborg, G. 1999. Effect of nasal salmon calcitonin on calcium and bone metabolism. Dan. Med. Bull. 46:118–126.

Thiagarajan, D.G., M.R. Bennink, L.D. Bourquin, and F.A. Kavas. 1998. Prevention of precancerous colonic lesions in rats by soy flakes, soy flour, genistein, and calcium. Am. J. Clin. Nutr. 68:1394S–1399S.

Thompson, D.D., H.A. Simmons, C.M. Pirie, and H.Z. Ke. 1995. FDA Guidelines and animal models for osteoporosis. Bone 17:125S–133S.

Thompson, L.U., P. Robb, M. Serraino, and F. Cheung. 1991. Mammalian lignan production from various foods. Nutr. Cancer. 16:43–52.

Tikkanen, M.J., K. Wahala, S. Ojala, V. Vihma, and H. Adlercreutz. 1998. Effect of soybean phytoestrogen intake on low density lipoprotein oxidation resistance. Proc. Natl. Acad. Sci. 95:3106–3110.

Tonkelaar, Isolde den, L. Keinan-Boker, P. Van't Veer, C.J.M. Arts, H. Adlercreutz, J.H.H. Thijssen, and P.H.M. Peeters. 2001. Urinary phytoestrogens and postmenopausal breast cancer risk. In Cancer epidemiology. Biomarkers & Prevention 10:223–228.

Tovar-Palacio, C., S.M. Potter, J.C. Hafermann, and N.F. Shay. 1998. Intake of soy protein and soy protein extracts inflluences lipid metabolism and hepatic gene expression in gerbils. J. Nutr. 128:839–842.

Tsai, K., K. Huang, P. Chieng, and C. Su. 1991. Bone mineral density of normal Chinese women in Taiwan. Calcif. Tissue Int. 48:161–166.

Tsai, K., S. Twu, P. Chieng, R. Yang, and T. Lee. 1996. Prevalence of vertebral fractures in Chinese men and women in urban Taiwanese communities. Calcif. Tissue Int. 59:249–253.

Tsai, K.S. 1997. Osteoporotic fracture rate, bone mineral density, and bone metabolism in Taiwan. J. Formos Med. Assoc. 96:802–805.

Tsuchida, K., S. Mizushima, M. Toba, and K. Soda. 1999. Dietary soybean intake and bone mineral density among 995 middle-aged women in Yokohama. J. Epidemiol. 9:14–19.

Tucci, J.R. 1998. Osteoporosis update. Med. Health Rhode Island 81:169–73.

Uehara, M., Y. Arai, S. Watanabe, and H. Adlercreutz. 2000. Comparison of plasma and urinary phytoestrogens in Japanese and Finnish women by time-resolved fluoroimmunoassay. Biofactors 12:217–225.

Uehara, M., A. Ohta, K. Sakai, K. Suzuki, S. Watanabe, and H. Adlercreutz. 2001. Dietary fructooligosaccharides modify intestinal bioavailability of a single dose of genistein and daidzein and affect their urinary excretion and kinetics in blood of rats. J. Nutr. 131:787–795.

Uesugi, T., T. Toda, K. Tsuji, and H. Ishida. 2001. Comparative study on reduction of bone loss and lipid metabolism abnormality in ovariectomized rats by soy isoflavones daidzin, genistin, and glycitin. Biol. Pharm. Bull. 24:368–372.

Ullah, A., and A.M. Shamsuddin. 1990. Dose-dependent inhibition of large intestinal cancer by inositol hexaphosphate in F344 rats. Carcinogenesis (New York). 11:2219–2222.

Upmalis, D.H., R. Lobo, L. Bradley, M. Warren, F.L. Cone, and C.A. Lamia. 2000. Vasomotor symptom relief by soy isoflavone extract tablets in postmenopausal women: A multicenter, double-blind, randomized, placebo-controlled study. Menopause 7(4):236–242.

Urban, D., W. Irwin, M. Kirk, M. A. Markiewicz, R. Myers, M. Smith. H. Weiss, W. E. Grizzle, and S. Barnes. 2001. The effect of isolated soy protein on plasma biomarkers in elderly men with elevated serum prostate specific antigen. J. Urol. 165:294–300.

U.S.Department of Agriculture/Iowa State University Isoflavone Database. 1999. Available at http://www.nal.usda.gov/fnic/foodcomp/Data/isoflav/isfl_tbl.pdf.

Van der Schouw, Y.T., A. Pijpe, C.E.I. Lebrun, M.L. Bots, P.H.M. Peeters, W.A. Van Staveren, S.W.T. Lamberts, and D.E. Grobbee. 2002. Higher usual dietary intake of phytoestrogens is associated with lower aortic stiffness in postmenopausal women. Arterioscler. Thromb. Vasc. Biol. 22:1316–1322.

Vidal, O., L-G Kindblom, and C. Ohlsson. 1999. Expression and localization of estrogen-receptor-β in murine and human bone. J. Bone Miner. Res. 14:923–929.

Wagner, J.D., W.T. Cefalu, M.S. Anthony, K.N. Litwak, L. Zhang, and T.B. Clarkson. 1997. Dietary soy protein and estrogen replacement therapy improve cardiovascular risk factors and decrease aortic cholesteryl ester content in ovariectomized cynomolgus monkeys. Metabolism 46:698–705.

Walsh, B.W., L.H. Kuller, R.A. Wild, S. Paul, M. Farmer, J.B. Lawrence, A.S. Shah, and P.W. Anderson. 1998. Effects of raloxifene on serum lipids and coagulation factors in healthy postmenopausal women. JAMA 279(18):1445–1451.

Wang, H.J. 1997. Human gut microfloral metabolism of soybean isoflavones. M.S. thesis. Iowa State Univ. Library, Ames.

Wang, H.J., O. Lapcík, R. Hampl, M. Uehara, N. Al-Maharik, K. Stumpf, H. Mikola, K. Wähälä, and H.Adlercreutz. 2000. Time-resolved fluoroimmunoassay of plasma daidzein and genistein. Steroids 65: 339–48.

Wang, H-J., and P.A. Murphy. 1994. Isoflavone content in commercial soybean foods. J. Agric. Food Chem. 42:1666–1673.

Wang, H. J., and P.A. Murphy. 1996. Mass balance of isoflavones study in soybean processing. J..Agric. Food Chem. 44:2377–2383.

Wang, M-F., S. Yamamoto, H-M. Chung, S-Y. Chung, S. Miyatani, M. Mori, T. Okita, and M. Sugano. 1995. Antihypercholesterolemic effect of undigested fraction of soybean protein in young female volunteers. J. Nutr. Sci. Vitaminol. 41:187–195.

Wang, W., and C.M. Higuchi. 2000. Dietary soy protein is associated with reduced intestinal mucosal polyamine concentration in male Wistar rats. J. Nutr. 130:1815–1820.

Wangen, K.E., A.M. Duncan, B.E. Merz-Demlow, X. Xu, R. Marcus, W.R. Phipps, and M.S. Kurzer. 2000. Effects of soy isoflavones on markers of bone turnover in premenopausal and postmenopausal women. J. Clin. Endocrinol. Metab. 85:3043–3048.

Wangen, K.E., A.M. Duncan, X. Xu, and M.S. Kurzer. 2001. Soy isoflavones improve plasma lipids in normocholesterolemic and mildly hypercholesterolemic postmenopausal women. Am. J. Clin. Nutr. 73:225–231.

Washburn, S., G.L. Burke, T. Morgan, and M. Anthony. 1999. Effect of soy protein supplementation on serum lipoproteins, blood pressure, and menopausal symptoms in perimenopausal women. Menopause 6:7–13.

Wasnich, R.D. 1996. Vertebral fracture epidemiology. Bone 18 (3 Suppl.):179S–183S.

Watanabe, S., K. Terashima, Y. Sato, S. Arai, and A. Eboshida. 2000. Effects of isoflavone supplement on healthy women. Biofactors 12:233–241.

Watanabe, S., M. Yamaguchi, T. Sobue, T. Takahashi, T. Miura, Y. Arai, W. Mazur, K. Wahala, and H. Adlercreutz. 1998. Pharmacokinetics of soybean isoflavones in plasma, urine and feces of men after ingestion of 60 g baked soybean powder (Kinako). J. Nutr. 128:1710–1715.

Watkins, T.R., K. Pandya, and O. Mickelsen. 1985. Urinary acid and calcium excretion. Effect of soy versus meat in human diets. *In* C. Kies (ed.) Nutritional bioavailability of calcium. Am. Chem. Soc., Washington, DC.

Weaver, C.M. 1998. Calcium requirements: The need to understand racial differences. Editorial. Am. J. Clin. Nutr. 68:1153–1154.

Wilcox, G., M.L. Wahlqvist, H.G. Burger, and G. Medley. 1990. Oestrogenic effects of plant foods in postmenopausal women. Br. Med. J. 310:905–906.

Williams, J.K., and T.B. Clarkson. 1998. Dietary soy isoflavones inhibit in-vivo constrictor responses of coronary arteries to collagen-induced platelet activation. Coron. Artery Dis. 9:759–764.

Wilson, L. A., P.A. Murphy, and P. Gallagher. 1991. Japanese soyfoods: Markets and processes. 1: 1–63. Iowa State Univ., Ames, Ctr. for Crops Utilization.

Winter, J., L. H. Moore, V. R. Jr. Dowell, and V. D. Bokkenheuser. 1989. C-ring cleavage of flavonoids by human intestinal bacteria. Appl. Environ. Micriobiol. 55:1203–1208.

Wiseman, H., J.D. O'Reilly, H. Adlercreutz, A.I. Mallet, E.A. Bowey, I.R. Rowland, and T.A.B. Sanders. 2000. Isoflavone phytoestrogens consumed in soy decrease F_2-isoprostane concentrations and increase resistance of low-density lipoprotein to oxidation in humans. Am. J. Clin. Nutr. 72:395–400.

Wright, C.D.P., N.J. Garrahan, M. Stanton, J-C Gazet, R.E. Mansell, and J.E. Compston. 1994. Effect of long-term tamoxifen therapy on cancellous bone remodeling and structure in women with breast cancer. J. Bone Miner. Res. 9:153–159.

Wu, A.H., F.Z. Stanczyk, S. Hendrich, P.A. Murphy, C. Zhang, P. Wan, and M.C. Pike. 2000. Effects of soy foods on ovarian function in premenopausal women. Br. J. Cancer 82(11):1879–1886.

Wu, A.H., R.G. Ziegler, A.M.Y. Nomura, D.W. West, L.N. Kolonel, P.L. Horn-Ross, R.N. Hoover, and M.C. Pike. 1998. Soy intake and risk of breast cancer in Asians and Asian Americans. Am. J. Clin. Nutr. 68(Suppl.):1437S–1443S.

Wynder, E.L., Y. Fujita, R.E. Harris, T. Hirayama, and T. Hiyama. 1991. Comparative epidemiology of cancer between the United States and Japan: A second look. Cancer 67:746–763.

Xu, L., A. Lu, X. Zhao, X. Chen, and S.R. Cummings. 1996. Very low rates of hip fracture in Beijing, People's Republic of China, the Beijing Osteoporosis Project. Am. J. Epidemiol. 144:901–907.

Xu, X., K. Harris, H.–J. Wang, P. Murphy, and S. Hendrich. 1995. Bioavailability of soybean isoflavones depends upon gut microflora in women. J. Nutr. 125:2307–2315.

Xu, X., H.-J. Wang, L. R. Cook, P. A. Murphy, and S. Hendrich. 1994. Daidzein is a more bioavailable soymilk isoflavone to young adult women than is genistein. J. Nutr. 124:825–832.

Xu, X, H.-J. Wang, P. A. Murphy, and S. Hendrich. 2000. Neither background diet nor type of soy food affects short-term isoflavone bioavailability in women. J. Nutr. 130:798–801.

Yamaguchi, M., and E. Sugimoto. 2000. Stimulatory effect of genistein and daidzein on protein synthesis in osteoblastic MC3T3-E1 cells: Activation of aminocyl-tRNA synthetase. Mol. Cell Biochem. 214:97–102.

Yamakoshi, J., M.K. Piskula, T. Izumi, K. Tobe, M. Saito, S. Kataoka, A. Obata, and M. Kikuchi. 2000. Isoflavone aglycone-rich extract without soy protein attenuates atherosclerosis development in cholesterol-fed rabbits. J. Nutr. 130:1887–1893.

Yamazaki, I., A. Shino, Y. Shimizu, R. Tsukuda, Y. Shirakawa, and M. Kinoshita. 1986a. Effect of ipriflavone on glucocorticoid-induced osteoporosis in rats. Life Sci. 38:951–958.

Yamazaki, I., A. Shino, and R. Tsukuda. 1986b. Effect of ipriflavone on osteoporosis induced by ovariectomy in rats. J. Bone Min. Metab. 3:205–210.

Yan, L., J.A. Yee, M.H. McGuire, and G.L. Graef. 1997. Effect of dietary supplementation of soybean on experimental metastasis of melanoma cells in mice. Nutr. Cancer 29:1–6.

Yang, G.Y., and A.M. Shamsuddin. 1995. IP_6-induced growth inhibition and differentiation of HT-29 human colon cancer cells: Involvement of intracellular inositol phosphates. Anticancer Res. 15:2479–2487.

Yasuda, T., S. Mizunuma, Y. Kano, K. Saito, and K. Ohsawa. 1996. Urinary and biliary metabolites of genistein in rats. Biol. Pharm. Bull. 19:413–417.

Zhang, J.-X., G. Hallmans, M. Landstrom, A. Bergh, J.-E. Damber, P. Aman, and H. Adlercreutz. 1997. Soy and rye diets inhibit the development of Dunning R3327 prostatic adenocarcinoma in rats. Cancer Lett. 114:313–314.

Zhang, Y. 2000. Bioavailability and biological effects of the isoflavone glycitein and isoflavone glucuronides: role of glucuronide in human natural killer cell modulation in vitro. Ph. D. Dissertation. Iowa State University Library, Ames, IA.

Zhang, Y., T.T. Song, J.E. Cunnick, P.A. Murphy, and S. Hendrich. 1999a. Daidzein and genistein glucuronides in vitro are weakly estrogenic and activate human natural killer cells in nutritionally relevant concentrations. J. Nutr. 129:399–405.

Zhang, Y., G-J Wang, T.T. Song, P.A. Murphy, and S. Hendrich. 1999b. Differences in disposition of the soybean isoflavones, glycitein, daidzein and genistein in humans with moderate fecal isoflavone degradation activity. J. Nutr. 129: 957-962. Erratum (2001) J. Nutr 131:147–148.

Zheng, Y.L. 2000. Ethnicity and diet habits: Influence on bioavailability of soybean isoflavones in women. M.S. thesis. Iowa State Univ. Library, Ames.

Zheng, W., Q. Dai, L.J. Custer, X.O. Shu, W.Q. Wen, F. Jin, and A.A. Franke. 1999. Urinary excretion of isoflavonoids and the risk of breast cancer. Cancer Epidemiol. Biomarkers Prevention 8:35–40.

Zhou, J.R., P. Mukherjee, E.T. Gugger, T. Tanaka, G.L. Blackbum, and S.K. Clinton. 1998. Inhibition of murine bladder tumorigenesis by soy isoflavones via alterations in the cell cycle, apoptosis, and angiogenesis. Cancer Res. 58:5231–5238.

SUBJECT INDEX